ENCYCLOPEDIA
OF
SPECTROSCOPY
AND
SPECTROMETRY

ENCYCLOPEDIA
OF
SPECTROSCOPY
AND
SPECTROMETRY

Editor-in-Chief

JOHN C. LINDON

Editors

GEORGE E. TRANTER

JOHN L. HOLMES

ACADEMIC PRESS

A Harcourt Science and Technology Company

San Diego   San Francisco   New York   Boston
London   Sydney   Tokyo

Copyright © 2000 by
ACADEMIC PRESS

ACADEMIC PRESS
A Harcourt Science and Technology Company
24–28 Oval Road
London NW1 7DX, UK
http.hbuk.co.uk/ap/

ACADEMIC PRESS
525 B Street, Suite 1900,
San Diego, CA 92101–4495, USA
http://www.apnet.com

ISBN 0-12-226680-3   100186564X

A catalogue record for this Encyclopedia is available from the British Library

Library of Congress Catalog Card Number: 98–87952

Access for a limited period to an on-line version of the Encyclopedia of Spectroscopy and Spectrometry
is included in the purchase price of the print edition.
This on-line version has been uniquely and persistently identified by the Digital Object Identifier (DOI)

10.1006/rwsp.2000

By following the link

http://dx.doi.org/10.1006/rwsp.2000

from any Web Browser, buyers of the Encyclopedia of Spectroscopy and Spectrometry
will find instructions on how to register for access.

Typeset by Macmillan India Limited, Bangalore, India
Printed and bound in Great Britian by The University Printing House, Cambridge, UK.
00 01 02 04 05 CU 9 8 7 6 5 4 3 2 1

Editors

EDITOR-IN-CHIEF

John C. Lindon
Biological Chemistry
Division of Biomedical Sciences
Imperial College of Science, Technology and Medicine
Sir Alexander Fleming Building
South Kensington
London SW7 2AZ, UK

EDITORS

George E. Tranter
Glaxo Wellcome Medicines Research
Physical Sciences Research Unit
Gunnells Wood Road
Stevenage
Hertfordshire SG1 2NY, UK

John L. Holmes
University of Ottawa
Department of Chemistry
PO Box 450
Stn 4, Ottawa, Canada KIN 6N5

Preface

This encyclopedia provides, we believe, a comprehensive and up-to-date explanation of the most important spectroscopic and related techniques together with their applications.

The *Encyclopedia of Spectroscopy and Spectrometry* is a cumbersome title but is necessary to avoid misleading readers who would comment that a simplified title such as the "Encyclopedia of Spectroscopy" was a misnomer because it included articles on subjects other than spectroscopy. Early in the planning stage, the editors realized that the boundaries of spectroscopy are blurred. Even the expanded title is not strictly accurate because we have also deliberately included other articles which broaden the content by being concerned with techniques which provide localized information and images. Consequently, we have tried to take a wider ranging view on what to include by thinking about the topics that a professional spectroscopist would conveniently expect to find in such a work as this. For example, many professionals use spectroscopic techniques, such as nuclear magnetic resonance, in conjunction with chromatographic separations and also make use of mass spectrometry as a key method for molecular structure determination. Thus, to have an encyclopedia of spectroscopy without mass spectrometry would leave a large gap. Therefore, mass spectrometry has been included. Likewise, the thought of excluding magnetic resonance imaging (MRI) seemed decidedly odd. The technique has much overlap with magnetic resonance spectroscopy, it uses very similar equipment and the experimental techniques and theory have much in common. Indeed, today, there are a number of experiments which produce multidimensional data sets of which one dimension might be spectroscopic and the others are image planes. Again the subject has been included.

This led to the general principle that we should include a number of so-called spatially-resolved methods. Some of these, like MRI, are very closely allied to spectroscopy but others such as diffraction experiments or scanning probe microscopy are less so, but have features in common and are frequently used in close conjunction with spectroscopy. The more peripheral subjects have, by design, not been treated in the same level of detail as the core topics. We have tried to provide an overview of as many as possible techniques and applications which are allied to spectroscopy and spectrometry or are used in association with them. We have endeavoured to ensure that the core subjects have been treated in substantial depth. No doubt there are omissions and if the reader feels we got it wrong, the editors take the blame.

The encyclopedia is organized conventionally in alphabetic order of the articles but we recognize that many readers would like to see articles grouped by spectroscopic area. We have achieved this by providing separate contents lists, one listing the articles in an intuitive alphabetical form, and the other grouping the articles within specialities such as mass spectrometry, atomic spectroscopy, magnetic resonance, etc. In addition each article is flagged as either a "Theory", "Methods and Instrumentation" or "Applications" article. However, inevitably, there will be some overlap of all of these categories in some articles. In order to emphasize the substantial overlap which exists among the spectroscopic and spectrometric approaches, a list has been included at the end of each article suggesting other articles in this encyclopedia which are related and which may provide relevant information for the reader. Each article also comes with a "Further Reading" section which provides a source of books and major reviews on the topic of the article and in some cases also provides details of seminal research papers. There are a number of colour plates in each volume as we consider that the use of colour can add greatly to the information content in many cases, for example for imaging studies. We have also included extensive Appendices of tables of useful reference data and a contact list of manufacturers of relevant equipment.

We have attracted a wide range of authors for these articles and many are world recognized authorities in their fields. Some of the subjects covered are relatively static, and their articles provide a distillation of the established knowledge, whilst others are very fast moving areas and for these we have aimed at presenting up-to-date summaries. In addition, we have included a number of entries which are retrospective in nature, being historical reviews of particular types of spectroscopy. As with any work of this magnitude some of the articles which we desired and commissioned to include did not make it for various reasons. A selection of these will appear in a separate section in the on-line version of the encyclopedia, which will be available to all purchasers of the print version and will have extensive hypertext links and advanced search tools. In this print version there are 281 articles contributed by more than 500 authors from 24 countries. We have persuaded authors from Australia, Belgium, Canada, Denmark, Finland, France, Germany, Hungary, India,

Israel, Italy, Japan, Mexico, New Zealand, Norway, Peru, Russia, South Africa, Spain, Sweden, Switzerland, The Netherlands, the UK and the USA to contribute.

The encyclopedia is aimed at a professional scientific readership, for both spectroscopists and non-spectroscopists. We intend that the articles provide authoritative information for experts within a field, enable spectroscopists working in one particular field to understand the scope and limitations of other spectroscopic areas and allow scientists who may not primarily be spectroscopists to grasp what the various techniques comprise in considering whether they would be applicable in their own research. In other words we tried to provide something for everone, but hope that in doing so, we have not made it too simple for the expert or too obscure for the non-specialist. We leave the reader to judge.

John Lindon
John Holmes
George Tranter

Acknowledgements

Without a whole host of dedicated people, this encyclopedia would never have come to completion. In these few words I, on behalf of my co-editors, can hope to mention the contributions of only some of those hard working individuals.

Without the active co-operation of the hundreds of scientists who acted as authors for the articles, this encyclopedia would not have been born. We are very grateful to them for endeavouring to write material suitable for an encyclopedia rather than a research paper, which has produced such high-quality entries. We know that all of the people who contributed articles are very busy scientists, many being leaders in their fields, and we thank them. We, as editors, have been ably supported by the members of the Editorial Advisory Board. They made many valuable suggestions for content and authorship in the early planning stages and provided a strong first line of scientific review after the completed articles were received. This encyclopedia covers such a wide range of scientific topics and types of technology that the very varied expertise of the Editorial Advisory Board was particularly necessary.

Next, this work would not have been possible without the vision of Carey Chapman at Academic Press who approached me about 4 years ago with the excellent idea for such an encyclopedia. Four years later, am I still so sure of the usefulness of the encyclopedia? Of course I am, despite the hard work and I am further bolstered by the thought that I might not ever have to see another e-mail from Academic Press. For their work during the commissioning stage and for handling the receipt of manuscripts and dealing with all the authorship problems, we are truly indebted to Lorraine Parry, Colin McNeil and Laura O'Neill who never failed to be considerate, courteous and helpful even under the strongest pressure. I suspect that they are now probably quite expert in spectroscopy. In addition we need to thank Sutapas Bhattacharya who oversaw the project through the production stages and we acknowledge the hard work put in by the copy-editors, the picture researcher and all the other production staff coping with very tight deadlines.

Finally, on a personal note, I should like to acknowledge the close co-operation I have received from my co-editors George Tranter and John Holmes. I think that we made a good team, even if I say it myself.

John Lindon
Imperial College of Science, Technology and Medicine
London
22 April 1999

Guide to Use of the Encyclopedia

Structure of the Encyclopedia

The material in the Encyclopedia is arranged as a series of entries in alphabetical order.

There are 4 categories of entry:

- Historical Overview
- Theory
- Methods and Instrumentation
- Applications

To help you realize the full potential of the material in the Encyclopedia we have provided the following features to help you find the topic of your choice.

1. Contents lists

Your first point of reference will probably be the main alphabetical contents list. The complete contents list appearing in each volume will provide you with both the volume number and the page number of the entry.

Alternatively you may choose to browse through a volume using the alphabetical order of the entries as your guide. To assist you in identifying your location within the Encyclopedia a running headline indicates the current entry. Furthermore, a "reference box" is provided on the opening page of each entry so that it is immediately clear whether it is a theory, methods & instrumentation, applications, or historical entry, and which of the following areas of spectroscopy it covers.

- Atomic Spectroscopy
- Electronic Spectroscopy
- Fundamentals in Spectroscopy
- High Energy Spectroscopy
- Magnetic Resonance
- Mass Spectrometry
- Spatially Resolved Spectroscopic Analysis
- Vibrational, Rotational, & Raman Spectroscopies

Example:

NUCLEAR OVERHAUSER EFFECT   1643

Nuclear Overhauser Effect

Anil Kumar and **R Christy Rani Grace**,
Indian Institute of Science, Bangalore, India

MAGNETIC RESONANCE
Theory

You will find "dummy entries" in the following instances:

1. where obvious synonyms exist for entries. For example, a dummy entry appears for ESR Imaging which directs you to EPR Imaging where the material is located.
2. where we have grouped together related topics. For example, a dummy entry appears for Arsenic, NMR Applications which leads you to **Heteronuclear NMR Applications (As, Sb, Bi)** where the material is located.

3. where there is debate over whether an entry title begins with the application of a technique, or with the technique itself. For example, a dummy entry appears for Raman Spectroscopy in Biochemistry which directs you to Biochemical Applications of Raman Spectroscopy where the material is located.

Dummy entries appear in both the contents list and the body of the text.

Example:

If you were attempting to locate material on the application of spectroscopic techniques in astronomy via the contents list the following information would be provided.

Astronomy, Applications of Spectroscopy   *See* Interstellar Molecules, Spectroscopy of; Stars, Spectroscopy of.

The page numbers for these entries are given at the appropriate location in the contents list.

If you were trying to locate the material by browsing through the text and you looked up Astronomy then the following would be provided.

Astronomy, Applications of Spectroscopy

See **Interstellar Molecules, Spectroscopy of; Stars, Spectroscopy of.**

Alternatively if you looked up Stars the following information would be provided.

STARS, SPECTROSCOPY OF 2199

Stars, Spectroscopy of

AGGM Tielens, Rijks Universiteit, Groningen,
The Netherlands

Copyright © 1999 Academic Press

ELECTRONIC SPECTROSCOPY
Applications

Further to aid the reader to locate material the main alphabetical Contents list is followed by a list of the entries grouped within their relevant subject area. (Subject areas follow each other alphabetically). Within each subject area the entries are further broken down into those covering historical aspects, theory, methods & instrumentation, or applications. The entries are listed alphabetically within these categories, and their relevant page numbers are given.

2. Cross References

To direct the reader to other entries on related topics a "see also" section is provided at the end of each entry

Example:

The entry Nuclear Overhauser Effect includes the following cross-references:

See also: **Chemical Exchange Effects in NMR; Macromolecule–Ligand Interactions Studied By NMR; Magnetic Resonance, Historical Perspective; NMR Pulse Sequences; NMR Relaxation Rates; Nucleic Acids Studied Using NMR; Proteins Studied Using NMR Spectroscopy; Structural Chemistry Using NMR Spectroscopy, Organic Molecules; Structural Chemistry Using NMR Spectroscopy, Peptides; Structural Chemistry Using NMR Spectroscopy, Pharmaceuticals; Two-Dimensional NMR Methods.**

3. Index

The index appears in each volume. Any topic not found through the Contents list can be located by referring to the index. On the opening page of the index detailed notes on its use are provided.

4. Colour plates

The colour figures for each volume have been grouped together in a plate section. The location of this section is cited at the end of the contents list.

5. Appendices

The appendices appear in volume 3.

6. Contributors

A full list of contributors appears at the beginning of each volume.

Contributors

Adams, Fred
Department of Chemistry
University of Instelling Antwerp
University Pleim 1, B-2610, Antwerp, Belgium

Aime, S
University of Torino
Department of Chemistry
via Giuria 7, 10125, Torino, Italy

Andersson, L A
Vassar College
Poughkeepsie, Box 589, NY 12604-0589, USA

Ando, Isao
Department of Polymer Chemistry
Tokyo Institute of Technology
Meguro Ku, Tokyo, 152, Japan

Andrenyak, David M
University of Utah
Center for Human Toxicology
Salt Lake City, Utah 84112, USA

Andrews, David L
School of Chemical Sciences
University of East Anglia
Norwich, NR4 7TJ, UK

Andrews, Lester
Chemistry Department
University of Virginia
McCormick Road,
Charlottesville, VA 22901, USA

Appleton, T G
Department of Chemistry
The University of Queensland
Brisbane, Queensland 4072, Australia

Arroyo, C M
USA Medical Research Institute for Chemical
Defense, Drug Assessment Division
Advanced Assessment Branch
3100 Ricketts Point Road, Aberdeen Proving
Ground, Maryland, MD 21010, USA

Artioli, Gilberto
Dipartimento di Scienze della Terra
Universita degli Studi di Milano
via Botticelli 23, I-20133, Milan, Italy

Ashfold, Michael N R
University of Bristol
School of Chemistry
Bristol, BS8 1TS, UK

Aubery, M
Laboratorie de Glycobiologie et Reconnaissance
Cellulaire
Université Paris V - UFR Biomédicale
45, rue des Saints-Péres, 75006, Paris, France

Baer, Tom
Department of Chemistry
University of North Carolina
Chapel Hill, NC 27599-3290, USA

Bain, A D
Department of Chemistry
McMaster University
1280 Main Street W., Hamilton,
Ontario L8S 4M1, Canada

Baker, S A
Beltsville Human Nutrition Research Center
U.S. Department of Agriculture
Food Composition Lab
Beltsville, MD 21054, USA

Baldwin, Mike
Mass Spectrometry Facility
University of California
San Francisco, CA 94143-0446, USA

Bateman, R
Micromass LTD
Manchester, M23 9LZ, UK

Batsanov, Andrei
Department of Chemistry
University of Durham
South Road, Durham, DH1 3LE, UK

Beauchemin, Diane
Department of Chemistry
Queen's University
Kingston, ONT K7L 3N6, Canada

Bell, Jimmy D
The Robert Steiner MR Unit, MRC Clinical
Sciences Centre
Imperial College School of Medicine
Hammersmith Hospital
Du Cane Road, London, W12 0HS, UK

Belozerski, G N
Post Box 544, B-155, 199155, St. Petersburg,
Russia

Bernasek, S L
Princeton University
Department of Chemistry
Princeton, NJ 05844, USA

Berova, Nina
Columbia University
Department of Chemistry
New York, NY 10027, USA

Berthezene, Y
Hospital L. PradelUMR CNRS 5515
Dept de Imagerie Diagnostique et Therapeutique
BP Lyon Montchat, F-69394,
Lyon 03, France

Boesl, Ulrich
Institut für Physikalische und Theoretische
Chemie
Technische Universität München
Lichtenbergstrasse 4,
D-85748, München, Germany

Bogaerts, Annemie
Department of Chemistry
University of Antwerp
Universiteitsplein 1, B-2610, Wilrijk, Belgium

Bohme, D
Department of Chemistry & Centre for Research in
Earth & Space Science
York University
North York, Ontario M3J 1P3, Canada

Bonchin, Sandra L
Los Alamos National Laboratory
Nuclear Materials Technology-Analytical
Chemistry
NMT-1, MS G740, Los Alamos,
NM 87545, USA

Bowie, John H
The University of Adelaide
Department of Organic Chemistry
South Australia 5005, Australia

Brand, Willi A
Max-Planck-Institute for Biochemistry
P.O. Box 100164, 07701, Jena, Germany

Braslavsky, Silvia E
Max Planck-Institut für Strahlenchemie
Postfach 101365,
D-45470, Mülheim an der Ruhr, Germany

Braut-Boucher, F
Laboratorie de Glycobiologie et Reconnaissance
Cellulaire
Université Paris V - UFR Biomédicale
45, rue des Saints-Peres,
75006, Paris, France

Brittain, H G
Center for Pharmaceutical Physics
10 Charles Road, Milford,
NJ 08848, USA

Brumley, W C
National Exposure Research Laboratory
US EPA, Division of Environmental Science
PO Box 93478, Las Vegas,
Nevada 89193, USA

Bryce, David L
Dalhousie University
Department of Chemistry
Halifax, Nova ScotiaCanada

Bunker, Grant
Department of Physics
Illinois Institute of Technology
3101 S. Dearborn, Chicago, IL 60616, USA

Burgess, C
Rose Rae
Startforth
Barnard Castle, Durham, DL12 9AB, UK

Buss, Volker
University of Duisberg
Department of Theoretical Chemistry
D-47048, Duisberg, Germany

Callaghan, P T
Department of Physics
Massey University
Palmerston North, New Zealand

Calucci, Lucia
Dipartimento di Chimica e Chimica Industriale
via Risorgimento 35, 56126, Pisa, Italy

Cammack, Richard
Division of Life Sciences, Kings College
University of London
Campden Hill Road, London, W8 7AH, UK

Canè, E
Universita di Bologna
Dipartimento di Chimica Fisica e Inorganica
Viale Risorgimento 4,
40136, Bologna, Italy

Canet, D
Laboratorie Methode RMN
Universite de Nancy 1
FU CNRS E008, INCM,
F-S4506, Vandoeuvre, Nancy, France

Carter, E A
University of Bradford
Chemical and Forensic Sciences
Bradford, BD7 1DP, UK

Caruso, Joseph A
University of Cincinnati
Department of Chemistry
Cincinnati, OH 45221-0037, USA

Cerdan, Sebastian
Instituto de Investigaciones Biomedicas,
C.S.I.C
c/ Arturo Duperier 4, 28029,
Madrid, Spain

Chakrabarti, C L
Chemistry Department
Carlton University
Ottawa, Ontario K1S 5B6, Canada

Chen, Peter C
Department of Chemistry
Spelman College
Spelman Lane, Atlanta, Georgia 30314-4399,
USA

Cheng, H N
University of Delaware
Department of Chemistry
Newark, DE 19176, USA

Chichinin, A I
Institute of Chemical Kinetics and Combustion
Institutskaya 3, 630090, Novosibirsk, Russia

Claereboudt, Jan
University of Antwerp
Department of Pharmaceutical Sciences
Universiteitsplein 1, B-2610, Antwerp, Belgium

Claeys, Magda M
University of Antwerp
Department of Pharmaceutical Sciences
Universiteitsplein 1, B-2610, Antwerp, Belgium

Colarusso, Pina
National Institues of Digestive and Diabetes and
Kidney Diseases, National Institues of Health
Laboratory of Chemical Physics
Bethesda, MD 20892, USA

Conrad, Horst
Fritz Haber Institute of the
Max Planck Gessellschaft
Faradayweg 4-6, D14195, Berlin, Germany

Cory, D G
Department of Nuclear Engineering
MIT
Cambridge, MA 02139, USA

Crouch, Dennis J
University of Utah
Center for Human Toxicology
Salt Lake City, Utah 84112, USA

Cruz, Fatima
Instituto de Investigaciones Biomedicas, C.S.I.C
c/ Arturo Duperier 4, 28029,
Madrid, Spain

Curbelo, Raul
Digilab Division
Bio-Rad Laboratories
237 Putnam Avenue, Cambridge, MA 02139, USA

Dåbakk, Eigil
Foss Sverige
Turebergs Torg 1, Box 974, SE 191 92,
Sollentuna, Sweden

Davies, M C
The University of Nottingham
Laboratory of Biophysics and Surface Analysis,
School of Pharmaceutical Siences
University Park, Nottingham,
NG7 2RD, UK

Dawson, P H
Iridian Spectral Technologies Ltd
Industry Partnership Facility [M5O]
1200, Montreal Road,
Ottawa, K1A 0R6, Ontario Canada

Demtroder, W
Fachbereich Physik
Universtität Kaiserslautern
D-6750, Kaiserslautern, Germany

Di, Qiao Qing
University of Florida
College of Pharmacy
P.O. Box 100485, Gainesville,
FL 32610, USA

Dirl, Rainer
Centre for Computational Material Science
Institut für Theoretische Physik
Tu Wien, Wiedner Hauptstrabe 8-10,
A-1040, Vienna, Austria

Dixon, Ruth M
MRC Biochemical and Clinical Magnetic
Resonance Unit
Department of Biochemistry
South Parks Road, Oxford, OX1 3QU, UK

Docherty, John
Institute of Biodiagnostics
436 Ellice Avenue, Winnipeg,
Manitoba R3B 1Y6, Canada

Dong, R Y
Department of Physics and Astronomy
Brandon University
Brandon, Manitoba R7A 6A9, Canada

Douglas, D
University of British Columbia
Department of Chemistry
2036 Main, Mall,
Vancouver, BC, V6T 1Z1, Canada

Dua, Suresh
The University of Adelaide
Department of Chemistry
South Australia 5005, Australia

Dugal, Robert
The Canadian Pharmaceutical Manufacturers
Association
Doping Control Laboratory
Ottawa, Canada

Durig, J R
University of Missouri-Kansas City
5100 Rockhill Road, Kansas City,
Missouri 64110-2499, USA

Dworzanski, Jacek
Center for Micro Analysis and Reaction Chemistry
University of Utah
110 South Central Campus Drive, Room 214,
Salt Lake City, Utah 84112, USA

Dybowski, Cecil R
University of Delaware
Department of Chemistry
Newark, DE 19716, USA

Eastwood, DeLyle
Air Force Institution of Technology
MS AFIT/ENP
Wright-Patterson AFB,
OH 45433-7765, USA

Edwards, H G M
Chemistry and Forensic Sciences
University of Bradford
Bradford,
West Yorkshire BD7 1DP, UK

Eggers, L
Gerhard-Mercator-Universität
Institut für Physikalische und Theoretische
Chemie
D-47048, Duisburg, Germany

Emsley, James W
Department of Chemistry
University of Southampton
Highfield, Southampton, UK

Endo, I
Department of Physics
Hiroshima University
1-3-1 Kagamiyama,
Higashi Hiroshima, 739, Japan

Ens, W
Department of Physics
University of Manitoba
Winnipeg,
Manitoba R3T2N2, Canada

Farley, J W
University of Nevada
Department of Physics
Las Vegas, NV 89154, USA

Farrant, R D
Physical Sciences
GlaxoWellcome R & D
Gunnels Wood Road, Stevenage, SG1 2NY, UK

Feeney, J
National Institute for Medical Research
Medical Research Council
The Ridgeway, Mill Hill, London, NW7 1AA, UK

Fennell, Timothy R
Chemical Industry Institute of Toxicology
6 Davis Drive, PO Box 12137, Research Triangle
Park, North Carolina NC 27709-2137, USA

Ferrer, N
Serv. Cientif. Tecn. University of Barcelona
Lluis Sole Sabaris 1, E-08028, Barcelona, Spain

Fisher, A J
Department of Physics and Astronomy
University College London
Gower Street, London, WC1E 6BT, UK

Flack, H D
University of Geneva
Laboratory of Crystallography
24 Quai Ernest Ansernet,
CH 1211, Geneva, Switzerland

Flytzanis, Chr.
Laboratorie d'Optique Quantique
CNRS - Ecole Polytechnique
F-91128, Palaiseau, Cedex France

Foltz, Rodger L
Center for Human Toxicology
University of Utah
20 S 2030 ERM 490,
Salt Lake City, UT 84112-9457, USA

Ford, Mark
University of York
Department of Chemistry
Heslington, York Y010 5DD, UK

Friedrich, J
Lehrstuhl für Physik, Weihenstephan
Technische Universität München
D-85350, Freisling, Germany

Fringeli, Urs
Insitute of Physical Chemistry
University of Vienna
Althanstrasse 14/UZA II,
A-1090, Vienna, Austria

Frost, T
GlaxoWellcome R & D
Temple Hill, Dartford, Kent DA1 5AH, UK

Fuller, Watson
Keele University
Department of Physics
Keele, Staffs ST5 5BG, UK

Futrell, J H
Department of Chemistry & Biochemistry
University of Delaware
Newark, Delaware 19716, USA

Geladi, Paul
Department of Chemistry
Umeå University
SE 901 87, Umeå, Sweden

Gensch, Thomas
Katholieke Universiteit of Leuven
Department of Organic Chemistry
Molecular Dynamics and Spectroscopy
Celestijnenlaan 200F, B-3001,
Heverlee, Belgium

Gerothanassis, I P
Department of Chemistry
University of Iannina
GR-45110, Iannina, Greece

Gilbert, A S
19 West Oak, Beckenham,
Kent BR3 5EZ, UK

Gilchrist, Alison
University of Leeds
Department of Colour Chemistry
Leeds, LS2 9JT, UK

Gilmutdinov, AKh
Kazan Lenin State University
Department of Physics
Kazan, 420008, Russia

Gorenstein, D G
Sealy Centre for Structural Biology
Medical Branch
University of Texas
Galveston, Texas 77555-1157, USA

Grace, R Christy Rani
Department of Physics
Indian Institute of Science
Bangalore, India

Green-Church, Kari B
Department of Chemistry
Louisiana State University
Baton Rouge, LA 70803, USA

Greenfield, Norma J
Department of Neuroscience and Cell Biology,
Robert Wood Johnson Medical School
University of Medicine and Dentistry of
New Jersey
675 Hoes Lane, Piscataway,
NJ 08854, USA

Grime, G
Department of Materials
University of Oxford
Parks Road, Oxford, UK

Grutzmacher, Hans
Universität Bielefeld
Fakultat für Chemie
Postfach 100131,
D-33501, Bielefeld, Germany

Guillot, G
Unite de Recherche en Resonance,
Magnetique Medicale, CNRS URS 2212
Bat.220 Universite Paris-Sud
91405, ORSAY, Cedex France

Hallett, F R
Department of Physics
University of Guelph
Guelph, Ontario N1G 2W1, Canada

Hannon, A C
ISIS Facility, Rutherford Appleton Laboratory
Didcot, Oxon OX11 0QX, UK

Harada, Noboyuki
Tohoku University
Institute of Chemical Reaction Science
Sendai, 980 77, Japan

Hare, John F
SmithKline Beecham Pharmaceuticals
The Frythe, Welwyn,
Herts, AL6 9AR, UK

Harmony, Marlin D
Department of Chemistry, Marlott Hall
University of Kansas
Lawrence, Kansas 66045, USA

Harrison, A G
Chemistry Department
University of Toronto
80 St George Street, Toronto,
Ontario M5S 3H6, Canada

Hawkes, G E
Department of Chemistry,
Queen Mary and Westfield College
University of London
Mile End Road, London, E1 4NS, UK

Hayes, Cathy
Department of Botany
Trinity College
Dublin 2, Eire

Heck, Albert J R
Bijvoet Center for Biomolecular Research,
Utrecht University
Department of Chemistry and Pharmacy
Sorbonnelaan 16, 3584 CA, Utrecht,
The Netherlands

Herzig, Peter
Institut für Physikalische Chemie
Universität Wien
Währingerstraße 42,
A-1090, Wien, Austria

Hess, Peter
Physikalisch-Chemisches Institut
Universität Heidelberg
Im Neuenheimer Feld 253,
D-69120, Heidelberg, Germany

Hicks, J M
Department of Chemistry
Georgetown University
Washington DC, 20057, USA

Hildebrandt, Peter
Max-Planck-Institut für Strahlenchemie
Postfach 101365,
D-45413, Mülheim/Ruhr, Germany

Hill, Steve J
Department of Environmental Science
University of Plymouth
Drake Circus, Plymouth PL4 8AA, UK

Hills, Brian P
Institute of Food Research
Norwich Laboratory
Norwich Research Park, Colney,
Norwich NR4 7UA, UK

Hockings, P D
SmithKline Beecham Pharmaceuticals
Analytical Sciences Department
The Frythe, Welwyn,
Herts, AL6 9AR, UK

Hofer, Tatiana
Universität Kaiserslautern
Fachbereich Chemie der
D-67663, Kaiserslautern, Germany

Hoffmann, G G
Hoffmann Datentechnik
Postfach 10 06 31,
D-46006, Oberhausen, Germany

Holcombe, James A
Department of Chemistry
University of Texas
Austin, Texas7871-1167, USA

Holliday, Keith
University of San Francisco
Department of Physics
2130 Fulton Street, San Francisco,
CA 94117, USA

Holmes, John L
Department of Chemistry
University of Ottawa
PO Box 450, Stn 4, Ottawa, K1N 6N5, Canada

Homer, J
Chemical Engineering and Applied Chemistry,
School of Engineering and Applied Science
Aston University
Aston Triangle, Birmingham, B4 7ET, UK

Hore, P J
Physical and Theoretical Chemistry Laboratory
University of Oxford
South Parks Road, Oxford, OX1 3QZ, UK

Huenges, Martin
Technische Universität Munchen
Institut für Organische Chemie and Biochemie -
Leharul II
Lichtenbergatrahe 4,
D-85747, Garching, Germany

Hug, W
Institut de Chimie Physique
Universite de Fribourg
CH-1700, Fribourg, Switzerland

Hunter, Edward P L
Physical and Chemistry Properties Division (838)
Physics Building (221), Room A 113
NIST, PHY A 111, Gaithersburg,
Maryland 20899, USA

Hurd, Ralph
GE Medical Systems
47697 Westinghouse Drive, Fremont,
California 94539, USA

Imhof, Robert E
Department of Physics and Applied Physics
Strathclyde University
Glasgow, G4 0NG, UK

Jackson, Michael
National Research Council Canada
Institute for Biodiagnostics
435 Ellice Avenue, Winnipeg,
Manitoba R3B 1Y6, Canada

Jalsovszky, G
Chemistry Research Centre, Institute of Chemistry
Hungarian Academy of Sciences
PO Box 17, H-1525, Budapest, Hungary

Jellison, G E
Oak Ridge National Laboratory
Solid State Division
POB 2008, Oak Ridge, Tennessee, TN 37831, USA

Jokisaari, J
Department of Physical Sciences
University of Oulu
P O Box 3000, Oulu, FIN-90Y01, Finland

Jonas, J
School of Chemical Sciences
University of Illinois
Urbana, Illinois, 61801, USA

Jones, J R
Department of Chemistry
University of Surrey
Guildford, Surrey GU2 5XH, UK

Juchum, John
University of Florida
College of Pharmacy
P.O.Box 100485, Gainsville, FL 32610, USA

Katoh, Etsuko
National Institute of Agrobiological Resources
2-1-2, Kannondai, Tsukuba, Ibaraki 305-0856,
Japan

Kauppinen, J
University of Turku
Department of Applied Physics
FIN-20014, Turku 50, Finland

Kessler, Horst
Institut für Organische Chemie und Biochemie
Technische Universität München
Lichtenbergstrabe 4,
D-85747, Garching, Germany

Kettle, S F
School of Chemical Sciences
University of East Anglia
Norwich, NR4 7TJ, UK

Kidder, Linda H
National Institues of Digestive and Diabetes and
Kidney Diseases, National Institutes of Health
Laboratory of Chemical Physics
Bethesda, MD 20892, USA

Kiefer, Wolfgang
Institut für Physikalische Chemie
Der Universität Wurzburg
Am Hubland,
D-97074, Wurzburg, Germany

Kiesewalter, Stefan
Universität Kaiserslautern
Fachbereich Chemie der
D-67663, Kaiserslautern, Germany

Kimmich, Rainer
Universität Ulm
Sektion Kernresonanzspektroskopie
D-89069, Ulm, Germany

Klinowski, J
Department of Chemistry
University of Cambridge
Lensfield Road,
Cambridge, CB2 1EW, UK

Koenig, J L
Case Western Reserve University
Department of Macromolecular Science
10900 Euclid Avenue, Cleveland,
Ohio 44106-7202, USA

Kolemainen, E
Department of Chemistry
University of Jyvaskyla
Jyvaskyla, FIN-40351, Finland

Kooyman, R P H
University of Twente
Department of Applied Physics
Enschede, NL 7500 AE, The Netherlands

Kordesch, Martin E
Department of Physics and Astronomy
Ohio University
Athens, Ohio 45701, USA

Kotlarchyk, M
Department of Physics
3242 Gosnell, Rochester Institute of Technology
85 Lomb Memorial Drive, Rochester,
NY 14623-5603, USA

Kramar, U
Institute of Petrography and Geochemistry
University of Karlsruhe
Kaiserstrasse 12,
D-76128, Karlsruhe, Germany

Kregsamer, P
Atominstitut der Osterreichischen Universitaten
Stadionallee 2, 1020, Wien, Austria

Kruppa, Alexander I
Institute of Chemical Kinetics and Combustion
Novosibirsk-90, 630090, Russia

Kuball, Hans-Georg
Universität Kaiserslautern
Fachbereich Chemie der
D-67653, Kaiserslautern, Germany

Kumar, A
Department of Physics and Sophisticated
Instruments Facility
Indian Institute of Science
Bangalore, 560012,
Karnataka, India

Kushmerick, J G
The Pennsylvania State University
Department of Chemistry
University Park, PA 16802-6300, USA

Kvick, Ake
European Synchrotron Radiation Facility
BP 220, Avenue des Martyrs, F-38043,
Grenoble, France

Laeter, J Rde
Curtin University of Technology
Bentley, Western Australia 6102, Australia

Latosińska, Jolanta N
Insitute of Physics
Adam Mickiewicz University
Umultowska 85, 61-614, Poznań, Poland

Leach, M O
Clinical Magnetic Resonance Research Group
Institute of Cancer Research
Royal Marsden Hospital
Sutton, Surrey SM2 5PT, UK

Lecomte, S
CNRS-Université Paris VI
Thiais, France

Leshina, T V
Russian Academy of Sciences
Institute of Chemical Kinetics and Combustion
Novosibirsk-90, Russia

Levin, Ira W
National Institues of Digestive and Diabetes and
Kidney Diseases, National Institutes of Health
Laboratory of Chemical Physics
Bethesda, MD 20892, USA

Lewen, Nancy S
Bristol Myers Squibb, 1 Squibb Dr., Bldg. 101 Rm
B18, New Brunswick, NJ 08903, USA

Lewiński, J
Department of Chemistry
Warsaw University of Technology
Noakowskiego 3, PL-00664, Warsaw, Poland

Lewis, Neil
National Institutes of Digestive and Diabetes and
Kidney Diseases, National Institute of Health
Laboratory of Chemical Physics
Bethesda, MD 20892, USA

Leyh, Bernard
F.N.R.S. and University of Leige
Department of Chemistry (B6)
B.4000, Sart Tilman, Belgium

Lias, S
Physical and Chemical Properties Division (838)
Physics Building (221), Room A 113
NIST, PHY A 111, Gaithersburg,
Maryland 20899, USA

Lifshitz, Chava
Department of Physical Chemistry
The Farkas Centre for Light-induced Processes
The Hebrew University of Jerusalem
Jerusalem, 91904, Israel

Limbach, Patrick A
Louisiana State University
Department of Chemistry
Baton Rouge, LA 70803, USA

Lindon, John C
Biological Chemistry
Division of Biomedical Sciences
Imperial College School of Science, Technology
and Medicine
Sir Alexander Fleming Building
South Kensington, London SW7 2AZ, UK

Linuma, Masataka
Hiroshima University
Department of Physics
1-3-1 Kagamiyama, Higashi,
Hiroshima, 739, Japan

Liu, Maili
The Chinese Academy of Sciences
Wuhan Institute of Physics and Mathematics
Laboratory of Magnetic Resonance and Atomic
and Molecular Physics
Wuhan, 430071,
Peoples' Republic of China

Lorquet, J C
Departement of Chemie
Universite de Liege
Sart-Tilman B6 (Batiment. B6), B-4000, Liege 1,
Belgium

Louer, D
Groupe Crystallochimie, Chemical Solids and
Inorganic Molecules Laboratory
University of Rennes
CNRS UMR 6511 Ave. Gen. Leclerc,
F-35042, Rennes, France

Luxon, Bruce A
Sealy Center for Structural Biology
University of Texas Medical Branch
Galveston, Texas 77555, USA

Maccoll, Allan
10, The Avenue, Claygate, Surrey KT10 0RY, UK

Macfarlane, R D
Chemistry Department
Texas A & M University
College Station, TX 77843-3255, USA

Maerk, T D
Institut fuer Ionenphysik
Leopold Franzens Universitaet
Technikerstr. 25, A-6020, Innsbruck, Austria

Magnusson, Robert
Department of Electronic Engineering
University of Texas
Arlington, Texas TX 76019, USA

Mahendrasingam, A
Keele University
Physics Department
Staffordshire, ST5 5BG, UK

Maier, J P
Institut für Physikalische Chemie
Universitat Basel
Klingelbergstrasse, CH 4056, Basel, Switzerland

Makriyannis, A
School of Pharmacy
University of Connecticut
Storrs, CT 06269, USA

Malet-Martino, Myriam
Universite Paul Sabatier
Groupe de RMN Biomedicale, Laboratorie des
IMRCP (UMR CNRS 5623)
31062, Toulouse, Cedex, France

Mamer, O
Mass Spectrometry Unit
McGill University
1130 Pine Avenue West, Montreal,
Quebec H3A 1A3, Canada

Mandelbaum, Asher
Technion-Israel Institute of Technology
Department of Chemistry
Technion, Haifa 32000, Israel

Mantsch, H H
National Research Council of Canada
Institute for Biodiagnostics
435 Ellice Avenue, Winnipeg, R3B 1Y6, Canada

Mao, Xi-an
The Chinese Academy of Sciences
Wuhan Institute of Physics and Mathematics,
Laboratory of Magnetic Resonance and Atomic
and Molecular Physics
Wuhan, 430071, Peoples' Republic of China

March, Raymond E
Department of Chemistry
Trent University
Peterborough, Ontario K9J 7B8, Canada

Marchetti, Fabio
Universita di Camerino
Dipartmento di Scienze Chemiche
via S. Agostino 1, 62032, Camerino MC, Italy

Mark, Tilmann D
Loepold Franzens Universitat
Institut für Ionenphysik
Technikerstrasse 25, A-6020, Innsbruck, Austria

Marsmann, H C
University Gesemthsch. Paderborn
Fachbereich Chem.
Warburger Str. 100,
D-33095, Paderborn, Germany

Martino, Robert
Universite Paul Sabatier
Groupe de RMN Biomedicale
Laboratore des IMRCP (UMR CNRS 5623)
31062, Toulouse Cedex, France

Maupin, Christine L
Michigan Technological University
Department of Chemistry
Houghton, MI 4993, USA

McClure, C K
Department of Chemistry
Montana State University
Bozeman, MT 59171, USA

McLaughlin, D
Kodak Research Laboratories
Eastman Kodak Co.
Rochester, New York NY 14650, USA

McNab, Iain
Department of Physics
University of Newcastle
Newcastle-upon-Tyne, NE1 7RU, UK

McNesby, K L
6735 Indian River Drive, Citrus Heights,
CA 95621, USA

Meuzelaar, H L C
Center for Micro Analysis & Reaction Chemistry
University of Utah
110 South Central Campus Drive, Room 214,
Salt Lake City, Utah 84112, USA

Michl, Josef
Department of Chemistry & Biochemistry
University of Colorado
Boulder, CO 80309-0215, USA

Miklos, Andras
Institute of Physical Chemistry
University of Heidelberg
Im Neuenheimer Feld 253,
D-69120, Heidelberg, Germany

Miller, S A
Princeton University
Department of Chemistry
Princeton, NJ 08544, USA

Miller-Ihli, Nancy J
U.S. Department of Agriculture
Food Composition Laboratory
Building 161, Rm. 1, BARC-East, Beltsville,
MD 20705, USA

Morris, G A
Department of Chemistry
University of Manchester
Oxford Road, Manchester, M13 9PL, UK

Mortimer, R J
Department of Chemistry
Loughborough University
Loughborough, Leics LE11 3TU, UK

Morton, Thomas H
Department of Chemistry
University of California
Riverside, CA 92521-0403, USA

Muller-Dethlefs, K
University of York
Department of Chemistry
Heslington, York YO1 5DD, UK

Mullins, Paul G M
SmithKline Beecham Pharmaceuticals
Analytical Sciences Department
The Frythe, Welwyn, Herts, AL6 9AR, UK

Murphy, Damien M
Cardiff University
National ENDOR Centre, Department of Chemistry
Cardiff, CF1 3TB, UK

Nafie, L A
Department of Chemistry
Syracuse University
Syracuse, New York 13244-4100, USA

Naik, Prasad A
Room #201, R & D Block "D"
Centre for Advanced Technology
Indore, 452013, Madhya Pradesh, India

Nakanishi, Koji
Department of Chemistry
Columbia University
New York NY 10027, USA

Nicholson, J K
Biological Chemistry, Division of Biomedical Sciences
Imperial College of Science, Technology & Medicine
Sir Alexander Fleming Building,
South Kensington, London,
SW7 2AZ, UK

Nibbering, N M M
Institute of Mass Spectrometry
University of Amsterdam
Nieuwe Achtergracht 129, 1018 WS,
Amsterdam, The Netherlands

Niessen, W M A
hyphen MassSpec Consultancy
De Wetstraat 8, 2332 XT, Leiden,
The Netherlands

Nobbs, Jim
University of Leeds
Department of Colour Chemistry
Leeds, LS2 9JT, UK

Norden, B
Department of Physical Chemistry
Chalmers University of Technology
S-41296, Gothenburg, Sweden

Norwood, T
Department of Chemistry
University of Leicester
Leicester, LE1 7RH, UK

Olivieri, A C
Facultad de Ciencias Biochimicas y Farmaceuticas, Departamento Quimica Analitica
Universita Nacional Rosario
Suipacha 531,
RA-2000, Rosario, Santa Fe, Argentina

Omary, Mohammed A
University of Maine
Department of Chemistry
Orono, Maine ME 04469, USA

Parker, S F
Rutherford Appleton Laboratory, ISIS Facility
Oxon, Didcot OX11 0QX, UK

Partanen, Jari O
University of Turku
Department of Applied Phyics
FIN 20014,
Turku, Finland

Patterson, Howard H
Department of Chemistry
University of Maine
Orono, Maine ME 04469, USA

Pavlopoulos, Spiro
University of Connecticut
Institute of Materials Science and School of
Pharmacy
Storrs, Connecticut 06269, USA

Pettinari, C
Università degli Studi di Camerino
Scienze Chimiche
Via S. Agostino 1, 62032, Camerino, MC, Italy

Poleshchuk, O K
Department of Inorganic Chemistry
Tomsk Pedagogical University
Komsomolskii 75, 634041, Tomsk,
Russian Federation

Rafaiani, Giovanni
Università di Camerino
Dipartimento di Scienze Chemie
Via S Agostino 1, 62032, Camerino MC, Italy

Ramsey, Michael H
Imperial College of Science Technology and
Medicine
TH Huxley School of Environmental, Earth
Science and Engineering
London, SW7 2BP, UK

Randall, Edward W
Department of Chemistry, Queen Mary &
Westfield College
University of London
Mile End Road,
London, E1 4NS, UK

Rehder, D
Institute of Inorganic Chemistry
University of Hamburg
Martin-Luther-King Platz 6,
D-20146, Hamburg, Germany

Reid, David G
Smithkline Beecham Pharmaceuticals
Analytical Sciences Department
The Frythe, Welwyn, Herts, AL6 9AR, UK

Reid, Ivan D
Paul Scherrer Institute
CH-5232, Villigen PSI, Switzerland

Reynolds, William F
Lash Miller Chemical Laboratories
University of Toronto
80 George Street, Toronto,
Ontario M5S 1A1, Canada

Richards-Kortum, Rebecca
Dept. of Elec. and Computer Engin.
University of Texas at Austin
Austin, TX 78712, Inter-Office C0803, USA

Riddell, Frank G
Department of Chemistry
University of St Andrews
The Purdie Building, St Andrews, Fife KY16 9ST,
Scotland

Riehl, J P
Michigan Technological University
Department of Chemistry
Houghton, MI 49931, USA

Rinaldi, Peter L
Department of Chemistry
University of Akron
Akron, Ohio OH 44325-3061, USA

Roberts, C J
The University of Nottingham
Department of Pharmaceutical Sciences
University Park, Nottingham, NG7 2RD, UK

Rodger, Alison
University of Warwick
Department of Chemistry
Coventry, CV4 7AL, UK

Rodger, C
University of Strathclyde
Department of Pure & Applied Chemistry
Glasgow, G1 1XL, UK

Roduner, Emil
Universität Stuttgart
Physikalisch-Chemisches Institut
Pfaffenwaldring 55, D-70550, Stuttgart, Germany

Rost, F W D
45 Charlotte Street, Ashfield, NSW 2131, Australia

Rowlands, C C
Cardiff University
National ENDOR Centre, Department of Chemistry
Cardiff, CF1 3TB, UK

Rudakov, Taras N
7 Reen Street, St James, WA 6102, Australia

Salamon, Z
University of Arizona
Department of Biochemistry
Tuscon, AZ 85721, USA

Salman, S R
Chemistry Department
University of Qatar
PO Box 120174, Doha, Qatar

Sanders, Karen
University of Warwick
Department of Chemistry
Coventry, CV4 7AL, UK

Sanderson, P N
Protein Science Unit
GlaxoWellcome Medicines Research Centre
Gunnels Wood Road, Stevenage,
Herts, SG1 2NY, UK

Santini, Carlo
Università di Camerino
Dipartimento di Scienze Chemiche
Via s. Agostino 1, 62032, Camerino MC, Italy

Santos Gómez, J
Instituto de Estructura de la Materia, CSIC
28006, Madrid, Spain

Schafer, Stefan
Institute of Physical Chemistry
University of Heidelberg
Im Neuenheimer Feld 253,
D-69120, Heidelberg, Germany

Schenkenberger, Martha M
Bristol-Myers Squibb
Pharmaceutical Research Institute
1 Squibb Dr., New Brunswick,
NJ 08903, USA

Schrader, Bernhard
Institut für Physikalische und Theoretisch Chemie
Universität Essen
Fachbereich 8, D-45117, Essen, Germany

Schulman, Stephen G
Department of Medicinal Chemistry
College of Pharmacy
University of Florida
Gainesville, FL 32610-0485, USA

Schwarzenbach, D
University of Lausanne
Institute of Crystallography
BSP Dorigny,
CH-1015, Lausanne, Switzerland

Seitter, Ralf-Oliver
Universität Ulm
Sektion Kernresonanzspektroskopie
89069, Ulm Germany

Seliger, J
Department of Physics
Faculty of Mathematics and Physics
University of Ljubljana
Jadranska, 19, 1000,
Ljubljana Slovenia

Shaw, R Anthony
National Research Council Canada
Institute for Biodiagnostics
435 Ellice Avenue, Winnipeg,
Manitoba R3B 1Y6, Canada

Shear, Jason B
Department of Chemistry
University of Texas
Austin, TX 78712, USA

Sheppard, Norman
School of Chemical Sciences
University of East Anglia
Norwich, NR4 7TJ, UK

Shluger, A L
Department of Physics and Astronomy
University College London
Gower Street, London, WC1E 6BT, UK

Shockcor, J P
Bioanalysis and Drug Metabolism
Stine-Haskell Research Center
Dupont-Merck, P O Box 30, Elkton Road, Newark,
Delaware DE 19714, USA

Shukla, Anil K
University of Delaware
Department of Chemistry and Biochemistry
Newark,
DE 19176, USA

Shulman, R G
MR Center, Department of Molecular Biophysics
Yale University
New Haven, Connecticut 06520, USA

Sidorov, Lev N
Physical Chemistry Division
Chemistry Department
Moscow State University
119899, Moscow, Russia

Sigrist, Markus
Swiss Federal Insitute of Technology (ETH)
Insitute of Quantum Electronics, Laboratory for
Laser Spectroscopy and Environmental Sensing
Hoenggerberg,
CH-8093, Zurich, Switzerland

Simmons, Tracey A
Department of Chemistry
Louisiana State Universty
Baton Rouge, LA 70803, USA

Smith, W E
University of Strathclyde
Department of Pure & Applied Chemistry
Glasgow, G1 1XL, UK

Smith, David
The University of Keele
Department of Biomedical Engineering and
Medical Physics, Hospital Centre
Thornburrow Drive, Hartshill,
Stoke-on-Trent ST4 7QB, UK

Snively, C M
Department of Macromolecular Science
Case Western Reserve University
Cleveland, OH 44106, USA

Somorjai, J
Institute for Biodiagnostics
435 Ellice Avenue, Winnipeg,
R3B 1Y6, Canada

Španel, Patrick
Keele University
Department of Biomedical Engineering and
Medical Physics
Thornburrow Drive, Hartshill,
Stoke-on-Trent ST4 7QB, UK

Spanget-Larsen, Jens
Department of Life Sciences and Chemistry
Roskilde University
POB 260, DK-4000, Roskilde, Denmark

Spiess, H W
Max Plank Institute of Polymer Research
Postfach 3148, D-55021, Mainz, Germany

Spinks, Terence
PET Methodology Group, MRC Clinical Sciences
Centre, Royal Postgraduate Medical School
Hammersmith Hospital
Du Cane Road,
London, W12 0NN, UK

Spragg, R A
Perkin-Elmer Analytical Instruments
Post Office Lane, Beaconsfield,
Bucks HP9 1QA, UK

Standing, K G
Department of Physics
University of Manitoba
Winnipeg, Manitoba R3T 2N2, Canada

Steele, Derek
The Centre for Chemical Sciences
Royal Holloway
University of London
Egham, Surrey TW40 0EX, UK

Stephens, Philip J
University of Southern California
Department of Chemistry
Los Angeles, CA 90089-0482, USA

Stilbs, Peter
Royal Institute of Technology Physical Chemistry
S-10044, Stockholm, Sweden

Streli, C
Atominstitut of the Austrian Universities
Stadionallee 2, 1020, Wien, Austria

Styles, Peter
John Radcliffe Hospital
MRL Biochemistry & Chemical Magnetic
Resonance Unit
Headington, Oxford,
OX3 9DU, UK

Sumner, Susan C J
Chemical Industry Institute of Toxicology
6 Davis Drive, PO Box 12137, Research Triangle
Park, North Carolina
NC 27709-2137, USA

Sutcliffe, L H
Institute of Food Research
Norwich NR4 7UA, UK

Szepes, Laszlo
Eotvos Lorand University
Department of General and Inorganic Chemistry
Pazmany Peter Satany 2, 1117,
Budapest Hungary

Taraban, Marc B
Institute of Chemical Kinetics and Combustion
Novosibirsk-90, 630090, Russia

Tarczay, Gyorgy
Eotvos University
Department of General and Inorganic Chemistry
Pazmany Peter S. 2,
Budapest H-1117, Hungary

Taylor, A
Robens Institute of Health and Safety
Trace Elements Laboratory
University of Surrey
Guildford, Surrey GU2 5XH, UK

Tendler, S J B
University of Nottingham
Department of Pharmaceutical Sciences
University Park, Nottingham, NG7 2RD, UK

Terlouw, J
McMaster University
Department of Chemistry
ABB-455, 1280 Main Street West,
Hamilton, ON L8S 4M1, Canada

Thompson, Michael
University of London
Birkbeck College, Department of Chemistry
Gordon House, 29 Gordon Square,
London, WC1H 0PP, UK

Thulstrup, Erik W
Department of Chemistry & Life Sciences
Roskilde University (RUC)
Building 17.2, PO Box 260,
DK-4000, Roskilde, Denmark

Tielens, A G G M
Kapteyn Astronomical Institute
PO Box 800, 9700 AV, Groningen, The Netherlands

Tollin, Gordon
University of Arizona
Department of Biochemistry
Tucson, AZ 85721, USA

Traeger, John C
Department of Chemistry
La Trobe University
Bundoora, Victoria 3083, Australia

Tranter, George E
Glaxo Wellcome Medicines Research Centre
Gunnels Wood Road, Stevenage,
Herts, SG1 2NY, UK

Trombetti, A
Università di Bologna
Dipartimento Chimica Fisica e Inorganica
Viale Risorgimento 4, I-40136, Bologna, Italy

True, N S
Department of Chemistry
University of California Davis
Davis, CA 95616, USA

Ulrich, Anne S
Institute of Molecular Biology
University of Jena
Winzerlaer Strasse 10, D-07745, Jena, Germany

Utzinger, Urs
The University of Texas at Austin
Texas USA

Van Vaeck, Luc
Department of Chemistry
University Instelling Antwerp
University Pleim 1, B-2610, Antwerp, Belgium

Vandell, Victor E
Louisiana State University
Department of Chemistry
Baton Rouge, LA 708023, USA

Varmuza, Klaus
Department of Chemometrics
Technical University of Vienna
A-1060, Vienna Austria

Veracini, C A
Dipartimento Chemica
University of Pisa
I-56126, Pisa, Italy

Viappiani, Christiano
Dipartimento di Scienze Ambientali
Universita degli Studi di Parma
viale delle Scienze, I-43100, Parma, Italy

Vickery, Kymberley
Chalmers University of Technology
Department of Physical Chemistry
S-412 96, Gothenburg, Sweden

Wagnière, Georges H
Physikalisch-Chemisches Institut der Universitat
Zurich
Winterhurerstrasse 190,
CH-8057, Zurich, Switzerland

Waluk, Jacek
Institute of Physical Chemistry
Polish Academy of Sciences
01-224 Warszawa, Kasprzaka, 44/52, Poland

Wasylishen, Roderick E
Department of Chemistry
Dalhousie University
Halifax, Nova Scotia B3H 4J3, Canada

Watts, Anthony
Department of Biochemistry
University of Oxford
Oxford, OX1 3QU, UK

Webb, G A
University of Surrey
Department of Chemistry
Guildford, GU2 5XH, UK

Weiss, P S
Penn State University
Department of Chemistry
University Park,
Pennsylvania PA 16802-6300, USA

Weller, C T
School of Biomedical Sciences
University of St Andrews
St Andrews, KY16 9ST, UK

Wenzel, Thomas J
Department of Chemistry
Bates College
Lewiston, Maine ME 04240, USA

Wesdemiotis, Chrys
Chemistry Department
University of Akron
Akron, OH 44325-3601, USA

Western, Colin M
University of Bristol
School of Chemistry
Bristol, BS8 1TS, UK

White, R L
Department of Chemistry
University of Oklahoma
Norman, Oklahoma OK 73019-0390, USA

Wieser, Michael E
University of Calgary
2500 University Drive NW, Calgary,
Alberta T2N 1N4, Canada

Wilkins, John
Unilever Research
Colworth, Sharnbrook Beds. MK44 1LQ, UK

Williams, P M
The University of Nottingham
Laboratory of Biophysics and Surface Analysis,
School of Pharmaceutical Sciences
University Park, Nottingham, NG7 2RD, UK

Williams, Antony J
Wobrauschek P Atominstitut der Osterreichischen
Universitaten
Stadionallee 2, 1020, Wien, Austria

Wlodarczak, G
Laboratorie de Spectroscopie Hertzienne
URA CNRS 249, Univ. de Lille 1,
F59655, Villeneured'Aacg Cedese, France

Woźniak, Stanisław
Mickiewicz University
Warsaw, Poland

Young, Ian
Robert Steiner MR Unit
Royal Postgraduate Medical School
Hammersmith Hospital
Ducane Road, London, W12 0HS, UK

Zagorevskii, Dimitri
Department of Chemistry
University of Missouri-Columbia
Columbia MO 65201, USA

Zoorob, Grace K
Biosouth Research Laboratories Inc.
5701 Crawford Street, Harahan,
LA 70123, USA

Zwanziger, J W
Department of Chemistry
Indiana University
Bloomington, Indiana 47405, USA

Contents

Volume 1

A

B

G

H

Volume 2

I

N

Volume 3

O

X

Y

Z

Entry Listing by Subject Area

Subject areas follow each other alphabetically in this list. Within each subject area entries are categorised into those covering historical aspects, theory, methods and instrumentation, or applications. The entries are listed alphabetically within these categories.

Fundamentals of Spectroscopy

Theory

Methods & Instrumentation

High Energy Spectroscopy

Theory

Magnetic Resonance

Mass Spectrometry

Historical Overview

Theory

Methods & Instrumentation

Applications

Optical Frequency Conversion

Christos Flytzanis, Laboratoire d'Optique Quantique du C.N.R.S., Palaiseau, France

ELECTRONIC SPECTROSCOPY
Methods & Instrumentation

General aspects

The introduction of intense coherent light sources such as lasers has drastically and irreversibly modified the field of optics in all its aspects, spectroscopy in particular. The emergence of nonlinear optics strikingly illustrates this revolution. In turn, the impact of nonlinear optics on the development of coherent light sources, since the very inception of lasers, has also been tremendous and to an extent that one can hardly now draw a demarcation line between them. Nonlinear optical processes, besides being at the very heart of laser action, are also increasingly integrated with laser systems to improve the spatiotemporal characteristics of the optical beams and pulses, to control their intensities and most importantly to provide access to frequency ranges that are still inaccessible by primary laser sources. This latter aspect, which concerns nonlinear optical processes where radiation is generated at frequencies that are different from those of the incident radiation and usually goes under the denomination of *optical frequency conversion*, has had wide and profound implications in the evolution of spectroscopy and remains central to research and development in nonlinear optics and laser physics. In this article we will sketch and summarize the main schemes of optical frequency conversion with some indications regarding their performances, achievements and tendencies.

Nonlinear optics concerns effects in the response of a material medium that are nonlinear in the optical field intensity. By response we refer to the radiation sources that are set up inside the medium because of the forced motion of bound and unbound charges (actually electrons). There are different ways to classify such effects, depending on the emphasis one wishes to put on a particular aspect of the nonlinear optical response and the validity of certain criteria. Thus as long as the field intensity is below the cohesive field intensity E_c of a sample, which essentially keeps electrons bound to their atomic core (typically of the order $10^{10} \, \mathrm{V cm^{-1}}$) and concerns the reversible behaviour of bound electrons, the most common and straightforward classification is according to the powers of the electric field intensity E of the applied electromagnetic field. Within this context and referring to the Fourier decomposition of the incident field intensity one can also introduce the distinction between frequency-preserving and frequency non-preserving processes, the latter being the basis of the optical frequency conversion. Notwithstanding the complexity of the nonlinear optical effects the development here follows some rather well-classified patterns and goals. For higher intensities than those of the cohesive field E_c the electrons can be forced out of the field of the atomic core and the classification of their response in powers of the electric field intensity loses its utility, as also does the distinction between frequency-preserving and frequency non-preserving effects. Here different approaches, in particular non-perturbative ones, are being pursued, some of them still tentative, as are also the experimental schemes that are being used; we expect in this regime of extreme nonlinear optics new and exciting developments, in particular regarding X-ray generation, as our understanding and description of the effects progresses.

Most of this article is devoted to summarizing the main achievements and state-of-art in optical frequency conversion within the regime where the perturbative approach is valid, namely $E < E_c$, and only at the very end we will take up the regime of extreme nonlinear optics, namely $E > E_c$. The progress in the field of frequency conversion will continue as it is closely connected with progress in the performances of lasers, optical materials (linear and nonlinear ones) and improvements in the efficiency and understanding of the underlying nonlinear optical processes. Although some specific aspects may be modified in the near future, shifting the emphasis to new directions, the main patterns outlined here are expected to remain relevant.

Nonlinear polarization sources – conceptual background

To keep track of the origin and interconnection of the nonlinear optical processes we recall that these are related to the different polarization terms in the power series expansion of the polarization P induced in the medium by an incident electric field E which in the dipolar approximation takes the form

$$P = P^{(1)} + P^{(2)} + P^{(3)} + \dots$$
$$= \chi^{(1)} E + \chi^{(2)} E^2 + \chi^{(3)} E^3 + \dots \quad [1]$$

where the coefficient $\chi^{(n)}$ is the dipolar susceptibility of order n, a tensor of rank $(n+1)$, that measures the deformability of bound electron clouds to external fields as long as the latter do not exceed the cohesive field E_c intensity which is typically in the range of 10^{10}–10^{11} V cm^{-1} for atoms and molecules. Introducing the Fourier analysis of these polarization terms and fields one can also write

$$P_i^{(1)}(\omega) = D_1 \, \chi_{ij}^{(1)}(\omega) \, E_j(\omega) \quad [2]$$

$$P_i^{(2)}(\omega_1 + \omega_2) = D_2 \, \chi_{ijk}^{(2)}(\omega_1, \omega_2) \, E_j(\omega_1) \, E_k(\omega_2) \quad [3]$$

$$P_i^{(3)}(\omega_1 + \omega_2 + \omega_3) = D_3 \, \chi_{ijkl}^{(3)}(\omega_1, \omega_2, \omega_3)$$
$$\times E_j(\omega_1) \, E_k(\omega_2) \, E_1(\omega_3) \quad [4]$$

for the linear, second and third order terms, respectively, and similarly for the higher order ones. D_n are degeneracy factors equal to the number of distinct permutations of the applied field Fourier components (modes) involved in the interaction; with this convention the value of $\chi^{(n)}$ for $\omega_i \to 0$ is independent of the chosen frequency path. Far from resonances in the transparency range of the materials the magnitude of $\chi^{(n)}$ is of the order of $1/E_c^{n-1}$ or 10^{-7} esu and 10^{-14} esu for $\chi^{(2)}$ and $\chi^{(3)}$ respectively.

The precise behaviour of the $\chi^{(n)}$, which is a tensor of rank $(n+1)$, and in particular their frequency dependence can be obtained by quantum-mechanical perturbation approaches. These coefficients possess some general symmetry properties related to the invariance laws such as causality, intrinsic symmetrization, time and space symmetries. The latter can lead to a drastic reduction of the number of independent components of the susceptibility tensor of a given order/rank. In particular, we recall that all even order dipolar susceptibilities $\chi^{(2n)}$ vanish for a medium with inversion symmetry, for instance for a gas, a liquid or an amorphous solid. Other considerations such as the Kleinman relations lead to some simplifications and reduction of the number of independent components in certain cases. Their quantum-mechanical expressions show that these coefficients can be enhanced by single and multiple resonances or can be greatly reduced because of destructive interference of such resonances. In **Figure 1** we indicate some schemes for frequency conversion. These coefficients are generally complex and both their magnitude and phase are relevant in assessing the energy transfer efficiency either among the electromagnetic field components involved in the nonlinear interaction process or between the electromagnetic field and matter. In the previous development, Equations [1]–[4], which is valid in the dipolar approximation we neglected nonlocal contributions that involve spatial or temporal derivatives of the electric fields or equivalently quadrupolar and magnetic dipolar contributions. These are in general, but no always, weaker than the electric dipolar ones and in certain cases introduce complications with boundary conditions; although interesting at a fundamental level they are not of much use for efficient frequency conversion and will not be considered here.

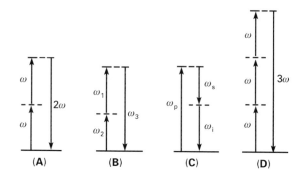

Figure 1 Main nonlinear frequency conversion schemes. (A) Second harmonic, (B) sum frequency, (C) parametric emission and (D) third harmonic generation.

Nonlinear propagation

The nonlinear polarization sources generate electromagnetic fields inside the medium that extract and transfer energy from the incident fields. These new fields can have markedly different spatiotemporal features from the incident ones. In particular, fields at new frequencies are generated such as harmonics of the incident frequencies or combinations thereof with conversion efficiencies close to one in certain cases. The assessment of the behaviour and efficiency of the frequency conversion process proceeds from the inhomogeneous propagation equation that contains the nonlinear polarization terms in Equation [1] as sources. Taking into account that the intense coherent light sources used in frequency conversion deliver pulses or wave packets whose envelope has a finite spatiotemporal extension one may set any of the nonlinear polarization terms in the form

$$P_{NL}(\boldsymbol{r}, t) = \mathrm{Re} \left\{ \mathscr{P}_{NL}(\boldsymbol{r}, t) \, e^{i(\boldsymbol{k}_\Sigma \cdot \boldsymbol{r} - \omega t)} \right\} \quad [5]$$

where $\mathscr{P}_{NL}(\boldsymbol{r}, t)$ is its envelope that varies slowly over the carrier period $T = 2\pi/\omega$ and $\boldsymbol{k}_\Sigma$ is the vector sum of the wave vectors of all field components that interact to set up the nonlinear polarization source (Eqn [5]). The electric field generated in the same frequency ω can be written in the form

$$\boldsymbol{E}(\boldsymbol{r}, t) = \mathrm{Re} \left\{ \hat{e} \, A(\boldsymbol{r}, t) e^{i(\boldsymbol{k} \cdot \boldsymbol{r} - \omega t)} \right\} \quad [6]$$

where $\boldsymbol{k}$ is related to ω through the dispersion relation

$$k^2 = \omega^2 (\hat{e} \, \varepsilon(\omega) \, \hat{e})/c^2 = \omega^2 \, n_\omega^2/c^2 \quad [7]$$

and in general $\boldsymbol{k} \neq \boldsymbol{k}_\Sigma$; $A(\boldsymbol{r}, t)$ is the envelope of the electric field amplitude and varies slowly over the wavelength $\lambda = 2\pi/k$ and over the period $T = 2\pi/\omega$ and $\varepsilon(\omega) = 1 + 4\pi\chi^{(1)}(\omega)$.

With the slowly varying envelope approximation (SVEA) which is valid for almost all presently realistic cases one derives the equation

$$\frac{\partial A}{\partial z} + \frac{1}{2ik} \Delta_\perp A - \frac{k''}{2i} \frac{\partial^2 A}{\partial \tau^2} = \frac{ik}{2\varepsilon(\omega)} \mathscr{P}_{NL} \quad [8]$$

with $\tau = t - z/v_g$ where the group velocity is defined by $v_g = -(\partial k/\partial \omega)^{-1}$, $k'' = (\partial^2 k/\partial \omega^2) = -(\partial v_g/\partial \omega)/v_g^2$ is its dispersion, $\Delta_\perp = \partial^2/\partial x^2 + \partial^2/\partial y^2$ is connected with diffraction and $\Delta k = k_\Sigma - k$ is the wave vector mismatch in the direction of propagation. For simplicity we have neglected absorption loss which can be taken into account by including an *ad hoc* term αA in the left-hand side of Equation [8] where α is the linear absorption coefficient. In general one is actually faced with a set of coupled equations which, however, can be simplified if one assumes that the incident (pump) beams are undepleted (parametric approximation). The solution can be obtained numerically but in several specific but useful cases one can obtain analytical solutions as well that elucidate several aspects of more general and complex configurations. The role played by the different terms in Equation [8] is evident (diffraction, group velocity dispersion, pulse broadening, etc.) and one can also make provisions to include the effect of anisotropy to account for the beam walk off and other related effects; the paraxial approximation is a good starting approach. The case of focused Gaussian beams has been extensively studied in harmonic generation as it has important implications on the phase matching configuration.

Several aspects regarding the efficiency of the frequency conversion processes can be grasped by restricting the analysis within the stationary and plane wave regime in an isotropic medium in which case Equation [8] reduces to

$$\frac{\partial A}{\partial z} = i \frac{2\pi k}{\varepsilon(\omega)} \mathscr{P}_{NL} e^{i\Delta k z} = i \frac{4\pi}{n\lambda} \mathscr{P}_{NL} e^{i\Delta k z} \quad [9]$$

which can actually be used even for pulses in the range of few nanoseconds. Again linear absorption at the frequency ω has been neglected in Equation [9] but can be easily accounted for as previously indicated.

It is evident from Equation [9] that other factors being equal the efficiency of the conversion process increases with frequency and the strength of the nonlinear polarization source; we also notice in Equation [9] that the conversion starts with a phase jump of $\pi/2$ with respect to the incident field at the boundary. The most drastic impact on the conversion efficiency, however, stems from the presence of the factor $\exp(i\Delta k z)$ in Equation [9]. To appreciate its impact let us assume that the incident 'pump' beam remains undepleted so that $\mathscr{P}_{NL}$ can be assumed to remain unaffected during the conversion process. The solution of Equation [9] is then simply

$$A = \frac{4\pi}{n_\omega \lambda} \mathscr{P}_{NL} e^{i(\pi + \Delta kz)/2} \sin c(\Delta kz/2) \qquad [10]$$

or in terms of the transmitted intensity at $z = L$.

$$I = (cn_\omega/2\pi)|A|^2 \sim (\mathscr{P}_{NL} L/n_\omega \lambda)^2 \sin c^2(\Delta kz/2)$$
$$[11]$$

where $\sin cx = \sin x/x$ and we assumed that $A(0) \approx 0$ which is equivalent to neglecting the reflected beam; a more rigorous approach shows that the latter is reduced by a factor $(\Delta k/k)^2$ with respect to the transmitted beam. It is quite evident that $A(z)$ is an oscillating function of z with a period $l_c = \pi/\Delta k$, the coherence length, with its amplitude scaling as $(1/\Delta k)$ as long as $\Delta k \neq 0$: the larger the wave vector mismatch Δk the smaller the amplitude A and the coherence length l_c and the conversion efficiency is dramatically reduced. For high conversion rates one must either achieve $\Delta k = 0$ namely $k_\Sigma = k$ along the propagation direction or artificially compensate for Δk. Both approaches are presently used for the most commonly exploited case of second harmonic generation but for more complex cases only the former is exploited.

Once the phase matching is achieved one can easily see from Equations [10] and [11] that $|A|$ will grow as z (or equivalently $I \sim z^2$) at the initial stage of the conversion process but this cannot continue indefinitely because of the compound impact of pump depletion and back reaction of the created beam on the incident pump beams. This problem has been extensively studied in the case of second harmonic and sum frequency generation where analytical solutions in terms of elliptic integrals can be obtained. In principle, conversion close to 100% for the fundamental to the harmonic can be achieved over a propagation length of the order of a few interaction lengths $L_i = (n_1^2 n_2^2 / 8\pi^3 \omega^2 \chi_e^{(2)} I_p)^{1/2}$; more precisely for the simplest case of second harmonic generation in the phase matched configuration one finds

$$I_2 = I_1 \tan^2 h(L/L_i) \qquad [12]$$

where I_1 is the intensity of the fundamental. Similar considerations apply for the sum frequency generation, difference frequency generation and other second order parametric processes. These considerations for second order processes are also qualitatively

corroborated by the Manley–Rowe conservation relations for the photon fluxes valid in the transparency range of any medium. The previous considerations valid for plane wave have also been numerically extended to Gaussian and Bessel beams.

Wave vector mismatch

From the previous general considerations it is quite evident that among the different factors that affect the frequency conversion efficiency the wave vector mismatch $\Delta k = k_\Sigma - k$ has by far the most dramatic effect and the achievement of phase matching, namely the cancellation of Δk, is an indispensable condition for a nonlinear beam configuration to be considered for frequency conversion. Parenthetically we point out here that the phase mismatch is a manifestation of nonlocality in electromagnetic propagation; in the phase matched configuration this nonlocality is suppressed and the interaction effectively becomes local. Because of the key role played by the phase matching in all frequency conversion processes we shall outline and classify below the main phase matching schemes. These fall in either of the following two classes:

(i) compensation of the refractive index dispersion, for instance by exploiting the natural or artificial birefringence of the nonlinear optical material to compensate for the refractive index mismatch so that one can achieve $k_\Sigma = k$ or equivalently $\Delta k = 0$.

(ii) artificial compensation of the wave vector mismatch by introducing an appropriate spatial (or eventually spatiotemporal) modulation of the nonlinear polarization source with a grating period $\Lambda = 2ml_c = 2m\pi/\Delta k$ where $m = 1, 3, 5$, etc.

Dispersion compensation

This is the first experimentally implemented scheme for phase matching. It is the most widely used because it is flexible and can be extended to a wide class of nonlinear interactions. It relies on the fact that the two polarization eigenstates of an electromagnetic wave propagating in a given direction inside a birefringent medium experience different refractive indices (different phase velocities). One may then chose the propagation configuration (polarization states, crystal orientation, etc.) of the incident and created beams so that the birefringence compensates the dispersion. This also sets the main restriction on the use of this method the others being the crystalline symmetry, the transparency range, etc.

Because of these restrictions, at present only linearly birefringent (or optically anisotropic) media

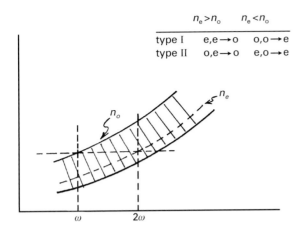

Figure 2 Phase-matching in a birefringent crystal by dispersion compensation, showing the two types of phase-matching for second order processes.

are used for phase matching over restricted spectral regions. For second-order effects one distinguishes the type I and II phase matching configurations. We recall that in such media for each propagation direction there are two initially orthogonal linear polarization eigenstates with different refractive indices, the one direction that is angle independent (o, ordinary) and the other direction that is angle dependent (e, extraordinary); the principle of phase matching is depicted in **Figure 2**, where the two types are defined. Natural linear birefringence is the most widely used but in some cases externally induced linear birefringence can also be used for finer tuning (for instance electrooptic or Pockels effect).

Several well-characterized crystals for frequency doubling or frequency sum up to 0.25 μm which corresponds to the fourth harmonic of the YAG laser frequency (1.053 μm), are proposed by crystal manufacturers with outstanding conversion efficiencies. This upper frequency limitation around 0.25 μm results from the onset of absorption above this frequency and concomitant heating in practically all efficient doubling crystals; there are similar limitations towards the mid- and far-IR spectrum because of strong absorption bands there due to IR active lattice modes (phonons, vibrations) and one resorts here to other nonlinear processes and coupling schemes.

In principle, circular birefringence, natural (rotatory power or optical activity) or artificial (Faraday effect), can also be used but circular birefringence is in general very weak in compensating dispersion and no useful results have been obtained yet.

For frequency conversion above 0.25 μm the previous schemes cannot be used because as previously stated in all presently available nonlinear crystals

absorption sets in above this frequency with concomitant heating of the crystal that can provoke beam instabilities and crystal damage. Here one proceeds instead with multiple (odd order) harmonic generation in gases, actually a two-component gas mixture, and exploits the anomalous dispersion to find coincidences of the refractive index for the fundamental and its multiple odd order harmonic; by controlling the relative concentration of the two gas components the dispersion can be 'tuned' to reach such a coincidence. The conversion efficiency is certainly lower than in crystals (<10%) and one needs long gas cell dimensions but in contrast to the previous cases one can now use very high beam intensities and in addition approach resonances to enhance the conversion efficiency; an additional advantage is the inherent isotropy of the medium. The beam focusing configuration plays a key role here and has been extensively studied for Gaussian beams and to some extent Bessel beams.

Wave vector mismatch compensation

This is actually the first proposed phase matching scheme but its implementation in practice had to wait the development of appropriate artificial crystal growth techniques; it has emerged as the most efficient and robust technique for the development of compact devices that can be integrated with a laser, a situation that could not be envisaged with the previously described class of phase matching schemes. This technique, which nowadays goes under the denomination of quasi-phase-matched (QPM), has now been applied for second harmonic generation but its extension in other situations involving second-order effects (frequency sum/difference) can also be considered.

In its simplest version the QPM technique for second harmonic generation (SHG) consists in alternatively reversing the polar axis direction in successive thin layers of the same thickness d of a crystalline material. This leads to a reversal of sign of $\chi^{(2)}$ in successive layers of the layered material and one can easily see that if $d = ml_c = m\pi/\Delta k$ then the conversion efficiency increases with the number of layers. This actually amounts to periodically modulating $\chi^{(2)}$ in the nonlinear optical source $\mathcal{P}^{(2)} = \chi^{(2)}EE$ in Equation [9] with a wave vector that precisely matches and cancels Δk in the propagation direction; note that the linear refractive index is unaffected and remains the same as in the bulk crystal while I_2 grows as z^2 (**Figure 3**). In a certain sense the quasiphase-matched systems can also be viewed as a nonlinear photonic crystal.

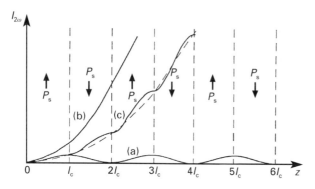

Figure 3 Wave vector compensation by quasi-phase matching (a) non-phase matched, (b) quasi-phase matched and (c) phase matched.

This technique has already been applied with outstanding results using certain ferroelectrics (domain reversal either by poling or doping) and holds much promise for the development of compact integrated lasing and doubling devices. It can also be used with poled polymers but their transparency range is not as extended as in the ferroelectrics. Attempts are being made to use this technique with cubic heteropolar semiconductors, ex:GaAs, but the use of molecular beam epitaxy techniques here to make the layered structure is costly and time consuming because of the required layer thickness. The technique can also be extended to cascading of SHG processes for higher frequency conversion within the transparency range of the material.

Nonlinear optical materials

Along with the phase matching problem the choice of the nonlinear material for frequency conversion (essentially SHG) has a major impact on the efficiency and robustness of the process. To the extent that cascading second-order effects can effectively and practically cover all cases of third-order processes as well, attention nowadays is being concentrated on non-centrosymmetric materials, inorganic or organic, with particular emphasis on oxygen polyhedra based inorganics such as $LiNbO_3$, $BaBO_4$, (BBO), $LiIO_3$, KDP, ADP, KTP, $KNbO_3$ and derivatives by substitution or specific semiconductors such as $AgGaGe_2$, $ZnGeP_2$ etc., certain organic crystals (urea, MAP, POM, MNA, NPP) or poled polymers. In view of the widespread uses of such materials not only in frequency conversion but also in light modulation and other optoelectronic applications in devices, and the need of their artificial growth, many factors must be taken into consideration when making the choice of a particular material, their

nonlinear properties per se not being the sole criterium. Aspects such as optical anisotropy and transparency range, multiphoton losses, threshold of optical breakdown, photochemical stability and mechanical and thermal properties are important in the choice of the appropriate material, as are the material growth technique, doping and ageing considerations, processability, interfacing and packaging and other aspects since for a wide class of materials the nonlinear figures of merit are of comparable magnitude. **Table 1** gives indicative values of the nonlinear optical coefficients of some crystals.

Cavity enhanced frequency conversion

The efficiency of frequency conversion processes can be substantially enhanced and the characteristics of the generated beams drastically improved by an appropriate choice of the geometry of the interaction configuration. In this context the insertion of the frequency conversion process in a cavity or guide can have a major impact. Different schemes have been used for second-order processes, such as intracavity frequency doubling, external cavity frequency doubling, quasi-phase-matched doubling and optical parametric oscillation. In all these cases one essentially exploits the fact that the power inside a cavity can be much higher than outside and the effective interaction length very long and this enhances the efficiency of the conversion process; together with other factors these substantially improve the generated beam quality. These schemes are briefly described below.

Intracavity SHG

The doubler crystal is inserted into the laser cavity with the output coupler replaced by a mirror that is highly reflective/transmittive in the fundamental/harmonic respectively. With appropriate choice of the transmittivity and coupling parameter the fundamental can be completely converted into the harmonic. High quality nonlinear optical crystals are required here and specific optical arrangements must be inserted to eliminate spatial hole-burning, amplitude fluctuations, etc. Clearly the ultimate goal here could be to compound the lasing gain and doubling processes in one and the same crystal, which requires appropriately doped nonlinear optical crystals such as rare earth doped ferroelectrics or dye doped nonlinear organic crystals or poled polymers. Some interesting results have already been achieved in this direction.

Table 1 Second-order nonlinear coefficients of some crystals

Crystal	Symmetry	n_0	d_{21}	d_{14} $(10^{-12}$ m V$^{-1})$ [a]	d_{33} $(10^{-12}$ m V$^{-1})$ [a]	d_{eff}	6	I_{ob} (GW cm^{-2}) [b]	Transm. (μ) [c]
LiNbO$_3$	3m	2.232	−2.1	0	−27	5.1	70	10	0.35–5
BaBO$_4$	3m	1.655	−2.3	0	0	1.9	16	14	0.2–2.6
LiIO$_3$	6m	1.857	0	0	4.5	1.8	13	2	0.34–4
KDP	$\bar{4}$2m	1.493	0	0.37	0	0.35	1	5	0.18–1.8
ADP	$\bar{4}$2m	1.509	0	0.47	0	0.39	1.2	6	0.18–1.5
AgGaGe$_2$	$\bar{4}$2m	2.594	0	33	0	28	81	0.3	0.78–18
ZnGeP$_2$	$\bar{4}$2m	3.073	0	69	0	70	292	0.05	0.74–12
KTP	mm2	1.737	0	0	8.3	3.2	47	15	0.35–4.5
KNbO$_3$	m2	2.119	0	0	−19.5	−11	312	7	0.4–5.5
					d_{11}				
Urea					12			5	0.2–1.4
MAP		1.508	16.8					3	0.5–2.5
POM		1.663		9.2				2	0.5–1.7
MNA		2.0			168		1000		0.48–2.0
NPP					85			0.05	0.48–2.0

[a] Divide each value by 4.2×10^{-4} to convert into esu.
[b] Optical breakdown.
[c] Transmission range.

External cavity SHG

For low gain lasers there is an advantage in using an external optical cavity configuration instead of the intracavity one as this also permits one to independently optimize the laser and doubling cavities. The main problems in this scheme are the requirements of frequency matching between the laser frequency and the doubling cavity resonance on one hand, and that of 'impedance matching' on the other. Excellent performances have been achieved here with monolithic cavities fabricated from a single piece of doubler crystal with mirrors directly coated on the crystal surfaces (other features have also been integrated to improve the efficiency and quality of the process). Quasi-phase-matched configurations along the lines previously discussed have also been considered here.

Optical parametric oscillators (OPOs)

This is an alternative solution to a tunable laser source in the optical frequency range. Here a powerful single frequency radiation (pump) is parametrically converted into equally powerful coherent radiation tunable over a wide optical frequency range (signal and idler). In its simplest version, which is also the most useful one, it is based on the optical parametric amplification of 'noise' photons of frequencies ω_1 and ω_2 provided by the dissociation of a pump photon of frequency ω_p such that $\omega_p = \omega_1 + \omega_2$, the selection of a particular frequency pair being made through the phase-matching condition

$k_p = k_1 + k_2$. Thus any one of the previous phase-matching schemes, combined with additional features regarding the cavity configuration, mirrors etc., can be used to select a particular frequency for the output radiation (or its complementary) and also to improve the output beam quality. The present surge in the development of OPOs stems from the improvements in pump beam quality and nonlinear crystal performances.

Optical parametric oscillators with continuous wave (CW) or pulsed regime operation are presently available with outstanding performances for the most demanding spectroscopic studies; particularly impressive has been the development of OPOs operating on the few-femtoseconds pulse regime.

Stimulated scattering processes – frequency shifters

The previous nonlinear optical processes for frequency conversion are coherent and essentially instantaneous as they do not involve any dynamics of material excitations and all energy transfer and storage only occur within the electromagnetic field modes; material resonances eventually serve to enhance the efficiency of the process.

There is a whole class of nonlinear optical processes involving real multiphoton transitions inside the medium where the incident beams are converted into beams of different frequencies at the same time as the material system undergoes a transition between two different energy levels and here the dynamics (relaxation) of this transition play a key role. Several such schemes can be conceived and exploited for frequency conversion, and they can be classified in two categories: stimulated scattering processes and up-conversion lasing processes. **Figure 4** shows some characteristic configurations. A characteristic feature of all these processes is that they exhibit an exponential growth with the spatial extension of the process as in the case of lasing processes and in contrast on the previously discussed frequency conversion schemes that exhibit a power law dependence on the interaction distance. In fact in all these schemes the generated beams are built up on spontaneous processes and can be viewed as generalized lasing processes that exploit multiphoton pumping schemes.

Stimulated Raman frequency shifter

This is the simplest and most used scheme of stimulated scattering process in which an incident wave at frequency ω is converted into a scattered wave at frequency ω_s, the difference in photon energy $\hbar(\omega_i - \omega_s) = \pm\hbar\omega_R$, being taken up (Stokes component) or supplied (anti-Stokes component) by the nonlinear medium, which undergoes a transition between two energy levels E_a and E_b such that $\omega_R = |E_a-E_b|/\hbar$. Various types of such transitions can be involved, such as electronic ones, vibrations, rotations etc., namely the same ones that give rise to spontaneous light scattering, the extreme case being that of Brillouin and Rayleigh scattering with negligible or no frequency shift. Actually the spontaneous process provides the 'seeding photons' from which the stimulated process builds up. The amplitude of the generated beam and the gain in the stationary regime, can be derived from Equation [9] after insertion of the appropriate nonlinear polarization source,

$\mathcal{P}^{(3)} = \chi_R^{(3)} E_L E_L^* E_s$ where $\chi_R^{(3)}$ is the $|\omega_i - \omega_s - \omega_R|$ - resonant part of $\chi^{(3)}(\omega_L, -\omega_L, \omega_s)$ or $\chi_R^{(3)} = i\chi''$ and χ'' is positive/negative for the Stokes/anti-Stokes component. The process is in fact automatically adjusted to satisfy the phase matching condition. Clearly the Stokes component has a gain and grows exponentially with the interaction distance as in the case of lasing action.

In the stimulated Raman Stokes scattering process the frequency shift ω_R is characteristics of the material medium, and by changing the medium new coherent sources with different down-shifted frequencies can be obtained. If the conversion to the Stokes wave is appreciable, generation of multiple Stokes waves $\omega-\omega_R$ can take place by a cascading process, the first one acting as the pump for the next one and so on. Although the Raman jumps are discrete, by using an initially tunable source one can have a tunable output as well one shifted en bloc by the Raman frequency. Note that because the process involves a material transition (excitation) its dynamics are conditioned by that transition. The previous considerations essentially concern the stationary regime where the pulse length is larger than the characteristic times of the material resonance, otherwise the transient regime must be used with much lower conversion efficiencies and other complications. This restricts the use of stimulated Raman frequency conversion technique to a certain pulse length regime.

Although the direct anti-Stokes process shows no gain from the outset, the generated Stokes field can, by interacting with the pump field, generate the anti-Stokes field through a phase matched four-wave mixing process. Here again one can have multiple anti-Stokes coherent generation through cascading processes. Because the phase-matching condition must be satisfied at each step the anti-Stokes output is generated in the form of cones around the direction of the Stokes output, which coincides with the direction of maximum gain or equivalently that of largest interaction distance between pump and Stokes. Actually the build up of the anti-Stokes field at the expense of the Stokes via a four-wave interaction process that requires the phase-matching to be satisfied is the simplest case of an 'optical balance'. In an 'optical balance' either configuration of two nonlinear optical processes sharing the same multiphoton resonance can be enhanced at the expense of the other by exploiting a phase matching condition: this is precisely the case of the Stokes/anti-Stokes generation.

Although the stimulated Raman conversion techniques have been superseded by others they are still useful in many instances, in particular because of their easy and economical implementation. One typically reaches 20% conversion efficiencies, or

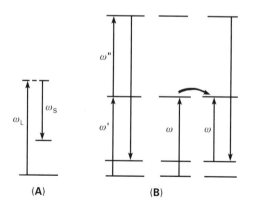

Figure 4 Main schemes of nonlinear stimulated processes for frequency conversion. (A) Stimulated Raman and (B) up-conversion lasing schemes.

even higher, but then the pump depletion must be taken into account.

Upconversion lasers

This is the generic denomination of a large class of stimulated lasing emission schemes between two widely separated energy levels E_a and E_b, with $(E_b - E_a)/\hbar$ in the UV or higher ($E_b > E_a$), where population inversion is reached by appropriate real multiple step transitions involving several photons of energy $\hbar\omega_i$ lower than $E_b - E_a$, usually provided by a powerful infrared pump. The key point here is that intermediate metastable states are strongly populated and used as a reservoir for the pump energy. By appropriate choice of the metastable states with regard to the transition oscillator strength and relaxation times a population inversion can be achieved between two states with widely spaced energies E_a and E_b or $E_b - E_a > \hbar\omega_i$ and subsequent lasing emission between them occurs when inserted in a cavity.

To the extent that the pump photon energy is stored at the intermediate metastable levels, which are then strongly populated, phase matching considerations are in general irrelevant in these schemes as in all lasing schemes, although in some cases 'optical balance' configurations may take place and population build-up may be suppressed if a phase matching condition is reached that results in a suppression of the population inversion.

There is a wide range of processes that can be exploited for up-conversion lasing, some of them involve transitions and levels within one atom or molecule but most frequently these schemes evolve through cooperative processes that involve two atoms or molecules. Many of these schemes in fact rely on the cooperative transitions originally predicted for impurity atom (or ion) pairs in crystals and subsequently extended in gases in the so-called laser-assisted van der Waals collisions and other similar schemes.

These processes in doped crystals, glasses, fibres or gases have been used to provide up-conversion schemes where IR laser radiation is converted into UV radiation with good efficiency. The main drawback of several of these schemes is the need of low temperature instrumentation because of the crucial role played by the line widths, relaxation and collision processes.

Terahertz radiation

Much effort is presently concentrated to extend the nonlinear frequency conversion techniques to cover the far-IR or terahertz region, roughly the spectral region 0.1–10 THz, where along with the XUV and X-ray regions to be discussed below there is a lack of coherent sources. In the past attention was concentrated on the frequency difference generation in noncentrosymmetric crystals the extreme case being the optical rectification effect. In several other cases a frequency down-conversion through a succession of stimulated Raman processes such as spin-flip or vibrational–rotational in cascade in various gases has been exploited but the efficiency drastically drops with the generated frequency unless one compensates this reduction with intermediate resonant enhancement. The best results were obtained with a combination of stimulated vibrational and rotational scattering in molecular gases which, however, cannot lead to compact far-IR sources.

Recent efforts to generate THz radiation have been essentially based on the optical rectification effect in ferroelectrics or the generation of transient photogenerated currents in semiconductors, both with femtosecond pulses; some variants of these approaches, also termed terahertz time-domain spectroscopy (THz-TDS), have produced very promising results and open the way to the development of compact coherent sources that emit pulses in the THz-region with durations close to the single terahertz optical cycle. Many technical problems, however, still remain before these schemes can be implemented in reliable devices for tunable THz-radiation.

Extreme nonlinear optics – X-ray generation

In the previously discussed frequency conversion schemes the laser field amplitude E_L is well below the cohesive field intensity E_c so that the perturbative approach, as exemplified by the power series development Equation [1], can be used to single out the appropriate nonlinear optical process for frequency conversion. In particular the order of the nonlinear process that contributes is well identified. If E_L exceeds E_c, which is typically of the order of 10^{10} V cm^{-1}, one may expect photoionization to take place, liberating electrons from the neutral atoms with concomitant depletion of the frequency conversion and irreversible damage to the material through optical breakdown. With presently available commercial amplified fs-lasers the intensities can easily be in the range of 10^{17} W cm^{-2} and with appropriate arrangements the pulse duration can be as short as the light cycle, typically 2–3 fs in the visible. The photo-ionization is initiated by the suppression of the Coulomb potential through the strong electric laser field, allowing the electrons to tunnel out of the atomic core within the oscillation period of the laser field. The full ionization of the atom can, however, be avoided if the intense light pulse is short enough so that the electrons are not fully liberated from the atomic core. Rather, the quasi-free electron is accelerated in the laser field and subsequently may re-encounter and recombine with the parent ion, emitting a photon, or collide with the surrounding atoms, causing electron ejection and photon emission through the inverse bremsstrahlung process. The two processes are laser polarization state dependent (linear and circular respectively). Both processes can lead to compact coherent X-ray sources although still many technical problems are to be surmounted before such sources become reliable for spectroscopic studies or other applications.

In this regime of extremely intense optical fields where the effective interaction time is close to the single light period and irreversible damage is avoided, the power series expansion [1] breaks down as the different terms acquire comparable magnitude. Non-perturbative approaches have been developed to treat this regime, predicting the generation of X-ray pulses from gas targets illuminated with intense ultrashort laser pulses. Several pulse and beam characteristics have a key impact on the efficiency of the X-ray generation process. Phase matching plays an important role in the process as also does the pulse duration and pulse phase. Because of the symmetry of the interaction, odd order harmonics are present in the emitted spectrum with a strength that drastically decreases above a given order.

The results obtained with rare gases and also some metallic surfaces are very promising and open the way to a generation of compact coherent XUV and X-ray radiation with easily accessible femtosecond pump sources.

Future prospects

The progress in frequency conversion techniques has been tremendous since the first observation of second harmonic generation in the early 1960s. This article gives only a rough picture of this progress which still continues. At present one can state without exaggeration that starting from a high quality primary laser that is tunable in a narrow region of the optical range and using the previously discussed optical frequency conversion schemes one can cover the whole spectral range from 10–0.1 μm with outstanding characteristics regarding the generated radiation, such as frequency tunability and stability, beam intensity and quality, pulse duration and shape. These schemes are now being implemented in compact and robust devices that greatly facilitate and widen their uses in different areas and in spectroscopy in particular. The

introduction of all-solid-state pump sources will greatly facilitate this trend.

This trend will continue unabated as the demand for tunable and versatile coherent light sources is continuously growing in many areas. In the near future important breakthroughs are expected in the two extreme frequency ranges, i.e. the far-IR and THz spectral ranges and the XUV and X-ray ranges. The expected applications in these frequency ranges are very important as are also the fundamental studies that such sources will generate.

List of symbols

A = amplitude, d = thickness of crystalline material, D_n = degeneracy factors, E_a, E_b = energy levels, E_c = cohesive electric field intensity, E_L = laser field amplitude, I_1 = intensity of the fundamental, k_Σ = vector sum of the wave vectors, l_c = coherence length, L_i = interaction length, P = polarization induced by an incident electric field, T = carrier period, v_g = group velocity, α = linear absorption coefficient, $\chi^{(n)}$ = dipolar susceptibility of order n, Λ = grating period, ω = frequency, ω_s = scattered wave frequency, $\hbar$ = Planck's constant/2π.

See also: **Laser Applications in Electronic Spectroscopy; Laser Magnetic Resonance; Laser Microprobe Mass Spectrometers; Laser Spectroscopy Theory; Nonlinear Optical Properties; Nonlinear Raman Spectroscopy, Theory; X-Ray Spectroscopy, Theory.**

Further reading

Armstrong JA, Bloembergen N, Ducuing J and Pershan P (1962) *Physics Review* **127**: 1918.

Auston DA, Glass AM and Ballman AA (1972) *Physics Review Letters* **28**: 897.

Bordui F and Fejer MM (1993) *Annual Reviews Material Science* **23**: 321.

Digiorgio V and Flytzanis C (eds) (1994) *Nonlinear Optical Materials. Principles and Applications.* Amsterdam: North-Holland.

Franken PA, Hill AE, Peters CW and Weinreich G (1961) *Physics Review Letters* **7**: 118.

Giordemaine J (1962) *Physics Review Letters* **8**: 19.

Grischkowsky D, Keiding S, van Exter M and Fattinger C (1990) *Journal of the Optical Society of America* **B7**: 2006.

Krausz F, Brabec T, Schnürer M and Spielman C (1998) *Optics and Photonics News*, July.

Maker PD, Terhune RW, Nisenoff M and Savage CM (1962) *Physics Review Letters* **8**: 21.

Miles RB and Harris SE (1971) *Applied Physics Letters* **19**: 385.

Mittleman D (1998) *Laser Focus World*, May issue, p 191.

Rabin H and Tang CL (eds) *Quantum Electronics: A Treatise*, Volumes 1 and 2. New York: Academic Press.

Reintjes J (1984) *Nonlinear Optical Parametric Processes in Liquids and Gases.* New York: Academic Press.

Shen YR (ed) (1977) *Nonlinear Infrared Generation.* Berlin: Springer Verlag.

Shen YR (1984) *Principles of Nonlinear Optics*, p 46. New York: Wiley.

Simon U and Tittel FK (1997) In *Experimental Methods in the Physical Sciences*, Vol 29C.

Yablonovitch E, Flytzanis C and Bloembergen N (1972) *Physics Review Letters* 29: 865.

Yariv A (1990) *Quantum Electronics*, Ch 9. New York: Wiley, New York: Academic Press.

Optical Spectroscopy, Linear Polarization Theory

Josef Michl, University of Colorado, Boulder, CO, USA

ELECTRONIC SPECTROSCOPY
Theory

The degree of linear polarization of light arriving at a detector from a sample contains information on (i) the anisotropy of molecular optical properties and (ii) the molecular orientation distribution. In the following, we outline a general theoretical framework for extracting this information, equally useful for electronic and vibrational spectroscopy. We apply it to processes that involve the interaction of one photon or two photons (successively, as in photoluminescence and photoinduced dichroism; or simultaneously as in two-photon absorption and Raman scattering) with an isotropic or partially aligned solution of molecules that interact with light one at a time. We emphasize uniaxial assemblies (e.g. solutes in an electric field, in nematic liquid crystals, in stretched polymers, or in membranes, and solutes 'photoselected' with linearly polarized light).

Probabilities of optical events

Amplitudes of molecular optical events are proportional to off-diagonal matrix elements of interaction operators between the wavefunctions of the initial, final, and possibly also intermediate states of the molecule, $|0\rangle$, $|f\rangle$, and $|j\rangle$ respectively. These operators are projections of molecular transition vector and tensor operators onto the polarization directions of photons created or annihilated in the event (**Table 1**). The amplitudes depend on the wavenumber of the light used and can be real or complex. The probability of an optical event W is proportional to the square of the absolute value of its amplitude (**Table 2**). The proportionality constant is of no

consequence in polarization spectroscopy, which deals with ratios of probabilities observed for the same collection of molecules under different conditions of light polarization. In the ratios, the proportionality constant cancels (except in strongly birefringent solvents, for which corrections are necessary).

Polarized light

We label the laboratory system of axes X, Y, Z. Linearly polarized light (electromagnetic radiation) is characterized by a real unit vector ε^U oriented along the direction U of its electric field, and perpendicular to the propagation direction. Circularly polarized light is characterized by a complex unit vector. For right-handed light (Z component of photon angular momentum equal to $-\hbar$) propagating in the positive direction of Z, the polarization vector is $(\varepsilon^X + i\varepsilon^Y)/\sqrt{2} = \varepsilon^+$, and for left-handed light, it is $(\varepsilon^X - i\varepsilon^Y)/\sqrt{2} = \varepsilon^-$. Circularly polarized light is essential in chiroptical spectroscopy, and important in two-photon and Raman spectroscopy.

Molecular transition moments and spectra

In electronic and vibrational spectroscopy we can neglect both molecular dimensions relative to the wavelength and the effects of the magnetic field of light relative to those of its electric field (electric dipole approximation). Then, the interaction of a single U-polarized photon with a molecule is described by the projection of the electric dipole moment vector operator M (**Table 3**) into ε^U (photon creation) or ε^{U*} (photon annihilation). Creation

Table 1 Amplitudes of molecular optical events[a]

1. One-photon events (photon polarization: U)

Annihilation	Creation
$\varepsilon^{U*} \cdot M(0f)$	$\varepsilon^{U} \cdot M(f0)$

2. Successive two-photon events (photon polarizations: U, V)

Annihilation + annihilation	Annihilation + creation
$\varepsilon^{U*} \cdot M(0j) M(j' f) \cdot \varepsilon^{V*}$	$\varepsilon^{U*} \cdot M(0j) M(j' f) \cdot \varepsilon^{V}$

3. Simultaneous two-photon events (photon polarizations: U, V)

Annihilation + annihilation	Annihilation + creation
$\varepsilon^{U*} \cdot T(j,f) \cdot \varepsilon^{V*}$	$\varepsilon^{U*} \cdot \alpha'(j,f) \cdot \varepsilon^{V}$

[a] M is the electric dipole transition moment vector; T is the two-photon absorption tensor; α' is the Raman scattering tensor; and ε^{U} is a unit vector in the direction of light polarization U.

Table 2 Probabilities of molecular optical events

1. One-photon events (tensor O, photon polarization: U)

Absorption	Emission
$[M(0f)]^{U*}[M(0f)]^{U}$	$[M(0f)]^{U}[M(0f)]^{U*}$

2. Successive two-photon events (tensor $^{(4)}O$, photon polarizations: U, V)

Photoinduced dichroism	Photoluminescence
$[M(0j)]^{U*}[M(j' f)]^{V*}$	$[M(0j)]^{U*}[M(j' f)]^{V}$
$[M(0j)]^{U}[M(j' f)]^{V*}$	$[M(0j)]^{U}[M(j' f)]^{V*}$

3. Simultaneous two-photon events (photon polarizations: U, V)

Two-photon absorption	Raman
$[T(0f, \tilde{v}_1)]^{UV*}[T(0f, \tilde{v}_1)^*]^{UV}$	$[\alpha'(0f, \tilde{v}_1)]^{UV}[\alpha'(0f, \tilde{v}_1)^*]^{UV*}$

and/or annihilation events involving two photons are described by projections of tensor operators of rank two, etc. (**Tables 1–3**). The off-diagonal matrix elements of M between molecular states, $M(0f)$, are the (electric dipole) transition moments. The analogous matrix elements of tensor operators T and α' (**Table 3**) are the two-photon absorption tensor $T(0f, \tilde{v}_1)$ and the Raman scattering tensor $\alpha'(0f, \tilde{v}_1)$.

The one-photon events in question (**Table 1**) are (i) the annihilation of a photon of wavenumber $\tilde{v}_1$ by the initial state 0 to yield an excited state f of energy $\tilde{v}_1$ (ordinary absorption), and (ii) the creation of a photon of wavenumber $\tilde{v}_1$ by an excited state f to yield a lower-energy state 0 (luminescence). The simultaneous two-photon events that we consider are (i) annihilation of two photons of wavenumbers $\tilde{v}_1$ and $\tilde{v}_2$ by the initial state 0 to yield an excited state of energy $\tilde{v}_1 + \tilde{v}_2$ (two-photon absorption), and (ii) annihilation of photon of wavenumber $\tilde{v}_1$ and creation of a photon of wavenumber $\tilde{v}_2$ to yield a (usually vibrationally) excited state of energy $\tilde{v}_1 - \tilde{v}_2$ above the initial state 0 (Raman scattering). The successive two-photon events that we consider are (i) annihilation of a photon of wavenumber $\tilde{v}_1$ that converts the initial state 0 to an intermediate state j followed later by the annihilation of a photon of wavenumber $\tilde{v}_2$ that converts the intermediate state

to the final state f (photoinduced dichroism), and (ii) annihilation of a photon of wavenumber $\tilde{v}_1$ that converts the initial state 0 to an intermediate state j followed later by the creation of a photon of wavenumber $\tilde{v}_2$ that converts the intermediate state to the final state f, often another (or even the same) vibrational level of the initial state 0 (photoluminescence). For both of these, the intermediate state j may develop into state j' during the interval between the first and the second event: the molecule may undergo internal conversion or intersystem crossing to another electronic state, it may vibrationally relax, it may change its conformation or chemical structure, etc. Especially important types of change are molecular rotation and the transfer of excitation energy to another molecule of the same kind, which may be differently oriented. For both compound events the tensor operator M is the direct product of the electric dipole moment operator M with itself, $M = MM$ (**Table 3**).

Optical transitions are associated with line shapes. The overall spectrum of a molecule is given by a sum of contributions provided by transitions from the initial state 0 to all possible final states f plotted against $\tilde{v}_1$, $\tilde{v}_2$, or in two dimensions against $\tilde{v}_1$ and

Table 3 Definition of transition moment vectors and tensors[a]

Electric dipole transition moment vector[b]

$$M(0f) = \langle f | -|e| \sum_{l=1}^{n} r_l + |e| \sum_{k=1}^{N} Z_k R_k | 0 \rangle$$

Two-photon absorption tensor[c]

$$T(0f, \tilde{v}_1) = \sum_j [M(0j)M(jf)/(\tilde{v}_j - \tilde{v}_1) + M(jf)M(0j)/(\tilde{v}_j - \tilde{v}_2)]$$

Raman scattering tensor[d]

$$\alpha'(0f, \tilde{v}_1) = \langle f | \sum_j [M(0j)M(0j)/(\tilde{v}_j + \tilde{v}_1) + M(0j)M(0j)/(\tilde{v}_j - \tilde{v}_1)] | 0 \rangle$$

[a] For transitions from an initial vibronic state 0 to the final state f. Z_k is the atomic number of nucleus k, e is the charge of the electron, n is the number of electrons, N is the number of nuclei, r_l is the position vector of the lth electron, and R_k is the position vector of the kth nucleus.

[b] In vibrational transitions, 0 and f belong to the same electronic state and differ in the vibrational part, whereas in electronic transitions, they describe different electronic states.

[c] The sum is over all electronic states of the molecule j; $\tilde{v}_1 + \tilde{v}_2 = \tilde{v}_f$, where $\tilde{v}_f$ is the wavenumber of excitation to the final state f.

[d] Here, 0 and f on the outside refer to the initial and final vibrational wavefunctions, $\tilde{v}_1$ is the excitation wavenumber, v_j is the wavenumber of excitation to state j, the integration in the evaluation of $M(0j)$ is only over the electronic degrees of freedom, and the summation is over all electronic states of the molecule j.

$\tilde{v}_2$. In vibrational spectra, the lines are usually sufficiently narrow that the contributions of the individual transitions do not overlap. It is then easy to measure the ratio of intensities with which any one line appears for various choices of photon polarization U (or U and V). When contributions from transitions provided by two or more excited states f overlap, as often happens in electronic spectra, it is sometimes still possible to derive the ratios of spectral intensities for different choices of U (or U and V) using the stepwise reduction procedure.

Polarized intensity ratios equal the ratios of the probabilities listed in **Table 2**, and reveal the macroscopic optical anisotropy of a sample. To derive molecular optical anisotropy, the molecular orientation distribution in the laboratory system of axes needs to be considered.

Description of molecular alignment

Polarization spectroscopy is most commonly performed on three types of samples: (i) isotropic fluid or rigid solutions (only for two-photon and higher-order processes); (ii) uniaxial partially aligned solutions possessing a unique direction (Z), with all directions perpendicular to it equivalent (any orthogonal pair can be chosen for X and Y); and (iii) crystals, which contain molecules all oriented in the same way or in a small number of distinct ways. Isotropic samples are the easiest to study, since they do not change the state of polarization of light that propagates through them. This is also true of uniaxial samples if the electric vector of the light is either parallel or perpendicular to Z (other orientations are best avoided). High-symmetry crystals, e.g. cubic, also satisfy this condition, but most low-symmetry crystals do not. Crystals are otherwise generally easier to handle since no or little orientation averaging is involved, but their spectra are often complicated by strong intermolecular interactions, absent in dilute solutions. In the following, we treat dilute isotropic or uniaxial samples with light polarization direction U (and/or V) along X, Y or Z.

To describe molecular orientation in a large assembly, we choose an arbitrary set of molecular axes x', y', z' associated rigidly with the molecular framework (the primes indicate the arbitrariness of the choice). The orientation of a molecule in the sample is then described by values for the three rotational degree of freedom (e.g., the three Euler angles α', β', γ') that convert the laboratory directions X, Y, Z into the molecular directions x', y', z', and denote them collectively by Ω'. The probability that a molecule has an orientation between Ω' and $\Omega' + d\Omega'$ is $f(\Omega')d\Omega'$. The orientation distribution function $f(\Omega')$ is normalized and permits the calculation of the orientation average of any angle-dependent quantity $a(\Omega')$:

$$\int f(\Omega')\,\mathrm{d}\Omega' = 1$$
$$\int a(\Omega')f(\Omega')\,\mathrm{d}\Omega' = \langle a \rangle \qquad [1]$$

In uniaxial samples, $f(\Omega')$ has the same value for all angles α' and depends only on β' and γ. In the special case of molecules for which all angles of rotation about their own z' axis are equally likely, $f(\Omega')$ also has the same value for all angles γ.

A redundant but convenient set of quantities that specify the orientation of the molecular x', y', z' axes relative to the laboratory X, Y, Z axes are the nine angles between the two sets. In uniaxial samples the angles $\eta_{x'}$, $\eta_{y'}$ and $\eta_{z'}$ between x', y', z', respectively, and the Z axis are sufficient, since rotation of the sample about Z makes no observable difference. Only two of them are independent:

$$\cos \eta_{x'} = -\sin \beta' \cos \gamma'$$
$$\cos \eta_{y'} = \sin \beta' \sin \gamma'$$
$$\cos \eta_{z'} = \cos \beta'$$
$$\cos^2 \eta_{x'} + \cos^2 \eta_{y'} + \cos^2 \eta_{z'} = 1 \qquad [2]$$

There are several equivalent choices of statistical parameters for the description of molecular alignment. We use one of these (orientation factors), and give formulae for conversion to two others.

Orientation factors

Orientation factors are orientation averages of products of the direction cosines $\cos \eta_u$ of the molecular axes with respect to the unique sample axis Z. For one- and two-photon events only the second-order factors $[\mathbf{K}]_{uv}$ and fourth-order factors $[\mathbf{L}]_{stuv}$ are needed (the indices s, t, u, v can acquire one of the values x', y' or z', and their order is immaterial):

$$[\mathbf{K}]_{uv} = \langle \cos \eta_u \cos \eta_v \rangle \qquad [3]$$

$$[\mathbf{L}]_{stuv} = \langle \cos \eta_s \cos \eta_t \cos \eta_u \cos \eta_v \rangle \qquad [4]$$

Like the angles $\eta_{x'}$, $\eta_{y'}$ and $\eta_{z'}$ themselves, the orientation factors are redundant:

$$[\mathbf{K}]_{x'x'} + [\mathbf{K}]_{y'y'} + [\mathbf{K}]_{z'z'} = 1 \qquad [5]$$

$$[\mathbf{L}]_{x'x'uv} + [\mathbf{L}]_{y'y'uv} + [\mathbf{L}]_{z'z'uv} = [\mathbf{K}]_{uv} \quad [6]$$

and there are only $2j + 1$ independent orientation factors of order j.

The orientation factors are elements of orientation tensors $\mathbf{K}$ (rank 2), $\mathbf{L}$ (rank 4), etc. The freedom available in the arbitrary choice of the axes system x', y', z' in the molecular framework can be used to diagonalize the tensor $\mathbf{K}$. The principal axes of this tensor are called the molecular orientation axes x, y, and z. The eigenvalues are the molecular orientation factors $K_x = K_{xx}$, $K_y = K_{yy}$ and $K_z = K_{zz}$ ($K_{yz} = K_{zx} = K_{xy} = 0$), where $K_{uv} = [\mathbf{K}]_{uv}$. The axes are labelled such that $K_z \geq K_y \geq K_x$. The z axis is the effective molecular orientation axis and is the direction in the molecular frame that is aligned best with Z, while the x axis is the direction that makes the largest angle with Z. In a plot of K_z against K_y, all orientation distributions are represented by points that lie in the 'orientation triangle', defined by its vertices (K_z, K_y): $(1,0)$, molecular axis z aligned perfectly with Z, x and y equivalent; $(1/2, 1/2)$, the yz plane aligned perfectly with Z, y and z equivalent; and $(1/3, 1/3)$, x, y, and z all equivalent. The sides of the triangle represent special orientation distributions: $K_y = K_x$, all those with x and y equivalent (rod-like alignment, all angles γ equally likely); $K_z = K_y$, all those with y and z equivalent (disc-like alignment); $K_y = 1 - K_z$, $K_x = 0$, those with yz plane along Z (all Euler angles γ equal to 90°).

If the molecular framework possesses planes or axes of symmetry, the orientation axes x, y, z lie in them or perpendicular to them. For example, in molecules of C_{2v} or D_{2h} symmetry the positions of

the x, y, z axes are fully determined and only $1+j/2$ nonzero independent orientation factors of order j remain. Using $L_{stuv} = [\mathbf{L}]_{stuv}$ and choosing the three independent fourth-order factors as $L_z = L_{zzzz}$, $L_y = L_{yyyy}$, and $L_x = L_{xxxx}$, we have for $L_{uv} = L_{uvuv}$ (with u and v in any order),

$$L_{uv} = [(K_u - L_u) + (K_v - L_v) - (K_w - L_w)]/2 \quad [7]$$

For examples of the values of molecular orientation factors K and L for selected orientation distributions, see **Table 4**.

Saupe orientation matrices

The molecular orientation factor matrices $\mathbf{K}$ and $\mathbf{L}$ provide the simplest expressions for polarized spectral intensities but acquire awkward values for isotropic samples, $K_u = 1/3$ and $L_u = 1/5$. This shortcoming is removed by transformation to Saupe matrices:

$$S_{uv} = \tfrac{1}{2}(3K_{uv} - \delta_{uv})$$
$$\begin{aligned} S_{stuv} = \tfrac{1}{8}[35L_{stuv} &- 5(K_{st}\delta_{uv} + K_{su}\delta_{tv} + K_{sv}\delta_{tu} \\ &+ K_{tu}\delta_{sv} + K_{tv}\delta_{su} + K_{uv}\delta_{st}) \\ &+ (\delta_{st}\delta_{uv} + \delta_{su}\delta_{tv} + \delta_{sv}\delta_{tu})] \quad [8] \end{aligned}$$

where δ_{ij} is the Kronecker delta: $\delta_{ij} = 1$ if $i = j$ and $\delta_{ij} = 0$ if $i \neq j$. In isotropic solution, $S_{uv} = S_{stuv} = 0$.

In the molecular orientation system of axes x, y, z, the second-order Saupe matrix is diagonal

Table 4 Orientation factors for some uniaxial orientation distributions

	A	B	C	D	E	F	G	H	I	J
K_x	0	$(1-K_z)/2$	1/5	3/10	1/3	1/3	1/3	1/5	K_x	0
K_y	0	$(1-K_z)/2$	1/5	3/10	1/3	1/3	1/3	2/5	$(1-K_x)/2$	1/2
K_z	1	K_z	3/5	2/5	1/3	1/3	1/3	2/5	$(1-K_x)/2$	1/2
L_x	0	$3(1-2K_z+L_z)/8$	3/35	6/35	1/5	1/9	1/3	3/35	L_x	0
L_y	0	$3(1-2K_z+L_z)/8$	3/35	6/35	1/5	1/9	1/3	9/35	$3(1-2K_x+L_x)/8$	3/8
L_z	1	L_z	3/7	9/35	1/5	1/9	1/3	9/35	$3(1-2K_x+L_x)/8$	3/8
L_{xy}	0	$(1-2K_z+L_z)/8)$	1/35	2/35	1/15	1/9	0	2/35	$(K_x-L_x)/2$	0
L_{xz}	0	$(K_z-L_z)/2$	3/35	1/14	1/15	1/9	0	2/35	$(K_x-L_x)/2$	0
L_{yz}	0	$(K_z-L_z)/2$	3/35	1/14	1/15	1/9	0	3/35	$(1-2K_x+L_x)/8$	1/8

For these orientation distributions, other orientation factors K and L vanish. (*A*) perfect alignment of z axis; (*B*) rod like; (*C*) photoselected with Z-polarized light (absorption z-polarized); (*D*) photoselected with natural light propagating along Z (absorption xy-polarized); (*E*) random; (*F*) x,y,z axes all at magic angle (54.7°) to Z; (*G*) any one of x,y,z axes aligned with Z, with equal probabilities; (*H*) photoselected with natural light propagating along Z (absorption x-polarized), or photoselected with Z-polarized light (absorption yz-polarized); (*I*) disc-like; (*J*) perfect alignment of yz plane, y and z equivalent.

$(S_{yz} = S_{zx} = S_{xy} = 0)$. The vertices of the orientation triangle are (S_{zz}, S_{yy}): $(-1/2, 1)$, $(1/4, 1/4)$, and $(0, 0)$.

If a uniaxial alignment of a set of local axes Z'' relative to Z is described by $S_{Z''Z''}$ and a uniaxial alignment of a set of axes z relative to each Z'' is described by S_{zz}'', the uniaxial alignment of the axes z with respect to Z is given by $S_{zz} = S_{zz}'' S_{Z''Z''}$.

Order parameters

The use of Wigner matrices as an orthonormal basis set for the expansion of $f(\Omega)$ has the advantage of providing a nonredundant set of expansion coefficients. The coefficients define the order parameters A and B, related to the orientation factors by

$$A_0^{(0)} = 1$$
$$A_0^{(2)} = (1/2)(3K_z - 1)$$
$$A_1^{(2)} = -(3/2)^{1/2}K_{xz}$$
$$B_1^{(2)} = (3/2)^{1/2}K_{yz}$$
$$A_2^{(2)} = (1/2)(3/2)^{1/2}(K_x - K_y)$$
$$B_2^{(2)} = -(3/2)^{1/2}K_{xy}$$
$$A_0^{(4)} = (1/8)(35L_z - 30K_z + 3)$$
$$A_1^{(4)} = -(1/4)5^{1/2}(7L_{xzzz} - 3K_{xz})$$
$$B_1^{(4)} = (1/4)5^{1/2}(7L_{yzzz} - 3K_{yz})$$
$$A_2^{(4)} = (1/4)(5/2)^{1/2}[7(L_y - L_x) - 6(K_y - K_x)]$$
$$B_2^{(4)} = -(1/2)(5/2)^{1/2}(7L_{xyzz} - K_{xy})$$
$$A_3^{(4)} = (1/4)35^{1/2}(3L_{xyyz} - L_{xxxz})$$
$$B_3^{(4)} = (1/4)35^{1/2}(3L_{xxyz} - L_{yyyz})$$
$$A_4^{(4)} = (1/8)(35/2)^{1/2}(L_x + L_y - 6L_{xy})$$
$$= (1/8)(35/2)^{1/2}[4(L_x + L_y) + 3(1 - L_z)$$
$$- 6(K_x + K_y)]$$
$$B_4^{(4)} = (1/2)(35/2)^{1/2}(L_{xyyy} - L_{xxxy}) \qquad [9]$$

For the x, y, z axes, $A_1^{(2)} = B_1^{(2)} = B_2^{(2)} = 0$. Orientation triangle vertices are $[A_0^{(2)}, A_2^{(2)}]$: $[1,0]$, $[0, -(1/4)(3/2)^{1/2}]$, and $[0,0]$. In isotropic solutions, $A_0^{(0)} = 1$ and the other order parameters vanish.

Polarized intensities for aligned samples

Expressions for polarized spectral intensities are of the form $\varepsilon^U \cdot \mathbf{O}(0f) \cdot \varepsilon^U$ for one-photon processes and of the form $\varepsilon^U \varepsilon^V \cdot {}^{(4)}\mathbf{O}(0f) \cdot \varepsilon^U \varepsilon^V$ for two-photon processes (**Table 2**). The nature and possible complex conjugation of the unit vectors ε^U and ε^V and the nature of the second-rank tensor $\mathbf{O}(0f)$ and the

fourth-rank tensor ${}^{(4)}\mathbf{O}(0f)$ depend on the polarization of the photons and the specific measurement in question. Knowledge of the nine elements of $\mathbf{O}(0f)$ or the 81 elements of ${}^{(4)}\mathbf{O}(0f)$ thus permits the calculation of event probability for any choice of polarizations U and V.

If all molecules were lined up with their z' axis along Z, y' axis long Y, and x' axis along X, both $\mathbf{O}(0f)$ and ${}^{(4)}\mathbf{O}(0f)$ would have the same elements in the laboratory and molecular system of axes, $[\mathbf{O}(0f)]^{ZZ} = [\mathbf{O}(0f)]_{z'z'}$, etc. When the molecular axes are rotated to a general orientation Ω', the tensor elements in laboratory axes become linear combinations of those in molecular axes:

$$[\mathbf{O}(0f)]^{UV} = \Sigma_{uv}[\mathbf{D}^{UV}]_{uv}[\mathbf{O}(0f)]_{uv}$$
$$[{}^{(4)}\mathbf{O}(0f)]^{STUV} = \Sigma_{stuv}[{}^{(4)}\mathbf{P}^{STUV}]_{stuv}[{}^{(4)}\mathbf{O}(0f)]_{stuv}$$
$$[10]$$

The measured elements of the optical tensor in the laboratory XYZ system are related to its elements in molecular $x'y'z'$ system by ensemble averaging:

$$[\langle \mathbf{O}(0f) \rangle]^{UV} = \Sigma_{uv}[\langle \mathbf{D}^{UV} \rangle]_{uv}[\mathbf{O}(0f)]_{uv} \qquad [11]$$

$$[\langle {}^{(4)}\mathbf{O}(0f) \rangle]^{STUV} = \Sigma_{stuv}[\langle {}^{(4)}\mathbf{P}^{STUV} \rangle]_{stuv}$$
$$\times [{}^{(4)}\mathbf{O}(0f)]_{stuv} \qquad [12]$$

For light electric field along or perpendicular to Z, we need two $\langle \mathbf{D}^{UV} \rangle$ and seven $\langle {}^{(4)}\mathbf{P}^{STUV} \rangle$:

$$\langle \mathbf{D}^{ZZ} \rangle, \; \langle \mathbf{D}^{YY} \rangle = \langle \mathbf{D}^{XX} \rangle,$$
$$\langle {}^{(4)}\mathbf{P}^{ZZZZ} \rangle, \; \langle {}^{(4)}\mathbf{P}^{YYYY} \rangle = \langle {}^{(4)}\mathbf{P}^{XXXX} \rangle,$$
$$\langle {}^{(4)}\mathbf{P}^{YZYZ} \rangle = \langle {}^{(4)}\mathbf{P}^{XZXZ} \rangle,$$
$$\langle {}^{(4)}\mathbf{P}^{ZYZY} \rangle = \langle {}^{(4)}\mathbf{P}^{ZXZX} \rangle,$$
$$\langle {}^{(4)}\mathbf{P}^{XYXY} \rangle = \langle {}^{(4)}\mathbf{P}^{YXYX} \rangle,$$
$$\langle {}^{(4)}\mathbf{P}^{++++} \rangle, \; \langle {}^{(4)}\mathbf{P}^{+-+-} \rangle \qquad [13]$$

Equations [11] and [12] separate the information on anisotropic molecular optical properties ($\mathbf{O}(0f)$ and ${}^{(4)}\mathbf{O}(0f)$) from that on orientation distribution and polarizer orientation ($\langle \mathbf{D}^{UV} \rangle$ and $\langle {}^{(4)}\mathbf{P}^{STUV} \rangle$), which can be expressed through the parameters chosen to describe the orientation distribution: the orientation factors [3], [4], the Saupe matrix elements [8], or the order parameters [9].

For uniaxial samples, the two required $\langle \mathbf{D} \rangle$ tensors are

$$\langle \mathbf{D}^{ZZ} \rangle = \mathbf{K}$$
$$\langle \mathbf{D}^{YY} \rangle = \langle \mathbf{D}^{XX} \rangle = (\mathbf{1} - \mathbf{K})/2 \qquad [14]$$

where $\mathbf{1}$ is the unit tensor. The five (in the xyz system of axes, two) orientation factors K describe these tensors fully. The tensors $\langle {}^{(4)}\mathbf{P} \rangle$ also contain the nine independent factors L:

$$[\langle {}^{(4)}\mathbf{P}^{ZZZZ} \rangle]_{stuv} = L_{stuv}$$
$$[\langle {}^{(4)}\mathbf{P}^{YYYY} \rangle]_{stuv} = (1/8)\{\delta_{st}\delta_{uv}(2\delta_{su} + 1)$$
$$\times (1 - K_{st} - K_{uv}) + (\delta_{st} - \delta_{uv})[2\delta_{st}(\delta_{s,|u \times v|} - 3)K_{uv}$$
$$- \delta_{uv}(2\delta_{|s \times t|,u} - 3)K_{st}] + [2\delta_{su}\delta_{tv} + \delta_{sv}\delta_{tu}) - 1]$$
$$\times K_{|s \times t|,|u \times v|} + 3L_{stuv}\}$$
$$[\langle {}^{(4)}\mathbf{P}^{YZYZ} \rangle]_{stuv} = (1/2)(\delta_{su}K_{tv} - L_{stuv})$$
$$[\langle {}^{(4)}\mathbf{P}^{ZYZY} \rangle]_{stuv} = (1/2)(\delta_{tv}K_{su} - L_{stuv})$$
$$[\langle {}^{(4)}\mathbf{P}^{XYXY} \rangle]_{stuv} = (1/8)\{\delta_{st}\delta_{uv}(2\delta_{su} - 1)$$
$$\times (1 - K_{st} - K_{uv}) + (\delta_{st} - \delta_{uv})[\delta_{st}(2\delta_{s,|u \times v|} - 1)K_{uv}$$
$$- \delta_{uv}(2\delta_{|s \times t|,u} - 1)K_{st}] + [3(3\delta_{su}\delta_{tv} - \delta_{su} - \delta_{tv})$$
$$- (3\delta_{sv}\delta_{tu} - \delta_{sv} - \delta_{tu})]K_{|s \times t|,|u \times v|} + L_{stuv}\}$$
$$[\langle {}^{(4)}\mathbf{P}^{++++} \rangle]_{stuv} = (1/4)\{\delta_{st}\delta_{uv}(2\delta_{su} - 1)$$
$$\times (1 - K_{st} - K_{uv}) + (\delta_{st} - \delta_{uv})[\delta_{st}(2\delta_{s,|u \times v|} - 1)K_{uv}$$
$$- \delta_{uv}(2\delta_{|s \times t|,u} - 1)K_{st}] + [2\delta_{su}\delta_{tv} + \delta_{sv}\delta_{tu}) - 1]$$
$$\times K_{|s \times t|,|u \times v|} + L_{stuv}\}$$
$$[\langle {}^{(4)}\mathbf{P}^{+-+-} \rangle]_{stuv} = (1/4)\{\delta_{st}\delta_{uv}(1 - K_{st} - K_{uv})$$
$$+ (\delta_{st} - \delta_{uv})[\delta_{uv}K_{st} - \delta_{st}K_{uv}] + [2(\delta_{su}\delta_{tv}$$
$$+ \{\delta_{sv} - \delta_{tu}\}^2) - 1]K_{|s \times t|,|u \times v|} + L_{stuv}\}$$
$$[\langle {}^{(4)}\mathbf{P}^{UVUV} \rangle]_{stuv} = [\langle {}^{(4)}\mathbf{P}^{UVUV} \rangle]_{uvst} \qquad [15]$$

where the cross-product of two indices is defined in a manner similar to the vector product of basis vectors ($|x \times x| = 0$, $|x \times y| = z$, $|x \times z| = y$, etc.) and a K with at least one subscript equal to zero vanishes.

For general orientation distributions, Equations [11] and [12] still apply, but $\langle \mathbf{D} \rangle$ and $\langle {}^{(4)}\mathbf{P} \rangle$ then also contain general orientation factors (averages of direction cosines with axes other than Z).

One-photon processes

Polarized absorption and emission

The same formulae apply in both cases, provided that the emission is excited in an isotropic fashion (e.g.,

chemiluminescence). Measurements on isotropic samples provide the same result for light polarized along any direction U and for unpolarized light, and useful information is obtained only if the samples are aligned. The probability of absorption and emission of linearly U-polarized light is proportional to $[\langle \mathbf{O}(0f) \rangle]^{UU}$, with $\mathbf{O}(0f) = M(0f)M(0f)$ (**Table 2**). From Equations [11] and [14], for a uniaxial sample the dichroic ratio d_f is

$$d_f = E^Z(0f)/E^Y(0f)$$
$$= [M(0f)]^Z[M(0f)]^Z/[M(0f)]^Y[M(0f)]^Y$$
$$= \Sigma_{uv}K_{uv}[M(0f)]_u[M(0f)]_v/\Sigma_{uv}(1/2)$$
$$\times (\delta_{uv} - K_{uv})[M(0f)]_u[M(0f)]_v \qquad [16]$$

where the indices u and v run over x', y' and z' and $E^U(0f)$ is proportional to the contribution of the fth transition to the observed U-polarized absorbance (or emission intensity). When the components of the transition moment $M(0f)$ are expressed in the molecular orientation axis system x, y, z,

$$d_f = 2\{K_z[M(0f)]_z^2 + K_y[M(0f)]_y^2$$
$$+ K_x[M(0f)]_x^2\}/(1 - K_z)[M(0f)]_z^2$$
$$+ (1 - K_y)[M(0f)]_y^2 + (1 - K_x)[M(0f)]_x^2\} \qquad [17]$$

Simplifications occur in the presence of molecular symmetry. If it is sufficient (e.g. C_{2v}, D_{2h}) to define the location of the x, y, z axes, and forces $M(0f)$ to lie in one of these axes, say u, the result for a purely polarized (nonoverlapping) transition is that anticipated from the proportionality of absorption probability to the cosine squared of the angle between U and $M(0f)$,

$$d_f = 2K_u/(1 - K_u)$$
$$K_u = d_f/(d_f + 2) \qquad [18]$$

Thus, a measurement of d_f for a u-polarized transition in a molecule of this symmetry determines K_u and only two such measurements for distinct axes u and v are needed to determine $\mathbf{K}$. If the orientation distribution is such that two molecular axes are equivalent, say $K_x = K_y$, a single measurement suffices (e.g., that of d_z, which yields K_z, and $K_x = K_y = (1 - K_z)/2$).

Total u-polarized absorption spectra $A_u(\tilde{\nu})$, $u = x, y, z$, can be obtained in special cases, but not in general, since three such spectra are possible but only two linearly independent spectra E^Z and $E^Y = E^X$ can be measured. For instance, when $A_u(\tilde{\nu})$ is negligible in a spectral region, as low-energy out-of-plane polarized absorption is in aromatic hydrocarbons,

$$A_z(\tilde{\nu}) = [(1 - K_y)E^Z(\tilde{\nu}) - 2K_y E^Y(\tilde{\nu})]/(K_z - K_y)$$
$$A_y(\tilde{\nu}) = [2K_z E^Y(\tilde{\nu}) - (1 - K_z)E^Z(\tilde{\nu})]/(K_z - K_y)$$
$$[19]$$

In molecules of lower symmetry, the direction cosines $\cos \phi_u^f$ of $M(0f)$ in an x', y', z' frame can be determined in favourable circumstances from relations such as

$$\Sigma_{uv} \cos \phi_u^f K_{uv} \cos \phi_v^f = d_f/(d_f + 2) \qquad [20]$$

which simplifies in the x, y, z system of axes to

$$\Sigma_u[K_u - d_f/(d_f + 2)] \cos^2 \phi_u^f = 0 \qquad [21]$$

Photoselection

Light polarized along Z, or, less effectively, unpolarized or circularly polarized light propagating along Z, selects molecules for excitation by the orientation of their absorbing transition moment. This process is known as photoselection, and because of it Equations [16]–[21] do not apply to emission by samples excited by a beam or another anisotropic light source. Explicit expressions can be written for the orientation distribution of the molecules excited from an isotropic solution and those left behind. They are simple only if the depletion of the ground state is small, but the remaining molecules can be aligned much more highly if the depletion is significant.

The results for molecules photoselected from an isotropic sample with Z-polarized light are particularly useful in two cases: (i) the molecule is of high enough symmetry that each $M(0f)$ must lie in one of the orientation axes x, y, z, but the transitions may overlap, and (ii) only one transition $(0 \rightarrow f)$ absorbs at $\tilde{\nu}_0$ (Table 4). In the former case,

$$K_u(\tilde{\nu}_0) = [1 + 2r_u(\tilde{\nu}_0)]/5$$
$$L_{uv}(\tilde{\nu}_0) = (1 + 2\delta_{uv})[1 + 2r_u(\tilde{\nu}_0) + 2r_v(\tilde{\nu}_0)]/35 \qquad [22]$$

where r_u is the fraction of absorption polarized along u.

Using the orientation factors Equation [22], Equations [16]–[21] provide expressions for polarized absorption and emission by molecules photoselected from isotropic samples and represent a special case of the treatment given below, which handles both one-photon events simultaneously and describes photoinduced absorption and luminescence from all uniaxial samples.

Two-photon processes

Polarized intensities of all successive and simultaneous two-photon events on isotropic or partially aligned optically inactive samples are given by Equation [12], using the same tensor $\langle {}^{(4)}\mathbf{P} \rangle$, with different choices for the tensor ${}^{(4)}\mathbf{O}$ (Table 2). For successive events, measurements with circularly polarized light do not offer any additional information and will not be considered, whereas for simultaneous two-photon processes they are indispensable in certain cases.

For uniaxial samples and molecules whose symmetry dictates the orientation axes x, y, z, there are only three independent orientation factors L and Equation [15] simplifies:

$$[\langle {}^{(4)}\mathbf{P}^{ZZZZ} \rangle]_{stuv} = \delta_{|s \times t|, |u \times v|} L_{stuv}$$
$$[\langle {}^{(4)}\mathbf{P}^{YYYY} \rangle]_{stuv} = (1/8)[\delta_{st}\delta_{uv}(2\delta_{su} + 1)$$
$$\times (1 - K_{st} - K_{uv}) + (\delta_{su}\delta_{tv} + \delta_{sv}\delta_{tu})$$
$$\times K_{|s \times t|, |u \times v|} + 3\delta_{|s \times t|, |u \times v|} L_{stuv}]$$
$$[\langle {}^{(4)}\mathbf{P}^{YZYZ} \rangle]_{stuv} = (1/2)(\delta_{su}\delta_{tv} K_{tv} - \delta_{|s \times t|, |u \times v|} L_{stuv})$$
$$[\langle {}^{(4)}\mathbf{P}^{ZYZY} \rangle]_{stuv} = (1/2)(\delta_{su}\delta_{tv} K_{su} - \delta_{|s \times t|, |u \times v|} L_{stuv})$$
$$[\langle {}^{(4)}\mathbf{P}^{XYXY} \rangle]_{stuv} = (1/8)[\delta_{st}\delta_{uv}(2\delta_{su} - 1)$$
$$\times (1 - K_{st} - K_{uv}) + (3\delta_{su}\delta_{tv} - \delta_{sv}\delta_{tu})$$
$$\times K_{|s \times t|, |u \times v|} + \delta_{|s \times t|, |u \times v|} L_{stuv}$$
$$[\langle {}^{(4)}\mathbf{P}^{++++} \rangle]_{stuv} = (1/4)[\delta_{st}\delta_{uv}(2\delta_{su} - 1)$$
$$\times (1 - K_{st} - K_{uv}) + (\delta_{su}\delta_{tv} + \delta_{sv}\delta_{tu})$$
$$\times K_{|s \times t|, |u \times v|} + \delta_{|s \times t|, |u \times v|} L_{stuv}]$$
$$[\langle {}^{(4)}\mathbf{P}^{+-+-} \rangle]_{stuv} = (1/4)[\delta_{st}\delta_{uv}$$
$$\times (1 - K_{st} - K_{uv}) + (\delta_{su}\delta_{tv} - \delta_{sv}\delta_{tu})$$
$$\times K_{|s \times t|, |u \times v|} + \delta_{|s \times t|, |u \times v|} L_{stuv}]$$
$$[\langle {}^{(4)}\mathbf{P}^{UVUV} \rangle]_{stuv} = [\langle {}^{(4)}\mathbf{P}^{UVUV} \rangle]_{uvst} \qquad [23]$$

Unlike one-photon measurements, two-photon measurements are useful even on isotropic samples. For these, the orientation factors K and L have the values given in Table 4. The results then are especially simple and this practically important case will be described separately.

Photoinduced absorption and emission (static)

Identical formulae apply in both cases. When excited molecules do not rotate or transfer excitation energy to differently oriented molecules between the initial absorption event and the later absorption (photoinduced dichroism) or emission (photoluminescence) event, their orientation factors are time independent. Negligible ground-state depletion is assumed in the first step.

When $M(0j)M(j'f)M(0j)M(j'f)$ is substituted for $^{(4)}O$ is Equation [12], only 36 of the 81 terms in the quadruple sum remain. The contribution to the overall event probability that is provided by a combination of transition $0 \to j$ involving a U-polarized photon in the first step with transition $j' \to f$ involving a V-polarized photon in the second step is proportional to $F^{UV}(0j, j'f)$:

$$F^{UV}(0j, j'f) = \Sigma_{(su),(tv)}[2(2 - \delta_{su} - \delta_{tv}) + \delta_{su}\delta_{tv}]$$
$$\times [\langle {}^{(4)}\mathbf{P}^{UVUV} \rangle]_{stuv}[M(0j)]_s[M(j'f)]_t[M(0j)]_u[M(j'f)]_v$$
$$[24]$$

where each of the unordered pairs (su) and (tv) can acquire the values $(x'x')$, $(y'y')$, $(z'z')$, $(y'z')$, $(z'x')$, and $(x'y')$. In molecules of high enough symmetry for transition moments to lie in the x, y, z axes, only the nine terms with $s = u$ and $t = v$ contribute:

$$F^{UV}(0j, j'f) = \Sigma_{st}[\langle {}^{(4)}\mathbf{P}^{UVUV} \rangle]_{stst}$$
$$\times [M(0j)]_s^2[M(j'f)]_t^2 \qquad [25]$$

If several transitions are present, their contributions can be combined into purely s-polarized excitation spectra $A_s(\tilde{v}_1)$ and purely t-polarized photoinduced absorption or emission spectra $B_t(\tilde{v}_2)$:

$$F^{UV}(\tilde{v}_1, \tilde{v}_2)/F^{U'V'}(\tilde{v}_1, \tilde{v}_2)$$
$$= \Sigma_{st}[\mathbf{S}^{UV}]_{st}A_s(\tilde{v}_1)B_t(\tilde{v}_2)/\Sigma_{st}[\mathbf{S}^{U'V'}]_{st}A_s(\tilde{v}_1)B_t(\tilde{v}_2)$$
$$[26]$$

where

$$[\mathbf{S}^{ZZ}]_{st} = L_{st}$$
$$[\mathbf{S}^{YY}]_{st} = (1/8)[(1 + 2\delta_{st})(1 - K_s - K_t) + 3L_{st}]$$
$$[\mathbf{S}^{YZ}]_{st} = (1/2)(K_t - L_{st})$$
$$[\mathbf{S}^{ZY}]_{st} = (1/2)(K_s - L_{st})$$
$$[\mathbf{S}^{XY}]_{st} = (1/8)[(3 - 2\delta_{st})(1 - K_s - K_t) + L_{st}] \qquad [27]$$

If only a single u-polarized transition $0 \to j$ contributes at $\tilde{v}_1$ and a single v-polarized transition $j' \to f$ at $\tilde{v}_2$,

$$F^{UV}(\tilde{v}_1, \tilde{v}_2)/F^{U'V'}(\tilde{v}_1, \tilde{v}_2) = [\mathbf{S}^{UV}]_{uv}/[\mathbf{S}^{U'V'}]_{uv} \qquad [28]$$

The 'stepwise reduction' procedure may yield the ratio $[\mathbf{S}^{UV}]_{uv}/[\mathbf{S}^{U'V'}]_{uv}$ even in the case of overlapping transitions. Overlapping phosphorescent emissions from the three triplet sublevels can be separated at very low temperatures when their populations do not equilibrate and lifetimes differ.

Photoinduced absorption and emission in rigid isotropic solution (static)

If the starting orientation distribution is random (ordinary photoselection), the orientation factors are known (**Table 4**). Equation [24] then yields for $F^{\parallel}$ (both polarization directions equal) and $F^{\perp}$ (polarizations mutually perpendicular):

$$F^{\parallel}(\tilde{v}_1, \tilde{v}_2)/F^{\perp}(\tilde{v}_1, \tilde{v}_2) = [\Sigma_{jf}r_j(\tilde{v}_1)q_f(\tilde{v}_2)$$
$$\times (1 + 2\Sigma_u\cos^2 \phi_u^f \cos^2 \phi_u^j)]/[\Sigma_{jf}r_j(\tilde{v}_1)q_f(\tilde{v}_2)$$
$$\times (2 - \Sigma_u\cos^2 \phi_u^f \cos^2 \phi_u^j)] \qquad [29]$$

where $r_j(\tilde{v}_1)$ is the fraction of the absorption at $\tilde{v}_1$ due to the jth transition and $q_f(\tilde{v}_2)$ is the fraction of absorption or emission at $\tilde{v}_2$ due to the fth transition. The degree of anisotropy R is

$$R(\tilde{v}_1, \tilde{v}_2) = (F^{\parallel} - F^{\perp})/(F^{\parallel} + 2F^{\perp})$$
$$= \Sigma_{jf}r_j(\tilde{v}_1)q_f(\tilde{v}_2)(3\Sigma_u \cos^2 \phi_u^f \cos^2 \phi_u^j - 1)/5$$
$$[30]$$

If symmetry constrains transition moments to the x, y, z axes,

$$F^{\parallel}(\tilde{v}_1, \tilde{v}_2)/F^{\perp}(\tilde{v}_1, \tilde{v}_2) = [\Sigma_{uv}(1 + 2\delta_{uv})$$
$$\times A_u(\tilde{v}_1)B_v(\tilde{v}_2)][\Sigma_{uv}(2 - \delta_{uv})A_u(\tilde{v}_1)B_v(\tilde{v}_2)] \qquad [31]$$

$$R(\tilde{v}_1, \tilde{v}_2) = (3/5)[\Sigma_u r_u(\tilde{v}_1)q_u(\tilde{v}_2) - (1/3)] \qquad [32]$$

In a molecule of any symmetry, when the jth and fth transitions do not overlap with others,

$$R(j, f) = (3\cos^2 \phi - 1)/5 = (2/5)P_2(\cos \phi) \qquad [33]$$

where ϕ is the angle between the transition moments $M(0j)$ and $M(j'f)$ and P_2 is the second Legendre polynomial. The limiting values are reached when ϕ equals 0 (parallel transitions, $R = 2/5$) or 90° (perpendicular transitions, $R = -1/5$). Molecules with a threefold or higher symmetry axis may have degenerate states. If one transition is polarized parallel to the high-order axis and the other in the plane perpendicular to it, $R = -1/5$. If both are polarized in this plane, $R = 1/10$.

Photoinduced absorption and emission in fluid isotropic solution (dynamic)

If the excited molecule rotates during its lifetime, or transfers its excitation to another differently oriented molecule, the polarization of the photoinduced absorption or luminescence becomes a function of time τ that has elapsed after the initial excitation event. We assume that the initial orientation distribution is isotropic and that the optical properties of the molecule are known from static studies. The steps required are an evaluation of the rotational correlation functions and their interpretation in terms of a suitable model for diffusion or energy transfer.

If symmetry constrains the directions of transition moments and the principal axes of the orientation tensor to the symmetry axes x, y, z, the results are the simplest. If the first photon is absorbed in a transition purely polarized along u and the second is absorbed or emitted in a transition purely polarized along v, the degree of anisotropy shows a double exponential decay in time:

$$R(u, u, \tau) = (3/10)\{[2/3 + (D - D_u)/\Delta]$$
$$\times \exp[-2(3D + \Delta)\tau] + [2/3 - (D - D_u)/\Delta]$$
$$\times \exp[-2(3D - \Delta)\tau]\}$$
$$R(u, v, \tau) = (3/10)\{[1/3 + (2D - D_u - D_v)/\Delta]$$
$$\times \exp[-2(3D + \Delta)\tau] + [1/3 - (2D - D_u - D_v)/\Delta]$$
$$\times \exp[-2(3D - \Delta)\tau]\}$$
$$[34]$$

where the second equation applies if $u \neq v$, and $D = (1/3)\Sigma_u D_u$, $\Delta = [(1/2)\Sigma_{uv}(3\delta_{uv} - 1)D_u D_v]^{1/2}$, and D_u are the diagonal elements of the diffusion tensor. In lower-symmetry cases, up to five simultaneous exponential decays are expected, not counting the decay of the excited state itself.

If the diffusion tensor is isotropic, only a single exponential decay of R is observed even in low-symmetry molecules, and the rotational relaxation time is

$\tau_R = 1/6D$:

$$R(\tilde{v}_1, \tilde{v}_2, \tau) = R(\tilde{v}_1, \tilde{v}_2)\exp(-6D\tau) \quad [35]$$

where $R(\tilde{v}_1, \tilde{v}_2) = R(\tilde{v}_1, \tilde{v}_2, 0)$ is the degree of anisotropy at time $\tau = 0$ after excitation, given for purely polarized absorbing and emitting transitions by Equation [33]. In this limit, the degree of anisotropy $\bar{R}(\tilde{v}_1, \tilde{v}_2)$ in a steady excitation experiment on the same sample is

$$\bar{R}(\tilde{v}_1, \tilde{v}_2) = R(\tilde{v}_1, \tilde{v}_2)/(1 + 6D\tau_0) \quad [36]$$

where τ_0 is the excited state lifetime.

Two-photon absorption and Raman scattering

In uniaxial samples, seven distinct combinations of photon polarization are important. The observed intensities are given by I^{ZZ}, $I^{YY} = I^{XX}$, $I^{YZ} = I^{XZ}$, $I^{ZY} = I^{ZX}$, $I^{XY} = I^{YX}$, and two intensities involving circularly polarized light propagating along Z, I^{CON} and I^{DIS} (or I^{SYN} and I^{ANTI}). In a measurement of I^{CON} (I^{DIS}) the tip of the electric vector viewed along Z rotates in the same (opposite) sense for both photons, regardless of whether they propagate parallel or antiparallel to each other. In a measurement of I^{SYN} (I^{ANTI}) the two photons are of the same (opposite) handedness regardless of the sense of their propagation. If the photons propagate in the same direction, SYN = CON, ANTI = DIS. If they propagate in opposite directions, SYN = DIS, ANTI = CON.

In Raman spectroscopy, all seven measurements are available. In two-photon spectroscopy, all seven are accessible only if the two photons are taken from different beams, and otherwise only $I^{ZZ}(\tilde{v}_1, \tilde{v}_1)$, $I^{YY}(\tilde{v}_1, \tilde{v}_1) = I^{XX}(\tilde{v}_1, \tilde{v}_1)$, and $I^{CON}(\tilde{v}_1, \tilde{v}_1)$ can be measured.

Up to a proportionality constant, the contribution of the fth transition to the observed intensity $I(0f, \tilde{v}_1)$ is again given by Equation [12], with proper choice of the tensor $^{(4)}\mathbf{O}$ (Table 2). $\mathbf{T}(0f, \tilde{v}_1)$ is symmetric if the two photons have the same wavenumber and $\alpha'(0f, \tilde{v}_1)$ is symmetric for nonresonant Raman. The elements of $\langle^{(4)}\mathbf{P}^{UVUV}\rangle$ for circularly polarized light differ:

Two-photon absorption:

$$\langle^{(4)}\mathbf{P}^{CON}\rangle = \langle^{(4)}\mathbf{P}^{++++}\rangle \quad \langle^{(4)}\mathbf{P}^{DIS}\rangle = \langle^{(4)}\mathbf{P}^{+-+-}\rangle$$

Raman:

$$\langle {}^{(4)}\mathbf{P}^{CON} \rangle = \langle {}^{(4)}\mathbf{P}^{+-+-} \rangle \qquad \langle {}^{(4)}\mathbf{P}^{DIS} \rangle = \langle {}^{(4)}\mathbf{P}^{++++} \rangle$$

$$[37]$$

Simplification occurs if the molecule has high enough symmetry to dictate the positions of the x, y, z axes. Then, only two orientation factors K and three L are independent, the tensors $\langle {}^{(4)}\mathbf{P}^{UVUV} \rangle$ have 21 nonzero elements of the types $[\langle {}^{(4)}\mathbf{P}^{UVUV} \rangle]_{sstt}$, $[\langle {}^{(4)}\mathbf{P}^{UVUV} \rangle]_{stst}$, and $[\langle {}^{(4)}\mathbf{P}^{UVUV} \rangle]_{stts}$, and the quadruple sum in Equation [12] is reduced to three double sums. The two-photon absorption cross-section is proportional to

$$
\begin{aligned}
I^{UV}(0f, \tilde{v}_1) = \Sigma_{st} \{ &[\mathbf{R}_F{}^{UV}]_{st} [\mathbf{T}(0f, \tilde{v}_1)]_{ss} [\mathbf{T}^*(0f, \tilde{v}_1)]_{tt} \\
+ &[\mathbf{R}_G{}^{UV}]_{st} [\mathbf{T}(0f, \tilde{v}_1)]_{st} [\mathbf{T}^*(0f, \tilde{v}_1)]_{st} \\
+ &[\mathbf{R}_H{}^{UV}]_{st} [\mathbf{T}(0f, \tilde{v}_1)]_{st} [\mathbf{T}^*(0f, \tilde{v}_1)]_{st} \}
\end{aligned}
$$

$$[38]$$

and for Raman the result is identical except that $\alpha'(0f, \tilde{v}_1)$ replaces $\mathbf{T}(0f, \tilde{v}_1)$. The matrices $\mathbf{R}_F{}^{UV}$, $\mathbf{R}_G{}^{UV}$, and $\mathbf{R}_H{}^{UV}$ are analogous to the matrix $\mathbf{S}^{UV}$ introduced in Equation [27]:

$$\mathbf{R}_F{}^{UU} = \mathbf{R}_G{}^{UU} = \mathbf{R}_H{}^{UU} = \mathbf{R}^{UU}$$

$$\mathbf{R}_F{}^{UV} = \mathbf{R}_F{}^{VU} = \mathbf{R}_H{}^{UV} = \mathbf{R}_H{}^{VU} \quad \text{if } U \neq V$$

$$\mathbf{R}_F{}^{SYN} = \mathbf{R}_H{}^{ANTI} = 2\mathbf{R}_F{}^{XY} = 2\mathbf{R}_H{}^{XY}$$

$$\mathbf{R}_F{}^{ANTI} = \mathbf{R}_G{}^{ANTI} = \mathbf{R}_G{}^{SYN} = \mathbf{R}_H{}^{SYN}$$

$$[\mathbf{R}^{ZZ}]_{st} = [\mathbf{R}^{ZZ}]_{ts} = (1/3)(3 - 2\delta_{st})L_{st}$$

$$[\mathbf{R}^{YY}]_{st} = [\mathbf{R}^{YY}]_{ts} = (1/8)[1 - K_s - K_t + (3 - 2\delta_{st})L_{st}]$$

$$[\mathbf{R}_F{}^{YZ}]_{st} = [\mathbf{R}_F{}^{YZ}]_{ts} = (1/6)[\delta_{st}K_s - (3 - 2\delta_{st})L_{st}]$$

$$[\mathbf{R}_G{}^{YZ}]_{st} = (1/6)(3 - 2\delta_{st})(K_t - L_{st})$$

$$[\mathbf{R}_G{}^{ZY}]_{st} = (1/6)(3 - 2\delta_{st})(K_s - L_{st})$$

$$[\mathbf{R}_F{}^{XY}]_{st} = [\mathbf{R}_F{}^{XY}]_{ts} = (1/24)[(4\delta_{st} - 3) \\ \times (1 - K_s - K_t) + (3 - 2\delta_{st})L_{st}]$$

$$[\mathbf{R}_G{}^{XY}]_{st} = [\mathbf{R}_G{}^{XY}]_{ts} = (1/24)[(9 - 8\delta_{st}) \\ \times (1 - K_s - K_t) + (3 - 2\delta_{st})L_{st}]$$

$$[\mathbf{R}_F{}^{ANTI}]_{st} = [\mathbf{R}_F{}^{ANTI}]_{ts} = (1/12)(3 - 2\delta_{st}) \\ \times (1 - K_s - K_t + L_{st})$$

$$[39]$$

The presence of molecular symmetry also simplifies the molecular tensors $\mathbf{T}(0f, \tilde{v}_1)$ and $\alpha'(0f, \tilde{v}_1)$. They are commonly written as a sum of a spherically

symmetric isotropic part $\mathbf{T}^{(0)}$ or $\alpha'^{(0)}$, a symmetric traceless part $\mathbf{T}^{(s)}$ or $\alpha'^{(s)}$, and an antisymmetric part $\mathbf{T}^{(as)}$ or $\alpha'^{(as)}$:

$$
\begin{array}{ccc}
(0) & (s) & (as) \\
\begin{pmatrix} a & 0 & 0 \\ 0 & a & 0 \\ 0 & 0 & a \end{pmatrix} & \begin{pmatrix} b & c & d \\ c & e & f \\ d & f & -b-e \end{pmatrix} & \begin{pmatrix} 0 & g & h \\ -g & 0 & i \\ -h & -i & 0 \end{pmatrix}
\end{array}
$$

$$[40]$$

If the axes x, y, z are symmetry-determined, Equation [38] simplifies, as the tensor has either (i) only two symmetrically disposed possibly unequal off-diagonal elements (forms $\mathbf{T}^{(s)} + \mathbf{T}^{(as)}$ or $\alpha'^{(s)} + \alpha'^{(as)}$), or (ii) only up to three possibly different diagonal elements (forms $\mathbf{T}^{(0)} + \mathbf{T}^{(s)}$ or $\alpha'^{(0)} + \alpha'^{(s)}$).

Two-photon absorption in isotropic solution

While measurement in an isotropic solution does not differentiate among the molecular x, y, z, axes, it does permit a separation of the F, G, and H parts of the intensity. Three of the four distinct measurements are linearly independent:

$$
\begin{aligned}
I^{\parallel}(0f, \tilde{v}_1) &= I^{ZZ}(0f, \tilde{v}_1) = I^{YY}(0f, \tilde{v}_1) \\
&= (1/30)(2\delta_F + 2\delta_G + 2\delta_H) \\
I^{\perp}(0f, \tilde{v}_1) &= I^{YZ}(0f, \tilde{v}_1) = I^{ZY}(0f, \tilde{v}_1) = I^{XY}(0f, \tilde{v}_1) \\
&= (1/30)(-\delta_F + 4\delta_G - \delta_H) \\
I^{SYN}(0f, \tilde{v}_1) &= (1/30)(-2\delta_F + 3\delta_G + 3\delta_H) \\
I^{ANTI}(0f, \tilde{v}_1) &= (1/30)(3\delta_F + 3\delta_G - 2\delta_H)
\end{aligned}
$$

$$[41]$$

where the rotational invariants of the tensor are defined by

$$
\begin{aligned}
\delta_F &= \Sigma_{st}[\mathbf{T}(0f, \tilde{v}_1)]_{ss}[\mathbf{T}(0f, \tilde{v}_1)^*]_{tt} \\
\delta_G &= \Sigma_{st}[\mathbf{T}(0f, \tilde{v}_1)]_{st}[\mathbf{T}(0f, \tilde{v}_1)^*]_{st} \\
\delta_H &= \Sigma_{st}[\mathbf{T}(0f, \tilde{v}_1)]_{st}[\mathbf{T}(0f, \tilde{v}_1)^*]_{ts}
\end{aligned}
$$

$$[42]$$

If $\mathbf{T}(0f, \tilde{v}_1)$ has the form $\mathbf{T}^{(0)}$, $\delta_F = 3\delta_G$ and $\delta_H = \delta_G$. For the form $\mathbf{T}^{(s)}$, $\delta_F = 0$ and $\delta_H = \delta_G$, and for $\mathbf{T}^{(as)}$, $\delta_F = 0$ and $\delta_H = -\delta_G$. Circularly polarized light is needed to determine all three invariants. When both photons are taken from the same beam, only $I^{\parallel}(0f, \tilde{v}_1)$ and $I^{SYN}(0f, \tilde{v}_1)$ can be measured. However, $\mathbf{T}(0f, \tilde{v}_1)$ is then symmetric, $\delta_H = \delta_G$, and the two remaining invariants can be determined.

Raman scattering in isotropic solution

The x, y, z axes cannot be distinguished, and three rotational invariants of $\alpha'(0f, \tilde{v}_1)$ result, usually defined as the isotropic part $\bar{\alpha}^2$, symmetric anisotropy γ_s^2, and antisymmetric anisotropy γ_{as}^2:

$$\bar{\alpha}^2 = (1/9)\Sigma_{st}[\alpha'(0f, \tilde{v}_1)]_{ss}[\alpha'(0f, \tilde{v}_1)^*]_{tt}$$
$$\gamma_s^2 = (1/2)\Sigma_{st}|[\alpha'(0f, \tilde{v}_1)]_{ss} - [\alpha'(0f, \tilde{v}_1)]_{tt}|^2$$
$$+ (3/4)\Sigma_{s\neq t}|[\alpha'(0f, \tilde{v}_1)]_{st} + [\alpha'(0f, \tilde{v}_1)]_{ts}|^2$$
$$\gamma_{as}^2 = (3/4)\Sigma_{st}|[\alpha'(0f, \tilde{v}_1)]_{st} - [\alpha'(0f, \tilde{v}_1)]_{ts}|^2 \quad [43]$$

These are related to δ_F, δ_G and δ_H through

$$\bar{\alpha}^2 = (1/9)\delta_F$$
$$\gamma_s^2 = (3/4)(\delta_G + \delta_H) - (1/2)\delta_F$$
$$\gamma_{as}^2 = (3/4)(\delta_G - \delta_H) \quad [44]$$

and have particularly simple values for the tensors $\alpha'^{(0)}$, $\alpha'^{(s)}$, and $\alpha'^{(as)}$:

$$\alpha'^{(0)}: \quad \bar{\alpha}^2 = (1/3)\delta_G, \gamma_s^2 = \gamma_{as}^2 = 0$$
$$\alpha'^{(s)}: \quad \gamma_s^2 = (3/2)\delta_G, \bar{\alpha}^2 = \gamma_{as}^2 = 0$$
$$\alpha'^{(as)}: \quad \gamma_{as}^2 = (3/2)\delta_G, \bar{\alpha}^2 = \gamma_s^2 = 0 \quad [45]$$

The observable intensities are proportional to

$$I^{\parallel}(0f, \tilde{v}_1) = I^{ZZ}(0f, \tilde{v}_1) = I^{YY}(0f, \tilde{v}_1)$$
$$= \bar{\alpha}^2 + (4/45)\gamma_z^2$$
$$I^{\perp}(0f, \tilde{v}_1) = I^{YZ}(0f, \tilde{v}_1) = I^{ZY}(0f, \tilde{v}_1)$$
$$= I^{XY}(0f, \tilde{v}_1) = (1/15)\gamma_s^2 + (1/9)\gamma_{as}^2$$
$$I^{SYN}(0f, \tilde{v}_1) = (2/15)\gamma_s^2$$
$$I^{ANTI}(0f, \tilde{v}_1) = \bar{\alpha}^2 + (1/45)\gamma_s^2 + (1/9)\gamma_{as}^2 \quad [46]$$

Unless $\gamma_{as}^2 = 0$ (nonresonant Raman), measurement with circularly polarized light is needed to determine the three invariants. The results are usually expressed as the depolarization ratio $I^{\perp}(0f, \tilde{v}_1)/I^{\parallel}(0f, \tilde{v}_1)$. For nonresonant Raman scattering this equals 3/4 for $\alpha' = \alpha'^{(s)}$, is lower otherwise, and reaches zero if $\alpha' = \alpha'^{(0)}$ (totally symmetric vibration). In resonant Raman, γ_{as}^2 can be different from zero, and the depolarization ratio can exceed 3/4 ('anomalous polarization').

List of symbols

A,B = order parameters; D_u = diagonal elements of the diffusion tensor; $\mathbf{D}$ = rank 2 orientation transformation tensor; $\hbar$ = Planck constant/2π; I = light intensity; $\mathbf{K}$ = orientation tensor, rank 2; K_{uv} = orientation factor, rank 2; L_{stuv} = orientation factor, rank 4; $\mathbf{L}$ = orientation tensor, rank 4; M = electric dipole moment operator; $\mathbf{M}$ = tensor operator = MM; $\mathbf{O}$ = one-photon event tensor; $^{(4)}\mathbf{O}$ = two-photon event tensor; P_2 = second Legendre polynomial; $^{(4)}\mathbf{P}$ = rank 4 orientation transformation tensor; $q_f(\tilde{v}_2)$ = fraction of absorption or emission at $\tilde{v}_2$ due to the fth transition; r_u = fraction of absorption polarized along u; R = degree of anisotropy; $\mathbf{R}$ = simplified form of orientation transformation tensor for Raman and two-photon absorption spectroscopy; $r_j(\tilde{v}_1)$ = fraction of absorption $\tilde{v}_1$ due to the jth transition; s, t, u, v = indices taking values x, y, z or x', y', z'; $\mathbf{S}$ = simplified form of orientation transformation tensor for fluorescence and transient dichroism spectroscopy; S_{uv}, S_{stuv} = Saupe matrix elements; $\mathbf{T}$ = two-photon absorption tensor; $U V$ = photon electric field vector direction (polarization); x, y, z = molecular orientation axes; x', y', z' = arbitrary molecular axes; X, Y, Z = laboratory axes; α', β', γ = Euler angles; $\bar{\alpha}^2$ = isotropic part of α'; α' = Raman scattering tensor; γ_s^2 = symmetric anisotropy; γ_{as}^2 = antisymmetric anisotropy; δ_F, δ_G, δ_H = rotational invariants of the two-photon absorption tensor; δ_{ij} = Kronecker delta; ε^U = real unit vector along U; ε^+, ε^- = polarization vector for left-, right-handed light; $\eta_{x'}$, $\eta_{y'}$, $\eta_{z'}$ = angles of x', y', z' with Z; $\tilde{v}$ = wavenumber; τ = time after initial excitation event; τ_0 = excited-state lifetime; τ_R = rotational relaxation time; ϕ = angle between $M(0j)$ and $M(j'f)$; ϕ_u^f = direction cosines of $M(0f)$ in x', y', z'; $\Omega' = \alpha', \beta', \gamma$.

See also: **Chiroptical Spectroscopy, General Theory; Electromagnetic Radiation; Fluorescence Polarization and Anisotropy; IR Spectroscopy, Theory; Linear Dichroism, Applications; Rayleigh Scattering and Raman Spectroscopy, Theory; Symmetry in Spectroscopy, Effects of.**

Further reading

Clarke D and Grainger JF (1971) *Polarized Light and Optical Measurement.* Oxford: Pergamon Press, Oxford.

Dörr F (1971) In: Lamola AA (ed) *Creation and Detection of the Excited State*, Vol 1, Chapter 2, pp 53–122. New York: Marcel Dekker.

Krause S (ed) (1981) *Molecular Electro-Optics.* New York: Plenum Press.

Michl J and Thulstrup EW (1995) *Spectroscopy with Polarized Light*, 2nd edn. New York: VCH Publishers.

Thulstrup EW and Michl J (1989) *Elementary Polarization Spectroscopy*. New York: VCH Publishers.

Zannoni C (1979) In: Luckhurst GR and Gray GW (eds) *The Molecular Physics of Liquid Crystals*, Chapter 3. New York: Academic Press.

ORD and Polarimetry Instruments

Harry G Brittain, Center for Pharmaceutical Physics, Milford, NJ, USA

ELECTRONIC SPECTROSCOPY
Methods & Instrumentation

Introduction

Chiral molecules interact with electromagnetic radiation in exactly the same fashion as do achiral molecules in that they will exhibit optical absorption, have a characteristic refractive index, and can scatter oncoming photons. Optically active compounds are also capable of an additional interaction with light whose electric vectors are circularly polarized. One particular manifestation of this property is an apparent rotation of the plane of linearly polarized light upon passage through a medium containing the optically active agent. When this property is measured at a single wavelength, the phenomenon is commonly referred to as polarimetry. The wavelength dependence of polarimetric response is termed optical rotatory dispersion (ORD). Charney, who has provided the most readable summary of the history and practice associated with chiroptical spectroscopy, has provided a general introduction to optical activity. A number of other monographs have been written which concern various applications of chiroptical spectroscopy, ranging from the very theoretical to the very practical, and these texts contain numerous references suitable for those unfamiliar with the field.

Polarization properties of light

An understanding of the polarization properties of light is essential to polarimetry and optical rotatory dispersion, and considerable insight can be gained by considering the characteristics of electric vectors. Unpolarized light propagating along the z-axis will contain electric vectors whose directions span all possible angles within the x–y plane. Linearly polarized light represents the situation where all the transverse electric vectors are constrained to vibrate in a single plane. The simplest way to produce linearly polarized light is by dichroism, where the polarization is obtained by passage of the incident light beam through a material that totally absorbs all electric vectors not lying along a particular plane. The other elements suitable for the production of linearly polarized light are crystalline materials that exhibit optical double refraction, and these are the Glan, Glan–Thompson, and Nicol prisms.

As with any vector quantity, the electric vector describing the polarization condition can be resolved into projections along the x and y axes. For linearly polarized light, these will always remain in phase during the propagation process unless passage through another anisotropic element takes place. Attempted passage of linearly polarized light through another polarizer (referred to as the analyzer) results in the transmission of only the vector component that lies along the axis of the second polarizer. If the incident angle of polarization is orthogonal to the axis of transmission of the analyzer, then no light will be transmitted.

Certain crystalline optical elements have the property of being able to alter the phase relationships existing between the electric vector projections. When the vector projections are rendered 90° out of phase, the electric vector executes a helical motion as it passes through space, and the light is now denoted as being circularly polarized. Since the helix can be either left- or right-handed, the light is referred to as being either left- or right-circularly polarized. It is preferable to consider linearly polarized light as being the resultant formed by combining equal amounts of left- and right-handed circularly polarized components, the electric vectors of which are always exactly in phase. When the phase angle between the two vector components in any light beam lies between 0 and 90°, the light is denoted as being elliptically polarized. The production of a 90° phase shift is termed quarter wave retardation, and an optical element that effects such a change is a quarter wave-plate. Passage of linearly polarized light through a quarter wave place produces a beam of circularly polarized light, the sense of which depends on whether the phase angle has been either advanced or retarded by 90°. The passage of circularly polarized light through a quarter wave-plate will produce linearly polarized light, whose angle is rotated by 90° with respect to the original plane of linear polarization.

Circular birefringence (optical rotation)

The study of molecular optical activity can be considered as beginning with the work of Biot, who demonstrated that the plane of linearly polarized light would be rotated upon passage through an optically active medium and designed a working polarimeter capable of quantitatively measuring the effect. Mitscherlich introduced the use of calcite prisms in 1844, and Soleil devised the double-field method of detection in 1845. Since these early developments, many advances have been made in polarimetry and a large number of detection schemes are now possible. An extensive summary of methods is available in Heller's comprehensive review.

The phenomenon of optical activity is determined by the relative indices of refraction for left- and right-circularly polarized light within the medium under consideration. In an optically inactive medium, the refractive indices of left- and right-circularly polarized light are equal. Since the two polarization senses would remain in phase at all times during passage through the medium, the resultant vector leaving the medium would be unchanged with respect to the vector that entered the medium. In an optically active medium, the refractive indices for left- and right-circularly polarized light are no longer equal, so the components are no longer coherent when they leave the medium. These phenomena are illustrated in **Figure 1**. When viewed directly along the direction of the oncoming beam, the resultant vector appears as a rotation of the initial plane of polarization (see **Figure 2**). Optical activity is therefore a manifestation of circular birefringence.

In principle, the measurement of optical rotation is extremely simple, and a suitable apparatus is shown in schematic form in **Figure 3**. The incident light is collimated and plane-polarized, and allowed to pass

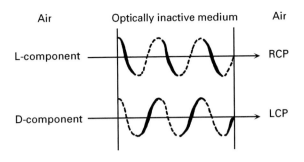
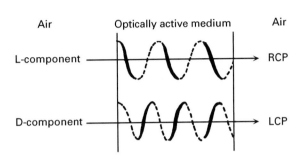

Figure 1 Behaviour of the right-circularly polarized (RCP) and left-circularly polarized (LCP) components of linearly polarized light as they pass through an optically inactive medium (upper diagram) or thorough an optically active medium (lower diagram).

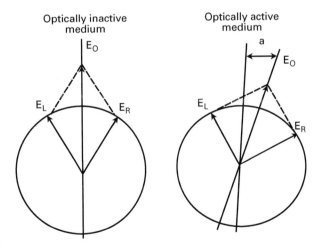

Figure 2 Phase relations associated with the electric vectors of linearly polarized light as it is passed through various media. For optically inactive media, recombination of left-circularly polarized (E_L) and right-circularly polarized (E_R) components yields linearly polarized light whose resultant electric vector (E_O) is unchanged with respect to the incident axis. For optically active media, recombination of the E_L and E_R components yields a resultant E_O vector rotated from this incident axis by the angle α.

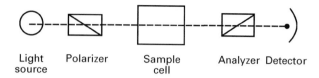

Figure 3 Block diagram of a simple polarimeter. Monochromatic light from the source is linearly polarized by the initial polarizer, and then allowed to pass through the sample medium. The angle of polarization associated with the light leaving the medium is determined by rotating the analyzer polarizer to the new null position. In automatic operation, the observed angle of rotation is determined using a photoelectric determination of the null points.

through the medium of interest. In most common measurements, the medium consists of the analyte dissolved in an appropriate solvent. The plane of the incident light is specified, and then the angle of rotation is defined with respect to this original plane. This is carried out by first determining the orientation of polarizer and analyzer for which no light can be transmitted (the null position). The medium containing the optically active material is then introduced between the prisms, and the analyzer is rotated until a null position is again detected. The observed angle of rotation is taken as the difference between the two null angles.

The measurement of the null angle as a minimum in the transmitted light is experimentally difficult to observe when conducted visually. For this reason, a more efficient detection mechanism was developed, which is commonly known as the 'half-shade' technique. The half-shade is a device that transforms the total extinction point into an equal illumination of two adjacent fields. This mode of detection was superior for manual polarimeters, since the human eye is much more adept at balancing fields of transmitted light than at detecting a minimum in light intensity. In most cases, setting an appropriate device in front of a simple analyzing prism, such as a Lippich prism, brings about the half-shade effect.

With the development of photoelectric devices, the manual detection of null positions in polarimetry became superseded by instrumental measurements of the endpoint. As would be anticipated, measurements can be made far more easily and accurately using photoelectric detection of the null position. Early versions of automated polarimeters used the half-shade method, but the two light intensities were measured using photomultiplier tubes. The position of the analyzer was rotated until the difference in signals detected by the two detectors reached a minimum. Other methods have made use of modulated light beams, and variations on the method of symmetric angles. A wide variety of detection

methods have been developed, enabling accurate measurements of optical rotation to be made on a routine basis. The influence of polarimeter design on the observed signal-to-noise characteristics has been discussed by a number of investigators.

The velocity of light (v) passing through a medium is determined by the index of refraction (n) of that medium:

$$v = c/n \qquad [1]$$

where c equals to the velocity of light in vacuum. For a non-chiral medium, the refractive index will not exhibit a dependence on the sense of the polarization state of the light. For a chiral medium, the refractive index associated with left-circularly polarized light will not normally equal the refractive index associated with right-circularly polarized light. It follows that the velocities of left- and right-circularly polarized light will differ on passage through a chiral medium. Since linearly polarized light can be resolved into two in-phase, oppositely-signed, circularly polarized components, the components will not longer be in phase once they pass through the chiral medium. Upon leaving the chiral medium, the components are recombined, and linearly polarized light is obtained whose plane is rotated (relative to the original plane) by an angle equal to half the phase angle difference of the circular components. This phase angle difference is given by:

$$\beta = \frac{2\pi b'}{\lambda_o}(n_L - n_R) \qquad [2]$$

In Equation [2], β is the phase difference, b' is the medium path length (in cm) λ_o is the vacuum wavelength of the light used (in cm), and n_L and n_R are the refractive indices for left- and right-circularly polarized light, respectively. The quantity $(n_L - n_R)$ defines the circular birefringence of the chiral medium, and is the origin of what is commonly referred to as optical rotation. The observed rotation of the plane-polarized light is given in radians by:

$$\phi = \frac{\beta}{2} \qquad [3]$$

The usual practice is to express rotation in terms of degrees, and in that case Equation [2] becomes:

$$\alpha = \frac{1800\, b}{\lambda_o}(n_L - n_R) \qquad [4]$$

In Equation [4], b represents the medium path length in *decimeters*, which is the conventional unit. The optical rotation (α) of a chiral medium can be either positive or negative depending on the sign of the circular birefringence. It is not practical to measure the circular birefringence directly, since the magnitude of $(n_L - n_R)$ exhibits typical magnitudes of between 10^{-8} and 10^{-9}.

The optical rotation exhibited by a chiral medium depends on the optical path length, the wavelength of the light used, the temperature of the system, and the concentration of dissymmetric analyte molecules. If the solute concentration (c) is given in terms of grams per 100 mL of solution, then the observed rotation (an extrinsic quantity) can be converted into the specific rotation (an intrinsic quantity) using:

$$[\alpha] = \frac{100\,\alpha}{b\,c} \qquad [5]$$

The molar rotation [M], is defined from:

$$[M] = \frac{(\text{FW})\,\alpha}{b\,c} \qquad [6]$$

$$= \frac{(\text{FW})\,[\alpha]}{100} \qquad [7]$$

where FW is the formula weight of the dissymmetric solute. When the solute concentration is given in units of molarity C, Equation [6] becomes:

$$[M] = \frac{10\,\alpha}{b\,C} \qquad [8]$$

The temperature associated with a measurement of the specific or molar rotation of a given substance must be specified. Thermal volume changes or alterations in molecular structure (as induced by a temperature change) are capable of producing detectable changes in the observed rotations. In the situations where solute–solute interactions become important at high concentrations, it may be observed that the specific rotation of a solute is not independent of concentration. It is acceptable practice, therefore, to obtain polarimetry data at a variety of concentration values to verify that a true molecular parameter has been measured.

The verification of polarimeter accuracy is often not addressed on a routine basis. The easiest method to verify the performance of a polarimeter is to use quartz plates that have been cut to known degrees of

retardation. These plates are normally certified to yield a specified optical rotation at a specific wavelength. A reasonable criterion for acceptability is that the observed optical rotation should be within $\pm 0.5\%$ of the certified value. The quartz plates are available from a variety of sources, including manufacturers of commercial polarimeters or houses specializing in optical components.

Another possibility of verify the accuracy of polarimetry measurements is to measure the optical rotation of a known compound. Owing to the stability of their rotatory strengths once dissolved in a fluid medium, steroids are probably most suitable for this purpose. Chafetz has provided a compilation of the specific rotation values obtained for a very extensive list of steroids. When possible, data have been reported for alternate solvents, as well as for wavelengths other than 589 nm.

Optical rotation measurements are most commonly used to confirm the enantiomeric identities of resolved enantiomers. When reference standards of totally resolved materials are available, polarimetry can be used to determine the enantiomeric purity of samples of defined composition.

Optical rotatory dispersion

Aside from the intrinsic molecular contribution, the most important parameter in determining the magnitude of optical rotation is the wavelength of the light used for the determination. Generally, the magnitude of the circular birefringence increases as the wavelength becomes shorter, so specific rotations increase in a regular manner with a decrease in wavelength. This behaviour persists until the light is capable of being absorbed by the chiral substance, whereupon the refractive index exhibits anomalous behaviour. The variation of specific or molar rotation with wavelength is termed optical rotatory dispersion (ORD).

Biot observed that the optical rotation of tartaric acid solutions was a function of the wavelength used for the determination. When measured outside of absorption bands, an ORD spectrum (as illustrated in one half of **Figure 4**) will consist either of a plain positive or plain negative dispersion curve. When measured inside of an absorption band, the ORD will exhibit anomalous dispersion, which is referred to as a Cotton effect. A positive Cotton effect consists of positive ORD at long wavelengths, and negative ORD at shorter wavelengths (see **Figure 4**). In the simplest measurement of ORD, the fixed wavelength source of **Figure 3** is replaced by a tunable source (such as a xenon arc combined with a

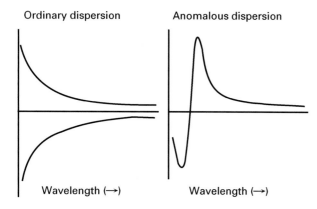

Figure 4 Optical rotatory curves, as would be obtained for simple dispersion (both positive and negative curves are shown) and anomalous dispersion (illustrating a positive Cotton effect).

monochromator), and the angle of rotation is automatically determined as the wavelength is swept.

The anomalous dispersion observed in ORD spectra arises since the refractive index of material is actually the sum of a real and imaginary part:

$$n = n_o + i\,k \qquad [9]$$

where n is the observed refractive index at some wavelength, n_o is the refractive index at infinite wavelength, and k is the absorption coefficient of the substance. It is evident that if $(n_L - n_R)$ does not equal zero, then $(k_L - k_R)$ will not equal zero either. The relation between the various quantities was first conceived by Cotton, and is illustrated schematically in **Figure 5**. The effect of the differential absorption $(k_L - k_R)$ (i.e. circular dichroism) is to render the incident linearly polarized light into an emergent beam that is elliptically polarized, for which the major axis of the ellipse will be rotated with respect to the incident vector.

The earliest ORD spectrometer was the non-recording instrument designed by Rudolph, consisting merely of a manual polarimeter through which light of various wavelengths could be passed. The optical rotation was measured at each incident wavelength, and the ORD spectrum obtained by plotting the observed values. In an improvement, the incident polarizer was rocked through a small angle by an electric motor at low frequency, the transmitted light measured by a photoelectric cell, and the analyzer rotated until a null position was again achieved.

Recording ORD spectrometers have been developed which either use the Faraday effect to obtain both modulation and balance, or use servo-driven analyzers to determine the null position. A block diagram illustrating the latter mode of operation is

given in **Figure 6**. Comparison of **Figure 3** and **Figure 6** reveals that the only real difference between an ordinary polarimeter and such an ORD spectrometer is the additional servo-mechanisms and associated feedback circuits which automatically and continuously determine the null positions as the incident wavelength is changed. Currently available ORD instrumentation is, however, invariably based on adapted circular dichroism spectrometers with their enhanced performance through phase modulation. The calibration of ORD spectrometers is verified using the same standards as used for the calibration of ordinary polarimeters.

The earliest investigative work involving chiral organic molecules was entirely based on ORD methods, since little else was available at the time. One of

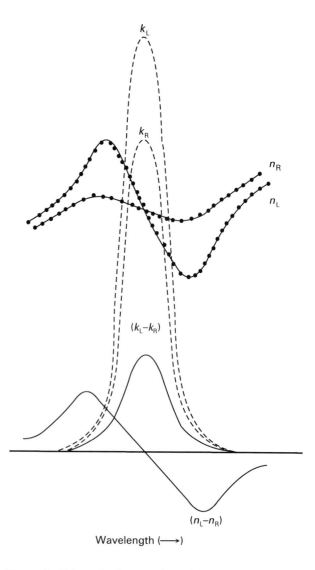

Figure 5 Schematic diagram of the Cotton effect, illustrating the effects of circular birefringence and circular dichroism within an isolated absorption band.

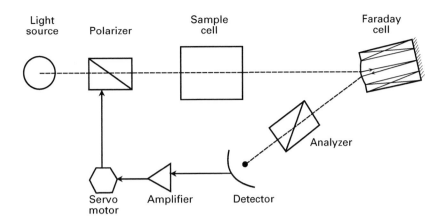

Figure 6 Block diagram of a servo-driven ORD spectropolarimeter. Monochromatic light from the source is linearly polarized by the initial polarizer, and then allowed to pass through the sample medium. Further modulation is effected by the Faraday cell, and then the angle of polarization associated with the light leaving the medium is determined by rotating the analyzer polarizer to the new null position. The observed angle of rotation is determined using a photoelectric determination of the null points.

the largest data sets collected to date concerns the chirality of ketone and aldehyde groups, which eventually resulted in the deduction of the octant rule. The octant rule was an attempt to relate the absolute stereochemistry within the immediate environment of the chromophore with the sign and intensity of the ORD Cotton effects. To apply the rule, the circular dichroism and/or ORD within the n–π* transition and 300 nm is obtained, and its sign and intensity noted. The rule developed by Djerassi and co-workers states that the three nodal planes of the n- and π*-orbitals of the carbonyl group divide the molecular environment into four front octants and four back octants. A group or atom situated in the upper-left or lower-right rear octant (relative to an

observer looking at the molecule parallel to the C=O axis) induces a positive Cotton effect in the n–π* band. A negative Cotton effect would be produced by substitution within the upper-right or lower-left back octant. This has been illustrated in **Figure** 7 for the particular example of a generic 3-hydroxy-3-alkyl-cyclohexanone.

Although exceptions to the octant rule have been shown, the wide applicability of the octant rule has remained established. The ability to deduce molecular confirmations in solution on the basis of ORD spectra data has proven to be extremely valuable to synthetic and physical organic chemists, and enabled investigators of the time to develop their work without requiring the use of more heroic methods.

See also: **Biomacromolecular Applications of Circular Dichroism and ORD; Chiroptical Spectroscopy, Emission Theory; Chiroptical Spectroscopy, General Theory; Chiroptical Spectroscopy, Oriented Molecules and Anisotropic Systems; Circularly Polarized Luminescence and Fluorescence Detected Circular Dichroism; Light Sources and Optics; Luminescence, Theory; Nonlinear Optical Properties; Vibrational CD Spectrometers; Vibrational CD, Applications; Vibrational CD, Theory.**

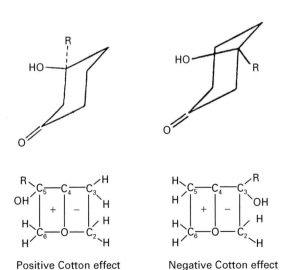

Figure 7 Application of the octant rule, applied to the enantiomers of 3-hydroxy-3-alkyl-cyclohexanone.

List of symbols

b = path length in decimetres; b' = pathlength in cm; c = velocity of light *in vacuo*; c = concentration in grams per 100 mL; C = molarity concentration; E_L = left circularly polarized electric vector; E_R = right circularly polarized electric vector; E_0 = resultant electric vector; FW = formula weight; k_L/k_R = absorption coefficient for left/right circularly polarized light;

k = absorption coefficient; $[M]$ = molar radiation; n = index of refraction; n_L/n_R = index of refraction for left/right polarized light; n_0 = refractive index at infinite wavelength; α = optical rotation in degrees; $[\alpha]$ = specific rotation; β = phase angle of difference; ϕ = optical rotation in radians; λ_0 = vacuum wavelength of light; v = velocity of light in medium.

Further reading

Barron L (1982) *Molecular Light Scattering and Optical Activity*. Cambridge: Cambridge University Press.

Caldwell DJ and Eyring H (1971) *The Theory of Optical Activity*. New York: Wiley-Interscience.

Chafetz L (1993) Implementing changes in polarimetry. *Pharmacopeial Forum* 19: 6159–6162.

Charney E (1979) *The Molecular Basis of Optical Activity*. New York: Wiley.

Crabbe P (1972) *ORD and CD in Chemistry and Biochemistry*. New York. Academic Press.

Djerassi C (1960) *Optical Rotatory Dispersion: Applications to Organic Chemistry*. New York: McGraw Hill.

Heller W (1972) Optical Rotation - Experimental Techniques and Physical Optics. In: Weissberger A and Rossiter BW (eds) *Physical Methods of Chemistry*, Vol I, Chapter 2. New York: John Wiley.

Kankare J and Stephens R (1986) The influence of optical design on the signal-to-noise characteristics of polarimeters. *Talanta* 33: 571–576.

Kirk DN (1986) The chiroptical properties of carbonyl compounds. *Tetrahedron* 42: 777–818.

Mason S (1982) *Molecular Optical Activity and the Chiral Discrimination*. Cambridge: Cambridge University Press.

Moffitt W, Woodward RB, Moscowitz A, Klyne W and Djerassi C (1961) Structure and the optical rotatory dispersion of saturated ketones. *Journal of the American Chemical Society* 83: 4013–4018.

Purdie N and Brittain HG (1994) *Analytical Applications of Circular Dichroism*. Amsterdam: Elsevier Science.

Snatzke G (ed) (1967) *Optical Rotatory Dispersion and Circular Dichroism in Organic Chemistry*. London: Heyden.

Velluz L, Legrand M and Grosjean M (1965) *Optical Circular Dichroism: Principles, Measurements, and Applications*. Weinheim: Verlag Chemie.

Viogtman E (1992) Effect of source 1/f noise on optical polarimeter performance. *Analytical Chemistry* 64: 2590–2595.

Yeung ES (1985) Signal-to-noise optimization in polarimetry. *Talanta* 33: 1097–1100.

ORD Spectroscopy of Biomacromolecules

See **Biomacromolecular Applications of Circular Dichroism and ORD.**

Organic Chemistry Applications of Fluorescence Spectroscopy

Stephen G Schulman, Qiao Qing Di and **John Juchum**, University of Florida, Gainesville, FL, USA

ELECTRONIC SPECTROSCOPY

Applications

Introduction

The fluorescence of organic molecules consists of the emission of light by molecules, which have previously absorbed visible or ultraviolet radiation. The measurement of fluorescence often permits very low analyte detection limits (10^{-15}–10^{-4} mol dm^{-3}) and is widely employed in quantitative analysis, especially as a detection and quantitation method in liquid chromatography. Most applications are derived from the relationship between analyte concentration and fluorescence intensity and are, therefore, similar in concept to other spectrochemical methods of analysis. However, as well as spectral intensity, other features of fluorescence spectral bands of organic molecules, such as position in the electromagnetic spectrum, emission lifetime and excitation spectrum are exquisitely sensitive to the molecular environment and to molecular structure and, therefore, are also analytically useful, especially for probing the environment of the fluorophore.

This article will deal with the nature of organic molecular fluorescence, its dependence upon molecular structure, reactivity and interactions with the environment and its utility in the trace analysis of organic compounds.

The origin of molecular fluorescence

The electronic excitation of molecules occurs as the result of the absorption of near ultraviolet or visible light.

Subsequent to excitation, the loss of excess vibrational energy, known as vibrational relaxation, takes place in about 10^{-12} s, the excess energy being lost to inelastic collisions with solvent molecules.

There is also an efficient radiationless pathway for the demotion of the excited molecule from higher to lower electronically excited states called internal conversion. In aliphatic molecules which have a high degree of vibrational freedom, vibrational relaxation and internal conversion may return the excited molecule to the ground electronic state radiationlessly within 10^{-12} s after excitation, in which case fluorescence is not observed.

However, in aromatic molecules, the degree of vibrational freedom is restricted. In this case, the excited molecule may, radiationlessly, arrive in the lowest vibrational level of the lowest electronically excited singlet state. Subsequently (after 10^{-11}–10^{-7} s), it may return to the ground electronic state by emitting near ultraviolet or visible fluorescence.

With very few exceptions, fluorescence always originates from the lowest excited singlet state. This means that only one fluorescence band may be observed from a given molecule, even though it will usually have several absorption bands. Therefore, the observation of several fluorescence bands in a solution of pure sample, suggests the occurrence of a chemical reaction in either the ground or excited state, resulting in two or more fluorescent species. Alternatively, the purity of the sample must be questioned. Because the fluorescence spectrum of any one organic compound can demonstrate only one fluorescence band, a band which is usually broad and lacking features, the fluorescence spectrum does not reveal the detailed information about molecular structure that

NMR, IR or mass spectrometry does. Nevertheless, fluorescence spectra can give information about the molecular environment that is unobtainable by other methods.

Chemical structural effects upon fluorescence

Most fluorescence spectra arise from functionally substituted aromatic molecules. Consequently, the compounds of interest in this article are those derived from organic compounds that possess aromatic rings, such as benzene, naphthalene or anthracene, or their heteroaromatic analogues pyridine, quinoline, acridine, etc. The fluorescence spectra of these substances may often be understood in terms of the electronic interactions between the simple aromatic structures and their substituents.

Chemical structure and fluorescence intensity

The intensity of fluorescence observable from a given molecular species depends upon the probability of light absorption and the probability of fluorescence. The molar absorptivity is a macroscopic manifestation of the probability of light absorption by molecules in the optical path. For most aromatic molecules, the π, π^* absorption bands lying in the near-ultraviolet and visible regions of the spectrum have molar absorptivities of 1×10^3–1×10^5 dm^3 mol^{-1} cm^{-1} so that the appropriate choice of the transition to excite can influence the intensity of fluorescence by about two orders of magnitude. If the absorbance of the potentially fluorescing species, at the wavelength of excitation, is below 0.02 the intensity of fluorescence is proportional to the molar absorptivity at that nominal wavelength. At high absorbances the sample is not equally illuminated along the optical path, giving rise to 'self-shadowing' effects.

More important is the quantum yield or efficiency of fluorescence, which may affect the intensity of fluorescence over about four orders of magnitude and may determine whether fluorescence is at all observable. The quantum yield of fluorescence is dependent upon the rates of processes competing with fluorescence for the deactivation of the lowest excited singlet state.

Aromatic molecules that contain lengthy aliphatic side-chains generally tend to fluoresce less intensely than those without the side-chains because of greater opportunity for vibrational deactivation (the 'loose-bolt' effect). In unsubstituted aromatic molecules, the rigidity of the aromatic ring results in lower probabilities of vibrational deactivation and hence, higher quantum yields.

The fluorescence of organic molecules is quenched (diminished in intensity) by heavy atom substituents such as $-As(OH)_2$, Br and I and by certain other groups such as $-CHO$, $-NO_2$ and nitrogen in six-membered heterocyclic rings (e.g. quinoline). These substituents cause mixing of the spin and orbital motions of the valence electrons. Spin–orbital coupling obscures the distinct identities of the singlet and triplet states and, thereby, enhances the probability or rate of singlet $\rightarrow$ triplet intersystem crossing. This process favours population of the lowest triplet state at the expense of the lowest excited singlet state and thus decreases the fluorescence quantum yield. Consequently, aromatic arsenites, nitro compounds, bromo and iodo derivatives, aldehydes, ketones and N-heterocyclics tend to fluoresce very weakly or not at all.

Chemical structure and position of the fluorescence

The energies of the ground and excited states of fluorescing molecules are affected by molecular structure. This is reflected in the positions of the fluorescence maxima in the spectrum.

$$E = h\nu = hc/\lambda \qquad [1]$$

According to Equation [1], where E is the energy, ν the frequency and λ, the wavelength of fluorescence and h and c are, respectively, Planck's constant and the velocity of light, the greater the separation between the ground and excited states the greater will be the frequency and the shorter will be the wavelength of fluorescence. This separation depends upon the energy difference between the highest occupied and lowest unoccupied molecular orbitals and the repulsion energy between the electronic configurations corresponding to the ground and excited states. In aromatic hydrocarbons, the greater the degree of linear annulation the closer together will be the highest occupied and lowest unoccupied orbitals. Consequently, benzene fluoresces at shorter wavelengths than naphthalene which fluoresces at shorter wavelengths than anthracene. Phenanthrene, which is angularly annulated, emits at wavelengths between those of naphthalene and anthracene. In functionally substituted aromatic molecules the substituents with lone electron pairs, e.g. $-NH_2$, $-OH$, will have highest occupied orbitals, higher in energy than those of the unsubstituted hydrocarbons while the substituents with vacant π^* orbitals, e.g. $-CHO$, $-CO_2H$, will have π^* orbitals lower in energy than those of the unsubstituted hydrocarbons. This means that in sub-

stituted aromatic molecules fluorescence will be at wavelengths longer than in the unsubstituted molecules. This is so regardless of whether the substituents are electron donating or electron withdrawing.

Influence of the chemical and physical environment on fluorescence spectra

The solvent

The solvents in which fluorescence spectra are observed play a role secondary only to molecular structure in determining the spectral positions and intensities with which fluorescence bands occur and occasionally determine whether fluorescence is observed.

The electronic transition accompanying excitation entails a change in electronic charge distribution. If the excited state is more polar than the ground state a more polar solvent will stabilize the excited state more than the ground state and cause the fluorescence to shift to longer wavelengths relative to that observed in a less polar solvent. If, however, the ground state is more polar than the excited state, which is rarely the case, the fluorescence will tend to shift to shorter wavelengths upon going to a more polar solvent.

Hydrogen bonding in the lowest excited singlet states occasionally results in a decrease in fluorescence quantum yield upon going from hydrocarbon to hydrogen bonding solvents. Many arylamines and phenolic compounds demonstrate this behaviour which is due to internal conversion enhanced by coupling of the vibrations of the molecule of interest to those of the solvent.

Molecules having the lowest excited singlet states of the n, π^* type rarely fluoresce in hydrocarbon solvents because the n, π^* singlet state is efficiently deactivated by intersystem crossing. However, in polar, hydrogen bonding solvents, such as ethanol or water, these molecules often become fluorescent. This results from the stabilization of the lowest singlet π, π^* state relative to the lowest n, π^* state by hydrogen bonded interaction. If the interaction is sufficiently strong, the lowest π, π^* state drops below the n, π^* state in the strongly solvated molecule, becoming the lowest excited singlet state and permitting intense fluorescence. Quinoline and 1-naphthaldehyde, for example, do not fluoresce in cyclohexane but do so in water.

The influence of pH

The spectral shifts accompanying protonation or dissociation of basic or acidic functional groups depend upon whether the functional group undergoing protonation or dissociation is directly coupled to the aromatic system and whether it gains or loses electronic charge upon going from the ground to the excited state. The higher the positive charge on electron-attracting groups and the higher the negative charge on electron-donating groups, the lower, in general, will be the energy of fluorescence. Thus, the protonation of electron-withdrawing groups, such as carboxyl, carbonyl and pyridinic nitrogen, causes shifts of fluorescence spectra to longer wavelengths while the protonation of electron-donating groups, such as the amino group, produces spectral shifts to shorter wavelengths. The protolytic dissociation of electron-donating groups, such as hydroxyl, sulfhydryl or pyrrolic nitrogen, produces spectral shifts to longer wavelengths while the dissociation of electron-withdrawing groups, such as carboxyls, shifts the fluorescence spectra to shorter wavelengths.

In some molecules, the presence of nonbonded electrons obviates the occurrence of fluorescence. However, protonation of the functional group possessing the non-bonded electron pair raises the n,π^* lowest excited singlet state above the lowest excited singlet π,π^* state and thereby allows fluorescence to occur. Benzophenone, for example, does not fluoresce as the neutral molecule but does so, moderately intensely, as the cation in concentrated sulfuric acid.

An interesting aspect of acid–base reactivity of fluorescent molecules is derived from the occurrence of protonation, and dissociation, during the lifetime of the lowest excited singlet state and is occasionally observed in the pH dependence of the fluorescence spectrum.

The lifetimes of molecules in the lowest excited singlet state are typically 10^{-10}–10^{-7} s. Typical rates of proton transfer reactions are $\leq 10^{10}$ s^{-1}. Consequently, excited state proton transfer may be much slower, much faster, or competitive with radiative deactivation of the excited molecules. If excited state proton transfer is much slower than fluorescence, the relative fluorescence intensity will vary with pH exactly the same way as does the absorbance, reflecting only the ground-state acid–base equilibrium. If excited state proton transfer is much faster than fluorescence, the fluorescence intensity will vary with pH in a way that reflects the acid–base equilibrium in the lowest excited singlet state. Equilibrium in the excited state is a rare phenomenon and will not be dealt with further here.

If the rate of proton transfer, in the excited state, is comparable to the rates of photophysical deactivation of excited acid and conjugate base, the variations of the fluorescence intensities of acid and conjugate base, with pH, will be governed by the kinetics of the excited state proton transfer reactions and the fluorescence of acid and conjugate base will be observed over a wide pH range.

The influence of high solute concentrations

Several types of excited state solute–solute interaction are common at high solute concentrations. The aggregation of excited solute molecules with unexcited molecules of the same type may result in a new excited molecule called an excimer, which may either not luminesce or may luminesce at lower frequency than the monomeric excited molecule. Because excimer formation takes place in the excited state it is sometimes demonstrable as a shifting of the fluorescence spectrum. However, after fluorescence, the deactivated polymer, which is unstable in the ground state, rapidly decomposes. Hence, the absorption spectrum does not reflect the presence of the excited state complexes. Occasionally, excited state complex formation may occur between two different solute molecules. The term 'exciplex' has been coined to describe a heteropolymeric excited state complex. Excimer and exciplex formation are usually observed only in fluid solution because diffusion of the excited species is necessary to form the excited complexes. One concentration effect that is observed in molecules in fluid or rigid media is resonance energy transfer. Energy transfer entails the excitation of a molecule which, during the lifetime of the excited state, transmits its excitation energy to a nearby molecule. The probability of resonance energy transfer decreases as the inverse sixth power of the distance between donor and acceptor and can occur between molecules which are separated by up to 100 nm. Because the mean distance between molecules decreases with increasing concentration, energy transfer is favoured by increasing the concentration of the acceptor. For energy transfer to occur between two dissimilar molecules, the fluorescence spectrum of the energy donor must overlap the absorption spectrum of the energy acceptor.

Fluorescence may be diminished in intensity or eliminated due to the deactivation of the lowest excited singlet state of the analyte by interaction with other species in solution. This is called quenching of fluorescence. Mechanisms of quenching appear to entail internal conversion, intersystem crossing, electron-transfer and photodissociation as modes of deactivation of the excited fluorescer–quencher complexes.

Quenching processes may be divided into two broad categories. In dynamic or diffusional quenching, interaction between the quencher and the potentially fluorescent molecule takes place during the lifetime of the excited state. As a result, the efficiency of dynamic quenching is governed by the rate constant of the quenching reaction, which is usually typical of that for a diffusion-controlled reaction, the lifetime of the excited state of the potential fluorescer and the concentration of the quenching species. Interaction between quencher and fluorophore results in the formation of a transient excited complex which is non-fluorescent and may be deactivated by any of the usual radiationless modes of deactivation of excited singlet states. Because interaction occurs only after excitation of the potentially fluorescing molecule, the presence of the quenching species has no effect on the absorption spectrum of the fluorophore. Many aromatic molecules, for example, the quinolines and acridines, are dynamically quenched by halide ions such as Cl^-, Br^- and I^-. Static quenching is characterized by complexation in the ground state between the quenching species and the molecule which, when excited alone, should fluoresce. The complex is generally not fluorescent and, as a result, the ground state reaction diminishes the intensity of fluorescence of the potentially fluorescent species. The quenching of the fluorescence of o-phenanthroline by complexation with iron(II) is an example of static quenching.

Applications

Native fluorescence of organic compounds

Numerous organic compounds are intrinsically fluorescent and so may be assayed directly, eliminating any need of derivatization or labelling. This section lists some examples of types of compounds which possess native fluorescence and some factors which exert influence on that fluorescence. The assay of organic compounds by fluorescence spectroscopy is covered in greater detail by Guilbault and that of organic natural products by Wolfbeis. Simple fluorescence spectra of 2000 compounds have been published by Sadtler Research Laboratories.

Three of the twenty common amino acids exhibit fluorescence. These are phenylalanine (Phe), tyrosine (Tyr) and tryptophan (Trp). Phenylalanine, as the name implies, consists of alanine with a benzene ring attached. It is weakly fluorescent at $\lambda_f = 282$ nm ($\lambda_{ex} = 260$ nm) and cannot be detected in the presence of Tyr or Trp. Tyrosine is phenolic and fluoresces ($\lambda_{ex} = 275$ nm; $\lambda_f = 303$ nm) with much greater intensity than Phe just as phenol fluoresces far more intensely than benzene. The phenolic group, which ionizes at pH above 10, introduces the need to control pH in the assay. The phenolate form of

Tyr fluoresces with far less intensity than the non-ionized form and redshifts to 345 nm. Tryptophan has a quantum yield very close to that of Tyr but absorbs far better and so fluoresces ($\lambda_{ex} = 287$ nm; $\lambda_f = 348$ nm) more intensely. The nitrogen-containing indole moiety in Trp becomes protonated at low pH thus decreasing its fluorescence. The fluorescence of Trp is also quenched at high pH and so Trp fluoresces best over a range in pH from 4 to 9. The fluorescence properties of these amino acids as mentioned apply only to the free amino acids in solution. Such properties can change significantly when they become part of a protein.

The indole moiety seen in tryptophan is a common structure in nature and is seen in abundance in alkaloids. Indole in water fluoresces at $\lambda_f = 352$ nm ($\lambda_{ex} = 275$ nm). Many simple natural derivatives of indole fluoresce in the 330–350 nm range and can be maximally exited at 270–290 nm. The 'indole alkaloid' is a major class of alkaloids, which include a number of well-known drugs, legal as well as illicit. The ergot alkaloids are in this class. The famous hallucinogen lysergic acid diethylamide (LSD) at pH 7 fluoresces at 365 nm with an excitation maximum at 325 nm. Bromolysergic acid diethylamide is not hallucinogenic. It is brominated at the 2 position on the indole moiety. It fluoresces ($\lambda_{ex} = 315$ nm; $\lambda_f = 460$ nm) at pH 1 with far less intensity than LSD.

Other classes of alkaloids, which exhibit fluorescence, include the quinoline and isoquinoline alkaloids. Quinoline is weakly fluorescent. The antimalarial drug quinine includes a methoxy substituent on the 6 position of the quinoline moiety and fluoresces very intensely. Quinine, in sulfuric acid solution, is often used as a standard in fluorescence spectroscopy for determining a quantum yield. Its fluorescent properties are sensitive to pH. At pH 2 it has an excitation maximum of 347 nm with fluorescence at 448 nm. At pH 7 the peaks shift to absorb at 331 nm and emit at 382 nm. In hydrochloric acid solution, absorption is unaffected but fluorescence intensity is quenched greatly by the halide anions.

Isoquinoline alkaloids can be subdivided into two groups. Those alkaloids which preserve the isoquinoline moiety and those which have a reduced ring structure, the tetrahydroisoquinolines. Papaverine, a smooth muscle relaxant, possesses the isoquinoline moiety. In chloroform, papaverine fluoresces at $\lambda_f = 347$ nm ($\lambda_{ex} = 315$ nm). The addition of some trichloroacetic acid to this protonates the isoquinoline moiety and causes a shift in absorption and fluorescence peaks ($\lambda_{ex} = 415$ nm; $\lambda_f = 452$ nm). Berberine is an isoquinoline alkaloid which has a quaternary nitrogen with hydroxide as the standard counterion. In ethanol it has excitation maxima of

432 and 352 nm and fluoresces at 548 nm. A change of solvent to DMF causes a shift of the excitation maximum to 380 nm and emission to 510 nm. Berberine deposited on a TLC plate shows excitation maxima of 433 and 353 nm and fluoresces at 510 nm. 13-Methoxyberberine in ethanol has one excitation peak, at 433 nm, and its emission peak shifts to 562 nm.

Tetrahydroisoquinoline alkaloids have a benzenoid fluorophore rather than isoquinoline and so fluoresce at shorter wavelengths. The opiate alkaloids are of this class of alkaloids. Morphine and codeine differ only at the 3 position. Morphine has a hydroxy substituent, making it phenolic, and codeine has a methoxy substituent. In water at pH 1 both compounds fluoresce at 350 nm. Both are excited at 285 nm. Codeine has an additional excitation peak at 245 nm. They can be assayed in admixture because morphine loses its fluorescence under basic conditions by way of phenolate formation but codeine retains its fluorescence at high pH.

Caffeine is a member of the xanthine alkaloids. Caffeine in water fluoresces at $\lambda_f = 303$ nm ($\lambda_{ex} = 270$ nm). Caffeine is 1,3,7-trimethylxanthine. Xanthine in water at pH 1 fluoresces at $\lambda_f = 435$ nm ($\lambda_{ex} = 275$ nm). The fluorophore of caffeine and xanthine is that of purine ($\lambda_{ex} = 300$ nm; $\lambda_f = 360$ nm at pH 13).

Purine is, likewise, the fluorophore associated with the purine bases adenine and guanine. Adenine at pH 1 fluoresces at $\lambda_f = 380$ nm ($\lambda_{ex} = 265$ nm). Adenosine and its various phosphates all fluoresce at $\lambda_f = 390$ nm ($\lambda_{ex} = 272$ nm) in 5 M sulfuric acid. Guanine at pH 1 fluoresces at $\lambda_f = 350$ nm ($\lambda_{ex} = 275$ nm) and at pH 11 at $\lambda_f = 360$ nm ($\lambda_{ex} = 275$ nm). Guanosine and GMP (guanosine 5′-phosphate) at pH 1 fluoresce at $\lambda_f = 390$ nm ($\lambda_{ex} = 285$ nm).

The other bases associated with DNA and RNA are the pyrimidines: cytosine, thymine and uracil. They also exhibit fluorescence. Cytosine in water fluoresces at $\lambda_f = 313$ nm ($\lambda_{ex} = 267$ nm) but CMP (cytidine 5′-phosphate) in water fluoresces at $\lambda_f = 330$ nm ($\lambda_{ex} = 248$ nm). Thymine in water at pH 7 fluoresces at $\lambda_f = 320$ nm ($\lambda_{ex} = 265$ nm) but TMP (thymine 5′-phosphate) fluoresces at $\lambda_f = 330$ nm ($\lambda_{ex} = 248$ nm). Uracil in water at pH 7 fluoresces at $\lambda_f = 309$ nm ($\lambda_{ex} = 258$ nm) but UMP fluoresces at $\lambda_f = 320$ nm ($\lambda_{ex} = 248$ nm). The quantum yields of all the bases, nucleosides and nucleotides are extremely low so extraordinary conditions must be applied to the fluorescence analysis of DNA, RNA and their component parts.

Examples of compounds possessing native fluorescence listed so far have had a benzene ring or a nitrogen-containing heterocycle as the fluorophore.

There is also a great, number of oxygen-containing heterocyclic compounds that fluoresce. Coumarins and flavonoids are the two largest classes of oxygen heterocycles. Coumarins fluoresce more intensely under basic conditions where flavones fluoresce weakly. In 30% sulfuric acid solution, flavones fluoresce intensely and coumarins do not.

Coumarin is not normally fluorescent but hydroxy substitution in any position except 8 leads to intense fluorescence at room temperature. In methanol, 3-hydroxycoumarin fluoresces at $\lambda_f = 372$ nm ($\lambda_{ex} = 316$ nm), 4-hydroxycoumarin fluoresces at $\lambda_f = 357$ nm ($\lambda_{ex} = 300$ nm), 6-hydroxycoumarin fluoresces at $\lambda_f = 431$ nm ($\lambda_{ex} = 341$ nm) and 7-hydroxycoumarin fluoresces at $\lambda_f = 392$ nm ($\lambda_{ex} = 333$ nm). Other oxygen-containing heterocycles are also of interest. Cannabinols in ethanol fluoresce at $\lambda_f = 318$ nm ($\lambda_{ex} = 280$ nm). Tocopherols (vitamin E) in ethanol fluoresce at $\lambda_f = 340$ nm ($\lambda_{ex} = 295$ nm).

Other vitamins and coenzymes exhibit fluorescence. Riboflavin (vitamin B_2) is a three-ring heterocycle which in water at pH 7 fluoresces at $\lambda_f = 565$ nm ($\lambda_{ex} = 370$ or 440 nm). The various forms of vitamin B_6 all have native fluorescence: pyridoxine ($\lambda_{ex} = 340$ nm; $\lambda_f = 400$ nm), pyridoxal ($\lambda_{ex} = 330$ nm; $\lambda_f = 385$ nm) and pyridoxamine ($\lambda_{ex} = 335$ nm; $\lambda_f = 400$ nm). Vitamin B_{12}, as cyanocobalamin, has a porphyrin moiety (cf. haem and chlorophyll) and fluoresces at $\lambda_f = 305$ nm ($\lambda_{ex} = 275$ nm). All D vitamins fluoresce when treated with strong acid but that is due to a degradation product. Vitamin D_2 (calciferol) has been reported exhibit native fluorescence in ethanol at $\lambda_f = 420$ nm ($\lambda_{ex} = 348$ nm). Calciferol's fluorescence is due to a rigid conjugated triene rather than an aromatic ring. Vitamin A (retinol) also has no aromatic ring but rather has an extended conjugation of π-bonds which leads to fluorescence. Retinol in ethanol fluoresces at $\lambda_f = 470$ nm ($\lambda_{ex} = 325$ nm) and in pentane–hexane it fluoresces at $\lambda_f = 513$ nm ($\lambda_{ex} = 321$ nm).

Fluorescent derivatization

Not all substances are fluorescent. Non-fluorescent substances may be analysed by indirect methods of which there are several.

(1) Some organic compounds, themselves non-fluorescent, can be converted into fluorescent compounds by a chemical reaction with another organic compound which is itself also non-fluorescent.

o-Phthalaldehyde (OPA), itself non-fluorescent, is one of the most widely used reagents for the assay of amines, amino acids, peptides, amino carbohydrates, etc. It reacts with the primary amino group, in the presence of a thiol (usually 2-sulfanylethanol), to give strongly fluorescent condensation products. The fluorescence is measured at ~455 nm with the excitation wavelength at ~340 nm. The assay can be conducted in the nanomole range.

For example, this method has been employed for analysis of carbamate pesticides in surface water after the pesticides are hydrolysed in strong base to yield methyl amine and phenols.

New reagents such as naphthalene-2,3-dicarboxaldehyde (NDA), 1-phenylnaphthalene-2,3-dicarboxaldehyde (ϕNDA), and anthracene-2,3-dicarboxaldehyde (ADA) have been synthesized as improved OPA/thiol type reagents. While similar in many respects, there are important differences in these isoindole products. For example, the products formed with NDA, ϕNDA or ADA are considerably more stable than the corresponding OPA derivatives and possess substantially higher fluorescence quantum efficiencies and minimal interference compared with the latter.

(2) Some non-fluorescent organic compounds can react with fluorescent dyes to give fluorescent products which usually show altered fluorescence properties with regard to the free dye.

4-(Aminosulfonyl)-7-(1-piperazinyl)-2,1,3-benzoxadiazole (ABD-PZ), (maximum wavelength 565 nm with excitation at 413 nm), has been synthesized as a fluorescent reagent for determination of carboxylic acids. It reacts with carboxylic acids in the presence of diethyl phosphorocyanidate (DEPC) to produce fluorescent adducts with fluorescence at longer wavelengths. For example, the maximum wavelength of fluorescence of arachidic acid labelled with ABD-PZ is 580 nm with excitation at 440 nm. This method has been applied to a reversed-phase HPLC column for fatty acid mixture analysis. The detection limits for eight fatty acids are in the 10–50 fmol range.

When the piperazinyl group in ABD-PZ is substituted by a chiral group such as the 3-aminopyrrolidinyl group, it becomes a chiral derivatization reagent (D-ABD-APy or L-ABD-APy). This chiral derivatization reagent reacts with carboxylic acid enantiomers to form diastereomers for fluorescence detection. The diastereomers derived from antiinflammatory drugs and N-acetylamino acids are efficiently resolved by a reversed-phase column after they react with D-ABD-APy. The detection limit, for example, of ABD-APy-Naprofen on HPLC chromatograms is 30 fmol.

(3) Fluorometric enzyme assay involves a reaction catalysed by an enzyme, in which the product must show different fluorescence properties compared with those of the substrate. One example is the fluorometric peroxygenase assay for lipid hydroperoxides in meats and fish. In the reaction, catalysed by pea peroxygenase, the lipid hydroperoxide is reduced to an equimolar amount of alcohol during hydroxylation of the substrate, 1-methylindole, which shows no fluorescent product in the absence of the peroxygenase. The maximum wavelengths of excitation and emission for the hydroxylated product, 3-hydroxy-1-methylindole, appear at 410 and 485 nm, respectively, in n-butanol. The detectabilities of hydroperoxides are in the range of 25–150 nmol and α-tocopherol, an antioxidant at levels equivalent to those in meats and fish, did not affect the peroxygenase reaction. The method enables determination of total lipid hydroperoxides in sample homogenates without extracting total lipids from meats and fish.

(4) Fluoroimmunoassay involves attaching a fluorescent-labelled antibody to its specific antigen or vice versa, making use of a complexation reaction between the antigen and the antibody, for fluorescence detection at nanogram and lower levels. The specific affinity reactions may include the following: enzyme–substrate, hormone–receptor, neurotransmitter–receptor, pharmacological agent–receptor, etc. Such fluoroimmunoassays are divided into two categories – heterogeneous assays, which involve physical separation of the assay mixture before detection, and homogeneous assays, in which no separation steps are required. The most common fluorescent labels employed for fluoroimmunoassay are fluorescein isothiocyanate (FITC), which emits apple-green fluorescence (520 nm) when excited by ultraviolet or, preferably, by blue light (494 nm), and tetramethylrhodamine isothiocyanate (TRITC), which emits orange fluorescence (575 nm) when excited by ultraviolet or, preferably, by green light (550 nm).

One example is the fluoroimmunoassay for the routine detection of buprenorphine in urine samples of persons suspected of Temgesic® abuse. Buprenorphine antibody is labelled with pseudobuprenorphine, the dimer of buprenorphine. In this case, pseudobuprenorphine has a higher affinity for the antibody than that of FITC–norbuprenorphine. Pseudobuprenorphine shows an intense blue fluorescence with maximum at 435 nm when excited at 326 nm. The minimum detectable dose of buprenorphine by the fluoroimmunoassay is calculated to be 20 ng mL^{-1}.

List of symbols

c = velocity of light; E = energy; h = Planck's constant; λ_{ex} = excitation wavelength; λ_f = fluorescent wavelength; ν = frequency.

See also: **Biochemical Applications of Fluorescence Spectroscopy; Fluorescence Microscopy, Applications; Fluorescent Molecular Probes; Fluorescence Polarization and Anisotropy; Inorganic Condensed Matter, Applications of Luminescence Spectroscopy; UV-Visible Absorption and Fluorescence Spectrometers; X-Ray Fluorescence Spectrometers; X-Ray Fluorescence Spectroscopy, Applications.**

Further reading

Baeyens WRG, de Keukeleire D, Korkidis K (ed) (1991) *Luminescence Techniques in Chemical and Biochemical Analysis*. New York: Marcel Dekker.

Brand L and Johnson ML (1997) (ed) *Fluorescence Spectroscopy*. San Diego: Academic Press.

Eastwood D and Love LJC (1988) (ed) *Progress in Analytical Luminescence*. Philadelphia: ASTM.

Goldberg MC (ed) (1989) *Luminescence Applications in Biological, Chemical, Environmental, and Hydrological Sciences*. Washington, DC: American Chemical Society.

Guilbault GG (ed) (1990) Assay of organic compounds. In: *Practical Fluorescence*, 2nd edn, pp 231–366. New York: Marcel Dekker.

Lakowicz JR (ed) (1991) *Topics in Fluorescence Spectroscopy*, Vol 1–5. New York: Plenum Press.

Lumb MD (ed) (1978) *Luminescence Spectroscopy*. London and New York: Academic Press.

Mason WT (ed) (1993) *Fluorescent and Luminescent Probes for Biological Activity: a Practical Guide to Technology for Quantitative Real-time Analysis*. London, San Diego: Academic Press.

Rendell D (1987) *Fluorescence and Phosphorescence Spectroscopy*. Published on behalf of ACOL, London by Wiley, Chichester and New York, 1987.

Sadtler Research Laboratories (1974–1976) *Fluorescence Spectra*, Chapter 3, Vol 1–8. Philadelphia.

Schulman SG (1977) *Fluorescence and Phosphorescence Spectroscopy: Physicochemical Principles and Practice*. New York: Pergamon Press.

Schulman SG (ed) (1985–1988) *Molecular Luminescence Spectroscopy: Methods and Applications*, Part I–II. New York: Wiley.

Soper SA, Warner IM and McGown LB (1998) Molecular fluorescence, phosphorescence and chemiluminescence. *Analytical Chemistry* 70: 477R–494R.

Winefordner JD, Schulman SG and O'Haver TC (1972) *Luminescence Spectrometry in Analytical Chemistry*. New York: Wiley-Interscience.

Wolfbeis OS (1985) The fluorescence of organic natural products. In: Schulman SG (ed) *Molecular Luminescence Spectroscopy: Methods and Applications*, Part I, Chapter 3, pp 167–370. New York: Wiley.

Organic Chemistry Applications of NMR Spectroscopy

See **Structural Chemistry Using NMR Spectroscopy, Organic Molecules.**

Organometallics Studied Using Mass Spectrometry

Dmitri V Zagorevskii, University of Missouri-
Columbia, Columbia, MO, USA

MASS SPECTROMETRY
Applications

Mass spectrometry is playing a significant role in organometallic chemistry. One of the most important applications of mass spectrometry is the determination of the molecular mass and elemental composition of metal compounds, and the identification of their structure. Mass spectrometry was first applied to the analysis of relatively volatile metal complexes using electron impact and chemical ionization techniques. Further development of ionization techniques has made it possible to analyse and to study the structure of a wide range of organometallics, including nonvolatile, ionic, multiply charged, polymetallic, and high-molecular-mass derivatives. Some of these ionization techniques allow identification of metal-containing intermediates directly from the condensed phase, providing beneficial information about reaction mechanisms.

Mass spectrometry is a unique method that allows study of the reactivity of isolated metal-containing ions in the gas phase in the absence of solvent. A number of fundamental thermochemical characteristics of organometallic molecules and ions, such as ionization energies, proton affinities, electron affinities and metal–ligand bond dissociation energies, can be determined from mass spectrometry experiments.

One of the most promising applications of mass spectrometry to the chemistry of metal compounds is in the investigation of the reactivity of metal-containing ions and molecules in the gas phase. Information about transformations of molecules on metal centres can be provided by these experiments. This is especially profitable in the study of mechanisms of reactions involving elusive intermediates (catalytic processes, interstellar chemistry, etc.) that cannot be isolated or characterized by 'traditional' spectral methods.

Basics of the mass spectral analysis of metal complexes

The following types of molecular characterization and recognition of metal complexes can be performed using a combination of mass spectrometric methods. The determination of the molecular mass of the analyte is a common first step in the mass spectral analysis of an unknown (organometallic) compound. A variety of ionization techniques can be employed to obtain molecular ions or charged adducts with a specific ionizing reactant. The isotope distribution in the molecular ion cluster (most metal atoms have a well-recognized isotope pattern) allows a thorough preliminary evaluation of the kind and number of metal atoms present. High-resolution experiments (accurate mass measurement) on molecular ions or adducts provide information about the elemental composition of the compound of interest, in many instances replacing relatively expensive and time-consuming traditional elemental analyses.

Detailed information about the structure of metal complexes may be obtained from the reactivities of their ions, including dissociation processes and ion-molecule reactions. A general problem in the elucidation of the structure of organometallic compounds is the determination of ligands surrounding the central metal atom(s). A common approach to this problem is to break metal–ligand bond(s) by allowing the ion to dissociate. This dissociation may occur upon ionization in (molecular) ions having an excess of internal energy. A sequential loss of metal–ligand bonds also can be achieved by activation of the ion. The observed mass loss leads to the deduction of the elemental composition of the ligand. Similar information can be obtained from ligand substitution ion–molecule reactions. Note however, that the latter method is limited to fragment ions having a coordinatively unsaturated metal atom. Molecular ions are usually inert to neutral reactants.

The recognition of atom connectivity in ligands is not always a direct task and a combination of experimental data should be considered. Reactions involving losses of radical and neutral molecules as well as migration of groups to the metal atom are the most informative for the recognition of the structure and location of substituents in the ligand. The most abundant processes are those involving atoms and groups located in α and β positions of the substituent in ligands. It is useful to compare the mass spectral behaviour of the unknown compound with the reactivity of similar derivatives of well-established structure.

Some isomeric organometallic complexes may be recognized using mass spectral methods. The difference in the mass spectra of isomers usually comes from the interaction of the metal atom with electronegative groups or unsaturated bonds in the substituent. These interactions increase the bond strength between the metal atom and the ligand, resulting in a preferential loss of other ligands. If an electronegative group is involved in such an interaction, it may migrate to the metal atom. The extent of the latter reaction depends on how close the migrating group is to the metal atom. Both effects can be illustrated by the electron impact ionization-induced fragmentation of complexes [1], having a hydroxyl group in the *exo* or *endo* position relative to the metal atom. The intensities of peaks due to the loss of the unsubstituted cyclopentadienyl ring and the migration of OH to the Fe atom were higher for the *endo* isomer (**Table 1**) in which the hydroxy group is in close proximity to the metal atom.

Intramolecular interaction of the substituent with the metal atom makes it possible to distinguish *ortho*- from *meta*- and *para*-ferrocenylbenzenes, $C_5H_5FeC_5H_4C_6H_4R$ (R = NH_2, COOH, $COOCH_3$, $COCH_3$, etc.). The dissociation of molecular ions (P^+) of *ortho* isomers gave rise to a strong peak due to the loss of the unsubstituted cyclopentadienyl ring, whereas their *meta* and *para* analogues produced no or very low intensity $[P-C_5H_5]^+$ ions.

The interaction of the metal atom may also induce loss of a radical from the substituent. This process is usually more intense if the lost group is in the *exo* position (**Table 1**).

Another effect that may result in differentiation of isomers is a metal-catalysed loss of neutral molecules from ligands. The isomer having an electronegative group in closer proximity to the metal atom usually loses this group more readily as a part of a neutral molecule. For example, the electron-impact ionization mass spectrum of $C_5H_5FeC_5H_4CH(OH)CH_3$ displays a stronger peak corresponding to the loss of H_2O than the peak of the same mass in the EI mass spectrum of $C_5H_5FeC_5H_4CH_2CH_2OH$.

Metal-containing ions are useful reactants for identification of organic compounds. The formation of metal adducts is especially advantageous when traditional methods of ionization (electron-impact (EI) and chemical ionization (CI)) do not result in stable molecular ions or protonated species. Chemical ionization with metal and metal-containing ions provides high selectivity and sensitivity to specific types of analytes (unsaturated and functionalized hydrocarbons, peptides, crown ethers, polymers, etc.) and can be successfully used in GC-MS experiments. Regiospecific and stereospecific dissociation of metal adducts allows isomers to be distinguished. For example, the stereochemistry of the hydrogenated D_8-naphthalene, $C_{10}D_8H_4$, was deduced from the reactivity of the adduct with gas-phase Co^+ ions. Its collisional activation resulted in losses of H_2, $2H_2$, D_2 and $2D_2$, but no elimination of HD had been observed. These results were consistent with *cis* orientation of all hydrogen atoms in the organic molecule.

Interpretation of mass spectra of metal compounds

Two formal approaches are used for the interpretation of mass spectra of organometallic compounds. The concept of valence-change has been applied to rationalize and to predict the type of fragmentation for a variety of σ complexes and metal chelates, whereas the concept of charge localization on the metal atom was more advantageous for the interpretation of mass spectra of π complexes of transition metals.

The valence-change concept makes the following assumptions:

(1) The metal atom contains an even number of electrons in the parent neutral molecule.
(2) Stabilization of molecular ions occurs when the metal atom can increase its oxidation state (OS). Fragmentation of these complexes involves a loss of even-electron species (molecules).
(3) If the metal atom has only one stable OS then molecular ions have a low intensity and their fragmentation will probably involve a loss of a radical followed by losses of neutral molecules.
(4) If the metal atom can easily reduce OS then the dissociation of (relatively unstable) molecular ions will involve a predominant loss of two radicals.

Table 1 Relative intensities (relative to the molecular ion, P) of some ions in the mass spectra of *exo*-[1] and *endo*-[1]

Isomer		$[P-C_5H_5]^+$	$C_5H_5FeOH^+$	$[P-OH]^+$
Exo:	R¹=H R²=OH	0.1	0.1	1.6
Endo:	R¹=OH R²=H	0.6	0.33	1.7

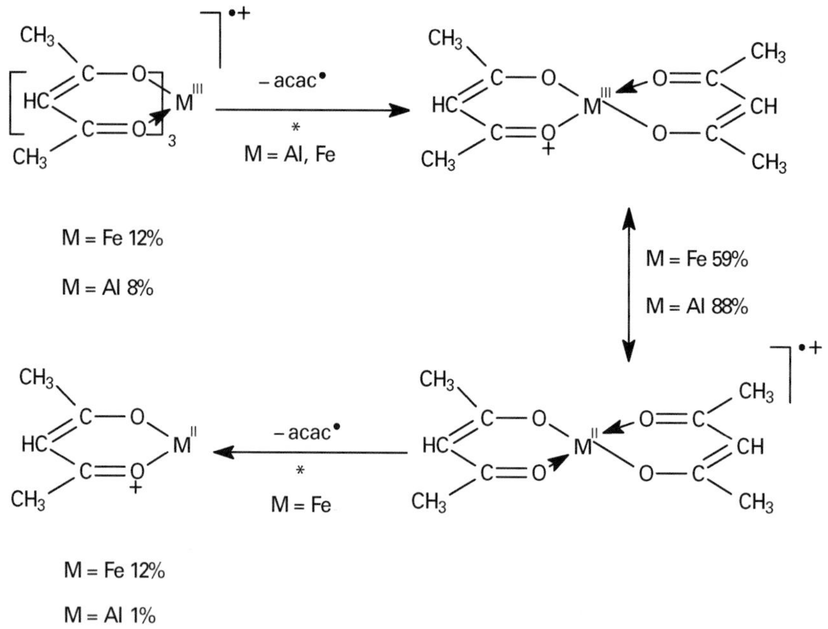

Scheme 1 Comparative fragmentation pathways of the Al(III) and Fe(III) acetylacetonates. Reproduced with permission of John Wiley & Sons Limited from Lacey MJ and Shannon JS (1972) Valence-change in the mass spectra of metal complexes. *Organic Mass Spectrometry* **6**: 931–937.

The valence-change effect is illustrated by the behaviour of aluminium (one stable OS: III) and iron (two stable OS: III and II) trisacetylacetonates, $M(acac)_3$ (**Scheme 1**). Both complexes produced modest molecular ions in the EI mass spectra with the dominant loss of one β-diketonate ligand producing $M(acac)_2^+$ ions in which metal atoms retained their stable OS III. The Fe-derivative also underwent a significant loss of the second ligand, giving rise to ions having the metal atom in the OS II. The abundance of this process in the EI mass spectrum of $Al(acac)_3$ was very low owing to the low stability of Al(II).

The dissociation of complexes having two metals with two stable oxidation states follow the above rules. The most abundant reactions result in ions containing the metal atom in stable oxidation states and ions having two different (stable) OS can be easily formed (**Scheme 2**). In spite of much formalism, this concept has made a positive contribution to the interpretation and prediction of mass spectra of inorganic and organometallic compounds.

The most successful interpretation of mass spectra of transition metal π complexes is based on the concept of charge localization on the metal atom. The ionization energies of the majority of organometallic

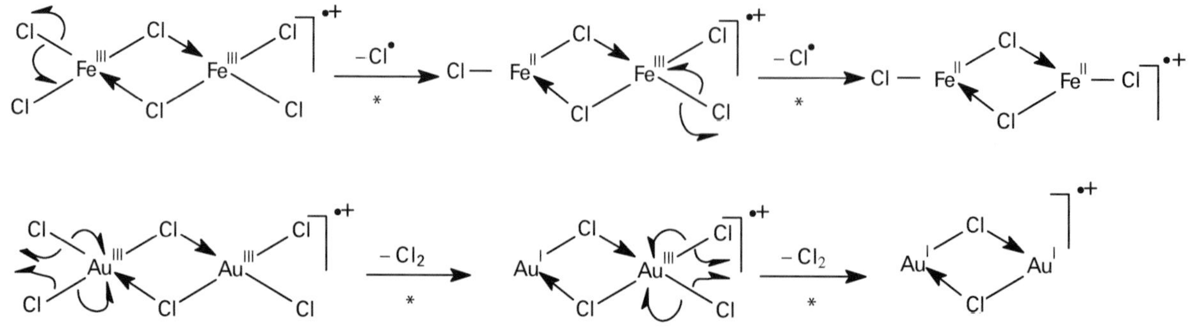

Scheme 2 Comparative ion fragmentations of Fe_2Cl_6 and Au_2Cl_6. Reproduced with permission of John Wiley & Sons Limited from Lacey MJ and Shannon JS (1972) Valence-change in the mass spectra of metal complexes. *Organic Mass Spectrometry*, **6**: 931–937.

molecules are lower than the ionization energies of the ligands. They differ only slightly from the ionization energies of the free metal atoms. Accordingly, the ionization of transition metal complexes probably involves the removal of an electron from the metal atom. As a result, on the decomposition of molecular ions of transition metal–hydrocarbon compounds, the positive charge usually remains on the metal-containing fragment. The other consequence of charge localization on the metal atom is that the central metal atom controls the dissociation of organometallic ions. Unlike ionized metal-free organic molecules, their metal complexes lose even-electron neutrals rather than radicals. Also, the mechanisms of similar reactions in the coordinated and metal-free organic molecules are often different. Participation ('catalysis') of the metal atom in transformations of the ligand regulates these mechanisms as illustrated in Scheme 3. The loss of ketene from the ionized acetamide involves a 4-centre cyclic transition state, whereas its adduct with Cr^+ ions dissociates via a (formally) 6-centre intermediate.

Strong support for charge localization on the metal atom is provided by the reactivity of organic molecules upon their interaction with metal ions. Similarly to the unimolecular dissociation of coordinated ligands, most of these reactions result in the loss of even-electron neutrals formed via direct interaction of leaving group(s) with the metal atom.

The formation of neutral metal-containing species upon fragmentation of organometallic ions does not contradict the initial charge localization at the metal atom. The positively charged metal atom is easily attacked by electronegative groups (as anions) present in ligands, forming stable metal-containing molecules:

$$C_5H_5FeC_5H_4COCl^+ \rightarrow C_5H_5Fe^{II}Cl + C_5H_4CO^+ \quad [1]$$

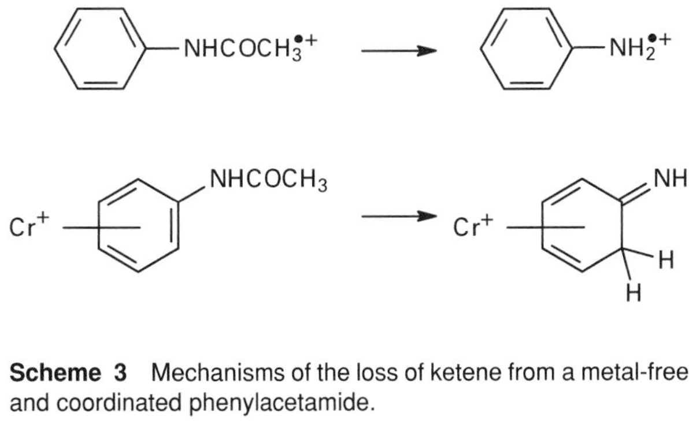

Scheme 3 Mechanisms of the loss of ketene from a metal-free and coordinated phenylacetamide.

This effect is common in the interaction of metal ions with functionalized organic reactants:

$$Mg^{\bullet+} + ClCH_2CH_2Cl \rightarrow Mg^{II}Cl_2 + C_2H_4^{\bullet+} \quad [2]$$

Thermochemistry of organometallic molecules and ions

Mass spectrometry can be used to determine fundamental thermochemical properties of organometallic molecules such as ionization energy (IE), electron affinity (EA) and proton affinity (PA). IEs can be obtained by determining ionization thresholds of electron induced ionization (adiabatic IE) or photoionization (vertical IE). Electron exchange between a positively charged ion and a neutral molecule (electron transfer bracketing) allows the estimation of adiabatic IEs. In this approach, various reference compounds are introduced and the observation of (direct or reverse) electron transfer reactions indicates which molecule has the lower IE. If it is possible to establish accurately a pressure for the neutral reactant, then the electron transfer equilibrium can be measured to give IE values with high accuracy. This method requires the use of reference compounds having well-established ionization energies.

Similar experiments, involving electron transfer between an anion and neutral molecule, yield relative or absolute EAs. The method has been used to determine relative free energies for electron attachment for a variety of metallocenes and β-diketonate molecules. The results for tris(hexafluoracetylacetonate) and tris(acetylacetonate) showed a distinctive metal effect on EAs in the order Cr < Fe < Co < Mn. Measurements of electron attachment energies provide an important component of the thermochemical cycles involving oxidation/reduction of metal complexes. The latter may be used to obtain values of average heterolytic metal–ligand gas-phase bond energies and homolytic metal–ligand bond energies and to understand the thermochemistry of metal ion–solvent interaction.

The analysis of substituent effects in metal complexes is one of the goals of electron transfer equilibrium studies. For example, alkyl substitution in metallocenes predictably decreases their IEs. At the same time it has been shown that the gas-phase EAs can be increased by a larger alkyl substituent in the hydrocarbon ligand (Figure 1).

Anion transfer reactions to/from metal complexes are sources for anion affinities of organometallic molecules. To illustrate, the hydroxide affinity of $(CO)_5Fe$ has been determined by measurement of the

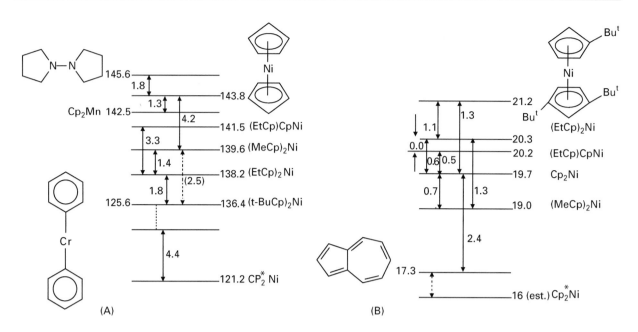

Figure 1 Electron transfer equilibrium ladders showing free energies of ionization (A) and electron attachment (B). Cp* denotes the pentamethylcyclopentadienyl ligand. Reproduced with permission of the American Chemical Society from Richardson DE, Ryan MF, Khan MDNI and Maxwell KA (1992) Alkyl substituent effects in cyclopentadienyl metal complexes; trends in gas-phase ionization and electron attachment energetics of alkylnickelocenes. *Journal of the American Chemical Society*, **114**: 10482–10485.

equilibrium constant for hydroxide transfer exchange between $(CO)_5Fe$ and SO_2. This value was used to estimate the heat of formation of $(CO)_4FeCOOH^-$. Ion–molecule reactions of these ions and their collision-induced dissociation gave rise to a variety of negatively charged species having a coordinatively unsaturated metal atom. Study of their reactivity is a good source for obtaining thermochemical characteristics of elusive metal complexes.

Proton transfer equilibrium measurements and proton transfer bracketing methods are the sources for proton affinity values of organometallic complexes. The determination of the site of protonation, i.e. metal atom versus a ligand, is a fundamental dilemma of any study on the protonation of metal complexes. It was demonstrated, for example, that $Fe(CO)_5$ was protonated exclusively at the metal

atom, whereas the results for the proton transfer to ferrocene can be explained by the formation of a metal-protonated compound [2a] and a ring-protonated [2b] form. The observation of hydrogen (deuterium) atom exchange between the cyclopentadienyl rings in $[(C_5H_5)_2Fe]D^+$ and $[(C_5D_5)_2Fe]H^+$ ions was rationalized by the agostic interaction with the metal atom (structure [2c]).

Chemistry of metal–ligand bonds

Knowledge of metal–ligand bond energies is fundamental information for organometallic chemistry. It is essential for the understanding of catalytic reaction mechanisms, which often involve the cleavage or the formation of these bonds. Mass spectrometry offers a series of experimental methods for determining absolute and relative bond strengths between a positively (or negatively) charged metal centre and organic (or inorganic) ligands. Direct determination of metal–ligand bond dissociation energies can be performed by measuring appearance energies (AEs) of the molecular and fragment ions. The best results are obtained from AEs of ions produced by metastable dissociation of mass-selected precursors. The kinetic energy release distribution (KERD) during metastable dissociation of ions is another source for the quantitative characterization of metal–ligand bonds. The experimental results obtained by this method require theoretical calculation to extract the information on the enthalpy change for the observed processes. Photoelectron–photoion coincidence is also a useful source for determining the thermochemistry of metal-containing ions.

A large number of both relative and absolute bond energies in metal-containing ions have been measured by the method that depends upon the abundances of products formed by competitive ligand loss (the kinetic method). The metal–ligand bond enthalpies can be determined from the metastable and collision-induced dissociation of LM^+L' ions, where L and L′ are different molecules. The general trend in metal–ligand bond dissociation energies is that the larger alkyl derivatives are more strongly bound to the metal atom than are their smaller homologues. A study of systems containing three and four ligands at the metal atom, L_2M^+L' and $L_2M^+L'_2$ reveals information about relative molecular pair and molecular triplet metal ion affinities. A particular advantage of the kinetic and some other methods is that they can probe ions containing thermally unstable ligands whose chemistry is difficult to study in the condensed phase.

A variety of other methods for obtaining metal–ligand bond dissociation characteristics employ activation of the ion of interest to initiate its dissociation. In addition to activation by collisions with a target gas, the dissociation of metal–containing ions of interest can be induced by photoexcitation, colliding them into a surface, or bombarding them with electrons.

Ion–molecule ligand exchange reactions are convenient for obtaining relative and absolute metal–ligand bond strengths. Using this approach, the affinities of molecules for a variety of 'bare' and ligated metal ions have been determined. From these data the order of relative softness of the Lewis acids was $H^+ < Al^+ << Mn^+ \leq FeBr^+ < Co^+ \leq C_5H_5Ni^+ < Ni^+ < Cu^+$. The metal ion affinities obtained represented relative metal–ligand binding energies, which can also be measured for two-ligand systems, $D^0(M^+\text{-}2L)$, as well as for negatively and doubly charged positive ions. An important application of ligand exchange

reactions is the determination of relative metal ion affinities for molecules of biological origin, e.g. peptides and amino acids.

Monitoring of the thresholds of endothermic reactions by the guided ion beam technique is one of the methods used to obtain the most accurate bond energies. Similar measurements can be performed employing (Fourier transform) ion cyclotron resonance (FT-ICR) spectroscopy. General types of endothermic reactions studied to obtain bond dissociation energies are shown in Equations [3]–[5] (M^+ is a 'bare' or ligated metal ion):

$$M^+ + RX \longrightarrow MX^+ + R \quad [3]$$
$$M^+ + RX \longrightarrow MX + R^+ \quad [4]$$
$$MY^+ + RX \longrightarrow MX^+ + RY \quad [5]$$

If the thermochemistry of the reactants (M^+, MY^+, RX) and one of the reaction products (R, R^+, RY) is well established, then using the experimentally measured threshold energy of reaction one can calculate M^+–X and M–X bond strengths. Important organometallic intermediates, such as metal hydrides, carbynes, carbenes and metal-alkyls, have been characterized by the above methods and their ionic and neutral heats of formation were deduced from metal–ligand bond energies. An extensive study of exothermic and thermoneutral ion–molecule reactions provide complementary and in many cases unique information on the thermochemistry of metal-containing species. All of these results have made a great contribution to the understanding of metal–ligand bonding and of periodic trends in transition metal–ligand bonds, and to the evaluation of the multiplicity of metal–ligand bonds and other fundamental topics of organometallic chemistry.

The results on metal–ligand bond dissociation energies (strengths) obtained by different methods are usually in good agreement. For example, the Fe^+–CO bond dissociation energies determined from KERD, ion beam and FT-ICR experiments (32 ± 5, 31.3 ± 1.8 and 37 ± 6 kcal mol^{-1}, respectively) were very similar and close to the theoretically calculated value of 30.3 kcal mol^{-1}, whereas the appearance energy measurements for FeCO$^+$ and Fe$^+$ ions in the EI mass spectrum of Fe(CO)$_5$ gave significantly higher values (between 41.5 and 51.4 kcal mol^{-1}). Good agreement between two or more methods supported by the results of high-level theoretical calculations is usually a sufficient criterion for which number should be used in thermochemical calculations (33 kcal mol^{-1} is probably a good estimate for Fe^+–CO). However, a critical analysis of the experimental techniques, understanding their limitations and sources of errors, as well as knowledge of trends in changing bond dissociation energies is highly recommended before a specific thermochemical value is accepted.

Transformations of ligands on charged metal centres

Mass spectrometry is widely used for studying reaction mechanisms involving metal-containing reaction intermediates. A majority of these studies involve the investigation of transformations of organic molecules on ligated or 'bare' metal ions. Reactions [1]–[5] provide a few examples from a large number of processes involving bond cleavage within ligands. Particular interest in this field of research is focused on intrinsic mechanisms of metal ion-induced C–H, C–C and C–heteroatom bond activation reactions. The results of these studies help the understanding and modelling of the elementary stages of important homogeneous and heterogeneous catalytic processes, metal ion biochemistry, synthesis of electronic and ceramic materials, metal–solvent interactions, reactions taking place in interstellar systems, etc.

A series of experiments on the reactivity of metal ions with nitriles (RCN) led to the discovery of the remote functionalization mechanism. The initial interaction of the metal ion involves coordination at the nitrile group. The insertion of the metal atom into a C–H or C–C bond occurs only after the alkyl chain becomes long enough (at least 3 or 4 methylene groups) to interact with a remote bond. The dissociation of the metal–hydride (Scheme 4) or metal–alkyl intermediate results in a loss of H_2, alkene or alkane molecules depending on the structure of the hydrocarbon group R.

Other important reactive metal-containing intermediates (e.g. metal–benzene complexes) and processes (e.g. decarboxylation of ketones by metal ions) of practical interest have been characterized using various mass special methods and provide insight into the mechanisms of organometallic reactions.

Elusive neutral organometallics generated from ions

Neutralization–reionization mass spectrometry (NRMS) is a unique mass spectral technique that allows the generation of *neutral* species from their charged counterparts. The major application of NRMS is to produce unstable reaction intermediates that cannot be isolated or characterized by other means to yield new, previously unknown, molecules and radicals. The method has been used successfully

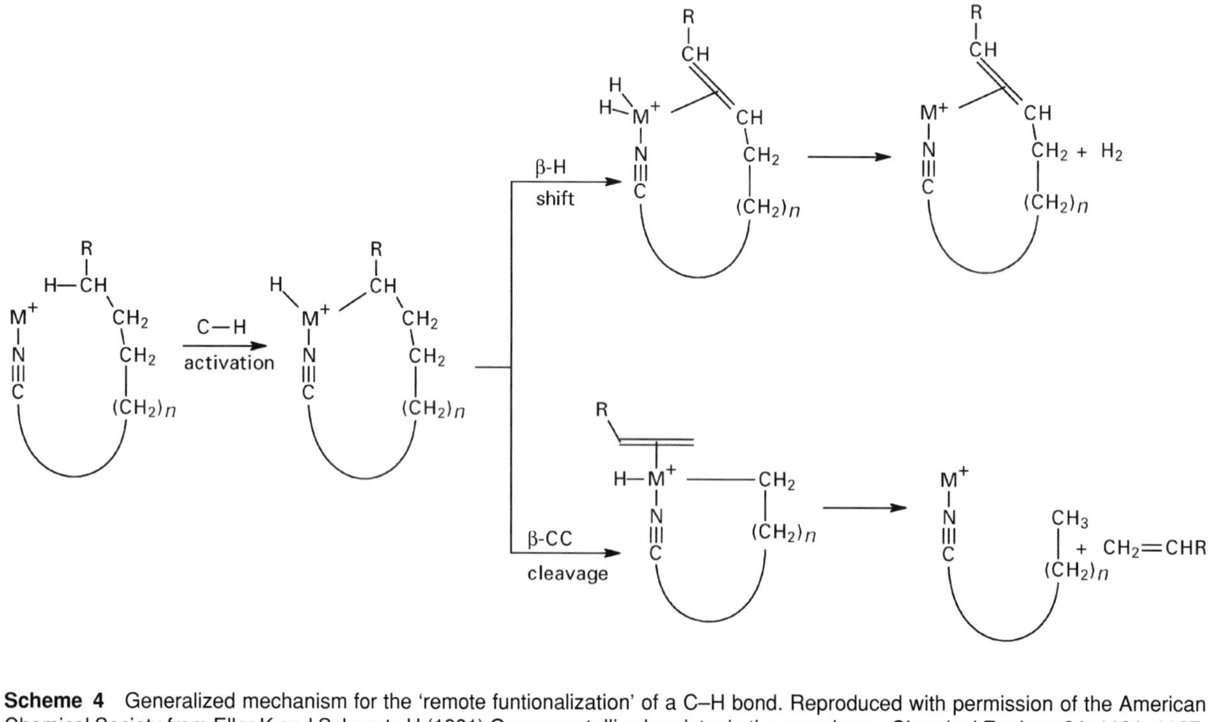

Scheme 4 Generalized mechanism for the 'remote funtionalization' of a C–H bond. Reproduced with permission of the American Chemical Society from Eller K and Schwartz H (1991) Organometallic chemistry in the gas phase. *Chemical Reviews* **91**: 1121–1127.

to generate a variety of organometallic species. A wide range of elusive metal-containing molecules (AuF, PrF, CH_2SiH_2, $C_5H_5Rhacac$) and radicals (NiCCH, $C_5H_5FeC_5H_4CO$, $(C_5H_5)_2Zr$ and others) have been generated and characterized for the first time using the NRMS method. The observation of these species in the gas phase suggests their possible formation in other than gas-phase experimental conditions, at least as short-lived reaction intermediates. This information is used by chemists to confirm and evaluate reaction mechanisms, including elementary steps in catalytic transformations on metal centres. **Scheme 5** represents a common sequence in the NRMS experiment for the generation of elusive organometallic species. Complementary information about the intrinsic stability of molecules and radicals can also be obtained by identifying the neutral products of ion fragmentation using the collision-induced dissociative ionization method followed by the detection of its positively charged counterpart. This technique was used, for example, for the observation of the neutral ferrocenyloxy radical originating from the dissociation of ferrocenylbenzoate to benzoyl ion (see **Scheme 5**).

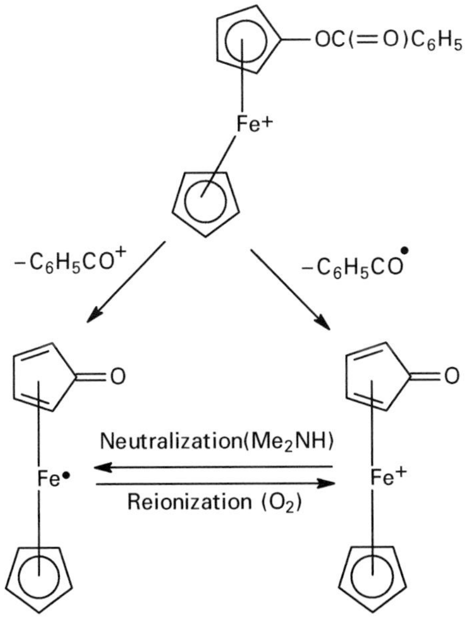

Scheme 5 Generation of ferrocenyloxy radical from the ionized ferrocenylbenzoate.

See also: **Fragmentation in Mass Spectrometry; Ion Energetics in Mass Spectrometry; Ion Molecule Reactions in Mass Spectrometry; Metastable Ions; Neutralization-Reionization in Mass Spectrometry; Photoelectron-Photoion Coincidence Methods in Mass Spectrometry (PEPICO); Photoionization and Photodissociation Methods in Mass Spectrometry; Proton Affinities.**

Further reading

Bruce MI (1968) Mass spectra of organometallic compounds. In: Stone FGA and West R (eds) *Advances in Organometallic Chemistry*, Vol 6, pp 273–333. New York: Academic Press.

Cais M and Lupin MS (1970) Mass spectra of metallocenes and related compounds. In: Stone FGA and West R (eds) *Advances in Organometallic Chemistry*, Vol 8, pp 211–333. New York: Academic Press.

Charalambous J (ed) (1975) *Mass Spectrometry of Metal Compounds*. London: Butterworths.

Freiser BS (1994) Selected topics in organometallic ion chemistry. *Accounts of Chemical Research* 27: 353–360.

Freiser BS (ed) (1996) *Organometallic Ion Chemistry*. Dordrecht: Kluwer.

Eller K and Schwarz H (1991) Organometallic chemistry in the gas phase. *Chemical Reviews* 91: 1121–1177.

Lacey MJ and Shannon JS (1972) Valence-change in the mass spectra of metal complexes. *Organic Mass Spectrometry* 6: 931–937.

Litzov MR and Spalding PH (1973) *Mass-Spectrometry of Inorganic and Organometallic Compounds*. Amsterdam: Elsevier.

Marks TJ (ed) (1990) *Bonding Energetics in Organometallic Chemistry*, Washington, DC: American Chemical Society.

Mass Spectrometry (Specialist Periodic Report Vols 1–10) (1971–1989). London: The Chemical Society.

Muller J (1972) Decomposition of organometallic complexes in the mass spectrometer. *Angewandte Chemie, International Edition in English* 11: 653–665.

Squires RR (1987) Gas-phase transition-metal negative ion chemistry. *Chemistry Reviews* 87: 623–646.

Zagorevskii DV and Holmes JL (1994) Neutralization-reionization mass spectrometry applied to organometallic and coordination chemistry. *Mass Spectrometry Reviews* 13: 133–154.

Oriented Molecules Studied By Chiroptical Spectroscopy

See Chiroptical Spectroscopy, Oriented Molecules and Anisotropic Systems.

Osmium NMR, Applications

See Heteronuclear NMR Applications (La–Hg).

Oxygen NMR, Applications

See Heteronuclear NMR Applications (O, S, Se, Te).

³¹P NMR

David G Gorenstein and **Bruce A Luxon**, University of Texas Medical Branch, Galveston, TX, USA

MAGNETIC RESONANCE
Applications

Introduction

Although ³¹P NMR spectra were reported as early as 1951, it was the availability of commercial multinuclear NMR spectrometers in about 1955 that led to the application of ³¹P NMR as an important analytical tool for structure elucidation. Early spectrometers generally required neat samples in large nonrotating tubes (8–12 mm OD). By the mid-1960s NMR became more sensitive and the availability of higher field electromagnets and signal averaging spurred rapid growth in the number of reported ³¹P spectra and the publication of the first monograph devoted to this field.

With the introduction of Fourier-transform (FT) and high-field superconducting-magnet NMR spectrometers in about 1970, ³¹P NMR spectroscopy expanded beyond the study of small organic, organometallic and inorganic compounds to biological phosphorus-containing compounds as well. The latest multinuclear FT NMR spectrometers have reduced if not eliminated the serious limitation to the widespread utilization of phosphorus NMR, which is the low sensitivity of the phosphorus nucleus (6.6% at constant field compared to ¹H NMR; magnetogyric ratio γ, 10.839×10^7 rad T^{-1} s^{-1}; NMR frequency at 4.7 T, 80.96 MHz). Today, routinely, millimolar (or lower) concentrations of phosphorus nuclei in as little as 0.3 mL of solution are conveniently monitored. The ³¹P nucleus has other convenient NMR properties making it suitable for FT NMR: spin $\frac{1}{2}$ (which avoids problems associated with quadrupolar nuclei), 100% natural abundance, moderate relaxation times (providing relatively rapid

signal averaging and sharp lines), and a wide range of chemical shifts (>2000 ppm).

In this article the interpretation of various ³¹P NMR spectroscopic parameters, particularly chemical shifts and coupling constants will be described. A major emphasis will be placed on newer developments in ³¹P NMR methods which have considerably expanded the utility of this important spectroscopic probe in organic and biological structure determination.

Practical considerations

Today's commercial NMR spectrometers cover the ³¹P frequency range from 24 to 323 MHz (on a 800 MHz ¹H NMR spectrometer). Generally, for small, phosphorus-containing compounds, high signal-to-noise and resolution requirements dictate use of as high a magnetic field strength as possible since both sensitivity and chemical-shift dispersion increase at higher operating frequency (and field). However, consideration must be given to field-dependent relaxation mechanisms such as chemical shift anisotropy which can lead to substantial line broadening of the ³¹P NMR signal at high fields. Indeed, especially for larger biomolecules, sensitivity is often poorer at very high fields because of considerably increased line widths. The latest probe designs provide a remarkable improvement in signal-to-noise. Indirect detection 2D NMR experiments have further improved sensitivity.

Typical acquisition times for the 1D ³¹P free induction decay (FID) following a 90° radiofrequency pulse are 1–8 s depending on the required resolution (dictated by the line width of the signal, $1/\pi T_2^*$, where T_2^* is the time constant for the FID). Waiting longer than $2T_2^*$ will generally not improve the signal-to-noise ratio (S/N). Additional consideration for optimization of the S/N must be given to the time it takes for the ^{31}P spins to return to thermal equilibrium after a 90° radiofrequency pulse, which is roughly 3 times the spin–lattice relaxation time (T_1). If $T_1 \approx T_2^*$, as would be true for small phosphorus-containing molecules where magnetic field inhomogeneity and paramagnetic impurities do not lead to any additional line broadening, then a waiting period between pulses of $3T_2^*$ provides a good compromise between adequate resolution and signal sensitivity. If $T_1 > T_2^*$, as is often the case in larger biomolecular systems, then waiting only $3T_2^*$ does not allow the magnetization to return to equilibrium and an additional delay must generally be introduced so that the total time between pulses is ~$3T_1$. This wait can be substantially shortened if the Ernst relationship is used to set the pulse flip angles to < 90°. At low field, 60–70° pulses, 4 to 8 k data points and 2.0–5.2 s recycle times are generally used. The spectra are generally broadband ^{1}H decoupled.

The ^{31}P spectra are generally referenced to an external sample of 85% H_3PO_4 or trimethylphosphate which is ~3.46 ppm downfield of 85% H_3PO_4. Note that throughout this review the IUPAC convention is followed so that *positive values are to high frequency* (low field). One should cautiously interpret reported ^{31}P chemical shifts because the early literature (pre-1970s) and even many later papers use the opposite sign convention.

Quantification of peak heights

The intensity of a resonance can be measured in several ways: (1) peak heights and areas obtained from the standard software supplied by the spectrometer manufacturer, (2) peak heights measured by hand, (3) peaks cut and weighed from the plotted spectrum, and (4) peaks fitted to a Lorentzian line shape. For flat baselines, intensity measurements are generally straightforward. However, in the event of curved baselines the measurements are somewhat uncertain and manual measurements are generally more reliable than intensity values obtained from computer software.

It is often necessary that experiments be carried out without allowing time for full recovery of longitudinal magnetization between transients because of the limited availability of spectrometer time or of the limited lifetime of the sample. Because of variations in T_1 between different phosphates and variation in

the heteronuclear NOE to nearby protons, care should be made in interpretation of peak area and intensities. Addition of a recycle delay of at least $5 \times T_1$ between pulses and gated decoupling only during the acquisition time to eliminate the ^{1}H–^{31}P NOE largely eliminates quantification problems.

^{31}P chemical shifts

Introduction and basic principles

The interaction of the electron cloud surrounding the phosphorus nucleus with an external applied magnetic field B_0 gives rise to a local magnetic field. This induced field shields the nucleus, with the shielding proportional to the field B_0 so that the effective field, B_{eff}, felt by the nucleus is given by

$$B_{eff} = B_0(1 - \sigma) \qquad [1]$$

where σ is the shielding constant. Because the charge distribution in a phosphorus molecule will generally be far from spherically symmetrical, the ^{31}P chemical shift (or shielding constant) varies as a function of the orientation of the molecule relative to the external magnetic field. This gives rise to a chemical-shift anisotropy that can be defined by three principal components, σ_{11}, σ_{22} and σ_{33} of the shielding tensor. For molecules that are axially symmetrical, with σ_{11} along the principal axis of symmetry, $\sigma_{11} = \sigma_\parallel$ (parallel component), and $\sigma_{22} = \sigma_{33} = \sigma_\perp$ (perpendicular component). These anisotropic chemical shifts are observed in solid samples and liquid crystals, whereas for small molecules in solution, rapid tumbling averages the shift. The average, isotropic chemical shielding σ_{iso} (which would be comparable to the solution chemical shift) is given by the trace of the shielding tensor or

$$\sigma_{iso} = \frac{1}{3}(\sigma_{11} + \sigma_{22} + \sigma_{33}) \qquad [2]$$

and the anisotropy $\Delta\sigma$ is given by

$$\Delta\sigma = \sigma_{11} - \frac{1}{2}(\sigma_{22} + \sigma_{33}) \qquad [3]$$

or, for axial symmetry,

$$\Delta\sigma = \sigma_\parallel - \sigma_\perp \qquad [4]$$

Theoretical 31P chemical shift calculations and empirical observations

Three factors appear to dominate ^{31}P chemical shift differences $\Delta\delta$, as shown by

$$\Delta\delta = -C\Delta\chi_X + k\Delta n_\pi + A\Delta\theta \qquad [5]$$

where $\Delta\chi_X$ is the difference in electronegativity in the P–X bond, Δn_π is the change in the π-electron overlap, $\Delta\theta$ is the change in the σ-bond angle, and C, k, and A are constants.

As suggested by Equation [5], electronegativity effects, bond angle changes, and π-electron overlap differences can all potentially contribute to ^{31}P shifts in a number of classes of phosphorus compounds. While these semiempirical isotropic chemical-shift calculations are quite useful in providing a chemical and physical understanding for the factors affecting ^{31}P chemical shifts, they represent severe theoretical approximations. More exact *ab initio* chemical-shift calculations of the shielding tensor are very difficult although a number of calculations have been reported on phosphorus compounds. Whereas the semiempirical theoretical calculations have largely supported the importance of electronegativity, bond angle, and π-electron overlap on ^{31}P chemical shifts, the equations relating ^{31}P shift changes to structural and substituent changes unfortunately are not generally applicable. Also, because ^{31}P shifts are influenced by at least these three factors, empirical and semiempirical correlations can only be applied to classes of compounds that are similar in structure. It should also be emphasized again, that structural perturbations will affect ^{31}P chemical shift *tensors*. Often variations in one of the tensor components will be compensated for by an equally large variation in another tensor component with only a small net effect on the isotropic chemical shift. Interpretation of variations of isotropic ^{31}P chemical shifts should therefore be approached with great caution.

Within these limitations, a number of semiempirical and empirical observations and correlations, however, have been established and have proved useful in predicting ^{31}P chemical-shift trends. Indeed, unfortunately, no single factor can readily rationalize the observed range of ^{31}P chemical shifts (**Figure 1**).

Bond angle effects Changes in the σ-bond angles appear to make a contribution ($|A|$, Equation [5]) to the ^{31}P chemical shifts of phosphoryl compounds, although electronegativity effects apparently predominate. Empirical correlations between ^{31}P chemical shifts and X–P–X bond angles can be found, although success here depends on the fact that these correlations deal with only a limited structural variation: in the case of phosphate esters, it is the number and chemical type of R groups attached to a tetrahedron of oxygen atoms surrounding the phosphorus nucleus. For a wide variety of different alkyl phosphates (mono-, di-, and triesters, cyclic and acyclic neutral, monoanionic, and dianionic esters), at bond angles < 108° a decrease in the smallest O–P–O bond

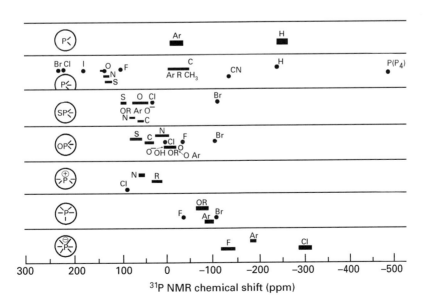

Figure 1 Typical ^{31}P chemical shift ranges for phosphorus bonded to various substituents in different oxidation states. (P– indicates the P_4 molecule.)

angle in the molecule generally results in a deshielding (downfield shift) of the phosphorus nucleus.

Torsional angle effects on ^{31}P chemical shifts Semi-empirical molecular orbital calculations and *ab initio* gauge-invariant-type molecular orbital, chemical-shift calculations suggested that ^{31}P chemical shifts are also dependent on P–O ester torsional angles which has been shown to be of great value in analysis of DNA structure (see below). The two nucleic acid P–O ester torsional angles, ζ (5'-O–P) and α (3'-O–P) are defined by the (5'-O–P–O-3') backbone dihedral angles. These chemical-shift calculations and later empirical observations indicated that a phosphate diester in a B_I conformation (both ester bonds gauche(–) or –60°) should have a ^{31}P chemical shift 1.6 ppm upfield from a phosphate diester in the B_{II} conformations (α = gauche(–); ζ = trans or 180°).

^{31}P signal assignments

If the proton spectra of the molecule has been previously assigned, then 2D ^{31}P–^{1}H heteronuclear correlation NMR spectroscopy can generally provide the most convenient method for assigning ^{31}P chemical shifts in complex spectra. Whilst application of these experiments to DNA is clear, the 2D methods will of course equally apply to organophosphorus compounds as well.

Conventional 2D ^{31}P–^{1}H heteronuclear shift correlation (HETCOR) NMR spectroscopy, the 2D long-range COLOC (correlation spectroscopy via long range coupling) experiment and indirect detection (^{1}H detection) HETCOR experiments can be used to assign multiple ^{31}P signals in complex spectra such as those of oligonucleotide duplexes. Additional 2D heteronuclear *J* cross-polarization hetero TOCSY (TOCSY = total correlation spectroscopy), 2D heteronuclear TOCSY-NOESY (NOESY = nuclear Overhauser effect spectroscopy), and even a 3D hetero TOCSY-NOESY experiment can be used if additional spectral dispersion, by adding a third frequency dimension, is desirable. This may prove to be extremely valuable for ribo-oligonucleotides where very little ^{1}H spectral dispersion in the sugar proton chemical shifts is unfortunately observed.

Generally these 2D experiments correlate ^{31}P signals with coupled ^{1}H NMR signals. Assuming the ^{1}H NMR spectra have been assigned, these methods allow for direct assignment of the ^{31}P signals. The HETCOR measurements, however, suffer from poor sensitivity as well as poor resolution in both the ^{1}H and ^{31}P dimensions, especially for larger biomolecular structures. The poor sensitivity is largely due to the fact that the ^{1}H–^{31}P scalar coupling constants are

generally about the same size or smaller (except for organophosphorus molecules with directly bonded hydrogens) than the ^{1}H–^{1}H coupling constants. Sensitivity is substantially improved by using a heteronuclear version of the 'constant time' coherence transfer technique, referred to as COLOC and originally proposed for ^{13}C–^{1}H correlations.

An example of a 2D HETCOR spectrum of the self-complementary 14-base-pair oligonucleotide duplex d(TGTGAGCGCTCACA)$_2$, is shown in **Figure 2**. The cross-peaks represent scalar couplings between ^{31}P nuclei of the backbone and the H3' and H4' deoxyribose protons. Assuming that the chemical shifts of these protons have been assigned (by ^{1}H–^{1}H NOESY and COSY spectra) the ^{31}P signals may be readily assigned (COSY = homonuclear chemical shift correlation spectroscopy).

Coupling constants

Directly bonded phosphorus coupling constants $^1J_{PX}$

One bond P–X coupling constants (J_{PX}) have generally been rationalized in terms of a dominant Fermi-contact term

$$J_{PX} = \frac{A a_P^2 a_X^2}{1 + S_{PX}^2} + B \qquad [6]$$

where A and B are constants, a_P^2 and a_X^2 are percentage s character on phosphorus and atom X, respectively, and S_{PX} is the overlap integral for the P–X bond. Because the Fermi-contact spin–spin coupling mechanism involves the electron density at the nucleus (hence the s-orbital electron density), an increase in the s character of the P–X bond is generally associated with an increase in the coupling constant. The percentage s character is determined by the hybridization of atoms P and X, and as expected sp^3-hybridized atoms often have $^1J_{PX}$ larger than p^3 hybridized atoms. Thus $^1J_{PH}$ for phosphonium cations of structure PH$_n$R$_{4-n}^+$ with sp^3 hybridization are ~500 Hz, whereas $^1J_{PH}$ for phosphines PH$_n$R$_{3-n}$ with phosphorus hybridization of approximately p^3 are smaller, ~200 Hz. Furthermore, as the electronegativity of atom X increases, the percentage s character of the P–X bond increases, and the coupling constant becomes more positive. In many cases, however, these simple concepts fail to rationalize experimental one-bond P–X coupling constants (**Table 1**) because other spin–spin coupling mechanisms can also contribute significantly to the coupling constant. For tetravalent phosphorus, a very good correlation

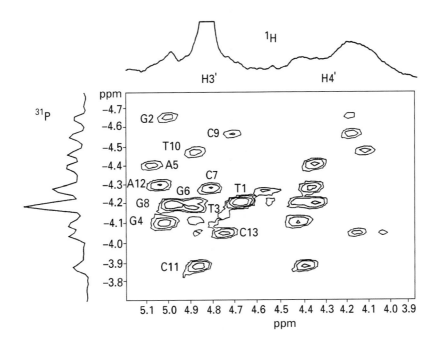

Figure 2 Pure absorption phase ³¹P–¹H heteronuclear correlation spectrum of tetradecamer duplex d(TGTGAGCGCTCACA)₂ at 200 MHz (¹H). ³¹P chemical shifts are reported relative to trimethyl phosphate which is 3.456 ppm downfield from the 85% phosphoric acid. Reproduced with permission.

is found between $^1J_{PC}$ and the phosphorus 3s–carbon 2s bond orders, the percentage s in the P–C bonding orbital in going from alkyl to alkenyl to alkynyl (sp³ → sp² → sp), and $^1J_{PC}$. Calculations and empirical observations on trivalent phosphorus compounds are *not* successful however, and suggest that the Fermi-contact contribution only dominates tetravalent phosphorus compounds.

One-bond P–H coupling constants appear always to be positive and vary from about +120 to +1180 Hz. Other heteroatom one-bond P–X coupling constants vary over a similar wide range and can be either positive or negative. The expected range of values is given in **Table 1**.

Two bond coupling constants: $^2J_{PX}$

Two-bond $^2J_{PX}$ coupling constants may be either positive or negative and are generally smaller than one-bond coupling constants (**Table 2**). The $^2J_{PCH}$ and $^2J_{PCF}$ constants are stereospecific and a Karplus-like dihedral dependence to the two-bond coupling constant (H or F)–C–P–X (X = lone pair or heteroatom) has been found. Thus in the *cis*- and *trans*-phosphorinanes, the $^2J_{PC}$ constants are 0.0 and 5.1 Hz in the cis- and trans-isomers, respectively.

Three-bond coupling constants, $^3J_{PX}$

Three-bond coupling constant, $^3J_{PX}$, through intervening C, N, O, or other heteroatoms are generally

Table 1 One-bond phosphorus spin–spin coupling constants $^1J_{PX}$

Structural class (or structure)	^1J (Hz)[a]	Structural class (or structure)	^1J (Hz)[a]
P(II)		P(IV) (continued)	
—PH₂⁻	139	\P(O)H/	460–1030
\PH/	180–225	—P(M)F/	1000–1400
\P—C/	0–45	(M=O, S)	
\PF/	820–1450	\P(S)H/	490–650
\P—P/	100–400	P(V)	
P(IV)		\\|−PH/|	700–1000
\+−PH/	490–600	\\|P—F/|	530–1100
\−PC/	50–305	PF₅	938
+P(CH₃)₄	+56	P(VI)	
		PF₆⁻	706

[a]For structural classes, only absolute value for *J* is given.

Table 2 Two-bond phosphorus spin–spin coupling constants $^2J_{PX}$

Structural class (or structure)	$^3J\ (HZ)^a$	Structural class (or structure)	$^3J\ (HZ)^a$
P(III)		P(IV) (Continued)	
>PCH	0–18	—P⁺CH	12–18
P(CH₃)₃	+2.7	—P⁺CC	0–40
>PCF	40–149	P⁺(C₂H₅)₄	–4.3
P(CF₃)₃	85.5	—POC	–6
>PNH	13–28	P(V)	
>PCC	12–20	>PCH (|)	10–18
P(C₂H₅)₃	+14.1	>PCF (|)	124–193
>POC	10–12	P(VI)	
>PXP	70–90	—PCF (|)	130–160
(X=S, C) P(IV)			
>P(O)CH	7–30		
P(O)(CH₃)₃	–12.8, –13.4		

^{a}For structural classes, only the absolute value for J is given.

<20 Hz (**Table 3**). The dihedral-angle dependence of vicinal $^3J_{POCH}$ coupling $^3J_{PCCH}$ and $^3J_{PCCC}$ has been demonstrated. The curves may be fitted to the general Karplus equation

$$J(\phi) = A\cos^2\phi + B\cos\phi + C \qquad [7]$$

where ϕ is the dihedral angle and A, B and C are constants for the particular molecular framework. Caution is recommended when attempting to apply these Karplus equations and curves to classes of phosphorus compounds that have not been used in

establishing these relationships because separate correlations and values for the constants A, B, and C in Equation [7] probably exist for each structural class. In all cases, a minimum in these Karplus curves is found at ~90°.

Applications to nucleic acid structure

The Karplus-like relationship between HCOP and CCOP dihedral angles and $^3J_{HP}$ and $^3J_{CP}$ three-bond coupling constants, respectively, has been used to determine the conformation about the ribose–phosphate backbone of nucleic acids in solution. Torsional angles about both the C3′–O3′ and C5′–O5′ bonds in 3′,5′-phosphodiester linkages have been determined from the coupled ¹H and ³¹P NMR spectra.

Within the limitations just described for the general application of the Karplus relationship, the best Karplus relationship for the nucleotide H3′–P coupling constants appears to be

$$J = 15.3\cos^2(\theta) - 6.1\cos(\theta) + 1.6$$

Table 3 Three-bond phosphorus spin–spin coupling constants $^3J_{PX}$

Structural class (or structure)	$^3J\ (HZ)^a$	Structural class (or structure)	$^3J\ (HZ)^a$
P(III)		P(IV) (Continued)	
>POCH	0–15	>P(O)OCH	0–13
P(OCH₃)₃	10.8–11.8	P(O)(OCH₃)₃	10.2–11.4
>PCCH	10–16	>P(M)CCH	14–25
>PNCH	3–14	(M=O, S)	
P[N(CH₃)₂]₃	8.8–9.0	—POCC	
P(IV)		P(V)	
—P⁺OCH	7–11	>PCCH (|)	20–27
—P⁺CCH	15–22	>POCH (|)	12–17
—PSCH	16–20		

^{a}For structural classes, only the absolute value for J is given.

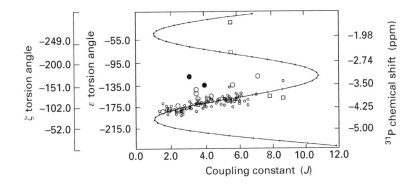

Figure 3 Plot of ³¹P chemical shifts for duplex oligonucleotide sequences (O) and an actinomycin D bound d(CGCG)₂ tetramer complex (□) with measured $J_{H3'-P}$ coupling constants (●, phosphates in a tandem GA mismatch decamer duplex which shows unusual, slowly exchanging signals). Also shown are the theoretical ε and ζ torsion angles (solid curve) as a function of the coupling constant derived from the Karplus relationship (ε) and the relationship $\zeta = -317 - 1.23\varepsilon$. ³¹P chemical shifts are reported relative to trimethyl phosphate. Reproduced with permission.

From the H3′–C3′–O–P torsional angle θ, the C4′–C3′–O–P torsional angle ε ($= -\theta - 120°$) may be calculated.

The $J_{H3'-P}$ coupling constants in larger oligonucleotides cannot generally be determined from the coupled 1D ³¹P or ¹H spectra because of spectral overlap. 2D J-resolved long-range correlation pulse sequences can be used to overcome this limitation. The Bax–Freeman selective 2D J experiment with a DANTE (delays alternating with nutations for tailored excitation) sequence for a selective 180° pulse on the coupled protons can be readily implemented on most spectrometers. This is particularly useful for measuring phosphorus–H3′ coupling constants in duplex fragments, which can vary from ~1.5 to 8 Hz in duplexes as large as tetradecamers. There is a strong correlation ($R = -0.92$) between torsional angles C4′–C3′–O3′–P (ε) and C3′–O3′–P–O5′ (ζ) in the crystal structures of various duplexes. Thus both torsional angles ε and ζ can often be calculated from the measured P–H3′ coupling constant.

Coupling constants of both 5′ protons are analysed in order to determine conformations about the C5′–O bond. Unfortunately, these β torsional angles have in practice been generally unobtainable even in moderate-length duplexes. Selective 2D J-resolved spectra generally fail for H4′, H5′, or H5″ coupling to ³¹P because the spectral dispersion between these protons is so limited. However, with either ¹³C labelling or even natural abundance ¹³C methods, it is possible to measure not only the ¹H–³¹P but also the ¹³C–³¹P coupling constants. Analysis of the 2D multiplet pattern, especially the 'E. COSY' pattern of the ¹H–¹³C HSQC spectrum, has allowed extraction of many carbon (C3′,C4′,C5′) and proton (H3′,H4′,H5′,H5″) coupling constants to phosphorus. The larger line

widths of longer duplexes limit measurement of the small coupling constants.

As shown in **Figure 3**, the Karplus relationship provides for four different torsional angle solutions for each value of the same coupling constant. Although all four values are shown in **Figure 3**, the limb which includes ε values between 360° and –270° is sterically inaccessible in nucleic acids. As shown in **Figure 3**, nearly all of the phosphates for normal Watson–Crick duplexes fall along only a single limb of the Karplus curve. Thus, for 'normal' B-DNA geometry, there is an excellent correlation between the phosphate resonances and the observed torsional angle, while phosphates that are greatly distorted in their geometry must be more carefully analysed.

It is clear from **Figure 3** that ³¹P chemical shifts and coupling constants provide probes of the conformation of the phosphate ester backbone in nucleic acids and nucleic acid complexes. It is important to remember that ³¹P chemical shifts are dependent on factors other than torsional angles alone. As noted above, ³¹P chemical shifts are very sensitive to bond angle distortions as well. It is quite reasonable to assume that backbone structural distortions as observed in unusual nucleic acid structures also introduce some bond angle distortion as well. Widening of the ester O–P–O bond angle indeed is expected to produce an upfield shift, while narrowing of this bond angle causes a downfield shift, and it is possible that this bond angle effect could account for the anomalous shifts. Indeed, very large ³¹P chemical shift variations (~3–7 ppm) are observed in transfer RNA and hammerhead RNA phosphates, and are probably due to bond angle distortions in these tightly folded structures.

Generally the main-chain torsional angles of the individual phosphodiester groups along the oligonucleotide double helix are responsible for sequence-specific variations in the ³¹P chemical shifts. In duplex B-DNA, the gauche(−), gauche(−) (g⁻, g⁻; ζ, α) (or B_I) conformation about the P–O ester bonds in the sugar phosphate backbone is energetically favoured, and this conformation is associated with a more shielded ³¹P resonance. In both duplex and single stranded DNA the trans, gauche(−) (t, g⁻; ζ, α) (or B_II) conformation is also significantly populated.

The ³¹P chemical shift difference between the B_I and B_II phosphate ester conformational stages is estimated to be 1.5–1.6 ppm. As the result of this sensitivity to the backbone conformational state, ³¹P chemical shifts of duplex oligonucleotides have been shown to be dependent both upon the sequence and the position of the phosphate residue.

The possible basis for the correlation between local helical structural variations and ³¹P chemical shifts can be analysed in terms of deoxyribose phosphate backbone changes involved in local helical sequence-specific structural variations. As the helix winds or unwinds in response to local helical distortions, the length of the deoxyribose phosphate backbone must change to reflect the stretching and contracting of the deoxyribose phosphate backbone between the two stacked base pairs. To a significant extent, these changes in the overall length of the deoxyribose phosphate backbone 'tether' are reflected in changes in the P–O ester (as well as other) torsional angles. These sequence-specific variations in the P–O (and C–O) torsional angles may explain the sequence-specific variations in the ³¹P chemical shifts.

³¹P NMR of protein complexes

³¹P NMR spectroscopy has proven to be very useful in the study of various protein complexes. **Table 4** provides an indication of the range of ³¹P chemical shifts and the titration behaviour of various phosphoprotein model compounds. Two examples of such studies are described below.

Ribonuclease A

Secondary ionization of a phosphate monoester produces approximately a 4 ppm down field shift of the ³¹P signal. Thus the pH dependence of the ³¹P signal of various phosphate monoesters bound to proteins can provide information on the ionization state of the bound phosphate ester. For example, pyrimidine nucleotides, both free in solution and when bound to

Table 4 Chemical shifts and pH titration data for representative model compounds

Compound	Chemical shift (ppm)[a]	Titratable[b]	pK_a
Phosphomonoesters			
Phosphoserine	4.6	+	5.8
Phosphothreonine	4.0	+	5.9
Pyridoxal phosphate	3.7	+	6.2
Pyridoxamine phosphate	3.7	+	5.7
Flavin mononucleotide	4.7	+	~6.0
Phosphodiesters			
RNA, DNA, phospholipids	0 to −1.5	−	
Diphosphodiesters			
Flavin adenine dinucleotide	−10.8 to −11.3	−	
Phosphotriesters			
Dialkyl phosphoserine	0 to −3.0		
Phosphoramidates			
N^3-Phosphohistidine	−4.5	−	
N^1-Phosphohistidine	−5.5	−	
Phosphoarginine	−3.0	+	4.3
Phosphocreatine	−2.5	+	4.2
Acyl phosphates			
Acetyl phosphate	−1.5	+	4.8
Carbamyl phosphate	−1.1	+	4.9

[a] All chemical shifts are reported with respect to an external 85% H_3PO_4 standard; upfield shifts are given a negative sign.

[b] Titrability: + indicates that changes are observed in the chemical shift on changes in pH: for phosphomonoesters this change is 4 ppm; for phosphoramidates 2.5 ppm; for acyl phosphates 5.1 ppm; − indicates no change observed.

bovine pancreatic ribonuclease A (RNase A), demonstrate this point. The ³¹P chemical shift of free solution cytidine 3′-monophosphate (3′-CMP) follows a simple titration curve, and the ionization constant derived form the ³¹P shift variation agrees with potentiometric titration values. The ³¹P chemical shift titration curve for the 3′-CMP·RNase A complex, however, cannot be analysed in terms of a single ionization process. Two inflections observed in this titration indicated two ionizations with $pK_1 = 4.7$ and $pK_2 = 6.7$.

These results suggest that the nucleotide binds at around neutral pH in the dianionic ionization state. Thus the 3′-CMP·RNase A complex ³¹P resonance is shifted upfield less than 0.3 ppm from the free 3′-CMP between pH 6.5 and 7.5, whereas monoprotonation of the free dianion results in a 4 ppm upfield shift. Furthermore, the addition of the first proton to the nucleotide complex ($pK_2 = 6.0$–6.7) must occur mainly on some site other than the dianionic phosphate because the ³¹P signal is shifted upfield by only 1–2 ppm. The addition of a second proton

$(pK_1 = 4.0–5.7)$ to the complex shifts the ^{31}P signal further upfield so that at the lowest pH values, the phosphate finally appears to be in the monoanionic ionization state.

On the basis of X-ray and ^{1}H NMR studies, it is known that the nucleotides are located in a highly basic active site with protonated groups histidine-119, histidine-12 and, probably, lysine-41, quite close to the phosphate. This suggests that pK_1 is associated with ionization of a protonated histidine residue which hydrogen bonds to the phosphate. This highly positive active site, which is capable of perturbing the pK of the phosphate from 6 to 4.7, must have one or more hydrogen bonds to the phosphate over the entire pH region. Yet at the pH extrema, little if any perturbation of the ^{31}P chemical shift is found. Apparently, the ^{31}P chemical shift of the phosphate esters is largely affected by the protonation state and not by the highly positive local environment of the enzyme.

Two-dimensional exchange ^{31}P NMR of phosphoglucomutase

Phosphoglucomutase (PGM) catalyses the interconversion of glucose 1-phosphate and glucose 6-phosphate. The enzyme has 561 residues on a single polypeptide chain with molecular weight 61 600 Da. Catalysis proceeds via a glucose 1,6-bisphosphate intermediate where the formation and breakdown of this intermediate results from two phosphate transfer steps involving a single enzymic phosphorylation site, Ser-116. A metal ion is required for activity and the most efficient metal ion is the physiological activator, Mg^{2+}. The phosphate transfer steps are shown below.

$$E_P + Glc1P^* \underset{k_{-1}}{\overset{k_1}{\rightleftharpoons}} E_D \cdot Glc1P^*6P \underset{k_6}{\overset{k_{-6}}{\rightleftharpoons}} E_{P^*} + Glc6P$$

$$[8]$$

$$E_{P^*} + Glc1P \underset{k_{-1}}{\overset{k_1}{\rightleftharpoons}} E_D \cdot Glc1P6P^* \underset{k_6}{\overset{k_{-6}}{\rightleftharpoons}} E_P + Glc6P^*$$

$$[9]$$

E_P and E_D are the phospho and dephospho forms of the enzyme, respectively, Glc1P is glucose-1-phosphate and Glc6P is glucose-6-phosphate.

Metal-free PGM and complexes with a variety of metal ions, substrates, and substrate analogues have been studied by ^{31}P NMR. Under conditions where the enzyme is inactive, each of the three enzyme-bound intermediates in the above scheme can be studied.

Exchange processes can be detected by 2D ^{31}P NMR in addition to the conventional 1D methods. The 2D exchange experiment (NOESY) described by Ernst and co-workers involves three 90° pulses. Nuclei are frequency labelled by a variable delay time (t_1) separating the first and second pulses. The mixing time is between the second and third pulses, and the detection of transverse magnetization as a function of time (t_2) follows the third pulse. During the mixing time, nuclei labelled in t_1 with a frequency corresponding to one site are converted by the exchange processes to a second site and evolve in t_2 with the frequency of the second site, giving rise to cross-peaks in the 2D spectrum. A 2D ^{31}P exchange spectrum of PGM shows cross-peaks indicating exchange between bound Glc6P and free Glc6P and between the two bound phosphorus sites, indicating transfer through free E_P involving a full catalytic cycle (see Eqn [8]).

Medical applications of ^{31}P NMR

In vivo ^{31}P NMR and ^{31}P magnetic resonance imaging are also important applications of this nucleus. ^{31}P signals from inorganic phosphate, adenosine triphosphate, adenosine diphosphate, creatine phosphate, and sugar phosphates can be observed in whole-cell preparations, intact tissues, and whole bodies and can provide information on the viability of the cells and tumour localization. Low sensitivity continues to be a problem in widespread application of these techniques. Additional details can be found in several of the entries in the Further reading section.

Conclusions

^{31}P NMR has become an indispensable tool in studying the chemistry and reactivity of phosphorus compounds, as well as in studying numerous biochemical and biomedical problems. Newer NMR instrumentation has enormously enhanced the sensitivity of the experiment and allowed 2D NMR studies to provide new means of signal assignment and analysis. Through 2D and 3D heteronuclear NMR experiments it is now possible to unambiguously assign the ^{31}P signals of duplex oligonucleotides and other phosphate esters. Both empirical and theoretical correlations between measured coupling constants, ^{31}P chemical shifts, and structural parameters have provided an important probe of the conformation and dynamics of nucleic acids, protein complexes, and small organophosphorus compounds.

List of symbols

a^2 = percentage s character; J = coupling constant; S = overlap integral; t_1 = delay time; t_2 = observe time in 2D NMR; T_1 = spin–lattice relaxation time; T_2^* = time constant for the FID; $\Delta\delta$ = chemical shift difference; $\Delta\theta$ = change in the σ-bond angle; $\Delta\chi_X$ = electronegativity difference in the P–X bond; σ = shielding constant; σ_{iso} = isotropic shielding constant; $\sigma_{\parallel}$ = parallel component of shielding constant; $\sigma_{\perp}$ = perpendicular component of shielding constant; $\sigma_{11}, \sigma_{22}, \sigma_{33}$ = components of shielding tensor; ϕ = dihedral angle.

See also: **Cells Studied By NMR; *In vivo* NMR, Applications, ³¹P; NMR Pulse Sequences; Nuclear Overhauser Effect; Nucleic Acids Studied Using NMR; Nucleic Acids and Nucleotides Studied Using Mass Spectrometry; Parameters in NMR Spectroscopy, Theory of; Perfused Organs Studied Using NMR Spectroscopy; Proteins Studied Using NMR Spectroscopy; Two-Dimensional NMR Methods.**

Further reading

Burt CT (1987) *Phosphorus NMR in Biology*, pp. 1–236. Boca Raton, FL: CRC Press.

Crutchfield MM, Dungan CH, Letcher LH, Mark V and Van Wazer JR (1967) Topics in phosphorus chemistry. In: Grayson M and Griffin EF (eds) *Topics in Phosphorous Chemistry*, pp. 1–487. New York: Wiley (Interscience).

Gorenstein DG (1984) *Phosphorus-31 NMR: Principles and Applications*, pp. 1–604. Orlando, FL: Academic Press.

Gorenstein, DG (1992) Advances in P-31 NMR. In: Engel, R (ed) *Handbook of Organophosphorus Chemistry*, pp. 435–482. New York: Marcel Dekker.

Gorenstein DG (1994) Conformation and Dynamics of DNA and Protein–DNA Complexes by ³¹P NMR, *Chemical Reviews* **94**: 1315–1338.

Gorenstein DG (1996) Nucleic Acids: Phosphorus-31 NMR. In: Grant DM and Harris RK (eds) *Encyclopedia of Nuclear Magnetic Resonance*, pp. 3340–3346. Chichester: Wiley.

Karaghiosoff K (1996) Phosphorus-31 NMR. In: Grant DM and Harris RK (eds) *Encyclopedia of Nuclear Magnetic Resonance*, pp. 3612–3618. Chichester: Wiley.

Mavel G (1973) *Annual Reports on NMR Spectroscopy* 5B: 1–350.

Quin LD and Verkade JG (1994) *Phosphorus-31 NMR Spectral Properties in Compound Characterization and Structural Analysis*, p. 1. New York: VCH.

Tebby JC (1991) *Handbook of Phosphorus-31 Nuclear Magnetic Resonance Data*, p. 1. Boca Raton FL: CRC Press.

Verkade JG and Quin LD (1987) *Phosphorus-31 NMR Spectroscopy in Stereochemical Analysis; Organic Compounds and Metal Complexes*, pp. 1–455. Deerfield Beach, FL: VCH.

Palladium NMR, Applications

See **Heteronuclear NMR Applications (Y–Cd).**

Parameters in NMR Spectroscopy, Theory of

GA Webb, University of Surrey, Guildford, UK

MAGNETIC RESONANCE
Theory

Introduction

High-resolution NMR provides spectra that consist of a number of lines and bands whose frequency, relative intensity and shape may be analysed to yield molecular parameters. The NMR parameters in questions are the nuclear shielding, σ, which describes the shielding of the nucleus from the applied magnetic field by the surrounding electrons and gives rise to chemical shifts; J, which relates to nuclear spin–spin coupling and depends upon relative nuclear orientations; and the times T_1 and T_2 which refer to the relaxation processes encountered by the nuclei excited in the NMR experiment.

Both the nuclear shielding and spin–spin coupling interactions are interpreted within the framework of quantum chemistry, whereas a quasi-classical form of mechanics is usually adopted to describe the nuclear relaxation interactions.

Nuclear shielding (chemical shifts)

For an NMR experiment the basic resonance condition is given as

$$\omega = \gamma B_0 \qquad [1]$$

where B_0 is the applied magnetic field in which the experiment is performed, γ is the magnetogyric ratio of the nucleus in question and ω is the angular frequency of the radiation producing the NMR transition. From this expression it follows that all nuclei with a given value of γ, e.g. protons, will produce a single absorption in the NMR spectrum. In such a situation NMR spectroscopy would not be of much chemical interest. In reality the expression for the resonance condition needs to be modified to include the fact that the value of the magnetic field experienced by the resonating nuclei is usually less than B_0 owing to shielding of the nucleus in a molecule by the surrounding electrons. Thus the expression for the resonance condition becomes

$$\omega = \gamma B_0 (1 - \sigma) \qquad [2]$$

where σ is the nuclear shielding. In NMR experiments the resonance frequencies are normally reported relative to that of a given nucleus in a standard molecule added to the experimental sample as a reference. The shielding difference, or chemical shift δ, is then defined as the difference in shielding between the given nucleus in the reference compound, σ_{ref}, and that of the nucleus of interest, σ_{sample}. Namely

$$\delta = (\sigma_{\mathrm{ref}} - \sigma_{\mathrm{sample}})$$

From which it follows that a shift of resonance to high frequency, denoted by an increase in δ, corresponds to a decrease in σ_{sample}.

In seeking a molecular interpretation for σ it is important to realize that the nuclear shielding is represented by a second-rank tensor. Many NMR experiments are performed on nonviscous solutions, or sometimes on gaseous samples, in which case rapid, and random, molecular motion ensures that the nuclear shielding experienced is the scalar corresponding to one-third of the trace of the tensor.

NMR measurements taken on the solid and liquid crystal phases can yield values for the individual components of the shielding tensor and its anisotropy, $\Delta\sigma$. For linear and symmetric-top molecules,

$$\Delta\sigma = \sigma_{\parallel} - \sigma_{\perp} \qquad [3]$$

where $\sigma_{\parallel}$ refers to the shielding component along the major molecular axis and $\sigma_{\perp}$ is that in the direction perpendicular to it. For less symmetrical molecules,

$$\Delta\sigma = \sigma_{\alpha\alpha} - \frac{1}{2}(\sigma_{\beta\beta} + \sigma_{\gamma\gamma}) \qquad [4]$$

where the σ_{ii} are the principal tensor components taken in accordance with the convention $\sigma_{\alpha\alpha} > \sigma_{\beta\beta} > \sigma_{\gamma\gamma}$.

The first report on the theory of nuclear shielding appeared in 1950; since then many reports have appeared of attempts to calculate shieldings, most of them within the framework of molecular orbital

(MO) theory. Some of the earlier results, particularly those based upon semiempirical MO methods, are at best indicative of shielding trends in series of closely related molecules. In general these are unsuitable for predictive purposes. In recent years this situation has changed dramatically and *ab initio* MO calculations of nuclear shielding are routinely providing satisfactory results. In principle, quantum chemistry can provide a full account of all molecular properties. In practice, various approximations are introduced into the calculations to make them tractable. Such approximations tend to produce limitations on the results obtained from the calculations. For example, calculations at the Hartree–Fock (HF) level involve a single determinant for a rigid isolated molecule; consequently, the effects of electron correlation, variations in geometry and media influences on nuclear shielding are ignored. Normally such effects are considered separately as are possible relativistic effects on the shielding of heavier nuclei. Calculations of molecular magnetic properties, such as nuclear shielding, can suffer from all of these limitations and an additional one known as the gauge problem. This arises from the use of perturbation theory to describe the rather small contribution to the total electronic energy of the molecule provided by the applied magnetic field in the NMR experiment. The magnetic perturbation is described by the orbital angular momentum operator. Since this operator is not invariant with respect to translations, its influence depends upon the position at which it is evaluated. Consequently, the result obtained for the calculated nuclear shielding depends upon the choice of origin for the calculation. This theoretical artefact has to be dealt with before comparison takes place between experimental and theoretical shielding data.

One way to combat the gauge problem in nuclear shielding calculations is to employ large basis sets in calculations using the coupled Hartree–Fock (CHF) approach. If smaller basis sets are employed, the shielding results obtained are gauge-dependent unless the gauge origin is taken to be at the nucleus in question; these are referred to as common-origin calculations. An example of ^{13}C nuclear shielding calculations of this type is provided by buckminsterfullerene, C_{60}: all of the carbon atoms are equivalent in this molecule and thus symmetry arguments can be used to reduce the number of integrals to be evaluated. If a relatively modest basis set, such as 6-13G*, is used in the calculation of the nuclear shielding, then about ten days of CPU time is required on a DEC 8400 computer. Consequently, it seems unlikely that common-origin nuclear shielding calculations will become widely affordable.

An alternative to using large basis sets to overcome the gauge problem is to introduce gauge factors either into the atomic orbitals of the basis set or into the MOs of a CHF calculations of nuclear shielding. The inclusion of gauge factors in the atomic orbitals used gives rise to the gauge-included atomic orbital (GIAO) method. In contrast the IGLO (individual gauges for localized orbitals) method employs individual gauge origins for different localized molecular orbitals. Both the GIAO and IGLO methods are referred to as local origin variants of the CHF method.

An alternative to the CHF calculations of second-order magnetic properties is to use the random-phase approximation within the equations of motion procedure. This has developed into a method using localized MOs with local origins (LORG). The LORG method results in a localization of the MOs used and provides a pathway to the decomposition of the calculated nuclear shielding into individual local bond and bond–bond contributions. Thus the LORG and IGLO methods of calculating nuclear shieldings are analogous to each other. The results of some ^{13}C shieldings and their anisotropies, produced by IGLO, LORG and GIAO calculations, are given in **Table 1**.

The results given are obtained by the use of medium-sized basis sets; e.g. sets of triple zeta quality with a set of d polarization functions for the heavy atoms. In general, the calculated and experimental results are in satisfactory agreement. In comparing the relative merits of the GIAO, IGLO and LORG methods, it appears that the GIAO procedure is the more efficient in terms of the convergence of the shielding value with respect to the size of basis set used. However, the IGLO and LORG calculations

Table 1 Comparison of some ^{13}C shieldings and their anisotropies $\Delta\sigma$ (in ppm) produced by IGLO and GIAO calculations and experimental values

Molecule and $\Delta\sigma$	IGLO	LORG	GIAO	Experimental
CH_4	196.7	196.0	193.0	195.1
HCN	72.9	77.3	74.8	82.1
$\Delta\sigma$	306	301	304.8	316.3 ± 1.2
C_2H_2	116.4	122.3	118.3	117.2
$\Delta\sigma$	243.3	235.0	241.0	240 ± 5
CO	−6.0	–	−21.3	1.0
$\Delta\sigma$	420.0	–	439.0	405.5 ± 1.4
C_2H_6	183.5	184.7	181.2	180.0
$\Delta\sigma$	13.5	8.0	11.3	–
CH_3OH	157.2	145.7	134.6	136.6
$\Delta\sigma$	–	60.5	77.3	63.0
H_2CO	−3.8	4.0	2.6	−8.0
$\Delta\sigma$	183.8	183.0	196.5	–

produce shielding contributions that may be attributed to specific molecular regions. Since the GIAO, IGLO and LORG calculations can all be performed with different levels of basis set quality, the results obtained are found to be dependent upon the choice of basis set. An example of the dependence of ^{13}C shieldings calculated by the GIAO method upon the choice of geometry and basis set is provided in **Table 2.**

The use of experimental geometries is indicated by NOOPT (none optimized geometry); also included are optimized geometries from *ab initio* MO calculations using 4-31G and 431G** basis sets, whereas the ^{13}C shieldings are calculated with 3-21G, 4-31G and 4-31G** basis sets. The best agreement with the experimental shieldings is given by the calculations using experimental molecular geometries and the

4-31G** basis set; in this case the average least-squares error for the set of molecules studied is 2.7 ppm.

In general, GIAO, LORG and IGLO calculations are capable of producing shieldings for nuclei from the first and second long rows of the periodic table to within about 3 or 4% of an element's shielding range. Thus these calculations can be used for predictive purposes as well as providing some information on the molecular electronic factors that determine the extent of nuclear shielding and its variations.

Another method of tackling the gauge problem in nuclear shielding calculations is to employ individual gauges for atoms in molecules (IGAIM). This procedure differs from the GIAO, IGLO and LORG procedures in that the gauge origins in IGAIM are

Table 2 Calculated and observed isotropic ^{13}C chemical shifts (in ppm from CH_4) of the resonant nuclei (*C) using NOOPT/4-31G, 4-31G/4-31G, NOOPT/3-21G, NOOPT/4-31G** and 4-31G**/4-31G** basis sets

Molecule	NOOPT/4-31G	4-31G/4-31G	NOOPT/3-21G	NOOPT/4-31G**	4-31G**/4-31G**	Experimental
CH_4	0.0	0.0	0.0	0.0	0.0	0.0
C_2H_6	4.2	4.8	2.9	5.0	5.5	8.0
C_2H_4	131.2	127.4	119.2	125.4	121.5	125.4
*$CH_3CH_2CH_3$	16.1	15.4	13.5	16.6	16.3	17.7
CH_3*CH_2CH_3	16.7	13.8	13.7	18.6	16.4	18.2
cis-*$CH_3CH=CHCH_3$	12.4	12.9	10.6	12.3	12.6	12.7
cis-CH_3*$CH=CHCH_3$	130.8	130.1	119.0	126.3	124.9	125.9
trans-*$CH_3CH=CHCH_3$	19.9	18.6	17.4	19.5	18.4	19.4
trans-CH_3*$CH=CHCH_3$	134.2	129.6	121.2	129.3	124.5	127.2
cyclo-C_3H_6	−2.0	−2.6	−2.6	−1.4	−2.2	0.1
cyclo-C_6H_{12}	29.5	25.5	24.9	30.3	26.2	29.9
C_6H_6	133.2	132.1	119.7	129.2	127.3	130.0
*$CH_2=CHCH_3$	123.7	128.3	112.8	118.2	115.1	117.5
$CH_2=$*$CHCH_3$	138.7	138.0	124.9	133.7	131.2	137.8
$CH_2=CH$*CH_3	17.5	22.9	15.2	17.1	19.3	20.8
*$CH\equiv CCH_3$	74.5	74.6	69.2	69.6	69.0	69.0
$CH\equiv$*CCH_3	79.2	80.4	71.7	74.1	74.5	82.0
$CH\equiv C$*CH_3	5.0	4.0	4.3	4.9	3.9	4.0
Toluene (C-1)	141.3	140.2	126.3	137.7	136.4	140.8
Toluene (C-2)	133.3	132.5	119.9	128.9	128.1	131.4
Toluene (C-3)	134.5	133.6	120.6	130.6	129.6	132.2
Toluene (C-4)	130.2	129.3	117.1	126.1	125.1	128.5
Toluene (CH_3)	21.9	22.1	19.6	21.1	21.5	24.3
*$CH\equiv C-C\equiv CH$	68.4	69.8	63.6	65.1	66.0	69.3
$CH\equiv$*$C-C\equiv CH$	71.9	73.0	65.4	66.5	66.8	71.0
cyclo-$C_3H_4(CH_2)$	–	−3.7	–	–	–	3.0
cyclo-$C_3H_4(=C)$	–	117.1	–	–	–	108.0
Averaged least-squares error (ppm)	3.16	3.24 (without cyclo-C_3H_4)	7.34	2.72	3.20	–
Correlation coefficient	0.998	0.995	0.997	0.998	0.999	
Slope	0.96	0.97	1.06	1.00	1.02	

determined by properties of the charge density in real space rather than by the behaviour of the chosen basis functions in the Hilbert space of the molecular wavefunction. The results of some IGAIM ^{13}C shielding calculations are given in **Table 3**, where they are compared with the results obtained from conventional CHF calculations using the same basis set.

In the CHF calculations, the common gauge origin is placed at the nucleus whose shielding is being deduced. **Table 3** shows that the IGAIM results are in much better agreement with experiment than those produced by the CHF calculations. The absence of electron correlation effects from HF calculations is most noticeable in cases of 'electron-rich' molecules containing, for example, multiple bonding and lone pair electrons. It is possible to enhance the local origin methods for calculating nuclear shieldings by including some electron correlation effects. The GIAO method has been extended by means of manybody perturbation theory (MBPT). The results of some GIAO and GIAO-MBPT calculations of ^{17}O shieldings are compared in **Table 4**, where the effect of including electron correlation is seen to lead to an increase in the calculated values of the shieldings. This increase usually results in a closer agreement between the calculated and experimental shieldings.

Electron correlation effects have been included in the IGLO method by means of a nonperturbative multiconfiguration extension to give the MC-IGLO method. **Table 5** shows the results of some MC-IGLO calculations of ^1H, ^{13}C, ^{17}O, ^{19}F and ^{31}P nuclear shieldings in comparison with comparable results from experiment and from the self-consistent fixed (SCF)-IGLO method.

The effects of electron correlation on the calculated nuclear shieldings are shown to be small for methane and phosphine but much more significant for fluorine, carbon monoxide and ozone, which are 'electron rich' molecules. For the central oxygen atom of ozone, the effect of including electron correlation in the shielding calculations is to produce an increase by over 2000 ppm.

The LORG method of calculating nuclear shieldings has been combined with the second-order polarization propagator (SOPPA) technique to produce the second-order LORG or SOLO procedure. The results of some LORG and SOLO ^{15}N shielding calculations are compared with experiment and with some IGLO results in **Table 6**.

The conjugated heterocycles chosen for the study represent cases where electron correlation effects are predicted to be significant. In general, the inclusion of electron correlation leads to an increase in the calculated nitrogen shieldings and usually to an improved agreement with the experimental results.

A comparison of the LORG and SOLO data given in **Table 6** shows root mean square errors of 49.2 and 19.9 ppm, respectively.

Table 3 Comparison of absolute ^{13}C shieldings in (ppm) calculated by the IGAIM method and the CHF procedure, using the same 6-31 G**(2d, 2p) basis set, with experimental values taken as thermal averages at 300 K in the limit of zero gas density

Molecule	IGAIM	CHF	Experimental
CH_4	197.4	198.5	195.1
HCN	79.9	89.5	82.1
C_2H_2	119.3	127.0	117.2
C_2H_4	66.4	73.4	64.5
C_2H_6	186.3	192.3	180.9
C_3H_4 (C-1)	119.5	130.2	115.2
C_3H_4 (C-2)	−34.8	−22.4	−29.3
C_6H_6	61.5	82.1	57.9
CO	−7.4	−11.9	1.0
CO_2	57.9	78.9	58.8
CS_2	−41.1	51.9	−8.0
CSO	21.9	78.2	30.0
CH_3NH_2	167.0	173.9	158.3
CH_3OH	148.0	155.8	136.6
CH_3F	130.2	140.0	116.8
CF_4	86.0	122.3	64.5
HCOOH	32.2	50.2	23.7

Table 4 Comparison of some ^{17}O shieldings (in ppm) produced by GIAO and GIAO-MBPT calculations and experimental values

Molecule	GIAO	GIAO-MBPT	Experimental
H_2O	323.18	339.79	357
H_2O_2	139.01	150.88	134
CO	−113.47	−54.06	−40.1 ± 17.2
H_2CO	−471.40	−345.02	−312.1
CH_3OH	341.55	354.41	344.9
CO_2	200.37	236.37	243.4
OF_2	−471.13	−465.53	−473.1
NNO	107.54	192.12	200.5

Table 5 Comparison of some ^1H, ^{13}C, ^{17}O, ^{19}F and ^{31}P shieldings (in ppm) produced by SCF-IGLO and MC-IGLO calculations and experimental values

Molecule	Nucleus	SCF-IGLO	MC-IGLO	Experimental
CH_4	C	193.8	198.4	198.7
	H	31.22	31.13	30.61
PH_3	P	583.4	598.2	594.4
	H	29.43	29.65	29.28
F_2	F	−165.3	−204.3	−192.8
CO	C	−23.4	13.4	3.0
	O	−83.9	−36.7	−42.3
O_3	O (central)	−2730.1	−657.7	−724.0
	O (terminal)	−2816.7	−1151.8	−1290.0

Table 6 Comparison of some ^{15}N shieldings (in ppm) produced by IGLO, LORG, and SOLO calculations on some conjugated heterocycles and experimental values

Molecule	IGLO	LORG	SOLO	Experimental values
Sym-Triazine	−41	−33	−28	−39
Pyrimidine	−71	−58	−45	−51
Pyridine	−104	−94	−72	−73
Pyrazine	−121	−136	−102	−90
Sym-Tetrazine	−221	−213	−159	−141
Pyridazine	−240	−235	−197	−156
1,2,4-Triazine				
N-3		−76	−42	−54
N-2		−171	−151	−134
N-1		−255	−207	−178

Density functional theory (DFT) is an alternative to HF methods for describing molecular electronic structure. Electron correlation effects are explicitly included in DFT calculations. Coupled DFT (CDFT), together with the IGLO method, has been used in some nuclear shielding calculations and some results for ^{13}C, ^{15}N, ^{17}O, ^{19}F and ^{31}P are shown in **Table 7**. For comparison purposes, the results of some GIAO calculations, not including electron correlation, and experimental results are given. The CDFT results are seen to be in much better agreement with experiment than are those from the GIAO calculations.

Spin–spin couplings

Many NMR signals appear as multiplets, the structure of which arises from spin–spin coupling interactions with other nuclei in the molecule. The separation between adjacent members of a multi-

Table 7 Comparison of some ^{15}C, ^{15}N, ^{17}O, ^{19}F and ^{31}P shieldings in (ppm) produced by GIAO and CDFT calculations and experimental values

Molecule	Nucleus	GIAO	CDFT	Experimental
PN	P	−15.8	42.1	53
	N	−409.4	−347.3	−349
P_2H_2	P	−294.2	−190.9	−166
CO	C	−8.0	−0.3	1
	O	−61.3	−63.4	−42.3
NNO	N (terminal)	89.0	97.0	99.5
	N(central)	−2.0	5.9	11.3
	O	219.4	185.4	200.5
H_2O_2	O	191.5	157.2	133.9
N_2	N	−80.0	−69.3	−61.0
N_2CO	C	14.2	−12.3	−1
	O	−406.2	−362.6	−312.1
F_2	F	−181.4	−197.8	−193.8

plet can give the value of J, the spin–spin coupling interaction between the spin coupled nuclei. For nuclei whose spin is $\frac{1}{2}$, the relative signal intensities of the members of a given first-order multiplet are given by the factors of a binomial expansion. If A and B are the two spin-$\frac{1}{2}$ coupled nuclei then the NMR signal for A will consist of a multiplet with $n + 1$ lines due to spin–spin coupling to n equivalent B nuclei, provided the chemical shift between A and B is large relative to J_{AB}.

As for σ, the value of J depends upon the chemical environment of the nuclei concerned. Hence values of J are of use in molecular structure determinations. Unlike the case for nuclear shieldings, values of J are independent of the magnitude of the applied magnetic field used in the NMR experiment; thus the gauge problem does not arise when considering quantum-chemical calculations of J. Nuclear spin–spin couplings arise from indirect interactions between the spin, I, of neighbouring nuclei. The spin orientation information is transmitted from one nucleus to the other by means of both bonding and nonbonding electrons encountered on the spin coupling pathway.

Values of J are usually given in Hz as is apparent from the following definition of the energy, E_{AB}, of the coupling interaction between nuclei A and B:

$$E_{AB} = hJ_{AB}I_AI_B \qquad [5]$$

As in the case of nuclear shielding, J_{AB} is a scalar quantity; an estimate of the anisotropy of the corresponding second-rank tensor may be forthcoming from measurements on oriented samples. The theoretical aspects of spin–spin coupling are based upon three types of electron-coupled interactions between the electrons and nuclei of the molecule concerned. Normally the largest of these is the contact (C) interaction between the electron and nuclear spins; the second one is a magnetic dipolar (D) interaction between the electron and nuclear spins; finally there is the orbital (O) interaction between the magnetic field produced by the orbital motion of the electrons and the nuclear magnetic dipole.

Accurate calculations of spin–spin couplings provide a challenge to the theoretician. Reliable results are difficult to obtain for molecules of chemical interest, because spin–spin couplings rely upon subtle aspects of molecular electronic structure. Consequently, a deeper understanding of the relationships between spin–spin couplings and molecular structure could considerably enhance the application of high-resolution NMR spectroscopy to the elucidation of molecular electronic structure. At present the theoretical analysis of spin–spin couplings is advancing in two different directions. For small molecules with light atoms, i.e. those up to the second row, highly accurate *ab initio* MO calculations are being applied. Alternatively, rather simple semiempirical calculations are used to provide some understanding of possible relationships between physical phenomena and experimental data.

At the HF level, the C contribution to spin–spin couplings is the most difficult to evaluate accurately owing to the poor description provided of the electron spin densities at the coupled nuclei. Consequently, it becomes necessary to include electron correlation effects to provide accurate calculations of spin–spin couplings. Many-body perturbation theory can be used to introduce some electron correlation into calculations of the C contribution to spin–spin couplings. Using this approach for some first-row hydrides, where the C contribution is expected to dominate, satisfactory agreement is found between calculated and observed values of one-bond couplings. However, the calculated values of ^2J(H–H) are much too large, which suggests that electron correlation effects beyond second order are important in determining the magnitudes of spin–spin couplings.

The use of multiconfiguration linear response (MCLR) theory is another approach to the calculation of spin–spin coupling interactions. As is usual for *ab initio* MO calculations, the results obtained are found to be basis set dependent. In general, satisfactory agreement with the available experimental data is achieved.

Other *ab initio* MO calculations of spin–spin couplings include those based upon polarization propagator methods, e.g. RPA, SOPPA and the coupled cluster single and double polarization propagator approximation (CCSDPPA). These three methods have been used to calculate the C contributions to the values of ^1J(C–H) and ^2J(H–H) for methane as functions of bond length variation in the region of the equilibrium geometry, as shown in **Figures 1** and **2**, where S_1 represents the symmetric stretching coordinate. In the case of the CCSDPPA result, about 91% of the correlation contribution to the value of ^1J(C–H) is recovered, whereas the corresponding figure for the SOPPA calculation is about 79%. For the calculations on ^2J(H–H), the corresponding recoveries are 88% and 79% for the CCSDPPA and SOPPA methods, respectively.

Semiempirical MO calculations of spin–spin couplings are often used in conjunction with conformational analysis studies. In general, the investigations are based upon a dihedral angle dependence of the $^3J(^{13}\text{C}-^1\text{H})$ values. However, calculations of longer-range couplings can also play a role in understanding molecular structure.

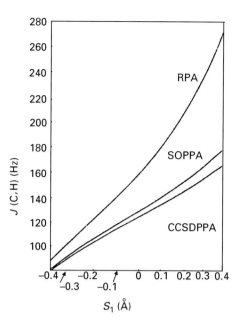

Figure 1 Dependence of the contact contribution to ^1J(C–H) on the symmetric stretching coordinate S_1 of methane. Results are given at the RPA, SOPPA and CCSDPPA levels of theory.

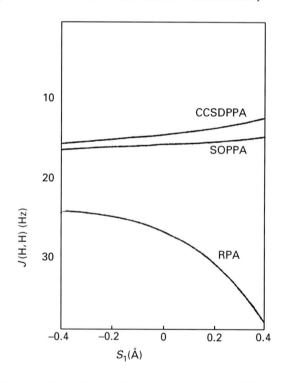

Figure 2 Dependence of the contact contribution of ^2J(H–H) on the symmetric stretching coordinate S_1 of methane. Results are given at the RPA, SOPPA and CCSDPPA levels of theory.

Self-consistent perturbation theory (SCPT) semi-empirical calculations have been used in a study of the effects of the oxygen lone pair electrons on ^1J(C–C) values in furan derivatives. The results show that the effects of the lone pairs on the spin–spin couplings, and the changes due to protonation, are similar to those resulting from the lone pair electrons on the nitrogen atom in imines.

Nuclear spin relaxation

The time taken for nuclear spin relaxation to occur constitutes the third type of chemically interesting NMR parameter. Since NMR is normally observed in the radiofrequency region of the electromagnetic spectrum it involves rather low-energy transitions; consequently spontaneous emission tends to be of negligible importance for NMR relaxation. Nuclear spin relaxation may be characterized by two relaxation times, T_1 and T_2. The spin–lattice relaxation time, T_1, relates to the exchange of nuclear magnetization in a direction parallel to that of the applied magnetic field. T_2, the spin–spin relaxation time applies to the exchange of magnetization in directions perpendicular to that of the applied magnetic field.

The ideal NMR line shape is Lorentzian and its full width at half-height, $W_{1/2}$, is controlled by T_2:

$$W_{1/2} = \frac{1}{\pi T_2} \qquad [6]$$

For nonviscous liquids, T_1 and T_2 are usually equal; thus comments made about T_1 apply equally to T_2.

A number of mechanisms may contribute to nuclear spin relaxation times. These mechanisms operate in chemically distinct ways, such that the identification of which particular mechanism(s) is operative can be of chemical interest. For any mechanism to be operative in producing spin relaxation it must produce an oscillating magnetic field at the nuclear site. The frequency of this local magnetic field must be equal to the resonance frequency of the nucleus to be relaxed. If this situation occurs, then a relaxation transition may be induced.

The microdynamic behaviour of molecules in fluids is attributed to Brownian motion, and the frequency distribution of the components of the local fluctuating magnetic field is expressed by a power spectral density. The component of this spectral density at the resonance frequency is responsible for nuclear relaxation. The magnitude of this component, taken together with the energy of interaction between the nuclear spin system and the molecular motions, determines the value of T_1.

In discussing nuclear relaxation phenomena it is normally assumed that the motional narrowing limit applies:

$$\omega_0{}^2 \tau_0{}^2 \ll 1$$

where ω_0 refers to the resonance frequency and τ_0 is the correlation time characterizing the appropriate molecular motion. For the motional narrowing limit to apply, the molecules in question must be tumbling rapidly; this implies small molecules in a low-viscosity medium and a relatively high temperature. As shown in **Figure 3**, under these conditions T_1 becomes frequency independent and equal to T_2. Larger molecules may not satisfy the motional narrowing limit, for example macromolecules, in which case T_1 and T_2 are almost certain to be unequal and to have different frequency dependences.

Provided the extreme narrowing conditions are satisfied, then the left-hand side of **Figure 3** is the appropriate one for further discussion of the various mechanisms that contribute to T_1.

Nuclear magnetic dipole relaxation interactions may occur with other nuclei, or with unpaired electrons. These processes usually dominate the relaxation of spin-$\frac{1}{2}$ nuclei. Both intra- and intermolecular interactions may contribute to dipole–dipole nuclear relaxation times. The value of T_1 due to the intramolecular dipole–dipole process is proportional to the sixth power of the internuclear separation. Consequently, this process becomes rather inefficient in the absence of directly bonded magnetic nuclei. However, it follows that a measurement of T_1 can be provide an estimate of internuclear separation that can be of chemical interest. The nuclear Overhauser effect (NOE) depends upon the occurrence of dipole–dipole relaxation processes and can similarly provide an estimate of internuclear separation.

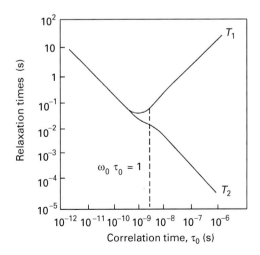

Figure 3 Schematic representation of the nuclear relaxation times T_1 and T_2 as functions of the correlation time τ_0.

The large magnetogyric ratio of the proton coupled with its common molecular occurrence ensures that dipole–dipole interactions with protons frequently dominate the relaxation of other spin-$\frac{1}{2}$ nuclei such as ^{13}C and ^{15}N.

The electron has a magnetogyric ratio that is more than 600 times larger than that of the proton; thus, if unpaired electrons are present their dipole–dipole interaction with a given nucleus normally controls the relaxation of that nucleus. Consequently, paramagnetic centres may be introduced to override nuclear–nuclear relaxation processes in certain cases, for example to reduce embarrassingly long relaxation times and to remove NOEs in cases where they are not required.

Nuclei with a spin $I > \frac{1}{2}$ have electric quadrupole moments in addition to the magnetic dipole moments required for the NMR experiment. The quadrupole moment may interact with a local electric field gradient to provide a very efficient nuclear relaxation process, and thus broad NMR signals. The value of T_1 for the quadrupolar relaxation process depends critically upon the electronic environment of the nucleus in question. This is demonstrated by the NMR line widths of about $10\,Hz$ for ^{35}Cl in NaCl and about $10\,kHz$ for ^{35}Cl in CCl_4. In the former example, the electronic environment of the chloride ion is approximately spherical, thus there is only a small field gradient, at best, at the site of the chlorine and the line width is controlled by the less efficient dipole–dipole process. For covalently bonded CCl_4, the large field gradients at the chlorine nuclei give rise to rapid quadrupolar relaxation.

Spin–rotation interactions may also produce nuclear relaxation. These arise from interactions between nuclear magnetic moments and rotational magnetic moments of the molecules containing the nuclei in question. A direct transfer occurs of nuclear spin energy to the molecular motion. This contrasts with the dipole–dipole and quadrupole mechanisms, which operate via an indirect energy transfer. The value of T_1 due to spin–rotation interactions decreases as the temperature increases, which is in contrast to the other nuclear relaxation mechanisms. Hence the observed temperature dependence of T_1 may be used to demonstrate the contribution or absence of spin–rotation interaction processes to the nuclear relaxation. Spin–rotation relaxation is most likely to be dominant for small molecules tumbling rapidly at high temperatures. Thus it is likely to be of particular importance for vapour-phase studies.

Anisotropy of the nuclear shielding tensor may also contribute to nuclear relaxation. Brownian motion can modulate the nuclear shielding tensor and thus provide a fluctuating magnetic field. The corresponding relaxation times depend inversely upon the square of the applied magnetic field and the square of the shielding anisotropy. Thus this relaxation process is likely to be of most importance at very high magnetic field strengths and for heavier nuclei, which tend to have very large shielding anisotropies, e.g. ^{195}Pt and ^{199}Hg. The fact that T_1 values for this process depend upon the strength of the applied magnetic field provides a means of determining the contribution or absence of nuclear shielding anisotropy to the relaxation of a given nucleus.

If chemical exchange or internal rotation causes the spin–spin coupling interaction between two nuclei to become time dependent, then scalar relaxation of the first kind can occur. Scalar relaxation of the second kind relates to the case where the relaxation rate of a coupled nucleus is fast compared with $2\pi J$. Coupling to a quadrupolar nucleus can give rise to this relaxation mechanism. For scalar coupling relaxation to be operative, it is generally important that the resonance frequencies of the coupled nuclei be similar. This is, perhaps, the least common of the nuclear spin relaxation processes considered.

List of symbols

B_0 = applied magnetic field strength (flux density); E_{AB} = energy of couplings interaction between nuclei A and B; J = spin coupling constant; T_1 = spin–lattice relaxation time; T_2 = spin–spin relaxation time; $W_{1/2}$ = full width at half-height of NMR line; δ = chemical shift; ω = angular frequency of applied radiation; $\Delta\sigma$ = shielding anisotropy; σ = nuclear shielding parameter; τ_0 = molecular correlation time.

See also: **^{13}C NMR, Parameter Survey; Chemical Shift and Relaxation Reagents in NMR; Gas Phase Applications of NMR Spectroscopy; NMR in Anisotropic Systems, Theory; NMR Principles; NMR Relaxation Rates; Nuclear Overhauser Effect.**

Further reading

Abragam A (1961) *The Principles of Nuclear Magnetism.* Oxford: Clarendon Press.

Ando I and Webb GA (1983) *Theory of NMR Parameters.* London: Academic Press.

Contreras RH and Facelli JC (1993) Advances in theoretical and physical aspects of spin–spin couplings. In: Webb GA (ed) *Annual Reports on NMR*, Vol 27, p 255. London: Academic Press.

de Dios AC (1996) Ab initio calculations of the NMR chemical shift. *Progress in NMR Spectroscopy* 29: 229.

Specialist Periodical Reports on NMR, published annually by the Royal Society of Chemistry, Webb GA (ed), contain chapters dealing with all aspects of NMR Parameters. Latest edition is Vol 29 (1999).

Webb GA (1978) Background theory of NMR parameters. In: Harris RK and Mann BE (eds) *NMR and the Periodic Table*, p 49. London: Academic Press.

Webb GA (1993) An overview of nuclear shielding calculations. In: Tossell JA (ed) *Nuclear Magnetic Shieldings and Molecular Structure*, p 1. Dordrecht: Kluwer.

Peptides and Proteins Studied Using Mass Spectrometry

Michael A Baldwin, University of California, San Francisco, CA, USA

MASS SPECTROMETRY
Applications

Thirty years ago it was impossible to ionize and analyse even a small peptide by mass spectrometry unless it was first made volatile by derivatization, such as acetylation and/or permethylation. In recent years 'soft ionization' methods have made mass spectrometric analysis of peptides and proteins a routine activity. Such methods employed for ionization and analysis of peptides and proteins have included field desorption (FD) from a heated emitter by high electric fields, direct chemical ionization (DCI) by the interaction of a hot plasma with a solid sample, fast atom bombardment (FAB) involving bombardment of an analyte solution with high energy xenon atoms or caesium ions, plasma desorption (PD) using nuclear fission fragment bombardment of a sample on a solid support such as nitrocellulose, electrospray ionization (ESI) by evaporation of charged droplets of analyte solution, and matrix-assisted laser desorption/ionization (MALDI) by laser irradiation of crystals of a matrix doped with analyte. Several of these are still in limited use but the almost universal utility of ESI and MALDI for the analysis of macromolecules of virtually unlimited mass range with extreme sensitivity has caused these two methods to supplant all other techniques, so only these methods will be discussed further.

At its simplest level, MS measures molecular masses. With calibration it can also determine quantities on a relative or absolute scale for 'pure' compounds and, with varying degrees of success, for components in a mixture. The analysis of complex mixtures such as a protein digest may require coupling with a separative method such as chromatography (GC-MS or LC-MS) or electrophoresis, either off-line (SDS-PAGE) or on-line (CE-MS). Further experiments can provide detailed structural information, e.g. peptides can be sequenced by collision-induced dissociation (CID) of their molecular ions and tandem MS (MS/MS). MS may also be used in conjunction with chemical modification or enzymatic digestion of a protein to aid its identification and/or sequence analysis. In practice, the diverse techniques available for ionization and mass analysis allow experiments to be optimized to answer very specific questions.

Mass spectrometry

Sample preparation and ionization methods

Optimization of sample preparation depends upon the nature of the sample, the information required and the type of mass spectrometer available. It is desirable to minimize salts and detergents, and if buffers are unavoidable these should be volatile whenever possible, e.g. ammonium formate or ammonium bicarbonate. In general MALDI is more tolerant of impurities than ESI. It may be essential to remove salts and detergents by dialysis, precipitation, absorption/elution from beads or a membrane, or absorption onto a small column and elution into the mass spectrometer. Achieving such separations without substantial losses is frequently complicated by limited amounts of material, sample aggregation, hydrophobicity and binding to surfaces.

Most peptides and proteins contain readily protonated basic sites, suitable for positive ion MS. Analytes are ionized directly from liquid solution for ESI but from the solid state for MALDI, consequently sample handling is fundamentally different for these alternative methods. In ESI-MS, liquid is usually introduced in a continuous stream, ideal for direct coupling with reversed phase high performance liquid chromatography (RP-HPLC), which is applicable to the separation of most peptide mixtures and many proteins. However, trifluoroacetic acid (TFA), widely used to optimize separations by RP-HPLC, can inhibit ionization in ESI-MS. Solvent systems developed for LC-MS replace TFA by formic acid, alternatively low flow rates from capillary HPLC columns can be supplemented with solvents more compatible with ESI-MS. An alternative to an externally pumped system is provided by nanospray, in which a small quantity of sample solution is placed in a capillary tube drawn to a fine tip. Liquid flows out at ~20–50 nL min⁻¹ under the combined influence of capillary action and an applied electric field, allowing each sample to be studied for an hour or more, which assists studies on mixtures such as protein digests. Fortunately, peak intensity in ESI or nanospray is largely independent of flow rate; consequently low flow rates efficiently conserve samples that are difficult to isolate and purify.

For MALDI-MS the analyte as a pure compound or a mixture is co-crystallized with a matrix that absorbs laser radiation and promotes ionization. Matrix materials ideal for peptides and proteins are aromatic acids such as sinapinic acid, 2,4-dihydroxybenzoic acid, and α-cyano-4-hydroxycinnamic acid, each having slightly different ionization characteristics. Published protocols for optimization of sample preparation try to achieve multiple, evenly distributed, small crystals. It is often necessary to remove salts by washing the crystals with water after they have been deposited.

Measurement of molecular mass

MS separates ions according to mass/charge (m/z). Peptides and proteins ionized by ESI under acidic conditions acquire multiple charges, z being roughly proportional to m, with m/z in the range 500–1500. In practice a distribution of charges gives multiple peaks in the mass spectrum, the spacing of which allows z to be calculated. The raw data for a pure compound can be deconvoluted to a zero-charge profile of the molecular mass, although this is more difficult for mixtures. An advantage of multiple peaks is the statistical improvement in mass accuracy. Multiple charging allows the m/z range of the mass spectrometer to be modest, even for large proteins. Mass analysers for ESI are mostly quadrupoles and ion traps with m/z ranges of 2000–3000, but orthogonal acceleration TOFs, hybrid quadrupole-TOFs, sector instruments and FTICRs of higher mass range are all available with ESI sources.

MALDI attaches only a single charge or a small number of charges to a peptide or protein,

consequently the m/z range for a suitable mass analyser must be much greater. Potentially a linear TOF instrument, with or without a reflectron, has unlimited mass range. Mass separation is based on ion velocity; slower ions take longer to arrive at the detector, therefore mass range is limited only by the observation time. In practice, factors such as detector design may inhibit the effective observation of the most massive species, but a mass range of several hundred thousand daltons is attainable.

MS methods for analysing peptides and proteins have two different operating regimes, which can be called low mass and high mass. Low mass describes the range where individual isotopic contributions to the overall molecular ion signal can be resolved as separate peaks. This is 1–2 kDa for a quadrupole of modest performance, perhaps 5 kDa for a high performance MALDI-TOF or ESI orthogonal-acceleration TOF, and significantly higher for FT-ICR. This regime which applies to peptides rather than proteins gives narrow peaks. With internal calibration 'monoisotopic' masses can be measured to 5–20 ppm for the ions containing the lowest mass isotopes, including ¹H, ¹²C, ¹⁴N and ¹⁶O. Only for the smallest species is this sufficient for an unambiguous isotopic assignment but it frequently differentiates between alternative isobaric species (ions of the same nominal mass). For multiply charged ions, the spacing of adjacent peaks within an isotopic cluster is equal to the reciprocal of the charge ($1/z$), thus z can be determined from a single peak. This is useful for complex spectra with multiple peaks that would otherwise be difficult to assign. In the high mass regime the isotopic clusters are not resolved and the 'average' molecular mass is obtained. Here mass spectrometer resolving power has less effect on overall mass accuracy, although any factor that broadens or distorts peak envelopes will introduce errors. This can include small covalent modifications such as methionine oxidation (+16 Da) or addition of a cation such as sodium (+23 Da) rather than a proton. These should be clearly resolved for small proteins of perhaps 20 kDa, but not for large proteins of say 100 kDa with inherently broad peaks. The best mass accuracy likely to be achieved with standard instrumentation is approximately 0.1–0.3 Da at 10 000 Da or 1–3 at 100 000 Da.

FT-ICR with a high field magnet represents a divergence from the above statement as this can have extraordinarily high resolving power. **Figure 1** shows the resolved isotopic cluster for the +49 charge state of bovine serum albumin (molecular mass 66.4 kDa), measured with a resolving power of 370 000 using a 11.5 T magnet. Thus, with such an instrument, almost any sample can give isotopic resolution.

Sensitivity of detection

Soft ionization usually gives molecular ions but not fragment ions, thus ion current is concentrated into a single peak or isotopic cluster. Although the ion yield of the ionization methods is relatively low (~1 ion per 1000 neutral molecules), MS is highly sensitive. Less than 100 ions are sufficient to define a mass spectrometric peak, i.e. ~10^5 molecules or 0.1 attomole. To exploit this inherent sensitivity it is necessary to integrate the entire ion signal, rather than scan a spectrum in which only a small fraction of the ions is monitored while most go unobserved. This is achieved by MALDI-TOF as each laser shot forms a packet of ions which are accelerated into the mass analyser to ultimately arrive at the detector. MALDI also has the advantage that a discrete quantity of sample on the target is available for analysis for as long as the experimenter chooses to select a new region to investigate or until the sample is exhausted, a dried spot from 1 μL of sample being sufficient for several thousand laser shots. By contrast ESI is used mostly with scanning instruments and spectra are recorded during the limited time the analyte enters the ionization region. This is relatively inefficient and ESI has been regarded as less sensitive than MALDI. A new generation of TOF instruments compatible with ESI integrate the signal, and nanospray is more like MALDI as sample is retained throughout the experiment, providing a substantial sensitivity enhancement. MALDI and nanospray both provide detection limits in the low femtomole region or better.

Additional techniques

Direct analysis of mixtures versus LC-MS

Because MALDI gives predominant singly charged molecular ions with few fragments, analysis of multi-component mixtures such as protein digests is readily achieved. Each peak corresponds to a separate peptide and can be selected for 'post-source decay' or PSD. However, some components in a mixture may not compete effectively for the available charges and may be weak or absent, e.g. tryptic peptides terminating in lysine rather than arginine. ESI is less suitable for direct analysis of mixed peptides as each component gives several multiply charged peaks that cause complex spectra. However, ESI is ideal for LC-MS and is less discriminatory as components elute separately, giving more comprehensive coverage of the original protein. Although unimportant for identification of a protein in a database, this is essential to find protein modifications or mutations. Automation is available from some instruments manufacturers for MS-MS analysis on each molecular species eluting

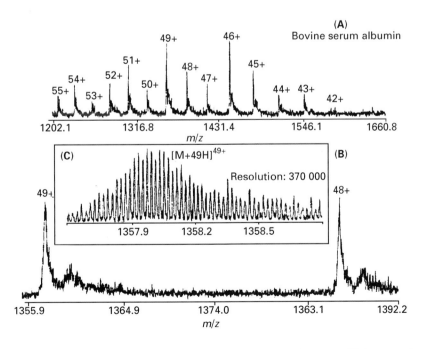

Figure 1 (A) ESI-FTICR spectrum of bovine serum albumin recorded using an 11.5 T magnet; (B) An expansion of the 49+ and 48+ charge states; (C) A further expansion of the 49+ charge state recorded at 370 000 resolving power showing isotopic separation. Reproduced with permission of Elsevier Science from Gorshkov MV, Tolić LP, Udseth HR, et al. (1998) Electrospray ionization – Fourier transform ion cyclotron resonance mass spectrometry at 11.5 Tesla: instrumental design and initial results. *Journal of the American Society for Mass Spectrometry* **9**: 692–700.

from the chromatograph. Disadvantages of LC-MS include added cost and complexity of the instrumentation and additional time spent equilibrating columns and waiting for components to elute.

Peptide sequencing by CID

Peptides and proteins can be sequenced by Edman chemistry at levels down to 1–10 pmol, depending on the number of residues to be determined. This requires a free amino terminus and is ineffective for modified amino acids unless appropriate standards are available. Through the use of CID and tandem mass spectrometers, MS has established itself as a more sensitive, faster alternative to Edman sequencing, although it cannot handle an intact undigested protein. Unlike Edman sequencing, fragment data can be obtained on each component of a mixture without separation. The efficiency of peptide sequencing by MS depends greatly upon the type of instrument available. Tandem mass spectrometers for CID include multisector instruments, triple quadrupoles (QQQs) and hybrid quadrupole-sectors. The less expensive and easier to operate QQQ is widely used. A molecular ion selected in Q1 fragments in Q2 through low energy collisions with a gas, then fragment ions are analysed in Q3. The same experiment, and even MS^n, can be carried out in a relatively modest ion trap, or in an FT-ICR with higher resolution

but at considerably greater expense. Hybrid QQ-TOFs offer substantially superior performance to QQQs, giving high sensitivity, high resolution data. A MALDI-TOF instrument equipped with a reflectron is equally capable of MS-MS by PSD.

Interpretation of spectra from CID-MS-MS of peptides of up to ~20–25 residues has been well documented. Peptides have a repeating linear backbone with sidechains defining the constituent amino acids. Backbone fragmentation of a singly charged peptide ion gives two species, an ion and a molecule. Retention of the proton by the N-terminal fragment gives an **a**, **b** or **c** ion, whereas C-terminal ions are classified as **x**, **y** or **z** (**Figure 2**). The most common cleavage at the amide bonds gives **b** or **y** ions; tryptic peptides with a C-terminal basic residue generally exhibit predominant **y** ions. Because ionization occurs by proton addition, this gives an ion with no odd electrons, which can be more stable than a corresponding radical cation. Subsequent cleavage of a backbone bond is associated with transfer of a hydrogen radical to prevent the thermodynamically unfavourable formation of a radical cation and a neutral radical. In forming a **b** ion this hydrogen moves to the neutral, whereas a **y** ion gains one hydrogen in addition to that added during ionization; thus **y** ions are sometimes designated as **y″** or **y+2**. Ion masses are calculated as the sum of the amino acid residues involved plus 1 H for **b** ions or plus

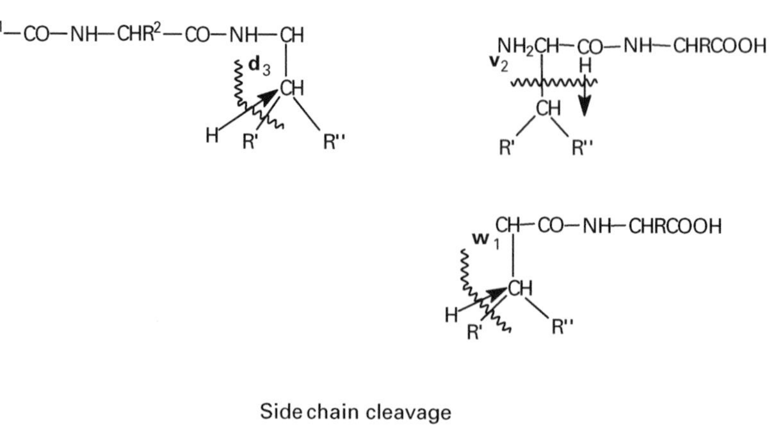

Backbone cleavage

Side chain cleavage

Figure 2 Collision induced fragmentation scheme for peptides and proteins. The initial ionizing proton is not shown and proton transfers are not shown for backbone cleavages (see text).

H_3O for y ions. Theoretically, cleavage can occur at each amide bond, giving a series of ions defining the amino acid sequence. In **Figure 2**, subscripts attached to the ion types identify which bonds have broken to form the fragment ions, e.g. for a peptide of n amino acids, the C-terminal ionic fragment formed by loss of the most N-terminal amino acid is designated y_{n-1}. Some ions are formed by further cleavages at the sidechains, giving peaks referred to as d, v and w ions, some of which can identify specific amino acids and differentiate isomeric amino acids such as leucine and isoleucine. The d ions represent loss of a group from the β-carbon of an a ion, w ions are formed by the equivalent loss from a z ion, and v ions represent loss of the intact sidechain at the α-carbon of a y ion. Multiple bond cleavages also give internal fragments, including individual amino acids that appear in the low mass region of the spectrum as immonium ions $^+NH_2=CHR$ that are valuable for diagnostic purposes. Note that a_1 is the immonium ion for residue 1.

Interpretation of MS-MS spectra of multiply charged ions from ESI-MS is complicated by the different charge states possible amongst the fragment ions, unless high mass resolution is available. The most comprehensive fragment ion spectra are generally obtained from high energy CID, such as from sector instruments. Programs exist for both the straightforward prediction of spectra and the more difficult interpretation of experimentally obtained spectra.

Chemical derivatization

Many straightforward chemical reactions can enhance the quality and utility of MS data from peptides and proteins. Before the advent of ESI or MALDI, polar groups in biomolecules were often derivatized to increase volatility, e.g. by permethylation or silylation. This could provide additional information on the number of replaceable hydrogens of a given type. Such procedures are still useful. Acetylation with acetic anhydride adds 42 Da for each free amino group, confirming whether the amino terminus is free or blocked, and can distinguish isobaric glutamine from lysine, the latter becoming acetylated. An equimolar mixture of perdeutero and protonated reagent gives double peaks separated by 3 Da for N-terminal but not C-terminal fragment ions. Esterification with acetyl chloride/methanol adds 14 Da per carboxylic acid and provides similar information about the C-terminus and the location of glutamate or aspartate residues. Trypsin digestion in $H_2^{16}O/H_2^{18}O$ differentially labels the C-termini of all resulting peptides, except the original protein C-

terminus. All of these techniques employing stable heavy isotopes enhance the information content in MS/MS as the N- and C-terminal fragments are readily distinguishable. Other derivatizations to improve MS/MS spectra include the addition of a permanent positive charge at one or other terminus, usually the N-terminus, which directs the fragmentation and aids spectral interpretation.

Protein disulfide bonds can be reduced with dithiothreitol and alkylated with a reagent such as iodoacetic acid before digestion, adding 58 Da per cysteine. Although MS is replacing Edman sequencing to a significant degree, Edman chemistry is used for ladder sequencing, in which phenyl thiocyanate is included at each cycle with the normal Edman reagent, phenylisothiocyanate. This blocks the N-terminus of a small fraction of the analyte molecules and prevents further cleavage, giving mixed products differing from each other by single amino acids. The sequence is read directly from the MALDI spectrum of the unseparated mixture.

Applications

Quality control of synthetic peptides and recombinant proteins

MS analysis has become a routine aid in the purification of synthetic peptides and recombinant proteins and plays an essential role in quality control of materials required to be of high purity. Following cleavage from the solid-phase resin, high quality peptides are purified by RP-HPLC. Fractions collected from an analytical run can be surveyed by MS to identify the elution profile of the desired product, with a minimum of impurities. Fractions may be dried down and used directly or they may act as a guide to fraction collection for a larger scale separation. The presence of unwanted side products such as those formed by amino acid deletions, incomplete removal of protecting groups and chemical modifications should be immediately apparent from the measured masses. If careful attention was paid to the sequence of amino acids loaded onto the synthesizer, the observation of the desired molecular mass should be sufficient to confirm the anticipated product. If necessary, MS/MS can be used to confirm the sequence.

MS and MS/MS are particularly useful for identifying and locating heavy isotopes. For example, **Figure 3** shows a portion of the MS/MS spectrum of three versions of a 14-residue peptide prepared for an NMR study, two of which contain ^{13}C labels at a carbonyl and an α-carbon in two alanine residues. Mass differences between successive a and b ions allow the positions of these residues to be determined precisely. As b_6 is at m/z 656 for all three species including the unlabelled control, no ^{13}C labels are present in the first six residues. However, a_7 for compound (C) is 1 Da higher than the equivalent ion for (A) or (B), therefore the α-carbon of the seventh residue in (C) is ^{13}C. Similar logic allows each of the other labels to be identified.

A number of potential chemical modifications can cause recombinant proteins to differ from the desired product. Cysteine-containing proteins may show a time-dependent shift of both chromatographic retention time and mass as aerobic oxidation causes disulfide formation (−2 Da per disulfide). Methionine oxidation to the sulfoxide is quite common and is readily identified (+16 Da). N-terminal glutamine may eliminate ammonia to form pyroglutamic acid (−17 Da), especially if stored in acidic solution. Harder to detect may be deamidation of asparagine to form a succinimide intermediate that is then hydrolysed to aspartate, or its isomer isoaspartate (+1 Da). Posttranslational modifications such as glycosylation occurring in mammalian systems are rarely observed in proteins expressed in bacteria but processes such as phosphorylation are not unknown. Enzyme impurities from the expression system may be responsible for numerous reactions, including the total degradation of the desired product. Carboxypeptidases and aminopeptidases can result in unexpected trimming of the intact sequence. N-terminal methionine is quite often observed to be partly or completely absent (−131 Da). This list of potential variants is far from complete but it gives an indication of the role that MS can play in their identification.

Protein identification by in-gel digestion and database searching

The closing decade of the twentieth century witnessed the initiation of a major concerted programme to sequence the human genome, and genomes for several other organisms are already completed. This effort is yielding a vast array of information about genes, but this will be the tip of the iceberg compared with the unanswered questions relating to proteins, including cellular and tissue-specific variations in levels of expression, posttranslational modifications, and their associations to form functional multimolecular units. MS will play an essential role in the elucidation of this information, often referred to as proteomics. Techniques are now available for the analysis of the major proteins in specific cell types. At present the most productive methods link 2D electrophoresis with high sensitivity MS and database searching, sometimes referred to as 'mass fingerprinting'.

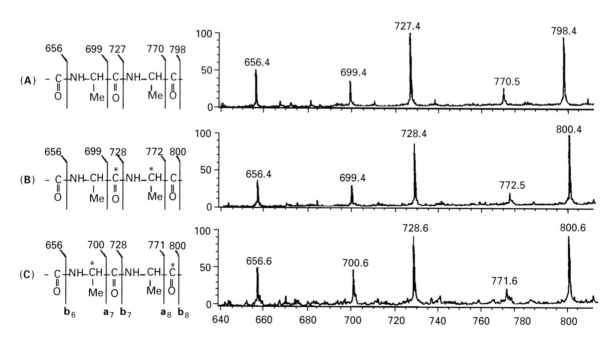

NH$_2$-MKHMAGA*A*AAGAVV-NH$_2$

Figure 3 Partial CID-MS/MS spectrum obtained on a tandem 4-sector mass spectrometer for a 14-residue synthetic peptide. (A) without ^{13}C labels; (B) and (C) with ^{13}C labels as indicated by asterisks.

As many as 2000 proteins from a cell digest may be separated on a 2D gel as discrete spots stained with Coomassie blue (100 ng sensitivity), silver (1–10 ng) or a fluorescent dye (<1 ng). These proteins may be difficult to elute from the gel in high yield for direct analysis by MS. Furthermore, if it could be determined, even the precise atomic composition would be insufficient information to identify a protein. However, cutting the spot from the gel, digesting the protein with an enzyme such as trypsin, eluting the peptides and analysing them by MS can provide sufficient information to unambiguously identify any known protein in a standard protein database, particularly if ion masses are measured accurately. Tryptic peptides are advantageous as they all terminate in a readily protonated basic residue, lysine or arginine. They can either be analysed in the intact mixture, usually by MALDI or nanospray, or separated and analysed by LC-MS. Several programs are freely available via the World Wide Web for database searches. As yet the databases are incomplete and many proteins remain unidentified, but ultimately all possible proteins will be predictable for an entire genome. As databases increase in size it will become necessary to increase the number of peptides identified or further improve the accuracy of mass measurement. Alternatively, sequence information from CID and MS/MS will greatly increase the reliability of identification.

Identification of genetic mutations

Most genetic mutations are identified by molecular biological techniques including gene sequencing. However, mass spectrometry of protein digests was a valuable additional method even before the development of ESI and MALDI. The molecular mass of an intact protein can now be determined by ESI or MALDI, which compared with a normal protein or the predicted sequence from a c-DNA will indicate the presence of a mutation. The precise location and identity can be confirmed by digestion and further analysis of the peptides by LC/MS or LC/MS/MS. The identification of haemoglobin variants represents one of the best known examples of mutational analysis by MS. Most patients are heterozygous, i.e. blood samples contain both the normal and variant forms of the protein. With a molecular mass in the region 15–16 kDa, monomeric haemoglobin is a relatively small protein and the differences between normal and variant forms are easily measured with high precision. Mass spectrometry complements the normal electrophoretic detection of the numerous haemoglobin variants responsible for a number of serious diseases including sickle-cell anaemia.

Posttranslational modifications

Gene sequences specify the initial amino acid sequences of proteins as expressed by the ribosomes. However, complex enzymic reactions in the cell can greatly modify the chemical composition of mature proteins. There are numerous possible modifications such as trimming of the expressed sequence to remove signal peptides, cysteine oxidation to cystine and addition of functionalities such as phosphates, sulfates, oligosaccharides and lipids. Understanding these modifications may be fundamental to understanding the biological action of a protein. To some extent they may be predictable from the existence of a consensus sequence such Asn-X-Thr or Asn-X-Ser for glycosylation of asparagine, but the extent of the modification may range from 0–100% of protein molecules. Also, because a process such as glycosylation is carried out by a complex cascade of enzymes, the glycoforms at a single site are frequently diverse in both structure and molecular mass. MS of a glycopeptide or glycoprotein will reveal the presence of the oligosaccharide by an increase in mass, but the potential heterogeneity may make it difficult to resolve the various glycoforms. Furthermore, mass alone will not distinguish the many isomeric forms of the individual monosaccharide units, their sequences or any branching patterns. The sugars can also be modified, e.g. by presence of phospholipids as in the glycosyl phosphatidylinositol-anchored proteins.

In order to determine the mass of the amino acid sequence of the protein alone the oligosaccharides can be removed chemically or with a glycosidase enzyme such as Endo-H or peptide N-glycosidase F (PNGase F). The released sugars can also be studied although their complete characterization may be more difficult than that of the peptide or protein. Tandem MS may identify the glycopeptides in a complex peptide mixture as CID gives low mass ions characteristic of the monosaccharide units, including m/z 163 (hexose), 204 (N-acetylhexose) and 292 (sialic acid). These masses are ideal for LC-MS-MS precursor ion scans to identify the glycopeptides and define their masses. Methods have been developed for the detection and analysis of N-linked and O-linked oligosaccharides and the more recently discovered but widespread N-acetyl glucosamine modification of serine (and threonine), which apparently plays a regulatory role similar to that of phosphorylation.

Figure 4 illustrates the complexity that can arise due to glycosylation. A 75-residue glycopeptide derived from the prion protein by digestion with endopeptidase Lys-C contained a single glycosylation site. The oligosaccharides attached to this site were highly heterogeneous, giving the complex ESI spectrum in **Figure 4A**, with the deconvoluted spectrum in **Figure 4B**. Treatment with PNGase F removed the sugars completely, leaving just the amino acid chain. This gave a simplified spectrum from which a molecular mass of 8607.8 was determined. Subtraction of the mass of the deglycosylated peptide from those of the prominent glycopeptides gave the masses of the sugars, which could be related to partial structures for the glycoforms determined previously.

Modifications such as phosphorylation are chemically simpler and are more easily identified by MS, although the addition of 80 Da can be due to either a phosphate (HPO_3) or a sulfate (SO_3). Serine and threonine phosphates are acid sensitive and may not survive LC/MS whereas tyrosine phosphates are generally more robust. Sulfate is also labile and may not be observed in positive ion MALDI but it usually can be identified by ESI or negative ion MALDI. The importance of phosphorylation in signalling and the control of cellular events has greatly increased interest in identification of protein phosphorylation sites. A valuable technique to locate phosphopeptides in a protein digest uses LC-ESI-MS-MS precursor ion scans of m/z 79 in the negative ion mode to identify all ions forming the characteristic fragment PO_3^-. The extent of phosphorylation may be much less than stoichiometric, so the same peptides may occur with and without phosphate. The identified phosphopeptides can be sequenced by MS-MS to identify the specific residues carrying the phosphate groups. A current challenge is to increase the sensitivity for detection and identification of such modifications in complex mixtures of proteins extracted from cells and tissues.

Disulfide bond analysis by MS is usually straightforward for proteins but is challenging for small

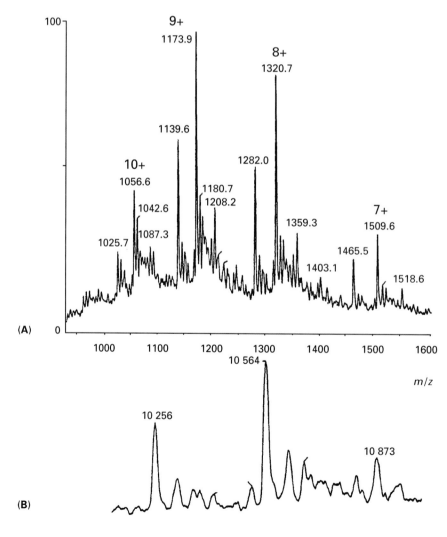

Figure 4 (A) ESI-MS spectrum of a 75-residue glycopeptide from the prion protein showing multiple peaks due to oligosaccharide heterogeneity. (B) Deconvolution of the spectrum in (A) showing molecular masses rather than *m/z*.

cysteine-rich proteins and peptides such as the cono-toxins which may contain as many as 6 cysteine residues within a peptide of only 20–30 amino acids. For a protein an effective strategy is to carry out chemical or enzymic digestion and peptide mapping on the native material after reduction with dithiothreitol and alkylation of the cysteines using a reagent such as iodoacetic acid. The addition of a carboxymethyl group to a cysteine increases the mass by 58 Da, so cysteine-containing peptides are readily detected by comparison with a digest of the unmodified protein. A protein digest with the disulfide bonds intact will result in some peptides linked by the disulfides. If these are not identifiable on the basis of mass alone, they may be separated by HPLC, reduced and then analysed as two separate peptides.

Noncovalent interactions and analysis of higher order structure

Peptides and proteins do not carry out their biological functions in isolation. Their active forms may be associated with metal ions, small molecules, macromolecules including proteins and DNA, and they may form complex multiprotein assemblies. Biological processes occur in aqueous solution in the presence of salts and other ions whereas MS is used to probe substances isolated *in vacuo*, conditions that could be incompatible. Despite this, the use of MS to study these interactions is growing rapidly. MALDI, which ionizes directly from the solid phase, does not lend itself to studies that mirror solution conditions, but ESI, which abstracts ions directly from solution by rapid evaporation of nebulized droplets, can monitor noncovalent associations very successfully.

The most successful studies have been on systems involving electrostatic rather than hydrophobic interactions. The metal-binding properties of a peptide (or protein) can be monitored directly from aqueous solution of perhaps $10\ \mu M$ peptide containing the metal cation salt at various concentrations. The pH is controlled by the use of a weak buffer such as 1–10 mM ammonium acetate. An equilibrium in solution between free peptide and the complex can be preserved during the rapid drying of the microdroplets, occurring within 0.5 ms. Low micromolar dissociation constants derived from the peak intensities are similar to those obtained by conventional methods. Although removal of the dielectric shielding effect of water might encourage random binding of cations to negative amino acids, strong specific binding can be clearly distinguished from weak nonspecific effects. Peptides homologous to some ATP-ases containing His-X-His or His-X-X-His bound either Cu^{2+} or Ni^{2+}

but not Zn^{2+}, whereas analogous peptides with a single histidine bound only Cu^{2+}. By contrast a peptide corresponding to a region of the Alzheimer's precursor protein having the motif His-X-His-X-His was seen to bind all three divalent cations.

Figure 5 shows partial mass spectra at two time points for a metal ion transfer reaction between metallothionen coordinated with 7 zinc ions and carbonic anhydrase. The conversion of the apo form to the holo form was monitored over a wide pH range for up to 120 min. It was also seen that zinc ion uptake was associated with addition of a hydroxyl radical or water molecule. Such an experiment gives direct information about the stoichiometry of reaction intermediates and complexes and can yield kinetic and thermodynamic data.

Complexes between the trp repressor (TrpR) and its specific operator DNA were monitored in a competition experiment. When TrpR was mixed with an equimolar mixture of DNA containing two consensus sequences separated by 2, 4 or 6 base pairs, 1:1 protein:DNA complexes formed only with DNA having the 4-bp spacer. The dimerization of the AIDS

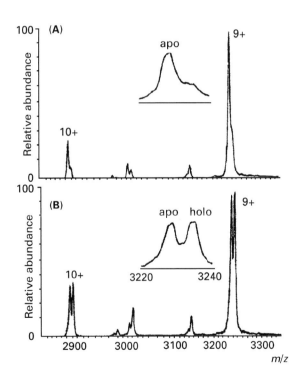

Figure 5 A metal transfer reaction between metallothionen-Zn_7 and carbonic anhydrase at pH 7.5 monitored by ESI-MS showing the major peaks for carbonic anhydrase, (A) after 3.5 min, and (B) after 43.5 min. Reproduced with permission of Cambridge University Press from Zaia Z, Fabris D, Wei D, Karpel RL and Fenselau C (1998) Monitoring metal ion flux in reactions of metallothionen and drug-modified metallothionen by electrospray mass spectrometry. *Protein Science* **7**: 2398–2404.

protease has also been demonstrated by ESI but hydrophobic interactions such as this are stabilized by the dielectric of the solvent and by the presence of salts. This stabilization is lost on transfer to the gas phase, necessitating very mild conditions for the nebulization, evaporation and transfer of such delicate complexes into the vacuum system.

ESI-MS studies at neutral pH of proteins in their native state usually results in the attachment of far fewer protons compared with ionization from acidified solution. Consequently m/z values are frequently 5000 or more and may range up to 10 000. Such species are beyond the normal working range of many mass spectrometers including most quadrupoles and ion traps, but they can be studied with TOF or high field magnetic instruments.

Hydrogen–deuterium exchange

MS is obviously well suited to establishing the primary amino acid sequence and posttranslational modifications of peptides and proteins and it also has a role in determining the quaternary structure, i.e. the noncovalent association of subunits in multiprotein assemblies. What is less apparent is that hydrogen–deuterium exchange allows secondary and tertiary structure to be probed. Other techniques commonly used for such biophysical studies include optical spectroscopy, fluorescence, circular dichroism, light scattering, NMR and X-ray crystallography. Only the last two of these can give a detailed 3D molecular structure, the others give integrated views that obscure individual features. MS requires much less material than any other method and can follow the dynamics of rapid protein folding and unfolding.

The amide hydrogens along the protein backbone and in some sidechain positions are labile and undergo rapid exchange with solvent protons unless they are involved in hydrogen bonding. Secondary structural elements such as helices and sheets are maintained by hydrogen bonds, as are many aspects of tertiary structure. The rate of exchange of such protected hydrogens is usually reduced dramatically. Usually a peptide or protein can be deuterated fully by lengthy exchange in D_2O, each deuteron increasing the mass by 1 Da, so that the total increase gives a measure of the number of exchangeable hydrogen atoms. The accessibility of these atoms is assessed by measuring the rate of back exchange with H_2O. Above pH 3 the rate of hydrogen exchange for freely accessible hydrogens is proportional to $[OH^-]$, i.e. is approximately 10 000 times faster at physiological pH than at pH 3. After carrying out the exchange reaction at pH 7.4, the reaction can be effectively frozen by taking aliquots at different times and rapidly

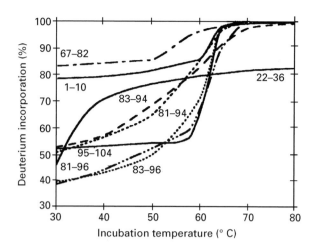

Figure 6 Deuterium incorporation into specific segments of cytochrome C as a function of temperature, showing thermal melting at 58–64°C of α-helical regions. Reproduced with permission of Cambridge University Press from Zhang Z and Smith DL (1993) Determination of amide hydrogen exchange by mass spectrometry: A new tool for protein structure elucidation. *Protein Science* **2**: 522–531.

dropping both the pH and the temperature. The aliquots can then be analysed by ESI-MS to determine the degree of exchange. Alternatively the exchange process occurs in a flowing stream in which the reagents mix for a controlled period of time before being quenched by mixing with an acidified solution then injected directly into the ESI source.

Peptides can be sequenced by MS-MS to determine the precise location of the deuterons. However it is usually essential to employ intact proteins rather than shorter peptides to address questions concerning secondary and tertiary structure, but the location of the deuterons in a protein requires digestion to peptides after exchange. Fortunately the non-specific protease pepsin is most active at the low pH values at which further exchange is minimized. Samples of the protein that have been subjected to back exchange can be digested quickly using a high enzyme:substrate ratio, then separated and analysed rapidly by LC-MS using fast flow rates through chilled, highly porous HPLC columns. Deuterium incorporation into specific segments of cytochrome C as a function of temperature is illustrated in **Figure 6**. This shows strong evidence for thermal denaturation with a sharp transition at 58–64°C that affects only the most structured regions known to be involved in α-helices.

Conclusions

MS is a critically important method for the identification and characterization of peptides and proteins. This role will continue to grow as proteomic studies spur the demand for high throughput protein analysis and characterization. Many mass spectrometric techniques lend themselves to rapid analysis using automation and robotics, particularly MALDI for which it is already easy to deposit hundreds or even thousands of samples on a single target plate and to programme the mass spectrometer to collect a spectrum for each one. One serious factor limiting further exploitation is the availability of suitable computerized data systems to analyse and summarize the enormous quantities of data such automation can yield. However, it is likely that the concerted efforts of the academic community, mass spectrometer manufacturers and major users such as multinational pharmaceutical companies, will successfully solve such problems quite rapidly.

See also: **Atmospheric Pressure Ionization in Mass Spectrometry; Biochemical Applications of Mass Spectrometry; Chromatography-MS, Methods; Laboratory Information Management Systems (LIMS); Nucleic Acids and Nucleotides Studied Using Mass Spectrometry; Proteins Studied Using NMR Spectroscopy; Time of Flight Mass Spectrometers.**

Further reading

Burlingame AL, Boyd RK and Gaskell SJ (1998) Mass spectrometry. *Analytical Chemistry* 70: 647R–716R.

Burlingame AL, Carr SA and Baldwin MA (1999) *Mass Spectrometry in Biology and Medicine.* Totowa, NJ: Humana Press.

Chapman R (1996) *Protein and Peptide Analysis by Mass Spectrometry. Methods in Molecular Biology*, Vol 61. Totowa, NJ: Humana Press.

Cole RB (1997) *Electrospray Ionization Mass Spectrometry: Fundamentals, Instrumentation and Applications.* New York: Wiley.

Cotter RJ (1997) *Time-of-Flight Mass Spectrometry: Instrumentation and Applications in Biological Research*, ACS Symposium Series 549. Washington DC: American Chemical Society.

Ens W, Standing KG and Chernushevich IV (1998) *New Methods for the Study of Biomolecular Complexes.* NATO ASI Series. Dordrecht: Kluwer Academic.

Loo JA (1997) Studying noncovalent protein complexes by electrospray ionization mass spectrometry. *Mass Spectrometry Reviews* 16: 1–23.

McCloskey JA (1990) Mass spectrometry. *Methods in Enzymology* 193.

Perfused Organs Studied Using NMR Spectroscopy

John C Docherty, National Research Council of Canada, Winnipeg, Manitoba, Canada

MAGNETIC RESONANCE

Applications

Introduction

Studies of isolated, perfused organs have been invaluable in furthering our knowledge of metabolism and function. This experimental system allows the direct observation of the effects of an intervention on an organ in the absence of neurohumoral control mechanisms. Experimental conditions are simpler to control, and to modify, in isolated organs than in the whole animal. Isolated organs also maintain important intercellular interactions that do not take place in isolated cells, subcellular organelles or cellular homogenates. Nuclear magnetic resonance (NMR) spectroscopy of these organs has greatly increased the versatility and power of such studies. NMR spectroscopy allows data to be collected nondestructively throughout the experimental protocol in discrete time segments. The types of data that can be acquired include information on energy status, ion homeostasis, substrate utilization and enzyme activities (**Table 1**). Most studies have been performed using isolated hearts. Liver, kidney and skeletal muscle have been used to a lesser extent. In addition to furthering knowledge of biochemistry and physiology, such studies have provided a basis for the development of *in vivo* NMR spectroscopy as a tool for assessing normal and pathophysiological function in humans. This article will describe the uses of ^{31}P, ^{23}Na, ^{13}C, ^{1}H, ^{19}F, ^{87}Rb and ^{7}Li NMR spectroscopy in cardiac physiology and pathophysiology. The application of these techniques to studies of isolated liver and kidney will also briefly be described.

Studies of isolated hearts

NMR spectroscopy has been performed on hearts isolated from a wide variety of species including dog, pig, sheep, rabbit, turkey, ferret, guinea-pig, rat and mouse. Hearts from the larger animals have been studied in horizontal-bore magnets with a clear bore of 20–30 cm and operating at field strengths of 4.7–7.0 T. Rodent hearts have generally been studied in vertical-bore magnets with a clear bore of approximately 8 cm and operating at field strengths of 4.7, 8.7, 9.4 and 11.75 T. In most studies perfusion is performed with a modified Krebs–Henseleit buffer with the approximate composition (mM) NaCl 118, KCl 4.7, $MgSO_4$ 1.2, $NaHCO_3$ 25 and with $CaCl_2$ varied between 1.2 and 2.5 mM depending upon the species. Ethylenediaminetetraacetic acid (EDTA) is often added at a concentration of 0.5 mM to chelate heavy-metal contaminants. For studies involving ^{31}P spectroscopy, phosphate is normally omitted from the perfusate to allow identification and quantification of the intracellular inorganic phosphate (P_i) peak. For other studies, perfusate may be supplemented with phosphate at a final concentration of 1.2 mM. The perfusate is aerated with 95% O_2/5% CO_2 to achieve a pH of 7.4, a $p_{O_2} > 600$ mmHg and a p_{CO_2} of 35–45 mmHg. Glucose (11 mM) is often supplied as the sole substrate source, although in many situations a mixture of glucose and pyruvate (or other substrates) is used. For studies of hearts from the larger species, the perfusate is supplemented with serum albumin to minimize the oedema that results from perfusion with crystalloid solutions. The use of crystalloid perfusate results in coronary flows ~3 times those observed under conditions where whole blood or buffers supplemented with washed red cells are used.

Hearts can be perfused in two modes; the working heart preparation and the isovolumic (or Langendorff) preparation. Both methods allow assessment of cardiac mechanical function throughout the experimental protocol. For both preparations, the ascending aorta is cannulated and retrograde perfusion of the aorta is initiated. In Langendorff

preparations, perfusion is continued in this manner and the coronary arteries are continuously perfused throughout the protocol. Perfusion is performed under conditions of constant pressure (60–80 mmHg) or of constant flow by means of a pump. A compliant water-filled balloon is inserted into the left ventricle and connected to a pressure transducer in order to measure left ventricular pressure. The balloon is inflated to achieve a relevant end-diastolic pressure (generally in the region of 10 mmHg). In the working heart preparation, following stabilization in the Langendorff mode, the left atrium is cannulated and perfusion is continued through this chamber. The perfusate enters the left ventricle and is ejected into the aorta. Perfusion of the coronaries occurs during diastole when the aortic valve closes. Cardiac function is assessed on the basis of cardiac output (measured with flow probes) and aortic pressures. Owing to the physical constraints imposed by working within a magnet, most MR spectroscopy studies are performed using the Langendorff preparation. Temperature regulation is achieved using water-jacketed perfusion lines, by immersing the heart in the perfusion buffer and by means of a flow of warm air within the bore of the magnet.

Hearts isolated from rodents can be contained within commercially available NMR tubes (20–30 mm) and make use of commercially available broad-band or nuclei-specific NMR probes. Studies on hearts from larger mammals generally require an organ bath that incorporates a custom-built NMR coil within its structure or a surface coil attached to the left ventricular wall.

^{31}P NMR spectroscopy

^{31}P NMR spectroscopy is widely used for studies of isolated hearts. Using the endogenous ^{31}P signal arising from the tissue, it is possible to obtain information about the energy status of the heart and also to determine the intracellular pH (from the chemical shift of P_i). Assessment of extracellular pH is also possible using phosphonates that are confined to the extracellular space (e.g. phenylphosphonic acid or methylphosphonic acid). The heart metabolizes substrates (fatty acids, ketones, lactate, glucose, etc.), with the resultant energy being stored in the high-energy phosphate compound adenosine triphosphate (ATP). Most of this ATP is formed by mitochondrial oxidative phosphorylation. The phosphocreatine shuttle is responsible for transferring the energy from this mitochondrial ATP to sites of energy expenditure at the myofibrils and sarcolemma. **Figure 1** shows a typical ^{31}P spectrum obtained from an isolated guinea-pig heart. The phenylphosphonic

Table 1 NMR-visible nuclei relevant to the study of perfused organs

Nucleus	Information obtained
^1H	Levels of lactate and creatine. Changes in lipid
^7Li	Congener of Na^+. Measure Na^+ fluxes
^{13}C	Substrate selection. Citric acid cycle activity
^{19}F	Measure intracellular Ca^{2+} using fluorinated Ca^{2+} probes
^{23}Na	Measure intracellular Na^+ levels
^{31}P	Assess energy status
	Measure intracellular pH (from chemical shift of P_i)[a]
	Measure enzyme kinetics – saturation transfer for creatine kinase reaction
^{87}Rb	K^+ congener. Measure K^+ fluxes

[a]P_i = inorganic phosphate.

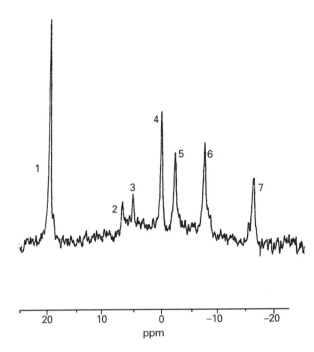

Figure 1 ^{31}P NMR spectrum of a guinea-pig heart perfused with Krebs–Henseleit solution. The spectrum was acquired at 8.7 T using a broad-band probe tuned to 145.8 MHz. Peak assignments are: 1, phenylphosphonic acid (external reference); 2, phosphomonoesters; 3, inorganic phosphate (P_i); 4, phosphocreatine (PCr); 5, γ-phosphorus of adenosine triphosphate (ATP); 6, α-phosphorus of ATP; 7, β-phosphorus of ATP. The spectrum was acquired in 2.5 min by summing 72 free induction decays (FIDs) with a 35 μs pulse and a repetition time of 2 s. Prior to Fourier transformation the FID was subjected to exponential multiplication with a 20 Hz line broadening factor.

acid, which acts as an external standard, is contained within a capillary tube placed alongside the heart and contained entirely within the coil of the NMR probe. ^{31}P NMR can detect phosphorus-containing compounds that are present in the fluid phase at concentrations of 0.6 mM or greater. The compounds visible by this technique are inorganic phosphate (P_i), phosphomonoesters (in this case sugar phosphates), phosphocreatine (PCr) and adenosine triphosphate (ATP). Adenosine diphosphate (ADP) is not visible because most of this nucleotide is protein bound within the cardiomyocytes.

Spectra are routinely collected with a repetition time of approximately 2 s, permitting the acquisition of data with adequate signal-to-noise in 2–5 min. This leads to 10–20% saturation of the PCr signal ($T_1 \approx 3$ s in rat heart at 8.7 T). The free induction decays are normally subjected to Fourier transformation following exponential multiplication using an appropriate line broadening (5–20 Hz). In many situations, alterations in the high-energy phosphate content of the heart over the course of an experiment are expressed as changes relative to the starting level.

Absolute quantification of metabolite levels can be achieved by use of an appropriate external reference (e.g. phenylphosphonic acid in **Figure 1**) and correction for the partial saturation of the PCr signal. ATP contents are determined from the integral of the β ATP peak; the α and γ ATP peaks overlap with resonances from other molecular species. The PCr and β ATP phosphates are 100% NMR-visible under aerobic conditions. The free ADP concentration can be calculated from the creatine kinase equilibrium equation,

$$K_{eq} = \frac{[\text{Cr}][\text{ATP}]}{[\text{PCr}][\text{ADP}][\text{H}^+]} \qquad [1]$$

where total creatine is normally determined biochemically and free [Cr] is determined from the difference between total [Cr] and [PCr]. Literature values for K_{eq} are ~10^9. The free energy of hydrolysis of ATP may be calculated from

$$\Delta G_{ATP} = \Delta G_0 + RT \ln \frac{[\text{ADP}][\text{P}_i]}{[\text{ATP}]} \qquad [2]$$

where ΔG^0 is taken to be -30.5 kJ mol^{-1}. It is often suggested that ΔG_{ATP} more accurately reflects the energetic capabilities of tissue than does a determination of the levels of high-energy phosphates.

^{31}P NMR can also be used to determine intracellular pH from the pH-dependent chemical shift of P_i using the following formula based on the Henderson–Hasselbach equation:

$$\text{pH} = \text{p}K + \log\left[(\delta - 3.27)/(5.69 - \delta)\right] \qquad [3]$$

where pK = pK_2 of inorganic phosphate (6.75). This technique yields an intracellular pH of 7.10–7.20 when the heart is perfused with buffer at pH 7.4.

^{31}P NMR spectroscopy has been applied to questions relating both to normal and to pathophysiological conditions. ^{31}P NMR spectroscopy has been used in the normal heart to investigate the regulation of cardiac energy supply in response to increased demand. This study showed that there is no simple equilibrium between the phosphorylation potential and the mitochondrial redox state and that other factors are involved in coordinating energy supply and demand.

^{31}P spectroscopy has also been used to study the mechanisms responsible for myocardial ischaemia–

reperfusion injury. During ischaemia, blood (or perfusate) flow is restricted or totally occluded, resulting in an insufficient supply of oxygen to support oxidative metabolism. Anaerobic metabolism, in the form of glycolysis, is stimulated but is not adequate to maintain the energy balance. This leads to a depletion of high-energy phosphates. In addition, intracellular acidosis develops as a result of ATP hydrolysis and the accumulation of acidic end products of glycolytic metabolism. ^{31}P spectroscopy can follow the time course of changes in intracellular pH and high-energy phosphates during ischaemia and reperfusion (**Figure 2**). The effects of drug interventions on these profiles can provide insights into the mechanisms responsible for any observed cardioprotection conferred by the drug. Such studies also provide essential information on the roles of high-energy phosphate depletion and intracellular acidosis in ischaemia–reperfusion injury.

^{23}Na NMR spectroscopy

Ionic concentration gradients exist across cell membranes and are responsible for maintaining the resting membrane potential. Intracellular and extracellular Na^+ are approximately 10 mM and 140 mM, respectively. The converse is true for K^+, with an intracellular concentration of 130–140 mM and an extracellular concentration of 4–5 mM. Sodium enters the cells of excitable tissue during the up stroke of the action potential and potassium leaves

the cell during the repolarization phase. The gradients are maintained by the operation of a Na^+K^+ ATPase (the sodium pump) that exchanges intracellular Na^+ for extracellular K^+. In the heart these ionic gradients can be disrupted by factors that prevent full activity of the sodium pump such as ischaemia or drugs (e.g. the cardiac glycosides related to digitalis). The ability to measure intracellular Na^+ levels in the intact heart makes ^{23}Na NMR spectroscopy a very powerful technique for assessing the role of altered Na^+ homeostasis in disease states.

Intracellular water represents about half the total water of the intact heart, the exact proportion being dependent on the species and perfusion conditions. This fact and the low intracellular concentration mean that, of the total Na^+ signal from the heart, less than 3% originates from the intracellular Na^+. Several studies have used double- or triple-quantum filtering techniques to discriminate this small intracellular Na^+ signal from the dominant extracellular Na^+ signal. Most studies, however, make use of non-cell-permeant paramagnetic reagents to 'shift' the extracellular peak and allow quantification of the intracellular Na^+ signal. These shift reagents are anionic chelates of lanthanide ions that do not cross membranes and thus are excluded from the intracellular space. The original agent used for this purpose, dysprosium bis(triphosphate), $Dy(PPP)_2^{7-}$ possesses the largest paramagnetic shift of any such complex for Na^+ or K^+. However, this reagent is quite sensitive to Ca^{2+} and Mg^{2+} and the shifts produced are greatly

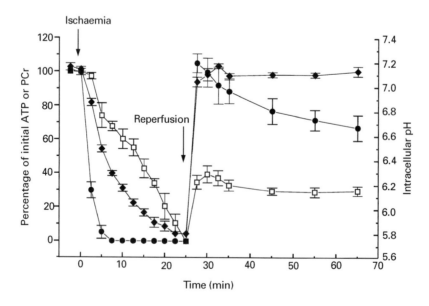

Figure 2   Time course of changes in intracellular pH (♦), ATP(□) and PCr (•) in a rat heart subjected to 25 min of total global ischaemia followed by 30 min of reperfusion. pH was determined from the chemical shift of P_i. Changes in ATP and PCr are expressed as percentage change from the basal levels measured prior to ischaemia. Total global ischaemia was achieved by stopping all flow of perfusate to the heart.

reduced by the presence of these ions. This fact precludes the use of $Dy(PPP)_2^{7-}$ in the intact heart, which requires both Ca^{2+} and Mg^{2+} for full functional integrity. The triethylenetetraminehexaacetic acid chelate of dysprosium, $Dy(TTHA)^{3-}$, produces smaller shifts but is much less sensitive to the effects of Ca^{2+} and Mg^{2+}. For studies in intact hearts that use $Dy(TTHA)^{3-}$, perfusates containing the shift reagent must be supplemented with Ca^{2+} to offset the Ca^{2+}-chelating properties of the shift reagent. This is well tolerated by the heart and results in adequate mechanical function. $Dy(TTHA)^{3-}$ at 5 mM causes a significant shift in the extracellular peak but also results in considerable line broadening, which still makes it somewhat difficult to fully resolve the small intracellular peak without specific processing strategies to maximize the resolution (**Figure 3**). Increasing the concentration of the shift reagent will cause a larger shift; however, the benefit is offset by an increase in line broadening.

The most recent shift reagent to be introduced is the tetraazacyclododecane-1,4,7,10-tetrakis(methylenephosphonate) chelate of thulium, $Tm(DOTP)^{5-}$ (**Figure 4**). This shift reagent also chelates Ca^{2+} and the perfusate must be supplemented with Ca^{2+} to maintain adequate mechanical function. At 4–5 mM $Tm(DOTP)^{5-}$ causes a significant shift in the extracellular Na^+ signal with very little line broadening. This latter property of the shift reagent makes it possible to perform interleaved ^{31}P and ^{23}Na NMR spectroscopy (in conjunction with a switchable NMR probe) in the presence of $Tm(DOTP)^{5-}$. Such studies are not possible in the presence of $Dy(TTHA)^{3-}$ owing to the excessive line broadening effects.

This strategy has been successfully applied to studies on isolated rat hearts to determine the involvement of the Na^+–H^+ exchanger in ischaemia-reperfusion. This sarcolemmal protein exchanges one Na^+ for one H^+. It is thought that this exchanger may contribute to myocardial ischaemia–reperfusion injury. During ischaemia, intracellular acidosis develops and this activates the Na^+–H^+ exchanger. This leads to an increase in intracellular Na^+ as intracellular H^+ is exchanged for extracellular Na^+. The increased intracellular Na^+ may activate the Na^+–Ca^{2+} exchanger, with intracellular Na^+ exchanging with extracellular Ca^{2+}. The end result is an increase in intracellular Ca^{2+}, which may be a major factor in the deleterious effects of ischaemia–reperfusion injury. ^{31}P and ^{23}Na NMR experiments performed in the presence of $Tm(DOTP)^{5-}$ provided data on intracellular pH and the Na^+ content. Inclusion of a relatively specific inhibitor of the Na^+–H^+ exchanger (ethyl isopropyl amiloride) in the perfusate partially attenuated the changes in pH and Na^+ and significantly

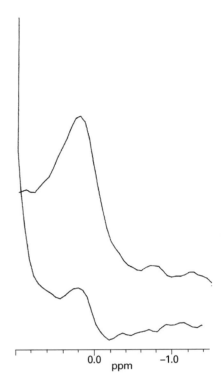

Figure 3 ^{23}Na NMR spectra of a rat heart in the presence of 5 mM $Dy(TTHA)^{3-}$. Spectra were acquired at 8.7 T using a broadband probe tuned to 95.25 MHz. The lower trace is a spectrum acquired during normal perfusion. The addition of shift reagent to the perfusate has shifted the large extracellular Na^+ peak 2 ppm downfield and has also caused a 0.2 ppm shift in the smaller intracellular Na^+ peak. The upper trace is a spectrum acquired following 25 min of total global ischaemia. The intracellular Na^+ peak has grown substantially, reflecting the intracellular Na^+ accumulation that occurs during ischaemia. Resolution of the peaks was enhanced using Gaussian multiplication with line broadening of –25 Hz and GB parameter of 0.15.

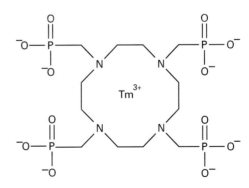

Figure 4 $Tm(DOTP)^{5-}$.

decreased mechanical dysfunction following ischaemia–reperfusion. This provided good evidence for the involvement of the Na^+–H^+ exchanger in ischaemia–reperfusion injury and confirmed the presumed mechanism of action of the drug. In most studies only relative changes in Na^+ levels are reported rather than intracellular concentrations. This is in large part due to the need to make assumptions regarding the visibility of the Na^+ NMR signal under various experimental conditions.

^{13}C NMR spectroscopy

Most studies on isolated hearts use glucose as the sole energy source. Normally the heart would be exposed to a variety of substrates including glucose, pyruvate, lactate, acetoacetate and a mixture of fatty acids. ^{13}C NMR spectroscopy has been used to demonstrate that fatty acids and acetoacetate are the preferred substrates for the heart under normal physiological conditions. It is also important to determine how substrate selection and the efficiency with which the heart metabolizes these substrates are altered under pathological conditions. The citric acid cycle is the central pathway for energy production in the heart and it is critical to determine how the flux of metabolites through this pathway is altered in diabetes and cardiomyopathy and during reperfusion of the ischaemic myocardium. Citric acid cycle flux has been determined indirectly by measuring the enrichment of ^{13}C into glutamate from α-ketoglutarate by the action of aspartate aminotransferase. For these studies, hearts are provided with substrate, or substrate mixtures, highly enriched with ^{13}C at specific carbon atoms (e.g. [1-^{13}C]glucose, [1,2-^{13}C]acetate, [3-^{13}C]lactate, etc.). The contribution of selected substrates to overall citric acid cycle activity may then be determined by isotopomer and multiplet analyses of ^{13}C enrichment in glutamate. Such studies have been performed under steady-state and non-steady-state conditions. Analyses are most usually performed by high-resolution spectroscopy on trichloroacetic acid extracts of hearts perfused with ^{13}C-enriched substrates, although useful data can be obtained by performing spectroscopy on intact beating hearts.

^{1}H NMR spectroscopy

The abundance of ^{1}H in water forms the basis for most magnetic resonance imaging. However, ^{1}H NMR spectroscopy has not found such universal applicability to studies of isolated hearts. This technique has been used to quantify the total content of creatine, which may provide a useful index of tissue viability. ^{1}H NMR spectroscopy has also been used to determine lactate levels during ischaemia and following interventions designed to modulate metabolism. The lactate methyl group and lipid methylene groups resonate at 1.3 ppm. These resonances can be differentiated and quantified using spin-echo spectral editing techniques.

^{19}F NMR spectroscopy

The most important use of ^{19}F NMR spectroscopy in studies of isolated hearts is to measure intracellular Ca^{2+} levels. This is based on the use of fluorinated derivatives of calcium chelators. The extracellular Ca^{2+} concentration is ~1.2 mM. The intracellular Ca^{2+} concentration at diastole is less than 100 nM. This increases to several hundred nM during cardiac excitation. This elevated Ca^{2+} (or calcium transient) is responsible for contractile activity at each heartbeat. Efficient relaxation at each beat depends upon the intracellular Ca^{2+} being restored to diastolic levels. Most of the cytosolic Ca^{2+} enters the cell through voltage-regulated Ca^{2+} channels during the plateau phase of the action potential or is released from the intracellular organelle, the sarcoplasmic reticulum. Diastolic Ca^{2+} levels are restored by active pumping of the Ca^{2+} back into the sarcoplasmic reticulum and by activation of the sarcolemmal Na^+–Ca^{2+} exchanger, exchanging intracellular Ca^{2+} for extracellular Na^{2+}. Thus, the level of Ca^{2+} within the cardiac cell is tightly controlled under normal physiological conditions. If the intracellular level of Ca^{2+} rises significantly above normal physiological limits, consequences may be deleterious as a result of activation of Ca^{2+}-dependent proteases and phospholipases and also due to mitochondrial damage.

Intracellular Ca^{2+} has been measured by ^{19}F NMR spectroscopy of intact hearts loaded with the 5,5′-difluoro derivative of 1,2-bis(o-aminophenoxy)ethane-N,N,N′,N′-tetraacetic acid (5F-BAPTA) (**Figure 5**). 5F-BAPTA is loaded into the heart as the cell-permeant acetoxymethyl (AM) ester. Esterases within the cardiomyocyte hydrolyse the AM ester to the free acid, which, being charged and therefore unable to cross the cell membrane, is trapped within the cell. The calcium-bound and calcium-free 5F-BAPTA species undergo slow exchange resulting in two NMR-visible peaks. The intracellular Ca^{2+} concentration is

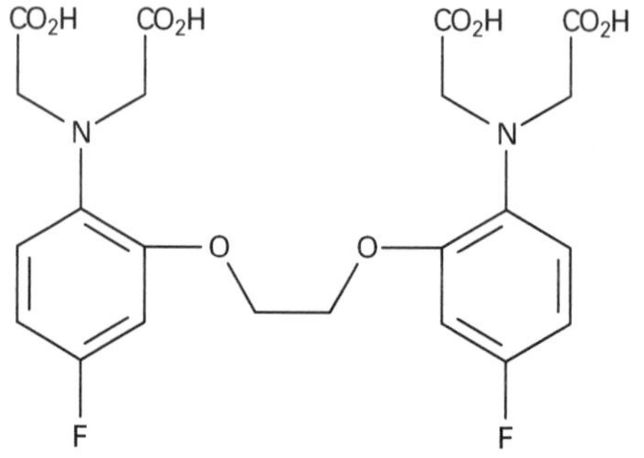

Figure 5 5F-BAPTA.

calculated using the relation

$$[Ca^{2+}] = K_d \frac{[B]}{[F]} \qquad [4]$$

where K_d is the dissociation constant of Ca^{2+}–5F-BAPTA (literature values of 537, 500, 635 and 285 nM), [B] is the area under the Ca^{2+}– 5F-BAPTA peak and [F] is the area under the 5F-BAPTA peak.

A K_d of approximately 500 nM makes 5F-BAPTA an ideal probe for measuring intracellular Ca^{2+} in the physiologically relevant range (100 nM to 1 μM). Unfortunately, a K_d of 500nM also causes 5F-BAPTA to provide excellent buffering to intracellular Ca^{2+} levels. This causes a decrease in mechanical function of the heart. When 5F-BAPTA is present within cardiomyocytes at sufficiently high concentration (300 μM) to provide adequate signal to noise in the ^{19}F NMR spectrum (> 10:1), developed pressure is reduced by 75–80%. This results from an increase in diastolic pressure and a decrease in systolic pressure. Thus, although studies using 5F-BAPTA can provide valuable information regarding changes in intracellular Ca^{2+}, it must always be borne in mind that these results were acquired under conditions of severely compromised cardiac function.

A BAPTA derivative with a higher K_d overcomes the problem of buffering of intracellular Ca^{2+}. The recently developed analogue, 1,2-bis(2-amino-5,6-difluorophenoxy)ethane-N,N,N′,N′-tetraacetic acid (TF-BAPTA) (Figure 6) has a K_d of 65 μM. Loading of intact hearts with this derivative causes <10% decrease in mechanical function. This high K_d causes TF-BAPTA to be a less accurate probe for measuring basal Ca^{2+} levels but does make it more suitable for measuring intracellular Ca^{2+} under pathophysiological conditions (e.g. ischaemia) where $[Ca^{2+}]$ may rise above 2 μM. The Ca^{2+}-bound and Ca^{2+}-free TF-BAPTA species are in intermediate–fast exchange. This leads to a single resonance with its chemical shift position being dependent upon the extent of Ca^{2+} binding (analogous to the pH-dependent shift in the P_i peak). Binding of Ca^{2+} to TF-BAPTA does not alter the chemical shift of the fluorine in the 6 position of TF-BAPTA. The fluorine in the 5 position shifts downfield upon binding of Ca^{2+}. The chemical shift difference between the 5 and 6 fluorines is used to determine intracellular $[Ca^{2+}]$ (with corrections for the effects of pH and $[Mg^{2+}]$). TF-BAPTA accumulates in the sarcoplasmic reticulum (SR) and this property has been used to determine $[Ca^{2+}]$ in this subcellular organelle in intact rabbit hearts. A 5-F resonance at 5 ppm (with the 6-F resonance set at 0 ppm) corresponds to a time-averaged basal

Figure 6 TF-BAPTA.

cytosolic $[Ca^{2+}]$ of 600 nM. A second 5-F resonance at 14 ppm corresponds to a SR $[Ca^{2+}]$ of 1.5 mM. This technique may prove to be invaluable in assessing the role of alterations in SR Ca^{2+} handling in various pathological conditions.

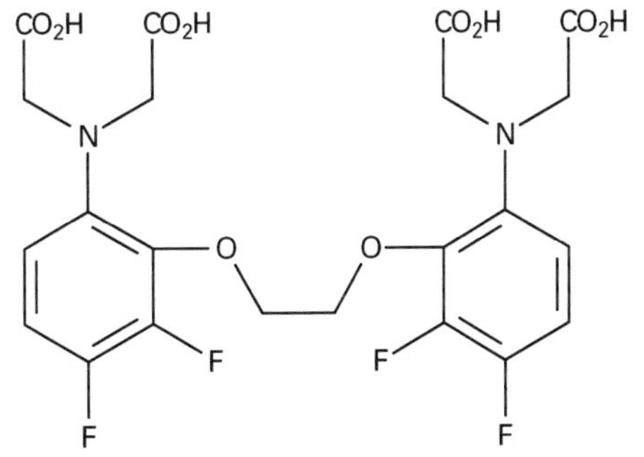

^{87}Rb and ^{7}Li NMR spectroscopy

^{23}Na spectroscopy is useful for measuring steady-state levels of intracellular sodium but is not suitable for measuring fluxes of this cation. The contribution of various ion channels, exchangers and pumps to the total movement of ions across the sarcolemmal membrane can be assessed by measuring ion fluxes in the presence and absence of selective inhibitors of these membrane proteins. It is possible to acquire this type of information using electrophysiological techniques. These data would be complemented by the use of techniques that can measure ion fluxes in the intact heart. ^{87}Rb and ^{7}Li, congeners of K^+ and Na^+, respectively, have been used to assess fluxes of K^+ and Na^+ in intact hearts and vascular tissue. For ^{87}Rb spectroscopy, 20% of the perfusate K^+ content can be replaced with Rb^+ with no effects on function. ^{87}Rb NMR spectroscopy can be used to determine rate constants for the uptake of Rb^+ on switching from Rb^+-free to Rb^+-supplemented perfusate. The converse, switching from Rb^+-supplemented to Rb^+-free perfusate, can be used to determine the rate constant for Rb^+ washout. This approach has been used to demonstrate that under normal conditions the bulk of Rb^+ uptake occurs through the Na^+K^+ ATPase rather than the Na^+–K^+–2Cl$^-$ exchanger or K^+ channels. ^{7}Li NMR spectroscopy has been used in similar types of experiments to study Na^+ channel activity in intact hearts. For these studies an excellent signal could be achieved by substituting a modest amount of perfusate Na^+ with Li^+ (15 mM). Similar to the ^{87}Rb studies, the kinetics of release of Li^+ could be determined in washout experiments. These studies demonstrated that Li^+ efflux from cardiomyocytes is predominantly through Na^+ channels.

Studies of isolated livers

Livers isolated from rats or mice may conveniently be studied by NMR spectroscopy. The organs are perfused through the portal vein with solutions similar to those described for isolated heart studies (supplemented with albumin). The methods described above for ^{31}P and ^{23}Na NMR spectroscopy have been successfully applied to the isolated liver. The ^{31}P NMR spectrum of the isolated liver differs from the heart spectrum in that it lacks the PCr peak observed in spectra obtained from hearts. This technique has been utilized to examine various aspects of ethanol metabolism in the liver, including the effects of chronic ethanol exposure on subsequent acute ethanol exposure and hypoxia. These studies revealed that chronic ethanol exposure caused an adaptation in the liver such that it becomes more resistant to acute ethanol exposure and also to hypoxia.

^{15}N NMR spectroscopy has been used to study the urea cycle in isolated liver. ^{15}N was provided to the liver in the form of ^{15}NH$_4$Cl or [^{15}N]alanine in the presence and absence of unlabelled lactate or ornithine. Proton-decoupled spectra were obtained from the intact liver, from the perfusing medium or from tissue extracts and yielded peaks corresponding to glutamine, arginine, urea, citrulline, glutamate, alanine and ammonia. Such studies may prove to be useful in the *in vivo* liver as a means of assessing the effects of disease states on urea cycle activity.

Studies of isolated kidneys

Kidneys isolated from rodents are suitable for study by NMR spectroscopy following cannulation of the renal artery and perfusion with an albumin-supplemented Krebs–Henseleit solution. Combined ^{31}P and ^{23}Na NMR spectroscopy has been used to determine the energetic cost of Na$^+$ transport in the kidney. Similar multinuclear techniques have been used in studies investigating the factors influencing renal function during the progression from pre-hypertension to hypertension in a spontaneously hypertensive rat model. ^{31}P NMR spectroscopy is being applied to studies into the viability of kidneys used for transplant. At present, donor availability is the limiting factor for transplant programmes. As the use of non-heart-beating donors increases, there is a greater need for a rapid, reliable noninvasive technique for assessing organ viability. ^{31}P NMR spectroscopy is showing promise in this regard.

List of symbols

[A] = concentration of species A; K_d = dissociation constant; K_{eq} = equilibrium constant; p_{A} = (partial) pressure of species A; P$_i$ = inorganic phosphate; δ = chemical shift; ΔG^0 = free energy change under standard conditions; ΔG_{ATP} = free energy of hydrolysis of ATP.

See also: **Cells Studied By NMR; Chemical Shift and Relaxation Reagents in NMR; ^{13}C NMR, Methods;** *In Vivo* **NMR, Applications, Other Nuclei;** *In Vivo* **NMR, Applications, ^{31}P; ^{31}P NMR.**

Further reading

Balaban RS (1989) MRS of the kidney. *Investigative Radiology* 24: 988–992.

Barnard ML, Changani KK and Taylor-Robinson SD (1997) The role of magnetic resonance spectroscopy in the assessment of kidney viability. *Scandinavian Journal of Urology and Nephrology* 31: 487–492.

Deslauriers R, Kupryianov VV, Tian G *et al* (1996) Heart preservation: magnetic resonance studies of cardiac energetics and ion homeostasis. In: Dhalla NS, Beamish RE, Takeda N and Nagano M (eds) *The Failing Heart*, pp 463–487. Philadelphia: Lippincott-Raven.

Elgavish GA (1993) Shift reagent-aided ^{23}Na nuclear magnetic resonance spectroscopy. In: Pohost GM (ed) *Cardiovascular Applications of Magnetic Resonance*, pp 371–391. Mount Kisco, NY: Futura.

Evanochko WT and Pohost GM (1993) ^{1}H NMR studies of the cardiovascular system. In: Schaefer S and Balaban RS (eds) *Cardiovascular Magnetic Resonance Spectroscopy*, pp 185–193. Norwell, MA: Kluwer Academic.

Ingwall JS (1993) Measuring sodium movement across the myocardial cell wall using ^{23}Na NMR spectroscopy and shift reagents. In: Schaefer S and Balaban RS (eds) *Cardiovascular Magnetic Resonance Spectroscopy*, pp 195–213. Norwell, MA: Kluwer Academic.

Kusuoka H, Chacko VP and Marban E (1993) Measurement of intracellular Ca^{2+} in intact perfused hearts by ^{19}F nuclear magnetic resonance. In: Pohost GM (ed) *Cardiovascular Applications of Magnetic Resonance*, pp 393–401. Mount Kisco, NY: Futura.

Malloy CR, Sherry AD and Jeffrey FMH (1993) ^{13}C nuclear magnetic resonance methods for the analysis of citric acid cycle metabolism in heart. In: Pohost GM (ed) *Cardiovascular Applications of Magnetic Resonance*, pp 261–270. Mount Kisco, NY: Futura.

Ugurbil K and From AHL (1993) Nuclear magnetic resonance studies of kinetics and regulation of oxidative ATP synthesis in the myocardium. In: Schaefer S and Balaban RS (eds) *Cardiovascular Magnetic Resonance Spectroscopy*, pp 63–92. Norwell, MA: Kluwer Academic.

PET, Methods and Instrumentation

TJ Spinks, Hammersmith Hospital, London, UK

SPATIALLY RESOLVED
SPECTROSCOPIC ANALYSIS
Methods & Instrumentation

Introduction

This article discusses the ways in which the design of instrumentation for positron emission tomography (PET) has evolved to provide data with ever greater accuracy and precision. Even though a PET tracer may be intrinsically capable of providing highly specific biochemical and physiological information, this is of little use if the radiation detection system is inadequate. The central aims of PET instrumentation are to increase the number of unscattered photons detected, making maximum use of the amount of tracer it is permissible to administer, and to resolve with greater accuracy their point of origin. A complementary aim is to reduce statistical noise in the data. In addition, developments in the measurement of tracer in the blood are important. A central component of quantification in PET is the combined use of tracer input to the tissue (arterial blood) and tissue uptake (determined from the image). Later in the article a brief discussion will be made of the methods used to obtain physiological, biochemical and pharmacological parameters using different mathematical models of the biological system under investigation.

Detector materials and signal readout

The overall aim in the search for new detectors is to maximize photon stopping power (efficiency) and the signal generated and to minimize response time. The latter property is required to cut down losses due to dead time or the time during which the detection system is 'busy' dealing with one interaction and thereby misses others that might occur. Theoretical predictions can be made of these properties for different materials, but good detectors have generally been found by a process of trial and error. The dominant detector material in PET today is bismuth germanate ($Bi_4Ge_3O_{12}$, or BGO for short), although sodium iodide (NaI(Tl), activated with thallium) is still quite widely used. A comparison of the principal properties of these substances and those of a promising new material that is being developed (lutetium orthosilicate, LSO) is shown in **Table 1**. LSO is very attractive because it has a similar density to BGO but also has about five times its scintillation efficiency and a very much shorter scintillation decay time. In

Table 1 Characteristics of some scintillation detectors used in PET

Property	Sodium iodide (NaI(Tl))	Bismuth germanate (BGO)	Lutetium orthosilicate (LSO)
Density (g cm^{-3})	3.7	7.1	7.4
Effective atomic number	51	75	66
Scintillation efficiency (% of NaI(Tl))	100	15	75
Scintillation decay time (ns)	230	300	40
Hygroscopic?	Yes	No	No

consequence, its energy resolution and timing resolution are higher.

The method by which the response of a detector is recorded and analysed (the 'readout') has undergone several developments, and refinements are still being made. Most PET tomographs at present utilize the BGO block detector, a device introduced in the mid-1980s. This was revolutionary in the sense that it enabled a multiple-ring tomograph to be produced at a reasonable cost. Earlier designs had utilized one-to-one coupling of crystal and photomultiplier (PMT), but the multiplicity of coincidence circuits had restricted commercial scanners to at most two detector rings. Furthermore, because of the desire to reduce detector size, the dimensions of readily available PMTs made one-to-one coupling impractical. The schematics of the block detector are shown in **Figure 1**.

A BGO crystal is divided into a number of elements (in modern scanners 8×8 with dimensions 4×4 mm^2 face $\times$ 30 mm depth) which are viewed by four relatively large, standard PMTs. The cuts between the elements contain light-reflecting material and are made to different depths across the crystal. This partial separation controls the amount of scintillation light reaching the PMTs and leads to a ratio of signals characteristic of each detector element. The x and y coordinates of the interaction are calculated by the formulae given in **Figure 1**. A ring of such block detectors would constitute eight rings of closely packed detector elements and multiple rings

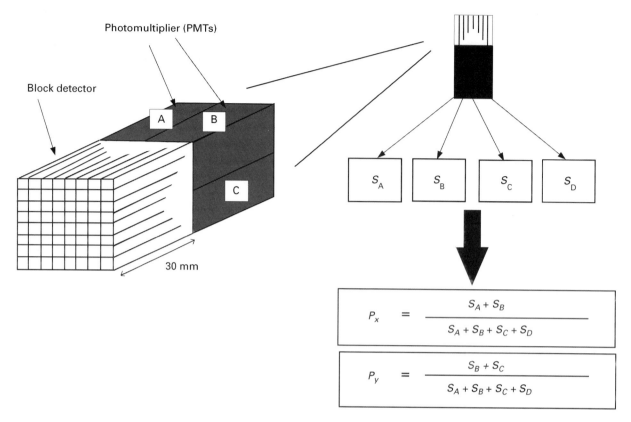

Photomultiplier (PMTs)

Block detector

A

B

C

30 mm

S_A

S_B

S_C

S_D

$$P_x = \frac{S_A + S_B}{S_A + S_B + S_C + S_D}$$

$$P_y = \frac{S_B + S_C}{S_A + S_B + S_C + S_D}$$

Figure 1 Principle of the block detector.

of blocks can be placed adjacent to each other to give the desired field of view (FOV) (within financial constraints). With this arrangement, the degree of multiplexing of signals is greatly reduced. On the other hand, the block detector has poorer dead-time properties than an individual crystal–PMT pairing. When one element is hit by a photon, all other elements are effectively 'dead' until the original inter-action has been analysed. In some experimental systems avalanche photodiodes (diode detectors with internal amplification) are used instead of PMTs. Their availability and performance still does not compete with those of PMTs, but developments continue and their great advantage is their very compact size and ability to view small detector elements.

Full collection of scintillation light from a photon interaction results in a charge pulse from the PMT output whose height is proportional to the energy deposited. The timing of a given event is determined by the point at which the pulse crosses a certain voltage level. The statistical nature of pulse generation and the range of pulse heights encountered give rise to a spread or uncertainty in this timing. An ingenious electronic device known as the constant fraction discriminator greatly reduces this spread by shaping the pulse so that triggering is made at a constant fraction of the pulse height. The rise and fall of

scintillation light determine timing resolution τ (how accurately the time of a single interaction can be determined) and integration time (how much time is required to measure the scintillations produced and hence energy deposited). For BGO, the timing resolution is about 6×10^{-9} s (6 ns) and the integration time about 10^{-6} s (1 μs). The time spectrum of two detectors (the measured time differences between the arrival of two events) shows a peak superimposed on a constant background (for a given activity). The peak corresponds to true coincidences and the background to random coincidences. The detection of coincidence events involves a convolution of the timing resolutions of each detector and the coincidence timing window is thus set at twice the timing resolution (i.e. 2τ). If this window is made narrower, the random events will fall linearly but the true events will decrease more rapidly. It is observed experimentally that a window of 12 ns is optimum for BGO detectors.

Even if a beam of monoenergetic 511 keV photons is incident on a crystal, the different scattering and absorption interactions occurring lead to different amounts of energy being deposited and result in output signals with a range of intensities. **Figure 2** is a sketch of the energy spectrum from a BGO detector. The photopeak at the right represents events that

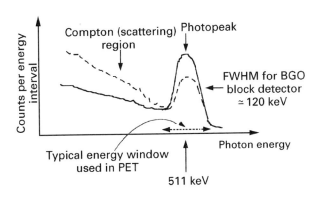

Figure 2 Shape of energy spectrum from a BGO detector for monoenergetic 511 keV photons (——) and those from a scattering medium (– – –).

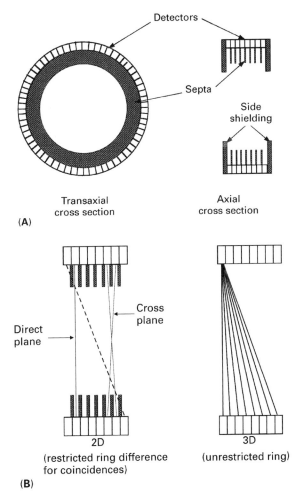

Figure 3 (A) Arrangement of inter-ring septa in multi-ring scanners. (B) Axial cross section of a multiple-ring tomograph illustrating inter-ring coincidence combinations for 2D and 3D modes.

have been totally absorbed (photoelectric effect), while the broad continuum corresponds to partial energy deposition (Compton scattering). If the source is within a scattering medium such as the body, the scattering region of the spectrum will be enchanced and the photopeak depressed. The width of the photopeak or the energy resolution (full width at half maximum, FWHM) is about 120 keV for a BGO block detector. (It is only about half this for an uncut block owing to better light transferrence to the PMT.) In a similar way as for the timing spectrum, the acceptance of events is restricted by setting an energy window of about $2 \times$ FWHM (about 350–650 keV). This scheme attempts to minimize detection of incident scattered (lower energy) photons but, although BGO has high stopping power, its scintillation efficiency and hence energy resolution is relatively poor and a fraction of scattered photons are inevitably accepted. Methods for subtracting these events are discussed later in this article.

2D and 3D acquisition modes

The great advantage or annihilation coincidence detection is the automatic or 'electronic' collimation that it provides, but scattered photons are an ever-present complication, the effects of which need to be minimized. The first PET tomographs consisted of 'area' detectors such as the large sodium iodide crystals of gamma cameras that were already a standard device for radioisotope imaging. However, it became clear that such scanners had a high sensitivity to scattered photons and random coincidences. Subsequent designs were based on rings of individual detectors with tight collimation using lead shielding. The first commercial scanner comprised a single ring of NaI(Tl) crystals with variable lead collimators in front of the detectors and heavy shielding on either

side of the ring. Multiple-ring designs arose from the inevitable demand for a larger axial FOV and in these a form of side shielding known as septa, consisting of lead or tungsten annuli, were inserted between the rings as shown in **Figure 3A**. In such an arrangement, data are acquired as a number of contiguous transaxial 'planes' or 'slices' and it is, for this reason, termed the '2D mode'. Each plane consists of data from coincidences either within an individual ring or in closely adjacent rings. The example given in **Figure 3B** shows a maximum ring difference (rd_{max}) of 1, but, as axial detector width became narrower to improve resolution, greater values of rd_{max} were used in order to maintain efficiency. However, it can be seen that for large ring differences coincidence counts would be severely attenuated by the septa (as shown by the dotted line). The approach to statistical image noise has emphasized the importance of maximizing detection efficiency. The restriction of data to '2D' slices tends to go against this and so there have been increasing moves during the 1990s to acquire data without septa inserted and with all possible coincidence lines-of-response (LORs) operational. Such a scheme is naturally known as the '3D mode'. Objections to 3D PET scanning stem, of course, from the original desire to reduce the effects of scattered and random coincidences, but the proponents of the method point to the significant increase in the efficiency of detection of unscattered true coincidences. For example, with a tomograph having eight crystal rings and a 2D rd_{max} of 1 (**Figure 3B**), the total number of inter-ring combinations is 22 (8 direct $+ 2 \times 7$ cross), whereas for 3D the number is 64 (8×8). This threefold increase is further enhanced (by about a factor of 2) owing to 'shadowing' of the crystals by the septa.

This simple analysis is appropriate at low count rates when random events are negligible, but at higher rates the advantage must be defined in terms of improvement in statistical noise as embodied in the parameter noise-equivalent counts (NEC). It is found that there is an NEC gain with the 3D mode over the whole range of count rates. For brain studies this varies from about a factor of 5 at low rates (for example, in receptor binding studies with ^{11}C-labelled compounds) to about 3 at high rates (such as are encountered in blood flow studies with ^{15}O-labelled water).

Data handling and image reconstruction

In addition to the reticence concerning the 3D mode on physical grounds, there are practical challenges. Current hardware designed to process coincidence events runs at rates sufficiently high to accommodate the tracer doses that can be administered, but a principal problem is one of data transfer, storage and reconstruction. 2D data are 'compressed' into a number of slices, while 3D data are conventionally acquired without any (or relatively little) compression. The increase in the number of (individual) LORs approaches an order of magnitude and these have to be backprojected during reconstruction and stored. However, despite these perceived disadvantages, the storage and archiving media (multi-gigabyte hard disks, DAT tape, optical disks etc.) and ever faster processors available today have lessened the practical burdens of 3D scanning.

A feature of data acquisition that is becoming of greater interest in PET is that of list mode. In this, the events are not stored over preselected time frames in the sinogram data matrices but instead each event is stored separately to disk. If a large number of counts are acquired over a short time, this is not necessarily an efficient method, but for studies following the uptake and clearance of tracer for 2 hours or more, it becomes much more efficient. However, even for shorter scanning times, list mode provides very high temporal ($\simeq 1$ ms) resolution and the ability to utilize physiological gating, or separation of the data into specific phases, for example of the cardiac and respiratory cycles. Furthermore, list mode acquisition would be of great advantage in the correction for patient movement, which is still an area of development.

Readily available dedicated processing hardware can reconstruct a set of images from the largest 3D PET data volume in about 10 minutes but the ability of tomographs to acquire data in 3D mode somewhat preceded the development of appropriate reconstruction algorithms. The specific problem with filtered backprojection lay in the variation of the point response function (PRF). Reconstruction of 2D slices relies on the fact that the response to a point source (efficiency of detection) remains constant over the slice. This requirement needs to be met because the filtering process represents a convolution with the measured projection data. In the 3D mode the PRF is not constant over the FOV, being a maximum in the centre and falling steadily (axially) towards the edge (**Figure 4A**). A way of overcoming this difficulty, and one that is now commonly used, is the reprojection method. In this (**Figure 4B**), 2D images are first reconstructed from direct plane (single ring) data and then 'missing' projections are created by forward projection (images to views). The missing projections are those that would have been acquired if additional detector rings had been present and these complement the data to provide an invariant PRF over the (original) FOV.

Alternative tomograph designs

Tomographs consisting of multiple rings of individual crystals are the most widely used. **Figure 5** shows the CTI/Siemens model 966 (covers removed) which is the most sensitive PET scanner yet constructed. However, there are a number of other designs in operation.

Tomographs based on planar sodium iodide detectors

Large-area planar sodium iodide (NaI(Tl)) detectors, which are used routinely in gamma cameras for imaging of single-photon tracers, have seen a revival in

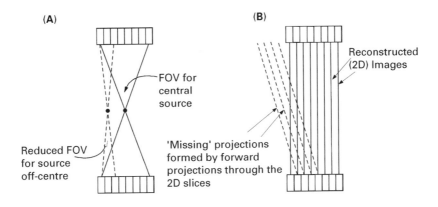

Figure 4 (A) Variation of point response function (PRF) in a multiple-ring tomograph. (B) Creation of 'missing' projections for the reprojection reconstruction algorithm.

PET over the last few years. A commercial design consists of six planar detectors arranged in a hexagon and operated without septa (3D mode). The crystal is viewed by an array of PMTs and the point of photon interaction is determined, in a similar way to the block detector, by comparing signals between adjacent tubes. This device takes advantage of the high light output (**Table 1**) of NaI(Tl) and possesses a similar spatial resolution to BGO systems. On the other hand, its stopping power (efficiency) is significantly less than that of BGO. Indeed, the standard detector thickness used for single-photon imaging (at about 100 keV) of about 10 mm is increased to 25 mm for the 511 keV photons in PET. One advantage of this system is that large NaI(Tl) crystals can be produced at relatively low cost, giving a large FOV. This aspect and the increasing availability of the glucose analogue tracer [^{18}F]fluorodeoxyglucose (FDG, see later for clinical diagnostic imaging) have been responsible for its commercial success.

There is also increasing interest in 'dual-use' systems for both single-photon emission tomography (SPET) and PET, consisting of double-headed gamma cameras (operated with and without multihole lead collimators, respectively). The significantly lower efficiency for PET compared with purpose-built tomographs needs to be borne in mind, but such systems could be useful for specific diagnostic tests (e.g. detection of tumour metastases) and their flexibility for general nuclear medicine use is attractive.

Partial-ring tomographs

As alluded to above, the expense of multiple-ring BGO tomographs has led to the development of lower-cost systems, principally those employing NaI(Tl) planar detectors. An alternative to these is the commercially available partial-ring scanner. In this device, two banks of BGO detector blocks (about 1/3

Figure 5 A commercial PET scanner.

of a complete ring) rotate around the body (at 30 revolutions per minute) and the output of data is achieved via optical coupling. Voltage supplies are provided through slip rings.

Tomographs based on multiwire proportional chambers

The multiwire proportional chamber (MWPC) is a device used extensively in high-energy particle physics experiments that has been modified in different ways for use in PET. The principle of operation is the detection of an electron (i.e. an electron avalanche) on planes of closely spaced fine wires (~1 mm apart) held at a high electric potential. Cathode and anode wires are arranged orthogonally to each other, thus providing the spatial localization, the wire spacing being the basic determinant of resolution. The electrons are produced in various ways, such as interaction of photons in thin sheets of lead or photoionization of a gas in the wire chamber by ultraviolet light from a barium fluoride (BaF_2) scintillator.

Tomographs for experimental studies

A number of research centres have designed and implemented smaller-diameter tomographs (BGO, NaI(Tl) and multiwire chamber detectors have all been used) specifically for the scanning of small animals. Spatial resolution close to 1 mm has been achieved. Although this does not compete with the resolution of autoradiography or dissection, the time–activity curve can be followed in a single animal, which is a great advantage in many studies, such as the investigation of new tracers and the testing of models of disease processes. It is anticipated that important new information will also be forthcoming in the development of new pharmaceuticals.

Data correction procedures

Normalization

Variations in the fabrication of detectors and their geometrical arrangement in a tomograph inevitably lead to variations in efficiency for different LORs. For example, the detectors at the edge of a BGO block have lower efficiency than those at the centre and the LORs crossing the centre of the FOV will have a different solid angle of detection to those at the edge. If these effects are not corrected for, systematic errors (of both high and low spatial frequency) will occur in the reconstructed image. The process of correction is known as normalization. The basic data for normalization are acquired by exposing each LOR to the same activity, for example in the form of a thin planar source or a line source scanning across the FOV.

Attenuation

The principles of attenuation and its correction in PET are outlined in an associated article. Here, some specific practical examples are given. Most PET tomographs utilize sources of ^{68}Ge (half-life about 9 months), which are stored in shields within the gantry of the tomograph and moved into the FOV by remote control. In turn, a blank scan (empty FOV) and a transmission scan (patient in position, usually before tracer administration) are acquired. The ratio between these two provides the attenuation

correction factors for each LOR. The logarithm of these ratios can also be backprojected to yield a transmission image (tissue density map). As for emisssion data, scattered radiation (see below) contaminates the data and one method of reducing this is electronic 'windowing'. This is illustrated in **Figure 6**. The transmission source in this case is a rotating rod whose position is encoded. If a photon from the rod is scattered in the object and measured, the event will be rejected because the resulting LOR (LOR2) does not pass through the rod. Unscattered events (e.g. LOR1) are accepted.

A continuing theme of PET is improvement in efficiency, and this is so for transmission scanning. Constraints of time and detector performance mean that transmission data can be suboptimal and this has led to the recent implementation of the single-photon transmission technique. This is similar to the process of X-ray computerized tomography and takes advantage of the fact that about two orders of magnitude more single-photon events are collected than coincidence events. A working mechanism employs a ^{137}Cs point source (single-photon: energy 662 keV, half-life 30 years) rotating in a helical tube around the subject. However, as for the rotating ^{68}Ge source in **Figure 6**, detection of scattered radiation will also contaminate the data in this case and lead to inaccurate correction. Recourse to the windowing method cannot be made for ^{137}Cs, but another way to compensate for scatter is to create a transmission image that is then segmented into regions of similar density, and correct attenuation factors are assigned accordingly.

Scattered radiation

The inclusion of scattered events in the projection data does not significantly affect the spatial resolution or 'sharpness' of the reconstructed image but it will lessen the contrast between different regions and cause inaccuracies in the measurement of activity concentration. The distributions of scatter due to a line source placed axially in the centre of a water-filled cylinder (20 cm diameter) are shown in **Figure 7** for 2D and 3D acquisition modes. The central peak corresponds to the position of the source and the broad continuum ('wings') on either side is due to scatter, which decreases with increasing angle. The scatter fraction (SF, integral of the 'wings' divided by the total events) for brain scanning is 10–15% with septa and 30–40% without septa; even larger values obviously occur in body scanning.

A number of methods have been used to correct for scatter, particularly for the 3D scanning mode. Correction schemes have broadly been based on the measured spatial or energy distribution (spectrum)

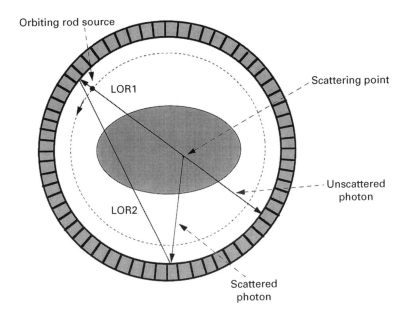

Figure 6   Mechanism for attenuation correction using a rotating [68]Ge rod with 'windowing'.

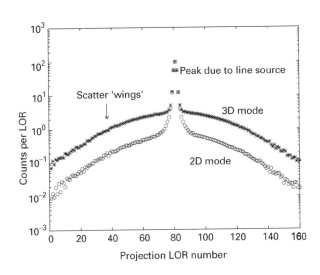

Figure 7   Scatter distributions for a line source in the centre of a cylinder (2D and 3D); peaks are normalized to the same height.

scatter and so the initial calculation is an overestimate. A shortcoming of the simple kernel above is that the shape of the scatter distribution does not stay constant with the position of the source. As it moves to one side of the object, the distribution becomes more and more asymmetric. Strictly, in this case, an integral transform rather than a convolution should be employed since the kernel shape is position dependent. However, quite accurate scatter corrections for the head have been demonstrated, even in the 3D mode, with an invariant kernel.

A correction method based on the photon energy spectrum utilizes the varying proportions of scattered and unscattered events for different ranges or energy windows. This technique had its origins in single-photon tomography. Some PET tomographs have the ability to acquire data in two energy windows (over the photopeak and over part of the Compton continuum). A simple form of the dual-window correction is based on the assumption that the ratios are constant throughout the object. The dual-window method has been shown to give similarly accurate corrections (within about 6%) to the convolution method in 3D brain scanning but has slightly poorer noise characteristics.

The simplest method of scatter correction consists of fitting a function (typically a Gaussian) to the 'wings' (pure scatter) outside the object. This has the advantage of being based on a direct measurement of scatter for each object. On the other hand, assumptions must be made about the shape of the distribution *inside* the object. Again, accurate results are

of scattered events or, more recently, on calculation from first principles of the probability of scattering through different angles.

The first scatter correction methods, which are still employed, treated the scatter distribution as a convolution of the projection data with a function or 'kernel', the shape of which was derived from curves such as that in **Figure 7**. The simplest form of the kernel is $\alpha\exp(-\beta|x|)$, where α and β are positive constants and $|x|$ is the absolute distance along a projection. In general, convolution is carried out iteratively because the measured projections contain obtained for the head, but for the chest a simple Gaussian is not necessarily a good choice of function.

The speed of current computers has made it feasible to calculate scatter distributions analytically from first principles for each set of data. This is carried out by taking the uncorrected tracer (emission) image and calculating the scatter that would arise from selected points given the transmission image. The probabilities of scattering through given angles are known precisely from physical principles. In practical implementations, a relatively coarse grid of points is selected. This saves time and utilizes the fact that the scatter distribution is smooth and is amenable to interpolation. The beauty of this technique is that it is makes few assumptions. Convolution and dual-window methods work well in a relatively uniform object such as the head but have more difficulty, for example, in the chest where there are abrupt density changes. A big challenge for PET is to obtain accurate quantification in 3D body scanning and it is likely that analytical methods such as this will prove the most valuable.

Dead time and random coincidences

As the activity in the body increases, the problems of electronic dead time and registration of random coincidences, leading to a reduction in efficiency and an increase in statistical noise, become ever more pressing.

Dead time correction schemes have ranged from those that are founded on an intimate knowledge of the electronic circuitry of the scanner to those based on empirical curve-fitting using test objects. All commercial scanners provide automatic dead time correction. The variation of true and random coincidences with activity is shown in **Figure 8A**. The peak and steady falloff in the trues curve at high rates is described as paralysable dead time behaviour and is typical of PET tomographs. In this case, if an event occurs during the analysis of a previous event ('busy' period) the effect is a successive lengthening of the dead time. Theoretically it can be shown that measured (N_m) and corrected (N_0) count rates are related by

$$N_{\mathrm{m}} = N_0 \exp(-N_0 \tau_{\mathrm{d}}) \qquad [1]$$

in which case the measured rate can eventually go to zero (or be 'paralysed') for very high activities (τ_{d} is the system dead time). The fundamental determinant of dead time is the rate of single events that are striking the detectors. Tomographs continuously record this rate and use it as a basis of the calculation of dead

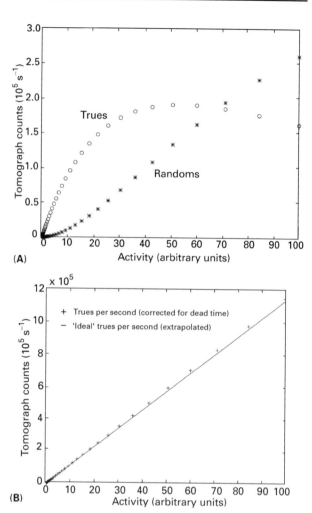

Figure 8 (A) The variation of measured trues and randoms with activity in the FOV. (B) The correspondence between 'ideal' trues rates (extrapolated from low activities) and dead time-corrected trues.

time correction factors. It should be noted that, for coincidence counting, the overall dead time is the *product* of the dead times of opposing detectors. Generally speaking, correction is accurate to within 5% for the range of counting rates encountered in *in vivo* studies (**Figure 8B**), but it needs to be stressed that the magnitude of the correction factor should not be too large and that scanning should not be carried out near to or beyond the peak of the 'trues' curve.

The steep rise in random events with activity is, similarly to dead time, due to the product of the rates of single events on opposing detectors. Without dead time, trues rise linearly with activity, whereas randoms rise quadratically (the dead time for trues and randoms is the same because they are counted by the same coincidence circuits). This behaviour of randoms means that judicious administration of activity should be adhered to and/or good shielding of

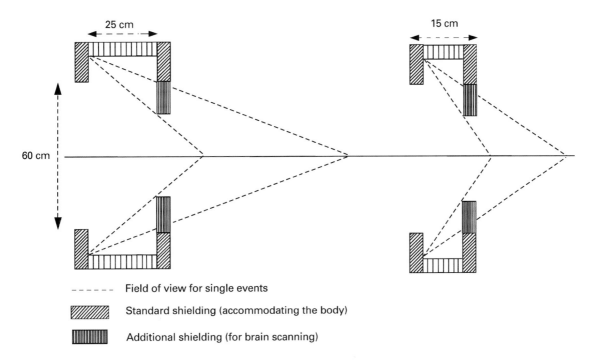

- - - - - - Field of view for single events

▨▨▨▨▨ Standard shielding (accommodating the body)

▮▮▮▮▮ Additional shielding (for brain scanning)

Figure 9 The change in FOV for single events for different detector ring diameters and axial lengths.

activity outside the FOV should be provided. The move towards 3D PET and the desire to accommodate any part of the body and to increase the axial FOV pose problems in this regard. The dilemma can be illustrated by simple geometry (**Figure 9**). The largest axial FOV for a multi-ring tomograph is about 25 cm (left of **Figure 9**). With standard side shielding giving an aperture of 60 cm diameter, the FOV for single photons extends about 75 cm beyond the coincidence FOV. With the insertion of additional shielding appropriate for the brain (an aperture of 35 cm), the single-photon FOV is reduced significantly. The more common axial detector length of about 15 cm clearly gives a much reduced singles FOV, which for brain scanning barely extends beyond the side shielding. Providing effective shielding for body studies is not easy, but a way of processing the random events in order to lessen their effect is to apply some form of smoothing. As for scatter, the distribution of randoms is of a broad, low-frequency nature and thus amenable to smoothing. The future alternative to this is to use detectors with faster response, narrower coincidence window and proportionately lower randoms.

Spatial resolution effects

The spatial spread of a point source in an image gives rise to a phenomenon known as spillover, in which activity in one region affects that measured in an adjacent region. A source of decreasing size in a 'background' of lower activity concentration will appear to have decreasing activity. This is purely an effect of finite resolution and not of efficiency and is often referred to as the partial volume effect (that is, the object only partially fills the detector resolution field-of-view and is 'mixed' with its surroundings); it is also expressed by saying that the recovery coefficient of the object is less than unity. This is a difficult problem, but one correction technique that has been applied in the brain relies on the much higher resolution of a magnetic resonance image to provide accurate anatomical data.

Coincidence detection gives good uniformity of resolution along an LOR, but for circular ring systems the resolution in the radial direction gradually worsens owing to the interaction of photons with detectors at increasingly oblique angles (**Figure 10**). This effect is magnified as the ring diameter decreases. The desire for smaller diameters (and hence less expensive systems) requires a remedy for this non-uniformity. A number of methods of correction are being tested, such as the dual use of photodiodes and PMTs on either ends of a detector, but no one method has yet gained general acceptance.

Models of tracer kinetics

Even if all physical corrections have been applied and the image is an accurate representation of tracer distribution, the further big challenge in PET is to derive biochemical, physiological and pharmacological

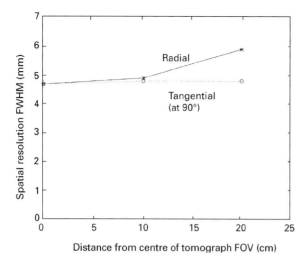

Figure 10 Variation of spatial resolution in radial and tangential (at right angles) directions with distance from the centre of the tomograph FOV.

parameters from the data. This is carried out by diverse mathematical models of the biological system. What may be termed a conventional approach is to treat the system as a number of separate compartments within each of which the tracer is uniformly distributed. The passage of tracer between compartments is described by rate constants that have dimension per unit time (time^{-1}). An example of such a system is shown in **Figure 11**, which divides the volume under investigation into blood (plasma) and free and fixed tracer in tissue. In PET research blood activity is usually measured either by taking discrete samples or by on-line monitoring in a detector placed by the side of the patient. A refinement of this is analysis of the blood into different radioactive components or metabolites because the tracer itself is broken down in its passage through tissue.

A compartmental model describes the system by a number of differential equations from which the rate constants are estimated. A common application of the model in **Figure 11** is in the dynamics of glucose utilization. The most widely used tracer in PET is an analogue of glucose known as FDG (^{18}F-labelled fluorodeoxyglucose). It is transported into the cell and undergoes the biochemical process of phosphorylation similarly to glucose, but it then remains 'trapped' in the tissue. Because of this, good quality

images of glucose utilization can be obtained. Over the period of a study (an hour or so), there is negligible release from the fixed compartment (k_4 in **Figure 11** is negligible) and this simplifies the model. In this case, the metabolic rate for glucose (MRGl) is given by

$$\mathrm{MRGl} = \frac{C_\mathrm{p} k_1 k_3}{\mathrm{LC}(k_2 + k_3)} \qquad [2]$$

where C_p is the (natural) glucose concentration in the plasma and LC is a constant that describes the difference in transport and phosphorylation rates between FDG and glucose. Examples of functional images of MRGl (two slices through the brain) are displayed in **Figure 12** (left) along with magnetic resonance (MRI) images showing anatomical detail (right) and coregistered PET and MRI images (centre). Coregistration covers a number of techniques used to overlay functional and anatomical images which, for example, make use of specific anatomical markers or the minimization of the variance between the two images.

Other types of tracer model do not seek to impose a compartmental structure but instead determine a combination of kinetic components which best fit the data. For example the technique of spectral analysis views the tissue response (C_tiss) as a convolution between the plasma/blood input (C_p) and a large but finite range of so-called basis functions of the form $\gamma \exp(-\delta t)$ where t is time:

$$C_\mathrm{tiss}(t) = C_\mathrm{p}(t) \otimes \sum_{j=1}^{j=n} \gamma_j \exp(-\delta_j t) \qquad [3]$$

When the γ_j are fixed and the δ_j are constrained to be zero or positive, it transpires that typically only two to four positive δ_j are given for the time–activity curve on each image pixel. The resulting parameters are used to generate the impulse response function (the tissue time–activity curve resulting from a unit pulse input at $t = 0$). The intercept of this function (for each pixel) at $t = 0$ gives an image of the clearance of tracer from blood to tissue (denoted K_1) and its integral gives the volume of distribution (concentration in tissue relative to blood, denoted V_d).

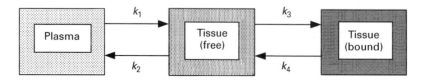

Figure 11 Three-compartment tracer model.

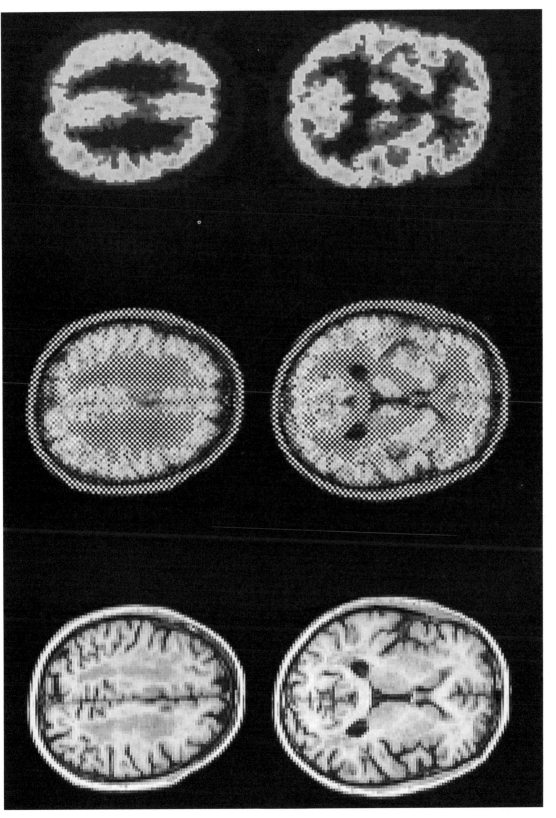

Figure 12 PET functional images of glucose metabolic rate (MRGI) (right), MRI images (magnetic resonance images, left) showing anatomical detail, and coregistered (overlaid) PET/MRI (centre). (See Colour Plate 42).

Frequency analysis is an example of a more objective way of extracting kinetic information from PET time–activity data. Other forms of this are factor, principal components and cluster analyses. These attempt to define pixels in the projection or image data that have similar kinetic characteristics. Their fundamental aim is the derivation of more specific and objective images of function rather than purely radioactivity concentration, a task that is enhanced by the superior physical properties of positron annihilation coincidence detection.

List of symbols

C_p = plasma concentration, plasma response; C_{tiss} = tissue response; k_{1-4} = rate constants; K_1 = tracer clearance (blood to tissue); MRGl = metabolic rate for glucose; N_0 = corrected count rate; NEC = noise-equivalent counts; N_m = measured count rate; rd_{max} = maximum scanner ring difference; t = time; V_d = volume of distribution; δ_j and γ_j = parameters of basis functions; τ = timing resolution; τ_d = system dead time.

See also: **MRI Applications, Biological; MRI Applications, Clinical; MRI Instrumentation; PET, Theory; SPECT, Methods and Instrumentation; Structural Chemistry Using NMR Spectroscopy, Inorganic Molecules; Scattering Theory; SPECT, Methods and Instrumentation; Two-Dimensional NMR, Methods.**

Further reading

Barrett HH and Swindell W (1981) *Radiological Imaging: The Theory of Image Formation, Detection, and Processing*. San Diego: Academic Press.

Bendriem B and Townsend DW (eds) (1998) *The Theory and Practice of 3D PET*. Dordrecht, Boston, London: Kluwer Academic.

Carson RE, Daube-Witherspoon ME, Herscovitch P (eds) (1998) *Quantitative Functional Brain Imaging with Positron Emission Tomography*. San Diego: Academic Press.

Casey ME and Nutt R (1986) A multicrystal two dimensional BGO detector system for positron emission tomography. *IEEE Transactions on Nuclear Science* NS-33: 570–574.

Cho ZH and Farukhi S (1977) New bismuth germanate crystal — a potential detector for the positron camera application. *Journal of Nuclear Medicine* 18: 840–844.

Knoll GF (1979) *Radiation Detection and Measurement*. Chichester: Wiley.

Melcher CL and Schweitzer JS (1992) Cerium-doped lutetium orthosilicate: a fast, efficient new scintillator. *IEEE Transactions on Nuclear Science* 39: 502–505.

Murray IPC, Ell PJ and Strauss HW (eds) (1994) *Nuclear Medicine in Clinical Diagnosis and Treatment*. Edinburgh: Churchill Livingstone.

Myers R, Cunningham V, Bailey D and Jones T (eds) (1996) *Quantification of Brain Function Using PET*. San Diego: Academic Press.

Phelps M, Mazziotta J and Schelbert H (eds) *Positron Emission Tomography and Autoradiography: Principles and Applications for the Brain and Heart*. New York: Raven Press.

Schwaiger M (ed) (1996) *Cardiac Positron Emission Tomography*. Boston: Kluwer Academic.

Spinks TJ, Jones T, Bailey DL *et al* (1992) Physical performance of a positron tomograph for brain imaging with retractable septa. *Physics in Medicine and Biology* 37: 1637–1655.

PET, Theory

TJ Spinks, Hammersmith Hospital, London, UK

SPATIALLY RESOLVED
SPECTROSCOPIC ANALYSIS
Theory

Introduction

The acronym PET can be used to stand for both positron-emitting tracers and positron emission tomography; it is a representation of the radioactive isotopes used and the methods by which their distribution is visualized (tomography derives from the Greek tomos or 'cut' – i.e. an imaged slice through the body). PET is the most sensitive and specific method of studying molecular interactions and pathways in the living organism and is assuming ever greater importance in medical diagnosis and research, in the understanding of biochemistry and physiology in health and disease, and in the development of drugs. Most of the world's PET centres are in North America, Europe and Japan but many other countries are making plans for the installation of PET facilities. Most of the centres are dedicated to diagnosis, particularly in heart disease and cancer, but there are also a number of centres associated with large medical research institutes that concentrate purely on research. This article will deal with the theoretical aspects of PET: (a) isotope production, (b) radiation interactions and detection, (c) data acquisition and image formation, and (d) properties of the image of radioisotope distribution.

Physics of the positron

A positron (positively charged electron) is a particle of so-called 'antimatter' that cannot coexist for long with the 'ordinary' matter of which we and all that surrounds us is made. Such particles were postulated in the 1930s by the physicist Paul Dirac, who pictured the vacuum as a 'sea' of electrons in negative energy levels that could be excited into positive energy levels by the absorption of quanta of energy. Although this concept was not readily accepted by most physicists, the existence of the positron was demonstrated experimentally by Anderson three years after the theoretical prediction. It was observed that a photon, of energy greater than or equal to twice the rest mass energy of the electron, in the field of the nucleus could give rise to the simultaneous appearance of a positron and an electron. This is known as the pair-production process. The positrons used in PET, however, arise from the disintegration of atomic nuclei that are unstable because they have an 'excess' of positive charge.

Production of positron emitters

Positron-emitting atoms do not normally exist in nature. The radionuclides used for PET are usually produced by a cyclotron, which, by harnessing powerful electric and magnetic fields, accelerates charged particles (such as protons, deuterons or alpha particles) to high energies (about 2–5% of the speed of light); these then bombard stable atoms in a target to give rise to radioactive isotopes. **Table 1** gives examples of the reactions used to produce the principal radionuclides used in PET (^{11}C, ^{15}O, ^{18}F and ^{13}N). It can be seen from column 3 that there are generally more protons than neutrons in each of the product nuclei. This 'excess' charge is released during nuclear disintegration (beta decay) by the emission of a (positively charged) positron (or in a smaller fraction of cases by the capture of an orbiting electron). Two principal characteristics of the positron emitters in **Table 1** that are responsible for their success as *in vivo* radiotracers are that (a) they are radioisotopes of major

body elements (or in the case of ^{18}F serve as *in vivo* analogues) and (b) they have short half-lives. These properties enable them to label biological molecules without altering the biochemical action and to be injected into a patient or normal volunteer in usable quantities with an acceptably low radiation dose. The physical characteristics of detection of these tracers, which will be described below, provide additional reasons for their pre-eminence in nuclear medicine. However, the short half-lives demand that the scanner (tomograph) and cyclotron are in close proximity. Such a necessity has given rise to the impression that PET is an expensive technique, but an increasing number of clinical centres are obtaining tracers from shared central cyclotron facilities and lower-cost scanner designs are commercially available.

Positron annihilation

Positrons are emitted from nuclei of a given isotope with a range of energies up to a characteristic maximum 'end point' energy E_{max} (**Table 2**), the mean energy being roughly one-third E_{max}. Positrons lose their energy by Coulomb interactions with atomic electrons, following a tortuous path until they are brought to rest within a precisely defined range (dependent on their energy and the effective atomic number of the medium). The ranges for mean and maximum energies in soft tissue are given in **Table 2**. When the energy of the positron is close to zero, the probability of interaction with an electron is highest. From direct interaction or after the formation of a transient system with an electron known as positronium, two photons, each of energy 511 keV (the rest mass energy of the electron or positron), are emitted in opposite directions with the disappearance (annihilation) of both particles. The 'back-to-back' photon emission arises from the conservation of momentum. However, there is only precisely 180° between the photon directions if the net momentum is zero at annihilation. The small residual momentum of the positronium system leads to an angular spread of about ± 0.3°. Positron range and the angular spread determine the physical limits of spatial resolution in PET.

Annihilation coincidence detection

Simultaneous detection of the annihilation photons provides significantly greater efficiency and improved uniformity of spatial resolution than detection of individual photons. These points are illustrated schematically in **Figure 1**. Consider a point source positron emitter in air moving across the channel between the two detectors D_1 and D_2 (**Figure 1A**). The channel is conventionally termed a

Table 1 Production and characteristics of principal isotopes used in PET

Positron emitting product	Stable element	Nuclear reaction	Number of protons (p) and neutrons (n) in the product nucleus	Half-life of product (min)	Stable nucleus after positron emission
^{11}C	Nitrogen(^{14}N)	$^{14}N(p,\alpha)^{11}C$	6p, 5n	20.4	^{11}B
^{18}F	Oxygen(^{18}O)	$^{18}O(p,n)^{18}F$	9p, 9n	109.8	^{18}O
^{15}O	Nitrogen(^{14}N)	$^{14}N(d,n)^{15}O$	8p, 7n	2.03	^{15}N
^{13}N	Carbon (^{12}C)	$^{12}C(d,n)^{13}N$	7p, 6n	10.0	^{13}C

Table 2 Positron ranges in soft tissue for the principal positron emitters

Positron emitter	Positron energy (MeV)		Positron range in soft tissue (mm)		Contribution to resolution (mm FWHM)
	Maximum (E_{max})	Mean	Maximum	Mean	
^{18}F	0.635	0.250	2.6	0.61	0.2
^{11}C	0.970	0.386	4.2	1.23	0.3
^{13}N	1.200	0.491	5.4	1.73	0.4
^{15}O	1.740	0.735	8.4	2.97	1.2

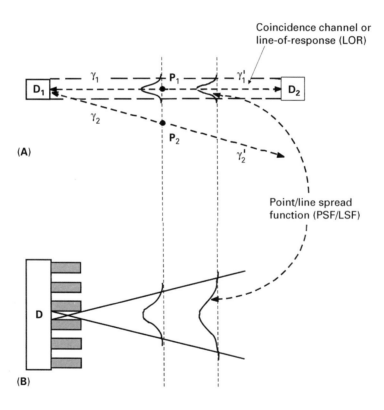

Figure 1 (A) Annihilation coincidence and (B) single-photon detection.

'line-of-response' (LOR) and the detectors are connected such that a 'count' is only registered when a photon interaction is recorded in each detector at the same time (or within the time resolution of the detector, see below). Only when the source is within the LOR (e.g. position P_1) do annihilation photons such as γ_1 and γ_1' have a chance of being detected as a coincident pair. With the source at position P_2 (e.g. photons γ_2 and γ_2') there is no possibility of a coincidence event. Therefore, coincidence counting confers a natural collimation of the radiation, often referred to as electronic collimation and, in addition, the resolution varies relatively little for different positions along the LOR. The variation in counts obtained by passing a point or (orthogonal) line source across the LOR is termed the point or line spread function (PSF, LSF) and is conventionally characterised by its full-width at half-maximum (FWHM). The PSF obtained in this way represents the intrinsic resolution of the detector pair, in other words, the best achievable.

To obtain spatial localization with a source emitting only single photons (γ rays from nuclear disintegration), a physical (lead) collimator has to be placed between source and detector D (**Figure 1B**). The solid angle subtended by the collimator aperture shows that the resolution in this case will vary with distance from the detector. The aperture can be made smaller to give a spatial resolution as high as desired, but this will be at the expense of detection efficiency and so some compromise is needed. Single-photon emission tomography (SPECT) systems normally consist of large-area detectors, whereas most modern PET systems consist of thousands of coincidence detector pairs and the detection efficiency in PET is about 100 times that of SPECT.

Photon interactions and attenuation in scattering media

The previous discussion of coincidence detection becomes more complicated when the source is within a scattering medium (such as body tissue). At the energy of the annihilation photons (511 keV), the possible interactions (with atomic electrons) are photoelectric (total) absorption and coherent and incoherent (Compton) scattering. Scattering refers to change of direction without (coherent) or with (incoherent) energy loss. Compton scattering is overwhelmingly predominant at 511 keV in the body (the probability of photoelectric absorption and coherent scattering can be considered negligible) and is thus, practically speaking, totally responsible for the removal of a photon from a particular LOR. This is termed attenuation and is illustrated in **Figure 2A**. If the scattering medium is uniform, it will have a constant attenuation coefficient (μ), dependent on the effective atomic number, defined as

$$\mathrm{d}N = \mu N_x\, \mathrm{d}x \qquad [1]$$

where $\mathrm{d}N$ are the number of photons scattered within a distance x to $x + \mathrm{d}x$ and N_x is the number of unscattered photons at x. Integrating Equation [1] gives the following formula for the number of photons (A_T) left

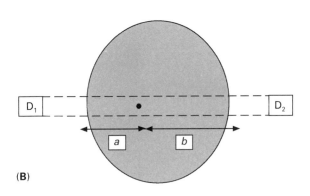

Figure 2 Photon attenuation in a given LOR due to Compton scattering.

unscattered after a distance T along the LOR:

$$A_T = A_0 \exp[-\mu T] \qquad [2a]$$

If the attenuation medium is not uniform, then Equation [2a] is modified to give

$$A_T = A_0 \exp[-\sum_i \mu_i \Delta T_i] \qquad [2b]$$

where μ_i = the attenuation coefficient within a small element ΔT_i.

Equation [2] describes what is known as 'narrow beam' attenuation because if one is just interested in the LOR between two detectors, a scattered photon is completely lost. For an array of detectors (as in a PET scanner), however, it can be imagined that the scattered photon might be detected in another LOR; this is considered later.

The larger the angle (Θ) through which a scattered photon is deflected (**Figure 3**), the greater is its loss of energy. From simple kinematics (conservation of

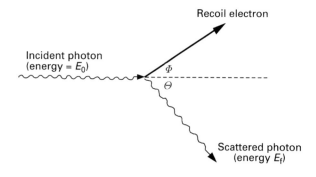

Figure 3 Mechanism of photon (Compton) scattering.

Table 3 Examples of scattering angles and energies in Compton scatter of annihilation photons

Scattering angle Θ (°) (see **Figure 3**)	Energy of scattered photon (keV)	Probability of scatter (%) ($0° = 100\%$)
30	451	68.7
60	341	31.5
90	256	18.8
180 (Back-scattering)	170	18.5

energy and momentum), the relationship between the initial photon energy E_0 and final energy E_f is

$$E_f = \frac{E_0}{1 + E_0(1 - \cos\Theta)/m_0c^2} \quad [3a]$$

where m_0c^2 is the rest mass energy of the electron (m_0 = electron rest mass; c = the speed of light). Since the rest mass energy is equal to 511 keV, this equation simplifies for PET to

$$E_f = \frac{E_0}{2 - \cos\Theta} \quad [3b]$$

The maximum energy that can be transferred to an electron in this interaction is when the photon is deflected back along its original path ('backscattering'), that is when $\Theta = 180°$. In this case E_f (minimum) = $E_0/3$ = 170 keV. Some other examples of scattering angles and energies are given in **Table 3**. However, Equation [3] only gives the resultant energy for a given angle; the probability of scattering through a particular angle is described by a much more complicated expression known as the Klein–Nishina formula. The relative probabilities of scattering are shown in the third column of **Table 3** (taking that for 0° as 100%). It can be seen that, even for a relatively large scattering angle, the photon still retains a significant fraction of its energy but that the probability of scattering falls quite quickly with increasing angle.

To obtain an accurate measurement of the regional distribution of isotope concentration in the body, the primary correction is that for attenuation. The basic principles of attenuation correction, relatively straightforward in PET, are as follows. Consider the attenuation lengths a and b on either side of the point source in **Figure 2B**. From Equation [2], attenuation along these two paths will be proportional to exp[−μa]

and exp[−μb]. For coincidence counting, the total attenuation will be proportional to the product of these two factors: exp[−μ($a+b$)]. This expression is independent of the position of the source along the LOR and indeed would be so if the source were outside the object. If such an external source is measured with and without the (inactive) object in the LOR, the ratio of the measurements gives the attenuation correction factor along that line. For single-photon detection, it should be noted that attenuation is dependent on depth in the object and this makes correction more complicated. Specific methods of attenuation correction in PET are dealt with in the article on Instrumentation and Methods for PET.

Detection of annihilation photons

Most PET scanners consist of individual detector elements arranged in a number of adjacent coaxial rings surrounding the patient. Each element is connected in coincidence with a number of other elements in both the same ring and any number of other rings, and modern scanners consist of thousands of detectors and millions of LORs. Specific scanner configurations are given in the article covering Instrumentation and Methods.

Detector materials must have a high atomic number to maximize their attenuation of annihilation photons and must also produce a measurable response. The great majority of detectors used in PET are scintillators, which respond to the absorption of photon energy by the emission of visible light. This occurs when electrons fall from excited energy levels to the ground state, a process often facilitated by the inclusion of a small amount of impurity (or activator) into the scintillation crystal. The light output rises rapidly to a peak and then falls with a characteristic 'decay time' and is converted into an electrical pulse by a photomultiplier tube (PMT) and subsequently amplifi.ed. The speed of response of the detector determines the width of this pulse and the precision with which the time of the interaction can be measured – the timing resolution, denoted by τ. In addition, the detector electronics take a finite time to process each event, during which other events go unrecorded. This time is known as dead time and corrections have to be made for accurate quantification. Based on the timing resolution, a coincidence time window is set within which events in two opposing detectors are regarded as constituting a coincidence event. This can be either a true event or a random (chance) event, depending on whether the photons came from the same or different annihilations. These collectively are known as prompt events (events are commonly termed 'prompts', 'trues' and 'randoms'). A distinction can be made between trues and randoms by counting the number of events occurring when the time window for one detector is delayed (by about 100 ns) relative to the other. A coincidence recorded in the 'delayed circuit' cannot be a true coincidence and it is assumed that the 'delayed events' are equal to the randoms recorded in the undelayed ('primary') circuit. Trues are therefore determined by subtracting the randoms from the prompts.

An alternative way of calculating random events is by recording the total rate of single photons striking each detector. If these singles rates for detectors 1 and 2 are S_1 and S_2, then the rate of random events (R_{12}) between the two detectors is given by

$$R_{12} = S_1 \cdot S_2 \cdot T \qquad [4]$$

where T, the coincidence time window, is twice the timing resolution (i.e 2τ). The distribution of randoms is quite uniform over the field-of-view (FOV) of the scanner but if not subtracted will impair the quantification of regional radiotracer concentration and reduce contrast in the image.

Spatial sampling and resolution

One of the continuing aims in PET is to improve spatial resolution or the clarity of definition of isotope distribution in the body. The physical limit of resolution is dictated by positron range and non-collinearity of annihilation photons as outlined above. The ranges of positrons in the body for the most important isotopes are given in **Table 2**. These appear to be rather large, but it should be borne in mind that the contribution of positron range to resolution in the image is reflected in the average range and that positrons travel in all directions and not just orthogonally to an LOR. The net contribution to resolution (FWHM of the point spread function) is given in **Table 2**. The physical limitation imposed by the small angular spread around the 180° ('back-to-back') photon emission leads to an additional

'blurring' of resolution independent of positron energy but increasing with the diameter of the detector array. For a scanner of diameter 80 cm (common for imaging of humans) the fundamental limit imposed by these effects is about 2 mm, whereas for a diameter of 20 cm the limit is less than 1 mm.

To image the distribution of a radiotracer, the active volume must be sampled as finely as possible. In other words, measurements must be made along a number of LORs, as closely spaced as practicable, both across the object and at different angles. The basic property of a tomographic system with good resolution is the ability to distinguish changes in tracer concentration. Another way of expressing this is that the system has a good spatial frequency response. This may be envisaged analytically by presenting a 'bar pattern' of alternating white and black (active/inactive) stripes to the imaging system. As the width of the bars is reduced, there will come a point when the imager will no longer be able to reproduce the pattern; the input (object) frequency will no longer produce a faithful output (image). Mathematically this is expressed in terms of the modulation transfer function (MTF), which gives the fraction of signal amplitude that a system will transfer to the image at each spatial frequency. A broader MTF function will give a sharper resolution.

The requirement of an imaging system is expressed formally by the sampling theorem, which states that the highest frequency that can reliably be measured (known as the Nyquist frequency) is equal to $1/(2\Delta d)$, where Δd is the sampling distance (the spacing between adjacent LORs). One of the principal aims in the design of a PET tomograph is to provide as high a degree of sampling as possible. However, a fundamental limit (apart from positron range and photon noncollinearity) is imposed by the detector width (the intrinsic detector resolution as defined above). Better sampling schemes will deliver an image resolution ever closer to the intrinsic resolution, but this cannot be exceeded without additional 'post-processing' of the image. Early tomographs of the late 1970s/early 1980s with relatively large (~25 mm) detectors, which were not densely packed, optimized their sampling by incremental linear and angular motion. Most tomographs in use today consist of circular rings of detectors of width 6 mm or less. For these, adequate linear and angular sampling is achieved without any motion, although in some designs a rotatory motion (known as 'wobble') is incorporated. This has largely been abandoned because the data volumes were increased typically by a factor of 4 without a great improvement in resolution.

The organization of data acquisition in a circular ring of detectors is shown in **Figure 4**. Each detector

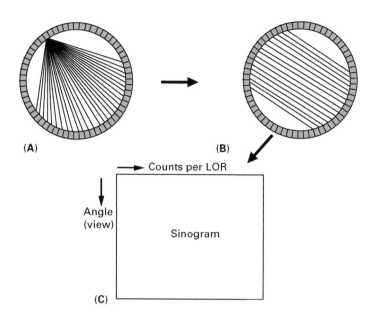

Figure 4 Geometrical relation between LORs and the sinogram data matrix.

is connected in a coincidence circuit with a number of detectors on the other side of the ring (**Figure 4A**) (The more general case of multiple rings is discussed in the article on Instrumentation and Methods. The LORs that are parallel to each other are grouped together to form projections or views of the object at each angle (**Figure 4B**). The counts recorded in each of these LORs form one row of a data matrix called a sinogram (**Figure 4C**), each row corresponding to an angle of view. In a recent design there are 576 detectors in the ring and each detector is in coincidence with 288 other detectors. The number of views in the sinogram in this case is $576/2 = 288$ (each separated by an angle $180°/288 = 0.625°$) and the number of LORs in each view is 288.

Formation of an image can be achieved by a number of different methods, but all of them involve back-projection of the views across the FOV. For practical purposes the position of an annihilation event along an LOR is indeterminate. Some detectors, notably barium and caesium fluorides, are capable of modest time-of-flight resolution (FWHM ~5 cm) by measuring the time difference of the photon interactions, but this has not proved to give significant advantages because of their lower efficiency.) The raw process of image formation conventionally divides the 'image space' into a matrix of square boxes or pixels (picture elements) and places 'counts' in each pixel proportional to the counts recorded in the particular LOR and the area of overlap of LOR and pixel. This process is illustrated for a point source in **Figure 5A**. Only a small number of projections, each consisting of a peak corresponding

to the source position, are shown for clarity. The back-projected profiles for each projection will intersect at the position of the point source, but the image will be a poor representation of the original because of the 'background' imposed. This may be described mathematically by saying that the high spatial frequency components of the source have been attenuated and low frequencies enhanced. This process is reversed by frequency filtering in which the Fourier transform is first applied to the projection data to give the magnitude of each component spatial frequency and a filter is applied with increasing weight given to higher frequencies. This filter is accordingly termed a ramp in frequency space. In real space, the form of the filter is as sketched in **Figure 5B**, which shows alternating positive and negative oscillations decreasing in intensity from its centre. When this is convolved with the projections and back-projection is performed, the effect is that the positive and negative components cancel each other out, so removing the low-frequency 'blur' and restoring the high-frequency nature of the object in the filtered image.

Filtered back-projection is the most commonly used method of image formation in PET because Fourier transforms can be calculated rapidly with modern computers. However, the method has its drawbacks and these are intimately allied to statistical variations or noise in the projection data. As the number of counts recorded by the system decreases, so the statistical uncertainty increases, as described by Poisson statistics, and 'star artefacts' (remnants of the back-projection process) become increasingly apparent. Filtering causes each pixel in the image to be

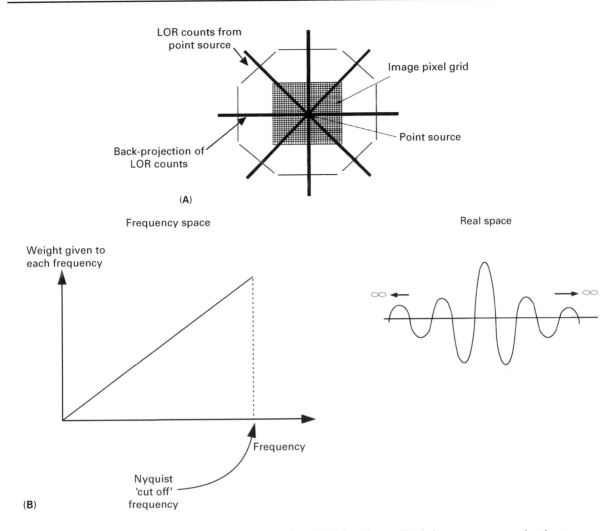

Figure 5 (A) Formation of an image by filtered back-projection. (B) Shape of ramp filter in frequency space and real space.

correlated to some degree with every other pixel and poor statistics enhances this.

Inherently superior methods of image reconstruction are the so-called iterative methods in which the image is successively corrected to be consistent with the projection data to within desired error limits. These techniques involve iterative forward-projection and back-projection (between image and projections) and a number of different algorithms have been developed for the correction step at each iteration. Such methods have much more flexibility than filtered back-projection because models for the statistical nature of the data and the physical processes involved in data acquisition can be incorporated. For example if the projection data are assumed to obey Poisson statistics, the likelihood of the reconstructed image can be maximized according to this, giving the so-called EM-ML (expectation maximization–maximum likelihood) algorithm. In contrast to filtered back-projection, which produces a 'one-off' solution, iterative algorithms gradually converge to the desired

solution but the number of iterations needs to be carefully assessed to optimize the signal-to-noise ratio in the image and regional quantification of tracer.

Detection efficiency and noise

The efficiency of detection of annihilation photons obviously depends on the geometry of the tomograph and the ability of the detector material to absorb the photon energy (detector 'stopping power'). Considering a single ring of detectors of radius r and width (axial thickness) t, decreasing the radius increases the solid angle of detection (with r^2) but the detector volume decreases with r. Therefore overall efficiency increases with r. However, as r decreases, the solid angle for detection of random and scattered events increases, and clearly r must be large enough to provide the desired FOV (for head or body). In addition, spatial resolution worsens in the radial direction as r decreases. This implies that some compromise must be reached to balance these factors.

Although the 'raw' efficiency, or number of true coincidences acquired, is a basic determinant of the quality of a PET scanner, the complicating factors of scattered and random events and dead time have to be brought into the analysis. Details of the distribution of scattered events and correction methods can be found elsewhere, but for the present purposes it can be stated that scattered radiation (or 'scatter' for short) produces a relatively flat 'background' on the projection and image data, impairing contrast and reducing quantitative accuracy. The quantity of scatter detected, the scatter fraction (SF), is expressed simply in terms of the total true (unscattered + scattered) events (T_{tot}) and the scattered events (S) by

$$\text{SF} = S/T_{tot} \qquad [5]$$

A similar reduction in contrast and quantification would result if random events were not subtracted from the data. As discussed above, subtraction of randoms is usually carried out 'on-line' during data acquisition and, although accurate quantitatively, imposes a statistical penalty on the net counts. If the numbers of prompt and random events are designated by P and R, then

$$T_{tot} = P - R \qquad [6]$$

Assuming that Poisson statistics apply to these counts (e.g. the standard error of $P = \sqrt{P}$), the standard error of T_{tot} is

$$\sqrt{\left[(\sqrt{P})^2 + (\sqrt{R})^2\right]} = \sqrt{(P + R)} = \sqrt{(T_{tot} + 2R)} \qquad [7]$$

Thus the standard error of T_{tot} is larger than $\sqrt{(T_{tot})}$ and increases with the rate of random events. Any procedure that reduces randoms, such as better shielding of extraneous radiation or reduction in the coincidence time window (Eqn [4]), will lower this uncertainty. The resultant standard error is conventionally described by the noise equivalent count (NEC) which is defined as follows. The fractional error of T_{tot} is $\sqrt{(T_{tot} + 2R)}/T_{tot}$. If a number (NEC) of hypothetical counts are collected (free of background) the fractional error is $(\sqrt{NEC})/NEC = 1/\sqrt{NEC}$. Equating this to the fractional error of T_{tot} gives

$$\text{NEC} = \frac{T_{tot}^2}{T_{tot} + 2R}$$

However, this expression is modified to take into account (a) the subtraction of scatter and (b) the fact that random events are spread fairly uniformly over the whole FOV of the PET tomograph, whereas unscattered true events are confined to the limits of the object. If the fraction of the FOV subtended by the object is f then the final expression for NEC is

$$\text{NEC} = \frac{[T_{tot}(1 - \text{SF})]^2}{T_{tot} + 2fR} \qquad [8]$$

Another assumption implicit in this is that the subtraction of scattered events does not lead to an increase in noise. This would be so if, for example, a mathematical function were used to describe the distribution of scatter, but is not quite true in all methods.

NEC provides an overall factor for the determination of statistical quality, but it is of importance to investigate noise (statistical 'ripple') in the reconstructed image. As mentioned above in the discussion of reconstruction, image noise is greater than expected purely from the Poisson statistics of the projection data, owing to the filtering process. If there are an adequate number of projection angles and sampling points (LORs) per projection, then it can be shown that the noise/signal ratio (% root mean square) for a uniformly labelled object is given by

$$\text{N/S}(\%) = \frac{120\{N_{re}\}^{3/4}}{N^{1/2}} \qquad [9]$$

where N is the total number of counts (trues) acquired and N_{re} is the number of 'resolution elements' contained within the object. The resolution element is defined as a region of dimensions (sampling distance)^2 or $(\Delta d)^2$. If the object is a disc (or a slice through a cylinder) of diameter 200 mm, the resolution FWHM = 6 mm and 10^6 counts are acquired, N/S is 19%. This is about 6 times what would be expected purely from Poisson statistics. Furthermore, if the resolution decreased to 3 mm FWHM, the number of counts N would have to increase by a factor of 8 to keep the N/S per resolution element constant. These simple calculations illustrate the importance of efficiency in PET. Resolution might be technically improved by using narrower detectors, but the advantage will be lost if efficiency of detection is not increased. This point is of fundamental importance and is discussed more fully in the article on Instrumentation and Methods.

List of symbols

A_T = number of photons left unscattered after distance T along the LOR; A_0 = original number of photons; c = speed of light; E_0, E_f = initial and final energy of scattered photon; f = fraction of FOV subtended by the object; m_0 = rest mass of the electron; N = total number of counts (trues); N_{re} = number of resolution elements; N_x = number of unscattered photons at x; P = number of prompt events; r = detector radius; R = number of random events, detector ring radius; S = total number of scattered events; t = detector width (axial thickness); T = distance along LOR, coincidence time window; T_{tot} = total number of true (unscattered + scattered) events; Δd = sampling distance = spacing between adjacent LORs; Θ = scattering angle of photon; μ = attenuation coefficient; τ = timing resolution.

See also: **Fourier Transformation and Sampling Theory; PET, Methods and Instrumentation; Scattering Theory; Statistical Theory of Mass Spectra.**

Further reading

Barret HH and Swindell W (1981) *Radiological Imaging: The Theory of Image Formation, Detection, and Processing.* San Diego: Academic Press.

Bendriem B and Townsend DW (eds) (1998) *The Theory and Practice of 3D PET.* Dordrecht: Kluwer Academic.

Herman GT (1980) *Image Reconstruction from Projections: The Fundamentals of Computerized Tomography.* New York: Academic Press.

Jones T (1996) The imaging science of positron emission tomography. *European Journal of Nuclear Medicine* **23**: 807–813.

Kinahan PE and Rogers JG (1989) Analytic 3D image reconstruction using all detected events. *IEEE Transactions on Nuclear Science* **NS-36**: 964–968.

Knoll GF (1979) *Radiation Detection and Measurement.* New York: Wiley.

Murray IPC, Ell PJ and Strauss HW (eds) (1994) *Nuclear Medicine in Clinical Diagnosis and Treatment.* Edinburgh: Churchill Livingstone.

Phelps M, Mazziotta J and Schelbert H (1986) *Positron Emission Tomography and Autoradiography: Principles and Applications for the Brain and Heart.* New York: Raven Press.

Shepp LA and Vardi V (1982) Maximum likelihood reconstruction for emission tomography. *IEEE Transactions on Medical Imaging*, **MI-1**: 113–122.

Webb S (ed) (1988) *The Physics of Medical Imaging.* Bristol: Institute of Physics Publishing.

Pharmaceutical Applications of Atomic Spectroscopy

Nancy S Lewen and **Martha M Schenkenberger**,
Bristol-Myers Squibb, New Brunswick, NJ, USA

ATOMIC SPECTROMETRY

Applications

Introduction

The United States Federal Food and Drug Administration (FDA) regulations require the complete characterization of drug compounds. Since most pharmaceutical agents are organic compounds, much of this characterization involves various chromatography-based analytical techniques, as well as NMR, IR and various physical testing methods (such as DSC, TGA and XRD). The field of atomic spectroscopy has not traditionally played a major role in the characterization of pharmaceutical products, but on closer inspection it is clear that absorption and emission spectroscopic techniques can play a valuable role in the process of drug development and the quality of the product that finally reaches consumers.

From drug synthesis to quality control (QC) monitoring of over-the-counter medications, metals are found in all phases of the drug development process. Many metal-based products are used as imaging agents, and metals are used in the synthesis of drug substances, as excipients in tablets, capsules and liquids. In addition, trace metals can arise from the equipment used to manufacture a drug substance or compound. Because of the prevalence of metals associated with the drug development and manufacturing process, various atomic and emission-based techniques are often used to help fully characterize pharmaceutical products. The wide variety of pharmaceutical dosage forms and matrices, such as tablets, capsules, injectables, liquids, effervescing compounds, ointments and creams, makes the development of analytical methods and the analysis of samples a challenging and interesting process. In this article we will describe the types of situations in the pharmaceutical industry where an analyst is likely to use atomic spectroscopy to solve the analytical problem and meet regulatory requirements. The pharmaceutical development process described will be based on the regulations and requirements in the USA.

Techniques of interest in the analysis of pharmaceutical products

The need for the determination of metallic constituents or impurities in pharmaceutical products has, historically, been addressed by ion chromatographic methods or various wet-bench methods (e.g. the USP heavy metals test). As the popularity of atomic spectroscopy has increased, and the equipment has become more affordable, spectroscopy-based techniques have been routinely employed to solve analytical problems in the pharmaceutical industry. Table 1 provides examples of metal determinations in pharmaceutical matrices, using spectroscopic techniques, and the reasons why these analyses are important. Flame atomic absorption spectrometry (FAAS), graphite furnace atomic absorption spectrometry

(GFAAS), inductively coupled plasma-atomic emission spectroscopy (ICP-AES – also referred to as inductively coupled plasma-optical emission spectroscopy, or ICP-OES) and inductively coupled plasma-mass spectrometry (ICP-MS) are all routinely utilized in pharmaceutical applications. While there are other techniques of note available, such as microwave induced plasma (MIP) or direct coupled plasma (DCP), they have not been routinely used in the pharmaceutical industry, and will, therefore, not be discussed here. The theories involved in the use of FAAS, GFAAS, ICP and ICP-MS may be found in other articles of this Encyclopedia.

The first atomic spectroscopic techniques to see increased usage in the pharmaceutical field were FAAS and GFAAS. Among the current instrumental techniques available, they are among the most inexpensive, and have seen considerably more usage in all fields of endeavour, thus availing the pharmaceutical analyst of a vast array of knowledge upon which to draw and develop analytical methods. Because of the relatively low cost of the instrumentation, as well as its ease of use, QC laboratories in the pharmaceutical industry are more likely to have this type of atomic spectroscopy equipment than any other type. The speed and sensitivity of FAAS for elements such as Na, K and Li make it superior to wet-bench techniques. Examples of pharmaceutical products which require Na, K or Li determinations are nafcillin sodium (an antibiotic), oral solutions of potassium chloride (an electrolyte replenisher) and Lithane® (a psychotropic drug).

Table 1 Examples of metals that are determined in pharmaceutical analyses

Element	Reason for assay/therapeutic area of use		Suggested analytical technique
Ag	1.	Determination of geographical origin of illicit drugs	1. Graphite furnace AA
	2.	Complex formation with drug for indirect determination (e.g. tetracycline)	2. ICP-AES
	3.	Monitor Ag content of material (e.g. antiseptic creams, ophthalmic solutions)	3. ICP-AES, graphite furnace AA
Al	1.	Monitor Al in antihaemophilia preparations, which are sometimes precipitated with aluminium hydroxide.	1. Graphite furnace AA
	2.	Determine geographical origin of illicit drugs	2. Graphite furnace AA
	3.	Monitor Al concentratiosns in dialysis solutions.	3. ICP-AES
Au	1.	Monitor Au concentration in arthritis drugs	1. ICP-AES
B	1.	Monitor for presence of B in regrents used in synthesis	1. ICP-AES, ICP-MS
	2.	Monitor for leaching of B from glass vials, containers	2. ICP-AES, ICP-MS
Ba	1.	Determination of geographical origin of material (e.g. illicit drugs)	1. ICP-MS
	2.	Monitor Ba content in materials (e.g. used for diagnostic imaging)	2. Flame AA, ICP-AES
Bi	1.	Indirect determination of cocaine	1. ICP-AES
	2.	Monitor Bi content in materials (e.g. antacid products)	2. ICP-AES
Br	1.	Determination of geographical origin of material (e.g. illicit drugs)	1. ICP-MS
	2.	Monitor for presence of reagents used in synthesis, or as part of the compound	2. ICP-MS
Ca	1.	Determination of Ca in calcium supplements and vitamins	1. ICP-AES, graphite furnace AA
	2.	Monitor Ca impurities in magnesium oxide (often used as an excipient in pharmaceutical preparations)	2. Flame AA
	3.	Determination of geographical origin of material (e.g. illicit drugs)	3. ICP-MS

Table 1 *Contd.*

Element	Reason for assay/therapeutic area of use	Suggested analytical technique
Cd	1. Monitor Cd in dialysis solutions	1. Graphite furnace AA
	2. Monitor heavy metals content of medicinal plants or herbal drugs	2. Graphite furnace AA, flame AA
	3. Determination of geographical origin of material (e.g. illicit drugs)	3. ICP-MS.
	4. Monitor trace metals content in materials (e.g. penicillin G)	4. ICP-AES
Co	1. Complexing agent for indirect determination of drug (e.g. salicylic acid, lidocaine)	1. Flame AA
	2. Monitor Co in dialysis solutions	2. Graphite furnace AA
	3. Monitoring trace metals content in materials (e.g. penicillin G)	3. ICP-AES
	4. Determination of B-vitamins	4. HPLC-FAAS
Cr	1. Complexing agent for indirect determination of drug (e.g. thioridazine, amitriptyline, imipramine, orphenadrine)	1. Flame AA
	2. Determine geographic area of origin of illicit drugs	2. Graphite furnace AA
	3. Monitor trace metals content in materials (e.g. penicillin G)	3. ICP-AES
	4. Monitor Cr content in vitamins	4. Graphite furnace AA
Cs	1. Monitor for presence of reagents used in synthesis	1. Flame AA
Cu	1. Complexing agent for indirect determination of drug (e.g. lincomycin, isonicotinic acid hydrazid, ethambutol hydrochloride, neomycin, streptomycin)	1. Flame AA
	2. Monitor Cu in dialysis solutions	2. Flame AA, graphite furnace AA
	3. Moinitor heavy metals content in medicinal plants	3. Graphite furnace AA
	4. Determine Cu concentrations in vitamins	4. ICP-AES
	5. Monitor Cu in herbal drugs	5. Flame AA
	6. Monitor trace metals content in materials (e.g. penicillin G)	6. ICP-AES
	7. Determination of synthetic route and geographical origin of material (e.g. illicit drugs or to prevent patent infringement)	7. ICP-MS
Fe	1. Monitor Fe in dialysis solutions	1. Flame AA
	2. Monitor Fe contamination in magnesium oxide (often used as excipient in pharmaceutical preparations)	2. Flame AA
	3. Monitor Fe concentrations in vitamins	3. ICP-AES
	4. Determination of geographical origin of illicit drugs	4. ICP-MS
	5. Determination of trace metals content of materials (e.g. penicillin G)	5. ICP-AES
	6. Monitor Fe concentrations in imaging agents.	6. ICP-AES, flame AA
Gd	1. Monitor Gd content in imaging agents (e.g. Prohance®)	1. ICP-AES
Hg	1. Monitor Hg content of materials (e.g. antiseptic solutions and creams, ophthalmic solutions)	1. Flame AA
	2. Monitor heavy metals content of materials	2. ICP-MS
I	1. Determination of geographical origin of illicit drugs	1. ICP-MS
In	1. Monitor trace metals concentration in final drug substance	1. ICP-MS
K	1. Monitor for presence of reagents used in synthesis	1. Flame AA
	2. Monitor salt counter-ion concentration	2. Flame AA, ICP
Li	1. Monitor for presence of reagents used in synthesis	1. Flame AA, ICP-MS
	2. Monitor Li concentration in drug (e.g. lithium-based psychotropic drugs for treatment of manic/depressive disorder)	2. Flame AA
Mg	1. Determine Mg concentrations in vitamins	1. ICP-AES
	2. Determination of geographical origin of illicit drugs	2. ICP-MS
	3. Monitoring magnesium stearate content or magnesium oxide (used as lubricant, sorbent, respectively, in pharmaceuticals)	3. ICP-AES, flame AA
Mn	1. Detemination of geographical origin of illicit drugs.	1. Graphite furnace AA
	2. Monitor trace metals content of materials (e.g. penicillin G)	2. ICP-AES, ICP-MS
Mo	1. Monitor trace metals content of materials (e.g. penicillin G)	1. ICP-AES
Na	1. Determination of synthetic route and geographical origin of material (e.g. to prevent patent infringement; illicit drugs)	1. Flame AA, ICP-MS
	2. Monitor salt counter-ion concentration of salt content (e.g. in diagnostic agents, in electrolyte replenishing solutions, in cathartics)	2. Flame AA, ICP-AES
Ni	1. Monitor Ni in dialysis solutions	1. Graphite furnace AA
	2. Determination of geographical origin of illicit drugs	2. Graphite furnace AA

Table 1 *Contd.*

Element	Reason for assay/therapeutic area of use	Suggested analytical technique
	3. Monitor trace metals content of materials (e.g. penicillin G)	3. ICP-AES, ICP-MS, graphite furnace AA
P	1. Determine P concentration of vitamins	1. ICP-AES
	2. Determination of geographical origin of illicit drugs	2. ICP-MS
	3. Determination of constituents of materials (e.g. alendronate sodium)	3. ICP-AES
Pb	1. Determination of Pb in calcium supplements	1. ICP-AES, graphite furnace AA
	2. Monitor Pb in dialysis solutions	2. Flame AA
	3. Monitor heavy metals content in medicinal plants.	3. Graphite furnace AA
	4. Determination of geographical origin of illicit drugs	4. ICP-MS
	5. Monitor trace metals content of materials (e.g. penicillin G)	5. ICP-AES
Pd	1. Determination of residual catalyst in pharmaceuticals (e.g. fosinopril, semisynthetic penicillin)	1. ICP-MS, graphite furnace AA
	2. Determination of geographical origin of illicit drugs	2. ICP-MS
Pt	1. Speciation of Pt-containing compounds (cisplatin, transplatin, carboplatin, JM-216)	1. HPLC-ICP-MS, graphite furnace AA
	2. Monitor for residual catalysts	2. ICP-MS, graphite furnace AA
Rh	1. Monitor for residual catalysts used in synthesis of pharmaceuticals	1. ICP-MS
Sb	1. Determination of synthetic route or geographical origin of material (e.g. for illicit drugs, or to prevent patent infringement)	1. ICP-MS
	2. Monitor Sb content in materials (e.g. final drug substances)	2. ICP-MS
Se	1. Monitor for presence of reagents used in synthesis	1. Graphite furnace AA, ICP-MS
	2. Monitor Se concentration in vitamins	2. Graphite furnace AA, ICP-MS
	3. Monitor Se concentration in anti-fungal and anti-seborrhoeic products	3. Graphite furnace AA, ICP-MS
Si	1. Determination of geographical origin of illicit drugs	1. ICP-MS
	2. Monitor for Si contamination from silicone-based compounds used in packaging processes	2. ICP-MS, ICP-AES
	3. Monitor for silica gel (used to prevent caking or as a suspending agent)	3. ICP-AES
Sn	1. Monitor for presence of reagents used in synthesis	1. Graphite furnace AA, ICP-MS
	2. Monitor heavy metals content of materials	2. Graphite furnace AA, ICP-MS
Sr	1. Determination of geographical origin of illicit drugs	1. Graphite furnace AA, ICP-MS
Ti	1. Determine Ti concentration in sunscreens (titanium dioxide is often used in sunscreens)	1. Flame AA
	2. Monitor trace metals content of materials (e.g. penicillin G)	2. ICP-AES
Zn	1. Monitor heavy metals content in medicinal plants	1. Graphite furnace AA
	2. Determine Zn concentration in vitamins	2. ICP-AES
	3. Determination of geographical origin and synthetic route of material (e.g. illicit drugs or to prevent patent infringement)	3. ICP-MS
	4. Monitor trace metals of content of materials (e.g. penicillin G)	4. ICP-AES
	5. Monitor Zn content of materials (e.g. insulin, antibiotics, sunscreens)	5. ICP-AES, flame AA

The speed of FAAS is, undeniably, a tremendous asset of the technique. Sample analysis times of less than 1 min per sample enables the analyst to process numerous samples in a given day by FAAS. FAAS is particularly useful when analysing a trace level analyte in the presence of another metal whose concentration is very high. This situation is encountered when analysing products which incorporate a metal into the drug substance, such as Platinol® (an oncology agent), Prohance® (an imaging agent) and Myochrysine® (a product used for the treatment of rheumatoid arthritis). These products contain high concentrations of platinum, gadolinium and gold, respectively. These elements have very rich spectra, with numerous spectral lines, which may overlap with the spectral lines of the analyte elements. In such cases, FAAS is a better choice for trace metals determinations than ICP-AES, since coincident line overlap is not a problem with the former technique, but presents a considerable problem for the latter.

GFAAS is also commonly found in pharmaceutical company laboratories, owing to the affordability of this spectroscopic instrumentation. GFAAS is ideally suited for the analysis of samples which are available only in small quantity, because it requires considerably less sample for a given analysis than FAAS or any of the plasma-based techniques (e.g. 20 µL per determination, versus 3 mL per determination). Additionally, GFAAS has the ability to remove the sample matrix before atomization of the sample for analyte determination, thus affording the analyst great versatility in the analysis of samples which are composed of rich organic matrices. As with FAAS, GFAAS is also well suited to those pharmaceutical applications where a low concentration analyte is determined in the presence of a high concentration metal.

Second in popularity to atomic absorption based techniques for applications in the pharmaceutical industry is ICP-AES. This instrumentation affords the analyst greater flexibility, with a wider dynamic range and a broader range of elements which can be analysed in a single run. It is often employed to simultaneously determine metals such as Al, Cr, Fe, Mn, Ni, Zn, P, B, Pd and Pt in pharmaceutical matrices. Though FAAS and GFAAS may also be used to monitor these elements, ICP-AES can scan all of these elements in a single analysis (either by scanning, or by the use of a simultaneous unit). In addition, ICP-AES can monitor multiple wavelengths for each element for confirmation of its presence, making it an attractive alternative to either FAAS or GFAAS. The wide linear range of ICP-AES is quite useful in the analysis of pharmaceutical samples, owing to the time saved in developing methods. In addition, the need to make multiple dilutions of a sample is eliminated, as is the need to run multiple standard concentrations within an analysis.

Depending on the stage of development of a pharmaceutical product, the decision for selecting ICP-AES over an AAS technique may simply be the amount of sample available for a set of analyses. The advent of axial ICP-AES systems, with their increased sensitivity, makes ICP-AES an excellent choice where large amounts of sample are not available. The axial ICP-AES system allows the analyst to use considerably less sample than in the past, while achieving the same detection limits and minimum quantifiable limits.

Owing to its ability to monitor multiple wavelengths for a given analyte, and its wide linear range, ICP-AES is well suited for identity testing. An identity test is one in which the analyst is only confirming or denying the presence of a given analyte. In some cases, a compound may have a sufficiently high concentration of a given metal, making it possible to monitor the metal to determine if the compound is authentic. Monitoring of multiple wavelengths is often used to positively confirm the identity of the analyte metal, thus fulfilling the needs of an identity test.

ICP is seeing more use as a sample introduction system for various hyphenated techniques. New to the pharmaceutical industry is the use of inductively coupled plasma-mass spectrometry (ICP-MS). ICP-MS offers excellent versatility and sensitivity to the analyst, and greatly complements any pharmaceutical

atomic spectroscopy laboratory. The sensitivity of the technique and its scanning capabilities make it an ideal choice for the analysis of pharmaceuticals in the early stages of development, when sample material may be in extremely short supply, as the chemists try to optimize and change the synthesis. ICP-MS has been used in our laboratories as an alternative to the United States Pharmacopeia (USP) heavy metals test, providing more accurate, element-specific results for several very toxic metals. The USP test requires a minimum of 1 g of material to perform the nonspecific sulfate ashing procedure. In comparison, the ICP-MS procedure requires only 25 mg and provides element-specific information on 14 different metals. Additionally, ICP-MS is able to examine different isotopes of a given metal present in a sample. This can be quite useful when studying imaging agents, which may be formulated with radioisotopes as part of the desired active ingredient. ICP-MS is very useful in the analysis of trace metals in a matrix containing another metal at high concentrations. As noted before, coincident lines may cause problems for ICP-AES determinations in these cases, and the sensitivity or need for individual lamps may slow or preclude the use of FAAS or GFAAS for these determinations as well. With ICP-MS, the high concentration metal may be 'skipped', while several analyte metals present at trace concentrations may be examined in a single analysis.

All of the techniques discussed have been used to analyse pharmaceutical products which have no chromophores and cannot be analysed by traditional UV-based chromatographic systems. In these cases, metallic complexes are formed with the compounds of interest and then indirectly determined by FAAS, GFAAS, ICP-AES or ICP-MS analysis. This approach can provide valuable information in a short time, one of the chief advantages of spectroscopic techniques when compared with a chromatographic technique, which may take several minutes to an hour per sample analysis. In addition to the situations where a pharmaceutical product is complexed with a metal before analysis by these techniques, FAAS, ICP-AES and ICP-MS are also used in concert with various chromatographic techniques, such as LC-ICP-AES, LC-ICP-MS, IC-ICP-AES, IC-ICP-MS, LC-FAAS and IC-FAAS. The coupling of chromatographic systems with FAAS, ICP-AES and ICP-MS instruments has provided the pharmaceutical analyst with tools which can be used to speciate metallic constituents in drug products, to achieve even lower detection limits, and to examine the different isotopes of metallic constituents present in a sample. Indeed, the sensitivity, flexibility and speed of each of these techniques prove to be valuable in the pharmaceutical industry.

The plasma-based techniques can also serve as detectors for laser ablation (LA) and electrothermal vaporization (ETV). These techniques are well-suited for the analysis of solid samples. ETV can also be used to analyse liquid and slurry samples. Both techniques use small quantities of material and, when interfaced with ICP-MS, are quite sensitive. A cool plasma accessory can also be interfaced with the ICP-MS. This allows for the removal or minimization of interferences caused by the formation of molecular species in the plasma, permitting the determination of Li, Na, Ca, K, Fe and Cr which can not be analysed successfully by conventional ICP-MS. Such analyses exhibit the same sensitivity as afforded by FAAS.

How is an analytical technique selected in a pharmaceutical laboratory?

The stages of the drug development process – some background

The role that atomic spectroscopy plays in the pharmaceutical industry may be directly linked to the various stages of the drug development process. To understand how these techniques might be encountered it is important to examine, in closer detail, what happens during each step of the drug development process.

From the time a potential new drug candidate is identified to the time that it reaches the market it undergoes considerable testing and evaluation. It is imperative that the testing and evaluation of a new drug candidate be completed as quickly as possible, since the pharmaceutical company's patent on a drug has a finite life. The patent gives the pharmaceutical company exclusive rights to the production and sale of the drug once it is approved by the FDA. Once the drug goes off patent, other pharmaceutical companies are allowed to produce a generic form of the

drug. The sale of these generic forms can have a great impact on the sales of the originator company product. One of the goals of drug development is to maximize the length of time in which the company has exclusive marketing rights. It is not uncommon for the sales of a major pharmaceutical product to exceed $1 billion a year at the time the exclusivity period expires. This translates to ~$100 million per month or more. Thus, each month the company can reduce from the development cycle can literally be worth hundreds of millions of dollars.

SmithKline-Beecham's product, Tagamet®, illustrates this well. The earnings from the sale of Tagamet® were £484 million in 1994 (last year of exclusivity). The earnings in 1995, the first full year Tagamet® was off patent, were £286 million, a drop of almost 41%. Therefore, it is beneficial to the company to reduce the time and expense required to get a drug product through the discovery and development phases to market. The steps and the goals of each phase in the development process are quite specific and well defined by the FDA, and are summarized in **Table 2**.

Preclinical (discovery) testing

The earliest stage of drug development is the preclinical or discovery phase. During this phase, a potential drug candidate is identified, and work begins on developing an optimal synthesis. Preliminary assessments are made regarding the safety and biological activity of the potential drug candidate in laboratory and animal studies. At this stage of development, the synthetic chemist has very little experience with the molecule and may utilize exotic catalysts to produce the first few grams of the materials. Analyses of preliminary batches of the drug candidate and key intermediates are performed to ensure the preliminary safety data is reflective of the drug candidate and not impurities generated by the synthetic process. Since

Table 2 The phases of the drug development process

Step in the drug development process	Goal/objective of the step in the process
Preclinical (discovery) phase	Assess the drug's safety and biological activity in the laboratory and animal studies
Phase I clinical trials (IND phase)	Establish the safety and bioavailability of the compound in humans. This is typically done in studies using healthy volunteers
Phase II clinical trials (IND phase)	Determine proof-of-principle for the drug's mode of action. Monitor for any possible side effects and evaluate the drug's effectiveness. This study uses patient volunteers who have the disease / condition for which the drug is targeted
Phase III clinical trials (IND phase)	Establish dose form and dosage strength for registrational filing. Continue to monitor possible side effects and adverse reactions. Verify the effectiveness of the drug in the targeted patient population. This study involves many more patient volunteers than the Phase II study
FDA review/approval process	The FDA reviews the new drug application (NDA). If it is fully approved, the drug may proceed to market. If the NDA is not fully approved, the FDA may require additional testing, answers to questions or it may reject the application

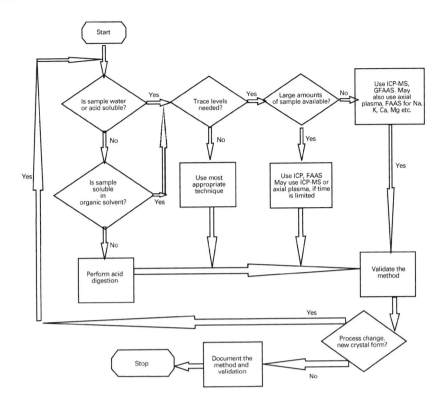

Figure 1 Flow chart of method development decision-making process. Reproduced with permission of the editor from *Atomic Spectroscopy: Pharmaceutical Applications of Atomic Spectroscopy* **12**(9): 14–23 (1997) published by Advanstar Communications.

patents have finite lifetimes, time is of the essence, especially at this early stage, when the drug has not yet been evaluated in man. During this phase of development the salt and/or crystal form of the drug substance may not have been selected. As a result, samples analysed can vary in solubility properties, pH, purity, etc. Analyses that are typically performed include counterion, trace metals (from equipment sources, e.g. stainless steel) and trace catalyst determinations.

The flow chart given in **Figure 1** illustrates the thought processes involved in the selection of a given analytical technique for the determination of metals in pharmaceutical related samples.

The atomic spectroscopist is typically involved in supporting a new potential drug candidate before the final salt and/or crystal form have been selected. Several forms of the drug substance (salt forms, and/or polymorphs) are considered during the discovery phase and are generated in small laboratory batches via several synthetic pathways or crystallization procedures. The atomic spectroscopy laboratory plays an important role in the selection of the final form by assaying these samples for trace metals, salt counter-ions and trace catalysts used in the syntheses. Once a final form has been selected, testing continues to support the optimization of the synthetic process.

The selection of an appropriate analytical technique is highly dependent upon the time constraints that pharmaceutical companies set for the complete development of a drug product. As the costs of developing drugs has risen, the push within the pharmaceutical industry has been to reduce the time from discovery to the clinical studies as much as possible. This is driven by the fact that somewhere between one in seven and one in ten potential drug candidates in development actually make it to market. Therefore, results on preclinical samples are usually required in a short time (a few days, or hours) so that refinements to the synthetic process can be made, if necessary. The technique that is chosen for the assay must be rapid, meet the sensitivity requirements and generally consume small quantities of material. To expedite the analysis, it is prudent to perform as many determinations as possible in one assay. ICP-AES and ICP-MS are well suited for this type of determination; however, they are not ideal for all trace metals. The elements Na and K must be analysed by FAAS unless they are present at concentrations high enough for ICP-AES.

Once the final form has been selected and the synthesis refined, methods are developed and validated for metals that are present in reagents used in the synthesis, metals that may arise from the equipment used in the synthesis and metals that are incorporated into the active ingredient in the final drug product. Validated methods are often required for synthetic intermediates as well. In addition, the FDA requires that a method be validated for the determination of heavy metals (i.e. lead, mercury, etc.) in the final drug substance. The USP heavy metals test requires one gram of sample for each determination. This method is non-specific and is based on a sulfate ashing of the sample, followed by a colorimetric comparison with a lead solution standard. Since this much material is usually not available in this phase of drug development, ICP-MS has been demonstrated to be an excellent technique for the determination of heavy metals in early development drug candidates. ICP-MS offers element-specific information and utilizes substantially less sample.

In addition to support of the drug substance, analyses are performed on starting materials and any raw materials used in the last step of the synthesis. Small batches of the drug substance which will be used in animal toxicology and pharmacokinetic studies must also be analysed. Once a synthetic process and final form have been selected, an investigational new drug (IND) application is filed with the FDA.

Clinical (IND) phase

The IND contains information regarding the drug's composition and synthesis and lists all specifications that have been set for the drug substance. Specifications are set for many tests, which may include trace metals. All subsequent batches of the drug that will be used in clinical studies must meet these specifications before their release. The IND contains information regarding animal toxicology study data and protocols for clinical trails. The IND clinical study protocol for a new drug candidate consists of three clinical phases (**Table 2**).

Optimization and refinement of the synthetic process continues during the IND phase. The synthetic chemists scale up the synthesis to produce kilogram size batches. Support of this stage of drug development is similar to that performed to support synthesis optimization on small laboratory batches during the preclinical phase. Validated methods must be refined as the synthesis is refined, because even the slightest change in the synthesis can have a profound effect on whether a previously validated method will continue to be adequate for the trace metal determination. The use of a different solvent or reagent can be sufficient to invalidate a method.

During the preclinical stage of drug development, speed and sample consumption are typically the most important factors when selecting a technique; however, as the compound moves through the clinical phase of development there are other factors to consider. First, the atomic spectroscopist must consider the analytes of interest and the sensitivity that is required. Speed is still an important issue; however, sample consumption is less of a critical factor, since batch sizes of several hundred grams to several kilograms are routinely being produced. One must consider whether the method will be transferred to a QC laboratory since the instrumentation within their laboratory will often dictate which spectroscopic technique is used. If a QC laboratory will be performing the analysis, then usually either FAAS or GFAAS will be preferred since this instrumentation is typically found in QC laboratories, owing to the lower cost, compared with plasma-based instrumentation. This poses a challenge when one requires the sensitivity of GFAAS but must dissolve the drug substance in an organic solvent that is too viscous for GFAAS systems to handle. ICP-AES or ICP-MS would be the ideal alternative, but, most QC laboratories cannot afford such instrumentation.

Once the development of the drug candidate passes into the clinical Phase II and III studies, the demand for bulk substance increases. The synthesis is scaled up in the pilot plant to batch sizes ranging from ten to several hundred kilograms, and eventually to final production size batches of the final drug substance. The atomic spectroscopist will sometimes be called upon to help troubleshoot the process during the scale-up. Troubleshooting samples come in a variety of forms: discoloured drug substance or intermediate; scrapings from the equipment used in the synthesis; reagents used in the synthesis; filters used in the synthesis; liquid streams from the processing or slurries that were produced owing to a malfunctioning of the equipment. Sometimes the chemist will have an idea as to why the process failed and can help narrow down the investigation for the analyst. The cause of a process excursion can range from the use of a reagent contaminated with metals to equipment failure, such as a lubrication oil or coolant leak or the corrosion of the stainless steel equipment by the reaction by-products. ICP-MS is an excellent tool for assessing the problem quickly by performing qualitative or semi-quantitative scans of the periodic table. If these scans indicate that any metals are present at concentrations high enough for concern (several parts per million), alternative techniques, such as ICP-AES or FAAS, are used to confirm and quantitate their presence in the sample. Usually, but not always, sample consumption is not of great concern, but the speed of the technique is critical, since the chemist cannot proceed with the processing of the batch(es) until the source of the problem is identified. In analysing oils, contaminated filters, discoloured drug substance or intermediates, the analyst will often focus on the possible presence of wear metals from lubricating oils (Al, Cr, Cu, Fe, Pb, Sn and Mo), metals from coolant contamination (Na, K and B) and metals found in stainless steel (Ni, Cd, Pb, Al, Fe, Cr, Cu, Mn and Zn). ICP-AES is sensitive enough for most of these metals and, because it is capable of multielement analyses, it is also rapid enough to satisfy the short turn-around-times required for processing these samples. Na and K must be assayed by FAAS unless they are present at high enough concentrations for ICP-AES.

LA-ICP-MS can be used to quickly analyse solid samples, such as filters or scrapings from the equipment. A minute amount of sample is consumed and analysis is fast, as no sample preparation is required. LA-ICP-MS is especially useful when analysing solid samples with distinct discolourations, since the laser can be focused on the area of interest to increase sensitivity. Each discoloured area can be ablated and assayed separately to determine its metallic composition. Often, these areas are caused by contamination from an oil or coolant that has leaked from the equipment. This affords the spectroscopist great selectivity over a conventional dilute-and-shoot method in which the small discoloured areas cannot be analysed separately.

Before the introduction of cool plasma ICP-MS, qualitative ICP-MS scans did not provide accurate information on Li, Na, K, Ca, Cr and Fe (owing to spectral interferences or the element being easily ionized). Therefore, ICP-AES or FAAS assays were required for accurate information on these elements. The advent of cool plasma ICP-MS makes it possible to quickly analyse all metals using only one spectroscopic technique, which is important when only a small amount of sample is available.

NDA phase (Phase IV)

In the final stage of drug development, a new drug application (NDA) is filed with the FDA. Animal and clinical studies continue during the NDA phase. Stability tests of the drug substance and product continue, including studies of the commercial formulation in the market packaging. The spectroscopy laboratory supports this stage of development by performing analyses that are included in the specifications that have been set for the drug substance and product, using the methods filed with the NDA. The spectroscopist may see samples during this phase that are generated when the process of the drug substance or product is transferred to a new production facility. The steps that are taken in selecting and using a spectroscopic technique are the same as those described in the previous section.

Table 3 Examples of pharmaceutical compounds which contain metals

Metal	Examples of uses in pharmaceutical compounds
Na, Mg, Ca, K	Used in various excipients, in vitamins, in dialysis solutions and Eye Stream®, a liquid used for irrigating eyes, contains Na, Mg and K
Pt	Used in several oncology drugs: Paraplatin®, Platinol® and Cisplatin®
Zn	Used in Insulin and in Cortisporin® ointment (a steroid–antibiotic ointment)
Li	Used in Lithobid® and Cibalith-S®, both of which are used for the treatment of manic-depressive psychosis
Al	Often used in antacid preparations, such as AlternaGel™ or Mylanta®
Ag	Used as a topical antimicrobial for the treatment of burns in Silvadene® cream, 1%
Au and Na	Used in Myochrysine® injection, for the treatment of rheumatoid arthritis
Fe	Active ingredient in Chromagen®, a drug used in the treatment of anaemia. Also, sometimes used in pigments for printing tablets or capsules
Se	Selsun Blue®, a dandruff shampoo
Mn	Active ingredient in LumenHance®, an imaging agent

The last stage in this phase consists of FDA inspections and a review of the NDA. During these inspections, the auditors may examine instrument calibration records and previous batch results to ensure that they were collected under Good Laboratory Practices (GLP) and/or Good Manufacturing Practices (GMP). Based on the inspection, the FDA will either approve the new drug candidate, require additional testing, request answers to questions and concerns they have or reject the drug product. Some examples of pharmaceutical compounds that contain metals are given in **Table 3**.

The role that the atomic spectroscopy laboratory plays in the drug development process is an important one. It helps ensures the safety and quality of the drug products that are approved by the FDA.

See also: **Atomic Absorption, Methods and Instrumentation; Atomic Absorption, Theory; Atomic Emission, Methods and Instrumentation; Biomedical Applications of Atomic Spectroscopy; Forensic Science, Applications of Atomic Spectroscopy; Hyphenated Techniques, Applications of in Mass Spectrometry; Inductively Coupled Plasma Mass Spectrometry, Methods; Inorganic Chemistry, Applications of Mass Spectrometry.**

Further reading

Ali SL (1983) Atomic absorption spectrometry in pharmaceutical analysis. *Journal of Pharmaceutical and Biomedical Analysis* 1: 517–523.

Lewen N, Schenkenberger M, Larkin T, Conder S and Brittain H (1995) The determination of palladium in

Fosinopril sodium (Monopril) by ICP-MS. *Journal of Pharmaceutical and Biomedical Analysis* 13: 879–883.

Lewen N, Schenkenberger M, Raglione T and Mathew S (1997) The application of several atomic spectroscopy techniques in a pharmaceutical analytical research and development laboratory. *Spectroscopy* 12: 14–23.

Ma TS (1990) Organic elemental analysis. *Analytical Chemistry* 62: 78R–84R.

(1992) *Physician's Desk Reference*, 46th edn. Medical Economics Data, a division of Medical Economics Company. 1992.

Rousselet F and Thuillier F (1979) Atomic absorption spectrometric determination of metallic elements in pharmaceutical products. *Progress in Analytical Atomic Spectroscopy* 1: 353–372.

Schulman SG and Vincent WR (1984) Atomic spectroscopy (in pharmaceutical analysis). *Drugs Pharmaceutical Science* 11: 359–399.

Taylor A, Branch S, Crews HM and Halls DJ (1993) Atomic spectroscopy update – clinical and biological materials, foods and beverages. *Journal of Analytical Atomic Spectrometry* 8: 79R–149R.

United States Pharmacopeial Convention (1975). *The United States Pharmacopeia*, nineteenth revision.

http://www.searlehealthnet.com/pipeline.html

http://www.allp.com/drug _dev.html

Photoacoustic Spectroscopy, Applications

Markus W Sigrist, Swiss Federal Institute of Technology (ETH), Zurich, Switzerland

> **ELECTRONIC SPECTROSCOPY/
> VIBRATIONAL, ROTATIONAL &
> RAMAN SPECTROSCOPIES**
> **Applications**

Introduction

In conventional absorption spectroscopy the measurement of absorption is transferred to a measurement of the radiation power transmitted through the sample. On the contrary, in photoacoustic spectroscopy, the absorbed power is determined directly via its heat and hence the sound produced in the sample. Photoacoustics, also known as optoacoustics, was pioneered by AG Bell, in 1880. The photoacoustic (PA) effect concerns the transformation of modulated or pulsed radiation energy, represented by photons, into sound. In general, two aspects have to be considered: first, the heat production in the sample by the absorption of radiation; and secondly, the resulting generation of acoustic waves. Closely related to the PA effect are photothermal (PT) phenomena which are caused by the original heating via absorption of radiation. While the PA effect is detected via acoustic sensors such as microphones, hydrophones or piezoelectric devices, the PT phenomena are sensed via the induced changes of the refractive index of the media by probe beam deflection, thermal lensing or, also, PT radiometry. Both PA and PT spectroscopy are widely used today in many applications. Experimental aspects are outlined in a separate article while this article discusses the main characteristics of this spectroscopic tool. The great potential is illustrated with examples from applications on solids, liquids and gases as well as in life sciences.

Spectroscopic applications

PA and PT phenomena are widely used for numerous non-spectroscopic applications such as the determination of thermal diffusivity, non-destructive testing of materials (in particular the probing of sub-surface defects) by thermal wave imaging, time-resolved studies of de-excitation processes or on biological photoreceptors, studies of phase transitions, etc. Here, only spectroscopic applications are considered that demonstrate the main characteristics and the potential of photoacoustic spectroscopy (PAS). In the following, illustrative examples are presented for solids, liquids, gases, biological and medical samples.

Studies on solids

A main advantage of PAS applied to solids is the fact that no elaborate sample preparation is required and unpolished sample surfaces pose no problems. Since the PA signal is proportional to the *absorbed* energy, even spectra of strongly scattering samples, e.g. powders, can easily be measured. However, it should be mentioned that owing to the complex nature of the signal generation involving interstitial gas expansion, etc., PA studies on powders are usually only qualitative. Another advantage is the high sensitivity that is achieved because the PA detection is a null method for measuring absorption. Hence, absorbances as low as 10^{-7} can be detected. For modulated radiation a simple theoretical model has been developed which is based on the fact that the acoustic signal is due to a periodic heat flow from the solid to the surrounding gas, as the solid is cyclically heated by the absorption of the chopped light. Six different cases are distinguished, depending on the optical and thermal properties of the solid samples. This allows the unique feature of measuring totally opaque materials which is impossible by conventional transmission measurements. Hence PAS is technique to study weak bulk and surface absorption in crystals and semiconductors, to evaluate the level of absorbed energy in thin films, to measure the spectra of oxide films in metals, various powders, organic materials, etc. and also to investigate multi-layered samples.

An early example is shown in **Figure 1** for the insulator Cr_2O_3. Spectrum (A) depicts the normalized PA spectrum of Cr_2O_3 powder in the 200 to 1000 nm region. In comparison spectrum (B) shows an optical absorption spectrum obtained on a 4.4 µm thick bulk crystal, taken parallel and vertical to the crystal c-axis whereas spectrum (C) represents a diffuse reflection spectrum of Cr_2O_3 powder. The advantage of PAS is obvious in that the two crystal-field bands of the Cr^{3+} ion at 460 and 600 nm are almost as clearly resolved in the PA spectrum of the powder as they are in the crystal spectrum, and substantially better resolved than in the diffuse reflectance spectrum. It should be noted, however, that the theoretical description of the PA effect in strongly scattering media is not straightforward and quantitative data are therefore difficult to determine from such spectra.

Another example concerns adsorbates on the surfaces of solids. PAS is expected to be rather sensitive to surface adsorption, especially if the substrate is transparent or highly reflective in the wavelength region in which the adsorbate absorbs. Both sinusoidal modulation of the incident laser beam and pulsed lasers have been used for this purpose. An interesting version is the modulation of the laser beam polarization to suppress the background signal that originates from substrate absorption. A fraction of only 0.005 of a monolayer of ammonia (NH_3) adsorbed on a cold silver substrate in ultrahigh vacuum was detected. An example is shown in **Figure 2** where the PT signal is recorded as a function of time as ammonia is slowly admitted to the system and condenses

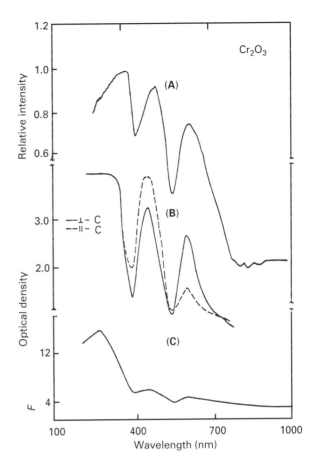

Figure 1 (A) Normalized PA spectrum of Cr_2O_3 powder, (B) optical transmission spectrum of a 4.4 µm thick Cr_2O_3 crystal, (C) diffuse reflectance spectrum of Cr_2O_3 powder. All spectra were taken at 300 K. Reproduced with permission of Academic Press from Rosencwaig A (1977). In: Pao Y-H (ed) *Optoacoustic Spectroscopy and Detection*. New York: Academic Press.

on the silver substrate. The signal of a microbalance as indicator of molecular coverage is monitored simultaneously. Later studies were aimed at investigating the kind of adsorption in more detail, e.g. to differentiate between chemisorption and physisorption, by combining the high spectral resolution and high sensitivity offered by pulsed laser PAS.

In other studies, the wide free spectral range offered by Fourier transform infrared (FTIR) spectroscopy combined with the step-scan methods has been increasingly applied in conjunction with PA detection for infrared spectral depth profiling of laminar and otherwise optically heterogeneous materials. IR spectra that are often unavailable by use of other techniques become accessible from samples that are strongly absorbing or even opaque, from strongly light-scattering samples and from samples in situ. The scheme is also applied as an analytical tool for chemical characterization and quantification, e.g. of

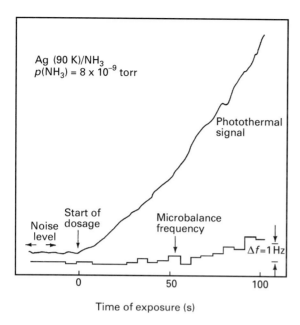

$Ag (90 K)/NH_3$
$p(NH_3) = 8 \times 10^{-9}$ torr

Photothermal
signal

Noise
level

Start of
dosage

Microbalance
frequency

$\Delta f = 1$ Hz

0   50   100

Time of exposure (s)

Figure 2 Photothermal signal and microbalance record versus exposure time as ammonia molecules are slowly adsorbed on a silver substrate. The maximum coverage is 0.8 monolayers, the ammonia partial pressure in the system is 8×10^{-9} torr. The noise level (left) indicates the signal from the clean substrate. Reproduced with permission of Elsevier from Coufal H, Trager F, Chuang T and Tam A (1984) *Surface Science* 145: L504.

polychlorinated biphenyls (PCBs) in industrial waste management such as PCB contamination of soils.

Finally, the available spectral range for PA studies on solids has been extended to the X-ray region by using hard X-rays from synchrotron radiation. As an example, the X-ray absorption near Cu K-edge regions has been measured on copper (Cu), Cu alloys (brass) and Cu compounds (CuO, Cu_2O and $CuInSe_2$) with a PA detector and compared with the usual X-ray absorption (10 μm thick Cu and brass foils and < 50 μm thick powdered samples of CuO, Cu_2O and $CuInSe_2$ put on Scotch tape were used as specimens). It was found that the energy peak values derived from the PA spectra agree with those deduced from optical density spectra, suggesting that the heat production processes are also reflected in the absorption spectra. A more detailed insight is obtained by dividing the PAS data by the optical density data, i.e. by forming the ratios PAS:$\log(I_0/I_t)$, which are proportional to the heat production efficiency. In **Figure 3** these ratios are plotted for Cu, Cu_2O and $CuInSe_2$ versus the photon energy near the K-edge of Cu. The results clearly indicate differences between X-ray absorption and PA spectra and hence imply a spectral variation of the heat production efficiency. Obviously, the heat production process is also different in Cu_2O compared with the other Cu compounds.

Studies on liquids

Experimental and theoretical PA and PT studies on liquids comprise a wide absorption range from 'transparent' to opaque liquids.

For investigations on weakly absorbing media a flash-lamp-pumped dye laser with pulse energies of 1 mJ was used as excitation source and a submersed piezoelectric transducer for detecting the generated acoustic signals. The high sensitivity permits, e.g. the recording of the water spectrum in the visible range where accuracies of other techniques such as long-path absorption measurements are often limited. Another example concerns the study of weak overtones of the C–H stretch absorption band of hydrocarbons up to the 8th harmonic. In **Figure 4** the absorption band of the 6th harmonic at 607 nm of benzene dissolved in CCl_4 is plotted for different dilution ratios of benzene. With increasing dilution, the absorption peak is obviously blue-shifted and both the line width and the line asymmetry decrease. These and other results demonstrate that PAS permits the measurement of minimum absorption coefficients of 10^{-6} cm^{-1}, corresponding to absorbed laser pulse energies of only 1 nJ.

Another field of interest concerns analytical investigations on pollutants in liquids. Detection limits in the sub-ppb range were achieved by PAS, e.g. for carotene or cadmium in chloroform or for pyrene in heptane. More recently, pesticides in aqueous solutions have attracted interest. Different experimental arrangements with pulsed or CW pump lasers and various PA and PT lens detection schemes were used in these studies. Limits of detection are down to below 10^{-6} cm^{-1}, corresponding to ppb concentrations. An example is presented in **Figure 5** where the calibration curves for the detection of the dinitrophenol herbicide DNOC in aqueous solutions are compared with the untreated standard solution. The techniques used involved PT techniques, namely PT deflection spectroscopy (PDS, **Figure 5A**), thermal lensing (TL, **Figure 5B**), PT interferometric spectroscopy (PIS, **Figure 5C**), PAS (**Figure 5D**) and a conventional spectrophotometer (Cary 2400, **Figure 5E**). Obviously, the detection limit of the spectrophotometer in the low ppb (μg kg^{-1}) range is exceeded by the PA and PT methods. In particular, TL and PDS appear superior in the determination of environmental pollutants. It should be noted that both US and EU standards require detection limits of 0.1 μg L^{-1} for pesticides in drinking water.

On the other end of the scale are opaque or strongly absorbing liquids. PA spectroscopy offers the great advantage that absorption coefficients that are two to three orders of magnitude higher than is

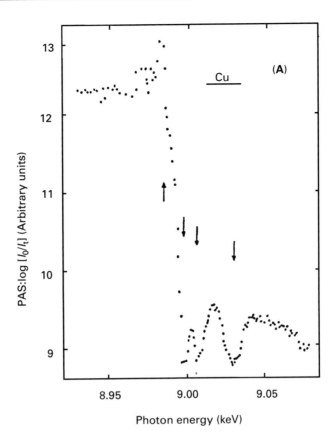

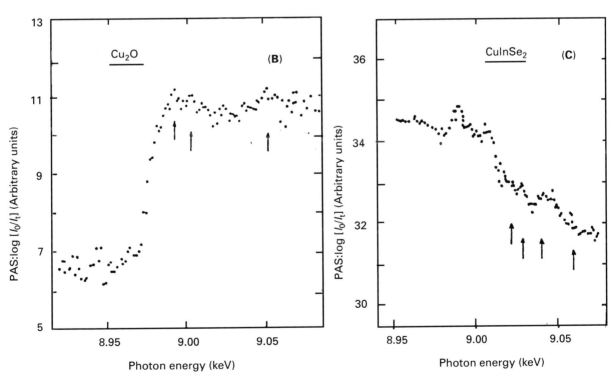

Figure 3 X-ray PA spectra normalized with optical transmission spectra (PAS : $\log I_0/I_t$), where I_0 and I_t denote the incident and transmitted intensity, respectively, at the K-edge region for different copper compounds. (A) Pure Cu, (B) Cu_2O and (C) $CuInSe_2$. Reproduced with permission of IGP AS, Trondheim, Norway from Toyoda T, Masujima T, Shiwaku H and Ando M (1995) *Proceedings of the 15th International Congress on Acoustics*, Vol I, 443.

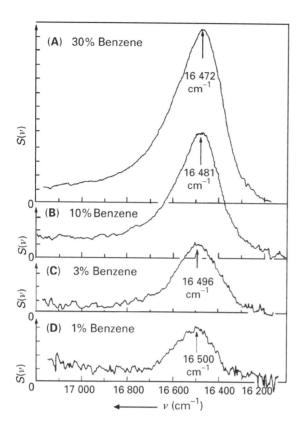

Figure 4 PA spectra of the 6th harmonic absorption of the C–H bond of benzene dissolved in carbon tetrachloride (CCl_4) in arbitrary linear units, as the volume dilution ratios indicate. The positions of the absorption peaks are given. Reproduced with permission of the Optical Society of America (OSA) from Tam AC, Patel C and Kerl R (1979) *Optics Letters* **4**: 81.

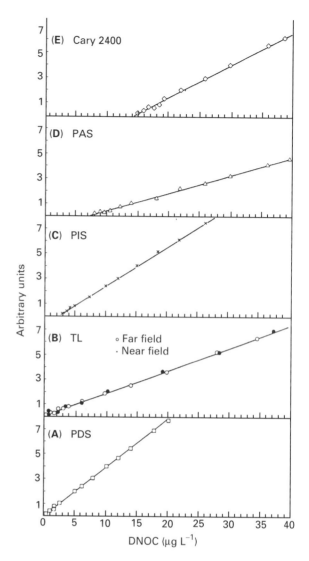

Figure 5 Calibration curves for the dinitrophenol herbicide DNOC in aqueous solution when using different techniques: (A) PDS: Photothermal deflection spectroscopy, (B) TL: thermal lensing, (C) PIS: Photothermal interferometric spectroscopy, (D) PAS: a photoacoustic spectroscopy, (E) a conventional spectrophotometer Cary 2400. Reproduced with permission of SPIE from Faubel W (1997) Detection of pollutants in liquids and gases. In: Mandelis A and Hess P (eds) *Life and Earth Sciences*. Progress in Photothermal and Photoacoustic Science and Technology, Vol III, Chapter 8. Bellingham: SPIE.

accessible by conventional transmission spectroscopy can be determined without difficulties. Various schemes have been proposed for this case including the optothermal window. As example, the *trans*-fatty acid (TFA) content of margarine was determined using a CO_2 laser and the optothermal window. Good agreement with alternative techniques such as FTIR, gas–liquid chromatography and thin-layer chromatography was obtained.

Studies on gases

Early PAS studies on gases had already demonstrated the high sensitivity that is achieved with a rather simple setup and have subsequently favoured further developments in trace gas monitoring. In comparison with conventional optical absorption measurements, PAS offers the following main advantages: (i) only short pathlengths are required which enables measurements at wavelengths outside of atmospheric transmission windows, (ii) the microphone as detector represents a simple room-temperature device with a wavelength-independent responsivity, (iii)

scattering effects are less important, and (iv) the dynamic range comprises at least five orders of magnitude. Measurements are generally performed with the gas either contained in or flowed through a specially designed PA cell. Typically, a minimum detectable absorption coefficient α_{min} of the order of 10^{-8} cm^{-1} atm^{-1}, corresponding to ppb (10^{-9}) concentrations, i.e. densities of μg m^{-3}, is achieved with laser-based setups. At the cost of dynamic range this limit can be lowered further to the <100 ppt range by

Table 1 List and detection limits of selected gas species monitored by laser PAS under interference-free conditions

Species	Type of laser	Spectral region (μM)	Detection limit [ppb]
Formic acid	Dye	220[a]	140
Sulfur dioxide	Dye	290–310[a]	0.12
Formaldehyde	Dye	303.6[a]	50
Nitrogen dioxide	Kr+	406.8[a]	2
Methane	DF	3.8	Few
Nitrous oxide	DF	3.8	Few
Carbon monoxide	$PbS_{1-x}Se_x$	4.6	40
Nitric oxide	CO–SFR	5.3	<0.1
Phosgene	CO	5.45	Few
Acetaldehyde	CO	5.66	3
Carbon disulfide	CO	6.48	0.01
Ethane	CO	6.7	1
Pentane	CO	6.8	0.1
Trimethyl amine	CO	6.93	10
Dimethyl sulfide	CO	6.95	3
Acetylene	CO	7.2	1
Hydrazines	CO_2	9–11	<10
Freons	CO_2	9–11	< 4
Explosives	CO_2	9–11	0.2–25
Ammonia	CO_2	9.22	0.4
Ethanol	CO_2	9.46	17
Ozone	CO_2	9.50	13
Methanol	CO_2	9.68	5
Ethylene	CO_2	10.53	0.3
Sulfur hexafluoride	CO_2	10.59	0.01
Vinyl chloride	CO_2	10.61	20

[a] In nm.

operating the PA cell intracavity. **Table 1** lists some gaseous compounds, laser sources used and (extracavity) detection limits achieved.

In practice, one usually deals with multicomponent samples. The analysis is done on the basis of the individual spectra and measurements performed at properly selected wavelengths to reduce absorption interferences. Apart from the PA signal amplitude the PA phase yields additional information for the analysis. A broad tuning range of the laser source and a narrow line width are advantageous for obtaining a high selectivity which is further enhanced for species with well structured spectra.

Most PA studies on trace gases have been devoted to laboratory investigations on collected air samples of different origin such as vehicle exhausts or industrial emissions. If the temporal evolution of the gas composition is of interest, the air is flowed continuously through the PA cell and the laser is switched repeatedly between appropriate wavelengths that are characteristic for the absorption of

the gases to be recorded. In addition to laboratory analyses field studies yielding temporally and spatially resolved data on ambient trace gas concentrations and on their distributions are required to obtain a profound knowledge of atmospheric chemistry as well as of emission processes. Unlike lidar or long-path absorption measurements in the open atmosphere, PA schemes are not suited for remote studies but are applied to in situ measurements for the simultaneous monitoring of various compounds. Apart from non-laser based PA gas sensors only a few mobile laser PA systems have been operated so far. Examples include a balloon-borne system equipped with a spin-flip Raman CO laser PA spectrometer that was employed for recording stratospheric diurnal NO concentration profiles. Furthermore, a waveguide CO_2 laser PA system was applied successfully to in situ measurements at a power plant where the commonly used scheme for the reduction of the emission of nitric oxides (NO_x) by injection of ammonia (NH_3) into the combustion process was to be tested. This required a reliable, fast and selective monitoring of NH_3 down to the 1 ppm level under rough measurement conditions. Studies of this type at power-, incineration- or industrial-plants play a key role for the evaluation of the pollution, for the emission control and the testing of remedial strategies.

During the last few years a fully automated CO_2 laser PA spectrometer, which is installed in a trailer, has been developed and which is operated unattended for longer time periods. The computer control ensures a proper laser wavelength selection with a long-term frequency stability of 10^{-3} cm^{-1} for ~70 laser transitions between 9.2 and 10.8 μm using $^{12}C^{16}O_2$ or for ~65 transitions between 9.6 and 11.4 μm with a $^{13}C^{16}O_2$ laser tube. The resonant PA cell is connected to a gas flow system and the air to be analysed is pumped continuously through the cell at atmospheric pressure and with a flow rate of typically 0.5–1 L min^{-1}. The air is not pretreated by any means except for measurements in dusty environments where a micropore filter is inserted into the gas stream at the air inlet. The humidity and CO_2 monitor in the air stream allow independent measurements of water vapour and CO_2 concentrations for comparison. Furthermore, the trailer is equipped with meteorological devices for wind and solar irradiance measurements. Hitherto, we have applied this mobile system to industrial stack emission sensing and ambient air monitoring in urban and rural environments. The stack emission measurements demonstrated the good time resolution and the high selectivity that are achieved in multicomponent analyses. In certain cases it could even be differentiated between isomers, e.g. between o- and m-dichlorobenzene, among various other compounds, mostly VOCs, at ppm concentration levels. The selectivity is determined by the compound to be measured, and the tuning characteristics and bandwidth of the laser source.

The air in urban environments often contains numerous pollutants with rather high and varying concentrations. More recently, the mobile system was used in a harsh and noisy environment at the exit of a freeway tunnel to record gases emitted by road traffic during one week. The polluted air, filtered by a Teflon dust filter with a porosity of 1 μm, was flowed continuously through the PA cell and the laser was tuned sequentially to wavelengths characteristic for ammonia (NH_3), ethylene (C_2H_4) and CO_2 absorption as well as to reference wavelengths with no appreciable absorption by these compounds. Concentration profiles of these three species could thus be recorded almost simultaneously with a time resolution of 10 min. As **Figure 6** shows, the temporal concentration data are clearly correlated with the independently monitored CO concentration and the traffic density. Rather high concentrations are recorded, particularly for ammonia which are most probably caused by the majority of those cars that are equipped with catalytic converters. Based on the gas concentrations and the calculated air flow through the tunnel the corresponding emission factors (mass of an exhaust component per vehicle and kilometre) were determined. These factors are 15 mg km^{-1} for ammonia, 26 mg km^{-1} for ethylene,

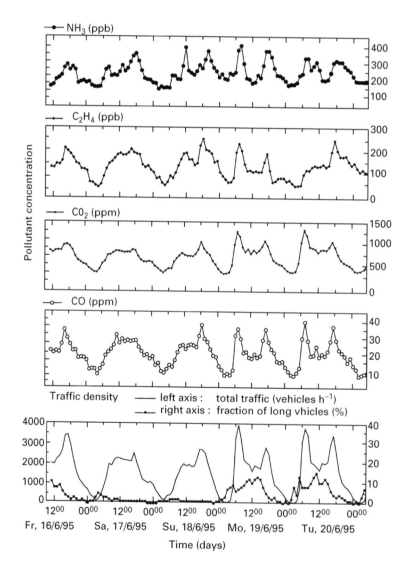

Figure 6 Temporal concentration profiles of four air pollutants during 5 days taken at the exit of a freeway tunnel. The NH_3, C_2H_4 and CO_2 were measured photoacoustically, CO was recorded with a commercial IR gas analyser. Total traffic density and fraction of long vehicles are recorded at the bottom. Reproduced with permission of the American Chemical Society from Moeckli M, Fierz M and Sigrist M (1996) *Environmental Science and Technology* **30**: 2864.

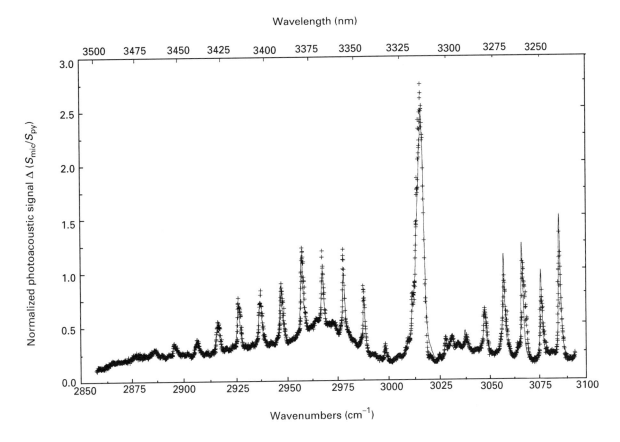

Figure 7 Spectrum of a gas mixture containing methane, methanol, ethanol, isopentane, benzene and toluene, all at ppm concentrations, buffered to 960 mbar total pressure with synthetic air (80% N_2, 20% O_2). The measured spectrum (+) was taken photoacoustically with a difference frequency laser spectrometer based on an optical parametric oscillator (OPO) with a line width of 0.2 cm^{-1}. Excellent agreement is obtained with the superimposed fitted spectrum (——) using the HITRAN database for methane and the previously recorded reference spectra for the other substances. Reproduced with permission of Elsevier from Bohren A and Sigrist M (1997) *Infrared Physics and Technology* **38**: 423.

5.2 g km^{-1} for CO and 201 g km^{-1} for CO_2. Such data are valuable for the estimation of the total annual emission of certain compounds from road traffic and their fractions of the total load in certain area.

Ambient monitoring in rural air requires detection schemes with very high sensitivity. So far, studies of PA sensing have concentrated on a few gases. An example concerns the in situ recording of ambient NH_3, H_2O vapour and CO_2 in a heath in the central Netherlands for several months in 1989. The mobile system was applied to measurements in a rural location in central Switzerland where H_2O vapour, CO_2, NH_3, O_3 and C_2H_4 were recorded simultaneously with a time resolution of 10 min by using nine carefully selected CO_2 laser transitions.

A key issue in analysing multicomponent samples is the detection selectivity, which is strongly influenced by the tuning characteristics and line width of the source as well as by the measuring conditions themselves (e.g. reduced gas pressure). Recent laser developments can enhance the performance substantially.

A continuously tunable narrowband high pressure CO_2 laser has, for example, enabled the analysis of a mixture of six CO_2 isotopes. In one study, presented in **Figure 7**, a mixture of six hydrocarbons at ppm concentrations has been analysed by an all-solid-state laser PA spectrometer.

Studies in life sciences

The inherent high light scattering and the often strongly varying depth structure render biological and medical samples rather difficult for investigations with conventional spectroscopic tools. As various researchers, however, have demonstrated, PA and PT techniques can successfully by applied to media such as skin tissue, blood or plants. *In vivo* studies were performed on human skin with specially designed PA cells that allowed the study on living skin by avoiding the noise induced by pulsating blood. In particular, the absorption of UV light by protein (α-keratin) and the application of sunscreens to protect skin from UV damage were studied spectroscopically by the PA

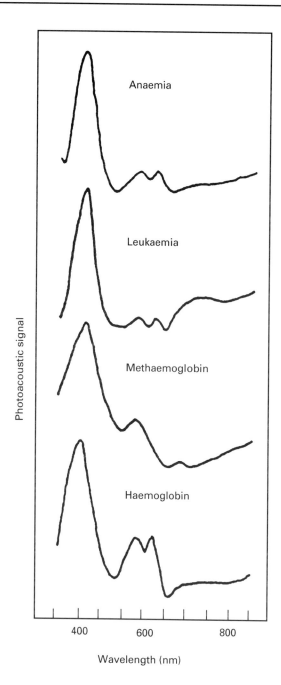

Figure 8 Photoacoustic spectra of haemoglobin and of the blood from anaemia, leukaemia and methaemoglobin patients. Reproduced with permission of Springer from Pan Q, Qui S, Zhang S, Zhang J and Zhu S (1987) Springer Series in Optical Sciences, Vol 58, 542.

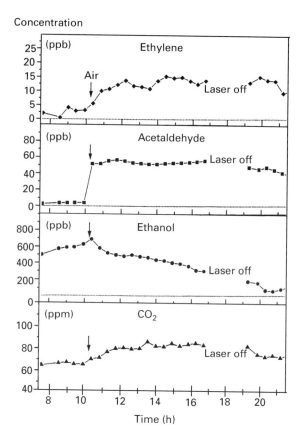

Figure 9 Temporal evolution of the emission of different gases (ethylene, acetaldehyde, ethanol and CO_2) from a cherry tomato subjected to a change from anaerobic to aerobic conditions at $t = 10.2$ h (indicated by the arrow). The emission of a reference fruit kept under aerobic conditions was subtracted to obtain the data in this figure. Reproduced with permission of SPIE from Harren F and Reuss J (1997). Applications in plant physiology, entomology, and microbiology; gas exchange measurements based upon spectral selectivity. In: Mandelis A and Hess P *Life and Earth Sciences*. Progress in Photothermal and Photoacoustic Science and Technology, Vol III, Chapter 4. Bellingham: SPIE.

technique. The dependence on the kind of sunscreen and the amount applied (in $\mu g\ cm^{-2}$) were investigated as well as the rate of penetration into the skin and the time of residence in different skin layers.

The ability of PAS to detect specific compounds present in a highly diffusive medium is a great advantage for spectroscopic studies on blood. The protein haemoglobin is responsible for the O_2–CO_2 exchange between blood and tissue. It exhibits three absorption bands, at 415, 540 and 580 nm, that are caused by the tetraporphyrin cycle bound to its amino acid skeleton. In **Figure 8** the typical haemoglobin spectrum is compared with blood spectra of patients suffering from anaemia, leukaemia and methaemoglobin. The deviations from normal blood are clearly visible and can complement other diagnostic results. The noncontact character of PA investigations and the fact that only minor sample amounts are required is another advantage that permits further studies on the same sample by alternative techniques.

It should be noted that no complicated preparation, treatment or any purification processes are required before measurements. This is advantageous also for studies on living plants. Photosynthetic activities of plants and the influence of environmental stress have been investigated by various research groups. As example, a PA apparatus has been developed to measure the oxygen evolution rate directly at a single leaf of a living plant. Environmental factors such as the effects of water stress, temperature extremes, varying light flux and gaseous pollutants were studied in details. A study on plant physiology is presented in **Figure 9**. The measurements were performed with a CO-laser intracavity PA arrangement. To enhance the detection specificity selective trapping was applied by leading the incoming air stream over different temperature levels of a cold trap. In the experiment, three cherry tomatoes were kept under pure nitrogen in a cuvette for 10 h before switching back to an air flow. This re-exposure obviously caused significant changes of the production rates of the different compounds. This example demonstrates that the PA technique is well suited for the study of fast responses of plant tissues to changing ambient conditions.

Another area of interest where PA spectroscopy is a valuable tool concerns food science. Examples include the determination of the iron content in milk powder concentrate, moisture in instant skim milk powder, stem in ground pepper or the detection of adulterated powdered coffee. An example of a deliberate adulteration in spices concerns the contamination of ground red paprika spice by red lead (Pb_3O_4) which enhances the colour of paprika but also adds to its total weight. PA studies on ground sweet red paprika have demonstrated that PA spectroscopy can be recommended as a method for rapid and gross screening for Pb_3O_4 adulterant. The current limit of detection of 2% w/w) is, however, above the internationally adopted maximum permissible level and inferior to that of established techniques like atomic absorption spectroscopy (AAS) or inductively-coupled plasma spectroscopy (ICPS).

See also: **Environmental Applications of Electronic Spectroscopy; IR Spectrometers; Laser Applications in Electronic Spectroscopy; Photoacoustic Spectroscopy, Theory; Photoacoustic Spectroscopy, Applications; Photoacoustic Spectroscopy, Methods and Instrumentation; Surface Studies By IR Spectroscopy; Zeeman and Stark Methods in Spectroscopy, Applications.**

Further reading

Almond DP and Patel PM (1996) *Photothermal Science and Techniques*. London: Chapman & Hall.

Bialkowski SE (1996) *Photothermal Spectroscopy Methods for Chemical Analysis*, Chemical Analysis Series, Vol 134. New York: Wiley.

Bicanic D (ed.) (1998) Photoacoustic, Photothermal and related methods as problem solvers in agricultural and environmental sciences. Special Issue of *Instrumentation Science and Technology* 26 (2&3).

Hess P (ed) (1989) *Photoacoustic, Photothermal and Photochemical Processes in Gases*. Topics in Current Physics, Vol 46. Berlin: Springer.

Hess P (ed) (1989) *Photoacoustic, Photothermal and Photochemical Processes at Surfaces and in Thin Films*, Topics in Current Physics, Vol. 47. Berlin: Springer.

Mandelis A (ed) (1982) *Principles and Perspectives of Photothermal and Photoacoustic Phenomena*, Progress in Photothermal and Photoacoustic Science and Technology, Vol I. New York: Elsevier.

Mandelis A and Hess P (eds) (1997) *Life and Earth Sciences*, Progress in Photothermal and Photoacoustic Science and Technology, Vol III. Bellingham, Washington: SPIE.

Meyer PL and Sigrist MW (1990) Atmospheric pollution monitoring using CO_2-laser photoacoustic spectroscopy and other techniques. *Reviews of Scientific Instruments* 61: 1779.

Pao Y-H (ed) (1977) *Optacoustics Spectroscopy and Detection*. New York: Academic Press.

Rosencwaig A (1978) Photoacoustic spectroscopy. In: *Advances in Electronics and Electron Physics*, Vol 46, Chapter 6.

Sell JA (ed.) (1989) *Photothermal Investigations of Solids and Liquids*. San Diego: Academic Press.

Sigrist MW (1986) Laser generation of acoustic waves in liquids and gases. *Journal of Applied Physics* 60: R83–R121.

Sigrist MW (1994) Air monitoring by laser photoacoustic spectroscopy. In: Sigrist MW (ed) *Air Monitoring by Spectroscopic Techniques*, Chemical Analysis Series, Vol 127, Chapter 4. New York: Wiley.

Tam AC (1983) Photoacoustics: spectroscopy and other applications. In: Kliger DS (ed) *Ultrasensitive Laser Spectroscopy*, Chapter 1. New York: Academic Press.

Tam AC (1986) Application of photoacoustic sensing techniques. *Reviews of Modern Physics* 58: 381–431.

Zharov VP and Letokhov VS (1986) *Laser Optacoustics Spectroscopy*, Springer Series in Optical Sciences, Vol 37. Berlin: Springer.

Photoacoustic Spectroscopy, Methods and Instrumentation

Markus W Sigrist, Swiss Federal Institute of
Technology (ETH), Zurich, Switzerland

ELECTRONIC SPECTROSCOPY
Methods & Instrumentation

Introduction

Since the discovery of the photoacoustic (PA) effect by Bell in 1880, who used the Sun as radiation source, a foot-operated chopper for modulation and an earphone as acoustic detector, the PA effect has found numerous applications as a sensitive and rather simple technique for determining optical, thermal and mechanical properties of all kinds of samples. This article focuses on methods and instrumentation employed in *spectroscopic* applications. Since *photothermal* (PT) spectroscopy is discussed elsewhere in the encyclopedia, PT schemes are only briefly mentioned here, whereas emphasis is put on instrumentation used in *photoacoustic* spectroscopy.

Although the technique of photoacoustic spectroscopy has existed for more than a century it is particularly the advent of lasers as radiation sources with high spectral brightness that has initiated a renaissance of the PA effect. In the meantime a great variety of experimental schemes have been developed which render the PA method a very versatile spectroscopic tool.

Photoacoustic and photothermal schemes

Experimental arrangements

As the photoacoustic (and related photothermal (PT)) phenomena comprise a large diversity of facets there exist various detection techniques which rely on the acoustic or thermal disturbances caused by the absorbed radiation. The selection of the most appropriate scheme for a given application depends on the sample, the sensitivity to be achieved, ease of operation, ruggedness, and any requirement for non-contact detection, e.g. in aggressive media or at high temperatures and/or pressures. **Figures 1–3** present the most typical arrangements applied for solid, liquid and gaseous samples, respectively. Experimental schemes for PA studies on solid samples include the measurement of the generated pressure wave either directly in the sample with a piezoelectric sensor for the pulsed regime, or indirectly in the gas which is in contact with the sample by a microphone. These most widely used setups are depicted in **Figures 1A** and **1B**. The indirect detection of the generated acoustic wave in the gas phase with the microphone is inevitable if a direct contact with the sample is not readily possible, e.g. for samples such as powders, gels or grease. The properties of piezoelectric transducers and microphones as pressure sensors are discussed below.

If the use of pressure sensors is not appropriate, because, for example, a piezoelectric transducer cannot be attached to the sample or measurements need to be done at high temperatures which hinders the application of microphones, noncontact techniques are to be applied. As shown in **Figure 1C** the induced spatial and temporal gradient of the refractive index can be sensed in this case by monitoring the deflection of a probe beam either within the (transparent) sample or directly above the (plane) sample surface (PT beam deflection or so-called mirage effect). Changes of the surface reflectivity or slight deformations of the surface (PT beam displacement) can also be detected in a noncontact manner by a probe beam. Finally, as depicted in **Figure 1D**, variations of the thermal radiation from the surface can be monitored with an infrared detector (PT radiometry). This method is of particular interest for measurements at elevated temperatures owing to the increased radiation intensity according to the Stefan–Boltzman law. Still other techniques include pyroelectric detection in thin films, thermal lensing and interferometric methods.

The typical experimental arrangement for absorption spectroscopy in weakly absorbing liquids is shown schematically in **Figure 2A**. The beam of a pulsed tunable laser is directed through the PA cell that contains the sample under study. The generated acoustic waves are detected by a piezoelectric transducer with fast response time. Usually, only the first peak of the ringing acoustic signal is taken and further processed. Pulse-to-pulse variations of the laser power are accounted for by normalizing the

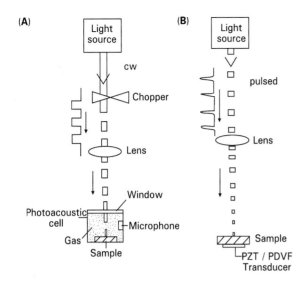

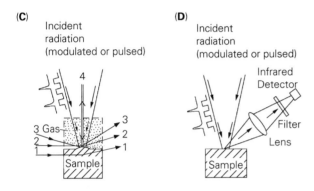

Figure 1 Typical experimental arrangements used for photoacoustic (PA) and photothermal (PT) studies on solids. As indicated one differentiates between modulated (⎍⎍⎍) or pulsed (⎍⎍⎍) incident radiation. (A) Indirect PA detection by microphone in the gas phase. (B) Direct PA detection with PZT transducer or PVDF foil attached to solid. (C) PT sensing of the gradient of the refractive index with probe beam deflection in the (transparent) sample (probe beam 1), or above the sample surface (Mirage effect, probe beam 2). Monitoring of the generated surface displacement (probe beam 3) or of the change of surface reflectivity (probe beam 4). (D) PT radiometry senses the induced change of the IR radiation that is radiated off the sample surface.

piezoelectric signal with the laser power measured with the power meter after the cell.

Another area of interest is the measurement of opaque or strongly absorbing liquids. A simple open PA cell called an optothermal window was developed for this purpose as displayed in **Figure 2B**. It essentially consists of an uncoated ZnSe window to which an annular lead zirconate titanate (PZT) piezotransducer is glued from the bottom. The excitation beam from a laser passes unobstructed through this window and is absorbed by a droplet of the sample deposited on the other side of the ZnSe disc. The generated heat diffuses into the disk which

expands. The induced stress is then recorded by the PZT transducer. Unlike in conventional transmission spectroscopy where the cell thickness is the restricting factor in dealing with strongly absorbing samples, the magnitude of the optothermal window signal depends solely on the product between the absorption coefficient and the thermal diffusion length whereby the latter can be adapted via the modulation frequency.

The typical setup for gas phase measurements is shown in **Figure 3A**. A tunable laser with narrow line width, or a conventional (broadband) radiation source followed by optical filters, is used. In general, amplitude-modulated (or sometimes pulsed) radiation is directed through the PA cell. The acoustic sensor is usually a commercial electret microphone or a condenser microphone. These devices are easy to use and sensitive enough for trace gas studies with very low absorptions. Often, the detection threshold is neither determined by the microphone responsivity R_{mic} itself nor by the electrical noise but rather by other sources (absorption by desorbing molecules from the cell walls, window heating, ambient noise, etc.). However, if this latter background is known from reference measurements, the ultimate detection sensitivity is determined solely by fluctuations of the radiation intensity, and by microphone and amplifier noise. The frequency dependence of R_{mic} is usually rather small and the temperature dependence may have to be taken into account in special cases only. If modulated radiation is employed the microphone signal is fed to a lockin amplifier locked to the modulation frequency. Since, according to theoretical

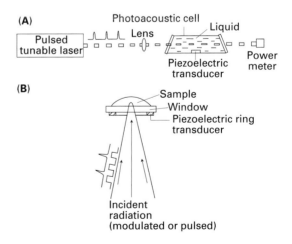

Figure 2 Typical experimental arrangements used for PA detection in liquids. (A) PZT detection of acoustic wave generated by pulsed radiation in weakly absorbing liquid. (B) Optothermal window setup applied for studies on strongly absorbing or opaque liquids with modulated or pulsed radiation.

considerations, the microphone signal amplitude is proportional to the absorbed power for weakly absorbing media, the average radiation power is recorded simultaneously by a power meter for normalization.

If pulsed radiation is employed, the microphone bandwidth is often not sufficient to resolve the temporal shape of the generated acoustic pulses. However, common microphones can still be used even for nanosecond laser pulses because the length of a single acoustic pulse is essentially determined by the transit time of the acoustic wave across the beam radius. Normalized PA amplitudes are obtained by dividing the microphone signal peaks by the corresponding laser pulse energy that is recorded by a sensor such as a pyroelectric detector. Averaging over several pulses improves the signal-to-noise ratio. Another approach for pulsed radiation consists of using an acoustic resonator with high Q-factor as the gas cell, recording the microphone signals in the time domain but analysing the PA signal amplitudes after Fourier transformation in the frequency domain. The excited cell resonances then appear in the PA frequency spectra.

An important issue for many applications concerns the calibration of the entire PA or PT detection system. Since the PA signal depends on many factors that are not known with sufficient accuracy, a straightforward calibration is often achieved by employing a reference sample with known absorption.

As an example, certified gas mixtures (trace gas diluted in a nonabsorbing buffer gas) or well characterized dye solutions in the case of liquids are used. The situation is more difficult with solid and biological samples, particularly layered media, powders, gels or tissue. In such cases, quantitative data are difficult, if not impossible, to obtain. But even qualitative instead of quantitative spectra are often valuable, especially when other spectroscopic techniques fail owing to opaqueness or strong scattering of the sample.

It should be emphasized that numerous different versions and modifications of these general schemes have been presented in the literature. In particular, combinations of conventional methods, such as Fourier-transform IR (FT-IR) or gas chromatography (GC), with PA detection have been reported. Some types of PA detection schemes are also implemented in commercial spectrometers.

In the following, the different components of PA spectrometers are briefly discussed.

Radiation sources

In commercial PA spectrometers, incoherent sources such as lamps are employed in combination with filters or with an interferometer. Devices equipped with a small light bulb, with either a chopper or direct current modulation as modulated radiation source and appropriate filters to avoid absorption

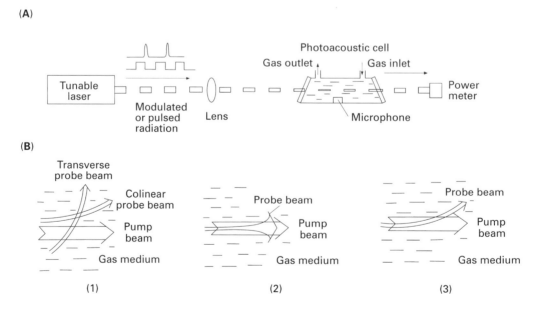

Figure 3 Typical experimental PA and PT arrangements used for gas monitoring with tunable laser sources. (A) PA detection with conventional microphone in resonant gas cell for modulated cw radiation or in nonresonant cell for pulsed radiation. (B) Noncontact refractive index sensing schemes with displaced colinear or transverse probe beam (PA deflection, 1), thermal lensing (2), or colinear probe beam (PT deflection, 3).

interferences with other species, are used as compact gas sensors, e.g. for indoor CO_2 monitoring.

However, since the generated PA signal is proportional to the absorbed (and thus to the incident) radiation power, powerful radiation sources, particularly lasers offering high spectral brightness, are advantageous for achieving high detection sensitivity and selectivity in spectroscopic applications. In the UV and visible spectral range, excimer and dye lasers have been employed, whereas in the midinfrared (fundamental or mid-IR) wavelength range line-tunable CO_2 and CO lasers dominate the applications. Diode lasers have so far only rarely been employed in PA spectroscopy owing to their limited power. This situation may, however, change with the ongoing developments in this field. On the one hand, near infrared diode lasers with sufficient power for PAS are available for monitoring overtones and combination bands of molecular fundamental absorptions. On the other hand, current efforts focus on the implementation of widely tunable narrowband all-solid-state laser devices in the mid-IR region for accessing the (much stronger) fundamental absorptions. Optical parametric oscillators (OPOs) and difference frequency generation (DFG) in nonlinear crystals are certainly of great interest for compact spectrometers. Furthermore, recent developments in quantum cascade lasers look very promising in this respect.

Modulation schemes

Modulation schemes can be separated into the modulation of the incident radiation and the modulation of the sample absorption itself. The first technique includes the most widely used amplitude modulation (AM) of continuous radiation by mechanical choppers, electrooptic or acoustooptic modulators as well as the modulation of the source emission itself by current modulation or pulsed excitation. In comparison to AM, frequency (FM) or wavelength (WM) modulation of the radiation may improve the detection sensitivity by eliminating the continuum background caused by a wavelength-independent absorption, e.g. of the cell windows, known as window heating. This type of modulation is obviously most effective for absorbers with narrow line width and most easily performed with radiation sources whose wavelength can rapidly be tuned within a few wavenumbers. Pulsed excitation is often applied for liquids but is also of interest for gaseous samples because it permits time gating and the excitation of acoustic resonances.

In certain cases the modulation of the absorption characteristics of the sample itself is advantageous. In gas studies the Stark or Zeeman effect has been employed, i.e. by applying a modulated electric or magnetic field to the sample. The result is a suppression of the continuum background and an enhancement of detection selectivity in multicomponent samples because, for example, Stark modulation only affects molecules with a permanent electric dipole moment like ammonia (NH_3) or nitric oxide (NO) while other, possibly interfering molecules, are not affected. Finally, combinations of both amplitude and sample absorption modulation have been successfully applied, e.g. for the sensitive detection of ammonia in the presence of absorbing water vapour and carbon monoxide.

Photoacoustic cell designs

The PA cell serves as a container for the sample under study and for the microphone or some other device for the detection of the generated acoustic wave. An optimum design of the PA cell represents a crucial point when background noise ultimately limits the detection sensitivity. In particular, for trace gas applications many cell configurations have been presented including acoustically resonant and nonresonant cells, single- and multipass cells, as well as cells placed intracavity. Nonresonant cells of small volume are mostly employed for solid samples with modulated excitation or for liquids and gaseous samples with pulsed laser excitation. As a unique example, a small-volume cell equipped with a 'tubular' acoustic sensor consisting of up to 80 single miniature microphones has been developed. These microphones are arranged in eight linear rows with ten microphones in each row. The rows are mounted in a cylindrical geometry parallel to the exciting laser beam axis and located on a circumference around the axis. This configuration is thus ideally adapted to the geometry of the generated acoustic waves.

Resonant cells, in combination with modulated excitation, are normally applied for gas monitoring. These cells are operated on longitudinal, azimuthal, radial, or Helmholtz resonances. The signal enhancement by the Q-factor (usually >100) is often advantageous. Resonance frequencies lie in the kHz range resulting in resonance widths of a few Hz. Furthermore, the gas handling for the cell can be designed in such a way that the gas inlets and outlets are located at pressure nodes of the acoustic resonance which allows measurements in flowing gas with flow rates of the order of 1 L min^{-1} without increasing the noise level.

Finally, cells developed for special purposes have been suggested, such as windowless cells equipped with acoustic baffles to reduce the influence of the ambient noise or heatable cells for studying liquid samples with low volatility.

Detection sensors

As mentioned above the acoustic disturbances generated in the sample are detected by some kind of pressure sensor. In contact with liquid or solid samples these are piezoelectric devices such as lead zirconate titanate (PZT), $LiNbO_3$ or quartz crystals with a typical responsivity R in the range of up to V bar^{-1} or thin polyvinylidene-difluoride (PVF_2 or PVDF)-foils with lower responsivity. These sensors offer fast response times and are thus ideally adapted for pulsed photoacoustics.

For studies in the gas phase, commercial microphones are employed. These include miniature electret microphones such as Knowles or Sennheiser models with typical responsivities R_{mic} of 10–20 mV Pa^{-1} as well as condenser microphones, e.g. Brüel & Kjær models with typical R_{mic} of 100 mV Pa^{-1}. Usually R_{mic} depends only weakly on frequency. The electret microphones produced for hearing aids exhibit a rather flat frequency response between, say, 20 Hz and 20 kHz whereas the bandwith Δv of condenser microphones extends to frequencies of 100 kHz. All these microphones are thus well suited for typical modulation frequencies in the 100 Hz to kHz range. For pulsed applications, the general relation between R_{mic} and Δv, as well as the occurrence of external noise implies a reduction of the signal-to-noise ratio for very large bandwidths so that miniature electret microphones are often appropriate detectors also in this case. The detection sensitivity can be enhanced by adding the signals of several microphones. In such a configuration the signal increases with the number of microphones used, whereas the microphone random noise decreases with the square root of their number. Since electret microphones are small and cheap, a number of them can be arranged in a still compact geometry. A further improvement of sensitivity is expected from the insertion of an electrical filter that cuts the low-frequency components below, say, 1 kHz of the signal because these components contribute less to the increase of the signal-to-noise ratio than the higher-frequency components. Finally, an adaption of the frequency response of the microphone preamplifier and amplifier stages to that of the microphone is advantageous to fully exploit all the sensed frequency contributions except noise components at frequencies not contributing to the acquired signal.

If there is a need for noncontact detection, refractive index sensing, notably thermal lensing and both PA and PT deflection, are employed. These methods use a pump beam and a probe beam (HeNe or diode laser) in either colinear or transverse arrangement as shown in **Figure 3B**. In comparison to the conventional PA method with pressure sensors, these schemes offer similar sensitivity but require a somewhat more sophisticated setup and imply a more difficult calibration.

List of symbols

R = responsivity; R_{mic} = microphone responsivity; Δv = bandwidth.

See also: **Environmental and Agricultural Applications of Atomic Spectroscopy; Environmental Applications of Electronic Spectroscopy; IR Spectrometers; Photoacoustic Spectroscopy, Applications; Photoacoustic Spectroscopy, Theory; Surface Studies By IR Spectroscopy; X-Ray Absorption Spectrometers.**

Further reading

Almond DP and Patel PM (1996) *Photothermal Science and Technique. London*: Chapman & Hall.

Mandelis A (ed) (1992) Principles and perspectives of photothermal and photoacoustic phenomena. In: *Progress in Photothermal and Photoacoustic Science and Technology*, Vol 1. New York: Elsevier.

Meyer PL and Sigrist MW (1990) Atmospheric pollution monitoring using CO_2-laser photoacoustic spectroscopy and other techniques. *Review of Scientific Instruments* 61: 1779–1807.

Pao, Yoh-Han (ed) (1977) *Optoacoustic Spectroscopy and Detection*. New York: Academic Press.

Rosencwaig A (1980) Photoacoustics and photoacoustic spectroscopy. In *Chemical Analysis Series*, Vol 57. New York: Wiley.

Sigrist MW (1986) Laser generation of acoustic waves in liquids and gases. *Journal of Applied Physics* 60: R83–R121.

Sigrist MW (1994) Air monitoring by laser photoacoustic spectroscopy. In: Sigrist MW (ed) *Air Monitoring by Spectroscopic Techniques*, Chemical Analysis Series, Vol 127, Chapter 4. New York: Wiley.

Tam AC (1983) Photoacoustics: spectroscopy and other applications In: Kliger DS (ed) *Ultrasensitive Laser Spectroscopy*, Chapter 1. New York: Academic Press.

Tam AC (1986) Applications of Photoacoustic Sensing Techniques. *Review of Modern Physics* 58: 381–431.

Zharov VP and Letokhov VS (1986) *Laser Optoacoustic Spectroscopy*. Springer Series in Optical Sciences, Vol 37. Berlin: Springer.

Photoacoustic Spectroscopy, Theory

András Miklós, **Stefan Schäfer** and **Peter Hess**,
University of Heidelberg, Germany

Copyright © 1999 Academic Press

ELECTRONIC SPECTROSCOPY/
VIBRATIONAL, ROTATIONAL &
RAMAN SPECTROSCOPIES
Theory

Introduction

The phenomenon of the generation of sound when a material is illuminated with nonstationary (modulated or pulsed) light is called the photoacoustic (PA) effect. Photoacoustic spectroscopy (PAS) is the application of the PA effect for spectroscopic purposes. It differs from conventional optical techniques mainly in that, even though the incident energy is in the form of photons, the interaction of these photons with the material under investigation is studied not through subsequent detection and analysis of photons after interaction (transmitted, reflected or scattered), but rather through a direct measurement of the effects of the energy absorbed by the material. Since photoacoustics measures the transient internal heating of the sample, it is clearly a form of calorimetry as well as a form of optical spectroscopy.

In the following discussion the complex PA effect will be divided into three steps:

Heat release in the sample material due to optical absorption.

Acoustic and thermal wave generation in the sample material.

Determination of the PA signal in a PA detector.

A quantitative analysis of the PA signal is possible only when all three steps can be described quantitatively. Up to now this could be achieved only in a few special cases. Excitation of rovibrational levels leads to a complete transformation of the absorbed radiation into heat, whereas for vibronic excitation competing channels exist such as emission of radiation and photoinduced reactions. In this latter case, additional information on these competing channels is needed for the theoretical description.

A quantitative treatment of acoustic and thermal wave generation is usually possible only for simple excitation geometries, as can be realized by laser excitation, using the basic equations of fluid mechanics and thermodynamics.

The PA signal finally detected with a calibrated microphone for quantitative analysis is also influenced by the shape and nature of the photoacoustic cell, which impose boundary conditions on the evolution of the generated acoustic waves. However,

only for highly symmetric resonant setups with high Q factors does a theoretical analysis seem to be possible. This means that for most experimental arrangements used in PAS a quantitative signal analysis cannot be achieved and more or less drastic approximations have to be introduced into the signal analysis.

Heat release in the sample material due to optical absorption

The interaction of photons with the material may produce a series of effects (**Figure 1**). If any of the incident photons are absorbed by the material, internal energy levels (rotational, vibrational, electronic) within the sample are excited. The excited state may lose its energy by radiation processes, such as spontaneous or stimulated emission, and by nonradiative

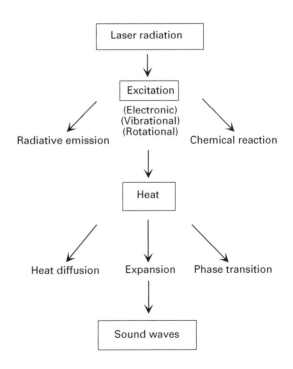

Figure 1 Elementary processes occurring during PA signal generation. The absorbed photon energy is partly transformed into heat and acoustic energy.

deactivation, which channels at least part of the absorbed energy into heat. In a gas this energy appears as kinetic energy of the gas molecules, while in a solid it appears as vibrational energy of ions or atoms (phonons). In the case of vibrational excitation of gas molecules, radiative emission and chemical reactions do not play an important role, because the radiative lifetime of vibrational levels is long compared with the time needed for collisional deactivation at ordinary pressures and the photon energy is too small to induce reactions. Thus the total absorbed energy is released as heat. However, in the case of electronic excitation, the emission of radiation and chemical reaction processes may compete efficiently with collisional deactivation. Chemical reactions may also contribute to the release of heat, and thus they may increase the PA effect. If photodissociation occurs, for example, the local increase of the number of molecules and the thermalization of the recoil energy of the fragments generates a local pressure and temperature rise.

The heat release due to optical absorption in the sample material can be modelled by a rate equation. If it is assumed that the thermalization of the absorbed photon energy can be described by a simple linear relaxation process, the heat released per unit volume and time can be determined by solving the rate equation. If the near-resonant vibration–vibration (V-V) and the vibration–translation (V-T) relaxation are the fastest processes (as, for example, in many gases at atmospheric pressure), the heat power density will be proportional to the absorption coefficient and to the incident light intensity. In cases of very short laser pulses or high light intensity, optical saturation may occur. Then the heat production will be a nonlinear function of the light intensity and absorption coefficient. The timescales of the temporal evolution of the processes involved are summarized in **Figure 2**. The heat release may be delayed in gas mixtures if the excess energy of the excited molecule can be channelled by collisions to a long-lifetime transition of another species.

Acoustic and thermal wave generation in the sample material

Sound and thermal wave generation can be theoretically described by classical disciplines of physics such as fluid mechanics and thermodynamics. The governing physical laws are the energy, momentum and mass conservation laws, given in the form of the heat-diffusion, Navier–Stokes and continuity equations, respectively. The physical quantities characterizing the PA and photothermal (PT) processes are the temperature T, pressure P (mechanical stress in case of

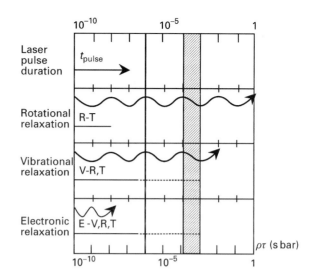

Figure 2 Timescales for the radiative emission and relaxation processes. The shaded area indicates the typical response time of a PA resonator equipped with a microphone. The thick line represents the acoustic transit time. The wavy lines depict the radiative emission and the horizontal lines the range of relaxation processes characterized by the relaxation time τ.

solids), density ρ and three components of the particle velocity vector v. As five equations are not enough to determine the above six quantities, a sixth equation is added, the thermodynamic equation of state in the form of $\rho = \rho(P, T)$. As the changes of ρ, P and T induced by light absorption are usually very small compared to their equilibrium values, the equations can be linearized by introducing the deviations from the equilibrium values as new variables, by neglecting the products of the small variables and by regarding the equilibrium values as constants. Moreover, the velocity vector v can always be separated into two components u and w, where curl $u = 0$ and div $w = 0$. As the heat-diffusion and continuity equations are coupled only by the nonrotational component u to the Navier–Stokes equation, the w component of the particle velocity can be omitted. The governing equations may be written then as follows:

$$\frac{\partial \theta}{\partial t} + \frac{\gamma - 1}{\beta} \operatorname{div} \boldsymbol{u} - \kappa_{\mathrm{V}} \nabla^2 \theta = \frac{H(\boldsymbol{r}, t)}{\rho C_{\mathrm{V}}} \qquad [1]$$

$$\frac{\partial}{\partial t}\left(p - \frac{\beta}{K_{\mathrm{T}}}\theta\right) + \frac{\operatorname{div}\boldsymbol{u}}{K_{\mathrm{T}}} = 0 \qquad [2]$$

$$\rho \frac{\partial \boldsymbol{u}}{\partial t} + \operatorname{grad} p - \frac{4\eta}{3}\nabla^2 \boldsymbol{u} = 0 \qquad [3]$$

where $\theta = T - T_0$, $p = P - P_0$, γ, β, κ_{V}, C_{V}, H, K_{T} and η are the new temperature and pressure variables, the adiabatic coefficient, the heat expansion coefficient, the thermal diffusivity and heat capacity at constant volume, the density of the deposited heat power, the isothermal compressibility and the dynamic viscosity, respectively. For solid materials the Navier–Stokes equation is replaced by the wave equation of the longitudinal elastic waves.

The heat power density H, released in the material as the result of all nonradiative de-excitation processes, appears as the source term on the right hand side of the heat-diffusion equation. The spatial size and shape of the source volume depend on the light-beam geometry and on the absorption length in the material. Similarly, the time dependence of the heat source is determined by the time evolution of the light excitation and by the relaxation processes in the material. A photoacoustic effect can be generated by modulated radiation as well as by pulsed radiation. As the theoretical treatments of the two cases are different, they will be discussed separately.

Modulated PAS

In this case the intensity (or the wavelength) of the incident light beam is modulated by an angular frequency ω to generate the acoustic signal. As the modulation frequency is usually in the audio frequency range, the time delay between heat release and light intensity may be negligible. In this case the source term has the same time dependence as the light intensity. Assuming an $\exp(i(\omega t - kr))$ dependence of the variables, θ, p and u, two independent plane wave solutions of Equations [1] and [3] can be derived: a thermal wave and a sound wave. The wave-lengths of the two plane waves can be determined from the corresponding eigenvalues of the wave vector k, taking into account that the length $|k| = k$ of the wave vector (called wavenumber) is inversely proportional to the wavelength λ ($k = 2\pi/\lambda$). The eigenvalues of the wavenumber for the thermal and sound wave may be given as $k^2_{th} \cong -i\omega/\kappa_P$ and $k^2_s \cong \omega^2/c^2(1 - i\omega v/c^2)$, respectively, where κ_P, $v = 4\eta/3\rho$ and c are thermal diffusivity at constant pressure, the effective kinematic viscosity and the sound velocity, respectively. As the orders of magnitude of κ_P and c are 10^{-5}–10^{-7} m^2 s^{-1} and 10^2–10^4 m s^{-1} respectively, the wavelength of the thermal wave is much shorter than that of the sound wave. That is, two types of waves are simultaneously generated: a very strongly damped thermal wave with submillimetre wavelength, and a slightly damped sound wave with wavelengths in the centimetre to metre range. The thermal wave corresponds more or less to an isobaric thermal expansion, i.e. the changes of the temperature and density are much larger than that of the pressure. Because of the large damping coefficient,

this wave cannot propagate far away from the heated region; it appears only in the neighbourhood of the exciting light beam. In the sound wave a quasi-adiabatic state change propagates with the velocity of sound; here the orders of magnitude of the relative changes in pressure, temperature and density are the same. The amplitude of the periodic pressure change is proportional to the time-varying (AC) component of the released heat power density and inversely proportional to the modulation frequency.

As the average of the intensity of a modulated light beam is nonzero, the heat energy in the illuminated volume will rise continuously. Therefore, the temperature will slowly increase and the density will decrease until the heat deposition rate is equal to the loss rate due to heat conduction. This process is also governed by the heat-diffusion equation. For a closed cell the average density is constant; therefore a pressure rise will occur. This DC component of the released heat power density changes the thermodynamic state of the material, in particular in very small PA detectors ($\approx$ cm^3).

The amplitude of the periodic temperature change (θ) is also proportional to the AC component of the heat power density and inversely proportional to the modulation frequency. This means that the PA and PT effects are inherently coupled; the heat deposited by the interaction of radiation with the material generates a localized temperature rise, a thermal wave and a propagating sound wave. The first two effects will result in a periodically pulsating temperature distribution.

Pulsed PAS

In the case of pulsed excitation, the absorption of photons ceases when the laser pulse is over, but the relaxation processes will continue until the full thermalization of the absorbed energy is achieved. Therefore, the duration of the thermal pulse is always longer than that of the light pulse. Nevertheless, the heat pulse may be regarded as instantaneous acoustic excitation if the characteristic time of the acoustic event is much larger than the duration of the heat pulse. Two quantities characterize the acoustic process, namely the transit time τ_s of the sound through the heated volume and the response time of the PA detector. The value of τ_s is usually in the microsecond range (**Figure 2**), because the laser beam diameter usually does not exceed a few millimetres in PAS. The PA response time depends on the period of the eigenmodes of the detector and is usually in the millisecond range (**Figure 2**). As the pulse duration of most pulsed lasers is in the nanosecond range, and in many cases, such as V-V and V-T

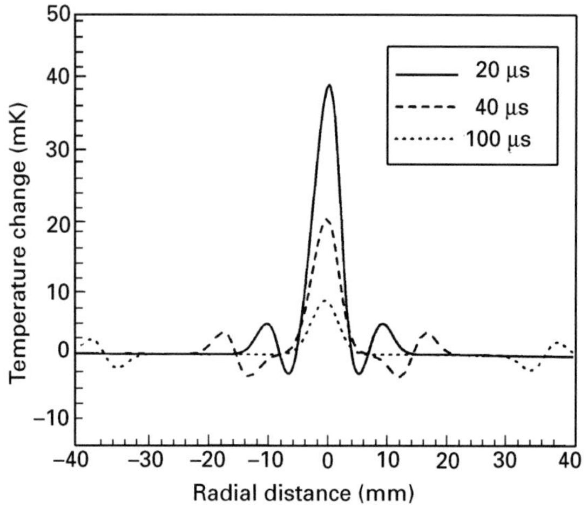

Figure 3 Modelling of the gas temperature distribution as a function of the radial distance from the light excitation source. The diameter of the light beam was 6 mm and the pulse duration 20 ns. The three curves represent three different time delays after illumination.

relaxation in gases, the relaxation time is also in the nanosecond to microsecond range, the source term of Equation [1] may be regarded as a Dirac-delta pulse. The governing equations (Eqns [1] to [3]) can be solved by the Green's function technique, taking into account the Green's functions of both the thermal and acoustic problems. The solution is composed of a slowly broadening quasi-Gaussian temperature distribution and an outward propagating sound pulse of duration $\geq 2\tau_s$. After a short time the sound pulse will be separated from the thermal distribution, allowing the measurement of both features separately (**Figure 3**). The spatial symmetry of the solution depends on the shape and size of the heated region. In strong absorbers, the heated region is usually small compared to the sound wavelength; therefore mostly spherical sound waves are produced. In weakly absorbing gases, cylindrical acoustic waves are generated. The outward-propagating primary wavefront can be detected by appropriate high-frequency pressure sensors. As the achievable sensitivity and signal-to-noise (S/N) ratio are small, the direct detection of the primary waves has no practical significance. In practice the acoustic excitation takes place inside a PA detector, and the time evolution of the acoustic signal is strongly influenced by the properties of this detector.

Determination of the PA signal in a PA detector

The actual solutions of the governing equations (Eqns [1] to [3]) depend on the boundary conditions determined by the type and geometry of the PA detector. In gas-phase photoacoustics, the PA detector consists of a cavity and a microphone to monitor the acoustic signal. From an acoustic point of view, the PA detector is a linear acoustic system that responds in a characteristic way to an excitation. The acoustic properties of the PA detector can be determined independently using acoustic modelling techniques or by measuring them in an acoustic laboratory. Once the acoustic properties are known, the response of the PA detector for any kind of PA excitation can be determined by calculation. Although PA detectors can usually be used with both modulated and pulsed excitation, the theoretical description of the two cases will be presented separately.

PA detectors excited by modulated light

A simple PA detector consists of a cavity and a microphone to monitor the acoustic signal (**Figure 4A**). Even such a simple system has acoustic resonances. If the modulation frequency is much smaller than the lowest resonance, the PA detector or PA cell operates in a nonresonant mode. Such a PA cell is frequently called a 'nonresonant' cell in the literature. In this case the sound wavelength is much larger than the cell dimensions and thus the sound cannot propagate. The average pressure in the detector will oscillate with the modulation frequency. The amplitude of the oscillation may be determined by integrating Equations [1] to [3] over the volume of the detector. The pressure amplitude will be inversely proportional to the volume of the cell. The photoacoustically generated pressure can be approximated by the expression

$$p(\omega) = \frac{(\gamma - 1)\alpha l W_{\mathrm{L}}}{\mathrm{i}\,\omega\,V_{\mathrm{cell}}} \qquad [4]$$

where α, l, W_{L}, ω, V_{cell} and γ denote the absorption coefficient of the material at the light pass length, the incident light power, the modulation frequency, the cell volume and the adiabatic coefficient of the material, respectively. In the case of a small cell and low modulation frequency, the signal may be quite large. The PA signal has a 90° phase lag with respect to the light intensity. Unfortunately, the noise also increases with decreasing frequency and volume; thus the S/N ratio will usually decrease.

As mentioned, the acoustic and thermal processes are inherently coupled. Until the cell dimensions are much larger than the size of the pulsating thermal distribution, the simple model described above can be applied. The thermal wave is usually not completely

+

damped at the walls of the cell in the case of a very small cell and very low modulation frequency. Then both the average temperature and the amplitude of the temperature oscillation will be influenced by the heat conduction through the cell walls to the environment. As in a closed cell the pressure is proportional to the temperature, the PA signal will also depend on the heat conduction through the walls. Since the theoretical modelling of this effect is practically impossible, such small cells and very low modulation frequencies allow only a qualitative signal analysis.

A special case should be mentioned here, one of the oldest arrangements among PA detectors. Termed the 'gas-microphone cell' it is used for investigating samples of condensed materials (**Figure 4B**). It usually consists of a small cylinder equipped with a sample holder, a microphone and a window. The gas in the cell is nonabsorbing; only the sample and the backing material (in the case of an optically thin sample) absorb the incident radiation. The PA signal is produced in an indirect way; the thermal waves generated in the solid sample are transmitted to the gas, and the periodic heat expansion of the gas above the sample surface acts as a piston and produces pressure oscillations in the closed cell volume. As the penetration depth of the thermal wave is usually much smaller than the diameter of the light beam, one-dimensional theoretical modelling is possible.

In resonant photoacoustics, an acoustic resonator with an optimized geometry such as a cylinder or sphere is used as the gas cell. Such an acoustic system has several eigenresonances, whose frequencies depend on the geometry and size of the cavity. For a lossless cylinder, the resonance frequencies of the different eigenmodes can be calculated from the equation

$$f_{n,m,n_z} = \frac{c}{2} \sqrt{\left(\frac{\chi_{m,n}}{R}\right)^2 + \left(\frac{n_z}{L}\right)^2} \qquad [5]$$

where c, R and L are the sound velocity of the gas in the cavity, the radius and the length of the cylinder, respectively. The indices m, n and n_z may take the values 0, 1, 2, etc. The quantity $\chi_{m,n}$ is the nth zero of the derivative of the mth-order Bessel function divided by π.

When only one index is nonzero, the eigenmodes separate into longitudinal ($n_z \neq 0$), radial ($n \neq 0$) and azimuthal ($m \neq 0$) modes (**Figure 5**). In the other mixed eigenmodes the spatial distribution of the sound pressure is much more complicated.

The modulation frequency may be tuned to one of the eigenresonances of the PA detector. Such 'resonant PA cells' or detectors have found widespread application in PA trace gas measurements. Since the eigenmodes of a lossless cavity are orthogonal, the sound pressure field in a lossy cavity can be approximated in the form of a series expansion of the eigenmodes, where the amplitudes of the terms depend on frequency. In fact, each amplitude function is a resonance curve characterized by the corresponding resonance frequency and quality factor (Q factor). The first term of the series expansion determines the sound pressure below the lowest resonance. To aid understanding of the behaviour of a resonant PA cell a computer simulation is shown in **Figure 6**, for which the frequency dependence of the PA signal at one end of a closed cylinder filled with a strongly absorbing gas was calculated. It can be seen that several resonances (the odd-numbered longitudinal ones) lie on a nonresonant curve. For well-separated sharp resonances, only two terms of the series expansion are necessary to determine the sound pressure around a certain resonance. These are the first nonresonant term, which is independent of the spatial coordinates but depends inversely on the frequency, and the resonant term, corresponding to the selected eigenmode. This term depends on the overlap of the acoustic mode pattern and the light beam distribution, on the position of the microphone and on the

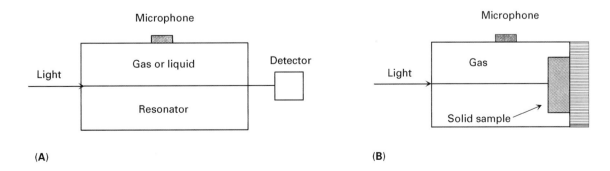

Figure 4 PA setups for monitoring the PA signal with a microphone in a resonator: (A) gas or liquid, (B) solid sample in a gas microphone cell.

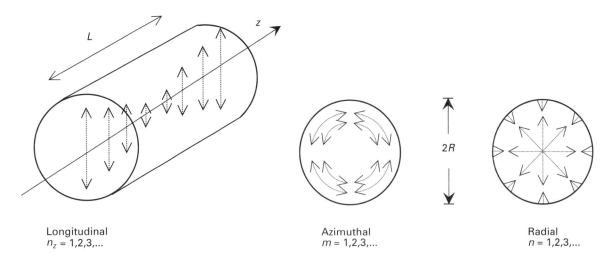

Longitudinal
$n_z = 1,2,3,...$

Azimuthal
$m = 1,2,3,...$

Radial
$n = 1,2,3,...$

Figure 5 Schematic representation of the longitudinal, azimuthal and radial acoustic modes in a cylindrical resonator.

quality factor of the eigenresonance. In the case of high Q factors ($Q > 100$), the contribution of the nonresonant term can be neglected and the PA signal amplitude at the resonance frequency ω_j and at the position r_M of the microphone can be calculated as

$$p(r_M, \omega_j) = \frac{(\gamma - 1)\, l\, U_j\, p_j(r_M) Q_j}{V_{cell}\, \omega_j\, D_j} \alpha\, W_L \qquad [6]$$

where l, p_j, U_j, Q_j, D_j, V_{cell}, α and W_L are the length of the light path within the cell, the pressure distribution of the jth eigenmode of the cell, the overlap integral of the light intensity distribution with p_j, the Q factor and the normalization factor of the jth eigenmode, the cell volume, the absorption coefficient and the incident light power, respectively. As the quantities in the first term of Equation [6] are independent of the light power and the absorption coefficient, the first term may be regarded as a characteristic setup quantity, called the 'cell constant', of the PA arrangement.

Since the resonance frequency depends on the speed of sound, a temperature drift of the gas inside the cavity causes a shift of all resonance frequencies. In modulated PAS this problem can be solved by temperature stabilization, by continuous monitoring of the gas temperature or by synchronizing the modulation frequency to the resonance peak using appropriate electronics.

As the modulation frequency of a CW light source may be arbitrary, a given PA detector can operate in both the nonresonant and resonant modes. The main advantage of resonant operation is the amplification of the PA signal by the Q factor of the resonator if the modulation frequency of the incoming light is properly tuned to the selected acoustic resonance ('acoustic amplifier').

If the modulation frequency is so low that the corresponding wavelength is much longer than the dimensions of the detector, Equation [3] may be used to calculate the PA signal. Nonresonant operation is also possible by tuning the modulation frequency away from a resonance. In such cases the PA signal can be estimated by substituting $Q = 1$ into Equation [6], but a better accuracy may be achieved by taking into account the nonresonant term and at least the neighbouring eigenmodes in the series expansion.

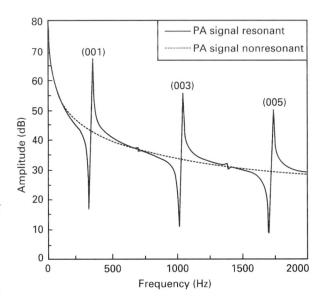

Figure 6 Modelling of the frequency-dependent PA signal. The excited modes are the first (001), third (003) and fifth (005) longitudinal modes of the cavity. The dotted line shows the nonresonant contribution.

PA detectors excited by light pulses

The absorption of a light pulse generates a primary acoustic pulse inside the PA detector. This pulse acts as a broadband acoustic source for the PA detector, exciting all eigenmodes simultaneously. If the acoustic transit time is much shorter than the period of the detected eigenmode of the resonator, the excitation of the PA signal can be regarded as instantaneous (**Figure 2**). In this case the slow PA detector responds to the excitation similarly to the way in which a ballistic pendulum or galvanometer responds to force or charge pulses; a sudden rise of the PA signal followed by a slow decay can be observed. The amplitude of the first period of the sound pressure oscillation will be proportional to the released heat energy. Thus, a pulsed PA detector can be used for absolute (calorimetric) measurement of that part of the absorbed light energy that was converted to heat. As calibrated microphones are available, the absolute measurement of the photoacoustically generated sound pressure is possible. The solutions of Equations [1] to [3] can again be given in the form of a series expansion of the orthogonal eigenmodes, but in this case the amplitudes will depend not only on the frequency but also on time. A PA cell for pulsed operation is designed for optimal excitation of a selected eigenmode. This mode should be well separated from the neighbouring ones and should have a high Q factor. As the PA response in the time domain shows a very complicated behaviour, it is much better to evaluate the PA signal by converting the time signal to the frequency domain using Fourier transformation. A part of the frequency spectrum measured in a cylindrical cell, optimized for the first radial mode, is shown in **Figure 7**. The PA signal amplitude at the peak of the selected resonance can be determined from the theory as

$$p(\mathbf{r}_{M}) = \frac{(\gamma - 1)\, lU_j\, p(\mathbf{r}_{M})}{V_{\text{cell}}\, D_j}\, \alpha\, E_{L} \qquad [7]$$

where E_{L} is the pulse energy. As the fast Fourier transform (FFT) algorithm applied for calculating the spectrum delivers the average spectrum of the signal over the recorded time window, the amplitude of the resonance peak has to be corrected in order to obtain the value of $p(\mathbf{r}_{M})$. The ratio of the corrected PA amplitude and laser pulse energy depends only on the product of several geometry factors and the absorption coefficient α of the absorbing component. In contrast to Equation [6], the PA signal amplitude $p(\mathbf{r}_{M})$ in Equation [7] does not depend on the Q factor of the cavity, which cannot be calculated

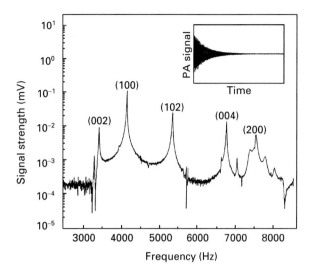

Figure 7 Time-dependent PA signal recorded by a microphone (inset) and corresponding frequency spectrum of a cylindrical resonator. In the displayed frequency range, the second longitudinal (002), first radial (100), combination (102), fourth longitudinal (004) and second radial (200) modes of the resonator are detected.

with high accuracy theoretically. Thus, pulsed PAS is an absolute method for measuring the absorption coefficient. Since α is given as the product of the number density N and the absorption cross section σ of the absorbing molecules, pulsed PAS can be applied for both spectroscopic studies (known N) and trace gas analysis (known σ).

Summary

The theory of PAS is sufficiently complicated that only an overview could be presented. In the theoretical description of a given PAS experiment, a separated treatment of the three physical processes, as presented here, is often not possible. Moreover, important quantities of the theory, such as the Q factor and the overlap integral U_j, are very difficult to keep under control. Small changes in the experimental adjustment (e.g. microphone position or beam focusing) may cause considerable changes, so that the agreement between theory and experiment may be degraded. As the PA signal does not depend on the Q factor in pulsed PAS, and the Q factor, which is needed only for calculating the correction factor, can be derived from the measured spectrum, pulsed PAS is more suitable for quantitative measurements than is modulated PAS. On the other hand, the probability of optical saturation is quite high in pulsed PAS, since the pulse energy of the available laser sources is usually in the millijoule range and the corresponding instantaneous power in the kilowatt to megawatt range. Therefore, the linear dependence of the PA signal on the pulse energy must always be checked in pulsed PAS.

List of symbols

c = sound velocity; C_V = heat capacity at constant volume; D_j = normalization factor of jth eigenmode; E_L = pulse energy; H = heat power density; $k = |\mathbf{k}|$; $\mathbf{k}$ = wave vector; K_T = isothermal compressibility; L = cavity length; l = light path length; P = pressure; $p = P - P_0$; $Q_j = Q$ factor of jth eigenmode; $\mathbf{r}$ = position vector; R = cavity radius; T = temperature; U = overlap integral; $\mathbf{u}$, $\mathbf{v}$ = particle velocity vector; W_L = incident light power; α = absorption coefficient; β = thermal expansion coefficient; γ = adiabatic coefficient; η = dynamic viscosity; $\theta = T - T_0$; κ_V, κ_P = thermal diffusivity at constant volume, pressure; λ = wavelength; ν = effective kinematic viscosity = $4\eta/3\rho$; ρ = density; τ_s = transit time of sound; $\chi_{m,n}$ = the nth zero of the derivative of the mth-order Bessel function $\times$ $(1/\pi)$; ω = angular frequency of modulation.

See also: **Laser Spectroscopy Theory; Light Sources and Optics; Photoacoustic Spectroscopy, Applications.**

Further reading

Diebold GJ (1989) Application of the photoacoustic effect to studies of gas phase chemical kinetics. In: Hess P (ed) *Photoacoustic, Photothermal and Photochemical Processes in Gases*, pp 125–170, Vol 46 in Topics in Current Physics. Berlin: Springer.

Fiedler M and Hess P (1989) Laser excitation of acoustic modes in cylindrical and spherical resonators: theory and applications. In: Hess P (ed) *Photoacoustic, Photothermal and Photochemical Processes in Gases*, pp 85–121, Vol 46 in Topics in Current Physics. Berlin: Springer.

Hess P (1992) Principles of photoacoustic and photothermal detection in gases. In: Mandelis A (ed) *Progress in Photothermal and Photoacoustic Science and Technology*, Vol 1, pp 153–204. New York: Elsevier.

Morse PM and Ingard KU (1986) *Theoretical Acoustics*. Princeton, NJ: Princeton University Press.

Pao YH (1977) *Optoacoustic Spectroscopy and Detection*. New York: Academic Press.

Rosencwaig A (1980) *Photoacoustics and Photoacoustic Spectrosopy*. New York: Wiley.

Schäfer S, Miklós A and Hess P (1997) Pulsed laser resonant photoacoustics/applications to trace gas analysis. In: Mandelis A and Hess P (eds) *Progress in Photothermal and Photoacoustic Science and Technology*, Vol 3, pp 254–289. Bellingham, WA: SPIE.

Schäfer S, Miklós A and Hess P (1997) Quantitative signal analysis in pulsed resonant photoacoustics. *Applied Optics* **36**: 3202–3211.

West GA, Barrett JJ, Siebert DR and Reddy KV (1983) Photoacoustic spectroscopy. *Review of Scientific Instruments* **54**: 797–817.

Zharov VP and Letokhov VS (1986) *Laser Optoacoustic Spectroscopy*. Berlin: Springer.

Photoelectron Spectrometers

László Szepes and **György Tarczay**, Eötvös University, Budapest, Hungary

HIGH ENERGY SPECTROSCOPY
Methods & Instrumentation

This article is centred on one of the most important molecular spectroscopic applications of photoionization. First, conventional photoelectron spectrometers based on the pioneering work of Turner, Vilesov and Siegbahn are reviewed, the basic building elements are shown and the principles of operation are discussed. Besides the conventional photoelectron experiment, threshold analysis and photoelectron-photoion coincidence spectroscopy as well as conceptually new techniques related to laser photoionization are discussed briefly.

Conventional photoelectron spectrometers

Photoelectron spectroscopy is a molecular spectroscopic method which is based on photoionization. If an atom or molecule (M) is irradiated with photons of larger energy than the ionization energy of the particle ionization may occur. The fundamental process and its energetics are given by the following equations:

$$M + h\nu = M_i^+ + e^- \qquad [1]$$

$$h\nu = \mathrm{IE}_i + \mathrm{KE}_i \qquad [2]$$

where IE_i is the i-th ionization energy of the system studied and KE_i is the kinetic energy of the ejected electron. Complementary to mass spectrometry, where the fate of the ion is investigated, in the case of photoelectron spectroscopy the other 'reaction product', the electron, is the source of the information. As a consequence, photoelectron spectroscopy normally requires a monoenergetic radiation source, a sample inlet system, a target chamber where photon–atom/molecule interaction occurs, an electron kinetic energy analyser, a detector and a recording system. A further requirement comes from the fact that electrons lose energy through collisions with gas molecules, thus electrons can only be handled in high vacuum where the mean free path is of the same order as the characteristic geometric dimension of the spectrometer. This is normally achieved at a pressure equal to or less than 10^{-5} mbar (10^{-3} Pa). A block diagram of a typical experimental arrangement is shown in **Figure 1**.

Two types of photoelectron spectrometers are distinguished on the basis of the energy of the radiation sources. Energy needed to eject electrons from the valence shell (above 6 or so eV) corresponds to photon wavelengths in the vacuum ultraviolet (VUV) region of the electromagnetic spectrum. This branch of instruments is known as VUV photoelectron spectrometers, normally abbreviated as UPS. More energetic X-ray photons, commonly in the range of 1000–2000 eV, are used for core electron ionizations. This type of photoelectron experiment is called X-ray photoelectron spectroscopy, XPS, also named ESCA (electron spectroscopy for chemical analysis).

As far as operation is concerned there is no principle difference between the above two methods. However, the diverse fields of application – gas-phase, molecular studies on the valence shell versus solid-phase surface studies based on core electron ionizations – make sense of this distinction.

The term 'conventional photoelectron spectrometers' designates those UPS instruments which are technically, based on the invention of Turner and Vilesov and operate with an energy resolution between 5 and 30 meV. As far as the main building units are concerned, an XPS spectrometer has a similar experimental arrangement, but lower energy resolution (0.5–1 eV) is obviously achieved.

Light sources

A photon source used in photoelectron (PE) spectroscopy must fulfil at least two basic requirements. First, the incident radiation must be monochromatic; the second requirement is high photon intensity. In order to achieve sufficient electron flux at the detector a source with a typical intensity of 10^{10}–10^{12} photons s^{-1} is needed. The most commonly used source in the XPS technique is the X-ray tube, while in valence shell PE spectroscopy resonance lamps are applied.

In the X-ray tube, radiation is generated by the bombardment of the anode with high-energy (10–15 keV) electrons. The schematic view of the widely used dual anode source is seen in **Figure 2**. In this set-up the two filaments, which are at ground potential, can be heated separately. Electrons from the hot cathode hit the corresponding positive anode face. The emerging characteristic X-ray radiation leaves the source through a thin aluminium window. As the two anode faces are covered by two different metal layers, the photon energy can be changed without a break. There are many metals which can be used as anodes (see

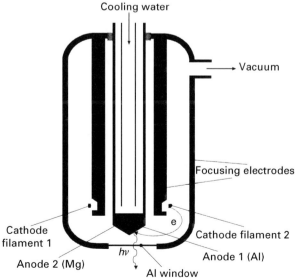

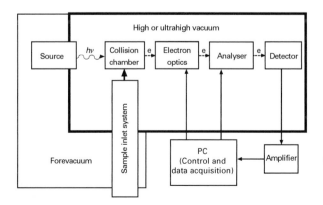

Figure 1 Block diagram of a photoelectron spectrometer.

Figure 2 A dual anode X-ray tube.

Table 1) but in most cases Mg or Al are used due to their narrow line width. With the use of a monochromator it is possible to filter out the satellite lines as well as the continuous radiation (the Bremsstrahlung) and to decrease the line half width to about 0.2 eV.

Ultraviolet resonance radiation may be produced by spark, direct current (d.c.) or microwave/radio frequency discharge through a rare gas. The most common type in use is the low-pressure helium d.c. discharge lamp (**Figure 3**). The discharge is initiated by applying a high voltage (~5 kV) across the discharge capillary. After the ignition period the voltage falls to 600–700 V and the gas pressure can be optimized for the generation of He(I) (21.21 eV), or He(II) radiation (40.81 eV). (The Roman numerals denote the emitting species which can be a neutral atom (I) or a singly (II) or doubly (III) ionized atom). Since there is no transmitting material in this energy region, the beam has to pass from the cathode area into the collision chamber through a windowless collimating capillary. To prevent self-absorption and to keep the gas out of the high vacuum side, differential pumping of the lamp is necessary. The line half width of this radiation is about 1 meV.

Besides helium some other noble (or atomic) gases can also be used. Their resonance energies are summarized in **Table 2**.

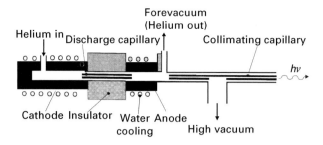

Figure 3 Low-pressure helium d.c. discharge lamp.

For some special studies (autoionization, predissociation, etc.) it is favourable to use a tunable source, and it is indispensable for threshold PE spectrometer experiments. The simplest way to generate tunable radiation is the use of a many-line or a continuous radiation source with a monochromator. A many-line source is usually a low-pressure hydrogen lamp while the other can be continuous helium or synchrotron radiation. Synchrotron radiation is produced by electron accelerators or storage rings as a consequence of radial acceleration of the electrons in a magnetic field. The main advantage of synchrotron radiation is its high intensity and broad-range tunability.

Sample inlet systems

Although both XPS and UPS techniques can be adapted for the investigation of either condensed or gas-phase samples, the former technique is almost exclusively applied for surface analysis while UPS is considered mainly as a gas-phase electron structure elucidation method.

The ideal solid sample is approximately 1 cm square with a mirror-smooth surface. Lumpy samples can also be measured directly while powders can be stuck onto a plastic tape or pressed into a pellet.

Regardless of the physical appearance of the solid sample, it is usually pretreated. The simplest treatment is washing the surface with organic solvents. The main drawback of this method is that the composition of the surface may change during the treatment: on the one hand, some ions may be washed out from the surface layer, on the other hand, the solvent, or even elemental carbon, may be adsorbed on the surface. These solvents, together with adsorbed water and gases can be removed in the instrument by baking the sample in ultrahigh vacuum conditions. The most sophisticated surface cleaning method is electron or ion bombardment, and laser vapourization can also be used. These techniques are also appropriate for the etching of the surface layer-by-layer, providing the possibility for depth analysis.

Table 1 Some X-ray lines used in X-ray photoelectron spectroscopy

Anode	Transition	Energy (eV)	Line width (eV)
Be	K	108.9	5.0
Ni	L_α	851.5	2.5
Cu	L_α	929.7	3.8
Mg	K_α	1253.6	0.7
Al	K_α	1486.6	0.85
Zr	L_α	2042	1.7
Ti	K_α	4510	2.0
Cr	K_α	5417	2.1
Cu	K_α	8048	2.6

Table 2 Resonance energies of some discharge lamps

Gas	Line	Energy (eV)
He	He(I)α	21.2175
	He(I)β	23.0865
	He(II)α	40.8136
Ne	Ne(I)α	16.6704, 16.8476
Ar	Ar(I)α	11.6233, 11.8278
Xe	Xe(I)	8.4363, 9.5695
H	Lymanα	10.1986
	Lymanβ	12.0872
	Lymanγ	12.7482

In the case of a gas-phase sample a large pressure gradient must be sustained between the interaction region and the rest of the system in order to obtain a sufficient electron current and to avoid the scattering of the electrons on the sample. The easiest way to achieve this gradient is the application of a collision chamber which may or may not be differentially pumped. In this setup the sample and the UV light pass into the chamber through a capillary or a bore and the electrons exit through a third slit directed toward the electron optics and the analyzer. The geometry of the chamber is not significant but its surface must be well cleaned, smoothed and uncharged. Local surface charges or the adsorption of dipolar molecules on the surface will shift the spectrum and decrease the resolution. These effects are most easily minimized by surface coatings of gold or colloidal graphite.

It is also possible to investigate less volatile solid samples in the gas phase. In this case the sample is evaporated close to (or in) the collision chamber. The sample can be heated by circulating hot liquid, by the heat of the discharge lamp or by a resistance oven. The temperature usually attainable by these methods is about 300°C. The temperature can be increased further up to 1000°C by special oven designs. In these constructions the disturbing magnetic and electric fields of the resistance heater are reduced by the use of noninductively wound and shielded wires. Even higher temperatures (2000–2500°C) can be achieved by laser vapourization.

Short-lived reaction products prepared *in situ* can also be studied. The sample molecules can undergo reaction shortly before ionization. Monomolecular reactions are commonly induced by UV photolysis or pyrolysis, while the feeding of highly reactive radicals into the sample vapour can cause bimolecular reactions. A well-tried variant of the latter method is the fluorine atom reaction where atoms generated from fluorine molecules in a microwave discharge outside the instrument are injected into the sample just before the ionization chamber. These methods qualify PE spectroscopy for the investigation of highly reactive molecules, transients and radicals not stable under normal conditions.

Analysis of energetic electrons

Basically, there are three ways of measuring the kinetic energy of electrons. The first is to measure the time needed to traverse a known distance. In the second method a retarding potential is applied to the electrons to be analysed. The third method is based on the deflection of the electrons in an electrostatic or magnetic field.

The time-of-flight (TOF) of an energetic electron over a distance of a few centimetres is very short, consequently, electronics with a response time of the order of nanoseconds are needed. This sort of kinetic energy analysis has not gained widespread application in the conventional techniques.

In retarding-field analysers, only those electrons are permitted to reach the detector which have higher energy than the retarding potential. This type of energy analysis can be performed by placing a grid in front of the detector and varying its potential. Many variations of retarding-field analysers are known; cylindrical and spherical arrangements are the most common. Analysers based on this principle yield a stepwise integral energy distribution curve in which a rise in the detected electron current indicates a group of energetically different electrons. The energy distribution can be obtained by differentiating this plot of detector signal versus retarding potential. The advantages of this type of analyser are an almost equal transmission of electrons of all energies and a large acceptance angle and consequently high sensitivity. One drawback of this method is that the examination of low-energy electrons is sometimes difficult. Furthermore, any contamination of the grids may give rise to surface potential variations which result in the loss of resolution. Retarding-field analysers can be very easily constructed but problems related to their performance limit their use for high-resolution studies.

Finally, the energies of electrons can be determined by passing them through an electrostatic or magnetic field where the deflection of the electron paths is a function of their energy. Since it is generally easier to produce a uniform electric field than a uniform magnetic field, electrostatic analysers have come to predominate in photoelectron spectroscopy. Two types of dispersive analysers are generally distinguished. In deflectors, electrons follow equipotential lines during analysis, while in mirrors electrons cross equipotential lines on moving between two electrodes.

The following part is restricted to the description of some frequently used electrostatic analysers, namely the *radial cylindrical*, the *hemispherical* and the *cylindrical-mirror* analysers.

One of the most important parameters of an analyser is the energy resolution. It may be defined as a measure of the ability of an analyser to resolve two adjacent peaks in the spectrum separated by ΔE. The resolution for deflection analysers at energy E is given by an equation of the form

$$\Delta E/E = a\omega + b(\Delta\alpha)^2 + c(\Delta\beta)^2 \qquad [3]$$

where a, b and c are constants characteristic of the particular analyser; ω is the entrance and exit slit width; and $\Delta\alpha$ and $\Delta\beta$ are the angular deviation of the electron beam in the plane of deflection and the perpendicular plane, respectively.

Figure 4 shows a 127° electrostatic cylindrical analyser, one of the most frequently used in gas-phase molecular photoelectron spectroscopy. Electrons produced in the ionization chamber pass through the entrance slit (S_1) and enter the analyser where they are deflected by a radial electric field produced by the electric potentials V_1 and V_2 placed on the concentric cylindrical electrodes. According to charged particle optics, focusing is achieved after an angular deflection of 127°. If E is the energy of electrons following the central path transmitted through the analyser and is related to the electrical potential V by $E = qV$, then the potentials to be applied to the outer and inner electrodes are

$$V_2 = V + 2V \ln(R_2/R) \qquad [4]$$

and

$$V_1 = V + 2V \ln(R_1/R) \qquad [5]$$

where R_2 and R_1 are the radii of the outer and the inner electrodes while R is the mean radius.

The majority of commercial electron spectrometers – either UPS or XPS instruments – are equipped with some form of spherical deflector analyser. The 180° spherical sector – shown in **Figure 5** – is often used because of its compactness and good resolving power.

In order to transmit electrons of energy $E = qV$ along a circular path of radius R, electrical potentials applied to the outer (V_2) and inner (V_1) electrodes are given by

$$V_2 = V[2(R/R_2) - 1] \qquad [6]$$

and

$$V_1 = V[2(R/R_1) - 1] \qquad [7]$$

where R_2 is the radius of the outer hemisphere while R_1 is that of the inner one.

For a hemispherical deflector, the coefficient c in the expression of resolution is equal to zero. This means that exact focusing is achieved in the plane perpendicular to the plane of deflection.

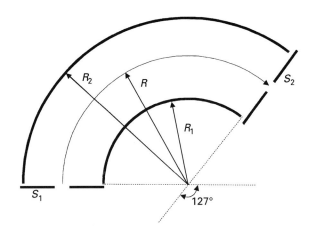

Figure 4 The 127° radial cylindrical analyser R_1, R, R_2 are the radius of the inner electrode, the midradius, and the radius of the outer electrode, respectively, S_1 and S_2 are the entrance and exit slits, respectively.

The cylindrical mirror analyser (CMA) is shown in **Figure 6**. In this analyser, electrons emerging from the ionization region (located on the axis) pass through an annular slit into the space between two concentric cylinders. Two arrangements may be distinguished for this analyser system: axial focusing and slit-to-slit focusing. In the former case, electrons of energy E are deflected so that after leaving the exit slit they are focused along the axis. Here, focusing occurs in both the deflection plane and the perpendicular plane. A source of modest size is expected for this design. Further parameters of operation are: the injection angle is 42.3°; the inner cylinder is earthed together with the ionization region, and the potential on the outer cylinder is

$$V = 1.3 \, E \ln(R_2/R_1) \qquad [8]$$

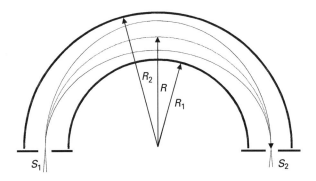

Figure 5 The hemispherical analyser. R_1, R, R_2 are the radius of the inner electrode, the midradius, and the radius of the outer electrode, respectively. S_1 and S_2 are the entrance and exit slits, respectively.

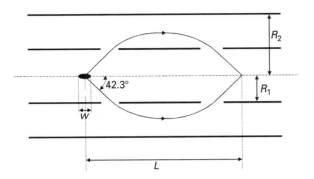

Figure 6 The axial focusing cylindrical mirror analyser. R_1 and R_2 are the radius of the inner and the outer cylinder, respectively; w is the axial extent of the source, L is the source–detector distance.

The resolution is approximately

$$\Delta E/E = 1.09(w/L) \qquad [9]$$

where w is the axial extent of the source and L is the distance from the source to the detector.

If the source is not small and well defined, slit-to-slit focusing can be chosen. In this arrangement energy-resolving slits in the inner electrode are used and the image occurs on the surface of the inner cylinder. This design yields large signals but focusing is obtained in the deflection plane only. Slit-to-slit focusing has been successfully used in high temperature UPS studies where small signals are expected.

CMA systems consisting of two axial-focusing analysers in series have been applied for XPS instruments. This combination gives an improved resolution at the cost of sensitivity.

In all electrostatic-deflection analysers the width of the bandpass ΔE is proportional to the transmitted energy E. That is why the absolute energy resolution can be improved by preretardation of the electrons prior to energy analysis. Most spectrometers employ this technique using appropriate electron optical elements.

Detection and registration

An electron current of less than 10^{-14} A, which is typical in photoelectron spectroscopy, can be detected with an electron multiplier. The multipliers used earlier consisting of Cu–Be dynodes have now been replaced by channeltrons. This variant of the electron multipliers is a curved glass or ceramic tube representing a continuous dynode system based on its semiconducting inner surface. The resistance between the two ends of channeltron is about 10^9 Ω and a potential of about 3 kV is applied across it in operation. The attainable gain is about 10^8. Channeltrons have the advantage of small size, low cost and ruggedness. Multichannel plates operating by the same principle are also in use. In this case, electrons within a range of energies are collected simultaneously at the focal plane of a suitably constructed analyser.

Following detection, data are collected and processed by a PC.

Operational considerations

In photoelectron spectroscopy one must compromise between energy resolution and the intensity of the detected signal (S) relative to the noise (N). (Both quantities expressed in counts per second.) The resolution may be enhanced at a lower signal count rate and vice versa. The maximum attainable signal-to-noise ratio is

$$(S/N)_{\text{max}} = S^{1/2}t^{1/2} \qquad [10]$$

where t is the data collection time. The above relationship indicates that theoretically any desired resolution and signal-to-noise ratio can be achieved if sufficient data collection time is provided.

Practically, there are two modes of operation of PE spectrometers. The first is scanning the voltage V applied to the analyser electrodes. In this case, the energy resolution, $\Delta E/E$, is constant and consequently peak widths vary across the spectrum. Alternatively, if the analyser voltages are fixed while the retarding potential is scanned by the electron optics, the peak width does not change throughout the spectrum. Finally, it has to be mentioned that an accurate electron energy cannot be determined from the voltage applied to the analyser because an energy shift will be present due to contact potentials within the analyser and other part of the spectrometer. Therefore, energy calibration by introducing an inert gas together with the sample is necessary for precise ionization energy determinations.

Threshold photoelectron spectroscopy

As described above, conventional photoelectron spectrometers operate at fixed photon energy while the electron energy is scanned. If a continuous light source is available the photon energy may be scanned while electrons of essentially zero kinetic energy, called threshold electrons, are detected. This is the operational principle of threshold photoelectron spectroscopy (TPES). Threshold electron detection uses a simple technique which discriminates strongly against energetic electrons while electrons formed with zero, or near zero kinetic energy are collected very effectively with small electric fields. In practice, threshold electrons are selected by a steradiancy analyser which consists of a stainless steel collimated hole system with a specific length-to-diameter ratio. During the analysis, threshold electrons are directed to pass through the tubes while those with appreciable energy are likely to hit the tube walls and be lost. Considering the fact that energetic electrons with initial velocity vector directed toward the detector are not discriminated by the analyser, the energy resolution using a discharge lamp is not better than 2–4 meV. This type of photoelectron spectroscopy is suited for experiments with synchrotron radiation sources as well as for photoelectron–photoion coincidence (PEPICO).

Photoelectron–photoion coincidence spectroscopy

In VUV photoelectron spectroscopy, the kinetic energy spectrum of electrons is measured which provides information about the ionic states of atoms and molecules. However, very limited information is given about the fate of the excited species formed in the photoionization process. By the use of photoelectron–photoion coincidence (PEPICO) spectroscopy a detailed analysis of the dissociation of molecular ions is possible: the fragmentation of a particular ionic state – identified through the photoelectron energy – can be directly studied. So, if ions are detected in coincidence with electrons of a given kinetic energy, the unimolecular decay of molecular ions in selected energy states can be investigated. There are two types of PEPICO instruments. In one case a light source of fixed energy, usually He(I), is used together with a dispersive electron energy analyser. Two modes of operation of this experimental setup are feasible: (i) the mass spectrum is scanned at a fixed electron energy, or (ii) a particular ion is chosen and the PE spectrum in coincidence is measured. According to the other approach, a continuum light source in conjunction with a vacuum monochromator is used and threshold electrons are collected in coincidence with parent and fragment ions, so that the ion internal energy is given by $h\nu$–IE. In a typical PEPICO arrangement, which proved to be extremely powerful in the field of dissociation dynamics, threshold electrons not only serve to identify the formation of an ion of known internal energy but also provide a start signal for measuring the time-of-flight (TOF) of the ion. The ion TOF distribution contains all dynamical information such as the kinetic energy released in polyatomic dissociations, metastable ion lifetimes and collision dynamics of ion–molecule interactions.

High-resolution photoionization techniques

Since the 1980s, some conceptually new photoionization techniques have appeared in the field of gas-phase investigations whose resolution is better than that of the conventional UPS technique by some orders of magnitude. Two technical novelties preceded the appearance of these techniques: the introduction of supersonic jets in photoelectron spectroscopy and the availability of high-energy lasers.

Supersonic jets have been used from the early 1950s in the fields of beam studies. According to this technique, the sample vapour is adiabatically expanded from several atmospheres into a high vacuum through a nozzle which has a diameter between 50 and 300 μm. As a result, a molecular beam with low translational (~1 K), rotational (~10 K) and vibrational (~100 K) temperatures is produced. This means that, according to the Boltzmann distribution, the non-ground rotational and vibrational states will be barely populated and the sample molecules are obtained in a well-defined energetic state. Feeding a noble gas to the sample beam as a carrier enhances this cooling effect. At an early stage this method was carried out by installation of huge pumps. Nowadays, the use of a pulsed nozzle is more common. In this case the nozzle opens repeatedly for some 100 μs, and the pump has to be large enough to evacuate the chamber between two pulses only. The diverging molecular beam can be collimated or skimmed. For this purpose, separate pumping of the additional chamber (between the nozzle and the skimmer) is also needed.

The next step toward the new techniques was the availability of high-power UV lasers. In a typical setup the UV laser beam is generated as follows: an excimer or a YAG laser produces monochromatic photons, the photon energy is then tuned by a dye laser and at the last stage the selected frequency may be doubled by a crystal (e.g. by β-barium borate, BBO). The largest photon energy attainable by this method is ~6.5 eV. Higher energies can be reached by frequency mixing in gaseous media. Four-wave mixing and the third harmonic generations are the two typical approaches to this problem. These relatively new techniques extend the laser photon energy into the VUV region up to 19 eV.

Resonance enhanced multiphoton ionization

The resonance enhanced multiphoton ionization (REMPI) technique was originally developed as a sensitive method for the investigation of excited states. In the REMPI experiment a laser excites the molecule, then either the same laser (one-colour REMPI or 1+1 REMPI) or another with a different photon energy (two-colour or 1+1' REMPI) ionizes the molecule from the excited state. If the photon density in the well-focused laser beam is high enough, nonresonant two-photon excitation may occur and then, similar to the previous case, the same laser, or a second one, can ionize the molecule (2+1 and 2+1' REMPI). By scanning the wavelength of the first laser and measuring the amount of ions formed, information can be derived about the excited states.

Since excitation and ionization are usually carried out by pulsed lasers the obvious detection method is TOF-mass spectrometry. Formerly, linear TOF tubes were in use but they have now been replaced by reflectron TOF tubes. These tubes are folded at the middle and the ions are reflected backward by positively charged rings in the fold (see **Figure** 7). This type of TOF detection has two advantages. Firstly, ions of the same mass and charge but different kinetic energy will hit the detector at the same time. Secondly, the parent ions and the ions formed by fragmentation in the drift (or in the accelerating) region can be separated by tuning the reflectron voltage.

Beside ions, the detection of electrons is possible. If electrons are analysed according to their kinetic energy the result is a new type of photoelectron spectroscopy, called REMPI-PES. This technique usually uses TOF analysers. Prolongation of the flight time of

the electrons, and consequently enhanced effectiveness of separation, can be achieved by the use of an external magnetic field.

Information about the ionic states can be gained more easily by scanning with the second (ionizing) laser and monitoring the total ion intensity. This method is called REMPI-photoionization efficiency (REMPI-PIE) spectroscopy. The REMPI-PIE spectrum is a staircase function, and the conventional photoelectron spectrum can be obtained from its differentiation.

Zero kinetic energy spectroscopy and mass-analysed threshold ionization

In 1984 Müller-Dethlefs and Schlag pioneered a new laser photoionization method called zero kinetic energy (ZEKE) spectroscopy. The original idea was to detect only the strictly zero kinetic energy electrons in the following way. After the laser pulse the zero kinetic energy electrons remain in the ionization chamber for some microseconds. During this time the nonzero kinetic energy electrons have left this region. After the delay the zero kinetic energy electrons are extracted by a pulse of electric field ($\sim 0.1-1$ V cm^{-1}) toward the flight tube and the detector (**Figure** 7). The latter is gated for the time of arrival of these electrons. However, practice has shown that an apparatus operating along these principles is unable to prevent the detection of electrons having a relatively small but still measurable kinetic energy (the so-called near-threshold electrons) which has a detrimental effect on the resolution. As it turned out, the mechanism described above – although it had been supposed since the late 1980s – only accurately describes

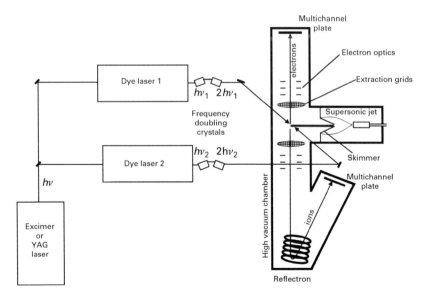

Figure 7 Schematic illustration of an experimental setup capable of performing REMPI, ZEKE and MATI measurements.

the process related to the electron removal from anions (electron detachment spectroscopy). In 1988 Reiser showed that when neutral molecules are ionized, the applied pulsed field extracts the electrons from long-lived Rydberg states formed by laser excitation. (This is the so-called pulsed field ionization). Current ZEKE instruments, utilizing the recognition of this mechanism, employ a weak electric field pulse (usually opposite in sign to the extracting field) just after the laser pulse, followed by a delay of some microseconds and then the extracting pulse. The first pulse removes the near-threshold electrons and also those in the highest-lying Rydberg states. The main extracting pulse then ionizes the remaining molecules in next-to-highest states; therefore the measured ionization energy will be somewhat smaller than the real value. The exact ionization energies can be obtained by repeated measurements at different fields followed by extrapolation to zero field. More commonly, the energies are corrected using the following equation:

$$\delta E \approx 4\sqrt{F} \qquad [11]$$

where F is the applied field. This method provides photoelectron spectra with a resolution of 0.1–0.5 cm^{-1} (~0.05 meV).

The extension of this technique to ions is somewhat complicated. The difficulty is due to the higher mass of the ions. In order to distinguish between ions formed by the electric field pulse and those formed by the laser pulse, longer pulses are needed. This spectroscopy is called mass-analysed threshold ionization (MATI). The main benefit of this method is the possibility of recording the spectra of various fragment ions together with that of the parent ion.

Technically, laser excitation to Rydberg states can be performed either directly or, similarly to the REMPI technique, via an excited state, using two lasers. The advantage of the first method is that we need no prior information about the excited states. The advantage of the second method is that investigation of chemically impure samples can be carried out since one can select of the molecule of interest by varying the excitation laser energy.

List of symbols

E = energy; I = extracting field; IE = ionization energy; KE = kinetic energy; L = distance from source to detector; N = noise; R = mean radius; R_1 = radius of inner electrode; R_2 = radius of outer electrode; S = signal; V_1 = potential applied to inner electrode; V_2 = potential applied to outer electrode; w = axial extent of source; ΔE = resolution; $\Delta\alpha$ = angular deviation of electron beam in plane of deflection; $\Delta\beta$ = angular deviation of electron beam in perpendicular plane; ω = entrance and exit slit width.

See also: **Laser Applications in Electronic Spectroscopy; Laser Spectroscopy Theory; Light Sources and Optics; Multiphoton Excitation in Mass Spectrometry; Multiphoton Spectroscopy, Applications; Optical Frequency Conversion; Pharmaceutical Applications of Atomic Spectroscopy; Photoelectron Spectroscopy; Photoionization and Photodissociation Methods in Mass Spectrometry; Pyrolysis Mass Spectrometry, Methods; Time of Flight Mass Spectrometers; X-Ray Spectroscopy, Theory; Zero Kinetic Energy Photoelectron Spectroscopy, Applications; Zero Kinetic Energy Photoelectron Spectroscopy, Theory.**

Further reading

Baer T (1979) State selection of photoion-photoelectron coincidence. In: *Gas Phase Ion Chemistry.* New York: Academic Press.

Briggs D (ed). (1977) *Handbook of X-ray and Ultraviolet Photoelectron Spectroscopy.* London: Heyden.

Brundle CR and Baker AD (eds) (1977) *Electron Spectroscopy: Theory, Techniques and Applications.* Vol 1–4 London: Academic Press.

Eland JHD (1974) *Photoelectron Spectroscopy.* London: Butterworth.

Hollas JM (1997) Photoelectron and related spectroscopies; lasers and laser spectroscopy. In: *Modern Spectroscopy, Third Edition.* Chichester: Wiley.

Kamke W (1993) Photoelectron-photoion coincidence studies of clusters. In: *Cluster Ions.* Chichester: Wiley.

Martensson N, Baltzer P, Brühwiler PA *et al* (1994) A very high resolution electron spectrometer. *Journal of Electron Spectroscopy and Related Phenomena* 70: 117–128.

Müller-Dethlefs K and Schlag W (1998) Chemical applications of zero kinetic energy (ZEKE) photoelectron spectroscopy. *Angewandte Chemie, International Edition in English.* 37: 1346–1374.

Powis I, Baer T and Ng CY (eds) (1995) *High Resolution Laser Photoionization and Photoelectron Studies.* Chichester: Wiley.

Rabalais JW (1977) *Principles of Ultraviolet Photoelectron Spectroscopy.* New York: Wiley.

Schlag EW (1998) *ZEKE Spectroscopy.* Cambridge: Cambridge University Press.

Schlag EW, Peatman WB and Müller-Dethlefs K. (1993) Threshold photoionization and ZEKE spectroscopy: a historical perspective. *Journal of Electron Spectroscopy and Related Phenomena* 66: 139–149.

Photoelectron Spectroscopy

John Holmes, University of Ottawa, Ontario, Canada

Photoelectron spectroscopy

The Encyclopedia includes a specialized article entitled 'Zero Kinetic Energy Photoelectron Spectroscopy' that describes a modern development of a long-established technique.

The principle upon which photoelectron spectroscopy (PES) is based is simple. If a molecule is excited by a high-energy photon in the ultraviolet region of the spectrum that has sufficient energy to ionize the molecule, the excited species will eject electrons. PES is the analysis of the kinetic energies of the ejected electrons. For a given excitation energy, the energy distribution of the ejected electrons reflects the distribution of accessible energy levels of the excited (ionized) molecule.

$$E_{\text{ion}} = h\nu - E_{\text{electron}} = h\nu - \tfrac{1}{2} m_e v_e^{\,2}$$

The first experiments in this field were performed in Russia in 1961 and in England in 1962.

With the measurement of highly resolved electron energy distributions, the orbitals from which the electrons are lost can be identified as well as vibrational progressions for the various excited ionic states.

The most commonly used UV photons are those from the He(I) line $(1s^2 2p^1 \rightarrow 1s^2)$ from a helium discharge lamp, at 58.43 nm, corresponding to an energy of 21.22 eV, well above the ionization energy of organic compounds. For a thorough review of the technique and its results, the book by Turner, a founding father of the subject, should certainly be consulted.

List of symbols

E_X = energy of species X; h = Planck constant; m_e = electron mass; v_e = electron velocity; ν = photon frequency.

See also: **Photoelectron Spectrometers; Photoelectron Spectroscopy, Applications; Photoelectron Spectroscopy, Theory.**

Further reading

Brundle CR (1993) UPS at the beginning. *Journal of Electron Spectroscopy and Related Phenomena* 66: 3–17.

Eland JHD (1983) *Photoelectron Spectroscopy; an Introduction to UV Photoelectron Spectroscopy in the Gas Phase*, 2nd edn. London: Butterworths.

Turner DW, Baker AD, Baker C and Brundle CR (1970) *Molecular Photoelectron Spectroscopy*. London: Wiley Interscience.

Photoelectron–Photoion Coincidence Methods in Mass Spectrometry (PEPICO)

Tomas Baer, University of North Carolina, Chapel Hill, NC, USA

Introduction

Photoelectron–photoion coincidence (PEPICO) is a method for energy selecting ions and studying their reaction dynamics. This subfield of mass spectrometry provides information about the mechanism and dissociation dynamics of ions. It is also a subfield of reaction kinetics because it provides a simple method for investigating reaction rates with energy-selected ions. Mass spectrometry is a useful analytical tool for investigating the structure of molecules. It is based on the principle that a molecule such as A-B-C-D-E-F once ionized in the source of a mass spectrometer, breaks apart into the ionic fragments A-B-C$^+$, C-D-E-F$^+$, A-B$^+$, etc. By piecing together the various units whose masses have been measured, it is possible to reconstruct the original molecule. Such an approach works very well if the molecule is well behaved and breaks apart in a manner suggested above. However, it is common for ions to rearrange to new structures prior to dissociation. In those cases, the fragment ions observed in the mass spectrometer may not have any simple relationship to the original molecule. An example is acetol, which reacts at low energies in the following manner:

$$CH_3C(O) - CH_2OH \begin{cases} CH_3CO^+ + CH_2OH^\bullet \quad [1] \\ C_2H_5O^+ + HCO^\bullet \quad [2] \end{cases}$$

The first dissociation path yields a product ion that is clearly related to the original acetol molecule. However, the second dissociation path, the loss of HCO$^\bullet$, is not possible from the original structure. Indeed, a rather major isomerization of the acetol ion must precede the loss of HCO$^\bullet$. Isomerization reactions often dominate the dissociation dynamics at low energies, when the ion is formed with an energy just above the dissociation limit. At higher energies, reaction channels that are dominant near threshold are often superseded by other dissociation paths. Thus, learning about reaction mechanisms and their variation with the ion internal energy becomes an important part of understanding the final mass spectrum. Information of this sort is provided by the PEPICO technique.

Methods for ion energy selection

Review of the methods

A number of methods for ion energy selection have been developed. They can be divided into two categories. In one, ions are first produced in the ground (or near ground) state and then excited by photons, usually in the form of laser light. Such experiments have been carried out in ion cyclotron resonance (ICR) mass spectrometers, in sector instruments, and in laser-based time-of-flight (TOF) instruments. The advantage of this method is that laser methods can be

used which provide good intensity and good energy resolution. Furthermore, it is possible to investigate very fast rate processes. The disadvantage is that parent ions are sometimes difficult to produce with little excess energy so that the final ion energy may be uncertain.

The other method involves direct ionization of the molecule to the ion energy of interest. This can be done optically by photoelectron–photoion coincidence (described below), or by charge exchange, which is a form of chemical ionization. Common chemical ionization methods include charge transfer and proton transfer.

$$Xe^{\bullet+} + CH_3OH \rightarrow Xe + CH_3OH^{\bullet+} \quad [3]$$

$$CH_4^{\bullet+} + CH_3OH \rightarrow CH_3^\bullet + CH_3OH_2^+ \quad [4]$$

Both methods produce ions with little internal energy. Charge transfer between a rare gas atom and a molecule deposits all of the reactant ion energy into the molecular ion as internal energy, with small amounts going into the translations. Charge transfer with ions of various ionization energies has thus been used to energy select ions. The major shortcoming of this method is that only a limited number of atomic ions can be used for this purpose.

Energy selection by photoelectron photoion coincidence

The photoionization process   Ionization by photons takes place by both direct ionization and by autoionization.

$$AB + h\nu \rightarrow AB^{\bullet+} + e^- \quad [5]$$

$$AB + h\nu \rightarrow AB^* \rightarrow AB^{\bullet+} + e^- \quad [6]$$

The processes differ in that the transition probability in the former is determined by the ionization continuum, whereas it is governed by the transition to the excited neutral state, AB*, in the latter. Of interest to the PEPICO experiment is the decay of these autoionizing resonances and their effect on the ability to select electrons of a given energy. It has been found that, because of autoionization, ions can be prepared in Franck–Condon gap regions that are normally not accessible by direct photon excitation from either the ground neutral or the ground ionic state.

The principle of energy selection by PEPICO When a molecule absorbs a vacuum ultraviolet (VUV) photon with an energy ($h\nu$) above the molecule's ionization energy (IE), the ejected electron can have energy that ranges from 0 to IE – $h\nu$. Thus, the ion is produced with an internal energy given by

$$E_{ion} = h\nu - IE - E_{el} + E_{thermal} \qquad [7]$$

where E_{el} is the energy carried away by the electron and $E_{thermal}$ is the thermal energy of the molecule prior to ionization. It is evident from Equation [7] that an ion of internal energy E_{ion} is associated with an electron of a given kinetic energy. Thus, by detecting only ions that are collected in coincidence with an electron of a given energy, it is possible to selectively investigate energy-selected ions. This is done by placing a small electric field across the ionization region so that electrons and ions are accelerated in opposite directions. The electrons are passed through an appropriate energy analyser and collected by an electron multiplier. In the meantime, the much heavier ion requires much longer to travel down the drift tube of a TOF mass spectrometer. The time difference between the arrival time of the electron and the ion is the ion TOF. Thus, the basic information is contained in the ion TOF distribution in which the energy analysed electron provides the start signal and its corresponding ion provides the stop signal. **Figure 1** shows a schematic of a typical threshold PEPICO apparatus. The start and stop signals in **Figure 1** are usually sent to a time-to-pulse height converter (TPHC) followed by a multichannel pulse height analyser that stores and displays the ion PEPICO spectrum.

Two types of PEPICO experiments have been carried out. In one the light source has a fixed energy (e.g. the He(I) source at 21.2 eV) and the electron energies are selected by a dispersive analyser such as a hemispherical analyser. In the more versatile approach (**Figure 1**), the light source energy is tunable and the electrons of interest have zero energy. Thus the ion energy is selected by varying the photon energy while the electron energy analyser remains fixed to pass initially zero energy, or threshold electrons. The major advantages of the latter approach are (a) the better energy resolution possible with threshold electron detection, (b) the much higher collection efficiency for threshold than for energetic electrons, (c) the ability to select ions in Franck–Condon gap regions (due to autoionizing resonances), and (d) the lower level of false coincidence signals. The major disadvantages are (a) the cost and complexity of the tunable VUV source and (b) the problem with discrimination against energetic electrons whose initial velocity vector is directed towards the electron multiplier. The latter problem is suitably solved only with the use of pulsed synchrotron radiation light sources.

Electron energy analysis

Energetic electrons

When a light source of fixed energy is used, the electron energy selection is made by a dispersive energy analyser such as a hemispherical analyser. The resolution of such an analyser is given by $(d/r)E_e$ in which d is the diameter of the input aperture, r is the average radius of the inner and outer sphere in between which the electrons pass, and E_e is the energy of the electron as it passes through the analyser. Although such analysers can be run with a resolution of 10 meV, this is not really practical for coincidence experiments because of a number of factors. If an electric field of ε V cm^{-1} is applied across the ionization region in order to extract ions, then the electrons will experience a voltage drop of εd_{ph} across the photon beam of width d_{ph} thus limiting the resolution to εd_{ph}. In order to avoid this, the ionization region must be kept field free and the ions extracted by a pulsed electric field applied whenever an electron is detected.

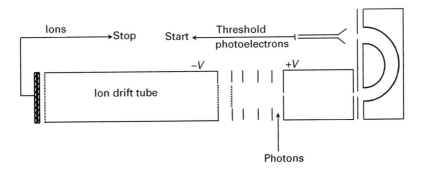

Figure 1 Schematic diagram of a threshold photoelectron photoion coincidence experiment. The electron resolution is determined primarily by the length of the electron drift tube and the size of the apertures.

Threshold electrons

The principle of threshold electron detection is totally different from the energy analysis of energetic electrons. A threshold has no initial kinetic energy, and thus must be extracted from the ionization region with an electric field. Because such an electron can be accelerated directly towards the detector, it can be collected with near 100% efficiency. On the other hand, energetic electrons will have a distribution of initial ejection angles which, as a first approximation, can be assumed to be isotropic. The application of a small electric field will not be sufficient to bend the electron trajectories towards the detector; thus most of the energetic electrons will be lost. Only those electrons whose initial velocity is directed toward the detector can be collected. (These are the electrons that are collected in a fixed photon energy PEPICO experiment.) It is possible to collect threshold electrons with good discrimination against energetic ones by simply passing them through small apertures in a long pipe. No real energy analysis is required. An important benefit for the PEPICO experiment is that a constant electric field can be applied to the ionization region, which gently extracts the threshold electrons and at the same time extracts the ions.

The electron energy resolution, or the electron collection efficiency, $R_{el}(E_0)$, as function of the electron's initial energy can be derived for an analyser that consists of a single acceleration region and a long drift region of length l and with apertures of diameter, d. If we assume a point source for the electrons within this electric field, the electron collection efficiency is given by

$$R_{el} = 1 - \left[1 - \left(\frac{E_g}{E_0} + 1 \right) \sin^2 \phi_c \right]^{1/2}$$
$$E_0 > E_g \tan^2 \phi_c \qquad [8]$$

$$R_{el} = 1 \qquad E_0 \leq E_g \tan^2 \phi_c \qquad [9]$$

where E_g is the energy gained by the electron as it is accelerated out of the ionization region, E_0 is the electron's initial kinetic energy, and ϕ_c is the planar critical collection angle, which for large l is approximately $d/2l$.

A resolution of 25 meV is readily attained with electric fields of 10 V cm^{-1}, an electron drift tube of 10 cm and apertures of 3 mm. The real power of this method is realized when the light source is pulsed with high repetition rates, as is possible with radiation from a storage ring of a synchrotron. In that case, an additional energy analysis, TOF, can be used to improve the resolution the resolution to a few meV.

Counting statistics — real and false coincidences

If an electron of the appropriate energy is collected but its corresponding ion is (for whatever reason) not collected, then it is possible that some other, totally unrelated ion can provide the stop signal. Such a coincidence event is called a 'false' coincidence because the electron and ion come from different precursor molecules. In a DC extraction field, these false coincidences appear at random times and thus provide a uniform background to the coincidence TOF mass spectrum. These can be easily distinguished from the real coincidences, which appear at a TOF that is determined by the ion's mass. However, if the ions are extracted with a voltage pulse, then both real and false ions will have the same TOF and are correspondingly more difficult to distinguish.

The number of false coincidence events depends upon the total ionization rate. If the total ionization rate is 10^6 ions s^{-1}, then an electron and ion pair are born every microsecond. Because most electrons are of the wrong energy, they are rejected so that the observed electron count rate is much smaller than 10^6. However, all ions can in principle be detected and, since it takes about 5 μs to extract an ion from the ionization region, it is clear that there are many ions that are able to provide stop pulses. A total ionization rate of 10^6 is not unreasonably high for a fixed light source PEPICO experiment because the helium resonance lamp is very intense and thus generates a very large ion signal.

PEPICO experiments work best with a continuous light source that is relatively weak, and with very high collection efficiencies for both electrons and ions. A pulsed source such as a 10 Hz pulsed VUV laser would not work because each laser pulse would produce many electrons and ions within a 10 ns interval. Thus, it is impossible to distinguish which electron is associated with which ion.

One of the unique features of PEPICO is the ability to determine the collection efficiencies and thus the absolute total ionization rate, N_T. The ion and electron collection efficiencies (E_i and E_e) can be calculated from the coincidence count rates (N_c) and the observed ion and electron count rates (N_i and N_e). They are given by

$$\text{Ion collection efficiency} = E_i = N_c / N_e \qquad [10]$$

$$\text{Electron collection efficiency} = E_c = N_c/N_i \quad [11]$$

Once the efficiencies are known, the total ion production rate is given by $N_T = N_i/E_i$, or N_e/E_e. These calculations make one important assumption, which is that the electrons and ions extracted from the ionization region originate from the same ionization volume.

The ion TOF distribution

The information content of the PEPICO technique lies in the ion TOF distribution. Not only does the TOF distribution disperse ions according to their masses, but it also provides information about the kinetic energy released in a dissociation, and about the dissociation rate of the ion if it is in the range of 10^4–10^7 s^{-1}. Examples of such data are discussed below.

The ion breakdown diagram

A breakdown diagram is a plot of the fractional abundances of ion masses as a function of the ion internal energy. An example is shown in **Figure 2** for the case of chromium hexacarbonyl. When the photon energy is above the molecule's ionization energy but below the first dissociation limit, the TOF distribution shows only one peak, the parent ion. Once the ion internal energy exceeds the dissociation limit, the parent ion signal disappears and in its place the first fragment ion appears, $Cr(CO)_5^+$, which is associated with the loss of a CO molecule. The cross-over energy at which parent and fragment ion intensities are equal provides a very precise means for determining this first dissociation limit.

As the photon energy is increased, the excess energy is partitioned between the translational, rotational and vibrational energies of the fragments. The rotational and vibrational energy remaining in the $Cr(CO)_5^+$ increases with photon energy until it exceeds the next dissociation limit for the loss of a second CO fragment. Because in the energy partitioning most of the energy remains in the vibrational modes of the larger fragment, the $Cr(CO)_5^+$ signal decreases to nearly zero while the $Cr(CO)_4^+$ ion takes its place. This continues up to the final loss of the last CO unit. Because of energy partitioning, each new onset is more gradual than the previous one. It must be pointed out that this breakdown diagram is very peculiar and not at all typical. It is characterized by the following successive dissociation paths:

$$Cr(CO)_6^{\bullet+} \rightarrow Cr(CO)_5^{\bullet+} \rightarrow Cr(CO)_4^{\bullet+} \rightarrow Cr(CO)_3^{\bullet+}$$
$$\rightarrow Cr(CO)_2^{\bullet+} \rightarrow Cr(CO)^{\bullet+}$$

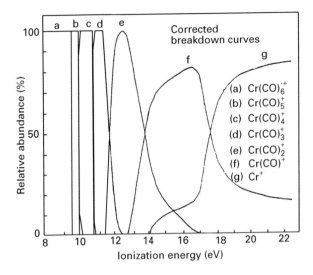

Figure 2 A breakdown diagram of chromium hexacarbonyl ions obtained by PEPICO. This ion dissociates by a sequential mechanism. The heats of formation of the various ions can be determined from the energy onsets of their formation. Reproduced with permission from Das PR, Nishimura T and Meisels GG (1985) Fragmentation of energy selected hexacarbonylchromium ion. *Journal of Physical Chemistry* **89**: 2808–2812.

Most organic ions dissociate not in sequential manners but in a combination of parallel and sequential reactions such as those for nitrobenzene:

$$C_6H_5NO_2^+ \rightarrow C_6H_5O^+ + NO^\bullet \quad AE = 10.98 \text{ eV} \quad [12]$$
$$\hookrightarrow C_5H_5^+ + CO \quad AE = 11.30 \text{ eV} \quad [13]$$
$$\hookrightarrow C_4H_3^+ + C_2H_2O \quad AE = 11.40 \text{ eV} \quad [14]$$
$$\rightarrow C_6H_5O^\bullet + NO^+ \quad AE = 11.04 \text{ eV} \quad [15]$$
$$\rightarrow C_6H_5^+ + NO_2^\bullet \quad AE = 11.14 \text{ eV} \quad [16]$$

Breakdown diagrams such as those in **Figure 2** provide an overview of the dissociation dynamics for ions as a function of the internal energy. Generally, fragments produced by low-energy rearrangement reactions appear first, while the direct bond cleavage products appear at higher energies. An example is the loss of $NO_2^\bullet$ from the nitrobenzene ion at 11.14 eV. Lower-energy dissociation paths involving the production of $NO^\bullet$ and NO^+ involve rearrangements.

An important aspect of the breakdown diagram is the cross-over energy for the first dissociation limit where the parent and daughter ion signals are equal. For samples at a temperature, T, this energy is located below the 0 K dissociation onset. However, there is no ambiguity in identifying this energy in the breakdown diagram if the sample temperature and its vibrational frequencies are known. To convert the 298 K cross-over energy to the cross-over energy expected for reactants initially at 0 K, one simply

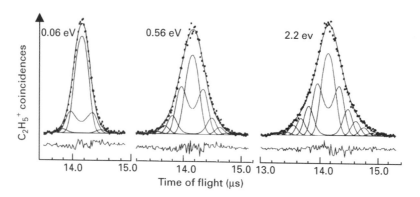

Figure 3 The PEPICO TOF distribution of $C_2H_5^+$ ions from energy selected $C_2H_5I^{+\bullet}$ ions at various energies above the dissociation limit for I• loss. The solid lines that fit the experimental points are single energy release distributions. From these data, the whole distribution of product translational energies could be determined. Reproduced with permission from Baer T, Büchler U and Klots TCE (1980) Kinetic energy release distributions for the dissociation of internal energy selected $C_2H_5I^{+\bullet}$ ions. *Journal de Chimie Physique* **77**: 739–743.

adds the median thermal energy to the 298 K cross-over energy.

Kinetic energy release in the dissociation

Because the timing in PEPICO experiments is very precise, all of the broadening in TOF peaks can be associated with the kinetic energy of the parent or the fragment ions. Consider two ions that are ejected along the TOF axis but in opposite directions with initial velocity, $\pm v_0$ where the positive sign signifies the direction toward the ion TOF tube. The negative-going ion will be decelerated and turn around and come back to its original point of formation but with its velocity vector now converted from − to +. From this point, the ion will have the identical TOF to the ion whose initial velocity vector was positive. Thus, the TOF difference between + and − ions is just the 'turn-around' time of the negative-going ion. This time is readily calculated from Newton's equation, $F = ma = q\varepsilon$, where q is the ion charge, m is the ion mass and ε is electric field, and is given by $2v_0 m/q\varepsilon$, where again v_0 is the initial velocity of the fragment ion. In a polyatomic molecule, much of the energy in excess of the dissociation limit remains in the vibrational modes. In that case, the ion fragments with a *distribution* of translational energies that is often well modelled by a Maxwell–Boltzmann distribution, as predicted by the statistical theory of unimolecular decay. If the three-dimensional distribution in the centre of mass is a Maxwell–Boltzmann type, the TOF peak will appear Gaussian, and the average kinetic energy release (KER) will be given by

$$\langle \mathrm{KER} \rangle = \frac{M}{(M-m)\,m} \frac{3}{16 \ln 2} (q\varepsilon)^2 (\mathrm{FWHM})^2 - \frac{m}{M-m} \langle E_{\mathrm{th}} \rangle \qquad [17]$$

where M and m are the masses of the parent and fragment ions, and FWHM is the TOF peak width. The first term is a result of kinetic energy release in the dissociation, while the second term subtracts the width due to the thermal motion of the parent ion. An example of this sort of statistical distribution is shown in **Figure 3** for the case of $C_2H_5I^+$ dissociation at three energies above the dissociation limit. As the ion energy increases, the peak widths broaden. The solid lines in this figure are TOF distributions expected for single kinetic energy releases of $9n^2$ ($n = 2, 3, 4,..$) meV. Because of apertures, ions with significant off-axis velocities will be stopped. Thus, each TOF peak for ions with KER greater than $n = 2$ will appear as a doublet, which accounts for the forward and backward ejected ions.

One of the most useful applications of peak widths in the TOF distributions is in distinguishing an ionization from a dissociative ionization process. This is particularly useful in the case of photoionization of clusters. An example of such a study is the case of $ArCO^+$, which can arise from a variety of process in a molecular beam expansion of Ar and CO gases, including those shown in Equations [18]–[20].

$$\mathrm{ArCO} + h\nu \rightarrow \mathrm{ArCO}^{\bullet +} + \mathrm{e}^- \qquad [18]$$

$$\mathrm{Ar_2CO} + h\nu \rightarrow \mathrm{ArCO}^{\bullet +} + \mathrm{Ar} + \mathrm{e}^- \qquad [19]$$

$$\mathrm{Ar(CO)_2} + h\nu \rightarrow \mathrm{ArCO}^{\bullet +} + \mathrm{CO} + \mathrm{e}^- \qquad [20]$$

If one is interested in the appearance energy or the thermochemistry of the $ArCO^+$ ion, it is necessary to know by which process this ion is created. In each of the three reactions, the product ion mass is the same.

However, the first reaction can be distinguished from the dissociative ionization processes because its TOF peak is expected to be narrow, whereas the latter two reactions will impart kinetic energy to their fragments and thus result in broadened TOF peaks. This is illustrated in **Figure 4**. The narrow peak with a width of 34 ns is due to reaction [18], while the broad peak with a width of 160 ns is due to the dissociative processes. By this approach, an onset of 13.03 eV for the ArCO⁺ ion from ArCO could be measured. In addition, by measuring the width of the broad peak as a function of the cluster ion energy, it was possible to extrapolate the kinetic release in the second reaction down to the threshold (as shown in **Figure 5**).

A similar study found that $Ar_n^{\bullet+}(n>2)$ ions produced in a supersonic expansion of argon originated exclusively by dissociative photoionization from higher-order clusters. Only in the case of Ar_2 could ions of that mass be produced without the accompanying dissociation. This was obvious from the broad peak widths, which showed no evidence of any narrow components for $Ar_n^{\bullet+}$ other than for $n=2$.

Ion dissociation rate measurements

The dissociation rates of metastable ions can be measured by PEPICO because the ion TOF distribution is very sensitive to the position of dissociation in the acceleration region. Ions that dissociate rapidly gain the full kinetic energy in the acceleration region, whereas ions that dissociate some distance into the acceleration region end up with less kinetic energy

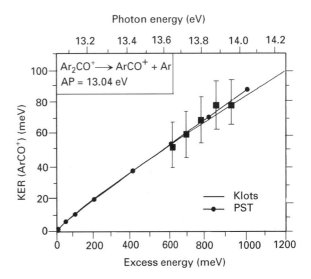

Figure 5 The derived kinetic energy release from the energy-selected $Ar_2CO^{\bullet+}$ ions as a function of the trimer ion internal energy. The solid line is a calculated kinetic energy release based on the statistical theory of dissociation: phase space theory (PST) or the version of PST due to C.E. Klots. AP is the threshold energy for ArCO⁺ formation. The onset leads to a heat of formation of the trimer ion. Reproduced with permission from Mahnert J, Baumgartel H and Weitzel KM (1997) The formation of ArCO⁺ ions by dissociative ionization of argon/carbon monoxide clusters. *Journal of Physical Chemistry* **107**: 6667–6676.

because it is partitioned between the ion and neutral fragments. Typical ion TOF distributions of products formed from metastable $C_6H_6^{\bullet+}$ ions are shown in **Figure 6**. The benzene ion dissociates via four major paths at low ion energies:

$$C_6H_6^{\bullet+} \begin{cases} \rightarrow C_6H_5^+ + H^\bullet \\ \rightarrow C_6H_4^{\bullet+} + H_2 \\ \rightarrow C_4H_4^{\bullet+} + C_2H_2 \\ \rightarrow C_3H_3^+ + C_3H_3^\bullet \end{cases} \quad [21]$$

Only the latter two fragments are shown in **Figure 6** because the mass difference between the parent ion and the H and H_2 loss channels are not sufficient to be resolved in this low-resolution TOF spectrum. However, what is evident is the asymmetry of the $C_4H_4^{\bullet+}$ and $C_3H_3^+$ ion TOF distributions. The peak shapes can be modelled (solid lines in **Figure 6**) knowing the mass of the ions, the electric field, the length of the acceleration and drift distances, and the ion dissociation rate, which is an adjustable parameter. As the ion energy is increased from 14.85 eV to 15.32 eV, the rate constant increases from 0.16 to ~1.2 μs^{-1}.

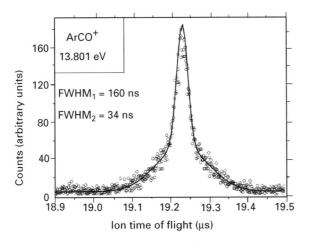

Figure 4 The PEPICO TOF distribution of $ArCO^{\bullet+}$ ions from various precursor cluster ions. The sharp peak is due to the direct ionization of ArCO dimers, whereas the broad peak with a width of 160 ns is due to dissociative ionization of Ar_2CO trimers. Reproduced with permission from Mahnert J, Baumgartel H and Weitzel KM (1997) The formation of ArCO⁺ ions by dissociative ionization of argon/carbon monoxide clusters. *Journal of Physical Chemistry* **107**: 6667–6676.

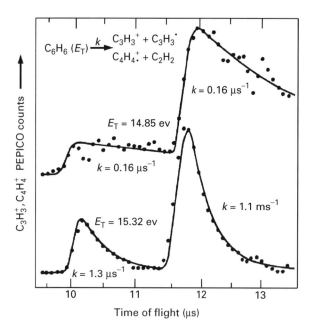

Figure 6   (k = rate constant; E_T = total ion energy) The PEPICO TOF distribution of $C_3H_3^+$ and $C_4H_4^+$ ions from energy-selected $C_6H_6^+$ ions. The asymmetric TOF distributions are a result of the slow reaction of the metastable ions. The solid lines are calculated distributions using the mean ion dissociation rate as an adjustable parameter. Reproduced with permission from Baer T, Willet GD, Smith D and Phillips JS (1979) The dissociation dynamics of internal energy selected $C_6H_6^+$. *Journal of Chemical Physics* **70**: 4076–4085.

Such data collected at a number of ion energies leads to a $k(E)$ vs E curve shown in **Figure 7**. The solid lines in this figure are the results of a statistical theory calculation using the Rice–Rampsperger–Kassel–Marcus (RRKM) formulation. Studies such as this one on many molecules have shown that the RRKM theory is extremely useful and accurate in predicting the dissociation rate constants of ionic reactions. It is also very useful for determining the onset of dissociation. This is of interest both for fundamental reasons as well as for a very practical one. Dissociation onsets are often used to determine thermochemical properties of ions and free radicals. Consider the reaction

$$AB + h\nu \rightarrow A^+ + B^\bullet \qquad [22]$$

in which the stable molecule is dissociatively ionized to the ion A^+ and a free radical, $B^\bullet$. By energy conservation, the heats of formation of the three species in Equation [22] are related to the minimum energy of

dissociation (DE) by

$$DE = \Delta H_f(A^+) + \Delta H_f(B^\bullet) - \Delta H_f(AB) \qquad [23]$$

Thus, if any two of the three heats of formation are known, the third can be calculated from the measured dissociation energy, DE. Many heats of formation derived using this approach have been reported. However, the validity of this method depends upon a rapid dissociation of the molecular ion. What is actually measured is the appearance energy, AE, of the product ions which is always greater than DE. Ions prepared just above their dissociation threshold often fragment very slowly (e.g. $k < 10^4$ s^{-1}) so that no A^+ ions will be observed since ions are collected in just a few microseconds after their formation. It is not until the ion energy reaches a level at which the dissociation rate constant, k, exceeds 10^4 s^{-1} that A^+ ions are observed. An effective means for circumventing this problem is to measure the dissociation

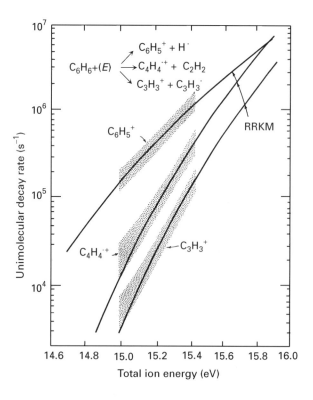

Figure 7   The dissociation rate constants of $C_6H_6^{\bullet+}$ ions as a function of the ion internal energy. The solid lines are calculated rate constants using the statistical theory of unimolecular decay (RRKM). Reproduced with permission from Baer T, Willet GD, Smith D and Phillips JS (1979) The dissociation dynamics of internal energy selected $C_6H_6^+$. *Journal of Chemical Physics* **70**: 4076–4085.

rate constant as a function of the ion internal energy and to extrapolate this $k(E)$ curve to the dissociation onset by use of the statistical theory of unimolecular decay. The extrapolations in **Figure 7** predict that the true onsets for these dissociation channels for loss of H·, C_2H_2, and C_3H_3· lie at 13.7, 14.32, and 14.47 eV, respectively. It is apparent that these onsets are well below the data in **Figure 7**, which were all collected above 15 eV because below this energy no fragment ions were observed. This is an example of the 'kinetic shift' that has shifted the onset towards higher energy because the dissociation rate constant at threshold is too slow.

Dissociation rate measurements are also very useful for determining the mechanism of ionic dissociation reactions. If the ion isomerizes prior to dissociation, the rate constant will be slower than predicted by the RRKM theory. Often it has proved possible to determine experimentally the energy of the isomerized structure by modelling the $k(E)$ curve with the RRKM theory. The reliability of such studies is greatly enhanced by the use of *ab initio* molecular orbital theory, which provides valuable input for the RRKM theory calculations.

List of symbols

d = diameter of analyser input aperture; diameter of analyser drift region aperture; d_{ph} = photon beam width; DE = dissociation energy; E_0 = electron initial kinetic energy; E_e = electron collection efficiency; E_e = energy of electron as it passes through the analyser; E_g = electron energy gain on acceleration from ionization region; E_i = ion collection efficiency; h = Planck constant; IE = ionization energy; k = dissociation rate constant; KER = kinetic energy release; l = length of analyser drift region; M, m = masses of parent and fragment ions, respectively; N_c = coincidence count rate; N_e = electron count rate; N_i = ion count rate; N_T = total ionization rate; q = ion charge; r = average radius of inner and outer spheres between which electrons pass in the analyser; R_{el} = electron collection efficiency; ΔH_f = heat of formation; ε = electric field strength; v_0 = initial ion velocity; ϕ_c = analyser planar critical collection angle.

See also: **Ion Dissociation Kinetics, Mass Spectrometry; Ion Energetics in Mass Spectrometry; Ioniza-**tion Theory; Metastable Ions; Photoionization and Photodissociation Methods in Mass Spectrometry; Statistical Theory of Mass Spectra.

Further reading

Baer T (1986) The dissociation dynamics of energy selected ions. *Advances in Chemical Physics* **64**: 111–202.

Baer T and Hase WL (1986) *Unimolecular Reaction Dynamics: Theory and Experiments*. New York: Oxford University Press.

Baer T and Mayer PM (1997) Statistical RRKM/QET calculations in mass spectrometry. *Journal of the American society of Mass Spectrometry* **8**: 103–115.

Baer T, Peatman WB and Schlag EW (1969) Photoionization resonance studies with a steradiancy analyzer. II. The photoionization of CH_3I. *Chemical Physics Letters* **4**: 243–247.

Baer T, Willett GD, Smith D and Phillips JS (1979) The dissociation dynamics of internal energy selected C_6H_6·. *Journal of Chemical Physics* **70**: 4076–4085.

Baer T, Buchler U and Klots CE (1980) Kinetic energy release distributions for the dissociation of internal energy selected $C_2H_5I^+$ ions. *Journal de Chemie Physique* **77**: 739–743.

Berkowitz J (1979) *Photoabsorption, Photoionization, and Photoelectron Spectroscopy*. New York: Academic Press.

Dannacher J, Rosenstock HM, Buff R *et al* (1983) Benchmark measurement of iodobenzene ion fragmentation rates. *Chemical Physics* **75**: 23–35.

Das PR, Nishimura T and Meisels GG (1985) Fragmentation of energy selected hexacarbonylchromium ion. *Journal of Physical Chemistry* **89**: 2808–2812.

Eland JHD (1972) Predissociation of triatomic ions studied by photoelectron photoion coincidence spectroscopy and photoion kinetic energy analysis. *International Journal of Mass Spectrometry and Ion Processes* **9**: 397–406.

Mahnert J, Baumgartel H and Weitzel KM (1997) The formation of $ArCO^+$ ions by dissociative ionization of argon/carbon monoxide clusters. *Journal of Physical Chemistry* **107**: 6667–6676.

Ng CY (1998) State-selected and state to state ion-molecule reaction dynamics by photoionization methods. In Farrar JM and Saunders WHJ (eds) *Techniques for the Study of Ion–Molecule Reactions*, 417–488. New York: Wiley.

Stockbauer R (1977) A threshold photoelectron photoion coincidence mass spectrometer for measuring ion kinetic energy release on fragmentation. *International Journal of Mass Spectrometry and Ion Physics* **25**: 89–101.

Photoionization and Photodissociation Methods in Mass Spectrometry

John C Traeger, La Trobe University, Bundoora, Victoria, Australia

MASS SPECTROMETRY

Methods & Instrumentation

The formation of ions in a mass spectrometer ion source can be accomplished by a variety of ionization methods. Photoionization of a molecule or atom involves the absorption of a vacuum ultraviolet (VUV) or soft X-ray photon. Depending on the wavelength (λ) of the photon, whose energy E is given by the Planck–Einstein relationship,

$$E(\text{eV}) = \frac{hc}{\lambda} = \frac{1239.85}{\lambda(\text{nm})} \qquad [1]$$

both ionization and subsequent fragmentation of the ionized molecule may occur. It is also possible to induce fragmentation of an initially formed ion by absorption of a single photon, or by the sequential absorption of a number of photons of equal or different wavelengths, in a photodissociation process. One particularly useful advantage of using a photon beam for ionization of a neutral, or dissociation of an ion, is that the amount of energy deposited depends only on the wavelength, which can be precisely measured and controlled. For this reason, the techniques of photoionization and photodissociation mass spectrometry are primarily used for fundamental energetic, kinetic and structural studies, although there are various analytical applications that can also benefit from these particular experiments. Because mass spectrometers used for photoionization and photodissociation experiments are not widely available, they are generally designed and constructed in the laboratory engaging in the particular studies.

A major source of energetic information is the photoionization efficiency (PIE) curve in which the relative numbers of photoions produced per number of photons transmitted is measured as a function of photon wavelength (energy). A typical PIE curve is shown in **Figure 1**.

Apart from yielding accurate thermochemical data, the PIE is also a convenient means of studying the energy dependence of a range of gas-phase unimolecular processes, which include those involving weakly bound clusters and ion–neutral complex intermediates. The detection of near zero-energy

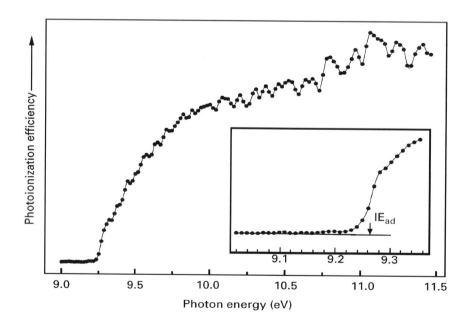

Figure 1 Photoionization efficiency curve with (inset) an expanded threshold region for the molecular ion from cyclopentanone. The photoions detected below the adiabatic ionization energy IE_{ad} (indicated by the arrow) are due to hot bands.

photoelectrons coincident with photoion formation (threshold PEPICO (photoelectron–photoion coincidence)) provides a direct measure of the internal energy of the initially formed ion; for many systems similar information can be obtained from the corresponding first differential PIE curve.

Photodissociation experiments involve the photon-assisted decomposition of a preformed ion beam with subsequent detection of the resulting fragment ions. To observe photodissociation with a mass spectrometer it is necessary that the photoabsorption process produces an excited state of the precursor ion above its lowest dissociation threshold. The dissociation should also be relatively fast compared to any other competing process, such as relaxation. The photodissociation spectrum of an ion, which is a measurement of the photodissociation efficiency as a function of photon energy, is related to the photoelectron spectrum for the corresponding neutral molecule and is comparable to the ion absorption spectrum. However, unlike photoabsorption experiments, photodissociation is a more sensitive technique for obtaining structural information because any small changes are not superimposed on a relatively large signal. There is also less interference from other contaminating species, such as ion–molecule reaction products.

Photodissociation spectra of excited ionic species that predissociate contain more structural information than those resulting from a direct dissociation. Unfortunately, there are few ions that undergo suitable predissociation processes. Time-resolved photodissociation, which involves the monitoring of fragment ion build-up following photoexcitation, provides a direct measure of the bond cleavage kinetics, from which bond strength information can be obtained. These experiments, in conjunction with Rice–Ramsperger–Kassel–Marcus (RRKM) calculations, can also help to determine dissociation thresholds.

Photoionization

Sample introduction

The simplest method for introducing a sample into an ion source is via an effusive beam through a pinhole or needle valve. For some experiments it is advantageous to cool the sample prior to introduction to minimize problems associated with the thermal population of vibrationally and rotationally excited species. Because of condensation problems, this method is not applicable to molecules with relatively high freezing points. However, the use of a supersonic molecular beam for sample introduction can result in a substantial internal cooling effect, with rotational and vibrational temperatures below 20 K

being achieved for polyatomic molecules. This significantly reduces the interference from low-energy tailing, called hot band structure, that is usually observed in the pre-threshold region of a room-temperature PIE curve (see **Figure 1**). Various free radicals and transient molecules have been studied by photoionization, although they often require special methods of preparation and introduction. Because these species are usually present at very low concentrations, it is preferable to have access to a high-intensity photon source. A molecular beam can also facilitate the production of clusters for investigation by photoionization.

Photon sources

Several different laboratory light sources can be used for VUV studies. These usually involve a gas discharge, which may produce either a continuum or a discrete-line emission spectrum. Although the output of light is generally much greater for a line spectrum, these cannot be used for very high-resolution studies because PIE measurements are then restricted to the finite number of emitted photon energies available. A commonly used photon source is the hydrogen many-lined spectrum (pseudocontinuum), which is shown in **Figure 2**. This is produced by a simple DC discharge in hydrogen at a relatively low pressure (<1 kPa), and produces usable photons in the energy range of 7–14 eV.

Care must be exercised with the data processing when using such a discrete-line source as it is possible for false structure to be generated in the PIE curve, particularly in the low photon intensity regions (e.g. 9.3–9.6 eV) of the lamp.

Two continuum sources that have been used for photoionization studies are the argon continuum (8–12 eV) and the Hopfield helium continuum (12–21 eV). Both of these involve emission from a transient diatomic molecule generated via a pulsed high-voltage, high-pressure (>10 kPa) rare gas discharge. As for the hydrogen pseudocontinuum, a typical photon intensity is approximately 10^9–10^{10} s^{-1}, which is substantially lower than the photon intensity obtained with either an electron monochromator or a conventional electron ionizing source.

Experiments involving continuous emission at energies above 21 eV require the use of synchrotron radiation, in which light is emitted from rapidly moving electrons accelerated in a circular path. The spectral distribution and intensity available from the electron storage ring vary depending on a number of different operating parameters. However, usable photon energies typically extend from the visible into the X-ray region, with intensities significantly greater

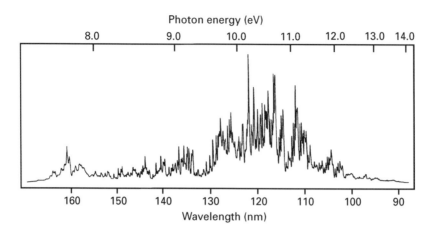

Figure 2   The many-lined molecular hydrogen pseudocontinuum.

than those of comparable gas discharge lamps. The high intensity of synchrotron radiation has proved particularly successful for photoionization studies involving low concentrations of free radicals.

The use of lasers as a photoionization light source has been limited by the availability of a suitable tunable VUV laser. The fundamental problem in developing such a high-energy laser is that the laser action becomes increasingly difficult to sustain as the photon energy is increased. Consequently, most laser-based photoionization experiments have involved multiphoton techniques in which the energies of several photons are combined. A commonly used method for the ionization of molecules is resonance-enhanced multiphoton ionization (REMPI), in which the laser frequency is tuned to an excited neutral electronic state. This greatly enhances the cross section for ionization. One photon is involved in the excitation process while a second photon, which may have a different wavelength (colour), is used to ionize the excited molecule.

Monochromators

For energetic studies it is necessary to select a given photon wavelength from the light source with a monochromator. Although there are numerous types available, it is preferable to use a design in which the entrance and exit slits remain in a fixed position relative to the mass spectrometer. The image at the exit slit should also stay in focus. Two monochromators that essentially satisfy these requirements are the Seya–Namioka and the near-normal-incidence type. The Seya–Namioka monochromator, in which wavelength selection is accomplished by the simple rotation of a diffraction grating about its centre, unfortunately suffers from an astigmatic and curved image at the exit slit, which result in a decrease in

both intensity and resolution. Despite its more complex scanning mechanism, the near-normal-incidence monochromator is better suited for high-resolution studies. Photoionization experiments with mono-chromator resolutions better than 0.002 nm have been reported, although a typical operating resolution is 0.1 nm, which corresponds to 0.008 eV for a photon with an energy of 10 eV. Wavelength calibration is normally done using atomic lines generated in the emission spectrum of the lamp.

The concave diffraction gratings used in these monochromators are usually optimized to maximize the light intensity over a selected operating wavelength region. This includes the selection of a blaze angle and a particular overcoating of the ruled surface. As shown in **Figure 3**, an aluminium surface with a thin film of MgF_2 or LiF provides the highest reflectance for photon energies below 12.0 eV. Metal coatings, such as platinum, are best used for experiments that extend to higher energies. The dispersion (resolution) of a grating depends on both its radius of curvature and the ruled line density. Typical values for these parameters are 1 m and 1200 grooves mm^{-1} respectively. Ghosting effects and the amount of scattered light can be reduced with a grating that has been holographically ruled.

If a high-pressure gas discharge is being used as the VUV light source, it is preferable to isolate the monochromator from the lamp. Apart from the problems associated with interfacing the lamp and monochromator to the vacuum environment of a mass spectrometer, self-absorption of light by the discharge gas greatly reduces the available photon intensity. In some cases the installation of an optical window at the entrance slit can be used to avoid this, although the choice of window material is somewhat restricted. A LiF filter can be used for photon energies up to 12.0 eV, although for photoionization

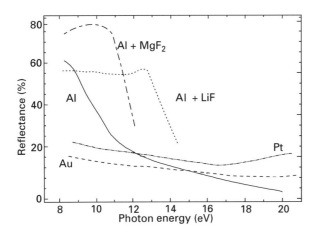

Figure 3 Vacuum ultraviolet reflectance curves for different grating surfaces.

studies above this sharp cut-off energy the monochromator needs to be windowless. This requires the use of substantial vacuum pumping systems to maintain a satisfactory pressure differential between the lamp and the monochromator, especially when a wide entrance slit is being used. Because the storage ring of a synchrotron light source is operated under a high vacuum, it does not suffer from this particular interfacing problem. However, the wide energy range of the generated light can result in contamination of a PIE curve by second-order or higher-order diffraction, requiring analytical removal during the data reduction. Alternatively, for low-energy studies an appropriate filter, such as a LiF window, can be used to effectively remove this interference.

Photon detectors

When conducting PIE experiments it is essential to monitor the total amount, or a representative sample, of light that is being used for the ionization process. The simplest device involves counting the photoelectrons emitted from either a simple irradiated metal plate, or, for low levels of light, via a discrete or continuous dynode (channeltron) electron multiplier. However, both methods require calibration of the photocathode spectral response. Electron multipliers have also been developed in conjunction with special photosensitive cathodes for operation well into the vacuum ultraviolet region. Because they are relatively insensitive to light of wavelengths greater than 300 nm, which is approximately the limit of solar radiation reaching the earth's surface, they are often referred to as 'solar blind' photomultipliers. Again, the photoelectric yield is wavelength dependent and must be calibrated.

An alternative approach to photon detection is to use a phototransducer in conjunction with a conventional photomultiplier. This involves measuring the fluorescence produced following irradiation with VUV light. A particularly convenient and widely used phosphor is sodium salicylate, which has a high fluorescence efficiency and nearly constant quantum yield over a wide photon energy range (5–35 eV). Periodic replacement of this phosphor is required as its low-wavelength efficiency deteriorates with time in a typical vacuum environment.

Mass analysers

Several different types of mass analyser have been used for photoionization studies, including quadrupole, magnetic sector and time-of-flight (TOF) mass spectrometers. The technique of time-resolved photoionization mass spectrometry uses a quadrupole ion trap to store ions for a selected time prior to detection, which greatly increases the threshold sensitivity for slow fragmentation processes. Many experiments require extended periods of data acquisition, so that a high mass stability is vital. In addition, because photoionization produces only small numbers of ions in the ion source, it is important to ensure that the mass analyser has a high ion transmission efficiency. The probability of photoionization can be enhanced by increasing the sample pressure in the ion source. However, to minimize absorption corrections and collision-induced effects, such as ion–molecule reactions and collisional activation, maximum pressures are generally restricted to below 0.1 Pa.

The quadrupole mass filter is an inexpensive and compact device in which the ion source operates at ground potential, making it convenient for interfacing to a monochromator. The ion optics for entry to the quadrupole RF/DC field are relatively straightforward, while it is a simple procedure to alter the mass resolution and maximize the ion count rate for a particular experiment.

A magnetic mass analyser is considerably more expensive and physically larger than its quadrupole counterpart. In normal operation the ion source operates at a high positive voltage, so that the monochromator interface region must be carefully designed to ensure that photoelectrons ejected from either the photodetector or any metal surfaces are not accelerated into the ion source to produce spurious ionization. For maximum ion transmission it is necessary to use an ion-optical lens, such as an electrostatic quadrupole, to focus the ions from a wide area into the narrow entrance slit of the mass spectrometer. The magnetic mass spectrometer, unlike a quadrupole mass filter, is able to detect metastable ions, which makes it particularly useful for the study of unimolecular fragmentation processes.

The TOF mass spectrometer is well suited to the mass analysis of ions produced from a pulsed source. In contrast to scanning mass analysers, the TOF detects all ions sampled from the ion source. For this reason it is widely used for coincidence PEPICO measurements and various laser ionization experiments. The pulsed nature of synchrotron radiation also lends itself to TOF photoionization studies. Peak shapes for different ions in a TOF mass spectrum can provide valuable information about their kinetic energy distributions.

Ion detectors

Various ion multipliers have been used to detect mass-selected photoions. These include the discrete dynode electron multiplier, the continuous dynode channeltron and the Daly detector, which uses a combined scintillator/photomultiplier to monitor the fluorescence produced from secondary electrons generated by ion bombardment of a high-voltage conversion electrode (**Figure 4**). Because of the small numbers of ions generated in photoionization experiments, pulse counting, rather than analogue, techniques are usually employed. The digital method of detection is more immune to noise, although it is preferable to use a multiplier that has a low background pulse count rate, typically less than one pulse every 10 s. Stray external magnetic fields, such as those encountered with a magnetic mass analyser, can impair the multiplier performance and require appropriate shielding.

Data processing

Most photoionization experiments require extended data acquisition to increase both the ion count statistics and the signal-to-noise ratio to acceptable levels, particularly in the threshold region where the photoion count rate can be extremely low. It is not unusual for some experiments to last for 24 hours or more. However, as the reduction in random noise only increases as the square root of the accumulated counts, there is a practical limit to the improved quality of data obtained using this technique.

The PIE curve provides the basis for obtaining precise gas-phase thermochemical information, such as bond energies, proton affinities, and cationic and neutral heats of formation. Because the ionization process is governed by the Franck–Condon principle, the most probable process will involve a vertical transition. Depending on the geometry change following ionization, this may not always correspond to the adiabatic (minimum) ionization energy (IE), as shown in **Figure 5**.

Because the precursor molecules have a thermal energy distribution, ionization will not all occur from the neutral ground state, resulting in ions being produced below the true ionization threshold energy (**Figure 5**). This is observed as pre-threshold hot band structure in the PIE curve and will typically extend over a range of 0.1 to 0.2 eV for a polyatomic molecule at room temperature (see **Figure 1**).

When variable-energy photons are used for ionization it is possible for autoionization to occur. This is

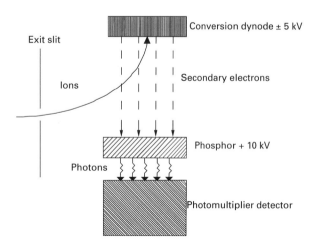

Figure 4 Schematic representation of a Daly scintillator/photomultiplier ion detector.

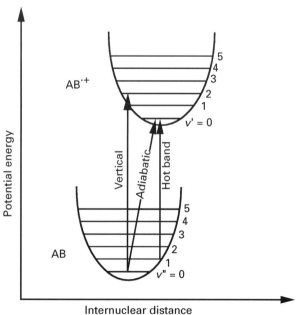

Figure 5 Single ionization of a molecule AB, showing the adiabatic, the vertical and a hot band transition.

the result of ionization from an excited neutral Rydberg state above the ionization threshold. Autoionization processes are observed as peaks in the PIE curve and can often complicate the assignment of the adiabatic ionization energy. **Figure 6** shows the effect of autoionization in the PIE curve for xenon. The structural profiles are similar to those observed in comparable photoabsorption experiments.

Apart from the higher resolution relative to similar electron ionization experiments, photoionization has an additional advantage in that there is a finite cross section at threshold, making it easier to detect the actual ionization onset. This also applies to unimolecular fragmentation reactions and is a result of the different threshold laws for the two ionization processes. An ionization efficiency curve produced by photoionization is comparable to a first derivative of its electron-ionization counterpart.

The appearance energy (AE) is the minimum energy required to photoionize and fragment the molecule AB in the reaction shown in Equation [2] and, in the absence of any excess internal energy, can provide useful thermochemical information about the reactant and products.

$$AB + h\nu \rightarrow AB^{\bullet+} + e^- \rightarrow A^+ + B^\bullet + e^- \quad [2]$$

However, the AE will not provide a true thermochemical value if there is any reverse activation energy (E_{rev}) associated with the decomposition. Furthermore, if the rate of fragment ion production is low for photon energies just above threshold, there

will be a kinetic shift (E_{kin}). This is the energy in excess of the thermochemical threshold, including any reverse activation energy, for the ion to be detectable by the mass spectrometer. These effects are illustrated in **Figure 7**.

A kinetic shift can be minimized by increasing the detection sensitivity or by trapping the dissociating precursor ion and extending the time prior to mass analysis. An additional complication will arise when there is competition with another more favourable fragmentation process. This can also cause an increase in the observed AE and is often referred to as a competitive shift.

Unless reaction [2] is carried out at 0 K, the products will not be formed at any well-defined equilibrium thermodynamic temperature because the fragmentation is a unimolecular process with no means for the products to thermalize with their surroundings. Provided that there is no excess internal energy, the experimental AE for reaction [2] at a given temperature is given by

$$\begin{aligned} AE = {} & \Delta H_f^0(A^+) + \Delta H_f^0(B^\bullet) + \Delta H_f^0(e^-) \\ & - \Delta H_f^0(AB) - \Delta H_{cor} \end{aligned} \quad [3]$$

where ΔH_{cor} is a thermal energy correction term to convert the products to the same temperature as the neutral precursor. ΔH_{cor} varies with different fragments, but is typically in the 10–30 kJ mol^{-1} range at 298 K.

For threshold studies it is convenient to consider the ejected electron at rest at all temperatures, i.e. $\Delta H_f^0(e^-) = 0$. This stationary electron convention is

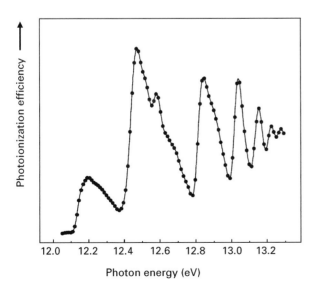

Figure 6 Photoionization efficiency curve for Xe$^+$ showing the extensive autoionization structure above the ionization energy of 12.13 eV.

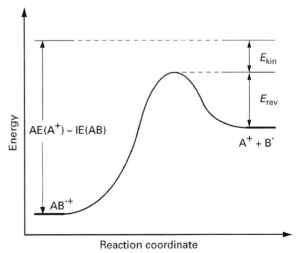

Figure 7 A unimolecular reaction involving both a reverse activation energy and a kinetic shift.

most commonly used for cationic heats of formation derived from mass spectrometry experiments. However, there are many other data compilations that invoke a thermal energy convention, in which the electron has a translational energy commensurate with the reaction temperature. At 298 K this corresponds to a difference of 6.2 kJ mol^{-1}. Consequently, when using such information it is important to know which particular convention has been adopted.

Photodissociation

Sample introduction

The precursor ions used in a photodissociation experiment are usually generated via a mass spectrometer. This minimizes complications that can arise when ion–molecule reactions occur in the ion source. The most common method of initial ion production is electron ionization as this invariably produces larger numbers of primary ions than either photoionization or collisional activation techniques. The ionizing electron energy is kept as low as practicable to minimize the initial internal energy of the precursor ions, which are then mass filtered before introduction into the photodissociation region of the instrument. Variation of the ionizing energy can be used to prepare ions with different internal energies and also facilitates the determination of the electronic state(s) involved in the photoinduced transition(s). Because of space-charge effects the precursor ion concentration is considerably lower than that of a comparable beam of neutral molecules.

Light sources

A high-intensity source of photons is required to produce detectable numbers of photofragment ions from a mass-selected ion beam. For this reason, most experiments have used a laser for the photodissociation process. Continuous lasers generally produce the highest average power outputs, with an excellent stability and a very narrow photon energy bandwidth (~10^{-5} eV). They may be used to pump a dye laser to produce light from the near-infrared to the visible region of the spectrum. The use of a pulsed excimer laser to pump the dye laser can extend the wavelength range to cover the 1.5–3.5 eV energy region, although this requires the use of over twenty different dye solutions. Frequency doubling and etalon bandwidth narrowing techniques can be used to further enhance the laser performance. Accurate wavelength calibration presents a problem for high-resolution studies. One procedure used in the visible region of 1.7–2.5 eV is to simultaneously record the

iodine absorption as a reference, producing calibrations to better than 10^{-6} eV. For near UV studies, optogalvanic lines provide one of the few convenient methods available.

The interaction between the photon and ion beams can be achieved by either a crossed or a coaxial configuration. The crossed arrangement produces a better definition of the interaction region and is used when absolute cross section and angular distribution measurements are required. However, a better overlap of the two beams, and hence improved sensitivity, results from an arrangement in which both beams travel in parallel along the interaction region. This also permits the use of a Doppler tuning technique to improve the photon energy resolution and to reduce the problems associated with the ion beam thermal energy.

Mass analysers

Photodissociation experiments are usually performed with either a beam or an ion-trap mass spectrometer. The beam instrument is a tandem mass analyser in which the first stage is used to mass select a given precursor ion. This ion beam is passed through a laser interaction region and the resulting photofragments are then energy or mass analysed by a second stage. A schematic diagram of a triple-quadrupole mass spectrometer with a coaxial configuration is shown in **Figure 8**.

Apart from restricting the laser–ion interaction time, fast ion beams, particularly those produced with a magnetic sector instrument, essentially restrict the lifetimes of ions that can be detected to less than 10 μs. This is not a limitation for an ion cyclotron resonance (ICR) or quadrupole ion storage (QUISTOR) trap, where ions can be stored almost indefinitely. However, the presence of relatively high concentrations of neutral species may result in interfering ion–molecule reactions taking place in the cell. The use of an ion trap with an external ion source can help to remove these competing processes and also allows photodissociation studies of precursor ions formed by ionization techniques other than electron ionization.

Data processing

As in threshold photoionization experiments, photofragment ion concentrations are extremely low and it is necessary to utilize signal-enhancement techniques. Although phase-sensitive detection has been used, it is more usual to employ signal-averaged pulse counting. For photodissociation with a pulsed laser, an appropriate gating system synchronized

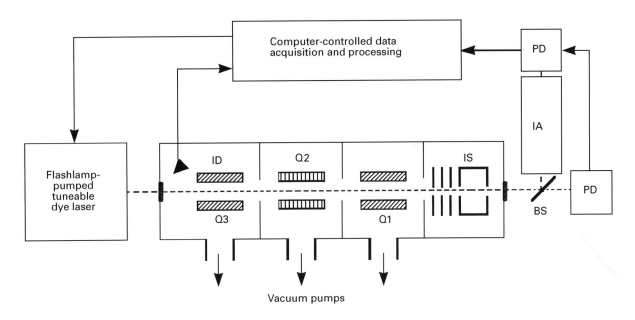

Figure 8 Schematic diagram of a triple-quadrupole laser-induced photodissociation mass spectrometer operated in a coaxial configuration. IS is the ion source; Q1 is the precursor ion mass filter; Q2 is the ion photodissociation region; Q3 is the photofragment ion mass filter; ID is the off-axis ion detector; PD is a photon detector; BS is a beam splitter; and IA is an iodine absorption cell.

with each laser pulse can further improve the signal-to-noise ratio.

List of symbols

c = speed of light; E = energy; h = Planck's constant; ΔH_{cor} = thermal energy correction term; ΔH_f^0 = standard enthalpy of formation; λ = wavelength.

See also: **Ion Dissociation Kinetics, Mass Spectrometry; Ion Energetics in Mass Spectrometry; Ionization Theory; Multiphoton Excitation in Mass Spectrometry; Photoelectron–Photoion Coincidence Methods in Mass Spectrometry (PEPICO); Spectroscopy of Ions; Time of Flight Mass Spectrometers.**

Further reading

Becker U and Shirley DA (1996) *VUV and Soft X-ray Photoionization.* New York: Plenum Press.

Berkowitz J (1979) *Photoabsorption, Photoionization, and Photoelectron Spectroscopy.* New York: Academic Press.

Bowers MT (1984) Ions and light. *Gas Phase Ion Chemistry,* Vol 3. New York: Academic Press.

Dunbar RC (1979) Ion photodissociation. In: Bowers MT (ed) *Gas Phase Ion Chemistry,* Vol 2, pp 181–220. New York: Academic Press.

Dunbar RC (1996) New approaches to ion thermochemistry via dissociation and association. In: Adams NG and Babcock LM (eds) *Advances in Gas Phase Ion Chemistry,* Vol 2, pp 87–124. London: JAI Press.

Futrell JH (1986) *Gaseous Ion Chemistry and Mass Spectrometry.* New York: Wiley.

Moseley JT (1985) Ion photofragment spectroscopy. *Advances in Chemical Physics* 60: 245–298.

Ng CY (1983) Molecular beam photoionization studies of molecules and clusters. *Advances in Chemical Physics* 52: 263–362.

Ng CY (1991) *Vacuum Ultraviolet Photoionization and Photodissociation of Molecules and Clusters.* Singapore: World Scientific.

Reid NW (1971) Photoionization in mass spectrometry. *International Journal of Mass Spectrometry and Ion Physics* 6: 1–31.

Samson JAR (1967) *Techniques of Vacuum Ultraviolet Spectroscopy.* New York: Wiley.

Tecklenburg RE and Russell DH (1990) An evaluation of the analytical utility of the photodissociation of fast ion beams. *Mass Spectrometry Reviews* 9: 405–451.

van der Hart WJ (1989) Photodissociation of trapped ions. *Mass Spectrometry Reviews* 8: 237–268.

Plasma Desorption Ionization in Mass Spectrometry

Ronald D Macfarlane, Texas A&M University,
College Station, TX, USA

MASS SPECTROMETRY
Methods & Instrumentation

Californium-252 is the key ingredient in a mass spectrometric method known as ^{252}Cf-plasma desorption (^{252}Cf-PD). The ^{252}Cf prefix to 'plasma' distinguishes this method from other mass spectrometric methods that also use a plasma to generate ions. The method uses the energy released in nuclear fission to produce mass spectra of a wide variety of materials ranging from refractory inorganic matrices to samples of biological materials.

A high-energy fission fragment from the spontaneous nuclear fission of ^{252}Cf, penetrating a sample of insulin, penicillin or an unknown protein, decomposes most of what is in its path, but occasionally, an intact molecular ion of the sample is ejected from the surface of the fission track into the vacuum of a mass spectrometer where its molecular mass is measured. This article outlines the essential features of the method including the nuclear properties of ^{252}Cf, the interaction of fission fragments with solids, ion emission from fission tracks in solids, the design of the mass spectrometer, sample preparation, and how data are acquired. Some applications have been selected that have utilized some of the unique features of the method.

A brief history

The development of mass spectrometry for the study of molecules of biological origin did not progress until the problem of forming intact gas phase molecular ions was solved. Volatilization of the molecules by heating led to decomposition. Field desorption ionization introduced in the late 1960s was the first solution to the volatilization–decomposition problem. The discovery of the ^{252}Cf-PD method in the 1970s expanded the spectrum of molecules that could be studied by mass spectrometry. One of the first applications was determining the molecular mass of pharmaceuticals and natural products. The structures of these species were determined by a combination of other spectroscopic techniques, and the molecular mass of the intact molecule was used to verify the atomic composition of the structure that was deduced. In several cases, the molecular mass did not

correlate with the proposed structure and a re-examination of the data from the other spectroscopic measurements led to the final structure. Some of the pharmaceuticals currently in use had their molecular masses first determined by ^{252}Cf-PD. These include vancomycin, amphotericin, thiostreptin, and bleomycin, and the marine toxins, palytoxin, tetrodotoxin and the red-tide toxin.

With the introduction of fast-atom bombardment (FAB) in 1982, and matrix-assisted laser desorption/ionization (MALDI) and electrospray ionization (ESI), most of the biomedical applications have been directed towards these methods. The ^{252}Cf-PD method has been found to have wide applicability, including the study of refractory materials, catalysts, semiconductors and frozen gases. Electronics capable of measuring the timing of events with subnanosecond resolution (the time it takes for a single photon to travel 1 cm) is used by this method as well as event-by-event data acquisition using the computer to make decisions at the molecular level, the basis of correlation mass spectrometry, a unique feature of ^{252}Cf-PD.

Properties of californium-252

Radioactive decay properties

Californium-252 is produced in a high-flux nuclear reactor by multiple neutron capture from ^{239}Pu. It has a half-life of 2.6 years and decays by alpha-particle emission (97%) and spontaneous fission. In the fission process, two high-energy ions (fission fragments) are emitted in opposite directions (180°). (This feature of fission is utilized in the mass spectrometer.) Most of the energy released in the fission process appears in the kinetic energy of the fission fragments, 90–130 million electron volts (MeV). The high kinetic energy coupled with a high charge (15–25 times the electron charge) is responsible for what happens when a fission fragment interacts with a solid. An example of one of the more abundant fission fragments is the technetium ion with a kinetic energy of 100 MeV and charge of +20.

The californium-252 source

Sources of ^{252}Cf specifically designed for ^{252}Cf-PD applications are commercially available. A carrier-free sample of ^{252}Cf is electroplated onto a thin nickel foil in the form of a small circular spot with a diameter of 3 mm and is then sealed with a thin nickel foil to contain the ^{252}Cf. (Fission fragment tracks formed in the source eject ^{252}Cf atoms from the source; a process called self-transfer.) A source used in ^{252}Cf-PD contains about 0.5 μg of ^{252}Cf and, when located to within a few millimetres of a sample, will deliver 2000–3000 fission fragments into the sample per second. The useful life of the source in a ^{252}Cf-PD mass spectrometer is of the order of 4 years.

Properties of the fission track

In a ^{252}Cf-PD measurement, fission fragments form a fission track in the sample to be analysed. The range

of the fission fragment is of the order of 10 μm. If the sample is thinner, the fission fragment passes through the sample depositing 100 eV nm^{-1} of energy in the form of electron excitation and ionization. **Figure 1** depicts the time evolution of the fission track for a 100 MeV Tc ion passing through a thin film and emerging with an energy of 70 MeV. Positive ions and electrons are formed creating a cylindrical high-temperature plasma that lasts for 10^{-15} s. If the material is a conductor, the electrons and positive ions (holes) recombine as shown on the right side of **Figure 1** and the energy is dissipated as heat. In the initial stages of the track development, atomic and molecular ions and photons are ejected from both ends of the fission track. After a period of a few nanoseconds, the track region returns to close to its original state with some evidence of dislocation of some of the atoms in the matrix.

If the matrix is an insulating material, the evolution of the fission track follows a different course.

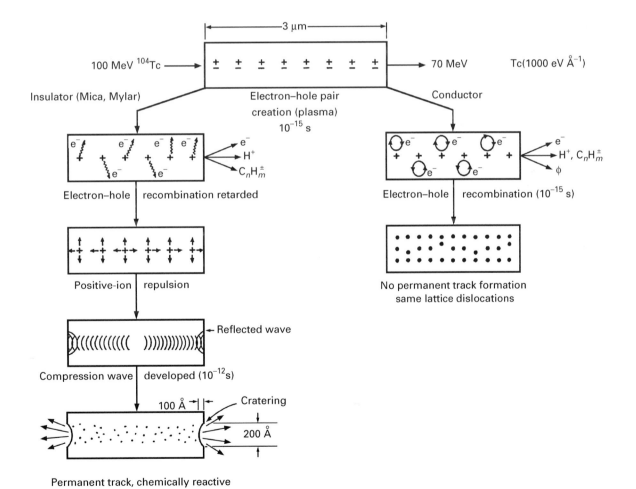

Figure 1 Time evolution of a fission track in an insulator and conductor. A typical PD MS experiment will have an insulator as the energy deposition medium. Protonated molecules from a biological matrix are probably formed during the last stage when material ablation from the surface takes place. The craters depicted here have been observed using scanning tunnelling microscopy.

The electrons ejected from the fission track in the primary ionization are much less mobile and, as a result, the cylinder of positive charge expands due to Coulomb repulsion resulting in the development of a pressure pulse that propagates to the ends of the fission track. The high-pressure gradient generated at the surface of the sample results in ablation of material from the surface, leaving a crater approximately 20 nm in diameter and 10 nm deep. These craters have been studied in detail by scanning tunnelling microscopy. In the ^{252}Cf-PD measurement, the molecules to be analysed are located on the surface of the matrix, and those that are located close to the fission track will be desorbed from the surface within a few trillionths (10^{-12}) of a second after the fission fragment passed through the sample. A permanent damage track is formed in the insulating matrix that is chemically reactive and can be etched to increase the size of the track for microscopic analysis. It is not known when in the sequence of events species are ejected that are detected that contribute to the mass spectrum. For penicillin, 1500 neutral molecules, on average, are ejected from a single fission track and, typically, 1 track in 10 produces an intact penicillin molecular ion.

Subnanosecond chemistry and the fission track

Ions that are formed as a result of excitation of the sample by fission fragments are detected in the ^{252}Cf-PD measurement. They are formed in various stages of the evolution of the fission track, from the surface, and in the gas phase immediately above the surface (called the selvedge region). The high density of ejected species close to the surface supports gas phase chemical reactions and the ejected clusters of molecules in an excited state are also a source of ions that comprise the mass spectrum. **Figure 2** summarizes the processes that can occur leading to the formation of gas phase ions from a perspective that depicts the surface region around the track. The primary fission track with a diameter of 0.1 nm is at the centre of the track. Emanating from the track is the region excited by secondary electrons extending to 6 nm diameter. The range of vaporization extends to 20 nm.

It is likely that the species contributing to the formation of molecular ions of penicillin or insulin are located in the peripheral regions of the fission track. The molecular ions that are detected include radical cations and anions of redox species such as chlorophyll, protonated and deprotonated molecules

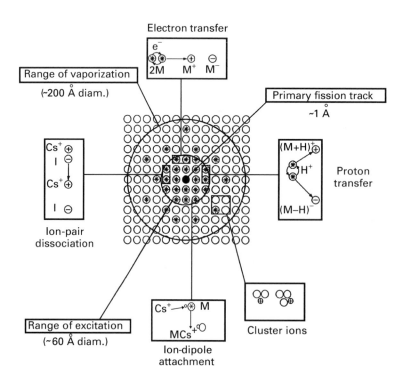

Figure 2 Depiction of the surface of a nascent fission track showing the various modes of ionization processes taking place based on the nature of the gas phase ions observed using different matrices.

such as in the case of insulin, and cations formed from the addition of Na ions to the molecule. The mechanism of the formation of these molecular ions is an open question. There is strong evidence that the important molecular-ion-producing chemistry is taking place in the gas phase, in the selvedge region.

Spectrum measurement

Sample preparation

Most of the molecules of the species to be analysed (analyte) that contribute to the mass spectrum are located in the surface region of the sample. By using adsorption at a solution–solid interface, a uniform layer of the analyte molecules can be made on the surface of a substrate. The substrate is chosen depending on the affinity of the analyte. For example, nitrocellulose is an effective matrix for adsorbing proteins from solution. Hydrophobic substrates, and cationic and anionic substrates have also been used. The sequence for sample preparation is outlined in **Figure 3**.

A thin metallized polymer film is stretched over a sample holder providing mechanical support for the sample to be analysed. (In the ^{252}Cf-PD measurement, the conducting layer is used to establish an electric field for accelerating the ejected ions into the mass spectrometer.) The adsorbing substrate is then deposited as a thin layer on the metallized surface. If nitrocellulose is used, it is electrosprayed from an acetone solution onto the surface. A solution of the analyte is then deposited onto the adsorbing layer. The solution is then removed, and the adsorbed layer

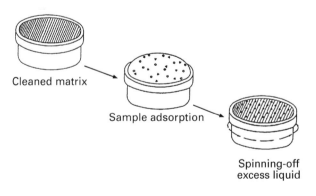

Cleaned matrix

Sample adsorption

Spinning-off excess liquid

Figure 3 The steps involved in the preparation of a sample for PD MS analysis. A matrix with a high adsorption affinity for the analyte is first deposited on the surface of a thin support film. A solution containing the analyte is deposited on the matrix layer and analyte molecules are adsorbed on the surface. The solution is then removed and the matrix–analyte layer washed to remove impurities. Reproduced from Macfarlane, R.D. (1988) Trends in Analytical Chemistry, **7(5)**, 179–183 with permission from Elsevier Science.

washed to remove species that do not adsorb to the surface. The sample is then mounted into the mass spectrometer and analysed.

The mass spectrometer

The arrangement of ^{252}Cf source and sample foil in the mass spectrometer is shown in **Figure 4**. The sample is positioned 4 mm in front of the ^{252}Cf source and behind a high-transmission metal grid at ground potential. In high vacuum, a voltage is then applied to the sample foil. When a fission fragment from the ^{252}Cf source passes through the sample foil ions are ejected from the surface of the sample foil and are accelerated through the grid with the same energy to charge ratio into a 1 m long tube called the flight tube. An ion detector, located at the end of the flight tube generates an electronic pulse every time an ion hits the detector.

The mass of an ion is determined by measuring how long it takes for the ion to travel the distance through the flight tube to the detector, since all ions have been accelerated through a common electric field and have the same energy. (Some ions, particularly protein molecular ions, are multiply charged. In these cases, mass to charge ratio is measured and the energy of the accelerated ion is the voltage + charge product.) The travel time is referred to as the time-of-flight (TOF). To measure the TOF of an ion, a 'time zero' marker must be generated. The complementary fission fragment, the companion of the fission fragment passing through the sample foil, provides this marker. Travelling in the opposite direction, it is detected by an ion detector located behind the ^{252}Cf source that first identifies it as a fission fragment and not an alpha-particle by the amplitude of the electronic pulse that is generated, and then sends a timing pulse to an electronic module called a time interval digitizer. For the fission track depicted in **Figure 4**, five ions were ejected from an adsorbed layer of insulin molecules. This particular track, which released a protonated insulin molecule, two fragment ions from the β-chain, a sodium ion and a hydrogen ion were identified as contributors to the mass spectrum of the insulin sample under study. For illustration, what happens to the data generated by these five ions as the analysis passes into the data acquisition phase, is followed in the subsequent sections.

Time-of-flight measurement and spectrum

The time interval digitizer (TID) is a fast electronic clock that has a resolution of 80 ps. The electronic pulse from the fission detector travels to the fast clock through a wire at the speed of light and arrives at the

TID while the ions are still travelling down the flight tube. The hydrogen ion arrives first at the ion detector at the end of the flight tube because it has the lowest mass and highest velocity. The electronic signal generated by this ion is sent to the TID which records the time lapsed since the arrival of the time-zero marker from the fission fragment. In this particular case, the lapsed time was determined to be 977.260 ns as shown in **Figure 5**. The clock keeps running and as the other three ions arrive, their flight times are also recorded in the same data set. After the maximum time has lapsed, the clock is reset for the next event and the data set is sent to an interrogation module, an interface between the clock and a computer, for the initial procession step. This scenario is repeated 2000 times a second as 2000 fission fragments pass through random sites in the insulin monolayer, generating the ions whose TOF values will comprise the mass spectrum for this sample.

Event-by-event analysis

The interrogation module is the link between experimentalist and the computer. It can be programmed to provide a variety of functions. Some of these options are indicated in the flow chart shown in **Figure 6**. The first option is the sorting of those ions used in obtaining a mass calibration, converting TOF values into m/z values. In the experiment used in the illustration, a decision was made to use the sodium and hydrogen ions as calibration ions, and it was determined by previous measurements that the TOF values for these ions were in time windows of 980 ± 10 ns and 1880 ± 10 ns respectively. TOF values were stored 'as is' as a separate file. Counting the number of ions detected for a particular event, truncation of the significant figures for a TOF value, and detecting whether particular sets of species of ions are formed in the same event are some of the options that have been used in ^{252}Cf-PD studies. Examining the ions formed from a single fission track and interacting with the data set at the molecular level is a unique feature of the ^{252}Cf-PD method.

Generating a mass spectrum

What comes out of the interrogation interface is stored in a computer. For the insulin measurement being used for illustration, the TOF values for the calibration ions are identified by the interrogation module and stored in a separate computer file which sorts by TOF value and records the frequency distributions of TOF values. All other TOF values are truncated to 1 ns resolution and stored in a file that will be used to generate the mass spectrum over the recorded mass range. The sequence of events is depicted in **Figure 7**.

The TOF spectrum is a composite of contributions from 10^6 to 10×10^6 fission fragments passing through the sample foil in a typical measurement requiring 10–100 min of running time. After the measurement is completed, the accumulated TOF spectra for the calibration ions is displayed as shown in **Figure 7** and the mean TOF values are determined. A calibration equation is calculated using the known masses of the calibration ions. The accumulated TOF spectrum for the range of TOF values set for the time interval digitizer (e.g. 32 µs) is then processed. A portion of the TOF spectrum is shown in the lowest

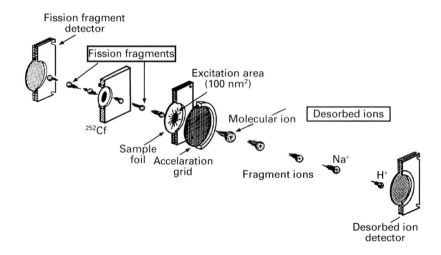

Figure 4 The geometrical arrangement of the essential components of a PD MS system. In the event portrayed here, five ions were ejected from the fission track. Subsequent measurement and processing of the data for these five ions are illustrated in **Figures 5**, **6** and **7**.

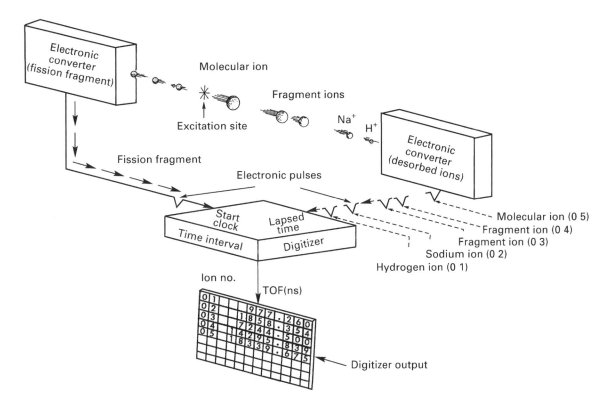

Figure 5 Conversion of ion detection into electronic pulses that are transmitted to a fast clock. The measured time-of-flight for the five ions displayed at the digitizer output is then transferred to a computer for processing.

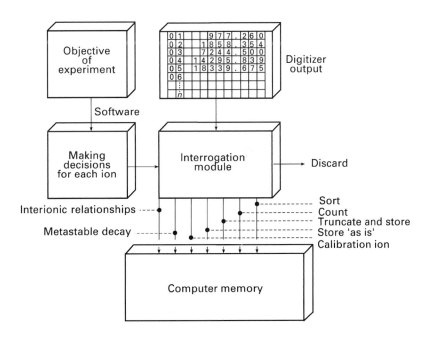

Figure 6 Features of event-by-event analysis. The data from the digitizer are sorted according to the objective of the experiment. A standard PD MS analysis involves identifying the calibration ions and storing the data with 78 ps resolution. The data for other ions are truncated to 1 ns resolution and stored. Examples of other software-controlled options are also listed.

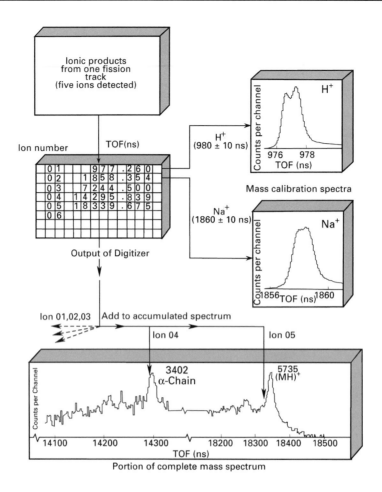

Figure 7 Example of the ion sorting process. The software has identified ions 1 and 2 in the event depicted in **Figure 4** as H+ and Na+ calibration ions and has stored these data in separate files. It then truncates the TOF values for all five ions and stores these data in a separate file that will be used to generate the full mass spectrum. This process is repeated 2000 times per second over a period of a few hours. The mass spectrum is a composite of the accumulation of the ions emitted from 10^6 to 10×10^6 fission tracks.

block in **Figure 7**. Peaks in the spectrum are identified, and mean TOF values for each peak determined. These TOF values are converted into m/z ratios and are tabulated. From this analysis, it was determined that ion number 5 depicted in **Figure 4** was from a protonated insulin molecule, and ion numbers 3 and 4 were fragment ions from the β-chain of a neighbouring insulin molecule close enough to be ejected by the same fission track. It is possible that the two insulin molecules were ejected from the surface as part of a cluster of molecules and, in the region just above the surface, processes took place within the cluster resulting in the formation of the protonated insulin molecule with the protonated second insulin ion acquiring enough internal excitation to dissociate into a protonated fragment of the β-chain and, presumably, a neutral fragment of the α-chain. The formation of these two ions from the same fission track is an example of a correlation phenomenon and is the consequence of two insulin

molecules being within nanometres of each other on the surface of the matrix.

Characteristics of a mass spectrum

Common features for all matrices

Although ions from the fission track are emitted from both ends, generally only the ions emitted from one end are used to produce a mass spectrum. Some of the ions emitted are not structurally related to the sample molecules being studied but are to high-energy processes that are characteristic of a high-temperature plasma. These ions include H+, the dominant ion in the spectrum, $H_2^{\bullet+}$, H_3^+, electrons, H−, and a set of positive and negative hydrocarbon ions in the m/z range 12–100. These ions originate from impurity molecules on the surface of the sample and are formed in high-energy gas phase processes near the surface.

Involatile organic samples

Most of the studies carried out for organic samples have involved involatile, highly polar biological molecules because this class of compound was not generally amenable to mass analysis prior to the development of ^{252}Cf-PD. Peptides, proteins, synthetic oligonucleotides are some of the classes of compound that were characterized by this method. These compounds produce protonated molecules and ions formed by alkali metal ion attachment to the molecule. In addition, many of these species form deprotonated negative molecular ions as well. By measuring the m/z values of these species, the M_r of the parent molecule can be identified. In addition to the protonated molecule, an extensive pattern of fragment ions is also formed due to the dissociation of ions that have acquired a high level of internal excitation in the desorption/ionization process. This feature of ^{252}Cf-PD, the identification of M_r and acquisition of an extensive high-excitation fragmentation pattern in a single measurement is unique amongst the desorption/ionization mass spectrometric methods.

Involatile inorganic species

The ^{252}Cf-PD mass spectrum of metal halides and oxides consists of a family cluster ions of these compounds extending to over m/z 10000, produced by the ejection of small domains of the crystal lattice in the region around the fission track. In addition, cluster ions are also observed that do not correlate with the composition of the crystal lattice, indicating that some of the cluster ions are involved in gas phase reactions in the desorption plume. One of the unique applications of ^{252}Cf-PD is the elucidation of the composition of large transition metal cluster compounds with M_r values approaching 10^5.

Correlations and ion multiplicity

While most of the ^{252}Cf-PD measurements have involved the recording of mass spectra, a higher level analysis is made possible by the event-by-event feature of the data acquisition, as described above. In this section, two examples are given that make use of this feature. The evolution of the fission track and the processes leading to ion formation are complex and variable. Each fission track is unique. Some tracks generate a large number of ions while others produce none. The generation of a protonated insulin molecule is a rare event, observed in one in a hundred fission tracks. To learn more about those tracks which produce this ion, the interrogation module can be programmed to determine how many other ions

are formed in the event that produced the protonated insulin molecule. This experiment is outlined in **Figure 8**.

A time window is set up in the interrogation module encompassing the TOF values for the protonated insulin molecule. The data file formed for each event is interrogated to determine the number of ions detected. Two scenarios are portrayed, one in which four ions are detected but none with the TOF of insulin. The multiplicity of this event (4) is stored in a data file. In the second event, 10 ions were detected, and one was within the TOF window of insulin. The multiplicity of this event was then stored in a separate data file. After 10^6 fission fragments had passed through the sample, two multiplicity histograms were generated; the blocks shown in **Figure 8**. By recording data at the event-by-event level, it was possible to determine that some fission tracks evolve in a manner that enhances the ion emission probability and these tracks have a greater probability for generating the large protonated insulin molecule.

^{252}Cf-PD as a microprobe for heterogeneity at the nanometre level using correlation analysis

Most of the samples that have been studied by ^{252}Cf-PD have been chemically homogeneous, pure samples of a biological molecule or inorganic matrix. Under these circumstances, if two or more ions are detected in a single fission fragment event, desorption/ionization is correlated because they were formed from the same desorption plume. The plume contains approximately 1500 identical molecules that were in close proximity in the sample matrix and within the excitation volume of the fission track. However, if the sample is not chemically homogeneous, microscopic heterogeneity at the nanometre level can be measured by studying the composition of the desorption plume at the event-by-event level. If the sample consists of two components that are homogeneous at the nanometre level, ions from both components are ejected from the same fission track. Event-by-event analysis will show that when ions from one of the components are desorbed it is highly likely that ions from the other component will be desorbed from the same track because both components are in close proximity at the nanometre level. If the components exist in separate domains separated by more than 10 nm within the matrix, fission tracks formed in one domain only desorb ions characteristic of that domain. Although the total spectrum is a composite of ions from both domains, the event-by-event analysis makes it possible to determine what fraction of the two matrices reside with the 10 nm dimensions of the fission track.

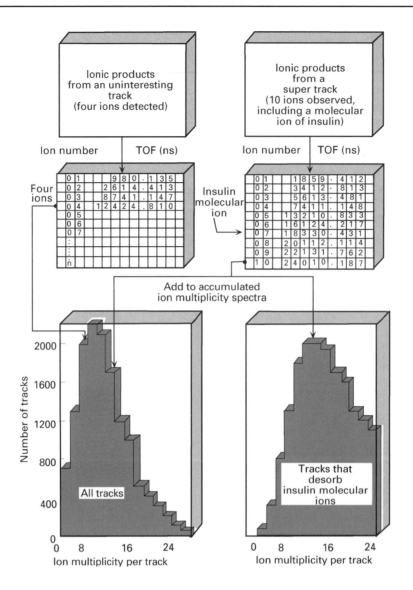

Figure 8 Example of a correlation measurement. The experiment was to determine how many ions are emitted from a fission track that produces a protonated insulin molecule. The interrogation module was programmed to count the number of ions detected for each event where an insulin molecular ion was detected, and store that sum in a separate file. The ion multiplicity for all other events was then stored in a separate file. The conclusion of this study was that tracks that desorb large molecular ions also desorb a much larger number of other ions than average. Other types of correlation measurements include sorting events where one or two ions of a particular type are emitted from the same track.

A historical perspective (1999)

The ²⁵²Cf-PD method has been largely supplanted by MALDI and ESI MS for the analysis of complex biological molecules because these methods are more efficient and widely applicable. The method has stimulated considerable interest in the field of particle-induced desorption and understanding of the mechanisms of the desorption/ionization process. Scanning tunnelling microscopy has been used to characterize the craters produced by fission fragments and high-energy heavy ions from particle accelerators have been used to study the influence of charge, mass and energy on the desorption/ionization process. High-energy cluster ions, including C_{60} fullerene ions, are being used to study the chemistry and physics of particle-induced energy deposition and transfer and ion emission from solids.

See also: **Fast Atom Bombardment Ionization in Mass Spectrometry; Proton Microprobe (Method and Background); Time of Flight Mass Spectrometers.**

Further reading

Da Silveira EF, Duarte SB and Schweikert EA (1998) Multiplicity analysis: a study of secondary particle correlation. *Surface Science* **408**: 28–42.

Della Negra S and Le Beyec Y (1983) Secondary ion emission from surfaces of solids bombarded by high energy heavy ions. Applications in analytical mass spectrometry and studies with beams from accelerators. *Nuclear Science Applications* **1**: 569–590.

Demirev P (1995) Particle-induced desorption in mass spectrometry. Part I Mechanism and processes. *Mass Spectrometry Reviews* **14**: 279–308.

Demirev P (1995) Particle-induced desorption in mass spectrometry. Part II Effects and applications. *Mass Spectrometry Reviews* **14**: 309–326.

Eriksson J, Rottler J and Reimann CT (1998) Fast-ion-induced surface tracks in bioorganic films. *International Journal of Mass Spectrometry and Ion Processes* **175**: 293–308.

Fritsch H-W, Schmidt L, Koehl P, Jungclas H and Duschner H (1993) Application of 252-Cf-PDMS in dental research. *International Journal of Mass Spectrometry and Ion Processes* **126**: 191–196.

Jonsson GP, Hedin AB, Hakansson PL *et al* (1986) Plasma desorption mass spectrometry of peptides and proteins adsorbed on nitrocellulose. *Analytical Chemistry* **58**: 1084–1087.

Jungclas H, Koehl P, Schmidt L and Fritsch H-W (1993) Quantitative matrix assisted plasma desorption mass spectrometry. *International Journal of Mass Spectrometry and Ion Processes* **126**: 157–161.

Macfarlane RD (1983) Californium-252 plasma desorption mass spectrometry – Large molecules, software, and the essence of time. *Analytical Chemistry* **55**: 1247A–1283A.

Macfarlane RD, Hill JC, Jacobs DL and Geno PW (1989) 252-Cf-Plasma desorption mass spectrometry – Past and present. *Advances in Mass Spectrometry* **11**: 3–21.

Matthaus R, Moshammer R, von Hayn G, Wien K, Della Negra S and Le Beyec Y (1993) Secondary ion emission from various metals and semiconductors Si and GaAs induced by MeV ion impact. *International Journal of Mass Spectrometry and Ion Processes* **126**: 45–58.

Roepstorff P and Sundqvist B (1986) Plasma desorption mass spectrometry of high molecular weight biomolecules. In: Gaskell, SJ (ed) *Mass Spectrometry in Biomedical Research*, pp 269–285. Chichester, UK: Wiley.

Sundqvist B and Macfarlane RD (1985) 252-Cf-Plasma desorption mass spectrometry. *Mass Spectrometry Reviews* **4**: 421–460.

Van Stipdonk MJ and Schweikert EA (1995) High energy chemistry caused by fast ion–solid interactions. *Nuclear Instruments and Methods in Physics Research B* **96**: 530–535.

Van Stipdonk MJ, Schweikert EA and Park MA (1997) Coincidence measurements in mass spectrometry. *Journal of Mass Spectrometry* **32**: 1151–1161.

Wien K (1989) Fast heavy ion induced desorption. *Radiation Effects and Defects* **109**: 137–167.

Platinum NMR, Applications

See **Heteronuclear NMR Applications (La–Hg).**

Polarimeters

See **ORD and Polarimetry Instruments.**

Polymer Applications of IR and Raman Spectroscopy

CM Snively and **JL Koenig**, Case Western Reserve University, Cleveland, OH, USA

VIBRATIONAL, ROTATIONAL & RAMAN SPECTROSCOPY

Applications

The two techniques of Raman and infrared (IR) spectroscopy have some similarities, yet are quite different in a number of ways. They provide complementary vibrational information. The reader is referred to the articles on the fundamentals of these techniques for details. Raman and IR will here be collectively referred to as VS (Vibrational Spectroscopy), except where details of a particular technique warrant discussion. It should also be noted that most of the techniques mentioned can also be used with near-IR radiation, but specific examples will not be cited.

Among the large number of applications of VS, applications to polymeric systems are especially interesting because of the wide variety of chemical structures and physical ordering that is present in polymer systems. The article is arranged as follows. First, the application of VS to the determination of chemical properties of polymeric systems will be illustrated. In particular, the use of VS as an identification tool for complex polymeric systems and the application of VS to the various chemical reaction processes will be detailed. Then the application of VS to the determination of polymer structure on a wide range of length scales will be explained, with particular emphasis on the determination of stereoregularity, chain conformation and crystallinity. The remainder of the article will focus on the study of dynamic properties of polymers such as diffusion and rheological properties and current topics such as millisecond time-resolved and microimaging applications.

Spectroscopic considerations unique to polymers

In contrast to small-molecule compounds, in polymer molecules the atoms are all linked together to form long chains. The presence of such long chains causes additional vibrational modes to be present that do not exist in small-molecule analogues. These arise owing to the vibrations of the chain as a whole. This topic is best treated using classical physics and normal coordinate analysis, which is beyond the scope of this article. A more indepth discussion of

these topics can be found in the books by Koenig listed in Further reading. In addition, the long chains can possess ordering along the chain as well as between neighbouring chains. This quality is also unique to polymers and is responsible for many of the physical properties that make polymers the material of choice for a wide variety of applications.

Several properties of polymers complicate their analysis via VS. First, a problem unique to transmission IR spectroscopy is that polymers are very strong absorbers of IR radiation. Therefore, in order to be within the linear region of Beer's law, an extremely thin polymer film must be used in transmission. A good rule of thumb is to keep the thickness below 5 μm. While the production of such thin films is possible in the laboratory, it must be remembered that most commonly encountered polymer systems are much thicker than this. As a result, the most commonly used industrial IR techniques are reflectance techniques such as attenuated total reflectance (ATR) or reflection–absorption spectroscopy (RAS), which have much smaller effective optical path lengths, typically on the order 1 μm or below.

A problem unique to Raman spectroscopy is the fact that most polymers fluoresce strongly when exposed to laser radiation. This problem can be reduced by using Fourier transform and resonance Raman techniques. Because of this and other difficulties associated with Raman spectroscopy, the quality of Raman spectra of polymers is typically less than that of IR spectra. Therefore, it is not surprising that a quick search of the literature reveals many more quantitative studies of polymers using IR than using Raman.

Using a combination of IR and Raman spectroscopies can yield valuable information. IR spectroscopy is sensitive to any chemical groups that possess a significant dipole moment, such as C–H, and C=O, which are commonly found in polymer side groups. Raman spectroscopy is more sensitive to highly polarizable groups such as C–C and C=C, which are commonly found in the polymer chain backbone. Thus, when used together, IR and Raman spectroscopy can be used to gain more information than is available from either of the individual techniques.

Complete structure determination

Vibrational spectroscopy can be used in the complete chemical and physical structure determination of polymers. Four size scales of structure and orientation found in polymers are covered. The most basic is the chemical identity of the chains, including chemical groups present, monomer sequences and stereoregularity. This is followed by studies of the local conformation of individual chains and interactions between chains, and finally orientation induced by the application of macroscopic forces.

Chemical identity of polymer chains

The techniques used for identification of the basic chemical structure of polymer chains are similar to those used for small molecules. The difference is that there is not a single chemical structure present; rather there is a distribution of chemical structures. The semirandom statistics involved in the polymerization reaction produce a distribution of molecular masses. Therefore, an average molecular mass is reported. Although chromatography and light scattering are most commonly used for molecular mass determination, IR can provide complementary chemical structure information. A number-average relative molecular mass M_n can be obtained by determining the number of end groups present for a given amount of polymer by

$$M_n = \frac{2}{E_{\text{end}}}$$

where E_{end} is the number of mass equivalent weights of end groups per gram of polymer, as determined from a simple Beer's law treatment. This method works if the structure of the end groups and the nature of the polymerization process are known.

Another unique property of polymerization and copolymerization reactions is the presence of a distribution of chemical connectivities in the final polymer, which come about from the orientation of monomer addition. This is commonly referred to as regiochemistry or regioisomerism. In a typical monosubstituted vinyl polymerization, the monomer has two options of addition to the growing chain, commonly referred to as head or tail addition, depending on whether the propagating chain attacks the monomeric carbon with the pendant group or the unsubstituted carbon, respectively. The ratio of head-to-head, head-to-tail, etc. groups present in the final polymer can be determined by comparing the ratios of spectral bands associated with these linkages. In a

similar manner, copolymer systems can be analysed to determine the statistics of the polymerization process. NMR is the most commonly used technique for this type of analysis, but not all polymers can be analysed using this method. VS can be used in these cases and also to verify the NMR results.

Perhaps the most common application of VS in the determination of chemical makeup in polymeric systems is the identification of components in complex polymer mixtures. Polymeric products are rarely composed of a single component. There are always additives present that aid in processing, appearance, adhesion, chemical stability or other properties important to the function of the final product. In an industrial setting, it is important to be able to determine both the identity and quantity of polymers and additives in a specific formulation for quality control purposes. This can be a fairly routine operation if tools such as spectral libraries are utilized. In this method, a computer search algorithm compares a spectrum with a catalogue of standard spectra to determine the identity of the compound or compounds present. Advanced statistical techniques, such as partial least squares (PLS) and principal-component analysis (PCA), are also often used to identify known and unknown components in polymeric systems. The details of these methods are described elsewhere in the Encyclopedia.

In certain polymerization processes, the final polymer may contain side branches. The most notable example of this is polyethylene, which commonly contains short branches that result from chain transfer reactions during the polymerization process. For each type of branch, up to six carbons in length, the methyl rocking band around the 900 cm^{-1} has a unique frequency. By comparing the intensities of these branch peaks to peaks associated with the main chain, the relative amount of each type of branch can be determined.

Owing to the length of typical polymer chains, the arrangements of side groups along the chain becomes important in the determination of the chemical properties of the polymer. The polymerization process permanently locks in this arrangement of side groups. The persistence of a specific pattern along the entire chain is known as stereoregularity. Stereoregular effects appear in vibrational spectra as spectral peak splitting or other changes in band shape. An example is shown in **Figure 1**, which shows the IR spectra from the three pure stereoisomers of polystyrene. As can be seen, the bands around 1070, 550 and 900 cm^{-1} are quite different for each stereoisomer, revealing that the local molecular environments are different in each case.

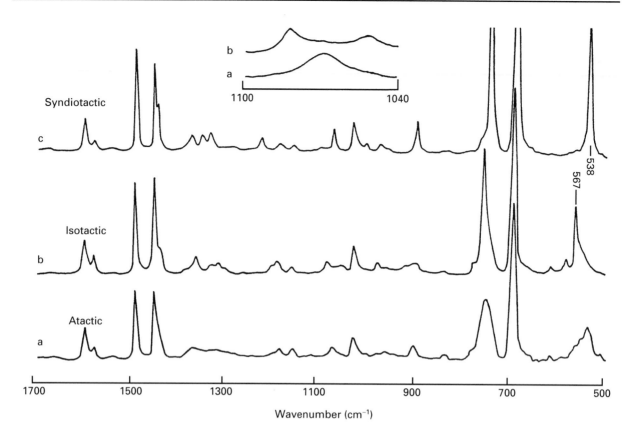

Figure 1 Infrared spectra of the pure stereoisomers of polystyrene.

Orientation of a single polymer chain

The next highest level of order in polymer chains is the presence of rotational isomers, which result from stable configurations of the polymer chain. The presence of *cis* and *trans* configurations and the relative ratios of each can be determined using VS because each of these structures possess a unique spectral band. Owing to the complementary nature of IR and Raman spectroscopy, a wealth of information about the chain conformation can be determined. Raman is particularly useful in this application owing to its enhanced sensitivity to the local environments around carbon–carbon bonds in the chain backbone. Using a simple thermodynamic treatment, this technique can be extended to determine quantitatively the activation energy barrier between the two conformers. The kinetic rate constant, k, can be determined from spectral information using the expression:

$$k = \frac{A_1/a_1}{A_2/a_2}$$

where A_1 and A_2 are the absorbances of the bands associated with each conformer, and a_1 and a_2 are the corresponding extinction coefficients of the bands. By acquiring spectra at a range of different temperatures, the activation energy, ΔG, can be determined from the well-known relation

$$\Delta G = RT \ln k$$

where R is the gas constant and T is the temperature.

Stable crystalline forms arise from the propagation of a particular sequence of conformations along the entire length of the chain, resulting in a helical arrangement. The most commonly used method for analysing crystalline materials is X-ray diffraction, which can detect long-range crystalline order, like that found in semicrystalline polymers such as poly(vinylidene fluoride) and polyethylene. VS probes the local molecular environment, and therefore has the advantage over X-ray methods that it can detect the short-range order present in the amorphous phase. The presence of helical order shows up in a predictable manner in both IR and Raman spectra.

VS has been applied to polystyrene, which can possess several localized crystalline forms. In this particular case, order results from a planar zigzag arrangement of the chains that pack together to form structures that possess either trigonal or orthorhombic symmetries. As subtle as this ordering may seem, it can be detected effectively using VS. This can be seen in **Figure 2**, which shows that the pure crystalline forms of syndiotactic polystyrene have distinct IR spectra. This particular example illustrates the spectral subtraction technique used to determine the spectra of pure crystalline forms of polymers. The spectrum from a completely amorphous sample is subtracted from the spectrum of a partially crystalline sample to yield the 'pure' crystalline spectrum. This is repeated for all crystalline forms. Using these spectra, the crystal forms present in any sample can then be determined. This method of spectral subtraction is commonly used to isolate the spectral contributions of individual components to determine which structures are present.

Interactions between chains

To achieve desirable properties, polymers are often blended to form stable mixtures. The formation of successful polymer blends depends on the presence of favourable interactions, such as hydrogen bonding, between chains. These strong interactions show up in the vibrational spectra as the appearance or disappearance of peaks or as peak shifts. The most commonly studied blend systems involve polymers with carbonyl groups, which show different carbonyl bands for hydrogen-bonded and nonbonded groups. In these studies, the degree of compatibility can be determined by monitoring the spectral shifts and relative peak intensities of the bonded and nonbonded forms for a range of concentrations and temperatures. With the proper mathematical treatment, thermodynamic and kinetic parameters for these processes can be determined. The details are described in depth in the book by Coleman *et al.* (see Further reading). This technique has also been used to probe the interactions between polymers that do not contain such obvious interacting groups, such as the classic totally miscible pair polystyrene and poly(phenylene oxide).

Orientation induced by processing

The production of polymer products typically involves processes in which the polymer is raised above its glass transition temperature (T_g) and forced into some desired shape. The most common of these methods include injection moulding and blow moulding and depend on the desired geometry of the final product. During this process, the highly flexible polymer chains in different parts of the mould orient according to the local shear and elongational stresses. These stresses cause orientation of both the crystalline and amorphous regions of the polymer and can produce ordered regions with undesirable anisotropic mechanical properties in the final products. This orientation can be measured using a variety of methods involving transmission, ATR or RAS depending on the sample geometry and transparency. These reflectance methods are especially useful for determining the orientation at or near the surface of the finished product. The three-dimensional orientation of the chains near the surface can be quantified in this manner.

The most commonly used method for quantifying the extent of orientation in polymeric systems is the determination of the dichroic ratio. Two spectra are collected: one with radiation polarized parallel to a reference direction and one perpendicular to this direction. The reference direction is most commonly chosen along the direction of orientation. The ratio of these two spectra, often called the dichroic ratio spectrum, can then be used to characterize the orientation in the system. Dichroic ratios greater

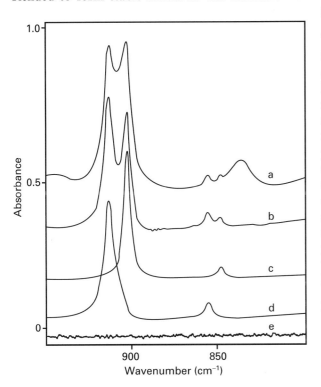

Figure 2 Infrared spectra of polystyrene with (curve a) a mixture of crystalline and amorphous forms, (curve b) a mixture of crystalline forms only, (curve c) one 'pure' crystalline form obtained by spectral subtraction, (curve d) another 'pure' crystalline form obtained by spectral subtraction and (curve e) subtraction result of curves b − c − d. Reproduced with permission of John Wiley & Sons Limited from Musto P, Tavone S, Guerra G and DeRosa C (1997) Evaluation by Fourier transform infrared spectroscopy of the different forms of syndiotactic polystyrene samples. *Journal of Polymer Science B* **35**: 1055–1066.

than unity indicate orientation along the reference direction, while those less than unity indicate orientation in the orthogonal direction. A value of 1 is indicative of a lack of orientation.

It should also be noted that the presence of orientation in samples can give rise to errors in quantitative studies. The presence of orientation changes the measured absorbance values, which affects the results of quantitative analysis. One way around this is to calculate the equivalent absorbance value with no orientation, the so-called structure factor A:

$$A = \frac{A_x + A_y + A_z}{3}$$

where A_x, A_y, and A_z represent the absorbances obtained with x, y, and z polarization, respectively. The structure factor is then used in place of absorbance in quantitative studies.

The behaviour of polymers can also be followed using a time-resolved technique known as rheooptical spectroscopy or dynamic IR linear dichroism (DIRLD). In this technique, a strain is applied to a polymer sample, and the spectral changes are monitored over time. This strain is usually either a sinusoid or a step function, these being the easiest to reproduce experimentally. With the inclusion of stress and strain gauges, the spectral changes can be directly related to the mechanical properties. This technique has been used to directly study strain-induced conformational changes. A more advanced application is the study of the mechanical behaviour of polymer blends. When a miscible blend is examined, it can be seen that the spectral bands corresponding to different chains respond in phase, implying that they are molecularly mixed and act as one unit. The spectral responses have different phases for an immiscible blend, which shows that each polymer functions alone.

Chemical property determination

Polymerization reactions

Perhaps the most obvious chemical application is the monitoring of polymerization reactions involved in the production of polymers. Like most chemical reactions, the polymerization process shows up very well in vibrational spectra as the simultaneous disappearance and appearance of spectral bands. For example, in the polymerization of a typical vinyl monomer, carbon–carbon double bonds are broken and carbon–carbon single bonds are formed. This can be seen clearly in the spectra as the disappearance of the vinyl stretch band around 1500 cm^{-1} and the simultaneous appearance of carbon–carbon single-bond stretching bands and aliphatic carbon–hydrogen stretching bands. This technique can be used to determine both the kinetic order of the polymerization reaction and the kinetic rate constants.

On-line monitoring has also received much recent interest in this area. This application allows spectra to be collected during the polymerization reaction without the necessity of stopping the reaction or performing the experiment under restrictive laboratory conditions. With the advent of low-loss fibreoptics, it is possible to monitor a polymerization reaction in a variety of harsh conditions far away from the spectrometer. This is particularly useful for large production settings commonly found in the polymer industry. An example of the quality of spectra that can be obtained using this method is shown in **Figure 3**, which shows several spectra from different times during a copolymerization reaction between styrene and 2-ethylhexyl acrylate. As can be seen, the quality of the spectra is quite good despite the drastic experimental conditions.

Several types of reactions involving polymers themselves have been studied using VS, including cross-linking, vulcanization, and degradation. The same methods are used for these processes as are used to study polymerization reactions.

Time-dependent phenomena and spatially resolved studies

Vibrational spectroscopy is perhaps the most frequently used technique for the study of diffusion in polymeric systems because it provides a rapid way to quantitatively describe this phenomenon. The two areas of this type of research include the diffusion of small molecules into polymers and polymer–polymer interdiffusion. The easiest and most commonly used technique for this purpose is ATR. In this experiment, a polymer film is placed in contact with an ATR crystal, and the diffusing species is placed on top of the polymer. As diffusion progresses, the diffusing species moves closer to the ATR crystal and shows up in the spectrum obtained as an increase in the diffusant specific spectral band. This spectral change can be used to determine the diffusion coefficient of the system with the appropriate diffusion equation. This technique is limited to IR because of the optics and sample geometry required.

A more modern approach to this problem is to use microscopic techniques, such as an IR microscope or Raman microprobe, which directly monitor the movement of diffusing species. In this technique, the

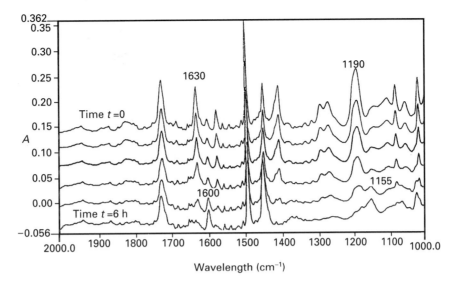

Figure 3 *In situ* infrared spectra measured during the progress of a copolymerization reaction using a fibreoptic probe. Reproduced with permission of John Wiley & Sons Limited from Chatzi EG, Kammona O and Kipanssides (1997) Use of a midrange infrared optical-fiber probe for the on-line monitoring of 2-ethylhexyl acrylate/styrene emulsion copolymerization. *Journal of Applied Polymer Science* **63**: 799.

contact method is used, in which a thin film of polymer is placed into edge-on contact with the diffusing species. After diffusion has been allowed to progress for a certain period of time, the diffusion process is stopped by quenching, and spectra are collected along the diffusion direction to yield spatially resolved concentration information. As in the ATR method, this data are fitted to the appropriate diffusion equation to yield a diffusion coefficient. This method is more difficult to apply than ATR because of the difficulty in sample preparation, but it has the advantage of utilizing a simplified diffusion equation.

A recent trend is the incorporation of two-dimensional detectors into IR and Raman microscopes. For IR studies, focal plane array detectors are used, while charge-coupled device (CCD) detectors are used for Raman studies. When combined with the appropriate hardware, these systems are capable of collecting images of high spatial and spectral resolution in a matter of a few minutes. This allows systems that vary in time to be monitored *in situ*, and in real time depending on the rapidity of the process under study. This technique has the advantage of producing spatially resolved, chemically specific spectral images, which aids in the visualization of the process, in addition to providing high-fidelity quantitative information. An example of the quality of data that can be obtained is shown in **Figure 4**, which shows a diffusion profile obtained using an IR microscope equipped with a 64×64 element focal plane array detector. The normalized absorbance values of the diffusant peak are plotted against diffusion distance and fitted to a Fickian diffusion profile, which is also shown. This

data was extracted from an image that was acquired in less than 3 minutes.

The determination of the spatial distribution of chemical species in polymeric systems is perhaps the most basic and most commonly encountered use of microspectroscopy. This technique is frequently used for the identification of defects in finished polymer products and for the identification of phase-separated regions of polymer blends. Polymer laminate films, which typically consist of layers between 2 and 10 μm thick, are also frequently studied using this technique. Raman techniques are typically more useful than IR

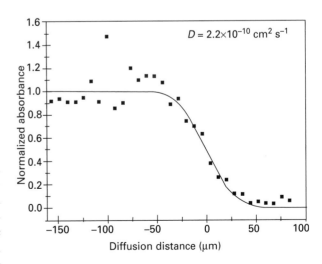

Figure 4 Diffusion profile obtained from an infrared image along with the fit to the diffusion equation. D = diffusion coefficient.

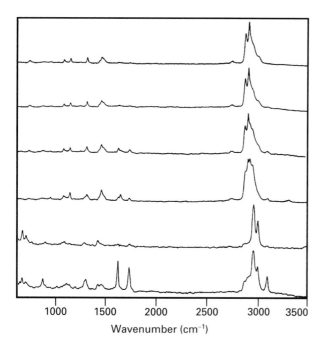

Figure 5 Raman spectra taken at 10 μm increments from a multilayer polymer laminate film. Reproduced with permission of Elsevier Science Limited from Xue G (1997) Fourier transform Raman spectroscopy and its application for the analysis of polymeric materials. *Progress in Polymer Science* **22**: 313–406.

techniques owing to the higher spatial resolution attainable. This is shown in **Figure 5**, which shows Raman spectra obtained from 10 μm regions of a laminate film. As can be seen, the identity of each layer can be distinguished from the others by the spectral features. Another technique, known as confocal Raman microscopy, allows the acquisition of spectra from thin regions, typically around 1 μm, through the depth of the sample. This method has advantages over other methods because it can be used as a non-destructive quality control test to determine the thickness and chemical identity of the individual layers of a thin film.

An additional technique that can obtain spatially resolved spectral information is step-scan photo-acoustic IR spectroscopy. Spectra can be obtained from different depths of a layered sample quickly and with higher spatial resolution (in some cases less than 1 μm) than the diffraction-limited optics of IR microscopes are capable of obtaining.

Mechanical property determination using Raman spectroscopy

An interesting application of Raman spectroscopy is in the determination of the modulus of pure crystalline forms of a polymer. Raman is very sensitive to the longitudinal acoustic vibrational modes of simple polymer chains. Using a combination of normal coordinate analysis and experiment, the modulus of pure crystalline forms of simple straight-chain polymers, such as polyethylene and poly(oxymethylene) have been determined. Similar techniques have been used to study the pressure-dependent band shifts that occur when a polymer sample is placed under stress. This technique is complementary to mechanical analysis because it gives insight into what is occurring at the molecular level during mechanical deformation.

List of symbols

a_1, a_2 = extinction coefficients of bands associated with conformers 1, 2; A = structure factor; A_1, A_2 = absorbances of bands associated with conformers 1, 2; $A_{x,y,z}$ = absorbances obtained with x,y,z polarized radiation; k = kinetic rate constant; M_n = number-average relative molecular mass; R = gas constant; T = temperature; T_g = glass transition temperature; ΔG = activation energy.

See also: **ATR and Reflectance IR Spectroscopy, Applications; IR Spectral Group Frequencies of Organic Compounds; Nuclear Quadrupole Resonance, Instrumentation; Photoacoustic Spectroscopy, Applications; Rayleigh Scattering and Raman Spectroscopy, Theory.**

Further reading

Bark LS and Allen NS (eds) (1982) *Analysis of Polymer Systems.* London: Applied Science.

Bower DI and Maddams WF (1989) *The Vibrational Spectroscopy of Polymers.* New York: Cambridge University Press.

Coleman, MM, Graf JF and Painter PC (1991) *Specific Interactions and the Miscibility of Polymer Blends.* Lancaster: Technomic.

Fawcett AH (ed) (1996) *Polymer Spectroscopy.* New York: Wiley.

Griffiths PR and DeHaseth JA (1986) *Fourier Transform Infrared Spectrometry.* New York: Wiley.

Koenig JL (1980) *Chemical Microstructure of Polymer Chains.* New York: Wiley.

Koenig JL (1992) *Spectroscopy of Polymers.* Washington DC: American Chemical Society.

Painter PC, Coleman MM and Koenig JL (1982) *The Theory of Vibrational Spectroscopy and Its Application to Polymeric Materials.* New York: Wiley.

Siesler HW and Holland-Moritz K (1980) *Infrared and Raman Spectroscopy of Polymers.* New York: Dekker.

Spells SJ (ed) (1994) Characterization of Solid Polymers. New York: Chapman & Hall.

Zbinden R (1964) *Infrared Spectroscopy of High Polymers.* New York: Academic Press.

Porosity Studied By MRI

See **MRI of Oil/Water in Rocks.**

Positron Emission Tomography, Methods and Instrumentation

See **PET, Methods and Instrumentation.**

Positron Emission Tomography, Theory

See **PET, Theory.**

Potassium NMR Spectroscopy

See **NMR Spectroscopy of Alkali Metal Nuclei in Solution.**

Powder X-Ray Diffraction, Applications

Daniel Louër, Université de Rennes, CNRS, France

HIGH ENERGY SPECTROSCOPY
Applications

Introduction

X-ray powder diffraction is a nondestructive technique widely used for the characterization of micro-crystalline materials. The method has been traditionally applied for phase identification, quantitative analysis and the determination of structure imperfections. In recent years, applications have been extended to new areas, such as the determination of moderately complex crystal structures and the extraction of three-dimensional microstructural properties. This is the consequence of the higher resolution of modern diffractometers, the advent of high-intensity X-ray sources and the development of line-profile modelling approaches to overcome the line overlap problem arising from the one-dimensional data contained in a powder diffraction pattern. The method is normally applied to data collected at room temperature. Nevertheless, it is also used with data collected *in situ* as a function of an external constraint (temperature, pressure, electric field, atmosphere, etc.), offering a useful tool for the interpretation of chemical reaction mechanisms and materials behaviour. Various kinds of microcrystalline materials may be characterized from X-ray powder diffraction, such as inorganic, organic and pharmaceutical compounds, minerals, catalysts, metals and ceramics. For most applications, the amount of information which can be extracted depends on the nature and magnitude of the microstructural properties of the sample (crystallinity, structure imperfections, crystallite size), the complexity of the crystal structure and the quality of the experimental data (instrument performance, counting statistics).

Line profile parameters

The observed diffraction line profiles are distributions of intensities $I(2\theta)$ defined by several parameters. The most commonly used measure for the reflection angle is the *position* $2\theta_0$ of the maximum intensity (I_0). It is related to the lattice spacing d of the diffracting hkl plane and the wavelength λ by Bragg's law.

$$\lambda = 2d\sin\theta$$

The *dispersion* of the distribution, or diffraction line broadening, is measured by the full width at half the maximum intensity (FWHM) or by the integral breadth (β) defined as the integrated intensity (I) of the diffraction profile divided by the peak height ($\beta = I/I_0$). Line broadening arises from the convolution of the spectral distribution with the functions of instrumental aberrations and sample-dependent effects (crystallite size and structure imperfections).

Since Fourier-series methods play an important role in X-ray diffraction by imperfect solids, the coefficients (A_n, B_n) of the Fourier series used to represent a line profile are also characteristics of the line broadening. The line *shape factor* is described by the ratio ϕ of the FWHM to the integral breadth ($\phi = \mathrm{FWHM}/\beta$). There are alternative shape factors according to the analytical functions Φ, given in **Table 1**, commonly used for modelling individual diffraction lines, e.g. the mixing factor η for the pseudo-Voigt and the exponent m for the Pearson VII. The *area* is defined by the integrated intensity I of the diffraction line. It is related to the atomic content and arrangement in the unit cell, to the amount of diffracting sample and to angle-dependent factors (Lp). It is proportional to the square of the structure factor amplitude $|F_{hkl}|$:

$$F_{hkl} = \Sigma_j N_j f_j \exp 2\pi\mathrm{i}(hx_j + ky_j + lz_j)$$
$$\times \exp\left(-B_j \sin^2\theta/\lambda^2\right)$$

where N_j is the site occupation factor, f_j is the atomic scattering factor, B_j is the isotropic atomic displacement (thermal) parameter, h, k and l are the Miller indices, and x_j, y_j and z_j are the position coordinates of atom j in the unit cell. **Table 2** lists the specific applications of the powder diffraction method according to the line profile parameters.

Diffraction geometry and data collection

For most applications it is essential that the powder diffraction data be collected appropriately. Therefore, it is of prime importance to spend time optimizing the adjustment of the diffractometer, the quality of the radiation employed and the randomization of

Table 1   Some flexible line-profile functions $\Phi(x)$ used to model powder diffraction line profiles (L and G denote the Lorentzian and Gaussian functions, respectively)

Name	Function	Shape factor
Pseudo-Voigt (p-V)	$C_1[\eta L + (1-\eta)G]$	*Mixing factor η*
	(η, C_1: adjustable parameters)	$\eta = 1$ Lorentzian shape
		$\eta = 0$: Gaussian shape
		$\eta > 1$: super-Lorentzian shape
Pearson VII (PVII)	$C_2(1+Cx^2)^{-m}$	*Exponent m*
	(m, C, C_2: adjustable parameters)	$m = 1$: Lorentzian shape
		$m = \infty$: Gaussian shape
		$m < 1$: super-Lorentzian shape
Voigt (V)	$C_3 \int L(x')G(x-x')\mathrm{d}x'$	$\phi = \mathrm{FWHM}/\beta$
	(C_3: adjustable parameter)	$\phi = 0.636\,6$: Lorentzian shape
		$\phi = 0.939\,4$: Gaussian shape

Table 2 Main applications of X-ray powder diffraction

Diffraction line parameter	Applications
Peak position	Unit-cell parameter refinement
	Pattern indexing
	Anisotropic thermal expansion
	Homogeneous stress
	Phase identification
Line intensity	Phase abundance
	Chemical reaction kinetics
	Crystal-structure determination and refinement (whole pattern)
	Search/match (d–I)
	Space-group determination ($2\theta_0$–absent I_{hkl})
	Preferred orientation
Line width and shape	Instrumental resolution function
	Microstructure
Line-profile broadening	Microstructure (crystallite size, size distribution, lattice distortion, structure mistakes, dislocations, composition gradient), crystallite growth kinetics

the crystallites in the sample. There are several designs for X-ray powder diffractometers (reflection or transmission modes), each of them having advantages and disadvantages. **Figure 1** shows an optics commonly used with conventional divergent-beam X-ray sources, based on the Bragg–Brentano parafocusing geometry. The beam converges on the receiving slit after diffraction by the sample. The geometry is characterized by two circles, the goniometer circle with a constant radius R and the focusing circle with a radius dependent on θ. It uses a flat sample which lies tangent to the focusing circle. The main advantage of this reflection geometry is that no absorption correction has to be made if an 'infinitely' thick sample is used. Disadvantages are that by using a flat sample, preferred orientation effects are increased and at low angles the illuminated area can become larger than the sample. Preferred orientation effects can be reduced by using side-loaded sample holders. The most frequent angular errors arise from a shift of the zero-2θ position and a displacement δ of the specimen from the goniometer axis of rotation ($\Delta 2\theta = 2\delta \cos\theta/R$). Transmission geometry, often combined with a position-sensitive detector (PSD), can be employed with thin flat samples or capillaries. Main advantages are the small amount of sample used and, with capillaries, the reduction of preferred orientation. Transmission optics are ideal for 'transparent' materials, containing only light atoms, but for highly absorbing materials, patterns are difficult to measure. X-ray synchrotron radiation presents some important advantages over conventional X-ray sources. The excellent angular collimation combined with the wavelength tunability and high brightness of the synchrotron source make it ideal for many types of X-ray diffraction experiments. Moreover, with parallel-beam optics, sample-displacement aberrations in diffraction lines are completely eliminated.

Although a basic requirement of the Bragg law is the use of monochromatic radiation, the doublet $K\alpha_1$–$K\alpha_2$ from copper is the most popular wavelength

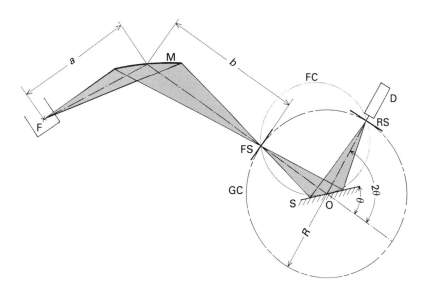

Figure 1 Optics of a conventional focusing powder diffractometer with monochromatic X-rays (Bragg–Brentano geometry with reflection specimen): F line focus of X-ray tube, M incident-beam monochromator, a short focal distance, b long focal distance, FS focal slit, S flat specimen, GC goniometer circle, O goniometer axis, R goniometer radius, FC focusing circle, θ Bragg angle, 2θ reflection angle, RS receiving slit, D detector.

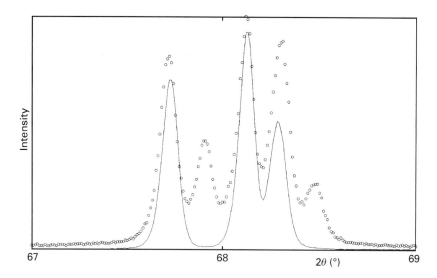

Figure 2 The quartz cluster of reflections 212, 203 and 301 collected with Cu$K\alpha_1$ radiation (continuous line) and with the Cu$K\alpha_1$–$K\alpha_2$ doublet (circles).

used with laboratory diffractometers. However, for the more demanding applications it is desirable to remove the $K\alpha_2$ component with a monochromator located in the incident beam (**Figure 1**). There is some reduction in the intensity of $K\alpha_1$, but the number of reflections in the pattern is halved and thereby the degree of line overlap is reduced. **Figure 2** shows the improvement in resolution achieved with the monochromatic Cu$K\alpha_1$ radiation (λ = 1.5406 Å) with respect to the $K\alpha_1$–$K\alpha_2$ doublet.

The performances of powder diffractometers are determined by the precision on peak position measurements and the instrument resolution function (IRF), normally expressed by the angular dependence of the FWHM obtained with a reference material. The most widely used standard reference materials (SRMs) for diffractometer characterization are those from the National Institute of Standards and Technology (NIST). For instance, powders of Si (SRM 640b, a = 5.430 940 ± 0.000 035 Å) and fluorophlogopite mica (SRM 675, interlayer spacing d_{001} = 9.981 04 ± 0.000 07 Å) are proposed as d-spacing standards, while LaB$_6$ (SRM 660) is recommended as a line-profile standard. With laboratory X-ray diffractometers errors on peak positions can be less than 0.01°(2θ) and the IRF has typically a minimum around 0.06°(2θ) at intermediate angles, increasing to twice this value at ~130°(2θ) as a consequence of spectral dispersion. The best instrumental resolution (~0.01–0.02°2θ) is obtained with the synchrotron parallel beam optics with a crystal analyser mounted in the diffracted beam. This is of particular interest in some applications, such as the study of complex crystal structures.

Pattern modelling

Pattern modelling techniques are used in most current applications. The intensity at point x_i in the calculated pattern is given by

$$y_{\mathrm{cal}}(x_i) = \Sigma_k\, I_k \Phi(x_i - x_k) + b(x_i)$$

where I_k is the integrated intensity of reflection k, Φ is one of the normalized profile functions given in **Table 1** and $b(x_i)$ is the background contribution. The summation is over all reflections contributing to the intensity at point x_i. Parameters defining the model are refined until the quantity

$$S = \Sigma\, w(x_i)[y_{\mathrm{obs}}(x_i) - y_{\mathrm{cal}}(x_i)]^2$$

is a minimum, the summation being over all data points in the diffraction pattern with $w(x_i)$ being the appropriate weighting factor. Although a visual inspection of the difference curve between observed and calculated patterns is the best way to judge the quality of the fit over the whole angular range, numerical factors are used to assess the quality of the final refinement. These are listed in **Table 3**.

Pattern decomposition

This is a systematic procedure for decomposing a powder pattern into its component Bragg reflections without reference to a structure model and, thereby,

Table 3 Some numerical criteria of fit used in pattern-fitting methods.

R-profile	$R_p = \dfrac{\sum \lvert y_i(\text{obs}) - y_i(\text{cal})\rvert}{\sum y_i(\text{obs})}$
R-weighted profile	$R_{wp} = \left[\dfrac{\sum \lvert w_i[y_i(\text{obs}) - y_i(\text{cal})]\rvert^2}{\sum w_i[y_i(\text{obs})]^2}\right]^{1/2}$
R-structure factor	$R_F = \dfrac{\sum \lvert I_{hkl}(\text{'obs'})^{1/2} - I_{hkl}(\text{cal})^{1/2}\rvert}{\sum I_{hkl}(\text{'obs'})^{1/2}}$
R-Bragg factor	$R_B = \dfrac{\sum \lvert I_{hkl}(\text{'obs'}) - I_{hkl}(\text{cal})\rvert}{\sum I_{hkl}(\text{'obs'})}$

line-profile parameters ($2\theta_0$, I_0, I, FWHM, β, shape factors) are extracted. **Figure 3** shows the fitting of pseudo-Voigt functions to the individual lines of the pattern of a ZnO sample displaying diffraction line broadening. However, for extracting integrated intensities over the complete pattern, for a subsequent structural analysis, an approach incorporating constrained peak positions, according to the refined unit-cell dimensions and to the space group, allows the generation of only space-group allowed intensities.

The Rietveld method

In the Rietveld method an observed and a calculated powder diffraction pattern are compared and the difference is used to refine the atomic coordinates of the structure model. The calculated intensity at point x_i is given by the equation

$$y_{\text{cal}}(x_i) = s\Sigma_k \, m_k(Lp)_k \lvert F_k\rvert^2 P_k \Phi(x_i - x_k) + b(x_i)$$

where s is a scale factor, m_k is the reflection multiplicity and P_k is a function to deal with the preferred orientation of the crystallites. They are two groups of refined parameters arising from the structure model (atomic coordinates x_j, y_j and z_j, atomic displacement parameters, unit-cell dimensions) and the instrumental model (angular dependence of the profile parameters, FWHM and shape factors, 2θ-zero position, preferred orientation, etc.). Recommendations for Rietveld-refinement strategies have been formulated by the Commission on Powder Diffraction of the International Union of Crystallography. **Figure 4** shows a typical final Rietveld plot obtained from powder data collected with the capillary method. The precision of a structure refinement depends on many factors, e.g. the number of parameters to be refined, data quality (preferred orientation, counting statistics, anisotropic line broadening), the contrast between atoms and the size of the unit cell. Moreover, structure refinement from X-ray diffraction data is strongly influenced by the fall-off of the scattering atomic factors f_j with d^{-1} (or 2θ). This is in contrast with neutron-diffraction data, for which the scattering length does not fall off significantly over the range of observations. Another important difference between neutron and X-ray

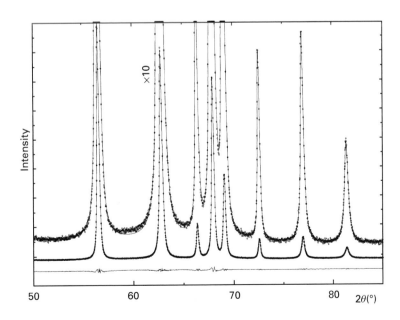

Figure 3 Fitting of a part of the pattern of a sample of nanocrystalline ZnO, using pseudo-Voigt functions for modelling individual diffraction lines, CuKα_1 radiation, R_{wp} = 1.2%. The observed intensity data are plotted as circles and the calculated pattern is shown as a continuous line. The lower trace is the difference curve. The ×10 scale expansion shows the fit in the line-profile tails.

diffraction is that the relative scattering powers of atoms for neutrons and X-rays are significantly different. With neutrons there are pronounced differences in the scattering lengths of neighbouring elements. In the case of X-rays, there is a monotonic variation in X-ray scattering factors and hence light atoms are weak X-ray scatterers. For instance, in the case of $U(UO_2)(PO_4)_2$ the X-ray powder data are dominated by the U atom, the ratio of the scattering factors of U to O being greater than 11.5, while with neutrons oxygen is a relatively strong scatterer and then the ratio of the appropriate scattering lengths is 1.45. This property explains the major role played by neutron powder diffraction in determining oxygen content and position in high T_c cuprates in the presence of heavy atoms such as Ba, Hg, Tl and Bi.

Qualitative and quantitative analyses

Phase identification

Qualitative phase identification is traditionally based on a comparison of observed data with interplanar spacings d and relative intensities I_0 compiled for crystalline materials. The Powder Diffraction File (PDF), edited by the International Centre for

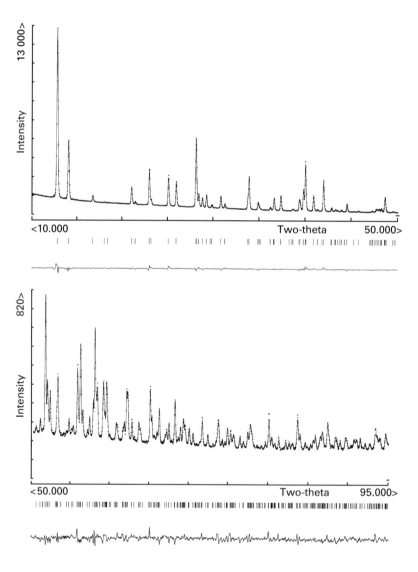

Figure 4 Example of a typical Rietveld plot. The powder data of $LiB_2O_3(OH).H_2O$ were collected with the Debye–Scherrer (capillary) geometry using monochromatic $CuK\alpha_1$ radiation and a curved position-sensitive detector. The observed data are plotted as points and the calculated pattern as a continuous line. The lower trace is a plot of the difference of observed minus calculated. The vertical markers indicate the positions of calculated Bragg reflections. The intensity scale is magnified for the high-angle part, where the intensity of observed diffraction lines is low. Reprinted with permission of the International Union of Crystallography from Louër D, Louër M and Touboul M (1992) Crystal structure determination of lithium diborate hydrate, $LiB_2O_3(OH)·H_2O$, from X-ray powder diffraction data collected with a curved position-sensitive detector. *Journal of Applied Crystallography* **25**: 617–623.

Diffraction Data (ICDD), contains powder data for more than 106 000 (sets 1–48) substances, including approximately 38 000 calculated patterns from the Inorganic Crystal Structure Database (ICSD). The Boolean search program supplied by the ICDD with the database offers great flexibility in phase identification and characterization. In a more recent search/match procedure, complete digitized observed diffraction patterns are used, instead of simply a list of extracted ds and Is. To decide whether or not the pattern contains the data of a particular PDF entry, its data are compared with parts of zero intensity in the pattern for the unknown substance. The method can be used for multiphase patterns. As each phase is identified, the diffraction lines can be removed and the procedure is repeated by using the remaining regions of the pattern. The method is very efficient, even for the identification of minor phases if some known chemical constraints are introduced into the search.

Quantitative phase abundance

Quantitative phase analysis is the determination of the amounts of different phases present in a sample. The powder diffraction method is widely used to determine the abundance of distinct crystalline phases, e.g. in rocks and in mixtures of polymorphs, such as zirconia ceramics. The principle of the method is straightforward; the integrated intensity (I) of the diffraction lines from any phase in a mixture is proportional to the mass of the phase present in the sample. One analytical approach is based on a reference-intensity ratio (RIR) defined as the integrated intensity of the strongest reflection for the phase of interest to the strongest line of a standard (usually the 113 reflection of corundum) for a 1:1 mixture by weight. The measurement and use of RIRs is straightforward for materials for which the intensity does not vary with composition and preferred orientation. Because of the potential health hazard of respirable crystalline silica, X-ray powder diffraction is also used for detecting, identifying and quantifying the crystalline and amorphous silica of all types of samples from airborne dusts to bulk commercial products. SRMs 1878 (α-quartz) and 1979 (cristobalite) from NIST are certified with respect to amorphous content for analysis of silica-containing materials in accordance with health and safety regulations.

An extension of the Rietveld method is its application to multiphase samples for the determination of phase abundance. There is a simple relationship between the individual scale factors determined in a Rietveld analysis and the weight fractions (W_i) of the phase concentration in a multicomponent mixture:

$$W_i = s_i(ZMV)_i/\Sigma_j s_j(ZMV)_j$$

where s_i, Z_i, M_i and V_i are the scale factor, the number of molecules per unit cell, the mass of the formula unit and the unit-cell volume of phase i and the summation is over all phases present. A requirement of the method is that the crystal structure is known for each phase in the mixture. The use of all reflections in the selected angular range is a great advantage, since the uncertainty on phase abundance is reduced by minimizing preferred orientation effects. In general, phase-analysis results obtained with X-ray data are inferior to those obtained from neutron data. This is related to residual errors arising from an imperfect preferred orientation modelling and from microabsorption effects.

Ab initio structure determination

Among the most recent advances of the powder method is the determination of crystal structures from powder diffraction data. It is an application for which the resolution of the pattern is of prime importance. A series of successive stages are involved in the analysis, including the determination of cell dimensions and identification of the space group from systematic reflection absences, the extraction of structure factor moduli $|F_{hkl}|$, the solution to the phase problem to elaborate a structure model and, finally, the refinement of the atomic coordinates with the Rietveld method.

Pattern indexing

The purpose of pattern indexing is to reconstruct the three-dimensional reciprocal lattice of a crystalline solid from the radial distribution of lengths $d^*(=1/d)$ of the diffraction vectors. The basic equation used for indexing a powder diffraction pattern is obtained by squaring the reciprocal-lattice vectors d_{hkl}^* ($=ha^*+kb^*+lc^*$), expressed in terms of the basis vectors of the reciprocal lattice (a^*, b^*, c^*) and hkl Miller indices,

$$Q(hkl) = h^2 Q_A + k^2 Q_B + l^2 Q_C + 2kl Q_D$$
$$+ 2hl Q_E + 2hk Q_F$$

where $Q(hkl) = 1/d^2$, $Q_A = a^{*2}$, $Q_B = b^{*2}$, $Q_C = c^{*2}$, $Q_D = b^* c^* \cos\alpha^*$, $Q_E = c^* a^* \cos\beta^*$, $Q_F = a^* b^* \cos\gamma^*$; a^*, b^*, c^* are the linear parameters and α^*, β^*, γ^* the angles of the reciprocal unit cell. This quadratic form corresponds to the triclinic crystal symmetry. Only four parameters are required for a monoclinic cell, three for an orthorhombic cell, two for tetragonal and hexagonal cells, and one for a cubic cell. Indexing a powder pattern consists in finding the linear and angular dimensions of the unit cell, from which a set of Miller indices hkl can be assigned to each observed line Q_{obs}, within the experimental error on the observed peak positions. Automatic procedures are available to index a powder diffraction pattern. They are based on three main approaches, regardless of symmetry: the zone-indexing method, the index-permutation method and the successive-dichotomy method. The use of pattern-indexing methods requires a high precision on peak positions ($|\Delta 2\theta| < 0.03°2\theta$). The assessment of the reliability of an indexed pattern is carried out with the de Wolff figure of merit M_{20}:

$$M_{20} = Q_{20}/2 \langle \Delta Q \rangle N_{cal}$$

where Q_{20} corresponds to the 20th observed line, $\langle \Delta Q \rangle$ is the average absolute discrepancy between $Q_{k\,obs}$ and the nearest $Q_{k\,cal}$ value and N_{cal} is the number of distinct calculated Q values smaller than Q_{20}, not including any systematic absences if they are known. M_{20} is greater when $\langle \Delta Q \rangle$ is small and N_{cal} is as close as possible to 20. A related figure of merit, F_N, introduced for the evaluation of powder data quality, is also used. It is reported as 'value ($\langle \Delta 2\theta \rangle$, N_{cal})', where $\langle \Delta 2\theta \rangle$ is the average angular discrepancy. A solution with M_{20} greater than 20 is generally correct, although some geometrical ambiguities or the presence of a dominant zone can mask the true solution. An additional check of the reliability of the solution consists in indexing all measurable peak positions in the pattern, from which systematic absent reflections can be detected and, then, possible space groups are proposed. With synchrotron parallel-beam optics, higher figures of merit are obtained, since angular precision and resolution (more lines are observed) are considerably improved. There is no particular problem for indexing, from conventional X-ray data, patterns of materials with moderate cell volumes and, for instance, patterns of monoclinic compounds with volumes up to 3000 Å^3 can be normally handled.

Structure solution

Following the determination of the unit cell and space group assignment, integrated intensities (I_{hkl}) (or structure factor amplitudes) are extracted with a pattern-decomposition technique. For clusters of overlapping lines an equipartition of the overall intensity is generally applied. Therefore, intensity data sets contain a limited number of unambiguously measured reflections. An estimation of the proportion of statistically independent reflections can be calculated from an algorithm based on line width and reflection proximity. It is a useful indicator which depends on the selected angular range. The methods for solving the phase problem used with single-crystal data (Patterson methods and direct methods) are generally applicable with powder data. Computer programs have been adapted to the powder diffraction case. In general, only fragments of the structure are found from these methods; the remaining atoms are then obtained from subsequent Fourier calculations. However, with good data and a favourable proportion of unambiguous reflections, the success of the direct methods can be very high and even complete models (non-H atoms) may be revealed from one calculation. Considerable effort has been devoted to the development of new approaches for the treatment of powder-diffraction data. They include the Monte Carlo and the simulated annealing approaches, the maximum-entropy method, the atom–atom potential method and genetic algorithms. Computer-modelling approaches operate in direct space. Trial crystal structures are generated independently of the observed powder diffraction data. The suitability of each structure model is assessed by comparison between the calculated and observed diffraction patterns and is quantified using an appropriate profile R-factor (see **Table 3**). The direct-space methods present the advantage of avoiding the critical stage of extracting the individual intensities from pattern decomposition. They are suitable for organic molecules, but a detailed knowledge of the expected molecule (bond lengths and torsional angles) is a basic requirement. An example is the structure determination of a metastable form of piracetam, $C_6H_{10}N_2O_2$, whose lifetime at room temperature is only 2 h, solved from powder data collected with a PSD (**Figure 5**). Structure determination from powder data is used in varied fields of materials science, including inorganic, organometallic and organic chemistry, mineralogy and pharmaceutical science. Although the analysis is generally applicable to moderately complex structures, it has been successful for solving structures containing, for instance, 29 atoms in the asymmetric unit, e.g. $Ga_2(HPO_3)_3.4H_2O$, Ba_3AlF_3, $Bi(H_2O)_4(OSO_2CF_3)$, or having a high unit-cell volume, e.g. 7471 Å^3 for $[(CH_3)_4N]_4Ge_4S_{10}$. With the ultra-high resolution available with X-ray synchrotron radiation, determination of more complex structure can now be undertaken, particularly with materials containing light atoms.

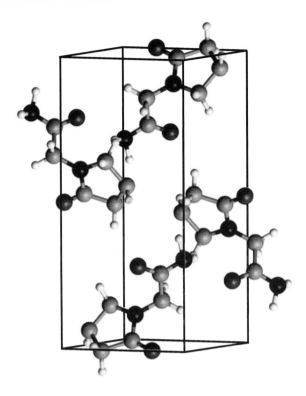

Figure 5 View of the crystal structure of a metastable phase of piracetam, $C_6H_{10}N_2O_2$, solved from the atom–atom potential method and refined with the Rietveld method. The crystal symmetry is monoclinic, space group $P2_1/n$ (the unit cell is shown with the a axis horizontal and b axis vertical) (crystal data are reported in *Acta Crystallographica* **B51**: 182, 1995). Powder diffraction data were collected from a conventional X-ray source.

For specific applications, the tunability of synchrotron radiation sources allows the X-ray wavelength to be changed readily, and this can be exploited to enhance the contrast between close elements in the periodic table. These studies are termed anomalous (resonant) scattering experiments. The atomic scattering factor for X-rays is defined as

$$f = f_0(\theta) + f'(E) + \mathrm{i}f''(E)$$

where f_0 varies only with $\sin\theta/\lambda$ and f' and f'' are the energy-dependent real and imaginary parts of the anomalous contribution. By selecting a wavelength close to the absorption edge of an element the scattering factor may change by a few electrons. By comparing powder data collected near-edge and off-edge, the technique can be used to determine the distribution of cations with similar atomic numbers over crystallographically distinct sites. Applications can be extended to mixed-oxidation states of an absorbing element if these are ordered within a crystal structure, e.g. Eu^{2+} and Eu^{3+} in Eu_3O_4 or Ga^+ and Ga^{3+} in $GaCl_2$.

Diffraction line broadening analysis

Microstructural imperfections (lattice distortions, stacking faults) and the small size of crystallites (i.e. domains over which diffraction is coherent) are usually extracted from the integral breadth or a Fourier analysis of individual diffraction line profiles. Lattice distortion (microstrain) represents departure of atom position from an ideal structure. Crystallite sizes covered in line-broadening analysis are in the approximate range 20–1000 Å. Stacking faults may occur in close-packed or layer structures, e.g. hexagonal Co and ZnO. The effect on line breadths is similar to that due to crystallite size, but there is usually a marked *hkl*-dependence. Fourier coefficients for a reflection of order *l*, $C(n,l)$, corrected from the instrumental contribution, are expressed as the product of real, order-independent, size coefficients $A^S(n)$ and complex, order-dependent, distortion coefficients $C^D(n,l)$ $[=A^D(n,l)+\mathrm{i}B^D(n,l)]$. Considering only the cosine coefficients $A(n,l)$ $[=A^S(n).A^D(n,l)]$ and a series expansion of $A^D(n,l)$, $A^S(n)$ and the microstrain $\langle e^2(n)\rangle$ can be readily separated, if at least two orders of a reflection are available, e.g. from the equation

$$A(n,l) = A^S(n) - A^S(n)2\pi^2 l^2 n^2 \langle e^2(n)\rangle$$

The inverse of the initial slope of the size coefficients $A^S(n)$ versus the Fourier harmonic number n (or $L = n/\Delta s$, where Δs is the range of the intrinsic profile in reciprocal units) is a measure of the Fourier apparent size, ε_F, defined as an area-weighted crystallite size. The second derivative of $A^S(n)$ is a measure of the crystallite size distribution in a direction perpendicular to the diffracting plane *hkl*, obtainable when strains are negligible. Fourier coefficient $C(n)$ are also related to the integral breadth, expressed in reciprocal units, β^* $[=\Delta s/\Sigma_n|C(n)|]$. For strain-free materials the reciprocal of this quantity is the integral-breadth apparent size, ε_β, defined as a volume-weighted thickness of crystallite in the direction of the diffraction vector. For a spherical crystallite the direction of the diffraction vector is unimportant and there is a simple relation between the diameter D and ε_β $(=3D/4)$ or ε_F $(=2D/3)$. The ideal ratio $\varepsilon_\beta/\varepsilon_F$ is 1.125. Any departure from this ratio is due to the presence of a distribution of crystallite sizes. Similar expressions have been derived for other crystallite shapes, e.g. the cylinder, for which the two limiting cases are acicular and disk-like forms. Anisotropic models require precise determination of apparent

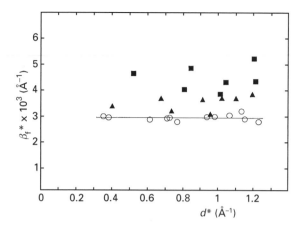

Figure 6 Example of a Williamson–Hall plot for a sample of nanocrystalline ZnO powder exhibiting size and stacking-fault diffraction line broadening according to the *hkl* values. (O) reflections unaffected by mistakes (*hk*0 and *hkl* with *l* even, $h - k = 3n$), (▲) first reflection set affected by stacking faults (*hkl* with *l* odd, $h - k = 3n \pm 1$), (■) second reflection set affected by stacking faults (*hkl* with *l* even, $h - k = 3n \pm 1$).

sizes in various crystallographic directions. Representative examples of average spherical and cylindrical crystallite shapes are found in CeO_2 and ZnO loose powders.

These methods can be applied to more diffraction lines when profile-fitting techniques are employed, provided the observed data are precisely modelled. A more-detailed (three-dimensional) characterization of the microstructure can then be obtained. In the Fourier approach, line profiles must be reconstructed analytically, from the line-profile parameters extracted from pattern-decomposition techniques, prior to the physical interpretation. With the integral-breadth method, the use of Voigt functions in the modelling of observed h and instrumental g profiles represents a definitive advantage, since the integral breadth β_f of the intrinsic f profile can be precisely determined. Whatever the diffraction-line-broadening analysis employed, an essential preliminary step in the analysis is to examine a (Williamson–Hall) plot giving the variation of β_f (expressed in reciprocal units) as a function of d^*. Although this plot is not used for quantitative characterization, it gives an overview of the nature of the broadening due to sample imperfections and orients the subsequent analysis. **Figure 6** shows the *hkl*-dependence of the integral breadths for a sample of nanocrystalline ZnO. These plots can be varied according to the effects at the origin of line broadening and to their anisotropic nature. Since perfect modelling of line profiles are required in the Rietveld method, the technique can also, in principle, be used to extract microstructural properties. However, unless line broadening is isotropic, to obtain

meaningful information the procedure must be able to handle the various possible sources of broadening present in the pattern, such as crystallite size, strain, stacking faults, dislocation density, and the residual errors related to the structure and preferred orientation models must be minimized.

Dynamic and non-ambient diffraction

Time and temperature dependent X-ray diffraction includes the measurement of a series of diffraction patterns as a function of time, temperature or other physical constraint. The time required for collecting data decreases considerably with the availability of fast detectors, such as PSDs, and the brightness of the X-ray source. In principle, line-profile parameters can be extracted for each pattern and interpreted in structural (peak position and integrated intensity) and microstructural (breadths and line shapes) terms. Consequently, the structural and microstructural changes taking place as a function of the external constraint (temperature, atmosphere, pressure, electric field, etc.) are displayed from the successive patterns. The method affords the possibility of establishing the pathways during chemical solid-state reactions, such as phase transition and thermal decomposition, and to determine the kinetics of processes, e.g. the crystallization of nanocrystalline solids or phase transformations. Patterns can be collected on a time scale of a few minutes with conventional X-rays but the high brightness of synchrotron radiation makes it possible to measure diffraction data in very short time periods. A representative application is the investigation of fast and self-propagating solid combustion reactions on a subsecond time-scale. *In situ* powder diffraction can also be combined with other complementary techniques applied simultaneously, such as EXAFS (extended X-ray absorption fine structure). The two structural probes may provide long-range order information (powder diffraction) and short-range order details (EXAFS). For instance, *in situ* combined X-ray diffraction and EXAFS have been used for the study of the formation of heterogeneous catalysts.

List of symbols

A_n, B_n = Fourier coefficients; A^D = distortion coefficient; A^S = size coefficient; B = isotropic atomic displacement; d = interlayer spacing; D = crystallite diameter; e = microstrain; f = atomic scattering factor; F = structure factor; h,k,l = Miller indices; I = intensity; Lp = angle-dependent factor; m = Pearson exponent; M_{20} = de Wolff figure of merit;

m_k = reflection multiplicity; N = site occupation factor; R = goniometer circle radius; R_p = profile factor; s = scale factor; W = weight fraction; Z = number of molecules per unit cell; β = integral breadth; δ = specimen displacement; ε = apparent crystallite size; η = pseudo-Voigt mixing factor; θ = diffraction angle; λ = wavelength; ϕ = shape factor; Φ = line-profile function.

See also: **Inorganic Compounds and Minerals Studied Using X-Ray Diffraction; Materials Science Applications of X-Ray Diffraction; Neutron Diffraction, Theory; Small Molecule Applications of X-Ray Diffraction; X-Ray Absorption Spectrometers.**

Further reading

Bish DL and Pose JE (eds) (1989) Modern powder diffraction. *Reviews in Mineralogy* 20: 1–369.

Harris KDM and Tremayne M (1996) Crystal structure determination from powder diffraction data. *Chemistry of Materials* 8: 2554–2570.

Jenkins R and Snyder RL (1996) *Introduction to X-ray Powder Diffractometry.* New York: Wiley.

Langford JI and Louër D (1996) Powder diffraction. *Reports on Progress in Physics* 59: 131–234.

McCusker LB, Von Dreele RB, Cox DE, Louër D and Scardi P (1999) Rietveld refinement guidelines. *Journal of Applied Crystallography* 32: 36–50.

Parrish W (1995) Powder and related techniques: X-ray techniques. In Wilson AJC (ed) *International Tables for Crystallography*, Vol. C, 42–79. Dordrecht: Kluwer Academic Publishers.

Smith DK (1997) Evaluation of the detectability and quantification of respirable crystalline silica by X-ray powder diffraction methods. *Powder Diffraction* 12: 200–227.

Young RA (ed) (1995) *The Rietveld Method.* Oxford: IUCr-Oxford University Press.

Product Operator Formalism in NMR

Timothy J Norwood, Leicester University, UK

MAGNETIC RESONANCE
Theory

The introduction of the product operator formalism by Ernst and co-workers can be regarded as one of the milestones in the development of NMR. Unusually, this is not because it constitutes an advance in either theory or technique, for it is neither, but because it makes it possible for many of the users of NMR to understand the experiments they carry out. It is necessary to use quantum mechanics to explain the behaviour of nuclear magnetism and a density operator to describe the state of the system. While this approach is rigorous, it can often appear abstract and lacking in physical intuition to those who do not have a background in quantum mechanics. Furthermore, calculations often involve cumbersome matrix operations. While there is an alternative in the physically intuitive net-magnetization vector model, this classical approach is only really useful for explaining very simple experiments such as the spin echo, and breaks down when it comes to explaining phenomena such as coherence transfer and multiple-quantum coherence that are ubiquitous in modern NMR spectroscopy. The

product operator formalism provides a half-way house between the two approaches; it retains both the rigour of the density matrix and the physical intuition and ease of use of the vector model. While with the density matrix it is usual to follow the evolution of the system as a whole during the course of an experiment, with the product operator formalism the fate of individual components can easily be traced. The product operator formalism can equally well be used either quantitatively to calculate the amplitude of the magnetization following particular coherence transfer pathways in the course of an experiment or qualitatively to sketch out those pathways. The product operator formalism has seen extensive use both as an education tool for explaining how pulse sequences work and as a research tool used as an aid for designing new ones. Its main limitation is that it is really only suitable for describing weakly coupled spin systems. When multiple-quantum coherence is under consideration, it can be usefully supplemented with raising and lowering operators.

Theory

Origin and definitions

The density operator $\sigma(t)$ can be expanded into a set of orthogonal base operators $\{B_s\}$:

$$\sigma(t) = \sum_{s=1}^{n^2} b_s(t) B_s \qquad [1]$$

Product operators are one of many possible sets of base operators that can be chosen; the choice usually depends on the application. Product operators are based upon the angular momentum operators I_x, I_y and I_z of individual spins, and for spin-$\frac{1}{2}$ nuclei are defined by

$$B_s = 2^{(q-1)} \prod_{k=1}^{N} (I_{k\alpha})^{a_{ks}} \qquad [2]$$

where N is the number of spin-$\frac{1}{2}$ nuclei in the spin system, k is the spin index, $\alpha = x$, y or z, and q is the number of operators in the product. The coefficient a_{ks} has a value of 1 for spins in the product and 0 for all other spins.

For a system consisting of a single uncoupled spin k there are four possible product operators ($\frac{1}{2})E$, I_{kz}, I_{kx} and I_{ky}, where E is the unity operator, which need not be considered further. The three remaining operators and their corresponding representations in the net-magnetization vector model are given in **Figure 1**. The operator I_{kz} corresponds to the k-spin z magnetization found at thermal equilibrium as a result of the Boltzmann distribution between the two spin states α and β. The operators I_{kx} and I_{ky} correspond to x and y components of k-spin magnetization in the xy plane. These two operators can also be described as the x and y components of the in-phase magnetization or as the in-phase single-quantum coherence of spin k. The importance of making these qualifications will become apparent below. Neither I_{kx} nor I_{ky} is present at thermal equilibrium but they may be present after the application of a radiofrequency pulse to the equilibrium magnetization. The operators I_{kx} and I_{ky} are also important because they are the only operators that are directly observed in an NMR experiment; it is from these operators that the signal observed in the free induction decay arises.

Most systems of interest consists of more than one spin. The product operators for a two-spin system are listed in **Table 1**. Where applicable, representations of characteristic examples of these operators are given in the net-magnetization vector

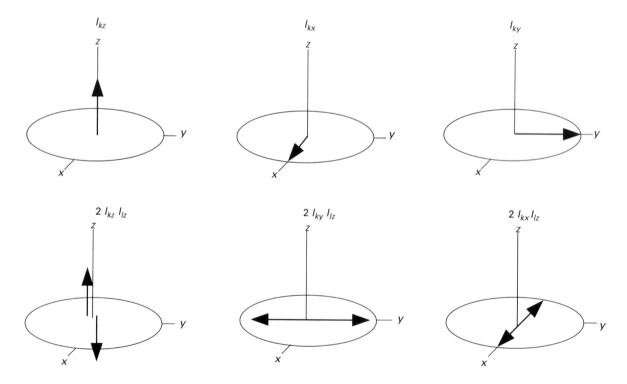

Figure 1 Vector model representations of one-spin and two-spin product operators arising from a spin k scalar coupled to a spin l. In the case of $2I_{kz}I_{lz}$ the l-spin vectors will be similarly arranged.

Table 1 Product operators for two spin-$\frac{1}{2}$ nuclei

Product operator	Description
$\frac{1}{2}E$	E = unity operator
I_{kx}, I_{ky}	In-phase x magnetization and y magnetization, respectively, of spin k
I_{lx}, I_{ly}	In-phase x magnetization and y magnetization, respectively, of spin l
I_{kz}, I_{lz}	Longitudinal magnetization of spins k and l, respectively
$2I_{kx}I_{lz}, 2I_{ky}I_{lz}$	x and y components, respectively, of k-spin magnetization antiphase with respect to l
$2I_{kx}I_{lx}, 2I_{kx}I_{ly}, 2I_{ky}I_{lx}, 2I_{ky}I_{ly}$	Components of two-spin coherence between spins k and l
$2I_{kz}I_{lz}$	Longitudinal two-spin order between k and l

model in **Figure 1**. It should be noted that there are two methods for differentiating between spins. Different subscripts, k and l in the operator $2I_{kz}I_{lz}$, are commonly used to denote different spins in homonuclear systems, while in heteronuclear systems the operators themselves may be denoted by different letters, I and S in the operator $2I_zS_z$. In the latter case I is usually taken to denote ^{1}H. In the case of a doublet, longitudinal two-spin order arises when the equilibrium magnetization of one of its two transitions has been inverted. Antiphase magnetization arises when the two components of the doublet point in opposite directions in the xy plane. The longitudinal component of the operator indicates the scalar coupling that separates the two components of the multiplet with opposite phases. The four operators denoting two-spin coherence consist of linear combinations of zero-quantum and double-quantum coherence; these will be considered below.

Evolution

The evolution of product operators can be calculated in a similar way to that of the density matrix, by a series of transformations of the type

$$\exp\{-i\phi B_r\}B_s\exp\{i\phi B_r\} = \sum_t b_{ts}(r, \phi)B_t \qquad [3]$$

where ϕB_r, corresponds to the relevant Hamiltonian. This takes the form $(\omega_k\tau)I_{kz}$ for chemical shift evolution of a spin k for a period τ, $(\pi J_{kl}\tau)2I_{kz}I_{lz}$ for scalar coupling evolution between two weakly coupled spins k and l for a period τ, and $(\beta)I_{kx}$ for a radiofrequency pulse applied to a spin k producing a rotation of β about the x axis. In each case the bracket has been added to indicate the angle through which the affected operators will evolve. Since all terms in the free precession Hamiltonian commute,

the evolution of individual chemical shifts and scalar couplings in multispin systems can be treated separately and in any order. In general, each process causes a rotation in a subspace, (**Figure 2**) corresponding to the evolution of one operator into another. In the case of a $\beta_x{}^\circ$ pulse applied to the equilibrium magnetization of a spin k, a component of the magnetization may be rotated onto the y axis:

$$I_{kz} \xrightarrow{\ \beta I_{kx}\ } I_{kz}\cos\beta - I_{ky}\sin\beta \qquad [4]$$

A vector description of this process is given in **Figure 3**. However, a $\beta_x{}^\circ$ pulse has no effect on I_{kx}:

$$I_{kx} \xrightarrow{\ \beta I_{kx}\ } I_{kx} \qquad [5]$$

It is important to use the correct sign conventions, which are given in **Figure 2**. If a pulse acts on a product of several operators, its overall effect can be determined by calculating its effects on each individual operator and then multiplying out the results. For example

$$
\begin{aligned}
2I_{kz}I_{lz} \xrightarrow{\ \beta(I_{kx}+I_{lx})\ } \ & 2(I_{kz}\cos\beta - I_{ky}\sin\beta) \\
& \times (I_{lz}\cos\beta - I_{ly}\sin\beta) \\
\equiv \ & 2I_{kz}I_{lz}\cos^2\beta \\
& - 2I_{kz}I_{ly}\cos\beta\sin\beta - 2I_{ky}I_{lz}\sin\beta \\
& \times \cos\beta + 2I_{ky}I_{ly}\sin^2\beta
\end{aligned}
$$
$$[6]$$

Where a pulse affects all of the nuclear spins under consideration the shorthand '$\beta_x{}^\circ$' is often written over the arrow.

Chemical shift evolution may be expressed in a similar fashion to that of radiofrequency pulses. For example, the chemical shift evolution of the operator I_{ky} at a frequency of ω_k over a period τ can be written as

$$I_{ky} \xrightarrow{\ \omega_k\tau I_{kz}\ } I_{ky}\cos\omega_k\tau - I_{kx}\sin\omega_k\tau \qquad [7]$$

A vector model representation of this process is given in **Figure 3**.

Scalar coupling evolution results in the interconversion of components of in-phase and antiphase magnetization. In the case of a component of in-phase k-spin magnetization I_{ky}, evolution due to a scalar coupling to a spin l during a period τ can be

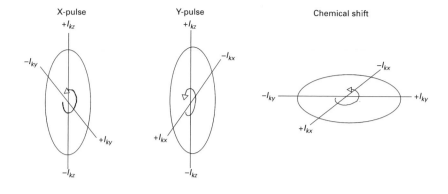

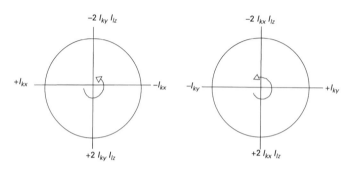

Figure 2 Sign conventions for product operator evolution under radiofrequency pulse, chemical shift and scalar coupling evolution.

described by

$$I_{ky} \xrightarrow{\pi J_{kl}\tau 2I_{kz}I_{lz}} I_{ky}\cos\pi J_{kl}\tau - 2I_{kx}I_{lz}\sin\pi J_{kl}\tau \quad [8]$$

This is represented pictorially in **Figure 3**. If the spin k is scalar coupled to a further spin m, the effects of evolution due to the two scalar couplings can be calculated sequentially:

$$I_{ky} \xrightarrow{\pi J_{kl}\tau 2I_{kz}I_{lz}} \xrightarrow{\pi J_{km}\tau 2I_{kz}I_{mz}}$$

$$\begin{aligned}
&I_{ky}\cos\pi J_{kl}\tau\cos\pi J_{km}\tau \\
&- 2I_{kx}I_{mz}\cos\pi J_{kl}\tau\sin\pi J_{km}\tau \\
&- 2I_{kx}I_{lz}\sin\pi J_{kl}\tau\cos\pi J_{km}\tau \\
&- 4I_{ky}I_{lz}I_{mz}\sin\pi J_{kl}\tau\sin\pi J_{km}\tau
\end{aligned}$$

$$[9]$$

The term $4I_{ky}I_{lz}I_{mz}$ corresponds to a y component of doubly antiphase k-spin magnetization; it is antiphase with respect to both l and m. It is important to note that if a component of magnetization has become antiphase with respect to one spin, scalar coupling evolution with respect to another spin will not make it in-phase again; only a further period during which evolution occurs due to the first spin's scalar coupling will have that effect.

Multiple-quantum coherence

The components of transverse magnetization considered above are all what are known as single-quantum coherences. Single-quantum coherences are phase coherences between states differing in overall magnetic quantum number by ±1. Phase coherences between states that do not fulfil this criterion are known as multiple-quantum coherence. Multiple-quantum coherences cannot be created by the application of a single nonselective radiofrequency pulse to the equilibrium magnetization and neither can they be detected directly since they have no net magnetization associated with them. Nevertheless, multiple-quantum coherences play an important role in many NMR experiments. The product operators $2I_{ky}I_{lx}$, $2I_{kx}I_{ly}$, $2I_{ky}I_{ly}$ and $2I_{kx}I_{lx}$ all describe linear combinations of multiple-quantum coherences; however, their precise meaning only becomes apparent when they are rewritten in terms of raising and lowering operators:

$$I_{kx} = \left(\frac{1}{2}\right)(I_k^+ + I_k^-) \quad [10]$$

$$I_{ky} = \left(\frac{1}{2i}\right)(I_k^+ - I_k^-) \quad [11]$$

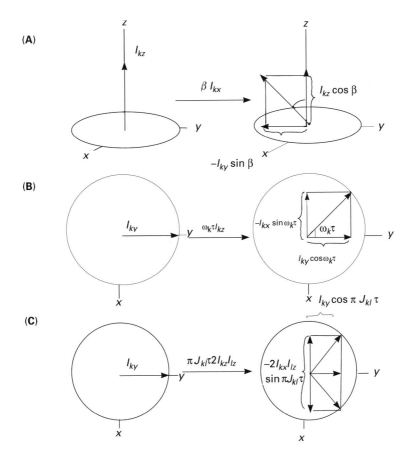

Figure 3 Vector model representations of the evolution of k-spin magnetization. (A) The effect of a β_x° pulse on z magnetization. (B) The effect of chemical shift evolution for a time τ on y magnetization. (C) The effect of scalar coupling evolution for a time τ on y magnetization.

The operators I_k^+ and I_k^- are counterrotating components corresponding to $+1$ and -1 quantum coherence, respectively; the former precesses at a frequency of $+\omega_k$ and the latter at a frequency of $-\omega_k$. Re-writing $2I_{ky}I_{lx}$ in terms of raising and lowering operators we find that

$$2I_{ky}I_{lx} = \left(\frac{1}{2i}\right)(I_k^+ I_l^+ + I_k^+ I_l^- - I_k^- I_l^+ - I_k^- I_l^-) \quad [12]$$

Clearly, $2I_{ky}I_{lx}$ consists of a linear combination of $+2$, 0, 0 and -2 quantum coherence. The other three operators can be re-written in a similar fashion. Pure zero-quantum coherence (ZQC) and double-quantum coherence (DQC) can be isolated by taking linear combinations of these four product operators:

$$\text{ZQC}_x = \left(\frac{1}{2}\right)(2I_{kx}I_{lx} + 2I_{ky}I_{ly}) = \left(\frac{1}{2}\right)(I_k^+ I_l^- + I_k^- I_l^+)$$

$$[13]$$

$$\text{ZQC}_y = \left(\frac{1}{2}\right)(2I_{ky}I_{lx} - 2I_{kx}I_{ly}) = \left(\frac{1}{2i}\right)(I_k^+ I_l^- - I_k^- I_l^+)$$

$$[14]$$

$$\text{DQC}_x = \left(\frac{1}{2}\right)(2I_{kx}I_{lx} - 2I_{ky}I_{ly}) = \left(\frac{1}{2}\right)(I_k^+ I_l^+ - I_k^- I_l^-)$$

$$[15]$$

$$\text{DQC}_y = \left(\frac{1}{2}\right)(2I_{kx}I_{ly} + 2I_{ky}I_{lx}) = \left(\frac{1}{2i}\right)(I_k^+ I_l^+ - I_k^- I_l^-)$$

$$[16]$$

The spins involved in a coherence, in this case k and l, are often referred to as the active spins. The evolution of these x and y components of multiple-quantum coherences can be calculated in the same fashion as single-quantum coherence (I_{kx} and I_{ky}) with several important provisos:

(1) The precessional frequency of a multiple-quantum coherence is a linear combination of those of its active spins. A double-quantum coherence with active spins k and l will evolve at $(\omega_k + \omega_l)$, while the corresponding zero-quantum coherence will evolve at $(\omega_k - \omega_l)$

(2) Multiple-quantum coherences do not exhibit scalar couplings between their active spins. Their scalar couplings to passive spins are linear combinations of those of their individual active spins to the passive spin concerned. Thus a double-quantum coherence between spins k and l would exhibit a scalar coupling constant of $(J_{km} + J_{lm})$ to a passive spin m; the corresponding zero-quantum coherence scalar coupling constant would be $(J_{km} - J_{lm})$.

Consequently, the chemical shift evolution of the y component of a zero-quantum coherence with active spins k and l during a period τ can be written as

$$ZQC_y \xrightarrow{(\omega_k I_{kz} + \omega_l I_{lz})\tau} ZQC_y \cos(\omega_k - \omega_l)\tau$$
$$- ZQC_x \sin(\omega_k - \omega_l)\tau \quad [17]$$

and the scalar coupling evolution of the corresponding double-quantum coherence to passive spin m can be described by

$$DQC_y \xrightarrow{(\pi J_{km} 2 I_{kz} I_{mz} + \pi J_{lm} 2 I_{lz} I_{mz})\tau} DQC_y \cos(J_{km} + J_{lm})\pi\tau$$
$$- DQC_x 2 I_{mz}$$
$$\times \sin(J_{km} + J_{lm})\pi\tau$$
$$[18]$$

Applications

The product operator formalism has been used to analyse most liquid-state NMR experiments since it was first introduced in 1983. The examples given here have been chosen either because of the way in which they exemplify the manipulation of product operators or because the pulse sequences concerned play an important role in contemporary NMR spectroscopy. The spin echo is considered because of its wide use and the opportunity it presents to undertake a relatively simple product operator calculation in full without resort to any of the short cuts that will be used to simplify later calculations. COSY (correlation spectroscopy) is also a widely used experiment and allows the introduction of the concept of coherence transfer, and its double-quantum filtered variant exemplifies some important points about how

multiple-quantum coherence is handled in product operator calculations. The HSQC (heteronuclear single-quantum coherence) pulse sequence is widely used as a basis for many heteronuclear experiments, particularly for structural studies of proteins in solution, and DEPT (distortionless enhanced polarization transfer) is one of the mainstays of perhaps the largest group of NMR users, organic chemists.

Spin echo

The spin echo is important both as an experiment in its own right for measuring transverse relaxation and as a building block of numerous other pulse sequences. The spin echo pulse sequence is

$$90^{\circ}_x - \tau - 180^{\circ}_y - \tau - \text{Acquisition}$$

The properties of this pulse sequence can be determined using the product operator formalism. Its effects on the equilibrium magnetization of a spin k with a scalar coupling to a spin l are calculated below. As an exercise the intermediate steps in the calculation are given in full (see Eqn [19] opposite).

It can be seen from the above that the product operators for even very simple systems can proliferate rapidly. Consolidation reduces the 16 terms present at the end of the calculation to two. It can be seen from the result that overall there is no evolution due to chemical shift (and by implication the effects of magnetic field inhomogeneities), which makes the pulse sequence useful for measuring transverse relaxation. However, scalar coupling evolution continues unaffected. In reality, it is rarely necessary to undertake a calculation of such complexity; the results of standard pulse sequences such as the spin echo are well known, so it is only necessary to write

$$I_{kz} \xrightarrow{90^{\circ}_x} - I_{ky} \xrightarrow{\tau, 180^{\circ}_y, \tau} - I_{ky} \cos(2\pi J_{kl}\tau)$$
$$+ 2 I_{kx} I_{lz} \sin(2\pi J_{kl}\tau) \quad [20]$$

In the remaining examples given below, the product operators will only be given at key stages during an experiment and any not following the selected coherence transfer pathway (and which are not observed) will be omitted.

COSY

The COSY (correlation spectroscopy) experiment has become one of the mainstays of modern NMR spectroscopy: it is also one of the simplest

$$I_{kz} \xrightarrow{90^\circ_x} -I_{ky} \xrightarrow{\omega_k\tau I_{kz}} -I_{ky}\cos(\omega_k\tau) + I_{kx}\sin(\omega_k\tau)$$

$$\downarrow \pi J_{kl}\tau 2I_{kz}I_{lz}$$

$$-I_{ky}\cos(\omega_k\tau)\cos(\pi J_{kl}\tau) + I_{kx}\sin(\omega_k\tau)\cos(\pi J_{kl}\tau)$$
$$+2I_{kx}I_{lz}\cos(\omega_k\tau)\sin(\pi J_{kl}\tau) + 2I_{ky}I_{lz}\sin(\omega_k\tau)\sin(\pi J_{kl}\tau)$$

$$\downarrow 180^\circ_y$$

$$-I_{ky}\cos(\omega_k\tau)\cos(\pi J_{kl}\tau) - I_{kx}\sin(\omega_k\tau)\cos(\pi J_{kl}\tau)$$
$$+2I_{kx}I_{lz}\cos(\omega_k\tau)\sin(\pi J_{kl}\tau) - 2I_{ky}I_{lz}\sin(\omega_k\tau)\sin(\pi J_{kl}\tau)$$

$$\downarrow \omega_k\tau I_{kz}$$

$$-I_{ky}\cos^2(\omega_k\tau)\cos(\pi J_{kl}\tau) - I_{kx}\sin(\omega_k\tau)\cos(\pi J_{kl}\tau)\cos(\omega_k\tau)$$
$$+2I_{kx}I_{lz}\cos^2(\omega_k\tau)\sin(\pi J_{kl}\tau) - 2I_{ky}I_{lz}\sin(\omega_k\tau)\sin(\pi J_{kl}\tau)\cos(\omega_k\tau)$$
$$+I_{kx}\cos(\omega_k\tau)\cos(\pi J_{kl}\tau)\sin(\omega_k\tau) - I_{ky}\sin(\omega_k\tau)^2\cos(\pi J_{kl}\tau)$$
$$+2I_{ky}I_{lz}\cos(\omega_k\tau)\sin(\pi J_{kl}\tau)\sin(\omega_k\tau) + 2I_{kx}I_{lz}\sin^2(\omega_k\tau)\sin(\pi J_{kl}\tau)$$

$$\downarrow \pi J_{kl}\tau 2I_{kz}I_{lz}$$

$$-I_{ky}\cos^2(\omega_k\tau)\cos^2(\pi J_{kl}\tau) - I_{kx}\sin(\omega_k\tau)\cos^2(\pi J_{kl}\tau)\cos(\omega_k\tau)$$
$$+2I_{kx}I_{lz}\cos^2(\omega_k\tau)\sin(\pi J_{kl}\tau)\cos(\pi J_{kl}\tau) - 2I_{ky}I_{lz}\sin(\omega_k\tau)\sin(\pi J_{kl}\tau)\cos(\omega_k\tau)\cos(\pi J_{kl}\tau)$$
$$+I_{kx}\cos(\omega_k\tau)\cos(\pi J_{kl}\tau)\sin(\omega_k\tau)\cos(\pi J_{kl}\tau) - I_{ky}\sin(\omega_k\tau)^2\cos^2(\pi J_{kl}\tau)$$
$$+2I_{ky}I_{lz}\cos(\omega_k\tau)\sin(\pi J_{kl}\tau)\sin(\omega_k\tau)\cos(\pi J_{kl}\tau) + 2I_{kx}I_{lz}\sin^2(\omega_k\tau)\sin(\pi J_{kl}\tau)\cos(\pi J_{kl}\tau)$$
$$+I_{kx}I_{lz}\cos^2(\omega_k\tau)\cos(\pi J_{kl}\tau)\sin(\pi J_{kl}\tau) - 2I_{ky}I_{lz}\sin(\omega_k\tau)\cos(\pi J_{kl}\tau)\cos(\omega_k\tau)\sin(\pi J_{kl}\tau)$$
$$+I_{ky}\cos^2(\omega_k\tau)\sin^2(\pi J_{kl}\tau) + I_{kx}\sin(\omega_k\tau)\sin^2(\pi J_{kl}\tau)\cos(\omega_k\tau)$$
$$+2I_{ky}I_{lz}\cos(\omega_k\tau)\cos(\pi J_{kl}\tau)\sin^2(\omega_k\tau) + 2I_{kx}I_{lz}\sin(\omega_k\tau)^2\cos(\pi J_{kl}\tau)\sin(\pi J_{kl}\tau)$$
$$-I_{kx}\cos(\omega_k\tau)\sin(\pi J_{kl}\tau)\sin(\omega_k\tau)\sin(\pi J_{kl}\tau) + I_{ky}\sin^2(\omega_k\tau)\sin^2(\pi J_{kl}\tau)$$

$$\equiv -I_{ky}\cos(2\pi J_{kl}\tau) + 2I_{kx}I_{lz}\sin(2\pi J_{kl}\tau) \qquad [19]$$

---

two-dimensional experiments:

$$90^\circ_x - t_1 - 90^\circ_\phi - \text{Acquisition}$$

where $\phi = x, y$. The value of ϕ determines whether the $\sin(\omega_k t_1)$ or the $\cos(\omega_k t_1)$ modulated component of the data is observed; it is necessary to measure both if absorptive phase-sensitive spectra are to be obtained. The coherence transfer pathways detected in each case are given below. The magnetization of a spin k coupled to a spin l is considered. When $\phi = y$:

$$I_{kz} \xrightarrow{90^\circ_x} -I_{ky} \xrightarrow{t_1} -I_{ky}\cos\omega_k t_1 \cos\pi J_{kl}t_1$$
$$+2I_{kx}I_{lz}\cos\omega_k t_1 \sin\pi J_{kl}t_1$$
$$\downarrow 90^\circ_y$$
$$-I_{ky}\cos\omega_k t_1 \cos\pi J_{kl}t_1$$
$$-2I_{kz}I_{lx}\cos\omega_k t_1 \sin\pi J_{kl}t_1$$

$$[21]$$

and when $\phi = x$:

$$I_{kz} \xrightarrow{90^\circ_x} -I_{ky} \xrightarrow{t_1} -I_{kx}\sin\omega_k t_1 \cos\pi J_{kl}t_1$$
$$+2I_{ky}I_{lz}\sin\omega_k t_1 \sin\pi J_{kl}t_1$$
$$\downarrow 90^\circ_x$$
$$-I_{kx}\sin\omega_k t_1 \cos\pi J_{kl}t_1$$
$$-2I_{kz}I_{ly}\sin\omega_k t_1 \sin\pi J_{kl}t_1$$

$$[22]$$

These equations provide an introduction to the concept of coherence transfer, the process by which one coherence is transformed into another. In each case a component of k-spin magnetization is transformed into l-spin magnetization. In the first equation the coherence transfer process, brought about by the second 90° pulse, is

$$2I_{kx}I_{lz} \xrightarrow{90^\circ_y} -2I_{kz}I_{lx} \qquad [23]$$

Coherence transfer plays a central role in many NMR experiments and usually involves antiphase magnetization. The components of in-phase magnetization present at the end of the schemes above have evolved at ω_k during t_1 and will evolve at ω_k during the acquisition time t_2: consequently they will give rise to a diagonal COSY peak with the coordinates (ω_k, ω_k). However, the antiphase magnetization will have evolved at ω_k during t_1, but since it underwent coherence transfer from k to l at the second 90° pulse it will evolve at ω_l during the acquisition period. Consequently it will give rise to an off-diagonal peak in the COSY spectrum at the coordinates (ω_k, ω_l). The components of magnetization giving rise to the diagonal and off-diagonal peaks are 90° out of phase; I_{ky} and $2I_{kz}I_{lx}$ in the case of the $\cos(\omega_k, t_1)$ modulated components. This leads to diagonal peaks that are always dispersive when the off-diagonal peaks are phased to be absorptive, and vice versa.

Double-quantum filtered COSY

Double-quantum filtered COSY is now the most widely used version of the COSY experiment. It has two main advantages over its precursor: diagonal and off-diagonal peaks can be phased to be absorptive simultaneously, and singlets are removed from the spectrum. The pulse sequence is

$$90^\circ_x - t_1 - 90^\circ_\phi - 90^\circ_x - \text{Acquisition}$$

where $\phi = x, y$ to select both $\sin(\omega_k, t_1)$ and $\cos(\omega_k, t_1)$ modulated components of the data. The selected pathway followed by the magnetization of a spin k with a single scalar coupling partner l is given below. When $\phi = x$:

$$I_{kz} \xrightarrow{90^\circ_x} -I_{ky} \xrightarrow{t_1} +2I_{kx}I_{lz}\cos\omega_kt_1\sin\pi J_{kl}t_1$$

$$\downarrow 90^\circ_x$$

$$\left(\frac{1}{2}\right)\left(2I_{ky}I_{lx}+2I_{kx}I_{ly}\right)\cos\omega_kt_1\sin\pi J_{kl}t_1$$

$$\downarrow 90^\circ_x$$

$$\left(\frac{1}{2}\right)\left(2I_{kz}I_{lx}+2I_{kx}I_{lz}\right)\cos\omega_kt_1\sin\pi J_{kl}t_1$$

$$[24]$$

and when $\phi = y$:

$$I_{kz} \xrightarrow{90^\circ_x} -I_{ky} \xrightarrow{t_1} +2I_{ky}I_{lz}\sin\omega_kt_1\sin\pi J_{kl}t_1$$

$$\downarrow 90^\circ_y$$

$$\left(\frac{1}{2}\right)\left(2I_{ky}I_{lx}+2I_{kx}I_{ly}\right)\sin\omega_kt_1\sin\pi J_{kl}t_1$$

$$\downarrow 90^\circ_x$$

$$\left(\frac{1}{2}\right)\left(2I_{kz}I_{lx}+2I_{kx}I_{lz}\right)\sin\omega_kt_1\sin\pi J_{kl}t_1$$

$$[25]$$

The second 90° pulse actually creates $2I_{ky}I_{lx}$ or $2I_{kx}I_{ly}$, depending on ϕ. However, phase cycling (or magnetic field gradient pulses) select only that component of the operator corresponding to double-quantum coherence. Using Equations [13]–[16], these operators can be resolved into components of zero-quantum and double-quantum coherence. For example, in the case of $2I_{ky}I_{lx}$:

$$2I_{ky}I_{lx} = \left(\frac{1}{2}\right)\left(2I_{ky}I_{lx}+2I_{kx}I_{ly}\right)+\left(\frac{1}{2}\right)\left(2I_{ky}I_{lx}-2I_{kx}I_{ly}\right)$$

$$[26]$$

where the first linear combination on the right of the equation consists of pure double-quantum coherence and the second pure zero-quantum coherence. The diagonal and off-diagonal peaks now both arise from components of antiphase magnetization with the same phase.

HSQC

The HSQC (heteronuclear single-quantum coherence) pulse sequence is an important tool in heteronuclear ^{1}H–^{15}N and ^{1}H–^{13}C NMR spectroscopy, particularly for larger molecules such as proteins. The pulse sequence is given below:

$$I \quad 90^\circ_x - \left(\frac{1}{4J_{IS}}\right) - 180^\circ_y - \left(\frac{1}{4J_{IS}}\right) - 90^\circ_y - \frac{t_1}{2} -$$

$$S \qquad\qquad\qquad 180^\circ_y \qquad\qquad\qquad 90^\circ_y$$

$$180^\circ_y - \frac{t_1}{2} - 90^\circ_y - \left(\frac{1}{4J_{IS}}\right) - 180^\circ_y - \left(\frac{1}{4J_{IS}}\right) - \text{Acquisition}$$

$$90^\circ_\phi \qquad\qquad 180^\circ_y \qquad\qquad \text{B.B. decoupling}$$

where $\phi = x, y$ in alternate experiments. HSQC is used both in its own right and as a building block for more sophisticated experiments. It produces two-dimensional spectra correlating the chemical shift of ^{1}H (I) in the F_2 dimension with that of the bonded heteronucleus (S) in the F_1 dimension. To maximize sensitivity, ^{1}H magnetization is both initially excited and detected. In the scheme given below, the magnetogyric ratios of both nuclei are assumed to have the same sign, and hence to rotate with the same sense in response to a radiofrequency pulse:

$$I_z \xrightarrow{90^\circ_x(I)} -I_y \xrightarrow{\frac{1}{4J_{IS}},180^\circ_y(I,S),\frac{1}{4J_{IS}}} +2I_xS_z$$

$$\xrightarrow{90^\circ_y(I,S)} -2I_zS_x$$

$$\downarrow \frac{t_1}{2},180^\circ_x(I),\frac{t_1}{2}$$

$$+2I_zS_x\cos(\omega_S t_1) \quad + \quad 2I_zS_y\sin(\omega_S t_1)$$

$$\xrightarrow{90^\circ_y(I,S) \ \text{(OR)}} \quad 90^\circ_x(S),90^\circ_y(I)$$

$$-2I_xS_z\cos(\omega_S t_1) \quad + \quad 2I_xS_z\sin(\omega_S t_1)$$

$$\downarrow \frac{1}{4J_{IS}},180^\circ_y(I,S),\frac{1}{4J_{IS}}$$

$$-I_y\cos(\omega_S t_1) \quad \text{(OR)} \quad I_y\sin(\omega_S t_1) \qquad [27]$$

Since the length of each spin echo is $(1/2J_{IS})$, in-phase magnetization present at the begining of one will become completely antiphase by the end and vice versa. The scheme splits into two after the evolution period to show how both sine and cosine t_1-modulated data are acquired in consecutive experiments by changing the phase ϕ of the second $90^\circ(S)$ pulse; this is necessary to obtain F_1 spectra that are both phase sensitive and absorptive.

DEPT

The DEPT (*distortionless enhanced polarization transfer*) pulse sequence is probably one of the most widely used of all NMR experiments. It produces ^{13}C subspectra edited according to the number of protons bonded to each ^{13}C. The DEPT pulse sequence is

$$^1\text{H} \qquad 90^\circ_x - \left(\frac{1}{2J_{CH}}\right) - 180^\circ_x - \left(\frac{1}{2J_{CH}}\right) - \theta^\circ_y$$

$$^{13}\text{C} \qquad\qquad\qquad\qquad 90^\circ_x \qquad\qquad 180^\circ_x$$

$$- \left(\frac{1}{2J_{CH}}\right) - \text{B.B. decoupling}$$

$$\text{Acquisition}$$

The experiment is repeated with θ set to 45°, 90° and 135° and linear combinations of the resulting data are taken to obtain the edited subspectra. The pulse sequence contains staggered ^{1}H and ^{13}C spin echos. This structure results in the cancellation of all chemical shift evolution by the start of acquisition, and consequently nothing is lost by omitting chemical shift evolution from the product operator description. The analysis for a ^{13}CH group is given below:

$$I_z \xrightarrow{90^\circ_x(I)} -I_y \xrightarrow{\left(\frac{1}{2J_{CH}}\right)} +2I_xS_z \xrightarrow{90^\circ_x(S)}$$

$$-2I_xS_y \xrightarrow{\left(\frac{1}{2J_{CH}}\right)} -2I_xS_y$$

$$\downarrow \theta^\circ_y(I),180^\circ_x(S) \qquad\qquad\qquad [28]$$

$$-2I_zS_y\sin\theta$$

$$\downarrow \left(\frac{1}{2J_{CH}}\right)$$

$$+S_x\sin\theta$$

The $90^\circ(S)$ pulse converts the antiphase magnetization present after the first period of free precession into heteronuclear two-spin coherences. This corresponds to a linear combination of heteronuclear zero-quantum and double-quantum coherence. Since both are selected and both evolve in the same way under the conditions of the experiment, there is no need to separate them and consider their evolution separately.

As noted above, all chemical shift evolution cancels out and neither coherence will exhibit significant scalar coupling evolution during the subsequent $(1/2J_{CH})$ evolution period. The $\sin\theta$ dependence of the coherence that is ultimately observed on the angle of the subsequent ^{1}H pulse is important, as will become evident below. This pulse converts part of the coherence into antiphase ^{13}C single-quantum coherence which is observed after it has become in-phase.

The product operator description for a ^{13}CH$_2$ group is the same except for the last three steps:

$$I_z \xrightarrow[+4I_xS_xI_z]{90^\circ_x(I)\left(\frac{1}{2J_{CH}}\right),\,90^\circ_x(S)} -2I_xS_y \xrightarrow{\left(\frac{1}{2J_{CH}}\right)}$$

$$\xrightarrow{\theta^\circ_y(I),180^\circ_x(S)} 4I_zS_xI_z\sin\theta\cos\theta \quad [29]$$

$$\downarrow \left(\frac{1}{2J_{CH}}\right)$$

$$-S_x\sin\theta\cos\theta$$

In this case the two-spin coherences generated by the $90^\circ(S)$ pulse both exhibit large scalar couplings to the remaining proton in the ^{13}CH$_2$ group. These are

$(J_{CH} - J_{HH})$ and $(J_{CH} + J_{HH})$ for the zero-quantum and double-quantum coherences, respectively. However, since scalar couplings are ineffective between equivalent spins, this reduces to J_{CH} in each case. Consequently, the operator evolves to become completely antiphase with respect to the remaining proton in the $^{13}CH_2$ group during the subsequent $(1/2J_{CH})$ period. This time the observed magnetization has a $\sin\theta\cos\theta$ dependence on the subsequent ^1H pulse.

For a $^{13}CH_3$ group:

$$I_z \xrightarrow{\; 90^\circ_x(I)\, \left(\frac{1}{2J_{CH}}\right),\, 90^\circ_x(S) \;} -2I_x S_y \xrightarrow{\; \left(\frac{1}{2J_{CH}}\right) \;}$$

$$+ 8I_x S_y I_z I_z \xrightarrow{\; \theta^\circ_y(I),\, 180^\circ_x(S) \;} 8I_z S_y I_z I_z \sin\theta\cos^2\theta \quad [30]$$

$$\downarrow \left(\tfrac{1}{2J_{CH}}\right)$$

$$-S_x \sin\theta\cos^2\theta$$

The two-spin coherences generated by the 90°(S) pulse exhibit scalar couplings of J_{CH} to the two remaining protons in the $^{13}CH_2$ group and consequently evolve to become antiphase with respect to both of them during the subsequent $(1/2J_{CH})$ period. As a result, the observed magnetization will exhibit a $\sin\theta\cos^2\theta$ dependence. Since each group has a different dependence on the $\theta°$ pulse, it is possible to isolate the subspectra arising from each group by acquiring data with three values of θ and taking the appropriate linear combinations of the results.

List of symbols

B_s = orthogonal base operator; E = unity operator; I_x, I_y, I_z = angular momentum operators; J = coupling constant; k, l, m = spin indices; N = number of spin-½ nuclei; q = number of operators in product; S_x, S_y, S_z = angular momentum operators; t_1 = evolution time; t_2 = acquisition time; θ = pulse angle; $\sigma(t)$ = density operator; τ = duration of evolution; ϕ = pulse phase; ω = angular frequency of spin (nucleus).

See also: **Two-Dimensional NMR Methods; ^{13}C NMR, Methods; NMR Principles; NMR Pulse Sequences; Proteins Studied Using NMR Spectroscopy.**

Further reading

Ernst RR, Bodenhausen G and Wokaun A (1987) *Principles Of Nuclear Magneic Resonance in One and Two Dimensions.* Oxford: Clarendon Press.

Kessler H, Gehrke M and Griesinger C (1988) Two-dimensional NMR spectroscopy: background and overview of the experiments. *Angewandte Chemie International Edition in English* **27**: 490–536.

Sorensen OW, Eich GW, Levitt MH, Bodenhausen G and Ernst RR (1983) Product operator formalism for the description of NMR pulse experiments. *Progress in NMR Spectroscopy* **16**: 163–192.

Proteins Studied Using NMR Spectroscopy

Paul N Sanderson, GlaxoWellcome Research and Development, Stevenage, UK

> **MAGNETIC RESONANCE**
> **Applications**

Introduction

The study of proteins by NMR spectroscopy has gained great impetus in recent years, providing a focus for the proliferation of many new complex NMR experiments and, quite possibly, justification for the purchase of more high-field spectrometers than any other field within NMR. The first ^{1}H NMR spectrum of a protein was published in 1957; this accurately reflected the amino acid composition but had neither the sensitivity nor resolution to yield further information. In the last two decades, however, many protein structures have been characterized and greater insight into the activity of proteins has been obtained from protein NMR studies. These achievements became possible as a result of the concurrent development of higher field (≥ 500 MHz) superconducting magnets, powerful computational hardware and software, complex multidimensional heteronuclear NMR experiments and isotopic labelling techniques. The quest for structural knowledge has been driven by the recognition that functions of biologically active proteins (such as enzymes, hormones and receptors) are fundamentally dependent on their three-dimensional structure. The development of NMR techniques, in parallel with X-ray crystallography, to obtain greater structural information from increasingly complex protein systems has resulted in increased understanding of biological processes. NMR also has a role in the characterization of protein interactions via, for example, titrations that map the binding surface of a protein through specific chemical shift changes. These binding studies, which are discussed in a separate article, benefit from prior characterization of solution structure by NMR and it is the structural aspects of protein NMR that are the focus of the present article. The theory and practical aspects of many areas of NMR spectroscopy which are of importance to protein NMR studies are described in detail elsewhere in this Encyclopedia; these are consequently mentioned only briefly here. Protein NMR has been thoroughly documented in many books and reviews and for more detailed discussion the reader is directed towards the Further reading section.

What proteins are suitable for study by NMR?

Full structural characterization of proteins in solution is possible for proteins of up to ~100 amino acids using homonuclear proton NMR. For larger proteins, of up to ~30 kDa, isotopic enrichment with ^{15}N and ^{13}C is required. There is no clear 'cut-off' in terms of protein size; each protein has to be considered on its own merits. Often, the likelihood of a successful structure determination only becomes clear after considerable protein purification and preliminary one-dimensional (1D) ^{1}H NMR data have been obtained.

Proteins must be non-aggregated and monomeric, or at least not present as large heteromultimers, under conditions of NMR measurements. Aggregated proteins give increased line widths and thus reduced spectral resolution and sensitivity. Internal mobility and the presence of multiple interconverting conformations also influence resonance line width.

Protein structure

Proteins are composed of a linear sequence of L-amino acid residues linked via amide bonds. The 20 naturally occurring amino acids are distinguished by the chemical nature of their side-chains. The amide bonds, or peptide linkages, are essentially planar and provide structural rigidity to the protein backbone, the only freedom to rotate being around the bonds to the α carbon. The angles of rotation are called, in IUPAC nomenclature, phi (ϕ) for the N–Cα bond [C(O)$_{i-1}$–N$_i$–Cα_i–C(O)$_i$] and psi (ψ) for the Cα–C(O) bond [N$_i$–Cα_i–C(O)$_i$–N$_{i+1}$] (**Figure 1**). If these two torsion angles are known for each amino acid the conformation of the whole polypeptide backbone is defined.

The linear sequence of amino acids represents the *primary structure* of the protein. Local regions within this sequence can adopt stable, defined, *secondary structure* such as α helices or β sheets. The packing of these secondary structural elements into compact domains gives rise to the protein's *tertiary structure*, such that distant regions of the peptide chain can be spatially close together. Multimeric

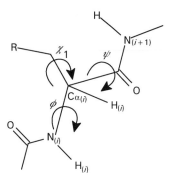

Figure 1 Stylized representation of a portion of a polypeptide chain, indicating the nomenclature of the backbone torsion angles (ϕ and ψ) and the side-chain torsion angle (χ_1) for the bonds emanating from the α carbon of an amino acid residue(i).

proteins are composed of several polypeptide chains arranged together in a *quaternary structure*. Adoption of the correct tertiary and quaternary structure is usually essential for biological function of a protein and a central dogma of biochemistry is that knowledge of a protein's structure leads to greater understanding of its activity.

Isotopic labelling of proteins

The range of proteins that can be studied by NMR and the nature of the structural information available have been extended through the biosynthetic incorporation of ^{15}N and ^{13}C isotopes. Data on isotopes encountered most frequently in protein NMR are given in **Table 1**.

Most proteins for structural analysis are prepared in cultures of bacteria or yeast that have been genetically modified to overexpress the protein of interest. Typically, a primary expression of 20–100 mg of protein is required for a structural NMR study. *Escherichia coli* (*E. coli*) bacteria are preferentially used as they can be grown rapidly in large quantities on chemically defined media which can be supplemented with ^{15}N-labelled ammonium salts and ^{13}C-

labelled compounds, such as glucose, acetate or glycerol, as required. It is essential to exclude all sources of natural abundance nitrogen and carbon from growth media to maximize incorporation of isotopic label. The production of ^{15}N-labelled protein is reasonably straightforward and relatively inexpensive, provided it can be expressed in *E. coli*. Additional incorporation of ^{13}C-label is more costly, but necessary for full analysis of larger proteins. The activity of expressed proteins should be validated to ensure correct folding and show that isotope incorporation has not impaired function.

Mammalian proteins produced in bacteria or yeast may not undergo correct post-translational processing, i.e. disulfide cross-linking, protein folding, phosphorylation or glycosylation. This problem may be overcome by expression in baculovirus systems, in insect cells or in mammalian cell lines such as Chinese hamster ovary (CHO) cells. Mammalian cells will not grow on minimal media, so their growth media must contain appropriately labelled amino acids, which can be obtained, for example, from hydrolysates of labelled algae.

Isotopic labelling with deuterium (^{2}H) can be used to provide spectral editing. For example, specific incorporation of a deuterium atom into methylene or methyl groups of amino acids can be used to obtain detailed structural and dynamic information. Dramatic increases in proton NMR resolution for larger proteins can be achieved through random labelling of the protein with deuterium, at levels between 50 and 85%, by growth on substrates in which the ratio of ^{1}H to ^{2}H is controlled.

Preparation of protein samples for NMR

A typical sample for protein NMR will contain 1 mM protein, of at least 95% purity, in 0.5 mL of aqueous solution. The final stages of purification may include desalting, buffer exchange, ^{2}H$_2$O exchange and concentration by lyophilization (if the protein is sufficiently stable), vacuum centrifugation or ultrafiltration. Sample volumes can be as small as 100 µL (for 2.5 or 3 mm diameter tubes) if only limited amounts of protein are available.

Data collection can take several days, during which the protein must remain stable; therefore oxidation, hydrolysis and microbial contamination must be minimized. Samples are not usually degassed, as protein solutions tend to 'froth' and paramagnetic broadening by oxygen can usually be ignored. Cysteine and methionine residues are susceptible to oxidation, so spontaneous formation of

Table 1 Properties of isotopes most commonly encountered in protein NMR

Isotope	Natural abundance (%)	Resonance frequency at 14.0926 T (MHz)	Relative sensitivity
^{1}H	99.98	600.00	1.0000
^{2}H	0.015	92.10	0.00965
^{13}C	1.108	150.86	0.0159
^{15}N	0.37	60.80	0.00104
^{31}P	100	242.88	0.0663

non-native intra- or intermolecular disulfide bonds should be prevented by addition of low levels of dithiothreitol or β-mercaptoethanol. The growth of microorganisms can be prevented by sodium azide.

The quality of NMR spectra of proteins can be strongly influenced by sample pH, ionic strength, buffer, concentration and temperature; optimum conditions are generally determined empirically via 1D spectra. For ^{1}H NMR, solutions are generally buffered, at 10–50 mM, with phosphate (which can cause precipitation or aggregation of some proteins) or with deuterated buffer salts (e.g. Tris). Deuterated reducing agents, cation chelators and proteolytic enzyme inhibitors may also be incorporated as necessary.

Obtaining NMR data from proteins

Protein NMR spectra are usually recorded in aqueous solutions to obtain data from the exchangeable amide protons. In these spectra the solvent water signal is at least 10^5 times more intense than the protein protons of interest and must be suppressed to allow detection of protein signals at an acceptable signal-to-noise ratio. The most common method for suppressing the water signal is presaturation, in which continuous low level irradiation is applied at the frequency of the water resonance. This is the most effective method if the effects of exchange between solvent and solute are not important. Many alternative methods that do not reduce the intensity of solute signals in exchange with the water are available; these include the use of pulsed-field gradients. Full discussion of solvent suppression techniques can be found in a separate article.

One-dimensional NMR data

Initial assessment of the suitability of a protein sample and preliminary solution optimization can be achieved via 1D ^{1}H NMR experiments; these should also be used to assess the consistency of a sample before and after lengthy acquisitions. Broad signals may indicate protein aggregation and suggest that further experiments are not worth pursuing on that sample. The presence (or absence) of protein structure can be assessed from the distribution of signals; typically, upfield shifted methyl signals ($\delta 0$), and downfield shifted alpha proton ($\delta 5$–6) and NH proton ($> \delta 9$) signals indicate that a protein is 'structured'. These regions are highlighted in a ^{1}H NMR spectrum of lysozyme in **Figure 2**. For a denatured (or random coil) protein, such signals are absent from these regions.

Multidimensional heteronuclear NMR experiments

Having established optimum sample conditions, all subsequent assignment and structural data are obtained from multidimensional NMR experiments. For smaller proteins, this can be achieved using two-dimensional (2D) homonuclear proton NMR experiments: correlation spectroscopy (COSY) and total correlation spectroscopy (TOCSY) to give scalar through-bond coupling data, and nuclear Overhauser effect spectrometry (NOESY) to provide 'through space' information. For larger proteins, resonance overlap is considerable, increased line width results in cancellation of COSY cross peaks and much signal loss occurs during long TOCSY mixing times. Full assignment and NOE analysis can, in these cases, be achieved by spreading data into three or four dimensions to increase signal resolution and by further spectral simplification through ^{15}N and ^{13}C isotope editing.

In heteronuclear correlation experiments, magnetization transfer between protons and heteronuclei can be via either heteronuclear single quantum coherence (HSQC) or heteronuclear multiple quantum coherence (HMQC) pathways. The HSQC sequence gives rise to narrower lines, but uses more pulses and requires a longer phase cycle than the HMQC. Thus, HSQC is used for 2D experiments where the highest resolution is required and HMQC is preferred for 3D sequences in which the experimental time is limited.

Labelling with ^{15}N alone can be sufficient to overcome spectral overlap for proteins of up to 20 kDa and, for these proteins, virtually complete resolution can often be achieved for the backbone amide groups in 2D-^{1}H–^{15}N HSQC experiments. These experiments are very robust and can be used to determine amide proton exchange rates or chemical shift temperature coefficients. For high protein concentrations ^{1}H–^{15}N HSQC data sets can be acquired rapidly; typically within 10 min for a 2 mM sample or 2–3 hr for a 0.2 mM protein sample. Consequently, this experiment has become the mainstay of NMR approaches to monitor the binding of ligands to ^{15}N-labelled proteins through titration experiments.

Information on side-chain resonances or sequential connectivities can be obtained by converting the 2D-[1]H–[15]N heteronuclear correlation experiment into a 3D experiment by adding either a TOCSY or a NOESY step. These 3D experiments can be considered as a series of 2D homonuclear experiments in which each is edited by a different ^{15}N frequency. Thus, having established amide ^{1}H/^{15}N pairs via the HSQC, the TOCSY-HMQC correlates alpha protons

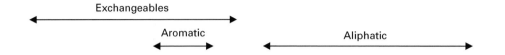

Chemical shift ranges for random coil peptide protons

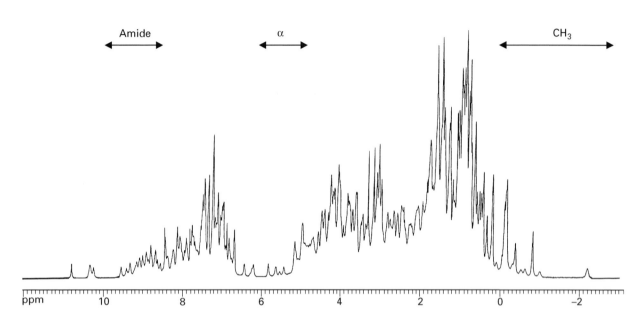

Figure 2 One-dimensional ^{1}H NMR spectrum of lysozyme in aqueous solution obtained with presaturation of the water signal, illustrating the range of proton chemical shifts expected for a random coil, or denatured, protein. The positions of upfield shifted methyl signals and downfield shifted alpha and amide proton signals that are indicative of a non-random, ordered protein structure are also shown.

(and in favourable cases side-chain protons) to these pairs and residue types can be assigned. Sequential assignment can then follow in a manner analogous to the 2D procedure (see below) via the NOESY-HMQC spectrum. Several other 3D experiments have been devised to facilitate assignments of ^{15}N-labelled proteins. One such, which is of particular use for identifying helical regions of a protein, is the HMQC-NOESY-HMQC experiment, which allows identification of [1]H–[1]H NOEs between residues with degenerate amide proton resonances.

For larger, double-labelled proteins, assignments are made via a range of three- and four-dimensional triple resonance experiments in which sequential assignments are facilitated via magnetization transferred through ^{13}C-couplings. These couplings are usually at least as large as the ^{13}C line width for proteins of up to 30 kDa and are reasonably independent of the protein backbone conformation. A summary of the magnitudes of coupling constants utilized in correlation experiments is shown in **Figure 3**. These experiments each focus on a particular coupling network (their nomenclature usually reflects this in a logical manner) and several different experiments may perform the same correlation, but via a different route. Residue-specific experiments can extend assignments from the peptide backbone along amino acid side-chains, for example through to the guanidino group of arginine residues.

Sequence-specific assignments

NMR spectra of small proteins can be fully assigned in a systematic manner by a sequential assignment procedure using well-resolved 2D homonuclear spectra and a knowledge of the amino acid sequence of

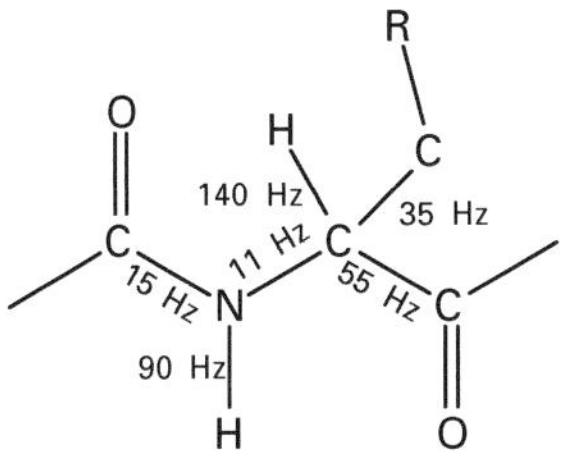

Figure 3 Schematic representation of the peptide backbone, showing the magnitudes of the one-bond coupling constants that are utilized in multidimensional heteronuclear correlation experiments to provide sequential connectivities.

the protein. This procedure depends on the fact that for all of the sterically allowed values of the torsion angles ϕ, ψ and χ_1 (**Figure 1**) NOEs will be observed for at least one of the interproton distances between NH, αH or βH of adjacent amino acids. Hence, spin systems of individual amino acids are first identified via spin–spin couplings in COSY and TOCSY spectra and are then connected to neighbouring amino acids across the peptide linkage via 'through-space' correlations in NOESY spectra. The latter are often obtained from a small region of the 2D NOESY spectrum, the 'fingerprint region'. This comprises amide proton frequencies in the directly detected dimension (F2) and α (and side-chain) proton frequencies in the indirectly detected dimension (F1) and contains both intra- and interresidue NH–αH connectivities, allowing sequential assignments to be made simply by 'walking' from peak to peak. The early stages of spin system identification are facilitated by the fact that many amino acids, either individually or in groups, give rise to characteristic peak patterns in COSY and TOCSY spectra and this feature provides the basis for automated assignment routines.

Protein structure information from NMR parameters

Conformation-dependent data, primarily in the form of NOEs and scalar coupling constants, are obtained from essentially the same experiments (NOESY and COSY respectively) as the assignments. Additional structural information can also be extracted from chemical shifts, relaxation times and amide exchange data. It should be emphasized here that all of these NMR parameters are population-weighted and time averaged; thus, as molecular systems are inherently dynamic in nature, they do not in general represent single, precise values of interatomic distances and angles. Nevertheless these NMR parameters, which

are considered below in terms of the structural information they represent, provide useful restraints to allow calculation of protein structures and, in the case of proteins that cannot be crystallized, access to structural information that cannot be obtained by other means.

NOEs

A cross peak in a NOESY experiment indicates dipolar cross relaxation between two nuclei that are spatially close to one another. The cross peak intensity is dependent upon the inverse sixth power of the distance between the two nuclei. Thus, for two protons, i and j separated by a distance r, which give rise to a NOESY cross peak, the intensity of that cross peak I is proportional to r^{-6}.

This simplified relationship assumes that the protein can be considered as a rotating rigid body in which correlation times for all proton pairs are the same, and are equal to the correlation time for the overall tumbling of the protein. For a more thorough analysis, account must be taken of, amongst other factors, internal mobility and spin diffusion. These are dealt with in greater detail in a separate article.

Interproton distances can, therefore, be determined from unambiguously assigned, well-resolved, high signal-to-noise NOESY data, by analysis of cross peak intensities. These may be obtained by volume integration and can be converted into estimates of interproton distances, using the equation above, for NOESY data at short mixing times. In this method, each proton pair is considered in isolation and NOESY cross peak intensities are compared with a reference cross peak from a proton pair of fixed distance, such as a geminal methylene proton pair or aromatic ring protons. A problem inherent to this approach is that the fixed distance is usually smaller than the unknown distance; this usually leads to systematic underestimation of the latter.

An alternative method of analysis of NOESY data, which is usually sufficient for resolved peaks with a digital resolution much greater than the intrinsic line width and coupling constants, is to measure the maximum peak amplitude or to count the number of contours. NOESY cross peaks can then be classified as strong, medium or weak and can be translated into upper distance restraints of around 2.5, 3.5 and 5.0 Å respectively. The lower distance constraint is usually the sum of the van der Waals radii (1.8 Å for protons). This simple approach is reasonably insensitive to the effects of spin diffusion or non-uniform correlation times and can usually lead to definition of the global fold of the protein, provided a sufficiently large number of NOEs have been identified.

Greater accuracy can be achieved by methods that involve calculation of a full relaxation matrix from the NOESY data to generate interproton distances. A model protein structure can then be iteratively refined by back calculation until differences in the empirical and calculated data are minimized. The resulting distances can be used as restraints for further refining the protein structure by distance geometry or molecular dynamics methods.

Coupling constants

Geometric information, particularly for the bonds around the peptide backbone (**Figure 1**), can be obtained from vicinal spin–spin coupling constants. The magnitude of the coupling constant J is dependent upon the dihedral angle θ as well as the nature and orientation of substituents in a manner that is defined by the Karplus relationship, which has the general form:

$$^3J = A\cos^2\theta - B\cos\theta + C$$

For a given coupling constant there are up to four valid solutions of the Karplus equation although knowledge of protein torsion angles (from protein structure databases) can be used to discount unlikely values. For example, the backbone torsion angle ϕ is usually negative, except in the case of asparagine, aspartate and glycine residues.

The $^3J_{NH,\alpha H}$ coupling constant, which is dependent upon the dihedral angle: [HN_i–N_i–$C\alpha_i$–$H\alpha_i$] ($\theta = \phi - 60°$ for L-amino acids), is commonly used for assessing secondary structure. Thus, a sequence of small (< 5 Hz) values indicates an α helix, whereas extended β structures have large (> 9 Hz) values that reflect the *trans* relationship of the NH and αH protons. Intermediate values are indicative of nonstandard structure or conformational averaging. The $^3J_{NH,\alpha H}$ values can be measured, in exceptional cases for small proteins, from high digital resolution 1D spectra but are more commonly obtained from 2D DQF-COSY spectra. If ^{15}N-labelled protein is available they can be extracted from ^{15}N-filtered correlation experiments and, in cases of signal overlap or insufficient digital resolution, from cross peak intensities in quantitative J-correlation experiments.

The $^3J_{\alpha H,\beta H}$ vicinal coupling constants can be determined from Exclusive COSY (E-COSY) type experiments and, together with intraresidue NOEs, can be used to obtain stereospecific assignments of β-methylene protons and side-chain χ angle restraints (**Table 2**).

Table 2 NMR parameters that define conformations about the Cα–Cβ bond in amino acids

Parameter			
Conformation	g^-	t	g^+
χ_1	60°	180°	−60°
$d_{NH,H\beta2}$	3.5–4.0 Å	2.5–3.4 Å	2.2–3.1 Å
NH–Hβ_2 NOE	Weak	Strong/medium	Strong
$d_{NH,H\beta3}$	2.5–3.4 Å	2.2–3.1 Å	3.5–4.0 Å
NH–Hβ_3 NOE	Strong/medium	Strong	Weak
Hα–Hβ_2 NOE	Strong	Strong	Weak
Hα–Hβ_3 NOE	Strong	Weak	Strong
$^3J_{H\alpha,H\beta2}$	< 5 Hz	< 5 Hz	> 10 Hz
$^3J_{H\alpha,H\beta3}$	< 5 Hz	> 10 Hz	< 5 Hz
$^3J_{N,H\beta2}$	~ 5 Hz	~ 1 Hz	~ 1 Hz
$^3J_{N,H\beta3}$	~ 1 Hz	~ 1 Hz	~ 5 Hz

Chemical shifts

The chemical shift of a resonance reflects the chemical environment of the atom that gives rise to it. This is determined mostly by covalent bonding and to a smaller extent by the non-bonded environment. In unstructured peptides each amino acid exists in an ensemble of conformations and the random coil chemical shift represents the population-weighted mean value of these environments. The chemical shifts for amino acids in denatured proteins are close to random coil values; however, for structured proteins, many resonances are far removed from their random coil position (**Figure 2**). Even greater changes in proton chemical shift can result from the proximity of a proton to an aromatic ring (the ring current effect) or to a paramagnetic centre, as found in haem proteins such as haemoglobin and cytochromes.

The availability of chemical shift assignments for many proteins of defined structure has allowed changes in chemical shift from random coil values to be related to secondary structure and, consequently, to be used predictively. Thus, upfield shifts of ~0.3 ppm are characteristic for α protons in α helices, whereas α protons in β sheets experience downfield shifts of ~0.3 ppm. Similar effects are observed for ^{13}C chemical shifts of α carbons, which are shifted downfield by ~3 ppm in α helices and upfield by ~1.5 ppm in β sheets. A sequence of similar changes can therefore help in the initial characterization of regions of secondary structure. Chemical shifts can also be used for refining protein tertiary structure; ^{13}C shifts, in particular, are sensitive to backbone geometry and can therefore help define backbone torsion angles.

Relaxation times

The introduction of ^{15}N and/or ^{13}C labels into a protein facilitates the study of dynamic properties and, in particular, localized intramolecular motions. This arises because relaxation of these nuclei is usually dominated by dipole–dipole interactions with the directly bonded proton and this relaxation is dependent upon internuclear distance (which is fixed) and the rotational correlation time, which is only uniform throughout a 'rigid' protein. Proteins, however, usually contain regions that have greater flexibility, such as surface loops, which have different local correlation times that are reflected in heteronuclear relaxation times.

Amide proton exchange and other exchange effects

Reduced rates of exchange of amide protons with the bulk solvent water indicate reduced solvent accessibility and potential involvement in hydrogen bonds. Almost all amide protons in regions of regular protein secondary structure (except for those near the edges) are hydrogen-bonded. A corollary of this is that fast amide exchange rates generally imply the absence of 'structure'. Measurement of amide proton exchange rates by following the time-course of the disappearance of signals in COSY, TOCSY or ^{1}H–^{15}N HSQC spectra therefore provides supportive evidence of secondary structure.

Other exchange phenomena that manifest themselves in NMR spectra include *cis–trans* isomerization of proline residues, aromatic ring 'flipping' and the rotation of primary amides of asparagine and glutamine.

Deriving protein structures from NMR data

Extensive computational calculations are necessary to translate the information contained within NMR data into a protein structure. The quality of the structure obtained is dependent on the accuracy and, to a greater extent, quantity of the NMR data. As a rule of thumb, at least ten long-range NOE restraints

Table 3 Characterization of protein secondary structure from NMR data

Parameter[a]	α Helix[b]	β Sheet[c]
ϕ [C(O)$_{i-1}$–N$_i$–Cα_i–C(O)$_i$]	–57°	–139°
$^{3}J_{NH,\alpha H}$	< 4 Hz	> 9 Hz
$d_{\alpha N}(i,i)$ (NOE intensity)	2.6 Å (strong)	2.8 Å (strong)
$d_{\alpha N}(i,i+1)$ (NOE intensity)	3.5 Å (weak)	2.2 Å (very strong)
$d_{\alpha N}(i,i+2)$ (NOE intensity)	4.4 Å (weak)	–
$d_{\alpha N}(i,i+3)$ (NOE intensity)	3.4 Å (medium)	–
$d_{\alpha \beta}(i,i+3)$ (NOE intensity)	2.5–4.4 Å (medium)	–
$d_{\alpha N}(i,i+4)$ (NOE intensity)	4.2 Å (weak)	–
$d_{NN}(i,i+1)$ (NOE intensity)	2.8 Å (strong)	4.3 Å (weak)
$d_{NN}(i,i+2)$ (NOE intensity)	4.2 Å (weak)	–
$d_{\alpha\alpha}(i,j)$ (NOE intensity)	–	2.3 Å (very strong)
$d_{\alpha N}(i,j)$ (NOE intensity)	–	3.2 Å (medium)
$d_{NN}(i,j)$ (NOE intensity)	–	3.3 Å (medium)
NH exchange rate	Slow	Slow
$^{1}H\alpha$ Chemical shift change[d]	~ –0.3 ppm	~ +0.3 ppm
$^{13}C\alpha$ Chemical shift change[d]	~ +3 ppm	~ –1.5 ppm

[a] $d_{xy}(i, i+n)$ refers to the distance between proton x in residue i and proton y in the residue n positions from the C-terminus of residue i. In the case of the β sheet j refers to the cross-strand partner.
[b] A 3_{10} helix differs from an α helix in that the NH of residue i is hydrogen bonded to the carbonyl of residue i–4 in the α helix and of residue i–3 in the 3_{10} helix. The consequences of this for differences in NMR data are small but are, most notably, a small decrease in $d_{\alpha N}(i, i+2)$, an increase in $d_{\alpha N}(i, i+4)$ to the point where a NOE is not observed and a small decrease in $^{3}J_{NH,\alpha H}$ for the 3_{10} helix.
[c] The values given above for a β sheet are for an antiparallel β sheet. Equivalent values for a parallel β sheet are essentially the same except for the interstrand distances, in particular $d_{\alpha\alpha}(i, j)$ is much larger (4.8 Å).
[d] $^{1}H\alpha$ and $^{13}C\alpha$ chemical shift changes are relative to random coil values.

are required for each amino acid residue to generate a 'reasonable' protein structure.

Regions of protein secondary structure are identified via characteristic short-range (i.e. ≤ 5 residues apart) NOEs, coupling constants, amide proton exchange and chemical shift data (**Table 3**). Longer-range NOEs then define the tertiary structure, which can be refined further against all available data.

Calculation of protein structures from NMR data requires conversion of the NMR data into distances and angles, usually in the form of allowed ranges, as described above. These are incorporated as restraints into protein structure calculations, such that deviation from these values incurs an energetic penalty. In these calculations, distances and angles within a protein structure are optimized using a combination of distance geometry, molecular dynamics and simulated annealing procedures.

Distance geometry calculations aim to optimize all interatomic distances within a protein on the basis of the experimental restraints and are often used to define the global fold of a protein. This may then be refined using restrained molecular dynamics simulations in which structures evolve over time under the influence of a force field that contains potential energy terms for both covalent and non-bonded interactions and includes NMR restraints. A variation on molecular dynamics is simulated annealing, which differs in that normally 'prohibited' potential energy barriers can be crossed, allowing regions of a molecule to 'pass through' one another, thereby sampling larger regions of conformational space; this technique does not require a defined starting structure. After optimization of the NMR restraints by any or all of these techniques the potential energy of the resulting structure is minimized by molecular mechanics calculations to determine the lowest energy structure. The accuracy of the protein structure can be further increased by refinement against coupling constants, chemical shifts (both ^{1}H and ^{13}C), relaxation time rates ($T_1 : T_2$) ratios and residual dipolar couplings.

Structural calculations usually result in an ensemble of protein structures that must be assessed to determine how well they satisfy the initial restraints. Violations in excess of 1 Å may indicate that regions of the structure are ill defined. NOESY spectra can be back calculated on the basis of the structures and compared with the experimental data to identify potential errors. Comparison of calculated and experimental NOEs can also lead to an 'R factor' which gives an indication of the quality of the structures in a manner analogous to that used in X-ray crystallography. An alternative is to measure the RMS deviation of the ensemble of structures from the average structure. This measure should be used with care as it may indicate a high level of precision in the structures but not give a true indication of their accuracy.

Other applications of protein NMR

Protein dynamics

Static 3D protein structures cannot always explain biological processes or point the way, for example,

to rational drug design; the dynamic properties of a protein may be of equal functional importance. Internal motions within proteins were recognized by early NMR experiments, and NMR spectroscopy has developed into a powerful technique for the study of dynamics, from the picosecond motions of bond vectors to millisecond motions. For labelled proteins, information about backbone dynamics and motions of nitrogen-containing side-chains can be obtained *via* ^{15}N relaxation, which is dependent upon reorientation of the ^{15}N–^{1}H bond vectors. A more complete description can result from the additional use of ^{13}C relaxation data. Side-chain dynamics can be further characterized *via* deuterium relaxation by incorporation of a single deuterium atom into methyl or methylene groups and determining the attenuation of intensities in ^{1}H–^{13}C-correlation experiments.

Protein folding

A major advantage of NMR over X-ray crystallography is the ability to characterize the structure and dynamics of unfolded and partially folded states of proteins. These are of importance in protein folding and many cellular processes; indeed, many proteins or domains are intrinsically unstructured and only become structured upon binding other molecules. In these states, proteins exhibit rapid fluctuations between a range of conformations and insight into these processes can be gained from NMR, either through studying equilibrium states or through direct monitoring of kinetic folding events. Stabilized intermediates or fully denatured states at equilibrium can be characterized by essentially the same techniques as structured proteins, although chemical shift dispersion, with the exception of ^{15}N and ^{13}C-carbonyl resonances, is usually poor. The kinetics of protein folding can be monitored *via* the time-course of hydrogen–deuterium exchange in 2D experiments, by pulse labelling or by stopped-flow techniques. For ^{15}N-labelled proteins, folding can also be monitored during the course of a single 2D-HSQC experiment by analysis of line shape changes.

Concluding remarks

There are other areas of protein NMR that have not been considered here but are covered elsewhere in the Encyclopedia. These include the binding of ligands and the study of membrane-associated proteins in detergent micelles or phospholipid bilayers using solid state magic-angle spinning techniques.

The present article has focused on structural aspects of protein NMR. Determination of a de novo protein structure by NMR takes longer, in general, than by X-ray crystallography – provided the protein can be crystallized. However, NMR is the only technique for obtaining high resolution structural data from proteins that cannot be crystallized and for which appropriate concentrations of non-aggregated protein can be achieved. Limitations of molecular size can be overcome, as with other biochemical and structural techniques, by considering protein fragments, or domains. Structural data for large proteins can thus be obtained for domains individually and the linkage and assembly of domains can then be determined. The extension of protein NMR to larger molecules may be further facilitated by measurement of residual anisotropic interactions, which give structural restraints that are orientational rather than distance based.

To summarize, NMR spectroscopy can provide a wealth of structural information about proteins and, leading from this, a greater insight into protein interactions and dynamic processes.

List of symbols

I = cross peak intensity; J = coupling constant; r = distance between two protons; T_1, T_2 = relaxation times; ϕ = dihedral angle defined by H_i–N_i–$C\alpha_i$–$H\alpha_i$; ψ = dihedral angle defined by $H\alpha_i$–$C\alpha_i$–$C = 0$; χ_1 = dihedral angle defined by $H\alpha_i$–$C\alpha_i$–$C\beta$–R.

See also: **Labelling Studies in Biochemistry Using NMR; Laboratory Information Management Systems (LIMS); Macromolecule–ligand Interactions Studied By NMR; Magnetic Field Gradients in High Resolution NMR; NMR Data Processing; NMR Pulse Sequences; NMR Relaxation Rates; Nuclear Overhauser Effect; Parameters in NMR Spectroscopy, Theory of; Solvent Suppression Methods in NMR Spectroscopy; Structural Chemistry Using NMR Spectroscopy, Peptides; Two-Dimensional NMR, Methods.**

Further reading

Cavanagh J, Fairbrother WJ, Palmer AG and Skelton NJ (1996) *Protein NMR Spectroscopy: Principles and Practice*, 587 pp. San Diego: Academic Press.

James TL and Oppenheimer NJ (eds) (1989) *Methods in Enzymology,* Vol 176 (Nuclear Magnetic Resonance, Part A: Spectral Techniques and Dynamics) 530 pp and Vol 177 (Nuclear Magnetic Resonance, Part B: Structure and Mechanism) 507 pp. San Diego: Academic Press.

James TL and Oppenheimer NJ (eds) (1994) *Methods in Enzymology*, Vol 239 (Nuclear Magnetic Resonance, Part C) 813 pp. San Diego: Academic Press.

Leach AR (1996) *Molecular Modelling – Principles and Applications*, 595 pp. Harlow, UK: Longman.

Nature Structural Biology (NMR I Supplement) (1997) Vol 4, 841–866.

Nature Structural Biology (NMR II Supplement) (1998) Vol 5, 492–522.

Reid DG (ed) (1997) *Methods in Molecular Biology: Protein NMR Techniques*, 419 pp. New Jersey: Humana Press.

Roberts GCK (ed) (1993) *NMR of Macromolecules – A Practical Approach*, 399 pp. Oxford: Oxford University Press.

Wüthrich K (1986) *NMR of Proteins and Nucleic Acids*, 292 pp. New York: John Wiley & Sons.

Proton Affinities

Edward PL Hunter and **Sharon G Lias**, National Institute of Standards and Technology, Gaithersburg, MD, USA

MASS SPECTROMETRY
Theory

What is a proton affinity?

The 'proton affinity' and the related quantity 'gas-phase basicity', are defined thermodynamic quantities that enable us to assign numeric values to the tendency of a molecule to accept a proton in the gas phase, or conversely, the tendency of a positive ion to donate a proton. That is 'proton affinities' and 'gas-phase basicities' provide quantitative measures of the acid–base properties of positive ions and the corresponding conjugate-base neutral species in the gas phase. These quantities have proved to be of scientific interest because of what they tell us about acid–base chemistry in the absence of a solvent, and of practical interest because of the applications of gas-phase proton transfer reactions in the analytical technique of chemical ionization mass spectrometry. This article will be concerned with defining these quantities, describing how they are determined experimentally, and discussing the current status of the collective database of proton affinity/gas-phase basicity data. For a discussion of the implications of those data in organic chemistry, the reader is referred to the available reviews of that subject, particularly to the work of R. W. Taft.

The proton affinity of species M, $PA_T(M)$, is defined as the negative of the enthalpy change, ΔH_T^0, of the hypothetical gas-phase reaction at temperature T:

$$M(g) + H^+(g) \rightarrow MH^+(g) \qquad [1]$$

$$\begin{aligned} PA_T(M) &\equiv -\Delta H_T^0 \text{ (Eqn [1])} \\ &= \Delta_f H_T^0(M) + \Delta_f H_T^0(H^+) - \Delta_f H_T^0(MH^+) \end{aligned}$$
$$[2]$$

while the gas-phase basicity, $GB_T(M)$, is the negative of the Gibbs free energy change of the same reaction, ΔG_T^0, and therefore related to the proton affinity through the standard thermochemical expression:

$$\begin{aligned} GB_T(M) &\equiv -\Delta G_T^0 \text{ (Eqn [1])} \\ &= PA_T(M) + T\Delta S_T^0 \text{ (Eqn [1])} \end{aligned} \qquad [3]$$

The entropy change of Equation [1], ΔS_T^0, can be expressed in terms of absolute entropies of the species involved:

$$\begin{aligned} \Delta S_T^0 &\text{ (Eqn [1])} \\ &= S_T^0(MH^+) - S_T^0(M) - S_T^0(H^+) \end{aligned} \qquad [4]$$

$$= \Delta S_{p,T}(M) - S_T^0(H^+) \qquad [5]$$

where $\Delta S_{p,T}(M)$ is the entropy of protonation of M:

$$\Delta S_{p,T}(M) = S_T^0(MH^+) - S_T^0(M) \qquad [6]$$

Thus, the relationship between the gas-phase basicity, proton affinity and entropy of protonation is given by

$$GB_T(M) = PA_T(M) + T[\Delta S_{p,T}^0(M) - S_T^0(H^+)] \quad [7]$$

In practice, proton affinity and gas-phase basicity values are used to predict the occurrence of bimolecular proton transfer reactions in the gas phase:

$$MH^+(g) + B(g) \rightarrow BH^+(g) + M(g) \qquad [8]$$

(In the text that follows, all references to compounds in reactions or in thermochemical equations refer to species in the gas phase, although a specific designator, (g), is omitted in the interest of improving readability.) At any particular temperature, the species, MH^+ or BH^+, with the lower gas-phase basicity will transfer a proton to the conjugate base of the other (B or M), and the exact difference in the gas-phase basicities can be measured by determining the equilibrium constant, K_{eq}, of Equation [8].

$$K_{eq} = \frac{[BH^+][M]}{[MH^+][B]} \qquad [9]$$

$$\begin{aligned} -RT \ln K_{eq} &= \Delta G_T^0 \text{ (Eqn [8])} \\ &= \Delta H_T^0 \text{ (Eqn [8])} \\ &\quad - T\Delta S_T^0 \text{ (Eqn [8])} \\ &= GB_T(M) - GB_T(B) \end{aligned} \qquad [10]$$

Similarly, the enthalpy change of Equation [8] is equal to the difference in the proton affinities of the two species,

$$\Delta H_T^0 \text{ (Eqn [8])} = PA_T(M) - PA_T(B) \qquad [11]$$

and the entropy change is

$$\Delta S_T^0 \text{ (Eqn [8])} = \Delta S_{p,T}^0(B) - \Delta S_{p,T}^0(M) \qquad [12]$$

Although the occurrence or nonoccurrence of the proton transfer depends, at any given temperature, on the relative gas-phase basicity values, rather than the proton affinities, entropy changes associated with Equation [8] are typically (although not always) small, and proton affinities are, in practice, often consulted to predict the occurrence or nonoccurrence of particular proton transfer reactions.

According to common usage, unless it is stated otherwise proton affinity values are assumed to refer to 298 K, unlike electron affinities and ionization energies, which are specifically referred to 0 K. Therefore, it is useful to question how absolute values of proton affinities and protonation entropies vary with temperature. Differentiating and integrating Equation [2] with respect to temperature gives:

$$PA_{T_2}(M) - PA_{T_1}(M)$$
$$= \int [C_p(H^+) + C_p(M) - C_p(MH^+)] dT \qquad [13]$$

where the C_p are the molar heat capacities at constant pressure of the parenthetically indicated species, and the integration is carried out from T_1 to T_2. At room temperature and above, $C_p(H^+)$ is taken as the classical value of $(5/2)R$, while $C_p(MH^+)$ will be close to, but greater than, $C_p(M)$. Thus, for example, the difference in the value of the absolute proton affinity of M at 298 K and 600 K will be less than $(5/2) R$ (600 K–298 K) = 6.2 kJ mol^{-1}.

The relative proton affinities, $PA_T(M) - PA_T(B)$, of a pair of molecules, M and B in Equation [8], are essentially temperature independent, i.e.

$$-\Delta H_{T_1}^0 \ (Eqn \ [8]) = PA_{T_1}(M) - PA_{T_1}(B)$$
$$\approx -\Delta H_{T_2}^0 \ (Eqn \ [8]) = PA_{T_2}(M) - PA_{T_2}(B) \qquad [14]$$

This can be shown more formally by differentiating and integrating Equation [11] with respect to temperature,

$$[PA_{T_2}(M) - PA_{T_2}(B)] - [PA_{T_1}(M) - PA_{T_1}(B)]$$
$$= \int [C_p(BH^+) + C_p(M) - C_p(MH^+) - C_p(B)] dT$$
$$\qquad [15]$$

and noting that, because of the structural similarities of reactants and products, the heat capacity terms of Equation [15] will essentially cancel. When a relative proton affinity is derived from a van't Hoff analysis of a proton transfer equilibrium over a suitable temperature range, it is safe to assume that ΔH_T^0 (Eqn [8]) is independent of temperature over that range. The above discussion suggests that the temperature independence of ΔH_T^0 (Eqn [8]) can be safely assumed throughout the range 298 K $\leq T \leq$ 600 K. The feature is a generally observed phenomenon for reactions in which the number of reactants and products is the same, as is the case for proton

transfer reactions. Similar considerations also apply to relative protonation entropies, i.e.

$$\Delta S_{T_1}^0 \ (Eqn \ [8]) = \Delta S_{p,T_1}(M) - \Delta S_{p,T_1}(B)$$
$$\approx \Delta S_{T_2}^0 \ (Eqn \ [8]) = \Delta S_{T_2}(M) - \Delta S_{T_2}(B) \qquad [16]$$

These considerations are important in establishing the validity of comparing data obtained at different temperatures, as must be done in evaluating gas-phase basicity and proton affinity data obtained under different experimental conditions.

How are proton affinities and gas phase basicities determined?

The proton affinity of any species can easily be derived from Equation [2], provided that enthalpies of formation of all the relevant species, M, H$^+$ and MH$^+$, are known. Actually, in practice, there are relatively few species for which this condition is fulfilled. Usually, reliable data on the enthalpy of formation of MH$^+$ are lacking, since direct determination of this quantity requires either that the neutral precursor, MH, exists and is sufficiently stable that its ionization energy can be experimentally determined, or that the appearance energy of MH$^+$ formed in the dissociation of a larger molecule, MNH, can be determined:

$$MNH \rightarrow MH^+ + N + e^- \qquad [17]$$

Obviously, neutral analogues (MH) exist for only a very few protonated molecular species (MH$^+$). For example, although CH_5^+ (protonated methane) is a commonly used proton-donor species, the proton affinity of methane cannot be derived using this approach (Eqn [2]), since the molecule CH$_5$ does not exist as a stable species, nor is CH_5^+ formed as a fragment ion from any known dissociation of an organic molecular ion.

Most of the existing quantitative data on proton affinities and gas-phase basicities have been derived from determinations of equilibrium constants for proton transfer reactions (Eqn [8]) in the gas phase (Eqns [9] and [10]). Note that the measurement of a series of equilibrium constants at a single temperature will generate relative, rather than absolute, values for gas-phase basicities. In order to determine the enthalpy change of Equation [8] – that is, the relative proton affinities of M and B (see Eqn [11]) – values for the entropy changes of the reaction must be obtained, through measurements of the equilibrium constant as a function of temperature (van't Hoff plot), through statistical-mechanical estimations or by *ab initio* calculations. Absolute values must then be assigned to the relative proton affinity scales using data for molecules whose position in the scale has been established, and for which absolute values of enthalpies of formation of both M and MH⁺ are known from other measurements (Eqn [2]).

Equilibrium constants for gas-phase proton transfer reactions can be determined as follows. A mixture, of known composition, of gases M and B is introduced into a mass spectrometric instrument designed to allow multiple collisions of ions before sampling (that is, designed so that ion–molecule reactions can be observed). Most such measurements have been carried out using one of three types of mass spectrometer that operate in very different pressure regimes. An ion cyclotron resonance (ICR) spectrometer typically observes such equilibria at very low pressures ($\sim 10^{-4}$ Pa), but at long times, so that there have been a sufficient number of collisions that an equilibrium can be established. Other measurements are done at higher pressures (100–1000 Pa) using a high-pressure mass spectrometer or a flow tube apparatus such as a flowing afterglow instrument.

If one or both of the parent molecular ions, $M^{\bullet +}$ or $B^{\bullet +}$, is a species that contains a labile proton that can be transferred to M or B, then Equation [8] may be observed in the system. When neither $M^{\bullet +}$ nor $B^{\bullet +}$ serves as a source of protons, the proton transfer reaction (Eqn [8]) may be initiated by the addition of a bath gas of some compound (methane, for example) that generates ions that transfer protons to the two subject compounds. If the reverse (endothermic) proton transfer reaction is fast enough to be observed on the timescale of the particular experiment, it is possible that a thermodynamic equilibrium can be established, and the equilibrium constant determined simply by observing the equilibrium ratio of the ion concentrations, [MH⁺] and [BH⁺]; the composition of the mixture [M]/[B] does not change, since the neutral compounds are present in great abundance relative to the ions. Each measurement provides a value for the Gibbs free energy change of Equation [8] at a single temperature – that is, a value for the difference in gas basicities of molecules M and B (see Eqn [10]). There exist large interlocking scales of relative gas phase basicities (at a single temperature) determined by carrying out such measurements on series of molecules.

In certain cases, it is not possible to establish a proton transfer equilibrium, for example, if the subject compound, M, is unstable, or if MH⁺ undergoes a fast reaction with M, or a reaction other than proton transfer with B. In these instances, other strategies have been adopted to determine relative gas-phase basicities or proton affinities. The simplest such approach is called the 'bracketing technique'; the ion MH⁺ is generated and the occurrence or non-occurrence of proton transfer with a series of molecules, B_1, B_2, etc., is observed. Reference compounds are chosen whose position in the relative scale of gas basicities is known.

$$MH^+ + B_1 \rightarrow \text{no proton transfer} \qquad [18]$$

$$MH^+ + B_2 \rightarrow B_2H^+ + M \qquad [19]$$

Under the assumption that proton transfer will be observed only if the reaction is associated with a negative value of the Gibbs free energy, the basicity of M is taken to be between the basicities of B_1 and B_2. In a variation of this approach proposed by Bouchoux, Salpin and Leblanc, called the 'thermokinetic method', trends in the reaction efficiency ($k_{Rn}/k_{Collision}$) with reactants of varying gas basicity are examined, and a correlation is used to predict the relative gas basicity.

Another approach, proposed by McLuckey, Cameron and Cooks, that is often used when compounds M and B are of low volatility is based on the observation of the collision-induced dissociation of proton-bound dimer ions, $M \cdot H^+ \cdot B$, formed in association reactions:

$$M \cdot H^+ \cdot B + X \rightarrow MH^+ + B + X \qquad [20a]$$

$$\rightarrow BH^+ + M + X \qquad [20b]$$

A semiquantitative relationship between the ratios of the two product ions and the relative proton affinities has been developed, and can be used to derive relative values of the proton affinities provided that the entropy changes associated with processes [20a] and [20b] are similar.

Quantitative information about relative proton affinities can also be obtained through the determination of the energy barrier associated with endothermic proton transfer reactions through an Arrhenius treatment of the temperature dependence of the rate coefficients, although this approach has rarely been used. In addition, determinations of the equilibrium constants of association reactions

$$AH^+ + B \rightarrow ABH^+ \qquad [21]$$

can provide values for enthalpies of formation of the product ion, ABH^+, provided the enthalpies of formation of AH^+ and B are known; if the enthalpy of formation of AB is also known, its proton affinity can be derived.

The 1998 scale of gas-phase basicities and proton affinities

Beginning in 1971, when the first determinations of gas-phase proton transfer equilibria were published, several extensive scales of relative gas-phase basicities were generated in different laboratories. With the exception of a small number of entropy-change measurements made in the laboratory of Paul Kebarle at the University of Alberta, most of these initial studies were carried out at a single temperature; in most cases, entropy changes were estimated from statistical-mechanical considerations, usually making the simplifying assumption that the complete expression derived from the partition functions could be adequately approximated using the ratio of the rotational symmetry numbers (σ) of M and MH$^+$:

$$\Delta S_p^0(M) = R \ln[\sigma(M)/\sigma(MH^+)] \qquad [22]$$

Using such estimated entropy changes, the experimentally determined interlocking scales of relative basicities were converted to scales of proton affinities. By the early 1980s, values of proton affinities for about 800 molecules had been reported, but comparing data from different laboratories was not always straightforward because different researchers chose different primary standards to assign absolute values to the proton affinities, and proton affinity values assigned to these reference standards were not always internally consistent. Therefore, in 1984 the Ion Energetics Data Center at the National Bureau of Standards (now the National Institute of Standards and Technology) carried out a comprehensive evaluation of available data to put all data on the same basis, and provide an internally consistent scale.

That evaluated scale (sometimes referred to as 'the NBS scale') proved to be sufficiently useful that it continues to be cited years after its publication. However, in the intervening years, a large amount of new data has appeared in the literature, so the so-called 'NBS scale' (which will be referred to here as the '1984 scale') is seriously out of date, missing information on about 900 compounds. In addition, since 1991 several important publications (both experimental and theoretical) have presented data indicating that portions of the thermochemical scale, as evaluated in 1984, are in need of re-evaluation. Furthermore, recent publications from two laboratories (Mautner and Sieck at the National Institute of Standards and Technology, and Szulejko and McMahon at the University of Waterloo) provided extensive data sets in which equilibrium constants had been determined as a function of temperature, i.e. experimental entropy change determinations had become available.

The results of these recent experimental proton transfer equilibrium studies indicated that portions of the 1984 gas-phase basicity/proton affinity scale were constricted. For example, in the 1984 scale the difference between the proton affinities of isobutene and ammonia was given as 33.5 kJ mol^{-1}, while according to the newer results, this interval is actually about 50 kJ mol^{-1}. Also, these experimental results indicated that the portion of the 1984 scale representing gas-phase basicities/proton affinities higher than that of ammonia (a portion of the scale that was not 'anchored' by data for a primary standard) was seriously constricted.

Furthermore, according to the 1993 theoretical calculation carried out by Smith and Radom of the proton affinity of isobutene (one of the primary standards for the 1984 scale),

$$(CH_3)_2C = CH_2 + H^+ \rightarrow (CH_3)_3C^+ \qquad [23]$$

the value for the proton affinity of isobutene used in the 1984 evaluation was too high by about 16 kJ mol^{-1}; this implied that the previously accepted enthalpy of formation of the $(CH_3)_3C^+$ ion must be in error by this amount. New experimental determinations of that enthalpy of formation carried out by Baer and colleagues and by Traeger soon gave evidence that the theoretical prediction was correct. These experimental and theoretical results corroborated the new equilibrium constant data that indicated that the interval between the proton affinities of isobutene and ammonia was greater by 17 kJ mol^{-1} than previously accepted. All these results had profound implications for the evaluation of the central portion of the thermochemical scales, which had been 'anchored' by taking the proton affinity of isobutene as reference standard.

For these reasons, the present authors undertook a re-evaluation of the entire corpus of data on gas-phase basicities and proton affinities, which now includes data for about 1740 compounds. This effort involved an evaluation of the entire interrelated thermochemical scale. Users of proton affinity data are sometimes confused when a re-evaluation results in a change in the value assigned to a particular molecule, and inquire whether a particular proton affinity value has been re-determined; it is important to understand that, with the exception of the few standards used to 'anchor' the scales, a compound-by-compound evaluation of the scale of gas-phase basicities (or proton affinities) is not possible, and when the entire scale is expanded, contracted or shifted (owing to the appearance of newer, more reliable data), values assigned to individual compounds will change, even when the original experimental determinations (of relative gas-phase basicities/proton affinities) remain the only source of information about the proton affinity of the compound in question. Although users of the data attach importance to absolute values assigned to particular proton affinity values, it is usually the relative values that are of importance in designing experiments involving proton transfer reactions; it is of primary importance to ascertain that the values being used, be they relative or absolute, are internally consistent.

The evaluation of such a body of interrelated thermodynamic data involves first an evaluation of the thermochemical scales for internal consistency in the three parameters, ΔG^0 (at different temperatures), ΔH^0 and ΔS^0. Final values assigned for the proton affinities and entropy changes must be consistent with what is known about the thermochemistry of M and MH$^+$. The lengths of segments of the scale linking different primary standards (compounds for which Eqn [2] can be used to derive an absolute proton affinity value) must of course match the known interval between the 'known' proton affinity values.

The 1998 evaluation of these data took as a starting point an examination of several extensive scales of gas-phase basicities obtained in different laboratories and using different kinds of instrumentation. The existence of such extensive independent data sets, each of which contained numerous checks on internal consistency, was invaluable in establishing confidence in the final evaluated scale. The gas-phase basicity scales were chosen as the starting point for the evaluation, rather than the newly published experimentally determined proton affinity scales (based on van't Hoff plots determined in high-pressure mass spectrometers over the temperature range ~450–650 K), because there was poor agreement between the relative proton affinity scales determined in the different studies. (There are several plausible explanations for the lack of agreement between these data sets determined in different laboratories, including pyrolysis of ions in the high-temperature region (> 600 K), clustering of neutral molecules to the ions in the low-temperature region (< 500 K), or poor characterization of the relative and absolute pressures of the reacting compounds in the high-pressure mass spectrometer ion sources, as demonstrated by Grimsrud and colleagues.)

These internally consistent scales of gas-phase basicity data on which the evaluation was primarily based included three data sets obtained at 600 K using high-pressure mass spectrometry, and a very extensive scale determined several years ago at 'ambient' temperature using ion cyclotron resonance spectrometry. These independent determinations of the extensive thermochemical ladder were, with a few minor exceptions, in excellent agreement over the central and upper parts of the energy range, except that the older data yielded scales that appeared to be slightly 'contracted' relative to the recent data, most likely as a result of problems in temperature measurements in the early experiments. A careful evaluation, involving comparisons with analogous experiments where the temperature was carefully measured, permitted a 'standardization' of the various data sets by making appropriate corrections for temperature.

From the evaluated composite gas-phase basicity scale, there was established a 'backbone' scale based on data for some 25 compounds, which included selected primary standards as well as a few other compounds for which entropy changes could be reliably assigned. Primary standards are compounds for which an absolute proton affinity can be established using known enthalpies of formation (Eqn [2]). Through the use of this 'backbone' scale, the established scale of relative gas-phase basicities could be tied to absolute proton affinity values. With the 'backbone' scale established, all other published gas basicity data (for some 1700 additional compounds) were related to that scale at appropriate temperatures. Then values for the entropy of protonation (Eqn [6]) were assigned, compound-by-compound, to generate the final complete scale of proton affinity values. Finally, the resulting data were examined to verify internal consistency and 'reasonableness' of all the proton affinity and entropy-change values.

In carrying out the evaluation, the scale of proton affinity values produced by Smith and Radom through *ab initio* calculations was an invaluable aid. That scale, of proton affinity values for 31 molecules covering an energy range of about 500 kJ mol^{-1}, effectively spanned most of the experimental scale reported from equilibrium constant determinations, and provided an independent check on the

Table 1 'Backbone' scale of proton affinities and gas-phase basicities

Molecule	1998 scale			Ab initio	1984 scale	
	GB_{298}	PA_{298}	$\Delta S_{p,298}$	PA_{298}	GB_{298}	PA_{298}
$(CH_3)_3N$	918.1±6.0	948.9±6.0	5.6	951.6[a]	909	942
Pyridine	898.1±6.0	930±6.0	2.0	929.8[a]	892	924
$(CH_3)_2NH$	896.5±6.0	929.5±6.0	−2.0	931.7[a]	890	923
$C_2H_5NH_2$	878±6.0	912±6.0	−5.1	941.0[a]	872	908
CH_3NH_2	864.5±6.0	899±6.0	−7.0	901.0[a]	861	896
NH_3	819	853.6	−6.4	853.6[a]	818	853.5
$CH_2=C=O$[c]	793.6±3.0	825.3±3.0	2.4	825.0[a]	793	828
CH_3COCH_3	782.1±6.0	812±6.0	8.7	811.9[a]	790	823
$(CH_3)_2C=CH_2$[d]	775.6±1.2	802.1±1.4	20	802.1[a]	784	820
$(CH_3)_2O$	764.2±6.0	792±6.0	16.5	792.0[a]	771	804
C_2H_5CN	763±6.0	794.1±6.0	4.7	794.3[a]	770	806
$C_6H_5CH_3$	756.3±6.0	784±6.0	16	–	761	794
$CH_2=CHCN$	753.7±6.0	784.7±6.0	4.9	784.7[a]	761	794
$HCOOCH_3$	751.5±6.0	782.5±6.0	5	782.2[a]	757	790
CH_3CN	748±6.0	779.2±6	4.3	780.0[a]	756	788
CH_3CHO[e]	736.5±1.6	768.5±1.6	1.5	770.2[a]	747	781
CH_3OH	724.5±6.0	754.3±6.0	9	754.3[a]	728	761
$CH_3CH=CH_2$[f]	722.7±3.0	751.6±3.0	12	744.3[a]	718	751
$CH_2=O$[g]	683.3±1.1	712.9±1.1	9.5	711.8[a]	687	718
H_2S[h]	673.8±5.3	705.0±5.3	4.3	707.7[a]	681	712
H_2O[i]	660.0±3.0	691.0±3.0	5	688.4[a]	665	697
CS_2	657.7±6.0	681.9±6.0	28	681.9[a]	672	699
$CH_2=CH_2$[j]	651.5	680.5±1.7	11.5	681.9[a]	651	680
CO[k]	562.8±3.0	594.0±3.0	4.2	539.0[a], 593.1[b]	562	594
CO_2[l]	515.8±2.0	540±2.0	26	539.3[a], 541.0[b]	520	548

[a] Smith and Radom (1993).
[b] Kormornicki and Dixon (1992).
[c] $\Delta_f H(CH_3CO^+)$ from appearance energies in methyl ketones.
[d] $\Delta_f H((CH_3)_3C^+)$ from appearance energy determinations (Keister *et al* 1993).
[e] $\Delta_f H(CH_3CHOH^+)$ from appearance energy in C_2H_5OH.
[f] $\Delta_f H(CH_3)_2CH^+)$ from appearance energies in 2-halopropanes.
[g] $\Delta_f H(CH_2OH^+)$ from appearance energy in CH_3OH.
[h] $\Delta_f H(H_3S^+)$ from appearance energy in Van der Waals dimer $(H_2S)_2$.
[i] $\Delta_f H(H_3O^+)$ from appearance energy in Van der Waals dimer $(H_2O)_2$.
[j] $\Delta_f H(C_2H_5^+)$ from adiabatic ionization energy of the ethyl radical, and from appearance energy in C_2H_5I.
[k] $\Delta_f H(HCO^+)$ from appearance energy in HCOOH.
[l] $\Delta_f H(CO_2H^+)$ from appearance energy in HCOOH and other carboxylic acids.

evaluation. Theoretical proton affinity values from this study were available for the 'backbone' scale compounds, and served to verify the evaluations.

Table 1 shows the results of the 1998 evaluation of the gas-phase basicity and proton affinity scales as exemplified by the 'backbone' scale. The results shown include the evaluated proton affinity, gas-phase basicity and entropy of protonation values, the *ab initio* value reported by Smith and Radom and, for comparison, the proton affinity value cited in the

1984 evaluation. Primary standards that anchor the scale are given in italic and the footnotes give an indication of the type of experiment that yielded a value for the enthalpy of formation of the protonated molecule, MH+ (see Eqn [2]) used in establishing the absolute proton affinity value. Note that the values adopted for ammonia are essentially the same as those recommended in the 1984 evaluation, and that the portion of the scale above ammonia is expanded by about 8–9% compared to the values assigned in that evaluation. The dramatic change in the proton affinity and gas-phase basicity of isobutene results in a general lowering of values assigned to compounds below isobutene in the scale. Below ethylene in the scale, the changes are less dramatic. It should be mentioned that the scale of gas basicities is not yet well established in the low-basicity region (below the basicities of H_2S and H_2O), where there are major inconsistencies in the data reported from different laboratories.

In **Table 1**, the standard uncertainties assigned to the primary anchor molecules (italic entries) are the usual root-sum-of-squares combination of individual uncertainties associated with relevant enthalpies of formation and the uncertainty of some key measurement, such as an ionization or an appearance energy. The uncertainties assigned to all the other molecules are based on our best judgment using all the relevant information and a general knowledge of and experience with interlocking thermochemical scales.

List of symbols

$BG_T(M)$ = gas-phase basicity of species M at temperature T; C_p = molar heat capacity at constant pressure; k = reaction rate constant; K_{eq} = equilibrium constant; $PA_T(M)$ = proton affinity of species M at temperature T; $S_T^0(M)$ = absolute entropy of species M at temperature T; T = temperature; $\Delta G_T^0(Eqn[n])$ = Gibbs free energy change of reaction (given as Eqn $[n]$) at temperature T; $\Delta H_T^0(Eqn[n])$ = enthalpy change of reaction (given as Eqn $[n]$) at temperature T; $\Delta_f H^0(M)$ = enthalpy of formation of species M at temperature T; $\Delta S_T^0(Eqn[n])$ = entropy change of reaction (given as Eqn $[n]$) at temperature T; $\Delta S_{p,T}(M)$ = entropy of protonation of species M at temperature T; σ = rotational symmetry number.

See also: Chemical Ionization in Mass Spectrometry; Ion Dissociation Kinetics, Mass Spectrometry; Ion Molecule Reactions in Mass Spectrometry.

Further reading

Bouchoux G, Salpin J-Y and Leblanc D (1996) A relationship between the kinetics and thermochemistry of proton transfer reactions in the gas phase. *International Journal of Mass Spectrometry and Ion Processes* 153: 37–48.

East ALL, Smith BJ and Radom L (1997) Entropies and free energies of protonation and proton-transfer reactions. *Journal of the American Chemical Society* 119: 9014–9020.

Hunter EP and Lias SG (1997) Proton affinity evaluation. In: Mallard WG and Linstrom PJ (eds) *NIST Standard Reference Database Number 69*. Gaithersburg, MD: National Institute of Standards and Technology (http://webbook.nist.gov).

Hunter EP and Lias SG (1998) Evaluated gas phase basicities and proton affinities of molecules: an update. *Journal of Physical Chemistry Reference Data* 27: 413–656.

Kebarle P (1997) Ion thermochemistry and solvation from gas phase ion equilibria. *Annual Review of Physical Chemistry* 28: 445–476.

Kebarle P, Yamdagni R, Hiroaka JK and McMahon TB (1976) Ion molecule reactions at high pressure: recent proton affinities, gas phase acidities and hydrocarbon clustering results. *International Journal of Mass Spectrometry and Ion Physics* 19: 71–87.

Keister JW, Riley JS and Baer T (1993) The *tert*-butyl ion heat of formation and the isobutene proton affinity. *Journal of the American Chemical Society* 115: 12 613–12 614.

Kormornicki A and Dixon DA (1992) Accurate proton affinities: *ab initio* proton binding energies for N_2, CO, CO_2, and CH_4. *Journal of Chemical Physics* 97: 1087–1094.

Lau YK (1979) PhD thesis, University of Alberta.

Lias SG, Liebman JF and Levin RD (1984) Evaluated gas phase basicities and proton affinities of molecules; heats of formation of protonated molecules. *Journal of Physical Chemistry Reference Data* 13: 695–808.

McGrew DS, Knighton WB, Bognar JA and Grimsrud EP (1994) Concentration enrichment in the ion source of a pulsed electron beam high pressure mass spectrometer. *International Journal of Mass Spectrometry and Ion Physics* 139: 47–58.

McLuckey SA, Cameron D and Cooks RG (1981) Proton affinities from dissociations of proton bound dimers. *Journal of the American Chemical Society* 103: 1313–1317.

Meot-Ner (Mautner) M and Sieck LW (1991) Proton affinity ladders from variable temperature equilibrium measurements. 1. A re-evaluation of the upper proton affinity range. *Journal of the American Chemical Society* 113: 4448–4460.

Smith BJ and Radom L (1993) Assigning absolute values to proton affinities: a differentiation between competing scales. *Journal of the American Chemical Society* 115: 4885–4888.

Szulejko J and McMahon TB (1991) A pulsed electron beam, variable temperature high pressure mass spectrometric re-evaluation of the proton affinity difference between 2-methylpropene and ammonia. *International Journal of Mass Spectrometry and Ion Processes* 109: 279–294.

Szulejko J and McMahon TB (1993) Progress towards an absolute, proton affinity scale. *Journal of the American Chemical Society* 115: 7839–7848.

Taft RW (1983) Protonic acidities and basicities in the gas phase and in solution: substituent and solvent effects. *Progress in Physical Organic Chemistry* 14: 248–350.

Taft RW (1975) Gas phase proton transfer equilibria. In: Caldin EF and Gold V (eds) *Proton Transfer Reactions*, pp 31–78. New York: Wiley.

Traeger JC (1996) The absolute proton affinity for isobutene. *Rapid Communications in Mass Spectrometry* **10**: 119–121.

Williamson DH, Knighton WB and Grimsrud EP (1996) *International Journal of Mass Spectrometry and Ion Physics* **154**: 15–24.

Yamdagni R and Kebarle P (1976) Gas phase basicities and proton affinities of compounds between water and ammonia and substituted benzenes from a continuous ladder of proton transfer equilibrium measurements. *Journal of the American Chemical Society* **98**: 1320–1324.

Proton Microprobe (Method and Background)

Geoff W Grime, University of Oxford, UK

HIGH ENERGY SPECTROSCOPY
Methods & Instrumentation

Although the techniques of ion beam analysis (IBA) have been used for many years with broad beams (5–10 mm diameter), it is only in the 1990s that the technology for focusing high-energy ion beams has developed to the point where the spatial resolution is comparable to that of other forms of probe beam microanalysis (around 1 μm). The interactions most commonly used require protons with energy of 1–4 MeV for optimum sensitivity, and in this form the instrument is commonly known as the proton or nuclear microprobe.

Using a combination of analytical techniques, the nuclear microprobe can provide simultaneous multielemental analysis over the entire Periodic Table with a spatial resolution of 1 μm, a minimum detection limit of 1–100 ppm depending on the conditions and a quantitative accuracy of 5–20% depending on the type of analysis. Although the penetration depth of MeV protons can be in the region of 100 μm in some materials, the nuclear microprobe is a surface-biased technique since signals are detected preferentially from the near surface region ($\leqslant$ 10 μm depth).

Focusing systems

The high momentum of MeV ions which gives them their advantage for proton-induced X-ray emission (PIXE) analysis also makes them difficult to focus. In particular, the standard cylindrical magnetic lenses used in electron focusing columns are too weak to be used with MeV ions, and alternative technologies must be sought. Many different arrangements of apertures and electromagnetic fields have been proposed, but the system which has shown consistently the best performance is the magnetic quadrupole multiplet.

Quadrupole lenses have four magnetic poles arranged symmetrically N–S–N–S around the beam axis (**Figure 1A**). They have a strong focusing action on charged particle beams but, because of the antisymmetry, a single quadrupole lens converges the beam only in one plane and diverges in a plane normal to this. For this reason, two or more quadrupoles of alternating polarity are required to form a point focus. Combinations of two, three and four lenses have been used, as shown in **Figure 1B**.

Like conventional glass lenses, quadrupole lenses suffer from significant angle-dependent aberrations which increase the beam diameter and, together with the other parameters of the experiments, these set a limit to the smallest usable beam diameter that can be achieved. This dependence can be loosely expressed as follows:

$$d \propto \sqrt[n]{\frac{\delta R f(C)}{Y B \Omega}}$$ [1]

where R is the smallest count rate that is required to perform the experiment with good enough statistics in a reasonable time, δ is the energy spread of the beam, Y is the normalized yield of the analytical reaction being used, B is the brightness of the beam from the accelerator, Ω is the solid angle of the

(A)

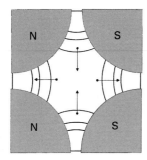

(B)

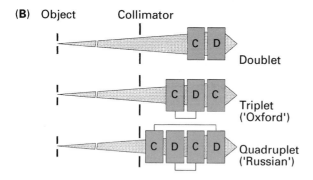

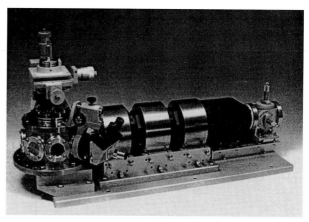

Figure 2 Photograph of the end-stage of a commercially available microbeam focusing system using a triplet of magnetic quadrupoles. The beam enters the system from the right and passes through the collimator aperture, magnetic deflection coils to sweep the beam across the sample and the three quadrupole magnets before entering the target chamber where the sample is mounted on a micrometer controlled positioning stage. Courtesy *Oxford Microbeams Ltd.*

Figure 1 (A) Cross-section of a quadrupole magnet used as a focusing lens for charged particles. Particles travelling in the vertical plane experience a force directed toward the axis; in the horizontal plane the force is directed away from the axis. (B) Combinations of quadrupoles which have been used for microbeam applications. In the triplet configuration of quadrupole lenses used in the Oxford nuclear microprobe the quadrupoles are excited alternately converging–diverging–converging and the first two magnets have the same excitation (for ease of alignment). Also shown is the object aperture (typically $50 \times 10 \ \mu m$) and the collimator aperture used to control the divergence of the beam (and hence the aberrations).

detector and $f(C)$ is some function of the aberration coefficients of the lens system, which in turn depend upon the precise shape and magnitude of the magnetic field in the system; n varies between 2 and 4, depending on the type of the dominant aberration. From this it can be seen that for the best spatial resolution we must accept a low count rate with a high-yield reaction using a beam with small energy spread and high brightness and a detector with a large solid angle. In practice, there is little that can be done to improve these parameters. However, the aberration coefficients of the lens can be minimized to a certain extent by the configuration and precision of construction and alignment of the lens.

Apart from the intrinsic chromatic and spherical aberration present even in a perfect quadrupole field, numerical ray-tracing studies of quadrupole probe-forming systems show that the major contributions to the beam broadening are from small imperfections in the construction and alignment of the lenses, especially departures from exact fourfold symmetry in the poles of the quadrupole magnet and errors in the relative rotational alignment of the lenses. The lenses used in the most successful design are cut from a single piece of high-quality magnet iron to minimize errors due to the assembly of individual components and are mounted on tables permitting precise mechanical alignment to micrometer accuracy (**Figure 2**). Using this system, a spatial resolution of 1 μm can be achieved routinely with a beam current of around 150 pA of 3 MeV protons, sufficient for PIXE analysis. These lenses are available commercially and are now installed in a number of facilities world-wide. Other facilities differ only in scale or details of the layout, and the general principles remain the same.

Analytical reactions used in nuclear microscopy

Analytical reactions

Equation [1] indicates that for optimum spatial resolution, high-yield analytical reactions are required. Ion beam analysis (BA) techniques are discussed elsewhere in this volume and here it is shown that the highest cross-section reactions are PIXE and Rutherford backscattering (RBS), which together cover almost the whole of the periodic table. Nuclear reaction analysis has in general a much lower cross-section and is not routinely used for high-resolution microbeam applications.

Table 1 Comparison between microPIXE and electron probe microanalysis

	MicroPIXE	Electron probe microanalysis
Primary particle beam	1–3 MeV protons from a nuclear accelerator focused to 1–10 μm with beam current 0.1–10 nA	10–100 keV electrons focused to 10–100 nm with beam current 1 nA–1 μA
Minimum detection limit	1–100 ppm depending on beam energy, sample matrix, elemental overlaps, etc. Limited by count rate	100 ppm–1% depending on beam energy, sample matrix, elemental overlaps, etc. Limited by background
Beam penetration depth	50–100 μm	1–3 μm
Analysis depth	Depends on absorption of emerging X-rays; can be >50 μm	Similar to penetration depth
Spatial resolution	Equal to beam diameter	Determined by spreading in sample (typically 0.5–2 μm depending on sample thickness)
Quantitative accuracy	Moderate (depends on sample homogeneity). Standardless analysis possible	Good
Other options available	Light elements using RBS and NRA. In-air analysis using external beam facility	High-resolution imaging with Z contrast
Scale of equipment	Medium to large facility (possible shared access to accelerator)	Compact instrument
Availability	< 50 world-wide	Common

RBS = Rutherford back scattering.
NRA = Nuclear reaction analysis.

Table 1 presents a comparison of nuclear microbeam analysis using PIXE (microPIXE) with the analogous electron probe microanalysis.

Other imaging techniques

Other non-analytical techniques are also available using MeV ions which can be used to give imaging contrast for a sample.

Secondary electron imaging MeV ions impinging on a surface liberate copious quantities of electrons, and these may be measured using a detector such as a channel electron multiplier to generate images of the sample. The energy of the emitted electrons is relatively low (typically < 1 keV) and so they are easily obstructed by surface irregularities. Detecting the electrons at a low (grazing) angle to the surface enhances this effect and allows clear topographic images to be obtained.

Secondary electrons resulting from proton bombardment do not carry any compositional information about the sample, and the spatial resolution is limited by the beam diameter of the ion beam to 1 μm or greater, which is significantly worse than the same technique used in the scanning electron microscope, but the technique can be valuable in helping to relate elemental distributions to the physical structure of the sample.

Scanning transmission ion microscopy (STIM) MeV ions in matter lose energy gradually by collisions with atomic electrons, so by measuring the energy of particles passing through thin (typically <30 μm) samples, a measure of the electron density along the ion path can be obtained. This can be used as a technique for mapping density fluctuation in samples thin enough to transmit the beam. STIM can be carried out in two modes: with the detector directly in the beam behind the sample (bright-field STIM), the contrast occurs solely due to proton energy loss in passing through the sample; with the detector mounted off-axis behind the sample (dark-field STIM), scattered particles are detected and contrast is due to a combination of small-angle scattering and energy loss in the sample. With bright-field STIM, each transmitted proton is a signal (i.e. the yield Y in Equation [1] is 100%) and so the beam current can be reduced to around 1000–2000 particles s^{-1} (~0.1 fA) to avoid detector damage and saturation of the acquisition electronics. This is normally carried out by reducing the beam defining apertures in the final lens, and so an added benefit of bright-field STIM is that the beam diameter is reduced and STIM imaging can be carried out at spatial resolutions of the order of 100 nm. For many applications, dark-field STIM is often used to map the sample to locate regions of interest prior to analysis, and this is normally carried out simultaneously with the PIXE and RBS mapping. A STIM image of two mouse red blood cells infected with malaria parasites on a thin plastic film is shown in **Figure 3**.

Ion beam induced charge (IBIC) Charged particles passing through the junctions of active semiconductor devices create electron–hole pairs which can be

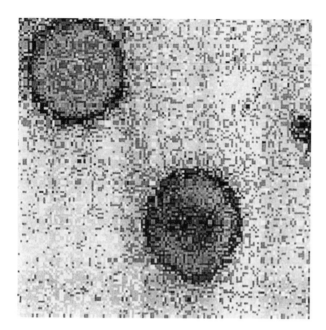

Figure 3 STIM images of mouse red blood cells (5 μm in diameter), both infected with malaria parasites. The bottom one is well developed and shows the parasite coiled around the inside wall of the cell. Denser or thicker regions of the sample are indicted by lighter colours. Courtesy Professor F. Watt, National University of Singapore.

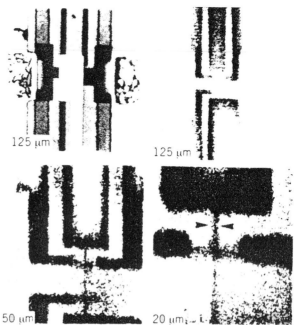

Figure 4 Optical (top left) and IBIC images of a gallium arsenide transistor at three different magnifications (length of side of image area indicated on each area). The IBIC images show the intensity of charge collected between the p-type and n-type contacts, with the highest charge shown as darker. In the highest magnification image, the arrows indicate a 0.8 μm depletion region, showing the resolution that can be achieved using this technique. Reproduced from Breese MBH, Grime GW, Watt F and Blaikie RJ (1993) *Vacuum* **44**: 175, with permission.

detected as charge pulses on the electrodes of the device. Thus scanning a semiconductor device using a microbeam gives the possibility of mapping the active regions of the device either for fault finding or for investigating sensitivity to radiation. This technique is more commonly used with electron beams but the use of MeV ions gives the advantage of the long range, which means that junctions may be studied which are covered by metallization or passivation layers. Like STIM, each particle creates a signal and so it is a high-yield technique with the possibility of high-resolution imaging. **Figure 4** shows IBIC images of a gallium arsenide transistor.

External beam milliprobe

One consequence of the long range of MeV ions is that they can be brought out of the vacuum system through a suitable exit foil into air, so that large or vacuum-incompatible objects can be analysed. Scattering in the foil and in the air means that the resolution is degraded and, according to the design of

Table 2 Sample preparation techniques for nuclear microbeam analysis[a]

Type of sample	Preparation technique
Biological cells	Cryo-fixation on to thin (<1 μm) plastic films
Biological tissue sections	Cryo-fixation followed by cryo-sectioning to a thickness of 5–10 μm and freeze-drying on thin (<1 μm) plastic films
Environmental particles, other particulates and powders	Small particles on filters: analysis *in situ* on filter membrane.
	Small particles in bulk: dispersion in dilute resin solution which can be cast into a thin film
	Large particles: adhesion to conducting adhesive pad
Mineralogical samples, metals, bone, etc.	Resin embedding and polishing as for electron probe analysis. Insulating samples may need carbon coating

[a] Because IBA is predominantly an inner shell/nuclear technique with most of the signal coming from many atomic layers beneath the surface, there is no strong dependence on surface condition (although ideally the surface to be analysed should be flat). Because of this, many samples can be analysed with no further preparation (especially in the external beam). Other types of sample require some preparation, as summarized in this table.

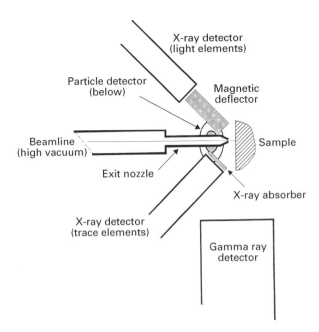

Figure 5 Schematic plan view of the external beam analysis facility at Oxford. The beam emerges from the vacuum of the beamline through a 300 μm diameter hole sealed with a thin (8 μm) plastic foil. X-rays emitted from the sample are detected by two detectors mounted at 45° on either side of the beam. One detector is fitted with an X-ray absorber to kill intense low-energy X-ray lines from elements in the sample matrix to ensure good detection limits for the trace elements while the other detector has no absorber and is fitted with a magnetic deflector to ensure that recoiling high-energy protons do not reach the detector. A detector for gamma rays is mounted at 90° to the beam. Below the beamline, a detector for recoiling protons allows the total amount of charge falling on the sample to be monitored. Not shown in this diagram are a video microscope which uses a mirror to view the front surface of the sample during analysis and a low power alignment laser to assist in positioning the sample for analysis.

the exit aperture, spatial resolutions from 30 μm up to 1 mm can be achieved. The external beam facility used at Oxford University is shown schematically in **Figure 5**. PIXE is the technique most commonly used with external beams, although RBS can be used (with degraded depth resolution because of scattering from air molecules) and nuclear reaction analysis using gamma ray detection can be used to measure light elements because in general the beam currents are higher than in the microbeam. External beam milliprobes are of particular value for analysing archaeological or historical artifacts that it is not possible to sample or hydrated (or even living) biological samples.

Practical aspects

Sample preparation Sample preparation for nuclear microscopy is similar to that required for electron microscopy with the benefit that because of the much longer range of protons in matter (typically 30–100 μm), there is no requirement for ultrathin samples, and in many cases samples can be analysed with little or no sample preparation. **Table 2** summarizes the main types of sample preparation that may be required for nuclear microscopy.

One important consideration for microPIXE analysis is the need to avoid sample treatments which may introduce trace element contamination at the ppm levels detectable using PIXE. This is a particular problem for biological sample preparation, where fixing and staining agents are normally used to preserve the structure of the sample and render visible the regions of interest. For microPIXE analysis of biological samples, the optimum preparation technique is cryo-fixation and sectioning. The structure of the sample must then be determined by a technique such as staining of adjacent sections or STIM imaging of the sample.

Unlike electron microscopy, there is no strong requirement to have a conducting sample, since the high energy of the proton beam will ensure that the beam will not be deflected by the relatively modest potentials encountered at the impact point. However, sample charging can be a problem for PIXE analysis, since secondary electrons present in the system can be accelerated to the positively charged impact points and create an enhanced bremsstrahlung background, which degrades the sensitivity. To reduce this effect, insulating samples (e.g. minerals, bone) are usually coated with a thin conducting film, such as carbon.

See also: **High Energy Ion Beam Analysis; NMR Microscopy; X-Ray Emission Spectroscopy, Applications; X-Ray Emission Spectroscopy, Methods.**

Further reading

Breese MBH, Jamieson DN and King PJC (1996) *Materials Analysis Using a Nuclear Microprobe.* New York: Wiley.

Llabador Y and Moretto Ph (1996) *Nuclear Microprobes in the Life Sciences.* Singapore: World Scientific.

Watt F and Grime GW (1987) *Principles and Applications of High Energy Ion Microbeams.* Bristol: Adam Hilger.

Pyrolysis Mass Spectrometry, Methods

Jacek P Dworzanski and **Henk LC Meuzelaar**,
University of Utah, Salt Lake City, UT, USA

> **MASS SPECTROMETRY**
> **Methods & Instrumentation**

Pyrolysis indicates decomposition caused by thermal energy and, as a process of covalent bond dissociation and rearrangement brought about by heat, represents a form of thermolysis. However, to indicate the relatively high temperature of the process, namely approaching the temperature of a fire, the use of the term 'pyrolysis' – that originates from the Greek: pyr, fire – is commonly used. Primary bond-scission occurring during such processes may generate reactive species which can be involved in further reactions called secondary pyrolysis processes or recombination reactions. The products of pyrolysis are composed of fragment molecules, called pyrolysate, and are generated using a family of devices for performing pyrolysis, namely, pyrolysers. An analytical technique based on direct coupling of a pyrolyser with a mass spectrometer that allows the detection and analysis on-line of that portion of the pyrolysate which has adequate vapour pressure to reach the detector has been named pyrolysis-mass spectrometry (Py-MS).

It is assumed that, under appropriate conditions, thermal degradation products reflect to a large extent the original structure of the pyrolysed material. Therefore, pyrolysis is utilized as a form of sample pretreatment and is considered as one of the basic approaches during mass spectrometric investigations of complex materials, because both techniques can be fully integrated.

The observation of primary reaction products of high activation energy processes requires both high temperatures and short residence times of the pyrolysate in the hot zone. Flash vacuum pyrolysis performed in proximity to the ion source of a mass spectrometer in many cases satisfies these criteria; however, the design of the pyrolyser plays an important role in the overall performance of the Py-MS

system. Reduction of secondary reactions, which tend to decrease both the character and reproducibility of the pyrolytic products, may be achieved by providing short residence times of the pyrolysate in the hot pyrolysis region. The ultimate objective is to obtain precisely controlled and defined temperature/time profiles of the samples.

The basic steps used to obtain structural information about complex samples by Py-MS are summarized in **Figure 1**.

Pyrolysis techniques

Pyrolysers can be divided into two main categories on the basis of their mode of operation, i.e. the continuous type, where the sample is supplied to a furnace preheated to the final temperature, and pulse mode reactors in which the sample is introduced into a cold furnace which is then heated to the final pyrolysis temperature. In the analytical pyrolysis of solid and some liquid materials mainly pulse mode pyrolysers are used and the following sections will focus on a few of the most popular pyrolysis techniques utilizing this mode of operation. However, for pyrolytic studies of liquid and gaseous samples continuous pyrolysers are applied.

Direct probe pyrolysers

Samples to be analysed may be introduced into an MS ion source using heated probe units that allow the material to be deposited directly on straight, folded or coiled filaments or ribbons. Alternatively, a quartz tube or crucible can be placed inside a heater coil and subjected to a selected rate of temperature increase by resistive or inductive heating. The filament or coil is positioned close to the ion source to

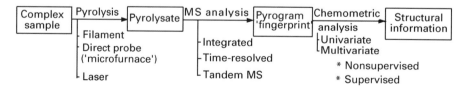

Figure 1 Basic approach options to computerized pyrolysis mass spectrometry.

release pyrolysis products generated under high vacuum conditions.

In principle, any type of magnetic or quadrupole mass spectrometer can be utilized for the analytical pyrolysis of organic materials, if a direct introduction system capable of producing a desired temperature/time profile is available. For example, direct insertion probes (DIPs) and direct exposure probes (DEPs) are widely used for sample introduction and such probes are supplied with control units that allow heating and temperature programming of the sample up to 500–800°C. Therefore, such modules should be considered as the most readily available probes for Py-MS studies.

The temperature rise time (TRT), i.e. the time required for the pyrolyser temperature to be increased from its initial to the final temperature, can be chosen in the range from several milliseconds to several minutes. Simultaneously, the temperature time profile (TTP), representing temperature as a function of time for a particular pyrolysis experiment, may be easily programmable. Pyrolysis may be carried out at a fast rate of temperature increase, e.g. $10\,000$ K s^{-1} in the case of flash pyrolysis, or the sample can be heated at a controlled rate over a temperature range in which pyrolysis occurs using a stepwise, linear or ballistic heating approach characterized by a total heating time (THT) of several minutes or even hours.

To illustrate the relationships among parameters that determine the heating profile for a pyrolytic experiment, a graphical representation of a TTP for an isothermal pyrolysis taking place at the equilibrium temperature is presented in **Figure 2**.

Time-resolved recording of pyrolysate spectra, sometimes referred to as a linear programmed thermal degradation mass spectrometry (LPTDMS), by using a slow heating rate adds additional information (**Figure 3A**) and enables one to perform pyrolysis close to the ion source (**Figures 4** and **5**). However, considerable contamination of the ion

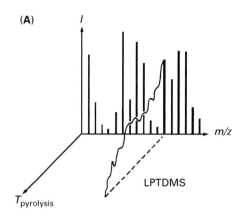

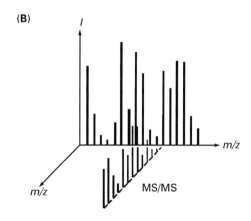

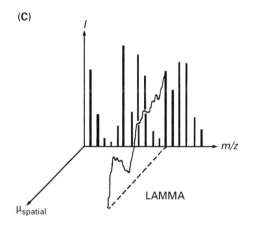

Figure 3 Third dimension in pyrolysis mass spectrometry approaches: (A) linear programmed thermal degradation mass spectrometry [LPTDMS – third dimension = temperature]; (B) collisionally activated dissociation of 'parent' ions coupled with scanning of product ions using tandem mass spectrometry [MS/MS – third dimension = spectrum of product ions]; (C) laser microprobe mass analyser [LAMMA – third dimension = spatial resolution].

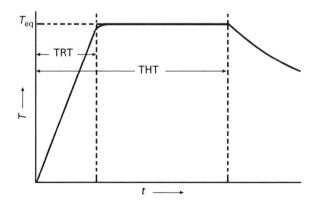

Figure 2 Schematic representation of the parameters that determine the heating profile in filament pyrolysis, namely temperature rise time (TRT), equilibrium temperature (T_{eq}) and total heating time (THT).

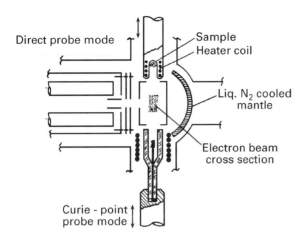

Figure 4 Typical instrumental configurations for pyrolysis electron-impact ionization mass spectrometry: direct insertion probe pyrolysis mode (upper) and Curie-point pyrolysis mode (lower). Reproduced by permission of Elsevier Science from Meuzelaar HLC, Windig W, Huff SM and Richards JM (1986). *Analytica Chimica Acta* **190**: 119–132.

source, together with increased charring and the occurrence of secondary reactions, make this approach troublesome in many cases.

Curie-point pyrolysers

Curie-point pyrolysis takes advantage of fast inductive heating of ferromagnetic materials placed in the high frequency (0.5–1.0 MHz) electromagnetic field generated inside the inductive coil surrounding the pyrolytic reaction tube. This produces fast heating of solid or liquid materials coated on – or gases flowing around – a ferromagnetic sample carrier in the shape of a filament, foil or tube.

When the high frequency field is switched on, the ferromagnetic carrier inductively heats to its Curie-point temperature (T_c), at which the metal loses its ferromagneticity, becomes paramagnetic and the heating effect decreases drastically owing to strongly reduced energy absorption from the high frequency field in the coil. This thermostatic effect ensures control of the final pyrolysis temperature (T_{eq}) determined by a sharply defined transition temperature that is specific for the composition of the alloy used, at the point where the residual energy absorption by eddy currents is balanced by the loss of heat through radiation and conduction.

The TRT for a pyrolyser reactor temperature (**Table 1**) is determined by the shape and diameter of the carrier, the composition of the alloy, the strength of the high frequency field and its frequency (**Figure 6**). Since under practical Py-MS conditions pyrolysis may take place before the wire reaches the final temperature, the heating rate is generally thought to have a critical influence on the pyrolysis patterns. However, other factors, such as the choice of the filament cleaning technique and of the solvent, as well as factors that govern either transfer of the pyrolysate to the ion source or influence ionization conditions, appear to determine the reproducibility of Py-MS more strongly than the temperature/time profile, especially with respect to long-term reproducibility.

Careful attention should be paid to the purity of the surfaces of the pyrolysis filament as well as of the

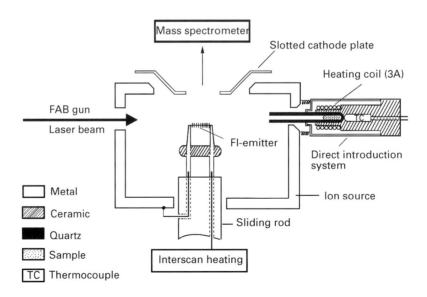

Figure 5 Schematic drawing of the experimental setup for pyrolysis field ionization mass spectrometry. Reproduced by permission of the American Chemical Society from Schulten H-R, Simmleit N and Muller R (1987). *Analytical Chemistry* **59**: 2903–2908.

Table 1 Temperature rise time (TRT) and Curie-point versus alloy composition

Alloy composition (%)			Curie-point (°C)	Quoted TRTs (ms)	
Fe	Ni	Co		Fisher-Varian (1500 W)	Phillips (30 W)
0	100	0	358	300	1300
61.7	0	38.3	400	40	500
50.6	49.4	0	510	150	700
42.0	41.0	16.0	600	70	500
29.2	70.8	0	610	130	1150
33.0	33.0	33.0	700	90	1350
100	0	0	770	110	2100

Reproduced by permission of John Wiley & Sons from Grob RL (ed) (1995) *Modern Practice of Gas Chromatography*, 3rd edn. New York: John Wiley & Sons.

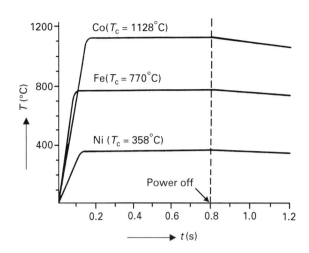

Figure 6 Temperature/time profiles and Curie-point temperatures for pure Ni, Fe, and Co wires (diameter 0.5 mm) when using a 1.5 kW, 1.1 MHz high frequency power supply for pyrolysis/mass spectrometry studies. Reproduced by permission of Elsevier Science from Meuzelaar HLC, Haverkamp J and Hileman FD (1982), *Pyrolysis Mass Spectrometry of Recent and Fossil Biomaterials; Compendium and Atlas*. Amsterdam: Elsevier Science.

glass reaction tube. Although Curie-point wires are inexpensive and are usually discarded after use, glass or quartz reaction tubes should be carefully cleaned before their next use. The choice of the filament cleaning method may influence the pyrolysis pattern. For example, prolonged heating in water-saturated hydrogen, as a reductive cleaning technique, removes oxide layers on the filament surface which otherwise could have an oxidative or catalytic effect and change the emissivity of the filament surface. However, this cleaning technique can cause severe hydrogen absorption by the metal and this could conceivably influence the pyrolysis process by promoting catalytic hydrogenation reactions. Several other factors, such as substitution of a wire for a ferromagnetic tube, or wrapping the sample inside a metal foil, produce pronounced changes in the pyrolysis mass spectra of most compounds.

A pyrolysate composed of a multicomponent mixture, under ideal conditions, should allow the pyrolysis products to reach the ionization zone without any loss, degradation or recombination of products during transfer. However, in practice these conditions are never fulfilled owing to the dynamic character of the whole process, which is governed by a complex set of kinetic parameters related to every chemical reaction taking place during the pyrolysis stage as well as to heat and mass transfer conditions, and interactions with walls. In fact, pyrolytic products represent a mixture of compounds characterized by a broad range of relative molecular masses and vapour pressures. As a result, some pyrolysis products tend to remain on the filament in the form of macromolecular, nonvolatile chars and consequently are inaccessible for further analysis by mass spectrometry. It is well established that the amount of char formed is inversely proportional to the heating rate and can be minimized by avoiding excessively slow heating rates. Another group of compounds is volatile enough to escape from the hot pyrolysis carrier but condenses on the walls of the reaction chamber and transfer lines. Although pyrolysis directly in the ion source is possible, in practice it is difficult to avoid strong contamination of the ion source under these conditions. Therefore, the glass reaction tube surrounding the ferromagnetic carrier serves as a trap for relatively nonvolatile pyrolysis products which otherwise could contaminate the ion source. In addition, these tubes help to obtain maximum signal intensity by producing a forward oriented beam of volatile pyrolysis products which may directly enter the expansion chamber or the ion source, and the ions formed are mass analysed or studied using collisionally induced dissociation (**Figure 3B**).

To broaden the pressure–time profile in the ion source, and thus to enable the recording of a sufficient number of mass spectra generate a representative averaged mass spectrum of the pyrolysate, an expansion chamber is usually placed between a pyrolytic reactor and the ion source. The expansion chamber should be heated and have chemically inert walls to avoid interaction with pyrolysis products. Quartz or gold-plated expansion chambers, heated to about 150–200°C are frequently used to provide a buffer volume (**Figure 7**). However, the removal of the expansion chamber and positioning of the reaction tube directly in front of the ion source, together with slowing the heating rate of the wire, allow for

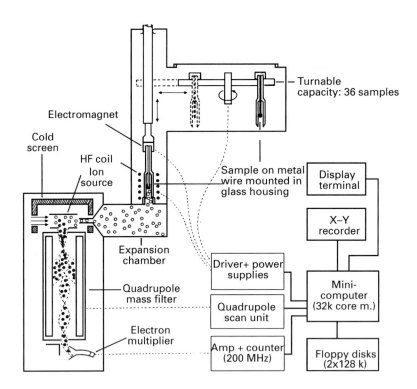

Figure 7 Scheme of an automated Curie-point Py-MS system with a dedicated minicomputer. Samples are selected by a pickup arm from the turntable and positioned inside the high frequency coil. The pyrolysate diffuses via a buffer volume into the mass spectrometer where the molecules are ionized and mass-analysed. Reproduced by permission of Plenum Press from Wieten G, Meuzelaar HLC and Haverkamp J (1984). Analytical pyrolysis in clinical and pharmaceutical microbiology. In: Odham G, Larsson L and Mardh P-A (eds) *Gas Chromatography/Mass Spectrometry Applications in Microbiology*. New York: Plenum Press.

sufficient scanning of the pressure–time profile by the quadrupole mass spectrometers (**Figures** 5 and 8).

In the case of gas phase pyrolytic reactions the pyrolysis module works as a flow reactor with gaseous molecules passing along the preheated filament into a mass spectrometer. This technique allows quantitative detection of products with half-lives greater than 1 ms.

Laser pyrolysers

Laser pyrolysers achieve very high TRTs together with the possibility of characterizing very small surfaces. With a spatial resolution as high as 0.5 μm (Figure 3C) and sample quantities in the range of picograms such instruments permit analysis of very small objects, for example a single bacterial cell. A high power ($\sim 10^9$ W cm^{-2}) beam pulse from a commercially available laser microprobe mass analyser (LAMMA) is provided by Nd:YAG laser and focused on the sample by a microscope objective.

Short pulses (~ 15 ns) create a plasma of positive and negative ions which are mass analysed by a time-of-flight (TOF) mass spectrometer. Although the mechanism of laser-induced ionization and volatilization of solids is not well known, it is established that the duration and shape of the laser pulse affect

the nature of the volatilization process. The region of direct impact of the laser on a sample is associated with high temperatures and consequently only small atomic and molecular fragments are emitted from this region. In the region immediately adjacent to the area of direct laser beam impact, a temperature gradient governed by a complex set of pyrolysis reactions will occur, resulting in emission of higher relative molecular mass products.

Analytical systems based on pyrolysis-mass spectrometry

Analytical systems based on Py-MS are intended to perform pyrolytic reactions and to analyse the composition of the resultant pyrolysate by mass analysis of ions formed through the ionization of its components molecules. The structural complexity of the spectra recorded is usually considerable, owing to the formation of fragments with the same nominal mass from originally different components and the absence of reliable separation procedures. In addition, the observed complexity originates from secondary fragmentation processes, hence various techniques have been used that minimize these

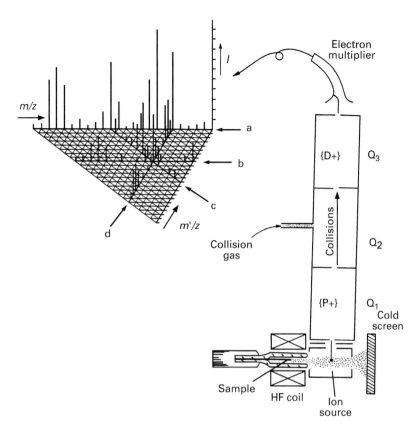

Figure 8 Curie-point pyrolysis MS-MS system: pyrolysis products diffuses into the ion source where the ions are formed by electron-impact (EI) ionization and analysed by a triple quadrupole mass analyser (Q1–Q3). Different modes of ion analysis: (a) EI spectrum of a pyrolysate; (b) neutral loss spectrum; (c) single ion spectrum; and (d) collisionally induced dissociation spectrum of a 'parent' ion {P+}.

effects. These include low-voltage electron ionization (LVEI) with 10–15 eV electrons instead of the standard 70 eV; chemical ionization (CI) with different reagent ions under vacuum or at atmospheric pressure and field ionization (FI). Despite the complexity of the mass spectra recorded, here referred to as mass pyrograms, even low-resolution spectra frequently provide structural information through the presence of characteristic ions or ion series. However, to provide clarity, many other techniques can be used, e.g. time/temperature resolved thermal degradation/pyrolysis MS, high-resolution mass spectrometry (HRMS) and tandem in space or tandem in time mass spectrometry (MS^n, $n \geq 2$) following collisionally activated dissociation of selected ions.

A schematic overview of some of the most frequently reported experimental arrangements of pyrolysis zones with respect to ionization regions is given in **Figure 9**. In configurations A–C the reaction zones are under vacuum whereas in D and E pyrolytic regions are kept at near-ambient pressures and for F at high pressure. Examples of integral zones include pyrolysis-field desorption in

the ion source of a mass spectrometer (**Figure 10**), laser pyrolysis/photolysis/desorption/ionization mass analyser (LAMMA) and, to some extent, in-source pyrolysis chemical ionization MS. This type of configuration allows the detection of large, and frequently highly informative, molecular or fragment ions from nanogram quantities of a sample. The use of field-desorption ionization coupled with HRMS, as well as field ionization techniques which have been developed to produce a high intensity of molecular ions and to reduce fragmentation of polar compounds, are well suited for the molecular characterization of polymeric building blocks. This approach is characterized by a very short residence time of the pyrolytic sample in the reaction zone (10^{-11} s), thus allowing for detection of short-lived primary products.

The other way to avoid losses of higher relative molecular mass components and less volatile thermal fragments, which are unable to enter the ion source owing to the condensation, is to place a pyrolysis probe very close to the ionization source, thus providing minimal separation between reaction and

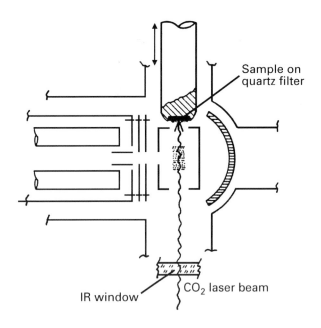

Figure 11 Instrumental configuration for laser pyrolysis EI ionization mass spectrometry. Reproduced by permission of Elsevier Science from Meuzelaar HLC, Windig W, Huff SM and Richards JM (1986). *Analytica Chimica Acta* **190**: 119–132.

Figure 9 Schematic representation of six on-line pyrolysis-mass spectrometry configurations. (⁎) Ionization zone; (•) pyrolysis zone.

ionization zones. Typical examples of this type of configuration are shown in **Figures 4, 5, 8 and 11**; however, vacuum TG-MS systems also fulfil the criteria of the configuration in **Figure 9B**. The so-called in-source pyrolysis produces compounds that can be ionized either under EI, CI or FI conditions. The application of such techniques for analysis of bacteria, lignocellulosic materials and polysaccharides has been documented, indicating additional advantages of this technique derived from the slower heating rate and the absence of mixing in an expansion chamber that represent the type of configuration shown in **Figure 9C** and includes a specially designed, fully automated Curie-point Py-MS system with an expansion chamber (**Figure 7**) but frequently represents the result of adapting an existing gas inlet manifold for on-line reaction studies.

The temperature-resolved data acquisition mode of operation provides additional layers of information. The evaluation of time-resolved mass spectral data of temperature-programmed pyrolyses include the shape of the total ion current profile, variations of the average relative molecular mass of volatilized pyrolysis products as well as the selection of characteristic mass signals, typical for distinct pyrolysis intervals. In addition, the differentiating key signals can be easily found and used for further investigation, e.g. temperature-resolved profiles of single ion intensities or specific sets of ion profiles can be selected to study pyrolysis processes of distinct sample components. Such components can be selected using chemometric techniques, e.g. factor or principal component analysis of time-resolved series of spectra combined with the 'numerical extraction' of chemical components, as revealed by the calculated mass spectra.

The identity of mass peaks or assignment of chemical structures to mass signals is complicated by the presence of isobaric fragment ions and isomeric structures of molecular ions. However, the presence of isobaric ions with a different composition can be investigated by HRMS, and pyrolysis product identification may further be supported by using MS-MS and/or GC-MS methods. Nevertheless, the combination of various ionization techniques, HRMS and, especially, MS-MS (**Figure 8**) can provide the extra

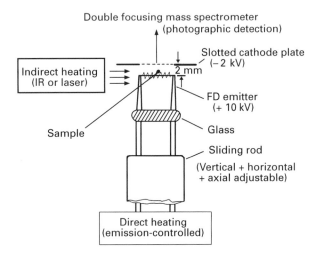

Figure 10 Field desorption (FD) pyrolysis in the ion source of a mass spectrometer. The place of pyrolysis and ionization is identical. Reproduced by permission of Elsevier Science from Schulten H-R (1977) Pyrolysis field ionization and field desorption mass spectrometry of biomacromolecules, microorganisms, and tissue material. In: Jones CER and Cramers CA (eds) *Analytical Pyrolysis*. Amsterdam: Elsevier Science.

specificity required for the analysis of complex materials.

If the pressure drop across the orifice or short capillary tube in a mass spectrometer sampling system exceeds a critical value of about 2.5, the flow velocity will reach the speed of sound and the flow beyond the exit expands into a supersonic 'free' jet, thereby producing a molecular beam (MB). This process is characterized by extreme collisional as well as internal energy (rotational and vibrational) cooling caused by isentropic and adiabatic expansion of a gas after crossing the sampling orifice (**Figure 9D**). A typical molecular-beam mass spectrometer (MBMS) sampling system coupled to a pyrolysis vapour generator is shown in **Figure 12**. It consists of a conical, extractive sampling probe, a skimmer to collimate the molecular beam and a line-of-sight MB inlet into the ion source of the mass spectrometer. Ions formed by EI ionization are directly mass analysed – or additionally studied using collisionally induced dissociation – with quadrupole mass filters or other mass spectrometric analysers.

The remaining two configurations (**Figures 9E and 9F**) are characterized by inherently lower molecular conductances and are usually referred to as pyrolysis GC-MS or represent a form of mass

spectrometric monitoring of high pressure reactors, including industrial-scale units.

Reproducibility and data analysis procedures

Short-term reproducibility of pyrolysis-mass spectra recorded during Py-MS studies is usually good and variations of peak intensities to within 5% are readily achieved. However, interlaboratory and long-term reproducibilities are not well evaluated. Some of the factors that influence long-term reproducibility and which should be carefully evaluated in interlaboratory comparisons are presented in **Table 2**.

Progress in the design of mass spectrometers and the availability of online computer systems has allowed the integration of Py-MS data acquisition with multivariate mathematical data reduction methods into a single analysis technique. Such an approach combines rapid analysis capability with expert system or pattern recognition based data evaluation (**Figure 13**).

Applications of Py-MS methods

Natural products

Natural products in general, and organic natural products in particular, are characterized by a high degree of complexity, thus requiring highly specialized analytical methods which tend to be labour intensive, lengthy and expensive. However, in the early 1960s the search for extraterrestrial life, as part of the new space probe programmes, prompted the development and application of pyrolytic methods to complex biomaterials and microorganisms. The combination of pyrolysis and mass spectrometry was found to be an especially powerful tool

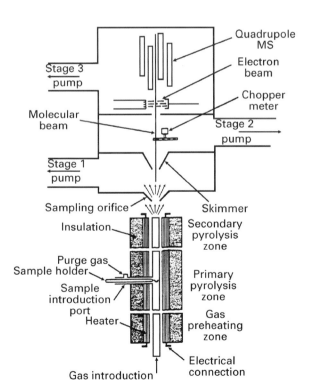

Figure 12 Schematic of a pyrolysis vapour generator coupled to a molecular-beam mass spectrometer sampling system. Reproduced by permission of the American Chemical Society from Evans RJ and Milne TA (1987) *Energy and Fuels* **1**: 123–137.

Table 2 Factors that possibly influence long-term reproducibility

Sample preparation	Filament heating	Product transfer	Mass analysis
Cleaning of sample carrier	Temperature rise time	Inlet temperature	Ionization
Solvent/suspending liquid	Equilibrium temperature	Residence time	Extraction
Sample size	Total heating time	Surface activity	Transmission

Reproduced by permission of Elsevier Science from Meuzelaar HLC, Haverkamp J and Hileman FD (1982). *Pyrolysis Mass Spectrometry of Recent and Fossil Biomaterials; Compendium and Atlas*, Amsterdam: Elsevier Science

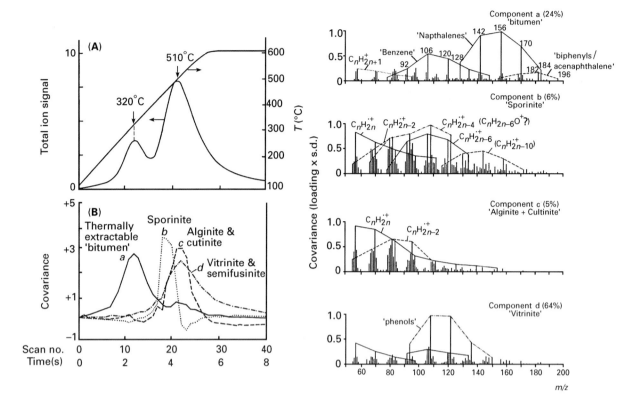

Figure 13 Time-resolved Curie-point pyrolysis MS total ion signal profile of a coal sample obtained at a heating rate of 100 K s⁻¹ (A) and the 'deconvolution' of the second maximum (at 510°C) into at least three components (b–d) by means of factor analysis (B). Tentative interpretation of components a–d was based on comparison of numerically extracted spectra with the actual spectra of maceral concentrates. Numbers in parentheses show the percent of the total variance for each component. Reproduced by permission of Plenum Press from Meuzelaar HLC, Yun Y, Chakravarty T and Metcalf GS (1992) Computer-enhanced pyrolysis mass spectrometry: a new window on coal structure and reactivity. In: Meuzelaar HLC (ed) *Advances in Coal Spectroscopy*. New York: Plenum Press.

to produce structural information about the molecular building blocks of many natural products that are solid, high relative molecular mass, insoluble, homo- or heteropolymeric materials. Moreover, advantages of Py-MS methods include sensitivity, specificity and speed. However, it should be noted that inorganic substances, such as salts and metal cations, may have a catalytic influence on the formation of organic pyrolysis products and that highly heterogeneous, cross-linked materials will not yield 100% pyrolysate but will 'disproportionate' into volatile matter and char. For example, it was shown that alkali metals affect the thermal activation of polysaccharides, which can result in a total loss of oligomer information.

To make quantitative estimates of the portion of sample analysed, it is advisable to determine the amount and approximate composition of the pyrolysis residue. Generally, volatilization tends to decrease with the increasing polarity of a material – owing to the presence of additional intermolecular forces. As a consequence, pyrolysates of complex

natural materials, e.g. soil particles, will under represent functional groups containing polarized bonds. To overcome these problems, various on-line and off-line derivatization techniques have been developed. For example, addition of a tetramethylammonium hydroxide (TMAH) solution to a sample deposited on a pyrolytic wire allows a wide range of O–, N–, S– and P–containing polar compounds to be analysed owing to the largely increased volatility of methylated products formed during the pyrolytic methylation step. However, in some cases, by using Py-MS systems with integrated reaction and ionization zones, high molecular mass oligomers can be recorded, as illustrated in **Figure 14**. In this case a series of ions representing 1,6-anhydro-oligosaccharides, ranging from monomer to dodecamer, obtained by in-source pyrolysis chemical ionization of cellulose can be observed.

In other cases, highly specific derivatives may be produced under pyrolytic conditions. For example, diketopiperazines (DKPs) are formed during the thermal decomposition of oligopeptides that range

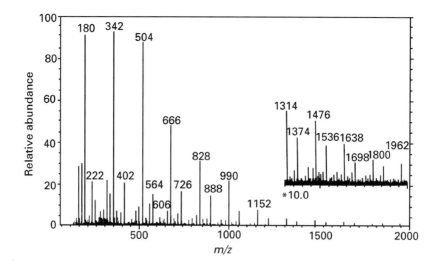

Figure 14 In-source pyrolysis ammonia chemical ionization mass spectrum of cellulose, showing pseudomolecular ions [MNH$_4$]$^+$ of a series of 1,6-anhydro-oligosaccharides, ranging from the monomer (m/z 180) to the dodecamer (m/z 1962). The spectrum was obtained using a Pt–Rh filament probe and an E–B-type sector instrument. Reproduced by permission of Elsevier Science from Boon JJ (1992) Analytical pyrolysis mass spectrometry: new vistas opened by temperature-resolved in-source PYMS. *International Journal of Mass Spectrometry and Ion Processes* **118/119**: 755–787.

from two to six amino acids. In **Figure 15A** a pyrolysis mass spectrum of product ions derived from the parent ion at m/z 260, obtained during Py-MS of phenylalanyl-leucine is presented, whereas **Figure 15B** shows general fragmentation pathway of these dipeptide-derived ions. The results showed a variability in the relative intensities of the various DKPs which depends on the amino acids involved in the cyclization process and their position in the peptide. Identification of pure dipeptides from DKP formation observed in Py-MS is rapid and involves no derivatization or complicated sample preparation procedures.

Bacteria

Classification and identification of microorganisms based on rapid and specific methods for determining the chemical composition have proved to be a valuable approach in microbiology. The identification of many chemical constituents that show a degree of specificity suitable for chemotaxonomic and/or diagnostic purposes made it clear that improvements in methods of sample preparation play an important role in fully utilizing the high speed and specificity offered by MS.

Py-MS is commonly applied for bacterial fingerprinting, i.e. classification of characteristic signal patterns followed the first reports published in the 1960s on the applicability of analytical pyrolysis techniques to clinical and pharmaceutical microbiology. Application of Py-MS as an independent tool for the char-

acterization and identification of bacteria represents an alternative approach that may provide important information for the discrimination of closely related microbial strains that is difficult to obtain by other techniques. Hence, the objectives of most projects were focused on the differentiation and classification of bacterial strains at either the genus, species or subspecies level and subsequent identification of unknown strains. The structural information which can be obtained from the pyrograms and the quantitative nature of the data allow them to be successfully used not only for both the chemical characterization of whole cells and cellular components and also in quality control and screening.

Important advantages of Py-MS techniques are the small sample size required, the high sample throughput, and the ready compatibility of the data with computerized evaluation. To classify the Py-MS profiles of various bacteria an expert system was constructed using multivariate statistical analysis. However, various aspects should be taken into consideration during application of this technique in microbiology, i.e. instrument and biological. Standardization of instrumental conditions is essential to achieve an acceptable level of long-term and interlaboratory reproducibility while standardized culturing and sampling are required to minimize biological heterogeneity among the samples. Sampling directly from the solid medium, and thus avoiding further manipulations, proved to be superior to methods involving the washing of harvested bacteria to remove culturing media, often followed by lyophylization

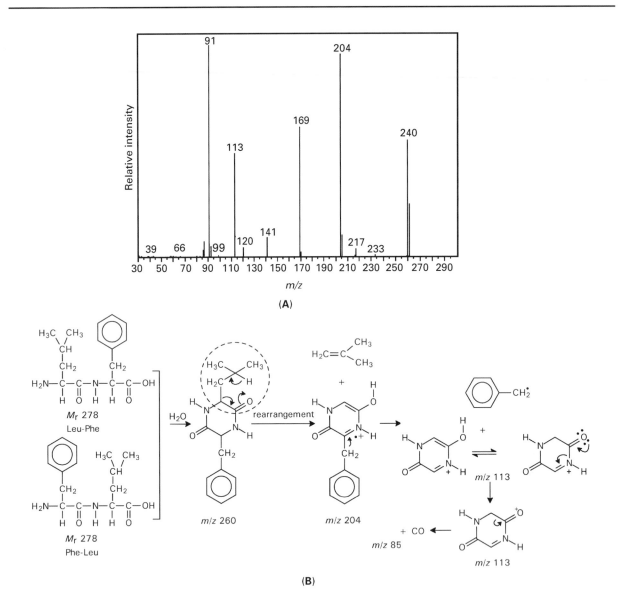

Figure 15 Pyrolysis mass spectrum of product ions of m/z 260 of phenylalanyl-leucine [Phe-Leu] (A) and a general fragmentation pathway of Phe-Leu and leucyl-phenylalanine [Leu-Phe] (B). Reproduced by permission of the American Society for Mass Spectrometry from Noguerola AS, Murugaveri B, Voorhees KJ (1992), *Journal of the American Society for Mass Spectrometry* **3**: 750–756.

and resuspension of cells in sterile water. In addition, the pyrolytic derivatization approach allows one to utilize chemical derivatization for some groups of less volatile compounds to convert them into a suitable form for MS analysis.

In the 1990s the search for characteristic pyrolysis products of complex biomolecules with the use of GC-MS and MS-MS instruments has firmly established the usefulness of analytical pyrolysis techniques when used in a biochemical marker detection, rather than in a 'fingerprinting' mode. Either direct pyrolysis followed by analysis of thermal fragmentation products or, alternatively, chemical derivatization followed by analysis of the derivatized products

can be used to transform nonvolatile polar and macromolecular components into characteristic chemical markers with sufficient volatility for MS analysis. The success of analytical pyrolysis was based on its abilities to detect compounds that are unique for particular groups of microorganisms.

For example, diacylpropenodiols, originating as pyrolytic products of bacterial phospholipids (PLs), under the standard conditions of EI ionization form molecular ions suitable for the evaluation of the acyl residue composition of PLs (number of carbon atoms and double bonds). In addition, the types of fatty acids involved in the composition of PLs may be determined using fragment ions of the general

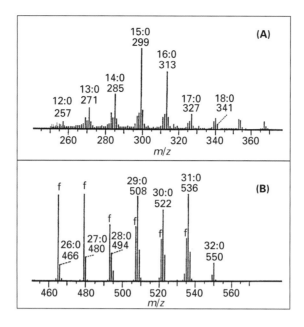

Figure 16 Curie-point pyrolysis EI mass spectrum of *B. anthracis* showing (A) fragment ions of a series [R–C=O + 74]⁺, representing the contribution of particular types of fatty acid residues to the overall profile of bacterial fatty acids, and (B) molecular ions of dehydrated diacylglycerols (diacylpropenodiols) originating from cellular phospholipids.

formula $[R-C=O + 74]^+$ (**Figure 16**). To perform detailed structural determinations of individual fatty acids, pyrolytic in situ hydrolysis-methylation of whole cells with TMAH was applied. This procedure, performed on a micro-scale on the surface of a pyrolytic wire, was shown to produce results which were qualitatively and quantitatively identical, or near-identical, to those from conventional methods. However, the Py-MS approach allows one to shorten the time needed to obtain this type of chemotaxonomic information by a few orders of magnitude. In addition, this method provides additional information, owing to the possibility for simultaneous analysis of other cell components, e.g. mycolic and mycocerosic acids from mycobacteria or of dipicolinic acid methyl ester from bacterial spores in addition to a plethora of pyrolytic products of proteins, sugars and nucleic acids.

Synthetic polymers

Bond cleavages that occur as a result of thermal excitation of all vibrational modes of the polymer lead to the formation of macroradicals that undergo secondary reactions via intra- or intermolecular mechanisms. Investigations of polymer decomposition mechanisms using Py-MS have been widely used to characterize the polymer microstructure and the accumulated knowledge applied to unknown polymers and to investigations of polymer decomposition products.

Plastics and rubbers are mainly copolymers or blends of different polymers, and commercial products contain other additives, namely, plasticizers, antioxidants, pigments, cross-linking agents, flame retardants and many other components that are used to obtain desired physical properties of polymeric products. Hence, the Py-MS technique can be used as a method to quickly examine an unknown polymer sample to determine its composition, thus providing a rapid means of identifying copolymers, blends and additives in industrial rubbers and plastics. The investigation of polymer decomposition mechanisms allows for the qualitative and quantitative characterization of unknown polymers and Py-MS should be considered as a useful tool for studies of polymer micro-structure. In **Figure 17** the results of temperature-programmed pyrolysis-MS analysis of the 'sheen' on anti-static matting are presented – showing the capabilities of a Py-MS system to perform rapid structural investigations of polymers.

In spite of the complexity of the starting material, Py-MS allows one to analyse even microgram quantities of analytes thus providing the means for the investigation of heterogeneous products and for direct analysis, circumventing the need for dissolving or pre-separation of components. In addition, different parts of heterogeneous materials may be studied separately and pyrogram patterns may be used to compare related materials. Thus, the Py-MS technique can also be used for rapid investigation of

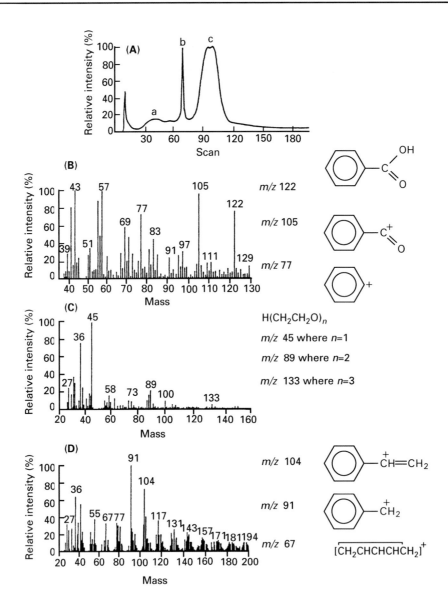

Figure 17 Temperature-programmed pyrolysis-mass spectrometry analysis of the sheen on anti-static matting: (A) temperature-resolved total ion current signal profile of pyrolysis products obtained; (B) mass spectrum of component a; (C) component b; (D) component c. Reproduced by permission of Elsevier Science from Mundy SAJ (1993), *Journal of Analytical and Applied Pyrolysis* **25**: 317–324.

polymer sample to identify copolymers, blends, and selected additives for forensic applications.

Flash desorption of oligomers relies on kinetic competition between evaporation and thermal decomposition. Molecular ion signals from the MS have been used to determine the average molecular masses and distribution of oligomers, whereas expert systems can be used to establish mechanisms for the thermal degradation of polymers, e.g. to determine the relationships between polymer structure and the corresponding Py-MS spectra.

See also: **Atmospheric Pressure Ionization in Mass Spectrometry; Biochemical Applications of Mass Spectrometry; Chemical Ionization in Mass Spectrometry; Chemical Structure Information from Mass Spectrometry; Forensic Science, Applications of Mass Spectrometry; Hyphenated Techniques, Applications of in Mass Spectrometry; Laser Microprobe Mass Spectrometers; MS–MS and MS^n; Peptides and Proteins Studied Using Mass Spectrometry; Quadrupoles, Use of in Mass Spectrometry.**

Further reading

Boon JJ (1992) Analytical pyrolysis mass spectrometry: new vistas opened by temperature-resolved in-source

PYMS. *International Journal of Mass Spectrometry and Ion Processes* **118/119**: 755–787.

Brown SD and Harper AM (1993) Multivariate analysis of time-resolved pyrolysis mass spectral data. In: Wilkins CL (ed) *Computer-Enhanced Analytical Spectroscopy*, Vol 4, pp 135–163. New York: Plenum Press.

Irwin WJ (1982) *Analytical Pyrolysis. A Comprehensive Guide*. New York: Marcel Dekker.

Meuzelaar HLC, Haverkamp J and Hileman FD (1982) *Pyrolysis Mass Spectrometry of Recent and Fossil Biomaterials; Compendium and Atlas*. Amsterdam: Elsevier Science.

Schulten H-R and Lattimer RP (1984) Applications of mass spectrometry to polymers. *Mass Spectrometry Reviews* **3**: 231–315.

Schulten H-R and Leinweber P (1996) Characterization of humic and soil particles by analytical pyrolysis and computer modeling. *Journal of Analytical and Applied Pyrolysis* **38**: 1–53.

Simmleit N and Schulten H-R (1989) Analytical pyrolysis and environmental research. *Journal of Analytical and Applied Pyrolysis* **15**: 3–28.

Simmleit N, Schulten H-R, Yun Y and Meuzelaar HLC (1992) Thermochemical analysis of U.S. Argonne premium coal samples by time-resolved pyrolysis-field ionization mass spectrometry. In: Meuzelaar HLC (ed) *Advances in Coal Spectroscopy*, pp 295–339. New York: Plenum Press.

Snyder AP (1990) Acrylic compound characterization by oxidative pyrolysis, atmospheric pressure chemical ionization-tandem mass spectrometry. *Journal of Analytical and Applied Pyrolysis* **17**: 127–141.

Voorhees KJ (ed) (1984) *Analytical Pyrolysis. Techniques and Applications*. London: Butterworths.

Voorhees KJ, Harrington PB, Street TE, Hoffman S, Durfee SL, Bonelli JE and Firnhaber CS (1990) Approaches to pyrolysis/mass spectrometry data analysis of biological materials. In: Meuzelaar HLC (ed) *Computer Enhanced Analytical Spectroscopy*, Vol 2, pp 259–275. New York: Plenum Press.

Wampler TP (ed) (1995) *Applied Pyrolysis Handbook*. New York: Marcel Dekker.

Wieten G, Meuzelaar HLC and Haverkamp J (1984) Analytical pyrolysis in clinical and pharmaceutical microbiology. In: Odham G, Larsson L and Mardh P-A (eds) *Gas Chromatography/Mass Spectrometry Applications in Microbiology*, pp 335–380. New York: Plenum Press.

Quadrupoles, Use of in Mass Spectrometry

PH Dawson, Iridian Spectreal Technologies Ltd., Ottawa, Ontario, Canada
DJ Douglas, University of British Columbia, Vancouver, Canada

> **MASS SPECTROMETRY**
> **Methods & Instrumentation**

Quadrupole mass spectrometers (often referred to as quadrupole mass filters because of the way they operate) are the most successful example of the class of mass spectrometers called 'dynamic'. Their performance depends upon a dynamic interaction of ions with time-varying electric fields. Most classes of spectrometers are 'static' and use either the interaction of fixed magnetic and electric fields or 'time-of-flight'.

The quadrupole mass spectrometer derives its name from the nature of the electric potential which is quadrupolar in the direction transverse to ion injection, i.e. dependent on the square of the distance from the centre of the field. The field is achieved by using four parallel rods as schematically illustrated in **Figure 1**.

This article begins with a brief history. It then explains the principles of operation of this mass spectrometer. The idealized view of its operation has to be tempered by consideration of some real-world situations that influence performance, such as the finite length of the field, the inevitability of fringing fields and field imperfections. Observations of performance and its limitations are illustrated. The applications section deals with residual gas analysis, gas chromatography and liquid chromatography mass spectrometry (GC-MS and LC-MS), collision induced dissociation using triple quadrupoles (MS/MS) and inductively coupled plasma mass spectrometry (ICP-MS) used for elemental analysis.

A history of development

The possibility of using quadrupole radiofrequency fields for mass analysis was first suggested in 1953 by Paul and Steinwedel and in a US government report by Post. The first practical implementation was by Paul and co-workers in 1958. This work became the foundation for the field. Wolfgang Paul was awarded a share of the Nobel Prize for physics in 1989 for the development of the ion trap technique that also used quadrupole fields.

Early quadrupole mass filters were very limited in mass range and resolution but their physical simplicity and the absence of a magnet made them attractive for upper atmosphere and space applications. Major development occurred in the 1960s inspired by this application. This same period also coincided with a rising demand for residual gas analysers because ultrahigh vacuum technology began to have routine

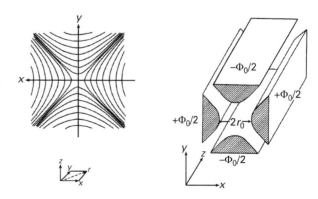

Figure 1 A schematic illustration of a quadrupole mass spectrometer. Ions that are to be filtered to identify the presence of a particular mass are injected in the direction of the axis of the instrument. If the combination of radiofrequency and direct voltages applied to the rods is correctly chosen, only ions of one particular mass to charge ratio (m/z) will be successfully transmitted to the detector.

application, first in research laboratories and later in production equipment, especially in semiconductor processing. Gradually the quadrupole mass filter became the dominant instrument for this application and remains so today.

Quadrupole performance slowly improved but in the 1970s new applications to organic analysis and particularly the implementation of the combination of GC and MS placed increasing emphasis on better understanding of how the instruments worked in order to overcome their limitations. Computer simulations of performance became important. Combined with detailed experimental analysis, these led to a much improved knowledge of real-world quadrupoles with fringing fields and field imperfections. An important advance came with the application of phase space dynamics for calculating quadrupole performance. The new advances were incorporated in the classic textbook of the field written by Dawson and various collaborators and published in 1976. This book was re-issued by the American Institute of Physics in 1995 as a 'paper-back classic'.

In the 1980s and 1990s, the limits to quadrupole performance have been pushed back by precision manufacture and careful source design. A mass range of up to 2000 or more is commonly achieved with unit mass resolution. For LC-MS applications the mass range may reach 4000. There have been equally significant improvements in trace analysis capability based on a combination of sensitivity and more perfect peak shapes. High-performance quadrupole mass spectrometer manufacture has come to demand very high precision.

Principles of operation

The perfect field

In a perfect quadrupole mass filter field, motion in the x and y (transverse) directions is independent. There is no field in the axial direction and motion is unchanged along the axis. Both x and y motion are governed by the Mathieu equation (see the Further reading section for the derivation of these equations);

$$d^2u/d\xi^2 + (a_u - 2q_u \cos 2\xi)\, u = 0$$

i.e. where u represents x or y and where

$$a_x = -a_y = 8zU/mr_o^2\omega^2, \quad q_x = -q_y = \frac{4zV}{mr_o^2\omega^2}$$

the applied voltages across the quadrupole are direct voltage U and an alternating radiofrequency voltage of $V \cos \omega t$ and the minimum separation between opposite pairs of electrodes is $2r_o$. The charge to mass ratio of the ion is z/m. Time is expressed by $\xi = \omega t/2$, where t is in seconds. For ion transmission the ions must have finite amplitude of oscillation in both x and y directions so that they do not strike the rods, i.e. the trajectories are both 'stable'.

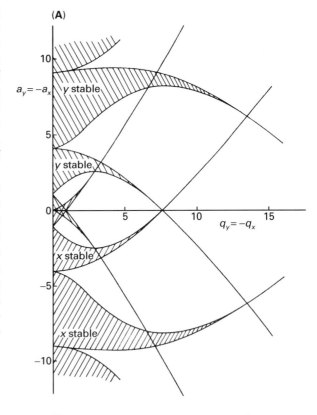

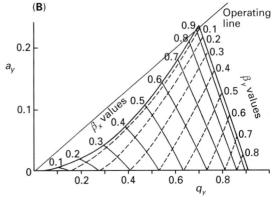

Figure 2 Zones for stable trajectories in both x and y transverse directions expressed in terms of the parameters a and q which are related to the m/z ratios of the ions. (A) General zones of stability, (B) a detail of the zone commonly used. The iso-beta lines are related to frequencies of ion oscillation.

Combinations of a and q values that give stable motion are shown in **Figure 2**. There are several areas of simultaneous stability. The one near the origin is commonly used but the higher zones have been examined both theoretically and experimentally especially in the search for peak shapes with more abrupt fall-off at the edges for applications in trace analysis. The quadrupole uses a sinusoidal alternating field. In principle any alternating field is possible but the higher harmonics may lead to complexity in the behaviour of the ions. The RF frequency is generally in the range of 1–2 MHz.

Mass selection is obtained by choosing a ratio of q/a such that only a narrow region of the stability zone near the apex ($a = 0.23699$ and $q = 0.706$) is intersected by the operating line using $q/a =$ constant. For a given RF and DC voltage, ions of different mass to charge ratio (m/z) appear at different points along the line. Mass scanning is carried out by altering the values of U and V while maintaining their ratio constant to bring ions of different m/z into the tip of the stability region. This would give a spectrum of constant resolution ($M/\Delta M$). In practice a resolution that increases with mass is preferred and the ratio q/a is adjusted electronically throughout the mass scan in order to achieve more or less constant peak width. An alternative to voltage scanning would be to scan the frequency but this is rarely done because of technical difficulties in covering a large mass range.

This simple description implicitly assumes that the length of the quadrupole is infinite so that all ions with stable trajectories are differentiated from those with unstable trajectories. In practice, resolution may be limited by the length of the field. It is found that $R_{max} = n^2/h$, where R_{max} is the maximum attainable resolution and n is the number of RF cycles the ions spend in the field. The parameter h depends upon the source and on the fringing fields. A value of 25 is not unusual; a value of 10 would be an excellent performance. High resolution is favoured for ions of low axial velocity (lower energy ions but, fortunately, also higher mass ions). One is limited, however, in reducing axial ion energies by the detrimental influence of fringing fields.

The real world: transmission and fringing fields

In the real world, there are inevitably fringing fields at the entrance to and the exit from the mass filter. Ion motion in these fields can be very complex and the motion in each of the three coordinate directions becomes coupled. Very low energy ions may be reflected or even trapped in the fringing fields. However, a good theoretical approximation in most cases

is to assume a linear increase of the x and y direction fields as the entrance is approached. The fringing fields extend over a distance comparable to the filter radius. The y-direction (a,q) values in the fringing field correspond to intrinsically unstable ion trajectories. If too much time is spent in the fringing fields, ion amplitudes will increase and the effective aperture of the spectrometer will be reduced.

The finite diameter of the field (r_0) means that ions are only 'accepted' for transmission when they enter the field with a small initial transverse displacement from the axis and small transverse velocities. The combination of displacement and transverse velocities that are possible defines the 'acceptance' of the instrument. The acceptance and so the overall sensitivity becomes smaller as the resolution is increased. The acceptance is calculated using phase space dynamics.

The sensitivity of the mass spectrometer is best expressed in terms of a phase space diagram as shown in **Figure 3**. This shows the initial combinations of

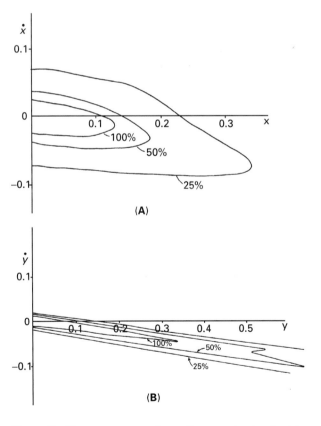

Figure 3 Phase space acceptance diagrams showing the initial conditions of transverse displacement and velocities that lead to ion transmission at 100% and 50% of the initial phases of the RF field. (A) x-direction, (b) y-direction. This is for the centre of a peak and for ions spending two RF cycles passing through the fringing fields. The displacement from the axis is measured in units of r_0 and the velocity in terms of r_0/ξ.

transverse position and transverse velocity that will result in ion transmission. In practice, this depends upon the initial phase of the field when the ion first experiences the field. The Figure shows the x and y 'acceptances' for transmission in 100% of the initial phases, 50%, and so on. At higher resolutions the acceptance area decreases.

At a given resolution, combining the x and y acceptances together gives an indication of sensitivity for different numbers of RF cycles spent in the fringing fields giving a diagram such as that in **Figure 4**. This shows that fringing fields can be advantageous if properly chosen. There have been many attempts to tailor fringing fields for optimum performance, such as using retardation of ions after they enter the field or by adding RF only sections at the ion entry (the 'delayed DC ramp'), or by using specially shaped electrodes.

If the relative ion transmission is measured versus resolution for a particular mass number, there will be a curve such as that in **Figure 5**. If the source is

evenly illuminated, at low resolution the source emittance will be less than the quadrupole acceptance and the transmission will vary little with resolution. Also the peaks will tend to be flat topped. At some point the acceptance will become limiting and transmission will tend to fall with the square of the resolution (taking fringing fields into account). Finally, at the length limitation of the quadrupole the transmission drops abruptly with resolution. Curves (a) and (b) illustrate different ion energies. In some cases, the source emittance may be rather diffuse or may not be evenly illuminated, then quite different transmission versus resolution behaviour will be observed.

The real world: field imperfections

There are inevitably other departures from the perfect fields and these become more and more important at high resolution. Displacement of one or more rods from the ideal position is the simplest case. This leads to higher order terms in the expression for the electric potential. One rod displaced would give predominantly third-order or hexapole correction terms. Opposite rods displaced gives fourth-order or octopole terms. These imperfections limit the resolution that is attainable, i.e. even if n is increased, the resolution will not increase further. There have been various theoretical and more limited experimental studies of these effects. The field faults may also cause badly shaped or even split peaks because of nonlinear resonances in the ion oscillations at certain critical (a, q) values.

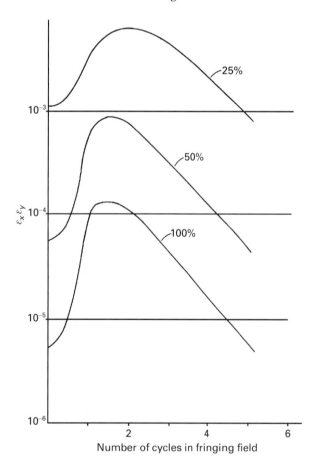

Figure 4 Combined phase space acceptance areas for x and y directions as function of the length of the fringing field (expressed as the number of RF cycles that the ions spend within the fringing field). This illustrates how sensitivity may vary depending on ion velocity.

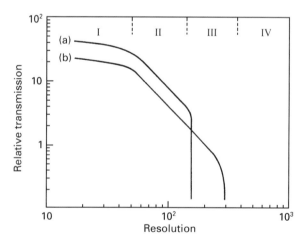

Figure 5 A typical example of transmission efficiency versus resolution for an instrument with a well-defined source emittance. There are three regions of transmittance as resolution is increased. (I) Sensitivity is source limited, (II) sensitivity is filter acceptance limited, (III) resolution is length limited (number of cycles in the field) and (IV) resolution is limited by field imperfection.

Other field faults may arise from nonparallel rods (bending or bowing) or from errors in the electrical waveform. It is common to substitute round rods in the quadrupole for the ideal hyperbolic ones. Round rods are easier to precision-manufacture. If the diameter and positioning of the rods is correctly matched, field faults are minimized and only sixth-order distortions are produced. These are not expected to significantly influence performance. However, there is controversy over this choice of round and hyperbolic rods with many assertions being made but little solid experimental evidence. One confusing factor is that other field faults (e.g. from rod positioning) may be more serious when using round rods.

Observations of limits to performance

Figure 6 illustrates some of the data from an extensive examination of performance of a particular quadrupole. These data demonstrate most of the features discussed above. In addition, there is an ultimate limit to achievable resolution set by the perfection of the quadrupole field.

Similar performance data have been generated for operation using other regions of the stability diagram such as near $a = 0$, $q = 7.5$ or $a = 3$, $q = 3$. These regions may have interest for specialized application.

Applications

Residual gas analysis

One of the first uses of quadrupole mass filters was for residual gas analysis in high-vacuum chambers. An electron beam ionizes the background gas and a mass spectrum is recorded with the quadrupole. This allows determination of the composition of the background gas and, after calibration, the partial pressures of each of the components, such as N_2, O_2, H_2O, etc. Because only light gases are generally involved, the mass range of the quadrupole need only be 100–200 m/z. The compact and rugged construction of a quadrupole with purely electronic scanning make quadrupoles particularly attractive for this application. Often the quadrupole is mounted on a flange which is simply bolted onto the vacuum chamber. Sometimes a pressure reduction stage is used. Modern systems have computer controlled scanning and allow the possibility of searching libraries to match an unknown spectrum.

GC-MS

The largest fraction of quadrupole mass spectrometers sold today are used as detectors for GC to identify

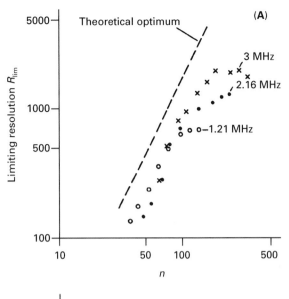

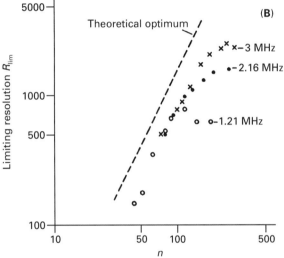

Figure 6 Some examples of performance measurement for a particular quadrupole mass filter showing how the limiting resolution varied with the number of RF cycles in the field. The ultimate resolution reached was dependent on frequency and mass number.

and quantify trace levels of organic compounds. Environmental analysis and drug testing of athletes, for example, rely extensively on GC-MS. Organic compounds separated by a gas chromatograph elute into the ion source of a quadrupole mass filter. Ions are formed either by electron impact (EI) or chemical ionization (CI). In EI, positive molecular and fragment ions are usually formed. The resulting mass spectrum gives a fingerprint of the compound. Unknown compounds can be identified by searching a library of spectra. In CI, analytes react with a reagent ion present in excess to produce either positive or negative molecular adduct ions, usually with minimal fragmentation. The combination of the retention time on the chromatograph and the appropriate molecular mass is often sufficient to identify trace analytes.

Quadrupole mass filters have become the standard for GC-MS because they are easily interfaced to computers, scan rapidly on a time scale compatible with peaks eluting from a GC, require only medium vacuum (10^{-5} mbar), are compact, and are of comparatively low cost. Gas chromatography is restricted to relatively volatile compounds with moderate molecular masses and so the m/z range of a quadrupole used as a detector for GC is usually ~500–1000. A complete EI spectrum can be obtained on ~10 pg of an analyte. If only ions of one m/z are monitored ('single ion monitoring'), with CI, the detection limits can be lowered to low femtogram levels. Alternatively a few selected m/z values (say four) corresponding to the major peaks in the spectrum of a targeted compound can be monitored by switching the quadrupole between these m/z values without scanning intervening regions ('multiple ion monitoring'). The ability to switch a quadrupole from full scans to single ion monitoring to improve detection limits is an advantage over other methods where a complete spectrum must be acquired (TOF, ion trap ICR). A quadrupole system dedicated to GC/MS can be quite compact, often smaller than the GC itself.

LC-MS

To separate and detect less volatile, more polar, more labile or higher molecular mass compounds, the GC is replaced by a LC. A number of ion sources have been used for LC-MS but two dominate today: atmospheric pressure chemical ionization (APCI) and electrospray ionization (ESI). In APCI, solvent and analytes eluting from the LC are sprayed into a heated tube at atmospheric pressure where they rapidly vaporize. Ions of the solvent are formed in a corona discharge. Typically in positive-ion mode these are protonated. These ions then transfer charge to analytes to produce molecular ions. In ESI the solution eluting from the LC is passed through a metal capillary that has a high voltage applied to it (3000–5000 V). Charged droplets emerge from the capillary tip at atmospheric pressure and lose solvent through evaporation, leading to the formation of gas phase ions characteristic of the ions in solution. Compounds that are present as simple ions in solution (M^+, M^- or MH^+, MH^-) give the same ion in the gas phase. Protein ions with molecular masses 5000–100 000 acquire multiple charges to produce ions with m/z ratios of <4000 (**Figure 7**) that can still be analysed by a quadrupole. Molecules with

intermediate molecular masses produce ions with a few charges, depending on the number of basic residues and their pKa values. Conventional ESI operates best at flow rates of 1–10 µL min^{-1}. LC flow rates are usually considerably higher (~1 mL min^{-1}) so the output of the LC is often split, with a fraction of the flow going to the ESI source.

APCI and ESI produce ions at atmospheric pressure. These are transferred into the vacuum system of a quadrupole mass filter with two or more stages of differential pumping. The ability to operate a quadrupole at moderate pressures of ~10^{-5} torr means only modest, lower cost vacuum pumps are required. ESI and APCI generally produce protonated molecular ions. As in GC-MS, for targeted compound analysis the quadrupole is operated in 'single ion' mode to monitor the m/z of interest. Fragment ions can be formed by applying high electric fields to the ions in the ion sampling region. These fields accelerate the ions through the locally formed high density of gas causing collision-induced dissociation. If the system is operated in this mode a full scan over the spectrum of an analyte can be obtained. Alternatively multiple ion monitoring can be used for a few m/z values of interest and detection limits of a few picograms of organic compounds are possible.

A major application of LC-MS is to identify proteins. The molecular mass of the protein is

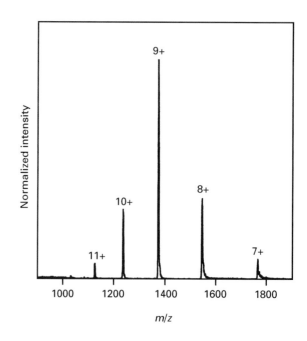

Figure 7 The mass spectrum of the protein cytochrome c (M_r 12 200) obtained with an ESI ion source and quadrupole mass filter. The isotopic structure of the peaks is not resolved. Each peak is identified by the number of protons attached to the protein.

determined by ESI-MS. A proteolytic enzyme such as trypsin is then used to cleave the protein into peptides. These are separated by LC and their molecular masses determined. These molecular masses, the molecular mass of the protein and specifying the residues cleaved by the enzyme can be used to search libraries to identify an unknown protein. If this is insufficient, some sequence information on the peptides can be obtained by tandem MS (see below).

Peptides often produce ions with two to four charges. Although quadrupoles are normally operated at unit resolution, sufficient to resolve peaks differing by one m/z, the resolution can be increased to separate the isotopic peaks of multiply charged ions. The spacing of these peaks allows the determination of the charge state directly (e.g. triply charged ions have isotopic peaks spaced by 0.33 m/z). Quadrupoles have demonstrated sufficient performance to resolve isotopic peaks of up to +4 ions in the range $m/z < 2000$. This is usually more than sufficient for peptide analysis. A higher mass range is required for LC-MS and this has pushed quadrupole performance to new limits. Current systems have an m/z range of 2000–4000.

Triple quadrupole mass spectrometers (MS/MS)

In tandem mass spectrometry (MS/MS) a first mass analyser selects an ion from a mixture, the ion is fragmented by collision, and a second mass analyser produces a spectrum of the fragment ions. MS/MS is used to determine ion structure and to detect and quantify targeted compounds in complex mixtures.

MS/MS can be carried out with a triple quadrupole system such as that shown in **Figure 8**. A first mass analysing quadrupole, Q1 mass selects a 'precursor' ion from the ESI source. The ion enters the collision cell with energies typically 10–500 eV. Here collisions with a neutral gas such as N_2 or Ar at a pressure 10^{-4} to 10^{-2} mbar transfer translational energy to the internal energy of the ion. It then undergoes unimolecular reaction to produce fragment or 'product' ions (collision-induced dissociation). Ions are confined to the collision cell by a quadrupole, Q2, operated with only a radiofrequency voltage between the poles. A broad range of ions with $q < 0.9$ have stable trajectories and are transmitted to the exit of the collision cell. They are then mass analysed in quadrupole Q3. The system in **Figure 8** shows an additional quadrupole Q0. This also operates in RF only mode and acts as an ion guide to transport ions from the skimmer to the first mass analyser.

Early triple quadrupoles were unable to efficiently extract higher mass ions from Q2 and at the same time to attain unit resolution in Q3 because of high fragment ion energies. A solution to this problem was found when it was recognized that by increasing the pressure in the collision cell, product ions have additional collisions with neutrals. This causes them to lose radial and axial kinetic energy, i.e. to 'cool', and to move to the centre of the RF quadrupole. They are then well within the acceptance of Q3 and the transmission increases. In addition, product ions emerge from Q2 with energies and energy spreads of only about 1 eV. With a modern triple quadrupole system, at least 50% of the precursor ions that are transmitted by Q1 can be converted to fragment ions that are transmitted through Q3 with unit resolution or better. Resolution of the isotopic peaks of up to +4 ions has been demonstrated (**Figure 9**). The high collision pressure also minimizes any ion focusing effects which could lead to variable transmission. Hexapole or octopole fields have also been used to confine ions in the collision region.

A triple quadrupole has many scan modes. In 'product ion' scans, Q1 mass-selects an ion from a

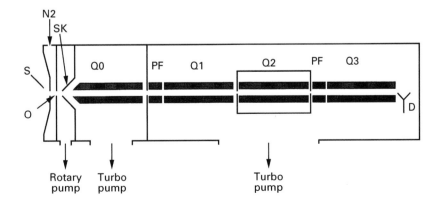

Figure 8 A triple quadrupole mass spectrometer system with an electrospray ion source. S, electrospray source; N2, nitrogen curtain gas; O, ion sampling orifice; SK, skimmer; Q0, RF-only quadrupole ion guide; PF, delayed DC ramp 'prefilters'; Q1 mass analysing quadrupole; Q2, RF quadrupole enclosed in a collision cell; Q3 mass analysing quadrupole; D, ion detector.

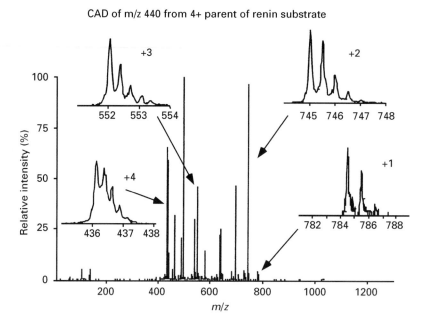

Figure 9 Mass spectra of multiply charged fragment ions of the peptide renin substrate tetradecapeptide. The precursor was the $(M+4H^+)^{4+}$ ion at m/z 440. The insets show the resolution of the isotopic peaks for the +1 to +4 ions. Reproduced with permission of The American Chemical Society from Thomson BA, Douglas DJ, Corr JJ, Hager JW, Jolliffe CL (1995) *Analytical Chemistry* **67**: 1696–1704.

mixture, it is fragmented in Q2 and a mass spectrum of product ions is obtained by scanning Q3. This scan mode is useful to obtain structural information of an ion such as sequence information of a peptide. In 'precursor ion' scans, Q3 is fixed on a particular m/z value and Q1 scans through all the ions produced by the source. This scan is useful to identify those ions in the source that contain a particular functional group. In 'neutral loss' scans, Q1 and Q3 scan together with a constant difference in m/z that corresponds to loss of a given neutral group. For example, if the loss corresponds to Cl_2 (70 amu) this scan could identify ions that contain two or more chlorine atoms such as polychlorinated dioxins. For targeted compound analysis, the system can be run in 'single reaction monitoring' mode. Here Q1 is fixed at the m/z of the precursor ion of a targeted compound and Q3 is set to a m/z value of a major fragment ion of that compound. There is a massive discrimination against other compounds. If still greater selectivity is required, multiple reaction monitoring can be done. Here Q1 is set to the m/z value of the precursor and Q3 'peak hops' to several (say four) m/z values of fragments that come from the targeted compound. If the intensity ratios of the fragments are correct the compound is identified. If there is an interference on one of the fragments, the remaining fragments can be used to identify and quantify the compound. The ability to independently

scan Q1 or Q3 under computer control makes a triple quadrupole MS/MS system a flexible tool for trace analysis. It has become the workhorse for LC-MS/MS and GC-MS/MS analysis.

An example of the use of a triple quadrupole MS/ MS system to identify a protein is given here. The enzyme telomerase rebuilds the ends of chromosomes (telomeres) when cells divide. It consists of one RNA and two protein subunits of molecular mass 43 kDa and 123 kDa. The 123 kDa protein was separated on a 2D gel, extracted and digested to produce a mixture of peptides. Q1 of a triple quadrupole was scanned to produce a spectrum of all the peptides, shown in **Figure 10A**. To identify these peptides, Q1 was set to transmit a 2 m/z mass window and tandem mass spectra were collected with a 0.2 m/z step size. The fragment ion spectrum of a doubly charged peptide at m/z 830.4 is shown in **Figure 10B**. The doubly charged peptide ion fragments to singly charged ions to give fragments with m/z greater than the precursor. This mass spectrum along with esterification of the peptide allowed unambiguous assignment of the amino acid sequence. The sequences of eight of the peptides in the mass spectrum were determined. These amino acid sequences were then used to make DNA probes that led to identifying the gene containing the complete sequence of the protein. The total amount of

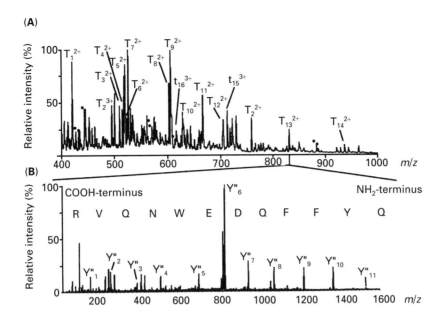

Figure 10 (A) Mass spectrum of the peptides obtained from digesting the 123 kDa subunit of telomerase. Peptides were not separated by chromatography. Peptides that were sequenced fully or partially are marked by T or t, respectively. (B) Tandem mass spectrum of the peptide at m/z 830.4. Reproduced with permission of The American Association for the Advancement of Science from Lingner J, Hughes TR, Shevchenko A, Mann M, Lundblad V, and Cech TR (1997) *Science* **276**: 561–567.

protein available for this experiment was in the low picomole range.

ICP-MS

Another major application area for quadrupoles is in ICP-MS systems for trace element analysis. A schematic of a system is shown in **Figure 11**. This ion source is an induction plasma in argon at atmospheric pressure with a temperature of 5000–7000 K contained in a torch. Samples are introduced to the plasma as aerosols, usually solutions that are sprayed. At the high plasma temperature dissolved solutes are vaporized, atomized and ionized. Most

elements of the Periodic Table are present in the plasma as singly charged atomic ions (the degree of ionization is typically 90% or more). The plasma expands through an orifice about 1 mm in diameter into a region at a pressure of a few mbar, and the centreline flow then passes through a skimmer into a region at a pressure of about 10^{-4} torr. In this region ions are extracted from the rarefied plasma, pass through ion lenses and then are mass-analysed in a quadrupole.

ICP-MS with a quadrupole gives simple mass spectra that are easy to interpret. **Figure 12** for example shows the mass spectrum at unit resolution of some transition metals at a concentration of 100 ng mL^{-1}. By scanning a quadrupole over the range ^{7}Li$^+$ to ^{238}U$^+$, over 70 elements can be determined in a 1 min scan. Alternatively the quadrupole can peak hop to selected isotopes or elements, to improve the duty cycle. Detection limits are typically in the 10 pg mL^{-1} region for elements in solution. Isotopic information is inherent in ICP-MS. With a quadrupole the precision on isotope ratios is typically 0.2% with a measurement time of a few minutes. This is insufficient for many geological dating applications but is more than adequate for many isotopic tracing experiments in nutrition and other studies. In addition, it greatly facilitates quantification of trace elements by isotope dilution.

Ideally the ICP would produce only singly charged atomic ions of each element. However, it also

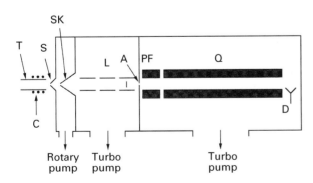

Figure 11 A quadrupole ICP-MS system. T, torch; S, sampler; SK, skimmer; L, ion lenses; A, differential pumping aperture and lens; PF, delayed DC ramp 'prefilter'; Q, quadrupole mass filter; D, ion detector.

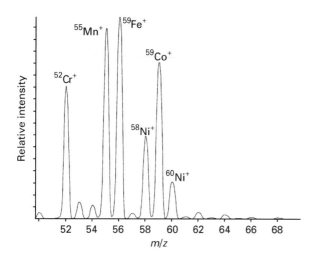

Figure 12 Mass spectrum of transition metals each present at 100 ng ml^{-1} in solution. The peak at m/z 56 also includes a contribution from $^{40}Ar^{16}O^+$. Small peaks at m/z 63–68 indicate contamination by Cu and Zn at concentrations of few ng mL^{-1}.

produces some molecular ions such as ArO$^+$ which interfere with $^{56}Fe^+$, or Ar$_2^+$ which interferes with $^{80}Se^+$. Such interferences are most common at m/z < 80. To separate these interferences requires a resolution that is beyond the capabilities of quadrupoles operated conventionally, although the use of alternative stability regions is being investigated. The interferences are not prohibitive because in many cases an alternative isotope can be found that is free of interference. As with many applications it is the comparatively low cost and electronic control of quadrupoles that make them attractive for ICP-MS.

List of symbols

$a = 8zU/m\omega^2r_o^2$; f = RF Frequency; m = mass of ion; n = number of RF cycles; $q = 4zU/m\omega^2r_o^2$; r_o = separation between electrodes; R_{lim} = limiting resolution; R_{max} = maximum resolution; t = time; $u = x$ or y direction; U = direct voltage; V = voltage; z = charge of ion; $\xi = \omega t/z$; $\omega = z\pi f$.

See also: **Atmospheric Pressure Ionization in Mass Spectrometry; Biomedical Applications of Atomic** Spectroscopy; Chemical Ionisation in Mass Spectrometry; Chromatography-MS, Methods; Hyphenated Techniques, Applications of in Mass Spectrometry; Inductively Coupled Plasma Mass Spectrometry, Methods; Mass Spectrometry, Historical Perspective; MS-MS and MS^n; Photoacoustic Spectroscopy, Applications.

Further reading

Bruins AP (1994) Atmospheric pressure ionization mass spectrometry. *Trends in Analytical Chemistry* **13**: 37–43; 81–90.

Busch KL, Glish GL and MacLuckey SA (1998) *Mass Spectrometry/Mass Spectrometry: Techniques and Applications of Tandem Mass Spectrometry*. Weinheim: VCH.

Cole RB (ed) (1997) *Electrospray Ionization Mass Spectrometry*. New York: Wiley.

Dawson PH (ed) (1976) *Quadrupole Mass Spectrometry and its Applications*. Amsterdam: Elsevier: (Reissued as a paperback Dawson PH (ed) *Quadrupole Mass Spectrometry and its Applications* (1995) New York: American Institute of Physics Press.

Dawson PH (1980) Mass filter design and performance. *Advances in Electronics and Electron Physics* **53**: 153.

Dawson PH and Bingqi Yu (1984) Performance comparison in conventional and higher stability regions. *International Journal of Mass Spectrometry and Ion Processes* **56**: 41.

Douglas DJ and Ying J-F (1996) High resolution ICP mass spectra with a quadrupole mass filter. *Rapid Communications in Mass Spectrometry* **10**: 649–652.

Du Z, Olney TH and Douglas DJ (1997) Inductively coupled plasma mass spectrometry with a quadrupole operated in the third stability region. *Journal of the American Society for Mass Spectrometry* **8**: 1230–1236.

Heumann K (1982) Isotope dilution mass spectrometry for micro- and trace-element determination. *Trends in Analytical Chemistry* **1**: 357–361.

Houk RS (1994) Elemental and isotopic analysis by inductively coupled plasma mass spectrometry. *Accounts of Chemical Research* **27**: 333–339.

Paul W, Reinard HP and von Zahn U (1958) *Zeitschrift für Physik* **152**: 143–182.

Yost RA and Enke CG (1979) Triple quadrupole mass spectrometry for direct mixture analysis and structure elucidation. *Analytical Chemistry* **51**: 1251A–1264A.

Quantitative Analysis

T Frost, Glaxo Wellcome, Dartford, UK

FUNDAMENTALS OF SPECTROSCOPY
Methods & Instrumentation

Spectroscopic techniques are particularly useful for quantitation. They offer speed and great flexibility in instrumentation. UV spectrophotometry is widely used for quantitation using data at the maximum absorbance of a chromophor. Fluorescence spectrometry is also widely used for quantitation as it provides greater selectivity and sensitivity than UV spectrophotometry. The use of FT-IR spectrometry for quantitation is less common. The use of NIR for quantitation is an area of great activity and innovation. In addition, NMR spectroscopy can be used for quantitation. Here the criteria for success are very different from those for absorption or fluorescence spectroscopy. NMR quantitation is the subject of a separate article in this encyclopedia.

Quantitation can be carried out using either absorption or reflectance measurements and the laws governing these two types of measurements will be discussed first.

Absorption of light

The absorption of light by a compound depends on its chromophor, the wavelength of the light and the thickness of the sample. Bouguer derived the relationship between absorption and the thickness of the sample. The integrated form of the equation is shown in Equation [1] where I_o is the intensity of the incident radiation and I the intensity of the transmitted radiation. The factor a related to the absorptivity of the chromophor and b is a measure of the sample thickness.

$$\log(I_o/I) = ab \quad [1]$$

Beer's law states that the absorption of monochromatic radiation by a sample is proportional to the concentration of the sample. The law is defined by Equation [2] where a' is a constant and c is the concentration.

$$\log(I_o/I) = a'c \quad [2]$$

Equations [1] and [2] are combined to give the Bouguer–Beer law shown in Equation [3] where ε is the molar absorptivity, i.e. the absorbance of a one molar solution of the compound. Note that ε is specific for each compound at particular wavelength.

$$\log(I_o/I) = \varepsilon bc \quad [3]$$

The term $\log(I_o/I)$ is the absorbance, A, of the sample and Equation [3] is presented more simply in Equation [4], often referred to as Beer's law.

$$A = \varepsilon bc \quad [4]$$

Spectrophotometers measure the ratio of the intensity of the incident and transmitted radiation, which is known as the transmittance, T.

$$T = I/I_o \quad [5]$$

Absorbance is then related to transmittance as shown in Equation [6].

$$A = -\log T \quad [6]$$

Logarithms to the base 10 are usually used.

Reflection of radiation

The use of reflection techniques is becoming common for quantitation. Taking measurements from the surface of a sample avoids the need for sample preparation and thus provides a very rapid method of analysis. One drawback of the use of reflectance measurements is that it is necessary to assume that the surface is representative of the sample as a whole. Reflectance measurements are commonly used in the NIR and FT-IR regions.

In reflectance measurements, $\log(1/R)$, where R is the reflectance of the sample, is proportional to concentration. The proportionality constant is not as universal as in absorption. The constant will depend on factors such as the particle size of the sample and the moisture. The constant is thus unique for each sample and this makes quantitation using reflectance techniques very challenging.

Fluorescence

Fluorescence quantitation can be described by Equation [7]

$$F = (I_o)(\Phi_f)(1 - e^{\varepsilon bc}) \qquad [7]$$

where F is the fluorescence intensity of the sample and I_o is the intensity of the incident light and Φ_f is the fluorescence quantum yield. The quantity $1-e^{\varepsilon bc}$ derives from Beer's law, Equation [4], and relates to the fraction of light absorbed by the fluorescing species. Equation [7] can be expanded to Equation [8]

$$F = I_o\Phi_f\left(1 - \left[1 - \varepsilon bc + \frac{(-\varepsilon bc)^2}{2} - \frac{(-\varepsilon bc)^3}{3} + \cdots\right]\right) \qquad [8]$$

At low concentrations with absorbance less that ~ 0.05, Equation [8] can be reduced to Equation [9] and fluorescence is linearly related to concentration.

$$F = I_o\Phi_f\, \varepsilon bc \qquad [9]$$

The approximations made to derive this equation must be remembered and at high concentrations fluorescence will not vary linearly with concentration and a curved calibration graph will be obtained.

Quantitative techniques

Data at a single wavelength

The simplest method of quantitation is to use data from a single wavelength. Ideally measurements should be taken at the wavelength of an absorption maximum because this avoids errors due to differences in wavelength calibration. This method of quantitation is usually used for absorption measurements rather than reflectance techniques which tend to require data from more than one wavelength. The absorbance of a solution of known concentration, the standard, is measured and compared to the absorbance of the sample whose concentration needs to be determined. Thus, if A_{sam} is the absorbance of the sample and A_{std} is the absorbance of the standard and C_{std} is the concentration of the standard the concentration of the sample, C_{sam} is given in Equation [10].

$$C_{sam} = A_{sam} \times C_{std}/A_{std} \qquad [10]$$

Two standards should be prepared and the responses compared to ensure they have been prepared correctly. Typically, the two responses should agree within 1 per cent. The mean response from the two standards can then be used to predict the concentration of the samples.

If the concentration of the samples varies over a wide range, a calibration curve, constructed from standards of varying concentration, can be used. The responses of each standard should be checked to ensure that they agree and the regression coefficient of the line should be close to 1.

Often the molar absorptivity of the compound under investigation is so well characterized that the use of a standard is not necessary and published values of the absorbance of a 1 per cent solution in a 1 cm cell, A_1^1 can be used. The use of A_1^1 values is particularly common in the pharmaceutical industry.

The drawback of using data at one wavelength is that the wavelength may not be specific for the compound of interest. The sample may contain other compounds that absorb at the same wavelength and interfere with the measurement. Thus quantitation using data at a single wavelength is limited to solutions of simple compounds. The technique is most commonly used in the ultraviolet and visible regions of the spectrum.

Derivative spectroscopy

Derivative spectroscopy is useful in quantitative analysis both as a quantitative technique in its own right and for preprocessing data prior to analysis by some of the chemometric techniques described below.

A common problem in spectroscopic quantitation is that the band may overlap with a broad interfering band. The interfering band may arise from the sample matrix and give a sloping baseline making quantitation difficult. This is illustrated in **Figure 1** where the left-hand spectrum shows a band on a sloping baseline. The first derivative removes the sloping baseline provided it is linear. First derivative spectra are not always ideal for quantitation and often a second derivative spectrum, shown on the right of **Figure 1**, is used. The second derivative will remove any baseline curvature provided it can be fitted to a quadratic equation with respect to wavelength.

A first derivative spectrum is found by differentiating the absorbance spectrum and shows the rate of change of the slope of the absorbance spectrum. The derivative spectrum is found by calculating $dA/d\lambda$, where A is the absorbance and λ is the wavelength.

Derivative spectra are normally calculated off-line on the digital spectral data. The use of the Savitzky–Golay routine for calculating smoothed derivatives is common but other methods are available. Instrumental parameters need to be optimized to obtain successful derivatives. In particular, the data interval used to record the spectrum, the scan speed and the interval used to calculate the derivative need to be optimized.

The spectrum must first be scanned using a suitable data interval. If the data interval is too large then vital high frequency features will be smoothed out when the spectrum is recorded and the power of derivative spectroscopy will be lost. For derivative work it is usually best to choose a relatively small data interval to scan the spectrum. Any noise recorded by the small interval can be smoothed in the calculation of the derivative by varying the derivative interval. The interval used for the calculation of the derivative will influence the spectrum. Some trial and error may be necessary to find the correct interval.

Derivatives for quantitation When Beer's law, Equation [4], is differentiated with respect to wavelength only ε, the molar absorptivity, will vary with wavelength. The concentration C and the path length b are constant. The first derivative is given in Equation [11].

$$\frac{dA}{d\lambda} = \frac{d\varepsilon}{d\lambda}Cb \qquad [11]$$

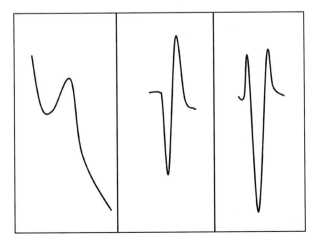

Figure 1 Left: A typical absorption band; Centre: The first derivative of the absorption band; Right: The second derivative of the absorption band.

The derivative amplitude, $dA/d\lambda$, will be proportional to concentration and can be used for quantitation. Differentiating the first derivative spectrum forms derivative spectra of higher order. The second and fourth order derivatives are generally used for quantitation. **Figure 1** shows first and second derivative spectra.

Derivative spectroscopy can be used for analysis of mixtures when two components have different bandwidths. The technique is particularly useful when one compound has a spectrum with sharp features and another component has broad features. The sharp band can be emphasised at the expense of the broad band.

The use of derivative spectroscopy for direct quantitation is limited to compounds where the spectra show major differences. Quantitation can be carried out using data at single wavelengths or the whole spectrum can be used in some of the chemometric techniques described later.

Chemometric techniques

Chemometric techniques have found widespread use in spectroscopic quantitation. The techniques are used when a single wavelength, that is specific for the compound of interest, cannot be found. Wavelengths have to be found where the species contributing to the spectrum have different absorption or reflectance. Mathematical approaches can then be used to unravel the contribution to the spectrum of the compounds of interest and hence deduce their contribution.

In a mixture of compounds the absorbance at a particular wavelength can be considered to arise from a linear combination of the absorbances from the individual components. If two components contribute to the spectrum we can expand Beer's law to reflect the individual contribution of each component as shown in Equation [12], where the subscripts 1 and 2 refer to the two components.

$$A = \varepsilon_1 bc_1 + \varepsilon_2 bc_2 \qquad [12]$$

If we know the molar absorptivities, ε_1 and ε_2, for each component and the path length, b, Equation [12] still cannot be solved for the concentrations, c_1 and c_2, because there are too many variables. To solve for the concentrations of the two components it is necessary to have data at two wavelengths where the molar absorptivities of the two components are different both from each other at both wavelengths. Two simultaneous equations can be constructed and solved for the two concentrations. If more than two components contribute to the spectrum it is necessary to find extra wavelengths for each component where the molar absorptivities differ. Analogous equations can be written for reflectance measurements.

Rather than use simultaneous equations it is usual to resort to regression techniques. These fall into three main techniques multiple linear regression (MLR), principal component regression (PCR) and partial least squares (PLS). All three techniques have found widespread use in spectroscopic quantitation using both absorption and reflectance techniques.

Multiple linear regression requires careful choice of wavelengths to find a quantitation model that is robust. PCR and PLS overcome problems of wavelength selection by using the full spectrum. PLS is often considered to give superior results to PCR. Full discussions of the algorithms and details of how to apply the techniques can be found in the monographs on chemometrics listed under Further reading.

Multiple linear regression

Multiple linear regression can be used to solve for the constants in Equation [13], which can be described as a general equation for any n components as shown in Equation [13].

$$A = \varepsilon_1 bc_1 + \varepsilon_2 bc_2 + \varepsilon_1 bc_1 + \varepsilon_3 bc_3 + \cdots + \varepsilon_n bc_n \quad [13]$$

Matrix algebra is used to simplify the mathematics and Equation [13] is described in matrix terms in Equation [14]. This is a general equation for any spectroscopic data, be it absorption, reflection or derivative data.

$$\mathbf{y} = \mathbf{Xb} + \mathbf{e} \quad [14]$$

The vector, $\mathbf{y}$, is a column vector of spectroscopic data for each sample at one wavelength. The matrix $\mathbf{X}$ contains the concentrations of the samples. If an intercept is included in the model, then the first column of $\mathbf{X}$ must be a column of 1s. The vector $\mathbf{e}$ is a column of residuals associated with lack of fit in the model. The vector $\mathbf{e}$ contains the errors that are minimized in the regression analysis.

The matrix solution for $\mathbf{b}$ in Equation [14] is given in Equation [15].

$$\mathbf{b} = (\mathbf{X'X})^{-1}\mathbf{X'y} \quad [15]$$

where $\mathbf{X'}$ is the transpose of $\mathbf{X}$ and $(\mathbf{X'X})^{-1}$ is the inverse of the covariance matrix $\mathbf{X'X}$. The inversion of the covariance matrix is not always possible and this is the main drawback to MLR.

The practical steps involved in MLR are as follows. A set of calibration samples is selected and the concentration of the component of interest is determined by a reference technique. The spectroscopic data and the concentrations from the reference technique are fed into Equation [15] to give an equation that can be used to predict the concentration of the component in further samples.

MLR can result in simple equations that can be used for quantitation. For example, the constituents of food, such as moisture and fat, can be determined using spectroscopic data at a few NIR wavelengths. The disadvantages of MLR are that it requires the operator to select the wavelengths. The selection of inappropriate wavelengths can result in poor models that are mathematically unstable.

PCR and PLS

PCR and PLS were developed to overcome the limitations of MLR. They use all the spectral data and so avoid the need for wavelength selection. PCR is essentially a mathematically more robust way of carrying out MLR. The regression is performed on the principal components of the data set. The principal components are determined from the data set with the specific aim that they will provide robust models. The principal components are linear combinations of the original measurements such that the first component explains the most variance in the data and subsequent components, all orthogonal, explain decreasing amounts of data variance.

The data set is broken down into loadings, which are analogous to abstract spectra, and scores, which describe the amount of each loading associated with each sample and are similar to abstract concentrations. Some of the loading spectra will describe noise or minor components and will not be relevant to the model and can be rejected. Determining which principal components can be rejected affects the robustness of the model.

A set of calibration samples is identified and the concentration of the components of interest is determined by a reference technique. The spectra are assembled into a data matrix often known as the calibration matrix. The data matrix is decomposed into the product of a score and a loading matrix using principal component analysis. Irrelevant principal components describing noise are rejected, typically those components of higher index.

Multiple linear regression is then used to relate the scores matrix to the matrix of concentrations determined by the reference technique.

The model is then complete and can be used to predict the concentration of the components of samples. The calibration set can be used many times provided the samples in the set have been chosen carefully. A good calibration set of samples will contain samples that arise from every source of variation likely to be encountered in further samples that the method will be used to quantify.

PCR takes no account of the concentration of the samples in the calibration set when finding the principal components and this has been considered a drawback. PLS was developed to use the concentration data when developing the mathematical model. PLS is thus claimed to be superior to PCR.

The practical application of PCR and PLS is very similar. Samples are selected for the calibration set and a reference technique used to determine the concentration of the component or components of interest. The spectra and concentrations then allow a calibration model to be set up for predictions of further concentrations.

Pretreatment of the spectroscopic data is usually carried out to improve the model. A first derivative can be applied to remove sloping backgrounds and offset errors. This is a very common method of pretreating NIR spectra. The data can be centred about its mean. Routines are available to smooth the data and remove spectroscopic scatter.

Once a set of calibration spectra have been set up the use of data pretreatment can be examined to find the optimum pretreatment. Similarly PCR and PLS can be compared to see which gives the best results.

Validation

Whatever method is chosen to perform the quantitation, whether a simple single wavelength assay or a complex chemometric method, the method should be validated to ensure it produces meaningful data. Before the method can be validated, the exact procedure should be documented so that the final method is used in the validation experiments.

Validation should look at the following parameters:

- Linearity
- Specificity
- Accuracy
- Precision
- Robustness

Linearity

The linearity of the method should be checked over and beyond the likely operating range of the method. Typically six concentrations of the standard solution over the range 25 to 150% of the normal working concentration should examined. A graph of absorbance against concentration can then be constructed. The intercept of the line should be examined to see how close to the origin the line crosses the X axis. A large intercept will indicate an offset error in the method or nonlinearity.

Specificity

The specificity of the method can be checked by running a sample without the compound of interest, if one is available. The method can also be compared with other methods of known specificity. The acceptable limit of interference from the sample matrix will vary with the sample and the use of the results from the analysis. Typically, an interference of < 2% would be expected but higher values can often be tolerated.

Accuracy

The accuracy of the method is a measure of how close to the true value the results from the method are. Accuracy is usually checked by preparing samples spiked with known amounts of the component of interest. The accuracy should be examined over a range that extends beyond the range of samples the method is likely to analyse. If the method is designed to measure samples at 100% of expected strength then it would be useful to check the accuracy at that value and at 80% and 120% of the expected strength. Results from the accuracy experiments should normally agree to within 2% of the true value.

When samples cannot be spiked with the component of interest then comparisons with a reference technique should be carried out.

Precision

Precision is a measure of the spread of results from a method. Precision can be broken down into subsections. The first measurement of precision relates to how repeatable the spectroscopic measurements are. Taking ten readings and looking at the coefficient of variation can assess the repeatability of the spectroscopic measurement. Normally the coefficient of variation should be < 2%.

The repeatability of the sample and standard preparation can be assessed by preparing ten samples and standards from the sample source and examining the coefficient of variation of the response.

Depending on where the method is used the precision of different operators, equipment and laboratories should also be examined. Interlaboratory precision can be determined by having different laboratories analyse the same sample.

Robustness

The robustness of the method should be examined by stressing the method parameters. Thus the effects of slight changes in wavelength, spectroscopic settings, sample preparation, etc. can be varied to see if any of the values are critical. The most effective way of doing this is to use experimental design so that parameters are varied in a planned way and statistics and response surfaces can be used to evaluate the outcome of the results.

List of symbols

a = constant; A = absorbance; A = absorptivity; b = sample thickness, path length; c, C = concentration; F = fluorescence intensity; I = intensity of transmitted radiation; I_o = intensity of incident radiation; R = reflectance; T = transmittance; ε = molar absorptivity; λ = wavelength; Φ_f = fluorescence quantum yield.

See also: **Dyes and Indicators, Use of UV-Visible Absorption Spectroscopy; Fourier Transformation and Sampling Theory; Multivariate Statistical Methods; UV-Visible Absorption and Fluorescence Spectrometers.**

Further reading

Clark BJ, Frost T and Russell MA (eds) (1993) *UV Spectroscopy: Techniques, Instrumentation and Data Handling.* Chapman and Hall.

Geladi P and Kowalski BR (1986) Partial Least-Squares Regression: A Tutorial. *Analytica Chimica Acta* **185**: 1–17.

Martens H and Naes T (1989) *Multivariate Calibrations.* Wiley.

Madden HH (1978) Comments on the Savitzky–Golay Convolution Method for Least Squares Fit Smoothing and Differentiation of Digital Data. *Analytical Chemistry* **50**: 1383–1386.

Miller JC and Miller JN (1993) *Statistics for Analytical Chemistry.* Ellis Horwood.

Osbornce BG, Fearn T and Hindle PH (1993) *Practical NIR Spectroscopy with Applications in Food and Beverage Analysis*, 2nd edn. Longman.

Stavitsky A and Golay MJE (1964) Smoothing and Differentiation of Data by Simplified Least Squares Procedures. *Analytical Chemistry* **36**: 1627–1639.

Radiofrequency Field Gradients in NMR, Theory

Daniel Canet, Université H. Poincaré, Vandoeuvre, Nancy, France

MAGNETIC RESONANCE
Theory

Any NMR experiment requires two magnetic fields, a static one denoted by B_0 which serves to polarize the nuclear spins (or to create two distinct energy levels in the case of spins $\frac{1}{2}$) and an alternating magnetic field B_1, also called the radiofrequency field (RF field) because of its usual frequency domain, whose purpose is to induce transitions. It can be mentioned that, in pulse NMR experiments, the RF field serves merely to rotate nuclear magnetization, or, in a more general way, to rotate spin operators involved in the description of the various populations or coherences that may exist or appear in the course of the experiment. In fact, these rotations occur in the so-called rotating frame, that is a frame which rotates at a constant angular velocity ($\omega_r = 2\pi\nu_r$, ν_r being the transmitter frequency) around the Z direction defined by B_0. We shall denote (x,y,z) as the rotating frame and (X,Y,Z), with $Z \equiv z$, as the laboratory frame. In the rotating frame, the RF field is stationary and oriented along x,y or any axis in the (x,y) plane according to its phase. Finally, during receive operations, the rotating frame has still to be considered. This is due to the particular scheme usually employed in NMR that detects the signal, not at its own frequency ν_0, but at a relative frequency $(\nu_0-\nu_r)$ by means of appropriate devices (mixers). Moreover, by means of the so-called quadrature detection scheme, two signals can be acquired simultaneously: $\cos2\pi(\nu_0-\nu_r)t$ and $\sin 2\pi(\nu_0-\nu_r)t$.

Most of the time, the experimentalist aims at uniform magnetic fields, either B_0 or B_1. For the former,

this need arises from the basic Larmor equation which provides the resonance frequency ν_0:

$$\nu_0 = \frac{\gamma B_0}{2\pi}(1-\sigma) \qquad [1]$$

(γ is the gyromagnetic ratio and σ the shielding coefficient, which determines the chemical shift of the considered nucleus). Thus a uniform B_0 field prevents line broadening or resonance spreading. For the latter, the need arises from the relation which provides the flip angle α (the angle by which a magnetization component or a spin operator rotates around the B_1 direction in the rotating frame) is

$$\alpha = \gamma B_1\tau \qquad [2]$$

(τ is the duration of the RF field application or more simply the duration of the RF pulse). A homogeneous B_1 field prevents a distribution of flip angles and, if the same coil is used for signal detection, a nonuniform receptivity by virtue of the reciprocity principle.

However, very early on it was recognized that some advantages may be drawn from using nonuniform magnetic fields. The simplest situation is a linear variation of the B_0 amplitude (B_0 still being oriented along Z) with respect to a spatial direction, say X. Let g_0 be the (uniform) slope of this variation, leading to an equation indicating how B_0 varies with respect to X

$$B_0(X) = B_{00} + g_0 X \qquad [3]$$

Combining Equations [1] and [3], it is apparent that the resonance frequency also depends linearly on X and therefore can provide information about any property which occurs that is spatially dependent. This simple feature led (1) to the capability of measuring self-diffusion coefficients, (2) to the various imaging methods, and (3) in a more subtle way, to the selection of coherence pathways.

Radiofrequency field gradients (RF gradients, or B_1 gradients) work in a different way (although strongly related) to B_0 gradients. In this case, it is the nutation frequency $\nu_1 = \gamma B_1/2\pi$, i.e. the frequency at which magnetization rotates in a plane perpendicular to B_1 (see Eqn [2]), which is spatially dependent; in other words, spatial encoding occurs via nutation and not via precession. It will be shown in this article that virtually all experiments that can be carried out with B_0 gradients can also be considered with B_1 gradients, sometimes with clear advantages, but unfortunately with some inconveniences which make the two approaches rather complementary. A common feature to these two types of gradients is the ability to *defocus* nuclear magnetization. This means that if a gradient of sufficient strength is applied for a sufficient time, it is capable of spreading out in the relevant plane all elementary magnetization vectors of a homogeneous sample so that the net result is zero.

Creation of B_1 gradients and their specificity with respect to B_0 gradients

The first problem of concern is the uniformity of gradients. As will be discussed below, gradient uniformity appears to be mandatory for imaging experiments, recommended for self-diffusion measurements, but not really necessary in pure spectroscopic applications (selection of coherences, solvent suppression, etc.). The best way for creating a uniform B_1 gradient is a flat coil which may comprise a single or several turns. For a circular single-turn coil, it can be shown that good uniformity is obtained in a region extending from $0.2\,r$ to $0.9\,r$ from the coil centre (r being the coil radius). The gradient strength is directly proportional to the B_1 field at the coil centre and thus depends on the coil quality factor and on the RF power it can hold. This is of course strongly related to the measurement frequency, low frequencies allowing

stronger gradients. The state-of-the-art, as far as uniform gradients are concerned, corresponds to gradients up to 800 mT m^{-1} at a measurement frequency of 100 MHz (obtained with a two-turn coil). It must be borne in mind that the gradient efficiency is in any event weighted (or enhanced) by the gyromagnetic ratio of the observed nucleus. As an example, **Figure 1** shows the experimental B_1 distribution in a 2 mm diameter sample obtained with a two-turn flat coil of 15 mm and 11 mm diameters respectively. Anyhow, the production of a B_1 gradient makes sense only if a detection coil with uniform receptivity exists within the probe. Because, most of the time, the detection coil is tuned at the same frequency as the gradient coil, this poses the leakage problem between the two coils. Ideally, they should be orthogonal. Achieving this orthogonality, although requiring fine mechanical adjustments, proves to be feasible. However, creation of RF gradients in the two complementary directions is actually a challenge. Let X be the direction of the RF gradient created by the specific coil and let Y be the axis of the detection coil (**Figure 2**). It is conceivable that the inherent inhomogeneity of the detection coil affords a B_1 gradient in the Z direction (if, for instance, it is of the saddle-shape design), although uniformity is far from being warranted. What about a B_1 gradient in the Y direction? If one imagines the installation in the probe of a second gradient coil, identical to the first one and orthogonal to it, its axis would be collinear to the detection coil with evidently a full leakage. This solution cannot be obtained in practice at present and whenever RF gradients in the X and Y directions appear to be necessary, the best way for tackling this problem is to rely on a single coil and to rotate the sample. This is clearly a major drawback of B_1 gradients vs. B_0 gradients which can be created independently in the three spatial directions without perturbing the RF system. On the other hand, RF gradients offer some distinct advantages over B_0 gradients which are summarized in **Table 1**.

In a general way, two points in favour of B_1 gradients must be emphasized: (i) a simple instrumentation (at least much simpler than the one currently employed for B_0 gradients); (ii) their double function, since they perform both spin excitation and spatial labelling. This latter feature may lead to a considerable simplification of experiments, usually requiring magnetic field gradients.

Self-diffusion measurements

In order to illustrate the above statement, let us consider the B_1 gradient experiment used for measuring

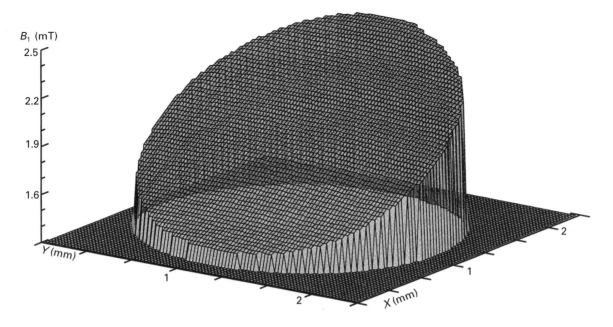

Figure 1 Three-dimensional diagram demonstrating the excellent uniformity of a RF gradient created by a two-turn flat coil (respective diameters: 15 and 11 mm).

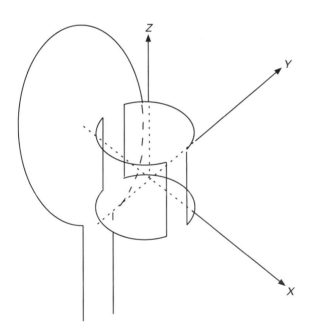

Figure 2 Sketch of a probe possessing B_1 gradient capabilities. The single-turn coil serves for generating a RF gradient in the X spatial direction. The saddle coil (in principle electrically orthogonal to the latter) is used essentially for detection purposes. Possibly it can produce a RF gradient in the Z direction.

Table 1 Properties of B_1 vs. B_0 gradients

	B_1 gradients	B_0 gradients
Rise and fall times	< 1 µs	≥ 100 µs
Eddy currents	None	Important, can be greatly attenuated by self shield
Perturbation of the static magnetic field and lock system	None	Strong
Effects of magnetic susceptibility variations across the sample	None	Important, must be compensated for
Effects of short T_2	Reduced by a factor of two	Full
Maximum strength (1997)	800 mT m⁻¹	10 T m⁻¹
Spatial directions available	1–2	3
RF deposition	High (improper for medical applications)	Low (limited to short RF pulses)

self-diffusion coefficients. The classical method is the PFGSE (pulsed field gradient spin echo) experiment which involves two pulses of B_0 gradient on both sides of the 180° RF pulse in a spin echo experiment. Here, with B_1 gradients, the experiment is especially simple and robust. It is sketched in **Figure 3** and starts with a B_1 gradient pulse $(g_1)_x$ along the x direction of the rotating frame of duration δ, separated by a 'diffusion interval' Δ from a second gradient pulse of duration identical to the first one (δ), immediately followed by a read pulse which probes the longitudinal magnetization (the read pulse may be incorporated into the second gradient pulse). This experiment can be understood as follows. The first gradient pulse is assumed to totally defocus nuclear magnetization, while the

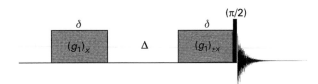

Figure 3 The simplest experiment for measuring self-diffusion coefficients by RF gradient pulses (shaded rectangles). $+x$ and $-x$ denote the RF phases. The simple phase cycling ($\pm x$), corresponding to two successive experiments whose results are coherently added, eliminates any transverse component of magnetization. The ($\pi/2$) pulse is a standard read pulse.

second one would act in a refocusing fashion provided that no motion occurred during Δ. Because of translational diffusion, the refocusing effect is not complete and some attenuation is observed at a rate depending on the self-diffusion coefficient D. The simple phase alternation $(g_1)_{\pm x}$ of the second gradient pulse eliminates any transverse magnetization contribution so that what is left is longitudinal magnetization whose amplitude decays according to (provided that $\Delta \gg \delta$):

$$\exp(-\Delta/T_1)\exp(-\gamma^2 g_1^2 \Delta \delta^2 D) \qquad [4]$$

where T_1 is the longitudinal relaxation time and g_1 the gradient strength. In practice, this decay is monitored by the last $\pi/2$ pulse (which converts the longitudinal magnetization into observable transverse magnetization) as a function of δ^2 and a semilogarithmic plot (for instance) provides D. An important point is that the additional attenuation due to relaxation involves T_1 and not T_2 as in the PFGSE experiment (in many systems of interest T_1 has a relatively high value while T_2 may become very small). In these latter cases, when employing B_0 gradients, stimulated echo experiments (more complicated and more instrumentally demanding) must be available to avoid an unacceptable attenuation by relaxation phenomena. Another point in favour of B_1 gradients is the variation of magnetic susceptibility inside heterogeneous samples which is responsible for internal B_0 gradients at interfaces. This feature obviously affects B_0 gradient experiments and requires the application of compensatory external B_0 gradient pulses. Such complications are by nature absent in B_1 gradient experiments.

NMR imaging by radiofrequency field gradients

An early application of B_1 gradients in conjunction with spatial discrimination (imaging) is the so-called

chemical shift imaging method; it is a two-dimensional experiment which produces chemical shift information in one dimension and spatial information (in the form of spin density) along the second dimension. This amounts to a one-dimensional experiment as far as spatial information is concerned. The experimental arrangement as well as the pulse sequence are very simple. It consists of using a surface coil (widely employed in biomedical applications of NMR), located upon the object under investigation and which therefore produces an inhomogeneous RF field along its axis. Applying a RF pulse of incrementable duration t_1 (this implies a series of experiments for each value of t_1) and acquiring the NMR signal immediately afterwards (according to the time variable t_2) generates a signal $S(t_1, t_2)$ modulated in t_1 according to the spin density distribution and in t_2 according to the chemical shifts of the different species existing within the object. Because these modulations are in the form of sine or cosine functions, a double Fourier transform with respect to t_1 and t_2 yields the result indicated above. However, because the B_1 gradient (originating from the inhomogeneity of the surface coil) is far from being uniform and because the receptivity of this coil (which serves as well for detection purposes) is by nature strongly non uniform, the results along the spatial dimension are obviously not quantitative. Nevertheless, the method has been widely used and is found very useful for obtaining the distribution of, for example, phosphorus metabolites. There are two instrumental prerequisites for obtaining quantitative spatial information: (i) a detecting coil with homogeneous receptivity and (ii) a coil delivering a uniform B_1 gradient. Clearly this corresponds to the arrangement of **Figure 2**. In practice, for making the experiment viable, the acquisition of spatial information must also be accomplished in a very short time (in a fraction of a second), as this is done by the read gradient in classical NMR imaging with B_0 gradients. These ideas led to the development of an experimental procedure adapted to B_1 gradients, which mimic the spatial encoding by a B_0 read gradient. The problem here is that acquisition of the NMR signal cannot take place while an RF field is on. The solution is indeed very simple and consists of applying the B_1 gradient in the form of short pulses separated by short intervals devoted to acquisition of the NMR signal. In practice a B_1 gradient pulse is followed by the acquisition of a single data point and the process is repeated until the NMR signal vanishes due to relaxation phenomena. The experiment is sketched in **Figure 4** and can be analysed as follows. Omitting

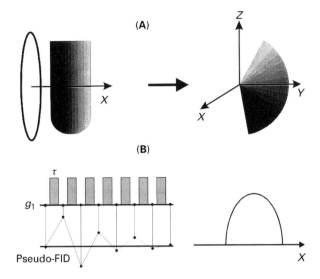

Figure 4 (**A**) Left: Different elementary slices subjected to the decreasing RF field produced by the gradient coil. Right: corresponding nutations assuming that nuclear magnetization was initially at thermal equilibrium. (**B**) Left: The RF gradient pulse train with interleaved detection windows leading to the acquisition of a pseudo-FID. Right: the Fourier transform of this pseudo-FID yields a profile representative of the object shape (see A).

relaxation effects and defining, for the lth data point, the time t as $l\tau$, where τ is the duration of each individual gradient pulse, we can write the amplitude of the NMR signal (which appears in the form of a 'pseudo-free induction decay', pseudo-FID) as

$$S(k) = \int_{\text{object}} \rho(X) \sin(2\pi X k) \, \mathrm{d}X \qquad [5]$$

where $\rho(X)$ is the spin density at abscissa X (the gradient is assumed to act along the X spatial direction) and where the variable k, according to usual practice, is substituted to the variable t

$$k = (2\pi)^{-1} \gamma g_1 t \qquad [6]$$

(g_1 is the gradient amplitude). It appears that there exists a Fourier relationship between S and ρ which can therefore be deduced from the following integral (inverse Fourier transform)

$$\rho(X) = \int S(k) \sin(2\pi k X) \, \mathrm{d}k \qquad [7]$$

yielding the object profile along X.

A two-dimensional (X,Y) image, possibly after a slice-selection procedure along the Z direction, can be reconstructed from a series of profiles obtained at different orientations resulting from the sample rotation around Z. Moreover, every type of contrast used in MRI can also be considered with RF gradients. These contrasts can originate from relaxation times, diffusion coefficients and, most importantly, chemical shifts. This latter possibility is illustrated by **Figure 5** which shows separate images of two components in a solvent mixture, sorted according to their chemical shift.

RF gradients in pure high-resolution NMR spectroscopy

It is now well established that the judicious use of B_0 gradient pulses in a complicated multipulse (and possibly multidimensional) NMR experiment can lead to invaluable improvements in the quality of spectra, eventually lifting barriers which were thought to be impossible to overcome. This is especially true for the suppression of huge peaks such as those of solvent or of protons bound to a majority isotope (e.g. ^{12}C vs. ^{13}C in natural abundance). Another attractive feature concerns the suppression of phase-cycling procedures which can be replaced by appropriate gradient pulses

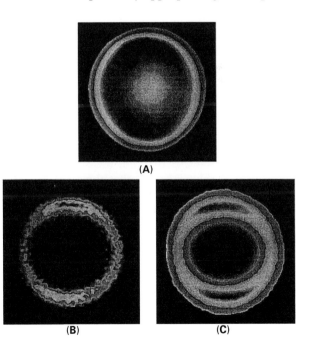

Figure 5 (**A**) Full image of a polymer rod that has been immersed in a solvent mixture (isooctane, ethanol, toluene). (**B,C**) Chemical shift selected images of isooctane and toluene, respectively, showing the different degrees of penetration.

in a single experiment thus considerably reducing measuring times. Basically, the benefit of using gradients in high-resolution NMR relies on the possibility of defocusing all magnetizations and, by the end of the experiment, selectively refocusing the desired magnetization(s) or coherence(s). (Throughout, gradient strengths and gradient pulse durations are supposed to entail a complete defocusing, i.e. a complete disappearance of the NMR signal.) Actually refocusing can be achieved in two ways: (i) by applying a gradient pulse of the same strength and same duration with an inverse polarity, thus producing the same precession in the reverse direction and (ii) by applying exactly the same gradient pulse after a 180° RF pulse which changes the sign of magnetization and amounts to reversing the gradient sign. Now, if the 180° pulse is selective, the only magnetization which survives (i.e. which has been refocused) is the one corresponding to the bandwidth of the 180° selective pulse. These simple considerations show how it is possible to manipulate gradients in order to preserve one resonance while destroying all the others. In fact, gradients can accommodate much more complicated schemes involving coherences of different orders and selection of a given coherence pathway. It can be recalled that coherence orders stem from the state of the spin system in the course of the experiment. For instance, in the case of two weakly coupled nuclei of spin-$\frac{1}{2}$, denoted A and X, the coherence orders are as follows (I_+ and I_- are the classical raising and lowering operators): $p = 0$, longitudinal magnetization or zero quantum coherence (represented by e.g. $I_+^A I_-^X$); $p = \pm 1$, single quantum coherences corresponding to observable transverse magnetization; $p = \pm 2$, double quantum coherences (represented by e.g. $I_+^A I_+^X$).

Recognizing that a coherence of order 0 is not affected by B_0 gradients, that defocusing of a double quantum coherence is twofold, the defocusing of a one-quantum coherence and that the sense of defocusing is given by the sign of the coherence order, it can easily by recognized that the coherence pathways leading to a final observable signal must satisfy the simple equation

$$\sum_i p_i g_i = 0 \qquad [8]$$

where p_i and g_i are respectively the coherence order and the gradient strength relevant to the ith interval of the considered experiment. This equation is in fact valid for homonuclear experiments; in a heteronuclear experiment, gyromagnetic ratios must be introduced as multiplying factors. At this point, it must be emphasized that the above discussion applies to B_0

gradients, i.e. to precession (or rotation) in the transverse (x,y) plane. If one assumes that these B_0 gradients are sufficiently strong so that the natural precession (due to chemical shift) is negligible, a full equivalence with B_1 gradients can be established by recognizing that the latter correspond to the same type of rotation in the vertical plane of the rotating frame. Consequently, this equivalence amounts to flipping the vertical plane into the xy plane and, after the B_1 gradient application, flipping it back to its initial position. This is achieved by clusters of the type

$$(\pi/2)_x (g_1)_y (\pi/2)_{-x}; \ (\pi/2)_{-x} (g_1)_{-y} (\pi/2)_x;$$
$$(\pi/2)_{-y} (g_1)_x (\pi/2)_y; \ (\pi/2)_y (g_1)_{-x} (\pi/2)_{-x} \qquad [9]$$

where $(\pi/2)$ stands for a standard hard pulse (homogeneous) and (g_1) for the application of a RF gradient of sufficient strength and duration (to produce a complete defocusing); RF phases are indicated as subscripts. By means of this equivalence, the numerous experiments involving B_0 gradients can easily be transposed with B_1 gradients and will not be discussed further.

The emphasis will rather be put on two simple and illustrative experiments which take advantage of the specific features afforded by RF gradients. In principle they could be converted into B_0 gradient experiments by using the reciprocal of Equation [9]; however, due to the inherent shortcomings of B_0 gradients (especially rise and fall times) and to the complexity of the resulting sequences, this is not practicable. The first example deals with solvent peak suppression. Among several possibilities allowed by B_1 gradients, the one using a train of short gradient pulses appears somewhat specific and difficult to convert into a B_0 gradient version. It works in the manner of a DANTE (delays alternating with nutations for tailored excitation) sequence. It can be recalled that the train of short, homogeneous RF pulses (each corresponding to a small flip angle) of a DANTE sequence is equivalent to a selective pulse because the effect of pulses is cumulative only for on-resonance magnetization. Magnetizations corresponding to other resonance frequencies are tipped back to the z axis (to their equilibrium position) in the course of the pulse train. If, instead of homogeneous pulses, RF gradient pulses are used, selectivity is of course retained, but their effect is to defocus that magnetization which is on-resonance and thus to suppress the relevant signal. The efficiency of this procedure is illustrated in **Figure 6** where a sufficient number of cycles (n) effectively leads to complete on-resonance peak suppression.

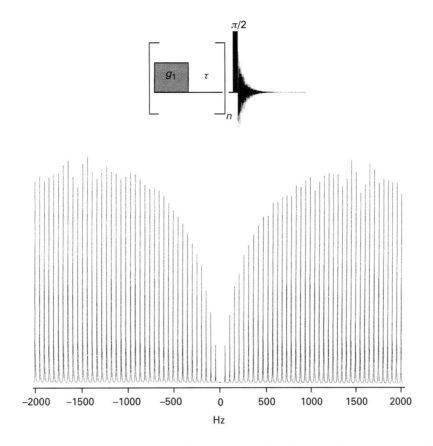

Figure 6 Suppression profile resulting from the sequence shown at the top of the Figure and obtained by repeating the experiment for a series of transmitter frequencies. When the signal is on-resonance, it is selectively defocused by the train of RF gradient pulses (g_1). τ is chosen so that the first sideband (2π precession) is outside the spectrum of interest.

Another appealing feature of RF gradient pulses is their ability to select coherences, and this will be discussed here for systems of spin-$\frac{1}{2}$ nuclei. A p-spin high-pass filter is indeed achieved by the following cluster of two B_1 gradient pulses of respective phases x and y:

$$(g_1)_x (rg_1)_y; \quad r = p/(p-1) \qquad [10]$$

rg_1 denoting a second gradient pulse that is r times longer than the first one. r is actually set according to the desired order of filtering, meaning that one-spin systems are rejected for $r = 2$, one- and two-spin systems are rejected for $r = 3/2$ (**Figure 7**) and so on. Before going into detail for a particular spin system, it can be anticipated that the filtering properties should rest on antiphase coherences rather than on multiple quantum coherences as might be the case with B_0 gradients. To make this statement clearer, it can be recalled that an antiphase A doublet (one line positive, the other negative) of a two-spin AX system is represented by the operator product $2I_x^A I_z^X$ (provided the

two A lines are along the x axis of the rotating frame). Now, B_1 gradients acting in vertical planes (e.g. x,z or y,z) are prone to operate on such antiphase coherences rather than on multiple quantum coherences (e.g. of the type $2I_x^A I_y^X$). In order to clarify this point, consider the A coherences which can exist after an evolution interval following a standard $\pi/2$ pulse: I_x^A, I_y^A, $2I_x^A I_z^X$, $2I_y^A I_z^X$ and let θ be the angle of nutation produced by g_1 for a given spatial location, acknowledging that, at the outcome, an average over all possible values of θ must be performed (for sufficient gradient strength and duration: $\langle \sin \theta \rangle = \langle \cos \theta \rangle = 0$, $\langle \sin^2 \theta \rangle = \langle \cos^2 \theta \rangle = 1/2$). From these considerations, it is easy to derive the way in which the various coherences transform under the gradient cluster $(g_1)_x(2g_1)_y$:

$$I_x^A \xrightarrow{(g_1)_x(2g_1)_y} 0$$

$$I_y^A \xrightarrow{} 0$$

$$2I_x^A I_z^X \xrightarrow{} 0$$

$$2I_y^A I_z^X \xrightarrow{} \left(2I_y^A I_z^X\right)\big/4 + \left(2I_z^A I_y^X\right)\big/4$$

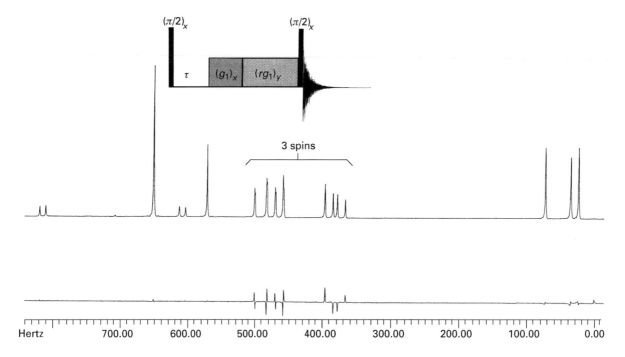

Figure 7 Reference (top) and 3 spin-filtered (bottom) spectra. The latter has been obtained with the sequence shown at the top of the Figure (*r* 3/2; see text). The time τ is chosen so that antiphase coherences can develop; the last $\pi/2$ pulse is for purging purposes.

Clearly, all quantities cancel (including those corresponding to a single spin system) except the antiphase coherence $2I_y^A I_z^X$, scaled down by a factor 4 and which leads to the corresponding quantity transferred to the X nucleus. Thus, in addition to its filtering properties, this gradient cluster affords transfer capabilities suitable for two-dimensional correlation spectroscopy without the need of phase cycling procedures.

List of symbols

B_0 = static magnetic field; B_1 = alternating magnetic field; D = self-diffusion coefficient; g_1 = gradient amplitude; I_+ and I_- = raising and lowering operators; p_i = coherence order; r = coil radius; S = signal; T_1 = longitudinal relaxation time; T_2 = transverse relaxation time; α = flip angle; γ = gyromagnetic ratio; δ = chemical shift; Δ = diffusion interval; θ = angle of nutation; ν_0 = resonance frequency; ν_r = transmitter frequency; ρ = spin density; σ = shielding coefficient; ω_r = notating frame angular velocity.

See also: **Diffusion Studied Using NMR Spectroscopy; Magnetic Field Gradients in High Resolution NMR; MRI Theory; NMR Microscopy; NMR Principles;** **NMR Pulse Sequences; NMR Spectrometers; Product Operator Formalism in NMR; Two-Dimensional NMR Methods.**

Further reading

Bosch CS and Ackerman JJH (1992) Surface coil spectroscopy. *NMR Basic Principles and Progress* 27: 3–44.

Bottomley PA (1992) Depth resolved surface coil spectroscopy. *NMR Basic Principles and Progress* 27: 67–102.

Callaghan PT (1993) *Principles of Nuclear Magnetic Resonance Microscopy.* Oxford: Oxford University Press.

Canet D (1997) Radiofrequency field gradient experiments. *Progress in NMR Spectroscopy* 30: 101–135.

Kimmich R (1997) *NMR Tomography, Diffusometry, Relaxometry.* Berlin: Springer.

Kormoroski RA (1993) Non medical applications of NMR imaging. *Analytical Chemistry* 65: 1068A–1077A.

Maffei P, Mutzenhardt P, Retournard A *et al* (1994) NMR microscopy by radiofrequency field gradients. *Journal of Magnetic Resonance* A107: 40–49.

Norwood TJ (1994) Magnetic field gradients in NMR: friend or foe? *Chemical Society Reviews* 23: 59–66.

Price WS (1996) Gradient NMR. *Annual Reports on NMR Spectroscopy* 32: 51–142.

Price WS (1998) NMR imaging. *Annual Reports on NMR Spectroscopy* 35: 139–216.

Radiofrequency Spectroscopy, Applications

See **Microwave and Radiowave Spectroscopy, Applications.**

Raman and Infrared Microspectroscopy

Pina Colarusso, **Linda H Kidder**, **Ira W Levin** and
E Neil Lewis, National Institutes of Health, Bethesda,
MD, USA

> **VIBRATIONAL, ROTATIONAL &
> RAMAN SPECTROSCOPIES**
> **Methods & Instrumentation**

Introduction

Vibrational Raman and infrared microspectroscopy
are analytical tools that characterize both the chemistry and the physical structure of materials. These
methods combine two separate approaches, vibrational spectroscopy and light microscopy, for elucidating chemical systems at the micron and the
submicron levels. Whereas vibrational spectroscopy
probes the details of molecular composition, light
microscopy reveals sample morphology based on the
variations in optical properties. Vibrational micro-
sopy builds on both of these techniques, and thus
provides chemically selective visualizations of microscopic samples. Domains within complex samples,
ranging from diamond inclusions in a mineral to malignant cells in a tissue biopsy, may be examined
with Raman and infrared techniques. The efficacy
and flexibility of vibrational microspectroscopy is
highlighted by its wide adoption in diverse areas
such as geology, medicine, forensic science, and industrial process control.

Since the theoretical background and practical implementation of vibrational spectroscopy are described elsewhere in this encyclopedia, the emphasis
here will be on the extension of Raman and infrared
spectroscopy to the microscopic realm. It should be
noted that many of the methods described here are
applicable to other microspectroscopic methods, especially those based on optical phenomena such as
fluorescence.

The balance between spectroscopy and microscopy
distinguishes one vibrational microscopic technique
from another. At one extreme, spectroscopic and microscopic analyses are carried out separately. In a
typical arrangement, the sample is examined through
a microscope under white-light illumination, and
then either Raman or infrared spectra are recorded at
one or more selected points. Approaches that more
fully integrate the acquisition of morphological and
spectroscopic data have also been developed. Mapping techniques record spectra at successive points or
lines within the sample; the spectra are combined to
provide a view of the sample at specific vibrational
frequencies. Imaging techniques, by contrast, record
spectra simultaneously for contiguous points within
a given sample area.

Mapping and imaging methods allow for sample
visualizations over sequential wavelength intervals.
The data sets are represented by a three-dimensional
'image cube' having two spatial and one spectral dimension (see **Figure 1**). Depending on the orientation, two-dimensional slices along a particular axis
yield either a set of image planes stacked as a function of wavelength or a group of spatially resolved
spectra. Vibrational spectroscopic mapping and imaging combine the morphological and chemical analyses of microscopic systems and reveal trends that
are often difficult to extract from bulk or isolated
single-point measurements.

Instrumentation

A standard Raman or infrared microspectrometer
consists of an excitation source, a compound microscope, a spectrometer, and a detector. As in bulk
techniques, the design of microspectroscopic
experiments is guided by the sample composition as
well as demands for frequency response, sensitivity,
data acquisition rates, and spectral resolution. Factors specific to vibrational microspectroscopy include
the spatial resolution and the optical throughput between the microscope and the spectrometer.

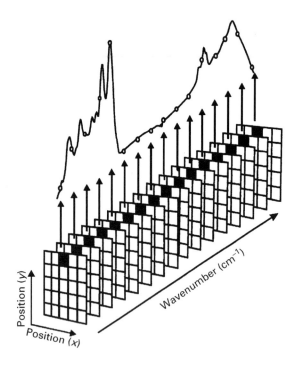

Figure 1 Mapping and imaging data can be depicted as an 'image cube' having two spatial and one spectral dimensions.

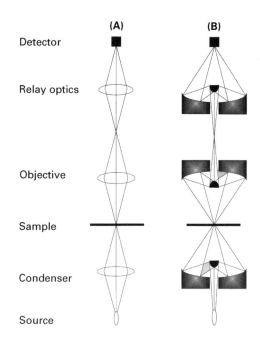

Figure 2 Schematics of (A) refractive and (B) reflective microscopes.

Sources

Raman microspectroscopy is usually implemented from the ultraviolet to near-infrared wavelength regions, from about 0.3 to 1.5 μm. Excitation with sources such as Ar⁺, HeNe, and solid-state lasers is standard. Infrared microspectroscopy, by contrast, is usually carried out between 0.78 and 25 μm with broadband sources such as quartz lamps or ceramic globars. More recently, excitation with synchrotron radiation has also been demonstrated for infrared measurements.

The compound microscope

The compound microscope is central to vibrational microspectroscopy. The optical components and light paths for both refractive and reflective microscopes are shown in **Figure 2**. As illustrated, a compound microscope contains a light source, a condenser, an objective, various apertures, and an ocular. The sample is first visualized under white light to aid in observation and alignment. The condenser illuminates the sample, and the transmitted light is collected by the objective. Light also can be reflected (or scattered) from the sample; in this configuration, which is known as epi-illumination, the objective also serves as the condenser. In either case,

an inverted and usually enlarged image of the sample is formed at the back focal plane of the objective. This intermediate image is either relayed through collection optics to a video camera or through the ocular, which further magnifies the image for viewing. Microscopes tailored for operation at ultraviolet and infrared wavelengths usually contain a parfocal white-light path for sample observation.

Following the preliminary microscopic examination, the spectroscopic source is introduced on to the sample. In Raman microspectroscopy, the probed region is defined by the size of the impinging laser beam. For infrared applications, the radiation is localized within a given area in part by placing an aperture between the source and the sample. An image of the irradiated area, corresponding to the Raman or infrared signal, is formed behind the objective, and is diverted to the spectrometer for analysis. The optical path of the vibrational signal is configured such that it focuses in the same plane as the white-light image; the parfocal light paths ensure that identical sample regions are examined in both procedures, while maximizing the optical throughput.

Microscope lenses Microscope lenses are complex assemblies that are designed to balance the requirements for magnification, focal length, and light collection with various factors that degrade image fidelity (note that the optics in **Figure 2** are simplified for clarity). Refractive and reflective optics exhibit aberrations that cause image blurring and deformation. These include effects such as spherical aberrations and astigmatism. Refractive optics also exhibit chromatic aberrations which causes light of different wavelengths to focus at separate points.

The highest quality images are generally obtained with light microscopes that contain refractive glass elements. Visible Raman microspectroscopy, in particular, benefits from the mature optical technology that has been developed for visible wavelengths. Refractive optics are available for vibrational microspectroscopic applications at other wavelengths; examples include quartz for the ultraviolet and germanium for the infrared. Reflective optics can also be tailored for different wavelength intervals and typically consist of glass substrates coated with a metallic layer. Since they do not exhibit chromatic aberration, reflective optics are particularly useful when large wavelength intervals are covered, as in mid-infrared microspectroscopy.

A standard optic in a reflective microscope is the Schwarzchild (also referred to as Cassegrainian) lens; it can be used as both a condenser and an objective. As depicted in **Figure 2B**, the Schwarzchild lens is centrally obscured, which reduces the amount of light collected compared with a refractive optic of similar dimensions. Since the optic is reflective, chromatic aberration is not a concern, but the spherical design can lead to image warping.

It is standard to label objectives with the numerical aperture, NA, which is a measure of the light-gathering capability of an optic:

$$NA = n \sin \phi_{\max} \qquad [1]$$

where n is the refractive index of the medium between the sample and objective and $\phi_{\max}$ is the half-angle of the maximum cone of light collected by the lens (see **Figure 3**). In a given apparatus, the NA of the objective, relay optics, and spectrometer aperture are matched for optimum light collection.

Vibrational objectives typically operate in air, which has a refractive index near unity; the highest NA value obtained in air is about 0.95 using refractive visible objectives. By changing the surrounding medium, and thus n, it is possible to obtain values greater than 0.95. Immersion objectives operating under media such as water or oil achieve NA values greater than 1. For mid-infrared applications, the maximum NA for reflective objectives is about 0.65. For optimal image formation, it is important to follow the manufacturer's specifications for thickness

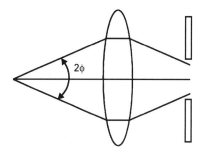

Figure 3 The numerical aperture is defined as $n \sin \phi$, where n is the refractive index of the surrounding medium and ϕ is the half-angle of the cone of light collected by the optic.

and refractive index of slides and coverslips (if applicable).

Spatial resolution Diffraction effects influence the spatial resolution of a vibrational microspectroscopic measurement. To illustrate consider, for example, monochromatic light from a point source as it propagates through the objective. The image that forms at the focal plane of the objective is not a point, but rather a diffraction pattern consisting of alternating light and dark concentric circles. The bright central disc in the pattern is known as the Airy disc, with a radius r_A given by

$$r_A \approx 0.61 \, \lambda / NA_{obj} \qquad [2]$$

where λ is the wavelength of light and NA is the numerical aperture of the optic.

Two incoherent point sources of equal brightness lying in a plane perpendicular to the objective are just resolved when

$$R \approx 1.22 \, \lambda / (NA_{obj} + NA_{cond}) \qquad [3]$$

where R is the distance between the two points and NA_{obj} and NA_{cond} are the numerical apertures of the objective and condenser, respectively. Equation [3] is known as the Rayleigh criterion, and is a standard measure of the lateral spatial resolution. For an epi-illumination measurement, Equation [3] is recast as

$$R \approx 0.61 \, \lambda / NA_{obj} \qquad [4]$$

From Equations [3] and [4], it is seen that the optimum spatial resolution is obtained when objectives with high NA are used for measurements at short-wavelengths.

The spatial resolution criteria given in Equations [3] and [4] apply for diffraction-limited microscope objectives in the far-field limit. In recent years, near-field techniques have been devised that exceed the diffraction limit by scanning a tapered light source, with a spot size less than the probe wavelength, in close proximity to the sample.

Spectrometers

Dispersive elements and interferometers are widely used in vibrational microspectroscopy. As in bulk measurements, microscopic Raman studies are carried out with grating monochromators, spectrographs, or Fourier transform spectrometers, although Fourier transform instruments are usually limited to applications in the near-infrared spectral region. Infrared microspectroscopy, by contrast, is almost exclusively a Fourier transform technique.

At visible and near-infrared wavelengths, vibrational microspectroscopy is also implemented with solid-state filters, particularly in imaging applications. Electronically driven devices such as acousto-optic and liquid crystal filters provide the tunability and moderately narrow passbands required for spectroscopic imaging. These high-speed filters contain no moving parts and can be custom-built to provide spectral resolutions of less than 10 cm^{-1} over large spectral ranges. Acousto-optic tunable filters (AOTFs) consist of a piezoelectric transducer bonded to a birefringent crystal such as TeO_2. When an RF frequency is applied to the crystal, an acoustic wave is generated that diffracts light over a narrow spectral interval. The AOTF passband is modified by varying the input RF frequency.

Another useful solid-state device is an interference filter fabricated from a set of liquid crystals. The birefringent properties of a liquid crystal tunable filter (LCTF) can be varied by applying an external voltage across a crystal axis. A filter is constructed from a series of polarizers and liquid crystals. A particular passband is selected by tuning the individual liquid crystal elements.

Thin-layer interference filters with passbands between 18 and 50 cm^{-1} are also applied in microspectroscopic imaging. These devices can be tuned over large wavenumber ranges by varying the angle of incidence. Broader wavelength coverage may be obtained with a series of filters, which can be placed in a device such as a filter wheel.

Detectors

The detectors used in vibrational microspectroscopy operate by transducing either the capture of a photon or a minute change in temperature into an electrical response. Both photon and thermal detectors are used individually or configured into an array. The single-point detectors that comprise the array detector are known as picture elements or pixels. **Table 1** lists the wavelength ranges and operating temperatures of several array detectors that are used in Raman and infrared microspectroscopy and imaging.

For the visible to near-infrared region, charge-coupled device (CCD) detectors have been widely adopted for single-point, mapping, and imaging Raman microspectroscopy. These photosensitive arrays, in most cases, are monolithic silicon devices that can be fabricated with millions of individual pixels. The wavelength response of CCDs generally declines in the red, cutting off at about 1.1 μm, the band-gap of silicon. For this reason, CCDs have limited application in near-infrared microspectroscopy.

For infrared microspectroscopy, single-element detectors are used for point and mapping measurements. More recently, array detectors have been applied for spectroscopic imaging in the infrared. In infrared focal plane arrays, the monolithic silicon design used in CCDs is replaced by a hybrid construction. In a hybrid detector, photon detection occurs in a semiconductor layer (indium antimonide, mercury cadmium telluride, and doped-silicon are typical detector materials), while the readout and amplification stages are carried out in a silicon layer. The two layers are electrically connected at each pixel through indium 'bump-bonds'. Other innovations such as microbolometer arrays also show promise for spectroscopic imaging applications.

Raman microspectroscopy

Point microscopy

Raman microspectroscopy is routinely implemented using epi-illumination. As shown in **Figures 4A–4E**, the objective directs the laser excitation onto the

Table 1 Properties of various array detectors

	CCD	InGaAs	Pt:Si	InSb	HgCdTe (MCT)	Si:As	Microbolometer
Wavelength range (μm)	300–1.1	0.9–1.7	1–5.7	0.5–5.4	0.8–12.5	1–25	8–14
Operating temperature(K)	77	300	77	77	77	<10	300

sample and collects the light scattered from the surface. Often fibre optics are used for safe and convenient optical coupling of the microscope to the laser and spectrometer. In point microspectroscopy (**Figure 4A**), the Raman signal from a small spot on the sample is dispersed by a grating spectrometer and focused onto a CCD detector. The constituent wavelengths are mea-sured by one or more pixels along a long narrow strip on the CCD. The undesired Rayleigh scattered light, corresponding to the laser excitation wavelength, is removed by placing one or more holographic or di-electric notch rejection filters between the microscope and spectrometer. Various grating designs, such as dual-stage and subtractive monochromators, also may be used to remove the laser excitation, but are less optically efficient than instruments that incorporate notch rejection filters and single-stage monochromators.

Confocal Raman microscopy

Optical sectioning of a sample may be achieved with confocal Raman microscopy. This technique rejects out of focus light by introducing pinhole apertures into the optical train of the microscope. A confocal scheme is shown in **Figure 5**, where one aperture is placed between the laser and the objective, and another is placed at the image plane of the objective. The apertures block the light scattered from regions outside the focal plane of the objective. Crisp images of thin optical sections may be obtained by mapping the sample point-by-point. Furthermore, a three-dimensional image of the sample may be constructed by carrying out measurements at different sample depths.

Mapping

Mapping techniques scan the excitation source or the sample in a raster or line pattern (see **Figures 4B and 4C**). In a raster scan, spectra may be recorded at successive points by moving either the sample or source on a motorized platform. Mapping may also be implemented by irradiating the specimen with a line source. In one method, a line is produced by rapidly sweeping the laser beam back and forth with a mirror powered by a piezoelectric transducer. It is also possible to focus the laser beam into a line with cylindrical optics. High-precision designs for mapping apparatus minimize mechanical instabilities and positioning errors.

For point mapping, the Raman signal is recorded as in single-point measurements. When the excitation is along a line, the Raman signal is collected through the objective, directed onto a grating and dispersed on to a CCD. The two-dimensional detector then records the spatial information along one axis and wavelength data along the other. An image cube is gradually built up as the line is moved across the full sample area.

Imaging

As shown in **Figures 4D** and **4E**, point and line excitation are replaced by wide-field irradiation in both spatial encoding and direct imaging methods. In a common arrangement, a given sample area is illuminated by defocusing the laser through a beam expander prior to the objective. The sample illumination is not perfectly uniform in this configuration because the intensity distribution of the laser beam (typically Gaussian) is preserved. Other properties of monochromatic coherent excitation may also affect the image quality. Despite the experimental concerns, wide-field illumination is useful when dealing with fragile specimens, since the power density is reduced, thus minimizing damage from thermal or photolytic processes.

Hadamard transform imaging Spatial encoding methods such as Hadamard transform imaging also can be used for the spectroscopic visualization of samples. More recently, developments in digital microarray technology will likely provide a convenient new approach for spatial encoding from the mid-infrared to the ultraviolet. In one common arrangement, the entire sample area is irradiated with wide-field, epi-illumination (**Figure 4D**). Part of the Raman signal emanating from the sample is blocked with a mask containing a series of apertures. The spatially filtered signal is focused on to an entrance slit of a monochromator, which disperses the signal across a two-dimensional array detector. The slit preserves one image axis while the Hadamard mask is used to encode the other axis. Subsequent measurements are carried out with the mask in different positions. Each measurement corresponds to the Raman signal from the unmasked points on the sample along one spatial axis over the entire spectral range of interest. The experiment is designed such that the number of independent measurements equals the number of points on the sample. The spatially dependent images are then converted to spectroscopic images through a Hadamard transform. Unlike the other methods mentioned in this article, Hadamard spectroscopic imaging systems are limited to research activities and are not yet commercially available.

Direct imaging Raman microscopic imaging can also be carried out by imaging the sample through an interferometer or filter (shown in **Figure 4E**). The sample is irradiated with wide-field, epi-illumination and then modulated or filtered images are recorded

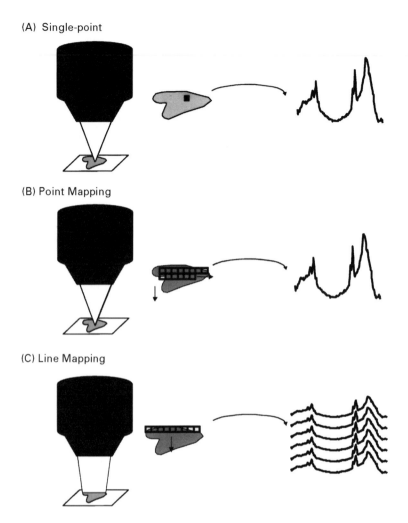

(A) Single-point

(B) Point Mapping

(C) Line Mapping

Figure 4 Several different approaches employed in vibrational microspectroscopy. (A) point measurements: vibrational spectra are obtained at individual, selected points within the sample. (B) Point mapping: spectra are recorded at successive spots along within the sample. (C) Mapping is also carried out with line excitation. (D) Spatial encoding methods such as Hadamard transform imaging: a physical mask blocks part of the signal from reaching the detector. A series of images is obtained with the mask in different positions, and then the data are converted to wavelength-dependent images through a Hadamard transform. (E) Direct imaging: the spectroscopic image is obtained by recording the signal from all points on the sample simultaneously over a narrow spectral interval. A series of images at discrete wavelengths is recorded to provide spectrscopic information for each pixel.

with a CCD. Spectroscopic information is obtained by recording sequential images over a range of optical retardations in the case of the interferometer or a range of frequencies in the case of the tunable filter. In this way, spectra corresponding to various points within the sample are obtained at each detector pixel. Since the image is captured in its entirety, the results are not prone to mechanical positioning errors that may occur in point or line mapping methods. The image quality, in fact, is primarily limited by diffraction. When image shifts occur in direct imaging, they are usually predictable and can be compensated by either experimental or computer procedures.

The basic principles of Raman microscopy and imaging are encapsulated by **Figure 6**, which depicts a Raman image and a spectrum of a 1 µm diameter polystyrene bead. The data were obtained with a mi-

croscope coupled to a CCD array; individual images were recorded through an AOTF, which was successively tuned between 747 and 1363 cm^{-1} from the exciting 647 nm Kr^+ laser line. The Raman image shown was recorded at 1000 cm^{-1}, corresponding to a symmetric aromatic ring vibration in polystyrene. The bright areas correspond to areas rich in polystyrene, thus revealing the distribution of beads in the sample. A spectrum for every pixel is available because images were recorded over a series of wavelengths. **Figure 6** illustrates a spectrum obtained from a single pixel in one of the bead centres.

Infrared microspectroscopy

Infrared microspectroscopy is based on either transmission or reflection measurements. Transmission is

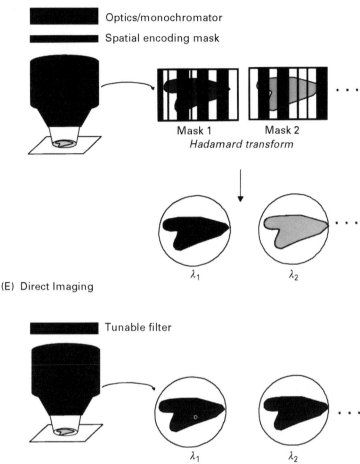

(D) Hadamard Imaging

Optics/monochromator

Spatial encoding mask

Mask 1 Mask 2
Hadamard transform

λ_1 λ_2

(E) Direct Imaging

Tunable filter

λ_1 λ_2

Figure 4 (*Continued*)

the simplest implementation; the maximum sample thickness is sample and wavelength dependent, however, and is limited to ~15–20 μm for mid-infrared measurements. Reflection measurements are indispensable for samples that are thick or opaque. Various microscope objectives have been tailored for measurements based on the corresponding bulk infrared techniques, such as diffuse reflectance, attenuated total reflectance, and grazing angle reflectance.

Optical effects such as stray light or scattering are treated by placing one or more apertures in the optical train of the infrared microscope. For transmission measurements, one aperture is placed between the condenser and the sample and another is placed between the objective and the image plane. The two apertures are matched in size in order to optimize the signal. For reflectance measurements, only one aperture is used. This type of arrangement is widely employed in infrared spectroscopy, and is known as 'redundant aperturing'.

Single-point measurements

Most single-point infrared microspectroscopy is implemented with Fourier transform spectrometers, for which the well known multiplex and throughput advantages generally apply. The multiplex advantage arises because all of the input radiation is detected over the entire scan time; however, it applies only when the dominant source of noise is the detector. When the detection is shot-noise limited, as with CCDs, the multiplex advantage does not hold. The throughput advantage for interferometers arises from the use of circular apertures. Dispersive instruments, by contrast, require narrow exit and entrance slits, particularly for higher spectral resolution. Furthermore, a circular aperture provides a more convenient geometry for coupling a microscope to the spectrometer.

Trace amounts of sample can be analysed with infrared microspectroscopy. The technique, for

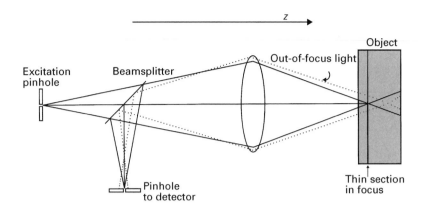

Figure 5 Confocal microscopy: pinhole apertures placed in the optical train of the microspectrometer lead to rejection of out-of-focus light.

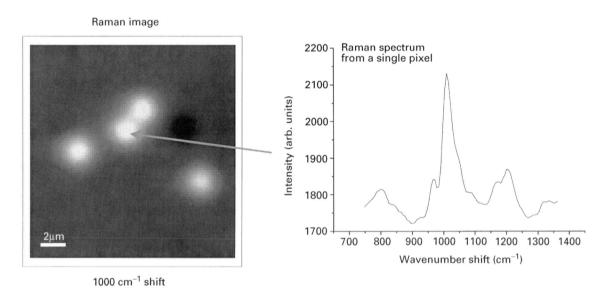

Figure 6 A Raman spectroscopic image of 1 μm diameter polystyrene beads. The spectroscopic image on the left exhibits bright regions that correspond to the Raman signal at 1000 cm^{-1} shift, a symmetric aromatic ring vibration. A series of images was recorded between 747 and 1363 cm^{-1} shift; a spectrum is then available for every spatial resolution element in the image. The trace on the right corresponds to the spectrum of a single pixel from the centre of a single polystyrene bead.

example, is a standard tool in forensic science. **Figure 7** illustrates fibre spectra that were key evidence in a criminal investigation. A cotton fibre fragment, which was recovered from the nose of a bullet, was matched spectroscopically to fibres from the vest of a police officer who had been shot. Fibre samples represent a system that is difficult to examine with bulk techniques, but which is amenable to vibrational microscopic analysis.

Mapping and imaging

Infrared mapping and imaging approaches are applied as in the corresponding Raman approaches, with the appropriate choices for source, spectrometer, and detector. For example, a combination of an acousto-optic filter and an InSb focal plane array can be used to record frequency-dependent images in the near-infrared. In the mid-infrared, direct images are recorded with a step-scan interferometer coupled to an infrared focal plane array detector. To illustrate, consider a transmission measurement in which a step-scan interferometer modulates the output from a blackbody source. The infrared radiation is directed on to the sample with the condenser. The transmitted infrared radiation is collected by the objective and then focused on to the array detector. Over the course of the scan, the movable mirror in the

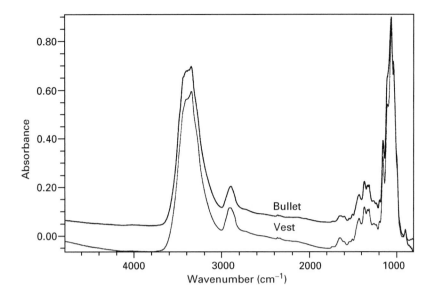

Figure 7 Infrared microspectra of cotton fibres collected in a forensic application. The spectrum of a fibre retrieved from a bullet (top trace) is seen to match the spectrum of a fibre from the police officer's vest (bottom trace). Data courtesy of John A. Reffner, Spectra-Tech, Inc. and Ronald P. Kaufman, Maine State Police Crime Laboratory.

interferometer is paused for several milliseconds at every step in order to record the image. At the completion of a single scan, a complete interferogram is generated at each detector pixel. The interferograms are then converted to frequency by standard Fourier transform processing.

As an example of the Fourier transform infrared imaging technique, **Figure 8** depicts an infrared spectroscopic image of human breast cells which was recorded with a 64×64 mercury–cadmium–telluride (MCT) array. The data set then encompasses 4096 separate interferograms, one for each pixel. Spectra

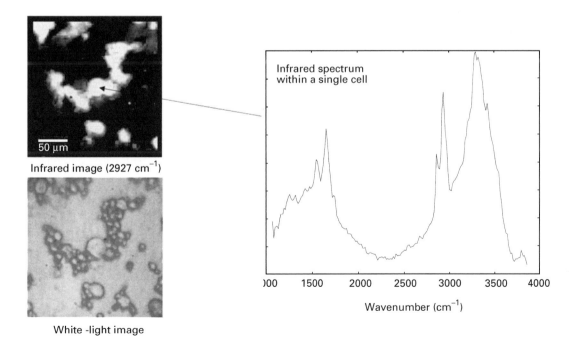

Figure 8 An infrared spectroscopic image of human breast cells. The spectroscopic image on the left exhibits contrast based on the intensity of the absorption band centred at $2927\,cm^{-1}$ (antisymmetric CH_2 stretch), which contains contributions from both the lipid and protein fractions within the cell. Since images are obtained over a contiguous wavenumber interval, it is possible to construct a spectrum for every image pixel. The spectrum extracted from one of the cells is shown on the right.

are obtained through standard Fourier transform procedures. The resulting data cube contains the set of image planes stacked as a function of frequency. Each image exhibits contrast based on the differences in infrared spectral response, which in turn reflects the variation in the chemical composition within the sample. **Figure 8** depicts one such image plane at 2927 cm^{-1}, which corresponds to a protein and lipid infrared molecular marker (CH$_2$ antisymmetric stretch). In this way, the distribution of biochemical species can be visualized across the sample. More detailed chemical information can be obtained by examining the infrared spectra that are associated with each pixel in the image.

Sampling considerations in image generation

The Nyquist theorem specifies that a sinuisoidal function in time or distance can be regenerated with no loss of information as long as it is sampled at a frequency greater than or equal to twice per cycle. The Nyquist theorem must be considered in direct imaging applications because the signal is sampled by the discrete pixel elements in an array. Consider a diffraction limited arrangement with a lateral spatial resolution R_L. If the total magnification M_{tot} is a product of the magnifications of the microscope objective M_{obj} and the projection lens M_{proj}, the Nyquist theorem requires

$$M_{tot}R_L \geq 2p \qquad [5]$$

where p is the pixel size. That is, the sampling interval must be at least twice the highest spatial interval. If the smallest resolvable feature is 5 µm, then each detector pixel must sample intervals that are ≤ 2.5 µm. As long as Equation [5] is obeyed, the spatial fidelity of the microscopic image is preserved and sampling artifacts are avoided. It follows that oversampling does not provide any additional information; this is also known as empty magnification.

See also: **IR Spectrometers; Raman Spectrometers; Scanning Probe Microscopes.**

List of symbols

n = refractive index; NA = numerical aperture; M = magnification; p = pixel size; r_A = radius of the Airy disc; R = distance between two incoherent point sources; R_L = lateral spatial resolution; λ = wavelength of light; ϕ_{max} = half-angle of the maximum cone of light collected by the lens.

Further reading

Humecki HJ (ed) (1995) *Practical Guide to Infrared Microspectroscopy.* New York: Marcel Dekker.

Inoué S and Oldenburg R (1995) Microscopes. In: van Stryland EW, Williams DR and Wolfe WL (eds) *Handbook of Optics*, Vol II, pp 17.1–17.52. New York: McGraw-Hill.

James J and Tanke HJ (1991) *Biomedical Light Microscopy.* Dordrecht: Kluwer Academic Publishers.

Katon JE and Summer AJ (1992) IR microspectroscopy. *Analytical Chemistry* **64**: 931A–940A.

Katon JE (1996) Infrared microspectroscopy: a review of fundamentals and applications. *Micron* **27**: 303–314.

Laserna JJ (ed) (1996) *Modern Techniques in Raman Spectroscopy.* Chichester: Wiley.

Lewis EN, Treado PJ, Reeder RC *et al* (1995) Fourier transform spectroscopic imaging using an infrared focal-plane array detector. *Analytical Chemistry* **67**: 3377–3381.

Messerschmidt RG and Harthcock MA (eds) (1998) *Infrared Microspectroscopy: Theory and Applications.* New York: Marcel Dekker.

Puppels GJ, de Mul FFM, Otto C *et al* (1990) Studying single living cells and chromosomes by confocal Raman spectroscopy. *Nature* **347**: 301–303.

Reffner JA (1998) Instrumental factors in infrared microspectroscopy. *Cellular and Molecular Biology* **44**: 1–7.

Rieke GH (1994) *Detection of Light: from the Ultraviolet to the Submillimeter.* Cambridge: Cambridge University Press.

Turrell G and Corset J (eds) (1996) *Raman Microscopy: Developments and Applications.* London: Academic Press.

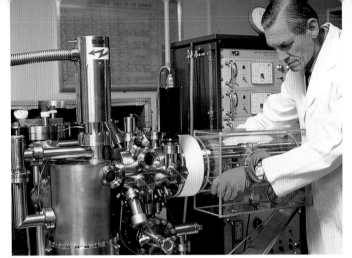

Plate 43 X-ray photoelectron spectrometer. This is an analytical instrument, used mainly in metallurgy. It consists of an electron source, a sample stage and X-ray detectors. A stream of electrons is accelerated toward and focussed on the sample. When an electron strikes an atom in the target, it may cause the ejection of an electron in the atomic shell. If this is replaced from an electron in a higher-energy shell, a photon of a specific wavelength is emitted, normally in the X-ray region. This may be detected and analysed, giving an indication of the identity and quantity of given elements in the sample. *See Photoelectron Spectrometers*. Reproduced with permission from Science Photo Library.

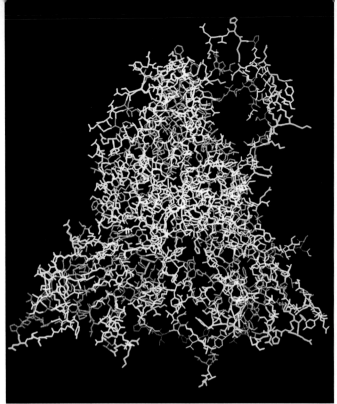

Plate 44 (above) Common cold virus protein: computer graphics representation of protein comprising 1 of the 60 faces of the icosahe-dral capsid (casing) of rhinovirus-14, a virus causing the common cold. This image is of the atomic backbone of the protein. Rhinovirus belongs to the picornavirus (small RNA viruses) group, which also includes the enteroviruses and the agent of foot and mouth disease in cattle. They are all relatively small (24–35 nanometres diameter), non-enveloped, with their genetic information held in the form of RNA. Rhinovirus infects the nose and throat, and is spread in droplets from talking, coughing and sneezing. *See Proteins Studied Using NMR Spectroscopy*. Reproduced with permission from Science Photo Library.

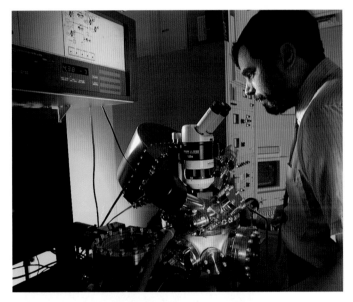

Plate 45 (left) A researcher at the BOC Group Technical Centre, Murray Hill, New Jersey, USA, using an advanced Electron Spectroscopy For Chemical Analysis (ESCA) unit. ESCA provides qualitative and quantitative analysis of the chemistry of elements present in the outermost layers of solid materials. The unit is used in the development of thin-film coatings, medical sensors, molecular sieves and catalysts. *See Photoelectron Spectrometers; Photoelectron Spectroscopy*. Reproduced with permission from Science Photo Library.

Plate 46 (above) Photoacoustic Multi-gas Monitor. *See Photoacoustic Spectroscopy, Methods and Instrumentation*. Reproduced with permission from INNOVA AirTech Instruments.

Plate 47 (right) The molecular structure of Bovine Seminal Ribonuclease, as determined using X-ray diffraction. The disulphide bridges are shown in yellow. The structure was drawn from the protein database entry drawn. *See Proteins Studied Using NMR Spectroscopy*. Supplied by Dr G. G. Hoffmann.

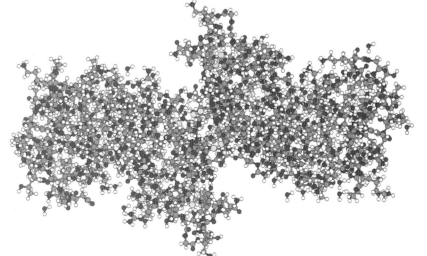

Plate 48 A photoacoustic spectroscopy (PAS) measuring cell, shown with the germanium window half removed. The gas analysis is based on the same principle as conventional infra-red (IR) monitoring except that here the amount of IR light absorbed is measured directly by determining the amount of sound energy emitted upon the absorption. *See Photoacoustic Spectroscopy, Methods and Instrumentation*. Reproduced with permission from INNOVA AirTech Instruments.

Plate 49 Transaxial planes through the brains of three subjects representing glucose utilisation obtained by application of the radiotracer fluoro-deoxy-glucose (FDG) PET tomography. The top of the images are the front of the brain and the person is looked upon from below. The values of the images are colour coded according to a linear scale red being the highest value. The middle panel shows the distribution in a healthy person. The left panel is from a person with Huntington's disease illustrating a loss of tracer uptake at the centre as a consequence of local nervous tissue loss. The right panel is from a person who had similar features of disordered movement (cholera), but based on different brain pathology. The image in this person shows however a clear increase in tracer uptake in the central regions (called: the striatum) of the brain. *See PET, Methods and Instrumentation*. Reproduced with permission from Prof. Leenders.

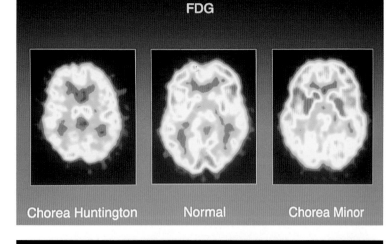

Plate 50 3-D computerized reconstruction of FDG uptake measured by PET in a healthy volunteer. Part of the right upper and frontal brain is "cut out". The left panel shows the tracer uptake in a rest condition. The right panel shows the uptake in the same person but now during intake of a hallucinogen. It can be seen that the frontal cortex region in the brain has a higher (more red) glucose uptake in the activated condition. *See PET, Methods and Instrumentation*. Reproduced with permission from Prof. Leenders.

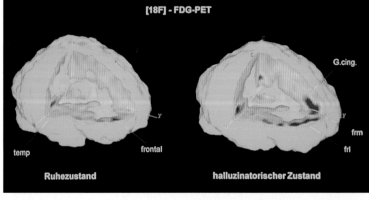

Plate 51 Transaxial planes through the human brain after administration of the of the radiotracer fluoro-dopa indicating dopa decarboxylase capacity as measured by PET. The left panel shows a healthy volunteer whilst right panel shows a patient with Parkinson's disease. It can be seen that the tracer uptake measured with PET in the patient is markedly reduced within the regions of the basal ganglia. *See PET, Methods and Instrumentation*. Reproduced with permission from Prof. Leenders.

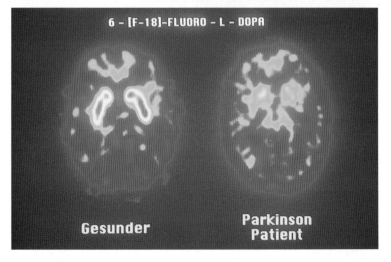

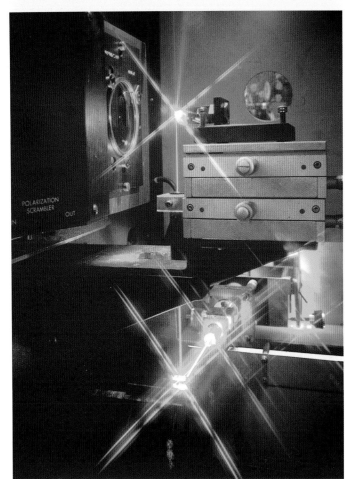

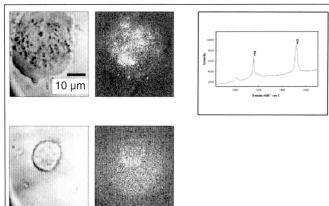

Plate 53 (above) Direct Raman imaging, using the **b**-carotene line at 1552cm^{-1}, of live corpus luteum cells. This provides a rapid insight into the distribution of different molecules within the cell (old cell below, young above). *See Raman and IR Microspectroscopy*. Reproduced with permission from Renishaw plc/Prof. D. N. Batchelder *et al*, Dept. Physics and Astronomy, University of Leeds.

Plate 52 (above) Laser optical bench system used as an excitation source for a Raman spectrometer. Raman spectroscopy provides essentially the same sort of information about molecular structure and dimensions as do infrared and microwave spectroscopy. The laser is an ideal Raman source: it provides a narrow, highly mono-chromatic beam of radiation which may be focused accurately into a small sample. *See Raman Spectrometers; FT-Raman Spectroscopy, Applications; Industrial Applications of IR and Raman Spectroscopy; Polymer Applications of IR and Raman Spectroscopy; IR and Raman Spectroscopy of Inorganic, Coordination and Organometallic Compounds*. Reproduced with permission from Science Photo Library.

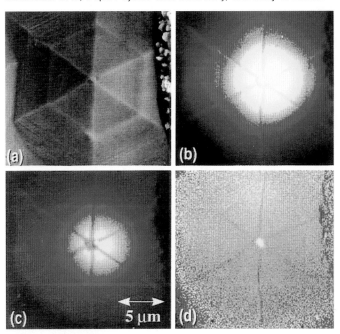

Plate 54 (above) Gallium Nitride Impurity Doping Images. *See Raman and IR Microspectroscopy*. Reproduced with permission from Renishaw plc.

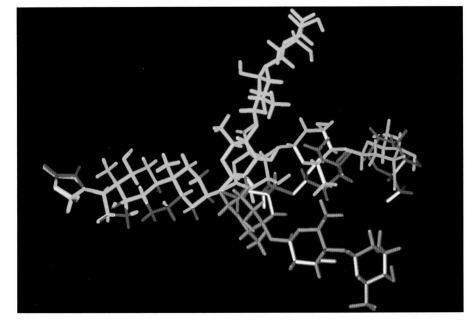

Plate 55 (left) A computer generated molecular model of the cardiac drug digoxin showing the molecular structure as derived by different techniques. The green structure is that determined in the solid state using X-ray diffraction. The red structure is calculated using a molecular dynamics approach. The cyan and magenta structures are the two possible structures determined in solution using NMR spectroscopy. *See Structural Chemistry Using NMR Spectroscopy, Pharmaceuticals*. Reproduced with permission from John Lindon.

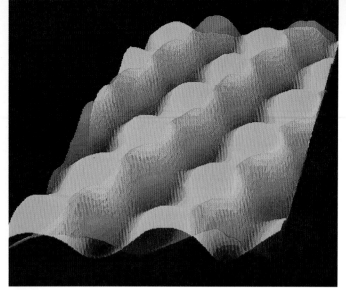

Plate 56 A composite of two images of the GaAs (110) surface. The orange features obtained at positive sample bias are the Ga atoms, while green features obtained at a negative sample bias are the As atoms. Feenstra RM, unpublished results.

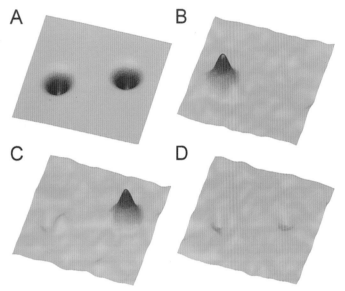

Plate 58 Vibrational spectroscopic imaging of C_2H_2 and C_2D_2. (A) Constant current STM image of a C_2H_2 molecule (left) and C_2D molecule (right). The d2/dV^2 images of the same area recorded with a bias voltage of (B) 358 mV, (C) 266 mV, and (D) 311 mV, with a 10 mV modulation. All images are 48 Å x 48 Å with 1 nA DC tunnelling current. Reproduced with the permission of the American Association for the Advancement of Science from Stipe B.C. et al. (1998) Science **280**: 1732–1735.

Plates 56–59 See Scanning Probe Microscopes.

Plate 60 (right) Sir Isaac Newton. He is shown using a prism to decompose light. See Electromagnetic Radiation. Reproduced with permission from Mary Evans Picture Library.

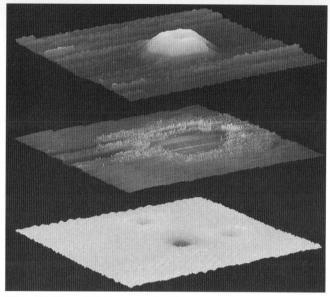

Plate 57 Three STM images of a Ni_3 cluster adsorbed on a MoS_2 basal plane at 4K. All three images show a 60 Å x 60 Å area and are plotted as three-dimensional representations with the same aspect ratio and with the same angle of view. The images were acquired with sample biases of: +2 V (upper), +1.4 V (middle), and –2 V (lower). Reproduced with the permission of the American Chemical Society from Kushmerick JG and Weiss PS (1998) Journal of Physical Chemistry **B102**: 10094-10097.

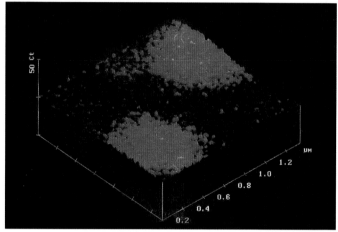

Plate 59 Fluorescence emmission near-field scanning optical microscope image (1.3 m m x 1.3 x m m) of a photosynthetic membrane fragment. Reproduced with permission of the American Chemical Society from Dunn RC et al. (1994) Journal of Physical Chemistry **98** 3094-3098.

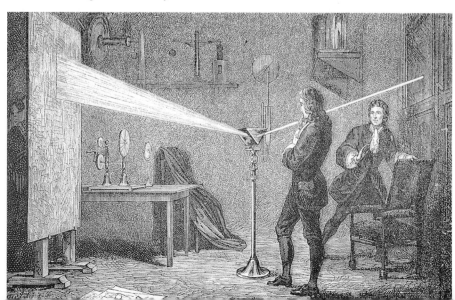

Raman Optical Activity, Applications

Günter Georg Hoffmann, Universität Essen,
Germany

> **VIBRATIONAL, ROTATIONAL &
> RAMAN SPECTROSCOPIES**
> **Applications**

Introduction

Raman optical activity (ROA), which is most often measured as the circular intensity difference (CID, the difference between a Raman spectrum excited with right circularly polarized light and that excited with left circularly polarized light), is a very small effect which can be easily obscured by artifacts. It can also be reported as the dimensionless circular intensity difference (CID or better DCID, $[I^R-I^L]/[I^R+I^L]$). The first ROA spectrometers were normal scanning Raman spectrometers modified by adding devices to modulate the polarization of the exciting laser beam. Measurements done on this instrumentation were plagued by the named artifacts and took up to 40 h to record a spectrum. Therefore relatively few spectra were measured before the advent of a new generation of spectrometers. Two main technical advances characterize the construction of these sensitive instruments: firstly the development of multichannel detectors culminating in backthinned CCDs which are practically ideal light detectors (with quantum efficiencies of about 80% at the wavelength of peak sensitivity), and secondly the development of the holographically manufactured line filter. The first improvement of shortened the time of measurement to about 20 min for pure samples and allowed the measurement of dilute samples in a few hours, the second allowed the replacement of multiple monochromators by a single polychromator with much higher optical throughput. Later the use of a backscattering arrangement was added to further improve the spectra. As these newer spectra are of much higher quality, mainly the relatively recent literature is covered in this article. For historic data, the interested reader is directed to earlier reviews given in the Further reading section.

In a typical modern ROA instrument, a linearly polarized laser beam is directed through a Pockels cell, which renders the beam circularly polarized. With a frequency of a few hertz the exciting radiation is switched between right circularly and left circularly polarized and focused onto the sample. In the latest spectrometers the Raman effect is observed in the so-called 180° or backscattering arrangement. This method of observation is achieved by drilling a hole into a plane mirror through which the exciting laser beam may pass. The mirror then reflects the backscattered part of the excited radiation through a highly efficient holographically manufactured notch filter to remove the Rayleigh radiation. The Raman radiation, having passed through a Lyot depolarizer to prevent polarization artifacts, can then be processed by a polychromator that consists of a slit and a single grating to finally reach the detector, a backthinned CCD chip.

As in normal Raman spectroscopy, two methods of detection are possible when not using the backscattering arrangement: through a linear polarizer horizontal to the scattering plane (those spectra are called depolarized) or through a vertical polarizer (polarized spectra).

To be useful for structural determinations, ROA spectra have to be compared with calculated spectra. Models confined to special functional groups or semiempirical calculations have been used with limited success for this purpose. During the last few years *ab initio* studies have become possible. These calculations, especially the time consuming ones with large basis sets or even with the inclusion of configurational interaction, allow for useful derivation of absolute configuration or even conformation of the molecules under study when compared with experimental data.

Stereochemistry of small chiral molecules

One of the simplest chiral molecules one can imagine is bromochlorofluoromethane [1]. As a liquid which until now has only been obtained as an enriched but not pure substance, no crystal structure suitable to deduce the absolute configuration could be obtained. It is therefore a great success for ROA, together with *ab initio* calculations, that it is possible to deduce the absolute conformation of this molecule by comparing experimental and calculated spectra. **Figure 1** shows the depolarized Raman and ROA spectra of CHFClBr [1] (36% enantiomeric excess) together with the calculated spectrum of the (*R*)-enantiomer. Seven of the nine ROA band calculations arrive at the correct sign.

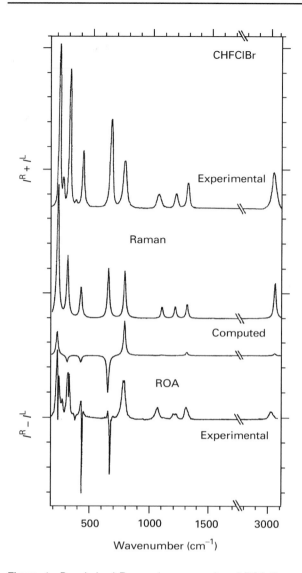

Figure 1 Depolarized Raman (upper curve) and ROA (lower curve) spectra of (–)-CHFClBr together with the calculated spectra (middle curves) for the (R)-isomer. Reproduced with permission of Wiley-VCH from Costante J, Hecht L, Polavarapu PL, Collet A and Barron LD (1997) *Angewandte Chemie International Edition in English* **36**: 885–887. Copyright 1997 Wiley-VCH Verlag GmbH.

[1]

The well-known molecule (2R,3R)-(+)-tartaric acid [2] has been studied both as its d_0 and d_4 isotopomer. As small molecules are less demanding concerning computation time, the molecule could be studied at three theoretical levels: *ab initio* calculations using the 6-31G, 6-31G* and DZP basis sets were performed and compared with the experimental

spectra. The molecule was found to exist as a single conformer in aqueous solution, i.e. the conformation with the carboxylic groups *trans* to each other.

[2]

A comparison between VCD (vibrational circular dichroism) and ROA spectra of the four structurally similar molecules *trans*-pinane [3], *cis*-pinane, α- and β-pinene together with theoretical calculations show that the ratios of the respective parent technique (IR or Raman spectroscopy) to its chiral counterpart are by a factor of 3 in favour of ROA. This advantage is, unfortunately, cancelled by the somewhat more difficult method of measurement. Quantitative comparisons enlighten the supposition that VCD and ROA are highly complementary nonredundant stereochemical techniques.

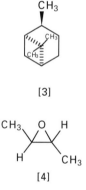

[3]

[4]

The experimental and calculated (6-31G* basis set) spectra of (R)-(+)-dimethyloxirane [4] in the 200–1500 cm^{-1} region agree for the majority of bands. Depolarized, polarized and 'magic angle' Raman and ROA spectra have been measured. By setting the transmission axis of the analyser at the 'magic angle' of ± 35.26° to the vertical, only contributions from the electric dipole–magnetic dipole polarizability are present in the ROA spectra. As expected, calculations at higher theoretical levels give better results for the calculation of normal modes, but for the calculation of the polarizability derivatives they do not show any remarkable advantage. This is very important for practical reasons, as one can use the less time-consuming calculations for the computation of ROA.

Methyl torsion Raman optical activity of *trans*-2,3-dimethyloxirane has been compared with that of *trans*-2,3-dimethylthiirane [5]. Torsion of the two methyl groups in these molecules can occur as in-phase and out-of-phase combinations. The former can be observed as a very weak and broad Raman band at ~200 cm^{-1} in the dimethyloxirane with positive ROA of medium size, whereas it is observed as a shoulder at ~220 cm^{-1} in the spectrum of the dimethylthiirane with large positive ROA. Contrary to the dimethyloxirane spectra, the dimethylthiirane spectra also contain the out-of-phase torsion: it shows up as a medium size Raman band at ~245 cm^{-1} with zero or small negative ROA. This is in good agreement with *ab initio* calculations and even with the old inertial model for methyl torsions.

From the Raman optical activity of *trans*-2,3-dimethylthiirane the absolute configuration has been determined. Its experimental spectrum is very similar to that calculated (*ab initio*, basis set 6-31G*) for the 2R,3R-isomer [5] with best agreement in the skeletal vibration region.

The vibrational spectrum of (S)-3-methylcyclopentanone [6] has been calculated on the 6-31G and on the 6-31G** level, whereas Raman optical activity has only been predicted on the 6-31G**/6-31G and on the 6-31G level. The calculations with the larger basis set did not show any improvements.

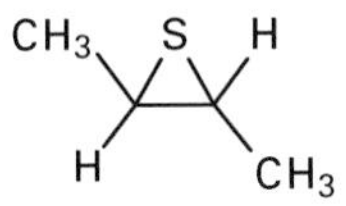

[5]

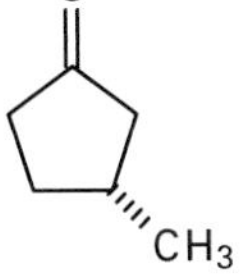

[6]

The ROA spectrum of (+)-3-methylcyclohexanone [7] shows some couplets, which gave rise to much discussion about their origin. Assignments of those vibrations were based on semiempirical calculations with limited reliability. The latest spectrum, however, together with an *ab initio* vibrational analysis, attributes, for example, the couplet at 494 and 518 cm^{-1} to a ring bending mode and an antisymmetric in-plane C=O bending motion coupled to the bending motion of the methyl group relative to the ring, and correctly assigns the (R)-configuration to the molecule.

An extensive study has focused on terpenes with structures related to pinane, camphor [8], and limonene [9]. The in-phase dual circular polarization (DCP$_1$) ROA spectra and the normalized CIDs are presented for fourteen compounds. Correlations between ROA features and structural elements of the molecules are discussed. The study clearly demonstrates the advantage of the backscattering measurements, which for theoretical reasons should be about three times as intensive as the normal right-angle incident circular polarization (ICP) measurements.

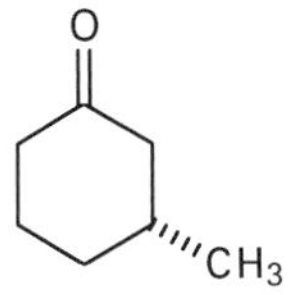

[7]

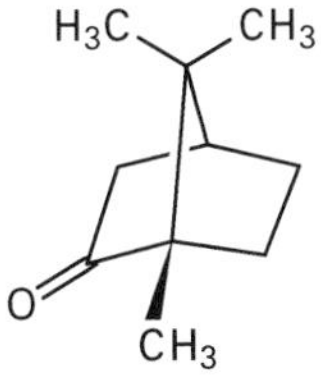

[8]

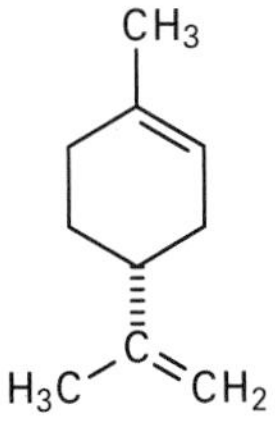

[9]

The Raman and ROA spectra of the hydrochloride salts of four medically applied ephedrine molecules in aqueous solution have been compared in the 700–1700 cm^{-1} region. The salts of (1S,2R)-ephedrine [10], (1S,2R)-norephedrine [11], (1S,2S)-pseudoephedrine [12] and (1S,2S)-norpseudoephedrine [13] show some bands which with a high degree of

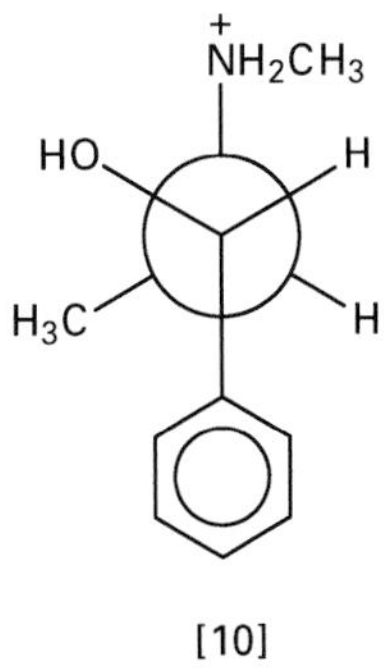

[10]

probability can be used as configurational ROA markers. For example, a band near 840 cm⁻¹ (strongly positive for [11–13]) reflects the chirality at C-1 and probably arises from the symmetric CCO stretch, while a band near 910 cm⁻¹, which is strongly negative in [11 and 12] arises from the *anti*symmetric CCO stretch at C-1. The chirality of the –C*HCH₃(NH₂R⁺) moiety at C-2 may be reflected by the bands of the antisymmetric methyl deformation at 1450 cm⁻¹, as they are oppositely signed in the two (1S,2R)-ephedrines (both positive) and the two (1S,2S)-pseudoephedrines (both negative).

Crystals

Optical activity in crystals is not only the effect of a single molecule but contains also the effects of a large array of similar entities. This is why the crystal

[11]

ROA is about one order of magnitude larger and one can measure vibrations which cannot be observed in solutions or neat liquids. But one has to be careful not to observe artifacts produced by the linear birefringence of the crystals. An interesting class of crystals are the cubic crystals of the sodium halogenates. These are composed of achiral subunits that form a

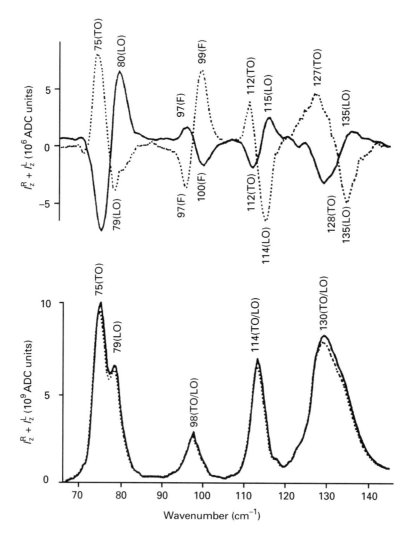

Figure 2 Depolarized right-angle scattering Raman and ROA spectra for the lattice vibrations of (+)-NaBrO₄ (—) and (−)-NaBrO₄ (· · ·), both 1.8 cm⁻¹ resolution. Reproduced with permission of John Wiley & Sons from Lindner M, Schrader B and Hecht L (1995) *Journal of Raman Spectroscopy* **26**: 877. Copyright 1995 John Wiley & Sons.

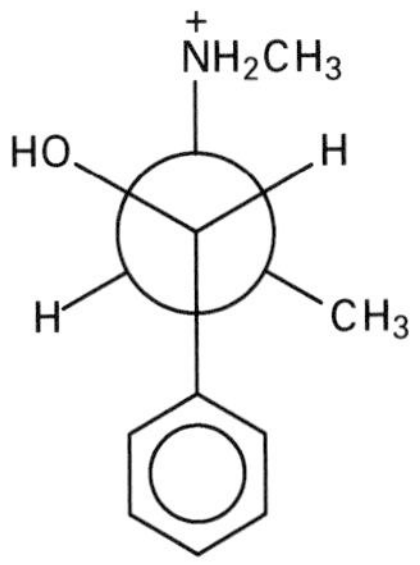

$\overset{+}{NH_2CH_3}$

[12]

$\overset{+}{NH_3}$

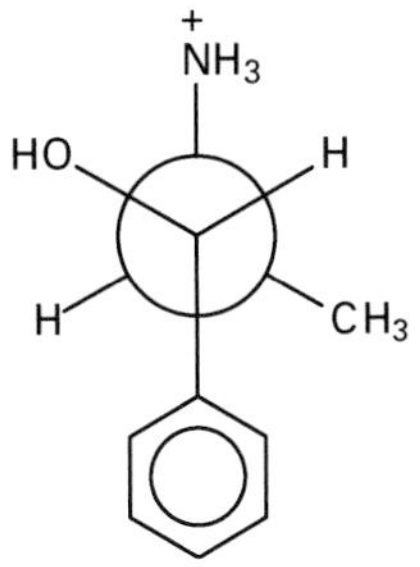

[13]

chiral array and therefore crystallize in enantiomorphic crystals, belonging to space group T^4 ($P2_13$). In the ROA spectra of sodium chlorate and bromate, longitudinal and transverse optical F phonons could be resolved. As an example, the depolarized right angle scattering Raman and ROA spectra of sodium bromate are shown in **Figure 2**. At a resolution of 1.8 cm^{-1} the non-degenerate longitudinal (LO) and doubly degenerate transverse (TO) optical modes can be observed separately. The observed signs can be attributed to one of the two possible chiral arrays.

Biochemical applications

One of the great advantages of Raman spectroscopy over IR spectroscopy is the ready applicability of water as a solvent. This is of course very important in biochemistry, where the compounds to be studied are, preferably, to be solved in water, as that is their natural medium. Nearly all biochemical compounds are chiral, so it is reasonable to study them not only by Raman spectroscopy but also by ROA, where additional features can be observed or resolved.

Carbohydrates

High quality spectra of carbohydrates can be measured in saturated aqueous solution, exhibiting clear bands that can be readily attributed to the orientation of OH substituents, anomeric preference and the conformation of exocyclic CH_2OH groups. As 15 monosaccharides (e.g. arabinose, glucose, xylose,

galactose and mannose) are examined, the conclusion that the ROA features reflect local stereochemical details seems to be based on enough material to be reliable.

In di- and polysaccharides ROA can probe the type and conformation of the glycosidic link. Early investigations on D-maltose [14] show a glycosidic couplet at 890–960 cm^{-1} (centred at about 917 cm^{-1}) similar to that of D-glucose [15]. It is concluded that for the $\alpha(1\rightarrow4)$ glycosidic link the couplet is positive at lower and negative at higher wavenumbers.

ROA observed at lower wavenumbers is also valuable for the determination of configuration. Compared with D-galactose, the disaccharide D-maltose and the polysaccharide laminarin [16] show a large ROA couplet centred at about 430 cm^{-1}. The configuration giving rise to a couplet that is negative at low, but positive at high wavenumbers is a β-glycosidic link (e.g. D-cellobiose, laminarin), whereas an α-glycosidic link (D-maltose) is positive at low, negative at high wavenumbers.

An extension of the investigations to the disaccharides D-maltose, D-maltose-O-d_8, D-cellobiose, D-isomaltose, D-gentobiose, D-trehalose and α-D-cyclodextrin confirms these results.

In cyclodextrins (CDs), conformational flexibility is studied by ROA. Using maltoheptose, β-cyclodextrin [17] and its derivatives with two- resp. three-times O-methylated sugar rings, an order of increasing flexibility of the CD ring could be established. Increases in couplet signal strength centred at ~915 cm^{-1} are interpreted in terms of a reduction in conformational flexibility of the cyclodextrin ring. Maltoheptose (linear) is expected to be the most flexible of the compounds, as found in the

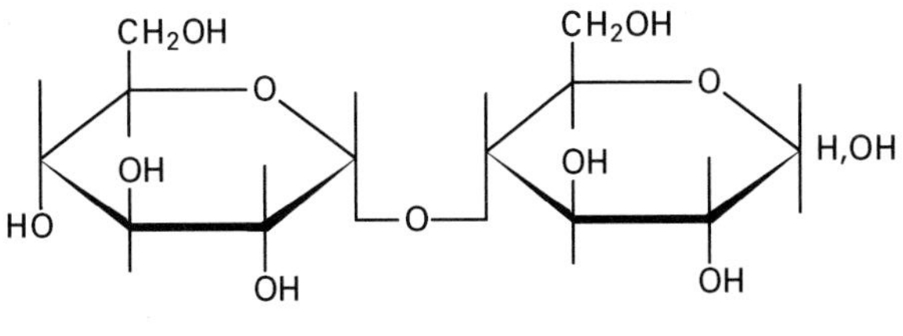

[14]

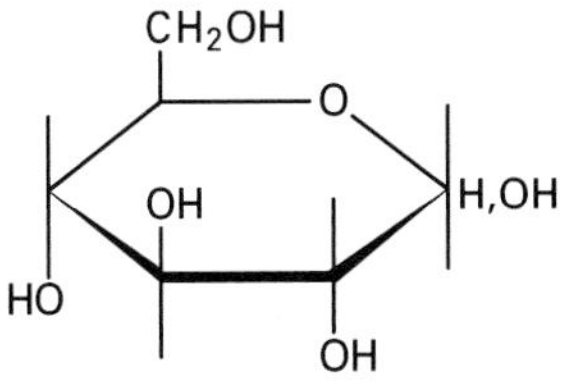

[15]

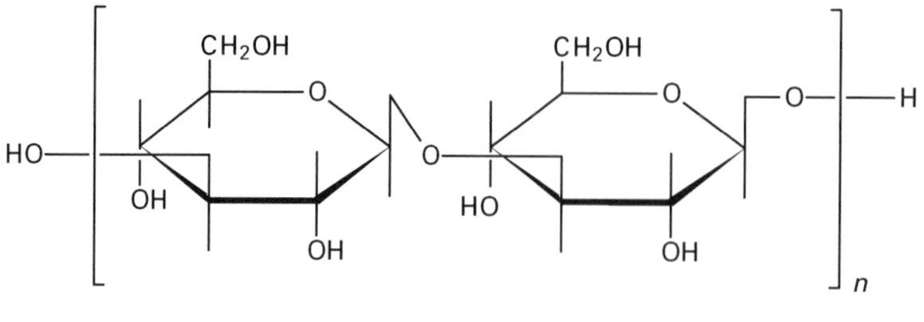

[16]

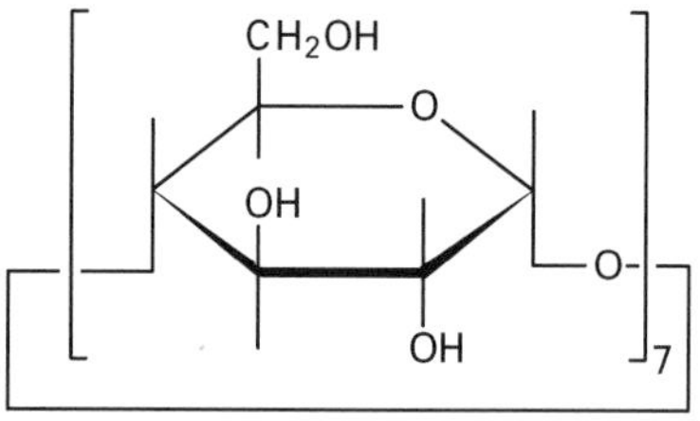

[17]

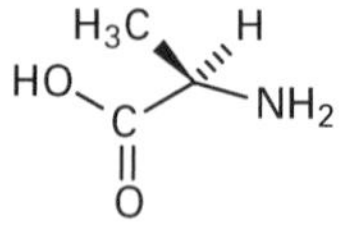

[19]

investigation. Good complexation of a guest in the CDs also lowers flexibility.

Other polysaccharides studied by ROA are laminarin [16] and pullulan [18]. Laminarin, which is a $\beta(1{\rightarrow}3)$ linked polymer of D-glucose, adopts a triple helix conformation. This is concluded by comparison with the ROA of the compound that corresponds to laminarin's dimer subunit, D-laminaribose. Pullulan is a linear polymer of D-glucose, consisting of D-maltotriose units connected through $\alpha(1{\rightarrow}6)$ glycosidic links. It adopts a random coil structure in aqueous solution, as can be deduced by the similarity of its ROA spectrum to that of D-maltotriose.

Amino acids

ROA spectra of L-(S)- as well as D-(R)-alanine [19] together with a detailed *ab initio* vibrational analysis of their zwitterionic structure at the 6-31G and the

6-31G* level have been reported. Though in the Raman spectra dependence on pH is clearly visible, the ROA spectra are not influenced by H^+ concentration. It is concluded from the fact that VCD spectra *do* respond to pH variation that ROA is more directly coupled to chirality and thus can yield more reliable information about absolute configuration.

The simple amino acids L-serine, L-cysteine, L-valine, L-threonine and L-isoleucine have been studied in aqueous solution by backscattering, using the spectral range from 600–1600 cm^{-1}. Bands originating in C_α^*H deformation and the symmetric CO_2^- stretch provide the best signatures for stereochemical studies. The ROA of the former vibration is positive for all acids studied and that for the latter negative. Isoleucine is an exception as it shows a nearly featureless ROA spectrum. The reason may be that it exists as an equal mixture of two rotamers in water.

Peptides and proteins

ROA spectra have proven to be of great value to study the secondary structure of proteins. By examining the

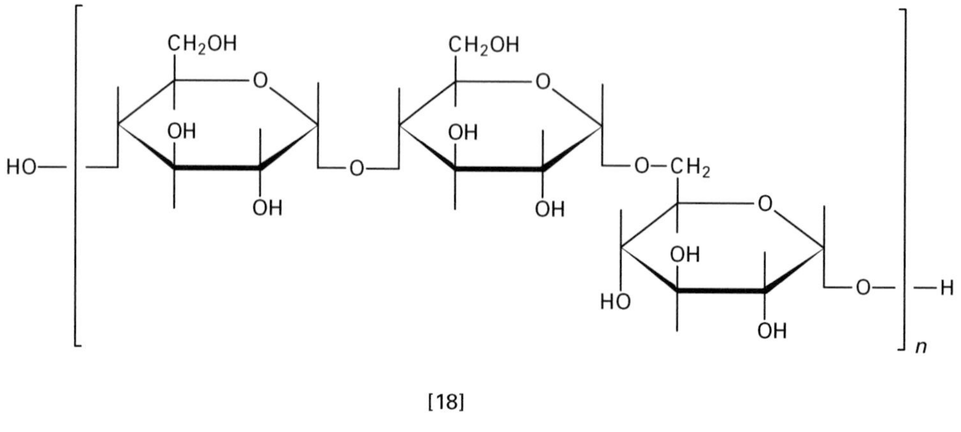

[18]

[20]

[21]

four different regions that contain the stretching vibrations of the backbone of proteins (the extended amide III region, the sidegroup and amide II region, and the amide I region), their α helix, β sheet, reverse turn and random coil content can quite readily be determined. This approximate division is shown in the lysozyme part of **Figure 4** (see below). Of course some sidegroups can also be observed by their Raman active vibrations, e.g. lysozyme and α-lactalbumin show tryptophan bands.

As a model for the peptide backbone N-acetyl-N'-methyl-L-alaninamide (NANMLA) [20] was subjected to a detailed ROA analysis. Its experimental spectrum was compared with the calculated spectra of nine different conformations. Both geometry optimizations and vibrational analyses using *ab initio* methods and the basis set 6-31G* were conducted on these conformers. As a result of these calculations the four conformers C_5, $C_{7,eq}-C_5$, $C_{7,eq}$ and α_R were found to play an important role in the room temperature equilibrium. These four conformers are characterized by torsion angles of $\phi = -157, -85, -99$ and $-60°$, and of $\psi = 159, 136, 79$ and $-40°$, respectively; here ϕ is the OC–NH–C*–CO backbone torsional angle and ψ is the HN–C*–CO–HN angle. Though the comparison between experimental and calculated spectra does not give unambiguous results, it gives strong hints that the predominant conformer in H_2O, D_2O and chloroform solution is the $C_{7,eq}-C_5$ conformer and that the α_R conformer is a minor component in H_2O and chloroform solutions, whereas the $C_{7,eq}$ conformer could be present in small amounts in D_2O solution.

The dipeptide L-alanyl-L-alanine [21] shows bands similar to the spectrum of L-alanine. In water, bands between 850 and 950 cm^{-1} can be attributed to the C^α–N stretch, symmetric CO_2 bend and the C^α–C(O) stretch. In addition, the dipeptide shows a negative ROA band at ~1270 cm^{-1} and a large positive ROA at ~1340 cm^{-1}. These vibrations occur in all known protein ROA spectra. According to newer investigations, the Raman band associated with the former effect is not only the amide III vibration but a superposition of three modes, of which only the C_C–H^I deformation (which is the first of two possible orthogonal C_C–H deformations) contributes to the ROA. The positive ROA at ~1340 cm^{-1} arises from

the N–H in-plane deformation coupled to the C_N–H^{II} deformation.

Looking at the ROA spectra of α helical and several unordered poly(L-lysine) preparations of increasing relative molecular mass (M_r), one not only finds bands at ~931 and 943 cm^{-1}, but one also finds a strong sharp positive ROA band at ~1340 cm^{-1}, which belongs to loops with a local order of a 3_{10} helix connecting regions of α helices. This feature, which is quite small at lower M_r (26 000), becomes more prominent at increasing M_r and becomes the dominant feature in a 268 000 Da protein, thus showing the 3_{10} helix content growing at the expense of the α helix.

Bovine serum albumin, which has a high α helix but no β sheet content, shows a strong positive band at 1340 cm^{-1}, probably connected to surface loop features forced mainly by its seventeen disulfide linkages. This prevents coupling of modes in this spectral region to other parts of the molecule and now the localized 'amide III' modes strongly resemble the vibrations in L-alanyl-L-alanine. The large globular protein ovalbumin, which has only a few disulfide bridges, lacks this large ROA feature.

α-Lactalbumin contains α helices as well as β sheet regions, and therefore shows bands at the corresponding wavenumbers in the 1020–1060 cm^{-1} part of the C^α–N stretch region. Large changes are induced in the extended amide III region as the hydrogen on nitrogen is replaced by deuterium, simplifying the spectra and thus allowing for easier identification of the remaining features.

It has been shown that ROA can detect residual amounts of structure in mainly unfolded proteins. Investigations on hen egg-white lysozyme and bovine ribonuclease A show that both proteins lose their natural structure to quite a large extent on reducing all the disulfide bonds (shown in **Figure 3** for ribonuclease A) in citrate buffer. The Raman and ROA spectra of the native proteins are shown in **Figure 4**. Lysozyme and ribonuclease A are known to contain both large α helix and β sheet regions, but on denaturation by reduction the two proteins behave differently. Whereas unfolded lysozyme loses nearly all of its structure at room temperature and only regains about 20% of its α helix content when cooled to 2°C, ribonuclease A still has about 50% of its secondary structure at room temperature.

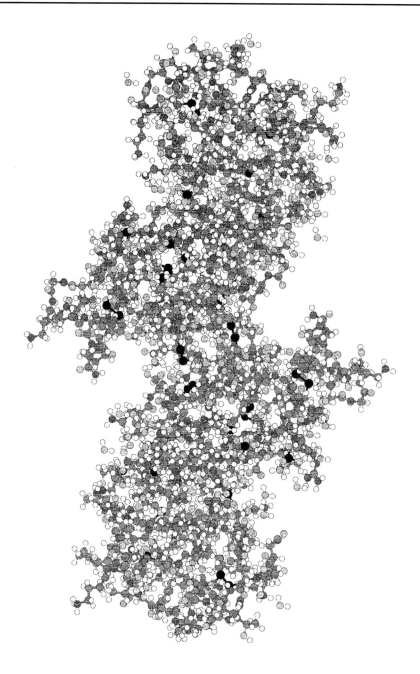

Figure 3 Bovine seminal ribonuclease, disulfide bridges shown in black. Structure drawn with HyperChem 5.0 from PDB entry 1BSR of Mazzarella L, Capasso S, Demasi D, Di Lorenzo G, Mattia CA and Zagari A (1993) *Acta Crystallographica Section D* **49**: 389; The Protein Data Bank: Bernstein FC *et al* (1977) *Journal of Molecular Biology* **112**: 535–542.

A further study of native hen egg-white lysozyme, using the pH 5.4 buffer solution spectra of the molecule in the temperature range 2–50°C, revealed a new cooperative transition at ~12°C. A detailed analysis shows the protein's adoption of a more rigid structure, where tertiary loops and tryptophan side-groups lose their residual mobility.

The human α_1-acid glycoprotein orosomucoid shows ROA bands that are characteristic of a high β-sheet content. It is to be expected from this single reported example that the ROA spectra of glycoproteins will yield valuable information about the conformation of the molecule, especially upon the mutual influence of the different protein and carbohydrate parts.

Another human protein that has been studied by ROA methods is the human serum albumin. Most prominent features in its ROA spectrum are bands

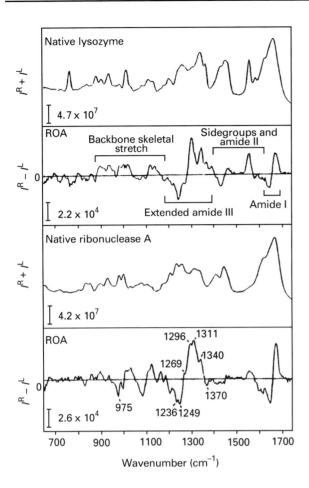

Figure 4 Backscattered Raman and ROA spectra of native hen egg-white lysozyme (top pair) and of native bovine ribonuclease A (bottom pair), both in acetate buffer at pH 5.4. Reprinted with permission from Wilson G, Hecht L and Barron LD (1996) *Biochemistry* **35**: 12 518–12 525. Copyright 1996 American Chemical Society.

that arise from the large α helix content and the already discussed strong positive band at ~1340 cm^{-1} which is characteristic for loop structures with local order of a 3_{10} helix. The latter band decreases to ~40% on reducing the pH to 3.4, where the mole-

[22]

cule assumes its F state, which is a mildly denaturated form created by dissociation of the two halves of the heart-shaped molecule. These findings may be explained by a loss of rigidity of the molecule and are in accordance with X-ray analyses of the crystalline state and derived structure proposals for the F state.

Nucleosides, nucleotides and nucleic acids

Pyrimidine nucleosides have been studied in backscattering. Unlike in normal Raman spectroscopy, where the sugar moiety only gives rise to weak signals, ROA detects signals of comparable intensity as well from the sugar as from the interaction of the sugars with the bases, but there are no signals from nearly achiral vibrations localized in the planar base rings. Of great diagnostic value are signals arising from the stretching at the C(1′)–N(1) glycosidic bond. These signals (~1200, 1230, 1380 and 1400 cm^{-1}) have opposite signs for α- and β-thymidine [22].

Of the three nucleotides adenylic acid, uridylic acid and cytidylic acid the three single-stranded polyribonucleotides as well as two double-stranded nucleotides were examined by ROA spectroscopy. Three regions were especially valuable for the analysis: 1550–1750 cm^{-1} (called the base stacking region), 1200–1550 cm^{-1} (sugar–base coupling region) and 950–1150 cm^{-1} (sugar–phosphate backbone region). As the name suggests, the first region reflects base stacking. The second shows the orientation of the sugar and base rings. The third reflects the conformation of the sugar ring, even though the backbone may also be involved. Typical spectra in H$_2$O as well as in D$_2$O are shown in **Figure 5**. These spectra of polyadenylic acid show two large negative peaks in the sugar base region in the ROA. As expected, the base-dependent signatures lose much of their intensity on exchange with D$_2$O, whereas the sugar phosphate backbone region remains virtually unaltered.

Other applications

Apart from probing molecular conformation and configuration, ROA has been employed in a couple of other interesting tasks.

The applicability of ROA for the determination of enantiomeric excess in mixtures of chiral enantiomers has been exploited. Using the test compound α-pinene an accuracy of 0.1% enantiomeric excess (ee) was achieved.

The possibility of observing Raman optical activity from chiral surfaces has been mentioned and a theory for the generation of second-harmonic optical activity (SHOA) from chiral surfaces and interfaces has been derived.

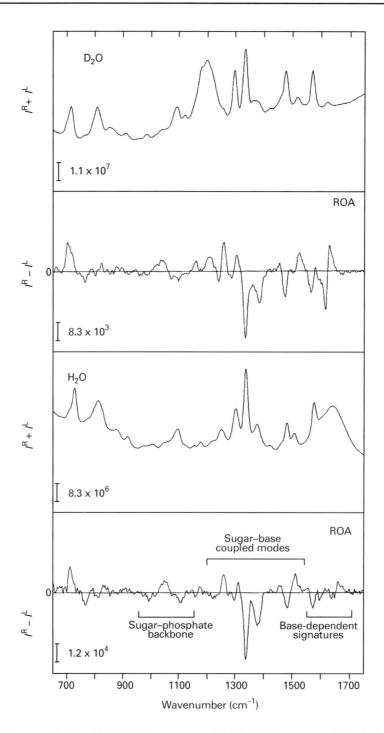

Figure 5 Backscattered Raman (I^R+I^L) and ROA (I^R–I^L) spectra of poly(rA) in H_2O (bottom) and D_2O (top). Reprinted with permission from Bell AF, Hecht L and Barron LD (1997) *Journal of the American Chemical Society* **119**: 6006–6013. Copyright 1997 American Chemical Society.

Even theoretical aspects may be attacked by ROA spectroscopy. In benzene derivatives containing heteroatoms the question arises as to whether Rydberg transitions centred on the heteroatom are involved in the production of Raman spectra. Using a comparison between the polarized and depolarized ROA spectra, this contribution of Rydberg transitions could be detected in l-phenylethanol and l-phenylethylthiol, as large deviations from the factor of 2 in the ratio of polarized to depolarized spectra were found.

Protein folding is an area of much recent activity. It has been shown by ROA that water acts as lubricant in helix coil unfolding. Extremely fast conformational fluctuations seem to occur with a frequency of about 10^{12} s^{-1}. These turnover rates, which are close to the theoretical limit of kinetics, could only be observed owing to the very short time scale (~10^{-14} s) of Raman techniques.

The very high quality of ROA spectra makes it possible even to obtain difference ROA spectra of substances at two different temperatures. By this method the premelting of poly(rA)·poly(rU) was monitored. The compound, which has an A-type double helical structure, shows qualitatively the same ROA spectrum at both 20 and 45°C, while the intensities of the bands are lowered. From the fact that the difference spectrum is similar to the ROA spectra at both temperatures, it is concluded that the same average structure is maintained throughout all temperatures in the range examined. The investigation demonstrates the usefulness of the new technique to probe the dynamics of nucleic acids in aqueous solution.

List of symbols

I^L = intensity of left circularly polarized light; I^R = intensity of right circularly polarized light, ϕ and ψ = torsional angles.

See also: **Biochemical Applications of Raman Spectroscopy; Carbohydrates Studied By NMR; Chiroptical Spectroscopy, Emission Theory; Chiroptical Spectroscopy, General Theory; FT-Raman Spectroscopy, Applications; Hydrogen Bonding and other Physicochemical Interactions Studied By IR and Raman Spectroscopy; IR and Raman Spectroscopy of Inorganic, Coordination and Organometallic Compounds; IR Spectral Group Frequencies of Organic Compounds; Matrix Isolation Studies By IR and Raman Spectroscopies; Non-linear Raman Spectroscopy, Applications; Non-linear Raman Spectroscopy, Instruments; Non-linear Raman Spectroscopy, Theory; Nucleic Acids and Nucleotides Studied Using Mass Spectrometry; Nucleic Acids Studied Using NMR; Polymer Applications of IR and Raman Spectroscopy; Proteins Studied Using NMR Spectroscopy; Raman and IR Microspectroscopy; Raman Optical Activity, Spectrometers; Raman Optical Activity, Theory; Raman Spectrometers; Stereochemistry Studied Using Mass Spectrometry; Vibrational, Rotational and Raman Spectroscopy, Historical Perspective.**

Further reading

Barron LD and Hecht L (1994) Vibrational Raman optical activity: from fundamentals to biochemical applications. In: Nakanishi K, Berova N and Woody RW (eds) *Circular Dichroism – Principles and Applications*, pp 179–215. New York: VCH.

Barron LD, Hecht L and Polavarapu PL (1989) Polarized Raman optical activity in methyl antisymmetric deformation: influence of heteroatom Rydberg orbitals. *Chemical Physics Letters* 154: 251–254.

Barron LD, Hecht L and Wilson G (1997) The lubricant of life: a proposal that solvent water promotes extremely fast conformational fluctuations in mobile heteropolypeptide structure. *Biochemistry* 36: 13 143–13 147.

Bell AF, Hecht L and Barron LD (1997) Vibrational Raman optical activity of pyrimidine nucleosides. *Journal of the Chemical Society, Faraday Transactions* 93: 553–562.

Bell AF, Hecht L and Barron LD (1997) Vibrational Raman optical activity as a probe of polyribonucleotide solution stereochemistry. *Journal of the American Chemical Society* 119: 6006–6013.

Bell AF, Hecht L and Barron LD (1997) New evidence for conformational flexibility in cyclodextrins from vibrational Raman optical activity. *Chemistry – A European Journal* 3: 1292–1298.

Costante J, Hecht L, Polavarapu PL, Collet A and Barron LD (1997) Absolute configuration of bromochlorofluoromethane from experimental and ab initio theoretical vibrational Raman optical activity. *Angewandte Chemie International Edition in English* 36: 885–887.

Hecht L and Barron LD (1994) Rayleigh and Raman optical activity from chiral surfaces. *Chemical Physics Letters* 225: 525–530.

Hecht L, Philips A and Barron LD (1995) Determination of enantiomeric excess using Raman optical activity. *Journal of Raman Spectroscopy* 26: 727–732.

Hoffmann GG (1995) Vibrational optical activity (VOA). In: Schrader B (ed) *Infrared and Raman Spectroscopy – Methods and Applications*, pp 543–572. Weinheim: VCH.

Lindner M, Schrader B and Hecht L (1995) Raman optical activity of enantiomorphic single crystals. *Journal of Raman Spectroscopy* 26: 877–882.

Polavarapu PL (1990) Ab initio vibrational Raman and Raman optical activity spectra. *Journal of Physical Chemistry* 94: 8106–8112.

Qu X, Lee E, Yu G-S, Freedmann TB and Nafie LA (1996) Quantitative comparison of experimental infrared and Raman optical activity spectra. *Applied Spectroscopy* 50: 649–657.

Teraoka F, Bell AF, Hecht L and Barron LD (1998) Loop structure in human serum albumin from Raman optical activity. *Journal of Raman Spectroscopy* 29: 67–71.

Wilson G, Hecht L and Barron LD (1996) Residual structure in unfolded proteins revealed by Raman optical activity. *Biochemistry* 35: 12 518–12 525.

Wilson G, Hecht L and Barron LD (1997) Evidence for a new cooperative transition in native lysozyme from temperature-dependent Raman optical activity. *Journal of Physical Chemistry, B* 101: 694–698.

Yu G-S, Freedman TB and Nafie LA (1995) Dual circular polarization Raman optical activity of related terpene molecules: comparison of backscattering DCP$_I$ and right-angle ICP spectra. *Journal of Raman Spectroscopy* 26: 733–743.

Raman Optical Activity, Spectrometers

Werner Hug, University of Fribourg, Switzerland

VIBRATIONAL, ROTATIONAL &
RAMAN SPECTROSCOPIES
Methods & Instrumentation

Raman optical activity (ROA) or, more precisely, spontaneous vibrational Raman optical activity scattering is, like vibrational circular dichroism (VCD), a spectroscopic method that directly probes the chirality, or handedness, of molecular vibrations. ROA and VCD therefore have an obvious stereochemical potential. That such phenomena could yield structural information not otherwise available was realized long before their measurement became feasible and the first observations of ROA and VCD date back only a quarter of a century. For ROA, measurement was preceded by a detailed theoretical analysis and, perhaps inevitably so in view of the experimental difficulties, some false claims of its observation.

Considerable progress on the collection of ROA data has been made since, but while at present such data can be reliably recorded for many samples it would be far from true to claim that the experimental situation is satisfactory. The currently most successful instruments are based on a ROA backscattering configuration first described at the start of the 1980s and characterized in ROA shorthand as 'ICP'. However, the field of ROA instrumentation is in full motion, and at this time it is not obvious which of several competing ROA variants will eventually emerge as the dominant method for routine ROA data collection.

We can conveniently divide the methodical advances, demonstrated or proposed, in the measurement of ROA into two categories. The first, and most extensively implemented, simply reflects the progress in general Raman instrumentation: multichannel detection systems based on backthinned CCD (charge coupled device) technology, and high luminosity spectrographs. The second consists of refinements specific to ROA. On one hand these are light gathering and data acquisition techniques which have no counterpart, or are of no interest, in ordinary spontaneous Raman spectroscopy. A typical example is a dual lens light collection system dating from the late 1970s. On the other hand are new ROA variants described in the late 1980s and early 1990s which have opened up new ways to collect ROA data, but so far it has not been shown convincingly that one of them may be better suited than the originally proposed techniques to conquer the two

great enemies of ROA measurements: statistical noise and systematic spurious scattering differences.

ROA variants

The reason why there is a whole set of different possibilities for measuring ROA arises from the fact that, as in any light scattering experiment, diffused light can be observed at different angles to the propagation direction of the incident, or exciting light, and that the intensity of the observed light depends on the polarization of the incident light and the eventual use, and the nature of, a polarization analyser for the scattered light. In all cases, however, a difference in the interaction of right and left circularly polarized light – chiral light – with chiral molecules of the sample is observed, i.e. there is what the chemist calls a diastereoisomeric interaction difference.

Basic measurement arrangements

The scattering geometries of practical importance are right angle, or 90° scattering and collinear, or 0° and 180° scattering. The 0° and 180° scattering geometries yield totally different ROA information, and the magnitude of their ROA signals differs strongly. In general, ROA signals measured at 180° are larger than those measured at 0°, but this can depend on the particular vibrational Raman band under observation. For 90° scattering, ROA signal strengths are somewhere in between. Although the signal strength is lower at 90° than at 180°, backward scattering is not always more advantageous than right-angle scattering.

For any chosen scattering angle, ROA can be observed if either the incident light, or the scattered, polarization analysed and detected light, or both, are circularly polarized. In accordance with this one uses the terms ICP (incident circularly polarized), SCP (scattered circularly polarized) and DCP (dual circularly polarized). Two DCP variants obviously exist: in-phase (DCP_I) and out-of-phase (DCP_{II}), where in-phase means that if the incident light is right circularly polarized then the right circularly polarized component of the scattered light is detected, while in an out-of-phase experiment one would observe the left circularly polarized component. Moreover, in ICP the scattered light can, or cannot, be polarization analysed with a linear polarization analyser, and in SCP the incident light can be natural or linearly polarized. Of the large number of ensuing experimental configurations **Table 1** shows those for which there are solid arguments that they will play a definite role in ROA instrumentation, either because of inherent advantages they have in measuring ROA or because they yield specific information not otherwise available.

Comparison of data from different configurations

In the general non-resonant case there are three contributions to the observed ROA spectra: $\alpha G'$, the isotropic ROA invariant stemming from the electric dipole–magnetic dipole optical activity tensor, which is also responsible for the anisotropic invariant γ^2, and the anisotropic invariant δ^2 due to the quadrupole transition tensor. The anisotropic invariants are also often written as $\gamma^2 = \beta(G')^2$ and $\delta^2 = \beta(A)^2$.

Relevant quantities of interest, e.g. differential scattered intensities ΔI, are obtained from the formulae in **Table 1** by an appropriate choice of K. The difference between the intensity of right (R) and left (L) circularly polarized scattered light in a backscattering SCP experiment with unpolarized (u) exciting light from Equation [8] is

$$\Delta I = I_R^u - I_L^u = \frac{4K}{c}[24\gamma^2 + 8\delta^2] \qquad [10]$$

$$K = \frac{1}{90}\left(\frac{\omega^2 \mu_0 E^{(0)}}{4\pi R}\right)^2 \qquad [11]$$

where ω is the angular frequency of the scattered light, $E^{(0)}$ is the magnitude of the electric field at the site (origin) of the scattering molecule, and R is the distance to the observer.

The way in which the three invariants are associated is different for different scattering geometries. Thus, data measured for forward, backward and right-angle scattering cannot be directly compared with each other, even qualitatively. On the one hand, from a purely analytical point of view this is unfortunate. If very small samples only are available, then 90° is the only choice, as ROA requires fair sized scattering volumes and avoidance of skew angle reflections and refractions of the scattered light by the walls of the sample cell, while the measurement of dilute solutions is best performed by 180° scattering because of its higher values for the ratio $\Delta I/I$. On the other hand, from the point of view of the amount of information which can be extracted from ROA measurements this fact is fortunate, and there is, in my opinion, therefore a strong case for instruments which measure ROA under more than one scattering angle.

Why ROA measurements are difficult

Signal-to-noise considerations

Before lasers were available the recording of a Raman optical activity spectrum would have been impossible. To appreciate the signal-to-noise (S/N) problem it suffices to consider the typical depolarized 90° ICP variant. From Equations [1] and [6] in **Table 1**, neglecting the often small quadrupole contribution, one has

$$\frac{\Delta I}{I} \approx \frac{4\gamma^2}{c\beta^2} \qquad [12]$$

Table 1 ROA variants most likely to play a role in future ROA instrumentation

Scattering angle (°)	Type	Polarization of exciting light	Polarization analyser detected light	Average quantity		Difference (right minus left) quantity	
90	ICP	◯	‖	$K\,6\beta^2$	[1]	$4\,(K/c)\,[6\gamma^2-2\delta^2]$	[6]
			⊥	$K\,[45\,\alpha^2+7\beta^2]$	[2]	$4\,(K/c)\,[45\alpha G'+7\gamma^2+\delta^2]$	[7]
180	ICP	◯	None	$K\,[90\,\alpha^2+14\,\beta^2]$	[3]	$4\,(K/c)\,[24\gamma^2+8\delta^2]$	[8]
	SCP	Unpolarized	◯				
	DCP$_I$	◯	◯	$K\,12\,\beta^2$	[4]		
0	ICP	◯	None	$K\,[90\,\alpha^2+14\,\beta^2]$	[3]	$4K/c\,[180\alpha G'+4\gamma^2-4\delta^2]$	[9]
	SCP	Unpolarized	◯				
	DCP$_I$	◯	◯	$K\,[90\,\alpha^2+2\,\beta^2]$	[5]		

All formulae are for the far from resonance limit only. ◯: circularly polarized. ‖, ⊥: linearly polarized parallel and perpendicular to the scattering plane respectively. Other symbols explained in the text.

The value of Equation [12] is therefore at most a few times the ratio of a magnetic to an electric dipole transition moment. The largest observed values of $\Delta I/I$ are of the order of 5×10^{-3}, and values of 10^{-3} or less are more common.

In the shot noise limited case, where I is represented by N events, the statistics of a Poisson distribution hold. With $I_\parallel^R \approx I_\parallel^L \approx I$ the rms deviation of $\overline{\Delta N}$ is given by $\sqrt{2N}$. Thus, to recover $\overline{\Delta N} = a\overline{N}$ with a given S/N ratio one has

$$\overline{\Delta N} = a\overline{N} = (\text{S/N})\sqrt{2N}$$
$$\overline{N} = 2\frac{(\text{S/N})^2}{a^2} \qquad [13]$$

For a typical situation in ROA one may have $a = 5 \times 10^{-4}$, S/N = 10, which corresponds to some 10^9 detected photons per spectral element of resolution for what might be, in ordinary Raman spectroscopy, an unremarkable depolarized Raman band. This number is put in perspective by comparing it with the value of 5.5×10^4 detected photons per second obtained in backscattering at 4 cm^{-1} resolution with a modern commercial Raman instrument, equipped with a CCD detector and a 15 mW He-Ne laser, for the peak height of the intense polarized 992 cm^{-1} band of benzene.

With modern slow-scan CCD detectors, read-out noise in a well-designed ROA spectrometer is negligible in comparison with photon shot noise. Yet, flicker, or $1/f$ noise, clearly plays its role. Unfortunately, to date no serious attempt to identify and reduce its influence has been made. What is well known since the first recording of entire ROA spectra in the mid 1970s is that dust can have a deleterious effect. In particular in microcapillaries, where sample volumes are small and convection essentially absent, single grains of dust sometimes appear to get trapped and gently oscillate in the modulated focused laser beam. Though not understood at that time, it probably was one of the first observations of the optical tweezer effect. Thus, careful sample preparation is important.

Systematic spurious scattering differences

Also called artefacts, spurious scattering differences are arguably the biggest and best discussed nuisance in ROA spectroscopy. They can easily be understood and appreciated without any elaborate theoretical treatment, and the two optical solutions to the problem discussed in the section on light collection optics are based on simple practical considerations.

The easiest viewpoint is to think of a perfect, artefact-free instrument, and then to try to find means to systematically spoil its performance. How best to degrade instrument performance can depend on the ROA variant but the basic ideas are the same. We will, therefore, consider only the ICP backscattering case and for the sake of simplicity assume a fully polarized Raman band, but the insight gained is not limited to this particular situation. The sample is assumed to be achiral. Thus, our perfect instrument registers no scattering difference if modulation between right and left circularly polarized exciting light occurs.

Initially, we now assume the polarization of the exciting light remains perfect in the scattering zone, but we introduce a circular polarization analyser, which transmits only one circular polarization component, somewhere between the scattering zone and the detector. As the scattered light is purely left and right circularly polarized, huge scattering differences are obviously observed for the two modulation periods. Unfortunately, the elements for a partial circular polarization analyser can be present in a ROA instrument: birefringence in the walls of the scattering cell and in the optics, and polarization-dependent diffraction efficiency of the grating of the spectrograph, among other things.

In a second step we spoil the circular polarization of the exciting light in the scattering zone by placing a quarter-wave plate between it and the circular polarization modulator. The light in the scattering zone now becomes linearly polarized in orthogonal planes for the two modulation periods, as is also true for the scattered light. Again, large scattering differences will be observed due to the polarization-dependent efficiency of mirrors and gratings in the light collection and analysing system. A partial quarter-wave plate ahead of the scattering zone can easily be formed in a ROA instrument by birefringence in the windows of the scattering cell and in lenses, as well as by imperfections in the circular polarization modulator.

The simultaneous measurement of several Raman bands and the general increase in measurement speed have, in practical terms, been important for artefact reduction, perhaps more so than optical advances because of their easier implementation. Taken together these two advances allow one to check and adjust the baseline before, and if need be, during a ROA measurement.

Instrument configurations

General building blocks

All ROA instruments share the common features depicted schematically in **Figure 1** regardless of the

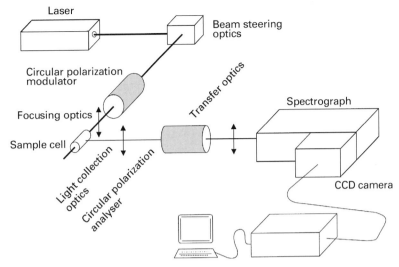

Figure 1 Building blocks of Raman optical activity spectrometers. Either a circular polarizer (in ICP), analyser (in SCP) or both (in DCP) may be present.

particular ROA variant they measure. Many of these building blocks are also part of ordinary Raman instrumentation, but for ROA there are specific, often more exacting, demands.

Lasers ROA requires laser powers of 100 to 1000 mW at the sample, with the upper limit determined by what a sample can stand without serious degradation rather than by what would be desirable for the experiment. In view of the light losses by polarizers, modulators, beam splitters, lenses and mirrors, about 1.5 W of single line output power from the laser is desirable.

The wavelength of the laser has to meet two contradictory requirements. It should be as short as possible so that ROA intensities are maximized, but long enough so that absorption by the sample and the well-known fluorescence problem of Raman spectroscopy are avoided. The standard exciting wavelengths used to date are the 488 nm and, particularly, the 514.5 nm argon ion laser line. Attempts have been made to use the 647.1 nm krypton ion laser line but they proved unsuccessful. From an inspection of Equation [11] it transpires that, for a fixed intensity of the exciting light, the intensity of the scattered light varies as the fourth power of $1/\lambda$. For the ROA difference intensities, however, another factor of $1/\lambda$ is hidden in the expressions of the tensor invariants αG, γ^2 and δ^2 because ROA samples the variation of the vector potential of the light wave over the molecular dimension. On going from 514.5 to 647.1 nm the average Raman intensity therefore decreases by a factor of 2.5, but the ROA difference

intensity drops by a factor of well over 3. Our new ROA spectrometer uses the 532 nm frequency doubled $Nd:YVO_4$ line. Average and difference intensities will be reduced by factors of 1.14 and 1.18 as compared with the 514.5 nm Ar^+ line, values which appear to be acceptable.

Laser stability is of crucial importance, power as well as beam pointing stability. Modern research argon ion lasers no longer present noticeable problems in this respect, but early models did. Frequency doubled $Nd:YVO_4$ lasers are expected to show even better performance.

Scattering cells Ordinary Raman scattering cells are generally unsuitable for ROA work, which requires cells of high optical quality without any strain induced birefringence in the windows through which the exciting light enters and the observed scattered light leaves. Moreover, care has to be taken to avoid the collection of Raman scattered light reflected from oddly oriented cell surfaces. High-quality fluorescence cells selected for their absence of birefringence are suitable for ROA work, but more sophisticated designs have also been developed. For right-angle scattering, microcapillary cells with sample volumes as low as 0.5 µL have been successfully used. Rotating cells that permit higher laser powers might be an option in backscattering if the optical problems can be solved.

The dispersive system The earliest ROA spectrometers were of the scanning type and used Czerny–Turner double monochromators. It took weeks to record a single ROA spectrum. The scanning technology has now been completely superseded by multichannel instruments with spectrographs.

One particular ROA instrument, which dates from the late 1970s, fully exploited concave aberration corrected holographic grating technology to maximize throughput, and set a standard for the following decade. Its spectrograph is depicted in **Figure 2A**. At an average resolution of about 9 cm^{-1}, typical for ROA solution spectra, the field stop, represented by the entrance slit, has a surface area S of 0.25×17 mm = 4.25 mm^2 at an average input focal-ratio of 6.33 with essentially no vignetting. The resulting étendue G is so large that special light collection optics, described in later sections, are required to effectively fill it and amounts to

$$G = \pi S \, \sin^2 \alpha \approx \frac{\pi S}{(2\text{focal-ratio})^2} = 0.083 \text{ mm}^2 \text{ sr} \quad [14]$$

Sin α is the numerical aperture NA of the system, and we have made the usual approximation that $2 \, NA \approx 1/(\text{focal-ratio})$. The larger input than output focal length reduces the size of the slit image in the output focal plane to about half the length of the input slit. The larger kind of the currently available CCDs for spectroscopic use is therefore well suited for such a spectrograph design. In its original application, before such CCDs became available, an image intensifier was used in the output focal plane of the spectrograph, and a further, anamorphic image size reduction was performed between the image intensifier and the solid state detector.

Spectrograph technology made another advance with the introduction of planar volume holographic transmission phase gratings in the 1990s. Such gratings are typically used in combination with holographic notch filters. The latest versions of these filters provide a Rayleigh light suppression of the order of 10^6 and allow the observation of Raman bands down to 100 cm^{-1} shifts. Commercial spectrographs use a simple back-to-back arrangement of two photographic single-lens reflex camera lenses with the grating placed in between (**Figure 2B**). The short focal length of these lenses, coupled with their high speed, keeps object and image size small while maintaining throughput. However, back-to-back configurations of photographic lenses are notorious for their vignetting problems and it does indeed reach 30% at the corners of a relatively small 6.6×27 mm CCD detector. Still, throughput is impressive. At 9.5 cm^{-1} resolution a slit size of 0.1×6.6 mm is realized at a focal-ratio of about 1.8, which corresponds to an étendue of 0.16 mm^2 sr, 1.9 times the value of the concave grating spectrograph. In addition, holographic transmission gratings are claimed to exhibit close to 100% diffraction efficiency for s-polarized and up to 60% for p-polarized light, essentially double the average values of concave gratings, but precise data have not been forthcoming by the manufacturer.

A considerable disadvantage of such spectrographs is the curvature of the image of the entrance slit on the detector. In ROA this can limit the data acquisition rate.

Detectors and their electronics The first detectors used in multichannel Raman instruments were silicon intensified target tubes preceded by an image intensifier. Their large carry-over of information from one illumination period to the following severely limited their usefulness for ROA spectroscopy, where modulation takes place between right and left circularly polarized exciting or detected light. Thus, the first viable, shot noise limited multichannel ROA instrument used self-scanned diode array technology, coupled via high-speed ($f/1.3$) anamorphic optics to a high sensitivity, high gain (10^6) multistage image intensifier, with critical phosphor screens selected for their rapid decay times. While technically highly successful, the design was too elaborate to be widely adopted. The ready availability of backthinned CCDs with average quantum efficiencies of up to 70% for green to orange light has finally provided an easy solution to the detector problem. The high quantum efficiency more than makes up in ROA for the lack of shot noise limited performance, with CCD read-out noise of the order of 20 detected

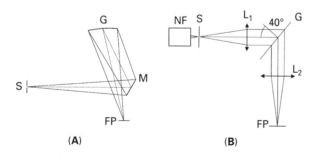

Figure 2 (A) Concave aberration corrected holographic grating spectrograph. S: entrance slit. M: folding mirror. G: concave grating 1500 mm^{-1}, 110 mm diameter, f_{in} = 658 mm, average f_{out} = 356 mm. FP: output focal plane 9 × 50 mm. (B) Holographic transmission grating spectrograph. S: entrance slit. L$_1$: input lens, f = 85 mm, f/1.8. G: grating 53 mm × 64 mm, 2400 mm^{-1}. L$_2$: output lens, f = 85 mm, f/1.4. FP: output focal plane 6.6 × 27 mm (limited by vignetting and detector size; radius of curvature 39 mm of image of straight slit). NF: holographic notch filter assembly (in ROA integrated in light collection optics).

photons or less. Still, to keep it well below photon shot noise, extremely low modulation speeds, of the order of a second per half-cycle, are presently used in ROA. The S/N degradation by flicker noise is bound to be an unavoidable consequence. Dark current with its associated shot noise can be kept low by cooling the detector to at least −70°C. Carry-over is inherently low and can be completely suppressed by rapid charge clearing cycles.

A detail of underestimated importance to ROA is data digitization. It is mandatory that precautions are taken to ensure that differences of less than 1 in 10^4 for small Raman bands are reliably recovered by 14 to 16 bit AD converters by repeated digitization; if not, ROA difference spectra are simply swamped. The misunderstanding of this problem, particularly in recent high-throughput, low noise instruments has doubtless contributed to the abandonment of right-angle scattering. Systematic dither, i.e. small variations of the illumination time of the sample for consecutive modulation periods, compensated numerically in the data acquisition system, should solve this problem in future ROA instruments.

Data acquisition and treatment Two important additional requirements need to be met compared with ordinary Raman spectroscopy. These are the need to switch, synchronized to an optical modulator, the acquisition between a period for right and one for left circularly polarized light, and to recover, in addition to the average scattered intensity I, the difference intensity ΔI, where ΔI is often less than 100 ppm of I. Modern PC-based data acquisition systems with a 32 bit word-length have no difficulty with this, and the vast amount of available memory capacity even allows storage of individual modulation cycles for entire acquisition runs. This facilitates checking data for corruption by cosmic ray events on the detector, as well as for sample degradation and similar problems. Cosmic ray events occur at a rate of about $2 \ \text{min}^{-1} \ \text{cm}^{-2}$ of CCD surface. Typically, at one second per half-cycle and a standard $6.6 \ \text{mm} \times 27 \ \text{mm}$ CCD detector, the data for every 17th half-cycle have to be discarded. Unfortunately, cosmic ray interference is not yet systematically eliminated in current ROA instruments.

Specific features

Circular polarization modulators Circularly polarized light is generated by passing linearly polarized light through a $\lambda/4$ retarder (**Figure 3**). Switching between right and left circular light occurs if the orientation of the fast and slow axis is interchanged. The exciting laser light is monochromatic and highly collimated. Fast switching longitudinal KDP ring-

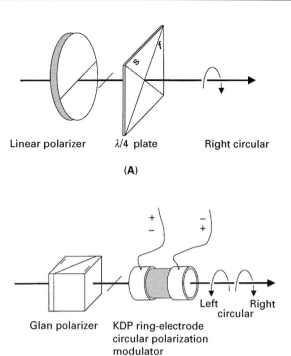

Figure 3 (A) Basic circular polarizer. (B) Practical implementation most often used in ROA.

electrode modulators with their even retardation over the whole aperture have therefore been the preferred choice in ICP instruments. They have the inconvenience of requiring a quarter-wave voltage of about ± 1700 V for 532 nm light, suffer from residual stress-induced birefringence, and their retardation is strongly temperature dependent. Liquid crystal retarders, which require just a few volts of modulating voltage and are available with temperature compensation electronics, will doubtless be used in the future. Their slow switching speed can be accommodated by commuting them during the read-out time of the CCD detector.

Circular polarization analysers The circular polarization analysers used in SCP and DCP are modulators in reverse (**Figure 4**). Their practical implementation is complicated by the fact that the light passing through them is neither monochromatic nor, as it emanates from a scattering zone with finite dimensions, is it collimated. Mechanically rotated compensated zero-order and achromatic crystal quarter-wave plates have been the choice for the retarder. To date, in the one operating DCP instrument a true zero-order polymer quarter-wave plate with a much larger acceptance angle is now being used. Small as such advances might appear, they can be decisive for instrument performance.

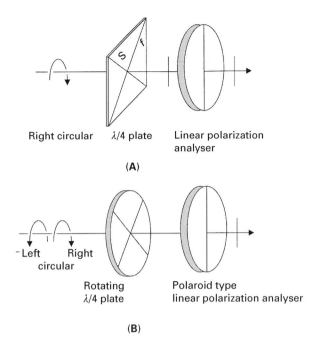

Figure 4 (A) Basic circular polarization analyser. (B) Practical implementation currently used in ROA; future instruments may use LC retarders and thin film polarizers.

In view of the aperture size and the required acceptance angle the linear polarization analysers employed in SCP and DCP are presently of the dichroic variety. Given the advances in thin film technology, thin film polarizers with better transmission characteristics will appear in future ROA instruments. Their use might solve the intensity problem SCP suffers in comparison with ICP and DCP by allowing the simultaneous recording of right and left circularly polarized scattered light. Yet, in order to be on a par with ICP, such SCP instruments will require a spectrograph with twice the étendue of a comparable ICP instrument, but should enjoy the benefit of reduced flicker noise. With holographic transmission grating spectrographs the size of the étendue is no longer an issue. This is the justification for having included SCP in **Table 1**.

DCP$_I$ backscattering (**Figure 5**) is somewhat in a class of its own. The discarded polarization component corresponds to DCP$_{II}$ and carries no ROA information in the non-resonant case. For depolarized Raman bands ($\alpha^2 = 0$) DCP$_I$ should slightly outperform ICP backscattering with respect to shot noise, but it is for polarized bands where it is expected to shine. As seen from Equation [5] (**Table 1**), isotropic scattering does not contribute to the average scattering intensity. Unfortunately, this is also one of the prime reasons why DCP$_I$ is so artefact prone. Tiny differences in the leakage of isotropic scattering through the circular polarization analyser for the right and left circular modulation periods simply tend to swamp the ROA difference spectrum. DCP$_I$ will only become of general usefulness if active stabilization techniques for the circularity of the exciting light are developed.

Light collection and transfer optics High light-collection efficiency and avoidance of artefacts are prime concerns in ROA. Fortunately, well-designed light collection optics can combine them both.

In ICP right-angle scattering the perfect arrangement would be circular light collection around the sample. This is hard to realize, but a similarly advantageous practical approach consists in using two light-collection lenses placed under 90° to each other (**Figure 6**). It is the only light collection system capable of filling the étendue of high luminosity spectrographs with small sample volumes. Though

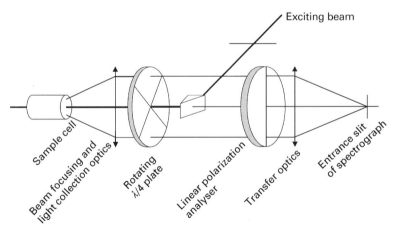

Figure 5 DCP$_I$ backscattering arrangement. A single quarter-wave plate can function in a circular polarizer and analyser configuration. True zero-order retarders may be placed ahead of the light collection lens, albeit with loss of precision. The plate is rotated within 200 ms in the appropriate orientation and stopped during the data acquisition half-cycle.

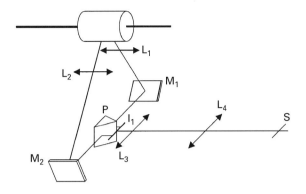

Figure 6 Dual lens light collection system for right-angle scattering. The two light collection lenses L_1 and L_2 form a joint intermediate image I_1. Their optical axes intersect at 90° at the sample. M_1, M_2: 22.5° mirrors. L_3: field lens. L_4: lens which projects I_1 onto the entrance slit S of the spectrograph.

backscattering has the advantage of approximately double the value of $\Delta I/I$, right-angle scattering remains in this respect a far superior technique.

In forward and backward scattering, ICP, SCP and DCP, a regular light collection cone about the axis of the exciting light is the equivalent of circular light collection in right-angle scattering and provides similar benefits. The approach was demonstrated with the help of a fibre optics cross-section transformer in the first ROA backscattering instrument. Care has to be taken to keep focal-ratio degradation low. If this is not done, and if the fibre optics are not carefully matched to the remaining optics, then the light losses can be disastrous, and in later backscattering instruments the use of fibre optics was therefore dropped. Yet, there appears to be no alternative to fibres when filling the large étendue of modern spectrographs in collinear scattering is required, and future ROA instruments will doubtless revert to their use.

Artefact suppression is only substantial in the two lens system if the two light-collection branches are well balanced. Similarly, light collection in a regular cone in backscattering makes sense only if a direction perpendicular to the axis of the cone is not distinguished in some other way. Unfortunately, for ICP, the polarization dependence of the dispersive grating of the spectrograph does precisely that. For holographic transmission gratings, for example, a 50% higher diffraction efficiency is quoted for s polarization than for p polarization. The action of the grating therefore is akin to that of a low efficiency polarization analyser. Other optical elements, such as the 45° mirror in **Figure** 7, can add to or subtract from this effect. Depolarization of the scattered light therefore is an absolute necessity in ICP. A Lyot depolarizer placed directly after the sample cell into the divergent scattered light was used in the first instrument, complementing the effect of the fibre optics, and has remained the standard method ever since to depolarize quasi-monochromatic Raman light in all ICP ROA backscattering instruments.

Instrument performance and the future

The rapid advance in the measurement of ROA is evident from the spectra recorded by the latest backscattering instruments equipped with backthinned CCDs and transmission phase gratings. To date, such spectra have been published for two ICP instruments operating at the University of Glasgow and for a DCP_1 instrument at Syracuse University. Depending on the earlier multichannel instrument taken for comparison, quoted increases in measurement speed vary from 5 to 100. Within the statistical noise limit, artefact control, achieved by nulling with achiral samples

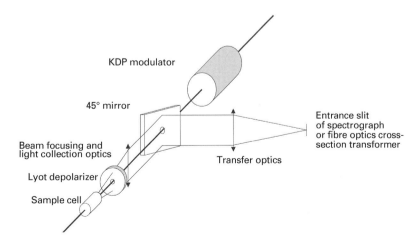

Figure 7 Conventional ICP backscattering arrangement with a Lyot depolarizer in divergent light. Filling the étendue of a spectrograph requires a fibre optics cross-section transformer as used in the original design.

by, for example, slightly rotating the plane of polarization of the incident light (DCP_I) or the axes of the Lyot depolarizer (ICP), is essentially complete.

A SCP instrument constructed by the author at the University of Zürich uses even more advanced optics (**Figure 8**). It represents a further increase, by a factor of 4.4 (compared with the DCP_I instrument) and 7.8 (ICP instrument), in light gathering and throughput capability (**Table 2**). Yet, its main attraction is

dramatically reduced flicker noise and self-balancing of artefacts. This is shown in the very first ROA spectrum recorded on this instrument (**Figure 9**). Deliberately, no artefact nulling was performed, the scattering cell was not selected or individually calibrated, and the liquid crystal circular analyser was not stabilized, in contrast, for example, to the high degree of stabilization used for the retarder in ICP experiments. Likewise, no sample filtration was

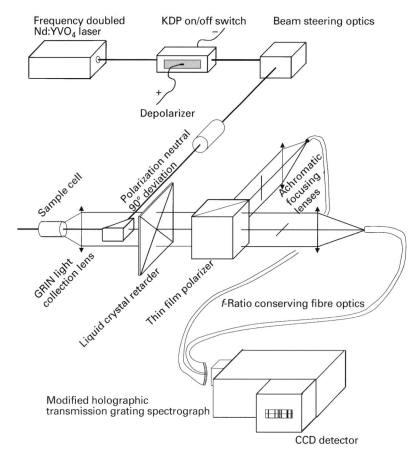

Figure 8 A new generation SCP backscattering instrument. The right and left circular components are measured simultaneously, with the two channels being interchanged by switching the liquid crystal retarder. Sample illumination during read-out is suppressed by the KDP switch. The curved fibre optics output forms the entrance slit of the spectrograph and yields a straight line image on the detector. The notch filter is incorporated in the light collection/transfer optics.

Table 2 Comparison of luminosity of current backscattering instruments as measured by the number of detected electrons per CCD column per joule of exciting energy at the sample; the values refer to the height of the depolarized 1436 cm^{-1} band of α-pinene. Dispersion and CCD efficiency is similar for all instruments

Instrument	Simultaneous R/L detection	Path length cell (mm)	CCD column size (mm^2)	Exciting wavelength (nm)	Slit width (μm)	Detected charge (electrons J^{-1})
ICP (Glasgow)	No	5	0.027 × 6.9	514.5	83	5.6 × 10^6 [a]
DCP_I (Syracuse)	No	10	0.024 × 7.9	514.5	83	10 × 10^6 [b]
SCP (Zürich)	Yes	6	0.026 × 6.65	532	71 [c]	44 × 10^6 [d]

[a] L Hecht, private communication.
[b] DCP_I value multiplied by 7/6 for comparison.
[c] Slit width of the SCP instrument is the equivalent value calculated for the standard 85 mm focal length input lens of the spectrographs used in the other instruments.
[d] Limited by the sample cell.

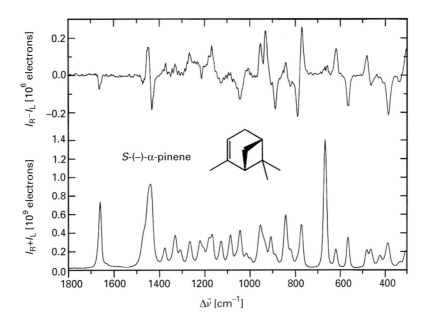

Figure 9 ROA spectra of (–)-α-pinene recorded with the SCP instrument of **Figure 8**. The exciting energy was 21 J at the sample, yielding approximately 10^9 electrons (= detected photons) for the peak height of the 1436 cm^{-1} band (see S/N discussion).

performed and no sample handling precautions were adopted. As a consequence, bright flashes of light due to grains of dust passing through the scattering zone of the exciting beam focused into the sample cell illuminated the walls of the laboratory during recording of the spectra. Ordinarily, such conditions would have totally invalidated any ROA measurement, and very large offsets were indeed observed in the two individual branches of the instrument. Yet, upon summing the data, offsets and artefacts essentially cancelled except at very low wavenumber shifts. The detector and electronics were probably temporarily overloaded at these wavelengths, and some Raman scattered light may also have been collected from the quartz windows of the sample cell, which is not yet optimized for the awesome light gathering power of the non-imaging optics of the instrument. These problems are amenable to straightforward solutions.

The spectrum in **Figure 9** is clearly not yet perfect but it is of acceptable quality, particularly if the short illumination time (300 s) and low laser power (70 mW at the sample) are also taken into account. Together with the above-cited ICP and DCP$_I$ data it lends strong support to the view that ROA instrumentation has finally progressed to the point where this powerful chiroptical method will become a generally useful analytical tool.

List of symbols

c = speed of light; $E^{(0)}$ = electric field at the scattering molecule; f = frequency or focal length; G = étendue;

I = scattered light intensity; K = defined by Equation [11]; N = number of events; NA = numerical aperture; R = distance between scattering molecule and observer; S = surface area; S/N = signal to noise ratio; α = half angle of light cone; α^2 = isotropic Raman invariant; $\alpha G'$ = isotropic ROA invariant due to the optical activity tensor; β^2 = anisotropic Raman invariant; $\gamma^2 = \beta(G')^2$ = anisotropic ROA invariant due to the optical activity tensor; $\delta^2 = \beta(A)^2$ = anisotropic ROA invariant due to the quadrupole tensor; μ_0 = permeability of the vacuum; ω = angular frequency.

See also: **Fibre Optic Probes in Optical Spectroscopy, Clinical Applications; Light Sources and Optics; Raman Optical Activity, Applications; Raman Optical Activity, Theory; Raman Spectrometers; Vibrational CD Spectrometers.**

Further reading

Barron LD and Hecht L (1994) Vibrational Raman optical activity: from fundamentals to biochemical applications. In: Nakanishi K, Berova ND and Woody RW (eds) *Circular Dichroism: Principles and Applications*, pp 179–215. New York: VCH.

Barron LD and Hecht L (1994) Recent developments in Raman optical activity instrumentation. *Faraday Discussions* 99: 35–47.

Barron LD and Hecht L (1996) Recent developments in Raman optical activity of biopolymers. *Applied Spectroscopy* 50: 619–629.

Greulich KO and Monajembashi S (1996) Laser microbeams and optical tweezers: how they work and why they work. In: *Optical and Imaging Techniques for Biomonitoring*, Vol 2628 of Proceedings of SPIE – International Society of Optical Engineering, 116–127. Bellingham: SPIE Press.

Hug W (1982) Instrumental and theoretical advances in Raman optical activity. In: Lascombe J and Huong PV (eds) *Raman Spectroscopy, Linear and Nonlinear*, pp 3–12. Chichester: Wiley-Heyden.

James JF and Sternberg RS (1969) *The Design of Optical Spectrometers*. London: Chapman & Hall.

Nafie LA (1996) Vibrational optical activity. *Applied Spectroscopy* 50: 14A–26A.

Nafie LA and Zimba CG (1987) Raman optical activity and related techniques. In: Spiro TG (ed) *Biological Applications of Raman Spectroscopy, Vol 1, Raman Spectra and The Conformation of Biological Macromolecules*, pp 307–343. New York: John Wiley & Sons.

Vargek M, Freedman TB and Nafie LA (1997) Improved backscattering dual circular polarization Raman optical activity spectrometer with enhanced performance for biomolecular applications. *Journal of Raman Spectroscopy* 28: 627–633.

Raman Optical Activity, Theory

Laurence A Nafie, Syracuse University, New York, NY, USA

VIBRATIONAL, ROTATIONAL & RAMAN SPECTROSCOPIES
Theory

Raman optical activity (ROA) is defined as the difference in Raman scattering intensity for right minus left circularly polarized light. Along with optical rotational and circular dichroism, ROA is a form of natural optical activity. All forms of optical activity can be defined as the differential interaction of a molecule with right versus left circularly polarized radiation. Only chiral molecules exhibit natural optical activity and, for such molecules, the mirror image of the molecule cannot be superimposed on itself. The most common form of ROA is vibrational ROA. Vibrational ROA is also one of two form of vibrational optical activity. The other form is vibrational circular dichroism (VCD), which is the difference in the IR absorption of a molecule for left versus right circularly polarized radiation for a vibrational transition. VCD and ROA are complementary, non-redundant forms of vibrational optical activity in the same way that IR absorption and Raman scattering are complementary forms of ordinary vibrational spectroscopy.

There are many forms of ROA, depending on the choice of polarization modulation, scattering geometry and proximity of the exciting laser radiation to resonance with excited electronic states in the molecules. A general theory of ROA can be written from which all special cases can be derived. The first division of the theory is between circular polarization (CP) ROA and linear polarization (LP) ROA. To date

only different forms of CP ROA have been measured experimentally. There are four forms of both CP and LP ROA. For CP ROA, the original and most common form is called incident circular polarization (ICP) ROA in which only the state of the incident laser radiation is modulated between right and left circular polarization (RCP and LCP). Analogously, the other three forms are called scattered circular polarization (SCP) ROA, where only the polarization of the Raman scattered radiation is sampled for RCP and LCP content, dual circular polarization one (DCP$_I$), where both the incident and scattered polarization states are modulated in-phase, and dual circular polarization two (DCP$_{II}$), where both states are modulated out-of-phase. Similar definitions have been provided for the four forms of LP ROA.

The two principal resonance limits of the theory of ROA are the far-from-resonance (FFR) limit, the original form of the theory of ROA, and the single-electronic-state (SES) limit, for the case of strong resonance between a single excited electronic state and the incident laser radiation. In the case of FFR ROA, *ab initio* calculations have been carried out for direct comparison with experiment. The SES theory is so simple that the complete SES-ROA spectrum can be predicted from the parent resonance Raman spectrum and the electronic circular dichroism spectrum of the resonant electronic state.

ROA is a new spectroscopic tool that has been applied with a high degree of success to the study of the structure of chiral molecules in solution. Areas of application include proteins, nucleic acids, carbohydrates, natural products, pharmaceuticals and other kinds of molecules of biological or therapeutic significance.

Polarized light scattering

One of the essential properties associated with the scattering of light by molecules is the polarization of state of the light. Changes in the polarization state affect the nature and information content of the scattered light. This holds for Rayleigh scattering, where the frequency of the scattered radiation is unchanged from the exciting laser radiation, and for Raman scattering, where the scattered radiation differs from the incident laser radiation by a vibrational energy change in the molecule. The intensity of light scattering for any experiment can be expressed in terms of the general scattering tensor $\tilde{a}_{\alpha\beta}$ and the polarization vectors for the incident and scattered radiation $\tilde{e}_{\alpha}^{i}$ and $\tilde{e}_{\alpha}^{d}$, respectively, and is given by

$$I(\tilde{\boldsymbol{e}}^{d}, \tilde{\boldsymbol{e}}^{i}) = 90K \left\langle \left| \tilde{\boldsymbol{e}}_{\alpha}^{d*} \tilde{a}_{\alpha\beta} \tilde{\boldsymbol{e}}_{\beta}^{i} \right|^{2} \right\rangle \qquad [1]$$

In this equation, K is a constant, given below, that depends on among other things on the intensity of the incident laser radiation. The angular brackets designate an average over all angles of orientation of the molecule to the laboratory frame of reference. This is needed for liquid, solution or gaseous samples where there is no unique molecular axes relative to the laboratory axes. The polarization vectors have one Greek subscript and the scattering tensor has two. For repeated Greek subscripts, summation over the Cartesian directions x, y and z is implied. Hence, Equation [1] has nine terms within the vertical brackets, and these brackets designate the absolute value of the complex quantities within the brackets. The tilde above a quantity, such as a polarization vector or a scattering tensor, indicates that this quantity can be complex. The star superscript for the polarization vector of the scattered light designates complex conjugation. The constant K is given by

$$K = \frac{1}{90} \left(\frac{\omega^{2} \mu_{0} \tilde{\boldsymbol{E}}^{(0)}}{4\pi R} \right)^{2} \qquad [2]$$

where ω is the angular frequency of the scattered light, μ_{0} is the magnetic permeability $\tilde{E}^{(0)}$ is the electric field strength of the incident laser radiation, and R is the distance from the scattering to the detector. The general scattering tensor is given through its lowest-order tensors as

$$\tilde{a}_{\alpha\beta} = \tilde{\alpha}_{\alpha\beta} + \frac{1}{c} \left[\varepsilon_{\gamma\delta\beta} \boldsymbol{n}_{\delta}^{i} \tilde{G}_{\alpha\gamma} + \varepsilon_{\gamma\delta\alpha} \boldsymbol{n}_{\delta}^{d} \tilde{G}_{\gamma\beta} \right.$$
$$\left. + \frac{1}{3} \omega (\boldsymbol{n}_{\gamma}^{i} \tilde{A}_{\alpha,\gamma\beta} - \boldsymbol{n}_{\gamma}^{d} \tilde{A}_{\beta,\gamma\alpha}) \right] \qquad [3]$$

where the first tensor is simply the polarizability tensor that is responsible for ordinary Raman (and Rayleigh) scattering. The four tensors in square brackets are the ROA tensors. The first two are magnetic dipole–electric dipole ROA tensors and the second two are electric quadrupole–electric dipole ROA tensors. The vectors $\tilde{\boldsymbol{n}}_{\alpha}^{i}$ and $\tilde{\boldsymbol{n}}_{\alpha}^{d}$ are the propagation vectors for the incident and scattered light, respectively, and $\varepsilon_{\alpha\beta\gamma}$ is the unit antisymmetric tensor that is +1 for even permutations of the order x, y, z, −1 for odd permutation of this order, and zero if any two directions are the same.

The Raman polarizability tensor is given by

$$\tilde{\alpha}_{\alpha\beta} = \frac{1}{\hbar} \sum_{j \neq m,n} \left[\frac{\langle m|\hat{\boldsymbol{\mu}}_{\alpha}|j\rangle \langle j|\hat{\boldsymbol{\mu}}_{\beta}|n\rangle}{\omega_{jn} - \omega_{0} + i\Gamma_{j}} + \frac{\langle m|\hat{\boldsymbol{\mu}}_{\beta}|j\rangle \langle j|\hat{\boldsymbol{\mu}}_{\alpha}|n\rangle}{\omega_{jm} - \omega_{0} + i\Gamma_{j}} \right] \qquad [4]$$

where $\hbar$ is Planck's constant divided by 2π, and the summation is over all excited electronic states, j, except the initial and final states, n and m, respectively. The states n and m differ by a vibrational quantum of energy. The denominators contain frequency terms, and ω_{jn} is the angular frequency difference between the state j and n. The terms $i\Gamma_{j}$ are imaginary terms proportional to the width of the electronic state j, and hence inversely proportional to its lifetime. The first term in Equation [4] is called the resonance term since the frequency difference between the jn-transition frequency and the laser frequency vanishes at the resonance condition. The quantities in angular brackets are quantum mechanical matrix elements with electric dipole moment operators μ_{α} given by

$$\hat{\boldsymbol{\mu}}_{\alpha} = \sum_{k} e_{k} r_{k\alpha} \qquad [5]$$

which is simply the summation over the charge and position in the α direction of all particles k, in the molecule, electrons and nuclei.

The matrix element in Equation [4] involving the operator $\hat{\mu}_{\beta}$ describes the interaction of the molecule with the incident radiation while the matrix elements with the operator $\hat{\mu}_\alpha$ describes the interaction of the molecule with the scattered radiation. The matrix element products in each term can be read from right to left in a time-ordered sense, and hence the resonance term describes the molecule interacting first with a laser photon and subsequently creating a scattered photon, whereas the non-resonance terms reverses the natural order of the those two events.

The four ROA tensors differ from the Raman polarizability tensor by substitution of a higher-order operator for an electric dipole operator in Equation [4]. The two operators needed for ROA are the magnetic dipole moment operator and the electric quadrupole moment operator given respectively by

$$\hat{\boldsymbol{m}}_\alpha = \frac{1}{2}\sum_k \frac{e_k}{m_k}\varepsilon_{\alpha\beta\gamma}r_{k\alpha}p_{k\gamma} \qquad [6]$$

$$\hat{\Theta}_{\alpha\beta} = \frac{1}{2}\sum_k e_k(3r_{k\alpha}r_{k\alpha} - r_k^2\delta_{\alpha\beta}) \qquad [7]$$

The relationships between the operator substitutions and the resulting ROA tensors in Equation [4] is given by

$$\alpha_{\alpha\beta} \quad \tilde{G}_{\alpha\beta} \quad \text{when} \quad \hat{\mu}_\beta \quad \hat{m}_\beta \qquad [8]$$

$$\alpha_{\alpha\beta} \quad \tilde{G}_{\alpha\beta} \quad \text{when} \quad \hat{\mu}_\alpha \quad \hat{m}_\alpha \qquad [9]$$

$$\alpha_{\alpha\beta} \quad \tilde{A}_{\alpha,\beta\gamma} \quad \text{when} \quad \hat{\mu}_{\beta\gamma} \quad \hat{\Theta}_\beta \qquad [10]$$

$$\alpha_{\alpha\beta} \quad A_{\beta,\alpha\gamma} \quad \text{when} \quad \hat{\mu}_{\alpha\gamma} \quad \hat{\Theta}_\alpha \qquad [11]$$

The expressions given above provide the theoretical formalism for the description of all forms of polarized Raman scattering through first-order in the magnetic dipole and electric quadrupole interaction of light with matter. This is sufficient to describe the various forms of ROA within the assumptions given above.

ROA observables

By specifying the desired polarization states of the incident laser radiation and the scattered radiation, it is possible to construct theoretical expressions to describe the various ROA observables that can be measured in the laboratory. If we restrict our attention to circular polarization ROA experiments, the expressions can be obtained from pairs of intensity expressions that differ only in the change in the circular

polarization state of one or both the light beams from right circular to left circular, or vice versa.

The fundamental ROA observables are classified by polarization and scattering angle, ξ. There are four different forms of CP ROA given by

ICP ROA: $\quad \Delta I_\alpha(\xi) = I_\alpha^R(\xi) - I_\alpha^L(\xi) \qquad [12]$

SCP ROA: $\quad \Delta I^\alpha(\xi) = I_R^\alpha(\xi) - I_L^\alpha(\xi) \qquad [13]$

DCP$_\text{I}$ ROA: $\quad \Delta I_\text{I}(\xi) = I_R^R(\xi) - I_L^L(\xi) \qquad [14]$

DCP$_\text{II}$ ROA: $\quad \Delta I_\text{II}(\xi) = I_L^R(\xi) - I_R^L(\xi) \qquad [15]$

These different forms of ROA are illustrated in **Figure 1**. In the case of ICP- and SCP-ROA, the polarization states α is any fixed value. The standard

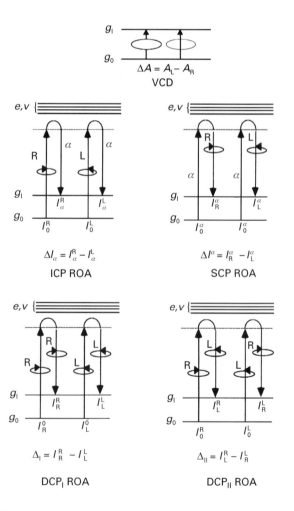

Figure 1 Energy level diagram showing the transitions and polarization states associated with the definition of VCD and the various forms of ROA.

choices are unpolarized, linearly polarized parallel to the scattering plane (depolarized) or linearly polarized perpendicular to the scattering plane (polarized). The standard scattering angles are 90° (right-angle scattering) 180° (backscattering), and 0° (forward scattering).

As an example of an ROA observable we present in **Figure 2** the backscattering DCP-ROA and Raman spectra of neat (−)-β-pinene in the region between 200 and 1700 cm⁻¹. The stereospecific structure of this molecule is given in the figure. The opposite enantiomer, the mirror-image molecule, would yield an ROA spectrum in which the signs of all the bands would be reversed, i.e. a 'mirror-image' ROA spectrum. Note that the ROA spectrum is approximately three orders of magnitude smaller than the corresponding Raman spectrum. Each parent Raman band has associated with it an ROA band of a particular sign and magnitude. There is no correlation between strong Raman bands and strong ROA bands.

General theory of ROA

The general, complete theory of ROA embraces all possible polarization experiments, scattering geometries and degress of resonance Raman intensity enhancement. Because of this generality, the level of theory is complex and too lengthy to describe in the present context. Instead, we provide a verbal description of the formalism and refer the interested reader to a comprehensive review by Nafie and Che (1994) of the theory and measurement of ROA.

ROA and Raman intensity are proportional to the square of a tensor quantity, as expressed in Equation [1]. For Raman scattering only the square of the polarizability is needed, whereas ROA intensity arises from the product of the polarizability and an ROA tensor. The ROA tensor are approximately three orders of magnitude smaller than the polarizability, and hence an ROA spectrum is approximately three orders of magnitude smaller than its parent Raman spectrum. As noted above, the Greek subscripts of the tensor refer to the molecular axis system. However, for both Raman and ROA, linear combinations of products of tensors can be found that do not vary with the choice of the molecular coordinate frame. Such combinations are called invariants. All Raman intensities from samples of randomly oriented molecules can be expressed in terms of only three invariants, called the isotropic invariant, the symmetric

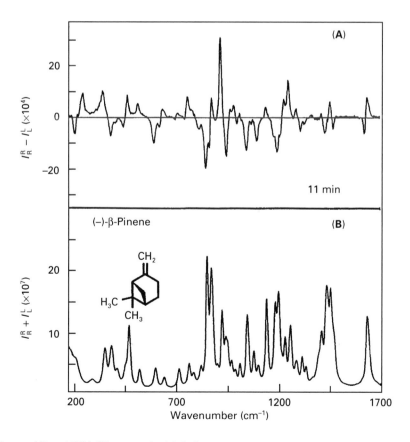

Figure 2 DCP₁ Raman (A) and ROA (B) spectra for (−)-β-pinene.

anisotropy and the antisymmetric anisotropy, given by

$$\alpha^2 = \frac{1}{9}\text{Re}[(\tilde{\alpha}_{\alpha\alpha})^{\text{s}}(\tilde{\alpha}_{\beta\beta})^{\text{s}*}] \quad [16]$$

$$\beta_{\text{s}}(\alpha)^2 = \frac{1}{2}\text{Re}[3(\tilde{\alpha}_{\alpha\beta})^{\text{s}}(\tilde{\alpha}_{\alpha\beta})^{\text{s}*} - (\tilde{\alpha}_{\alpha\alpha})^{\text{s}}(\tilde{\alpha}_{\beta\beta})^{\text{s}*}] \quad [17]$$

$$\beta_a(\alpha)^2 = \frac{1}{2}\text{Re}[3(\tilde{\alpha}_{\alpha\beta})^{\text{a}}(\tilde{\alpha}_{\alpha\beta})^{\text{a}*}] \quad [18]$$

where the symmetric and anti-symmetric forms of the polarizability tensors are given by

$$(\tilde{\alpha}_{\alpha\beta})^{\text{s}} = \frac{1}{2}[(\tilde{\alpha}_{\alpha\beta}) + (\tilde{\alpha}_{\beta\alpha})] \quad [19]$$

$$(\tilde{\alpha}_{\alpha\beta})^{\text{a}} = \frac{1}{2}[(\tilde{\alpha}_{\alpha\beta}) - (\tilde{\alpha}_{\beta\alpha})] \quad [20]$$

For ROA there are ten invariants, five associated with the Roman (font) tensors, $[\alpha G, \beta_{\text{s}}(\tilde{G})^2, \beta_{\text{a}}(\tilde{G})^2, \beta_{\text{s}}(\tilde{A})^2$ and $\beta_{\text{a}}(\tilde{A})^2$ and five with the Arial tensors, $[\alpha G, \beta_{\text{s}}(\tilde{G})^2, \beta_{\text{a}}(\tilde{G})^2, \beta_{\text{s}}(\tilde{A})^2$ and $\beta_{\text{a}}(\tilde{A})^2]$. All of the different ROA experiments can be expressed in terms of these invariants. The ROA intensity for each experiment is expressed as a linear combination of some or all of the ten invariants. Although sets of experiments can be devised to isolate all three ordinary Raman invariants, only six distinct combinations of ROA invariants can be isolated.

Far from resonance theory of ROA

The theory of ROA simplifies drastically in the limit, where the exciting laser radiation is far from the lowest allowed excited state in the molecule, and the interaction of the light with the molecule is approximately the same for both the incident and the scattered radiation. This symmetry reduces the number of Raman invariants from three to two, the isotropic and (symmetric) anisotropic invariants, and the number of ROA invariants from ten to three. The relationships that reduces these thirteen Raman and ROA invariants to only five are

$$\beta_{\text{a}}(\alpha)^2, \beta_{\text{a}}(\tilde{G})^2, \beta_{\text{a}}(\tilde{A})^2, \beta_{\text{a}}(\tilde{G})^2, \beta_{\text{a}}(\tilde{A})^2 = 0 \quad [21]$$

$$\beta_{\text{s}}(\alpha)^2 = \beta(\alpha)^2 \quad [22]$$

$$\alpha G = -\alpha\mathsf{G} = \alpha G' \quad [23]$$

$$\beta_{\text{s}}(\tilde{G})^2 = -\beta_{\text{s}}(\tilde{\mathsf{G}})^2 = \beta(G')^2 \quad [24]$$

$$\beta_{\text{s}}(\tilde{A})^2 = \beta_{\text{s}}(\tilde{\mathsf{A}})^2 = \beta(A)^2 \quad [25]$$

where the following definition of the real and imaginary parts of complex tensor has been used

$$\tilde{T} = T - iT' \quad [26]$$

The equations for the two Raman invariants and three ROA invariants are

$$\alpha^2 = \frac{1}{9}\alpha_{\alpha\alpha}\alpha_{\beta\beta} \quad [27]$$

$$\beta(\alpha)^2 = \frac{1}{2}(3\alpha_{\alpha\beta}\alpha_{\alpha\beta} - \alpha_{\alpha\alpha}\alpha_{\beta\beta}) \quad [28]$$

$$\alpha G' = \frac{1}{9}\alpha_{\alpha\alpha}G'_{\beta\beta} \quad [29]$$

$$\beta(G')^2 = \frac{1}{2}(3\alpha_{\alpha\beta}G'_{\alpha\beta} - \alpha_{\alpha\alpha}G'_{\beta\beta}) \quad [30]$$

$$\beta(A)^2 = \frac{1}{2}\omega_0\alpha_{\alpha\beta}\varepsilon_{\alpha\gamma\delta}A_{\gamma,\delta\beta} \quad [31]$$

where the FFR polarizability and optical activity tensors are given by

$$\tilde{\alpha}_{\alpha\beta} = \frac{2}{\hbar}\sum_{j\neq n}\frac{\omega_{jn}}{\omega_{jn}^2 - \omega_0^2}\text{Re}\left[\langle n|\hat{\mu}_\alpha|j\rangle\langle j|\hat{\mu}_\beta|n\rangle\right] \quad [32]$$

$$G'_{\alpha\beta} = \frac{-2}{\hbar}\sum_{j\neq n}\frac{\omega_0}{\omega_{jn}^2 - \omega_0^2}\text{Im}\left[\langle n|\hat{\mu}_\alpha|j\rangle\langle j|\hat{\mu}_\beta|n\rangle\right] \quad [33]$$

$$A_{\alpha\beta\gamma} = \frac{2}{\hbar}\sum_{j\neq n}\frac{\omega_{jn}}{\omega_{jn}^2 - \omega_0^2}\text{Re}\left[\langle n|\hat{\mu}_\alpha|j\rangle\langle j|\hat{\Theta}_{\beta\gamma}|n\rangle\right] \quad [34]$$

Using these invariants, we can write intensity expressions for ROA and Raman that cover all possible polarization modulations and scattering geometries in the FFR approximation. These expressions are:

$$I(R) - I(L) = \frac{8K}{c}[D_1\alpha G' + D_2\beta(G')^2 + D_3\beta(A)^2] \quad [35]$$

$$I(R) + I(L) = 4K[D_4\alpha^2 + D_5\beta(\alpha)^2] \quad [36]$$

The values of the constants D_1 through to D_5 are given in **Table 1**.

Most of the ROA spectra measured to date have been for one of three kinds of experiments, right-angle depolarized ICP ROA, backscattering unpolarized ICP ROA, and backscattering DCP$_I$ ROA. The early work on ROA was almost exclusively right-angle depolarized ICP ROA where the ROA and Raman intensity expressions from **Table 1** are

$$I_z^R(90°) - I_z^L(90°) = \frac{8K}{c}[3\beta(G')^2 - \beta(A)^2] \quad [37]$$

$$I_z^R(90°) - I_z^L(90°) = 4K[3\beta(\alpha)^2] \quad [38]$$

The corresponding polarized ICP ROA experiment included isotropic ROA invariants and was more difficult to measure without interference from polarization artifacts. In the late 1980s, the virtues of backscattering ROA were implemented on a routine basis. Two forms of backscattering ROA, each having their own theoretical or experimental advantages, have been used extensively. They are unpolarized backscattering ICP ROA, given by

$$I_u^R(180°) - I_u^L(180°) = \frac{8K}{c}[12\beta(G')^2 + 4\beta(A)^2] \quad [39]$$

$$I_u^R(180°) + I_u^L(180°) = 4K[45\alpha^2 + 7\beta(\alpha)^2] \quad [40]$$

and backscattering DCP$_I$ ROA, given by

$$I_R^R(180°) - I_L^L(180°) = \frac{8K}{c}[12\beta(G')^2 - 4\beta(A)^2] \quad [41]$$

$$I_R^R(180°) + I_R^L(180°) = 4K[6\beta(\alpha)^2] \quad [42]$$

From these expressions, the advantages of ICP$_u$ and DCP$_I$ ROA in backscattering are apparent. The ROA invariants have a multiplicative advantage of 4 in backscattering and are additive rather than subtractive between the magnetic-dipole and electric-quadrupole ROA invariants. Comparing right-angle ICP$_z$ and DCP$_I$ Raman, both represent depolarized scattering with the backscattering stronger by a factor of 2. It can also be seen that the ROA expressions

for backscattering ICP$_u$ and DCP$_I$ ROA are identical even though the ICP$_u$ Raman intensity represents classical polarized Raman scattering and DCP$_I$ Raman intensity the corresponding depolarized scattering. The additional Raman intensity in ICP$_u$ Raman $4K[45\alpha^2 + \beta(\alpha)^2]$ carries no ROA intensity since this additional strongly polarized Raman intensity corresponds to DCP$_{II}$ ROA which has not intensity in the FFR approximation as shown in **Table 1**. Thus backscattering DCP$_I$ ROA discriminates against DCP$_{II}$ ROA by analysing the circular polarization of the scattering light, whereas both of these intensities are present in ICP$_u$ ROA which has no such discrimination.

Another interesting property of these expressions is the possibility of isolating the ROA spectra for the magnetic-dipole and electric-quadrupole ROA invariants. By proper experimental normalization of the depolarized Raman spectra given in Equations [38] and [42], the corresponding and suitably scaled ICP$_z$(90°) and DCP$_I$ (180°) ROA spectra can be added and subtracted to yield these invariants. This has been accomplished for the molecule (+)-*trans*-pinane as shown in **Figure 3**.

Of the two backscattering ROA schemes predominantly in use these days, the ICP$_u$ approach enjoys an advantage of experimental simplicity, while the DCP$_I$ enjoys an advantage of higher ROA intensity per Raman intensity, particularly in regions of strongly polarized Raman bands where no such intensity enters the DCP$_I$ Raman spectrum.

Table 1 Values of Raman and ROA invariant coefficients for the far-from-resonance ROA and Raman intensity expressions in Equations [41]

$\xi(°)$	Form	Raman (4K)		ROA (8K/c)		
		α^2	$\beta(\alpha)^2$	$\alpha G'$	$\beta(G')^2$	$\beta(A)^2$
0	ICP$_u$	45	7	90	2	−2
	SCP$_u$	45	7	90	2	−2
	DCP$_I$	45	1	90	2	−2
	DCP$_{II}$		6			
90	ICP$_p$	$\frac{45}{2}$	$\frac{7}{2}$	$+\frac{45}{2}$	$+\frac{7}{2}$	$+\frac{1}{2}$
	ICP$_d$				+3	−1
	ICP*	$\frac{45}{3}$	$\frac{10}{3}$	$+\frac{45}{3}$	$+\frac{10}{3}$	
	SCP$_p$	$\frac{45}{2}$	$\frac{7}{2}$	$+\frac{45}{2}$	$+\frac{7}{2}$	$+\frac{1}{2}$
	SCP$_d$		3		+3	−1
	SCP*		$\frac{10}{3}$	$+\frac{45}{3}$	$+\frac{10}{3}$	
	DCP$_I$	$\frac{45}{4}$	$\frac{13}{4}$	$+\frac{45}{2}$	$+\frac{13}{2}$	$-\frac{1}{2}$
	DCP$_{II}$	$\frac{45}{4}$	$\frac{13}{4}$			
100	ICP$_u$	45	7		+12	+4
	SCP$_u$	45	7		+12	+4
	DCP$_I$		6		+12	+4
	DCP$_{II}$	45	1			

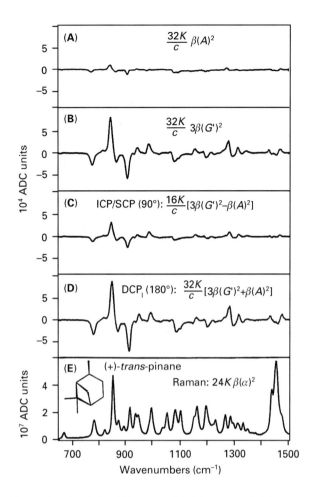

Figure 3 Depolarized Raman and ROA spectra of (+)-*trans*-pinane, showing the decomposition of right-angle depolarized ICP and DCP$_I$ ROA into their magnetic-dipole and electric-quadrupole anisotropic ROA invariants.

Single electronic state theory of resonance ROA

When the frequency of the incident laser radiation in a Raman scattering experiment is in resonance with a single electronic state (SES), it is well known that strong enhancement of the Raman scattering occurs. This is because the denominator of the resonant term in the polarizability expression, Equation [4] approaches zero and the value of the polarizability can increase by several orders of magnitude. This resonance condition brings simplifying conditions to the theory of Raman scattering as it becomes known as resonance Raman (RR) scattering. Under conditions of strong resonance, the non-resonant terms can be dropped and the contributions of all other electronic states can be dropped as too small to consider. The Raman polarizability in Equation [4] for a 0 to 1 vibrational transition in the ground electronic state, g_0

to g_1, becomes

$$(\tilde{\alpha}_{\alpha\beta})_{g_1,g_0} = \frac{1}{\hbar} \sum_\nu \frac{\langle g_1|\hat{\mu}_\alpha|e\nu\rangle\langle e\nu|\hat{\mu}_\beta|g_0\rangle}{\omega_{e\nu,g_0} - \omega_0 + \mathrm{i}\Gamma_{e\nu}} \quad [43]$$

If the transition moment of the resonant electronic state is taken to lie in the z-direction, the general set of three Raman invariants and ten ROA invariants reduces effectively to only one Raman invariant and one ROA invariant, as

$$\alpha^2 = \frac{1}{9}\left|(\tilde{\alpha}_{zz})_{g_1,g_0}\right|^2 \quad [44]$$

$$\beta_s(\alpha)^2 = \left|(\tilde{\alpha}_{zz})_{g_1,g_0}\right|^2 \quad [45]$$

$$\alpha G = -\alpha\mathsf{G} = \frac{1}{9}\mathrm{Im}\left[(\tilde{\alpha}_{zz})_{g_1,g_0}(\tilde{G}_{zz})^*_{g_1,g_0}\right] \quad [46]$$

$$\beta_s(\tilde{G})^2 = \beta_s(\tilde{\mathsf{G}})^2 = \mathrm{Im}\left[(\tilde{\alpha}_{zz})_{g_1,g_0}(\tilde{G}_{zz})^*_{g_1,g_0}\right] \quad [47]$$

$$\beta_s(\tilde{A})^2 = \beta_s(\tilde{\mathsf{A}})^2 = 0 \quad [48]$$

$$\beta_a(\alpha)^2 = \beta_a(\tilde{G})^2 = \beta_a(\tilde{\mathsf{G}})^2\beta_a(\tilde{A})^2 = \beta_a(\tilde{\mathsf{A}})^2 = 0 \quad [49]$$

In addition, it can be shown the Raman invariant is proportional to the square of the electronic absorption strength for the resonant electronic state, and the ROA is proportional to the product of the electronic circular dichroism (CD) and the electronic of this state through the relationships,

$$\left|(\tilde{\alpha}_{zz})_{g_1,g_0}\right|^2 = (1/\hbar)|(\vec{\mu})^0_{eg}|^4 U(\omega_0) \quad [50]$$

$$\mathrm{Im}\left[(\tilde{\alpha}_{zz})_{g_1,g_0}(\tilde{G}_{zz})^*_{g_1,g_0}\right]$$
$$= (1/\hbar)|(\vec{\mu})_{eg}|^2\mathrm{Im}\left[(\vec{\mu}^0_{ge}\cdot(\vec{m})^{0*}_{eg}\right]U(\omega_0) \quad [51]$$

where $\vec{\mu}^\nu_{eg}$ and $\vec{m}^\nu_{eg}$ are the electric-dipole and magnetic dipole transition moments, respectively, between the ground and resonant electronic states, respectively. In the case of resonance ROA in the SES limit, the most efficient form of ROA is backscattering DCP$_I$ where the intensity expressions are:

$$I^R_R(180°) - I^L_L(180°) = \frac{96K}{c}\mathrm{Im}[(\tilde{\alpha}_{zz})_{g_1,g_0}(\tilde{G}_{zz})^*_{g_1,g_0}] \quad [52]$$

$$I_{\mathrm{R}}^{\mathrm{R}}(180°) + I_{\mathrm{R}}^{\mathrm{L}}(180°) = 24K|(\tilde{\alpha}_{zz})_{g_1,g_0}|^2 \qquad [53]$$

From these relationship emerges a deep connection between RROA in the SES limit and the electronic CD of the resonant electronic state. Since the anisotropy ratio, g_{eg} is defined as the ratio of the CD intensity to the parent intensity, the following expression is found

$$\frac{I_{\mathrm{R}}^{\mathrm{R}}(180°) - I_{\mathrm{R}}^{\mathrm{L}}(180°)}{I_{\mathrm{R}}^{\mathrm{R}}(180°) + I_{\mathrm{R}}^{\mathrm{L}}(180°)} = -\left(\frac{4}{c}\right) \frac{\mathrm{Im}\left[(\vec{\mu}_{ge}^{0} \cdot (\vec{m})_{eg}^{0}\right]}{|(\vec{\mu})_{eg}|^2} = -g_{ge} \qquad [54]$$

The minus sign arises from the definition of ROA being right minus left circular polarization intensities, whereas the corresponding definition for CD is left minus right. Resonance ROA promises to open up new applications for ROA in the same way that resonance Raman spectroscopy extended the reach of Raman spectroscopy, particularly for biological applications.

Simple models of ROA

Before the development of molecular orbital approaches to the calculation of ROA, a number of simple models of ROA were advanced to provide a conceptual basis for understanding ROA spectra. In various ways, these models arise from considering the polarizability of the molecule $\alpha_{\alpha\beta}$ as the sum of local polarizability units, such as those associated with bonds or atoms. While the polarizability and its individual local components are independent of the location of the origin of the molecule, the magnetic dipole and electric quadrupole ROA tensors do depend on this origin. In the FFR approximation, we can express these tensors in terms of local contributions as

$$\alpha_{\alpha\beta} = \sum_i \alpha_{i,\alpha\beta} \qquad [55]$$

$$G'_{\alpha\beta} = \sum_i G'_{i,\alpha\beta} + \sum_i \tfrac{1}{2}\omega\varepsilon_{\beta\gamma\delta}R_i\alpha_{i,\alpha\delta} \qquad [56]$$

$$A_{\alpha\beta\gamma} = \sum_i A_{i,\alpha\beta\gamma} - \sum_i \tfrac{1}{2}[3\boldsymbol{R}_{i,\beta}\alpha_{i,\alpha\delta} \\ + 3\boldsymbol{R}_{i,\gamma}\alpha_{i,\alpha\beta} - 2\boldsymbol{R}_{i,\delta}\alpha_{i,\alpha\delta}\delta_{\beta\gamma}] \qquad [57]$$

where $R_{i,\alpha}$ is the location of the ith polarizability unit relative to the origin. In the application of ROA models it is assumed that only the local polarizability units

are non-zero and that contributions from the local ROA tensors, $G'_{i,\alpha\beta}$ and $A_{i,\alpha\beta\gamma}$, are zero. The simplest of these models is the two-group model which describes the ROA from the two local symmetric polarizability groups in a molecule that are twisted in a chiral sense with respect to one another. ROA arises from the interference of the independent Raman scattering from each of these two groups. Only limited success has been achieved in the use of these models to understand the details of ROA spectra. For a quantitative understanding, one must use *ab initio* molecular orbital methods.

Ab initio calculations of ROA

The first calculations of ROA using *ab initio* molecular orbital methods have been carried out recently. This has been achieved by starting with expressions for the polarizability and Rayleigh optical activity tensors in the zero-frequency limit of the FFR approximation as

$$\alpha_{\alpha\beta} = \frac{2}{\hbar} \sum_{j\neq n} \frac{\mathrm{Re}\left[\langle n|\hat{\boldsymbol{\mu}}_\alpha|j\rangle\langle j|\hat{\boldsymbol{\mu}}_\beta|n\rangle\right]}{\omega_{jn}} \qquad [58]$$

$$G'_{\alpha\beta} = -\frac{2\omega_0}{\hbar} \sum_{j\neq n} \frac{\mathrm{Im}\left[\langle n|\hat{\boldsymbol{\mu}}_\alpha|j\rangle\langle j|\hat{\boldsymbol{m}}_\beta|n\rangle\right]}{\omega_{jn}^2} \qquad [59]$$

$$A_{\alpha\beta\gamma} = \frac{2}{\hbar} \sum_{j\neq n} \frac{\mathrm{Re}\left[\langle n|\hat{\boldsymbol{\mu}}_\alpha|j\rangle\langle j|\hat{\Theta}_{\beta\gamma}|n\rangle\right]}{\omega_{jn}} \qquad [60]$$

where the distinction between the initial and final states is not needed. ROA and Raman intensities can be obtained from these tensors by calculating their variation with the normal coordinates of vibrational motion. The method by which these tensors have been calculated is the field perturbation approach. The summation over all the excited states j can be avoided by substituting field perturbed wavefunctions in first-order perturbation theory for their nonperturbed counterparts as

$$\alpha_{\alpha\beta} = 2 \sum_{j\neq n} \mathrm{Re}\left[\langle n|\hat{\boldsymbol{\mu}}_\alpha|n'(E_\beta)\rangle\right] \qquad [61]$$

$$G'_{\alpha\beta} = -2\hbar\omega_0\mathrm{Im}\left[\langle n'(E_\alpha)|n'(B_\beta)\rangle\right] \qquad [62]$$

$$A_{\alpha\beta\gamma} = 2\langle n|\Theta_{\beta\gamma}|n'(E_\alpha)\rangle \qquad [63]$$

where E_α and B_α are components of the electric and magnetic fields, and the prime on the wavefunction indicates the first derivative with respect to the field. In **Figure 4** we show the results of a comparison of *ab initio* ROA calculations for the molecule L-alanine in aqueous solution. A high degree of correspondence has been achieved between theory and experiment. This demonstrates that ROA can be calculated with success for small chiral molecules of biological significance.

Applications to biological molecules

Experimental measurements of ROA were first achieved in the mid-1970s. Since that time ROA spectra of many classes of molecules of biological significance have been published. The include terpenes, amino acids, sugars, carbohydrates, peptides, proteins, and nucleic acids. Through these studies, ROA can be seen to be a sensitive probe of the stereo-conformational detail of these molecules in their native environments.

As described in this article, the theory of ROA is rich in content, offering experimentalists many options in the measurements of ROA spectra. These include the wavelength of the exciting radiation and its proximity to allowed transitions to excited electronic states in the molecules. Also of importance is the polarization modulations scheme and the scattering geometry. Recent studies have established that the optimum polarization and scattering conditions for biological applications are either unpolarized ICP or in-phase DCP in backscattering geometry.

The theoretical understanding of ROA is well in hand. What remains to be demonstrated is the ability to calculate ROA intensities accurately across a wide range of theoretical limits, including the most general cases, for most molecules of biological interest. To date, only the simplest level of theory has been used

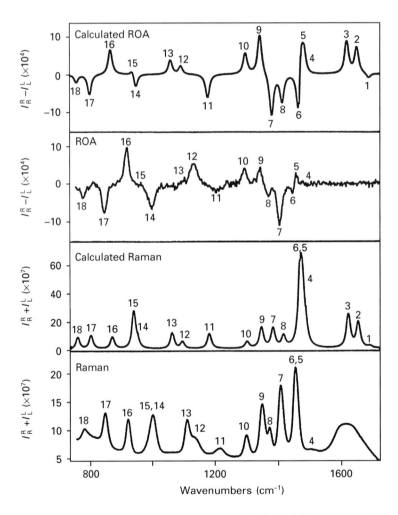

Figure 4 Comparison of the experimentally measured and the *ab initio* calculated DCP$_I$ Raman and ROA spectra of L-alanine in aqueous solution.

for *ab initio* calculations of relatively simple biomolecules. Beyond the zero-frequency limit of the FFR approximation are the dynamic frequency limit, the near resonance conditions, and the strong resonance conditions involving more than one electronic state. As demonstrated above, the case of strong resonance with a single electronic state is trivial in that the ROA spectrum is completely predicted from the resonance Raman spectrum and the electronic CD of the resonance electronic state.

The problem of extending ROA calculations to molecules of increasing complexity will accompany the steady increase in the power of computational calculations made possible by advances in the speed and memory capacity of computers.

With the realization of improvements in the measurements and theoretical calculation of intensities, ROA will assume a place of special importance among our spectroscopic probes of the structure and dynamics of molecules of biological interest.

List of symbols

$\tilde{a}_{\alpha\beta}$ = general scattering tensor; D_1–D_5 = constants; $\tilde{e}_\alpha^i$ = polarization vector for incident light; $\tilde{e}_\alpha^s$ = polarization vector for scattered light; $\tilde{E}^{(0)}$ = electric field strength of incident radiation; g_{eg} = anisotropy ratio; $\hbar$ = Planck's constant/2π; I = intensity of light scattering; K = a constant; n = initial electronic state; $\tilde{n}_\alpha^i$, $\tilde{n}_\alpha^d$ = propagation vectors for incident and scattered light, respectively; m = final electronic state; $(\vec{m})_{eg}^v$ = magnetic dipole transition moment; R = distance from the scattering to the detector; $R_{i,\alpha}$ = location of the ith polarizability relative to the origin; α^2 = the isotropic invariant; $\tilde{\alpha}_{\alpha\beta}$ = Raman polarizability tensor; $\beta_a(\alpha)^2$ = antisymmetric anisotropy; $\beta_s(\alpha)^2$ = symmetric anisotropy; $\varepsilon_{\alpha\beta\gamma}$ = the unit antisymmetric tensor; ξ = scattering angle; μ_0 = magnetic permeability; $(\vec{\mu})_{eg}^v$ = electric dipole transition moment; ω = angular frequency.

See also: **Biochemical Applications of Raman Spectroscopy; Chiroptical Spectroscopy, Emission Theory; Chiroptical Spectroscopy, General Theory; Chiroptical Spectroscopy, Oriented Molecules and Anisotropic Systems; Electromagnetic Radiation; ORD and Polarimetry Instruments; Raman Optical Activity, Applications; Raman Optical Activity, Spectrometers; Raman Spectrometers; Scattering Theory; Vibrational CD Spectrometers; Vibrational CD, Applications; Vibrational CD, Theory.**

Further reading

Barron LD (1982) *Molecular Light Scattering and Optical Activity*. Cambridge: Cambridge University Press.

Barron LD and Hecht L (1994) Vibrational Raman optical activity: from fundamentals to biochemical applications. In: Nakanishi K, Berova ND and Woody RW (eds) *Circular Dichroism: Principles and Applications*. New York: VCH.

Barron LD, Hecht L, Bell AF and Wilson G (1996) Recent developments in Raman optical activity of biopolymers. *Applied Spectroscopy* 50: 619–629.

Nafie LA and Che D (1994) Theory and measurement of Raman optical activity. In: Evans M and Kielich S (eds) *Modern Nonlinear Optics, Part 3*, Vol 85, pp 105–149. New York: Wiley.

Nafie LA, Yu G-S, Qu X and Freedman TB (1994) Comparison of IR and Raman forms of vibrational optical activity. *Faraday Discussions* 99: 13–34.

Nafie LA, Yu G-S and Freedman TB (1995) Raman optical-activity of biological molecules. *Vibrational Spectroscopy* 8: 231–239.

Nafie LA (1995) Circular polarization spectroscopy of chiral molecules. *Journal of Molecular Structure* 347: 83–100.

Nafie LA (1996) Vibrational optical activity. *Applied Spectroscopy* 50: 14A–26A.

Nafie LA (1996) Theory of resonance Raman optical activity: The single-electronic state limit. *Chemical Physics* 205: 309–322.

Nafie LA (1997) Infrared and Raman vibrational optical activity: theoretical and experimental aspects. *Annual Reviews in Physical Chemistry* 48: 357–386.

Raman Spectrometers

Bernhard Schrader, Universität Essen, Germany

> **VIBRATIONAL, ROTATIONAL &**
> **RAMAN SPECTROSCOPIES**
> **Methods & Instrumentation**

Synopsis

Raman spectrometers are quite different from 'ordinary' spectrometers. In Raman spectra the very weak Raman lines are accompanied by the extremely strong Rayleigh line. The stray light of it produces a background in the spectrometer which may be more intense by orders of magnitude than the Raman lines. Therefore, a Raman spectrometer has to combine the elimination of the Rayleigh line with the spectral dispersion and isolation of the Raman lines. Additionally, the necessary resolving power of Raman spectrometers has to be considerably higher compared with 'ordinary', e.g. infrared spectrometers.

This article describes the elements of Raman spectrometers for routine analyses which are available commercially. Instruments designed only for special research are not covered. Only spectrometers for 'classical' (linear) Raman scattering are mentioned, not those for observing resonance Raman scattering (RRS), surface-enhanced Raman scattering (SERS) and all nonlinear Raman techniques; they are described elsewhere in this Encyclopedia.

Rayleigh and Raman scattering, the Raman spectrum

Raman spectra are complementary to infrared spectra. Both are composed of lines (or bands of lines) which are images of the vibrations of molecules. The intensity of Raman lines represents the change of the molecular polarizability by a vibration, while the intensity of infrared lines represents the change of the molecular dipole moment (actually the square of the change of the molecular polarizability or the molecular dipole moment). The Raman lines are accompanied by the 'Rayleigh line' at the wavelength of the exciting radiation; its intensity is proportional to the square of the molecular polarizability. In addition, the intensity of this 'unshifted' line is enhanced further by the exciting radiation which is directly scattered at the surfaces of the particles of powders or at the windows of the sample cuvettes.

When monochromatic exciting radiation of light quanta $h\nu_0$ hits a molecule, an *elastic scattering* process, i.e. *Rayleigh scattering* of quanta having the energy $h\nu_0$ has the highest probability. The *inelastic scattering process*, during which vibrational energy $h\nu_s$ is exchanged with the molecules, has a much lower probability, and is called *Raman scattering*. It emits quanta of energy $(h\nu_0 \pm h\nu_s)$. At ambient temperature most molecules are in their vibrational ground state. According to Boltzmann's law, a much smaller number are in their vibrationally excited state. Therefore, Raman scattering of quanta of energy $(h\nu_0 - h\nu_s)$ has much higher probability than the Raman scattering of quanta of energy $(h\nu_0 + h\nu_s)$. While studying fluorescence spectra Stokes, in 1852, postulated that the wavelength of light produced by photoluminescence, fluorescence or phosphorescence, is usually longer than that of the exciting light. In analogy, Raman lines are referred to as *Stokes lines* and *anti-Stokes lines*. Stokes lines are caused by the quanta of lower energy. Since their intensities are higher than those of anti-Stokes lines, only they are usually recorded as Raman spectrum. The line intensities are usually drawn over the Raman shift, ν_s, in wavenumbers. The intensity ratio of Stokes and anti-Stokes lines of the same Raman shift allows, by employing the Boltzmann equation, the determination of the temperature of the sample under illumination.

The line width $\Delta\nu_s$ of the infrared and Raman lines is about the same. However, the necessary resolving power in the infrared spectrum for isolating an infrared line of width $\Delta\nu_s = 10$ cm^{-1} at a wavelength of 10 μm equivalent to a wavenumber of $\nu_s = 1000$ cm^{-1} is $R_{IR} = \nu_s/\Delta\nu_s = 100$. In the Raman spectrum excited with radiation of wavelength of 500 nm equivalent to $\Delta\nu_0 = 20\,000$ cm^{-1} the same vibration, at $\nu_s = 1000$ cm^{-1}, is recorded at an absolute wavenumber of $(\nu_0 - \nu_s) = 20\,000 - 1000 = 19\,000$ cm^{-1}. Therefore, the necessary resolving power is $R_{RA} = 19\,000/10 = 1900$, this is larger by a factor of 19 than that necessary for recording the infrared line. This shows that Raman spectrometers have to supply a considerably large resolving power than infrared spectrometers.

The intensity of the Raman lines is given by the square of the change of the molecular polarizability by the vibrations. There is an analogy between the molecular polarizability and the molecular volume:

vibrations that modulate the molecular volume are *Raman active*.

Figures 1A and **1B** show a potential energy diagram of a molecule with the vibrational ground state and one vibrational excited state. The direct transition can be observed (**Figure 1A**) by an absorption of a light quantum from the infrared spectral range. There is another way of exciting vibrational states (**Figure 1B**). The molecule is illuminated with light quanta of higher energy, from the visible or near-infrared range. These quanta are, on the one hand, scattered elastically, producing the Rayleigh line. On the other hand, the inelastic scattering process emits a light quantum of the energy of the exciting radiation which is reduced by the vibrational energy. Therefore, infrared absorption and Raman scattering may produce—by different mechanisms—molecules in exactly the same vibrational excited state. The molecular symmetry may 'allow' or 'forbid' the activity in the Raman or infrared spectrum. An example is the *rule of mutual exclusion*: If a molecule has a centre of symmetry, then infrared active vibrations are forbidden in the Raman spectrum and vice versa.

For the observation of 'classical' Raman spectra exciting radiation from the visible part of the spectrum is used. When molecules have electronic energy levels, which can be excited by light quanta from the visible range of the spectrum, the molecules may undergo a transition into an electronic excited state. As a consequence the molecule emits fluorescence radiation (**Figure 1C**). Since this process has a quantum yield much larger than the Raman effect (of the order of 1 versus 10^{-6}–10^{-11}) fluorescence is quite strong, much stronger than the Raman lines. Therefore, Raman lines are then overlaid by the strong (quasi-continuous) fluorescence radiation. This process may hinder the observation of Raman lines. The fluorescing molecules may be impurities, which can be removed by purification (by distillation, sublimation, or recrystallization). However, the fluorescing molecules may be normal constituents of the sample. This is true for all living cells or tissues. They have a chemical 'machinery', composed of enzymes, which are absorbing visible radiation. 'Purification' by destroying the fluorescing molecules will destroy the cells or tissues. The only way to prevent superimposing by fluorescence is excitation of Raman spectra with light quanta having a low energy which is not sufficient to excite electronic states (**Figure 1D**), this is done by using the Nd:YAG laser radiation at 1064 nm.

Figure 2 shows the Stokes and anti-Stokes part of the Raman spectrum of coumarin, excited by the Ar-ion laser line at 514.5 nm. Three abscissa scales are drawn: the wavelength scale, the absolute wavenumber scale and the Raman shift, the energy difference between the exciting energy and the energy of the light quanta, scattered by the Raman effect. This is the only scale which is usually used to draw Raman spectra.

The components of Raman spectrometers

A Raman spectrometer analyses the radiation scattered by molecules, when they are illuminated with monochromatic exciting radiation. The scattered radiation is composed of the strong Rayleigh line and the very weak Raman lines. The Rayleigh line has a radiant power that may exceed that of the Raman lines by about 10^6 up to 10^{15}. The electric signal S produced by a radiation detector is proportional to

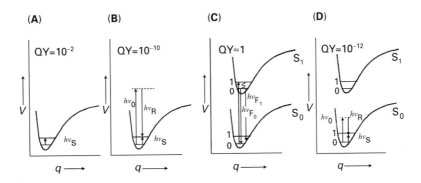

Figure 1 Observation of the excitation of a vibrational state in the electronic ground state S_0 by (A) infrared absorption; (B) Raman scattering; excitation in the visible range ($\lambda = 488$ nm); (C) absorption of the exciting radiation with subsequent fluorescence, (D) Raman scattering, excitation in the near-infrared range, ($\lambda = 1064$ nm), the energy of the exciting light quanta is only 46% of that of (B), V = potential energy, QY = order of magnitude of the quantum yield, q = normal coordinate (describing the vibrational motion), S_0 = electronic ground level, S_1 = electronic excited level. Reproduced from Schrader B (ed) (1995) *Infrared and Raman Spectroscopy*. Weinheim: VCH Publishers, with permission of VCH.

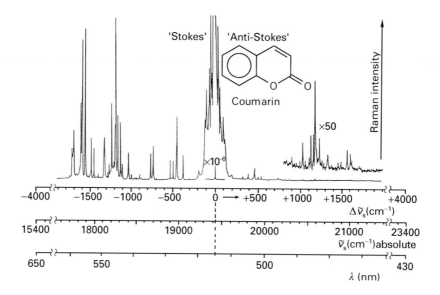

Figure 2 Raman spectrum of coumarin, excited with the radiation of the Ar^+ laser at $\lambda = 514.53$ nm equivalent to $\nu = 19\,430$ cm^{-1}. Reproduced from Schrader B (ed) (1995) *Infrared and Raman Spectroscopy*. Weinheim: VCH Publishers, with permission of VCH.

the radiant power Φ (W), received by the detector. A Raman spectrometer has to facilitate recording of the Raman lines with a high signal-to-noise ratio (S/N) sufficiently resolved.

The radiant power transmitted Φ by any optical instrument is given by

$$\Phi = LG\tau$$

Here, L describes the *radiance* of the radiation source, power per area and solid angle (W cm^{-2} sr^{-1}), G is the *optical conductance*, solid angle times area (sr cm^2), and τ represents the *transmission* of the whole system. In order to record Raman spectra with a large S/N, all factors, L, G and τ, have to be maximal. L is optimized by appropriate sample arrangements. G is maximal when the spectrometer uses a maximal area of the essential elements (the prism or grating or the beam splitter of an interferometer and the entrance aperture). A maximal value of τ is guaranteed by a proper design of the instrument, especially by antireflection coating of all glass/air interfaces and by a maximal reflectivity of all mirrors. As an approximation the optical conductance of a pair succeeding elements of a spectrometer constructed by apertures, lenses or mirrors having an area F_i, $i = 1$ and 2 at distance of a_{12} is given by

$$G \approx F_1 F_2 / a_{12}^2$$

The flux through a spectrometer is appropriately described by

$$\Phi_\nu = L_\nu G_\nu (\Delta\nu^2)\tau$$

Here the subscript ν stand for 'per wavenumber'. L_ν, the spectral radiance, is a property of the radiation source, G_ν the spectral optical conductance and $\Delta\nu$ the bandwidth of the instrument (in wavenumber units), τ is the overall transmission factor of the entire instrument.

In every well-constructed optical instrument the optical conductance of all pairs of succeeding elements should be the same. The overall optical conductance of any instrument is given by the smallest optical conductance of any succeeding pair of elements.

The practical resolving power is given by the ratio of the wavenumber (or wavelength) and the bandwidth. Its upper limit is the theoretical resolving power, R_0 determined by the properties of the dispersive elements, the number of grating rules in a grating spectrometer or the pathlength difference of the interfering rays in an interferometer (**Figures 3** and **4**).

The ratio of the spectral optical conductance of an interferometer, compared to a grating spectrometer (the Jacquinot advantage) is given by

$$G_\nu^I / G_\nu^G \approx 2\pi f/h$$

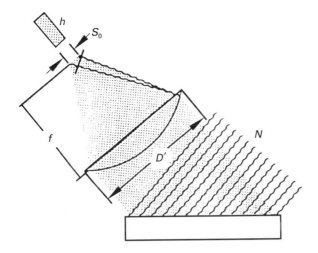

Figure 3 The main components of a grating spectrometer: N is the number of interfering rays, given by the number of rules; S_0 is the halfwidth of the diffraction pattern of the collimator lens with diameter D' and focal length f, which determines the 'optimal' slit width, h is the slit length. Reproduced from Schrader B (ed) (1995) *Infrared and Raman Spectroscopy*. Weinheim: VCH Publishers, with permission of VCH.

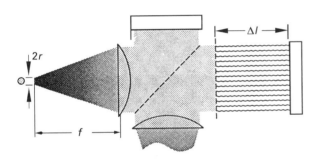

Figure 4 The significant features of an interferometer: Δl displacement of the moving mirror, $2r$ diameter of the Jacquinot stop. Reproduced from Schrader B (ed) (1995) *Infrared and Raman Spectroscopy*. Weinheim: VCH Publishers, with permission of VCH.

with f the focal length of the collimator and h, the slit height of the grating spectrometer. For spectrometers with the same beam area at the interferometer and the grating this factor amounts to about 500. However, common interferometers generally have a smaller beam area than grating instruments, therefore the Jacquinot advantage of actual instruments is of the order of 100. This can be compensated by using array detectors with grating polychromators, employing the multichannel advantage.

Light sources

Raman spectroscopy began in 1928 by using the lines of a mercury discharge lamp at 435.8 or 404.7 nm as the exciting radiation. Since 1960 lasers have been available as ideal monochromatic sources of exciting radiation. The ruby laser (694.0 nm), HeNe-laser (632.8 nm), Ar^+ laser (488.0 and 514.5 nm), the GaAs diode laser (780 nm) and the Nd:YAG laser (1064 nm) are mainly employed. The light flux necessary for recording of Raman spectra is of the order 10 up to 1000 mW.

All laser lines in the visible range of the spectrum, especially the Ar^+ laser lines, but also the line at 780 nm, may excite fluorescence spectra, overlaying the Raman spectra. Excitation by radiation with longer wavelengths reduces the danger of fluorescence. With the exciting radiation of 1064 nm the minimal probability of fluorescence is reached.

Sample arrangements

'Classical' Raman arrangements use the Raman radiation which is emitted at an angle of about 90° relatively to the direction of the exciting radiations (90° arrangement). This is the straightforward arrangement to illuminate the entrance slit of a grating spectrometer by the Raman radiation excited by a focused laser beam in a liquid. However, when interferometers having a circular entrance aperture are employed it has been proven to be superior to analyse the Raman intensity emitted at about 180° to the exciting radiation (180° or back-scattering, arrangement).

In order to record Raman spectra of a sufficiently large S/N in an acceptable time the spectral radiance of the sample has to be maximal. Since Raman spectra are very weak and an excessive large power of the exciting radiation could destroy the sample, sample arrangements have to be designed very carefully. In particular, the ratio of the usable intensity of the Raman radiation at the entrance aperture of a spectrometer versus the intensity of the available exciting radiation, the figure of merit of the sample cell, has to be optimized. This can be done, on the one hand, by observing the Raman radiation produced from a large thickness of the sample, for instance with *end-on* capillary cells (**Figure 5D**) or, on the other hand, by employing different kinds of multiple reflection cells. The largest part of the exciting radiation in cells as in **Figure 5A** is not used for producing Raman radiation, scattered in a direction to the spectrometer. Since the exciting radiation just passes the sample once, more than 99% of the exciting radiation is lost. With spherical mirrors surrounding the sample cell this radiation is reflected back to the sample. Also, the Raman radiation, which is not taken up by the spectrometer is reflected back to the sample. This is done by cells as shown in **Figure 5B**. A 'notch filter'

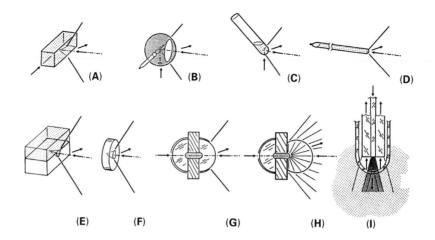

Figure 5 Sample arrangements for Raman spectroscopy: (A) rectangular cell; (B) spherical cell for liquids and powders which are in a melting point capillary at the centre of the sphere with a reflecting surface; (C) liquid in an NMR tube, the axis of which has angle of 45° relative to the axis of the entrance optics. The cuvettes (A)–(C) may be used in a 90° or a 180° arrangement; (D) light pipe cuvette; (E) arrangement for solids or surface layers: the sample is placed upon a block of aluminium or stainless steel with a polished conical indentation, providing a multiple reflection arrangement; (F) tablet with a cone-shaped bore; (G) arrangement for Raman spectroscopy of powders in 0° and 180° arrangement, the half-spheres have a reflecting surface which reflects the exciting and Raman radiation back to the sample increasing the usable Raman intensity; (H) same as (G), but one half-sphere is exchanged against a Weierstrass lens (Weierstrass 1856), which collects the radiation emitted into a solid angle of ~2π; (I) sample head for a two-way bundle of optical fibres for spectroscopy of liquids or powders. The head can be introduced directly into the sample, it is protected by a cover in order to prevent sticking and pyrolysis of the sample at the central fibre which transports the exciting radiation from the laser to the sample. Reproduced from Schrader B (ed) (1995) *Infrared and Raman Spectroscopy*. Weinheim: VCH Publishers, with permission of VCH.

just behind the entrance optics reflects the exciting radiation emerging in the direction as the Raman radiation recorded by the spectrometer, while it transmits the Raman radiation. When the overall reflectivity of the multiple reflection cell for the exciting radiation is ρ, the increase of the observable Raman intensity is given by $I = 1/(1 - \rho)$. With $\rho = 90\%$, the Raman intensity will be larger by a factor of 10, compared to the original intensity. Since also the intensity of the effective Raman radiation is increased by the multiple reflection cell, the intensity of the Raman spectrum increases further (to a maximal of I^2). **Figure 5** gives examples of different sample cells for liquids and powders which have been used successfully. Rectangular sample cells, which are supplied with most Raman spectrometers, have a very small figure of merit, since they do not employ multiple reflection enhancement or the observation of an increased effective thickness of the sample.

In **Figure 5I** an arrangement for investigation of remote samples by fibre optics is shown. The high transmittance of optical fibres in the NIR range makes possible the recording of Raman spectra of samples, which are located up to several hundred metres away from the spectrometer. The quartz fibres, developed for telecommunications, have a large transmission in the spectral range of Raman spectra excited in the NIR—more than 80% per km! One fibre transports the exciting radiation from the laser to the sample, a second fibre or fibre bundle transports the Raman radiation from the sample to the spectrometer. However, it cannot be avoided that the Raman lines of quartz are excited in both fibres. Therefore, the sample head has to be combined with a transmission filter which eliminates, on the one hand, the Raman line of the quartz from the exciting radiation and, on the other hand, the exciting radiation coming back from the sample on the way to the spectrometer (see the next section: Rayleigh filters). Some companies supply such sample heads which allow remote product or production control.

The laser radiation can be focused to a diameter of the order of a few multiples of its wavelength. Therefore, even by using the normal entrance optics of the spectrometer one can investigate Raman spectra of micro samples. However, in order to be able to exactly adjust the area from which the Raman spectrum is taken, common microscopes, able to adjust the sample by observation with visible light are modified for the excitation and observation of Raman spectra. Such microscopes may even allow confocal observation with 3-dimensional spatial resolution. It must be remembered that the optical conductance of microscopes is quite small, therefore much longer observation times are needed for Raman spectroscopy of micro samples. This cannot be compensated by a larger power of the laser radiation, since this may then overheat the sample.

Rayleigh filters

A spectrometer set to pass radiation of a particular wavelength band always has a small amount of stray radiation of other wavelengths. The Rayleigh line is stronger by 10^6–10^{12} than the Raman lines. In ordinary spectrometers the Rayleigh scattering produces stray radiation, which conceals the Raman spectrum completely. Therefore, the Rayleigh radiation has to be eliminated in order to record Raman lines with a maximal signal/background ratio. Different principles are employed for this purpose:

Absorption filters composed of solutions with appropriate absorption bands or glass filters with absorption edges are able to absorb the spectral band of the Rayleigh radiation. Especially interesting is the elimination of the exciting line by atomic absorption of the same element, which produces the exciting radiation. Rasetti, in 1930, used the low-pressure mercury arc to excite rotational Raman spectra of gases with the mercury resonance line at 253.6 nm. With a drop of mercury in the spectrograph he could completely absorb the Rayleigh radiation. Similar procedures use the resonance absorption of rubidium and caesium.

Interference line filters reflect the whole spectrum with a high reflectivity except for the spectral line where the filter has the largest transmittance. The Rayleigh line intensity can thus be reduced by reflection on the filter. Combinations of such filters in a row make up a very efficient Rayleigh filter.

The same filters are employed satisfactorily to reduce the unwanted plasma lines of the laser radiation for the excitation of the Raman spectra, also, Raman lines of the optical fibres transporting the exciting radiation to the sample can be eliminated using interference transmission filters.

Volume holographic optical elements developed for military uses, are successfully applied for Raman spectroscopy: a so-called notch filter reflects a laser line virtually completely, its intensity is reduced by >6 orders of magnitude, with a transmission of about 70% at the other wavelengths.

Holographic laser bandpass filters reflect the laser radiation by >90% at an angle of exactly 90° while reducing the intensity of all unwanted plasma or Raman lines (of an optical fibre) considerably.

A similar construction, a so-called Holoplex transmission grating, disperses the Raman spectrum for recording on a CCD (charge-coupled devices), which are very sensitive detector arrays. They can be fabricated with 2-dimensional dispersion allowing the instantaneous recording of several spectral channels on a 2-dimensional CCD detector as for an echelle spectrograph.

Dispersive mono- and polychromators, interferometers

When illuminated with monochromatic radiation a single monochromator usually shows continuous stray light of the order of 10^{-5} of the intensity of the monochromatic radiation. Therefore, 2 or 3 monochromators in series combined with additive dispersion reduce the stray radiation by about 10^{-10} or 10^{-15}, respectively. However, the intensity of the Raman lines is also reduced when passing a monochromator: Since every monochromator has a transmittance only of about 30%, this means that a double monochromator has only a transmittance of 9%, a triple monochromator of 2.7%. Such monochromators are usually very voluminous and expensive. However, they are widely used for recording of Raman spectra with single detectors (photomultipliers).

When array detectors are used to record simultaneously the whole spectrum, or parts of it, three monochromators are used in a special arrangement. The first two monochromators are combined in a subtractive arrangement. A diaphragm at the middle slit blocks out the Rayleigh line. The third grating instruments acts as a polychromator, it produces a spectrum directly upon the elements (pixels) of a CCD.

Interferometers record an interferogram of a spectrum. By applying the Fourier transformation, the original spectrum is calculated. Jacquinot has pointed out in 1954, that interferometers have a considerably higher optical conductance, by a factor of 100–500, compared to prism or grating spectrometers. Interferometers make use of *the multiplex advantage* when compared to dispersive spectrometers. By recording all channels simultaneously with the same single detector, the detector noise is therefore distributed over all spectral channels. It is not recorded at every spectral channel separately as for prism or grating spectrometers. This increases the *S/N* considerably.

For recording the weak Raman spectrum excited by the Nd:YAG laser line at 1064 nm, interferometers are successfully used. They have to be combined with very powerful Rayleigh filters. The quantum noise of every strong line is distributed over the whole interferogram. By Fourier transformation it is distributed as white noise over the whole spectrum. To avoid this *multiplex disadvantage* all strong lines have to be removed from the spectrum to be analysed.

Detectors

The first generation of Raman spectroscopists used photoplates for the recording of Raman spectra. Later, photomultipliers were used as very powerful single channel detectors at the exit slit of a monochromator recording the Raman spectrum sequentially.

They are now being replaced by metal oxide semiconductors where charge produced by light quanta is stored in a small area, a pixel. Arrays composed of many independent pixels store the charge pattern corresponding to the irradiation pattern. These arrays of 'charge-coupled devices', CCDs, can be linear or two-dimensional, thus storing spectra or images of the sample. Since the number of pixels is limited (usually to about 1024 in a row) spectra can be recorded either *completely, in low resolution*, or *in high resolution sequentially*. Combined with an echelle spectrograph several channels representing different spectral orders of grating can be recorded simultaneously in high resolution.

Since the spectral information 'seen' by the individual pixel is recorded simultaneously, an array with n pixels is equivalent to n spectrometers working separately or one spectrometer working n times. Simultaneous recording of n pixels provides the *multichannel advantage*.

Cooling reduces thermal noise of CCD detectors, so that integration times may be long, up to days. Thus, very faint Raman lines can be recorded.

Interferometers work with single detectors. For the NIR range InGaAs or Ge semiconductor detectors are used. They have to be extremely sensitive, since the intensity of the Raman lines decreases with the fourth power of its absolute frequency (the ν^4 factor). In order to reduce their thermal noise they are cooled by Peltier elements or liquid nitrogen.

Complete Raman spectrometers

Complete Raman spectrometers are produced by several companies. Due to limited space and to the fact that the market is changing continuously, only the names of the main producers can be given here: Andor, Bio-Rad, Bruker*, Dilor, Instruments S.A, Jobin-Yvon, Kaiser Optical Systems, Nicolet*, Ocean Optics, Perkin-Elmer*, Renishaw, Sentronik, Spex. The companies marked with a * supply Raman spectrometers with excitation at 1064 nm able to record 'fluorescence-free' Raman spectra. Most companies supply Raman microscopes.

List of symbols

a = distance; f = collimator focal length; F = area; G = optical conductance; G_v = spectral optical conductance; h = slit height; $h\nu_0$ = quantum of energy; L = radiance; L_v = spectral radiance; N = number of pixels; q = normal coordinate; QY = quantum yield; $2r$ = diameter of Jacquinot stop; R_0 = theoretical resolving power; R_{IR} = resolving power (infrared range); R_{RA} = resolving power (Raman range); S = electric signal; S/N = signal-to-noise ratio; $\Delta\nu$ = line width (bandwidth); Δl = displacement of mirrors; ν_0 = frequency; ν_s = frequency for inelastic scattering; ρ = overall reflectivity; τ = transmission; Φ = radiant power; Φ_v = flux.

See also: **Biochemical Applications of Raman Spectroscopy; FT-Raman Spectroscopy Applications; Hydrogen Bonding and other Physicochemical Interactions Studied By IR and Raman Spectroscopy; IR Spectral Group Frequencies of Organic Compounds; IR and Raman Spectroscopy of Inorganic, Coordination and Organometallic Compounds; Matrix Isolation Studied By IR and Raman Spectroscopies; Nonlinear Raman Spectroscopy, Applications; Nonlinear Raman Spectroscopy, Instruments; Nonlinear Raman Spectroscopy, Theory; Raman and IR Microspectroscopy; Rayleigh Scattering and Raman Spectroscopy, Theory; Surface-Enhanced Raman Scattering (SERS), Applications; Vibrational, Rotational and Raman Spectrocopy, Historical Perspective.**

Further reading

International Union of Pure and Applied Chemistry (1998) *Compendium of Analytical Nomenclature, Definite rules 1997.* Oxford: Blackwell Science.

Jacquinot P (1954) The luminosity of spectrometer with prisms, gratings or Fabry–Perot étalons. *Journal of the Optical Society of America* 44: 761–765.

Schrader B (ed) (1985) *Infrared and Raman Spectroscopy.* Weinheim: VCH Verlagsgesellschaft.

Schrader B (ed) (1989) *Raman/Infrared Atlas of Organic Compounds.* Weinheim: VCH Verlagsgesellschaft.

Schrader B and Moore DS (1997) Laser-based molecular spectroscopy for chemical analysis – Raman scattering processes (IUPAC Recommendations 1997). *Pure and Applied Chemistry* 69: 1451–1468.

Raman Spectroscopy in Biochemistry

See **Biochemical Applications of Raman Spectroscopy.**

Rayleigh Scattering and Raman Effect, Theory

David L Andrews, University of East Anglia, Norwich, UK

Copyright © 1999 Academic Press.

VIBRATIONAL, ROTATIONAL & RAMAN SPECTROSCOPIES
Theory

Rayleigh scattering, the commonplace phenomenon which accounts for the brightness of the sky (amongst many other familiar aspects of the world we inhabit) and the Raman effect, a weaker analogue seen only at high intensities, are closely similar processes in which light is scattered by atoms or molecules. The interactions each entails differ in that the Rayleigh process is technically *elastic* whilst its Raman counterpart is *inelastic* – all of the features in which the two processes significantly differ owe their origin to that fundamental difference in the energetics. Matter responsible for Rayleigh scattering neither loses nor gains energy thereby – and so the scattered light has the same frequency as the radiation from which it is produced. However, atoms or molecules engaged in Raman scattering either gain or lose energy in the process, so that the frequency of the emergent light differs from that impinging on them – by conservation of energy, the emergent light has either a lower or higher frequency, respectively, as a result. The two types of Raman process, known as Stokes and anti-Stokes, are illustrated schematically in the energy level or ladder diagrams of **Figures 1A** and **1B**; the Stokes process results in a molecular transition to a state of higher energy, its anti-Stokes counterpart is a transition to a state of lower energy. Rayleigh scattering processes are represented by **Figures 1C** and **1D**.

A simple picture widely used for didactic purposes portrays Rayleigh scattering in terms of the electric field of impinging radiation generating, through its interaction with the electron cloud of the scattering molecule, an outgoing field that oscillates at the same frequency. The Raman process is considered to be the generation of an emergent field modulated by molecular vibrations. However, theory cast at that

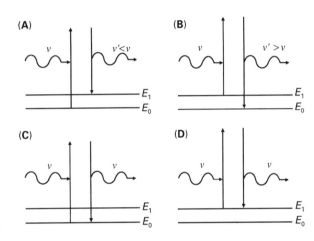

Figure 1 Energy level diagrams illustrating Raman and Rayleigh scattering, with incoming radiation on the left, scattered radiation emergent on the right. Only energy levels directly involved are depicted: (A) Stokes Raman transition, (B) anti-Stokes Raman transition; (C) and (D) Rayleigh scattering.

level is of severely limited value – it fails, for example, to address the relative magnitudes of the Stokes and anti-Stokes Raman signals; and it is not well suited to processes involving electronic transitions. To fully understand those aspects we have to look further into the theory of the fundamental interactions involved in the scattering of light.

Rayleigh scattering

The blue of the sky attests to more efficient Rayleigh scattering at the higher frequency, shorter wavelength end of the visible spectrum – the familiar red sky before dusk and at dawn manifesting the loss of bluer light to skies in other parts of the world. In fact the efficiency of scattering has a cubic dependence on the optical frequency, the scattered intensity having a fourth power dependence because the photon energy itself then enters into the equation. Although it is a moot point for Rayleigh scattering, these dependences actually relate to the frequency of the scattered rather than the incident light, a distinction which nonetheless becomes significant in the case of Raman scattering where the same power laws apply.

The mechanism for Rayleigh scattering entails the electronic polarizability α of the molecules of the sample. The polarizability itself is essentially a measure of how easily the molecular charge distribution is shifted through its interaction with electromagnetic radiation. At simplest, and in an isotropic system such as an atom, the polarizability is a quantity which represents the constant of proportionality between the electric field E of the radiation and the electric dipole moment it induces in the same direction, $\mu_x^{\text{ind}} = \alpha E$. A further guide to its nature can be gained from the polarizability volume, $\alpha' = \alpha/4\pi\varepsilon_0$ (where ε_0 is the vacuum permittivity), which casts this constant in units of volume and often yields a value similar in magnitude to the molecular volume. This correctly suggests that systems such as large atoms and aromatic molecules tend to have large polarizabilities. Nonetheless in all but the highest symmetry molecules the ease of charge displacement within the molecule varies with direction, so that in general the induced dipole moment is not parallel with the applied electric field, but slanted towards the direction of least resistance. Then the polarizability is a second rank tensor and we have;

$$\mu_i^{\text{ind}} = \sum_j \alpha_{ij} E_j \qquad [1]$$

where i and j represent Cartesian coordinates – for example, the dipole moment induced in the x-direction is determined by $\mu_x^{\text{ind}} = \alpha_{xx}E_x + \alpha_{xy}E_y + \alpha_{xz}E_z$.

It is the development of a fully quantum theoretical depiction of scattering at the molecular level which leads to the detailed structure of the electronic polarizability. Here, each Rayleigh (or Raman) scattering event is understood as involving the absorption of one photon of the incoming radiation, accompanied by the emission of one photon. It is important to recognize that the absorption and emission take place together in one concerted process; there is no measurable time delay between the two events. The energy–time uncertainty relation $\Delta E \Delta t \geq h/2\pi$ allows for each process to take place even when there is no energy level to match the energy of the absorbed photon, as indicated by the

absence of any level at the upper end of the arrows in **Figure 1**. In other words the absorption does not populate a real intermediate state, since it is accompanied by emission. Thus it is, for example, that Rayleigh scattering of visible light takes place even in transparent media.

Despite their widespread adoption and utility, energy diagrams such as **Figure 1** are potentially misleading for any such processes involving the concerted absorption and/or emission of more than one photon – for in this case they incorrectly suggest that emission takes place subsequent to absorption. All such processes are best described with the aid of time-ordered diagrams which symbolically represent such interactions as a series of photon absorptions and emissions, and which lead to a more correct theoretical representation. Both for Rayleigh and Raman scattering there are two possible sequences, depending on whether the absorption or the emission comes first. These two cases are illustrated by the time-ordered diagrams of **Figures 2A** and **2B**, in which the vertical line represents the successive states of the molecule, and the wavy lines photons, the sequence of interactions being read upwards. Thus in both diagrams the molecular progress from an initial state m to a final state n proceeds via an intermediate state r; in (A) the transition from m to r is accompanied by absorption of a photon $h\nu$ from the incident beam, and the transition from r to n by emission of a photon $h\nu'$; in (B) this ordering of absorption and emission is reversed. In reality these processes are not separable; the diagrams simply assist development of the theory. Therefore, although the overall process is subject to energy conservation, i.e. $E_m + h\nu = E_n + h\nu'$, energy need not be conserved in the individual absorption and emission stages. For

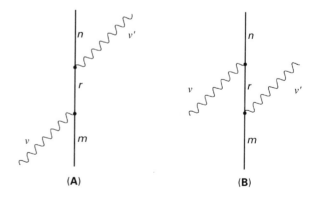

Figure 2 Time-ordered diagrams for light scattering, with time progressing upwards. Taken together, each applies to both Rayleigh and Raman processes; in the Rayleigh case the initial state m and the final state n of the molecule become the same, and $\nu' = \nu$.

this reason the state r is often referred to as a virtual state, and all possible energy levels must be taken into account.

Interpreting the time-ordered diagrams of **Figure 2** by the rules with which they are associated, and given that the states m and n can be identified with the electronic ground state with wavefunction φ_0 and energy E_0, the following result for the polarizability is obtained:

$$\alpha_{ij} = \sum_r \left[\frac{\langle \varphi_0 \mid \hat{\mu}_i \mid \varphi_r \rangle \langle \varphi_r \mid \hat{\mu}_j \mid \varphi_0 \rangle}{E_r - E_0 - h\nu - ih\Gamma_r} \right.$$
$$\left. + \frac{\langle \varphi_0 \mid \hat{\mu}_j \mid \varphi_r \rangle \langle \varphi_r \mid \hat{\mu}_i \mid \varphi_0 \rangle}{E_r - E_0 + h\nu - ih\Gamma_r} \right] \qquad [2]$$

where the two complete terms on the left and right of the plus sign correspond directly to **Figures 2A** and **2B** respectively. In Equation [2], φ_r is the wavefunction of state r with energy E_r and line width $\frac{1}{2}\Gamma_r$, and $\hat{\mu}_i$ is the ith component of the electric dipole moment operator. In frequency regions close to an optical absorption band, one of the states in the summation over r will be such that $E_r - E_0 \approx h\nu$, so that the first term of Equation [2] will have a small denominator, approximating to the line width factor $-ih\Gamma_r$, and the term as a whole – see **Figure 2A** – will overwhelm all else. However, in more common off-resonant circumstances the line width factor can be neglected in each denominator. Moreover for most electronic states the Dirac brackets $\langle \varphi_0, \mid \hat{\mu}_i \mid \varphi_r \rangle$ and $\langle \varphi_r, \mid \hat{\mu}_i \mid \varphi_0 \rangle$ are identical and can be expressed more concisely as components of the transition dipole moment μ^{r0}. Again for conciseness, defining $h\nu_{r0} = E_r - E_0$, Equation [2] then finally reduces to

$$\alpha_{ij} \approx \frac{2}{h} \sum_r \left(\frac{\nu_{r0}}{\nu_{r0}^2 - \nu^2} \right) \mu_i^{r0} \mu_j^{r0} \qquad [3]$$

Rayleigh scattering generally produces radiation with a changed polarization state; polarized incident light is to some extent depolarized by the process whilst unpolarized light is to some extent polarized. Both effects are normally characterized by a depolarization ratio defined as the intensity ratio of plane polarized components of the scattered light. For right-angled scattering of light polarized in the z-direction and incoming along the y-direction, as shown in **Figure 3**, the depolarization ratio ρ_l of light scattered in the x-direction is calculated as

$$\rho_l = I(z \to y) / I(z \to z) \qquad [4]$$

where the subscript on the ρ denotes linear polarization (synonymous with plane polarization). The value of ρ_l depends on the molecules responsible for the scattering and is directly expressible in terms of polarizability parameters. Specifically, if we define the polarizability mean $\overline{\alpha}$ and anisotropy γ through

$$\overline{\alpha} = \frac{1}{3} \sum_{i=x,y,z} \alpha_{ii}; \qquad \gamma^2 = -\frac{9}{2}\overline{\alpha}^2 + \frac{3}{2} \sum_{i=x,y,z} \sum_{j=x,y,z} \alpha_{ij}^2 \qquad [5]$$

then for scattering by a gas or liquid we find

$$\rho_l = \frac{3\gamma^2}{4\gamma^2 + 45\overline{\alpha}^2} \qquad [6]$$

with a value in the interval $(0, 0.75)$. The lower limit corresponds to scattering with full retention of linear polarization and corresponds to $\gamma^2 = 0$, a case which occurs only for molecules of very high symmetry. Although introduced here for right-angled scattering, the above result is in fact independent of scattering angle. In contrast the extent of polarization introduced by the scattering of unpolarized light is an angle-dependent quantity. For right-angled scattering where the effect is largest, the corresponding 'depolarization ratio' is given as

$$\rho_n = I(n \to y) / I(n \to z) \qquad [7]$$

(where the subscript of the ρ stands for natural), giving

$$\rho_n = \frac{6\gamma^2}{7\gamma^2 + 45\overline{\alpha}^2} \qquad [8]$$

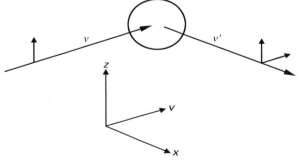

Figure 3 Scattering geometry for the usual measurement of depolarization ratios; incident radiation is z-polarized and light scattered at right-angles is resolved for its y- and z-polarization components.

For scattering at other angles θ (where $\theta = 0$ relates to forward scattering) the result can be written as

$$\rho_n(\theta) = \cos^2\theta + \rho_n \sin^2\theta \qquad [9]$$

The angle dependence of the polarization which Rayleigh scattering confers on unpolarized light is immediately evident on viewing a clear daytime sky through polarizing spectacles.

The Raman effect

The Raman effect was one of the first processes whose explanation, largely through the work of Placzek in 1934, exploited and vindicated the still nascent quantum theory. As for Rayleigh scattering, each scattering event involves concerted processes of photon absorption and emission, without the need for an energy level to match the absorbed photon. The difference is that as a result of this process, the scatterer undergoes an overall transition from one energy level to another, as depicted in **Figures 1A** and **1B**. The Raman effect is a very weak phenomenon; typically only one incident photon in $\sim 10^7$ produces a Raman transition, and observation of the effect thus requires a very intense source of light.

Raman scattering generally involves transitions amongst energy levels that are separated by much less than the photon energy of the incident light. The two levels, denoted by E_0 and E_1 in **Figure 1**, for example, are most often vibrational levels, whilst the energies of the absorbed and emitted photons are commonly in (or near to) the visible range – hence the effect provides the facility for obtaining vibrational spectra using visible light. In general the Stokes Raman transition from level E_0 to E_1 results in scattering of a frequency given by

$$\nu_s = \nu - \Delta E/h \qquad [10]$$

where $\Delta E = (E_1 - E_0)$, and the corresponding anti-Stokes transition from E_1 to E_0 produces a frequency

$$\nu_{AS} = \nu + \Delta E/h \qquad [11]$$

Thus each allowed Raman coupling generally produces two frequencies in the spectrum of scattered light, shifted to the negative and positive sides of the dominant Rayleigh line by the same amount, $\Delta\nu = \Delta E/h$. For this reason Raman spectroscopy is concerned with measurements of frequency shifts, rather than absolute frequencies. In most cases a number of Raman transitions can take place, involving various molecular energy levels, and the

spectrum of scattered light contains a range of frequencies shifted away from the irradiation frequency. In the particular case of vibrational Raman transitions, the shifts can be identified with vibrational frequencies.

Although each Stokes line and its anti-Stokes counterpart are equally separated from the Rayleigh line, they are not of equal intensity. This is because the intensity of each transition is proportional to the population of the energy level from which the transition originates; under equilibrium conditions the ratio of populations is given by the Boltzmann distribution. With the fourth-power dependence on the scattering frequency, the ratio of intensities of the Stokes line and its anti-Stokes partner in a Raman spectrum is given by

$$I_{AS}/I_S = \{(\nu + \Delta\nu)/(\nu - \Delta\nu)\}^4 (g_1/g_0)\exp(-h\Delta\nu/kT) \qquad [12]$$

where g_0 is the degeneracy of the ground state and g_1 that of the upper level – and hence the anti-Stokes line is almost invariably weaker in intensity. The dependence of this ratio on the absolute temperature T can, for example, be employed as a means of flame thermometry. However, since the Stokes and anti-Stokes lines give precisely the same information on molecular frequencies, it is usually only the stronger (Stokes) part of the spectrum that is recorded.

A development of detailed theory, again based on the time-ordered diagrams of **Figure 2**, establishes the dependence of Raman scattering on a transition tensor which takes on the same role as the polarizability in Rayleigh scattering. Once the usual Born–Oppenheimer separation of electronic and vibrational wavefunctions has been effected, then for the vibrational Raman transition $v \rightarrow v'$ involving a normal mode of vibration λ, this tensor takes the form

$$\alpha_{ij}^{v_\lambda' v_\lambda} = \sum_{r,v''(\mu)} \left[\frac{\langle \chi_0^{v_\lambda'} | \langle \varphi_0 | \hat{\mu}_i | \varphi_r \rangle | \chi_r^{v_\mu''} \rangle \langle \chi_r^{v_\mu''} | \langle \varphi_r | \hat{\mu}_j | \varphi_0 \rangle | \chi_0^{v_\lambda} \rangle}{E_r^{v_\mu''} - E_0^{v_\lambda} - h\nu - ih\Gamma_r^{v_\mu''}} \right.$$

$$\qquad [13]$$

$$\left. + \frac{\langle \chi_0^{v_\lambda} | \langle \varphi_0 | \hat{\mu}_j | \varphi_r \rangle | \chi_r^{v_\mu''} \rangle \langle \chi_r^{v_\mu''} | \langle \varphi_r | \hat{\mu}_i | \varphi_0 \rangle | \chi_0^{v_\lambda} \rangle}{E_r^{v_\mu''} - E_0^{v_\lambda} + h\nu - ih\Gamma_r^{v_\mu''}} \right]$$

where, for example, the vibrational wavefunction $\chi_r^{v_\mu''}$ denotes a state with a quantum number v'' in the vibrational mode μ, within the set of levels associated with electronic state r. Here also $E_r^{v_\mu''}$ and $\Gamma_r^{v_\mu''}$ relate to the total (electronic plus vibrational) energy and the damping, respectively, of that state. Away from resonance, in other words when using frequencies ν well removed from any optical absorption bands, then the vibrational energy contributions in each denominator term of Equation [13] can safely be neglected. Then, using the completeness relation of quantum mechanics, the $|\chi_r^{v_\mu''}\rangle \langle \chi_r^{v_\mu''}|$ sum can be effected to give

$$
\alpha^{v_\mu^{\prime}v_\lambda} = \sum_r \left[\frac{\langle \chi_0^{v'_\lambda} | \langle \varphi_0 | \hat{\mu}_i | \varphi_r \rangle \langle \varphi_r | \hat{\mu}_j | \varphi_0 \rangle | \chi_0^{v_\lambda} \rangle}{E_r - E_0 - h\nu - ih\Gamma_r} \right.
$$
$$
\left. + \frac{\langle \chi_0^{v'_\lambda} | \langle \varphi_0 | \hat{\mu}_j | \varphi_r \rangle \langle \varphi_r | \hat{\mu}_i | \varphi_0 \rangle | \chi_0^{v_\lambda} \rangle}{E_r - E_0 - h\nu - ih\Gamma_r} \right]
$$
$$
= \langle \chi_0^{v'_\lambda} | \alpha_{ij}(Q_\lambda) | \chi_0^{v_\lambda} \rangle \qquad [14]
$$

using Equation [2]. The result thus involves the dependence of the electronic polarizability on the nuclear coordinate Q_λ relating to the excited vibration. Although all molecules have a finite polarizability, that is not the case for $\alpha_{ij}^{v'_\lambda v_\lambda}$ – but no Raman signal will emerge when the latter is zero. Here a powerful symmetry rule emerges: any Raman-active vibration must transform under an irreducible representation spanned by components of the polarizability tensor (transforming as the quadratic variables x^2, xy, etc., or one of their combinations). Some of the broad implications of this are highlighted below.

To obtain the major selection rules for Raman scattering we can first expand Equation [14] in a Taylor series about the equilibrium configuration Q_e;

$$
\alpha_{ij}(Q_\lambda) = \alpha_{ij}(Q_e) + \left. \frac{\partial \alpha_{ij}}{\partial Q_\lambda} \right|_{Q_e} (Q_\lambda - Q_e) + \cdots \quad [15]
$$

and hence we have

$$
\alpha_{ij}^{v'_\lambda v_\lambda} = \alpha_{ij}(Q_e)\delta_{v'v} + \left. \frac{\partial \alpha_{ij}}{\partial Q_\lambda} \right|_{Q_e} \langle \chi_0^{v'_\lambda} | (Q_\lambda - Q_e) | \chi_0^{v_\lambda} \rangle + \cdots
$$
$$
[16]
$$

The first term on the right is non-zero only when the initial and final states are identical – which relates back to Rayleigh scattering. It is the second term which is significant for the Raman process and its detailed form establishes two rules governing Raman-allowed transitions, since both of its factors must then be non-zero. For the Dirac bracket to be non-zero dictates $v' = v \pm 1$, as in infrared absorption spectroscopy. For the polarizability derivative to be non-zero, the polarizability must change

during the vibration, as the molecule passes through its equilibrium configuration. This is the key selection rule for the Raman effect, illustrated for CO_2 in **Figure 4**.

It is immediately apparent that Raman transitions are governed by different selection rules from absorption or fluorescence. The case of CO_2 illustrates a general principle applicable to all centrosymmetric molecules, which is that only gerade vibrations (those which are even with respect to inversion symmetry) appear in the Raman spectrum, whilst only ungerade vibrations (odd with respect to inversion) show up in infrared absorption. This illustrates the so-called mutual exclusion rule for centrosymmetric molecules, which states that vibrations active in the infrared spectrum are inactive in the Raman, and vice versa. Even for complex polyatomic molecules lacking much symmetry, the intensities of lines resulting from the same vibrational transition may be very different in the two types of spectrum, so that in general there is a useful complementarity between the two methods. Generally it is the vibrations of the most polarizable groups which are strongest in the Raman spectrum, those of the most polar groups being strongest in the infrared, as nicely illustrated in the spectra of the drug acetaminophen (UK paracetamol; *p*-hydroxyacetanilide) shown in **Figure 5**.

Further information on the symmetry properties of Raman-active molecular vibrations can be obtained by measurement of the depolarization ratios of the lines in the Raman spectrum – see **Figure 3** and Equation [4]. Interpretation of the results here invokes Equation [17]:

$$
\rho_l = \frac{3\gamma'^2}{4\gamma'^2 + 45\overline{\alpha}'^2} \qquad [17]
$$

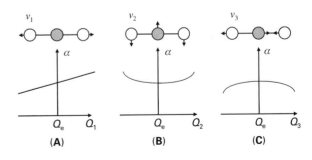

Figure 4 Variation of polarizability in the course of three normal modes of vibration of carbon dioxide: (A) symmetric stretch, (B) bending mode and (C) antisymmetric stretch. The slope $\partial\alpha/\partial Q_\lambda$ on crossing the vertical axis is non-zero only for the symmetric stretch, and hence only this vibration gives a Raman signal.

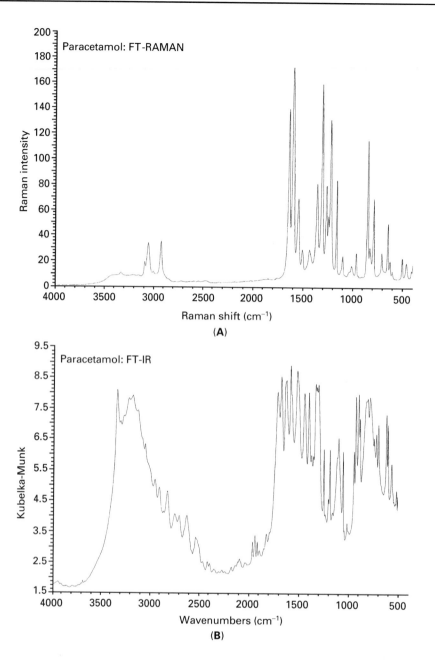

Figure 5 Fourier-transform spectra of paracetamol: (A) Raman, and (B) infrared. Stretch vibrations of the non-polar C–H groups, close to 3000 cm^{-1}, show up well in the Raman spectrum. In the infrared spectrum this whole region is dominated by stretching vibrations of the highly polar O–H and N–H groups, much broadened through association with hydrogen bonding. Reproduced with permission of Nicolet Instruments.

The prime on the mean and anisotropy parameters, $\bar{\alpha}'$ and γ', respectively, denote values obtained from the polarizability derivative – defined in the sense of Equation [5], but in components of the tensor $\partial\alpha_{ij}/\partial Q_\lambda$ rather than α_{ij} itself. In the case of gases and liquids, ρ_l is lower than $\frac{3}{4}$ for vibrations that are totally symmetric (vibrations transforming under the totally symmetric representation of the molecular point group), but exactly for $\frac{3}{4}$ for other vibrations that lower the molecular symmetry, since $\bar{\alpha}'$ is then zero.

Although the frequency of radiation used for the study of Raman scattering is generally well removed from any absorption band of the sample, to forestall problems associated with absorption and subsequent fluorescence, special features become apparent on irradiation at a frequency close to a broad and intense optical absorption band. Quite simply, the closer the approach, the greater is the intensity of the Raman spectrum. Spectra obtained under such conditions are known as resonance Raman spectra. In the case of large polyatomic molecules where any electronic absorption band may be due to localized absorption in a particular chromophore, the vibrational Raman lines which experience the greatest amplification are those of the appropriate symmetry involving vibrations of nuclei close to the groups responsible for the resonance.

Equation [13] correctly represents the Raman tensor even under resonance or pre-resonance conditions – and the resonance enhancement is clearly attributable to the fact that if there is an excited state for which $E_r^{v_\mu^n} - E_0^{v_\lambda}$ is close to $h\nu$, the first term of that equation has a denominator of greatly diminished magnitude. However, the subsequent development of theory leading to Equation [16] is no longer valid under such conditions; for example, the $v' = v \pm 1$ selection rule breaks down and overtones commonly appear. Other vibrations (those which transform like the rotations R_x, R_y and R_z) can also become active through changed selection rules, associated with the fact that the Raman tensor is no longer real and index-symmetric, but complex and non-symmetric. As a result the equations for the Raman depolarization ratio also require modification to the following form

$$\rho_l^{\text{res}} = \frac{3\gamma'^2 + 5\delta'^2}{4\gamma'^2 + 45\overline{\alpha}'^2} \qquad [18]$$

where

$$\delta'^2 = \frac{1}{4} \sum_{i=x,y,z} \sum_{j=x,y,z} \left| \alpha'_{ij} - \alpha'_{ji} \right|^2 \qquad [19]$$

is a measure of the degree of antisymmetry in the Raman tensor. One consequence of including this factor in Equation [18] is the possibility of depolarization ratios exceeding the normal upper bound of $\frac{3}{4}$ – in some cases indeed yielding an infinite result (complete depolarization).

Generally, the use of circularly polarized light in studies of Rayleigh or Raman scattering offers no further information beyond that provided by plane polarizations. However, optically active (chiral) compounds in the liquid or solution state respond differentially to circularly polarized light, according to its handedness, making it possible to obtain a spectrum showing a marginal difference in the Raman intensity $I^R - I^L$ as a function of scattering frequency. The extent of this differential for each molecular vibration is directly related to the detailed stereochemical structure responsible for the manifestation of chirality. In particular, the extent of differential scattering in a region of the spectrum associated with a particular group frequency can be interpreted in terms of the chiral environment of the corresponding functional group. In contrast to the theory developed here, the scattering entails not only electric dipole but also the much weaker magnetic dipole and electric quadrupole interactions.

At the high intensities now available from laser sources, numerous other variants of the Raman effect can be observed, many associated with optically nonlinear behaviour. For analytical purposes the most important of these come under the heading of coherent Raman spectroscopy, of which the process known as CARS (coherent anti-Stokes Raman spectroscopy) is the most common. Here, two beams are directed into the sample: one has a fixed frequency playing the role of ν and the other, a frequency ν' tunable across the Stokes range. As ν' tunes into each Stokes frequency ν_S a four-photon process occurs, essentially combining all the elements of **Figures 1A** and **1B**, and generating coherent emission at the corresponding anti-Stokes frequency ν_{AS}. The laser-like nature of this output facilitates its collection for spectroscopic analysis, and permits the analysis of microscopic samples.

List of symbols

E = electric field; E_m = energy of level m; g_0 = degeneracy of ground state; g_1 = degeneracy of upper level; h = Planck's constant; I = intensity; k = Boltzmann's constant; Q = Nuclear coordinate; t = time; T = absolute temperature; α = electronic polarizability; $\overline{\alpha}$ = mean polarizability; γ = polarizability anisotropy; Γ_r = damping of level r; ε_0 = vacuum permittivity; θ = scattering angle; λ = normal mode of vibration; μ^{ind} = induced electric dipole moment; μ^{r0} = transition dipole moment; ν = frequency of incident radiation; ν' = frequency of emitted radiation; ν_S = Stokes frequency; ν_{AS} = anti-Stokes frequency; ρ = depolarization ratio; ϕ_0 = wavefunction with energy E_0; χ = vibrational wavefunction.

See also: **Biochemical Applications of Raman Spectroscopy; Nonlinear Optical Properties; Raman Optical Activity, Applications; Raman Optical Activity, Spectrometers; Raman Optical Activity, Theory; Raman Spectrometers.**

Further reading

Andrews DL (1997) *Lasers in Chemistry*, 3rd edn, pp 128–149. Berlin: Springer-Verlag.

Barron LD (1982) *Molecular Light Scattering and Optical Activity*. Cambridge: Cambridge University Press.

Craig DP and Thirunamachandran T (1984) *Molecular Quantum Electrodynamics*. London: Academic Press.

Long DA (1977) *Raman Spectroscopy*. New York: McGraw-Hill.

Placzek G (1934) Rayleigh-Streuung und Raman-Effekt. In: Marx E (ed) *Handbuch der Radiologie*, Vol. 6, Part 2, pp 205–374. Leipzig: Akademische Verlag.

Raman CV and Krishnan KS (1928) A new type of secondary radiation. *Nature* **121**: 501.

Sheppard N (1990) Chemical applications of molecular spectroscopy – A developing perspective. In: Andrews DL (ed) *Perspectives in Modern Chemical Spectroscopy*, pp 1–41. Berlin: Springer-Verlag.

Regulatory Authority Requirements

See **Calibration and Reference Systems (Regulatory Authorities).**

Relaxometers

Ralf-Oliver Seitter and **Rainer Kimmich**,
Universität Ulm, Germany

MAGNETIC RESONANCE
Methods & Instrumentation

Purpose and classification of NMR relaxometers

Nuclear magnetic relaxation, that is, thermal equilibration of the spin systems with respect to longitudinal or transverse magnetization components, multiple-quantum spin coherences and longitudinal dipolar, quadrupolar or scalar order, comprises a vast variety of different experimental protocols and phenomena. In a typical relaxation experiment, one first produces a nonequilibrium population of the spin states, often combined with spin coherences. It is then a matter of the fluctuations of the spin couplings to induce spin transitions towards thermal equilibrium. 'Equilibrium' means (i) populations following Boltzmann's distribution, and (ii) completely vanishing spin coherences. Consequently there are three elements inherent in a typical relaxation experiment: *Preparation* of a nonequilibrium state of the spin systems; the variable *evolution* interval allowing for the induction of spin transitions; and the *detection* of the populations, longitudinal order, or coherences after the evolution interval. The time constants with which the 'observable' approaches equilibrium during the evolution interval are the relaxation times, such as the spin–lattice relaxation time T_1, the transverse relaxation time T_2, the rotating-frame relaxation time $T_{1\rho}$, the dipolar-order relaxation time T_d, and so on. In the following we will focus on T_1 in particular.

The spin–lattice relaxation rate of dipolar coupled homonuclear two-spin I systems, for instance, is given by

$$\frac{1}{T_1} = \left(\frac{\mu_0}{4\pi}\right)^2 \frac{3}{2}\gamma^4\hbar^2 I(I+1)\left[J^{(1)}(\omega) + J^{(2)}(2\omega)\right] \quad [1]$$

where μ_0 is the magnetic field constant, γ is the gyromagnetic ratio, and $J^{(i)}(\omega)$ is the intensity function of the Larmor frequency, $\omega = \gamma B_0$, depending on the flux density of the external magnetic field, B_0. The intensity function is given as the Fourier transform of the dipolar autocorrelation function $G_i(\tau)$,

$$J^{(i)}(\omega) = \int_{-\infty}^{\infty} G_i(\tau)\exp(-\mathrm{i}\omega\tau)\mathrm{d}\tau \quad [2]$$

The dipolar autocorrelation function is defined by

$$G_i(\tau) = \left\langle F^{(i)}(0) F^{(-i)}(\tau) \right\rangle \qquad [3]$$

where

$$F^{(1)} = \frac{1}{r^3} \sin\vartheta \cos\vartheta \exp(-i\varphi),$$

$$F^{(2)} = \frac{1}{r^3} \sin^2\vartheta \exp(-2i\varphi) \qquad [4]$$

The polar coordinates r, ϑ, φ define the internuclear vector of the spin system. That is, any molecular motion affecting these coordinates leads to fluctuations of the functions $F^{(i)}$, so that the spin–lattice relaxation rate (Eqn [1]) directly reflects these motions via the intensity and autocorrelation functions. The prominent goal of NMR relaxometry, hence, is to monitor the features of molecular dynamics. This is best done by recording the frequency dependence of spin–lattice relaxation.

Variation of the Larmor frequency means variation of the external flux density B_0. The range within which one can do that using conventional NMR spectrometers is very limited. The reason is that the flux density has a fixed value given by the magnet in use. Typical values range from 1 to 20 T. The radio frequency (RF) part of conventional NMR spectrometers is tuned to resonance in the particular flux density provided by the magnet. That is, there is no reasonable way to study frequency dependences using spectrometers with fixed flux densities, because the signal-to-noise ratio, $S/N \propto B_0^{3/2}$, and the RF bandwidth deteriorates the lower the flux density and the frequency become.

The solution of the problem is field-cycling NMR relaxometry. The essence of this technique is that the magnetic flux density relevant for relaxation is different from that during signal detection. The 'relaxation field' may be varied over several orders of magnitude while the 'detection field' is kept fixed at the highest possible value. That is, the RF console is tuned to this particular detection field. The design of a field-cycling relaxometer thus implies the possibility to switch the flux density rapidly and precisely between different levels with very good stability. A typical field cycle is shown in **Figure 1**.

Field-cycling (FC) NMR relaxometry

Principle

A typical field cycle consists of a preparation interval, an evolution interval and a detection interval. In

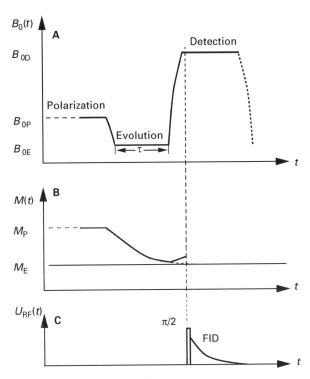

Figure 1 Typical cycle of a field-cycling NMR relaxometer serving for spin–lattice relaxation experiments: (A) External magnetic flux density $B_0 = B_0(t)$, (B) longitudinal magnetization $M = M(t)$ in the sample, (C) RF pulse with free-induction decay (FID) during the detection period. Note that the field level in the (short) detection period is much higher than in the (long) polarization interval. This ensures equivalent heat production in both intervals.

the preparation interval the magnetic flux density, B_P, is chosen as high as technically possible. The purpose is to polarize the sample so that the magnetization becomes as large as feasible.

A nonequilibrium state is then produced by switching the magnetic field to another level, namely to that of the evolution interval, $B_E \neq B_P$. If the magnetic field is switched fast enough, so that the magetization cannot follow the instantaneous equilibrium value given by Curie's law, we have a nonequilibrium situation at the beginning of the evolution interval. That is, the magnetization starts to relax towards the equilibrium value corresponding to the field level of the evolution interval.

The relaxation curve is then probed point-by-point by switching the field up to the detection field level and by acquiring the NMR signal using a RF pulse or an echo RF pulse sequence. The detection field is chosen as high as possible, and is the same for all experiments. This is the field to which the RF console is tuned. The amplitude of the NMR signal represents the magnetization at the end of the evolution interval, in which relaxation takes place. Thus varying the evolution interval, permits one to record the whole relaxation curve to which the spin–lattice relaxation time can be fitted. Assuming mono-exponential relaxation and neglecting losses during the switching intervals, the detected magnetization is given by

$$M(\tau) = M_E + [M_P - M_E]\exp[-\tau/T_1(B_E)] \quad [5]$$

where M_E and M_P are the Curie magnetizations corresponding to the magnetic flux density in the evolution and polarization intervals, respectively. Incrementing the evolution field level, B_E, in subsequent experiments permits one to measure such relaxation curves at field levels not accessible to conventional NMR. That is, the whole frequency dispersion of spin–lattice relaxation can be probed with a relaxometer tuned to a fixed resonance frequency. This implies that the signal-to-noise ratio remains the same for all Larmor frequencies.

Requirements and limitations

Signal-to-noise ratio According to Curie's law, the magnetization reached in the polarization field, B_P, is $M_P \propto B_P$. The induction signal increases proportionally to the detection field, B_D, whereas noise is proportional to the square root of the carrier frequency, i.e. the square root of the detection field. The signal-to-noise ratio is thus determined by

$$S/N \propto B_D^{1/2} M(\tau) \quad [6]$$

The magnetization at the end of the evolution interval obeys

$$M(\tau) \geq B_P \exp[-\tau/T_1(B_E)] \quad [7]$$

so that the signal-to-noise ratio is limited by

$$S/N \propto B_D^{1/2} B_P \exp[-\tau/T_1(B_E)] \quad [8]$$

For optimum signal-to-noise ratio, this suggests field levels B_D and B_P as high as technically feasible.

Feasibility of phase-sensitive detection An insufficient signal-to-noise ratio, as is likely with deuteron relaxation or with very dilute systems and small samples, can be improved with signal accumulation. This, however, requires a detection field stability permitting phase-sensitive detection. That is, after

having switched the field up to the detection level, resonance must be reproducibly reached in all cycles of the experiment with an accuracy of 10^{-5} before signal acquisition. This is a rather demanding condition requiring a technical solution like the battery driven power supply described below. By contrast, the accuracy of the polarization and evolution levels is much less critical. A few percent may be acceptable.

Thermal stability The *polarization field*, B_P, is applied as long as needed for reaching thermal equilibrium, typically a period $t_P \approx 5 \times T_1(B_P)$. The *detection field*, B_D, is applied only in the very short interval t_D needed for acquiring the free-induction decay. That is,

$$t_P \gg t_D \quad [9]$$

Apart from a few superconducting versions reported in the literature, the magnet coils of field-cycling relaxometers are resistive, so that Joule's heat produced in the course of a field cycle affects the thermal stability of the system. If the magnet coil gets unduly warm during the polarization interval, thermal drifts of the field levels are unavoidable. Actually, the cooling properties of the magnet coil form a crucial factor limiting the applications. Considering the inequality given in Equation [9] it becomes obvious that it is more favourable to restrict the polarization field to a level still compatible with thermal stability, whereas the detection field should be as high as the magnet power supply permits. This is the reason why the field levels in the scheme shown in **Figure 1** are chosen in such a way that $B_D \gg B_P$.

Accuracy of the evolution field level On a logarithmic field or frequency scale covering several orders of magnitude, an accuracy of better than 10% may be adequate. However, at low magnetic fields, relaxation times less than a millisecond may occur. That is, a stability of the evolution field level of less than 10% must be reached in a ring-down time of less than 1 ms after switching. The evolution field level must begin with a sharp edge in the time dependence of the magnetic flux density. This is another demanding specification which requires special current control measures (see the design described below). It is also an essential condition to check the switch interval from the polarization level to that of the evolution field with the aid of a suitable Hall probe and a fast high-resolution (e.g. 500 kHz/16 bit) analogue-to-digital converter (**Figure 2**).

Influence of the switching times Provided that the time dependence of the magnetic flux density

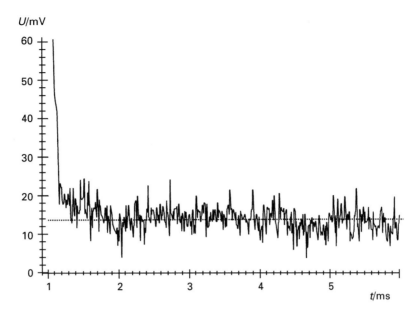

Figure 2 Precision of the initial stabilization of the evolution field. The plot represents the voltage transient recorded with a Hall probe (6 μT/mV) and a 16 bit analogue-to-digital converter. The field level corresponds to a proton Larmor frequency of 3.5 kHz. The field-cycling relaxometer is described below. This plot demonstrates that the evolution field can be stabilized within ±1 digital unit in a ring-down time less than 1 ms.

(see **Figure 1**) is reproducible in an experiment for a given evolution flux density, there is no influence on the measured spin–lattice relaxation time even when the switching intervals are comparable with the low-field relaxation times. However, the dynamic range of the magnetization variation in a relaxation experiment is reached only in full if the switching intervals are much shorter than the low-field relaxation times. That is, the best measuring accuracy is achieved if the relaxation losses during the switching intervals are negligible.

Lower limit of the evolution field When the evolution field approaches the earth field, i.e. the field when the magnet is switched off, a compensation of this zero-field with the aid of correction coils becomes necessary. In this way, flux densities less than 5×10^{-5} T can reliably be reached. Another limit is due to the local field within the sample. The local field arises because of secular spin interactions. It can therefore not be compensated by external correction coils. If the local field exceeds the external evolution field, two situations can arise. Firstly, the local field is approached by adiabatically switching off the polarization field. That is, the local magnetization vectors follow the instantaneous field direction which is finally given by the local field. In this case, dipolar or, in the case of quadrupole nuclei, quadrupolar order is produced. The relaxation time measured under such circumstances is the dipolar-order spin–lattice relaxation time. In the opposite case, the

polarization field is switched off nonadiabatically so that the magnetization partly becomes transverse to the local field. This is the situation of the so-called zero-field NMR spectroscopy probing the evolution of spin coherences in the local field.

Field homogeneity Fast field switching stipulates small magnet coils so that the field energy to be varied in a cyclic way is low. On the other hand, the field homogeneity of small magnets tends to be poor. For field-cycling relaxometry the condition is that the homogeneity within the sample should correspond to the stability (or reproducibility) of the detection field. That is, if the detection field can be reproduced with an accuracy of 10^{-5}, the field homogeneity in the sample need not to be better than 10^{-5} using suitable magnet designs this can be achieved easily.

Crucial specifications of field-cycling NMR relaxometers

From the above outline of the factors limiting field-cycling NMR relaxometry, it becomes obvious that the following specifications of a relaxometer are most crucial. (a) The polarization and detection fields should each be as high as compatible with the thermal stability of the magnet coil. This normally means that the detection field is much higher than the polarization field. (b) The evolution field must achieve stability within a few percent after a stabilization time of less than the shortest relaxation time to be measured. The total field switching intervals should preferably be shorter than the typical relaxation times of interest. (c) The detection field should be reproducible within 10^{-5} after a stabilization period of much less than the high-field relaxation time. This is the condition permitting phase-sensitive detection and, hence, automatic signal accumulation. (d) The field homogeneity in the sample should correspond to the reproducibility of the detection field, i.e. a relative field variation of less than about 10^{-5}.

Typical setup of a field-cycling relaxometer

Historically the first field-cycling instruments consisted of two magnets of varying flux densities. The sample was 'shuttled' between a resonant magnet and a magnet producing the variable evolution field. With modern relaxometers the field is varied much faster by controlling the current through the magnet coil electronically. This, of course, stipulates that the use of iron magnets is excluded. The magnets in use are resistive or even superconducting magnet coils with solenoid-like geometries.

The magnet current can be controlled with the aid of gate-turn-on thyristors (GTOs), metal-oxide semiconductor field-effect transistors (MOSFETs) or insulated gate bipolar transistors (IGBTs). The latter are most powerful with respect to maximum current and breakthrough voltage. A corresponding field-cycling setup is described in the following block diagram **Figure 3**.

Field-cycling magnet

In order to facilitate fast field transitions, the total field energy, which is to be changed in a field cycle, should be as small as possible. That is, the magnet should be very compact. It is also desirable to restrict the maximum induction voltages occurring in a field cycle. Therefore, the inductance should be small.

The magnet coil may be cut out of a solid copper or aluminium block. In this way, a current density distribution for optimum homogeneity can be provided. A less demanding but nevertheless operational construction is to wind the magnet from ordinary copper wires. The magnet coil of the system described here, for instance, is composed of six double-winding layer split solenoids, so that refrigerated cooling oil can be pressed through the 1 mm spacings between the winding layers. The copper wire used has a cross section of 2.34×1.35 mm^2. The four inner layers consist of 2×21 windings; the two outer layers with 2×18 windings have a 10 mm gap in the middle. The inductance is 3.2 mH; the room temperature ohmic resistance is $0.46\,\Omega$. The room temperature bore is 28 mm and the outer diameter is 70 mm. The axial length is 98 mm. A cross section is shown in **Figure 4**. The magnetic field in the sample

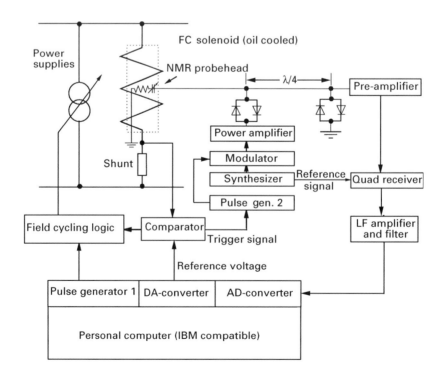

Figure 3 Block diagram of a typical field-cycling NMR relaxometer.

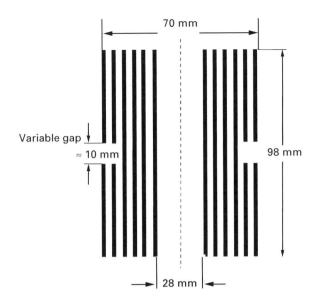

Figure 4 Cross section of an easy-to-make and efficiently cooled field-cycling magnet. The solid vertical lines represent the six double-layers of windings of the magnet. The cooling medium is pressed through between these double layers.

volume can be homogenized by adjusting the position of the two outermost layers correspondingly. The relative inhomogeneity of the flux density is much less than $\Delta B/B = 1 \times 10^{-4}$. The current/field ratio amounts to 188.73 A/T.

Digital control

The pulse sequence for the field cycle is generated with a personal computer equipped with a SMIS MR3020 pulse programmer board. The actual current through the magnet coil is sensed by a temperature compensated precision shunt. This current is compared with the set point given by the programmer board. The comparator unit needs an analogue reference voltage which is generated with the aid of a digital-to-analogue conversion board (Keithley Metrabyte DDA012) in the computer. For example, when the detection field level is reached, the comparator triggers the RF-pulse so that the signal acquisition is always performed at the same magnetic flux density within the accuracy of the voltage comparison (ca. 10^{-5}). The free-induction decay recorded in phase quadrature in order to permit phase-sensitive detection is then digitized using an 8 bit analogue-to-digital converter board in the computer.

Magnet power supply system

The magnet power supply is the most crucial part of a field-cycling relaxometer. It is a complicated system of commercial laboratory power supplies and certain measures for switching between different power supplies are optimized for different current ranges, and for connecting driver high-voltage capacitors to the magnet current circuit. The polarization and evolution fields above 4.2 mT are produced by a parallel combination of four modules of Kepco ATE 75-15M. The maximum current in these intervals is 60 A. The resolution of evolution fields below 4.2 mT is improved by using a special low-current power supply (< 800 mA) temporarily replacing the Kepco system.

The detection field is most demanding with respect to strength, stability and reproducibility. A series of 13 ordinary 12 V car batteries turned out to be best in these regards. In this way an extremely stable and reproducible current of about 300 A can be generated easily.

The transitions from high to low and from low to high magnetic fields are accelerated with the aid of a capacitor precharged to a voltage of 600 V. The polarity by which this driver capacitor is connected to the magnet current circuit depends on the direction of the transition.

For security, the system is surveyed by a unit checking the most important instrument parameters such as the cooling-oil temperature and pressure, and the storage capacitor voltage. In case the parameters deviate from the set point adjustment, the complete high-power part of the relaxometer is switched off within milliseconds.

In passive operation the time constant of the magnet coil is given by

$$\tau_{\mathrm{m}} = \frac{L}{R} \qquad [10]$$

where L and R are the inductance and the resistance of the coil, respectively. Employing semiconductor power switches as mentioned before permits one to actively control the field transition intervals. The following measures are in use.

- Reducing the flux density can be accelerated by dissipating the field energy in an ohmic damping resistor, R_{d}, connected to the magnet in series during the switching-down process. The magnet current then obeys the differential equation

$$-L\frac{\mathrm{d}I}{\mathrm{d}t} = I(R + R_{\mathrm{d}}) - U_0 \qquad [11]$$

where U_0 is the voltage induced by the current variation in the magnet coil. With the initial condition

$I(0) = I_{max} = U_0/R$ the solution becomes

$$I_d(t) = \frac{U_0}{R + R_d} + \left[\frac{U_0}{R} - \frac{U_0}{R + R_d}\right] \exp(-t/\tau_d) \quad [12]$$

with the new time constant

$$\tau_d = \frac{L}{R + R_d} \quad [13]$$

- The damping resistors can be efficiently supplemented by voltage suppressor diodes which cut off all voltages above a certain nominal value. The field energy is then dissipated in the diodes as well.
- The third possibility is to transfer the field energy temporarily to a capacitor. With opposite polarity, such a capacitor can be used to drive the field up as mentioned before. The principle is illustrated in **Figure 5**. The switches are single IGBT modules (Semikron SKM 400 GA) with maximum currents of 400 A at 1200 V. The switching rate can be changed by altering the gate resistor R_G. Induction-voltage peaks due to fast switching can be reduced by increasing R_G. The IGBT switches are controlled by a Semikron SKHI 20 module, which galvanically separates the input TTL signal from the gate. Additionally it contains a small surveillance unit for the IGBT (overvoltage and overtemperature protection), which feeds an error signal back to the Control Logic so that the cycle is immediately stopped in the fault case. During the *polarization period* the switch S_1 (IGBT 1) is closed, and G_1

drives the magnet current. Simultaneously the capacitor C is precharged to about 600 V. In the switching interval *polarization field → evolution field*, the digital-to-analogue converter defines a new set value for G_1. The switch S_1 opens, and the network operates as a series resonance circuit. That is, the magnet energy is transferred to the capacitor. After the desired evolution field level has been reached, the switch S_1 closes again. In the *evolution interval*, either G_1 or G_2 controls the magnet current depending on the adjusted field level. The two alternate current amplifiers are used in order to improve the field resolution, limited by that of the 12 bit digital-to-analogue converter. In the switching interval *evolution field → detection field*, the switch S_2 is closed, and the precharged capacitor provides the peak power for driving the current up to the detection field level. S_3 then closes after a 1 ms delay. When the level of the detection field is reached, S_2 opens and the batteries supply a current of 300 A required for the detection field. Finally, in the *switching-off interval*, the switch S_1 opens, and the magnet energy is transferred to the capacitor again. When the current has decayed to zero, S_1 is closed. Before starting a new polarization/evolution/detection field cycle, a settling time of 2 to 3 s is allowed.

The probehead

As the bore of field-cycling magnets are kept as small as possible, extremely compact designs of the RF probehead are unavoidable. Typical diameters range between 20 and 30 mm. All electrically conducting

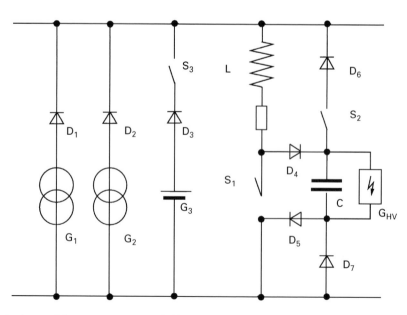

Figure 5 Simplified scheme of the magnet current switching circuit for field-cycling purposes.

parts must be thoroughly slit in order to prevent the induction of eddy currents during the field cycle. The materials and the dimensions must be chosen in a way that the ohmic resistance is as high as possible. On the other hand, one must keep in mind that every slit works like an antenna, which might pick up undesired RF from other sources. Multiple layer shielding with interdigitated slits turn out to be best.

Figure 6 schematically illustrates the 'heart' of the probe, the sample compartment with the RF coil and the temperature sensor. It is arranged closely below the tune and match capacitors, and is surrounded by the RF shielding. The outer diameter is 23 mm, the length of the sample coil 17 mm, with a maximal diameter of 12 mm for the sample containers. The coil is tuned either for proton or for deuteron resonance. That is, a 5 turn coil for 62 MHz or a 70 turn coil for 9.5 MHz, respectively. A glass tube supplies the air stream for heating or cooling of the sample. The temperature is controlled using a Pt100 resistor with an accuracy of ±1°C.

General specifications

The total transition and ring-down time from the polarization field to a value of the evolution field,

stable within ±10%, is less than 3 ms at all evolution field levels. Resonance in the detection field is reached within ±10^{-5}, in a transition time of 2.5 ms. The field switching rates involved are 750 T s^{-1}. Typical field values are listed in **Table 1**.

Typical spin–lattice relaxation dispersion curves

Field-cycling NMR relaxometry is a versatile and powerful method for investigating molecular dynamics over a large range of timescales. It has been applied to manifold materials which show broad distributions of molecular motions, for example proteins, liquid crystals, synthetic polymers, and liquids confined in porous materials. **Figure 7A** represents an example for the investigation of polymer dynamics. The T_1-dispersion curve in the double-logarithmic scale shows the typical slopes observed in polyethylene oxide melts above the critical molecular

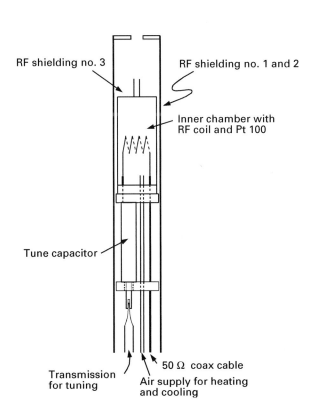

Figure 6 Cross section through a probehead for field-cycling NMR relaxometry.

(a)
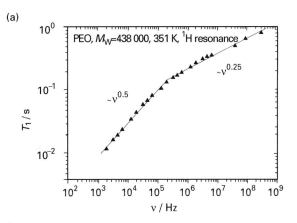

(b)
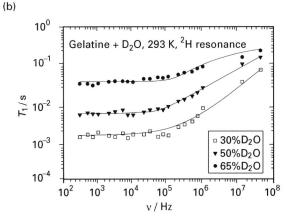

Figure 7 Typical ^{1}H and ^{2}H spin–lattice relaxation curves. The power laws in (A) are discussed in Kimmich, Fatkullin, Weber and Stapf (1994) (see Further reading). The solid lines in (B) correspond to a fit of 'reorientations mediated by translational displacements (RMTD) model described in Kimmich (1997) (see Further reading).

Table 1 Field values of the home-built relaxometer, expressed by the current and the Larmor frequencies of protons and deuterons.

Field	Current (A)	$v_L(^1H)$ (Hz)	$v_L(^2H)$ (Hz)	Rel. deviation
Polarization	55	12.4×10^6	1.904×10^6	2×10^{-4}
Evolution (max.)	28.8	6.51×10^6	1.0×10^6	3×10^{-4}
Evolution (min.)	0.009	2.0×10^3	300	0.2^b
Detection^a	267–273	$60–61.5 (\times 10^6)$	$9.2–9.5 (\times 10^6)$	1.5×10^{-5}

^{a} Depending on the state of charge of the batteries.
^{b} Accuracy of the earth field compensation.

mass $\propto v^{0.5}$ and $\propto v^{0.25}$. The data can be explained by the renormalized Rouse theory. **Figure 7B** shows the ^{2}H-dispersion of D_2O in gelatine. Using deuterated water helps to clarify the mechanisms of water relaxation in biological tissue.

List of symbols

B_0 = external magnetic flux density; B_D = detection field; B_E = magnetic flux density, evolution interval; B_P = magnetic flux density, preparation interval; $G_i(\tau)$ = dipolar autocorrelation function; $J^{(i)}(\omega)$ = intensity function of the Larmor frequency; M_E = Curie magnetization, evolution interval; M_P = Curie magnetization, preparation interval; S/N = signal-to-noise ratio; T_1 = spin–lattice relaxation time; T_d = dipolar-order relaxation time; $T_{1\rho}$ = rotating-frame relaxation time; T_2 = transverse relaxation time; γ = gyromagnetic ratio; μ_0 = magnetic field constant.

See also: **Liquid Crystals and Liquid Crystal Solutions Studied By NMR; Magnetic Field Gradients, in High Resolution NMR; NMR Relaxation Rates; NMR Spectrometers; Proteins Studied Using NMR Spectroscopy.**

Further reading

Kimmich R (1980) Field-cycling in NMR relaxation spectroscopy: Applications in biological, Chemical and polymer physics. *Bulletin on Magnetic Resonance* **1**: 195.

Kimmich R, Fatkullin N, Weber HW and Stapf S (1994) Nuclear spin–lattice relaxation and theories of polymer dynamics. *Journal of Non-Crystalline Solids* **172–174**: 689.

Kimmich R (1997) *NMR Tomography, Diffusometry, Relaxometry.* Springer-Verlag: Berlin.

Koenig SH and Brown RD (1990) Field-cycling relaxometry of Protein solutions and tissue: Implications for MRI. *Progress in NMR Spectroscopy* **22**: 487.

Noack F (1986) NMR field-cycling spectroscopy: Principles and applications. *Progress in NMR Spectroscopy* **18**: 171.

Noack F, Notter M and Weiss W (1988) Relaxation dispersion and zero-field spectroscopy of thermotropic and lyotropic liquid crystals by fast field-cycling NMR, *Liquid Crystals* **3**: 907.

Stapf S, Kimmich R and Seitter R.-O. (1995) Proton and deuteron field-cycling NMR relaxometry of liquids in porous glasses: Evidence for Lévy-Walk statistics. *Physical Review Letters* **75**: 2855.

Rhenium NMR, Applications

See **Heteronuclear NMR Applications (La–Hg).**

Rigid Solids Studied Using MRI

David G Cory, Massachusetts Institute of Technology, Cambridge, MA, USA

MAGNETIC RESONANCE
Applications

Introduction

The function of many rigid solids is tied to their structure and hence there are questions in the study of synthetic, natural and biomaterials that can be addressed by imaging methods. In fact, the first nuclear magnetic resonance (NMR) images of rigid solids were acquired by Peter Mansfield and co-workers in 1973, the same year that Paul Lauterbur introduced the field of NMR imaging. NMR imaging of liquids rapidly advanced to where very sophisticated methods are used for clinically important diagnosis in medicine along with related techniques for the study of microscopy and complex flow dynamics. Progress in the study of rigid solids by NMR imaging has been somewhat slower, due partially to the priorities given to health care issues, but also due to technical challenges introduced by the broad NMR resonance lines of rigid solids. From a spectroscopy point of view the difference between a liquid and rigid solid is tied to the orientational dependence of NMR parameters such as chemical shifts and dipolar couplings. Mobile liquids have sufficient rotational motion that orientational dependencies are averaged to their isotropic value (hence the dipolar interaction vanishes since it has no isotropic component). For imaging purposes the sharper line widths seen in liquids leads to higher spatial resolution with simple imaging methods, however, the tensor properties of the full interactions permit more interesting forms of contrast to be achieved for solids. The goal then in NMR imaging of solids has been to develop methods that provide high quality images in the presence of broad lines and which can also take advantage of the greater array of contrast mechanisms. These methods fall into two general classes, wide-line methods that achieve spatial resolution by overwhelming the broad resonance line with a stronger magnetic field gradient, and coherent averaging approaches that carry out the imaging experiment in an interaction frame where the line width is reduced.

Coherent and incoherent imaging

The basic ideas of NMR imaging are frequently presented as a simple result of the following three concepts:

1. the NMR resonance frequency is directly proportional to the applied magnetic field strength,
2. the NMR signal intensity is directly proportional to the spin concentration,
3. time-dependent linear magnetic field gradients can be applied along any arbitrary direction.

The simple idea is that a spatially varying magnetic field encodes the positions of the spins in their resonance frequencies, and thus the number of spins at any given location may be directly measured as the intensity of the NMR signal at the corresponding resonance frequency.

This real space picture of imaging corresponds best to an incoherent measurement where each spatial location is interrogated sequentially and the full image of the sample is built up from a sequence of such measurements. In such an approach the relative NMR signals from two different spatial locations have no phase (or coherent) relationship to one another. Incoherent imaging methods are occasionally employed in solid state imaging with the stray field imaging (STRAFI) method being the best known example (see below).

Most solid state imaging (and virtually all modern liquid state imaging) is carried out as coherent measurements where a specific phase relation is established between the signal from spins at different locations. It is this phase that is modulated throughout the course of the measurement. Such approaches have the advantage that the entire sample is measured simultaneously, leading to a multiplexing gain in the sensitivity. Coherent imaging is best introduced in a space reciprocal to real space, and discussed in terms of magnetization gratings.

Magnetic field gradients, gratings, and *k*-space

The presence of a spatially varying magnetic field across the sample results in spins at different locations precessing at different rates, and so after some time they rotate to be out of phase with each other. This establishes a relative phase shift that depends on the difference in the magnetic fields that the spins see and the evolution time. In most imaging, the spatially varying field is arranged to be a linear magnetic field gradient so that the difference in fields is directly proportional to the separation of the spins, with the proportionality constant being the strength of the magnetic field gradient (an experimentally controlled parameter).

Differential spin precession sets up a magnetization grating across the sample corresponding to a linear ramp of the phase of the nuclear spin (see **Figure 1**). Grating pictures of the spin dynamics emphasise the coherent nature of the imaging experiment, making it clear that there is a phase relation between any two given spins. The magnetization grating is most conveniently characterized by its corresponding wavenumber, k, which is,

$$k = 2\pi/\lambda \qquad [1]$$

where λ is the spatial period of the grating – identified in **Figure 1**.

As the spins evolve in the magnetic field gradient, they are wound up into progressively tighter spirals and the pitch of the spiral continuously decreases, corresponding to an increasing wavenumber. In a constant magnetic field gradient the wavenumber increases linearly with time,

$$k(t) = k(0) + \gamma \partial B_z/\partial Z t \qquad [2]$$

For simplicity the gradient has been written as though it varies along the z-axis, in practice this could be any direction, and γ is the gyromagnetic ratio.

The key relationship between the grating and the imaging measurement is that the NMR signal is the spin magnetization integrated over the sample. The grating thus is a spatial modulation of the spin density, $\rho\,(x, y, z)$, at a given spatial period. The signal as a function of the wavenumber, $s(k)$, is a Fourier component of the local spin density of the sample,

$$s(\boldsymbol{k}) = \int \rho(\boldsymbol{r})\exp(\boldsymbol{k} \cdot \boldsymbol{r})\mathrm{d}\boldsymbol{r} \qquad [3]$$

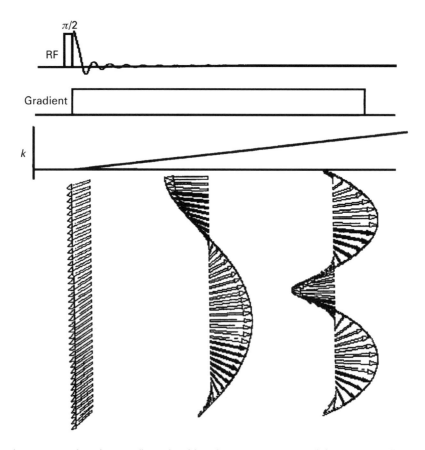

Figure 1 A schematic representation of a one-dimensional imaging measurement and the corresponding magnetization grating. Notice that the pitch of the spatial helix becomes finer over time resulting in measurements of progressively higher wavenumbers, or Fourier components of the spin density. In a multidimensional imaging measurement wave vectors in 2 or 3 dimensions are independently modulated.

where k is the reciprocal variable of the real space laboratory frame and r is a vector in 3D space.

In a coherent picture then the imaging experiment reduces to a series of measurements as a function of the wavenumber in three-dimensional space and taking the inverse Fourier transform of the NMR signal returns the real space image. At this level of approximation, the field of view and spatial resolution are given directly by the Nyquist sampling theorem.

There are a wealth of schemes for how to efficiently sample k-space in two or three dimensions. Since these are not unique to solid state studies the reader is referred to the descriptions in MRI theory.

The complications from rigid solids

The picture set up above is complete when the gradients are the sole source of spin evolution, there is no relaxation and the spins do not move. In the liquid state the gradients can often be made sufficiently strong that relaxation and chemical shifts are minor complications and the limits to resolution and image distortions arise from limited NMR sensitivity and molecular transport. For rigid solids, molecular transport is not an issue, but the short lifetime of the transverse spin magnetization is.

The transverse relaxation time of proton-rich, rigid solids can be as short as tens of microseconds resulting in only this very limited time to impose a grating across the sample and to measure a given Fourier component. As seen from Equation [2] a given wavenumber (and hence spatial resolution) can still be reached, but as the lifetime of the transverse spin magnetization decreases the gradient strength must correspondingly increase. To obtain a wavenumber of $2\pi/20\ \mu m$ in 10 μs a gradient of 1000 G cm^{-1} is needed. Typical gradients for small samples are of the order of 100 G cm^{-1} although gradients as strong as 100 000 G cm^{-1} have been employed in NMR.

The gradient strength is only a technical issue and large gradients are certainly available, their use in imaging, however, is limited by a corresponding loss in sensitivity. NMR is a very insensitive spectroscopy, with the most significant noise source being white noise from the receiver coil. As the gradient strength is increased the effective frequency bandwidth of the smallest volume element (voxel) also increases and the noise increases as the square root of the frequency spread (or gradient strength). Thus, as the resolution is made finer, and the voxel shrinks, the signal decreases with the number of spins in the voxel and the noise increases with the gradient strength. The overall result is that wide-line methods that rely upon large gradients to overwhelm the broad NMR line

width typically have low resolution and low signal-to-noise ratios.

Formally, since the gradient evolution and the natural evolution of the spin system (that due to chemical shifts and dipolar couplings, etc.) commute with each other, the NMR spectrum blurs the image as a convolution. Of course the spectrum is in frequency units, and the image is in spatial units, so one must convert between these, and the gradient strength provides the necessary scaling,

$$\text{image}(x, yz) = \rho(x, y, z) \otimes \text{spectrum}\left(\frac{2\pi\nu}{\gamma(\partial B_z/\partial z)}\right)$$

[4]

The comparison between liquid and solid state imaging is shown schematically in **Figure 2**.

Methods

While obtaining an image is relatively straightforward and can be accomplished with methods that are closely related to those successfully employed in liquid state studies, the broad line width of rigid solids will result in low resolution and sensitivity unless special care is taken. This can be accomplished in three ways,

1. wide-line methods can be employed but with multiple sampling to increase the limited sensitivity,
2. constant-time methods can be used that accept the loss in sensitivity, but recover resolution by

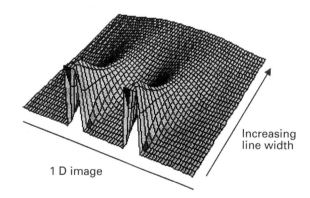

Figure 2 A comparison of a series of 1D images as a function of the NMR line width with a constant gradient. The NMR line width is varied from 0.1% to 100% of the separation between the two bands in the image. As the resolution degrades so also does the sensitivity. Coherent averaging recovers both by artificially narrowing the NMR resonance. The solid state and coherent averaging line widths can differ by a factor of 10^4.

using the fact that the gradient and internal Hamiltonians commute,

3. finally, the broad line width does not need to be accepted, it can be refocused through the use of coherent averaging resulting in a simultaneous increase in both resolution and sensitivity.

Wide-line/strong gradients

Wide-line methods accept the limitation to resolution imposed by the sample's line width and use strong gradients to achieve high spatial encoding. The most successful of these is stray field imaging (STRAFI) where the sample is rotated and translated through the strong magnetic field gradient that is found at the fringe of any superconducting solenoid magnet. The imaging process is incoherent since only a single sensitive slice of the sample is observed at any given time, and the reconstruction process is based on the filtered back projection algorithm. The STRAFI method is described in detail in the article on MRI using stray fields.

Although the STRAFI method gives up the sensitivity advantage from the multiplexing inherent in coherent imaging approaches, a smaller multiplexing advantage can be achieved since there is no need to wait for spin lattice relaxation between the measurement from one slice to the next. In addition, the gradient-imposed noise bandwidth of coherent imaging can be partially avoided through pulsed spin-locked detection, essentially a narrow-band excitation and filtering that locally has good sensitivity.

STRAFI, like most wide-line methods, is typically free of distortions that can plague coherent averaging based imaging methods, but may introduce its own artifacts if the mechanics for sample motion are not excellent. In addition, the method is inherently three-dimensional and so the imaging time is very long.

A challenge of stray field (and wide-line methods) is that the types of contrast that can be introduced are rather limited, and usually based on rotating frame relaxation times. Contrast arising from the commonly observed spectroscopic parameters, particularly the chemical shift cannot be achieved.

Constant time methods

As pointed out above, the gradient evolution and that from internal interactions commute at high fields, and so these act independently on the spin system. Emid and Creyghton pointed out that since these commute and since the gradient interaction is under experimental control, it is possible to set up the k-space sampling in such a fashion that the data

are taken at a constant time following the excitation pulse (see **Figure 3**). The internal interactions lead to their normal complex spin evolution, but this is identical from measurement to measurement and therefore can be removed from the convolution: the only thing that changes is the gradient; Equation [4] may thus be rewritten as,

$$\text{image}(x, y, z) = \text{fid}(t_s)\rho(x, y, z) \qquad [5]$$

where the value of the normalized free induction decay at the sampling time, $\text{fid}(t_s)$, is a complex number with magnitude less than unity. Notice that since the sampling time is kept constant, the free induction decay is not transformed into a spectrum by the inverse Fourier transform.

In constant-time methods the line width of the NMR resonance no longer blurs the image and the image resolution is solely determined by the sampling conditions for the coverage of k-space. The line width does, however, limit the sensitivity, and so

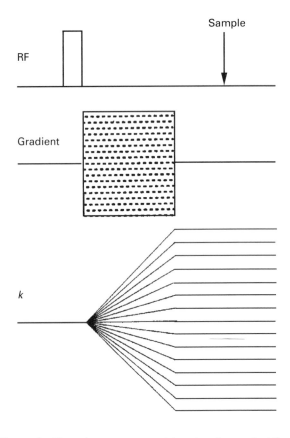

Figure 3 The pulse sequence and k-vectors for constant-time imaging. Notice that all of the data are collected at a constant time following the excitation pulse and yet the k-vector is still modulated from point-to-point by systematically varying the strength of the gradient.

constant-time methods make the normal trade-off of sensitivity for resolution. At the same time, however, they achieve a robust experimental setup that is easy to apply.

With the typical 100 G cm^{-1} gradients available from commercial small scale liquid state imaging equipment, constant-time methods can be used to observe rigid solids, but at very low S/N ratios and at low spatial resolution.

Constant-time methods are also used in liquid state imaging and in NMR microscopy to avoid distortions from chemical shifts and local variations in the bulk magnetic susceptibility.

Coherent averaging

One of the most significant advances in the development of high resolution NMR methods for solids is Waugh's demonstration that the complex spin evolution in strongly coupled rigid solids can be refocused. This result is most easily understood in an interaction frame picture that depends on the time independence of the dipolar Hamiltonian in a rigid solid, rather than focusing on the complex spin dynamics. The interaction frame Hamiltonian can be periodically modulated through mechanical rotation (to modulate the geometric dependence), or through a series of RF pulses (to modulate the spin dependence), and in such a fashion that the time-averaged value is zero. Provided that the NMR response is sampled within this frame then the interaction appears to vanish. Coherent in this case corresponds to the structure of the modulation that is responsible for line-narrowing, and not a spatial phase as in the imaging case. Coherent differentiates the steady experimentally controlled, driven modulation from a natural stochastic, incoherent process.

For imaging, coherent averaging means is that we are not left to suffer the blurring from the broad resonance of a typical solid, but we can 'artificially' narrow this to achieve both higher sensitivity and resolution. Indeed, the first solid state images that were acquired by Mansfield and co-workers in 1973 used coherent averaging precisely for this purpose. Coherent-averaging methods have been developed to average dipolar couplings, chemical shifts, susceptibility shifts, and all of these simultaneously.

Coherent averaging for imaging has taken four approaches:

- Magic angle sample-spinning – which has the advantage of permitting contrast based on the isotropic chemical shift. MAS by itself, however, rarely removes the dipolar broadening efficiently.
- Multiple-pulse sequences based on solid echoes – which are very efficient at refocusing all internal

interactions and can be combined with pulsed gradients to avoid image distortions.
- Multiple-pulse sequences based on magic echoes – are very similar to the solid echo-based methods, but have the advantage of longer windows for the gradients.
- Multiple-pulse sequences based on off-resonance excitation – where the effective interaction frame rotates at the magic angle and dipolar interactions can be made to vanish. Imaging in this interaction frame requires a combination of matched RF and d.c. gradients.

Magic angle sample spinning (MAS) is conceptually the simplest of these. Andrew demonstrated that a pair of dipolarly coupled spins are effectively decoupled from one another if the sample is spun at the 'magic-angle', the [111] direction – or the bisector of a cube. This rotation effectively removes the dipolar broadening from the system, and has the advantage of also removing the anisotropic chemical shift while leaving the isotropic shift for contrast generation. The challenge, of course, is that as the sample is rotating the imaging experiment must keep up with the sample and a variety of means have been engineered to synchronize the imaging experiment to the frame of reference rotating with the sample (see **Figure 4**). Once in this frame, the imaging methods are very similar to liquid state methods and the solid state aspects can, for the most part, be ignored.

Unfortunately, MAS alone, even at 20 kHz rotation rates, does not produce narrowing of the line width of a dipolarly coupled rigid solid. The challenge here is that the state of the nuclear spins are modulated by energy conserving mutual flip/flops (one spin flips to up while a neighbour flops to down) at a correlation time that is much faster than the modulation frequencies reached through sample rotation. So, MAS works well in special cases but in general must be combined with a multiple-pulse method.

Multiple-pulse methods achieve line narrowing by modulating the spin states quickly and symmetrically (see **Figure 5**). The promise of multiple-pulse approaches to solid-state imaging is both high resolution and good sensitivity, and that these can be combined with great flexibility in terms of contrast and image geometry. The challenge for multiple-pulse imaging is that the gradient evolution no longer commutes with the internal Hamiltonians as they are modulated, and so the spin dynamics of the gradient and internal Hamiltonians are no longer independent of each other. This can lead to pronounced image distortions, where the residual NMR linewidth depends on the gradient-induced frequency shift. The cleanest way to avoid such distortions is to apply the gradient

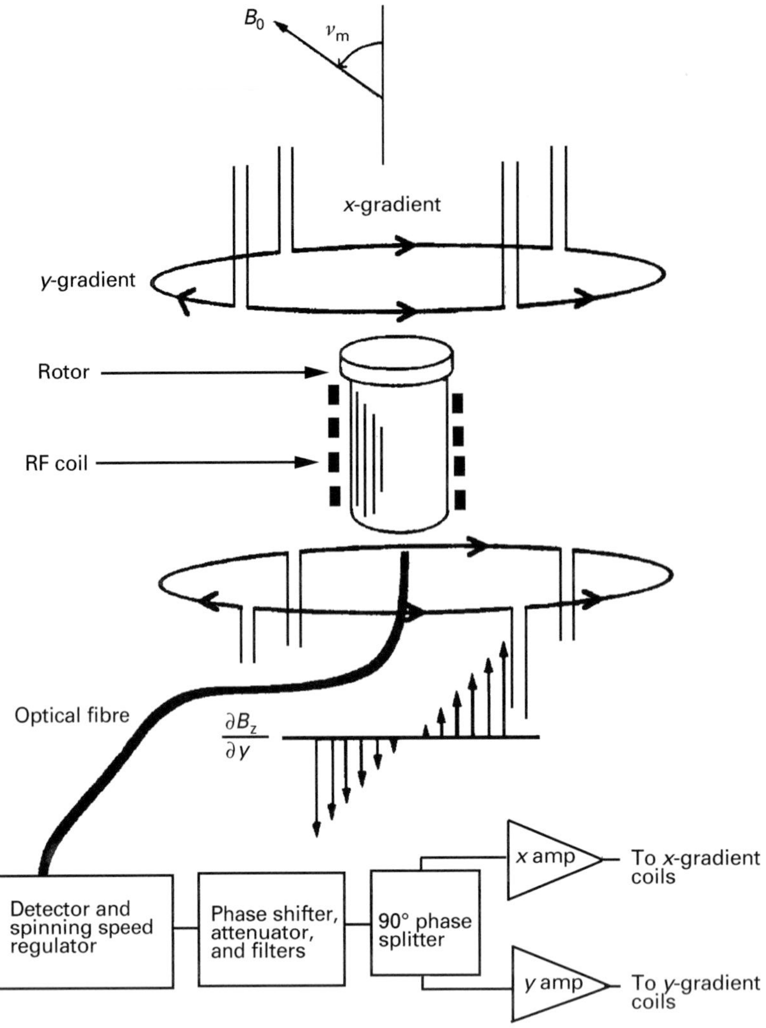

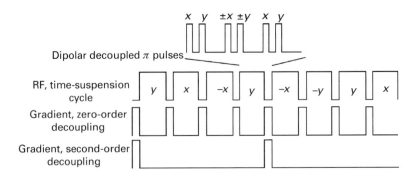

Figure 4 The experimental setup for MAS imaging. The sample rotates about the magic angle at a frequency of about 10 kHz, and the current through the gradient coils is modulated so that the gradient fields move with the sample. ν_m is the magic angle, $54.7°$.

Figure 5 The pulse sequence for a representative multiple-pulse, coherent averaging imaging measurement. The series of RF pulses lead to a refocusing of the dipolar and chemical shift interactions over the full cycle, and the gradients are added in selected windows. Even though the gradient and internal Hamiltonians do not commute, by restricting the gradients to selected windows an image which is substantially free of distortions can be collected. By placing the gradient pulses between each dipolar decoupled π pulse a larger effective gradient is achieved, but at a cost of some distortions, these are avoided by placing the gradients more sparsely where the gradient RF interaction is avoided, to second order.

in selected, symmetry related windows, which, however, requires specialized hardware to deliver strong fast gradient pulses.

Multiple-pulse imaging permits a surprisingly wide range of image contrast even though the internal Hamiltonian is suppressed during the imaging sequence. This, of course, removes the possibility of building contrast into the image during data sampling, and so most solid-state imaging methods are preceded with a contrast-generating sequence and perhaps slice selection if a two-dimensional image is desired.

Resolution and field of view

Resolution in NMR imaging of solids is primarily limited by the low sensitivity of NMR. Typical detection limits are of the order of 10^{15} spins in the solid state, so that minimum voxels are approximately $10^4 \, \mu m^3$ (or a cube 20 μm on edge). Along any one dimension the resolution can be somewhat higher, but the volume remains approximately constant. The various approaches to solid state imaging present additional limitations to the resolution.

STRAFI methods rely on sample translation and so the precision and reproducibility of the mechanical arrangement also limits the resolution, here to approximately $10 \, \mu m$. STRAFI methods are also set up to provide isotropic voxels in 3D-imaging, although much work has been done with 1D-STRAFI measurements where the question has been carefully tailored so they can be answered by studying profiles of the sample.

Constant-time methods trade-off significant sensitivity for resolution at a rate determined by the gradient strength. These methods typically work best for large objects where lower resolution is required and the gradient evolution time can be kept short.

Magic angle spinning-based imaging methods are very close to liquid state studies and the resolution limits normal in microscopy (better than 10 μm) apply here, however, spinning speeds are not sufficient to study most rigid solids. For rigid materials multiple-pulse coherent averaging is preferred and here the limitation relates to the challenges of sampling in the presence of multiple-pulse and pulsed gradients. In practical terms this permits about 128 volume elements to be acquired along a given axis and so the resolution is tied to the sample size.

This last point is more general, given sampling time and the lower efficiencies of larger detection coils imaging methods are typically limited to 512 elements along a side (or less). So the highest resolutions are achieved with small samples.

A related issue is the allowable size of the field of view. Solid-state imaging methods typically impose mechanical constraints on the sample size, such as to fit within the strong gradient in STRAFI (about 15 mm), or within the rotor for MAS methods (about 5 mm). Multiple-pulse methods work best in very high RF fields, and so the sample is limited to fitting within a relatively small RF coil (about 10 mm). Some single-sided imaging methods have been explored and are certainly feasible for soft materials, but are less well developed for rigid solids.

Availability of instruments and methods

To date, NMR imaging of rigid solids is still mainly limited to experts who typically build a significant portion of their own equipment. Constant-time methods are the most widely available and can be used for preliminary studies, but the low sensitivity make many studies prohibitively long. STRAFI probes are commercially available, but here also the time to acquire an image is long and the restricted means of generating contrast limits applications. Recently, magic angle sample spinning probes with a single magnetic field gradient along the spinner axis have become commercially available (for NMR spectroscopic studies of semisolids) but probes with 3D gradient sets must still be home-made. All multiple-pulse coherent averaging approaches to imaging are carried out in homebuilt probes, and used with or without an associated fast gradient pulser. A large part of the challenge in employing multiple-pulse methods is the required stability and precision of the RF which typically demands the attention of an experienced spectroscopist to set up.

Applications

Most of the effort in NMR imaging of rigid solids has been directed towards developing robust, high quality imaging methods, and few applications have been investigated in any detail. However, there are now a variety of good methods and it is an appropriate time for applications.

Proposed applications include the study of structure and morphology in synthetic polymers, an example being the selective imaging of one component of a multicomponent blend (see **Figure 6**). Such studies can be extended to the mapping of rigid versus mobile segments in a processed material (see **Figure 7**). There are a wealth of questions in materials processing and ageing that can be investigated by

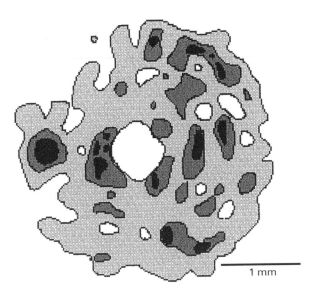

Figure 6 MAS image of the polybutadiene fraction of a poly-butadiene/polystyrene blend cast as a film from toluene. MAS alone at 5 kHz spinning is not sufficient to narrow the polystyrene NMR resonance. The in-plane resolution is 50×50 μm². Figure reproduced with permission from Cory DG, de Boer JC and Vee-man WS (1989) Magic angle spinning ¹H NMR imaging of poly-butadiene/polystyrene blends. *Macromolecules* **22**: 1618.

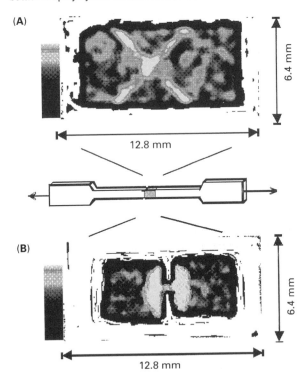

Figure 7 (A) T_{2e^-} magic-echo image of a piece of polycarbonate (see schematic) in which a crossed shearband has been created. (B) An image of a sample that has been stretched nearly to the breaking point: The regions with high intensity correspond to areas of low molecular mobility. Figure reproduced with permission from Traub B, Hafner S, Maring D and Spiess HW (1998) In: Blümler P, Blümich B, Botto B and Fukushima E, *Spatially Resolved Magnetic Resonance*, Weinheim: Wiley-VCH, p 191.

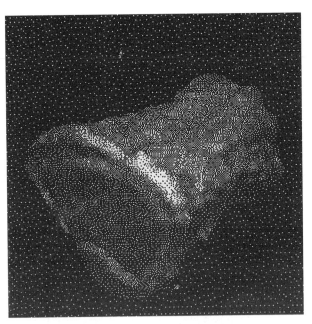

Figure 8 Three-dimensional constant-time image of a bovine femur showing both the compact bone and the interior bone marrow. The image resolution is $1.1 \times 1.1 \times 4.4$ mm³ and the image was acquired over a period of 5 h. Figure reproduced with permission from Balcon BJ (1998) In: Blümler P, Blümich B, Botto B, and Fukushima E (eds) *Spatially Resolved Magnetic Resonance*. Weinheim: Wiley-VCH, p. 84.

such methods, even though the resolution of the image is on the order of tens of micrometres.

There has also been interest in biomedical studies of bone, bone cements and the function of synthetic organs. An example of an image of bovine bone is shown in **Figure 8**. Progress has also been made in developing methods of following bone repair *in vivo*.

List of symbols

B = magnetic field strength; k = wave number; s = signal; t = time; γ = gyromagnetic ratio; λ = spatial period; ν = frequency; ρ = spin density.

See also: **High Resolution Solid State NMR, ¹H, ¹⁹F; MRI Instrumentation; MRI of Oil/Water in Rocks; MRI Theory; MRI Using Stray Fields; NMR Microscopy; NMR of Solids; NMR Principles.**

Further reading

Blümler P, Blümich B, Botto B and Fukushima E (1998) *Spatially Resolved Magnetic Resonance*. Weinheim: Wiley.

Blümler P and Blümich B (1994) Solid state NMR. *NMR Basic Principles and Progress* 30: 209.

Callaghan PT (1991) *Principles of NMR Microscopy*. Oxford: Oxford University Press.

Cory DG (1992) Solid state NMR imaging. *Annual Reports on NMR Spectroscopy*, **24**: 87–180.

Kimmich R (1997), *NMR-Tomography, Diffusometry, Relaxometry*. Berlin: Springer-Verlag.

Mansfield P and Grannell PK (1975) "Diffraction" and microscopy in solids and liquids by NMR. *Physical Review B* **12**: 3618–3634.

Rotational Resonance in Solid State NMR

See **Solid State NMR, Rotational Resonance.**

Rotational Spectroscopy, Theory

Iain R McNab, University of Newcastle upon Tyne, UK

**VIBRATIONAL, ROTATIONAL &
RAMAN SPECTROSCOPIES**

Theory

Synopsis

The rotational spectra of gas phase molecules is explained using the quantum theory of angular momentum. The spectra of simple rigid molecules are explained on the basis of selection rules and energy level expressions which depend upon the moments of inertia of the molecules. We consider how the theory needs to be extended to include effects of nonrigidity, and also small effects due to interactions between the angular momentum of the nuclear framework, electron spins, and nuclear spins. A calculation of relative intensities using spherical tensor techniques is given for a diatomic molecule using Hund's case *a*.

Introduction

The rotational spectroscopy of molecules typically (but by no means always) occurs in the microwave region of the spectrum. In solids, molecules are usually not free to rotate, and in liquids collisions normally render absorption featureless; we therefore consider only the rotational spectroscopy of gaseous molecules.

Spectroscopic assignments are often discussed in terms of selection rules which determine whether or not it is possible for a transition in a molecule to be caused by a particular interaction, such as the electric dipole interaction. The only true selection rules are for total angular momentum (F) and parity (+ or –). For complex molecules with many different angular momenta, it is more profitable to discuss spectra in terms of relative intensities rather than 'allowed transitions'. We can readily calculate the relative intensities for many types of transitions in different molecules using the quantum theory of angular momentum.

Rotational effects (fine structure) are also seen in high resolution electronic and vibrational spectra; additional structure due to the interaction of electrons with nuclear electric and magnetic moments may also be observed and is called hyperfine structure. The simplest rotational spectra are associated with diatomic molecules with no electronic orbital or spin angular momentum (i.e. singlet sigma states) and these are considered first.

The theory of rotational spectroscopy depends upon an understanding of the quantum mechanics of angular momentum. Useful results from the quantum theory of angular momentum are given in **Table 1** and useful results in spherical tensor notation are given in **Table 2**.

A full understanding of a rotational spectrum is achieved when we can calculate the rotational energy levels of the molecules and then calculate the frequencies and intensities of the transitions between them. The cheapness and availability of computers and subroutines for diagonalizing matrices means that most spectra are now assigned on the basis of calculated spectra iterating towards convergence with measured line positions and intensities; we discuss how this is accomplished.

Table 1 Useful results from the quantum theory of angular momentum

$$\boldsymbol{J}^2 \left| JKM \right\rangle = J(J+1) \left| JKM \right\rangle \hbar^2$$
$$J_c \left| JKM \right\rangle = K \left| JKM \right\rangle \hbar$$
$$J_z \left| JKM \right\rangle = M \left| JKM \right\rangle \hbar$$
$$J^\pm \left| JKM \right\rangle = J_x \pm iJ_y) \left| JKM \right\rangle = [J \mp M)(J \pm M+1)]^{1/2} \left| JKM \pm 1 \right\rangle \hbar$$

The lower result shows the operation of the shift (or ladder) operator in the space fixed axis system using the normal phase convention.

Angular momentum relationships are based on the commutation relationships

$$[J_x, J_y] = i\hbar J_z$$
$$[\boldsymbol{J}^2, J_z] = 0$$

Addition of two angular momenta:

$$\boldsymbol{J} = \boldsymbol{J}_1 + \boldsymbol{J}_2$$

and the allowed values of the quantum number J are given by

$$J = J_1 + J_2, J_1 + J_2 - 1,\dots \left| J_1 - J_2 \right| \tag{1}$$

Molecule fixed angular momenta are problematic, because the commutation relationships between different components are reversed. There are many ways of taking account of this, but the easiest is to calculate everything in space fixed axes and convert (where necessary) to molecule fixed axes by using Equations [13] and [14].

Table 2 Useful relationships for the use of spherical tensors in the calculation of matrix elements of angular momentum operators (courtesy of Prof. JM Brown, PTCL, Oxford University)

The three components of a first rank spherical tensor are defined as a linear combination of Cartesian components:

$$T^1_{p=1}(\boldsymbol{A}) = -\frac{1}{\sqrt{2}}(A_X + iA_Y)$$
$$T^1_{p=0}(\boldsymbol{A}) = A_Z \tag{2}$$
$$T^1_{p=-1}(\boldsymbol{A}) = \frac{1}{\sqrt{2}}(A_X - iA_Y)$$

where the superscript 'k' is the rank and the subscript 'p' is the component which can have any of $2k + 1$ values, $k, k-1, k-2,\dots-k$. Note that this transformation essentially corresponds to using J_z, J^+ and J^- in place of J_z, J_x and J_y.
The scalar product of 2 irreducible spherical tensor operators of rank k is defined by

$$T^k(\boldsymbol{A}).T^k(\boldsymbol{B}) = \sum_p (-1)^p T^k_p(\boldsymbol{A}) T^k_{-p}(\boldsymbol{B}) \tag{3}$$

The Wigner–Eckart theorem enables the M_J dependence of a matrix element to be factored out as follows:

$$\langle \gamma J M_J | T^k_p(\boldsymbol{A}) | \gamma' J' M_J' \rangle = (-1)^{J-M_J} \begin{pmatrix} J & k & J' \\ -M_J & p & M_J' \end{pmatrix} \langle \gamma J \| T^k(\boldsymbol{A}) \| \gamma' J' \rangle \tag{4}$$

where $\langle \gamma J \| T^k(\boldsymbol{A}) \| \gamma' J' \rangle$ is defined by the equation and is known as a reduced matrix element (it has no M dependence).
The Wigner 3-j symbol $\begin{pmatrix} J & k & J' \\ -M_J & p & M_J' \end{pmatrix}$ which appears in the definition is itself defined by the coupling equation

$$|J_1 J_2 JM\rangle = \sum_{M_1, M_2} |J_1 M_1\rangle |J_2 M_2\rangle (-1)^{-J_1 + J_2 - M} (2J + 1)^{1/2} \begin{pmatrix} J_1 & J_2 & J \\ M_1 & M_2 & -M \end{pmatrix} \tag{5}$$

J_1 and J_2 (with component quantum numbers M_1 and M_2, respectively) are coupled together to yield the resultant angular momentum $\boldsymbol{J}$ with component M. The 3-j symbol is zero unless the bottom row sum equals zero and the three arguments in the top row must add together according to Equation [1] in **Table 1**.

Continued

Table 2 (*Continued*)

Computation of reduced matrix elements $\langle \gamma J \parallel T^k(\boldsymbol{A}) \parallel \gamma'J'\rangle$ is accomplished by choosing values of components which give a simple value for the matrix element $\langle \gamma JM_J|T_p^k(\boldsymbol{A})|\gamma'J'M_J'\rangle$ and dividing the matrix element by

$$(-1)^{J-M_J}\begin{pmatrix} J & k & J' \\ -M_J & p & M_J' \end{pmatrix}$$

Two common examples are (a) $\langle J\Omega M|J_Z|J'\Omega'M'\rangle = M\langle J\Omega M|J'\Omega'M'\rangle = M\delta_{JJ'}\delta_{MM'}\delta_{\Omega\Omega'}$
hence $\langle J\Omega \parallel T^1(\boldsymbol{J})\|J\Omega\rangle = \delta_{JJ'}\delta_{\Omega\Omega'}[J(J+1)(2J+1)]^{1/2}$
(b) similarly, $\langle S\|T^1(\boldsymbol{S})\|S'\rangle = \delta_{SS'}[S(S+1)(2S+1)]^{1/2}$

Matrix elements of the tensor product of two tensor operators: Let $X_p^k(\boldsymbol{A}, \boldsymbol{B})$ be the tensor product of rank k of $T^{k_1}(\boldsymbol{A})$ and $T^{k_2}(\boldsymbol{B})$:

$$X_p^k(\boldsymbol{A}, \boldsymbol{B}) = (-1)^{k_1-k_2+p}[(2k+1)]^{1/2}\sum_{p_1,p_2}\begin{pmatrix} k_1 & k_2 & k \\ P_1 & P_2 & -P \end{pmatrix} T_{p_1}^{k_1}(\boldsymbol{A})T_{p_2}^{k_2}(\boldsymbol{B}) \tag{6}$$

Then the reduced matrix element of $X_p^k(\boldsymbol{A}, \boldsymbol{B})$ is related to those of $T^{k_1}(\boldsymbol{A})$ and d $T^{k_2}(\boldsymbol{B})$ by the general expression

$$\langle \gamma J \parallel X^k(\boldsymbol{A}, \boldsymbol{B}) \parallel \gamma'J'\rangle = [2k+1]^{1/2}(-1)^{k+J+J'}\sum_{\gamma'',J''}\begin{Bmatrix} k_1 & k_2 & k \\ J' & J & J'' \end{Bmatrix}\langle \gamma J \parallel T^{k_1}(\boldsymbol{A}) \parallel \gamma''J''\rangle\langle \gamma''J'' \parallel T^{k_2}(\boldsymbol{B}) \parallel \gamma'J'\rangle \tag{7}$$

where $\begin{Bmatrix} k_1 & k_2 & k \\ J' & J & J'' \end{Bmatrix}$ is a Wigner 6-j symbol.

The general formula can be simplified if the two operators commute, i.e. if $T^{k_1}(\boldsymbol{A})$ operates only on one part of a coupled system $(\boldsymbol{J}_1)$ and $T^{k_2}(\boldsymbol{B})$ on another $(\boldsymbol{J}_2)$ then the reduced matrix elements of $X_p^k(\boldsymbol{A}, \boldsymbol{B})$ are given by:

$$\langle \gamma J_1 J_2 J\|X^k(\boldsymbol{A}, \boldsymbol{B})\| \gamma'J_1{}'J_2{}'J'\rangle = \sum_{\gamma''}[(2J+1)(2J'+1)(2k+1)]^{1/2}$$

$$\times \begin{Bmatrix} J_1 & J_1' & k_1 \\ J_2 & J_2' & k_2 \\ J & J' & k \end{Bmatrix}\langle \gamma J_1\| T^{k_1}(\boldsymbol{A}) \| \gamma''J_1'\rangle\langle \gamma''J_2 \| T^{k_2}(\boldsymbol{B}) \| \gamma'J_2'\rangle \tag{8}$$

$\begin{Bmatrix} J_1 & J_1' & k_1 \\ J_2 & J_2' & k_2 \\ J & J' & k \end{Bmatrix}$ is a Wigner 9-j symbol.

A simplification of the above expression which is often encountered is for the scalar product $(X^k = X^0)$ of two commuting tensor operators:

$$\langle \gamma J_1 J_2 JM|T^k(\boldsymbol{A}).T^k(\boldsymbol{B})| \gamma'J_1{}'J_2{}'J'M'\rangle = (-1)^{J_1'+J_2+J'}\delta_{JJ'}\delta_{MM'}$$

$$\times \begin{Bmatrix} J & J_2 & J_1 \\ k & J_1' & J_2' \end{Bmatrix}\sum_{\gamma''}\langle \gamma J_1\| T^k(\boldsymbol{A}) \| \gamma''J_1'\rangle\langle \gamma''J_2 \| T^k(\boldsymbol{B}) \| \gamma'J_2'\rangle \tag{9}$$

If a single operator in a coupled scheme of angular momenta acts on only one part $(J_1$ or $J_2)$ then we have the reduced matrix element expressions:
(a) for operator $T^{k_1}(\boldsymbol{A})$ operating only on J_1:

$$\langle \gamma J_1 J_2 J \parallel T^{K_1}(\boldsymbol{A}) \parallel \gamma'J_1'J_2'J'\rangle = (-1)^{J_1+J_2+J'+k_1}\delta_{J_2 J_2'}[(2J+1)(2J'+1)]^{1/2}\begin{Bmatrix} J_2 & J_1 & J \\ k_1 & J' & J_1' \end{Bmatrix}\langle \gamma J_1 \parallel T^{k_1}(\boldsymbol{A}) \parallel \gamma'J_1'\rangle \tag{10}$$

(b) for operator $T^{k_2}(\boldsymbol{B})$ operating only on J_2:

$$\langle \gamma J_1 J_2 J \parallel T^{k_2}(\boldsymbol{B}) \parallel \gamma'J_1'J_2'J'\rangle = (-1)^{J_1'+J_2'+J+k_2}\delta_{J_1 J_1'}[(2J+1)(2J'+1)]^{1/2}\begin{Bmatrix} J_1 & J_2 & J \\ k_2 & J' & J_2' \end{Bmatrix}\langle \gamma J_2 \parallel T^{k_2}(\boldsymbol{B}) \parallel \gamma'J_2'\rangle \tag{11}$$

Continued

Table 2 (*Continued*)

For the scalar product of two operators which **both** act on part one ($\boldsymbol{J}_1$) of a coupled scheme, we have

$$\langle \gamma J_1 J_2 J M | T^{k_1}(\boldsymbol{A}).T^{k_1}(\boldsymbol{B}) | \gamma' J_1' J_2' J' M' \rangle$$
$$= \delta_{JJ'} \delta_{MM'} \delta_{J_1 J_1'} \delta_{J_2 J_2'} \sum_{\gamma'', J_1''} (-1)^{J_1 - J_1''} (2J_1 + 1)^{-1} \langle \gamma J_1 \parallel T^{k_1}(\boldsymbol{A}) \parallel \gamma'' J_1'' \rangle \langle \gamma'' J_1'' \parallel T^{k_1}(\boldsymbol{B}) \parallel \gamma' J_1' \rangle \quad [12]$$

For the relation between operators in space-fixed and molecule fixed co-ordinate systems, we use the notation that p gives the sPace fixed coordinate and q gives the moleQule fixed coordinate of a tensor operator. The relationship between the same tensor in the two coordinate systems is given by:

$$T_p^k(\boldsymbol{A}) = \sum_q D_{pq}^k(\alpha, \beta, \gamma)^* T_q^k(\boldsymbol{A}) \quad [13]$$

where $D_{pq}^k (\alpha, \beta, \gamma)^* = D_{pq}^k (\omega)^*$ is the complex conjugate of the kth rank rotation matrix $D^k(\omega)$. α, β, γ are the three Euler rotation angles that specify the relationship of the molecule fixed coordinate system to the space fixed coordinate system. The phase convention used here is opposite to that of Edmonds. The inverse relationship is

$$T_q^k(\boldsymbol{A}) = \sum_p D_{pq}^k(\omega) T_p^k(\boldsymbol{A}) = \sum_p (-1)^{p-q} D_{-p,-q}^k(\omega)^* T_p^k(\boldsymbol{A}) \quad [14]$$

The $D_{pq}^k (\omega)^*$ are simply related to the eigenfunctions of the symmetric top:

$$|JKM\rangle = \left[\frac{2J+1}{8\pi^2} \right]^{1/2} D_{MK}^J(\omega)^* \quad [15]$$

We now quote two additional useful relationships for tensor operators which can be quantized in both molecule- and space-fixed coordinate systems:

$$\langle JKM | D_{p,q}^k(\omega)^* | J'K'M' \rangle = (-1)^{K-M} [(2J+1)(2J'+1)]^{1/2} \begin{pmatrix} J & k & J' \\ -K & q & K' \end{pmatrix} \begin{pmatrix} J & k & J' \\ -M & p & M' \end{pmatrix} \quad [16]$$

$$\langle JK \parallel D_{\bullet,q}^k(\omega)^* \parallel J'K' \rangle = (-1)^{J-K} [(2J+1)(2J'+1)]^{1/2} \begin{pmatrix} J & k & J' \\ -K & q & K' \end{pmatrix} \quad [17]$$

Note that the Wigner 3-j, 6-j and 9-j symbols are simply numbers when evaluated for particular values of the angular momentum symbols. There are numerous tabulations and programs available which allow their evaluation and the program Mathematica can handle 3-j and 6-j symbols analytically.

Born–Oppenheimer approximation

The Born–Oppenheimer approximation assumes that the molecular wavefunction can be written in the form $\psi_{\text{total}} = \psi_{\text{electronic}} \psi_{\text{vibration}} \psi_{\text{rotation}}$ and therefore that the energies due to each type of motion are additive $E_{\text{total}} = E_{\text{electronic}} + E_{\text{vibrational}} + E_{\text{rotational}}$. This gives rise to electronic and vibrational quantum numbers which are often a good first order description of the energy levels and wavefunctions of a molecule. The application of the Born–Oppenheimer approximation leads naturally to the rigid rotor description of a molecule – we treat the rotation separately from the vibrational motion of the nuclei, which are considered later as a perturbation.

The rotational motion of a molecule is quantized in units of $\hbar$, and this free rotation can take place about three orthogonal axes which are associated with three distinct moments of inertia, I_A, I_B, I_C. The moments of inertia are, in fact, quantum mechanical expectation values of the moments of inertia in a

particular electronic and vibrational state of the molecule. We associate the moments of inertia with three rotational constants, A, B, C, which determine the energy levels of the molecule.

The Hamiltonian for rotation is $H_r = A J_A^2 + B J_B^2 + C J_C^2$. If the molecule has symmetry then we can take advantage of the symmetry to rewrite the Hamiltonian in a form which is easy to solve. The appropriate Hamiltonians and energy level expressions are given in **Table 3**.

Molecules without electronic or nuclear angular momentum

The simplest spectra belong to molecules with no electronic angular momentum and no nuclear spin angular momenta and we consider these first. The relevant classifications, Hamiltonians and energy level expressions are given in **Table 3**. The quantum mechanical treatment of spherical motion gives rise

Table 3 Rigid rotor classifications.

We classify rigid rotors according to their moments of inertia about three perpendicular axes. There is a convention (due to Mulliken) that the three moments of inertia are labelled ordered according to size by $I_A \leq I_B \leq I_C$ and hence $C \leq B \leq A$

	Moments of inertia	Hamiltonian and energy levels
Spherical tops:	$I_A = I_B = I_C$	$H_r = B\mathbf{J}^2$
		$E_r = BJ(J+1)$
Linear molecules:	$I_A = 0 \quad I_B = I_C$	$H_r = B(J_B^2 + J_C^2) = B\mathbf{J}^2$
		$E_r = BJ(J+1)$
Symmetric tops:		
Prolate	$I_A < I_B = I_C$	$H_r = AJ_A^2 + B(J_B^2 + J_C^2) = B\mathbf{J}^2 + (A-B)J_A^2$
		$E_r = BJ(J+1) + (A-B)K^2$
Oblate	$I_A = I_B < I_C$	$H_r = AJ_A^2 + B(J_B^2 + J_C^2) = B\mathbf{J}^2 + (C-B)J_A^2$
		$E_r = BJ(J+1) + (C-B)K^2$
Asymmetric top	$I_A \neq I_B \neq I_C$	$H_{rotation} = AJ_A^2 + BJ_B^2 + C_C^2 J$
		analytic solutions exist for only the first few levels, these solutions are given in **Table 4.**

to the eigenfunctions of the spherical top ψ_{rotation} which we denote by the ket $|JKM\rangle$; the letters inside the bracKET denote the quantum numbers used to describe the state. For other problems involving rotation it is convenient to use a basis of $|JKM\rangle$ states and express the eigenfunctions of each problem as a linear combination of them, that is:

$$|J\Gamma\rangle = \sum_K a_K^J |JKM\rangle$$

$$H_{\text{rotation}}|J\Gamma\rangle = E|J\Gamma\rangle$$

For a molecule, there are two projection quantum numbers of J which are simultaneously conserved, M and K. M is the space fixed projection of $\mathbf{J}$ and K is the molecule fixed projection of $\mathbf{J}$. Both M and K can take values from $+J$ to $-J$ in steps of one. In the absence of an electric or magnetic field the eigenvalues are independent of M and each eigenfunction gives rise to an eigenvalue which is $2J+1$ degenerate.

Spherical top molecules

The spherical top has the same simple energy level pattern as a linear molecule. However, a spherical top cannot have a permanent dipole moment, and therefore has no rotational spectrum.

Linear molecules (including diatomic molecules)

The linear molecule energy levels are given by $E = B.J(J+1)$, where J is the rotational quantum number for rotation perpendicular to the bond, and B is the rotational constant. The energy level structure is given in **Figure 2A**. A typical rotational spectrum (of a diatomic molecule) is shown in **Figure 1**.

To understand the spectrum we must be able to calculate the absorption frequencies of the lines and their relative intensities. The Beer–Lambert law

$$I(\omega) = I_0(\omega)\exp[-\gamma(\omega)l]$$

describes absorption in a thin sample (after length l is traversed) in terms of the absorption coefficient $\gamma(\omega)$ which is given by:

$$\gamma(\omega) = \frac{8\pi^3}{3hc}\omega\left(\frac{N_m}{g_m} - \frac{N_n}{g_n}\right)|\langle J\Gamma|\boldsymbol{\mu}|J'\Gamma'\rangle|^2 \, S(\omega, \omega_0)$$

where $S(\omega, \omega_0)$ is a normalized line shape function. Evaluation of $\langle J\Gamma|\boldsymbol{\mu}|J'\Gamma'\rangle$ yields selection rules for the basis functions (see below). For the linear molecule with no electronic angular momentum and no nuclear spin, we can calculate selection rules which tell us whether a particular transition is allowed or not, by understanding that we are adding a photon angular momentum of one to the angular momentum of the molecule. Therefore, by the formula for addition of angular momenta (see **Table 1**), the change in angular momentum allowed is $\Delta J = 0, \pm1$ and $\Delta J = 0$ cannot occur from $J = 0$. A complete treatment shows also that $\Delta K = 0$, $\Delta M = 0, \pm1$ and the dipole change occurs along the internuclear axis. The possible absorption energies are given by

$$\Delta E = hf = B[J'(J'+1) - J(J+1)]$$
$$= B[(J+1)(J+2) - J(J+1)] = 2B(J+1)$$

So we expect a series of lines separated by $2B$, as shown in **Figure 1**.

The relative intensities of the lines also depend upon the difference in population of the two levels concerned at the temperature considered, $N_m/g_m - N_n/g_n$ (g_n is the degeneracy of level n, etc.). For a sample in thermal equilibrium the relative populations are given by the Boltzmann distribution

$$N_m/g_m = (N_n/g_n)\exp[-(E_n - E_m)/kT]$$

and this factor accounts for the envelope of the spectral intensities shown.

In the case of a molecule with two identical nuclei (such as CO_2), relative line intensities are affected by the nuclear spins. The nuclear spin governs the intensities through the generalized Pauli exclusion principle; 'all wavefunctions are antisymmetric with respect to the interchange of fermions (half-integer spin particles) and symmetric with respect to the interchange of bosons (integer spin particles)' A general treatment uses the molecular symmetry group. For a linear molecule with zero spin nuclei the antisymmetric rotational levels are missing entirely. In the case of CO_2 the Σ_g^+ electronic state has only even J levels. If only one pair of identical nuclei have nonzero spins, then the ratio of the statistical

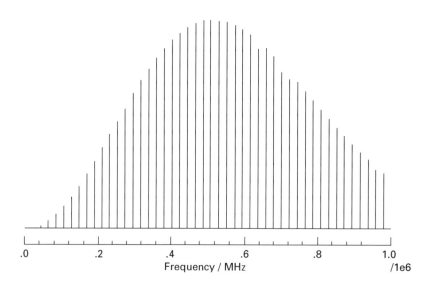

Figure 1 A simulated spectrum of BrF at 300 K. All simulated spectra were generated using Pgopher, courtesy of Dr. CM Western, University of Bristol.

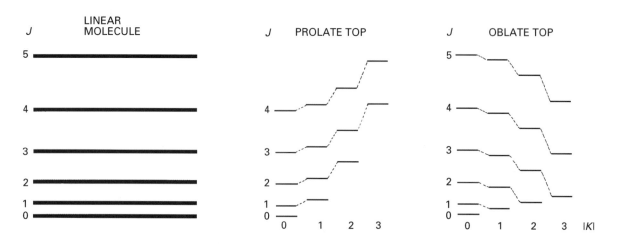

Figure 2 Energy levels of the linear molecule and prolate and oblate symmetric tops. Modified from Figures 3.1 and 3.2 of Kroto HW (1992) *Molecular Roatation Spectra*, pp 32–33, Canada: Dover Books.

weights of the symmetric and antisymmetric rotational levels is $(I+1)$: I for bosons or I: $(I+1)$ for fermions.

The rotational constant B is easily determined from the spectrum. For a diatomic molecule, B is given by the expression

$$B = \left\langle \frac{\hbar^2}{2\mu R^2} \right\rangle = \frac{\hbar^2}{2\mu} \left\langle \frac{1}{R^2} \right\rangle$$

where μ is the reduced mass, R is the bond length and the angled brackets show that an average value (expectation value) must be calculated. Back calculation yields the expectation value of $1/R^2$, but inverting this and taking the square root does NOT give the expectation value of R, even for a harmonic oscillator! For linear molecules with more than one atom, isotopic substitution can be used to find different B values, and hence the expectation values of $1/R^2$ for several different bonds can be found simultaneously.

Symmetric top molecules

We consider molecules to be symmetric tops if two of their moments of inertia are the same. It is useful to consider two cases separately, the 'prolate' and 'oblate' symmetric tops. A diagram of the lowest energy levels of both prolate and oblate symmetric tops is given in **Figure 2B**. The selection rules for a symmetric top are the same as for a linear molecule: $\Delta J = 0, \pm 1$ and $\Delta J = 0$ cannot occur from $J = 0$, $\Delta K = 0$, $\Delta M = 0, \pm 1$ and the dipole change occurs along the symmetry axis.

Asymmetric top molecules

The energy levels of an asymmetric top molecule cannot in general be expressed in a simple algebraic form. Those levels for which such simple solutions exist are given in **Table 4**. The complication arises because the functions $|JKM\rangle$ which we are using as our basis are NOT eigenfunctions for the asymmetric top. However, many asymmetric rotors are nearly a prolate or oblate top, and the perturbations to the energy levels in such cases may be quite small. **Figure 3** shows a schematic correlation diagram between the energy levels in each case. The levels of the asymmetric rotor are usefully LABELLED by the number $\tau = K_A - K_C$ which is NOT a quantum number, but has the useful property that its value increases with increasing energy of the levels. Note that levels of a given J do not cross.

The general problem of finding the allowed energies and transition intensities of the asymmetric rotor

Table 4 Energies of the rigid asymmetric rotor. Analytic expressions for the total rotational energy in terms of the rotational constants A, B and C. As the rotational transitions involving low J can often be observed, these relationships can be of considerable use in making a preliminary determination of molecular constants

$J_{K_{prolate}, K_{oblate}}$	$E(A, B, C)$
$0_{0,0}$	0
$1_{1,0}$	$A+B$
$1_{1,1}$	$A+C$
$1_{0,1}$	$B+C$
$2_{2,0}$	$2A+2B+2C+2[(B-C)^2+(A-C)(A-B)]^{1/2}$
$2_{2,1}$	$4A+B+C$
$2_{1,1}$	$A+4B+C$
$2_{1,2}$	$A+B+4C$
$2_{0,2}$	$2A+2B+2C-2[(B-C)^2+(A-C)(A-B)]^{1/2}$
$3_{3,0}$	$5A+5B+2C+2[4(A-B)^2+(A-C)(B-C)]^{1/2}$
$3_{3,1}$	$5A+2B+5C+2[4(A-C)^2-(A-B)(B-C)]^{1/2}$
$3_{2,1}$	$2A+5B+5C+2[4(B-C)^2+(A-B)(A-C)]^{1/2}$
$3_{2,2}$	$4(A+B+C)$
$3_{1,2}$	$5A+5B+2C-2[4(A-B)^2+(A-C)(B-C)]^{1/2}$
$3_{1,3}$	$5A+2B+5C-2[4(A-C)^2-(A-B)(B-C)]^{1/2}$
$3_{0,3}$	$2A+5B+5C-2[4(B-C)^2+(A-B)(A-C)]^{1/2}$
$4_{4,0}$	no analytic solution
$4_{4,1}$	$10A+5B+5C+2[4(B-C)^2+9(A-C)(A-B)]^{1/2}$
$4_{3,1}$	$5A+10B+5C+2[4(A-C)^2-9(A-B)(B-C)]^{1/2}$
$4_{3,2}$	$5A+5B+10C+2[4(A-B)^2+9(A-C)(B-C)]^{1/2}$
$4_{2,2}$	no analytic solution
$4_{2,3}$	$10A+5B+5C-2[4(B-C)^2+9(A-C)(A-B)]^{1/2}$
$4_{1,3}$	$5A+10B+5C-2[4(A-C)^2-9(A-B)(B-C)]^{1/2}$
$4_{1,4}$	$5A+5B+10C-2[4(A-B)^2+9(A-C)(B-C)]^{1/2}$
$4_{0,4}$	no analytic solution
$5_{4,2}$	$10A+10B+10C+6[(B-C)^2+(A-B)(A-C)]^{1/2}$
$5_{2,4}$	$10A+10B+10C+6[(B-C)^2+(A-B)(A-C)]^{1/2}$

is solved by setting up the Hamiltonian matrix and the dipole transition matrix (the 'D' matrix) using the $|JKM\rangle$ functions as a basis. The Hamiltonian matrix is not now diagonal but by diagonalizing the Hamiltonian matrix we find the eigenvalues and eigenvectors of the Hamiltonian that we have used. The eigenvalues can be used to extract the relative intensities of all allowed transitions from the 'D' matrix. The magic of this approach is that we *do not need to know the eigenfunctions themselves*. The diagonalization provides the mixture coefficients for the basis that we have chosen (a_K^J) without us having to specify the basis functions apart from their

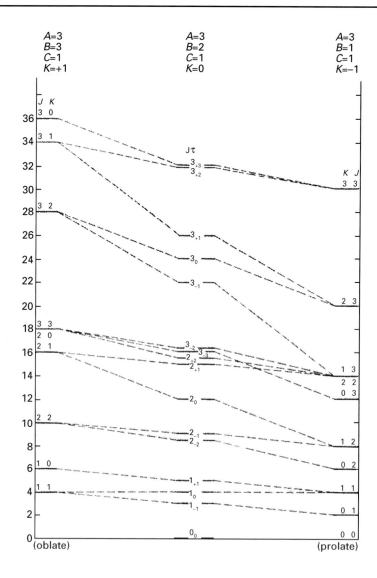

Figure 3 Correlation between the energy levels of prolate, oblate and asymmetric tops. Reproduced from Figure 2.11 of Gordy W, Smith WV and Trambarulo RF (1953) *Microwave Spectroscopy*, p. 110. New York: Wiley.

properties under various quantum mechanical operators, i.e. their matrix elements. The properties of the basis functions are given in **Table 1** and in spherical tensor form in **Table 2**. For more general problems where the molecule (and therefore the Hamiltonian) includes electron or nuclear spins, it is not possible to write an eigenfunction which describes the property in three-dimensional space, but it is still possible to set up and diagonalize the Hamiltonian and D matrices.

Vibration–rotation interaction

In some sense we can understand a spectrum if we can assign values of A, B, C which generate the spectral line positions using the selection rules. The best test of this fit is to examine if the tested and predicted transitions agree with one another. A spectrum has

not been correctly fitted unless the constants which have been determined can reproduce the experimental transition frequencies to within the experimental accuracy. From our initial study, we know that B is a measure of $\langle 1/R^2 \rangle$, averaged over the particular electronic and vibrational state involved. In order to fit the experimental data accurately it is usually necessary to consider a more general energy level expression. This is achieved by adding terms to the Hamiltonian which allow for the interaction of vibration and rotation. When a molecule rotates, the centrifugal forces distort the molecule and this changes the moments of inertia. In a linear molecule the effect can be seen as a slight decrease in the spacing between successive rotational lines as J increases. The effect for low J lines is usually less than 1 part in 10^4, but such changes are observable.

In linear molecules we account for the interaction by changing the rotational Hamiltonian to

$$H_{\text{rotation}} = B\boldsymbol{J}^2 - D\boldsymbol{J}^4 + H\boldsymbol{J}^6 + K\boldsymbol{J}^8 + L\boldsymbol{J}^{10} + \cdots$$

giving rise to the new eigenvalue expressions:

$$E_{\text{rotation}} = B[J(J+1)] - D[J(J+1)]^2 + H[J(J+1)]^3 \\ + K[J(J+1)]^4 + L[J(J+1)]^5 + \cdots$$

As many terms as necessary are used in order to fit the spectrum. Unfortunately such power series expressions are heavily correlated—this manifests itself when say, B and D are determined if a further power is added, the values of B, D and H are now determined, but the values of B and D are not the same as before.

For nonlinear molecules, the Hamiltonian including the effects of vibration–rotation interaction is more complicated, but well understood.

Molecules with electronic angular momentum

In this section we shall only consider the additional complications caused in diatomic molecules but similar considerations apply to more complicated molecules. So far we have considered molecules which have no angular momentum due to their electrons: that is no electronic angular momentum. The addition of spin angular momentum into the problem changes the energy level expressions and causes additional splittings of lines; the ground electronic state of O_2 is $^3\Sigma$ and therefore has two unpaired electron spins yielding a total spin of 1. Each rotational level with $J>1$ is now split into three and each line in the spectrum involving $J>1$ is split into six components, three of which are intense (hence the name 'triplet').

Orbital electronic angular momentum gives rise to non sigma electronic states, such as $^2\Pi$ states which have an orbital angular momentum of 1 and a spin angular momentum of $\frac{1}{2}$. NO is a molecule with a $^2\Pi$ ground electronic state and a simulated rotational spectrum of NO is shown in **Figure 4**.

One now has a choice of how to construct a basis set in which to set up the Hamiltonian, and the best basis is the one which most nearly diagonalizes the Hamiltonian matrix. However, nature does not care which basis we choose and a full calculation in any basis yields the same results; which basis is chosen is purely a matter of convenience. In diatomic molecules the choices of basis for the inclusion of spin most normally used were first considered by Hund, and are called Hund's cases a and b (see **Figure 5** for explanation).

We shall only consider Hund's case a. We are creating a basis function by coupling together the angular momenta in a particular order: L and S (the electronic and spin angular momentum of the electrons) are strongly coupled to the internuclear axis, and have projections Λ and Σ on this axis. The combination of $\Lambda + \Sigma = \Omega$. The angular momentum along the internuclear axis (r) is Ωr which is coupled with the angular momentum of the rotating nuclear

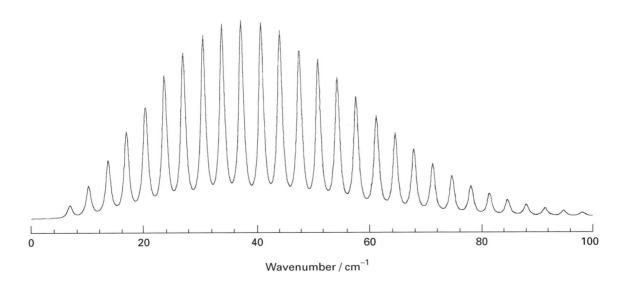

Figure 4 Simulated rotational spectrum of the NO ground state (X$^2\Pi$) – note that the first rotational line is 'missing' due to the electronic angular momentum $\Lambda = 1$ (Π-state); such missing lines aid the assignment of electronic states.

framework (R) to give the resulting total angular momentum (excluding nuclear spins) of J. In the case of the diatomic molecule, the projection quantum number K is called Ω in Hund's *case a* and the eigenfunctions are labelled as $|J\Omega M\rangle$ rather than $|JKM\rangle$. The full eigenfunction includes the dependence on electron spin and can be written as the simple product function $|S\Sigma\rangle|J\Omega M\rangle = |S\Sigma; J\Omega M\rangle$. The fact that we can simply multiply the two functions together in this way (one for electron spin, one for resultant angular momentum) makes calculation of some matrix elements extremely simple. The total eigenfunction, including the electronic and vibrational state involved is written $|v, \Lambda, S\Sigma; JKM\rangle$.

Extra terms in the Hamiltonian

The inclusion of electron spin into the problem increases the complexity of the Hamiltonian which must be considered. In general, we must now add the following terms to the rotational Hamiltonian, some of which may be zero in a particular problem

$$\mathrm{H}_{\text{fine structure}} = \mathrm{H}_{\text{spin-spin}} + \mathrm{H}_{\text{spin-orbit}} + \mathrm{H}_{\text{spin-rotation}}$$
$$+ \mathrm{H}_{\text{lambda doubling}}$$

Hyperfine interactions

If the nuclei in a molecule have angular momenta $I_n > 1$ then complications in the spectra may arise due to the electric quadrupole field of the nuclei. Such additional structure is commonly observed and arises because there is an interaction between the electric field gradient along the internuclear axis (due to the electrons), the electric quadrupole moment of the nucleus, and the angular momentum of the molecule. The energy change due to the quadrupole moment can be written:

$$E_{\text{quadrupole}} = -eqQf(I, J, F)$$

where $-eqQ$ is the nuclear quadrupole moment multiplied by the field gradient and $f(I, J, F)$ is an angular momentum coupling coefficient called Casimir's function. The effect of the quadrupole interaction on a spectral line is shown in **Figure 6**.

If a molecule has nonzero spin angular momentum, then magnetic hyperfine interactions may be observed between any nuclei with nonzero spin and between such nuclei and the electron spins. Such interactions cause the zero-field splittings in NMR.

The additional terms which must be considered due to magnetic hyperfine interactions are as follows:

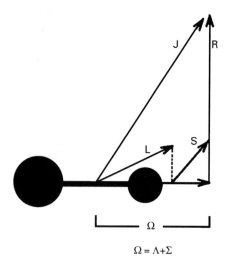

$$\Omega = \Lambda + \Sigma$$

Vector coupling diagram for Hund's case (a).

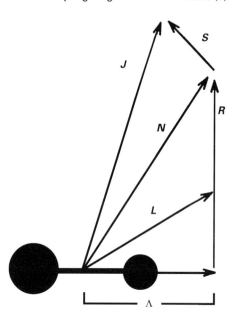

Vector coupling diagram for Hund's case (b).

Figure 5 Vector diagrams for Hund's coupling cases *a* and *b*. Reproduced from Figures iii.17 and iii.18 of Carrington A (1973) *Microwave Spectroscopy of Free Radicals*, pp 129–130. New York: Academic.

$$\mathrm{H}_{\text{hyperfine}} = \mathrm{H}_{\text{fermi-contact}} + \mathrm{H}_{\text{spin-dipole, nuclear dipole}}$$
$$+ \mathrm{H}_{\text{nuclear dipole-nuclear dipole}}$$

there are other possible terms, but they have not commonly been necessary in interpreting the spectra which have been observed. Typically, hyperfine interactions have magnitudes which lead to splittings in spectra between 1–1000 MHz.

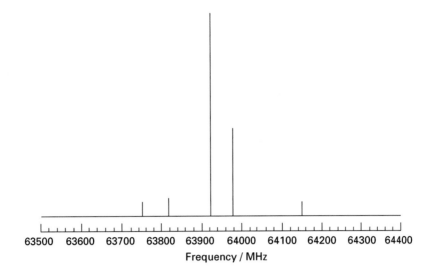

Figure 6 Additional structure in a single rotational line ($J = 3$–2) of BrF due to the quadrupolar constant $eqQ = 909.2$ MHz of the bromine nucleus ($I = 3/2$).

Frequencies and Intensities of transitions

To calculate the relative intensity of a particular transition we use the interaction Hamiltonian between the dipole moment of the molecule and the electric vector of the radiation: $H_{\text{interaction}} = -\boldsymbol{E}.\boldsymbol{\mu}$. We need to calculate the matrix elements of this interaction using the eigenfunctions for the states of interest. In the simplest cases we have considered our eigenfunctions are the functions $|JKM\rangle$ and we need to calculate the matrix elements $\langle \nu, \Lambda, S\Sigma; JKM| - \boldsymbol{E}.\boldsymbol{\mu} |\nu, \Lambda, S'\Sigma'; J'K'M'\rangle$; the intensity of the transition is proportional to the square of this matrix element. The procedure is to expand the scalar product in space fixed axes using Equation [3] – both $\boldsymbol{E}$ and $\boldsymbol{\mu}$ are 1st rank tensors (vectors):

$$\boldsymbol{E}.\boldsymbol{\mu} = T^1(\boldsymbol{E}).T^1(\boldsymbol{\mu}) = \sum_p (-1)^p T_p^1(\boldsymbol{E}) T_{-p}^1(\boldsymbol{\mu})$$

and we insert this into the matrix element

$$\langle \nu, \Lambda, S\Sigma; JKM| - \sum_p (-1)^p T_p^1(\boldsymbol{E}) T_{-p}^1(\boldsymbol{\mu})|\nu,$$
$$\Lambda, S'\Sigma'; J'K'M'\rangle$$

as the electric field does not operate on the basis states, we can take the electric field outside the matrix element:

$$-\sum_p (-1)^p T_p^1(\boldsymbol{E}) \langle \nu, \Lambda, S\Sigma; JKM|T_{-p}^1(\boldsymbol{\mu})|\nu,$$
$$\Lambda, S'\Sigma'; J'K'M'\rangle$$

To evaluate the matrix element we need to transform the tensor for the dipole moment into molecule fixed coordinates, which we can do using Equation [13]

$$T_{-p}^1(\boldsymbol{\mu}) = \sum_q D_{-pq}^1(\omega)^* \, T_q^1(\boldsymbol{\mu})$$

which gives us:

$$-\sum_{p,q} (-1)^p T_p^1(\boldsymbol{E}) \langle \nu, \Lambda, S\Sigma; JKM|D_{-p,q}^1(\omega)^* \, T_q^1(\boldsymbol{\mu})|\nu,$$
$$\Lambda, S'\Sigma'; J'K'M'\rangle$$

Now $T_q^1(\boldsymbol{\mu})$ only operates on ν, Λ and $D_{-p,q}^1(\omega)^*$ only operates on $|JKM\rangle$, and neither operator effect $|S\Sigma\rangle$ so we can separate this out to:

$$-\sum_{p,q} (-1)^p T_p^1(\boldsymbol{E}) \langle \nu, \Lambda|T_q^1(\boldsymbol{\mu})|\nu, \Lambda\rangle$$
$$\times \langle JKM|D_{-p,q}^1(\omega)^*|J'K'M'\rangle \langle S\Sigma|S'\Sigma'\rangle$$

$T_p^1 (E)$ is a constant number which tells us the intensity of the light and (through p) its polarization; $\langle S\Sigma | S'\Sigma' \rangle$ yields the condition that $S = S'$ and $\Sigma = \Sigma'$ if the expression is nonzero; $\langle \nu, \Lambda | T_q^1 (\mu) | \nu, \Lambda \rangle$ is constant for a particular state, and so (finally) it is the matrix elements $\langle JKM | D_{-p,q}^1(\omega)^* | J' K' M' \rangle$ (which are the elements of the 'D-matrix') which determine the relative intensities of the transitions and the selection rules. Using Equation [16] (**Table 2**) we evaluate the D-matrix elements to:

$$(-1)^{K-M}[(2J + 1)(2J' + 1)]^{1/2}$$
$$\times \begin{pmatrix} J & 1 & J' \\ -K & q & K' \end{pmatrix}$$
$$\times \begin{pmatrix} J & 1 & J' \\ -M & -p & M' \end{pmatrix}$$

The selection rules arise from the properties of the 3-j symbols which yield $\Delta J = 0, \pm 1$. The selection rule on K depends upon which components of the dipole moment are nonzero but, as $q = -1, 0, 1$ only, then $\Delta K = 0, \pm 1$. The selection rule on M depends on the polarization of the light. For z polarized light $p = 0$ and $\Delta M = 0$. For nonpolarized light $\Delta M = 0, \pm 1$.

In more complicated cases, where the eigenfunctions are linear combinations of the basis functions, the relative intensities are found by multiplying the D matrix by the eigenfunction coefficients.

The frequencies of transitions are found by setting up the requisite Hamiltonian matrix in the same basis and diagonalizing it, evaluation of the matrix elements of the Hamiltonian matrix is accomplished in the same fashion as for the example above. The eigenvectors and eigenvalues found together with the D-matrix enable a complete calculation of the spectrum to be made.

Summary

We have considered the theory necessary to understand rotational spectroscopy, concentrating on some simple cases. In a first approximation, spectra can be understood on the basis of rigid rotor models. Allowing for the fact that molecules are not rigid complicates spectra and introduces new molecular constants. We have seen that when electronic spin and orbital angular momenta are included in the
picture we need to extend our basis sets, and the utility of spherical tensors has been illustrated in calculating expressions which yield the relative intensities of rotational transitions.

List of symbols

a_K^J = mixture coefficients; B = rotational constant; D = dipole transition matrix; $-eqQ$ = nuclear quadrupole moment; E = energy; E = electric vector; $f(I,J,F)$ = angular momentum coupling coefficient (Casimir's function); F = total angular momentum quantum number; g = degeneracy; H = Hamiltonian; I = moment of inertia; I = intensity; I_0 = incident intensity; J = rotational quantum number; L, S = electronic and spin angular momenta; M, K = projection quantum numbers; R = bond length; $S(\omega, \omega_0)$ = normalized line shape function; $\gamma(\omega)$ = absorption coefficient; μ = reduced mass; ψ = wavefunction.

See also: **IR Spectroscopy, Theory; Laser Applications in Electronic Spectroscopy; Laser Spectroscopy Theory; Spectroscopy of Ions; Symmetry in Spectroscopy, Effects of.**

Further reading

Bernath PF (1995) *Spectra of Atoms and Molecules*, Chapter 6. New York: Oxford University Press.

Bunker PR (1979) *Molecular Symmetry and Spectroscopy*. London: Academic Press.

Dixon RN (1965) *Spectroscopy and Strucure*, Chapter 3. London: Methuen and Co.

Edmonds AR (1985) *Angular Momentum in Quantum Mechanics*, 2nd edn. Princeton: Princeton University Press. (contains tables of 3-j and 6-j coefficients).

Gordy W and Cook RL (1984), *Microwave Molecular Spectra; Techniques of Chemistry Volume XVIII* 3rd edn. New York: Wiley Interscience.

Herzberg G (1991) *Molecular Spectra and Molecular Structure; Volume I: Spectra of Diatomic Molecules*. Florida: Krieger Publishing Company.

Herzberg G (1991) *Molecular Spectra and Molecular Structure; Volume II: Infrared and Raman Spectra of Polyatomic Molecules*. Florida: Krieger Publishing Company.

Hirota E (1985) *High-Resolution Spectroscopy of Transient Molecules*. Berlin: Springer-Verlag.

Kroto HW (1991) *Molecular Rotation Spectra*. Toronto: Dover.

Townes CH and Schawlow AL (1975) *Microwave Spectroscopy*. New York: Dover.

Rubidium NMR Spectroscopy

See **NMR Spectroscopy of Alkali Metal Nuclei in Solution.**

Ruthenium NMR, Applications

See **Heteronuclear NMR Applications (Y–Cd).**

S

^{29}Si NMR

Heinrich C Marsmann, Universität Paderborn, Germany

MAGNETIC RESONANCE
Applications

Introduction

All group 14 elements – except germanium – share, from the viewpoint of the NMR spectroscopist, one rather important feature: they all have at least one isotope with a spin of $\frac{1}{2}$ that is a minor component of the isotopes of the element. The most important element in this group is, of course, carbon and most people are familiar with obtaining and interpreting 13C NMR spectra. Consequently a comparison with the situation of carbon might be helpful (**Table 1**).

An inspection of **Table 1** shows that ^{29}Si has a higher share of the isotopic mixture but the absolute value of the magnetic moment is slightly lower than that of ^{13}C. This leads to a lower resonance frequency. A complication arises from the fact that the spin and magnetic moment are antiparallel, leading to negative sign of γ.

The chemistry of silicon is very different to that of carbon and in organic compounds there is usually a lower silicon content than carbon. These shortcomings meant that silicon NMR spectroscopy had a slow start. After the first report by Lauterbur and co-workers in 1962 there had been only a few papers per year. But this has now changed dramatically and one collection of ^{29}Si chemical shifts now contains about 9000 data sets. One area of special growth is ^{29}Si NMR measurement on solids. Here applications range from crystalline silicates over amorphous silica gels to silicon-containing surfaces and coatings.

Measuring ^{29}Si NMR

Referencing

The established standard compound to calibrate ^{1}H and ^{13}C spectra is tetramethylsilane $(CH_3)_4Si$ (TMS). As this substance contains silicon also, it is natural to use it as standard for Si NMR. Although TMS is an inert substance, has a low boiling point and a rather short relaxation time, its chemical shift is in the middle of the shift range of other organosilicon compounds and peak misassignments are possible. Two strategies to circumvent this problem have been developed. The first is the use of secondary standards. **Table 2** lists a number of those used in the literature.

M_8Q_8, shown in **Table 2**, has a lower symmetry in the solid state and its resonances are split into several lines, but it is the common reference compound used as an internal standard for solid state Si NMR. The second tactic is to use no standard compound in the sample at all. The referencing is done externally relative to a separate sample that contains TMS in the same solvent as is used in the analysis sample.

Early reports of silicon NMR data employ the magnetic field definition of chemical shifts instead of

Table 1 Comparison of carbon and silicon isotopes.

	Natural abundance (%)	Spin	γ^a	v^b (MHz)	Other isotopes
^{13}C	1.108	1/2	6.7263	25.145004	^{12}C (98.89%)
^{29}Si	4.70	1/2	−5.3141	19.867184	^{28}Si (92.21%)
					^{30}Si (3.09%)

^{a} Gyromagnetic ratio.
^{b} In a magnetic field, where the nuclei of TMS resonate at 100 MHz.

Table 2 Common reference compounds for silicon NMR.

Formula	Abbreviation	Shift relative to TMS (ppm)		
$(CH_3)_4Si$	TMS	0.00		
$[(CH_3)_3Si]_4C$		3.6		
$[(CH_3)_3Si]_2$	HMDS	−19.79		
$[(CH_3)_3Si]_2O$	M_2	7.22		
$[(CH_3)_2SiO]_4$	D_4	−19.86		
$(CH_3O)_4Si$	TMOS	−78.22		
$(C_2H_5O)_4Si$	TEOS	−81.65		
$[(CH_3)_3SiO]_4Si$	M_4Q	8.62, −104.08		
$[(CH_3)_3SiOSiO_{3/2}]_8$	M_8Q_8	12.4,[a] − 108.6[a]		
		11.77,[b]	11.72,[b]	11.51[b]
		−108.36,[b]	−108.64,[b]	
		−109.36,[b]	−109.71[b]	

[a] In solution.
[b] In the solid state.

the now universally accepted frequency based one, resulting in a reversed sign for chemical shift data.

Pulse sequences

In principle there are two cases to distinguish: many, especially inorganic, compounds contain only ^{29}Si as a useful magnetic nucleus. Here, single pulse experiments are applicable only. Because of the rather slow relaxation in such compounds, a 30° pulse is used with a repetition rate of about 20 s.

On the other hand, silicon compounds with organic side-groups usually have resonances split into many lines by spin–spin couplings with the protons. These are normally removed by decoupling. Because of the negative gyromagnetic ratio of ^{29}Si the nuclear Overhauser effect (NOE) is also negative, leading to negative signals with an enhancement factor of $\eta_0 = -2.52$ if the relaxation is dominated by dipolar interactions with the protons. With few exceptions, the relaxation path is divided between the dipolar term and others – mostly the spin–rotation interaction. This results in much lower enhancement factors and even null signals. This can be overcome by:

1. Doping of the sample with a shiftless relaxation agent such as $Cr(acac)_3$ at a concentration of ~10^{-2} mol dm^{-3}. However, such a compound might react with the sample and it is also difficult to remove.
2. Inverse gated decoupling. Here the proton decoupler is switched off during a recovery time (three to five times the silicon relaxation time). The advantage of not contaminating the sample is offset by an ineffective use of spectrometer time.
3. The use of pulse programmes such as insensitive nuclei enhanced by polarization transfer (INEPT) or distortionless enhancements by polarization

transfer (DEPT). There are two advantages in using these programmes. The first is a signal enhancement by population transfer from the protons that depends on the number of coupling protons. A good review has been given by Blinka *et al*. The second advantage is the removal of the background signal caused by the silicon content of the probe. However, these techniques depend on the proton–silicon coupling constant and as these can cover a considerable range in organosilicon compounds this can lead to unexpected results. An example is shown in **Figure 1** where a ^{29}Si INEPT spectrum of an octasilsesquioxane is displayed. Because the siloxane skeleton consists of a cube with silicon on the vertices and oxygen on the edges of the cube, the cage can only rotate as a whole. This way the silicon relaxes almost solely by the spin–dipole path. The relaxation times are rather long and the signals very narrow.

The NMR of solids differs from that of liquids insofar as spatial interactions are not averaged out by Brownian motion. The NMR resonances are therefore so broad that chemical shifts and indirect spin–spin couplings are not resolved. There are different ways of handling solid state NMR problems. Here is not the place to cover all of them. The most common presently used method starts with a powder of crystalline or amorphous silicon compound. In the absence of nuclei with a strong magnetic moment such as ^{1}H or ^{19}F or ions with unpaired electrons, the source of broadening of the signals comes from the fact that the chemical shift is a tensor. This chemical shift anisotropy (CSA) can be removed by fast spinning (3000–15 000 Hz) around an angle of 54°44′ (magic-angle spinning, MAS). The CSA is then reduced to a series of sidebands spaced at the rotational frequency and with an intensity profile governed by the CSA. By rotating the sample fast enough, these sidebands can be moved out of the region of the chemical shifts of the sample. On the other hand, by rotating slowly, the CSA can be determined. The *direct* spin–spin coupling between ^{29}Si nuclei is small because the magnetic moment is small and the average distance between ^{29}Si nuclei is large because the abundance of ^{29}Si is just 4.7%. For most crystalline silicates, single resonance NMR spectroscopy under MAS is the best choice. Depending on the quality of the crystals, narrow peaks with line widths between 0.1 to 3 ppm can be obtained. In addition to single pulse spectroscopy, incredible natural abundance double quantum transfer experiment (INADEQUATE) or 2D correlation spectroscopy (2D-COSY) can be carried out to determine the connectivity of the silicon atoms.

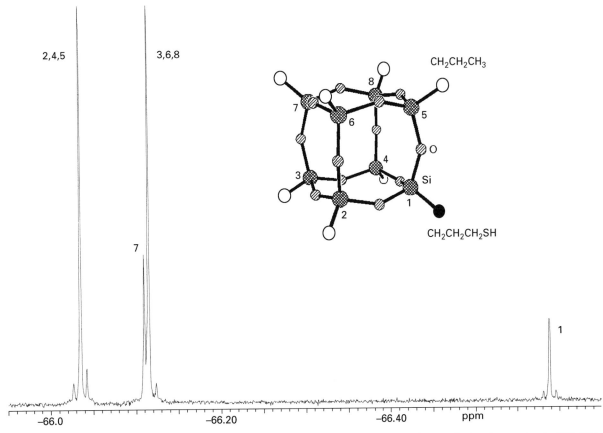

Figure 1 ²⁹Si NMR spectrum of (3-mercaptopropyl)hepta(propyl)octasilsesquioxane. The small signals are due to ²⁹Si–²⁹Si coupling.

Organosilicon compounds contain hydrogen and the strong magnetic moment of ¹H leads to very broad lines in the solid state by *direct* proton–silicon spin–spin interaction. This coupling is usually removed by high power decoupling of the protons. The spin coupling with protons offers the opportunity to obtain stronger signals by the cross-polarization (CP) technique. The connection between the proton and silicon spins is possible if both spins are rotating with the same speed around the magnetic field, i.e. $\gamma_H B_{1H} = \gamma_{Si} B_{1Si}$ (spin locking, Hartmann–Hahn condition). The most effective time for this spin locking is determined by the proton–silicon distance. Because of the higher incidence of protons the ¹H relaxation time is usually shorter, so that beside the inherently stronger signals the pulse repetition rate can be faster.

²⁹Si chemical shifts

General features

In principle, two kinds of silicon compounds are possible. The first kind, derived from divalent silicon [Si(II)] are normally thermodynamically unstable.

Some exist at high temperatures or may be trapped using cryogenic temperatures. However, a few are stable enough to be handled at room temperature, but the small number of examples does not allow a general definition of a specific range of chemical shifts. Nevertheless, derivatives with a high coordination number are strongly shielded. Thus, for instance, decamethylsilicocene, $[(CH_3C)_5]_2Si$, has a shift of −398 ppm, the highest shielding measured so far. Another situation where Si(II) seems to be stable is between two nitrogen atoms, as in [1]. The silicon is then rather deshielded with values of the chemical shift between +78.3 and +96.9 ppm.

The majority of the ^{29}Si NMR data reported involve derivatives of Si(IV). The most deshielded silicon (+268.7 ppm) has been reported for [2], the most shielded for SiI_4 (−351.7 ppm), but the majority of shifts are found between −200 and 150 ppm. As with the other heavier nuclei, the chemical shift depends primarily on the coordination number of the silicon in such a way that a low number of substituents around the silicon leads to deshielding (e.g. disilylenes: 13 to 175 ppm) and a high coordination number gives high negative numbers (e.g. sixfold coordination: −198 to −135 ppm). The regions of chemical shifts for some important classes of silicon compounds are depicted in **Figure 2**.

There have been several proposals to rationalize silicon chemical shifts. Most of them centre on the charge on the silicon atom. Ernst and co-workers observed that the chemical shifts follow a parabolic curve if plotted against the sum of the electronegativities of the substituents of the silicon. Thus, above the sum of electronegativities of 9.5, a further increase of the electronegativity leads to an increase of the shielding. However, a decrease of shielding is observed if this sum is below 9.5.

Radeglia and co-workers consider that not only the charge on the silicon determines the shielding but also its distribution. This is due to the paramagnetic contribution to the shielding constant. They define a reduced shielding constant σ^* on the basis of the valence shell p orbitals of the silicon:

$$\sigma^* = \frac{\sigma_p}{\sigma_{p_0}} = \frac{P_u \langle r^{-3} \rangle_P}{P_u \langle r^{-3} \rangle_{P_0}} = P_u^* R^* \qquad [1]$$

where the index 0 pertains to a situation in which all electrons are distributed evenly around the silicon. The terms P_u^* and R^* are then calculated to various degrees of sophistication. One approach uses Slater-type orbitals and Slater rules; R^* is then given by:

$$R^* = \frac{\langle r^{-3} \rangle_p}{\langle r^{-3} \rangle_{p_0}} = \left(1 + 0.35 \frac{f q_{Si}}{Z_0(Si)} \right)^3 \qquad [2]$$

with $Z_0(Si) = 4.15$, $f = 0.2135$ as an empirical factor and q_{Si} stands for the charge on the silicon atom:

$$q_{Si} = 4 - \sum_{i=A}^{D} h_i \qquad [3]$$

The bond polarity h_i, calculated from their electronegativities, is summed up over all four substituents (A to D) on the silicon. The term P_u takes elements of the bond order matrix around the silicon atom into account, using the electronegativity as a measure of electron distribution:

$$P_u^* = \frac{P_u}{P_{u_0}} = \frac{1}{2} \sum_i h_i - \frac{1}{6} \sum_{i>j} \sum h_i h_j \qquad [4]$$

The σ^* values so calculated do not give chemical shifts directly but have to be fitted empirically. Nevertheless, it is remarkable that the shape of the curve of ^{29}Si chemical shifts as a function of the number of substituents (**Figure 3**) is reproduced by such calculations. The literature and a discussion of this procedure and other aspects of silicon chemical shift interpretation has been given by Marsmann (see Further reading section). Empirical aspects of chemical shifts of different classes of some silicon compounds only are given briefly below.

Organosilanes

Silicon surrounded by four carbon atoms gives resonances between ~+100 and −107 ppm. As expected, shifts of silicon with aliphatic substituents are clustered around 0 ppm. Increased shielding is found if the ligands around the silicon atom have π bonds in the β position and thus the centre of shifts is moved

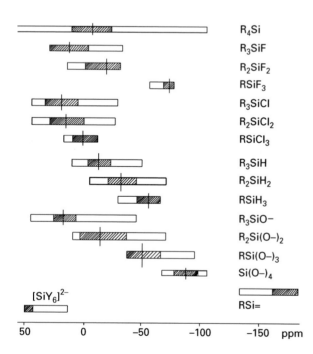

Figure 2 Ranges of chemical shift of selected classes of silicon compounds. (R is an arbitrary organic ligand, Y = F or OR). See also **Figures 5, 6** and **8**.

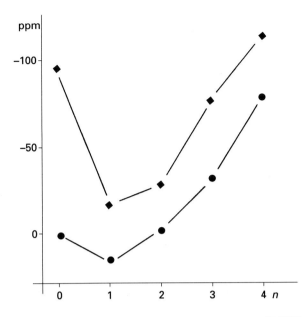

Figure 3 The effect of successive substitution on ²⁹Si NMR chemical shifts. (●) $(CH_3)_{4-n}Si(OCH_3)_n$; (◆) $H_{4-n}SiF_n$.

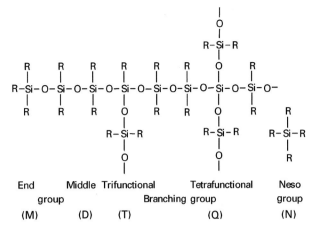

Figure 4 Building units of a hypothetical polysiloxane.

to −35 ppm if silicon is connected to four sp²-hybridized carbons. Very high shielding is observed for silicon in silacyclopropanes (~−60 to −40 ppm) and silacyclopropenes [3] (−87 to −106 ppm). The influence of π orbitals is discussed in the later case. The strong deshielding for the bridge silicon in silanorbornadienes [4] (76–98 ppm) is also attributed to a σ–π interaction.

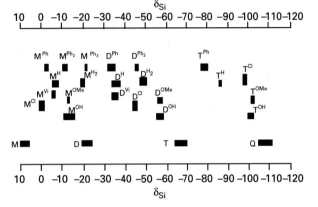

Figure 5 Shift ranges of building units in polysiloxanes. Reproduced with permission of John Wiley & Sons from Williams EA (1989) NMR spectroscopy of organosilicon compounds. In: Patai S and Rappoport Z (eds) *The Chemistry of Organic Silicon Compounds*. New York: John Wiley & Sons.

[3]   [4]

Siloxanes and silicates

One of the main interests in silicon NMR is because it allows distinction between the building units of a polysiloxane. A sketch of an imaginary polysiloxane to make the concept of building units and their nomenclature clearer is given in **Figure 4**. If the substituent R is not a methyl group, appropriate notation for that substituent is added as a superscript. A collection of chemical shift ranges of such building units has been published by Williams and is shown here as **Figure 5**.

Ring strain has a marked effect on the shifts. This can be seen for example, with dimethylsiloxanes in (**Table 3**). In polysiloxanes with different middle

groups, it is possible to distinguish the triad and even the pentad structures in the backbone of the polymer.

Approximately the same behaviour, but with a smaller range, is observed if R = OH or O⁻ in silicates. However, as they are now all Q units, they are distinguished by a superscript n ($n = 0$–4), indicating the number of non-connecting oxygen atoms around the silicon. For instance, Q^2 designates a middle group with two connecting and two non-connecting

Table 3 ²⁹Si chemical shifts of dimethylsiloxanes.

n in $[(CH_3)_2SiO]_n$	δ (ppm relative to TMS)
3	−9.12
4	−19.51
5	−21.93
6	−22.48

oxygens. In aqueous solution there are fast equilibria between ionized and unionized forms. This causes a dependency of the chemical shifts of the building units on the pH. As the acidity of the building units differs and the existence of individual silicate ions is pH-dependent, this makes the analysis of such aqueous silicate solutions very difficult and some resonances are still not assigned. To make things a little more manageable, the signal of the monosilicate $Si(OH)_4$ or its corresponding ions are taken as a secondary shift standard. A rough conversion into the TMS scale can be obtained by adding −71.3 ppm. The use of highly ²⁹Si-enriched material proved to be very successful and a number of individual ions could be identified by selective decoupling and ²⁹Si, ²⁹Si COSY experiments. It was found that silicate ions in solution consist mostly of small fused rings. A collection of the shift ranges of acyclic silicic acid ions and esters is given in **Table 4**. Ring strain again leads to deshielding, so that, for example, the signal of the middle groups of trimeric rings can be found close to the end groups and that of their trifunctional branching close to the region of middle groups. The use of Si NMR of solid silicates provides valuable information. The isotropic shifts in silicates follow similar rules to aqueous silicates. The low CSA value of <100 ppm makes the MAS spectra of silicates easy to observe. There are, however, some differences. In crystalline silicates, the shifts also depend on the crystallographic symmetry, which can be lower than in the dissolved state. The most famous example is that M_8Q_8, which gives two sharp lines in solution but splits into seven signals in the solid (**Table 2**). The Si–O–X angle dependence of shifts – not observable for solutes – is responsible for the spread of ~30 ppm for the disilicate from ~−70 to −100 ppm compared with a region of about 4 ppm for the end groups in aqueous silicates. It is possible to determine the Si–O–X angles on the basis of chemical shifts. One of the triumphs of ²⁹Si MAS NMR spectroscopy was the determination of the distribution of silicon and aluminium in aluminosilicates, e.g. zeolites. Silicon chemical shifts move ~5 ppm to higher field if an Si–O–Si bond is exchanged for an Si–O–Al, as can be seen from **Figure 6**. Although it is not possible to disprove a statistical distribution of Si and Al in the framework, it seems that the Al sites are highly ordered and that the Loewenstein rule is observed, i.e. that no direct Al–O–Al connections in aluminosilicates are stable. Another field for ²⁹Si MAS NMR is the characterization of glasses, silica-gels and their organically modified surfaces which are used in chromatography or catalysts. Here the silicate framework is amorphous, i.e. although each silicon is surrounded by four oxygen atoms there is no long-range order. Compared with the signals of well-crystallized

Table 4 Shift ranges (ppm) of some classes oligomeric silanes.

X	X_4Si	X_3Si-	$X_2Si<$	$XSi-$	$-Si-$
H	−95.6	−107 to −67	−124 to −79	−161 to −53	−165
CH₃	0.0	−2 to +28	−49 to −20	−88 to −78	−118 to −136
Cl	−18.5	−10 to +20	−10 to +15	−46 to −15	−79 to −81

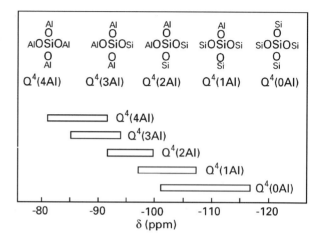

Figure 6 Chemical shifts of the building units in aluminosilicates. Reproduced with permission of John Wiley & Sons from Engelhardt G and Michel D (1987) *High Resolution Solid-State NMR of Solids and Zeolites*. New York: John Wiley & Sons.

silicates this kind of material gives very broad signals with almost no features. In silica-gels and similar materials the non-connecting oxygens are in the form of OH or OR group. This allows CP MAS NMR spectra to be obtained. **Figure 7** shows a CP MAS spectrum of a silica-gel with an organically modified surface.

Trimethylsilyl derivatives

The $(CH_3)_3Si-$ moiety is a very useful fragment in several kinds of chemistry. It is used in organic synthesis because it can direct reactions along a particular path or act as a protecting group. In addition, it can be split off by gentle methods. The exchange of the protons in OH or >NH groups for a $(CH_3)_3Si-$ group increases the volatility and the solubility in organic solvents. This has many applications in analytical chemistry. Schraml showed that the trimethylsiloxy group is useful as a NMR tag for the analysis of natural products such as sugars, lignins, etc. Therefore, silicon chemical shifts for these types of compounds are very common in the literature. About a quarter of all reports of silicon NMR shifts deal with them. The regions of chemical shifts for trimethylsilylated compounds are given in **Figure 8**.

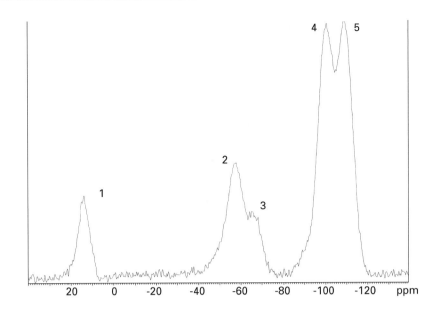

Figure 7 ²⁹Si MAS CP spectrum of a surface modified silica-gel. The signal strength is no indication of the concentration of the corresponding building group. Peak 1: (CH₃SiO–; peaks 2,3: T², T³ of the functional group RSi(O–)₃; peaks 4,5: Q³, Q⁴ of the silica-gel skeleton.

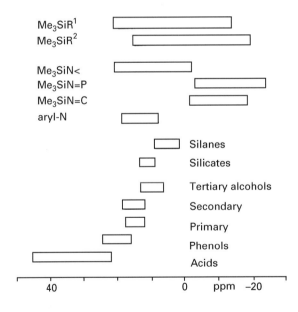

Figure 8 Regions of chemical shifts for trimethylsilylated compounds (R¹ = aliphatic R² = unsaturated organic residue).

The trimethylsilyl group bound to a carbon atom has relatively clear cut regions. If the carbon is an aliphatic one, most of the resonances appear between −1.6 and +3.8 ppm. If the carbon is substituted with electronegative elements, more positive values are possible. A sp²-hybridized carbon results in shifts to higher fields and the resonances are now clustered around −3.8 ppm. The silicon is more shielded if the carbon is a ring member (~−11 ppm).

In (CH₃)₃SiN– derivatives the spread of chemical shifts is small if the nitrogen is aliphatic (−3.3 to 17.3 ppm), while it is more diverse if the nitrogen has a double bond. If the nitrogen is in an aromatic ring system, a compact region with positive values is discernible (3 to 18 ppm). However, no clear pattern emerges for other environments of this type.

A wealth of data exist for the trimethylsiloxy group. The greatest spread of data is found for the trimethylsilyl esters of acids. There is a linear correlation between the silicon chemical shift and the strength of the acid. The overlapping regions for the shift values of alcohols makes further experiments necessary for an assignment. One procedure is the use of hydrogen–silicon couplings to the non-methyl part of the molecule. A review has been given by Schraml.

Polysilanes

An extensive review of this area has been presented by Williams. The structural principles of the polysilanes are the same as those of the alkanes. Thus, the rules applied for alkanes were extended to the silanes. This approach after Paul and Grant, was used by West and Stanislawski to derive Equation [5] for dimethylsilanes.

$$\delta_{Si(k)} = B + \sum_l A_l\, n_{kl} \qquad [5]$$

with A_l the chemical shift parameter for the lth atom from the kth position and n_{kl} the number of atoms present in that position. The term B is constant (8.5 ppm) and A_l is -25.8 ± 0.1 ppm for the α position, $+3.9 \pm 0.05$ ppm for the β position, $+1.2 \pm 0.09$ ppm for the γ position and $+0.2 \pm 0.01$ ppm for the δ position relative to the silicon atom considered. This works well for end, middle and trifunctional branching groups but not so well for tetrafunctional branching and otherwise substituted silanes. Hahn used a similar concept to describe the shifts of silanes:

$$\delta_{Si} = -96.02 - 2.43a - 3.73a^2 + Bb + C \qquad [6]$$

where -96.02 represents the idealized shift of monosilane, a the number of silicon atoms in the α position, b those in the β position, and B and C are the incremental shifts produced by the β and γ silicon atoms.

Silicon with a coordination number of three

A coordination number of three is found for compounds in which the silicon extends a p_π–p_π bond to a partner. The usual condition is that the two other ligands to the silicon are sterically demanding, such as a phenyl group with substituents in the *ortho* position. The assumption of a p_π–p_π bond in disilylenes is supported strongly by the chemical shift anisotropy data obtained from solid-state silicon NMR measurements which are comparable to that of a p_π–p_π carbon bond. Although little data exist for this class of silicon compound it seems that all of them are deshielded compared with TMS. At present, compounds with a Si=C fragment have shifts between 13 and 144 ppm, where a silicon attached to the carbon leads to the higher numbers. Compounds with a Si=Si moiety resonate between 52 and 65 ppm and in a Si=P fragment the silicon resonances are found between 148 and 176 ppm. The spread of shifts is surprisingly low for simple transition metal derivatives of silylenes. Shift values range from 83.6 ppm for $(t\text{-}C_4H_9S)_2Si\,Fe(CO)_4$ to 142.1 ppm for $ArHSi=MnCp(CO)_2$.

The effect of higher coordination numbers

Although the normal coordination number of silicon is four, the atom still acts as a Lewis acid, especially if the silicon is bound to electronegative ligands and hence coordination numbers of five and six are possible. However, because of equilibria it is not always possible to establish the degree of higher coordination in solution. In this respect, couplings to other magnetic nuclei are of help. Thus for silatranes [5] – the largest group of fivefold coordinated silicon compounds measured so far – the additional bond to nitrogen could be assured by the observation of a small ^{29}Si–^{15}N coupling (0.20–3.37 Hz). In addition, the difference between the shifts in solution and in the solid state is small (-1.0 to -5.1 ppm). It is estimated that the additional bond to the nitrogen results in an increased shielding of ~ -10 to -30 ppm. Shifts of silatranes have been observed between -107 and -59 ppm.

[5]

Diols and other oxygen-containing ligands can give rise to sixfold coordination. The mineral thaumasite displays resonances between -182.1 and -177.9 ppm. Complexes with diols have resonance lines between -197 and -135 ppm. Sixfold coordination is discussed here. The effect of putting two additional oxygen atoms in the coordination sphere is ~ -100 ppm upfield shift. The result is less pronounced if five oxygen atoms form the first ligand sphere around the silicon. Resonances are found between -131 to -120 ppm, giving an extra shift of ~ -50 ppm compared with tetraalkoxysilanes.

Fluorine is also capable of expanding the coordination sphere of silicon. Here the coupling between ^{29}Si and ^{19}F can be used to confirm the number of fluorines around the silicon. Fluxional behaviour is often observed. The hexafluorosilicate anion (SiF_6^{2-}) resonates at a very high field (-191.7 ppm). More common are fivefold coordinated compounds measured in solution as well as in the solid state. The extra fluorine results in shifts to low frequencies between -28 and -80 ppm; the larger effect is observed in difluorides.

Donating solvents such as hexamethylphosphoramide or dimethylpropylurea give rise to upfield shifts which have been interpreted as caused by the formation of pentacoordinated complexes. The solvents shifts seem to follow Gutmann's donor number for solvents.

Transition metal derivatives

This has been a fast growing area but, owing to the very diverse situation for a silicon bound to a transition metal ion, only very rough trends can be distinguished so far. The most positive shift values arise from non-classical transition metal complexes such as [2]. The other limit is formed by a number of platinum complexes that display values between −90 to −29 ppm. Within a group of the periodic table there seems to be a trend towards lower frequency on going to the heavier homologues. Some useful observations have been made by Pannell and Bassingdale for more conventional complexes where the silicon is connected to the metal by a single bond. Comparing the shift of the complex to that of a compound where the metal is exchanged by a methyl group, it is found that the silicon shifts are +40 ppm higher in the complex if the metal is Fe or Ru, but −13 to −39 ppm lower for Re.

Coupling constants involving ²⁹Si

All magnetic nuclei within a molecule interact, leading to line splittings or sometimes line broadening. There are two cases to consider:

1. The silicon interacts with an abundant isotope, e.g. ^{1}H, ^{19}F or ^{31}P. Then the silicon resonance is split according to the well-known rules for spin–spin interaction.
2. The silicon couples with another dilute spin such as ^{29}Si itself or, for example, ^{13}C or ^{195}Pt. The coupling then gives rise to doublets to the left and right of a strong centre line in the form of satellites.

For a complete collection of coupling constants, the relevant reviews should be consulted (see Marsmann, and Schraml and Bellama in the Further reading section). Some frequently encountered coupling situations for silicon are discussed below.

Silicon–hydrogen couplings

The coupling with hydrogen is also visible in proton NMR spectrum and this was used even before ^{29}Si NMR spectroscopy was feasible. Because of the many and complex splittings arising from silicon–proton interactions, the coupling is usually removed by decoupling in the spectra of organosilicon compounds. Of the three possible pathways, the Fermi contact term is the most efficient. Therefore, because of the negative gyromagnetic ratio of ^{29}Si, the sign of the coupling over one bond is negative, but absolute values only are given in the following. The magnitude of the coupling then depends on the s orbital density between the hydrogen and the silicon.

Semiempirical and empirical relationships have been developed by several authors, e.g.

$$^1J(\text{Si,H}) = 725\alpha_{\text{Si}}^2 + 15.9 \qquad [7]$$

$$^1J(\text{Si,H}) = 969.7\alpha_{\text{Si}}^2 + 5.9N_{\text{P}} - 41.0 \qquad [8]$$

$$^1J(\text{Si,H}) = 810\alpha_{\text{Si}}^2 + 4.3N_{\text{P}} - 1.40N_{\text{Me}} \qquad [9]$$

for compounds of the type H_nSiR_{4-n} (R = CH_3, C_6H_5; n = 1–3). Here α_{Si}^2 is the square of the s character of the bond used by the silicon, N_{P} is the number of phenyl groups and N_{Me} is the number of methyl groups bonded to the silicon. On a purely empirical basis, there is a rough correlation between the electronegativity of the other bonding partners of the silicon and the magnitude of the coupling constant J in such a way that highly electronegative substituents favour a large value for $J(\text{Si,H})$. The range for the magnitude of coupling constants over one bond, $^1J(\text{Si,H})$, is between 74.8 Hz (H_3SiK) and 371.7 Hz ($HSiF_3$). An exception is the value of 420 Hz for the transition metal complex $H_2Si[Mn(CO)_5]_2 \cdot 4py$.

Of the couplings over two bonds, $^2J(\text{Si,H})$, the most important is for the Si-CH₃ fragment. Values range from 6.0 to 9.5 Hz in most cases. An exception is $(CH_3)_3SiLi$ which has a value for $^2J(\text{Si,H})$ of 2.8 or 3.5 Hz.

The large values of ^2J found if the bonding goes over a transition metal have been interpreted as a sign of an agostic interaction, e.g. constants between 20 and 69 Hz for $^2J(\text{Si,Mn,H})$.

Silicon–fluorine couplings

For one-bond couplings, $^1J(\text{Si,F})$, typical values lie in the 167–488 Hz range, if only compounds with a coordination number of four for the silicon are considered. The decreased share of the s orbital per bond leads to lower values for the coupling in compounds with higher coordination numbers for silicon, e.g. 131–300 Hz for fivefold and 108–182 Hz for sixfold coordinated silicon compounds.

Silicon–phosphorus couplings

In silylphosphines, it was possible to rationalize the one-bond phosphorus–silicon coupling constants $^1J(\text{P,Si})$ according to an empirical equation:

$$^1J(\text{P,Si}) = 26.3 + 8u + 5v - Zw \qquad [10]$$

where 26.3 Hz is the value for the reference compound H_3SiPH_2, u the number of silyl groups, v the number of methyl groups substituting P–H and w is the number of methyl groups substituting the Si–H bonds. The empirical factor Z lies between 3.3 and 5. A silicon–phosphorus double bond gives a splitting of 149 to 155 Hz.

A body of data exists for two-bond silicon coupling to phosphorus with an intervening nitrogen. Very low values are observed if the nitrogen is sp^3-hybridized, but values of $^2J(Si,N,P)$ between 8 and 42 Hz are typical for nitrogen with a double bond.

Silicon–carbon and silicon–silicon couplings

Both ^{13}C and ^{29}Si count as magnetically dilute isotopes. Therefore couplings are found only in the form of small satellites with an intensity of ~0.5% (^{13}C) or 2.3% (^{29}Si) on both sides of a strong central line. The Fermi contact term should be the dominant pathway to transfer magnetization between silicon and carbon. Thus the interaction depends on the s-electron density on both nuclei. The following semiempirical equations have been discussed:

$$^1J(Si,C) = 555.5\alpha_c^{\,2}\alpha_{Si}^{\,2} + 18.2 \qquad [11]$$

$$^1J(Si,C) = -1227.77 P_{sSisC}^{\,2} + 26.0 \qquad [12]$$

where $\alpha_C^{\,2}$ and $\alpha_{Si}^{\,2}$ are the square of the s character in the bonds that connect the carbon and the silicon and P_{sSisC} is the bond order element pertaining to the Si–C bond.

Values for $^1J(C,Si)$ are between 44 and 107 Hz. The high end is characteristic of silicon bonded to substituents of high electronegativity and sp^2-hybridized carbon. For a silicon–carbon double bond a value of 83.7 Hz has been reported.

Silicon–silicon coupling constants over a single bond range between 23 and 186 Hz. For Si=Si double bonds, values between 155–158 Hz are characteristic. This is somewhat higher than the values found for diarylsilanes (~85 Hz). The $^2J(Si,Si,Si)$ in stress-free situations lies in the 3–13 Hz range, but in polycyclic silanes ranges between 14 and 24 Hz. Because ring strain seems to decrease one-bond coupling constants, both types of constants are difficult to distinguish in that class of silanes. Two-bond couplings over oxygen are generally small. $^2J(Si,O,Si)$ data are between 0.5 and 14 Hz. The lower end is typical of silsesquioxanes and the larger values are found for Si–O–Si fragments in the skeletons of trimethylsilyl silicates.

Spin–lattice relaxation

For an isotope with a negative gyromagnetic ratio it is even more important to know of the pathways by which the ^{29}Si dissipates its energy from the excited state to the surroundings (lattice). In principle there are five paths, and the total spin–lattice relaxation time can be calculated from the contributions arising from each one:

$$\frac{1}{T_1} = \frac{1}{T_1^{DD}} + \frac{1}{T_1^{SR}} + \frac{1}{T_1^{CSA}} + \frac{1}{T_1^{SC}} + \frac{1}{T_1^{E}} \qquad [13]$$

Here T_1^{DD} is the dipolar, T_1^{SR} the spin–rotation, T_1^{CSA} the chemical shift anisotropy, T_1^{SC} the scalar and T_1^{E} the electronic contribution to the spin–lattice relaxation. At the heart of all relaxation paths is the formation of fluctuating magnetic fields.

Owing to the strong magnetic moment of the electron, T_1^{E} is very efficient. The extreme variation of T_1 in solids is explained by the effect of paramagnetic ions or molecular oxygen. In solution, shiftless paramagnetic relaxation agents [e.g. $Cr(acac)_3$, 10^{-2} molar] are used to shorten T_1 to ~10 s. Molecular oxygen also acts as a relaxation agent. It was calculated that for a solution saturated with oxygen at 1 bar, T_1^{E} is between 35 and 57 s. If other relaxation pathways are to be measured or a small line width is desired, paramagnetic impurities have to be removed. Paramagnetic ions that leach from the sample tube are removed by rinsing them with a chelating reagent (e.g. EDTA) before use. Oxygen is removed by bubbling argon through the sample or by vacuum techniques.

In the absence of unpaired electrons, the most efficient pathways for relaxation are the dipolar and the spin–rotation mechanisms. Because the interaction with protons is the most dominant one in this respect, the discussion will be restricted to this aspect. The dipolar term T_1^{DD}

$$\frac{1}{T_1^{DD}} = \hbar^2 \tau_c \sum \gamma_H^2 \gamma_{Si}^2 r_{SiH}^{-6} \qquad [14]$$

then depends on the gyromagnetic ratios of the ^1H and ^{29}Si isotopes, the distance between them and a correlation time, τ_c (Eqn [14]). Owing to the larger radius of silicon compared with that of carbon, dipolar relaxation is less efficient than in ¹³C NMR. Another aspect of the longer Si–H bond distance is that hydrogens placed farther apart in the molecule or on solvents can now also contribute to the relaxation process. The correlation time τ_c is the mean time the molecule, of a part of it, rotates one radian. This process is temperature dependent (**Figure 9**).

If in organosilicon compounds, the signal splittings caused by the coupling with the protons are removed by decoupling, then a modulation of the signal intensity occurs that is connected with the dipolar term. If the silicon relaxes by dipolar interaction with the protons only, then the signal intensity is multiplied by a factor of −1.52 from the nuclear Overhauser effect (NOE):

$$\text{NOE} = \frac{\text{signal intensity with irradiation of the protons}}{\text{signal intensity without irradiation}}$$

$$= 1 + \frac{\gamma_H T_1}{12\gamma_{Si}T_1^{DD}} \qquad [15]$$

The Overhauser enhancement factor η is calculated as

$$\text{NOE} = 1 + \eta \qquad [16]$$

If the dipolar mechanism is the only pathway, then η becomes $\eta_0 = -2.52$. Otherwise the share of the dipolar relaxation on the whole relaxation process can be calculated by:

$$T_1^{DD} = -\frac{2.52T_1}{\eta} \qquad [17]$$

Therefore, in the unfortunate case of $\eta = -1$ the signal intensity is zero. The other major relaxation mechanism involves spin–rotation interaction. The rotation of a molecule or a part of it generates a ring current through the electrons connected to it. By the tumbling motion of the molecules, fluctuating magnetic fields are produced. The spin–rotation relaxation time T_1^{SR} is given by

$$\frac{1}{T_1^{SR}} = \frac{2\pi I k T}{h^2}\frac{2C_\perp^2 + C_\parallel^2}{3}\tau_j \qquad [18]$$

The correlation time τ_j is the angular momentum correlation time, the time the molecule changes its angular momentum, usually the time between collisions. The terms $C_\parallel$ and $C_\perp$ are components of the spin–rotation interaction tensor and I is the moment of inertia of the molecule. For small step diffusion, τ_c and τ_j are connected by the Hubbard relation:

$$\tau_c\tau_j = \frac{I}{6kT} \qquad [19]$$

Because for any given molecule in Equation [17] all data are constant except T and τ_j, this results in a temperature dependence which is the reverse of that for T_1^{DD} (**Figure 9**). At high temperatures the relaxation process is dominated by the spin–rotation contribution and at low temperatures by the dipolar part. In between is a maximum value for the total relaxation time T_1. Another consequence is that the intensity of proton decoupled ²⁹Si spectra is temperature dependent.

Typical values for the relaxation times for organosiloxanes are between 35 to 205 s. Lower values are found for aqueous solutions of silicates (0.3–4.6 s), higher ones for silicon compounds without protons, e.g. chlorosiloxanes (53–352 s).

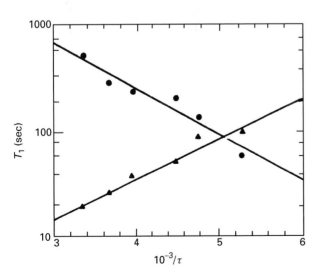

Figure 9 ²⁹Si Spin–lattice relaxation in TMS. (●) T_1^{DD}; (▲) T_1^{SR}; assuming $T_1^{CSA} \gg T_1^{DD}$. Reproduced with permission of the American Chemical Society from Levy GC, Cargioli JD, Juliano PC, and Mitchell TD (1973) *Journal of the American Chemical Society*: **95**: 3445.

List of symbols

B = magnetic flux density; J = coupling constant; T_1^{DD} = dipolar spin–lattice relaxation time; T_1^{CSA} = chemical shift anisotropy relaxation time;

T_1^E = electronic contribution to the spin–lattice relaxation time; T_1^{SC} = scalar relaxation time; T_1^{SR} = spin–rotation relaxation time; γ = gyromagnetic ratio; η = NOE enhancement factor; δ^* = reduced shielding constant; τ_c = correlation time; τ_j = angular momentum correlation time.

See also: **Chemical Shift and Relaxation Reagents in NMR; NMR of Solids; NMR Pulse Sequences; NMR Relaxation Rates; Nuclear Overhauser Effect; Parameters in NMR Spectroscopy, Theory of; Solid State NMR, Methods; Structural Chemistry Using NMR Spectroscopy, Inorganic Molecules.**

Further reading

Blinka TA, Helmer BJ and West R (1984) Polarization transfer NMR spectroscopy for silicon-29: the INEPT and DEPT techniques. *Advances in Organometallic Chemistry* 23: 193–218.

Denk M *et al* (1994) *Journal of the American Chemical Society* 116: 2691–2692.

Engelhardt G and Michel D (1987) *High Resolution Solid-State NMR of Silicates and Zeolites.* Chichester: John Wiley & Sons.

Gehrhus B *et al* (1996) *Journal Organometallic Chemistry* 521: 211–220.

Harris RK and Knight CTG (1983) *Journal of the Chemical Society, Faraday Transactions 2,* 79: 1525; 1539.

Horn H-G (1992) Spectroscopic properties of silicon-sulfur-compounds with at least one Si–S-bond – a review. *Journal für Praktische Chemie* 334: 201–213.

Jutzi P *et al* (1989) *Chem. Ber.* 122: 1629–1639.

Likiss PD (1992) Transition meta complexes of silylenes, silenes, disilenes and related species. *Chemical Society Reviews* 271.

Marsmann H (1981) *NMR Basic Principles Progress* 17: 65–235.

Pannell KH and Bassingdale AR (1982) *Journal of Organometallic Chemistry* 229: 1.

Schraml J (1990) ^{29}Si NMR spectroscopy of trimethylsilyl tags. *Progress in Nuclear Magnetic Resonance Spectroscopy* 22: 289–348.

Schraml J and Bellama JM (1976) ^{29}Si nuclear magnetic resonance. In: *Determination of Organic Structures by Physical Methods,* Vol 6, pp. 203–269. New York: Academic Press.

Uhlig F, Herrmann U and Marsmann H, ^{29}Si NMR database system. http://oc30.uni-paderborn.de/~chemie/fachgebiete/ac/ak_marsmann or http://platon.chemie.uni-dortmund.de/acii/fuhlig

Williams EA (1983) *Annual Reports on NMR Spectroscopy* 15: 235.

Williams EA (1989) Spectroscopy of organosilicon compounds. In: Patai S and Rappoport Z (eds) *The Chemistry of Organic Silicon Compounds,* Vol 1, pp 511–554. New York: John Wiley & Sons.

Williams EA and Cargioli JD (1979) *Annual Reports on NMR Spectroscopy* 9: 221.

Scandium NMR, Applications

See **Heteronuclear NMR Applications (Sc–Zn).**

Scanning Probe Microscopes

JG Kushmerick and **PS Weiss**, Pennsylvania State University, University Park, PA, USA

SPATIALLY RESOLVED SPECTROSCOPIC ANALYSIS
Methods & Instrumentation

The atomic resolution and spectroscopic capabilities of scanning probe microscopes (SPMs) have enabled elucidation of the great heterogeneity of surface sites including: defects, step edges, lattice impurities, adsorbates, and grown structures. One or more of these minority sites often function as the active sites for surface processes, and their individual investigation is thus required to gain insight into such processes. Such specific information cannot typically be acquired by spectroscopies that measure ensemble averages of the surface.

The scanning tunnelling microscope (STM) is the most suited, and the most developed of the various SPMs, to perform local spectroscopic measurements. Discussion of STM techniques will constitute the bulk of this article. It also has the most restricted range of accessible substrates in terms of conductivity and roughness. The atomic force microscope (AFM) has limited spectroscopic capabilities but can image a wider range of samples. The near-field scanning optical microscope (NSOM) has excellent spectroscopic, but limited spatial resolution. These latter two SPMs are discussed at the end of this article.

The basic working principles of the STM rely on the quantum mechanical properties of electrons. When an atomically sharp metal probe tip is brought within a few Å of a conducting or semiconducting surface, electrons can tunnel through the energy barrier between the probe tip and surface. By applying a constant DC bias voltage (V), a net tunnelling current (I) can be induced between the probe tip and the sample under study. Raster scanning the tip across the surface, through the use of piezoelectric transducers while maintaining a constant tunnelling current, images a surface of constant density of electronic states. The resulting image is a convolution of topographic and electronic properties of the sample surface.

The tunnelling current is exponentially dependent on the tip–sample separation and linearly dependent on the densities of tip and sample electronic states. The applied bias voltage determines which electronic states, on the both sides of the tunnelling junction, are being sampled, and thus allows acquisition of *spatially resolved spectra*. **Figure 1** is a schematic of the tunnelling potential barrier. Adjusting the bias

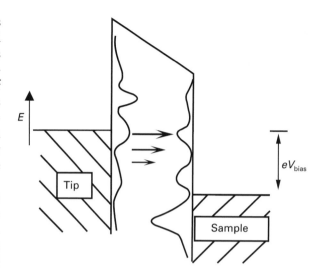

Figure 1 Energy level diagram for the tip–sample tunnelling gap, depicting electrons tunnelling from tip to sample. The local density of states for both the tip and sample as a function of energy is represented graphically.

voltage probes different electronic states, allowing the local density of states to be mapped as a function of energy. This is the basis of electronic spectroscopy with the STM.

Scanning tunnelling microscope

Experimental methods

Voltage-dependent imaging   One straightforward means to gain spectroscopic information from the STM is to acquire multiple images of the same area sequentially at different bias voltage, and thus different energies. These images are then overlaid. Each image shows the relative contributions from features at the particular energy relative to the Fermi level defined by the bias voltage (at energies eV, where V is the bias voltage and e is the charge on an electron). Another method is to acquire tunnelling current vs. bias voltage (I–V) characteristics at every imaged point, or at selected locations.

A dramatic example of this technique is the contrast reversal observed for some semiconductor surfaces. When voltage-dependent imaging of GaAs(110) is performed, only the Ga atoms are imaged when electrons are tunnelling into surface states, while the As atoms are imaged when electrons are tunnelling from filled surface states to the tip. **Figure 2** is a superposition of two images of the GaAs(110) surface. The orange image was obtained at positive sample bias, thus imaging the Ga atoms, while the green image was obtained with a negative sample bias, showing the As atoms. The cause of this contrast reversal is charge transfer from the Ga to the more electronegative As atoms, which results in the localization of the lowest lying empty states and highest lying filled states on the Ga and As atoms, respectively.

Voltage-dependent imaging can also be applied to understand bonding of adsorbate molecules to surfaces. **Figure 3** shows a Ni_3 cluster on MoS_2 at a temperature of 4 K, imaged at three different bias voltages. The cluster's contribution to the electronic structure varies dramatically as can be seen at these different energies. At +2 V sample bias (electrons tunnelling into the empty states on the surface) the Ni_3 cluster appears as a significant protrusion from the MoS_2 surface, indicating that it enhances the local density of empty electronic states at this energy. Similarly, we see that the cluster depletes the local density of filled states at –2 V sample bias (electrons tunnelling from the filled sample electron states to

Figure 3 Three STM images of a Ni_3 cluster adsorbed on a MoS_2 basal plane at 4 K. All three images show a 60Å × 60Å area and are plotted as three-dimensional representations with the same aspect ratio and with the same angle of view. The images were acquired with sample biases of: +2 V (upper), +1.4 V (middle), and –2 V (lower). Reproduced with the permission of the American Chemical Society from Kushmerick JG and Weiss PS (1998) *Journal of Physical Chemistry B* **102**: 10094–10097. (See Colour Plate 57).

the tip) since the cluster then appears as a depression in the surface. At a sample bias of +1.4 V the cluster is not directly apparent but a diffuse ring ~30 Å in diameter surrounding it is imaged. This ring is outside the atomic positions of the Ni_3 cluster imaged at +2 V (~ 16 Å diameter), and results from a perturbation of the MoS_2 surface electronic structure by the Ni_3 cluster. This ring is believed to be purely electronic in origin with little change in atomic positions of the substrate. Each image shows a different contribution of the Ni_3 cluster to the surface electronic structure, demonstrating how voltage-dependent imaging can measure electronic states as a function of both energy and position.

Voltage-dependent imaging can be a powerful technique but it does have some inherent problems. It is necessary to obtain many images in order to map the surface electronic structure as a function of energy. Thermal drift and piezoelectric creep can make overlay and comparison of successive images difficult. The fact that the constant current images acquired are a convolution of geometric and electronic effects further complicates the interpretation of observed features. The latter technique of recording complete or selected sets of *I–V* characteristics,

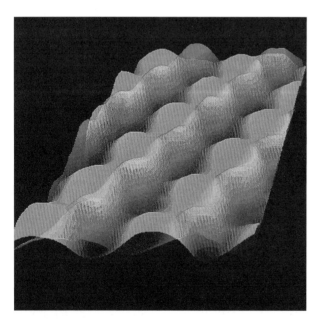

Figure 2 A composite of two images of the GaAs(110) surface. The orange features obtained at positive sample bias are the Ga atoms, while green features obtained at a negative sample bias are the As atoms. Feenstra RM, unpublished results. (See Colour Plate 56).

discussed in the next section, overcomes some of these problems.

Local current–voltage measurements Spectroscopic information over a large energy scale can be obtained by acquiring complete I–V curves at one or many locations. This can be accomplished by releasing feedback control while holding the probe tip steady and measuring the current with respect to applied bias voltage. Spectroscopy with the probe tip held in place allows scanning at both polarities and through zero bias.

Spectra are usually plotted as $(dI/dV)/(I/V)$ vs. V for comparison to other measurements of surface densities of states. This normalizes the spectra, removing the dependence on voltage and the exponential dependence on tip–sample separation. The derivative dI/dV can be numerically calculated, or a lock-in amplifier can be used to measure dI/dV phase sensitively, with a superimposed sinusoidal modulation on the bias voltage.

Synchronizing the feedback blanking and the applied voltage ramp enables acquisition of I–V curves at specified points in an image, thus mapping the energy and position dependence of the surface electronic structure in one image. **Figure 4** shows spectra acquired over Si atoms at three different locations in the unit cell of the Si(111)-(7 $\times$ 7) surface reconstruction. The three spectra show different electronic features due to the local bonding and environment of the Si atom probed.

Collection of data for the entire energy region of interest circumvents the need to construct a spectrum from several images. The dynamic signal range plays a role in determining how large an energy range can be scanned. The current goes to zero as the magnitude of the bias voltage decreases. Thus, features at high and low bias voltages can be hard to resolve in a single spectrum (recall that the preferred display of spectra, $(dI/dV)/(I/V)$ vs. V, has tunnelling current in the denominator).

Scanning tunnelling spectroscopy Closely related to voltage-dependent tunnelling is the modulation technique of scanning tunnelling spectroscopy (STS). This technique consists of superimposing a small sinusoidal modulation voltage on the constant DC bias voltage. The modulation frequency is typically chosen to be higher than the cut-off frequency of the feedback loop, resulting in a constant average tunnelling current. If the modulation frequency is too low the control electronics will attempt to compensate by adjusting the gap spacing. Alternatively, control of the tip height can be released during acquisition of spectra. By measuring the in-phase

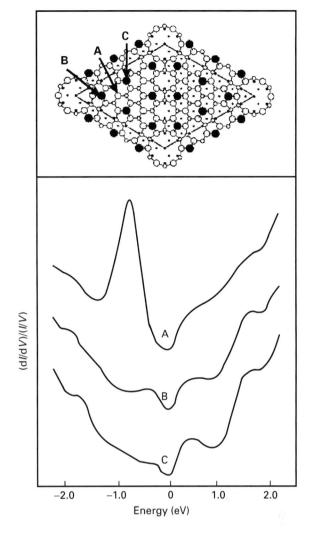

Figure 4 Atom-resolved spectra of the Si(111)-(7 $\times$ 7) surface obtained at positions indicated in schematic. Reproduced with permission of the American Institute of Physics from Wolkow R and Avouris PH (1988) *Physical Review Letters* **60**: 1049–1052.

tunnelling current modulation with a lock-in amplifier, dI/dV can be recorded simultaneously with the constant current image. Structure in dI/dV as a function of the applied bias voltage, can be attributed to structure in the local density of surface states.

Application of STS at constant average tunnelling current suffers from two disadvantages. The first is that the magnitude of the dI/dV signal scales as I/V, and thus becomes progressively small at low bias voltages. The second is that at low bias voltages the tip–sample separation reduces in order to maintain a constant average tunnelling current. If the density of states is too low the tip will come into contact and damage the surface. This form of STS, like voltage-dependent imaging, requires the acquisition of multiple images to map out electronic structure as a function of energy.

Barrier height spectroscopy Lack of direct knowledge of the effective height and width of the tunnelling barrier during spectroscopic measurements makes quantitative understanding of spectra difficult. Properties of the tunnelling barrier can however be investigated, giving information complementary to *I–V* spectroscopy. Barrier height spectroscopy consists of measuring the dependence of the tunnelling current on the tip–sample separation, at constant bias voltage. What is actually measured is the apparent barrier height, which is defined as

$$\Phi_{ap} = \left[\frac{1}{1.025} \cdot \frac{d(\ln G_t)}{dz}\right]^2$$

where G_t is the tunnelling conductance. Thus by measuring current as a function of tip–sample separation, G_t and Φ_{ap} can be calculated.

By applying a small modulation to the tip–sample separation (z) at constant bias voltage and constant tunnelling current, a lock-in amplifier can be used to measure dI/dz. The modulation frequency is chosen to be larger than the feedback loop bandwidth but smaller than resonant mechanical frequencies of the microscope. The dI/dz signal measured is directly related to the local surface work function and often provides useful contrast.

Another method of measuring the apparent barrier height is to record the tunnelling current directly as a function of tip–sample separation for a constant bias voltage. The tip–sample separation is reduced while the tunnelling current is measured. Although conceptually simple, there are experimental complications that must be taken into account. As the tip–sample separation becomes very small, the attractive forces between them tend to deform the tip and cause the actual separation to be smaller than expected. In fact, while point contact is typically used as the reference for tip–sample separation, this contact is usually realized with a jump to contact from small separation. It can also be difficult to maintain a constant bias voltage, as the tip–sample separation becomes very small, since the junction impedance can become comparable to that of the current preamplifier causing a significant decrease in the actual bias voltage. By measuring the voltage across the junction as well as the tunnelling current this problem can be overcome. **Figure 5** shows tunneling current and bias voltage as a function of tip–sample separation for a Au(110) sample and W tip (most probably covered with Au). The dramatic decrease in bias voltage at small tip–sample separations can be seen.

Inelastic electron tunnelling spectroscopy The majority of the tunnelling current consists of elastically tunnelling electrons. Inelastic pathways in which tunnelling electrons excite transitions can be used for recording local spectra. For a vibrational transition of a molecule contained in the junction this can occur above the threshold voltage of $|V| = \hbar\omega/e$ (**Figure 6**), where $\hbar\omega$ is the energy of a molecular

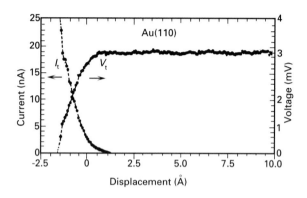

Figure 5 Tunnelling current (I_t), and bias voltage (V_t) during tip approach on Au(110). Reproduced with permission of the American Institute of Physics from Olesen L *et al.* (1996) *Physics Review Letters* **76**: 1485–1488.

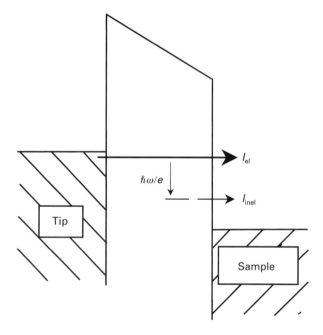

Figure 6 Energy level diagram for tip–sample tunnelling gap depicting both elastic (I_{el}) and inelastic (I_{inel}) tunnelling. Above the threshold voltage $|V| = \hbar\omega/e$ tunnelling electrons can excite a vibrational transition, for molecules in the tunnel junction. The densities of states of both tip and sample have been neglected for clarity.

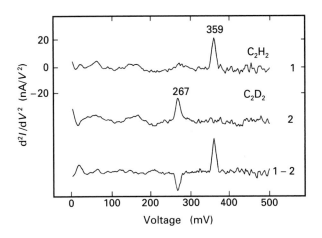

Figure 7 Inelastic electron tunnelling spectra of C_2H_2 (1) C_2D_2 (2) taken with the same STM tip and the difference spectra (1 − 2). Reproduced with the permission of the American Association for the Advancement of Science from Stipe BC *et al.* (1998) *Science* **280**: 1732–1735.

vibrational transition, and e is the charge of an electron. Inelastic pathways effectively increase the available number of final states for a tunnelling electron, thus producing a kink in the $I–V$ curve at $|V| = \hbar\omega /e$ for each vibrational transition excited. Using a similar modulation technique to that previously described d^2I/dV^2 vs. V can be measured directly with a lock-in amplifier (as the second harmonic of the modulation frequency). This has the benefit of transforming the kinks in $I–V$ to peaks and dips in d^2I/dV^2 vs. V, some of which may be assigned to molecular vibrations (of energy eV). Vibrational spectra can be obtained in this fashion for molecules in metal–insulator–metal sandwich tunnel junctions at low temperature. Limiting peak widths of the observable features also requires low temperatures as in tunnel diodes and sandwich tunnel junctions. **Figure 7** shows inelastic electron tunnelling spectra of acetylene (C_2H_2) and perdeuterated acetylene (C_2D_2), adsorbed on Cu(100) at 8 K, and the difference spectrum. The spectrum of acetylene has a peak at 359 meV corresponding to the C=H stretch. This peak is shifted to 267 meV for the C=D stretch in perdeuterated acetylene. Tuning the bias voltage to the energy of the vibrational mode and monitoring d^2I/dV^2 allows vibrational spectroscopic imaging. **Figure 8** demonstrates vibrational spectroscopic imaging of acetylene and perdeuterated acetylene of Cu(100).

Inelastic electron tunnelling spectroscopy with the STM allows unambiguous chemical identification of surface species, as demonstrated above. Electronic spectroscopy is also capable of differentiating between limited sets of adsorbates but does not as a rule enable such determinations. The vibrational spectra of isolated molecules also shed light on the chemical environment and bonding changes of minority

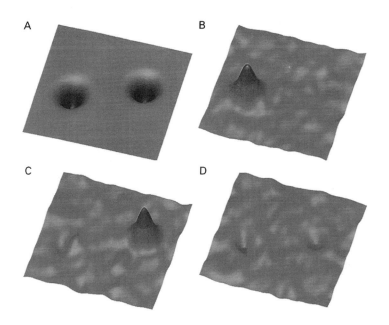

Figure 8 Vibrational spectroscopic imaging of C_2H_2 and C_2D_2. (A) Constant current STM image of a C_2H_2 molecule (left) and C_2D_2 molecule (right). The d^2I/dV^2 images of the same area recorded with a bias voltage of (B) 358 mV, (C) 266 mV, and (D) 311 mV, with a 10 mV modulation. All images are 48 Å × 48 Å with 1 nA DC tunnelling current. Reproduced with the permission of the American Association for the Advancement of Science from Stipe BC *et al.* (1998) *Science* **280**: 1732–1735. (See Colour Plate 58).

surface sites, and their critical role in surface processes, such as chemistry, corrosion, and film growth.

Photon emission   Electrons tunnelling inelastically between tip and surface can induce photon emission from the tunnel junction. On some surfaces, the inelastic tunnelling is enhanced by exciting surface plasmons, which decay by emitting a photon. Measuring emitted photons as a function of applied bias voltage yields information analogous to conventional inverse photon emission experiments. Unlike conventional inverse photon emission experiments occupied and unoccupied electronic states can be probed with photon-emission STM by scanning positive or negative bias voltages, respectively. Dispersing the emission with a spectrometer allows the spectral fingerprint of a feature to be measured. The spatial resolution of the STM allows the emission spectrum of isolated molecules to be recorded.

Instrumentation

Microscope Design considerations for spectroscopy with the STM are the same as for the STM in general. Due to the exponential dependence of tunnelling current on tip–sample separation, vibration isolation is of critical importance. The most demanding of the techniques mentioned above, inelastic electron tunnelling spectroscopy, requires special vibration isolation down to the level of ~ 0.001 Å over a limited bandwidth and operation at extremely low temperatures (ca. 4 K). Various designs for vibration isolation have been implemented, from placing the STM on top of a Viton stack, to mounting the instrument on a pneumatically suspended laser table enclosed in an acoustic isolation chamber. The STM itself should be constructed rigidly so as to yield the highest possible resonance frequencies. Shielding from electronic noise is also important, but is determined primarily by the design of the control electronics as will be discussed below.

Other aspects of the microscope design depend upon the intended experiments. *I–V* spectra can be obtained in air, if the system to be studied is not air-sensitive. Investigation of isolated adsorbates on metal single crystals requires ultrahigh vacuum, to enable sample preparation, and often cryogenic temperatures, to limit thermally activated diffusion. Cryogenic cooling also reduces thermal drift allowing an area to be studied repeatedly.

Electronics Low-noise electronics are important to maintain the stability of the probe tip and to avoid coupling of the control electronics to AC voltages powering the electronics or in nearby equipment.

Proper shielding and planning can reduce the electronic noise to sufficiently low levels that this is rarely a limiting factor.

Blanking the feedback loop, which is required for many of the spectroscopic techniques discussed, is typically accomplished through the use of a sample-and-hold circuit. In normal operation, the STM maintains a constant tunnelling current by driving the *z*-piezoelectric transducer with the error signal generated from the difference of the tunnelling current (converted to a voltage by the preamplifier) and a reference voltage. A sample-and-hold circuit interrupts the input to the feedback control loop and thus maintains a constant voltage to the piezoelectric transducer controlling tip–sample separation. The applied *z*-piezoelectric voltage can be held constant for up to several seconds with such a circuit. If longer blanking times are required, the use of a digital feedback blanking circuit can hold the voltage constant indefinitely. In addition, nearly constant drift can be corrected in microscopes where hold times are long compared to drift rates.

Microscope probe tip   Since the observed spectral features are a convolution of both the tip and the sample electronic density of states, understanding the role of the microscope probe tip in determining the observed spectra is critical to allowing interpretation of spectral features. A tip with a constant, preferably flat, density of states is typically desirable so that the sample's electronic structure dominates the spectral features observed. Alternatively a probe tip with a single sharp spectral feature can be useful in obtaining spectra.

Semiconductors have greatly varying densities of states and thus contributions from metal probe tips are less prominent. Metal surface state densities vary to a much smaller degree and are thus comparable to those of the tip states, making electronic spectroscopy of metals more complicated. To enable comparison between spectra obtained at various surface positions it is important that the tip structure, and thus density of states remains constant between measurements. Rearrangement of the tip apex can greatly affect the observed spectra and thus lead to spurious data interpretation.

Heteroatoms adsorbed to the tip can also play a large role in determining the observed spectra. Many studies have shown that the transfer of an adatom or molecule to the STM tip can affect the observed topography, e.g. yielding atomic resolution on an electronically flat close-packed metal substrate. The spectroscopic effects can be even larger. Special care must be taken when probing semiconductor surfaces. Tip–sample contact resulting in some

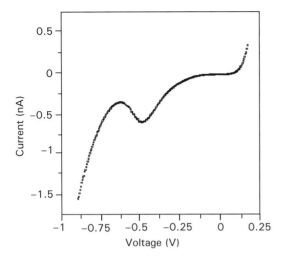

Figure 9 *I–V* scan of MoS_2 with a tungsten tip demonstrating negative differential resistance caused by the presence of MoS_2 on the tip. Kushmerick JG and Weiss PS unpublished results.

semiconductor material on the probe tip apex can give rise to odd effects, such as negative differential resistance (NDR). The tunnelling current between tip and sample normally increases with increasing bias voltage. If, however, there is a band gap on both the tip and sample, the tunnelling current can decrease with increasing bias voltage, due to the decrease in overlap of the density of states when the two band gaps are not aligned. **Figure 9** is an *I–V* curve exhibiting NDR of MoS_2, with a tungsten tip that came in contact with the surface. By backing away from the surface and field cleaning the tip, *I–V* curves without NDR are again obtained.

Negative differential resistance can also be observed for surfaces with localized trap states. A tunnelling electron can become localized for long times in these surface states, when they are in resonance with the Fermi level of the tip. Electrons so localized electrostatically repel other electrons causing a decrease in tunnelling current, referred to as a coulomb blockade. The voltage at which the NDR occurs is a measure of the energy of the localized trap state.

Other scanning probe microscopes

Atomic force microscope

It is possible to perform spectroscopy with SPMs other than the STM. The use of metal-coated cantilevers allows spectroscopic measurements with the AFM. The AFM maps surface topography by monitoring the attractive and/or repulsive probe–surface interactions as a surface is scanned by a cantilever. With a metal-coated cantilever, *I–V* curves can be obtained

for surfaces incompatible for STM study, such as metal structures covered with an insulating oxide layer. The insulating film acts as a tunnelling barrier of constant width, allowing spectroscopic measurements analogous to constant-separation *I–V* scans.

Functionalization of a cantilever by deposition of a molecular film allows the chemical forces between molecules to be probed. The chemical force (bonding) between a functionalized cantilever and surface is measured by monitoring cantilever deflection while the sample approaches, makes contact with, then is drawn back from the probe. The deflection of the cantilever is then converted to a force from the cantilever spring constant. **Figure 10** is a plot of force versus cantilever displacement curves for three combinations of tip and sample functionalization. The hysteresis in the curves is a measure of the adhesive interactions between probe

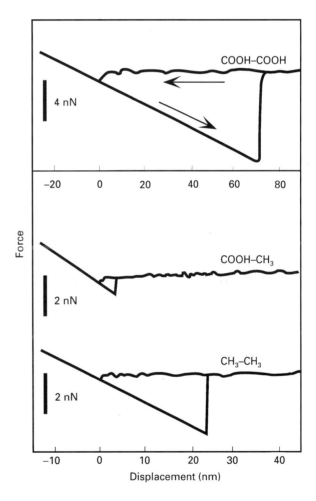

Figure 10 Force versus displacement curves recorded between functionalized atomic force microscope cantilever probes and surfaces. The adhesive interactions are strong for like–like interactions (COOH–COOH and CH_3–CH_3) but weak for interaction between unlike functional groups (COOH–CH_3). Noy A, Frisbie CD and Lieber CM, unpublished results.

and sample. It can be seen that the strongest adhesion is when both the cantilever and sample are functionalized with the –COOH (hydrophilic) groups. Functionalization of the tip and sample with (hydrophobic) –CH₃ groups also yields a strong attractive interaction, but adhesion between –COOH and –CH₃ is minimal due to their smaller interactions. Such measurements are sometimes called chemical force microscopy.

Near-field scanning optical microscope

The NSOM also enables localized spectroscopic measurements. This technique circumvents the diffraction-limited resolution of conventional optical microscopy by scanning a light source or collector in close proximity to the surface of interest. A metal-clad optical fibre typically serves as the probe, allowing light to be either emitted or collected from its apex which is free of any metal cladding. Other probes can also be used. By bringing the probe–sample separation into the near-field regime, the spatial resolution achieved is determined by the size of the unclad probe apex and can go well below the far-field limit of $\lambda/2$, where λ is the wavelength of light used. The NSOM can record absorption and fluorescence spectra, as well as measure the refractive index of surface and subsurface species. Spatial resolution of 12 nm has been achieved with visible light. While resolution is lower than that attainable with the STM, the ease of interpretation and familiarity of optical spectra make this technique attractive. Systems particularly suited to study with NSOM include unique biological structures such as the photosynthetic membrane shown in **Figure 11**.

If infrared absorption or Raman scattering is used as the contrast mechanism, vibrational spectra of samples can be obtained. The combination of the nanoscale spatial resolution of a scanned probe with the chemical specificity of vibrational spectroscopy allows *in situ* mapping of chemical functional groups with subwavelength spatial resolution. **Figure 12** is a shear force image of a thin polystyrene film along with a representative near-field spectrum of the

A

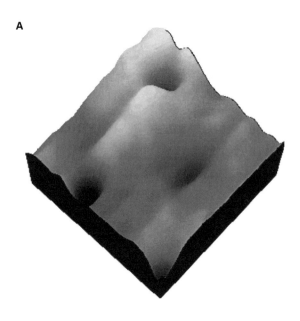

B

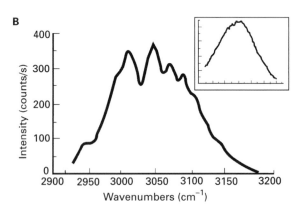

Figure 12 (A) A 3.1 × 3.1 µm shear force image of a thin polystyrene film deposited on a glass cover slip. The full-scale z-range is 62 nm. (B) Near-field IR transmission spectrum of a thin polystyrene film in the aromatic C–H stretching region. The inset is the laser output over the same spectral range in the absence of polystyrene absorption. Reproduced with permission of Stranick SJ, Richter LJ, Cavanagh RR and Michaels C, unpublished results.

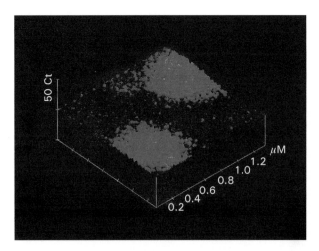

Figure 11 Fluorescence emission near-field scanning optical microscope image (1.3 µm × 1.3 µm) of a photosynthetic membrane fragment. Reproduced with permission of the American Chemical Society from Dunn RC *et al.* (1994) *Journal of Physical Chemistry* **98**: 3094–3098. (See Colour Plate 59).

aromatic C–H stretch region recorded in the microscope with a 1 second integration time per point.

Also suitable for study are dopants in semiconductors and local structures in semiconductor lasers.

Overview

Spectroscopies with SPMs are being developed rapidly. The ability to study isolated or small structures of adsorbates has allowed incredible insight into the rich chemistry of surfaces, particularly the defining roles that defect-sites play. The recent demonstrations of single molecule vibrational spectroscopies with the STM and NSOM have further opened up new avenues for investigation.

List of symbols

Å = Ångstrom; e = charge of an electron; E = energy; G_t = tunnelling conductance; I = tunnelling current; K = Kelvin; V = bias voltage; z = tip–sample separation; $\hbar$ = Plancks constant/2π; Φ_{ap} = apparent barrier height; ω = vibrational frequency.

See also: **Inorganic Compounds and Minerals Studied Using X-Ray Diffraction; Inorganic Condensed Matter, Applications of Luminescence Spectroscopy; Scanning Probe Microscopy, Applications; Scanning Probe Microscopy, Theory; Surface Plasmon Resonance, Applications; Surface Plasmon Resonance, Instrumentation; Surface Plasmon Resonance, Theory; Surface Studies By IR Spectroscopy.**

Further reading

Betzig E and Trautman JK (1992) Near-field optics: Microscopy, spectroscopy, and surface modification beyond the diffraction limit. *Science* 257: 189–195.

Bonnell DA (ed) (1993) *Scanning Tunneling Microscopy and Spectroscopy: Theory, Techniques, and Applications*. New York: VCH Publishers.

Chen CJ (1993) *Introduction to Scanning Tunneling Microscopy*. New York: Oxford University Press.

Feenstra RM (1994) Scanning tunneling spectroscopy. *Surface Science* 299/300: 965–979.

Sarid D (1991) *Scanning Force Microscopy: With Applications to Electric, Magnetic, and Atomic Forces*. Cambridge: Cambridge University Press.

Stroscio JA and Kaiser WA (1993) *Methods of Experimental Physics: Scanning Tunneling Microscopy*. New York: Academic Press.

Wiesendanger R (1994) *Scanning Probe Microscopy and Spectroscopy: Methods and Applications*. Cambridge: Cambridge University Press.

Scanning Probe Microscopy, Applications

CJ Roberts, MC Davies, SJB Tendler and PM Williams, The University of Nottingham, UK

> **SPATIALLY RESOLVED SPECTROSCOPIC ANALYSIS**
> **Applications**

The family of scanning probe microscopes (SPMs) have revolutionary imaging capabilities on a range of materials. For example, atomic resolution images of metal and semiconductor surfaces produced by the scanning tunnelling microscope (STM) or images of individual biomolecules in aqueous environments recorded by atomic force microscopy (AFM) are now routine in the literature. Perhaps, less well known, is the even greater potential of these and other probe-based techniques to produce spatially resolved spectroscopic information at the atomic or molecular level. Following a brief introduction to the principal SPMs available, this article will review as comprehensively as possible the wide ranging applications of spectroscopic SPM in semiconductor, material and life sciences.

Methods and instrumentation

The term SPM encompasses a family of surface-sensitive techniques, each based upon the interrogation at the nanometre level of a specific physical property by a sharp proximal probe. For example the original SPM, the STM measures local conductivity and the AFM local surface hardness. **Figure 1** provides a summary of the operation of the most popular types of SPM. Extensive discussions on the operations of the various forms of SPMs can be found in numerous reviews of the subject. Here we briefly highlight the three SPMs most readily applied to spectroscopic measurements, the STM, the AFM and the near-field scanning optical microscope (NSOM).

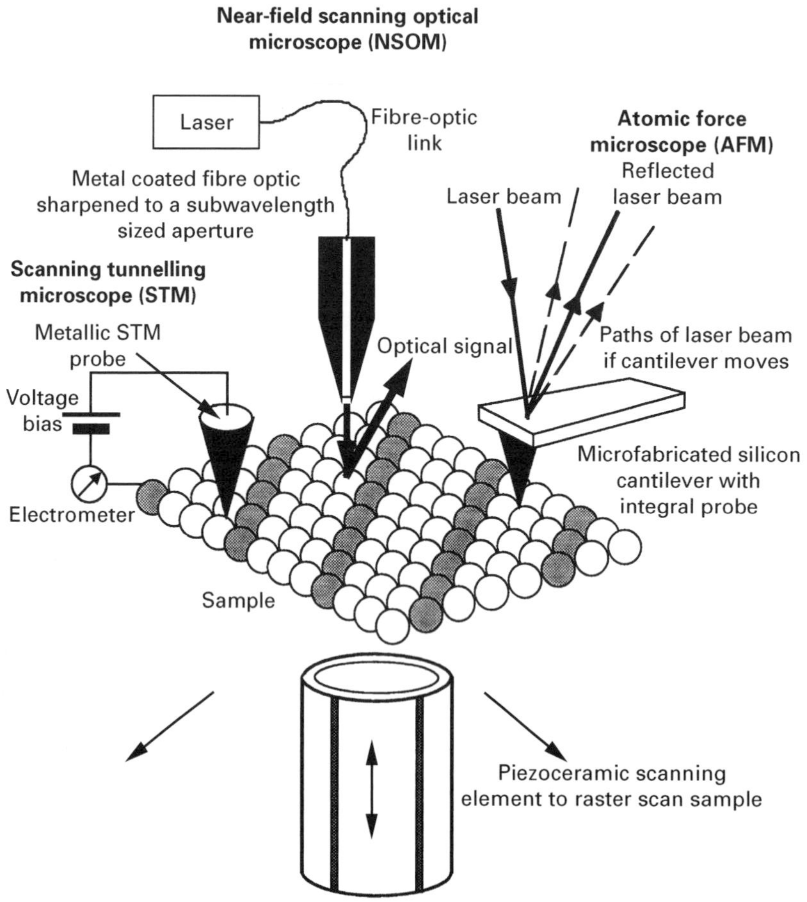

Figure 1 Schematic representation of the key components of a scanning tunnelling microscope (STM), a near-field scanning optical microscope (NSOM) and an atomic force microscope (AFM). In a STM, a sharp metallic probe is brought into close proximity with a conducting sample. A small bias voltage between probe and sample causes a tunnelling current to flow. This current is recorded as the sample is scanned beneath the probe. In NSOM, the sample is placed in the near-field region of a subwavelength-sized light source. The transmitted or reflected optical signal is used to form an image of the scanning sample. AFM microscopy relies upon the effect of repulsive and attractive forces between the probe and sample to bend a supporting cantilever. The bending of the cantilever, and hence the force, is extracted by monitoring the path of a laser beam reflected from the back of the cantilever.

STM

It was noted early in the use of STM that the appearance of a surface, particularly at atomic resolution, often changes dramatically with applied sample-tip bias voltage. This phenomenon results since electrons tunnelling between the tip and the surface do so between discrete electronic states and these states change with applied bias. In the extreme case of changes in tip bias polarity, either occupied (tip positive) or empty states (tip negative) in the sample are responsible for image contrast. Hence, it is possible to 'image' the location of specific bonding and antibonding orbitals and, with care, identify surface species by their electronic signature.

The first type of spectroscopy investigated by STM was current (I) versus gap distance (s). Here the tunnelling current is recorded as the metallic STM probe approaches a sample surface. The local barrier height (ϕ) of the surface can also be estimated from this I–s data using, $\phi = 0.952(\text{d} \ln I/\text{d}s)^2$ where s is in Å (note that ϕ is a strong function of tip shape, a generally unknown parameter). Electronic structure has also been studied by ramping the tip bias voltage (V) while recording the resultant tunnelling current. This measurement can be carried out at a single point or at each point in a topography scan, hence producing spatially resolved spectroscopic data. Plotting d ln I/d ln V versus V has been shown to be proportional to the density of states in the low voltage limit ($\phi > V$).

AFM

Since its inception by Binnig *et al.* in 1986, AFM has become an important and widespread tool for imaging surface topography with nanometre resolution. AFM is essentially a very sensitive profilometer, measuring surface topography using a sharp stylus, or probe, mounted on a soft spring, or cantilever (**Figure 1**). The ability of AFM to image samples in ambient or aqueous environments is particularly attractive in biomolecular and electrochemical studies.

By exploiting the local nature of an AFM probe and its pico-Newton force sensitivity considerably more information can be extracted from AFM than just surface topography. For example local tribology and force-probe–sample separation spectra. During such force probe–sample measurements the probe is moved towards the surface at constant velocity until it is brought into contact with the sample up to a predetermined point of maximum load. The direction of motion is then reversed and the probe is withdrawn from the sample surface. As the probe is withdrawn from the sample the probe may stick to the surface due to interactions between the probe and the sample. The magnitude of this 'sticking' force and its temporal evolution can reveal details of the type and dynamics of the forces occurring between probe and surface.

NSOM

NSOM represents one of the most promising optical techniques that aims to overcome the Abbé barrier, and yet retain most of traditional optical microscopy's utility. A NSOM typically illuminates a local area of a sample by transmitting laser light through a subwavelength sized aperture, as defined by the end of a tapered metal-coated optical fibre. An image is then formed by raster-scanning the aperture close to the sample surface and collecting either the transmitted or the reflected light (**Figure 1**). In such a regime, it is the aperture of the NSOM probe that determines the ultimate resolution.

Since NSOM utilizes optically based contrast it has the potential to exploit optical spectroscopies with resolution comparable to the probe dimensions, i.e. tens of nanometres. For example, steady state and time-resolved fluorescence spectroscopy and Raman spectroscopy have been demonstrated with NSOM. Despite some notable successes it is important to note that the combination of the difficulty in the interpretation of NSOM data and the often poor optical efficiency of NSOM probes, presently makes the application of NSOM the most challenging form of SPM.

Applications of spatially resolved SPM

The breadth of applications of SPM for spectroscopic measurements is considerable. In order to highlight key examples, the discussion is classified by the nature of the sample investigated, ranging from atomic scale studies on semiconductors and metals to the study of biomolecular interactions.

Semiconductors

The first atomically resolved scanning tunnelling spectroscopy (STS) data were obtained by Hamers *et al.* (1986) on Si(111)-(7 × 7), showing chemically inequivalent atoms within each unit cell. Since this time many semiconductors have been studied by STS studies carried out under ultrahigh vacuum. For example, dI/dV curves recorded from In atoms adsorbed onto the Si(111)-(7 × 7) surface show that covalent bonding between the In and Si surface states saturates the Si intrinsic metallic states. STS studies of Li on p-type Si(001) show strong negative differential resistance and the related existence of thermally activated electron traps. Spatially resolved STS has also distinguished inequivalent sites on a roughened Si(001) surface. Spectra recorded from terraces show bonding and antibonding states at +0.5 V and −0.5 V; however, Si atoms recorded from a step show a marked metallic character. It should be noted that a quantitative analysis of such spectra can be problematic due to the generally unknown nature of the tip's electronic structure and the change in shape of the tunnelling barrier with applied tip–sample voltage bias. Nevertheless, such sensitivity has been very successfully exploited for the study of semiconductor surfaces.

Although many methods exist for the optical characterization of semiconductor surfaces, when high spatial resolution is required, for example with an inhomogeneous surface, new techniques are required. Here NSOM has significant potential and has been employed to map photoluminescence intensity simultaneously with topography on quantum well structures and hence local carrier density. The data was shown to be consistent with a diffusion-based model and the existence of short-lifetime carriers at the quantum-well boundary. Low-temperature NSOM at 4.2 K has been used to show a vertical dependence of spectral shape in GaAs quantum dots. NSOM has also been employed to acquire Raman spectra from rubidium-doped regions of $KTiO_2PO_4$ sample, although along acquisition times and very small signal levels have limited progress. Despite this, NSOM Raman has also been successfully employed to map residual stresses in silicon wafers associated with deformation (**Figure 2**).

Metals

In comparison to semiconductor surfaces, metals display smaller corrugations in their density of states due to very delocalized bonding. Nevertheless, early STM studies revealed relatively high atomic corrugations on Au(111). It has since been shown that this 'super' resolution results from the nature of the electronic density of states on face-centred cubic metal surfaces.

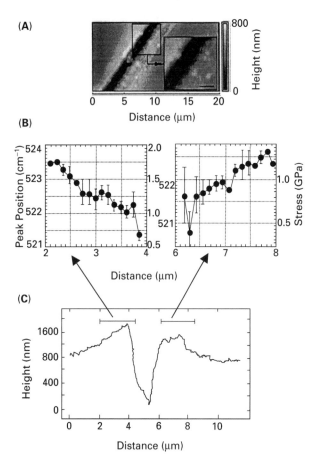

Figure 2 (A) Topography image recorded using NSOM showing a scratch on a silicon wafer surface. Inset is an enlargement of the area that was Raman mapped. The scale bar is 1 μm. An array of 26 by 21 spectra were recorded with step sizes of 154 nm and 190 nm in the X and Y directions, respectively. Each spectrum took 60 s to acquire giving a total image acquisition time of just over 9 h. In (B) the value of centre frequency of the silicon band was extracted and is shown as a function of distance across the scratch; the lateral position of the data points is shown on the topographic cross section (C). Reprinted with permission from Webster S, Batcheldes DN and Smith DA (1998) Submicron resolution measurement of stress in silicon by near-field Raman spectroscopy. *Applied Physics Letters* **72**, 1478–1480.

Elemental-specific contrast on metals using STS has been demonstrated for copper on W(110) and Mo(110) surfaces. Resonant tunnelling via surface and image states provide elemental identification. Theoretical treatment of elemental identification of adsorbates on metals indicates that up to 2 Å peaks should be present in images of electropositive elements on Pt(111) and that 0.35 Å depressions would result from electronegative oxygen atoms.

Low-temperature ultrahigh-vacuum STM has been used to perform atomically localized spectroscopic measurements on single Fe atoms adsorbed onto the Pt(111) surface. Using dI/dV spectra a resonance was found to occur in the adatom local densities of states that is centred 0.5 eV above the Fermi energy. This feature had a width of approximately 0.6 eV, and occurred only when the tip was within angstroms (laterally) of the centre of an Fe adatom. Following on from this work it was found that Fe adatoms strongly scatter metallic surface state electrons and hence are good building blocks for constructing atomic-scale barriers to confine electrons. 'Quantum corral' barriers constructed by individually positioning Fe adatoms using the STM tip reveal, via STS, discrete resonances inside the corrals, consistent with size quantization (**Figure 3**).

Superconductors

Cryogenic STS is an ideal tool for studying the electronic nature of superconductors and has produced local dI/dV spectra at the BiO cleavage planes of a bismuth cuprate superconductor at 4.2 K. The spectra confirm a large gap parameter associated with an apparently gapless density of states on the uppermost layer. Spatial variations of the gap parameter on a 100 Å scale were attributed to variations in BiO metallicity with two characteristic dI/dV spectral shapes over regions of metallic and nonmetallic BiO layers. Low-temperature STS has also been employed to probe the local effects of magnetic impurities on superconductivity. Tunnelling spectra obtained near magnetic adsorbates reveal the presence of excitations within the superconductor's energy gap that can be detected over a few atomic diameters around the impurity at the surface. These excitations are locally asymmetric with respect to tunnelling of electrons and holes. A model calculation based on the Bogoliubov–de Gennes equations can be used to understand the details of the local tunnelling spectra.

Polymers

New higher throughput NSOM probes have recently permitted the recording of Raman spectra from

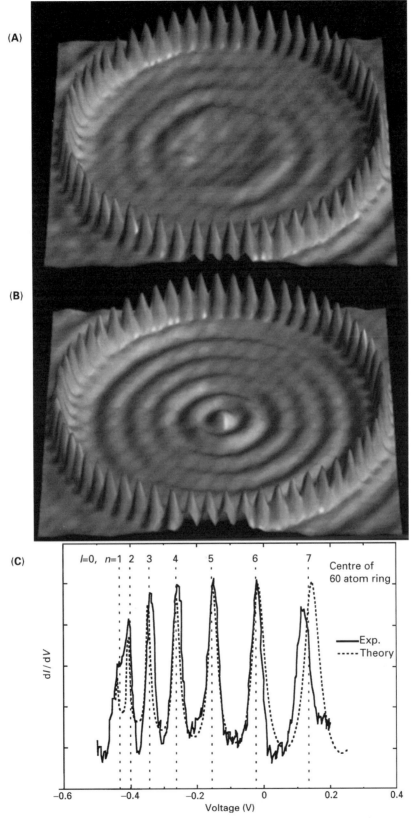

Figure 3 Perspective views of a 60 atom Fe ring recorded at tunnelling current of 1 nA and tip bias voltages of (A) 10 mV and (B) −10 mV. The quantum interference patterns inside the ring change with energy. The energy dependence of the lowest density of states at the centre of the ring is illustrated by the dI/dV spectra in (C). The sharp peaks in the spectra indicate sharp resonances in the lowest density of states. These data match theoretical results based upon the particle-in-a-box model very closely. Reprinted from *Physica D* **83**: Crommie MF, Lutz CP, Eigler DM and Heller EJ, Quantum corrals, pp 98–108, 1993 with kind permission of Elsevier Science – NL, Sara Burgerhartstraat 25, 1055 KV Amsterdam, The Netherlands.

polystyrene spheres labelled with different dyes and adsorbed on silver substrates. No significant differences in near-field and far-field Raman spectra were observed with the NSOM data, clearly demonstrating true chemical identification. Nonresonance Raman spectra of polydiacetylene crystals (**Figure 4**) demonstrate the feasibility of acquiring spatially resolved Raman spectra despite very low signal levels.

NSOM has also been used to probe the excitonic transitions in J-aggregates of 1,1'-diethyl-2,2-cyanine iodide grown in poly(vinyl sulfate) thin films. Fluorescence spectra recorded as a function of the NSOM tip position along individual aggregates show only slight variations and are very similar to the bulk aggregate spectrum. The absence of spectral broadening is assigned as evidence for a uniform, well-ordered molecular structure within the aggregates.

Biological systems

NSOM fluorescence image and spectrographs of recombinant *Escherichia coli* cloned to produce green fluorescence protein (GFP) show a difference in fluorescence distribution within individual bacteria. Fluorescence activity of GFP can thus be used as a convenient indicator of transformation. Improvements in NSOM probe–sample distance control have facilitated the fluorescence imaging of thick biological specimens, such as neurons, astrocytes and mast cells, which also fluoresce in the far-field and hence would normally reduce optical resolution. NSOM has also been used to provide high-resolution information on *in situ* interactions between proteins in biological membranes, in particular human red blood cells invaded by the malaria parasite, *Plasmodium falciparum*. During infection, the parasite expresses proteins that are transported to the cell membrane. Host and parasite proteins were selectively labelled in indirect immunofluorescence antibody assays, and simultaneous NSOM dual-colour excitation fluorescence maps produced.

Presently karyotypes of human metaphase chromosomes are used to detect genetic defects like deletions or translocations, where the chromosomes are treated by the trypsin–Giemsa protocol, to produce a typical banding pattern and imaged by optical microscopy. Because of the diffraction limit in optical microscopy, even the smallest visible band contains around 1 million base pairs. Improved resolution has been demonstrated using fluorescence NSOM on the treated chromosomes compared to conventional light microscopy.

Single molecule studies

Spatially resolved STS has been used to characterize the electronic structure of C_{60} molecules on a range of substrates including Au(001), Au(110), Au(111) and Al(111). Due to a lattice mismatch between the overlayer C_{60} and the substrate Au(100) surface, a uniaxial stress is applied resulting in several types of oblique lattices and modified electron charge density around the C_{60} molecules. Charge transfer from the substrate to the molecules and intermolecular bonding under stress were observed in STS data. STS also clearly differentiates inequivalent adsorption sites on Au(111) and Al(111). The STM tip has been used to locally excite single C_{60} molecules to luminesce with an emission spot size of 0.4 nm. Fullerenes have been

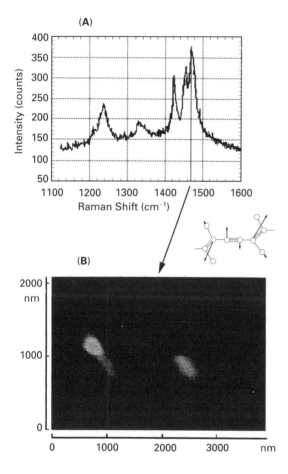

Figure 4 (A) A pre-resonant near-field Raman spectrum of a polydiacetylene microcrystal taken using 633 nm excitation, approximately 150 nm aperture and a 60 s exposure. (B) A 3 µm × 2 µm near-field Raman image of the polydiacetylene microcrystallites. A Raman spectra as in (A) is obtained at every point in the image, and the intensity of any peak may be chosen to produce a grey scale image, in this case the 1485 cm^{-1} feature. The vibration mode responsible for this line is shown.

employed as STM tips, showing improved performance when studying graphite surfaces.

A number of groups have employed NSOM to record fluorescence spectra from single dye molecules. Fluorescence spectra taken in the near-field showed no broadening due to long-range inhomogeneities as are apparent in far-field spectra. Using picosecond light pulses, time-resolved near-field fluorescence images of single sulforhodamine 101 dye molecules and rhodamine 6G have been recorded. Since metal surfaces near radiating dipoles influence fluorescence lifetimes, the fluorescence decay of single molecules is dependent on the relative position of the tip and the molecule. Polarization-sensitive NSOM has been used to resolve mesoscopic spectral inhomogeneities in small crystals of the dye 1,1-diethyl-2,2-cyanine iodide, where the crystals showed strong absorption perpendicular to their long axis and no absorption in the two orthogonal directions.

The sensitivity of fluorescence resonance energy transfer (FRET) has been extended to the single-molecule level by measuring energy transfer between single donor and acceptor fluorophores linked by a short DNA molecule using NSOM. Dual colour images and emission spectra combined with photodestruction dynamics have been used to determine the presence and efficiency of energy transfer. In contrast to ensemble measurements, dynamic events on a molecular scale are observable in single-pair FRET measurements because they are not cancelled out by random averaging.

The ability of AFM to directly measure discrete intermolecular forces as low as 10 pN was highlighted as long ago as 1993. Since then, a number of groups have exploited this ability, using AFM to determine the forces required to separate individual receptor-ligand molecules including avidin-biotin, cell adhesion proteoglycans, antibody–antigen and hydrogen bonding between nucleotide bases. In addition, the potential for mapping surface groups by their functionality using AFM has been exploited to spatially locate the adhesive and frictional interactions between hydrophobic and hydrophilic organic monolayers and biotin–streptavidin (see **Figure 5**).

Recently molecular dynamics has been used to model the disruption of biotin–streptavidin as it occurs in force adhesion experiments and to relate these forces to molecular structure and conformation. The force is calculated from the steepest slope in the free energy profile along the unbinding pathway. Interestingly the calculated rupture forces show a similar spread in values as is found in experimental data, suggesting that this spread is due to heterogeneity in the reaction pathway of biotin–streptavidin.

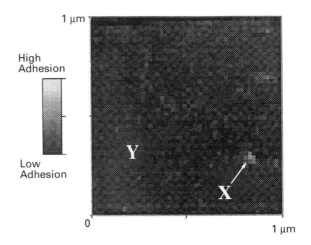

Figure 5 A spatially resolved force adhesion map recorded using an AFM probe coated in biotin on a 90% biotin-blocked streptavidin surface. Biotin–streptavidin are a ligand–receptor pair with very high affinity (10^{15} M^{-1}) often employed as a model system for molecular recognition studies. The contrast on the image corresponds to the amount of adhesion felt by the probe at the surface. Light areas represent high levels. If the biotin-coated probe contacts the surface in a region where free binding sites exist (i.e. streptavidin unblocked by free biotin) then adhesion based on the biotin–streptavidin specific molecular interaction would be expected. Position X in the image corresponds to an area of adhesion and Y a typical area of no interaction. Hence, X marks the spot of an open streptavidin binding pocket. Reproduced with permission of Gordon and Breach Publishers from Roberts CJ, Allen S, Chen X, Davies MC, Tendler SJB and Williams PM (1998). Measurement of intermolecular forces using force microscopy: Breaking individual molecular bonds. *Nanobiology* **4**: 163–175.

AFM has also been used to measure inter- and intra-chain forces in DNA nucleotides. The interaction forces between complementary 20 base pair lengths of single-stranded oligonucleotides ((ACTG)$_5$ and (CAGT)$_5$) and the forces required to stretch and break polydisperse homopolymers of inosine were measured. AFM has also been employed to study the interaction between d(T)$_{20}$ (dT = 2′-deoxyribosyl-thymine) and poly(dA) oligonucleotides (dA = 2′-deoxyribosyladenine). As before, rupture forces after binding were measured; however, the dependence of the rupture force with time of contact between the probe and sample was also observed. After 30 s contact, only low adhesion was observed. A maximum probe sample adhesion was observed after 2 min. This incubation-time dependence was interpreted as slow reorientation of partially hybridized oligonucleotide strands into more stable structures and was suggested as a means of studying double-helix annealing processes.

Single-molecule force spectroscopy on dextran filaments linked to a gold surface has been carried out using AFM by vertical stretching, the applied force being recorded as a function of the elongation. At low

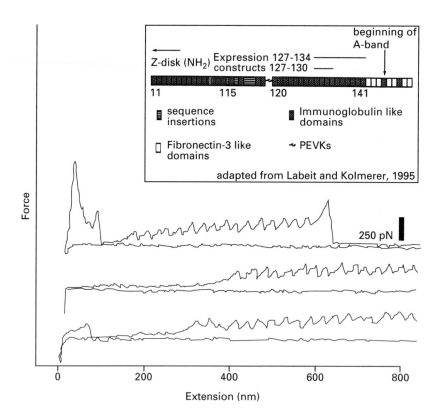

Figure 6 Three typical force extension curves obtained by stretching titin molecules. The curves show periodic structure on the retract portion of the data consistent with the unfolding of individual titin domains. Reproduced with permission from Rief M, Gautel M, Oesterhelt F, Fernandez JM, Gaub HE (1997) Reversible unfolding of individual titin immunoglobulin domains by AFM. *Science* **276**, 1109–1112.

forces the entropic deformation of dextran dominates and can be described by the Langevin function with a 6 Å Kuhn length. At elevated forces the strand elongation was governed by the twist of bond angles. At higher forces the dextran filaments underwent a distinct reversible conformational change. This ability to stretch and relax long-chain molecules has also been exploited to unfold individual domains of the giant sarcomeric protein of striated muscle, titin (**Figure 6**). At large extensions, the restoring force exhibited a sawtooth-like pattern. Measurements on recombinant titin immunoglobulin segments of two different lengths exhibited the same pattern and allowed the discontinuities to be attributed to the unfolding of individual immunoglobulin-like domains. The forces required to unfold individual domains ranged from 150 to 300 pN and depended on the pulling speed. Upon relaxation, refolding of immunoglobulin domains was observed.

Future trends

The resolution and adaptability of scanning probe techniques are increasingly being exploited to carry out spectroscopy measurements, particularly for electronic and optical surface properties. In addition, new variants of traditional scanning probe spectroscopic methods continue to be developed. For example, STM spectroscopy performed with magnetic probe tips has yielded new information about the spin-resolved nanoelectronic properties of magnetic nanostructures. Also, an adaption of STM to allow the probing of surface acoustic waves is proposed that reaches submicrometre resolution for the quantitative evaluation of elastic constants and studies of nanoscale structures. In AFM, measuring the oscillation amplitude of the probing AFM tip and phase-shift between the cantilever response as a function of the tip–sample distance, allows the analyses of the dynamic interaction of the AFM tip with the sample surface. This has been termed dynamic force spectroscopy and has been proposed as a new method of rapidly mapping probe–sample interactions. Advances in NSOM spectroscopic applications are presently centred on improving the optical efficiency of the probe, either through more precise control of fibre optic probe geometry or through the development of semiconductor-based tips not dissimilar to those in use in AFM.

Other SPM technologies for specific applications are also being developed. For example, magnetic resonance effects have been studied using modified AFMs. The sample is mounted on an AFM cantilever in a magnetic field gradient and exposed to a radiofrequency field which drives the spins into precession. The resultant periodic force is sensed in the normal way from the flexure of the AFM lever. This technique has also been extended to spatially map magnetic resonance data and to detect NMR effects. These data exemplify the reason for the continued rapid growth in SPM applications. The adaptability of the technique to address problems from a range of scientific disciplines and its ability to operate under conditions of vacuum to liquid from 4 K to 1000 K, will ensure this pace of SPM advancement continues for the foreseeable future.

List of symbols

I = current; s = gap distance (Å); V = voltage; ϕ = barrier height.

See also: **Scanning Probe Microscopes; Scanning Probe Microscopy, Theory.**

Acknowledgement

SJBT is a Nuffield Foundation Science Research Fellow.

Further reading

Allen S, Davies J, Davies MC *et al* (1996) *In situ* observation of streptavidin–biotin binding on an immunoassay well surface using SFM. *FEBS Letters* 390: 161–164.

Ambrose WP, Goodwin PM, Martin JC and Keller RA (1994) Alteration of single molecule fluorescence lifetime in near-field optical microscopy. *Science* 265: 364–367.

Betzig E and Chichester RJ (1993) Single molecules observed by near-field scanning optical microscopy. *Science* 262: 1422–1425.

Binnig G, Quate CF and Gerber C (1986) Atomic force microscope. *Physical Review Letters* 56: 930–933.

Berndt R, Gaisch R, Gimzewski JK, Reihl B, Schlittler RR, Schneider WD and Tschudy M (1993) Photon emission at molecular resolution induced by a scanning tunnelling microscope. *Science* 262: 1425–1427.

Crommie MF, Lutz CP and Eigler DM (1993) Spectroscopy of a single atom. *Physical Review B* 48: 2851–2854.

Crommie MR, Lutz CP, Eigler DM and Heller EJ (1993) Quantum corrals. *Physica D* 83: 98–108.

Grubmüller H, Heymann B and Tavan P (1996) Ligand binding: molecular mechanics calculation of the streptavidin–biotin rupture force. *Science* 271: 997–999.

Hallen HD, Larosa AH and Jahncke CL (1995) Near-field scanning optical microscopy and spectroscopy for semiconductor characterization. *Physica Status Solidi A* 152: 257–268.

Hamers RJ, Tromp RM and Demuth JE (1986) Surface electronic structure of Si(111)-(7×7) resolved in real space. *Physical Review Letters* 56: 1972–1975.

Hamers RJ (1996) Scanned probed microscopies in chemistry. *Journal of Physical Chemistry* 100: 13103–13120.

Higgins DA and Barbara PF (1995) Excitonic transitions in J-aggregates probed by NSOM. *Journal of Physical Chemistry* 99: 3–7.

Hoh JH, Cleveland JP, Prater CB, Revel JP and Hansma PK (1993) Quantized adhesion detected with the atomic force microscope. *Journal of the American Chemical Society* 114: 4917–4919.

Kazantsev DV, Gippius NA, Oshinivo D and Forchel A (1996) Spectroscopy of GaAs/AlGaAs microstructures with submicron spatial resolution using a near-field scanning optical microscope. *JETP Letters* 63: 550–554.

Lee GU, Kidwell DA and Colton RJ (1994) Sensing discrete streptavidin–biotin interactions with atomic force microscopy. *Langmiur* 10: 354–357.

Rief M, Oesterhelt F, Heymann B and Gaub HE (1997) Single molecule force spectroscopy on polysaccharides by atomic force microscopy. *Science* 275: 1295–1297.

Rugar D, Yannoni CS and Sidles JA (1992) Mechanical detection of magnetic resonance. *Nature* 360: 563–566.

Shao Z and Yang J (1995) Progress in high resolution atomic force microscopy in biology. *Quarterly Reviews in Biophysics* 28: 195–251.

Smith DA, Webster S, Ayad M, Evans SD, Fogherty D and Batchelder D (1995) Development of a scanning near-field optical probe for localized Raman spectroscopy. *Ultramicroscopy* 61: 247–252.

Wolf EL, Chang A, Rong ZY, Ivanchenko YM and Yu FR (1994) Direct mapping of the superconducting energy gap in single crystal $Bi_2Sr_2CaCuO_{8+X}$. *Journal of Superconductivity* 7: 355–360.

Zeisel D, Dutoit B, Deckert V, Roth T and Zenobi R (1997) Optical spectroscopy and laser desorption on a nanometer scale. *Analytical Chemistry* 69: 749–754.

Scanning Probe Microscopy, Theory

AJ Fisher, University College London, UK

SPATIALLY RESOLVED SPECTROSCOPIC ANALYSIS

Theory

Introduction

The three most important scanning probe techniques are

- Scanning tunnelling microscopy (STM);
- Scanning force microscopy (SFM, also known as atomic force microscopy, AFM);
- Scanning near-field optical microscopy (SNOM).

The three methods give different types of information, and require correspondingly different theoretical treatments. STM probes the electronic states of a surface; SFM probes the force (or force gradient) between a tip and a surface; while SNOM probes the electromagnetic field near a surface.

However, all three techniques share several common features. First, they measure local, not average, surface properties. Any theory must therefore include the local surface properties if it is to be useful. Second, they all lack a simple *inversion theorem*: in no case is it possible to infer directly physical properties of the system from the scanning probe results. Interpretation therefore has to proceed by an indirect 'interpretation cycle':

1. Build a model of the relevant local features (e.g. structure, excitations) of the system under study;
2. Develop a theory of the scanning probe experiment concerned;
3. Combine (1) and (2) to determine the predicted experimental signal from the model adopted;
4. Alter the model if the predictions and the experiment do not match.

In this article we shall examine what type of model of the physical system under study is appropriate under item (1) of the 'interpretation cycle' for each technique, and how a suitable theory of the experiment can be constructed for item (2).

The scanning tunnelling microscope: electronic spectroscopy

General considerations

The fundamental physical process in STM is the tunnelling of electrons between the tip and the sample under study, through the barrier formed by the vacuum between them (see **Figure 1**). The 'height' of this barrier in energy is approximately equal to the work functions of the tip or sample material. In the simplest possible one-dimensional model, we assume that the electron potential energy V takes a constant value V_0 through the tunnelling gap; the barrier height is therefore $(V_0 - E)$ where E is the electron energy. The electron wavefunctions then decay in the vacuum like $\exp(-\kappa z)$, where z is the coordinate normal to the surface and

$$\frac{\hbar^2 \kappa^2}{2m_e} = (V_0 - E) \qquad [1]$$

If $V_0 - E$ is $5\,\text{eV} = 8.01 \times 10^{-19}\,\text{J}$ then $\kappa = 1.15 \times 10^{10}\,\text{m}^{-1}$.

Tunnelling can occur from tip to sample and from sample to tip. If no bias is applied to the system (i.e. if the electrochemical potentials of the electrons in tip and sample far from the junction are equal) the rates of tunnelling in opposite directions are equal, and no net current flows. (Note that if the tip and sample have different work functions, a finite charge transfer will occur at zero bias to establish a dipole layer at the surfaces, and hence an electric field in the vacuum gap; it is the potential difference arising

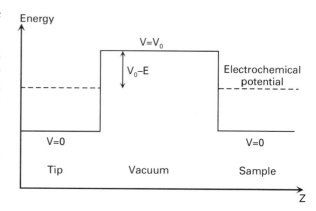

Figure 1 Schematic diagram of an STM junction at zero bias, illustrating the meaning of the symbols defined in the text.

from this field which equalizes the electrochemical potential in the two materials.)

Suppose now that a finite bias potential $\Delta\Phi$ is applied to the system (see **Figure 2**), of a sign which raises the electrochemical potential for electrons on the left of the junction by $|e|\Delta\Phi$ relative to those on the right. Over a range of energies, electrons are now more likely to tunnel from left to right than *vice-versa*, and a net current flows from right to left. If the difference in chemical potentials is small so that current is dominated by electrons with a single energy E, we can use the fact that the current is proportional to the tunnelling probability and hence to the absolute square of the wavefunction to deduce that it will vary with the tip-sample separation d like $\exp(-2\kappa d)$, with κ given by Equation [1]. Taking the value of κ we estimated earlier, we obtain the often-quoted rule of thumb that the tunnel current should reduce by roughly a factor of ten whenever the tunnel gap is increased by 1 Å = 10^{-10} m.

This is an approximate theory of the tunnelling process, but it says nothing about the contrast to be expected when the STM tip is moved across the surface. A better theory must take account of the atomistic structure of the tip and the surface, as well as a better theory of the tunnelling between them. In doing this, it is important to realize that the energy of the tunnelling electrons being used to probe the system is very similar to the energy of electrons in the bonding orbitals holding the atoms together. There is therefore a very close relationship between the tunnelling process, the electronic structure, and the atomic (or chemical) structure of the system.

Step (1) of the 'interpretation cycle' for STM must therefore involve a model of the atomic and electronic structure of the surface, including any adsorbates or surface defects. In practice this is most often obtained numerically using density-functional theory, in which the total energy of the electrons in the system is calculated from the electronic charge density, rather than from the full many-electron wavefunction. The Hartree–Fock method, which employs an approximate form for the many-electron wavefunction which neglects the correlations between the motions of the electrons, is also used. Such calculations are now relatively standard, and many can be found in the literature for surfaces of different types. Step (2) must involve a three-dimensional theory of electron tunnelling between the surface (represented in this way) and the tip; we now turn to this more difficult step.

Perturbation theory

The interpretation of many spectroscopies (for example, optical spectroscopy) proceeds by the identification of a well-defined 'perturbation' which is applied to the system when the experiment is performed. This is both convenient (because the response of the system to the perturbation is not too difficult to evaluate in terms of the matrix elements of the perturbation) and conceptually useful (because it allows a clear separation between the 'system' and the 'probe').

This is not straightforward in STM. There are two reasons: (i) the tip and the sample may be very close, and hence strongly coupled together, and (ii) even when this is not the case, there are mathematical difficulties in separating the Hamiltonian into parts describing a noninteracting tip and sample, because the kinetic energy operator for the electrons appears in both parts. Nevertheless problem (ii) has been solved, and it has been shown that a sensible perturbation theory can be constructed in which the appropriate matrix element is that of the electron current density operator, evaluated over a surface S separating the tip and the sample (see **Figure 3**). We write

$$M_{ts} = \frac{\hbar^2}{2im_e} \int_S d^2\boldsymbol{r} \left[\psi_t \nabla \psi_s - \psi_s \nabla \psi_t\right] \qquad [2]$$

where ψ_t is a one-electron state of the tip (in the absence of the sample) and ψ_s is a state of the sample (in the absence of the tip). Note that in order to derive this result, one has to assume that *both* states are valid solutions of the Schrödinger equation in the neighbourhood of the surface S; this implies that the potential for electrons at S must be equal to the vacuum potential. The transition rate for electrons from state t to state s (or *vice-versa*) can then be written

$$W_{ts} = \frac{2\pi}{\hbar} |M_{ts}|^2 \delta(E_t - E_s) \qquad [3]$$

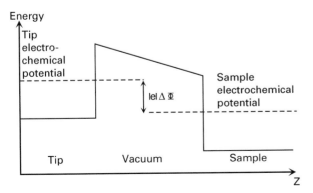

Figure 2   Schematic diagram of an STM junction at finite bias.

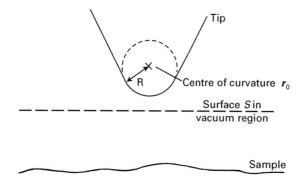

Figure 3 The important quantities in the Tersoff–Hamann model of STM. The matrix element is evaluated on the surface S; the conductance is proportional to the sample density of states at the tip centre of curvature r_0.

and the total current from tip to sample as

$$I = \sum_{ts} \frac{2\pi}{\hbar} |M_{ts}|^2 \delta(E_t - E_s) f_t(E_t)[1 - f_s(E_s)] \quad [4]$$

where $f_t(E)$ and $f_s(E)$ are the occupation probabilities for electron states with energy E in the tip and the sample respectively.

The most commonly used model in interpreting STM data is the Tersoff–Hamann model, in which the analysis is carried a step further. It is assumed that the tip wavefunction is an s-wave, and decays into the vacuum like

$$\psi_t(\boldsymbol{r}) = \Omega^{-1/2} \kappa R \exp(\kappa R) \frac{\exp(-\kappa|\boldsymbol{r} - \boldsymbol{r}_0|)}{\kappa|\boldsymbol{r} - \boldsymbol{r}_0|} \quad [5]$$

where Ω is a normalization volume, $\boldsymbol{r}_0$ is the centre of the curvature of the tip, R is the radius of curvature, and κ is as defined earlier. In this special case the integral in Equation [2] can be evaluated exactly (under the assumption that ψ_s obeys the free-space Schrödinger equation), and one finds in the limit of small bias that the differential conductance σ of the STM is

$$\sigma = \frac{32\pi^3}{\hbar} e^2 N_t(E_F) R^2 \kappa^{-4} \exp(2\kappa R)$$
$$\sum_s |\psi_s(\boldsymbol{r}_0)|^2 \delta(E_s - E_F) \quad [6]$$

Here $N_t(E_F)$ is the total tip density of states at the Fermi energy. This is a very simple and important result; it tells us that the tunnelling conductance measures the sample density of states at the Fermi energy, evaluated at the centre of curvature of the tip (i.e. some distance outside the sample surface). This is relatively straightforward to calculate, and easy to interpret in simple chemical terms. The model disregards all details of the tip; they are absorbed into the values of the constants R and Ω. The (usually unknown) structure of the end of the tip can therefore be disregarded, at the cost of sacrificing any information about the absolute value of the conductance. It is largely because of these advantages that the Tersoff–Hamann model is so popular.

The approximations leading to Equation [6] are valid only if there is no electric field in the vacuum. Nevertheless, the Tersoff–Hamann model is often used to interpret images taken at finite bias voltage $\Delta\Phi$, or even data from the 'spectroscopic' mode of the STM in which the tip position is held fixed and the bias varied. The density of states involved in Equation [6] is projected onto a 'window' of energies $\Delta E = e\Delta\Phi$, rather than onto a single energy. There is no theoretical justification for this, as the true states of the system are bound to be modified by the addition of such a bias voltage, but it has proved useful as a way of qualitatively rationalizing STM data provided the bias is not too large.

It is possible to extend perturbation theory beyond the Tersoff–Hamann model, for example by including tunnelling to or from states of non-zero angular momentum on the tip, or by using states explicitly calculated from a particular atomistic model to find the matrix element in Equation [2]. However, both these approaches require additional information about the geometry of the tip and the electronic states it supports. This is generally not available from experiment, as a tip will be modified by the forces acting in the course of the experiment (as discussed in more detail below); even if the tip is well-characterized before use (for example, by electron microscopy or field-ion microscopy), this information will become out-of-date once the experiment starts.

Another extension of this type of perturbation theory is to the case where there is some additional electronic order in the tip or the sample – for example, magnetic or superconducting order. In the case of magnetic order one is led to consider separate currents of spin-up and spin-down electrons, proportional to the spin-resolved components of the density of states. For a superconductor, the tunnel current depends on the quasiparticle density of states.

Beyond perturbation theory

Perturbation theory of this kind leads to an appealing picture of STM. Nevertheless it is not always justified; here we list some of the reasons why it may break down.

- A substantial redistribution of charge and potential takes place, so the effective one-electron Schrödinger equation is altered. This effect has been predicted theoretically when the tip-sample separation drops below about 3 Å; it tends to result in a lowering of the potential energy for an electron in the vacuum and a collapse of the tunnelling barrier.
- The electron tunnelling probability between tip and sample is not small. In practice this occurs only when the electron transport is no longer dominated by tunnelling – either because a physical contact or *nanojunction* has been formed between the two, or because the tunnel barrier has collapsed completely (see above). The signature of this state of affairs is that the STM conductance becomes of the order of the quantum of conductance, e^2/h.
- Although small, the tunnelling matrix element through the vacuum is not the smallest energy scale in the problem. This can occur when, for example, a highly insulating molecule is adsorbed on a surface; tunnelling through the molecule can then be just as difficult as tunnelling through the vacuum, so it is not appropriate to treat the vacuum tunnelling as a perturbation.

For all these problems, the theoretical cure is the same: one must perform a single coupled calculation for the electron states in the whole system (tip plus adsorbate – if any – plus sample) under a non-zero bias, allowing a current to flow.

However, this has proved to be very difficult without additional simplifications. In the elastic scattering quantum chemistry (ESQC) method developed by Joachim and Sautet, there is no self-consistency in the Hamiltonian for the electrons and only a relatively small basis set, giving very limited flexibility to the electron wavefunctions. In another approach, pioneered by the group of Tsukada, a more detailed numerical representation of the wavefunction is adopted: the wavefunctions are calculated on a mesh of points and full self-consistency is achieved between the wavefunctions and the electronic potential. The simplification in this case is that the wavefunctions far from the tunnel junction are those of a fictitious 'jellium' in which the positive charge of the nuclei is smeared out into a uniform background. In yet a third approach the conductance is calculated

in a non-perturbation manner between two localized states, rather than between the true bulk states of the tip and sample.

Other factors

There are also other factors that are known to be important in STM. One of these is the mechanical interaction between the tip and the sample; the forces that arise can distort the tip, with the result that the displacement of the tip apex is not the same as that recorded from the piezoelectric actuators controlling the tip (see **Figure 4A**). This effect was revealed by careful measurements of the *corrugation* of the STM image of adsorbates on metals as a function of the conductance.

A second effect is that of the tip electric field. This can be very large: fields above 10^9 V m^{-1} can be obtained when a potential difference of a few volts is

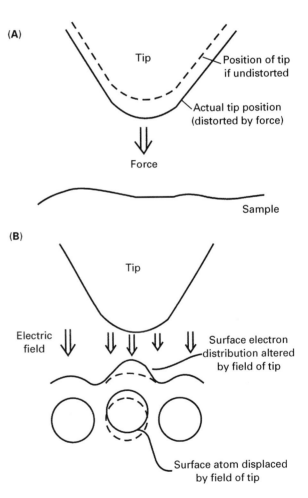

Figure 4 Two physical phenomena which alter STM images from those predicted by simple theory: (A) Tip–sample forces distort the actual change in separation from that measured by the piezoelectric transducers. (B) Electric fields from the tip cause motion of surface atoms and distortion of surface electron states.

dropped over a narrow tunnelling gap. This can have two results: first it distorts the atomic structure of the surface, causing movements of a few tenths of an angstrom in the surface atoms, and second it distorts the electronic structure, changing the tunnelling probability at different points on the surface (see **Figure 4B**). These effects have been shown to be important in images of the Si(001) surface.

A third complicating effect is the inelastic scattering of electrons from other excitations during tunnelling. These other excitations may be electronic; the most important example is surface plasmons, which may be found in either the tip or the sample. Scattering from surface plasmons produces electromagnetic disturbances near the tunnel gap which can result in the emission of electromagnetic radiation from the STM. The frequency distribution of the emitted light is then characteristic of the surface plasmon spectrum. The distortion of the surface plasmons can also have other, more subtle, effects since it determines whether or not the electron experiences the effect of the 'image interaction' outside a surface. This can in turn have a large effect on the electron potential and the tunnelling current.

Alternatively, the scattering excitations may be atomic vibrations. These may result in phonon-assisted sidebands around resonant tunnelling peaks, corresponding to the absorption or emission of phonons. In extreme cases the transfer of electronic energy to atomic motion may produce atom transfer between tip and sample, or even desorption; this is a form of DIET (desorption induced by electronic transitions) and may be used to break bonds selectively on surfaces.

The scanning force microscope: force spectroscopy

In order to interpret these experiments one needs to bear in mind the different types of forces that can act between the tip and the sample.

- At large distances the force most commonly present is the Van der Waals force. Between two atoms the Van der Waals force energy decays with separation z according to the well-known z^{-7} law, but for a sphere above a planar surface (one simple model for the tip–surface system) the decay is only as z^{-2}. This relatively slow fall-off tells us that in SFM, unlike STM, the large-scale structure of the tip is important.
- If the sample is an insulator, it may be locally charged. The interaction between these local 'patch charges' and the tip also decays like a power law in the tip–sample separation. The patch

charges are difficult to control; the highest-resolution SFM results are generally obtained on conducting samples.

- At smaller distances (of the order of 3–5 Å separation) local interactions between the closest atoms of the tip and sample start to become important. These include the onset of covalent bonding, and local electrostatic forces.
- As the tip-sample separation drops below the sum of the atomic radii of the atoms, the Pauli exclusion principle raises the energy of the overlapping electron distributions, producing a repulsive force. If the tip and sample are forced together beyond this point, atomic deformations (first elastic, then plastic) occur.

Models for the forces

Of these interactions, the Van der Waals attraction and the Pauli repulsion are universal; the presence of the others depends on the nature of the material. The combination of Van der Waals and Pauli interactions is often captured by the simple '6-12' Lennard-Jones interatomic potential

$$V(r) = \epsilon \left[\left(\frac{r}{\sigma} \right)^{-12} - \left(\frac{r}{\sigma} \right)^{-6} \right] \qquad [7]$$

in which the attractive r^{-6} term represents the Van der Waals force and the repulsive r^{-12} term the Pauli force. Simulations of generic interatomic interactions are often performed using this potential, although it cannot be expected to be realistic for anything other than interactions between the simplest rare-gas solids. More realistic calculations include approximate forms for the electrostatic and covalent interactions between the atoms, or (better still) find these forces directly from the electronic structure of the materials involved.

High-resolution SFM operation

With this in mind, let us examine the most common modes of SFM operation when high-resolution information about the surface is required.

- **Non-contact mode.** In this mode the tip is kept at a distance from the sample in the attractive part of the force–distance curve; usually it is then scanned across the sample, and the tip–sample distance adjusted to keep the cantilever displacement (and hence the force) constant. This procedure keeps the tip in the region where the tip–sample force is (relatively) well understood, but with the price that the force is determined by the cumulative effect of a large number of atoms – hence the resolution of individual atomic-scale features is seldom possible.

- **Contact mode.** Here, by contrast, the tip is allowed to penetrate into the repulsive regime of **Figure 5**. This has the advantage that one expects a large component of the force to be determined by a relatively small number of atoms near the tip apex, but the disadvantage that the force becomes dependent on complex atomic processes involving the irreversible deformation of the tip–sample junction. Images with apparent atomic resolution can be seen in contact mode on simple crystalline materials such as alkali halides, but the conclusion of careful simulations is that the atomic-scale features are not, in fact, correlated with the positions of atoms in the surface. This theoretical conclusion is reinforced by the failure to resolve atomic defects (known to be present on the surface) in experiments.

One might think that a technique intermediate between contact and non-contact modes could be devised simply by bringing the tip close to the surface, but not in contact with it. In fact this is very difficult because of the 'jump-to-contact' phenomenon: a static tip held above a surface by a SFM cantilever with a given force constant k_{cant} can be stable only as long as the force gradient from the tip–sample interaction is less than k_{cant} (see **Figure 5**). The force gradient of a Van der Waals interaction between a tip and a flat surface diverges as the separation between them is reduced, so this condition is always violated and the tip snaps into contact with the sample. If the tip is pulled off the surface, a similar jump out of

contact occurs (although between different values of tip–sample separation).

Since a very interesting range of tip–surface separations is rendered unavailable by the jump to contact, it would be desirable to eliminate it. To date, this has been done in two ways. First, a dynamical approach is used: the cantilever is vibrated above the surface with an amplitude of several hundred angströms, in such a way that its point of closest approach is only a few angströms from the surface. The difference from before is that the tip is accelerating rapidly away from the surface as it approaches; this suppresses the jump to contact. One way of expressing this is to say that the effective cantilever force constant is increased from k_{cant} to $k_{cant} + M_{tip}\omega^2$, where M_{tip} is the total mass of the vibrating tip and ω is the angular frequency of vibration. The tip is usually scanned while keeping the vibrational period constant; this corresponds approximately to a scan of constant force gradient. Atomic resolution has been obtained using this technique, initially on the Si(111)–7 × 7 surface but now also on others. It seems this resolution can be understood in terms of the interaction between the tip and the surface near the point of closest approach, but the theory is complicated because the vibration of the tip samples all the different regions of the potential surface described above during a cycle, so a unified model containing all of them must be used.

A second approach is to control the force on the tip directly, generally by means of a small magnet mounted on the back. This removes the need to model a complicated tip oscillation, but imposes stringent demands on the response and stability of the electronics controlling the force. Direct measurements of tip–sample potential curves have now been reported using this technique, but comparison with theory is still in its infancy.

Measurements of elastic properties

If local but not ultra-high-resolution measurements are required to probe the elastic properties of a hard material, there are advantages in using high-frequency measurements.

The scanning near-field optical microscope-optical spectroscopy

The theory of scanning near-field optical microscopy is somewhat similar to that of STM, with the transport of light (or photons) replacing the transport of electrical current (electrons). Instead of the Schrödinger equation, the Maxwell equations for the electromagnetic field must be solved near the tip and

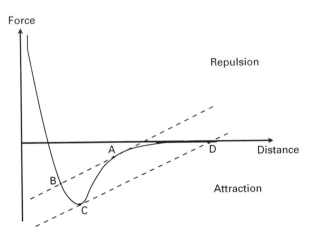

Figure 5 Schematic force–distance curve for an SFM experiment. On approach, the tip jumps from point A to point B; on retraction, it jumps from C to D. The dotted lines have a slope equal to the cantilever force constant k_{cant}.

the sample, taking into account the local electromagnetic properties of each medium. In some respects this is easier, because (in the absence of non-linear media) the Maxwell equations are truly linear and no self-consistency (of the type needed between effective one-electron wavefunctions and the potential) is needed. Also, since the characteristic wavelengths and decay lengths for optical photons are much larger than atomic dimensions, a continuum treatment of the tip and sample materials is almost always sufficient. On the other hand, the Maxwell equations require treatment of two coupled vector fields.

As in the STM case, the equations must in practice be solved numerically. Perturbation theory is seldom employed, and most calculations make a direct solution for the optical modes at a fixed frequency, either by a transfer matrix approach, or by using the Dyson equation to obtain the solution from that of an exactly soluble system (for example, free space).

Conclusions

STM appears to be the most subtle of the scanning probe methods, relying as it does on quantum-mechanical tunnelling. In fact the theory for this technique is the best developed of all the scanning probe family, but much progress remains to be made in accounting correctly for the nature of the tip and for tip–sample interactions. The theory of near-field optical microscopy is similar in spirit, and in some ways more straightforward. The understanding of SFM data is very incomplete, particularly for experiments with resolution on the atomic scale.

List of symbols

d = tip–sample separation; e = electronic charge; $f(E)$ = occupation probability for electron state with energy E; $\hbar$ = Planck's constant divided by 2π; k_{cant} = force constant of SFM cantilever; m_e = mass of electron; M_{tip} = mass of vibrating tip assembly; M_{ts} = matrix element connecting tip and sample states in STM; r_0 = centre of curvature of STM tip; R = radius of curvature; V = potential energy of electron; W_{ts} = transition rate between tip and sample states; κ = decay constant for electron wavefunctions ψ; σ = differential conductance; Φ = electrostatic potential; ω = angular frequency of vibration; Ω = normalization volume for tip wavefunction in the Tersoff–Hamann model.

See also: **Scanning Probe Microscopes; Scanning Probe Microscopy, Applications; Surface Plasmon Resonance, Applications; Surface Plasmon Resonance, Instrumentation; Surface Plasmon Resonance, Theory.**

Further reading

Briggs GAD and Fisher AJ (1999) Molecules on semiconductor surfaces: STM experiment and atomistic theory hand in hand. *Surface Science Reports* **246**: 1–81.

Blöchl PE, Joachim C and Fisher AJ (eds) (1993) *Computation for the Nanoscale.* Dordrecht: Kluwer Academic.

Chen J (1993) *Introduction to Scanning Tunnelling Microscopy.* Oxford: Oxford University Press.

Datta S (1995) *Electronic Transport in Mesoscopic Systems.* Cambridge: Cambridge University Press.

Israelachvili J (1992) *Intermolecular and Surface Forces,* 2nd edn. London: Academic Press.

Sautet P (1997) Images of adsorbates with the scanning tunnelling microscope: theoretical approaches to the contrast mechanism. *Chemical Reviews* **97**: 1097–1116.

Wiesendanger R (1994) *Scanning Probe Microscopy and Spectroscopy – Methods and Applications.* Cambridge: Cambridge University Press.

Wiesendanger R and Güntherodt H-J (eds) (1996) *Scanning Tunnelling Microscopy III,* 2nd edn. Berlin: Springer-Verlag.

Scanning Tunnelling Microscopes

See **Scanning Probe Microscopes.**

Scanning Tunnelling Microscopy

See **Scanning Probe Microscopy, Applications**

Scanning Tunnelling Microscopy, Theory

See **Scanning Probe Microscopy, Theory.**

Scattering and Particle Sizing, Applications

F Ross Hallett, University of Guelph, Ontario, Canada

Copyright © 1999 Academic Press

ELECTRONIC SPECTROSCOPY
Applications

The scattering of radiation has been used for many years as a noninvasive technique for the determination of particle sizes in suspension and in aerosols. Two fundamentally different experimental approaches are used. In the first approach, fluctuations in the scattered light intensity arising from the diffusive motion of the particles are monitored and analysed by autocorrelation, a procedure that yields the intensity autocorrelation function, a function that corresponds to the Fourier transform of the frequency spectrum. The intensity fluctuation approach, when applied to light scattering, has been given a wide variety of names, including quasi-elastic light scattering (QELS), intensity fluctuation spectroscopy (IFS), photon correlation spectroscopy (PCS) and the currently favoured dynamic light scattering (DLS). Diffusing wave spectroscopy (DWS) is a special type of DLS technique which provides estimates of particle sizes in concentrated, nearly opaque, suspensions. The second approach involves the measurement of the intensity of light, X-rays or neutrons scattered by the particles as a function of the scattering angle. Depending on the source of radiation these are named static light scattering (SLS), small angle X-ray scattering (SAXS) and small neutron scattering (SANS).

When the scattering particles are small, monodisperse and spherical, size information can be obtained from either SLS or DLS data by fitting to relatively simple mathematical expressions. Generally, however, the particles are polydisperse (i.e. characterized by a size distribution) and sometimes they can be multimodal, and/or nonspherical. These properties of the sample can severely complicate the analysis and recovery of size information can require the use of relatively complex and indirect data analysis methods. This is because the direct analysis falls into a class of mathematical operations known as 'ill conditioned' transforms that are unstable and severely affected by truncation errors and noisy data. Much of the more recent work in light scattering has centred on finding analysis procedures that minimize these problems. The presence of polydispersity introduces a further consideration. Since the amount of light scattered by a particle is proportional to r^n, where r is the particle radius and n can be as large as 6, there is the consequence that the larger particles of a distribution can scatter considerably more light than the smaller ones. This leads to z-average particle sizes and intensity weighted size distributions in DLS. Conversion of these to number averaged quantities, such as those obtained by analysis of electron microscopic images, requires knowledge of the particle's scattering factor or structure factor and its incorporation into the analysis. These factors can be very simple for very small uniform spheres and approach unity for point particles. When particular criteria are satisfied, approximately when particle diameters are significantly smaller than a wavelength, Rayleigh–Gans–Debye scattering factors can be used. However, once the particle's diameter is of the same order of magnitude or larger than the wavelength then Mie theory must be used to compute scattering factors. To date, Mie factors can be easily computed only for spherical or coated spherical systems.

Basic physics of elastic scattering

The scattering vector

During elastic scattering processes, the incident radiation is redirected at all scattering angles as a result of its encounter with the scatterers. By suitable use of pinholes or slits the light scattered at a particular angle, θ, from the main beam can be detected and analysed (see **Figure 1**). The momentum transfer, which results from such a process, is defined as the vector difference between the incident wave vector, k_0, and the scattered wave vector k_s. This vector difference, called the scattering vector (Q), has a magnitude that can be obtained from the geometry of the scattering arrangement. That is,

$$|\boldsymbol{Q}| = \frac{4\pi n_0}{\lambda} \sin(\theta/2) \qquad [1]$$

where n_0 is the refractive index of the medium surrounding the particles and λ is the wavelength (in a vacuum) of the incident light. It is clear from Equation [1] that Q has the dimensions of an inverse length. Indeed, Q^{-1} sets the length scale which will be probed by the light scattered at angle θ. At low angles, Q^{-1} is large and light scattered in this direction will contain information only on the gross dynamic and structural properties of the particles, whereas light scattered at a high angle contains this information at a finer scale and might, for example, carry information on the internal structure of the particle. **Figure 2** shows a comparison of scattering techniques, and their respective particle sizing ranges. DLS has a greater range than other scattering techniques because it is concerned with the distance $(2\pi/Q)$ that a particle diffuses before dephasing of the scattered light occurs. Note also that due to the relatively short wavelengths of X-rays and neutrons (typically 0.1 to 1 nm as compared with visible light, 400 to 700 nm), Q^{-1} is appropriate for particle sizing only when θ is very small. As a result, only small

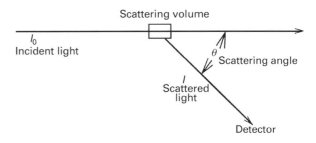

Figure 1 The scattering geometry.

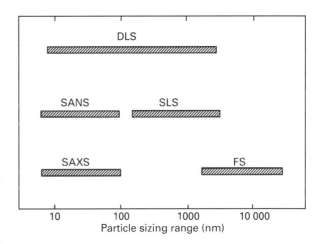

Figure 2 Particle sizing ranges for different scattering instruments. DLS: dynamic light scattering, SANS: Small-angle neutron scattering, SLS: static light scattering, SAXS: small-angle X-ray scattering, FS: Fraunhofer scattering.

angle X-ray or neutron spectrometers are of interest here.

Dynamic light scattering

DLS is an established technique for measuring the average size and the size distribution of particles in a suspension. The technique has advantages of being relatively fast, noninvasive, and requires minimal sample preparation (as compared with electron microscopy for example). But it does require low particle concentration. As well, dynamic light scattering results are often open to misinterpretation if one is unaware of the optical properties of the sample and the method of data analysis. The following discussion reviews some of the basic concepts of dynamic light scattering and outlines some of the pitfalls which are often encountered in data interpretation. A modification of DLS is DWS, which can be used to obtain approximate size information at high particle concentrations and is described later.

All DLS spectrometers use a laser as a source of coherent light. This means that all the light incident on the sample is in phase. At any instant the scattering particles will have a particular set of positions within the scattering volume. Because of these positions, the relative phases of the electric fields of the scattered wavelets will differ at the detector due to the differing optical pathlengths that they must travel. The instantaneous electric field at the detector is the superposition of the fields due to all the scattered wavelets and will, at time t, have a value $E(t)$. At the time, $t + \tau$, which is a very small delay time, τ, later than t, the particles, which are diffusing, will have new positions that are slightly removed from those at the earlier time. Superposition of the slightly shifted scattered wavelets will modify the total electric field at the detector to a new value. $E(t + \tau)$. Therefore, as time progresses, the electric field at the detector will fluctuate as the Brownian random walks that characterize the diffusion of the particles continue.

The main problem in data recovery in the DLS experiment is the extraction of quantitative information from a fluctuating signal. Since detectors are sensitive to light intensity (which is related to the square of the total electric field), small rapidly diffusing particles will yield rapidly fluctuating intensities, whereas larger particles and aggregates generate relatively slow fluctuations. In modern DLS spectrometers the rate of the fluctuations is measured by autocorrelation analysis of a real-time sequence of intensity data that has been digitized into time intervals (sometimes called sampling times) each of which are much shorter than a fluctuation time. The result is an intensity autocorrelation function, $g^{(2)}(\tau)$, where τ is an instrumentally delay time having a value $\tau = kt$, where k is the channel number of the autocorrelation function and t is equivalently the length of the sampling time or the delay time between one channel of $g^{(2)}(\tau)$ and the next. Ideally, the function (see **Figure 3**) has $\tau = 0$ limit equal to the normalized mean square intensity $\langle I(t)^2 \rangle / \langle I(t) \rangle^2 = 2$ and decays to an asymptotic limit (as τ approaches infinity) equal to the normalized mean intensity squared, $\langle I(t) \rangle^2 / \langle I(t) \rangle^2 = 1$. Practically, however, finite pinhole sizes and the detection optics reduce the $\tau = 0$ intercept from its ideal value.

The rate of decay of $g^{(2)}(\tau)$ is indicative of the typical fluctuation time of the scattering signal and of the rate of diffusion of the scatterers. Quantitatively, $g^{(2)}(\tau)$ can be related to the electric field autocorrelation function through the relation

$$g^{(2)}(\tau) = 1 + |g^{(1)}(\tau)|^2 \qquad [2]$$

and when the scatterers are spherical and monodisperse, $g^{(1)}(\tau)$ is related to the translational

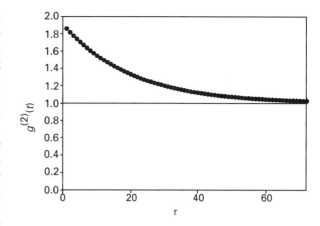

Figure 3 A normalized intensity autocorrelation function.

diffusion coefficient, D, by

$$g^{(1)}(\tau) = e^{-DQ^2\tau} \qquad [3]$$

Thus, for such a system, diffusion coefficients can rapidly be obtained by simple least-squares fits of Equation [3] to electric field autocorrelation functions recovered from the experimental intensity autocorrelation functions, or by linear fits to the natural logarithm of $g^{(1)}(\tau)$.

The more common situation is one where the solution contains a size distribution of scatterers. In this situation

$$g^{(1)}(\tau) = \int\limits_0^\infty G(\Gamma)\, e^{-\Gamma\tau}\, d\Gamma \qquad [4]$$

where $\Gamma = DQ^2$ and $G(\Gamma)$ is the distribution of decay rates that would be obtained from spherical particles whose radius distribution is given by $G(r)$. Since $G(\Gamma)\,d\Gamma = G(r)\,dr$ and the Stokes–Einstein relation is

$$D = \frac{k_B T}{6\pi\eta r} \qquad [5]$$

where k_B = Boltzmann's constant; T = temperature K and η = coefficient of viscosity then Equation [4] becomes

$$g^{(1)}(\tau) = \int\limits_0^\infty G(r)\, e^{-\frac{k_B T}{6\pi\eta r}Q^2\tau}\, dr \qquad [6]$$

In principle, a complete Laplace inversion of Equation [6] would yield the $G(r)$, the radius distribution of the scattering particles. However, this inversion is termed 'ill conditioned' because of mathematical instability and because unattainably high precision in the experimental data is required. Several alternative methods of various complexities have been developed and brief treatments of two of these are given here. The first and most common method of proceeding is called moment analysis or the method of cumulants. Any monomodal distribution can be described in terms of its moments. The object is to obtain these moments of $G(\Gamma)$ without actually performing the inversion. Specifically one attempts to obtain the first and second moments from which one can obtain the mean value and the variance of $G(\Gamma)$ respectively. In this approach the exponential $e^{-\Gamma t}$ in Equation [4] is expanded about the mean value $e^{-\bar{\Gamma}t}$. The result, after taking the natural logarithm of both sides, becomes a polynomial of the form

$$\ell n[g^{(1)}(\tau)] = -\bar{\Gamma}\tau + \frac{\mu_2}{2!}\tau^2 - \frac{\mu_3}{3!}\tau^3 + \cdots \quad [7]$$

Most spectrometers contain fitting routines which output the correlation time $\tau_c = \Gamma^{-1}$, the z-average diffusion coefficient and radius, calculated using Equation [5] and μ_2. Since

$$\mu_2 = \overline{\Gamma^2} \quad [8]$$

the variance of the distribution can be determined from

$$\text{var} = \overline{\Gamma^2} - \bar{\Gamma}^2 \quad [9]$$

Most modern spectrometers provide a visual display of the log-normal distribution that has the same mean radius and variance as obtained from the moments analysis of the data. This can be somewhat misleading if the true size distribution of the sample is not close to being log-normal and can be a serious misinterpretation if the sample is multimodal. Some spectrometers do, however, provide enhancements of the above procedure if a sample is suspected to be multimodal. Further, since no consideration has been given to the fact that large particles scatter more light than small ones, then all the results of moments analysis are intensity weighted or z-averages quantities.

The broader the size distribution, the more these quantities can differ from their corresponding number-average quantities.

A variety of mathematically more sophisticated procedures have been developed to approximate the inversion of Equation [6], but without presupposing the form of the size distribution. Most of these approaches are variants of a discrete method in which $g^{(1)}(\tau)$ is fitted to a sum of m exponential functions, each premultiplied by a weighting factor ω_m. The intent is to minimize the sum of squares

$$\left[\left(g^{(1)}(\tau) - \sum^m \omega_m \exp(-\Gamma_m\tau)\right)\right]^2 \quad [10]$$

with respect to the variables ω_m and Γ_m. As mentioned earlier, this is notoriously unstable, and if no constraints were applied it is essentially impossible to obtain a unique set of best fit parameters. One of the more common restraints is to specify, in advance of the fit, the range and values of a 'trial' set of Γ_m that are expected to span the range of Γs corresponding to the distribution of particle sizes present. The smallest and largest values of the trial Γ_m (i.e. $m = 1$, and $m = m$) can easily be estimated from the final and the initial slopes of $g^{(1)}(\tau)$. In the technique called exponential sampling, the remaining Γ_ms are exponentially spaced between these limiting values. Once all the Γ_m have been preset, only the ω_ms corresponding to the amplitudes, or relative weights of each of the exponentials remain to be determined by the fit. Even with this simplification there is rarely a single solution with a unique set of ω_m, especially if m is 50 or more. However, by constraining all ω_m to be positive [by using non-negative-least squares (NNLS) methods] a unique set of ω_m can usually be found. Since each Γ_m has a corresponding r_m, plots of ω_m versus r_m represent the radius distribution of the sample. If the data are of sufficient quality (often runs of several hours are necessary) then reliable and reproducible plots of the radius distribution, $G(r)$, can be obtained.

The amplitudes ω_m represent the amount of light scattered by each particle size r_m and represent an intensity weighted distribution of radii. To obtain number distributions one must include the relative scattering ability of each size in the distribution. This can be accomplished by including Rayleigh–Gans–Debye or Mie scattering factors in the analysis. The intensity weighted distribution and the number distribution for a set of phospholipid vesicles is shown in **Figure 4**.

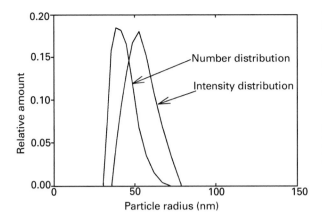

Figure 4 Intensity and number distributions obtained from the same DLS data from dioleoylphosphatidylglycerol vesicles prepared by extrusion through filters of 50 nm radius pore size.

Although a number of 'user friendly' DLS spectrometers and advanced analytical software packages are available from a number of manufacturers, and size distributions are rapidly and neatly displayed on the computer monitor, it is still important that the user perform a number of checks to ensure that these results are reliable. If one is safely operating in the single scattering regime (no multiple scattering effects), then the particle size distribution should remain the same if the sample concentration is doubled or halved. The presence of multiple scattering leads to multiple phase shifts in the scattered light which, in turn, lead to 'fictitious' particles with sizes smaller than the true size. On the other hand, if sample concentrations are too small, number fluctuations can lead to particles whose apparent size is erroneously large. The presence of dust particles in the sample must be scrupulously avoided for the same reason. Secondly, it is essential that experimental runs be sufficiently long that the measured autocorrelation function is a true statistical representation of the sample. Recovery of size distributions requires extremely accurate data and it is not uncommon that runs of several hours are necessary to achieve the required level of data quality. Better quality data and more data always leads to more confidence in the result. For this reason, new multi-angle DLS experiments have been shown to improve the resolution of measured size distribution over single angle determinations.

Samples must be diluted to the point where they are optically almost clear for proper DLS investigations. As a result standard DLS methods cannot be applied to more concentrated suspensions such as milk or paint unless they are diluted extensively. However, DWS can provide estimates of mean particle sizes in concentrated opaque suspensions. The apparatus used in DWS is the same as for DLS except that the incident light and the scattered light are usually conducted through single fibre optics. DWS depends on multiple scattering being so severe that the incident photons experience a random walk (diffusing wave) in the sample before being finally scattered into the returning fibre optic for detection. For back-scatter detection, the simplest form of the DWS electric field autocorrelation function is

$$g^{(1)}(\tau) = \exp\left[-\gamma(6\tau/\tau_0)^{1/2}\right] \qquad [11]$$

where $\tau_0 = (Dk_0)^{-1}$. The constant γ must be determined independently by calibration with known size scatterers or by transmission measurements. Once this is done, however, good estimates of D (and therefore r for spherical systems) have been obtained in a variety of dense scattering media.

Static light scattering, small angle X-ray and neutron scattering

Even though lasers are still commonly used as the source of incident radiation, static scattering does not require coherence. Therefore, other forms of radiation such as X-ray and neutrons, even when pulsed, are suitable sources. The scattering vector dependence of the intensity, $I(Q)$, of the radiation scattered by particles has the form

$$I(Q) = KNm^2 P(Q) S(Q) I_0 \qquad [12]$$

where I_0 is the incident radiation intensity, N is the number of particles in the scattering volume, m is the particle mass and K is an instrumental constant. The functions $S(Q)$ and $P(Q)$ are the interparticle scattering factor and the intraparticle scattering factor respectively. The interparticle scattering factor arises from interference between wavelets scattered from different particles. In a suspension this term contains information on the radial distribution function of the scatterers and on their interaction potential. In a more ordered system, where regular spacings are found, this term results in Bragg diffraction peaks when conditions for constructive interference are satisfied. In the case of light scattering, the avoidance of multiple scattering suspensions usually ensures that particle concentrations are sufficiently dilute that interactions are negligible and $S(Q) = 1$. However, since neutrons and X-rays are more penetrating and are scattered more weakly by particles, they are more commonly used at concentrations where interactions are important and where $S(Q) \neq 1$. However, measurements of $S(Q)$ will not be considered here.

The intraparticle scattering factor $P(Q)$, is related to the structural properties of the scatterer and is the term that contains particle size information. For extremely small noninteracting particles where $r \ll \lambda$, $P(Q)$ approaches unity, and the scattering is said to be isotropic. In this situation Equation [12] greatly simplifies and $I(Q) \propto m^2$. For uniform spherical or globular particles this means that $I(Q) \propto r^6$. If concentrations can be accurately determined by alternative means, then the radius of an unknown particle can be obtained by simple ratios, provided the unknown and the known particles have identical refractive indices.

In the regime where $Q^2 r^2 \ll 1$, the scattering is weakly dependent on Q, and $P(Q)$ can be described by the Guinier approximation.

$$P(Q) = \exp\left(\frac{Q^2 R_g^2}{3}\right) \quad [13]$$

For small angle scattering (e.g. small angle neutron and X-ray scattering) Equation [13] is further approximated by

$$P(Q) = 1 - \frac{Q^2 R_g^2}{3} \quad [14]$$

For larger particles, explicit Rayleigh–Gans–Debye (RGD) expression, for $P(Q)$ are known for several different particle shapes and are available in the literature. These RGD functions are valid if the condition

$$Q^2 r^2 \left(\frac{n^2}{n_0^2} - 1\right) < 1 \quad [15]$$

is satisfied (n is the refractive index of the particle and n_0 is the refractive index of the medium). For the case of scattering of vertically polarized incident light from uniform spherical particles, $P(Q)$ has the form

$$P(Q) = \left(\frac{n}{n_0} - 1\right)^2 \left[\frac{3}{u^3}(\sin u - u\cos u)\right]^2 \quad [16]$$

where $u = Qr$.

When particle sizes are too large to satisfy the RGD criterion then Mie theory must be used to evaluate scattering factors. The need for a more sophisticated theory arises because wavefronts of the incident light do not remain planar as they traverse large particles, thereby altering intraparticle interference and the angular dependence of the scattered light. For vertically polarized incident light, the angle dependence of the scattered light is contained in the S_{11} component of the Mueller matrix:

$$S_{11}(\cos\theta, n, n_0, r) = \sum_{j=1}^{\infty} \frac{2j+1}{j(j+1)}(a_j \pi_j + b_j \tau_j) \quad [17]$$

where

$$\pi_j = \frac{P_j^1 \cos(\theta)}{\sin(\theta)} \quad \text{and} \quad \tau_j = \frac{d}{d\theta} P_j^1 \cos(\theta) \quad [18]$$

and $P_j^1 \cos(\theta)$ is the associated Legendre polynomial. The scattering coefficients, a_j and b_j are known only for spherical systems such as uniform spheres, hollow shells and coated spheres.

Fraunhofer scattering has become a popular optical sizing tool for particles substantially larger than a micrometre in diameter. In this case each particle behaves similarly to a pinhole aperture, and the low angle scattered light is a superposition of Fraunhofer diffraction patterns. This technique is widely used for the analysis of commercial food-based emulsions and sauces.

Since X-rays and neutrons have relatively low scattering cross-sections for most particulate suspensions, RGD scattering factors apply extremely well in the analysis of SAXS and SANS data, even when the scattering particles are above the size limitations imposed by Equation [15]. Expressions equivalent to Equations [12] and [16] have been written down for X-ray and neutron scattering. In addition to particle size analysis, these techniques, SANS especially, offer the ability to use contrast variation methods to enhance the scattering from the structure or substructure of interest. For SANS studies of water-soluble particles, this can be accomplished by varying the $H_2O:D_2O$ ratio in the solvent mixture or by exchanging deuterium for hydrogen in the scatterer. For SAXS the contrast variation can be accomplished if heavy atoms can be strategically attached to the scatterer (**Figure 5**).

Effects due to polydispersity are also important in the analysis of static light scattering data. For a

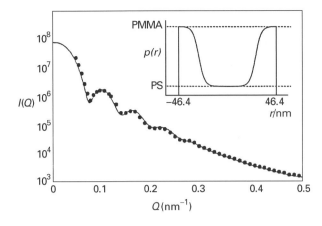

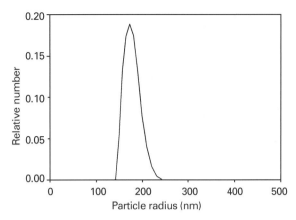

Figure 5 Experimental SAXS data from a composite latex with a polystyrene core and a poly(methyl methacrylate) shell. The number average diameter is 92.8 nm. The inset displays the electron density cross-section of the particle derived from contrast variation studies. Reproduced with permission of Hüthig & Wepf Verlag from Ballauff MB, Bolze J, Dingenouts N, Hickl P. and Pötschke D (1996) Small-angle X-ray scattering on latexes. *Macromolecular Chemistry and Physics* **197**: 3043–3066.

Figure 6 Number distribution of particle sizes from a sample of microfluidized milk obtained by the inversion of $I(Q)$ data from SLS.

polydisperse sample, Equation [12] becomes.

$$I(Q) = I_0 KN \int_0^\infty M^2(r) P(Q, n, n_0, r) G(r) \, \mathrm{d}r \quad [19]$$

Recovery of the distribution of radii $[G(r)]$ present in the sample requires the inversion of Equation [21] and as for dynamic light scattering (see Eqn [6]) this transform is ill conditioned. However, discrete methods analogous to those used for DLS have been applied successfully, thereby allowing size distributions to be obtained from SLS data. Depending on the particle size range, $P(Q, n, n_0, r)$ can be Guinier, RGD, or Mie, with the latter being the most universal. **Figure 6** demonstrates a size distribution of microfluidized milk obtained from the inversion $I(Q)$ from SLS data according to Equation [19] and assuming the sample behaves as uniform spherical particles. This is corroborated by electron microscopic studies in this case. However, knowledge of the shape and structure (e.g. sphere, coated sphere, rod, Gaussian coil, etc.) is not always available. If the wrong morphology is assumed then the inversion results can be meaningless. In addition, for SLS, $I(Q)$ data from particle sizes below ~ 100 nm in diameter is so close to being isotropic that inversion and recovery of size distributions becomes impossible. The opposite limitation arises in SANS and SAXS. Because of the significantly shorter wavelengths of neutrons and X-rays, the upper limit on diameter for particle sizing for these techniques is typically 100 nm (see **Figure 2**). Finally, as for DLS the need for extremely high quality data is just as vital for the successful recovery of particle size distributions, and all inversion routines work best when the distributions are unimodal and relatively narrow. To improve the recovery of broad size distributions and multimodal distributions, SLS measurements have sometimes been combined with sample fractionation techniques. The most successful of these devices, multiangle light scattering-field flow fractionation (MALS-FFF) spectrometers shows great promise for sizing complex particulate mixtures.

List of symbols

$E(t)$ = electric field at time t; $g^{(1)}(\tau)$ = electric field autocorrelation function; $g^{(2)}(\tau)$ = intensity autocorrelation function; I_0, I = intensity of incident and scattered light respectively; k_0, k_s = incident and scattered wave vector respectively; n_0, n = refractive index of the medium and of the particle respectively; Q = scattering vector; r = particle radius; R_g = radius of gyration; T = temperature (K); η = viscosity coefficient; θ = scattering angle; λ = wavelength of incident light; τ = time delay; ω_m = weighting factor.

See also: **Diffusion Studied Using NMR Spectroscopy; Electromagnetic Radiation; Fourier Transformation and Sampling Theory; Laser Applications in Electronic Spectroscopy; Laser Spectroscopy Theory; Light Sources and Optics; Neutron Diffraction, Instrumentation; Scattering Theory.**

Further reading

Ballauff M, Bolze J, Dingenouts, Hickl P and Pötschke D (1996) Small-angle X-ray scattering on latexes. *Macromolecular Chemistry and Physics* **197**: 3043–3066.

Barnes MD, Lermer N, Whitten WB and Ramsey JM (1997) A CCD based approach to high-precision size and refractive index determination of levitated microdroplets using Fraunhofer diffraction. *Review of Scientific Instruments* **68**: 2287–2291.

Berne B and Pecora R (1976) *Dynamic Light Scattering*. New York: John Wiley & Sons.

Brown W (ed) (1993) *Dynamic Light Scattering*. Oxford: Clarendon Press.

Brown W (ed) (1996) *Static Light Scattering, Principles and Development*. Oxford: Clarendon Press.

Bryant G, Abeynayake C and Thomas JC (1996) Improved particle size distribution measurements using multiangle dynamic light scattering 2. Refinements and applications. *Langmuir* **12**: 6224–6228.

Chu B (1991) *Laser Light Scattering*. Academic Press: Boston.

Dalgleish DG, West SJ and Hallett FR (1997) The characterization of small emulsion droplets made from milk proteins and triglyceride oil. *Colloids Surfaces A: Physiochemical and Engineering Aspects* **123–124**: 145–153.

De Vos C, Deriemaeker L and Finsy R (1996) Quantitative assessment of the conditioning of the inversion of quasi-elastic and static light scattering data for particle size distributions. *Langmuir* **12**: 2630–2636.

Filella M, Zhang J, Newman ME and Buffle J (1997) Analytical applications of photon correlation spectroscopy for size distribution measurements of natural colloidal suspensions: capabilities and limitations. *Colloids Surfaces, A: Physicochemical and Engineering Aspects* **120**: 27–46.

Glatter O and Kratky O (eds) (1982) *Small Angle X-Ray Scattering*. London: Academic Press.

Hallett FR, Craig T, Marsh J and Nickel B (1989) Particle size analysis: number distributions by dynamic light scattering. *Canadian Journal of Spectroscopy* **34**: 63–70.

Maret G (1997) Diffusing-wave spectroscopy. *Current Opinion in Colloid and Interface Science* **2**: 251–257.

Pike ER and McNally B (1997) Theory and design of photon correlation and light-scattering experiments. *Applied Optics* **36**: 7531–7538.

Schmitz KS (1990) *Introduction to Quasielastic Light Scattering by Macromolecules*. New York: Academic Press.

Strawbridge KB and Hallett FR (1994) Size distributions obtained from the inversion of $I(Q)$ using integrated light scattering spectroscopy. *Macromolecules* **27**: 2283–2290.

Wyatt PJ (1998) Submicrometer particle sizing by multiangle light scattering following fractionation *Journal of Colloid and Interface Science* **197**: 9–20.

Scattering Theory

Michael Kotlarchyk, Rochester Institute of Technology, NY, USA

ELECTRONIC SPECTROSCOPY
Theory

The term 'scattering' refers to an interaction between an incident radiation and a target material resulting in a redirection, and possibly a change in energy, of the incident radiation. In effect, there is an exchange of momentum and energy between the radiation and target. From a fundamental standpoint, the scattering of photons, neutrons, and charged particles from atomic and molecular systems needs to be treated quantum-mechanically. On the other hand, for electromagnetic radiation having wavelengths near or larger than those in the visible region of the spectrum, a classical wave approach is justified as well. In either case, the aim of any useful scattering theory is to provide the calculational framework for predicting the quantity most readily measured in the laboratory, namely the flux or intensity of radiation scattered into a detector. Ultimately, the connection between a scattering measurement and theory is made through a quantity known as the double-differential scattering cross-section.

Scattering cross-sections

A scattering cross-section, σ, is a quantity proportional to the rate at which a particular radiation–target interaction occurs. More specifically, if the incoming radiation is considered as being composed of quanta or 'particles' (for example, photons or neutrons), a cross-section is a scattering rate (number of scattering events per unit time) per unit incident radiation flux, where the latter is the number of incident particles striking the target surface per unit time per unit area. In cases where the radiation is being treated as a continuous classical wave, as in the case of long-wavelength electromagnetic radiation, scattering cross-sections are determined by dividing the power of the scattered wave by the intensity of the incident wave. Dimensionally, a cross-section represents an area, with the basic unit being the barn, which represents an area of 10^{-28} m^2. A scattering cross-section should not be interpreted as a true geometric cross-sectional area, but as an effective area that is proportional to the probability of interaction between the radiation and target.

In a laboratory setting, one measures differential scattering cross-sections. These are determined by placing a detector at a particular angular position at a substantial distance away from the scattering target. **Figure 1** shows a standard scattering geometry. The incident beam travels in the positive z-direction and one considers the radiation scattered into a differential solid angle $d\Omega$ at polar angle θ and azimuthal angle ϕ. One can then measure an angular differential scattering cross-section (in barn steradian^{-1}) given by

$$\frac{d\sigma}{d\Omega} = \frac{dR/d\Omega}{\Phi_{in}} \qquad [1]$$

where dR is the rate of scattering into solid angle $d\Omega$, and Φ_{in} is the incident flux. The most fundamental type of cross-section is the double-differential scattering cross-section, $d^2\sigma/d\Omega dE'$. The quantity $[d^2\sigma/(d\Omega\, dE')]\, d\Omega\, dE'$ is the number of particles, each with incident energy E, scattered (per unit time) into solid angle $d\Omega$ with energy between E' and $E' + dE'$,

divided by the flux of the incident beam. Once the double-differential cross-section is derived or measured, $d\sigma/d\Omega$ and σ can be calculated by integrating over the energy of the scattered radiation and solid angle.

Development of the double-differential cross-section

From a quantum mechanical standpoint, the calculation of the double-differential cross-section is based on time-dependent perturbation theory, which provides the general framework for calculating transition rates between quantum states.

A scattering event is a type of transition between two states of the combined system of incident particle + scattering medium. Take the system's unperturbed Hamiltonian H_0 to be

$$H_0 = H_S + H_R \qquad [2]$$

where H_S and H_R are the Hamiltonians of the scatterer and radiation, respectively, assuming no interaction between the two. The combined system is in an eigenstate of H_0 both before and after the scattering interaction takes place. Employing Dirac notation, let the kets $|E_i\rangle$ and $|E_f\rangle$ denote the initial and final energy eigenstates of the scatterer, i.e.

$$H_S |E_i\rangle = E_i |E_i\rangle \quad \text{and} \quad H_S |E_f\rangle = E_f |E_f\rangle \qquad [3]$$

If k is the wavenumber (2π/wavelength) of the incident beam, a quantum of the incident radiation is represented by an eigenstate $|k\lambda\rangle$ of definite momentum $\hbar k$ and polarization (or spin) λ (see **Figure 2**). The scattered quantum falls into state $|k'\lambda'\rangle$. These are also energy eigenstates, i.e.

$$H_R |k\lambda\rangle = E |k\lambda\rangle \quad \text{and} \quad H_R |k'\lambda'\rangle = E' |k'\lambda'\rangle \qquad [4]$$

Again, E and E' are the energies of the incident and scattered particles. For the composite system, the initial state $|m\rangle$ and final state $|n\rangle$ are given by the products

$$|m\rangle = |k\lambda\rangle |E_i\rangle \quad \text{and} \quad |n\rangle = |k'\lambda'\rangle |E_f\rangle \qquad [5]$$

which are eigenstates of H_0, so

$$H_0 |m\rangle = E_m |m\rangle \quad \text{and} \quad H_0 |n\rangle = E_n |n\rangle \qquad [6]$$

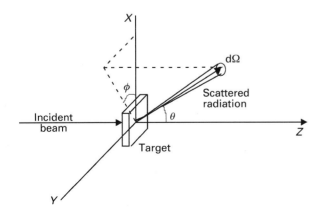

Figure 1 Standard scattering geometry for measuring differential scattering cross-section.

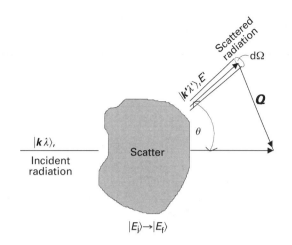

Figure 2 Setup for deriving a double-differential scattering cross-section

where the initial and final energies of the system are $E_m = E_i + E$ and $E_n = E_f + E'$, respectively.

Time-dependent perturbation theory gives rise to the following general result for the transition probability per unit time between initial state $|m\rangle$ and final state $|n\rangle$:

$$W(m \to n) = \frac{2\pi}{\hbar} |\langle n | \mathbf{T} | m \rangle|^2 g(E_n) \qquad [7]$$

Here, $g(E_n)$ denotes the density of final states for the system, where $g(E_n)dE_n$ is the number of final states between E_n and $E_n + dE_n$. $\langle n | \mathbf{T} | m \rangle$ is called the transition matrix or $\mathbf{T}$-matrix, and it is based on V, the interaction potential between the incident radiation and the scatterer, according to the expansion

$$\langle n | \mathbf{T} | m \rangle = \langle n | V | m \rangle \\ + \sum_I \frac{\langle n | V | I \rangle \langle I | V | m \rangle}{E_m - E_I} + \cdots \qquad [8]$$

Basically, the interaction potential induces scattering events and, hence, transitions between the initial and final states of the combined system. When the first term, i.e. matrix element $\langle n | V | m \rangle$, is the only one responsible for transitions, one says that the transition is first-order in the perturbation, and Equation [7] is referred to as Fermi's golden rule. Higher-order transitions require passing through some intermediate states, denoted here by I, between the specified initial and final states.

Once the transition rate is developed for a specific type of interaction, Equation [1] can be used to obtain the angular differential cross-section, $d\sigma/d\Omega$. For a target in a known initial state $|E_i\rangle$ and known

final state $|E_f\rangle$, one writes

$$\left(\frac{d\sigma}{d\Omega} \right)_{if} d\Omega = \frac{dR_{if}}{\Phi_{in}} = \frac{W(E_i, \boldsymbol{k}\lambda \to E_f, \boldsymbol{k}'\lambda')}{\Phi_{in}} \qquad [9]$$

In this equation, Φ_{in} now represents the flux of a single incident particle. It is usually the case, however that the scatterer is not initially in one of its pure eigenstates. Instead, the target is normally in thermal equilibrium at some known temperature T. In this case, the scatterer is in a 'mixed quantum state' and it is necessary to perform a summation over possible initial states, $|E_i\rangle$, weighted according to the Maxwell–Boltzmann probability distribution, $P_i = \exp(-E_i/k_B T)/\Sigma_j \exp(-E_j/k_B T)$, where k_B is Boltzmann's constant. The angular differential cross-section then becomes

$$\frac{d\sigma}{d\Omega} = \frac{\sum_i P_i W(E_i, \boldsymbol{k}\lambda \to E_f, \boldsymbol{k}'\lambda')}{\Phi_{in} d\Omega} \qquad [10]$$

The procedure for obtaining the double-differential scattering cross-section is to now sum over possible final states $|E_f\rangle$ of the scatterer, subject to the condition that the energy of the combined system of radiation + target is conserved. The latter is accomplished by attaching the Dirac delta function, $\delta(E_m - E_n)$, to each term in the sum. Finally, we can introduce

$$\hbar\omega = E - E' \qquad [11]$$

which is an important parameter that represents the energy transferred to the material medium by a quantum of the radiation. Putting these ideas together gives the following expression for the double-differential cross-section:

$$\frac{d^2\sigma}{d\Omega \, d\omega} \\ = \hbar \frac{d^2\sigma}{d\Omega \, dE'} \\ = \frac{\hbar \sum_{i,f} P_i W(E_i, \boldsymbol{k}\lambda \to E_f, \boldsymbol{k}'\lambda') \delta[\hbar\omega - (E_f - E_i)]}{\Phi_{in} d\Omega} \qquad [12]$$

Light scattering – an example

The purpose of this section is to illustrate, for a specific case, the derivation of the angular and double-differential scattering cross-section just outlined. The vehicle chosen is the scattering of visible light from a collection of N optically isotropic atoms in thermal equilibrium. As will be seen, the calculation culminates in a central quantity known as the dynamic structure factor, which not only appears in light scattering, but in the scattering of X-rays and thermal neutrons as well.

For optical frequencies, where the wavelength of light is large compared with the size of an atom, the interaction occurs through the coupling of the radiation to the electric dipole moment μ of each atom in the target material. The Hamiltonian of the atomic system in the presence of the electromagnetic (EM) field of the light beam is

$$H = H_S + H_R - \sum_l \boldsymbol{\mu}_l \cdot \boldsymbol{E}(\boldsymbol{R}_l) \qquad [13]$$

where H_S is the Hamiltonian of the atomic system in the absence of the EM field, H_R is the Hamiltonian of the pure light field, and the last term represents the sum of the interactions between the dipole moment of the lth atom in the system and the light's electric field vector, E, at the location $\boldsymbol{R}_l$ of that atom.

The present derivation of the scattering cross-section is based on a non-relativistic quantum electrodynamic approach. In this picture, the modes of the radiation field are quantized and the electric field is treated as a quantum-mechanical operator that annihilates or creates photons populating the various modes. The field operator is given by

$$\boldsymbol{E}(\boldsymbol{R}_l) = \sum_{k\lambda} \mathrm{i}\sqrt{\frac{\hbar c k}{2\varepsilon_0 L^3}}\, \epsilon_{k\lambda}[a_{k\lambda}\exp(\mathrm{i}\boldsymbol{k}\cdot\boldsymbol{R}_l)$$
$$- a_{k\lambda}^{\dagger}\exp(-\mathrm{i}\boldsymbol{k}\cdot\boldsymbol{R}_l)] \qquad [14]$$

At the initial stage of the calculation it is necessary to imagine that the radiation field is confined to a region of space having volume L^3; however, this volume drops out of the calculation shortly. The term $\epsilon_{k\lambda}$ is the unit polarization vector of photon state $|k\lambda\rangle$. $a_{k\lambda}$ and $a_{k\lambda}^{\dagger}$ are sometimes called the photon 'annihilation' and 'creation' operators. $a_{k\lambda}$ has the effect of lowering the number of photons $n_{k\lambda}$ in state $k\lambda$ to $n_{k\lambda}-1$, whereas $a_{k\lambda}^{\dagger}$ raises the number of photons from $n_{k\lambda}$ to $n_{k\lambda}+1$.

The transition rate for the lth atom is now given by Equation [7], where the interaction potential associated with the transition matrix is $\boldsymbol{\mu}_l \cdot \boldsymbol{E}(\boldsymbol{R}_l)$. If the atom is initially in state $|A\rangle$, then the combined initial state for the field + atom is $|m\rangle = |1_{k\lambda}, 0_{k'\lambda'}\rangle$

$|A\rangle$, indicating that there is one photon in incident mode $k\lambda$ and no photons in the scattered mode $k'\lambda'$. Likewise, the final state is $|n\rangle = |0_{k\lambda}, 1_{k'\lambda'}\rangle|B\rangle$. Because of the action of the operators $a_{k\lambda}$ and $a_{k\lambda}^{\dagger}$, along with the orthogonality property of the photon states, the terms in the perturbation expansion of the transition matrix can only be non-zero if the initial and final states differ by unity in one, and only one, mode of the field. A careful inspection shows that all terms in the perturbation expansion vanish except for the second-order contributions. These are terms that involve intermediate states, $|I\rangle$, of the combined system, which are sometimes referred to as virtual states. For single photon scattering, two types of virtual transitions can occur, as illustrated by the Feynman diagrams in **Figure 3**. In the first type of transition, we have the intermediate state $|I\rangle = |0_{k\lambda}, 0_{k'\lambda'}\rangle|J\rangle$, where one pictures that an intermediate atomic state $|J\rangle$ is created when the incident $k\lambda$-photon is absorbed; the scattered $k'\lambda'$-photon is then born, leaving the atom in final state $|B\rangle$. In the second type of transition, the intermediate state is represented by $|I\rangle = |1_{k\lambda}, 1_{k'\lambda'}\rangle|J\rangle$ where one imagines that the scattered $k'\lambda'$-photon appears before the incident $k\lambda$-photon is absorbed.

Because their existence is not real, transitions to and from the virtual states do not need to conserve energy. However, the energies of the initial state and the final state must match exactly, i.e. $E_A + \hbar c k = E_B + \hbar c k'$. Taking this into account, the transition matrix for atom l works out be the sum of two terms

$$\langle n | \mathbf{T} | m \rangle_l = -\frac{\hbar c}{2\varepsilon_0 L^3}\sqrt{kk'}\exp[\mathrm{i}(\boldsymbol{k}-\boldsymbol{k}')\cdot\boldsymbol{R}_l]$$
$$\times \sum_J \left(\frac{\langle B|\mu_{k'}^l|J\rangle\langle J|\mu_k^l|A\rangle}{E_A - E_J + \hbar c k} \right.$$
$$\left. + \frac{\langle B|\mu_k^l|J\rangle\langle J|\mu_{k'}^l|A\rangle}{E_A - E_J - \hbar c k'} \right) \qquad [15]$$

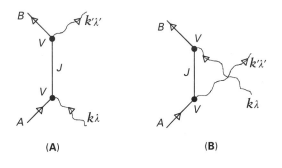

Figure 3 Feynman diagrams for the two types of virtual transitions associated with the scattering of a photon by an atom

where $\mu_k^l = \epsilon_{k\lambda} \cdot \mu_l$ and $\mu_{k'}^l = \epsilon_{k'\lambda'} \cdot \mu_l$. Details leading to this result can be found in the Further reading section.

Carrying out the calculation of the scattering cross-section also requires the incident flux associated with a single photon, which is

$$\Phi_{\text{in}} = c/L^3 \qquad [16]$$

and the density of states factor for a photon (of a specified polarization) scattered into solid angle $d\Omega$ about wave vector k'. If recoil of the atom is neglected, the expression is given by

$$g(E_n) = \left(\frac{L}{2\pi}\right)^3 \frac{k'^2}{\hbar c} \, d\Omega \qquad [17]$$

One now has the following expression for the angular differential cross-section for the scattering of light by the lth atom:

$$\left(\frac{d\sigma}{d\Omega}\right)_l = \frac{kk'^3}{(4\pi\varepsilon_0)^2} \left| \sum_J \left(\frac{\langle B \mid \mu_{k'}^l \mid J \rangle \langle J \mid \mu_k^l \mid A \rangle}{E_A - E_J + \hbar ck} \right. \right.$$
$$\left. \left. + \frac{\langle B \mid \mu_k^l \mid J \rangle \langle J \mid \mu_{k'}^l \mid A \rangle}{E_A - E_J - \hbar ck'} \right) \right|^2 \qquad [18]$$

This result is valid for either inelastic Raman scattering or elastic Rayleigh scattering. For the case of Raman scattering, $|A\rangle$ and $|B\rangle$ are different discrete electronic states of the atom, hence the frequency of the scattered photon is either less than or greater than the frequency of the incident radiation. A down-shifted frequency gives rise to the so-called Stokes spectral line, whereas an up-shifted frequency corresponds to the anti-Stokes line. In the case of Rayleigh scattering, the atom returns to its original state, so that $|B\rangle = |A\rangle$ and $k' = k$. The cross-section then becomes

$$\left(\frac{d\sigma}{d\Omega}\right)_l^{\text{Rayleigh}} = \frac{k^4}{(4\pi\varepsilon_0)^2} \left| \epsilon_{k\lambda} \cdot \tilde{\alpha}_l(k) \cdot \epsilon_{k'\lambda'} \right|^2 \qquad [19]$$

where we define the *atomic polarizability tensor* for atom l to be

$$\tilde{\alpha}_l(k) = \sum_J \left(\frac{\langle A \mid \mu_l \mid J \rangle \langle J \mid \mu_l \mid A \rangle}{E_A - E_J + \hbar ck} \right.$$
$$\left. + \frac{\langle A \mid \mu_l \mid J \rangle \langle J \mid \mu_l \mid A \rangle}{E_A - E_J - \hbar ck} \right) \qquad [20]$$

If the atomic charge distribution is spherically symmetric, then the polarizability is a scalar quantity and

$$\left(\frac{d\sigma}{d\Omega}\right)_l^{\text{Rayleigh}} = \frac{k^4}{(4\pi\varepsilon_0)^2} \left| \epsilon_{k\lambda} \cdot \epsilon_{k'\lambda'} \right|^2 |\alpha_l(k)|^2 \qquad [21]$$

The well-known k^4-dependence of the cross-section is responsible for pronounced Rayleigh scattering of light at the short-wavelength end of the visible spectrum.

Now consider the full system of N atoms in thermal equilibrium. Recall from Equation [15] that the transition matrix for a given atom contains a phase factor $e^{i(k-k')\cdot R_l}$ that depends on the position of that atom. At this point, let us introduce the momentum transferred to the material medium as

$$\hbar Q = \hbar k - \hbar k' \qquad [22]$$

Q is usually referred to as the wave vector transfer (see **Figure 2**). The angular cross-section for coherent scattering from all the atoms of the system is obtained by first summing the transition matrix element over l, then squaring the result, and finally averaging over the initial energy states associated with the thermal motion of the atoms at equilibrium temperature T. In the case of identical atoms, all with the same polarizability α, the cross-section reduces to

$$\left(\frac{d\sigma}{d\Omega}\right) = N \left(\frac{d\sigma}{d\Omega}\right)_0 S(Q) \qquad [23]$$

where $(d\sigma/d\Omega)_0 = [k^4/(4\pi\varepsilon_0)^2] \left| \epsilon_{k\lambda} \cdot \epsilon_{k'\lambda'} \right|^2 |\alpha|^2$ is the Rayleigh scattering cross-section for a single atom and

$$S(\boldsymbol{Q}) = \frac{1}{N}\sum_i P_i \langle E_i \mid \left| \sum_{l=1}^{N} \exp(\mathrm{i}\boldsymbol{Q}\cdot\boldsymbol{R}_l) \right|^2 \mid E_i \rangle$$

$$= \frac{1}{N}\left\langle \sum_{l,l'} \exp[-\mathrm{i}\boldsymbol{Q}\cdot(\boldsymbol{R}_l - \boldsymbol{R}_{l'})] \right\rangle \qquad [24]$$

is called the 'static structure factor'. The large angled bracket represents an ensemble average at temperature T. The function $S(\boldsymbol{Q})$ reflects the spatial correlations between pairs of particles in the scattering medium.

To write the double-differential cross-section, the expression for $(\mathrm{d}\sigma/\mathrm{d}\Omega)$ is summed over the final motional states with the inclusion of an energy-conservation factor, i.e.

$$\frac{\mathrm{d}^2\sigma}{\mathrm{d}\Omega\mathrm{d}\omega} = N\left(\frac{\mathrm{d}\sigma}{\mathrm{d}\Omega}\right)_0 S(\boldsymbol{Q},\omega) \qquad [25]$$

where

$$S(\boldsymbol{Q},\omega) = \frac{1}{N}\sum_{i,f} P_i \left| \langle E_f \mid \sum_{l=1}^{N} \exp(\mathrm{i}\boldsymbol{Q}\cdot\boldsymbol{R}_l) \mid E_i \rangle \right|^2$$

$$\times \delta\left(\omega - \frac{E_f - E_i}{\hbar}\right) \qquad [26]$$

is the aforementioned dynamic structure factor. By using an integral representation of the delta function

$$\delta\left(\omega - \frac{E_f - E_i}{\hbar}\right) = \frac{1}{2\pi}\int_{-\infty}^{+\infty} \mathrm{d}t \exp(-\mathrm{i}\omega t)$$

$$\times \exp[\mathrm{i}(E_f - E_i)t/\hbar] \qquad [27]$$

the structure factor can (with some work) be transformed into the following more useful form:

$$S(\boldsymbol{Q},\omega) = \frac{1}{2\pi}\int_{-\infty}^{+\infty} \mathrm{d}t \exp(-\mathrm{i}\omega t)$$

$$\times \left\langle \frac{1}{N}\sum_{l,l'} \exp[-\mathrm{i}\boldsymbol{Q}\cdot\boldsymbol{R}_l(0)] \exp[\mathrm{i}\boldsymbol{Q}\cdot\boldsymbol{R}_{l'}(t)] \right\rangle \qquad [28]$$

$S(\boldsymbol{Q},\omega)$ is determined by the spatial and temporal correlations between the various atoms in the scattering medium. Integrating the dynamic structure factor over ω produces the expression for the static structure factor, $S(\boldsymbol{Q})$.

Scattering of classical electromagnetic waves

Light scattering can also be treated in the framework of classical EM waves. Since Maxwell's equations accurately describe electromagnetic phenomena almost down to the atomic scale, one might ask whether it is really necessary to go through the detailed quantum treatment previously described. To answer this question, the classical approach is outlined below.

Consider Maxwell's equations for the electric and magnetic field vectors, $\boldsymbol{E}$ and $\boldsymbol{B}$, in a nonmagnetic scattering medium void of free charges and currents. Assigning an $\exp(-\mathrm{i}\omega_0 t)$ time-dependence to all fields, two of the equations (Faraday's law and Ampere's law) become the pair

$$\nabla \times \boldsymbol{E} - \mathrm{i}c k_v \boldsymbol{B} = 0 \qquad [29]$$

$$\nabla \times \boldsymbol{B} + \mathrm{i}\frac{n^2 k_v}{c}\boldsymbol{E} = 0 \qquad [30]$$

$k_v = \omega_0/c$ is the wavenumber of the light in vacuum and $n = (\varepsilon/\varepsilon_0)^{1/2}$ is the refractive index of the medium (ε and ε_0 are the permittivity of the medium and vacuum, respectively.) Eliminating $\boldsymbol{B}$ from the equations results in a single equation for the electric field:

$$\nabla \times \nabla \times \boldsymbol{E} - n^2 k_v^2 \boldsymbol{E} = 0 \qquad [31]$$

The triple product becomes $\nabla\times\nabla\times\boldsymbol{E} = \nabla(\nabla\cdot\boldsymbol{E})-\nabla^2\boldsymbol{E}$. However, Gauss's law for a charge-free region is $\nabla\cdot\boldsymbol{E} = 0$, so the result is the following equation for the field:

$$\nabla^2\boldsymbol{E} + n^2 k_v^2 \boldsymbol{E} = 0 \qquad [32]$$

For a homogeneous medium with uniform index of refraction $n = \bar{n}$, the solution to Equation [32] is a propagating plane-wave with modified wavenumber $k = \bar{n}k_v$. For scattering to occur, there must be local refractive-index fluctuations present:

$$\Delta n^2(\boldsymbol{r}, t) = n^2(\boldsymbol{r}, t) - \bar{n}^2$$

$$= \frac{\varepsilon(\boldsymbol{r}, t) - \bar{\varepsilon}}{\varepsilon_0} = \frac{\Delta\varepsilon(\boldsymbol{r}, t)}{\varepsilon_0} \qquad [33]$$

Equation [32] then becomes

$$\nabla^2 \boldsymbol{E} + k^2 \boldsymbol{E} = -\frac{\Delta\varepsilon(\boldsymbol{r}, t)}{\varepsilon_0} k_v^2 \boldsymbol{E} \qquad [34]$$

$\Delta\varepsilon(\boldsymbol{r}, t)/\varepsilon_0$ represents local fluctuations of the relative permittivity (dielectric constant) in the target. The task of classical light-scattering theory has been reduced to solving Equation [34]. The only role of quantum theory, therefore is to calculate the atomic polarizability α – in other words, the microscopic properties of the scattering medium. Once this is achieved, the polarizability can then be related to the dielectric constant through the well-known Clausius–Mossotti relation.

The scattered wave reaching location $\boldsymbol{r}$ is constructed from the Green's tensor $\tilde{G}(\boldsymbol{r} - \boldsymbol{r}')$, which is the solution to Equation [34] with the right-hand side, or source term, replaced by $\tilde{I}\delta(\boldsymbol{r} - \boldsymbol{r}')$ where $\tilde{I}$ is the unit dyad. The delta function represents a point scatterer situated at $\boldsymbol{r}'$. For a field point far from the scatterer, i.e. $r \gg r'$, the form of the Green's tensor is

$$\tilde{G}(\boldsymbol{r} - \boldsymbol{r}') = (\tilde{I} - \boldsymbol{e}_r \boldsymbol{e}_r) \frac{\exp[ikr]}{4\pi r} \exp[-i\boldsymbol{k}' \cdot \boldsymbol{r}'] \qquad [35]$$

where $\boldsymbol{e}_r$ is the unit vector in the direction of $\boldsymbol{r}$, and $\boldsymbol{k}' = k\boldsymbol{e}_r$ denotes the scattered wave vector. The scattered wave, $\boldsymbol{E}_s(\boldsymbol{r}, t)$, corresponds to the inhomogeneous solution to Equation [34]. It is constructed from the Green's tensor to be the following integral over points $\boldsymbol{r}'$ in the scattering volume:

$$\boldsymbol{E}_s(\boldsymbol{r}, t) = \int_V d^3r' \tilde{G}(\boldsymbol{r} - \boldsymbol{r}') \cdot \frac{\Delta\varepsilon(\boldsymbol{r}', t)}{\varepsilon_0} k_v^2 \boldsymbol{E}(\boldsymbol{r}', t) \qquad [36]$$

If scattering from the medium is sufficiently weak, one is justified in applying the so-called Rayleigh–Gans–Debye approximation. This simply says that the field inside the scatterer may be approximated to

be the same as the field of the incident light. That is, one makes the replacement $\boldsymbol{E}(\boldsymbol{r}', t) = \boldsymbol{E}_0 \exp[i(\boldsymbol{k} \cdot \boldsymbol{r} - \omega_0 t)]$, which leads to

$$\boldsymbol{E}_s(\boldsymbol{r}, t) = \frac{k_v^2}{4\pi\varepsilon_0} \frac{\exp[ikr - \omega_0 t]}{r} (\tilde{I} - \boldsymbol{e}_r \boldsymbol{e}_r)$$

$$\cdot \boldsymbol{E}_0 \int_V d^3r' \Delta\varepsilon(\boldsymbol{r}', t) \exp(i\boldsymbol{Q} \cdot \boldsymbol{r}') \qquad [37]$$

In the case of condensed phases, fluctuations in the dielectric constant may contain a slow time-dependence caused by slowly varying density fluctuations on length-scales of the same order of magnitude as the wavelength of light. The dielectric fluctuations may then be replaced by fluctuations in the number density $\Delta\rho(\boldsymbol{r}', t)$ according to

$$\Delta\varepsilon(\boldsymbol{r}', t) = \left(\frac{\partial\varepsilon}{\partial\rho}\right)_T \Delta\rho(\boldsymbol{r}', t) \qquad [38]$$

where $\Delta\rho(\boldsymbol{r}', t) = \rho(\boldsymbol{r}', t) - \bar{\rho}$ with $\bar{\rho}$ being the average density of the material. Furthermore, if one denotes the spatial Fourier transform of $\Delta\rho(\boldsymbol{r}', t)$ by

$$\Delta\rho(\boldsymbol{Q}, t) = \int_V d^3r' \Delta\rho(\boldsymbol{r}', t) \exp(i\boldsymbol{Q} \cdot \boldsymbol{r}') \qquad [39]$$

then the scattered field becomes

$$\boldsymbol{E}_s(\boldsymbol{r}, t) = \frac{k_v^2}{4\pi\varepsilon_0} \frac{\exp[i(kr - \omega_0 t)]}{r}$$

$$\times (\tilde{I} - \boldsymbol{e}_r \boldsymbol{e}_r) \cdot \boldsymbol{E}_0 \left(\frac{\partial\varepsilon}{\partial\rho}\right)_T \Delta\rho(\boldsymbol{Q}, t) \qquad [40]$$

This clearly shows that light scattering arises because of local density fluctuations in the medium.

The angular differential cross-section is calculated by taking $dP/d\Omega$, where dP is the time-averaged power scattered into solid angle $d\Omega$, and dividing by the intensity of the incident light. This is equivalent to writing $d\sigma/d\Omega = r^2 \langle |\boldsymbol{E}_s|^2 \rangle / |\boldsymbol{E}_0|^2$, so that

$$\frac{d\sigma}{d\Omega} = \frac{k_v^4}{(4\pi\varepsilon_0)^2} \left(\frac{\partial\varepsilon}{\partial\rho}\right)_T^2 \left|(\tilde{I} - \boldsymbol{e}_r \boldsymbol{e}_r) \cdot \boldsymbol{\varepsilon}\right|^2 \langle \Delta\rho(\boldsymbol{Q})\Delta\rho(-\boldsymbol{Q}) \rangle$$

$$[41]$$

where ϵ is the unit polarization vector of the incident beam. The time average $\langle \Delta\rho(Q)\Delta\rho(-Q)\rangle$ is statistically equivalent to an ensemble average. Since the number density can be expressed as a sum of delta functions, i.e.

$$\rho(\mathbf{r}') = \sum_{l=1}^{N} \delta(\mathbf{r}' - \mathbf{R}_l) \qquad [42]$$

its Fourier transform is simply

$$\rho(\mathbf{Q}) = \sum_{l=1}^{N} \exp[i\mathbf{Q}\cdot\mathbf{R}_l] \qquad [43]$$

so that

$$\Delta\rho(\mathbf{Q}) = \sum_{l=1}^{N} \exp[i\mathbf{Q}\cdot\mathbf{R}_l] - \bar{\rho}\delta(\mathbf{Q}) \qquad [44]$$

Therefore, with the exception of forward scattering along the direction of the incident light ($Q = 0$), the quantity $\Delta\rho(Q)$ may be replaced by $\rho(Q)$, and

$$\frac{1}{N}\langle\rho(\mathbf{Q})\rho(-\mathbf{Q})\rangle = \frac{1}{N}\left\langle\sum_{l,l'} \exp[-i\mathbf{Q}\cdot(\mathbf{R}_l - \mathbf{R}_{l'})]\right\rangle \qquad [45]$$

The latter expression is identical to that of the static structure factor $S(Q)$ previously identified in the quantum derivation. Thus, the cross-section $d\sigma/d\Omega$, is again given by $N(d\sigma/d\Omega)_0 S(Q)$, except the basic unit of scattering from the system is now

$$\left(\frac{d\sigma}{d\Omega}\right)_0 = \frac{k_v^4}{(4\pi\varepsilon_0)^2}\left(\frac{\partial\varepsilon}{\partial\rho}\right)_T^2 \left|(\tilde{I} - \mathbf{e}_r\mathbf{e}_r)\cdot\epsilon\right|^2 \qquad [46]$$

The double-differential cross-section is developed in the same way as before, and the result is identical to the expression given by Equation [25], which again involves the dynamic structure factor $S(Q, \omega)$. The latter can now also be written in terms of correla-tions involving Fourier components of the density fluctuations:

$$S(\mathbf{Q},\omega) = \frac{1}{2\pi N}\int_{-\infty}^{+\infty} dt \exp[-i\omega t]$$
$$\times\langle\rho(-\mathbf{Q},0)\,\rho(\mathbf{Q},t)\rangle \qquad [47]$$

Role of dynamic structure factors in spectroscopy

For light scattering, both the quantum and classical derivations of the double-differential cross-section result in an expression that is the product of $(d\sigma/d\Omega)_0$ and the dynamic structure factor, $S(Q, \omega)$. This decomposition of $d^2\sigma/d\Omega d\omega$ into two factors, the first being the angular cross-section from a single scattering unit and the second representing space and time-dependent structure within the system, is what makes the radiation useful as a spectroscopic tool. This separation occurs whenever the radiation couples weakly to the scattering medium. For example, the scattering of X-rays and thermal neutrons fall into this category. X-rays interact with the atomic electrons, and the basic scattering unit involves the classical electron radius $r_0 = (e^2/mc^2)/4\pi\varepsilon_0$ (e and m are the electron's charge and mass, respectively) and

$$\left(\frac{d\sigma}{d\Omega}\right)_0 = r_0^2\,|\epsilon_{k\lambda}\cdot\epsilon_{k'\lambda'}|^2 \left(\frac{k'}{k}\right)^2 \qquad [48]$$

The scattering of thermal neutrons occurs because of the interaction between the neutrons and the atomic nuclei of the target. The basic scattering unit in this case is given by

$$\left(\frac{d\sigma}{d\Omega}\right)_0 = b^2\left(\frac{k'}{k}\right)^2 \qquad [49]$$

where b is called the bound scattering length of an atomic nucleus.

The dynamic structure factor is a function of the two parameters Q and ω. Loosely speaking when the radiation imparts a momentum $\hbar Q$ and an energy $\hbar\omega$ to the system, it, in effect, probes the structure and dynamics of the system with a spatial resolution of

Q^{-1} and a temporal resolution of ω^{-1}. **Table 1** shows the regions of $(\boldsymbol{Q}, \omega)$-space and the corresponding resolutions in real-space and real-time accessible to photon and neutron spectroscopies using current instrumentation and measurement techniques.

One often expresses the dynamic structure factor in the form

$$S(\boldsymbol{Q}, \omega) = \frac{1}{2\pi} \int\limits_{-\infty}^{+\infty} dt \exp\left[-i\omega t\right] F(\boldsymbol{Q}, t) \quad [50]$$

with

$$F(\boldsymbol{Q}, t) = \left\langle \frac{1}{N} \sum_{l,l'} \exp\left[-i\boldsymbol{Q} \cdot \boldsymbol{R}_l(0)\right] \exp\left[i\boldsymbol{Q} \cdot \boldsymbol{R}_{l'}(t)\right] \right\rangle$$

$$[51]$$

$F(\boldsymbol{Q}, t)$ is known as the intermediate scattering function. The dynamic structure factor is just the temporal Fourier transform of $F(\boldsymbol{Q}, t)$. For scattering measurements that involve the frequency, or energy, domain, one determines $S(\boldsymbol{Q}, \omega)$ directly. Examples include optical mixing spectroscopy in light scattering, as well as crystal spectrometry and time-of-flight measurements in the case of thermal neutron scattering. On the other hand, the intermediate scattering function is directly measured in the time domain. Both photon correlation spectroscopy and neutron spin–echo spectroscopy measure $F(\boldsymbol{Q}, t)$.

Detailed discussions of $S(\boldsymbol{Q}, \omega)$ and $F(\boldsymbol{Q}, t)$ for various target systems appear in the Further reading section. To illustrate the type of information that is contained in these functions, consider the specific case of scattering from a fluid consisting of N identical, independently moving scattering centres (or particles). In this case, the scattering from the various particles adds incoherently and is determined by a simplified self intermediate scattering function

governed by the motion of a single particle:

$$F_S(\boldsymbol{Q}, t) = \langle \exp\left[-i\boldsymbol{Q} \cdot \boldsymbol{R}(0)\right] \exp\left[i\boldsymbol{Q} \cdot \boldsymbol{R}(t)\right] \rangle \quad [52]$$

The brackets represent a thermal average at temperature T.

From purely classical considerations, the scattering function always reduces, at least approximately, to the Gaussian form

$$F_S^{cl}(\boldsymbol{Q}, t) = \exp\left[-\tfrac{1}{2} Q^2 W(t)\right] \quad [53]$$

where $W(t)$ is a width function that corresponds to the mean-square displacement of the scattering particle as a function of time. The specific form of the width function depends on the type of motion exhibited by the particle. $W(t)$ for a particle (mass M) in an ideal gas ($V_0 = k_B T/M$) and for a diffusing particle (diffusion coefficient D) are given by

$$W(t) = \begin{cases} V_0^2 t^2 & \text{for ideal gas} \\ 2Dt & \text{for diffusing particle} \end{cases} \quad [54]$$

Figure 4 is a comparison of the behaviour of these functions, along with $W(t)$ for a particle moving through a typical liquid. Fourier transforming the $F_S^{cl}(\boldsymbol{Q}, t)$s gives the corresponding forms for the (self) dynamic structure factors:

$$S_S^{cl}(\boldsymbol{Q}, \omega) = \begin{cases} (2\pi Q^2 V_0^2)^{-1/2} \exp\left[-\omega^2/2V_0^2 Q^2\right] \\ \quad \text{for ideal gas} \\ \dfrac{1}{\pi}\left[\dfrac{DQ^2}{\omega^2 + (DQ^2)^2}\right] \\ \quad \text{for diffusing particle} \end{cases}$$

$$[55]$$

Table 1 Accessible regions of (Q,ω)-space and corresponding resolutions in real space and time for various types of spectroscopies[a]

Type of spectroscopy	Q (μm^{-1})	$\hbar\omega$ (eV)	Spatial resolution (μm)	Temporal resolution (s)
Optical mixing	$10^{-1} - 10^{1}$	$10^{-9} - 10^{-6}$	$10^{-1} - 10^{1}$	$10^{-10} - 10^{-6}$
Photon correlation	$10^{-1} - 10^{1}$	$10^{-15} - 10^{-9}$	$10^{-1} - 10^{1}$	$10^{-6} - 10^{0}$
X-ray	$10^{1} - 10^{5}$	$10^{1} - 10^{2}$	$10^{-5} - 10^{-1}$	$10^{-17} - 10^{-16}$
Thermal neutron	$10^{2} - 10^{5}$	$10^{-8} - 10^{1}$	$10^{-5} - 10^{-2}$	$10^{-16} - 10^{-6}$
Neutron spin-echo	$10^{1} - 10^{2}$	$10^{-9} - 10^{-2}$	$10^{-2} - 10^{-1}$	$10^{-14} - 10^{-7}$

[a] Ranges are approximate and to nearest order of magnitude.

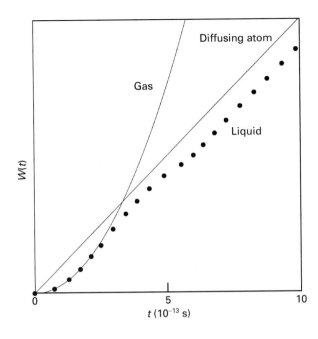

Figure 4 Mean-square displacement function for a particle in an ideal gas, a diffusing particle, and a particle moving through a typical liquid.

For a fixed value of Q, the functions $S_S^{cl}(Q, \omega)$ give the line shapes for the gas and diffusion cases to be Gaussian and Lorentzian, respectively.

The latter results neglect any quantum corrections to the self intermediate scattering functions. In the case of the ideal gas, the full quantum-mechanical expression for $S_S(Q, \omega)$ is obtained from the corresponding classical one simply by multiplying by two exponential factors, i.e.

$$S_S(\boldsymbol{Q}, \omega) = \exp\left[\hbar\omega/2k_B T\right] \exp\left[-\hbar^2 Q^2/8Mk_B T\right]$$
$$S_S^{cl}(\boldsymbol{Q}, \omega)$$

[56]

The first factor, $\exp(\hbar\omega/2k_B T)$, is known as the 'detailed balance factor' – it produces an asymmetry in the quantum-mechanical structure factor, whereas the classical one is an even function of ω. The second factor, $\exp(-\hbar^2 Q^2/8Mk_B T)$, can also be written as $\exp(-E_R/4k_B T)$, where $E_R = \hbar^2 Q^2/2M$ is the recoil energy of the target particle. Hence this exponential factor is known as the recoil factor. Equation [56] is exactly true only in the ideal gas case; however, it is also approximately valid for other scattering systems as well.

List of symbols

$|A\rangle$ = initial state of atom; $a_{k\lambda}$ = photon annihilation operator for mode $k\lambda$; $a_{k\lambda}^\dagger$ = photon creation operator for mode $k\lambda$; b = bound scattering length of atomic nucleus; $\boldsymbol{B}$ = magnetic field vector; $|B\rangle$ = final state of atom; c = speed of light in vacuum; D = diffusion coefficient; dP = time-averaged power scattered into solid angle $d\Omega$; dR = rate of scattering into solid angle $d\Omega$; $(d\sigma/d\Omega)_l$ = angular differential cross-section for scattering by the lth atom; $(d\sigma/d\Omega)_{if}$ = angular differential cross-section for target in specific initial state $|E_i\rangle$ and final state $|E_f\rangle$; $d\sigma/d\Omega$ = angular differential cross-section; $d^2\sigma/d\Omega dE'$ = double differential cross-section; $(d\sigma/d\Omega)_0$ = angular differential cross-section from single scattering unit; $d^2\sigma/d\Omega d\omega = d^2\sigma/d\Omega dE'$ multiplied by $\hbar$; $(d\sigma/d\Omega)_l^{Rayleigh}$ = angular differential cross-section for Rayleigh scattering by the lth atom; $d\Omega$ = differential solid angle; e = electronic charge; e_r = unit vector in direction of r; E = energy of incident radiation quantum; $\boldsymbol{E}$ = electric field vector; $\boldsymbol{E}_0$ = electric field vector for incident wave; E_A = initial energy of atom; E' = energy of scattered radiation quantum; E_f = energy of target medium after scattering; $|E_f\rangle$ = quantum state of target medium after scattering; E_i = energy of target medium before scattering; E_I = energy of intermediate state of radiation/target system; $|E_i\rangle$ = quantum state of target medium before scattering; E_J = energy of intermediate atomic state; E_m = energy of radiation/target system before scattering; E_n = energy of radiation/target system after scattering; E_R = recoil energy of target particle; $\boldsymbol{E}_s$ = electric field vector for scattered wave; $F(Q, t)$ = intermediate scattering function; $F_S(Q, t)$ = self intermediate scattering function; $F_S^{cl}(Q, t)$ = classical self intermediate scattering function; $g(E_n)$ = density of final states; $\tilde{G}$ = Green's tensor; $\eta = h/2\pi$, where h is Planck's constant; H_0 = Hamiltonian of incident radiation + scattering medium, with no interaction present; H_S = Hamiltonian of scattering medium; H_R = Hamiltonian of incident radiation; i = square-root of -1; $\tilde{I}$ = unit dyad; $|I\rangle$ = an intermediate state of radiation/target system; $|J\rangle$ = an intermediate atomic state; k = wavenumber of incident radiation; k_B = Boltzmann's constant; k_v = wavenumber of light in vacuum; $\boldsymbol{k}$ = wavevector of incident radiation; k' = wavenumber of scattered radiation; $\boldsymbol{k}'$ = wavevector of scattered radiation; $|k\lambda\rangle$ = quantum state of incident radiation quantum; $|k'\lambda'\rangle$ = quantum state of scattered radiation quantum; L^3 = volume of region in which radiation is confined; m = mass of electron; $|m\rangle$ = quantum state of radiation/target system before scattering; M = mass of scattering particle;

n = index of refraction of medium; $|n\rangle$ = quantum state of radiation/target system after scattering; $n_{k\lambda}$ = number of photons in mode $k\lambda$; $\bar{n}$ = average index of refraction of medium; N = number of atoms; P_i = Maxwell–Boltzmann probability distribution for target in different initial energy states $|E_i\rangle$; Q = wave vector transfer; r = position vector of field-point where radiation is detected; r' = position vector of source-point inside scattering medium; r_0 = classical electron radius; R_l = location of lth atom; $S(Q)$ = static structure factor; $S_S(Q, \omega)$ = self dynamic structure factor; $S_S^{cl}(Q, \omega)$ = classical self dynamic structure factor; $S(Q, \omega)$ = dynamic structure factor; t = time; T = absolute temperature; $\mathbf{T}$ = transition matrix; V = interaction potential between radiation and scatterer; V_0 = speed, $k_B T/M$; W = transition probability per unit time; $W(t)$ = width function or mean-square displacement function; $\tilde{\alpha}_l$ = atomic polarizability tensor for lth atom; α_l = scalar polarizability of lth atom; $\Delta\varphi(Q, t)$ = spatial Fourier transform of local number-density fluctuations in scattering medium; Δn = local fluctuation in refractive index; $\Delta\varepsilon$ = local fluctuation in permittivity; $\Delta\rho(r', t)$ = local fluctuation in number density in scattering medium; ε = electric permittivity of scattering medium; ε_0 = electric permittivity of vacuum; $\epsilon_{k\lambda}$ = unit polarization vector for mode $k\lambda$; $\bar{\varepsilon}$ = average permittivity of medium; θ = polar angle; λ = polarization of incident radiation; λ' = polarization of scattered radiation; μ_l = electric dipole moment of lth atom; $\mu_{k'}^l$ = product of $\epsilon_{k\lambda}$ and μ_l; $\mu_{k'}^l$ = product of $\epsilon_{k'\lambda'}$ and μ_l; $\bar{\rho}$ = average number density in scattering medium; $\rho(r', t)$ = local number density in scattering medium; $\rho(Q, t)$ = spatial Fourier transform of local number density in scattering medium; σ = scattering cross-section; Φ_{in} = incident flux; ϕ = azimuthal angle; ω = angular-frequency shift of radiation due to scattering; ω_0 = angular frequency of light.

See also: **Electromagnetic Radiation; Inelastic Neutron Scattering, Applications; Inelastic Neutron Scattering, Instrumentation; Light Sources and Optics; Neutron Diffraction, Theory; Rayleigh Scattering and Raman Spectroscopy, Theory; Scattering and Particle Sizing, Applications; X-Ray Spectroscopy, Theory.**

Further reading

Berne BJ and Pecora R (1976) *Dynamic Light Scattering.* New York: Plenum Press.

Bohren C and Huffman D (1983) *Absorption and Scattering of Light by Small Particles.* New York: John Wiley & Sons.

Chen SH and Kotlarchyk M (1997) *Interaction of Photons and Neutrons with Matter.* Singapore: World Scientific.

Chu B (1974) *Laser Light Scattering.* San Diego: Academic Press.

Foderero A(1971) *The Elements of Neutron Interaction Theory.* Cambridge, MA: M.I.T. Press.

Heitler W (1954) *The Quantum Theory of Radiation.* London: Oxford University Press.

Kerker M (1969) *The Scattering of Light and Other Electromagnetic Radiation.* New York: Academic Press.

Louisell WH (1973) *Quantum Statistical Properties of Radiation.* New York: John Wiley & Sons.

Lovesey SW (1984) *Theory of Neutron Scattering from Condensed Matter.* Oxford: Clarendon Press.

Mott NF and Massy HSW (1949) *The Theory of Atomic Collisions*, 2nd edn. Oxford: Oxford University Press.

Sector Mass Spectrometers

R Bateman, Micromass, Wythenshaw, Manchester, UK

MASS SPECTROMETRY

Methods & instrumentation

The magnetic sector is the oldest type of mass analyser. Experiments initiated in 1906 by JJ Thompson demonstrated the different deflections of various 'positive' rays in superposed crossed magnetic and electric fields. In 1912, using this equipment, the isotopes of neon were discovered, and over the next decade magnetic sector instruments built separately by Aston and by Dempster were used to identify most of the isotopes and determine their relative abundance. In 1931 Bainbridge added a velocity filter to Dempster's basic design to limit the energy distribution of ions entering the magnet. This refinement improved mass resolution and the relative mass measurement precision of the isotopes. This allowed the 'packing fraction' of the isotopes to be determined.

Industrial interest in mass spectrometry grew in the following decade and in 1942 the first commercial mass spectrometer, a magnetic sector instrument, was built by Consolidated Engineering Corporation and sold to the Atlantic Refining Company for the analysis of gasoline refining streams. Magnetic sector instruments are now only one of several different types of mass spectrometer manufactured commercially. Nevertheless magnetic sector mass spectrometers remain the instrument of choice for a number of important applications, in particular in the areas of target compound trace analysis, accurate mass measurement, isotope ratio measurement and fundamental ion chemistry studies.

Principal of operation

Ions with mass (m) and charge (ze) accelerated through an electrical potential difference (V) will have velocity (v) and kinetic energy (KE) where:

$$\text{KE} = zeV = \tfrac{1}{2}\,mv^2 \qquad [1]$$

Ions with charge (ze) moving through a magnetic field (B) with velocity (v) are subject to the Lorentz force (F) orthogonal to the direction of the field and direction of travel. Consequently ions travel with a circular trajectory with radius (r_m) in which the centripetal force is provided by the Lorentz force:

$$F = B\,ze\,v = m(v^2/r_m) \qquad [2]$$

Eliminating (v) from Equations [1 and 2]:

$$m/ze = B^2 r_m^2 / 2V \qquad [3]$$

In the SI system of units for B (tesla), r_m (metres), V (volts) and m/z (daltons per unit of electronic charge) Equation [3] becomes:

$$m/z = 4.786 \times 10^7 \cdot B^2 r_m^2 / V \qquad [4]$$

In its simplest form, the mass spectrometer transmits ions of a particular mass and charge from source to detector via a circular trajectory through a magnetic sector. Here the magnetic sector is a mass filter, and a mass spectrum can be recorded by scanning the magnetic field or accelerating voltage with serial detection of mass peaks. Alternatively, several detectors may simultaneously record several different ion masses, each taking a different trajectory. This principle may be extended to incorporate a continuous detector array in which a complete portion of the mass spectrum is recorded simultaneously. The magnet sector is usually constructed from a laminated iron-cored electromagnet with low inductance coils to allow fast scanning or switching. Superconducting magnets are not appropriate since they are not readily scanned. However, permanent magnets may be used for certain dedicated applications such as isotope ratio determinations.

Optics

An ion moving in a magnetic field is dispersed with respect to its momentum (ρ). The momentum of an ion is given by:

$$\rho = mv = (2m \cdot \mathrm{KE})^{1/2} \qquad [5]$$

Therefore ions with the same kinetic energy (KE) are, in effect, dispersed with respect to their mass. In addition, the shape of the magnetic sector can be designed to have ion directional focusing properties. A magnetic sector of a particular shape and size will have a particular combination of ion dispersion and directional focusing characteristics.

Single focusing

An arrangement which combines an ion source with an ion beam width defining slit, a magnetic sector with convergent directional focusing characteristics and an ion 'collector' slit positioned at the image point of the source slit is defined as single focusing. The magnetic sector directional focusing characteristics can be designed to a very high order, but its imaging properties will be limited by any spread in ion energy. Consequently such instruments are used where the energy spread of ions is low, for example with an electron impact ionization source, and where a high resolution is not required (**Figure 1**).

Stigmatic focusing

An inhomogeneous magnetic field has field components in more than one direction. If the magnetic field is inhomogeneous the focusing in the direction of dispersion is modified, and focusing in the direction orthogonal to the direction of dispersion is introduced. As the focal power in the direction of dispersion reduces, the focal power in the direction orthogonal to that increases, and when the focal power in these two directions become the same the sector has stigmatic focusing. This can provide more efficient ion transmission from source to detector,

although higher order focusing terms in the direction of dispersion are usually compromised and ultimate resolution reduced.

A homogeneous magnetic field in which the shape of the sector has been designed to have non-normal ion entry and/or ion exit will also modify focusing in the direction of dispersion and introduce focusing in the orthogonal direction. In a similar way such a magnetic sector can be designed to provide stigmatic focusing (**Figure 2**).

Mass dispersion and resolution

The mass dispersion coefficient (D_m) of a single focusing magnetic sector is proportional to the radius of curvature (r_m) of the ion beam trajectory in the magnetic field. The spatial separation (y) of two similar masses with mean mass (m) and mass difference (Δm) is related to the mass dispersion coefficient by:

$$y = D_m \cdot \Delta m / m \qquad [6]$$

The ion beam width (w_b) at the image position is related to the ion beam width defined by the source slit (w_s), the image lateral magnification (M) and the sum of the imaging aberration coefficients (α) by:

$$w_b = M \cdot w_s + \alpha \cdot r_m \qquad [7]$$

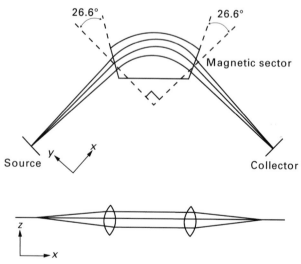

Figure 2 Single-focusing magnetic sector, with stigmatic focusing in 'Y' and 'Z' directions.

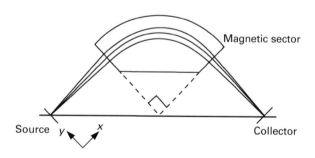

Figure 1 Single-focusing magnetic sector, with focusing in 'Y' direction only.

The mass resolving power ($m/\Delta m$) for a 'collector' slit width (w_c) is given by:

$$m/\Delta m = D_m/(w_b + w_c) = D_m/(M \cdot w_m + w_c + \alpha \cdot r_m)$$

[8]

Thus, the mass dispersion coefficient and the slit widths are the most significant parameters in setting the resolution, and the ratio of dispersion to magnification (D_m/M) is a key figure of merit in the design of an instrument. However, ultimate resolution is limited by the imaging aberrations.

Double focusing

It has been pointed out that the magnetic sector will disperse ions with respect to their momentum, and hence with respect to their mass if they are monoenergetic. However, ions will normally have a spread in kinetic energy, depending on the nature of the ion source, and this will broaden the image width. This usually becomes the limiting factor to achieving high resolution.

Momentum dispersion may be considered a combination of mass dispersion and energy dispersion. An electric sector will only disperse ions with respect to their energy, and so if an electric sector is combined with a magnetic sector the overall energy dispersion will be modified. A combination of magnetic sector and electric sector, which is directional focusing and in which the overall energy dispersion is zero, is said to be double focusing. Such a combination does not suffer the same image broadening due to spread in ion kinetic energy, and can achieve much higher resolution.

If the first sector has energy dispersion D_{e1}, and the second sector has energy dispersion D_{e2}, and image magnification of the second sector is M_2, then the overall energy dispersion (D_e) is

$$D_e = M_2 \cdot D_{e1} + D_{e2} = 0 \qquad [9]$$

If the electric sector precedes the magnetic sector, the double-focusing arrangement is said to be 'forward' geometry, and 'reverse' geometry if the electric sector follows the magnetic. Either arrangement is equally effective as a double focusing optical system. However, the electric sector can provide additional benefits, which are different for each 'geometry'.

Forward geometry Here the electric sector, which precedes the magnetic sector, can be used to further

advantage if designed to have a lateral magnification less than unity. This will increase the dispersion to magnification ratio (D_m/M) and the source slit will be larger than it otherwise would have been to achieve a required resolution. The larger source slit usually results in higher transmission and reduced susceptibility to contamination.

Reverse geometry Here the electric sector follows the magnetic sector, and therefore can no longer be used to increase the dispersion-to-magnification ratio. However, the electric sector can now be used as an energy filter to prevent transmission of ions with very different energies – such as the product ions from metastable ion decompositions. In a single-focusing system, or a 'forward' geometry double-focusing arrangement, these product ions appear as defocused artefact peaks in the spectrum when they are generated in the field-free region before the magnetic sector. If the decompositions occur in the magnetic field this usually results in an increase in background noise in the spectrum. In the 'reverse' geometry arrangement these are removed by the electric sector. The electric sector can also be used to further advantage by providing a means of analysing products from metastable ions resulting from ion decompositions in the field-free region between the magnetic and electric sectors. This type of analysis is considered in more detail in the Metastable ion analysis section.

Split forward and reverse geometry Here the electric sector is divided into two smaller electric sectors, positioned before and after the magnetic sector. Each electric sector can be smaller than the single electric sector in a 'forward' or 'reverse' geometry, and provided the overall energy dispersion is zero, the arrangement can still be double focusing. This arrangement can benefit both from an increased dispersion-to-magnification ratio, by appropriate design of the first electric sector, and from removal of artefact peaks and background noise – resulting from metastable ion decompositions – by virtue of the second electric sector. Again the final electric sector can provide the means for directly analysing the products from metastable ions dissociating in the field-free region immediately preceding that electric sector (**Figure 3**).

Parallel detection

Unlike the quadrupole type of mass filter, a magnetic sector may be designed to record the signal from several different masses simultaneously. This is referred to as parallel detection.

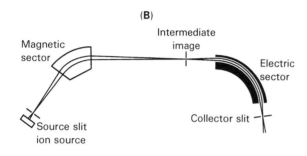

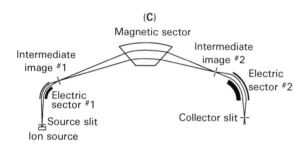

Figure 3 Modified 'Nier–Johnson'-type double-focusing magnetic sector mass spectrometer (A) with 'forward' geometry, (B) with 'reverse' geometry, (C) with split 'forward' and 'reverse' geometry.

Multiple collectors

Parallel detection provides a means of accurately recording the relative abundance of two or more different masses, since measurement of the ratio of these peak intensities is not susceptible to fluctuations or drift in the ionization source, or to rapidly changing sample concentration such as encountered in chromatography. Magnetic sector mass spectrometers designed to simultaneously transmit a number of different masses, each with a different trajectory, and incorporating multiple discrete detectors, are used to make the most accurate isotope ratio determinations. Instruments designed specifically for accurate isotope ratio determinations have included combinations of up to 16 Faraday and/or ion counting detectors, and isotope ratios may be determined to a precision of better than 10 ppm using such equipment.

Continuous array detectors

A magnetic sector mass spectrometer with a single detector may be used to record a mass spectrum by scanning and sequentially detecting mass peaks. The duty cycle for recording each mass in the spectrum is generally poor, and the higher the resolution or the wider the mass range the poorer the duty cycle. An array detector allows simultaneous acquisition over a range of masses thereby improving the duty cycle when used to record a spectrum. Array detectors employing high-density arrays of discrete charge-sensitive detectors or 'single ion' position sensitive detectors are very sensitive, although they are usually limited in size. This imposes a limit on the combination of mass resolution and mass range of simultaneous detection. For example, an array of 2048 discrete detectors can be arranged to simultaneously detect ions over a 10% mass range with a resolution of ~ 4000 (FWHM). A 'single ion' position-sensitive detector simultaneously recording over a similar mass range may achieve approximately twice that resolution, but cannot cope with two ions arriving within one measurement period. This imposes an upper limit on the total ion current onto the detector of between 10^5 and 10^6 ions/s, which in turn imposes a limit on the practical mass range for simultaneous detection.

Alternatively, a large-scale photosensitive emulsion plate detector may be used to simultaneously record a large mass range with acceptable mass resolution. Such detectors require the mass spectrometer design to allow simultaneous transmission of a wide range of masses and focusing on to a flat plane. The double focusing arrangement first described by Mattauch and Herzog in 1934 gives first-order double focusing over the entire length of the 'photo-plate'. Mass spectrometers based on this theory have been constructed to successfully record ions simultaneously over a 60:1 mass ratio (**Figure 4**).

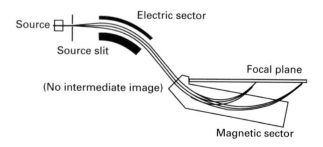

Figure 4 Modified 'Mattauch–Herzog'-type double-focusing magnetic sector with flat focal plane.

High resolution

The combination of magnetic and electric sectors to form a double-focusing arrangement includes sufficient degrees of freedom in the choice of design to allow higher order focusing to be achieved. A design constructed by Nier and Roberts in 1951, consisting of a 90° electric sector and 60° magnetic sector, arranged sequentially in a C-shaped geometry, was shown by Nier and Johnson to have first-order energy focusing and second-order directional focusing. Many modern designs of double-focusing instruments, in which all second-order directional- and energy-focusing terms are either zero or near zero, are essentially variations of the 'Nier–Johnson' geometry. Commercially manufactured instruments of this type achieve resolving powers in excess of 150 000 by the 10% valley definition (based on peak width at 5% height) or in excess of 350 000 by the FWHM definition (based on peak width at 50% height). They are used in particular in the petroleum industry for characterization of complex oil mixtures (**Figure 5**).

Accurate mass measurement

The capacity for high resolution enables double-focusing magnetic-sector mass spectrometers to be used for accurate mass measurement. Accurate mass is always determined by reference to one or more known masses simultaneously introduced into the mass spectrometer, and there are two different methods in common use.

Peak matching

This is the simplest and generally most accurate method of measuring the mass of a single peak. Here the magnetic field strength is held constant and the electric sector field and all other voltages are switched together such as to transmit in turn the known and unknown mass peaks. The difference between the two masses, ideally, is small. A narrow voltage scan is applied as each mass is selected to display the peak profile, and the voltages are adjusted until the two peak profiles are exactly superimposed. The unknown mass is calculated from the ratio of the two voltage settings and the known reference mass. A refinement of this method entails the use of two known reference masses, preferably 'bracketing' the unknown mass, and adjusting the switched voltages until all three peak profiles are exactly superimposed. This allows correction for certain instrumental offsets, and a mass accuracy of better than 1 ppm is routinely achievable.

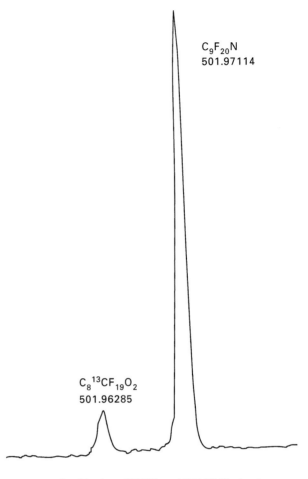

$C_9F_{20}N$
501.97114

$C_8{}^{13}CF_{19}O_2$
501.96285

Figure 5 Doublet from PFTBA and FOMBLIN showing a resolving power of 200 000 (10% valley definition), or 450 000 (full width at half height definition).

Alternatively voltage scanning through each of the three mass peaks while measuring the peak profiles with a high sampling rate analogue-to-digital converter, and computing the 'centre of mass' of each peak profile, will yield a similar mass measurement accuracy.

Magnetic scanning

The 'peak matching' method is not appropriate if it is required to quickly measure the mass of a large number of peaks over a wide mass range. Here it is more appropriate to scan the magnetic field instead. However, the magnetic field strength from an iron-cored magnet is difficult to scan precisely. Consequently, a reference mixture is simultaneously introduced into the ion source such as to give a large number of known reference mass peaks at regular intervals throughout the mass range of interest. It is important that these mass peaks do not 'interfere' with any other mass peaks in the spectrum. The known and unknown masses are 'digitized' and recorded into a data system, and from each peak centre the accurate mass of the unknown peaks may be calculated. Perfluorokerosene (PFK) is commonly used as a 'reference mixture' since its 'electron impact' spectrum exhibits peaks every 12 or 14 daltons from m/z 31 to beyond m/z 800. Furthermore, the peaks are mass deficient and may easily be resolved from most other organic compounds. This method can yield an average mass measurement accuracy of about 1 milli-dalton over this mass range.

High-resolution selective ion recording

A double-focusing magnetic-sector mass spectrometer can be used to select and record the response from target compounds at high resolution and with a high sensitivity. The high resolution enables chemical background masses to be eliminated and consequently allow a lower detection level to be achieved. Selected ion recording provides a much better duty cycle, and therefore sensitivity, than scanning.

The detection and quantification of polychlorinated dibenzo-p-dioxins, and in particular the 2,3,7,8-tetrachlorinated dibenzo-p-dioxin (2,3,7,8-TCDD), is a major application for double-focusing magnetic-sector mass spectrometers. Despite extensive clean-up procedures, samples still contain compounds such as polychlorinated biphenyls and benzyl phenyl ethers, which have the same nominal masses as the compounds of interest. The sample is 'spiked' with a known amount of the ^{13}C isotope labelled form of 2,3,7,8-TCDD, introduced via gas chromatography and recorded by 'high-resolution' mass spectrometry. The measurement is quantified by comparison of the native dioxin response to that from the ^{13}C-labelled form, and verified by confirmation of the ratio of the major isotopes of both the native and the ^{13}C-labelled dioxins. At 10 000 resolving power (10% valley definition) the detection level for 2,3,7,8-TCDD is about 1 femtogram, or 3 attomole (**Figure 6**).

Metastable ion analysis and MS/MS

The decomposition of a precursor or 'parent' molecular ion to a product or 'daughter' ion, while in flight through a single focusing magnetic sector mass spectrometer, can give rise to an artefact peak in the mass spectrum. However, such artefact peaks can be instructive in understanding the structure of the precursor ion. Methods have been developed for double-focusing instruments to eliminate such

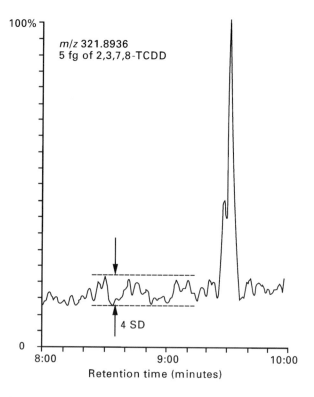

Figure 6 Detection of 5 femtogram of 2,3,7,8-TCDD with S/N greater than 10:1 from an injection of 1 μL onto a GC capillary column and 'selective ion recording' of four masses plus a 'lock mass' at 10 000 resolving power (10% valley definition).

artefact peaks from mass spectra, to specifically record daughter ions from a selected parent ion, and to detect specific parent–daughter ion transitions as a means of detecting target compounds or classes of compounds.

If a parent ion (mass m_p) fragments in a field-free region the daughter ion (mass m_d) retains the velocity of the parent, only modified by the addition of a velocity component as a result of energy released in the decomposition reaction. Hence, the ratio of the daughter ion kinetic energy to that of the parent is equal to ratio of their masses (m_d/m_p). The additional velocity component is randomly distributed and this is seen as a superimposed energy spread.

Single-focusing magnetic sector

If a parent ion fragments in the flight path between the source and the magnetic field, or the 'first field-free region', the daughter ion has an energy relative to that of the parent in proportion to the ratio of their masses (m_d/m_p). Ions originating in the ion source are accelerated to the same kinetic energy. Consequently, after traversing the magnetic sector, the daughter ion appears at a different mass (m^*) when viewed relative to the normal spectrum. The apparent mass is given by:

$$m^* = m_d \cdot (m_d/m_p) \qquad [10]$$

The daughter ion peak is often recognizable since it is usually broader as a result of its increased energy spread. The parent and daughter ions can often be deduced from m^* with reasonable confidence if the sample is pure, but not if a mixture. This can provide useful additional information or become a source of noise depending on the circumstances.

Double-focusing magnetic sector

If a parent ion fragments in the 'first field-free region' of a double-focusing mass spectrometer, the daughter ion, which will have a lower kinetic energy, is unlikely to be transmitted by the electric sector and therefore will not appear in the mass spectrum. If a parent ion fragments in the field-free region preceding the final sector the situation is more complex. The daughter ion can be transmitted through the final (magnetic) sector of a 'forward geometry' arrangement just as it would be in a single-focusing sector design, but will not be transmitted through the final (electric) sector of a 'reverse geometry' or 'split geometry' arrangement. In these latter cases, if the final electric sector only is scanned, the daughter ion will be transmitted when the electric field is at value equal to (m_d/m_p) of the value required to transmit 'full energy' ions originating from the source. The result is a spectrum of all the daughter ions of the parent ion transmitted by the magnetic sector. The electric sector scan is a spectrum of ion energies, and therefore will also exhibit the energy distribution of daughter ions, albeit at the cost of low 'mass' resolution. This is known as a 'mass analysed ion kinetic energy spectrum' (MIKES).

Transitions occurring in the 'first field-free region' of a double-focusing mass spectrometer can be studied by scanning the electric sector (E) and magnetic sector (B) fields synchronously such that the ratio B/E is constant, determined by the value of the selected precursor ion mass (m_p). This is known as 'B/E linked scanning'. The dissociation energy information is lost as a result of the double-focusing action of the combined electric and magnetic sectors but on the other hand the observed daughter ion 'mass' resolution is increased.

Different types of linked magnetic sector field (B) and electric sector field (E) scans can yield different information from transitions in the first field-free region, and from the penultimate field-free region

Table 1 Linked magnetic field (B) and electric field (E) scans for the study of metastable ion decompositions on double-focusing magnetic sector mass spectrometers

Type of analysis	'Field-free region' in which decomposition takes place	
	First	Penultimate (Final sector is electric)
Daughter ions (m_d = constant)	B/E = constant	B = constant (E scan only)
Parent ions (m_p = constant)	B^2/E = constant	B^2E = constant
Constant neutral loss ($m_p - m_d$ = constant)	$(B/E)^2(1-E')$ = constant	$B^2(1-E')$ = constant

$E' = E/E_o$, where E_o is the electric field for transmission of undissociated parent ions.

where the final sector is an electric sector. These include 'parent ion scans' in which spectra of all the parent ions from which a selected daughter ion (m_d) are obtained, and 'constant neutral loss scans' in which spectra of all the transitions in which the same neutral loss ($m_p - m_d$) occur. These are listed in **Table 1**.

Sector mass spectrometers with two or more sectors, and gas 'cells' in which 'high energy' ion–molecule collisions take place, are used to study ion and neutral chemistry. Tandem sector mass spectrometers with up to six sectors (two magnetic and four electric sectors in EBEEBE sequence) have been constructed to study ion–molecule interactions, multistep dissociations and neutralization–reionization processes.

List of symbols

B = magnetic field; D_{e1} = energy dispersion (first sector); D_{e2} = energy dispersion (second sector); D_m = mass dispersion coefficient; E = electric sector field; F = Lorentz force; KE = kinetic energy; m = mass; m_d = mass of daughter ion; m_p = mass of parent ion; m^*= apparent mass; M = magnification; r_m = radius; v = velocity; V = potential difference; w_b = ion beam width; w_c = collector slit width; w_s = source slit width; ze = charge; α = imaging aberration coefficient; ρ = momentum.

See also: **Ion Dissociation Kinetics, Mass Spectrometry; Ion Imaging Using Mass Spectrometry; Ion Molecule Reactions in Mass Spectrometry; Isotope Ratio Studies Using Mass Spectrometry; Mass Spectrometry, Historical Perspective; Metastable Ions; MS–MS and MS^n; Neutralization-Reionization in Mass Spectrometry.**

Further reading

Beynon JH (1960) *Mass Spectrometry and its Applications to Organic Chemistry*. Amsterdam: Elsevier.

Cooks RG, Beynon JH, Caprioli RM and Lester GR (1973) *Metastable Ions*. Amsterdam: Elsevier.

Duckworth HE (1958) *Mass Spectroscopy*. Cambridge: Cambridge University Press.

Enge H (1967) In: Septier A (ed) *Focusing of Charged Particles*, Chapter 4.2, p 203. New York: Academic Press.

Hintenberger H and Koenig LA (1959) Mass spectro-meters and mass spectrographs corrected for image defects. In: Waldron J (ed) *Advances in Mass Spectrometry*, Vol 1, pp 16–35. Oxford: Pergamon Press.

Jennings KR (1983) In: McLafferty FW (ed) *Tandem Mass Spectrometry*, Chapter 9.

McDowell CA (1963) *Mass Spectrometry*, New York: McGraw-Hill.

Milne GW (1971) *Mass Spectrometry: Techniques and Applications*. New York: Wiley Interscience.

Wollnik H (1987) *Optics of Charged Particles*. New York: Academic Press.

Selenium NMR, Applications

See **Heteronuclear NMR Applications (O, S, Se, Te).**

SIFT Applications in Mass Spectrometry

David Smith, Keele University, Stoke-on-Trent, UK
Patrik Španěl, Czech Academy of Science,
Prague, Czech Republic

MASS SPECTROMETRY
Applications

Introduction

The selected ion flow tube, SIFT, technique is a fast-flow tube/ion-swarm method for the study of the reactions of ions (positive or negative) with atoms and molecules under truly thermalized conditions over a wide range of temperature. It has been extensively used to study ion–molecule kinetics. Its application to atmospheric and interstellar ion chemistry by several eminent groups over a 20-year period has been crucial to the advancement and understanding of these interesting topics. Recently it has been developed as a very sensitive analytical technique for the detection and quantification of trace gases in air and in human breath down to the ppb level and in real time.

Principle of the SIFT

The SIFT technique was developed because the original fast-flow-tube method, the eminently productive flowing afterglow technique, was not suitable for the study of the reactions considered at that time to be involved in the complex ion chemistry of interstellar clouds. The SIFT (and the flowing afterglow) are flow-tube 'swarm' experiments. The 'ideal' swarm experiment involves the creation of an ensemble of reactant charged particles of number density n_1 in an inert buffer gas of number density n_2 such that $n_1 \ll n_2$. Then multiple collisions between the charged particles and the buffer gas ensure the randomization (Maxwellianization) of the charged particle velocities and the relaxation of the charged particle mean energies (which may be high initially) to those appropriate to the buffer gas temperature, T_g. Then the introduction of reactant neutral atoms or molecules at a number density, n_3, $(n_3 \ll n_2)$ initiates an ion neutral reaction. Thus the rate of change of n_1 as a function of n_3 can be determined, and the rate of coefficient, k, for the reaction can be calculated since:

$$\frac{dn_1}{dt} = -kn_1n_3 \Rightarrow n_1(t) = n_1(0)\exp\left(-kn_3t\right)$$

$$\Rightarrow \ln n_1(t) = \ln n_1(0) - kn_3t \qquad [1]$$

The k and the ion products of the reaction (determined at the defined temperature T_g) are the important parameters needed to model the ion chemistry of ionized media. In a flow system the reaction time t is the length of the reaction zone divided by the ion-flow velocity.

In the SIFT apparatus the ions are created in an ion source which is external to the flow tube. The ions are then extracted from the ion source, selected according to their mass-to-charge ratio using a quadrupole mass filter (see **Figure 1**) and injected into a flowing carrier gas (usually helium at a pressure of 0.5 Torr) via a small orifice (typically ~1 mm diameter). The carrier gas is inhibited from entering the quadrupole mass filter chamber by injecting it into the flow tube through a Venturi-type inlet at near-supersonic velocity in a direction away from the orifice (see **Figure 1**). In this way a swarm of a single-ion species thermalized at the same temperature as the carrier gas are convected along the flow tube (which is usually ~1 m long), sampled by a down-stream pinhole orifice, mass analysed and counted by a differentially pumped quadrupole mass spectrometer system.

To study a particular ion–molecule reaction, a reactant gas is introduced at a measured flow rate into the carrier gas. Then by measuring the count rates, I, of the reactant and product ions using the downstream mass spectrometer and relating them to the number destiny of the reactant gas in the carrier gas, the k for the reaction (following Eqn [1]) and the ion products can readily be determined (see **Figure 2**). Some SIFT apparatuses can be operated over the wide temperature range 80–600 K. Constructional details for a SIFT are given in some of the cited reviews.

No electrons are present in the carrier gas and therefore the SIFT medium is not a gaseous plasma medium like the flowing afterglow but rather a swarm of positive or negative ions in the carrier gas. Notwithstanding the great success of the flowing afterglow for the study of ion–neutral reactions, it has a serious limitation in that the primary reactant ions may react with their parent (source) gas which is usually present in the carrier gas, and this complicates the interpretation of the mass spectrometric data. This is avoided in the SIFT by the upstream external ion source/mass filter system. So, for example, if methane is introduced into the SIFT ion source, C^+, CH^+, $:CH_2^+$, CH_3^+ or $CH_4^{\bullet+}$ (and even CH_5^+) can be

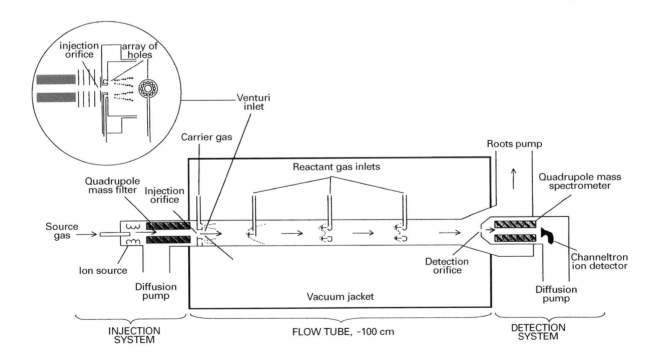

Figure 1 Schematic representation of a SIFT apparatus. One form of Venturi-type inlet is shown in the upper-left part of the diagram. The vacuum jacket facilitates operation at high and low temperatures.

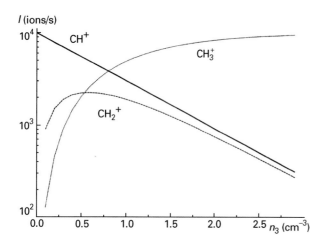

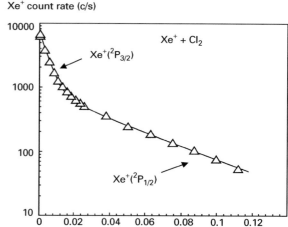

Figure 2 Dependence of the ion count rate on the H_2 number density for the reaction of CH^+ with H_2. The rate coefficient, k, is obtained from the slope of the linear decay plot following Equation [1]. The primary product ion is CH_2^+. The CH_3^+ ion is produced in the secondary reaction of CH_2^+ with H_2.

Figure 3 A typical SIFT decay (c/s = counts per second) obtained when two differently reacting species are present at the same mass. In this example, the excited ion $Xe^+(^2P_{1/2})$ reacts more slowly with Cl_2 than the ground state $Xe^+(^2P_{3/2})$ ion.

separately injected into the flowing carrier gas. Significantly, the CH_4 source gas is not present in the helium carrier gas and cannot therefore confuse kinetics studies.

The rate coefficients and ion products of the reactions with many gases of practically any positive or negative ion species can be studied using the SIFT, provided that the ions can be extracted from the ion source at a sufficient current ($\sim 10^{-9}$ A is the practical lower limit) and can be injected at a sufficiently low energy to avoid their fragmentation in collisions with the carrier gas. Even weakly bound species such as $H_3O^+(H_2O)_3$ ions have been injected without undue dissociation. Low-pressure and high-pressure election-impact sources and even flowing afterglow sources are routinely used to prepare a wide variety of positive and negative ions. For our recent development of the SIFT for trace gas analysis, a microwave discharge source is used (see below).

A fraction of the ions produced in SIFT ion sources may be electronically and vibrationally excited and after injection survive in the inert carrier gas. Their reactions can then be studied and if their reactivity is different from their ground state analogues the decay curves from which the rate coefficients are derived are nonlinear (see **Figure 3**). However, excited ions can often be relaxed to their ground states by the addition of a suitable 'quenching' gas.

Using SIFT apparatuses, the rate coefficients and products ions for a large number of positive ion–neutral and negative ion–neutral reactions have been studied in several laboratories around the world. They include the bimolecular reactions of doubly charged ions, electronically and vibrationally excited ions and cluster ions, and termolecular association reactions, some over a wide range of temperature. This has led to a greater understanding of the mechanisms, kinetics and energetics of such reactions and also to a clearer understanding of the ion chemistry of ionized media.

Ion chemistry of the terrestrial atmosphere and interstellar clouds

The SIFT is ideally suited to the study of the ion chemistries of the terrestrial atmosphere, TA, and interstellar clouds, ISC. These are ionized media in which gas phase ion–neutral reactions occur which produce the exotic ions and molecules observed in these regions. The challenge to ion chemists is to identify these reactions, and to this end the SIFT has made a major contribution.

Atmospheric ion chemistry

In the tenuous upper TA (above 100 km altitude; the ionospheric E- and F-regions) the ion chemistry is simple. Only bimolecular reactions occur between the precursor ions H^+, $He^{\bullet+}$, O^+, N^+, $O_2^{\bullet+}$ and $N_2^{\bullet+}$ (formed by photoionization) and the ambient neutrals O, O_2 and N_2, e.g.

$$He^{\bullet+} + N_2 \rightarrow N_2^{\bullet+} + He \qquad [2]$$
$$\rightarrow N^+ + N^{\bullet} + He$$

The k and the product ion distributions for such reactions are readily determined using the SIFT. Rocket-borne mass spectrometers have shown that NO^+ ions are a major species in the upper TA, even though neutral NO does not exist in measurable concentrations. Flowing afterglow and SIFT experiments have shown that these ions result primarily from the ion–molecule reaction:

$$O^{\bullet+} + N_2 \rightarrow NO^+ + N^{\bullet} \qquad [3]$$

The ion chemistry of the upper TA (summarized in **Figure 4**) proceeds to convert the energetic precursor ions to the less energetic, predominantly diatomic molecular ions, which can be neutralized by the ambient free electrons via the process of dissociative recombination as shown. It was also realized that metastable electronically excited ions of O^+ and $O_2^{\bullet+}$ are produced in the upper ionosphere and so it became important to study the reactions of these excited ions. The SIFT is well suited to these studies (excited ions are quenched in flowing afterglow plasmas). The chemistry of these excited species is represented in **Figure 4** by the thick lines.

The ion chemistry of the lower TA (below 100 km) is dominated by termolecular (three-body) reactions of both positive ions and negative ions. At altitudes between about 50 to 90 km, in the ionospheric D-region, most ionizing solar radiations have been filtered out except for L_α and L_β radiation, which can selectively ionize NO and $O_2(^1\Delta_g)$ molecules. So the initial ions in the positive ion chemistry are NO^+ and $O_2^{\bullet+}$.

SIFT and flowing afterglow studies have shown that both NO^+ and $O_2^{\bullet+}$ are relatively unreactive in bimolecular collisions with the major ambient atmospheric neutrals, but they both undergo termolecular (three-body) reactions forming weakly bonded association ions, an important initial reactions in the D-region being:

$$NO^+ + N_2 + N_2 \rightarrow NO^+ \cdot N_2 + N_2 \qquad [4]$$

These weakly bonded $NO^+\cdot N_2$ ions react efficiently with the ambient CO_2 and H_2O molecules in the 'switching reactions':

$$NO^+ \cdot N_2 + CO_2 \rightarrow NO^+ \cdot CO_2 + N_2$$
$$NO^+ \cdot CO_2 + H_2O \rightarrow NO^+ \cdot H_2O + CO_2 \qquad [5]$$

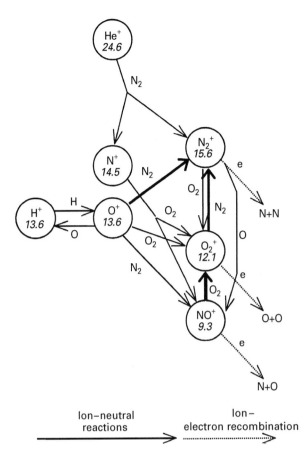

Figure 4 The ion chemistry of the upper terrestrial atmosphere. Only biomolecular ion–neutral reactions occur, and ion–electron reactions (dissociative recombination) maintain the ionization equilibrium.

Termolecular H_2O addition reactions build up the NO^+ hydrates $NO^+(H_2O)_{2,3}$. Finally the NO^+ is 'switched out' from the latter ions thus:

$$NO^+(H_2O)_3 + H_2O \rightarrow H_3O^+(H_2O)_2 + HNO_2 \quad [6]$$

Further termolecular reactions produce the hydrated hydronium ions $H_3O^+(H_2O)_n$, which are observed to be the major positive ions in this altitude region.

In the lowest altitude region of the TA the initial positive ions, e.g. $O_2^{\bullet+}$ and $N_2^{\bullet+}$, formed by cosmic ray ionization, are quickly converted to the dimer ions $O_4^{\bullet+}$ and $N_4^{\bullet+}$, which undergo switching reactions with the relatively abundant H_2O and CO_2 to finally produce $H_3O^+(H_2O)_n$ (see the left part of **Figure 5**). The chemistry does not stop here, because the many minor reactive species that exist in the lowest part of TA, such as the bases NH_3, CH_3CN and CH_3OH, undergo ligand switching with the hydrated hydronium ions producing 'mixed' cluster ions, e.g.

$$H_3O^+(H_2O)_n + CH_3CN \rightarrow H^+(H_2O)_n \cdot CH_3CN + H_2O \quad [7]$$

Switching reactions like [7] can be studied in the SIFT because of the facility to inject cluster ions without undue break-up.

The negative ion chemistry in the lower TA (see the right part of **Figure 5**) is initiated by the electron

attachment reactions:

$$O_2 + electron + M \rightarrow O_2^- + M$$
$$O_3 + electron \rightarrow O^- + O_2 \quad [8]$$

This chemistry follows a similar pattern to the positive ion chemistry in that the initial ions O^- and O_2^- are converted to O_3^- and O_4^- and these undergo switching reactions with H_2O and CO_2 to produce ions such as $O_2^- \cdot H_2O$, $O_2^- \cdot CO_2$ and CO_3^-. The further reactions of these ions with other minor neutral

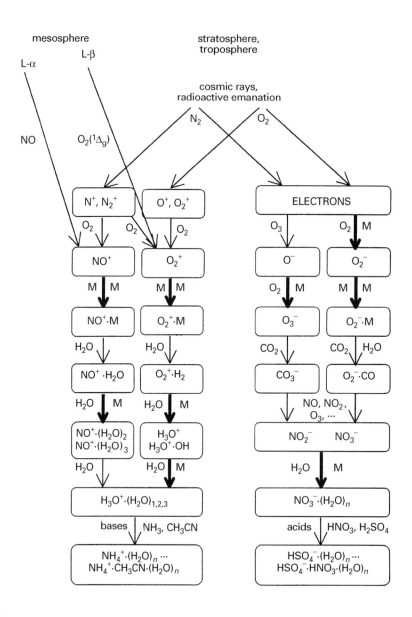

Figure 5 The ion chemistry of the lower atmosphere. The positive ion chemistry (left column) and the negative ion chemistry (right column) are dominated by termolecular reactions, finally producing the cluster ions in the lower boxes.

species such as NO_x molecules result in the very stable negative ion NO_3^-. This ion undergoes ter-molecular association reactions with H_2O producing $NO_3^-\cdot(H_2O)_n$ cluster ions which dominate the ionospheric D-region. At even lower altitudes the production of NO_3^- and its hydrates is followed by their reactions with acids, mostly HNO_3 and H_2SO_4 (produced from pollutant NO_x and SO_2 in the parallel neutral chemistry that is occurring in the lower TA). These reactions produce cluster ions like $NO_3^-(HNO_3)_{2,3}$ and $HSO_4^-(HNO_3)_n$. However, the large ambient concentration of H_2O shifts the equilibrium to the right in the generalized reaction:

$$HSO_4^-\ (H_2O)_{m-1}(acid)_n + H_2O \rightleftharpoons$$
$$HSO_4^-\ (H_2O)_m(acid)_{n-1} + acid \qquad [9]$$

producing the mixed cluster ions indicated. The SIFT has been of great value in the study of these and many other reactions of cluster ions.

Interstellar ion chemistry

One of the most interesting events in astronomy during the last few decades has been the discovery of many types of molecules, both ionized and neutral, in the diffuse and dense ISC that pervade the Milky Way galaxy. These ISC are at such enormous distances from the solar system that *in situ* probes, e.g. conventional mass spectrometers, cannot be exploited to analyse their composition. The only tool available to probe their physical conditions and chemical compositions is spectroscopy over the whole spectral range (from radio waves to gamma rays). The diffuse clouds of very low number density (about 10^2 cm^{-3}) consist mainly of H and H_2 together with C, N and

O atoms which play a central role in the gas phase ion chemistry. The dense clouds of much higher number density (10^4–10^3 cm^{-3}) consist mainly of H_2 and He together with minority C, N and O atoms.

In diffuse ISC it is stellar ultraviolet and galactic cosmic rays that largely create the ions, the important initial ions being C^+, H^+, $H_2^{\cdot+}$ and $He^{\cdot+}$. The dense ISC through which ultraviolet cannot penetrate because of the presence of micrometre-sized 'dust' grains, are ionized by galactic cosmic rays producing H^+, and $H_2^{\cdot+}$ and $He^{\cdot+}$. From these initial positive ions begins the gas phase ion chemistries illustrated in **Figures 6** and **7**. The SIFT is especially valuable for the study of interstellar ion chemistry and (together with the ion cyclotron resonance technique) has provided the major contribution to this field.

The primary H^+ ions can react with O atoms by 'accidental resonance' charge transfer producing $O^{\cdot+}$ ions, which react rapidly with H_2 producing OH^+. The $H_2^{\cdot+}$ primary ions react rapidly with H_2 producing H_3^+, which transfers a proton to O atoms also producing OH^+

$$H_3^+ + O \rightarrow OH^+ + H_2 \qquad [10]$$

The OH^+ ions react with H_2 to give $H_2O^{\cdot+}$, which then reacts with H_2 to give H_3O^+ in the sequence of H-atom abstraction reactions indicated in **Figure 6**. The closed-shell ion H_3O^+ ends this chain, since it does not react with H_2, but it does undergo dissociative recombination with electrons, producing OH and H_2O molecules. A similar sequence of H-atom abstraction reactions probably leads to the production of NH, NH_2 and NH_3 in ISC as is indicated in **Figure 6**. This sequence begins with N^+ ions produced from $He^{\cdot+}$ (via

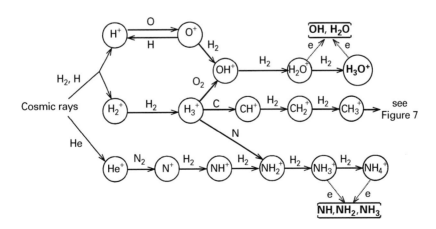

Figure 6 The initial steps in the ion chemistry of interstellar clouds leading to H_2O, NH_3 and CH_4 (see also **Figure 7**).

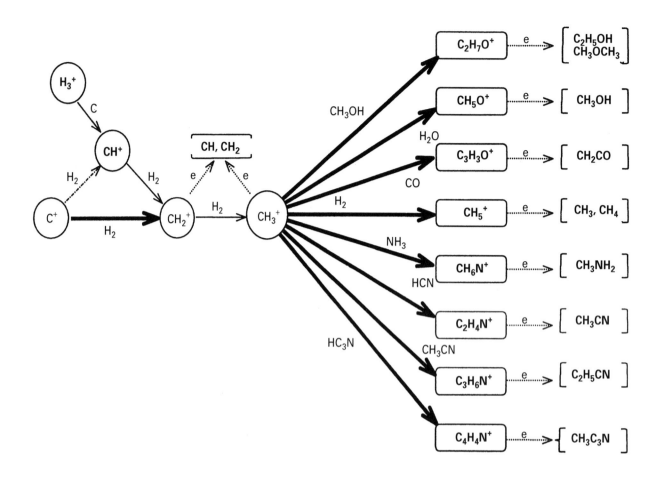

Figure 7 The production of polyatomic ions and neutral molecules in dense ISC following the radiative association reactions (thick arrows) of CH_3^+ with several known interstellar molecules and the subsequent dissociative recombination.

Eqn [2]). The reaction of H_3^+ with N atoms is also involved in NH_3 production. Similar sequences lead to small hydrocarbon molecules. These sequences of reactions can readily be studied using the selected ion injection facility of the SIFT.

The ion chemistry that results in the polyatomic hydrocarbon molecules observed in dense ISC can begin with slow radiative association reaction (see below) of C^+ with H_2, producing CH_2^+ ions which then react rapidly with H_2 to form CH_3^+. The last ion can also be produced by the sequence of reactions beginning with the proton transfer reaction of H_3^+ with C atoms producing CH^+ ions, whence:

$$CH^+ + H_2 \rightarrow CH_2^+ + H$$
$$CH_2^+ + H_2 \rightarrow CH_3^+ + H \qquad [11]$$

SIFT studies have shown that these bimolecular reactions are very fast, but that CH_3^+ reacts only slowly

with H_2 in a termolecular association reaction producing CH_5^+ thus:

$$CH_3^+ + H_2 + M \rightarrow CH_5^+ + M \qquad [12]$$

Such termolecular reactions proceed via the formation of loosely bound excited ion, e.g.

$$CH_3^+ + H_2 \rightleftharpoons (CH_5^+)^* \qquad [13]$$

which can be stabilized against decomposition at sufficiently high pressures in collisions with an inert third body M, leaving M internally or kinetically excited thus:

$$(CH_5^+)^* + M \rightarrow CH_5^+ + M^* \qquad [14]$$

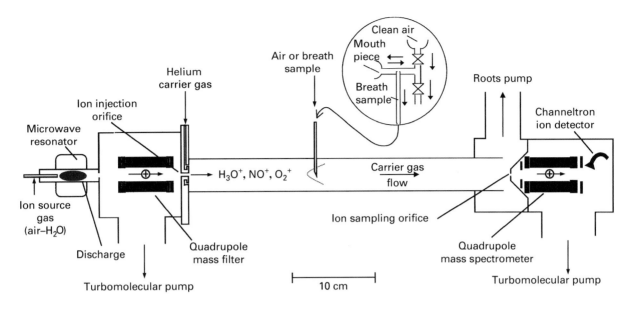

Figure 8 A SIFT apparatus configured for trace gas analysis, with the stable discharge ion-source for H_3O^+, NO^+ and O_2^+ ions, and a single air/breath sample inlet port.

The kinetics of such termolecular reactions are well understood, and many have been studied using the SIFT. However, they cannot occur in ISC because the pressures in these regions are very low. Instead, the analogous process of radiative association occurs. Continuing with the above example, the bimolecular reactions [13] can occur in ISC, and indeed it is promoted by the low ambient temperatures (as low as 10 K in some clouds), and if the dissociation lifetime of the excited $(CH_5^+)^*$ ion is long enough, then it may emit a photon which will stabilize it against dissociation:

$$(CH_5^+)^* \longrightarrow CH_5^+ + h\nu \qquad [15]$$

Radiative association reactions are very important in ISC chemistry and lead to the production of many of the observed polyatomic molecules in these regions. The CH_5^+ ions formed in reaction [15] are neutralized by electrons in ISC to form CH_4 (see **Figure 7**).

SIFT studies have shown that CH_3^+ ions readily undergo termolecular association reactions with many known interstellar molecular species, including CO, H_2O, HCN, NH_3, CH_3OH and CH_3CN. These ion–molecule associations must surely proceed via radiative association in dense ISC, and in this way complex molecules can be formed, as is indicated in **Figure 7**. Thus these SIFT studies have been crucial in indicating the importance of radiative association reactions in ISC.

Many of the molecular species detected in dense interstellar clouds are seen to contain the rare (heavy) isotopes of some elements (e.g. D, ^{13}C, ^{18}O, etc.). However, at first sight the abundance ratios of some molecules containing the rare and common isotopes (e.g. DCN/HCN) were very surprising because they are orders-of-magnitude greater than those expected from their cosmic isotopic ratios. Following extensive SIFT studies, it is now understood that this is due to the phenomenon of 'isotope fractionation' in gas phase ion–molecule reactions. This phenomenon is exemplified by the elementary reaction:

$$D^+ + H_2 \rightleftharpoons H^+ + HD \qquad [16]$$

It is a simple matter to determine the forward (exothermic) and reverse (endothermic) rate coefficients, k_f and k_r for such reactions using the SIFT technique (ideally suited for such studies because separate isotopic ions can be injected and the reaction temperature can be accurately controlled even down to 80 K). For reactions [16] it is observed that k_f increasingly exceeds k_r as the temperature is reduced. This is because the reverse reaction is endothermic by 39.8 meV (3.84 kJ mol^{-1}), by virtue of the zero-point-energy difference between H_2 and HD, and the ionization energies of H and D differ by 4 meV (0.38 kJ mol^{-1}). Hence in cold ISC the reverse reaction is effectively stopped whilst the forward reaction proceeds at the gas kinetic rate. This effectively ensures that much of the deuterium in dense interstellar

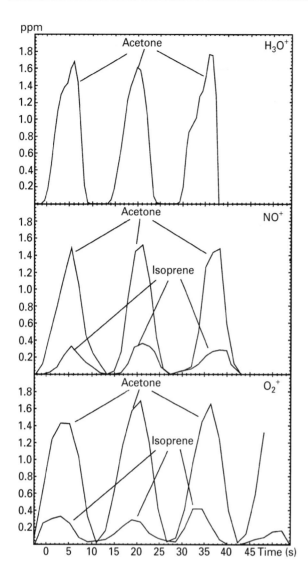

Figure 9 Concentration time profiles of the trace gases acetone and isoprene in single exhalations of breath obtained with the SIFT using H_3O^+, NO^+ and O_2^+ precursor ions. Concentrations are given in parts per million (ppm) of the breath.

clouds is contained in HD. Also important in ISC are the reactions :

$$H_3^+ + HD \rightleftharpoons H_2D^+ + H_2$$
$$CH_3^+ + HD \rightleftharpoons CH_2D^+ + H_2 \qquad [17]$$

in which D is fractionated into H_2D^+ and CH_2D^+. Thus the subsequent reactions of these deuterated ions (occurring, of course, in parallel with the H_3^+ and CH_3^+ reactions) result in the enrichment of deuterium in many interstellar molecules. Similarly, SIFT studies have shown that fractionation of the rare isotopes of heavier elements also occurs, notably of ^{13}C

into the very abundant interstellar molecule CO via the reaction of $^{13}C^+$ with ^{12}CO.

Trace gas analysis; SIFT/MS

The SIFT can be used for accurate determinations of the concentrations n_M of trace gases, M, in an air sample that has been introduced into the carrier gas, if the rate coefficients k are known for the reactions of a chosen injected precursor ion species with the M. Measurements of the count rates at the downstream mass spectrometer system of the precursor ion and each product ion, I_1 and I_p respectively, provide values of n_M for each trace gas if $I_p \ll I_1$. Then following Equation [1]:

$$n_M = \frac{I_p}{I_1 kt} \qquad [18]$$

Since currently I_1 may be typically 10^5 precursor ions per second, a typical k is 10^{-9} cm^3 s^{-1} and the reaction time t is of the order 10^{-2} s. Then for I_p at the low value of one product ion per second (a sensible detection limit), n_M in the carrier gas is 10^6 cm^{-3}. This is a fractional number density of 10^{-10} of that of the helium carrier atoms (usually 10^{16} cm^{-3}). Thus for an air or breath sample introduced into the carrier gas at a relative concentration of 1% of the carrier gas, trace gases at a partial pressure of 10 parts per billion (ppb) can be detected and quantified. Clearly, for large I_1, greater air sample concentrations, and longer integration times for I_p, the sensitivity can be improved and the detection limit lowered. The experimental configuration for this SIFT analytical method is indicated in **Figure 8**. A transportable version of the SIFT is now available commercially for use *in situ* for breath and environmental analyses.

This analytical method can be easily used for the detection and accurate quantification of several trace gases simultaneously in multicomponent vapour mixtures such as polluted air and human breath. This can be achieved in real time because of the short time response of the method ($t \sim 10^{-2}$ s), by rapidly switching the downstream mass spectrometer between chosen precursor and product ion masses. Thus the time profiles of the concentration of several trace gases in breath can be obtained from single exhalations, as are shown in **Figure 9**. Samples introduced into the apparatus directly from the atmosphere or from containers can be analysed by multiscanning the downstream mass spectrometer. A spectrum obtained by sampling the SIFT laboratory air is shown in **Figure 10**.

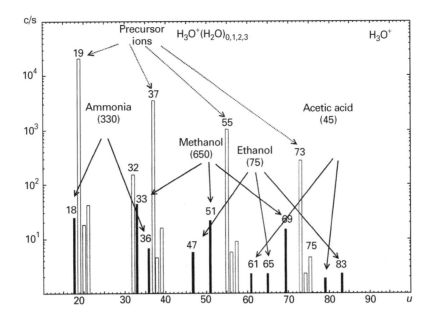

Figure 10 A mass spectrum obtained using H_3O^+, precursor ions following the introduction of laboratory air into the SIFT. Concentrations of the detected species (released from an adjacent laboratory) are given in parts per billion (ppb) in parentheses. u = ion mass-to-charge ratio, c/s = counts per second.

A vital point to note is that the mass spectrum of the product ions in much simpler than would be obtained using electron impact ionization of the air or breath sample, because chemical ionization is used. However, only a limited number of precursor ions can be used, which must not react at a significant rate with the major components of air or breath, N_2, O_2, CO_2, H_2O and Ar, but obviously must react efficiently with the trace gases in the sample. In this respect H_3O^+, NO^+ and $O_2^{\bullet+}$ are the prime candidates, and SIFT studies of a great number of the reactions of these three ions with a wide variety of organic and inorganic compounds have shown that H_3O^+ and NO^+ are of widest application, with $O_2^{\bullet+}$ being useful for fewer particular trace gases.

Recent SIFT studies have shown that H_3O^+ transfers a proton at the gas kinetic (maximum efficiency) rate to all molecules, M, that have proton affinities greater than H_2O molecules, and usually only a single product ion MH^+ is formed, which greatly simplifies the analysis of the product ion mass spectrum. However, two products are sometimes observed from these proton transfer reactions, as is the case for some aldehyde reactions, e.g.

$$H_3O^+ + C_4H_9CHO \rightarrow C_4H_9CHOH^+ (+H_2O)$$
$$C_5H_9^+ (+2H_2O) \quad [19]$$

NO^+ ions often react with organic molecules via hydride ion (H^-) transfer producing $(M-H)^+$ ions and HNO molecules e.g.

$$NO^+ + CH_3CHO \rightarrow CH_3CO^+ + HNO \quad [20]$$

They react with most carboxylic acids and with tertiary alcohols via hydroxide ion (OH^-) transfer producing $(M-OH)^+$ ions and HNO_2 molecules, and undergo rapid association reactions with most ketones producing $NO^+ \cdot M$ ions.

$O_2^{\bullet+}$ is particularly useful as a precursor ion for species of low proton affinity that do not react with H_3O^+ and of high ionization potential that do not react with NO^+. The reaction process is invariably charge transfer; it is particularly useful for quantifying NO, NO_2 and NH_3 in air:

$$O_2^{\bullet+} + NO \rightarrow NO^+ + O_2$$
$$O_2^{\bullet+} + NO_2 \rightarrow NO_2^+ + O_2$$
$$O_2^{\bullet+} + NH_3 \rightarrow NH_3^{\bullet+} + O_2 \quad [21]$$

The beauty of this SIFT/MS analytical technique is that all three suitable precursor ion species can be used on the same air or breath sample by switching

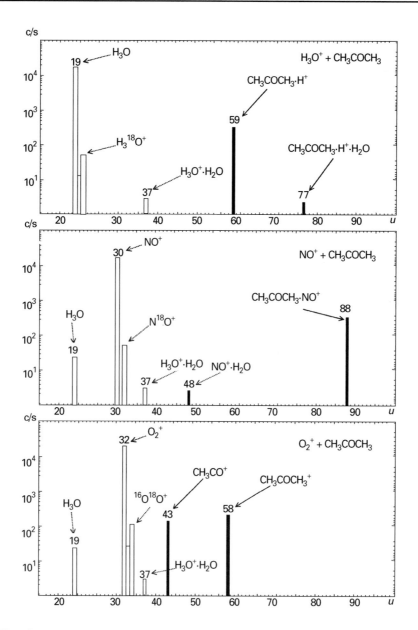

Figure 11 The SIFT analyses of an air–acetone mixture using H_3O^+, NO^+ and O_2^+ precursor ions injected sequentially into the carrier gas by switching the upstream mass filter (see **Figure 8**).

the upstream mass filter (see **Figure 8**). **Figure 11** shows the results obtained simply for an air–acetone sample. The use of two or three precursor ions ensures that few of the various trace gases in the sample are missed.

An extensive database is required of the rate coefficients and product ions of the reactions of H_3O^+, NO^+ and $O_2^{\bullet+}$ with many different types of molecules, and this is rapidly being compiled from SIFT experiments. With sufficient knowledge of the ion chemistry, very complex mass spectra can be analysed, such

as those shown in **Figure 12**, obtained for a sample of the volatiles from gasoline using both H_3O^+ and $O_2^{\bullet+}$ precursor ions. Note the identification of the aromatic and aliphatic hydrocarbons, each compound of mass M, being recognized as MH^+ ions in the H_3O^+ spectrum (proton transfer) and M^+ ions in the $O_2^{\bullet+}$ spectrum (charge transfer).

The potential of this SIFT/MS analytical method is enormous for human breath analysis and hence in physiology and biochemistry, for clinical diagnosis and therapeutic monitoring, and for measuring air

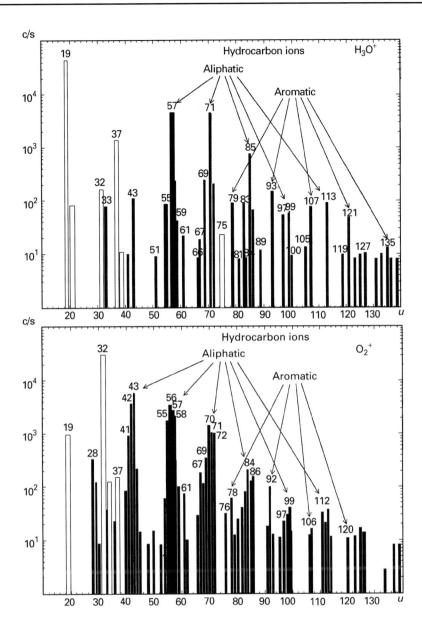

Figure 12 Mass spectra obtained in the SIFT for an air–gasoline vapour mixture using H_3O^+ and O_2^+ precursor ions. The component hydrocarbon molecules M, are protonated by H_3O^+ producing MH^+ ions, whereas charge transfer between O_2^+ and the M produces ions like M^+ and $(M–H)^+$. u = ion mass-to-change ratio, c/s = counts per second.

pollution and analysing food vapour emissions and food flavours. The medical and biochemical value is well illustrated by the analyses of the breath of six healthy individuals following the morning ingestion of a liquid protein meal. The breath concentrations of five species (ammonia, methanol, ethanol, acetone and isoprene) were obtained from only single breath exhalations before and some six hours after the meal. Shown in **Figure 13** are the rise in the ammonia levels as the protein is metabolized and the decrease in the acetone as the body is nourished.

SIFT/MS analyses of the breath of some 30 uraemic patients with end-stage renal failure have shown greatly elevated levels of ammonia compared to healthy subjects. The spectrum obtained of breath of a uraemic patient with diabetes (indicated by the elevated acetone level) who also smokes cigarettes (indicated by the presence of acetonitrile) is shown in **Figure 14**. Many further applications of this new method for trace gas analysis are in train including applications in agriculture (animal welfare) and grassland research.

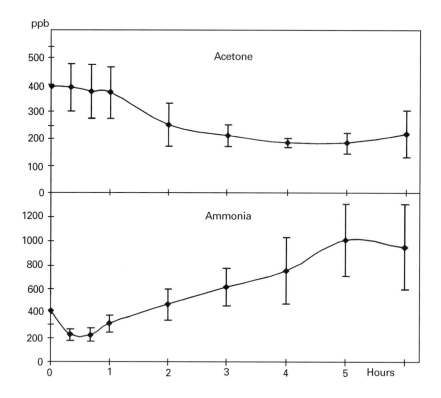

Figure 13 The mean concentrations of acetone and ammonia (ppb) obtained by the SIFT in single exhalations of breath from six volunteers following a protein meal taken at time 0 (see the text). The vertical bars represent the standard deviations of the six concentrations at each time.

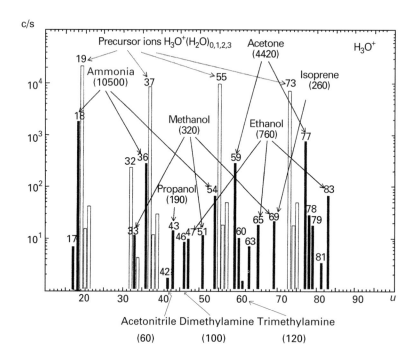

Figure 14 The SIFT spectrum obtained following the introduction of breath from a diabetic patient with end-stage renal failure who smokes cigarettes (see the text). The concentrations of the major breath gases are given in ppb in parentheses. u = mass. c/s = counts per second.

List of symbols

h = Plank constant; I = count rate; I_1 = count rate of precursor ion; I_p = count rate of product ion; k = rate coefficient; k_f = forward rate coefficient; k_r = reverse rate coefficient; n_M = concentration of trace gas M; n_1 = number density of reactant charged particles; n_2 = number density of buffer gas; n_3 = number density of reactant neutral particles; T_g = buffer gas temperature; t = reaction time; u = ion mass-to-charge ratio; ν = frequency.

See also: **Biochemical Applications of Mass Spectrometry; Chemical Ionization in Mass Spectrometry; Cluster Ions Measured Using Mass Spectrometry; Food Science, Applications of Mass Spectrometry; Ion Collision, Theory; Ion Molecule Reactions in Mass Spectrometry; Isotope Ratio Studies Using Mass Spectrometry; Isotopic Labelling in Mass Spectrometry; NQR, Applications; Proton Affinities; Quadrupoles, Use of in Mass Spectrometry.**

Further reading

Anicich VG (1993) Evaluated bimolecular ion–molecule gas phase kinetics of positive ions for use in modeling planetary atmospheres, cometary comae, and interstellar clouds. *Journal of Physical Chemistry and Reference Data* 22: 1469–1569.

Duley WW and Williams DA (1984) *Interstellar Chemistry*. London: Academic Press.

Farrar JM and Saunders WH Jr (eds) (1988), *Techniques for the Study of Gas-Phase Ion–Molecule Reactions*, New York: Wiley.

Ferguson EE, Fehsenfeld FC and Albritton DL (1979) Ion chemistry of the Earth's atmosphere. In: Bowers MT, (ed) *Gas Phase Ion Chemistry*, pp 45–81. New York: Academic Press.

Fontijn A and Clyne MAA (eds) (1983) *Reactions of Small Transient Species*. London: Academic Press.

Herbst E and Leung CM (1989) Gas phase production of complex hydrocarbons, cyanopolynes, and related compounds in dense interstellar clouds. *Astrophysical Journal, Supplement Series* 69: 271–300.

Smith D (1992) The ion chemistry of interstellar clouds. *Chemical Reviews* 92: 1473–1485.

Smith D and Adams NG (1987) The selected ion flow tube (SIFT): studies of ion–neutral reactions. *Advances in Atomic and Molecular Physics* 24: 1–49.

Smith D and Spanel P (1995) Swarm techniques. In: Dunning FB and Hulet RG (eds) *Experimental Methods in the Physical Sciences: Atomic, Molecular, and Optical Physics: Charged Particles*, pp 273–298. New York: Academic Press.

Smith D and Spanel P (1996) Ions in the terrestrial atmosphere and in interstellar clouds. *Mass Spectrometry Reviews* 14: 255–278.

Španěl P, Cocker J, Rajan B and Smith D (1997) Validation of the SIFT technique for trace gas analysis of breath using the syringe injection technique. *Annals of Occupational Hygiene* 41: 373–382.

Španěl P and Smith D (1996) Application of ion chemistry and the SIFT technique to the quantitative analysis of trace gases in air and on breath. *International Review of Physical Chemistry* 15: 231–271.

Španěl P and Smith D (1996) Selected ion flow tube: a technique for quantitative trace gas analysis of air and breath. *Medical and Biological Engineering and Computing* 34: 409–419.

Španěl P and Smith D (1997) SIFT studies of the reactions of H_3O^+, NO^+ and O_2^+ with a series of alcohols. *International Journal of Mass Spectrometry and Ion Processes* 167/168: 375–388.

Španěl P and Smith D (1999) Selected ion flow tube studies of the reactions of H_3O^+, NO^+ and O_2^+ with some chloroalkanes and chloroalkenes. *International Journal of Mass Spectrometry* 184: 175–181.

Španěl P and Smith D (1999) Selected ion flow tube-mass spectrometry: Detection and real-time monitoring of flavours released by food products. *Rapid Communications in Mass Spectrometry* 13: 585–596.

Wayne RP (1985), *Chemistry of Atmospheres*, Oxford: Clarendon.

Silver NMR, Applications

See **Heteronuclear NMR Applications (Y–Cd).**

Single Photon Emission Computed Tomography

See **SPECT, Methods and Instrumentation.**

Small Molecule Applications of X-Ray Diffraction

Andrei S Batsanov, University of Durham, UK

> **HIGH ENERGY**
> **SPECTROSCOPY**
> **Applications**

Single crystal X-ray diffraction is the main source of information on the geometrical structure of molecules and molecular solids, including bond distances (and hence bond orders), bond angles, shapes of coordination polyhedra, conformations of flexible molecules, as well as intermolecular contacts. It can always distinguish between configurational isomers (e.g. *cis* or *trans*), and often optical isomers (enantiomers), too. The enormous and fast-growing mass of X-ray structural data defies any attempt to review it even in the most general way. Numerous compen-diums and reference tables on these data have been published. In fact, X-ray crystallography provides the main bulk of data for any book on structural chemistry and the theory of the chemical bond. More recently, computerized structural databases have given easy access to all available X-ray crystal structures and sophisticated means of extracting from them necessary parameters, performing statistical analysis of the data and finding regularities (structural correlations). On a higher level of precision, it is possible now to determine not only the atomic positions, but also the entire map of electron density in a crystal, visualizing concentrations of bonding electrons and lone electron pairs, determining directly atomic charges, electrostatic potentials, etc. The analysis of the displacement parameters of atoms is also of much importance, both to improve the accuracy of molecular geometry determinations and to understand the dynamic behaviour of crystal structures (**Figure 1**).

Atomic structure

A crystal structure consists of a periodic pattern of atomic nuclei and a continuous (also periodic) distribution of electron density. As X-rays are scattered by electrons, it is possible to extract structure factors from the reflection intensities, and, by their Fourier transform, to calculate the electron density map, the peaks of which correspond to the centres of atoms. We can then approximate the structure by placing at these points isolated atoms of corresponding elements with ideal spherical symmetry. The positions (coordinates) of these atoms and their displacement parameters (see Atomic displacements below), can be refined by the least-squares technique so as to achieve the best

agreement with the observed intensities. Usually, the structure determination ends at this stage. Atomic co-ordinates are referred to the coordinate axes parallel to the edges of the unit cell and published in fractional form, i.e. as x/a, y/b and z/c, where a, b and c are the unit cell parameters. All kinds of geometrical parameters of the molecule in a crystal can be calculated from these coordinates. Shortest vectors between centres of atoms give bond lengths and, of course, the atomic connectivity, i.e. the order in which the atoms are linked. Usually, the bond order (multiplicity) can be estimated roughly from a bond length in a straightforward way. Angles between bonds indicate the state of hybridization of a given atom. The conformation of a sterically flexible molecule can be described in terms of dihedral angles between planar fragments, or torsion angles. For ring systems, an unequivocal way of describing the conformation is provided by the ring puckering coordinates.

X-ray diffraction provided the first determination ever (in 1912) of the covalent bond length (C–C 1.544 Å in diamond) and is to the present day the overwhelming source of information on 'molecular metrics'. The number of organic and organometallic crystal structures, deposited at the Cambridge Structural Database (CSD) by April 1998, exceeds 182 000 (cf. 67 000 in 1988). Of these, less than 1% were determined by neutron diffraction, all the rest by X-ray diffraction. All other spectroscopic methods that can give exact interatomic distances (gas-phase electron diffraction, microwave spectroscopy, etc.) yielded only hundreds of structure determinations. Furthermore, X-ray diffraction is unique among spectroscopic methods in that the number of experimental data (reflection intensities) exceeds the number of unknown variables (for each atom, three coordinates plus one isotopic or six anisotropic displacement parameters) by an order of magnitude. Currently, the reflections-to-variables ratio of 8 is regarded as the minimum acceptable for a good quality experiment. Therefore in determining the structural formula of a molecule and, to moderate precision, its geometry, the X-ray method needs no preconceived model of the structure, nor prior knowledge of the molecular symmetry. The number of possible reflections being proportional to the unit cell volume, and the latter to the

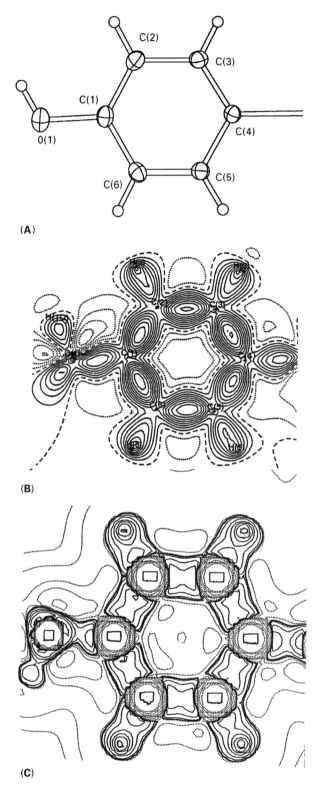

(A)

(B)

(C)

Figure 1 (A) Molecular structure of the hydroxyphenyl group in the crystal of 4-(methylamido)phenol, showing 50% displacement ellipsoids; maps of the deformation electron density (B) and Laplacian (C) in the same moiety (positive contours solid, negative dashed). Unpublished results of Yufit DS and Muir K, courtesy of the authors.

molecular volume, this favourable situation persists up to molecules of considerable size (200 nonhydrogen atoms and even more).

This makes the X-ray method a very cost-effective tool for identification of newly synthesized compounds or newly isolated natural products: one needs a single crystal of ~0.01 mg to learn the composition and structural formula, usually without destroying the sample itself. In the past, solution of the phase problem was the major difficulty, the only available means (besides trial and error) being the Patterson method, effective only for structures containing at least one heavy atom. The advent of efficient direct methods of phase determination in the 1970s and the enhancement of these methods by phase relations of higher orders in the 1980s (which eliminated the problems for space groups without glide planes and screw axes) reduced the solution of a wholly unknown structure from art to near routine, if a crystal of good quality can be obtained.

A primary (but not unfallable) estimate of the quality of structure determination is R-factor, showing the discrepancy between the experimentally observed structure factors of reflections (F_o) and those calculated from the determined structure (F_c), $R = \Sigma \left| |F_o| - |F_c| \right| / \Sigma |F_o|$. With modern precision of the data measurement, 'quality' structures have R of 0.02 to 0.06 (for observed reflections). R-factor based on F^2 (rather than F) became popular recently, being more meaningful for weak reflections. For the same data, $R(F^2)$ is roughly twice the magnitude of $R(F)$.

Bond distances and their precision

It was realized early on that bond lengths in molecular crystals depend mainly on the nature of the elements involved and on the bond order (the number of bonding electron pairs minus the number of antibonding ones), the crystal environment playing only a secondary role. Numerous tables and compendiums of bond length exist. **Tables 1** and **2** list some average ('standard') values, mainly from the most recent and comprehensive reviews by Kennard *et al.* based on the data from the CSD.

The measure of random errors in geometrical parameters, estimated standard deviations (esd), are routinely calculated in least-squares refinements. In good-quality studies of purely organic compounds, they may be as low as one or several thousandths of Å (for bonds not involving hydrogen atoms). Errors being in general in inverse relation to the atomic number, in organometallic structures the esd of the heavy atom coordinates may be as small as 0.0001 Å, but those of the lighter ligand atoms can be of the order of 0.01 Å or more, so the latter define

Table 1 Selected average bond distances in organic compounds (in Å, σ = 0.01–0.02 Å)

RH_2C–CH_2R	1.524	RHC=CHR	1.316
R_2HC–CHR_2	1.542	R_2C=CR_2	1.331
R_3C–CR_3	1.588	C=C(=C), in allenes	1.307
$C(sp^2)$–$C(sp^2)$ in:		C≡C	1.181
Biphenyls	1.490	C=O in ketones	1.210
Conjugated dienes	1.455	C–N peptide bond	1.332
$C(sp)$–$C(sp)$	1.377	C=N	1.279
C–C in phenyl rings	1.380	C≡N	1.144
$N(sp^3)$–$N(sp^3)$	1.454	C=S	1.671
$N(sp^2)$–$N(sp^2)$	1.401	R_3P=O	1.489
N=N	1.240	R_3P=S	1.954
RO–OR	1.469	Se–Se	2.340
RS–SR	2.048	Si–Si	2.359

X	$C(sp^3)$–X	$C(sp^2)$–X	X	$C(sp^3)$–X	$C(sp^2)$–X
F in RCH_2F	1.399	1.340	P^+R_3	1.800	1.793
Cl in RCH_2Cl	1.790	1.739	PR_2	1.855	1.836
Br	1.966	1.899	$P(=O)R_2$	1.813	1.801
I	2.162	2.095	SO_2R	1.786	1.763
OH	1.432	1.362	$S(=O)R$	1.809	1.790
$N^+(sp^3)$	1.499	1.465	SR	1.819	1.773
$N(sp^3)$	1.469	1.416	SeR	1.970	1.893
$N(sp^2)$	1.454	1.355	SiR_3	1.863	1.868

the esd of the metal–ligand *distances*. Differences in bond lengths, not exceeding 3 esd, are regarded as statistically insignificant. However, esd represent only the internal consistency of the data and say nothing of the various systematic errors inherent in the experiment (absorption, extinction, radiation decay of crystal, etc.), or of the model, which has the following limitations.

1. The X-ray diffraction method measures the distances between 'centres-of-gravity' of atoms' electron clouds, which need not necessarily coincide with the atomic nuclei. The difference is most pronounced for H (or D), its sole electron being shifted into the bond region; the X-ray estimates of $C(sp^2)$–H, N–H and O–H bond lengths are 0.93, 0.89 and 0.82 Å, vs. 1.08, 1.01 and 0.97 Å determined by neutron diffraction in the same compounds. The effect is also considerable for sp-hybridized C and N atoms in acetylene and cyano groups.

2. Spherically symmetrical 'isolated' atoms model chemically bonded atoms, which are far from spherical. On refinement, atomic positions shift so as to compensate for the disregarded bonding electrons and/or lone pairs. Thus, a two-coordinate oxygen atom is shifted by 0.007–0.013 Å from its 'neutron' position (**Figure 2**) and C–O bond lengths are overestimated by 0.003–0.005 Å.

3. The X-ray method measures neither instantaneous nor time-average bond lengths, but the distances between mean atomic positions, averaged over all the crystal and all the duration of the experiment. This is much the biggest source of bias (see below).

Table 2 Selected average metal–ligand distances in organometallic complexes (in Å, $\sigma = 0.03$–0.04 Å)

Metal	CO(terminal)	C(aryl)	C(alkyl)	C_5H_5 (centroid of)	PMe$_3$	Cl
				Ligands		
V	1.95	2.11		1.95	2.51	2.29
Cr	1.87	2.08		1.88	2.39	2.34
Mn	1.81	2.06	2.18	1.82	2.46	2.45
Fe	1.78	2.03	2.09	1.71	2.25	2.26
Co	1.78	1.93	2.01	1.70	2.22	2.27
Ni	1.77	1.92		1.75	2.20	2.34
Mo	1.98	2.19	2.25	2.01	2.46	2.41
Ru	1.90	2.09	2.18	1.89	2.31	2.42
Rh	1.85	2.01	2.09	1.90	2.27	2.38
W	2.00		2.19	2.01	2.49	2.41
Re	1.94		2.17	1.96	2.37	2.39
Pt	1.85	2.05	2.08		2.30	2.32

Intermolecular contacts

Distances between atoms not bonded directly to each other, are rationalized in terms of the close packing model. An atom is regarded as a hard sphere of certain radius, van der Waals (vdW) radius, and a molecule as a superposition of such spheres. In a crystal, these bodies are closely packed but cannot dent each other. It is implied that a vdW radius is specific for a given element and remains the same in any contact,

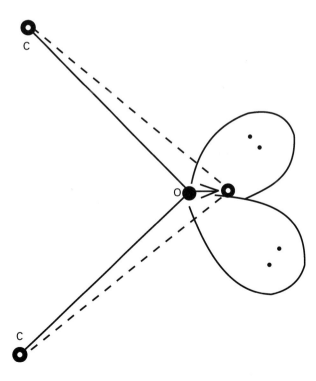

Figure 2 Ether group geometry determined by neutron (solid) and X-ray (dashed) diffraction; in the latter the O atom is shifted towards its lone-pair electrons.

so that any A···B contact distance equals the average of the A···A and B···B ones. Many systems of vdW radii have been developed. The pioneering one of Pauling (1939) used anion radii (which are close to the vdW radii) for the lack of direct data. The system of Bondi (1964), the most oft-quoted today, was based on experimental distances in a relatively small number of organic crystals and on some noncrystallographic data (gas kinetic collisions, critical densities, liquid properties).

However, because of the weakness and nonspecific nature of vdW forces, nonbonded distances vary more widely than bond lengths. By a statistical analysis of numerous structures, Rowland and Taylor derived a novel system of vdW radii (**Figure 3**, **Table 3**). Atoms of Cl, Br, I, S, Se and Te, forming one covalent bond, show a highly anisotropic shape in intermolecular contacts (**Figure 4**), which can be described as an ellipsoid compressed along the bond direction, while harder N, O and F atoms remain essentially spherical.

Nonbonded contacts considerably shorter than predicted by the vdW radii, are often indicative of specific interactions between molecules (hydrogen bonds, secondary coordination, charge-transfer, etc.) and are useful in analysing solid-state properties (e.g of organic conductors, so called 'organic metals').

It must be stressed that vdW radii have different meanings in molecular mechanics, where the sum $R(A) + R(B)$ corresponds to the minimum of the potential curve of the *two-atom* A···B interaction. Crystallographic vdW radii were derived to describe the distances of *closest* approach between *polyatomic* molecules, often of complex shapes, and are 8–20% shorter than the equilibrium radii. Only for inert gases, monoatomic in the solid state, do the two systems coincide.

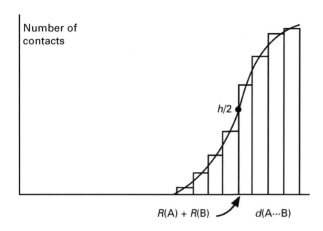

Figure 3 A typical histogram of nonbonded contact distances; the half-height point A corresponds to the sum of the van der Waals radii (after Rowland and Taylor).

Structural databases

The abundant data on X-ray crystal structures was made easily accessible with the development of computerized databases. Small-molecule organic and organometallic structures are covered by the Cambridge Structural Database (Cambridge, UK), which lists all compounds containing at least one 'organic' carbon (carbide, carbonate, CN or CO ligand carbon is not regarded as such). For each entry, the crystal lattice parameters and related data, atomic coordinates and (for more recent studies) ADP, bond distances and angles, molecular and structural formula are stored. The database is updated quarterly and widely distributed among academic and nonacademic users worldwide. Its software permits retrieval of entries according to the chemical class, full or partial compound name, composition and chemical connectivity, conditions of determination (pressure, temperature), as well as features of the three-dimensional structure. It also facilitates various statistical and numerical tests on the retrieved data, in order to find regularities of structural parameters and correlations between them.

Crystal structures of minerals and other inorganic substances are covered by the Inorganic Crystal Structure Database which is compiled by the University of Bonn (Germany). It also lists large numbers of molecular structures.

Absolute structure

Chiral molecules are those not superimposable with their mirror images, such as those containing a tetra-

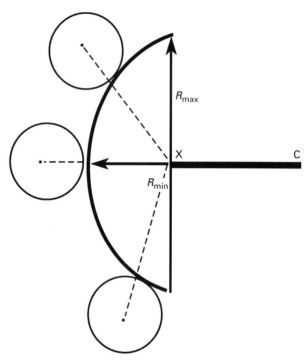

Figure 4 Anisotropic shape of an atom, forming one covalent bond. Van der Waals radius is at a minimum on the continuation of this bond and maximum normal to it.

Table 3 Nonbonded contact (or van der Waals) radii for crystals (Å)

Element	P	B	ZZ	RT	NF,max	NF, min
H	1.2	1.20	1.16	1.10	1.26	1.00
C	1.7	1.70	1.71	1.77		
N	1.5	1.55	1.50	1.64	1.60	1.60
O	1.4	1.52	1.29	1.58	1.54	1.54
F	1.35	1.47	1.40	1.46	1.38	1.30
Cl	1.8	1.75	1.90	1.76	1.78	1.58
Br	1.95	1.85	1.97	1.87	1.84	1.54
I	2.15	1.98	2.14	2.03	2.13	1.76
S	1.85	1.80	1.84	1.81	2.03	1.60

P, Pauling, 1939–1960; B, Bondi, 1964; ZZ, Zefirov and Zorkii, 1974–1989; RT, Rowland and Taylor, 1996 (see **Figure 3**); NF, Nyburg and Faerman, 1985–1987 (anisotropic system, see **Figure 4**).

hedral carbon (or other) atom with four different substituents (asymmetric centre). They are of great import in biochemistry: natural sugars, amino acids and hence proteins, etc. Molecules of the same chemical composition and molecular geometry, but differing by a mirror reflection (enantiomers, or optical isomers) are indistinguishable in their chemical behaviour (unless another chiral reagent is involved) and in most physical properties. Few physical methods can indicate the difference between enantiomers: optical rotation (ability to rotate the polarization plane of light, in the opposite sense for enantiomers), circular dichroism and the formation of crystals with chiral outer form. None, except X-ray diffraction, can determine the absolute configuration of an asymmetric atom, i.e the real configuration of the substituents in space (R or S, see **Figure 5** for explanation). The notation introduced by Fischer was based on an arbitrary assignment of the absolute configuration of (+)-glyceraldehyde, to which other optically active compounds could be related by virtue of their chemical origin from one another.

For normal X-ray scattering, centrosymmetrically related reflections with indices hkl and $\bar{h}\bar{k}\bar{l}$ always have equal intensities (Friedel's law), thus the diffraction pattern is always centrosymmetric, no matter whether the crystal itself has an inversion centre or not. However, for wavelengths slightly lower than the absorption edge of an atom, a resonant interaction of the X-rays with the inner electrons results in *anomalous scattering*, in which both the intensity and the phase of the radiation is changed. The atomic scattering factor (f_a) becomes a complex function:

$$f_a = f + \Delta f' + i\Delta f''$$

where $\Delta f'$ and $\Delta f''$ depend on the wavelength, and for a noncentrosymmetric crystal the hkl and $\bar{h}\bar{k}\bar{l}$ reflections differ in intensity. The effect is stronger for heavier atoms and longer wavelengths. Bijvoet (1951) first utilized this anomaly to determine the absolute configuration of the dextrorotatory (+)-tartaric acid in the form of its sodium-rubidium salt, using ZrK_α radiation. (By good luck, this coincided with Fischer's notation.) Today, it is possible to determine reliably the absolute configuration of a medium-sized molecule containing one atom of phosphorus or a heavier element, using MoK_α or CuK_α X-radiation. With the latter wavelength, a careful routine experiment can determine an absolute configuration from the anomalous scattering of oxygen atoms as the heaviest element.

Various methods are used to determine the absolute configuration of the crystal structure, which necessarily defines that of a molecule (although a chiral crystal may consist of nonchiral molecules). The intensities of a few dozens reflections, for which the effect is the strongest, can be calculated together with their inversion equivalent (Friedel pairs) and compared to the observed intensities. Least squares refinements of the 'solved' structure and then of its inverted equivalent, will give a lower R-factor for the correct enantiomer. The anomalous scattering corrections themselves ($\Delta f''$) or some coefficients linked to them can be included in the least-squares refinement as a variable. The best method is to refine the 'Flack' parameter during the least-squares procedure. This not only gives the absolute configuration of the crystal, but also (through the esd) a measure of the reliability of the assignment. In addition, it permits the identification and analysis of 'twins by inversion', a common problem faced by physicists trying to grow large enantiomerically pure crystals. Finally, if a structure contains more than one chiral centre, and one of them has a known absolute configuration, those of other(s) can always be determined, even if anomalous scattering is unobservable. For this purpose, an unknown chiral product can be cocrystallized with a known one and the structure of the complex solved by diffraction methods.

Atomic displacements

Atoms and molecules in crystals are not static, but oscillate at their potential minima and in some cases can jump between different minima. At room temperature, nonhydrogen atoms in a well-ordered organic crystal perform thermal oscillations with mean square amplitudes $\langle u^2 \rangle$ of 0.04–0.05 Å^2 (which means the r.m.s amplitudes of ~0.2 Å, comparable

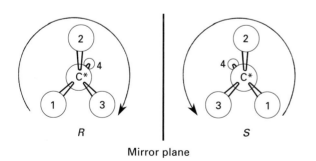

Figure 5 Notation of the absolute configuration of an asymmetric carbon atom. The substituents are numbered according to certain rules (e.g. beginning with the highest atomic weights, e.g. I > Br > Cl > F); the highest-number substituent pointing away, the numbers of the three others increase clockwise for the R configuration and anticlockwise for the S one.

with bond lengths). At 100 K (the practical limit for most cooling devices using liquid nitrogen) $\langle u^2 \rangle$ can be reduced to ~0.015 Å², but some (zero-point) oscillations persist even at 0 K. Generally, u is in inverse relation to the atomic mass; for outlying atoms of a molecule or some easily rotating substituents, it can be twice the average. Furthermore, atoms can be *disordered*, i.e. have different positions in different asymmetric units of the structure, which are presumed identical by definition of the symmetry of the crystal lattice. Both dynamic and static distortions of the lattice periodicity result in 'smearing' of the electron density, averaged in space over the entire crystal and in time over the entire duration of the experiment. The smearing is described either by atomic thermal parameters, or more precisely atomic displacement parameters (ADP), either in isotropic approximation (spherically averaged for each atom) or in anisotropic approximation, as a symmetrical third-order tensor with six independent parameters. Both representations imply harmonic oscillations; the real vibrations are always anharmonic to some extent, but it is really important to take this effect into account only in high-precision studies of electron density.

Visually, anisotropic ADPs are commonly presented as 'thermal ellipsoids' of a given probability, say, 50% (**Figure 6**). The atom's centre can be found inside the ellipsoid with this probability, and it has *equal* probability of reaching the ellipsoid in any direction. The principal axes of the ellipsoid are proportional to the components of $\langle u^2 \rangle$ in these directions, but other axes are not. The surface visualizing

$\langle u^2 \rangle$ components in every direction is not an ellipsoid but a quadric surface of a 'peanut' shape (**Figure 6B**).

Models of molecular vibrations

The coherent elastic (Bragg) scattering, which alone is measured in the normal X-ray diffraction study, can inform only about the average displacement of every single atom with respect to the crystal lattice. It carries no information whatsoever on how a displacement of one atom relates to that of another atom. In principle, such knowledge can be obtained from inelastic, or thermal diffuse scattering, but the latter is difficult both to measure (its peaks lying under the Bragg ones) and to interpret. Without this knowledge, the same ADPs can be interpreted in different ways, depending on whether the atoms oscillate in phase or out of phase (**Figure 7**).

However, vibrations of a molecular crystal are mainly external modes (oscillations of entire molecules). Intramolecular modes (except some torsional ones), which imply distorting a relatively rigid covalent-bond skeleton, contribute relatively little to the overall amplitudes. From this follows the rigid-bond criterion (Hirshfeld, 1975): if two atoms are linked by a covalent bond, their $\langle u^2 \rangle$ components along the bond direction must be almost equal. For accurately determined structures, this is true within 0.001 Å for X–X bonds (X = C, N, O), 0.003 Å for metal–ligand bonds and 0.005 Å for X–H or X–D bonds. The latter difference is in agreement with the $\langle u^2 \rangle$ of 0.006 Å, calculated from the C–H bond stretching frequency ν 2700–3300 cm^{-1}. Larger differences are indicative of molecular flexibility, as in crystalline Fe(III) complexes (spin-crossover effects), octahedral Cu(II) complexes (dynamic Jahn–Teller and pseudo-Jahn–Teller distortions) and binuclear Mn(II)–Mn(IV) complexes (valence disorder).

In an ideally rigid molecule, the same criterion applies to *all* interatomic vectors. The TLS model,

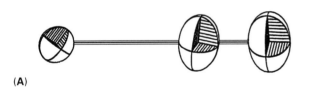

(A)

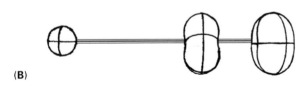

(B)

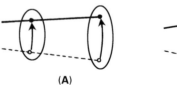

(A)

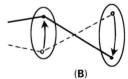

(B)

Figure 6 Atomic displacements in a metal–carbonyl moiety represented by thermal ellipsoids (A) and 'peanut-shaped' surfaces of $(\langle u^2 \rangle)^{1/2}$ (B).

Figure 7 In-phase vibrations (A) of a rigid group (CO, CN) and out-of-phase (B) of a flexible group (ethyl) can result in similar ADP ellipsoids.

introduced by Schomaker and Trueblood (1968), describes the motion of such a molecule with three tensors, representing respectively: translations along three principal axes, librations around these axes and screw motions (for noncentrosymmetric cases only). Subtracting the ADPs predicted by this model, from the actual ones, internal nonrigidity of the molecule can be revealed. Torsional motions of a (presumed rigid) group within a molecule can be analysed in the same way, determining torsional amplitudes and, from them, force constants and barriers to internal motion.

Correcting bond lengths

Rigidly bonded atoms usually perform oscillations along arcs (librations). The electron density thus smeared has its centre-of-gravity (taken for the average atomic position) inside the arc. Hence the X-ray method tends to underestimate bond lengths systematically, the more so (up to 0.03 Å) at higher temperatures (while the real bonds slightly lengthen with temperature). The libration of a C–C bond (~1.5 Å) by 5–6 Å results in a spurious shortening of 0.005 Å, and when by 10 Å, in a shortening of 0.025 Å (**Figure 8**).

If the ADPs from a standard least-squares are consistent with the rigid-body model, bond lengths can be corrected for the spurious shortening. Thus, in a nonrigid molecule of o-terphenyl, the mean C–C distance in the phenyl ring is 1.389 Å before correction and 1.398 Å after it, as compared with the r_g = 1.399 ± 0.002 Å from gas-phase electron diffraction. Nevertheless, any thermal motion correction is based on some preconceived idea about the nature of the motion and demands caution. The rigid-bond test can disprove molecular rigidity but not always prove it, as the atoms may vibrate along their connecting vector with the same $\langle u^2 \rangle$, but quite out of phase. The best way to reduce these errors is to diminish the vibrations themselves by cooling the crystal.

Disorder

From a diffraction study at one temperature one cannot distinguish between dynamic (thermal motion) and static (disorder) displacements. However, if cooling fails to reduce the ADPs, the latter are certainly of static nature, as in the case of ferrocene. The early X-ray studies at room temperature (1951–1956) of the monoclinic modification proved the molecule to possess the crystallographic C_i symmetry, implying the staggered conformation of the rings, in contrast with the eclipsed one in the gas phase. The persistence of a large ADP at 173 K could only be explained as a static disorder, for which various models were suggested. At 164 K, a phase transition occurred into an ordered triclinic structure with four symmetrically independent molecules, all of which adopted nearly (within 10°) eclipsed conformations. A statistical averaging of these molecular orientations results in a picture similar to that observed in the disordered phase. Thus the centrosymmetric conformation is likely to be spurious.

Erratic ADP

In a least-squares refinement, ADPs tend to become ultimate 'sinks' of all unaccounted for systematic errors, particularly absorption, anisotropic extinction, sometimes incorrect assignment of atomic type or nonstoichiometric occupancy of the position. Erratically elongated ADP ellipsoids, uncorrelated, with those of adjacent atoms, usually betray gross blunders in structure solution, e.g. incorrect crystallographic symmetry or lattice parameters.

Electron density

The second level of a crystal structure investigation is to go beyond the independent atom model and to resolve more subtle details of the electron density distribution. For such a study one needs to measure

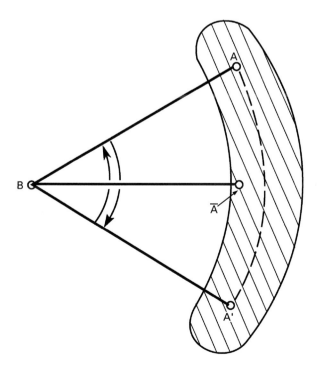

Figure 8 Systematic underestimation of the length of a librating A–B bond. The terminal atom moves along AA' arc; its average position $\bar{A}$ is found as the centre-of-gravity of the smeared electron density (hatched) and lies within the arc.

by an order of magnitude more experimental data (measuring all symmetrically equivalents of each reflection rather than just one, increasing the maximum 2θ angle) and to address carefully all possible sources of systematic errors, which can be irrelevant on the atomic approximation level. The mathematical formalisms and software for such studies are still evolving. Only a few hundreds of electron density studies of very uneven quality have been published since 1972 (the earlier works had been attempted with totally inadequate means), of these about 50 were with the multipole technique (see below).

Deformation maps

The earliest approach was to calculate the total electron density map by a Fourier synthesis and to subtract from its density of isolated atoms, centred at atomic nuclei positions (*promolecule*). These nuclei positions were either determined directly by a neutron diffraction experiment (X–N method) or calculated using only high-order X-ray reflections (X–X_{ho} method), making use of the fact that valence electrons scatter X-rays mainly at lower angles, while the inner electron shells remain practically unperturbed by chemical bonding. The resulting *deformation electron density* maps showed some well-expected features: peaks of electron density in the areas of covalent bonds and lone electron pairs, outward shift of bonding peaks in a cyclopropane ring ('bond bending', **Figure 9**) and so on. However, no peaks appear on bonds involving atoms with the valence shells more than half-filled, e.g. fluorine. In organometallic complexes, only insignificant peaks of electron density were found on the formally quadruple Cr–Cr bond in chromium acetate (0.1 e$Å^{-3}$) and the single Mn–Mn bond in $Mn_2(CO)_{10}$(0.05 e$Å^{-3}$), and none at all on the Fe–Fe bond in [$(C_5H_5)Fe(CO)_2$]$_2$. This reveals the inadequacy of the spherically averaged independent atom as the reference: the $(1s)^2(2s)^2(2p)^5$ atom F has an average of $\frac{5}{3}$ electrons on each p-orbital, which are subtracted from the bonding density. While in fact this atom contributes only one electron for bond formation, the excessive subtraction of $\frac{2}{3}$ e more than outweighs the accumulation of electrons in the bond area.

More informative deformation maps could be obtained by subtracting a correctly preoriented independent atom, or a reference atom in a corresponding state of hybridization. Features of one particular bond in a molecule can be most highlighted by subtracting the (calculated by MO methods) electron densities of the two fragments, differing from the molecule by the absence of only this bond.

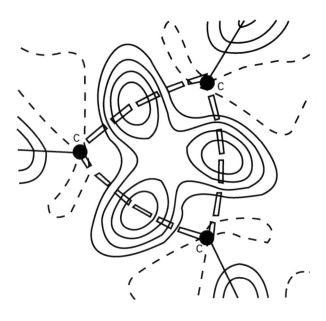

Figure 9 Deformation electron density map in a cyclopropane ring confirms the concept of 'bent bonds' (dashed lines).

More information can be extracted by topological analysis of the electron density map, particularly using the Laplacian (second gradient) of the electron density:

$$\nabla^2\rho(\mathbf{r}) = \partial^2\rho(\mathbf{r})/\partial x^2 + \partial^2\rho(\mathbf{r})/\partial y^2 + \partial^2\rho(\mathbf{r})/\partial z^2$$

which has maxima in the areas where electron density accumulates (i.e. on chemical bonds) and minima where it depletes.

Determining the phases of structure factors (especially for noncentrosymmetric structures) with the precision sufficient for a deformation density map is an extremely difficult task, as is combining X-ray and neutron data in a sensible way. The irrelevance of the valence electron density for high-order X-ray scattering is also only an approximation (it is correct for the spherically symmetric case only).

Multipole analysis

More promising is to describe the deformation electron density by a series of spherical harmonic density functions (multipoles), which can be included into least-squares refinement. The inner (core) electron shells of an atom are presumed and the κ parameter, which describes the isotropic expansion ($\kappa < 1$) or contraction ($\kappa > 1$) of the valence shell as a whole. Multipole parameters of higher orders describe deviations of the electron density from spherical symmetry. They can be related to the products of atomic orbitals: monopolar to ss, dipolar to sp, quadrupolar to pp, octupolar to 2p3d and hexadecapolar to 3d3d, and used to analyse atomic orbital populations more directly than density maps.

Studies of octahedral complexes $M(NH_3)_6^{3+}$, $M(CN)_6^{3-}$ (M = Cr, Co) and $Cr(CO)_6$ have confirmed the predictions of the ligand field theory: 66 to 77% of 3d electrons occupy the t_{2g} orbitals (field-stabilized) to the depletion of the destabilized e_g orbitals.

Attempts were made recently to develop a set of deformation parameters, transferable between chemically similar molecules. Their application is much less time-consuming than a full electron density study and may give more realistic least-squares refinement (reduction of displacement parameters) and better predictions of molecular properties than the atomic approximation.

All methods of electron density analysis are hindered by the difficulty of distinguishing between the real (static) deformation and smearing due to thermal motion.

Estimates of electric properties

It is possible to estimate directly the charges on atoms and groups of atoms in a crystal, by integrating the electron density over the corresponding areas, if physically sensible borders for them can be suggested. For a monoatomic ion, the dependence of the integral electron density inside a sphere centred on this atom as the function of the radius of this sphere has a distinct minimum, indicating the 'ion boundary'. For the anion in NH_4Cl, this gives the radius of 1.75 Å with 17.5 e inside the sphere, showing an incomplete charge transfer. In TTF·TCNQ, an organic complex with metallic conductivity, integration over parallelepiped-shaped boxes gave the degree of charge transfer of ~0.6 e from TTF to TCNQ, in agreement with other estimates. In more complex cases, the physical boundary of an ion can be determined as the surface of zero flux, on which the electron density gradient vanishes.

The electrostatic potential (energy needed to bring a unit positive charge from infinity to a given point) and its first and second derivatives (electric field and its gradient) are of great importance in understanding intermolecular interactions, molecular recognition and reaction mechanisms (as they define the path by which the reagent approaches the substrate). The electrostatic potential can be calculated from the experimental electron density or directly or directly from a high-precision set of structure factors.

List of symbols

e = electron charge; f_a = atomic scattering factor, comprising the parts independent from (f) and dependent on (real f' and imaginary f'') the wavelength; $i = \sqrt{-1}$; $\mathbf{r}$ = radius-vector; R = discrepancy factor between the observed structure factors of reflections (F_o) and those calculated from the structure (F_c), $R = \Sigma \left| |F_o| - |F_c| \right| / \Sigma |F_o|$; r_g = thermal average internuclear distance; $\langle u^2 \rangle$ = mean square amplitude of oscillations; κ = expansion–contraction parameter; ν = frequency; ρ = electron density; σ = estimated standard deviation (esd); ∇ = Laplacian.

See also: **Chiroptical Spectroscopy, General Theory; Laboratory Information Management Systems (LIMS); Microwave Spectrometers; Structure Refinement (Solid State Diffraction).**

Further reading

Allen FH, Bergerhoff G and Sievers R (eds) (1987) *Crystallographic Databases*. Chester: International Union of Crystallography.

Burgi H-B and Dunitz JD (eds) (1994) *Structure Correlation*, Vol 1, 2. Weinheim: VCH.

Coppens P (1997) *X-Ray Charge Densities and Chemical Bonding*. Oxford: Oxford University Press.

Domenicano A and Hargittai I (eds) (1992) *Accurate Molecular Structures, their Determination and Importance*. Oxford: Oxford University Press.

Dunitz JD (1979) *X-Ray Analysis and the Structure of Organic Molecules*. Ithaca: Cornell University Press.

Glusker JP, Lewis M and Rossi M (1994) *Crystal Structure Analysis for Chemists and Biologists*. Weinheim: VCH.

Kitaigorodskii AI (1994) *Molecular Crystals and Molecules*. New York: Academic Press.

Sodium NMR Spectroscopy

See **NMR Spectroscopy of Alkali Metal Nuclei in Solution.**

Solid State NMR of Macromolecules

See **High Resolution Solid State NMR, ^{13}C.**

Solid-State NMR Using Quadrupolar Nuclei

Alejandro C Olivieri, Universidad Nacional de
Rosario, Argentina

MAGNETIC RESONANCE
Applications

Compared with solution NMR, the study of solid samples presents certain peculiarities derived from the nuclei being fixed at their lattice positions in a crystal. This rigidity gives rise to the effect of various anisotropic interactions in NMR spectra, the most common of which are given **Table 1**. All these interactions produce orientation-dependent resonances, resulting in characteristic splittings and/or non-Lorentzian broadening of the solid-state NMR signals. The line broadening may range from a few ppm up to thousands of ppm, which helps to explain why much effort has been devoted to the development of line narrowing techniques. Since 1950, for example, Andrew and co-workers have pioneered a method known as magic-angle spinning (MAS), in which the sample is spun inside a suitable rotor, at an angle of 54.7° with respect to the external magnetic field B_0 (**Figure 1**). The expected effect of MAS is to average out the orientational dependence of NMR lines, rendering a solution-like spectrum where the relevant

Table 1 Anisotropic interactions affecting NMR spectra in the solid state

Interaction	*G* Tensor	*k*	*P*	Anisotropy	Asymmetry	Average value in solution
			Parameter^a			
Chemical shift	σ	$\gamma/2\pi$	B_0	$\Delta\sigma$	η	σ_{iso}
Direct dipolar coupling	D	D	S	$-3D$	0	0
Indirect dipolar coupling	J	1	S	ΔJ	$\eta_J{}^b$	J_{iso}
Quadrupole coupling^c	q	$\dfrac{e^2Q}{2S(2S-1)h}$	$P=I=S$	$3q_{zz}/2$	η^{θ}	0

^{a} The general form of the Hamiltonian is $h^{-1}H = k\,GP$.
^{b} The usual assumption is $\eta_J = 0$ and J coaxial with D.
^{c} The anisotropy is defined as $\Delta q = q_{zz} - (q_{xx} + q_{yy})/2$, and the asymmetry as $3(q_{yy}-q_{xx})/2\Delta q$, where q_{xx}, q_{yy} and q_{zz} are the principal components of the q tensor.
^{d} Q is the nuclear quadrupole moment and q_{zz} is the maximum value of the field gradient tensor q.

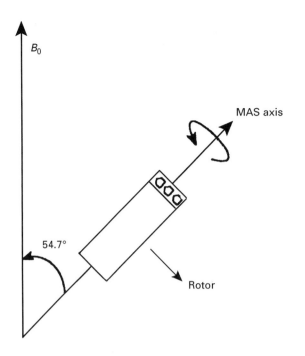

Figure 1 Scheme showing how MAS is implemented: the sample is placed inside a rotor, which spins in the kHz range about an axis inclined at 54.7° with respect to the external magnetic field B_0.

interactions are collapsed to their corresponding isotropic values (**Table 1**). Owing to the successful combination of MAS with sensitivity enhancement pulse sequences (most notably cross-polarization from abundant to dilute spins), solid state NMR has evolved into a technique with sensitivity and resolution comparable to its solution counterpart. It should be borne in mind, however, that MAS is successful provided two conditions are met. On one hand, the dependence of NMR transitions with the orientation should be of the form $(1 - 3\cos^2\theta)$, otherwise split and/or broad lines may result. On the other hand, the frequency of sample spinning should be comparable or higher than the anisotropy. If this latter condition is not attained, the spectrum will not only show an isotropic resonance, but also a number of spinning sidebands separated from the latter by integer multiples of the spinning frequency. The presence of such a sideband manifold may in itself constitute a nuisance, but in certain circumstances it provides a means to measure interesting molecular properties.

The present article has been divided in two main sections, which are briefly described below.

(1) When the observed nucleus is spin-$\frac{1}{2}$ and a neighbouring nucleus is quadrupolar, i.e. spin $>\frac{1}{2}$. In this case, the main interactions suffered by the nucleus are the chemical shift and the dipolar coupling to neighbouring nuclei. One would expect a simple, solution-like spectrum if high-speed MAS was applied. However, a subtle combination of the dipolar coupling with quadrupole interactions at neighbouring nuclei leads to the appearance of additional splittings, which cannot be averaged by MAS. Interesting information concerning molecular and structural properties of solid materials has been gathered from the study of these splittings.

(2) When the observed nucleus is itself quadrupolar. Here the quadrupole interaction is dominant, the others being usually very minor in comparison. Normal MAS speeds are much smaller than typical quadrupole couplings (which lie in the MHz range), producing an enormous number of sidebands, except in the case of the central transition of half-integer spins. In this latter case distinctive line broadening effects appear which cannot be completely removed by MAS. Methods to overcome these difficulties will be reviewed.

Quadrupole effects on spin-$\frac{1}{2}$ NMR spectra

Simple MAS spectra

In this section we explain the basis of the success of MAS in achieving spectral narrowing in solid-state NMR. When a single, observed spin-$\frac{1}{2}$ I nucleus is coupled to a single quadrupolar S nucleus, the complete Hamiltonian is:

$$H = H_Z(I) + H_Z(S) + H_{CS}(I) + H_{CS}(S) \\ + H_D(I,S) + H_J(I,S) + H_Q(S) \quad [1]$$

where $H_Z(I)$ and $H_Z(S)$ are the Zeeman terms for nuclei I and S respectively. They are given by:

$$h^{-1}H_Z(I) = -\nu_{0I}I_z \quad [2]$$

$$h^{-1}H_Z(S) = -\nu_{0S}S_z \quad [3]$$

where $\nu_{0I} = \gamma_I B_0$ and $\nu_{0S} = \gamma_S B_0$ are the nominal NMR frequencies. The forms of the other Hamiltonian terms appearing in Equation [1] are summarised in **Table 2**.

The usual assumption in computing NMR spectral resonances is to consider that the Zeeman terms are

Table 2 Hamiltonian terms affecting an I,S spin system and secular contributions to the NMR spectra.

Term	Interaction	Affected nucleus	Secular contribution to NMR spectra[a]
$h^{-1}H_{cs}(I)$	Chemical shift of I	I	$-\nu_{0I}[\sigma_{iso}(I) - \Delta\sigma(I)(1 - 3\cos^2\theta)/3]I_z$
$h^{-1}H_{cs}(S)$	Chemical shift of S	S	$-\nu_{0S}[\sigma_{iso}(S) - \Delta\sigma(S)(1 - 3\cos^2\theta)/3]S_z$
$h^{-1}H_D(I,S)$	Direct dipolar coupling [b,c]	I and S	$D\,I_zS_z(1 - 3\cos^2\theta)$
$h^{-1}H_J(I,S)$	Indirect dipolar coupling [c,d]	I and S	$J_{iso}I_zS_z - (\Delta J/3)I_zS_z(1 - 3\cos^2\theta)$
$h^{-1}H_Q(S)$	Quadrupole coupling[e]	S	$\dfrac{\chi}{4S(2S-1)}[3S_z^2\cos^2\theta - S^2 + \left(\dfrac{3}{4}\right)(S_+S_- + S_-S_+)\sin^2\theta]$

[a] The angle θ between the main tensor axes and the external magnetic field is characteristic of each interaction; the expressions shown in this table are valid only if the relevant tensors are axially symmetric and coaxial.

[b] $D = (\mu_0/4\pi)(\gamma_I\gamma_S h/4\pi^2 r_{I,S}^3)$ is the dipolar coupling constant ($r_{I,S}$ is the internuclear distance).

[c] The secular contributions for these interactions are valid for heteronuclear spin systems. For homonuclear spin systems, the flip-flop term containing $(I_+S_- + I_-S_+)$ is also secular.

[d] The usual assumption of J coaxial with D leads to an effective dipolar constant $D' = D - (\Delta J)/3$.

[e] The quadrupole coupling constant is defined as $\chi = e^2Qq_{zz}/h$.

dominant in Equation [1], and thus that first-order perturbation theory can be applied. Using the information given in **Table 2**, the NMR lines for an observed I nucleus are given by

$$\nu_I = \nu_{0I}[1 - \sigma_{iso}(I)] - m_S J_{iso} + [\Delta\sigma(I)\nu_{0I}/3 - m_S D'](1 - 3\cos^2\ ) \quad [4]$$

where $m_S = -S, -S+1, ..., S - 1, S$ are the spin components for nucleus S. Equation [4] is true provided the relevant tensors σ, D and J are axially symmetric and coaxial. Notice that the quadrupole interaction plays no role in Equation [4], owing to the fact that the latter does not contain I spin operators (**Table 2**). If MAS is applied at a sufficiently high speed, i.e. when the sample spinning frequency ν_r is larger than the effective anisotropy ($\Delta\sigma\nu_{0I}$–$3m_S D'$), the factor $(1 - 3\cos^2\theta)$ in Equation [4] becomes $(1 - 3\cos^2\xi)$ $[1 - 3\cos^2(54.7°)]/2$ (here ξ is the angle between the main axes of the interaction tensors and the sample spinning axis). Since $\cos^2(54.7°) = \frac{1}{3}$, Equation [4] predicts, under MAS, a J-coupled multiplet centred at the isotropic I chemical shift:

$$\nu_I = \nu_{0I}[1 - \sigma_{iso}(I)] - m_S J_{iso} \quad [5]$$

i.e. a solution-like spectrum. This constitutes the basis of the successful applications of MAS which led to the development of high-resolution solid-state NMR.

When the spinning speed is low, the result is well known: the isotropic resonance appears flanked by a number of sidebands, located at integer multiples of the spinning frequency. Suitable analysis of the intensities of these sidebands allows of the chemical shift parameters $\Delta\sigma$ and η for nucleus I.

Second-order effects

The simple, solution-like Equation [5] was found to apply in most solid-state NMR spectra. However, conference reports in 1979, and papers in the scientific literature soon after, showed both experimentally and theoretically that this was not the case when the quadrupole interaction at a neighbouring S nucleus is comparable to its Zeeman interaction. To interpret these results, first-order theory needs to be corrected with second-order effects. The latter are the result of the interplay of the tensors D and q, through non-secular terms of the corresponding $H_D(I,S)$ and $H_Q(S)$ Hamiltonians (specifically, the single quantum transition terms containing the operators $I_\pm S_z$ and $I_z S_\pm$). They lead to the appearance of terms having an orientational dependence other than $(1 - 3\cos^2\theta)$, which cannot be averaged out even by high-speed MAS. Second-order theory allowed the derivation of the following simple equation:

$$\nu_I = \nu_{0I}[1 - \sigma_{iso}(I)] - m_S J_{iso} + \left(\frac{3D'\chi}{10\nu_{0S}}\right) \times \left[\frac{S(S+1) - 3m_S^2}{S(2S-1)}\right] \quad [6]$$

From the second-order shift (the last term in the right-hand side of Equation [6]), the following six conclusions can be drawn.

(1) Second-order effects scale inversely with the applied magnetic field, in contrast to J_{iso} or chemi-

cal shift effects. Thus, experiments conducted at different fields provide evidence that the observed splittings are indeed due to quadrupole effects.

(2) The second-order shifts depend on m_S^2, and therefore their values occur in pairs. When $J_{iso} = 0$ (and ΔJ is also zero), Equation [6] predicts the I line as a 2:1 doublet for $S = 1$, and as a 1:1 doublet for $S = 3/2$ (**Figures 2A** and **2B**). Simple equations for the doublet splittings s (**Figures 2A** and **2B**) exist (Eqn [6]): $s = 9D\chi/10\nu_{0S}$ for $S = 1$, and for $s = 6D\chi/10\nu_{0S}$ for $S = 3/2$. Notice that for $S = 1$ the doublet is asymmetric (**Figure 2A**), causing s to have a definite sign (the convention is that s is positive if the smallest peak appears at higher frequencies).

For a finite J_{iso}, the I line is predicted to be a distorted J-multiplet (**Figures 2C** and **2D**). Since the outermost lines of the multiplet shift in opposite direction as compared with the innermost lines, the spectra are 'bunched' at one end, in a manner which also depends on the sign of χ the quadrupole coupling constant (**Table 2**).

It should be noticed that in all cases the lines are not single peaks but have a distinct powder pattern shape (**Figures 3A** and **4A**).

(3) The average of the second-order shifts over m_S is zero, and hence the isotropic $\sigma_{iso}(I)$ is obtained by averaging the multiplet line frequencies.

(4) The value of J_{iso} is obtained by averaging the multiplet line spacings, or from the central spacing if S is half-integer (**Figures 2C** and **2D**).

(5) The sign of s or the sense of spectral 'bunching' provides experimental access to the sign of χ, which is difficult to obtain from other techniques. An exception is the 1:1 doublet for $S = 3/2$ and $J_{iso} = 0$, in which case the information on the sign of χ is lost.

(6) If the molecular geometry and the quadrupole parameters are known, the observation of distorted multiplets may allow the determination of the asymmetry in the J tensor ΔJ.

When the assumptions of axial symmetry and co-axiality among the tensors are relaxed, the following general equation is obtained:

$$\nu_I = \nu_{0I}[1 - \sigma_{iso}(I)] - m_S J_{iso}$$
$$+ \left(\frac{3D'\chi}{20\nu_{0S}}\right)\left[\frac{S(S+1) - 3m_S^2}{S(2S-1)}\right]$$
$$\times (3\cos^2\beta^D - 1 + \eta_Q \sin^2\beta^D \cos 2\alpha^D) \qquad [7]$$

Equation [7] incorporates not only χ but also η_Q, as well as two angles (β^D and α^D) that fix the mutual orientation of D and q. As expected, Equation [6] is a special case of Equation [7] when $\beta^D = 0$, i.e. when D and q are coaxial.

Finally, when the value of χ is larger than the NMR frequency ν_{0S} second-order theory breaks down, and one needs to resort to complete full-matrix Hamiltonian calculations, with the results shown in **Figures 3B** and **4B** for $S = 1$ and $S = 3/2$, respectively (in both cases $J_{iso} = 0$). As can be appreciated, for low values of the ratio (χ/ν_{0S}) the predictions of Equation [6] are in agreement with the full calculations.

It is important to note that the effects described by Equation [6] are only observed in rigid solids. Both in solution and in highly mobile solid phases, random molecular motions average out all anisotropic contributions, leaving only Equation [5] (a further motional effect may be a fast quadrupole relaxation on nucleus S, which would erase the multiplet structure of the I signal).

Experimental examples and applications

Tables 3 and 4 summarize nuclear, molecular and structural parameters as well as the spectral appearance for some studied spin pairs giving rise to second-order quadrupole effects on the I line, as described above. In the case of ^{13}C, ^{14}N, the ratios (χ/ν_{0S}) are low and have therefore been studied mainly

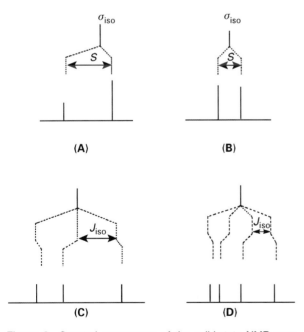

Figure 2 Spectral appearance of the solid-state NMR spectrum of a spin-$\frac{1}{2}$ nucleus (I), including second-order quandrupole effects from S when: (A) $S = 1$, $J_{iso} = 0$; (B) $S = \frac{3}{2}$, $J_{iso} = 0$; (C) $S = 1$, $J_{iso} \neq 0$; (D) $S = \frac{3}{2}$, $J_{iso} \neq 0$. In all cases Equation [6] applies, with χ positive. The frequency axes increase from right to left.

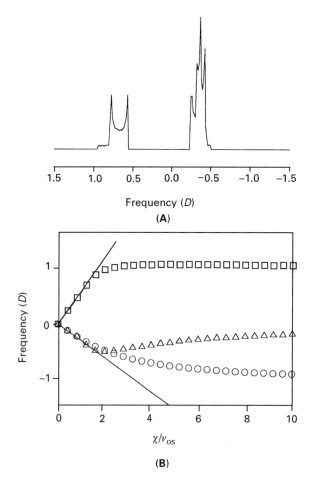

Figure 3 (A) Powder pattern line shape of an I nucleus coupled to a quadrupolar ($S = 1$) nucleus when $(\chi/\nu_{0S}) = 1$, $J_{\text{iso}} = 0$. The frequency axis is in units of the dipolar coupling constant D. (B) Frequencies (in units of D) of the three lines expected for an I,S pair ($S = 1$) as a function of the ratio χ/ν_{0S}. The line positions marked with symbols have been obtained by full-matrix Hamiltonian calculations. The solid lines are the values given by Equation [6].

on the basis of the simple Equation [6] (and its extension to non-symmetric q tensors); in most cases $J_{\text{iso}} = 0$ and $D' = D$ (see **Table 2**). The theoretical equations have been used to (1) predict the spectral appearance once the molecular geometry and the quadrupole parameters are known, (2) to derive approximate values of χ (including sign) from the spectra and (3) to aid in spectral assignment, since the affected carbons appear as characteristic doublets.

Most studies on ^{13}C, ^{14}N second-order effects were done using relatively low field solid-state NMR spectrometers. The advent of high-field instruments has displaced this interesting phenomenon to a rather unfortunate second place: at 7.05 T the effects are rarely seen, unless favourable circumstances occur. An interesting example is provided by a recently studied metal cyanide polymer, in which the molecular geometry suggests that all relevant tensors σ, D and q are axially symmetric and coaxial. The ^{13}C solid-state

MAS NMR spectrum at 7.05 T and high spinning speed shows three asymmetric doublets (corresponding to non-equivalent cyanide sites), from which values of χ in the range -1.9 to -2.5 MHz have been derived (**Figure 5**). The sideband shapes are also doublets with characteristic line shapes, and have been successfully simulated using second-order theory (**Figure 6**). This simulation also provided, as a by-product, $\Delta\sigma = 350$ ppm for the ^{13}C chemical shift tensor.

Another spin-1 nucleus causing second-order effects is ^{2}H (**Table 3**). ^{13}C MAS NMR spectra of solid deuterated organic molecules show distorted triplets, as expected from Equation [6] when $J_{\text{iso}} \neq 0$. In this case, the small value of χ for ^{2}H is compensated by a large dipolar coupling constant D (**Table 4**).

In the case of ^{13}C nuclei coupled to chlorine (**Table 3**), the values of (χ/ν_{0S}) for ^{35,37}Cl are such

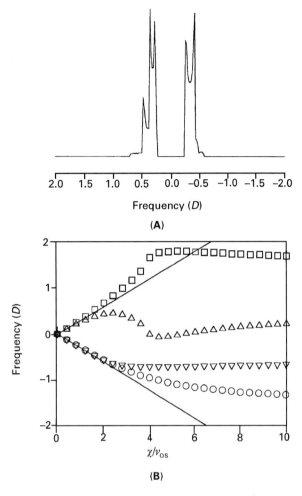

Figure 4 (A) Powder pattern line shape of an I nucleus coupled to a quadrupolar ($S = 3/2$) nucleus when $\chi/\nu_{0S} = 1$, $J_{iso} = 0$. The frequency axis is in units of the dipolar coupling constant D. (B) Frequencies (in units of D) of the four lines expected for an I,S pair ($S = 3/2$) as a function of the ratio (χ/ν_{0S}). The line positions marked with symbols have been obtained by full-matrix Hamiltonian calculations. The solid lines are the values given by Equation [6].

that 1:1 doublets are observed even at high fields, as described by Equation [6] when J_{iso} is negligible. On the other hand, in ^{119}Sn spectra of chlorostannic compounds, distorted quartets have been observed which allowed the determination of the parameter ΔJ (**Table 4**). Notice that axial symmetry and coaxiality of D and q are plausible assumptions for the X–Cl bond. Since the nuclear properties of both chlorine isotopes are similar (**Table 3**), only average effects are observed in the spectra, except in the special circumstances of very high spectral resolution.

When the quadrupolar nucleus is bromine, J_{iso} effects are important, and ^{13}C NMR spectra of C–Br carbons appear as asymmetrically distorted quartets (**Figure 7**). Further, the ratio (χ/ν_{0S}) is in this case large (**Tables 3** and **4**), and second-order theory cannot be applied. Thus, full-matrix calculations were used to account for the observed spectra, as well as a so-called 'inverse' first-order theory, in which the Zeeman term is considered as a small perturbation

on the quadrupole Hamiltonian. In any case, knowing both the C–Br distance and Br quadrupole coupling constants, and assuming axial symmetry of all tensors around the C–Br bond, allows one to derive approximate values of J_{iso} and ΔJ from the spectra (**Table 4**). As with chlorine, the separate effects of both bromine isotopes are difficult to distinguish (**Table 3**).

Finally, cases involving ^{31}P coupled to metals should be mentioned. Spectra which involve coupling of ^{31}P to several quadrupolar metals, e.g. ^{63,65}Cu ($S = 3/2$), ^{55}Mn ($S = 5/2$), ^{59}Co ($S = 7/2$) and ^{93}Nb ($S = 9/2$) have been found to consist of dis-torted J-multiplets, as expected from Equation [6]. The pair ^{31}P, ^{63,65}Cu has been extensively studied in a series of phosphine-Cu(I) complexes. Since the ratio (χ/ν_{0S}) is low, second-order theory allowed the easy calculation of χ and ΔJ from the spectra (**Table 4**). It is interesting to note that solid-state NMR is one of the few techniques which allows one to measure ΔJ,

Table 3 Nuclear properties for studied pairs of nuclei as regards second-order effects on spin-$\frac{1}{2}$ spectra.

I.S.Pair	v_{0I} (MHz)[a]	Natural abundance of I (%)	v_{0S} (MHz)[a,b]	S[b]	Natural abundance of S(%)[b]	$10^3 Q(S)$ (barn)
$^{13}C,^{14}N$	75.43	1.11	21.67	1	99.6	20.1
$^{13}C,^{2}H$	75.43	1.11	46.05	1	0.015[c]	2.86
$^{13}C,^{35,37}Cl$	75.43	1.11	29.40	$\frac{3}{2}$	75.5	−81.1
			24.47	$\frac{3}{2}$	24.5	−63.9
$^{119}Sn,^{35,37}Cl$	111.82	8.58	d	d	d	d
$^{13}C,^{79,81}Br$	75.43	1.11	75.16	$\frac{3}{2}$	50.5	331
			81.02	$\frac{3}{2}$	49.5	276
$^{31}P,^{63,65}Cu$	121.44	100	79.52	$\frac{3}{2}$	69.1	−211
			85.18	$\frac{3}{2}$	30.9	−195

[a] At $B_0 = 7.05$ T, for which $v(^1H) = 300$ MHz.
[b] When two isotopes occur, the first entry corresponds to the nuclear properties for the lighter isotope.
[c] Enriched samples.
[d] See the $^{13}C,^{35,37}Cl$ case.

Table 4 Typical values of I,S distances, scalar, dipolar and quadrupole coupling constants, and spectral appearance of I spectra owing to second-order quadrupole effects.

I,S pair	$r_{i,s}$ (pm)	D (kHz)	ΔJ (kHz)	J_{iso} (kHz)	Range of $\lvert\chi\rvert$ (MHz)	Spectral appearance of I
$^{13}C,^{14}N$	110–150	0.6–1.6	$\Delta J \ll D$	~0	0.5–5	2:1 Doublet
$^{13}C,^{2}H$	100	3.6	$\Delta J \ll D$	0.02	0.1–0.3	Distorted triplet
$^{13}C,^{35,37}Cl$	170–180	0.5–0.6	$\Delta J \ll D$	~0	60–80	1:1 Doublet
$^{119}Sn,^{35,37}Cl$	220–240	0.3–0.4	−0.4 to −0.8	0.2–0.4	60–80	Distorted quartet
$^{13}C,^{79,81}Br$	180–190	1.2–1.3	~0.5	0.1–0.2	450–500	Distorted quartet
$^{13}P,^{63,65}Cu$	220–240	0.8–1.0	~0.6	1–2	10–100	Distorted quartet

a parameter of somewhat elusive experimental accessibility.

Self-decoupling

As discussed above, second-order effects are only observed in rigid solid samples. Random molecular motions in solution or in highly mobile solids produce two phenomena: (1) anisotropic dipolar and quadrupolar interactions are averaged to zero (**Table 1**), and (2) fast longitudinal relaxation is induced on the quadrupolar nucleus S, leading to the collapse of all coupling interactions (both dipolar and scalar). The latter result is known as self-decoupling, and is responsible, for example, for why ^{13}C nuclei in solution do not normally appear as J-coupled when bonded to ^{14}N or $^{35,37}Cl$ nuclei. An interesting situation arises when the solid-state motion is anisotropic: the relevant interactions do not completely disappear, but are scaled down, depending on the extent of the motion. Only self-decoupling would be able to erase the expected splittings in this case.

Appropriate examples are provided by sodium chloroacetates. Both $ClCH_2COONa$ and $Cl_2CHOONa$ show the ^{13}C–Cl signal as the expected

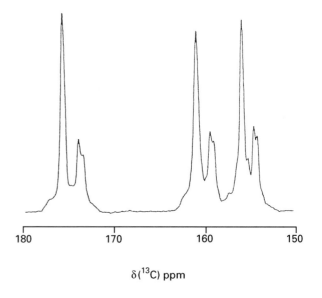

$\delta(^{13}C)$ ppm

Figure 5 Solid-state ^{13}C MAS spectrum of a sample of the polymer $[\{(CH_3)_3Pb\}_4Ru(CN)_6]_\infty$ obtained at a nominal frequency of 75.43 MHz ($B_0 = 7.05$ T) and a spinning speed of 4.3 kHz, by summing all relevant sidebands. There are three different cyanide sites in the solid, each giving a characteristic (negative) splitting s.

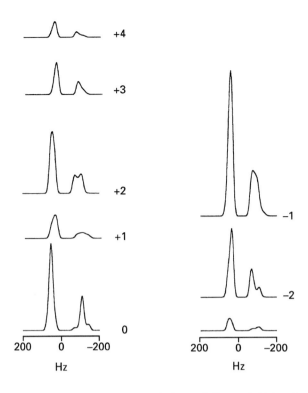

Figure 6 Simulated shapes of the isotropic line and sidebands (as numbered) for the sample of **Figure 5**. The simulations were done assuming $\chi(^{14}N) = -2.3$ MHz and $\Delta\sigma(^{13}C) = 350$ ppm.

1:1 doublet and 1:2:1 triplet, respectively (provided second-order theory applies) (**Figures 8A** and **8B**). However, $Cl_3CCOONa$ shows a narrow single peak even at low temperatures (**Figure 8C**), owing to a fast rotation of the Cl_3C group around the C–C bond. That the collapse of the expected 1:3:3:1 quartet in the latter case is due to self-decoupling was confirmed by independently measuring the ^{35}Cl longitudinal relaxation times in all three salts by using nuclear quadrupole resonance. As expected, the ^{35}Cl T_1 in $Cl_3CCOONa$ is significantly shorter than in other two compounds.

Quadrupole effects on NMR spectra from nuclei with spin $> \frac{1}{2}$

MAS spectra

From the information given in **Table 2**, the following expression for the S NMR signal can be obtained when the observed nucleus is itself quadrupolar:

$$\nu_S = \nu_{0S}[1 - \sigma_{iso}(S)] - m_I J_{iso}$$
$$+ \left[\frac{3\chi(m_S + \frac{1}{2})}{4S(2S-1)} - \frac{1}{3}\Delta\sigma(S)\nu_{0S} \right] (1 - 3\cos^2\theta) \quad [8]$$

In general, the spectra are dominated by the quadrupole interaction (i.e. the term containing χ in Eqn [8]). Although the relevant term in Equation [8] also contains the usual factor $(1 - 3\cos^2\theta)$, the ability of MAS to average out the quadrupole effects depends critically on the spinning speed. Typical values of χ lie in the MHz range; thus, at experimentally accessible spinning speeds (which rarely exceed tens of kHz) the spectra will have the signal intensity distributed over an enormous number of very weak sidebands. There is an exception to this rule: when $m_S = -\frac{1}{2}$ in Equation [8], the first-order quadrupole effect is zero, and MAS should yield simple spectra. Thus, efforts have been directed to the study of the central transition $(-\frac{1}{2}, +\frac{1}{2})$ in half-integer spin systems. Even when the first-order effect is zero for the latter transition, second-order quadrupolar effects remain which are not completely removed by MAS. The expression for the NMR central transition of half-integer quadrupolar nuclei with axially symmetric q, when high-speed MAS is applied, is

$$\nu_S = \nu_{0S}[1 - \sigma_{iso}(S)] - m_I J_{iso} + \nu_{0S}\Delta\sigma_{qs}$$
$$+ \frac{1}{10}\left(\frac{\chi^2}{\nu_{0S}}\right)\frac{9[4S(S+1) - 3]}{[16S(2S-1)]^2}$$
$$\times (35\cos^4\xi - 30\cos^2\xi + 3) \quad [9]$$

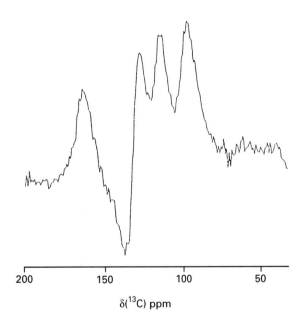

Figure 7 Quaternary-only solid-state ^{13}C NMR spectrum of 1,4-dibromobenzene at 50.33 MHz (4.7 T), showing the C–Br signal as a distorted quartet produced by interaction with $^{79,81}Br$. The negative peak at ~135 ppm is an artifact of the pulse sequence.

spectrum with respect to the isotropic chemical shift, and (3) the occurrence of sidebands (notice that, according to **Table 5**, $\chi^2/\nu_{0S} \ll \chi$ and hence typical spinning speeds will lead to a reasonably low number of sidebands).

In general, appropriate spectral simulations based on Equation [9] are required to retrieve the relevant

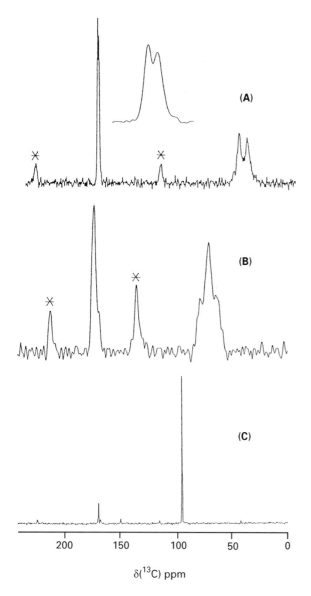

Figure 8 Solid-state ^{13}C NMR spectra of sodium chloroacetates at 75.4 MHz (7.05 T). (A) ClCH$_2$COONa at room temperature (the insert shows an expansion of the carboxyl region; notice that the effect of coupling to Cl is not limited to directly bonded carbons), (B) Cl$_2$CHCOONa at 158 K and (C) Cl$_3$CCOONa at 163 K. Asterisks denote spinning sidebands.

where

$$\nu_{0S}\, \Delta\sigma_{qs} = \frac{3}{10}\left(\frac{\chi^2}{\nu_{0S}}\right)\frac{4S(S+1)-3}{[4S(2S-1)]^2}$$

is the so-called isotropic second-order quadrupolar shift and ξ is the angle between q_{zz} and the rotor axis. Equation [9] will conceivably lead to three effects (**Figure 9**): (1) broad powder pattern line shapes, (2) a shift $\Delta\sigma_{qs}$ of the centre-of-gravity of the

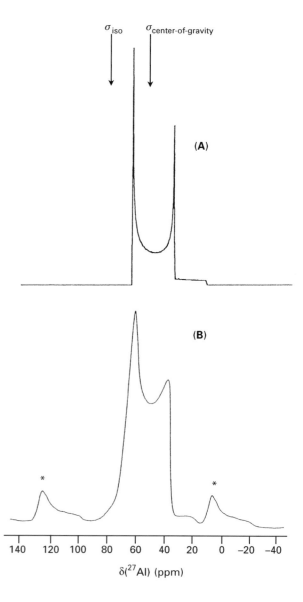

Figure 9 (A) Typical line shape of an observed quadrupolar nucleus S, showing the second-order quadrupole shift $\Delta\sigma_{qs}$, and the relative position of the centre-of-gravity with respect to the isotropic chemical shift σ_{iso}. (B) ^{27}Al solid-state MAS NMR spectrum of Sr$_8$(AlO$_2$)$_{12}$·Se$_2$ at 78.15 MHz (7.05 T). Asterisks denote sidebands. Reproduced with permission of Elsevier Science Publishers from Weller MT, Brenchley ME, Apperley DC and Davies NA (1994) Correlations between ^{27}Al magic-angle spinning nuclear magnetic resonance spectra and the coordination geometry of framework aluminates. *Solid State Nuclear Magnetic Resonance* **3**: 103–106.

Table 5 Nuclear and field gradient properties for some quadrupolar nuclei

Nucleus	ν_{0S} (MHz)[a]	S	Natural abundance (%)	$10^3 Q(S)$ (barn)	Range of $|\chi|$ (MHz)
^{23}Na	79.35	$\frac{3}{2}$	100	100.6	1–4
^{27}Al	78.17	$\frac{5}{2}$	100	140.3	0.5–1
^{11}B	96.25	$\frac{3}{2}$	80.42	40.59	2–5
^{17}O	40.27	$\frac{5}{2}$	0.037[b]	−25.58	1–5
^{51}V	78.86	$\frac{7}{2}$	99.76	−52	0.5–10
^{7}Li	116.59	$\frac{3}{2}$	92.58	−40.1	0.03–0.05

[a] At B_0 = 7.05 T, for which $\nu(^1\text{H})$ = 300 MHz.
[b] Enriched samples.

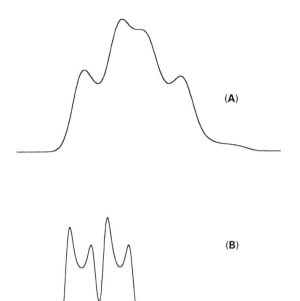

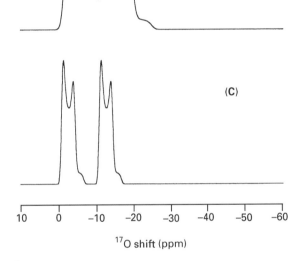

information concerning χ (and η_Q in non-symmetric cases), σ_{iso} and the chemical shift parameters $\Delta\sigma(S)$ (and η). Notice that the broadness of the lines may lead to substantial overlap, thereby complicating the spectral interpretation. (**Figure 10**) shows the theoretical effect in a spectrum with two overlapping lines, for typical parameters of ^{17}O in minerals. Complete separation of peaks is not achieved even at very high magnetic fields.

Figure 10 Second-order quadrupolar spectra expected for two ^{17}O lines (S = 5/2) located at values of σ_{iso} of 0 and 10 ppm, assuming χ = 2 MHz and axial symmetry for both sites. The external magnetic fields are: (A) 7.05, (B) 11.75 and (C) 17.62 T, corresponding to ^{1}H NMR frequencies of 300, 500 and 750 MHz, respectively. The line shapes have been convoluted with a Gaussian broadening.

Double rotation NMR and other techniques

A significant increase in the fundamental knowledge of quadrupole effects in NMR spectroscopy has taken place in the last decade. Various techniques have been developed to diminish the effects produced by the orientational dependence of Equation [9], or to separate chemical shift from pure quadrupole effects. The most successful seems to be double-rotation (DOR), in which the sample spins simultaneously around two different 'magic' angles: 54.7 and 30.6°. The latter angle has the property that, for $\xi = 30.6°$, the factor $[35 \cos^4\xi - 30 \cos^2\xi + 3]$ in Equation [9] is zero, leaving a simple spectrum where only the chemical shift and the isotropic second-order quadrupole shift remain. Further distinction of these shifts can be made by recording spectra at different magnetic fields. **Figure 11** shows an example of the dramatic reduction in line width which is attained by application of DOR to a solid sample.

Other relevant methods are (1) quadrupole nutation spectroscopy, a two-dimensional technique which allows the projection of the conventional spectrum in one dimension, and only quadrupolar information in the second frequency axis, (2) dynamic-angle spinning (DAS), in which the sample is spun sequentially rather than simultaneously (as in DOR) about two different 'magic' axes and (3) satellite transition spectroscopy (SATRAS), which monitors NMR transitions other than the central $(-\frac{1}{2}, +\frac{1}{2})$ under high-speed MAS, allowing the measurements

of both the correct isotropic chemical shift and the quadrupole coupling in a single experiment.

Studied nuclei

The interest in studying quadrupolar nuclei with NMR is to combine useful chemical shift correlations with information concerning the quadrupole coupling constant. The value of χ depends on the nucleus itself (though the quadrupole moment Q), and on the maximum electric field gradient q_{zz}. The latter is a function of both the symmetry and density of the

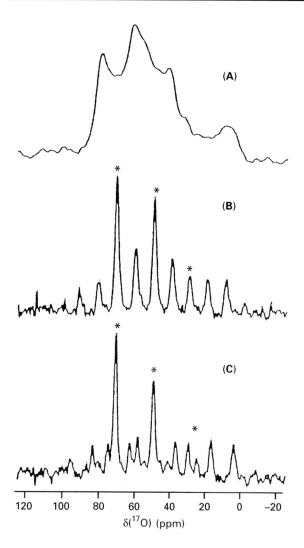

Figure 11 Solid-state ¹⁷O NMR spectra of a sample of the mineral diopside, CaMgSi$_2$O$_6$: (A) using only high-speed MAS, (B) using DOR with a speed of 540 Hz around the second magic-angle of 30.6°, and (C) as in (B), but with a speed of 680 Hz. There are three different oxygen sites in the crystal structure, corresponding to the three ¹⁷O lines marked with asterisks. The latter are identified in spectra (B) and (C) since their positions are not affected by the spinning speed. Reproduced with permission of Macmillan Magazines Ltd. from Chmelka BF, Mueller KT, Pines A, Stebbins J, Wi Y and Zwaziger JW (1989) Oxygen-17 NMR in solids by dynamic-angle spinning and double rotation. *Nature* **339**: 42–43.

electron distribution, i.e. on the molecular structure. For reasons discussed above, the focus has been restricted on the central transition of half-integer nuclei (which make up almost one-third of all NMR active nuclei). In this regard, ²³Na and ²⁷Al are the most studied, but interest has also been paid to ¹¹B, ¹⁷O, ⁵¹V and ⁷Li (see **Table 5** for nuclear and field gradient properties).

A great deal of attention has been paid to ²⁷Al, owing to its importance in the preparation and dealumination of zeolites, and in the study of inorganic materials and catalysts. Since the values of χ are relatively low (**Table 5**), the second-order quadrupole shift is small and therefore relatively narrow signals are obtained. Furthermore, the chemical shift range spanned by distinctly coordinated aluminium sites is large (**Table 6**), allowing not only the distinction of Al sites, but also environments within like sites which may differ in bond distances and angles or in hydrogen bonding. In fact, useful correlations between ²⁷Al chemical shift or quadrupole parameters and structural features have been found, such as (1) $\sigma_{iso}(^{27}Al)$ varies linearly with Al–O–Al angles in aluminates and with Al–O–Si angles in aluminosilicates, and (2) $\chi(^{27}Al)$ is linearly correlated with the distortion $|\psi| = \Sigma|\tan(\alpha_i) -109.48°|$ of the AlO$_4$ tetrahedron.

By studying other quadrupolar nuclei, interesting structural details have been obtained on a number of inorganic solids of undoubted technical importance, such minerals, zeolites, catalysts, ceramics, glasses and cements (**Table 6**). Owing to the technological significance of the studied materials, it is likely that this area of solid-state NMR will experience great progress in the years ahead.

List of symbols

B_0 = magnetic flux density; D = dipolar coupling constant; D' = effective dipolar coupling constant; h = Planck's constant; I = spin-$\frac{1}{2}$ nucleus; J = coupling constant; q = field gradient tensor; Q = nuclear quadrupole moment; s' = doublet splitting;

Table 6 Chemical shift range in solid materials studied by solid-state NMR of quadrupolar nuclei

Nucleus	Chemical shift range (ppm)	Usual reference	Studied materials
²³Na	−50–50	NaCl (1 M)	Na oxides and salts, zeolites
²⁷Al	150–200 (tetracoord.) 20–80 (pentacoord.) −20–50 (hexacoord.)	AlCl$_3$ (1 M)	Zeolites, molecular sieves, catalysts, Al oxides, aluminates, Al glasses and ceramics
¹¹B	−2–2 (tetracoord.) 10–30 (tricoord.)	BF$_3$·Et$_2$O	B glasses, oxides, borates
¹⁷O	0–500	H$_2$O	Oxides, oxoanions, metal carbonyls
⁵¹V	−100–1000	VOCl$_3$	V oxides, vanadia catalysts, vanadates
⁷Li	−5–5	LiCl (1 M)	Li glasses, oxides

S = quadrupolar nucleus; γ = magnetogyric ratio; θ = angle between main tensor axes; δ = chemical shift; τ_1 = relaxation time; ξ = angle between main axes of interaction tensors and sample spinning axis; χ = quadrapole coupling constant.

See also: **High Resolution Solid State NMR, ¹³C; NMR in Anisotropic Systems, Theory; NMR of Solids; Solid State NMR, Methods; Structural Chemistry Using NMR Spectroscopy, Inorganic Molecules.**

Further reading

Abragam A (1989) *Principles of Nuclear Magnetism.* Oxford: Oxford University Press.

Alarcón SH, Olivieri AC, Carss SA and Harris RK (1994) Effects of $^{35}Cl/^{37}Cl$, ^{13}C residual dipolar coupling on the variable-temperature ^{13}C CP/MAS NMR spectra of solid, chlorinated sodium acetates. *Angewandte Chemie, International Edition in English*: 33: 1624–1625.

Davies NA, Harris RK and Olivieri AC (1996) The effects of interplay between quadrupolar, dipolar and shielding tensors on magic-angle spinning NMR spectra: shapes of spinning sidebands. *Molecular Physics* 87: 669–677.

Fyfe CA (1983) *Solid State NMR for Chemists.* Ontario: CFC Press.

Grondona P and Olivieri AC (1993) Quadrupole effects in solid-state NMR spectra of spin-$\frac{1}{2}$ nuclei: a perturbation approach. *Concepts in Magnetic Resonance* 5: 319–339.

Harris RK (1986) *Nuclear Magnetic Resonance Spectroscopy. A Physicochemical View.* New York: Longman Scientific & Technical.

Harris RK (1996) Nuclear spin properties & notation. In: Grant DM and Harris RK (eds) *Encyclopedia of Nuclear Magnetic Resonance* Vol 5, pp 3301–3314. Chichester: Wiley.

Harris RK and Olivieri AC (1992) Quadrupole effects transferred to spin-$\frac{1}{2}$ magic-angle spinning spectra of solids. *Progress in Nuclear Magnetic Resonance Spectroscopy* 24: 435–456.

Lucken EAC (1969) *Nuclear Quadrupole Coupling Constants.* London: Academic Press.

Mason J (ed) (1987) *Multinuclear NMR.* New York: Plenum Press.

Mehring M (1983) *High Resolution NMR in Solids*, 2nd edn. Berlin: Springer-Verlag.

Wasylishen RE and Fyfe CA (1982) High resolution NMR of solids. *Annual Reports in NMR Spectroscopy* 12: 1–80.

Solid State NMR, Methods

JW Zwanziger, Indiana University, Bloomington, USA
HW Spiess, Max-Planck-Institut für Polymerforschung, Mainz, Germany

MAGNETIC RESONANCE
Methods & Instrumentation

As in NMR of liquid samples, solid state NMR probes the magnetic interactions of atomic nuclei. These interactions yield detailed information about the local structure and dynamics of the sample, including the bonding types and geometry, the site–site connectivity patterns, and the spatial characteristics and timescales of atomic and molecular motions. All kinds of solids can be studied with NMR, including single crystals and powders, disordered materials such as glass and rubber, and metals and superconductors. Although not as high as in liquid state NMR, spectral resolution is still extraordinary (parts per million or better) but sensitivity is not. Sample volumes of order 100 μL are typical.

The magnetic interactions probed in solid state NMR include those studied in the liquid state, beginning with the Zeeman interaction between the nuclear spin and the applied magnetic field. This induces precession at the Larmor frequency ω_0, which is defined by the nucleus and the strength of the external field. Fields as high as 17.5 T are in use, yielding proton Larmor frequencies of 750 MHz. Internal interactions observed include the chemical shift, and, in favourable cases, scalar couplings. Additionally, magnetic dipole and electric quadrupole interactions, which are observable only indirectly in liquid state NMR spectra, can be detected as frequency shifts in the solid state. In metals, the major spectral observable is the Knight shift. Because all these interactions depend sensitively on the local bonding geometry, they can be used to measure dynamic properties of the sample, either directly through spectral changes as a function of experimental parameters, or indirectly through the nuclear spin relaxation time.

The primary difference between solid state and liquid state NMR is one of timescale. The atomic dynamics of the sample define a natural internal timescale, denoted τ. The motion of interest might be, for example, the rotational tumbling of molecules in a liquid, the reorientation of segments in a polymer, or the hopping of ions in a solid electrolyte. Clearly τ can range from picoseconds to seconds or more. As the observed nucleus moves to different locations or orientations, its NMR spectrum changes. This occurs both because the different sites may differ chemically, and also because the observable interactions are orientation-dependent, the sense of orientation being defined by the external magnetic field. The different orientations define a range of frequencies $\Delta\omega$ centred on the Larmor frequency. If $\Delta\omega\tau \ll 1$, then a given nucleus samples many local environments during the NMR experiment. All interactions that depend on molecular orientation are in this way averaged to zero, and the spectrum is 'liquid-like'. 'Solid-like' spectra result when $\Delta\omega\tau \gg 1$, for then the anisotropic portions of the interactions remain. These include for example the above-mentioned magnetic dipole and electric quadrupole terms. Typical examples of both regimes are shown in **Figure 1**.

Figure 1 also shows the principal difficulty encountered in solid state NMR spectra: the additional information provided by the anisotropic interactions can seriously congest the spectrum, making interpretation difficult. Our aim in this article is to outline the current principal methods by which solid state NMR spectra can be acquired in interpretable form. Rather than giving an exhaustive account of the current developments in the field, we present the most important techniques in the context of the physical and chemical problems that they can help to solve.

Resolving chemically distinct sites

The most frequent application of solid state NMR, as in the liquid state, is resolution of chemically distinct sites in a material. However, as **Figure 1** shows, the anisotropy observed in solid spectra typically create so much spectral congestion that assignment is difficult. Moreover, as **Figure 1** also shows, the anisotropically broadened lines exhibit a variety of step and singularity features which, while informative in their own right, further obstruct a rapid assessment of the types and relative concentrations of distinct sites. The most important method for improving resolution in solid state NMR is magic angle spinning (MAS). This method, so-called because the sample is rotated about an axis inclined at the magic angle of 54.74° with respect to the

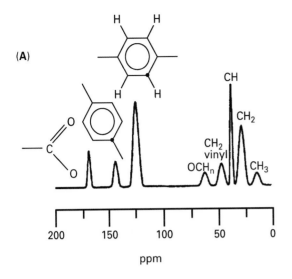

(A)

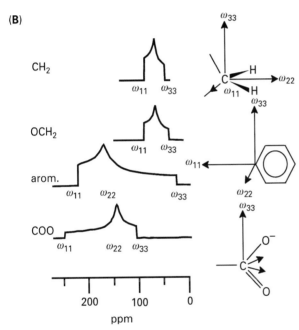

(B)

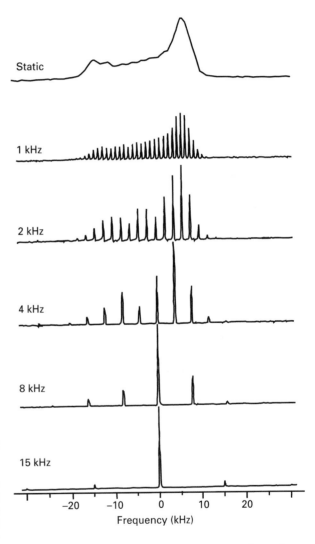

2nd rank spherical tensors, i.e. like d-orbitals. Recall that the d_z^2 orbital has an angular node; this is in fact at 54.74°. Thus an interaction which transforms in the same way can be averaged to zero by spinning about an axis located at this node.

To achieve effective averaging, the rotation frequency ω_r in MAS must be of the order of, or greater than, the spread of interaction frequencies: $\omega_r/\Delta\omega \gtrsim 1$. The resulting resolution enhancement is dramatic, as shown in **Figure 2**.

Figure 1 (A) Typical chemical shifts of carbon in different functional groups. The line widths indicate the ranges observed; in a liquid sample, the actual line width will typically be much smaller. (B) Chemical shift powder patterns of carbons in the same functional groups. Such shapes are observed in powdered solids. The complex shapes, with steps and singularities, arise from the nontrivial orientation dependence of the chemical shift interaction, which is averaged to zero in a liquid but is observable in solids. The powder patterns are shown separately here for convenience; in real samples they overlap, making interpretation difficult. This problem is addressed by techniques such as magic angle spinning. Figure adapted from Schmidt-Rohr K and Spiess HW (1994) *Multidimensional Solid-State NMR and Polymers.* London: Academic Press, 1994.

Figure 2 The effect of magic angle spinning. The figure shows ^{31}P spectra of $Na_4P_2O_7 \cdot 10H_2O$, as a function of rotor frequency. Note the extreme line-narrowing achieved, while still in a powdered solid. This illustrates that the symmetry of the chemical shift interaction is such that the full isotropic averaging of the liquid state is more than necessary to suppress the anisotropy; rotation about a single axis is in this case sufficient. The small splittings show that crystallographically, as well as magnetically, different sites can be resolved. Figure adapted from Schnell I, Diploma Thesis, Johannes-Gutenberg-Universität Mainz, 1996; see also Kubo A and McDowell CA (1990) *Journal of Chemical Physics* **92**: 7156.

magnetic field, can enhance the resolution by more than 2 orders of magnitude. It arises because the dominant anisotropies transform under rotations as

For nuclei like ^{13}C, ^{31}P, and ^{29}Si, which have modest chemical shift ranges, spinning frequencies of 5–10 kHz are often sufficient, and are well within range of typical commercial MAS probes. ^1H spectroscopy is particularly challenging in solid state NMR, in contrast to liquids, because of the strong magnetic dipole coupling between protons. This interaction gives a proton line width typically in the range of 20–50 kHz. Commercial MAS NMR probes are now available with spinning frequencies as high as 35 kHz, and so in many cases even ^1H solid state NMR can be accomplished with the MAS technique.

While MAS can provide significant resolution enhancement, it enhances sensitivity only insofar as the signal from broad resonances is concentrated into narrower resonances. For naturally low-abundance nuclei like ^{13}C (1% naturally occurring), this increase may be insufficient. For dilute spins in the presence of an abundant species with good sensitivity (such as protons), i.e. nearly all organic solids, double resonance methods may be used to achieve an additional gain in sensitivity. Coupled with MAS, these techniques are collectively referred to as CP-MAS (CP = cross-polarization). In CP-MAS, magnetization is first excited using the abundant species (typically ^1H), and then transferred to the dilute species (^{13}C or ^{15}N, say) by simultaneously irradiating both nuclei, at their respective Larmor frequencies. Then, the dilute spin is detected, often with decoupling of the abundant spin. All this is carried out in the presence of MAS, to obtain good resolution of the resulting spectrum. The theoretical sensitivity gain is the ratio of the Larmor frequencies of the two species, for example, 4 for the ^1H–^{13}C pair. In practice, protons often have significantly shorter relaxation times than their CP partners, so this method also allows for shorter recycle delays in pulsed NMR, and thus more rapid acquisition of the spectrum. A disadvantage of CP-MAS is that the CP efficiency is a function of proximity of the abundant and dilute species, so that CP-MAS spectra cannot be assumed to be quantitative reflections of the abundances of the resolved sites.

While MAS and CP-MAS are often sufficient to resolve chemical sites for nuclei like ^{13}C, ^{31}P and ^{29}Si, this is not the case for other nuclei such as ^{27}Al, ^{17}O and ^{11}B. The reason is that the first group of nuclei have spin $\frac{1}{2}$, while the second have higher spin. Nuclei with spin greater than $\frac{1}{2}$ are subject to electric quadrupole effects, in addition to chemical shift, and these effects present a significant additional source of line-broadening. Because the quadrupole anisotropy in the presence of a strong magnetic field does not transform simply like a 2nd rank tensor, it cannot be

removed completely by MAS alone. During the last 10 years significant progress has been made in devising methods to average quadrupole interactions in addition to chemical shift anisotropy. Double rotation (DOR) and dynamic angle spinning (DAS) both make use of spinning the sample about a time-dependent axis. DOR is a direct extension of MAS, and makes use of a complex rotor-within-a-rotor device. In DAS, the sample spinning axis is hopped between two angles during the experiment. Both DOR and DAS require mechanically sophisticated probes. A third method, multiple-quantum magic angle spinning (MQ-MAS), uses just a MAS probe, but a complex pulse sequence to excite and detect triple and higher order coherences during the MAS experiment. It is mechanically the simplest to implement, but uniform excitation of different chemical sites is difficult with existing pulse sequences. Despite the limitations for each method mentioned above, they have yielded impressive resolution advances for quadrupolar nuclei (**Figure 3**), similar to what is achieved with MAS for nuclei such as ^{13}C.

The methods described above yield greatly enhanced resolution, and as such help to determine the types and amounts of distinct sites in material. This resolution gain comes at the price of discarding the information available from the interaction anisotropies. This information typically relates to the local site symmetry, for example, distortions in the bond angles, number of nearest neighbours, and so forth. Both high-resolution and interaction anisotropies may be obtained by taking advantage of a second spectral dimension, using so-called 'separation of interactions' experiments. An example is shown in **Figure 4**, where it is seen that the anisotropically broadened resonances are sorted according to their isotropic shift. In this way each resonance may be examined in isolation, and the anisotropy parameters determined with no congestion from neighbouring bands. There are many ways to implement such experiments; an elegant approach for simple chemical shift correlations is to spin the sample at an angle other than the magic angle during the first part of the experiment, followed by a hop to the magic angle and subsequent signal acquisition. In this way the anisotropic interactions 'label' the detected signal, and a double Fourier transform gives the type of spectra shown in **Figure 4**.

Determining the connectivity between sites

Once the types of sites in the material have been determined, using for example the techniques discussed

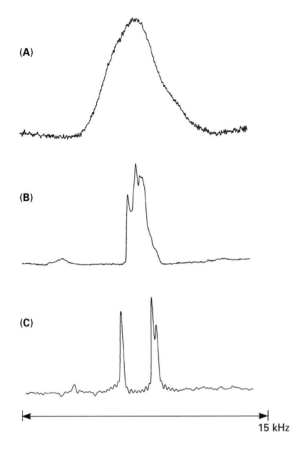

Figure 3 ^{87}Rb NMR spectra of RbNO$_3$. (A) The static spectrum, (B) The effect of magic angle spinning. The symmetry of strong quadrupole interactions is such that spinning about a single axis alone is not sufficient to remove all the anisotropy broadening. (C) Results of multiple-quantum magic angle spinning, one of several methods currently available to obtain high-resolution spectra of quadrupolar nuclei like ^{87}Rb. In this spectrum, the three crystallographically distinct rubidium sites are resolved, and the total line-narrowing is comparable to that achieved by MAS alone for spin-$\frac{1}{2}$ nuclei like ^{13}C and ^{31}P (see **Figure 2**). Figure adapted from Brown S, D. Phil. Thesis, Oxford University, 1998, and Brown S and Wimperis S (1997) *Journal of Magnetic Resonance* **128**: 42–61.

above, the second step in determining the material structure can be considered, namely what the connectivity between these sites is. In liquid state NMR connectivities are primarily determined through scalar couplings, a through-bond interaction mediated by the electrons. However, this interaction is very small compared to the anisotropies of the magnetic interactions, and so is hard to probe in solids in any but the most well-ordered samples. On the other hand, magnetic dipole interactions, i.e. the through-space effects of the nuclear magnetic moments on each other, can be substantial. This interaction varies with distance as r^{-3}, and so is of particular use in determining local structure. As noted above, dipole couplings are averaged to zero in 'liquid-like' spectra, but in solids they have easily observable effects

on the resonance line shapes. Such interactions are also observed *indirectly* in liquids, through the nuclear Overhauser effect where they appear as second-order interactions and thereby survive the averaging due to the molecular motion. Knowledge of the through-space connectivities does not give directly a map of the bonding network but, when combined with knowledge of the material composition and chemistry, can yield much about the bonding pattern.

For example, the CP-MAS experiment described above can already be used to obtain some degree of through-space information. The magnetization transfer, from ^{1}H to ^{13}C say, is mediated by the magnetic dipole interactions. By varying the duration of the transfer time (usually called the contact pulse), sites can be distinguished by their transfer efficiency. For example, primary and secondary carbons can be selectively excited, relative to tertiary and quaternary carbons due to their greater proximity to protons. This is done simply by using a short contact pulse. Much more elaborate spectral editing schemes yield more accurate results, and are more flexible.

While variants of CP-MAS are particularly suitable for exploring proximities in heteronuclear systems with protons as one partner, other experiments can be performed to probe other heteronuclear systems, and homonuclear couplings. All make use of the dipole coupling as the mechanism for encoding distance information.

In general, one wants to combine the distance measurement with some kind of resolution enhancement, in order to determine which sites are close to which. This is not always possible. For example, in many inhomogeneous solids, only one (broad) resonance will be observed, even with MAS or similar techniques. A well-studied example is sodium in a glass. Because techniques like MAS average the dipole coupling to zero, if they do not provide sufficient resolution enhancement, they should not be used. Then, one studies a static sample. The spatial distribution of species in a static sample can be estimated, by measuring the decay properties of spin echoes.

In spin-echo experiments, an excitation pulse is followed at some time τ later by a refocussing pulse. At time τ, after the second pulse, an echo will typically form. The typical use of this experiment for measuring distances is to estimate the so-called second moment (M_2) of the resonance line. Interactions on a local scale, e.g. the chemical shift and quadrupole interactions, and interactions involving isolated pairs of spins, can be refocussed by using suitably chosen pulses. However, dipole coupling to a bath of partners cannot. Therefore, the echo cannot be

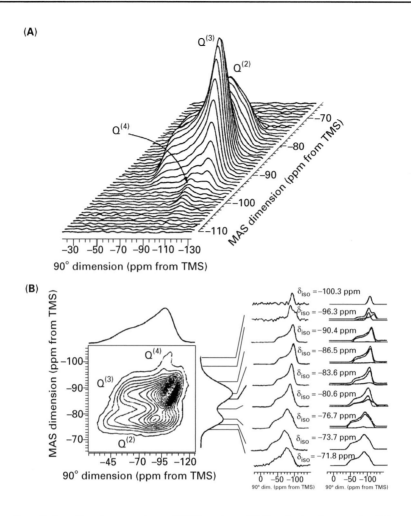

(A)

$Q^{(3)}$
$Q^{(2)}$
$Q^{(4)}$

MAS dimension (ppm from TMS)

−70
−80
−90
−100
−110

−30 −50 −70 −90 −110 −130
90° dimension (ppm from TMS)

(B)

MAS dimension (ppm from TMS)

$Q^{(4)}$
$Q^{(3)}$
$Q^{(2)}$

−100
−90
−80
−70

−45 −70 −95 −120
90° dimension (ppm from TMS)

$\delta_{iso} = -100.3$ ppm
$\delta_{iso} = -96.3$ ppm
$\delta_{iso} = -90.4$ ppm
$\delta_{iso} = -86.5$ ppm
$\delta_{iso} = -83.6$ ppm
$\delta_{iso} = -80.6$ ppm
$\delta_{iso} = -76.7$ ppm
$\delta_{iso} = -73.7$ ppm
$\delta_{iso} = -71.8$ ppm

0 −50 −100   0 −50 −100
90° dim. (ppm from TMS)   90° dim. (ppm from TMS)

Figure 4 A separation of interactions-type spectrum of ^{29}Si in a glass. (A) By spinning the sample off the magic angle during the first part of the experiment, and on the magic angle in the second, a two-dimensional spectrum is generated that has a high-resolution dimension correlated with the anisotropies of the individual sites. (B) Here, in a glass, each site itself shows a distribution of environments, which can be mapped out quantitatively by taking slices through the two-dimensional spectrum. In this way, bond angle distributions for example, even in complex materials, can be determined, often with superior precision as compared to diffraction-based methods. Figure adapted from Zhang P et al. (1996) Journal of Non-Crystalline Solids **204**: 294–300.

refocussed indefinitely, but only up to a characteristic time which is a measure of properties of the bath of nuclei coupled to the studied species. The decay constant of the spin-echo envelope is proportional to M_2, which is given essentially by summing over r_{ij}^{-6}, where the r_{ij} are internuclear distances. Because of the exponent −6, this experiment gives short-range information. It is valuable in assessing qualitative features of the distribution of species in inhomogeneous materials. An important extension of this experiment is called SEDOR, for spin-echo double resonance, in which an additional refocussing pulse is applied to a second nuclear species, and the echo behaviour with and without this secondary pulse are compared. In this way the mixing of different species in an inhomogeneous solid may be assessed.

Similar to SEDOR, but appropriate for isolated pairs of spins, is the rotational echo double resonance

method, or REDOR. In REDOR, the combined dynamics of the isolated two-spin system and the sample rotation serve to generate echos at the rotor period. These echoes can be dephased by application of a pulse to one of the coupled partners. The amount of dephasing caused by this additional pulse is a measure of the coupling strength, and hence proximity, of the spins in the pair. This experiment is most applicable to doubly labelled samples, e.g. biopolymers enriched at selected sites with ^{13}C and ^{15}N.

When the spectrum of the material consists of resolved resonances, much more detailed information on the nuclear distances can be derived than is possible with the spin-echo techniques outlined above. Magnetic dipole coupling is still the interaction to probe but, when the various sites are resolved, experiments can be used that give signals *only* if two distinct sites are near enough to each other to have a significant interaction. Clearly this yields much more detailed information than when the sites serve primarily to generate a background bath. Several types of signals can be generated and measured in this context, but the most precise are the so-called double-quantum coherences. Isolated nuclei (here we have in mind only spin-$\frac{1}{2}$, such as ^{1}H or ^{13}C) do not have enough energy levels to support quantum number changes greater than unity, and therefore also cannot support coherences greater than unity. If two such spins are coupled, however, the composite system can support 2-quantum coherence. Experiments can be designed that are selective only for 2-quantum coherence, thus yielding a connectivity map of the sites that are close enough spatially to couple in this way.

Generating such a connectivity map for a solid requires additionally a resolution-enhancement technique, such as MAS. Such techniques, as discussed previously, suppress precisely the spin–spin interactions that the connectivity map is meant to reveal. Therefore, to combine multiple-quantum experiments with MAS, a pulse sequence which counteracts the averaging effect of MAS, thereby restoring the dipole coupling, must be implemented during excitation and reconversion of the 2-quantum coherence. A variety of such dipolar recoupling sequences currently exist, of varying levels of performance and complexity.

The resulting two-dimensional spectra give a connectivity map that can be traced in much the same way as is routinely done for liquid samples (**Figure 5**).

It must be remembered, of course, that the signals observed reflect coupling through space, not through chemical bonds, so additional information about the chemistry must be used to interpret such spectra. Nevertheless, this method is fast becoming routine, as it requires only standard solids NMR instrumentation.

Dynamics in solids

The dynamics of atoms in solids may be probed directly, through their effects on the NMR spectra, and indirectly, through the nuclear spin relaxation. Because the NMR signal is observed only after the nuclear magnetization has been perturbed from its equilibrium state, relaxation is a standard feature of all NMR experiments. Two primary relaxation processes are usually identifiable. The first is the relaxation of the total magnetization back to its thermal equilibrium value; this occurs on a timescale denoted T_1. The second is the timescale for relaxation of quantum coherences in the spins, and is denoted T_2. In liquids, due to the strong decoupling resulting

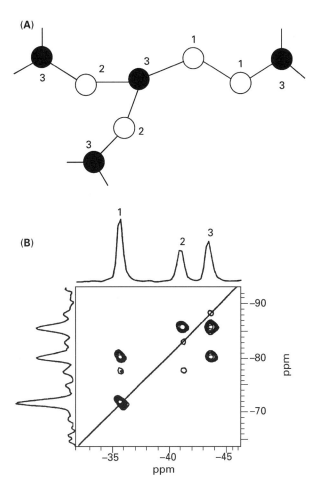

Figure 5 Double-quantum correlation spectrum of ^{31}P in a solid phosphate. At top (A) is the molecular fragment derived from assignment of the spectrum (oxygen atoms not shown), and includes phosphate chain branch points (site 3), branch connections (site 2) and chain phosphates (site 1). The spectrum (B) gives signals symmetric across the diagonal, for pairs of sites that are close enough in space to be coupled. Thus from such a spectrum the spatial proximity of resolved sites can be traced, as is done routinely in liquid NMR using scalar couplings (a through-bond interaction). Figure adapted from Feike M *et al.* (1996) *Journal of the American Chemical Society* **118**: 9631–9634.

from the molecular motion, these two processes occur on similar timescales. In solids, however, they are usually very different, with T_2 ranging typically from 10^{-4} to 10^{-2} s, and T_1 from 10^{-3} to 10^3 s. Relaxation occurs because fluctuations in the surroundings of a spin induce transitions within the spin quantum states. Therefore, measurements of relaxation times are indirect probes of the dynamics in the solid. However, identifying what sort of fluctuation is operative is usually very difficult, unless one type of interaction is clearly dominant (conduction electrons in a metal or superconductor is a good example). Otherwise, the best that can be done is to estimate the temperature and magnetic field dependence of the relaxation to be expected from candidate fluctuation modes, and to compare the predictions with the data.

An easier qualitative assessment of dynamics can often be obtained from resonance line shapes. As noted above, the key distinction between solid-like and liquid-like NMR spectra is the timescale of the atomic motions, compared to the frequency spread of the detected interactions. As the rate of a dynamic process increases, say as a function of temperature, it can be followed through changes in the line widths of the nuclei involved. These changes can be substantial, as a resonance goes from solid-like at low temperatures to liquid-like at high temperatures, with the line width decreasing by orders of magnitude.

The line width strategy is particularly effective if no additional line-narrowing is needed to interpret the low-temperature spectra; however, it is often the case that these spectra will be so congested that additional techniques such as MAS must be applied. In that case, the line-narrowing caused by heating the sample is much less dramatic. In this case, two-dimensional spectroscopy can again be very helpful. For organic solids and polymers, the wideline separation of interactions (WISE) experiment is a convenient qualitative measure of the relative site dynamics. This experiment combines the good resolution found in ^{13}C spectra under MAS, with the strong inter- nuclear coupling of protons. The latter feature makes proton spectra particularly good indicators of motional narrowing due to dynamics: broad proton resonances (30–50 kHz) are seen in static samples, and narrow (< 1 kHz) for mobile sites. The WISE experiment works by adding an additional evolution time to the CP-MAS sequence, between the proton excitation and the contact pulse to the carbons, and using relatively slow sample spinning. In this way, a two-dimensional spectrum is obtained, with proton resonances sorted by the carbon sites to which they are bonded (**Figure 6**). One can immediately see, therefore, which carbon sites are mobile (narrow associated proton resonances) and which are static (broad proton resonances).

More detailed information on dynamics is available from so-called exchange experiments. This class of two-dimensional technique provides a correlation between spectral components, which exchange during a mixing period. The exchange may occur because of a chemical transformation during the mixing time, resulting in a new frequency due to a new chemical environment, or because of site reorientation. Since the nuclear spin interactions are orientation-dependent, if the molecular unit changes its orientation during the mixing time, the involved nuclear spins will exhibit altered NMR frequencies, which can be correlated to their initial values.

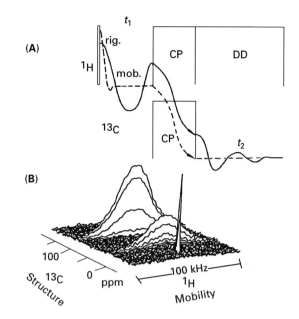

Figure 6 Two-dimensional wideline separation of interactions (WISE) spectrum of polystyrene-poly(dimethyl siloxane) diblock copolymer (PS-b-PDMS). In the first part of the experiment (A) proton magnetization is allowed to evolve, and then transferred to carbon in the second part of the experiment. In this way the proton spectra of individual carbon sites are sorted by the shift of each site, and the result is a wide-line proton spectrum in one dimension, and a high-resolution MAS carbon spectrum in the other. In this example, (B), the PDMS is seen to be quite mobile: it gives the carbon signal near 0 ppm, and the proton spectrum for this site is very sharp, indicating significant motional narrowing. The PS peaks, on the other hand, give very broad proton resonances, showing that the PS part of this block copolymer is essentially static at this temperature. With this experiment, quick qualitative assessments of relative local mobility in organic solids and polymers can be made. Figure adapted from Schmidt-Rohr K and Spiess HW (1994) *Multidimensional Solid-State NMR and Polymers*. London: Academic Press.

Because the orientation dependences of NMR interactions are well known, it is often a straightforward matter to relate the observed exchange spectrum to the underlying molecular motion that gave rise to it. In this way, very detailed information on microscopic molecular dynamics can be obtained.

Deuterium NMR spectra provide particularly clear examples of the above approach. For the deuterium nucleus, the quadrupole interaction is dominant, by far and in a C–D bond, is aligned with the C–D bond itself. Therefore, changes in time of the deuterium quadrupole orientation give a direct reflection of the orientational dynamics of the C–D bond itself. **Figure 7** shows the two-dimensional exchange spectrum of deuterated dimethyl sulfone $((CD_3)_2SO_2)$. The strong diagonal ridge is a typical deuterium NMR spectrum, and arises from molecules that did not exchange during the mixing time. The pattern of ellipses off the diagonal arises due to

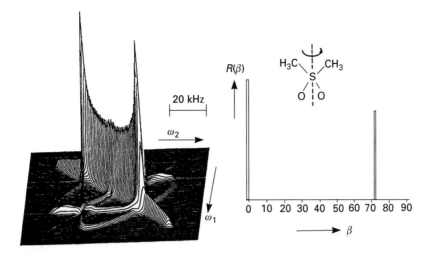

Figure 7 Two-dimensional ^{2}H exchange spectrum of deuterated dimethylsulfoxide. The strong diagonal ridge reflects molecules that have not reoriented during the mixing time, while the pattern of ellipses off the diagonal shows those that have. The ellipses arise due to the orientational dependence of the ^{2}H quadrupole interaction, the dominant anisotropy here. The distribution of jump angles is shown on the right, and sharply peaked at zero (static molecules) and 72°, the included angle of the C–D bonds as the entire molecule executes hops about its symmetry axis. With this type of experiment, slow to moderate dynamics of molecules and polymers can be followed in atomic-level detail. Figure adapted from Schmidt-Rohr K and Spiess HW (1994) *Multidimensional Solid-State NMR and Polymers*. London: Academic Press.

deuterium nuclei with one orientation, and hence one frequency, at the start of the experiment, and a second orientation, hence frequency, after the mixing time. This pattern is consistent with 180° jumps of the molecules about their symmetry axis. The exchange experiment can be applied to other nuclei as well, such as ^{13}C, although it can be harder to relate the spin interaction orientation to a molecular frame of reference.

Summary

In this article we have attempted to provide a brief overview of modern techniques and their applications in solid state NMR. Far from being exhaustive, we hope instead to have informed the reader about the types of problems that can be investigated fruitfully with this approach, using what have become standard methods. There are many other more specialized techniques, suitable for particular problems, which are described in the current literature. The following bibliography is meant to provide a starting point for newcomers to the field. The 'Further reading' section provides entry into the technical primary literature.

List of symbols

M_2 = second moment; r_{ij} = internuclear distances; T_1 = relaxation time of total magnetization;

T_2 = relaxation time of quantum coherences; τ = timescale; ω_0 = Larmor frequency; ω_r = rotation frequency.

See also: **^{13}C NMR Methods; Chemical Exchange Effects in NMR; Chiroptical Spectroscopy, Orientated Molecules and Anisotropic Systems; Heteronuclear NMR Applications (Ge, Sn, Pb); Heteronuclear NMR Applications (O, S, Se, Te); Liquid Crystals and Liquid Crystal Solutions Studied By NMR; Magnetic Resonance, Historical Perspective; NMR Data Processing; NMR in Anisotropic Systems, Theory; NMR Relaxation Rates; NMR Spectroscopy of Alkali Metal Nuclei in Solution; ^{31}P NMR; Parameters in NMR Spectroscopy, Theory of; Product Operator Formalism in NMR; Relaxometers; Xenon NMR Spectroscopy.**

Further reading

Abragam A (1961) *Principles of Nuclear Magnetism.* Oxford: Clarendon Press.

Blümich B (ed) (1994) Solid State NMR I–IV, vols. 30–33 of *NMR Basic Principles and Progress.* Diehl P, Fluck E, Günther H, Kosfeld R, and Seelig J (eds) Berlin: Springer-Verlag.

Fukushima E and Roeder SBW (1981) *Experimental Pulse NMR: A Nuts and Bolts Approach.* London: Addison-Wesley.

Fyfe C (1983) *Solid State NMR for Chemists.* Guelph: CFC Press.

Harris RK and Grant DM (1996) *Encyclopedia of Nuclear Magnetic Resonance.* Chichester: Wiley.

Mehring M (1983) *High Resolution NMR in Solids,* 2nd edn. Berlin: Springer-Verlag.

Slichter CP (1983) *Principles of Magnetic Resonance,* 3rd edn. Springer: Berlin.

Schmidt-Rohr K and Spiess HW (1994) *Multidimensional Solid-State NMR and Polymers.* London: Academic Press.

Traficante DD (ed) *Concepts in Magnetic Resonance, An Educational Journal.* New York: Wiley.

Solid-State NMR, Rotational Resonance

David L Bryce and **Roderick E Wasylishen**,
Dalhousie University, Halifax, Nova Scotia, Canada

MAGNETIC RESONANCE
Applications

Introduction

One of the primary goals of solid-state NMR spectroscopists has been to develop techniques that yield NMR spectra of solid samples with resolution approaching that observed for samples in isotropic liquids. Rapidly spinning samples about an axis inclined at the magic angle ($arc\cos(1/\sqrt{3}) = 54.7356°$) relative to the applied static magnetic field has been found to be highly effective in this regard. In addition, high-power decoupling of abundant spins (e.g. ^1H) eliminates heteronuclear spin–spin coupling interactions (direct dipolar and indirect J-coupling) involving the abundant spins when dilute spins are examined. The availability of commercial NMR instrumentation that permits users to apply these two techniques has contributed to spin-$\frac{1}{2}$ NMR becoming a routine method for examining a wide range of solid materials. Finally, cross-polarization (CP) from abundant spins to dilute spins has been important in improving the sensitivity of the dilute-spin NMR experiment.

Ironically, it is sometimes desirable to selectively reintroduce interactions which are effectively averaged in the magic-angle-spinning (MAS) experiment. Dipolar coupling, for instance, may be recovered in the form of the direct dipolar coupling constant (R_{DD}) between isolated spin pairs. The value of R_{DD} is of interest due to its simple relationship with the distance separating two spins, r_{12} (Eqn [1])

$$R_{DD} = \frac{\mu_0}{4\pi} \frac{\gamma_1 \gamma_2}{\langle r_{12}^3 \rangle} \frac{\hbar}{2\pi} \qquad [1]$$

where μ_0 is the permeability of free space, and γ_i are the magnetogyric ratios of the nuclei under consideration.

Rotational resonance (RR) is a MAS NMR technique which selectively restores the dipolar interaction between a homonuclear spin pair, thus allowing the determination of the dipolar coupling constant, R_{DD}, and hence, the internuclear distance. Historically, the RR phenomenon was discovered by Andrew and co-workers in a ^{31}P NMR study of phosphorus pentachloride, which consists of PCl_4^+ and PCl_6^- units in the solid state. This group noticed that when the rate of sample spinning matched the difference in resonance frequencies of the nonequivalent phosphorus centres, their peaks broadened and the rate of cross-relaxation was enhanced.

It is now known that if the RR condition is satisfied, direct dipolar coupling is restored selectively to a homonuclear spin pair. That is, if the sample spinning rate is adjusted to a frequency, ν_r, such that

$$n\nu_r = (\nu_1^{iso} - \nu_2^{iso}) \qquad [2]$$

where n is an integer, generally 1–3, and ν_1^{iso} and ν_2^{iso} are the isotropic resonant frequencies of spins 1 and 2 respectively, then the two nuclei are said to be in RR. As a result, dipolar coupling between the nuclei is restored (via the 'flip-flop' term in the dipolar Hamiltonian), and 'line broadening' of the resonances at ν_1^{iso} and ν_2^{iso} is observed (see **Figure 1**). Additionally, a rapid oscillatory exchange of Zeeman

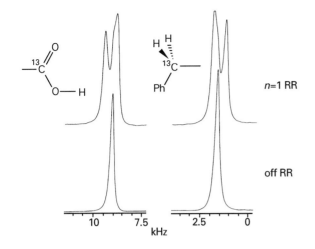

Figure 1 Effect of RR on the ^{13}C NMR line shape of Ph^{13}CH$_2$^{13}COOH. Spectra acquired at 4.7 T (50.3 MHz). Top spectrum at $n = 1$ RR, $\nu_{rot} = 7207$ Hz. Bottom spectrum off RR, $\nu_{rot} = 10\,000$ Hz.

magnetization occurs. In fact, it is this exchange of magnetization rather than the line shape which is usually monitored in order to determine the dipolar coupling constant. In order to generate an exchange curve which may be analysed and simulated, one of the two resonances involved must be inverted selectively; the intensity difference of the peaks is then monitored as function of time.

Obviously, it is highly desirable to develop techniques capable of recovering weak dipolar coupling constants from high-resolution NMR spectra obtained under MAS conditions. The focus of the present discussion will be to provide an overview of the basic RR scheme. First, the theory of RR will be outlined, followed by a discussion of the most important experimental techniques employed to measure dipolar coupling constants under conditions of RR. Finally, some examples that illustrate the applications and limitation of the techniques will be described.

Theory

Restoring the dipolar interaction: a theoretical approach

The most important interaction in NMR results from the application of a large external magnetic field, B_0, to the sample. Termed the Zeeman interaction, its effect on the normally degenerate nuclear spin energy levels is to cause them to split. The Zeeman levels are perturbed by local fields generated by the motion of electrons in the vicinity of the nucleus. These induced local fields are proportional to the applied field. In frequency units, the Hamiltonian operator which accounts for both the Zeeman interaction and this chemical shielding (CS) interaction is

$$h^{-1}\hat{\mathcal{H}}_{Z,\,CS} = -\sum_i \nu_i \hat{I}_{zi} \qquad [3]$$

where

$$\nu_i = \frac{\gamma_i B_0 (1 - \sigma_i^{iso})}{2\pi} \qquad [4]$$

and σ^{iso} is the isotropic chemical shielding constant.

The interaction of interest in RR is the dipolar interaction, an orientationally dependent through-space spin–spin coupling, which leads to a perturbation of the CS-perturbed Zeeman energy levels. For a homonuclear two-spin system, the truncated dipolar Hamiltonian operator is given by the following:

$$
\begin{aligned}
h^{-1}\hat{\mathcal{H}}_{DD} = &-\left(\frac{\mu_0}{4\pi}\right)\gamma^2\hbar\langle r_{12}^{-3}\rangle \\
&\times \left[\hat{I}_{z1}\hat{I}_{z2} - \frac{1}{4}\left(\hat{I}_{1+}\hat{I}_{2-} + \hat{I}_{1-}\hat{I}_{2+}\right)\right] \\
&\times (3\cos^2\theta - 1) \qquad [5]
\end{aligned}
$$

Here, $\hat{I}_+$ and $\hat{I}_-$ are the raising and lowering operators and θ is the angle between the applied magnetic field and the internuclear vector, r_{12}. The factor containing the raising and lowering operators is sometimes referred to as the 'flip-flop' term.

The final interaction that must be considered is the indirect spin–spin coupling interaction, which is mediated by the intervening electrons. The indirect spin–spin Hamiltonian, $\hat{\mathcal{H}}_J$, is often ignored because it is frequently considerably smaller than $\hat{\mathcal{H}}_{DD}$.

Up until this point, we have implicitly assumed time independence of the interactions and their corresponding Hamiltonian operators. This assumption is valid for a rigid stationary sample. However, when the sample is spun rapidly, each of the internal Hamiltonians becomes time-dependent. For example, $\hat{\mathcal{H}}_{Z,CS}$ becomes time-dependent when there is chemical shielding anisotropy due to the fact that the orientations of the chemical shielding tensors relative to the applied magnetic field change as the sample rotates. Then Equation [3] becomes

$$\hbar^{-1}\hat{\mathcal{H}}_{Z,CS}(t) = -\sum_i \gamma_i B_0 \big[(1 - \sigma_i^{iso})$$
$$+ (\sigma_{i,zz} - \sigma_i^{iso})\xi(t) \big] \hat{I}_{zi} \qquad [6]$$

where $(\sigma_{i,zz} - \sigma_i^{iso})$ is a measure of the orientation dependence of the chemical shielding and $\xi(t)$ represents the time dependence of the interaction:

$$\xi(t) = C_1 \cos(\omega_r t) + S_1 \sin(\omega_r t) + C_2 \cos(2\omega_r t)$$
$$+ S_2 \sin(2\omega_r t) \qquad [7]$$

Here, C_1, C_2, S_1, and S_2 are constants that depend on the nature of the interaction (i.e. CS, dipolar) and ω_r is the rotor angular frequency.

Summing the CS-perturbed Zeeman Hamiltonian given in Equation [6] with the time-dependent dipolar Hamiltonian gives the total Hamiltonian

$$\hbar^{-1}\hat{\mathcal{H}}(t) = -\sum_i \gamma_i B_0 \big[(1 - \sigma_i^{iso}) + (\sigma_{i,zz} - \sigma_i^{iso})\xi(t) \big] \hat{I}_{zi}$$
$$- \left(\frac{\mu_0}{4\pi}\right) \gamma^2 \hbar \langle r_{12}^{-3} \rangle$$
$$\times \xi(t) \left[\hat{I}_{z1}\hat{I}_{z2} - \frac{1}{4} (\hat{I}_{1+}\hat{I}_{2-} + \hat{I}_{1-}\hat{I}_{2+}) \right] \qquad [8]$$

The parameter $\xi(t)$ completely describes the time-dependence of a rotating solid.

In order to understand some of the essential features of the RR experiment, it is convenient to assume negligible chemical shielding anisotropy. Under these conditions, ω_1 and ω_2, the CS-perturbed Zeeman angular frequencies, are independent of time. In addition, it is convenient to use the spherical tensor notation to describe the direct dipolar interaction. Thus, the total Hamiltonian is

$$\hbar^{-1}\hat{\mathcal{H}}(t) = \omega_1 \hat{I}_{z1} + \omega_2 \hat{I}_{z2} + A_d(t)T_{20} \qquad [9]$$

where

$$A_d(t) = -\sqrt{6}(2\pi R_{DD}) \sum_{m=-2}^{2} d_m \exp(-i m\omega_r t) \qquad [10]$$

Here, $A_d(t)$ represents the spatial dependence of the dipolar Hamiltonian, and d_m are Fourier

components of the dipole–dipole coupling which depend on Euler angles defining the crystallite orientation with respect to the rotor frame. They are time-independent. The spin part of the truncated dipolar Hamiltonian is

$$T_{20} = \frac{2}{\sqrt{6}} \left[\hat{I}_{z1}\hat{I}_{z2} - \frac{1}{4} (\hat{I}_{1+}\hat{I}_{2-} + \hat{I}_{1-}\hat{I}_{2+}) \right] \qquad [11]$$

To transform the total Hamiltonian into the doubly rotating frame of reference defined by the Zeeman interactions, the propagator is

$$U^{int} = \exp\big[i(\omega_1 \hat{I}_{z1} + \omega_2 \hat{I}_{z2})t \big] \qquad [12]$$

The Zeeman terms and the $\hat{I}_z$ terms of the dipolar Hamiltonian are unaffected by this rotation since it is about the z-axis, and so the desired transformation is:

$$T_{20}^{int} = (U^{int})T_{20}(U^{int})^{-1} \qquad [13]$$

The result of the transformation gives a periodic interaction frame dipolar Hamiltonian,

$$\hbar^{-1}\hat{\mathcal{H}}_{DD}^{int}(t)$$
$$= -2(2\pi R_{DD}) \sum_{m=-2}^{2} d_m \Bigg(\hat{I}_{z1}\hat{I}_{z2} \exp(-im\omega_r t)$$
$$- \frac{1}{4} \left\{ \begin{array}{l} \hat{I}_{1+}\hat{I}_{2-} \exp[i(\omega_\Delta^{iso} - m\omega_r t)] \\ + \hat{I}_{1-}\hat{I}_{2+} \exp[-i(\omega_\Delta^{iso} + m\omega_r t)] \end{array} \right\} \Bigg) \qquad [14]$$

where $\omega_\Delta^{iso} = \omega_1 - \omega_2$. If we re-express the rotational resonance condition [2] in angular frequency units as $n\omega_r = \omega_\Delta^{iso}$, and if $|R_{DD}\tau_r| \ll 1$, where $\tau_r = v_r^{-1}$ is the rotor period, then the time-independent terms vanish and the time average of Equation [14] over one rotor period is

$$\overline{\hbar^{-1}\hat{\mathcal{H}}_{DD}} = \frac{1}{2}(2\pi R_{DD})(d_m \hat{I}_{1+}\hat{I}_{2-} + d_{-m}\hat{I}_{1-}\hat{I}_{2+}) \qquad [15]$$

where $m = \pm 1, \pm 2$. The result of this exercise is that at rotational resonance, parts of the 'flip-flop' term do not average to zero and will therefore contribute to the MAS NMR spectrum.

Some qualitative results of an approximate theoretical treatment of rotational resonance are useful to examine. For $n = 1$ RR, the splitting of each peak is given by $R_{DD}/(2\sqrt{2})$, or $\sim 0.35\ R_{DD}$. For the $n = 2$ case, the splitting is $R_{DD}/4$. The splitting decreases as the order of the RR increases. More rigorous treatments also indicate that the splitting decreases as the chemical shielding anisotropy increases. It is important to note that the observed line widths for homonuclear spin systems are not strictly independent of spinning speed; for a spin pair with differing isotropic chemical shifts, the line widths take on a ω_r^{-2} dependence at high spinning speeds.

When the RR condition is satisfied, a rapid exchange of Zeeman magnetization occurs in addition to dipolar broadening. We will not present a complete theoretical description of the origins of this exchange, but rather present some of the important approximate results of such a treatment. It is convenient to define

$$\Lambda^2 = \left(T_2^{ZQ}\right)^{-2} - 4\left|\overline{\omega_B^{(n)}}\right|^2 \quad [16]$$

where T_2^{ZQ} is the zero-quantum relaxation time constant, and $\overline{\omega_B^{(n)}}$ are the resonant Fourier components associated with the flip-flop term of the dipolar Hamiltonian for RR of order n. To monitor the exchange of magnetization, one plots $\langle \hat{I}_{z1} - \hat{I}_{z2} \rangle$ as a function of time. In the limit of very fast dephasing where T_2^{ZQ} is relatively short and thus $\Lambda^2 \gg 0$, the decay of magnetization is exponential:

$$\langle \hat{I}_{z1} - \hat{I}_{z2} \rangle(t) \approx \exp\left(-T_2^{ZQ}\left|\overline{\omega_B^{(n)}}\right|^2 t\right) \quad [17]$$

In the case of very slow dephasing where T_2^{ZQ} is relatively long and $\Lambda^2 \ll 0$, the exchange of magnetization oscillates as it decays:

$$\langle \hat{I}_{z1} - \hat{I}_{z2} \rangle(t) \approx \exp\left(-\frac{t}{2T_2^{ZQ}}\right) \cos\left(\left|\overline{\omega_B^{(n)}}\right| t\right) \quad [18]$$

In practice, the parameters which influence the observed magnetization exchange curve include R_{DD}, T_2^{ZQ}, the magnitude of the principal components of the chemical shielding tensors, the relative orientation of the CS tensors with respect to r_{12}, and the J-coupling constant.

A pictorial representation of rotational resonance and the exchange of Zeeman magnetization

At this point, it is instructive to provide a qualitative picture of the rotational resonance phenomenon. The energy level diagram for an isolated homonuclear two-spin system where the two nuclei have resonance frequencies ν_1^{iso} and ν_2^{iso} is shown in **Figure 2**. Transitions 1 and 2 correspond to the two isotropic peaks, which would be observed in a MAS NMR spectrum. The difference between the isotropic chemical shifts (or, alternatively, the energies) is, according to the diagram, equivalent to the angular frequency ω_Δ^{iso}. As shown earlier, rotational resonance occurs when an integer multiple of the spinning frequency is equivalent to ω_Δ^{iso}. In terms of the diagram, it is convenient to think of the mechanical rotation of the sample as supplying the necessary energy for zero-quantum coherence between the two intermediate energy levels. The fact that these two states are linked by mechanical rotation ensures that the dipolar interaction will be recoupled, and that exchange of Zeeman magnetization will occur rapidly.

Figure 3 illustrates the exchange experiment, in which one of the transitions is selectively inverted, thus creating a nonequilibrium situation in which spins must relax so that the equilibrium Boltzmann populations are re-established. If we consider the diagram on the left to reflect the excess populations in arbitrary units as determined by the Boltzmann distribution, a selective inversion of transition 1 will result in the population distribution shown on the right. Transition 1 is inverted while the intensity of transition 2 remains unperturbed. Techniques for accomplishing this experimentally will be discussed in the next section. Once the inversion has been carried out, the diagram on the right shows a difference of five population units between the two intermediate energy levels. Rotational resonance provides the zero-quantum coherence necessary for an exchange

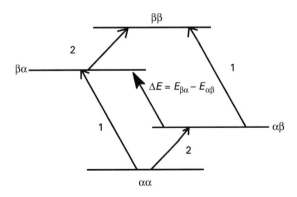

Figure 2 Simplified energy level diagram for two spin-½ nuclei with different isotropic chemical shifts. The two transitions are labelled '1' and '2', and their difference is greatly exaggerated. The energy of the zero-quantum transition is indicated, which corresponds to the mechanical energy supplied at RR. Here, J-coupling is ignored and dipolar coupling is not shown.

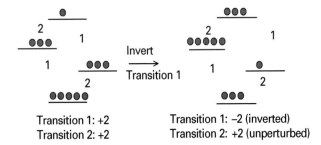

Transition 1: +2
Transition 2: +2

Transition 1: –2 (inverted)
Transition 2: +2 (unperturbed)

Figure 3 Energy level and population distribution diagrams for two spin-$\frac{1}{2}$ nuclei. The circles indicate the excess population in arbitrary units relative to the least populated level. On the left, an equilibrium Boltzmann-type distribution is represented. Both transitions would show a signal of relative intensity +2. Upon inversion of transition 1 (at right) the populations related to this transition are switched, while the net difference in population for transition 2 is unchanged. Transition 1 would now show an inverted signal with relative intensity –2.

of Zeeman magnetization between these two levels. The zero-quantum relaxation which dampens this exchange is described by the time constant T_2^{ZQ}.

Experimental techniques

Pulse sequences and cross-polarization

The basic rotational resonance experiment can be as simple a single-pulse excitation, with the rate of MAS adjusted to satisfy the RR condition. A $\pi/2$ pulse followed by acquisition of the free induction decay (FID) will generate a spectrum with significant broadening of the two resonances concerned. Many typical applications involve ^{13}C in the presence of ^{1}H, and benefit from standard CP techniques. **Figure 1** shows an example of the line broadening observed for the $n = 1$ RR condition for the ^{13}C–^{13}C spin pair in Ph^{13}CH$_2$^{13}COOH, with CP.

A typical pulse sequence for carrying out the RR experiment with selective inversion of a particular transition and CP is shown in **Figure 4**. Note that a flipback pulse is applied to the rare spin channel to store the magnetization along the z axis before carrying out the inversion. The efficiency of CP becomes sensitive to the spinning rate, particularly as ω_r increases. One technique which attempts to circumvent this problem is known as variable amplitude cross-polarization (VACP), where the spin-locking pulses vary in amplitude.

The goal of the RR experiment is the extraction of the homonuclear dipolar coupling constant, R_{DD}. This can be done by carrying out lineshape simulations. However, in general this is not done because a Zeeman magnetization exchange experiment is more

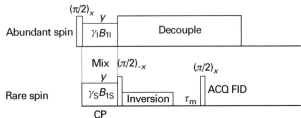

Figure 4 Pulse sequence for carrying out the RR experiment (see text). In the case of the magnetization exchange experiment, CP of the rare spins is followed by a flipback pulse on the rare spin channel, selective inversion of a particular resonance, and a variable delay before acquisition.

informative, and more sensitive to the magnitude of R_{DD}, as will be shown.

The exchange experiment

To generate an exchange curve, one of the two resonances involved must be inverted selectively. By whichever technique a selective inversion is carried out, it is important that the other resonances not be perturbed. The most frequently used inversion techniques in RR experiments are a long, soft pulse or an asynchronous DANTE (delays alternating with nutation for tailored excitation) sequence.

In cases where the CS anisotropy at one or both of the sites is comparable to the isotropic chemical shift difference between them, difficulties arise in carrying out the inversion with selectivity. Total sideband suppression pulse sequences combined with their time-reversed counterparts may be used to overcome the difficulties associated with large chemical shift anisotropies. Regardless of what technique is used to establish the initial condition of maximum polarization difference, the next step in the experiment is to allow the exchange of Zeeman magnetization for a variable time, τ_m (see **Figure 4**), before applying a $\pi/2$ acquisition pulse. The equilibration of magnetization between the two sets of spins is described by the approximate Equation [17] or [18], depending on the system.

Applications and limitations

As mentioned previously, the primary goal of the rotational resonance experiment is to determine the dipolar coupling constant, R_{DD}, from which the internuclear distance, r, may be calculated. Carbon–carbon separations as large as 6.8 Å have been successfully determined, which corresponds to measuring a coupling as small as 24 Hz. Occasionally, dihedral angle measurements have also been carried out using RR. At higher order rotational resonances (i.e. $n = 3$ or $n = 4$), where CS anisotropy is more likely to be comparable to ν_r, the lineshapes and exchange curves are more sensitive to the orientation of the chemical shielding tensors. In general, when simulations (of either line shapes or exchange curves) are performed, they depend on R_{DD} and, to varying degrees, on the magnitudes of the principal components of the chemical shielding tensors, their orientations with respect to the internuclear vector and with respect to each other, the magnitude of the J-coupling, and the zero-quantum transverse relaxation time constant, T_2^{ZQ}.

Lineshape simulations

In order to effectively determine a dipolar coupling constant based on a lineshape simulation, the chemical shielding tensors and their orientations must be known, as well as the J-coupling constant, and T_2^{ZQ}. To determine the principal components of the chemical shielding tensors, a MAS NMR spectrum acquired on a singly-labelled compound in the slow-spinning regime may be used to emulate the powder pattern provided the isolated spin approximation is valid. In some cases it is also possible to determine the CS tensor components from a spectrum of the stationary sample. Determining the *orientations* of the CS tensors is a more involved process, although in some cases careful assumptions and clues from local symmetry may be helpful. In practice, a value for T_2^{ZQ} is usually estimated from the observed line widths off the RR condition.

$$(T_2^{ZQ})^{-1} = \pi[\nu_{1/2}(1) + \nu_{1/2}(2)] \qquad [19]$$

In many cases, not all these parameters are known for the specific spin system under investigation. Therefore, two techniques that may be employed when a lineshape simulation is desirable are (i) a simulation based on known chemical shielding tensors (σ), J-coupling constants, and T_2^{ZQ} values, where only R_{DD} is varied; (ii) use of a calibration with respect to similar compounds, where R_{DD} (or r itself) can be extracted analytically from the observed splitting of a resonance.

For example, the calibration method (ii) has been employed for a series of ^{13}C-labelled retinals containing vinylic and methyl carbons, shown in **Figure 5**. The three isotopomers were ^{13}C-labelled at the (10,20), (11,20), and (12,20) positions. It must be emphasized that the required parameters (σ, J, T_2^{ZQ} and r) were known independently from X-ray and previous NMR studies. T_2^{ZQ} was estimated using

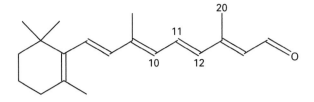

Figure 5 Structure of the retinal studied using RR, with the labelled carbons indicated. See text for details. (Reprinted with permission of the American Chemical Society from Verdegem PJE, Helmle M, Lugtenburg J and de Groot HJM (1997) (*Journal of the American Chemical Society*, **119**: 169–174).

Equation [19]. The goal of the calibration was to be able to employ a simple, analytic equation relating r to the observed broadening of the vinylic peaks at RR. To accomplish this, the 'ideal' splitting presented above, $R_{DD}/(2\sqrt{2})$, was plotted against the observed splitting, $\Delta\omega$. Simulations showed that $\Delta\omega$ could be reliably reproduced, independent of the actual shape of the line. The resulting equation,

$$\frac{R_{DD}}{\sqrt{8}} = 1.15 \left(\frac{\Delta\omega}{2\pi}\right) + 7 \ \ (Hz) \qquad [20]$$

shows that the approximate theory fits well with experimental results in this case, and allows for a very straightforward determination of r from the observed splitting. The major advantage of using line shape simulations to extract the dipolar coupling constant, in general, is that the spectrometer time involved is less than that for the corresponding magnetization exchange experiment. For molecules similar to the retinal in **Figure 5**, where r is unknown, the analytical empirical Equation [20] can provide the information after a simple 1D NMR experiment.

In spite of the results of the preceding example, the lineshape simulation method has rarely been used in practice, mainly because the R_{DD} values are too small to result in splittings. In such cases, the exchange curve method discussed below is the standard technique for extracting the dipolar coupling constant under RR conditions.

Exchange curves and simulations

By far the most common method for deriving structural information under RR conditions is through the analysis and simulation of a magnetization exchange curve. Once a suitable state of polarization difference has been achieved between the two sets of spins, 1 and 2, the delay time, τ_m, is varied before applying a $\pi/2$ observe pulse and acquiring the spectrum (see **Figure 4**). Separate NMR experiments must be performed to generate each point on a magnetization exchange plot.

In order to extract the dipolar coupling constant or structural information, the observed magnetization decay must be simulated. Qualitative and relative distance information is more readily available than quantitative information since the exchange curve depends on the same parameters that the RR line shape depends on. Two common procedures for extracting R_{eff} are: (i) comparison of the exchange curve with a series of exchange curves of model compounds for which r is known, and (ii) complete simulation of the exchange curve, where σ, J, and T_2^{ZQ} are known (or estimated).

The magnetization due to naturally abundant NMR-active spins in the sample must be considered. This is done by subtracting the natural-abundance spectrum from that of the labelled sample. Failure to make such a correction could lead to an overestimation of r and an underestimation of T_2^{ZQ}.

A recent example of the application of RR to a structural problem will serve to illustrate its utility as a comparative tool. **Figure 6** shows two peptide fragments in different conformations. This compound models the peptide Aβ1-42, a constituent of the amyloid plaques characteristic of Alzheimer's disease. Rotational resonance MAS NMR was used in a qualitative fashion by Costa and co-workers to determine whether the amide conformation in the solid state was 'cis' or 'trans'. From previous

experiments, model compounds served to give the chemical shielding tensor orientations of the carbonyl carbon. The orientation of the internuclear vector connecting the two labelled carbon atoms with respect to the chemical shielding tensor of the carbonyl carbon is drastically different for the two conformers shown in **Figure 6**. Note that in the *trans* conformation, the internuclear vector lies nearly along the σ_{11} component, while in the *cis* conformation it lies nearly along σ_{22}. The dipolar coupling for the two conformations should, however, be nearly identical. Hence, the variable of interest in this experiment is the CS tensor orientation. Experimental Zeeman magnetization exchange curves were generated and matched to simulated curves (**Figure 7**). It was found that theory matched experiment only when a *trans* geometry was assumed. The $n = 2$ RR experiment was used in this case because at higher spinning speed (i.e. $n = 1$), the orientations of the CS tensors become less influential in determining the course of the magnetization exchange.

This example shows that RR experiments can be used for more than simply extracting the dipolar coupling constant and determining an accurate value for r_{12}. In fact, the basic RR technique is probably

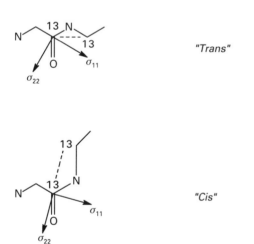

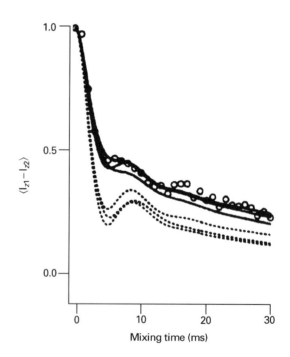

Figure 6 Fragments of the peptide β34-42 showing the *cis* and *trans* conformations. Also indicated is the orientation of the carbonyl carbon chemical shielding tensor, with σ_{33} perpendicular to the plane. Note the different orientations of the C–C internuclear vector with respect to the CS tensor components. Reprinted with permission of the American Chemical Society from Costa PR, Kocisko DA, Sun BQ, Lansbury PT Jr and Griffin RG (1997) *Journal of the American Chemical Society*, **119**: 10487–10493)

Figure 7 Zeeman magnetization exchange plot for the peptide β34-42 fragments shown in **Figure 6**. The open circles are experimentally determined data points; the solid lines result from simulations assuming a *trans* geometry; the dotted lines result from simulations assuming a *cis* geometry. Reproduced with permission of the American Chemical Society from Costa PR, Kocisko DA, Sun BQ, Lansbury PT Jr and Griffin RG (1997) *Journal of the American Chemical Society* **119**: 10487–10493.

Table 1 A summary of some homonuclear dipolar recoupling techniques.

Name	Acronym	Principle
Rotational resonance	RR	Recouples when the difference in chemical shift frequencies is an integer multiple of the MAS speed.
Dipolar recovery at the magic angle	DRAMA	In its simplest form, a pair of x and $-x$ π pulses separated by a delay, τ, results in an observable dipolar broadening (Tycko R and Dabbagh G (1991) Double-quantum filtering in magic-angle-spinning NMR spectroscopy: an approach to spectral simplification and molecular structure determination. *Journal of the American Chemical Society* **113**: 9444–9448).
Simple excitation for the dephasing of the rotational-echo amplitudes	SEDRA	Synchronously applied pulses lead to signal dephasing for dipolar coupled spins (Gullion T and Vega S (1992) A simple magic angle spinning NMR experiment for the dephasing of rotational echoes of dipolar coupled homonuclear spin pairs. *Chemical Physics Letters* **194**: 423–428).
Radio frequency driven dipolar recoupling	RFDR	Rotor-synchronized π-pulses reintroduces flip-flop term (Bennett AE, Ok JH, Griffin RG and Vega S (1992) Chemical shift correlation spectroscopy in rotating solids: radio frequency-driven dipolar recoupling and longitudinal exchange. *Journal of Chemical Physics* **96**: 8624–8627).
Unified spin echo and magic echo	USEME	Spin-echo and magic-echo sequences are applied to recover the dipolar interaction (Fujiwara T, Ramamoorthy A, Nagayama K, Hioka K and Fujito T (1993) Dipolar HOHAHA under MAS conditions for solid-state NMR. *Chemical Physics Letters* **212**: 81–84).
Combines rotation with nutation	CROWN	Dipolar dephasing occurs due to applied RF pulses (Joers JM, Rosanske R, Gullion T and Garbow JR (1994) Detection of dipolar interactions by CROWN NMR. *Journal of Magnetic Resonance* **A106**: 123–126).
Double quantum homo-nuclear rotary resonance	2Q1-HORROR	RF field applied at half the rotation frequency in conjunction with RF pulses (Nielsen NC, Bildsøe H, Jakobsen HJ and Levitt MH (1994) Double-quantum homonuclear rotary resonance: efficient dipolar recovery in magic-angle-spinning nuclear magnetic resonance. *Journal of Chemical Physics* **101(3)**: 1805–1812).
Melding of spin-locking dipolar recovery at the magic angle	MELODRAMA	Rotor-synchronized 90° phase shifts of the applied spin-locking field (Sun B-Q, Costa PR, Kocisko D, Lansbury PT Jr and Griffin RG (1995) Internuclear distance measurements in solid state nuclear magnetic resonance: Dipolar recoupling via rotor synchronized spin locking. *Journal of Chemical Physics* **102**: 702–707).
Rotational resonance in the tilted rotating frame	R2TR	Application of an RF field allows selective recoupling when the chemical shift difference is small (Takegoshi K, Nomura K and Terao T (1995) Rotational resonance in the tilted rotating frame. *Chemical Physics Letters* **232**: 424–428).
Sevenfold symmetric radio-frequency pulse sequence	C7	Seven phase-shifted RF pulse cycles lead to dipolar recoupling (Lee YK, Kurur ND, Helmle M, Johannessen OG, Nielsen NC and Levitt MH (1995) Efficient dipolar recoupling in the NMR of rotating solids. A sevenfold symmetric radiofrequency pulse sequence. *Chemical Physics Letters* **242**: 304–309).
Dipolar recoupling with a windowless multipulse irradiation	DRAWS	Windowless DRAMA sequence (Gregory DM, Wolfe GM, Jarvie TP, Sheils JC and Drobny GP (1996) Double-quantum filtering in magic-angle-spinning NMR spectroscopy applied to DNA oligomers. *Molecular Physics* **89(6)**: 1835–1850).
Rotational resonance tickling	R^{2}T	Ramped RF field during the variable delay removes the T_2^{ZQ} dependence (Costa PR, Sun B and Griffin RG (1997) Rotational resonance tickling: accurate internuclear distance measurement in solids. *Journal of the American Chemical Society* **119**: 10821–10830).
Adiabatic passage rotational resonance	APRR	MAS speed varied during CP mixing to achieve more complete polarization transfer (Verel R, Baldus M, Nijman M, van Os JWM and Meier BH (1997) Adiabatic homonuclear polarization transfer in magic-angle-spinning solid-state NMR. *Chemical Physics Letters* **280**: 31–39).
Supercycled POST-C5	SPC-5	Fivefold symmetric pulse sequence leads to homonuclear dipolar recoupling (Hohwy M, Rienstra CM, Jaroniec CP, Griffin RG (1999) *Journal of Chemical Physics* **110**: 7983–7992).

better suited to qualitative distance measurements such as in the example given.

It is necessary to make a general comment regarding the influence of molecular motion on the measurement of dipolar coupling constants. In the context of solid-state NMR, it is not r_{12} which is directly measured, but rather the dipolar coupling constant. Molecular librations and vibrations will cause a certain degree of averaging of the dipolar interaction and thus R_{DD}. The net result of the motional averaging of the dipolar coupling is that the calculated distances, r, will be too large.

Finally, it is important to recognize that the dipolar coupling constant measured in any NMR experiment also has, in principle, a contribution from the anisotropy in the indirect spin–spin coupling, ΔJ. That is, only an effective dipolar coupling constant, R_{eff} can be measured where

$$R_{eff} = R_{DD} - \Delta J/3 \qquad [21]$$

The last term in equation [21], $\Delta J/3$, is generally ignored.

Other homonuclear recoupling methods

Restoring the dipolar coupling between both heteronuclear and homonuclear spin pairs is of great interest. Rotational resonance applies strictly to homonuclear spin pairs, and **Table 1** provides a brief overview of some of the other techniques available for recovering the dipolar coupling and extracting R_{eff} for homonuclear spin pair from high-resolution MAS spectra.

Conclusions

At present, RR is best suited for use as a qualitative probe into molecular structure rather than a quantitative one. In most experiments which have been done using the basic RR technique, the internuclear distances were known beforehand as a result of other investigations. Further developments of related RR techniques (such as rotational resonance tickling) may prove to be more useful in obtaining quantitative results. Still, the standard RR experiment is an excellent one for confirming distances between homonuclear spin pairs in a proposed structure.

List of symbols

$A_d(t)$ = spatial dependence of the dipolar Hamiltonian; B_0 = external applied magnetic field; d_m = Fourier components of the dipole–dipole coupling; h = Planck constant; $\hbar$ = Planck constant divided by 2π; $\bar{\mathcal{H}}_{DD}$ = average dipolar Hamiltonian; $\hat{\mathcal{H}}_{DD}$ = direct dipolar Hamiltonian operator; $\hat{\mathcal{H}}_J$ = indirect spin–spin coupling Hamiltonian; $\hat{\mathcal{H}}_{Z,CS}$ = chemical shielding perturbed Zeeman Hamiltonian operator; $\hat{I}_-$ = lowering operator; $\hat{I}_+$ = raising operator; $\hat{I}_i$ = spin angular momentum operator for spin i; $\hat{I}_{zi}$ = z-component of the spin angular momentum operator for spin i; J = indirect spin–spin coupling constant; n = order of the rotational resonance; r_{12} = distance between spins 1 and 2; internuclear vector; R_{DD} = direct dipolar coupling constant (in Hz);

R_{eff} = observed dipolar coupling constant; t = time; T_{20} = spin term in the spherical tensor representation of the dipolar Hamiltonian; T_2^{ZQ} = zero-quantum relaxation time constant; U = propagator; γ_i = magnetogyric ratio of spin i; ΔJ = anisotropy of the indirect spin–spin interaction; θ = angle between the applied field and the internuclear vector; Λ^2 = dephasing parameter; μ_0 = permeability of free space; ν_r = rotor frequency in Hz; ν_i, ν_i^{iso} = isotropic resonant frequency of nucleus i (in Hz); ν_{rot} = rotor frequency (in Hz); $\nu_{1/2}$ = line width at half-height (in Hz); $\xi(t)$ = time-dependence of the NMR interactions as a result of sample rotation; σ_i = chemical shielding tensor of spin i; σ_{ii} = principal component of the chemical shielding tensor (i = 1, 2, 3); σ^{iso} = isotropic chemical shielding constant; τ_m = variable mixing time; $\omega_B^{(n)}$ = resonant Fourier components; ω_i = CS-perturbed Zeeman angular frequency of spin i (in rad s^{-1}); ω_r = rotor frequency (in rad s^{-1}); ω_Δ^{iso} = difference in isotropic angular frequencies of spins 1 and 2.

See also: **Chemical Exchange Effects in NMR; High Resolution Solid State NMR, ¹³C; High Resolution Solid State NMR, ¹H, ¹⁹F; NMR in Anisotropic Systems, Theory; NMR of Solids; NMR Pulse Sequences; NMR Relaxation Rates; Solid State NMR, Methods.**

Further reading

Andrew ER, Bradbury A, Eades RG and Wynn VT (1963) Nuclear cross-relaxation induced by specimen rotation. *Physics Letters* 4: 99–100.

Garbow JR and Gullion T (1995) Measurement of internuclear distances in biological solids by magic-angle-spinning ¹³C NMR. In Beckmann N (ed), *Carbon-13 NMR Spectroscopy of Biological Systems*, pp. 65–115. New York: Academic Press.

Griffiths JM and Griffin RG (1993) Nuclear magnetic resonance methods for measuring dipolar couplings in rotating solids. *Analytica Chimica Acta* 283: 1081–1101.

Peersen OB and Smith SO (1993) Rotational resonance NMR of biological membranes. *Concepts in Magnetic Resonance* 5: 303–317.

Raleigh DP, Levitt MH and Griffin RG (1988) Rotational resonance in solid-state NMR. *Chemical Physics Letters* 146: 71–76.

Smith SO (1993) Magic angle spinning NMR methods for internuclear distance measurements. *Current Opinion in Structural Biology* 3: 755–759.

Smith SO (1996) Magic angle spinning NMR as a tool for structural studies of membrane proteins. *Magnetic Resonance Review* 17: 1–26.

Webb GA, Recent advances in solid-state NMR are reviewed annually in: *Nuclear Magnetic Resonance: Specialist Periodical Reports*. Cambridge: The Royal Society of Chemistry.

Solvent Suppression Methods in NMR Spectroscopy

Maili Liu and **Xi-an Mao**, Wuhan Institute of Physics and Mathematics, Chinese Academy of Sciences, Wuhan, PR China

MAGNETIC RESONANCE
Methods & Instrumentation

Introduction

In order to get useful information from NMR spectroscopy of biofluids or biomolecules (proteins, DNA, RNA, carbohydrates, amino acids), it is often necessary to measure ^1H NMR spectra in aqueous solutions. In these samples, the concentration of solvent water protons is about 110 M and is about 10^5 times higher than that of the molecules of interest which are usually in the mM concentration range or less. Such a huge excess of water spins can cause many problems for NMR measurements. Firstly, the receiver gain of the spectrometer must be set to a low value to avoid the water signal overloading the receiver. In this circumstance, the analogue-to-digital converter (ADC) will be filled by the water resonance and many of the small signals from the molecules of interest will be below 1 bit of the ADC resolution and hence will not be digitized adequately. Secondly, the water resonance may obscure many solute peaks, resulting in the loss of molecular structural information that could make the spectrum useless. Thirdly, the strong water resonance can cause radiation damping, which provides another relaxation mechanism and shortens the relaxation time of water and hence broadens the water peak. Therefore for ^1H NMR spectroscopy to be useful in aqueous solution, it is clearly necessary to attenuate the water signal.

There has been a continued interest in developing new methods for solvent suppression in NMR spectroscopy of biomolecules and biofluids and many methods have been proposed. These fall into five categories (1) presaturation, (2) nonexcitation, (3) pulsed field gradient (PFG)-based methods, (4) filtering methods and (5) post-acquisition data processing. Among these methods, soft pulse presaturation is the most frequently used and it is also easy to be incorporated into one- (1D) and multidimensional (nD) pulse sequences. The other methods can also be found in some applications, such as the study of exchangeable protons. In the past decade, the use of PFG to enhance the suppression efficiency and to develop new methods has been a popular area of

study in NMR spectroscopy. Another advantage of using PFGs is the reduction of radiation damping. The important criteria for a good suppression method are the efficiency, selectivity and phase and baseline properties of the resulting spectrum.

Owing to the limitation of space, this article focuses on the introduction of general principles of the solvent suppression methods, and emphasizes some of the important developments. The reader is encouraged to refer to the original literature and recent reviews for the fundamentals and details of the methods.

Solvent presaturation

The presaturation method (PR) normally consists of a low-power, soft (long or continuous-wave) pulse at the solvent resonance. It is the simplest and the most widely used method for solvent suppression. The method can be found in a large number of 1D- and nD-NMR pulse sequences for measuring the ^1H NMR spectra of biofluids or biomolecules in aqueous solutions. The general 1D presaturation pulse sequence is shown in **Figure 1A**, where PR_x is the saturation pulse applied along the x-axis of the rotating frame. The duration of the PR pulse is normally equal to the preacquisition or relaxation delay. It has been found that a long PR pulse with a fixed phase can lock part of the water magnetization along the direction of the RF field. The locked resonance can be reduced by a so-called phase-shifted PR method as shown in **Figure 1B**. The pulse sequence of the phase-shifted presaturation method has two PR pulses with a duration ratio of 9:1 (PR_x:PR_y) and a $\pi/2$ phase shift. It had been reported that in conventional 2D COSY [homonuclear chemical shift correlation spectroscopy] and NOESY [2D NOE spectroscopy] experiments on 2 mM lysozyme in a 90% H_2O–10% D_2O solution, a water signal suppression of a factor of 10^6 was achieved using the phase-shift PR method. **Figure 1C** shows a pulse sequence known as NOESYPRESAT, which can be considered as a combination of the phase-shifted PR

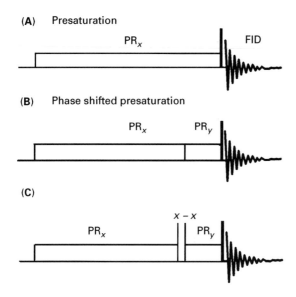

Figure 1 Water suppression pulse sequences, where the bar symbol represents a 90° pulse: (A) presaturation, (B) phase-shifted presaturation (the ratio of the duration of PR_x and PR_y pulses is 9:1), (C) NOESYPRESAT.

and conventional NOESY approaches and which can further improve the spectral phasing and baseline. It also provides a more effective suppression in the wide base of the water resonance.

In HPLC–NMR or other applications with solvent mixtures, it is often necessary to suppress two solvent peaks at the same time. This is normally achieved by applying soft pulses or continuous irradiation at both solvent resonances or by fast switching the frequency offset between the two signals. With advances in NMR software, it is now possible to automatically search the solvent frequencies and to define suppression pulse offsets accordingly without any reduction in the suppression efficiency.

The presaturation method can cause a loss of signal intensity of exchangeable protons. This is the result of saturation transfer from the solvent resonance that is caused by the chemical exchange during the period of the saturation pulse. Generally, care must be taken when using the method to study systems containing labile protons. On the other hand, since the amount of the saturation transfer can be controlled by the (saturation) pulse length, the approach provides a facile method for the assignment of labile protons and for the study of their exchange rate with water.

Solvent resonance nonexcitation

A well-established nonexcitation method is the 'Jump-Return (JR)' sequence. It is also known as the

[1, –1] method. The sequence consists of a pair of 90° hard pulses that are separated by a delay τ and have a π phase shift,

$$90°_x - \tau - 90°_{-x} - \text{FID} \qquad [1]$$

where FID = free-induction decay. The principle of the sequence can be described using the product operator approach. The first $90°_x$ RF pulse generates transverse magnetization of $-I_y$. During the delay period τ, the transverse magnetization evolves at the relative frequency (ω) with respect to the transmitter

$$-I_y \xrightarrow{\omega\tau} -I_y \cos(\omega\tau) + I_x \sin(\omega\tau) \qquad [2]$$

The last $90°_{-x}$ puts the $-I_y$ term on the right side of Equation [2] back along the z-axis of the rotating frame and the I_x term remains unaffected. The sequence thus gives rise approximately to a sine-shaped excitation profile of $\sin(\omega\tau)$. The excitation bandwidth (SW_b) depends on the delay τ and is $SW_b = 1/\tau$. The maximum magnitude of the transverse magnetization can be obtained at the relative frequencies of $1/4\tau$ and $-1/4\tau$, respectively, and the two excitation bands have a 180° phase shift between them. The on-resonance solvent magnetization will be put along the z-axis and remains unchanged but the signals at the relative frequencies of $\pm 1/2\tau$ will be inverted. By increasing the number of the RF pulses from two, it is possible to make the excitation bands flatter and the suppression region narrower. One of the improved versions of this pulse sequence is the [1–3–3–1] approach. A disadvantage of the general method is that it is complicated by a phase roll over the spectrum and by the effects of radiation damping. However, the saturation transfer effect remains at a minimum in these methods because of the very short overall duration, typically a few milliseconds, of the pulse sequences.

One other variation of the nonexcitation approaches is the combination of selective and nonselective subsequences. It consists of a soft pulse of angle β and a hard pulse of angle $-\beta$. The response to the soft pulse is limited to a narrow range around the solvent resonance, which is cancelled subsequently by the hard pulse. The sequence thus nearly provides a flat and phased excitation profile with a gap at the solvent frequency. On the other hand, the inherent asymmetry of the method makes it sensitive to hardware imperfections.

Solvent suppression using pulsed field gradients

Because of the advances in probe technology, PFGS have been widely used in high-resolution NMR spectroscopy for coherence selection, magnetization destruction and molecular diffusion coefficient measurement. Many of the water suppression techniques and conventional pulse sequences can be, and have been, modified using PFGs to improve suppression and to increase the spectral quality. It is rare to find a newly proposed pulse sequence not including PFGs. Reviews on the usage of PFGs in water suppression and the other specific topics are available in the recent literature.

A PFG pulse applied along the magnetic field (z-axis) direction causes the transverse magnetization (coherence) to rotate with an additional phase of

$$\phi(r) = n\gamma\delta G(r) \qquad [3]$$

where n is the coherence quantum order, γ is the gyromagnetic ratio of the spin, r is the position of the spin in the gradient direction (the z-axis in this case), and $G(r)$ and δ are the gradient strength and duration, respectively. This position-dependent dephasing can be reversed by applying a PFG pulse of the same strength $[\delta G(r)]$ but in the opposite direction. The dephasing and rephasing properties of the PFG pulses can then be used for solvent resonance suppression.

A simple and efficient pulse sequence using PFGs to dephase the water resonance is the RAW (randomization approach to water suppression, **Figure 2A**) method, in which the transverse magnetization of solvent generated by a selective 90° pulse is destroyed immediately by the following strong PFG pulse. The selective pulse can be a Gaussian-shaped soft pulse to provide a narrowband excitation and to minimize the off-resonance excitation. Other selective pulses could be used as well. The scheme of 'sel. 90°-G_z' can be used preceding a preparation pulse in most 2D experiments for solvent suppression. To prevent the longitudinal recovery of the water resonance during the PFG pulse, the G_z pulse can be replaced by a scheme of a composite 180° pulse sandwiched by a pair of bipolar PFG pulses, -G_z-90°$_x$-180°$_y$-90°$_x$-G_{-z}-.

The WATERGATE (water suppression by gradient-tailored excitation) method has proved popular recently because of its high efficiency and short duration compared with the methods using PR or selective nonexcitation. The method resembles a spin-

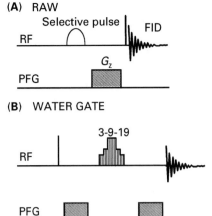

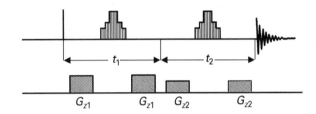

Figure 2 Water suppression pulse sequences using PFG to selectively dephase the solvent resonance, where the bar and open symbols represent 90° pulses. (A) Randomization approach to water suppression (RAW) sequence. (B) WATERGATE (the composition of the 3-9-19 pulse train and its variations are listed in **Table 1**). (C) Double WATERGATE echo method. It is recommended that different echo times ($t_1 \neq t_2$) and different gradient strengths ($G_{z1} \neq G_{z2}$) are used.

echo sequence with a selective refocusing pulse flanked by two symmetrical gradient pulses as shown in **Figure 2B**. Transverse coherences are dephased by the first gradient and can be rephased by the second gradient, provided they experience a 180° rotation by the selective pulses or the pulse train, denoted by W. This can be a 180° hard pulse sandwiched by a pair of 90° selective pulses. More commonly, it uses a frequency-selective pulse train of 3α-τ-9α-τ-19α-τ-19α-τ-9α-τ-3α (denoted by 3-9-19 or W3 for short), where $62\alpha = 180°$ and τ is a short delay that is used to control the null-inversion points ($\pm 1/\tau$ Hz). When either kind of selective refocusing pulse is used, the spectral resonances experience a 180° rotation and will be rephased by the second gradient pulse, whilst the net flip-angle at the water resonance frequency approaches zero and thus the water signal will be dephased by both gradient pulses. When selective pulses are used, the duration of the pulse is about 10 ms if a narrower suppression bandwidth is desired. WATERGATE with a W3 pulse train normally takes less than 5 ms, and the saturation of exchangeable proton resonances is not too serious. Another advantage of using the W3 pulse train is that the null-points can be easily modified for off-resonance water suppression. The disadvantage of using the hard pulse train is that the peak elimination region is wider than when using presaturation and thus any resonances close to the solvent peak will be suppressed. New pulse trains with a narrower noninversion region have been introduced recently. The improvement is achieved by using four (W4) or five (W5) pairs of hard pulses in the pulse train instead of three pairs as in the original W3 sequence. The experimental results (**Figure 3**) indicate that when more element pulses are used in the pulse train, the non inversion region becomes narrower, and the spectral profile becomes wider and flatter, but this is balanced by some sacrifice of suppression efficiency. The composition of different hard pulse trains and parameters are listed in **Table 1**.

The suppression efficiency may be improved by using the double gradient-echo method **Figure 2C**. Although the method provides much better suppression and phase properties, the intensities of any peaks near the solvent will be reduced because the profile of suppression has a squared form.

For comparison of PFG and PR based suppression methods, **Figures 4A** and **4B** show 500 MHz ^{1}H NMR spectra of human blood plasma measured using the pulse sequences of WATERGATE (**Figure 2C**) and of NOESYPRESAT (**Figure 1C**) at 30°C, respectively. The low-field region was enlarged and plotted as an inset for both spectra. The experiments were carried out under identical conditions with the exception of the different pulse sequences. The pulse train of W5 was used as the refocusing pulse for the WATERGATE and the inversion bandwidth was set to 3000 Hz (τ = 1/3000 s). A 2 s PR$_x$ and a 100 ms PR$_y$ low-power (γB_1 = 60 Hz; where B_1 is the RF magnetic field) pulse were used for solvent saturation in NOESYPRESAT. Both methods provide a high efficiency of solvent suppression; however, the resonances of labile protons (marked by arrows) were observable in **Figure 4A**

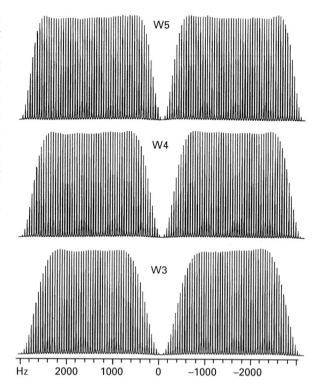

Figure 3 Experimental excitation profiles of the NMR pulse sequences using W3, W4 and W5 methods. The experiments were carried out at 500 MHz. The bandwidth was set nominally to 3000 Hz (τ = 1/3000 s).

but suppressed in **Figure 4B** by the saturation transfer effects.

Solvent suppression using relaxation, diffusion or multiple quantum filters

Spin and molecular properties that distinguish water from solute signals could be used for the water resonance suppression. The most frequently used properties are the longitudinal relaxation time (T_1, also known as the spin–lattice relaxation time), the molecular diffusion coefficient (D) and double quantum or multiple quantum coherence filters.

Table 1 WATERGATE pulse sequence parameters

Pulse type	Pulse train composition (deg)[b]	Phasing for null-points[c] at k/τ	Phasing for nullpoints[c] at $(2k+1)/\tau$
W1	90, 90	ϕ-θ [d]	ϕ-ϕ
W2	45, 135, 135, 45	$(\phi)_2$-$(\theta)_2$	ϕ-$(\theta)_2$-ϕ
W3[a]	20.8, 62.2, 131.6, 131.6, 62.2, 20.8	$(\phi)_3$-$(\theta)_3$	ϕ-θ-$(\phi)_2$-θ-ϕ
W4	10.4, 29.4, 60.5, 132.8, 60.5, 29.4, 10.4	$(\phi)_4$-$(\theta)_4$	ϕ-θ-ϕ-$(\theta)_2$-ϕ-θ-ϕ
W5	7.8, 18.5, 37.2, 70, 134.2, 134.2, 70, 37.2, 18.5, 7.8	$(\phi)_5$-$(\theta)_5$	ϕ-θ-ϕ-θ-$(\phi)_2$-θ-ϕ-θ-ϕ

[a] Original pulse train for WATERGATE sequence.
[b] Each pulse element is separated by a period τ.
[c] k = 0, 1, 2, 3,....
[d] $\theta = \phi + \pi$.

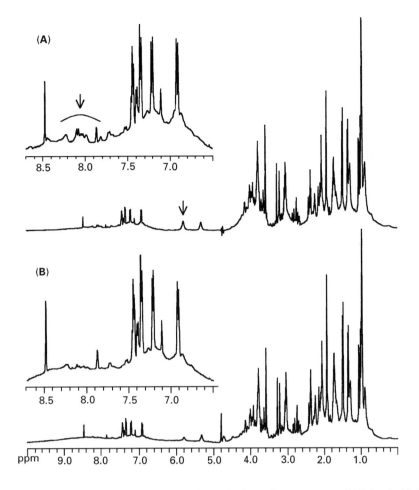

Figure 4 500 MHz ^{1}H NMR spectra of human blood plasma obtained using pulse sequences of (A) the double WATERGATE echo method and of (B) NOESYPRESAT under identical conditions. The low-field region is expanded and plotted as the inset. The labile proton resonances (marked by arrows) are observable in (A) but suppressed in (B) by saturation transfer effects.

In biological samples, the longitudinal relaxation times of spins in a protein or other large molecule are often in the region of tens or hundreds of milliseconds, while those of solvent water are 2 to 3 s. Such a large difference in relaxation times makes it possible to suppress the water signal using an inversion recovery scheme (**Figure 5A**). A π pulse inverses the magnetization of both solvent and solutes. The remaining transverse magnetization is dephased by the PFG pulse, which also blocks the relaxation pathway associated with the radiation damping effect. The observation pulse is applied when the longitudinal magnetization of water becomes null, after a time $T_1^w \ln 2$, where T_1^w is the longitudinal relaxation time of water. The spins in larger molecules with smaller T_1s relax much faster and get closer to their equilibrium magnitude at the time of $T_1^w \ln 2$. For small solute molecules with similar longitudinal relaxation times, the π pulse can be replaced by a selective pulse applied at the solvent resonance. Another variation is the use of a series of selective pulses with small flip

angles, each of the selective pulses being followed by a PFG pulse. The major advantage of this T_1 filter method is its higher selectivity since it provides less attenuation of the resonances close to that of the solvent.

It is also possible to attenuate the water resonance in biofluids and other aqueous samples by the addition of a reagent that causes a significant reduction of the water T_2. The broadened water peak can then be attenuated by measuring the spectrum using a spin-echo pulse sequence. This can be achieved for biofluid samples by adding substances containing exchangeable protons that cause water to exchange at an intermediate rate and induce a line broadening. Commonly, guanidinium chloride is used. Alternatively, addition of a paramagnetic ion would also be effective, although care must be taken not to broaden the solute resonances in this case.

Figure 5B shows the diffusion coefficient (D) filtering method. The method utilizes the difference in molecular mobility between water and solutes such

(A) T_1 Filter

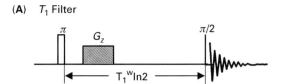

(B) Diffusion filter

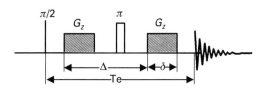

(C) DFQ-COSY

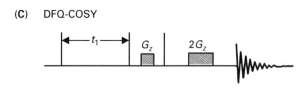

Figure 5 Solvent suppression pulse sequences based on filtering methods. Method (A) uses a T_1 filter to discriminate resonances of solvent and solutes. The difference in molecular diffusion coefficients is used in method (B). T_e is the spin-echo time. (C) Double-quantum filtering COSY, which uses the fact that there is no J-coupling between the two equivalent protons in water molecules and thus it cannot be excited to higher quantum coherence. The PFG pulses in (A) and (B) are used to attenuate radiation damping effects and dephase any transverse magnetization. They are used for the desired coherence selection in (C).

as macromolecules. In this experiment, the signal magnitude (M) is attenuated according to

$$M = M_0 \exp\left[-D(\gamma\delta\Delta G_z)^2(\Delta - \delta/3)\right] \qquad [4]$$

where Δ is the time interval between the leading edges of the two PFG pulses, and G_z and δ are the PFG pulse strength and duration, respectively. The diffusion coefficient of water is more than 10 times that of most larger biological molecules, and thus the water signal will be attenuated, resulting in a spectrum of larger or less mobile molecules. Since D is a molecular property, the attenuation to different spins in a molecule will be the same. This should be useful for quantitative measurement. The diffusion filtering method has now been implemented into most conventional 2D pulse sequences used for diffusion coefficient measurement and for the editing of complex NMR spectra.

The multiple-quantum filtering (MQF) method is an efficient approach to suppression of any resonance that does not experience a specific defined

quantum order during the pulse sequence. Since water protons give rise to a single peak in an NMR spectrum and cannot be excited to a quantum order higher than unity, the water resonance cannot pass a MQF and will be eliminated from the spectrum. However, when the MQF is built up based on ^1H–^1H spin coupling, the resulting spectrum is expected to have a dispersive line shape on the coupling splittings, and thus the homonuclear MQF is commonly used for 2D NMR experiments. For example, the pulse sequence of double quantum filtered (DQF) COSY is shown in **Figure 5C**, in which PFG pulses are used to enhance the selectivity. A heteronuclear MQF NMR spectrum often has in-phase line shape, but generally the lower natural abundance of heteronuclei will reduce the sensitivity.

Postacquisition data processing

As discussed in previous sections, a variety of solvent suppression methods are available. However, even when a high-quality method is used, it is quite common for there to be phase or baseline distortion (or both) in the resulting spectrum, especially in the region of the solvent resonance. In other cases, although the water peak is suppressed, it still remains as the largest peak in the spectrum. This will cause serious ridges (known as 't_1 noise') in the F_1 dimension of multidimensional NMR experiments and will strongly affect the cross peaks close to it. These disadvantages can be overcome by postacquisition data processing. It has been a standard approach to eliminate a (solvent) signal using a frequency filtering method on most modern NMR spectrometers. Some NMR machine manufacturers also provide software packages of linear prediction and maximum entropy to enhance the quality of NMR spectra and to extract more information from the spectra. However, most off-line data processing methods are focused on automatic baseline correction and convolution, and filtration or subtraction of solvent signal from time- or frequency-domain data sets. The water signal is commonly considered as being a Lorentzian line shape. In some cases it can be treated as a Gaussian function or a weighted Lorentzian–Gaussian function, and can be subtracted from the spectrum or, more commonly, from the time-domain data.

Radiation damping effects on water suppression

When water is used as solvent in ^1H NMR spectroscopy, radiation damping is not avoidable, but its effects can be minimized if a proper solvent suppression method is applied. So far radiation damping has been regarded as a negative effect on ^{1}H NMR experiments when water is used as the solvent, although some positive use of radiation damping has been proposed. Radiation damping is a dynamic process similar to the longitudinal relaxation, both leading the magnetization toward the equilibrium state. Physically, radiation damping is caused by the interaction of the FID current with the magnetization itself, and is characterized by a time constant T_{rd}, defined by $2(\mu_0\gamma\eta QM_0)^{-1}$, where μ_0 is the vacuum permeability, η is the filling factor of the sample in the probe, Q is the quality factor of the detection coil. The large values of η and Q of the probe and huge magnetization of water make the radiation damping time very short (in the region of milliseconds), much shorter than its true relaxation time T_1 at low concentration (in the region of seconds). As a result, radiation damping interferes with water suppression in many ways.

Among the four experimental strategies discussed above, some could be seriously affected by radiation damping, while others may not be. Because the RF irradiation tends to null the total magnetization, radiation-damping effects will be correspondingly removed. It has been shown that during continuous-wave irradiation, the decay of the magnetization is dependent on T_1, T_2 and the inhomogeneous contributions from both the static field B_0 and the RF field B_1, but is independent of radiation damping. The PR method has proven to be a reliable method for water suppression, but with the disadvantage of saturation transfer for exchanging systems. As for the PFG method, if only z-gradients are used, radiation damping could occur. Since z-gradients can destroy the transverse magnetization only, the remaining longitudinal magnetization, if it is in the $-z$ direction and is still very strong, can easily evolve into a transverse magnetization under the influence of a radiation damping field. As a result, there will be a strong water signal in the spectrum. Care should also be taken when the refocusing PFG is used, since after the refocusing, the transverse magnetization will behave as if PFG were not applied. It should be pointed out that only the positive longitudinal magnetization does not lead to radiation damping. Thus, in order to prevent radiation damping from occurring, PFG experiments should be carefully designed.

If neither PR nor PFG is used, radiation damping may bring about serious problems in water suppression. Because the water magnetization is not attenuated by hard pulses, radiation damping would make the [1, −1] sequence, the inversion-recovery sequence and the DQF-COSY sequence useless, as far as water suppression is concerned. For the simple inversion-recovery method for example, after the spin inversion, the water magnetization returns to the z-direction more quickly than the spins in large molecules because of the short radiation damping time. Therefore, it is not possible to use the simple inversion-recovery sequence to measure the relaxation times, nor as a relaxation filter, unless PFG is utilized. In fact, the [1, −1] method, the inversion-recovery method and MQF method have been improved significantly by PFG modifications.

Summary

Solvent suppression has been one of the rich areas in biological NMR research. The driving force comes from *in vivo* and *in vitro* proton NMR spectroscopy, protein structure determination, metabolic and toxicological studies of biofluids, and medical and functional magnetic resonance imaging. New methods could emerge from the following two fundamental techniques:

1. development of fast switching B_0 and B_1 gradient coils. A high-quality gradient facility is essential for all types of NMR experiment, including solvent suppression. The use of switching gradients means that transverse magnetization is rapidly dephased, and less signal attenuation caused by diffusion and magnetization transfer results. A properly self-shielded gradient coil can minimize the eddy-current effect which can cause phase distortion throughout the spectrum. The use of magic angle PFGs has proved to be very efficient in the suppression of the solvent resonance and for reducing artefacts in 2D experiments.
2. development of precise and flexible phase- and amplitude-controlled excitation pulses. This facilitates the design of pulse sequences that could produce an exact excitation profile for the solvent peak. Since HPLC–NMR has become more and more routine, the development of double and multiple solvent-peak suppression methods will become important.

Whatever the fundamental technique a new method is based on, it should provide excellent baseline and phase properties and high suppression efficiency, have a short duration to avoid suppression transfer and be easy to be implemented into 1D and nD pulse sequences.

List of symbols

B_1 = RF magnetic field; B_0 = static magnetic field; D = diffusion coefficient; $G(r)$ = gradient strength;

I_x and I_y = transverse magnetizations; M = signal magnitude; n = coherence quantum order; Q = quality factor; r = spin position; T_1 = longitudinal relaxation time; T_1^w = water longitudinal relaxation time; T_{rd} = radiation damping time constant; β = pulse angle; γ = gyromagnetic ratio; δ = gradient pulse duration; Δ = time interval; η = sample filling factor; μ_0 = vacuum permeability; τ = pulse delay period; ω = relative frequency.

See also: **Biofluids Studied By NMR; Chromatography-NMR, Applications; Diffusion Studied Using NMR Spectroscopy; Magnetic Field Gradients in High Resolution NMR; NMR Data Processing; NMR Principles; NMR Pulse Sequences; NMR Relaxation Rates; NMR Spectrometers; Product Operator Formalism in NMR; Proteins Studied Using NMR Spectroscopy; Two-Dimensional NMR, Methods.**

Further reading

Altieri AS, Miller KE and Byrd RA (1996) A comparison of water suppression techniques using pulsed field gradients for high-resolution NMR of biomolecules. *Magnetic Resonance Review* 17: 27–82.

Gueron M, Plateau P and Decorps D (1991) Solvent signal suppression in NMR. *Progress in NMR Spectroscopy* 23: 135– 209.

Gueron M and Plateau P (1996) Water signal suppression in NMR of biomolecules. In: Grant DM and Harris RK (eds), *Encyclopedia of Nuclear Magnetic Resonance*, pp 4931–4942. Chichester: Wiley.

Mao X-A and Ye C-H (1997) Understanding radiation damping in a simple way. *Concepts in Magnetic Resonance* 9: 173–187.

Moonen CTW and Van Zijl PC (1996) Water suppression in proton MRS of humans and animals. In: Grant DM and Harris RK (eds), *Encyclopedia of Nuclear Magnetic Resonance*, pp 4943–4954. Chichester: Wiley.

Sonically Induced NMR Methods

John Homer, Aston University, Birmingham, UK

MAGNETIC RESONANCE
Methods & Instrumentation

Conventionally, nuclear magnetic resonance (NMR) spectroscopy may be viewed as proceeding by photon-stimulated transitions between energy levels of certain nuclei that are quantized due to the influence of a strong homogeneous polarizing magnetic field. However, the required quanta ($h\nu$) of energy necessary to induce transitions need not be restricted in origin to electromagnetic radiation, but may be derived from the phonon, which is the acoustic analogue of the photon. An acoustic wave, which is manifest as sinusoidal pressure variations with a characteristic frequency $\nu(\omega/2\pi)$, can be considered to be composed of a beam of phonons each carrying energy quanta of $h\nu$. The pressure wave and phonon approaches may be invoked to explain various acoustic phenomena. For example, phonons can be considered to be capable of facilitating both experimentally stimulated and naturally occurring nuclear relaxation transitions in solids. On the other hand, the wave approach proves convenient for describing nuclear relaxation processes, and also cavitation, in liquids. Ultrasound can be used experimentally to stimulate NMR transitions, modify nuclear relaxation processes, narrow the resonance bands derived from solids and modify the spectra of solutes dissolved in liquid crystals.

Principles

Phonons and nuclear spin-lattice relaxation

Before considering the various possibilities of experimentally manipulating nuclear spin systems with sound it is beneficial to summarize the role of phonons in naturally occurring nuclear relaxation processes.

Debye's theory of the specific heats of solids depends on the existence of a high number of standing, high frequency, elastic waves that are associated with thermal lattice vibrations. Central to his approach is the proposal that in a solid the phonon spectral density (ρ) increases continuously, and with a direct dependence on the square of the frequency (ω) to a cutoff frequency (Ω, at about 10^{13} Hz) above which the phonon density vanishes: for a solid continuum containing N atoms in a sample of volume V, the proportionality constant is $6V/v^3$, where v is the velocity of propagation. At a typical nuclear

Larmor frequency (ν_0) of around 10^9 Hz there will be a significant phonon spectral density that could facilitate nuclear spin-lattice relaxation. It emerges that the transition probability for such a direct process is very low. However, phonons at frequencies other than ν_0 can contribute to relaxation through indirect processes. Of those possible, one can involve two phonons that could be involved sequentially in absorption or emission, but the probability of this happening is extremely low simply because the numbers of pairs of phonons having energies that combine to match the required transition energy is very low. An alternative indirect process is the so-called Raman process in which a phonon of frequency ν interacts with the nuclear spins to cause either absorption or emission with the accompanying emergent phonons having frequencies of $\nu-\nu_0$ and $\nu+\nu_0$, respectively. As absorption can only occur when ν lies between ν_0 and Ω this process is very much less efficient than emission for which ν can have any value up to Ω: despite other processes being capable, in principle, of affecting relaxation, the indirect Raman emission process can, therefore, be viewed as being responsible for spin-lattice relaxation in solid samples.

In the case of liquid lattices, the difficulty in adequately characterizing their structures renders them unsuitable for treatment by quantum mechanics. Accordingly, liquid lattices are often treated classically by considering the effects of molecular rotations and translations, with characteristic correlation times, τ, on time-dependent magnetic and electric fields that may influence relaxation processes. Accordingly, it becomes convenient to depart from the phonon description of sound and adopt a classical view of this as the sinusoidal propagation of a pressure wave through a medium.

Ultrasound

Sound having frequencies above about 18 kHz is traditionally called ultrasound, and in addition to its frequency is characterized by its intensity and velocity of propagation through a medium. As a matter of convenience ultrasound is usually categorized as diagnostic ultrasound (as used for imaging, with frequencies in the megahertz region) or power ultrasound (as used for cleaning, welding etc., with frequencies in the kilohertz range).

The passage of ultrasound through liquids can result in some spectacular phenomena, particularly those resulting from acoustic cavitation. Simplistically, when an acoustic wave passes through a liquid it produces compression and rarefaction of the liquid on successive cycles. On the rarefaction cycle the liquid experiences reduced pressure and, provided a suitable nucleation centre is available, a small cavity will form in the liquid. As migration of entrapped vapour into the cavity depends on the surface area of the interface of the cavity with the liquid this is greater on the rarefaction (expansion) cycle and so the cavity will grow on successive rarefaction cycles. The cavity will either become stable or, over a few acoustic cycles, unstable. In the latter case, the cavity collapses catastrophically with the generation of extreme local temperatures and pressures and the emission of shock waves. If this process occurs near a solid surface, so that there is an imbalance in the force field, a microjet of liquid, starting from the region of the cavity that is most remote from the solid surface, is ejected along the internal symmetry axis of the cavity and strikes the solid surface at velocities in the order of hundreds of m s^{-1}. Evidently, ultrasound can be used below the cavitational threshold to pressure modulate the molecular motion in liquids or above the cavitational threshold to subject both liquids and solids to shock waves, liquids to extreme local temperatures and pressures, and solids to the violent impact of microjets.

Experimentation

Although ultrasound can be generated in a variety of ways, it often proves most convenient to derive it from piezoelectric transducers by applying alternating voltages across opposite silvered faces of discs of the piezoelectric materials (for example, lead zirconium titanate) that have been suitably cut to generate either longitudinal or transverse waves from their surfaces. An alternating signal, applied at the natural resonance frequency of the piezoelectric crystal, causes the surfaces of the crystal to expand and contract at the resonance frequency so that corresponding sinusoidally varying pressure waves are generated and propagated in the desired direction through chosen media.

For many NMR purposes, acoustic waves are often transmitted to liquids using metal (such as titanium) horns that are coupled to one of the piezoelectric crystal faces, or more simply by enabling intimate contact between the crystal and the liquid: to provide acoustic transmission through solids the transducer can be bonded directly to an optically flat surface of the solid.

In the first low frequency (20 kHz) ultrasound/NMR experiments, the ultrasound was delivered to samples in a conventional iron magnet NMR spectrometer using apparatus similar to that shown in **Figure 1**. The titanium alloy horn used was sufficiently long (77 cm) to enable the piezoelectric device

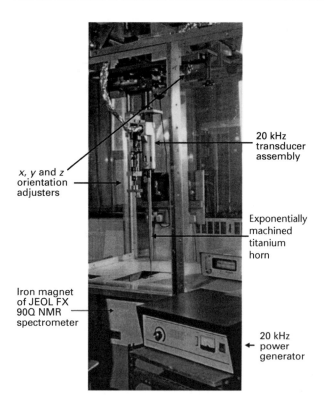

Figure 1 Early 20 kHz SINNMR apparatus.

Labels on Figure 1:
- 20 kHz transducer assembly
- x, y and z orientation adjusters
- Exponentially machined titanium horn
- Iron magnet of JEOL FX 90Q NMR spectrometer
- 20 kHz power generator

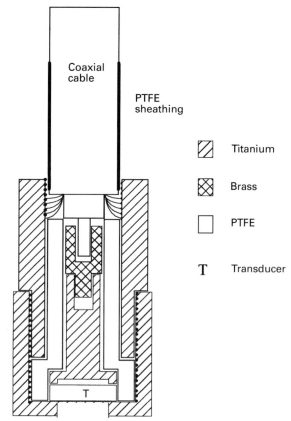

Labels on Figure 2:
- Coaxial cable
- PTFE sheathing
- Titanium
- Brass
- PTFE
- T Transducer

Figure 2 Schematic diagram of a transducer assembly capable of generating high intensity ultrasound in the megahertz region. Reproduced with permission from RL Weekes, Ph.D. thesis, Aston University, 1998.

to be remote from the NMR detector region as the latter is sensitive to pick-up of extraneous a.c. signals. The horn was machined to provide exponential reduction in its diameter both to provide a coupling tip capable of fitting inside a conventional NMR tube and also to provide mechanical amplification of the ultrasound. Evidently, this equipment with its physically large horn is not suited to insertion down the bore of cryomagnets. Consequently, devices have been produced that are particularly suited to operation with MHz ultrasound and which can be easily inserted into the top of NMR sample tubes in cryomagnets: such a device is illustrated schematically in **Figure 2**. Essentially, this facilitates electrical contact with the piezoelectric transducer by way of compressional contacts and enables the latter to be brought into intimate contact with liquid samples about 1 cm above the active NMR coil region where little pick-up by the latter from the former is experienced: very recently, similar devices to that shown in **Figure 2** have been constructed that allow acoustic irradiation of NMR samples from underneath. By using this general approach NMR/ultrasound experiments have become possible at acoustic frequencies up to 10 MHz. It is worth noting that many high frequency piezoelectric transducer discs have an impedance that can be matched to the output of readily available, and relatively inexpensive, transceiver devices that can, therefore, be used to drive the transducers without extensive modification.

Naturally, when contemplating high acoustic frequency/NMR experiments in cryomagnets it was a matter of priority to resolve two questions. First, does the introduction of ultrasound into the bore of a cryomagnet cause the latter to quench? Extensive experiments have shown that even with acoustic intensities approaching 500 W cm^{-2} the magnets do not quench, although naturally caution is recommended when undertaking such experiments. Second, what is the limit of acoustic frequency at which cavitation can be induced? Using the self-indicating colour-sensitive Weissler reaction as dosimeter it has been shown that cavitation can readily be achieved at frequencies up to 10 MHz, without the involvement of undertones of the transducer natural frequency.

Applications

By extrapolation of the role of lattice phonons in nuclear relaxation processes in solids it is not a great step to appreciate that the application of ultrasound to both solids and liquids may be used to manipulate phenomena of interest to NMR spectroscopists. Although not yet an area of considerable activity the relatively few examples of the combined use of NMR and ultrasound that will now be described indicate that further such studies will prove profitable.

Acoustic nuclear magnetic resonance (ANMR)

If a solid is irradiated with ultrasound it produces a phonon spectral density that is much larger than the spectral density arising from natural lattice vibrations at the irradiation frequency. If acoustic irradiation is at the Larmor frequency of dipolar nuclei, the rate of stimulated transitions is increased by a factor of about 10^{11}, but this is probably insufficient to make the detection of acoustically stimulated transitions detectable. However, for quadrupolar nuclei the ultrasonic transition rates can be increased by a further factor of about 10^4 and enable the observation of net acoustic energy by ANMR despite the competition from relaxation transitions. The selection rules for allowed transitions due to quadrupolar coupling are $\Delta m = \pm 1$ (but not $m = -\frac{1}{2} \leftrightarrow +\frac{1}{2}$) and ± 2. Accordingly, ANMR (which is not susceptible to skin depth problems) can be detected at both ν_0 and $2\nu_0$. Although ANMR can, with difficulty, be detected directly it is usual to detect its effect through the additional saturation of NMR signals that are detected normally through stimulation by RF irradiation.

Although there are many examples of ANMR experiments on solid samples there has been considerable debate as to whether similar experiments are possible using liquid samples. This debate appears to have been resolved by relatively recent work on ^{14}N ($\nu_0 = 6.42$ MHz at a magnetic field of 2.1 T) ANMR saturation experiments on acetonitrile and N,N-dimethylformamide using acoustic frequencies ranging from about 1 MHz to 10 MHz. Only at an acoustic frequency corresponding to the nuclear Larmor frequency was saturation of the ^{14}N signal observed, using an acoustic intensity of about 2.5 W cm^{-2}.

Acoustically induced nuclear relaxation

If solids are irradiated with ultrasound having a frequency below the nuclear Larmor frequency, reference to the discussion of the Raman phonon relaxation process indicates that relaxation emission transitions should be favoured and that spin-lattice relaxation times might be reduced. Correspondingly, the acoustic modulation of normal molecular motion in liquids might result in the reduction of T_1. The earliest indication that T_1 can be reduced by the application of ultrasound derived from work on an aqueous colloidal sol of As_2S_3 when, in the 1960s, a reduction in T_1 was noted. Since that observation, detailed investigations have been undertaken on the effects of ultrasound, at various frequencies and intensities, on the values of T_1 for ^{1}H, ^{13}C and ^{14}N in several liquids and liquid mixtures. Importantly, it has been established that the normal values of T_1 in liquids can be reduced by irradiation with ultrasound. Although the acoustic frequency used (1–6 MHz) appeared to have little effect on the observations, it was found that as the acoustic intensity was increased the value of T_1 decreased by up to 60% of its natural value, and that the extent of the decrease appeared to correlate roughly with the molecular environment of the nucleus studied. As, for the small molecules studied, the nuclei were all in their extreme narrowing limit (short correlation times) a possible explanation for the reduction in the values of T_1 is that their correlation times were increased by the application of ultrasound. This is not inconsistent with the rarefaction–compression effects of the acoustic pressure wave, which may be considered to impose on the molecules a motion that corresponds to a dominant translational correlation time of the order of the inverse of the acoustic frequency ($\sim 10^{-6}$ s). It was also observed that as the intensity of the ultrasound was increased further the values of T_1 increased from the minimum value achieved. Although several explanations of this increase in T_1 are possible, it is most likely that it arose as a result of the rather crude apparatus used, causing heating of the samples and a normal increase in T_1. Recently, further investigations of the acoustic reduction in the values of T_1 have been conducted using improved apparatus. The now reproducible results show that T_1 for liquid samples can indeed be progressively reduced, to a limiting value, by the systematic increase in intensity of the acoustic field. **Figure 3** shows typical plots of the signal-to-noise ratio of the ^{13}C quaternary carbon of 1,3,5-trimethylbenzene in cyclohexane with increasing acoustic intensity at 2 MHz and reflects the progressive reduction of T_1. If the explanation for the reduction in the values of T_1 for nuclei in the liquid phase is in fact that the ultrasound imposes a dominant translational correlation time on small molecules, an exciting possibility, currently being investigated, is that the choice of a suitable acoustic frequency could be used to modify the correlation times of large biomolecules and hence

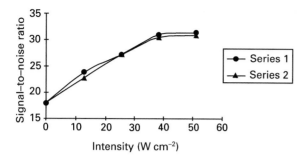

Figure 3 Dependence on 2 MHz ultrasound intensity of the ^{13}C signal-to-noise ratio for the quaternary carbon of 1,3,5-trimethyl-benzene in a 1:1 molar mixture with cyclohexane. Reproduced with permission from AL Weekes, Ph.D. thesis, University of Aston, 1998.

reduce the associated values of T_1 and speed up their study by NMR. Such an approach, however, has accompanying problems, not least of which is the possibility that the application of ultrasound may cause conformational changes to the macromolecules studied. Similar changes have been observed during studies of N,N-dimethylacetamide where increasing intensity of 20 kHz ultrasound was found to induce free rotation about the N–C=O single bond and cause averaging of the two N-methyl ^1H chemical shifts to a single value.

It has also been demonstrated by other workers that ultrasound can reduce T_1 for a gadolinium chloride solution, and they, like the originators of the technique, have suggested that the approach might find use in magnetic resonance imaging: this exciting possibility remains to be investigated.

Although less work has been done on the acoustic reduction of T_1 in solids than in liquids, it has been established that, as for liquids, the natural values of T_1 in solids can be reduced by the application of ultrasound. By coupling 20 kHz ultrasound to a sample of trisodium phosphate dodecahydrate in an open mesh nylon sack immersed in a liquid, the normal value of the ^{31}P T_1 was reduced from 7.1 s (obtained from MAS NMR measurements) to 2.1 s. Subsequently, similar reductions (by a factor of about two) have been observed for the values of T_1 for ^{13}C in diamonds to which high frequency piezoelectric transducers were bonded directly.

Ultrasound and the NMR of liquid crystals

Due to the anisotropy in the molecular magnetic and electrical properties of liquid crystals they can, when in their nematic mesophase, be orientated by the application of external magnetic and electric fields. In the context of NMR this enables liquid crystals containing low concentrations of dissolved solutes to

be orientated by the magnet polarizing field. One beneficial consequence of this is that the solutes themselves become orientated and yield NMR spectra which show dipolar spin coupling splittings and which are quite different from their usual isotropic liquid state spectra. These, for example, enable the solute molecules structures to be determined. If a liquid crystal, in its nematic mesophase, is located in a magnetic field alone the molecular director adopts a reasonably well defined orientation with respect to the direction of the applied field. If, in a constant magnetic field, the orientation of the liquid crystal director can be changed, the appearance of the spectra of dissolved solutes should be changed. The possibility of using ultrasound to manipulate the director orientation of both thermotropic and lyotropic liquid crystals in appropriate NMR magnets has been investigated. The ^2H spectrum of benzene-d_6 dissolved in the nematic liquid crystal ZLI-1167, that normally aligns with its director perpendicular to the directions of the applied magnetic fields, was studied at about 30°C below the clearing temperature, both with and without the application of ultrasound along the bore of a cryomagnet. In the absence of ultrasound, the normal ^2H spectrum, composed of a pair of sharp quadrupolar split resonances, was observed. When the sample was irradiated with 2 MHz ultrasound the ^2H resonances broadened as shown in **Figure 4**. A temperature gradient within the sample should result in tails between the inner edges of the two resonances the minimum separation between which corresponds to the sample being close to the clearing temperature. On the other hand, a gradient in the director orientation should result in outer trailing edges to the resonances, as observed. Analyses of several such spectra led to the conclusion that the ultrasound, possibly through acoustic streaming, can result in a dispersion of the normal director orientation with a maximum induced change of about 20° from the normal direction.

SINNMR spectroscopy

It is well known that the width of the naturally broad resonances arising from static solids can be reduced by the coherent averaging processes that result from rapidly spinning samples about so-called magic angles. For dipolar nuclei the magic angle is set at 54° 44′ in the MAS NMR technique, while for quadrupolar nuclei an additional magic angle is used in the double orientation rotor, or DOR, method in order to minimize second order quadrupolar broadening effects. This is not necessary for liquids where the normally rapid and random molecular motion

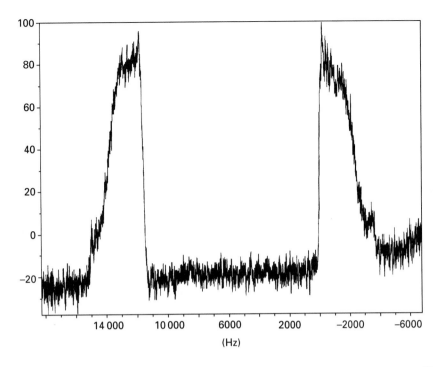

Figure 4 ^{2}H NMR spectrum of benzene-d_6 in liquid crystal ZLI-1167 irradiated with 2 MHz ultrasound at ~ 20 W cm^{-2}. Reproduced with permission from SA Reynolds, Ph.D. thesis, Aston University, 1997.

leads naturally to narrow lines with isotropic characteristics. Evidently, if particulate matter could be made to mimic the motional characteristics of (large) molecules in the liquid phase, and undergo rapid random motion, it should be possible to narrow the resonances from the solids through an incoherent motional process. If this situation could be achieved without spinning the sample, the production of spinning sidebands, that confuse MAS NMR and DOR spectra, could be avoided.

There are several possible ways of inducing the necessary incoherent motion of particles to facilitate their resonance line narrowing. An obvious way is to produce a fluidized bed in the NMR sample tube, but this has been tried without success. An alternative is to utilize the effects of Brownian motion where molecular bombardment of fine particles in suspension can cause their incoherent motion. The latter approach (ultrafine particle NMR) has been demonstrated using very small particles (nm size) that were perceived to be necessary to respond appropriately to Brownian motion. The necessity to use extremely small particles for this type of experiment is open to question because many experiments in the author's laboratory have shown that micrometre sized particles, suspended in density matched liquids, respond to Brownian motion and yield resonances that are significantly reduced relative to those from static solids.

Another way of inducing appropriate incoherent motion of suspended particles, and producing narrow

resonances from solids, is to irradiate the suspension with ultrasound. Whilst this idea relies on the consequences of several phenomena such as streaming, cavitation and shock waves, it may be considered initially from the viewpoint of early theoretical treatments of a single particle subject to an acoustic field. These showed that, for a physically anisotropic particle, the application of an acoustic field will drive the particle to one of three equilibrium orientations of which that with the long particle axis parallel to the direction of the acoustic field is the most stable. Having achieved the most stable equilibrium configuration any motional perturbation from this will be followed by a very rapid return to the equilibrium orientation, i.e. the particle will rotate rapidly through some relatively small angle in a more or less coherent fashion. Obviously this is not what is required to narrow the resonance lines arising from the solid. However, when an assembly of many particles is subject to acoustic irradiation the effects of cavitation producing microjet impact on particles will cause them to rotate, as will shock waves. The sequential effects of these phenomena, together with the effects of induced interparticle collisions, can be adequate to cause rapid incoherent rotation of the particles. These principles are implicit in the sonically induced narrowing of the NMR spectra of solids (SINNMR) technique that was first demonstrated in 1991.

There are so many interdependent parameters (such as support liquid density and viscosity, particle size, and acoustic frequency and intensity) that govern the success of the SINNMR experiment that it can by no means yet be considered a routine analytical tool. Nevertheless, continuing extensive investigations are slowly revealing the key features of the experiment and some of the findings are worthy of particular note.

The technique was placed on a fairly reproducible basis, using 20 kHz ultrasound delivered via a long titanium alloy acoustic horn to suspensions contained in a normal NMR high resolution tube in an iron magnet based spectrometer. Work on resin-coated (to prevent chemical reaction) 'particles' of aluminium and its alloys resulted in the production of ^{27}Al SINNMR spectra which revealed resonances of full width at half maximum height (FWHM) as low as 350 Hz: this compares most favourably with the FWHM of about 700 Hz obtained using MAS NMR and the FWHM of about 9000 Hz for a static sample of aluminium. The fact that the SINNMR resonances showed Knight shifts typical of the metallic species appears to provide an unequivocal proof of the validity of the SINNMR experiment. Subsequent studies of the ^{23}Na and ^{31}P spectra of trisodium phosphate dodecahydrate provided valuable insight into the SINNMR experiment. ^{31}P measurements of relaxation times using both MAS NMR and SINNMR revealed that the acoustically induced particle correlation times are of the order of 10^{-7} s, which is quite fast enough to cause the motional narrowing of the NMR spectra of the suspended solids. Interestingly, it was shown that the correlation times of the particles reduced as their size increased in support media of decreasing density and viscosity. Somewhat disappointingly, however, these detailed studies revealed that under the conditions employed only about 2% of the solid sample participated in the SINNMR narrowing. Such a low efficiency of SINNMR would appear to render it a poor competitor to MAS NMR and DOR. Nevertheless, the potential value of SINNMR has been established through successful narrowing of ^{11}B, ^{27}Al, ^{29}Si and ^{23}Na resonances in a range of materials, including glasses. In view of this a concerted thrust has been initiated to both improve the sensitivity of the technique and reduce its complexity.

The most recent work on SINNMR has resulted in the development of the first dedicated acoustic/NMR probehead. This accepts special SINNMR sample tubes that contain a piezoelectric transducer (interchangeable so that a range of acoustic frequencies from 1 to 10 MHz can be used) at its base. This configuration permits the ultrasonic irradiation of particles in less dense support liquids so that the particles can be levitated by the acoustic field and

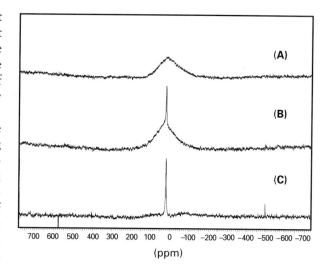

Figure 5 The ^{23}Na NMR spectra of particulate trisodium phosphate dodecahydrate suspended in a bromoform/chloroform mixture (A) without acoustic irradiation and (B) and (C) irradiated with 2 MHz ultrasound at ~30 and 50 W cm^{-2} respectively. Reproduced with permission from AL Weekes, Ph.D. thesis, Aston University, 1998.

induced to undergo incoherent motion in the NMR coil region. The use of ultrasound in the MHz range should produce smaller cavities and facilitate the study of much smaller particles than those used in the 20 kHz experiments, because the latter generates cavities of about 90 μm and microjetting from these is only effective with particles whose size is greater than the cavity dimensions. The reproducibility and efficiency of the SINNMR experiment has been improved using this dedicated apparatus and the preliminary results are most encouraging: typical SINNMR spectra are shown in **Figure 5**. When designing the dedicated SINNMR probehead and vessel mentioned above, particular attention was devoted to avoiding significant heating of the sample. As a result of this the apparatus can be used, not only for SINNMR studies, but for the reduction in the values of T_1 for liquid samples. This is now routinely possible for small molecules and the possibility of inducing similar changes in macromolecules is now the subject of intensive investigation.

List of symbols

N = number of atoms; T_1 = nuclear spin–lattice relaxation time; v = velocity of propagation; V = sample volume; Δm = change in magnetic quantum number: transition selection rule; v_0 = Larmor frequency; $v(\omega/2\pi)$ = characteristic frequency; ρ = phonon spectral density; ω = frequency; Ω = cut-off frequency.

See also: **Heteronuclear NMR Applications (As, Sb, Bi); Heteronuclear NMR Applications (B, Al, Ga, In, Tl); Heteronuclear NMR Applications (Ge, Sn, Pb); Heteronuclear NMR Applications (La–Hg); Heteronuclear NMR Applications (O, S, Se, Te); Heteronuclear NMR Applications (Sc–Zn); Heteronuclear NMR Applications (Y–Cd); Liquid Crystals and Liquid Crystal Solutions Studied By NMR; NMR Principles; NMR Relaxation Rates; NMR of Solids; Solid State NMR, Methods.**

Further reading

Abragam A (1989) *Principles of Nuclear Magnetism*. New York: Oxford University Press.

Beyer RT and Letcher SV (1969) *Physical Ultrasonics*. London: Academic Press.

Emsley JW and Lindon JC (1975) *NMR Spectroscopy in Liquid Crystal Solvents*, Oxford: Pergamon Press.

Homer J (1996) Ultrasonic irradiation and NMR. In: Grant DM and Harris RK (eds) *Encyclopedia of Nuclear Magnetic Resonance*, Vol 8, pp 4882–4891. Chichester: Wiley.

Homer J, Patel SU and Howard MJ (1992) NMR With Ultrasound, *Current Trends in Sonochemistry*, Cambridge: Royal Society of Chemistry Special Publication No. 116.

Homer J, Paniwnyk L and Palfreyman SA (1996) Nuclear Magnetic Resonance Spectroscopy Combined with Ultrasound. *Advances in Sonochemistry* **4**: 75–99.

Suslick KS and Doktycz SJ (1990) Effects of Ultrasound on Surfaces and Solids, *Advances in Sonochemistry* **1**: 197–230.

SPECT, Methods and Instrumentation

John C Lindon, Imperial College of Science, Technology and Medicine, London, UK

SPATIALLY RESOLVED SPECTROSCOPIC ANALYSIS
Methods & Instrumentation

Introduction

The SPECT technique is part of the armoury of nuclear medicine for the diagnosis of pathological conditions. Unlike other imaging techniques such as X-ray tomography or even some uses of magnetic resonance imaging (MRI), this technique can be used to identify functional abnormalities rather than anatomical disturbances.

SPECT involves the injection into the body of a radioactive pharmaceutical product (a radionuclide) such as technetium-99m or thallium-201 which decays with the emission of gamma rays. Usually the radiopharmaceutical is a protein with the radioactive atom attached and the molecule is designed to have the desired absorption properties for the tissue to be imaged. Some are accumulated into heart muscle and are used for cardiac imaging, whilst others penetrate the brain and are used for studying brain function. Yet others can be targeted to the lungs. Thus a healthy tissue will take up a known amount of the SPECT agent and this then appears as a bright area in a SPECT image. If a tissue is abnormal it is possible that uptake of the radiopharmaceutical will be amplified or depressed according to circumstances and then this will appear as a more intense spot or a dark area, respectively, on the image. This can be interpreted by a nuclear medicine expert in terms of the suspected pathological state.

Like X-ray tomography, SPECT imaging requires the rotation of a photon detector array around the body to acquire data from multiple angles. Using this technique, the position and concentration of the radionuclide distribution can be determined. Because the emission sources (in this case injected radiopharmaceuticals) are inside the body, this is more difficult than for X-ray tomography, where the source position and strength are known because the X-ray source is outside the body. It is necessary to compensate for the attenuation experienced by emission photons from injected tracers in the body and contemporary SPECT machines use mathematical reconstruction algorithms to generate the image taking this into account.

SPECT imaging has lower attainable resolution and sensitivity then positron emission tomography (PET). The radionuclides that are used for SPECT imaging emit a single gamma ray photon (usually about

140 keV), whereas in PET the positron emission results in two high-energy gamma ray 511 keV photons.

Methods and instrumentation

The technique requires the detection of the gamma rays emitted by the distribution of the radiopharmaceutical in the body. These gamma rays have to be collimated for detection by a gamma ray camera. The collimators contain thousands of parallel channels made of lead with square, circular or hexagonal cross-sections through which the gamma rays pass. These typically weigh about 25 kg and are about 5 cm thick with a length and breadth of about 40 by 20 cm. Models for special high-energy studies can be much more substantial and can weigh up to 100 kg. Such a collimator is termed a parallel-hole collimator and has a resolution which increases with distance from the gamma ray source. The resolution can be altered by using channels of different sizes. By going to smaller channels there is a trade-off in sensitivity. This has to be borne in mind during patient studies as it has an effect on scan times. Other types of collimator have been developed and these include converging hole collimators. Collimators are positioned above a very delicate single crystal of sodium iodide which is the heart of a gamma camera. This type of collimator/camera arrangement is called an Anger camera after its inventor.

The gamma rays emitted by the radiopharmaceutical in the body can be scattered by electrons within molecules in the body. This is known as Compton scattering and some such scattered photons are thus lost to the Anger camera because of the deflections caused. Second, the gamma rays can cause a photoelectron effect within an atom in the body (promotion of an electron to a higher orbital or even release of the electron) and again this gamma photon will be lost to the detection process.

Usually, Compton scattering is the most probable cause of attenuation in a SPECT image. Conversely, it is possible for a Compton scattered photon to be deflected into the Anger camera's field of view and in this case there is no information available on where the gamma photon originated and hence no spatial information on the location of the radiopharmaceutical. This process leads to loss of image contrast. A typical Anger camera equipped with a low-enegy collimator detects only about 1 in 10^4 gamma photons emitted by the radiopharmaceutical and a modern Anger camera has an intrinsic resolution of between 3 and 9 mm.

A gamma ray which passes through the collimator assembly will hit the sodium iodide crystal and generate a light photon which interacts with a grid of photomultiplier tubes behind and which collect the light for further processing. SPECT images are produced from these light signals. Sensitivity has been improved by the introduction of multi-camera SPECT systems. A triple-camera SPECT system equipped with high-resolution parallel-hole collimators can produce a resolution of 4–7 mm. Finally, other types of collimator such as the so-called pinhole type have been designed for imaging small organs such as the thyroid gland or limb extremities and for studies on laboratory animals.

The signals for the detected photons are reconstructed into an image using algorithms originally based on those used for X-ray tomography but which allowed for photon attenuation and Compton scattering. A typical method would be the filtered back-projection approach.

A number of experimental parameters have to be optimized in order to obtain the best SPECT image. These include attenuation, scatter, linearity of detector response, spatial resolution of the collimator and camera, system sensitivity, minimization of mechanical movements, image slice thickness, reconstruction matrix size and filter methods, sampling intervals and system deadtime. In a hospital, calibrating and monitoring these functions are usually performed by a Certified Nuclear Medicine Technician or a medical physicist.

Applications

SPECT is used routinely to help diagnose and stage the development of tumours and to pinpoint stroke, liver disease, lung disease and many other physiological and functional abnormalities. Although SPECT imaging resolution is not as high as that of PET, the availability of new SPECT radiopharmaceuticals, particularly for the brain and head, and the practical and economic aspects of SPECT instrumentation make this mode of emission tomography particularly attractive for clinical studies of the brain.

See also: **MRI Theory; PET, Methods and Instrumentation; PET, Theory; Zero Kinetic Energy Photoelectron Spectroscopy, Theory.**

Further reading

Brooks DJ (1997) PET and SPECT studies in Parkinson's disease. *Baillieres Clinical Neurology* 6: 69–87.
Corbett JR and Ficaro EP (1999) Clinical review of attenuation-corrected cardiac SPECT. *Journal of Nuclear Cardiology* 6: 54–68.

Dilworth JR and Parrott SJ (1998) The biomedical chemistry of technetium and rhenium. *Chemical Society Reviews* **27**: 43–55.

Germano G (1998) Automatic analysis of ventricular function by nuclear imaging. *Current Opinion in Cardiology* **13**: 425–429.

Hom RK and Katzenellenbogen JA (1997) Technetium-99m-labeled receptor-specific small-molecule radiopharmaceuticals: recent developments and encouraging results. *Nuclear Medicine and Biology* **24**: 485–498.

Krausz Y, Bonne O, Marciano R, Yaffe S, Lerer B and Chisin R (1996) Brain SPECT imaging of neuropsychiatric disorders. *European Journal of Radiology* **21**: 183–187.

Kuikka JT, Britton KE, Chengazi VU and Savolainen S (1998) Future developments in nuclear medicine instrumentation: a review. *Nuclear Medicine Communication* **19**: 3–12.

Powsner RA, O'Tuama LA, Jabre A and Melhem ER (1998) SPECT imaging in cerebral vasospasm following subarachnoid hemorrhage. *Journal of Nuclear Medicine* **39**: 765–769.

Ryding E (1996) SPECT measurements of brain function in dementia; a review. *Acta Neurologica Scandinavica* **94**: 54–58.

Spectroelectrochemistry, Applications

RJ Mortimer, Loughborough University, UK

ELECTRONIC SPECTROSCOPY
Applications

Introduction

Spectroelectrochemistry encompasses a group of techniques that allow simultaneous acquisition of spectroscopic and electrochemical information *in situ* in an electrochemical cell. Electrochemical reactions can be initiated by applying potentials to the working electrode, and the processes that occur are then monitored by both electrochemical and spectroscopic techniques. Electronic (UV-visible) transmission and reflectance spectroelectrochemistry has proved to be an effective approach for studying the redox chemistry of organic, inorganic and biological molecules, for investigating reaction kinetics and mechanisms, and for exploring electrode surface phenomena. In this article a selection of representative examples are presented, the emphasis being on the applications of transmission electronic (UV-visible) spectroelectrochemistry to the study of redox reactions and homogeneous chemical reactions initiated electrochemically within the boundaries of the diffusion layer at the electrode–electrolyte interface.

Organic systems

Many organic systems exhibit redox states with distinct electronic (UV-visible) absorption spectra and are therefore amenable to study with spectroelectrochemical techniques.

o-Tolidine

The technique of transmission spectroelectrochemistry, using an optically transparent electrode (OTE), was first demonstrated in 1964 using o-tolidine, a colourless compound that reversibly undergoes a 2-electron oxidation in acidic solution to form an intensely yellow coloured species (Eqn [1]). This system soon became a standard for testing spectroelectrochemical cells and new techniques.

Figure 1 shows absorbance spectra, for a series of applied potentials, recorded in an electrochemical cell employing an optically transparent thin-layer electrode (OTTLE). Curve a was recorded after application of +0.800 V vs saturated calomel electrode (SCE), which under thin-layer electrode

$$^+H_3N-\text{(ring)}-\text{(ring)}-NH_3^+ \underset{+2e\ +2H^+}{\overset{-2e\ -2H^+}{\rightleftharpoons}} {}^+H_3N=\text{(ring)}=\text{(ring)}=NH_2^+$$

[1]

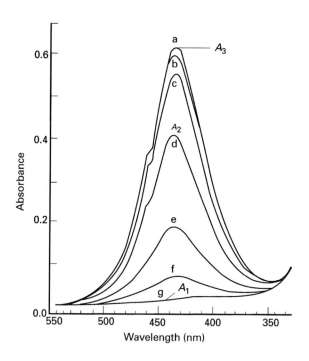

Figure 1 Thin-layer spectra of 0.97 mM o-tolidine, 0.5 M ethanoic acid, 1.0 M HClO$_4$ for different values of $E_{applied}$. Cell thickness 0.017 cm. Potential vs SCE: (a) 0.800 V, (b) 0.660 V, (c) 0.640 V, (d) 0.620 V, (e) 0.580 V, (f) 0.600 V, (g) 0.400 V. Reprinted with permission from DeAngelis TP and Heineman WR (1976) *Journal of Chemical Education* 53: 594–597. © 1976 Division of Chemical Education, American Chemical Society.

conditions causes complete electrolytic oxidation of o-tolidine to the yellow form ([O]/[R] > 1000, where O represents the oxidized form and R the reduced form). Curve g was recorded after application of +0.400 V, causing complete electrolytic reduction ([O]/[R] < 0.001), with the intermediate spectra corresponding to intermediate values of $E_{applied}$. The absorbance at 438 nm reflects the amount of o-tolidine in the oxidized form, which can be calculated from the Beer–Lambert law.

Determination of E^0, the reversible electrode potential, and n, the number of electrons in the o-tolidine redox reaction, can be determined from the sequence of spectropotentiostatic measurements (**Figure 1**). For a reversible system,

$$O + ne \rightleftharpoons R \qquad [2]$$

the [O]/[R] ratio at the electrode surface is controlled by the applied potential according to the Nernst equation:

$$E_{applied} = E^0 + \frac{RT}{nF} \ln\left(\frac{[O]}{[R]}\right)_{surface} \qquad [3]$$

In an OTTLE cell, on application of a new potential, the concentrations of O and R in solution are quickly adjusted to the same values as those existing at the electrode surface. Thus, at equilibrium:

$$\left(\frac{[O]}{[R]}\right)_{surface} = \left(\frac{[O]}{[R]}\right)_{solution} \qquad [4]$$

The Nernst equation in a thin-layer cell can then be written as:

$$E_{applied} = E^0 + \frac{RT}{nF} \ln\left(\frac{[O]}{[R]}\right) \qquad [5]$$

For the o-tolidine spectra, 438 nm is used as the monitoring wavelength and the ratio [O]/[R] is determined from the Beer–Lambert law:

$$\frac{[O]}{[R]} = \frac{(A_2 - A_1)/\Delta\varepsilon b}{(A_3 - A_2)/\Delta\varepsilon b} = \frac{A_2 - A_1}{A_3 - A_2} \qquad [6]$$

where A_1 is the absorbance of the reduced form, A_3 is the absorbance of the oxidized form, A_2 is the absorbance obtained at an intermediate applied potential, $\Delta\varepsilon$ is the difference in molar absorptivity between O and R at 438 nm and b is the light path length in the thin-layer cell. Thus the Nernst equation can be expressed as

$$E_{applied} = E^0 + \frac{RT}{nF} \ln\frac{(A_2 - A_1)}{(A_3 - A_2)} \qquad [7]$$

Figure 2 gives $E_{applied}$ vs log([O]/[R]) for the data from **Figure 1**. The plot is linear as predicted from Equation [7], the slope being 0.031 V, which corresponds to an n value of 1.92, with an intercept of 0.612 V vs SCE.

Methyl viologen

Mechanistic information is often available from spectroelectrochemical measurements. To illustrate the acquisition of semiquantitative information using a rapid scan spectrometer (RSS), the reduction of methyl viologen (the 1,1′-dimethyl-4,4′-bipyridilium dication) under semi-infinite linear diffusion conditions is presented. Methyl viologen (MV^{2+}) undergoes two consecutive one-electron reductions to the radical cation (MV$^{\bullet+}$) and neutral species (MV^0) in an EE mechanism. In acetonitrile at an OTE coated with a

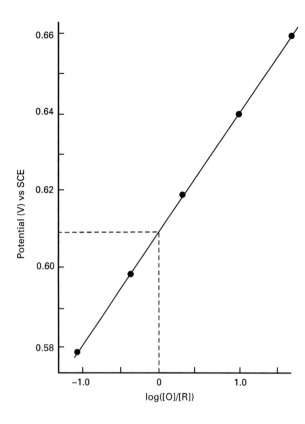

Figure 2 Plot of $E_{applied}$ vs log([O]/[R]) from spectra in **Figure 1**. Reprinted with permission from DeAngelis TP and Heineman WR (1976) *Journal of Chemical Education* **53**: 594–597. © 1976 Division of Chemical Education, American Chemical Society.

tin oxide film, both waves appear reversible with peak potentials $(E_p)_1 = -0.36$ V and $(E_p)_2 = -0.76$ V vs Ag/AgCl (Eqn [8]).

$$MV^{2+} \quad H_3C-^+N \bigcirc \bigcirc N^+-CH_3$$

$$-e^- \Updownarrow + e^-$$

$$MV^{\bullet+} \quad H_3C-^+N \bigcirc \bigcirc N-CH_3 \qquad [8]$$

$$-e^- \Updownarrow + e^-$$

$$MV^0 \quad H_3C-N \bigcirc \bigcirc N-CH_3$$

If the electrode is stepped some 0.200 V more negative than $(E_p)_1$ during a chronoamperometric experiment, absorbance spectra taken by RSS show two absorbance bands at λ_{max} equal to 390 and 602 nm. Interestingly, if the experiment is repeated with the

potential stepped slightly beyond $(E_p)_2$, the spectra taken are qualitatively identical to those obtained at $(E_p)_1$. This can be interpreted as due to the equilibrium between the three methyl viologen redox species in the diffusion layer, which greatly favours the radical ion MV$^{\bullet+}$, since $K_{eq} \gg 10^3$ for the reaction:

$$MV^{2+} + MV^0 \underset{k_b}{\overset{k_f}{\rightleftharpoons}} 2\,MV^{\bullet+} \qquad [9]$$

Analysis of the spectroelectrochemical working curves for this mechanism shows that when the radical ion is being monitored spectrally, the slopes of the A vs $t^{1/2}$ plots obtained by chronoamperometric reductions at potentials of the first and second waves, respectively, should be in a ratio of 1:1.20. This ratio assumes that the electrode reaction at both waves occurs at the diffusion-controlled rates, and that the three species are in thermodynamic equilibrium in the diffusion layer. The ratio for methyl viologen is 1.21 at $\lambda_{max} = 620$ nm. The larger ratio of 1.79 at $\lambda_{max} = 390$ nm is believed to be caused by band overlap from MV^0, which absorbs near the 390 nm band of MV$^{\bullet+}$.

If the chronoamperometric electrolysis is continued beyond several seconds, the rate of growth of the absorbance at the shorter wavelength of 390 nm decreases considerably owing to the formation of a dimer that absorbs near the longer-wavelength band.

Pyrene reduction

Reduction of the polycyclic aromatic pyrene serves as another excellent example of an EE mechanism where follow-up chemical reactions complicate the overall mechanism (**Figure 3**).

The one-electron reduction to the radical anion produces a ground doublet state with allowed transitions expected in the visible region of the spectrum. Spectra taken by RSS during chronoamperometric reduction at a potential 0.200 V more negative than E_p of the first wave ($E_p = -2.06$ V vs SCE in acetonitrile–TEAP (tetraethylammonium perchlorate) showed only a major band with a wavelength maximum at 492 nm in the visible region (**Figures 3A and 3C**). Reduction at E_p of the second wave produced spectra with wavelength maxima at 455 and 520–530 nm (**Figure 3A**, curve b). These maxima are similar to those for a spectrum obtained from the chemical reduction of pyrene and attributed to the dianion, except that the long-wavelength band at 602 nm reported earlier is absent. There is doubt, however, that this spectrum is the dianion because the second wave is irreversible; a new oxidation

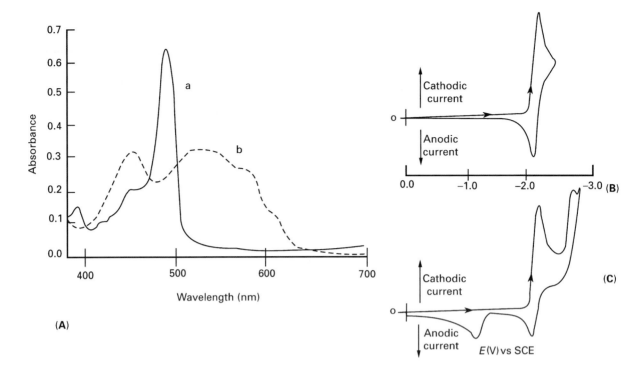

Figure 3 Spectra and cyclic voltammograms for the reduction of pyrene. (A) Curve a, spectrum of monoanion radical; λ_{max} 492, 446, 385 nm; curve b, spectrum obtained from reduction at the second wave for pyrene (see (C)); λ_{max} 455 and 520–530 nm. (B) Cyclic voltammogram for the reduction of pyrene to the monoanion radical in acetonitrile–TEAP at a tin oxide OTE. (C) Cyclic voltammogram for the reduction of pyrene. After reduction at the second wave, a new oxidation wave more positive than for the oxidation of the radical appears. Reprinted by courtesy of Marcel-Dekker, Inc. from Kuwana T and Winograd N (1974) Spectroelectrochemistry at optically transparent electrodes. I. Electrodes under semi-infinite diffusion conditions. In: Bard AJ (ed) *Electroanalytical Chemistry. A Series of Advances*, Vol 7, pp 1–78. New York: Marcel-Dekker.

wave more positive in potential than the wave for the oxidation of the radical anion appears (**Figure 3C**); and no spectrum due to the free radical appears during chronoamperometric reduction at E_p of the second wave. In any EE mechanism where the waves are sufficiently separated that the equilibrium constant for the disproportionation reaction is large, the equilibrium between the three species (pyrene, radical anion and dianion) would favour the presence of the radical anion in the diffusion layer. The supposed absence of rapid electron exchange between pyrene and dianion to form the radical anion suggests an EEC mechanism in which the dianion undergoes a fast homogeneous chemical reaction to a species more stable than the radical. A likely candidate is the monoanion formed through protonation.

Inorganic systems

There is a wide range of inorganic systems amenable to study by the spectroelectrochemical approach. In particular, transition metal complexes, with their rich redox state-dependent electronic spectra, have been intensively studied.

Hexacyanoferrate(III/II)

The hexacyanoferrate(III/II) (ferricyanide/ferrocyanide) system in aqueous solution is a well known electrochemically reversible redox couple (Eqn [10]).

$$[Fe^{III}(CN)_6]^{3-} + e \rightleftharpoons [Fe^{II}(CN)_6]^{4-} \qquad [10]$$

Furthermore, as the hexacyanoferate(III) ion is brilliant yellow in colour and the hexacyanoferrate(II) ion is only very pale yellow, this redox couple is particularly suited as a model system for electronic (UV-visible) absorbance spectroelectrochemical studies.

Figure 4 shows UV-visible absorption spectra recorded in a spectropotentiostatic experiment in an OTTLE cell on reduction of hexacyanoferrate(III) at a sequence of applied potentials. Curve a is at +0.50 V vs SCE reference electrode, where the redox system is in the oxidized state ($[Fe^{III}(CN)_6]^{3-}/[Fe^{III}(CN)_6]^{4-} > 1000$). Curve h is at +0.00 V vs SCE, where the redox system is in the reduced state

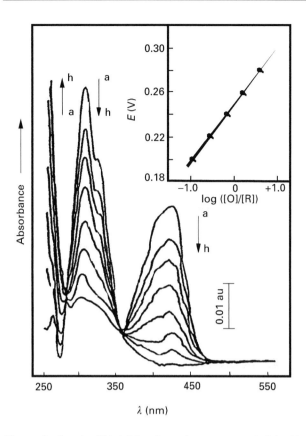

Figure 4 *In situ* UV-visible absorption spectra of 2.0 mM $K_3Fe(CN)_6$ in aqueous 1 M KCl at a sequence of applied potentials vs Ag/AgCl: (a) 0.50 V, (b) 0.28 V, (c) 0.26 V, (d) 0.24 V, (e) 0.22 V, (f) 0.20 V, (g) 0.17 V and (h) 0.00 V. Inset shows the plot of $E_{applied}$ vs log ([O]/[R]): ● at 312 nm and ▲ at 420 nm. Reprinted from Niu J and Dong S (1995) *Electrochimica Acta* **40**: 823–828, © 1995, with permission from Elsevier Science.

$([Fe^{III}(CN)_6]^{4-}/[Fe^{III}(CN)_6]^{3-} > 1000)$, while the intermediate spectra correspond to intermediate values of applied potentials. The inset plot in **Figure 4** demonstrates the reversibility of this system in accordance with Equation [7].

$[Tc^{III}(diars)_2Cl_2]^+$

The complex $[Tc^{III}(diars)_2Cl_2]^+$ (diars = [1]) provides another example of a reversible redox couple for which the spectropotentiostatic method has been applied.

As(CH₃)₂
As(CH₃)₂

[1]   Diars

Figure 5 shows a thin-layer cyclic voltammogram for this system and **Figure 6** gives a series of spectra

for a spectropotentiostatic experiment. Each spectrum was recorded 5 min after potential application so that $([O]/[R])_{solution}$ is at equilibrium with the electrode potential. Spectrum h is the oxidized form, whereas spectrum a is the reduced form. A Nernst plot from the spectra in **Figure 6** is shown in **Figure 7** ($E^0 = -0.091$ V vs SSCE, $n = 0.99$).

Polypyridylruthenium(II) complexes

The prospect of developing new materials of relevance to the emerging field of molecular electronics, modelling electron-transfer processes in biological systems and producing new electroactive and photoactive catalysts has led in recent years to considerable interest in transition metal polypyridyl complexes. Two recent examples of the application of the OTTLE spectroelectrochemical technique to the study of these fascinating systems are described here.

Identification of mixed-valence states in polynuclear polypyridylruthenium(II) complexes Mixed-valence complexes provide an ideal way of studying electron transfer – the most fundamental process in chemistry – under controlled conditions. Polynuclear complexes

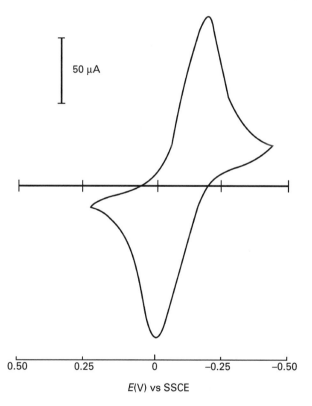

Figure 5 Thin-layer cyclic voltammogram at 2 mV s⁻¹ of 0.87 mM $[Tc^{III}(diars)_2Cl_2]^+$, 0.5 M TEAP in DMF. (SSCE = Sodium chloride saturated calomel electrode.) Reprinted with permission from Hurst RW, Heineman WR and Deutsch E (1981) *Inorganic Chemistry* **20**: 3298–3303. © 1981 American Chemical Society.

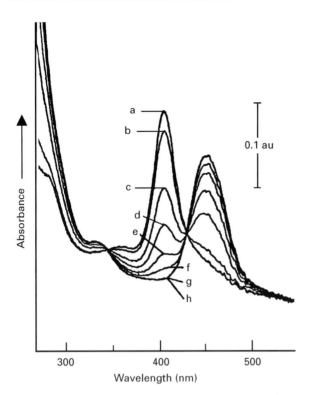

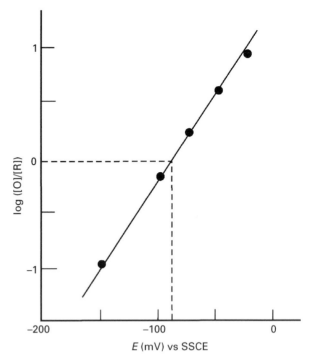

Figure 6 Spectra recorded during an OTTLE spectropotentiostatic experiment on 0.87 mM $[Tc^{III}(diars)_2Cl_2]^+$, 0.5 M TEAP in DMF. Applied potentials vs SSCE: (a) −0.250 V; (b) −0.150 V; (c) −0.100 V; (d) −0.075 V; (e) −0.050 V; (f) −0.025 V; (g) 0.100 V; (h) 0.250 V. Reprinted with permission from Hurst RW, Heineman WR and Deutsch E (1981) *Inorganic Chemistry* **20**: 3298–3303. © 1981 American Chemical Society.

Figure 7 Nernst plot for spectropotentiostatic experiment on 0.87 mM $[Tc^{III}(diars)_2Cl_2]^+$, 0.5 M TEAP in DMF. Data at 403 nm from **Figure 6** are used. Reprinted by courtesy of Marcel-Dekker, Inc. from Heineman WR, Hawkridge FM and Blount HN (1984) Spectroelectrochemistry at optically transparent electrodes. II. Electrodes under thin-layer and semi-infinite diffusion conditions and indirect coulometric titrations. In: Bard AJ (ed) *Electroanalytical Chemistry. A Series of Advances*, Vol 13, pp 1–113. New York: Marcel-Dekker.

containing polypyridylruthenium(II) moieties are of particular interest for the study of mixed valency because of their kinetic inertness in both the +II and +III oxidation states, generally reversible electrochemical behaviour, and good π-donor ability which allows interaction with bridging ligand orbitals. Spectroelectrochemical measurements can be used to probe electrogenerated mixed-valence states in such complexes. A recent example (**Table 1** and **Figure 8**) is the controlled-potential oxidation of the [2,2] species of the complex $[\{Ru(bipy)_2\}_2(\mu\text{-OMe})_2][PF_6]_2$ in an OTTLE cell. Oxidation of the [2,2] species to the mixed-valence [2,3] state results in the collapse of the metal-to-ligand charge transfer (m.l.c.t) bands at 589 and 364 nm and the generation of a new transition at ~1800 nm ($\varepsilon = 5000$ dm^3 mol^{-1} cm^{-1}), which disappears on further oxidation to the Ru^{III} state. The observations that this transition is not solvatochromic and that the half-width of the peak is much narrower than the value predicted from Hush theory for vectorial intervalence charge-transfer bands both point to a class III (Robin and Day fully delocalized) mixed-valence state.

Electronic properties of hydroquinone-containing ruthenium polypyridyl complexes Ruthenium polypyridyl complexes bound to hydroquinone/quinone moieties are expected to yield information on the behaviour of hydroquinone-type compounds in biological processes. Furthermore, ruthenium(II)–hydroquinone complexes involving O and N bonds are likely to absorb well into the visible region and therefore have potential as dyes in sensitized solar cells. A recent example in the application of spectroelectrochemistry to the study of hydroquinone-containing ruthenium polypyridyl complexes is the oxidation of $[Ru(bipy)_2(HL^0)]^+$ (H_2L^0 = 1,4-dihydroxy-2,3-bis(pyrazol-1-yl)benzene) (**Figure 9**).

The spectral changes associated with the first two-electron oxidation step are reversible, and unstable long-lived intermediates are not present, as indicated by the clear isobestic points at 327, 398, 446 and 614 nm (**Figure 9**). After the first two-electron oxidation the m.l.c.t. band at 490 nm blue shifts to approximately 416 nm, and a new feature appears at 700 nm for $[(Ru(bipy)_2(HL^0)]^{2+}$. The presence of

Table 1 Electronic spectral data for the dinuclear complex [{Ru(bipy)$_2$}$_2$(μ-OMe)$_2$][PF$_6$]$_2$ in CH$_2$Cl$_2$ at 240 K

Oxidation state	λ_{max} (nm) ($10^{-3} \varepsilon$(dm^3 mol^{-1} cm^{-1}))
[2,2]	572 (12), 420 (sh), 359 (15), 293 (79), 242 (58)
[2,3]	1 800 (5), 480 (9), 340 (12), 292 (94), 242 (57)
[3,3]	580 (6), 380 (sh), 248 (64)

Reprinted with permission from Bardwell DA, Horsburgh L, Jeffrey JC *et al* (1996) *Journal of the Chemical Society, Dalton Transactions*, 2527.

significant absorption features between 400 and 500 nm in the spectrum of the oxidized compound suggests that in the complex the metal centre is still in the ruthenium(II) state, consistent with interpretation from electrochemical data. The oxidized complex is therefore most likely the analogous ruthenium(II)–quinone species. After oxidation of the hydroquinone to quinone, the Ru^{II} → bipy(π_2^*) m.l.c.t. shifts to the blue as a result of the stabilization of the t$_{2g}$ level when the σ-donating ability of the ligand is decreased. Further oxidation results in the irreversible loss of the intense feature between 700 and 800 nm and of the band at 416 nm and the generation of a yellow complex likely to be a

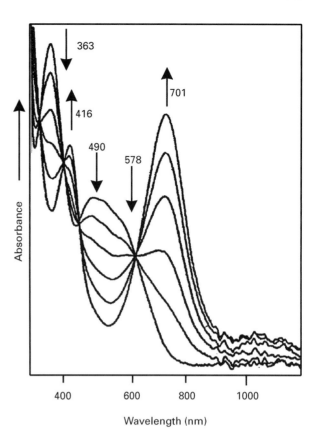

Figure 9 Spectroelectrochemical oxidation of [Ru(bipy)$_2$(HL^0)]$^+$ (H$_2$L^0 = 1,4-dihydroxy-2,3-bis(pyrazol-1-yl)benzene) as a function of time between 0 and 20 min. Reprinted with permission from Keyes TE, Jayaweera PM, McGarvey JJ and Vos JG (1997) *Journal of the Chemical Society, Dalton Transactions*, 1627–1632.

complex in which the pyrazole is bound to the ruthenium in a monodentate fashion.

Biological systems

Numerous biological redox systems have been studied by the spectroelectrochemical approach, including cytochromes, myoglobin, photosynthetic electron transport components, spinach ferrodoxin, blue copper proteins, retinal, and vitamin B$_{12}$ and its analogues. Two classic examples are presented here.

Vitamin B$_{12}$

Vitamin B$_{12}$ (cyanocob(III)alamin) is an example of a quasi-reversible redox system that exhibits slow heterogeneous electron-transfer kinetics. Cyclic voltammetry alone suggests that the reduction of vitamin B$_{12}$ is a single two-electron process at $E_{pc} = -0.93$ V vs SCE to the Co(I) redox state (**Figure 10A**). However, thin-layer spectroelectrochemistry using a

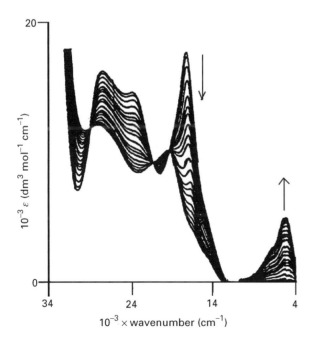

Figure 8 Successive electronic spectra of the dinuclear complex [{Ru(bipy)$_2$}$_2$(μ-OMe)$_2$][PF$_6$]$_2$ in propylene carbonate at 240 K recorded during electrochemical oxidation to the mixed-valence Ru^{II}Ru^{III} state, showing the disappearance of the Ru^{II} → m.l.c.t. bands and the appearance of the near-IR band. Reprinted with permission from Bardwell DA, Horsburgh L, Jeffrey JC *et al* (1996) *Journal of the Chemical Society, Dalton Transactions*, 2527–2531.

Hg–Au minigrid OTTLE in a spectropotentiostatic mode reveals that reduction takes place via two consecutive one-electron steps (**Figures 11** and **12**). **Figure 11** shows thin-layer spectra for the reduction to B_{12r}, which occurs in the potential range −0.580 to −0.750 V, and **Figure 12** shows the spectral changes for the further reduction to B_{12s}, which occurs in the range −0.770 to −0.950 V. Nernst plots for these two reduction processes (using Eqn [7] above) give values of $E_1 = -0.655$ V, $n = 1$ and $E_2 = -0.880$ V, $n = 1$, respectively. The two one-electron reduction processes are clearly shown by the plot of absorbance at 363 nm vs potential in **Figure 10B**, the first one-electron reduction occurring in a region with no apparent cathodic current (**Figure 10A**).

Cytochrome *c* Often biological macromolecules will not undergo direct heterogeneous electron transfer with an electrode. Instead, mediator titrants are used that exchange electrons heterogeneously with the electrode and homogeneously with the macro-

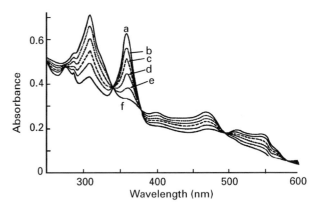

Figure 11 Thin-layer spectra for reduction of vitamin B_{12} to B_{12r} in a solution of 1 mM vitamin B_{12}, Britton–Robinson buffer pH 6.86, 0.5 M Na_2SO_4. To obtain the spectra, the potential was stepped in 0.5 mV increments and maintained at each step for 3–5 min until spectral changes ceased. Applied potentials vs SCE: (a) −0.550 V; (b) −0.630 V; (c) −0.660 V; (d) −0.690 V; (e) −0.720 V; (f) −0.770 V. Reprinted by courtesy of Marcel-Dekker, Inc. from Heineman WR, Hawkridge FM and Blount HN (1984) Spectroelectrochemistry at optically transparent electrodes. II. Electrodes under thin-layer and semi-infinite diffusion conditions and indirect coulometric titrations. In: Bard AJ (ed) *Electroanalytical Chemistry. A Series of Advances*, Vol 13, pp 1–113. New York: Marcel-Dekker.

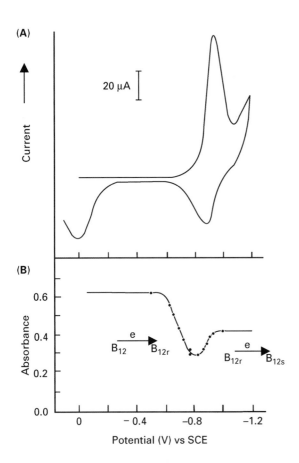

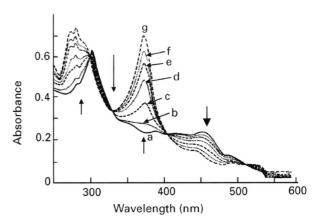

Figure 10 (A) Thin-layer cyclic voltammogram of 1 mM vitamin B_{12}, Britton–Robinson buffer pH 6.86, 0.5 M Na_2So_4. (B) Plot of absorbance at 368 nm vs potential, recorded at effectively ∼ 0.003 mV s⁻¹, from spectra in **Figures 11** and **12**. Reprinted by courtesy of Marcel-Dekker, Inc. from Heineman WR, Hawkridge FM and Blount HN (1984) Spectroelectrochemistry at optically transparent electrodes. II. Electrodes under thin-layer and semi-infinite diffusion conditions and indirect coulometric titrations. In: Bard AJ (ed) *Electroanalytical Chemistry. A Series of Advances*, Vol 13, pp 1–113. New York: Marcel-Dekker.

Figure 12 Thin-layer spectra for reduction of vitamin B_{12r} to B_{12s} in a solution initially of 1 mM vitamin B_{12}, Britton–Robinson buffer pH 6.86, 0.5 M Na_2SO_4. To obtain the spectra, the potential was stepped in 0.5 mV increments and maintained at each step for 3–5 min until spectral changes ceased. Applied potentials vs SCE: (a) −0.770 V; (b) −0.820 V; (c) −0.860 V; (d) −0.880 V; (e) −0.900 V; (f) −0.920 V; (g) −1.000 V. Reprinted with permission from Rubinson KA, Itabashi E and Mark Jr HB (1982) *Inorganic Chemistry* **21**: 3771–3773. © 1982 American Chemical Society.

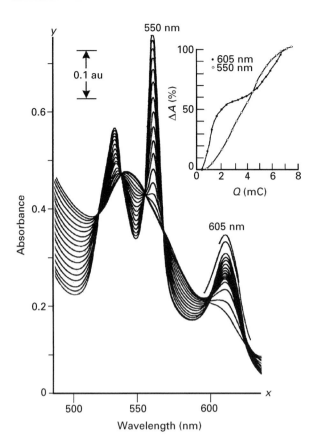

Figure 13 Spectrocoulometric titration of cytochrome c (17.5 μM) and cytochrome c oxidase (6.3 μM) by reduction with electrogenerated methyl viologen radical cation (MV$^{\bullet+}$) at a SnO$_2$ OTE. Each spectrum was recorded after 5×10^{-9} equivalents of charge (0.5 mC) were passed. Spectra correspond to titration from totally oxidized to totally reduced forms. The final two spectra around 605 nm were recorded after excess MV$^{\bullet+}$ was present. Inset shows titration curves at 550 and 605 nm. Reprinted with permission from Heineman WR, Kuwana T and Hartzell CR (1973) *Biochemical and Biophysical Research Communications* **50**: 892–900.

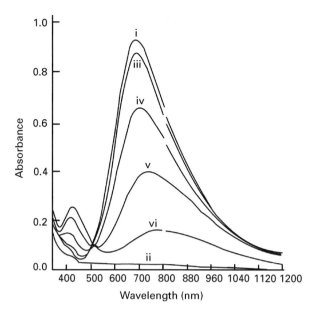

Figure 14 Spectra of iron hexacyanoferrate films on ITO-coated glass at various potentials [(i) +0.50 V (PB, blue); (i) –0.20 V (PW, transparent); (iii) +0.80 V (PG, green); (iv) +0.85 V (PG, green); (v) +0.90 V (PG, green); (vi) +1.20 V (PX, yellow)] vs SCE with 0.2 M KCl + 0.01 M HCl as supporting electrolyte. Reproduced with permission from Mortimer RJ and Rosseinsky DR (1984) *Journal of the Chemical Society*, Dalton Transactions, 2059–2061.

molecules. **Figure 13** gives spectra obtained for the reduction of a mixture of the haem proteins cytochrome c and cytochrome c oxidase, both initially in the fully oxidized state.

Each spectrum was recorded after the coulometric addition of 5×10^{-9} equivalents of reductant, the methyl viologen radical cation (MV$^{\bullet+}$) electrogenerated at a SnO$_2$ OTE. The reaction sequence is an EC catalytic regeneration mechanism:

$$\text{Electrode} : MV^{2+} + e \rightleftharpoons MV^{\bullet+}$$

$$\text{Solution} : MV^{\bullet+} + \text{cyt } c \text{ (ox)} \rightleftharpoons MV^{2+} + \text{cyt } c \text{ (red)}$$

[11]

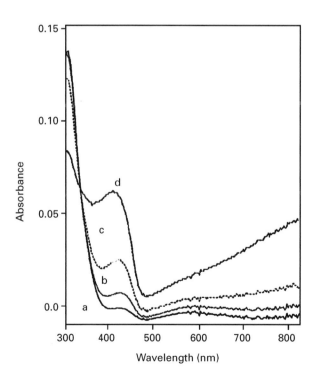

Figure 15 Spectra of poly(m-toluidine) films on ITO in 1 M hydrochloric acid at (a) –0.20 V, (b) +0.10 V, (c) +0.20 V, (d) +0.30 V vs SCE. Reproduced with permission from Mortimer RJ (1995) *Journal of Materials Chemistry* **5**: 969–973.

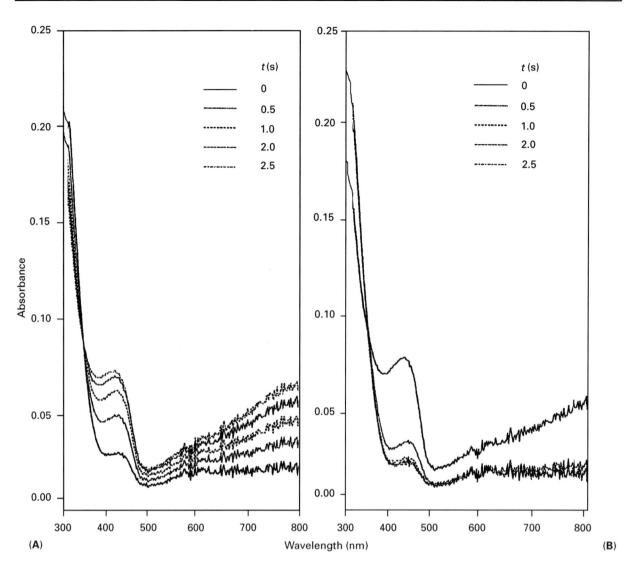

Figure 16 Spectra recorded at times indicated after potential switching of poly(*m*-toluidine) films on ITO in 1 M hydrochloric acid (A) Potential step −0.20 to +0.40 V vs SCE. (B) Potential step +0.40 to −0.20 V vs SCE. Reproduced with permission from Mortimer RJ (1995) *Journal of Materials Chemistry* **5**: 969–973.

In solution, one MV$^{\bullet+}$ species can reduce a single haem site in cytochrome *c* or one of two in the oxidase. The absorbance increase (**Figure 13**) at 605 nm corresponds to the reduction of the two haem components of cytochrome *c* oxidase; the increase at 550 nm corresponds to the reduction of the haem in cytochrome *c*. Study of plots of absorbance change vs coulometric charge (see inset of **Figure 13**) indicate that MV$^{\bullet+}$ initially reduces one of the haem groups in cytochrome *c* oxidase, then the haem in cytochrome *c*, before it reduces the second haem of the oxidase.

Modified electrodes

Immobilization of chemical microstructures onto electrode surfaces has been a major growth area in

electrochemistry in recent years. Compared to conventional electrodes, greater control of electrode characteristics and reactivity is achieved on surface modification. Potential applications of such systems include the development of electrocatalytic systems with high chemical selectivity and activity, coatings on semiconducting electrodes with photosensitizing and anticorrosive properties, electrochromic displays, microelectrochemical devices for the field of molecular electronics and electrochemical sensors with high selectivity and sensitivity.

Spectroelectrochemical measurements, both *ex situ* and *in situ*, are frequently used in the characterization of modified electrodes. In the case of *in situ* spectroelectrochemical measurements, the modified electrode can be considered to be analogous to an OTTLE, the redox active layer being physically or

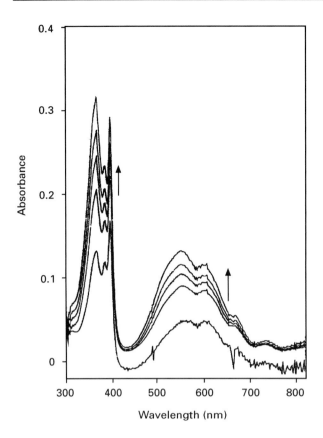

Figure 17 Spectra recorded at –0.90 V vs SCE during the 2nd, 4th, 6th, 8th and 10th cyclic voltammograms for an ITO/Nafion electrode in 0.1 mM 1,1'-dimethyl -4,4'-bipyridilium dichloride +0.2 M KCl (pH 5.5). The vertical arrows indicate absorbance increase with scan number. For a comparable experiment in the absence of Nafion, the maximum absorbance was <0.01. Reproduced with permission from Mortimer RJ and Dillingham JL (1997) *Journal of the Electrochemical Society* **144**: 1549–1553.

chemically confined to the electrode surface. Electronic spectroelectrochemistry sees significant use in the study of electrodes modified with electrochromic surface films, for which some examples are given below.

Characterization of electrochromic materials

Chemical species that can be electrochemically switched between different colours are said to be electrochromic. Electrochromism results from the generation of different visible-region electronic absorption bands on switching between redox states. The colour change is commonly between a transparent ('bleached') state and a coloured state, or between two coloured states. In cases where more than two redox states are electrochemically available, the electrochromic material may exhibit several colours and be termed polyelectrochromic. Likely applications of electrochromic materials include their use in controllable light-reflective or

light-transmissive devices for optical information and storage, antiglare car rear-view mirrors, sunglasses, protective eyewear for the military, controllable aircraft canopies, glare-reduction systems for offices, and 'smart windows' for use in cars and in buildings.

Prussian blue Prussian blue (PB; iron(III) hexacyanoferrate(II)) thin films can be switched to Prussian white (PW) on electrochemical reduction and to Prussian yellow (PX) on oxidation via the partially oxidized Prussian green (PG). For all these electrochromic redox reactions, there is concomitant ion ingress/egress in the films for electroneutrality. The spectra of PX, PG, PB and PW are shown in **Figure 14**, together with two intermediate states between blue and green. The intense blue colour in the $[Fe^{III}Fe^{II}(CN)_6]^-$ chromophore of PB is due to an intervalence charge-transfer (CT) absorption band centred at 690 nm. The yellow absorption band in PX corresponds with that of $[Fe^{III}Fe^{III}(CN)_6]$ in solution, both maxima ($\lambda_{max} = 425$ nm) coinciding with the (weaker) $[Fe^{III}(CN)_6]^{3-}$ absorption maximum. On increase from +0.50 V vs SCE to more oxidizing potentials, the original PB peak shifts continuously to longer wavelengths with diminishing absorption, while the peak at 425 nm steadily increases, owing to the increasing $[Fe^{III}Fe^{III}(CN)_6]$ absorption. The reduction of PB to PW is by contrast abrupt, with transformation to all PW or all PB without pause, depending on the potential that is set. In the cyclic voltammogram of a PB-modified electrode, the broad peak for PB $\rightleftharpoons$ PX in contrast with the sharp PB $\rightleftharpoons$ PW transition emphasizes the range of compositions involved. This difference in behaviour, supported by ellipsometric measurements, indicates continuous mixed-valence compositions over the blue-to-yellow range in contrast with the presumably immiscible PB and PW, which clearly transform one into the other without intermediacy of composition.

Conducting polymers Chemical or electrochemical oxidation of numerous resonance-stabilized aromatic molecules including pyrrole, thiophene, aniline, furan, carbazole, azulene and indole produces electronically conducting polymers. In their oxidized forms, such conducting polymers are 'doped' with counteranions (*p*-doping) and possess a delocalized π electron band structure, the energy gap between the highest occupied π electron band (valence band) and the lowest unoccupied band (the conduction band) determining the intrinsic optical properties of these materials. The doping process (oxidation) introduces polarons (in polypyrrole, for example, these are radical cations delocalized over

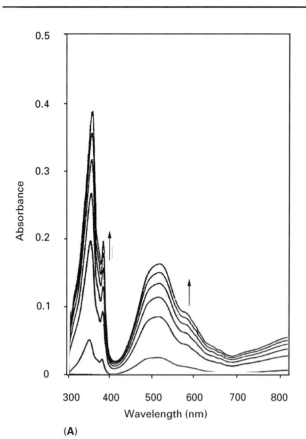

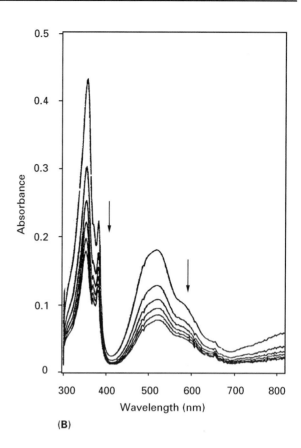

Figure 18 (A) Spectra recorded at $t = 0$, 10, 20, 30, 40 and 50 s in response to a potential step from +1.00 to –0.90 V vs SCE for an ITO/Nafion/1,1'-di-n-hexyl-4,4'-bipyridilium electrode in 0.1 mM 1,1'-di-n-hexyl-4,4'-bipyridilium dibromide +0.2 M KCl (pH 5.5). The vertical arrows indicate absorbance increase with time. (B) Spectra recorded at $t = 0$, 10, 20, 30, 40 and 50 s in response to a potential step from –0.90 to +1.00 V vs SCE for an ITO/Nafion/1,1'-di-n-hexyl-4,4'-bipyridilium electrode in 0.1 mM 1,1'-di-n-hexyl-4,4'-bipyridilium dibromide +0.2 M KCl (pH 5.5). The vertical arrows indicate absorbance decrease with time. Reproduced with permission from Mortimer RJ and Dillingham JL (1997) *Journal of the Electrochemical Society* **144**: 1549–1553.

ca. four monomer units), which are the major charge-carriers. Reduction of conducting polymers with concurrent counteranion exit removes the electronic conjugation, to give the 'undoped' (neutral) electrically insulating form.

All conducting polymers are potentially electrochromic in thin-film form, redox switching giving rise to new optical absorption bands in accompaniment with transfer of electrons/counteranions. Good examples are the polymers of aniline, o-toluidine and m-toluidine, which are easily prepared as thin films by electrochemical oxidation from aqueous acid solutions of the appropriate monomer. The electrical and electrochromic properties of such polyanilines depend not only on oxidation state but also on the protonation state, and polyanilines are in fact polyelectrochromic (transparent yellow to green to dark blue to black), the yellow–green transition being durable to repetitive colour switching. Spectra for a poly(m-toluidine)-modified electrode are illustrated in **Figure 15**.

The two low-wavelength spectral bands observed are assigned to an aromatic $\pi-\pi^*$ transition ($\leq$330 nm) related to the extent of conjugation between the adjacent rings in the polymer chain, and to radical cations formed in the polymer matrix ($\leq$440 nm). With increase in applied potential, the $\leq$330 nm band absorbance decreases and the $\leq$440 nm increases (**Figure 15**), the isobestic point indicating that the two species have the same chemical stoichiometry with differences only in electrons. Beyond +0.30 V, the conducting region is entered; the $\leq$440 nm band decreases as a broad free carrier electron band ~800 nm is introduced. Response times for the yellow–green transition following a potential step can be determined using a diode array spectrophotometer (**Figure 16**).

Viologens in Nafion In addition to being important mediator titrants, 1,1'-disubstituted-4,4'-bipyridiliums (viologens) are a major group of electrochromic materials. Electrochromism occurs in bipyridiliums

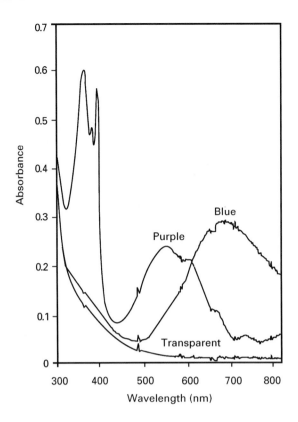

Figure 19 Spectra recorded at +0.50 V (blue), −0.20 V (transparent) and −0.90 V (purple) vs SCE for an ITO/PB/ Nafion/methyl viologen electrode in 0.1 mM 1,1′-dimethyl-4,4′-bipyridilium dichloride +0.2 M KCl (pH 5.5). Reproduced with permission from Mortimer RJ and Dillingham JL (1997) *Journal of the Electrochemical Society* **144**: 1549–1553.

because, in contrast to the bipyridilium dications, the radical cations formed on electroreduction have a delocalized positive charge, coloration arising from an intramolecular electronic transition. Suitable choice of nitrogen substituents to attain the appropriate molecular orbital energy levels can, in principle, allow colour choice of the radical cation. For short alkyl chain length, 1,1′-dialkyl-4,4′-bipyridiliums, both the dication and radical-cation states, are soluble in water and any electrochromic device (ECD) using such bipyridiliums would have the limitation of a low write–erase efficiency. One solution to this problem involves electrostatic binding of bipyridilium dications into anionic polyelectrolyte films. When a Nafion-modified electrode is immersed in an aqueous solution of 1,1′-dialkyl-4,4′-bipyridilium (alkyl = methyl, ethyl, n-propyl, n-butyl, n-pentyl, n-hexyl), the 1,1′-dialkyl-4,4′-bipyridilium accumulates in the anionic polyelectrolyte such that its concentration is considerably higher than that in the bulk solution. **Figure 17** illustrates the uptake of 1,1′-dimethyl-4,4′-bipyridilium into a Nafion film monitored by *in situ* spectral measure-

ments. The coloured form in each case is purple, from the presence of monomeric (blue, λ_{max} 600 nm) and dimeric (red, λ_{max} 500 nm) viologen radical cations. For such viologen-incorporated Nafion films, the electrochromic response times are in excess of 60 s for both coloration and bleaching and independent of viologen size. **Figure 18** shows absorbance spectra measured every 10 s in response to a potential step between the oxidized and reduced forms, for the case of the 1,1′-di-n-hexyl-4,4′-bipyridilium system. The longer response time for the oxidation (bleaching) reflects the slower diffusion of radical-cation dimers through the Nafion film compared to that of the monomeric radical cations.

Five-colour polyelectrochromicity, by application of an outer Nafion layer (with subsequent electrostatic incorporation of the methyl viologen system) to an inner layer of PB, is possible (**Figure 19**). The transparent/purple viologen dication/radical cation electrochromicity operates in the potential region where the PB is in its (reduced) transparent state; the bilayer electrode system thus exhibits yellow/green/blue/transparent/purple colours.

List of symbols

A = absorbance; b = path length; e = electron; E = electrode potential; E^0 = reversible electrode potential; E_p = peak potential; E_{pa} = anodic peak potential; E_{pc} = cathodic peak potential; F = Faraday constant (96 485 C mol^{-1}); k = rate constant; K_{eq} = equilibrium constant; n = number of electrons; O = oxidized form; R = reduced form; R = gas constant (8.315 J K^{-1} mol^{-1}); t = time; T = temperature in kelvin; ε = molar extinction coefficient; λ_{max} = wavelength of maximum absorbance.

See also: **Colorimetry, Theory; Dyes and Indicators, Use of UV-Visible Absorption Spectroscopy; Ellipsometry; Spectroelectrochemistry, Methods and Instrumentation.**

Further reading

Heineman WR and Jensen WB (1989) Spectroelectro-chemistry using transparent electrodes – an anecdotal history of the early years. *ACS Symposium Series* **390**: 442–457.

Heineman WR and Kissinger PT (1984) Large-amplitude controlled-potential techniques. In: Kissinger PT and Heineman WR (eds) *Laboratory Techniques in Electro-analytical Chemistry*, Chapter 3. New York: Marcel Dekker.

Heineman WR, Hawkridge FM and Blount HN (1984) Spectroelectrochemistry at optically transparent electrodes. II. Electrodes under thin-layer and semi-infinite diffusion conditions and indirect coulometric titrations. In: Bard AJ (ed) *Electroanalytical Chemistry. A Series of Advances*, Vol 13, pp 1–113. New York: Marcel-Dekker.

Kuwana T and Heineman WR (1976) Study of electrogenerated reactants using optically transparent electrodes. *Accounts of Chemical Research* 7: 241–248.

Kuwana T and Winograd N (1974) Spectroelectrochemistry at optically transparent electrodes. I. Electrodes under semi-infinite diffusion conditions. In: Bard AJ (ed) *Electroanalytical Chemistry. A Series of Advances*, Vol 7, pp 1–78. New York: Marcel-Dekker.

Mortimer RJ (1994) Dynamic processes in polymer modified electrodes. In: Compton RG and Hancock G (eds) *Research in Chemical Kinetics*, Vol 2, pp 261–311. Amsterdam: Elsevier.

Mortimer RJ (1997) Electrochromic materials. *Chemical Society Reviews* 26: 147–156.

Murray RW (1984) Chemically modified electrodes. In: Bard AJ (ed) *Electroanalytical Chemistry. A series of Advances*, Vol 13, pp 191–3\68. New York: Marcel-Dekker.

Niu J and Dong S (1996) Transmission spectroelectrochemistry. *Reviews in Analytical Chemistry* 15: 1–171.

Spectroelectrochemistry, Methods and Instrumentation

Roger J Mortimer, Loughborough University, UK

Copyright © 1999 Academic Press

ELECTRONIC SPECTROSCOPY
Methods & Instrumentation

Introduction

Spectroelectrochemistry encompasses a group of techniques that allow simultaneous acquisition of electrochemical and spectroscopic information *in situ* in an electrochemical cell. A wide range of spectroscopic techniques may be combined with electrochemistry, including electronic (UV-visible) absorption and reflectance spectroscopy, luminescence spectroscopy, infrared and Raman spectroscopies, electron spin resonance spectroscopy and ellipsometry. Molecular properties such as molar absorption coefficients, vibrational absorption frequencies and electronic or magnetic resonance frequencies, in addition to electrical parameters such as current, voltage or charge, are now being used routinely for the study of electron transfer reaction pathways and the fundamental molecular states at interfaces. In this article the principles and practice of electronic spectroelectrochemistry are introduced.

Cell design

In electronic UV-visible spectroelectrochemistry an optical beam traverses an optically transparent electrode (OTE) in one of the three configurations shown in **Figure 1**.

In transmission spectroelectrochemistry the optical beam is directed perpendicularly through the OTE and the adjacent solution either under semi-infinite linear diffusion conditions (**Figure 1A**) or in a thin-layer cell where diffusion is restricted (**Figure 1B**). Internal reflection spectroscopy (IRS) involves introducing the optical beam through the rear side of an OTE at an angle greater than the critical angle so that the beam is totally reflected (**Figure 1C**). In IRS spectral changes are observable owing to the small penetration of the electric field vector into the solution. Both transmission and reflection spectroscopy have been coupled with numerous electrochemical excitation signals to generate a variety of spectroelectrochemical techniques.

Optically transparent electrodes

As the working electrode in a spectroelectrochemical experiment, an OTE needs to have both wide optical and potential 'windows', a sufficiently low resistance for good electrode potential control, good stability and surface reproducibility. **Table 1** summarizes the optical and resistance data of some OTEs which are typically thin conducting films on substrate surfaces or minigrids (electroformed mesh).

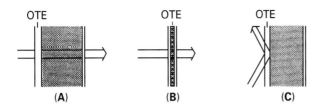

Figure 1 Schematic diagram of spectroelectrochemical techniques at an optically transparent electrode (OTE). (A) Transmission spectroelectrochemistry; (B) transmission spectro-electrochemistry with an optically transparent thin-layer electrode (OTTLE) cell; (C) internal reflection spectroscopy (IRS). Reprinted by courtesy of Marcel Dekker, Inc. from Heineman WR, Hawkridge FM and Blount HN (1984) Spectroelectrochemistry at optically transparent electrodes. II. Electrodes under thin-layer and semi-infinite diffusion conditions and indirect coulometric titrations. In: Bard AJ (ed) *Electroanalytical Chemistry. A Series of Advances*, Vol 13, pp 1–113. New York: Marcel-Dekker.

Thin conducting films Any conductor becomes transparent if it is made sufficiently thin, and in OTE design a compromise is usually made between resistance and transmission values (**Table 1**). Most OTEs are prepared by vapour deposition or cold sputtering of a thin film of metal such as platinum or gold or a 'doped' oxide such as tin oxide (Nesa) or indium oxide (Nesatron) on a transparent substrate such as glass, quartz or plastic. OTEs based on thin conducting films were first introduced in 1964 with the use of an antimony-doped tin oxide film on a glass sub-

strate (Nesa glass) in a transmission spectroelectro-chemical study of the electrooxidation of o-tolidine. Tin-doped indium oxide (ITO) is presently the most common conducting film used for UV-visible spectroelectrochemical studies.

The 'doped' semiconductor oxides, as used in liquid-crystal displays, are particularly attractive owing to their wide potential window (+1.2 to −0.6 V vs saturated calomel electrode (SCE) between solvent oxidation and reduction. Furthermore, the absence of surface oxidation/reduction currents (since these surfaces are already oxidized) is another advantage over platinum and gold OTEs. **Figure 2** shows typical spectra of n-type tin oxide films on various substrates. The optical absorption by the free carriers in the 'doped' tin oxide is in the infrared region, giving transparency in the visible region. Film thickness can be accurately calculated from the interference patterns that are observed in the spectra.

Minigrids Minigrid electrodes consist of electroformed wire mesh (40–800 wires cm^{-1}) of Au, Pt, Ni, Ag or Cu. The light is transmitted through the microscopic holes between the wires of the minigrid, which in operation functions as a planar electrode after electrolysis has proceeded for sufficient time that the diffusion layer depth becomes large compared to the wire and hole dimensions. The minigrid transmittance varies from 22% to 82%, depending upon the number of wires per centimetre. Since light passes through the holes in the minigrid, the optical window is essentially unlimited. The electrochemical

Table 1 Optical transmission and electrochemical data on various optically transparent electrodes (OTEs)

Type of OTE	Transmission range	Resistance (Ω sq^{-1})
Pt film (vapour deposited)	220–near IR, 10–40%	15–25
Hg–Pt film (electrode-posited Hg)	220–near IR, 10–30%	10–25
Au film (vapour deposited)	220–near IR, 10–80%	5–20
Sb-doped indium oxide (Nesa)	360–near IR on glass, 70–85%	5–20
	240–near IR on quartz, 50–85%	
Sn-doped indium oxide (Nesatron)	As Sb-doped indium oxide	5–20
Au, Hg–Ni and Hg–Au minigrids	UV–visible–IR, 22–80%	< 0.1

Reprinted with permission from Kuwana T and Heineman WR (1976) Study of electrogenerated reactants using optically transparent electrodes. *Accounts of Chemical Research* 7: 241–248 © 1976 American Chemical Society.

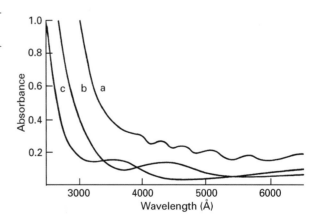

Figure 2 Transmission spectra of SnO$_2$ coatings on various substrates: (curve a) glass, 3 Ω sq^{-1}; (curve b) Vicor, 6 Ω sq^{-1}; (curve c) quartz, 20 Ω sq^{-1}. Reprinted by courtesy of Marcel Dekker, Inc. from Kuwana T and Winograd N (1974) Spectroelectrochemistry at optically transparent electrodes. I. Electrodes under semi-infinite diffusion conditions. In: Bard AJ (ed) *Electroanalytical Chemistry. A Series of Advances*, Vol 7, pp 1–78. New York: Marcel-Dekker.

properties of the gold minigrid are similar to those of the gold film OTE. The negative potential limit of gold and nickel minigrids can be extended by around 0.4 V by deposition of a thin mercury film, which has a high overpotential to hydrogen evolution.

Transmission spectroelectrochemical cells

Two types of cell geometry can be defined on the basis of the electrolyte solution thickness adjacent to the electrode (**Figure 1A** and **1B**).

Semi-infinite linear diffusion conditions The rate of an electrochemical process depends not only on electrode kinetics but also on the transport of species to/from the bulk solution. Mass transport can occur by diffusion, convection or migration. Generally, in a spectroelectrochemical experiment, conditions are chosen in which migration and convection effects are negligible. The solution of diffusion equations, that is the discovery of an equation for the calculation of oxidized form [O] and reduced form [R] concentrations as functions of distance from electrode and time, requires boundary conditions to be assumed. Usually the electrochemical cell is so large relative to the length of the diffusion path that effects at walls of the cell are not felt at the electrode. For semi-infinite linear diffusion boundary conditions, one can assume that at large distances from the electrode the concentration reaches a constant value.

Semi-infinite linear diffusion spectroelectrochemical cells In the semi-infinite linear diffusion spectroelectrochemical cell geometry (**Figure 1A**), the cell is analogous to a conventional electrochemical cell, the electrode being in contact with solution much thicker than the diffusion layer adjacent to the electrode. Semi-infinite linear diffusion spectroelectrochemical cell design requires that electrolysis products generated at the counterelectrode should not interfere with the absorbance measurement and that complete deoxygenation should be easily achieved. **Figure 3** shows the classic sandwich-cell design, with a thin film OTE as working electrode. The reference electrode and counterelectrode and the side arms for degassing are positioned so that the cell may be placed with the surface of the OTE in a horizontal plane. A Luggin capillary places the reference electrode near to the surface of the OTE for minimization of solution resistance in the control of the working electrode potential.

These cells are normally used for experiments (chronoamperometry and chronocoulometry) in which large-amplitude steps are applied in order to carry out an electrolysis in the diffusion region. For

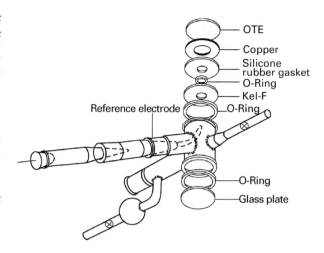

Figure 3 Sandwich cell for transmission measurement under semi-infinite linear diffusion conditions. Reprinted by courtesy of Marcel-Dekker, Inc. from Kuwana T and Winograd N (1974) Spectroelectrochemistry at optically transparent electrodes. I. Electrodes under semi-infinite diffusion conditions. In: Bard AJ (ed) *Electroanalytical Chemistry. A Series of Advances*, Vol 7, pp 1–78. New York: Marcel-Dekker.

the case of a general reduction reaction,

$$O + ne \rightleftharpoons R \qquad [1]$$

the absorbance change can be described by considering a segment of solution of thickness dx and cross-sectional area A_{elec} (**Figure 4**). If species R of molar absorption coefficient ε_R, is the only species absorbing at the monitored wavelength, then the differential absorbance upon passage of light through this segment is

$$dA = \varepsilon_R[R](x,t)\,dx \qquad [2]$$

The total absorbance is then

$$A = \varepsilon_R \int_0^\infty [R](x,t)\,dx \qquad [3]$$

If R is a stable species, the integral in Equation [3] is the total amount of R produced per unit area and is equal to Q/nFA_{elec}, where Q is the charge passed in electrolysis, n is the number of electrons and F is the Faraday constant. Since Q is given by the integrated Cottrell equation, which describes the

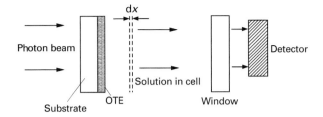

Figure 4 Schematic view of the experimental arrangement for transmission spectroelectrochemistry.

chronoamperometric response,

$$Q = \frac{2nFA_{\text{elec}}[O]D_O^{1/2}t^{1/2}}{\pi^{1/2}} \qquad [4]$$

we have

$$A = \frac{2\varepsilon_R[O]D_O^{1/2}t^{1/2}}{\pi^{1/2}} \qquad [5]$$

which shows that the absorbance should be linear with $t^{1/2}$. Analysis of the slopes of A vs $t^{1/2}$ plots are useful in mechanistic studies where coupled homogeneous reactions follow the initial electrode reaction.

Optically transparent thin-layer electrochemical cells The optically transparent thin-layer electrode (OTTLE) cell, first reported in 1967, consists of a sandwich structure with a minigrid OTE working electrode (**Figure 5**). The assembly is placed in a cup of solution containing both the counterelectrode and reference electrode and is filled either by capillary action or by applying nitrogen pressure to give a thin (<0.2 mm) solution layer confined next to the OTE. These cells can easily be constructed in the laboratory using ordinary microscopy slides, 100 μm Teflon adhesive tape spacers, a minigrid and epoxy resin.

A large ohmic potential drop is often present in OTTLE cells owing to the nonuniform current distribution within the thin-layer cavity caused by the large distance between the working electrode and counterelectrode. This is not a problem, as experiments generally involve exhaustive electrolyses, where any ohmic drop can be 'out-waited'. The OTTLE cell design enables the techniques of thin-layer electrochemistry, cyclic voltammetry, controlled potential coulometry and UV-visible spectros-

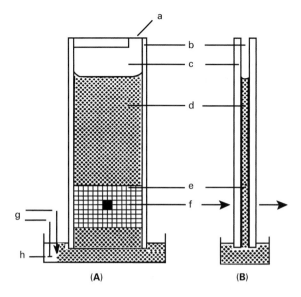

Figure 5 Optically transparent thin-layer electrode (OTTLE) cell: (A) front view; (B) side view. (a), Point of suction application to change solution; (b) Teflon tape spacers; (c) microscope slides (1 × 3 in.); (d) solution; (e) transparent gold minigrid electrode; (f) optical path of spectrometer; (g) reference and counter electrodes; (h) solution cup. Epoxy resin holds the cell together. Reprinted with permission from DeAngelis TP and Heineman WR (1976) *Journal of Chemical Education* **53**: 594–597. © 1976 American Chemical Society.

copy to be performed in one unified experiment using a small quantity of solution.

Spectropotentiostatic measurements with an OTTLE cell Determination of E^0, the reversible electrode potential, and n, the number of electrons in a redox reactions, can be performed in a sequence of spectropotentiostatic measurements with an OTTLE cell. Generally, for the reversible system given in Equation [1], the [O]/[R] ratio at the electrode surface is controlled by the applied potential according to the Nernst equation:

$$E_{\text{applied}} = E^0 + \frac{RT}{nF}\ln\left(\frac{[O]}{[R]}\right)_{\text{surface}} \qquad [6]$$

In an OTTLE cell, on application of a new potential, the concentration of O and R in solution are quickly adjusted to the same values as those existing at the electrode surface, giving at equilibrium:

$$\left(\frac{[O]}{[R]}\right)_{\text{surface}} = \left(\frac{[O]}{[R]}\right)_{\text{solution}} \qquad [7]$$

The Nernst equation is a thin-layer cell can then be generally expressed as

$$E_{\text{applied}} = E^0 + \frac{RT}{nF}\ln\left(\frac{[O]}{[R]}\right) \qquad [8]$$

In order to obtain E^0 and n values, a series of potentials are sequentially applied to the thin-layer cell containing a test solution. The redox couple is incrementally converted from one oxidation state into another by the applied potentials, resulting in different ratios of $[O]/[R]$ that can be determined spectrally. Each applied potential is maintained until electrolysis ceases, so that the equilibrium value of $[O]/[R]$ is established as defined by the Nernst equation. The E^0 and n values can then be obtained by a Nernst plot made by the values of E_{applied} and the corresponding ratios of $[O]/[R]$. In practice, the range of applied potentials is selected to span E^0 of the redox couple, so that spectra for complete reduction/oxidation and intermediate values of $[O]/[R]$ can be obtained. Selecting a wavelength (usually the maximum wavelength) of O as the monitoring wavelength, for example, the ratio $[O]/[R]$ determined by recording the *in situ* absorbance changes at a certain applied potential is produced using the Beer–Lambert law,

$$\frac{[O]}{[R]} = \frac{(A_i - A_R)/\Delta\varepsilon b}{(A_O - A_i)/\Delta\varepsilon b} = \frac{(A_i - A_R)}{(A_O - A_i)} \qquad [9]$$

where A_R is the absorbance of the reduced from, A_O is the absorbance of the oxidized form, A_i is the absorbance obtained at an intermediate applied potential, $\Delta\varepsilon$ is the difference in molar absorptivity between O and R at the selected wavelength and b is the light path length in the thin-layer cell. Substituting Equation [9] into Equation [8], the Nernst equation expressed by absorbance is obtained:

$$E_{\text{applied}} = E^0 + \frac{RT}{nF}\ln\frac{(A_i - A_R)}{(A_O - A_i)} \qquad [10]$$

The intercept and slope of the straight line of E_{applied} vs $\ln(A_i - A_R)/(A_O - A_i)$ then give E^0 and n, respectively. Values of E^0 and n for numerous reversible redox couples in aqueous, nonaqueous and molten salt solvents have been determined from Nernst plots of such spectropotentiostatic experiments.

Spectroelectrochemical cells for modified electrode studies

Immobilization of chemical microstructures onto electrode surfaces has been a major growth area in electrochemistry in recent years. In their characterization using *in situ* UV-visible absorption spectroelectrochemical measurements, a modified electrode can be considered to be analogous to an OTTLE, the redox active layer being physically or chemically confined to the electrode surface. Spectroelectrochemical cell design is often simple, a rectangle of OTE being mounted transverse to optical beam direction in a conventional 1 cm cuvette. The counterelectrode, placed opposite the working electrode, is typically a loop of platinum wire through which the light beam can pass. A machined polytetraethylene lid with appropriate holes is used to hold the Luggin capillary from the reference electrode above the light path and the working electrode and counterelectrode in place.

Internal reflection spectroelectrochemical cells

A typical cell configuration is shown **Figure 6**. This cell has most of the electrode area masked and

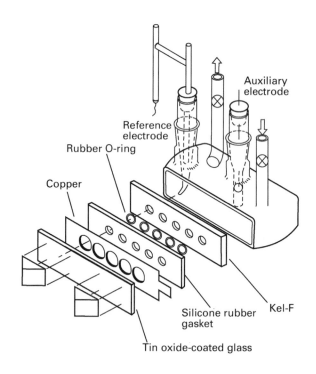

Figure 6 Internal reflection spectroscopy (IRS) cell and various attachments. Reprinted by courtesy of Marcel Dekker, Inc. from Kuwana T and Winograd N (1974) Spectroelectrochemistry at optically transparent electrodes. I. Electrodes under semi-infinite diffusion conditions. In: Bard AJ (ed) *Electroanalytical Chemistry. A Series of Advances*, Vol 7, pp 1–78. New York: Marcel-Dekker.

exposes only the region where the light beam is incident to the electrode–solution interface. The cell shown allows for five reflections, but the number can be selected for the requirements of the experiment.

Electrochemical control and techniques

Modern electrochemical research encompasses a wide range of techniques, including those based on potential sweep, potential or current step, use of hydrodynamic electrodes, impedance measurements and electrolysis. Electrode processes are investigated at the working electrode, under potentiostatic or galvanostatic control, with a counterelectrode being used to complete the electrical circuit. Historically, in controlled potential experiments, the counterelectrode was also the reference electrode, with the double function of passing current and acting as a reference potential for controlling the potential of the working electrode. Nowadays, three-electrode systems are routinely used where the current passes from the working electrode to a counterelectrode (of larger area than the working electrode), the separate reference electrode serving purely as a reference potential and not passing current. Voltammetric, step and coulometric methods are the most frequently used electrochemical techniques in UV-visible spectroelectrochemistry.

Voltammetric methods

Electrochemical techniques in which a potential is imposed upon an electrochemical cell and the resulting current is measured are termed voltammetric methods. Numerous methods have been developed, with variation in the type of potential waveform impressed on the cell, the type of electrode used and the state of the solution in the cell (quiescent or flowing). Voltammetry has proved to be very useful for analysing dilute solutions, both quantitatively and qualitatively, for inorganic, organic and biological components, measuring thermodynamic parameters for redox systems, and studying the kinetics of coupled homogeneous chemical reactions. In spectroelectrochemistry, linear-sweep and cyclic voltammetric methods are generally used in quiescent solutions. In linear-sweep voltammetry, for the general redox reaction in Equation [1] above, the potential of the working electrode is swept from a value E_1, at which O cannot undergo reduction, to a potential E_2, at which the electron transfer is driven rapidly. In cyclic voltammetry, a triangular waveform is applied; once the potential reaches E_2, the direction of the sweep is reversed and the electrode potential is scanned back to E_1.

Potentiostatic control The heart of modern electrochemical instrumentation is the potentiostat, which has control of the voltage across the working electrode–counterelectrode pair; it adjusts this voltage in order to maintain the potential difference between the working and reference electrodes (which it senses through a high-impedance feedback loop) in accord with the programme supplied by a function generator. The instrumentation requirements for thin-layer spectroelectrochemistry are not as exacting or expensive as some other types of spectroelectrochemistry. Since the large ohmic potential drop precludes very fast measurements, relatively inexpensive (slow) potentiostats with a good digital voltmeter are usually adequate.

Step techniques

In step techniques, a potential or current step is instantaneously applied to the working electrode. Following this perturbation to the electrochemical system, current, charge or potential is monitored versus time.

Chronoamperometry Chronoamperometry involves the study of the variation of the current response with time under potentiostatic control. Generally the working electrode is stepped from a potential at which there is no electrode reaction to one corresponding to the mass-transport-limited current, and the resulting current–time transient is recorded. In double-step chronoamperometry, a second step inverts the electrode reaction and this method is useful in analysing cases where the product of the initial electrode reaction is consumed in solution by a coupled homogeneous chemical reaction.

Chronocoulometry Chronocoulometry is similar to chronoamperometry except that the current is integrated and the variation of charge with time is studied. The advantages of integration are that the signal increases with time, facilitating measurements towards the end of the transient, when the current is almost zero; integration is effective in reducing signal noise; it is relatively easy to separate the capacitive charge from the faradaic charge.

In an OTTLE cell, coulometry is generally performed by application of a potential that causes complete electrolysis of the electroactive species. Electronic integration of the resulting current gives the total charge consumed by the electrode process, which can be related to the number of moles and electrons involved in the redox reaction by Faraday's law. It is important to carry out a second experiment in the absence of the electroactive species, to allow subtraction of the 'blank' charge for charging of the electrode/electrolyte interface and any background redox reactions involving the solvent and electrode.

Chronopotentiometry Generally constant-current chronopotentiometry is employed, in which the constant current applied to the working electrode causes the electroactive species to be reduced at a constant rate. The potential of the electrode moves to values characteristic of the redox couple and varies with time as the [O]/[R] concentration ratio changes at the electrode surface. In the case of a reduction, after the concentration of O drops to zero at the electrode surface, the flux of O becomes insufficient to accept all the electrons being forced across the electrode–electrolyte interface. The potential, at this transition time, then rapidly changes towards more negative values until a new, second reduction can start.

Spectroscopic measurements

UV-visible spectral measurements under electrochemical control can often be made using a conventional spectrometer. For thin-layer spectroelectrochemistry experiments, a sufficiently large sample spectrometer compartment is required to accommodate the OTTLE cell.

Rapid scan and diode array spectrometers

For kinetic measurements, analysis of complete time-resolved spectra is possible using a rapid-scan spectrometer (RSS) interfaced with a microcomputer. With a RSS it is possible to record a 1000-point 450 nm wide spectrum in the range 240–800 nm in about 5 ms, although signal averaging is generally necessary to obtain the required sensitivity. Although RSS instruments have been employed extensively and with great success, the instrumental design is now obsolete and usually instruments are used that employ diode arrays in combination with a polychromator. In these 'optical multichannel analysers', the whole spectrum is spatially dispersed by a polychromator and then imaged onto the detector, which consists of an array of tiny photodiodes.

Reflectance spectroscopy measurements

A schematic diagram of the typical apparatus required for reflectance spectroscopy is given in **Figure 7**. The optical components consist of a highly stabilized intense light source, frequently a mercury or mercury/xenon arc, a monochromator, a

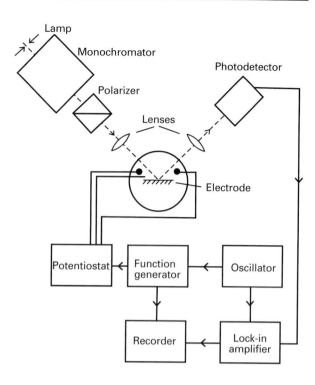

Figure 7 Block diagram of typical apparatus for reflectance spectroscopy Reprinted with permission from Robinson J (1984) Spectroelectrochemistry. *Electrochemistry Specialist Periodical Reports*. Vol 9, pp 101–161. London: Royal Society of Chemistry.

polarizer, the electrochemical system, a photodetector (photomultiplier or photodiode) and appropriate collimating and focusing lenses. The electrode potential is periodically modulated by either a square or a sinusoidal waveform and the small changes in reflectivity so caused are detected with a lock-in amplifier. For kinetic measurements where the shape of the reflectance–time transient is required, the lock-in amplifier is replaced with a signal averager. The requirements for the cell and electrode design are identical to those of ellipsometry.

List of symbols

A = absorbance; A_{elec} = electrode area; b = path length; D = diffusion coefficient; e = electron; E = electrode potential; E^0 = reversible electrode potential; F = Faraday constant (96 485 C mol^{-1}); n = number of electrons; O = oxidized form; Q = charge; R = gas constant (8.315 J K^{-1} mol^{-1}); R = reduced form; t = time; T = temperature in kelvin; x = distance from electrode.

See also: **Colorimetry, Theory; Dyes and Indicators, Use of UV-Visible Absorption Spectroscopy; Ellipsometry; Light Sources and Optics; Spectroelectrochemistry, Applications.**

Further reading

Bard AJ and Faulkner LR (1980) *Electrochemical Methods: Fundamental and Applications*, Chapter 14. New York: Wiley.

Fisher AC (1996) *Electrode Dynamics*. Oxford: Oxford University Press.

Heineman WR and Jensen WB (1989) Spectroelectrochemistry using transparent electrodes –an anecdotal history of the early years. *ACS Symposium Series* **390**: 442–457.

Heineman WR and Kissinger PT (1984). In: Kissinger PT and Heineman WR (eds) *Laboratory Techniques in Electroanalytical Chemistry*, Chapter 3. New York: Marcel Dekker.

Heineman WR, Hawkridge FM and Blount HN (1984) Spectroelectrochemistry at optically transparent electrodes. II. Electrodes under thin-layer and semi-infinite diffusion conditions and indirect coulometric titrations. In: Bard AJ (ed) *Electroanalytical Chemistry. A Series of Advances*, Vol 13, pp 1–113. New York: Marcel Dekker.

Kolb DM (1988) UV-visible reflectance spectroscopy. In: Gale RJ (ed) *Spectroelectrochemistry*, Chapter 4. New York: Plenum Press.

Kuwana T and Heineman WR (1976) Study of electrogenerated reactants using optically transparent electrodes. *Accounts of Chemical Research* **7**: 241–248.

Kuwana T and Winograd N (1974) Spectroelectrochemistry at optically transparent electrodes. I. Electrodes under semi-infinite diffusion conditions. In: Bard AJ (ed) *Electroanalytical Chemistry. A Series of Advances*, Vol 7, pp 1–78. New York: Marcel Dekker.

Mortimer RJ (1994) Dynamic processes in polymer modified electrodes. In: Compton RG and Hancock G (eds) *Research in Chemical Kinetics*, Vol 2, pp 261–311. Amsterdam: Elsevier.

Murray RW (1984) Chemically modified electrodes. In: Bard AJ (ed) *Electroanalytical Chemistry. A Series of Advances*, Vol 13, pp 191–368. New York: Marcel-Dekker.

Niu J and Dong S (1996) Transmission spectroelectrochemistry. *Reviews in Analytical Chemistry* **15**: 1–171.

Robinson J (1984) Spectroelectrochemistry. *Electrochemistry Specialist Periodical Reports*, Vol 9, pp 101–161. London: Royal Society of Chemistry.

Sawyer DT, Sobkowiak A and Roberts JL Jr (1995) *Experimental Electrochemistry for Chemists*, 2nd edn, pp 284–286. New York: Wiley.

Spectroscopy of Ions

John P Maier, University of Basel, Switzerland

MASS SPECTROMETRY
Methods & Instrumentation

Ions and radicals are transient species which are not readily accessible to conventional techniques for spectroscopic characterization. There are essentially three problems to be overcome–the production in sufficient concentration, the availability of a sensitive technique enabling their IR or electronic spectra to be recorded and the ability to identify the observed spectral features. The involvement of mass-selection not only leads to the solution of the last problem, but enables methods based on particle detection – fragment ions, electrons and photons – to be incorporated. The aim of the spectroscopic studies is, on the one hand, to provide a fingerprint of the species by its vibrational or electronic spectrum, enabling its identification in various terrestrial and space environments, and on the other hand, the spectroscopic analysis leads to information on geometric structures, force fields and fundamental interactions.

Ions are known to be important entities in space, for example in comets and interstellar clouds and their electromagnetic spectrum is the means not only to identify them but also to gain physicochemical data on their surroundings. The same applies in the nonintrusive monitoring in the laboratory, such as of plasmas and flames. It is the purpose of this article to give some examples of the studies on mass-selected species, both of their electronic and IR spectra. This has been achieved in the 1990s in experiments measuring the electronic absorption spectra of mass-selected cations, anions and neutral radicals in an inert neon matrix at ≈ 5 K and more recently in the gas phase for anions following electron photo-detachment. IR spectroscopy of ionic complexes, prototypes of fundamental interactions in chemical and biochemical phenomena, and intermediates of ion–molecule reactions, can also be achieved with a mass-selected technique involving vibrational predis-sociation spectroscopy in an ion trap.

Absorption spectroscopy in neon matrices

Technique

This method combines mass spectrometry and matrix isolation. The study of transients trapped in the neutral environment of a rare gas is an established spectroscopic approach. It suffers, however, from selectivity; though sufficient concentrations of elusive radicals and ions can be obtained in rare gas matrices, spectral overlap as a result of the simultaneous presence of many such species usually restricts the interpretation. This is overcome by growing matrices with mass-selected ion beams.

Figure 1 shows the essential features of the instrument developed. A number of ion sources have been used, hot-cathode discharge for anions and cations as well as a cesium sputter one for anions. These have been designed to generate copious amounts of specific ions. The extracted ions are passed through an electrostatic deflector into a quadrupole mass spectrometer before codeposition with excess neon to form a matrix at ≈ 5 K. Several stages of differential pumping are incorporated in the instrument to isolate the ion source from the cryosurface. Mass-selected ion currents in the nA range are usually required for a successful measurement and the kinetic energies of the ions are chosen to be in the 50–150 eV range. The deposition takes about 2 h resulting in a thin neon matrix (≈ 150 μm) over about 1 cm^2 area. A typical ion density is 10^{15} to 10^{16} cm^{-3}. The neutrality in the matrix is assured by the presence of counter-ions generated *in situ* either from impurities present or via the species deposited themselves.

The absorption spectrum of the mass-selected molecules is measured by a technique which enables the

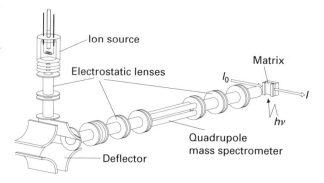

Figure 1 Apparatus used for the measurement of absorption spectra of mass-selected cations, anions and neutral radicals in neon matrices.

whole length of the thin matrix to be interrogated. The species and neon are condensed on a copper substrate and the light is passed through slits into the side of the matrix and traverses it parallel to the metal surface. By this means, absorption path lengths of 1–2 cm are achieved. For measurements in the IR, only a reflection configuration can be used, leading to two orders of magnitude lower sensitivity.

A major aim of this technique is to identify and locate the characteristic electronic transition of carbon chain-like species. This is for two reasons: (1) to be able to consider their relevance in astrophysical phenomena in view of their spectral features and (2) to plan gas-phase experiments with this knowledge.

Electronic spectra of cations

As an example, the polyacetylene cations HC_nH^+ are considered. They are readily produced in a hot-cathode discharge source fed with a mixture of acetylene diluted with helium. Ion–molecule reactions lead to polymerization. When a particular species is selected according to the number of carbon atoms in the chain, with currents in the nA range, and codeposited with excess neon, the electronic absorption spectra can be measured. **Figure 2** shows the characteristic strong transitions for the species

with an even number of carbon atoms. The polyacetylene cations have an open-shell electron configuration with $X\ ^2\Pi$ ground states and the band systems apparent correspond to $\pi*-\pi$ electron excitation, i.e. to electronic transitions of $^2\Pi-X\ ^2\Pi$ symmetry.

In the spectra observed, the first band at lower energy is the most intense and the origin transition. The peaks lying to shorter wavelength involve the excitation of vibrational modes in the upper electronic state. Under the conditions of the experiment, rotational motion is eliminated because the species are held rigid in the surrounding neon lattice. The low ambient temperature of 5 K constrains the population to the vibrations $v = 0$ level in the ground electronic state from which all transitions then originate. In addition, the geometry change on π-electron excitation is relatively small; hence the origin bands dominate and the relative simplicity of the spectra (**Figure 2**). The vibrational pattern and the types of normal modes excited indicates that these are transitions of linear (or quasilinear) polyacetylene chains.

The electronic transition shifts towards the IR by regular increments (≈ 100 nm) for each acetylene unit. This is a characteristic feature of carbon chain molecules and can be simply modelled by an electron

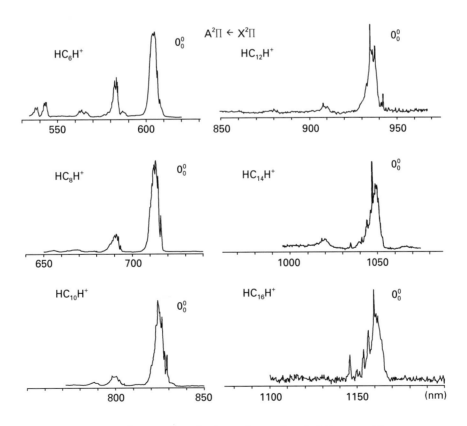

Figure 2 Electronic absorption spectra of mass-selected polyacetylene cations in 5 K neon matrices.

in a box treatment. In addition to the inverse energy dependence on the length of the chain, the oscillator strength of these transitions grows. This is a contributing factor for the successful detection of the longer species in the laboratory even though the attainable ion current of the mass-selected species is decreasing. It is also one of the attractive features for consideration of such species in astrophysical phenomena. Finally, with the location of the transition in a neon matrix, the corresponding measurements in the gas phase have become a realistic proposition. Typical shifts for the electronic transitions of the cations (**Figure 2**) should be in the 100–200 cm^{-1} range to the blue on passing from the neon matrix to the gas phase.

Electronic spectra of anions

In an analogous way to the measurement of the electronic absorption spectra of cations, those of anions can be obtained. A sputter source has been used to generate the pure carbon anions. In this, a graphite rod is bombarded with cesium ions. The carbon species are formed by sputtering and gas phase processes and the ions produced are extracted for mass-selection. Sufficient ion concentrations have been attained for the spectroscopic studies of carbon anions in the C_4^- to C_{20}^- range. The carbon species are unusual among anions in that they have large

electron detachment energies (3–5 eV) and possess one or more bound excited electronic states. This is illustrated in **Figure 3** by the observed electronic spectra of the C_{2n+1}^- anions detected after mass-selected deposition.

Up to four band systems are discernible (for $n = 4$, 5) and these are the various $^2\Pi - X\ ^2\Pi$ transitions arising by $\pi - \pi$ electron excitation. The arrow placed on the wavelength scale (**Figure 3**) indicates the electron detachment threshold in the gas phase. Thus, the highest lying excited electronic state is stabilized with respect to the isolated state as a result of solvation by the neon atoms. The vibrational structure is again relatively simple, indicating a chain structure for the carbon skeleton. In addition to the frequencies of the fundamentals, which can be inferred from the electronic spectra—these are mainly the totally symmetric stretching modes and some bending ones excited in double quanta—antisymmetric, IR active modes have also been observed with this approach. This followed the recording of the infrared spectrum after the mass-selected deposition by a reflection arrangement.

Electronic spectra of mass-selected neutral radicals

The technique has been extended to the spectroscopic study of mass-selected neutral species. In the case of

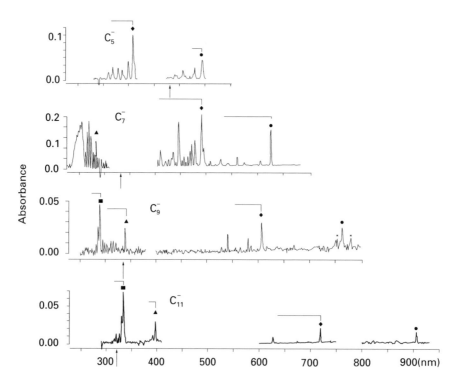

Figure 3 Absorption spectra of mass-selected carbon anions in neon matrices at 5 K showing several $^2\Pi - X\ ^2\Pi$ electronic transitions.

the carbon molecules this is best accomplished by selection of the corresponding anions and subsequent detachment of the electron either during or after growth of the matrix using a broad band photon source. By this means the long sought electronic spectra of the linear carbon chains C_{2n+1} ($n = 2$–7) could be identified. Some of these are shown in **Figure 4**. The electronic transition corresponds to $\pi* - \pi$ excitation. The carbon chains with odd numbers of atoms are closed–shell species and the symmetry of the observed band systems is $^1\Sigma_u^+$–$X\ ^1\Sigma_g^+$. In contrast, the chains with even numbers of carbon atoms are paramagnetic with a $X\ ^3\Sigma_g^-$ ground state. Their transitions have been detected for C_{2n} ($n = 2$–5) by this method. The band systems of the linear forms of the larger species are not apparent in the spectra, providing direct experimental evidence for a change of geometrical structure above C_{10}. In contrast, the linear chains C_{2n+1} persist beyond $n = 8$. Thus, the potential of the technique of combining mass and matrix-isolation spectroscopies for the characterization of neutral species in addition to the ionic ones is clear.

Electronic spectra of anions in the gas-phase

The electronic spectra of carbon chain anions have now also been measured in the gas phase incorporating a mass-spectrometric technique. Because the energy region of the electronic transition is known from the measurements on the mass-selected species in neon matrices (see above), the gas-phase study is made that much easier. The sensitive approach adopted involves selection of the respective anion, laser excitation and photodetachment. A schematic outline of the apparatus is given in **Figure 5**.

The anions are conveniently prepared in a pulsed d.c. discharge source containing a few per cent of acetylene in argon expanded in a supersonic free jet with a backing pressure of about 9 bar. The argon carrier gas ensures clustering and cooling of the formed anions. The beam is passed through a skimmer into a time-of-flight mass spectrometer. The ion of interest, e.g. C_7^-, is identified by its flight time after which the excitation and photodetachment takes place. In the case of C_7^-, the A $^2\Pi$ – X $^2\Pi$ transition (cf. **Figure 3**) is sought. A tunable laser is scanned in this wavelength region while a second laser photon of fixed energy (532 nm) causes the photodetachment whenever the first photon is in resonance with the electronic transition. The total energy of the two photons absorbed exceeds the electron affinity of C_7.

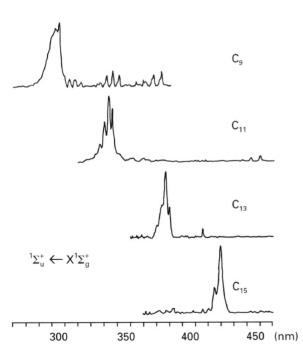

Figure 4 Electronic absorption spectra ($^1\Sigma_u^+ - X^1\Sigma_g^+$) of neutral carbon chains observed after mass-selected codeposition of their anions with neon at 5 K and detachment of the electrons by photon illumination.

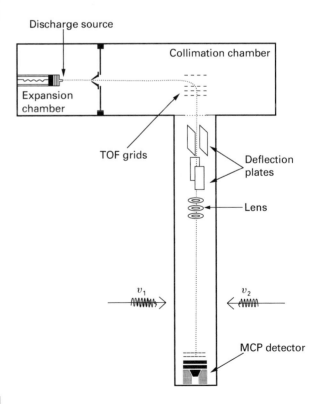

Figure 5 Experimental arrangement used to observe the electronic transitions of mass-selected carbon anions in the gas-phase.

C_7^- A $^2\Pi_u$–X $^2\Pi_g$

Figure 6 The A $^2\Pi_u$ – X $^2\Pi_g$ electronic transition of C_7^- measured in the gas phase using a two-colour photon excitation and electron detachment approach.

The detection is usually of the neutral mass-selected species (C_7).

In **Figure 6** is seen the recorded spectrum of the A $^2\Pi$ – X $^2\Pi$ transition of C_7^-. The striking feature is the narrow line width of the bands, attributed primarily due to the production of cold anions (20–40 K) in the supersonic discharge source. The analysis of the vibrational structure is straightforward – it is consistent with a transition of a molecule linear in both the ground and excited electronic state. The excitation of the three totally symmetric stretching modes in progressions and as combinations in the upper electronic state is apparent. The absence of bands corresponding to bending modes is associated with a relatively rigid structure of C_7^-. Similar measurements in the gas phase have now been realized for carbon anions in the C_5^- to C_{11}^- range. Such data then allow a direct comparison with astronomical observations, for example, on the diffuse interstellar bands, a long-standing puzzle.

Infrared spectra of ionic complexes

Approach

The aim is to measure IR spectra in the gas phase of ionic complexes. As illustration, the H_2–HCO^+ species is considered. Such ionic complexes are intermediates in ion–molecule reactions and have been detected by mass spectrometry in the earth's stratosphere. The forces involved in their binding are ion(induced) dipole, and are involved in biological phenomena. The binding energies are generally intermediate between those of van der Waals species and

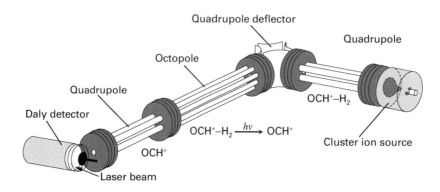

Figure 7 The setup of the instrument used to measure infrared spectra of ionic complexes via vibrational predissociation spectroscopy.

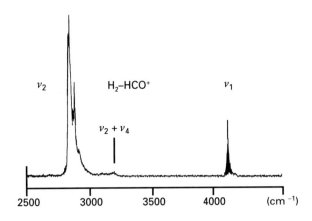

Figure 8 Infrared spectrum of the mass-selected H_2–HCO^+ ionic complex recorded with the apparatus shown in **Figure 7**.

those involving hydrogen bonds. In the case of H_2–HCO^+, the binding energy is ≈ 1400 cm^{-1}, whereas it is merely 150 cm^{-1} for He-HCO^+, another of the species studied.

There are three concepts in the experiment, the schematic outline of which is given in **Figure 7**. The first is the production and selection of the complexes. This can be achieved only at low temperatures and consequently a supersonic expansion combined with electron impact ionization has been used. To produce H_2–HCO^+, a 15:1 mixture of H_2/CO at a total pressure of 4–5 bar is passed through a pulsed orifice and the complex is formed in the expansion reaction where 70 eV electron impact produces the ions. The latter are extracted via a skimmer into a quadrupole mass spectrometer and the chosen ion is injected into an octopole, and confined for the desired time. The

second part involves the excitation of a vibrational transition of the ionic complex. Radiation from a tunable infrared laser passes down the middle of the octopole and intercepts the species. The photon energy is chosen so that it exceeds the binding energy of the ionic complex. When the infrared transition is excited, in due course the complex dissociates. Because the vibrational predissociation is slow for such ionic complexes – the transfer of energy from the mode excited is inefficient, e.g. ν_1(H–H) in H_2–HCO^+, into the fragmentation channel (HCO^+ + H_2) – there is usually no spectral broadening.

The fact that fragment ions are produced leads to the third underlying principle of the experiment, namely the sensitive detection of the absorption process by counting the resulting fragment ions. Thus, in the octopole both the ionic complex (H_2–HCO^+) as well as the product ion (HCO^+) are constrained, but the final quadrupole (**Figure 7**) is tuned to transmit only the fragment ion. The IR spectrum of an ionic complex is observed by monitoring the intensity of the fragment ions as a function of the laser frequency.

Vibrational spectrum of H_2–HCO^+

Figure 8 shows an IR spectrum for the complex in the 2500–4500 cm^{-1} region. The two strong bands are the ν_1 (H–H stretch) and ν_2 (C–H stretch) fundamentals as well as combination bands involving the bending levels (in the spectrum only the $\nu_2 + \nu_4$ band is indicated—others lie adjacent to the ν_2 peak). Both ν_1 and ν_2 frequencies are lower than the values for the isolated units indicating a weakening of the H–H

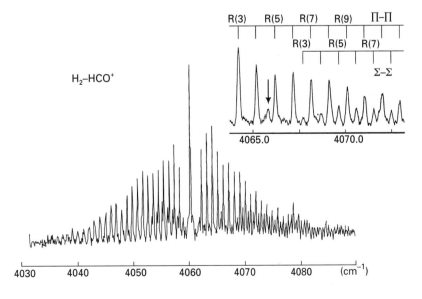

Figure 9 Rotationally resolved ν_1 band (H–H stretch) of the H_2–HCO^+ ionic complex.

and H–C bonds in the ionic complex. Thus it can be seen that this approach enables vibrational spectra of mass-selected ionic complexes to be recorded, like the spectra of stable molecules, although the concentrations of the ionic species are minute in comparison.

When the resolution is increased (0.02 cm^{-1}), rotational structure on some of the bands becomes apparent. This is particularly striking for the ν_1 band (4060 cm^{-1}) of the H–H stretching motion (**Figure 9**). The rotational structure, with Σ–Σ and Π–Π components, is consistent with a T-shaped semirigid symmetric top. Assuming that the HCO$^+$ unit has similar bond distances as in the free species, then the rotational constant evaluated from the analysis of the rotational structure implies a H$_2$ to HCO$^+$ distance of ≈ 175 pm.

Similar studies have now been accomplished on a number of ionic complexes ranging from R–HCO$^+$, R–N$_2$H$^+$ to species such as R–NH$_4^+$, with R = He, Ne, Ar and H$_2$. This leads to an insight and understanding of the interactions occurring between neutral species and ions as well as of their geometrical structure.

Solvation of ionic cores

As the measurement of the IR spectra is based on mass-selected species, the number of ligands surrounding the ionic core can readily be varied. This is illustrated in **Figure 10** for the case of the HCO$^+$ surrounded by an increasing number of argon atoms where the changes of the ν_1 transition energy are followed. The IR spectrum can be recorded for each mass-selected entity by monitoring the dominant photofragmentation product ion. The characteristic shifts and patterns in the IR spectrum lead to models of solvation and representation of structural changes.

The addition of the first argon atom results in a shift of 274 cm^{-1} in the ν_1 frequency (H–C stretch) relative to that in the free HCO$^+$ ion. This large change indicates that the argon atom is bound at the hydrogen end. The analysis of the rotational structure points to a linear, rigid geometry. The argon atom is located about 213 pm from the ion core. As further argon atoms are added, the ν_1 frequency shifts in the opposite direction, towards higher energy, in a fairly systematic fashion; by the time 12 argon atoms are solvating the ionic core, the red shift relative to the free HCO$^+$ fundamental is back to 156 cm^{-1}. As can be seen from **Figure 10**, the incremental shifts are largest for 2–5 argon atoms, then there occurs a small red shift, while in the range 6–12 atoms the blue shift is rather small. When 13

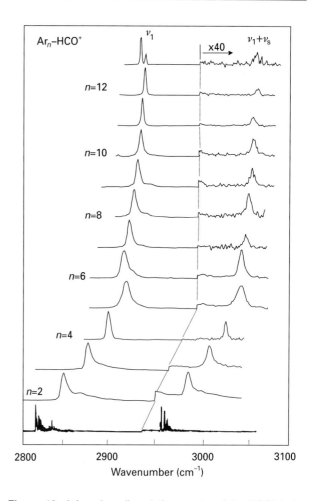

Figure 10 Infrared predissociation spectra of the HCO$^+$ ionic core solvated by specific number of argon atoms.

argon atoms are attached, the ν_1 band splits into two components.

These spectroscopic signatures can be rationalized in terms of a simple structural model. The first argon atom occupies a linear proton-bound position. The next 2–5 atoms form the first primary solvation ring at one end. The addition of 5–11 argon atoms forms the second ring at the other end. The marked splitting with 13 atoms, as well as a distinct drop in the binding energy, is associated with the beginning of a second solvation shell and the presence of at least two isomers.

See also: **Cluster Ions Measured Using Mass Spectrometry; Ion Dissociation Kinetics, Mass Spectrometry; Ion Energetics in Mass Spectrometry; Ion Molecule Reactions in Mass Spectrometry; Ion Structures in Mass Spectrometry; Multiphoton Excitation in Mass Spectrometry; Photoionization and Photodissociation Methods in Mass Spectrometry.**

Further reading

Bieske EJ and Maier JP (1993). Spectroscopic studies of ionic complexes and clusters. *Chemical Reviews* 93: 2603–2621.

Herbst E (1990) The chemistry of interstellar space. *Angewandte Chemie, International Edition* (English) 29: 595–608.

Kroto HW (1981) The spectra of interstellar molecules. *International Reviews in Physical Chemistry* 1: 309–376.

Maier JP (ed) (1989) *Ion and Cluster Ion Spectroscopy and Structure*. Amsterdam, Elsevier.

Maier JP (1992) Electronic spectroscopy – ions in space and on earth. *Chemistry in Britain* 437–440.

Maier JP (1997) Electronic spectroscopy of carbon chains. *Chemical Society Reviews* 26: 21–28.

Weltner W and Van Zee RJ (1989) Carbon molecules, ions and clusters. *Chemical Reviews* 89: 1713–1747.

Spin Trapping and Spin Labelling Studied Using EPR Spectroscopy

Carmen M Arroyo, US Army Medical Research Institute of Chemical Defense, Aberdeen Proving Ground, MD, USA

MAGNETIC RESONANCE

Applications

The EPR/spin label technique

Spin labelling is used to monitor physical, biophysical or biochemical properties of substances, proteins, lipids, and cell membranes. This is achieved by introducing a persistent free radical (such as a nitroxide) into the system, which is sensitive to this physical surrounding. Such spin probes are ideally suited to investigate the dynamic aspects of molecular interactions. A spin probe study should proceed in two steps. The first is an analysis of the EPR spectra, yielding physical parameters about the orientation and the motion of the spin probe in the host system. The second step is the interpretation of these parameters in terms of relevant molecular models for the formulation. The section on spin labels will use as an illustrative example the results obtained when various perfluorinated polyether (PEPE) materials containing reactive antivesicant agents were labelled with stearic acid spin probes. These spin-labelled ointment formulations were exposed to different concentrations of vesicant (blistering) agents. The attachment of the spin label at a particular site on an ointment formulation provides a unique means of probing the physical environment in the formulation. If the probe molecule is placed at or near the reactive component of the formulation, then the magnetic resonance experiments may be used to determine the penetration profile. However, there have been many examples of spin labelling in other fields, including liquid crystals and cell membranes.

EPR/spin label technique for antivesicant agents

Initial screening was performed using the electron paramagnetic resonance (EPR)/spin-label technique by inserting stable organic nitroxide free radicals, known as spin labels, into a heterogeneous complex system consisting of polytetrafluoroethylene (PTFE) particles and fluorinated oils of the proposed formulation. The EPR spectra of the spin labels are very sensitive to the rate at which the label is able to reorient after a magnetic field (B_o) is applied. Thus, knowledge of this functional dependence allows the evaluation of the degree of mobility and permeability permitted in the environment of the label. Ointment formulations of reactive topical skin protectants (rTSPs) or topical skin protectants (TSPs) based on PFPE (i.e. fomblin®) were prepared and spin labelled. Four N-oxyl-4-4'-dimethyloxazolidine derivatives of stearic acid, 5-NS, 7-NS, 12-NS and 16-NS, were used as spin probes. The spin-labelled vehicle, fomblin® and the vehicle containing chloroamide [chlorinated glycoluril; 1,3,4,6-tetrachloro-7,8-diphenyl-2,5-diiminoglycoluril (S-330), an antivesicant] were exposed to various concentrations of 2-chloroethyl ethyl sulfide (half-sulfur mustard). The physical insight into the molecular motion is defined by the order parameter (S) which depends on the frequency and amplitude. The EPR experiments detect only the motion and orientation of the –N•–O group, so the order parameter is related to the x, y, z coordinates. Therefore, S depends on the depth of penetration of the paramagnetic group into the vehicle (fomblin®) and on the chemical composition of the reactive antivesicant under investigation. The net change of the viscosity of the vehicle and the chemical composition were seen to affect the penetration profile.

The value of S obtained from the EPR data provides an analytical tool for comparison of the formulation of topical skin protectants and reactive topical skin protectants. The changes in the interior of this heterogeneous system of the labelled-TSP or labelled-rTSP (controls) were compared with exposed labelled-TSP/or rTSP. The spin-labelled formulations were exposed to the vesicant agents in a dose/time-dependent manner. The results show that the EPR/spin-labelling technique provides an analytical tool for determining the resistance of rTSP to the breakthrough and fluxes of vesicant agents.

The formulated candidates were labelled with stearic acid spin probes. Four kinds of N-oxyl-4,4′-dimethyloxazolidine derivatives of stearic acid (**Figure 1**) were used as spin probes. The nitroxide group (–N•–O) is attached at various positions along the fatty acid chain to situate the nitroxide groups at different depths in the hydrophobic interior of this heterogeneous complex system of TSP or rTSP. These probes are amphiphilic with polar regions which anchor the probes at the hydrophilic interface and the large, rigid steroid frameworks embedded in the lipid chain region. The largest hyperfine splitting (32 G) occurs when the magnetic field is parallel to the long axis of the fatty acid (and hence perpendicular to the plane of a well-ordered multilayer system). These steroid probes lack the rigidity of the group to which the doxyl moiety is attached. It is the behaviour of the lipid to which the –N•–O group is attached that is important, so the order matrix is usually transformed to the molecular coordinate system, with z as the long axis of the spin label. It is in the molecular coordinate system that the S-matrix has axial symmetry. The outermost EPR lines move in and the line widths become narrower as the doxyl group moves from C-5 to the C-7, C-12 and C-16 positions.

The mixture of the formulation labelled with the spin probe was placed in a tissue cell sample holder (illustrated in **Figure 2**) for EPR measurement at room temperature. Once the EPR spectra of the control samples were recorded and characterized, titration experiments were performed on the same

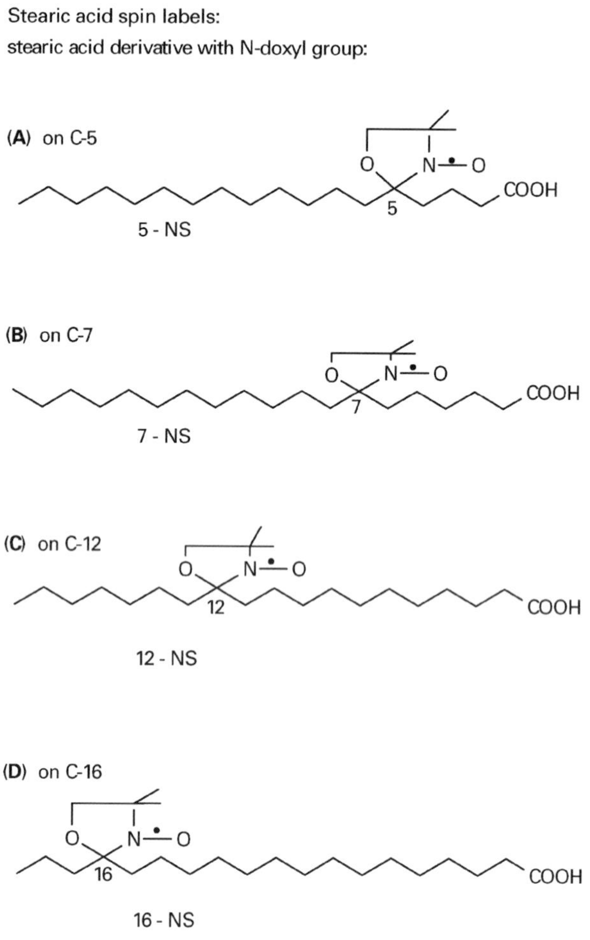

Stearic acid spin labels:
stearic acid derivative with N-doxyl group:

(A) on C-5

5 - NS

(B) on C-7

7 - NS

(C) on C-12

12 - NS

(D) on C-16

16 - NS

Figure 1 Chemical structure of the stearic acid spin labels.

control with the particular vesicant agent under study. The vesicant agent, an oily liquid, was carefully distributed over the surface area using a plastic spatula. This dispersion process was performed without disturbing the surface of the labelled formulation. The EPR quartz flat cell was maintained at room temperature in a chemical fume hood for an hour to allow venting of the volatile agent and, after one hour, the EPR spectra were monitored as described for the control samples.

The EPR of the labelled-formulation control exhibits two low-field peaks as illustrated in **Figure 3**, which shows the EPR spectrum of a vehicle (fomblin®-RT-15) incubated with 5-doxyl stearic acid. The spectrum resembles that of an immobilized spin probe. The value of S of the strongly immobilized probe can be calculated from the spectra using the equation:

$$S = \frac{(T_{\parallel} - T_{\perp})}{3[(T_{zz} - a^1)/2]}$$

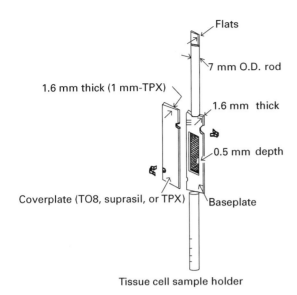

Tissue cell sample holder

Figure 2 EPR cavity cell for semisolid samples.

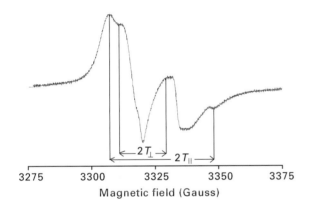

Magnetic field (Gauss)

Figure 3 EPR spectrum of a formulation incubated with 5-doxyl stearic acid. The spectrum is typical of an immobilized spin-probe where $T_{\parallel}$ is the outer hyperfine splitting ($A_{\parallel}$) and $T_{\perp}$ is the inner hyperfine splitting ($A_{\perp}$).

where $T_{\parallel}$ is the outer hyperfine splitting, $T_{\perp}$ is the inner hyperfine splitting, a^1 is $(T_{\parallel}+2T_{\perp})/3$, and the tensor $T_{zz} = 32$ G. A change in S could be interpreted as representing a net change in the breakthrough and flux of the vesicant agent.

Some general principles should be followed in such formulation studies. (1) The use of experimentally determined 'low' probe–lipid ratios is a requirement. A sufficient quantity of the spin probe should be incorporated into the formulation to permit the recording of an EPR trace with a reasonable signal-to-noise level. (2) The spectral parameters $T_{\perp}$ ($A_{\perp}$), $T_{\parallel}$ ($A_{\parallel}$) and S derived from the EPR spectra should be plotted as a function of the vesicant concentration

under study. If titration experiments indicate optimal conditions of probe–lipid ratios, then S may be used as a measure of formulation fluidity. (3) It must, however, be remembered that S is a function of both the motion of the probe and the polarity of the environment of the probe. (4) The probe spectral perturbation could be monitored by initially loading the formulation with probe and continuously measuring the EPR spectra to identify 'low' probe concentrations for formulation systems that destroy the EPR signal of the spin label with time. The decrease in probe concentration could be estimated from a double integration of the EPR trace. The decrease in the various spectral parameters could be plotted as a function of time.

The –N•–O moiety is attached directly to the hydrocarbon, as illustrated in **Scheme 1**. The probe is incorporated into the lipid portion of the perfluorinated grease and produces an EPR spectrum of an immobilized spin-label probe. An idealized cross section of this heterogeneous system is illustrated in the schematic representation I. The fatty spin-label (A) intercalates with the long hydrophobic chain normal to the plane of the matrix. The rigid spirane structure of the doxyl group places the z-axis (arrow) of the nitroxide parallel to the extended lipid chain. Thus, the direction of maximum splitting will be observed when the magnetic field (B_o) is normal to the plane of the matrix (i.e. along the arrow A); the splitting becomes progressively smaller as the matrix is rotated in the magnetic field, and the minimum splitting occurs when the matrix is parallel to the magnetic field.

In principle, the polarity profile could be determined by measuring the three-hyperfine splitting constants T_{xx}, T_{yy} and T_{zz} for the –N•–O group at various positions along the lipid chain. In practice, none of these parameters can be measured directly because of the partial motion average of the electron–nuclear dipolar interaction. An indirect method of obtaining the polarity profile is to estimate $T_{\perp}$ and $T_{\parallel}$ with the spin labels in randomly oriented samples. The –N•–O group at the C-5 position in the lipid matrix has been reported to be in an environment that is more polar than the C-12 and C-16 positions which are in a more hydrocarbon-like environment (**Scheme 1**).

Humidity is a very important factor in such studies since vesicant agents hydrolyse at a relatively rapid rate. The polarity profile can be studied using lipid films supported on glass wool in a hydration chamber. These samples can be equilibrated at 100% relative humidity or dehydrated over phosphorus pentoxide (P_2O_5); the removal of water would effectively abolish the polarity profile. The main point is

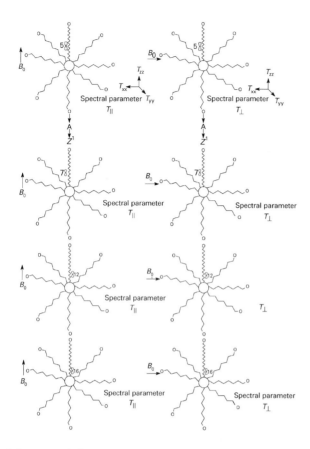

Scheme 1 A fatty acid spin probe incorporated into a very complex heterogeneous system that consists of a particular probe (0.2–10 μm) and a lipid matrix. The axis of motional averaging Z^1 is the normal on the matrix surface. The arrow (i.e. along the arrow A) indicates the direction of the nitroxide z'-axis.

that the defined parameter S provides an operationally relative change at some depth in this complex heterogeneous system. Ion transport across lipid films can be studied by measuring the maximum and minimum splitting for the $(-N^\bullet-O)$ group at various positions along the lipid chain. In addition, a penetration profile into the lipid matrices can be measured.

The order parameter S depends on the depth of penetration of the paramagnetic group $(-N^\bullet-O)$ into the homogeneous hydrophobic vehicle, the viscosity of the fluorochemical material and the chemical properties of the vesicant agent under investigation. Therefore, the net change in viscosity of the vehicle and the chemical composition affect the penetration profile. The probability that the paramagnetic group of the probes will be located at different depths in this complex heterogeneous system provides a unique tool to observe the permeation process at a molecular level. In addition, if we assume that the labelled and the unlabelled lipids (base oil of the formulation) intercalate to different depths in the

lipid matrix, then to account for the large effect at the C-5 position, the vesicant agent must penetrate to at least the C-2 position to change the environment of the formulation. This penetration profile can be improved by using spin labels bound in the polar head group region, by filling in all points from C-3 to C-16, and by performing X-ray diffraction experiments on the same samples to obtain a more accurate estimate of the overall matrix thickness. Nevertheless, the essential features are clear, stearic acid spin-labels provide a means of estimating system heterogeneity, because they are randomly distributed in the plane of the matrix, position themselves so that the α-carbon of the fatty acyl chain is aligned with the head group region of the lipid matrix, and faithfully reflect the degree of orientational order and dynamics of the lipid molecules. The observed decay of S can be related to the permeability properties of the system, because it is known that the EPR signal of the nitroxide group disappears almost instantaneously when the paramagnetic moiety is located at the polar interface of a lipid matrix. This observation is reinforced by the results obtained with the stearic acid spin probes (**Figure 4**). **Figure 4** shows representative EPR spectra of fomblin® HC/ 04 (relative molecular mass = 1500; viscosity index: 60) labelled with 12- and 7-doxyl stearic acids in the presence of a chlorinated glycoluril (S-330) control and exposed to different concentrations of half-mustard. It is well known that compounds containing a chloroamide group are generally preferred. S-330 provides an efficient antivesicant, which is characterized by its high efficiency and long protective times. One possible mechanism is that S-330 produces sufficient chloride ions to prevent the formation of the sulfonium ion intermediate, which is the active form of sulfur mustard. The graph in **Figure 4** illustrates the decay of the EPR integral signal (signal intensity in relative units) as a function of H-MG concentration. The decay of the EPR signal was dependent on the depth of penetration of the paramagnetic group into this heterogeneous matrix. The increase of H-MG concentration affects the penetration profile as shown in **Figure 4** for the four different spin-probes.

Spin probes only indirectly reflect the ordering of the host system. What information about the molecular architecture of this heterogeneous system can be deduced from these experiments? These labelled-formulations exhibit very little anisotropy (i.e. their magnitude and sign depend on the orientation of the radical with respect to the applied magnetic field) when hydrated with distilled water. Addition of non-electrolytes such as sucrose has no effect, but the addition of salt results in a large increase in EPR spectral anisotropy. This effect is thus one of charge,

Fomblin HC/04 + S330 + 12-doxyl control

Fomblin HC/04 +S330 + 12-doxyl + 30 µL H-MG

Fomblin HC/04 + S330 + 7-doxyl labelled control

Fomblin HC/04 + S330 + 7-doxyl + 20 µL H-MG

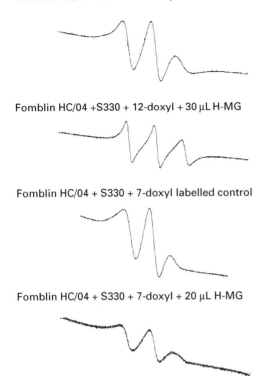

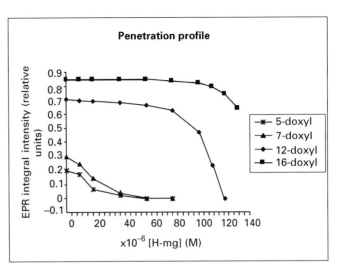

Figure 4 EPR spectra of a labelled reactive topical skin protectant with 12- and 7-doxyl in selected controls and exposed samples. The graph shows a plot of EPR signal integral (relative units) obtained from the labelled formulation as a function of H-MG concentrations for spin-label probes, 5-, 7-, 12-, and 16-doxyl stearic acids.

not of osmotic or vapour pressure. It has been shown to be owing to the cation and to depend on the cation valence. The order of effectiveness in promoting spectral anisotropy is $NaCl = KCl = LiCl < MgCl_2$, whereas the chloride, thiocyanate and sulfate salts of the same cation were equal in effectiveness. The cations act by reducing the surface charge density of the heterogeneous complex system, which consists of small particles and the lipid matrix. With the reduction in repulsion between groups, the ointment formulation can contract and achieve a higher degree of order. A net charge of the groups leads to molecular repulsion, an expansion of the formulation, and a decrease in EPR spectral anisotropy.

Vesicant agents of the 'mustard' type present some special features in their reaction mechanism(s). It is known that the mechanism involves, as a first step, the formation of a cyclic sulfonium ion and the release of chloride (Cl^-). This mechanism is illustrated for the cases of half-mustard and sulfur mustard in **Scheme 2**. In addition, these vesicant agents are persistent agents, depending on pH and moisture. The mustard is hydrolysed to form HCl and thiodiglycol. Ointment formulation-incorporated spin probes are sensitive to the polarity and fluidity of

their local environment. Cation binding, pH alterations, and the action of many substances perturb labelled formulations to yield characteristic changes in their respective EPR spectra. An adequate interpretation of EPR spectral changes in terms of the structure of the host matrix requires that alterations in fluidity and/or polarity be distinguished from changes in probe–probe interaction (e.g. dipole–dipole and electron–electron exchange broadening).

The spectral alterations noted upon addition of vesicant agents to a labelled formulation are caused by changes in (1) motion of the probe, (2) the polarity of the environment of the probe, (3) alteration in fluidity, and/or (4) the permeability profile. If the vesicant agent induces changes that involve radical interactions, the magnitude of the spectral alteration that depends on probe concentration will disappear.

EPR/spin trapping

EPR spin trapping techniques have successfully been applied to determine and identify free radical intermediates in biology and toxicology. Spin trapping allows one to determine if short-lived free radicals

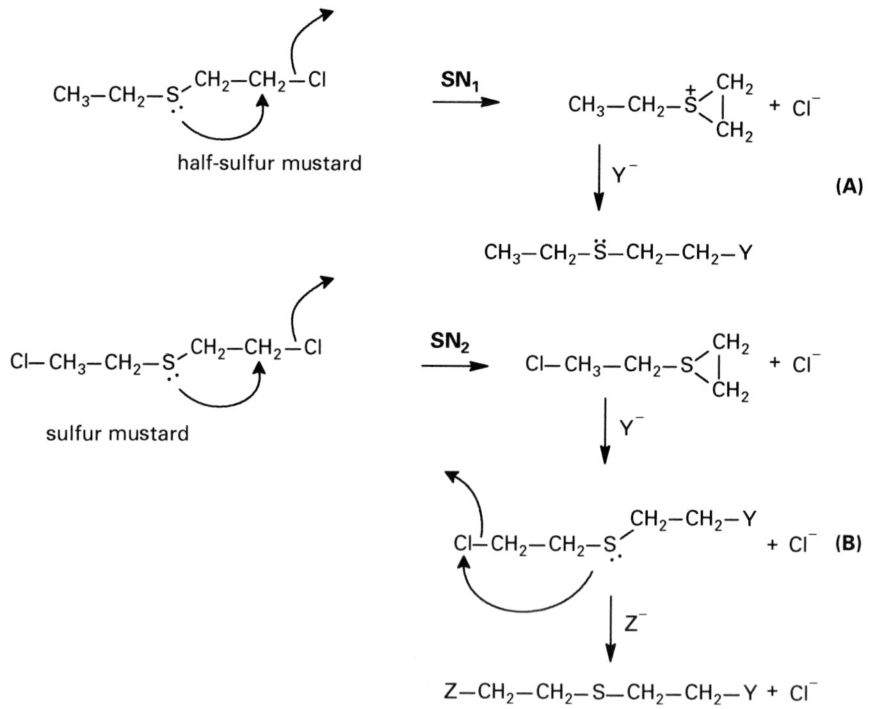

$$CH_3-CH_2-\overset{..}{S}-CH_2-CH_2-Cl \quad \xrightarrow{SN_1} \quad CH_3-CH_2-\overset{+}{S}\overset{CH_2}{\underset{CH_2}{\big\vert}} + Cl^-$$

half-sulfur mustard

$$\downarrow Y^- \qquad \textbf{(A)}$$

$$CH_3-CH_2-\overset{..}{S}-CH_2-CH_2-Y$$

$$Cl-CH_3-CH_2-\overset{..}{S}-CH_2-CH_2-Cl \quad \xrightarrow{SN_2} \quad Cl-CH_3-CH_2-S\overset{CH_2}{\underset{CH_2}{\big\vert}} + Cl^-$$

sulfur mustard

$$\downarrow Y^-$$

$$Cl-CH_2-CH_2-\overset{..}{S}\overset{CH_2-CH_2-Y}{} + Cl^- \quad \textbf{(B)}$$

$$\downarrow Z^-$$

$$Z-CH_2-CH_2-S-CH_2-CH_2-Y + Cl^-$$

Scheme 2

are involved as reaction intermediates by scavenging the reactive radical to produce more stable nitroxide radicals. This technique involves reaction of the initially-generated radical (itself either too short-lived or of too low a concentration to directly detect) with an added organic compound, known as spin trap, to generate stable radical adducts from whose EPR spectra information about the original radical may be obtained.

Two kinds of spin traps have been developed, nitrone and nitroso compounds. **Figure 5** illustrates the most common commercially available traps. Nitrones are the spin traps of choice for the study of oxygen-centred radicals. The most popular nitrone traps identified in **Figure 5** have a β-hydrogen that can provide considerable information about the radical trapped. However, some information is lost using these nitrones because the trapped radical adds to a carbon adjacent to the nitrogen. Nitroso compounds can provide more information than nitrones as the radical to be trapped adds to the nitroso nitrogen, and, therefore, more information about the hyperfine splitting parameters is obtained.

In vivo spin trapping of oxygen-centred radicals

Nitrones have emerged as the most popular spin traps for biological applications, and out of several nitrone spin traps, the cyclic 5,5-dimethyl-1-pyrroline N-oxide (DMPO) has received most attention, since it yields distinct and characteristic adducts with superoxide radical anion ($O_2^{\cdot-}$) and hydroxyl radical (•OH). The use of DMPO as a probe for oxyradical generation in biology systems is not without limitations as high concentrations of DMPO have been suggested to have serious toxic effects on biological tissue. A low concentration of DMPO (10 mM) was used to detect free radical generation in hearts with ischaemia/reperfusion insult. **Figure 6** shows a scheme of a typical spin trapping experiment in an isolated heart. In the effluent immediately after reperfusion, DMPO–OOH, the superoxide spin adduct of DMPO, was obtained. DMPO at a 10 mM concentration range did not interfere with the left ventricular (LV) function during the control perfusion period. Enzyme leakage from hearts also supported non-toxicity findings of DMPO at 10 mM, confirming that the DMPO superoxide adduct is genuine evidence of the generation of superoxide upon reperfusion and is not an artificial generation owing to the cytotoxicity of DMPO.

Furthermore, application of the spin trapping technique in intact animals require an understanding of the stability of the spin traps and the spin adducts *in vivo*. A new class of α-phosphorus-containing DMPO analogues, 5-(diethoxyphosphoryl)-5-methyl-1-pyrroline N-oxide (DEPMPO) has been

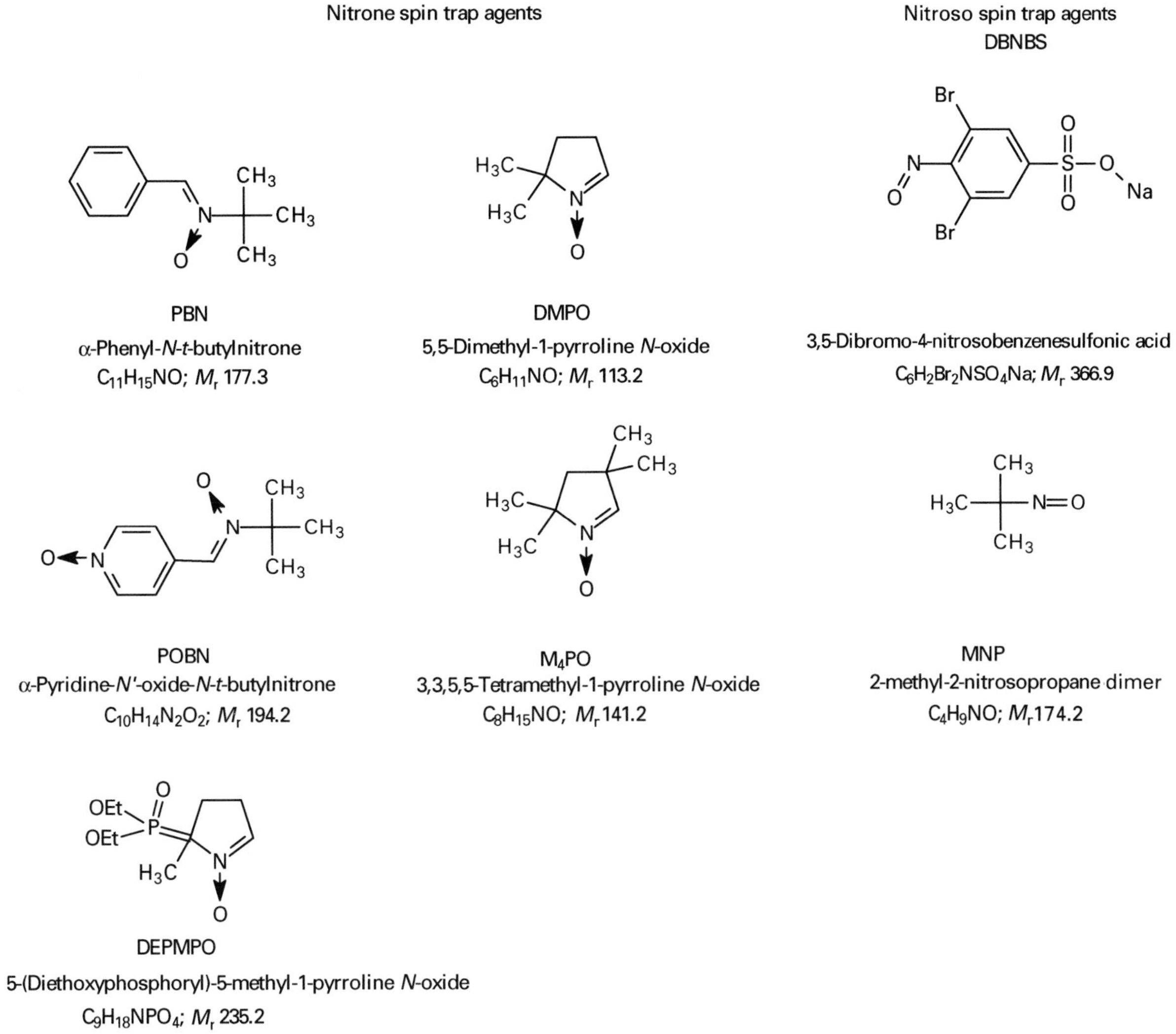

Nitrone spin trap agents

Nitroso spin trap agents
DBNBS

PBN
α-Phenyl-*N-t*-butylnitrone
$C_{11}H_{15}NO$; M_r 177.3

DMPO
5,5-Dimethyl-1-pyrroline *N*-oxide
$C_6H_{11}NO$; M_r 113.2

3,5-Dibromo-4-nitrosobenzenesulfonic acid
$C_6H_2Br_2NSO_4Na$; M_r 366.9

POBN
α-Pyridine-*N'*-oxide-*N-t*-butylnitrone
$C_{10}H_{14}N_2O_2$; M_r 194.2

M₄PO
3,3,5,5-Tetramethyl-1-pyrroline *N*-oxide
$C_8H_{15}NO$; M_r 141.2

MNP
2-methyl-2-nitrosopropane dimer
C_4H_9NO; M_r 174.2

DEPMPO
5-(Diethoxyphosphoryl)-5-methyl-1-pyrroline *N*-oxide
$C_9H_{18}NPO_4$; M_r 235.2

Figure 5 Names, structural formulae and relative molecular mass of the most common spin trapping agents commercially available.

synthesized and characterized for the generation of superoxide during the reperfusion of ischaemic isolated hearts. DEPMPO can trap and form a stable adduct for both •OH as DEPMPO–OH, and $O_2^{\div}$, as DEPMPO–OOH, giving EPR spectra that are characteristic of each. Thus, unlike DMPO, DEPMPO can be used to distinguish between superoxide dependent and independent mechanisms that lead to the hydroxyl radical. DEPMPO is a good candidate for trapping radicals in functioning biological systems, and represents an improvement over the commonly used trap DMPO.

A very important aspect of spin trapping is to positively identifying the radicals under study. This assignment requires a knowledge of how certain features of the structure of the trapped radical influence the EPR spectrum of the spin adduct.

Sometimes it is very difficult to determine unambiguously the precise structure of the spin adduct from the EPR signal obtained. Isotopic substitution EPR experiments are recommended in an attempt to identify the observed adducts. The strategy is that the unpaired electron in a radical interacts with the nucleus of the atom it orbits, and the spin of the nucleus determines the number of lines or peaks in the spectrum. For example, ^{13}C has a nuclear spin of $\frac{1}{2}$ while ^{12}C has no spin. An unpaired electron, which is associated with atoms having no spin, will exhibit an EPR spectrum containing only a single line. The spin of the nucleus influences the resonance of the unpaired electron so that the EPR resonance splits into two or more lines. The number of EPR resonance observed is equal to $2I + 1$, where I is the nuclear spin. A practical

Spin trapping method of isolated heart

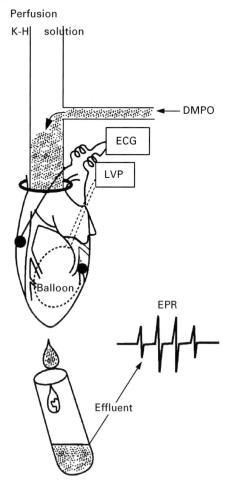

Figure 6 Scheme of a typical spin trapping experiment in an isolated heart. Coronary effluents are collected and immediately frozen in liquid N_2 to prevent spin adduct decay. Frozen samples are thawed just before EPR measurement and EPR spectra are recorded using 160 or 250 μL flat cells. Typical spectrometer conditions are as follows: magnetic field 3350 G, power 8 mW, response 0.3 s, modulation amplitude 1.0–1.25 G, receiver gain 6.5×10^3, sweep time 2 min, modulation frequency 100 kHz and temperature 23–28°C.

example using this concept follows. When human monocyte cells were exposed to 2-chloroethyl ethyl sulfide (H-MG) in the presence of the nitroso spin trap MNP, an EPR signal consisting mainly of a primary triplet was observed (**Figure 7A**). The complexity of the EPR spectrum suggested the trapping of a hydrogen-abstraction radical of polypeptide molecules. To verify the assignment, ^{13}C-labelled (at the C_2 amino acids) of human tumour necrosis factor-alpha (TNF-α) were reacted with H-MG in the presence of MNP. The resulting EPR spectrum (**Figure 7B**) was detected and simulated using five different hyperfine coupling

constants, indicating the presence of a MNP-^{13}C-centred adduct. Therefore, the, spin trapping technique, properly applied, leaves no doubt about the nature of the radical or the intensity and duration of its production in biological systems.

EPR/Spin trapping in toxicology

The involvement of reactive intermediate species (RIS) in chemical agent toxicity could be also investigated using EPR/spin trapping techniques. For example, toxic oedemagenic gases, such as phosgene ($OCCl_2$), bis(trifluoromethyl)disulfide (TFD) and perfluoroisobutylene (PFIB) can be studied by spin trapping. Two types of apparatus that have proved useful for these experiments are shown in **Figure 8**. They consist of a 'tube' which connects via a 7/25 tapered ground-glass joint to a Wilmad EPR 'flat cell' (**Figure 8A**) and a 'U' tube which also with a tapered ground-glass joint to a Wilmad 'flat cell' (**Figure 8B**). In a typical experiment, one positions the tube or the 'U' tube vertically and a solution of the spin trap is placed in one chamber. The chambers are stopped with rubber septa through which long (~#8) syringe needles are inserted. The gas radical producers are then passed through the spin trap solution for 15–20 min. For deoxygenating purposes, bubbling purified nitrogen or argon gas through the solution is sufficient. The excess of gas can escape through a small syringe needle, as indicated in **Figure 8**. When gassing is complete the system is stopped and the contents of the tubes and sample cell are thoroughly mixed and shaken into the EPR flat cell, which is inserted into the microwave cavity of the EPR spectrometer. Relatively simple modifications of this basic experimental design allow the study of other toxic chemicals. Highly reactive intermediate species of phosgene have been identified by adding ^{13}C-labelled phosgene into a PBN solution in benzene (**Figure 9**).

EPR/spin trapping can be used in the study of reactive intermediate species in a wide range of biological and toxicological systems. This technique allows the identification of radical species [and hence their mechanism(s) of production], and the indirect and real-time observation and quantification of radical reactions. To summarize, spin trapping is a powerful technique for the indirect EPR observation of many reactive free radicals. More ideas for dealing with this technique are well documented and discussed in reviews by Janzen, Anderson Evans, Mason, Thornalley and Buettner.

(A) Experimental

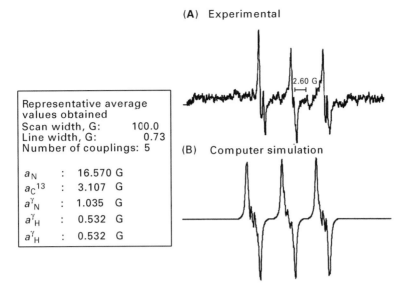

Representative average
values obtained
Scan width, G: 100.0
Line width, G: 0.73
Number of couplings: 5

a_N	:	16.570 G
$a_C{}^{13}$	:	3.107 G
a^γ_N	:	1.035 G
a^γ_H	:	0.532 G
a^γ_H	:	0.532 G

(B) Computer simulation

Figure 7 (A) EPR spectrum of MNP-adducts observed when THP-1 (monocyte cells) suspensions were exposed to H-MG (2×10^{-4}M) in the presence of MNP (7 mg). The magnetic field was set at 3350 G, microwave power 10 mW, modulation amplitude 8 G, microwave frequency 9.474 GHz. (B) EPR spectrum of the MNP adduct of ^{13}C-labelled serine $^{13}C_2$ amino acid generated by H-MG via a hydrogen atom abstraction mechanism. Computer simulation using the EPR parameters is given in the box.

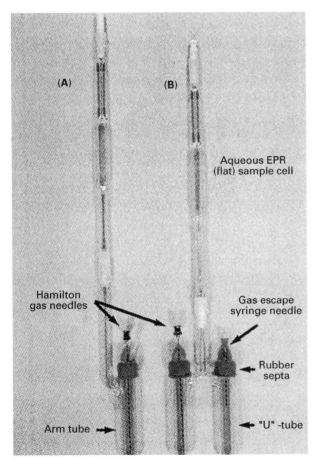

Figure 8 Typical apparatus for gas aqueous or dielectric solvent EPR/spin trapping experiments.

(A) Experimental

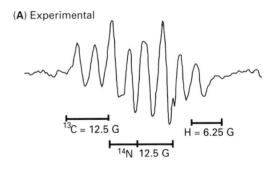

(B) Computer simulated

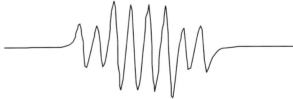

Figure 9 (A) Spin adducts formation of PBN-^{13}COCl in benzene. Receiver gain 1.25×10^5; modulation amplitude 1.0 G. (B) Computer simulation of (A) using $a_N = 12.4$ G; $a_H = 6.25$ G and $a_{13C} = 12.5$ G with a line width of 2.5 G. Insert: spectral parameters for ^{13}C-phosgene obtained from the observed nitrone spin adducts.

List of symbols

B_0 = magnetic flux density; I = nuclear spin; S = order parameter; $T_\parallel$ = outer hyperfine splitting; $T_\perp$ = inner hyperfine splitting.

See also: **Chemical Applications of EPR; EPR, Methods; EPR Spectroscopy, Theory.**

Further reading

Anderson Evans C (1979) Spin trapping. *Aldrichimica Acta* 12: 23–29.

Arroyo CM and Janny SJ (1995) EPR/Spin label technique as an analytical tool for determining the resistance of reactive topical skin protectant (rTSPs) to the breakthrough of vesicant agents. *Journal of Pharmacological and Toxicological Methods* 33: 109–112.

Arroyo CM, Von Tersch RL and Broomfield CA (1995) Activation of alpha-human tumour necrosis factor (TNF-α) by human monocytes (THP-1) exposed to 2-chloroethyl sulphide (H-MG). *Human & Experimental Toxicology* 14: 547–553.

Arroyo CM and Keeler JR (1997) Edemagenic gases cause lung toxicity by generating reactive intermediate species. In: Baskin SI and Salem H (eds) *Oxidants, Antioxidants, and Free Radicals*, Vol 17, pp 291–314. Washington, DC: Taylor & Francis.

Buettner GR (1987) Spin trapping: ESR parameters of spin adducts. *Free Radical Biology Medicine* 2: 259–303.

Berliner LJ (1976) *Spin Labelling – Theory and Application.* New York: Academic Press.

Frejaville C, Karoui H, Tuccio, Le Moigne F, Culcasi M, Pietri, Lauricella R and Tordo P (1995) 5-(Diethoxy-phosphoryl)-5-methyl-1-pyrroline N-oxide: A new efficient phosphorylated nitrone for the *in vitro* and *in vivo* spin trapping of oxygen-centered radicals. *Journal of Medicinal Chemistry* 38(2): 258–265.

Janzen EG (1980) A critical review of spin trapping biological systems. In: Pryor WA (ed) *Free Radicals in Biology*, Vol IV, pp 115–154. New York: Academic Press.

Mason RP, Stolze K and Morehouse KM (1987) Electron spin resonance studies of the free radical metabolites of toxic chemicals. *British Journal of Cancer* 8: 163–171.

Mason RP, Hanna PM, Burkitt MJ and Kadiiska MB (1994) Detection of oxygen-derived radicals in biological systems using electron spin resonance. *Environmental Health Perspective* 102(10): 33–36.

McConnell HM and McFarland BG (1970) Physics and chemistry of spin labels. *Quarterly Reviews of Biophysics* 3(1): 91–136.

Pantini G and Antonini A (1988) Perfluoropolyethers for cosmetics. *Drugs & Cosmetic Industry*, September 1988.

Speck JC (1959) Polychloro-7,8-disubstituted-2,5-diiminoglycoluril for use as an antivesicant. *United States Patent Office Patented*, No 2 885 305, May 5, 1959.

Thornalley PJ (1986) Theory and biological applications of the electron spin resonance technique of spin trapping. *Life Chemistry Reports* 4: 57–112.

Stark Methods in Spectroscopy, Applications

See **Zeeman and Stark Methods in Spectroscopy, Applications.**

Stark Methods in Spectroscopy, Instrumentation

See **Zeeman and Stark Methods in Spectroscopy, Instrumentation.**

Stars, Spectroscopy of

AGGM Tielens, Rijks Universiteit, Groningen, The Netherlands

| ELECTRONIC SPECTROSCOPY |
| Applications |

Introduction

Spectroscopy is the key to unlocking the information in starlight. Stellar spectra show a variety of absorption lines which allow a rapid classification of stars in a spectral sequence. This sequence reflects the variations in physical conditions (density, temperature, pressure, size, luminosity) between different stars. The strength of stellar absorption lines relative to the continuum can also be used in a simple way to determine the abundances of the elements in the stellar photosphere and thereby to probe the chemical evolution of the galaxy. Further, the precise wavelength position of spectral lines is a measure of the dynamics of stars and this has been used in recent years to establish the presence of a massive black hole in the centre of our galaxy and the presence of planets around other stars than the Sun.

Stellar classification

All stellar spectra show absorption lines due to a variety of species. For the Sun, these were first discovered by Joseph von Fraunhofer in the early 1800s. A sample of stellar spectra is shown in **Figure 1**. The patterns in these lines allow stellar spectra to be grouped in a classification scheme. Depending on the spectral characteristics, stars are designated by a letter from the sequence O, B, A, F, G, K and M. This spectral sequence is summarized in **Table 1**. Since temperature controls ionization and excitation of the atoms, this spectral classification basically reflects a temperature sequence. This temperature sequence is also obvious from the colours of stars.

The strength of an absorption line is a measure of the opacity in the line compared with that in the continuum. For A stars, which have surface temperatures around 10 000 K, the $n = 2$ level of hydrogen can be excited and, because H is by far the most abundant element in (almost) all stars, the opacity and hence the visual region of the spectrum are dominated by absorption out of these levels giving rise to prominent Balmer lines (**Figure 1**). For higher stellar surface temperatures (e.g. O stars), the fraction of the H atoms excited to the $n = 3$ level – which provide the continuum opacity in the visible – increases more rapidly than that in the $n = 2$ level. As a result, the strength of the lines relative to the continuum decreases (**Figure 1** and **Table 1**). For the hottest stars, H is completely ionized and Thompson scattering by free electrons now dominates the continuum opacity. Lines of helium, which has a higher ionization potential, are still present. Of course, these He lines, which originate from levels much higher in energy than the ionization potential of hydrogen, require high stellar surface temperatures for their excitation. For stars much cooler than A, the H^- ion provides the continuum opacity. Because the population of the levels leading to Balmer absorption of hydrogen becomes very small in such cool stars, the strength of the H lines decreases relative to the continuum. Various trace elements with lower energy

Table 1 Spectral types

Spectral class[a]	T_{eff}[b] (K)	Colour	Spectral characteristics
O	>30 000	Bluish white	Relatively few lines. He+ lines dominate
B	10 000–30 000	Bluish white	More lines; neutral He lines dominate; hydrogen Balmer lines developing
A	7 500–10 000	White	Very strong hydrogen Balmer lines, decreasing later; Ca+ line appears
F	6 000–7 500	White	Hydrogen Balmer lines and ionized metal lines declining; neutral metal lines increasing
G	5 000–6 000	Yellow	Many metal lines; lines of Ca+ strong; neutral metal lines continue to increase
K	3 500–5 000	Reddish	Molecular bands appear; neutral metal lines dominate
M	2 500–3 500	Red	Neutral metal lines strong; molecular bands dominate

[a] Each spectral class is subdivided into subclasses ranging from 0 to 9 with 0 the hottest and 9 the coolest type in the class.

[b] Approximate photospheric temperature range for main sequence stars.

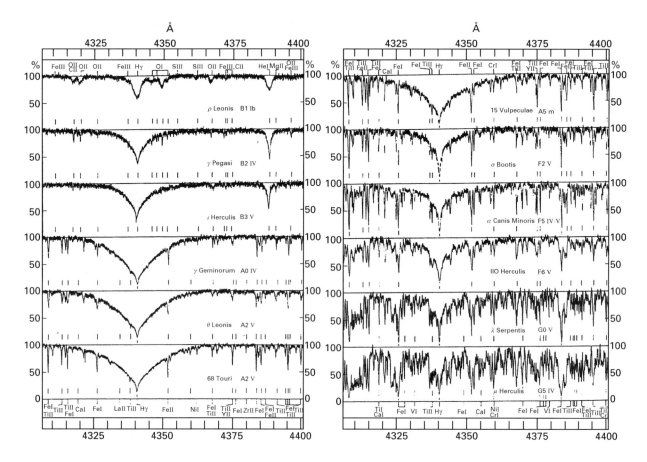

Figure 1 Line intensities in the 4300–4400 Å region of spectra of representative stars of spectral types B–G. The more prominent lines are labelled at the top and bottom. Note how the strength of the Balmer Hγ line increases in strength relative to the continuum from top to bottom in the left-hand column and then decreases again in the right-hand column. At the same time, metal lines increase in strength, first from ionic and then from neutral species. Finally, note the difference in line width between different luminosity classes (see top left-hand side).

levels now take over the spectrum (**Figure 1** and **Table 1**). For the coolest K and M stars, most of the metals are neutral and molecules can survive in the stellar photosphere. Lines from these species will dominate the spectral appearance.

In addition to this spectral class, stars are also characterized by a luminosity parameter. This luminosity classification is made on the basis of the width of spectral lines. **Table 2** summarizes this classification. The width of spectral lines increases as the gas pressure increases. This so-called pressure broadening is due to the perturbation of atomic energy levels by other, nearby species. The physically largest stars have the lowest surface densities and pressures. Lines from these stars are therefore broader than from smaller stars (**Figure 1** and **Table 2**). This difference in size, which results in a difference in stellar luminosity, has led to the naming scheme from supergiants to dwarfs.

Stellar classification of a spectrum is actually based upon the intensity ratio of pairs of lines which

Table 2 Luminosity classes[a]

Luminosity class	Star type
I	Supergiants
II	Bright giants
III	Giants
IV	Subgiants
V	Main sequence[b]
VI	Subdwarfs
VII	White dwarfs

[a] Along this sequence, the width of spectral lines increases from supergiants to white dwarfs.
[b] Main sequence stars are also known as dwarfs.

are sensitive to temperature or luminosity. The lines used depend on the appropriate spectral type. This provides a straightforward way to classify stellar spectra and to determine rapidly the physical conditions in the stellar photosphere, including density, temperature, pressure, luminosity and size.

While these spectral and luminosity classes can be used to classify most stars, there is in addition a bewildering collection of stars with spectra which deviate in various respects. Generally, these variations reflect differences in the elemental abundances in the photospheres of these stars. Most stars have abundances very similar to the Sun. However, deep in the interior of each star, nucleosynthesis converts hydrogen and helium into heavier elements such as carbon and nitrogen. In some stars, these freshly synthesized elements can be exposed in the stellar photosphere either owing to the effects of extensive mixing of deeper layers with the surface or because much of the stellar envelope has been lost in a stellar wind. **Table 3** contains a sample of such special stars and their spectroscopic characteristics.

The carbon stars are rare variable giants with temperatures similar to those of classes G and M. Their spectra are characterized by strong lines from carbon-bearing molecules (CN, C_2). These stars form when nucleosynthetically processed material, which is enriched in carbon owing to helium burning, is dredged up from the interior. When the abundance of carbon becomes larger than that of oxygen in the photosphere, all the oxygen is locked up in carbon monoxide and carbonaceous molecules rather than oxides dominate the composition. The spectra of these stars also show lines due to technetium (Tc), which is radioactive with a half-life of about 1 million years. Clearly, this element was recently formed in the interior and dredged up to the surface. Hence the presence of Tc in stellar spectra directly attests to the importance of nucleosynthesis in stellar interiors. So-called S stars are thought to be an intermediate stage in the stellar evolution from M to C stars where the abundances of carbon and oxygen approximately balance. ZrO bands now dominate the spectra (**Table 3**). Wolf Rayet stars are very luminous and hot stars with weak hydrogen lines and very strong helium lines. Their spectra show very wide emission features due to ionized He, C, N and O, originating in a wind from the star. Depending on type, they have excess carbon (WC) or excess nitrogen (WN) in their photosphere (**Table 3**). Some O, B and A stars (hot emission line stars) show hydrogen emission lines originating in a stellar wind.

Elemental abundances

The determination of stellar abundances is one of the main applications of stellar spectroscopy. The strength of photospheric absorption lines can provide information on the relative abundances of the elements in the photosphere of the star. Generally, this is done by measuring the strength of spectral lines relative to the continuum, the so-called equivalent width. The relationship between the equivalent width and the number of absorbers is called the curve of growth in stellar spectroscopy. For weak lines, the equivalent width is directly proportional to the number of absorbers. When the intrinsic strength of a line is larger or the number of absorbers is larger, the centre of the line saturates and the equivalent width becomes almost independent of the number of absorbing particles. For very strong lines or very large number of absorbers, absorption in the wings of the line become important and the equivalent width of the line will increase proportionally to the square root of the number of absorbers present. Stellar spectra contain many lines of a given element with known intrinsic strength. These can be used to construct an empirical curve of growth for that element. Comparison of such curves of growth for different elements yields then the relative elemental abundances. **Figure 2** shows an example for iron and titanium lines in the Sun. Similarly, we can compare the curves of growth for other stars with that for the Sun and determine elemental abundances for these stars relative to solar.

This semiempirical method is fairly straightforward but does assume that all the lines and the continuum involved are formed in the same region. Moreover, these line formation regions would have to have similar physical conditions in all stars. This

Table 3 Additional spectral types

Name	Spectral class	Spectral characteristics
Carbon stars	C	Strong CN bands and C_2 bands
Heavy metal stars	S	ZrO bands
Wolf Rayet stars	WN	N^{2+} and N^{3+} emission lines
	WC	Ionized carbon and oxygen lines
Hot emission line stars	Of, Be, Ae	Bright hydrogen emission lines

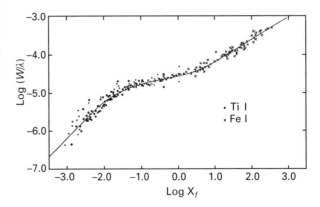

Figure 2 Empirical curve of growth for solar Fe I and Ti I lines. The *y*-axis is the equivalent width (line strength relative to the continuum) and the *x*-axis is based on the oscillator strength of the transition.

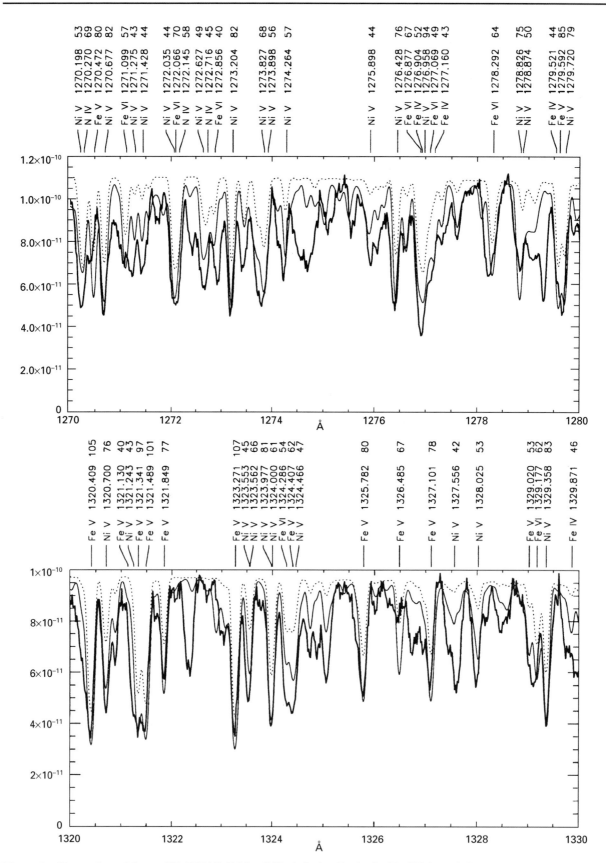

Figure 3 Observations of the star BD+75°325'; (thick solid line) obtained by the Goddard high-resolution spectrometer on board the Hubble space telescope, showing numerous iron and nickel lines in various ionization stages. These data are compared with model atmosphere spectra. The thin solid line is the best fit model for an iron abundance of 4×10^{-4} and the dotted line is for solar abundances (4×10^{-5}). A solar iron-to-nickel ratio has been assumed in both models. Reproduced with permission from Lanz T, Hubeny I and Heap SR (1997) *Astrophysical Journal* **485**: 843.

is not always justified. Furthermore, cool stars have very crowded spectral regions where line overlap is a severe problem. In these cases, detailed modelling is a prerequisite for the determination of accurate stellar abundances. Sophisticated techniques have been developed which model in detail the physical structure of the stellar photosphere and its interaction with light. These models solve the equation of statistical equilibrium, regulating the individual level populations, the equation for hydrostatic equilibrium, governing the stellar pressure structure, and the radiative transfer equations describing the absorption and emission of light in the stellar photosphere. Comparison of models calculated for a variety of abundances with the observations allows the determination of the elemental abundances. In general, good agreement between models and observations can be obtained (**Figure 3**).

Analyses of this kind have shown that nearly all stars have very similar elemental compositions. The spectra of some stars, however, reveal much lower abundances than in the Sun. These so-called subdwarfs are metal-poor by factors up to 500. These compositional variations are correlated with the mass, age and dynamics of the stars within the Milky Way. The stars that formed first have the lowest elemental abundances. As those stars evolved – the more massive ones more rapidly than the less massive ones – they 'polluted' the interstellar medium with the nucleosynthetic products formed in their interiors either through a gentle wind (low-mass stars) or through a violent supernova explosion (massive stars). In this way, later generations of stars are formed from gas with higher elemental abundances. It is this elemental enrichment that drives the evolution of the Milky Way and other galaxies.

Stellar dynamics

In addition to spectral and luminosity classification and abundance determination, stellar spectra also provide the radial motion of the absorbing gas through the Doppler shift. High-resolution spectra can thus provide information on stellar outflows in K and M giants, carbon stars, Of, Be and Ae emission line stars and Wolf Rayet stars. In general, the velocity information available in stellar spectra can be used to probe the dynamics of stars in the galaxy. For example, this technique has been used to trace the dynamics of stars in the centre of our galaxy. The derived velocity law implies a supermassive object in the centre of the galaxy with 3×10^6 solar masses. This provides strong evidence for the presence of a massive black hole in the centre of the Milky Way.

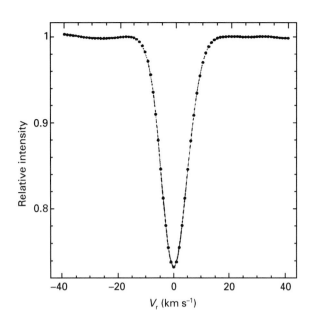

Figure 4 Typical cross-correlation function used to measure the radial velocity. This figure represents the mean of the spectral lines of the star 51 Peg. The position of the Gaussian function fitted (solid line) is a precise measurement of the Doppler shift to an accuracy of about 15 m s^{-1}. The width of the cross-correlation function reflects the star's rotational velocity. Reproduced with permission of Macmillan Magazines Ltd. from Mayor M and Queloz D (1995) *Nature* **378**: 355.

By monitoring radial velocities over a long period, stellar spectroscopy can also be used to search for stellar companions, be they binary stars, brown dwarfs (stellar-like companions which are not massive enough to start hydrogen burning in their interiors) or planets. In recent years, new techniques have been developed to search for planets orbiting solar-type stars using Doppler shifts. In order to eliminate systematic wavelength shifts, the spectra are calibrated either by passing the stellar light through iodine gas or by using stable, fibre-fed spectrometers with simultaneous Th–Ar wavelength calibration. In long-term monitoring programmes, radial velocities of stars can then be measured with an accuracy of about 15 m s^{-1}, using a cross-correlation technique which concentrates the Doppler information for some 5000 stellar absorption lines (**Figure 4**). For comparison, the wobble introduced in the Sun's radial motion by Jupiter is about 13 m s^{-1}. The handful of planetary companions found so far in this way have masses in the range 0.5–10 Jupiter masses and orbits with solar system dimensions. In this way, stellar spectroscopy has allowed us to complete the 'Copernican' revolution.

See also: **Atomic Spectroscopy, Historical Perspective; Cosmochemical Applications Using Mass**

Spectrometry; Environmental and Agricultural Applications of Atomic Spectroscopy; Interstellar Molecules, Spectroscopy of.

Further reading

Gustafsson B (1989) Chemical analyses of cool stars. *Annual Review of Astronomy and Astrophysics* 27: 701–756.

Jaschek C and Jaschek M (1987) *The Classification of Stars*. Cambridge: Cambridge University Press.

Kudritzki RP and Hummer DG (1990) Quantitative spectroscopy of hot stars. *Annual Review of Astronomy and Astrophysics* 28: 303–345.

Mihalas D (1970) *Stellar Atmospheres*. San Francisco: Freeman.

Yamashita Y, Nariai K and Norimoto Y (1977) *An Atlas of Representative Stellar Spectra*. Tokyo: University of Tokyo Press.

Statistical Theory of Mass Spectra

JC Lorquet, Université de Liège, Belgium

MASS SPECTROMETRY
Theory

The model

Ionization via 70 eV electronic impact brings about a large number of Franck–Condon transitions to many electronic states of the ion. The molecular ion then undergoes internal conversions to its lowest electronic state owing to the existence of very fast and efficient radiationless transitions which themselves result from the presence of numerous crossings between potential energy surfaces. **Figure 1** shows a calculated set of potential energy curves for the $F_2^{\bullet+}$ molecular ion.

This picture gives some idea of the complexity to be expected, for example, in the case of the isoelectronic ion $CH_3CH_3^{\bullet+}$. For polyatomic molecular ions, the pattern of surface crossings is extremely complicated because the density of electronic states is much higher than that detected by photoelectron spectroscopy and because of the number of nuclear degrees of freedom. In a polyatomic system, the sequence of radiationless transitions results from the presence of conical intersections and is usually over after about 10^{-13} s. Once in the electronic ground state, vibrational energy is assumed to be redistributed (randomized) throughout the different vibrational degrees of freedom on a timescale short with respect to the reaction lifetime. When it fragments, the vibrationally excited molecular ion has forgotten its

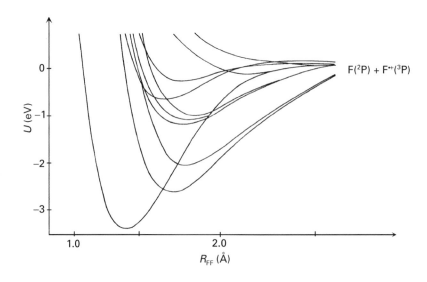

Figure 1 Potential energy curves for the $F_2^{\bullet+}$ ion.

initial conditions. It undergoes a series of competing, consecutive, unimolecular reactions that can be described by a statistical theory. The theory is known under two acronyms, viz., RRKM (after Rice, Ramsperger, Kassel and Marcus) and QET (for quasi-equilibrium theory, as suggested by Rosenstock, Wahrhaftig and Eyring).

Probably the most convincing argument for the validity of a statistical approach comes from the independence of the fragmentation pattern with respect to the way energy is delivered. Except in rare cases, there is no correlation between the breakdown diagram determined from photoion–photoelectron coincidence (PIPECO) experiments and the photoelectron spectrum. In addition, when the molecular ion (e.g., $C_2H_4^{\bullet+}$) can be prepared by a collision process (e.g., $C_2H_2^{\bullet+} + H_2$ or $C^+ + CH_4$), the fragmentation ratios (e.g., $C_2H_4^{\bullet+}$: $C_2H_3^+$: $C_2H_2^{\bullet+}$) at a given internal energy are usually found to be equal within the experimental error to those obtained when the system is prepared by photon impact (see **Figure 2**).

These observations imply that the dissociation rate constants are smooth, monotonically increasing functions of the internal energy E of the decaying molecular ion alone, irrespective of the way energy is delivered to the molecule, and do not depend on the electronic and vibrational quantum numbers that specify a particular molecular state. The criterion of validity of a statistical treatment (viz., that all of the properties of the system be completely characterized by a single parameter, its internal energy) is thus fulfilled. Exceptions to this pattern of behaviour exist, but they are rare. Many authors have attempted to develop a 'mode-selective chemistry', i.e. to promote reaction involving a chosen bond by excitation of that bond alone. These efforts have met with very limited success for unimolecular reactions, except in the case of van der Waals complexes where the coupling between the high-frequency intramolecular modes and the low-frequency intermolecular mode is weak.

Phase space

In principle, the evolution of a molecular ion on its lowest potential energy surface is governed by the equations of classical (or quantum) mechanics. Consider a molecular ion made up of N atoms. Its evolution is known when we know the values of the $3N$ coordinates and of the $3N$ conjugated momenta (related to generalized velocities) as a function of time. This information can be graphically represented as a trajectory in a $6N$-dimensional hyperspace, called the phase space.

Randomness is not accounted for by the deterministic classical equations. Indeed, at low internal energies, the nuclear trajectories appear regular (quasi-periodic). The molecular ion can then be described as a set of $(3N-6)$ harmonic or anharmonic but separable oscillators to which, if relevant, one, two or three overall rotations have to be added. Only a small fraction of the available phase space is then visited. However, at higher internal energies, the usual model of a collection of $(3N-6)$ independent oscillators breaks down. The nuclei then carry out extremely complicated trajectories and the fraction of phase space visited increases dramatically. If, for all trajectories characterized by a given energy, this fraction reaches a value of 100%, the system is declared 'ergodic' and statistical mechanics then generates exact equations. In technical terms, one has reached a state of microcanonical equilibrium. Even when the limit of 100% is not reached, useful equations can be derived from a statistical treatment, and this is the origin of the denomination 'quasi-equilibrium theory' adopted by Rosenstock and colleagues.

There are essentially two explanations that account for the success of the statistical approach.

Efficient intramolecular vibrational energy redistribution (IVR)

The potential energy surface of the ground electronic state of the molecular ion is assumed to be so anharmonic that the various vibrational degrees of freedom are strongly coupled. IVR is then expected to be rapid compared with the timescale of the reaction. When it follows photon excitation or electron impact, IVR is a sequential process. First, after a time of the order of 10^{-13} s, energy is redistributed among the optically active modes. The second step (for which a timescale of the order of about 10^{-11} s can be proposed) consists of an energy exchange from symmetric to antisymmetric modes by anharmonic coupling (e.g. Fermi resonances). Efficient IVR implies that each individual nuclear trajectory has visited most parts of the available phase space before dissociating. This case is represented schematically in **Figure 3A**. The isolated molecular ion is then expected to reach spontaneously the ergodic limit under collision-free conditions. IVR is known to be inefficient in the case of van der Waals complexes (and probably also in the case of weakly bonded species) because of the disparity between the high intramolecular frequencies and the low intermolecular ones.

Randomness of the initial conditions

For thermal reactions, the success of the statistical approach can be explained by invoking the

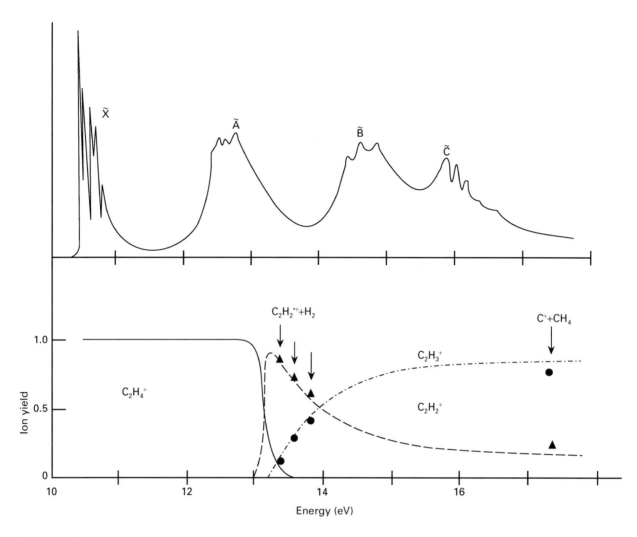

Figure 2 Top: photoelectron spectrum of ethylene. Bottom: breakdown graph of ethylene determined by photoion–photoelectron coincidence measurements. Triangles and dots: yields of $C_2H_2^{\bullet+}$ and $C_2H_3^+$ ions, respectively, obtained by collision experiments. The ground and excited electronic states of the $C_2H_4^{\bullet+}$ ion are denoted by $\tilde{X}$, $\tilde{A}$, $\tilde{B}$ and $\tilde{C}$.

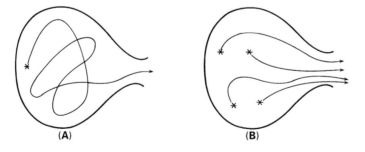

Figure 3 (A) A single ergodic trajectory visits a substantial part of phase space. (B) A swarm of nonergodic trajectories originating from widely different initial conditions fills up the entire phase space.

impossibility of specifying the initial conditions if energy is delivered by molecular collisions. As shown in **Figure 3B**, even if each individual trajectory does not meet the requirements of ergodicity, an average over a large number of randomly distributed initial conditions will lead to a quasi-uniform sampling of the

available phase space. For mass spectrometric experiments, this argument is apparently irrelevant. However, it should be noted that the initial conditions are also poorly defined, even if the experiment is carried out under collision-free conditions. As mentioned in the first section, the initial conditions are determined by the conversion of electronic into vibrational energy via a cascade of radiationless transitions to the ground state of the molecular ion. As suggested by **Figure 1**, the pattern of surface crossings can be extremely complicated in a polyatomic system. Each process ends up at a different phase-space cell of the reacting ground-state ion. In addition, autoionization (a process in the course of which internal energy of the ionic core is transferred to a Rydberg electron, which is then ejected) generates a molecular ion characterized by its particular initial distribution of the vibrational energy. All of these processes lead to a pattern where the initial conditions are scattered at various positions in phase space. The averaging over these widely different initial conditions ensures the validity of a statistical approach.

The transition state

The concept of transition state plays an essential role in the theory because it provides a simple, compact, but approximate description of the dynamics of the reaction. It can be defined in various ways. Essentially, one assumes that there exists some critical configuration of the reacting molecule, called the transition state, such that once the molecule has reached this configuration, it will irreversibly proceed on to products. (The term activated complex is sometimes used, but it is best to restrict its use to bimolecular reactions.)

The transition state, defined as it is as a point of no return, is often associated with the top of a potential energy barrier; that is, it is defined as a saddle point with a negative curvature of its potential energy surface along the reaction coordinate and hence with a single imaginary frequency. However, many molecular ions dissociate with no reverse activation energy barrier.

The transition state is not a stable molecule but a fictitious entity obtained after removal of the reaction coordinate, thus having one degree of freedom less than the reactant. Its best definition is that of a dividing surface in phase space. Reactant and products are assumed to be separated by a surface whose crossing is assumed to be irreversible (i.e., having crossed it once, the trajectories terminate as products without recrossing the surface backwards). The reaction is modelled as a flux through this dividing surface and its rate is measured by counting the number of times the surface is crossed per second (see **Figure 4**).

The RRKM-QET equation

Consider a microcanonical ensemble of molecular ions with an energy between E and $E + \delta E$. The total

number of quantum states available to members of the ensemble is denoted $\rho(E)\delta E$, where the function $\rho(E)$ is called the density of states. Of these, a certain fraction corresponds to transition states. The rate constant $k(E)$ is proportional to the ratio of these two quantities. Many unknown parameters defining the transition state cancel out in the final expression, which reads simply

$$k(E) = \sigma \int_0^{E-E_0} \rho^{\ddagger}(E')\, dE'/h\rho(E) = \sigma N^{\ddagger}(E-E_0)/h\rho(E)$$

[1]

where $N^{\ddagger}$ denotes the number of accessible quantum states of the transition state, i.e. those having a potential energy E_0 and a translational kinetic energy ε in the reaction coordinate, with the remainder ($E - E_0 - \varepsilon$) in the bound degrees of freedom, and σ is the reaction path degeneracy, i.e. the number of equivalent paths leading from the reactant to the products. (The quantities affixed with a double dagger refer to the transition state, which has one degree of freedom less than the reactant.)

At threshold (i.e. when $E = E_0$), $N^{\ddagger}(0) = 1$. Thus, the minimum rate is in principle equal to $1/h\,\rho(E_0)$. However, when the internal energy E is equal to, or even is slightly less than the barrier, then the reaction proceeds via tunnelling and a transmission probability κ is then inserted into Equation [1]. Isotope effects are then to be expected. When the shape of the barrier can be described by an inverted parabola, a simple expression can be derived for the transmission

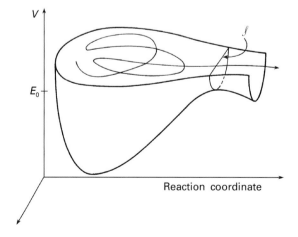

Figure 4 Potential-energy surface for a unimolecular fragmentation, with the dividing surface s separating the reactant from the products.

coefficient:

$$\kappa(E) = \frac{\exp[2\pi(E-E_0)/h\nu^{\ddagger}]}{1+\exp[2\pi(E-E_0)/h\nu^{\ddagger}]} \qquad [2]$$

where $\nu^{\ddagger}$ is the modulus of the imaginary frequency of the saddle point. Expressions for more complicated, unsymmetrical barriers are also available.

Practical calculations

There exists a computer algorithm (developed by Beyer and Swinehart) for counting all the possible combinations of the vibrational quantum numbers that are consistent with a specified value of the internal energy, even for a collection of anharmonic (but independent) oscillators.

Alternatively, analytical expressions for the energy-level densities can be obtained by inverse Laplace transformation of the corresponding partition function. This works well for the rotational degrees of freedom. For example, for a single rotor having a symmetry number equal to σ, one has simply (except at very low energies):

$$\rho(E) = \sigma^{-1}(BE)^{-1/2} \qquad [3]$$

where B is the rotational constant.

However, for a collection of oscillators, the set of energy levels is not dense enough and the energy-level density has to be numerically calculated by the steepest-descent method implemented by Forst. Nevertheless, an approximate closed-form expression involving an empirical correction for the effect of the zero-point vibrational energy has been developed by Whitten and Rabinovitch. For a system of n oscillators having frequencies ν_i, one has:

$$\rho(E) = \frac{(E+aE_z)^{n-1}}{(n-1)!\prod_{i=1}^{n} h\nu_i} \qquad [4]$$

where E_Z is the total zero-point energy and a is an empirical parameter that usually ranges between 0.7 and 0.98. This formula leads to satisfactory results for the calculation of the denominator of Equation [1]. For its numerator, a direct count of the sum of states $N^{\ddagger}(E-E_o)$ is preferable, because at lower energies the vibrational levels are usually too widely spaced for a classical or semiclassical approximation to work.

The numerical values of the frequencies and moments of inertia needed for an actual calculation can be obtained from commonly used quantum-chemistry programs. This is the only possibility for the frequencies of the transition state.

The role played by the overall rotations raises a difficult problem. The constraint of angular momentum conservation prevents certain rotational degrees from getting involved in randomization. Very often, the transition state is approximated as a symmetric top, with two equal moments of inertia. It is then usually assumed – possibly incorrectly – that the degenerate two-dimensional external rotation is unable to couple significantly with the vibrational degrees of freedom, whereas the rotation about the reaction coordinate is able to exchange energy with them. However, although unavailable for randomization, the rotational energy stored in the degenerate external two-dimensional rotation modifies the radial potential and gives rise to an effective potential characterized by a centrifugal barrier.

More elaborate statistical models

It has been found useful to introduce new concepts and methods, particularly in the study of reactions for which no potential barrier is encountered along the reaction coordinate, a case often encountered in the dissociation of molecular ions. In such a case, the conservation of the orbital angular momentum during the reaction process has important consequences.

Microcanonical variational transition state theory (VTST)

If the transition state is defined as a structure denoting unstable equilibrium between the reactant and the products, then any point of its phase space having a nonzero velocity in the forward direction will react. However, this criterion is oversimplified: the condition is necessary but not sufficient. After having left the transition state, some trajectories may return to cross it again. Hence, the rate constant calculated by the transition state theory is an overestimate of the exact value. (This conclusion has been challenged, however, when energy is selectively delivered to the system, as in laser light excitation.) Leaving the controversial cases aside for deeper scrutiny, it follows that the best choice for the transition state is the one that minimizes the calculated rate constant. Therefore, in VTST, the numerator of Equation [1] is written as $N^{\ddagger}[E-V_{eff}(R^{\ddagger})]$, where $V_{eff}(R)$ denotes the effective potential for motion along the reaction coordinate R (usually, the sum of the actual and a centrifugal potential), and where $R^{\ddagger}$

denotes the position of the dividing surface that minimizes $N^{\ddagger}$.

Transition state switching

The rate constant does not necessarily admit a unique minimum. Thus, several transition states may simultaneously exist along the reaction coordinate with characteristics depending on the energy and angular momentum. For example, a tight transition state may be followed by an orbiting complex. The former determines the magnitude of the rate constant, whereas the latter controls the translational, rotational and vibrational energy distributions of the products.

Bottlenecks

This term is often encountered, but is used with two different connotations. First, it may describe an impediment to IVR (resulting, for example, from the presence of a heavy atom or from a disparity in the frequencies). Alternatively, it may describe a throttling of the reactive flux (i.e., a new kind of transition state) due, for example, to a region of strong curvature of the reaction path resulting, for example, from a detour around a conical intersection between two potential-energy surfaces.

The statistical adiabatic channel model (SACM)

Troe and Quack have defined reaction channels as a set of potential-energy curves obtained by connecting the vibration–rotational states of the reactant to those of the pair of products along the reaction coordinate. (This procedure is referred to as an adiabatic correlation.) Barriers arise when imposing conservation of the orbital angular momentum during the reaction process to the transitional modes (i.e. the vibrational bending modes that correlate with rotations or orbital motion of the fragments). A channel is said to be open when its potential energy curve never exceeds the available internal energy E. No reference to a transition state is made in this theory. The rate constant now depends on two parameters, viz., the internal energy E and the total angular momentum J. Its expression is remarkably similar to the RRKM-QET equation:

$$k(E, J) = \frac{N^*(E, J)}{h\rho(E, J)} \qquad [5]$$

where $N^*(E, J)$ is now defined as the total number of open channels.

The orbiting transition state phase space theory (OTS/PST)

The unimolecular reactions of polyatomic ions are assumed to be governed by the long-range part of the potential. The principle of microscopic reversibility has been used by Klots to express the rate constant, not in terms of the properties of a transition state, but in terms of the cross section for association of fragments. The latter is then evaluated by the Langevin theory, which is based on the consideration of a long-range attractive potential (e.g. $-\alpha q^2/2R^4$). Alternatively, an orbiting transition state located at a centrifugal barrier has been reintroduced by Chesnavich and Bowers. In both cases, the density of states is no longer evaluated as a simple convolution between the vibrational and rotational parts, but by introducing restrictions resulting from the conservation of angular momentum. This theory often provides too large estimates for the rate constant.

Effective number of states

Information theory shows that if the assumption of complete energy randomization breaks down, then Equation [1] remains valid provided it is interpreted as a ratio between an *effective* number of states divided by an *effective* density of states. Both the numerator and the denominator are then reduced (but not necessarily to the same extent) by a so-called entropy deficiency factor. A mechanism of cancellation of errors arises, which accounts for the success of the simple theory.

Nonadiabatic reactions

Nonadiabatic reactions (i.e. those involving a change in the electronic state or structure) occur more frequently in mass spectrometry than is commonly thought. Surprisingly enough, statistical methods are found to be useful even in these cases.

Kinetic energy release distributions (KERDs)

The internal energy in excess of the thermodynamic dissociation threshold is partitioned among the translational, rotational and vibrational degrees of freedom of the pair of fragments. In the case of molecular ions, the translational kinetic energy release distribution (KERD) can be measured and its study is a precious source of information on the dynamics of the reaction. Indeed, it is a stronghold of the more elaborate statistical theories presented in the previous paragraph because the usual RRKM-QET theory is in principle not appropriate for such a study. The reason is that transition state theory concentrates on the behaviour of the system up to the dividing surface, whereas the product energy disposal is determined by the potential felt by the fragments as they separate. However, Marcus, Wardlaw and Klippenstein have proposed an extension that describes the evolution of the transitional modes along the reaction coordinate and that, at the same time, conserves angular momentum.

Phase space orbiting transition state theory works much better for the calculation of KERDs than for that of the rate constants, thereby demonstrating that the former are controlled by the long-range part of the potential whereas the latter are governed by its shorter range. In addition, Klots has introduced a set of effective temperatures to parametrize the observed distributions. The SACM has also demonstrated its usefulness in the case of weakly bonded species.

The maximum entropy method and the associated surprisal theory are an outgrowth of information theory. They involve a comparison between the actual shape of the KERD and the hypothetical, most statistical, so-called 'prior distribution'. Two precious pieces of information can be derived from this comparison: (i) an identification of the constraint that operates on the dynamics and prevents it from being statistical and (ii) the magnitude of the entropy deficiency which can be related to the fraction of phase space effectively sampled by the transition state. Values of 75–80% have been obtained in the case of the halogenobenzene ions.

Concluding remarks

It has often been naggingly remarked that the RRKM-QET theory can fit anything and predict nothing. To counter this criticism, many authors have multiplied skilful consistency checks (study of isotope effects, preparation of the ion via a bimolecular reaction or via charge reversal in addition to electron or photon impact, time-resolved studies all the way from the millisecond to the nanosecond timescales, etc.) and have removed arbitrariness via *ab initio* calculations of frequencies. However, it should be realized that ability to fit the experiments by no means implies that the theory is exact and that its basic assumptions (full energy randomization and existence of a good transition state) are fulfilled. It has been seen that Equation [1] cannot be grossly in error because of a mechanism of cancellation of errors. In contradistinction, KERDs (for which the cancellation of errors does not work because they basically depend on the numerator only) provide a much better way to test the validity of the

assumption of complete energy randomization than a study of fitted rate constants.

List of symbols

B = rotational constant; E = internal energy; E_Z = zero-point energy; h = Planck constant; J = total angular momentum; $k(E)$ = rate constant; $N^{\ddagger}$ = number of accessible states of the transition state; $R^{\ddagger}$ = position of dividing surface that minimizes $N^{\ddagger}$; $V_{eff}(R)$ = effective potential for motion along reaction coordinate R; κ = transmission probability; $\nu^{\ddagger}$ = modulus of imaginary frequency of saddle point; ν_i = oscillator frequency; $\rho(E)$ = density of states; σ = symmetry number.

See also: **Fragmentation in Mass Spectrometry; Ion Dissociation Kinetics, Mass Spectrometry; Metastable Ions; Photoelectron-Photoion Coincidence Methods in Mass Spectrometry (PEPICO).**

Further reading

Baer T and Hase WH (1996) *Unimolecular Reaction Dynamics. Theory and Experiments.* Oxford: Oxford University Press.

Baer T (1996) The calculation of unimolecular decay rates with RRKM and *ab initio* methods. In: Baer T, Ng CY and Powis I (eds) *The Structure, Energetics and Dynamics of Organic Ions*, pp 125–166. Chichester: Wiley.

Baer T and Mayer PM (1997) Statistical Rice–Ramsperger–Kassel–Marcus quasiequilibrium theory calculations in mass spectrometry. *Journal of the American Society for Mass Spectrometry* 8: 103–115.

Chesnavich WJ and Bowers MT (1979) Statistical methods in reaction dynamics. In: Bowers MT (ed) *Gas Phase Ion Chemistry*, Vol 1, pp 119–151. New York: Academic Press.

Derrick PJ and Donchi KF (1983) Mass spectrometry. In: Bamford CH and Tipper CFH (eds) *Comprehensive Chemical Kinetics*, Vol 24, pp 53–247. Amsterdam: Elsevier.

Gilbert RG and Smith SC (1990) *Theory of Unimolecular and Recombination Reactions.* Oxford: Blackwell.

Illenberger E and Momigny J (1992) *Gaseous Molecular Ions.* Darmstadt: Steinkopff.

Lifshitz C (1989) Recent developments in applications of RRKM-QET. *Advances in Mass Spectrometry* **11A**: 713–729.

Lifshitz C (1992) Recent developments in applications of RRKM-QET. *Advances in Mass Spectrometry* **12**: 315–337.

Lorquet JC (1994) Whither the statistical theory of mass spectra? *Mass Spectrometry Reviews* **13**: 233–257.

Lorquet JC (1996) Non-adiabatic processes in ionic disso-
ciation dynamics. In: Baer T, Ng CY and Powis I (eds)
*The Structure, Energetics and Dynamics of Organic
Ions*, pp 167–196. Chichester: Wiley.

Wardlaw DM and Marcus RA (1988) On the statistical
theory of unimolecular processes. *Advances in Chemical Physics* 70: 231–263.

Stereochemistry Studied Using Mass Spectrometry

Asher Mandelbaum, Technion–Israel Institute of
Technology, Haifa, Israel

MASS SPECTROMETRY

Applications

Stereochemical effects in mass spectrometry have
been used for configurational assignment in numerous organic systems. The unsurpassed sensitivity of
mass spectrometry and the possibility of interfacing
the mass spectrometer with a variety of separating
devices (most commonly gas chromatography–mass
spectrometry (GCMS) and liquid chromatography–
mass spectrometry (LCMS)), make it a most useful
tool for structural assignment of minor constituents
of complex mixtures, including stereochemical information. The mass spectral stereochemical effects
may also be an important and useful tool in structural studies of gas-phase ions and in mechanistic investigations of their fragmentation processes.

Mass spectral stereochemical effects occur in the
molecular radical cations $M^{\bullet+}$, obtained usually on
electron ionization (EI), in protonated molecules
MH^+ upon chemical ionization (CI) and other soft
ionization techniques (fast atom bombardment
(FAB), electrospray, thermospray), and, to a lesser
extent, in negative ions. These effects are observed in
the normal mass spectra and also by tandem mass
spectrometric techniques (MS/MS).

The different mass spectral behaviour of stereoisomers (this term will not include enantiomers in
this article) may be due either to the different thermochemical nature of the chemistry of their gas-
phase ions, or it may result from the different
kinetics of their reactions. These two aspects will be
dealt with in the following sections.

Thermochemical considerations

Electron ionization

Small, often insignificant or within the experimental
error, differences (in most reports below 0.1 eV)

have been observed in the ionization energies of
stereoisomers in a number of systems. The appearance energies of some specific fragment ions often
show a more pronounced dependence on the configuration of stereoisomers. Lower energies have been
reported for $[M-R]^+$ ions obtained from stereoisomers with higher enthalpies of formation in various dialkylcycloalkanes and in other more complex
heterocyclic and polycyclic systems. The difference
between the appearance energies could be quantitatively correlated with the difference in the enthalpies
of formation of the stereoisomers in several systems.
The lower appearance energies of the thermochemically less stable stereoisomers have been attributed to
the release of steric strain in the course of the fragmentation.

Since the early days of mass spectrometry it has
been proposed, that the release of steric strain results, in many cases, in lower abundances of the molecular ions and in higher abundances of fragments
in the EI mass spectra of the thermochemically less
stable isomers in numerous systems. The differences
between the mass spectra of stereoisomers in those
systems are usually not large, and there are numerous exceptions which cast doubt on the reliability of
this approach in real problems of configurational
assignments.

Chemical ionization

In contrast to the ionization energies, considerably
different proton affinities and gas-phase basicities
have been reported for a number of stereoisomeric
difunctional compounds, which have an impact on
their behaviour upon chemical ionization. For
example, the gas-phase basicities of *cis*- and *trans*-2-
amino-3-hydroxybicyclo[2.2.2]octanes 1c and 1t
(**Scheme 1**) are 926 and 909 kJ mol^{-1}, respectively.

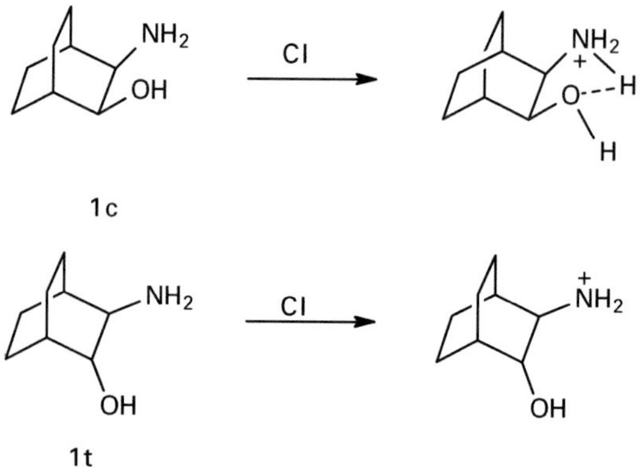

Scheme 1

The difference is attributed to the different interfunctional distance in the two stereoisomers (2.31 and 3.50 Å respectively) resulting from their different dihedral angles about the HOC–CNH$_2$ bond (30° and 150°). The effective intramolecular hydrogen bond (often termed proton bridging), which is possible in the MH$^+$ ion of the *cis*-amino alcohol 1c (**Scheme 1**), results in the higher gas-phase basicity of this stereoisomer.

A smaller difference (4–5 kJ mol^{-1}) has been reported between the stereoisomeric *cis*- and *trans*-4-phenyl-1-alkoxycyclohexanols 2c and 2t (**Scheme 2**), and it has been attributed to proton bridging between the alkoxylic oxygen atom and the phenyl ring, which is possible only in the MH$^+$ ions of 2c (**Scheme 2**).

Easy distinction between stereoisomers containing two or more basic sites, may be achieved upon CI using selected reagent gases, based on the remarkable difference in their proton affinities and gas-phase basicities, due to the effectiveness of proton bridging. Thus, diesters with remote alkoxycarbonyl groups (e.g. fumarates, mesaconates, *trans*-1,3 and 1,4-cyclohexane dicarboxylates) give rise to low abundance MH$^+$ ions and to abundant [M+NH$_4$]$^+$ adduct ions upon NH$_3$-CI, because of the higher proton affinity of ammonia. On the other hand, the *cis* analogues with adjacent ester groups (e.g. maleates, citraconates, *cis*-1,3- and 1,4-cyclohexane dicarboxylates), which may undergo intramolecular proton bridging on protonation, afford abundant MH$^+$ ions under NH$_3$-CI (**Scheme 3**).

Stabilization of the MH$^+$ ions by intramolecular hydrogen bonding may be used as a simple tool in the configurational assignment of stereoisomeric diols, diethers and other difunctional analogues. Stereoisomers with adjacent basic functions afford abundant proton-bridged MH$^+$ ions under CI conditions. Counterparts with remote functions, which cannot be stabilized by proton bridging, undergo fragmentation processes resulting in less abundant MH$^+$ ions in their CI mass spectra. Typical examples are shown in **Scheme 4**, and many others have been reported.

The above stabilization effect is limited to systems such as 4 and 5 (**Scheme 4**), containing at least one functional group (e.g. OH, OR) which undergoes ready dissociation upon CI. Other systems, e.g. diesters, with functional groups that are stable under CI conditions, behave in a different manner. Protonated esters of alkanoic or cycloalkanoic acids exhibit a high degree of stability upon isobutane-CI. Protonation occurs at the carbonylic oxygen atom,

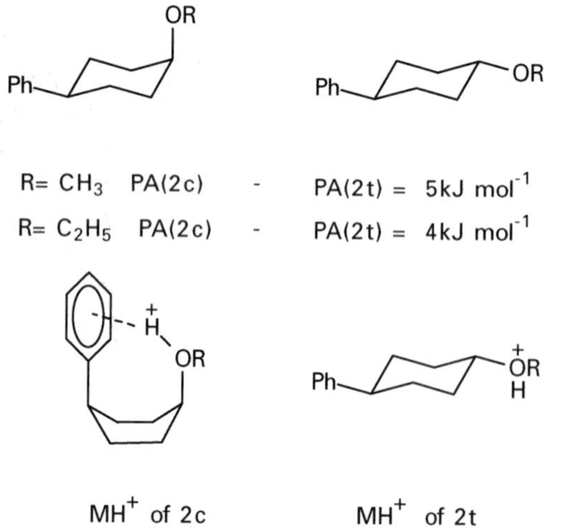

R= CH$_3$ PA(2c) – PA(2t) = 5kJ mol^{-1}
R= C$_2$H$_5$ PA(2c) – PA(2t) = 4kJ mol^{-1}

MH$^+$ of 2c   MH$^+$ of 2t

Scheme 2

cis-3   MH$^+$

trans-3

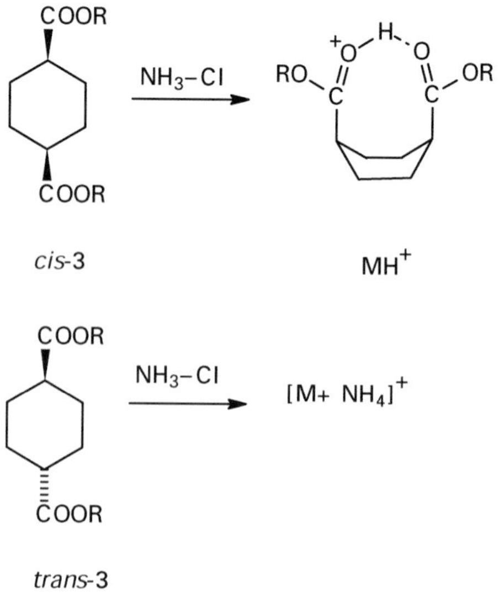

Scheme 3

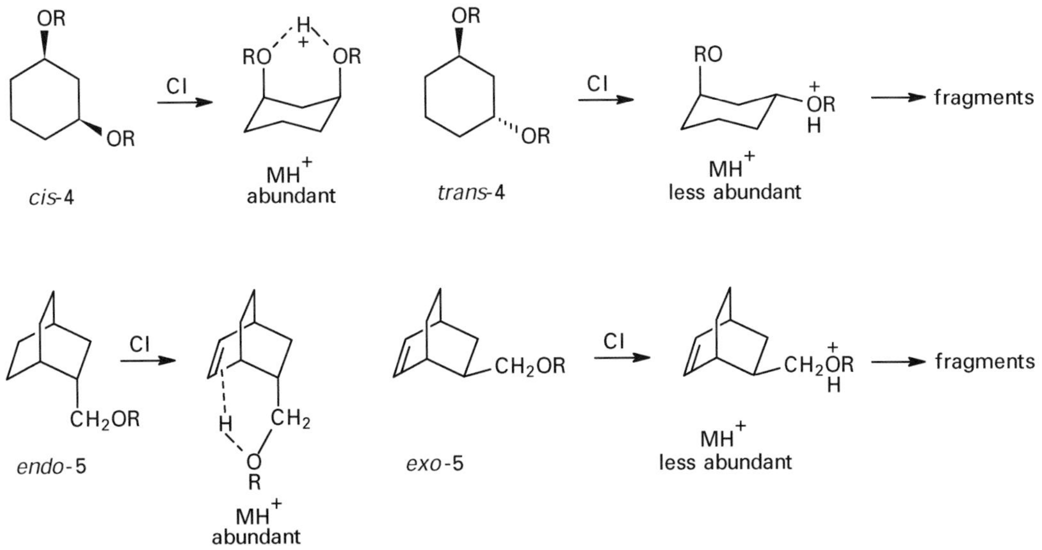

Scheme 4

and the 1,3-proton transfer to the alkoxy oxygen, that would enable alcohol elimination, is a symmetry forbidden process. The nonbridged MH$^+$ ions of diesters, with the two remote noninteracting alkoxycarbonyl groups (e.g. *trans*-6) (**Scheme 5**), are stable (and consequently highly abundant) under isobutane-CI. On the other hand, stereoisomers with adjacent ester groups (e.g. *cis*-6) (**Scheme 5**), undergo facile proton transfer between the two alkoxycarbonyls *via* proton-bridged intermediates, resulting the efficient elimination of ROH (**Scheme 5**). In the latter systems, the thermochemically stabilized proton-bridged species are kinetically destabilized, resulting in less abundant MH$^+$ ions.

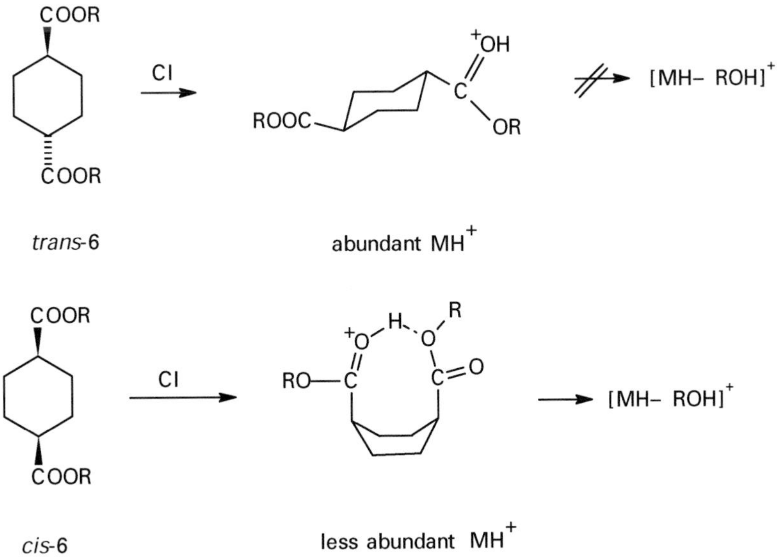

Scheme 5

Kinetic considerations

A wide variety of fragmentation processes of the $M^{•+}$ and MH^+ ions obtained upon EI and CI involve formation of new bonds. Such unimolecular rearrangements take place via cyclic transition states. Molecular ions of stereoisomers often differ in the accessibility of the transition structures for fragmentation processes, which involve bond formation or concerted rupture of several bonds. Relatively large differences in the energy of activation of such processes may result in extreme cases in total suppression of particular pathways in one of the stereoisomers. A large number of systems have been reported that show stereospecific fragmentation behaviour upon EI and CI. A limited number of typical examples will be given in the following.

Hydrogen transfer

Elimination of H_2O from cycloalkanols and related processes One of the early cases of stereospecific fragmentation processes of gas-phase ions was the EI-induced dehydration of stereoisomeric 4-sustituted cyclohexanols. *Trans*-4-*t*-butyl- and -4-arylcyclohexanols 7t (**Scheme 6**) afford highly abundant $[M-H_2O]^{•+}$ ions, which are much less abundant in the mass spectra of the *cis* isomers 7c (**Scheme 6**). Similar stereospecificity has been also observed in the elimination of methanol and acetic acid from the corresponding methyl ethers and acetates. Deuterium labelling studies have shown that the H-atom from position 4 is abstracted in the course of the elimination of H_2O from the *trans*-alcohols, but not to an appreciable extent in the case of the *cis* isomers. The low energy H-transfer from the tertiary

(and benzylic when R=aryl) position 4 to the hydroxy group is the stereospecific step in this fragmentation. The cyclic transition state of this H-transfer is attainable only in 7t (**Scheme 6**) in boat conformation. The high stereospecificity of this process indicates retention of the original structure in the molecular radical cation.

A similar stereospecific behaviour has also been observed in the stereoisomeric 3-arylcyclohexanols and in their methyl ethers and acetates, but not in the 2-aryl analogues, which undergo cleavage of the C_1–C_2 bond prior to fragmentation, resulting in the loss of stereochemical information. Many additional systems have been shown to undergo EI-induced stereospecific eliminations of water and other neutral molecules, which may be applied in structural studies.

It is noteworthy that stereoisomeric cycloalkanols and their derivatives exhibit a lower degree of stereospecificity in their dissociation upon CI. Protonation occurs at the oxygen atom in these materials, and the concurrent elimination, that takes place by a simple cleavage of the C–O bond, is governed by thermochemical factors.

Cycloalkylidene acetates The geometrically isomeric *E*- and *Z*- disubstituted cycloalkylidene acetates (e.g. E-8 and Z-8) exhibit a highly specific EI-induced loss of one of the two substituents R and R′ at the two homoallylic positions, as shown in **Scheme 7**. The specific H-transfer from the allylic position adjacent to the carbonyl toward the carbonylic oxygen atom has been proposed as explanation for the stereospecificity of this process. These results enable easy distinction and configurational assignment of stereoisomers in analogous systems

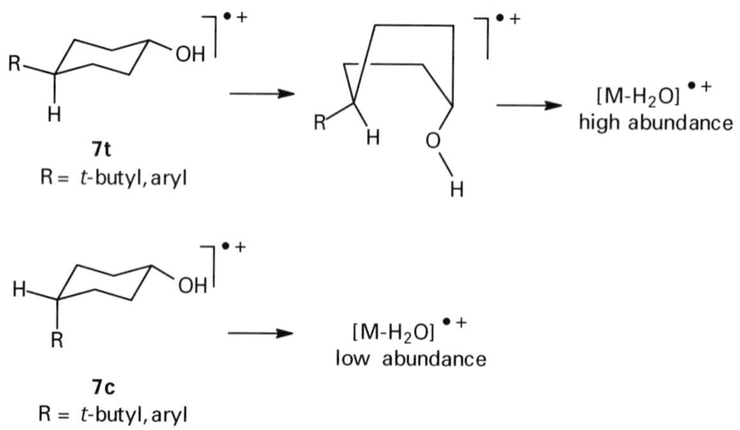

Scheme 6

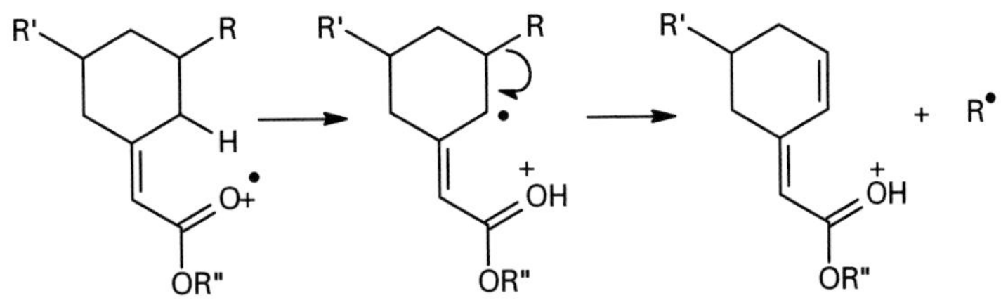

Scheme 7

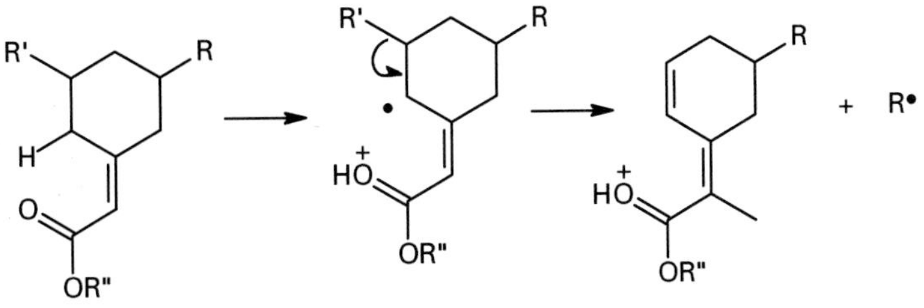

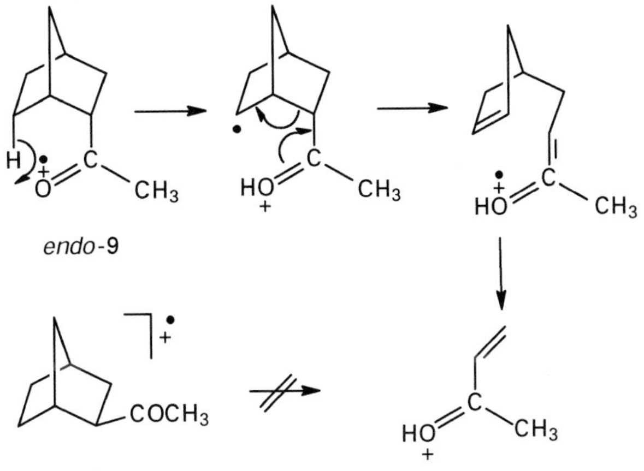

Scheme 8

(including acyclic analogues), which is not straightforward by other spectroscopic methods. They also indicate that the rotation about the double bond in the $M^{\bullet+}$ ions must be slower than the H-transfer in this process.

McLafferty rearrangement and related processes The stereospecificity of the McLafferty rearrangement is exemplified by the different behaviour of *endo-* and *exo-*acetylnorbornanes *endo-*10 and *exo-*10 upon EI, shown in **Scheme 8**. The *m/z* 71 ion is abundant in the EI-mass spectrum of *endo-*10, but absent in that of *exo-*10. McLafferty rearrangement

is the initial step in the formation of the *m/z* 71 ion with the involvement of the H-atom from position 6, which is in the proper distance from the carbonylic oxygen only in the *endo-* isomer.

Steric interactions in the cyclic transition structures for a hydrogen transfer result in different abundance of the butene radical cation fragment, obtained from the diastereoisomeric unsaturated alcohols *erythro-*10 and *threo-*10. The higher energy of the transition state of *erythro-*10, due to the steric interaction of the methyl and R groups, results in a lower abundance of the fragment, as compared with the stereoisomeric *threo-*9 (see **Scheme 9**).

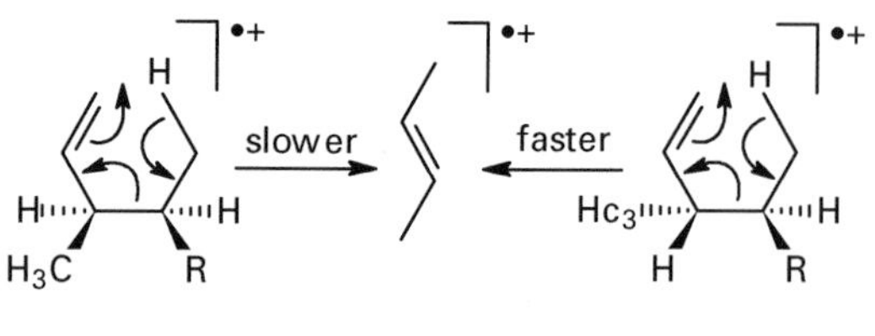

erythro-10   m/z 56   threo-10

Scheme 9

Alcohol elimination from MH⁺ ions diesters upon CI and CID It has been previously mentioned that MH⁺ ions of diesters, with the two remote noninteracting alkoxycarbonyl groups, are stable and consequently highly abundant under isobutane-CI. On the other hand, stereoisomers with adjacent ester groups undergo a facile proton transfer between the two alkoxycarbonyls via proton bridged intermediates, resulting in the efficient elimination of ROH (**Scheme 5**). This behaviour enables easy distinction between stereoisomeric diesters by CI mass spectrometry, and also by CID measurements of MH⁺ ions. CID mass spectra of the m/z 173 MH⁺ ions of diethyl maleate Z-11 and fumarate E-11 exhibit an entirely different behaviour (**Scheme 10**): that of Z-11 shows an abundant m/z 127 [MH–EtOH]⁺ ion which is absent in the spectrum of E-11, while that of E-11 exhibits abundant m/z 145 [MH–C₂H₄]⁺ and m/z 117 [MH–2C₂H₄]⁺ ions, which do not appear in the CID spectrum of Z-11.

These distinctive features of the CID spectra also enable structural assignments and quantitative relative abundance estimates of protonated maleate and fumarate ions, which are formed by mass spectral fragmentation of higher systems. For example, this method was used to determine that the retro-Diels–Alder (RDA) fragmentation of cis- and trans-2,3-diethoxycarbonyl-5,6,7,8-dibenzobicyclo[2.2.2.]octanes 12 under i-C₄H₁₀-CI and CH₄-CI conditions is highly stereospecific, giving rise to protonated diethyl maleate and fumarate, respectively (see **Scheme 11**). This behaviour is consistent with a single-step concerted mechanism, analogous to the ground state RDA process occurring in neutral molecules in the condensed phase. On the other hand, analogous dissociation is nonstereospecific in endo-, exo- and trans-2,3-diethoxycarbonyl-5,6-benzobicyclo[2.2.2]-octanes 13, indicating involvement of a step-wise mechanism in this system (**Scheme 12**).

Retro-Diels–Alder (RDA) fragmentation

The retro-Diels–Alder (RDA) fragmentation is highly stereospecific in a variety of bi-, tri-, tetra- and pentacyclic systems. Two examples are shown in **Scheme 13**. The diene radical cations are the most abundant species observed in the EI mass spectra of the cis isomers 14c and 15c, and practically absent in the trans counterparts 14t and 15t. The very high efficiency of this process in the cis isomers and the high stability of the molecular ions of the trans analogues suggest that the RDA dissociation of gas-phase ions takes place by a concerted mechanism, which exhibits symmetry conservation characteristics similar to those of the ground-state RDA process in neutral species.

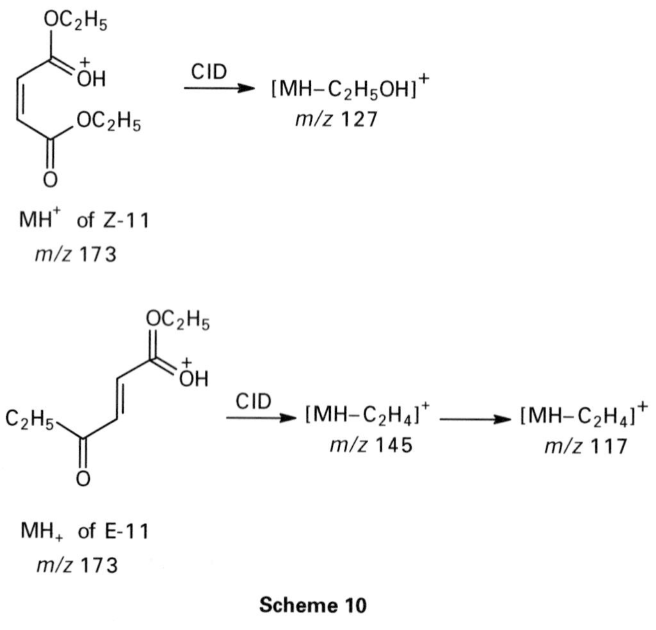

MH⁺ of Z-11
m/z 173

MH₊ of E-11
m/z 173

Scheme 10

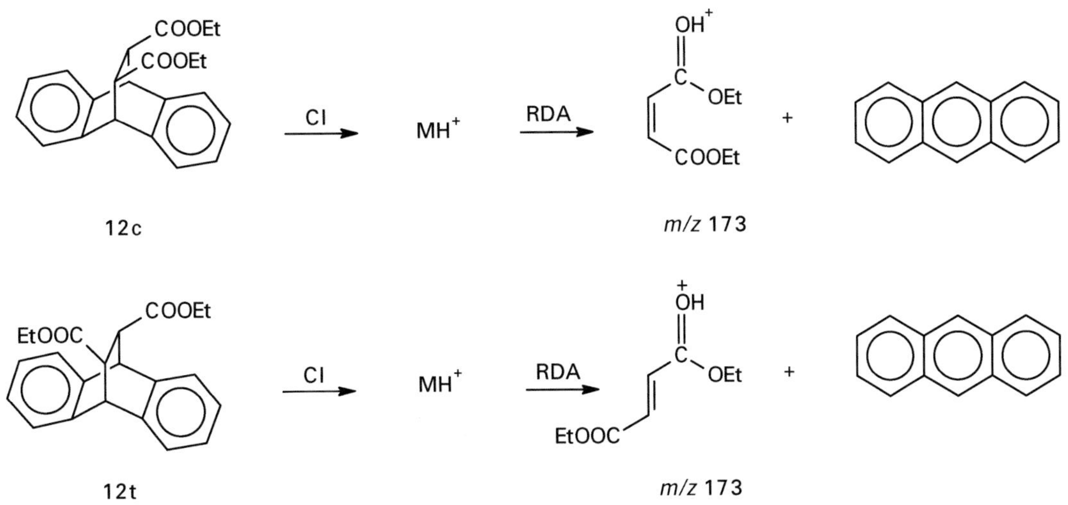

Scheme 11

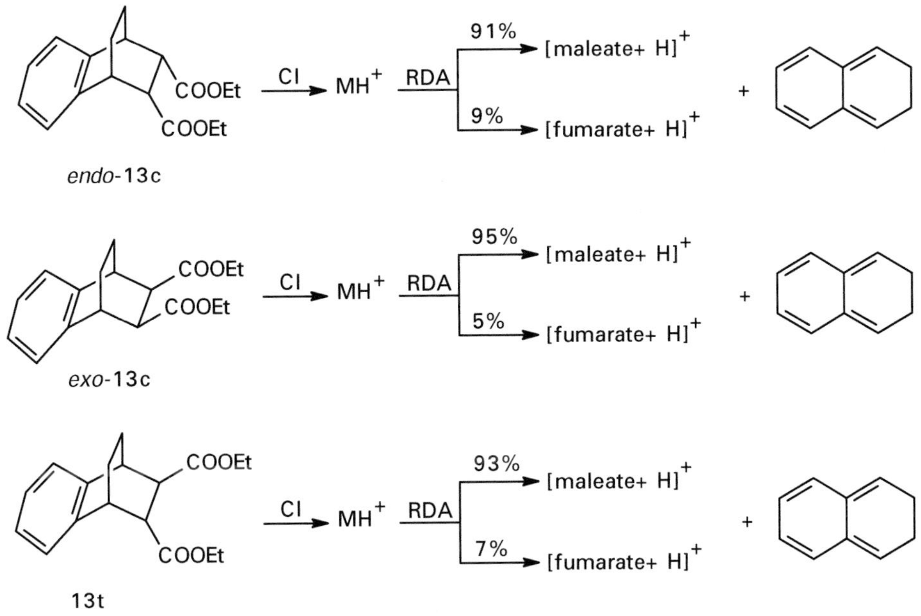

Scheme 12

Similar behaviour has been also observed upon CI. Protonated dienes are formed only from *cis* isomers. Protonated dienophiles have been also observed in certain cases under CI, again only in *cis* isomers.

In contrast with the above behaviour, there have been reports on the occurrence of EI-induced RDA fragmentation in compounds with a *trans* junction of the cyclohexane and the adjacent rings (e.g. 16 and 17, **Scheme 14**). These results are indicative of a step-wise dissociation in these systems, in which the allylic bonds cleaved in the course of this process are either highly (as in 16) or lightly (as in 17) substituted. It has been proposed that the nonstereospecific step-wise behaviour of systems such as 16, results from the low energy requirement for the cleavage of the fully substituted allylic 9–14 bond, which presumably is the initial step of the step-wise RDA dissociation. In systems such as 17, the low substitution pattern of the bonds cleaved in the course of the RDA process may result in a relatively high energy of activation of the concerted pathway and in the consequent preference of the step-wise mechanism.

The partial stereospecificity of the RDA process in system 18 (**Scheme 15**) has been the subject of a thorough examination. Theoretical calculations and critical energy measurements led to the conclusion

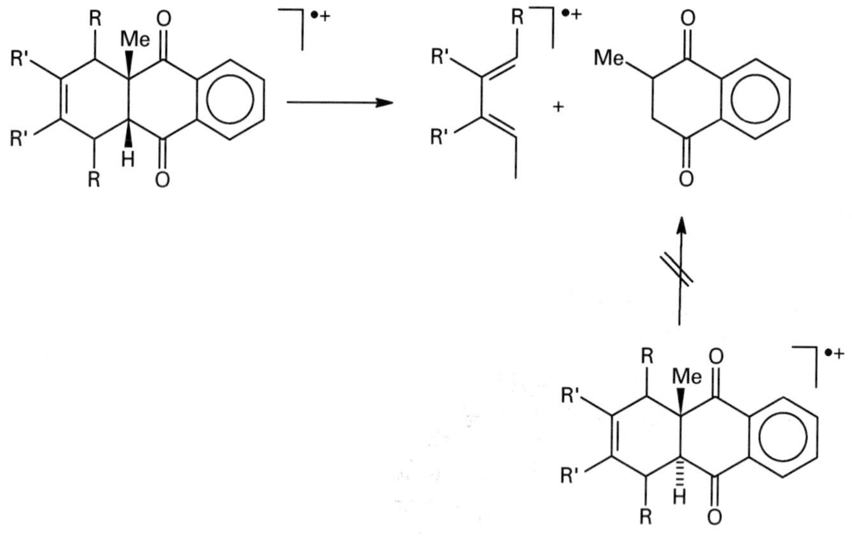

R= Me,Et; R'= Me,H; RR'= (CH$_2$)$_n$ (n = 3,4,5)

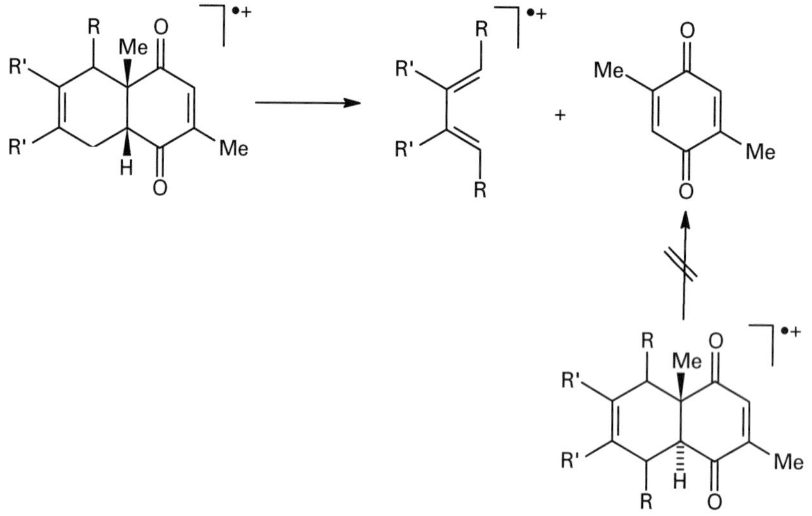

R= Me,Et; R'= Me,H; RR'= (CH$_2$)$_n$ (n = 3,4,5)

Scheme 13

that the RDA fragmentation of both stereoisomers is a step-wise process involving a common distonic intermediate, but the energy barriers are different for the stereoisomers (**Scheme 15**).

Anchimeric assistance

In many systems, variations in the abundances of certain fragment ions in the mass spectra of stereoisomers have been ascribed to anchimeric assistance. The loss of bromine from the molecular ions of stereoisomeric 1,2-dibromocyclopentanes and dibromocyclohexanes 19 (**Scheme 16**) is strongly affected by their configuration. The ion abundance ratio [M–Br]$^+$/M$^{\bullet+}$ is higher for the *trans* isomer by a factor of 10 in the dibromocyclopentane and 42 in the dibromocyclohexane system at 70 eV, and the difference between the stereoisomers increases at lower ionization energies. A similar effect has also been observed in acyclic diastereoisomeric vicinal dibromoalkanes. Anchimeric assistance of one bromine atom in the expulsion of the other from the molecular ion is proposed as the explanation of the stereospecificity of this process

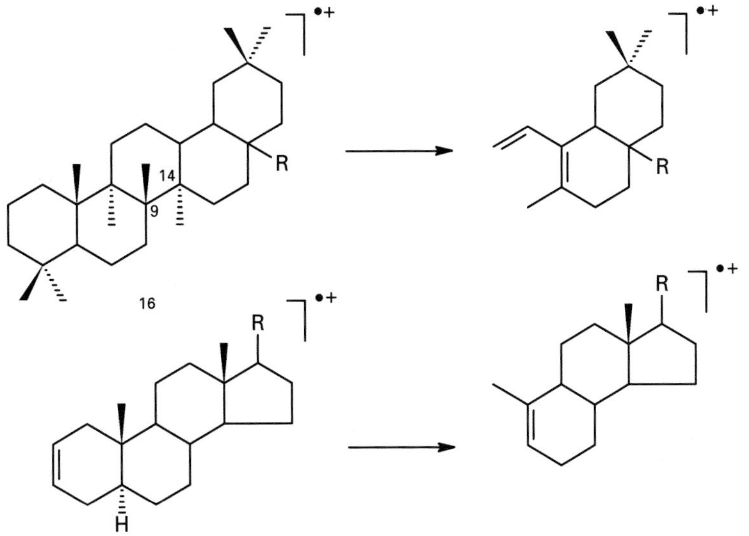

Scheme 14

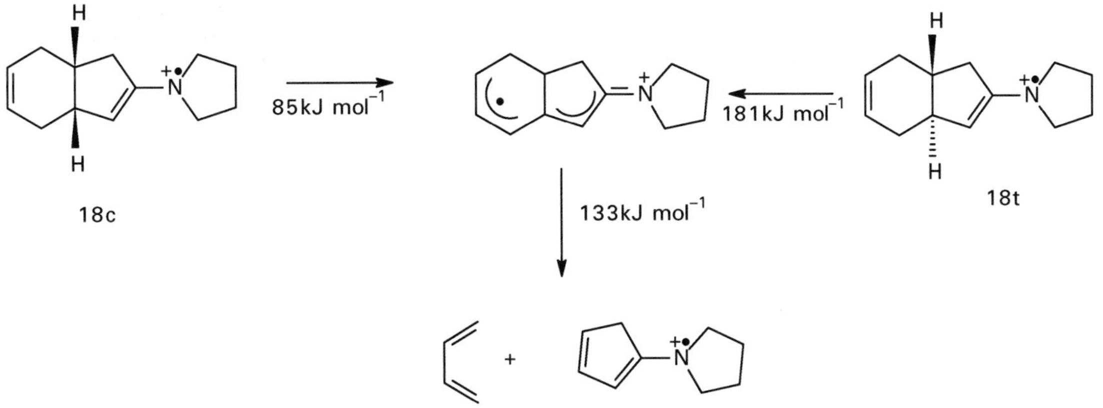

Scheme 15

(**Scheme 16**). This proposed mechanism finds support in the stereospecific behaviour of the two *trans*-1,2-dibromo-4-*t*-butylcyclo-hexanes (a)-20t and (e)-20t (**Scheme 17**). The stereoisomer (a)-20t, with the two axial Br atoms in the antiperiplanar conformation, gives rise to a much more abundant [M–Br]⁺ ion than the diequatorial analogue (e)-20t.

The greater extent (by a factor of 2–10) of elimination of acetic acid from the MH⁺ ions of *trans*-1,2- and 1,3-diacetoxycyclopentanes and -cyclohexanes upon CH₄-CI and isobutane-CI, as compared with the *cis* isomers, has been also interpreted in terms of anchimeric assistance. The carbonylic oxygen of the nonprotonated acetoxy group in the *trans* isomers (e.g. 21t in **Scheme 18**) assists in the elimination of

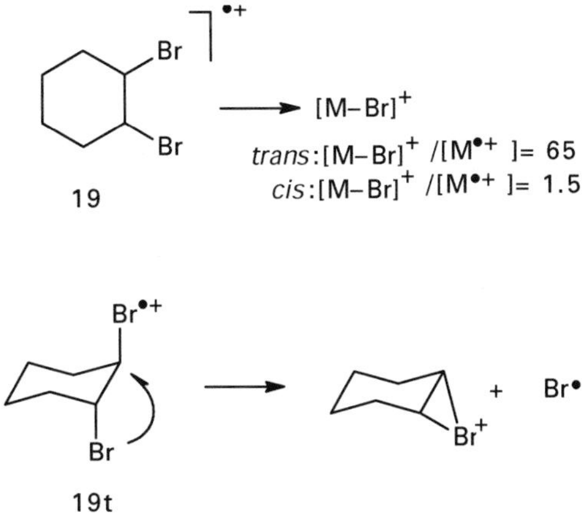

$trans$:[M–Br]⁺ /[M•⁺]= 65
cis:[M–Br]⁺ /[M•⁺]= 1.5

Scheme 16

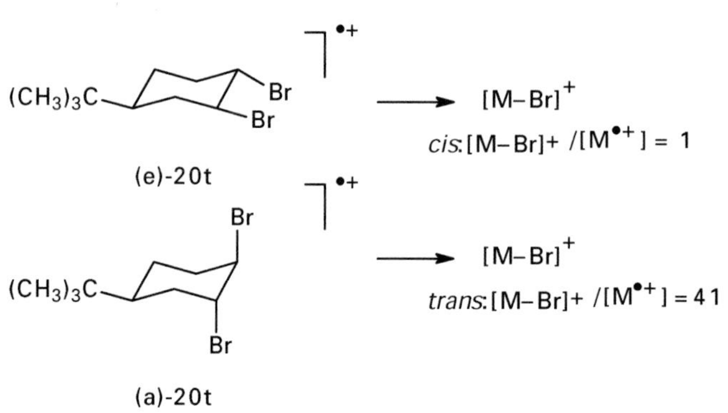

Scheme 17

CH$_3$COOH, affording the stabilized cyclic fragment-ion.

Numerous additional examples of a distinctive mass spectral behaviour of stereoisomers, which have been attributed to anchimeric assistance both under EI and CI conditions, have been proposed in the literature. In some cases, particularly upon CI, additional factors may be responsible for the distinctive behaviour. For instance, the higher abundance of the [MH–CH$_3$COOH]$^+$ ion in the CI mass spectra of the *trans*-diacetate 21t could also be attributed (at least in part) to the greater stability of the internally hydrogen bonded MH$^+$ ion of the *cis*-isomer.

Direct evidence for the operation of anchimeric assistance has been found in the CI behaviour of stereoisomeric 1,4-dialkoxycyclohexanes 22 (**Scheme 19**). The *trans*-diethers 22t afford very abundant [MH–ROH]$^+$ ions upon CI in contrast to the corresponding *cis* counterparts, suggesting anchimeric assistance in the elimination of alcohol from the MH$^+$ ions of the *trans*-diethers. Collision induced dissociation (CID) measurements of the [MH–ROH]$^+$ ions, obtained from various suitably deuterium labelled stereoisomeric 1-ethoxy-4-methoxy-cyclohexanes, indicated formation of symmetrical bicyclic ethyl and methyl oxonium ions by an anchimerically assisted alcohol elimination from the *trans*-diethers (the elimination of methanol is shown in **Scheme 19**). On the other hand, the CID measurements show that the *cis* isomers afford isomeric [MH–ROH]$^+$ ions, in which positions 2 and 3 (as well as 1 and 4, and 5 and 6) are not equivalent. These two results, namely the symmetrical structure and the high abundance of the [MH–ROH]$^+$ ions in the CI mass spectrum of the *trans*-diether 22t, in contrast to the non-symmetrical monocyclic structure and low abundance of these ions in the *cis* counterpart, are suggested as direct evidence for anchimeric assistance in the gas-phase ion dissociation process in that system. *Ab initio* calculations support the anchimerically assisted elimination observed in 22t. The energy difference between the anchimerically assisted and non-assisted elimination mechanisms in this system is not large (~ 2–3 kcal mol^{-1}).

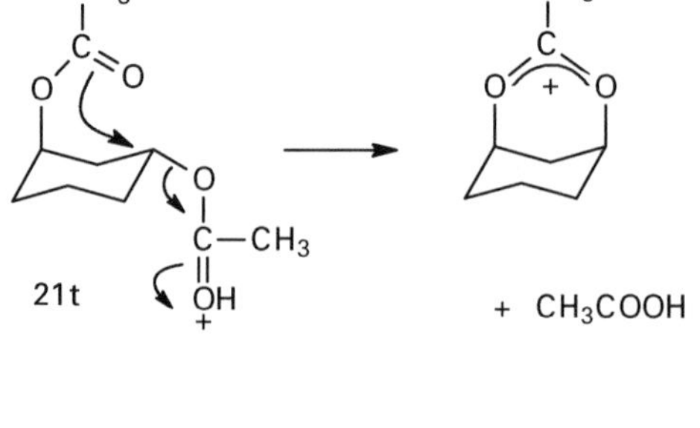

Stereoelectronic effects

One of the early reported cases of distinctive mass spectra of stereoisomers was the EI-induced behaviour of deacetylcyclindrocarpol 23a and of its epimer at C-19, 23b (**Scheme 20**). The more pronounced loss of the hydrogen atom from position 19 of the molecular ion of 23b, as compared with that of 23a (10.6% versus 1.7%), was attributed to the antiperiplanar relationship of the 19-C–19-H bond and the p-orbital of the adjacent nitrogen atom in 23b, in contrast to the epimer 23a.

The pronounced different behaviour of the stereoisomeric bicyclic carbamates 24a and 24b (**Scheme 21**) also indicates the occurrence of a stereoelectronic effect. The *trans* isomer 24a, with the axial methoxycarbonylmethyl group, affords the much more abundant [M–CH$_2$COOCH$_3$]$^+$ ion

Scheme 18

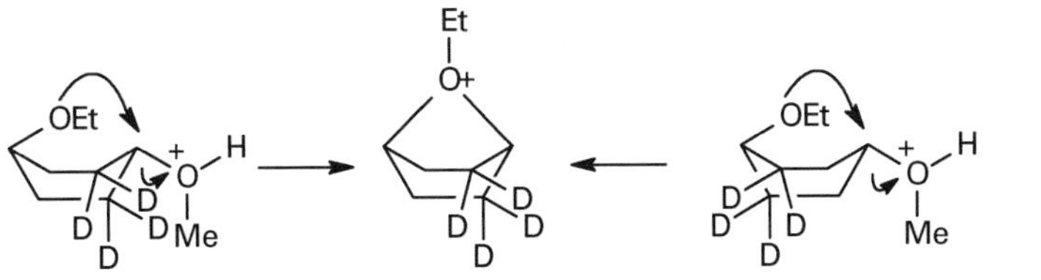

MH+ of 3,3,5,5-d₄-22t   symmetrical   MH+ of 2,2,6,6-d₄-22t
                          [MH-MeOH]+ ion

MH+ of 3,3,5,5-d_4-22t   symmetrical   MH+ of 2,2,6,6-d_4-22t
                          $[MH-MeOH]^+$ ion

Scheme 19

possibly due to the stereoelectronic assistance of the π-orbital at the adjacent carbamate group. Similar stereoelectronic effects, resulting in distinctive bond dissociation processes of stereoisomers, have been observed in several other heterocyclic systems.

Ion–molecule reactions

Ion–molecule reactions, studied using the ion cyclotron resonance (ICR) technique, show considerable stereospecificity in a number of systems, which enable distinction between stereoisomers. An early example of such a process is the gas-phase acetylation of *endo*- and *exo*-norborneol, shown in **Scheme 22**. The fourfold lower reactivity of the *endo*-isomer *endo*-25, as compared with the epimeric *exo*-25, is attributed to the steric hindrance in the approach of the bulky triacetyl cation to this stereoisomer.

Similar stereoselective behaviour has also been observed in the analogous gas-phase acetylation of *cis*- and *trans*-1-decalones and of the acetates of *cis*- and

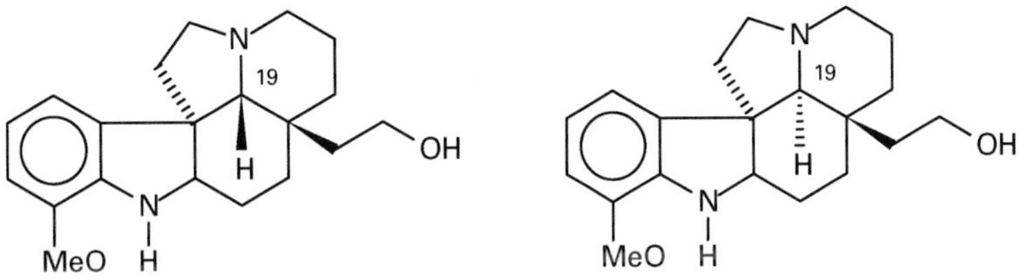

23a

$[Ma-H]^+$ 1.7%

23b

$[M-H]^+$ 10.6%

Scheme 20

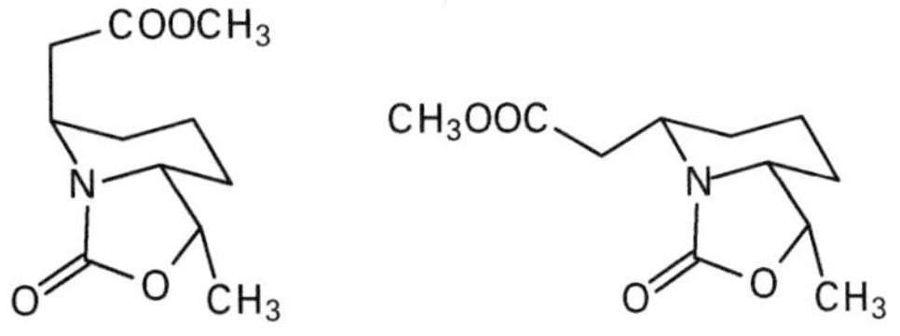

24a

$$\frac{[M-CH_2COOCH_3]^+}{M^{\bullet+}} \quad 137$$

24b

6.3

Scheme 21

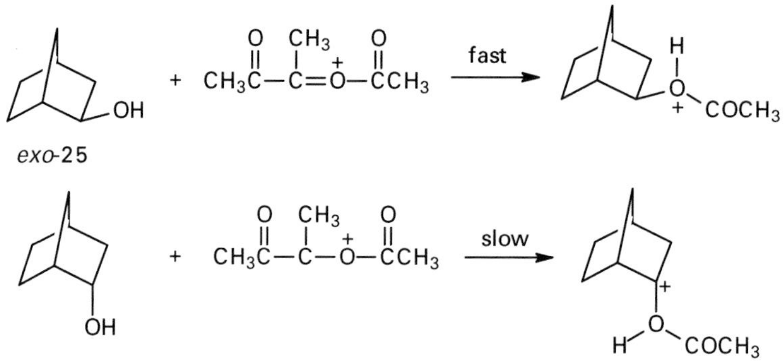

Scheme 22

trans-4-*t*-butylcyclohexanols. The reaction is faster with the less hindered *trans* isomers in both cases.

Enantiomers: chiral recognition

Unimolecular fragmentation processes studied by MS are of achiral nature, and consequently insensitive to chirality differences. A key procedure for chiral recognition using MS is to add a chiral component to the process and detect the possibly different diastereoisomeric interactions under a variety of conditions (CI, FAB, electrospray ionization (ESI), Fourier transform ion cyclotron resonance (FTICR), CID).

The formation and dissociation behaviour of the gas-phase protonated dimers (or higher clusters) of dialkyl tartarates under CI conditions is one of the early and most widely studied examples of chiral recognition by mass spectrometry. The protonated dimer consisting of two enantiomeric molecules and that consisting of two molecules of the same enantiomer are diastereoisomeric species, and as such they may exhibit distinctive behaviour, usually of a quantitative nature. Isotope labelling of one of the enantiomers is necessary in order to observe the different behaviour of the two protonated dimers.

Chemical ionization using chiral reagent gases (e.g. 1-amino-2-propanol or 2-amino-1-propanol) has been shown to induce distinctive behaviour between enantiomers. Enantiomers could also be differentiated via S_N2 reactivity with chiral reagents in ion–molecule reactions. The resulting diastereoisomeric products were distinguished by tandem mass spectral techniques (MS/MS).

Host–guest interactions using chiral hosts have been relatively widely investigated under a variety of mass spectral conditions as tools for chiral recognition in numerous systems. The use of chiral matrices

in secondary ion mass spectral measurements (FAB ionization) is a promising way for distinguishing between enantiomers.

Conformational studies of biopolymers

Electrospray ionization (ESI) has become a powerful method for the investigation of thermally labile polar biopolymers. The attachment of a large number of protons to the basic sites of proteins in the course of the ESI experiment affords a spectrum of multiply charged ions in the gas phase, which allows mass analysis using mass spectrometers with relatively low mass-to-charge ranges. With the steadily increasing use of this technique it has become apparent that more information may be obtained from the results of ESI analysis than just the molecular weight of the protein.

Careful measurements of ESI mass spectra of proteins show that changes in the solvent or in the pH of the examined solution may have an effect on the charge-state distributions. These different distributions have been interpreted as being the result of differences in the conformation of the proteins in the examined solutions. For example, ESI spectra of lysozyme, obtained from a 100% aqueous solution at pH 5.0 and from a solution containing 50% acetonitrile and 1% formic acid, show charge state maxima at 9+ and 12+, respectively. These different charge-state distributions are interpreted in terms of the degree of folding of the proteins in the original solutions. The protein molecules are unfolded to a greater extent in the organic than in the aqueous solution.

The unfolding of the molecules exposes sites or protonation, which were buried in the folded native conformation, resulting in the higher charge states in the ESI mass spectrum obtained from the 50% acetonitrile–1 % formic acid solution.

Hydrogen/deuterium exchange measurements with the aid of ESI mass spectrometry have also been used successfully to explore conformational changes of proteins in solution. Elements of secondary structure that involve internal hydrogen bonding in the core of the protein are protected against hydrogen exchange, whereas regions exposed to the solvent undergo hydrogen exchange more readily.

See also: **Chemical Ionization in Mass Spectrometry; Fast Atom Bombardment Ionization in Mass Spectrometry; Fragmentation in Mass Spectrometry; Ion Molecule Reactions in Mass Spectrometry; Ion Structures in Mass Spectrometry; MS–MS and MS[n]; Peptides and Proteins Studied Using Mass Spectrometry; Proton Affinities.**

Further reading

Green MM (1976) Mass spectrometry and the stereochemistry of organic molecules. *Topics in Stereochemistry* 9: 35–110.

Harrison AG (1992) *Chemical Ionization Mass Spectrometry*, 2nd edn, pp 172–185. Boca Raton, Florida: CRC Press.

Mandelbaum A (1977) Application of mass spectrometry to stereochemical problems. In Kagan H (ed) *Stereochemistry*, vol. 1, pp 137–180. Stuttgart: Georg Thieme Publishing.

Mandelbaum A (1983) Stereochemical effects in mass spectrometry. *Mass Spectrometry Reviews* 2: 223–284.

Meyerson S and Weitkamp AW (1968) Stereoisomeric effects on mass spectra. *Organic Mass Spectrometry* 1: 659–668.

Robinson CV (1996) Protein secondary structure investigated by electrospray ionization. In: Chapman JR (ed.) *Protein and Peptide Analysis by Mass Spectrometry*, pp 129–139, Totowa, New Jersey: Humana Press.

Splitter J and Turecek F (1994) *Application of Mass Spectrometry to Stereochemical Problems*. New York: VCH.

Sawada M (1997) Chiral recognition detected by fast atom bombardment mass spectrometry. *Mass Spectrometry Reviews* 16: 73–90.

Turecek F (1991) Stereoelectric effects in mass spectrometry. *International Journal of Mass Spectrometry and Ion Processes* 108: 137–164.

Turecek F (1987) Stereochemistry of organic ions in the gas phase. *Collection Czech. Chemical Communications* 52: 1928–1984.

Stray Magnetic Fields, Use of in MRI

See **MRI Using Stray Fields.**

Structural Chemistry Using NMR Spectroscopy, Inorganic Molecules

GE Hawkes, Queen Mary and Westfield College, London, UK

MAGNETIC RESONANCE
Applications

A very wide range of experiments is available to the NMR spectroscopist for inorganic structural analysis. The full armoury of single- and multi-dimensional NMR techniques used for the structural analysis of organic compounds may be used for inorganic systems, with the added dimension of the multinuclear approach. X-ray structural analysis of crystalline compounds has long been fundamental to inorganic structure determination and it is now clear that solid-state NMR, particularly with the magic angle spinning (MAS) technique, is capable of providing complementary data making the synergistic combination of X-ray and MAS-NMR methods very powerful indeed. In reviewing applications of NMR methods to inorganics, it is important to consider the insight that NMR provides into dynamic aspects of the molecular structures; intermolecular exchange processes in solution, and intramolecular fluxionality in both solution and solid state can be delineated, and very often thermodynamic and kinetic parameters derived which can be interpreted in mechanistic terms. It is often the case, particularly for transition metal compounds, that the molecule is paramagnetic. While the unpaired electron(s) can cause undesirable effects on the NMR spectra, e.g. excessive broadening of resonances, it may often be the case that the paramagnetic *shift* at ligand resonances may be useful in resolving erstwhile overlapping signals, or in providing information on the distribution of unpaired spin density throughout the molecule. In addition to the intrinsic information from such paramagnetic molecules, the addition of a paramagnetic compound (e.g. a lanthanide chelate) to a solution of a diamagnetic compound can induce paramagnetic shifts and changes in line widths in resonances of the latter compound (shift and relaxation agents). These changes can be valuable in providing structural information or in enhancing contrast in magnetic resonance images.

The multinuclear approach

In any structural problem, the first step is to decide which NMR experiments will provide the most definitive information in the shortest time. Usually, although not exclusively, solution-state spectra are measured, and if the inorganic compound includes an organic ligand, then ^1H and ^{13}C spectra are essential. The choice of additional spectra from other nuclei must be directed using the same criteria – information content and economy of time. In this it is necessary to know the inherent sensitivity of the isotope, and the value of its spin quantum number (I). Isotopes with $I = 1/2$ generally give rise to narrower resonances ('high-resolution' nuclei) whereas those with $I > 1/2$ often yield much broader resonances (quadrupolar nuclei) and spectral information (chemical shifts and scalar couplings) may be obscured. Some NMR-active isotopes of potential interest to the inorganic chemist are listed in **Table 1**, and this list is by no means exhaustive since there are considerably more than 100 NMR-active stable isotopes across the periodic table.

Careful consideration should be given to the ease of observation of the NMR spectrum of a given isotope. Generally, the ease of observation will increase with increase in the magnetic field strength of the spectrometer, but it is possible that at the higher end of the available magnetic field strength range (≥ 14 T), the resonances of certain nuclei, particularly the heavier metals such as platinum, mercury or lead, may become broadened by the influence of chemical shift anisotropy (see below), thereby abrogating the beneficial effects of the higher field. Another consideration is the combination of low natural abundance and small nuclear magnetic moment (low resonance frequency) giving a low intrinsic receptivity, as for ^{57}Fe. In such cases the situation may be partially alleviated by resorting to isotopic enrichment, as illustrated in **Figure 1**, which shows two ^{57}Fe resonances from a 20 mM solution of the superstructured haem model compound $^{57}Fe^{II}PocPiv(1,2\text{-diMeIm})$ (CO) [1] (94.5% enriched in ^{57}Fe). Even at this high level of enrichment and high magnetic field, the spectrum still required about 20 h of instrument time for the spectral accumulation. The spectrum does illustrate the extreme sensitivity of the ^{57}Fe chemical shift to fairly

Table 1 NMR properties of some isotopes for the inorganic chemist

Isotope	I	Abundance[a]	Receptivity[b]	Q (10^{-28} m²)[c]	Ξ (MHz)[d]	Reference[e]
^{1}H	1/2	99.985	5.7×10^3		100.0	SiMe$_4$
^{2}H	1	0.015	8.2×10^{-4}	2.7×10^{-3}	15.4	
^{6}Li	1	7.4	3.6	-8×10^{-4}	14.7	Li$^+$aq
^{7}Li	3/2	92.6	1.5×10^3	-4.5×10^{-2}	38.9	Li$^+$aq
^{11}B	3/2	80.4	7.5×10^2	3.6×10^{-2}	32.1	BF$_3$·Et$_2$O
^{13}C	1/2	1.1	1.00		25.1	SiMe$_4$
^{17}O	5/2	0.037	6.1×10^{-2}	-2.6×10^{-2}	13.6	H$_2$O
^{19}F	1/2	100	4.7×10^3		94.1	CCl$_3$F
^{23}Na	3/2	100	5.3×10^3	0.12	26.5	Na$^+$aq
^{27}Al	5/2	100	1.2×10^3	0.15	26.1	Al(H$_2$O)$_6^+$
^{29}Si	1/2	4.7	2.1		19.9	SiMe$_4$
^{31}P	1/2	100	3.8×10^2		40.5	85% H$_3$PO$_4$
^{51}V	7/2	99.8	2.2×10^3	0.3	26.3	VOCl$_3$
^{57}Fe	1/2	2.2	4.2×10^{-3}		3.2	Fe(CO)$_5$
^{59}Co	7/2	100	1.6×10^3	0.4	23.6	Co(CN)$_6^{3-}$
^{71}Ga	3/2	39.6	3.2×10^2	0.11	30.5	Ga(H$_2$O)$_6^{3+}$
^{103}Rh	1/2	100	0.18		3.2	
^{109}Ag	1/2	38.2	0.28		4.7	Ag$^+$aq
^{113}Cd	1/2	12.3	7.6		22.2	CdMe$_2$
^{119}Sn	1/2	8.6	25.2		37.3	SnMe$_4$
^{139}La	7/2	99.9	3.4×10^2	0.2	14.1	La^{3+}aq
^{183}W	1/2	14.4	5.9×10^{-2}		4.2	WO$_4^{2-}$aq
^{195}Pt	1/2	33.8	19.1		21.4	Pt(CN)$_6^{2-}$
^{199}Hg	1/2	16.8	5.4		17.9	HgMe$_2$
^{207}Pb	1/2	22.6	11.8		20.9	PbMe$_4$

[a] The natural abundance of the isotope.
[b] A rough guide to the ease of observation of the NMR spectrum, relative to ^{13}C.
[c] Approximate values for the quadrupole moment.
[d] The resonance frequency in a magnetic field strength that gives the ^{1}H resonance of SiMe$_4$ at exactly 100 MHz.
[e] Commonly accepted chemical shift reference standard material.

remote structural effects since the two ^{57}Fe resonances are believed to arise from the presence of the two atropisomers (α and β) due to restricted rotation of the pivaloylamido picket. In considering NMR spectra of the quadrupolar nuclei, in addition to the question of sensitivity, there is the question of resolution of chemical shifts, since chemical shift differences within a spectrum may be obscured by relatively large line widths (W_Q) exhibited by the resonances due to quadrupolar relaxation. As a very rough guide to the line width problem, the line width may increase with the square of the quadrupole moment, hence less broadening is expected in spectra of ^{2}H or ^{6}Li:

$$W_Q \propto \left(\frac{e^2 q_{zz} Q}{h} \right)^2 \left(1 + \frac{1}{3} \eta^2 \right) \qquad [1]$$

where e is the electronic charge, h Planck's constant, q_{zz} the largest component of the electric field gradient and η the asymmetry parameter for q. However, since the electric field gradient at the nucleus, caused by the surrounding electron distribution, is also important, this may counter the effect of a larger quadrupole moment as for the ^{51}V spectrum of the product of partial hydrolysis of VO(NO$_3$)$_3$ (**Figure 2**, which also includes the ^{17}O spectrum). ^{17}O enrichment can often be achieved starting with the relatively inexpensive source H$_2^{17}$O, and the relatively narrow lines often exhibited by ^{17}O resonances can provide a wealth of structural data as shown in **Figure 3** for the aqueous isopolytungstate solution (enriched to 5% ^{17}O), where the ^{17}O chemical shifts are sensitive to a variety of structural features, particularly the metal–oxygen bond lengths.

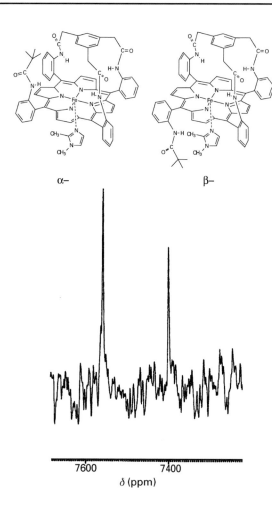

α–   β–

Figure 1 19.58 MHz ^{57}Fe NMR spectrum of the ^{57}Fe^{II} PocPiv(1,2-diMeIm)(CO) adduct in CD$_2$Cl$_2$ solution at 298 K. Reproduced with permission from Gerothanassis IP, Kalodimos CG, Hawkes GE and Haycock PR, *Journal of Magnetic Resonance* 1998, **131**: 163–165.

The NMR parameters

Chemical shifts

For samples of inorganic molecules in solution, each chemically distinct site for an atom in the molecule will result in a distinct *isotropic* chemical shift for its nucleus in the NMR spectrum. What is important from the structural point of view is that these chemical shifts are resolved in the spectrum, and this in turn will be determined by the sensitivity of the chemical shift to structural changes. Some nuclei are more sensitive than others, and this is usually represented by the reported chemical shift range of the nucleus. ^{1}H chemical shifts in inorganic compounds typically span a range ~20 ppm, whereas heavier isotopes often exhibit much greater chemical shift ranges, e.g. hundreds of ppm for ^{17}O, ^{19}F, ^{29}Si and ^{31}P, and this may run to thousands of ppm for heavy

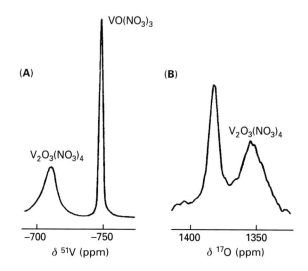

Figure 2 NMR spectra at 294 K of the VO(NO$_3$)$_3$ – H$_2$O (mole ratio 1 : 0.3) system in MeNO$_2$: (A) 105.1 MHz ^{51}V spectrum and (B) 54.2 MHz ^{17}O spectrum. Reproduced with permission from Hibbert RC, Logan N and Howarth AW, *Journal of the Chemical Society, Dalton Transactions* 1986, 369–372.

metals such as ^{195}Pt or ^{199}Hg. Each isotropic chemical shift is in fact an average of three principal chemical shift values. The chemical shift is determined by the interaction of the electron distribution with the spectrometer magnetic field, hence a nucleus in an asymmetric electronic environment (the general case) will experience a change in chemical shift with the orientation of the molecule in the magnetic field. The chemical shift is thus represented as a second-rank tensor (3 × 3 values) and it is possible to find a molecule-fixed Cartesian coordinate system which diagonalizes this tensor to give the three principal components ($\delta_{11}, \delta_{22}, \delta_{33}$). In any solution sample the molecules undergo rapid random motion, including rotation, and as a result these components average to a single isotropic chemical shift (δ_{iso}):

$$\delta_{iso} = (\delta_{11} + \delta_{22} + \delta_{33})/3 \qquad [2]$$

Although it is not possible to obtain values for the independent components from solution-state spectra, it is possible in certain cases to obtain some information such as the chemical shift anisotropy $\Delta\delta$ (see below):

$$\Delta\delta = \delta_{33} - (\delta_{11} + \delta_{22})/2 \qquad [3]$$

where δ_{33} is the component furthest removed from the average δ_{iso}.

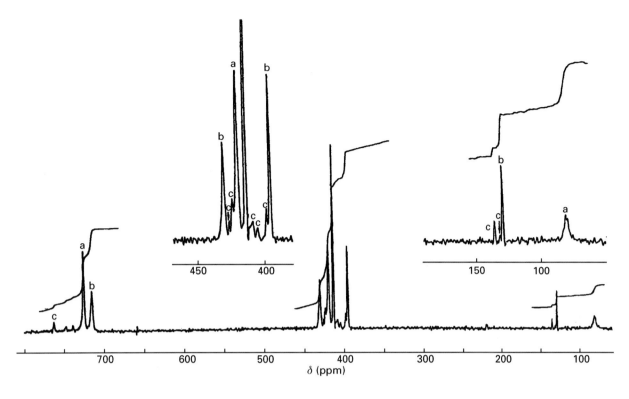

Figure 3 54.2 MHz ^{17}O NMR spectrum of isopolytungstate at 353 K, pH 1.1. Species: a, α-$[H_2W_{12}O_{40}]^{6-}$, metatungstate; b, α-$[HW_{12}O_{40}]^{7-}$; c, ψ'-metatungstate; probably β-$[HW_{12}O_{40}]^{7-}$. Reproduced with permission from Hastings JJ and Howarth OW, *Journal of the Chemical Society, Dalton Transactions* 1992, 209–215.

MAS-NMR spectra of the solid state can provide values for the components δ_{ii} as shown in **Figure 4**. The *static* spectrum (**Figure 4A**) of the powder sample is the superposition of different chemical shifts resulting from all possible orientations of the molecules in the discrete particles of the sample, and the components δ_{ii} are as indicated. Usually such broad lines would mask any resolution of distinct chemical shifts. The MAS spectra (**Figures 4B–D**) consist of a centre-band resonance at the isotropic chemical shift (δ_{iso}) and side bands spaced at the rotation frequency, and offer two advantages over static spectra. The first advantage is that within the centre band there may be chemical shift resolution (in this case there is only one ^{31}P environment and therefore only one isotropic chemical shift) and the second is that the total integrated intensity of the spectrum is within the relatively sharp lines and so the sensitivity of the observation is greatly enhanced in comparison with that of the static spectrum. The intensity pattern of the spinning side bands roughly follows the static spectrum, and these intensities can be used to obtain values for the components δ_{ii}. To illustrate the utility of such measurements, it has been shown that the ^{31}P isotropic shifts and chemical shift tensor components (δ_{ii}) from a series of phosphido-bridged iron complexes $Fe_2(CO)_6(\mu\text{-}X)(\mu\text{-}PPh_2)$ gave an excellent correlation with the crystallographically determined Fe–P–Fe bond angles. In a related study on iron complexes with asymmetrical bridging carbonyl ligands Fe$\cdots$CO–Fe the ^{13}C chemical shift anisotropy and the component δ_{33} (associated with the C–O bond axis) both correlated with the difference in the two Fe–C distances. The isotropic ^{13}C chemical shift is not a reliable indicator of the metal–carbonyl group bonding, and typically cannot be used to distinguish unequivocally between terminal and bridging carbonyl groups (e.g. M–CO *vs* M–CO–M). However, as shown in **Table 2**, the chemical shift anisotropy values are distinctive of the carbonyl group bonding.

Table 2 ^{13}C chemical shift anisotropy values for terminal and bridging carbonyl groups

	$\Delta\delta$ (ppm)		
	Co terminal	μ_2-CO	μ_3-CO
$(C_5H_5)_2Fe_2(CO)_4$	444	138	–
$Rh_6(CO)_{16}$	390	–	194

Data reproduced with permission from Gleeson JW and Vaughan RW, *Journal of Chemical Physics* 1983, **78**: 5384–5392.

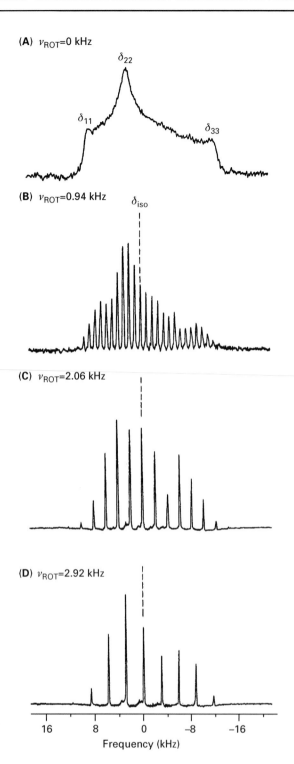

(A) ν_{ROT}=0 kHz

δ_{22}

δ_{11}

δ_{33}

(B) ν_{ROT}=0.94 kHz

δ_{iso}

(C) ν_{ROT}=2.06 kHz

(D) ν_{ROT}=2.92 kHz

16   8   0   −8   −16

Frequency (kHz)

Figure 4 119.05 MHz solid-state ^{31}P NMR spectrum of diethyl phosphate spinning at the magic angle. ν_{ROT} indicates the spinning frequency. Reproduced with permission from Herzfeld J and Berger AE, *Journal of Chemical Physics* 1980, **73**: 6021–6030.

Coupling constants

The two important parameters to consider are the internuclear scalar couplings and the internuclear

dipolar couplings. Scalar or spin–spin coupling constants are observed in both solution- and solid-state spectra and are usually considered to be transmitted *via* bonding electrons. The magnitude of the scalar couplings varies dramatically with the isotopes concerned, the nature and number of intervening bonds, coordination number, oxidation state, etc. Interproton couplings are typically small (< 20 Hz) but couplings between heavier isotopes may be fairly large, up to about 10 000 Hz, for example, for the one-bond ^{31}P–^{199}Hg coupling in Ph_3PHgX_2, X = $OCOCH_3$ or $OCOCF_3$. The splitting patterns induced by the scalar couplings are used to determine the number of interacting nuclei and the magnitudes of the couplings may be interpreted in terms of bonding and conformation. Internuclear dipolar couplings are direct through space interactions between magnetic nuclei and values may be as large as 50 000 Hz between protons, being dependent upon the inverse third power of the separation (r^{-3}). The splitting caused by this effect is dependent on the orientation of the internuclear vector in the spectrometer magnetic field and for molecules in solution undergoing rapid molecular tumbling the effect averages to zero. Therefore, in solution spectra there is no obvious direct effect on one-dimensional spectra due to dipolar couplings. However, they are responsible for indirect effects, such as nuclear relaxation and the nuclear Overhauser effect (see below). In solid-state NMR, dipolar interactions provide one mechanism for line broadening, particularly with hydrogen present in the molecule. MAS (e.g. using 4 mm o.d. rotors at 15 kHz) is used to reduce the dipolar interactions and for the observation of nuclei other than hydrogen this is often used in conjunction with high-power ^{1}H decoupling. For observation of solid-state ^{1}H spectra it is often necessary to use the combination of MAS with a suitable pulse sequence (CRAMPS; combined rotation and multiple pulse spectroscopy). If the dipolar interactions are relatively weak then MAS alone may be sufficient to allow resolution of chemical shifts, as shown in **Figure 5** for some metallo-hydride complexes. While much attention has been focused on methods for reducing or eliminating the effects of the dipolar interactions in solid-state NMR, the presence of the dipolar interaction could be useful in showing the spatial proximity of atoms in a structure. Several multiple-pulse two-dimensional experiments have been proposed, including trains of pulses, synchronized in time with the MAS rotation period. These sequences refocus the dipolar interaction, as illustrated in **Figure 6** for the ^{31}P double-quantum–single-quantum correlation spectrum of a polycrystalline powder sample of $Cd_3(PO_4)_2$. There are six crystallographically distinct

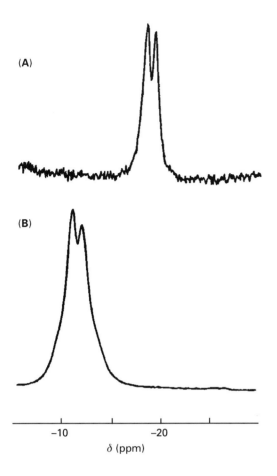

(A)

(B)

δ (ppm)

Figure 5 300 MHz solid-state ^{1}H MAS-NMR spectra: (A) $H_2Os_3(CO)_{10}$, MAS rate 8.1 kHz; (B) $H_2FeRu_3(CO)_{13}$, MAS rate 9.5 kHz. Reproduced with permission from Aime S, Barrie PJ, Brougham DF, Gobetto R and Hawkes GE, *Inorganic Chemistry* 1995, **34**: 3557–3559.

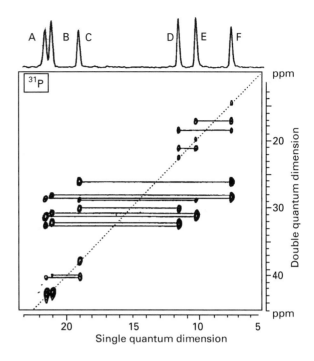

Figure 6 202.5 MHz solid-state ^{31}P MAS-NMR spectrum of $Cd_3(PO_4)_2$. The two-dimensional spectrum shows the single-quantum–double-quantum dipolar correlations Reproduced with permission from Dollase WA, Fecke M, Förster H, Schaller T, Schell I, Sebald A and Stevernagel S, *Journal of the American Chemical Society* 1997, **119**: 3807–3810.

phosphorus sites in the structure and six resolved ^{31}P resonances in the one-dimensional spectrum. The contours link pairs of phosphorus sites which have a measurable dipolar interaction (are in spatial proximity) and the more intense correlations indicate greater proximity in the structure.

Relaxation times

The principal relaxation times measurable for resolved resonances from solution state samples are the spin–lattice relaxation time (T_1) and the spin–spin relaxation time (T_2). Spin echo methods may be used to measure T_2 values and these are often useful in defining chemical exchange rate processes. T_1 values are readily obtained from the inversion–recovery experiment and can be directly used to provide structural information. For nuclei with spin $I = 1/2$ in diamagnetic molecules in solution there are two principal mechanisms which contribute to the rate of the relaxation (T_1^{-1}), namely the dipole–dipole inter-

action and the chemical shift anisotropy mechanism. The dipole–dipole interaction occurs between magnetic nuclei which are in close spatial proximity in the molecule and if the population distribution of nuclei across the energy levels of one site is disturbed away from its equilibrium value (usually the populations are equalized by a second radiofrequency field) then this is reflected as a change in the intensity of the resonance from the other site. This is the so-called nuclear Overhauser effect (NOE) and is widely used as a structural tool; particularly the interproton NOE is used in a qualitative or quantitative manner to estimate distances between protons in biomolecules, and thereby serve to define conformation. Such interproton NOEs will similarly be useful for determination of structure in organometallic species, and in addition the structural inorganic chemist will be able to utilize the homonuclear NOE between other nuclei which are present at a high level of abundance, for example the ^{31}P–^{31}P NOE might be particularly useful. The heteronuclear NOE might also be expected to provide useful conformational information where sensitivity permits, and ^{13}C–^{1}H and ^{31}P–^{1}H are obvious candidates, and ^{6}Li–^{1}H has also been used. The relaxation rate of a nucleus due to its dipole–dipole interaction (T_1^{-1}) with a neighbouring nucleus at a distance r is proportional to r^{-6}.

This distance dependence has been put to a number of uses, in particular in structural studies of molecular hydrogen complexes. If hydrogen is bound as two distinct M–H groups then the separation between the hydrogens will be greater than if the hydrogen is bound as molecular hydrogen M–(H_2). Therefore, for the molecular hydrogen case the hydrogens will experience a stronger mutual dipole–dipole interaction and the rate of spin–lattice relaxation will be greater. Ideally the experiment should be calibrated by measuring the relaxation rate for a pair of nuclei at a known separation in the same molecule, perhaps for nuclei in an organic ligand.

The rate of nuclear spin–lattice relaxation due to the chemical shift anisotropy mechanism $(T_1^{CSA})^{-1}$ depends upon the magnitude of the chemical shift anisotropy $(\Delta \delta)$ and the strength of the spectrometer magnetic field (B_0) squared. This dependence of the rate on B_0^2 provides a means of estimating a value for $\Delta \delta$ if the mechanism is important (significant value for $\Delta \delta$) and if the measurement of the rate of relaxation can be made at several different field strengths (B_0). This may be applied to ^{13}C relaxation for the carbonyl groups of organometallic complexes (see **Table 2**). A related study on metal carbonyl complexes made use of the quadrupolar relaxation rate shown by those nuclei with spin $I > 1/2$, here ^{17}O with $I = 5/2$. The relaxation rate of such quadrupolar nuclei is dominated by the contribution, $(T_1^Q)^{-1}$, from the quadrupolar mechanism, and in favourable cases it is possible to determine the quadrupole coupling constant (QCC; *cf.* Eqn [1]). **Table 3** shows values for the ^{17}O QCC for the carbonyl groups of some metallo-carbonyl complexes, and again the derived parameter is seen to be diagnostic of the type of carbonyl group.

Dynamic processes

NMR spectroscopy is a most powerful method for the investigation of dynamic processes occurring at the molecular level. In particular, both intramolecular and intermolecular chemical exchange processes in solution may be investigated and for inorganic compounds these processes include ligand exchange, conformational changes, rearrangements and fluxional processes. There are various NMR parameters which may be used to monitor the dynamic process, and the particular set of NMR experiments to be used may depend in part on the order of magnitude for the rate coefficient of the process. The rate process may be termed either 'slow' or 'fast', but these labels really depend on the NMR parameter being used to monitor the process. For many years the most common method to study rate processes both qualitatively and quantitatively has been the band shape method. Here the NMR parameters may be chemical shifts (measured in frequency units) and/or coupling constants. For a two-site exchange process $A \leftrightarrow X$ the spectrum will be affected when the rate is within the limits

$$T_2 \geq k_r^{-1} \geq \left(\tfrac{1}{2} \pi^2 \Delta \nu^2 T_2 \right)^{-1} \qquad [4]$$

where $\Delta \nu$ is the difference in resonance frequency (between sites A and X) or the coupling constant being averaged. Exchange processes in the fast exchange limit $(k_r \geq 10^4 \, Hz)$ may contribute to the rotating frame relaxation time $(T_{1\rho})$, and measurement of $T_{1\rho}$ as a function of the strength of the spin-lock field can give a value for the rate coefficient. More recently, magnetization transfer experiments, both one- and two-dimensional, have been used to explore multi-site slow exchange situations. In these experiments the population distribution between the nuclear energy levels for one or more of the sites is disturbed from the equilibrium. This can be by equalization of the populations (cf. saturation as described above for the NOE) or by inversion of the populations by a selective 180° radiofrequency pulse for the 1D experiment or a non-selective pulse for the 2D experiment. The chemical exchange can then transmit the disturbance throughout the exchanging system. However, since spin–lattice relaxation is always occurring in order to restore the equilibrium nuclear distribution, then this method is applicable when the rate coefficient $k_r \geq T_1^{-1}$. In the two-dimensional experiment, which is exactly the same as the 2D NOESY experiment, the advantage is that it is possible to obtain a very clear picture of the magnetization transfer pathways in addition to being able to quantify the rate coefficients. This is illustrated in **Figure 7**, where the off-diagonal contours link chemical shifts of pairs of slowly exchanging methyl groups among the six distinct methyls of the 2,4,6-tris(3,5-dimethylpyrazol-1-yl)pyrimidine (tdmpzp) ligands.

Table 3 ^{17}O quadrupole coupling constant values for terminal and bridging carbonyl groups

	QCC (MHz)		
	Co terminal	μ_2-CO	μ_3-CO
$(C_5H_5)_2Fe_2(CO)_4$	1.47	3.3	–
$Rh_6(CO)_{16}$	2.02	–	0.09

Data reproduced with permission from Hawkes GE and Randall EW, *Journal of Magnetic Resonance* 1986, **68**: 597–599.

Sensitivity enhancement by polarization transfer

One-dimensional experiments

The solution-state polarization transfer experiments described here all depend upon the existence of a resolved scalar coupling between a sensitive nucleus (e.g. ^{1}H, ^{31}P) and an insensitive nucleus (e.g. ^{57}Fe, ^{109}Ag). The one-dimensional experiments are based upon the so-called INEPT or DEPT pulse sequences and involve the initial creation of anti-phase magnetization for the more sensitive nucleus; this is effectively the selective inversion (of populations) for part of the multiplet of the sensitive nucleus. This has the effect of enhancing the population differences across the energy levels of the coupled, less sensitive nucleus, thus making the observed resonances more intense. Polarization (population differences) has thus been transferred from the more sensitive to the less sensitive nucleus. For a single acquisition the sensitivity improvement for spin $I = 1/2$ nuclei is of the order of the ratio of the resonance frequencies; hence using the ^{31}P polarization to drive the ^{109}Ag populations results in a sensitivity enhancement factor ~ 8.6 compared with 'single pulse' observation of the ^{109}Ag spectrum. There is a second benefit to using the polarization transfer sequences in that typically the spectrum must be accumulated over a period of time and each individual acquisition sequence should be separated by a 'relaxation delay'. The intensity of the observed resonance is derived from the nonequilibrium populations of the more sensitive nucleus, and it is often the case that relaxation times for ^{1}H and ^{31}P are shorter than for the metal nuclei, and therefore the accumulation sequence with polarization transfer can be repeated with greater frequency than the 'single pulse' observation. An example of the sensitivity enhancement is shown in **Figure 8** where the ^{109}Ag–$\{^{31}$P$\}$ INEPT experiment provides considerably improved signal/noise ratio over the normal acquisition, with about one sixth of the number of scans accumulated with INEPT. There is one disadvantage common to all 1D and 2D polarization transfer experiments and this is the need to have prior knowledge of the magnitude of the scalar coupling constant (J, Hz) since there are delays in the various pulse sequences which are related to J. It is

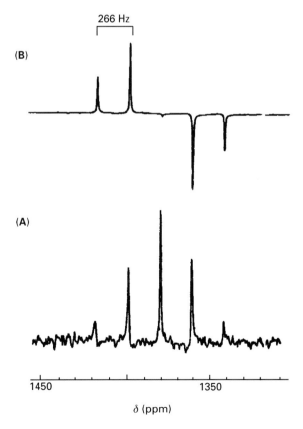

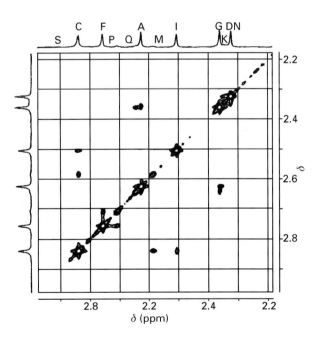

Figure 7 Methyl region of the 400 MHz ^{1}H two-dimensional EXSY NMR spectrum of [ReBr(CO)$_3$(tdmpzp)] in CDCl$_2$CDCl$_2$ solution at 296 K. Reproduced with permission from Gelling A, Noble DR, Orrell KG, Osborne AG and Šik V, *Journal of the Chemical Society, Dalton Transactions* 1996, 3065–3070.

Figure 8 13.97 MHz ^{109}Ag NMR spectra of [Ag(dppe)$_2$]NO$_3$ in CDCl$_3$ solution at 300 K: (A) single pulse acquisition (accumulation of 12 111 scans); (B) ^{109}Ag–$\{^{31}$P$\}$ INEPT experiment (accumulation of 2048 scans). Reproduced with permission from Berners Price SJ, Brevard C, Pagelot A and Sadler PJ, *Inorganic Chemistry* 1985, **24**: 4278–4281.

often possible to obtain a value for the coupling constant from the ^1H (or ^{31}P) spectrum, but in other cases a guess must be made for J, and the experiment possibly repeated for a range of assumed J values.

Two-dimensional experiments

Two-dimensional experiments provide heteronuclear shift correlations with the spectrum detected at the frequency of the more sensitive nucleus, and involve a double polarization transfer from sensitive to less sensitive to sensitive nucleus. The sensitivity improvement over the direct 'single pulse' observation of the less sensitive nucleus is even more dramatic than for the one-dimensional methods; here it is $\sim R^{5/2}$, where R is the ratio of the resonance frequencies, and for the ^{31}P–^{109}Ag example used above the sensitivity improvement is ~218. These experiments are often used to facilitate the observation of the spectrum of the less sensitive nucleus. Several experiments may be used and, when the one bond coupling is known the choice is between HMQC (heteronuclear multiple quantum coherence) and HSQC (heteronuclear single quantum coherence). There are advantages and disadvantages to both experiments; for example, the HMQC pulse sequence has fewer pulses than the HSQC, making the former experiment less susceptible to instrumental imperfections. However the resulting HMQC two-dimensional plot includes homonuclear coupling for the more sensitive nucleus (e.g. ^1H or ^{31}P), reducing the intensities of the correlation peaks, and the line widths are determined by the relaxation rate of the multiple quantum coherence which may be faster than that of the single quantum coherence leading to broader lines in the HMQC plot. An example is the ^{109}Ag–^{31}P correlation shown in **Figure 9**.

Paramagnetic systems

The incidence of paramagnetism in inorganic molecules and materials is fairly common, and is due to the presence of unpaired electrons, usually associated with a metal centre. This paramagnetism will often lead to large chemical shifts of the NMR-active nuclei in the sample and may also induce severe line broadening of the resonances. Certainly paramagnetic metal centres in a variety of biomolecules may result in spectra dispersed on the chemical shift scale over hundreds of ppm, compared with tens of ppm for diamagnetic analogues. Such effects are fairly common, for example, in the ^1H and ^{13}C NMR spectra of a wide range of natural haem systems containing iron, and in model haem and porphyrin systems wherein the paramagnetic shifts may be related, in

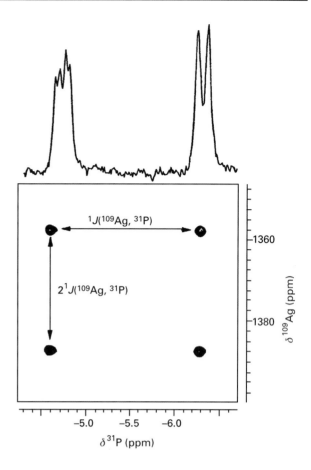

Figure 9 A ^{109}Ag–^{31}P HSQC correlation experiment with inverse (^{31}P) detection (^{109}Ag at 23.3 MHz, ^{31}P at 202 MHz) of a silver–chiral ferrocene complex in the presence of an excess of the isonitrile $CNCH_2(CO_2Me)$. This shows a single ^{109}Ag resonance. Reproduced with permission from Lianza F, Macchioni A, Pregosin P and Rüegger H, *Inorganic Chemistry* 1994, **33**: 4999–5002.

part, to the distribution of inpaired electron spin density throughout the molecule. The paramagnetism of inorganic complexes may be used to good effect in other areas. Lanthanide metals complexed with a range of organic ligands have been used for a number of years as shift or relaxation reagents. The shift (relaxation) reagent, when added to a solution of a diamagnetic compound, may form a weak complex with the diamagnetic molecule and result in paramagnetic changes in the chemical shifts (relaxation rates) of the nuclei in the substrate. The magnitudes of these changes in shift or relaxation rate depend upon the geometry of the weak complex and so may be analysed to give information about the structure of the diamagnetic compound. A second area of application of paramagnetic organometallic complexes is as contrast agents for use in MRI experiments. When the complex is introduced *in vivo*, if there is a differential distribution of the agent between normal tissue and a lesion, then the agent will induce a differential in the relaxation rates of the water protons in these regions. Since the contrast in the MR image can be tuned to the relaxation properties of the water protons, then enhanced contrast in the image is obtained.

List of symbols

B_0 = magnetic field strength; e = electronic charge; h = Planck's constant; I = spin quantum number; J = coupling constant; k_r = rate coefficient; q_{zz} = largest component of electric field gradient; Q = quadrupole moment; r = distance between nuclei; R = ratio of resonance frequencies; T_1 = spin–lattice relaxation time; T_2 = spin–spin relaxation time; $T_{1\rho}$ = rotating frame relaxation time; W_Q = line width; δ_{ii} = component of chemical shift; δ_{iso} = isotropic chemical shift; $\Delta\delta$ = chemical shift anisotropy; ν = resonance frequency; η = asymmetry parameter; Ξ = resonance frequency in a magnetic field strength that gives the ^1H resonance of $SiMe_4$ at exactly 100 MHz.

See also: **Chemical Exchange Effects in NMR; Chemical Shift and Relaxation Reagents in NMR; Halogen NMR Spectroscopy (excluding ^{19}F); Heteronuclear NMR Applications (As, Sb, Bi); Heteronuclear NMR Applications (B, Al, Ga, In, Tl); Heteronuclear NMR Applications (Ge, Sn, Pb); Heteronuclear NMR Applications (La–Hg); Heteronuclear NMR Applications (O, S, Se, Te); Heteronuclear NMR Applications (Sc–Zn); Heteronuclear NMR Applications (Y–Cd); High Resolution Solid State NMR, ^{13}C; High Resolution Solid State NMR, ^1H, ^{19}F; Inorganic Compounds and Minerals Studied Using X-ray Diffraction; NMR of Solids; NMR Relaxation Rates; NMR Spectroscopy of Alkali Metal Nuclei in Solution; Nuclear Overhauser Effect; ^{31}P NMR; ^{29}Si NMR; Solid State NMR, Methods; Solid State NMR, Rotational Resonance; Solid State NMR Using Quadrupolar Nuclei; Structural Chemistry Using NMR Spectroscopy, Organic Molecules; Two-Dimensional NMR Methods.**

Further reading

Aime S, Botta M, Fasano M and Terreno E (1998) Lanthanide(III) chelates for NMR biomedical applications. *Chemical Society Reviews* **27**: 19–29.

Bertini I and Luchinat C (1986) *NMR of Paramagnetic Molecules in Biological Systems.* Menlo Park, CA: Benjamin/Cummings.

Brey WS (ed) (1988) *Pulse Methods in 1D and 2D Liquid-phase NMR.* New work: Academic Press.

Gielen G, Willem R and Wrackmeyer B (eds) (1996) *Advanced Applications of NMR to Organometallic Chemistry.* Chichester: Wiley.

Mann BE (1974) ^{13}C NMR chemical shifts and coupling constants of organometallic compounds. *Advances in Organometallic Chemistry* **12**: 135–213.

Mann BE (1991) The Cinderella nuclei. *Annual Reports on Nuclear Magnetic Resonance Spectroscopy* **23**: 141–207.

Mason J and Jameson C (eds) (1987) *Multinuclear NMR.* New York: Plenum Press.

Orrell KG, Šik V and Stephenson D (1990) Quantitative investigations of molecular stereodynamics by 1D and 2D NMR methods. *Progress in Nuclear Magnetic Resonance Spectroscopy* **22**: 141–208.

Orrell KG (1999) Dynamic NMR spectroscopy in inorganic and organometallic chemistry. *Annual Reports on Nuclear Magnetic Resonance Spectroscopy* **37**: 1–74.

Pregosin PS (ed) (1991) *Transition Metal Nuclear Magnetic Resonance.* Amsterdam: Elsevier.

Sandström J (1982) *Dynamic NMR Spectroscopy.* London: Academic Press.

Sievers RE (ed) (1973) *Nuclear Magnetic Resonance Shift Reagents.* New York: Academic Press.

Willem R (1988) 2D NMR applied to stereochemical problems. *Progress in Nuclear Magnetic Resonance Spectroscopy* **20**: 1–94.

Structural Chemistry Using NMR Spectroscopy, Organic Molecules

Cynthia K McClure, Montana State University,
Bozeman, MT, USA

MAGNETIC RESONANCE
Applications

Nuclear magnetic resonance spectroscopy is one of the most powerful tools that chemists use to determine the structure of compounds. Generally, NMR spectroscopy is the technique that most chemists, especially organic chemists, use first and routinely in structural analysis. In organic compounds, this non-destructive spectroscopic analysis can reveal the number of carbon and proton atoms and their connectivities, the conformations of the molecules, as well as relative and absolute stereochemistries, for example. The recent advent of pulsed field gradient (PFG) technology for NMR spectrometers has allowed the routine acquisition of sophisticated one-dimensional (1D) and two-dimensional (2D) NMR spectra in relatively short periods of time on complex organic molecules. This in turn has revolutionized organic structure determination such that deducing the three-dimensional structure of compounds takes a fraction of the time it used to. Mention of relevant 2D experiments that can aid in structure determination will be made in the appropriate sections herein.

This article is geared toward the analyses of small organic compounds, and will cover the following topics: practical tips in sample preparation; basic principles of one-dimensional ^{1}H and ^{13}C NMR spectroscopy and their use in organic structure determination, including chemical shifts, coupling constants and stereochemical analyses; and the application of more sophisticated 1D and 2D experiments to structure elucidation. Examples of structural analyses of organic compounds via NMR methods are ubiquitous in the literature such that it is impractical to mention more than just a few of them here. Therefore, the reader is encouraged to peruse the organic chemistry literature to find structural analyses of the specific types of organic compounds of interest. This article will deal mainly with generalities of organic compound structure elucidation, although several relevant examples will be presented.

General practical considerations

Deuterated solvents are utilized with FT NMR spectrometers to provide an internal lock signal to compensate for drift in the magnetic field during the experiment. The more common solvents used for organic compounds are $CDCl_3$, CD_3CN, CD_3OD, acetone-d_6, benzene-d_6, DMSO-d_6 and D_2O. Since all deuterated solvents contain some protonated impurities (e.g. $CHCl_3$ in $CDCl_3$), one should choose a solvent that will not interfere with the NMR peaks of interest from the sample. Tetramethylsilane (TMS) is usually added to the sample as an internal standard for both proton and carbon spectra, being set at 0.0 ppm in both cases. However, the small protonated solvent impurities also make good standards as the chemical shifts of these peaks are published in many texts and are reported relative to TMS.

Protons provide the highest sensitivity for NMR observations, and therefore only small quantities of sample are needed (1–10 mg in 0.5 mL of solvent for an FT instrument). ^{13}C NMR has a much lower sensitivity than proton NMR due to the low natural abundance of ^{13}C (1.1%) compared with ^{1}H (100%), and the fact that the energy splitting and hence the resonance frequency for carbon is approximately one quarter that of proton. Thus, for a spectrometer whose ^{1}H frequency is 300 MHz, the frequency for ^{13}C is 75.5 MHz. To obtain a carbon NMR spectrum in a timely manner, one needs to use either more sample than for a ^{1}H NMR spectrum (>20 mg of a compound with MW ≈ 150–300 g mol^{-1}), or a higher field spectrometer.

^{1}H NMR

As mentioned earlier, ^{1}H NMR is a very valuable method for obtaining information regarding the molecular structure of organic compounds with any number of protons. The electronic environment, as well as near neighbours and stereochemistry, can be determined by analysing the chemical shifts and spin–spin couplings of protons. The relative number of protons can be determined by direct integration of the areas under the peaks (multiplets), as the number of protons is directly proportional to the area under the peaks produced by those protons. To obtain accurate integrations, however, the relaxation delay needs to be at least 5 times the longest T_1 in the sample.

Proton chemical shift

Chemical shifts are diagnostic of the electronic environment around the nucleus in question. Withdrawal of electron density from around the nucleus will deshield the nucleus, causing it to resonate at a lower field (higher frequency or chemical shift). Higher electron density around a nucleus results in shielding of the nucleus and resonance at higher field (lower frequency or chemical shift (δ)). Therefore, basic details of the molecular structure can be gleaned from analysis of the chemical shifts of the nuclei. Factors that affect the electron density around the proton in question include the amount of substitution on the carbon (i.e. methyl, methylene, methine), the inductive effect of nearby electronegative or electropositive groups, hybridization, conjugation interactions through π bonds, and anisotropic (ring current) effects. Tables of proton chemical shifts can be found in various texts, such as those listed in Further reading.

As alkyl substitution increases on the carbon that possesses the proton(s) in question, the deshielding increases due to the higher electronegativity of carbon compared with hydrogen (e.g. $CHR_3 > CH_2R_2 > CH_3R$), producing a downfield shift of the resonances (methine most downfield, methyl most upfield). The deshielding effect of electron-withdrawing groups depends directly upon the electronegativity of these groups, and upon whether their effects are inductive (less effective) or through resonance (more effective). This deshielding effect falls off rapidly with increasing number of bonds between the observed proton and the electronegative group. One can, therefore, estimate chemical shifts of alkyl protons by analysing the amount of carbon substitution and the effects of nearby electron-withdrawing groups. A fairly accurate calculation of chemical shifts for methylene protons attached to two functional groups ($X-CH_2-Y$) is possible by using Shoolery's rule, where the shielding constants for the substituents, Δ_i, are added to the chemical shift for methane. Tables of these shielding constants can be found in most texts on NMR spectroscopy.

To some extent, hybridization also influences the electron density around the proton in question by electronegativity effects. With increasing s character in a C–H bond, the electrons are held closer to the carbon nucleus. The protons consequently experience less electron density and are, therefore, more deshielded. This reasoning applies very well to protons attached to sp^3 rather than sp^2 carbons. For sp (acetylenic) protons, however, anisotropic effects are the dominating factors.

Electron-donating or electron-withdrawing groups directly attached to aromatic or alkene sp^2 carbons greatly affect the chemical shifts of aromatic or vinyl protons via π bond interactions (resonance). Thus, vinyl protons on the β-carbon of an α,β-unsaturated carbonyl system are further downfield (more deshielded) than the proton on the α-carbon due to resonance, and the opposite holds true for the β-proton(s) of a vinyl ether, as shown in **Figure 1**. In aromatic systems, electron-withdrawing groups deshield the protons ortho and para to it relative to unsubstituted benzene, while a group that is electron-donating by resonance will shield the ortho and para protons such that they resonate at a field higher than unsubstituted benzene ($\delta7.27$). Empirical methods for estimating the chemical shifts of protons on substituted alkenes and benzene rings have been developed (see Further reading). It should be realized, however, that the anisotropies of aromatic and alkenyl systems are also responsible for the larger than expected downfield shifts of the protons. The large downfield shift of aldehyde protons ($\sim\delta9.5$) is due in large part to the anisotropic shielding/deshielding effect (called the cone of shielding/deshielding in carbonyls), as seen in alkenes and aromatic compounds.

Shielding and deshielding effects via anisotropy caused by ring currents can also affect protons not directly attached to the alkene, alkyne, carbonyl or aromatic systems. A good example of this in shown in **Figure 2**. The calculated chemical shift of the methine proton H_a in the absence of any ring current effects is $\delta4.40$, while the observed chemical shift is

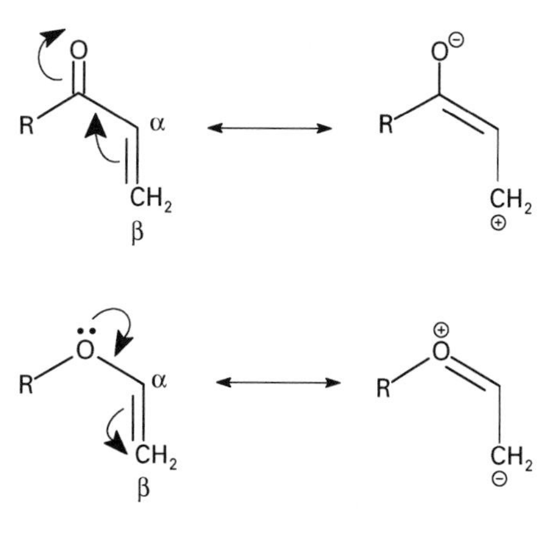

Figure 1 Shielding and deshielding effects due to resonance.

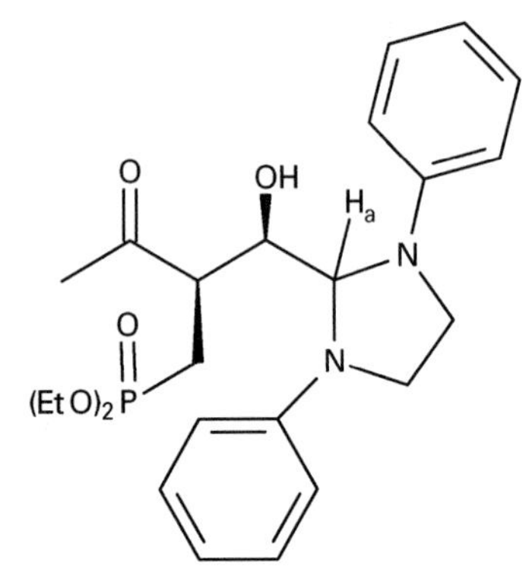

Figure 2 Proton H_a is deshielded by ~1 ppm due to the ring current of the nearby phenyl group.

δ5.44. The low energy conformation of the molecule (from molecular modelling) has one of the phenyl rings very near the proton H_a. Therefore, it appears that the ring current of this phenyl group is deshielding this proton by ~1 ppm.

In organic molecules possessing protons with very similar chemical shifts, such as steroids or carbohydrates, it would be advantageous to be able to simplify the spectrum by eliminating all spin–spin splittings, thereby allowing the determination of resonance frequencies by only chemical shift effects.

An improved method has been developed to produce this type of 'chemical-shift spectrum', and is illustrated in **Figure 3**. Overlapping resonances are resolved into singlets, and this allows for a more straightforward structural assignment of the resonances. Near neighbours, coupling constants and relative stereochemistries can be determined by other spectral editing techniques and experiments (see below).

Through-bond coupling: determination of near neighbours and stereochemistry

The analysis of through-bond spin–spin coupling (scalar or *J* coupling) allows for ready determination of the number of neighbouring protons, as well as the relative stereochemistry in certain cases. See the texts listed in Further reading for more in-depth discussions of spin–spin coupling. In short, spin–spin couplings occur between magnetically nonequivalent nuclei (here, protons) through intervening bonding electrons, and decreases with increasing number of intervening bonds. Protons that are chemically equivalent (interchangeable by a symmetry operation) are magnetically equivalent if they exhibit identical coupling to any other nucleus not in that set. However, protons with the same chemical shift do not split each other even when the coupling constant

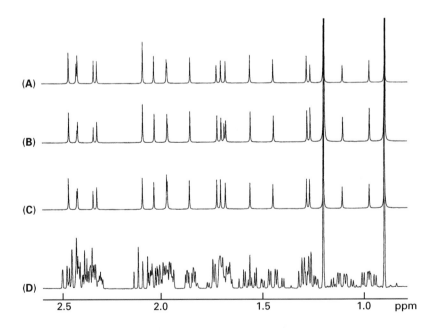

Figure 3 Chemical shift spectra of 4-androsten-3,17-dione obtained from (a) the reflected *J* spectrum; (b) the purged *J* spectrum (the additional response near δ1.7 is from the residual water signal); and (c) the *z*-filtered *J* spectrum. The conventional ¹H spectrum is shown in (d). Reprinted with permission from Simova S, Sengstschmid H and Freeman R (1997) Proton chemical-shift spectra. *Journal of Magnetic Resonance* **124**: 104–121.

between them is non-zero. Rapid rotation about a C–C single bond, such as with a CH_3 group, results in an average environment for each methyl proton, and hence, equivalence.

Interacting protons with very different chemical shifts are weakly coupled if the difference in chemical shifts between the coupled protons, $\Delta\delta$, is large compared to the coupling constant J, i.e. $\Delta\delta/J > 10$. The multiplets resulting from this weak coupling are considered 'first-order' patterns, and can be interpreted easily. The multiplicity is governed by the $(2nI + 1)$ rule, where n is the number of *magnetically equivalent* coupled protons and I is the spin of the nucleus. In first-order systems, the multiplicities and peak intensities of coupled protons can be predicted using Pascal's triangle. For example, a proton split by two magnetically equivalent neighbours will be a triplet with peak intensities of 1:2:1. The frequency difference between the lines of the multiplet is the coupling constant, J, reported in Hz, and is invariant with changes in the strength of the magnetic field. The recent greater accessibility to higher NMR field strengths has enabled the interpretation of most proton spectra as first-order. In symmetrical spin systems, these simple rules do not apply and a more rigorous analysis is needed.

The Pople spin notation system is generally utilized to indicate the degree of difference among nuclei. Thus, in a two spin system, AX indicates a molecule with two nuclei where the chemical shift difference is much larger than the coupling between them (weakly coupled system, first-order analysis possible), whereas AB indicates a molecule containing two strongly coupled nuclei with similar chemical shifts. An A_2BB' notation indicates a set of two equivalent nuclei (A) interacting with two nuclei (B, B') that are chemically, but not magnetically, equivalent.

For proton NMR, the most diagnostic couplings are 2-bond (^2J, geminal), 3-bond (^3J, vicinal), and 4-bond (^4J, W-type) couplings. Geminal couplings can be quite large, but may not be evident due to the symmetry associated with the carbon and protons in question. As mentioned above, the lack of geminal coupling is due to the identical chemical shifts of the protons involved. Vicinal 3-bond proton–proton couplings tend to be the most useful when determining stereochemistry, although coupling beyond three bonds can be important in systems with ring strain (small rings, bridged systems) or bond delocalization, as in aromatic and allylic systems.

For simple organic molecules, pattern recognition of multiplets can simplify structure determination. For example, the presence of an upfield triplet due to

three protons and a more downfield quartet due to two protons with the same coupling constant is most probably due to an ethyl group ($X–CH_2CH_3$). Therefore, it is useful to look for common patterns. Many preliminary assignments can be made in a standard 1D ^1H spectrum due to the reciprocity of coupling constants ($J_{AB} = J_{BA}$) With more complex patterns due to coupling to several magnetically non-equivalent protons, interpretation can be done via first-order analysis only if no two of the spins within an interacting multispin system have $\Delta\delta/J \leq 6$. Multiplets such as doublet of doublets (dd), doublet of triplets (dt), triplet of doublets (td), doublet of quartets (dq), doublet of doublets of doublets (ddd), etc., can usually be analysed by first-order techniques, especially if the spectrum was run at a fairly high magnetic field. A very useful and practical guide to first-order multiplet analysis that utilizes either a systematic analysis of line spacings or inverted splitting trees to determine the couplings is listed in Further reading.

Measurement of coupling constants can usually be done directly from the 1D spectrum with well resolved multiplets, or aided by simple 1D homonuclear decoupling experiments where irradiation of one of the *weakly* coupled nuclei (i.e. nuclei with very different chemical shifts) simplifies a multiplet by eliminating that spin–spin interaction. Several two dimensional techniques also help to determine the coupling network, and determine and assign coupling constants. A 2D COSY (correlated spectroscopy) spectrum is a homonuclear experiment, and provides a map of the proton–proton J-coupling network in the molecule. The spectrum contains a set of auto-correlated peaks along the diagonal ($\omega_1 = \omega_2$), which is the original spectrum. For those spins that exchange magnetization due to J-coupling, $\omega_2 \neq \omega_1$ and off-diagonal peaks appear. The diagonal peaks that correspond to J-coupled spins are connected by symmetrical pairs of off-diagonal peaks. In general, strongly coupled protons are handled better in a COSY experiment than with conventional 1D homonuclear decoupling. However, in molecules that have overlapping resonances, it can be difficult to accurately assign the cross peaks. New computer programs are being developed to provide automated processing and assignment of the data (see Further reading). Further simplification of the spectrum can be attained by utilizing a DQF-COSY (double quantum filtered COSY) experiment, where singlets are essentially eliminated from the spectrum. Coupling constants can be attained from a COSY spectrum, but it is not a trivial process. The J values measured from a COSY spectrum also tend to be slightly larger than the actual J value.

The two-dimensional experiment, homonuclear *J*-resolved spectroscopy, is utilized to accurately measure the scalar coupling constants. This method can readily resolve overlapping signals, as well as strongly coupled systems. From the contour plot of the spectrum, multiplets are resolved along the *y*-axis (Hz), and the coupling constants are read directly along this axis. Projection of the multiplet onto the *x*-axis (δ-axis) provides a single resonance line for each distinct spin system without the effects of coupling (i.e. is proton-decoupled), and accurate values of δ (ppm) can be attained.

Stereochemical assignments

Accurate stereochemical assignments are generally only possible in rigid or ring systems where free rotation about carbon–carbon bonds is hindered or not possible. As mentioned above, vicinal, three-bond couplings (^3J) can be quite diagnostic of the stereochemical relationship between the coupling protons. The Karplus equation (Eqn [1]) can predict the vicinal coupling constant $^3J_{H-C-C-H}$ with reasonable accuracy if the H–C–C–H dihedral angle is known. Thus, dihedral angles near 0° or 180° have the largest coupling constants, while a dihedral angle of 90° has a coupling constant near 0 Hz.

$$^3J = A\cos^2\phi + C \quad (\phi = 0-90°)$$
$$^3J = A'\cos^2\phi + C' \quad (\phi = 90-180°)$$

[1]

Use of this relationship in alkenes and ring (or bridged) systems works very well to predict stereochemistry. For alkenes, *trans* coupling is in the range of 12–18 Hz, and *cis* coupling is 6–12 Hz. See the texts listed in Further reading for tables listing coupling constants in various alkenyl systems. *Trans*, di-axial protons on a six-membered carbocyclic ring have a dihedral angle of ~180° and a *J* value of 6–14 Hz (typically 8–10 Hz), whereas protons oriented axial–equatorial or equatorial–equatorial with dihedral angles near 60° have coupling constants of 0–5 Hz (typically 2–3 Hz). In five-membered rings, vicinal *trans* and *cis* coupling constants may be similar in magnitude due to the reduced conformational flexibility of the ring relative to six-membered rings. Five-membered rings will adopt a twist conformation to relieve eclipsing interactions. For example, in the oxazolidinones in **Figure 4**, it was found that $^3J_{4-5\ cis} = 7.2$ Hz, and $^3J_{4-5\ trans} = 5.9$ Hz.

Where no coupling is possible because of a quaternary centre or if the coupling constants fall on the

Figure 4 Oxazolidinones exhibiting coupling constants of $^3J_{4-5\ cis} = 7.2$ Hz and $^3J_{4-5\ trans} = 5.9$ Hz.

borderline between two possible orientations nuclear Overhauser effect (NOE) measurements may need to be taken in order to definitively establish the relative stereochemistry. The use of pulsed field gradients (PFGs) in the acquisition of NOE enhancement spectra (called GOESY) now allows one to avoid the need to compute difference NOE spectra, as was done in the past. With NOE difference spectra, it was hard to avoid subtraction artifacts, and thus difficult to obtain accurate NOE values, especially small (<1%) enhancements. Utilizing PFGs, the only resonances now seen in the spectrum are those from spins which are cross-relaxing with the irradiated spin.

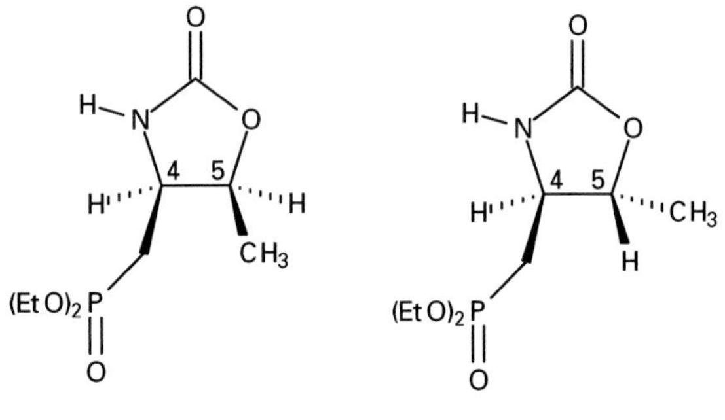

The stereochemical assignments for the oxazolidinones in **Figure 4**, were confirmed by the NOE enhancements of 13% between H4 and H5 in the *cis* isomer, and only 2% between H4 and H5 in the *trans* isomer. There was also an NOE enhancement of 5% between the CH_2 α to the phosphonate and H5 in the *trans* isomer. In the examples shown in **Figure 5**, the vicinal coupling constants were all small (0–3 Hz), thus indicating that all the methine protons were in axial–equatorial or equatorial–equatorial relationships. The NOE measurements (GOESY experiment) indicated in **Figure 5** enabled final assignments of the relative stereochemistries.

^{13}C NMR

Information regarding the number and types of carbons in an organic compound can be provided by carbon NMR spectroscopy. Since the chemical shift range is greater for carbon than for proton, a greater dispersion of signals is seen. Different functional groups that contain at least one carbon, such as

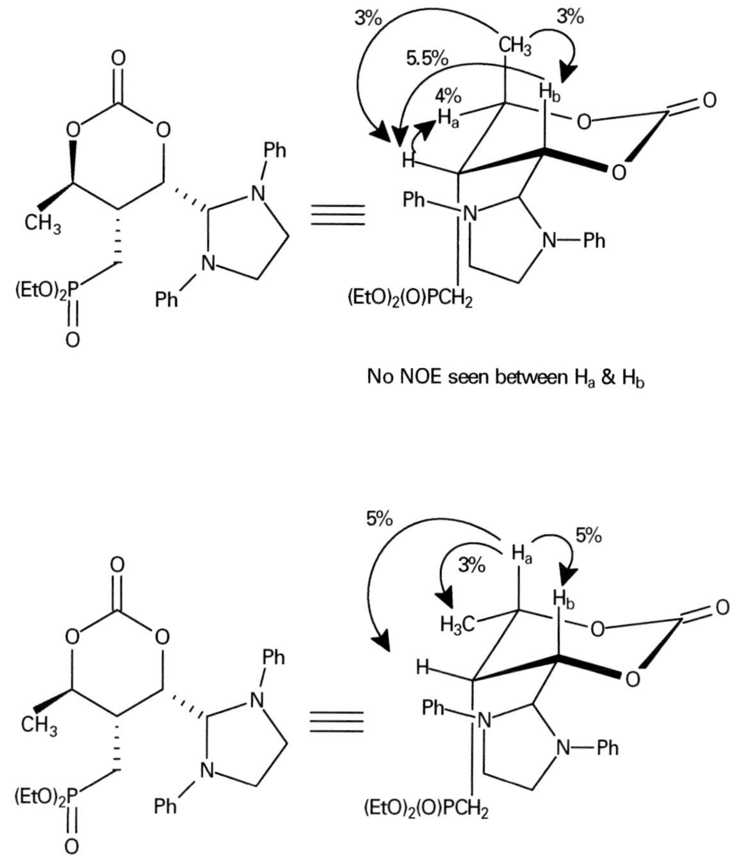

No NOE seen between H_a & H_b

Figure 5 NOE enhancements (GOESY experiments) of two diastereomeric cyclic carbonates to confirm relative stereochemistries.

ketone, ester and amide carbonyls, alkenes, alkanes, alkynes, nitriles, imines, etc., can generally be readily distinguished by ^{13}C NMR spectroscopy. The chemical shifts of aromatic and alkene carbons, however, are in the same chemical shift range, and at times cannot be differentiated. This is in contrast to aromatic and alkene protons which exhibit different chemical shift ranges.

In general, the factors that affect the chemical shifts of carbons are the same as for protons (i.e. electron density around the nucleus in question, and anisotropy effects). Carbon chemical shifts can be readily calculated from tables of shift effects found in many texts. However, unlike protons attached to sp^2 carbons, sp^3 carbons attached to sp^2 carbons exhibit only a small shift difference. There are also few good substituent parameters available for calculating the chemical shifts of alkene carbons bearing polar groups, unlike the calculation of ^1H NMR chemical shifts near polar groups. However, in systems where resonance is present, some predictions can be made of relative shift differences in the carbons (see **Figure 1**).

Carbon–proton connectivities can be determined using several methods. The number of protons directly attached to the carbon in question will split the carbon resonance according to the $2nI + 1$ rule seen in proton NMR. There tends to be, however, much overlap of the multiplets in fully proton-coupled carbon spectra, sometimes such that it is very difficult to distinguish between the various multiplets. Routine carbon spectra are therefore measured fully proton decoupled for simplicity. Information regarding the exact number of protons attached to the carbons can be acquired from APT, DEPT or INEPT experiments. In APT spectra, the carbons bearing an odd number of protons (CH, CH_3) can be distinguished from carbons with no or two attached protons (quaternary C, CH_2). DEPT and INEPT experiments can distinguish between all four types of carbons (primary, secondary, tertiary and quaternary). Heteronuclear 2D J-resolved spectroscopy can also be used to obtain the multiplicities of the carbons, as well as $^1J_{C-H}$.

A complete mapping of the protons to the carbons they are attached to is possible via a HETCOR (heteronuclear chemical shift correlation) experiment. This method correlates the peaks of the proton spectrum of a compound with the peaks of its carbon spectrum. A contour plot of the spectrum has a cross peak at the intersection of the vertical line drawn from a carbon resonance (plotted along the x-axis) with the horizontal line drawn from a proton peak (plotted along the y-axis). However, this method is relatively insensitive as it suffers from the low natural abundance of ^{13}C atoms in the molecule. Utilization of the inverse detection method, HMQC (heteronuclear correlation through multiple quantum coherence), can alleviate this problem as ^{13}C responses are observed in the ^1H spectrum. Normally, ^1H–^{13}C coupling information is included in the ^1H dimension, although proton decoupling from carbon is possible. Quaternary carbons, however, will not be present in a HMQC spectrum. A method to aid in assignments of quaternary carbons, as well as carbon and proton connectivities, is the HMBC (heteronuclear multiple bond correlation) experiment. In these spectra, cross peaks are observed connecting the ^{13}C signals to ^1H signals two or more bonds away.

The INADEQUATE experiment is designed to map out the entire carbon skeleton of a molecule by providing carbon–carbon connectivities, and offers great possibilities for organic structure determination. However, it suffers severely from the low natural abundance of covalently bound ^{13}C–^{13}C pairs. Several groups have recently offered modifications of the pulse sequence to try to overcome this limitation. It remains to be seen, however, if these new programmes will produce the higher sensitivities needed for this experiment to become a routine analytical procedure.

Even without the newer modifications to the pulse sequence, the INADEQUATE experiment can be an invaluable method for structure determination in some cases, as illustrated in the example in **Figure 6**.

The rearrangement of the cage hydrocarbon diazonium ion [1] in the presence of water could lead to three possible alcohol products [2], [3] and [4]. Only two alcohol products were obtained from this reaction in isolated yields of 38% and 7%. The structures of these two products could only be determined by the 2D INADEQUATE experiment due to the hydrocarbon nature of their structures, and therefore, the lack of any distinguishing details in the proton spectra. Each compound also had the same number of methylene, methine and quaternary carbons, thus precluding the utilization of structure determination by simple 1D ^{13}C spectroscopy.

The INADEQUATE and APT spectra for the major product are shown in **Figure 7**. In an INADEQUATE spectrum, the pairs of adjacent carbons, and hence the connectivity, can be mapped out similarly to a COSY spectrum. The major difference here is that the original spectrum is not on the diagonal in an INADEQUATE spectrum (as in a COSY spectrum), but is in the x-axis direction (= normal ^{13}C frequencies) along the line $v_1 = 0$ (residual single quantum signals). The y-axis is the frequency v_1, the double quantum frequency that is the sum of the frequencies of the two coupled nuclei referenced to a transmitter frequency at zero. The peaks arising from two coupled nuclei (here adjacent carbons) with shifts v_a and v_b have coordinates of $((v_a + v_b), X)$, where X is the frequency of the carbon in a single quantum coherence spectrum (1D spectrum). At the double quantum frequency (v_1) for each pair of adjacent carbons, doublets will occur at the coordinates of $((v_a + v_b), v_a)$ and $((v_a + v_b), v_b)$. Thus, the midpoint between each pair of signals lies on a line with slope of 2, and helps to distinguish the real peaks from artifacts. The (C7–C1) – (C1–C7), (C14–C8) – (C8–C14) and (C3–C5) – (C5–C3) peak pairs are illustrated on the spectrum (B) in **Figure 7**.

In the example illustrated in **Figure 6**, it may be noted that compound [3] would require the grouping

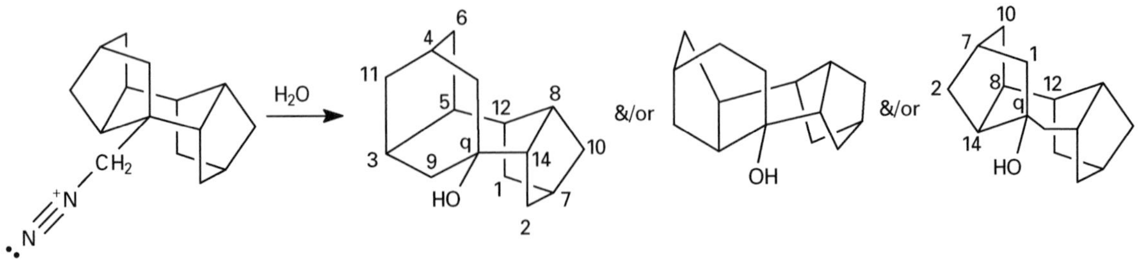

Figure 6 Possible products from the rearrangement of structure [1] in water. Bratis AD, Bruch MD and Murray RK Jr unpublished results.

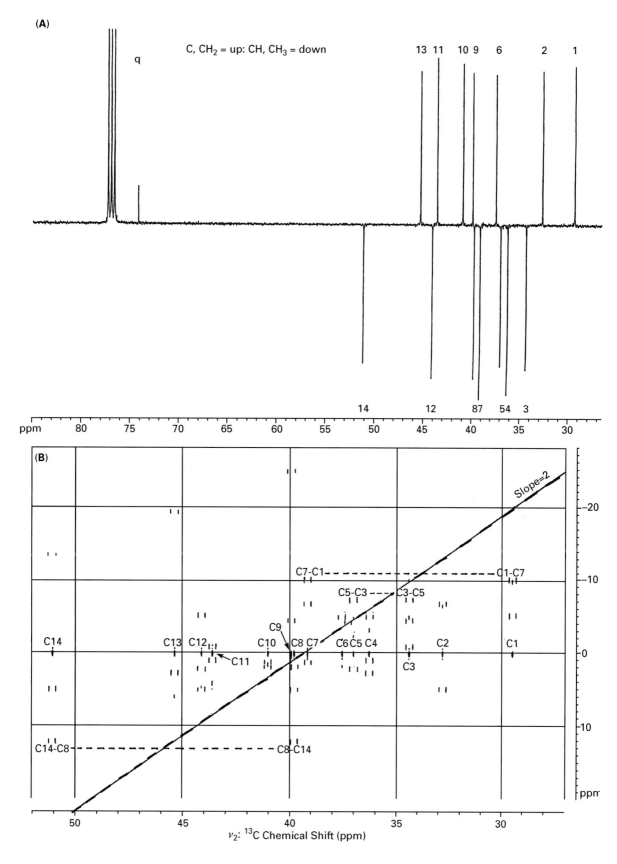

Figure 7 (A) APT ^{13}C spectrum of compound [2]; (B) INADEQUATE spectrum of compound [2], with the connectivities of C1 to C7, (see **Figure 6**) C8 to C14 and C3 to C5 shown. Bratis AD, Bruch MD and Murray RK Jr unpublished results.

Table 1 Comparison of calculated proton distances, dihedral angles and estimated *J* couplings, with *J* couplings and NOE results for compound [6]

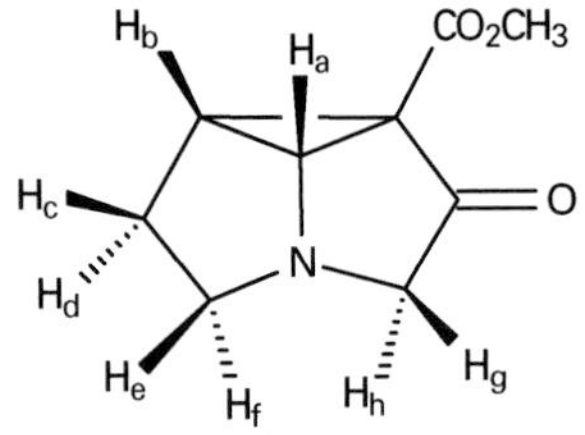

	Mechanics			Experimental	
Protons	Distance (Å)	Angle (°)	*J* coupling (Hz, estimated)	*J* coupling (Hz)	NOE
H_a–H_b	2.71	7.1	8	4.9	yes
H_b–H_c	2.47	13	7.5	6.5	yes
H_b–H_d	2.92	108	1	0	ND
H_c–H_d	1.81	109	10–20	14.0	yes
H_c–H_e	2.37	32	6	7.0	yes
H_c–H_f	2.77	87	<1	0	ND
H_d–H_e	3.09	154	8	NA	ND
H_d–H_f	2.48	35	6	6.2	ND
H_e–H_f	1.79	107	10–20	12.3	yes
H_g–H_h	1.8	108	10–20	18.0	yes

NA = not available; ND = no NOE detected.

Table 2 HETCOR data for compound [6]

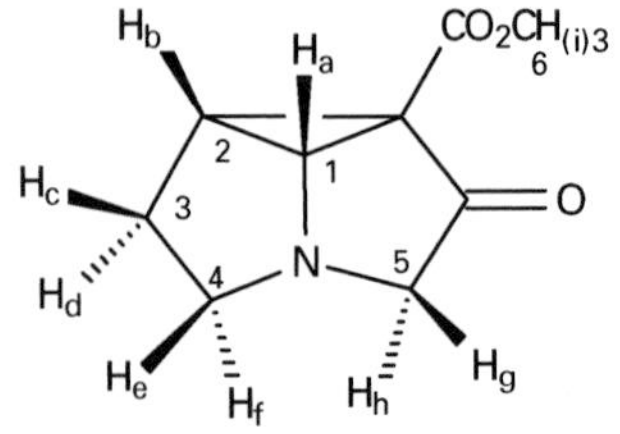

Carbon	^{13}C δ	Proton(s)	^1H δ
C-1	63.7	H_a	4.23
C-2	43.7	H_b	3.06
C-3	26.5	$H_{c,d}$	2.25, 1.99
C-4	64.3	$H_{e,f}$	3.32, 2.79
C-5	68.8	$H_{g,h}$	3.64, 2.84
C-6	53.1	H_i	3.75

–CH_2CH_2– to be present, which is not seen in the INADEQUATE spectrum. For the major compound to be structure [4], the connectivity of carbon 1 would have to be to carbon 7 and the quaternary carbon, q. The connectivity found for carbon 1 was to carbon 7 and the methine carbon 12. Therefore, only structure [2] fully supports all the spectral data for the major isolated product from this rearrangement.

Putting it all together

The following is an outline of the basic procedure that a practising organic chemist follows when deducing the structure of an organic compound via NMR spectroscopy. First, standard 1D ^1H and ^{13}C spectra are acquired and analysed. If needed, a 'chemical shift spectrum' can provide a straightforward assignment of the δ value of all the resonances, including overlapping multiplets. Proton spin–spin couplings and near neighbours are determined either directly from the 1D ^1H spectrum, or assisted by homonuclear decoupling experiments. If these experiments are not conclusive due to overlapping resonances, changing the solvent or utilization of a shift reagent can on occasion resolve the overlapping multiplets. A map of the *J*-coupling network in the molecule is available through 2D COSY or TOCSY experiments. New programs to assist in assignment of the cross-peaks in complicated COSY or TOCSY spectra are now becoming available. Utilization of the DQF-COSY experiment eliminates all the singlets from the spectrum. The homonuclear 2D *J*-resolved spectrum allows for the separation of overlapping resonances and, therefore, accurate measurement of the coupling constants and chemical shifts. A number of ^{13}C experiments, such as APT, DEPT, INEPT, INADEQUATE, and heteronuclear 2D *J*-resolved experiments, can be run to assist in determining proton–carbon attachment, carbon–

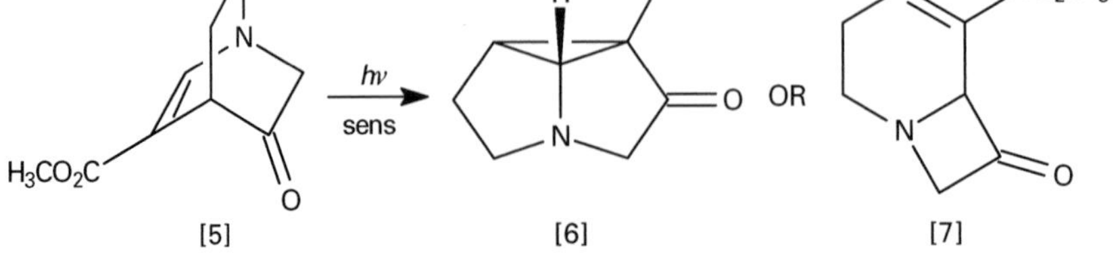

[5]

[6] Oxa-di-pi-methane rearrangement product

[7] 1,3-Acyl shift product

Figure 8 Possible products from the photolysis of compound [5]. Kiessling AJ and McClure CK unpublished results.

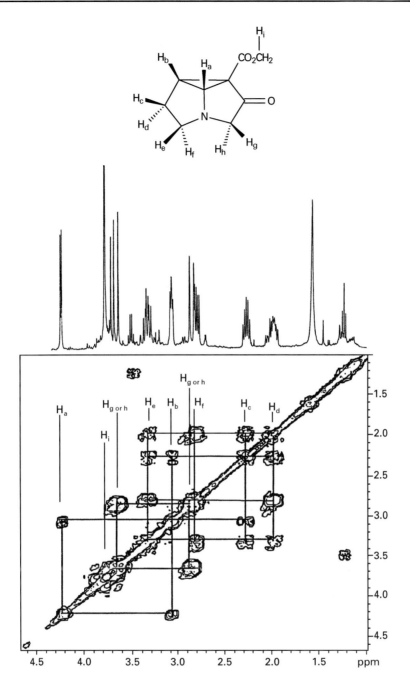

Figure 9 COSY spectrum of compound [6].

carbon attachments for carbon skeleton determination, and values of $^1J_{C-H}$. A 2D HETCOR spectrum will confirm the proton–carbon connectivities. NOE, GOESY and ROESY experiments can assist in determining stereochemical information where coupling constants cannot.

The following is an example that illustrates how several of the experiments listed above, as well as molecular mechanics calculations, can be used to deduce the structure of the compound produced by a photochemical rearrangement. The photochemical irradiation of the bicyclic compound [5] was run under sensitized conditions (acetophenone). Two possible products, [6] and [7] are theoretically possible, and are shown in **Figure 8**. The 1,3-acyl shift product [7] normally arises from photolysis under non-sensitized conditions, but can be formed from certain compounds under sensitized photolysis conditions. The oxa-di-pi-methane rearrangement product [6] was the desired compound. From 1D ^1H and ^{13}C NMR spectra, the only product isolated from the photochemical rearrangement did not appear to

(A)

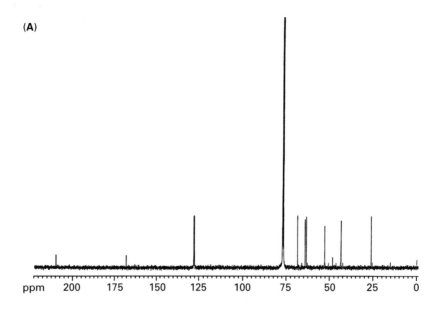

(B)

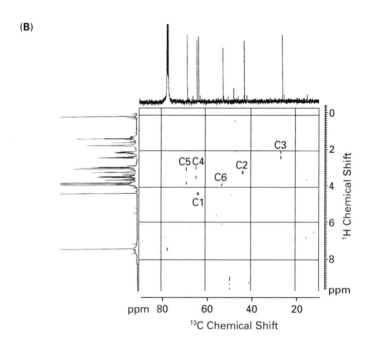

Figure 10   (A) ¹³C spectrum of compound [6] (see **Figure 2**) in CDCl₃/benzene-d₆; (B) HETCOR spectrum of compound [6].

contain an olefin, as would be seen in the α,β-unsaturated ester [7]. Thus, the rearrangement most likely went via the oxa-di-pi-methane rearrangement, and not by the 1,3-acyl shift mechanism.

Further proof that the structure of the photoproduct was indeed [6] is as follows. From the standard COSY spectrum (**Figure 9**), all the proton–proton coupling networks could be established. The proton responsible for the peak at δ4.22 (d, J = 4.8 Hz) was

coupled only to a proton at δ3.06 (dd, J = 6.5, 4.9 Hz), which in turn was coupled to only one other proton at δ2.25. Of the protons H_a–H_f, only proton H_a was expected to be coupled to only one other proton, namely H_b. The chemical shift of δ4.22 was also reasonable for H_a. The multiplet at δ3.06 was therefore assigned to H_b, and was coupled to one other proton at δ2.25. This other proton could be either H_c or H_d.

In order to assist in the assignments, the tricyclic structure [6] was submitted to molecular mechanics calculations to estimate the dihedral angles between the protons, and thus approximate the coupling constants using Equation [1] (see **Table 1**). According to these calculations, H_b and H_c had a dihedral angle of ~13° and thus an estimated coupling constant of 7.5 Hz, while H_b and H_d had a dihedral angle of ~110° and an estimated coupling constant of 1 Hz. The measured J value between signals at $\delta 3.06$ and $\delta 2.25$ was 6.5 Hz, closely matching the calculated coupling constant between H_b and H_c. Therefore, H_c was assigned to the signal at $\delta 2.25$. This multiplet (ddd) exhibited three coupling constants of 14.0, 7.0 and 6.9 Hz. The large coupling of 14 Hz would be consistent with geminal coupling to proton H_d. The calculations predicted that in addition to H_b, H_c would couple to H_e with a dihedral angle of 32° and an estimated coupling constant of 6 Hz. H_f was predicted to be nearly orthogonal to H_c, and thus have little or no coupling to H_c. From this data, H_d was assigned to the multiplet at $\delta 1.99$ and H_e to the multiplet at $\delta 3.32$. From the COSY spectrum, the multiplet at $\delta 1.99$ was further coupled to the multiplet at $\delta 2.79$, which was assigned to H_f. The multiplet (dd) at $\delta 2.79$ had two coupling constants of 12.3 and 6.2 Hz. The large coupling constant was geminal coupling with H_e. The other coupling constant was consistent with the calculation of the dihedral angle of 35° between H_f and H_d. The protons H_g and H_h were coupled only to each other, and the exo proton H_g was assigned as the downfield doublet of doublets at $\delta 3.65$.

The distances between the protons of the proposed structure were also calculated by molecular mechanics and are summarized in **Table 1**. The photoproduct was submitted to NOE experiments to verify the spatial relationships. The signals assigned to H_a, H_b and H_e yielded meaningful data, and the NOE results are shown in **Table 1**. Results of the NOE experiments are in agreement with the proposed structure, where H_a, H_b, H_c and H_e are shown to be in a *cis* relationship.

The ^{13}C and HETCOR spectra in **Figure 10A** and **Figure 10B** respectively, further verified the proposed structure [6]. No carbon signals were detected in the alkene region of the spectrum, consistent with the lack of alkene protons. The only carbonyl peak detected was at $\delta 207.6$, consistent with the ketone in [6]. The HETCOR data is summarized in **Table 2**.

See also: ^{13}C NMR Methods; ^{13}C NMR Parameter Survey; Chemical Exchange Effects in NMR; Chemical Shift and Relaxation Reagents in NMR; Chromatography-NMR, Applications; Enantiomeric Purity Studied Using NMR; Magnetic Field Gradients in High Resolution NMR; NMR Data Processing; NMR Pulse Sequences; Nuclear Overhauser Effect; Structural Chemistry using NMR Spectroscopy, Peptides; Structural Chemistry Using NMR Spectroscopy, Pharmaceuticals; Two-Dimensional NMR Methods.

Further reading

Bourdonneau M and Ancian B (1998) Rapid-pulsing artifact-free double-quantum-filtered homonuclear spectroscopy. The 2D-INADEQUATE experiment revisited. *Journal of Magnetic Resonance* 132: 316–327.

Bruch MD (ed) (1996) *NMR Spectroscopy Techniques*, 2nd edn. New York: Marcel Dekker.

Derome AE (1987) *Modern NMR Techniques for Chemistry Research*. Oxford: Pergamon.

Hoye TR, Hanson PR and Vyvyan JR (1994) A practical guide to first-order multiplet analysis in ^1H NMR spectroscopy. *Journal of Organic Chemistry* 59: 4096–4103.

Lambert JB, Shurvell HF, Lightner DA and Cooks RG (1998) *Organic Structural Spectroscopy*. New York: Macmillan.

Sengstschmid H, Heinz S and Freeman R (1998) Automated processing of two-dimensional correlation spectra. *Journal of Magnetic Resonance* 131: 315–326.

Silverstein RM, Bassler GC and Morrill TC (1998) *Spectrometric Identification of Organic Compounds*, 6th edn. New York: John Wiley.

Simova S, Sengstschmid H and Freeman R (1997) Proton chemical-shift spectra. *Journal of Magnetic Resonance* 124: 104–121.

Stonehouse J, Adell P, Keeler J and Shaka AJ (1994) Ultra-high-quality NOE spectra. *Journal of the American Chemical Society* 116: 6037–6038

Structural Chemistry Using NMR Spectroscopy, Peptides

Martin Huenges and Horst Kessler, Technische Universität München, Garching, Germany

MAGNETIC RESONANCE
Applications

Introduction

The majority of biological processes depend directly on peptides and proteins. The sequence of most peptides and all proteins are encoded genetically and the polypeptides are post-translationally modified, processed and transported to their specific location in the cell. The wide range of possible chemical structures (owing to the combination of functional groups of their amino acid residues), especially in their three-dimensional dynamic conformation, allows peptides and proteins to play many different roles in biological processes, such as hormone/receptor interactions, cellular adhesion and cellular recognition, transport mechanisms between cell compartments or through membranes, and the processing of almost all chemical compounds, including peptides and proteins, to name only a few.

Although the conformation and dynamics of peptides and proteins are encoded in their sequence, they are not yet reliably predictable based on it. The determination of the dynamic 3D structure therefore was, and still is, of utmost importance for the interpretation and artificial modulation of their functions.

The challenges posed by peptides and proteins have strongly stimulated the development of modern NMR spectroscopy. Most of the multidimensional NMR techniques now available designed and applied initially to peptides and proteins.

We will discuss in this article first some general problems which arise in the NMR spectroscopy of peptides. Then, NMR techniques for signal assignment and extraction of conformational parameters will be described, followed by a short excursion into structure determination using NMR parameters. The final part will include the analysis of peptide dynamics based on NMR data.

General problems with peptides

Peptides are composed of a linear, branched or cyclic array of amino acid residues (**Figure 1**). The peptide chain is defined and numbered from the N to the C terminus. The α carbon atoms of the amino acids are linked by peptide bonds. The bonds to the α carbon atom are described by their bond angles $\phi(N-C^\alpha)$, $\psi(C^\alpha-CO)$ and $\chi_1(C^\alpha-C^\beta)$. The usually planar peptide bonds prefer the *trans*-configuration ($\omega = 180°$) as shown in **Figure 1** for the Phe-Gly and Gly-Val bonds. Only in the case of Xaa-Pro pairs are *cis*- and *trans*-conformations of similar energy. In **Figure 1** the Val-Pro bond is in the *cis*-configuration. The barrier between *cis* and *trans* peptide bonds is between 16 and 20 kcal mol^{-1} which leads to a slow exchange between the conformations on the NMR time-scale at room temperature. At higher temperature the exchange between the two states occurs at an increased rate, leading to a coalescence of the two signal sets in NMR spectra.

Rotations around ϕ, ψ and χ_1 are fast on the NMR time-scale. At a result it is not straightforward to distinguish between a single preferred conformation and a rapid equilibrium between several conformations (see below). Linear peptides, approximately up to dodecamers, are normally very flexible and do not exist in or prefer a single conformation, although sometimes a slight preference for a distinct set of structures is observed. However, cyclization and/or sterically demanding substitution restrain the conformational space and often allow the identification of a preferred conformation.

The equilibrium between the different conformations of a peptide can be very sensitive to the environment. Hence, different conformations can be found in a single unit cell of a crystal, between different crystals, between crystal and solution as well as between 'free' and receptor bound peptides. 'Free' conformations are embedded in the solvent whose chemical and physical properties can induce drastic changes when different solvents are used. The evidence for such conformational changes often is indirect (solvent induced signal shifts), but a careful analysis of the whole 3D structure can also lead to the detection of such exchange processes (see e.g. antamanide). Direct observation of solvent induced conformational shifts is possible in the case of the *cis/trans* isomerization of an alkylated peptide bond,

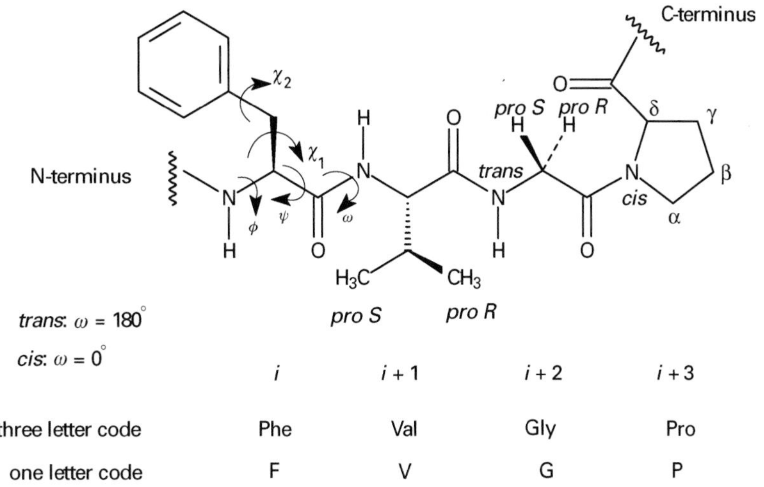

Figure 1 Schematic representation of the tetrapeptide sequence – FVGP – used to illustrate the nomenclature.

since rotation around this bond is slow enough to be resolved on the NMR time-scale. For example, when cyclosporin A is dissolved in CDCl₃, benzene or THF one strongly dominating conformation is observed that contains a NMeVal⁹-NMeVal¹⁰ *cis* amide bond. The corresponding conformation with the same amide in the *trans-* conformation is populated at less than 5%. In more polar solvents, such as CD₃CN and MeOH, a number of coexisting conformations are found, whereas a single but very different conformation is found when the peptide is bound to the receptor. However, in cases of peptides with un-modified CONH peptide bonds *cis*-conformations are only very rarely observed.

Generally, it is recommended that proof of conformational homogeneity, i.e. the dominance of a single or a few conformation(s) under given conditions, be obtained before beginning a detailed NMR analysis. Criteria for preferred conformations are:

- Large chemical shift dispersion within the set of Hᴺ and Hᵅ signals.
- Large shift difference between diastereotopic protons such as Hᵅ protons of Gly, or of Hᵝ protons in the side-chains of Phe, Tyr, His, Trp, Ser, Cys and the two β methyl groups of Val, for example.
- Strong differences in Hᴺ–Hᵅ coupling constants of different residues and Hᵅ–H$^{\beta proR}$/Hᵅ–H$^{\beta proS}$ coupling constants in each side-chain.
- Pronounced differences in NOE intensities.
- Appearance of long-range NOEs between protons of non-neighboured residues.

Only if these criteria are met is it worthwhile initiating a careful conformational analysis. In general, conformational restraints, such as cyclization,

binding to a receptor or complexation with metal ions are necessary to fulfil these conditions.

If the NMR data indicate a flexible structure, a structural discussion is not meaningful since the bio-active conformation of interest, i.e. the conformation the peptide bound at the receptor, is selected out of a large ensemble of alternative conformations.

Assignment of signals

A prerequisite for the extraction of conformational parameters is the assignment of each signal in the spectrum to a specific spin system, and the assignment of these spin systems to specific residues in the peptide chain. The pulse sequences discussed here are shown in **Figure 2**.

Assignment of spin systems

The COSY (correlation spectroscopy) experiment yields information about connectivity between nuclei. COSY cross peaks can be expected for each resolved scalar coupling between nuclei that are connected by two or three bonds. An unambiguous identification of individual amino acid spin systems can be complicated by the overlap of signals in the vicinity of the diagonal of spectrum, overlap of cross peaks for the long side-chains of Arg, Lys, Pro and Leu and the frequently insufficient signal intensities of resonances that are coupled to many neighbouring nuclei (e.g. Hᵞ of Leu couples to eight vicinal neighbours). A list of other pulse techniques used in structural studies of peptides is given in **Table 1**.

The TOCSY experiment is the most efficient way to obtain complete assignments of spin systems (**Figure 3**). TOCSY is often called HOHAHA.

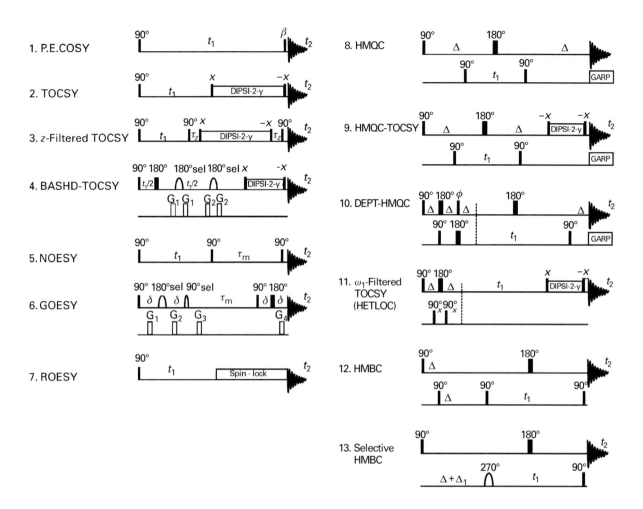

Figure 2 Pulse sequences used in this article. All experiments can be acquired in the phase-sensitive mode, using either TPPI or the States method. Sequences 12 and 13 are best displayed in the magnitude mode because of phase modulation owing to homonuclear couplings in F_2. All inverse correlations, except for sequences 12 and 13, can be preceded by a BIRD (bilinear rotating decoupling) pulse sandwich to allow for a fast repetition rate of scans. For simplicity gradients are not given here in most cases.

The duration of the mixing period determines the efficiency of transfer. If a sufficiently long mixing time is chosen, correlations of the whole proton spin systems are found. TOCSY experiments with short mixing times reveal mainly correlations between directly coupled nuclei.

The spin systems of the naturally occurring amino acid residues can be divided into two groups. While the side-chains of Ala, Arg, Gln, Glu, Gly, Ile, Leu, Lys, Met, Thr and Val exhibit unique spin systems and therefore can be identified with relative ease, the residues in the second group, including Asn, Asp, Cys, His, Phe, Ser, Trp and Tyr, all have similar AMXY proton spin systems. Nevertheless, the aromatic residues of the second group can be unambiguously assigned using H^β–H (ring) NOE cross peaks, whereas the Ser-spin system can be easily distinguished from all other possible AMXY spin systems in COSY experiments because of its weak H^β–$H^{\beta'}$

coupling. Trp is easy to identify via a heteronuclear HMQC experiment owing to its characteristic downfield shift of the β carbon signal.

Heteronuclear spectroscopy also reduces problems with signal overlap because of the large chemical shift dispersion of ^{13}C nuclei. The most popular heteronuclear correlation experiments for this purpose are HMQC and HMQC-TOCSY experiments (e.g. **Figure 4**). The latter contains information similar to the homonuclear ^{1}H-TOCSY, but, as for all heteronuclear experiments, its sensitivity is lower owing to the lower gyromagnetic ratio and low natural abundance of the ^{13}C nucleus.

HMQC sequences can be combined with DEPT editing, allowing for the editing of multiplicities in heteronuclear correlations. The resulting experiments, such as HDQC (heteronuclear double quantum correlation), HTQC (heteronuclear triple quantum correlation) and HQQC (heteronuclear

Table 1 Pulse techniques necessary for structural studies of peptides

Technique	Purpose
P.E.COSY	COSY with simplified multiplet structure. P.E.COSY allows for the accurate measurements of homonuclear coupling constants
TOCSY = HOHAHA	Assignment of spin systems. If a long mixing time is used, TOCSY gives total correlation between all nuclei in a spin system
z-filtered TOCSY	TOCSY sequence that leads to cross peaks with pure phases. z-filtered TOCSY needs long measurement times owing to the random variation of the z-filter in 6 to 12 steps between 110 µs and 20 ms
NOESY	NOESY gives distance information about nuclei that are separated by less than 500 pm in space
ROESY	ROESY is ideally suited for the observation of nuclear Overhauser effects for medium-sized peptides at low field strengths
HMQC	HMQC correlates the shifts of protons with a directly bound heteronucleus. Very sensitive
HMQC-TOCSY	HMQC with subsequent TOCSY transfer to coupled protons
DEPT-HMQC	DEPT-edited HMQC, which allows for the distinction of CH, CH_2 and CH_3 groups. Exclusive selection of these multiplicities is possible with the related HDQC, HTQC and HQQC techniques
ω_1- filtered TOCSY = HETLOC	Extraction of coupling constants to proton-bearing heteronuclei. Because magnetization is distributed among a large number of spins, this method is rather insensitive
HQQC	Assignment of methyl groups in crowded spectra, when folding is not feasible
HMBC	Assignment of carbons and protons and determination of long-range coupling constants
Selective HMBC	Useful variation of HMBC. For peptides, the selective pulse is usually applied to the carbonyl carbons

quadruple quantum correlation, make it possible to exclusively excite CH, CH_2 or CH_3 groups. This results in a simplified assignment procedure.

An alternative way of overcoming problems with overlapping resonances in crowded spectral regions is to apply band-selective excitations. Band-selective pulses can be used to selectively excite a desired spectral region in one or more dimensions. The reduction of spectral width in one or more dimensions improves the digital resolution attainable in the chosen dimension, and thus helps to reduce ambiguities in the resonance assignment procedure. As a welcome side-effect it also shortens the measuring time of the experiment.

Resolution can further be improved substantially by semi-selective homonuclear decoupling during both the acquisition and the evolution dimensions. This can be achieved in the acquisition dimension by use of homonuclear shaped pulse decoupling in combination with the time-shared decoupling mode during data acquisition and in the evolution dimension by application of a semi-selective refocusing pulse together with a non-selective refocusing pulse in the centre of the evolution period.

An example of an experiment implementing these techniques is the BASHD- (band selective homonuclear decoupled) TOCSY experiment. Band selection in the evolution dimension is achieved by the excitation sculpting method. The key element of this method is a double pulse field gradient spin echo (DPFGSE) that leads to pure phase spectra with flat baselines. This cluster of pulses rephases only the selected magnetization affected by the 180° pulses and avoids any evolution of the J-coupling during this period. The combination of selective pulses and pulse field gradients to select the desired coherence pathway results in pure phase spectra largely devoid of artefacts. This principle can also be extended to any existing homonuclear and heteronuclear selective NMR experiment, as demonstrated by semi-selective 2D TOCSY, ROESY, HSQC and HSQC-TOCSY experiments.

Sequential assignment

Sequential assignment of residues in a peptide chain requires correlation across the peptide bond that separates the proton spin systems of adjacent residues. This sequential information can be provided by dipolar couplings using NOESY or ROESY experiments, or by heteronuclear scalar couplings using HMBC experiments.

When only homonuclear proton experiments can be used (e.g. for reasons of sensitivity), NOESY or ROESY experiments are necessarily the method of choice. Short-range NOE signals, such as those observed between amide and aliphatic protons, are usually also observed for sequentially adjacent residues. Among these the H^N_i–H^N_{i+1} and H^α_i–H^N_{i+1} connectivities are especially important for establishment of the sequence (see **Figure 5**). In practice, however, ambiguities can be encountered in the sequential assignment step owing to overlap of cross peaks, particularly if the peptide contains multiple residues of one type, or if long-range NOE signals are also found in the H^α–H^N region of the NOESY or ROESY spectrum, as in the case of folded peptides and small proteins. Sequential assignment of these molecules therefore can be ambiguous and requires a

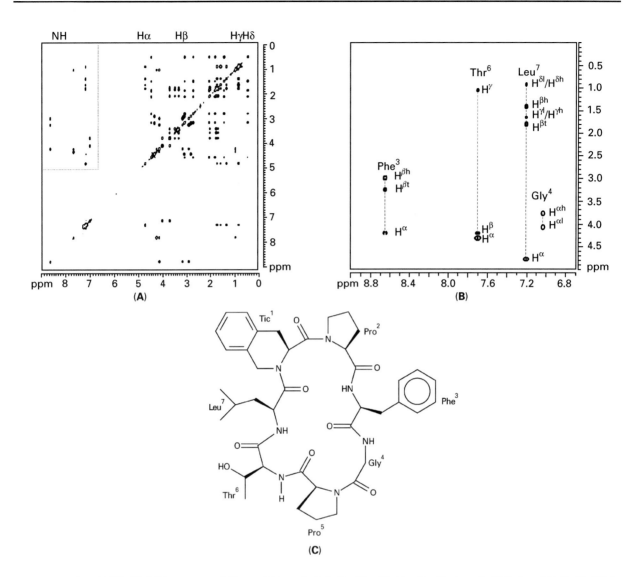

Figure 3 (A) 500 MHz TOCSY spectrum of 20-mmol-L⁻¹ *cyclo* (-Tic-Pro-Phe-Gly-Pro-Pro-Thr-Leu- in DMSO at 300 K with TPPI applied in the F_1 dimension. Mixing time was 80 ms. A number of relayed connectivities can be observed. This spectrum is typical for small cyclic peptides. Linear peptides of this size would exhibit much less chemical shift dispersion of H^N and H^α protons. The indicated part of the spectrum is expanded in (B) which is the fingerprint region (F_2: amide protons, F_1: aliphatic protons); coupled nuclei are connected by dashed lines and assigned to the respective residues. (C) Sequence of *cyclo* (-Tic-Pro-Phe-Gly-Pro-Pro-Thr-Leu-). Tetrahydroisochinolin (Tic) is an unnatural proline like amino acid which lacks the amide proton.

tedious and time-consuming analysis of the NOESY or ROESY spectra. In those cases, a significantly increased resolution in the H^α–H^N region of the ROESY spectrum can be achieved with a BASHD-ROESY pulse sequence, incorporating band selection and homonuclear decoupling in the H^α region of the spectra. Band selection in the evolution dimension is performed with the DPFGSE technique as described above.

This NOE-based approach requires previous knowledge of the peptide sequence. The known sequential position of a residue with a unique or characteristic spin system can be used as a first 'anchor point', from which sequencing in both directions can

be carried out. In cases where the sequence is unknown, a different strategy must be used. In this case, each spin system must be assigned individually to its type of amino acid *before* a sequential assignment can be achieved. Obviously this approach is restricted to relatively small molecules (up to 30 residues).

It is also possible to determine the sequence of a peptide using scalar coupling to the carbonyl carbon. This is best done with standard or selective HMBC experiments, where only carbonyl carbons are excited and indirectly detected during t_1. The sequential assignment is then unambiguous and does not require any knowledge about the conformation of the peptide (**Figure 6**). A clear distinction between the

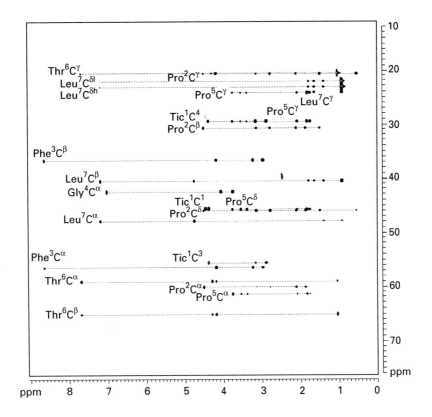

Figure 4 500 MHz HMQC-TOCSY spectrum of *cyclo* (-Tic-Pro-Phe-Gly-Pro-Pro-Thr-Leu-) in DMSO at 300 K. Mixing time was 80 ms. In addition to the directly bound protons the entire spin system can be observed. Coupled nuclei are connected by dashed lines and assigned to their respective residues.

proof of the complete assignment and the conformational analysis is then possible. Such a sequential assignment might also be used as an independent proof of the existence of a specific peptide bond, as for example formed by cyclization.

A complete assignment also includes the stereospecific assignment of diastereotopic groups such as methylene protons or geminal methyl groups in Val or Leu. Such additional information can significantly increase the quality of a 3D structure, which is especially important in the case of small peptides, where typically only a small number of long-range NOEs is available for the conformational analysis. Diastereotopic assignment will be discussed in more depth below.

Extraction of conformationally relevant parameters

NOE effects

Nuclear cross relaxation in liquids is caused by mutual spin flips in pairs of dipolar-coupled spins, induced by motional processes. Cross-relaxation efficiency depends on the spatial distances between the relaxing nuclei and leads to a transfer of magnetization between the spins. It causes intensity changes known as nuclear Overhauser effects (NOE). NOEs, as well as ROEs, can only be observed between nuclei that are separated in space by less than 500 pm. The NOE can be rationalized as heat flow from a non-equilibrium state to another neighbouring spin. In the NOE (or ROE) experiment such deviation from the Boltzmann equilibrium of spin states is created via specific pulsing and the efficiency of the heat flow to neighbouring nuclei (NOE build-up) is measured via the induced intensity changes of their NMR signals. The efficiency of dipolar relaxation is a function of the field strength (represented by ω_0) and the motion (rate of relaxation referred to the external magnetic field) of the molecule, described by the molecular correlation time τ_c.

The intensity of a cross peak appearing in a NOESY spectrum contains information about the relative distances between the two nuclei that contribute to the cross peak. It can be shown that the

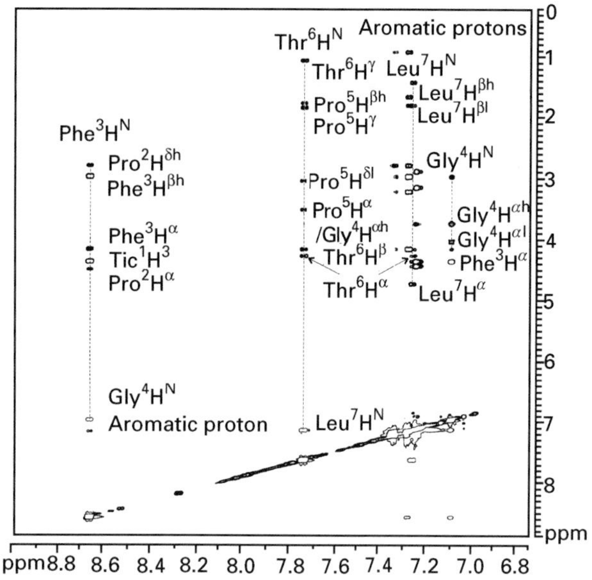

Figure 5 Part of the 500 MHz ROESY spectrum of *cyclo* (-Tic-Pro-Phe-Gly-Pro-Pro-Thr-Leu-) in DMSO at 300 K. Correlations between the amide protons and H^α protons of one residue and the H^α protons of the preceding residue are essential for the sequential assignment. Cross peaks between neighbouring amide protons are an important source of sequential information. Cross peaks not assigned belong to aromatic protons.

cross-relaxation rate σ that determines the NOE transfer is obtained from Equation [1].

$$\sigma = W_2 - W_0 = \frac{\gamma^4 h^2 \tau^c}{4\pi^2 10 r^6}\left(\frac{6}{1 + 4\omega^2\tau_c^2} - 1\right) \quad [1]$$

where r is the internuclear distance and W_2 and W_0 are the transition probabilities for the double-quantum and zero-quantum transition respectively. At a given ω the variables that determine the size of σ are the correlation time τ_c and the interproton distance $r^6{}_{ij}$. It should be noted that for specific combinations of τ_c and ω the second term becomes zero or negative. The build up of cross peak intensity in a multispin system is given by $A(\tau_m) = \exp\{-R\tau_m\}$:

$$A_{ij}(\tau_m) = \left\{-R_{ij}\tau_m + \frac{1}{2}\sum R_{ik}R_{ij}\tau_m^2 + \ldots\right\}A_{ii}(0) \quad [2]$$

where $A(t_m)$ is the cross peak intensity as a function of the mixing time τ_m, R the relaxation matrix and R_{ij} the relaxation rate between spins i and j. For sufficiently short mixing times the quadratic term and those of higher order in τ_m can be ignored. The cross peak intensity is then directly proportional to the

cross-relaxation rate and thus to the inverse sixth power of the distance between the nuclei (when the molecule behaves as a rigid body, τ_c = constant):

$$A_{NOE} \propto \sigma_{ij} \propto \frac{1}{r_{ij}^6} \quad [3]$$

The intensities of NOE effects depend on the size of W_2 and W_0. As can be seen for Equation [1] the NOE vanishes when $W_2 = W_0$, which occurs approximately when the inverse correlation time τ_c^{-1} is of the order of the Larmor frequency ω_0.

Similar relationships can be derived for ROESY. In this case, the cross peaks are generated by cross relaxation of transverse magnetization. In the rotating frame, σ is given by Equation [4]

$$\sigma = u_2 - u_0 = \frac{\gamma^4 h^2 \tau}{4\pi^2 10 r^6}\left(\frac{3}{1 + 4\omega^2\tau^2} + 2\right) \quad [4]$$

where u_2 and u_0 are the transition probabilities for the double-quantum and zero-quantum transitions in the rotating frame, respectively.

It is important to note that ROE effects, in contrast to NOE effects, are always positive and never vanish.

NOESY as well as ROESY experiments can both provide distance information. However, there are some important differences in their application and in the evaluation of the resulting spectra.

The usefulness of either of these techniques depend strongly on the time-scale of the motional processes that cause the cross relaxation. We have to distinguish three cases: (a) the fast-motion limit (extreme narrowing limit) with a short correlation time $\tau_c \ll \omega_0^{-1}$ (positive NOEs) (**Figure 7**). This applies for small molecules in non-viscous solutions. (b) The slow-motion limit (spin-diffusion limit) with a long correlation time $\tau_c \gg \omega_0^{-1}$ (negative NOEs) which applies to large macromolecules such as proteins at the maximum currently used magnetic field strengths. (c) For intermediate sized molecules only small, or even no NOE effects at all are observed. This is the case for peptides with a relative molecular mass of 500 to 1000 Da at resonance frequencies ω_0 of 300 MHz. Differences of internal mobility, for example via the rotation of side-chains, can then lead to the appearance of both positive and negative NOEs in the same NOESY spectrum, making it impossible to evaluate molecular distances from these data. If only small NOE effects are observed, the ROESY techniques should be used.

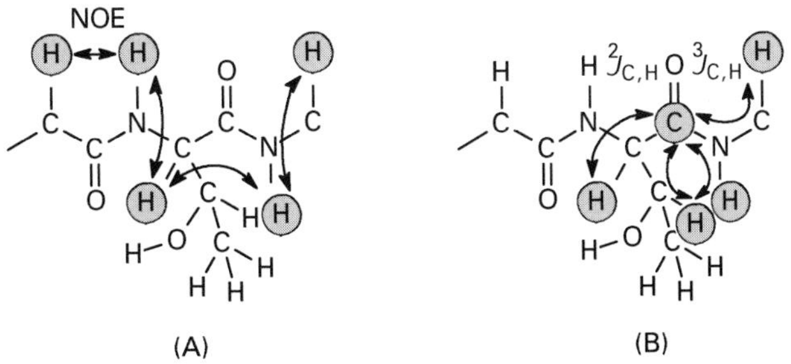

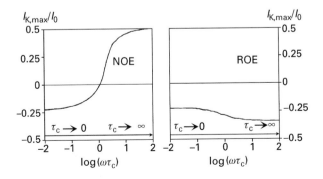

Figure 6 Polypeptide segment with the possible interresidue connectivities between spin systems. (A) Sequential short distance NOEs for peptides. (B) Observable through-bond couplings, useful for the sequencing of peptides.

Two problems have to be considered in the evaluation of ROESY spectra. First, the offset dependence of the spin-lock field introduces intensity variations into the spectrum. Peak intensities will have to be corrected to take this effect into account when distances are to be calculated. The cross peak intensity as a function of offset from the transmitter (when NOE effects are neglected) is given by Equation [5].

$$A(\gamma B_1) = A \, \sin^2\theta_k \, \sin^2\theta_l \qquad [5]$$

where $\theta_{k,l} = \arctan(\gamma B_l/\Omega_{k,l})$ and $\Omega_{k,l}$ are the offsets of spins k and l from the transmitter. The use of the compensated ROESY sequence leads to a higher intensity for peaks at the edge of the spectrum compared with the standard ROESY. In this case the peak intensity is given by Equation [6].

$$A(\gamma B_1) = A \, \sin\theta_k \, \sin\theta_l \qquad [6]$$

Figure 7 Dependence of the maximum NOE and ROE cross peak intensities $I_{k,max}$ (standardized on a diagonal signal I_0) on ω and τ_c for very short mixing times.

Second, undesired TOCSY peaks appear because some nuclei that are spin coupled experience similar fields during the application of the spin-lock and fulfil the Hartmann–Hahn condition. Since the TOCSY peaks are phase shifted by 180° with respect to the ROESY peaks, they can easily be recognized. However, the superposition of contributions from direct and indirect transfer results in a decrease of cross peak intensity and therefore in distances which are too long. When only lower boundaries are used as restraints in MD calculations this would lead to lower restraints and a less well-defined structure but would not induce wrong results. In addition, different internal correlation times, such as the above-mentioned different flexibility of the molecule have a smaller influence in ROESY than in NOESY spectra.

NOESY spectra are preferred in the slow-motion limit but never near the transition from positive to negative NOEs ($W_2 \approx W_0$, $\sigma \to 0$) because the different internal mobility induces larger errors in distances. In such cases, lowering the temperature (to slow down molecular rotation) is recommended.

Evaluation of NOESY and ROESY spectra

Dipolar cross-relaxation rates, and thus distances, can be determined through NOESY or ROESY experiments using various approaches. The measurement of build-up rates involves the recording of several NOESY spectra with different mixing times. To ensure equal conditions, the measurements should be made in succession. The integrals of cross peaks are determined, and the volumes are plotted as a function of mixing time:

$$A(\tau_m) = 1 - \exp(-\sigma\tau_m) \qquad [7]$$

The derivative of Equation [7] at an extrapolated mixing time of zero yields the rate of build-up of the cross peak. This initial build-up rate is directly proportional to the cross-relaxation rate.

$$dA \frac{(\tau_m = 0)}{d\tau_m} = \sigma \qquad [8]$$

The simplest and most common approach is the measurement of a single NOESY or ROESY spectrum with a short mixing time. At short mixing times the NOE build-up is in the linear range. Under this condition, it can be assumed that only direct enhancements σ_{ij} contribute to the cross peak intensity $a_{ij}(\tau_m)$. The evaluation of a single NOESY spectrum can be done by either integration of the cross peaks or in a more qualitative manner by visual inspection of the spectrum. The second approach is often used in the case of NOESY spectra of proteins where an insufficient signal-to-noise ratio and extensive overlap prevents the accurate integration of cross peaks. In large molecules 'spin diffusion', i.e. a rapid flow of magnetization from one nucleus via another nucleus to a third one, is the most severe problem. Only very short mixing times can be used and a complete treatment of the relaxation matrices is recommended. In the first approximation the NOEs are only used qualitatively in proteins: cross peaks are then classified according to several semi-quantitative categories – usually strong, medium, weak – which correspond to distance ranges. This approach is not recommended for peptides since the integration of cross peaks should lead to considerably more accurate distances and spin diffusion is not so efficient in peptides as it is in large molecules. The so-called ISPA (isolated spin pair approximation) is closer to reality for molecules which are close to the $\tau_c \approx \omega_0$ condition.

Owing to the fact that absolute values of correlation times are usually not available, interproton distances cannot be directly calculated. Distances are instead obtained by calibration of the cross peak intensities against an internal distance standard, usually the distance between diastereotopic geminal protons (178 pm) or aromatic protons of Tyr (242 pm). Assuming isotropic tumbling and rigid-body model for all parts of the molecule, Equation [9] is then used to calculate all interproton distances:

$$r_{ij} = r_{ref} \left(\frac{a_{ref}}{a_{ij}} \right)^{-6} \qquad [9]$$

Usually, the NOE enhancements for the structural and conformational analysis of peptides are

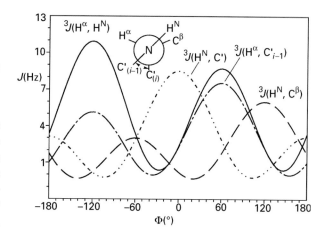

Figure 8 The Karplus curve for four coupling constants about the ϕ dihedral angle of an L-amino acid. There are four possible dihedral angles for a given coupling constant. Utilizing a combination of all four coupling constants it is usually possible to narrow down the choice to a single angle. Reproduced with permission of Wiley-VCH from Eberstadt *et al* (1995) *Angewandte Chemie, International Edition in English* **34**: 1671–1695.

extracted from 2D NOESY spectra. The GOESY experiment, a 1D version of the NOESY experiment, uses selective excitation of separated signals and yields accurate measurements even for tiny enhancements. The DPFGSE NOE technique (see above) achieves better sensitivity by not discarding one of the coherence transfer pathways (in contrast to the GOESY technique), while spectra have the same characteristics as GOESY spectra. Therefore, the DPFGSE NOE technique is to be preferred.

Determination of coupling constants

Many J coupling constants illustrate a clear dependency on dihedral angles and therefore are an important source of conformational information. This relationship is particularly distinct for ^3J couplings. The model for the relationship between bond angles and the coupling constant most often used is that proposed by Karplus (**Figure 8**).

$$^3J = A \cos^2\theta + B \cos\theta + C \qquad [10]$$

The equation holds for almost all coupling constants (**Table 2**). Only the coefficients A, B and C have to be adjusted, depending on the type of the two coupled nuclei and their environment. However, even if the coefficients have been determined, the multiple angles that fulfil Equation [10] (up to four values can be obtained) remain a problem.

Despite these problems, the application of coupling constants, in addition to proton–proton distances, in modern conformational analysis of peptides is indispensable. In principle, the coupling constants shown in **Table 2** can be used to determine the ϕ, ψ and χ_1 angles.

For the determination of homonuclear or heteronuclear three-bond coupling constants, three fundamentally different approaches are used: (1) direct measurements of splittings caused by J-couplings, (2) the so-called E.COSY-signal patterns in which the splittings can be measured as shift differences of signals and (3) special experiments which lead to modulation of signal intensities via J (**Table 3**). A frequent requirement for the latter is a ^{15}N-enriched sample, e.g. for the J-modulated (^{15}N, ^{1}H)-COSY experiment or the HNHA experiment. Whereas labelled compounds are routinely used for proteins and nucleic acids they are expensive for peptides and therefore rarely used.

Determination of coupling constants from the shape of the signal. All these techniques have to correct or compensate for the partial overlap of multiplet lines by either using additional parameters which depend on the shape of the signals or by fitting a model to the overlapped experimental signal. However, this procedure requires that the line width is still smaller than the coupling constant, and furthermore that the signals have a good signal-to-noise ratio.

The determination of coupling constants from an antiphase (e.g. COSY) cross peak yields values which are inherently too large owing to the reciprocal signal cancellation of the antiphase pattern. Kim and Prestegard have described an especially simple method for the determination of coupling constants for AX spectra. The splitting of the maxima in the absorptive and in the dispersive signal is measured and the J coupling is calculated via an extensive cubic equation (**Figure 9**). Only two calculations are required for the COSY spectra in which the phase are 90° shifted. This procedure is especially useful for the determination of coupling constants that cannot easily be extracted from peaks in E.COSY spectra (see below), e.g. for H^N–H^α cross peaks. No additional spectrum has to be recorded.

The simulation of the line shape using a linear combination of reference signals is based on the simulation of a complex nonresolved multiplet, starting out from a library of experimental multiplets. This technique has found widespread application in the determination of heteronuclear long-range coupling constants from HMBC cross peaks of peptide samples with ^{13}C in natural abundance. The major advantage is the relatively high sensitivity of the HMBC experiment because of inverse detection, as well as the fact that the identity of heteronuclear couplings directly ensues from the assignment of the 2D cross signals. In principle, long-range coupling constants can be determined directly from the cross peaks which represent the active coupling. However, since the long-range heteronuclear coupling constants are approximately of the same magnitude as $^2J_{H,H}$ and $^3J_{H,H}$ coupling constants (1–10 Hz), the rather small heteronuclear antiphase coupling constant $^nJ_{C,H}$ cannot be read directly because of overlapping and reciprocal cancellation of the numerous multiplet lines. Keeler *et al.* have developed an elegant procedure which allows the determination of the heteronuclear long-range coupling

Table 2 Coupling constants used to determine ϕ, ψ and χ_1 angles in peptides

Angle	Coupling constant
ϕ	$^3J_{HN,H^\alpha}$, $^3J_{C^\beta,HN}$ $^3J_{H^\alpha,CO(i-1)}$
ψ	$^3J_{H^\alpha N(i-1)}$
ϕ and ψ	$^1J_{C^\alpha,H^\alpha}$
χ_1	$^3J_{H^\alpha,H^\beta}$, $J_{CO,\ H^\beta}$, J_{N,H^β}

Table 3 Experiments that are used for determining the most important coupling constants in peptides

Coupling	Technique
H^N–H^α	Direct reading or Kim–Prestegard method
H^α–H^β	E.COSY techniques
Couplings within the proline pyrrolidine ring	E.COSY techniques
^{13}CO–H^β	HMBC (qualitative), Keeler method
H^N–C^β	HETLOC

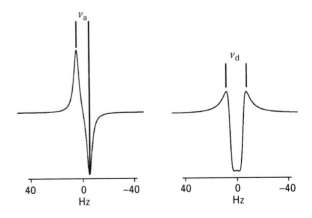

Figure 9 Determination of the separation of the signal maxima of an antiphase doublet for the absorptive (ν_a) and dispersive (ν_d) component. Based on these two parameters the coupling constant J can be calculated. Reproduced with permission of Wiley-VCH from Eberstadt *et al* (1995) *Angewandte Chemie, International Edition in English* **34**: 1671–1695.

constant from the line shape even in such cases (**Figure 10**). The difference between a cross section in a HMBC spectrum and the corresponding, reconstructed proton multiplet (e.g. from a 1D spectrum) is the heteronuclear coupling constant of interest. In practice, the proton multiplet will show some overlap in the 1D spectrum, and therefore a TOCSY spectrum with a pure phase (e.g. with a z-filter) has to be recorded. The homonuclear reference multiplet now differs from the HMBC multiplet only by the absence of the heteronuclear antiphase coupling and the signal amplitude. This can be simulated by scaling of the reference signal in the time domain with the term $\sin(\pi J_{\text{trial}} t)$. The amplitude and J_{trial} can now be varied by a nonlinear optimization until the deviation between the two spectra reaches a minimum. At this point J_{trial} should be equal to the coupling constant of interest.

This method has the advantage that a large number of coupling constants can be determined from a single HMBC spectrum, which usually contains a large number of long-range correlations. However, the quality of the heteronuclear spectrum and especially the signal-to-noise ratio is of crucial importance for the convergence of the optimization. Therefore, the method is time consuming with respect to both recording of the spectra as well as their processing, but it yields accurate values for the coupling constants between heteronuclei.

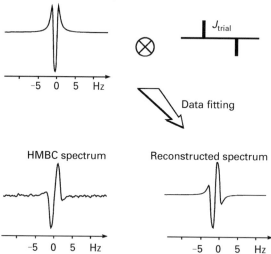

Homonuclear reference spectrum

Data fitting

HMBC spectrum

Reconstructed spectrum

Figure 10 Schematic of the Titman–Keeler method for the evaluation of heteronuclear coupling constants from HMBC spectra. The synthetic spectrum is computed from the homonuclear reference spectrum and a chosen heteronuclear coupling constant J_{trial}. This spectrum is then compared with the actual HMBC spectrum and J_{trial} is iteratively varied until a good fit is obtained. Reproduced with permission of Wiley-VCH from Kessler H and Seip S (1994) In: Croasmun WR and Carlson RMK (eds) *Two-Dimensional NMR Spectroscopy*, pp 619–654.

Assuming that the staggered rotamers are predominantly populated (**Figure 11**), qualitative considerations together with accurately determined homonuclear coupling constants are often sufficient for the diastereotopic assignment of methylene protons. The χ_1 angle can be set to $-60°$ if the two $^3J_{H^\alpha,H^\beta}$ and $^3J_{H^\alpha,H^{\beta'}}$ coupling constants are small (both ~3 Hz). If one strong and one weak coupling is observed, χ_1 can be either 60 or 180°. To differentiate these two cases, stereospecific assignment of the H^β protons is required. This is possible with the aid of qualitative heteronuclear J-couplings (e.g. between ^{13}CO and H^β or ^{15}N and H^β) and NOE or ROE cross peak intensities to the different H^β protons.

Determination of coupling constants by the E.COSY principle The E.COSY (exclusive correlation spectroscopy) principle yields a simplified cross peak multiplet, since only the 'connected transitions' are excited. This means that the signal intensity in an A, M cross peak from a three-spin system AMX can only be found in those parts of the multiplet pattern where the spin states of the third nucleus X have been conserved. To obtain such an E.COSY pattern, a mixing of spin states of the X nucleus (e.g. by the application of a non-sensitive 90° pulse) must be avoided. The coupling between M and X can then be extracted from the passive coupling of the A, M cross peak as the shift of two in-phase multiplets, which are separated in the indirect dimension and, therefore, have no interfering influence on each other. The only requirement is that the splitting in F_1 (the coupling between A and M) must be larger than the line width (**Figure 12**).

The E.COSY technique with the highest sensitivity is P.E.COSY (primitive E.COSY), where the retention of the spin states of the passive spin is achieved using a small flip angle of the mixing pulse and subtraction of the dispersive diagonal via a reference spectrum. The resulting cross peaks contain strong signal intensities for connected transitions but vanishing intensities for non-connected transitions.

In heteronuclear spectroscopy E.COSY patterns can be easily obtained if no 90° pulse is applied to the heteronucleus (i.e. the states α and β are not mixed). ω_1-Hetero-filtered (HETLOC) experiments are the method of choice for the determination of long-range coupling constants between protons and ^{13}C or ^{15}N nuclei in natural abundance that carry a directly connected proton (**Figure 13**). For the determination of the χ_1 angle the coupling $^3J_{H^\beta}$ can be determined by using a heteronuclear ω_1-half-filter (X-half-filter) at the beginning of a NOESY or TOCSY experiment (before the t_1 delay). Only protons which are directly coupled to the magnetically active

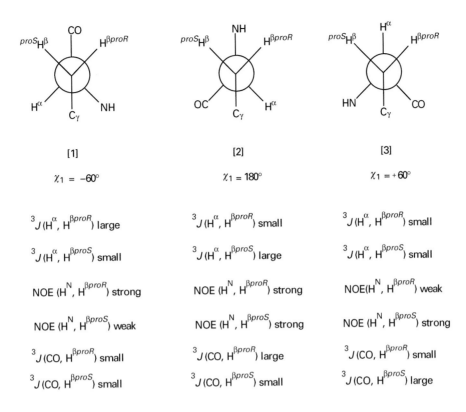

Figure 11 The three staggered conformers about the χ_1 angle of residues with β-methylene protons (see text for details).

heteronucleus (^{13}C or ^{15}N) are selected, while ^{12}CH or ^{14}NH protons are suppressed. Obviously, no heteronuclear decoupling can be performed during the acquisition. The delay Δ, is adjusted according to the value of the heteronuclear coupling constant for the ω_1-half-filter ($\Delta = 1/2J_{H,X}$) resulting in in-phase magnetization at the end of the two delays. The subsequent TOCSY sequence affects only the ^{13}C-coupled protons and transfers the magnetization through the entire spin system. In ω_1 the signals are split by the large value of the $^1J_{X,H}$ coupling (e.g. $^1J_{N,H} = 90$ Hz) in ω_2 by the desired long-range heteronuclear coupling constant. However, by this method heteronuclear $^nJ_{H,X}$ couplings can only be determined for heteroatoms which bear a directly bound proton. The latter causes the required large splitting in ω_1 via $^1J_{X,H}$. Fortunately, using the sensitivity of an inverse detection experiment many heteronuclear coupling constants can be determined from a single spectrum for molecules in natural isotopic abundances.

Structure determination of peptides

The utilization of NMR data for the determination of the three-dimensional structure of peptides involves the use of computer simulations. The methods can be broken down into two general categories: molecular mechanics/dynamics (MM or MD) and distance geometry (DG) calculations.

MM and MD use a force field to describe the molecule and estimate the potential energy of the given conformation. The standard force field contains a term for distortion of bond lengths, bond angles and dihedral angles plus non-bonded terms for Coulombic interactions and a Lennard-Jones description of the attraction/repulsion of atoms. The application of experimental restraints is achieved by simply introducing an additional term, a so-called penalty function. This penalty function serves to minimize the differences between the calculated values and experimental data.

The second general approach, DG calculations, utilizes a description of a molecule based solely on distances. Bond lengths, bond angles and torsion angles are converted into ranges of allowed distances according to the molecular constitution. Distances which satisfy these ranges are chosen randomly to create a distance matrix. The diagonalization of this matrix then produces Cartesian coordinates. The experimental distances are compared with the distances generated from DG calculations based on the covalent structure; if the experimental distances are tighter, they replace those on the consideration of the covalent geometry.

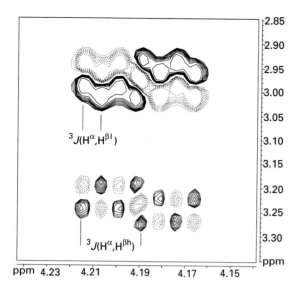

Figure 12 H$^\alpha$, H$^\beta$ region of the 500 MHz P.E.COSY spectrum of cyclo(-Tic-Pro-Phe-Gly-Pro-Pro-Thr-Leu-) in DMSO at 300 K. The enlarged view is of the Phe^3 H$^\alpha$, H$^\beta$ cross peaks. The displacement of the multiplet patterns can be used to determine the passive couplings with high accuracy. H$^{\beta l}$ means the low field β proton and H$^{\beta h}$ the high field proton.

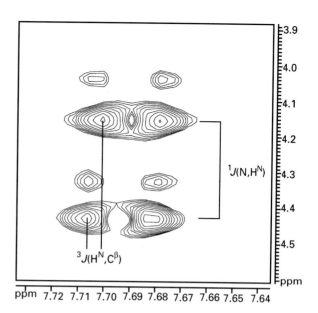

Figure 13 Part of the 500 MHz HETLOC (ω_1-filtered TOCSY) spectrum of cyclo(-Tic-Pro-Phe-Gly-Pro-Pro-Thr-Leu-) in DMSO at 300 K. The enlarged view is of the Thr^7 H^N, H$^\beta$ cross peak. The coupling constant is extracted from the separation of the cross peak components in the better resolved acquisition dimension (F^2).

The most important parameters for the determination of a three-dimensional structure are NOE-derived distances, bond angles derived from J-coupling constants and temperature dependences of H^N proton chemical shifts. Distances from NOESY or ROESY spectra are directly used as restraints for the calculations. Recently, coupling constants have also been used in computational structure refinement. The penalty function employed in this case is directly based on the Karplus equation. If more than one coupling for a single dihedral angle is available the restraints from coupling constants are quite useful at the level of structure refinement. A limited number of experimental NOE values for peptides normally requires special care. In small peptides more or less all of the protons are on the surface, in contact with the solvent, and thus there are only distances to one side whereas in proteins there are many protons in the core region that are completely surrounded by other protons. Hence, for peptides it is indispensable to collect as much experimental data as possible. This means that accurate, quantitative NOE data as well as correct treatment of J coupling constants are required.

The direct application of the temperature coefficients in structure refinement is problematic, since, while a temperature coefficient may indicate that an H^N proton is shielded from the solvent (for example by being involved in a hydrogen bond), it does not allow for identification of the acceptor of that hydrogen bond. Therefore, temperature coefficients,

unfortunately, are frequently used only to confirm the final structure. The analysis of the radial distribution function (rdf) of the solvent around an amide proton shows, in our experience, distinct peaks when this proton is solvent exposed. The size and sharpness of these rdfs correlate directly with the size of the temperature gradient.

Conformational analysis of a peptide normally begins with the simple assumption of only a single dominating conformation using restrained MD under vacuum, beginning with various starting structures (to prevent structural bias, it is, however, recommended to begin with DG calculations to create the first crude starting structures for the MD). The resulting MD structure is further refined by recalculations of the molecule within an explicit solvent box. This can contain H_2O, DMSO, $CHCl_3$, CH_3OH or others, depending on the solvent used for the measurement. The best procedure uses a truncated octahedron to allow periodic boundary conditions for an almost spherical box but also cubic boxes may be used. The quality of the final structure is finally checked by a long trajectory MD calculation (100 ps or more), without all experimental restraints but within the solvent. If all restraints are fulfilled (within about 10 pm), and the same result is always obtained regardless of the selected starting structure, it can be concluded that a 'single' satisfying conformation exists (a 'single conformation' may still include some flexibility, of the order of ±20° for torsions). There might be other conformations which also fulfil the experimental data, especially when the system is underdetermined owing to an insufficient number of constraints, but this should be apparent if different starting structures lead to different results.

If one part of the molecule turns out to be well defined while another part shows larger deviations from the experimental constraints, it can be assumed that the latter part is undergoing intramolecular motion that is fast on the chemical shift time-scale. Short distances r contribute most strongly to the observed NOE intensities (because of the r^6 dependence), and hence not all distance constraints derived from NOEs can be fulfilled at the same time if an equilibrium among several conformations exists. Given that there is a sufficient number of restraining parameters one can try to analyse this equilibrium by using time-dependent NOEs and/or time-dependent coupling constants, making the assumption that the constraints are not fulfilled at each simulation step, but rather only over a whole trajectory. This allows for analysis not only of the flexibility but also of the detailed nature of the molecular processes involved.

In addition, ensemble calculations may be used to analyse flexible structures. However, these calculations are time consuming, difficult to analyse and cannot directly include solvents. Hence, this procedure is only used in rare cases.

Side-chain mobility is analysed mainly by assuming a rapid equilibrium between three staggered conformations. The populations of the conformations can then be derived from the homonuclear and heteronuclear coupling constants using Pachler's equation. The observed coupling J_{obs} then results as the average over all three rotamers, J_i, weighted with their respective population P_i.

$$J_{obs} = \sum_i P_i J_i \quad (i = 1 - 3) \qquad [11]$$

$$P_{[1]} = \frac{J(H^\alpha, H^{\beta proR}) - J_{sc}}{J_{ap} - J_{sc}} \qquad [12]$$

$$P_{[2]} = \frac{J(H^{\beta proR}, CO) - J_{sc}}{J_{ap} - J_{sc}} \qquad [13]$$

$$P_{[2]} = \frac{J(H^\alpha, H^{\beta proS}) - J_{sc}}{J_{ap} - J_{sc}} \qquad [14]$$

$$P_{[3]} = \frac{J(H^{\beta proS}, CO) - J_{sc}}{J_{ap} - J_{sc}} \qquad [15]$$

For homonuclear couplings, the antiperiplanar coupling J_{ap} is 13.6 Hz and the synclinal coupling J_{sc} is 2.6 Hz while for heteronuclear ^{1}H–^{13}C couplings J_{ap} is 8.5 Hz and J_{sc} is 1.4 Hz. It should be noted that homonuclear coupling constants and NOE effects alone do not always yield an unambiguous diastereotopic assignment of the β-methylene protons. This means that incorrect values for the dominant bond angle (χ_1) may be obtained. This is especially the case if MD calculations are performed only under vacuum. In such cases, a globular structure is often predicted for the molecule, which only opens into a realistic conformation when the solvent is explicitly included in the calculations. Especially as peptides have a large surface which can interact with the solvent, it is essential to perform MD calculations in explicit solvents.

Relaxation parameters and molecular dynamics

NMR spectroscopy is uniquely capable of comprehensively characterizing the internal motions of peptides in solution at the atomic level over time-scales ranging from picoseconds to hours. NMR techniques used for the study of dynamics include relaxation rate measurements, dynamic NMR and line shape analysis, magnetization transfer experiments, NOESY and ROSEY and amide proton exchange measurements.

For diamagnetic peptides in isotropic solvents, the primary mechanism of nuclear magnetic relaxation of protonated ^{13}C nuclei and of ^{15}N nuclei at natural abundance is the dipolar interaction with the directly bound protons. At high magnetic fields, chemical shift anisotropy (CSA) also contributes to the relaxation of the heteronuclei. The rates of these relaxation processes are governed by both the internal motions and the overall rotational motion of the molecule. Consequently, characterization of ^{13}C and ^{15}N heteronuclear relaxation can provide information about internal dynamics of peptides on time-scales faster than the rotational correlation time.

The $T_{1\rho}$ times of protons have been measured to study conformational exchange on the microsecond to millisecond time-scale. However, the complex interaction with surrounding protons, which is strongly dependent on the molecular geometry, may lead to artefacts in the interpretation of the data.

The overall tumbling rate, an important parameter for NMR spectroscopy, can best be determined by measuring T_1 relaxation times. To determine the overall correlation time τ_c at least two different field strengths are required. In most cases, the function (T_1 versus τ_c) allows for two possible values of τ_c. Usually the true τ_c can be selected based either on reasonable estimates as a function of relative molecular masses or after consideration of additional relaxation rates such as T_2 relaxation times or heteronuclear NOEs (**Figure 14**).

Specific models for internal motions can be used to interpret heteronuclear relaxation, such as restricted diffusion and site-jump models. However, model-free formal methods are preferable, at least for the initial analysis, since available experimental data generally are insufficient to completely characterize complex internal motions or to uniquely determine a specific motional model. The model-free approach of Lipari and Szabo for the analysis of relaxation data has been used for proteins and even for peptides. It attempts to reproduce relaxation rates by a weighted product of spectral density functions with different correlation times τ_i. The weighting factors are identified as order parameters S_i^2 for the molecular rotational correlation time τ_c and optional further local correlation times τ_i. The term $(1 - S_i^2)$ would then be proportional to the amplitude of the corresponding internal motion. However, the Lipari–Szabo approach is based on the assumption that molecular and local correlation times are not coupled, i.e. they should be distinct enough (e.g. differing by at least a factor of 10 in time) to allow for this separation. However, in small molecules the rates of these different processes are of the same order of magnitude, and the requirements of the Lipari–Szabo approach may not be fulfilled. Molecular dynamics simulation provide a complementary approach for the interpretation of relaxation measurements.

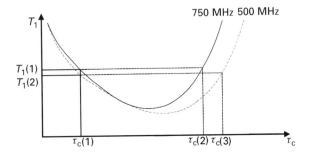

Figure 14 Dependence of T_1 on τ_c at two different field strengths. The different T_1 times [$T_1(1)$ and $T_1(2)$] clearly correspond to $\tau_c(1)$; the alternative values for τ_c can be ruled out.

List of symbols

A = cross peak intensity; B = magnetic flux density; J = coupling constant; P_i = population of rotamer i; R_{ij} = relaxation rate between spins i and j; r = internuclear distance; u_2, u_0 = probability of double- and zero-quantum transitions, respectively, in the rotating frame; W_2, W_0 = transition probability for double- and zero-quantum transitions, respectively; γ = gyromagnetic ratio, σ = cross-relaxation rate; τ_1, τ_2 = correlation times; τ_c = correlation time; τ_m = mixing time; ϕ, ψ, ω = peptide backbone angles; χ = bond angles of peptide side-chains; ω_0 = Larmor frequency.

See also: **NMR Pulse Sequences; Nuclear Overhauser Effect; Proteins Studied Using NMR Spectroscopy; Solvent Suppression Methods in NMR Spectroscopy; Structural Chemistry Using NMR Spectroscopy, Organic Molecules; Structural Chemistry Using NMR Spectroscopy, Pharmaceuticals; Two-Dimensional NMR Methods.**

Further reading

Eberstadt M, Gemmecker G, Mierke DF and Kessler H (1995) Scalar coupling constants – their analysis and their application for the elucidation of structures. *Angewandte Chemie, International Edition in English* **34**: 1671–1695.

Evans JNS (1995) Biomolecular NMR Spectroscopy. Oxford: Oxford University Press.

Kessler H and Seip S (1994) NMR of Peptides. In: Croasmun WR, and Carlson RMK (eds) *Two-Dimensional NMR Spectroscopy*, pp 619–654. Weinheim: VCH.

Kessler H and Schmitt W (1996) Peptides and polypeptides. In: Grant DM and Harris RK (eds.) *Encyclopedia of Nuclear Magnetic Resonance*, pp 3527–3537. Chichester: John Wiley & Sons.

Lipari G and Szabo A (1982) Model-free approach to the interpretation of nuclear magnetic resonance relaxation in macromolecules. *Journal of the American chemical society* **104**: 4546–4559.

Neuhaus D and Williamson MP (1989) *The Nuclear Overhauser Effect in Structural and Conformational Analysis*. Weinheim: VCH.

Parella T (1996) High quality 1D spectra by implementing pulsed-field gradients as the coherence pathway selection procedure. *Magnetic, Resonance in chemistry* **34**: 329–347.

van Gunsteren WF and Berendsen HJ (1990) Computer simulation of molecular dynamics: methodology, applications and perspectives in chemistry. *Angewandte Chemie, International Edition in English* **29**: 992.

Wüthrich K (1986) *NMR of Proteins and Nucleic Acids*. New York: Wiley.

Structural Chemistry Using NMR Spectroscopy, Pharmaceuticals

Alexandros Makriyannis and **Spiro Pavlopoulos**,
University of Connecticut, Storrs, CT, USA

MAGNETIC RESONANCE
Applications

Scope and applications

Information on the structure of drug molecules and their interactions with their therapeutic sites of action is of critical importance in the design and development of new drugs. Of all the analytical methods, nuclear magnetic resonance spectroscopy (NMR) is the most exquisitely suited to provide such experimental results. The field of NMR is advancing continuously to include new pulse sequences and methods as well as progressively larger field instruments and improved probes. This fast progress in NMR methods and technologies has served to expand dramatically its applications in drug research. Indeed NMR, used jointly with X-ray crystallography and computational/graphical approaches, has revolutionized structure-based drug design.

Currently the availability of a plethora of multidimensional/multinuclear NMR methods allows us to extract information on the structures and dynamic behaviours of a wide range of drug molecules of up to 30 kDa in size. These include the small and medium-sized traditionally used therapeutic drugs, to higher molecular weight peptides, proteins, nucleotides, nucleic acids and polysaccharide biotechnology products.

Progressively, more detailed structural and dynamic information has become available because of our increased ability to measure more effectively the basic NMR parameters used in structural analysis, namely, proton and carbon chemical shifts, coupling constants, relaxation parameters (T_1, T_2) and the exceedingly valuable nuclear Overhauser effect. Such measurements are, in turn, used to obtain information on the three-dimensional structure of molecules as well as their conformational properties and dynamic behaviour. Additionally, the new solution NMR methods allow researchers to study the interactions of a drug molecule with its site of action on the biopolymer (enzyme, receptor, nucleic acid, etc.). Such studies can lead to insights regarding the bioactive conformation of a flexible drug molecule, which constitutes invaluable information for drug design. Here we shall discuss the most commonly exploited

experiments for extracting the individual NMR parameters mentioned above, and also how such parameters are utilized to obtain structural information.

Conformational analysis of small molecules

Information on the structural properties of small drug molecules in solution can be obtained from a number of NMR parameters including ^{1}H and ^{13}C chemical shifts, [1]H–[1]H and [1]H–[13]C scalar coupling constants, ^{1}H nuclear Overhauser effects (NOEs), as well as relaxation measurements. Here, the conformational analysis of CP55,940 (**Figure 1**) is used to illustrate the most common experiments encountered in studying small molecules (<1000 Da) in solution. This synthetic compound is structurally related to Δ^9-tetrahydrocannabinol (Δ^9-THC), a psychoactive component of marijuana, and has received much attention because it was used as the high affinity radioligand during the discovery and characterization of the G-protein coupled cannabinoid receptor (CB1). The elucidation of the conformational properties of this compound and its congeners provides information on the steric requirements for a productive interaction at the cannabinoid receptor active site.

Double quantum filtered correlation (DQF-COSY) and total correlation (TOCSY) spectroscopy

These experiments provide information on ^{1}H chemical shifts and [1]H–[1]H scalar coupling. Spectral assignments are made initially by an analysis based on integrated peak areas and chemical shifts in the one-dimensional spectrum. Subsequently, they are speci-fically assigned by analysis of scalar or spin–spin coupling connectivities observed by [1]H–[1]H double quantum filtered correlation spectroscopy (DQF-COSY). This is a two-dimensional experiment where the information is spread onto a plane in which the diagonal is equivalent to the one-dimensional spectrum, and the scalar coupling is manifested as an off-diagonal crosspeak between the two

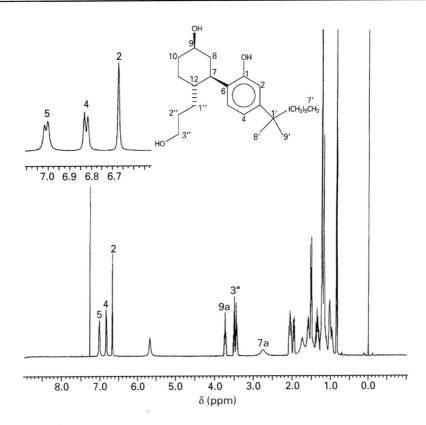

Figure 1 The one-dimensional ^{1}H spectrum of CP55,940 with an expanded scale of the aromatic region.

resonances in question. Thus even though a resonance may not be visible in a one-dimensional spectrum due to overlap, its position may be identified from a crosspeak in a two-dimensional experiment. Protons that are part of a spin system often give rise to a pattern of diagonal peak–crosspeak connectivities that can be traced from a 'starting point' in the spin system to an 'end point'. It is the presence of such connectivity patterns that makes this experiment such a powerful tool in assignment.

For example, the ^{1}H NMR spectrum of CP55,940 is shown in **Figure 1**. A logical starting point for assignment was the resonance at δ 3.74 that was assigned to H9a because the size of the peak as measured by integration was consistent with one proton and because this aliphatic resonance is expected to be deshielded. In the DQF-COSY spectrum (**Figure 2**) vicinal coupling to H10e, H8e, H10a and H8a is observed. Assignment of H10e and H10a resonances was made with the support of the crosspeak connectivities of H9a with H10a, H10e, H8a and H8e. The DQF-COSY spectrum clearly shows three components (H10e, H8e and H11e) under the multiplet at δ 2.06, in which three related strong geminal ^2J couplings, H10e/a ($F_1 = \delta\ 2.09$, $F_2 = \delta\ 1.38$), H8e/a ($F_1 = \delta\ 2.05$, $F_2 = \delta\ 1.53$) and H11e/a ($F_1 = \delta\ 1.98$, $F_2 = \delta\ 1.13$) can be discerned.

This is a prime example of the improvement in spectral resolution of two-dimensional experiments over one-dimensional experiments.

Complete assignment of a whole spin system may be limited because of severe spectral overlap. To overcome this, DQF-COSY data are often used in conjunction with total correlation spectroscopy (TOSCY). This experiment results in a transfer of magnetization across an entire spin system and consequently crosspeaks may be observed between each resonance of a spin system. Thus it is possible to determine whether a particular overlapped region of the spectrum contains all unidentified members of a chemical spin system.

^{1}H chemical shifts and scalar coupling constants can be measured directly from one-dimensional spectra if the peaks are well resolved, or, if spectra are too complex, they may be measured from DQF-COSY spectra crosspeaks. However, such measurements are often inaccurate and so are used as a basis of simulating the observed 1D spectrum to obtain more accurate values. The measurements are used as a starting point and are systematically altered until the stimulated spectrum best matches the observed spectrum. ^{1}H–^{1}H scalar coupling constants are especially useful in providing information on the dihedral angle within a HC–CH system and are thus one

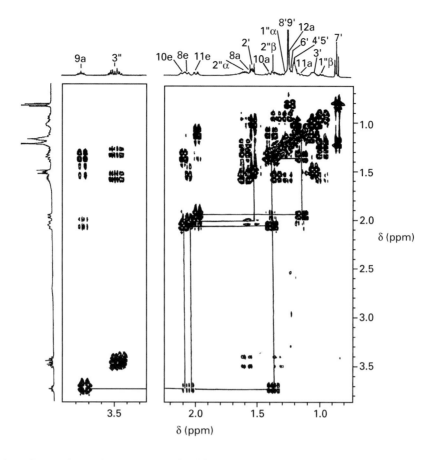

Figure 2 Contour plots of expanded regions from the DQF-COSY spectrum of CP55,940. The lines highlight the connectivities between the H8, H9a, H10 and H11 resonances.

Nuclear Overhauser effect spectroscopy (NOESY)

The nuclear Overhauser effect (NOE) is another important NMR parameter used in conformational analysis because the magnitude of the NOE is inversely proportional to the sixth power of the interproton distance in space ($I_{NOE} \propto r^{-6}$). NOE spectroscopy (NOESY) is two-dimensional experiment that may be run routinely in which the NOE is manifested as a crosspeak between two resonances indicating that the two protons are near in space.

For example, in the case of CP55,940, two NOE crosspeaks were assigned to the spatial coupling of H5 with H8a and H12a (**Figure 3**). Such crosspeaks are congruent with a conformation in which the planes of the two rings are almost perpendicular, with the Ph–OH oriented toward the α face of the cyclohexyl ring. An NOE crosspeak between the phenolic hydroxyl proton and the adjacent aromatic H2 indicates that these two protons are spatially near each other and thus coupled through a dipole–

dipole interaction (**Figure 3**). Such a result indicates that in its preferred conformation, the Ph–OH proton points away from the cyclohexyl ring and towards the H2 proton.

The full analysis of NOESY and DQF-COSY spectra of other analogues, plus computational studies, further showed that this was typical for all congeners of CP55,940 and that the dimethylheptyl chain adopts one of four preferred conformations, in all of which the chain is almost perpendicular to the phenol ring. The most biologically active conformations were such that all hydroxyl groups were oriented towards one face of the cyclohexyl ring system (**Figure 4**), a feature that may be an important requirement for cannabimimetic activity.

Structure of drug macromolecules

An increasing number of therapeutic drugs are composed of proteins or peptides and knowledge of their three-dimensional structures has helped in the design of structurally modified variants with improved biological activities and pharmacokinetic properties.

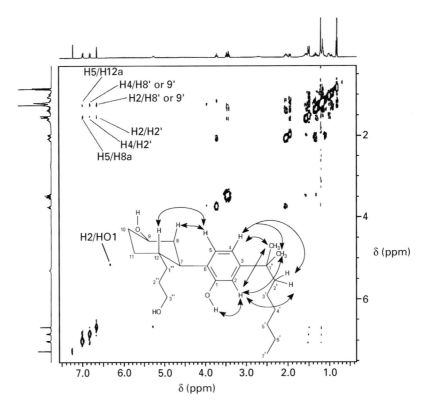

Figure 3 Contour plot of the 500 MHz CP55,940 NOESY spectrum in CDCl₃. The NOE interactions for CP55,940 are indicated with arrows.

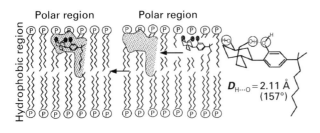

Figure 4 Representation of the active biological conformation of CP55,940. According to the current hypothesis, the ligand preferentially partitions in the membrane bilayer where it assumes an orientation and location allowing for a productive collision with the active site.

Such a case is insulin, which was first extracted from pancreas tissue, used in a patient in 1922, and its structure first determined in 1972. It has a molecular weight of 5.8 kDa and consists of a 21 amino acid peptide (chain A) that is connected to a 30 amino acid peptide (chain B) by two disulfide bonds. A third intra-subunit disulfide bond exists in chain A.

The structure of insulin has been probed using X-ray crystallography, while NMR spectroscopy was used to determine its three-dimensional structure in solution. As with other similar-sized molecules, standard two-dimensional ¹H homonuclear experi-

ments can adequately provide such structural information.

The use of two-dimensional experiments such as DQF-COSY, TOCSY and NOESY to assign and determine the three-dimensional conformation of peptides and proteins is well established. Briefly, the method used for the assignment of amino acid chains is not dissimilar from that of assigning small molecules. Each amino acid residue gives rise to a characteristic spin pattern that can be identified using the complementary DQF-COSY and TOCSY spectra, where connectivities between all protons within a spin system are observed in the TOCSY, and connectivities between neighbouring protons are observed in the DQF-COSY. An example of the DQF-COSY spectrum for the des-pentapeptide insulin monomer is shown in **Figure 5**. Of the 46 amino acids in the monomer, non-overlapped spin systems for four valine, two gylcine, one threonine and one alanine residue were distinguished. The latter two were sequence specifically assigned as threonine-8 in the A chain and alanine-14 in the B chain as these were the only alanine and threonine residues in the sequence. In addition, a further nine AMX spin systems could be clearly distinguished that were subsequently assigned to amino acids with the aid of the TOCSY and NOESY spectra. A combination of DQF-COSY

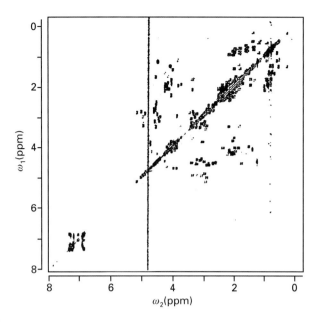

Figure 5 The 500 MHz ¹H DQF-COSY spectrum of a short-ened analogue of insulin, des-(B26–B30)-pentapeptide insulin in D_2O. The spectrum was recorded on a Bruker AM 500 spectro-meter interfaced with an ASPECT 3000 computer in a phase-sensitive mode. 2048 complex data points were acquired in the t_2 dimension, with a total of 512 free induction decays (FID) collected for transformation in the t_1 dimension. Each FID was acquired with a total of 128 scans.

and TOCSY data were required in order to delineate the remaining overlapped amino acid spin systems.

Once the spin systems arising from individual ami-no acids have been identified, the correct sequence of amino acids is determined from NOE data, acquired using the NOESY experiment. For molecules that fall within a particular molecular weight range (1000–3000 Da), the magnitude of the NOE is small, and a modification that is referred to as rotat-ing Overhauser enhancement spectroscopy (ROESY) must be employed. Typically, a series of NOESY (and/or ROESY) spectra are collected in different solvents such as D_2O, H_2O or DMSO, with mixing times ranging from 50 to 600 ms. Different mixing times are required to gain an estimate of the magni-tude of the NOE and subsequently an estimate of the distance between the protons. Care must be taken at longer mixing times, as it is possible to observe an indirect magnetization transfer between protons that are further apart than 5Å via a third proton that is appropriately positioned between them. To avoid this spin diffusion effect, it is preferable to acquire NOESY experiments with the shortest mixing times possible that will result in good quality spectra.

The determination of the amino acid sequence is based on the fact that NOEs will always be observed between particular protons from neighbouring resi-dues regardless of the secondary structure of the pro-tein. For example the proton attached to the α carbon (Hα) of an amino acid will always be within approximately 4.5 Å of the proton attached to the nitrogen (NH) of the neighbouring amino acid that is attached to the carboxyl end. In the NOESY spec-trum, the NH resonances tend to occur within a par-ticular spectral region, as do the Hα resonances. Thus, there is a particular region of crosspeaks between them in which a continuous connectivity pattern can be distinguished that begins with an NOE between the first and second residue and ends with an NOE between the terminal residue and the one preceding it (**Figure 6**). Such patterns reveal the sequence of amino acids in the protein. Similar pat-terns can also be distinguished between other groups of resonances (e.g. correlations between NH protons of neighbouring residues, Hβ protons and NH pro-tons of neighbouring residues) that can be used to confirm the sequence or resolve ambiguities. Breaks in this continuous pattern do occur in cases where resonances overlap or in some instances due to the structure of the amino acid chain, for example when a proline is present. However, once various lengths of polypeptides have been identified it is usually pos-sible to surmise the order in which the various lengths are connected. This is achieved by consider-ing different possibilities until a process of elimina-tion arrives at the arrangement that best fits the data.

Once the spectra have been assigned, data that contain structural information can be extracted. Nuclear Overhauser effects between non-neighbour-ing residues are the most revealing source of confor-mational information, and scalar coupling constants are important in providing information on torsion angles. All such conformational parameters are used as constraints in computational calculations that arrive at the three-dimensional conformation of the protein. For small molecules, such as the non-classi-cal cannabinoids, it is possible to infer the conforma-tion from NMR data without the aid of a computer, given that there are a relatively small number of NOEs generated and there are limited conformation-al possibilities from which to choose. However, for peptides, proteins and other macromolecules such as DNA, where a large number of measured NMR parameters must be taken into consideration, compu-tational methods are an essential tool in the determi-nation of the vast conformational possibilities. The computational techniques utilized seek to systemati-cally adjust the position of all nuclei in the molecule, so that all distances and bond angles derived from NOE data and coupling constants are satisfied. At the same time, the structure must not exceed set

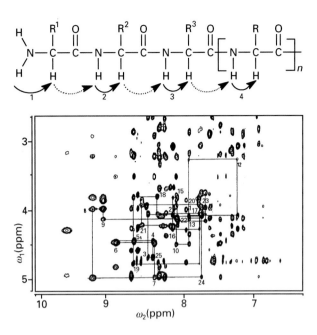

Figure 6 The fingerprint region of a 500 MHz ^{1}H NOESY spectrum which contains NOEs between NH and Hα resonances of des-(B26–B30)-pentapeptide insulin. The expected intra-residue and inter-residue NOEs that form the highlighted sequential pattern of B chain backbone NH and Hα resonances are represented as solid and dashed arrows, respectively.

physical limits for bond lengths, bond angles, torsion angles, van der Waals contacts and Coulombic interactions between atoms. The challenge in such methods is to ensure that all possible conformations are sampled while not allowing the molecule to exceed the set physical limits. Various algorithms and methods for calculation exist; however, the most common are the distance geometry and/or restrained molecular dynamics methods.

Using this basic methodology, the three-dimensional conformation of insulin was determined, and a significant amount of structure–activity information was gained by the study of insulin analogues and insulins purified from different sources.

Structural analysis of drug-binding domains

Macromolecules such as proteins and nucleic acids form the sites at which drugs interact. Knowledge of three-dimensional conformations assists in the design of analogues that are more potent and have improved pharmacokinetic properties. Furthermore, the structural analyses of protein receptors and enzymes adds to the knowledge of biological systems, and therefore assists in identifying novel

types of therapeutic agents. Because of their large molecular weights, most macromolecular therapeutic targets cannot be studied using exclusively ^{1}H homonuclear methods.

The advent of three-dimensional and heteronuclear pulse techniques has greatly expanded the ability to study macromolecules of up to 30 kDa in size. Three-dimensional techniques are a natural progression of the two-dimensional experiments. The pulse sequence is altered so that a vertical domain is introduced and the information is spread into a third dimension, so that the spectrum now is projected into a cube instead of a plane. The diagonal of the cube is equivalent to the one-dimensional spectrum, and crosspeaks that may be overlapped in a two-dimensional spectrum can be resolved in the third dimension.

A case in point is the structure determination of the insulin receptor substrate-1 (IRS-1). Insulin binds to a membrane-bound receptor that is a ligand-activated protein tyrosine kinase. Upon insulin binding there is an autophosphorylation of several tyrosine residues on the cytosolic side of the receptor. This enhances the tyrosine kinase activity of the insulin receptor towards other substrates and is required for signal transduction. A cascade of events is initiated, the first of which is the phosphorylation of IRS-1. This occurs when IRS-1 binds to the insulin receptor via a specific domain of the protein that is termed the phosphotyrosine binding (PTB) domain.

The structure of this domain was determined while interacting with a tyrosine-containing peptide derived from a receptor. As such, this study is also an example of the use of NMR to study the interactions between molecules. The three-dimensional NMR experiment was coupled with heteronuclear techniques to increase the level of resolution of the spectra. For larger proteins such as IRS-1, homonuclear experiments are limited because proton resonances become broader, and the efficiency of magnetization transfer between protons is decreased. With the advent of recombinant DNA technology, this problem can be overcome by producing the proteins to contain isotopic labels, such as ^{13}C, ^{15}N or ^{2}H, either at specific sites or uniformly throughout the protein. The magnetic properties of these nuclei offer significant advantages over protons. Carbon and nitrogen nuclei resonate over a much larger spectral width or range of frequencies, and as such are less likely to suffer from overlap. Also their scalar couplings to each other and to protons are higher in magnitude, and a more efficient magnetization transfer takes place than for ^{1}H–^{1}H couplings. Thus, for macromolecules, heteronuclear experiments that correlate ^{15}N to ^{13}C nuclei or to protons offer significant gains in sensitivity and resolution compared to homonuclear experiments. The most often used heteronuclear experiments are the heteronuclear single quantum coherence experiment (HSQC) and the related heteronuclear multiple quantum coherence experiment. These experiments allow the measurement of one- and two-bond heteronuclear couplings (and homonuclear ^{13}C–^{13}C couplings). They are most often combined with traditional two-dimensional experiments such as NOESY and TOCSY to yield a three-dimensional experiment. For example, in the case of an HSQC-NOESY spectrum of a protein, two of the axes represent the heteronuclei such as ^{15}N and the protons which are directly attached to the nitrogen nuclei, while the third axis contains chemical shifts of protons which share an NOE effect with the amide proton. This offers a significant increase in resolution compared to a traditional two-dimensional NOESY. A large array of these types of three-dimensional, heteronuclear-edited experiments have been designed to extract structural information in various situations.

Using these methods, the structure of the PTB domain of the IRS-1 protein was found to be similar to phosphopeptide-binding regions of several other proteins. Once the structural details were known, the different binding specificities could be compared and rationalized based on the interactions with their substrates.

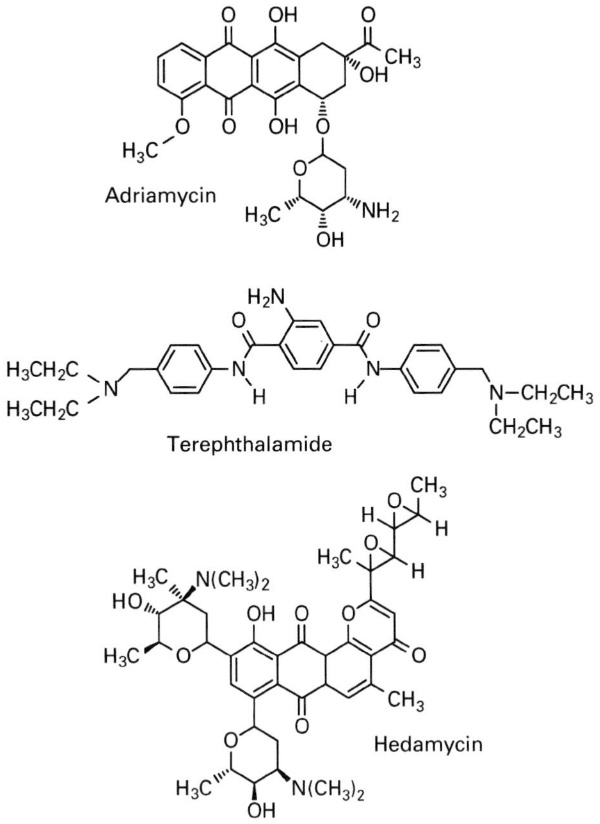

Drug interactions with macromolecular targets

NMR spectra are capable of supplying information about molecular interactions in solution. When a drug interacts with a receptor in a reversible manner, a number of effects may be observed in the spectra due to the exchange of the molecules between free and bound states. These effects will be discussed in relation to studies of antitumour antibiotics binding to short sequences of oligonucleotides. Compounds such as adriamycin (**Figure 7**) are currently in use as chemotherapeutic agents, and an understanding of factors involved in binding specificity may lead to more effective drugs to combat cancer. Chemical exchange effects have been exploited to great effect in obtaining information about these interactions.

The free and bound states of the molecules represent two different chemical environments in which participating molecules may be found. Thus, the same nucleus of a particular molecule may be characterized by different values for NMR parameters, such as chemical shift, when located in each different environment and give rise to different sets of resonances. The ability to measure the parameters that

Figure 7 Chemical structures of intercalating and minor groove DNA binding ligands.

characterize each environment is dependent on the rate at which the nucleus exchanges between them. The exchange rate therefore has a significant effect on the appearance of NMR spectra, and the exchange rate is dependent upon the affinity of the drug for the receptor. The kinetics of the interaction can be examined by acquiring a series of spectra in which the concentrations of the reactants or the temperature are modulated.

Figure 8 shows an expansion of the aromatic region of a series of ^1H NMR spectra of the oligonucleotide duplex d(GGTAATTACC)$_2$ to which has been added increasing amounts of a terephthalamide derivative (**Figure 7**). It is clear that the addition of the ligand causes significant perturbations of the free DNA resonances, and these indicate that the molecules are interacting. At ligand: DNA ratios between 0:1 and 1:1 there is a mixture of DNA molecules that are bound and unbound. The DNA resonances observed are averages of resonances arising from each of these states, and this is characteristic of fast exchange between the bound and unbound state due to a low affinity of the ligand for the binding site. There is a point in the titration in which the DNA

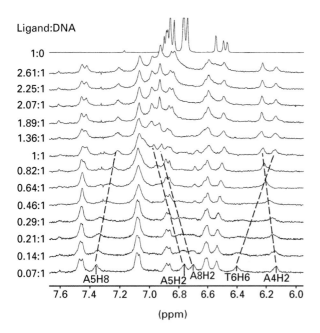

Figure 8 Expanded aromatic regions from the 300 MHz ¹H NMR spectra of complexes between a terephthalamide derivative and d(GGTAATTACC)₂, recorded at 10 °C. Increasing ligand concentrations cause perturbations to chemical shifts of nuclei located at or near the binding site. All perturbations are upfield except for adenine H2 resonances that are perturbed downfield and are located on the floor of the minor groove.

resonances are no longer perturbed, and this identifies the point at which all binding sites have been occupied and there are only bound DNA molecules present in solution. By noting the resonances that are most perturbed and the direction of the perturbation (upfield or downfield), it was determined that the ligand was bound at the 'ATTA' binding site. The observed perturbations were consistent with the ligand being inserted edge on into the minor groove. This places protons on the floor of the groove in the same plane as the aromatic drug which are then deshielded due to ring current effects. Protons above and below the plane of the ring experience shielding ring current effects.

An example of a ligand that has a high affinity for the binding site and exhibits slow exchange characteristics is shown in **Figure 9**, where the antitumour antibiotic hedamycin (see **Figure 7**) was titrated into a solution of the oligonucleotide duplex d(CACGTG)₂. Upon addition of hedamycin, the free DNA resonances diminish in intensity and new peaks appear in the spectrum. These new peaks do not correspond to chemical shifts of the free duplex or ligand resonances and increase in intensity with increasing ligand concentration. This suggests that they arise from the bound form of these molecules,

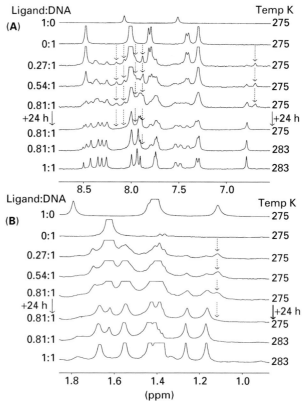

Figure 9 Expanded regions from the 300 MHz ¹H NMR spectra of complexes between hedamycin and d(CACGTG)₂, showing (A) aromatic resonances and (B) methyl resonances. Spectra up to 0.8:1 ligand:DNA ratio were recorded at 2 °C and the complex allowed to equilibrate for 24 h at this temperature. Subsequent spectra were recorded to 10 °C as the resonances were sharper. The dotted arrows highlight resonances that disappear after the 24 h equilibration period.

and that slow exchange conditions exist due to the high affinity of the ligand for the binding site. In this particular case it became apparent that time-dependent changes in the spectra were taking place. Allowing the mixture to equilibrate for 24 h resulted in certain resonances disappearing from the spectrum and sharpening of the remainder of the resonances. Given that hedamycin was subsequently shown to intercalate and alkylate, the transient peaks may represent reaction intermediates prior to alkylation where the chromophore is intercalating reversibly to sites other than the most favoured binding site, and the sharpening of the spectra is caused by alkylation of the DNA by the ligand.

In cases where the binding affinity is high, or the binding is irreversible, a detailed model of the interaction between the molecules can be constructed based on intermolecular NOEs. A NOESY spectrum of the 1:1 complex of hedamycin with d(CACGTG)₂ is shown in **Figure 10**. Assignment of the resonances

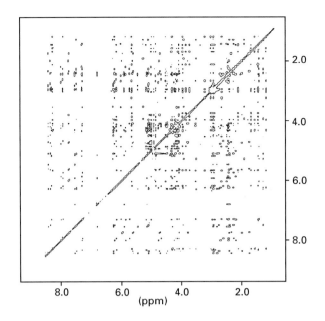

Figure 10 Contour plot of the 500 MHz ^{1}H NOESY spectrum of the 1:1 hedamycin:DNA complex recorded at 10 °C with a mixing time of 300 ms. Spectra were acquired on a Bruker 500AMX spectrometer with 2048 complex data points in the t_2 dimension and a total of 512 free induction decays (FID) collected for transformation in the t_1 dimension. Each FID was acquired with a total of 128 scans.

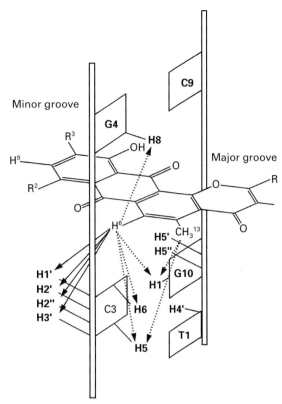

Figure 11 Schematic representation of NOEs observed between the central GC basepairs and the hedamycin chromophore. Contacts to major groove protons are shown as dotted arrows and contacts to the minor groove are shown as solid arrows.

was achieved using a combination of COSY, TOCSY and NOESY spectra. As in the case of proteins, the NOESY spectra of oligonucleotides yield characteristic crosspeak patterns that allow the sequential assignment of residues. Once the spectrum had been assigned and intramolecular NOEs arising from the DNA duplex and hedamycin had been eliminated, a total of 61 intermolecular NOEs from each of the oligonucleotide strands to protons located on the chromophore, epoxide chain and sugar groups of hedamycin were identified. These intermolecular NOEs identified the binding site and allowed the orientation of the ligand within the binding site to be determined.

For example, a number of intermolecular NOEs, summarized in **Figure 11**, showed that the molecule was intercalated between the central CG basepairs. Similarly, intermolecular contacts showed that the sugar groups of the molecule were located in the minor groove (**Figure 12**) and the epoxide chain in the major groove. Furthermore, the epoxide chain was shown to be in an appropriate location, so that the terminal carbon was capable of alkylating to the N7 of the guanine. Direct evidence for this was observed in that the guanine H8 proton became labile (i.e. disappeared when spectra were acquired in D$_2$O) following complexation.

Another example involving proteins is the interaction of the PTB domain of the IRS-1 protein with a tyrosine phosphorylated peptide in which the peptide was found to bind in a surface-exposed pocket of the PTB domain. More specifically, the peptide was bound along one strand of the β sheet structure of the PTB domain and interacted with an α helix. Hydrogen bonding, van der Waals contacts and hydrophobic interactions stabilized the interaction. The peptide was found to be in a type I β turn, and the N-terminal residues of the peptide were in an extended conformation that formed an additional strand of the PTB domains β sheet.

Combinatorial methods

Traditional methods of drug discovery involved a search amongst a range of diverse compounds that were derived from nature. This was a long and arduous process in which advances were due more often to serendipity rather than scientific thought and technique. The means to study the three-dimensional conformation of drugs and their receptors opened the door to a more rational approach in which compounds could be synthesized to better interact with a

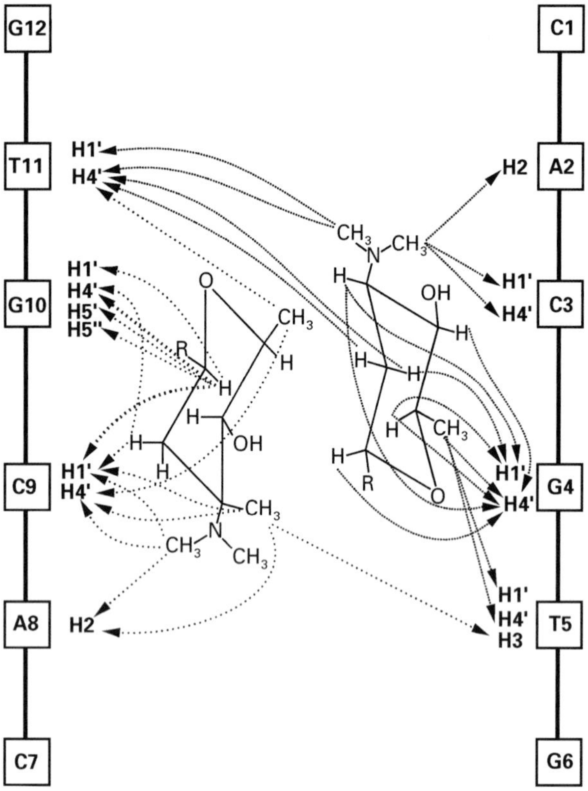

Figure 12 Schematic representation of contacts between the sugar rings of hedamycin and the DNA duplex. Contacts are observed only to minor groove protons and each ring is associated predominately with one DNA strand.

receptor. Along with the better understanding of drug/receptor interactions came improvements in biochemical techniques to isolate receptors and test compounds for their affinity towards these receptors. The ability to test thousands of compounds in a relatively short amount of time using high throughput screening methods resulted in pressure to produce large numbers of diverse compounds to be tested.

Analysis of solid-state intermediates

Combinatorial chemistry is a term used to describe the production of a large number (thousands) of diverse compounds, and different methods are employed to achieve this. One such method is solid-phase synthesis where the chemistry occurs on a solid support that may be packed into a column. The solid support may be a material such as crosslinked polystyrene and the compounds to be produced generally consist of a number of basic pharmacologically relevant structures (pharmacophores) that are to be linked in different ways. The method requires that one of the pharmacophores be linked to the solid support, then each subsequent pharmacophore can be delivered to the column with appropriate reagents, allowed to react, and then washed off the column. By varying the order in which the pharmacophores are delivered, different compounds can be produced.

This method relies heavily on the use of NMR to assess the success of each reaction step. As a non-destructive technique, a sample of the resin may be removed, placed in the NMR machine and then returned to the column. However, the fact that the compound is attached to the resin results in broad resonances if solution NMR techniques are utilized. This difficulty is circumvented by using solid-state NMR methods, including magic angle spinning (MAS). These experiments lead to narrow linewidths and high quality spectra, so that even two-dimensional experiments can be performed.

Combinatorial methods in drug discovery

One of the most interesting applications of NMR in drug research is in the field of high throughput screening. One such recently described approach bridges combinatorial chemistry with biochemical screening and was named 'SAR by NMR'. The first step in this method is to identify the ability of ligands to bind to a target preparation of the molecular therapeutic target (e.g. an enzyme) in solution. Thus, batches of ligand mixtures (e.g. one batch containing 20 compounds) are allowed to interact with the enzyme preparation in solution while following changes in the protein ^{15}N and ^{1}H NMR frequencies.

In this manner, batches that produce a response can be identified quickly and the compounds within each batch further tested to identify specific compounds that bind to adjacent but different sites on the protein. These ligands are then further optimized using rational design techniques to improve the binding at each respective site. The second generation ligands are then linked together to produce a compound that has a higher affinity than either of the two lead compounds. Using this method it was possible to identify two ligands that bound with micromolar affinities to the FK506 binding protein, which is involved in immunosuppression when activated. Linking these two individual ligands resulted in a compound that bound to FK506 binding protein with a nanomolar affinity. Preliminary use of this technique, which is applicable only to biomolecules of less than 30 kDa, indicates that it can have wide-ranging usefulness and is a potentially valuable tool in drug research.

List of symbols

$F_1 = \omega_1 = y$-axis frequency domain obtained by Fourier transformation with respect to t_1; $F_2 = \omega_2 = x$-axis frquency domain obtained by Fourier transformation with respect to t_2; t_1 = evolution time in a 2D pulse sequence; t_2 = detection period in a 2D pulse sequence; T_1 = spin–lattice (longitudinal) relaxation time; T_2 = spin–spin (transverse) relaxation time.

See also: **High Resolution Solid State NMR, ^{13}C; High Resolution Solid State NMR, ^{1}H, ^{19}F; Macromolecule-Ligand Binding Studied By NMR; NMR Pulse Sequences; Nuclear Overhauser Effect; Nucleic Acids Studied Using NMR; Proteins Studied Using NMR Spectroscopy; Small Molecule Applications of X-ray Diffraction; Structural Chemistry Using NMR Spectroscopy, Organic Molecules; Structural Chemistry Using NMR Spectroscopy, Peptides.**

Further reading

Anderson RC, Stokes JP and Shapiro MJ (1995) Structure determination in combinatorial chemistry: Utilization of magic angle spinning HMQC and TOCSY NMR spectra in the structure determination of wang-bound lysine. *Tetrahedron Letters* 36: 5311–5314.

Boelens R, Ganadu ML, Verheyden P and Kaptein R (1990) Two-dimensional NMR studies on despentapeptide-insulin. Proton resonance assignments and secondary structure analysis. *European Journal of Biochemistry* 191: 147–153.

Brange J, Ribel U, Hansen JF *et al.* (1988) Monomeric insulins obtained by protein engineering and their medical applications. *Nature* 333: 679–682.

Craik DJ (1996) *NMR in Drug Design*, Boca Raton: CRC Press.

Davis SN and Granner DK (1996) Insulin, oral hypoglycemic agents, and the pharmacology of the endocrine pancreas. In: Hardman JG, Limbird LE, Molinoff PB, Ruddon RW and Gilman AG (eds) *The Pharmacological Basis of Therapeutics*, 9th edn, pp 1487–1517. New York: McGraw-Hill.

Pavlopoulos S, Bicknell W, Craik DJ and Wickham G (1996) Structural characterization of the 1:1 adduct formed between the antitumor antibiotic hedamycin and the oligonucleotide duplex d(CACGTG)$_2$ by 2D NMR spectroscopy. *Biochemistry* 35: 9314–9324.

Shuker SB, Hajduk PJ, Meadows RP and Fesik SW (1996) Discovering high affinity ligands for proteins: SAR by NMR. *Science* 274: 1531–1534.

Wilson SR (1997) Introduction to combinatorial libraries: Concepts and terms. In: Wilson SR and Czarnick AW (eds) *Combinatorial Chemistry, Synthesis and Applications*, pp 1–23. New York: Wiley Interscience.

Xie XQ, Melvin LS and Makriyannis A (1996) The conformational properties of the highly selective cannabinoid receptor ligand CP-55,940 *The Journal of Biological Chemistry* 271: 10640–10647.

Zhou MM, Huang B, Olejniczak ET *et al.* (1996) Structural basis for IL-4 receptor phosphopeptide recognition by the IRS-1 PTB domain. *Nature Structural Biology* 3: 388–393.

Structure Refinement (Solid State Diffraction)

Dieter Schwarzenbach, University of Lausanne, Switzerland
Howard D Flack, University of Geneva, Switzerland

HIGH ENERGY SPECTROSCOPY
Methods & Instrumentation

Crystallographic structure refinement is generally understood to be the last step in the determination of a crystal structure by diffraction methods. The usual procedure of a crystal structure analysis includes collection of X-ray or neutron diffraction intensities, data reduction yielding structure factor amplitudes, the solution of the crystallographic phase problem yielding approximate structural parameters and finally refinement of these parameters to obtain a best fit of the observed structure factor amplitudes with the amplitudes calculated from the optimized model. The methods used to accomplish these successive steps depend on the type of compound and crystal. It is convenient to distinguish between small structures containing up to 100 or 200 symmetrically independent atoms, and macromolecular structures. The former are solved with atomic resolution, often routinely and nearly automatically, using standard and well-tested program packages. Modern efficient data collection apparatus employing area detectors and adequate computing power have enabled small-molecule crystallography to become an analytical technique with turn-around times measured in days or even fractions of a day. Complete small-molecule structure determinations are normally carried out on single crystal data, but the number of successful structure solutions from synchrotron powder diffraction data is steadily increasing. The methodology applied to macromolecular structures has evolved quite independently. Corresponding software packages and algorithms are distinct to a large extent from those applied to small structures, but small-molecule crystallography has started to incorporate some of the techniques applied to macromolecules and the dialogue is open. The present article is devoted mainly to methods of small-molecule structure refinement against single-crystal diffraction data, offered by widely distributed program packages such as SHELX-97, CRYSTALS, XTAL and applied to chemical crystallography.

Model fitting

The interpretation of measured quantities using a model derived from theory is a universal process in the physical sciences. The model is expressed by mathematical equations and contains constant numbers and parameters whose values are not fixed by the theory. The aim is then to find values for these variable parameters that best reproduce the observed quantities. The quality of the fit is judged according to a criterion that defines the distance between observed and modelled quantities. Model fitting is thus equivalent to the minimization of a multiparameter function expressing a distance criterion. Observations are always carried out to a limited precision, characterized by the standard uncertainty akin to the standard deviation or square root of the variance of a statistically distributed quantity. An important question to be answered is then the derivation of the standard uncertainties of the adjusted parameters of the model. This question is complicated by the fact that models are invariably imperfect: since they are obeyed only approximately, they cannot in principle obtain a perfect fit even for perfectly precise data. A crystallographic example is the use, in all models, of the kinematic scattering theory, which is correct only in the limiting case of vanishing diffracted intensity. Such model deficiencies are often referred to as 'systematic errors'. They may be reduced by performing experiments more closely related to the theoretical model. A further problem derives from applying corrections to the observations. Thus, observed diffraction intensities are corrected for background scattering, Lorentz and polarization factors and absorption effects. The resulting pseudo-observations (squared structure amplitudes on a relative scale) are then used as observations in the model fitting. Their values are in general correlated since an uncertainty in a correction and also systematic errors will affect all of them. This correlation is neglected in nearly all refinement software.

The crystallographic model

The standard *procrystal model* assumes that the electron distribution in the crystal is very nearly equal to a superposition of previously known rigid atomic density distributions, which are smeared by harmonic lattice vibrations. The structure factor $F(h, k, l)$ with integer Miller indices h, k, l for N atoms per unit cell, located at the positions x_n, y_n, z_n then becomes

$$F(h, k, l) = \sum_{n=1}^{N} p_n f_n T_n \exp[2\pi i(hx_n + ky_n + lz_n)] \quad [1]$$

The atomic scattering factors f_n are the Fourier transforms of the spherical atomic electron distributions. They are considered as known from quantum-chemical calculations. The site occupation parameters p_n may assume values different from unity if the structure is disordered. The Debye–Waller factors T_n allow for the atomic thermal motions. They are functions of the atomic displacement parameters $U^{i,j}$. Omitting the atom index n and representing the Miller indices and lengths of the reciprocal lattice vectors by h_i and a_i^*, respectively:

$$T = \exp\left[-2\pi^2 \sum_{i,j} U^{i,j} a_i^* a_j^* h_i h_j\right] \quad [2]$$

There results nine adjustable parameters per atom, i.e. three positional coordinates and six displacement parameters, and sometimes an additional site occupation parameter, although the point symmetry of an atom in a special position may reduce the number of adjustable parameters. The model value of the net intensity $I(h, k, l)$, Bragg peak minus background, is

$$I(h, k, l) = K \, Lp \, T \, y \, |F(h, k, l)|^2 \quad [3]$$

where K is a scale factor, L_p is the Lorentz-polarization correction, T is the transmission factor allowing for X-ray absorption, and y is an extinction correction allowing for some deficiencies in the kinematic scattering theory. The observed net intensities are routinely corrected for Lp and somewhat less routinely for T. The resulting corrected observations $|F_{obs}|^2$ and the corresponding model quantities $|F_{calc}|^2$ thus become

$$|F_{obs}(h, k, l)|^2 \rightarrow K\,y\,|F(h, k, l)|^2 = |F_{calc}(h, k, l)|^2 \quad [4]$$

where y depends on one or more parameters, and K is also adjustable. Taking the square root of both sides gives the model in terms of structure amplitudes $|F|$. Each observed quantity is accompanied by an estimation of its uncertainty, usually derived from counting statistics and/or the spread of multiple and symmetry-equivalent observations around their average value. The model may of course be augmented at will by additional parameters (twinning ratios, enantiomorph-polarity parameter, charge density or anharmonic motion parameters, etc.), but the above-mentioned are included in every refinement software.

The principle of least squares

By far the most important distance criterion between observed and modelled quantities is the least-squares deviance

$$D = \sum_{h,k,l} w(h, k, l) \, \{|F_{obs}(h, k, l)|^2 - |F_{calc}(h, k, l)|^2\}^2 \quad [5]$$

or analogously in terms of $|F|$, $w(h, k, l)$ being a weighting factor. The problem of structure determination, including the crystallographic phase problem, is thereby formulated in terms of an optimization criterion: find the minimal deviance D by varying all the parameters of the model defined by Equations [1], [2] and [4]. The solution would be unique and could be easily obtained if the equations for $|F_{calc}(h, k, l)|^2$ describing the model were linear. However, they are highly nonlinear, and a unique solution is not assured (although solutions of ordered structures at atomic resolution are indeed most probably unique). The optimization for nonlinear model equations becomes an iterative process, starting with approximate parameter values obtained by the methods of structure determination. For this reason, the structure solution and the structure refinement steps are usually dealt with separately.

In the following, we denote the M variable parameters of a structure by p_m ($1 \leq m \leq M$), the J observations $|F_{obs}(h, k, l)|^2$ or $|F_{obs}(h, k, l)|$ by O_j ($1 \leq j \leq J$) and the corresponding model quantities by C_j. The usual iterative Gauss–Newton method for approaching the optimal value of D, starting from approximate values $p_m^{(k)}$ for the adjustable parameters obtained in the (k)th iteration, is to linearize the model equations for C_j. The deviance D (see Eqn [5]) to be minimized with the improved parameter values $p_m^{(k+1)}$ then becomes

$$D = \sum_{j=1}^{J} w_j \, \{O_j - C_j[p_1^{(k+1)} \cdots p_M^{(k+1)}]\}^2 \quad [6]$$

$$C_j[p_1^{(k+1)} \cdots p_M^{(k+1)}] \approx C_j[p_1^{(k)} \cdots p_M^{(k)}]$$
$$+ \sum_{m=1}^{M} (\partial C_j / \partial p_m)_k \, [p_m^{(k+1)} - p_m^{(k)}] \quad [7]$$

The partial derivatives are evaluated with the parameters $p_m^{(k)}$. This leads to the matrix equation $\mathbf{N}[\mathbf{p}^{(k+1)} - \mathbf{p}^{(k)}] = \mathbf{v}$ with elements $N_{m,n}$ of the *normal matrix* $\mathbf{N}$ and v_m of the vector $\mathbf{v}$.

$$N_{m,n} = \sum_{j=1}^{J} w_j (\partial C_j / \partial p_m)_k (\partial C_j / \partial p_n)_k,$$
$$v_m = \sum_{j=1}^{J} w_j [O_j - C_j^{(k)}] (\partial C_j / \partial p_m)_k \quad [8]$$

The process starts with the parameters $p_m^{(0)}$ obtained with the methods of structure solution. Algorithms for solving Equation [8] are described in the literature. If the number of observations is much larger than the number of parameters, e.g. $J/M \approx 10$, and if the structure is well-ordered, the diagonal terms $N_{m,m}$ are large compared to the off-diagonal terms, and $\mathbf{N}$ is easily invertible. A refinement may be started with the scale factor and the positional parameters. The displacement parameters are then found automatically starting from values with reasonable orders of magnitude or from zero, and convergence is rapid. However, the observations may not be particularly sensitive to certain structural features of interest. The normal matrix then risks being nearly singular, the results of matrix inversion may be dominated by rounding errors, and convergence of the iterative adjustment is not guaranteed and may be meaningless. Examples of ill-conditioned problems are the parameters of weakly scattering hydrogen atoms in the presence of heavy atoms, or those of split atoms in disordered structures. In such cases, experience shows that it is in the use of a more appropriate model, constraints and restraints (see below), that refinement can be achieved rather than in resort to more advanced numerical algorithms.

The Gauss–Newton algorithm is not the only method capable of minimizing the deviance (Eqn [5]). The deviance of ill-conditioned nonlinear least-squares problems may possess several, often shallow, minima. There is then no guarantee that a minimum found starting from a given trial structure is the lowest one. The method of *simulated annealing* may be suitable for such problems as it permits us to leave a local minimum in search of another. It is frequently used in macromolecular crystallography.

Estimation of uncertainty

A result of a measurement is not complete without a quantitative statement of its uncertainty. Likewise, the value of a quantity derived from the measurement must be accompanied by its uncertainty. The uncertainty reflects the lack of exact knowledge of the value owing to random and systematic effects including deficiencies in the model. The quantitative measure of uncertainty is called standard uncertainty. It is an estimate of the standard deviation σ (i.e. the positive square root of the variance) of the probability density function of the quantity.

The method of least squares allows standard uncertainties to be obtained for all adjusted parameters. For any weights w_j, and for any probability density function of the observations, linear least squares is an unbiased estimator if the weights are independent of the observations, and if the model is perfect, i.e. represents physical reality correctly for some values of the parameters. In particular, the *Gauss–Markov theorem* states that minimal variances of the estimates are obtained if the weights are chosen equal to the inverse variances of the observations, $w_j = \sigma_j^{-2}$. For these, and only for these weights, the inverse of the normal matrix N^{-1} is an unbiased estimate of the variance–covariance matrix of the model parameters. In particular, the square roots of the diagonal terms of N^{-1} are the standard uncertainties of the adjusted parameters. We have tacitly assumed in the foregoing that the observations are statistically independent, i.e. that their covariances are zero. The Gauss–Markov theorem applies in fact also to correlated observations. Generally valid expressions for the variance–covariance matrix of the

model parameters for weights other than minimum-variance weights are rarely implemented in refinement software. As a rule, crystallographic software implements estimated minimum-variance weights, $w_j = u_j^{-2}$, where u_j is the standard uncertainty of O_j. The use of reliable values for these standard uncertainties is of utmost importance. They may be estimated from the spread of multiply measured or symmetry-equivalent Bragg intensities about their average values, the intensity fluctuations of periodically measured check reflections and Poisson statistics of count rates. The latter may be estimated from the observed O_j, the model C_j or a combination of these quantities. When other weights, e.g. unit weights, are used, the apparent standard uncertainties of the model parameters obtained from N^{-1} will be meaningless.

The goodness-of-fit S is a measure of the extent to which the calculated model values C_j agree with the observations O_j. For miminum-variance weights, S is calculated as

$$S^2 = (J - M)^{-1} D_{\min} = (J - M)^{-1} \sum_{j=1}^{J} u_j^{-2}(O_j - C_j)^2$$

[9]

where $D_{\min}$ is the deviance at convergence, and the u_j are the standard uncertainties of the O_j. J and M have been defined above. The expectation value of S for a perfect model is $\langle S \rangle = 1.0$. If the probability density functions of the observations are Gaussian, $D_{\min}$ is distributed like χ^2 with $J - M$ degrees of freedom. In practice, values for S near unity are rarely obtained even for fairly estimated u_j values. Apart from the possibility of inappropriate measurement procedures, this is explained by imperfections in the model. For example, we may recall that the standard procrystal model neglects the effects of chemical bonding and anharmonic thermal motions, that the kinematic theory of diffraction may be a questionable approximation to reality, and that X-ray absorption effects may have been incompletely accounted for. It is common practice among crystallographers to multiply the standard uncertainties of the parameters obtained from N^{-1} by S. Although this is equivalent to the unrealistic assumption that a lack of fit is due entirely to a constantly proportional underestimation of the standard uncertainties of the observations, it does have the advantage of increasing the standard uncertainties of the parameters obtained from N^{-1}. Moreover, other elaborate schemes of uncertainty manipulation are in use for attaining $S \approx 1$.

It may be important to stress at this point that standard uncertainties of parameters optimized with crystallographic programs depend on the method of refinement. Thus, if a parameter has been fixed because it was highly correlated with another parameter, thereby leading to an ill-conditioned normal matrix, the resulting standard uncertainties refer to an effective model where the fixed parameter assumes its true value which is previously known. Adjusting correlated parameters by alternatingly fixing one at its current value while refining the other is an unacceptable practice: it masks the fact that the parameters may not be both obtainable from the available data, and their standard uncertainties will be severely underestimated. Another example is block-diagonal refinement: in order to save storage space and computer time, the normal matrix is decomposed into a set of smaller matrices by neglecting off-diagonal terms outside the latter. With increasing computer power, this practice has been progressively abandoned but may still be attractive for large structures. The resulting standard uncertainties may, however, be underestimated, particularly if one or more of the neglected matrix elements assume large values. Wherever possible, the standard uncertainties should be calculated from an unrestricted full normal matrix.

Maximum likelihood

Maximum likelihood is a more general statement of the optimization problem than least squares. Instead of optimizing a distance criterion such as the deviance D, the problem may be stated in the following manner: 'Given a model and a set of observations, what is the likelihood of observing those particular values, and for what values of the parameters of the model is that likelihood a maximum?' [*International Tables for Crystallography*, Vol. C (1992) p 605. Dordrecht: Kluwer Academic.] The answer depends on the probability density functions ρ of the observations O_j. Assuming the observations to be uncorrelated and the model quantities C_j to be the expectation values of the O_j, the probability of observing the set of the O_j is the likelihood L,

$$L(p_1 \cdots p_M) = \prod_{j=1}^{J} \rho[O_j - C_j(p_1 \cdots p_M)] \qquad [10]$$

L, or more easily the logarithm $\ln L$, is maximized by varying the parameters p_m. Simple algebra gives for the gradient of $\ln L$

$$\partial \ln L / \partial p_m = (1/2) \sum_{j=1}^{J} [\zeta_j^{-1} \mathrm{d}\phi(\zeta_j)/\mathrm{d}\zeta_j] \, [\partial \zeta_j^2 / \partial p_m]$$

$$\phi(\zeta_j) = \ln \rho(\zeta_j); \quad \zeta_j = (O_j - C_j)/\sigma_j \qquad [11]$$

where σ_j is the square root of the variance of the function $\rho(\zeta_j)$, or a measure for the width of $\rho(\zeta_j)$. The term $\partial \zeta_j^2 / \partial p_m$ is the gradient of Equation [6] with $w_j = \sigma_j^{-2}$. For a Gaussian function, $\rho(\zeta) = (2\pi\sigma^2)^{-1/2} \exp(-\zeta^2/2)$, maximum likelihood is identical with minimum-variance least squares. If $\rho(\zeta)$ is not Gaussian, Equation [11] shows that maximum likelihood is equivalent to minimizing a least-squares deviance, where the minimum-variance weights are multiplied after each cycle by a function of the current deviates $O_j - C_j$. For a Cauchy function, $\sigma(\zeta) = \pi\sigma(1 + \zeta^2)^{-1}$, the modified weights are $w_j = [\sigma_j^2 + (O_j - C_j)^2]^{-1}$. Strongly discordant data are thus down-weighted because they are much more probable for a Cauchy function than for a Gaussian.

The probability density functions of the observations are generally unknown, but the Gauss–Markov theorem ensures that least-squares is always an acceptable estimator. However, the results of least squares are strongly influenced by discordant observations, so-called outliers. The *robust-resistant* techniques use weight-modification functions of $O_j - C_j$ which progressively down-weight outliers. These functions implicitly define probability functions ρ. They may alternatively be interpreted as an appreciation of the reliability of certain measurements. This approaches the frequently used option to simply omit discordant observations because they are judged to be unreliable.

Constraints

It is usually necessary, and often desirable, to impose relations between the parameters of a model. The problem to be solved is then a minimization of Equation [6] while imposing subsidiary conditions. A *constraint* satisfies such a condition exactly. A *restraint* (soft constraint) is a condition accompanied by a measure of uncertainty or a weight, and accordingly satisfies the condition only approximately.

Most crystallographic refinements require parameter constraints due to site symmetries. For example, the position of an atom on a mirror plane is defined by two independent coordinates only. The displacement tensor of such an atom is defined by four, rather than six, $U^{i,j}$ values, since one of its eigenvectors must be perpendicular to the plane. Modern refinement programs find and apply such constraints automatically for all possible site symmetries. In some space groups the origin of the coordinate system cannot be fixed with respect to the symmetry elements and thus must be defined by a constraint between positional parameters. Certain programs detect this condition and apply the constraint automatically, others do not. If site occupation parameters are refined, a constraint assuring electroneutrality may be required. The above types of constraint are usually implemented in the software by the method of parameter elimination.

Certain special features to be imposed on a model may be expressed by more complicated constraint equations. We note as an example the assumption of a rigid molecule with prescribed dimensions whose position and orientation are to be refined. The position may be described by the coordinates of the centre of mass and the orientation by three Euler angles with respect to a unitary coordinate system. The atomic coordinates and thus the structure factor, Equation [1], are expressed as functions of these six parameters. The latter may then be adjusted to optimize the deviance. A similar procedure can be used to constrain the atomic displacement parameters of a molecule to rigid-body movements described by a translation tensor, a libration tensor and a translation/libration-correlation tensor (TLS model). This model neglects intramolecular vibrations.

Restraints

Restraints are used to specify certain features of the structural model which are approximately known. For example, typical values of the bond lengths such as C–C or C–H are well-known; the quality of a structure determination is indeed judged by, among other properties, the plausibility of bond lengths. Bond lengths in well-ordered small-molecule structures are, as a rule, accurately determined from the diffraction data. This is not the case for those macromolecular structure determinations that do not attain atomic resolution. It may not be the case for disordered structures characterized by split-atom positions and superpositions of molecules or structural fragments in various orientations, such as found in structures of fullerenes, C_{60} and C_{70}, or in structures containing highly symmetric ions such as ClO_4^-. Powder diagrams may also contain insufficient information for the determination of all atomic positions. In such cases, the prior knowledge of bond lengths may permit a successful structure refinement.

Bond lengths are not known with absolute precision, nor does a bond between two atoms assume exactly the same value in all structures. Bond lengths should therefore not be fixed to a particular value by a constraint. Instead, they are introduced in the form of additional observational equations. The distance

between atoms m and n is

$$d_{calc} = |(x_m - x_n)\boldsymbol{a} + (y_m - y_n)\boldsymbol{b} + (z_m - z_n)\boldsymbol{c}| \approx d_{obs}$$

[12]

where d_{obs} has an associated standard uncertainty u expressing the confidence to be placed in this information. This pseudo-observation is used like all other observations in Equation [6], [7] and [8], with a weight $w = u^{-2}$. A residual $|d_{obs} - d_{calc}|$ of the order of u shows the distance information to be useful. A large residual, on the other hand, indicates a contradiction between distance and diffraction information which must be taken seriously: one of the two types of information or the corresponding weights may be erroneous. Simply increasing the weight of the restraint equation at the expense of the observations may be most inadequate, even if it leads to a closer agreement of the bond length with its expectation.

Other geometrical restraints may be applied to bond angles and torsion angles. Groups of atoms may be restrained to occupy a plane by fixing one or several torsion angles to be 0° or 180°. *Similarity restraints* are used to impose two or more bonds to be of the same length without specifying the value of the bond length. This permits the restraint of the hexagonal symmetry of a phenyl ring, or the icosahedral symmetry of a C_{60} molecule. Restraints are thus far more flexible than rigid-molecule constraints, and may be applied to only part of a molecule. Distance and angle information alone, without diffraction data, has been used to optimize atomic coordinates in framework structures. Examples are refinements of alumosilicate frameworks in zeolites starting with unit cell data as determined from powder diagrams, and assuming diverse space groups belonging to subgroups of the holohedral lattice symmetry (the highest symmetry compatible with the metric of the translation lattice).

The vibrationally rigid bond criterion expresses the expectation that the amplitudes of intramolecular vibrations should be much smaller than those of the translational and librational movements of a molecule. The mean-square displacements of two strongly bonded atoms along their bond should accordingly be approximately equal. Accurate structure determinations confirm this criterion for C–C bonds, but not for metal–ligand bonds in transition-metal complexes. Vibrationally rigid bonds may be imposed by rigid-bond restraints on the anisotropic displacement parameters. If all interatomic distances in a molecule are specified to be rigid, its thermal motions are restrained to rigid-body translation and libration (TLS), except if the molecule is linear or planar. With respect to a constrained TLS-refinement mentioned above, restraints offer the advantage that partially rigid molecules or structural groups may be defined.

Shift-limiting restraints are used to improve the convergence of iterative least-squares cycles. Strongly correlated parameters, such as displacement and site-occupation parameters refined against a low-resolution data set, result in an ill-conditioned pseudosingular normal matrix. The minimum of the deviance with respect to such parameters may be very shallow, successive shifts may erratically oscillate in opposite directions, or the refinement may diverge. A shift-limiting restraint ensures that the shift of the parameter value is of the order of the associated uncertainty u, or in terms of Equation [7] $p_m^{(k+1)} - p_m^{(k)} \approx 0$ with weight $w = u^{-2}$. Such a restraint simply adds w to the corresponding diagonal term of the normal matrix. This clearly leads to a better conditioned and more easily invertible matrix. After convergence, all shift-limiting restraints should be removed and a final cycle should be calculated to obtain acceptable standard uncertainties for the parameters. Shift-limiting restraints are a simplified version of the *Levenberg–Marquardt method* in which the diagonal terms of the normal matrix are multiplied by a number which approaches 1.0 at convergence.

In general, any function of the structural parameters may serve as a restraint. An important example in macromolecular crystallography is *molecular dynamics*. In addition to the diffraction data, an empirical potential energy expression E_{pot} is minimized, which contains terms for bond lengths, bond angles, torsion angles and nonbonded interactions. The function minimized is then the sum of the deviance, Equation [5], and the energy, $D + E_{pot}$. Molecular dynamics is a generalization of the distance and angle restraints discussed above where the weights are derived from the expression for the energy.

Judging the results

The quality of the fit between observed and calculated structure amplitudes is commonly characterized by the agreement factors

$$R1 = \sum_{h,k,l} \left| |F_{obs}| - |F_{calc}| \right| / \sum_{h,k,l} |F_{obs}|$$

$$wR2 = \left\{ \sum_{h,k,l} w(|F_{obs}|^2 - |F_{calc}|^2)^2 / \sum_{h,k,l} w|F_{obs}|^4 \right\}^{1/2}$$

[13]

The conventional index $R1$ usually includes only the stronger reflections; a common selection is $|F_{obs}| > 4u_F$ or $|F_{obs}|^2 > 2u_{F2}$, where u_F and u_{F2} are the standard uncertainties of $|F_{obs}|$ and $|F_{obs}|^2$, respectively. The goodness of fit S defined by Equation [9] is a less appropriate measure since some software adjusts the weights to bring this value close to 1.0. Values of $R1$ up to 0.07 are considered to be acceptable; for very accurate studies, all agreement factors are of the order of 0.02 or lower.

The optimal values of the parameters and their associated standard uncertainties are meaningful only in relation to the model employed in the final least-squares cycle. All constraints and restraints used should be reported. Note that the positions and displacement parameters of hydrogen atoms are often derived from the parameters of the heavier atoms and subsequently suitably restrained or constrained; corresponding bond lengths, such as C–H, are then features of the model and not experimental evidence.

In the final cycle, the best estimate of minimum-variance weights, $w = u^{-2}$, should be used, where $u = u_{F2}$ or u_F for refinements against $|F|^2$ or $|F|$, respectively. The standard uncertainties of the refined parameters obtained from the inverse normal matrix are meaningful only if the standard uncertainties of the observations are meaningful. It should also be borne in mind that even in careful studies, model errors are unavoidable. These include imperfect corrections for absorption and extinction, as well as the standard procrystal approximation neglecting chemical bonding and anharmonic motions. In addition, the uncertainties of molecular parameters such as bond lengths are often computed neglecting correlations between the model parameters. Therefore, published uncertainties of refinement results are approximate and tend to be underestimates. They can be improved only through a careful estimation of u_{F2} or u_F and a reduction of model errors.

List of symbols

a, b, c = lattice vectors defining unit cell; a_i^* = length of ith reciprocal lattice vector; $C_j = j$th calculated model quantity, either $|F_{calc}(h,k,l)|$ or $|F_{calc}(h,k,l)|^2$; D = deviance, distance criterion in model fitting; d_{calc} = calculated interatomic distance; d_{obs} = observed interatomic distance; f_n = atomic scattering factor of nth atom; $F(h, k, l)$ = structure factor with Miller indices h, k, l; $|F(h, k, l)|$ = structure amplitude with Miller indices h, k, l; $|F_{calc}(h, k, l)|$ = calculated structure amplitude on relative scale; $|F_{obs}(h, k, l)|$ = observed structure amplitude on relative scale; h, k, l = Miller indices; i = square root of -1; $I(h,k,l)$ = intensity of Bragg reflection; J = number of observations; K = scaling factor; k = index of kth refinement interaction (e.g. $p_m^{(k)}$); L = maximum likelihood function; Lp = Lorentz-polarization correction; M = number of adjustable parameters; N = normal matrix with elements $N_{m,n}$; O_j = jth observed quantity, either $|F_{obs}(h,k,l)|$ or $|F_{obs}(h,k,l)|^2$; p_m = mth adjustable parameter; p_n = site occupancy factor of nth atom; $R1$ = reliability index with respect to $|F(h,k,l)|$; $wR2$ = weighted reliability index with respect to $|F(h,k,l)|^2$; S = goodness of fit, $S = \{D/(J-M)\}^{1/2}$; TLS = rigid-body thermal displacement description; T = transmission factor or Debye–Waller factor; T_n = Debye–Waller factor of nth atom; u = standard uncertainty; $U^{i,j}$ = anisotropic atomic displacement tensor element; v = vector of normal equations with elements v_m; w_j = weight of jth observation; x_n, y_n, z_n = positional coordinates of nth atom; y = extinction correction; ϕ = logarithm of probability density function; $\rho(\zeta)$ = probability density function; σ = standard deviation, square root of variance measure for width; ζ = random variable with mean 0.0 and variance 1.0.

See also: **Powder X-Ray Diffraction, Applications; Small Molecule Applications of X-Ray Diffraction.**

Further reading

Albinati P, Becker PJ, Boggs PT *et al* (1992) Refinement of structural parameters. In: Wilson AJC (ed) *International Tables for Crystallography*, Vol. C, pp 593–652. Dordrecht: Kluwer Academic.

Brünger AT (1988) Crystallographic refinement by simulated annealing. In: Issacs NW and Taylor MR (eds) *Crystallographic Computing 4*, pp 126–140. Oxford: Oxford University Press.

Brünger AT (1991) A unified approach to crystallographic refinement and molecular replacement. In: Moras D, Podjarny AD and Thierry JC (eds) *Crystallographic Computing 5*, pp 392–408. Oxford: Oxford University Press.

Hall SR, King JDS and Stewart JM (1995) LSLS. In: *Xtal 3.4 User's Manual*. Lamb, Perth: University of Western Australia.

Press WH, Flannery BP, Teukolsky SA and Vetterling WT (1992) *Numerical Recipes: The Art of Scientific Computing*. New York: Cambridge University Press.

Prince E (1994) *Mathematical Techniques in Crystallography and Materials Science*. New York: Springer-Verlag.

Schwarzenbach D, Abrahams SC, Flack HD *et al* (1989) Statistical descriptors in crystallography. Report of the International Union of Crystallography Subcommittee on Statistical Descriptors. *Acta Crystallographica, Section A* **45**: 63–75.

Schwarzenbach D, Abrahams SC, Flack HD, Prince E and Wilson AJC (1995) Statistical descriptors in crystallography. II. Report of a working group on expression of uncertainty in measurement. *Acta Crystallographica, Section A* **51**: 565–569.

Sheldrick GM (1997) SHELXL-97. In: *The SHELX-97 Manual*. Gottingen: University of Göttingen.

Watkin DJ (1994) Lead article: The control of difficult refinements. *Acta Crystallographica, Section A* **50**: 411–437.

Watkin DJ, Prout CK, Carruthers JR and Betteridge PW (1996) *CRYSTALS Issue 10*. Oxford: Chemical Crystallography Laboratory, University of Oxford.

Sulfur NMR, Applications

See **Heteronuclear NMR Applications (O, S, Se, Te).**

Surface Induced Dissociation in Mass Spectrometry

SA Miller and **SL Bernasek**, Princeton University,
NJ, USA

MASS SPECTROMETRY

Methods & Instrumentation

Analytical tandem mass spectrometry (MS/MS) relies on the ability to activate and dissociate ions in order to identify or obtain structural information about an unknown compound. The most common means of ion activation in tandem mass spectrometry is collision-induced dissociation (CID). CID uses gas phase collisions between the ion and neutral target gas to cause internal excitation of the ion and subsequent dissociation. Surface-induced dissociation (SID) is analogous to the CID experiment except a surface is substituted for the collision gas as shown in **Figure 1**. SID offers several advantages as a means

of ion activation in tandem mass spectrometry, including fine control of the internal energy deposition, efficient translational to vibrational energy conversion, large average internal energy deposition, and applicability to both small organic and large biological molecules. Since 1985, low energy polyatomic SID and its accompanying processes have been the subject of much research and continue to gain interest in the mass spectrometry community as an alternative means for ion activation, to study interesting ion/surface reactions, and to cause selective surface modification.

(A)

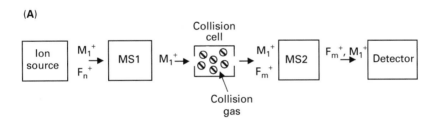

(B)

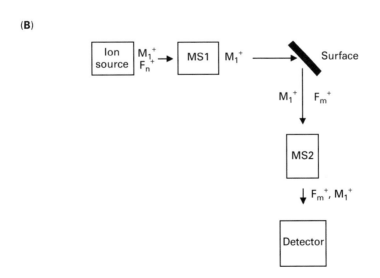

Figure 1 Simplified schematic of (A) a CID tandem mass spectrometer and (B) a SID tandem mass spectrometer where MS1 and MS2 are the mass analysers, M_1^+ is the parent or projectile ion, F_n^+ are fragment ions of M_1^+ generated in the ion source, and F_m^+ are fragment ions formed as a result of (A) CID and (B) SID.

Table 1 Low energy polyatomic ions/surface collision processes

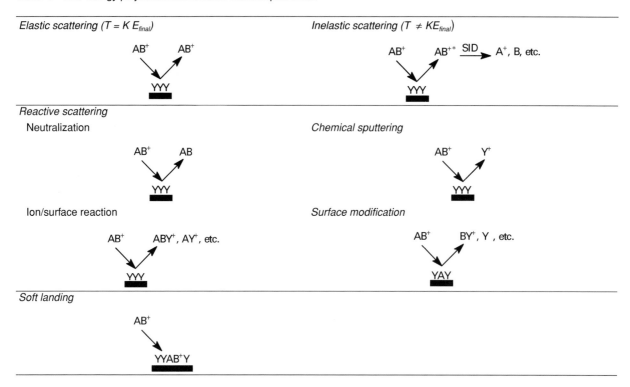

Low energy polyatomic ion/surface collisions

When an ion collides with a surface, several processes may occur including elastic, inelastic, and reactive scattering or soft landing. **Table 1** lists possible events which occur for low energy polyatomic ion collisions with organic covered surfaces. Note that 'low energy' refers to the component of kinetic energy normal to the surface which is usually in the range of 1 to 100 eV. The parent or projectile ion, AB^+, is accelerated (up to a few hundred eV) towards a substrate which has been chemically modified with an organic overlayer, Y, such as a self-assembled monolayer (SAM) surface (**Figure 2**). Since the projectile ion has a low normal component of collision energy, it interacts only with the outer layer of the adsorbate and not directly with the substrate. The ion will collide with the organic surface and may scatter from the surface without losing any kinetic energy. Generally, the polyatomic ion scatters from the surface with a loss of kinetic energy which is distributed into internal energy of AB^+ and of the organic overlayer Y as described in the following equation,

$$Q = T - \mathrm{KE_{final}} = V + \varepsilon_{\mathrm{surface}} \qquad [1]$$

where Q is the inelasticity of the collision event, T and $\mathrm{KE_{final}}$ are the initial and final kinetic energy of the projectile ion AB^+, V is the total vibrational energy of the ion AB^+, and $\varepsilon_{\mathrm{surface}}$ is the internal energy deposited in the surface (radiative losses are ignored). If AB^+ becomes sufficiently activated, it will dissociate into fragment ions which are characteristic of the structure and chemical composition of the projectile ion.

Several processes are representative of reactive scattering, including neutralization, chemical sputtering, ion/surface reaction, and surface modification. Most reactive scattering processes occur through electron transfer from the surface to the ion. If no other processes occur after the electron transfer, then the only product formed is the neutralized projectile ion, AB, and no ions are detected. Chemical sputtering occurs when sufficient energy is deposited in the surface following the electron transfer and collision event causing adsorbate ions to be liberated from the surface. The incident ion or a fragment ion can incorporate an atom or group of atoms from the surface to form an ion/surface reaction product. Ion/surface reactions may also cause chemical modification of the adsorbate through atom or molecule exchange reactions. One must distinguish between surface modifications that result from damage which occurs to an adsorbate during ion/surface collisions (for example, chemical sputtering) and a selective chemical modification that results from an ion/surface exchange reaction. Finally, specially chosen projectile ions can be soft landed into the adsorbate, specifically in fluorocarbon and hydrocarbon SAM surfaces, upon collision at very low energies (~1–10 eV). Remarkably, the deposited intact projectile ion remains in the surface as an ion protected by the adsorbate layer. This matrix-isolated ion has been shown to survive for several days in vacuum and a few days in ambient lab air.

Surface-induced dissociation instrumentation

The fundamental requirement for performing an SID experiment is a tandem mass spectrometer. Several tandem mass spectrometers have been developed using various combinations of mass analysers and geometries to perform SID. **Table 2** lists several SID systems (by no means an exhaustive list), which have been organized by on-axis (or in-line) and off-axis geometry. On-axis geometry refers to a tandem mass spectrometer where the exit slit or aperture of the first mass analyser is in-line with the entrance slit or aperture of the second mass analyser. In the off-axis geometry, there is no common axis between the exit and the entrance of the first and second mass analyser. The on-axis and off-axis groups can be further divided into instrument collision geometry, that is, large incident angle/low collision energy and glancing incident angle/high collision energy. The incident angle is defined here as being referenced to the surface normal. Finally, a special group of mass spectrometers which trap ions have also been used to perform SID experiments. Quadrupole ion traps and Fourier transform ion cyclotron resonance mass spectrometers perform the tandem mass spectrometry experiment in a single mass analyser, i.e. the tandem mass spectrometry experiment is performed in one analyser with each MS stage separated in time rather than separated in space as for a two-analyser system.

A simplified example of an SID tandem mass spectrometer is shown in **Figure 1B**. This system utilizes two mass analysers (MS1 and MS2) perpendicular to one another to mass select the projectile ion (MS1) and analyse the scattered product ions (MS2). Ions are generated in the ion source, accelerated, and focused into MS1 where the projectile ion, M_1^+, is mass selected. M_1^+ is then further accelerated or decelerated (depending on the desired collision energy) and focused onto the surface. The product ions scattered from the surface are accelerated toward MS2 and the SID mass spectrum is then collected by scanning MS2 and measuring the ion current at the detector.

There are several design considerations which must be addressed when constructing an SID

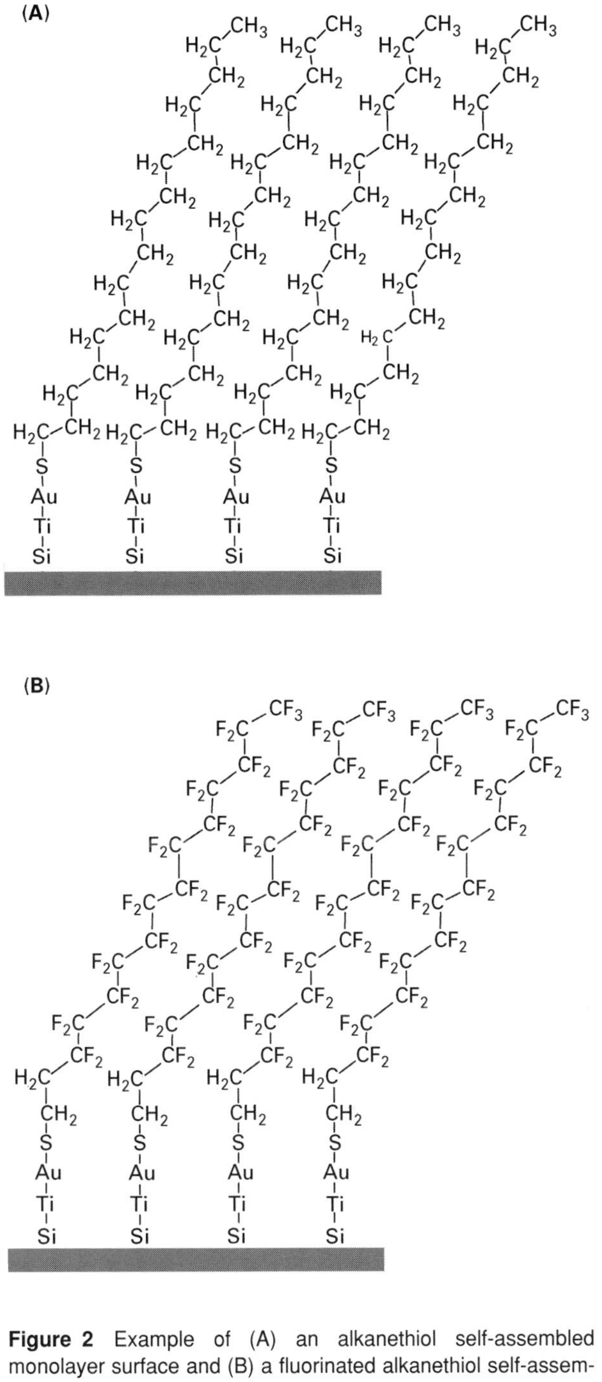

Figure 2 Example of (A) an alkanethiol self-assembled monolayer surface and (B) a fluorinated alkanethiol self-assembled monolayer surface. Typical substrates are glass or silicon with a thin Ti layer (~100 Å) followed by a layer of gold (~1000 Å). Reproduced with permission from Miller SA, PhD Thesis, Purdue University, 1997.

instrument, including vacuum conditions and instrument functionality. The first consideration is determining the vacuum requirement which will be sufficient for performing an SID experiment. For most SID instruments, the vacuum is typically in the range of 10^{-7} to 10^{-10} torr (1.3×10^{-5} to 1.3×10^{-8} Pa) and is dictated by the type of surface to be used in the experiment. If a clean and well characterized single crystal is used as the SID target, then ultrahigh vacuum (UHV) conditions must be met ($<10^{-9}$ torr). For an organic covered surface, such as a self-assembled monolayer surface, less stringent vacuum conditions ($<10^{-6}$ torr) will suffice.

The type of ion source used in an SID experiment depends on the projectile ion under study. For instance, if volatile, small organic compounds are of interest then an electron ionization (EI) and/or chemical ionization (CI) source will be sufficient. However, if one wishes to study biological compounds then a spray or desorption ionization method will be required, such as electrospray (ESI), atmospheric pressure chemical ionization (APCI), desorption chemical ionization (DCI), or matrix-assisted laser desorption ionization (MALDI).

As seen in **Table 2**, several combinations of mass analysers have been used in SID instruments, each with their own advantages and disadvantages. The type(s) of mass analysers used in an SID system depends to some extent on the class of projectile ions

Table 2 Low energy ion/surface collision instruments.

On-axis

Large incident angle/low collision energy
  1. Multisector
  2. Tandem quadrupole
  3. Hybrid sector/quadrupole

Glancing incident angle/high collision energy
  1. Hybrid sector/time-of-flight
  2. Hybrid sector/quadrupole
  3. Multi-Wien filter
  4. Multisector

Off-axis

Large incident angle/low collision energy
  1. Tandem quadrupole
  2. Tandem time-of-flight
  3. Hybrid sector/time-of-flight
  4. Hybrid sector/quadrupole
  5. Hybrid Wien filter/quadrupole

Trapping instruments
  1. Paul quadrupole ion trap
  2. Fourier transform ion cyclotron resonance mass spectrometers.

to be studied, as with the method of ionization, but more so on the overall performance mass analyser. One must consider the resolution, sensitivity, size, ruggedness, ease-of-use, cost, and the compatibility of one mass analyser with another as well as the information sought in the SID spectra. The authors refer the reader to other articles in this text that discuss advantages and disadvantages of the various mass analysers.

The extent of fragmentation in an SID experiment is highly dependent on the component of kinetic energy normal to the surface. Thus, having control of the collision energy (kinetic energy of the projectile) is necessary. The range of ion kinetic energy will depend on the collision geometry of the instrument. For example, if glancing incident angles are used then higher overall collision energies will be needed (hundreds of eV to keV), while more normal collisions will require a lower ion kinetic energy (1–100 eV). It has been shown that nearly identical SID spectra can be obtained from both glancing and large incident angle collisions as long as the component of kinetic energy normal to the surface is similar. While the angle of incidence is not that important analytically, subtle fundamental information can be obtained by controlling both the kinetic energy of the projectile and the angle of incidence. More fundamental information about the SID process and other ion/surface collision processes can be determined by measuring the energy of the scattered ions as well as their angular distribution. Hence, it is of interest to control the scattering angle independently of the incident angle and to have the ability to measure the kinetic energy of the scattered ions.

Surface science techniques are useful for monitoring changes in the structure and composition of the adsorbate before, during, and after ion/surface collisions. Common techniques include high resolution electron energy loss spectroscopy (HREELS), low energy electron diffraction (LEED), Auger electron spectroscopy (AES), secondary ion mass spectrometry (SIMS), reflection absorption infrared spectroscopy (RAIRS), X-ray photoelectron spectroscopy (XPS), and ultraviolet photoelectron spectroscopy (UPS). While little ion structural information is gained using these methods, changes in the adsorbate are of interest when studying ion/surface reactions, especially those which lead to surface modification.

A few examples of SID instruments are shown in **Figures 3** to **6** which incorporate some or all of the considerations discussed above. **Figure 3** is an off-axis large incident angle/low collision energy hybrid sector/quadrupole SID instrument of the BEEQ configuration (B = magnetic sector, E = electrostatic

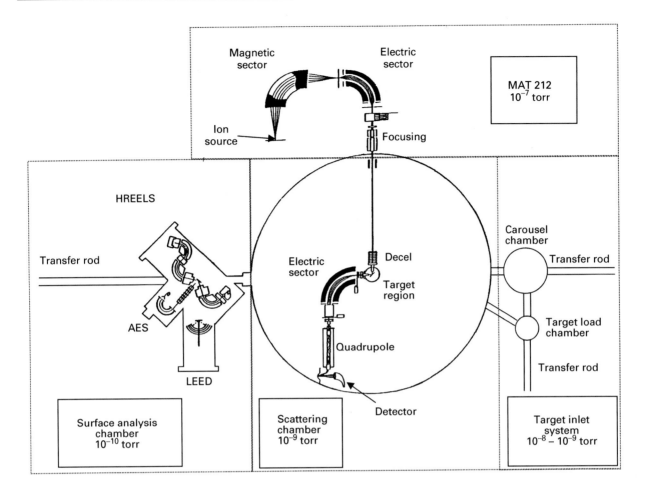

Figure 3 Overview of a hybrid BEEQ tandem mass spectrometer. Reproduced with permission from Miller SA, PhD Thesis, Purdue University, 1997.

sector, and Q = quadrupole mass filter). This system has the ability to change the scattering angle independently of the incident angle, to measure the kinetic energy of the scattered product ions with the post-collision electrostatic sector, and to do traditional surface analysis in a separate UHV chamber. **Figure 4** presents the SIMION calculated ion trajectories for a custom-built, on-axis SID device used on a commercial four sector tandem mass spectrometer. (Dahl DA and Delmore JE (1989) *SIMION PC/PC2 version 4.02*, Idaho National Engineering Laboratory, EG and G Idaho Inc.) The mass analysed projectile ion beam enters from the left of the device, strikes a target, and the scattered ions are extracted to the right side of the device into the final mass analyser. The SID device can be used in two different modes including a single deflection and dual deflection mode. It is clear in the SIMION calculation that the dual deflection mode (**Figures 4B** and **4C**) produces a more focused ion

beam. An advantage of this SID system is that the SID device can be exchanged with a commercial CID device and direct comparisons of SID to high energy CID collisions can be observed. **Figures 5** and **6** show two types of off-axis instruments that maintain large incident angle/low collision energy geometry but use different tandem mass analysers. The tandem time-of-flight system in **Figure 5** has the ability to use several ionization techniques making it amenable to projectile ions ranging from small organics to large biological compounds. This system also employs delayed extraction technology to improve the resolution in the SID product ion mass spectra. A unique feature of the tandem quadrupole system in **Figure 6** is that it is contained on a single 8″ conflat flange and is easily attached to a standard UHV surface analysis system. Most currently used SID systems utilize off-axis large incidence angle/low collision energy geometry due to enhanced overall performance compared to on-axis systems.

(A)

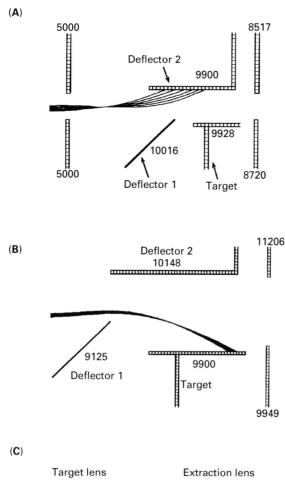

(B)

(C)

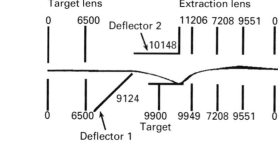

Figure 4 SIMION ion trajectory calculations of an on-axis SID device with (A) precursor ion deflection in single deflection mode, (B) precursor ion deflection in double deflection mode, and (C) precursor and product ion trajectories in double deflection mode. Numerical labels represent the voltages applied to the electrodes. Reproduced with permission from Durkin DA, Schey KL, (1998) Characterization of a new in-line SID mode and comparison with high energy CID. *International Journal of Mass Spectrometry and Ion Processes* **174**: 63–71.

Surface-induced dissociation

The characteristics associated with the SID process are listed in **Table 3** and emphasize the analytical utility of SID as a method of ion activation. The first

Table 3 Characteristics of the SID process.

1.	Internal energy (V) deposition is variable depending on the collision energy (T)
2.	Internal energy deposition can be made large, more than 10 eV is readily accessible
3.	Internal energy distribution is narrow, typically 4 eV FWHM, and is approximately Gaussian in shape
4.	The efficiency of converting translational energy to internal energy is dependent on the nature of the surface and is approximately 20–30% for fluorocarbon surfaces and 10–20% for hydrocarbon surfaces
5.	The SID efficiency varies over a wide range 1–75% but is typically more than 10% for most polyatomic ion/organic covered surface pairs.
6.	Fragment ions are similar to those noted in CID experiments
7.	SID is proving to be applicable to large biomolecules

characteristic of SID listed in **Table 3** is the ability to control the amount of fragmentation, or internal energy, by adjusting the collision energy. **Figure 7** displays the SID mass spectrum resulting from the collision of p-methyl phenetole molecular ion (136 Th) with a fluorocarbon self-assembled monolayer (F-SAM) surface at two different collision energies. This example illustrates the effect that collision energy has on the extent of fragmentation. At a 15 eV collision energy, only the molecular ion and one fragment ion, m/z 108 (loss of C_2H_4), are observed. When the collision energy is increased to 30 eV, fragmentation occurs giving rise to a high energy fragment ion, m/z 29 ($C_2H_5^+$ and/or CHO^+) as the base peak and *complete loss* of the molecular ion at m/z 136. The fragment ions in **Figure 7B** are the same as those observed in the electron ionization mass spectrum of p-methyl phenetole but with different ion intensities.

A better fundamental understanding of the SID process is obtained by quantitating the partitioning of the transitional energy of the ion into internal energy of the ion, internal energy of the surface, and the kinetic energy of the scattered ions. In Equation [1], partitioning of the projectile kinetic energy in an ion/surface collision was described. Rearranging this equation to indicate energy conservation,

$$T = \text{KE}_{\text{final}} + V + \varepsilon_{\text{surface}} \qquad [2]$$

the ion transitional energy is the sum of the final kinetic energy of the scattered products, the total internal energy of the projectile ion and the energy absorbed by the surface. Despite the fact that experimentally determined values for KE_{final} and V

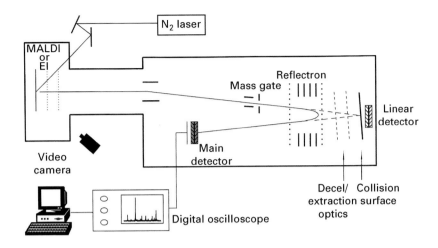

Figure 5 Overview of a tandem time-of-flight SID instrument. Reproduced with permission from Riederer Jr, DE, PhD, Department of Chemistry, University of Missouri, Columbia.

are dependent on the nature of the surface and projectile ion, general conclusions can be made about the distribution of the projectile ion kinetic energy. For example, the kinetic energy of the scattered product ions represents a small fraction of T, typically less than 10%. The value obtained for V is approximately 10–30% and has been shown to be linear over a wide range of collision energies. The remaining energy, 60–80%, is absorbed by the surface.

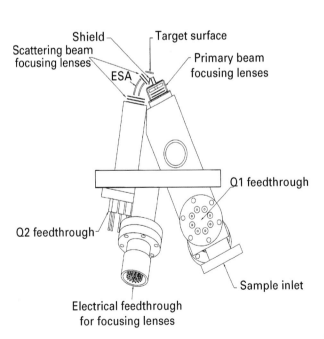

Figure 6 Overview of a tandem quadrupole SID instrument with a cross-sectional view of the target and ion focusing region. The ion optical element labelled ESA represents an electrostatic analyser. Reproduced with permission from Phelan LM, PhD Thesis, Princeton University, 1998.

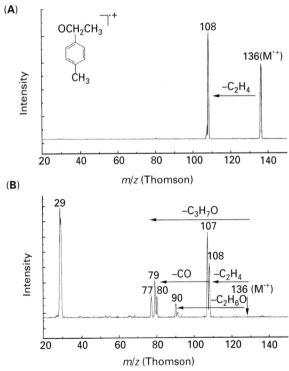

Figure 7 Surface-induced dissociation ion spectra of the p-methyl phenetole molecular ion (m/z 136) upon collision with an F-SAM surface at (A) 15 eV and (B) 30 eV. Reproduced with permission from Cooks RG and Miller SA, Collision of ions with surfaces. In: Jennings KR (ed) *NATO ASI Series*, Dordrecht, The Netherlands, Kluwer Academic Publishers, (in press).

Conversion of ion transitional energy into internal energy (T→V)

A quantitative variable used to judge the effectiveness of a collision ion activation technique is the translational to internal energy conversion efficiency or $T{\to}V\%$. The internal energy absorbed from a collision event is not single-valued, rather a distribution of internal energies is observed. Internal energy distributions can be measured using a thermometer molecule method which relies on a distinct set of dissociation pathways of known critical energy for a particular ion. The intensities of the fragment ions in the SID mass spectrum of the thermometer ion and the critical energies required to generate each fragment ion are used to calculate the internal energy distribution ($P(\varepsilon)$ versus internal energy, ε) deposited in the collision event. Metal carbonyls are often used due to their simple fragmentation pathway by consecutive losses of CO and well known critical energies. **Figure 8A** depicts the SID product ion spectrum resulting from 30 eV collisions of $W(CO)_6^{\bullet+}$ (352 Th) with an 8 µm thick liquid perfluoropolyether (PFPE) surface, $F[CF(CF_3)CF_2O]_{27(avg)}CF_2CF_3$. The internal energy distribution is then calculated from the mass spectrum in **Figure 8A** using the thermometer molecule method and the result is shown in **Figure 8B**. The distribution is approximately Gaussian in shape with a full width at half maximum (FWHM) of 4.4 eV and an average internal energy of 5.5 eV. The $T{\to}V$ conversion efficiency is calculated to be 18%. Although the values calculated in **Figure 8B** are for one particular ion/surface system, it is a general characteristic of SID to produce internal energy distributions which are approximately Gaussian in shape with FWHMs of approximately 4 eV.

Another method used to determine SID internal energy distributions is the deconvolution method. This method relies on the fact that the mass spectrum, breakdown curve (normalized fragment ion intensities as a function of internal energy), or internal energy distribution can be calculated if two of the three are known. For example, the mass spectrum can be calculated by convoluting the breakdown curve with the internal energy distribution. Since the benzene molecular ion has been well studied and breakdown curves have been measured, the internal energy distribution can be determined by deconvoluting the breakdown curve and the recorded SID mass spectrum of the benzene molecular ion. The deconvolution method has reported slightly higher $T{\to}V$ conversion efficiencies than the thermometer molecule method for similar surfaces. Small variations in the $T{\to}V$ conversion efficiency for a given

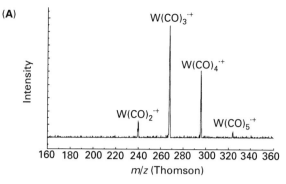

(A)

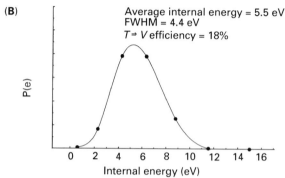

(B)

Average internal energy = 5.5 eV
FWHM = 4.4 eV
$T{\to}V$ efficiency = 18%

Figure 8 (A) Surface-induced dissociation mass spectrum produced from 30 eV collisions of $W(CO)_6^{\bullet+}$ (*m/z* 352) with an 8 µm thick liquid perfluoropolyether surface. (B) Calculated internal energy distribution for the spectrum in (A) using the thermometer molecule method. **Figure 8A** reproduced with permission from Pradeep T, Miller SA and Cooks RG (1993) Surface-induced dissociation from a liquid surface *Journal of American Society for Mass Spectrometry* **4**: 769–773.

surface have been noted with changes in the chemical composition of the projectile ion (see **Table 4**).

SID internal energy distributions are very different from those observed in CID. CID internal energy distributions are typically broad in nature and with uncharacteristic shapes. The narrow internal energy distribution found in SID gives a finer control over the energy that is deposited into an ion and thus more control over the extent of fragmentation than CID.

As a result of the high $T{\to}V$ conversion efficiencies in SID, large amounts of internal energy can be readily deposited into the projectile ion. For example, experiments using the benzene molecular ion have shown the ability of SID to impart more than 20 eV of internal energy. This large internal excitation and the high $T{\to}V$ conversion efficiency are due to the fact that in SID, the projectile ion cannot 'miss' its target, that is to say, the ion will always experience 'head-on' or small impact parameter collisions, as opposed to CID where ions are more likely to undergo larger impact parameter collisions causing less excitation.

Surface effects

Surface effects in ion/surface collision processes are very prominent and continue to be actively investigated. For example, surfaces covered with hydrocarbon pump oil were found to be more effective in reducing neutralization (or increasing secondary ion yield, that is, the ratio of the total secondary ion current to the primary ion current measured at the surface) than clean metal surfaces. In addition, it was also found that the organic nature of the surface affected the energetics of the ion/surface collision, especially the $T \rightarrow V$ conversion efficiency.

Table 4 Effect of target nature on the estimated translation-to-vibrational energy conversion[a]

Ion	Target (Co.Energy, eV)	$T \rightarrow V$(%)	
Cr(CO)$_6$	F-SAM (25)	19	
W(CO)$_6$	F-SAM (30)	18	Fluorine
W(CO)$_6$	F-SAM (50)	20	targets
W(CO)$_6$	PFPE (50)	18	
Ferrocene	F-SAM (20–50)	24 (20)[b]	
Benzene	F-SAM (10–70)	28	
Cr(CO)$_6$	D-SAM (25)	12	
Ferrocene	D-SAM (20–80)	15 (13)[b]	
Cr(CO)$_6$	Fc-SAM (25)	11	
n-Butylbenzene	Fc-SAM(25)	12	Hydrocarbon
Cr(CO)$_6$	SS (25)	11	like targets
W(CO)$_6$	SS (25–100)	13	
Fe(CO)$_5$	SS (25–60)	13	
Et$_4$Si	SS (10–100)	12	
Benzene	H-SAM (10–70)	17	
W(CO)$_6$	Alkenyl-terminated SAM	12	
	30–70		Other SAMs
Ferrocene	Amino-terminated SAM (25)	13	
Ferrocene	Cyano-terminated SAM (25)	12	
Ferrocene	Si(100) (20–90)	13	Single crystal
Pyridine	Ag(111) (32)	20	targets

[a] Where PFPE is perfluoropolyether, Fe is ferrocene, D is deuterated, and SS is a stainless steel surface with an adventitious hydrocarbon overlayer.
[b] Denotes that the ($T \rightarrow V$%) has been corrected for initial internal energy from the ionization event.

Prior to using organic covered surfaces, SID was performed on clean metal surfaces. The SID efficiency (ratio of the sum of total fragment ion current to the projectile ion current measured at the surface) of these ion/surface systems was often very poor, <1%. For example, collisions of the ferrocene molecular ion with a clean Si(100) surface at 10 eV result in an SID efficiency of 10^{-2}. This can be rationalized by examining the thermochemistry of an electron transfer reaction for the projectile ion and the surface. For example, a typical ionization energy for an organic compound is 10 eV and that of a clean metal surface is on the order of 3 or 4 eV (known as the work function of the surface). When the ion approaches a clean metal surface, the enthalpy of the electron transfer reaction (from the surface to the projectile ion) is very exothermic, 6–7 eV, and the neutralization of the projectile ion will be very efficient. However, when an organic layer covers the metal substrate, it changes the thermochemistry of the surface from a work function of a few eV to one that is of the order of the neutralization energy of the projectile ion. This change in thermochemistry reduces the exothermicity and extent of neutralization.

Self-assembled monolayer surfaces (SAMs) of n-alkane thiols on gold have become a surface of choice for studying the SID process. These surfaces offer an array of variability in the chemical functionality of the surface, they are stable in vacuum and air, can be well ordered, characterizable, and are relatively simple to fabricate. The most common of the SAM surfaces used in SID studies are the hydrocarbon SAM (H-SAM) and fluorocarbon SAM (F-SAM) surfaces shown in **Figure 2**. Continuing the thermochemical argument for the electron transfer process above, it would be expected that a fluorocarbon surface would yield higher SID efficiencies due to the larger ionization energy for fluorocarbon molecules, ~13 eV, making the electron transfer reaction endothermic for typical organic projectile ions. Note that although the reaction is endothermic, the translational energy of the projectile ion can be used to drive endothermic reactions. Studies have shown that fluorocarbon surfaces have higher total secondary ion yields than hydrocarbon surfaces, with some reports showing the F-SAM surface to have efficiencies of 55–75% as contrasted with 1–35% for H-SAM surfaces. The variation in secondary ion yields for a given SAM surface correlates with the neutralization energy of the projectile ion, that is, the overall thermochemistry of the electron transfer reaction.

The nature of the surface is an influential factor in the $T \rightarrow V$ conversion efficiency for a particular ion/surface pair. **Table 4** lists several thermometer molecules that have been used to determine the $T \rightarrow V$

conversion efficiency for several hydrocarbon and fluorocarbon surfaces. Fluorocarbon surfaces are more efficient at converting translational energy to internal energy than hydrocarbon surfaces due to a more rigid structure in the F-SAM surface and more massive end-groups. **Figure 9** illustrates the greater effectiveness of the F-SAM surface at dissociating the pyrazine molecular ion than a perdeuterated SAM (D-SAM) surface at the same ion kinetic energy of 25 eV. The F-SAM surface also exhibits a greater signal-to-noise ratio and SID efficiency as a result of reduced neutralization. Note that the pyrazine molecular ion is reactive with the D-SAM surface producing the ion/surface reaction product, $[M+CD_3]^+$, and the ion/surface reaction fragment $[M+D-HCN]^+$. It is interesting to note that the $T \rightarrow V$ conversion efficiency of the liquid PFPE film and an F-SAM surface are almost identical (similar chemical composition) but their physical state of matter is different. There is very little difference in SID and ion/surface reactions between the liquid PFPE and the F-SAM surface for a large variety of projectile ions, supporting the hypothesis that the ion only interacts with the outer layer of the surface.

Applications

Surface-induced dissociation produces structurally diagnostic ions similar to those found in the analogous gas phase CID experiment. For example, **Figure 10A** shows the SID mass spectrum of the pyrene molecular ion at a collision energy of 100 eV with an F-SAM surface. The spectrum displays significant fragmentation by diagnostic C_2 losses. **Figure 10B** displays the CID of pyrene at 100 eV under single collision conditions with Ar and only produces two minor fragment ions. Note that this comparison of SID to CID is made based on specific experimental conditions, namely single collision events for CID. CID is often performed under multiple collision conditions which imparts a larger average internal energy deposition yielding more diagnostic fragment ions. One must be careful when comparing SID and CID since the experimental conditions can greatly bias the observed results. Nonetheless, both ion activation methods provide useful information about ion structure.

Fragment ions produced by SID have been used to distinguish gaseous isomeric ions and to elucidate the structure of biomolecules. The unique characteristics of SID, narrow internal energy distribution and variable control of internal energy, make it a powerful tool for studying isomeric ions. Before low energy ion/surface collisions were used as a means of ion activation, isomeric ion distinction was performed

by mass spectrometric techniques such as CID, charge exchange, and kinetic energy release measurements, all of which used high energy collisions. Isomers of $C_5H_6^+$, $C_6H_6^{\bullet+}$, $C_6H_6^{2+}$, $C_2H_3S^+$, $C_2H_4O^{\bullet+}$, and $C_3H_4^{\bullet+}$ have been distinguished by low energy SID. It is worth noting that ion/surface reaction products can serve as a complementary tool when studying isomers in the gas phase due to the often selective nature of ion/surface reactions.

Activation and dissociation of large biomolecule ions is currently the focus of many investigators interested in analytical applications of SID. As noted in the introduction, the need to activate and dissociate increasingly larger molecules is a constant challenge in tandem mass spectrometry. SID offers the advantage of large $T \rightarrow V$ conversion efficiencies and narrow internal energy distributions. While neutralization in SID can pose a problem, improvements are made by choosing a surface which minimizes the extent of neutralization. Thus, fluorocarbon surfaces are often used in SID biomolecule applications. SID mass spectra of peptides yield results consistent with low energy CID, that is, the formation of peptide backbone fragments such as a-, b-, and y-type ions. In addition to backbone cleavages, some peptides exhibit side chain specific cleavage ions, d- and w-type, in low energy SID experiments which are normally characteristic of high collision energy (keV) CID data. **Figure 11** illustrates the diagnostic fragments resulting from collisions of the doubly protonated melittin ion with an F-SAM surface at 110 eV. The spectrum consists of mainly backbone cleavage fragments (a-, b-, and y-type) with nearly a complete y-type series from y_2 to y_{25}. A complete primary sequence of the melittin ion, with the exception of two residues, can be determined from this single SID mass spectrum. These experiments are also lending insight to the energetics and fragmentation mechanisms of peptides due to the controllable nature of the internal energy deposition characteristic of the SID method.

Surface-induced dissociation mechanisms

In most cases, polyatomic fragment ions noted in SID are similar to those found in CID and can be explained by unimolecular fragmentation pathways. This general observation has led to the prediction of a two-step SID mechanism involving the collision which causes internal excitation followed by a delayed fragmentation some distance and time away from the collision event. This concept for the mechanism of SID of polyatomic *organic* ions is fairly well accepted, although one recent study shows evidence for dissociation at the surface interface in a prompt

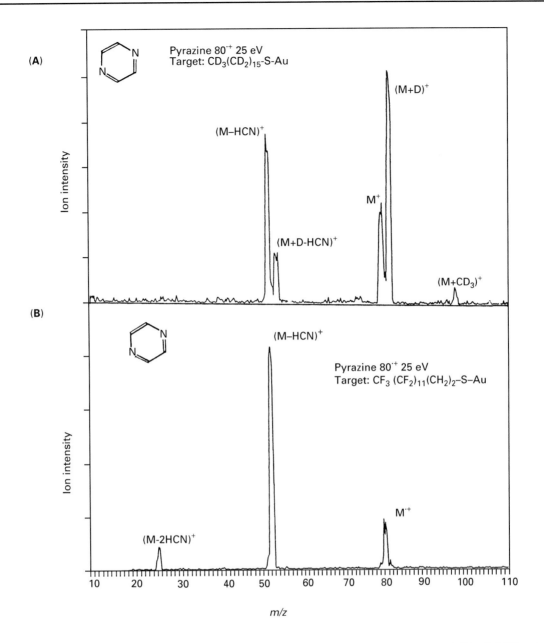

Figure 9 Product ion spectra resulting from 25 eV collisions of the pyrazine molecular ion (*m/z* 80) with (A) a D-SAM surface and (B) an F-SAM surface. Reproduced with permission from Winger BE, Laue H-J, Horning *et al*, (1992) Hybrid BEEQ tandem mass spectrometer for the study of ion/surface collision processes *Review of Scientific Instruments* **63**: 5613–5625.

or 'shattering' mechanism for a specific ion/surface pair.

SID mechanisms are typically interrogated by monitoring the kinetic energy or velocity of the scattered product ions. If delayed unimolecular fragmentation is occurring, then the velocity of the scattered fragment ions will be the same as the scattered projectile ion. If a prompt or shattering mechanism is occurring then the scattered product ions will leave the surface with different velocities or rather the same kinetic energy. At this early stage of investigation, it is not

clear if the prompt dissociation mechanism is only specific for certain ion/surface pairs or if both mechanisms commonly occur but to varied extents.

Other ion/surface processes

The examples discussed thus far have been relatively 'clean' examples of SID without the added complexity of peaks resulting from chemical sputtering and ion/surface reactions. There is a growing literature based on observations of chemical sputtering, ion/surface reactions, and ion/surface reactions which

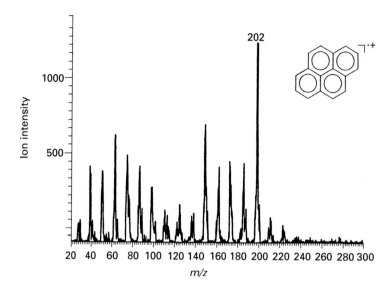

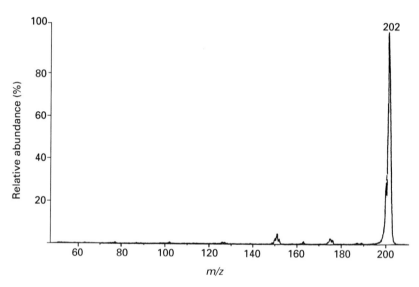

Figure 10 (A) Surface-induced dissociation spectrum of the pyrene molecular ion (*m/z* 202) with a stainless steel surface at a collision energy of 100 eV. (B) Collision-induced dissociation mass spectrum of the pyrene moelcular ion (*m/z* 202) with Ar under single collision condiitons. Note that the *m/z* axis is not aligned or the same scale between (A) and (B). Adapted with permission fron Riederer Jr DE, PhD Thesis, Purdue University, 1993.

lead to selective surface modification in low energy ion/surface collisions. In order to obtain a better understanding of SID, it is important to have an understanding of all of the ion/surface collision processes shown in **Table 1**.

Ion/surface reactions The organic covered surface can be thought of as a chemical reagent in a reaction that uses the projectile ion as the second chemical reagent. **Figure 12** shows the SID mass spectrum for collisions of $(CH_3)_2$ $SiNCS^+$ with an F-SAM surface at 50 eV. Peak assignments can be found in **Table 5**.

This spectrum displays a wealth of peaks originating from various ion/surface collision processes including SID, ion/surface reaction, and chemical sputtering. The number and abundance of ion/surface reaction products (especially Si containing product ions) for this ion/surface pair may appear to be remarkable for what would intuitively be thought of as a relatively inert fluorocarbon surface. Indeed, it has been found that the F-SAM surface is very reactive toward organometallic, metallic, and a few organic ions.

Two distinct ion/surface reaction mechanisms have been elucidated. One ion/surface reaction mechanism

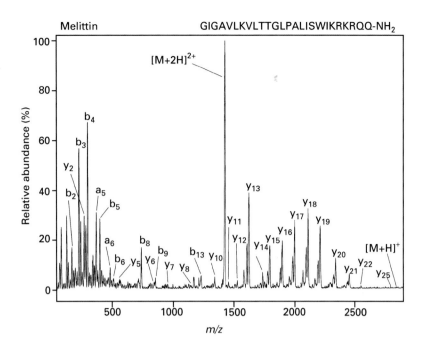

Figure 11 Surface-induced dissociation spectrum of doubly protonated melittin obtained at a 110 eV collision with a fluorinated SAM surface. Reproduced with permission from Dongré AR, Somogyi Á and Wysocki VH (1996) Surface-induced dissociation: an effective tool to probe structure, energetics and fragmentation mechanisms of protonated peptides *Journal of Mass Spectrometry* **31**: 339.

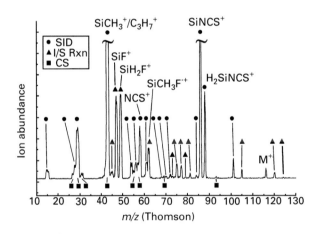

Figure 12 50 eV ion/surface collision spectrum of $(CH_3)_2SiNCS^+$ on an F-SAM surface. Note that the ion abundance scale has been expanded. For reference, the $M^+/SiNCS^+$ ratio is 0.05 and the $M^+/SiCH_3^+$ ratio is 0.04. Peak assignments are noted in **Table 5**. Reproduced with permission from Miller SA, Luo H, Jiang X, Rohrs HW and Cooks RG (1997) Ion/surface reactions, surface-induced dissociation, and surface modification resulting from hyperthermal collisions of $OCNCO^+$, $OCNCS^+$, $(CH_3)_2SiNCO^+$, and $(CH_3)_2SiNCS^+$ with a fluorinated self-assembled monolayer. *International Journal of Mass Spectrometry and Ion Processes* **160**: 83–105.

occurs by direct abstraction or oxidative addition of an atom or group of atoms from the surface by the projectile ion. This type of mechanism is most often observed for metal-containing ions and fluorocarbon

surfaces. For example, collisions of W^+ with an F-SAM surface result in ion/surface reaction products of the form, WF_n^+ ($n = 1–5$). These reaction products have been shown to occur via the direct abstraction process. The second mechanism occurs by a two-step process of (i) an electron transfer from the surface to the projectile ion and (ii) an associative ion/molecule reaction between the neutralized projectile and a chemically sputtered ion from the surface. This mechanism is typically observed for hydrocarbon projectile ions and hydrocarbon surfaces. An example of ion/surface reaction which occurs via an associative ion/molecule reaction is shown in **Figure 13**. The neutralized d_6-benzene, C_6D_6, reacts with a sputtered proton from the H-SAM surface to form the $[M+H]^+$ product ion. This example also illustrates the advantage of using an isotopically labelled projectile ion. Peak assignments for SID and ion/surface reaction products are less ambiguous than if the $C_6H_6^{\bullet+}$ molecular ion was used as the projectile. It is worth noting that both ion/surface reaction mechanisms have been indirectly supported by the corresponding gas phase ion/molecule experiments.

Chemical sputtering Chemical sputtering has been shown to be a very sensitive (submonolayer) surface analysis technique when using appropriate sputtering reagents in low energy ion/surface collisions. The chemical sputtering peaks originating from the

F-SAM surface in **Figure 12** are of low abundance but can be significantly enhanced by using the xenon radical cation which is an efficient charge transfer reagent for fluorocarbon surfaces. For example, **Figure 14** depicts the chemical sputtering spectrum of a fresh F-SAM surface and the liquid PFPE surface. Abundant chemical sputtering peaks representative of the chemical composition of each surface are labelled in the figure. This experiment is also sensitive toward hydrocarbon impurities in the F-SAM surface at m/z 27 ($C_2H_3^+$), m/z 29 ($C_2H_5^+$), m/z 39 ($C_3H_3^+$), m/z 41 ($C_3H_5^+$), and m/z 43 ($C_3H_7^+$). As noted earlier, the F-SAM and liquid PFPE surface produce very similar SID mass spectra. However, chemical sputtering experiments can differentiate the two, due to the characteristic oxygenated fluorocarbon peak at m/z 47, CFO^+, in the liquid PFPE spectrum. These experiments also suggest that the liquid/vacuum interface microscopic structure is similar to that of the F-SAM surface and the projectile ion only interacts with the outer portion of the surface at these low energies. Note that chemical sputtering is a low energy (reactive) collision which differs from the more common secondary ion mass spectrometry technique that uses keV ion beam (momentum transfer) collisions to analyse surface composition.

Surface modification   Selective chemical modification of surfaces is also a consequence of low energy

ion/surface collisions. For example, one of the many ion/surface reaction products seen in **Figure 12** is of particular interest, $CH_3FSiNCS^+$ (see also **Table 5**). This reaction product may be formed by the following group transfer reaction,

$$(CH_3)_2SiNCS^+ + F-CF_2\text{-(surface)}$$
$$\rightarrow CH_3FSiNCS^+ + CH_3-CF_2\text{-(surface)}$$

where $F-CF_2$-(surface) represents the outer CF_3 head group on an F-SAM surface. This reaction would yield a chemically modified surface which would be difficult to fabricate using common surface preparation methods. Xenon chemical sputtering has shown that the $(CH_3)_2SiNCS^+$ projectile ion modifies the F-SAM surface by incorporating NCS, methyl, and silyl groups into the fluorocarbon chain. Other examples of chemical surface modification involving group transfer reactions or transhalogenation in low energy ion/surface collisions include the interaction of $OCNCS^+$, $SiCl_4^{\bullet+}$, $CH_2Br_2^{\bullet+}$, and CH_2Br^+ with F-SAM surfaces.

Soft landing   Finally, soft landing of polyatomic ions into self-assembled monolayers can be achieved using appropriate projectile ions with low ion kinetic

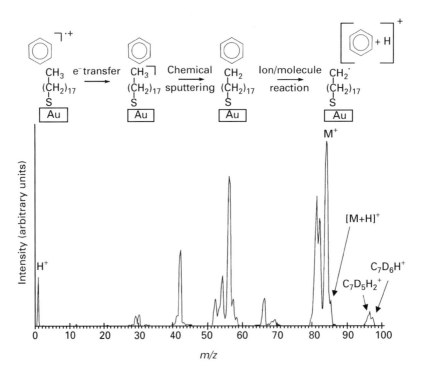

Figure 13   Product ion spectrum resulting from collisions of the d_6-benzene molecular ion (m/z 84) with an H-SAM surface at 30 eV. The inset represents the associative ion/surface reaction mechanism for the unlabelled benzene molecular ion.

Table 5 Peak assignments for $(CH_3)_2SiNCS^+$ from an F-SAM surface at a collision energy of 50 eV.

Mass	SID	Mass	Ion/surface reaction	Mass	Chemical sputtering
m/z 101	$CH_3SiNHCS^{\bullet+}$	m/z 124	F_2SiNCS^+	m/z 93	$C_3F_3^+$
m/z 88	H_2SiNCS^+	m/z 120	$CH_3FSiNCS^+$	m/z 81[a]	$C_2F_3^+$
m/z 86	$SiNCS^+$	m/z 105	$FSiNCS^+$	m/z 69	CF_3^+
m/z 84	$C_2H_6SiNC^+$	m/z 81[a]	$F_2SiCH_3^+$	m/z 57[a]	$C_4H_9^+$
m/z 72	$C_2H_6SiN^+$	m/z 79	F_2SiCH^+	m/z 55[a]	$C_4H_7^+$
m/z 70	$C_2H_4SiN^+$	m/z 77	$NCSF^+$ and $FSi(CH_3)_2^+$	m/z 43[a]	$C_3H_7^+$
m/z 68	$C_2H_2SiN^+$	m/z 75	$FSiC_2H_4^+$	m/z 31	CF^+
m/z 61[a]	$HSiS^+$	m/z 73	$FSiC_2H_2^+$	m/z 29[a]	$C_2H_5^+$
m/z 58	NCS^+ and $Si(CH_3)_2^{\bullet+}$	m/z 62	$FSiCH_3^{\bullet+}$	m/z 27	$C_2H_3^+$
m/z 57[a]	$SiC_2H_5^+$	m/z 61[a]	$FSiCH_2^+$		
m/z 56	$SiC_2H_4^{\bullet+}$	m/z 49	H_2SiF^+		
m/z 55[a]	$SiC_2H_3^+$	m/z 47	SiF^+ and FCO^+		
m/z 54	$SiC_2H_2^{\bullet+}$	m/z 45	$NCF^{\bullet+}$		
m/z 43[a]	$SiCH_3^+$				
m/z 29[a]	SiH^+				
m/z 28	Si^+				
m/z 15	CH_3^+				

[a] Indicates isobaric masses which can have more than one assignment. See **Figure 12** for a scattered product ion spectrum. Reproduced with permission from Miller SA, Luo H, Jiang X, Rohrs HW and Cooks RG (1997) *International Journal of Mass Spectrometry and Ion Processes* **160**: 83–105.

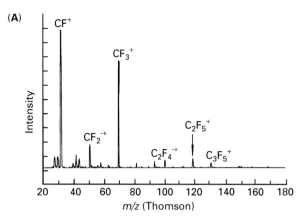

(A)

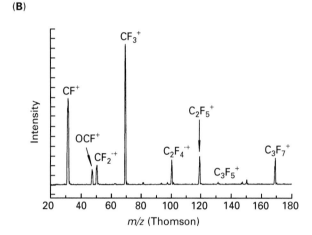

(B)

Figure 14 90 eV $Xe^{\bullet+}$ chemical sputtering spectra for (A) and F-SAM surface and (B) an 8 μm thick liquid perfluoropolyether surface. Reproduced with permission from Pradeep T, Miller SA, Rohrs HW, Feng B and Cooks RG (1995) Chemical sputtering of F-SAM and liquid PFPE. *Materials Research Society Symposium Proceedings* **380**: 93–98.

energies. The projectile ion used in **Figure 12**, $(CH_3)_2SiNCS^+$, can be soft landed in a F-SAM surface by lowering the ion kinetic energy below 10 eV. The soft-landed ion can be subsequently liberated into the gas phase by $Xe^{\bullet+}$ chemical sputtering. By using chemical sputtering experiments and temperature programmed desorption, the soft-landed species has been shown to survive as an ion in the F-SAM matrix for several days in vacuum and a few days in ambient air. Although these results have only recently been published, this approach may have exciting implications for both fundamental and applied studies.

List of symbols

KE_{final} = final kinetic energy; $P(\varepsilon)$ = internal energy distribution; Q = inelasticity of collision; T = initial kinetic energy (ion translational energy); V = total vibrational energy; ε = internal energy; $\varepsilon_{surface}$ = internal energy deposited in surface.

See also: **Fragmentation in Mass Spectrometry; Ion Collision Theory; Ion Energetics in Mass Spectrometry; Ion Molecule Reactions in Mass Spectrometry; Ion Structures in Mass Spectrometry; Ion Trap Mass Spectrometers; MS–MS and MS^n; Quadrupoles, Use of in Mass Spectrometry; Sector Mass Spectrometers; Time of Flight Mass Spectrometers.**

Further reading

Ada ET, Kornienko O and Hanley L (1998) Chemical modification of polystyrene surfaces by low-energy polyatomic ion beams. *Journal of Physical Chemistry B* **102**: 3959.

Busch KL, Glish GL and McLuckey SA (1993) *Mass Spectrometry/Mass Spectrometry: Techniques and*

Applications of Tandem Mass Spectrometry. New York: VCH Publishers.

Cooks RG, Ast T, Pradeep T and Wysocki VH (1994) Reactions of ions with organic surfaces. *Acc. Chemistry Research* 27: 316.

Cooks RG and Miller SA. *Collisions of Ions with Surfaces.* In: Jennings KR (ed) *NATO ASI Series.* Dordrecht: Kluwer (in press).

Cooks RG, Ast T and Mabud MA (1990). Collisions of polyatomic ions with surfaces. *International Journal of Mass Spectrometry and Ion Processes* 100: 209.

Dongre AR, Somogyi A and Wysocki VH (1996) Surface-induced dissociation: an effective tool to probe structure, energetics and fragmentation mechanisms of protonated peptides. *Journal of Mass Spectrometry* 31: 339.

Hanley L (ed) (1998) Special issue on polyatomic ion-surface interactions. *International Journal of Mass Spectrometry and Ion Processes* 174: 1.

Hayward MJ, Park FDS, Phelan LM, Bernasek SL, Somogyi A and Wysocki VH (1996) Examination of

sputtered ion mechanisms leading to the formation of $C_7H_7^+$ during surface induced dissociation (SID) tandem mass spectrometry (MS/MS) of benzene molecular cations. *Journal of the American Chemical Society* 115: 8375.

Mabud MA, Dekrey MJ and Cooks RG (1985) Surface-induced dissociation of molecular ions. *International Journal of Mass Spectrometry and Ion Processes* 67: 285.

Miller SA, Luo H, Cooks RG and Pachuta SJ (1997) Soft-landing of polyatomic ions at fluorinated self-assembled monolayer surfaces. *Science* 275: 1447.

Morris MR, Riederer DE Jr., Winger BE, Cooks RG, Ast T and Chidsey CED (1992) Ion/surface collisions at functionalized self-assembled monolayer surfaces. *International Journal of Mass Spectrometry and Ion Processes* 122: 181.

Vekey K, Somogyi A and Wysocki VH (1995) Internal energy distribution of benzene molecular ions in surface-induced dissocation. *Journal of Mass Spectrometry* 30: 212.

Surface Plasmon Resonance, Applications

Zdzislaw Salamon and **Gordon Tollin**,
University of Arizona, Tucson, AZ, USA

SPATIALLY RESOLVED SPECTROSCOPIC ANALYSIS
Applications

As described in the article on the theory of surface plasmon resonance, surface plasmons create a surface-bound evanescent electromagnetic wave which propagates along the surface of an active medium (usually a thin metallic film), with the electric field intensity maximized at this surface and diminishing exponentially on both sides of the interface. As a consequence of this property, the phenomenon has been utilized extensively in studies of surfaces and of thin dielectric films deposited on the active medium. Although numerous other optical techniques have also been applied to such systems (e.g. ellipsometry, interferometry, spectrophotometry, and microscopy the surface plasmon resonance (SPR) method has some important advantages over all other optical techniques, as follows. The method utilizes a relatively simple optical system, it has a superior sensitivity, and the complete system of measurement is located on the side of the apparatus that is remote from the sample, and thus there is no optical interference from the bulk medium. Furthermore, the surfaces of the sample need no extra treatment to increase reflectivity, because this is achieved by using total reflectance. Additional benefits include the fact that there are three parameters of the resonance that can readily be measured, thereby yielding much more information about the sample and changes within it than the simple interferometric step height used in other sensitive optical techniques. Finally, the recently developed new variant of SPR, which involves both plasmon and waveguide modes producing coupled plasmon waveguide resonance (CPWR), expands spectroscopic sensitivities and capabilities even further, allowing the measurement of anisotropies in both the refractive index and extinction coefficient.

Current interest in the properties of surfaces and surface coatings arises partly from increased applications of thin film devices, including the large field of integrated optics, which uses surface guided waves, and partly from recent developments in biosensors. In addition, many surface phenomena depend on molecular interactions occurring at dielectric interfaces. Surface chemistry applications include charge transfer interactions, acid–base chemistry, chemical bond formation, and van der Waals forces, all of which participate in the processes of adhesion, catalysis, lubrication, corrosion, contamination and packaging. Surface and interfacial phenomena are also of particular importance in all areas of biology, where molecular phenomena occurring at lipid membrane interfaces or during protein–protein interactions are key events in living systems.

This article focuses on the application of the two major properties of the surface plasmon (SP) evanescent wave (i.e. its identity as a surface-bound and surface-unique electromagnetic phenomenon), to the characterization of the properties of surfaces, interfaces, and thin films. It describes the two principal experimental modes, kinetics and spectroscopy, which are currently used in these SPR applications.

Principles of SPR applied to thin films

As described in the article on the theory of surface plasmon resonance, plasmons are transverse magnetic (TM; p-polarized) evanescent waves generated at the expense of light energy under total internal reflection conditions, which propagate along an active medium surface (usually silver or gold) with their field amplitudes decaying exponentially perpendicular to the metal surface. They obey the well-known dispersion relation, and can only be excited when matching energy and momentum conditions between photons and surface plasmons are fulfilled, as follows:

$$K_{SP} = K_{ph} = (\omega/c)\varepsilon_0^{1/2} \sin \alpha_0 \qquad [1A]$$

where:

$$K_{SP} = \omega/c \, (\varepsilon_1\varepsilon_2/\varepsilon_1 + \varepsilon_2)^{1/2} \qquad [1B]$$

is the longitudinal component of the SP wave vector, K_{ph} is a component of the light wave vector parallel to the active medium surface, ω is the frequency of the surface plasmon excitation wavelength, c is the velocity of light *in vacuo*, ε_0, ε_1, and ε_2 are the complex dielectric constants for the incident, surface active, and dielectric (or emerging) media, respectively, and α_0 is the incident coupling angle.

As can be seen, the resonance condition stated in Equation [1A] can be fulfilled by either changing the incident angle, α, at a constant value of photon energy, $h\omega$ (i.e. maintaining a constant value of the light wavelength, $\lambda = \lambda_0$; see **Figures 1A** and **1C**, curve 1), or varying λ at a constant value of the incident angle, $\alpha = \alpha_0$ (see **Figures 1B** and **1D**, curve 1). This means

that each photon of energy $h\omega$ allows the excitation of exactly one SP mode at a distinctive value of the incident angle, α, within a specified metallic layer (silver or gold). Furthermore, any alteration in the optical parameters (described by the complex dielectric constant, ε, in Equation [1B], which contains the refractive index, r, and extinction coefficient, k) of the metal–emerging medium interface, will affect the K_{SP} value and therefore change the resonance characteristics, as indicated by Equation (1A). Thus, deposition of any thin dielectric film (sensing layer) on a metallic surface introduces changes in the optical parameters of the metal–emerging medium interface, thereby causing a shift of the SP wave vector to a larger value:

$$K'_{SP} = K_{SP} + \Delta K_{SP} \qquad [2]$$

According to Equation [1A], such a shift of the SP wave vector moves the resonance to either a higher incident angle, α_1, at a constant value of exciting light wavelength, λ_0, or a longer exciting light wavelength, λ_1, at a constant value of the incident angle α_0, as shown in all parts of **Figure 1** (curves 2).

The spectra presented in **Figure 1** demonstrate that the angular (or wavelength) position and shape of the resonance curve is very sensitive to the optical properties of both the metal film and the emergent dielectric medium adjacent to the metal surface. This property is one of three major attributes of SPR which constitute the foundation for the application of surface plasmons (via their interactions with the medium in which they propagate) as an optical probe of the properties of any material in contact with an active layer capable of generating this effect. The other two attributes can be stated as follows: (i) The evanescent electromagnetic field generated by the free-electron oscillations is strongly enhanced as compared with the electromagnetic field of the exciting light, and reaches its maximum at the metal–emerging dielectric medium interface. The enhancement of the field at the metal–dielectric interface magnifies the spectral features of the interface, making optical measurements with the SP electromagnetic waves possible. (ii) The evanescent electromagnetic field generated in the film decays exponentially with penetration distance into the emerging medium, i.e. the depth of penetration into the dielectric material in contact with an active metal layer extends only to a fraction of the light wavelength used to generate the plasmons. This makes the phenomenon sensitive *only* to the metal–dielectric interface region, without any interference from the bulk volume of the dielectric material or any medium that is in contact with it.

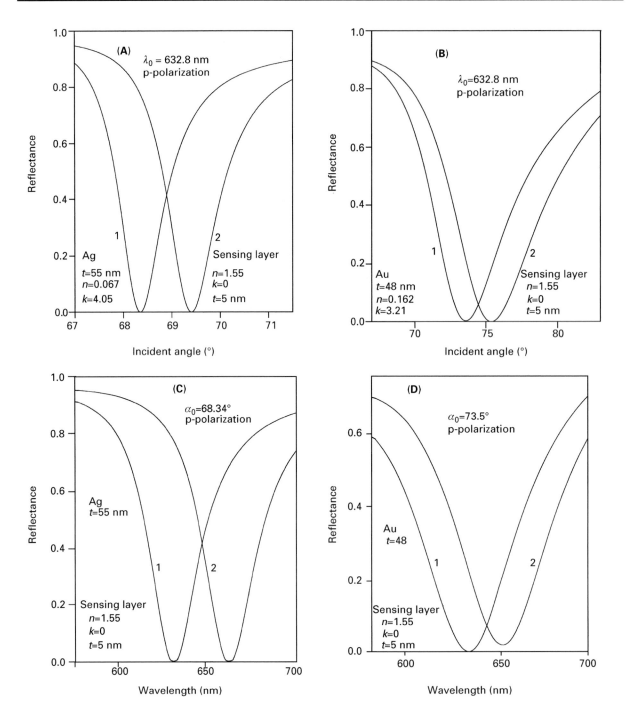

Figure 1 Influence on the theoretical SPR spectra generated either with silver (panels A and C) or gold (panels B and D) films (curves 1 in all panels), and represented by reflectance vs. either incident angle (panels A and B), or excitation light wavelength (panels C and D), of a thin sensing layer deposited on the metallic surface (curves 2 in all panels). The incident and emerging media are glass (n_g = 1.5151) and water (n_w = 1.33), respectively.

As a consequence of these characteristics, SPR is ideally suited to probe a few nanometres from the metal surface, a distance well below the wavelength of the light used to generate the plasmons. In essence, the surface plasmon technique can be likened to a multiple-beam interferometer, with one narrow

reflected fringe, and is therefore capable of similar sensitivity. In contrast, however, it allows optical measurements to be made of dimensions that are one-thousandth of a wavelength or less, which is well below the sensitivity limits of the other optical techniques (e.g. ellipsometry, spectrophotometry, and various forms of microscopy) used in studies of surfaces and thin films.

Analysis of SPR spectra

As illustrated above, SPR spectra can be presented either in the form of reflectance (R) as a function of incident angle (α):

$$R = f(\alpha)_{\lambda_0} \qquad [2A]$$

or as a function of the excitation light wavelength (λ):

$$R = f(\lambda)_{\alpha_0} \qquad [2B]$$

Such spectra incorporate information about the following three parameters, refractive index, n, extinction coefficient, k, and thickness, t, which describe the optical properties of both the metal and the dielectric layers. The influence of these parameters on the angular (or wavelength) *position*, the angular (or wavelength) *width*, and the *depth* of the SPR spectrum is completely contained in the characteristic matrix of the thin film assembly, which allows a determination of their values.

Structural analysis of thin dielectric films

As demonstrated in **Figures 2** and **3**, the above-mentioned three optical parameters of a thin dielectric film deposited on a metal surface are well separated in their effects on the SPR spectra. Therefore, the experimental spectra interpreted in the context of the characteristic matrix of the assembly allow a unique evaluation of n, k, and t. The evaluation procedure is based on fitting a theoretical resonance curve to the experimental one. As also indicated in **Figures 2** and **3**, the CPWR method provides a means for determining the optical parameters using both TM (p) and transverse electric (TE) (s) polarizations of the excitation light, resulting in two values of the refractive index (n_p and n_s) and two values of the extinction coefficient (k_p and k_s). These parameters can then be used to calculate the anisotropy of n and k, thereby describing the degree of both molecular order (by the anisotropy in n) and orientation of chromophoric groups attached to the molecules comprising the thin film (by the anisotropy in k). Such information, taken together with the film thickness (t), provides insights into the microscopic structure of the film. Furthermore, the optical parameters can also be employed to calculate the mass of a deposited thin layer (see next section for details).

It has to be emphasized that all of these characterizations can be obtained using a single device (in this case, CPWR), containing a metal film covered with a dielectric layer, and using a measurement method that involves only a determination of reflected light intensity under total reflection conditions as a function of either incident angle or light wavelength.

Determination of thin film mass

As noted above, the values of the n, k, and t parameters of a thin film layer deposited on the surface of a metal film contain information about the amount of material in the layer. There are two different ways of calculating the adsorbed mass from the refractive index value. The first approach is based on the assumption that the refractive index increment, dn/dC, is independent of the concentration, C, of the adsorbed substance. If this is so, the surface density, D_s, i.e. the amount of material per unit surface area, can be evaluated by the following expression:

$$D_s = t\left[(n - n_2)/(dn/dC)\right] \qquad [3]$$

where n and n_2 are refractive indices of the adsorbed thin film and the emerging medium, respectively, and dn/dC is the refractive index increment of the adsorbed substance. Equation [3] can of course only be used if the refractive index increment is known, and depends on the assumption that the value of the increment is constant over the concentration range of the adsorbed material.

The second method of mass calculation is based on the Lorentz–Lorenz relation which can be presented in the most general case, i.e. when the deposited layer contains a mixture of substances, by the following equation:

$$(n^2 - 1)/(n^2 + 2) = A_1 N_1 + A_2 N_2 + \dots \qquad [4]$$

where A_i and N_i are the molar refractivity and the number of moles of substance per unit volume, respectively. For a pure substance, a mass density, D, defined as mass per unit volume of adsorbed material, can be directly related to the refractive index by the following equation:

$$D = MN = M/A\left[(n^2 - 1)/(n^2 + 2)\right] \qquad [5]$$

For an adsorbed layer of thickness t formed from such a pure substance, the above equation can be used to calculate the adsorbed mass (M) in µg per cm², as follows:

$$M = Dt = 0.1 \, M/A \, \{t[(n^2 - 1)/(n^2 + 2)]\} \qquad [6]$$

where the thickness is expressed in nanometres. Such a simple mass calculation becomes more complicated when a surface layer is formed from a mixture of substances, as it often is in real measurements. This complexity can, however, still be dealt with, depending upon the specific experimental conditions.

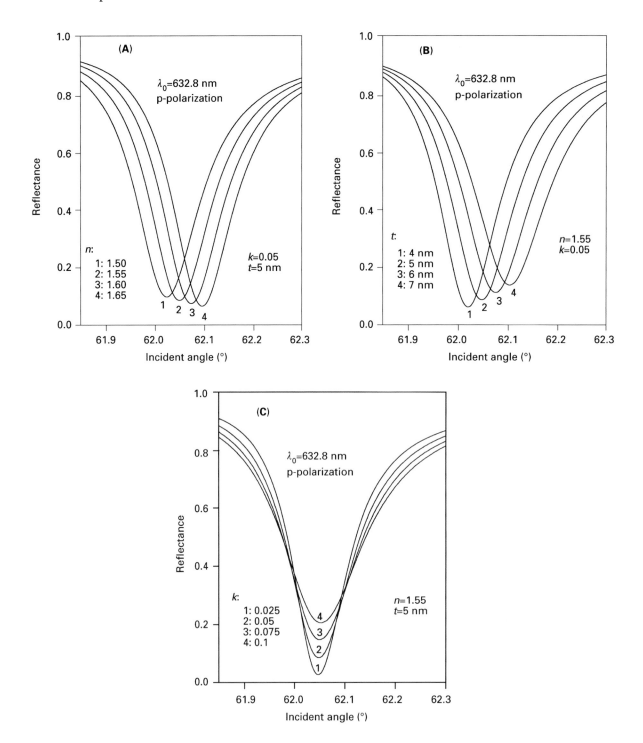

Figure 2 Changes in the *p*-polarization component of a theoretical SPR spectrum generated with a CPWR device, comprising a silver layer coated with a 460 nm SiO_2 film, caused by alterations in either the refractive index (A), the thickness (B), or the extinction coefficient (C), of a light-absorbing dielectric sensing film deposited on the silica film. The incident and emerging media are both as described in **Figure 1**.

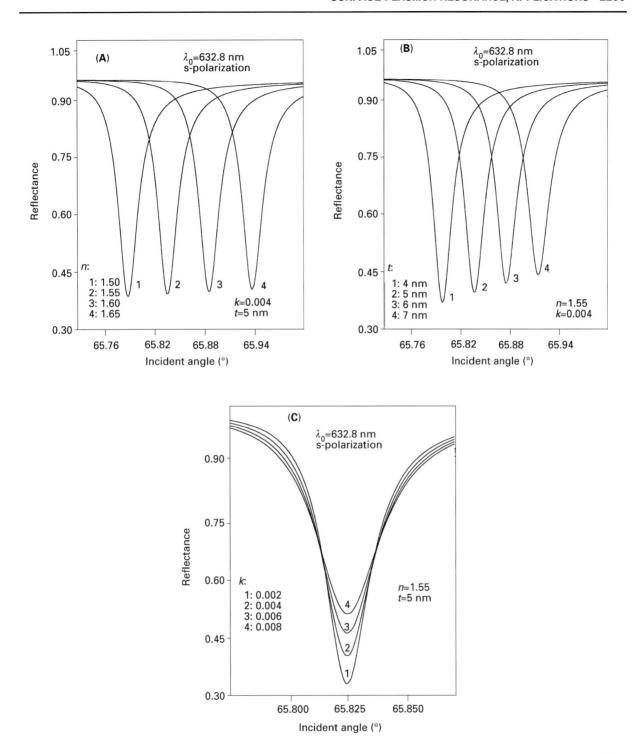

Figure 3 Changes in the *s*-polarization component of a theoretical SPR spectrum obtained with the CPWR device described in **Figure 2**, by alterations in either the refractive index (A), the thickness (B), or the extinction coefficient (C), of a light-absorbing dielectric sensing film deposited on the silica. The incident and emerging media are the same as in **Figure 1**.

Applications of the SPR technique

The SPR measurement can be performed, as with other optical measurements, in two modes. Used in a spectroscopic mode, i.e. by measuring and analysing the entire resonance spectrum, SPR provides a tool for the characterization of the microstructural properties of thin films, including thickness, mass distribution (packing density) within the film, and degree of order (orientation) of molecules and of chromophore groups attached to the molecules within the film. In addition, when combined with emission measurements, i.e. emission excited by the surface plasmon electromagnetic field, it can also provide a *molecular orientation distribution*.

When used in a kinetic mode i.e. by measuring and analysing the changes of reflectance as a function of time at $\alpha = \alpha_0 = $ const., and $\lambda = \lambda_0 = $ const.:

$$R = f(time)_{\alpha_0, \lambda_0} \qquad [7]$$

this experimental approach provides a means for real-time analysis of molecular interactions.

Steady-state and kinetic SPR measurements

As a consequence of the above-mentioned characteristics, SPR is ideally suited to studying both structural and mass changes of thin films, including molecular interactions occurring at surfaces and interfaces. These can be examined using either steady-state or time-resolved SPR spectroscopy, and can be applied to a wide range of materials. The time-resolved mode expands the capability of SPR techniques to allow probing of the dynamics of structural and mass alterations of thin films.

Although the phenomenon initially has been utilized extensively by physical scientists in studies of the properties of surfaces and thin films, with a major goal of creating thin film-based opto-electronic devices for optics, spectroscopy, laser optics, solar energy conversion, and space technology, the implementation of this technology to chemical, and especially to biological processes, has proven to be one of the most dynamic and fruitful application areas during recent years. The molecular mechanisms underlying biological and chemical processes are dependent on direct interactions between (bio)molecules. These interactions are often preceded by specific binding between two or more molecules, and for binding to occur the molecules must be able to come close enough to each other to make contact. Such binding may take place between the partners in solution, or with at least one partner attached to a biological surface (e.g. a lipid membrane or a chromosome) or to a physical surface. In situations in which the interactions occur at a surface, there must be a mechanism, such as lateral diffusion in the surface plane, that permits them to come close enough for interaction. These binding processes will not only result in alterations in the mass and composition of the membrane or other surface, but may also cause structural changes within such entities as well.

The simplest case of a biomolecular (or chemical) interaction is the association of two molecules, X

and Y, to form a complex XY. This process is characterized by rate constants for association, k_{ass}, and dissociation, k_{diss}, and an equilibrium (binding or binding affinity) constant, $K_B = k_{ass}/k_{diss}$. Steady-state SPR measurements, by determining mass density changes occurring in molecular assemblies accompanying binding interactions, allow an evaluation of the binding constants K_B. Kinetic SPR measurements enable direct observation in real time of such binding events, resulting in determination of association and dissociation rate constants (as discussed in the next section), as well as the dynamics of binding-induced structural changes.

The ability of SPR to probe both kinetic and thermodynamic processes, as well as to provide microstructural information, make it a very important component of the experimental methodology available to probe molecular interactions occurring at surfaces. Furthermore, it allows some of the limitations of other techniques to be overcome. For example, other methods often require one of the partners to be labelled in some way in order to allow it to be detected. Fluorescent probes, radioactive labels, and attachment of independently detectable molecules (e.g. enzymes) have all been used for this purpose. These suffer from the drawback that they may interfere with the binding of the labelled partner to the unlabelled one, or cause unwanted structural perturbations. SPR observations can be based solely on the dielectric properties of molecules, or their intrinsic light absorption characteristics, and thus require no specific labelling.

Association and dissociation rate constant measurements

The most straightforward application of SPR technology is the measurement of binding kinetics, i.e. association and dissociation rate constants. This usage is based on the following requirements: First, the binding of one molecule to another, in which one of the molecules is immobilized on the surface of an SPR device, must produce a change in the refractive index, n, of the interface which results in an alteration of the SPR spectrum. This condition is only valid when the SPR exciting light wavelength is outside of the absorption spectral region of both interacting molecules, and when the interaction does not induce any structural changes in the interface. Second, the changes in n are assumed to be proportional to the surface concentration of bound molecules, C, i.e. $dn/dC = $ constant over the whole concentration range. This assumption can give rise to some error, especially at high concentrations where the proportionality might not be fulfilled. Third, the changes in

n cause *only* a shift of the resonance curve without any alterations in its shape, which is a simplification of the theoretical influence of n on the resonance curve as determined by the characteristic matrix of the thin film assembly. Under these assumptions, changes in the n-value can easily be measured by the shift of the SPR resonance curve. This can be greatly simplified by monitoring the reflectance taken at one specific point of the resonance curve as a function of time, as described by Equation [7]. Such a simple optical measurement allows one to directly probe the molecular interaction by monitoring association and dissociation processes in real time.

Additional applications of SPR

Additional applications of the SPR phenomenon include using the surface plasmon electromagnetic waves to excite emission of surface-bound chromophores, to enhance Raman spectra (surface-enhanced Raman spectroscopy), and as 'surface-bound light' in optical microscopy.

Sensitivity of various types of surface plasmon resonances

The sensitivity, S, of an SPR measurement can be defined as the change in reflectance, measured either at a specified angle, α_1, or a specified wavelength, λ_1 within the range of the resonance curve, divided by the change in one of the optical parameters, i.e.

$$\text{either} \quad S_{\alpha 1} = [dR_{(\alpha)}/dn \; dk \; dt]_{\alpha_1} \quad\quad \text{[8A]}$$

$$\text{or} \quad S_{\lambda 1} = [dR_{(\lambda)}/dn \; dk \; dt]_{\lambda_1} \quad\quad \text{[8B]}$$

respectively. As can be seen from the explicit function of R given by the characteristic matrix of a thin film assembly, the reflectance is not only a function of the optical parameters of the sensing layer, but also depends on the optical parameters of the incident and emerging media, as well as those of the metal film that generates the surface plasmons. In addition, the refractive indices and extinction coefficients of these media are related to one another by the complex form of Snell's Law, which complicates the function R even further. In general, however, changes in the experimental value of R are generated by two factors: the shift of the position and the change of the shape of the resonance spectrum. The latter parameter is usually described either by the sharpness of the SPR spectrum, i.e. its half-width, or by the slope of the reflectance function, and

characterizes the resolution in either the resonance angle or the resonance wavelength.

Both of these factors affecting sensitivity are dependent upon the field distributions within the resonant structure. The extent of the resonance angle shift is dependent upon the fraction of the resonant mode that lies within the sensing layer; the higher the fraction, the greater the sensitivity. This is why tight confinement of the surface plasmon evanescent wave to the sensing layer, which results in maximizing the fraction of the evanescent field in this region, favours high sensitivity. The resolution in the resonance angle (or wavelength) shift (setting aside instrumental considerations) is determined by the width of the resonance, which is defined by the losses of electromagnetic field energy within the system. The absorption by the metallic layer is the dominant factor in the SPR spectral width. Scattering of the electromagnetic wave causes further losses and increases with increased irregularities in the film structure and surface roughness, and with lower exciting wavelengths.

In summary, the overall sensitivity of the SPR device depends on both the metallic layer material (e.g. gold or silver), and the type of surface plasmon resonance being measured. The increase in the evanescent field is much smaller (about 2-fold) with gold than silver, which translates into about a 4-fold smaller overall sensitivity for an SPR device based on gold. On the other hand, gold layers are usually more stable in practical use than are silver films. The different types of surface plasmon resonances show different distributions of the evanescent electromagnetic field within the resonator device resulting in widely varying sensitivities. Long-range surface plasmon resonance has about 2.5-fold higher overall sensitivity than conventional SPR, whereas CPWR shows an even higher increase in sensitivity: about 3.5-fold (for p-polarization) and about 8-fold for s-polarization, as compared to conventional SPR. Therefore, the final design of an SPR device is usually a compromise between the different factors influencing overall sensitivity, durability, and any other requirements of the device for a specific practical application.

List of symbols

A_i = molar refractivity; c = velocity of light; C = concentration of bound molecules; D = mass density; D_s = surface mass density; k = extinction coefficient; k_{ass}, k_{diss} = rate constant for association and dissociation; K_{sp} = longitudinal component of the SP wave vector; K_{ph} = photon component of wave vector; K_B = binding affinity equilibrium constant; M = adsorbed mass; n = refractive index;

N_i = number of moles of substance per unit volume; R = reflectance; S = sensitivity; t = thickness; α_0 = incident coupling angle; ε = dielectric constant; ω = frequency.

See also: **Biochemical Applications of Fluorescence Spectroscopy; Biomacromolecular Applications of UV-Visible Absorption Spectroscopy; Chiroptical Spectroscopy, General Theory; Chiroptical Spectroscopy, Orientated Molecules and Anisotropic Systems; Ellipsometry; Surface Plasmon Resonance, Instrumentation; Surface Plasmon Resonance, Theory; Surface-enhanced Raman Scattering (SERS), Applications.**

Further reading

Garland PB (1996) Optical evanescent wave methods for the study of biomolecular interactions. *Quarterly Reviews of Biophysics* 29: 91–117.

Harrick NJ (1967) *Internal Reflection Spectroscopy*. New York: Interscience.

Kovacs G (1982) Optical excitation of surface plasmon-polaritons in layered media. In: Boardman AD (ed) *Electromagnetic Surface Modes*, pp. 143–200. New York: Wiley.

Macleod AH (1986) *Thin Film Optical Filters*. Bristol, U.K.: Adam Hilger.

Raether H (1977) Surface plasma oscillations and their applications. *In*: Hass G, Francombe M and Hoffman R (eds) *Physics of Thin Films*, 9, pp. 145–261. New York: Academic Press.

Salamon Z and Tollin G (1998) Surface plasmon spectroscopy: A new biophysical tool for probing membrane structure and function. *In*: Chapman D and Haris P (eds) *Biomembrane Structure*. Amsterdam: IOS Press.

Salamon Z, Macleod AH and Tollin G (1997) Surface plasmon spectroscopy as a tool for investigating the biochemical and biophysical properties of membrane protein systems. II: Applications to biological systems. *Biochimica et Biophysica Acta* 1331: 131–152.

Salamon Z, Macleod AH and Tollin G (1997) Coupled plasmon-waveguide resonators: A new spectroscopic tool for probing proteolipid film structure and properties. *Biophysical Journal* 73: 2791–2797.

Salamon Z, Brown MF and Tollin G (1999) Plasmon resonance spectroscopy: probing molecular interactions with membranes. *Trends in Biochemical Sciences* 24: 213–219.

Wedford K (1991) Surface plasmon-polaritons and their uses. *Optical and Quantum Electronics* 23: 1–27.

Surface Plasmon Resonance, Instrumentation

RPH Kooyman, University of Twente, Enschede, The Netherlands

SPATIALLY RESOLVED SPECTROSCOPIC ANALYSIS
Methods & Instrumentation

In view of its simple instrumentation and its high surface sensitivity, surface plasmon resonance (SPR) and, more recently, SPR microscopy gains an increasing significance to numerous problems concerned with the study of interactions occurring near to or at surfaces. Applications can be found in the optical behaviour of metals, the study of Langmuir–Blodgett films and self-assembled monolayers, the interactions of proteins with interfaces, or redox reactions at interfaces. A fast-increasing field is the development of sensitive chemooptical sensors, intended to quantitatively and selectively monitor the presence of prespecified and chemical species, which can be in the range from a molecular mass of ~200 Da to 150 kDa. The main reason for the high sensitivity of SPR to surface phenomena should be attributed to the high local electromagnetic field strengths brought about by surface plasmons (SPs).

Several excellent monographs have been published on SPR theory, whereas an overview on applications is the subject of another article in this Encyclopedia. This chapter will be concerned with experimental techniques for a variety of SPR applications.

Basic requirements

Although SPs can be excited in an electron beam we will only consider the common case where they are generated by means of light.

SPs can be excited in any material where free charge carriers exist; in the majority of practical cases this means that a metal layer deposited on a dielectric material is used as a medium to carry SPs. It turns out that the particular SP properties strongly depend on the nature of this deposited layer: if the metal is present as an island-like structure with patches of the order of a few tens of a nm, *radiative* plasmons will be excited resulting in large enhancements of the applied (optical) field strengths. This effect is mainly used to enhance surface responses in vibrational spectroscopy, such as Raman and infrared spectroscopy. *Nonradiative* SPs can be excited if the metal layer is applied as a thin (~50 nm) homogeneous layer. These plasmons have the peculiar property that the associated electromagnetic field is evanescent in character, i.e. the wave vector k_x parallel to the interface is (partly) real, whereas that perpendicular to the interface is imaginary. The generation of SPs is an elastic scattering process, implying that both energy and momentum are conserved. Consider the system as depicted in **Figure 1A**, where a metal layer is sandwiched between two media a and p, which is the commonly used Kretschmann configuration. In view of Snell's law, which states that through this whole system k_x remains constant, this has the consequence that SPs can only be excited at the interface a/m when light enters the metal layer through another interface m/p, *under the condition that $\varepsilon_p > \varepsilon_a$*, where ε_p, ε_a denote the respective dielectric constants. This is further illustrated in **Figure 1B** where the two lines a and p represent the light dispersion relations in the media a and p, respectively, and the curve represents the SP dispersion relation. A further necessary condition is that the exciting light has to be p-polarized, because light polarized along the metal interface cannot exist.

This short theoretical description contains all the information needed for setting up a basic SPR experiment.

Choice of metal support

An approximate expression for the SPR wave vector can be found in the literature:

$$k_x = (2\pi/\lambda)(\varepsilon_m \cdot \varepsilon_a/(\varepsilon_a + \varepsilon_m)^{1/2}) \qquad [1]$$

where λ is the wavelength *in vacuo*, and ε_m is the metal dielectric constant at the wavelength used. Generally, k_x of the incoming light can be adapted to that required by Equation [1] by inclining the beam relative to the interface normal. The following relation holds (see also **Figure 1A**):

$$k_x = (2\pi/\lambda) \cdot \varepsilon_p^{1/2} \cdot \sin\theta \qquad [2]$$

However, in practice the maximum angle θ that can be chosen is around 80°, implying that in a number of cases the dielectric constant of the support, ε_p, has to be selected such that within this material a light beam can have a *practical* k_x matching that of Equation [1]. If medium a is a water solution (refractive index $n_a \sim 1.33$; note that $n = \varepsilon^{1/2}$) then BK7 glass ($n_p \sim 1.52$) is an appropriate material. For media with higher refractive index SF10 glass ($n_p \sim 1.7$) can be used. (The abbreviations BK7 and SF10 refer to a nomenclature common in lens-making technology.) Convenient shapes of the dielectric support are prisms with apex angle 60 degrees or hemicylinders, which minimize the reflection loss of light when entering the support (cf. **Figure 1A**).

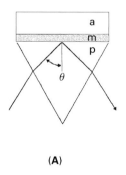

(A)

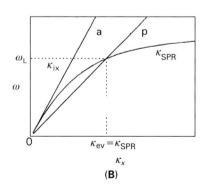

(B)

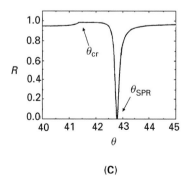

(C)

Figure 1 The SPR experiment. (A) a metal layer m is sandwiched between two dielectric media a and p. The direction *x* is defined parallel to the layer structure; (B) dispersion relation of SPs; see text; (C) a typical SPR curve when medium a has a refractive index $n_a = 1$. The angle θ_{cr} corresponds to the critical angle for the a/p interface.

Choice of metal layer

A typical application of SPR spectroscopy is to monitor a layer growth on the metal layer. Such a process can be modelled as a changing ε_a, which experimentally translates into a changing resonance angle θ_{SPR}. **Figure 1C** depicts a reflectance curve obtained from a SPR experiment. From the figure it is clear that in order to obtain maximum sensitivity to variation in ε_a the slope $dR/d\theta$ has to be as large as possible. For a given ε_a this depends in a complicated way both on the value of $\varepsilon_m(\lambda)$ and on the thickness of the metal layer. Generally, it can be said that smaller real and imaginary parts, $\mathrm{Re}(\varepsilon_m)$ and $\mathrm{Im}(\varepsilon_m)$, result in smaller resonance halfwidths, whereas the metal layer thickness is an important parameter determining the minimum reflectance. From **Table 1**, which gives an impression of $\varepsilon_m(\lambda)$ for some metals, it can be concluded that in the visible wavelength range silver is expected to exhibit the narrowest resonances. This is indeed found experimentally; however, in many situations where experiments are performed in a water or air environment, silver tends to undergo unwanted interactions with its environment, making this a less attractive material. In this respect gold is a much more stable material and thus this material has become the standard metal for SPR purposes, despite its resonance width that at $\lambda = 633$ nm is approximately three times larger than that of silver.

For a given metal the optimum layer thickness depends on the wavelength used, as illustrated in **Figure 2**; also the remarks on the importance of the value of $\varepsilon_m(\lambda)$ are apparent from the figure. For any wavelength within the visible range a gold or silver layer thickness can be found corresponding to a vanishing minimum reflectance; for the often used helium-neon laser wavelength $\lambda = 633$ nm this is approximately 47 nm for gold and 49 nm for silver.

Practical aspects As already mentioned, in order to be able to excite nonradiative SPs it is important to have available smooth, homogeneous metal layers

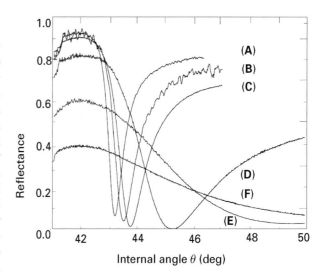

Figure 2 SPR reflectance curves for a bare gold layer with thickness 44 nm, measured with different wavelengths: (A) 676 nm; (B) 647 nm; (C) 633 nm; (D) 568 nm; (E) 514 nm; (F) 488 nm.

with a well-defined layer thickness. An appropriate way to prepare these is to sputter or to evaporate the metal at a high rate (~ 1 nm s^{-1}) in a vacuum chamber (10^{-6} mbar) on the substrate of choice. In order to increase the adhesion between metal and underlying dielectric, the support is often precoated with a 2 nm Ti or Cr layer. It is not mandatory to coat directly on the prism; alternatively, a flat substrate, such as a microscope glass cover slip, can be used as a substrate. After coating, this plate is optically connected to the prism using a matching oil; care has to be taken that, in order to avoid spurious reflections, plate, oil and prism have the same refractive index in the wavelength region of interest.

Although a simple incandescent light source can be used for SPR, the use of a small HeNe or diode laser is far more convenient, in view of their well-defined wavelength and high degree of collimation.

SPR instrumentation

Rotation stage

The most straightforward method to measure SPs is depicted in **Figure 3**, where the prism/gold assembly is placed on top of a rotation stage (angular resolution typically 0.01 degree). With such an arrangement, all the features of a SPR curve can be accurately determined. The light detectors consists of simple large area photodiodes whose output is fed into a low-noise current-to-voltage amplifier. The excitation laser beam is polarized such that both s- and p-polarized light enter the prism. The use of a

Table 1 Real and imaginary parts of dielectric constant for some metals

Metal	λ (nm)	ε_{re}	ε_{im}
Aluminium	600	−29.8	7
	700	−46.6	22
	900	−55.5	30
Silver	500	−8.23	0.3
	700	−21.3	0.7
	900	−38.7	1.3
Gold	600	−8.37	1.2
	750	−18.2	1.2
	900	−28.5	1.8

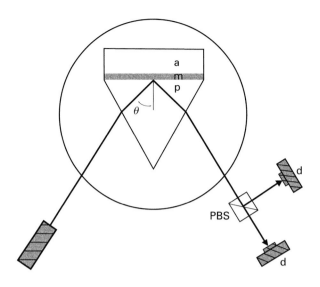

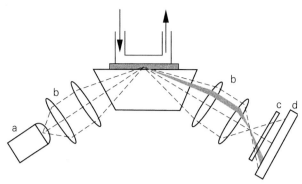

Figure 4 Use of focused beam. a: laser diode; b: focusing optics; c: neutral density filter; d: diode array. A flow cuvette is placed on top of the metal layer. Reproduced with permission from Sjölander S (1991) *Analytical Chemistry* **63**: 2338. © American Chemical Society.

Figure 3 Rotation stage SPR setup. d: detectors; PBS; polarizing beamsplitter.

polarizing beamsplitter serves two purposes: one detector measures only the reflected s-polarized light whose intensity is only affected by laser intensity fluctuations and angular dependent refraction losses, whereas the other detector in addition measures SPR effects. The ratio between the two outputs then provides a signal proportional to the net SPR response. In order to obtain an absolute angular read-out with sufficient accuracy the critical angle can be measured; this angle is very accurately known for a certain $\varepsilon_p/\varepsilon_a$ interface, and is an easily discernible feature in a SPR curve (cf. **Figure 1C**).

The obvious disadvantage of this setup is that it is difficult to monitor fast changes in the SPR curve. To be able to monitor time-dependent changes, the orientation of the rotation stage relative to the light beam is often set such that the reflectance is about halfway between maximum and minimum. A changing ε_a is then approximately linear in the monitored changed reflectance. However, one has to assume that a change of ε_a leaves the width of the SPR curve unchanged.

Use of a focused beam

A more elaborate setup is shown in **Figure 4**. Here, SPs are excited by a focused diode laser beam such that the beam waist is slightly displaced from the top metal/dielectric interface. In this way, a certain distribution of angles is present at the interface. The reflected beam is imaged onto an array consisting of a large number (50–100) of photodiodes. The angular range corresponding to resonance will exhibit a low intensity; consequently, the output of the diode array

is a measure of the angular dependent reflectance. Although the distance between individual diodes is relatively large ($\sim 20\ \mu m$) the use of numerical interpolation techniques makes it possible to obtain an angular resolution of ~ 1 millidegree. Because no moving parts are involved, this system is capable of monitoring relatively fast-changing SPR characteristics: the temporal resolution will now be determined by the read-out rate of the diode array. A drawback of this system is that obtaining an *absolute* angular read-out is not as straightforward as in the case of the rotation stage, in view of the limited angular range that is applied to the metal/dielectric interface.

Photothermal detection

When a nonradiative SP decays it generates heat at the metal/dielectric interface. This observation can be fruitfully exploited to detect the presence of SPs. In a *photoacoustic* approach, the prism/metal assembly is placed in an airtight chamber. SPs are excited with an intensity-modulated light beam; the resulting periodic heat flow from metal to dielectric as a result of plasmon decay causes a periodic pressure variation in the dielectric which can be detected by a microphone in the sample chamber. A significant difference as compared with optical detection is that SPR is now detected as a *maximum* in response. This method is mainly used in fundamental studies where one is interested in SP decay. The method is less suited in situations where the dielectric of interest consists of a fluid, such as water.

In a related approach, the heat flow is detected optically (*photothermal deflection spectroscopy*). A representative setup is shown in **Figure 5**. The prism/metal assembly is placed upon a rotation stage, and SPs are excited in the usual way. Heat produced by the decaying plasmon results in a gradient of the refractive index immediately above the metal layer. Consequently, the propagation direction of the light beam of the probe laser will be deflected and this can be measured, e.g. by using a position-sensitive detector. By modulating the SPR excitation beam and synchronous detection of the detector output, a differential signal is obtained whose angular dependence is directly related to the SPR curve. Using a HeNe laser as the excitation source an angular resolution of ~0.01 degree can be obtained; contrary to the setups where the reflectance is measured, this resolution can be improved by employing higher laser powers. It has been demonstrated that this deflection method is also useful in water solutions.

Sensor configurations

As already mentioned, an important application of SPR is in the field of chemical sensors. Such a SPR sensor system should be simple, compact, and relatively inexpensive, while retaining the angular sensitivity. Some representative SPR sensor systems will now be discussed.

Vibrating mirror device In the system as depicted in **Figure 6** the angle under which the light from a laser diode enters the interface is made time-dependent by means of a mirror, vibrating at a frequency of ~50 Hz. If the optical system in the excitation path is designed properly, the light spot during an angular scan of ~5 degrees is stationary on the interface to within 0.2 mm, while the beam divergence is kept within 0.02 degrees. Although this setup can be used to monitor the complete SPR curve, it is dedicated to determine only the angle of minimum reflectance θ_{SPR}, which for the majority of SPR applications is the main parameter of interest. This can be conveniently accomplished as follows: during one cycle of the vibrating mirror the beam traverses the reflectance minimum twice. The time span Δt between these two minimum is measured using appropriate electronics. If θ_{SPR} changes, Δt changes accordingly.

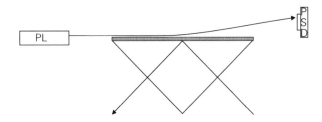

Figure 5 Basic photothermal detection setup. The position of the deflected beam of a probe laser PL is measured by a position-sensitive detector PSD.

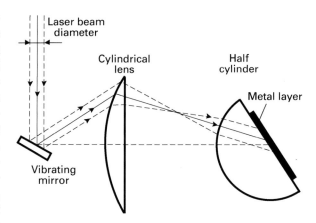

Figure 6 Vibrating mirror setup. a: vibrating mirror; b: cylindrical lens. Reproduced with permission from Lenferink ATM (1991) *Sensors and Actuators B (Chemical)* **3**: 261. Copyright 1991, with kind permission from Elsevier Science Ltd, The Boulevard, Langford Lane, Kidlington, OX5 1GB, UK.

Connected to a personal computer, this provides a versatile means to detect a SPR minimum with an angular resolution of 1 millidegree. The time resolution of such a setup will be determined by the vibration frequency of the mirror. (Note that vibrating mirrors operating at frequencies up to 10 kHz are available). For some applications the difficulty of obtaining an absolute angular read-out will be a disadvantage.

Use of diffraction-gratings In the foregoing configurations a SP was excited using a prism. However, plasmons can also directly be produced on a grating surface coated with a thin metal layer. The condition for SPR to occur is determined by the grating periodicity (typical value 2000 lines mm⁻¹) and the angle of incidence of the light beam, whereas the value of the reflectance minimum is solely determined by the grating depth (optimum value ~50 nm). Any of the above described setups can be used to detect SPs in this configuration. An advantage of this approach is that gold-coated gratings can be very easily and inexpensively manufactured as disposable replicas, once a holographic master grating has been produced. An important disadvantage is that the light beam has to enter the dielectric/metal interface from the dielectric side, implying that this is only a valuable approach for transparent media.

Fibre-optic devices In situations where the sensing surface should be remote from the signal processing equipment (such as in a hostile environment or in a living organism) the use of optical fibres can provide a practical solution. Apart from the trivial use of a fibre to transport the light to and from a separate

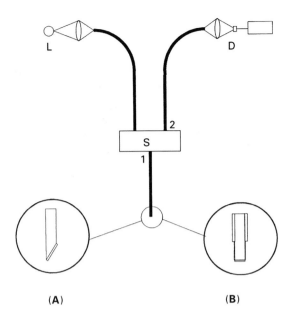

Figure 7 Use of fibre optics. L: light source; D: detector; S: fibre splitter/combiner; 1, 2: output ports. (A): single mode fibre; (B): multimode fibre (cf. text).

prism/metal assembly, it has also been demonstrated that the fibre can also be used as an *intrinsic* sensing interface, where part of the fibre surface replaces the prism. This concept is schematically shown in **Figure 7**. Light is fed into a fibre connected to an optical splitter/combiner S. The output port 1 of S is connected to a probe fibre, which is decladded at the distal end. After interaction with this end interface (see below) the light is reflected and again enters S, where it is transmitted to output port 2. A fibre connected to this port transports the light to a detector.

Several possibilities exist for the design of the decladded fibre tip. In **Figure 7A** the tip of a monomode fibre is cleaved at a specific angle and is subsequently coated with a metal layer of appropriate thickness. As the wave vector (cf. Eqn [2]) of the propagating light in a monomode fibre is well defined the cleaving angle can be calculated such that excitation of SPs is possible in the ε_a region of interest. Again, SP excitation is detected by a decreased reflected intensity. In order to obtain optimum response it is advisable to use polarization-preserving fibres. In **Figure 7B** another approach is depicted, involving the use of a multimode fibre. Here, the circumference of the tip is metal coated; additionally a mirror is deposited on the distal end to minimize reflection losses. For a multimode fibre a (discrete) range of wave vectors simultaneously propagates; in order to be able to detect changing characteristics of the metal/dielectric interface it is therefore necessary to use a broadband light source.

The SPR information will now be contained in the wavelength-dependent reflectance. An advantage over the monomode configuration is that a broader range of SP wavevectors can be excited.

SPR microscopy

The foregoing discussion has ignored the possibility to obtain spatially resolved information from a SPR experiment. However, it is known that SPs are collective electron oscillations *with a limited coherence length L_x*, implying that two regions in the metal with a mutual distance larger than L_x are capable of supporting SPs which are mutually independent. This phenomenon can be exploited to image structures on top of a metal layer that have a distribution of different ε_a: if the angle of light incidence is such that one particular ε_a corresponds to resonance, then regions with another ε_a will exhibit larger reflectance. An example of such an experiment can be seen in **Figure 8**, where the spatially resolved reflectance of an inhomogeneous monomolecular layer, consisting of molecules oriented either tilted or perpendicularly to the surface is depicted.

Of course, in such an experiment one aims to obtain maximum lateral resolution, while retaining the vertical (thickness) resolution. A general rule is that L_x decreases for increasing resonance halfwidths (cf. **Figure 1C**). However, as was pointed out in an

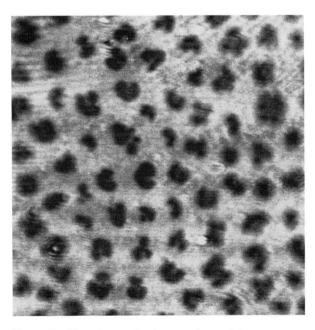

Figure 8 SP microscopic image of an inhomogeneous monolayer. Dimensions of image 0.2×0.2 mm^2. Thickness difference of the two types of domains is less than 0.4 nm. Lateral resolution approximately 3 μm.

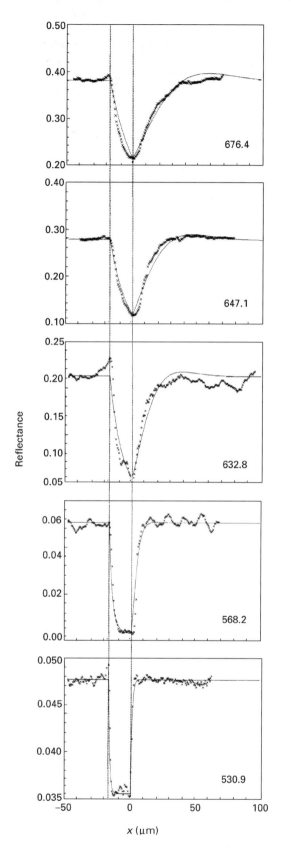

Figure 9 Lateral resolution in SPR microscopy. The vertical lines denote the physical width of a 2.5 nm SiO₂ ridge. Number insets correspond to the various wavelengths used.

earlier section, such an increasing SPR halfwidth simultaneously deteriorates vertical resolution. Therefore, for each particular situation a balance has to be sought between these two contradictory conditions. This is exemplified in **Figure 9** where SPR results are shown of a 2.5 nm SiO₂ ridge on gold for various wavelengths. It is clearly seen that at the shortest wavelength used the lateral resolution is the highest (~ 2 μm), but the slope is the lowest (compare also **Figure 2**), indicating that it is difficult to obtain sufficient intensity contrast to detect sub nm height differences, which is usually not a problem with standard SPR experiments.

Instrumentation

The first SPR microscopy experiments were done using a scanning focused beam. However, a straightforward approach, by imaging a collimated beam, proved to be much faster and instrumentally much simpler while at both the lateral and vertical resolution are comparable or better. Therefore, only this last approach will be discussed in more detail.

Figure 10 gives an overview of a representative setup. A prism/gold layer assembly and an optional cuvet system are placed on a rotation stage (angular increments 0.01 degree). SPs can be excited using light from a HeNe or an argon/krypton ion laser. This last light source has the advantage of providing a large number of high power wavelengths over the whole visible range. After having passed through a Pockels cell and a spatial filter, the light spot

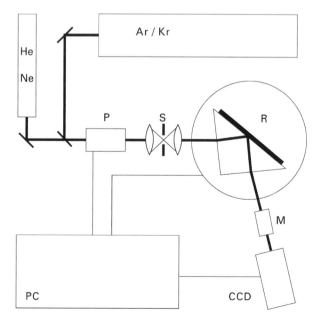

Figure 10 SPR microscope setup. P: Pockels cell; S: spatial filter; R: rotation stage; M: objective; CCD: video camera.

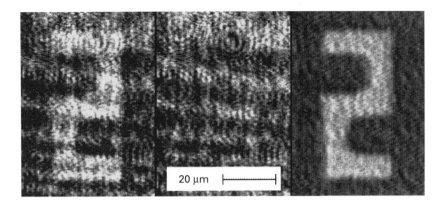

Figure 11 Improvement of image quality by dividing s- and p-polarized responses. From left to right: p response; s response; p/s response.

entering the prism has an area of ~1 cm². The intensity profile of the reflected beam is recorded with a microscope consisting of an objective and CCD video camera. It is important that the video camera has a response linear in the light intensity (see below). Depending on the light intensity and required magnification, objectives with focal distances between 5 and 50 mm are used. The Pockels cell and rotation stage are controlled by a microcomputer and the obtained images are also stored in this same computer by using a frame-grabber card.

It also proves possible to use the aforementioned gratings in a SPR microscopic experiment. Rotation of such a plate about the normal is equivalent to changing the angle of incidence of the exciting light. Compared to the standard prism setup this configuration is very compact and can easily be integrated in a conventional light microscope.

Improvement of image quality

Apart from the diffraction limit of the use of imaging lens there are a number of factors determining the eventual quality of a SPR microscopy image. As already mentioned, the use of shorter wavelengths generally improves lateral resolution, but simultaneously results in lower intensity contrasts for a given height difference. With such relatively low contrasts the use of a digital frame-grabber (which has a typical dynamic range of 2^8) can result in *quantization effects during image acquisition*, which become apparent if a low-contrast image is digitally amplified. To avoid this effect an option is to add a number of images with an effective dynamic range substantially larger than that of the frame-grabber used.

Integrating a number of images also results in averaging. This can be particularly important if shot noise is present when working with low light levels. Such an averaging will result in a signal-to-noise

(SNR) improvement with a factor of $n^{1/2}$ where n is the number of added images. Another possible way to increase SNR is to increase laser power but this option is of limited value in view of the resulting destructive heating effects on the sample.

Another experimental problem is *lateral nonhomogeneity of the incoming laser beam intensity profile*. This can be partially solved by spatial filtering as indicated in **Figure 10**; however, if there are any unwanted, spurious reflections in the light path between the spatial filter and CCD camera, then again a beam nonhomogeneity will occur owing to the relatively large coherence length of the laser light used. An appropriate method is the use of a Pockels cell with simultaneous digital image processing. Such a cell can be configured such that the application of a voltage results in transmission of either p- or s-polarized light, without moving any part in the light beam. In the SPR microscope the Pockels cell is employed to acquire two images using p- and s-polarized light, respectively. The image with p-polarization contains the SPR microscopic image together with the contrast generated by spatial nonhomogeneities, whereas the image obtained with s-polarization only contains the same unwanted nonhomogeneities. Because the CCD camera output is linear in intensity, the ratio of the two images is a true representation of the SPR-related reflectance variations over the imaged surface. An example of the result of such an approach can be seen in **Figure 11**. Although lateral nonhomogeneities are still visible in the ratioed SPR image, the resolution is approximately 2 μm; for this image the difference in reflectance between the two regions is ~0.01.

Combination with other microscopies

Since the birth of scanning probe microscopy several attempts have been undertaken to merge this approach with SP microscopy. In **Figure 12** a schematic diagram is given where a SPR microscope is combined with a scanning tunnelling microscope (STM). The hemisphere, serving as a prism, is metal-coated, and SPs are excited in the usual way. The metallic STM tip is brought close to the surface (distance ~5 nm). The area of interaction between the tip and the surface is imaged on a photodiode. The tip is allowed to scan laterally over the surface and the photodiode output is monitored simultaneously. It was found that small corrugations on the surface which were detected by the tunnelling current of the STM were equally well monitored by the SPR reflectance. In this way a lateral resolution of ~5 nm could be demonstrated *in a SPR image*. Similar conclusions could be made for a dielectric atomic force microscopy (AFM) tip. Even better results can be obtained if the conically scattered SPR radiation is monitored rather than the specularly reflected light.

A discussion on the contrast mechanism in both setups is far beyond the scope of the present article; a pragmatic remark is that, in most cases, an AFM type tip is preferable in view of the fact that in the latter case no conductivity conditions have to be imposed on the samples of interest.

List of symbols

k_x = wave vector; L_x = coherence length; n = refractive index; n = number of added images; R = reflectance; Δt = time span; ε = dielectric constant; λ = wavelength *in vacuo*; θ = angle of light; θ_{SPR} = angle of minimum reflectance.

See also: **Fibre Optic Probes in Optical Spectroscopy: Clinical Applications; Fourier Transformation and Sampling Theory; Light Sources and Optics; Photoacoustic Spectroscopy, Applications; Photoacoustic Spectroscopy, Theory; Scanning Probe Microscopes; Scanning Probe Microscopy, Applications; Scanning Probe Microscopy, Theory; Surface Plasmon Resonance, Applications; Surface Plasmon Resonance, Theory.**

Further reading

Agranovich VM and Maradudin AA (1982) *Surface Polaritons.* Amsterdam: North-Holland Publishing Company.

Kretschmann E (1971) Die bestimmungen optischer Konstanten von Metallen durch Anregung von Oberflächen plasmaschwingungen. *Zeitschrift Physica* **241**: 313–321.

Lawrence CR and Geddes NJ (1997) SPR for Biosensing. In: Kress-Rogers E (ed) *Handbook of Biosensors and Electronic Noses.* Boca Rata, FL: CRC Press.

Raether H (1988) *Surface Plasmons on Smooth and Rough Surfaces and on Gratings.* Berlin: Springer Verlag.

Rothenhäusler B and Knoll W (1988) Surface plasmon microscopy. *Nature* **332**: 615–616.

Specht M, Pedarnig JD, Heckl M, and Hansch TW (1992) Scanning plasmon near-field microscope. *Physical Review Letters* **68**: 476–478.

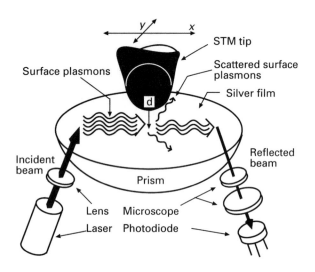

Figure 12 Scheme of a combined SP microscope and a scanning probe microscope. The tip can either be a STM or an AFM tip. Reproduced with permission from Specht M (1992) *Physical Review Letters* **68**: 476–479. American Physical Society.

Surface Plasmon Resonance, Theory

Zdzislaw Salamon and **Gordon Tollin**,
University of Arizona, Tucson, AZ, USA

SPATIALLY RESOLVED
SPECTROSCOPIC ANALYSIS
Theory

The physics of surface plasmons propagating along a metal/dielectric interface has been studied intensively, and their fundamental properties have been found to be in good agreement with theoretical concepts based upon the plasma formulation of Maxwell's theory of electromagnetism. The phenomenon has been utilized extensively by physical scientists in studies of the properties of surfaces and thin films. Current interest in the properties of thin surface coatings stems partly from increased applications to thin film devices and, in particular, to recent developments in biosensor devices. This article focuses on the characterization of the surface plasmon resonance phenomenon, with emphasis on the conditions of optical excitation of plasmon resonance and the theoretical analysis of different types of surface resonances.

Description of surface plasmons

The concept of surface plasmons originates from the plasma formulation of Maxwell's theory, where the free electrons of a metal (or a conductive electron gas) are treated as a high density liquid (plasma). Plasma oscillations in metals are collective longitudinal excitations of the conductive electron gas, and plasmons are the quanta representing these charge-density oscillations. Such oscillations can exist in the bulk media, and can also be localized on an interface between a metallic and a dielectric surface, along which they propagate as waves. In the latter case they are called surface plasmons (SP), or surface polaritons. The spreading electron density fluctuations generate a surface-localized electromagnetic wave which propagates along the plane interface between the metal/dielectric media, with the electric field normal to this interface and vanishing exponentially with penetration distance from the interface. These characteristics of the electromagnetic field are the same as those describing the guided surface waves (also known as evanescent waves) generated optically under total internal reflection conditions when all the incident light is reflected at the boundary of the incident and emerging media. Under such conditions the electric and magnetic fields do not, however, stop abruptly at the boundary. Rather they penetrate a distance into the emerging medium in the form of a surface wave. Although the existence of guided surface electromagnetic waves has been theoretically predicted from Maxwell's equations and investigated during the first decade of the 20th century, it is only since 1960 that they have attracted the interest of the experimentalist and the term 'surface plasmon' has been used. This is partly due to the fact that methods have been developed which enable the optical excitation and detection of surface-bound electromagnetic waves (see below).

It has been shown that Maxwell's equations have solutions resulting in the generation of surface plasmon electromagnetic waves only when the following conditions are fulfilled: (1) one of the adjacent media (i.e. the surface active medium which generates surface plasmon waves) has a negative value for the real part of its complex dielectric constant ε, (2) a component of the wave vector along the interface between these two media, K, satisfies an equation which involves the dielectric constants of both media. These conditions are discussed in more detail in the following sections.

Optical excitation of surface plasmons

The condition that the surface active medium has a negative dielectric constant results in several experimental requirements which have to be fulfilled in order to be able to generate surface plasmon waves. First, not all materials can be utilized as surface active media; gold and silver are the best examples of materials which can support surface plasmons. In addition, the electromagnetic wave in the surface active medium is an evanescent wave under all circumstances. Therefore, in order to satisfy boundary conditions in the surface active and dielectric media, which require that the tangential components of the electrical and magnetical fields be continuous across the boundary (i.e. that a component of the wave vector along the interface must be the same in both media), surface plasmons can only be optically generated by an evanescent wave whose wave vector matches that of the evanescent surface plasmon electromagnetic wave K_{sp}. The latter is given by the following equation:

$$K_{SP} = \omega/c(\varepsilon_1\varepsilon_2/(\varepsilon_1 + \varepsilon_2))^{1/2} \qquad [1]$$

where ω is the frequency of the surface plasmon wave, c is the velocity of light *in vacuo*, and ε_1 and ε_2 are the complex dielectric constants for the surface active and dielectric (emerging) media, respectively. The complex dielectric constant ε is directly related to the complex index of refraction $N = c/v = n - ik$, by the following relation:

$$\varepsilon = \varepsilon' + i\varepsilon'' = N^2 \qquad [2]$$

where the real part of the complex dielectric constant is $\varepsilon' = n^2 - k^2$ and the imaginary part is $\varepsilon'' = 2nk$, n is the refractive index, k is the extinction coefficient, and v is the velocity of light in the dielectric medium.

These constraints require that surface plasmons cannot be directly excited by incident light, and produce a resonance condition for the wave vector of the evanescent wave which excites the plasmons. Furthermore, indirect SP excitation can only be achieved by an evanescent wave generated by p-polarized incident light under total internal reflection conditions. This occurs when a beam of light propagating through an incident medium (e.g. a prism) of higher refractive index (n_0) meets an interface with a second (emerging) medium of lower refractive index (n_2) at incident angles α larger than the critical angle for total reflection α_c, given by $\alpha_c = \sin^{-1}(n_2/n_0)$. Since the prism has a dielectric constant, $\varepsilon_0 = n_0^2$ (i.e. $k_0 = 0$) the following relation must be satisfied: $\varepsilon_0 > \varepsilon_2$ (or $n_0 > n_2$), where ε_2 is the dielectric constant and n_2 is the refractive index of the emergent dielectric medium). Although metal-coated diffraction gratings may be used instead of prisms, they require a greater complexity of fabrication without additional benefits, and so they have not been widely used and will not be discussed further.

Despite being totally reflected, the incident beam generates an evanescent electromagnetic field that penetrates a small distance, the order of a wavelength, into the second medium, where it propagates parallel to the plane of the interface. This electromagnetic field can be used to measure the optical properties of interfaces and thin films in various ways. The two main types of applications of optically generated evanescent waves are those based on waveguiding systems, and those used to excite surface plasmon resonance (SPR).

In waveguiding techniques the measured interface (or thin film) is placed in the evanescent region of a guided mode propagating in a dielectric waveguide structure. The optical properties of the interface (or thin film) affect the propagation characteristics of the evanescent surface wave causing changes in the resonant waveguide mode. These changes can be measured by a variety of optical techniques including attenuated total reflection (ATR) spectroscopy, total internal reflectance fluorescence (TIRF), where the evanescent wave is employed to excite fluorescence from molecules in the interface, or interferometry.

In order to generate SPR using an evanescent wave which is produced during an internal total reflection of the light from a prism whose base is coated with a thin metal film, the following two conditions have to be fulfilled. First, a component of the incident light vector parallel to the prism/metal interface, K_{ph} which is described by the following equation:

$$K_{ph} = (\omega/c)\,\varepsilon_0^{1/2} \sin\alpha \qquad [3]$$

must be identical with the surface plasmon wave vector, K_{SP} (see Eqn [1]):

$$K_{ph} = K_{SP} \qquad [4]$$

The value of K_{ph} can be adjusted to match that of the surface plasmon wave by changing either ω, i.e. the frequency (or wavelength) of the excitation light, or α, i.e. the incident angle (see Eqn [3]). Secondly, because the oscillations of the free electrons in a metal film can only occur along the normal to the plane of the metal surface, only p- (or TM, transverse magnetic) polarization of the incident light is effective in generating surface plasmons. At the resonance condition (Eqn [4]), the incident light is coupled into a SP wave travelling along and bound to the outer active (metal) surface, and the phenomenon is known as surface plasmon resonance (SPR). The SP wave is nonradiative, and can either decay into photons of the same frequency ω if coupling by the surface roughness takes place, or be converted to heat.

In practical terms there are two configurations, both based on the ATR technique available to optically excite SPR at the metal/dielectric (or emerging medium) interface. In the first, the Kretschmann configuration, the prism is in direct contact with the surface active (metal) medium. In the second, the Otto configuration, the prism is separated by a thin layer of a dielectric (inactive) medium at a distance of approximately one wavelength of excitation light from the metal film. The practical consequences of using these configurations to excite SPR are described below.

Analysis of surface plasmon resonances excited by light

Although the SPR phenomenon can be accurately described in physical terms as propagating oscillations of free electrons at a metal surface, there is a simpler and more general approach which has been used to describe light propagation through optically anisotropic layered materials whose properties vary only along the layer normal. This is a standard mathematical tool used to describe the optical properties of multilayered thin-film devices, and the SPR phenomenon can be seen as a straightforward result of the application of such thin-film electromagnetic theory. The analysis applies Maxwell's equations to describe the propagation of a plane electromagnetic wave through a multilayer assembly of thin dielectric films, and is based on the following properties of the structure. The film is considered thin when the phase differences between the various waves in the assembly are constant with time. This condition invariably holds for films which have thicknesses of not more than a few wavelengths. A second requirement is that the thin-film materials are characterized by a complex refractive index, which in the optical region is numerically equal to the optical admittance in free space units. This is defined by the ratio of the total tangential electric (B) and magnetic (C) field amplitudes of the electromagnetic wave ($Y = C/B$). Additional constraints are that the tangential components of the electric and magnetic field vectors of the electromagnetic wave are continuous across the interface between any two thin films. Also, that in any thin film the amplitude reflection coefficient or reflectivity, sometimes known as the Fresnel reflection coefficient (r), defined as the ratio of the amplitudes of the incident and reflected electric field vectors, at any plane within the layer is related to that at the edge of the layer remote from the incident wave, r_{m}, by $r = r_{\mathrm{m}}e^{-2i\delta}$, where δ is the phase thickness of that part of the layer between the far boundary, m, and the plane in question. Solution of Maxwell's equations describing the propagation of a plane, monochromatic, linearly polarized, and homogeneous electromagnetic field within a multilayer thin-film assembly with the above mentioned attributes results in a relationship which connects the total tangential components of the electric and magnetic field amplitudes at the incident interface with the total tangential components of electric and magnetic field amplitudes which are transmitted through

the final interface. This result is of prime importance in describing the optical properties of thin films and forms the basis of almost all calculations. It has the following standard matrix notation, known as the characteristic matrix of the assembly:

$$\begin{bmatrix} B \\ C \end{bmatrix} = \prod_{j=1}^{s} \begin{bmatrix} M_j \end{bmatrix} \begin{bmatrix} 1 \\ y_{j+1} \end{bmatrix} \qquad [5]$$

where s is the number of layers deposited on the incident medium; y_j, gives the characteristic oblique admittance of the jth layer (y_{j+1} for the emerging medium): $y_j = y_j/\cos \alpha_j = (n - ik)_j/\cos \alpha_j$ for p-polarized incident light, or $y_j = Y_j \cos \alpha_j$, for an s-polarized electromagnetic wave, and $\mathbf{M_j}$ is known as the characteristic matrix of the jth thin layer and has the following form:

$$\mathbf{M_j} = \begin{bmatrix} \cos \delta_j & i(\sin \delta_j)/y_j \\ i\, y_j \sin \delta_j & \cos \delta_j \end{bmatrix} \qquad [6]$$

where $\delta_j = (2\pi Y_j/\lambda)t_j \cos \alpha_j$, gives the phase thickness of layer j in the thin-film assembly; $(Y_0 \sin \alpha_0) = (Y_j \sin \alpha_j)$, is the complex Snell's Law which relates α_j to α_0, the angle of incidence in the incident medium; t_j is the thickness of the jth layer, and α_0 and α_j are the incident angles of light of wavelength λ for the incident medium (a prism in the SPR system) and the jth layer, respectively.

There are two important conclusions which can be deduced from Equation [5]. First, the characteristic matrix of an assembly of j layers is simply the product of the individual matrices taken in the correct order. Second, sufficient information is included in Equations [5] and [6] to allow the full analysis of the electromagnetic field generated at each interface of a multilayer thin-film assembly, thereby yielding transmittance, absorbance, and reflectance for both p- and s-polarizations. Furthermore, the optical admittance presented at the incident interface by the system of layers and emerging medium is the product of the characteristic matrix of the assembly.

The optical admittance parameter has been introduced into thin-film optics with one specific aim, namely to visualize optical phenomena occurring within such systems by means of a graphical representation of the optical events known as the admittance diagram. Although this is one of a class of diagrams known collectively as circle diagrams, it is particularly powerful and attractive and therefore it is used extensively in thin-film optics.

In the case of SPR which is generated optically with the ATR technique, the reflectance of a multi-layer system, R, defined as the ratio of the energy reflected at the surface of such a structure to the energy which is incident, is an especially important parameter and can be calculated from the following relation:

$$R = rr^* \qquad [7]$$

where:

$$r = (Y_0\, B - C)/(Y_0\, B + C) = (Y_0 - Y)/(Y_0 + Y) \qquad [7a]$$

is the amplitude reflection coefficient (reflectivity or Fresnel reflection coefficient), and:

$$r^* = \{(Y_0\, B - C)/(Y_0\, B + C)\}^* = \{(Y_0 - Y)/(Y_0 + Y)\}^* \qquad [7b]$$

is the complex conjugate of r. Y_0 is the admittance of the incident medium (which in the case of the present application to SPR is a non-light-absorbing glass prism, i.e. with $k = 0$, and therefore the Y_0 value becomes real and equal to the refractive index of the incident medium, n_0).

Equations [5], [6], [7], [7a] and [7b] comprise a full set of mathematical tools to examine optically excited surface plasmon resonance. Such an analysis can be applied to the following three types of resonances (see the next section): (i) conventional surface plasmon resonance (SPR), (ii) the resonance associated with long-range surface plasmons (LRSPR), and (iii) the resonance associated with coupling of plasmon resonances in a thin metal film with waveguide modes in a dielectric overcoating, known as coupled plasmon waveguide resonance (CPWR). Although in all three cases the excitation of surface plasmons is based on coupling light photons to plasmons by the ATR technique, these resonance systems differ in their detailed thin-film structural designs, as described in the next section.

Variety of surface plasmon resonances

Conventional surface plasmon resonance

In the most straightforward case, for which the hypotenuse of the prism is coated with a single high-performance metal (Ag or Au) layer (Kretschmann coupling), one can generate surface plasmons on the outer surface of the metal, as indicated in **Figure 1**. This shows that for either a 55 nm Ag or a 48 nm Au layer, as a consequence of an enormous increase in the intensity of the evanescent electromagnetic field, which is produced as a function of either the incident angle, α, with $\lambda = \lambda_0 = $ constant (panel A), or the wavelength, λ, with $\alpha = \alpha_0 = $ constant (panel B). This very large increase of electric field amplitude, as compared to that of the incident light, with the characteristics of a sharp resonance and which can only be obtained with p-polarized excitation light, is a result of a resonant generation of free metal electron oscillations, and is known as conventional surface plasmon resonance (SPR). As anticipated in the previous sections, **Figure 1** demonstrates the dependency of SPR on the optical parameters of the metal films, resulting in a much stronger evanescent electric field obtained with silver than with gold. The easiest means of experimental detection of this phenomenon is by measuring the changes in intensity of the totally reflected light (R) which has been used to generate plasmons in the ATR arrangement (**Figure 2**). In both cases, i.e. for R calculated either as a function of α with $\lambda = \lambda_0 = $ constant (**Figure 2A**), or as a function of λ with $\alpha = \alpha_0$ = constant (**Figure 2B**), one obtains a resonance curve resembling in shape those shown in both panels of **Figure 1**. As demonstrated above, in conventional SPR the evanescent electromagnetic field reaches its maximum intensity on the outer metal surface and decays very rapidly with distance into the emerging dielectric medium. This effect is demonstrated in **Figure 3A**. In addition, **Figure 3B** illustrates the sensitivity of the SPR phenomenon to the thickness of the metal film.

Long-range surface plasmon resonance (LRSPR)

The second type of resonance, long-range surface plasmon resonance (LRSPR), is generated in the same way as conventional SPR, but uses a thinner metal layer which is surrounded by dielectric media that are beyond the critical angle so that they support evanescent waves. An example of a calculated LRSPR reflectance versus incident angle curve (**Figure 4A**) demonstrates a very narrow resonance (curve 1) as compared to conventional SPR (curve 2). The distribution of the resonantly generated evanescent electric field intensity along the normal to the film planes, presented in **Figure 4B**, clearly indicates two important differences between this type of resonance and conventional SPR: (1) LRSPR involves two surface bound waves on both the inner

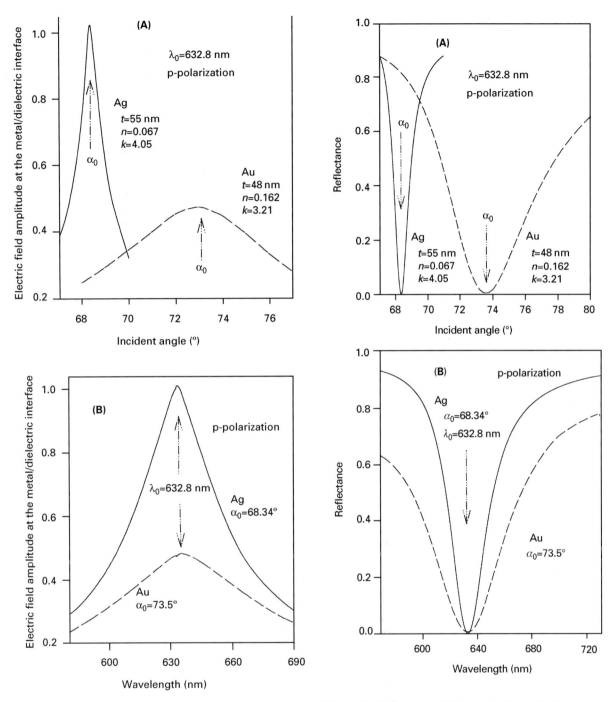

Figure 1 Resonance spectra of the two most frequently used metal films, i.e. silver (solid line) and gold (dashed line), with the indicated optical parameters, presented as the total evanescent electric field amplitude (normalized at its largest value) generated by surface plasmons on the outer surface of the metal film, versus either the incident angle (α, panel A) obtained with p-polarized (transverse magnetic) light of constant wavelength ($\lambda = 632.8$ nm), or light wavelength (panel B) obtained at $\alpha = \alpha_0 =$ constant. α_0 is the incident angle at which the resonance excited with light of $\lambda = \lambda_0 =$ constant reaches its maximum. The calculation has been done assuming a glass prism as an incident medium ($n_g = 1.5151$), and water as an emerging medium ($n_w = 1.33$), both at $\lambda = 632.8$ nm.

Figure 2 Reflectance SPR spectra, i.e. reflectance versus either the incident angle (α), with a constant value of the wavelength (λ_0) of the surface plasmon excitation light (panel A), or the wavelength, λ, at a constant value of α_0 (panel B), obtained with silver and gold films. Other experimental conditions and symbols as in **Figure 1**.

and outer surfaces of the metal layer; and (2) it is characterized by a much higher evanescent electric field at the outer metal surface, which results in a sharper resonance curve, as shown in **Figure 4A**.

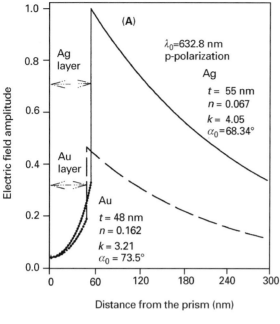

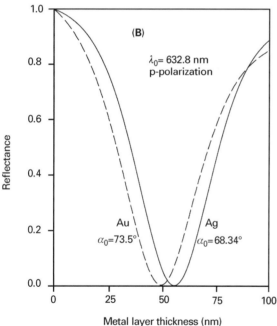

Figure 3 (A) Calculated amplitude of the evanescent electric field (normalized at its largest value) generated within a metallic film (shown by closed circles), obtained with both silver (solid line) and gold (dashed line) at constant values of α_0, as a function of the distance from the glass prism/metal interface. (B) Influence of metal film thickness on the SPR spectra obtained with an excitation wavelength $\lambda_0 = 632.8$ nm for silver (solid line) and gold (dashed line) films. Other conditions as in **Figure 1**.

Coupled plasmon-waveguide resonance (CPWR)

The third type of surface resonance involves even more complex assemblies in which surface plasmon resonances in a thin metal film are coupled with guided waves in a dielectric overcoating, resulting in excitation of both plasmon and waveguide resonances (CPWR). A coupled plasmon-waveguide resonator contains a metallic layer (the same as in a conventional SPR assembly), which is deposited on either a prism or a grating and is overcoated with either a single dielectric layer or a system of dielectric layers, characterized by appropriate optical parameters so that the assembly is able to generate surface resonances upon excitation by both p- and s-polarized light components (**Figure 5A**). The addition of such a dielectric layer (or layers) to a conventional SPR assembly plays several important roles. First, it functions as an optical amplifier which significantly increases electromagnetic field intensities at the dielectric surface in comparison to conventional SPR, as illustrated by **Figures 5B and 5C**. This results in an increased sensitivity and spectral resolution (the latter due to decreased resonance linewidths, as shown in **Figure 5A**). Secondly, it enhances spectroscopic capabilities (due to excitation of resonances with both p- and s-polarized light components), which results in the ability to directly measure anisotropies in refractive index and optical absorption coefficient in a thin film adsorbed onto the surface of the overcoating. Thirdly, the dielectric overcoating also serves as a mechanical and chemical shield for the thin metal layer in practical applications.

Coupled long-range plasmon-waveguide resonance (CLRPWR)

This type of surface resonance can be obtained with a resonator which combines both the long-range surface plasmon and coupled plasmon-waveguide resonators into one device. The resulting resonance spectra are similar in shape and intensity to those obtained with CPWR devices.

Detection of surface plasmon resonances

As discussed above, SP excitation generates surface electromagnetic modes bound to and propagating along the interface between a metal and a dielectric medium. Although these modes differ considerably from plane electromagnetic waves by having a pronounced dispersion (i.e. energy and momentum are not linearly connected by the speed of light), they demonstrate all other properties common to plane waves such as diffraction and interference. Therefore, the SP modes can, in principle, be detected by the same techniques as for plane electromagnetic

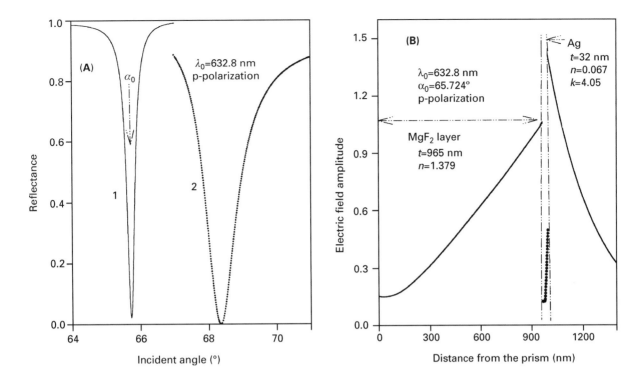

Figure 4 (A) Calculated long-range SPR spectrum represented by reflectance versus incident angle (curve 1), obtained with p-polarized excitation light ($\lambda_0 = 632.8$ nm) and a glass prism coated with a MgF$_2$ layer (thickness = 965 nm, and refractive index = 1.379) on top of which a 32 nm thick silver film has been deposited. α_0 indicates the incident angle at which resonance achieves its maximum. The emerging medium is water with $n_w = 1.33$. Curve 2 illustrates the conventional SPR spectrum obtained with a 55 nm thick silver layer deposited directly on a glass prism (see **Figure 2A**). (B) Amplitude of the evanescent electric field (normalized to the largest value of the conventional SPR electric field as presented in **Figure 3A**) along the normal to the layer plane, calculated for the thin film design described in panel A and using $\lambda_0 = 632.8$ nm as an excitation wavelength at $\alpha = \alpha_0$ (see panel A).

waves, with two additional constraints, namely, that these are surface bound and are primarily nonradiative modes. Taking these special properties into account, together with the requirements which have to be fulfilled in order to excite SP modes, the most direct means of detecting these modes is by analysing the totally reflected light used to excite SPR.

The reflected light can be examined by analysis of either the transformation of light polarization using ellipsometry techniques, or the alteration of light intensity by applying reflectometry methods. The state of polarization is characterized by the phase, δ, and amplitude, E_0, of light polarized parallel (p-polarization) and normal (s-polarization) to the plane of incidence. The difference in polarization state between the incident and reflected light (denoted i and r, respectively) is described by the parameters ψ and Δ, which are defined as follows:

$$\tan \psi = (E_{0p}^r / E_{0s}^r)/(E_{0p}^i / E_{0s}^i) \qquad [8]$$

$$\Delta = (\delta_p^r - \delta_s^r) - (\delta_p^i - \delta_s^i) \qquad [9]$$

where $\tan \psi$ and Δ are the changes in the amplitude ratio and the phase difference, respectively, on reflection. The variables ψ and Δ often referred to as ellipsometrical angles, are related to the ratio between the overall complex reflection (Fresnel) coefficients of the respective light components, r_p, and r_s, by the following equation:

$$r_p / r_s = \tan \psi e^{i\Delta} \qquad [10]$$

The Fresnel coefficients of the interface, r_p and r_s, can, with the aid of Maxwell's theory (as shown in the preceding section) be expressed as functions of the wavelength of the light, λ, the incident angle, α, and the optical properties of the reflecting system.

The reflectometry technique, which is based on measurement of reflected light intensity under ATR conditions, is experimentally a much simpler method than ellipsometry. It is therefore used much more frequently, especially in various sensor applications. As shown above, in this methodology, at a particular

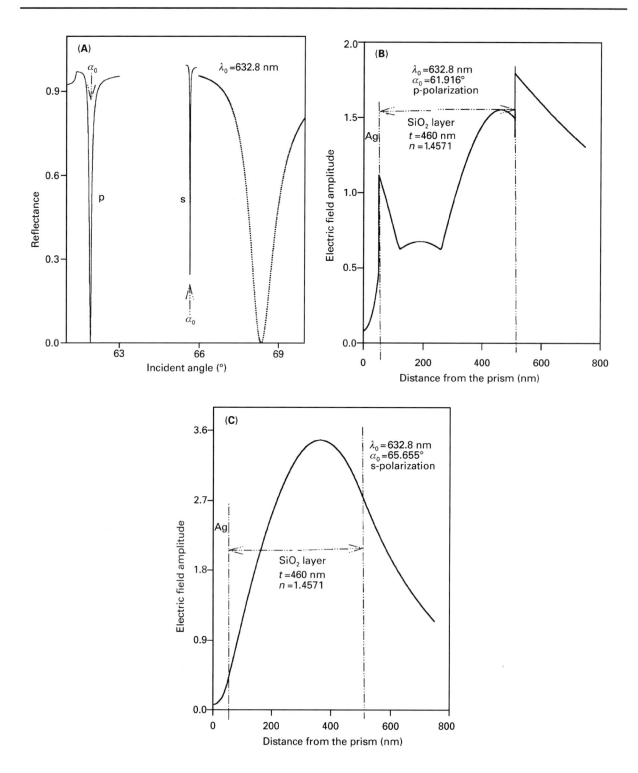

Figure 5 (A) Coupled plasmon-waveguide resonance (CPWR) spectra presented as reflected light intensity versus incident angle, calculated with $\lambda_0 = 632.8$ nm as an excitation wavelength for both p- and s-polarizations, and a glass prism ($n_g = 1.5151$) coated with a 55 nm thick silver layer overcoated with a 460 nm SiO_2 film. The emerging medium is water ($n_w = 1.33$). The curve plotted with solid points illustrates the conventional SPR spectrum obtained with a 55 nm thick silver layer (see **Figure 2A**). α_0 indicates the incident angle at which the resonance achieves its maximum. (B) and (C) Amplitudes of the evanescent electric fields, as a function of the distance from the glass prism/metal interface, within a silver layer, an SiO_2 film, and the emerging medium for p- (panel B) and s- (panel C) polarized light of $\lambda_0 = 632.8$ nm, calculated at $\alpha = \alpha_0$ (see panel A), and normalized to the largest value of the conventional SPR electric field as shown in **Figure 3A**.

incident angle the light wave vector matches the wave vector of the plasmon, fulfilling resonance conditions for plasmon generation (see Eqn [4]). During the resonance interaction, energy is transferred from photons to plasmons, so that the effect of plasmon excitation can be observed as a sharp minimum of the reflectance when either the angle of light incidence is varied at the same light wavelength, or the light wavelength is varied at the same incident angle, thus defining an SPR spectrum. In both instances, this spectrum reflects the resonance in absorption of incident photons.

SPR can also be detected with a fluorescence technique known as total internal reflectance fluorescence (TIRF) employed with waveguided systems. The application of TIRF to monitor SPR is based on the following properties of the surface modes. First, there is a possibility that the nonradiative SP modes can, under specific conditions, decay into light, therefore allowing emission techniques to be utilized to detect the resonance. As noted above, nonradiative surface plasmons can decay into photons of the same frequency if coupling by the surface roughness takes place. The intensity of emitted light images the SPR occurring at silver and gold surfaces producing an emission resonance curve similar to that of the reflectance resonance curve obtained under ATR conditions. Secondly, the surface bound electric field generated by SP modes, which can be much higher than that of the plasmon excitation light (see **Figures 1, 3A, 4B, 5B and 5C**), can be used as an efficient source to excite fluorescence emission of molecules adsorbed at the SPR active surface. This property of the surface plasmon electromagnetic field allows the monitoring of resonance by using fluorescent labelling of molecules adsorbed (or immobilized) on the active surface of the SPR producing medium. In both cases the measurement of fluorescence emission intensity as a function of either incident angle, α, with excitation wavelength, $\lambda = \lambda_0$ = constant, or λ with $\alpha = \alpha_0$ = constant, will generate an excitation emission curve. Such an excitation emission resonance curve will reflect either the SPR absorption resonance spectrum, as usually measured with ATR (see above), or the combination of the absorption spectrum of a fluorescent label and the SPR phenomenon, for measurement of emission intensity versus λ, at $\alpha = \alpha_0$ = constant.

List of symbols

B = tangential electric field amplitude; c = velocity of light in vacuo; C = tangential magnetic field amplitude; E_0 = amplitude of light; k = extinction coeffi-

cient; K = wave vector; K_{ph} = wave vector of light component parallel to interface; K_{SP} = evanescent wave vector; $\mathbf{M}_j$ = matrix of jth layer; n = refractive index; N = complex index of refraction; r = Fresnel reflection coefficient; r^* = complex conjugate of r; R = reflectance; s = number of layers; $\tan \psi$ = change in amplitude ratio; t_j = thickness of jth layer; v = velocity of light in dielectric medium; y_i = oblique admittance; Y_0 = admittance of incident medium; α = incident angle; α_c = critical angle for total internal reflection; δ = phase thickness; Δ = change in phase difference; ε' = real part of dielectric constant; ε'' = imaginary part of dielectric constant; λ = wavelength; ω = frequency of surface plasmon wave.

See also: **ATR and Reflectance IR Spectroscopy, Applications; Ellipsometry; Fluorescent Molecular Probes; Inorganic Condensed Matter, Applications of Luminescence Spectroscopy; Organic Chemistry Applications of Fluorescence Spectroscopy; Surface Plasmon Resonance, Applications; Surface Plasmon Resonance, Instrumentation.**

Further reading

Garland PB (1996) Optical evanescent wave methods for the study of biomolecular interactions. *Quarterly Reviews of Biophysics* 29: 91–117.

Harrick NJ (1967) *Internal Reflection Spectroscopy*. New York: Interscience.

Kovacs G (1982) Optical excitation of surface plasmon-polaritons in layered media. In: Broadman AD (ed) *Electromagnetic Surface Modes*, pp 143–200. New York: Wiley.

Macleod AH (1986) *Thin Film Optical Filters*. Bristol: Adam Hilger.

Raether H (1977) Surface plasma oscillations and their applications. In: Hass G, Francombe M and Hoffman R (eds) *Physics of Thin Films*, Vol. 9, pp 145–261. New York: Academic Press.

Salamon Z, Brown MF and Tollin G (1999) Plasmon resonance spectroscopy: probing molecular interactions within membranes. *Trends in Biochemical Sciences* 24: 213–219.

Salamon Z, Macleod AH and Tollin G (1997) Coupled plasmon-waveguide resonators: A new spectroscopic tool for probing proteolipid film structure and properties. *Biophysical Journal* 73: 2791–2797.

Salamon Z, Macleod AH and Tollin G (1997) Surface plasmon resonance spectroscopy as a tool for investigating the biochemical and biophysical properties of membrane protein systems. I: Theoretical principles. *Biochimica et Biophysica Acta* 1331: 117–129.

Wedford K (1991) Surface plasmon-polaritons and their uses. *Optical and Quantum Electronics* 23: 1–27.

Surface Studies By IR Spectroscopy

Norman Sheppard, University of East Anglia, Norwich, UK

VIBRATIONAL, ROTATIONAL & RAMAN SPECTROSCOPIES
Applications

The investigation of surfaces, and of molecular layers adsorbed on surfaces, by electromagnetic radiation has been carried out principally by infrared spectroscopy. This is because of the high sensitivity of present-day Fourier transform infrared (FT-IR) spectrometers; the capability for IR spectroscopy to obtain data from mixed-phase samples with gas/solid, liquid/solid, or gas/liquid interfaces; and because of the availability of very large databases relating the positions and relative strengths of infrared absorptions to structural features of organic and inorganic molecules. As described below, the sampling techniques used differ substantially whether the systems under investigation involve finely divided samples (powders or porous solids) or whether the surface involved is flat. In the early history of the subject, spectral features relating to adsorbed layers and other surface phenomena could only be detected if very high area finely divided samples were used so that the radiation beam could pass through many interfaces. However, since the advent of FT-IR spectrometers, infrared sensitivity has so much improved that nowadays a measurable spectrum can be produced from even a single monolayer on a flat surface.

After reviewing the experimental techniques involved, we survey the principal applications of the infrared method under the headings surface characterization, physical adsorption, and chemisorption and catalysis.

Experimental techniques

High-area, finely divided, surfaces

Surfaces, because of their unsaturated surface fields, normally require to be cleaned from contamination derived from the atmospheric environment before systematic research can be carried out on them. For finely divided samples of high area it is adequate to mount them in a high-vacuum enclosure ($\sim 10^{-6}$ mbar) provided with infrared-transparent windows and the means for treating the sample in oxygen or hydrogen at elevated temperatures. The samples themselves are most often studied in transmission, usually in the form of pressed discs derived from powders. These are prepared using a hydraulic press in a manner similar to that used for the standard potassium bromide pressed-disc sampling procedure for IR spectroscopy. The pressure required for coherent disc formation is greater for the commonly studied oxide layers than for the softer potassium bromide, but the discs so prepared remain porous for adsorption studies. Alternatively, a powdered sample can often be made to cohere on an infrared-transparent disc, or on a fine metal grid, through sublimation or by deposition from a solvent.

Finely divided metal samples require that the opaque particles are separated from each other for transmission purposes. Usually this is done by distributing (supporting) them on the surface of an oxide which is transparent over relevant regions of the spectrum. Such samples are prepared by depositing metal salts from solution on the oxide particles followed by evaporation of the solvent; a disc is then pressed from the mixed powder and inserted in the IR vacuum cell; finally, the salt is reduced in hydrogen at appropriate temperatures so as to form metal particles distributed over the surface of the oxide support. Very high area powders of silica and alumina, of areas between 200 and 300 m^2 g^{-1}, are commercially available and are frequently used as metal supports. They have the advantages that they are largely infrared-transparent down to ~ 1300 or ~ 1100 cm^{-1} respectively, and hence permit the study of many group-characteristic absorptions from organic adsorbates. Silica is a more catalytically inert support than alumina.

Samples prepared as described above can be good models for working catalysts of either the oxide or metal types and many infrared studies of surface phenomena are undertaken in conjunction with catalytic investigations. Loose powders can alternatively be studied by diffuse reflection, with the advantage for kinetic studies that surface reactions, rather than diffusion processes, are more likely to be rate determining than is the case with the fine-pored pressed discs.

Low-area flat surfaces

Adsorbents in the form of flat surfaces are of very low area and normally ultra-high vacuum (UHV) conditions ($\sim 10^{-10}$ mbar) are required in order to preserve them from contamination. Where the substrate is transparent, infrared spectra of the surface layers can be obtained either by transmission or by reflection; when the substrate is non-transmitting, as in the case of metals, then reflection is normally used. Experiments involving flat surfaces allow the application of polarized radiation for the purpose of obtaining information about the orientation of the adsorbed molecules with respect to the surface. For transparent substrates the used of radiation polarized in or perpendicular to the plane of incidence, in combination with the measured angle of incidence, can determine the direction of the dipole change with respect to the surface associated with each group-characteristic vibration. The orientations of even flexible molecules with respect to the surface can be deduced from such measurements.

In the case of metals the effect of the free response of the conduction-electrons to a charge above the surface can be modelled in terms of an image of opposite sign at the same distance below the surface as is shown in **Figure 1**. In the infrared context it is the dipole moment change associated with the vibration that interacts with the radiation. **Figure 1** shows that a component of such a dipole change that is parallel to the surface is cancelled out by its image, whereas a component perpendicular to the surface is doubled in magnitude. Hence only modes with perpendicular dipole components are IR allowed; in general these are those modes of vibration that are symmetrical with respect to all the symmetry elements associated with the surface complex. For example, a CO molecule adsorbed perpendicular with respect to the surface will give absorption bands from the νCO or νCM (M = metal) bond-stretching modes but not from the OCM bending modes. Such considerations constitute the metal surface selection rule (MSSR), which is widely used for the determination of molecular orientation or, if this is known, as an aid in the assignment of vibrational modes. For work with metals, reflection–absorption infrared spectroscopy(RAIRS) uses near-grazing incidence in order to maximize the strength of the electric vector of the incident infrared radiation that is perpendicular to the surface.

UHV is normally required when studying low-area flat surfaces (exceptionally this would not be a requirement if the adsorbate, such as a surfactant, is capable of displacing surface impurities) and this requires sophisticated equipment. Also, the high sensitivity needed for the measurement of spectra from single monolayers requires the use of FT-IR spectrometers with selective photoconductive infrared detectors; the mercury/cadmium telluride detector which covers the major range of the spectrum down

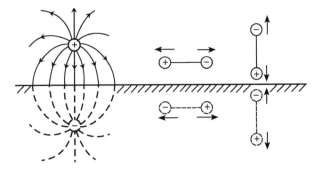

Figure 1 Charges and their images near metal surfaces; the origin of the metal surface selection rule (MSSR).

to ~700 cm^{-1} is widely used. **Figure 2** illustrates a typical experimental arrangement for RAIRS on a metal surface under UHV conditions. Spectroscopic work carried out on single crystals with known types of adsorption sites, such as are readily available for metals, are of great use in interpreting the more complex spectroscopic phenomena obtained from finely divided samples. Individual particles of the latter can exhibit facets with a variety of atomic arrangements and adsorption sites which can be studied one-at-a-time on single crystals. **Figure 3** shows the different atomic arrangements, and hence adsorption sites, of the (111), (100) and (110) faces of a face-centred cubic metallic lattice. UHV facilities also permit complementary spectroscopic methods involving particles such as electrons (as in high-resolution electron energy loss spectroscopy) or diffraction methods (as in low-energy electron diffraction) to be employed in order to characterize the same system further.

Adsorption on metal electrodes, which can be cleaned in solution by electrode reactions, is also studied by RAIRS. There is added interest in the effects of the variable electrode potential on the spectra and structures of the adsorbed species.

The surfaces of infrared-transparent materials that are available in the form of shaped and polished crystals, such as silicon or germanium, can be studied with good sensitivity by using attenuated total internal reflection (ATR) in conjunction with multiple reflection procedures.

Sum frequency generation (SFG) is a recent spectroscopic development in which two laser beams, one in the visible region and the other of variable frequency in the infrared region, generate infrared-modulated signals in the visible region at the sum of the two frequencies. As the signals come only from the interface and not from the bulk, this technique is being exploited in high-pressure catalyst work and for surfactant research.

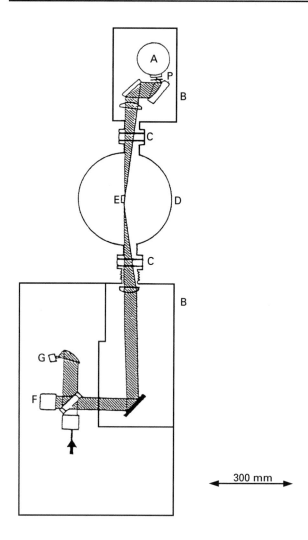

Figure 2 The optical arrangement of an FT-IR spectrometer for reflection–absorption (RAIRS) work in ultrahigh vacuum (UHV). A, detector; B, KBr lens: C, KBr window; D, UHV chamber; E, sample; F, Michelson interferometer; G, Globar source; P, grid polarizer. Reprinted from Chesters MA (1986) *Journal of Electron Spectroscopy and Related Phenomena* **38**:123 Copyright (1986), with permission from Elsevier Science.

Surface characterization

The long-range patterns of surface atomic arrangements are principally monitored by low-energy electron diffraction (LEED). Whereas in principle the top layers of a lattice have different frequencies and hence wavenumbers from those of the bulk lattice, the associated absorptions often fall within a spectral region dominated by the latter and are hence difficult to identify. Transition metal oxides are exceptional in that the variable valency associated with the metallic element can lead to the generation of surface M = O groups (M = metal) that give absorptions of notably higher wavenumber than those of the lattice modes.

Infrared contributions to our knowledge of surfaces are mostly short-range in type and involve the identification of different types of site through the adsorption of probe molecules chosen for this purpose. CO is a well known probe of surface sites on metal surfaces. As discussed in the Chemisorption section below, its νCO bond-stretching vibration has distinct wavenumber ranges for adsorption on linear (on-top), twofold and threefold bridging sites. Although linear and twofold sites can occur on each of the surfaces shown in **Figure 3**, the threefold one is specifically characteristic of (111) surfaces and can be used to identify such facets on metal particles. Distinctions can sometimes be made between twofold CO sites on different facets. The wavenumbers of CO absorptions can also be used to characterize surface cation sites of different charge (different formal oxidation states) on transition metal oxides as shown in **Figure 4** for a partially reduced Ni oxide sample. For hydrocarbon adsorption the characteristic spectrum of ethylidyne (CH_3C) also plays a useful role in identifying (111) facets on finely divided metals.

One of the earliest discoveries of surface infrared spectroscopy was that oxide surfaces, such as those of SiO_2 or Al_2O_3, retain chemisorbed OH groups after the removal of water molecules adsorbed from the atmosphere. These can only be removed by high-temperature treatment and are presumably generated by the reaction of ambient water molecules with otherwise free valencies on the surface of the oxide lattice, according to the reaction $O^{2-} + H_2O \rightarrow 2OH^-$ or its covalent equivalent. In the case of alumina, for example, individual absorptions amongst a multiplicity of OH bond-stretching absorptions can be identified with linear, two- and threefold adsorption sites, for each of two types of surface aluminium atoms which in the bulk lattice have four- or six-fold coordinations, i.e. are in formal IV or VI oxidation states. Silica has only four silicon coordination and correspondingly simpler νOH spectrum consisting

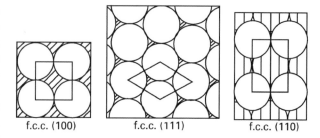

f.c.c. (100)   f.c.c. (111)   f.c.c. (110)

Figure 3 The arrangements of atoms, and the resulting adsorption sites, on the (100), (111) and (110) surfaces of a face-centred-cubic metal.

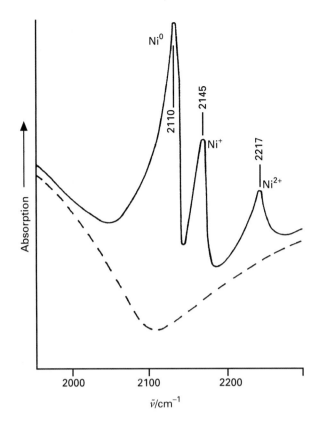

Figure 4 The IR spectrum of CO adsorbed on partially reduced NiY zeolite showing the oxidation-state dependence of νCO. Reproduced with permission of John Wiley & Sons Ltd. from Davydov AA (1984) *Infrared Spectroscopy of Adsorbed Species on the Surfaces of Transition Metal Oxides*, Copyright John Wiley & Sons Ltd.

mainly of absorptions of pairs of free and hydrogen-bonded OH groups on adjacent sites.

An acidic oxide such as alumina exhibits relatively inert OH groups, strongly acidic OH groups that are capable of proton donation (Brønsted acidity), plus aluminium ions that act as electron-deficient sites (Lewis acidity). The relative proportions of these on such oxide surfaces are analysed using the infrared spectrum of adsorbed pyridine. The spectra are measured in the 1700–1500 cm^{-1} region. Pyridine hydrogen-bonded to the weaker surface OHs gives a weakly perturbed spectrum; that interacting with the strongly acidic OHs forms the pyridinium ion by proton transfer which has a characteristic additional absorption at 1540 cm^{-1}; that interacting with metal-coordination sites shows a change in wavenumber of a skeletal absorption near 1600 cm^{-1}.

Basic oxides frequently adsorb carbon dioxide from the atmosphere to give surface carbonates according to the reaction $O^{2-} + CO_2 \rightarrow CO_3^{2-}$, a process readily monitored by infrared spectroscopy. With the well known exceptions of gold and plati-

num, metal surfaces adsorb oxygen from the atmosphere leading, in the cases of aluminium and iron, to the production of multilayer 'passivating' oxide films or (with the participation of absorbed water) thick films of rust, respectively. Infrared spectroscopy can monitor these surface corrosion processes.

Physical adsorption

In this section we consider the spectroscopic study of the association of molecules with surfaces by intermolecular forces ranging from van der Waals to strong hydrogen bonding.

Figure 5 shows the absorption bands in the νCH bond-stretching region from methane adsorbed on porous silica glass. In the gas phase the triply degenerate νCH mode at 3019 cm^{-1} is infrared active and is to be identified with the strong band from the adsorbed species at ~3006 cm^{-1}. On the surface an additional feature has appeared in the spectrum at 2899 cm^{-1} which is readily identified as the gas-phase forbidden νCH 'breathing' mode, known from gas-phase Raman spectroscopy to occur at 2917 cm^{-1}. The one-sided surface forces have distorted the original tetrahedral shape of the methane molecule so as to cause this mode to become active. The considerable breath of the ~3006 cm^{-1} absorption of the surface species, notably less than that of the gas-phase vibration–rotation band, was interpreted in terms of quasi-free rotation of the molecule about a single axis perpendicular to the silica surface.

The spectrum of **Figure 6** shows the interaction of the very acidic surface OH group on zeolite HY with adsorbed ethene. The low wavenumber, broad profile, and intensification of the shifted νOH absorption upon ethene adsorption indicate a hydrogen bond of considerable strength, comparable to that between water molecules, even although the bonding is only to the π-electrons of the adsorbed ethene. This complex

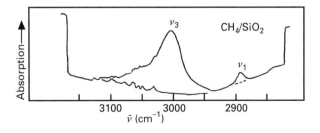

Figure 5 The IR spectrum of CH_4 adsorbed on high-area porous silica glass in the νCH bond-stretching region showing the presence of a gas-phase forbidden absorption. Reproduced with permission of the Royal Society from Sheppard N and Yates DJC (1956) *Proceedings of the Royal Society of London, Series A* **238**: 69.

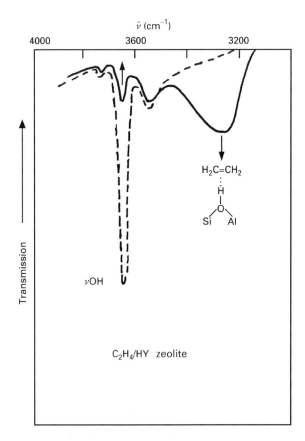

Figure 6 The IR spectrum of ethene adsorbed on the acid OH groups of HY zeolite. Solid line, ethene adsorbed; dashed line, background. Reproduced with permission of the Royal Society of Chemistry from Liengme BV and Hall WK (1966) *Transactions of the Faraday Society* **62**: 3229.

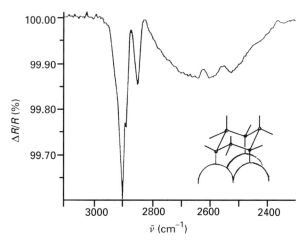

Figure 7 The IR spectrum of cyclohexane adsorbed on a Pt(111) surface. The broad absorption near 2600 cm⁻¹ is from a form of hydrogen bonding between axial CH bonds and surface Pt atoms. Reprinted from Chesters MA and Gardener P (1990) *Spectrochimica Acta, Part A* **46**: 1011, Copyright (1990), with permission from Elsevier Science.

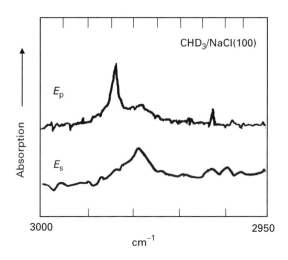

Figure 8 The IR spectrum from CHD₃ adsorbed at 33 K on NaCl(100). E_p and E_s refer to radiation polarized in and perpendicular to the plane of incidence, respectively. Reprinted with permission from Davis KA and Ewing GE (1997) *Journal of Chemical Physics* **107**: 8073. Copyright 1997, American Institute of Physics.

is clearly an intermediate in the higher temperature formation of the carbenium ion $C_2H_5^+$.

Hydrogen bonds are normally considered to form between acidic XH groups and electron-rich bases. However, surface infrared spectroscopy, in conjunction with HREELS, has shown that such bonds can also occur between electron-rich CH bonds of paraffins and electron-deficient sites on metal surfaces. **Figure 7** shows the spectrum of cyclohexane adsorbed on the (111) surface of platinum. The very broad band centred at ~2620 cm⁻¹ is from a proportion of CH bonds of the adsorbed cyclohexane in a hydrogen-bonded type of environment. As the separation between the three parallel axial CH bonds on one side of the cyclohexane molecule is almost exactly the separation between adjacent Pt atoms on a threefold site of the (111) surface, it is clear that the hydrogen bond is of the type CH···Pt.

Figure 8 is a spectrum taken at 33 K in the νCH region of CD₃H adsorbed on the (100) face of the face-centred-cubic lattice of NaCl. It is seen that there are two well resolved absorption bands, one sharp and the other broad, resulting from CH bonds that are oriented differently with respect to the surface. One of these occurs with the incident light polarized in the plane of incidence but is eliminated when the light is polarized perpendicular to this; the other is present in both spectra. The former band hence has its νCH vibrational dipole change perpendicular to the surface, whereas the direction of the latter has both parallel and perpendicular components. Considerations of relative intensities, taking into account the angle of incidence, show that the broader low-wavenumber band is from CH bonds that are at ~70° with respect to the surface. It is hence concluded that the parent CH_4 molecules are adsorbed with three of their four CH bonds on the surface.

Even in the absence of the special effects of hydrogen bonding, bandwidths of 10 cm⁻¹ or more are common from adsorbed species on polycrystalline substrates owing to interactions with sites that differ in their detailed environments. Absorptions obtained from adsorbed species on single-crystal planes with uniform and well defined sites can, in contrast, be very sharp with bandwidths of less than 1 cm⁻¹. In the case of methane itself, and of a number of other molecules such as CO and CO_2, the resolution of the spectra on alkali metal halide single-crystal surfaces are such that even the fine-structure splitting caused by the vibrational couplings of more than one molecule in the surface unit cell can readily be resolved.

Chemisorption and catalysis

The quantitative and energetic aspects of the chemisorption of molecules on surfaces have long been investigated but, until in the 1950s it became possible to obtain infrared spectra, the actual structures of the surface species could only be a matter for speculation. The spectra show that in fact finely divided adsorbents give absorptions from several different surface species, and that the nature of the latter can vary as a function of coverage. Simpler spectra are obtained on single-crystal surfaces of known atomic arrangements. However, even so, the deductions of the structures of the chemisorbed species can be difficult because of uncertainties related to the effects of surface bonding on the spectra of the attached groups. The usual group-characteristic wavenumber ranges can no longer be assumed to be reliable because of the electron-donating or -withdrawing properties of the surface atoms and also, when there is multiple bonding to the surface, because of strains associated with cyclic bonding features. The procedure adopted is to use the spectra to suggest possible alternative structures for the adsorption complexes, and then to look for molecular analogues of known structures whose spectra can be obtained for comparative pattern-recognition purposes. This approach is well exemplified by the results obtained for chemisorption on metal surfaces, an area much studied because of the ready availability of single crystals of metals which can be cut so as to display particular surface planes.

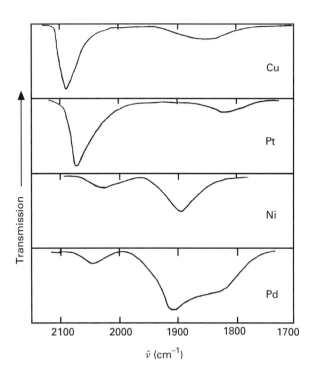

Figure 9 The IR spectra of CO adsorbed on the silica-supported metals Cu, Pt, Ni and Pd. Absorptions above 2000 cm⁻¹ are from linear (on top) CO bonded to one metal atom; those below this value are from CO bridge-bonded to two or three metal atoms. Reprinted with permission from Eischens RP, Pliskin WA and Francis S A (1954) *Journal of Chemical Physics*, **22**: 1786. Copyright 1954 American Institute of Physics.

Figure 9 shows high-coverage spectra obtained from CO chemisorbed on the silica-supported metals Cu, Pt, Ni and Pd. The several metals show different proportions of absorption bands above and below 2000 cm⁻¹ which are characteristic of adsorption on linear (on-top) and bridged sites, respectively. These structural assignments were deduced by comparison with the spectra of metal carbonyls. The spectral ranges attributable to such surface species are as follows: linear, 2120–2000 cm⁻¹; twofold bridge, 2000 – ~1870 cm⁻¹; and threefold bridge ~1900–1800 cm⁻¹. These ranges apply whichever crystal face is involved. Within each range the characteristic absorptions increase in wavenumbers with increasing coverage. This is caused by strong vibrational coupling within the array of parallel molecules on the surface, mostly of a dipolar nature related to the exceptional strength of the νCO absorptions. The mixture of linear and bridged CO species found from the spectra from the finely-divided samples is caused by adsorption on different sites, usually different facets, on the metal particles.

Figure 10 shows νCO spectra at full coverage from chemisorption on an Rh(111) single-crystal electrode. These are plotted as a function of the

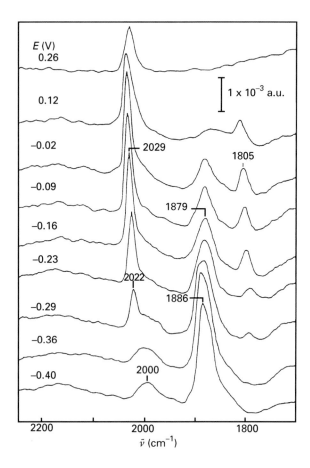

Figure 10 The IR spectra of CO adsorbed at full coverage on an Rh(111) electrode in 0.1 M NaClO$_4$ at various electrode potentials. Reprinted from Chang SC and Weaver MJ (1990) *Surface Science* **238**: 142, Copyright (1990), with permission from Elsevier Science.

electrode potential (with respect to a standard Ag/AgCl electrode) in 0.1 M NaClO$_4$ and show the interest of this additional variable in electrode work. It is seen that at the lowest electrode potential the spectrum is dominated by the absorption at 1886 cm^{-1} from a bridged species but at higher potentials, before desorption sets in, a linear species becomes dominant, absorbing at 2029 cm^{-1}. More generally, work on electrodes shows that, at a given coverage, negative potentials favour bridged species over linear species and that the wavenumbers of νCO absorptions from linear species increase in value with increasingly positive electrode potentials – a milder version of the dependence of νCO on metal oxidation state reported above.

Hydrocarbons on metal surfaces provide greater challenges in spectral interpretation and we choose the example of ethene chemisorbed on different metal surfaces. Here the relevant model compounds are inorganic binuclear or trinuclear metal clusters with the hydrocarbon ligand of interest and additional

CO ligands occupying the positions of the metal atoms of the surface complex. One of the unexpected aspects of the adsorption of ethene is that (111) faces of many metals are covered by the dissociative ethylidyne species CH$_3$CM$_3$ (M = metal) near room temperature. Its spectrum was attributed to this structure by comparison with the spectrum of the model compound (CH$_3$C)Co$_3$(CO)$_9$, considered as a possibility because electron diffraction had shown that the CC bond of the adsorbed species is perpendicular to the surface. This example shows the importance of the metal–surface selection rule (MSSR). For this species, as a ligand or as a surface complex, the modes of vibration are fully separable into those with dipole changes either perpendicular or parallel to the surface (parallel or perpendicular to the CC bond, respectively). Only the former modes are active under the MSSR but both sets are active in the infrared spectrum of the model compound. **Figure 11** compares the infrared spectrum of the ethylidyne species on the Pt(111) surface with that of the model compound. The bands marked with asterisks in the spectra of the model compound (in order of decreasing wavenumber, νCH$_3$ symmetrical stretch, δXH$_3$ symmetrical bend and νCC stretch) are those which give dipole changes perpendicular to the surface; the other doubly degenerate modes give dipole changes parallel to the surface (νCH$_3$ asymmetric stretch, δCH$_3$ asymmetric bend and CH$_3$ rock). The positions of the 'missing' modes of the surface species, indicated by arrows, have been identified in the HREEL spectrum of the same system where the selection rules are more relaxed.

It has been shown by spectroscopy that at low temperatures ethene adsorbs on Pt(111) as the

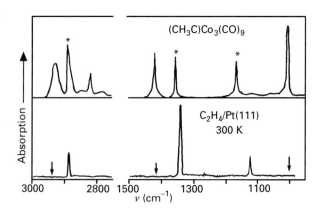

Figure 11 A comparison of the IR spectrum from ethene adsorbed on Pt(111) at room temperature with that of the model compound (CH$_3$C)Co$_3$(CO)$_9$. Asterisks indicate absorptions of the model compound allowed in the spectrum of the adsorbed species by the metal-surface selection rule; arrows indicate other bands observed by HREELS. Courtesy Chesters MA.

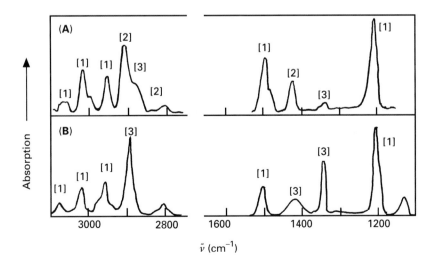

Figure 12 The structures of the principal adsorbed species from the adsorption of ethene on metal surfaces; [1] π-complex; [2] the di-σ species; [3] ethylidyne (CH_3C).

MCH_2CH_2M (di-σ adsorbed) species with a cyclic C_2M_2 skeleton and that this transforms into ethylidyne on warming to near room temperature. The (111) faces of other metals, notably Pd and Cu, show low-temperature spectra from another less strongly perturbed H_2CCH_2 adsorbed species in which there is bonding from a single metal atom to the π-electron distribution of the C=C double bond. Its spectrum is closer to that of ethene itself but with those modes which involve CC stretching occurring at lower wavenumbers. The structures of these three species are shown in **Figure 12**. **Figure 13** shows two spectra from ethene adsorbed on a silica-supported Pt sample, of the type used in catalysis, at 180 K and at room temperature. On these are indicated absorptions from the above three species with the MSSR-allowed modes still dominant. It is seen that the spectra from the catalyst sample are comprehensively accounted for in terms of the species that had been identified one-at-time on single-crystal surfaces.

For the purpose of catalysis, the structure of the surface-adsorbed reactant should be sufficiently perturbed in order to promote reactivity, but not so strongly adsorbed that it cannot be removed by reaction. Below the temperature for the onset of catalysis, controlled by the energy of activation, the reactive species will be one or more of the chemisorbed species. The spectra of such species will weaken or disappear when catalysis commences while less reactive species are retained. In the case of ethene hydrogenation over metal catalysts, the order of reactivity in the presence of hydrogen is [1] > [2] > [3], with [3], the ethylidyne species, being very slow to be removed. By room temperature over Pt/SiO_2, when the di-σ species has all been converted to ethylidyne, it is clear that the π-species, [1], is the catalytically active one. On Pt single crystals this mainly occurs on non-close-packed planes, and it may be inferred that catalytic reduction occurs on rougher, non-(111), surfaces of the metal particles. In a similar manner, it has been shown by single-crystal spectroscopy that the reactive species in the reduction of nitrogen to ammonia over the Fe catalyst (the Haber process) is a di-σ species involving the NN molecule chemisorbed to two Fe atoms which dissociates to adsorbed N atoms during catalysis.

The transition metal oxides form the other principal class of catalysts. These differ from the metals in that they have both acid and base sites in the same surface (the metal and oxygen atoms/ions, respectively) and react differently according to which of these properties is dominant. **Figure 14** shows the infrared spectrum from the heterolytic dissociation of hydrogen on polycrystalline ZnO to given surface

Figure 13 The IR spectra of ethene adsorbed on silica-supported Pt (A) at 180 K and (B) at room temperature, labelled according to the structural assignments of the absorption bands. Reprinted with permission from Mohsin SB, Trenary M and Robota H *Journal of Physical Chemistry*, (1988) **92**: 5229 and (1991) **95**: 6657. Copyright 1988,1991, American Chemical Society.

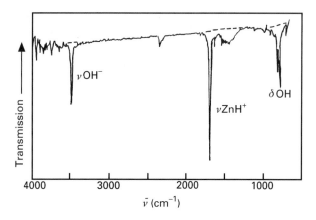

Figure 14 The IR spectra of hydrogen adsorbed on ZnO. Reproduced with permission of the Royal Society of Chemistry from Hussain G and Sheppard N (1990) *Journal of the Chemical Society, Faraday Transactions* **86**: 1615.

HZn$^+$ and OH$^-$ groups. Oxides have mostly been studied in high-area form to date, including the zeolites whose acidic activity occurs on well defined sites within the pores of the crystalline material. Flat single crystals of oxides are difficult to clean in UHV because of their insulating properties. 'Single-crystal' spectroscopic work on oxides is increasing being carried out on thin films grown epitaxially on metal surfaces.

Other vibrational spectroscopic techniques for surfaces

Raman spectroscopy provides valuable complementary vibrational information to IR spectroscopy but its applications to adsorbed molecules has been principally limited to the study of finely divided samples for reasons of reduced sensitivity. Exceptionally, flat monolayers of long-chain surfactant have given Raman spectra using multireflection techniques. Surface-enhanced Raman spectroscopy (SERS) gives greatly enhanced sensitivity but only for molecules adsorbed on the roughened surfaces of the coinage metals, particularly silver.

High-resolution electron energy loss spectroscopy (HREELS), with its higher sensitivity but lower resolution, has played a strongly complementary role to IR in the study of molecules adsorbed on single-crystal metal surfaces.

Inelastic neutron scattering (INS) and inelastic electron tunnelling spectroscopy (IETS) have found more limited applications to the study of the adsorption of molecules on high-area surfaces.

See also: **ATR and Reflectance IR Spectroscopy, Applications; High Resolution Electron Energy Loss Spectroscopy, Applications; Inelastic Neutron Scattering, Applications; Inelastic Neutron Scattering, Instrumentation; IR Spectroscopy, Theory; Raman and IR Microspectroscopy; Surface-Enhanced Raman Scattering (SERS), Applications.**

Further reading

Bell AT and Hair ML (1980) *Vibrational Spectroscopies for Adsorbed Species*, ACS Symposium Series 137. Washington, DC: American Chemical Society.

Clark RJH and Hester RE (eds) (1988) *Spectroscopy of Surfaces*, Advances in Spectroscopy, Vol 16. New York: Wiley.

Davydov AA (1984) *Infrared Spectroscopy of Adsorbed Species on the Surfaces of Transition Metal Oxides.* New York: Wiley.

Sheppard N and De La Cruz C (1996, 1998) Vibrational spectra of hydrocarbons adsorbed in metals. *Advances in catalysis*, Part I, 41: 1–112; Part II, 42: 181–313.

Sheppard N and Nguyen TT (1978) The vibrational spectra of CO chemisorbed on the surfaces of metal catalysts. In: Clark RJH and Hester RE (eds) *Advances in Infrared and Raman Spectroscopy*, Vol. 5. London: Heyden.

Suëtaka W (1995) *Surface Infrared and Raman Spectroscopy – Methods and Applications.* New York: Plenum Press.

Willis RF (ed.) (1980) *Vibrational Spectra of Adsorbates*, Springer Series in Chemical Physics 15. Berlin: Springer-Verlag.

Yates JT Jr and Madey TE (1987) *Vibrational Spectroscopy of Molecules on Surfaces.* New York: Plenum Press.

Surface-Enhanced Raman Scattering (SERS), Applications

WE Smith and **C Rodger**, University of Strathclyde, Glasgow, UK

VIBRATIONAL, ROTATIONAL & RAMAN SPECTROSCOPIES
Applications

Surface enhanced Raman scattering (SERS) was first demonstrated by Fleischmann and colleagues in 1974. In a study of the adsorption of pyridine at a silver electrode, they noted that the Raman scattering was considerably stronger when the surface of the electrode was roughened. Jeanmaire and Van Duyne and Albreicht and Creighton reported that the Raman scattering from pyridine adsorbed on a roughened surface was enhanced by a factor of 10^6 compared to the equivalent concentration of pyridine in solution. This huge increase in signal stimulated a great interest in the technique and it remains one of its main advantages. The technique has been applied in many fields, including surface science, medicinal chemistry and analytical chemistry. Several books and reviews have been written: early developments were surveyed by Furtak and Reyez and Laserna has produced an informative overview indicating the potential to develop a powerful quantitative and qualitative analytical methodology. Chang and Furtak have written a comprehensive book on the subject. Articles directed towards specific applications include one by Vo Dinh targeted at chemical analysis and two by Nabiev and colleagues and Cotton and colleagues targeted at biological and medicinal applications.

The mechanism of the surface enhancement

The nature of the mechanism that produces SERS is still the subject of debate. Two main mechanisms of enhancement are now most commonly proposed. These are electromagnetic enhancement and charge transfer or chemical enhancement. Electromagnetic enhancement does not require a chemical bond between the adsorbate and the metal surface. It arises from an interaction between surface plasmons on the metal surface and the adsorbed molecule. Chemical or charge transfer enhancement requires a specific bond between the adsorbate and the metal plus energy transfer between the metal and the adsorbate during the Raman scattering process. There is evidence for both mechanisms. The predominant view appears to be that both may occur.

Electromagnetic enhancement

On smooth surfaces, surface plasmons exist as waves of electrons bound and confined to the metal surface. However, on a roughened metal surface, the plasmons become localized and are no longer confined and the resulting electric field can radiate both in a parallel and in a perpendicular direction. When an incident photon falls on the roughened surface, excitation of the plasmon resonance of the metal may occur, causing the electric field to be increased both parallel and perpendicular to the surface. The adsorbate is bathed in this field and the Raman scattering is amplified. This mechanism has been studied and reviewed by Weitz, Moskovits and Creighton.

Since SERS has been obtained from molecules spaced off the surface, the existence of enhancement from this type of mechanism is well established.

Charge transfer enhancement

The enhancement from the charge transfer mechanism is believed to result from resonance Raman scattering from new resonant intermediate states created by the bonding of the adsorbate to the metal. The adsorbate molecular orbitals are broadened into resonance by interaction with electrons in the conduction band. Resonance states whose energies lie near the Fermi energy are partially filled, while those lying well below are completely filled. Otto has provided much evidence of the existence of this effect. He showed that there was a specific first layer and has extensively reviewed the field. Campion reported direct experimental evidence linking new features in the electronic spectrum of an adsorbate to SERS, under conditions where electromagnetic enhancements were unimportant. He noted that it was difficult to observe charge transfer only because the electromagnetic effects had to be accounted for and removed. This problem was overcome by conducting SERS on an atomically flat, smooth single-crystal surface where the electromagnetic effects were small and well understood. He adsorbed pyromellitic dianhydride (PMDA) on to copper(III) and observed an enhancement of a factor of 30. In addition, a low-energy band in the electronic spectrum from the adsorbed PMDA was observed that was absent in the solution PMDA spectrum.

Selection rules

Selection rules have been derived for electromagnetic SERS enhancement. The advantage of electromagnetic enhancement is that, since no new chemical species is formed on the surface, the selection rules can be based on the properties of the molecular adsorbate rather than on an ill-understood surface complex. In its simplest form, assuming no specific symmetry rules, the most intense bands are those where a polarization of the adsorbate electron cloud is induced perpendicular to the metal surface. However, more detailed selection rules can be obtained when the molecule has symmetry elements. Creighton and Moskovits have independently reviewed the principles.

Nature of the substrate

The active substrates are usually made from a limited number of metals. Silver, gold and copper are the most commonly used SERS-active metals, although the use of lithium is well established. These substrates were chosen because their surface plasmons exist in or close to the visible region. Ideally, the excitation from the laser should coincide with the plasmon resonance frequency of the particular surface created and conditions such that the efficiency of absorption of the light is reduced and the efficiency of scattering increased. Silver is the most commonly used substrate, although gold is often used particularly in the near infrared. The original experiments used electrochemistry and this is a good method of obtaining a suitable surface. The scale of the roughness required is between about 40 and 250 nm for visible excitation with silver. SERS of pyridine obtained using an electrode setup results in certain bands appearing strongly in the pyridine spectra and the relative intensity and absolute intensity is dependent on voltage. The maximum enhancement is believed to be when the Fermi level matches the energy of the π orbital of pyridine. The electrode working surface can be difficult to reproduce and is prone to annealing with time in certain environments. However, sensitive qualitative analysis is feasible.

Colloidal suspensions are particularly attractive as they can be prepared in a one-pot process and are inexpensive. Reliable SERS analysis is possible as a fresh surface is available for each analysis. Many different methods of colloid preparation have been re-

ported. Some groups always use freshly prepared colloid for their experiments, but recent emphasis has been on obtaining reproducible, monodisperse colloid that is stable for several months. In particular, colloid prepared by the citrate reduction of silver can be produced in almost monodisperse form and with a lifetime of up to one year or more. The particle size of these colloids varies. In one standard preparation of citrate-reduced silver colloid, a transmission electron microscopy study indicated that the particles were approximately 36 nm in their longest dimension and were small hexagonal units (**Figure 1**). Photoelectron correlation spectra of the suspension indicated that the average particle size approximated to a sphere was 28 nm.

Metal colloidal particles adsorbed upon or incorporated into porous membranes such as filter papers, gels, beads, polymers, etc. have been developed as SERS-active substrates. Although these substrates are claimed to be reproducible, they are not widely used, probably because they involve complicated preparative procedures and are susceptible to contamination and self-aggregation. Ruled gratings can be used to give good reproducibility and abraded surfaces, although not so reproducible, they are attractive because of their simplicity of preparation. Numerous researchers have reported that immobilization of the colloidal particles as ordered arrays on films gives reproducible and sensitive SERS sensors.

Surface enhanced resonance Raman scattering

Surface enhanced resonance Raman scattering (SERRS) is obtained by using a molecule with a

Figure 1 Transmission electron microscopy image of silver colloid, × 250.

chromophore as the adsorbate and tuning the excitation radiation to the frequency of the chromophore. The effect was originally reported by Stacey and Van Dyne in 1983. The enhancement obtained is very much greater than that of either resonance Raman or SERS, enabling very sensitive analysis and low detection limits to be achieved. Although SERRS is best considered as a single process, it arose experimentally from the combination of two previously studied effects, namely resonance scattering and surface enhanced Raman scattering (SERS). It is a unique process and different effects can be obtained depending on the nature of the chromophores used and the choice of laser excitation. **Figure 2** illustrates the main choices. In **Figure 2A**, the molecular chromophore (curve a) is chosen not to coincide with the maximum of the plasmon resonance (curve b). If laser excitation at the molecular absorption maximum is used, the maximum contribution to the overall effect from resonance enhancement would be expected. With the arrangement in **Figure 2A** and with the excitation at the molecular resonance maximum, it has been reported that for azo dyes there is reduced sensitivity to surface enhancement mechanisms, providing a signal that is less sensitive to the nature of the surface and that has a recognizable molecular 'fingerprint' related to the resonance spectrum, making this arrangement better for quantitative analysis.

The second possible arrangement illustrated in **Figure 2A** is where the laser excitation is set off the frequency of the adsorbate resonance and at the maximum of the plasmon resonance (2). For resonance experiments on the molecule alone, this would be described as a preresonant condition and often SERRS undertaken in this way is written as SE(R)RS. More orientation information is to be expected and additional bands have been observed and assigned as due to mechanisms of surface enhancement. However, in this preresonant condition, the selectivity of resonance still applies. Thus, it is possible to pick out individual molecules in the presence of a matrix of interferents, but the effect will now be more dependent on the angle of the adsorbate to the surface. For many surface studies this is a key point and consequently this experimental process may be preferred for surface analysis.

Figure 2B gives an alternative case in which the molecular chromophore (curve a) coincides with the surface plasmon maximum (curve b). Similar considerations will apply, but a greater increase in sensitivity is likely at the resonance and plasmon resonance maximum frequency.

Hildebrandt and Stockburger carried out an extensive study on SERRS of Rhodamine 6G in order to explore the enhancement mechanisms involved. They

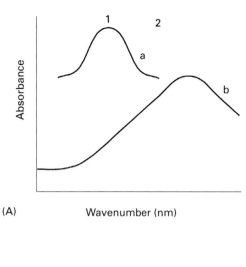

(A)   Wavenumber (nm)

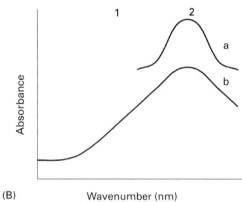

(B)   Wavenumber (nm)

Figure 2 Illustration of the different arrangements for SERRS: curve a is the molecular absorbance and curve b is the plasmon absorbance. In (A) the molecular and plasmon absorbances do not coincide. Position (1) represents excitation at the molecular maximum and (2) represents excitation at the plasmon maximum. In (B) the molecular and plasmon maxima coincide. Position (1) represents excitation away from molecular and plasmon maxima and (2) represents excitation at the absorbance and plasmon maximum.

reported that two different types of adsorption sites on the colloid surface were responsible for the enhancement experienced: an unspecific adsorption site that had high surface coverage on the colloid surface resulted in an enhancement factor of 3000 and could be explained by a classical electromagnetic mechanism; a specific adsorption site was only activated in the presence of certain anions (Cl^-, I^-, Br^-, F^- and SO_4^{2-}). This specific site had a low surface coverage (approximately 3 per colloidal particle); however, an enhancement of 10^6 was claimed to result. This enhancement was believed to be due to a charge transfer mechanism. This study was continued into the near-infrared region and extended to include gold colloid and gold and silver colloid supported on filter papers. The enhancement experienced from anion activation with silver colloid was stronger by a factor of 47 in the near-infrared region compared with the visible region. The authors concluded that this phenomenon could be accounted for by the charge transfer transition being shifted towards the red for Rhodamine 6G, increasing the resonance effect in the near-infrared region.

Advantages and disadvantages of SERS/SERRS

SERS incorporates many of the advantages of Raman spectroscopy in that visible lasers can be used so that flexible sampling is possible, and there is little signal from water so that *in situ* examination of such surfaces as those of colloidal particles in aqueous suspension or of electrode surfaces under solvent can be carried out. The greatest advantages are the sensitivity that can be obtained and the selectivity of the signals. Since SERS will detect compounds down to a level of about 10^{-9} M, adsorbates at monolayer coverage or less can be studied easily. Experiments on pyridine are a classic example. At well below monolayer coverage, pyridine is believed to lie with the plane of the ring almost flat on the metal surface. Under these conditions, there is very little intense Raman scattering since the main polarizability changes in the molecule are parallel to the surface. As the surface density of pyridine increases, the molecule is forced into a more vertical position and the signal begins to appear quite rapidly. This forms a good probe of when monolayer coverage occurs. Further, the existence of selection rules means that an indication of the nature of the surface processes can be obtained.

There are a number of key limitations on the method. First, to obtain a large effect, SERS can be used only for adsorbates on a limited number of metal surfaces in correctly prepared (roughened) form. Second, the very large surface enhancement coupled to the need for a specific molecule to be adsorbed on the surface makes the technique prone to interference. Contaminants that give strong surface enhancement can be detected in much lower concentration than the adsorbate studied, leading to problems in identification. The additional complexity that the intensity of the bands depends on only partially understood selection rules and can change depending on the angle of the molecule to the surface and the degree of packing makes it difficult to assign bands. Finally, there is a tendency in SERS for photodecomposition to occur on the surface. Characteristic broad signals that have been reported as being due to specific

surface adsorbates are probably pyrolysed species on the surface. Notwithstanding these problems, SERS is unique in providing a fascinating insight into the adsorption mechanisms of molecules on suitable surfaces *in situ*.

The technique of SERRS might be assumed to have some of the same disadvantages as SERS and more limitations, but in fact SERRS is proving to be a much more effective technique for analytical science. The major advantage of SERRS is that, if correctly applied, the chromophore signal dominates. Since related spectra are obtained by resonance from solution, the spectra on the surface can easily be recognized, and since the Raman signal from the chromophore is enhanced more than any other molecule this particular species is very readily identified at the surface. Thus, in contrast to the difficulty in assigning signals in SERS in some cases, the signal assignment in SERRS is often simple and reliable. Further, and rather surprisingly, a fluorescence quenching mechanism occurs on the surface so that both fluorescent and nonfluorescent dyes give good SERRS. Provided the molecule is attached to the surface, there is little fluorescence background. In fact, it is often useful to establish the fluorescence background against the strong SERRS signals in order to measure the degree of adsorption and desorption from the surface. Thus, a wide range of chromophores is available. Further, the technique requires very low laser powers and consequently the photodegradation common in SERS is seldom a problem.

The characteristic spectra routinely observed with SERRS permit the identification of mixtures without the need for preseparation. Munro and colleagues have reported the analysis and characterization of 20 similar monoazo dyes, all of which produced unique characteristic spectra that in turn permitted the simultaneous analysis and detection of five dyes presented in a crude mixture. They addressed the problems associated with reproducibility and have focused much attention on improving and standardizing the production of the silver colloid generally used to obtain SERS. They concluded that, with careful attention to detail, a relative standard deviation (RSD) of 5% was routinely obtained.

Applications of SERRS and SERS

The advantages reported above have been exploited in numerous research fields, including the following.

Biochemistry SERRS methodology can be modified in order to provide a biocompatible environment for biological materials. The identification of water-soluble porphyrins and their photostability and interaction with roughened metal surfaces have been reported. For copper chlorophyllin, spectra obtained by using different excitation frequencies permitted a better understanding of a complex system. SERRS enabled the identification of novel chromophores in eye lenses without preseparation of the crude mixture. Finally the oxidation and spin state of proteins such as myoglobin and cytochrome P450 can be probed, and the use of a fluorescent low-concentration protein solution to study labelled tyrosines has been reported.

Medicinal chemistry SERRS has been used to detect the antitumour drug mitoxantrone and its interaction with DNA *in situ*. The adsorption of the drug complex onto a colloidal surface did not destroy or interfere with the native structure. Therefore, bonding information from the complex was extracted from the SERRS spectra. Selective detection on DNA at ultra-low detection by SERRS has been developed.

Surface chemistry SERRS has been used to probe electrode surfaces *in situ* to extract structural information and to provide quantification. The *in situ* SERRS detection of compounds such as 2,4,6-trinitrobenzene sulfonic acid covalently bound to tin oxide was observed when the chemically modified surface was coated with silver. The spectra collected provided rapid and sensitive structural information that was semiquantitative. SERRS has also been used to follow reactions occurring at well below monolayer coverage at roughened metal surfaces.

Polymer science The ability to probe surfaces and boundaries using *in situ* SERS has been exploited extensively in polymer chemistry to characterize the surface of polymers for comparison with the bulk properties, and to determine the molecular geometry, orientation of polymer side groups adjacent to the metal surface and information on bonding, for example of polymer–metal composites such as adhesives and coatings.

Forensic science Modern Raman spectrometers connected to microscopes enable the examination of small amounts of material such as single fibres. The sensitivity and selectivity of SERRS can be exploited in forensic science by determining the nature of the dye mixture *in situ* from a single fibre, from an ink or from a lipstick smear.

Corrosion science Studies of corrosion inhibitors, particularly for copper using SERS of the inhibitor adsorbed on the roughened metal surface, have been used to selectively identify the species. The limitation of requiring a roughened metal surface of a particular metal can be overcome by applying colloid to a smooth nonactive surface, but this field has yet to be exploited.

Practical uses of SERRS have been developed. It has been used to prepare a robust disulfide pH indicator by coupling pH-sensitive dyes – methyl red, cresol violet and 4-pyridinethiol – to cystamine, which adsorbs strongly to the roughened metal surface, forming monolayer coverage of the complex with colloidal silver and allowing strong SERRS to be recorded. Changes in the pH result in changes in the chromophores of the dyes that were easily detected by SERRS. As the SERRS spectrum obtained from the complex was pH sensitive, it was possible to obtain quantitative pH determination.

Another example involves the exploitation of the sensitivity of the technique to analysis of trace amounts of nitrite in fresh and sea waters: sulfanilamide was added to the water sample and reacted with any nitrite present, forming a diazonium salt that was then coupled with ethylenediamine to produce an azo dye that was then detected by SERRS. An enhancement of a factor of 10^9, a relative standard deviation of 10–15% and a limit of detection of picograms were reported. This method was superior to existing colorimetric and chemiluminescence techniques used to analyse the water samples for nitrite. Ultrasensitive detection of metal ions has been reported. A limit of detection at the nanogram level was claimed. The metal ions nickel or cobalt and a ligand, a mixture of 2-pyridinecarboxyaldehyde and 2-pyridinehydrazone or 1,10-phenolanthroline, form a complex on the roughened metal surface that is then detected by SERRS.

Single molecule detection

Rhodamine 6G has been used extensively as a model dye to probe the nature of the SERRS effect. It is an extremely strong fluorophore when excited by visible radiation. Hence normal Raman is not observed except with near-infrared excitation. However, when SERRS is used the dye adsorbs strongly to the roughened metal surface and consequently this strong fluorescence is quenched and an extremely strong, enhanced Raman signal is observed. **Figure 3** illustrates the resonance Raman and SERRS spectra collected from Rhodamine 6G. Attomolar levels (10^{-18} M) of detection have been reported for this system, which is approaching single molecule detection. The fluorescence-quenching properties of surface enhancement coupled with the additional sensitivity obtained from SERRS have been exploited by several researchers. Rhodamine 6G adsorbs very effectively on the roughened silver

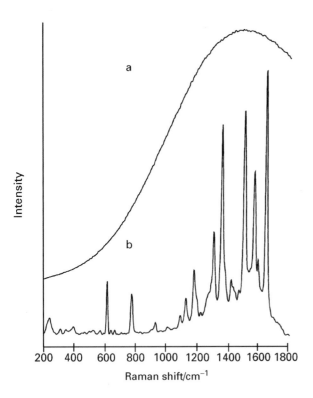

Figure 3 Curve a: Solution spectrum from a 10^{-6} M Rhodamine 6G solution using 514.5 nm excitation, demonstrating the predominance of fluorescence over resonance Raman scattering. Curve b: SERRS spectrum taken from a suspension of aggregated silver colloid to which 150 μL of a 10^{-8} M Rhodamine 6G solution has been added using 514.5 nm excitation.

surface. However, the detection of single adsorbates of dopamine or phthalazine on colloidal clusters, with a limit of detection at picogram levels, illustrates that ultrasensitivity of this technique for other adsorbates is possible.

The ability of SERRS to detect one molecule has recently been demonstrated by three groups. Nie has isolated colloidal particles with rhodamine adsorbed onto glass slides and obtained spectra from the individual particles. The particles are preselected for size to ensure that the surface plasmon of the single particle is in the visible region. Kneipp has used near-infrared anti-Stokes scattering and statistical methods to demonstrate that single molecules can be observed in suspension and Graham and colleagues have shown that one molecule of DNA labelled with a covalently attached fluorescein dye can be detected in the interrogation volume of suspended and aggregated colloid.

Conclusion

In summary, surface enhancement results in a huge enhancement in Raman scattering and the ability to observe Raman signals at very low concentrations.

The resulting spectra provide molecular information and are unique to individual molecules. Sample preparation is simple and it is possible to undertake analysis *in situ* under water or in air or in vacuum. Since there are new selection rules and the effect is dependent on the metal used and the degree of surface roughness, there is a wealth of surface information to be obtained from SERS provided the limitations in terms of contamination and photodecomposition are remembered. The main problems are the limited number of surfaces to which the method can be applied and difficulties in interpreting the spectra.

Additional advantages can be obtained from the use of SERRS. It is more sensitive and the resulting spectrum can be related back to the molecular resonance spectrum, making for more confidence in assignments. Fluorescence is quenched, signals from the adsorbate are much more intense than from contaminants and there is less dependence on the exact nature of the surface. This makes for unique applications for SERRS, which include simpler, more sensitive and more selective quantitative analysis and single molecule detection.

See also: **Biochemical Applications of Raman Spectroscopy; Dyes and Indicators, Use of UV-Visible Absorption Spectroscopy; FT-Raman Spectroscopy Applications; IR and Raman Spectroscopy of Inorganic, Coordination and Organometallic Compounds; MRI of Oil/Water in Rocks; Polymer Applications of IR and Raman Spectroscopy.**

Further reading

Campion A, Ivanecky JE, Child CM and Foster M (1995) *Journal of the American Chemical Society* 117: 11 807.

Chang RK and Furtak TE (1982) *Surface Enhanced Raman Scattering*. New York: Plenum Press.

Cotton TM and Chumanov G (1991) *Journal of Raman Spectroscopy* 22: 729.

Creighton JA (1988) In: Clark RJH and Hester RE (eds) *Spectroscopy of Surfaces*. Chichester: Wiley.

Fleischmann M, Hendra PJ and McQuillan AJ (1974) *Chemical Physics Letters* 26: 163.

Furtak TE and Reyez J (1980) *Surface Science* 93: 351.

Laserna JJ (1993) *Analytica Chimica Acta* 283: 607.

Moskovits M (1982) *Journal of Chemical Physics* 77: 4408.

Nabiev I, Chourpa I and Manfait M (1994) *Journal of Raman Spectroscopy* 25: 13.

Vo Dinh T, Alak A and Moody RL (1988) *Spectrochimica Acta* 43B: 605.

Weitz DA, Moskovits M and Creighton JA (1986) In: Hall RB and Ellis AB (eds) *Chemistry and Structure at Interfaces, New Laser and Optical Techniques*, p 197. Florida: VCH.

Symmetry in Spectroscopy, Effects of

SFA Kettle, University of East Anglia, Norwich, UK

FUNDAMENTALS OF
SPECTROSCOPY
Theory

In this brief article an attempt is made to review those aspects of group theory that may be of concern to a nonspecialist reader. It is written on the assumption that the reader may at some point decide to study a particular aspect of the subject in more detail. However, specialist articles are not always the most accessible and so a particular effort has been made to explain key points in nonmathematical terms. Hopefully, this will provide an insight which will be helpful should the going get tough! For simplicity, reference will always be made to 'molecules'. This should in no case be taken to exclude other species.

Modern spectroscopy makes extensive use of group theory. It does so at several related points. Perhaps most important is in the interpretation of the data that it produces. However, spectral interpretation does not stand alone. In order that the data may be interpreted, the states to which they relate must also be subject to a group theoretical description. In that the states are normally product functions of some sort, the individual functions must themselves have a group theoretical basis. Then, the spectral data correspond to transitions that are made allowed by virtue of some physical process – electric dipole transitions are perhaps the most widely studied but the continuing growth of NMR and EPR methodologies, *inter alia*, is making magnetic dipole transitions of increasing importance; it is also relevant when optical activity is the subject of discussion. The physical process also has to be susceptible to a group theoretical description.

There is an alternative approach to the topic which is of no less validity. This is the study of commuting operators. The statement that two operators commute is equivalent to the statement that you can, in principle at least, make simultaneous measurements of the physical properties associated with each of them. The set of operators that make up the point group of a molecule commute with the Hamiltonian operator and so one can have knowledge of energies which correspond to symmetry-distinct levels. If one can recognize the existence of a valid operator then it can be used in the classification of energy levels; there is no requirement that one can implement in some way the associated physical process. The permutation operator and time-reversal operators

fall into this category, but so too do the operations of the point groups of spectroscopic importance. So, it can prove convenient to extend the concept of the group of operations which commute with the Hamiltonian beyond the simple point group operators and to incorporate others with them.

But, to begin at the beginning. To the majority of users the applications of group theory to spectroscopy start with the simple point groups. The symmetry of a molecule is expressed by the statement that there exist within it certain symmetry elements, such as rotation axes or mirror planes, which relate equivalent parts of a molecule to each other. To apply such an approach there is an implicit assumption that the molecule is a rigid body locked into its lowest potential energy arrangement. It is commonplace to state that it is not the symmetry elements that are of importance but, rather, the corresponding operations. More accurate still is to state that it is the corresponding operators that are of importance, since it is they which commute with the Hamiltonian. It is also important to recognize that there is no requirement of a 1:1 relationship between symmetry element and operator. There is only a single identity (leave alone) operator – often denoted E – but for any molecule there is an infinite number of C_1 axes (C_n means rotation by $360/n$). There is a 1:2 relationship between C_n, $n > 2$, axes and the corresponding C_n operators. Other symmetry elements/operators that will commonly be encountered are mirror planes/mirror plane reflections, denoted σ. These admit of a classification into three types. First, σ_v, where the v stands for vertical and indicates that the mirror plane contains the rotational axis of highest n (i.e. if the axis is vertical, so too is the mirror). Second, σ_h, where the h stands for horizontal and indicates that the mirror plane is perpendicular to the rotational axis of highest n (i.e. if the axis is vertical, the mirror is horizontal). Thirdly, σ_d, where the d is the first letter of the word diagonal. These mirror planes bisect the angle between two twofold axes and also have an axis of at least twofold symmetry lying in them. Care is needed, however, because the use of the symbol σ_d is subject to abuse. If a set of mirror planes is correctly labelled σ_d in one point group they may well be labelled the same way in a subgroup, even though the latter does not contain the two-fold axes of the parent. Other symmetry elements/operators are a centre of symmetry/inversion. These are commonly denoted i (the same symbols are usually used for symmetry element and operator, although different fonts may be used to distinguish them). The final symmetry elements/operators are denoted S_n. They correspond to a rotation about $360/n$ followed either by reflection in a plane perpendicular to the axis or by inversion in the centre of mass of the molecule. Either definition may be used; the sets resulting are in a 1:1 correspondence. The former definition is perhaps the more popular but the latter is perhaps the more useful (it is convenient to have a simple way of determining parity in a descent in symmetry from spherical to point group). Each set of symmetry operators comprise a group which is given a unique label, usually a combination in some way of the labels used for the operations that it contains. An exception to this is found for those groups which have a symmetry such that the coordinate axes are all symmetry-related. They have individual names (I_h, O_h, T_d are the labels of the most important icosahedral, octahedral and tetrahedral groups).

The combination of the group operators gives rise to a group multiplication table and the characters of sets of matrices which multiply isomorphically to the group operators are collected in character tables. These character tables are the starting point for many applications and are widely available, most commonly as appendices in relevant books. A typical character table is given in **Table 1**.

The table of characters is square; each row corresponds to a different irreducible representation of the group. Each irreducible representation has a unique label, given at the left-hand side, such labels are widely used. The label E (with or without suffixes or primes) indicates double degeneracy. Not used here, but the label T (or sometimes F) indicates triple degeneracy. For emphasis, the characters are divided into four sets which will be seen to have a very simple relationship with each other. At the right-hand side of the table are given two columns of basis functions. These are functions which may be used to generate the set of characters in their row but, more usefully, they provide information which can be useful in a spectroscopic analysis. For instance, the electric dipole operators have the same symmetry species as the coordinate axes (here, $A_2'' + E'$). One important use of character tables is to decompose reducible representations into their irreducible components. Reducible representations are produced by most real-life applications of group theory to molecular problems, when a sum of irreducible representa-

Table 1 Table of characters

D_{3h}	E	$2C_3$	$3C_2$	σ_h	$2S_3$	$3\sigma_d$		
A_1'	1	1	1	1	1	1		z^2: X^2+y^2
A_2'	1	1	−1	1	1	−1	R_z	
E'	2	−1	0	2	−1	0	(T_x, T_y)	(x,y)
								$(1/\sqrt{2}\,[x^2-y^2],xy)$
A_1'	1	1	1	−1	−1	−1		
A_2''	1	1	−1	−1	−1	1	T_z	z
E''	2	−1	0	−2	1	0	(R_y, R_x)	(zx, yz)
?	6	0	−2	2	2	−2		

Table 2 Direct product table for the D_{3h} group

D_{3h}	A_1'	A_2'	E'	A_1''	A_2''	E''
A_1'	A_1'	A_2'	E'	A_1''	A_2''	E''
A_2'	A_2'	A_1'	E'	A_2''	A_1''	E''
E'	E'	E'	$(A_1'+A_2'+E')$	E''	E''	$(A_1'+A_2'+E'')$
A_1''	A_1''	A_2''	E''	A_1'	A_2'	E'
A_2''	A_2''	A_1''	E''	A_2'	A_1'	E'
E''	E''	E''	$(A_2'+A_2''+E'')$	E'	E'	$(A_1'+A_2'+E')$

tions is generated and the details of the sum have to be determined. For instance, immediately beneath the character table above is a reducible representation which has $2A_2' + E' + E''$ components. There are simple systematic ways of decomposing reducible representations based on the orthonormality of the irreducible representations. This can be seen in that the product of pairs of characters of any two different irreducible representations, summed over every operation of the group, sum to zero. On the other hand, the squares of characters of any irreducible representation, summed over all of the operations of the group, sum to the number of operations in the group, 12 in the group above.

One cannot go far in spectroscopy without encountering product functions. The individual component functions of product functions have their symmetry properties described by one of the irreducible representations in the appropriate character table. In order to determine the symmetry species of the product function a table of direct products is needed, which enables this to be determined from the irreducible representations of the component functions. Such direct product tables may be obtained from the corresponding character table. That corresponding to the D_{3h} group above is shown as **Table 2**.

The symmetry of **Table 2** arises, of course, because it results from the multiplication of numbers. Such direct product tables can be used to determine the symmetry species of product functions, irrespective of the number of functions; they can be included in sequence. So, the symmetry of the first overtone or combination bands in vibrational spectroscopy is immediately obtained from a direct product table.

However, care has to be taken in the case of overtones of degenerate vibrations. The case of the first overtone illustrates the problem. For the overtone of a doubly degenerate mode there are just three overtone functions, roughly aa, ab and bb. Four are given in the table above. Similarly, for the overtone of a triply degenerate mode the overtone functions, roughly, are aa, ab, bb, bc, cc and ca, six in total, compared with the nine that appear in a direct product table. The reason is that the direct product can be divided into a symmetric and an antisymmetric part (with respect to interchange of the component functions). For overtones, only the symmetric part is relevant, the antisymmetric functions self-cancel (the functions above are rough because they are not all properly symmetric).

Direct product tables also have importance in determining the selection rules of spectroscopy. The statement that a particular transition is allowed is a statement that a transition moment integral is nonzero. Selection rules are general statements about which of all of the possible transition moment integrals can be nonzero. It is group theory which is used to determine which integrals can be nonzero.

An integral may be regarded as the sum of an infinite number of tiny component contributions, one from each of the infinite number of tiny volume elements of space. But for any given tiny volume of space there are $(n-1)$ symmetry-related tiny volume elements (where the number of symmetry operations in the group is n; n is often called the 'order' of the group). When all of the symmetry-related volume elements make equal (in magnitude and sign) contributions to the integral their sum does not vanish and the integral (which includes the contributions from all of such sets) is nonzero. If $n/2$ of the symmetry-related tiny volume elements are positive and $n/2$ are negative then their combined contribution to the integral is zero. Since this pattern is repeated for each and every set of symmetry-related tiny volume elements, the integral itself is zero. But can one make statements about the phase relationships between the symmetry-related members of sets of tiny volume elements?

Character tables such as that above are statements about the phase relationships between functions in the space spanned by the character table. For instance, it is easy to see that any A_2' function has equal positive and negative contributions – multiply each A_2' character by the number of regions of space to which it refers (the numbers at the head of each column) – and so an integral over all space is symmetry-required to be zero. Only integrals of A_1' symmetry (in the D_{3h} point group) can be nonzero. In general, only integrals which transform as the totally

symmetric irreducible representation (which always has characters of +1) of a group can be nonzero. Now, this irreducible representation always occurs on the leading diagonal of the table of direct products and nowhere else in this table. This means that it occurs whenever the direct product is one formed between an irreducible representation and itself. It is this matching which is the basis of simple selection rules. For instance, the often-stated requirement for infrared activity – that to be spectrally active a vibrational mode has to transform like a coordinate axis. This arises because in the triple direct product derived from the symmetry species of:

(excited state) (operator) (ground state)

the vibrational ground state is totally symmetric – the molecule is assumed to be nonvibrating (actually, it would almost be equally valid to say that it is assumed that only nondegenerate vibrations are excited in the ground state – this is the way the group theory works out). For the corresponding integral to be nonzero, the symmetry species of the excited state (which is the symmetry of the mode excited) has to be the same as that of the operator. It is an electric dipole transition and so the symmetry of the operator is that of an electric dipole – and this is the same as that of a coordinate axis.

All of this has assumed a rigid, isolated, molecular species. If we are interested in molecular rotations the theory clearly needs extension – it covers bulk rotations (these find a place in the list of basis functions in the character table above) but there is nothing evident in the theory which would enable the classification of rotational energy levels. Equally, the zero point vibrational energy – and its destructive effect on the assumed molecular geometry – has conveniently been neglected. In fact, there is a validity in this last step. It arises from the fact that the vibrations of a molecule are such that it explores a multidimensional space (each molecular vibration represents an independent variable and so the greater the number of vibrations, the greater the number of dimensions that the molecule can explore vibrationally). This multidimensional space has an inherent symmetry. Thus, if a molecule has (when rigid) a mirror plane of symmetry, then if it is distorted there will be another configuration which is of equal energy to the first but the mirror image of it. That is, all of the symmetry operations of the rigid molecule have a relevance to the multidimensional potential surface. In fact, if one works through the theory, one ends up with a group which is isomorphic to that of the rigid molecule. It is for this reason that the use of the rigid-molecule group gives valid results – it is isomorphic to the correct group. This theory might be called the theory of nearly-rigid molecules; molecules that make but small deviations away from an equilibrium geometry. What of the other extreme – molecules which, whilst retaining their molecular identity, are nonetheless internally very mobile. One has to deal with the molecular symmetry group (as opposed to the molecular point group) of a molecule. Consider what has become a classic example, the molecule CH_3BF_2, which has a rather free rotation about the C–B bond axis. Consider the molecule in a configuration in which it has no point group symmetry other than the identity. Appropriate is an arrangement in which, viewed down the C–B axis, one F almost eclipses an H. Suppose that the H is slightly to the left. Two arrangements of equal potential energy can be generated by rotating one of the two other hydrogens into the position occupied by the first. Three equienergetic arrangements in all. Three more can be obtained by rotating the second F to take the place of the first. Now we have six equivalent arrangements. Six more can be obtained if we go back to the beginning but now place the H in an equivalent position but now to the right. The (physically feasible) interconversions which relate these 12 arrangements constitute the molecular symmetry groups of CH_3BF_2. The definition of 'physical feasibility' is at the heart of the definition of the molecular symmetry group of a molecule and, in contrast to the definition of a molecular point group, has sometimes proved somewhat controversial. In part, at least, this has resulted from attempts to define such groups in a compact way. For instance, inversion of the positions of all particles in the centre of mass of the molecule – a physically highly unfeasible operation – can usefully be included!

Next, the assumption of an isolated molecule. When in solution, a molecule will be subject to an infinity of different solvent environments, unless there is some reason that a particular solvent is 'frozen' around the dissolved species. Such cases are rare. The infinity of environments will cause a broadening of most spectral transitions but there is little more that can be said in general. Much more can be said when the environment is that of a crystalline solid. For a solid, formally, one is not concerned with a point group but, rather, with a space group. These are the space groups of classical crystallography but care has to be taken in that classical crystallography may make use of a nonprimitive unit cell (body or face centred). For spectroscopic purposes it is essential that only primitive unit cells are used (a unit cell, by pure translations generates the entire crystal. If one doubles the size of the unit cell, by body-centring it, for instance, then the number of translation oper-

ations is halved). Even so, it is to be noted that there is no unique definition of a unit cell. If the spectroscopy is such that the surface of the crystal is not of importance then the crystal faces are ignored and the crystal is, effectively, infinite. An important simplification arises because for the vast majority of spectroscopic measurements performed on a crystal, the wavelength of the incident radiation is much greater than a typical primitive lattice translation vector. This enables the use of a so-called factor group in which the entire set of translation operations are incorporated into the identity operation. The set of operations which remain are isomorphic to one of the 32 crystallographic point groups; this isomorphism is exploited by the use of the character table of the appropriate one of the 32 crystallographic point groups in the subsequent analysis. Pictorially, one can think of there being a coupling between all of the molecules within a (primitive and, of course, arbitrary) unit cell, but in reality the formal analysis covers all unit cells. The method of analysis can be called the 'unit cell' or 'factor group' models (they differ in approach, not results); the splittings that occur on 'molecular' features are referred to as 'factor group', 'correlation field' or 'Davydov' splittings. If coupling between molecules does not have to be invoked, the molecular environment in the crystal, the site, may well lead to splittings in (molecular) degenerate modes – site symmetries are commonly lower than molecular. The 'site group' model is then appropriate and the splittings are known as 'site splittings'. The irreducible representation of the group of all translations does not enter any of these analyses. However, they become important in vibrational spectroscopy, for example, if anharmonic vibrations are considered (and they well may be, involving for instance, coupling with lattice modes). In such cases the Brillouin zone, representing the k vectors which define the irreducible representations of the group of all translations, has to be invoked.

Related to the spectroscopy of crystals is the spectroscopy of surfaces and, particularly, the spectroscopy of species adsorbed on crystal surfaces. For perfectly conducting metals, there is an important selection rule in that such surfaces image any electric dipole within an adsorbed molecule. When such dipoles are perpendicular to the surface the dipoles reinforce; when they are parallel to the surface they cancel. This gives rise to the so-called 'surface selection rule', that it is only possible to observe by electric dipole spectroscopy those modes which involve dipole moment changes perpendicular to the surface. This requirement can be expressed group theoretically by use of the so-called diperiodic groups in two dimensions.

Often not too far away from an application of group theory in spectroscopy is the fact that functions derived from atomic orbitals are under study. The study of the group theory of spherical symmetry is fascinating, an infinitesimal rotation about any axis being a symmetry operation. The corresponding infinitesimal rotation operators have important commutation relationships which provide a basis for the theory of angular momentum, for instance. The descent from spherical to molecular symmetry usually means a reduction in degeneracy. For a species corresponding to an angular momentum j the character, χ corresponding to a rotation of $\theta°$ is given by:

$$\chi_j(\theta) = \frac{\sin\frac{1}{2}(2j+1)\theta}{\sin\frac{1}{2}\theta}$$

This equation is equally applicable when j is halfintegral but then it is necessary to work in the so-called double group, in which the factor of $\frac{1}{2}$ in front of the θ means that a rotation of 720° is required to give the identity. Corresponding to each operation in the 0–360° sector there is another in the 360–720°. A C_2 rotation still corresponds to a 180° rotation but it now takes four such operations in succession to generate the identity. Perhaps not surprisingly, C_2 in a double group closely resembles C_4 in a 'normal' group. So, although the group C_{2v} is Abelian (all characters |1|), the C_{2v} double group has degenerate representations – and, indeed, its character table is isomorphic with that of the C_{4v} 'normal' group.

An important application of group theory in spectroscopy arises from the recognition that direct products are at the heart of the topic. Consider the case where two degenerate irreducible representations combine to give a reducible representation (which they invariably do). Of course, the irreducible representations could describe either operators or functions. The irreducible components of the reducible representation will be linear combinations of functions of the type that we have already met, aa, ab, ac and bc, etc. (these functions have the disadvantage that they occur in a direct product of an irreducible representation with itself; however, we seek to illustrate the principle, not to give a general example). The coefficients which relate the basis functions of the starting irreducible representations to the final combinations of product functions do not depend on the particular problem in hand. The so-called vector coupling or 'Clebsch–Gordon' coefficients are universal and so tables of them are available. However, care is needed in their use. This is because for any degenerate irreducible representation an infinite number of choices of basis function is of equal validity and can be used. But the coefficients just introduced are based on a specific choice of basis functions; these same functions must be used in the application of tables of coupling coefficients or errors will surely result (and sometimes the basis functions chosen are not the most obvious – the reason why will become evident in the next paragraph). There are several important outcomes from the use of coupling coefficients. The existence of coupling coefficients means that the final functions are related. If one final (product) function, which could be the intensity of a spectral transition, were available, either by calculation or by measurement, then the values of all of the others follow, based on a so-called 'reduced matrix element' (which is written mathematically with two bars on each side of the operator). This is the Wigner–Eckart theorem. Related to this is the so-called replacement theorem. If the relative values of a set of integrals are determined using coupling coefficients then the same pattern holds for a different set of integrals involving basis functions from the same irreducible representations. The two sets of integrals are proportional to each other. This can be very useful if, for example, one only has an approximate expression for an operator or wavefunction.

It is evident that the more symmetry, the greater the utility of coupling coefficients. Put another way, the greater the inherent degeneracies, the greater their use. The logical consequence of this is that they will be of greatest use in spherical symmetry (and the descent from spherical symmetry to point group symmetry has been described above). In spherical symmetry, that which we have called vector coupling coefficients become replaced by 3j symbols (j as in j–j coupling) and higher extensions, the 6j and 9j symbols. These nj symbols have many advantages. First, they are defined in such a way that they are basis-independent. Secondly, they summarize in compact form what would otherwise be lengthy summations. Those practised in the art use them wherever they can!

See also: **Tensor Representations.**

Further reading

Butler PH (1981) *Point Group Symmetry Applications.* New York: Plenum.

Harter WG (1993) *Principles of Symmetry, Dynamics and Spectroscopy.* New York: Wiley.

Heine V (1960) *Group Theory in Quantum Mechanics.* London: Pergamon.

Piepho SB and Schatz PN (1983) *Group Theory in Spectroscopy.* New York: Wiley.

Tsukerblat BS (1994) *Group Theory in Chemistry and Spectoscopy.* London: Academic Press.

Wolbarst AB (1977) *Symmetry and Quantum Systems.* New York: Van Nostrand.

T

Tandem Mass Spectrometry

See **Hyphenated Techniques, Applications of in Mass Spectrometry; MS–MS and MS^n.**

Tantalum NMR, Applications

See **Heteronuclear NMR Applications (La–Hg).**

Technetium NMR, Applications

See **Heteronuclear NMR Applications (Y–Cd).**

Tellurium NMR, Applications

See **Heteronuclear NMR Applications (O, S, Se, Te).**

Temperature Measurement Using Vibrational Spectroscopy

See **Flame and Temperature Measurement Using Vibrational Spectroscopy.**

Tensor Representations

Peter Herzig, Universität Wien, Vienna, Austria
Rainer Dirl, TU Wien, Vienna, Austria

> **FUNDAMENTALS OF SPECTROSCOPY**
> **Theory**

Introduction

Tensor representations, synonymous for product representations and their decomposition into irreducible constituents, are useful concepts for the treatment of several problems in spectroscopy. Important examples are the classification of the electronic states in atoms and the derivation of selection rules for infrared absorption or the vibrational Raman or hyper-Raman effect in crystals. In the first case the goal is to reduce tensors which are defined as products of one-particle wave functions, while in the second case tensors for the dipole moment, the electric susceptibility or the susceptibilities of higher orders have to be reduced according to the irreducible representations of the relevant point groups.

Basics of group theory

Group postulates

A set of elements $g_1, g_2,...$ forms a group **G** if there is a composition law defined for the elements such that the following conditions are satisfied:

$$g_i \, g_j \; = \; g_k, \qquad g_k \in \mathbf{G} \qquad [1]$$

$$(g_i \, g_j) g_k \; = \; g_i (g_j \, g_k) \qquad [2]$$

$$g_i \, e \; = \; e \, g_i = g_i \qquad [3]$$

$$g_i \, g_i^{-1} \; = \; g_i^{-1} \, g_i = e \qquad [4]$$

Equation [1] represents the binary composition law, Equation [2] states the associative law, Equation [3] shows the existence of an identity element, and Equation [4] shows that to each $g \in \mathbf{G}$ there exists a unique inverse element.

Conjugate elements and classes

For any given group element $g_i \in \mathbf{G}$ the product $g g_i g^{-1}$ can be formed for any arbitrary $g \in \mathbf{G}$ and is called the *conjugate element* of g_i by g. This defines an equivalence relation which subdivides the group **G** into mutually disjoint subsets where the latter, symbolized by $[g_i] = \{g g_i g^{-1} | \, g \in \mathbf{G}\}$, are called classes of conjugate elements.

Products of groups

Direct product of groups Consider two groups **L** and **M** and let $\mathbf{L} \times \mathbf{M}$ be the product set which consists of all ordered pairs $\{(l_j, m_k) | l_j \in \mathbf{L}, m_k \in \mathbf{M}\}$. The set $\mathbf{L} \times \mathbf{M}$ defines a *direct product* group, if its composition law is defined by:

$$(l_j, m_k) * (l_r, m_s) = (l_j l_r, m_k m_s) \qquad [5]$$

If **L** and **M** are finite groups, then the order $|\mathbf{L} \times \mathbf{M}|$ of the product group is given by $|\mathbf{L} \times \mathbf{M}| = |\mathbf{L}| \cdot |\mathbf{M}|$, where the symbol $|\mathbf{G}|$ denotes the order of some group **G**.

Kronecker product of groups The special case $\mathbf{L} = \mathbf{M} = \mathbf{G}$ allows one to define the product group $\mathbf{G} \times \mathbf{G}$ which contains as a special subgroup $\mathbf{G} \boxtimes \mathbf{G} = \{(g,g) | g \in \mathbf{G}\}$ that is isomorphic to the group **G**. The subgroup $\mathbf{G} \boxtimes \mathbf{G}$ is sometimes called the *Kronecker product* group or likewise *diagonal subgroup* of the direct product group $\mathbf{G} \times \mathbf{G}$.

Representations of groups

Matrix representations If for every element $g_i \in \mathbf{G}$ there is a corresponding element $g_i' \in \mathbf{G}'$, such that if $g_i \, g_j = g_k$ in **G**, it follows for the corresponding product of operations $g_i' g_j' = g_k'$ in **G'**, then the two groups are said to be *homomorphic*. A *representation* of a group is defined as a group of nonsingular square matrices homomorphic with the group. The number of rows (columns) of the matrices is called the *dimension* of the representation. Consider the product $g_i \, g_j = g_k$ in **G**. The analogous product of representation matrices is written as:

$$G(g_i) \, G(g_j) = G(g_k) \qquad [6]$$

Carrier spaces – linear operator representations A vector space V is called the carrier (or representation) space for the group **G** if there exists a homomorphic image of **G** of operators $O(\mathbf{G}) = \{O(g)|g \in \mathbf{G}\}$ that leave the space V invariant. The representation $O(\mathbf{G})$ is called the *linear* operator representation $L(\mathbf{G})$ of **G** in V if the operators $O(g) = L(g)$ are linear ones. If V is a unitary space and each operator $L(g) = U(g)$ leaves any scalar product invariant, then $U(\mathbf{G})$ is called a *unitary* operator representation of **G** in V.

For instance, the Hilbert space $L^2(\mathbb{R})^3$ is the carrier space for the 3-dimensional rotation group $O(3, \mathbb{R}) = SO(3, \mathbb{R}) \times \mathbf{C}_i$ where $\mathbf{C}_i = \{e, i\}$ and the symbol i denotes the inversion operation $(\vec{x} \longmapsto -\vec{x})$ in $\mathbb{R}^3$, respectively. If, using the *Euler* angles to characterize the group elements $g = (\omega) = (\alpha, \beta, \gamma) \in \mathbf{SO}(3, \mathbb{R})$ then the linear operators

$$U(\omega) = \exp(-i\alpha L_z)\exp(i\beta L_y)\exp(-i\gamma L_z) \quad [7]$$

$$\lfloor U(i)f\rfloor(\vec{x}) = f(-\vec{x}) \quad [8]$$

define a unitary operator representation of $O(3, \mathbb{R})$ on $L^2(\mathbb{R}^3)$, where $f \in L^2(\mathbb{R}^3)$ represents an arbitrary function of the Hilbert space in question and where the operators L_x, L_y, L_z denote the Cartesian components of the orbital angular momentum operator.

As another example, the Hilbert space $L^2(\mathbb{R}^3) \otimes \mathbb{C}^2$ is the carrier space for the direct product group $O(3, \mathbb{R}) \times SU(2)$ where the unitary operators $U(\omega, i^\sigma) = U(\omega)U(i^\sigma)$ with $\sigma = 0,1$ that are defined by Equations [7] and [8] represent the orbital part together with the spin part a unitary operator representation of $O(3, \mathbb{R}) \times SU(2)$ on the Hilbert space in question. The spin part reads:

$$V(\omega') = \exp(i\alpha' S_z)\exp(-i\beta' S_y)\exp(-i\gamma' S_z) \quad [9]$$

where the entries S_x, S_y, S_z denote the Cartesian components of the intrinsic (spin) angular momentum which are proportional to the *Pauli* matrices σ_j, respectively. Here, it should be noted that due to the fact that $SU(2)$ represents the universal covering group of $SO(3, \mathbb{R})$, the range of variation of the group parameters $(\omega') \in [0, 2\pi) \times [0, \pi) \times [0, 4\pi)$ describing $SU(2)$-elements differs from that of $(\omega) \in [0, 2\pi) \times [0, \pi) \times [0, 2\pi)$ describing $SO(3, \mathbb{R})$ elements. In fact, $\mathrm{Ker}_\varphi = \{(0,0,0), (0,0,2\pi)\}$ where $\varphi(SU(2)) = SO(3, \mathbb{R})$ represents the nontrivial kernel of the homomorphism. Apart from this, we have

$$W(\omega, i^\sigma, \omega') = U(\omega, i^\sigma) \otimes V(\omega') \quad [10]$$

where $W(\omega, i^\sigma, \omega')$ denotes an arbitrary element of the unitary operator representation $U(O(3, \mathbb{R}) \times SU(2))$ acting on the Hilbert space $L^2(\mathbb{R}^3) \otimes \mathbb{C}^2$. In this context it should be noted that $L^2(\mathbb{R}^3)$ is used to describe the quantum mechanical motion of a spinless particle in 3 dimensions, whereas $L^2(\mathbb{R}^3) \otimes \mathbb{C}^2$ is used as the carrier space to describe the quantum mechanical motion of a particle with spin $\frac{1}{2}$ in 3 dimensions.

Basis of a representation Let $\mathcal{H}$ be the carrier space of the group **G** whose elements $g \in \mathbf{G}$ are represented by unitary operators $U(g) \in U(\mathbf{G})$, which is typical when applying group theoretical methods to quantum mechanical problems. Assume that dim $\mathcal{H} = n$ and consider a set of n linearly independent functions φ_i defined in this configuration space with the property:

$$U(g)\varphi_k = \sum_{i=1}^n G(g)_{ik}\, \varphi_i \quad [11]$$

This set of functions defines a basis for the representation G. It is clear from the last equation that the representation matrix for some particular group operation is completely defined if a basis is given and vice versa. It is therefore convenient to label the representation according to the basis, e.g. $G_\varphi(g)$ in order to stress the basis dependence of multidimensional matrix representations. Note in particular, if $\{\varphi_j|j = 1, 2, 3, \ldots n\}$ defines an orthonormal basis, then the corresponding matrix representation $G_\varphi(\mathbf{G}) = \{G_\varphi(g)|g \in \mathbf{G}\}$ of **G** must be a unitary one.

Reducibility of representations, irreducible representations Multidimensional matrix representations of groups are not unique and, if defined via bases of some carrier spaces, sensitively depend on the chosen basis. Any nonsingular linear transformation of the basis $\{\varphi_j|j = 1, 2, 3, \ldots n\}$ to a new basis say $\{\psi_j|j = 1, 2, 3, \ldots n\}$, leads to the following well-known transformation formulae:

$$\psi_k = \sum_{i=1}^n C_{ik}\, \varphi_i \quad [12]$$

$$G_\psi(g) = C^{-1} G_\varphi(g)C \quad [13]$$

where C is an invertible matrix with coefficients C_{ij} defining a similarity transformation. Note in passing, if $\{\varphi_j\}$ and its counterpart $\{\psi_j\}$ are orthonormal bases of the **G**-invariant carrier space $\mathcal{H}$, then C

can be chosen unitary without any loss of generality and the matrix representations $G_\varphi(\mathbf{G})$ and $G_\psi(\mathbf{G})$ are unitary too.

A representation is *reducible* if there is a matrix C which converts all matrices of a representation into the same block-diagonal form. When such a matrix is found the reducible representation is decomposed into a number of representations of dimensions smaller than the original dimension. Representations found in the reduction process that cannot be reduced further are called *irreducible representations*. There are a number of theorems for irreducible representations valid in particular for *finite* groups which shall be given without proof:

1. The number of irreducible representations which are nonequivalent, i.e. not related by a similarity transformation, is equal to the number of classes in the group.
2. The sum over the squares of the dimensions of the irreducible representations is equal to the number of elements in the group $\mathbf{G}$ (*order* of the group, $|\mathbf{G}|$).
3. There is an orthogonality relation between the different irreducible representations of a group (denoted by a Greek letter as a left superscript):

$$\sum_g {}^\alpha G(g)^*_{ij}\, {}^\beta G(g)_{kl} = (|\,\mathbf{G}\,| / |{}^\alpha G\,|)\delta_{\alpha\beta}\,\delta_{ik}\,\delta_{jl} \quad [14]$$

where $|{}^\alpha G|$ is the dimension of the irreducible representation α.

Characters Let the matrices ${}^\alpha G(g)$ form a (reducible or irreducible) representation of a group. The trace of matrix ${}^\alpha G(g)$ is called the *character* of the operation g_i in the representation ${}^\alpha G(g)$:

$$ {}^\alpha \chi(g) = \sum_i {}^\alpha G(g)_{ii} \quad [15]$$

An orthogonality relation corresponding to Equation [14] also exists for the characters of irreducible representations:

$$\sum_g {}^\alpha \chi(g)^{*\,\beta}\chi(g) = |\,\mathbf{G}\,|\,\delta_{\alpha\beta} \quad [16]$$

Symmetrization of states One of the most important applications of group theoretical methods in quantum mechanics consists of the task to construct so-called *symmetrized* states which by definition not

only form an orthonormalized basis of **G**-invariant carrier space but also have a peculiar transformation law with respect to the group **G**. Let $\mathcal{H}$ be a **G**-invariant carrier space which for the sake of simplicity implies that $U(\mathbf{G})$ forms a unitary operator representation on $\mathcal{H}$ and that there exists an orthonormalized basis $\Phi = \{{}^\alpha\phi^s_j | \alpha \in A_G, s = 1, 2, \dots m_\alpha, j = 1, 2, \dots |{}^\alpha G|\}$ with the following properties:

$$\left\langle {}^\alpha\phi^s_j, {}^\beta\phi^t_k \right\rangle = \delta_{\alpha\beta}\,\delta_{st}\,\delta_{jk} \quad [17]$$

$$U(g)\, {}^\alpha\phi^s_j = \sum_{k=1}^{|{}^\alpha G|} {}^\alpha G(g)_{kj}\, {}^\alpha\phi^s_k \quad [18]$$

Here it should be remarked that the symbol $\langle \phi, \psi \rangle$ denotes the scalar product in $\mathcal{H}$ and that A_G symbolizes the index set of the labels which characterize the irreducible **G**-representations, and that the index $s = 1, 2, \dots m_\alpha$ indicates the frequency of the irreducible representations ${}^\alpha G$ in $\mathcal{H}$, respectively.

Usually, the symmetrization of states is carried out by means of a specific projection method which mainly relies upon the properties of the underlying *group algebra*. Without going into any details, we merely state the form and properties of the so-called *units* of the group algebra. The units of the group algebra are represented in the Hilbert space $\mathcal{H}$ by the following operators:

$$\mathsf{P}^\alpha_{jk} = \frac{|{}^\alpha G|}{|\,\mathbf{G}\,|} \sum_g {}^\alpha G(g)^*_{jk}\, U(g) \quad [19]$$

whose properties are well known. Note that P^α_{jj} are projection operators, whereas P^α_{jk} with $j \neq k$ are shift operators. For further details the reader is referred to the relevant literature. However in this context it is worth emphasizing that there exist some modifications of this method. In particular, it is possible to construct complete sets of commuting operators instead of the units of the group algebra. This method deserves extra attention since it circumvents the representation dependence of the units by solving simultaneous eigenvalue equations.

Spherical harmonics – irreducible SO(3, $\mathbb{R}$)-representations The most prominent functions of mathematical physics with applications in many areas of physics and related disciplines are the so-called *spherical harmonics* which form a basis of the Hilbert space $L^2(S)$ where the symbol S denotes the unit sphere. Spherical harmonics are eigenfunctions of the angular momentum operators L^2 and L_z, respectively.

For integer values of l the spherical harmonics take the following form where, in particular, the Condon–Shortley phase convention has been adopted:

$$Y_l^m(\theta,\phi) = N_{lm}\,P_{lm}(\theta)\exp(im\phi) \qquad [20]$$

$$N_{lm} = i^{m+|m|}\left\{\frac{(2l+1)(l-|m|)!}{4\pi(l+|m|)!}\right\}^{1/2} \qquad [21]$$

$$P_{lm}(\theta) = \frac{1}{2^l l!}\sin^{|m|}\theta\,\frac{d^{l+|m|}(\cos^2\theta-1)^l}{d(\cos\theta)^{l+|m|}} \qquad [22]$$

The group elements $(\omega)\in \mathbf{SO}(3,\mathbb{R})$ are represented in the specific Hilbert space $L^2(S)$ in principle by the same unitary operators as given in Equation [7] though the corresponding Hilbert spaces are different. The matrix elements of the $(2l+1)$-dimensional irreducible representations of $\mathbf{SO}(3,\mathbb{R})$ are given by:

$$U(\omega)Y_{lm} = \sum_{m'} R^l_{m'm}(\omega)Y_{lm'} \qquad [23]$$

$$R^l_{m'm}(\omega) = \exp\{-i(m'\alpha+m\gamma)\}d^l_{m'm}(\beta) \qquad [24]$$

$$\begin{aligned}d^l_{m'm}(\beta) =\ &\{(l+m')!(l-m')!(l+m)!(l-m)!\}^{1/2}\\ &\times\sum_k(-1)^{k+m'-m}\\ &\times\frac{\cos^{2l-m'+m-2k}(\beta/2)\sin^{2l+m'-m}(\beta/2)}{(l-m'-k)!(l+m-k)!(m'-m+k)!k!}\end{aligned}$$
$$[25]$$

Equation [24] is also valid for the half-integral values j so that direct products of representations needed for the coupling of arbitrary angular momenta can be calculated. However, one should be aware that for half-integral values j the spherical harmonics due to their definition cannot serve as basis for the corresponding $\mathbf{SU}(2)$-representations.

Direct products of representations

Let $\mathcal{H}_{12}^{\alpha\beta} = \mathcal{H}_1^\alpha \otimes \mathcal{H}_2^\beta$ be the carrier space of the direct product group $\mathbf{G}\times\mathbf{G}$ where the latter contains $\mathbf{G}\boxtimes\mathbf{G}$ as a diagonal subgroup. Let us assume that $\mathcal{H}_1^\alpha$ is the carrier space for the irreducible representations $^\alpha G$ and $\mathcal{H}_2^\beta$ is the carrier space for the irreducible representations $^\beta G$, respectively. In other words, we assume symmetrized basis functions $^\alpha\varphi_i$, $i=1,2,\ldots,|^\alpha G|$ of $\mathcal{H}_1^\alpha$ and equivalently symmetrized basis functions $^\beta\psi_j, j=1,2,\ldots,|^\beta G|$ of $\mathcal{H}_2^\beta$,

respectively. The group elements $(g_1,g_2)\in\mathbf{G}\times\mathbf{G}$ are represented by the unitary operators $U(g_1,g_2) = U_1(g_1)\otimes U_2(g_2)$ where $U_1(g_1)$ acts nontrivially on $\mathcal{H}_1$ only and similarly $U_2(g_2)$ on $\mathcal{H}_2$, respectively. Now, if all possible products

$$\Psi_{ij}^{\alpha,\beta} = {}^\alpha\varphi_i \otimes {}^\beta\psi_j \qquad [26]$$

are formed, then an orthonormal basis of the product space $\mathcal{H}_{12}^{\alpha\beta}$ is defined, since the factor bases $\{^\alpha\varphi_i\}$ and $\{^\beta\psi_j\}$ are assumed to be symmetrized ones. This allows one to define for the Kronecker product group $\mathbf{G}\boxtimes\mathbf{G}$ a unitary $|^\alpha G|\times|^\beta G|$-dimensional matrix representation (in general reducible) which is of the following form:

$$U(g,g)\,\Psi_{ij}^{\alpha,\beta} = \sum_{rs} G^{\alpha,\beta}(g)_{rs,ij}\Psi_{rs}^{\alpha,\beta} \qquad [27]$$

$$G^{\alpha,\beta}(g) = {}^\alpha G(g)\otimes {}^\beta G(g) \qquad [28]$$

This representation is called the *direct product* of the representations $^\alpha G$ and $^\beta G$. It can be decomposed into the so-called *direct sum* of its irreducible constituents, say $^\sigma G$, of $\mathbf{G}$ written symbolically as:

$$G^{\alpha,\beta}(g) = \sum_\sigma \oplus\, a_{\alpha\beta;\sigma}\,{}^\sigma G(g) \qquad [29]$$

$$\begin{aligned}a_{\alpha\beta;\sigma} &= |\mathbf{G}|^{-1}\sum_g {}^{\alpha\times\beta}\chi(g)\,{}^\sigma\chi(g)^*\\ &= |\mathbf{G}|^{-1}\sum_g {}^\alpha\chi(g)\,{}^\beta\chi(g)\,{}^\sigma\chi(g)^* \quad [30]\end{aligned}$$

The coefficients $a_{\alpha\beta;\sigma}$ indicate how often the irreducible representation $^\sigma G$ is contained in the direct product (*multiplicity* or *frequency*).

Clebsch–Gordan coefficients The canonical basis for the irreducible representation $^\sigma G$ in Equation [29] is given by the set of functions $^\sigma\Phi_k^q$, $k=1,2,\ldots,|^\sigma G|$, where the index $q=1,2,\ldots,a_\sigma$ indicates how many times $^\sigma G$ is contained in $^\alpha G\otimes {}^\beta G$. The functions $^\sigma\Phi_k^q$ can be expressed as linear combinations of the products $^\alpha\varphi_i\otimes {}^\beta\psi_j = \Psi_{ij}^{\alpha\beta}$:

$$^\sigma\Phi_k^q = \sum_{ij}\Psi_{ij}^{\alpha\beta}\langle\alpha i,\beta j\,|\,\sigma qk\rangle \qquad [31]$$

The linear-combination coefficients $\langle\alpha i,\beta j\,|\,\sigma qk\rangle$ are commonly called *Clebsch–Gordan coefficients*. The Clebsch–Gordan coefficients form a unitary matrix

$C^{\alpha\beta}$ whose columns are given by the following expressions and are satisfying specific transformation laws which allow on their systematic computation:

$$\left\{\vec{C}^{\alpha\beta}_{\sigma qk}\right\}_{ij} = \langle\alpha i, \beta j\,|\,\sigma qk\rangle \qquad [32]$$

$$G^{\alpha,\beta}(g)\vec{C}^{\alpha,\beta}_{\sigma qk} = \sum_m {}^\sigma G(g)_{mk}\vec{C}^{\alpha\beta}_{\sigma qm} \qquad [33]$$

Here, it is worth emphasizing that Clebsh–Gordan matrices $C^{\alpha\beta}$ are nonsymmetrically indexed, since their rows are labelled by the pairs (i, j) whereas their columns are labelled by the triplets (σ, q, k), respectively. The inverse form of Equation [31] is given by the following expression:

$$\Psi^{\alpha\beta}_{ij} = \sum_{\sigma qk} {}^\sigma\Phi^q_k\langle\sigma qk\,|\,\alpha i, \beta j\rangle \qquad [34]$$

3*nj* Symbols and atomic spectra

Clebsch–Gordan coefficients and Wigner's 3*j* symbols: coupling of two angular momenta

We now consider the coupling of two angular momenta. For this purpose we have to specify Equation [31] to the case $\mathbf{G} = \mathbf{SU}(2)$ which is here written as:

$$|\,(j_1 j_2)jm\rangle = \sum_{m_1, m_2} |\,j_1 m_1 j_2 m_2\rangle\langle j_1 m_1 j_2 m_2\,|\,jm\rangle$$

$$[35]$$

First, it should be noted that the states $|j_1, m_1\rangle = \phi^{\lambda_1}_{j_1, m_1}$ and similarly $|j_2, m_2\rangle = \phi^{\lambda_2}_{j_2, m_2}$ are written in a shorthand notation where the principal quantum numbers λ_1, λ_2 are suppressed. Accordingly, the irreducible representation labels are j_1 and j_2 for the uncoupled representations and j for the coupled representations. The index m ($-j \leq m \leq j$) numbers the different functions of the basis. For the products of the uncoupled functions $|j_1, m_1\rangle \otimes |j_2, m_2\rangle$ we simply write $|j_1 m_1 j_2 m_2\rangle$ for short. The multiplicity index of Equation [31] can be dropped here, because in the special unitary group SU(2) each irreducible representation occurs exactly once in the direct product. It should be noticed that a variety of different symbols are in use in the literature for the Clebsch–Gordan coefficients and different phase conventions exist. Because of

$$m_1 + m_2 = m \qquad [36]$$

$$j_1 + j_2 \geq j \geq |j_1 - j_2| \qquad [37]$$

the sum (Eqn [35]) is only formally a double sum since the condition (Eqn [36]) must hold and for given values j_1 and j_2 the range of variation of j-values is determined by the triangle inequality (Eqn [37]) respectively.

Finally, we emphasize that the Clebsch–Gordan coefficients for **SU(2)** are directly related to the Wigner's 3*j* symbols where the latter have the advantage of being more symmetric than the former.

$$\langle j_1 m_1 j_2 m_2\,|\,jm\rangle = (2j+1)^{1/2}(-1)^{-j_1+j_2-m}$$
$$\times \begin{pmatrix} j_1 & j_2 & j \\ m_1 & m_2 & -m \end{pmatrix} \qquad [38]$$

Calculation and symmetry properties of the 3*j* symbols

The following general formula for the calculation of 3*j* symbols is due to Racah:

$$\begin{pmatrix} a & b & c \\ \alpha & \beta & \gamma \end{pmatrix}$$
$$= (-1)^{a-b-\gamma}F(abc)\{(a+\alpha)!(a-\alpha)!(b+\beta)!$$
$$\times (b-\beta)!(c+\gamma)!(c-\gamma)!\}^{1/2}$$
$$\times \sum_k (-1)^k\{k!(a+b-c-k)!(a-\alpha-k)!$$
$$\times (b+\beta-k)!(-b+c+\alpha+k)!$$
$$(-a+c-\beta+k)!\}^{-1} \qquad [39]$$

$$F(abc) = \left\{\frac{(-a+b+c)!((a-b+c)!(a+b-c)!}{(a+b+c+1)!}\right\}^{1/2}$$

$$[40]$$

where the summation over k extends over all values for which all the factorials are defined. It is valid for $\alpha + \beta + \gamma = 0$ and $\Delta(abc) = 1$ (the latter being a compact notation for the triangle condition (Eqn [37]) otherwise the 3*j* symbols vanish by definition. Much simpler formulae exist for some of the special cases.

The 3*j* symbols have a high degree of symmetry. Many (but not all) of these symmetries can be seen by using a highly redundant notation which has been introduced initially by Regge:

$$\begin{pmatrix} a & b & c \\ \alpha & \beta & \gamma \end{pmatrix} = \begin{bmatrix} -a+b+c & a-b+c & a+b-c \\ a+\alpha & b+\beta & c+\gamma \\ a-\alpha & b-\beta & c-\gamma \end{bmatrix}$$

$$[41]$$

The following operations performed on the square-bracket Regge symbol define 72 symmetries of the $3j$ symbol:

(i) Permutations of the columns. The symbol is multiplied by $(-1)^{a+b+c}$ if the permutation is odd.
(ii) Permutations of the rows. As in (i) the symbol is multiplied by $(-1)^{a+b+c}$ if the permutation is odd.
(iii) Transposition about the main diagonal (like the transposition of an ordinary matrix).

Coupling of three angular momenta

If three angular momenta $\mathbf{J}_1, \mathbf{J}_2, \mathbf{J}_3$ have to be coupled to the total angular momentum $\mathbf{J}$ there are different ways leading to bases related to each other by a unitary transformation. One possibility is to start from the uncoupled state $|j_1 m_1 j_2 m_2 j_3 m_3\rangle$ and to perform first of all the coupling $\mathbf{J}_1 + \mathbf{J}_2 = \mathbf{J}_{12}$ and then $\mathbf{J}_{12} + \mathbf{J}_3 = \mathbf{J}$ leading to a coupled state denoted as $|((j_1 j_2)j_{12} j_3)jm\rangle$. Another possibility is the coupling scheme $\mathbf{J}_2 + \mathbf{J}_3 = \mathbf{J}_{23} = \mathbf{J}_1 + \mathbf{J}_{23} = \mathbf{J}$ with the corresponding wavefunction $|(j_1(j_2 j_3)j_{23})jm\rangle$. With these definitions the following relations between the two types of state functions hold:

$$|((j_1 j_2)j_{12} j_3)jm\rangle = \sum_{j_{23}} |(j_1(j_2 j_3)j_{23})jm\rangle V_{j_{23} j_{12}}^{(j)}$$

[42]

$$V_{j_{23} j_{12}}^{(j)} = \langle (j_1(j_2)j_3)j_{23})jm | ((j_1 j_2)j_{12} j_3)jm\rangle$$

[43]

The coefficients $V_{j_{23} j_{12}}^{(j)}$ which only depend on the values of the js but are independent of m, are called *recoupling coefficients* and can be expressed in terms of Clebsch–Gordan coefficients:

$$V_{j_{23} j_{12}}^{(j)} = \sum_{\text{all } m_i} \langle j_2 m_2 j_3 m_3 | j_{23} m_{23}\rangle \langle j_1 m_1 j_{23} m_{23} | jm\rangle$$
$$\times \langle j_1 m_1 j_2 m_2 | j_{12} m_{12}\rangle \langle j_{12} m_{12} j_3 m_3 | jm\rangle$$

[44]

Analogous to the introduction of the $3j$ symbols

(Eqn [38]) one defines $6j$ symbols:

$$V_{j_{23} j_{12}}^{(j)} = \{(2j_{23} + 1)(2j_{12} + 1)\}^{1/2}(-1)^{j_1 + j_2 + j_3 + j}$$
$$\times \begin{Bmatrix} j_1 & j_2 & j_{12} \\ j_3 & j & j_{23} \end{Bmatrix}$$

[45]

Equation [41] can thus be written as:

$$\begin{Bmatrix} j_1 & j_2 & j_3 \\ l_1 & l_2 & l_3 \end{Bmatrix}$$
$$= \sum_{\substack{\text{all } m_i \\ \text{all } \mu_i}} (-1)^s \left[\begin{pmatrix} j_1 & j_2 & j_3 \\ m_1 & m_2 & m_3 \end{pmatrix} \begin{pmatrix} j_1 & l_2 & l_3 \\ m_1 & \mu_2 & -\mu_3 \end{pmatrix} \right.$$
$$\times \left. \begin{pmatrix} l_1 & j_2 & l_3 \\ -\mu_1 & m_2 & \mu_3 \end{pmatrix} \begin{pmatrix} l_1 & l_2 & j_3 \\ \mu_1 & -\mu_2 & m_3 \end{pmatrix} \right]$$

[46]

where $s = l_1 + l_2 + l_3 + \mu_1 + \mu_2 + \mu_3$.

Calculation and symmetry properties of the 6j-symbols

With the abbreviation (Eqn [40]) Racah's formula for the $6j$ coefficients reads:

$$\begin{Bmatrix} a & b & c \\ d & e & f \end{Bmatrix} = F(abc)F(aef)F(dbf)F(dec)$$
$$\times \sum_k \left\{ \frac{(-1)^k (k+1)!}{(a+b+d+e-k)!(b+c+e+f-k)!(a+c+d+f-k)!} \right.$$
$$\times \left. \frac{1}{(k-a-b-c)!(k-a-e-f)!(k-b-d-f)!(k-d-e-c)!} \right\}$$

[47]

The $6j$ symbols are invariant under interchange of the columns. A further invariance exists with respect to the interchange of any two numbers from the top row with their counterparts in the bottom row, e.g.:

$$\begin{Bmatrix} j_1 & j_2 & j_3 \\ l_1 & l_2 & l_3 \end{Bmatrix} = \begin{Bmatrix} j_1 & l_2 & l_3 \\ l_1 & j_2 & j_3 \end{Bmatrix}$$

[48]

The $6j$ symbols are different from zero only if the following four triangle conditions are fulfilled:

$$\Delta(j_1 j_2 j_3) = \Delta(j_1 l_2 l_3) = \Delta(l_1 j_2 l_3) = \Delta(l_1 l_2 j_3) = 1$$

[49]

and if the sum of the side lengths of these triangles are integers. Further symmetries of the $6j$ symbols have been found but they are not discussed here.

9j and 12j symbols

The $9j$ symbols are required for the coupling of four angular momenta. An example is the coupling of two orbital and two spin angular momenta either within the Russel–Saunders (LS) or the jj coupling scheme. The $9j$ symbols can also be used as recoupling coefficients for the transformation from one of the two schemes into the other. They can be expressed in terms of $3j$ or $6j$ symbols. Analogously, the coupling between five angular momenta can be described by the $12j$ symbols for which formulae in terms of $6j$ and $9j$ symbols exist.

Property tensors and vibrational spectra of crystals

Property tensors

Like other physical objects, as, for example, elementary particles, atoms and molecules, crystals are characterized by their symmetry which, as has been known now for more than a century, has a determining influence on their physical properties. The underlying symmetry principle, often called 'Neumann–Minnigerode–Curie principle' can, for our purposes, be written as:

$$G_{\text{object}} \subseteq G_{\text{property}} \quad [50]$$

where G_{object} is the symmetry group of the object and G_{property} the symmetry group of the physical property. Therefore the symmetry operations for the crystal are also valid symmetry operations for its physical properties.

On the other hand, crystals are media that are anisotropic, which means that the application of certain causes onto the crystal (e.g. by an electric field) leads to a response or an effect (like the induced polarization) which depends on the orientation of the crystal. Both quantities, the electric field **E** and the electric polarization **P**, are vectors which, in general, point in different directions. For sufficiently low electric fields this leads to a linear relationship between the electric field and the electric polarization, which can be written as tensor equations as follows:

$$\mathbf{P} = \chi \mathbf{E} \quad [51]$$

$$P_x = \chi_{xx}\, E_x + \chi_{xy}\, E_y + \chi_{xz}\, E_z$$
$$P_y = \chi_{yx}\, E_x + \chi_{yy}\, E_y + \chi_{yz}\, E_z \quad [52]$$
$$P_z = \chi_{zx}\, E_x + \chi_{zy}\, E_y + \chi_{zz}\, E_z$$

$$P_i = \chi_{ij} E_j \quad [53]$$

Note that Equation [51] represents formally the tensorial relationship, while Equation [52] expresses this relation by the (Cartesian) components of **P** and **E** and some coefficients χ_{ij}, whereas Equation [53] states this relation by using Einstein's summation convention. Here, the coefficients χ_{ij} are the components of the electric susceptibility tensor which is a tensor of rank 2. The tensor χ is an example of what is usually called a *property tensor* or *matter tensor*. Strictly speaking, property tensors describe physical properties of the static crystal which belong to the totally symmetric irreducible representation of the relevant point group. Properties, however, that depend on vibrations of the crystal lattice are described by tensors which belong to the different irreducible representations. The corresponding tensors are then often designated as *tensorial covariants*.

If the electric field is not low enough for a linear relationship (Eqn [53]) to hold higher-order terms become important:

$$P_i = \chi_{ij}^{(1)}\, E_j + \frac{1}{2}\chi_{ijk}^{(2)}\, E_k E_j + \frac{1}{6}\chi_{ijkl}^{(3)}\, E_l E_k E_j + \cdots \quad [54]$$

While $\chi^{(1)}$ is important for the Raman effect, the electric susceptibilities of higher order ($\chi^{(2)}$, $\chi^{(3)}$, etc.) are tensors of ranks 3, 4, etc., and are necessary for the description of the hyper-Raman effect.

All tensors can be classified by their parity, i.e. by their behaviour under the symmetry operation of the inversion of the space. Tensors (like vectors) are said to have even parity if they remain unchanged under space inversion and have odd parity if they change sign. In a tensor equation such as Equation [54], the parities on both sides of the equations must be the same. Thus, since **E** and **P** have odd parity, the parity of the susceptibility tensor must be even. Instead of using the term 'parity' one usually designates tensors as *polar tensors* (tensors of even rank having even parity or tensors of odd rank having odd parity) or *axial tensors* (tensors of odd rank having even parity or tensors of even rank having odd parity).

A further criterion for the classification of tensors is their behaviour under time reversal. This is important when one is interested in magnetic properties, for instance in the Raman effect of magnetic crystals. All tensors we are dealing with in this section are assumed to be invariant under time reversal. The latter are also known in the literature as *i*-tensors.

In addition to the above-mentioned classification, a property tensor may have an *intrinsic symmetry*, i.e. a symmetry with respect to the interchange of certain indices which is determined by the symmetry properties of the tensors for the cause and for the effect.

Transformation properties of tensors of rank [*m*]

Here we consider the *Euclidean* vector space $\mathbb{E}^3 \simeq \mathbb{R}^3$ and assume that the components of an arbitrary vector $\mathbb{R}^3$ are its Cartesian components with respect to a fixed orthonormal basis where the former are denoted by $x_1 = x$, $x_2 = y$, $x_3 = z$, respectively. Accordingly a tensor of rank [*m*] is defined as follows:

$$\left\{ \mathsf{T}^{[m]} \right\}_{i_1, i_2, \ldots i_m} = x_{i_1} x_{i_2} \cdots x_{i_m} \qquad [55]$$

$$\mathsf{T}^{[m]'} = \mathrm{g}\, \mathsf{T}^{[m]} \qquad [56]$$

$$\mathrm{g}\mathsf{T}^{[m]} = \mathbf{M}^{[m]}(g)\, \mathsf{T}^{[m]} \qquad [57]$$

$$\mathbf{M}^{[m]}(g) = M(g) \otimes M(g) \otimes \cdots \otimes M(g) \qquad [58]$$

$$\left\{ \mathsf{T}^{[m]'} \right\}_{i_1, i_2, \ldots i_m} = M_{i_1 j_1}(g) M_{i_2 j_2} \cdots M_{i_m j_m} \left\{ \mathsf{T}^{[m]} \right\}_{j_1, j_2, \ldots j_m}$$
$$[59]$$

Here the entries $\{\mathsf{T}^{[m]}\}_{i_1, i_2, \ldots i_m}$ denote the Cartesian components of the tensor of rank [*m*] where $i_j = 1, 2, 3$ has to be taken into account. Equations [56] to [58] describe the transformation properties of an arbitrary tensor of rank [*m*] with respect to the orthogonal group $\mathbf{O}(3, \mathbb{R})$, respectively. The matrix group $\mathbf{M}^{[m]}$ is the *m*-fold tensor representation of $\mathbf{O}(3, \mathbb{R})$, where $M(g)$ is a real 3-dimensional matrix representation of $g \in \mathbf{O}(3, \mathbb{R})$, which implies that $\mathbf{M}^{[m]}$ defines a real 3^m-dimensional $\mathbf{O}(3, \mathbb{R})$-representation. Moreover, note that in Equation [59] Einstein's summation convention has been used. Finally, one has to distinguish carefully between *polar* and *axial* tensors of rank [*m*] since their inherent transformation properties with respect to $\mathbf{O}(3, \mathbb{R})$,

being stated below, are significantly different.

$$\mathrm{g}\, \mathsf{T}^{[m]}_{\mathrm{polar}} = \mathbf{M}^{[m]}(g)\, \mathsf{T}^{[m]}_{\mathrm{polar}} \qquad [60]$$

$$\mathrm{g}\, \mathsf{T}^{[m]}_{\mathrm{axial}} = (\Delta(g))^m\, \mathbf{M}^{[m]}(g)\, \mathsf{T}^{[m]}_{\mathrm{axial}} \qquad [61]$$

$$\Delta(g) = \det M(g) \qquad [62]$$

Here it is important to notice that $\Delta(g) = \det M(g) = +1$ for proper rotations and $\Delta(\mathrm{g}) = \det M(\mathrm{g}) = -1$ for improper rotations, like for the inversion or reflections. Thus, the factor $(\Delta(g))^m$ occurring in Equation [61] is the *m*-th power of this phase factor and confirms what has been stated in the preceding section as regards the transformation properties of tensors with respect to the space inversion.

Calculation of property tensors (tensorial covariants)

Property tensors of arbitrary rank [*m*] can be constructed by symmetry adapting their Cartesian components along the lines of the well-known projection method where the group $\mathbf{G}$ is assumed to be a finite subgroup of $\mathbf{O}(3, \mathbb{R})$. In other words, here the symmetry adaptation of the tensors is carried out for point groups and some nontrivial extensions like for magnetic point groups, which we discuss later. The procedure can be divided into several steps. First, the projection operators are defined, and secondly they are applied to the Cartesian tensors, and finally the shift operators are applied to the symmetrized tensors to obtain the corresponding partner tensors. The procedure is summarized as follows:

$$^{[m]}\mathsf{P}^{\gamma}_{ij} = |^{\gamma} G \,|\, |\, \mathbf{G}\, |^{-1} \sum_{g} {}^{\gamma}G(g)^*_{ij}\, \mathrm{g} \qquad [63]$$

$$^{[m]}\mathsf{X}^{\gamma}_{i} = {}^{[m]}\mathsf{P}^{\gamma}_{ii}\, \mathsf{T}^{[m]} \qquad [64]$$

$$^{[m]}\mathsf{X}^{\gamma}_{j} = {}^{[m]}\mathsf{P}^{\gamma}_{ji}\, \mathsf{T}^{[m]} \qquad [65]$$

Several remarks are necessary. The units $^{[m]}\mathsf{P}^{\gamma}_{ij}$ of the corresponding group algebra are provided with the superscript [*m*] in order to indicate that their representation is realized in the space of tensors of rank [*m*], respectively. The first step is to use Equation [64] in order to generate just one tensor for the $|^{\gamma}G|$-dimensional irreducible representation. In fact, the number of free parameters must coincide with the frequency of $^\gamma G$ in the reducible tensor representation $\mathbf{M}^{[m]}$. The other tensors of a multidimensional tensor basis are obtained from Equation [64] by applying the appropriate shift operators. The given formulae are valid for the calculation of polar and axial tensors, since one merely has to use either Equation [60] for polar tensors or Equation [61] together with Equation [62] for axial tensors, respectively. For the sake of clarity, the explicit forms of the calculated tensors the following notation is used in order to define tensors of rank 2 and rank 3 in matrix form:

$$\mathbf{T}^{[2]} = \begin{bmatrix} xx & xy & xz \\ yx & yy & yz \\ zx & zy & zz \end{bmatrix} \quad [66]$$

$$\mathbf{T}^{[3]} = \begin{bmatrix} xxx & xxy & xxz \\ xyx & xyy & xyz \\ xzx & xzy & xzz \\ - & - & - \\ yxx & yxy & yxz \\ yyx & yyy & yyz \\ yzx & yzy & yzz \\ - & - & - \\ zxx & zxy & zxz \\ zyx & zyy & zyz \\ zzx & zzy & zzz \end{bmatrix} \quad [67]$$

Since the components of property tensors are quantities that can be measured experimentally, they have to be real. The results obtained from Equations [64] and [65] are real, if real rotation matrices and real irreducible representations are used. The latter can always be achieved for the multidimensional representations. However, there are 10 crystallographic point groups which have pairs of one-dimensional irreducible representations complex conjugate to each other. In a case like this, one first computes pairs of conjugate complex tensors and then forms two real linear combinations for each conjugate pair as illustrated below for C_3 (or 3 in Hermann–Mauguin notation).

Polar tensors of rank 2 for group C_3 The characters for C_3 are shown in **Table 1** and the required rotation matrix for the operation C_3^+ can be taken from **Table 4** (the matrix for C_3^- is the square of the matrix for C_3^+). With these data the polar tensors of rank 2 can be calculated. Using the abbreviations $a = xx + yy$, $b = xy - yx$, $c = zz$, $d = \frac{1}{2}(xx - yy)$, $e = \frac{1}{2}(xy - yx)$, $f = xz$, $g = yz$, $h = zx$ and $j = zy$, the resulting tensors that are complex for the representations ^1E and ^2E are displayed in **Table 2**. Adding the

two tensors for ^1E and ^2E gives the first real tensor and multiplying the one for ^1E with $-i$ and the one for ^2E with i and adding them together gives the second real tensor for the 'physically irreducible representation' E, as it is sometimes called (see **Table 3**).

A note on magnetic point groups In order to describe consistently some properties of magnetic systems that are related to their symmetry, the concept of ordinary point groups was extended, to so-called *magnetic* point groups. In order to cope these types of problems a new operation has been introduced designated as *antisymmetry operation*. The latter does not affect the space coordinates but only reverses the sign of the magnetic moment at each point in space. Instead of ordinary point groups one has to deal with magnetic point groups (colour groups) in which some of the ordinary point-group operations appear in combination with the antisymmetry operation. To summarize, one can also construct symmetry adapted tensors for magnetic point groups, where the crucial difference consists of a

Table 1 Character table for the group $\mathbf{C}_3$

C_3 (Eqn [3])	E	C_3^+	C_3^-
A	1	1	1
^1E	1	$\varepsilon^\star$	ε
^2E	1	ε	$\varepsilon^\star$

$\varepsilon = \exp(2\pi i/3)$

Table 2 Tensors of rank 2 for the group $\mathbf{C}_3$

A	^1E	^2E
$\begin{bmatrix} a & b & 0 \\ -b & a & 0 \\ 0 & 0 & c \end{bmatrix}$	$\frac{1}{2}\begin{bmatrix} d+ie & e-id & f-ig \\ e-id & -d-ie & g+if \\ h-ij & j+ih & 0 \end{bmatrix}$	$\frac{1}{2}\begin{bmatrix} d-ie & e+id & f+ig \\ e+id & -d-ie & g-if \\ h+ij & j-ih & 0 \end{bmatrix}$

Table 3 Real Raman tensors for the group $\mathbf{C}_3$

A	E	
$\begin{bmatrix} a & b & 0 \\ -b & a & 0 \\ 0 & 0 & c \end{bmatrix}$	$\begin{bmatrix} d & e & f \\ e & -d & g \\ h & j & 0 \end{bmatrix}$	$\begin{bmatrix} e & -d & -g \\ -d & -e & f \\ -j & h & 0 \end{bmatrix}$

Table 4 Generators for the groups $\mathbf{D}_3, \mathbf{D}_{3v}, \mathbf{D}_{3d}$

C_{3z}^+	C_{2x}	σ_x	i
$\begin{bmatrix} -\frac{1}{2} & -\frac{\sqrt{3}}{2} & 0 \\ \frac{\sqrt{3}}{2} & -\frac{1}{2} & 0 \\ 0 & 0 & 1 \end{bmatrix}$	$\begin{bmatrix} 1 & 0 & 0 \\ 0 & -1 & 0 \\ 0 & 0 & -1 \end{bmatrix}$	$\begin{bmatrix} -1 & 0 & 0 \\ 0 & 1 & 0 \\ 0 & 0 & 1 \end{bmatrix}$	$\begin{bmatrix} -1 & 0 & 0 \\ 0 & -1 & 0 \\ 0 & 0 & -1 \end{bmatrix}$

certain modification of the transformation formulae (Eqns [60] to [62]).

Raman and hyper-Raman tensors for Laue class $\bar{3}m$

Here, the Raman and the hyper-Raman tensors for trigonal lattices shall be given. The property tensor for the Raman effect is the derivative of the polarizability with respect to the normal coordinate and is a polar second rank i-tensor which is symmetric if the resonant Raman effect or a degenerate ground state are excluded while the corresponding tensor for the hyper-Raman effect is a polar i-tensor of rank three the internal symmetry of which is dependent on the experimental conditions.

The matrix generators for the groups belonging to this Laue class, namely $\mathbf{D}_3$ (32), $\mathbf{C}_{3v}$ (3m) and $\mathbf{D}_{3d}$ (3m) are shown in **Table 4** and the representation matrices (in real form) for the generators in **Table 5** (the information for $\mathbf{D}_{3d}$ can easily be generated by forming the direct product $\mathbf{D}_{3d} = \mathbf{D}_3 \times \mathbf{C}_i$ where $\mathbf{C}_i = \{E, i\}$).

The obtained tensors (here given without any intrinsic symmetry) are displayed in **Table 6** where for the tensors of rank 2 the two numbers on the right-hand side of the matrices give the numbers of independent components of the tensor without any intrinsic symmetry and for the symmetric tensor, respectively. Summing these numbers for all irreducible representations must give 9 (the number of components of a general tensor of rank 2 without intrinsic symmetry) and 6 (the number of components of a symmetric tensor of rank 2). For the tensors of rank 3 the numbers of independent components without intrinsic symmetry, for intrinsic symmetry with respect to the interchange of two indices and for the totally symmetric tensor are indicated in a similar fashion. The corresponding sums over the irreducible representations must yield 27, 18 and 10 in this case taking into account the required intrinsic symmetries. In **Table 7** the Raman and hyper-Raman tensors for the Laue class $\bar{3}m$ are displayed.

Selection rules

In order to determine whether a transition between given initial and final states is allowed or forbidden, one only has to know whether the following scalar product is zero or not:

$$\langle \Psi_i^\alpha, O_k^\gamma \Psi_j^\beta \rangle = \int_{\mathbb{R}^3} d^3x \, \Psi_i^\alpha(\vec{x})^* O_k^\gamma \Psi_j^\beta(\vec{x}) \quad [68]$$

where the operator O_k^γ is the electric dipole moment, the electric susceptibility or the electric susceptibility

Table 5 Irreducible representations for the generators

$\mathbf{D}_3$	$\mathbf{C}_{3z}^+$	$\mathbf{C}_{2x}$	$\mathbf{C}_{3v}$	$\mathbf{C}_{3z}^+$	σ_x
A_1	1	1	A_1	1	1
A_2	1	-1	A_2	1	-1
E	$\frac{1}{2}\begin{bmatrix} -1 & -\sqrt{3} \\ \sqrt{3} & -1 \end{bmatrix}$	$\begin{bmatrix} 1 & 0 \\ 0 & -1 \end{bmatrix}$	E	$\frac{1}{2}\begin{bmatrix} -1 & -\sqrt{3} \\ \sqrt{3} & -1 \end{bmatrix}$	$\begin{bmatrix} -1 & 0 \\ 0 & 1 \end{bmatrix}$

Table 6 Tensors

$$\mathbf{P}_2 = \begin{bmatrix} a & 0 & 0 \\ 0 & a & 0 \\ 0 & 0 & b \end{bmatrix} \begin{matrix} 2 \\ 2 \\ 2 \end{matrix} \qquad \mathbf{Q}_2 = \begin{bmatrix} 0 & c & 0 \\ -c & 0 & 0 \\ 0 & 0 & 0 \end{bmatrix} \begin{matrix} 1 \\ 0 \\ 0 \end{matrix}$$

$$\mathbf{R}_2^{(1)} = \begin{bmatrix} d & 0 & 0 \\ 0 & -d & e \\ 0 & f & 0 \end{bmatrix} \begin{matrix} 3 \\ 2 \\ \end{matrix} \qquad \mathbf{R}_2^{(2)} = \begin{bmatrix} 0 & -d & -e \\ -d & 0 & 0 \\ -f & 0 & 0 \end{bmatrix} \begin{matrix} 3 \\ 2 \\ \end{matrix}$$

$$\mathbf{P}_3 = \begin{bmatrix} a & 0 & 0 \\ 0 & -a & b \\ 0 & c & 0 \\ \hline 0 & -a & -b \\ -a & 0 & 0 \\ -c & 0 & 0 \\ \hline 0 & d & 0 \\ -d & 0 & 0 \\ 0 & 0 & 0 \end{bmatrix} \begin{matrix} \\ \\ \\ \\ 4 \\ 2 \\ 1 \\ \\ \\ \end{matrix} \qquad \mathbf{Q}_3 = \begin{bmatrix} 0 & e & f \\ e & 0 & 0 \\ g & 0 & 0 \\ \hline e & 0 & 0 \\ 0 & -e & f \\ 0 & g & 0 \\ \hline h & 0 & 0 \\ 0 & h & 0 \\ 0 & 0 & i \end{bmatrix} \begin{matrix} \\ \\ \\ \\ 5 \\ 4 \\ 3 \\ \\ \\ \end{matrix}$$

$$\mathbf{R}_3^{(1)} = \begin{bmatrix} j+k+l & 0 & 0 \\ 0 & j & m \\ 0 & n & p \\ \hline 0 & k & m \\ l & 0 & 0 \\ n & 0 & 0 \\ \hline 0 & q & r \\ q & 0 & 0 \\ s & 0 & 0 \end{bmatrix} \begin{matrix} \\ \\ \\ \\ 9 \\ 6 \\ 3 \\ \\ \\ \end{matrix} \qquad \mathbf{R}_3^{(2)} = \begin{bmatrix} 0 & l & m \\ k & 0 & 0 \\ n & 0 & 0 \\ \hline j & 0 & 0 \\ 0 & j+k+l & -m \\ 0 & -n & p \\ \hline q & 0 & 0 \\ 0 & -q & r \\ 0 & s & 0 \end{bmatrix} \begin{matrix} \\ \\ \\ \\ 9 \\ 6 \\ 3 \\ \\ \\ \end{matrix}$$

Table 7 Raman (RT) and hyper-Raman tensors (HRT) for Laue class

Point group	Rep.	RT	Rep.	HRT
$\mathbf{D}_{3d}$ ($\bar{3}m$)	A_{1g}	$\mathbf{P}_2$	A_{1u}	$\mathbf{P}_3$
	A_{2g}	$\mathbf{Q}_2$	A_{2u}	$\mathbf{Q}_3$
	E_g	$\mathbf{R}_2^{(1)}, \mathbf{R}_2^{(2)}$	E_u	$\mathbf{R}_3^{(1)}, \mathbf{R}_3^{(2)}$
$\mathbf{D}_3$ (32)	A_1	$\mathbf{P}_2$	A_1	$\mathbf{P}_3$
	A_2	$\mathbf{Q}_2$	A_2	$\mathbf{Q}_3$
	E	$\mathbf{R}_2^{(1)}, \mathbf{R}_2^{(2)}$	E	$\mathbf{R}_3^{(1)}, \mathbf{R}_3^{(2)}$
$\mathbf{C}_{3v}$ (3m)	A_1	$\mathbf{P}_2$	A_1	$\mathbf{Q}_3$
	A_2	$\mathbf{Q}_2$	A_2	$\mathbf{P}_3$
	E	$\mathbf{R}_2^{(2)}, \mathbf{R}_2^{(1)}$	E	$\mathbf{R}_3^{(1)}, \mathbf{R}_3^{(2)}$

of second order, depending on whether infrared absorption, Raman or hyper-Raman scattering is to be investigated and Ψ_i^α and Ψ_j^β are the initial and final state, respectively. The integral (Eqn [68]) is different from zero only if the totally symmetric irreducible representation is contained in the direct product representation $^\alpha G^* \otimes {}^\beta G \otimes {}^\gamma G$. According to Equation [30] this is the case if the corresponding frequency number

$$a_{\gamma\beta;\alpha} = |\,G\,|^{-1} \sum_g {}^\alpha\chi(g)^{*\gamma}\chi(g)^\beta\chi(g) \qquad [69]$$

is greater than zero. Often, Ψ_j^α belongs to the totally symmetric irreducible representation in which case its characters are all equal to 1.

In order to obtain the characters for the (reducible) representation to which the operator O_k^γ belongs one only has to consider matrices of the form:

$$M(g) = \begin{bmatrix} \cos\phi_g & -\sin\phi_g & 0 \\ \sin\phi_g & \cos\phi_g & 0 \\ 0 & 0 & \pm 1 \end{bmatrix} \qquad [70]$$

where the positive sign holds for proper operations (pure rotations about the z axis) and the negative sign for improper operations (rotations combined with reflections about the xy plane).

For infrared absorption O_k^γ is the operator for the electric dipole moment which is a polar vector transforming like the Cartesian coordinates x, y, z. The corresponding character is given by the trace of the matrix (Eqn [70]).

$$\chi_{\text{IR}}(g) = 2\cos\phi_g \pm 1 \qquad [71]$$

For the Raman effect the operator is that for the electric polarizability which is a polar tensor of rank 2. Without any intrinsic symmetry the character is the square of the trace of the matrix (Eqn [70]). However, since we assume a symmetric tensor here, the character becomes:

$$\chi_{\text{R}}(g) = 2\cos\phi_g(2\cos\phi_g \pm 1) \qquad [72]$$

For the hyper-Raman effect we only consider the case of a totally symmetric polar tensor of rank 3. Under this assumption the character is:

$$\chi_{\text{HR}}(g) = 2\cos\phi_g(4\cos^2\phi_g \pm 2\cos\phi_g - 1) \qquad [73]$$

With this information and only using group character tables one can easily find out to which symmetry species ψ_i^α may belong in order that a transition is observed.

There is a general rule that follows from the symmetry considerations above, called *mutual exclusion rule*: 'In groups having a centre of symmetry the modes active in Raman scattering are inactive in infrared absorption and vice versa'. An explanation for this behaviour is that the property tensor for infrared absorption is a polar vector and for Raman scattering a polar tensor of rank 2. While the parity under space inversion of the former is odd and thus belongs to an *ungerade* representation of the point group, it is even for the latter and belongs to a *gerade* representation.

In Raman spectroscopy the relative intensity of the scattered light is straightforwardly calculated from the appropriate Raman tensor as follows: If $\mathbf{v}_i$ is a unit vector defining the polarization of the incident laser radiation and $\mathbf{v}_s$ a unit vector characterizing the polarization of the scattered light, then the following proportionality for the intensity I of the total scattered radiation holds:

$$I \sim |\,\mathbf{v}_i \cdot \boldsymbol{\chi} \cdot \mathbf{v}_s\,|^2 \qquad [74]$$

where the inner product has to be formed between the three tensor quantities on the right-hand side. With a suitable experimental arrangement the components of the Raman tensor can thus be measured independently.

See also: **Atomic Absorption, Theory; IR Spectroscopy, Theory; Nonlinear Raman Spectroscopy, Applications; Nonlinear Raman Spectroscopy, Instruments; Nonlinear Raman Spectroscopy, Theory; Rotational Spectroscopy, Theory; Symmetry in Spectroscopy, Effects of; Vibrational, Rotational and Raman Spectroscopy, Historical Perspective.**

Further reading

Altmann SL and Herzig P (1994) *Point-Group Theory Tables*. Oxford: Clarendon Press.

Brandmüller J, Illig D and Herzig P (1999) *Symmetry and Physical Properties of Matter. Rank 1, 2, 3 and 4 property tensors for the irreducible representations of the classical and magnetic, crystallographic and non-crystallographic point groups*. IVSLA Series, Vol. 2, Amsterdam: IOS Press.

Brandmüller J and Winter FX (1985) Influence of symmetry on the static and dynamic properties of crystals. Calculation of sets of the Cartesian irreducible tensors for the crystallographic point groups. *Zeitschrift für Kristallographie* **172**: 191–232.

Chaichian M and Hagedorn R (1998) *Symmetries in Quantum Mechanics. From Angular Momentum to Supersymmetry.* Bristol and Philadelphia: Institute of Physics Publishing.

Chen JQ (1989) *Group Representation Theory for Physicists.* Singapore: World Scientific.

Claus R, Merten L and Brandmüller J (1975) Light Scattering by Phonon-Polaritons. *Springer Tracts in Modern Physics* **75**: 1–237.

Condon EU and Odabaşı H (1980) *Atomic Structure.* Cambridge: Cambridge University Press.

Joshua SJ (1991) *Symmetry Principles and Magnetic Symmetry in Solid State Physics.* Bristol, Philadelphia and New York: Adam Hilger.

Poulet H and Mathieu JP (1976) *Vibration Spectra and Symmetry of Crystals.* New York: Gordon and Breach.

Rotenberg M, Bivins R, Metropolis N and Wooten JK (1959) *The 3-j and 6-j symbols.* Cambridge, Massachusetts: MIT Press.

Wigner EP (1959) *Group Theory and its Application to the Quantum Mechanics of Atomic Spectra.* New York: Academic Press.

Thallium NMR, Applications

See **Heteronuclear NMR Applications (B, Al, Ga, In, Tl).**

Thermospray Ionization in Mass Spectrometry

WMA Niessen, hyphen MassSpec Consultancy, Leiden, The Netherlands

MASS SPECTROMETRY

Methods & Instrumentation

Introduction

Thermospray ionization is a soft ionization technique, applicable with a thermospray interface for combining liquid chromatography and mass spectrometry (LC-MS). The thermospray interface was developed by Vestal and co-workers at the University of Houston (TX) and subsequently commercialized by Vestal in the company Vestec (Houston, TX). The thermospray interface was the first LC-MS interface, where the analyte ionization is an integral part of the introduction of the column effluent into the mass spectrometer. Between 1987 and 1992, thermospray interfacing and ionization was the most widely used strategy for LC-MS coupling. After 1992, its use diminished in favour of interfacing strategies based on atmospheric-pressure ionization, i.e. electrospray and atmospheric-pressure chemical ionization (APCI).

In a thermospray interface, a jet of vapour and small droplets is formed by heating the column effluent of an LC column or any other continuous liquid stream in a heated vaporizer tube. Nebulization takes place as a result of the disruption of the liquid by the expanding vapour that is formed upon evaporation of part of the liquid in the tube. A considerable amount of heat is transferred to the solvent prior to the onset of the partial inside-tube evaporation. This assists in the desolvation of the droplets in the lower pressure region. By applying efficient pumping directly at the ion source, up to 2 mL min^{-1} of aqueous solvents can be introduced into the MS vacuum system. The ionization of the analytes takes place by mixed mechanisms based on gas-phase ion–molecule reactions and ion evaporation processes. The reagent gas for ionization can be made either in a conventional way using energetic electrons from a filament or discharge electrode, or in a process called thermospray buffer ionization, where the volatile buffer dissolved in the eluent is involved. Thermospray interfacing and ionization as well as its applications in LC-MS have been reviewed in two excellent review papers by Arpino and in two extensive book chapters.

Thermospray interface

The thermospray interface is the result of a long-term research project between 1978 and 1984, aimed at the development of an LC-MS interface which is compatible with a flow-rate of 1 mL min⁻¹ of an aqueous mobile phase and is capable of providing both electron ionization and chemical ionization mass spectra. Initially, the two most important research topics were related to the ability to (1) achieve a very rapid heating and subsequent vaporization of the column effluent, and to (2) achieve sufficient vacuum conditions to successfully perform analyte ionization and mass analysis, while introducing large amounts of liquid vapour, i.e. the equivalent of 1 mL min⁻¹ aqueous solvents. The developments with respect to the various heating systems investigated for the rapid vaporization of the mobile phase from the LC column are summarized in **Table 1**. The highly complex setup of the first prototype, featuring laser vaporization of the mobile phase and an extensive vacuum system containing orthogonal quadrupole analysers, was simplified in subsequent interface designs. The heat required for the evaporation of the 1 mL min⁻¹ aqueous mobile phase could, instead of using an expensive laser system, also be provided with an electrically heated vaporizer capillary (see **Table 1**). In addition, the vacuum system could be significantly simplified by connecting a mechanical

rotary pump directly to the ion source block. The highly directed flow of the liquid vapour jet from the thermospray vaporizer considerably enhances the pumping efficiency of this pump.

A commercial thermospray interface consists of a direct-electrically heated vaporizer type, mostly fitted into a spray probe, a heated source block featuring a filament, a discharge electrode, a repeller electrode and an ion-sampling cone to the mass analyser, and the exhaust pump outlet. A schematic diagram of a typical thermospray system is shown in **Figure 1**. The temperature of the vaporizer tube must be accurately controlled in order to ascertain the partial liquid evaporation, required for successful thermospray ionization. Automatic compensation for changes in the solvent composition during gradient elution should be incorporated. A liquid nitrogen trap is positioned in the exhaust line between the source block and the rotary pump in order to avoid contamination of the pump oil by solvent used in LC-MS. Obviously, minor instrumental differences with respect to vaporizer design, temperature control and source block design are present between the various commercial systems. Two types of thermospray vaporizers have been in use, i.e. the Vestec-type vaporizer where the temperature control is based on measuring temperatures both at the stem near the solvent entrance and at the tip, where the nebulization is complete, and the Finnigan-type vaporizer where a thermocouple is spot-welded close to the inlet side at approximately one-quarter of the heated length.

A schematic representation of the thermospray nebulization process is shown in **Figure 2**. Initially, in the first part of the vaporizer tube, the liquid is heated until, at a certain stage, the onset of vaporization takes place. The vaporization process will start at the heated capillary walls and results in tearing of the liquid: bubbles are formed within the liquid.

Table 1 Characteristics of various heating systems investigated in the development of the thermospray interface

Heat supply	Heated length (mm)	Energy flux (W cm⁻²)	Total power (W)
CO₂ laser beam focused on liquid jet	0.3	30 000	25
Hydrogen flames to heat a copper cylinder at the capillary exit	3	5 000	50
Indirect electrically heated capillary	30	700	100
Direct electrically heated capillary	300	70	150

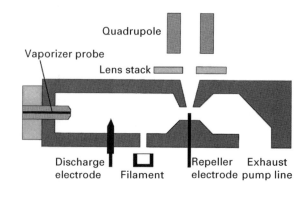

Figure 1 Schematic diagram of a thermospray interface and ion source.

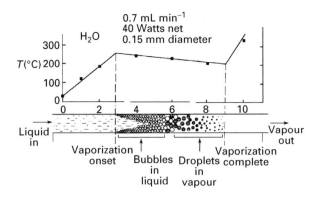

Figure 2 Schematic representation of the thermospray vaporization process. Reprinted with permission from Vestal ML and Fergusson GJ (1985) *Analytical Chemistry* **57**: 2373–2378, © 1985, American Chemical Society.

Upon continuing vaporization, the stage of bubbles in the liquid transforms to liquid droplets in a vapour. The temperature measured at the vaporizer tube wall over the length of the capillary is also shown in **Figure 2**. When complete solvent evaporation inside the tube would be achieved, a sharp increase of the capillary wall temperature would be observed, where the vapour is heated. However, optimum ionization conditions are achieved at nearly complete inside-tube vaporization. From this description of the nebulization process, it may be concluded that the contact time between the liquid and the analyte molecules dissolved in the liquid and the hot surface of the capillary is relatively short. This limits the extent of thermal decomposition of labile analytes.

Most thermospray interfaces have been fitted onto (triple) quadrupole mass analysers, although thermospray interfaces for magnetic sector instruments were commercially available as well.

Thermospray ionization modes

The thermospray interface can be used in various modes of ionization, depending on the settings of experimental parameters and the choice of the solvent composition. The thermospray nebulization process provides for a rapid and efficient means to partially evaporate the solvent mixtures introduced into the system by means of the production of small heated droplets. For clarity of the discussion, four ionization modes are distinguished here, i.e. two electron-initiated modes and two liquid-based ionization modes.

The two electron-initiated ionization modes are filament-on and discharge-on ionization. In these modes, the thermospray interface is used as a solvent introduction device, providing nebulization and soft transfer of analytes from the liquid to the gas phase. High-energy electrons are generated by means of a heated filament or at a corona discharge electrode. These electrons produce molecular ions of solvent molecules in the high-pressure (typically 1 kPa) source. In a series of ion–molecule reactions, the solvent molecular ions are converted to solvent-based reagent gas ions, i.e. protonated molecules and clusters, similar to the processes in a chemical ionization source. Protonated $[M+H]^+$ or deprotonated $[M-H]^-$ analyte molecules are produced in positive-ion and negative-ion mode, respectively, as a result of gas-phase proton-transfer reactions, while various other even-electron ionic species (adducts such as the ammoniated molecule $[M+NH_4]^+$, $[M+CH_3OH+H]^+$, or $[M+CH_3COO]^-$) may be produced as well.

The two liquid-based ionization modes are based on ion evaporation processes, initially proposed by Iribarne and Thomson. The mechanism can be summarized as follows: During thermospray nebulization, a superheated mist carried in a supersonic vapour jet is generated. Nonvolatile molecules are preferentially retained in the droplets, which are charged due to the statistical random sampling of the buffer ions in solution. As a result of continuous solvent evaporation from the droplets and repeated droplet breakdown by Rayleigh instabilities, a high local field strength is generated allowing charged species to desorb or evaporate from the droplets. These charged species comprise analyte molecules, present as preformed ions in solution, and buffer-solvent cluster ions, that rapidly equilibrate with the vapour in the ion source. Ion–molecule reactions may occur between the ions and neutrals in the source. A schematic illustration of the thermospray ionization mechanism is provided in **Figure 3**.

The ion evaporation mechanism in thermospray ionization has been criticized, by, among others, Röllgen and co-workers, who propose an alternative model, i.e. the charge residue or soft desolvation model. According to this model, the preformed ion of the nonvolatile analyte molecule is kept in a droplet, which decreases in size due to continuous solvent evaporation and repetitive Rayleigh instabilities until the droplet has become so small that it can be considered as a solvated ion. The ionization is thus the result of soft desolvation of the preformed analyte ions. Interestingly, the discussion between these two mechanisms reappears in the discussion on electrospray ionization, although in a slightly different manner.

Irrespective of the exact mechanism, ion evaporation or soft desolvation, it is important to pursue the generation of preformed ions in solution, i.e. by adjusting pH, and to reduce the influence of

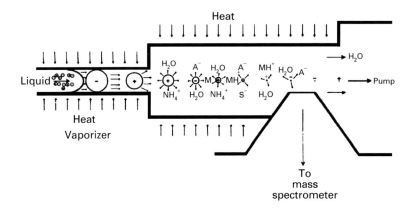

Figure 3 Schematic illustration of the liquid-based thermospray ionization modes. Reprinted with permission from Vestal ML (1983) *International Journal of Mass Spectrometry and Ion Physics* **46**: 193–196, © 1983, Elsevier Science.

competitive ions, i.e. to keep the ionic strength of other ions as low as possible. Surprisingly, the latter is not a common practice in thermospray ionization. The most widely applied liquid-based thermospray ionization mode appears to perform best in the presence of rather high (0.05–0.2 mol L^{-1}) concentrations of ammonium acetate or formate. Under these conditions, the evaporation of solvated ammonium ions generally will be more effective than the ion evaporation of preformed analyte molecules, especially because the latter are present in significantly lower concentrations. As a result, gas-phase ion-molecule reactions between ion-evaporated ammonium ions and neutral analyte molecules will significantly contribute to the ionization yield in most thermospray applications. Therefore, two liquid-based ionization strategies are indicated and discriminated here, i.e. one based on ion evaporation of preformed analyte ions in solution (thermospray ion evaporation mode), and one based on ion evaporation of solvent buffer ions followed by gas-phase ion-molecule reactions with neutral analyte species, efficiently transferred to the gas phase by means of nebulization and subsequent droplet evaporation, to produce protonated or deprotonated analyte ions (thermospray buffer ionization mode).

It must be emphasized that the resulting ions from either mechanism are the same. Therefore, it appears difficult to discriminate between the various mechanisms, especially because a mixed ionization mode, where various processes contribute to the final mass spectrum, is most likely.

Although the ion evaporation mechanism is the most popular view on the thermospray ionization mechanism, the ionization characteristics under typical operating conditions, and for most analytes, are best understood in terms of chemical ionization.

Although for particular compounds differences between the filament-on and discharge-on modes were observed, in general these two modes can be treated in the same way.

In both modes, analyte ionization is due to a gas-phase ion-molecule reaction between analyte molecules and reagent gas ions. The latter are generated from the solvent vapour in the high-pressure ion source by means of energetic electrons, basically similar to the generation of any other reagent gas in a source for chemical ionization.

The reagent gas composition is determined by the composition of the solvent mixture or mobile phase introduced. The reagent gas mass spectrum is often quite complex, containing several solvent cluster ions, but is generally dominated by the ionic species derived from the component in the solvent mixture with the highest proton affinity (in positive-ion mode) or the lowest gas-phase acidity (in negative-ion mode), although concentration effects may play a role as well. Proton affinities (gas-phase basicity) of common mobile-phase constituents are given in **Table 2**, while gas-phase acidities are given in **Table 3**. From these data it can be concluded that in positive-ion mode the reagent gas due to a 50:50 water-methanol mixture is dominated by methanol-related ions, e.g. at m/z 33, 65 and 97 due to $[(CH_3OH)_n + H]^+$ with $n = 1$, 2 and 3, respectively. After addition of ammonium acetate to this solvent mixture, the most abundant reagent gas ions are ammonium related ions, e.g. at m/z 50, 78 and 110, due to $[NH_4.CH_3OH]^+$, $[NH_4.CH_3COOH]^+$ and $[NH_4.CH_3OH.CH_3COOH]^+$, respectively.

In positive-ion mode, analyte ionization to a protonated molecule may take place when the proton affinity of the analyte exceeds that of the reagent gas. Typical values of proton affinities for a number of monofunctional analytes are given in **Table 2**. For multifunctional analytes, the proton affinity is roughly determined by the proton affinity of the functional group with the highest proton affinity. From **Table 2**, it may be concluded that the solvent mixture without ammonium acetate has a wider applicability range. In practice, however, ammonium acetate is added to the mobile phase in over 80% of the applications with filament-on or discharge-on modes, partly because of the need to apply a buffer in order to achieve reproducible retention times in LC. Next to protonated molecules [M+H]⁺ a variety of adduct ions may be generated. When the proton affinity of the analyte is within ~30 kJ mol⁻¹ of that of the reagent gas, adduct ions, e.g. [M+NH₄]⁺, may be found. Furthermore, a series of solvent cluster ions may be observed, generally with low intensity. Maeder elaborately studied the various ions observed in thermospray ionizations and proposed a general formula:

$$[M + A + xB - yC]^+$$

where M is the analyte molecule, A is the attached cation, e.g. the proton or ammonium ion, B is an attached solvent molecule, C is an eliminated molecule, e.g. water, and x and y take integer values of 0, 1, 2, ... The presence of adduct ions next to the protonated molecule may be useful to ascertain the molecular-mass determination.

In the negative-ion mode, analyte ionization to a deprotonated molecule [M–H]⁻ may take place when the gas phase acidity of the reagent gas exceeds that of the analyte, while similarly adduct ions may be observed as well. Typical values of the gas-phase acidity for a number of monofunctional analytes are given in **Table 3**.

These same ionization rules can be applied to predict the ionization behaviour of compounds in thermospray buffer ionization, where the analyte ionization is primarily dependent on the gas-phase ion–molecule reaction with ion-evaporated buffer ions, i.e. ammonium and acetate or formate.

In all instances, soft ionization of the analyte molecules is achieved, i.e. generally little fragmentation is observed. Obviously, there are a number of parameters other than the solvent composition that determine the ionization behaviour, e.g. analyte properties, temperatures, pressure. The temperature plays an important role because of its many influences on the ionization behaviour, but also on the production of ions due to thermal decomposition of thermolabile analytes. In most cases, thermal decomposition of analytes already takes place in the

Table 2 Proton affinities of some common mobile-phase constituents (PA_A) and of typical compound classes with one functional group (PA_M) in kJ mol⁻¹

Reagent gas	PA_A (kJ mol⁻¹)	Compound class	PA_M (kJ mol⁻¹)
Methane	536	Ethers, esters, ketones	630–670
Water	723	Polycyclic aromatic	710–800
Methanol	773	Hydrocarbons	
Acetonitrile	797	Carboxylic acids	<800
Ammonia	857	Carbohydrates	710–840
Methylamine	894	Alcohols	750–840
Pyridine	921	Thio	750–880
Dimethylamine	922	Amine, nitro	840–1000
		Peptides	880–1000

Table 3 Gas-phase acidities (ΔH_{acid}) of some common mobile-phase constituents and of typical compound classes with one functional group in kJ mol⁻¹

Reagent gas	ΔH_{acid} (kJ mol⁻¹)	Compound class	ΔH_{acid} (kJ mol⁻¹)
Ammonia	1657	Benzyl alcohol	1662
Water	1607	Toluene	1588
Methanol	1589	Alkyl alcohols	1560–1590
Acetonitrile	1528	Ketones, aldehydes	1530–1550
Acetic acid	1429	Anilines	1510–1540
Formic acid	1415	Thiols	1485–1510
Fluoroacetic acid	1394	Phenols	1400–1470
Trifluoroacetic acid	1323	Benzoic acid	1420

vaporizer tube, and the mixture of analyte-related molecules is subsequently ionized. The mass spectrum appears to show fragmentation, although some of the fragment ions observed cannot be explained from a mass spectrometric point-of-view, but rather are due to hydrolysis and subsequent ionization of the hydrolysis product.

The general lack of fragmentation under thermospray conditions has led to the more extensive application of MS/MS instrumentation as well as to research into the possibilities of collision-induced dissociation of ions in the ion source by means of high voltages on the repeller electrode. The latter showed nice perspectives in fundamental studies, with mass spectra quite similar to those observed in MS/MS but they proved to lead to a signal reduction that was too large for successful use in real-life applications.

Operation and optimization

The thermospray interface for LC-MS is generally considered as a difficult interface. This is due to the fact that for a proper operation the careful

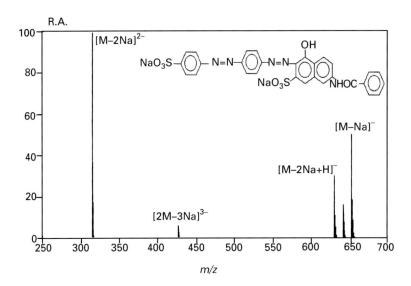

Figure 4 Negative-ion thermospray mass spectrum of the disulfonated azo dye. Direct Red 81 (mobile phase contains 10 mmol L^{-1} ammonium acetate).

optimization of a variety of mostly interrelated experimental parameters is required.

The performance of the interface is to a large extent determined by the quality of the spray, which in turn depends on the quality of the vaporizer, the solvent composition and the temperature control at the vaporizer. The temperature at the vaporizer depends on the type: with a Vestec-type vaporizer the stem and tip temperature are typically set at ~120°C and 220°C, respectively, while the vaporizer temperature of a Finnigan-type vaporizer is typically set at ~100°C.

Because the thermospray interface contains a dedicated ion source block, tuning and calibration of the source and analyser parameters is obligatory. Calibration and tuning cannot be performed with common calibrants like perfluorokerosene. Diluted solutions of polyethylene glycols are used in most cases, although a tuning and calibration based on clusters of ammonium acetate, ammonium trifluoroacetate and even simply water was proposed and used as well.

After tuning and calibration, the proper functioning of the interface can further be investigated by the injection of a number of standard compounds, e.g. adenosine and tertiary amines, as well as the compound(s) of interest. Subsequently, the system can be optimized to achieve the highest response or the best signal-to-noise ratio. Parameters to be studied are: the ammonium acetate concentration, the concentrations of the organic modifier and possible other mobile phase additives, the flow-rate, the optimum compound-dependent vaporizer temperature, the source block temperature, the repeller potential, as

well as the ionization mode. A lower flow-rate generally requires a lower vaporizer temperature, as does a higher content of the organic modifier. However, at a modifier content exceeding 40%, the thermospray buffer ionization mode is generally ineffective. Although the vaporizer temperature should be set in such a way that ~95% solvent vaporization inside the vaporizer is achieved, which in principle is primarily dependent on the flow-rate and the solvent composition, fine-tuning of the vaporizer temperature for a particular application may provide significant improvement of the performance. The analyte-related optimum of the vaporizer temperature may be sharp.

Applications

Thermospray ionization was especially applied between 1987 and 1992 in combination with LC-MS for a wide variety of compound classes, e.g. pesticides and herbicides, drugs and metabolites, alkaloids, glycosides and several other natural products, as well as peptides.

There are many studies available concerning the characterization of interface and ionization performance for the thermospray LC-MS analysis of pesticides, herbicides and insecticides, the improvement of detection limits and information content of the mass spectra. Compound classes most frequently studied are the carbamates, organophosphorous pesticides, triazine and phenylurea herbicides, chlorinated phenoxy acetic acids, and sulphonylureas.

Analytical strategies for the analysis of pesticides and herbicides in environmental samples, e.g. surface and tap water, are based on combined solid-phase extraction (SPE), LC separation and subsequent thermospray MS detection, often in a completely automated online system. Specific strategies have been developed for multiresidue screening as well as quantitative determination of pesticides and herbicides from specific compound classes. More recently, there is a growing interest in the determination and identification of pesticide and herbicide degradation products.

Environmental applications of LC-MS, not only pesticides and herbicides, but dyes, shellfish toxins, surfactants, organotin and other environmental contamination were recently reviewed in a multi-authored book, edited by Barceló.

Thermospray LC-MS has also found frequent application in the qualitative and quantitative analysis of drugs and their metabolites in biological fluids, like plasma and urine, and tissue extracts.

In the drug development area, thermospray ionization has found application in open-access approaches, as the technique allows the rapid determination of the molecular mass of a synthesized product without the need to optimize too many experimental variables. In this type of work, the thermospray interface is simply applied as an easy access to the MS.

Thermospray LC-MS, especially in combination with MS/MS was successfully applied in the characterization of drug metabolites. Metabolite screening strategies based on precursor-ion or neutral-loss scan modes in MS/MS were also proposed for the detection of both Phase I and Phase II metabolites. For the Phase II metabolites, neutral loss scan with 176 or 80 Da losses for glucuronide and sulfate conjugates, respectively, were proposed. However, this approach is successful for only some Phase II metabolites, because it was found that the Phase II metabolites often undergo thermally induced ammoniolysis, resulting in mass spectra of the aglycones.

The thermospray interface was the first LC-MS system that allowed reliable quantitative bioanalysis for a wide variety of compounds. Numerous examples were published in the literature. An excellent example is the automated analysis of bambuterol. The automated system, described by Lindberg and co-workers, contained a series of feedback steps in order to assure the various components of the system were operating properly during overnight, unattended analysis and to avoid the loss of valuable sample material. The same approach was applied to the quantitative bioanalysis of cortisol and related steroid compounds. In order to enhance the response of cortisol in thermospray ionization, the compound was derivatized to the 21-acetate using acetic

anhydride. This is a viable approach to slightly increase the proton affinity of an analyte to obtain improved ionization characteristics in thermospray buffer ionization.

Thermospray LC-MS was also frequently applied in the analysis of natural products, e.g in extracts from plants or cell cultures, of (modified) nucleosides, endogenous compounds such as prostaglandins, and some peptides. However, later it was demonstrated that alternative LC-MS strategies, e.g. based on electrospray ionization, were far more effective in the MS analysis of peptides.

Conclusion and perspectives

For a number of years (1987–1992), thermospray LC-MS was the most frequently applied interface for LC-MS. It has demonstrated its applicability in both qualitative and quantitative analysis in various application areas. With the advent of the more robust LC-MS interfaces, based on atmospheric-pressure ionization, the use of thermospray interfacing and ionization rapidly decreased. The newer technology pointed out the limitations of the thermospray system, e.g. in the analysis of thermolabile compounds, ionic compounds, high molecular-mass compounds, as well as in robustness and user-friendliness. Therefore, thermospray as an ionization and interface technique for LC-MS is now history. Thermospray nebulization will continue to be used, e.g. in nebulization for ICP-MS.

See also: **Chromatography-MS, Methods; Ionization Theory.**

Further reading

Arpino PJ (1990) Combined liquid chromatography mass spectrometry. Part II. Techniques and mechanisms of thermospray. *Mass Spectrometry Review* 9: 631–669.

Arpino PJ (1992) Combined liquid chromatography mass spectrometry. Part III. Applications of thermospray. *Mass Spectrometry Review* 11: 3–40.

Barceló D (ed) (1996) *Applications of LC-MS in Environmental Chemistry.* Amsterdam: Elsevier.

Blakley CR and Vestal ML (1983) Thermospray interface for liquid chromatography/mass spectrometry. *Analytical Chemistry* 55: 750–754.

Blakley CR, McAdams MJ and Vestal ML (1978) Crossed-beam liquid chromatograph–mass spectrometer combination. *Journal of Chromatography* 158: 261–276.

Conver TS, Shawn T, Yang J and Koropchak JA (1997) New developments in thermospray sample introduction for atomic spectrometry. *Spectrochimica Acta B* 52: 1087–1104.

Iribarne JV and Thomson BA (1976) On the evaporation of small ions from charged droplets. *Journal of Chemical Physics* 64: 2287–2294.

Iribarne JV and Thomson BA (1979) Field induced ion evaporation from liquid surfaces at atmospheric pressure. *Journal of Chemical Physics* 71: 4451–4463.

Lindberg C, Paulson J and Blomqvist A (1991) Evaluation of an automated thermospray liquid chromatography – mass spectrometry system for quantitative use in bioanalytical chemistry. *Journal of Chromatography* 554: 215–226.

Niessen WMA (1998) *Liquid Chromatography – Mass Spectrometry*, 2nd edn. New York: Marcel Dekker.

Röllgen FW, Nehring H and Giessmann U (1989) Mechanisms of field induced desolvation of ions from liquids. In Hedin A, Sundqvist BUR and Benninghoven A (eds), *Ion Formation from Organic Solids (IFOS V)*, pp 155–160. New York: Wiley.

Yergey AL, Edmonds CG, Lewis IAS and Vestal ML (1990) *Liquid Chromatography/Mass Spectrometry, Techniques and Applications*, pp 31–85. New York: Plenum Press.

Time of Flight Mass Spectrometers

KG Standing and **W Ens**, University of Manitoba, Winnipeg, Manitoba, Canada

MASS SPECTROMETRY
Methods & Instrumentation

The time of flight (TOF) mass spectrometer is perhaps the simplest type of mass analyser, at least in principle. The kinetic energy of an ion of mass m is given by $mv^2/2$, so a measurement of its speed v by timing the flight of the ion over a given path determines the mass when the kinetic energy is known, or when the spectrometer has been suitably calibrated. Such an instrument was first proposed (in 1946) to take advantage of the improvements in timing circuits developed in the Second World War and in succeeding years it developed a reputation as a device with fast response but low resolution when used with an electron impact ion source. The Bendix Corporation manufactured a commercial TOF instrument that achieved considerable popularity, but later the technique fell into disuse when quadrupole mass filters became common. This situation has changed dramatically in recent years, and the field is now one of the most active areas in mass spectrometry. This has come about partly because of improvements in electronics, but mainly because of the development of new methods of ionization, particularly matrix-assisted laser desorption/ionization (MALDI). In addition, interest has shifted to the measurement of compounds of larger mass, for which TOF methods are especially well suited.

Ionization methods

To define the 'start signal' for a TOF measurement, it is necessary to produce the ions as a series of short bursts. In some methods of ion production, this is achieved naturally, since the source itself is intrinsically pulsed. An early example of such a source is plasma desorption mass spectrometry (PDMS). In this technique, ions are produced by bombardment of the sample with particles of MeV energies, usually fission fragments, and the pulses are formed by individual bombarding particles arriving at the target. More recently, much of the activity in the field has stemmed from the widespread use of MALDI, as remarked above. MALDI is also an intrinsically pulsed source, where the ions are produced by irradiation of a sample with a beam from a pulsed laser. The laser provides the start pulse, so the coupling to a TOF instrument is a natural one.

On the other hand, ions produced in a continuous beam must be formed into pulses by an appropriate device, so an additional complication is introduced. Examples are electron ionization, secondary-ion mass spectrometry (SIMS), and most recently electrospray ionization (ESI). As mentioned above, the earliest commercial TOF instruments used electron-impact ionization, and the difficulty in producing short ion bursts with this method was mainly responsible for their limitations in mass resolution. However, new technology has enabled dramatically improved performance for continuous sources, particularly ESI, and TOF has been gaining popularity for such sources as well.

A simple model

An idealized TOF instrument is illustrated in **Figure 1**. Here particle or laser bombardment desorbs an ion of charge $+q$ and mass m at time $t = 0$ from a sample deposited on the target, a plane conducting surface ($z = 0$) that is held at potential $+V$. A parallel grid at $z = s$ is kept at ground potential so that there is a uniform electrostatic field directed along the z axis in the 'source region' between the target and the grid. If the ion starts out with zero velocity at the target, it is accelerated by the electric field and arrives at the grid with kinetic energy $\frac{1}{2}mv_z^2 = qV$ or velocity $v_z = \sqrt{(2qV)/m}$; its average velocity in the source region is half this value. The ion then passes through the ideal grid and travels with constant velocity through a drift region to the plane surface of a detector at $z = s + d$. Thus the total time of flight t is the sum of the time spent in the source region and the time spent in the drift region; i.e. $t = (2s + d)\sqrt{m/(2qV)}$. Measurement of the time of flight with a fast 'clock' determines the mass m, since the other parameters in the equation are known. Note that the time of flight is proportional to $\sqrt{m/q}$.

This geometry is close to that introduced by Macfarlane and Torgerson in their pioneering studies in PDMS in 1976, which marked a major step forward in the development of TOF techniques. As in the simple model, ions were ejected from an equipotential surface (by fission fragment bombardment in this case), so the spatial spreads that had limited the performance of earlier instru-ments were removed.

Features of TOF measurements

The advantages of the technique in the ideal case can be seen from the simplified description above:

- The mass range of the analyser is unlimited, since the clock can simply be allowed to run until the ion of interest arrives at the detector. The only limits on the mass range are imposed by the ion source and the detector.
- Parallel detection of all the ions over the complete mass range is straightforward, because the mass is determined by the measured arrival time at the detector, and the arrival times of all ions can be recorded.
- Defining slits are unnecessary.
- Because of the previous points, sensitivity is high and the instrument has a fast time response.

In the past, the main disadvantage of TOF systems has been poor mass resolution, because of the difficulty of producing an ion beam consisting of very short bursts, and because of the inevitable departures from the ideal case. However, TOF

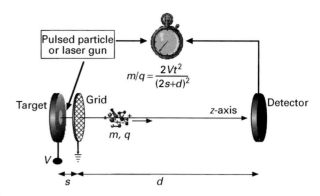

Figure 1 Schematic diagram of a simple idealized time-of-flight mass spectrometer.

resolution has been substantially increased recently by various technical improvements, as described below. Consequently, TOF instruments now often provide the optimum combination of resolution and sensitivity, particularly in cases where the whole mass spectrum is required. In contrast to the parallel detection capability of TOF instruments, most other types of mass spectrometer operate as mass filters, in which the mass spectrum is obtained by scanning through the mass range, one mass at a time. Thus, in these instruments there is a reciprocal relationship between mass resolution and sensitivity; resolution must be sacrificed to obtain high sensitivity.

Compensation for ion energy spreads with an electrostatic reflector

The most obvious defect of the simple model given above is its failure to take account of the initial energy that the ion possesses as it leaves the target. Variations in the initial energy may give rise to a considerable spread of times of flight, and thus to a deterioration in resolution. However, a modification to the instrument that alleviates this problem is the introduction of a reflector or ion mirror, as first proposed by Mamyrin (who called it a 'reflectron'). The simplest case is illustrated in **Figure 2**, where ions on a plane surface (the 'object plane') just outside the source region have a distribution of velocities. Ions travel freely from this surface until they enter a uniform retarding electrostatic field (an ion mirror). Like projectiles in the earth's gravitational field, ions follow parabolic paths within the mirror and leave it with the ion velocity component parallel to the mirror axis (v_z) reversed. The ion then travels freely to the detector. For $L = L_1 + L_2$, the time spent in free flight is L/v_z, and

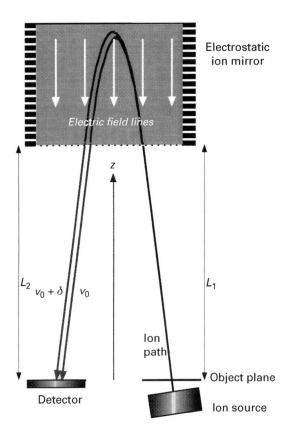

Figure 2 An illustration of the principle of ion mirror to compensate for velocity spreads. Two ion paths are shown for axial velocities $v_z = v_0$ and $v_z = v_0 + \delta$. The ion with higher velocity spends less time in the field-free region but more time in the mirror.

the time spent in the mirror is $2mv_z/qE$, where E is the magnitude of the retarding electric field. If $v_z = v_0 + \delta$, we can expand as a function of (δ/v_0) to give a total time of flight:

$$t = \left(\frac{2mv_0}{qE} + \frac{L}{v_0} \right) + \left(\frac{\delta}{v_0} \right) \left(\frac{2mv_0}{qE} - \frac{L}{v_0} \right) + \cdots \quad [1]$$

Setting $2mv_0/qE = L/v_0$, or $E = 2mv_0^2/qL$ removes the first-order term in δ/v_0. Thus the reflector eliminates the effect of a velocity variation δ to first order. Under this condition, the total time of flight $t = 2L/v_0$, so the ion spends equal amounts of time in the mirror and in free flight. Higher-order terms can be removed by the use of more complicated electric fields, for example by the two-stage mirror described by Mamyrin, but any advantage in doing so is often lost because the resolution may deteriorate because of other effects, particularly in the acceleration region, which must be considered separately.

Space and velocity focusing in the acceleration region

As pointed out above, variations in ion velocity on the object plane can be corrected to a large extent by the use of a reflector. However, effects during ion production or acceleration may give not only a spread in velocity on this plane but also a spread in time. The latter effects are not corrected by the mirror.

Some of these phenomena were discussed in a classic paper of 1955 by Wiley and McLaren in connection with their TOF studies of ions produced by electron ionization, which in general may have both an initial spatial spread and an initial velocity spread. They considered an ion source with two stages of acceleration in series, each with a uniform longitudinal electric field. They pointed out first that ions with a pure spatial spread in the acceleration region are subject to 'space focusing'; that is, there is some plane beyond the grid that ions of the same mass will reach at approximately the same time. This is because ions initially close to the grid acquire less energy in the acceleration region than more distant ions and therefore are overtaken by the latter after travelling some distance determined by their original position and the accelerating electric fields. In the special case of a single uniform field, the focal plane is a short distance $D = 2s_0$ beyond the grid if the ions originate an average distance s_0 inside it. If a two-stage acceleration region is used, the position of the focal plane can be adjusted by changing the ratio of the electric fields in the two regions. When the ions arrive at this plane, they will have a velocity spread because of the differing amounts of energy gained during acceleration, so the original spatial spread has been exchanged for a velocity spread.

This technique does not give perfect spatial focusing, even in the ideal case, but the usual limitation results from an initial velocity spread in addition to the spatial spread. The worst case involves two ions at the same z position but with velocities in opposite directions when the extraction pulse is applied. The ion in the negative z direction must first be turned around, and the two ions will arrive at the focal plane separated by this 'turn-around time'.

If the ions are produced in pulses (e.g. if the electron gun in an electron ionization source is pulsed), then some velocity compensation in the above situation is possible by the use of 'time-lag focusing', also proposed by Wiley and McLaren. In this technique, now usually called delayed extraction, a delay is introduced between ion production and the application of the accelerating fields, during which the ions drift freely. For simplicity, consider ions starting with zero spatial spread, i.e. with a pure velocity spread. When the accelerating field is applied, the ions will be separated in space according to their velocity. Those ions with higher initial v_z, will be closer to the end of the acceleration region, and will receive a smaller accelerating impulse. If the time delay and the amplitude of the accelerating voltage are adjusted properly, ions of the same mass will arrive at a focal plane at approximately the same time.

In the general case there may be both spatial and velocity spreads in the initial ion distribution, so some compromise is necessary to give optimum focusing. However, two currently popular ionization methods, ESI and MALDI, approximate the two limiting cases described above. Ions suitably injected from an ESI source have an appreciable spatial spread, but a very small velocity spread (see below). A pure velocity spread is approximated by the geometry normally used for MALDI, since the MALDI ions are ejected from an equipotential target by a very short laser pulse. In a simple linear TOF spectrometer, resolution can be optimized in both cases by using a two-stage acceleration region and setting the accelerating fields so that the focal plane coincides with the plane of the detector.

A modern TOF geometry

The best performance of TOF instruments is now achieved by a combination of an electrostatic reflector and the Wiley–McLaren focusing techniques, and this combination is the basis for most high-performance TOF systems.

By themselves, the Wiley–McLaren focusing methods are limited because the narrowest time distributions are achieved for short flight paths, but good mass resolution requires long flight paths to provide time dispersion between ions of different masses. The electrostatic mirror provides energy focusing but does not compensate for time spreads in the source region, so by itself it also offers limited improvement. However, the two methods are highly effective when used in combination. The ions from the source are focused into a flat ion packet near the source and coincident with the object plane of the mirror (see **Figure 2**). The mirror then images the ion packet onto the detector, greatly increasing the time of flight and therefore the time dispersion between species, without appreciably increasing the time spread. The ions at the first focal plane have a considerable velocity spread as mentioned above, but the mirror compensates to first order for velocity spreads in its object plane.

TOF measurements with a continuous beam

As remarked above, a continuous beam must be formed into pulses before it is introduced into the TOF spectrometer. This requirement is an extra complication that is not present if the beam is intrinsically pulsed. However, there are several cases in which mass analysis of ions produced in a continuous beam can benefit considerably from the features of TOF instruments. For example, electrospray ionization has been the most successful technique for producing ions from intact noncovalent complexes, but these ions are often formed with high mass-to-charge ratios, beyond the range of quadrupole mass filters; TOF imposes no limit on the mass-to-charge ratio (m/z) range. A second example involves coupling of separation techniques such as high-performance liquid chromatography with mass spectrometry. The sensitivity and fast time response of TOF instruments are well suited for such an application, but most separation techniques produce a continuous output. The same can be said of coupling TOF instruments with other types of mass spectrometer in order to perform MS/MS measurements as discussed below. For these reasons, there is clearly a need for an effective method for coupling continuous sources to TOF instruments.

A continuous beam can be injected into a TOF spectrometer in the longitudinal geometry of **Figure 1**, but only with very low efficiency. A more practical arrangement is illustrated in **Figure 3**; here a continuous beam of ions enters the TOF instrument perpendicular to its axis with low velocity and is injected into the flight path by the electrical pulses indicated in the figure. This technique takes advantage of the relative insensitivity of TOF measurements to spatial spreads in a plane perpendicular to the TOF axis. Such 'orthogonal injection' geometries were first introduced in the early 1960s, but acquired particular relevance when used with an electrospray source and an ion mirror by Dodonov. Limits on the injection efficiency and the resolution are set by the spatial and velocity spreads of the injected beam as described above. The properties of these instruments are therefore considerably improved if they are preceded by an ion guide running at relatively high pressure (up to ~10 Pa) to provide collisional translational cooling of the ions before they enter the TOF spectrometer. In this way, a beam is produced with a small energy spread, limited by thermal velocities, allowing effective spatial focusing as described above.

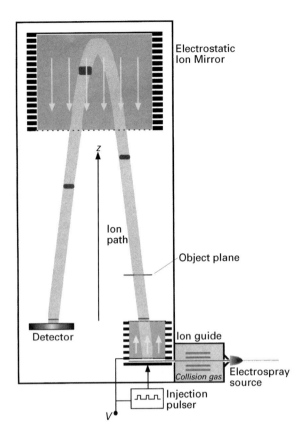

Figure 3 A schematic diagram of an orthogonal-injection TOF instrument with an electrospray ionization source. Collisional cooling is used in a quadrupole ion guide to produce a beam with a small energy spread, and a small cross section. The pressure in the ion guide is typically tens of millitorr; the main TOF chamber is under high vacuum, typically 10^{-7} torr. Ions are pulsed into the spectrometer at a repetition rate of several kilohertz; packets of ions for one mass are shown at several positions along the ion path.

Daughter-ion measurements in TOF spectrometers

The study of the products of ion breakup can often give useful information about the molecular structure of the parent ion. In the simplest measurement of this type, a metastable parent ion may suffer unimolecular decay as it passes down the flight tube (so-called 'post-source decay'). The velocities of the daughter ions or neutrals can be determined from their times of flight, but such a measurement in a linear TOF spectrometer yields little information, since the *velocities* of the decay products are approximately the same as the velocity of the parent ion, as a result of conservation of momentum. Thus a daughter ion cannot be distinguished by its time of flight in a simple linear instrument from the parent or from other daughters.

On the other hand, the kinetic energy of the parent ion is divided among its decay products, so daughter ions will have *energies* determined by their masses. In contrast to the situation in a linear TOF spectrometer, the total time of flight in a reflecting instrument depends on the ion's energy as well as its velocity, because the ratio of ion energy to charge determines the distance the ion penetrates the mirror and thus the time spent there. The mirror does not correct for the velocity spreads of the daughter ions as well as it does for the parent (assuming that the electric field in the mirror is set to the value appropriate for the parent ion). However, for optimum resolution the problem can be minimized by examining the daughter-ion spectrum in segments, with the mirror field set to an appropriate value for each segment. Alternatively, a mirror with a nonlinear electric field can be used.

Usually there are a number of different parent ions extracted from the sample, each giving rise to a daughter-ion spectrum. It is therefore necessary to separate these in order to identify the particular parent ion giving rise to an observed decay. This is normally done by deflecting all ions except the desired one out of the flight path by some form of 'ion gate', and examining the decay products of the parent ions one by one.

The post-source decay technique suffers from some limitations, such as the limited selectivity obtainable on the parent ion, modest mass accuracy of the daughter ions (compared to the accuracy achieved for parent ions), and the frequently incomplete information obtained from the metastable decay spectrum. These factors have stimulated the development of various tandem instruments, as discussed below.

Tandem instruments

Tandem mass spectrometers provide a more flexible means of studying molecular structure. Such devices combine a mass filter (MS1) to select the parent ion, a gas cell for collision-induced dissociation, and a second mass spectrometer (MS2) to analyse the daughter ions produced from breakup of the parent. A particularly suitable combination is a quadrupole mass filter (Q) as MS1, a quadrupole ion guide (q) as the collision cell (excited by RF only), and a reflecting TOF instrument as MS2. MS/MS experiments are often limited by the amount of sample available, and by the time available to obtain a spectrum, so the sensitivity and rapid response of TOF offers a significant advantage over scanning instruments for MS2. Because of the parallel detection, the sensitivity can be maximized without reducing mass resolution. On the other hand, MS1 is simply used as a mass filter, so parallel detection offers no advantage, and a quadrupole mass filter is a good choice because it can be efficiently coupled to the collision cell. The

QqTOF geometry thus offers an improved alternative to the popular triple-quadrupole instrument.

Detection methods

Detection in most types of mass spectrometer depends on measurement of the electrons or the secondary ions ejected from a surface by the impact of the ions of interest, usually after electron multiplication. A TOF detector must obviously have a fast time response. It must also have a large and flat active area, since the cross section of the ion beam at the detector is relatively large (usually several centimetres). Finally, to exploit the high mass range that TOF is capable of, the detector must be sensitive to ions with high m/z, which for a given energy have relatively low velocity.

The first two of the above demands are met effectively with microchannel plate (MCP) electron multipliers. These are flat arrays (of various dimensions) of micrometre-sized channels, each one acting as a very fast electron multiplier when a voltage gradient exists along it. Most high-resolution TOF instruments now use a microchannel plate for the first element of the detector.

The secondary emission coefficient decreases rapidly for decreasing velocity, and therefore for increasing m/z. For this reason, the ions are usually accelerated to relatively high energies before they are incident on the detector. Most MALDI/TOF instruments use 30 kV acceleration or more. Even so, for singly charged ions larger than about 50 kDa, the electron emission coefficient is considerably less than unity. Detection by electron emission still appears to be feasible in the high mass range because of the large number of ions desorbed in MALDI, although more efficient detection is possible by using a detector designed to take advantage of secondary ion emission at the cost of some loss of resolution.

The problem of detecting high masses does not occur when an ESI source is used, because it produces ions with much higher charge states, and therefore much higher energy (and lower m/z) for the same molecular mass. Intact molecular ions from brome mosaic virus with mass larger than 4.6 MDa (m/z ~25 000) have been observed by ESI/TOF using only 5 kV acceleration.

List of symbols

E = electric field strength of reflector; L_1, L_2 = field-free path lengths (see **Figure 2**); m = ion mass; q = ion charge; t = time; time of flight; v = ion velocity; V = electric potential; z = z coordinate; ionic charge; δ = variation in ion velocity.

See also: **Fragmentation in Mass Spectrometry; Ion Molecule Reactions in Mass Spectrometry; Ion Structures in Mass Spectrometry; Ionization Theory; Laser Applications in Electronic Spectroscopy; Metastable Ions; Plasma Desorption Ionization in Mass Spectrometry; Quadrupoles, Use of in Mass Spectrometry.**

Further reading

Chernushevich IV, Ens W and Standing KG (1999) Orthogonal-injection TOFMS for analyzing biomolecules. *Analytical Chemistry* 71.

Cotter RJ (1997) *Time-of-Flight Mass Spectrometry.* Washington DC: American Chemical Society.

Dodonov AF, Chernushevich IV and Laiko VV (1994) Electrospray ionization on a reflecting time-of-flight mass spectrometer). In Cotter RJ (ed) (1994) *Time-of-flight Mass Spectrometry*, pp 108–123. Washington DC: American Chemical Society.

Hillenkamp F, Karas M, Beavis RC and Chait BT (1991) Matrix-assisted laser desorption/ionization mass spectrometry of biopolymers. *Analytical Chemistry* 63: 1193A–1203A.

Macfarlane RD and Torgerson DF (1976) Californium-252 plasma desorption mass spectroscopy. *Science* 191: 920–925.

Mamyrin BA, Karataev VI, Shmikk DV and Zagulin VA (1973) The mass-reflectron, a new nonmagnetic time-of-flight mass spectrometer with high resolution. *Soviet Physics – JETP* 37: 45–48.

Spengler B (1997) Post-source decay analysis in matrix-assisted laser desorption/ionization mass spectrometry of biomolecules. *Journal of Mass Spectrometry* 32: 1019–1036.

Stephens WE (1946) A pulsed mass spectrometer with time dispersion. *Physical Review* 69: 691.

Wiley WC and McLaren IH (1955) Time-of-flight mass spectrometer with improved resolution. *Review of Scientific Instruments* 26: 1150–1157.

Tin NMR, Applications

See **Heteronuclear NMR Applications (Ge, Sn, Pb).**

Titanium NMR, Applications

See **Heteronuclear NMR Applications (Sc–Zn).**

Tritium NMR, Applications

John R Jones, University of Surrey, Guildford, UK

MAGNETIC RESONANCE
Applications

It is undoubtedly true that if tritium were not radioactive it would be one of the most widely used of all NMR nuclei. Its favourable properties – an isotope of hydrogen, one of the most important of all the elements, with a nuclear spin $I = \frac{1}{2}$ and the most sensitive of all NMR nuclei – counts but little with those who quake at the mention of the word 'radioactivity', let alone think of spinning radioactive samples. However, there are those, increasing in number, who believe that tritium is the most favoured of all the nuclei, combining the advantages of favourable radioactive properties (weak β⁻ emitter ($E_{avg} \sim 6$ keV), convenient half-life (12.3 years), ready detection by liquid scintillation counting with good efficiencies (typically $> 50\%$) with these positive NMR characteristics. Furthermore the technology for synthesizing and handling tritiated compounds has been in place for many years whilst the development of spectrometers operating at ever increasing fields means that less tritium is required for NMR detection. In addition there is virtually no natural abundance tritium concentration, unlike the situation that exists for stable isotopes, so that the dynamic range is enormous. It is this factor above all others that will lead to an expansion in the use of tritium and tritium NMR spectroscopy in the life sciences. Recent publications show that such possibilities are being increasingly appreciated.

Properties of the nucleus

As well as having a nuclear spin $I = \frac{1}{2}$ tritium has a high nuclear magnetic moment which is responsible for the magnetogyric constant being larger than for any other nucleus, as also is its sensitivity to NMR detection, 21% higher than that for ¹H. At 11.7 T,

at which field the ¹H NMR frequency is 500 MHz, the ³H NMR frequency will be 533.3 MHz.

Sample preparation and spectrum measurement

Before embarking on any ³H NMR work the personnel must become designated radiation workers, have the appropriate radiochemical facilities and become familiar with tritiation procedures. In this respect it is frequently useful for initial training to be given in appropriate deuteration studies although the corresponding tritium work will usually be carried out on a much smaller scale and the purification procedures will depend greatly on appropriate radio-chromatographic methods, as distinct from chromatographic methods. With appropriate rules and regulations in place a radiochemical laboratory need not be any more hazardous than an ordinary chemistry laboratory, particularly if the rule 'that only the minimum amount of radioactivity consistent with the requirements of the project' is used.

There are two separate units of radioactivity in use, the first being the curie (Ci) which is defined as an activity of 3.7×10^{10} disintegrations per second. This is a large unit, hence the frequent use of smaller subunits, the millicurie (10^{-3} Ci) and the microcurie (10^{-6} Ci). The second, and more recently introduced unit, is the becquerel (Bq). At one disintegration per second this is an extremely small amount of radioactivity. The conversions are

$$1 \text{ Bq} = 2.703 \times 10^{-11} \text{ Ci } (27.03 \text{ p Ci}),$$
$$1 \text{ Ci} = 3.7 \times 10^{10} \text{ Bq or 37 GBq}$$

Although there are a large number of methods available for preparing tritiated compounds the most widely used stem from the following categories:

- catalytic hydrogenation of an unsaturated precursor using $^3H_2$ gas;
- catalytic halogen–tritium replacement reactions;
- hydrogen isotope exchange reactions catalysed by acids, bases or metals;
- reductions using reagents such as sodium borotritide;
- methylation reactions using reagents such as tritiated methyl iodide.

Recently microwaves have been used to greatly accelerate the rates of many of these reactions whilst the development of microwave-enhanced solid state tritiation procedures offers considerable potential.

For ^3H NMR analysis 1–10 mCi of material dissolved in ~ 50–100 μl of a deuterated solvent is usually sufficient to obtain a spectrum of good signal-to-noise in a matter of 1–10 h, depending on whether the radioactivity is located at one site (a specifically labelled compound) or in several positions (a general labelled compound) – this assumes a spectrometer operating at 300 MHz for ^1H and 320 MHz for ^3H. For reasons of safety the radioactive samples are placed in narrow cylindrical tubes, sealed at the top, which themselves are placed in standard NMR tubes – this double containment procedure, initially introduced when much higher levels of radioactivity were required, provides a measure of safety as well as reassurance. Experience shows that ^3H NMR spectra are of two kinds. Firstly there are those in which the specific activity of the compound is less than ~1 Ci mmol^{-1} so that ^3H–^3H couplings are absent and the ^3H NMR (^1H decoupled) spectra consist of a series of single lines, which on integration give the relative incorporation of ^3H at each site. Nuclear Overhauser effects (NOEs) are small and differential effects even smaller so that there is no need to obtain NOE-supressed spectra. It should also be mentioned that there is no need to synthesize a tritiated organic standard – all the ^3H chemical shifts are obtained via the ^1H chemical shifts and the Larmor frequency ratio.

For compounds at high specific activity, e.g. prepared by the addition of $^3H_2$ gas across the unsaturated group of a precursor, there will be tritium–tritium couplings, the magnitude of which are similar to those of hydrogen, i.e. $J(^1H-^3H) = J(^1H-^1H) \times 1.066$. Small isotope effects are present but these do not complicate the interpretation of the spectra, on the contrary they can aid the analysis of the relative proportions of isotopomers present, e.g. RC $^3H_3$, RC $^1H^3H_2$ and RC $^1H_2{}^3H$. Tritium couplings to boron,

deuterium, carbon, fluorine and phosphorus are also in agreement with theory.

Applications of tritium NMR

New tritium labelling reagents

The development of ^3H NMR spectroscopy has made possible many new applications and in the process has stimulated research into the development of new labelling reagents and hence new/improved labelling procedures. One such area is that of tritide reagents. Essentially carrier-free LiB $^3H_4$ can now be obtained via the two-step sequence:

$$BuLi + {}^3H_2 \rightarrow Li^3H + Bu^3H$$
$$4Li^3H + BBr_3 \rightarrow LiB^3H_4 + 3LiBr$$

Similarly, carrier-free sodium triethylborotritide, a useful reagent for the stereo- and regiospecific reduction of carbonyl-containing compounds, can be synthesized in the following manner:

$$BuLi + NaOBu^t + {}^3H_2 \rightarrow Na^3H + LiOBu^t + Bu^3H$$
$$Na^3H + Et_3B \rightarrow NaEt_3B^3H$$

Tri-n-butyltin and lithium tri-s-butylborotritide are other useful reagents. Increasingly sophisticated tritium labelling technology is being developed as an alternative to the more traditional hydrogenation and catalytic dehalogenation reactions. The procedures will find wide application in the tritiation of molecules of biological importance. Thus N-tritioacetoxyphthalimide, a new high specific activity tritioacetylating reagent, has been used to label a number of acetylenes, ketones and alcohols whilst radical-induced tritiodeoxygenation reactions can lead to the synthesis of important heterocyclic compounds.

New more selective tritiation procedures

Hydrogen isotope exchange reactions are widely used not only to study reaction mechanisms but also for labelling compounds with either deuterium or tritium. The reactions may be catalysed by acids or bases under both homogeneous or heterogeneous conditions and frequently lead to generally labelled compounds. The same is true for transition metals. Recently considerable effort has been directed at developing more selective procedures – homogeneous rhodium trichloride has been shown to be very effective in introducing both deuterium and tritium into the *ortho*-aromatic positions of a wide range of pharmaceutically important compounds.

The well-known iridium catalyst [Ir(COD)(Cy$_3$P)-(Py)]PF$_6$, where COD = 1,5-cyclooctadiene and Py = pyridine, demonstrates excellent regioselectivity in isotopic exchange reactions of acetanilides and other substituted aromatic substrates. ^{3}H NMR spectroscopy is invaluable in identifying the site(s) of tritium incorporation – there are many instances where the broad signals in the corresponding ^{2}H NMR spectra are much less informative.

Another iridium catalyst that exhibits good regioselectivity in hydrogen isotope exchange reactions is the complex [IrH$_2$(acetone)$_2$(PPh$_3$)$_2$]BF$_4$. As in the previous studies the transient existence of a metallocyclic intermediate is indicated. Considerable interest has also been shown in the development of the high temperature solid state catalytic isotopic exchange (HSCIE) method developed by Myasoedev and colleagues. Although labelling is uniform in most instances, some regiocontrol can be exerted by careful temperature control.

Chiral methyl, stereochemistry and biosynthesis

The analysis of stereochemical problems in both chemistry and biochemistry has benefited greatly from the use of compounds that contain a methyl group with one atom each of ^{1}H, ^{2}H and ^{3}H. Such compounds exist as a pair of enantiomers, identified by *R* and *S*, and early work in this area will always be associated with the names of Arigoni and Cornforth. Recently a very efficient five-stage synthesis of chiral acetate has been reported (in which the penultimate reaction uses supertritide) with an enantiomeric purity of 100%.

(*S*)-Enantiomer   (*R*)-Enantiomer

In the past the determination of whether an unknown sample contained an excess of an (*R*)- or (*S*)-configured chiral methyl group relied on using a reaction in which one hydrogen is removed to generate a methylene group in which tritium is now unevenly distributed between the two methylene hydrogens. The condensation of acetyl coenzyme A with glyoxylic acid catalysed by the enzyme malate synthase, which exhibits a primary kinetic isotope effect k_H/k_D of 3.8, was the chosen reaction. Analysis of the tritium distribution, together with a knowledge of k_H/k_D and the steric course of the reaction, yields the required information – the configuration of the original chiral methyl group and an estimate of the enantiomeric excess.

^{3}H NMR spectroscopy can provide the necessary information directly; whether ^{3}H has ^{1}H or ^{2}H as a neighbour can be determined directly from the ^{1}H–^{3}H coupling and the ^{2}H isotope shift on the ^{3}H signal. The only problem with the ^{3}H NMR method is that it requires a few mCi of tritiated material, at least with current-day NMR spectrometers. With improvements in spectrometer design and the absence of 'natural abundance' tritium signals this may not always be the case. As it is, the method is direct, does not require any knowledge of the primary isotope effect and no chemical degradations are required.

There are many examples of enzymatic methyl-transfer reactions in biochemistry to which the chiral methyl/^{3}H NMR technology can be applied. One such example involves the important biological methyl donor *S*-adenosylmethionine. Combined with other studies the results show that the transfer of a methyl group to a variety of different nucleophiles all operate with inversion of methyl group configuration.

Substrate–receptor interactions

Most NMR studies in this area have used ^{13}C- or ^{15}N-labelled ligands, the synthesis of which is frequently more demanding than is the case for ^{3}H ligands. Furthermore, the sensitivity of both ^{13}C and ^{15}N nuclei to NMR detection is considerably less favourable than is the case for tritium. It is therefore somewhat surprising and at the same time disappointing that there are still relatively few examples of protein–ligand interaction studies based on the use of ^{3}H-labelled ligands. In an early study ^{3}H NMR spectroscopy was used to monitor the anomeric binding specificity of α- and β-maltodextrins binding to a maltose-binding protein whilst in another study ^{3}H NMR spectroscopy was used to measure the dynamic properties of tosyl groups in specifically ^{3}H-labelled tosylchymotrypsin. Preliminary details of a ^{3}H NMR binding study of a tritium-labelled phospholipase A$_2$ inhibitor to bovine pancreatic PLA$_2$ suggest that the tritium atoms are located within the hydrophobic pocket of the protein. In a more extensive study a number of high specific activity tritiated folic acids and methotrexates have been prepared and their complexes with *Lactobacillus casei* dihydrofolate reductase (DHFR) investigated. The ^{3}H NMR results confirm the presence of three pH-dependent different conformational forms in the complex DHFR·NADP⁺·folate, whereas both the binary and ternary methotrexate complexes (DHFR·MTX, DHFR·NADP⁺·MTX) were shown to exist as a single conformational state.

An interesting ^{3}H NMR study of the complex formed by [4-^{3}H]benzenesulfonamide and human carbonic anhydrase 1 reveals details that are widely different from those obtained when using a fluorinated inhibitor, highlighting the dangers of using fluorine as a 'substitute' isotope for one of the hydrogen isotopes.

Macromolecules

The methods that have been developed for tritiating small organic molecules do not lend themselves very readily to the tritiation of large macromolecules such as proteins although there are a few examples where Myasoedov's HSCIE procedure proved successful. It is not surprising therefore that very little work has been reported on, for example, the ^{3}H NMR of proteins. The polymer area, however, has seen more activity, mainly because it has been much easier to tritiate such compounds – hydrogenation with ^{3}H$_2$ gas of a suitable monomer followed by polymerization leads to a specifically tritiated product.

Many polymers are difficult to solubilize and the question has been asked several times whether in view of its good NMR characteristics it is possible to obtain satisfactory solid state spectra. The potential problems have recently been overcome partly by the development of zirconia rotors and partly by enclosing the tritium probe in a Perspex shield so that, in the event of an accident, radioactivity would be retained on a suitable filter. Magic-angle spinning at 17 kHz rotation provides spectra with line widths at half-height of the order of 120 Hz. This has been achieved without ^{1}H decoupling, this being a more difficult task than for solution studies. It is too early to say at this stage whether ^{3}H NMR spectroscopy of solids will develop into as widely used a technique as ^{13}C NMR spectroscopy. The main factor will undoubtedly be how far the current-day improvements in NMR sensitivity can be extended.

See also: **Biochemical Applications of Mass Spectrometry; Enantiomeric Purity Studied Using NMR; Isotopic Labelling in Mass Spectrometry; Labelling Studies in Biochemistry Using NMR; Macromolecule–Ligand Interactions Studied By NMR; Microwave Spectrometers; Solid State NMR, Methods; Stereochemistry Studied Using Mass Spectrometry; Structural Chemistry Using NMR Spectroscopy, Organic Molecules.**

Further reading

Andres H, Morimoto H and Williams PG (1990) Preparation and use of LiEt$_3$BT and LiAlT$_4$ at maximum specific activity. *Journal of the Chemical Society, Chemical Communications* 627.

Evans EA, Warrell DC, Elvidge JA and Jones JR (1985) *Handbook of Tritium NMR Spectroscopy and Applications*. Chichester: Wiley.

Floss HG and Lee S (1993) Chiral methyl groups: small is beautiful. *Accounts of Chemical Research* **26**: 116–122.

Junk T and Catallo WJ (1997) Hydrogen isotope exchange reactions involving C–H(D,T) bonds. *Chemical Society Reviews*, 401–406.

Kubinec MG and Williams PG (1996). Tritium NMR. In: Grant DM and Harris RK (eds) *Encyclopedia of NMR*, Vol 8, pp 4819–4839. Chichester: Wiley.

Saljoughian M, Morimoto H, Williams PG, Than C and Seligman SN (1996) *Journal of Organic Chemistry* **61**: 9625–9628.

Tungsten NMR, Applications

See **Heteronuclear NMR Applications (La–Hg).**

Two-Dimensional NMR, Methods

Peter L Rinaldi, University of Akron, OH, USA

MAGNETIC RESONANCE
Methods & Instrumentation

Introduction

While NMR has been a valuable tool for scientists who must understand the structures, reactions and dynamics of molecules, there have been two major advances since the 1960s, which, more than any other contributions, have kept this an exciting and rapidly evolving field. The first of these was the introduction of the Fourier transform NMR technique by Ernst and Anderson in 1966. This development helped to reduce problems associated with the biggest limitation of NMR, its poor sensitivity. It also set the stage for a second important development. The dispersion of NMR signals, and thus the complexity of molecules which can be studied is related to the magnetic field strength of the instrument. At a time when scientists were preparing evermore complicated structures, the incremental increases in the magnetic field strengths of commercially available instruments were growing smaller. However, the proposal of Jeener in 1971 and the first demonstration, by Muller, Kumar and Ernst in 1975, of multidimensional NMR spectroscopy resulted in a quantum leap in the capabilities of and the prospects for NMR. By dispersing the resonances into a second frequency dimension additional spectral dispersion could be achieved. The dispersion from a 2D experiment performed on a 1980 vintage 200 MHz spectrometer can match that obtained in the 1D spectrum from modern 800 MHz spectrometers. In 2D spectroscopy, the spectral dispersion increases as the square of the magnetic field strength. Furthermore, 2D experiments can have the unique characteristic of providing structural information based on the correlation of the frequencies at which peaks occur. This article deals with the background and practical aspects of obtaining 2D NMR data.

There are quite a few variations of 2D NMR experiments in which properties such as retention time (in liquid chromatography-NMR, LC-NMR), distances (imaging) or diffusion coefficients (diffusion ordered spectroscopy) are the variables along one or more axes in the spectra.

However, discussions in this article will be restricted to experiments in which two frequencies, related to NMR parameters, are plotted along the two axes of the spectra. Other forms of 2D NMR are discussed in other parts of this work. While this article can be read alone, it is useful to refer to other articles for details of various techniques (e.g. weighting, zero filling, sampling rates, complex versus real Fourier transforms, linear prediction, etc.) which are discussed here as they pertain to 2D NMR.

Fourier transform NMR spectroscopy

Figure 1A shows the time domain signal, called the free induction decay (FID), obtained by measuring the response of nuclear spins to an RF pulse. The FID is the sum of many exponentially decaying cosine waves, one for each resolvable singlet in the spectrum. In the example shown in this time domain spectrum, a single frequency is observed; from measuring its period, the frequency can be determined. A typical FID will contain the sum of many oscillating signal components, making it impossible to identify individual frequency components by visual inspection of the time-domain signal. By converting the time domain signal into a frequency domain signal, using a mathematical process called the Fourier transformation, a readily interpretable spectrum (**Figure 1B**) with peaks at discrete frequencies (one for each cosine wave in the original FID) can be obtained.

Each point in the time domain spectrum contains information about every frequency in the frequency domain spectrum. In a typical 1D spectrum up to 100k points are collected in the time domain, thus information about each peak is measured 100k times. The laws of signal averaging tell us that the signal (S) from n measurements increases linearly ($n \times S$), but that the noise (N) from n measurements

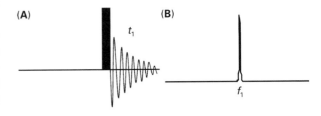

Figure 1 One-dimensional Fourier transform NMR data: (A) time domain free induction decay (FID) detected after a radiofrequency pulse; (B) frequency domain spectrum after Fourier transformation of the signal in (A).

increases as $n^{1/2}$ $(n^{1/2} \times N)$. Therefore, the signal-to-noise ratio (S/N) in the final spectrum improves as $n \times S$ $(n^{1/2} \times N) = n^{1/2} \times S/N$ as long as the signal is present throughout the FID. Consequently, the S/N in the final spectrum will be $(10^5)^{1/2} \sim 300$-fold better than that obtained in a single scanned spectrum. This improvement is known as the Felgett advantage. It is described here because it has some important consequences when n-dimensional experiments are performed. In practice, S/N gains in 1D NMR are lower than those predicted by the Felgett advantage, because the intensities of the signals decay exponentially during the signal acquisition period. However, in multidimensional NMR, short evolution and acquisition times are used to minimize the size of the data sets. Consequently, very little signal decay occurs and S/N improvement are close to those expected from theory.

Fundamentals of 2D NMR

General sequence for collection of 2D NMR spectra

Figure 2 contains a diagram of a 2D NMR pulse sequence called the NOESY (nuclear Overhauser enhancement spectroscopy) experiment. (NMR spectroscopists have been very liberal in their methods for selecting acronyms to name their experiments). This pulse sequence contains the four basic elements which are common to 2D NMR experiments: preparation, evolution (t_1) mixing and detection (t_2) times. The filled rectangular boxes represent 90° pulses, which are applied at the ^1H resonance frequency in this experiment. In general, these pulses can be at a variety of flip angles and can be applied at a variety of frequencies, depending upon the requirements of the experiment and the information desired.

The preparation period is used to put the nuclear spins into the initial state required by the experiment being performed. In this particular sequence the preparation period is a relaxation delay to allow the spins to return to their equilibrium Boltzmann distribution

among the energy levels. In some sequences, the preparation period might also contain a coherence transfer step (e.g. by an INEPT-type polarization transfer pulse sequence) to move NMR signal components from one nucleus to another in preparation for the evolution period. The evolution period is used to encode frequency information in the indirectly detected (t_1) dimension. The mixing period, which is present in some pulse sequences, is used to transfer magnetization from one nucleus (whose chemical shift information is encoded during t_1) to a second nucleus for detection during the acquisition period, t_2.

The NOESY sequence contains a delay during the evolution period to encode ^1H chemical shifts; however, some pulse sequences contain 180° refocusing pulses to remove chemical shift modulation or coupling to a second nucleus (if the pulse is at the frequency of that second nucleus), or combinations of pulses to remove selected signals or coupling interactions.

The key to the success of 2D NMR is the collection of a series of FIDs, while progressively incrementing the value of t_1. At the end of data collection a set of 100–1000 FIDs is obtained (the number of FIDs collected depends upon the desired resolution and spectral window in the t_1 dimension) as shown in **Figure 3A**. The intensities of these FIDs are modulated based on the length of the t_1 period and the precession of the coherence during t_1. If each of these FIDs is Fourier transformed (with respect to t_2), a series of spectra is obtained as shown in **Figure 3B**. Each spectrum contains signals which correspond to those found in the normal 1D spectrum of the nucleus which is detected. The intensity of a signal at a specific chemical shift varies from one spectrum to the next. Its intensity is modulated by the NMR interaction (J-coupling, chemical shift, multiple quantum coherence, etc.) which is in effect during t_1, and by the duration of the t_1 period. By plotting the intensities of the two peaks in **Figure 3B** as a function of t_1, the curves in **Figure 3C** are obtained. The modulation frequencies of these two curves are different because the detected signals in t_2 originate from different coherences which have different precession frequencies in t_1.

The intensity variations in these curves are reminiscent of the 1D FIDs. The obvious thing to do with these signals is to transpose the data matrix, and, at each frequency of f_2, Fourier transform the data with respect to t_1. The result is a spectrum with signal intensity variations as a function of two frequencies as shown in **Figure 3D**. The frequencies plotted along the f_2 dimension correspond to those which are detected during t_2 (i.e. ^1H chemical shift and J-coupling if the sequence in **Figure 2** is used). The frequency

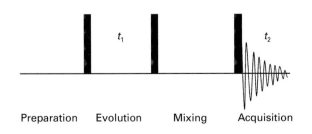

Figure 2 NOESY 2D NMR pulse sequence.

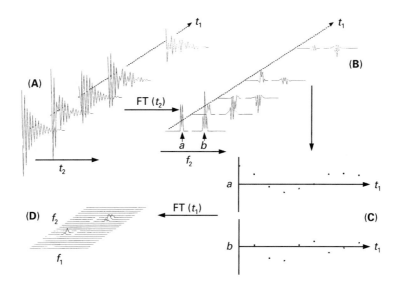

Figure 3 Schematic illustration of the process used to produce a 2D NMR spectrum.

plotted along the f_1 dimension corresponds to the precession properties of the coherences which are selected by the pulse sequence.

Presentation of 2D NMR data

Figure 4A shows a stacked plot of the COSY (correlation spectroscopy) spectrum from ethanol. Detailed fine structure is not resolved in this spectrum because of the greatly reduced digital resolution compared with that obtained in 1D NMR. This reduced digital resolution is not greatly detrimental,

and is necessary to keep the 2D data files to a manageable size (see below).

The stacked plot (**Figure 4A**) is not the most desirable way to present the data because it involves a lot of plotting time and background peaks are often hidden by those in the foreground. The preferred method of presenting 2D data is in the form of contours as shown in **Figure 4B**. For those unfamiliar with the generation of contour maps, planes are set at a range of intensity values above a user determined threshold in the spectrum. The contour map is generated by the intersection of the peaks with these

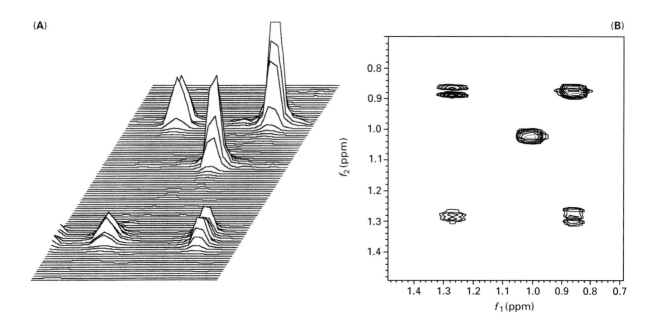

Figure 4 COSY 2D NMR spectrum of ethanol: (A) stacked plot; (B) contour plot with contours plotted at intensities of 2^n.

planes. The more intense peaks will intersect a larger number of planes, therefore peaks in **Figure 4B** which are defined by a larger number of contours are more intense than those defined by a small number of contours. In this display mode, peaks are not obscured and the printout is generated fairly rapidly. While it does take a significant amount of computer power to calculate a contour map, modern computers are capable of doing so in much less time than it takes to transmit the data to most plotters. Commercial software packages for manipulating NMR data permit the adjustment of the number and spacing of the contours so as to best display all the peaks in the spectrum. In cases where all the signal intensities are of the same order of magnitude, contour spacing can be small so that a large number of contours accurately defines the peak shapes. In cases where there is a considerable dynamic range of peak heights, contour spacing can be large to prevent the generation of a large number of contours around intense peaks.

Classes of 2D NMR experiments

Part of the power of 2D NMR comes from its ability to provide tremendous spectral dispersion; however, the structural information present from the correlation of frequencies is equally important. Organic chemists have been performing elegant syntheses for

many decades. They choose a target molecule for preparation, and based on their knowledge of chemical reactions select the proper reagents from their stockroom to carry out the chemical transformations necessary to obtain the desired product. Since the development of multidimensional NMR similar possibilities exist for studying molecular structure, reactivity and dynamics. The NMR spectroscopist first defines the nature of the information needed to solve a particular problem. It is then possible to go to the 'NMR stockroom', and choose from a variety of 'NMR reagents', which include pulses, delays, frequencies, RF phases, RF amplitudes, magnetic field gradients, etc. Using the right combination of these reagents, a spectrum can be produced with selected signals which contain the needed information, while removing other undesired signals which interfere with observation of the interesting signals and/or the interpretation of the data.

As an example, **Figure 5** shows a COSY pulse sequence (**Figure 5A**) and the 2D COSY spectrum of heptan-3-one with the 1D ^{1}H spectrum plotted across the top (**Figure 5B**). The 2D spectrum contains a series of peaks along the diagonal whose positions in f_1 and f_2 correspond to the positions of the peaks in the 1D spectrum. Off the diagonal, cross peaks exist which correlate the frequencies of proton pairs which are coupled to each other. In COSY

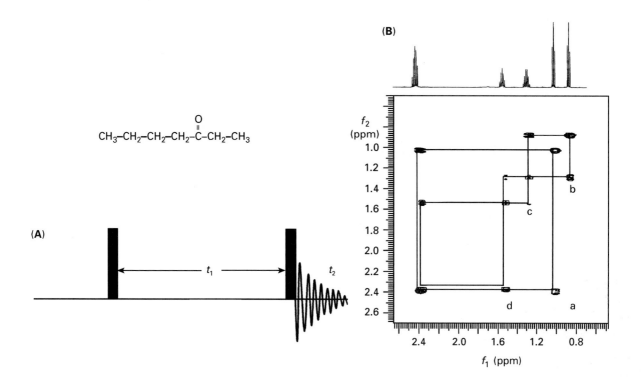

Figure 5 (A) COSY pulse sequence; and (B) COSY spectrum of heptan-3-one with its 1D ^{1}H spectrum plotted across the top.

spectra, these correlations indicate protons which are on adjacent carbons (or non-equivalent protons on the same carbon). Separate sets of cross peaks are observed for the ethyl (a) and butyl (b, c and d) fragments of the molecule. Because the protons on these two fragments are more than three bonds away from each other, there is no J coupling between protons on the two groups. Consequently, none of the resonances from protons on the butyl fragment contain correlations to the resonances of protons on the ethyl fragment.

In general, the NMR parameter plotted along the f_2 dimension is related to the signal detected during the t_2 time period. The NMR parameter plotted along the f_1 dimension is determined by the precession frequencies of the NMR coherences during the t_1 period. A specific sequence of pulses can be used to place the NMR coherence on selected nuclei (e.g. ^{1}H or ^{13}C) to encode their NMR properties during t_1; a second series of pulses and delays are then used to transfer that coherence to the detected nucleus based on an NMR interaction (e.g. by J coupling, dipolar coupling or internuclear relaxation). Cross peaks in the 2D NMR spectrum identify pairs of nuclei which share this interaction. Some common 2D NMR experiments are shown in **Table 1**, along with the NMR parameters plotted along the f_2 and f_1 dimensions, the interaction which produces the correlations, the structure information obtained from the spectrum, typical experiment times and some comments.

The experiments can arbitrarily be classified in four groups: homonuclear chemical shift correlation, heteronuclear chemical shift correlation, J-resolved, and multiple quantum experiments. The first five experiments in **Table 1** are homonuclear chemical shift correlation experiments. In one subset of homonuclear chemical shift correlation experiments, COSY and TOCSY-type experiments, the same chemical shifts (usually those of ^{1}H) are plotted along the f_1 and f_2 axes. The 2D spectrum contains peaks along a diagonal at the intersections of the chemical shifts of each nucleus. If there is J-coupling between two nuclei, then off-diagonal cross peaks connect the diagonal peaks to form a box as in the COSY spectrum of heptan-3-one described above. A second subset of homonuclear chemical shift correlation experiments (NOESY and ROESY) have an appearance identical to that of COSY-type experiments, but with off-diagonal cross peaks that indicate the proximity of two nuclei in space (usually the nuclei must be within 5 Å to produce NOESY cross peaks).

The second group of experiments produces 2D spectra with the chemical shifts of different nuclei along the two axes (e.g. ^{1}H along the f_1 axis and ^{13}C along the f_2 axis). A single cross peak is observed for each coupling interaction in HETCOR, COLOC, HMQC, HSQC and HMBC experiments. The latter three experiments are sometimes put in their own classification, and are called indirect detection experiments. HETCOR-type experiments, which involve detection of the ^{13}C signal during t_2, were the first

---

Table 1 Some common 2D NMR experiments and related information

Experiment name	f_2	f_1	NMR interaction	Structure information	Time (h)[a]/sample quantity (mg)	Comments
COSY	δ_H	δ_H	$^2J_{HH}$ & $^3J_{HH}$	H–C–H & H–C–C–H	0.25/1	Easy
Relayed COSY	δ_H	δ_H	$^2J_{HH}$ & $^3J_{HH}$	H's in a spin system	0.25/1	Easy
TOCSY	δ_H	δ_H	$^2J_{HH}$ & $^3J_{HH}$	H's in a spin system	0.25/1	Easy
NOESY	δ_H	δ_H	H–H dipole–dipole	r_{HH}, conformation	4–12/5	Usually 10–100 times weaker than COSY
ROESY	δ_H	δ_H	H–H dipole–dipole	r_{HH}, conformation	4–12/5	Usually 10–100 times weaker than COSY
HETCOR	δ_C	δ_H	$^1J_{CH}$	C–H	2–8/10	^{13}C detected
Long-Range HETCOR	δ_C	δ_H	$^2J_{CH}$ & $^3J_{CH}$	C–C–H & C–C–C–H	2–8/10	^{13}C detected
COLOC	δ_C	δ_H	$^2J_{CH}$ & $^3J_{CH}$	C–C–H & C–C–C–H	2–8/10	^{13}C detected
HMQC/HSQC	δ_H	δ_C	$^1J_{CH}$	C–H	1/5	^{1}H detected
HMBC	δ_H	δ_C	$^2J_{CH}$ & $^3J_{CH}$	C–C–H & C–C–C–H	1–4/5	^{1}H detected
HOESY	δ_C	δ_H	C–H dipole–dipole	r_{CH}	12/50	Extremely difficult
Homonuclear 2D-J	δ_H	J_{HH}	$^2J_{HH}$ & $^3J_{HH}$	Conformation	0.25/1	Easy
Heteronuclear 2D-J	δ_C	J_{CH}	$^2J_{CH}$ & $^3J_{CH}$	Conformation & no. of attached H	2–8/10	Moderate
2D-INADEQUATE	δ_C	$\delta_{Ca}+\delta_{Cb}$	$^1J_{CC}$	$^{13}C_a$–$^{13}C_b$	12–16/100	1 in 10^4 Molecules extremely difficult

[a] Typical experimental times for molecule with M_r = 500 and experiments performed on a 300–400 MHz spectrometer.

commonly used experiments to provide ^{1}H–^{13}C correlations. Later, after the performance of NMR instruments improved, the more sensitive and more challenging ^{1}H–^{13}C correlation experiments involving detection of the ^{1}H signal during t_2 became popular. In these experiments, the chemical shift of the nucleus X (usually ^{13}C) is indirectly detected in the t_1 dimension. Perhaps, if HMQC-type experiments were popular first, HETCOR-type experiments would now be called indirect detection experiments. The HOESY experiment is the heteronuclear version of the NOESY experiment, and contains cross peaks between the resonances of dissimilar nuclei if there is an NOE interaction between those nuclei.

The third class are J-resolved experiments. These produce spectra with the peaks at the frequencies along f_2, corresponding to the resonances observed in the 1D spectrum of the detected nucleus. In homonuclear 2D J-spectroscopy, the peaks are dispersed into the f_1 dimension based on homonuclear J-coupling. In heteronuclear 2D J-spectroscopy, the peaks are dispersed into the f_1 dimension based on heteronuclear J-coupling (e.g. detection of ^{13}C in f_2 and at the shift of each ^{13}C a multiplet, resulting from all of the resolved J_{CH}-couplings, is observed in f_1).

The fourth class of experiments is multiple quantum spectroscopy such as 2D-INADEQUATE. In these experiments, homonuclear shifts (such as those of ^{13}C) are plotted along the f_2 dimension. If two or more nuclei are J-coupled to each other, then they can be made to share a common multiple quantum precession frequency during the t_1 period. In the 2D-INADEQUATE experiment if C_A and C_B are coupled to each other, then the signals from each of these components will precess at a common double quantum frequency ($\nu_A + \nu_B$) in the f_1 dimension.

Usually, two or more experiments are run, where the correlations in each experiment provide a set of structure fragments. The combined fragments can then be fit together like the pieces of a puzzle, and, in most instances, the right combination of multidimensional experiments can provide complete information about the structure of an unknown molecule.

As an example, if HMQC and HMBC spectra were obtained from p-nitrotoluene, the HMQC spectrum would provide C–H connectivities, illustrated by the highlighted bonds in structure [1]; HMBC would provide information which relates the ^{1}H shifts with ^{13}C shifts of atoms two and three bonds away. Some of these correlations are illustrated in structure [2]. The combined information from the two experiments provides a complete structure of the molecule. While a complete description of these experiments is beyond the scope of this article, some

comments on experimental characteristics are worth noting. Experiments which involve ^{1}H detection are generally much more sensitive than those which involve ^{13}C detection, largely due to the higher γ of ^{1}H. Even though HETCOR and HMQC provide similar frequency correlations and identical structure information, the former involves ^{13}C detection and generally requires ~30 times more sample to produce a spectrum of the same quality.

Although many of the experiments use similar interactions to provide correlations, experiments which use smaller, long-range J-couplings require longer delays (usually ~1/2J) than experiments which use large one-bond J-coupling. During these longer delays, relaxation effects reduce the intensities of the signals which are finally detected during t_2. Consequently, experiments like HMBC produce spectra with poorer S/N than its counterpart, HMQC.

While the entries in the second and third columns of **Table 1** all refer to ^{1}H and ^{13}C, other combinations of NMR active nuclei can be used to perform most of these experiments. For example, the experiments in the first 5 rows of **Table 1** are ^{1}H–^{1}H homonuclear correlation experiments. These experiments will work just as well with ^{19}F–^{19}F homonuclear correlation experiments if there are a number of mutually coupled fluorine atoms in the structure to be studied. Likewise, ^{15}N could be substituted for ^{13}C in HMQC and HMBC experiments.

Experimental aspects of 2D NMR

Acquisition conditions

Instrument requirements Most instruments which have been installed in the 1990s are capable of performing all of the experiments shown in **Table 1**. The collection of 2D NMR spectra requires a stable instrument and a stable instrument environment. The exact requirements become more stringent at higher magnetic fields. For example, 600–800 MHz spectrometers generally require room temperature fluctuations less than $\pm 0.5°$, draughts should be minimized, and the magnet should be mounted on vibration isolation pads. In some instances it might be necessary to mount other mechanical equipment near the instrument (near is not used in an absolute sense since some buildings are more efficient at transmitting vibrations throughout the structure than others) on its own vibration isolation pads. All of the experiments shown in **Table 1** can be performed on standard two-channel (i.e. ^{1}H and X channels) spectrometers which have been installed since 1990, although HMQC and HMBC experiments might require special accessories in order to run these experiments conveniently.

Spectral resolution and data size   As mentioned above, a number of separate 1D FIDs are collected, each with a different value for t_1. In 1D NMR spectroscopy, 50–100k data points are collected and Fourier transformed to provide a spectrum. The exact number of points depends on the spectral window, expected line widths, and the desired digital resolution (usually 0.1–0.5 Hz per point) in the 1D spectrum. If this digital resolution were maintained in both dimensions of a 2D experiment, the file size could grow to many Gbytes and would be difficult to manipulate and store. Consequently, short cuts are used to minimize the sizes of 2D data files. The first of these shortcuts is to minimize the spectral windows to include only those regions which are expected to contain peaks of interest. For example, in a COSY experiment which contains cross peaks between the resonances of coupled protons, the spectral window is narrowed to exclude singlets and solvent resonances. It is usually worthwhile to collect 1D spectra which correspond to the windows in the two dimensions before attempting to run the 2D experiment. However, this may not be possible in some circumstances, e.g. if sample quantity is limited and the f_1 dimension is the ^{13}C chemical shift in an HMQC experiment.

The second shortcut is to drastically reduce the digital resolution in the 2D spectrum; typically the data is collected to provide 2–4 Hz per point digital resolution in the final spectrum. Typical t_2 times are 0.05–0.2 s, an order of magnitude smaller than those used in 1D NMR spectroscopy. The use of longer acquisition times has very little effect on data collection times. If a 1s relaxation delay is used, increasing t_2 from 0.05 to 0.2 s results in 15% longer experiment time and provides a four-fold increase in digital resolution in f_2 (and a four-fold increase in the size of the data file). The same is not true in the t_1 dimension.

Typically, 100–500 separate FIDs are collected to produce a 2D spectrum. To obtain a four-fold increase in digital resolution in f_1, four times as many FIDs must be collected, increasing the experiment time by more than four-fold (the t_1 period for the last FID will be significantly longer than for the first increment where $t_1 = 0$). If four transients are averaged for each of the 100 FIDs, the S/N in the resulting 2D spectrum would be comparable to that obtained in the corresponding 1D version of the experiment (i.e. only the first t_1 increment is collected) obtained by averaging 400 transients. However, in many cases this additional sensitivity is not required and the 2D experiment is longer than is required for signal detection.

Phase cycling for artefact suppression and coherence selection   There are a number of artefacts which can appear in 2D spectra, including peaks and ridges at the transmitter frequencies in f_1 and f_2, and mirror images of the real peaks on the opposite side of the spectrum. These can arise from a number of sources, including the fact that some spins experience imperfect pulses. Even with a properly functioning instrument, nuclei whose resonances lie near the edge of the spectral window can experience a significantly different flip angle compared with those nuclei whose resonances lie near the transmitter, as a consequence of resonance offset effects. Additionally, even those nuclei whose resonances fall near the transmitter experience a gradation of flip angles, depending on the position of the nuclei relative to the probe's transmitter coil (i.e. the nuclei in the portion of the sample near the middle of the tube experience a larger flip angle than nuclei in those portions of the sample near the top and bottom of the tube). These artefacts are reduced by using composite pulses in place of simple 180° pulses. To further reduce artefacts, the phases of 180° pulses are shifted by 180° in alternative transients; and the phases of 90° pulses are usually incremented by 90° in a sequence of four transients. A sequence with both a 90° and a 180° pulse requires the averaging of eight transients to obtain a spectrum resulting from all permutation of the two phase cycles. In experiments with many pulses, the number of transients required to cycle the phases of all pulses becomes extremely large (spectra in which the number of transients per FID is 64–256 are typical). Usually, artefacts from imperfections in one pulse are more prominent that those arising from imperfection in the other pulses in the sequence. In those cases, the phases which remove the most severe artefacts are cycled first. When setting up an experiment, it is necessary to know the number of transients needed to complete this minimum phase cycling, and to set up the experiment to collect an integral multiple of this number of transients.

Some of the 2D spectra result from cancellation experiments (i.e. coherence selection). The HMQC is an excellent example of experiments in this class. As described above, HMQC provides a 2D spectrum correlating the shifts of ¹H and directly bound ¹³C nuclei. In the ¹H spectrum of heptan-3-one, the peaks which are normally observed are those from ¹H bound to ¹²C (99% of the protons, **Figure 6A**); however, if the vertical scale of the spectrum is increased 100-fold a set of satellite resonances from ¹H atoms bound to ¹³C (1.1% of the signal, **Figure 6B**) are observed. To selectively detect the desired component from ¹H atoms bound to ¹³C, the pulse sequence in **Figure 7A** is used. If the phase, ϕ_1, of the 90° ¹³C pulse is applied along the +x- and −x-axes on odd and even transients, respectively, the sign of the undesired signals from ¹H bound to ¹²C are unaffected; however, the phases of the desired signals from ¹H bound to ¹³C are altered in odd and even transients. If the FIDs from odd transients (**Figure 6C**) are subtracted from those in even transients (**Figure 6D**) (by altering the phase of the receiver ϕ_3) the undesired signals cancel and the desired signals add (**Figure 6E**). Detection of the desired signals requires observation of small differences between two large signals. Minor variations in the state of the instrument or its environment will result in imperfect cancellation (as evident by the large residual centre signal in **Figure 6E**) and will produce large noise ridges which obscure the resonances of interest. With a 64-cycle sequence, the residual centre peak can be significantly reduced; however, once an adequate

S/N is achieved after a single transient, the experiment must still be run 64 times longer just to complete the phase cycling necessary to remove the artefacts. Furthermore, even when limited sample quantities result in the need to perform signal averaging, the residual signal intensity varies randomly from one pair of transients to the next, and adds like noise. The result is a ridge of noise along f_1 at the f_2 frequency of intense signals. This noise ridge often called t_1-noise, limits the ability to detect weak resonances in the spectrum.

Pulsed field gradients for coherence selection
Pulsed field gradients (PFGs), also known as gradient enhanced spectroscopy (GES), can be used to achieve coherence selection and minimize the need for extensive phase cycling. In PFG spectroscopy a large z-gradient is introduced along the sample's vertical axis (magnitude ~0.1–0.5 T m⁻¹ and duration ~1 ms); additional PFGs are introduced later in the sequence to selectively refocus the coherence components of interest and continue to destroy coherence components which are undesired. The spectrum in **Figure 6F** was obtained by collecting a single HMQC transient with the aid of PFG coherence selection. The residual centre peak from ¹H–¹²C fragments is completely suppressed in one transient. The first obvious advantage of PFG spectroscopy is that excellent coherence selection is obtained in a single transient. Under these circumstances, the number of transients per FID is

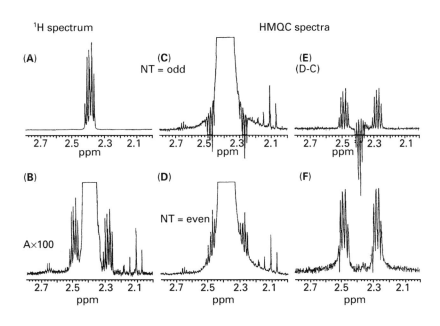

Figure 6 Methylene regions between 2.2 and 2.5 ppm from the proton spectra of heptan-3-one. (A and B) normal ¹H spectra; (C and D) HMQC spectra; (A) ¹H spectrum; (B) 100× vertical amplification of (A); (C and D) spectra obtained from collecting a single HMQC transient with phase cycling for odd and even transients, respectively; (E) is spectrum in (D) minus spectrum in (C); and (F) single transient from HMQC spectrum obtained with PFG coherence selection.

(A)

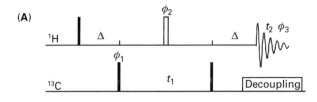

(B)

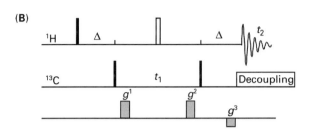

Figure 7 (A) HMQC pulse sequence with phase cycling for coherence selection; (B) PFG-HMQC pulse sequence. In these diagrams, the filled rectangles are 90° pulses, the unfilled rectangles represent 180° pulses and the shaded rectangles represent field gradient pulses.

determined by sensitivity requirements and not the need to complete a phase cycle. Additionally, in cancellation experiments, the suppression level is less sensitive to instrument instabilities; and because the large undesired signal component never reaches the receiver, instrument gain settings can be optimized for detection of the weak signals of interest.

Quadrature detection in f_1 and f_2 When collecting 1D spectra on modern instruments two detection channels are present which independently measure the signals 90° out of phase with respect to one another (**Figure 8A** below). Two FIDs, a real component and an imaginary component (which is 90° out of phase with respect to the real component of the signal), are saved; frequency information is obtained from the real component, and phase information (i.e.

whether the signal is to the left or right of the transmitter) is present in the imaginary component. A complex Fourier transformation produces a spectrum that shows peaks with the proper relationship with respect to the transmitter, depending on the relative phase of the imaginary component in **Figure 8A**.

In 2D NMR, it is not possible to use a second detector in the f_1 dimension. There are alternatives which provide the equivalent of phase sensitive detection in the f_1 dimension; two of these are the States method and time proportional phase incrementation (TPPI).

In the TPPI method a single data set with 512 t_1 increments is collected. In each successive t_1 increment the phase of the 90° pulse at the end of the t_1 period is incremented by 90° with respect to the phase of the corresponding pulse in the previous t_1 increment. (An equivalent experiment can be performed in which the phases of the pulses before the t_1 period are shifted by 90°). This is equivalent to changing the reference frame in f_1 so that the transmitter in the t_1 dimension appears to be shifted to one edge of the spectrum. After performing a real Fourier transformation, all peaks will appear to be shifted to one side of the transmitter in f_1. The main disadvantage of this technique is that phase distortions can appear for resonances in strongly coupled spin systems.

To obtain true quadrature detection two sets of data with real and imaginary components in t_1 must be obtained. In the States method, two set of 2D FIDs are collected and saved. Both sets of data might contain 256 FIDs (2 × 256) and the t_1 delays in the corresponding FIDs in the two data sets are identical. Their only difference is that the second set of FIDs is collected with the phase of the 90° pulse immediately after the t_1 period shifted by 90° compared with the phase of the same pulse in the first set of FIDs. The first set of FIDs contains the frequency

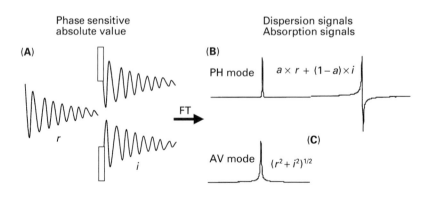

Figure 8 (A) Schematic illustration of real (r) and imaginary (i) components of the spectrum before FT ; (B) after FT the data can be represented in phase sensitive (PH) mode, or (C) an absolute value (AV) mode spectrum can be calculated.

modulation information in t_1 and the second set of FIDs contains the phase information in t_1, similar to quadrature detection in 1D NMR. After each of the FIDs in the two data sets are Fourier transformed, corresponding points from the two data sets are paired to form complex points in t_1 and a complex Fourier transformation is performed with respect to t_1. This latter method provides data sets which are identical in size (and digital resolution) to those obtained from the TPPI method with equivalent digital resolution. The States method of phase sensitive detection is usually preferred because artefacts are less problematical.

Establishing a steady state Ideally, a relaxation delay of $3–5 \times T_1$ should precede the cycle of pulses used to collect each transient. However, this would make experiments impractically long. Normally, 1–2 s relaxation delays are used, even though T_1 values, might be 10 s or more. For larger molecules with shorter T_1 values, relaxation delays of 100–500 ms are used. Under these conditions, incomplete relaxation occurs. Consequently, the intensities of the signals in the first t_1 increment are artificially enhanced relative to the same signals in later increments. This means that the first point in t_1 is offset, leading to a large zero frequency offset in the baseline at slices corresponding to the resonance frequencies of all peaks in t_2. To eradicate this problem, it is customary to perform 8–32 dummy scans, which are discarded, before collection of the data to be saved. During the dummy scans 'steady state' magnetization is established before data collection commences.

Data processing

Zero filling Most of the effort needed to produce quality 2D spectra from a working spectrometer occurs after the data is collected. With the many data processing techniques available, even poor data can frequently be worked up to produce a useful spectrum. Two basic operations which are common to 1D NMR are zero filling and mathematical weighting of the data to improve resolution or S/N. In 2D NMR it is almost always desirable to add zeros in the t_1 and t_2 dimensions so that the linear dimensions in each of these directions is 2–4 times larger than the data collected. For example, if the States method was used to collect 2×256 FIDs with 1024 points (in t_2) per FID, it is useful to zero fill the data so that Fourier transformation is performed on a $2 \times 1024 \times 2048$ matrix. The final displayed spectrum after phasing will be a 1024×1024 matrix. It is useful to optimize other processing conditions described below on the smaller data set without zero filling to shorten processing times for the many

iterations needed. After this is accomplished, zero filling should be used to produce the best quality spectrum for final display.

Absolute value versus phase sensitive display The selection between absolute value (AV) and phase sensitive (PH) display modes is governed by the nature of the data collected. The 1D NMR spectra obtained from modern instruments are always phase sensitive as illustrated in **Figure 8**. As described above, two FIDs are collected, a real (r) FID and an imaginary (i) FID with a 90° phase shift between them (**Figure 8A**). The phase of the imaginary component can be +90° or −90° with respect to the real component. When a complex FT is performed, the phase information [i.e. whether the imaginary component is a $+\sin(\omega t)$ or a $-\sin(\omega t)$ function] determines the direction of the frequency offset relative to the transmitter. After FT and phasing, if the PH display mode is chosen, two components of the data are obtained (**Figure 8B**). The real component is a pure absorption mode signal which is usually displayed, and the imaginary component is a dispersion spectrum which is either hidden from the operator or disposed. In cases where the spectrum is difficult to phase, the real and imaginary components are combined point by point to generate an AV mode signal as illustrated in **Figure 8C**. This display mode has the advantage that phasing is not required. However, because the resulting spectrum contains both absorptive and dispersive components the peaks are much broader. For this reason, the AV mode display is rarely used in 1D NMR.

In 2D NMR, the use of AV mode display is more common. If phase sensitive detection is not used and/or if the peaks in the spectrum are phase modulated, then it becomes necessary to use an AV mode display. In addition, in 2D NMR there are twice as many phase parameters to adjust, making the phase correction procedure somewhat cumbersome if the right experiment delays were not used to obtain the data. Under these circumstances cross peaks in the spectrum are dispersive rather than pure absorption, and it is necessary to display the data in an AV display mode.

Weighting In 1D NMR the selection of weighting functions is based on the desired tradeoff between resolution and S/N ratio. In 2D NMR, the selection of weighting functions is based on the nature of the experiment, the display mode (AV versus PH), the desired resolution and the desired S/N. An entire article could be written to describe the various weighting functions which have been developed. Rather than discuss all of these, it is useful to break them up into several groups based on the general effect they have on the spectrum and show a limited number of representative weighting functions. These are shown in **Figure 9** along with two types of FIDs which are commonly obtained in 2D NMR.

The groups shown in **Figure 9B** and **9C** are used to provide a smooth decay at the end of a truncated FID to minimize truncation artefacts in the spectrum. These functions are generally used when the FID is similar in shape to the one shown in **Figure 9D**. The use of an exponential decay function (**Figure 9A**) is not common in 2D NMR because it broadens the base of the peak and results in long ridges/tails at the base of intense signals. These tails would obscure other weak cross peaks that fall at the same frequency. It also produces severe line broadening under conditions which provide noticeable smoothing of the FID. If smoothing is desired, the functions in **Figure 9B** and **9C** accomplish this with minimal perturbation of the first half of the signal.

The group of weighting functions (**Figures 8E–G**) is used to provide resolution enhancement when the data must be displayed in AV mode. In addition, some 2D experiments such as HMBC produce FIDs which are echos, like the one shown in **Figure 9G**. Under these circumstances it is usually desirable to match the weighting function to the echo (by adjusting parameters which control the width and

displacement of the maximum of the weighting function) so that their maxima coincide and their initial buildup and later decay rates are matched. It is common to use different types of weighting functions in the t_1 and t_2 dimensions.

Linear prediction   Limited access to instrument time and the volume of data which must be collected require that short cuts, which adversely affect the appearance of the spectrum, must always be taken. Digital signal processing techniques can significantly enhance the appearance of a spectrum without increasing data collection times. In some cases, when instrument time is at a very high premium, it might even be desirable to deliberately reduce the experiment time below the minimum needed for a reasonable spectrum, knowing that processing techniques can be used to compensate for the lost data. Mathematically, it is possible to use the behaviour of a function during time t, during which a measurement is made, to predict the behaviour of the function if the measurement time had been extended by time t' (**Figure 10A**). As of writing, in multidimensional NMR, linear prediction is the most used and most useful of these mathematical methods. Essentially, the oscillatory behaviour of the signal intensity as a function of t_1 (at a specific f_2) is fitted to the sum of a series of cosine waves. Since the number of peaks present at a single f_2 in a 2D spectrum is relatively small, the sum of a relatively small number of frequency components is sufficient to stimulate the behaviour of the FID in t_1. The function can then be used to artificially synthesize values for the FID in t_1 as if a much larger number of t_1 increments had been collected. Usually the data is increased to 2–4 times the original size, and zero filling is applied to double the length of the data (e.g. if 2×256 t_1 increments were collected, linear prediction would be used to forward extend the data to 2×1024 and zero filling could be used to further increase the size in t_1 to 2×2048). This permits improvements in resolution comparable with what would be achieved from an experiment that is up to four times longer than the actual experiment time.

Linear prediction can also be used to remove experimental artefacts from data. For example, the intensity of a single FID in the middle of a 2D experiment (**Figure 10B**) could be distorted if the field were perturbed for some reason. If steady pulses were not applied at the beginning of the experiment, the first few points in t_1 might be more intense than they should be (**Figure 10C**). In the former case, with linear prediction, the behaviour of the FID on the either side of the distorted point could be used to approximate the correct value of the distorted point.

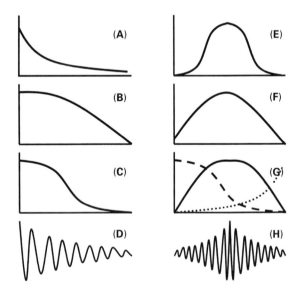

Figure 9 General shapes of various weighting function used to massage NMR data and typical FIDs: (A) exponential decay, (B) shifted sine function, (C) shifted Gaussian function, (D) typical FID in which acquisition has been truncated before the signal decays, (E) Gaussian function, (F) sine function (G) product of exponentially increasing and shifted Gaussian functions, and (H) echo signal FID.

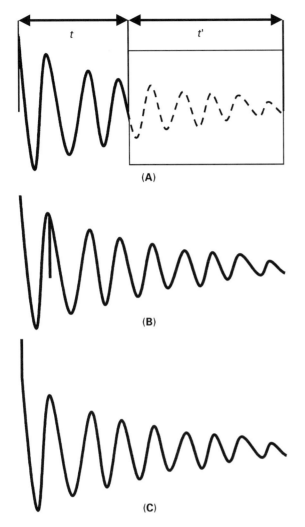

Figure 10 (A) Time domain NMR signal (—) detected (---) calculated using linear prediction. (B) FID with a distorted point in the middle. (C) FID with the first few points distorted by pulse breakthrough.

In the latter case, if the first two points are distorted, the behaviour of the function for points 3–10 could be used to back predict the proper value of the points in the beginning of the FID.

List of symbols

t_1 = evolution time; t_2 = detection time; T_1 = relaxation time; γ = gyromagnetic ratio.

See also: **Diffusion Studied Using NMR Spectroscopy; Macromolecule–Ligand Interactions Studied By NMR; Magnetic Field Gradients in High Resolution NMR; NMR Data Processing; NMR Pulse Sequences; Nuclear Overhauser Effect; Nucleic Acids Studied Using NMR; Product Operator Formalism in NMR; Proteins Studied Using NMR Spectroscopy; Solid State NMR, Methods; Structural Chemistry Using NMR Spectroscopy, Organic Molecules; Structural Chemistry Using NMR Spectroscopy, Peptides; Structural Chemistry Using NMR Spectroscopy, Pharmaceuticals.**

Further reading

Aue WP, Bartholdi E and Ernst RR (1976) Two-dimensional spectroscopy. Application to nuclear magnetic resonance. *Journal of Chemical Physics* **64**: 2229.

Berger S (1997) NMR techniques employing selective radio frequency pulses in combination with pulsed field gradients. *Progress in NMR Spectroscopy* **30**: 137.

Bovey FA and Mirau PA (1996) *NMR of Polymers.* Academic Press.

Cavanagh J, Fairbrother WJ, Palmer III AG and Skelton NJ (1996) *Protein NMR Spectroscopy Principle and Practice.* Academic Press.

Clore GM and Gronenborn AM (1991) Application of three- and four-dimensional heteronuclear NMR spectroscopy to protein structure determination. *Progress in NMR Spectroscopy* **26**: 43.

Croasmun WR and Carlson RMK (1994) *Two-Dimensional NMR Spectroscopy Applications for Chemists and Biochemists.* New York: VCH.

Ernst RR and Anderson WA (1966) Application of Fourier transform spectroscopy to magnetic resonance. *Review of Scientific Instruments* **37**: 93.

Freeman RA (1997) *A Handbook of Nuclear Magnetic Resonance* 2nd edn. Essex: Longman.

Griffiths PR (1978) *Transform Techniques in Chemistry.* New York: Plenum Press.

Jeener J (1971) *Abstracts AMPERE International Summer School,* Basko Polje, Yugoslavia.

Martin GE and Zektzer AS (1988) *Two-Dimensional NMR Methods for Establishing Molecular Connectivity.* New York: VCH.

Muller L, Kumar A and Ernst RR (1975) Two-dimensional carbon-13 NMR spectroscopy. *Journal of Chemical Physics* **63**: 5490.

Schmidt-Rohr K and Spiess HW (1994) *Multidimensional Solid State NMR and Polymers.* New York: Academic Press.

UV Spectroscopy of Biomacromolecules

See **Biomacromolecular Applications of UV-Visible Absorption Spectroscopy.**

UV Spectroscopy of Dyes and Indicators

See **Dyes and Indicators, Use of UV-Visible Absorption Spectroscopy.**

UV-Visible Absorption and Fluorescence Spectrometers

GE Tranter, GlaxoWellcome Medicines Research, Stevenage, Herts, UK

ELECTRONIC SPECTROSCOPY
Methods & Instrumentation

UV-visible absorption and fluorescence (together with phosphorescence) spectrometers work with light in the wavelength region extending from the far ultraviolet (175 nm) to beyond the red end of visible light (900 nm). Often, the highest specification absorption instruments are able to extend their range into the near-infrared region (NIR). Many of the elements of UV-visible absorption spectrometers likewise appear in the analogous fluorescence spectrometers and can be conveniently described alongside each other.

UV-visible absorption instruments

Typically, in a scanning UV-visible absorption spectrometer, light from a suitable source in transmitted through a monochromator (or filters in a low specification instrument) to yield light of the desired wavelength. This is then passed through the sample and thence to a detector (**Figure 1A**). As the monochromator in scanned through its wavelength range so a spectrum is measured from the detector's response. To improve spectral resolution and accuracy, the highest specification instruments have two, or even three, monochromators in series.

In contrast, diode array based instruments, popular as relatively low cost spectrometers, have a reverse arrangement by having the dispersion of wavelengths post-sample by a dispersive optic (e.g. a diffraction grating) which irradiates the diode array detector with the spectrum across its elements. Whilst commercially available diode array instruments are invariably of lower optical quality than the best scanning instruments, and are limited by their spectral resolution through the number of elements in their array, they do offer the advantage of rapid spectral acquisition as the complete spectrum across the wavelength range may be acquired almost simultaneously.

A similar 'reverse optics' arrangement, with a post-sample monochromator followed by a detector has infrequently been incorporated into scanning instruments, albeit rarely on a commercial level (**Figure 1B**). Reverse optics instruments, whether diode array or scanning, can be more prone to inducing sample photodecomposition and other photoreactive phenomena, as the full intensity of the light source, unattenuated by a monochromator, is incident on the sample.

As the light source and all of the optical components of an instrument will have a wavelength dependence it is necessary to acquire a 'background' spectrum in the absence of a sample with which to correct an acquired sample spectrum. If one wishes to correct simultaneously for the optical properties of a moiety in the sample, such as a solvent is solution studies or the vessel in which it is contained, then this 'reference' may replace the sample when acquiring the background. Although the background and sample spectra have to be acquired separately (although ideally at proximate times) in a 'single beam' instrument, many spectrometers have a 'dual beam' configuration that enables both to be acquired together (**Figure 1C**). In this case, the light is split into two equivalent beams prior to the sample. One beam passes through the sample, as in the single beam case, whereas the other passes through the reference. In most current instruments the two beams are then recombined onto a single optical path to the same detector. The resulting two signals are distinguished by alternately obscuring the beams through the use of 'choppers' (typically a rotating disk with apertures), rotating mirrors (which may be used to generate the two beams alternately) or a combination of the two. Nonetheless, for studies of the highest precision, a sequential background should be acquired to ensure complete correction for the sample optical path, which will inevitably differ slightly from that of the reference path in a dual beam arrangement.

The use of oscillating signals brings a further benefit in the ability to use AC rather than DC detection circuitry, where phase lock amplification can be used to best advantage, particularly if a dark period during which light is completely obscured from the detector is also introduced to provide a zero transmission level. This advantage is equally applicable in single beam configurations and therefore is to be found in virtually all scanning instruments. The drawback of employing oscillating signals is that rapid phenomena such as in kinetic studies may be immeasurable on the oscillation time-scale. In these cases a dual beam configuration with two separate detectors and continual illumination may be more appropriate.

Instruments are generally limited by the maximum absorption (measured in absorption units, AU) they can measure due to stray light (see later) and signal-to-noise concerns. Typical limits are, in practice, up to 1 AU for diode arrays and single monochromators, up to 2.5 AU for routine double monochromators and up to 4 AU for the highest specification double monochromator instruments. A once popular technique, now reappearing in the latest instrumentation, for aiding the precise absorbance determination of highly absorbing samples is to introduce an attenuator into the reference beam in a double beam configuration. In so doing, the relative absorbance of the sample to the reference in reduced to around 1 AU, so achieving near optimal conditions regarding signal-to-noise. However, stray light will continue to exert its effects.

As with all instrumentation, proper care and maintenance are essential for correct functioning. Notwithstanding this, UV-visible absorption spectrometers are very reliable if a few precautions are taken with their operation. In particular, instruments should sited in a vibration, vapour and dust free environment with low humidity and minimal temperature variation. For use in the UV region it is necessary to purge the instrument with dry evaporated nitrogen to exclude oxygen, which absorbs below 200 nm and will otherwise generate reactive ozone that damages the optical components. The discipline of continuously purging with nitrogen, whatever the wavelength

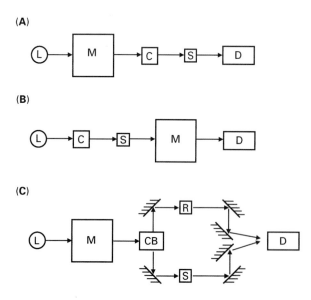

Figure 1 Conceptual diagrams of absorption spectrometer configurations: (A) single beam (B) single beam reverse optics and (C) dual beam. L = light source(s), M = monochromator(s), C = chopper, B = beam splitter, R = reference sample, S = test sample and D = detector.

region, protects the optical components from airborne contaminants and prolongs their life.

Light sources

The key requirements for a light source in UV-visible instruments are an adequate coverage of the spectral range with sufficient intensity together with stability of output. Deuterium arcs and tungsten filament lamps are frequently employed in tandem for the lower (180–350 nm) and higher (330–900 nm) wavelength region, respectively. Alternatives for the higher wavelength region include quartz halogen lamps, which are essentially a tungsten filament in a halogen atmosphere within a quartz envelope. These operate at higher temperatures than traditional tungsten filaments, the halogen prolonging the life of the lamp through reacting with vaporized tungsten to minimize blackening of the envelope.

For the complete wavelength region (175–1000 nm) high intensity xenon arcs may be employed. The output of such arcs, particularly those above 100 W, are of great value in fluorescence instruments where high intensity is a prerequisite, but are usually considered unnecessarily powerful for absorption spectrometers, particularly given their heat generation. Other alternatives, currently for specialist use, include tuneable lasers (which may obviate the need for monochromators) and pulsed arc lamps (which give a broad band of intense radiation, like steady-state arcs, but with much reduced heat generation).

Monochromators

The simplest of wavelength selectors is a set of optical filters in a movable mount such that an individual filter may be placed into the optical path. However, they are wholly inadequate for spectroscopy except for the most basic of instruments. Nonetheless, they may be fruitfully employed to prefilter light prior to a monochromator in order to reduce stray light and attenuate the incident radiation of the optical elements.

The simplest of monochromators consist of a rotatable diffraction grating or prism together with a number of mirrors to guide the beam from the entrance to the exit slit/aperture. As the grating or prism is rotated, so the wavelength of light issuing from the exit slit varies. The size of the slits through which the light is constrained, coupled with the dispersion of the monochromator's optics, determines the 'spectral bandwidth' (SBW) of the light produced. Light nominally of wavelength λ is better considered as having a distribution of wavelengths, the width of the distribution about λ being given by the SBW. Typically, single monochromators may achieve SBWs of 5–10 nm, whereas for accurate spectroscopy SBWs of the order of 1 nm or less may be appropriate. By coupling monochromators in series the SBW may be reduced to a more suitable figure and likewise improve stray light performance (see later), albeit with a commensurate loss in light intensity. Alternatively, SBW may be reduced by simply increasing the dimensions of a single monochromator such that the physical linear dispersion becomes larger. However, this can lead to impracticalities and problems with focusing of the beam. Whatever the design, the optimal configuration of components and the use of baffles to deflect and absorb unwanted light is of critical importance.

Diffraction gratings have the advantage of ease of manufacture and relatively constant dispersion with wavelength (i.e. the wavelengths are evenly spread out by the grating). Generally holographic gratings have a higher precision than ruled gratings, although the latter may give greater light throughput away from their central wavelength. With diffraction gratings the slits may be kept of fixed size to achieve a given SBW independent of the wavelength. In contrast, the dispersion of a prism is highly wavelength dependent, requiring the coupling of the wavelength drive (prism rotation) to the slit mechanisms. However, prisms do have useful transmission and polarization qualities that are of value in other optical instrumentation such as circular dichroism spectrometers.

Sample compartment

Solution studies are typically carried out using fused quartz rectangular cuvettes (cells) which are, for historical reasons, of 1 cm pathlength. However, many alternative variants are available, including those of cylindrical construction, thermostatted, flow cells, micro cells and a plethora of specialist types. In particular, pathlengths from 0.001 cm to 10 cm are readily available from commercial sources, allowing the study of a wide range of sample concentrations and quantities whilst avoiding problems of excessive or too little absorption. Whatever the cell construction, for reliable results cells must be located in a fixed, reproducible, orientation and position in the sample compartment. Many instrument manufacturers provide a wide range of attachments for controlling the sample, such as thermostatted cell holders, stirrers, sippers and cell autochangers.

Likewise, there are a vast array of attachments for solid samples, whether for transmission studies or for surface reflectance investigations. In particular, many attachments provide either a method of integrating total reflectance (as in integrating spheres to surround the sample) or to orientate the surface with respect to the light beam and detector so as to probe specular and diffuse reflectance.

Recent advances have been made in employing optical fibres to allow the study of samples remote from the spectrometer. Essentially a fibre optic redirects the light from the sample compartment to the external sample, with another returning the resultant beam back to the detector. By these means absorption can be monitored by either transmission or reflectance, using remote cells, surface probes and submersible probes.

Detectors

The detector in a standard UV-visible absorption spectrometer is most frequently a photomultiplier tube (PMT) or a silicon diode, the latter being extended to an array in diode array instruments. PMTs have a greater sensitivity and are employed in the most demanding research grade instruments. However, silicon diode devices are considerably smaller, cheaper and do not require the high voltages necessary with PMTs. Both PMTs and silicon diodes have wavelength dependencies which may also dictate their specific use. More recent detectors include charge coupled devices (CCD) and photomultiplier arrays, which will no doubt become more commonplace in the future.

Calibration

The two axes of an absorption spectrum, namely the wavelength (or correspondingly the energy, frequency or wavenumber) and the absorption (or transmission or intensity), dictate that these two scales of an instrument be calibrated. The wavelength scale calibration is typically accomplished by the use of either a series of line emissions from discharge lamps, the precise wavelengths of which have been tabulated, or through standard filters with known absorption spectra. For general convenience the filter method is the one of choice. The most common filters used are those of holmium oxide or didymium (a mixture of neodymium and praesodymium) oxide in glass. However, these can be difficult to produce consistently and can show variations of some ±4 nm in peak positions at the long wavelength end of their useful range (240–685 nm). Therefore, for accurate calibration, it is necessary to use filters provided with a table of determined peak positions from a reputable source such as a national laboratory for standards. As an alternative to glasses, solutions containing lanthanide ions have proved useful, with less variation but a corresponding decrease in convenience.

Historically, potassium dichromate has been the most extensively employed standard for calibrating the absorbance scale. However, in solution numerous species exist in a series of complex equilibria that are sensitive to pH, concentration and other environmental factors. Consequently, various organic and inorganic compounds have been investigated as alternatives, although many have other complicating features for calibration to the highest accuracy. Solutions also hold the problems of being a test not just of the instrument, but of the laboratory skills of the scientist preparing them. At present, the most practical method of calibration is through the use of neutral density filters, whose absorbance has been established by a reputable source.

Stray light

One of the main reasons for an apparent deviation from the Beer–Lambert law for absorption, excluding chemical phenomena specific to a sample, is the effect of stray light. In an ideal spectrometer, only light of the correct wavelength (within the spectral bandwidth window) that has impinged upon the sample would reach the detector and be monitored. Any additional sources of light detected in a real spectrometer may be thought of as 'stray light'. Broadly, there are five potential sources of stray light: (i) sample fluorescence/phosphorescence/luminescence etc, (ii) ambient light leakage into the instrument, (iii) transmission of light not through or from (in the case of reflectance) the sample, (iv) imperfections in the monochromator and light source and (v) imperfections in the detector optics.

The first of these, emission by the sample, when it does occur is invariably weak and would only cause problems in the most precise studies or extreme cases. As a molecular phenomenon specific to the sample it is not within the realms of 'instrumental' stray light and must be considered on a case-by-case basis.

The second two sources are manifestations of poor instrumental design; instruments should be light tight and the sample should be sufficiently masked in a blackened compartment to ensure that only light impinging on the sample reaches the detector. This latter condition is sometimes unfortunately overlooked by instrument manufacturers, who may, for example, introduce reflective components in the sample compartment, or cell holders that do not fully mask the cell to within its useable aperture and beyond the dimensions of the light beam.

Finally, the last two sources are, to some degree, unavoidable instrumental stray light. Nonetheless, they can be minimized through careful design and maintenance. Imperfections in the optical surfaces and compromises in the positioning of components in the monochromators, and elsewhere, give rise to unwanted reflections or dispersion. In particular, diffraction gratings are not perfect and furthermore, even in ideal circumstances, they generate repetitions of the wavelength range. Thus the choice of optimal component configurations, light baffles and component quality is crucial to the stray light performance.

Reverse optics instruments may similarly exhibit stray light introduced at the detector, post sample. In particular, diode array instruments may suffer through internal reflections in the optical surface covering the array, leading to apparent illumination of the incorrect array elements.

Polarization

All the optical components, particularly the diffraction grating or prism and the light source, cause the light beam to the polarized. For the study of isotropic samples, with no preferred orientation, via transmission methods this is of little consequence. This is not the case for non-isotropic materials, such as crystals and ordered solids, or reflection measurements where one encounters linear dichroic effects. To avoid polarization artefacts it is necessary to insert depolarizing optics at the appropriate positions in the optical path. However, care must be taken to choose a depolarizer that truly depolarizes at each wavelength, rather than one that simply gives a different, but specific, polarization at each individual wavelength (these are intended for use with white light applications). Suitable depolarizers are often based upon multiple scattering (*ala* frosted glass) which, in turn, may give additional stray light concerns.

Fluorescence spectrometers

Flourescence spectrometers can be divided into either lifetime or steady-state instruments, depending on whether they resolve the temporal behaviour of the emission (or more correctly the excited state), or not, respectively. In both cases there are strong similarities with single beam absorption instruments. Thus, much of the preceding sections is equally relevant to them. However, the levels of photons detected in fluorescence (or equally phosphorescence) are typically much lower than those in absorbance: in the former one is detecting the few photons that are emitted by the sample, in the latter one is detecting those of the light source attenuated by the number absorbed by the sample. As a consequence certain features are optimized differently for fluorescence.

Firstly, fluorescence is detected orthogonal to the direction of the excitation beam incident on the sample (**Figure 2**), so as to delineate the emission photons from those of the excitation beam and minimize those from Rayleigh and Raman scattering, although these always provide a residual level. Hence, for solution studies, special fluorescence cells are required

that have orthogonal faces optically transparent and flat. Due to the low levels of photons to be detected it is extremely important to exclude all sources of ambient light from the instrument.

To distinguish the wavelength dependencies of a sample's excitation and emission spectra, monochromators are placed in both the excitation and the emission optical paths. Again, the emission side monochromator and detector may be replaced with a fixed dispersive element (e.g. a diffraction grating) and a diode array. Likewise, in very basic instruments, filters may be substituted for monochromators. For instruments operating at a single excitation wavelength laser sources can be used to good effect.

The selection of excitation wavelength and detected emission wavelength may be independently controlled. Thus the excitation wavelength may be fixed and the emission wavelength scanned to give the emission spectrum, or vice versa to give the excitation spectrum. On many of the higher specification instruments it is possible to automatically scan both emission and excitation wavelengths to give an excitation–emission 2D map.

As the fluorescence is directly proportional to the number of photons absorbed by the sample (in the absence of inner-filter/self-shadowing effects of excessive absorption), it is advantageous to employ very high intensity light sources; xenon arcs are highly suitable. Additional excitation intensity may be achieved by greater spectral bandwidths employed on the excitation side, although this may

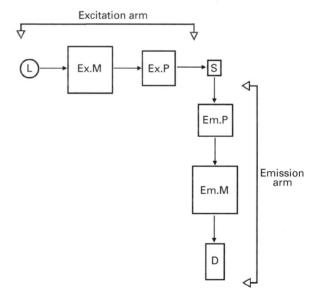

Figure 2 Conceptual diagram of a fluorescence spectrometer. L = light source, Ex.M = excitation monochromator(s), Ex.P = excitation polarizer, S = sample, Em.P = emission polarizer, Em.M = emission monochromator(s) and D = detector.

compromise the spectral resolution of the results. With the higher levels of light impinging on a sample, unwanted photoreactions can be a problem, as can heat generation in the sample, which accelerates all reactions. Alternatives are pulsed sources, which provide broad band radiation over short periods of time and thus may minimize some of the problems of steady-state arcs. Furthermore, in lifetime instruments they provide a means of determining the time lag between absorption and subsequent emission of photons by the sample – the emission lifetime. In this respect pulsed lasers and flash lamps such as hydrogen arcs are popular.

Similarly, the number of photons detected can be increased by modestly increasing the spectral bandwidth on the emission side. However, again this will have a corresponding effect on spectral resolution. In the most sensitive of instruments the detector, invariably a photomultiplier tube, is cooled to reduce noise and thus improve the signal-to-noise levels.

Calibration

Many fluorescence studies are carried out without recourse to correction for instrumental response variations with wavelength, or even with time. For some investigations this is adequate, but the reasons for not calibrating stem primarily from its difficulty rather than its irrelevance; for absolute measurements it is essential.

To correct excitation spectra it is necessary to determine the wavelength dependence of the excitation side of the instrument, with the emission side fixed. A common method is to employ a 'quantum counter'; essentially a compound whose absorption spectrum (at an appropriate concentration) is such that more than 99% of all the exciting photons are absorbed over a sufficiently wide wavelength range and whose emission spectrum and quantum yield are independent of the excitation wavelength over this range. The most frequently used quantum counter is rhodamine B in glycerol or ethylene glycol (at 3–8 g L^{-1}). This solution exhibits constant (to within 2%) fluorescence efficiency at 610–620 nm when excited in the range 350–600 nm, and only ±5% variation for excitation between 250–350 nm. Measurement of the apparent excitation spectrum of such a sample, monitored at an emission wavelength between 610 and 620 nm, allows direct determination of the wavelength dependence of the excitation side of the instrument.

The effectiveness of this method has led to instruments in which a quantum counter is incorporated into the design by diverting a portion of the excitation light to a separate quantum counter and detector. Whilst such a system importantly allows for correc-tion of any temporal variations whilst measurements on samples is ongoing, it does introduce a difference in the optical path from that of the true exitation beam, with a potential inaccuracy.

Determination of the wavelength dependence of the emission side of an instrument is more problematical. Ideally, light from a 'standard lamp', i.e. one whose calibrated spectral distribution is known, is directly introduced into the emission optical path from the sample compartment. Measurement of the apparent emission spectrum and comparison with the known true distribution of the lamp gives the wavelength characteristics of the emission side. One practical variation of this method is to employ the light from the excitation side of the instrument, for which the wavelength characteristics have already been determined via, say, a quantum counter. In order to direct this light into the emission optical path it is necessary to place a reference scatterer into the sample compartment. Such scatterers must have no appreciable wavelength dependence over the wavelength range of interest (mirrors, whilst achieving the redirection, have wavelength dependencies that make them poor choices). Common choices are flat cakes of magnesium oxide (MgO) or barium sulfate with potassium sulfate binder ($BaSO_4$ in K_2SO_4), which can be mounted in the sample position at 45°C to both the excitation and the emission optical paths.

As an alternative to standard lamps and their derivatives, there are numerous compounds whose absolute fluorescence spectra have been documented and may be employed to deduce the emission wavelength characteristics of the instrument. Nonetheless, the calibration of the excitation arm is still an essential procedure.

Fluorescence anisotropy and polarization

As in absorption spectroscopy, instrumental polarization effects can yield unwanted artefacts and therefore it is appropriate to introduce depolarizers into the optical path before and after the sample if the aim is to monitor the true unpolarized fluorescence spectrum. Unfortunately, this will reduce the light levels commensurately and is therefore frequently not pursued. Other methods, involving the use of polarizers set at 'magic angles' to minimize some unwanted polarisation effects have been devised, but are even less frequently employed.

However, deliberate polarization can be used to great advantage in probing the environment and motion of fluorescent molecules and groups in larger macromolecules. In this case, rotatable plane polarizers are inserted in the optical path just before and after the sample; spectra are acquired in the four possible combinations of the polarizers, each in the horizontal or vertical orientation relative to the 'horizontal' plane defined by the excitation and emission optical paths. By comparison of the four spectra the mobility of the fluorescent group during the lifetime of the exicted state can be deduced.

Fluorescence lifetime instruments

Lifetime instruments share most of the optical arrangement of steady-state instruments. Indeed there are commercial instruments that combine both into one versatile spectrometer. The essential optical difference is in the use of intense pulsed light sources, with an emission pulse width typically of the order of 1 ns or less. By coupling the detector and light source trigger with sophisticated electronics and post acquisition processing it is possible to correlate the time between the absorption and subsequent emission of photons by the sample. Essentially, the excitation pulse corresponds to the absorption profile in time. In time-correlated single photon counting methods the delay for the first photon to be subsequently detected is then recorded. This is repeated many thousands of times to give a statistical distribution from which the absorption time profile can be deconvoluted. Alternatively, if the lifetime is sufficiently long, as in phosphorescence, then the complete decay curve of emitted photons from a single exitation pulse can be directly monitored – the pulse excitation method. Finally, rather than employing a flash lamp to provide a pulse of excitation, the intensity of continuous excitation can be modulated and the phase lag of the resulting oscillations on emission intensity observed – the phase resolved method. Whichever method is adopted, in turn the data can be analysed in terms of the fluorescence or phosphorescence lifetimes of the molecular species involved.

Imaging instruments

The advent of imaging detectors, such as CCD cameras and more recently photomultiplier arrays, has prompted the development of monochromators that are able to spectrally disperse the individual pixels of an image, whilst preserving the spatial integrity of the image. Consequently, absorption and fluorescence instruments are beginning to be developed that are able to produce a spectroscopic image of a sample, each pixel of the image being a complete spectrum. It is apparent that such instruments will find growing use in the investigation of inhomogeneous material for which traditional methods are only able to give spatially averaged results.

See also: **Biochemical Applications of Fluorescence Spectroscopy; Biomacromolecular Applications of UV-Visible Absorption Spectroscopy; Dyes and Indicators, Use of UV-Visible Absorption Spectroscopy; Inorganic Condensed Matter, Applications of Luminescence Spectroscopy; Light Sources and Optics; Organic Chemistry Applications of Fluorescence Spectroscopy; X-Ray Fluorescence Spectrometers; X-Ray Fluorescence Spectroscopy, Applications.**

Further reading

Burgess C and Knowles A (eds) (1981) *Techniques in Visible and Ultraviolet Spectrometry*, Vol 1, Standards in absorption spectrometry. London: Chapman & Hall.

Miller JN (ed) (1981) *Techniques in Visible and Ultraviolet Spectrometry*, Vol 2, Standards in fluorescence spectrometry. London: Chapman & Hall.

Knowles A and Burgess C (eds) (1984) *Techniques in Visible and Ultraviolet Spectrometry*, Vol 3, Practical absorption spectrometry. London: Chapman & Hall.

Clark BJ, Frost T and Russell MA (eds) (1993) *Techniques in Visible and Ultraviolet Spectrometry*, Vol 4, UV spectroscopy. London: Chapman & Hall.

Mattis DA and Bashford CL (eds) (1987) *Spectrophotometry and Spectrofluorimetry, a Practical Approach*. Oxford: IRL.

Ingle JD Jr and Crouch SR (1988) *Spectrochemical Analysis*. Englewood Cliffs, NJ: Prentice Hall.

Silverstein RM, Bassler CG and Morrill TC (1974) *Spectrometric Identification of Organic Compounds*, 3rd edn. New York: Wiley.

Vanadium NMR, Applications

See **Heteronuclear NMR Applications (Sc–Zn).**

Vibrational CD Spectrometers

Laurence A Nafie, Syracuse University, NY, USA

> **VIBRATIONAL, ROTATIONAL & RAMAN SPECTROSCOPIES**
> **Methods & Instrumentation**

Introduction

Vibrational circular dichroism (VCD) is defined as circular dichroism (CD) in vibrational transitions in molecules. These transitions typically occur in the infrared (IR) region of the spectrum and hence a VCD spectrometer is an infrared spectrometer that can measure the circular dichroism associated with infrared vibrational absorption bands. CD is defined as the difference in the absorption of a sample for left versus right circularly polarized radiation. This difference is zero unless the sample possesses molecular chirality, either through its constituent chiral molecules or through a chiral spatial arrangement of non-chiral molecules.

A molecule, or an arrangement of molecules, is chiral if it is not superimposable on its mirror image. A chiral molecule possesses a handedness and can exist in either one form, an enantiomer, or its mirror image, the opposite enantiomer. A sample of chiral molecules can have varying chiral purity, referred to as enantiomeric excess. The percent enantiomeric excess (%ee) is defined as the percent excess of one enantiomer relative to the total sample. The %ee of a pure sample of only one enantiomer is 100%. If a sample is composed of an equal mixture of both enantiomers, the %ee is 0% and the sample is called racemic. Racemic samples of chiral molecules exhibit no CD spectra, or any other form of natural optical activity.

The first measurements of VCD were achieved nearly 25 years ago. The early instruments used for these measurements were relatively crude by today's standards, but they demonstrated that VCD was a natural phenomenon that could be used to study in more detail the structure and dynamics of chiral molecules. Subsequent improvements in VCD spectrometers included extending the wavelength of coverage from the region of hydrogen-stretching modes into the mid-infrared region where a greater variety of vibrational transitions could be studied. It also included the implementation of Fourier-transform (FT) methods for VCD measurement. This was a particularly important advance, since virtually all modern, commercially available infrared absorption spectrometers are now FT-IR spectrometers. With the advent of FT-VCD, it became possible to construct an efficient VCD spectrometer starting from a commercially available FT-IR spectrometer.

Within the past few years, accessory modules for the measurement of FT-VCD have become available from the manufacturers of several FT-IR spectrometers. In one case, that of the Bomem Chi-ra*lir*, the first stand-alone, factory-aligned FT-VCD spectrometer has become commercially available, opening the way for widespread applications of VCD spectroscopy.

The principal applications of VCD spectroscopy include measurements of the conformation, absolute configuration and enantiomeric excess of chiral molecules. Most of the molecules of interest for study with VCD are biological in origin. Many are molecules of pharmaceutical interest. The unique application of VCD is its ability to determine absolute configuration in conjunction with *ab initio* calculations. Remarkably close matches have been achieved between experimental VCD spectra and the corresponding spectra calculated from first principles using quantum-mechanical calculations.

General measurement principles

The measurement of VCD is quite simple in concept. A sample is placed in the VCD spectrometer and the polarization of the IR radiation passing through the sample is switched between left and right circularly polarized states. If the sample is chiral, a small difference in the intensity of the IR beam for left and right circularly polarized IR radiation occurs and is measured at the detector. **Figure 1** illustrates the definition of VCD with an energy-level diagram for molecular transitions between the zeroth and first vibrational sublevels of the ground electronic state, $g0$ and $g1$). The decadic absorbance, or IR intensity, of the sample for wavenumber frequency $\bar{\nu}$ (equal to the frequency of the radiation divided by the speed of light) is defined as

$$A(\bar{\nu}) = -\log_{10}[I(\bar{\nu})/I_0(\bar{\nu})] \qquad [1]$$

where $I(\bar{\nu})$ and $I_0(\bar{\nu})$ are the single-beam intensities at the detector with and without the sample present, respectively. The circular-polarization differential absorbance, or VCD intensity, is defined as

$$\Delta A(\bar{\nu}) = A_L(\bar{\nu}) - A_R(\bar{\nu}) \qquad [2]$$

The basic measurement layout is illustrated in **Figure 2**. Here radiation from an IR source is dispersed either by a diffraction grating or a Fourier-transform interferometer so that different wavelengths of the radiation can be distinguished. An infrared optical filter is placed in the beam to restrict

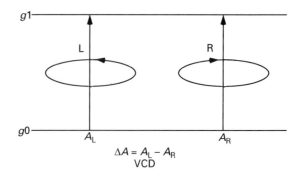

Figure 1 Energy-level diagram illustrating the definition of VCD as the difference in the absorbance of a molecule for left *versus* right circularly polarized IR radiation in a vibrational transition between states $g0$ and $g1$ in the ground electronic state.

the measurement to the spectral region of interest. This is followed by a linear polarizer to define a single state of polarization of the infrared beam. A photoelastic modulator (PEM) then modulates the polarization state of the beam between left (L) and right (R) circularly polarized states at a frequency in the tens of kilohertz range. Immediately afterwards, the sample is placed in the beam. The VCD of the sample creates an intensity modulation of the IR beam at the polarization-modulation frequency. The radiation is then focused on a detector after which further manipulations are carried out electronically to produce the final VCD spectrum. The detector signal is first amplified using a preamp and then divided into two pathways. One leads directly to the IR spectrum as in an ordinary infrared spectrometer. The other pathway is to a lock-in amplifier, referenced to the PEM, that demodulates the high-frequency polarization-modulation component of the signal and leads to the VCD spectrum as described in more detail below.

The precise way in which the detector signal is processed electronically depends on the kind of VCD spectrometer used. In the following sections, three different VCD spectrometer designs are discussed. The simplest of these is the dispersive VCD spectrometer, and we will use this design to illustrate the basic concepts associated with the electronic processing of VCD spectra. The two subsequent cases involving Fourier-transform VCD spectrometers are more complex, but share the same underlying conceptual basis as the dispersive VCD spectrometer.

Dispersive VCD spectrometers

In a dispersive VCD spectrometer, the IR source in **Figure 2** consists of a thermal or arc source of infrared radiation, a light chopper, and a grating monochromator. The infrared source of radiation is first focused on the entrance slit of the monochromator where it is spatially dispersed by the grating. A narrow band of wavelengths (wavenumber frequencies) emerges from the exit slit of the monochromator, and the spectrum is collected by turning the grating and scanning each point in the spectrum sequentially. Successive scans of the monochromator can be averaged to improve the signal-to-noise ratio.

The signal from the detector consists of two components. One is referred to as I_{DC}, which represents the IR single-beam transmission of the sample. In a dispersive VCD spectrometer, I_{DC} is modulated at the frequency of the light chopper and carries the information needed for the ordinary IR spectrum as indicated in **Figure 2**. The other component is I_{AC}, and it is modulated at the polarization-modulation frequency of the PEM, as well as the frequency of the light chopper. In terms of the transmission intensities at the detector for left and right circularly polarized radiation, I_L and I_R, these two components of the detector signal are given by

$$I_{DC}(\bar{\nu}) = (1/2)[I_R(\bar{\nu}) + I_L(\bar{\nu})] \qquad [3]$$

$$I_{AC}(\bar{\nu}) = (1/2)[I_R(\bar{\nu}) - I_L(\bar{\nu})] \sin \alpha_M(\bar{\nu}) \qquad [4]$$

where I_{AC} depends on the sine of the retardation angle of the PEM, which in turn varies sinusoidally at the PEM frequency $\bar{\nu}_M$:

$$\alpha_M(\bar{\nu}) = \alpha_M^0(\bar{\nu}) \sin 2\pi\nu_M t \qquad [5]$$

After some algebra, it can be shown that the ratio of the two intensity components in Equations [3] and [4] is proportional to the VCD intensity as

$$I_{AC}(\bar{\nu})/I_{DC}(\bar{\nu}) = 2J_1[\alpha_M^0(\bar{\nu})]1.15\Delta A(\bar{\nu}) \qquad [6]$$

where $J_1[\alpha_M^0(\bar{\nu})]$ is the first-order Bessel function and is a measure of the efficiency of the PEM setting for the wavenumber frequency specified.

In order to calibrate VCD measurements, one substitutes the sample with a multiple waveplate followed by a linear polarizer. The fast and slow axes of the multiple waveplate are aligned with the axes of the PEM, and the polarizer is set at 45 degrees from these axes. There are four positions of the multiple waveplate and the polarizer relative to the PEM, and these generate a family of four calibration curves. It can be shown that the intersections of these pseudo-VCD curves have the values

$$[I_{AC}(\bar{\nu})/I_{DC}(\bar{\nu})]_{CAL} = \pm 2J_1[\alpha_M^0(\bar{\nu})] \qquad [7]$$

Connecting all the positive intersection points, one obtains a spectral curve, which when divided into Equation [6] allows the isolation of the calibrated VCD intensity spectrum $\Delta A(\bar{\nu})$.

Examples of dispersive VCD and IR spectra for three closely related chiral molecules are presented in **Figure 3**. These spectra are illustrative of a number of basic concepts. All three sets of spectra are recorded in the region of carbon–hydrogen

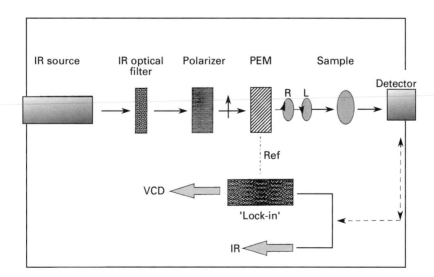

Figure 2 Diagram illustrating the basic optical layout and electronic pathways for the measurement of VCD. The diagram is applicable to both dispersive VCD spectrometers and FT-IR spectrometers that use a photoelastic modulator (PEM) as the source of the polarization modulation of the light beam between left (L) and right (R) circular states.

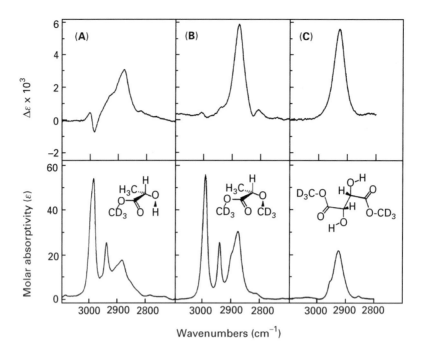

Figure 3   Dispersive IR and VCD spectra in the region of CH stretching vibrations for the molecules (A) (*S*)-methyl-d_3 lactate, (B) (*S*)-methyl-d_3 2-(methoxy-d_3)-propionate, and (C) di(methyl-d_3) D-tartrate illustrating the large positive VCD associated with the methine CH stretching mode in these molecules. The experimental conditions were 0.005 M or 0.01 M solutions in CCl$_4$ in a 1.00 cm fixed-pathlength cell. The resolution of the VCD spectra is 16 cm^{-1} and of the IR spectra is 4 cm^{-1}.

stretching vibrations. The VCD instrument was constructed in our laboratory at Syracuse University starting in 1975 and optimized over the years to include various kinds of improvements including computer control for automatic control and signal averaging. The source used was a xenon arc lamp and the detector was a liquid-nitrogen-cooled InSb detector, used for higher-frequency vibrations above 2000 cm^{-1}. The intensity scale is in molar absorptivity in units of M^{-1} cm^{-1} where the absorbance has been divided by the concentration in moles per litre and the pathlength in cm. The magnitudes of the VCD spectra are approximately four orders of magnitude smaller than the IR absorbance spectra. The VCD spectra all have a bias toward positive VCD intensity. The source of this bias is demonstrated by the series of three spectra, and it is shown to be the lone methine CH stretching mode. In **Figure 3A**, for (*S*)-methyl-d_3 lactate, four CH fundamental modes are present, two for the antisymmetric methyl stretching modes near 3000 cm^{-1}, one for the symmetric methyl stretching mode, with an additional Fermi component at lower frequency, near 2940 cm^{-1}, and the lone methine stretch near 2880 cm^{-1}. Converting this molecule to the deuteriomethoxy analogue, (*S*)-methyl-d_3 2-(methoxy-d_3)-propionate, further enhances the lone methine stretching mode relative to the methyl modes as

shown in **Figure 3B**. The interfering methyl group is eliminated in the case of **Figure 3C** for di(methyl-d_3) D-tartrate where the VCD of the methine can be observed free of other fundamental vibrational modes. The sign of the methine VCD is a marker for the absolute configuration of these and related molecules. The magnitude of the methine VCD is sensitive to the conformation of the molecule in the vicinity of the chiral centre, which for these relatively small molecules is essentially the whole molecule.

Although dispersive VCD spectrometers were the original kind of VCD instrument, they still retain some advantages over the newer Fourier-transform instruments. A relative advantage is present if only a limited spectral range is of interest. In that case a strong source and a narrow filter permit transmission intensities that are higher than could be maintained over a broader spectrum without saturating the detector.

Fourier transform VCD spectrometers

The first measurements of VCD using a Fourier-transform (FT-IR) spectrometer were published in 1979. The basic idea is to substitute the combination of light chopper and monochromator with an FT-IR spectrometer. In an FT-IR spectrometer, all wavelengths of the spectrum of interest are measured at once. The frequencies are distinguished from one another by the interferometer at the heart of the instrument. The infrared light from the source is divided by amplitude at a beamsplitter where one beam is sent to a fixed mirror and the other to a mirror that can change position. The two beams recombine at the beamsplitter and interfere with one another depending on the phase difference of the two light paths. Shorter wavelengths go in and out of phase more rapidly than longer wavelengths and, hence, the different wavelengths can be distinguished from one another by their interference rate or Fourier frequency. The Fourier interference frequency is analogous to the light chopper in the dispersive VCD instrument, but in the case of an FT-VCD instrument, each wavelength has its own 'chopper' frequency. No other changes are needed in the optical setup of the FT-VCD instrument, and hence **Figure 2** is applicable to this instrumental layout as well as that of the dispersive VCD instrument.

The intensity measured by the detector as a function of the moving mirror position, δ, is called an interferogram. The interferogram is a sum of all the intensities of the spectrum at each wavenumber frequency times their Fourier amplitude. Again, there are two intensity components at the detector. One is the ordinary interferogram associated with the single-beam transmission spectrum, and the other is the VCD interferogram that is modulated at the PEM frequency. These component interferograms are given by

$$I_{DC}(\delta) = \int_0^\infty I_{DC}(\bar{\nu}) \cos[2\pi\bar{\nu}\delta - \theta_{DC}(\bar{\nu})] d\bar{\nu} \quad [8]$$

$$I_{AC}(\delta) = \int_0^\infty I_{AC}(\bar{\nu}) \exp(-2V\bar{\nu}\tau) \cos[2\pi\bar{\nu}\delta - \theta_{AC}(\bar{\nu})] d\bar{\nu} \quad [9]$$

where V is the Fourier frequency and τ is the time constant of the PEM lock-in amplifier. The ordinary IR interferogram in Equation [8] contains a phase function, $\theta_{DC}(\bar{\nu})$, that must be determined before the interferogram can be Fourier transformed to yield $I_{DC}(\bar{\nu})$. This is evaluated by standard techniques. The VCD interferogram contains its own phase function and this phase is transferred from another VCD interferogram associated with a spectrum of only positive VCD intensities so that standard phase-correction algorithms can be used. Equation [9] also contains an exponential-decay function that decreases with higher wavenumber frequency. This function

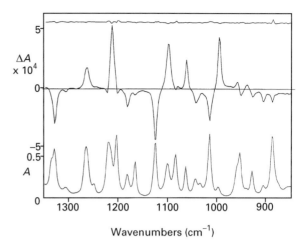

Figure 4 FT-IR and FT-VCD spectra of (–)-α-pinene in the mid-IR region. The experimental conditions were neat liquid with a pathlength of 75 μm a resolution of 4 cm⁻¹ and a collection time of 20 min per enantiomer. The final VCD spectrum was obtained from the subtraction of the VCD of the (+)-enantiomer from that of the (–)-enantiomer.

represents the effect of the time constant of the lock-in amplifier used to demodulate the VCD interferogram from the PEM modulation frequency.

Once Equations [8] and [9] have been Fourier transformed, Equations [6] and [7] can be used to isolate the VCD spectrum although both ratios now also include the exponential function of the lock-in time constant. However, this function vanishes when the calibration curve is divided into the ratio of the AC and DC intensities and does not enter the final VCD spectrum.

An example of an FT-VCD spectrum is presented in **Figure 4**. The IR and VCD spectra of (–)-α-pinene are between 1350 and 850 cm⁻¹. These spectra were measured on the Chiral*ir* VCD spectrometer from Bomem/BioTools. It employs a SiC glower source and a liquid-nitrogen-cooled HgCdTe (MCT) detector. Again we see the VCD spectrum is displayed on an intensity scale that is approximately four orders of magnitude smaller than the corresponding IR spectrum. Good correspondence is present between the peaks in the IR and VCD spectra, although some overlapping of bands is present. It is easy to see that some IR bands are positive and some are negative. According to the definition of VCD, the positive bands absorb left circularly polarized light more strongly than right circularly polarized light. There is no particular correlation between strong IR bands and strong VCD bands. The spectrum illustrates the relative strength of FT-VCD to measure spectra over a wide spectral range at high resolution in a relatively short period of time.

A second example of an FT-VCD spectrum is provided in **Figure 5**. The sample in this case is (+)-camphor and the spectral region is the higher-frequency range of CH-stretching vibrations near 3000 cm⁻¹. This spectrum was obtained using a step-scan VCD spectrometer based on an IFS 55 of Bruker Instruments and a VCD accessory bench aligned and optimized in our laboratory at Syracuse University. Step-scan operation offers the advantage of eliminating the decreasing exponential time-constant function associated with the VCD interferogram that disadvantages the higher-frequency region of vibrational transitions. Here a tungsten light source was used in conjunction with an InSb detector. This VCD spectrum is of much higher quality than the corresponding spectrum obtained with a dispersive VCD spectrometer. Step-scan FT-VCD measurement have been carried out as well in the OH and NH stretching regions between 3000 and 3700 cm⁻¹. In some respects collecting a step-scan VCD spectrum is similar to collecting a dispersive VCD spectrum. In each case the spectrum is scanned and averaged a limited number of times, typically two to four times, and a relatively long time constant can be employed with the PEM lock – since one is not trying to protect a band of Fourier frequencies. The principal difference between the two kinds of measurements is that the light level is not diminished by the reduction of slitwidth if higher resolution is desired, and all the light is used in the FT measurement rather than leaving most of it on the inside of the monochromator as in the case of the dispersive VCD measurement.

Polarization-division FT-VCD spectrometers

In 1989, a new kind of FT-VCD measurement was demonstrated, originally called polarization-modulation interferometry (PMI) and more recently called polarization-division interferometry (PDI). In this approach a polarizing beamsplitter is substituted for the normal amplitude-division beamsplittter. If a linearly polarized infrared beam, with a direction of polarization at 45 degrees relative to the polarization direction of the beamsplitter, is directed to this beamsplitter, then the beam is split into two orthogonally polarized beams. Upon recombination at the beamsplitter, the two beams combine but they do not interfere. The result of the movement of the moving mirror associated with one of the beams is that the polarization state of each wavelength of light cycles continuously through 360 degrees of relative phase retardation at its own Fourier frequency. The cycle starting with vertically polarized radiation

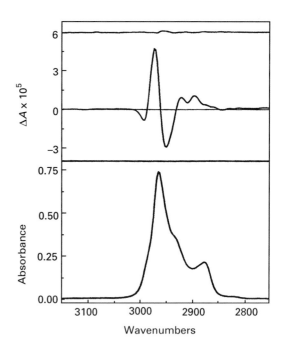

Figure 5   FT-IR and step-scan FT-VCD spectra of R-(+)-camphor in the CH-stretching region. The experimental conditions were a 0.6 M solution in CCl₄, a pathlength of 43 μm, a resolution of 16 cm⁻¹ and a collection time of 2 h per enantiomer. The final VCD spectrum was obtained from the subtraction of the VCD of the (−)-enantiomer from that of the (+)-enantiomer.

is vertical linear, right circular, horizontal linear, left circular and back to vertical linear. In the ideal case, there is no intensity modulation, only polarization modulation. In order to measure a conventional FT-IR absorption spectrum, one can insert a polarizer, say in the vertical position, and the beam is converted from polarization modulation to intensity modulation. Maximum intensity occurs when the beam is vertical linear and minimum when it is horizontal linear. Without the polarizer present, the interferometer is sensitive to linear dichroism in the sample oriented vertically or horizontally at the same Fourier phase as the absorption spectrum (cosine transform) and is sensitive to circular dichroism (VCD) out of phase relative to the absorption spectrum (sine transform). **Figure 6** illustrates the polarization cycles and the mode of operation of this instrument for absorption, linear dichroism and circular dichroism measurements.

The advantage of PDI-FT spectrometers is their independence of PEMs. A PEM has a limited range of wavelength coverage, and there are no PEMs commercially available that operate into the far IR. Yet, there are polarizing beamsplitters that have good efficiency in the far IR, and hence PDI-FT-VCD is the likely approach to extend VCD into this region of the spectrum. To date, the performance of

Absorbance, $A = -\log \frac{I}{I_0}$

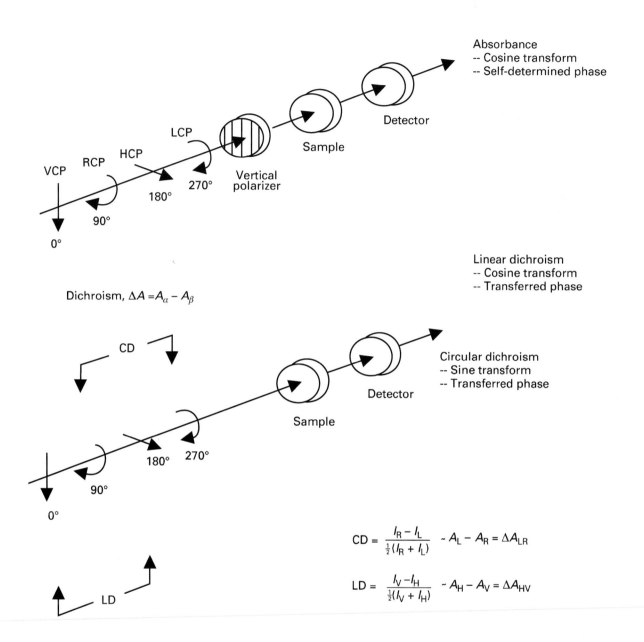

Dichroism, $\Delta A = A_\alpha - A_\beta$

$$CD = \frac{I_R - I_L}{\frac{1}{2}(I_R + I_L)} \sim A_L - A_R = \Delta A_{LR}$$

$$LD = \frac{I_V - I_H}{\frac{1}{2}(I_V + I_H)} \sim A_H - A_V = \Delta A_{HV}$$

Figure 6 Diagram illustrating the polarization sequence and measurement setups for FT-IR, FT-VCD and FT-VLD with a PDI-FT spectrometer. For absorption measurements, the placement of a vertical polarizer converts the train of polarization modulation to intensity modulation. The polarizer is removed for VCD and VLD measurements as illustrated.

PDI-FT-VCD spectrometers is somewhat below that of PEM-FT-VCD spectrometers in the region where they have been directly compared, the mid-IR region. Recently, an FT-VCD instrument was described that possessed both PDI capability and conventional polarization-modulation capability using a PEM. Referred to as double polarization modulation interferometry, this technique offers advantages of signal intensity compared to other single polarization modulation FT-VCD spectrometers.

Artifact suppression

VCD intensities are smaller than IR intensities by approximately four orders of magnitude. As a result they are subject to interference from optical imperfections in the instrument itself. The manifestations of these imperfections, which differ from instrument to instrument, are called artifacts. Artifacts arise from the combination of birefringence in the optics and a sensitivity to different states of linear polarization by the detector or by optical reflection surfaces. The birefringence arises from strain in windows and lenses. Birefringence alters the polarization state of the light and disturbs the balance between left and right circularly polarized light in the spectrometer. A PEM is an oscillating birefringent plate and its action creates the oscillating left and right circular polarization states in the first place. Stray birefringence in the optics, including within the PEMs, further alters the polarization states in undesirable, unknown ways. Once the symmetry between the left and right circular polarization states is broken, the instrument possesses some, perhaps small, degree of linear polarization modulation at the PEM frequency. If the detector or surface of some other optical element responds differentially to the linear polarization modulation, an artifact intensity is created that coexists with the VCD intensity.

There are two kinds of artifacts. One is independent of the sample and exists as a common background spectrum. It can be recorded in the absence of a sample or with any racemic or non-chiral sample, such as a solvent. Once recorded, it may be subtracted automatically from all future VCD spectra to remove this background signal from the measurement. The second kind of artifact is one that varies with the absorption spectrum of the sample or solvent. This is more difficult to remove. In fact, the only way currently known that it can be removed completely from a measurement is by subtraction of the VCD spectrum of the racemic mixture or the opposite enantiomer of the chiral sample. It is important that the racemic or enantiomer VCD measurement be made under the same conditions as the desired chiral measurement, namely, the same pathlength, concentration and cell position. From the standpoint of signal quality, it is more effective to record the VCD spectrum of the opposite enantiomer rather than the racemic mixture since in the former case, the subtraction adds additional VCD information while at the same time cancelling the common artifact spectrum. Unfortunately, a sample of the opposite enantiomer or the racemic mixture is not always available. For this reason, great care needs to be exercised to reduce the occurrence of both kinds of artifacts in the initial optical alignment of the VCD spectrometer.

In practice, it is found that reducing the constant background artifact also reduces the severity of the absorption-dependent artifact. It has also been found that using lenses instead of mirrors after the first polarizer in the optical train can reduce the background artifact. Mirrors possess both birefringence and sensitivity to different states of linear polarization. When used off-axis, as is usually the case, these effects are enhanced on the IR beam. Lenses, on the other hand, can be used on-axis and exhibit lower artifact-inducing effects. In addition to using lenses, the optical alignment should be purely axial and cylindrically symmetric so that a particular direction in space, beyond the direction of beam propagation, is not favoured in the alignment. The final alignment can be achieved by minor adjustment of the optics on a trial and error basis until a good instrument baseline is reached. If the baseline is relatively flat and close to zero across the spectrum, it is usually found that absorption-dependent artifacts are not a serious problem.

Absolute VCD intensity

An important aspect of instrumentation performance is absolute intensity calibration. The intensity-calibration procedure described above using the multiple waveplate and second polarizer has been the method of choice for the calibration of VCD spectra for many years. Nevertheless, the technique is prone to variation depending on the accuracy and care taken in the calibration measurement. If the multiple waveplate or the second polarizer is not positioned at the optimum angular orientation, a calibration spectrum is obtained that is not correct. Usually, the calibration spectrum is too small and the resulting calibrated VCD spectrum has intensities that are too large. Another common source of error is the aperture of the beam used in the calibration measurement relative to that used for the VCD measurement. The degree of polarization modulation in a PEM varies with aperture, decreasing from its centre. The calibration measurement determines the J_1 function of the PEM averaged over the beam profile, and the correct calibration is obtained only if the aperture of the VCD measurement matches the aperture of the calibration measurement.

In an effort to establish an intensity standard for VCD measurements, a number of laboratories have undertaken the measurement of the mid-infrared VCD spectrum of $(-)$-α-pinene. The results from several laboratories have been obtained to date and the results have been plotted in molar absorptivity units. In **Figure 7**, we present the absolute VCD measurements from three locations. The results are still preliminary, and though they show a variation of the order of 10%, these intensities and others appear to

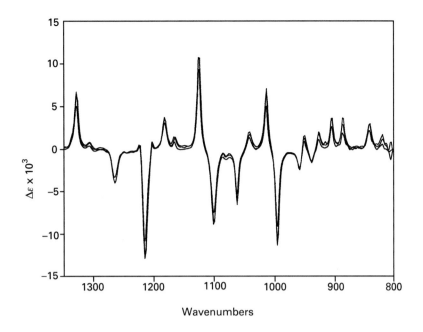

Figure 7 Measurements of the VCD spectrum of neat (+)-α-pinene from three different laboratories illustrating both the variation that can arise in the measurement of absolute intensities and the convergence of the measurements to a relatively narrow range of values.

be converging on a particular set of values. It is hoped that a set of accepted values with a small range of uncertainty will be available in the near future. With a set of absolute intensities in hand, the calibration of a VCD spectrometer could then be carried out by comparison with a standard set of spectra rather than by a calibration measurement subject to the operational errors discussed above.

Areas of application

There are three principal areas of application of VCD spectroscopy. The first, and simplest, is to use VCD spectra to measure the optical purity in terms of %ee of a sample or series of samples. The second application is to determine the absolute configuration of the molecules in the sample, and the third is to determine the solution-state conformations of molecules present in the sample.

Determination of enantiomeric excess

In the case of optical-purity measurements, the determination of %ee is based on the fact that the VCD intensity varies linearly from 100% to 0% with the %ee. The VCD spectrum obtains its maximum value for a chirally pure sample of a single enantiomer; it falls to half its value for a %ee value of 50% and

vanishes for the racemic solution where the %ee is 0%. This linear relationship is demonstrated in **Figure 8** where the VCD spectra of (−)-α-pinene for three different values of %ee are plotted for the region from 1150 to 1075 cm^{-1}. Here it is clear that the VCD in both bands decreases in value as the %ee is lowered from 100% to 95% to 90%. A partial least-squares analysis of the entire VCD spectrum of (−)-α-pinene from 1350 to 900 cm^{-1} for a wide range of %ee values leads to a degree of precision in predicting the %ee from the VCD spectrum of less than 1%. Similar accuracies have been achieved for other molecules. The only prior requirement for the determination of %ee is a high-quality VCD reference spectrum of a sample with known optical purity. From such a VCD spectrum, the VCD intensities for a pure sample at 100 %ee can be determined and all subsequent unknowns can be referenced to this measurement.

There are several advantages of VCD for the determination of optical purity that are not available in the more traditional methods. First, compared to the measurement of optical rotation, VCD intensity can usually be observed in most molecules at approximately the same level of intensity for the strongest bands in the spectrum. This intensity is approximately four orders of magnitude smaller than the IR absorbance spectrum. On the other hand, values of optical rotation can vary widely.

VCD intensities are not temperature sensitive. VCD spectra are composed of many bands and

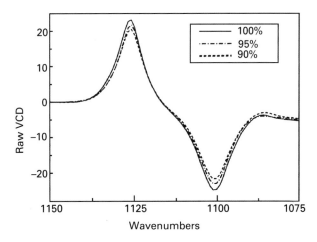

Figure 8 Three VCD spectra of (–)-α-pinene for decreasing values of optical purity, 100%, 95% and 90% enantiomeric excess. The experimental conditions were 70 μm pathlength, 8 cm⁻¹ resolution and 2 h of collection time for each sample.

many spectral points, each one of which is a determinant of the %ee relative to the same point in another sample. Averaging over all the points in the spectrum weighted by their importance to the spectrum, leads to an accurate overall determination. By contrast, optical rotation is a single, temperature-sensitive measurement. The multi-spectral aspect of VCD allows accurate results to be obtained even if there is more noise in a VCD spectrum than in a single optical-rotation measurement.

Compared to chiral chromatography, VCD spectra can be obtained without a physical separation of the two enantiomers. In some cases, enantiomeric pairs of molecules cannot be separated sufficiently with a column, and in this case VCD can be useful. Chiral columns are also expensive to develop and operate, and VCD can be used for routine measurement in a time that is less than that usually required for a physical separation and optical-purity measurement.

The prospects are bright for further improvements of VCD to determine the optical purity of samples. The use of VCD for this purpose is still in an early stage of development, and it is likely that improved measurement and analysis techniques will increase the accuracy of VCD %ee determinations to well below 1% for most samples.

Determination of absolute configuration

A powerful application of VCD is the determination of the absolute configuration of chiral molecules. VCD is more effective than electronic CD (ECD) in

this respect for at least two reasons. One is the number of transitions and the richness of a VCD spectrum compared to an ECD spectrum. There are many more possible transitions to consider in looking for ways to connect the CD spectrum to the absolute configuration. The second reason is that it is easier to calculate IR absorption and VCD spectra than it is to calculate UV absorption and ECD spectra. The latter require accurate descriptions of excited electronic state wavefunctions whereas a vibrational spectrum can be calculated only on the basis of the ground electronic state and its response to nuclear motion. This information is readily available when the equilibrium ground-state geometry is optimized and its molecular force field is determined.

There is widespread interest in the capability of VCD to determine absolute configurations of molecules because the method is free of the need to obtain a single crystal for X-ray diffraction measurements, the standard approach to the determination of absolute configuration of a chiral molecule. VCD measurements are typically carried out in solution or with neat liquids. Since many molecules are difficult to crystallize, VCD promises to be an effective way to determine stereo-specific structures in the absence of the availability of crystals.

Over the past several years, several laboratories have demonstrated that the absolute configuration of a molecule can be determined *de novo* without reference to any other measurement or information base. The method involves carrying out an *ab initio* calculation of the VCD of a particular enantiomer of the chiral molecule. This serves as a theoretical prediction of its VCD spectrum. Next, the VCD spectrum of the molecule is measured and compared to the theoretical prediction. If the sign pattern agrees, the absolute configuration is confirmed to be the one calculated. If the signs are opposite, then the absolute configuration of the molecule in the sample is opposite to the configuration used in the calculation.

An example of the determination of absolute configuration is shown in **Figure 9**. Here the FT-VCD spectrum of (*S*)-methyl lactate in the mid-infrared region is compared to the corresponding *ab initio* calculation using density-functional theory (DFT) and magnetic-field perturbation (MFP) at the 6-31G* basis set level. There are no adjustable parameters to the calculation, and the calculated intensities have been plotted using band shapes similar to the experimentally measured IR and VCD spectra. There is excellent agreement in sign and intensity for both the VCD and IR spectra and there is no doubt that the experimental spectrum was the *S*-enantiomer of this molecule and not the *R*-enantiomer. The theoretical programs used to carry out these calculations are

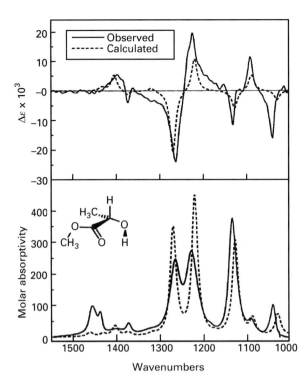

Figure 9 Comparison of the experimental and theoretical IR and VCD spectra of (S)-methyl lactate. The calculation was carried out using density-functional theory and the magnetic-field perturbation theory of VCD intensities. No adjustable parameters were used for the theoretical calculation other than choosing the bandshape for the vibrational transitions.

available commercially with the Gaussian 98 set of quantum-chemistry programs.

In addition to this new powerful approach to the determination of absolute configuration by VCD, it is also possible to gain information about the absolute configuration of members of a family of structurally related molecules by empirical correlation. Among the many bands present in a VCD spectrum, there are often one or more that serve as reliable markers of absolute configuration for a particular chiral centre. One of the best known examples, illustrated in **Figure 3**, is the methine CH stretch of amino acids and hydroxy acids. The S-enantiomer (L-amino acid) always exhibits a positive VCD band for reasons that are currently being explored using detailed quantum-chemistry calculations. Other examples of markers of absolute configuration abound, and such markers can typically be found for any set of structurally related molecules.

Stereo-conformational analysis

The final area of the application of VCD spectroscopy to be discussed here is stereo-conformational analysis. This is the most sophisticated level of VCD application. Here one is concerned about determining the conformation of chiral molecules in solution. This is the principal research application of VCD, and VCD has been used for this purpose for many years. Nearly all classes of chiral molecules have been explored using VCD. Most of these are molecules of biological significance such as amino acids, peptides, sugars, proteins, nucleic acids, and most classes of pharmaceutical molecules. Many excellent reviews have been written on the application of VCD to the study of these molecules. Most recently it has been demonstrated that quantum calculations can be used to determine the presence and relative population of various solution-state conformers of chiral molecules. Theoretically predicted VCD spectra of the most stable conformers of the chiral molecule are compared to the experimental VCD spectrum. A best fit is then sought starting from a Boltzmann distribution of the theoretically determined spectra. Deviations from the VCD spectrum predicted by the Boltzmann distribution are explained in terms of the influence of the solvent on the stability of the various conformers. In this way, new information is obtained about the solution-state structures that are present under particular conditions of solvent, temperature and concentration.

Future performance

The design and performance of VCD spectrometers have been advanced dramatically over the 25 years since the first measurements of VCD. Improvements are continuing today and the first commercially available VCD instrument was introduced only two years ago. Therefore, there is reason to expect advances to continue in the coming years making VCD of higher and higher quality accessible to those concerned about stereochemistry and molecular structure. With the advent of commercially available theoretical programs for the accurate simulation of VCD spectra, VCD spectroscopy is emerging as a powerful new tool for understanding the absolute structure and dynamics of chiral molecules in solution.

List of symbols

A = absorbance; g = ground electronic state vibrational sublevel; I = intensity; J_1 = first-order Bessel function; V = Fourier frequency; α_M = retardation angle; δ = mirror position; θ = phase function; ν = wavenumber frequency; τ = time constant of lock-in amplifier.

See also: **Biochemical Applications of Raman Spectroscopy; Biomacromolecular Applications of Circular Dichroism and ORD; Chiroptical Spectroscopy, Oriented Molecules and Anisotropic Systems; Chiroptical Spectroscopy, General Theory; ORD and Polarimetry Instruments; Raman Optical Activity, Applications; Raman Optical Activity, Spectrometers; Raman Spectrometers; Vibrational CD, Applications; Vibrational CD, Theory.**

Further reading

Ashvar CS, Stephens PJ, Eggimann T and Wieser H (1998) Vibrational circular dichroism spectroscopy of chiral pheromones: frontalin (1,5-dimethyl-6,8-dioxabicyclo [3.2.1]octane). *Tetrahedron: Asymmetry.* **9**: 1107–1110.

Devlin FJ and Stephens PJ (1997) *Ab Initio* prediction of vibrational absorption and circular dichroism spectra of chiral natural products using density functional theory: alpha-pinene. *Journal of Physical Chemistry* A. **101**: 9912–9924.

Freedman TB, Long F, Citra M and Nafie LA (1999) Hydrogen stretching vibrational circular dichroism spectroscopy: absolute configuration and solution conformation of selected pharmaceutical molecules. *Enantiomer* (in press).

Gigante DMP, Long F, Bodack L *et al* (1998) Hydrogen stretching vibrational circular dichroism in methyl lactate and related molecules. *Journal of Physical Chemistry A.* (submitted for publication).

Holzwarth G, Hsu EC, Mosher HS, Faulkner TR and Moscowitz A (1974) Infrared circular dichroism of carbon–hydrogen and carbon–deuterium stretching modes. Observations. *Journal of the American Chemical Society* **96**: 251–252.

Keiderling TA (1990) Vibrational circular dichroism. Comparison of technique and practical considerations. In: Ferraro JR and Krishnans K (eds) *Practical Fourier Transform Infrared Spectroscopy. Industrial and Laboratory Chemical Analysis*, pp 203–284. San Diego: Academic Press.

Lipp ED, Zimba CG and Nafie LA (1982) Vibrational circular dichroism in the mid-infrared using Fourier transform spectroscopy. *Chemical Physics Letters* **90**: 1–5.

Long F, Freedman TB, Tague TJ and Nafie LA (1997) Step-scan Fourier transform vibrational circular dichroism measurements in the vibrational region above 2000 cm^{-1}. *Applied Spectroscopy* **51**: 508–511.

McCann JL, Rauk A and Wieser H (1998) A conformational study of (1S,2R,5S)-(+)-menthol using vibrational circular dichroism spectroscopy. *Canadian Journal of Chemistry* **76**: 274–283.

Nafie LA (1996) Vibrational optical activity. *Applied Spectroscopy* **50** (5): 14A–26A.

Nafie LA (1997) Infrared and Raman vibrational optical activity: theoretical and experimental aspects: *Annual Review of Physical Chemistry* **48**: 357–386.

Nafie LA and Freedman TB (1998) Vibrational circular dichroism: an incisive tool for stereochemical applications. *Enantiomer* **3**: 283–297.

Nafie LA, Cheng JC and Stephens PJ (1975) Vibrational circular dichroism of 2,2,2-trifluoro-1-phenylethanol. *Journal of the American Chemical Society* **97**: 3842.

Nafie LA, Diem M and Vidrine DW (1979) Fourier transform infrared vibrational circular dichroism. *Journal of the American Chemical Society* **101**: 496–498.

Nafie LA, Lipp ED and Zimba CG (1981) Fourier transform infrared circular dichroism: a double modulation approach. In: Sakals J (ed) *Proceedings of the 1981 International Conference on Fourier Transform Infrared Spectroscopy*, pp 457–468. SPIE.

Nafie LA and Vidrine DW (1982) Double modulation Fourier transform spectroscopy. In: Ferraro JR and Basiles LJ (eds) *Fourier Transform Infrared Spectroscopy*, pp 83–123. New York: Academic Press.

Nafie LA (1988) Polarization modulation FTIR spectroscopy. In: Mackenzies MW (ed) *Advances in Applied FTIR Spectroscopy*, pp 67–104. New York: John Wiley & Sons.

Polavarapu PL and Deng ZY (1996) Measurement of vibrational circular-dichroism below ~ 600 cm^{-1} — progress towards meeting the challenge. *Applied Spectroscopy* **50**: 686–692.

Polavarapu PL (1997) Double Polarization modulation interferometry. *Applied Spectroscopy* **51**: 770–777.

Ragunathan N, Lee N-S, Freedman TB, Nafie LA, Tripp C and Buijs H (1990) Measurement of vibrational circular dichroism using a polarizing Michelson interferometer. *Applied Spectroscopy* **44**: 5–7.

Su CN, Heintz VJ and Keiderling TA (1980) Vibrational circular dichroism in the mid-infrared. *Chemical Physics Letters* **73**: 157–159.

Vibrational CD, Applications

Günter Georg Hoffmann, Hoffmann Datentechnik, Oberhausen, Germany

Introduction

Only a few methods are available for the determination of the absolute configuration of chiral molecules. The most common are synthesis by chemical degradation from a molecule with reliably known stereochemistry, the X-ray method of Bijvoet, or the measurement of electronic circular dichroism (ECD). The chemical method has often been used in the history of chemistry when no other methods were available, but it is too time consuming to be generally applicable. The X-ray method is surely the most important, as it gives starting points for all other methods, but it cannot be applied to molecules that are not crystallizable and it is often too time consuming and comparatively demanding. The measurement of ECD requires a suitable chromophore. If none of these methods is practicable, the measurement of vibrational circular dichroism is a good choice. Since vibrational circular dichroism (VCD) and Raman optical activity (ROA) are complementary techniques (a stereochemical problem that cannot be solved by one technique can most probably be solved by the other), the interested reader should also consult appropriate articles of ROA.

The history of VCD is marked by instrumental advances: the early instruments were useful only in the near IR, first for measuring O–H and N–H stretching vibrations and their overtones, advancing to the C–H stretching region, later covering the C=O stretching vibration, and finally intruding into the fingerprint region. Advances in theory are also clearly discernible: first, simple coupled oscillator models were compared to the experimental spectra, then semiempirical calculations were made, later Hartree–Fock (HF) *ab initio* calculations, and now mainly density functional theory (DFT) calculations.

Stereochemistry of small chiral molecules

In principle, the stereochemistry of a molecule can be determined by comparing the sign of a single band of one enantiomer with the calculated sign of that band. Unfortunately, the calculations are still not accurate enough for this method. One has to compare the enantiomer's spectrum in a larger spectral region with the calculated spectra of both configurations and then look for the best fit. In the early days of VCD spectroscopy, when the theory of VCD was not well developed, it was possible to derive 'chirality rules' for some classes of compounds.

The first publication of a single molecule VCD effect appeared in 1974. (S)-(+)- and (R)-(−)-2,2,2-tri-fluorophenylethanol [1] and (R)-(−)-neopentyl-1-d-chloride were examined using the C–H stretching vibration and its respective C–D analogue. As the signal-to-noise ratio of these first spectra was very low, the former compound was re-examined the following year by another group.

[1]

NMR data together with VCD data (O–H and C=O regions) were used to access the conformation of dimethyl tartrate and (2S)-(−)-malic acid dimethyl ester. A later investigation on the conformations of tartaric acid and its esters used *ab initio* calculations to find that the *trans*-COOH conformation with hydrogen bonding was the most stable. These results fit well with the VCD intensities in the C*–O stretching region of the examined compounds if charge flow along the C*–C* bond is assumed.

Using an empirical force field of the Urey–Bradley type, the infrared Raman, VCD and ROA spectra of chlorofluoroacetic acid and its anion were readily interpreted.

In an investigation of the O–H stretching vibration of the (R) enantiomer of 2,2-dimethyl-1,3-dioxoalane-4-methanol [2], the conformer containing an intramolecular hydrogen bridge showed a positive VCD effect at 3600 cm^{-1}. The free form of the alcohol showed no measurable VCD band.

Spectra of substituted allenes showed a correlation of the sign of the VCD for the asymmetrical stretching vibration of the C=C=C moiety (≈ 1950 cm^{-1}) with the absolute configuration. Thus for an (S) configuration the VCD was positive. Judging from two

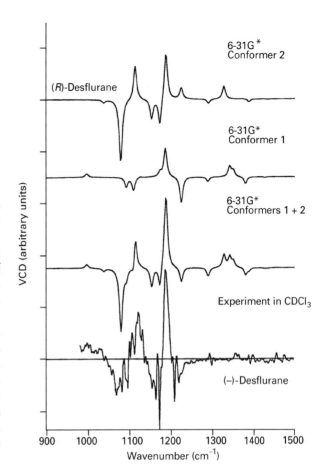

[2]

1-halo-3-*t*-butyl allenes, such a correlation also seems to exist in the C(X)–H stretching mode (≈ 3050 cm^{-1}).

Anisotropies from $+1.5 \times 10^{-4}$ to $+4.5 \times 10^{-4}$ were found for the VCD in the methine stretching mode of hydroxyacid methyl esters, whereas dimethyl-d_6-2,3-O-benzylidine-C-d_1-L-tartrate and (*S*)-methyl-2-chloropropionate showed only small VCD signals.

(*R*)-2,2'-Dihydroxy-1,1'-binaphthyl [3] was examined in the O–H stretching region and from 950 to 1700 cm^{-1}. For (1*R*, 5*R*, 6*R*)-(–)-spiro[4.4]nonane-1,6-diol [4], the theoretical VCD spectra were produced using vibronic coupling theory at the 6-31G level. A comparison of the crystal structure of the ketal of the compound with optically pure (+)-(5α-cholestan-3-one confirmed the results of the VCD determination.

[3]

[4]

Isoflurane [5] and desflurane [6] are relatively new fluorine-contaning anaesthetics. Unlike older anaesthetics such as diethyl ether or chloroform, they are chiral molecules; VCD spectra can be taken and, by comparison with theoretical spectra, their absolute configuration can be determined. This is especially

Figure 1 VCD of desflurane. The experimental spectrum shows the (–)-enantiomer (corrected, as the original label incorrectly read (+) due to a confusion); 0.2 M solution in CCl$_3$). The theoretical calculations were done on the (*R*) configuration. Reprinted with permission from Polavarapu PL, Cholli AL and Vernice G (1992) Determination of absolute configurations and predominant conformations of general inhalation anesthetics: desflurane. *Journal of Pharmaceutical Sciences* **82**: 791–793. © 1992 American Chemical Society.

valuable as the enantiomers have different biological activities; for example the (+)-isomer of isoflurane is nearly twice as effective in activating the potassium current as the (–)-isomer. Experimental and theoretical spectra are shown for desflurane in **Figure 1**. The same assignment has been made for both compounds: the (*R*) configuration for the (–)-isomer and accordingly the (*S*) configuration for the (+)-isomer. For each isomer of desflurane, two dominant conformations were found. In a reinvestigation of the compound using DFT methods (see **Table 1**) and large basis sets, the configurational assignment was confirmed, but three different conformers contributing to the experimental spectrum have been proposed. A study on a third volatile anaesthetic used the same high level of theory: for (–)-1,2,2,2-tetrafluoroethyl methyl ether the (*R*) configuration was been derived from the VCD spectrum and the *trans*-conformer was found to be dominant in CCl$_4$ solution.

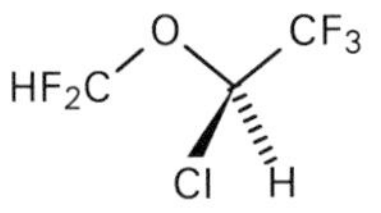

[5]

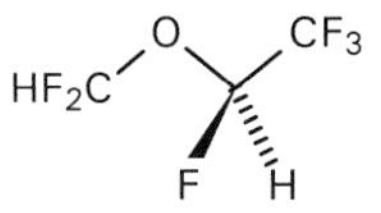

[6]

The enzymatic synthesis of (2R)-(+)-(^{2}H) cyclohexanone and *trans*-(2,6-^{2}H$_2$)cyclohexanone has been reported together with the CD spectrum and the VCD spectrum in the C–H and C–D region.

For (3R)-(+)-methylcyclohexanone [7] in the C–H stretching and deformation region, no temperature dependence of the VCD was found. This leads to the conclusion that only one conformation is present in solution. Four of its chiral deuterated isotopomers were also examined in the C–H and the C–D regions. In the first report of VCD in the CH$_2$ bending region, 3-methylcyclohexanone and (+)-*trans*-1,2-cyclopropane dicarboxylic acid dichloride were investigated. The spectra of (R)-(+)-3-methylcyclohexanone and (R)-(+)-3-methylcyclopentatone showed negative bands in the region of the overtones ($v = 4$) of the C–H stretch, which had a distinctly larger rotatory strength in the case of the cyclopentanone derivative. The very first investigation in the poorly accessible 370–620 cm^{-1} region was performed on (R)-3-methylcyclohexanone. The result was compared successfully with *ab initio* calculations.

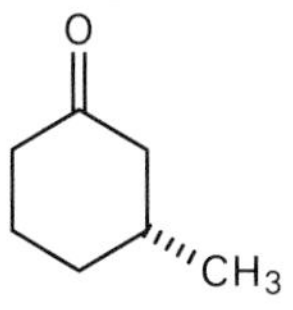

[7]

Small rings (three- and four-membered) have been of great interest among the VCD spectroscopists. The rings have a rigid structure and the compounds are small enough to allow the theoretical spectra to be calculated in reasonable time. Optically active cyclopropanes were studied in the C–H stretch and CH$_2$ bending regions, and deuterated compounds in

Table 1 Models/methods cited

FPC	Fixed partial charge
LMO	Localized molecular orbital
MFP	Magnetic field perturbation
APT	Atomic polar tensor
VCT	Vibronic coupling theory
EXC	Excitation scheme
HF	Hartree–Fock
DFT	Density functional theory

the respective regions. Spectra could readily be interpreted using the FPC model. The C=O region of a dimethyl ester and the C≡N stretching region of a dinitrile were also investigated. Here the coupled oscillator model could be used with advantage. The model failed for deformational modes.

In the VCD spectra of *trans*-2-phenylcyclopropane carboxylic acid and similar compounds, the symmetrical stretching vibration of the methylene group of the cyclopropane ring always showed a negative sign for the (1R, 2R) configuration.

With (S)-(+)-(1,2-^{2}H$_2$)cyclopropane in the gaseous phase, the region above 2000 cm^{-1} could only be resolved to 7.2 cm^{-1}, but from 900 to 1500 cm^{-1} a resolution of 1 cm^{-1} was reached for the first time, allowing the observation of P, Q, and R branches in the VCD spectrum.

The crystal structure of the triply bridged diborate ester tris(*trans*-1,2,-cyclopropanediyldimethylene) diborate [8] has been determined and its VCD spectrum has been measured from 3150 to 2750 cm^{-1} and from 1500 to 950 cm^{-1}.

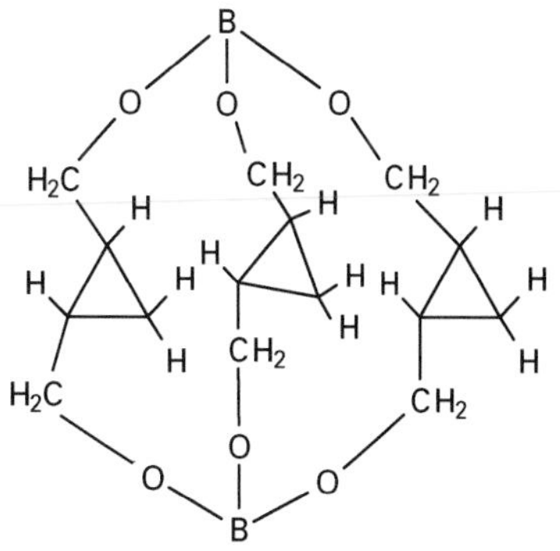

[8]

The symmetry of oxirane is lowered from C$_{2v}$ to C$_2$ by partial deuteration. The resulting (S,S)-(2,3-^{2}H$_2$) oxirane [9] exhibits two modes in the C–H as well as two in the C–D stretching region of the infrared spectrum, corresponding to two couplets in the VCD spectrum. Their intensities are affected by a ring current mechanism (C–H) and a Fermi resonance (C–D). The molecule has also been investigated in the gaseous phase.

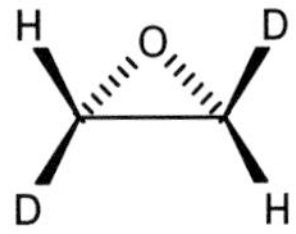

[9]

The VCD spectrum of (S)-(–)-epoxypropane [10] in the liquid and in the gaseous phase shows the splitting of the degenerate vibrational modes of the methyl group. Its analysis verified the VCD theory of the perturbed vibrational degenerate modes. Using a resolution of 1 cm^{-1}, the CD in the rotational–vibrational spectrum of (R)-(+)-methyloxirane has been measured with the result that the Q branch in some bands has the opposite sign to the R and P branches. This can be explained if methyloxirane (in spite of its chirality) is an approximate symmetrical top.

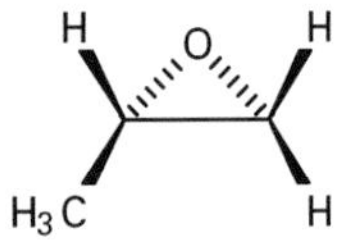

[10]

The absolute configuration of *trans*-2,3-dimethyloxirane [11] (2R,3R for the (+)-enantiomer) has been derived from the VCD spectra and *ab initio* calculations and is consistent with that determined by complexation chromatography.

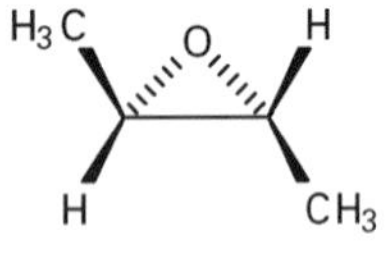

[11]

The vibrational circular dichroism of both enantiomers of methyloxirane [12] has been measured in CCl$_4$, in CS$_2$, and in the gaseous phase. The experimental spectra have been compared with a wide variety of theoretical calculations.

An extensive analysis of the experimental VCD of *trans*-2,3-dimethyloxirane [13] and its 2,3-d_2-isotopomer was published for 850–1650 cm^{-1}.

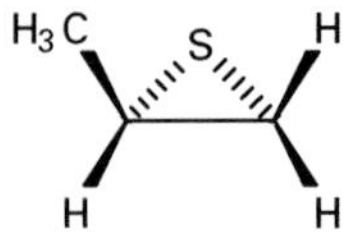

[12]

Comparing the experimental spectra with various calculations, the best results were obtained using the VCT model and the basis set 6–31G$^{*(0,3)}$.

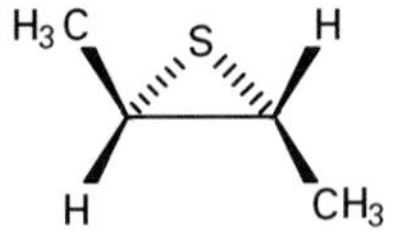

[13]

(2R)-2-Methylaziridine exists in solution as a mixture of the invertomers. According to *ab initio* calculations the ratio of (1R,2R)-2-methylaziridine (*trans*) [14] to (1S,2R)-2-methylaziridine (*cis*) should be 70 to 30. The experimental VCD spectrum is dominated by the effects of the *trans* isomer, as this is not only in excess but also shows greater rotatory strengths.

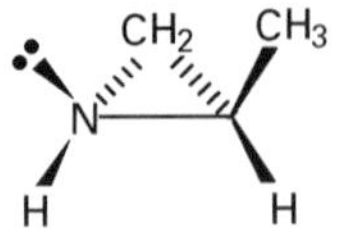

[14]

The heterocycles 1,2,- and 2,3,-dimethylaziridine were measured from 1500 to 1800 cm^{-1} and the experimental VCD spectra were compared with theoretical calculations using the VCT model.

Comparison of the experimental spectra of *trans*-1,2-dideuteriocyclobutane [15] with the FPC as well as with the LMO model shows the former to be considerably more reliable.

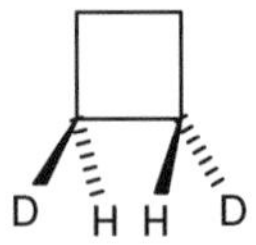

[15]

The synthesis, normal coordinate analysis and VCD spectrum have been reported for (2S,3S)-dideuteriobutyrolactone [16]. Comparison of the latter with *ab initio* calculations using the MFP method and basis set 6–31G** yielded good qualitative agreement.

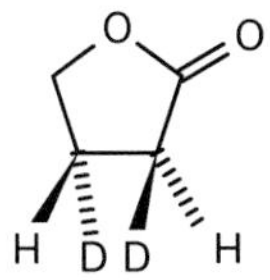

[16]

In a study on the VCD of (3*R*,4*R*)-dideuterio-cyclobutane-1,2-dione [17] the experimental spectra were compared with the calculated rotatory strengths using the MFP model, with good agreement.

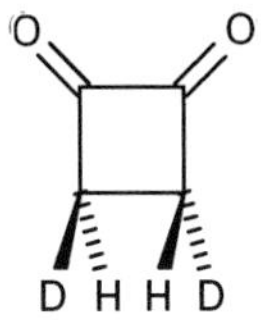

[17]

In the region of methyl deformational modes and in the C–H stretching region, the VCD of α-phenylethylamine, α-phenylethanol, α-phenylethylisocyanate, *p*-bromophenylethylamine and (*S*)-methyl mandelate [18] was examined. The bands near 1450 cm^{-1} were explained by interaction of the CH$_3$ deformational mode with an energetically neighbouring phenyl vibration.

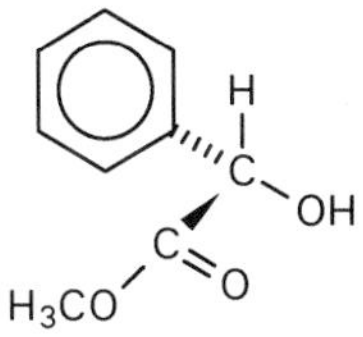

[18]

With 1-phenylethanol, 1-phenylethanethiol, 1-chloro-1-phenylethane, D-α-phenylglycine-N-*d$_3$* and (*S*)-methyl mandelate, the VCD of the methine stretching vibration is enhanced by ring currents. For methyl mandelate, a very large value of $\Delta\varepsilon = 5 \times 10^{-3}$ is found for the O–H stretching vibration.

α-Phenylethylamine, α-phenylethyl alcohol, α-phenylethyl isocyanate and methyl mandelate were measured in the 1625–860 cm^{-1} region. The VCD

effects, which occur at 1368 and 1182 cm^{-1} with phenylethylamine, were correlated with the stereochemistry of the molecules.

A simple chirality rule was derived for six phenyl-carbinols: orienting the fourth substituent to the back and arranging the remaining three substituents in the order OH–Ph–H clockwise, one finds a negative VCD band at 1200 cm^{-1}; orienting them counterclockwise results in a positive effect.

Deuterated phenylethanes were observed in the region from 3100 to 2000 cm^{-1}. All aliphatic C–H and C–D stretching vibrations could be assigned.

The measurement and the theoretical calculation of the VCD spectrum of 6,8-dioxabicyclo[3.2.1]octane [19] was presented together with a detailed *ab initio* normal coordinate analysis using the APT and the FPC models. In another study, mono- and dimethyl derivatives of 6,8-dioxabicyclo[3.2.1]octane were treated the same way, but compared with calculations of higher accuracy (see also the pheromone [34]). The determination of absolute configuration by VCD has been made for some *exo*-7-derivatives of 5-methyl-6,8-dioxabicylco[3.2.1]octane (R=H, OH, Br or CH$_3$). Using recurring patterns in the 1100–1400 cm^{-1} region, the chiral unit –C* (CH$_2$R)X–, with X=O or S was detected in rings of different size. The signs of these patterns corresponds to absolute configuration.

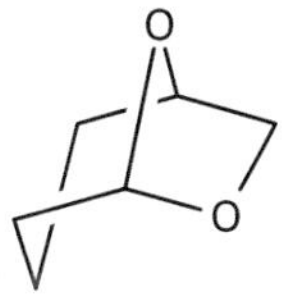

[19]

Crystals

The first sample in which VCD was detected unambiguously was a thin slice of a crystal of α-nickel sulfate hexahydrate. The compound crystallizes as tetragonal bipyramids in the narrow temperature range 31.5–53.3°C. The same paper, which was published in 1973, also reported the VCD of α-ZnSeO$_4$·6H$_2$O. As nickel sulfate is an achiral molecule, the VCD bands can be ascribed to vibrations of the chiral array of water molecules. Five main bands were found: at 2300 cm^{-1} (ν_2 + librations), at 4000 cm^{-1} (ν_3 or ν_2 + librations), at 4200 cm^{-1}, at 4350 cm^{-1}, and the first part of a strong negative band at 5100 cm^{-1} ($\nu_2 + \nu_3$), which was expected to be symmetrical. α-Nickel sulfate has been reinvestigated by the author's group and the latter band showed a sawtooth shape with the steeper descent

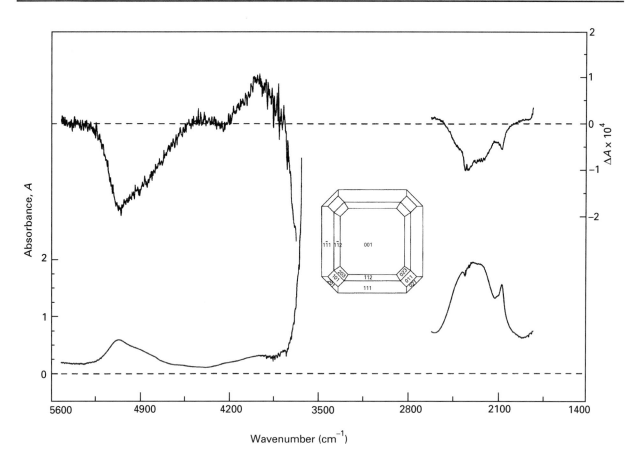

Figure 2 VCD (upper) and absorbance (lower) spectra of a thin crystal slice of α-nickel sulfate hexahydrate ($d = 63$ μm, cleaved parallel (001) as shown in inset). No spectrum could be taken of bands with absorbance > 2.

on the side of shorter wavelength (**Figure 2**). We also found a new band at 2070 cm^{-1}, which, owing to lower resolution, had formerly only been detected in the absorption spectrum and had been attributed to the first overtone of the ν_3 vibration of the SO$_4^{2-}$ ion (species F$_2$ at 1104 cm^{-1}).

Liquid crystals

Studies of the optical activity of liquid crystalline solutions in the infrared region were published one year before the publication of single-molecule VCD effects. Using a solution of 2 mol% d-carvone in a liquid-crystalline eutectic of the isomeric N-oxides of p-methoxy-p'-n-butylazobenzols, huge effects were observed from the liquid crystal forced into a helical arrangement (cholesteric state) by the chiral solute. The liquid crystal acts as a molecular amplifier and reliably allows the determination of absolute configuration using only tiny amounts of substance. Later the same year, the VCD of a solution of 2% (–)-menthol in N-(p-methoxybenzylidene)butylaniline was published. Again the effects were extraordinarily large.

Infrared circular dichroism can also be measured using an ATR arrangement consisting of a wire polarizer followed by a KRS5 half-cylinder as ATR element. The spectrum of a 1% solution of cholesteryl chloride in the liquid crystal ZLI-887 was recorded from 4000 to 400 cm^{-1} and compared favourably with a VCD spectrum recorded by the common-beam technique.

Organometallic compounds

Relatively few papers have been published on the VCD of organometallic complexes. This may be because the common organic ligands are achiral and chirality has to be sought in the arrangement of the ligands around the central atom. The very first spectra of organometallic compounds were of electron transitions of Pr^{3+}–tartrate complexes down to 2000 cm^{-1}. The next published spectra of real vibrational transitions featured the C–H region only. These studies were on tris(3-trifluoromethylhydroxymethylene-d-camphorato) complexes of europium and praseodymium. Studies on complexes with amino acids, ethylenediamine and acetylacetate followed.

The two diastereomeric complexes Δ- and Λ-bis(acetylacetonato)(L-alaninato)cobalt(III) give rise to VCD spectra that can be explained using the degenerate coupled oscillator model (antisymmetric C=O stretching at 1522 cm^{-1}) and the ring current mechanism (N–H stretching). The appearance of the latter is illustrated by the fact that while they give nearly equal absorption spectra, the VCD of the Δ-isomer is nearly ten times larger than that of the Λ-isomer.

The spectral data (C–H stretching) of five copper complexes of amino acids and (Δ)α′-tris(L-alaninato)cobalt(III) have been obtained. As for the parent amino acids at pH values favouring hydrogen bonds, one detects an enhancement of the methine band by ring current effects. This is even larger as a result of the better closure of the ring by the transition metal ion.

Complexes of the trivalent cobalt and chromium with ethylenediamine have also been examined by VCD. Again substantial enhancements of the VCD effects by ring currents are found for the N–H and C–H stretching vibrations.

The peptide *cyclo*-(Pro-Gly-)$_3$ forms complexes with different alkali and alkaline earth metals that show spectra with sensitivity to the conformation of the peptide; the arrangement of the carbonyl groups is especially of interest. The solvent plays the most important role in the development of conformation. The size of the ion-binding cavity formed by the carbonyl groups and the size and charge of the cation are only of secondary importance.

In studies of the interaction of two deoxyribo-oligonucleotides with divalent manganese ions, the resulting changes in the VCD spectra of d(GC)$_{20}$•d(GC)$_{20}$ and d(ATGCATGCAT)•d(AT-GCATGCAT) were interpreted in terms of structural changes.

Biochemical applications

Terpenes

The first VCD investigation on camphor [20] observed the C–H stretching in the principal region as well as in the first overtone and combination regions; the first overtone of the C=O stretching was also measured. Later reports showed VCD also in the mid IR and presented calculational results.

Inherently dissymmetric chromophores, meaning groups that do not gain their chirality only from the influence of their neighbourhood, have always been of interest to investigators studying optical activity, including VCD spectroscopists. The sequence of signs could be correctly predicted for 15 different molecules (including cyclohexanones, menthone

[20]

[21], pinenes, pulegone, and other natural products) featuring the CH$_2$–C$_2$H–C*H moiety.

[21]

Using a FT-VCD spectrometer, the spectra of (+)-3-methylcyclohexanone, (+)-carvone [22] and (−)-α-pinene [23] were observed in the mid-infrared region; a higher signal-to-noise ratio and twice as great anisotropy were obtained than with dispersive instruments. Matrix-isolated molecules feature even larger anisotropies. Accessing the band at about 2920 cm^{-1}, one finds values of 5.4×10^{-4} for (−)-α-pinene and -6.5×10^{-4} for (−)-β-pinene, which really are record values (excepting the huge value of 0.02 for methaemoglobin azide).

[22]

For the six monoterpenes (S)-(−)-limonene, (R)-(+)-limonene, (S)-(−)-perillyl alcohol, (S)-(−)-perillaldehyde, (R)-(+)-p-menth-1-ene and (R,R)-(+)-p-menth-1-en-9-ol, the VCD spectra of the second, third, and fourth overtones of the C–H stretching vibration have been published. The observed couplets can be attributed to a coupled vibration of the CH$_2$CH$_2$C*H fragment.

Other terpenes studied subsequently in the mid IR include nopinone and (−)-borneol, and a detailed

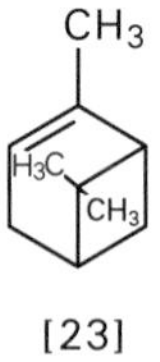

[23]

study has focused on the conformers of (+)-menthol
[24].

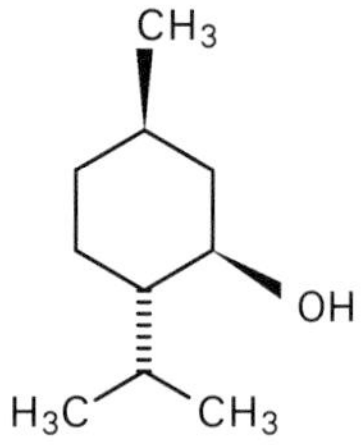

[24]

Carbohydrates

Sugars are very good candidates for the measurement of VCD as the more common ECD is dependent on a chromophore, which is almost always
absent in this class of natural compounds. The examination of the VCD spectra of six common sugars
revealed a chirality rule for the 1150 cm^{-1} band in
deuterated dimethyl sulfoxide. Later the FT-VCD
spectra of the carbohydrates D-fucose, D-arabinose,
D-ribose, D-galactose and D-glucose [25] and their
isotopomers deuterated at the hydroxyl group were
examined in the same solvent. Some useful correlations between structure and spectra are found, but
also some deviations.

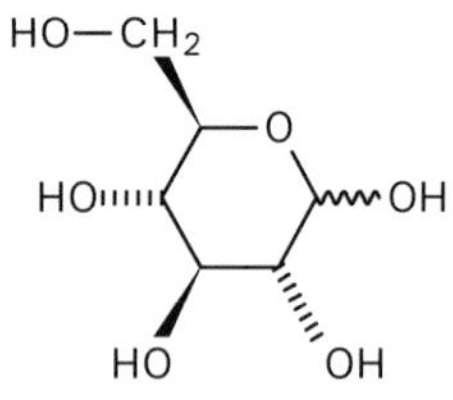

[25]

Cyclodextrins are water-soluble cyclic oligomers
of glucose, the most common of which are α-, β- and
γ-cyclodextrin with six, seven or eight glucose moieties, respectively. Owing to their conical shape with
a hydrophobic interior and a hydrophilic exterior,
they form water-soluble complexes with inorganic
or organic compounds. Comparison of the VCD
spectra of the α- and β-cyclodextrins with hydroxyl-

deuterated α-cyclodextrin, cyclodextrin–copper
complexes and cyclodextrin inclusion complexes
with methyl orange, methyloxirane, n-propanol and
substituted cyclohexanones sensitively monitors
structural changes in dimethyl sulfoxide-d_6.

Alkaloids

The spectrum of calycanthin in the C–H and N–H
stretching regions can be interpreted as due to the
coupling of the chromophore with the substituents.
This is in contrast to the common coupling of the
two chromophores in chiral dimers, which is commonly used to explain the electronic CD.

VCD investigation of a CCl$_4$ solution of (–)-
sparteine [26] – the alkaloid from lupin beans – and
comparison of the experimental results with calculations using the new EXC theory gave adequate
agreement for such a large molecule.

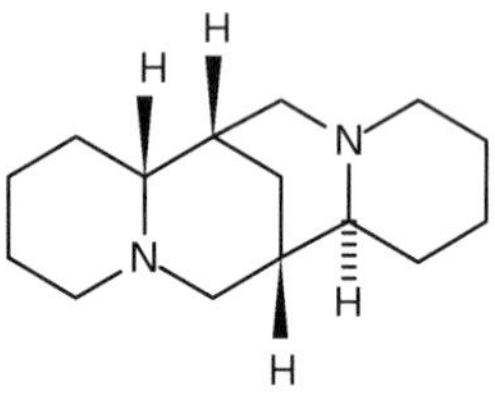

[26]

Vibrational CD in the O–H and N–H stretching
bands of the anticancer chemotherapeutic agent
taxol and two of its side-chain derivatives has been
measured and compared with calculations on taxol
fragments.

Steroids and their precursors

The simple coupled oscillator model, which can
readily be applied to large molecules with two identical oscillators, originates from electronic CD. An
example of the applicability of the model is given by
steroids carrying two carbonyl functionalities. Even
this simple model gives good results for the closely
related steroids 3,6-dioxo-5α-cholestane [27], 3,6-
dioxo-5β-cholic acid methyl ester, 3,7-dioxo-5β-
cholic acid methyl ester, 7,12-dioxo-5β-cholic acid,
3α-hydroxy-7,12-dioxo-5β-cholic acid, and 3-oxo-
5β-cholic acid with only one exception (the 3,7-
dioxo derivative).

(+)-5,6,7,8-Tetrahydro-8-methylindane-1,5-dione
[28] is an important precursor in the synthesis of estrone. The signs of its experimental VCD spectrum
in the 1400–850 cm^{-1} region can be reproduced adequately even using the small basis set 6–31G. In a
later paper the spectra of the target molecule, estrone

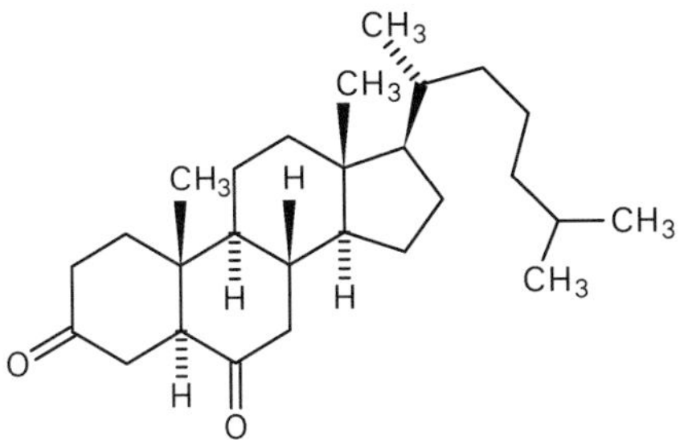

[27]

[29], were calculated with larger basis sets using HF and DFT methods.

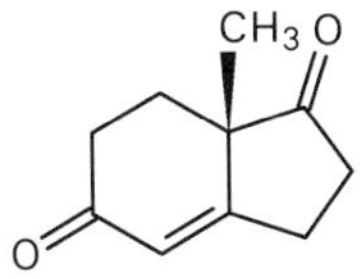

[28]

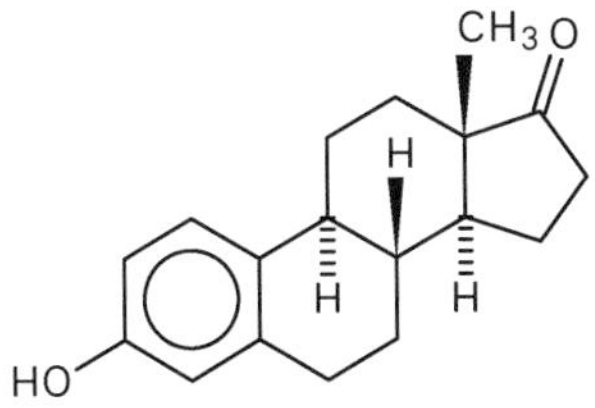

[29]

Amino acids

The simplest amino acid studied is (S)-(–)-glycine-C_α-d_1 [30]. Its weak VCD in the methine stretching at 2990 cm^{-1} was studied together with those of L-alanine and L-proline, which in contrast to [30] show a positive effect.

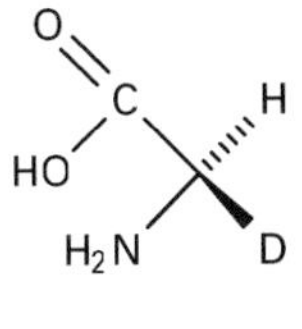

[30]

Nineteen different amino acids were examined using the C–H stretching vibrational region. Aided by these VCD spectra, shown explicitly only for L-valine-N-d_3, a chirality rule was deduced for the chiral methine. If it is supposed that the amino acids studied form an intramolecular ring in aqueous solution, the VCD of the C*–H stretching vibration will be enhanced so much by ring currents (**Figure 3**), that it will obscure all other vibrations in this region. A positive effect with a value of more than 10^{-4} cm^{-1} L mol^{-1} then indicates an L-amino acid.

The vibrational CD spectra of some L-amino acids have been recorded as a function of pH. A large positive bias has been found for the C–H stretching region at neutral or high pH, whereas at low pH the bias is absent and only very small VCD signals are observed. Again the large bias was attributed to ring currents that are possible in some conformations. Another study examined alanine and its deuterated isotopomers.

Seven N-acyl-N'-alkylamide derivatives of different amino acids were measured in CCl$_4$ and CHCl$_3$ using the spectral region 3600–3200 cm^{-1}. The local conformation of the amide moiety was determined as well as the hydrogen bridge bonds using the VCD spectra. Other derivatives studied include N-t-BOC-alanine and N-t-BOC-proline (BOC = butoxy-carbonyl).

Peptides and proteins

The determination of absolute configuration is not of importance in the study of peptides and proteins. Among the peptides that have been studied are poly-alanines, polyprolines, polylysines, polytryosines and poly(γ-benzyl-L-glutamate), as well as gramicidin S and other cyclic peptides; proteins examined include α-chymotrypsin, cytochrome c, haemoglobin, myoglobin, ribonuclease S and triose-phosphate isomerase. VCD spectroscopy is applied with advantage to access the secondary and tertiary structures of these biopolymers. Only a few typical examples are given here from studies published during the 1990s.

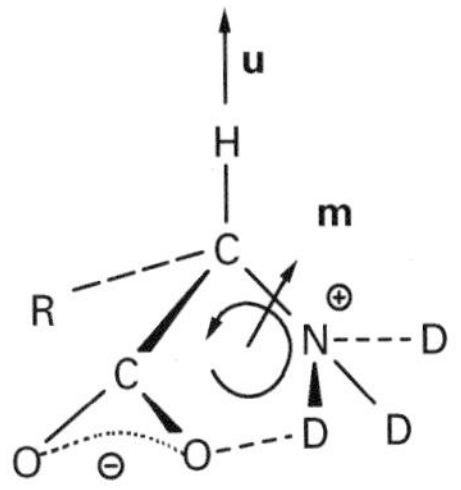

Figure 3 Ring current mechanism (positive VCD) in the C–H stretching vibration of an L-amino acid. Redrawn from Freedman TB, Balukjian GA and Nafie LA (1985) Enhanced vibrational circular dichroism via vibrationally generated electronic ring currents. *Journal of the American Chemical Society* **107**: 6213–6222.

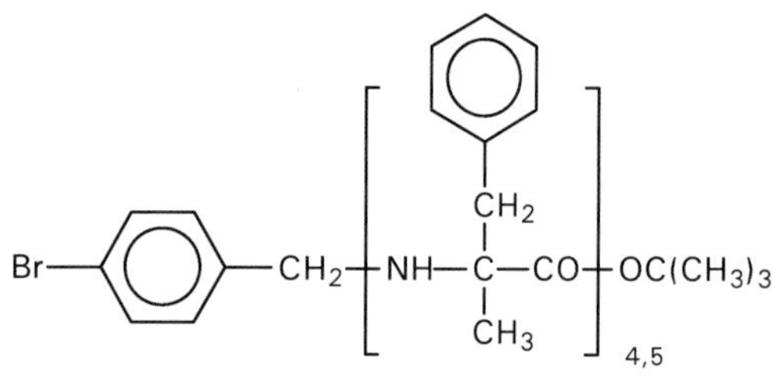

[31]

Using *ab initio* calculations of the model dipeptide CH_3–$CONH$–CH_2–$CONH$–CH_3, the VCD of the four most common secondary structures of proteins were calculated and compared successfully with experimental spectra of albumin, concanavalin A, $(Aib)_2Leu(Aib)_5$ and poly(L-lysine) (with Aib = α-amino isobutyric acid). The main structure of these proteins is the α-helical, β-sheet, 3_{10}-helical and poly(L-proline)II conformation.

The favoured screw sense of homo-oligopeptides of α-methylated phenylalanine and isovaline has been studied using p-BrBz-[D-(αMe)Phe]$_{4,5}$-OBu^t [31] and p-BrBz-[D-Iva]$_5$-OBu^t [32] in CDCl$_3$ solution. Analysis of their VCD spectra shows that the first two compounds are folded in a right-handed 3_{10}-helix, whereas the last pentapeptide forms a left-handed helix.

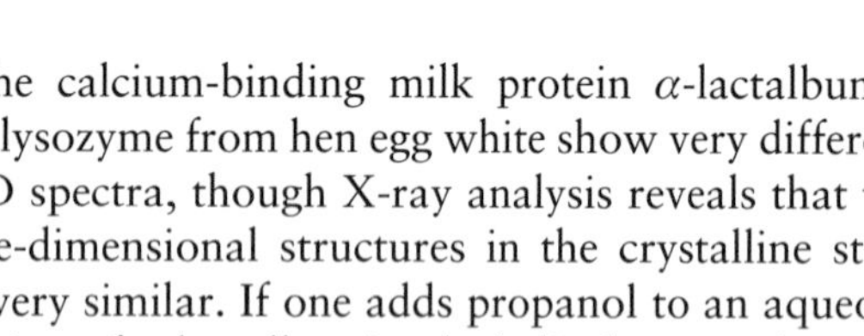

[32]

The calcium-binding milk protein α-lactalbumin and lysozyme from hen egg white show very different VCD spectra, though X-ray analysis reveals that the three-dimensional structures in the crystalline state are very similar. If one adds propanol to an aqueous solution of α-lactalbumin, the helical regions become enlarged and the spectra become similar.

Nucleotides and nucleic acids

Base-sequence-characteristic bands have been found in the VCD spectra of six different octadeoxynucleotides in buffered D_2O. These bands belong to the C=O and C–C stretching regions and do not have a counterpart in the absorption spectrum.

Polyribonucleic acids can be measured in aqueous solution using the windows at 1750–1550 cm^{-1}. In contrast to the monomers, which do not show VCD in this region, the dimers and higher polymers show bisignate VCD bands. The spectra have been calculated for the dimers ApA and CpC using the coupled oscillator model.

Other compounds

A study on the pharmaceutically applied ephedrines and pseudoephedrines derived valuable stereochemical information from the VCD spectra.

The model β-lactams 3-methyl- and 4-methylazetidine-2-one [33] readily form dimers in solution, as was clearly observed from the experimental VCD spectra and corresponding *ab initio* calculations.

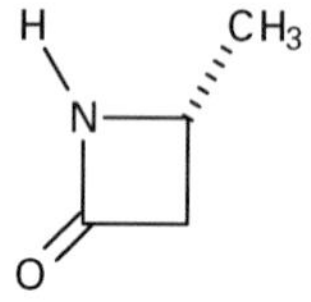

[33]

A very interesting field of research in the biological area is the chemistry of pheromones. These chemicals strongly attract animals of the same species but of opposite sex. Stereochemistry is essential for the effectiveness of these substances. As pheromones are often applied in the struggle against insect pests, methods are needed to test the chirality of the natural and synthetic pheromones. For frontalin [34] the pheromone of the southern pine beetle (*Dendroctonus frontalis*), the VCD and absorption spectra of two different conformers have been calculated and compared with the experimental spectrum (**Figure 4**). In conformation *a* the six membered ring is in the chair conformation, whereas in conformation *b* it is in the boat conformation. The (1R,5S) configuration and the energetically more stable *a* conformation were assigned to the (+)-isomer on this basis.

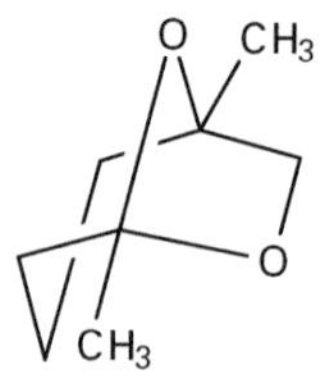

[34]

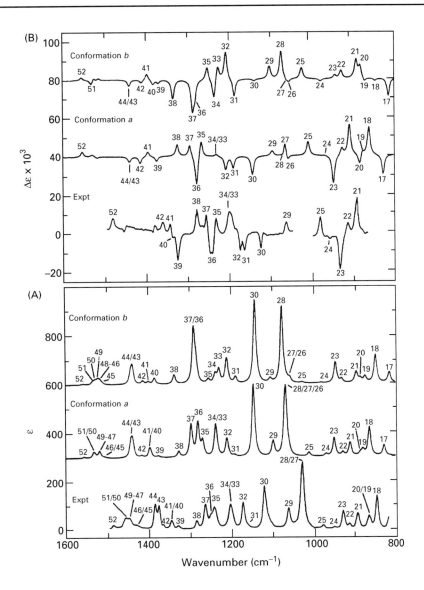

Figure 4 Experimental (in CCl$_4$) and theoretical (*ab initio* DFT, B3LYP/6-31G*) spectra of (1*R*, 5*S*)-(+)-frontalin: (A) absorption spectra, (B) VCD spectra. Reprinted with permission from Ashvar CS, Stephens PJ, Eggimann T and Wieser H (1998) Vibrational circular dichroism spectroscopy of chiral pheromones: frontalin (1,5-dimethyl-6,8-dioxabicyclo[3.2.1]octane). *Tetrahedron: Asymmetry* **9**: 1107–1110. © 1998 Elsevier Science B.V.

Synthetic polymers

Depending on the method of polymerization, a synthetic chiral polymer can be obtained from methyl methacrylate that has more or less extended isotactic regions. The VCD spectrum of the compounds is substantially more sensitive to its stereochemistry than is the normal IR spectrum.

Menthyl vinyl ether polymers of the diastereomeric menthols (+)-menthol, (+)-isomenthol and (+)-neomenthol have been synthesized and studied. While the menthyl and the neomenthyl derivatives both showed enhanced VCD features compared to the corresponding monomer, the VCD of the isomethyl derivative was found to stay virtually the same. Poly(menthyl vinyl ether) was studied in greater detail in later work.

Other applications

Magnetic VCD

If an achiral substance is put into a strong magnetic field, a VCD spectrum can be taken. Important information about molecules, such as the *g* value, can be obtained in this way.

Chiral detection

A very sensitive detector using the VCD of the O–H stretching vibration can be constructed using a solid-state laser that is circularly polarized by a photoelastic modulator. Using this, 2,2,2-Trifluoro-1-(9-anthryl)ethanol and benzoin have been separated on the microgram scale by column chromatography on a chiral stationary phase.

Kinetics

The signal-to-noise ratio of modern VCD spectrometers is now high enough to allow them to follow the course of a chemical reaction involving chiral molecules. Thus, the ratio of isomerization to stereomutation of 1.4±0.4 at 420°C was derived from study of the thermolysis of (1R,2R)-1-2-dideuteriocyclobutane, and the reaction constants of the racemization and isomerization of (2S,3S)-cyclopropane-1-^{13}C-1,2,3-d_3 were obtained at 407.0°C.

See also: **Biochemical Applications of Raman Spectroscopy; Biomacromolecular Applications of Circular Dichroism and ORD; Carbohydrates Studied by NMR; Circularly Polarized Luminescence and Fluorescence Detected Circular Dichroism; Induced Circular Dichroism; Magnetic Circular Dichroism, Theory; Nucleic Acids and Nucleotides Studied Using Mass Spectrometry; Organometallics Studied Using Mass Spectrometry; Polymer Applications of IR and Raman Spectroscopy; Proteins Studied Using NMR Spectroscopy; Vibrational CD Spectrometers; Vibrational CD, Theory.**

Further reading

Ashvar CS, Devlin FJ, Stephens PJ, Bak KL, Eggimann T and Wieser H (1998) Vibrational absorption and circular dichroism of mono- and dimethyl derivatives of 6,8-dioxabicyclo[3.2.1]octane. *Journal of Physical Chemistry A* 102: 6842–6857.

Bose PK and Polavarapu PL (1999) Vibrational circular dichroism of cyclodextrin complexes. *Journal of the American Chemical Society* 121.

Hoffmann GG (1995) Vibrational optical activity (VOA). In: Schrader B (ed) *Infrared and Raman Spectroscopy – Methods and Applications*, pp 543–572. Weinheim: VCH.

Keiderling TA (1994) Vibrational circular dichroism spectroscopy of peptides and proteins. In: Nakanishi K, Berova N and Woody RW (eds) *Circular Dichroism – Principles and Applications*, pp 597–521. New York: VCH.

Keiderling TA (1996) Vibrational circular dichroism – applications to conformational analysis of biomolecules. In: Fasman GD (ed) *Circular Dichroism and the Conformational Analysis of Biomolecules*, pp 555–598. New York: VCH.

McCann JL, Rauk A and Wieser H (1998) A conformational study of (1S,2R,5S)-(+)-menthol using vibrational circular dichroism spectroscopy. *Canadian Journal of Chemistry* 76: 274–283.

Nafie LA (1996) Vibrational optical activity. *Applied Spectroscopy* 50: 14A–26A.

Nafie LA (1997) Infrared and Raman optical activity: theoretical and experimental aspects. *Annual Review of Physical Chemistry* 48: 357–386.

Polavarapu PL (1998) *Vibrational Spectra: Principles and Applications with Emphasis on Optical Activity*. Amsterdam: Elsevier.

Rauk A and Freedmann TB (1994) Chiroptical techniques and their relationship to biological molecules, big or small. *International Journal of Quantum Chemistry* 28: 315–338.

Vibrational CD, Theory

Philip J Stephens, University of Southern California, Los Angeles, CA, USA

VIBRATIONAL, ROTATIONAL &
RAMAN SPECTROSCOPIES
Theory

Introduction

Circular dichroism (CD) can be observed in the vibrational transitions of chiral molecules: vibrational circular dichroism (VCD). An example of a VCD spectrum is shown in **Figure 1**, together with the corresponding unpolarized absorption spectrum. The sample is a 0.6 M solution of (1R,4R)-(+)-camphor in CCl_4. Here, we discuss the theoretical analysis of VCD spectra. The current state-of-the-art is illustrated in **Figure 1**, where VCD and absorption spectra of camphor, predicted within the harmonic approximation (HA) using *ab initio* density functional theory (DFT), are shown.

Theory

We restrict our discussion to the case of isotropic dilute solutions of randomly oriented molecules e.g. liquid solutions or amorphous solid solutions. (In practice, the vast majority of VCD experiments are carried out using liquids at room temperature.) Beer's Law applies:

$$A = \varepsilon c l \qquad [1]$$

$$\Delta A \equiv A_{\mathrm{L}} - A_{\mathrm{R}} = (\varepsilon_{\mathrm{L}} - \varepsilon_{\mathrm{R}})c l \equiv (\Delta\varepsilon)c l \qquad [2]$$

where A = absorbance, L and R denote left and right circularly polarized light, ΔA = circular dichroism, ε = molar extinction coefficient, c = concentration (mol L^{-1}) and l = pathlength (cm). The unpolarized absorption is

$$A \equiv \tfrac{1}{2}(A_{\mathrm{L}} + A_{\mathrm{R}}) = \tfrac{1}{2}(\varepsilon_{\mathrm{L}} + \varepsilon_{\mathrm{R}})\, c l \equiv \bar{\varepsilon} c l \qquad [3]$$

Semi-classical treatment of the interaction of molecules with electromagnetic waves leads to equations for ε and $\Delta\varepsilon$ in terms of molecular properties:

$$\bar{\varepsilon}(\nu) = \frac{8\pi^3 N \nu}{(2.303)3000hc}\sum_{g,k}\alpha_g D_{gk} f_{gk}(\nu_{gk},\nu) \qquad [4]$$

$$\Delta\varepsilon(\nu) = \frac{32\pi^3 N \nu}{(2.303)3000hc}\sum_{g,k}\alpha_g R_{gk} f_{gk}(\nu_{gk},\nu) \qquad [5]$$

$$D_{gk} = |\langle g|\boldsymbol{\mu}_{\mathrm{el}}|k\rangle|^2 \qquad [6]$$

$$R_{gk} = \mathrm{Im}[\langle g|\boldsymbol{\mu}_{\mathrm{el}}|k\rangle \bullet \langle k|\boldsymbol{\mu}_{\mathrm{mag}}|g\rangle] \qquad [7]$$

where $g \to k$ is a molecular excitation of frequency ν_{gk}, α_g is the fraction of molecules in state g, and $f(\nu_{gk},\nu)$ is a normalized line shape function (e.g. Lorentzian). D_{gk} and R_{gk} are the dipole strength and rotational strength of the excitation $g \to k$. $\boldsymbol{\mu}_{\mathrm{el}}$ and $\boldsymbol{\mu}_{\mathrm{mag}}$ are the electric and magnetic dipole moment operators:

$$\boldsymbol{\mu}_{\mathrm{el}} = -\sum_i e \boldsymbol{r}_i + \sum_\lambda (Z_\lambda e)\boldsymbol{R}_\lambda \equiv \boldsymbol{\mu}_{\mathrm{el}}^{\mathrm{e}} + \boldsymbol{\mu}_{\mathrm{el}}^{\mathrm{n}} \qquad [8]$$

$$\boldsymbol{\mu}_{\mathrm{mag}} = -\sum_i \frac{e}{2mc}\left(\boldsymbol{r}_i \times \boldsymbol{p}_i\right) + \sum_\lambda \left(\frac{Z_\lambda e}{2M_\lambda c}\right)$$
$$\times \left(\boldsymbol{R}_\lambda \times \boldsymbol{P}_\lambda\right) \equiv \boldsymbol{\mu}_{\mathrm{mag}}^{\mathrm{e}} + \boldsymbol{\mu}_{\mathrm{mag}}^{\mathrm{n}} \qquad [9]$$

Here, $-e$ and $Z_\lambda e$, $\boldsymbol{r}_i$, and $\boldsymbol{R}_\lambda$, $\boldsymbol{p}_i$ and $\boldsymbol{P}_\lambda$ are the charge, position and momentum of electron i and nucleus λ respectively. Equations [4] and [5] do not include the effects of the condensed-phase medium either on the molecular properties α_g, D_{gk}, R_{gk} and ν_{gk} or on the electromagnetic fields of the radiation: 'solvent effects'.

In the case of vibrational transitions, g and k are vibrational levels of the ground electronic state, G. Within the Born–Oppenheimer (BO) approximation:

$$\Psi_g(\boldsymbol{r},\boldsymbol{R}) = \psi_G(\boldsymbol{r},\boldsymbol{R})\chi_{Gg}(\boldsymbol{R}) \qquad [10]$$

$$\Psi_k(\boldsymbol{r},\boldsymbol{R}) = \psi_G(\boldsymbol{r},\boldsymbol{R})\chi_{Gk}(\boldsymbol{R}) \qquad [11]$$

where

$$H_{el}(\boldsymbol{r},\boldsymbol{R})\,\psi_G(\boldsymbol{r},\boldsymbol{R}) = W_G(\boldsymbol{R})\,\psi_G(\boldsymbol{r},\boldsymbol{R}) \qquad [12]$$

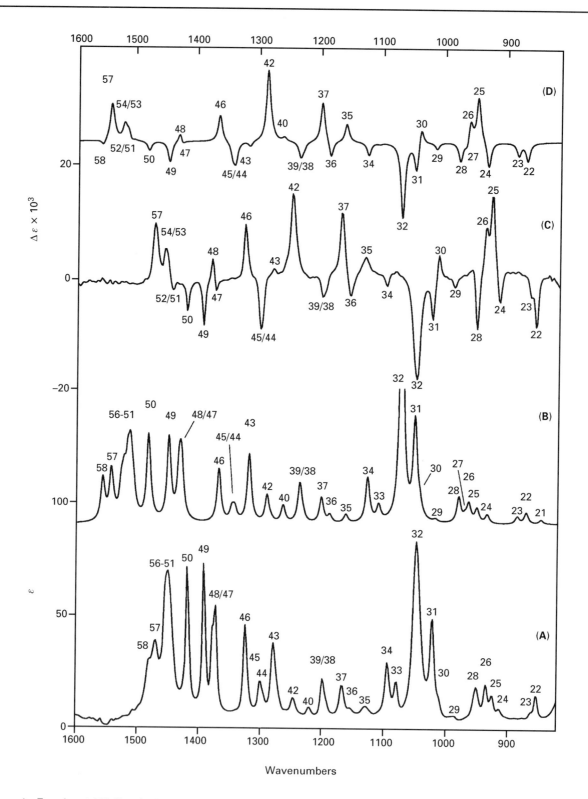

Figure 1 Experimental (A,C) and calculated (B,D) absorption (A, B) and VCD (C, D) spectra of (1R, 4R)-(+)-camphor. The resolution of the experimental spectra was 4 cm⁻¹. Calculated spectra were obtained using DFT, B3PW91 and 6-31G*. Band shapes are Lorentzian (half width at half height 4 cm⁻¹). Fundamental modes are numbered.

$$[W_G(\boldsymbol{R}) + T_n(\boldsymbol{R})]\chi_{Gv}(\boldsymbol{R}) = E_v\chi_{Gv}(\boldsymbol{R}) \qquad [13]$$

r and $\boldsymbol{R}$ denote electronic and nuclear coordinates respectively. H_{el} is the adiabatic 'electronic Hamiltonian':

$$H_{el} = T_e + V_{ee} + V_{en} + V_{nn} \qquad [14]$$

comprising the electronic kinetic energy and the Coulombic interactions of electrons and nuclei. Ψ_G and W_G are the wavefunction and energy of the ground electronic state. χ_{Gv} and E_v are the wavefunction and energy of the vibrational level v arising from vibrational motion on the potential energy surface (PES) $W_G(\boldsymbol{R})$.

For simplicity, we restrict discussion now by assuming that only the lowest vibrational level is populated and that the PES, W_G, is harmonic:

$$\begin{aligned} W_G &= W_G^o + \tfrac{1}{2}\sum_{\lambda\alpha\lambda'\alpha'}\left(\frac{\partial^2 W_G}{\partial X_{\lambda\alpha}\partial X_{\lambda'\alpha'}}\right)_o X_{\lambda\alpha}X_{\lambda'\alpha'} \\ &= W_G^o + \tfrac{1}{2}\sum_i k_i Q_i^2 \end{aligned} \qquad [15]$$

where W_G^0 is the energy of G at equilibrium, $\boldsymbol{R} = \boldsymbol{R}_0$; $X_{\lambda\alpha}$ is the displacement of nucleus λ ($\lambda = 1\ldots N$) along Cartesian axis α ($\alpha = x,y,z$); Q_i are normal coordinates, simultaneously diagonalizing the nuclear kinetic energy:

$$T_n = \tfrac{1}{2}\sum_i \dot{Q}_i^2 \qquad [16]$$

The force constants, k_i, determine the normal mode frequencies:

$$\nu_i = \frac{1}{2\pi}\sqrt{k_i} \qquad [17]$$

The vibrational states of this harmonic PES are of energy

$$E(v_1, v_2 \ldots v_{3N}) = \sum_i \left(v_i + \tfrac{1}{2}\right)h\nu_i \qquad [18]$$

$$(v_i = 0, 1, 2\ldots)$$

For six modes, corresponding to translational and rotational motions, k_i and ν_i are zero.

Within the HA, electric dipole transition moments are

$$\begin{aligned} \langle g|\boldsymbol{\mu}_{el}|k\rangle &\equiv \langle \psi_G\chi_{Gg}|\boldsymbol{\mu}_{el}|\psi_G\chi_{Gk}\rangle \\ &= \langle \chi_{Gg}|\langle\psi_G|\boldsymbol{\mu}_{el}|\psi_G\rangle|\chi_{Gk}\rangle \end{aligned} \qquad [19]$$

which, on expanding $\langle\psi_G|\boldsymbol{\mu}_{el}|\psi_G\rangle \equiv \boldsymbol{\mu}_{el}^G$

$$\boldsymbol{\mu}_{el}^G = \left(\boldsymbol{\mu}_{el}^G\right)_o + \sum_i \left(\frac{\partial\boldsymbol{\mu}_{el}^G}{\partial Q_i}\right)_o Q_i + \cdots \qquad [20]$$

leads to non-zero transition moments from the vibrational ground state (all $v_i = 0$) only for fundamental transitions involving one mode alone, i.e. to the states $v_i = 1$, $v_j = 0$ ($j \neq i$). The transition moment for the fundamental in mode i is

$$\langle 0|\boldsymbol{\mu}_{el}|1\rangle_i = \left(\frac{\partial\boldsymbol{\mu}_{el}^G}{\partial Q_i}\right)_o \left(\frac{\hbar}{4\pi\nu_i}\right)^{1/2} \qquad [21]$$

Equation [21] can be rewritten in terms of derivatives of the molecular electric dipole moment $\boldsymbol{\mu}_{el}^G$, with respect to the Cartesian displacement coordinates, $X_{\lambda\alpha}$. With

$$X_{\lambda\alpha} = \sum_i S_{\lambda\alpha,i}\, Q_i \qquad [22]$$

Equation [21] becomes

$$\langle 0|(\boldsymbol{\mu}_{el})_\beta|1\rangle_i = \left(\frac{\hbar}{4\pi\nu_i}\right)^{1/2}\sum_{\lambda\alpha} S_{\lambda\alpha,i}\, P_{\alpha\beta}^\lambda \qquad [23]$$

where

$$P_{\alpha\beta}^\lambda \equiv \left(\frac{\partial\left(\mu_{el}^G\right)_\beta}{\partial X_{\lambda\alpha}}\right)_o \qquad [24]$$

The second-rank molecular tensors, $P_{\alpha\beta}^\lambda$, are termed atomic polar tensors (APTs). Separating electronic and nuclear parts:

$$\begin{aligned} P_{\alpha\beta}^\lambda &= E_{\alpha\beta}^\lambda + N_{\alpha\beta}^\lambda \\ E_{\alpha\beta}^\lambda &= \left[\frac{\partial}{\partial X_{\lambda\alpha}}\left\langle\psi_G\left|(\mu_{el}^e)_\beta\right|\psi_G\right\rangle\right]_0 \\ N_{\alpha\beta}^\lambda &= (Z_\lambda e)\delta_{\alpha\beta} \end{aligned} \qquad [25]$$

We can further write

$$E_{\alpha\beta}^{\lambda} = 2\left\langle \left(\frac{\partial\psi_G}{\partial X_{\lambda\alpha}}\right)_0 \middle| (\boldsymbol{\mu}_{el}^e)_\beta \middle| \psi_G^0 \right\rangle \qquad [26]$$

The dipole strength of the fundamental excitation of mode i is then

$$D_i = \left(\frac{\hbar}{4\pi\nu_i}\right)\left|\left[\frac{\partial\left(\boldsymbol{\mu}_{el}^G\right)}{\partial Q_i}\right]_0\right|^X$$
$$= \left(\frac{\hbar}{4\pi\nu_i}\right)\sum_\beta \sum_{\lambda\alpha,\lambda'\alpha'}\left[S_{\lambda\alpha,i}P_{\alpha\beta}^\lambda\right]\left[S_{\lambda'\alpha',i}P_{\alpha'\beta}^{\lambda'}\right] \quad [27]$$

The formulation of magnetic dipole transition moments is unfortunately less straightforward. Compare the electronic contributions to the electric and magnetic dipole moments of G:

$$\left(\boldsymbol{\mu}_{el}^G\right)^e = \langle\psi_G|\boldsymbol{\mu}_{el}^e|\psi_G\rangle \qquad [28]$$

$$\left(\boldsymbol{\mu}_{mag}^G\right)^e = \langle\psi_G|\boldsymbol{\mu}_{mag}^e|\psi_G\rangle \qquad [29]$$

Considering only non-degenerate electronic ground states (in practice very few chiral molecules are exceptions) $\boldsymbol{\mu}_{el}^G$ and $\boldsymbol{\mu}_{mag}^G$ are qualitatively different because $\langle\Psi_G|\boldsymbol{\mu}_{mag}^e|\Psi_G\rangle = 0$ at all molecular geometries. That is, electrons make zero contribution to the adiabatic magnetic dipole moment. It follows that, in the case of magnetic dipole transition moments, the BO approximation leads to a non-physical result. The treatment of magnetic dipole transition moments requires more accurate vibronic wavefunctions. The vibrational states g and k must be written

$$\Psi_g = \psi_G\chi_{Gg} + \sum_{E\neq G}(\psi_E\chi_{Ee})c_{Gg,Ee}$$

$$\Psi_k = \psi_G\chi_{Gk} + \sum_{E\neq G}(\psi_E\chi_{Ee})c_{Gk,Ee} \qquad [30]$$

allowing for the admixture of BO functions of excited electronic states E into the ground state. This in turn permits non-zero vibrational transition moments of $\boldsymbol{\mu}_{mag}^e$ to be obtained; simply put, electronic magnetic dipole transition moments are

'stolen' by mixing of BO states. The reader is referred to the literature for the details. The final result is that

$$\left\langle 0|(\boldsymbol{\mu}_{mag})_\beta|1\right\rangle_i = \left(4\pi\hbar^3\nu_i\right)^{1/2}\sum_i S_{\lambda\alpha,i}M_{\alpha\beta}^\lambda \qquad [31]$$

where

$$M_{\alpha\beta}^\lambda = I_{\alpha\beta}^\lambda + J_{\alpha\beta}^\lambda$$
$$I_{\alpha\beta}^\lambda = \left\langle \left(\frac{\partial\psi_G}{\partial X_{\lambda\alpha}}\right)_0 \middle| \left(\frac{\partial\psi_G}{\partial H_\beta}\right)_0\right\rangle$$
$$J_{\alpha\beta}^\lambda = \frac{i}{4\hbar c}\sum_\gamma (Z_\lambda e)R_{\lambda\gamma}^0\varepsilon_{\alpha\beta\gamma} \qquad [32]$$

The tensors $M_{\alpha\beta}^\lambda$ are termed atomic axial tensors (AATs); $I_{\alpha\beta}^\lambda$ and $J_{\alpha\beta}^\lambda$ are the electronic and nuclear components. Here, $(\partial\psi_G/\partial X_{\lambda\alpha})_0$ is the same derivative which occurred already in Equation [26]. The electronic AAT, $I_{\alpha\beta}^\lambda$, is the overlap integral with the derivative $(\partial\psi_G/\partial H_\beta)_0$. This latter is defined via

$$H(H_\beta) = H_{el} + H'(H_\beta)$$
$$H'(H_\beta) = -\left(\boldsymbol{\mu}_{mag}^e\right)_\beta H_\beta$$
$$H(H_\beta)\psi_G(H_\beta) = W_G(H_\beta)\psi_G(H_\beta) \qquad [33]$$

That is: $\psi_G(H)$ is the wavefunction of G in the presence of a uniform external magnetic field, H, approximating the perturbation by the linear magnetic dipole interaction $H'(H)$.

The rotational strength of the fundamental excitation of mode i is then

$$R_i = \hbar^2\sum_\beta \sum_{\lambda\alpha,\lambda'\alpha'}\left[S_{\lambda\alpha,i}P_{\alpha\beta}^\lambda\right]\left[S_{\lambda'\alpha',i}M_{\alpha'\beta}^{\lambda'}\right] \qquad [34]$$

Computation

Within the HA, the prediction of a vibrational absorption spectrum amounts to the calculation of the harmonic normal mode frequencies, ν_i, and dipole strengths, D_i. The frequencies are obtained from the harmonic force field (HFF). With respect to Cartesian displacement coordinates, this is the Hessian $(\partial^2 W_G/\partial X_{\lambda\alpha}\partial X_{\lambda'\alpha'})_0$. Diagonalization (after mass-weighting) yields the force constants k_i; the frequencies, ν_i; and the normal coordinates, Q_i, i.e. the transformation matrices, $S_{\lambda\alpha,i}$. The dipole strengths depend in addition on the APTs; these require calculation of $(\partial\psi_G/\partial X_{\lambda\alpha})_0$.

The prediction of a VCD spectrum amounts likewise to the calculation of the harmonic frequencies and rotational strengths, R_i. All of the quantities required in predicting the absorption spectrum are again needed; in addition, the AATs must be calculated. Since $(\partial\psi_G/\partial X_{\lambda\alpha})_0$ is already required for the APTs, the AATs require additionally only $(\partial\psi_G/\partial H_\beta)_0$.

In sum: the prediction of both absorption and VCD spectra requires (i) $(\partial^2 W_G/\partial X_{\lambda\alpha}\partial X_{\lambda'\alpha'})_0$; (ii) $(\partial\psi_G/\partial X_{\lambda\alpha})_0$; (iii) $(\partial\psi_G/\partial H_\beta)_0$. The prediction of the VCD spectrum requires relatively little more than is needed for the absorption spectrum: specifically, $(\partial\psi_G/\partial H_\beta)_0$.

The calculation of molecular properties can be carried out at three distinct levels: (i) *ab initio*, (ii) semi-empirical, (iii) empirical. *Ab initio* methods have increased enormously in accuracy and efficiency in the last two decades and are the focus of our discussion here. *Ab initio* methods have developed in two directions: first, the level of approximation has become increasingly sophisticated and, hence, accurate. The earliest *ab initio* calculations used the Hartree-Fock/self-consistent field (HF/SCF) methodology, which is the simplest to implement. Subsequently, such methods as Møller-Plesset perturbation theory, multi-configuration self-consistent field theory (MCSCF) and coupled-cluster theory have been developed and implemented. Relatively recently, density functional theory (DFT) has become very popular, since it yields an accuracy much greater than that of HF/SCF while requiring relatively little additional computational effort.

The second dimension in which *ab initio* theory has progressed is that of derivative techniques. Many molecular properties of interest–including, as shown above, the HFF, APTs and AATs–can be expressed in terms of derivatives of energies and wavefunctions with respect to perturbations. Such derivatives can be evaluated using either numerical or analytical methods. For example, the energy gradients $(\partial W_G/\partial X_{\lambda\alpha})_0$ can be evaluated either by calculating W_G at $\boldsymbol{R}_0$ and $\boldsymbol{R}_0 + X_{\lambda\alpha}$ and using

$$\left(\frac{\partial W_G}{\partial X_{\lambda\alpha}}\right)_0 \approx \frac{W_G\left(\boldsymbol{R}_0 + X_{\lambda\alpha}\right) - W_G\left(\boldsymbol{R}_0\right)}{X_{\lambda\alpha}} \qquad [35]$$

or by formulating an equation for $(\partial W_G/\partial X_{\lambda\alpha})_0$ and then carrying out direct evaluation. Similarly, a Hessian matrix can be obtained by finite-differences of gradients or analytically. Analytical derivative methods are much more efficient. Much of the recent expansion in usage of *ab initio* quantum chemistry has resulted from advances in formulating and implementing analytical derivative techniques for an increasing diversity of molecular properties at an increasing number of theoretical levels.

At the present time, the simultaneous calculation of HFFs, APTs and AATs using analytical derivative *ab initio* methods has been implemented in three program packages: CADPAC, DALTON and GAUSSIAN. The levels of implementation are:

CADPAC	HF/SCF
DALTON	HF/SCF and MCSCF
GAUSSIAN	HF/SCF and DFT.

The accuracies of these methods are:

HF/SCF < MCSCF << DFT.

The computational effort is:

HF/SCF < DFT << MCSCF.

The ratio of accuracy to effort is:

DFT >> HF/SCF > MCSCF.

Thus, DFT is currently the most cost-effective methodology available.

An additional variable in *ab initio* calculations is the basis set. Two choices are to be made: (i) perturbation-independent or perturbation-dependent; (ii) size and composition. In calculating derivatives with respect to nuclear displacements, $X_{\lambda\alpha}$, one can adopt basis functions which either (a) are not or (b) are functions of nuclear position. The latter add computational complexity but vastly improve convergence of properties with increasing basis set size (i.e. decrease the errors associated with the use of basis sets of finite size). Modern computational packages use only nuclear-position-dependent basis sets. In the same way, derivatives with respect to magnetic fields can use basis functions which either (a) are not or (b) are functions of magnetic field. The standard choice for the latter are so-called London orbitals or gauge-invariant atomic orbitals (GIAOs). The use of GIAOs vastly reduces basis set error and is increasingly *de rigueur* in computation of magnetic properties (e.g. NMR shielding tensors). With regard to the implementation of AATs in CADPAC, DALTON and GAUSSIAN, we should add that DALTON and GAUSSIAN use GIAOs, while CADPAC does not.

With respect to basis set size we can simply note that (a) accuracy increases with increasing basis set size; (b) the rate of increase in accuracy is rapid at small sizes and less rapid at large sizes; (c) with respect to the ratio of accuracy to size, such basis sets as 6-31G* are generally regarded as optimal for calculations on organic molecules.

Finally, in DFT calculations there is the question of the density functional. The accuracy of DFT calculations varies greatly with the choice of functional. The exact functional gives exact results. Very crude functionals give very inaccurate results. Functionals used in the recent past can be grouped into three classes: (a) local; (b) non-local/gradient-corrected; (c) hybrid. Overall, the relative accuracy is

$$local < non\text{-}local < hybrid.$$

At this time, hybrid functionals are generally regarded as state-of-the-art. There are many: the original is 'B3PW91'; a popular-choice is 'B3LYP'. Undoubtedly, current functionals will be soon replaced by yet more accurate functionals.

Implementation

In **Figure 1** we compare predicted absorption and VCD spectra for camphor to experiment. Predictions are based on DFT calculations using the B3PW91 functional and the 6-31G* basis set. AATs are calculated using GIAOs. The calculations were carried out using the GAUSSIAN program. Spectra are simulated from frequencies, dipole strengths and rotational strengths assuming Lorentzian bandshapes of constant width (half width at half height 4 cm^{-1}).

Focussing first on the absorption spectrum, we observe an excellent one-to-one correspondence between predicted and experimental spectra. That is: we can assign the bands of the experimental spectrum to fundamental excitations in such a way that the pattern of frequencies and intensities is extremely similar to that predicted. One notices an overall shift of the predicted spectrum to higher frequency. This shift is attributable both to error in the calculated harmonic frequencies and to anharmonicity, which uniformly lowers experimental frequencies with respect to harmonic frequencies. Calculations on very small molecules, where harmonic frequencies are known, indicate that the two contributions are of the same sign and comparable in magnitude. It is also to be noted that almost all bands of the experimental spectrum can be assigned to fundamental transitions. Conversely, the number of overtone and combination bands observable is very small. It is clear that the HA is a very good approximation and that the neglect of anharmonicity in predicting spectra is not a serious deficiency.

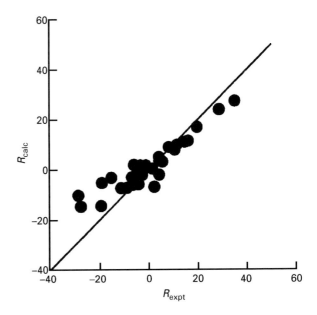

Figure 2 Comparison of calculated and experimental rotational strengths for (1*R*, 4*R*)-(+)-camphor. *R* values are in 10^{-44} esu^2 cm^2. The straight line has a slope +1.

We turn now to the VCD spectrum. Each VCD band corresponds to an absorption band. The assignment of the experimental absorption spectrum can thus be transferred directly to the experimental VCD spectrum. Comparison of predicted and experimental VCD intensities, fundamental by fundamental, can then be carried out. As seen in **Figure 1**, the agreement is qualitatively excellent. Quantitative comparison of calculated and experimental rotational strengths, the latter obtained by Lorentzian analysis of the experimental VCD spectrum, is shown in **Figure 2**.

Discussion

At the present time, calculation of absorption and VCD spectra within the harmonic approximation using DFT is computationally straightforward at the 6-31G* basis set level using GAUSSIAN 98 and standard parallel computers for organic molecules containing ≤ 200 atoms. For smaller molecules, larger basis sets can be used, yielding results of higher accuracy. For molecules with > 200 atoms one awaits further developments in computational algorithms and machine speed. The accuracy of DFT/6-31G* calculations using hybrid density functionals, illustrated in **Figure 1**, is already impressive. Further improvements in functionals, to be expected in the near future, will bring predicted spectra into even closer agreement with experiment.

There are two major deficiencies in the theoretical treatment of VCD described above: the neglect of anharmonicity and of solvent effects. Anharmonicity is relatively unimportant in the mid-IR spectral region, but becomes much more important at higher frequencies, e.g. in the C–H stretching region. Its inclusion for large molecules constitutes one of the remaining major theoretical challenges for the theory of VCD. Solvent effects are relatively unimportant in simple non-polar solvents such as CCl_4. However, they become much more important in solvents interacting much more strongly with solute molecules. Inclusion of solvent effects in such solvents, especially water, constitutes another major challenge for the theory of VCD.

We are here restricted to the theory of VCD. Its applications, past and future, are beyond the scope of our discussion. It is worth emphasizing, nevertheless, that the development of an accurate, computationally efficient methodology for the prediction of VCD spectra is of enormous consequence for the practical application of VCD spectroscopy to problems of stereochemistry in chiral molecules. In particular, the determination of absolute configuration in organic molecules containing ≤ 200 atoms is now practicable using VCD spectroscopy.

List of symbols

A = absorbance; c = concentration (mol L^{-1}); D = dipole strength; $-e$ = electronic charge; f = normalized line shape function; H_{el} = adiabatic electronic Hamiltonian; k = force constant; l = pathlength (cm); $M_{\alpha\beta}^{\lambda}$ = atomic axial tensor; p_i = electronic momentum; $P_{\alpha\beta}^{\lambda}$ = atomic polar tensor; P_{λ} = nuclear momentum; Q = normal coordinate; r_i = electronic position; R = rotational strength; R_{λ} = nuclear position; W = energy; $Z_{\lambda}e$ = nuclear charge; α_g = fraction of molecules in state g; ΔA = circular dichroism; ε = molar extinction coefficient; μ = dipole moment operator; v = frequency; Ψ = wavefunction.

See also: **Vibrational CD, Applications; Vibrational CD Spectrometers.**

Further reading

Cheeseman JR, Frisch MJ, Devlin FJ and Stephens PJ (1996) *Ab initio* calculation of atomic axial tensors and vibrational rotational strengths using density functional theory. *Chemical Physics Letters* **252**: 211–220.

Devlin FJ, Stephens PJ, Cheeseman JR and Frisch MJ (1997) *Ab initio* prediction of vibrational absorption and circular dichroism spectra of chiral natural products using density functional theory: camphor and fenchone. *Journal of Physical Chemistry* **101**: 6322–6333.

Devlin FJ, Stephens PJ, Cheeseman JR and Frisch MJ (1997) *Ab initio* prediction of vibrational absorption and circular dichroism spectra of chiral natural products using density functional theory: α-pinene. *Journal of Physical Chemistry* **101**: 9912–9924.

Hehre WJ, Schleyer PR, Radom L and Pople JA (1986) *Ab initio Molecular Orbital Theory*. New York: Wiley.

Laird BB, Ross RB and Ziegler T (eds) (1996) *Chemical Applications of Density Functional Theory*, ACS Symposium Series 629. ACS.

Schellman JA (1975) Circular dichroism and optical rotation. *Chemical Reviews* **75**: 323–331.

Stephens PJ (1985) Theory of vibrational circular dichroism. *Journal of Physical Chemistry* **89**: 748–752.

Stephens PJ (1987) Gauge dependence of vibrational magnetic dipole transition moments and rotational strengths. *Journal of Physical Chemistry* **91**: 1712–1715.

Stephens PJ and Lowe MA (1985) Vibrational circular dichroism. *Annual Reviews of Physical Chemistry* **36**: 213–241.

Stephens PJ, Cheeseman JR, Frisch MJ, Ashvar CS and Devlin FJ (1996) *Ab initio* calculation of atomic axial tensors and vibrational rotational strengths using density functional theory. *Molecular Physics* **89**: 579–594.

Yamaguchi Y, Osamura Y, Goddard JD and Schaefer HF (1994) *A New Dimension to Quantum Chemistry: Analytic Derivative Methods in Ab Initio Molecular Electronic Structure Theory*. OUP.

Vibrational, Rotational and Raman Spectroscopy, Historical Perspective

AS Gilbert, Beckenham, Kent, UK

VIBRATIONAL, ROTATIONAL & RAMAN SPECTROSCOPIES
Historical Overview

After a slow start in the nineteenth century, infrared (IR) spectroscopy saw a rapid increase in use and for a while, from 1945 onwards, was the most widespread method for determining the chemical structures of molecules. In the 1960s it became increasingly supplemented and overshadowed by other techniques, but recent developments in instrumentation have vastly improved its sensitivity and speed and enabled it to be applied to previously intractable problems. While research applications are many, there is overwhelming analytical and industrial usage.

The development of Raman spectroscopy has generally lagged behind that of IR owing to greater technical difficulties. The advent of the laser was the most important event in its history and has enabled many special and esoteric Raman experiments to be conceived. It is no longer confined mainly to the laboratory and is now used extensively in industry.

Beginnings: IR and Raman spectroscopy before the Second World War

The astronomer William Herschel, better known as the discoverer of the planet Uranus, first detected IR radiation in 1800. Using a glass prism to refract the rays of the sun, he observed a rise in temperature in a thermometer positioned beyond the red limit of the visible spectrum (**Figure 1**).

Over the next one hundred years the essential nature of IR radiation was slowly established, electrical methods of measurement were developed and a number of materials such as rock salt were found to be largely transparent to it and thus useful as refracting elements. Interest was mainly directed to the physics of the subject, with little attention given to any possibilities for application to chemistry.

Near the end of the nineteenth century, however, a number of workers observed that many specific classes of organic compounds absorbed in characteristic regions of the IR spectrum. A monumental compilation of such bands, mostly from his own measurements, was published by W.W. Coblentz in 1905. One of his spectra is shown in **Figure 2** and can be compared with a modern-day version of the same compound in **Figure 3**. At the time, the precise mechanism for absorption was unclear, but it was soon realized that it was derived from what could be considered to be intramolecular vibrations. By the 1930s, a fairly complete theory encompassing the relation of dipole moment change to band intensity, selection rules, molecular symmetry and anharmonicity was available for both vibrational and rotational motions.

In 1928, C.V. Raman announced the discovery of the effect that now bears his name. In fact he had already observed it a few years before as a weak 'residual fluorescence' from highly purified organic liquids. It is generally considered, however, that the effect had been predicted by A. Smekal. Raman was awarded the Nobel Prize, the second Indian to be so honoured. Although the Raman effect was found to be very weak, the relation of selection rules to symmetry differed, so that results could complement those from IR absorbance.

Instrumentation

Detection of IR radiation could be done by using bolometers or thermocouples, but point-by-point plotting of galvanometer readings made measurement of spectra somewhat tedious. IR-sensitive photographic film was available later and was occasionally used. Technical obstacles were many, however; for example the DC output from detectors could not be amplified and most galvanometers had long response times. Material for dispersion of IR radiation was scarce, rock salt of the quality and size suitable for prisms being difficult to obtain. Machines were single-beam and baselines were a major problem owing to the normal variations in laboratory heat background. Most instruments were built in-house, though commercial models were available, the first being introduced by Adam Hilger Ltd of the UK in 1913. It is illustrated in **Figure 4**.

The KBr disk method was not to be invented until 1952, so solids were usually sampled by mulling or reflection from whole crystals. The majority of samples studied were either liquids or gases.

Figure 1 Picture of William Herschel's experimental setup. Using blackened-bulb mercury thermometers, he observed a difference in temperature between one positioned at the red end of the visible and one positioned beyond. Originally from *Philosophical Translations*, Pt II **80**: 289 (1800). Photograph courtesy of Professor N. Sheppard, FRS.

By contrast, Raman spectra were usually easier to acquire so long as the sample was not coloured or turbid. Working in the visible region allowed simple spectrographs with silica/glass optics to be used and suitable photographic films were readily available, though long exposures were generally required. Discharge lamps, usually mercury vapour, provided a rather weak source of excitation radiation with a broad, diffuse background.

Studies between the wars

The benefits of theory soon allowed vibrational spectroscopy to move away from simple characterization

to deductions about the shape and symmetry of simple molecules. The observation that gaseous CO_2 yielded three fundamental modes of vibration, of which two were seen in the IR only and the other in the Raman only, immediately determined that it must be linear and symmetric. The tetrahedral structure of methane was confirmed by the lack of any observable pure rotational Raman spectrum. Such results gave valuable support to the developing theories of chemical bonding.

Knowledge of the actual frequency values enabled the calculation of the vibrational partition functions. Thermodynamic quantities of interest such as the heat capacity in the gas phase at different temperatures

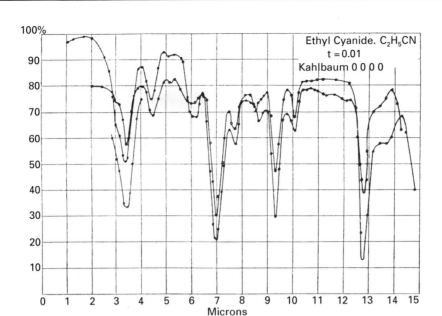

Figure 2 Point-by-point IR spectrum of ethyl cyanide. From Coblentz WW (1905) *Investigations of Infrared Spectra*. Washington DC: Carnegie Institution of Washington. Photograph courtesy of Professor N. Sheppard, FRS.

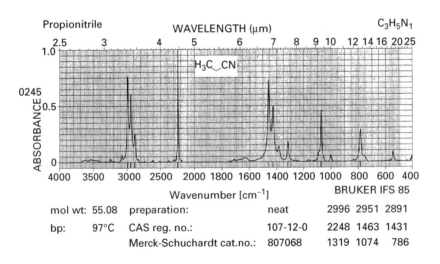

Figure 3 An FT-IR spectrum of ethyl cyanide. Reproduced with permission from Pachler KGR, Matlok F and Gremlich H-U (1998) *Merck FT-IR Atlas*. New York: VCH.

could then be estimated with great precision. In addition, the force constants of very simple molecules could be determined.

IR spectroscopy gave a dramatic illustration of the existence of hydrogen bonding, a suspected new type of molecular attraction. For instance, at high temperatures in the gas phase, formic acid yielded the expected single fundamental from O–H stretching at about 3570 cm^{-1}. But at lower temperatures and in the liquid, this band disappeared to be replaced by a new one at near 3080 cm^{-1}.

While most attention was naturally given to the IR region below 4000 cm^{-1}, some measurements were

made in the near IR (NIR), i.e. above 4000 cm^{-1}, often for no other reason than ease of experimentation. Overtones and combination bands were of course of interest in their own right but could also sometimes be used to make deductions about lower-frequency fundamental modes.

First flowering: IR spectroscopy to the 1970s

Developments in electronics during the 1930s made measurements of IR spectra somewhat easier.

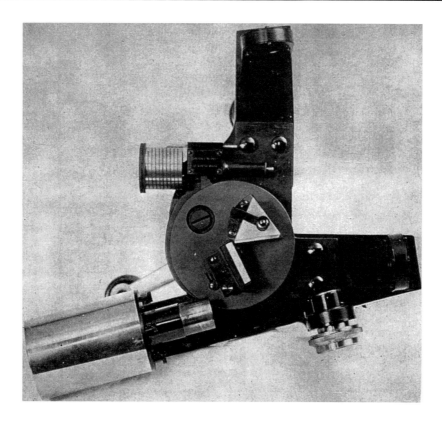

Figure 4 Plan view of a Hilger IR spectrometer made in 1918. A constant-deviation Wadsworth optical arrangement with a 60° rock salt prism was employed. The detector was a thermopile and the source a Nernst filament. From *Hilger Journal*, August 1955. Photograph courtesy Professor N. Sheppard, FRS, and reproduced with permission from Hilger Analytical.

Successful application to a number of tasks connected with the Second World War, such as monitoring the composition of Axis petroleum samples to determine origin, led to a considerable expansion in activities.

Considerable effort was put into organic group frequency correlations, and compilations soon became available for general use. These were utilized extensively for detailed structural characterization of organic compounds. The boon to organic chemists can be appreciated by considering the methods that had been available to them before. Elemental analysis could only give overall composition and ultraviolet–visible (UV-visible) spectroscopy little more than some idea of the degree or type of unsaturation. A limited number of chemical group types were detectable by chemical spot tests.

Moreover, IR spectroscopy was also applied to other work such as quantitative analysis and the study of physical interactions between molecules. The general rise of scientific activities in both universities and industry naturally led to a rise in demand for IR facilities and the 1940s saw the first commercial automatically recording spectrometers. Spectra were now produced routinely as continuous plots. Single-beam instruments were rapidly superseded by double-beam ones and gratings came into general use.

Dispersive double-beam spectrometers

In double-beam instruments compensation was, with one or two exceptions, by the optical null method whereby a servomotor drove a comb into the reference beam to minimize the signal difference between it and the sample beam as it was attenuated to varying degrees by absorbance. Movement of the comb was mechanically geared to a pen, which thus recorded the spectrum as a trace on a moving chart.

Sample and reference beams were alternately presented to the detector by means of mirrors and a chopper rotating at a frequency of several hertz, thus allowing selective amplification of the signal difference, which was fed to the servomotor. Thermal detectors continued to be used for many years but with developments, such as the pneumatic Golay cell, for greater sensitivity.

Good-quality large alkali halide crystals could now be grown on an industrial scale, so that fabrication of prisms and other optical components did not present any difficulties. In particular, use of synthetic KBr allowed spectra to be measured down to 400 cm^{-1}; pre-war natural NaCl (rock salt) prism spectrometers had a lower limit of about 650 cm^{-1} owing to the higher frequency absorption edge of the salt.

Figure 5 An early automatically recording IR spectrometer, the Hilger model D209 introduced in 1940. The spot from a mirror galvanometer was focused onto photographic film fixed to a rotating drum. Later during the Second World War, this machine was developed into the first commercial double-beam (ratio recording) IR spectrometer. Photograph courtesy of Professor N. Sheppard, FRS, and reproduced by permission of Hilger Analytical.

Rotation of the prism or grating was controlled by a cam to allow the spectral trace to be recorded in either constant wavelength or wavenumber. This cam was coupled to the slits to vary their width in order to keep the radiation reaching the detector roughly constant. Resolution therefore varied over the spectrum. Positional accuracy and repeatability were poor owing to backlash and wear in the mechanical linkages; most specifications quoted values around ± 0.5 cm^{-1}.

The inherent disadvantage of the optical null mechanism was the attenuation of the reference beam, so that accuracy and response time became worse as sample transmittance (%T) decreased. Linearity of response depended on how well the comb had been machined. Theoretically it was impossible for the equipment to measure 0%T. Scanning had to be slow to avoid excessive pen lag. Reliable quantitative analysis was difficult though possible with heed.

The amounts of sample required for good spectra of a solid was generally of the order of a milligram. By utilizing beam condensers and the best instruments, adequate spectra could be obtained from small samples of the order of a hundred nanograms.

Commercial instruments

Several companies built spectrometers; based in the USA were Perkin-Elmer, Beckman, Baird and Cary; in the UK, Grubb-Parsons, Hilger (**Figure 5**) and Unicam (later Pye-Unicam). Instruments were also built in France, Germany (both West and East),

Japan and the USSR. The commonest optical layout used was the Littrow arrangement which had the virtue of compactness. Low-cost instruments were introduced in the 1950s, some only with prisms though all were soon sold with gratings instead. Their relative cheapness and ease of use were major factors in the rapid expansion in the practice of IR spectroscopy. Research-grade models usually had a grating monochromator with a fore-prism. Two or more gratings with auto changeover were required to cover the whole range.

During the latter part of the period, prisms disappeared and more advanced features were built into dispersive spectrometers such as computer interfaces, principally for data acquisition only. Ratio recording became standard in place of the optical null. However, the Perkin-Elmer model 983, introduced in 1982, demonstrated the benefits that computerization could bring to instrument control. Grating rotation was carried out directly by a stepper motor under the command of a microprocessor. With no mechanical cam, position repeatability (± 0.005 cm^{-1}) was, in consequence, almost as good as that of FT-IR instruments.

Applications

A pre-eminent use already described was structure determination of organic compounds. The KBr disk technique allowed complete and unmarred spectra of solids to 400 cm^{-1}. Extensive collections of spectra (both literature and commercial) of organics, inorganics, pharmaceuticals and industrial chemicals now appeared.

A result of great topical interest in the late 1940s was the recognition that penicillin contained a fused four-membered lactam ring on the basis of several bands, including one at an unusually high value, near 1780 cm^{-1}, from the amidic carbonyl stretching mode.

Because of the distinctive fingerprint nature of the spectra of individual substances, a very common application was, and continues to be, compound verification in forensic identification and quality assurance to meet pharmacopoeial and other standards set by regulatory authorities.

The sensitivity of IR bands to changes in physical state led to numerous studies concerned with the thermodynamics of hydrogen bonding in solvents and the influence of solvent effects (donor/acceptor capacities). Phase transitions (polymorphism) in solids could easily be detected by changes in band intensity and position.

Inorganic and metal–organic compounds also received attention in structural studies. For instance, IR spectroscopy of metal carbonyl complexes could easily distinguish between terminal and bridging ligands.

Access to computing (mainframe) facilities from the 1950s onwards provided the means to analyse the fundamental mode frequencies of a polyatomic molecular structure on the basis of harmonic oscillation and obtain force constants. These could be transferred to other structures and the procedure reversed so as to estimate frequencies for cases where a spectrum was difficult to interpret. While fraught with difficulties, mainly because there were usually many more force constants to estimate than frequencies, such work did at least highlight the fact that many vibrational modes are strongly coupled to each other and that few bands can normally be assigned to specific bond motions.

Data handling was never more than very primitive. Although the largest commercial collections of spectra offered searching facilities, for example by matching of most prominent bands, these were not very successful in practice even for pure compounds.

The rapid spread of mass spectrometry (MS) and nuclear magnetic resonance (NMR) spectroscopy from the 1960s caused IR spectroscopy to be sidelined in many areas of analysis. NMR was much better at elucidating many fine details of molecular structure, while detection limits were lower using MS.

Renaissance: The (Fourier) transformation of IR spectroscopy

IR interferometers seem first to have been tried out for examining the very low signal levels of IR emission from astronomical sources. They were shortly also employed for far-IR spectroscopy where low source output made measurement extremely difficult using conventional spectrometers below 200 cm^{-1}.

There was sufficient interest for several far-IR instruments to be made available commercially from the late 1950s, but one drawback was the difficulty of numerically converting the interferograms to recognizable spectra. With only mainframe computers available, instrument-based digital computing was not feasible. RIIC/Beckman marketed a machine that stored data, after analogue-to-digital conversion, on magnetic core memory. The data were then reconverted back to analogue to be fed into an electronic wave analyser to generate the spectrum.

Ironically, these interferometers could not be used to advantage in the mid-IR region ($4000–400 \text{ cm}^{-1}$) in most circumstances. This was because, with scan speeds in minutes, signal levels were so large that the noise was well below the threshold of even the smallest bit of the best (16-bit) ADCs obtainable. However, by the early 1960s, Block Engineering of the USA had developed a fast scanning interferometer. Providentially, lasers (for accurate path difference referencing), minicomputers and the Cooley–Tukey fast Fourier transform (FFT) algorithm soon all appeared. A Block subsidiary, Digilab, exploited these devices to produce the first commercial mid-IR Fourier transform (FT-IR) spectrometer, the model FTS-14, in 1969.

They were followed during the 1970s by other companies, most of whom had had no previous experience in the IR field but made computers and data loggers, such as Nicolet and Bruker. The machines offered were principally research grade, though Digilab made an early and substantial entry into the specialist market for quality control of semiconductors.

Most manufacturers of dispersive instruments opted not to enter the new arena, presumably because of the costs involved in investment in the new technology, and in consequence left the business altogether as demand for dispersive instruments dried up. A major exception was Perkin-Elmer who, although starting late, had the resources to participate in a big way. Although offering research-grade instruments as well, they particularly targeted the analytical market, in which they were soon to be a major supplier.

At least twenty manufacturers worldwide have offered FT-IR spectrometers in recent years, several of them for niche markets such as dedicated gas analysis and space science.

Advantages of interferometry in the mid IR

Two factors were judged to be of major importance: The Fellgett (multiplex) and Jacquinot (throughput) advantages. The Jacquinot advantage, though large, and foremost at higher wavenumbers, was found, however, to be largely cancelled out by the much decreased performance (at the high modulation frequencies imposed by fast scanning interferometers) of the early pyroelectric detectors used, both factors having a roughly similar wavenumber dependency. Traditional thermal effect detectors such as thermocouples were far too slow.

As other factors (e.g. beam splitter versus grating efficiency) were of lesser importance, this meant that overall advantage in signal to noise (S/N) ratio of mid-IR FT instruments over dispersives was basically determined by multiplexing only. Strict comparisons were not possible, though, because FT instruments ran at constant resolution unlike dispersives where resolution varied somewhat across the spectrum.

Thus for spectra run at moderate resolution and over limited spectral range, FT machines with pyroelectric detectors were not strikingly better than dispersives. This was because the multiplex advantage (for FTIR) is no better than $\frac{1}{2}\sqrt{M}$, where M is the number of resolution elements.

However, for low signal levels a quantum detection, MCT (mercury cadmium telluride), was soon developed, that had much greater sensitivity at high modulation rates and thus did not trade off the Jacquinot advantage. This allowed FT spectrometers to tackle situations that dispersives could not even attempt and to go some way in competing with and complementing MS in trace analysis. Various advances in component design (e.g. sources) saw a considerable improvement in general performance during the 1970s.

The Connes advantage claimed for interferometers was that laser referencing would yield much greater wavelength accuracy and reproducibility. The latter was crucial for both signal averaging and spectral subtraction but, as the Perkin-Elmer company had shown, this could almost be matched by dispersives.

Photometric accuracy and reproducibility were actually worse than with the best ratio-recording grating instruments, owing mainly to sample reflection and emission back into the interferometer.

Fashion probably played a significant role in the cessation of dispersive spectrometer manufacture as the high S/N provided by FT machines was not required when sample was abundant and sample preparation took much longer than measurement.

Applications

Interferometers made far-IR spectroscopy possible, and enabled observation of lattice vibrations of crystals, low-frequency skeletal motions of organics and the stretching modes of heavy atoms. In the mid-IR region interest was not so much in new vibrations as in old ones in new and difficult situations, now accessible because of the high S/N available. Multitudes of hitherto indifferent samples suffering from low transmission were now directly amenable without resort to complex pre-preparation. Because data were obtained and stored in digital form, spectra could readily be manipulated and spectral subtraction to remove solvent bands from solution spectra, assess purity and reveal impurities became a popular activity.

High S/N, and in addition fast scanning, enabled events as rapid as fractions of seconds to be monitored. Thus FT-IR spectroscopy was applied with considerable success as a detection method for gas chromatography (GC) and manufacturers were obliged to offer suitable devices. However, optimizing the interface took a surprisingly long time; waiting on GC column technology was partly to blame. Some 20 years elapsed from the early 1970s before minimum identifiable limits, starting in micrograms, reached the mid-picogram level with the advent of cryogenic trapping methodology.

While many biological/biochemical samples had been studied by dispersive instruments, the difficulties of dealing with aqueous solutions was a severe limitation. Now, however, this general field became a major area of research attention for application of FT-IR in the 1980s. Of interest was the sensitivity of vibrational modes to subtle changes in molecular environment of the type crucial to biological mechanism and structure studies. Particular topics were how cell membrane conformations were affected by interactions with other chemicals or general physicochemical effects, and analyses of protein secondary structure.

Instrumentation

The GC–IR interface was the first of many specialized accessories that began to appear in increasing numbers during the 1980s. They included interfaces to various forms of chromatography (the combination being known as a hyphenated method), photoacoustic and diffuse reflectance cells and microscopes. Spectrometer design was influenced so as to avoid the inconvenience of changeover. Interferometers were therefore built with multiple beam ports to allow two or more accessories to be permanently connected. With such developments and the continuing miniaturization of computing hardware, the large floor-standing FT spectrometers of the 1970s were superseded by the modular benchtop machines of the 1980s.

The most significant development of the interferometer in recent years has been the step scanning

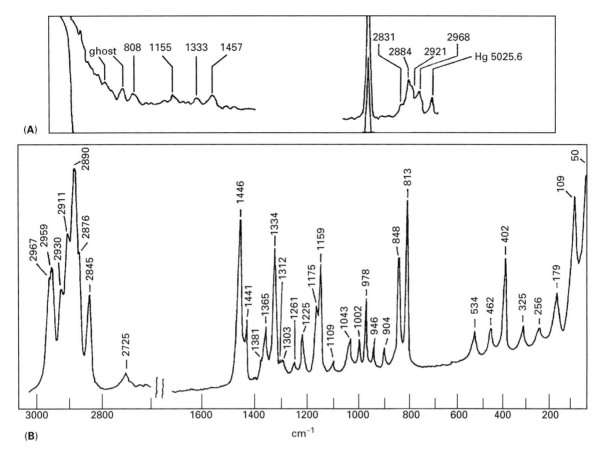

Figure 6 Comparison of the Raman spectra of isotactic polypropylene. (A) Mercury arc, 435.8 nm; densitometer trace of a photographic negative. (B) Laser, 632.8 nm; photoelectric recording. Reproduced by permission from Tobin MC (1971) *Laser Raman Spectroscopy*. New York: Wiley.

modification. This has allowed, among other things, the monitoring of fast processes on the timescale of single data point acquisition so that, for example, vibrational spectra of some excited states can be obtained.

Finally, the interferometer was not to be all-conquering, however, as simple filter spectrometers were found to be useful for many dedicated monitoring purposes. Tuneable IR lasers have also been applied to gas monitoring, but their general usefulness has been limited by the range of wavelengths that can be output.

The coming of the black box

The necessary provision of computers in FT-IR soon led to much more than simple data manipulation. After much unfulfilled promise and wasted effort, data transfer systems became fairly standardized and reliable by the late 1980s. With the universal adoption of the JCAMP format, spectra in digital form became truly portable and could become part of the business of laboratory information management systems (LIMS). More interestingly, they could be directly matched into large digital spectral databases

for automatic identification or structural analysis by expert systems.

In parallel, the 1980s also saw considerable development in the application of software for data analysis and instrument control. By this time, software was accounting for more than half the development cost of an FT-IR system.

Acronyms galore: The impact of the laser on Raman spectroscopy

Unlike the case for IR spectroscopy, there was little development in the immediate post-war years, so that for a long time practice was mainly confined to a few academic institutions.

An improved source, the mercury arc, a helical discharge lamp surrounding a cylindrical sample tube, yielded much-increased light energy, though probably equal in importance was the advent of photoelectric recording. The latter facility enabled the production of scanning instruments. Two companies, Cary and Hilger+Watts, offered models, the former fielding an image slicing device that attempted to overcome the optical incompatibility of the extended area of the mercury arc source and the narrow spectrometer slits.

Apart from the weakness of the spectra, the range of materials that could be examined was restricted by problems thrown up by the sample itself. Coloured samples absorbed source and scattered signal and possessed the propensity to fluoresce, which could easily drown out the Raman signal.

The advantages offered by the laser to Raman spectroscopy were recognized almost as soon as it had been demonstrated in 1960. It was to revolutionize the technique by providing dramatic increases in sensitivity (**Figure 6**) and the opportunity for unusual experiments by virtue of nonlinear effects that it could induce in many materials.

The continuous gas lasers were found to be most useful. The principal advantage of the laser in general over the mercury arc lay in the small point-source area that allowed larger flux throughput for a given spectrometer étendue.

Where formerly grams of material were needed, now milligram quantities, as solid powders, liquid or solutions in capillary tubes, were routinely amenable to examination.

The beckoning opportunities soon led several manufacturers (e.g. Spex, Jarrel-Ash, Coderg and Perkin-Elmer) to bring out new equipment, while older designs (Cary) were modified to accommodate the new type of source. Instruments were of course expensive compared to dispersive IR spectrometers as the optical requirements were far more stringent. Two coupled monochromators were necessary to cut down stray light from the exciting line. The Czerny–Turner layout was most common, the monochromators usually, but not always, being arranged for additive dispersion. **Figure 7** shows a very early photographic/mercury arc Raman spectrum of CCl_4 for comparison with what could be achieved with an automatically recording spectrometer of the late 1960s (**Figure 8**).

Photoelectric detection for Raman spectroscopy initially suffered from poor sensitivity in the red, but this was rectified during the 1980s when many cheap photomultiplier tubes became available. Rather more expensive were multichannel detectors, originally used for fast time-resolved experiments in the 1970s. Their take-up was slow but by the late 1980s very sensitive charge-coupled devices (CCDs) developed for astronomy were being incorporated in spectrographs for relatively mundane analytical work.

Complementing infrared

Compared to IR spectroscopy, the Raman technique was found to possess some advantages, one particular being the comparative ease in dealing with aqueous solutions. In the biochemical sphere this meant, for example, that ionization behaviour and pH change could be studied; even before the arrival of the laser, amino acids had been demonstrated to exist as zwitterions.

For a long time a major disadvantage was the common occurrence of interference from fluorescence, especially often encountered in biological material. A redeeming feature of Raman spectroscopy, however, was the many special experiments that could be performed.

(A)

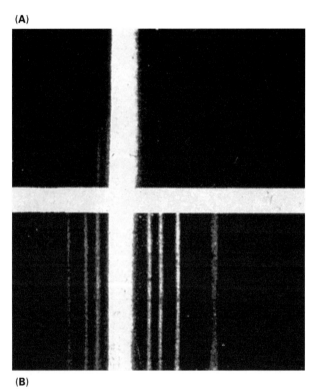

(B)

Figure 7 Mercury arc-excited Raman spectrum of carbon tetrachloride with photographic recording. (A) The spectrum of the mercury arc itself for reference. (B) The four Stokes lines (right side of the exciting line) and the weaker antiStokes (left side of the exciting line) lines, of which only three can be seen. From part of Plate 1, Raman CV and Krishnan KS (1929) The production of new radiations by light scattering. *Proceedings of the Royal Society (London)* **A122**: 23–35. Reproduced from a photograph courtesy of Professor N. Sheppard, FRS, and with permission of the Royal Society.

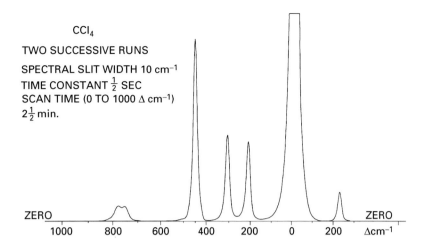

CCl₄

TWO SUCCESSIVE RUNS

SPECTRAL SLIT WIDTH 10 cm⁻¹

TIME CONSTANT ½ SEC
SCAN TIME (0 TO 1000 Δ cm⁻¹)
2½ min.

ZERO   ZERO

1000   800   600   400   200   0   200   Δcm⁻¹

Figure 8 Laser-excited Raman spectrum of carbon tetrachloride with photoelectric recording. Reproduced with the permission of Professor N. Sheppard, FRS.

Nonlinear effects and other esoterica

The high power densities available from lasers were found to be capable of inducing a number of strange (nonlinear) Raman effects, though many were of limited applicability to problems in chemistry. Typical examples included SIRS (stimulated inverse Raman scattering), RIKES (Raman induced Kerr effect spectroscopy) and the hyper Raman effect.

Probably the most useful was found to be CARS (coherent antiStokes Raman scattering), in which the signal could be detected largely free of background such as fluorescence. It was particularly beneficial in examining combustion processes in flames. Hopes that it might provide the complete answer to the fluorescence problem were to be largely unfulfilled owing mainly to its technical complexity.

Other interesting and useful discoveries did not require high-power excitation. Resonance Raman scattering (RRS) arises in certain samples when the laser excitation wavelength falls within an electronic absorbance band. RRS bands can be several orders of magnitude stronger than normal. One application was to employ small-molecule ligands exhibiting RRS as probes in the active sites of enzymes, thus avoiding major interference from the complex spectrum of the protein.

In 1974 news came that the Raman spectrum of pyridine was considerably enhanced when absorbed on a roughened silver substrate. This effect was named SERS (surface enhanced Raman scattering) and has since been observed from many other compounds and other metals. Electrochemistry was an early application. Inevitably, experimenters were to combine the effect with RRS to create SERRS, which in some circumstances can equal fluorescence for sensitivity of detection.

Instrumental developments

FT-Raman spectroscopy was introduced in 1986 and it is now available as a bolt-on to many FT-IR machines. Interestingly, interferometers might have been used earlier for Raman spectroscopy if the laser had not been invented, as their large circular aperture could have coped advantageously with the extended source area of the mercury arc. As it was, the multiplexing capability was needed to boost sensitivity so as to satisfactorily observe the weak spectra produced by a near-IR laser. The rationale was that fluorescence was largely eliminated. Thus, high-quality spectra of dyes, for instance, could be obtained that had formerly been impossible. Unfortunately, as overtones and combinations of H_2O vibrations possess significant absorbance in the near IR, spectra from aqueous solutions were affected.

An earlier development was the reintroduction of the spectrograph made possible by the availability of multichannel detectors. CCDs, with sensitivity equal to the traditional photomultiplier and up to 1024 channels, thus provided a considerable multiplexing advantage. They allowed the construction of small, rugged instruments able to acquire good-quality spectra very rapidly. Coupled to fibreoptic sampling devices, such spectrographs have recently found considerable use for process monitoring in industry.

A very useful accessory has been the microscope. Here there is a significant advantage over IR spectroscopy as spatial resolution is higher owing to the shorter wavelength of the source radiation.

Out of the orphanage: The rise of near-IR spectroscopy

The NIR region was for long neglected, a curiosity for users of many research-grade UV-visible and mid-IR spectrometers that had been provided with extended range capabilities. This was not unexpected, as absorption bands in the region originate from either uncommon electronic transitions in inorganic compounds or broad and heavily overlapped overtones and combinations of vibrational fundamentals. The latter are mostly derived from X–H stretching modes in organic compounds. In consequence, spectra, particularly of mixtures, are not easy to interpret.

Experimentally, however, the spectra are easy to observe, thick samples being tractable in either transmission or reflection without preparation. As the spectra seldom possess many narrow features that could be unduly affected by instrumental or other factors, robust methods for quantitation were possible. With the arrival of cheap instrumental computing from the 1980s onwards and the development of multivariate analysis methodology, NIR spectroscopy has undergone considerable expansion in use. It is now widely applied to automated, rapid and precise quantitative analyses in agriculture, industrial process control and noninvasive medical examinations. A typical and early example was determination of the protein content of grain and flour.

Most work is now done with dispersive and simple filter analysers. Though S/N is not usually a problem, FT-NIR and filter/CCD machines are becoming more popular.

Full circle: IR and Raman spectroscopy into the new millennium

Originally IR spectra were measured point by point; they then became continuous and now are again digital in nature. Spectrometers were single-beam, then double-beam, and now are almost all single-beam once more. With the introduction of spectrographs coupled to multichannel detectors, Raman spectroscopy is also in a sense back where it used to be, when photographic film effectively provided a multiplicity of channels. NIR CCDs are presently available that cover some of the range and the future possibilities of usable array detection right down into the mid IR could eventually spell a general return to dispersive techniques. Further into the future, fully tuneable IR lasers would remove even the need for a dispersing element for IR spectroscopy. If the rationale for interferometers, hitherto their advantage in throughput, is lost, then their disadvantages such as poor photometric accuracy and mechanical complexity could mean retention only in special circumstances.

Hand in hand with increasing sophistication of spectrometers and software, much of which runs transparently to the user, has gone an overall deskilling of operatives. This is inevitable (and usually desirable) given the workload and demands on modern analytical laboratories and industrial processes. There are obvious dangers, however, as the theory behind the methodology is beyond many scientific workers, as are the ramifications of many spectrometer function operations on the input data. The days are long gone when spectroscopists built and maintained their own machines, synthesized the chemicals they studied and contributed to advance of theory.

See also: **Chromatography–IR, Applications; Fourier Transformation and Sampling Theory; FT-Raman Spectroscopy, Applications; Hydrogen Bonding and other Physicochemical Interactions Studied By IR and Raman Spectroscopy; IR Spectral Group Frequencies of Organic Compounds; Industrial Applications of IR and Raman Spectroscopy; IR Spectrometers; IR and Raman Spectroscopy of Inorganic, Coordination and Organometallic Compounds; Near-IR Spectrometers; Nonlinear Raman Spectroscopy, Applications; Raman Spectrometers; Raman and IR Microspectroscopy.**

Further reading

Bellamy LJ (1975) *The Infra-red Spectra of Complex Molecules*. London: Chapman and Hall.

Brugel W (1961) *An Introduction to Infrared Spectroscopy*. London: Methuen.

Ferraro JR (1996) A history of Raman spectroscopy. *Spectroscopy* 11(3): 18–25.

Griffiths PR, Sloane HJ and Hannah RW (1977) Interferometers vs monochromators: separating the optical and digital advantages. *Applied Spectroscopy* 31(6): 485–495.

Herzberg G (1945) *Infra-red and Raman Spectra of Polyatomic Molecules*. New York: Van Nostrand.

Hibben JH (1939) *The Raman Effect and Its Chemical Applications*. New York: Reinhold.

Johnston SF (1991) *FTIR: A Constantly Evolving Technology*. Chichester: Ellis and Horwood.

Jones RN (1985) Analytical applications of vibrational spectroscopy – a historical review. In Durig JR (ed) *Chemical, Biological and Industrial Applications of Infrared Spectroscopy*. Chichester: Wiley.

Long DA (1988) Early history of the Raman effect. *International Reviews in Physical Chemistry* 7(4): 317–349.

White RG (1964) *Handbook of Industrial Infrared Analysis*. New York: Plenum.

Wood Products, Applications of Atomic Spectroscopy

See **Forestry and Wood Products, Applications of Atomic Spectroscopy.**

Xenon NMR Spectroscopy

Jukka Jokisaari, University of Oulu, Finland

MAGNETIC RESONANCE
Applications

Nuclear magnetic resonance experiments were originally planned for measuring nuclear magnetic moments. This explains why such exotic nuclei as ^{129}Xe (spin-$\frac{1}{2}$) and ^{131}Xe (spin-$\frac{3}{2}$) were the subjects of NMR investigations already as early as 1951, only about 5 years after the first successful NMR experiments at Harvard and Stanford. In the 1960s and 1970s, researchers were mostly interested in performing shielding and relaxation experiments of the two xenon isotopes (as well as of the other noble gas nuclei: ^{3}He, ^{21}Ne and ^{83}Kr) in gas, liquid and solid states. The expansion of xenon NMR spectroscopy started in the early 1980s when it was proposed that the physical properties of zeolites and clathrates, and in general, of porous solids, could be characterized by adsorbing xenon to the sample and recording the ^{129}Xe NMR spectrum. Since then, xenon NMR has been applied to derive information on (besides porous solids) various isotropic liquids and liquid mixtures, liquid crystals, proteins and membranes, myoglobin and haemoglobin, and polymers. The finding that ^{129}Xe can be spin-polarized by optical pumping, leading to a sensitivity increase up to a factor of 10^5, has widened the field of application to low surface area solids, human blood and perfused tissue, and imaging of organs of small animals, and recently, even of humans.

Properties of the nuclei

Xenon possesses nine stable isotopes, but only two of them, ^{129}Xe and ^{131}Xe, have a nonzero spin necessary for magnetic resonance. NMR properties of the isotopes are collected in **Table 1**. Xenon being a monatomic gas, the ^{129}Xe and ^{131}Xe NMR spectra recorded in isotropic solutions, for example, consist of a single resonance line. In anisotropic environments with nonzero static electric field gradient (EFG), the ^{131}Xe NMR spectrum is a triplet. Despite this simplicity, the spectra provide much information on the environment into which the xenon is introduced. This is due to the fact that the xenon shielding is extremely sensitive to the changes taking place in its local environment. Furthermore, as xenon is very inert it does not disturb the system into which it is taken to. In the literature, the change of the shielding of atomic xenon, arising from bulk and local effects, is often called the chemical shift. This is somewhat misleading. However, in this article chemical shift ($\delta = \sigma_0 - \sigma_{Xe}$, σ_0 and σ_{Xe} being the shielding of the reference gas, usually free xenon gas, and of xenon introduced into the environment under study, respectively) and shielding ($\sigma_{Xe} - \sigma_0 = -\delta$) are used in parallel.

The 32-fold sensitivity of ^{129}Xe compared to ^{13}C makes it an easy nucleus for NMR detection; the

Table 1 Properties of the NMR active xenon isotopes ^{129}Xe and ^{131}Xe

Isotope	Spin	Natural abundance (%)	Gyromagnetic ratio (10^7 rad T^{-1} s^{-1})	Quadrupole moment (10^{-28} m^2)	NMR frequency[a] (MHz)	NMR sensitivity[b]
^{129}Xe	$\frac{1}{2}$	26.44	−7.441	—	110.632	31.82
^{131}Xe	$\frac{3}{2}$	21.18	2.206	−0.12	32.795	3.318

[a] At the magnetic field $B_0 = 9.39$ T.

[b] Absolute sensitivity (product of the relative sensitivity and natural abundance) with respect to that of the ^{13}C isotope.

spectrum may be obtained on a single pulse from samples with ~1 atm or higher pressure of gas. This is a great advantage because often the spin–lattice relaxation time, T_1 is long, up to hundreds of seconds in solutions and up to over 100 min in the gas phase. In many applications, relaxation agents can be utilized to shorten T_1. The ^{131}Xe isotope possesses an electric quadrupole moment and thus its spin–lattice relaxation is dominated by the quadrupole interaction, the T_1 values are at the millisecond level allowing a relatively fast pulse repetition in accumulation.

A drastic improvement in the NMR sensitivity of ^{129}Xe can be achieved by the optical pumping method. In a conventional NMR experiment, the nuclear magnetization arises from the population difference due to Boltzmann distribution between energy states. For example, for ^{129}Xe at the magnetic field of 9.4 T and at thermal equilibrium at $T = 300$ K, the relative population difference between the $m = -\frac{1}{2}$ and $m = +\frac{1}{2}$ states is 9×10^{-6} = 9 ppm. When applying optical pumping with a high-powered laser, this figure can be increased up to >0.3, i.e. the magnetization increases by a factor of ~3×10^5. This increase is performed by placing xenon and nitrogen gas together with alkali-metal vapour (usually rubidium) in a glass cell and illuminating by circularly polarized laser light with a wavelength of 749.7 pm. During binary collisions, spin polarization is transferred from alkali-metal atoms to xenon atoms.

Xenon in gases

The shielding of pure xenon gas is often expressed as a virial expansion

$$\sigma(\rho, T) = \sigma_0 + \sigma_1(T)\rho + \sigma_2(T)\rho^2 + \sigma_3(T)\rho^3 \quad [1]$$

where σ_0 is the shielding constant of the atom *in vacuo*, ρ is density (given in amagat, the density of xenon under standard conditions at 298 K, ~2.5×10^{19} atoms cm^{-3}), and for the virial coefficients the following values have been reported (at 298 K): $\sigma_1 = -0.548 \pm 0.004$ ppm amagat^{-1}, $\sigma_2 = (-0.17 \pm 0.02) \times 10^{-3}$ ppm amagat^{-2}, and $\sigma_3 = (0.16 \pm 0.01) \times 10^{-5}$ ppm amagat^{-3}. The coefficients σ_1, σ_2 and σ_3 arise from two-, three- and four-body interactions, respectively. At low densities the shielding constant depends linearly on density, whereas at high pressures many-body collisions become important as well and cause deviation from linearity. The second virial coefficient arises from the Xe–Xe pair interactions (with the potential $V(r)$, r is the interatomic separation) and can be presented

in the form.

$$\sigma_1(T) = 4\pi \int_0^\infty \sigma(r) \exp[-V(r)/kT] \, r^2 \, dr \quad [2]$$

When xenon is mixed with another gas, G, the collisions Xe–G also contribute to the ^{129}Xe shielding. If only binary collisions are considered to be important, the shielding of xenon is given by

$$\sigma_{\text{exp}} = \sigma_0 + \sigma_1(\text{Xe–Xe})\rho_{\text{Xe}} + \sigma_1(\text{Xe–G})\rho_{\text{G}} \quad [3]$$

where ρ_{Xe} and ρ_{G} are the densities of Xe and G, respectively. Once σ_1 (Xe–Xe) is determined in pure xenon gas, the σ_1 (Xe–G) term can be solved from Equation [3]. Mixtures of xenon with other noble gases as well as with some small molecules, such as CO, N_2, O_2, CO_2, CH_nF_{4-n}, CH_4, SiF_4, SF_6, etc., have been studied. The very recent coupled Hartree–Fock calculations with gauge-including atomic orbitals on the shielding surfaces of the Xe–CO_2, Xe–N_2 and Xe–CO systems predict second virial coefficients in fair agreement with the corresponding experimental ones. The calculations revealed the shielding surfaces to be highly anisotropic.

In the early days of ^{129}Xe NMR, the spin–lattice relaxation time was assessed to be very long, and therefore a paramagnetic substance (Fe_2O_3) was used to shorten it in the first NMR experiments. The estimate of long T_1 was based on the dipolar interaction between two nuclei when they collide with each other. This model leads to the relaxation rate, $R_1 = 1/T_1$, which is inversely proportional to the mean collision time. Another interaction, more effective than the dipolar one, is the spin–rotation interaction, in which the nuclear spin couples with the angular momentum of a transient diatomic molecule formed during the collision. This model accounts for the experimental finding of R_1 being linearly dependent on the gas density; $R_1 = (5.0 \pm 0.5) \times 10^5 \rho$, where R_1 is given in s^{-1} and ρ in amagats. The T_1 value determined in hyperpolarized ^{129}Xe has been found to be dependent upon the magnetic field strength: 155 min (gas pressure 790 torr) and 185 min (896 torr) at 2.0 T, and 66 min (790 torr) and 88 min (896 torr) at 7.05 T, experimental error being ~5% and temperature 20 °C. The slight variation of T_1 at a constant magnetic field is assumed to arise from differences in the cell wall structure, whereas the field dependence was interpreted as a consequence of the less effective wall interaction at the higher magnetic field where the nuclear presession frequency is also higher.

The spin–lattice relaxation of the ^{131}Xe isotope is predominantly due to the interaction of the nuclear electric quadrupole moment with the electric field gradient (EFG) induced during binary collisions. Experiments have shown that also in this case the relaxation rate is linearly dependent upon the density: $R_1 = 0.039\,\rho$, where R_1 is in s^{-1} and ρ in amagats. Theory has given a similar relation with a proportionality coefficient of 0.046.

Xenon as a probe in liquid and solid environments

Isotropic liquids

For xenon dissolved in an isotropic liquid, the solvent effect on the shielding, σ_m, can be represented as

$$\sigma_m = \sigma_{exp} - \sigma_0 - \sigma_b = \sigma_a + \sigma_w + \sigma_E \quad [4]$$

where σ_{exp} is the experimental shielding constant, σ_0 is the shielding in the free atom (obtained by extrapolation of σ_{Xe} to zero pressure), σ_b arises from bulk susceptibility, σ_a from the magnetic anisotropy of the nearest neighbouring solvent molecules, σ_w from the van der Waals interactions, and σ_E is the shielding contribution caused by the permanent electric dipole of the solvent. Solvent-induced change of the ^{129}Xe shielding is about 250 ppm, as can be seen from **Table 2**. On the other hand, the ^{129}Xe gas-to-solution shifts, i.e. the change of the shielding compared to the shielding of free xenon, are over 330 ppm.

Various models have been developed for explaining the solvent-induced changes in the ^{129}Xe shielding. For example, it has been proposed, based on the reaction field theory of Onsager, that the medium shift is proportional to the function $f(n) = [(n^2 - 1)/(2n^2+1)]^2$ (this is called the van der Waals continuum model), where n is the refractive index of the solvent. Part of the experimental data indeed follows this relation, but most does not. An alternative model is provided with the pair interaction structureless approximation (PISA). The xenon–solvent dispersion energy, E_{dis}, calculated on this approximation is found to correlate better with the ^{129}Xe medium shift than $f(n)$.

One possible approach to gain insight into the solvent effects is to perform group contribution analysis. ^{129}Xe gas-to-solution shifts have been determined for pure n-alkanes, n-alkyl alcohols, n-alkyl carboxylic acids, di-n-alkyl ketones and cycloalkanes, and in solutions of lauric acid in n-heptane. It was found that the medium shifts corrected for solvent density are linearly dependent on the number of carbon atoms, except for the shortest members of the series of linear solvents (see **Figure 1**). Not only the structure of the environment but also temperature affects the ^{129}Xe shielding significantly. For example, in CD$_3$CN the shielding increases with increasing temperature (i.e. with decreasing density) at the rate of 0.30 ppm K^{-1}. This is 33 Hz K^{-1} at the magnetic field of 9.4 T. The position of the xenon resonance can be determined often with accuracy to better than 0.5 Hz, and consequently, ^{129}Xe shielding provides a good basis for a thermometer; accuracy may even be 0.02 K. A modified continuum

Table 2 Solvent effect on the ^{129}Xe shielding. Values are referenced to zero-pressure xenon gas.

Solvent	σ_m (ppm)	Solvent	σ_m (ppm)
Hexafluorobenzene	−85	Water	−196
Methanol	−148	Chlorobenzene	−202
Methyl chloride	−153	Bromobenzene	−219
Tetramethylsilane	−158	Carbon tetrachloride	−222
Ethanol	−165	Methyl iodide	−239
Fluorobenzene	−176	Iodobenzene	−248
Toluene	−190	Methylene iodide	−335

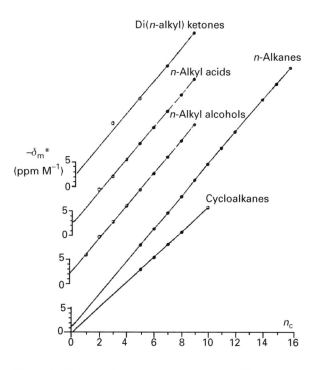

Figure 1 Molar medium effect on the ^{129}Xe gas-to-solution shifts, $-\sigma^*_m$ as a function of the number of carbon atoms, n_c. $-\sigma^*_m = -\sigma_m/\rho$, where ρ is density. Adapted with permission of the American Chemical Society from Luhmer M and Bartik K (1997) *Journal of Physical Chemistry A* **101**: 5278–5283.

model of van der Waals shifts has been presented to include also the effect of temperature. However, this model predicts temperature shifts in reasonable agreement with experiments only for the *n*-alkanes.

Although xenon NMR has been applied fairly widely to study physical properties of various materials, surprisingly little attention has been paid to its relaxation. The situation is, however, changing with the application of hyperpolarized ^{129}Xe; the longitudinal hyperpolarized magnetization decays with the spin–lattice relaxation time, T_1. The relaxation mechanisms of the ^{129}Xe isotope are exclusively due to interparticle (xenon–solute molecule) interactions. In pure xenon gas, the relaxation mechanism has been proposed to arise from spin–rotation (SR) coupling during atomic collisions or during the transient existence of diatomic molecules. The SR interaction may also partly explain the relaxation of ^{129}Xe in benzene; the xenon atom is located on the C_6 symmetry axis of benzene with a binding energy of 10.4 KJ mol^{-1}. The dominating interaction in protonated solvents, however, is the ^{129}Xe-^{1}H dipolar interaction; in benzene its contribution is over 50% and in cyclohexane over 90% of the total relaxation rate, $R_1 = 1/T_1$. The T_1 ranges from ~70 s to ~1000 s for xenon in typical isotropic solvents. In blood cells and plasma, the T_1 is 4.5 s and 9 s, respectively, whereas in blood foam it is 21 s (oxygenated) and 40 s (deoxygenated).

The relaxation of the quadrupolar ^{131}Xe nucleus is predominantly due to the interaction between the nuclear electric quadrupole moment and the fluctuating EFG at the nuclear site. The origin of the EFG contributing in a solution is, however, still partly an open question. Various models, both electrostatic and electronic, have been developed. The electrostatic models assume the EFG to be due to solvent molecules represented by point charges, point dipoles or quadrupoles, or a dielectric continuum. In the electronic approach, EFG is considered to be a consequence of the deformation of the spherical electron distribution of ^{131}Xe. The deformation arises from the collisions between xenon and solvent molecules. It is obvious (evidence is provided, for example, by ^{131}Xe NMR experiments in liquid–crystal solutions, and by first principles calculations) that neither of these approaches alone is sufficient. In typical isotropic solvents, the ^{131}Xe T_1 ranges from ~4 ms to ~40 ms.

Liquid crystals

Thermotropic liquid crystals (LC) are anisotropic liquids that possess a mesophase (a phase with crystal and liquid properties) within a certain temperature range. In a spectrometer magnet, LC molecules tend to orient to a common direction which defines the director of the liquid crystal. The director may orient either along the external magnetic field or perpendicular to it, depending upon the sign of the anisotropy of the diamagnetic susceptibility. When xenon is dissolved in a liquid crystal and its ^{129}Xe NMR spectrum is recorded at variable temperatures, a series of spectra, as shown in **Figure 2**, may be obtained. This kind of experiment provides information on phase transition, isobaric thermal expansion coefficient, liquid–crystal orientational order parameters and anisotropy of the ^{129}Xe shielding tensor ($\Delta\sigma_d$). The latter property arises from the fact that in a mesophase, the originally spherical electron cloud of xenon is deformed leading to an axially symmetric shielding tensor with nonzero $\Delta\sigma_d$. The above-mentioned quantities can be derived from Equation [5] by least-squares fitting:

$$
\begin{aligned}
\sigma_{\exp}(T) - \sigma_0 = [1 - \alpha(T - T_0)]\{\sigma_d[1 + 2c\tau_1{}^2(T)] \\
+ (2/3)P_2(\cos\phi)\Delta\sigma_d[S(T) \\
+ 2c\sigma_1(T)\tau_1(T)]\} \quad [5]
\end{aligned}
$$

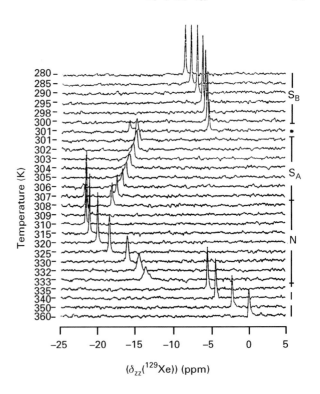

Figure 2 ^{129}Xe NMR spectra of natural xenon gas in a binary mixture of the Merck S1114 and EBBA liquid crystals. The shielding (in ppm) is referenced to that at 360 K. On the right are shown the various phases: I (isotropic), N (nematic), S_A (smectic A), and S_B (smectic B). Adapted with permission of Gordon and Breach Publishers from Jokisaari J, Diehl P and Muenster O (1990) *Molecular Crystals and Liquid Crystals* **188**: 189–196.

where $\sigma_{\text{exp}}(T) - \sigma_0$ is the shielding difference for xenon in liquid–crystalline and gaseous phases, α is the isobaric thermal expansion coefficient, T_0 is the reference temperature (for example, the isotropic–nematic or nematic–smectic A phase transition temperature), σ_{d} and $\Delta\sigma_{\text{d}}$ are the shielding constant and shielding anisotropy of xenon, $S(T)$, $\sigma_1(T)$ and $\tau_1 T$ are the conventional order parameter, translational–orientational order parameter and translational order parameter (the last two are present only in smectic phases), respectively, P_2 is the second Legendre polynomial, and ϕ is the angle between the external magnetic field and the liquid-crystal director, and the coefficient c describes how much the positional distribution function deviates from a uniform distribution.

As mentioned above, the ^{131}Xe nucleus possesses an electric quadrupole moment. In a liquid–crystalline solution the quadrupole moment interacts with the EFG at the nuclear site, and consequently, instead of a single resonance peak detectable in isotropic phases, a triplet with theoretical relative intensities of 3:4:3 is observed. (Generally, the multiplet consists of $2I$ resonance lines, where I is the spin of the nucleus.) An example is given in **Figure 3**. The quadrupole splitting, i.e. the separation of the resonance peaks in a spectrum, can be used for determining external EFGs, i.e. EFGs arising from the electric multipoles of LC molecules, EFGs arising from the deformation of the electron cloud of xenon when it collides with LC molecules and LC orientational order parameters.

Polymers

The physical and mechanical properties of polymeric systems are connected with their solid state morphology. NMR spectroscopy of the nuclear spins attached to a polymeric system is a very applicable means to gain insight into the microstructure as well as into the dynamics of the system. An alternative way is to make use of a probe, such as a xenon atom, which diffuses over the environment and gives information on the microscopic heterogeneity.

Since the xenon shielding is sensitive to the density of the surrounding medium, one may expect that it is affected by the glass transition of an amorphous polymer. Indeed this is the case, but, however, a more distinct change can be detected in the ^{129}Xe line width, as shown for poly(ethyl methacrylate) in **Figure 4**. ^{129}Xe NMR has proven to be particularly useful in studies of polymer blends whose components possess almost identical glass transition temperatures. Namely, for a phase-separated two-component blend, the ^{129}Xe spectrum consists of two resonance signals, while the homogeneous morphology of a miscible blend yields a single resonance. The application of thermal analysis techniques is restricted by the fact that the different glass transitions can only be detected if they differ at least by 20 K.

When xenon is adsorbed, for example, into a solid EPDM rubber (a terpolymer composed of ethylene, propylene and ethylidene norbornene) at least four distinct ^{129}Xe resonance signals can be observed indicating the presence of physically distinct domains; the intensity ratios may be used for the determination of the size of the domains, whereas the shielding differences reveal the variation of the destiny in the domains. When the rubber is cross-linked, the spectrum is clearly different from the one before

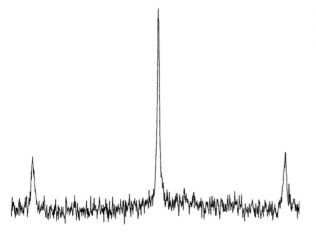

Figure 3 The ^{131}Xe triplet of xenon in the thermotropic Merck ZLI1167 liquid crystal. The frequency separation of the two outmost peaks is ~56 kHz. The intensity ratios are distorted because of experimental instabilities. Run parameters: ^{131}Xe resonance frequency 49.218 MHz ($B_0 = 14.1$ T), acquisition time ~7 min, $T = 325$ K. (Unpublished data from this laboratory.)

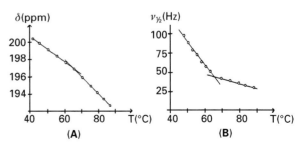

Figure 4 (A) ^{129}Xe chemical shift, and (B) ^{129}Xe line width at 24.79 MHz as a function of temperature for xenon adsorbed in poly(ethyl methacrylate), the glass temperature, T_g is 65 °C. Adapted (redrawn) with permission of the American Chemical Society from Stengle TR and Williamson KL (1987) *Macromolecules* **20**: 1430–1431.

cross-linking. As **Figure 5** shows, the cross-linking leads to the disappearance of the signal corresponding to the highest shielding of ^{129}Xe, i.e. the largest amorphous voids. This is consistent with the fact that cross-linking produces a more condensed polymer matrix.

One possibility for investigating microheterogeneity in polymers with the xenon probe is to apply the cross-polarization (CP) technique, in which polarization is transferred from polymer protons to ^{129}Xe. The necessary condition for ^{1}H–^{129}Xe CP is for the xenon atom to be trapped long enough near a proton for the dipolar coupling between the nuclei to be effective. **Figure 6** displays the normal single-pulse ^{129}Xe NMR spectrum, together with the ^{1}H–^{129}Xe CP spectrum of xenon in a polymer blend of a copolymer (a mixture of 2/3 polyethylene, PE, and 1/3 polypropylene, PP) dispersed in a polymer matrix. The latter

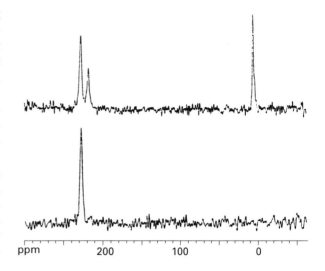

Figure 6 (Top) The conventional single-pulse ^{129}Xe NMR spectrum and (bottom) the ^{1}H–^{129}Xe CP spectrum of xenon in a polymer blend. The mixing time in the CP experiment was 3 ms. The signal at 0 ppm arises from free xenon gas, the signal at 216 ppm from xenon in copolymer and the signal at 226 ppm from xenon in PP. Note: the scale is the chemical shift scale, which is opposite to the shielding scale. Adapted with permission of Elsevier Science Ltd from Mansfeld M and Veeman WS (1994) *Chemical Physics Letters* **222**: 422–424.

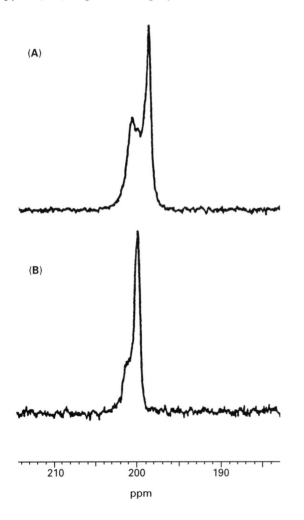

Figure 5 ^{129}Xe NMR spectra of xenon adsorbed in solid EPDM: (A) before, and (B) after crosslinking. Note: the scale is the chemical shift scale, which is opposite to the shielding scale. Adapted with permission of Springer-Verlag from Kennedy GJ (1990) *Polymer Bulletin* **23**: 605–606.

spectrum consists of a single resonance peak arising from xenon in the PP matrix where the translational mobility of the xenon atom is slow enough in order not to interrupt the dipolar coupling between the nuclei. Because of the CP between the PP protons and xenon adsorbed in the PP matrix, it is possible to obtain a correlation spectrum between the two spins, making it possible to identify the protons involved in the polarization transfer.

The efficiency of the polarization transfer depends upon the internuclear distance according to r^{-6}, and consequently, it is restricted to the nearest-neighbour protons of xenon. Thus CP experiments yield information only on the spatial proximity of distinguishable domains in a polymer. A much wider range of distance can be covered by two-dimensional (2D) exchange spectroscopy (EXSY). Its application is most useful in cases where the exchange rate of xenon between domains is slow compared to the chemical shift difference, and separate resonance signals from xenon in each domain can be observed. **Figure 7** displays results for ^{129}Xe 2D EXSY experiments on a model blend system of poly vinyl chloride (PVC) and poly vinyl methyl ether (PVME). The system consists of thin alternating layers (thickness 2–6 μm) of the two polymers. The EXSY experiments were performed with two mixing times, 0.8 s and 8 s. It is seen that during the shorter mixing time, xenon samples all the local environments in the

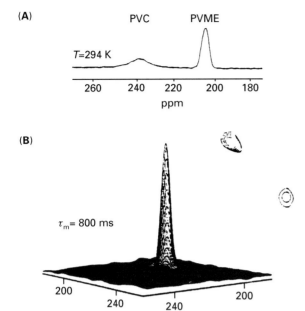

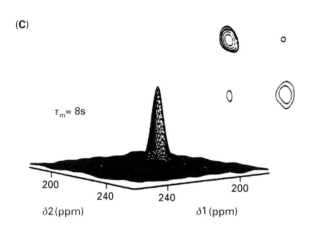

information on the size and shape of the pores of unknown structure through the ^{129}Xe shielding measurements. In principle this information is available but not very straightforwardly since the xenon shielding is affected not only by the two factors but also by xenon–xenon collisions, and the presence of strong absorption sites (SAS), paramagnetic species and adsorbed molecules, etc. The experimental shielding of xenon in zeolites and molecular sieves is usually represented in the form

$$\sigma_{\exp} = \sigma_0 + \sigma_{\mathrm{S}} + \sigma_{\mathrm{Xe}} + \sigma_{\mathrm{SAS}} + \sigma_{\mathrm{E}} + \sigma_{\mathrm{M}} \qquad [6]$$

where σ_0 is the shielding of the reference (usually a zero pressure gas), σ_{S} arises from the interaction of xenon with the pore walls, $\sigma_{\mathrm{Xe}} = \sigma_{\mathrm{XeXe}}\rho_{\mathrm{Xe}}$ is the shielding contribution of the xenon–xenon collisions, σ_{SAS} stems from Xe interaction with strong adsorption sites, σ_{E} in turn takes into account electric field effects (due to charge–compensating cations) and σ_{M} is the contribution of paramagnetic species.

Figure 8 shows the ^{129}Xe chemical shift as a function of the number of xenon atoms in different

Figure 7 ^{129}Xe spectrum (A), and ^{129}Xe 2D EXSY spectra (B and C) of xenon in the PVC/PVME model blend. τ_{m} is the mixing time. The insets show contour plots of the same data. Adapted with permission of Elsevier Science Ltd from Tomaselli M, Meier BH, Robyr P, Suter UW and Ernst RR (1993) *Chemical Physics Letters* **205**: 145–152.

PVC domains (this is indicated by the round shape of the diagonal peak), whereas no exchange between the PVC and PVME is taking place (there is no cross-peak in the spectrum). In contrast, during the mixing time of 8 s, exchange between the two phases is also seen.

Zeolites, molecular sieves and clathrates

Most attention has been drawn to the application of ^{129}Xe NMR to studies of zeolites, molecular sieves and clathrates. The main goal is to derive

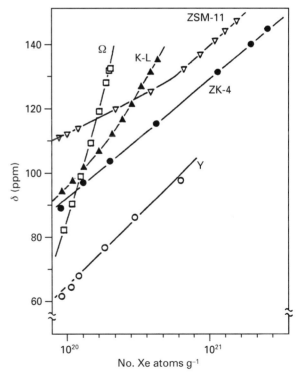

Figure 8 Dependence of the ^{129}Xe chemical shift (negative shielding) upon the number of xenon atoms adsorbed per gram of zeolite. Fraissard J (1996) In: Grant DM and Harris RK (eds) *Encyclopedia of Nuclear Magnetic Resonance Spectroscopy*, Vol. 5, pp. 3058–3064. Chichester: Wiley © John Wiley & Sons Limited. Reproduced with permission.

zeolites. The shielding values $(=-\delta)$, obtained by extrapolation to zero xenon number, range from ~ -110 ppm to ~ -60 ppm (in fact, in mordenite, which is not shown in the Figure, it goes down to -250 ppm), indicating the sensitivity of xenon shielding to zeolite structure.

No comprehensive theory exists for interpreting the experimental Xe shielding results in porous materials, although very significant progress in this direction has taken place recently; in particular, simulation calculations have improved our knowledge. In order to develop the method to a level giving as diversified information as possible of the shape and size of void space (and possibly cation distribution), it is important to investigate systems with varying properties and well-defined structure. The most simple systems are zeolites without charge compensating cations, and paramagnetic species when the two last shielding contributions in Equation [6] can be neglected. Very illustrative examples are the siliceous zeolite Si-ZSM-12 and molecular sieve AlPO$_4$-11. In the first approximation, both possess 1D channels with elliptical cross section. A closer inspection, however, reveals that the structure is more complex, consisting of series of cells, which have to be taken into account when interpreting experimental shielding data. The static ^{129}Xe spectrum in these two systems is a CSA (chemical shift anisotropy) powder pattern whose shape changes continuously from axially symmetric to asymmetric upon the xenon loading, as shown in **Figure 9**. Powder-like spectra have been observed for xenon only in a few zeolites. This may partly be due to misinterpretation of observed, slightly asymmetric line shapes; the xenon shielding is significantly affected by magnetic field inhomogeneity as well as by temperature gradients and fluctuations, and thus the CSA contribution, when small, has been masked by these instabilities during spectral recordings. The line shapes for xenon in Si-ZSM-12 and AlPO$_4$-11 have been explained by a dynamic model. In this model the shielding tensor is a dynamic, population-weighted average of tensors corresponding to three sites at which xenon has 0, 1 or 2 neighbouring xenons, and xenon samples rapidly fill the space available to it. When the elements of the shielding tensor are presented as a function of xenon loading, a linear dependence is observed at low loadings and thus the extrapolation to zero loading is straightforward and gives the shielding tensor elements corresponding to xenon with no neighbouring xenon atom.

The averaging of the xenon shielding tensor, and consequently, the derivation of structure information from the ^{129}Xe shielding data necessitates the

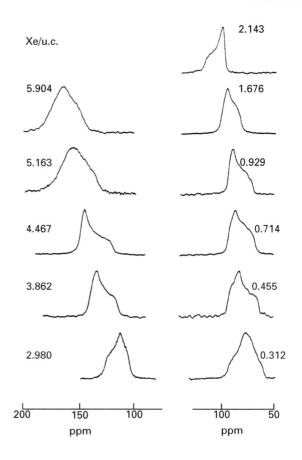

Figure 9 Static ^{129}Xe NMR spectra of xenon adsorbed in Si-ZSM-12 zeolite for different loading levels (xenon atoms per unit cell, Xe/u.c.). All spectra were recorded at 295 K. Adapted with permission of Springer-Verlag from Moudrakovski IL, Ratcliffe CI and Ripmeester JA (1996) *Applied Magnetic Resonance* **10**: 559–574.

knowledge of xenon dynamics. (One should emphasize here also that the dynamics of the zeolite framework affects significantly the averaging process. This is a consequence of the decrease of the effective potential barrier to intercellular jumps of xenon.) The observed line shape and shielding are averages over the shielding tensors of xenon sampling intracage volume and exchanging between cages, i.e. the mobility of xenon. For the first time, a CSA powder pattern was observed for xenon trapped in a clathrate where xenon is truly immobile. Xenon dynamics is available through spin–lattice relaxation time and diffusion measurements and 2D EXSY experiments.

2D EXSY has been applied, for example, to study xenon dynamics in the NaA zeolite whose structure is composed of large α-cages (inner diameter approximately 11.4 Å) and smaller β-cages (6.6 Å). At elevated temperatures and pressures the Xe atoms are distributed among the α-cages. This is seen in the ^{129}Xe 1D NMR spectrum which displays several distinct resonance signals (see **Figure 10**); xenon shielding decreases with increasing number of atoms in a cage. **Figure 10** also shows the results of the 2D EXSY experiments performed with three different mixing times. The emergence of cross-peaks arises from intercage motion of xenon during the mixing time. Performing the experiment at variable temperatures allows for the derivation of the rates of intercage motion as well as the adsorption and activation energies of the xenon atoms.

Applications of laser-polarized xenon

The ^{129}Xe magnetization can be enhanced compared to the thermal equilibrium magnetization by a factor of $\sim 10^5$ by utilization of optical pumping and spin-exchange between xenon and alkali-metal atoms. The resulting state is a nonequilibrium state, and the magnetization is often called hyperpolarized magnetization. One has to take into account the following facts when performing NMR experiments: (i) the hyperpolarized spins decay toward thermal equilibrium with the spin–lattice relaxation time T_1, and thus the hyperpolarization cannot be recovered by waiting for thermal equilibrium, (ii) when a radiofrequency θ pulse is applied to the spin systems, the longitudinal hyperpolarized magnetization decreases by $\cos \theta$. On the other hand, when applying repeated pulses the pulse repetition time can be short because there is no need to wait for the spin system to return to thermal equilibrium. In most applications the ^{129}Xe NMR spectrum of hyperpolarized xenon can be obtained on a single pulse. It is also possible to apply continuously flowing hyperpolarized ^{129}Xe.

Hyperpolarized ^{129}Xe can be utilized in two ways: firstly, by observing directly the ^{129}Xe NMR spectrum of xenon introduced to the environment under study, and secondly, by transferring hyperpolarization from xenon to other nuclei and recording their spectra. The polarization transfer takes place through cross-relaxation between spins without the need for irradiation of the spins. This technique has been denoted SPINOE (spin polarization induced nuclear Overhauser effect). The SPINOE may be either positive or negative depending upon whether right or left circularly polarized light is applied.

When hyperpolarized ^{129}Xe gas is dissolved, for example, into benzene, the ^{1}H NMR signal is enhanced because of SPINOE. This finding opens up new views for deriving information, for example, on Xe–protein interactions as well as on blood and other biological systems. The very high sensitivity of hyperpolarized ^{129}Xe makes it possible to investigate also low surface area (1–10 m^2 g^{-1}) nonporous solids

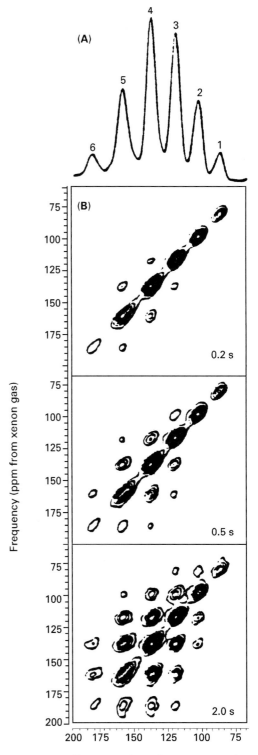

Figure 10 (A)^{129}Xe 1D NMR spectrum (on top of each peak is shown the number of Xe atoms in the cage) and (B) ^{129}Xe 2D EXSY spectra of xenon adsorbed in NaA zeolite at 523 K and 30 atm (1 atm = 101 325 Pa). The mixing times are: 0.2, 0.5 and 2.0 s as shown. Adapted with permission of Elsevier Science Ltd from Larsen RG, Shore J, Schmidt-Rohr K *et al* (1993) *Chemical Physics Letters* **214**: 220–226.

which otherwise, due to the low inherent sensitivity of NMR, are difficult if not impossible to investigate. On the other hand, magnetization transfer from hyperpolarized ^{129}Xe to other nuclei, e.g. ^{1}H and ^{13}C on a surface, allows for the direct observation of the magnetic resonances of these isotopes. Experiments have shown that enhancement of proton magnetization even by a factor of 20 may be achieved at low temperatures. The enhancement is more pronounced at low temperatures because of decreasing mobility of xenon atoms. An example of the ^{1}H magnetization enhancement is given in **Figure 11**.

Clathrates were the first systems investigated by ^{129}Xe NMR of natural xenon gas. The xenon atom is of the same size and shape as methane and it also forms a clathrate hydrate with water. The xenon shielding is much more sensitive (by a factor of about 30) than the ^{13}C shielding of methane to the

changes taking place in the structure of their environment. The use of hyperpolarized ^{129}Xe allows the investigation of large and small cages of clathrate hydrate; two distinct resonance signals are seen in the spectrum as shown in **Figure 12**. Much interest has been drawn to the application of hyperpolarized ^{129}Xe to materials, and in particular, to medical imaging (xenon gas is a safe general anaesthetic) and to spectroscopy in blood systems and in tissue, specifically the heart, lungs, brain and other organs. Recently, the technique was used to obtain human lung images. In this case, ~ 0.5 L of hyperpolarized ^{129}Xe is needed, necessitating higher power (100–120 W) lasers.

Xenon compounds

Xenon is known to covalently bond to fluorine, oxygen, nitrogen, carbon and to itself. In such circumstances, ^{129}Xe exhibits a large range of chemical

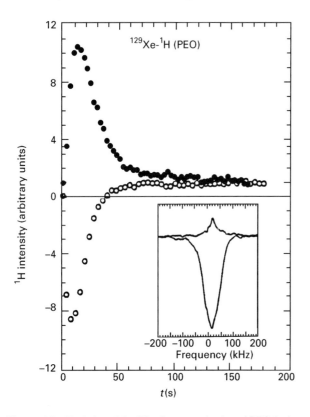

Figure 11 Evolution of the ^{1}H spin magnetization of PEO (polyethylene oxide)-coated Aerosil1130 as a function of time after exposure of the surface to hyperpolarized ^{129}Xe. The initial Xe pressure is 160 torr and the sample temperature is 130 K for positive SPINOE (●) and 125 K for negative SPINOE (○). The inset displays two single-shot ^{1}H spectra from a negative SPINOE run, one taken at the Boltzmann equilibrium for the unpolarized sample (positive peak) and the other taken at the time t_0, when the negative SPINOE enhancement has reached its maximum absolute value. Adapted with permission from Rõõm T, Appelt S, Seydoux R, Hahn EL, and Pines A (1997) *Physical Review B* **55**: 11604–11610.

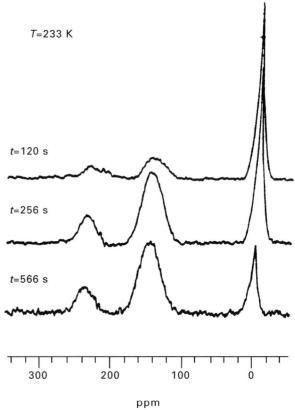

Figure 12 The ^{129}Xe NMR spectra of the formation of a xenon clathrate hydrate at 233 K and time *t* after admission of the xenon to the powdered ice sample. The signal at 160 ppm is attributed to xenon in the large tetrakaidecahedral cages and the one at 240 ppm to xenon in the smaller dodecahedral cages. Adapted with permission of the American Chemical Society from Pietrass T, Gaede HC, Bifone A, Pines A and Ripmeester JA (1995) *Journal of the American Chemical Society*, **117**: 7520–7525.

shifts, about 7500 ppm. **Figure 13** displays chemical shifts of some selected xenon compounds with different oxidation states. Also the ^{129}Xe shielding anisotropy may be large. For example, the theoretical estimate of the shielding anisotropy for XeF_2 is 5125–7185 ppm, for XeF_4 it is 3940 ppm and for $XeOF_4$ it is 365–1500 ppm. Spin–lattice relaxation time measurements have given an anisotropy of ~4700 ppm for XeF_2. In this particular case, the relaxation is dominated by the CSA and SR interactions and the T_1 values range from 150 to 430 ms depending upon magnetic field strength (CSA interaction depends upon the square of the magnetic flux density) and temperature (SR interaction depends linearly upon temperature). In general, the ^{129}Xe T_1 of various species in solution varies between 285 and 780 ms. Thus the relaxation is much faster than that of atomic xenon, and renders possible fast repetition rates in data accumulation. The one-bond spin–spin couplings of xenon with ^{19}F are relatively large and include a sizable relativistic contribution. The change of the absolute value of $^1J(^{129}Xe–^{19}F)$ can be used as a diagnostic tool to confirm the formal oxidation number of xenon as the coupling decreases in the order: Xe(II)> Xe(IV)>Xe(VI). Xenon couplings to other nuclei are smaller. No absolute sign determination has been made for any of the couplings to xenon. **Table 3**, shows absolute spin–spin coupling values of some selected xenon compounds.

List of symbols

c = distribution coefficient; E_{dis} = dispersion energy; I = nuclear spin; P_2 = second Legendre polynomial;

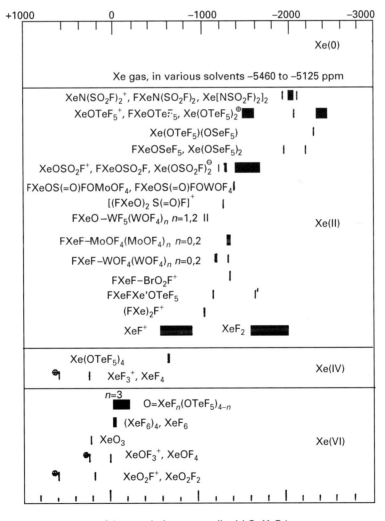

δ (ppm relative to neat liquid O=XeF$_4$)

Figure 13 ^{129}Xe NMR chemical shifts of a few selected xenon compounds. Adapted with permission from Jameson C (1987) The noble gases. In: Mason J (ed) *Multinuclear NMR*, Chapter 8, pp. 463–477. New York: Plenum Press.

Table 3 Absolute values of xenon spin–spin couplings for selected molecules.

Molecule	Coupling	Value (Hz)[a]
Xe(II)		
XeF_2	$^1J(^{129}Xe, ^{19}F)$	5644
$CH_3C≡CN—XeF^+$	$^1J(^{129}Xe, ^{19}F)$	6020
$C_6F_5C≡N—XeF^+$	$^1J(^{129}Xe, ^{19}F)$	6610
Xe(IV)		
XeF_5^-	$^1J(^{129}Xe, ^{19}F)$	1056–1082
cis-$F_2Xe(OTeF_5)_2$	$^1J(^{129}Xe, ^{19}F)$	3714
trans-$F_2Xe(OTeF_5)_2$	$^1J(^{129}Xe, ^{19}F)$	3503
XeF_4	$^1J(^{129}Xe, ^{19}F)$	3801–3900
Xe(VI)		
$OXeF(OTeF_5)_3$	$^1J(^{129}Xe, ^{19}F)$	1056–1082
$OXeF_3(OTeF_5)$	$^1J(^{129}Xe, ^{19}F)$	1127–1148
$XeOF_4$	$^1J(^{129}Xe, ^{19}F)$	1115–1131
$(XeF_6)_4$	$^1J(^{129}Xe, ^{19}F)$	330–331.7
$XeOF_4$	$^1J(^{129}Xe(II), ^{17}O)$	692–704
$CH_3≡N—XeF^+$	$^1J(^{129}Xe(II), ^{14}N)$	313
$C_6F_5Xe^+$	$^1J(^{129}Xe(II), ^{13}C)$	119
$Xe(OSeF_5)_2$	$^2J(^{129}Xe(II), ^{77}Se)$	130
$Xe(OTeF_5)_2$	$^2J(^{129}Xe(II), ^{125}Te)$	470
$HC≡NXeF^+$	$^3J(^{129}Xe(II), ^1H)$	24.7–26.8

[a] In some cases, the range of experimental values is given.

r = interatomic separation; R_1 = longitudinal relaxation rate; T_0 = reference temperature; T_1 = spin–lattice relaxation time; $V(r)$ = potential function for Xe–Xe pair interactions; α = isobaric thermal expansion coefficient; δ = chemical shift; $\Delta\sigma_d$ = shielding anisotropy of xenon; ρ = density; σ_0 = shielding constant of reference gas; σ_a = magnetic anisotropy shielding constant; σ_b = bulk susceptibility shielding constant; σ_{exp} = experimental shielding constant; σ_E = permanent electric dipole shielding constant; σ_M = paramagnetic species shielding constant; σ_S = surface induced shielding constant; σ_{SAS} = strong absorption sites shielding constant; σ_W = van der Waals interaction shielding constant; σ_{Xe} = shielding constant of xenon gas; τ = time; τ_M = mixing time; ϕ = angle between external magnetic field and liquid–crystal director.

See also: **Chemical Exchange Effects in NMR; Diffusion Studied Using NMR Spectroscopy; Gas Phase Applications of NMR Spectroscopy; Liquid Crystals and Liquid Crystal Solutions Studied By NMR; MRI Applications, Clinical; NMR in Anisotropic Systems, Theory; NMR Microscopy; NMR Relaxation Rates; Nuclear Overhauser Effect; Polymer Applications of IR and Raman Spectroscopy.**

Further reading

Albert MS, Cates GD, Driehuys B *et al* (1994) Biological magnetic resonance imaging using laser-polarized ^{129}Xe: *Nature (London)* 370: 199–201.

Barrie JP and Klinowski J (1992) ^{129}Xe NMR as a probe for the study of microporous solids: A critical review. *Progress in NMR Spectroscopy* 24: 91–108.

Dybowski C and Bansal N (1991) NMR spectroscopy of xenon in confined spaces: clathrates, intercalates, and zeolites. *Annual Review on Physical Chemistry* 42: 433–464.

Fraissard J and Ito T (1988) ^{129}Xe N.M.R. study of adsorbed xenon: A new method for studying zeolites and metal–zeolites. *Zeolites* 8: 350–361.

Jameson C (1987) The noble gases. In: Mason J (ed) *Multinuclear NMR*; pp 463–477. New York: Plenum Press.

Jokisaari J (1994) NMR of noble gases dissolved in isotropic and anisotropic liquids. *Progress in NMR Spectroscopy* 26: 1–26.

Miller JB (ed) (1995) Special issue on magnetic resonance studies of noble gases. *Applied Magnetic Resonance* 8: 337–595.

Raftery D and Chmelka BF (1994) Xenon NMR spectroscopy. *NMR Basic Principles and Progress*, Vol 30: pp. 112–158. Berlin: Springer-Verlag.

Ratcliffe CI (1998) Xenon NMR. *Annual Reports on NMR Spectroscopy* 36: 124–208.

Schrobilgen GJ (1996) Noble gas elements. In: Grant DM and Harris RK (eds) *Encyclopedia of Nuclear Magnetic Resonance*, Vol 5; pp 3251–3262. Chichester: Wiley.

Walker TG and Happer W (1997) Spin-exchange optical pumping of noble gas nuclei. *Review of Modern Physics* 69: 629–642.

X-Ray Absorption Spectrometers

Grant Bunker, Illinois Institute of Technology, Chicago, IL, USA

Introduction

The task of an X-ray absorption spectrometer is the precise and accurate measurement of the linear X-ray absorption coefficient of a substance. A principal use of such spectrometers is the measurement of X-ray absorption fine structure (XAFS) spectra of solids, liquids, and molecular gases. XAFS consists of modulations in the X-ray absorption coefficient in the vicinity of an X-ray absorption edge, which may extend more than one KeV beyond the edge. Applications and theory of X-ray Absorption Spectroscopy are covered elsewhere in this volume. This article is directed primarily to instrumental requirements for X-ray absorption spectroscopy over the energy range from several KeV X-ray photon energy to approximately 100 KeV, with emphasis on synchrotron radiation based instruments.

Absorption, fluorescence, and fluorescence excitation spectra

In the simplest case, the X-ray absorption coefficient (μ) of a homogeneous sample of thickness x is given by

$$\frac{I}{I_0} = \exp[-\mu(\varepsilon)x]$$

where I_0 and I respectively are the incident and transmitted X-ray intensities at X-ray photon energy ε. An analogous quantity in optical absorption spectroscopy is the product of the molar extinction coefficient and the concentration.

X-ray *absorption* spectrometers are distinct from X-ray *fluorescence* spectrometers (often referred to as 'X-ray spectrometers'), which measure the intensity of fluorescence radiation emitted by atoms in a specimen following their excitation by high energy photons, charged particle beams, or other interactions. X-ray fluorescence spectrometers are primarily used for measuring the elemental composition of samples, or measuring shifts in energy of fluorescence, to obtain chemical information about a sample.

A principal use of X-ray absorption spectrometers is in the measurement of X-ray absorption fine structure (XAFS) spectra, which provide quantitative information on the local structural and chemical environment within the region several ångstroms around selected atomic species in a sample. Absorption and fluorescence spectroscopies are complementary, and both can be used for spatially resolved spectroscopic mapping of samples. Fluorescence spectrometers also have considerable applicability for determining elemental composition in astrophysical research, planetary sciences, and numerous other areas.

There are many situations in which the X-ray absorption spectrum is most easily measured (indirectly) by monitoring the fluorescence produced following absorption of X-rays. One measures variations in the fluorescence intensity of a particular atomic species as the energy of incident photons is varied over an absorption edge of a selected element. This 'fluorescence excitation spectrum', distinct from the fluorescence spectrum, provides an indirect measurement of the X-ray absorption coefficient, albeit one that is subject to several well-known instrumental effects. If care is taken in sample preparation, systematic errors can be minimized. Fluorescence detection is the method of choice for dilute systems, because it provides an improved signal-to-noise ratio.

X-ray absorption spectra

Over the range from several KeV to 100 KeV, X-ray photons propagating through a sample are absorbed, scattered without loss of energy (elastic scattering), or scattered with loss of energy (inelastic scattering). X-rays require energies in excess of one MeV for positron–electron pair production to occur.

Cross sections vary significantly as a function of energy, as shown in **Figure 1A** and **B**. Absorption cross sections between the absorption edges decrease approximately as $1/\varepsilon^3$.

Absolute and relative measurements

Although it is essential to use samples of appropriate thickness and concentration in XAFS experiments, in practice it is seldom necessary to precisely determine the *absolute* absorption coefficient. The quantities of interest (e.g. inter-atomic bond lengths) are intrinsic to the material, while the absolute absorption coefficient depends on the thickness of the specimen, concentration of the element of interest, and other irrelevant factors. The standard methods of XAFS analysis treat the data in such a way that the structural parameters ultimately determined are invariant with respect to change of multiplicative scale factors, and additive background, provided it is a sufficiently smooth function of energy. To the extent that the X-ray scattering cross-sections vary slowly with energy, or are negligible compared to the photo electric cross-section, they do not affect the structure determination. It should be noted, however, that both the elastic and inelastic scattering cross-sections do have small contributions that vary near absorption edges and may need to be accounted for in some circumstances.

For some purposes the absolute cross-section is needed; for example, to determine the areal concentration of a particular atomic species in the sample, or to quantify the elemental composition of samples. The simplest approach is to measure the absorption coefficient with and without the sample inserted into the beam, as is done in a single beam optical spectrophotometer. Measurement precision can in principle be improved by modulation, i.e. rapidly performing differential measurements by rapidly inserting and removing the sample, for example by using a rotating disk with apertures that alternately contain a sample or a blank. This approach may be impractical for some samples, particularly if they require a special environment such as ultra high vacuum, high pressure, etc. In such cases it is possible to deflect the X-ray beam with a glancing incidence X-ray mirror, or by introducing an X-ray beam splitter.

Measurements of highest accuracy discriminate between the scattered, refracted, and transmitted beams. This can be made possible through the use of crystal analysers following the sample. If the X-ray beam is polarized and statistically non-isotropic it is important to account for orientation-dependent effects. Magic-angle spinning of the sample can be an effective means of averaging over anisotropies.

Requirements for X-ray absorption spectroscopy

X-ray absorption fine structure experiments require energy resolution on the order of several electron volts or less, to resolve modulations in the spectra. Spectra are intrinsically broadened by the 'core-hole lifetime broadening', which is an effect stemming from the relatively rapid ($<10^{-15}$ s) filling of the core-hole state (initial state vacancy) produced after the X-ray absorption event. Heisenberg's time–energy uncertainty relation $\Delta E \ \Delta t \geq h/2\pi$, implies that the rapid decay of the core-hole state broadens the energy spectrum with an energy level width ΔE inversely proportional to the lifetime Δt. The level width increases rapidly with atomic number.

Several stringent criteria must be met before an apparatus can be regarded as a suitable X-ray absorption spectrometer for XAFS. First, the device must have an appropriate energy resolution (several

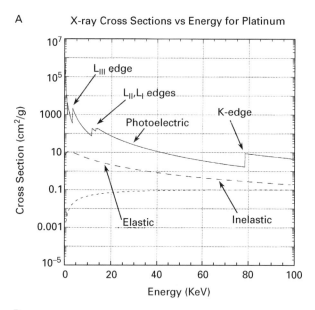

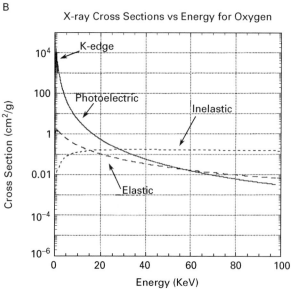

Figure 1 Absorption and scattering cross-sections for (A) platinum, and (B) oxygen.

eV or less); tunability (i.e. smooth, reliable scanning over an energy range of more than ~1000 eV range above selected absorption edges; and high flux generally >10^{10} photons s^{-1}). The harmonic content, i.e. contributions from high energy photons at multiples of the selected energy, should be limited to less than 0.1% of the fundamental. Beam intensity variation over a scan should be kept within ~20% depending on the linearity of detectors. **Figure 2** shows the layout of a typical instrument for transmission XAFS experiments.

Sources

The collimated beams, smooth energy spectrum, and high intensity of synchrotron radiation sources offer compelling advantages for XAFS experiments, compared to conventional fixed and rotating anode X-ray generators, although the latter are useful in some situations.

Historically, synchrotron radiation sources were associated with particle accelerators constructed for high energy physics experiments – so-called first generation sources. Electrons or positrons (anti-electrons) are accelerated in a closed orbit through an evacuated pipe at speeds exceedingly close to the speed of light:

$$\frac{v}{c} \sim 1 - \frac{1}{2\gamma^2}$$

where $\gamma = E/mc^2$ and E is the particle energy. These particles are used because particles of low mass radiate energy much more efficiently than those of high mass. Through the use of magnets the path of the particles is bent into a closed path of several hundred to one thousand metres circumference, through which the particles circulate at frequencies of hundreds of kilohertz to megahertz. Energy lost from the particle beam by synchrotron radiation is replenished through the use of radio-frequency cavities which apply a force to the particles along the direction of motion as they pass by. The particles travel in discrete 'bunches', and therefore the radiation produced has a pulsed structure that can be used to advantage for time-resolved experiments. The electrons circulate for many hours; scattering of the electrons from residual gas atoms in the ultra high vacuum environment in the ring, intra-bunch electron–electron interaction, quantum perturbations from spontaneous emission of photons, and other effects cause a slow loss of particles from the beam, which therefore must be periodically refilled. It is technically feasible to periodically replenish the beam and preserve nearly constant current in the ring.

All accelerating charged particles radiate energy, and if it were not for relativistic effects, the radiation produced by a synchrotron would be in the radio frequency spectrum. The relativistic motion causes the familiar dipole radiation lobes of an accelerating charge (seen in an inertial frame co-moving with the charge) to tilt forward along the direction of motion, in the observer's reference frame. The radiation pattern from a bend magnet is a horizontal fan of several milliradians angular width in the horizontal direction (in the orbital plane), and it is very well collimated in the vertical direction to an opening angle of order $1/\gamma$, where $\gamma = E/mc^2$, E is the particle beam energy, m is the rest mass of the electron, and c is the speed of light. For all bend magnets the broad spectrum (integrated over vertical opening angle) is described by a universal function

$$g_1(x) \equiv x \int_x^\infty K_{5/3}(t)\, \mathrm{d}t$$

where $K_{5/3}$ is the modified Bessel function of order $\frac{5}{3}$. The spectrum is parameterized by the 'critical energy'

$$\varepsilon_c = 3\frac{hc}{4\pi\rho}\gamma^3$$

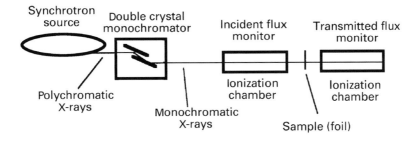

Figure 2 Schematic of a transmission XAFS experiment.

where ε is the X-ray energy, h is Planck's constant, and ρ is the bend radius [ρ_β (metres) $\sim 3.336\,E(GeV)/B(Tesla)$]. The root mean square angular width in the vertical direction can be approximated by $(0.57/\gamma)(\varepsilon_c/\varepsilon)^{0.43}$. **Figure 3** shows the universal spectral curve for bend magnets and planar wigglers (see below).

Second generation sources are constructed specifically to produce bend magnet radiation for experimental use. Third generation sources are optimized for the use of 'insertion devices' – magnetic structures inserted into the particle beam path that modify the trajectory so as to produce synchrotron radiation of the desired energy spectrum, spatial characteristics, and polarization. They are designed for low emittance, which is the product of the spread in momentum and the spread in position of the particle beam (the phase space volume). According to Liouville's theorem the emittance is conserved as the beam propagates through the dipole, quadrupole, and sextupole magnets used to guide and focus the particles.

Wigglers are arrays of alternating magnetic poles that apply an approximately sinusoidal magnetic field to the particles and cause their trajectory to oscillate. The spectrum and angular radiation pattern is similar to that of an array of bend magnets of alternating curvature, but with the advantages of higher flux owing to multiple magnetic poles, and the critical energy determined by experimental requirements rather than geometrical constraints.

Undulators are similar to wigglers in that they produce an alternating magnetic field that causes the particle trajectory to oscillate in an approximately sinusoidal manner. The tangent direction of the particle trajectory is kept within the intrinsic width of

the synchrotron radiation cone, which allows the X-rays emitted at each successive magnetic pole to interfere. This interference concentrates the energy into narrow energy bands and a narrow angular divergence in both horizontal and vertical directions. The amplitude of oscillation of the particles is characterized by the 'deflection parameter' $K = \gamma \delta_w$, where $\delta_w = \lambda_0/2\pi\rho_0$, λ_0 is the undulator period, and ρ_0 is the bend radius corresponding to the peak magnetic field. For small oscillations, there is a single peak in the spectrum at $2\gamma^2$ times the frequency of oscillation of the particle Ω_w, as measured in the inertial frame of the average particle velocity. Undulators in use for X-ray spectroscopy generally have sufficiently high fields that they have characteristics intermediate between an ideal undulator and wigglers. In this case the particle motion becomes relativistic even in its co-moving reference frame, and harmonics are generated. The X-ray frequency of the fundamental observed at an angle θ_0 is given approximately by

$$\frac{2\,\gamma^2\,\Omega_w}{1 + (K^2/2) + \gamma^2\,\theta_0{}^2}$$

This expression shows that the positions of peaks in the spectrum can be controlled by adjusting the deflection parameter K, by controlling the magnetic field presented to the particle beam. The energy width of undulator peaks is typically of the order of 100 eV, decreasing with the number of poles. The fluxes from even-order harmonics are of significantly lower amplitude, particularly on the undulator axis. **Figure 4** shows the spectrum from an APS type A undulator.

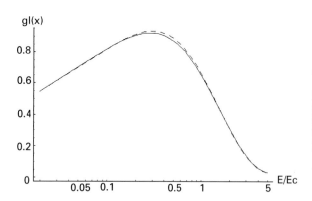

Figure 3 Synchrotron function $g\,\mathsf{l}\,(x)$ (solid) and simple approximation (dashes): $f(x) = 1.8\,x^{0.3}\exp(-x)$, where $x = \varepsilon/\varepsilon_c$. A more accurate approximation (not shown) is $g\,\mathsf{l}\,(x) = ax^b \exp(-cx)$ with $a = 1.71857$, $b = 0.281526$, $c = 0.968375$. The spectral photon flux (photons/sec/0.1% bandwidth $(\Delta\varepsilon/\varepsilon)$/mA beam current/mrad) integrated over the full vertical opening angle is $1.256 \times 10^7\,\gamma g\,\mathsf{l}\,[x]$, with $\gamma = E/mc^2$.

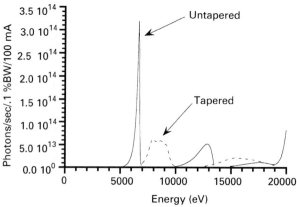

Figure 4 Integrated spectral flux for Advanced Photon Source (APS) undulator. The position of the peaks is adjustable by varying the undulator magnetic gap.

Optics

Monochromators

The desired energy bandwidth of approximately 1 eV is selected by allowing the beam to impinge at a selected angle θ onto a cooled single crystal of silicon, germanium, diamond, or other substance. The crystals reside on a goniometer inside a vacuum chamber or inert atmosphere to minimize ozone production and absorption, and the angle θ can be remotely scanned by a computer. X-rays that meet the Bragg diffraction condition $n\lambda = 2\,d_{hkl}\sin(\theta)$ are diffracted through an angle 2θ; the rest are absorbed. In this equation, λ is the X-ray wavelength, which is related to the photon energy $\varepsilon = hc/\lambda$; n is the harmonic number, and the spacing between diffracting atomic planes in the crystal for 'reflection' hkl is $d_{hkl} = a_0/(h^2+k^2+l^2)^{1/2}$, where a_0 is the lattice constant (0.5431 nm for Si). The crystals used are sufficiently perfect that they are well described by dynamical diffraction theory instead of kinematic theory. Some of the lower index allowed reflections are 111, 220, 311, 400, 331, 422, 333, 511, 440, and 531. Higher index crystals are used to obtain better energy resolution but at the expense of lower integrated reflectivity.

Normally, a parallel second crystal is placed after the first crystal to deflect the beam in a direction parallel to the incident beam direction, so that the X-ray beam angle is maintained constant as the energy is scanned. The first and second crystal faces can be formed from the same piece of silicon by cutting a channel in it, making a so-called channel-cut configuration. Alternatively, separate crystals can be mounted independently of each other. In such a double crystal Bragg monochromator, the beam is displaced by a distance $2H\cos(\theta)$ from the height of the incident beam, where H is the perpendicular separation of the crystals; consequently the beam height varies with energy, typically by less than a millimetre. The beam motion can be tracked by moving the sample and detectors under computer control; alternatively, in appropriately constructed monochromators, H can be adjusted to preserve a fixed beam height. A translation of the second crystal parallel to the first to keep the diffracted beam centred on the second crystal also may be beneficial. A fine adjustment of the relative orientation of first and second crystal is essential; this is accomplished with piezoelectric transducers or highly gear-reduced motors.

The 'rocking curve' – the reflectivity of the crystal versus θ for monochromatic X-rays – is approximately rectangular in shape, with a typical (energy dependent) width of 3–10 arc-seconds (25–50 microradians). The rocking curves have small, but long-range, tails that can degrade the energy resolution, and may distort X-ray absorption spectra in the near-edge region where there may be rapid changes in absorption with energy. The contribution of these tails can be reduced by using a second pair of crystals following the first pair, but at the cost of considerably greater instrumental complexity.

Focusing in the sagittal (usually horizontal) direction can be accomplished by bending the second crystal to an appropriate radius R, given by $2\sin(\theta)/R = (1/u+1/v)$, where u is the source to optic distance, and v is the optic to focal point distance. Substantial vertical (meridional) focusing cannot be achieved by bending the second crystal because doing so would cause the incidence angle of the beam on parts of the second crystal to fall outside of the rocking curve.

The high power density produced by undulator beams presents challenges for monochromator designers. The heat deposited in the first crystal creates a 'thermal bump' in the first crystal because the local heating creates thermal expansion in the silicon that degrades the rocking curves, and hence the resolution and throughput. The relevant parameter is the ratio of the thermal conductivity to the thermal expansion coefficient. Several approaches have been devised to deal with this problem. One is to cool the silicon with liquid nitrogen to a temperature around 100 K, which is beneficial because the thermal expansion coefficient is greatly reduced, and the thermal conductivity is increased. Another approach is to cut the crystal so the diffracting planes are at an inclined angle relative to the crystal face, so that the beam is spread over a larger area of the crystal. A third approach is to use diamond crystals instead of silicon, because the thermal conductivity of diamonds greatly exceeds that of silicon. A fourth approach is to use an X-ray mirror or synthetic multilayer as a pre-filter to reduce the power load on the first crystal.

Mirrors

X-ray mirrors are used for rejecting harmonics, focusing, power filtering, displacing the beam, and improving collimation of the beam incident on the first crystal in order to improve energy resolution.

At small angles of incidence (on the order of milliradians), X-rays are totally externally reflected from the surfaces of materials. This effect is the X-ray analogue of ordinary total *internal* reflection that is observed at visible wavelengths when looking from a dense medium into a less dense medium. For most materials, the index of refraction at X-ray energies is a complex number: $\tilde{n} = 1 - \delta - i\beta$, where $\delta = ne^2\lambda^2/$

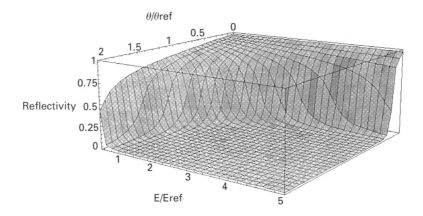

θ/θref

Figure 5 Mirror reflectivity as a function of angle and energy. θ_{ref} is the critical angle at (arbitrary) energy E_{ref}.

$2\pi mc^2$ and $\beta = \mu\lambda / 4\pi$. The real and imaginary parts describe dispersion and absorption, and are connected through a Kramers–Kronig transform. Here n is the number of dispersive electrons per volume of the material, e and m are the charge and mass of an electron, λ is the wavelength, and μ is the X-ray absorption coefficient. For elemental materials this reduces to $\delta = N(Z/A)\rho e^2\lambda^2/2\pi mc^2$ where N is Avogadro's number, Z is the atomic number, A is the atomic mass, and ρ is the mass density.

Total external reflection occurs at angles $\theta < \theta_c$, where the 'critical angle' $\theta_c = (2\delta)^{1/2}$. θ_c is approximately inversely proportional to the X-ray energy, as shown in **Figure 5**. This allows the experimenter to eliminate harmonics by selecting an angle so that the fundamental is reflected from the mirror, but the harmonics are not.

The reflectivity from a mirror can be expressed as a function of the reduced angle $\phi = \theta/\theta_c$ as

$$R(\phi) = \frac{(\phi - A)^2 + B^2}{(\phi + A)^2 + B^2}$$

where

$$2A^2 = \left[(\phi^2 - 1)^2 + \left(\frac{\beta}{\delta}\right)^2\right]^{1/2} + (\phi^2 - 1)$$

and

$$2B^2 = \left[(\phi^2 - 1)^2 + \left(\frac{\beta}{\delta}\right)^2\right]^{1/2} - (\phi^2 - 1)$$

For a given angle, materials or coatings of high atomic number have larger critical energies, which allows them to reflect X-rays at higher energies. The disadvantage of using high Z coatings are lower reflectivity and a less sharp cutoff of reflectivity against energy. This is a consequence of absorption in the material, the effect of which is shown in **Figure 6**. The product of energy and θ_c is an intrinsic property of the material coating: representative measured values (in KeV mrad, ±2%) are Si (31), Ni (59), Pd (62), Rh (67), Pt (82), Au (80).

X-ray mirrors are fabricated from polished glass (float glass, ultra low thermal expansion titanium silicate), silicon, silicon carbide, appropriate ceramics, or metal substrates. Typically they are tens to hundreds of centimeters in length in order to accept the vertical divergence of the beam. Surface roughness degrades the reflectivity, and accordingly the best mirrors are highly polished to as little as ~0.2 nm root mean square roughness. With present technology, ångstrom-level roughness and microradian rms slope errors are achievable over mirror lengths of more than 1 m. Meridional focusing in the vertical direction can be accomplished by bending a short mirror to an appropriate radius, given by $2/(R\theta) = (1/u+1/v)$, where u is the source to optic distance, and v is the optic to focal point

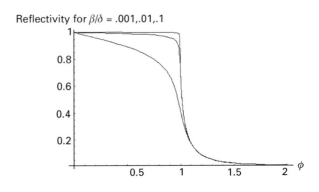

Reflectivity for β/δ = .001,.01,.1

Figure 6 Effect of X-ray absorption from mirror coatings on reflectivity. See text for explanation.

distance. Longer mirrors can be used provided the local radius of curvature at each point on the mirror satisfies this focusing equation.

Detectors

Ionization chambers are the most commonly used detectors for X-ray absorption spectroscopy. They have the virtue of being partially transparent so that the incident beam intensity can be monitored. Typically they consist of a pair of parallel conducting plates several cm in each dimension and separation approximately 1 cm, inside a gas-tight housing in which is placed an appropriate gas (e.g. helium, nitrogen, argon, krypton). The gas can be flowed through the chamber, or sealed inside under positive, negative or ambient pressure. A constant potential of hundreds to thousands of volts is applied between the plates using a high voltage power supply or battery, and the small current flowing between the plates through the gas between them is measured. The X-ray beam is allowed to pass between the plates, which ionizes the gas molecules, rendering the 'fill-gas' partially conductive. The current (microamperes to nanoamperes) is proportional to the photon flux, given approximately as $N\,(E/E_{cc})\,(1-\exp(-\mu_{fg}(E)l))$, where N is the number of photons per second of energy E, E_{cc} is the mean energy required to produce a charge carrier in the gas (typically around 32 eV), μ_{fg} is the absorption coefficient of the fill gas (or mixture), and l is the active length of the plates. The current is amplified with a transconductance (current to voltage) amplifier, and read by a computer with an analogue to digital converter; or it is converted to pulses with a voltage to frequency (V/F) converter and counted by a scaler.

Absorption within an ionization chamber is controllable by selection of fill-gas composition and pressure. For reliable absorption measurements, ionization chambers must be carefully constructed and operated at sufficiently high bias voltage that they are linear, i.e. the output current is linearly related to the absorbed photon flux. Under these conditions the chamber is said to be in the 'plateau region' of the flux versus bias voltage curve.

PIN diodes (positive-intrinsic-negative) are semiconductor devices that act essentially as solid state ion chambers. X-rays absorbed by the diodes create electron hole pairs that act as charge carriers. The electric field acting to separate them in the intrinsic (undoped) region is produced by the adjacent positively and negatively doped regions. The charge collected is amplified in the same manner as in an ionization chamber. A bias voltage can also be applied to alter the operational characteristics. PIN diodes are capable of excellent linearity but the X-ray thickness is not as easily experimentally controllable as it is for ionization chambers.

Ionization chambers with X-ray transparent plates made of aluminized plastic or thin metallic mesh are used for detection of fluorescence radiation. Detection of the fluorescence from dilute species requires a means of rejecting elastically scattered background from the sample. X-ray filters and slits used with ionization chambers are a standard method of rejecting background. Alternatively, arrays of solid state detector elements with appropriate electronics can provide useful energy resolution and adequate count rates for many purposes. Eliminating the scattered radiation with synthetic multilayer or crystal analysers is a promising approach for third generation synchrotron radiation sources.

List of symbols

E = particle beam energy; $I(I_0)$ = transmitted (incident) X-ray intensities; K = deflection parameter; u = source to optic distance; v = optic to focal point distance; $\gamma = E/mc^2$; ε = X-ray photon energy; ε_c = critical energy; θ = angle of incidence on an optical surface; λ_0 = undulator period; λ = photon wavelength; $\rho(\rho_0)$ = bend radius (bend radius at peak magnetic field).

See also: **Light Sources and Optics; X-Ray Fluorescence Spectrometers; X-Ray Fluorescence Spectroscopy, Applications.**

Further reading

Creagh DC and Hubbel JH (1987) Problems associated with the measurement of X-ray attenuation coefficients. I. Silicon. Report on the International Union of Crystallography X-ray Attenuation Project. *Acta Crystallographica* A43: 102–112.

Heald SM (1988) EXAFS with synchrotron radiation. In: Koningsberger D and Prins R (eds) *X-ray Absorption: Principles, Applications, Techniques of EXAFS, SEXAFS and XANES*. New York: John Wiley.

Koningsberger DC (1988) Laboratory EXAFS facilities. In: Koningsberger D and Prins R (eds) *X-ray Absorption: Principles, Applications, Techniques of EXAFS, SEXAFS and XANES*. New York: Wiley.

Knoll G (1989) *Radiation Detection and Measurement* 2nd edn. New York: Wiley.

Krinsky S (1983) Characteristics of synchrotron radiation and its sources. In: Koch EE (ed) *Handbook on Synchrotron Radiation*. Amsterdam: North-Holland.

Matsushita S and Hashizume H (1983) X-ray monochromators. In: Koch, EE (ed) *Handbook on Synchrotron Radiation*. Amsterdam: North-Holland.

X-Ray Diffraction in Materials Science

See Materials Science Applications of X-Ray Diffraction.

X-Ray Diffraction of Fibres and Films

See Fibres and Films Studied Using X-Ray Diffraction.

X-Ray Diffraction of Inorganic Compounds

See Inorganic Compounds and Minerals Studied Using X-Ray Diffraction.

X-Ray Diffraction of Minerals

See Inorganic Compounds and Minerals Studied Using X-Ray Diffraction.

X-Ray Diffraction of Powders

See Powder X-Ray Diffraction, Applications.

X-Ray Diffraction of Small Molecules

See Small Molecule Applications of X-Ray Diffraction.

X-Ray Emission Spectroscopy, Applications

George N Dolenko, Lermontova 35A/16, 664033
Irkutsk, Russia
Oleg Kh Poleshchuk, Tomsk Pedagogical University,
Tomsk, Russia
Jolanta N Latosińska, Adam Mickiewicz University,
Poznań, Poland

HIGH ENERGY SPECTROSCOPY
Applications

MO structure investigation

Any variations in the composition, structure, stereochemistry or coordination character of a molecule change its chemical properties and MO (molecular orbital) structure. MO changes are clearly observed by the fine structure of XFS (X-ray fluorescence spectroscopy). This makes it possible to relate some features of the chemical behaviour of compounds to their electronic structure and opens a way to various 'chemical property–electronic structure parameter' correlations which are frequently of help for explaining and predicting the chemical properties of compounds.

The transitions from valence atomic levels to vacancies in inner shells form X-ray valence emission lines, reflecting the structure of the valence levels or zones. Electron transitions between different inner levels form inner X-ray emission lines. The study of the fine structure of different valence emission lines of all the atoms in a compound allows detailed investigation of the structure of valence levels or zones. Research into the shifts of inner X-ray emission lines allows one to investigate effective charges on the corresponding atoms.

For example, consider the X-ray emission spectra of sulfur. The initial state of a sulfur atom for X-ray emission is that with a vacancy in the K or $L_{2,3}$ level. This vacancy is rapidly filled (within $10^{-16} - 10^{-14}$ s) as a result of transitions obeying the dipole selection rules, i.e. $2p \rightarrow 1s$ (Kα lines), $3p \rightarrow 1s$ (Kβ lines) or $3s \rightarrow 2p$, $3d \rightarrow 2p$ ($L_{2,3}$ lines) transitions. The energy released in this case is emitted from the atom as either an Auger electron or an X-ray quantum (**Figure 1**).

Whereas SKα are inner lines, SKβ and SL$_{2,3}$ are valence lines. The energies of atomic $np \rightarrow 1s$ transitions can be represented by the equations

$$h\nu = E_{\text{fin}} - E_{\text{init}} = E^{-np} - E^{-1s} \qquad [1]$$

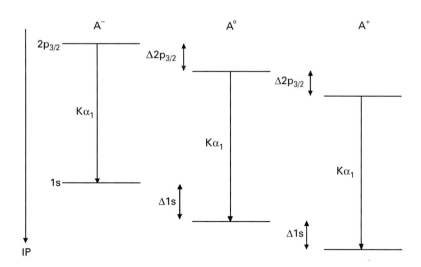

Figure 1 Scheme showing the change of one-electron energies of 1s, 2p levels and the energy of the A Kα line in the ions A$^+$ and A$^-$ with respect to the neutral atom A^0.

$$\Delta h\nu = \Delta E^{-\mathrm{np}} - \Delta E^{-1\mathrm{s}}$$
$$= \Delta E^{-\mathrm{np}} \approx \Delta(E^0 - \varepsilon_{\mathrm{np}}) = -\Delta\varepsilon_{\mathrm{np}} \qquad [2]$$

where $h\nu$ is the energy of the emission quantum, E_{fin} is the final energy of the system, E_{init} is the initial energy of the system, $E^{-\mathrm{nl}}$ is the energy of the system with an nl electron removed and $\varepsilon_{\mathrm{np}}$ is the one-electron energy of the np level.

Thus, in a one-electron approximation the distance between individual maxima in a spectral series is equal to the difference in one-electron energies of the corresponding atomic levels. In molecules the 3s, 3p and 3d electrons of the sulfur atom are involved in chemical bonding to form an MO system. In this case, the SKβ spectrum (S3p $\rightarrow$ S1s interatomic transitions), for example, corresponds to MO$_i \rightarrow$ S 1s transitions, and the distances between spectral maxima correspond to energy differences of the appropriate occupied molecular levels:

$$\Delta E_{ij}(\mathrm{SK\beta}) = \varepsilon(\mathrm{MO}_j) - \varepsilon(\mathrm{MO}_i) \qquad [3]$$

The intensity of X-ray emission lines is determined by the relation (for the SK series as an example):

$$I(\mathrm{Snp} \rightarrow \mathrm{S1s}) \sim N_{\mathrm{Snp}} E_1^3 \left| \langle \Phi_{\mathrm{Snp}} | r | \Phi_{\mathrm{S1s}} \rangle \right|^2 \qquad [4]$$

where N_{Snp} is the np level population of the sulfur atom, E_1 is the energy of Snp $\rightarrow$ S1s transitions, Φ_{Snp} is the wavefunction of sulfur np orbitals and Φ_{S1s} is the wavefunction of the sulfur 1s orbital.

For molecules this expression is transformed to the equation

$$I(\mathrm{MO}_i \rightarrow \mathrm{S1s}) \sim N_i E_1^3 \left| \langle \Psi^i | r | \Phi_{\mathrm{S1s}} \rangle \right|^2 \qquad [5]$$

where Ψ^i is the wavefunction of ith MO.

In the generally accepted MO LCAO model the wavefunction of any ith MO can be represented as an AO sum:

$$\Psi^i = \sum_j c_j^i \Phi_j \qquad [6]$$

where Φ_j is the wavefunction of jth AO and c_j^i are the coefficients.

Then, with account taken of the dipole selection rules, Equation [5] is transformed into

$$I(\mathrm{MO}_i \rightarrow \mathrm{S1s}) \sim N_i E_1^3 \left| \left\langle \sum_j c_j^i \Phi_j | r | \Phi_{\mathrm{S1s}} \right\rangle \right|^2$$
$$= N_i E_1^3 (c^i)^2 \left| \langle \Phi_{\mathrm{S3p}} | r | \Phi_{\mathrm{S1s}} \rangle \right|^2 \qquad [7]$$

where c^i is the coefficient of Ψ^i at the wavefunction of S3p AO (Φ_{S3p}). Thus, the ratio of partial intensities in the SKβ spectrum of a molecule is equal to that of the squares of the coefficients of the S3p wavefunction in the expansion of the corresponding MOs in terms of AOs:

$$I_i : I_j : I_k = (c^i)^2 : (c^j)^2 : (c^k)^2 \qquad [8]$$

The above is also true for other X-ray emission series. Important features of X-ray emission spectra are the comparative ease of interpretation and the possibility of investigating the electronic structure from the 'viewpoint' of any atom of the molecule investigated.

Example: electronic structure of the sulfate ion

Information concerning the electronic structure of molecules provided by XFS can be well illustrated with the sulfate ion as an example. The wavefunction of any ith valent MO of the sulfate ion can be described by the equation

$$\Psi^i = c_1^i \Phi_{\mathrm{S3s}} + c_2^i \Phi_{\mathrm{S3p}} + c_3^i \Phi_{\mathrm{S3d}} + c_4^i \Phi_{\mathrm{O2s}}$$
$$+ c_5^i \Phi_{\mathrm{O2p}} \qquad [9]$$

All possible X-ray fluorescence spectra of the sulfate ion are presented in **Figure 2** whereas **Figure 3** shows an MO diagram constructed from a full set of these spectra. 'Adjustment' to the scale of the ionization potentials [IP] of valence electrons is effected by subtracting the X-ray transition energies from the IPs of the corresponding inner levels (S1s, S2p, O1s) determined by the use of data of X-ray photoelectron spectroscopy:

$$E_i(\mathrm{SK\beta}) = \mathrm{IP}(\mathrm{S1s}) - \mathrm{IP}(\mathrm{MO}_i) \qquad [10]$$

$$E_j(\mathrm{SL}_{2,3}) = \mathrm{IP}(\mathrm{S2p}) - \mathrm{IP}(\mathrm{MO}_j) \qquad [11]$$

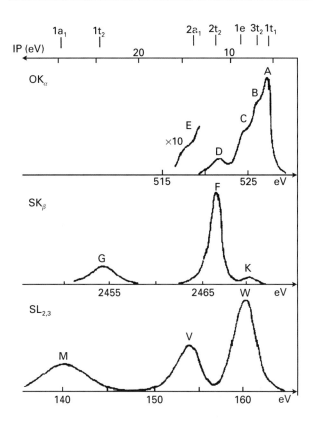

Figure 2 Full set of X-ray fluorescence spectra of the sulfate ion on an energy scale corresponding to the ionization potentials of valence electrons.

$$E_k(OK\alpha) = IP(O1s) - IP(MO_k) \qquad [12]$$

From these, the following equations are derived

$$
\begin{aligned}
IP(MO_i) &= IP(S1s) - E_i(SK\beta) \\
&= IP(S2p) + E(SK\alpha) - E_i(SK\beta) \qquad [13]
\end{aligned}
$$

$$IP(MO_j) = IP(S2p) - E_j(SL_{2,3}) \qquad [14]$$

$$IP(MO_k) = IP(O1s) - E_k(OK\alpha) \qquad [15]$$

All MOs with $c_5 \neq 0$ are displayed in the O Kα spectrum (O2p → O1s transitions), those with $c_2 \neq 0$ in the SKβ spectrum (S3p → S1s transitions) and those with $c_1 \neq 0$ and $c_3 \neq 0$ in the SL$_{2,3}$ spectrum (S3s → S2p and S3d → S2p transitions). From the spectra shown in **Figure 3** it follows that the highest occupied MO 1t$_1$ (maximum A in the OKα spectrum) consists of only O2p electrons; next, the 3t$_2$ and 1e levels (maxima B, C and W) significantly correspond to the π bond S3d – O2p; the 2t$_2$ level then follows (maxima D and F), which corresponds to the

strong S3p – O2p σ bond. The 2a$_1$ level (maxima E and V), consisting of the S3s AO and, possibly, the O2s with a small admixture of O2p AO, lies even deeper. Deep 1t$_2$ and 1a$_1$ MOs consisting mainly of the O2s AO are seen as low intensity long-wavelength maxima G and M respectively.

Consequently, much information on the MO structure of a chemical species under investigation can be derived from X-ray emission spectra.

Determination of the effective atomic charges

The concept of the effective charge on atoms in molecules is known to be fundamental in the field of theoretical chemistry. It assumes that the entire electron distribution of an investigated atom can be considered as a point charge that coincides with the coordinates of the nucleus. This simple and obvious form of the electron density distribution in the examined species is rather approximate: in real molecules the outer electron shells of atoms substantially lose their 'individuality' because of the delocalization of valence electrons of atoms. The approximation procedure (the replacement of the real distribution of the outer electron density of an atom by the point charge on its nucleus) is not simple in general and depends largely on the actual definition of effective charge. A number of calculations and experimental methods used for the determination of the latter are known. The effective charges on atoms (q_A, where A is an element) do not belong to the class of directly observed physical characteristics, and therefore the so-called 'experimental determination' of q_A values usually means the result of the interpretation of various experimental data in terms of the corresponding model with q_A as a parameter.

Shifts of the energy of inner nl levels of the A atom (ΔAnl), defined by X-ray photoelectron spectroscopy, are sensitive to q_A and the effective charges of all other atoms, and in terms of the so-called potential model can be written as

$$\Delta Anl = k(Anl)q_A + \sum_{i \neq A} \frac{q_i}{r_{iA}} + \Delta E_r \qquad [16]$$

where K(Anl) is the coefficient that is characteristic of the A nl level, $\sum_{i \neq A}(q_i/r_{iA})$ is the Madelung potential and E_r is the relaxation energy. Shifts of inner AKα_1 line (2p$_{3/2}$ → 1s electron transitions are more intense than for 2p$_{1/2}$ → 1s) are determined by the difference between the shifts of one-electron energies

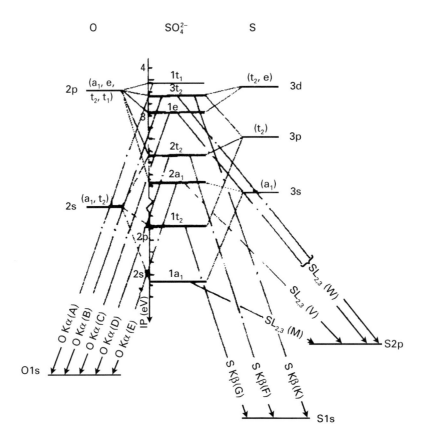

Figure 3 Scheme of the sulfate ion MOs obtained from the full set of X-ray fluorescence spectra.

of A1s and A2p$_{3/2}$ levels:

$$\Delta AK\alpha_{3/2} = (E^{-2p} - E^{-1s}) - (E_0^{-2p} - E_0^{-1s})$$
$$= \Delta E^{-2p} - \Delta E^{-1s} \approx \Delta A1s - \Delta A2p_{3/2} \quad [17]$$

where E_0^{-nl} is the total energy of the neutral A atom (A^0) containing a vacancy in the nl level.

The difference in the energy changes for different Anl levels is rather small and therefore the $\Delta AK\alpha_1$ values are determined by subtraction of two large and very similar quantities, $\Delta A1s$ and $\Delta A2p_{3/2}$ (the energy shift of the A1s level always prevails over that of A2p$_{3/2}$). As a result, the $\Delta AK\alpha_1$ values are normally about ten times smaller than X-ray photoelectron Anl shifts. However, AKα_1 shifts have an obvious advantage. The corresponding electron transitions are localized in the same potential well as that created by the Coulomb field of other atoms of the system investigated. Hence, the potential model for AKα shifts, analogous to Equation [16], can be written as

$$\Delta AK\alpha_1 = k_A q_A \quad [18]$$

or, in a more universal form

$$\Delta AK\alpha_1 = f(q_A) \quad [19]$$

Many authors have investigated the dependencies of Equation [19] for different inner X-ray emission lines and atoms by different methods. **Table 1** gives the AKα_1 shifts for some phosphorus-, sulfur- and chlorine-containing compounds and q_A values obtained by the correlation of Mulliken charges calculated by the SCF *ab initio* method using a 4-31G** basis set for a sufficiently large series of A-containing molecules with the experimental AKα_1 shifts.

Participation of the 3d atomic orbitals in L-emission spectra

From the X-ray L-emission spectra of S, Cl, P, within the framework of a MO method, one can estimate the 3d-population. The basic problem is the need to calculate the matrix elements of the transitions 2p → 3s and 2p → 3d. For Kβ spectra of elements in period 3 in the dipole approximation, transitions are permitted from levels involving 3pAO, from which

Table 1 Experimental AKα₁ shift and q_A values

A	Class of compounds	$\Delta AK\alpha_1$ (eV) relative to P_{red}, S_8, Cl_2	$q_A(e)$ in 4-31G** charge scale
P	R_3P	−0.01 − 0.25	−0.03 − 0.25
	R_4P^+	0.16 − 0.38	0.32 − 0.76
	$P(OR)_3$	0.48 − 0.52	0.97 − 1.05
	R_3PO	0.27 − 0.57	0.54 − 1.15
	R_3PS	0.17 − 0.24	0.34 − 0.48
	$(RO)_3PO$	0.69 − 0.76	1.39 − 1.53
	PO_4^{3-}	0.76 − 0.80	1.54 − 1.61
S	$=S$	−0.14 − −0.02	−0.22 − 0.09
	RSH	−0.08 − 0.00	−0.11 − 0.05
	R_2S	−0.09 − 0.14	−0.18 − −0.32
	R_3S^+	0.00 − 0.12	0.05 − 0.28
	RNSNR′	0.18 − 0.25	0.40 − 0.54
	R_2SO	0.36 − 0.42	0.75 − 0.87
	R_2SO_2	0.78 − 0.85	1.57 − 1.61
	SO_4^{2-}	1.00 − 1.20	2.00 − 2.40
Cl	RCl	−0.19 − 0.01	−0.28 − 0.02

the MO are constructed. The relative intensity of separate emission lines is proportional to the squares of the factors c_i^2. In L-emission spectra of atoms in period 3 the MO coefficients owing to dipole rules of selection ($\Delta l = \pm 1$), will be simultaneously displayed MO, which are constructed with participation of 3s AO and 3d AO. If the symmetry of a molecule is such that the MO of the systems can be constructed with simultaneous participation of 3s- and 3d AOs of a period 3 atom the intensity of the emission line will depend on the contributions of both the 3d and the 3s AOs to the appropriate MO. Thus estimations of c_i^2 values with 3d AO and 3s AO from X-ray spectra need a knowledge of matrix elements of transitions $|\langle 2p|r|3d\rangle|$, $|\langle 2d|r|3s\rangle|$. The determination of such matrix elements becomes complicated by the problem of choosing a 'good' 3d wavefunction. It is known that the atomic 3d functions are too diffuse and their electronic density, appropriate to them, is located far from the nuclei of atoms and to be unsuitable for participation in chemical bonds. In the case of X-ray transitions the important behaviour of 3d wavefunctions is not in the area of valence electrons but in the field of core electrons of atoms, in particular in the area of 2p AO. The 2p AO wavefunctions are located near the nucleus of an atom and have atomic character. It is possible to consider that 3d AO in this area also has atomic character. On this basis the estimation of matrix elements $|\langle 2p|r|3d\rangle|$ and $|\langle 2p|r|3s\rangle|$ is carried out to account for the intensity of X-ray atomic transitions.

The analysis of spectra of molecules and ions shows that the short-wave maximum W in L-spectra of S and Cl (**Figure 2**) is basically connected with an MO, in which there is a significant contribution from the 3d AO, while the contributions of the 3s AO to these MOs are insignificant. The maxima V and M are connected with an MO in which the 3s AO participates. Hence, from L-spectra it is possible to obtain experimental values of the relative intensity of various lines and to determine the contribution of AO and MO.

Estimations for the ion SO_4^{2-} and the molecule SF_6 give I_{3d}/I_{3s} values that correspond to theoretical results. The consideration of results from sulfur- and chlorine-containing compounds indicate that the participation of 3d AO in various MO becomes appreciable. The study of shifts to Kα lines over a range of molecules shows that the experimental relation I 3d/I 3s increases as the Kα shift grows, physically this is connected to the growth of a positive charge on an atom of sulfur. **Table 2** gives the relative experimental 3d occupations for sulfur in some compounds.

Application of SKα spectra to characteristic compounds

It is known that the energy of valence electrons of heteroatoms in periods 2 and 3 varies linearly with changes in the energy of their core electrons. The concept of an energy level of a hypothetical electron lone pair of a sulfur atom (hnS), whose energy depends only on the charge on the sulfur atom (qS), has been suggested. The position of this level in the SKβ spectrum was related to the values of the SKα shift, which are proportional to the net charge on the sulfur atom. The following relationship was

Table 2 Relative 3d sulfur population from X-ray spectral data

Molecule	I_{3d}/I_{3s}	$\Delta E K\alpha_{1,2}$ (eV)	q_s	$|\langle 2p|r|3d\rangle|/|\langle 2p|r|3s\rangle|$
$(CH_3)_2SO$	0.2	0.25	1.0	0.1
Cl_2SO	0.3	0.27	1.2	0.3
$(C_6H_5)_2SO$	0.4	0.31	1.2	0.3
SO_2	0.7	0.45	1.5	0.6
$(CH_3)_2SO_2$	0.8	0.75	1.9	0.9
SF_4	1.0	0.75	1.9	0.9
SO_3^{2-}	1.0	0.65	1.6	0.7
$(C_6H_5)_2SO_2$	1.2	0.82	2.1	1.1
SF_6	1.0	1.45	2.6	1.4
SO_4^{2-}	1.8	1.10	2.3	1.2

obtained by comparing the short-wave maximum in the SK spectra of saturated sulfides with the corresponding values of $\Delta(SK\alpha)$: $hn_S(K\beta)(eV) = E(n_S \rightarrow 1S_S) = 0.0056 \times 10^3 \Delta SK\alpha \ (eV) + 2468.37$, with $r = 0.973$, $s = 0.06$, $n = 26$.

Using the equation it is possible, knowing $\Delta SK\alpha$ values to predict the location of a hnS level in the SKβ spectra of any saturated sulfide compound. In fact, this level can be treated as a reference level in the analysis of the changes in the spectral structure caused only by orbital interactions devoid of the effect of charge changes on the sulfur atom. As an example one can consider the application for complex compounds with dimethyl sulfide.

It follows from **Figure 4** that the intensity of the short-wave maximum A, which in the SKβ spectra of sulfides corresponds to the transition from the nS level to the vacancy K of the sulfur atom, significantly decreases and is considerably shifted towards longer wavelength with respect to the hnS level (Kβ). This shift, $(\Delta n_S = E_A(SK\beta) - hn_S(SK\beta)$ (**Table 3**), characterizes quantitatively the bonding nature of the highest occupied molecular orbital. The observed shift towards longer wavelength indicates that the nS level interacts with the vacant levels of the acceptor, being mainly of the d type. The differences in shapes of the SKβ spectra of the complexes studied (**Figure 4**) can be explained by the presence of partly populated valence d orbitals, apart from the vacant ones, in Ti, in contrast to Sn and Sb.

Application of SK β and SKα spectra for rodano-group

It is known that the NCS-group in compounds can be coordinated with a metal in three ways: M–NCS (a), M–SCN (b) and M–NCS–M (c). Inorganic thiocyanates with coordination of type (a) have a characteristically large (negative) total electronic density on the sulfur atom, lower intensity of long-wave maxima and lower energy of a short-wave maximum (**Figure 5**). In organic isothiocyanates the $\Delta SK\alpha$ values vary in an interval -7400–5800 eV with the intense short-wave maximum A in the SKβ spectra

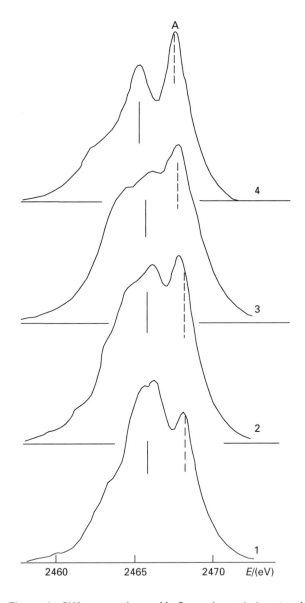

Figure 4   SKβ spectra of some Me$_2$S complexes. (—) centre of gravity; (---) $hn_S(K\beta)$. 1 = TiCl$_4$ · 2SMe$_2$; 2 = SbCl$_5$ · SMe$_2$; 3 = SnCl$_4$ · 2SMe$_2$; 4 = SMe$_2$.

(E_A) in the range $-2467.1 - 2467.8$ eV. In organic thiocyanates these parameters become $\Delta SK\alpha = 4.7 - 13.3$ eV 10^2, $E_A = 2468.1 - 2469.1$ eV (**Table 4**).

Table 3   Parameters determined from X-ray spectra of the sulfur atoms for some of the complexes studied

Compound	$\Delta(SK\alpha)^a$ 10^{-3}(eV)	q_s (e) in 4–31G** charge scale	$hn_s(K\beta)$ (eV)	$E_A(K\beta)$ (eV)	Δn_s (eV)
(CH$_3$)$_2$S	$-63(6)^b$	$-0.07(2)$	8.02(4)	8.1(1)	$-0.1(1)$
SnCl$_4$ · 2S(CH$_3$)$_2$	2(14)	0.05(2)	8.38(8)	8.2(1)	0.2(1)
SbCl$_5$ · S(CH$_3$)$_2$	17(7)	0.08(2)	8.47(5)	8.3(1)	0.2(1)
TiCl$_4$ · 2S(CH$_3$)$_2$	$-3(10)$	0.04(2)	8.35(6)	8.11(5)	0.24(8)

^{a} Relative to S$_8$.

^{b} The mean-square errors in the last significant digit, taken for 95% confidence interval by Student's criterion, are given in parentheses.

Table 4 X-ray spectral character of some thiocyanates and isothiocyanates

No.	Compound	$\Delta(SK\alpha)* 10^{-2}$ (eV)	q_s (e) in 4–31G** charge scale	$E_A(K\beta)$ (eV)	hn_s (Kβ) (eV) relative to 2467.0 eV	Δn_s (eV)	Coordination type
(I)	KNCS	−4.5(12)	−0.04(3)	0.2	1.1	−0.9	a
(II)	NH₄NCS	−9.2(6)	−0.13(2)	0.3	0.9	−0.6	a
(III)	Ba(NCS)₂	−3.6(10)	−0.02(2)	0.2	1.2	−1.0	a
(IV)	Sn(NCS)₂	−7.2(8)	−0.09(2)	0.3	1.0	−0.7	a
(V)	CuSCN	4.1(8)	0.13(2)	2.0	1.6	0.4	b
(VI)	CH₃SCN	4.7(8)	0.14(2)	1.8	1.7	0.1	b
(VII)	C₆H₅CH₂SCN	5.4(11)	0.16(3)	1.6	1.7	−0.1	b
(VIII)	C₆F₅SCN	9.4(8)	0.23(2)	1.9	1.9	0.0	b
(IX)	C₈H₄OPhSCN	13.3(9)	0.31(3)	2.1	2.1	0.0	b
(X)	C₆H₅SCN	5.0(8)	0.15(2)	1.7	1.7	0.0	b
(XI)	CH₃NCS	−4.7(8)	−0.04(2)	0.5	1.1	−0.6	a
(XII)	P(NCS)₃	0(2)	0.05(2)	0.2	1.4	−1.2	a
(XIII)	C₆F₅(NCS)₂	−7.4(6)	−0.09(2)	0.1	1.0	−0.9	a
(XIV)	C₈H₄OPhNCS	5.8(9)	0.16(2)	0.8	1.7	−0.9	a
(XV)	C₆H₅NCS	−5.3(7)	−0.05(2)	0.3	1.1	−0.8	a

The mean-square errors in the last significant digit, taken for 95% confidence interval by Student's criterion, are given in parentheses.
[a] Relative to S_8.

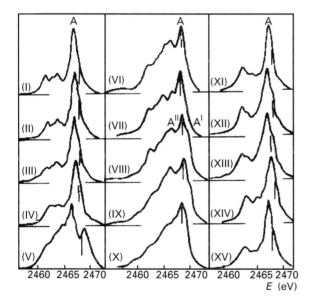

Figure 5 SKβ spectra of some inorganic and organic thiocyanates and isothiocyanates. The compound numbers are defined in **Table 4**. The energy levels of the hypothetical lone electron pair are marked by the vertical lines.

However, the ΔnS values of thiocyanates and isothiocyanates are divided almost unequivocally: ΔnS ≥ 0 for thiocyanates, ΔnS < 0 for isothiocyanates (in thiocyanates $3p_\pi$− electronic density of atom of sulfur enters into a strongly bonding MO).

From the presence in the SKβ spectra of thiocyanates (VI), (VII) of a single short-wave maxima A that is coincident with the level hnS, it follows that the level nS does not practically couple with the $\pi_{C\equiv N}$ orbitals. The existence of the advanced short-wave structure in SKβ spectra thiocyanates (VIII)–(X), beyond the hnS level, must relate to n_S–π_{Ar} interactions. Thus, the spectra of thiocyanates (VIII)–(X) indicates two orthogonal conformations of the molecules, one in which n_S–π_{Ar} interactions are absent giving rise to a peak coincident with the hnS level, and another in which the n_S and π_{Ar} couple to give two maxima, A' and A", corresponding to levels $n_S \pm \pi_{Ar}$.

Applications of Kα shifts for electronic density redistribution

It was of interest to use the data obtained for the investigation of the electron density redistribution on complexation between a donor molecule PCl₃ or of SPCl₃ (where one can define the effective charges on all atoms by their Kα shifts) and an acceptor molecule AlBr₃ containing no interfering atoms. The data presented in **Table 5** show that, in spite of a positive q_P growth on complexation, the total ligand electron density does not decrease (in the range of accuracy achieved) due to a strong dampening effect of the Cl atoms. This leads to a sufficient growth of

Table 5 Change of atomic charges and bond ionicities of free ligands on their complexation

Compound	ΔA Kα[a] (eV)			Ionicities of bonds[b] (%)		Change of effective atomic charge $(\delta q_A)^c$ of ligands on their complexation in 4-31G** charge scale			
	A = P	A = S	A = Cl	P–S	P–Cl	A = P	A = S	A = Cl	Σ δqi
PCl₃	0.373(7)	–	–0.09(1)	–	44(5)				
AlBr₃PCl₃	0.599(8)	–	–0.156(15)	–	72(7)	0.5(1)	–	–0.10(7)	0.2(2)
SPCl₃	0.430(8)	–0.10(2)	–0.11(2)	50(6)	51(6)				
AlBr₃SPCl₃	0.616(8)	–0.080(9)	–0.171(14)	68(7)	74(8)	0.4(1)	0.04(6)	–0.1(1)	0.1(2)

[a] The mean-square errors in the last significant digit, taken for 95% confidence interval by Student's criterion, are given in parentheses.
[b] The ionicity of the AB bond is equal to $(|q_A-q_B|/2)100\%$.
[c] The positive sign of δq_A corresponds to a decrease of the A atom electron density on complex formation.

the ionicity of all bonds of the ligands and acceptor. From **Table 5** it also follows that the positive charge on the central acceptor atom grows sufficiently on complexation while the electron density on acceptor geminal of atoms does not decrease.

List of symbols

c^i = coefficient of Ψ^i; E_0^{-nl} = total energy of neutral atom A with a vacancy at the nl level; E_{init} = initial energy of system; E_{fin} = final energy of system; E_r = relaxation energy; $h\nu$ = energy of emission quantum; I = intensity of X-ray emission lines; N_{Snp} = np level population of sulfur atom; q_A = effective charge on atom A; ε_{np} = one-electron energy of np level; Φ_{S1s} = wavefunction of the sulfur 1s orbital; Ψ^i = wavefunction of the ith MO.

See also: **X-Ray Emission Spectroscopy, Methods; X-Ray Fluorescence Spectrometers; X-Ray Fluorescence Spectroscopy, Applications.**

Further reading

Dolenko GN (1993) X-ray determination of effective charges on sulphur, phosphorus and chlorine atoms. *Journal of Molecular Structure* **291**: 23–57.

Dolenko GN, Litvin AL, Elin VP and Poleshchuk OKh, (1991) X-ray investigation of electron density redistribution on complexation. *Journal of Molecular Structure* **251**: 11–27.

Dolenko GN, Latajka Z and Ratajczak H (1995) X-ray spectral determination of the effective charges on P, S, and Cl atoms in chemical compounds with a nonempirical charge scale. *Heteroatom Chemistry* **6**: 553–557.

Dolenko GN, Poleshchuk OKh and Koput J (1998) Antimonium pentachloride electron density redistribution on complexation. *Heteroatom Chemistry* **9**: 543–548.

Mazalov LN and Yumatov VD (1984) *Electronnoe stroenie ekstragentov*. Novosobirsk: Nauka, 199 p.

Nogaj B, Poleshchuk OKh, Kasprzak J, Koput J, ElinVP, Dolenko GN (1997)Changes in electron density distribution resulting from formation of antimony pentachloride complexes studied by X-ray fluorescence spectroscopy. *Journal of Molecular Structure* **406**: 145–151.

Poleshchuk OKh, Nogaj B, Dolenko GN and Elin VP (1993) Electron density redistribution on complexation in non-transition element complexes. *Journal of Molecular Structure* **297**: 295–312.

Poleshchuk OKh, Nogaj B, Kasprzak J, Koput J, Dolenko GN, Elin VP, Ivanovskii AL (1994) Investigation of the electronic structure of SnCl₄L₂, TiCl₄L₂ and SbCl₅L complexes by X-ray fluorescence spectroscopy. *Journal of Molecular Structure* **324**: 215–222.

X-Ray Emission Spectroscopy, Methods

George N Dolenko, Lermontova 325A/16, 664033
Irkutsk, Russia
Oleg Kh Poleshchuk, Tomsk Pedagogical University,
Tomsk, Russia
Jolanta N Latosińska, Adam Mickiewicz University,
Poznań, Poland

Copyright © 1999 Academic Press

HIGH ENERGY SPECTROSCOPY
Methods & Instrumentation

General characteristics

Recently, intensive development of the theory of the electronic structure of chemical compounds has revealed a great need for physical experimental methods to which modern methods of quantum chemistry can be applied. Primary information provided by quantum chemical methods of electronic structure studies concerns the energy spectrum of molecules as well as the structures of wavefunctions of molecular orbitals (MOs). There is a need for physical methods which would allow direct measurement of MO energies as well as the determination of the degree of participation of different atomic electrons in chemical bonding. X-ray emission spectroscopy (XES), which provides both integral information about electronic structure (effective charges on atoms, atomic nl populations) and differential information (relative energies of occupied MOs and characteristics of their AO components), can be referred to these methods.

X-rays, discovered by Röentgen in 1895, are a form of electromagnetic radiation that occupy the spectral area between UV and γ radiation in the range of wavelength $\lambda = 10^{-3}$–10^2 nm, corresponding to energies $h\nu = 10$–10^6 eV ($\nu = c/\lambda$). X-ray spectroscopy is divided into X-ray emission and X-ray absorption spectroscopy and is subclassified into short wavelength ($\lambda \leq 0.2$ nm), long wavelength ($0.2 \leq \lambda \leq 2$ nm) and ultralong wavelength ($\lambda > 2$ nm) regions.

X-ray emission spectroscopy is used for the study of electronic structure and for the qualitative and quantitative analysis of substances. With the help of X-ray emission spectroscopy one may investigate all elements of the periodic table (with $Z > 2$) in compounds of any phase.

X-rays are divided into continuous and characteristic. Continuous X-rays occur as a result of the stopping of very fast charged particles (e.g. electrons) in the target substance and have a 'white' spectrum. Characteristic X-ray emission radiation is emitted by target atoms after their collisions with hot electrons (primary excitation) or with X-ray photons (secondary excitation, fluorescence radiation) and produces a line spectrum. These collisions may also remove an electron of any inner shell of a target atom. The resulting vacancy is filled by an electron transition from another inner or outer electron shell. As a consequence of this electron transition, energy is released which may be in the form of an X-ray quantum (**Figure 1**).

Characteristic X-ray emission spectra

Characteristic X-ray emission spectra consist of spectral series (K, L, M, N...), whose lines have a common initial state with the vacancy in the inner level. Labels of basic X-ray transitions are shown in **Figure 2**. All electron levels with the principal quantum number n equal to 1, 2, 3, 4, etc. are named as K, L, M, N etc. levels and denoted with corresponding Greek letters and digit indexes. The electron transitions which satisfy the dipole selection rules

$$\Delta l = \pm 1; \Delta j = 0, 1 \; (j = \pm \tfrac{1}{2}); \Delta n \neq 0 \qquad [1]$$

are most intense. The dependence of X-ray emission line energy on atomic number Z is defined by Moseley's law:

$$h\nu \sim (Z - \sigma)^2 \qquad [2]$$

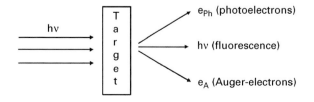

Figure 1 Scheme of radiation interaction with a substance.

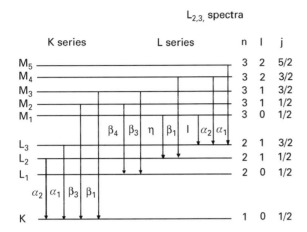

Figure 2 Scheme of the most important X-ray emission transitions; n, l and j are correspondingly the principal, orbital and total quantum numbers of K, L_1, L_2, L_3 levels, etc.

where Z is the atomic number and σ is the shielding constant, which varies from series to series. Therefore, any X-ray emission spectral line is the fingerprint of an element.

With X-ray emission excitation by electron bombardment (primary emission) all emission lines of the ith series appear when the X-ray tube voltage U exceeds the ionization potential of ith level (V_i). At higher U the intensity of all lines of the ith series, I_i, increases because the electrons penetrate deeper into the target substance and, therefore, the number of excited atoms in the target increases. In the $V_i < U < 3V_i$ region, the intensity obeys the rule $I_i \sim (U - V_i)^2$. With a further increase in U X-ray emission begins to be absorbed by the target atoms, therefore the increase in I_i is reduced. At $U \geq 11V_i$, I_i decreases because now most of the excited atoms are so deep in the target that their emitted radiation is absorbed by the target substance.

X-ray emission spectra are usually excited by X-ray photons because most chemical compounds are decomposed by electron bombardment. With X-ray emission spectra excitation by photons [secondary emission or fluorescence (XFS)] the fluorescent line intensity depends on the exciting photon energy $h\nu$ $I_i = 0$ if $h\nu < \mathring{a}V_i$. All lines of the ith series appear if $h\nu = \mathring{a}V_i$; however, I_i decreases little with further increase in $h\nu$. Therefore, to excite X-ray fluoresence one must use a target that contains a substance with intense characteristic X-ray lines whose energy just exceeds eV_i. Using the continous radiation of an X-ray tube with a target consisting mostly of heavy elements it is possible to excite X-ray fluorescence.

The intensity of a characteristic X-ray spectrum (both primary and fluorescent) depends on the

probability p_r of a radiation transition in the atom having the vacancy in the ith level. The value of p_r is determined by the total probability of photon emission when this vacancy is filled by outer electrons. However, with a probability p_A the vacancy may be filled by outer electrons without radiation as the result of the Auger-effect (see **Figure 1**). For the K series of medium and heavy elements $p_r > p_A$, for the light elements $p_r < p_A$. For all others series of any elements $p_r \ll p_A$. The ratio $f = p_r/(p_r + p_A)$ is called the yield of characteristic radiation.

However, X-ray characteristic lines appear because of single atom ionization; in X-ray emission spectra weaker lines are found to occur as a result of binary (or multiple) atom ionization when two (or more) vacancies are formed simultaneously in different electron shells. If, for example, one vacancy is formed in the K shell of atoms and filled by electrons belonging to the $L_{2,3}$ shell, atoms emit an $\hat{E}\alpha_{1,2}$ doublet. If another vacancy is formed simultaneously which too is filled by electrons from the $L_{2,3}$ shell, then the final state will have a binary ionization $L_{2,3}L_{2,3}$, and would correspond to the emission of radiation with energy exceeding that of the $\hat{E}\alpha_{1,2}$ doublet. As a result, in an X-ray emission spectrum a short wavelength $\hat{E}\alpha_{3,4}$ doublet, called a satellite of the main $\hat{E}\alpha_{1,2}$ doublet, would appear. Because of such processes of multiple ionization X-ray emission spectra may have a large number of satellites of the main lines. Usually, the satellite intensity is some orders of magnitude less than that of the main lines. However, if target atoms are bombarded by heavy ions with great energy, the probability of multiple atom ionization becomes higher than that of single ionization. Therefore, in this case the intensity of the main emission lines is essentially less than that of the satellites.

The continuous X-ray spectrum

The continuous X-ray radiation occurs because of electron deceleration in the target substance. Electron energy losses by radiation have quantum character, the emitted photon having an energy $h\nu$, which can not exceed the electron kinetic energy ε, i.e. $h\nu \leq \varepsilon$. The energy $h\nu_i = \varepsilon$ is called the quantum boundary of a continuous spectrum. The corresponding wavelength λ_i depends on the X-ray tube voltage charge U as follows:

$$\lambda_i(\text{nm}) = hc/(eU) = 1239.8/U(\text{V}) \qquad [3]$$

At $\lambda < \lambda_i$ the intensity of continuous radiation is absent. With an increase in λ from λ_i to $3\lambda_i/2$ the continuous radiation intensity increases and with further increases in λ it decreases.

X-ray sources

The most widespread source of X-rays is the X-ray tube. In an X-ray tube, electrons emitted from the cathode are accelerated by an electrical field and bombard the metal target (anode). Target atoms, excited by electron impact, and electrons losing their kinetic energy when decelerating in the anode substance, emit X-rays. The primary radiation of the X-ray tube consists of two parts – characteristic (line) and continuous radiation. As a result of the primary radiation impinging on a substance its atoms emit characteristic fluorescence (secondary) radiation.

Other sources of X-rays are radioactive isotopes which can directly emit X-rays or electrons or α particles. In the last two cases charged particles can bombard the target substance which then emits X-rays. The intensity of X-ray isotope sources is some orders of magnitude less than that of X-ray tubes; however the dimensions, weight and cost of X-ray isotope sources are less than that of X-ray tubes.

X-rays are also generated as synchrotron radiation. It can be selected by a crystal analyser and may be used as an X-ray source. The intensity of X-rays selected from synchrotron radiation is some orders higher than that from an X-ray tube.

The characteristic radiation of the X-ray tube is spread in space isotropically, whereas its continuous radiation has maximal intensity in directions in a plane perpendicular to the trajectory of electrons bombarding the target. The X-ray component of synchrotron radiation is polarized and spread only in the plane of the synchrotron ring.

Obtaining X-ray emission spectra

A schematic of an X-ray fluorescence instrument is presented in **Figure 3**. The X-ray tube is used as the source of primary radiation $h\nu_1$. The vacancies in inner shells of atoms of the substance investigated are formed as a result of primary radiation action. These vacancies are filled by other inner or outer electrons. This is accompanied by X-ray fluorescent photons $h\nu_2$ being emitted. This fluorescence radiation is spread out into the spectrum by means of a crystal analyser (or, for the ultrasoft X-ray region, by means of diffraction gratings) in accordance with Bragg's law

$$2d \sin \phi = n\lambda \qquad [4]$$

where n is the order of the spectrum, λ is the wavelength, d is the grating constant of the crystal analyser and ϕ is the angle of incidence of the collimated X-ray fluorescent beam on the specific set of parallel planes in the crystal from which the beam is diffracted (see **Table 1**).

The X-ray fluorescence spectrum is then registered on a photographic film or by Geiger, proportional or scintillation counters, semiconductor detectors, etc.

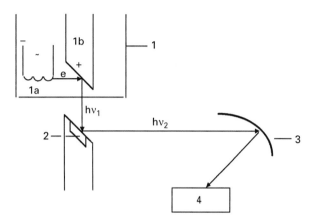

Figure 3 An X-ray fluorescence spectrometer. 1, X-ray tube; 1a, electron source; 1b, target; 2, substance investigated (secondary anode); 3, crystal analyser; 4, registration device; $h\nu_1$, primary radiation; $h\nu_2$, secondary radiation; and $h\nu_3$, registered radiation.

Table 1 Parameters of typical crystal analysers

Crystal	Reflecting plane	2d (nm)	Maximal solving ability ($\lambda/\Delta\lambda$)	Relative coefficient
KAP	001	2.7714	1400	8–18
Mica	002	1.9884	2 000	2–3
ADP	101	1.0659	10 000	1–10
EDDT	020	0.8808	–	–
PET	002	0.8726	8 000	10–20
Quartz	$10\bar{1}0$	0.8512	20 000	1–10
Quartz	$10\bar{1}\bar{1}$	0.671 53	10 000	2–14
Plumbago	002	0.6696	100	50–200
Ge	111	0.653 27	6 000	–
Si	111	0.6271	10 000	2–10
Calcite	211	0.6069	15 000	2–30
Quartz	$10\bar{2}0$	0.4912	30 000	0.4–3.3
LiF	200	0.4028	2 000	10
Ge	220	0.400	13 000	17–23
Si	220	0.383 99	29 000	1–6
Calcite	422	0.3034	64 000	0.4–0.9
LiF	220	0.2848	1 300	10–20
Quartz	$20\bar{2}3$	0.2806	90 000	0.3–0.9
Quartz	$22\bar{4}3$	0.2024	144 000	0.2–0.45
Calcite	633	0.202	122 000	0.3–0.6

The wavelengths and energies of the characteristic emission lines have been accurately measured and tabulated in handbooks, monographs and reference works for all chemical elements with $Z > 2$.

X-ray fluorescence analysis

X-ray fluorescence analysis (XFA) is based on the X-ray emission line's intensity dependence on the concentration of the appropriate element. XFA is widely used for the quantitative analysis of various materials, especially in black and colour metallurgy and geology. XFA is distinguished by rapidity and a high degree of automation. The detection limits depend on the element, matrix composition and spectrometer used and lie in the region 10^{-3}–$10^{-10}\%$. Defining any element (with $Z > 4$) is possible by means of XFA in both a solid and a liquid phase.

However, the fluorescence line intensity I_A of an investigated element A depends not only on its concentration C_A in the sample, but also on the concentration of other elements, C_i, because other elements promote both absorption and excitation of fluorescence of the element A (matrix effect). Moreover, the measured value I_A essentially depends on the sample surface, phase distribution, grain sizes, etc. Numerous methods have been developed to account for such effects. Most notable are the empirical methods of external and internal standards, using the background of scattered primary radiation and the method of dilution.

In the external standard method the unknown concentration C_A of the element A is determined by comparing the intensity I_A in the sample investigated with analogous I_{st} values of standards for which defined element concentrations C_{st} are known:

$$C_A = C_{st} I_A / I_{st} \qquad [5]$$

This method allows one to take into account corrections connected with the equipment used; however, the composition of the standard should be close to that of the investigated sample to precisely match the matrix effect.

In the internal standard method some amount $\Delta \tilde{N}_A$ of a defined element A is added to the sample investigated. This leads to an increase in the fluorescence intensity of ΔI_A. In this case:

$$C_A = I_A \Delta C_A / \Delta I_A \qquad [6]$$

This method is especially effective for the analysis of complex samples but needs the special requirements of sample preparation.

The use of the background of scattered primary radiation is based on the fact that the ratio $I_A : I_b$ (I_b is the background intensity) mainly depends on C_A and only weakly depends on the concentration of other elements, C_i.

In the method of dilution a great amount of a weak absorber or a small amount of strong absorber is added to the sample investigated. These additions should reduce the matrix effect. This method is effective for water solution analysis and for the analysis of complex samples when the internal standard method is inapplicable.

There are also models in which the measured intensity I_A is corrected on the basis of the intensities I_i and concentrations C_i of other elements. For example, C_A may be represented as:

$$C_A = a_{A0} + a_{A1} I_A + a_{a2} I_A{}^2 + I_A \sum_{m,i} m_{Ai} I_i$$
$$+ \sum_{m,i} b_{Ai} I_i + \sum_{m,i} d_{Ai} I_i^2 \ (i \neq A) \qquad [7]$$

a and b are values determined by the least-squares method with the help of I_A and I_i values measured in several standards with known concentrations C_A of element A. Such models are widely used for the serial analysis of many samples via computers.

X-ray microanalysis

X-ray microanalysis is a local analysis, fulfilled by means of microanalyser electron probe, for sample sites of ~ 1–3 μm². The electron probe is formed by electrostatic and magnetic fields to obtain a parallel electron beam with a diameter of ~ 1 μm. The analysis is via primary X-ray sample emission which is spread out into a spectrum by means of X-ray spectrometer. In this method corrections for the atomic number of the element, the absorption of its radiation in the sample, its fluorescence, and the characteristic spectra of other elements contained in the sample must be accounted for. Microanalysis is used for the investigation of two- and three-component systems such as mutual diffusion, crystallization processes, local variations of alloy structure, etc.

List of symbols

d = grating constant; I_A = fluorescence line intensity of an investigated element A; I_{st} = fluorescence intensity of a standard; n = order of spectrum;

p_A = probability of a vacancy being filled without emission; p_r = probability of photon emission when vacancy is filled; U = X-ray tube voltage; V_i = ionization potential of the ith level; Z = atomic number; ε = electron kinetic energy; λ = wavelength; δ = shielding constant; ϕ = angle of incidence.

See also: **X-Ray Absorption Spectrometers; X-Ray Fluorescence Spectrometers; X-Ray Fluorescence Spectroscopy, Applications.**

Further reading

Bearden JA (1967) X-ray wavelengths. *Review of Modern Physics* 39: 78–124.

Ehrhardt H (1981) *Röntgenfluoreszenzanalyse.* Leipzig: VEB Deutscher Verlag für Grundstoffindustrie, 250p.

Mazalov LN, Yumatov VD, Murakhtanov VV, Gelmukhanov FK, Dolenko GN, Gluskin ES and Kondratenko AV (1977) *Rentgenovskie Spectry Molekul.* Novosobirsk: Nauka, 331p.

Meisel A, Leonhardt G and Szargan R (1977) *Röntgenspektren und Chemische Bindung.* Leipzig: Geest & Portig, 320p.

Nemoshkalenko VV and Aleshin VG (1979) *Teoreticheskie osnovy rentrenovskoj emissionnoj spektroskopii.* Kiev: Naukova Dumka, 384p.

Siegbahn K, Nordling C, Fahlman A, Nordberg R, Hamrin K, Yedman J, Johansson G, Bergmark T, Karlsson SE, Lindgren I and Lindberg B (1967) *ESCA. Atomic, Molecular and Solid State Structure by Means of Electron Spectroscopy.* Uppsala: Nova acta Regiae Societatus Scientiarum Upsaliensis, 493p.

X-Ray Fluorescence Spectrometers

Utz Kramar, University of Karlsruhe, Germany

HIGH ENERGY SPECTROSCOPY
Methods & Instrumentation

X-ray fluorescence (XRF) spectrometers are widely used for the determination of elements with atomic numbers from 4 (beryllium) to 92 (uranium) at concentrations from 0.1 µg g^{-1} to high percentage levels. These elements can be analysed using characteristic K$_\alpha$-lines (KL$_{III}$) from 11.4 nm/0.1885 keV (Be K$_\alpha$) to 0.0126 nm/98.4 keV (U K$_\alpha$). Nevertheless, elements of higher atomic number (e.g. Cd L$_{III}$M$_V$ 0.3956 nm/3.133 keV; U L$_{III}$M$_V$ 0.09106 nm/13.61 keV) are often determined using their L-lines. The characteristic X-ray lines of these elements can be determined either with sequential or with simultaneous wavelength-dispersive spectrometers by Bragg diffraction, using the wave phenomena of X-rays or in energy-dispersive systems using their energy characteristic. Coherent and incoherent scattering of primary X-rays in the sample may cause increased background effects, and matrix-dependent absorption of the characteristic secondary X-rays may also cause severe matrix effects. Since XRF methods are routinely used, methods for correcting the matrix effects such as fundamental parameters have been developed and instruments with drastically improved peak to background ratios such as polarized X-ray fluorescence (PXRF) and total reflection X-ray fluorescence (TXRF) spectrometers have been designed during the 1990s. The principles of XRF spectroscopy and descriptions of the different kinds of instrumentation are given in an increasing number of monographs, books and reviews, with extensive data compilations.

Principles

If a target is irradiated with photons, or charged particles (electrons or ions) with energies exceeding the binding energy of the bound inner electrons, an electron from inner orbitals of the target atoms can be ejected.

$$E_\infty = 13.56 \cdot \frac{[Z_{eff}]^2}{n} \ (eV)$$

where Z_{eff} = effective atomic number and n = principal quantum number. If the total energy of the photon is transferred to the electron this interaction is called the photoeffect.

The resulting atom is unstable and regains its ground state by transferring an electron from a high-energy outer orbital to the vacancy in the inner electron shell. The energy difference between the initial and final energy state of the transferred electron is

released as a photon of the energy

$$E_{\text{photon}} = E_{\text{initial}} - E_{\text{final}}$$

In XRF spectroscopy photons are used to excite the characteristic elemental X-rays from the sample. Alternatively the incident photons can be scattered coherently (Rayleigh scattering) at an electron of the inner shell or incoherently at an electron of the outer shell (Compton scattering). In the second case the wavelength/energy of the scattered photon depends on the initial wavelength/energy of the original photon and the scattering angle Θ

$$E' = E/[1 + (1 - \cos\Theta)(E/E_e)]$$

where E' = photon energy after Compton scattering, E = initial energy of the photon and E_e = energy equivalent of the electron mass. The XRF spectrum of a sample is a mixture of the different characteristic X-rays emitted from the atoms in the sample and of coherent and incoherent scattered components of the primary radiation source. The task of XRF spectrometers is to separate the different spectral components, to determine their intensities and based on this to calculate the elemental concentrations. Typically, XRF spectrometers consist of a photon source for the excitation of the secondary X-rays, a sample support, an X-ray detection unit and a data evaluation unit.

X-ray sources

In most XRF spectrometers an X-ray tube is used as the photon source. Alternatively radioactive isotopes can be applied for the excitation of the characteristic X-rays from the sample. X-ray tubes can be used in both wavelength-dispersive and energy-dispersive systems. Due to their lower beam intensities, application of radionuclides is restricted to energy-dispersive systems.

X-ray tubes

In X-ray tubes, the X-rays are produced by the bombardment of matter with accelerated electrons. The X-ray tubes are built as a vacuum-sealed metal glass cylinder. The electrons are emitted from a heated tungsten filament which serves as the cathode and are accelerated by a high voltage applied between the filament and a metal anode. Two effects can occur if the accelerated electrons interact with the atoms of

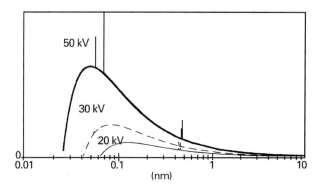

Figure 1 Primary spectrum of an X-ray tube with a rhodium anode.

the anode material. (1) An electron enters the electric field of an atomic orbital and is slowed down, and the loss of kinetic energy during slowing down is emitted as electromagnetic radiation, called bremsstrahlung. (2) An inner-shell electron of an atom is ejected if the kinetic energy exceeds the binding energy, and from this the characteristic X-rays of the anode material are emitted. Thus the spectrum emitted from the anode consists of the continuous bremsstrahlung and the characteristic X-ray lines of the anode material.

The maximum X-ray energy primarily emitted from the tube is determined by the applied acceleration voltage.

$$E_0 = e \cdot U_o$$

The radiation intensity is a function of the tube current and the applied high voltage (**Figure 1**).

$$I(\lambda)\mathrm{d}\lambda = kiZ\left(\frac{\lambda}{\lambda_0} - 1\right)\frac{1}{\lambda^2}\mathrm{d}\lambda$$

Excitation of the characteristic X-rays from the sample can be optimized by selecting the appropriate anode material, voltage and tube current (**Table 1**).

Table 1 Anode materials and application ranges of X-ray tubes used for X-ray fluorescence analysis

Anode material	Element range		Operating voltage (kV)
	K-spectra	L-spectra	
Cr	^{8}O–^{22}Ti	^{42}Mo–^{55}Cs	55
W	^{23}V–^{27}Co	^{56}Ba–^{72}Hf	55
Au	^{28}Ni–^{30}Zn; ^{40}Zr–^{92}U	^{73}Ta–^{75}Re	65/100
Mo	^{31}Ga–^{39}Y	^{76}Os–^{92}U	100
Rh	^{4}Be–^{56}Ba	^{42}Mo–^{92}U	60

In wavelength-dispersive XRF (WDXRF), TXRF and PXRF methods, efficiencies of primary-to-detected-secondary radiation are very low. Therefore tubes for these methods are generally operated at high power [~3 kW per kV up to 100 kV]. Because most of the tube power is converted to heat, these tubes have to be water cooled. Depending on the kind of application, different tube designs such as end-window (e.g. **Figure 9**), side-window (**Figure 2A**) and line focus are employed. When extremely high primary intensities are necessary rotating anodes are used (**Figure 2B**). In energy-dispersive systems radiation efficiency is several orders of magnitude higher than in WDXRF, therefore compact air-cooled low-power tubes (3–100 W) are used in most cases.

Radionuclide sources

In mobile EDXRF systems, where small dimensions are essential, sealed radionuclides are often used as primary radiation sources instead of X-ray tubes. The radionuclides have to be selected with respect to

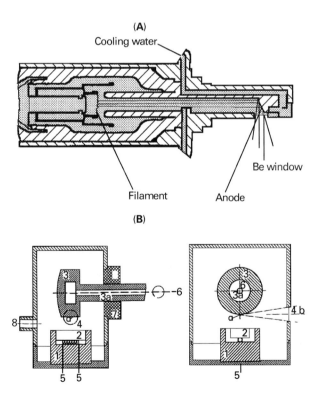

Figure 2 (A) Schematic sketch of a side window tube. (B) Rotating anode tube in two sectional views: (1) cathode unit; (2) filament; (3) cylindrical anode with (3a) rotary shaft; (4) window; (5) electrical connections; (6) cooling water connection; (7) sealing gasket; (8) vacuum flange. Reproduced with permission from Klockenkämper R (1997). *Total Reflection X-ray Fluorescence Analysis*. New York: Wiley.

their decay scheme, half-life and radiotoxicity (**Table 2**).

β⁻ Decay

From isotopes decaying in β⁻-mode, high-energy electrons ranging from several keV to MeV are emitted from the nucleus. The electrons interact with the matter and bremsstrahlung is produced. Most β⁻-sources are used for secondary target excitation.

Electron capture (EC)

The nucleus captures an electron of the innermost electron shell and a proton of the nucleus changes into a neutron. Due to this process a vacancy in the innermost electron shell occurs. By filling this vacancy with an electron from the outer shells, the characteristic X-rays of the daughter atom are emitted.

γ Decay

In most of the radioactive decays the daughter isotope is formed in an excited state. These nuclides are transferred into the ground state by emitting electromagnetic radiation, the γ-radiation.

Selection criteria

The choice of the radionuclide source depends on the type of sample, the elements to be determined and the detection technique. Generally a high specific activity of the radionuclide and emission energies suitable for the application are required. The most convenient isotopes for energy-dispersive XRF are those that decay exclusively by EC without emitting γ-radiation.

Radionuclides emitting high-energy photons (>150 keV), or decaying by β⁺ or high-energy β⁻, or those with short half-lives are not recommended for use in XRF systems. A selection of suitable radionuclides with suitable energies and high specific activity is given in **Table 2**.

X-ray dispersion and detection units

XRF spectrometers have two major objectives: (a) to determine the spectral distribution of the X-rays emitted from the sample; (b) the measurement of the intensity of the selected spectral component. In wavelength-dispersive XRF, the spectrum is dispersed into different wavelengths by Bragg diffraction at different crystals. Intensities are measured by electronic detectors. In energy-dispersive spectrometers both energy dispersion and intensity measurement are performed by electronic detectors; thus for

Table 2   Radionuclides used as primary sources in energy-dispersive X-ray fluorescence

Isotope	Decay mode	Half-life (years)	Energy (keV)	Elemental range		Recomm. activity (MBq)	Working life (years)
				K X-rays	L X-rays		
^{241}Am	α	433	59.5 Np L X-rays (12–22)	Zr–Ce	W–U	370–1 110	15
^{109}Cd	EC	1.26	88 Ag K X-rays (22–26)	Ti–Nb	Tb–U	111–740	5
^{57}Co	EC	0.74	122; 136 Fe K X-rays (6–7)	Ba–U		37–370	5
^{244}Cm	α	17.8	43; 99; 152 Pu L X-rays (12–23)	Ti–Se	Ce–Pb	370–3 700	10
^{55}Fe	EC	2.69	Mn K X-rays (5.9–6.5)	Si–V	Nb–Sn	185–1 850	5
^{153}Gd	EC	0.66	97; 103 Eu K X-rays (41–48)				
^{125}I	EC	0.17	35 Te K X-rays (27–32)			370–3 700	1
^{3}H	β⁻	12.43	E_{max}: 18.6 (beta)				
^{238}Pu	α	433	43 U L X-rays (11–22) U K X-rays (94–115)	Ti–Se	Ce–Pb	370–1 110	10

detectors used in energy-dispersive spectrometers extremely good energy resolutions are required.

Dispersing crystals

At the X-ray diffraction crystals the X-ray spectrum is dispersed into different wavelengths according to Bragg's law

$$n \cdot \lambda = 2d \sin \Theta$$

The wavelengths of characteristic lines used in X-ray spectrometry range from ~0.03 nm (Ba K) to ~10 nm (Be K). This range cannot be covered by use of a single diffraction crystal. The detectable wavelength and high-order reflections are limited by the relation between d and λ.

$$n\lambda \leq 2d$$

For example with LIF 200, the crystal with the broadest application range, the elements with atomic numbers <19 (K) are not detectable. At shorter wavelength, the dispersion between neighbouring lines decreases and they cannot be resolved from each other. The diffraction crystals have to be selected with respect to their reflectivity and to be suitable for the lines to be detected. Commonly used X-ray diffraction crystals and their application range are compiled in **Table 3**.

Table 3   Diffraction crystals commonly used in wavelength-dispersive X-ray fluorescence

Analyser crystal	Material	2d (nm)	Detectable elements		Efficiency
			K	L	
	Topaz	0.27	V–Ta	Ce–U	Average
LIF(220)	Lithium fluoride	0.29	V–Ta	Ce–U	High
LIF(200)	Lithium fluoride	0.4	K–La	Cs–U	High
Ge	Germanium	0.65	P–Zr		Average
PET	Pentaerythrite	0.87	Al–Ti	Kr–Xe	High
AdP	Ammonium dihydrogen phosphate	1.06	Mg		Low
TAP/TIAP	Thallium biphthalate	2.58	F–Na		High
OVO-55	W/Si multilayer	5.5	N–Si	Ca–Br	High
OVO-160	W/C multilayer	16	Be–O		?
OVO-C	V/C multilayer	12	C		?
OVO-B	Mo/B4C multilayer	20	Be–B		?
PbSD	Lead stearate decanoate	10	B–F		Average

X-ray detectors

In X-ray detectors the energy transported by the radiation is converted into forms that can be recognized visually or electronically. Generally the photons are absorbed by the detector material and energy transfer takes place by ionization. The number of ionizations N per photon is proportional to the energy E of the absorbed photon and depends on the average energy e necessary to produce an electron–ion pair in the detector material.

$$N = E/e$$

The energy resolution of a detector is determined by the statistical scatter of the number of ionizations per photon.

$$\sigma = \sqrt{(FN)}$$

with F = the Fano factor. Due to this relation it is possible to separate the radiation absorbed in the detector into the different energies.

Critical parameters for selection of an appropriate detector are: efficiency, energy resolution and deadtime, i.e. the pulse processing time of the detector. A measure for the energy resolution is the full width of half maximum (FWHM).

$$\mathrm{FWHM} = 2.36 \times \sqrt{(eFE)}$$

The detector efficiency depends on the radiation energy to be determined and the density and type of detector material. In XRF spectrometers three different type of X-ray detectors are used: gas-filled detectors, scintillation detectors and semiconductor detectors.

Gas proportional (flow-) counters

Gas-filled detectors are used for the determination of low-energy X-rays. An isolated thin wire is mounted in a metal cylinder filled with a suitable counting gas (e.g. 10% Ar, 90% CH_4). The wire is held at a high positive potential (1.2–2 kV) and the metal cylinder is grounded (**Figure 3**). A thin window (Be or Mylar) allows the radiation to enter the detector. The gas is ionized by the photons and the electrons produced are accelerated towards the anode and the ions towards the cathode. The energy necessary to produce an ion pair is 20–30 eV, depending on the type of counting gas. Additional ionizations are produced by collisions with other gas atoms. The electrons are collected at the anode, producing a short-term voltage drop. Applying an appropriate high voltage to the detector, the voltage drop is proportional to the radiation energy (**Figure 4**). High radiation doses degrade the counting gas. In addition, the thin detector windows, necessary for the detection of low-energy photons, can lead to a permanent loss of counting gas by diffusion. To avoid these problems the detectors are operated as

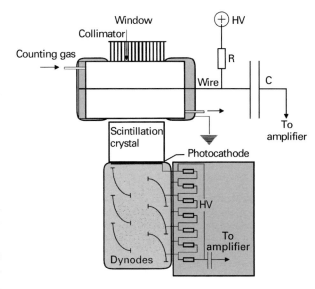

Figure 3 Schematic of gas proportional counter (flow counter) and scintillation detector operated as tandem detectors in wavelength-dispersive X-ray fluorescence spectrometers. R = resistor; C = capacitor.

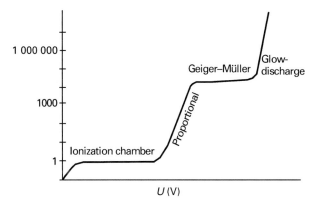

Figure 4 Gas-amplification characteristics of gas-filled detectors. Most flow counters for WDXRF are operated in the changeover region from proportional to Geiger–Müller.

flow counters, in which the counting gas is permanently exchanged.

Scintillation detectors

Scintillation detectors are used for the determination of the high-energy part of the X-ray spectrum. In scintillation detectors the material of the detector is excited to luminescence (emission of visible or near-visible light photons) by the absorbed photons or particles. The number of photons produced is proportional to the energy of the absorbed primary photon. The light pulses are collected by a photocathode. Electrons, emitted from the photocathode, are accelerated by the applied high voltage and amplified at the dynodes of the attached photomultiplier (**Figure 3**). At the detector output an electric pulse proportional to the absorbed energy is produced. The average energy necessary to produce one electron at the photocathode is approximately 300 eV. For X-ray detectors, in most cases NaI or CsI crystals activated with thallium are used. These crystals offer a good transparency, high photon efficiency and can be produced in large sizes.

Semiconductor detectors

Semiconductor detectors can be considered analogous to ionization chambers with the gas being replaced by the semiconductor material. The group 4 elements silicon and germanium are most widely used as semiconductor detectors when excellent energy resolution is required. Semiconductors have a small energy gap between their valence and conducting bands. At energies above absolute zero, thermal excitation will move some electrons from the valence band to the conductor band and holes or positive charges are left in the valence band. The quasi-free electrons in the conductor band move towards the anode if an electric field is applied. If ionizing radiation interacts with a semiconductor, electrons are transferred from the valence to the conducting band. The energy necessary to produce an ion pair is low (3–5 eV). Thus a large number of electrons, proportional to the absorbed radiation energy, can be collected at the anode by applying a high voltage of 1–2 kV to the detector, and statistical scatter is low. The current of the radiation-induced free electrons is superimposed by the temperature-dependent free charge density (leakage current) and electron recombination by acceptor-type impurities. Today silicon cannot be produced at purity levels that allows electron recombination by acceptor-type impurities to be neglected. In Si(Li) detectors acceptor-type impurities are compensated for by drifting the silicon with lithium as donator.

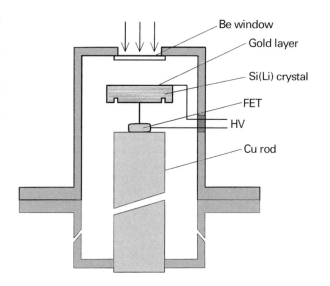

Figure 5 Schematic of a planar semiconductor detector. HV = high voltage; FET = field-effect transistor.

In order to reduce the thermal charge carrier generation (noise) to acceptable levels Si(Li) detectors and Ge detectors are operated at liquid nitrogen (LN$_2$) temperatures. LN-cooled detectors are mounted in a vacuum chamber, which is inserted in or attached to a LN$_2$ Dewar flask (**Figure 5**). The detector is in thermal contact with the LN$_2$ and the radiation enters the detector housing via a thin Be window. In mobile systems, and for applications requiring long-term unattended operation, detectors cooled by Joule–Thomson coolers, He-cycle refrigerators or Peltier elements are used, but all of them can show degraded energy resolution, with the Peltier exhibiting the most (~20%). Extremely compact room temperature detectors using semiconductors with higher energy gaps (CdZnTe, HgI$_2$, GaAs) have been developed in the last decade. Energy resolution of these detectors is poorer than for Si(Li) detectors but better than with proportional or scintillation detectors. The characteristic properties of the different detector types are compared in **Table 4**.

Table 4 Properties of radiation detectors used in XRF systems

Detector type	Material/filler gas	Ionization energy (eV)	Band gap (eV)	FWHM at 5.9 keV (eV)		Dead-time (μs)	Energy range (keV)	Remarks
				Theor.	Typical			
Proportional counter	Ar/methane	26.4		840	840	1	<1–10	
Scintillation	NaI(Tl)	300		3 150	3 150	0.2	3–100	
Semiconductor	Si(Li)	3.61	1.12	120	150	2–4	1–60	LN$_2$
	Ge	2.98	0.74	108	145		1–200	LN$_2$
	CdZnTe	~5	~1.5	135	285		>2	
	HgI$_2$	6.5	2.13	160	270		>2	

Wavelength-dispersive spectrometers

In wavelength-dispersive spectrometers the characteristic radiation emitted from the sample is dispersed into different wavelengths by Bragg diffraction and the resulting radiation is registered by electronic detectors. Since the wavelength selection is performed by diffraction of the radiation to different angles, only minor energy resolution of the radiation detectors is required. Nevertheless according to Bragg's law higher order diffraction radiations of the wavelength $\lambda/2$, $\lambda/3$, etc. are registered at the same angle as those of wavelength λ. These high-order signals cannot be separated by the diffraction crystals, but they are significantly different in their energies and can be separated using the energy-resolution capabilities of the detectors. **Figure 6** gives an example

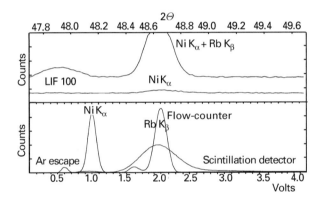

Figure 6 Wavelength and energy spectrum of Ni Kα in the presence of high Rb concentrations with open window (3 V) and window width set to 1.3 V to remove second-order radiation of Rb Kβ. Spectrum of reference rock MA-N, measured with WDXRF SRS 303 AS.

of elimination of second-order radiation by energy discrimination in WDXRF.

Thus, a wavelength-dispersive spectrometer consists of the excitation unit (X-ray tube with high-voltage supply), the sample chamber, the diffraction unit and the detector unit with amplifier, lower level and window discriminator, and registration unit (**Figure 7**). Generally the sample, the diffraction unit and the low-energy detector are mounted together in a vacuum chamber to avoid absorption of low-energy X-rays along the radiation path. For some applications, primary beam filters are used to optimize the tube spectrum. Wavelength-dispersive spectrometers are either designed as fixed (simultaneous) spectrometers or as sequential (scanning) spectrometers. Generally wavelength-dispersive spectrometers have a mass of > 400 kg and are operated in well established laboratories, equipped with cooling water facilities and high-current electric power supply. Nevertheless, mobile systems operated with low power tubes are also possible.

Sequential spectrometers

In most sequential XRF spectrometers a flat crystal is mounted on the central axis of a rotating goniometer and a proportional counter and a scintillation counter are mounted at the moving goniometer arm. The Bragg angle can be simply varied, rotating the crystal mount by Θ and the detectors by 2Θ. Primary and secondary collimators between sample and crystal and crystal and detectors are used to limit the beam divergence. Most common types of collimator consist of a pack of parallel plates of highly absorbing material (e.g. tungsten). Radiation (nearly) parallel to these plates can pass, whereas diverging radiation is absorbed by the plates. Narrow spacing

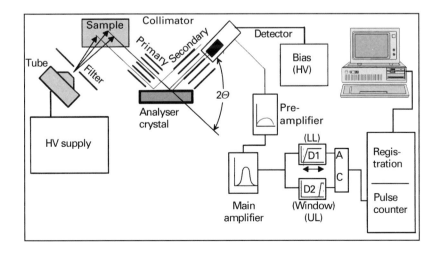

Figure 7 Schematic drawing of a typical sequential X-ray fluorescence setup. HV = high voltage; AC = anti-coincidence; D = discriminator; LL = lower level; UL = upper level.

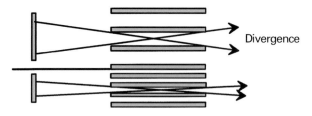

Figure 8 Schematic of coarse and fine collimator assignment.

of the plates results in optimum resolution but intensity is reduced by an amount equivalent to the space angle (**Figure 8**). Therefore most sequential spectrometers are equipped with coarse and fine spaced primary collimators.

Curved focusing scanning crystals are rarely used in wavelength-dispersive instruments, but are common in microprobes.

Simultaneous spectrometers

Simultaneous spectrometers are designed with fixed crystal arrangements. In modern simultaneous spectrometers up to 30 crystal–detector combinations are fixed at suitable angle positions for the peak and background measurements of the application. Generally, focusing crystal shapes are used. For fixed diffraction geometries the best focusing is achieved with logarithmically curved crystals (**Figure 9**).

Energy-dispersive spectrometers

In energy-dispersive systems the separation of the different components in the X-ray spectrum is performed exclusively by the detector and the subsequent electronic components. Thus good energy resolution, low electronic noise, low temperature drift and excellent linearity of the electronic components are required. Generally, an energy-dispersive spectrometer consists of a radiation source (X-ray tubes or radionuclides), sample support, pre- and main amplifier, pile-up rejector, analogue-to-digital converter and multichannel analyser.

As radiation sources, X-ray tubes or radionuclides are applied. The polychromatic radiation of the tube is scattered by the Compton or Rayleigh effect towards the detector, which results in a high bremsstrahlung background and line interferences caused by the characteristic lines of the anode material and the resulting Compton peak.

Primary beam filters are used routinely in energy-dispersive XRF (EDXRF) to optimize the tube spectrum for the specific application (**Figure 10**). Filters with a sharp absorption edge slightly above the characteristic tube lines are used for filtering out the high-energy bremsstrahlung and a quasi-monochromatic excitation spectrum is achieved (e.g. Rh tube and Pd filter). The characteristic lines can be removed by heavy-element filters without an absorption edge in the high-energy region (e.g. Rh tube and Cu filter) (**Figure 11**). The use of primary beam filters causes some intensity loss but this is overcompensated for by the background reduction.

Secondary targets are used to provide optimal excitation conditions for specific elements, but the drastic loss of intensity in the beam restricts their application to tubes with higher output powers or

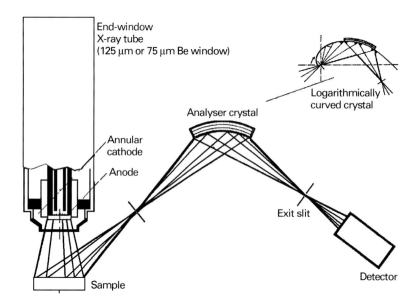

Figure 9 Schematic of a simultaneous WDXRF with end-window tube and logarithmically curved focusing analyser crystal. Redrawn from a sketch in a brochure from Bruker AXS AG.

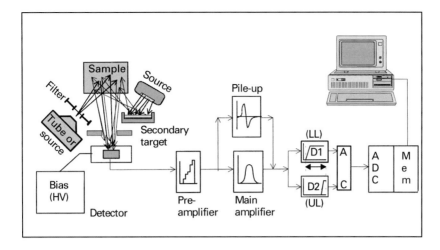

Figure 10 Schematic drawing of a typical EDXRF setup with excitation by tube or radionuclide with optional primary filter or secondary target excitation. ADC = analog to digital converter.

radionuclides of higher activities. Most radionuclides provide monoenergetic excitation, but the excitation energies are not tunable. For radionuclide sources a licence for handling and transportation is required.

In EDXRF systems, semiconductor detectors are used exclusively, apart from some mobile systems for low-energy application equipped with high-resolution proportional counters. Most systems are equipped with cooled Si(Li) or Ge detectors. The energy-proportional electric charge pulse is integrated by the preamplifier. Today state of the art preamplifiers are of the pulsed optical feedback type with the first active element, a field-effect transistor (FET), mounted in the detector cryostat in order to reduce electronic noise. The preamplifier output is converted to an energy-proportional voltage pulse of some microseconds duration, depending on the shaping time of the amplifier. Events arriving at the detector and amplifier during their signal processing period cannot be processed correctly. A pile-up rejector eliminates these pulses by inspecting the slopes of the pulses. The amplifier output signal is digitized by a subsequent analogue-to-digital converter (Wilkinson type) and collected in the memory of a multichannel analyser. Each channel is equivalent to a small energy interval and the channel content holds the counts of this energy interval, accumulated during the measuring time. All spectral components are registered simultaneously.

EDXRF systems are available as stationary systems, to be used for fast simultaneous multielement determinations in analytical laboratories, as compact equipment suitable for mobile field laboratories and as hand-held probes to be used directly on-site for screening analyses. The dimensions of the equipment range from several hundred kg for stationary systems

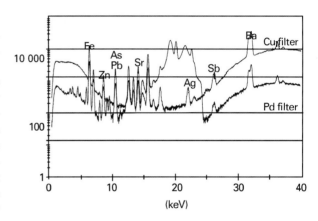

Figure 11 Energy-dispersive spectra of reference soil sample GXR-2, measured with Spectrace 5000 using a Rh anode tube and Pd and Cu primary beam filters to optimize excitation conditions.

down to 1.3 kg for the smallest radionuclide-excited hand-held probe.

Special instruments have been developed for trace element analysis in the picogram region in microsamples using total reflection of the X-ray beam (TXRF) and the sub-μg g^{-1} range in bulk samples using polarized X-rays.

Total reflection XRF (TXRF)

At radiation incident angles of less than the critical angle α_{crit}, the primary beam is totally reflected at the surface of a specimen.

$$\alpha_{\text{crit}} \approx \frac{1.65}{E} \sqrt{\frac{Z}{A}\rho}$$

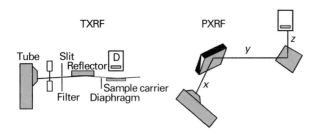

Figure 12 Schematic of TXRF and x–y–z geometry of PXRF.

where ρ = density, A = atomic mass and Z = atomic number. At angles $<\alpha_{crit}$ the X-rays can penetrate into the substrate only for a few nm. At these angles interaction of the X-rays with the total reflecting specimen is at a minimum. This effect is used in two types of TXRF instruments. In total reflection devices for trace element analysis, the sample is prepared as a thin film on a well polished quartz block. In conventional thin film methods the characteristic X-rays of elements in the sample are excited, but excitation and Compton scattering occurs within the material of the sample support as well. In TXRF the geometry is arranged to provide a total reflection of the primary X-ray beam (**Figure 12**). The low total reflecting incident angle provides an excellent interaction between the primary beam (absorption path ~1 mm per 1 μm sample thickness) and the sample, whereas absorption of the secondary radiation in the sample can be neglected in most cases. Generally, it is not necessary to use matrix correction methods as it is with other XRF methods. Once calibrated, samples of all the matrices can be analysed without recalibration and matrix corrections. Detection limits at the μg g⁻¹ level can be obtained from a few micrograms of solid sample and at the ng g⁻¹ level from a few microlitres of a liquid sample. For surface and thin layer analysis, instruments with angle variable sample positioning devices are used. With these instruments the angle-dependent intensity profile can be recorded as the basis for surface and thin layer analysis. For the determination of thin layer and surface contaminations, e.g. for wafer analysis, the measurement is performed at the critical angle with the beam grazing the surface. With these devices impurities of some 10^{11} – 10^{13} atoms cm⁻² can be determined.

Polarized X-ray fluorescence (PXRF)

In PXRF the primary X-ray beam is polarized by Barkla scattering or Bragg reflection. Under orthogonal conditions the scattered radiation is highly polarized. The polarized radiation is used to excite the characteristic X-rays of the elements in the sample. The secondary radiation is measured in orthogonal geometry to the polarization plane. Background, produced by Compton and Rayleigh scattering in the sample, is effectively reduced. Detection limits in bulk samples are improved by approximately a factor of 3 compared to normal EDXRF. Thus sensitivity and accuracy are equivalent (K–Mo, Hf–U) or better (Ag–Nd) than those obtained by WDXRF. Therefore, PXRF combines the capability of nondestructive multielement analysis with low detection limits and a minimum of sample preparation.

Sample preparation for XRF

With XRF solid as well as liquid samples can be analysed. Liquid samples have to fill an adequate sample container, with an X-ray transparent window (e.g. Mylar). For the determination of light elements the sample chamber has to be flushed with He instead of applying a vacuum. Solid samples can be analysed as pressed powder pellets, bulk powder samples, fused pellets or as pills (diluted with a flux, e.g. Spectromelt). For quantitative analyses a flat and homogenous sample surface is needed. Powder samples have to be pulverized to a fineness, where absorption in single grains can be neglected. Solid and liquid samples can be prepared either by thin film techniques, or as samples of infinite thickness. In thin film techniques the amount of sample in the irradiated spot has to be known. This can be established by adding an internal standard. A specimen can be regarded to be of infinite thickness if

$$D_\infty \geq 3 \cdot d_{1/2\,\mathrm{prim}} \quad \text{with} \quad d_{1/2\,\mathrm{prim}} = \frac{\ln 2}{\mu(E_{\mathrm{prim}})}$$

where $d_{1/2}$ = half absorption thickness and μ = mass attenuation coefficient.

Matrix correction methods

Thin specimens show a strictly linear dependency of the intensity of a characteristic line on the amount of the analyte. In thick specimens, primary radiation is absorbed along its path in the sample and secondary radiation on its way out. Additionally, secondary excitation by intense characteristic lines with energies above the absorption edge of the analyte can occur. These intensity variations (up to a factor of ~20) are strongly matrix dependent and have to be corrected. Different methods can be used.

Normalization of the Compton peak can be applied, if no absorption edge occurs between the Compton peak and the line of the analyte.

$$c_i = (a_i I_i + b_i) \frac{I_{C_{\text{ref}}}}{I_{C_{\text{sample}}}}$$

Empirical, semiempirical and theoretical matrix correction models are used for the determination of major components, where line intensities are influenced by absorption edges and enhancement. Routinely applied is the fundamental parameters model based on theoretical or experimentally determined α-coefficients:

$$c_i = [D + E \cdot I(\lambda_i)] \times [1 + \Sigma(c_j \cdot \alpha_{ij})]$$

or empirical corrections such as

$$c_i = \left(b_{0i} + b_{1i} I_i + b_{2i} I_i^2 + I_i \Sigma m'_{ij} \cdot I_j + \Sigma m''_{ij} \cdot I_j\right)$$

or under consideration of elemental concentrations

$$c_i = \left[b_{0i} + b_{1i}(I_{\text{F}})_i + b_{2i}(I_{\text{F}})_i^2\right] \times \left(1 + \Sigma k'_{ij} \cdot c_j + \frac{\Sigma k''_{ij} \cdot c_j}{1 + c_j}\right)$$

where I_{F} = fluorescence intensity.

Recent trends

All modern X-ray fluorescence systems are computer controlled and equipped with automatic sample changers. Different matrix correction models are included in the data evaluation software, and further developments on expert systems will reduce manual calibration work. With WDXRF, the detectability of light elements (down to Be) will be optimized, e.g. by using X-ray tubes and detectors with ultrathin windows and widely spaced (focusing) analyser crystals. In EDXRF, trends are for miniaturization, development and optimization of high-resolution room temperature detectors and extension of the application range towards the determination of light elements.

List of symbols

A = atomic mass; b_i = calibration coefficient for element i; c_i = concentration of element i; C_j = concentration of element j; d = d-spacing of a crystal; $d_{1/2}$ = half absorption thickness; D_∞ = saturation thickness; e = charge of an electron; E = initial energy of the photon; E_0 = maximum energy of accelerated electron; E_∞ = energy necessary to remove an electron from the orbital n; E' = photon energy after Compton scattering; E_e = energy equivalent of the electron mass or energy necessary to produce an electron–ion pair; E_{final} = final energy of the transferred electron; E_{inital} = initial energy of the transferred electron; E_{photon} = energy of the emitted photon; F = Fano factor; i = tube current; I = radiation intensity; I_{Cref} = intensity of reference Compton peak; I_{Csample} = intensity of sample Compton peak; I_{F} = intensity of fluorescence radiation; I_i = radiation intensity element; k = constant; k'_{ij} = coefficient for absorption correction; k''_{ij} = coefficient for enhancement effect; m'_{ij} = interelement coefficient for absorption and enhancement; m''_{ij} = coefficient for line interference; n = principal quantum number or order of diffraction; N = number of electrons or number of ionizations per photon; U_0 = acceleration voltage; Z = atomic number; Z_{eff} = effective atomic number; α_{crit} = critical angle of total reflection; α_{ij} = correction coefficient for interelement matrix effects; Θ = angle between primary and secondary radiation; μ = mass attenuation coefficient; λ = wavelength of radiation; ρ = density; σ = standard deviation.

See also: **Quantitative Analysis; Scanning Probe Microscopy, Theory; X-Ray Fluorescence Spectroscopy, Applications; X-Ray Spectroscopy, Theory.**

Further reading

Argawal BK (1979) *X-ray Spectroscopy*. Berlin: Springer.

Bertin EP (1978) *Introduction to X-ray Spectrometric Analysis*. New York: Plenum Press.

Buhrke VE, Jenkins R and Smith DK (1998) *Preparation of Specimens for X-ray Fluorescence and X-ray Diffraction Analysis*. New York: Wiley.

Hahn-Weinheimer P, Hirner A and Weber-Diefenbach K (1995) *Röntgenfluoreszenzanalytische Methoden. Grundlagen und praktische Anwendung in den Geo-, Material- and Umweltwissenschaften*. Wiesbaden: Vieweg, Braunschweig.

Heckel J (1995) Using BARKLA polarized X-ray radiation in energy dispersive X-ray fluorescence spectrometry. *Journal of Trace Microprobe Techniques* 13: 97–108.

Jenkins R, Gould RW and Gedcke D (1981) *Quantitative X-ray Spectrometry*. New York: Marcel Dekker.

Klockenkämper R (1997) *Total Reflection X-ray Fluorescence Analysis*. New York: Wiley.

Lachance GR and Claisse F (1995) *Quantitative X-ray Fluorescence Analysis*. New York: Wiley.

Whiston C (1987) *X-ray Methods. Analytical Chemistry by Open Learning*. Chichester: Wiley.

Williams KL (1987) *Introduction to X-ray Spectrometry*. London: Allen & Unwin.

X-Ray Fluorescence Spectroscopy, Applications

Christina Streli, P Wobrauschek and
P Kregsamer, Atominstitut of the Austrian
Universities, Wien, Austria

**HIGH ENERGY
SPECTROSCOPY**

Applications

Introduction

X-ray fluorescence spectrometry (XRF) has been applied during the 1970s to 1990s as a versatile tool to many analytical problems. The analysis of major, minor and trace elements in various kinds of samples can be performed qualitatively as well as quantitatively. The working principle is based on the excitation of the sample atoms by high-energy X-rays, followed by the emission of characteristic photons with a certain energy, well correlated to the atomic number Z of each element (Moseley's law). The determination of the energy (or wavelength) of the emitted photon allows qualitative analysis and the determination of the number of emitted characteristic photons allows quantitative analysis. The fundamental physical principle of X-ray fluorescence is described in the article about the theory of X-ray fluorescence spectroscopy. One of the features of XRF besides the accurate, rapid, multielement capacity, is that the analysis can be performed nondestructively. In fact, one has to consider that XRF is a surface-sensitive method, because of the energy of the excited and emitted radiation which is in the range 1–115 keV (Na to U K-radiation). The penetration depth of the primary radiation is some μm or so for low-Z elements and some 100 μm or so for heavy elements, depending also on the matrix or type of sample (solid, liquid and powder samples are common). In any case the surface of the object of investigation has to be representative of the entire volume, and thus requires a homogenous sample. Therefore, because of the sample preparation, the 'nondestructiveness' is lost. To perform XRF a spectrometer is required that consists of an excitation source, sample and a detection system, which can be either wavelength-dispersive or energy-dispersive.

The excitation is mostly performed by X-rays produced in and emitted by an X-ray tube. The spectral distribution of the emitted radiation is partly the bremsstrahlung, with a maximum energy corresponding to the applied voltage, and is partly superimposed by the characteristic lines of the respective anode material. The intensity of the emitted radiation depends on the atomic number of the target and the applied voltage. The intensity of the measured fluorescence signal depends on the intensity and energy of exciting photons hitting the sample atoms. Low-power X-ray tubes operating in the few W range and standard X-ray tubes dissipating up to 3 kW, as well as X-ray tubes with a rotating anode, up to 18 kW, are in use. Photons from radioisotopes are used for special applications, offering an excitation source independent of any power supply. The brightest excitation source is synchrotron radiation. It is emitted when bunches of electrons or positrons with energies in the GeV range are travelling along curved sections in a storage ring. An intensive continuous spectrum with a strong natural collimation in the forward direction is emitted. The radiation is continuous from the eV region to some hundreds of keV and linear polarized in the orbital plane.

Due to the different working principles of wavelength-dispersive (WD) and energy-dispersive (ED) XRF the applications differ strongly and it is necessary to work out the special techniques and their advantages and disadvantages.

Instrumentation and methodology

Wavelength-dispersive spectrometers

Two types of WD instruments are in use, the sequential spectrometer and the simultaneous spectrometer. The sequential spectrometer scans the radiation emitted by the sample by changing the angle sequentially. For different wavelength regions different analyser crystals must be used to fulfil Bragg's equation. A set of 6 to 8 crystals with various lattice spacings d is properly mounted and can be changed automatically. The simultaneous spectrometers consist of various combinations of analyser crystals and detectors, arranged around the sample at fixed angle settings. So each 'channel' is optimized to detect an individual wavelength corresponding to an element. Mostly these channels use focusing optics at the detector to increase the signal. These spectrometers are called simultaneous multielement spectrometers, but the number of simultaneously detected elements depends on the number of channels of the spectrometer. LiF, topaz and other natural crystals are used for the medium-Z elements. The use of synthetic multilayer structures (consisting of alternating layers of a high-Z and a low-Z material with a bilayer thickness of ~1–10 nm) as dispersive elements and measurement in an evacuated environment allows the efficient determination of low-energy characteristic radiation down to even Be–K lines ($\lambda = 11$ nm) if a flow counter with an ultrathin entrance window is used. The big advantage of the WD spectrometers is their excellent wavelength-to-energy resolution, especially in the low-energy region and the high count-rate range (10^6 cps) in operation.

Energy-dispersive spectrometers

An ED detector consists of a semiconductor crystal (Si, Ge) prepared as p-i-n diode, mounted under vacuum, generally operated at 77 K and cooled with liquid nitrogen (LN) and the necessary connected electronics. Therefore a large, heavy dewar (7–30 L of LN) is required for an ED detector. LN consumption is ~1 L a day. The crystal environment has to be a vacuum, and as an entrance window in front of the crystal generally Be is used. Be is chosen as it is available with 8–25 μm thickness, vacuum and light-tight and its absorption of low-energy photons is tolerable. If very low energy photons ($E < 1$ keV) are to be measured, an ultrathin (< 1 μm) entrance window is required. Generally, the energy resolution is much larger with ED detectors than with WD systems, so ED is more susceptible to line overlaps. Especially in the low-energy regions this leads to difficulties in the interpretation of the measured spectrum and mathematical procedures for spectrum deconvolution are required. To overcome the problem of LN cooling, new Peltier cooled detectors are available offering smaller size and lighter weight, but worse resolution. As common practice the value for the energy resolution is given at 5.9 keV and is in the range 130–180 eV for LN-cooled detectors, and 175–200 eV for Peltier cooled detectors.

ED spectrometers measure all photons coming from the sample simultaneously. On one hand this is an advantage, because many elements can be detected within a short time, but on the other hand it is a disadvantage, because the maximum count-rate is limited to 50–80 kcps. The processing of the measured signal from each photon requires a certain length of time and during that interval the system is not ready to process the signal from the next arriving photon. The result is a deadtime, which has to be corrected. The reason for the saturation is not the number of fluorescence photons from the sample mainly, but the exciting radiation being scattered by the sample and the sample carrier. Therefore special

techniques have been developed to reduce the scattered radiation.

Use of monochromatic radiation

The scattering can be drastically reduced, if monoenergetic radiation is used for excitation. Then only monoenergetic photons are arriving at the sample and can be scattered and so contribute to spectral background. Various methods of producing monoenergetic radiation are in use.

Filtered radiation The easiest way to reduce the number of photons not used for sample excitation is the insertion of a filter. It mainly absorbs the low-energy bremsstrahlung, but also a filter with Z as $Z_{anode}-1$ can be used to effectively reduce the characteristic K_β radiation from the anode.

Secondary target excitation The radiation from an X-ray tube is used to excite a suitable target and the fluorescence radiation from the secondary target is used to excite the sample. This method of achieving quasi-monochromatic radiation suffers from a tremendous loss of photon flux, but this can be compensated for by using higher current from high-power tubes.

Radioisotope excitation (RIXRF) A few radioisotope sources (with acceptable half-life) emit radiation in the X-ray region, such as Am-241 (59.5 keV), Cd-109 (Ag–K lines, 22 and 25 keV) and Fe-55 (Mn lines, 5.9 keV), and can be used for excitation. The advantage of radioisotope sources is their independence of a generator and power supply, which makes their use interesting for portable instruments, in field, on-line and even extraterrestrial applications. The disadvantage is the low photon flux in comparison to tube excitation.

Crystal monochromators and multilayer structures Crystal monochromators in the beam path of an X-ray tube allow the selection of either the characteristic line from the anode or the selection of an energy from the continuous spectrum. Mono-chromators with a high reflectivity as well as a large energy band width are usually preferred, because the product of these two parameters determines the photon flux on the sample. In comparison to crystal monochromators, the multilayer offers high reflectivity as well as larger band width ($dE/E = 10^{-2}$). Higher photon fluxes could be obtained. They are used with either X-ray tubes or synchrotron radiation.

XRF using a linear polarized beam

If linear polarized radiation is scattered, in the ideal case no scattering radiation is emitted in the direction of the polarization vector. This effect can be used to reduce the scattering from the sample itself. Barkla polarizers or Bragg polarizers can be used. The Barkla polarizer scatters the entire spectrum of the exciting radiation; the Bragg polarizer acts additionally as a monochromator. Both polarizers use the scattering of the unpolarized radiation through an angle of 90°. Using polarized radiation the spectral background is drastically reduced in comparison to nonpolarized radiation, but again losses in intensity occur.

Total reflection X-ray fluorescence analysis (TXRF)

TXRF is an EDXRF technique, utilizing the total external reflection of X-rays on the smooth plane surface of a reflector material, e.g. polished quartz. If a low divergent beam impinges on the reflector surface at an angle smaller than the critical angle for total reflection, most of the beam is reflected from the surface; only a small part penetrates into the reflector, causing scattered radiation. This leads to a reduced spectral background. The fluorescence signal is enhanced because the primary and also the reflected beam excite the sample, which is deposited on the reflector. Due to the small incidence angle the detector can be brought very close to the sample, so the detection efficiency is high. All these features lead to detection limits in the range of pg. Generally the samples have to be in liquid form. A droplet of 2–100 μL is pipetted onto the sample reflector and the liquid matrix is evaporated. Also, thin films or thin layers, as well as atoms implanted in a reflecting material such as Si-wafers, can be measured and thickness and depths determined.

Microfluorescence analysis (MXRF)

Microfluorescence analysis indicates the analysed area to be very small, leading to spatially resolved information of the sample composition. There are several methods of obtaining a beam with a small diameter, from a simple pinhole to highly sophisticated X-ray optical elements with focusing characteristics. One very effective method is the use of capillaries using the principle of total reflection of X-rays on the inner walls of a glass capillary. Small diameters down to 1 μm can be obtained with satisfactory intensity.

Synchrotron radiation induced XRF (SRXRF)

Synchrotron radiation, as described above, offers several advantages for use as an excitation source for XRF, especially the higher intensity, orders of magnitude greater than that offered by an X-ray

tube. In combination with a monochromator the exciting radiation can be tuned to the energy with the optimum value of the photoelectric cross section for the investigated sample. It is also possible to tune the energy below the absorption edge of a main element in the sample to excite an element at trace levels with $Z < Z_{\text{main element}}$. This method is called *selective excitation* and offers several advantages in trace element analysis. To perform microanalysis, focusing elements are inserted to produce high-intensity microbeams. Also TXRF can be done using SR as exciting radiation.

Trace element analysis

To detect elements at trace levels – μg g^{-1} (ppm), ng g^{-1}(ppb) or even pg g^{-1} (ppt) – with XRF, generally special techniques as well as special sample preparation methods have to be used. The relevant quantity for trace element analysis is the limit of detection (DL), which is given by the formula

$$\text{DL} = \frac{3 \cdot \sqrt{I_{\text{B}}/t}}{S}$$

I_{B} is the background intensity, t the measuring time and S the sensitivity (cps ppm^{-1} or cps ng^{-1}). Either increasing the sensitivity or reducing the background leads to a reduction of detection limits. Therefore, special techniques of XRF, mainly EDXRF techniques, are applied, like TXRF, MXRF or SRXRF, as methods for trace element analysis. Detection limits range from ng to fg or μg g^{-1} to pg g^{-1}.

Applications

The applicability of XRF is almost unlimited with respect to type of sample and concentration range, but depends very much on the chosen technique. It can be used for on-line analysis in production processes or in-field measurements of geological samples with portable instruments, or quality control, such as impurity determination on Si-wafer surfaces at ultratrace levels, and environmental investigations. Also sample preparation is an important factor influencing the applicability (see **Figure 1**). For fine art object investigation or precious objects from museums, absolute nondestructiveness is required, contrary to environmental sampling where homogenizing and pressing to a pellet is no problem.

To describe the various fields of application in detail they are divided into categories.

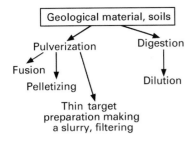

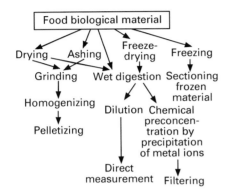

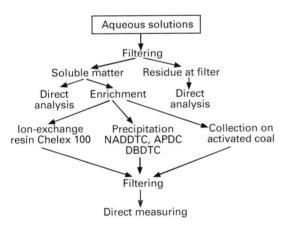

Figure 1 Overview of sample preparation techniques typical for XRF.

Metals and alloys

Process control in today's highly automated production facilities is strongly dependent upon fast, precise and accurate chemical analysis, and XRF has been found to be widely applicable in the metal industries. XRF of metallic samples includes several solvable problems, especially in the areas of sample preparation and modelling calculations to convert intensities into concentration data. In general, metallic samples do not need complicated sample preparation, but the analytical information is derived from a volume close to the surface which must be polished. XRF is applied to various kinds of alloys, such as Na–Mg alloys, and Al, Ti, ferrous, Ni, Cu, Zr, W and Au alloys, bronzes

and brass. Typically, simultaneous WD spectrometers with an automatic sample changer are used.

Mining and ore processing

In mining and ore processing XRF is used for quality and process control. The spectrometers used differ very much depending on their application. Laboratory spectrometers for quality control may be WD or ED systems whith tube excitation. The on-stream spectrometers are located at in-plant locations, can be WD and ED systems, with radioisotope excitation or X-ray tube excitation and equipped with flow cells. In-stream instruments can be installed in slurry streams, mainly equipped with radioisotope excitation and scintillation counters for single-element determination. Field instruments must be portable and battery operated. The big advantage of XRF over other analytical tools for that application is its simplicity and speed. The usefulness of X-ray instruments to selective mining is well established, the information being used for ore–waste sorting.

Cement analysis

Finishing cements and raw mixes typically contain Ca, Si, Al and O at high concentrations, plus Fe, K, Mg and Na at low concentrations. One of the major problems in accurate cement analysis is homogeneity, particularly in the case of raw mixes, where the source of raw material may be variable. Most of the elements to be determined are of low atomic number, hence the penetration of their characteristic lines will be of the order of a few micrometres only. Careful grinding and pelletizing will suffice, but the fusion bead technique with Li_3BO_4 is strongly recommended. Simultaneous multichannel WD systems are ideally suited for this kind of application.

Lubricating oil analysis

Raw or used oils are usually analysed for additive-element content including Ba, Zn, Mn, Ca, P and Cl, plus naturally occurring elements including S, N, Ni and Na. In blended stocks the concentration of these elements would typically lie in the range 0.01–2.5%. In a standard case the analysis will be performed in the liquid phase and under a He atmosphere. Large matrix effects are likely because of the variable concentration levels of relatively heavy elements in a very low average atomic number matrix.

Geology

XRF offers a rapid, accurate, low-cost method of analysis. Fully automated XRF spectrometers, sequential as well as simultaneous WD instruments, and ED instruments are in use. The analysis of geo-chemical samples often involves the analysis of samples having concentrations ranging from 0.0001 to 80%. Elements from Na to U are routinely determined. Detection limits depend very much on sample preparation and are in the range 20–1000 µg g^{-1} for low-Z elements, 5–10 µg g^{-1} for medium-Z elements and 1–20 µg g^{-1} for high-Z elements.

One spectacular application of EDXRF was the Mars Pathfinder which was designed to inspect the rocks on the surface of Mars. Radioisotope excitation was used and one of the new electrically cooled ED detectors mounted on a small vehicle. Surface analysis of rocks could be performed quickly and the data could be transferred to Earth.

Fine art and archaeological objects

In principle, with EDXRF, nondestructive analysis can be performed, which opens up the wide field of art and museum objects. Coins, bronzes, paintings, pottery, ceramics and ancient glasses can all be analysed. Paintings can be analysed pixel by pixel. With a spectrometer mounted on an *x–y* stage, a selected area can be analysed and the whole painting can be scanned. In particular, MXRF techniques are very well adapted to analysing specific points on figures or vessels, as well as lines on ancient documents.

Environmental analysis

This is probably the most versatile and important application as all kinds of biological material such as plants, roots, needles, food-stuff, algae and lichens as biomonitors can be analysed with XRF. Lichens offer the advantage of a natural sampler collecting aerosols without exchange with the substrate. See **Figure 2** as example of a sample lichen measured with an EDXRF spectrometer. Soil, river sediments, sewage sludge, dust, coal fly ash, car exhaust and fog condensate or the aerosols, sampled in impactor stages, are well suited for XRF. All kinds of liquids can be analysed: river water, sea water, snow, ice. However, special techniques for sample preparation should be applied especially when trace element analysis is required.

Microfluorescence analysis

MXRF allows spatial resolution of analysis. To perform it with sufficient intensity in the laboratory with X-ray tubes, special optics should be applied such as capillaries. Using capillaries, X-ray tube excitation with SR allows much higher intensities, and so lower detection limits can be obtained. Applications range from microelectronics and plating thickness,

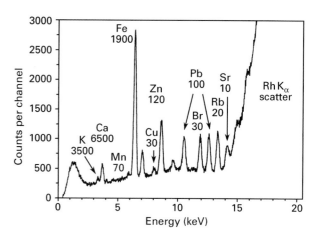

Figure 2 Spectrum of lichen sample. Pressed pellet, measured with a standard EDXRF spectrometer, Rh tube 30 kV, 0.2 mA, 500 s, Pd filter, Si(Li) detector; concentration values of respective elements given in µg g^{-1}.

maps of bone cross-sections, superconductor films, human hair, pig heart muscles, metal alloys, leaves, chinaware, environmental particles, tree rings and glass fragments.

Medicine

In vivo measurements as well as *in situ* measurements and analysis of malignant cells, as well as tissue sampling have been performed. *In vivo* measurements started with the determination of iodine in the thyroid and range from Cd in liver and kidney and Pt in kidneys and tumours, to Hg in the wrists and skulls of dentists, Pb in various near-surface bones, Cu in the eye and Fe in the skin. For the *in vivo* measurements, the use of polarized radiation offers big advantages. Hg and Pb can be analysed by their K-radiation thus giving information from even deeper tissues. Also, the analysis of biopsy samples, whole blood and blood serum (Se) can be performed using EDXRF. Trace element content in malignant and benign tissues was investigated, as well as lung tissue from different factory workers, showing different elements at higher concentrations corresponding to their profession. Various body fluids were analysed, and correlation between trace element content and diseases found. EDXRF, TXRF and SRXRF were used, depending on the task.

TXRF applications

TXRF extends EDXRF to a method for trace and ultratrace element analysis. A special feature is the small sample amount required, which is in the range µg–ng of a solid material and less than 100–10 µL of a liquid. Therefore, TXRF is a microanalysis method and samples can seldom be analysed as-received. Pretreatment is generally required to prepare the samples as solutions, suspensions, fine powders or thin sections. For a determination of ultratraces, the matrix of the sample should be separated and removed. All preparation techniques can be applied that have been tested with other methods of atomic spectrometry, e.g. AAS, inductively coupled plasma mass spectrometry. The quantification is generally very simple, because the sample forms a thin layer, so the thin-film approximation is valid. One element of known concentration has to be added as internal standard. Quantification can be performed after the determination of the sensitivity factors of all elements relative to the internal standard. Also, surface and thin-layer analysis, as well as analysis of atoms below a reflecting surface, can be performed by varying the angle of incidence in the region of total reflection. This angle-dependent intensity profile allows a qualitative differentiation between contaminations on the surface, in the layers, in implantations or in so-called residues after evaporation of liquid samples. Quantitative determinations can be made by applying an algorithm deduced from theory in combination with an external standard.

It is obvious that the sample preparation technique used influences the detection limits. **Table 1** shows this influence on various samples from different fields of application. **Table 2** gives an overview of applications of TXRF already analysed. **Figure 3** shows a spectrum of a water standard reference sample (NIST 1643c) obtained with a TXRF vacuum chamber, constructed at Atominstitut, Vienna. Generally, an excellent field of application of TXRF in trace element analysis can be seen in liquid samples. All kinds of liquids, ranging from different kinds of water to acids and oils, as well as body fluids, can be analysed. Environmental samples, like airborne particles, plant material or medical and biological samples such as tissue can be analysed directly on a reflector.

The main industrial application of TXRF is the surface quality control of Si-wafer material. Wafers offer the required flatness and are polished, so that they can be directly analysed by TXRF. Several commercial instruments have been developed as wafer analysers and some 100 instruments are utilized in the semiconductor industry. TXRF is capable of checking the contaminations brought in by different steps during the production process. The required sensitivity is now 10^9 atoms cm^{-2} for transition

Table 1 Influence of sample preparation on detection limits in TXRF (after Klockenkämper R (1997))

Sample	Drying	Freeze-drying	Chemical matrix separation	Open digestion	Ashing	Suspension	Solution	Pressure digestion	Freeze cutting
Rain, river water	0.1–3 ng mL^{-1}	20–100 pg mL^{-1}	3–20 pg mL^{-1}	1–3 ng mL^{-1}					
Blood, serum			Digest: 2–30 ng mL^{-1}	20–80 ng mL^{-1}	40–220 ng mL^{-1}				
Air dust, ash, aerosols				5–200 µg g^{-1}		10–100 µg g^{-1}	0.1–3 µg g^{-1}		
Air dust on filter					0.6–20 ng cm^{-2}			0.2–6 ng cm^{-2}	
Suspended matter				3–25 µg g^{-1}		10–100 µg g^{-1}			
Sediment						10–100 µg g^{-1}		15–300 µg g^{-1}	
Powdered biomaterial						1–10 µg g^{-1}		0.2–2 µg g^{-1}	
Fine roots				1–10 µg g^{-1}	Digest: 0.1–1 µg g^{-1}				
High-purity acids	5–50 pg mL^{-1}								
Tissue, foodstuff, biomaterial									0.5–5 µg g^{-1}
Mineral oil							1–15 µg g^{-1}		
Mussel, fish								0.1–1 µg g^{-1}	
High purity water		1 pg mL^{-1}							

Table 2 Applications of TXRF

Environment	
Water	Rain, river, sea, drinking water, waste water.
Air	Aerosols, airborne particles, dust, fly ash.
Soil	Sediments, sewage sludge.
Plant material	Algae, hay, leaves, lichen, moss, needles, roots, wood.
Foodstuffs	Fish, flour, fruits, crab, mussel, mushrooms, nuts, vegetables, wine, tea.
Various	Coal, peat.
Medicine/biology/pharmacology	
Body fluids	Blood, serum, urine, amniotic fluid.
Tissue	Hair, kidney, liver, lung, nails, stomach, colon.
Various	Enzymes, polysaccharides, glucose, proteins, cosmetics, biofilms.
Industrial/technical applications	
Surface analysis	Si-wafer surfaces, GaAs-wafer surfaces.
Implanted ions	Depth and profile variations.
Thin films	Single layers, multilayers.
Oil	Crude oil, fuel oil, grease.
Chemicals	Acids, bases, salts, solvents.
Fusion/fission research	Transmutational elements in Al Cu, iodine in water.
Mineralogy	
Ores, rocks, minerals, rare earth elements.	
Fine arts/archaeological/forensic	
Pigments, paintings, varnish.	
Bronzes, pottery, jewellery.	
Textile fibres, glass, cognac, dollar bills, gunshot residue, drugs, tapes, sperm, fingerprints.	

elements like Cr, Fe, Co, Ni, Cu and Zn. TXRF has an 'up-time' of 90% and is nondestructive. Surface mapping can be performed and differentiation between film type or particle type is possible. The detection limits can be improved by more than two orders of magnitude, if the impurities of the entire surface of the wafer are collected and preconcentrated prior to TXRF analysis. The native oxide layer is dissolved by HF vapour and the impurities remaining on the surface are collected by scanning the wafer with a drop of a suitable liquid. This method has the advantage of higher sensitivity, but nondestructiveness is lost.

It is also possible to measure the thickness of near-surface layers in the range 1–500 nm on reflecting substrates with TXRF. Single layers as well as multilayer samples can be analysed. Also, atoms implanted in the reflecting surface can be detected. The implantation depth as well as the depth profile can be determined.

TXRF can also be applied to fine art and museum objects. The sampling technique – a dry cotton bud can be used to rub off a small amount of paint – can only be applied during restoration, because the varnish has to be removed. For analysis the bud is dipped onto a sample carrier by a single tip. An amount of less than 100 ng is transmitted and can be analysed.

Application of SRXRF

The rapid development of the SR X-ray sources since about 1975 is starting to have an impact on X-ray analysis. Due to the features of SR, especially the small source size and therefore the high brilliance, the use of microprobes is obvious. There are several approaches to producing a microbeam; the simplest is to use a pinhole collimator, but more sophisticated systems use focusing optics. Monochromatic as well as continuous radiation is used. Because SR is linearly polarized in the orbital plane, scattering from the sample is reduced, leading to low detection limits. SRXRF is a trace element analytical method as well as an MXRF method. SRXRF is performed at several SR facilities, most prominently NSLS Brookhaven, HASYLAB Hamburg, SSRL Stanford, SLS Daresbury, Photon Factory Tsukuba, DCI Lure and ESRF Grenoble. The available spot sizes are in the region of 10 μm and the detection limits in the low pg to fg range. Interesting applications were found in the fields of geology (mineral inclusions can be analysed), as well as in biology and medicine (distribution of trace elements in bone, tooth, brain, hair or algae strands, tree rings and aerosol particles), giving interesting information. Also, application in archaeology is found, letters in different ancient papers being analysed to allow the differentiation of the ink used to help identify the workshop. Even extraterrestrial minerals and rocks have been analysed.

TXRF can also be done using SR as the excitation source (SR-TXRF) offering the advantage of higher photon flux and improved detection limits. Experiments are performed at SSRL, HASYLAB, Photon Factory and ESRF. The main application is the surface quality control of Si wafers. With SR-TXRF, detection limits on wafer surfaces of 10^7 atoms cm^{-2} have been obtained. At SSRL there is a beamline dedicated for routine wafer analysis. SR is the ideal source for the excitation of low-Z elements. It offers high intensity also in the low-energy region, in comparison to standard X-ray tubes. They do not emit enough intensity in the low-energy region; therefore the analysis of low-Z elements always lacks intensity

Table 3 Overview of various applications (from Török S and Van Grieken R (1994) X-ray spectrometry. *Analytical Chemistry* **66:** 186R–206R; Török S, Labar J, Injuk J and Van Grieken R (1996) *Analytical Chemistry* **68:** 467R–485R)

Field	Method	Field	Method
Archaeology		Se in soil	SR-TXRF
General	MXRF	Soil, marine sediments	WDXRF
Paintings	TXRF	Impurities in ice	TXRF
Obsidian	EDXRF	V, Ni, in oil, asphaltene	EDXRF
Medals	WDXRF	Bitumen solutions	EDXRF
Pottery	WDXRF	*Geology*	
Pigments	TXRF	Rocks	WDXRF
Biomedical	MXRF	Mineral grains	SRXRF
General	SR-TXRF	Soils, sediments	MXRF
Single-cell analysis	EDXRF, TXRF	Fossilized bone	WDXRF
Biopsy samples	EDXRF, RIXRF	Crysolite	EDXRF
Blood	EDXRF	Geological samples	WDXRF
Skin *in vivo*	RIXRF	Phosphate in rocks	EDXRF
Bone	EDXRF	Oxides, silicates, carbonates	WDXRF
Bone *in vivo*	RIXRF	Liquid petroleum products	EDXRF
Leaves	TXRF, WDXRF	Microlayer of Fe–Mn nodules	SRXRF
Hair	EDXRF	Au in micas	SRXRF
Vegetables	TXRF	*Materials science*	
Plants	WDXRF	Thin-film characterization	XRF, EDXRF
Lichens	WDXRF	Impurities on Si-wafers	TXRF, SRXRF
Moss	RIXRF	Multilayers	EDXRF
Cd in kidney	RIXRF	Superconductors	WDXRF
Mussel shells	WDXRF	Zirconium oxide	WDXRF
Cu, Se, Zn in kidney	SR-TXRF	Alumina	WDXRF
Marine bivalve shells	SRCXRF	Cu corrosions	EDXRF
Pt in tumour tissues	EDXRF	Ultrapure reagents	TXRF
Cu in human serum	EDXRF	Glasses	WDXRF, EDXRF
Pb in bones, serum, blood	RIXRF	Ferroalloys	EDXRF
Fe *in vivo*	RIXRF	High-purity Cu	EDXRF
Hg *in vivo*	RIXRF	GaAs-wafers	TXRF
Amniotic fluids	TXRF	Electrolytic solutions	RIXRF
Plankton in polluted lakes	EDXRF	Al_2O_3 thin films	WDXRF
Meadow moths	SRXRF	Plastic materials	EDXRF, WDXRF
Environmental		Ceramic materials	WDXRF
Aerosols	WDXRF, EDXRF, TXRF, SRXRF	Hf in Zr matrix	EDXRF
		Ga in polyurethane foam	WDXRF
Fly ash	SRXRF	P in PbO films	EDXRF
Rain water	TXRF	Cu, Sr, Bi film on MgO	RIXRF
River water	TXRF	Molybdate crystals	EDXRF
Sea water	TXRF	Ferrous alloy	WDXRF
Sediments, suspensions	EDXRF, TXRF	Ta in Ti–Ta alloys	WDXRF
Waste	EDXRF	Pb in houseware	RIXRF
Coal	EDXRF, WDXRF	Textile fibres	TXRF
Dust	EDXRF	W analysis	TXRF
Pb in dust	WDXRF	HTSC films	SRXRF

HTSC = High-temperature semiconductor.

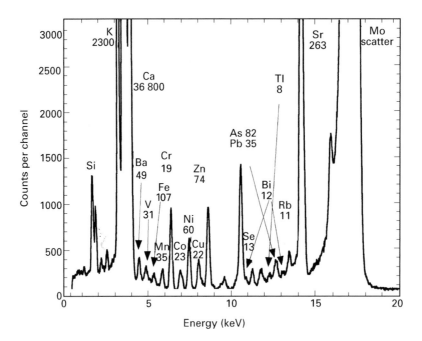

Figure 3 Spectrum of water sample, NIST 1643c standard reference material, 10 μL, dried on a quartz reflector, measured with the TXRF vacuum chamber of Atominstitut, Mo monochromatic excitation, 40 kV, 50 mA, 1000 s; concentration values of respective elements given in μg L^{-1}.

of the fluorescence lines. TXRF is also applied to the determination of low-Z elements using a special detector. Detection limits of 60 fg for Mg have been achieved using SR and have to be compared to 7 pg with windowless Si-anode tube (prototype) excitation.

SR-TXRF generally is a very fast growing field and the problem of reduced access to SR sources for routine analytical applications is becoming less severe due to the large number of dedicated facilities.

Table 3 gives an overview of applications and techniques published from 1994 to 1998.

Conclusions

Applications range from on-line analysis and in-field inspections to ultratrace analysis of semiconductor surfaces. There is almost no sample that cannot be analysed by XRF as long as elemental analysis is required. The achievable detection limits depend on the method used and range from μg g^{-1} (ppm) to pg g^{-1} (ppt). Nondestructive analysis can be performed but sometimes sophisticated sample preparation techniques are required. The elemental range (Be to U) depends on the excitation source as well as the detection system. Generally XRF can be seen as a work-horse for elemental analysis and is easy to use.

List of symbols

d = lattice spacing; DL = detection limit; I_B = background intensity; S = sensitivity; t = measuring time; Z = atomic number; λ = wavelength.

See also: **Environmental and Agricultural Applications of Atomic Spectroscopy; Environmental Applications of Electronic Spectroscopy; Geology and Mineralogy, Applications of Atomic Spectroscopy; Inorganic Compounds and Minerals Studied Using X-Ray Diffraction; IR and Raman Spectroscopy Studies of Works of Art; IR Spectroscopy Sample Preparation Methods; MRI of Oil/Water in Rocks; X-Ray Fluorescence Spectrometers.**

Further reading

Bertin EP (1978) *Introduction to X-ray Spectrometric Analyis.* New York: Plenum Press.

Carpenter DA (ed) (1997) Special Issue on Micro X-ray Fluorescence Analysis. *X-ray Spectrometry* 26(6):

Ellis A, Potts Ph, Holmes M, Oliver GL, Streli C and Wobrauschek P (1996) Atomic spectrometry update: X-ray fluorescence spectrometry. *Journal of Analytical Atomic Spectroscopy* 11: 409R–442R.

Ellis A, Potts Ph, Holmes M, Oliver GL, Streli C and Wobrauschek P (1997) Atomic spectrometry update: X-ray fluorescence spectrometry. *Journal of Analytical Atomic Spectrometry* **12**: 461R–490R.

Herglotz HK and Birks LS (eds) (1978) *X-ray Spectrometry*. New York: Marcel Dekker.

Holynska B (1993) Sampling and sample preparation in EDXRS. *X-ray Spectrometry* **22**: 192–198.

Iida A and Gohshi Y (1991) Trace element analysis by X-ray fluorescence. In: Ebashi S, Koch M and Rubenstein R (eds) *Handbook on Synchrotron Radiation*, Vol. 4, pp 307–349. Amsterdam: North Holland, Elsevier.

Jenkins R, Gould RW and Gedcke D (1981) *Quantitative X-ray Spectrometry*. New York: Marcel Dekker.

Klockenkämper R (1997) *Total-Reflection X-ray Fluorescence Analysis*. New York: Wiley.

Sparks CJ (1982) X-ray fluorescence microprobe for chemical analysis. In: Winick H and Doniach S (eds) *Synchrotron Radiation Research*, pp 459–509. New York: Plenum Press.

Török S and Van Grieken R (1994) X-ray spectrometry. *Analytical Chemistry* **66**: 186R–206R.

Török S, Labar J, Injuk J and Van Grieken R (1996) X-ray spectrometry. *Analytical Chemistry* **68**: 467R–485R.

Van Grieken R and Markowicz A (eds) (1993) *Handbook of X-ray Spectrometry*. New York: Marcel Dekker.

Wielopolski L and Ryon RW (eds) (1995) Workshop at the Denver X-ray conference on *in vivo* XRF measurement of heavy metals. *Advances in X-ray Analysis* **38**: 641.

X-Ray Spectroscopy, Theory

Prasad A Naik, Centre for Advanced Technology, Indore, India

> HIGH ENERGY
> SPECTROSCOPY
> Theory

X-ray is the region of the electromagnetic spectrum lying between gamma rays and extreme ultraviolet (XUV / EUV) corresponding to a wavelength range of about 0.1 to 100 Å. The radiation on the lower end of the XUV region, up to about 300 Å, is also sometimes referred to as X-ray. On the lower wavelength side, radiation of shorter wavelengths is termed X-ray if it is nonnuclear in origin. The wavelength of the radiation is related to the photon energy by the standard relation E (keV) = 12.4/λ(Å). In terms of energy, the X-ray region is roughly between 125 eV and 125 keV. Being electromagnetic radiation, X-rays can be reflected, refracted, scattered, absorbed, polarized etc. They also show interference and diffraction effects.

There are several sources of X-rays such as a Coolidge tube, vacuum sparks, hot-dense fusion plasmas, synchrotron, pinch devices, muonic atoms, beam-foil interaction, stellar X-ray emitters, solar flares, etc. The X-rays originating from all these sources can be broadly categorized into main types: (1) atomic inner shell transitions, (2) emission by free electrons, (3) X-rays from few electron systems. The basic spectroscopic aspects of the various types of X-rays are discussed in this article.

X-rays from inner shell transitions in atoms

X-rays are produced when an electron in an outer shell of an atom jumps to an inner shell to fill an electron vacancy. The difference in energy is emitted as an X-ray photon. The vacancy giving rise to such a transition can be produced by an energetic photon, bombardment of charged particles (e⁻, p, α ..), or by nuclear processes such as internal conversion, K-capture, etc. If a charged particle collision or a nuclear process produces the vacancy, the resulting X-ray emission is called primary. If the vacancy is produced by an X-ray photon, the subsequent emission is called secondary or fluorescence radiation.

In all these cases the singly ionized atom lowers its energy by emission of a photon of definite wavelength which is characteristic of the emitting atom. Hence, these X-rays are also called characteristic X-rays.

Characteristic X-rays

The most energetic X-ray emission comes when a vacancy in a K shell ($n = 1$) is filled by an outer electron. Removal of a 1s electron from a neutral atom raises it to the highest energy state represented by 1s1s⁻¹ or 1 $^2S_{1/2}$ or K_I. Removal of a 2s electron

$(2s2s^{-1})$ gives rise to L_1 state $(2\ ^2S_{1/2})$. Removal of a 2p electron from a filled 2p shell $(2p^5p^{-1})$ gives rise to L_{II} $(2^2p_{1/2})$ and $L_{III}(2\ ^2P_{3/2})$ states, respectively. Similarly, removal of a 3s,3p,3d electron from filled shells gives rise to $(3s3s^{-1},\ 3p^53p^{-1},\ 3d^93d^{-1})$. $M_I(3\ ^2S_{1/2})$, $M_{II}(3\ ^2P_{1/2})$, $M_{III}(3\ ^2P_{3/2})$, $M_{IV}(3\ ^2D_{3/2})$, $M_V(3\ ^2D_{5/2})$ states.

The selection rules applicable to optical dipole transitions also apply to X-ray transitions. The rules are $\Delta L = \pm1$, $\Delta j = 0,\pm1$. Intensity rules are also the same as those applicable to optical transitions. The transitions obeying selection rules are called normal transitions. Not all transitions allowed by selection rules are observed. On the contrary, some transitions, which are not allowed by selection rules, are sometimes observed. These are called forbidden transitions. The observed lines were initially given names as per their observed line intensities. The Siegbahn notation used to name various observed lines is given in **Table 1**. As this nomenclature is intensity-based

(e.g. $K_{\alpha1}$ more intense than $K_{\alpha2}$) and was adopted long before the origin of the lines was explained spectroscopically, this notation is somewhat confusing.

Moseley's law gives the frequency of line transition as $v(\text{cm}^{-1}) = KR(Z-\delta)^2$, where R is the Rydberg constant, and Z is the atomic number. For example, for K_α: $\delta = 1$ and $K = 3/4$, which gives $v(\text{cm}^{-1}) = (3/4)R(Z-1)^2$. Similarly, for $L_{\beta1}$, $v(\text{cm}^{-1}) = (1/2^2-1/3^2)R(Z-7.4)^2$.

The energy levels formed by single electron removal are similar to those of a hydrogen-like atom, obtained by replacing Z by $(Z-\sigma)$ and $(Z-s)$ in the Sommerfeld formula, where σ and s are screening constants. The energy of a level is given by the modified Sommerfeld formula as

$$T = R\,(Z-\sigma)^2/n^2 + R\,\alpha^2\,(Z-s)^4$$
$$(n/k - 3/4)/n^4 + O(\alpha^4, \alpha^6, \ldots)$$

Table 1 Siegbahn notation for various inner shell X-ray transitions

Shell structure	n	l	i	Optical notation	X-ray notation	K series	L series			M series				
1s 1s^{-1}	1	0	1/2	$^2S_{1/2}$	K_I									
					K_I					- : Forbidden by selection rules				
2s 2s^{-1}	2	0	1/2	$^2S_{1/2}$	L_I	-				[ ] : Forbidden but observed.				
2p^5 2p^{-1}	2	1	1/2	$^2P_{1/2}$	L_{II}	α_2				$\sqrt{}$: Allowed by selection rules but no name given				
2p^5 2p^{-1}	2	1	3/2	$^2P_{3/2}$	L_{III}	α_1								
							L_I	L_{II}	L_{III}					
3s 3s^{-1}	3	0	1/2	$^2S_{1/2}$	M_I	-	-	η	l					
3p^5 3p^{-1}	3	1	1/2	$^2P_{1/2}$	M_{II}	β_3	β_4	-	-					
3p^5 3p^{-1}	3	1	3/2	$^2P_{3/2}$	M_{III}	β_1	β_3	-	-					
3d^9 3d^{-1}	3	2	3/2	$^2D_{3/2}$	M_{IV}	$[\beta_5]$	$[\beta_{10}]$	β_1	α_2					
3d^9 3d^{-1}	3	2	5/2	$^2D_{5/2}$	M_V	$[\beta_5]$	$[\beta_9]$	-	α_1					
										M_I	M_{II}	M_{III}	M_{IV}	M_V
4s 4s^{-1}	4	0	1/2	$^2S_{1/2}$	N_I	-	-	γ_5	β_6	-	$\sqrt{}$	$\sqrt{}$	-	-
4p^5 4p^{-1}	4	1	1/2	$^2P_{1/2}$	N_{II}	$\beta_2(\gamma_2)$	γ_2	-	-	$\sqrt{}$	-	-	ξ_2	-
4p^5 4p^{-1}	4	1	3/2	$^2P_{3/2}$	N_{III}	$\beta_2(\gamma_1)$	γ_3	-	-	$\sqrt{}$	-	-	$\sqrt{}$	ξ_1
4d^9 4d^{-1}	4	2	3/2	$^2D_{3/2}$	N_{IV}	$[\beta_4]$		γ_1	β_{15}	-	$\sqrt{}$	γ_2	-	-
4d^9 4d^{-1}	4	2	5/2	$^2D_{5/2}$	N_V	$[\beta_4]$		-	β_2	-	-	γ_1	-	-
4f^{13} 4f^{-1}	4	3	5/2	$^2F_{5/2}$	N_{VI}	-	-	-	-	-	-	-	β_1	α_2
4f^{13} 4f^{-1}	4	3	7/2	$^2F_{7/2}$	N_{VII}	-	-	-	-	-	-	-	-	α_1
5s 5s^{-1}	5	0	1/2	$^2S_{1/2}$	O_I	-		γ_8	β_7	-	$\sqrt{}$	$\sqrt{}$	-	-
5p^5 5p^{-1}	5	1	1/2	$^2P_{1/2}$	O_{II}	δ_2	γ_4	-	-	$\sqrt{}$	-	-	$\sqrt{}$	-
5p^5 5p^{-1}	5	1	3/2	$^2P_{3/2}$	O_{III}	δ_1	γ_4	-	-	$\sqrt{}$	-	-	$\sqrt{}$	$\sqrt{}$
5d^9 5d^{-1}	5	2	3/2	$^2D_{3/2}$	O_{IV}	-	-	γ_6	β_5	-	$\sqrt{}$	$\sqrt{}$	-	-
5d^9 5d^{-1}	5	2	5/2	$^2D_{5/2}$	O_V	-	-		β_5	-	-	ε	-	-

where α is the fine structure constant, n is the total (principal) quantum number, and k is the Sommerfeld original azimuthal quantum number; $k = 1, 2$ and 3 for s,p and d electrons, respectively.

A pair of terms having the same n, s, L but different j (in the $^{2s+1}L_j$ Russel–Saunders notation) is called a spin-relativity doublet (earlier called a regular doublet). L_{II}–L_{III}, M_{II}–M_{III}, M_{IV}–M_V are examples of such doublets. The screening constant σ (same for a doublet) depends on s, p, d sub levels but s is the same for a spin-relativity doublet, almost independent of Z (s values: L_I:2.0, L_{II}–L_{III}: 3.5, M_I:6.8, M_{II}–M_{III}:8.5, M_{IV}–M_V:13).

A pair of terms having the same n, s and j but different L values is called as screening doublet (or irregular doublet). L_I–L_{II}, M_I–M_{II}, M_{III}–M_{IV} are examples of such screening doublets.

From the first term of the Sommerfeld formula, one obtains the screening doublet law, which states that the difference between the square roots of the term values of a given doublet is constant, i.e. independent of Z. This term also gives the irregular doublet law which states that the difference between term values of an irregular (screening) doublet is a linear function of Z.

The second term of the Sommerfeld formula gives the separation in energy for a spin-relativity doublet as proportional to the fourth power of the screened atomic number, i.e. $(Z–s)^4$. This is referred to as the regular doublet law. For example, for the L_{II}–L_{III} doublet (same σ), $\Delta v (cm^{-1}) = (R\alpha^2/16)(Z–3.5)^4$.

Satellite lines

These are weaker lines appearing on the shorter wavelength side of the normal (characteristic) lines. They were initially referred to as nondiagram lines because unlike the normal lines, these lines did not fit conveniently in the energy level diagrams of that time. Later, it was realized that the energies can also be predicted using energy level diagrams of multiply charged ions (**Figure 1**).

The satellite lines are due to additional electron vacancy in a doubly ionized atom. Due to the absence of a second electron, the energy levels shift to the higher energy side (relative to those of a singly ionized atom) due to reduced Coulomb screening. As a result, single electron transitions in such atoms (doubly ionized) are at a slightly higher energy compared to those in singly ionized atoms. As the probability of creation of two electron vacancies in an atom is much smaller than that of a single vacancy, the intensity of satellite lines is much less than that of normal lines.

Satellite lines are denoted as α', α'', α''' ... where the higher number of primes implies lower intensity. If K–M (i.e. K_I–M_{II},M_{III}) denotes a K_β transition, then KL–LM denotes the satellite line K_β''' which is due to an additional vacancy in the L shell (**Figure 1**).

Hypersatellite This is a special case of satellite lines wherein X-rays originate from atoms with two holes in the same inner shell. A K-hypersatellite line appears when an atom has initially two vacancies in its

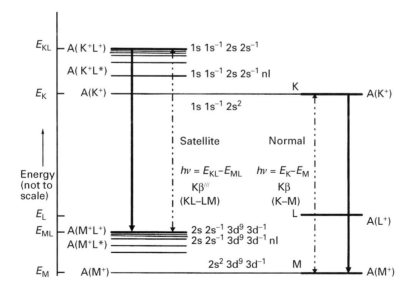

Figure 1 Energy diagrams for a K_β transition and $K\beta'''$ satellite transition. A() denotes the state of an atom. For example, A(M$^+$L*) denotes an atomic level where an M shell electron has been removed and an L shell electron is in an outer bound (excited) state. The energy difference E_{KL}–E_K is more than E_{ML}–E_M as more energy is required to remove an L shell electron in the presence of a K shell vacancy than in the presence of a M shell vacancy. Solid downward arrows show vacancy transitions, and dashed lines with double arrows show radiative transitions.

K shell. They are denoted by superscript H. For example, $K^H_{\alpha 1,2}$ is a hypersatellite of the $K_{\alpha 1,2}$ line (**Figure 2**). Due to the strong reduction of screening, there is a large energy difference between a normal line and its hypersatellite. Hypersatellites are more easily observed in heavy ion collision spectra or radioactive decay by K-electron capture.

Plasma satellites   These are low intensity structures, which can appear on either side of a parent line. They are equally spaced and correspond to an energy difference of $(h/2\pi)\omega_p$, where ω_p is the plasma frequency given by $\omega_p^2 = (N_e e^2/\varepsilon_0 m)$ where N_e is the electron density in the conduction band of the solid target material.

If an X-ray photon loses its energy in exciting a plasmon, one obtains a low energy plasmon satellite. On the other hand, if plasmons already exist in the solid at the time of X-ray emission, they can lead to high energy satellites.

Auger and allied processes

These processes compete with the radiative process in de-excitation of the atom. Thus they influence the energy-width of X-ray lines. Moreover, they transfer a vacancy from one level to another and thereby affect the intensity of X-ray lines. Many of these processes also lead to double ionization which (as discussed earlier) gives rise to satellite lines. Some of these processes are briefly described below.

Auger ionization   In this case, an inner shell vacancy is filled by an electron from an outer shell and the excess energy instead of being emitted as a photon, is consumed in ejecting a second electron (called an

Auger electron) from the atom. This is a radiation-less process leading to double ionization. A transition of an electron from an L shell to a K shell vacancy, accompanied by the ejection of an L shell electron, is represented as a KLL transition. The energy of the ejected electron in KLL transition is $E_{Auger} = E_K - E_{LL}$. Here E_K is the energy of the atom with one electron missing in the K shell and E_{LL} is the energy of the level formed by ejection of an L shell electron from a singly ionized atom with an L shell electron vacancy (**Figure 3A**).

Radiative Auger process   In this process, the Auger electron, instead of carrying away all the energy, receives only a part of it and the rest is emitted as an X-ray photon. The energy of the photon is given by $hv = E_{Auger} - E_{kin}$ where E_{Auger} is the energy of the electron in the Auger transition involving the same three levels (e.g. KLL), and E_{kin} is the kinetic energy of the ejected electron. The energy spectrum of the emitted X-rays is continuous, with a maximum energy $hv = E_{Auger}$, corresponding to zero kinetic energy of the ejected electron (**Figure 3B**).

Semi-Auger process   If, in the above radiative Auger process, the electron, instead of being ejected out of the atom, is transferred to some outer bound state, then the energy of the emitted X-ray photon would be discrete having the value $(E_K - E_L) - E_b$. Here, E_b is the difference in energy between the outer bound state and the energy level of the electron in an atom with a L shell vacancy (**Figure 3C**). This process is referred to as a semi-Auger process.

Coster–Kronig process   This is a special case of Auger ionization in which the vacancy transition is

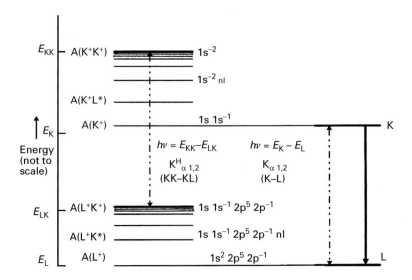

Figure 2 Energy diagram for $K^H_{\alpha 1,2}$ hypersatellite transition.

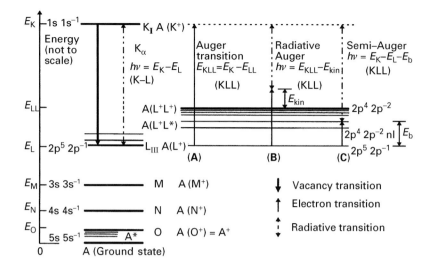

Figure 3 Energy diagrams for: (A) Auger transition, (B) radiative Auger transition, and (C) semi-Auger transition.

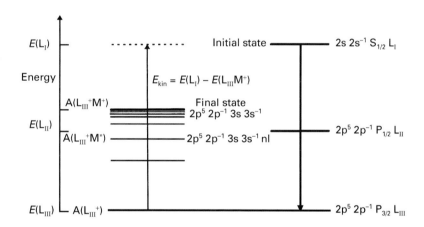

Figure 4 Energy diagram for LLM Coster–Kronig transition.

between two levels with the same principal quantum number (i.e. within the same shell; $\Delta n = 0$) and an electron from some outer shell (different n) is ejected (e.g. LLM transition: see **Figure 4**).

Super Coster–Kronig process This is an Auger transition wherein not only the vacancy transition is within the same shell ($\Delta n = 0$ as in a CK transition), but the electron is ejected from the same shell (e.g. MMM transition: see **Figure 5**).

Autoionization This process is similar to Auger ionization. In this case a vacancy is created by promoting an inner shell electron to an outer bound level, instead of being ejected from the atom. When this vacancy is filled by an outer shell electron, if the difference in energy exceeds the ionization potential of any electron, then that electron is ejected leading to a singly ionized atom (**Figure 6**).

The differences between auto ionization and Auger ionization are as follows: a) Auto ionization results from an electron vacancy produced by the excitation of a core electron to an outer bound level whereas Auger ionization results from an electron vacancy created by ejecting a core electron from an atom. b) Autoionization takes place in a neutral atom leading to a singly ionized atom whereas Auger ionization takes place in a singly ionized atom leading to a doubly ionized atom.

There is a quite high probability (at least for K shell ionization) of the order of 20%, that the first ionization producing the initial vacancy is simultaneously accompanied by the excitation or ionization of a second electron. If the second electron is excited to some bound state, the process is called shake-up. If the second electron is ejected, the process is called shake-off.

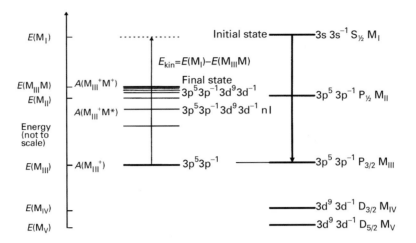

Figure 5   Energy diagram for MMM super Coster–Kronig transition.

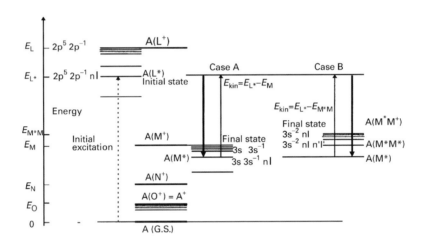

Figure 6   Energy diagram for auto-ionization. An L shell electron is excited to an outer nl shell. Subsequently, an M shell electron fills up the L shell vacancy. In case A, the excited electron in the nl shell leaves the atom, and in case B, another M shell electron leaves the atom while the nl shell electron remains a spectator electron.

The Auger processes compete with radiative decay. The probability of radiative decay is proportional to Z^4, whereas the probability of Auger ionization is constant, independent of Z. The fluorescence yield (ω) is defined as the ratio of the number of vacancies filled by X-ray emission to the total number of vacancies (filled by all processes). For K shell, ω_K = number of K X-ray photons/ (number of K X-ray photon + number of Auger electrons) = $Z^4/(Z^4 + \alpha)$, where $\alpha = 1.12 \times 10^6$. Hence, for low Z atoms, the fluorescence yield is low due to strong competition from Auger processes (for $Z < 33$, $\omega_K < 1/2$).

Because satellite lines are due to doubly ionized atoms, and since Auger processes dominate for low Z atoms, satellite lines are more prominent in the spectra of low Z atoms compared to high Z atoms.

X-ray emission by free electrons

From electromagnetic theory, any accelerating charge will radiate, with the intensity of emission being proportional to the square of the acceleration. Electrons, due to their smaller mass, undergo higher acceleration for a given force and hence emit more intense radiation. The spectrum of the emitted radiation called Bremsstrahlung, (braking radiation) depends on the nature and magnitude of the acceleration. We discuss here three main sources of X-ray based on electron acceleration (deceleration), namely (i) X-ray tubes (linear acceleration, monoenergetic electrons), ii) hot plasma sources (linear acceleration, Maxwellian distribution of energy) and iii) synchrotron sources (transverse acceleration). The X-ray spectra from these differ from each other.

X-ray tube

In this case, thermionically generated electrons are accelerated by a d.c. (or pulsed) potential (V) to several keV energy and are abruptly slowed down by impinging them on a solid target. Due to this deceleration, Bremsstrahlung radiation is emitted. The maximum emission is in the direction perpendicular to the acceleration. Accordingly, the target surface in an X-ray tube is kept at an angle to the electron beam.

The total intensity of the X-ray emitted is given by $I \propto ZV^2$.

The efficiency of X-ray conversion is given by ε = X-ray power/electron beam power. Experimentally, the value of ε is about $1.1 \times 10^{-9} Z(V+10.3Z)$ or $\sim 1.1 \times 10^{-9} ZV$ for large accelerating voltage. This means that the X-ray conversion efficiency is better for higher Z targets and at higher accelerating voltages. However, even for a high-Z material like tungsten ($Z = 74$) at a voltage of 50 kV, the efficiency is only about 0.4%. The rest of the e^- beam energy is spent mostly in heating the target. It is therefore necessary to have an X-ray tube anode with a high melting point in either a rotating or cooled (or both) configuration to prevent it from melting.

The X-ray energy distribution for a thin target is given by

$$I_\nu = I_o \quad \text{for } \nu \leq \nu_{max}$$
$$= 0 \quad \text{for } \nu > \nu_{max}$$

where $h\nu_{max}$ is the maximum energy of the X-ray photon, equal to the electron beam energy. This corresponds to the case where the electron loses its full energy in a single collision. It is also referred to as the Duane–Hunt limit.

The above spectral distribution is true for a thin target where the electron undergoes a single collision. However, in practice, the target is thick and the electron is scattered (decelerated) many times. As a result, the energy spectrum becomes

$$I_\nu = KZ \left(\nu_{max} - \nu \right) \quad \nu \leq \nu_{max}$$
$$= 0 \quad \nu > \nu_{max}$$

The spectral distribution, in terms of X-ray wavelength, is given by (**Figure 7**).

$$I_\lambda = c\, I_\nu / \lambda^2 \sim Kc^2 Z(\lambda - \lambda_{min})/\lambda^3\, \lambda_{min} \quad \text{for } \lambda \geq \lambda_{min}$$
$$= 0 \quad \text{for } \lambda < \lambda_{min}$$

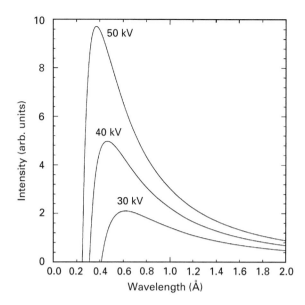

Figure 7 Bremsstrahlung emission spectrum from an X-ray tube.

Here $\lambda_{min}(\text{Å}) = 124\,000/V$. The peak of the X-ray emission is at $\lambda \cong (3/2)\lambda_{min}$. For example, for an accelerating voltage of 50 kV, $\lambda_{min} = 0.25$ Å and the spectral distribution has a peak at $\lambda = 0.37$ Å.

The Bremsstrahlung radiation from a thick target is only partially polarized. However, at wavelengths near λ_{min}, the radiation is strongly polarized in the plane containing the X-ray and the electron beam directions.

It may be noted that if the energy of the electron exceeds the binding energy of the inner shells of the target material, the impinging electron can also knock out core electrons from the target atoms, creating inner shell vacancies, leading to emission of characteristic X-rays of the target material. These X-ray lines are superimposed on the continuous Bremsstrahlung spectrum.

Hot plasma sources

These include sources such as laser produced plasmas, tokamak plasmas, pinch plasmas, solar flares, stellar X-ray emitters, etc. In such plasmas, the electron temperature (corresponding to a Maxwellian velocity distribution) can be a few hundreds of eV to several keV. On collision with plasma ions these energetic electrons undergo acceleration/deceleration and thereby emit Bremsstrahlung radiation. Electron–electron collisions do not emit any net radiation as the two colliding electrons undergo exactly equal and opposite accelerations. The radiation emitted by the two electrons is therefore equal in magnitude and opposite in phase. Hence, there is no net radiation emitted.

For a Maxwellian velocity distribution of the electrons in the plasma, the spectral distribution of the emitted Bremsstrahlung radiation is given by (**Figure 8**)

$$I_\lambda = C\, N_e\, \Sigma_i\, N_i\, z^2\, T_e^{-1/2}\, g_f \exp(-hc/\lambda\, k\, T_e)/\lambda^2$$

where N_e is the electron density, N_i is the ion density of charge i, and z is the average ion charge. The factor g_f is of the order of unity and it represents a departure of quantum mechanical calculations from the classical results. This factor is called the Gaunt factor. The peak of this spectrum is at $\lambda_p(\text{Å}) = hc/2k\, T_e \sim 6.2/T_e(\text{keV})$.

The spectral distribution of the X-ray Bremsstrahlung emitted from a hot plasma is shown in **Figure 8**, which shows a strong temperature dependence of the spectrum for $\lambda < \lambda_p$. This fact is often used for estimation of the electron temperature of the plasma.

It may be noted that, unlike the Bremsstrahlung emission from X-ray tubes, there is no short wavelength limit here as the electrons have all possible energies.

Recombination radiation In addition to Bremsstrahlung radiation, a hot plasma also emits recombination radiation. This radiation is emitted when a free electron is captured in a bound state of an ion. If E is the kinetic energy of a free electron and χ_n is the ionization potential of the energy level in which the electron is captured, the radiation is emitted with a photon energy of $hv = E + \chi_n$ (**Figure 9**). As the free

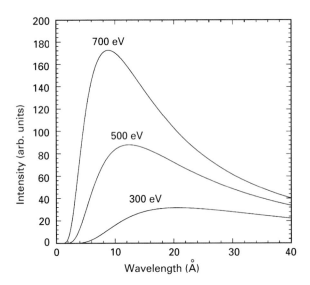

Figure 8 Bremsstrahlung emission spectrum from hot plasma.

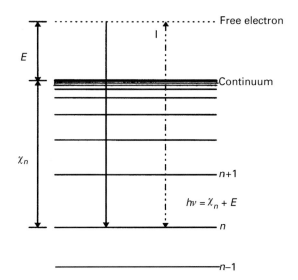

Figure 9 Energy diagram for recombination radiation.

electron has a continuous energy distribution, the emitted radiation spectrum is also continuous for $hv \geq \chi_n$. Further, since the recombination can occur in different energy levels of the ion, the overall spectrum is quasi-continuous showing discontinuities at energies equal to the ionization potential energies of various levels. The overall shape of the spectrum is similar to that of plasma Bremsstrahlung radiation shown in **Figure 8**.

Interestingly, whereas in an X-ray tube, the radiation is on the longer wavelength side of the Duane–Hunt limit, here the spectrum is on the shorter wavelength side of the ionization potential.

Cyclotron radiation If the hot plasma happens to be magnetized (like the tokamak/pinch plasma, X-ray binary stars, etc.), then the electrons also emit cyclotron radiation (also called magnetic Bremsstrahlung) due to their gyration around the magnetic lines of force. The acceleration is due to the Lorentz force acting on the moving electrons.

The radiation spectrum is a discrete line spectrum at an electron cyclotron frequency (ω_c) and its harmonics:
$\omega = n\omega_c$, where $\omega_c = eB/m \cong 1.76 \times 10^{11} B$ (T) rad s^{-1}.

Although the spectrum is independent of electron temperature, the total power radiated is proportional to both the electron temperature (T_e) and electron density (N_e) and is given by P_c (W/m^3) $\cong 4.4 \times 10^{-28} N_e$ (m^{-3})B(T)$^2 T_e$ (eV).

Synchrotron radiation sources In a synchrotron source, electrons move with relativistic speed ($v \sim c$). Although this speed is almost constant, when the electron trajectory is bent using bending magnets, the electrons undergo transverse acceleration and radiate energy. This radiation is called synchrotron radiation.

The power radiated is given by

$$S = (e^2 c/6\pi\varepsilon_0)\,\gamma^4/R^2$$

where $\gamma = E/m_0 c^2$ and R is the radius of the electron trajectory. For an electron beam of energy E(GeV) and a bending magnet field of B(T), the radius of the electron trajectory is given by $R = 3.33\,E/B$ m.

Since the motion of the electron is relativistic, the radiation (emitted perpendicular to the direction of acceleration) is highly concentrated in the direction of velocity within a narrow cone of nominal angular width $1/\gamma$.

Since the radiation is emitted in the form of a searchlight-like cone, an observer receives this radiation for a very short time period $\Delta t \sim 4R/3c\gamma^3$. As a result, the emitted radiation has a large bandwidth given by $\Delta\omega = 2\pi/\Delta t = 3\pi c\gamma^3/2R$. Thus the observed radiation is a continuous spectrum over a large frequency range.

Critical frequency is defined as the frequency which divides the synchrotron radiation power spectrum into two equal parts. This frequency is given by $\omega_c = 3c\gamma^3/2R$, and the corresponding critical energy is given by

$$E_c\,(keV) = 3hc\gamma^3/4\pi R = 0.665\,B\,(T)\,E^2\,(GeV)$$

Also, the critical wavelength is given by

$$\lambda_c = 2\pi c/\omega_c = 4\pi R/3\gamma^2 = 5.6\,R\,(m)/E^3\,(GeV)$$
$$= 18.6/B\,(T)\,E^2\,(GeV)$$

The synchrotron radiation from a bending magnet is linearly polarized when observed in the plane of the e^- orbit. Out of this plane, it is elliptically polarized with opposite helicity on either side of the plane.

Periodic magnetic structure In order to enhance the X-ray emission from the synchrotron, periodic magnetic structures are sometimes inserted in the linear sections of the synchrotron. This makes the electron notation sinusoidal in a horizontal plane. An important parameter characterizing the electron motion is the deflection parameter (K) given by $K = 93.4\,\lambda_u$ (m) B (T), where λ_u is the period of the magnetic structure.

Undulator For low magnetic field ($K<1$) the angular excursion of the electron is within the nominal $1/\gamma$ radiation cone. In this case, the electron beam breaks up into equally spaced bunches and the radiation from these bunches adds coherently (in phase). Such a magnetic structure is called an undulator. The emitted coherent radiation is at harmonics of $\lambda_L = (1+0.5\,K^2)\lambda_u/2\gamma^2$. The relative bandwidth of the nth harmonic is given by $\Delta\lambda/\lambda = 1/(nN)$, where N is the number of magnetic periods. Along the axis, only odd harmonics are observed. However, if a helical magnetic structure is used, the radiation will be circularly polarized and there will be no harmonics present in the on-axis radiation. The emission cone is further narrowed down to $\theta \sim 1/(\gamma\sqrt{N})$. The total radiation flux is N^2 times the flux due to a single bending magnet.

If such a structure is placed in an optical resonator or used as an amplifier for radiation of wavelength $\lambda = \lambda_L$, then it is referred to as a free electron laser.

Wiggler If the electron excursion angle exceeds $1/\gamma$ (i.e. when $K >> 1$) then the magnetic structure is called a wiggler. Here, the radiation from different sections of the electron trajectory (where the direction of electron motion makes an angle less than $1/\gamma$ with the axis) adds up incoherently. The total radiation flux is $2N$ times the flux due to a single bending magnet. The emission cone is several times larger than that of the synchrotron radiation.

The total power emitted by an undulator or wiggler is given by

$$P(kW) = 0.633\,E^2\,(GeV)\,B^2\,(T)\,L\,(m)\,I\,(A)$$

where L is the total length ($=\lambda_u N$) of the magnetic structure, and I is the beam current.

For a planar magnetic field in the vertical direction, the on-axis radiation is polarized in the horizontal plane, as in the case of a bending magnet. However, unlike the bending magnet case, for a periodic magnetic structure, the off-axis radiation is also plane-polarized in the horizontal plane. This is because the vertical component of polarization emitted in one half is cancelled by the radiation in the next half (out of phase) as both have the same direction of polarization. However, the horizontal polarization in the two halves being opposite in direction, the net radiation adds up.

X-rays from few electron systems

X-rays from a few electron systems such as hydrogen-like, helium-like, lithium-like ions are observed in hot plasmas (laser produced / tokamak / Z pinch), beam-foil experiments, heavy ion collisions, solar flares, etc. Muonic atoms also emit X-rays.

Ionic X-rays

Transitions in the highly charged ions are in the X-ray region and are interesting because they are relatively simple to interpret. They are also one of the few cases in atomic physics wherein high order multipole transitions are observed.

Hydrogen-like ions   These are ions having a single electron left. The energy levels of these ions are exactly the same as those of the hydrogen atom except that they are increased by a factor Z^2, where Z is the atomic number of the ion.

$$E\,(n,l\,) = Z^2\,E_{\mathrm{H}}(n,l\,)$$

The hydrogen-like series is composed of transitions of the type $(np)^2\,P_{3/2,1/2}-(1s)\,^2S_{1/2}$, where $n \geq 2$. Each line of the series is a doublet. The limit of this series is the highest energy X-ray line that can be emitted by a given element.

Unlike the hydrogen Lyman-α doublet ($\Delta v = 0.36$ cm^{-1}) which cannot be easily resolved due to Doppler broadening even in moderate temperature ($\sim 10^4$K) plasmas, the fine structure in high Z elements can be easily resolved even in high temperature ($\sim 10^6$K) plasmas. For example, in H-like calcium, the wavelengths of the Ly$_\alpha$ doublet are 3.018 Å and 3.24 Å, which can be easily resolved with a standard crystal spectrometer.

Helium-like ions   These are ions with only K shell electrons left. Here, except for the ground state which is a singlet ($^1S_0$), all the excited levels ($n \geq 2$) have both singlet and triplet states (**Figure 10**). The helium-like series is composed of transitions of the type (np 1s) $^1P_1-(1s^2)^1S_0$. The end point of the series is the ionization potential of the 1s electron in its $(1s^2)^1S_0$ ground state.

Types of transitions   The terminology for X-ray transitions in ions is the same as that of optical transitions in an atom. The three main types of transitions observed in an ionic line spectrum are: 1) resonance transitions, 2) intercombination transitions, and 3) satellite transitions.

Resonance transitions   These are transitions from an excited state to the ground state (**Figure 10**). The H-like series and He-like series discussed earlier are resonance transitions. The oscillator strengths of these transitions are higher than those of other types of transition. For the same reason, these lines can be strongly reabsorbed if they are emitted by a hot-dense plasma. They follow the normal selection rules applicable to optical transitions.

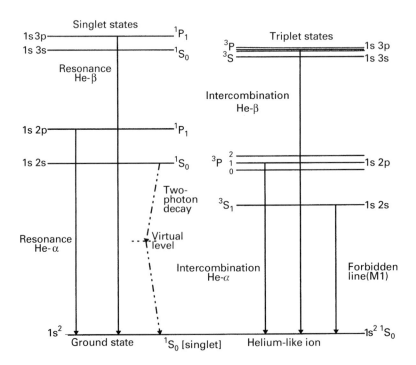

Figure 10   Energy level diagram for helium-like ions showing resonance, intercombination, and other transitions.

Intercombination transitions These transitions are similar to resonance transitions except that they are between states of different multiplicity (e.g. triplet to singlet). Corresponding to a resonance transition (1s2p) $^1P_1$–(1s^2) $^1S_0$ (which is a singlet–singlet transition), the intercombination transition (triplet–singlet) is (1s^{2}p)^{3}P–(1s^2) $^1S_0$ (**Figure 10**). These lines appear on the lower energy side of the resonance lines. Though $^3P_1$ – $^1S_0$ is a spin-forbidden transition, in high Z ions, the $^3P_1$ state decays by dipole transition through mixing with $^1P_1$ states.

Satellite lines These are weaker lines arising from doubly excited ions of higher ionization states. The satellite lines of H-like ions are due to He-like ions, those of He-like ions are due to Li-like ions and so on. For example, transitions in Li-like ions of the type 1s nln'l' → 1s^2 n'l' will appear as a satellite to 1s nl–1s^2 resonance transitions in He-like ions (**Figure 11**). The n'l' electron is a spectator electron. Due to the presence of this electron, Coulomb shielding decreases, which results in the transition occurring at a slightly lower energy than that of the resonance transition. Since two electrons are involved in such transitions (one active, one spectator), these lines are also referred to as dielectronic satellites.

The largest separation from a parent line occurs when the spectator electron is in the lowest *n* level (i.e. *n* = 2), when the Coulomb shielding effect is maximum. These satellites are referred to as 2s or 2p satellites. For example, the 1s 2s 2p – 1s^2 2s transition in Li-like ions will be a 2s satellite for the 1s 2p – 1s^2 (He-α) transition in He-like ions (**Figure 11**) Satellites of H-like ions in the Gabriel notation are

denoted by capital letters A, B,...J,... and those of He-like ions are denoted by lower case letters a, b, c,......u, v.

Muonic X-rays

When a negatively charged particle (μ^-, π^-, K$^-$ meson) replaces an electron in an atom, a mesonic atom is formed. For example, when a μ^- meson is brought to rest in a target, muonic atoms of the target element are formed by replacement of a valence electron by μ^-. The energy levels in a muonic atom are analogous to the electronic energy levels of an H-like ion except that the muon mass is higher (m_μ~207 m_e). The mesonic atom energy levels are related to those of the hydrogen atom by $E(n,l) = Z^2$ (M^*/m^*) $E_H(n,l)$, where M^* and m^* are the reduced masses of μ^- and the electron, respectively, and $E_H(n,l)$ denotes energy levels in hydrogen. The newly formed muonic atom is thus in a highly excited state and lowers its energy by ejecting electrons by successive Auger processes until the principal quantum number falls to less than 5 (in heavy atoms). At this point, the radiative transition probability becomes more prominent. The energy level differences become of the order of several keV. As a result, X-rays are emitted until the muonic atom reaches its ground state.

The average radius of the lowest (Bohr) orbit of the muonic atom is given by r~(m^*/M^*) (a_H/Z) which is considerably smaller than that of a normal atom. For example, for silver, r is 5×10^{-15}m, which is of the order of the nuclear size. As a result, the muon spends considerable time inside the nucleus. Consequently, the 1s level is strongly affected by the nucleus. This is corroborated by the fact that whereas the Balmer series (*n*→2) transition energies are found to be exactly $Z^2(M^*/m^*)$ times those of the H-atom, the Lyman series (*n*→1) transition energies are lower than expected.

List of symbols

E = photon energy; g_f = Gaunt factor; I = intensity; I = beam current; k = Sommerfeld quantum number; K = deflection parameter; L = length of magnetic structure; M^*, m^* = reduced mass of muon and electron; n = total quantum number; N = number of magnetic periods; N_e = electron density; N_i = ion density; P = total power; r = radius of Bohr orbit; R = radius of electron trajectory; R = Rydberg constant; S = power; T = term value; T_e = electron temperature; V = voltage; W = work function; z = ion charge; Z = atomic number; $\Delta\omega$ = bandwidth; α = fine structure constant; χ_n = ionization potential; ε = efficiency; γ = relativistic factor = E/m_0c^2; λ = wavelength; ν = frequency; θ = cone angle; σ, s = screening

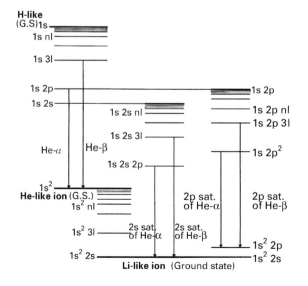

Figure 11 Energy diagram for 2s and 2p satellite transitions in lithium-like ions corresponding to He-α and He-β transitions in helium-like ions.

2498 X-RAY SPECTROSCOPY, THEORY

constants; ω = fluorescence yield; ω_c = electron cyclotron frequency; ω_p = plasma frequency.

See also: **Photoelectron Spectrometers; X-Ray Absorption Spectrometers; X-Ray Emission Spectroscopy, Applications; X-Ray Emission Spectroscopy, Methods; X-Ray Fluorescence Spectrometers; X-Ray Fluorescence Spectroscopy, Applications; Zero Kinetic Energy Photoelectron Spectroscopy, Theory.**

Further reading

Agarwal B.K (1991) *X-ray Spectroscopy: an Introduction,* Berlin: Springer-Verlag.

Azaroff LV (1974) *X-ray Spectroscopy,* New York: McGraw-Hill.

Bertin EP (1975) *Principles and Practice of X-ray Spectrometric Analysis,* New York: Plenum Press.

Bonnelle C and Mande C (1982) *Advances in X-ray Spectroscopy,* New York: Pergamon Press.

Craseman B (1985) *Atomic Inner Shell Physics.* New York: Plenum Press.

Herglotz HK and Birks, LS (1978) *X-ray Spectrometry.* New York: Dekker.

Janev RK, Presnyakov LP and Shevelko VP (1985) *Physics of Highly Charged Ions.* Berlin: Springer-Verlag.

Jenkins R (1976) *An Introduction to X-ray Spectrometry,* London: Heyden.

Kauffman RL and Richard P (1976) X-ray region. In: Williams D (ed) *Methods of Experimental Physics Vol. 13 Part A (Spectroscopy).* London: Academic Press.

Michette AG and Buckley CJ (eds) (1993) *X-ray Science and Technology.* London: IOP Publishing Ltd.

Thompson, M. Baker, MD Christie, A and Tyson JF (1985) *Auger Electron Spectroscopy (Chemical Analysis, Vol. 74)* New York: Wiley Interscience.

Williams KL (1987) *Introduction to X-ray Spectrometry* London: Allen and Unwin.

White HE (1986) *Introduction to Atomic Spectroscopy.* Singapore: McGraw-Hill Book Co.

Yttrium NMR, Applications

See **Heteronuclear NMR Applications (Y–Cd).**

Z

Zeeman and Stark Methods in Spectroscopy, Applications

Ichita Endo and **Masataka Linuma**, Hiroshima University, Japan

ELECTRONIC SPECTROSCOPY
Applications

Introduction

An atomic system is influenced by an external electric and magnetic field. Due to an interaction of the magnetic field with the magnetic moment of an atom, an electronic energy level in the atomic system is shifted in accordance with the formula of the Zeeman effect, while an external static electric field would polarize the atom, resulting in an energy shift referred to as the Stark effect. As the amount of energy shift depends on the magnetic quantum number of the level, the Zeeman and Stark effects resolve the otherwise degenerate energy levels into sublevels. From the pattern of level splitting we can assign the quantum numbers of the observed electronic level. The absolute value of separation between the split levels tells us about the magnetic moment and the polarizability for Zeeman and Stark spectroscopy, respectively.

A straightforward application of Zeeman spectroscopy is a magnetic-field determination using atoms with a known magnetic moment, and one of Stark spectroscopy is a measurement of an electric field using atomic levels with predetermined polarizability. Such measurements are useful when field-measuring probes based on other principles are either unusable or difficult to apply as in the astrophysical environment or in a plasma.

Indirect but important usage of Zeeman and Stark effects is found in fundamental physics researches: for example measurements of violation of symmetry in physical laws under time and space inversion known as the T-violation and the parity violation, respectively. Such measurements would eventually give

us clues of new physics beyond the standard model of unified electromagnetic and weak interactions.

We present here some selected topics in the application of Zeeman and Stark spectroscopy of atoms in the gas phase with special emphasis on parity-violation experiments.

Fundamental physics research

A steady state of an isolated atom is described by a quantum-mechanical state specified by the energy and the total angular momentum in accordance with the translation invariance in the time coordinate and the rotational symmetry of the space coordinates, respectively. If physical laws were completely invariant under the parity operation, i.e. space inversion, the wavefunction ψ of the atom remains exactly the same except for its sign; $P\psi$, where $P = \pm 1$. The parity quantum number P introduced in this way would also be a good quantum number in the atomic system if the forces acting on atomic electrons were due only to the classical electromagnetic interaction which is invariant under space inversion. Recent findings in fundamental physics, however, predict that parity is not conserved in the atomic system, though the amount of violation is extremely small.

When an electric field is applied to an atom, the space symmetry is destroyed so that an even-parity state is slightly contaminated by an odd-parity state and *vice-versa*. This makes the otherwise-forbidden E1 transition between the same parity levels to be observable; a Stark-induced transition. The interference term between the Stark-induced E1 transition amplitude and that due to the intrinsic parity violation changes sign when the direction of the electric field is reversed. Therefore, the parity nonconservation effect can be measured by comparing a small amount of change in the transition rates as we reverse the electric field. There have been several experiments based on this principle to acquire quantitative information of parity violation, the most precise experiments being laser spectroscopy of an atomic beam of Cs under an external electric and magnetic field.

Various attempts are being made to achieve higher accuracy in the parity nonconservation (PNC) measurements. One of the possibilities is to use a heavier atom and to observe a transition to a level with a relatively large amount of parity mixing. Rare earth atoms are deemed as good candidates, because they have many close-lying level pairs of opposite parity. However, there has been no experimental support for them to have a sizable enhancement in parity mixing.

Some examples of Zeeman and Stark spectroscopy of rare earth atoms are shown below. They were obtained in a series of studies aiming at finding the atomic states suitable for the PNC experiment.

In **Figure 1** Stark spectra of samarium atoms are shown. They were obtained by detecting the fluorescence from the level excited with the laser beam. The observed transition is from the ground state ($J = 0$) to the 1.9404 eV ($J = 1$) excited level, in which J is the electronic total angular momentum.

The strength of the electric field denoted by E is 0.0, 17.2 and 26.1 kV cm^{-1} for the upper, middle and lower part of the figure, respectively. The peaks labelled by open and solid circles correspond, respectively, to the transition from $|m| = 0 \rightarrow |m| = 0$ and from $|m| = 0 \rightarrow |m| = 1$, in which m is the magnetic quantum number. We see that each peak for ^{152}Sm and ^{154}Sm is split into two peaks of which the separation increases as the electric field is strengthened. In this case only the Stark effect on the upper level ($^7G_1$) is responsible for the splitting because the lower level has $J = 0$. The energy separation of the split is expressed by

$$\Delta W = \frac{3}{2}\alpha_2 J_u E^2 \qquad [1]$$

where α_2 is the tensor polarizability and J_u is the electronic total angular momentum of the upper level. The E^2-dependence of splitting of the ^{154}Sm peak for the same transition is shown in **Figure 2**. It is clearly seen that the energy interval of the splitting is proportional to E^2. The tensor polarizability is determined from the slope of the straight line: $\alpha_2 = -554.6$

± 1.3 kHz (kV)$^{-2}$ cm^2 for the data shown in **Figure 2**.

The Zeeman spectra for the transition between the 0.0363 eV level ($J = 1$) and the 1.9301 eV level ($J = 2$) of Sm are shown in **Figure 3**. The peaks labelled by squares and circles correspond to ^{154}Sm and ^{152}Sm. The applied magnetic field is 0, 115×10^{-4}, 224×10^{-4} and 352×10^{-4} T for the spectra shown in **Figures 3A, B, C** and **D** respectively. The open and solid symbols are for the σ and π components of the transition, respectively. The σ component is the transition associated with a change in magnetic quantum number, $\Delta m = \pm 1$, caused by a photon with its polarization perpendicular to the direction of the magnetic field. The π component is defined as the transition with $\Delta m = 0$. The energy interval, represented by the frequency, f, of Zeeman splitting, is given by the formula,

$$f = \mu_B \left(g_u \, m_u - g_l \, m_l\right)B + C \qquad [2]$$

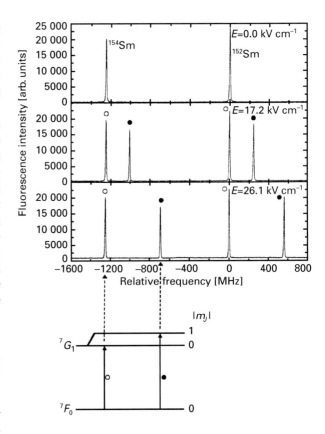

Figure 1 Stark spectra in an electric field E of 0.0, 17.2 and 26.1 kV cm^{-1} for ^{152}Sm and ^{154}Sm for the transition from the ground state with $J = 0$ to the 1.9404 eV level with $J = 1$. The energy levels responsible for these spectra are schematically shown in the lower part together with their electronic configuration.

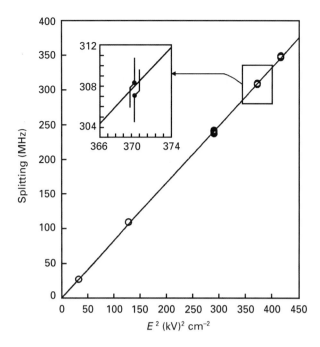

Figure 2 Stark splitting as a function of squared electric field E^2.

where μ_B is the Bohr magneton, m_u and g_u are the magnetic quantum number and the g-factor for the upper level, respectively, while m_l and g_l are those for the lower level. A constant C gives the original frequency without an external magnetic field. In the case of **Figure 3** the g-factor for the lower level is known to be zero so that the constants g_u, B and C are determined by least-squares fitting of Equation [2] to the relative frequencies among the Zeeman peaks.

Field measurements

If the g-factors (polarizabilities) are known in advance, it is possible to measure a static magnetic (electric) field by means of the Zeeman (Stark) effect. This is useful particularly in such situations as in hot plasma and in astronomical objects where the standard field-measuring probes, e.g. a nuclear magnetic resonance probe and a Hall probe, are unusable.

In plasma diagnostics, for example, Stark spectroscopy is used for determining the local electric field. Since the Stark splitting is large for the Rydberg levels, the excitation to the level with high principal quantum number n is used. A small amount of probe atoms mixed in the plasma are excited to metastable

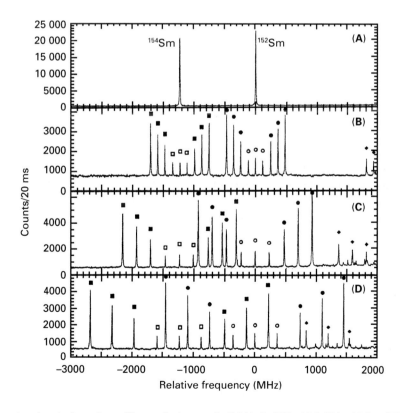

Figure 3 Zeeman spectra of samarium atoms. The applied magnetic field is 0, 115×10^{-4}, 224×10^{-4} and 352×10^{-4} T for (A), (B), (C) and (D) respectively. The peaks represented by squares and circles correspond, respectively, to the transition of ^{154}Sm and ^{152}Sm. The open and solid symbols represent the σ and π components of the transition, respectively.

states by an electric discharge. They are further pumped up to the Rydberg levels by a laser beam, followed by decay to intermediate levels due to collisional transitions. The fluorescence from the intermediate levels bears information on the electric field at the particular position in the plasma at which the laser is focused with its frequency being scanned. The electric field can also be evaluated from the ratio of the intensity of the forbidden transition induced by the Stark mixing to that of the allowed transition.

Another example is related to astronomical researches: the spectra of starlight usually reveal some emission and absorption lines. The position and the depth of the absorption lines tell us about the atomic species and their relative abundance in the cold outer gas of the star. If a strong field exists in a star, Zeeman and Stark splitting are identifiable in the spectra. From the pattern of absorption or emission lines, it is sometimes possible to determine the strength of magnetic and electric fields in astronomical objects.

List of symbols

E = electric-field strength; f = frequency of Zeeman splitting; J = electronic total angular momentum; m = magnetic quantum number; P = parity quantum number; W = energy separation of Stark splitting; α_2 = tensor polarizability; μ_B = Bohr magneton; ψ = atomic wavefunction;

See also: **Atomic Fluorescence, Methods and Instrumentation; Laser Applications in Electronic Spectroscopy; Laser Spectroscopy Theory; Zeeman and Stark Methods in Spectroscopy, Instrumentation.**

Further reading

Dalgarno A and Layzer D (1987) *Spectroscopy of Astrophysical Plasmas.* Cambridge: Cambridge University Press.

Demstroder W (1998) *Laser Spectroscopy*, 2nd edn. Berlin: Springer-Verlag.

Greenberg KE and Hebner GA (1993) Electric-field measurements in 13.56 MHz helium discharges. *Applied Physics Letters* **63**: 3282–3284.

Hanle W and Kleinpoppen H (1978) *Progress in Atomic Spectroscopy.* New York: Plenum Press.

Khriplovich IB (1991) *Parity Nonconservation in Atomic Phenomena.* Philadelphia: Gordon and Breach Science Publishers.

Kobayashi T, Endo I, Fukumi A *et al* (1997) Measurement of hyperfine structure constants, *g* values and tensor polarizability of excited states of Sm I. *Zeitschrift für Physik D* **39**: 209–216.

Shimoda K (1976) *High-Resolution Laser Spectroscopy.* Berlin: Springer-Verlag.

Svanberg S (1992) *Atomic and Molecular Spectroscopy*, 2nd edn. Berlin: Springer-Verlag.

Zeeman and Stark Methods in Spectroscopy, Instrumentation

Ichita Endo and **Masataka Linuma**, Hiroshima University, Japan

ELECTRONIC SPECTROSCOPY
Methods & Instrumentation

Introduction

The energy difference between the Zeeman and Stark sublevels is usually far smaller than the line width of the optical transition in atoms at normal temperature due to Doppler broadening. Doppler-free techniques are necessary for obtaining the values of *g*-factors and polarizabilities in optical spectroscopy.

Variations of coherent spectroscopy, such as level-crossing, quantum beat, and pulsed-field spectroscopy, are examples of Doppler-free techniques. They make use of the interference effect in the transition amplitudes of simultaneous excitation from a level to two closely separated higher levels.

Another technique widely used is atomic beam spectroscopy. In a gas jet ejected from a small orifice, the transverse motion of atoms is much reduced. The line width of the transition induced by a laser beam perpendicularly crossing the atomic beam can be narrow enough to resolve the Zeeman and Stark splitting.

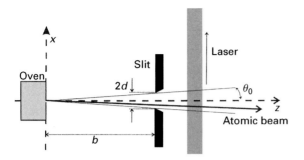

Figure 1 Schematic illustration of an atomic-beam technique to reduce Doppler broadening. Atomic vapour effuses from a small orifice of an oven. The angular divergence of atoms in the beam is limited to $\theta_0 = \tan^{-1} b/d$ by a slit whose aperture is $2d$ placed at a distance b from the orifice.

Atomic beam spectroscopy

The Doppler effect broadens absorption or emission lines from atoms in the gas phase at thermal equilibrium. Assume that an atom at rest is excited by a photon with wave vector $\boldsymbol{k}$ and de-excited to emit light with angular frequency ω_0. If the atom is moving at a velocity $\boldsymbol{v}$ the angular frequency of the emitted light is shifted to a value ω_0' according to the formula,

$$\omega_0' = \omega_0 - \boldsymbol{k} \cdot \boldsymbol{v} \qquad [1]$$

At thermal equilibrium, the velocities of atoms in the gas phase obey a Maxwellian distribution. This results in the broadened intensity profile around ω_0 as approximately represented by a Gaussian form,

$$I(\omega) = I_0 \exp\left[-\left(\frac{c(\omega - \omega_0)}{\omega_0 v_{\text{th}}} \right)^2 \right] \qquad [2]$$

where m/c is the velocity of light and $v_{\text{th}} = (2k_{\text{B}}T/m_{\text{a}})^{1/2}$ is the most-probable velocity of atoms with mass m_{a}, temperature T and Boltzmann constant k_{B}. The Doppler width defined by the full width at half maximum of the Gaussian profile is

$$\Delta\omega_{\text{D}} = 2\sqrt{\ln 2} \frac{\omega_0 v_{\text{th}}}{c} = \left(\frac{\omega_0}{c} \right) \sqrt{\frac{k_{\text{B}} T \ln 2}{m_{\text{a}}}} \qquad [3]$$

Let us consider a case where the atoms are effusing into a vacuum chamber, as shown in **Figure 1**, from an orifice of an oven filled with vapour at a temperature T. Let the atomic beam travel along the z-axis, while the laser beam is parallel to the x-axis. One can reduce the Doppler broadening by limiting the beam divergence with a slit with a small aperture $2d$ in the x direction at a distance b from the orifice. This makes the beam divergence in the x-z plane smaller than $\theta_0 = \arctan d/b$ and the Doppler broadening is reduced to $\Delta\omega_{\text{D}}' = \Delta\omega_{\text{D}} \sin \theta_0$.

An example of laser spectrometers for Zeeman and Stark spectroscopy using a collimated atomic beam is shown schematically in **Figure 2**. It consists of a continuous-wave (CW) tunable dye-laser system, a frequency calibration system, a vacuum chamber with a fluorescence detector, and a data-acquisition system. The interaction point of the atomic beam with the laser is inside the vacuum chamber.

A magnified view around the interaction point is illustrated in **Figure 3**. The oven made of molybdenum

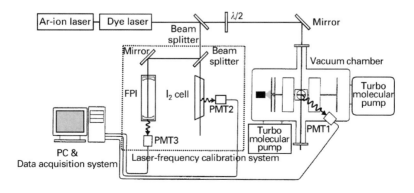

Figure 2 Typical setup for a laser spectrometer based on the atomic-beam method. The apparatus is composed of a continuous-wave (CW) tunable laser system, a laser-frequency calibration system, a vacuum chamber, and a data acquisition system. The fluorescence light from the excited I_2 molecules in a cell and the excited atoms in the vacuum chamber, and the transmitted light from a Fabry-Perot interferometer (FPI) are detected simultaneously with three photomultiplier tubes (PMTs). The signals are transformed to digital pulses event-by-event and introduced to the inputs of a multi-channel scaler (MCS).

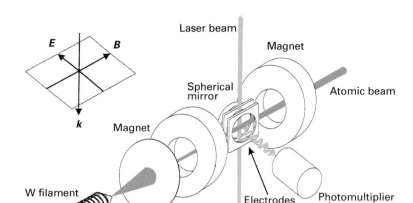

Figure 3 Magnified view around the interaction point of the atomic beam with the laser beam. The oven made of molybdenum is heated by a tungsten filament wound around it to eject a gas jet from the orifice with a diameter of 0.8 mm. The atomic beam is collimated with the slit to a diameter of about 4 mm at the interaction point, where the two electrodes for applying the electric field and a pair of Helmholtz coils to produce the magnetic field are installed. The photomultiplier tube (PMT) for detecting the fluorescence light from the atoms and a spherical mirror to collect light are shown.

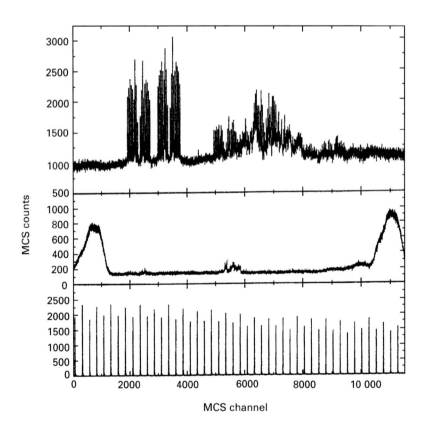

Figure 4 Typical set of raw data for Zeeman spectrum of samarium atoms with the natural isotopic abundance under the magnetic field of 167.38×10^{-4} T. The number of counts of detected photons per 20 ms is plotted against the MCS channels corresponding to the elapsed time from the starting point of the frequency sweep of the laser. The top part corresponds to the Zeeman spectrum in the transition from the level of $E = 0.184\,68$ eV$(J = 3)$ to the one of $E = 2.076\,5$ eV$(J = 3)$. In the middle and the bottom, the spectrum of $^{127}I_2$ and the spectrum of the transmitted light from the FPI are shown, respectively.

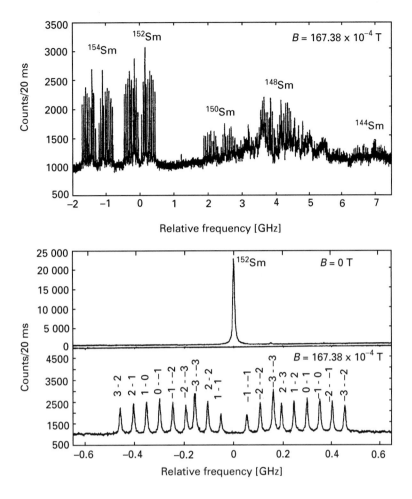

Figure 5 Zeeman spectra after calibration on the horizontal axis and peak assignments. The top part of the spectrum is the same as the one shown in **Figure 4**. Magnified spectra of just the area of the peaks for the ^{152}Sm atoms are shown in the middle and the lower parts, in which the magnetic field is 0 T and 167.38 × 10^{-4} T respectively. The label above each peak is to indicate the relevant change in the magnetic quantum number, m, to the one with m', associated with the optical transition.

is attached to an end plate of the vacuum chamber in which the pressure is kept to about 10^{-6} torr. The oven is heated by a tungsten filament wound around it to eject a gas jet from the orifice with a diameter of 0.8 mm. The temperature can be increased to about 1000 K and is monitored with a Pt-Rh thermocouple. The atomic beam is collimated by the slit and led to the interaction point, where the two electrodes made of BK7 glass plates coated with ITO(InSnO$_2$) on one side are installed to apply the electric field E. The magnetic field B in parallel with the atomic beam is applied by a set of Helmholtz coils. The fluorescence from the atoms is detected with a photomultiplier tube (PMT) which is cooled to reduce thermal noise. In order to increase the collection efficiency of the emitted photons, a spherical mirror is installed on the opposite side of the PMT.

Linearly polarized light from a laser is introduced to the inside of the vacuum chamber as shown in **Figure 2**. The CW dye-laser is capable of

continuously changing its frequency with time, sweeping over a certain frequency range. The laser polarization is adjusted with a half-wave plate when necessary. The signal from the PMT is fed to one of the inputs of a multi-channel scaler (MCS) where the number of counts in each time interval, corresponding to a small frequency segment, is recorded.

The spectra from molecular iodine, ^{127}I$_2$, together with the transmitted light through a Fabry-Perot interferometer (FPI), are recorded synchronously with the fluorescence from the excited atoms to give frequency marks separated by the free-spectral range (FSR) of the FPI.

In **Figure 4** a set of raw data obtained in Zeeman spectroscopy of Sm at $B = 167.38 \times 10^{-4}$ T is shown as an example. The uppermost part corresponds to the Zeeman spectra for samarium atoms with the natural isotopic abundance (^{144}Sm: 3.1%, ^{147}Sm: 15.0%, ^{148}Sm: 11.3%, ^{149}Sm: 13.8%, ^{150}Sm: 7.4%, ^{152}Sm: 26.7%, ^{154}Sm: 22.7%) in the transition from

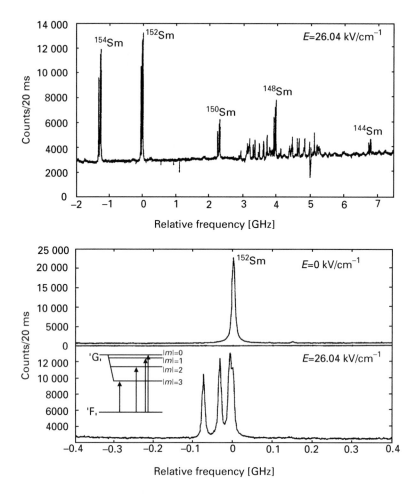

Figure 6 Stark spectra obtained by the analogous method to the one used for **Figure 5**. The transitions are the same as in **Figure 5**. The electric field of 26.04 kV cm⁻¹ is applied. The middle and lower graphs correspond to the spectra for the ¹⁵²Sm atoms under the electric field of 0 kV cm⁻¹ and 26.04 kV cm⁻¹ respectively.

the level of $E = 0.184\,68$ eV $(J = 3)$ to the one of $E = 2.076\,5$ eV $(J = 3)$ where J is the electronic total angular momentum. The spectrum of $^{127}I_2$ and that of the transmitted light from the FPI is shown in the middle and the lowest parts, respectively. After calibration on the horizontal axis and assignment of each peak, we obtain the spectra shown in **Figure 5** for which the Zeeman splitting is completely resolved. Combining the spectra measured with different magnetic field strengths, we can determine the g-factor for either the upper or lower level if one of them has been known in advance.

The Stark spectrum for the same transition at $E = 26.04$ kV cm⁻¹ is shown in **Figure 6**. Here again, the splitting is clearly seen thanks to the Doppler-free technique applied here.

Coherent spectroscopy

Although the line widths in the optical transitions observed in atoms in the gas phase at room temperature are larger than the spacing of the Zeeman and Stark splitting, the intervals of the sublevels themselves are scarcely altered by the thermal motion. In the coherent techniques, the relative energies between the sublevels are determined from the observation of the interference of amplitudes of coherent optical excitation followed by de-excitation.

Let us consider the case of two close-lying levels $|1\rangle$ and $|2\rangle$. It is possible that the atoms are simultaneously excited to these two levels from a common lower level $|i\rangle$ by a short-pulse laser with the pulse width of $\Delta t < \hbar / |E_2 - E_1|$, where $\hbar$ is the Planck constant divided by 2π. Assume, for simplicity, that the populations of levels 1 and 2 decay into another common lower level $|f\rangle$ with the same decay constant γ. The total intensity, $I(t)$, of fluorescence emitted from either level will vary with time according to the form

$$I(t) = C\,e^{-\gamma t}\,(A + B\cos\omega_{21}t) \qquad [4]$$

where A, B and C are constants depending on both the relevant atomic wave-functions and the experimental arrangement, and ω_{21} is given by $\omega_{21} = |E_2 - E_1|/\hbar$. This behaves as an exponential decay $\exp(-\gamma t)$ suffering from a sinusoidal modulation with the angular frequency ω_{21}, which is called a quantum beat.

In the case of the Zeeman (Stark) splitting, the Zeeman (or Stark) spectrum is reproduced in a Fourier transform of the time dependence in the fluorescence similar to Equation [4]. It is essential that the time response of the detection system is fast enough to observe oscillations with the characteristic period $2\pi/\omega_{21}$.

For other measuring techniques based on coherent excitation, see the 'Further reading' section for details.

List of symbols

b = distance of slit from orifice; B = magnetic field; $2d$ = slit aperture; E = electric field; I = intensity; k = photon wave vector; k_B = Boltzmann constant;

m = magnetic quantum number; m_a = atomic mass; T = temperature; v = atomic velocity; γ = decay constant; $\hbar$ = Planck constant divided by 2π; ω = angular frequency of emitted light.

See also: **Atomic Fluorescence, Methods and Instrumentation; Laser Applications in Electronic Spectroscopy; Laser Spectroscopy Theory; Zeeman and Stark Methods in Spectroscopy, Applications**.

Further reading

Demtroder W (1998) *Laser Spectroscopy*, 2nd edn. Berlin: Springer-Verlag.

Fukumi A, Endo I, Horiguchi T *et al.* (1997) Stark and Zeeman spectroscopies of $4f^6 6s 6p^7 G_{1-6}$ levels in Sm I under external electric and magnetic fields. *Zeitschrift für Physik D* **42**: 243–249.

Hanle W and Kleinpoppen H (1978) *Progress in Atomic Spectroscopy*. New York: Plenum Press.

Shimoda K (1976) *High-Resolution Laser Spectroscopy*. Berlin: Springer-Verlag.

Svanberg S (1992) *Atomic and Molecular Spectroscopy*, 2nd edn. Berlin: Springer-Verlag.

Zero Kinetic Energy Photoelectron Spectroscopy, Applications

K Müller-Dethlefs and **Mark Ford**, University of York, UK

HIGH ENERGY SPECTROSCOPY
Applications

Introduction

NO has been studied extensively by photoelectron spectroscopy. A study using vacuum ultraviolet photoelectron spectroscopy by Turner and co-workers can be compared with a ZEKE study through the $A^2\Sigma^+$ state, using a $1 + 1'$ photon experiment. As can be seen in **Figure 1** the resolution is improved by approximately three orders of magnitude; resolving the rotational structure of the NO cation. Benzene and paradifluorobenzene have been studied using time-of-flight photoelectron spectroscopy. These two systems have also been studied using ZEKE spectro-scopy. With benzene, rotational resolution has again been obtained, as shown in the comparison in **Figure 2**. The two techniques are compared for paradifluorobenzene in **Figure 3**. Both

of the above systems exhibit a breakdown in the Born–Oppenheimer approximation, and were useful indicators of the Herzberg–Teller, and Jahn–Teller effects.

ZEKE spectroscopy has been applied to a wide variety of molecular ions, clusters, van der Waals molecules, free radicals, reactive intermediates, and even to elusive transition states of chemical reactions. Examples of such typical applications of high-resolution ZEKE spectroscopy to molecules and clusters are given here. Compared to conventional photoelectron spectroscopy, ZEKE spectroscopy offers greatly increased spectral resolution, allowing the rotational structure of large molecular cations such as the benzene cation and the intermolecular vibrations of molecular clusters like phenol-water to be obtained.

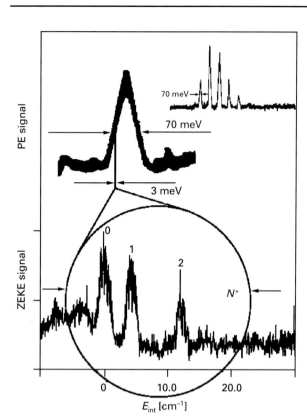

Figure 1 A comparison between conventional VUV PES and ZEKE spectroscopy on NO; with the latter technique rotational resolution is attained.

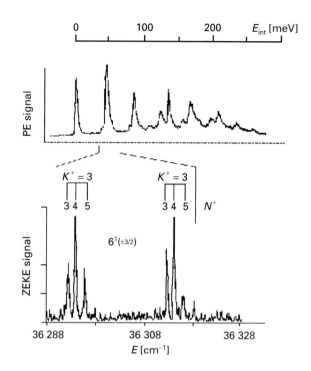

Figure 2 The rotationally resolved ZEKE spectrum of benzene compared with time-of-flight PES; again the resolution is improved by several orders of magnitude.

Smaller molecules

Iodine (I₂)

Iodine has been studied extensively by ZEKE spectroscopy, including $2 + 1'$ and $1 + 2'$ schemes carried out by Cockett and co-workers. In the first case, a number of centrosymmetric Rydberg excited states acted as resonant intermediate states, and in the second case, the valence $B^3\Pi_0$ state was the intermediate. These studies demonstrate how well ZEKE spectroscopy can give a detailed vibrationally resolved spectrum and how autoionization is unavoidable in the photoelectron spectroscopy of small molecules.

The $2 + 1'$ ZEKE spectra of I_2 exhibited non-Franck–Condon behaviour, having intense off-diagonal peaks in v^+, v, due to vibrational autoionization. **Figure 4** gives the spectra resulting from ionization through the band origin of the $[^2\Pi_{3/2}]_{core}$ 5d; 2_g state at about 62 600 cm⁻¹, and also through the first three vibrationally excited levels. This state was ionized into the lower spin-orbit state of the ion. Conversely **Figure 5** gives the spectra recorded through the first three vibrational levels of the $[^2\Pi_{1/2}]_{core}$ 5d;2_g

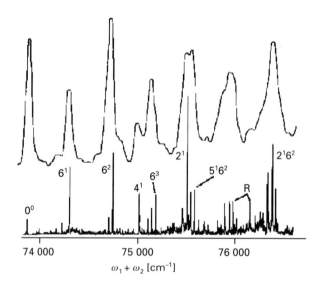

Figure 3 In *para*-difluorobenzene, the vibrational structure of the cation was not fully resolved until the introduction of ZEKE spectroscopy.

Rydberg state at about 68 000 cm⁻¹, which was ionized into the upper spin-orbit state of the ion. For the spectrum given in **Figure 4A**, which was through the origin, the $\Delta v = 0$ transition was most intense; the total transition energy to this level is 75 066 ± 2 cm⁻¹, which is the adiabatic ionization energy (to the ion in

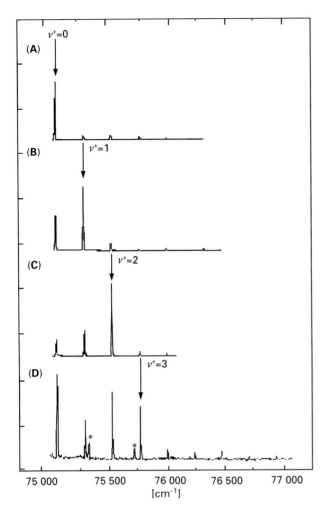

Figure 4 ZEKE spectrum of I_2 recorded through the $[^2\Pi_{3/2}]_{core}$ $5d;2_g$ state; the arrows indicate diagonal transitions, and the asterisks accidental resonances with A ← X transitions.

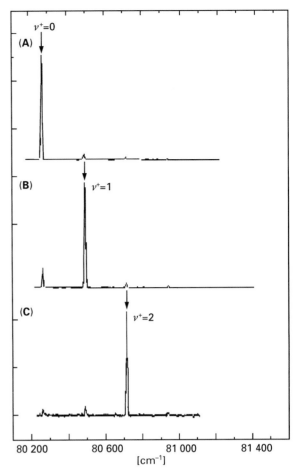

Figure 5 ZEKE spectrum of I_2 recorded through the $[^2\Pi_{3/2}]_{core}$ $5d;2_g$ state; the arrows indicate diagonal transitions.

its ground vibrational state). There is a weak vibrational progression from the band origin up to $v^+ = 4$. The dominance of the origin peak indicates a minimal change in geometry on ionization. This minimal change in geometry is expected after ionization from the (intermediate) Rydberg state; however, the progression does extend to higher v^+ than is expected merely on the basis of Franck–Condon factors. For the $v = 1$ intermediate state (**Figure 4B**) the non-Franck–Condon behaviour is even more pronounced: the stretching progression can be followed up to $v^+ = 7$; the $\Delta v = 0$ transition remains dominant; however, the peak is of about the same intensity as the $v^+ = 0$ peak. For the intermediate states $v = 2$ (**Figure 4C**) and $v = 3$ (**Figure 4D**) the progressions become longer. Although the spectra show a $\Delta v = 0$ propensity, a Franck–Condon envelope does not fit the intensity distribution.

The pattern for the vibrational propensity was found to be quite different in the upper spin-orbit state of the ion. In the spectrum recorded through the origin (**Figure 5A**) the most intense peak corresponds to the $\Delta v = 0$ transition (to the $v^+ = 0$ level of the ion). The corrected total transition energy to the origin is 80 266 ± 2 cm^{-1}; this again corresponds to the adiabatic ionization energy. This result, combined with the ionization energy for the lower $^2\Pi_{3/2}$ spin-orbit state, gives an improved spin-orbit splitting constant for I_2^+ in its ground electronic state of 5197 ± 4 cm^{-1}. For the upper spin-orbit component, the propensity for the $\Delta v = 0$ transition remains strong even as the vibrational level of the intermediate is increased. This corresponds to classic Franck–Condon behaviour.

The non-Franck–Condon intensities observed in the lower ($^2\Pi_{3/2}$) spin-orbit state spectrum were attributed to autoionization involving another Rydberg series which converges to a higher

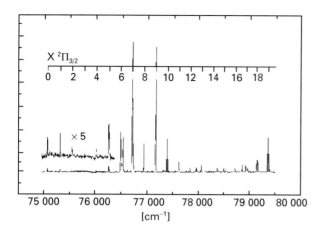

Figure 6 The long Franck–Condon forbidden progression exhibited in the ZEKE spectrum of the lower spin-orbit state recorded through $\nu = 15$ of the $B^3\Pi_u 0^+{}_u$ state.

vibrational state of the upper spin-orbit state in the ion. An interaction between the two Rydberg series, at the ionization threshold of the lower series, gave rise to autoionization from the higher series. No interaction occurs with the upper ($^2\Pi_{1/2}$) spin-orbit state, as there are no nearby Rydberg series.

The $1 + 2'$ photon study, via the valence $B^3\Pi_0$ state, extended the previous work. Long Franck–Condon progressions, arising from the valence character of the intermediate state, are evident in the ZEKE spectra of both spin-orbit components. In the lower spin-orbit component, the vibrational progression extends to at least $\nu^+ = 62$, and in the upper state as high as $\nu^+ = 34$. The spectrum in the range 75 000 to 80 000 cm^{-1} of the lower spin-orbit state, which was recorded via $\nu = 15$, is shown in **Figure 6**.

The vibrational progressions can be adequately simulated through the calculation of Franck–Condon factors; however the observed spin-orbit branching ratio, along with the intensity distribution, reflects a considerable contribution from both spin-orbit and field-induced resonant autoionization processes. Also, accidental resonances at the two-photon level with ion-pair states further perturb the distribution of peak intensities.

HCl

From the ZEKE spectra of hydrogen halides, the rotational-state distribution of the product ion provides a direct measure of the angular momentum of the outgoing electron; this is a sensitive probe of ionization dynamics. Autoionization occurs very readily in these molecules, via rotational, vibrational and electronic pathways, and is often evident in the spectra recorded. A further motivation in much of the

work on the hydrogen halides has been to investigate the artefacts of autoionization, particularly the role of rotational and spin-orbit autoionization processes. In the cation there is a π-vacancy in the ground electronic configuration. Hence, spin-orbit coupling is evident in the spectra.

HCl has been studied by both single-photon and two-colour multiphoton experiments. The studies focused on the vibrational ground state of the ion and observed a tendency to large changes in angular momentum on ionization, $|\Delta J| \leq 7/2$, indicating a preference for an outgoing d-partial wave. A preference for negative values of angular momentum transfer for both spin-orbit components was observed, which has been evident in many ZEKE spectra which exhibit rotational resolution; this is attributed to rotational autoionization. In the single-photon experiment by White and co-workers, anomalous branch intensities in the ZEKE spectrum were interpreted in terms of field- or dipole-induced mixing of Rydberg states converging on higher-ion rotational levels. Intensity anomalies were observed in the spin-orbit and rotational branching ratios of two-colour ZEKE spectra of de Lange and co-workers recorded via the $F^1\Delta_2$, $D^1\Pi_1$ and $f^3\Delta_2$ Rydberg states. The branching ratios were dependent on three experimental parameters:

(i) The delay time employed between excitation and ionization;
(ii) The magnitude of the bias electric field;
(iii) The magnitude of the applied pulsed electric field.

The results were rationalized on the basis of the increasing number of autoionization decay channels that become available to the high-n Rydberg states as each ionization threshold is reached. An analysis of the decay-dependence of the ZEKE spectra via the $F^1\Delta_2$ state provided evidence for a non-exponential decay of the high-n Rydberg states.

Studies on the other hydrogen halides have provided similar results; from HF, however, it was concluded that the s-channel dominates the photoionization process as opposed to the d-channel in HCl, HBr and HI.

Ammonia (NH₃)

The NH_3 molecule, when studied by Habenicht and co-workers, was the first polyatomic molecule studied by ZEKE spectroscopy for which full rotational resolution in the cation was achieved. Ionization was achieved by a $2 + 1'$ process, excited through the two-photon transition $\tilde{B} \leftarrow \tilde{X}$. The excitation

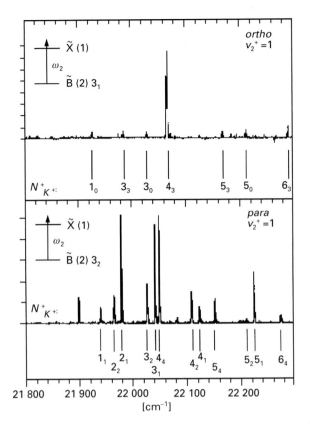

Figure 7 The symmetry-based selection rules applied to ZEKE spectroscopy are borne out by the different spectra obtained using the *ortho*- and *para*-nuclear spin states of ammonia.

spectrum for the 2^2_0 transition obtained by REMPI shows clearly resolved rotational structure with a Coriolis interaction giving *l*-type doubling. This allows for rotational-state selectivity in the intermediate state. The two rotational states corresponding to the *ortho*- ($J'_{K'} = 3_1$) and para- ($J'_{K'} = 3_2$) nuclear spin states of NH_3 were chosen as intermediate states for the ZEKE spectra. These two ZEKE spectra, recorded through the $\tilde{B}$-state, are given in **Figure 7**, taking the 2^2_1 vibronic transition. There is a clear difference between the spectra obtained for *ortho*- (top spectrum) and *para*-NH_3 (bottom spectrum). The ZEKE spectrum of *ortho*-NH_3 shows one strong transition into the ion rotational state with $N^+_{K+} = 4_3$, and other transitions with $K^+ = 0$ and 3, which are considerably weaker. On first sight this ZEKE spectrum appears similar to an atomic photoionization spectrum. On the other hand the ZEKE spectrum of *para*-NH_3 is much fuller giving the strongest lines observed for $K^+ = 1$ and $N^+_{K+} = 4_4$; also there are weaker transitions into $K^+ = 2$. These spectra are in very good agreement with the symmetry selection rules that apply to ZEKE transitions.

Larger molecules

Benzene (C_6H_6)

The neutral benzene molecule has a hexagonal, planar structure with D_{6h} symmetry; in the electronic ground state the electronic configuration is $a_{2u}^2 e_{1g}^4$. When benzene is ionized, one of the e_{1g} electrons is removed, leaving one e_{1g} electron unpaired. Thus, the cation has a doubly degenerate $^2E_{1g}$ electronic ground state. The Jahn–Teller theorem predicts that for any nonlinear polyatomic molecule in a degenerate electronic state, there exists a distortion of nuclear geometry along at least one non-totally symmetric normal coordinate that results in a splitting of the potential-energy function such that the potential minimum is no longer at the symmetrical position.

The structural distortions of the benzene cation have been discussed at length; quantum-chemical *ab initio* calculations predict three equivalent D_{2h} in-plane distortions corresponding to elongation, or compression along three of the twofold-symmetry axes of benzene. These give structures more stable than the hexagonal structure, with an experimentally determined stabilization energy of 266 cm^{-1}. This is approximately half the zero-point energy of the lowest-frequency Jahn–Teller active normal vibration.

For weak Jahn–Teller coupling, the stabilization energy for the distorted symmetry is much smaller than the zero-point energy of the Jahn–Teller active mode. Under collision-free conditions, the three equivalent D_{2h} structures of the cation would dynamically interconvert rapidly, and the ground state of the cation would still be described in the D_{6h} symmetry group. For strong Jahn–Teller coupling the cation would spend much time in one of the three structures, and would therefore be described in D_{2h}. The knowledge of the structure and the symmetry of the isolated benzene cation is desirable not only for testing quantum-mechanical model calculations: it also has a fundamental importance for organic chemistry. Through group-theoretical considerations, rotationally resolved ZEKE spectroscopy gives a clear and unambiguous determination of the symmetry of the cation. Thus, if the molecule were statically distorted to lower symmetry, transitions would appear in the rotationally resolved ZEKE spectra, which are forbidden in the D_{6h} structure. Thus, the observed rotational transitions are a sensitive and clear indication of the symmetry of the cation.

If one quantum of a Jahn–Teller active normal vibration in the benzene cation (these are the modes v_{6-9} with e_{2g} symmetry) is excited, the linear dynamic Jahn–Teller coupling leads to a splitting into two vibronic states with $j = \pm 1/2$ (E_{1g} symmetry) and

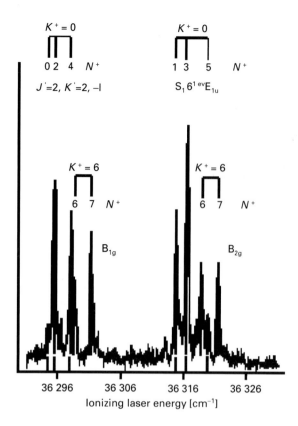

Figure 8 In the lower-energy band, recorded through the S_1 ^{161} E_{1u}, 2_2-l state, only even K are observed, whereas in the higher-energy band odd K are observed, indicating rigorously the symmetry of the vibronic state associated with each band.

$j = \pm 3/2$ ($B_{1g} \oplus B_{2g}$ symmetry) vibronic angular momentum. The $j = \pm 3/2$ states are further split by quadratic dynamic Jahn–Teller coupling, but the $j = \pm 1/2$ state remains doubly degenerate.

Detailed vibronic structure is seen in the low-resolution scan of the ZEKE spectrum of benzene, recorded via the 6^1 vibrational level in the S_1 state of the neutral. Bands with fundamental frequencies characteristic of the ν_6, ν_{16}, ν_4 and ν_1 vibrational modes are seen. However, no harmonic progressions can be observed, with the higher-energy portion of the spectrum exhibiting a highly irregular and dense system of vibronically active states. The active mode of lowest energy is ν_6, which is along the coordinate predicted for the Jahn–Teller distortion by *ab initio* methods. A key pair of bands, which appear at about 350 cm^{-1}, corresponding to the ion internal energy, are the B_{1g} and B_{2g} vibronic components of the ν_6 fundamental, shifted to lower energy by linear Jahn–Teller coupling. However, the relative ordering of B_{1g} and B_{2g} is unclear in this spectrum. The conservation of symmetry of the nuclear spin wavefunction restricts the possible transitions to these two vibronic states; thus, depending on the vibronic symmetry, only certain rotational progressions can be observed.

A high-resolution ZEKE scan of the B_{1g} and B_{2g} components of the 6^0 band recorded by exciting through the 6^1, $J'_{K'} = 2_2$, -l', S_1 state is shown in **Figure 8**. In this spectrum, only the $K^+ = 0$ projections of even N^+ are seen in the lower-energy vibronic component, whereas only those from odd N^+ are seen in the higher-energy vibronic component. This effect can be attributed to nuclear spin statistics, and indicates unambiguously that the lower-energy vibronic component has B_{1g} symmetry, and that the higher-energy vibronic component has B_{2g} symmetry. Thus the rotational structure in the ZEKE spectrum has established that the B_{2g} level in the quadratically split 6^1 ($j = 3/2$) levels of the benzene cation, lies above the B_{1g} level.

The cation is apparently distorted to a small extent by quadratic Jahn–Teller coupling in ν_6. It has been concluded from the rotational intensities in the ZEKE spectrum that the wells in the pseudorotation coordinate correspond to local B_{1g} electronic configurations (the elongated structure) whereas the saddle points are locally B_{2g} (compressed). From the coupling parameters that fit the vibronic structure in the coarse ZEKE spectrum, an energy difference between the stationary states of only 8 cm^{-1} is established. This is much less than the ν_6 zero-point energy of 413 cm^{-1}, indicating that the benzene cation is fluctuational, and therefore must be viewed in D_{6h} symmetry rather than in terms of the three D_{2h} structures with locally non-degenerate electronic configurations.

Toluene

The toluene molecule and its torsional states are classified according to its irreducible representations in the molecular symmetry (MS) group G_{12}, which is isomorphic to the point group D_{3h}. The problem of an unhindered, rigid methyl rotor attached to a rigid frame reduces to a one-dimensional Schrödinger equation with eigenfunctions, $\Psi_{\text{tor}}(\alpha) = (2\pi)^{-1/2} e^{\pm im\alpha}$ and eigenvalues, $E_M = Bm^2$, where B is an effective constant for the methyl-group rotation relative to the rigid molecular framework and $m = 0, \pm 1, \pm 2, \ldots$ is the rotational quantum number. In toluene, the torsional potential has 6-fold symmetry. A small torsional perturbation of the form $V(\alpha) = (V_6/2)(1 - \cos 6\alpha)$ leaves the $m = \pm 1, \pm 2, \pm 4$ and ± 5 states doubly degenerate but lifts the degeneracy of the $m = \pm 3$ and $m = \pm 6$ levels.

$\alpha = 0$ is defined as an eclipsed geometry, and $\alpha = \pi/6$ a staggered geometry. The rotor states are termed $0a_1'$, $1e''$, $2e'$, $3a_2''$, $3a_1''$, $4e'$, etc. Band intensities in the S_1-S_0 and cation-S_1 spectra reveal which $|m| = 3$ state lies lower in energy and thus the absolute phase of the torsional potential. For

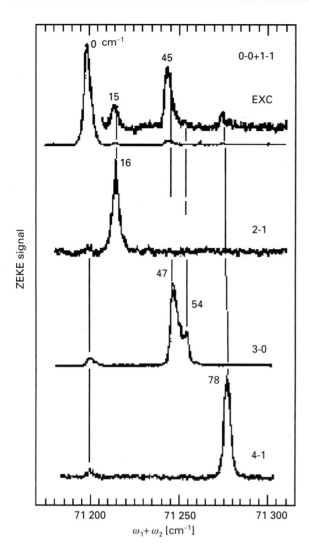

Figure 9 ZEKE spectra recorded through various torsional levels of the S_1 state in toluene. The label EXC indicates the torsional transition to the intermediate: $S_0 \leftarrow S_1$.

negative V_6 (potential minimum at $\alpha = \pi/6$), $3a_2''$ lies below $3a_1''$, whereas for positive V_6 (potential minimum at $\alpha = 0$), $3a_1''$ lies below $3a_2''$.

Weisshaar and co-workers recorded ZEKE spectra of toluene through different intermediate resonances of S_1 and they are presented in **Figure 9**. From the assignment it can be shown that there is a positive V_6 with $3a_1''$ lower in energy than the $3a_2''$ torsional state. The torsional states of the toluene cation (0, 15, 54 and 75 cm^{-1}) are not very different from the torsional states in the S_1 state (1, 15, 55 and 77 cm^{-1}), leading to the conclusion that the torsional barriers in the cation and in S_1 are quite similar.

For both $0a_1'-0a_1'$ and $3a_1''-0a_1'$ excitations of S_1, the ZEKE band at 46 cm^{-1} is stronger than the 54 cm^{-1} band. Using the assumption that torsion–electronic coupling allows all a–a torsional

transitions and forbids a–e transitions and the fact that the 47 cm^{-1} band is three times more intense than the 54 cm^{-1} band for ionization through the $3a_1''$ state of S_1, the 47 cm^{-1} band is assigned to the $3a_1''$–$3a_1''$ transition and the 54 cm^{-1} band to $3a_2''$–$3a_1''$. These assignments for the ZEKE spectra constitute a major step in understanding large-amplitude motions and the role of torsional–electronic couplings.

Weakly bound molecules

Phenol–methanol

The study of the PhOH–MeOH complex, by Müller-Dethlefs and co-workers, was carried out with a 1 + 1′ REMPI spectrum. This was difficult to interpret owing to the comparatively dense vibrational structure. Various vibrational levels in the S_1 state were used as intermediate states on the way to ionization. The ZEKE spectrum obtained by exciting *via* the S_1 vibrationless level, given in **Figure 10**, is striking with progressions of about ten quanta in a low-frequency vibrational mode of 34 cm^{-1}, denoted h_1, appearing in combination with components of an anharmonic progression of the intermolecular stretch of 278 cm^{-1}. The pattern suggests a rather substantial change of geometry upon ionization. The adiabatic ionization energy was derived as 63 207 ± 4 cm^{-1} which is a red shift from the S_0 state of 5421 ± 8 cm^{-1}, indicating a large increase in bond strength. The latter point is also exemplified by the large increase in the energy of the intermolecular stretch compared with 176 cm^{-1} in the S_1 state and 162 cm^{-1} in the S_0 state. Additionally, between the latter components, another set of progressions of the intermolecular mode of 34 cm^{-1} were seen, this time in combination with a third intermolecular mode of 52 cm^{-1}.

The ZEKE spectrum obtained *via* the S_1 state, with one quantum of the lowest-frequency intermolecular mode excited, showed the same vibrations but with a substantially changed Franck–Condon envelope which allowed the identification of a fourth intermolecular mode, denoted h_4, at 153 cm^{-1}. A slightly different envelope for the 34 cm^{-1} vibration was also obtained when exciting through the S_1 state with one quantum of the intermolecular stretch excited. The other two intermolecular modes of the Ph-MeOH cationic complex were identified from the ZEKE spectrum obtained *via* a combination band; their values being 76 cm^{-1} and 158 cm^{-1}.

I_2-Ar

Iodine–argon was one of the first complexes to be studied in the early jet spectroscopy experiments

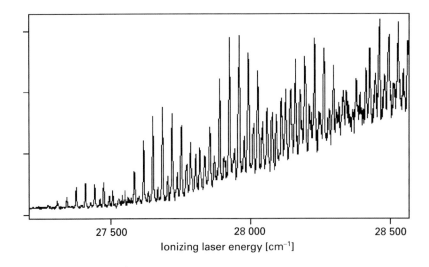

Figure 10 The striking vibrational progression seen in the ZEKE spectrum of phenol–methanol.

conducted by Levy and co-workers in the 1970s. Most of the work carried out on this complex since then has been concerned with the $\tilde{B}^3\Pi_{0^+} \leftarrow \tilde{X}^1\Sigma^+_g$ system studied using laser-induced fluorescence (LIF) spectroscopy. This state is not well-suited as a resonant intermediate in ZEKE spectroscopy, as it lies only 15 800 cm^{-1} above the electronic ground state. Since this study a number of grade ns and nd Rydberg excited states, based upon both spin-orbit states of the $\tilde{X}^2\Pi_{\Omega,g}I_2^+$-Ar core, have been characterized using 2 + 1 mass-resolved REMPI spectroscopy, by Cockett and co-workers. These Rydberg states lie between 53 000 and 69 000 cm^{-1} and are better suited as resonant intermediate states for ionization into the two spin-orbit components of the $\tilde{X}^2\Pi_{\Omega,g}$ ionic ground state.

Substantial differences in the binding energy were observed for the 6s, 5d and 6d Rydberg states of the I$_2$-Ar complex, which correlate with the degree to which the positive charge of the core is shielded from the argon atom by the Rydberg electron. More-penetrating Rydberg orbitals reduce the charge-induced dipole forces between iodine and argon. The observed binding energy increases are seen in the REMPI spectra as progressive increases in spectral red shifts and intermolecular van der Waals stretching frequencies.

An issue which had been the subject of considerable debate was whether I$_2$-Ar adopts a T-shaped or linear geometry. Initial speculation suggested that it should be linear by analogy with the known linear geometry of the ClF-Ar complex. However the first direct experimental evidence for the geometry of I$_2$-Ar emerged from a partially rotationally resolved B-X fluorescence excitation spectrum, which showed that I$_2$-Ar adopts a T-shaped geometry. It

was also suggested that a linear isomer was responsible for an observed fluorescence excitation continuum underlying the discrete B-X transitions.

The mass-resolved 2 + 1 REMPI spectrum recorded by monitoring the I$_2^{\bullet+}$-Ar mass channel is shown in **Figure 11**. The spectrum is composed of partially overlapping vibrational progressions arising from the $[^2\Pi_{3/2}]_c$ 5d; 2$_g$ (**Figure 11A**; recorded with circularly polarized light) and $[^2\Pi_{3/2}]_c$ 5d; 0^+_g (**Figure 11B**; recorded with linearly polarized light) Rydberg states of I$_2$-Ar. The vibrational structure for both states essentially arises from simultaneous excitation of both the I-I stretch (ν_1) and the I$_2$-Ar van der Waals stretch (ν_3). For the $[^2\Pi_{3/2}]_c$ 5d; 0^+_g state, the progression terminates abruptly at $\nu_3\nu_1$ which suggests that the complex dissociates at this point. The $\nu_3\nu_1$ band appears at an internal vibrational energy of 758 cm^{-1}, but the spectral red shift of the band origin dictates a lower limit to the zero-point dissociation energy for this state of 563 ± 5 cm^{-1}. Thus, it would appear that the coupling between the two vibrational modes is sufficiently weak to enable the complex to accommodate an excess 195 cm^{-1} of internal energy before it dissociates. This is consistent with the geometry of the complex being T-shaped.

The $[^2\Pi_{3/2}]_c$ 5d; 2$_g$ Rydberg state progression shown in **Figure 11A** has an additional complexity of structure when compared to the progression observed for the 0^+_g state. For both $\nu_1 = 0$ and $\nu_1 = 1$, each vibrational band appears significantly broader than observed for the 0^+_g progression, and on most of the bands, an apparent splitting can be resolved. However, for $\nu_1 = 2$ the doublet structure has disappeared and the peaks have adopted the narrower profile seen in the 0^+_g progression. In fact, the

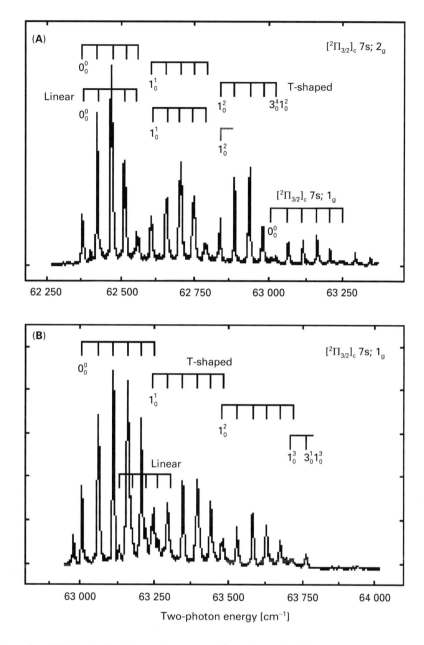

Figure 11 REMPI spectra of I_2-Ar showing the assignments attributed to each of the two structural isomers; (A) was recorded through the $[^2\Pi_{3/2}]_{core}$ 5d; 2_g state with circularly polarized light, and (B) was recorded through the $[^2\Pi_{3/2}]_{core}$ 5d; 0^+_g state with linearly polarized light.

doublet structure arises, not from any splitting of the peaks, but from two partially overlapping vibrational progressions with near-identical band origins. It appears from the spectrum that one of the progressions terminates at $\nu_1 = 1$ while the other continues at least as far as $\nu_1 = 2$.

Assuming that the point at which the progression terminates represents the point at which the complex dissociates, the conclusion drawn was that for the shorter progression, the onset of dissociation occurs at 463 cm^{-1} internal vibrational energy (compared with a calculated value for D_0 of about 503 cm^{-1}),

while for the longer progression, dissociation occurs at a lower limit of 655 cm^{-1}. On this basis, a provisional assignment was made of the shorter progression to the linear isomer, for which the van der Waals stretch might be expected to couple more efficiently with the I_2 stretch, and the longer progression to the T-shaped isomer.

Although the REMPI spectrum certainly provided a great deal of circumstantial evidence for the existence of two isomers, the assignment was to be greatly strengthened by extending the study to the ionic state by using ZEKE spectroscopy to probe

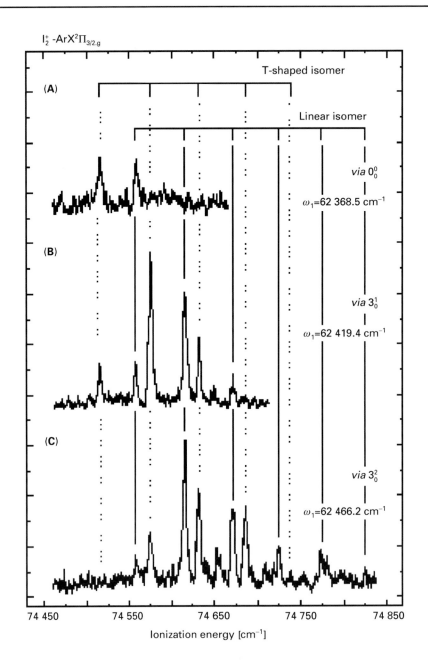

Figure 12   The ZEKE spectra of I_2-Ar recorded through the $[^2\Pi_{3/2}]_{core}$ 5d; 2_g state. (A) is through the overlapping (0^0_0) vibrational levels, (B) the (3^1_0) levels and (C) the (3^2_0) levels.

the $[^2\Pi_{3/2}]_c$ 5d; 2_g Rydberg state in a two-colour 2 + 1' ionization scheme, and recording isomer-specific ZEKE spectra. The ZEKE spectra of the ground electronic state of the ion recorded *via* several intramolecular vibrational levels in the $[^2\Pi_{3/2}]_c$ 5d; 2_g Rydberg state show, in each case, two well-separated vibrational progressions (**Figure 12**). As the level of vibrational excitation is increased in the intermediate Rydberg state, so the vibrational activity in the resulting ZEKE spectra increases.

The general propensity for diagonal transitions was consistent with the fairly small changes in geometry that occur on exciting from the Rydberg state to the ion. The experimental observations in this case were consistent with an assignment of the two overlapping Rydberg state progressions to two geometrical isomers. The measured difference in the ionization energies of 43 cm⁻¹ for the two isomers provided an indication of their relative stabilities in the ion, with the linear isomer being the more weakly bound.

List of symbols

B = constant for methyl-group rotation; j = vibronic angular momentum; J = total angular momentum with a projection of K in the molecular frame; m = rotational quantum number; n = principal quantum number; N = total angular momentum excluding spin; α = torsional angle; v = vibrational quantum number; ω = corresponding vibrational constant.

See also: **Photoelectron Spectrometers; Photoelectron Spectroscopy; Zero Kinetic Energy Photoelectron Spectroscopy, Theory.**

Further reading

Cockett M, Müller-Dethlefs K and Wright TG (1994) Recent applications and developments in ZEKE spectroscopy. *Royal Society of Chemistry Annual Reports Section C* **94**: Chapter 9, pp 327–373.

Habenicht W, Reiser G and Müller-Dethlefs K (1991) High resolution zero kinetic energy electron spectrum of ammonia. *Journal of Chemical Physics* **95**: 4809–4820.

Haines S, Dessent C and Müller-Dethlefs K Mass analysed threshold ionization of phenol-CO, intermolecular binding energies of a hydrogen bonded complex. *Journal of Chemical Physics* (submitted).

Lindner R, Müller-Dethlefs K, Wedum E, Haber K and Grant ER (1996) On the shape of $C_6H_6^+$. *Science* **721**: 1698–1702.

Müller-Dethlefs K (1995) High resolution spectroscopy with photoelectrons: ZEKE spectroscopy of molecular systems. In: Powis I, Baer T and Ng C-Y (eds) *High Resolution Laser Photoionization and Photoelectron Studies*, Chapter 2, pp 22–78. Wiley.

Müller-Dethlefs K and Cockett M (1998) *Nonlinear Spectroscopy for Molecular Structure Determination*, Chapter 7, pp 167–201. Oxford: Blackwell Science.

Müller-Dethlefs K and Schlag EW (1998) Chemical applications of zero kinetic energy (ZEKE) photoelectron spectroscopy. *Angewandte Chemie, International Edition English* **37**: 1347–1374.

Müller-Dethlefs K, Dopfer O and Wright TG (1994) ZEKE spectroscopy of complexes and clusters. *Chemical Reviews* **94**: 1845–1871.

Wang K and McKoy V (1995) High resolution photoelectron spectroscopy of molecules. *Annual Review of Physical Chemistry* **46**: 275–304.

Zero Kinetic Energy Photoelectron Spectroscopy, Theory

K Müller-Dethlefs and **Mark Ford**, University of York, UK

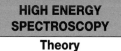

Introduction

Understanding of chemistry and the chemical bond is greatly influenced by molecular orbital theory. The power of the molecular orbital approach in providing an understanding of the structure and reactivity of molecules lies in its description of chemical bonds. The legitimacy of a molecular orbital description is attested to by the results of molecular electronic spectroscopy. In the single-electron molecular orbital picture, photoionization involves the excitation of an electron from a bound orbital into the ionization continuum. The energy from the photon is partitioned between the kinetic energy of the electron and the ionization energy as in Equations [1] and [2]. Thus by selecting the electron kinetic energy a spectrum can be recorded.

$$h\nu = E_{\text{ionization}} + E_{\text{kinetic}} \quad [1]$$

$$E_{\text{ionization}} = E_{\text{internal}}(\text{ion}) - E_{\text{internal}}(\text{neutral}) \quad [2]$$

To a first approximation the ionization energy is equal to the energy of the orbital from which the electron originates in the ground-state molecule. This is known as Koopmans' theorem, and allows each signal in the spectrum to be assigned to an orbital. Koopmans' theorem provides evidence that molecular orbitals are conceptually valid and has assumed a very important place in the development of our understanding of the electronic structure of molecules.

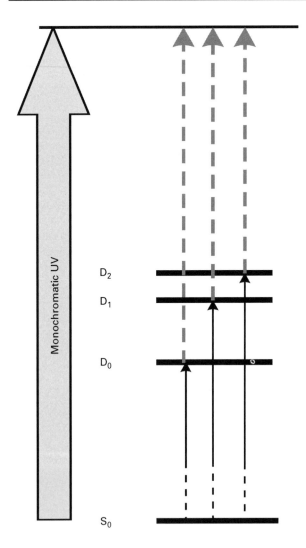

Figure 1 An ionization scheme for conventional PES.

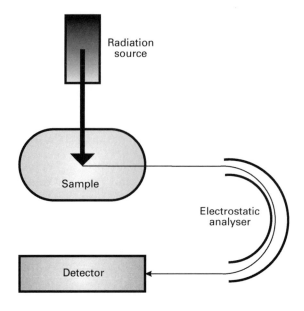

Figure 2 The setup for conventional photoelectron spectroscopy.

through the practical resolution of the geometrical or time-of-flight photoelectron analysers used to analyse the kinetic energy; in order to have a reasonable signal strength this is about 10 meV (80 cm^{-1}).

The energy of the radiation source used in photoelectron spectroscopy determines the configuration of the ion. If a high-energy source is used as in X-ray photoelectron spectroscopy (XPS) the electron is emitted from a core orbital, whereas ultraviolet photoelectron spectroscopy (UVPS) will only give signals from the valence orbitals of molecules with relatively low ionization energies.

ZEKE photoelectron spectroscopy

The method of ZEKE spectroscopy was invented by Müller-Dethlefs, in 1984, to bypass the resolution limitation. The technique uses a delayed electric field pulse to extract only the electrons ionized without kinetic energy, as shown in **Figure 3**. A signal is only seen when the radiation used is precisely that required to excite the electrons to the ionization threshold; thus, in theory, the resolution is limited only by the bandwidth of the radiation used in ionization. Dye lasers are used as the source of radiation in the ZEKE experiment, as these can have a very low bandwidth (of the order of 0.05 cm^{-1}). This allows the prospect of resolving rotational states of larger molecular cations. A graphic illustration of the greater resolving power of ZEKE spectroscopy can be seen in **Figure 4** in which the conventional photoelectron spectrum of I_2 is compared with a ZEKE spectrum.

Koopmans theorem is limited by its neglect of molecular orbital reorganization, and transitions between states of correct symmetry species that are allowed; hence it can be quite misleading. Thus a better picture is to say that, at low resolution, the signals observed correspond to different electronic states in the cation. At higher resolution, fine structure in the photoelectron spectrum corresponds to the vibronic levels of the cation; a broad band, with significant vibronic structure, indicates a large change in structure between the neutral molecule and the cation. This is due to the improved overlap with vibrational overtones, arising as a result of the Franck–Condon principle. An ionization scheme for conventional PES is given in **Figure 1**.

The setup for a conventional photoelectron spectrometer is illustrated in **Figure 2**. In modern spectrometers the sample is introduced in a molecular beam, with a very low temperature. The major restriction on the resolution of PES, however, arises

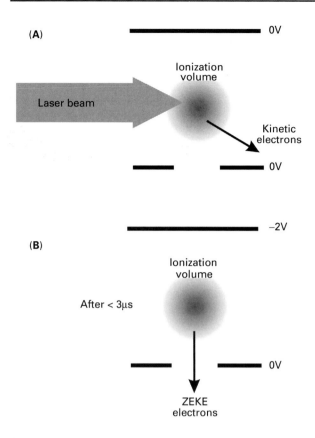

(A)

0V

Ionization
volume

Laser beam

Kinetic
electrons

0V

–2V

(B)

Ionization
volume

After < 3μs

0V

ZEKE
electrons

Figure 3 (A) A pulsed laser beam ionizes the sample, and kinetic electrons are scattered randomly. (B) A delayed electric-field pulse is used to extract ZEKE electrons from the ionization volume.

The photon energy from a dye laser, even after frequency-doubling, is normally insufficient to ionize a molecule; hence the experiment involves a multiphoton process, normally via a resonant intermediate state. An advantage from this resonant intermediate state selection is, for instance, that different vibrational levels of the intermediate state can be selected, thus allowing the Franck–Condon factors for ionization to change. In order to find the intermediate states a resonance-enhanced multiphoton ionization (REMPI) spectrum must first be recorded. This is usually done by incorporating a time-of-flight mass spectrometry apparatus into the ZEKE photoelectron spectroscopy apparatus. For the extraction of the heavier ions a high-voltage pulse (~1 kV) is used and the ion signal is detected by multichannel plates. The REMPI signal can be mass-selected, to ensure that it represents an excited state of the appropriate species; this property is very important when studying van der Waals clusters in molecular beams. The ionization scheme for a two-colour 1 + 1′ REMPI spectrum is shown in **Figure 5**.

With the exception of electron photodetachment from anions, it was soon realized that the ZEKE signal detected in a typical experiment does not actually arise from direct photoionization. With the application of an electric field to extract the ZEKE electrons there is a lowering of the ionization energy by the Stark effect, resulting in the ionization of high-n Rydberg states ($n > 100$) which are said to be within a 'magic region' lying about 5 to 10 cm^{-1} below each ion threshold, as indicated in **Figure 6**. As it turned out, the pulsed-field ionization of the Rydberg states sometimes leading to the name ZEKE-PFI, is in fact preferable to the 'true' ZEKE technique, since the neutral Rydberg states are less susceptible to stray electric fields in the apparatus than the 'true' ZEKE electrons.

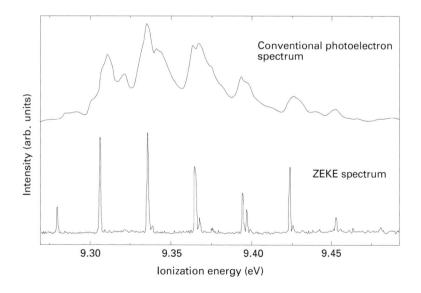

Conventional photoelectron
spectrum

ZEKE spectrum

Intensity (arb. units)

9.30   9.35   9.40   9.45

Ionization energy (eV)

Figure 4 The improvement in resolution gained by using ZEKE spectroscopy.

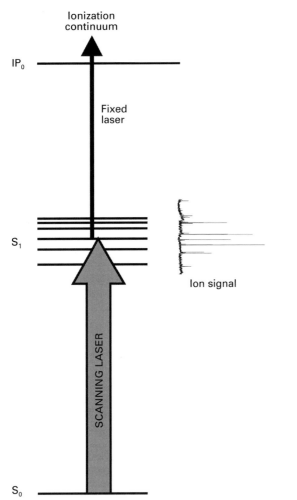

Figure 5 An ionization scheme for a 1 + 1′ REMPI experiment.

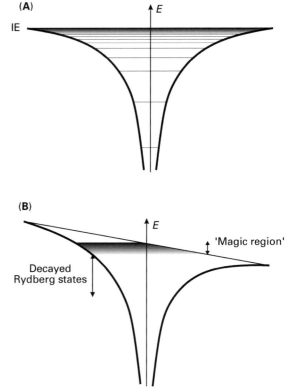

Figure 6 (A) the Rydberg states converge on the ionization energy. (B) the long-lived states in the 'magic' region are field-ionized by the extraction pulse.

The ionization scheme for ZEKE is given in **Figure 7**. In the case of ZEKE photodetachment, the only mechanism by which one can obtain a signal is the detection of free electrons with zero kinetic energy.

A great deal of research effort has been expended in trying to understand the nature of the Rydberg states in the 'magic region'. The origin of the exceptionally long lifetime of the ZEKE Rydberg states has been a matter of some considerable debate and although the discussion is continuously evolving, a degree of consensus has been reached concerning the principal contributory effects. The extended lifetime is attributed to a combination of the effects of small homogeneous fields (which typically originate from electronic equipment in the laboratory) and inhomogeneous electric fields (associated with regions of localized charge in the spectrometer, e.g. ions). The highly diffuse nature of high-lying Rydberg states renders them susceptible to l and m_l mixing through external perturbations. Stray DC electric

fields may cause substantial l mixing through the Stark effect, while inhomogeneous fields inducing m_l randomization arise from the presence of ions in low to medium concentrations. This slows down the rate of intramolecular relaxation considerably due to the conservation of angular momentum, as depicted in **Figure 8**. In the 'magic region' the Rydberg electron acquires a non-penetrating character and no longer interacts with the positive-ion core. Conversely, in the lower-lying Rydberg states the Rydberg electron still undergoes regular collisions with the core, which leads to intramolecular relaxation processes such as predissociation into neutral fragments and auto-ionization. This decay of the lower Rydberg states during the typical delay times used in ZEKE experiments is the underlying reason why peak widths observed in ZEKE spectroscopy are limited, even when high field strengths are used.

If laser-limited resolution is required, it is necessary to design sophisticated field-ionization schemes. The current resolution benchmark is the rotationally resolved ZEKE spectrum of benzene. From this it was demonstrated that the benzene cation is planar and adequately described in the D_{6h} molecular point group, despite being subject to Jahn–Teller

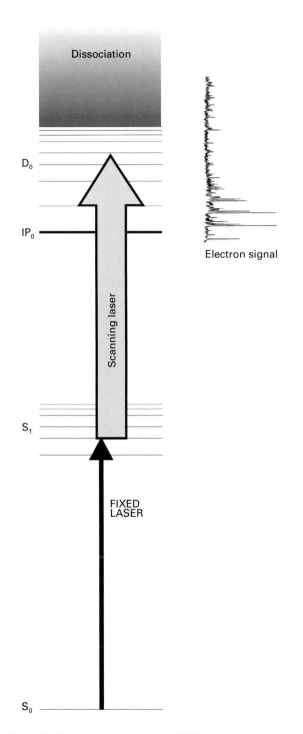

Figure 7 An ionization scheme for ZEKE spectroscopy.

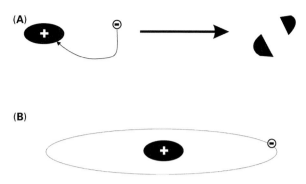

Figure 8 (A) Collisions between the Rydberg electron and the core in lower Rydberg states cause decay by intramolecular processes such as predissociation. (B) The nonpenetrating character of higher Rydberg states results in a longer lifetime.

comparable to that of Rydberg extrapolations but with less experimental effort.

Variation of the slope of the pulse allows the spectral resolution to be adjusted according to the laser bandwidth limitations and to the demands of the system under study (whether one requires vibrational or rotational resolution). **Figure 9** shows schematically the effect of pulse-slope risetime on the time-of-flight (TOF) of the corresponding electrons produced by PFI. A fast pulse generates all the signal within a narrow TOF distribution, whereas a slow pulse spreads the different 'slices' of Rydberg states into a broader TOF distribution. Thus, for a particular TOF gate a smaller spectral 'Rydberg slice' will be collected when the photon energy of the light source is scanned for a slow rather than for a fast risetime pulse.

There are many ways to improve the resolution by varying the pulse sequence. The use of an extraction pulse causes the $2l + 1$ degeneracy of a given state to be lifted, due to the Stark effect. Those with a negative m_l are raised in energy, and those with a positive m_l are lowered in energy; this has the effect of broadening the signal about the field-free ionization energy. However, it has been shown that under field ionization the Stark states shifted to the blue (with negative m_l values) have a longer lifetime than those shifted to the red. Hence the signal seen arises principally from states which have been red shifted. This can be accounted for using a multi-step staircase-like extraction pulse, to determine exactly the ionization energy under field-free conditions. It has also been observed that if a double extraction pulse is used, the second being of opposite polarity to the first, the Stark states which were blue shifted, and therefore were not ionized by the first pulse, kept the same orientation, and were red shifted with respect to the second pulse. The consequence of this fractional Stark-state selection is that a narrower slice of Stark

distortion. ZEKE spectroscopy has also been successfully applied to studies of the vibrational structure of large organic molecules, molecular and metal clusters, and hydrogen-bonded systems. One of the more routine strengths of the technique is that ionization energies can be determined with an accuracy

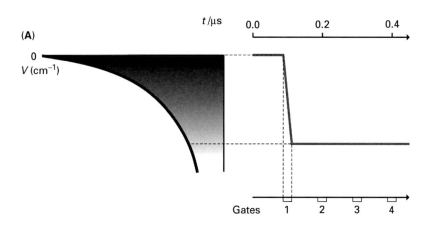

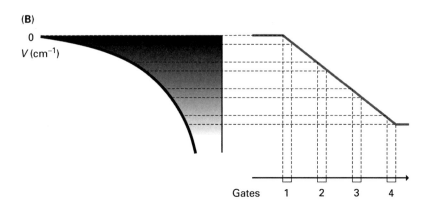

Figure 9 The effect of slope risetime on the resolution: in (A) the fast pulse gives a low resolution as a large slice of the Rydberg manifold is ionized; in (B) the slower pulse enables the detection of much smaller slices of the manifold, giving a higher resolution at the cost of signal strength.

states is selected, and a higher-resolution signal is obtained.

A typical experimental setup for a ZEKE experiment is shown schematically in **Figure 10**. It consists of a laser system and a vacuum apparatus which includes the molecular beam source, the extraction plates and a μ-metal-shielded flight tube with electron/ion detectors (dual multichannel plates) at each

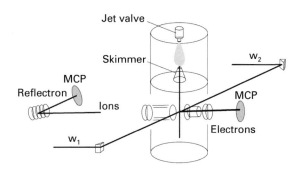

Figure 10 A schematic representation of the ZEKE apparatus.

end. In a typical two-colour experiment, both dye lasers (often frequency-doubled) are pumped simultaneously by an excimer laser or a Nd:YAG laser. The first dye laser excites a specific vibronic or rovibronic level of the intermediate state and the second laser ionizes the molecules or promotes them into long-lived Rydberg states ($n > 150$) converging to (ro)vibronic levels of the electronic ground state or an electronically excited state of the cation. After a delay time of a few μs, an extraction pulse is applied by either a simple electric pulsing device or by an arbitrary-function generator. The electrons are detected at the multichannel plates and their time of flight is recorded with boxcar integrators or a transient digitizer by setting narrow time gates (10–30 ns).

Intensities in ZEKE spectra

The selection rules for ZEKE transitions are governed by the usual principle that the transition-moment integral must be totally symmetric to give a non-zero intensity. For ZEKE transitions these can be affected by coupling to other Rydberg states. For rotational transitions one has to consider the possibility of transfer of angular momentum to the Rydberg electron; hence the selection rule for the quantum number representing the total angular momentum excluding spin is no longer $\Delta N = \pm 1, 0$. One model to account for the intensities of the rotational transition is the spectator model. In the spectator model it assumed that the Rydberg electron wave function is atomic hydrogen-like, with angular momentum l_0 and has no interaction with the core; hence the selection rules for N'', the total angular momentum of the Rydberg state, and N^+, the total angular momentum of the ion, are bound by the triangle condition $|N'' - N^+| \le l_0 \le |N'' + N^+|$. The intensities of ionizing transitions in the spectator model are dependent only on the transitions into the Rydberg state; a more complicated model is the compound model, where it is not assumed that the Rydberg state is initially fully decoupled from the core; hence further transfer of angular momentum can occur before the extraction pulse.

A further factor which affects the intensity of transitions is the role of autoionization from the Rydberg states. When a given Rydberg state is near the ionization threshold of another, lower-energy Rydberg series, the state has a shortened lifetime with respect to autoionization; as a consequence the intensity of higher-energy Rydberg series is usually depleted. In rotationally resolved spectra this is observed as a propensity for negative changes in angular momentum.

Mass-selected ZEKE spectroscopy

An extension of ZEKE spectroscopy is mass-analysed threshold ionization (MATI), 'photoelectron spectroscopy without photoelectrons'. This is effectively the same experiment; for every ZEKE electron produced, there must be a cation, and in MATI detection a signal is recorded from these ions. It is much harder to separate the ions produced from pulsed-field ionization of the ZEKE Rydberg states from the ever-present directly produced ions. Ions are much heavier than electrons and hence move more slowly, so a higher-voltage extraction pulse is required for the separation and the subsequent extraction and selection of the cations. The obvious advantage of this combination of ZEKE with mass spectrometry is the ability to select the cations on the basis of their mass.

The MATI signal also allows the study of fragmentation processes. It is interesting that at levels of ion internal energy at which a complex dissociates, the ZEKE spectrum can still be observed. When only

looking at the ZEKE signal, it is not obvious that such a fragmentation has occurred; however, by looking at the MATI signal, fragmentation can be observed, as the spectrum switches from the parent cation mass channel to a fragment mass channel. This gives a direct measure of the dissociation energy of the cation. Predissociation can also be observed directly by this technique, and the dissociation products observed, which is useful for obtaining a more complete rovibronic structure of molecules.

Another development in photoionization spectroscopy is the technique called photoinduced Rydberg ionization (PIRI). In this, the neutral molecule absorbs radiation to produce a high-n Rydberg state, as well as prompt ions. The prompt ions are separated using a delayed electric pulse, and the remaining Rydberg states, rather than being field-ionized, as in MATI, are photoexcited to form core-excited Rydberg states, which autoionize, and can be separated from the remaining Rydberg states. This technique effectively gives the absorption spectrum of the cation; also the problems arising from Stark shifts are no longer relevant as the molecules are not field-ionized.

List of symbols

E = energy; h = Planck's constant; l = orbital angular momentum of an electron with projection m on the laboratory Z-axis; n = principle quantum number; N = angular momentum; ν = frequency.

See also: **Photoelectron Spectrometers; Photoelectron Spectroscopy; Zero Kinetic Energy Photoelectron Spectroscopy, Applications.**

Further reading

Dietrich H-J, Müller-Dethlefs K and Baranov LY (1996) Fractional Stark state selective electric field ionization of very high-n Rydberg states in molecules. *Physical Review Letters* 76: 3530–3533.

Fischer I, Lindner R and Müller-Dethlefs K (1994) State to state photoionization dynamics probed by zero kinetic energy photoelectron spectroscopy. *Journal of the Chemical Society, Faraday Transactions* 90: 2425–2442.

Haines SR, Geppert WD, Chapman DM *et al* (1998) Evidence for a strong intermolecular bond in the phenol N_2 cation. *Journal of Chemical Physics* 109: 9244–9251.

Müller-Dethlefs K and Schlag EW (1991) High resolution zero kinetic energy (ZEKE) photo electron spectroscopy of molecular systems. *Annual Review of Physical Chemistry* 42: 109–136.

Müller-Dethlefs K (1995) High resolution spectroscopy with photoelectrons: ZEKE spectroscopy of molecular systems. In: Powis I, Baer T and Ng C-Y (eds) *High*

Resolution Laser Photoionization and Photoelectron Studies, Chapter 2, pp 22–78. Chichester: Wiley.

Müller-Dethlefs K (1995) Applications of ZEKE spectroscopy. *Journal of Electron Spectroscopy and related Phenomena* 75: 35–46.

Müller-Dethlefs K, Schlag EW, Grant ER, Wang K and McKoy BV (1995) ZEKE spectroscopy: High resolution spectroscopy with photoelectrons. *Advances in Chemical Physics* 90: 1–104.

Müller-Dethlefs K (1991) Zero kinetic energy electron spectroscopy of molecules – rotational symmetry selection rules and intensities. *Journal of Chemical Physics* 95: 4821–4839.

Reiser G and Müller-Dethlefs K (1992) Rotationally resolved zero kinetic energy photoelectron spectroscopy of nitric oxide. *Journal of Physical Chemistry* 96: 9–12.

Wright TG, Reiser G and Müller-Dethlefs K (1994) Good vibrations. *Chemistry in Britain* 30: 128–132.

Zinc NMR, Applications

See **Heteronuclear NMR Applications (Sc–Zn).**

Zirconium NMR, Applications

See **Heteronuclear NMR Applications (Y–Cd).**

Appendices

Appendices 1, 2, 4, 5, 6, 7, 8, 9, 10, 15, 16, 17, 18, 19, 20, 21, 22, 23, 24, 25, 26, 27, 28, 29, and 30 are reproduced with permission from Bruker Analytik, Rheinstetten, Germany.

Appendices 12, 13 and 14 are reproduced with permission from ROMIL Super Purity Solvents, Acids and Reagents catalogue.

Appendix 11 is reproduced with permission from Bard, A.J., Parsons, R. and Jordan, R. (eds) (1985) *Standard potentials in Aqueous Solution*. New York: Marcel Dekker Inc.

1. Periodic Table of Elements

IA	IIA	IIIA	IVA	VA	VIA	VIIA	VIII	VIII	VIII	IB	IIB	IIIB	IVB	VB	VIB	VIIB	O
1 **H** 1.008																	2 **He** 4.003
3 **Li** 6.941	4 **Be** 9.012											5 **B** 10.81	6 **C** 12.01	7 **N** 14.01	8 **O** 16.00	9 **F** 19.00	10 **Ne** 20.18
11 **Na** 22.99	12 **Mg** 24.31											13 **Al** 26.98	14 **Si** 28.09	15 **P** 30.97	16 **S** 32.06	17 **Cl** 35.45	18 **Ar** 39.95
19 **K** 39.10	20 **Ca** 40.08	21 **Sc** 44.96	22 **Ti** 47.90	23 **V** 50.94	24 **Cr** 52.00	25 **Mn** 54.94	26 **Fe** 55.85	27 **Co** 58.93	28 **Ni** 58.71	29 **Cu** 63.55	30 **Zn** 65.37	31 **Ga** 69.72	32 **Ge** 72.60	33 **As** 74.92	34 **Se** 78.96	35 **Br** 79.90	36 **Kr** 83.80
37 **Rb** 85.47	38 **Sr** 87.62	39 **Y** 88.91	40 **Zr** 91.22	41 **Nb** 92.91	42 **Mo** 95.94	43 **Tc** (99)	44 **Ru** 101.1	45 **Rh** 102.9	46 **Pd** 106.4	47 **Ag** 107.9	48 **Cd** 112.4	49 **In** 114.8	50 **Sn** 118.7	51 **Sb** 121.8	52 **Te** 127.6	53 **I** 126.9	54 **Xe** 131.3
55 **Cs** 132.9	56 **Ba** 137.3	57 **La** 138.9	72 **Hf** 178.5	73 **Ta** 180.9	74 **W** 183.9	75 **Re** 186.2	76 **Os** 190.2	77 **Ir** 192.2	78 **Pt** 195.1	79 **Au** 197.0	80 **Hg** 200.6	81 **Tl** 204.4	82 **Pb** 207.2	83 **Bi** 209.0	84 **Po** (210)	85 **At** (210)	86 **Rn** (222)
87 **Fr** (223)	88 **Ra** (226)	89 **Ac** (227)															

Lanthanides

58 **Ce** 140.1	59 **Pr** 140.9	60 **Nd** 144.2	61 **Pm** (147)	62 **Sm** 150.4	63 **Eu** 152.0	64 **Gd** 157.3	65 **Tb** 158.9	66 **Dy** 162.5	67 **Ho** 164.9	68 **Er** 167.3	69 **Tm** 168.9	70 **Yb** 173.0	71 **Lu** 175.0

Actinides

90 **Th** 232.0	91 **Pa** (231)	92 **U** 238.0	93 **Np** (237)	94 **Pu** (242)	95 **Am** (247)	96 **Cm** (248)	97 **Bk** (247)	98 **Cf** (251)	99 **Es** (254)	100 **Fm** (253)	101 **Md** (256)	102 **No** (254)	103 **Lr** (257)

2. Tables of SI and related units

SI base units

Quantity	Name	Symbol
length	metre	m
mass	kilogram	kg
time	second	s
electric current	ampere	A
thermodynamic temperature	kelvin	K
amount of substance	mole	mol
luminous intensity	candela	cd

Units in use with the international system

Name	Symbol	Value in SI units
minute	min	1 min = 60 s
hour	h	1 h = 60 min = 3600 s
day	d	1 d = 24 h = 86 400 s
degree	°	$1° = (\pi/180)$ rad
minute	'	$1' = (1/60)° = (\pi/10\ 800)$ rad
second	"	$1'' = (1/60)' = (\pi/648\ 000)$ rad
litre	L	$1\ L = 1\ dm^3 = 10^{-3}\ m^3$
tonne	t	$1\ t = 10^3\ kg$

cgs units with special names

Name	Symbol	Value in SI units
erg	erg	$1\ erg = 10^{-7}\ J$
dyne	dyn	$1\ dyn = 10^{-5}\ N$
poise	P	$1\ P = 1\ dyn\ s\ cm^{-2} = 0.1\ Pa\ s$
stokes	St	$1\ St = 1\ cm^2\ s^{-1} = 10^{-4}\ m^2\ s^{-1}$
gauss	Gs, G	$1\ Gs$ corresponds to $10^{-4}\ T$
oersted	Oe	$1\ Oe$ corresponds to $(1000/4\ \pi)\ A\ m^{-1}$
maxwell	Mx	$1\ Mx$ corresponds to $10^{-8}\ Wb$
stilb	sb	$1\ sb = 1\ cd\ cm^{-2} = 10^4\ cd\ m^{-2}$
phot	ph	$1\ ph = 10^4\ lx$

SI prefixes

Factor	Prefix	Symbol	Origin
10^{18}	exa	E	'$\dot{\varepsilon}\xi$', Greek 'six' (10^3 to sixth power is 10^{18})
10^{15}	peta	P	'$\pi\varepsilon\nu\tau\varepsilon$', Greek 'five' ($10^3$ to fifth power is 10^{15})
10^{12}	tera	T	'$\tau\varepsilon\rho\alpha\sigma$', Greek 'monster'
10^{9}	giga	G	'$\gamma\iota\gamma\alpha\sigma$', Greek 'giant'
10^{6}	mega	M	'$\mu\varepsilon\gamma\alpha\sigma$', Greek 'large'
10^{3}	kilo	k	'$\chi\iota\lambda\iota o\iota$', Greek 'thousand'
10^{2}	hecto	h	'$\dot{\varepsilon}\kappa\alpha\tau o\nu$', Greek 'hundred'
10^{1}	deca	da	'$\delta\varepsilon\kappa\alpha$', Greek 'ten'
10^{-1}	deci	d	'decima pars', Latin 'tenth'
10^{-2}	centi	c	'pars centesima', Latin 'hundredth'
10^{-3}	milli	m	'pars millesima', Latin 'thousandth'
10^{-6}	micro	μ	'$\mu\iota\kappa\rho o\sigma$', Greek 'little'
10^{-9}	nano	n	'nanus', Latin 'dwarf'
10^{-12}	pico	p	'pica', Latin 'pike, spear'
10^{-15}	femto	f	'femten', Danish, Norwegian 'fifteen'
10^{-18}	atto	a	'atten', Danish, Norwegian, 'eighteen'

SI derived units

Quantity	SI unit			
	Name	Symbol	Expression in terms of other units	Expression in terms of SI base units
Frequency	hertz	Hz		s^{-1}
Force	newton	N		$m\ kg\ s^{-2}$
Pressure, stress	pascal	Pa	$N\ m^{-2}$	$m^{-1}\ kg\ s^{-2}$
Energy, work, quantity of heat	joule	J	$N\ m$	$m^2\ kg\ s^{-2}$
Power, radiant flux	watt	W	$J\ s^{-1}$	$m^2\ kg\ s^{-3}$
Quantity of electricity, electric charge	coulomb	C		$s\ A$
Electr. potent., potent. diff., electromot. force	volt	V	$W\ A^{-1}$	$m^2\ kg\ s^{-3}\ A^{-1}$
Capacitance	farad	F	$C\ V^{-1}$	$m^{-2}\ kg^{-1}\ s^4\ A^2$
Electric resistance	ohm	Ω	$V\ A^{-1}$	$m^2\ kg\ s^{-3}\ A^{-2}$
Conductance	siemens	S	$A\ V^{-1}$	$m^2\ kg^{-1}\ s^3\ A^2$
Magnetic flux	weber	Wb	$V\ s$	$m^2\ kg\ s^{-2}\ A^{-1}$
Magn. flux density, magn. induct., magn. field	tesla	T	$Wb\ m^{-2}$	$kg\ s^{-2}\ A^{-1}$
Inductance	henry	H	$Wb\ A^{-1}$	$m^2\ kg\ s^{-2}\ A^{-2}$
Luminous flux	lumen	lm		$cd\ sr$
Illuminance	lux	lx	$lm\ m^{-2}$	$m^{-2}\ cd\ sr$
Dynamic viscosity	pascal second	Pa s		$m^{-1}\ kg\ s^{-1}$
Moment of force	metre newton	N m		$m^2\ kg\ s^{-2}$
Surface tension	newton per metre	$N\ m^{-1}$		$kg\ s^{-2}$
Heat flux density, irradiance	watt per square metre	$W\ m^{-2}$		$kg\ s^{-3}$
Heat capacity, entropy	joule per kelvin	$J\ K^{-1}$		$m^2\ kg\ s^{-2}\ K^{-1}$
Specific heat capacity, specific entropy	joule per kilogram kelvin	$J\ kg^{-1}\ K^{-1}$		$m^2\ s^{-2}\ K^{-1}$
Specific energy	joule per kilogram	$J\ kg^{-1}$		$m^2\ s^{-2}$
Thermal conductivity	watt per metre kelvin	$W\ m^{-1}\ K^{-1}$		$m\ kg\ s^{-3}\ K^{-1}$
Energy density	joule per cubic metre	$J\ m^{-3}$		$m^{-1}\ kg\ s^{-2}$
Electric field strength	volt per metre	$V\ m^{-1}$		$m\ kg\ s^{-3}\ A^{-1}$
Electric charge density	coulomb per cubic metre	$C\ m^{-3}$		$m^{-3}\ s\ A$
Electric displacement, electric flux density	coulomb per square metre	$C\ m^{-2}$		$m^{-2}\ s\ A$
Permittivity	farad per metre	$F\ m^{-1}$		$m^{-3}\ kg^{-1}\ s^4\ A^2$
Permeability	henry per metre	$H\ m^{-1}$		$m\ kg\ s^{-2}\ A^{-2}$
Molar energy	joule per mole	$J\ mol^{-1}$		$m^2\ kg\ s^{-2}\ mol^{-1}$
Molar entropy, molar heat capacity	joule per mole kelvin	$J\ mol^{-1}\ K^{-1}$		$m^2\ kg\ s^{-2}\ K^{-1}\ mol^{-1}$

3. Wavelength scale

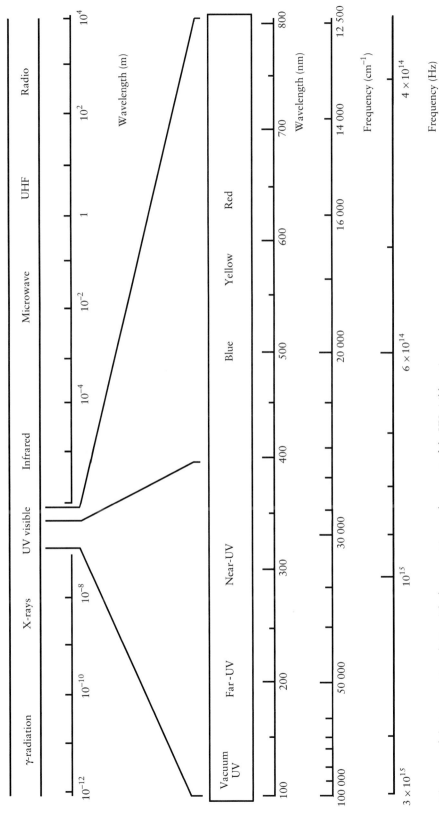

The range of electromagnetic radiation. The lower part is an enlargement of the UV-visible region.

4. Colour, wave length, frequency, wave number and energy of light

Colour	λ/nm	ν/Hz	$\bar{\nu}$/cm^{-1}	E/eV	E/kJ mol^{-1}
Infrared	1000	$3.00 \cdot 10^{14}$	$1.00 \cdot 10^4$	1.24	120
Red	700	$4.28 \cdot 10^{14}$	$1.43 \cdot 10^4$	1.77	171
Orange	620	$4.84 \cdot 10^{14}$	$1.61 \cdot 10^4$	2.00	193
Yellow	580	$5.17 \cdot 10^{14}$	$1.72 \cdot 10^4$	2.14	206
Green	530	$5.66 \cdot 10^{14}$	$1.89 \cdot 10^4$	2.34	226
Blue	470	$6.38 \cdot 10^{14}$	$2.13 \cdot 10^4$	2.64	254
Violet	420	$7.14 \cdot 10^{14}$	$2.38 \cdot 10^4$	2.95	285
Near Ultraviolet	300	$1.00 \cdot 10^{15}$	$3.33 \cdot 10^4$	4.15	400
Far Ultraviolet	200	$1.50 \cdot 10^{15}$	$5.00 \cdot 10^4$	6.20	598

5. Magnetic susceptibilities at 25°C

	κ	χ
Water	$-9.0 \cdot 10^{-5}$	$-9.0 \cdot 10^{-9}$
Benzene	$-7.2 \cdot 10^{-6}$	$-8.2 \cdot 10^{-9}$
Cyclohexane	$-7.9 \cdot 10^{-6}$	$-1.02 \cdot 10^{-8}$
Carbon tetrachloride	$-8.9 \cdot 10^{-6}$	$-5.4 \cdot 10^{-9}$
NaCl (s)	$-1.39 \cdot 10^{-5}$	$-6.4 \cdot 10^{-9}$
Cu (s)	$-9.6 \cdot 10^{-5}$	$-1.07 \cdot 10^{-9}$
S (s)	$-1.29 \cdot 10^{-5}$	$-6.2 \cdot 10^{-9}$
Hg (l)	$-2.85 \cdot 10^{-5}$	$-2.1 \cdot 10^{-9}$
$CuSO_4 \cdot 5H_2O$ (s)	$1.76 \cdot 10^{-4}$	$7.7 \cdot 10^{-8}$
$MnSO_4 \cdot 4H_2O$ (s)	$2.64 \cdot 10^{-3}$	$8.12 \cdot 10^{-7}$
$NiSO_4 \cdot 7H_2O$ (s)	$4.16 \cdot 10^{-4}$	$2.01 \cdot 10^{-7}$
$FeSO_4(NH_4)_2 SO_4 \cdot 6H_2O$ (s)	$7.55 \cdot 10^{-4}$	$4.06 \cdot 10^{-7}$
Al (s)	$2.2 \cdot 10^{-5}$	$8.2 \cdot 10^{-9}$
Pt (s)	$2.62 \cdot 10^{-4}$	$1.22 \cdot 10^{-8}$
Na (s)	$7.3 \cdot 10^{-6}$	$7.5 \cdot 10^{-9}$
K (s)	$5.6 \cdot 10^{-6}$	$6.5 \cdot 10^{-9}$

The CGS values of the susceptibility (per gram) are obtained by forming $1000 \chi / 4\pi$, and the value per gram mole from $1000 \chi M_r / 4\pi$.

6. Electronic configuration of elements

Z	Element	K	L		M			N				O				P			Q
		1s	2s	2p	3s	3p	3d	4s	4p	4d	4f	5s	5p	5d	5f	6s	6p	6d	7s
		1st period																	
1	H	1																	
2	He	2																	
		2nd period																	
3	Li	2	1																
4	Be	2	2																
5	B	2	2	1															
6	C	2	2	2															
7	N	2	2	3															
8	O	2	2	4															
9	F	2	2	5															
10	Ne	2	2	6															
		3rd period																	
11	Na	2	2	6	1														
12	Mg	2	2	6	2														
13	Al	2	2	6	2	1													
14	Si	2	2	6	2	2													
15	P	2	2	6	2	3													
16	S	2	2	6	2	4													
17	Cl	2	2	6	2	5													
18	Ar	2	2	6	2	6													
		4th period																	
19	K	2	2	6	2	6		1											
20	Ca	2	2	6	2	6		2											
21	Sc	2	2	6	2	6	1	2											
22	Ti	2	2	6	2	6	2	2											
23	V	2	2	6	2	6	3	2											
24	Cr	2	2	6	2	6	5	1											
25	Mn	2	2	6	2	6	5	2											
26	Fe	2	2	6	2	6	6	2											
27	Co	2	2	6	2	6	7	2											
28	Ni	2	2	6	2	6	8	2											
29	Cu	2	2	6	2	6	10	1											
30	Zn	2	2	6	2	6	10	2											
31	Ga	2	2	6	2	6	10	2	1										
32	Ge	2	2	6	2	6	10	2	2										
33	As	2	2	6	2	6	10	2	3										
34	Se	2	2	6	2	6	10	2	4										
35	Br	2	2	6	2	6	10	2	5										
36	Kr	2	2	6	2	6	10	2	6										
		5th period																	
37	Rb	2	2	6	2	6	10	2	6			1							
38	Sr	2	2	6	2	6	10	2	6			2							
39	Y	2	2	6	2	6	10	2	6	1		2							
40	Zr	2	2	6	2	6	10	2	6	2		2							
41	Nb	2	2	6	2	6	10	2	6	4		1							
42	Mo	2	2	6	2	6	10	2	6	5		1							
43	Tc	2	2	6	2	6	10	2	6	5		2							
44	Ru	2	2	6	2	6	10	2	6	7		1							
45	Rh	2	2	6	2	6	10	2	6	8		1							
46	Pd	2	2	6	2	6	10	2	6	10									
47	Ag	2	2	6	2	6	10	2	6	10		1							
48	Cd	2	2	6	2	6	10	2	6	10		2							
49	In	2	2	6	2	6	10	2	6	10		2	1						
50	Sn	2	2	6	2	6	10	2	6	10		2	2						

Electronic configuration of elements (continued)

Z	Element	K	L		M			N				O				P			Q
		1s	2s	2p	3s	3p	3d	4s	4p	4d	4f	5s	5p	5d	5f	6s	6p	6d	7s
51	Sb	2	2	6	2	6	10	2	6	10		2	3						
52	Te	2	2	6	2	6	10	2	6	10		2	4						
53	J	2	2	6	2	6	10	2	6	10		2	5						
54	Xe	2	2	6	2	6	10	2	6	10		2	6						
		6th period																	
55	Cs	2	2	6	2	6	10	2	6	10		2	6			1			
56	Ba	2	2	6	2	6	10	2	6	10		2	6			2			
57	La	2	2	6	2	6	10	2	6	10		2	6	1		2			
58	Ce	2	2	6	2	6	10	2	6	10	1	2	6	1		2			
59	Pr	2	2	6	2	6	10	2	6	10	2	2	6	1		2			
60	Nd	2	2	6	2	6	10	2	6	10	4	2	6			2			
61	Pm	2	2	6	2	6	10	2	6	10	5	2	6			2			
62	Sm	2	2	6	2	6	10	2	6	10	6	2	6			2			
63	Eu	2	2	6	2	6	10	2	6	10	7	2	6			2			
64	Gd	2	2	6	2	6	10	2	6	10	7	2	6	1		2			
65	Tb	2	2	6	2	6	10	2	6	10	8	3	6	1		2			
66	Dy	2	2	6	2	6	10	2	6	10	9	2	6	1		2			
67	Ho	2	2	6	2	6	10	2	6	10	10	2	6	1		2			
68	Er	2	2	6	2	6	10	2	6	10	11	2	6	1		2			
69	Tm	2	2	6	2	6	10	2	6	10	13	2	6			2			
70	Yb	2	2	6	2	6	10	2	6	10	14	2	6			2			
71	Lu	2	2	6	2	6	10	2	6	10	14	2	6	1		2			
72	Hf	2	2	6	2	6	10	2	6	10	14	2	6	2		2			
73	Ta	2	2	6	2	6	10	2	6	10	14	2	6	3		2			
74	W	2	2	6	2	6	10	2	6	10	14	2	6	4		2			
75	Re	2	2	6	2	6	10	2	6	10	14	3	6	5		2			
76	Os	2	2	6	2	6	10	2	6	10	14	2	6	6		2			
77	Ir	2	2	6	2	6	10	2	6	10	14	2	6	7		2			
78	Pt	2	2	6	2	6	10	2	6	10	14	2	6	9		1			
79	Au	2	2	6	2	6	10	2	6	10	14	2	6	10		1			
80	Hg	2	2	6	2	6	10	2	6	10	14	2	6	10		2			
81	Tl	2	2	6	2	6	10	2	6	10	14	2	6	10		2	1		
82	Pb	2	2	6	2	6	10	2	6	10	14	2	6	10		2	2		
83	Bi	2	2	6	2	6	10	2	6	10	14	2	6	10		2	3		
84	Po	2	2	6	2	6	10	2	6	10	14	2	6	10		2	4		
85	At	2	2	6	2	6	10	2	6	10	14	2	6	10		2	5		
86	Rn	2	2	6	2	6	10	2	6	10	14	2	6	10		2	6		
		7th period																	
87	Fr	2	2	6	2	6	10	2	6	10	14	2	6	10		2	6		1
88	Ra	2	2	6	2	6	10	2	6	10	14	2	6	10		2	6		2
89	Ac	2	2	6	2	6	10	2	6	10	14	2	6	10		2	6	1	2
90	Th	2	2	6	2	6	10	2	6	10	14	2	6	10		2	6	2	2
91	Pa	2	2	6	2	6	10	2	6	10	14	2	6	10	2	2	6	1	2
92	U	3	3	6	2	6	10	2	6	10	14	2	6	10	3	2	6	1	2
93	Np	2	2	6	2	6	10	2	6	10	14	2	6	10	4	2	6	1	2
94	Pu	2	2	6	2	6	10	2	6	10	14	2	6	10	5	2	6	1	2
95	Am	2	2	6	2	6	10	2	6	10	14	2	6	10	7	2	6		2
96	Cm	2	2	6	2	6	10	2	6	10	14	2	6	10	7	2	6	1	2
97	Bk	2	2	6	2	6	10	2	6	10	14	2	6	10	8	2	6	1	2
98	Cf	2	2	6	2	6	10	2	6	10	14	2	6	10	9	2	6	1	2
99	Es	2	2	6	2	6	10	2	6	10	14	2	6	10	10	2	6	1	2
100	Fm	2	2	6	2	6	10	2	6	10	14	2	6	10	11	2	6	1	2
101	Md	2	2	6	2	6	10	2	6	10	14	2	6	10	12	2	6	1	2
102	No	2	2	6	2	6	10	2	6	10	14	2	6	10	13	2	6	1	2
103	Lr	2	2	6	2	6	10	2	6	10	14	2	6	10	14	2	6	1	2

7. Properties of some important solvents

Solvent	Abbreviation	Formula	Molecular weight	F.p. (°C)	B.p. (°C)	Density	Dipole Mom.**	Susceptib. ×10⁶	Dielectr. Const.	Refract. Index	Chemical Shift δ ¹H (ppm)	δ ¹³C (ppm)
Acetic acid		$C_2H_4O_2$	60.0₅	16.7	117.9	1.049	1.2	0.551 (32°)	6.1	1.3719	2.1	21.1 / 177.3
Acetone	AC	C_3H_6O	58.1	−97.7	56.3	0.790	2.88	0.460	20.7	1.3587	2.2	30.2 / 205.1
Acetonitrile	AN	C_2H_3N	41.0₅	−44	81.6	0.777	3.92	0.534	37.5	1.3416	2.0	0.3 / 117.2
Benzene*		C_6H_6	78.1	5.5	80.1	0.879	0	0.699 (22°)	2.3	1.5011	7.4	128.7
t-Butyl alcohol	t-BuOH	$C_4H_{10}O$	74.1	25.5	82.2	0.789	1.66	0.534	1.8	1.3878	1.3	31.6 / 68.7
Carbon disulfide		CS_2	76.1	−111.6	46.2	1.270	0	0.532	2.6	1.6319		192.8
Carbon tetrachloride*		CCl_4	153.8	−23	76.7	1.584	0	0.691	2.2	1.4574		96.7
Chloroform*		$CHCl_3$	119.4	−63.5	61.1	1.480	1.01	0.740	4.8	1.4429	7.3	77.7
Cyclohexane		C_6H_{12}	84.2	6.6	80.7	0.774	0	0.627	2.0	1.4235	1.4	27.8
Cyclopentane		C_5H_{10}	70.1	−93.8	49.3	0.745		0.629	2.0	1.4065	1.5	26.5
Decaline		$C_{10}H_{18}$	138.2	−43	191.7	0.879		0.681		1.4758	0.9 to 1.8	24 to 44
Dibromomethane		CH_2Br_2	173.8	−52.6	97.0	1.5420	1.43	0.935	7.5	2.497	5.0	21.6
o-Dichlorobenzene*	ODCB	$C_6H_4Cl_2$	147.0	−17	180.5	1.306	2.50	0.748	9.9	1.5515	7.0 to 7.4	128 to 133
Diethylether		$C_4H_{10}O$	74.1	−116.2	34.5	0.7138	1.15		4.3	1.3526	1.2 / 3.5	17.1 / 67.4
1,2-Dichloroethane		$C_2H_4Cl_2$	99.0	−35.7	83.5	1.246	1.75		10.4	1.4421	3.7	51.7
1,2-Dichloroethylene Z		$C_2H_2Cl_2$	96.9	−80.0	60.6	1.284	1.90	0.679 (15°)	9.2	1.4490	6.4	119.3
1,2-Dichloroethylene E		$C_2H_2Cl_2$	96.9	−49.8	47.7	1.255	0	0.638 (15°)	2.1	1.4462	6.3	121.1
1,1-Dichloroethylene		$C_2H_2Cl_2$	96.9	−112.6	31.6	1.213	1.34	0.635 (15°)	4.6	1.4247	5.5	115.5 / 128.9
Dichloromethane		CH_2Cl_2	84.9	−95.1	39.8	1.315	1.60	0.733	8.9	1.4211	5.3	54.2
Dimethoxymethane		$C_3H_8O_2$	76.1	−105.2	42.3	0.866		0.611	2.6	1.3563	3.3 / 4.4	54.8 / 97.9
Dimethylacetamide	DMA	C_4H_9NO	87.1	−20.0	166.1	0.937	3.8		37.8	1.4356	2.1 / 3	34/38 / 169.6
Dimethylcarbonate*		$C_3H_6O_3$	90.1	3	90.5	1.069				1.3687	3.65	54.8 / 156.9
Dimethylether		C_2H_6O	46.1	−139	−24							59.4
N,N-Dimethylformamide	DMF	C_3H_7NO	73.1	−60.4	153.0	0.944	3.86		36.7	1.4282	2.9 / 8.0	31–36 / 161.7
Dimethylsulfoxide	DMSO	C_2H_6SO	78.1	−18.5	189.0	1.096	3.96		46.7	1.4773	2.6	39.5
Dioxane*	D	$C_4H_8O_2$	88.1	11.8	101.3	1.028	0.45	0.606 (32°)	2.2	1.4203	3.7	67.8

Properties of some important solvents (continued)

Solvent	Abbreviation	Formula	Molecular weight	F.p. (°C)	B.p. (°C)	Density	Dipole Mom.**	Susceptib. ×10⁶	Dielectr. Const.	Refract. Index	Chemical Shift δ ¹H (ppm)	δ ¹³C (ppm)
Ethyl acetate		$C_4H_8O_2$	88.1	-83.9	77.1	0.895	1.8	0.554	6.0	1.3698	1.2/2.0 4.1	14.3 60.1 170.4
Ethylene carbonate	EC	$C_3H_4O_3$	88.1	36.4	238	1.321	4.91		89.6	1.4250	4.4	65.0 155.8
Freon 12		CF_2Cl_2	120.9	-158	-29.8	1.18	0.5	0.642 (-30°)				
Freon 22		$CHClF_2$	86.5	-146	-40.8	1.491	1.4					
Formamide	F	CH_3ON	45.0	2.6	210.5	1.133	3.73	0.551	109	1.4475	7.2 8.1	165.1
Hexachloroacetone		C_3Cl_6O	264.8	-30	203	1.744				1.5112		123.7 126.4
Hexafluorodichloro-propane		$C_3F_6Cl_2$	220.9	-136	35	1.589						
Hexamethyl-phoshoramide*	HMPT	$C_6H_{18}N_3PO$	179.2	7.2	233	1.027	5.39	0.530	30.0	1.4588	2.4 2.6	36.6
Methanol	MeOH	CH_4O	32.0	-97.7	64.7	0.787	1.70		32.7	1.3265	3.5	49.3
Methyl chloride		CH_3Cl	50.5	-97.7	-24.1	0.916	1.87		12.6	1.3389		25.1
Morpholine		C_4H_9NO	87.1	-3.1	128.9	1.005		0.631	7.4	1.4573	2.6 3.9	46.8 68.9
Nitromethane		CH_3NO_2	61.0	-28.5	101.2	1.131	3.46	0.391 (25°)	35.9	1.3796	4.3	57.3
Pyridine		C_5H_5N	79.1	-41.6	115.3	0.978	2.2	0.611	12.4	1.5075	7.1 to 8.8	124 to 150
Quinoline		C_9H_7N	129.2	-14.9	237.1	1.098	2.2	0.729	9.0	1.6293	7 to 8.8	121 to 151
Sulfur dioxide		SO_2	64.1	-72.7	-10.0	1.434	1.6		17.6			
2,2-Tetrachloro-ethane		$C_2H_2Cl_4$	167.8	-43.8	146.2	1.578	1.3	0.856	8.2	1.4868	6.0	74.0
Tetrachloroethane		C_2Cl_4	165.8	-22.3	121.2	1.631		0.802 (15°)	2.3	1.5076		120.4
Tetrahydrofuran	THF	C_4H_8O	72.1	-66	66.0	0.889	1.75		7.6	1.4050	1.8 3.7	26.7 68.6
Tetramethylurea	TMU	$C_5H_{12}N_2O$	116.2	-1.2	176.5	0.969	3.47	0.634 (15°)	23.1	1.4459	2.8	38.5 165.6
Toluene		C_7H_8	92.1	-94.9	110.6	0.867	0.36	0.618	2.4	1.4969	2.3 7.2	21.3 125 to 138
Trichloroethylene*		C_2CHl_3	131.4	-73	87.2	1.476		0.734	3.4	1.4800	6.4	116.7 124.0
Water		H_2O	18.02	0	100	0.997	1.85	0.719	78.5	1.3329	4.8	

B.p.: boiling point   F.p.: freezing point   Reference of the shifts: ¹H and ¹³C (TMS)
** Dipole moment in Debye units: (1D = 10^{-18} e.s.u. cm $\triangleq 3.3356 \times 10^{-30}$ C m)
* Carcinogens or cancer suspect agents

8. Important acronyms in organic chemistry

A	adenine
AA	anisylacetone
AAO	acetaldehyde oxime
Ac	acetate
Ac	acetyl
ACAC (acac)	acetylacetone
ACTH	adrenocorticotropic hormone (corticotropin)
ADMA	alkyldimethylamine
ADP	adenosine 5′-diphosphate
Ala	alanine
Am	amyl
AMP	adenosine 5′-monophosphate
AN	acetonitrile
APS	adenosine 5′-phosphosulfate
Ar	aryl
Arg	arginine
Asn	asparagine
Asp	aspartic acid
ATA	anthranilamide
ATP	adenosine 5′-triphosphate
B	nucleoside base (adenine, cytosine, guanine, thymine, or uracil)
BA	benzyladenine
BaP (BAP)	benzo[a]pyrene
BBP	benzyl butyl phthalate
BHC	benzene hexachloride
Bn	benzyl (also Bz, BZL, or Bnz)
BN	benzonitrile
BON	β-oxynaphtoic acid
BPBG	butyl phthalyl butyl glycolate
BSA	bovine serum albumin
BTA	benzoyltrifluoroacetone
n Bu	n-butyl
t Bu	tert-butyl
Bz	benzoyl
p CBA	p-carboxybenzaldehyde
CD	cyclodextrin
CDP	cytidine 5′-disphosphate
CE	cyanoethyl
CMP	cytidine monophosphate
CoA	coenzyme A
Cp (or cp)	cyclopentadiene
12-Crown-4	1,4,7,10-tetraoxacyclododecane
CTA	citraconic anhydride
Cys	cysteine
DAA	diacetone acrylamide
DAP	dodecylammonium propionate
DCB	dicyanobenzene
DCEE	dichloroethyl ether
DDD	2,2′-dihydroxy-6,6′-dinaphthyl disulfide

DDH	1,3-dibromo-5,5-dimethylhydantoin
DDM	diphenyldiazomethane
DEA	N,N-diethylaniline
DEC	diethylaminoethyl chloride hydrochloride
DHA	dehydroacetic acid
Diglyme	diethylene glycol dimethyl ether
Diox	dioxane
DMA	dimethylacetamide
DMAA	N,N-dimethylacetoacetamide
DME	1,2-dimethoxyethane (glyme)
DMF	dimethylformamide
DML	dimyristoyl-lecithin
DMS	dimethylsiloxane
DMSO	dimethyl sulfoxide
DMT	dimethyl terephthalate
DNA	deoxyribonucleic acid
DNF	2,4-dinitrofluorobenzene
DOCA	deoxycorticosterone acetate
DPG	2,3 diphosphoglycerate
DPI	dipalmitoyl-lecithin
dpm	dipivaloylmethanato
DPPH	diphenylpicrylhydracyl
DSS	2,2 dimethyl-2-silapentane-5-salphonate (usually as the sodium salt)
DST	disuccinimidyl tartrate
DTBN	di-t-butyl nitroxide
EAA	ethyl acetoacetate
EAK	ethyl amyl ketone
EBA	N-ethyl-N-benzylaniline
EBBA	N-(p-ethoxybenzylidene)-p-butylaniline
EDC	ethylene dichloride
EDTA	ethylenediaminetetraacetic acid
EGS	ethylene glycol bis(succinimidyl succinate)
en	ethylenediamine
Et	ethyl
EVA	ethylene-vinyl acetate
FA	furfuryl alcohol
FAD	flavin adenine dinucleotide
Fl	flavin
FMA	fluoroscein mercuric acetate
G	guanine
GDP	guanosine 5′-diphosphate
gln	glutamine
Glu	glutamic acid
Gly	glycine
Glyme (glyme)	1,2-dimethoxyethane
HAB	4,4′- bis(heptyl)azoxybenzene
Hb	hemoglobin
Hex	hexane (or hexyl)
HFA	hexafluoroacetone

Important acronyms in organic chemistry (continued)

His	histidine
HMPA	hexamethylphosphoramide
HMPT	hexamethylphosphorous triamide
HOAB	p-n-heptyloxyazoxybenzene
HOAc	acetic acid
Hyp	hydroxyproline
IH	immobilized histamine
IHP	inositolhexaphosphate
Ile	isoleucine
IMP	inosine 5′-monophosphate
IPN	isophthalonitrile
KDP	potassium dihydrogen phosphate
LAP	leucine aminopeptidase
LDH	lactic dehydrogenase
Leu	leucine
Lys	lysine
M	metal
MA	maleic anhydride
MAA	menthoxyacetic acid
MBBA	N-(p-methoxybenzylidene)-p-butylaniline
MCA	monochloroacetic acid
Me	methyl
MeLeu	N-methylleucine
Met	methionine
MMH	methylmercuric hydroxide
MOM-	methoxymethyl-
MSA	methanesulfonic acid
NAD	nicotinamide adenine dinucleotide
NADH(P)	nicotinamide adenine dinucleotide (phosphate)
NAI	N-acetylimidazole
NCA	N-chloroacetamide
NM	nitromethane
NMA	N-methylolacrylamide
NMF	N-methylformamide
NTA	nitrilotriacetic acid
NU	nucleophile
OCBA	o-chlorobenzoic acid
OCT	o-chlorotoluene
ODCB	o-dichlorobenzene
P	polymer substituent
PAA	p-azoxyanisole
PAS	p-aminosalicylic acid
PBA	pyrene butyric acid
PBLG	poly(L-benzyl μ-glutamate)
PC	propylene carbonate
PCB	polychlorinated biphenyl
PCP	pentachlorophenol
PDMS	polydimethylsiloxane
PEG	polyethylene glycol

Ph	phenyl
Phe	phenylalanine
PMA	poly(methacrylic acid)
PMMA	poly(methyl methacrylate)
POC	cyclopentyloxycarbonyl
POM	poly(oxymethylene)
PPA	polyphosphoric acid
Pr	propyl
i Pr	isopropyl
Pro	proline
PS	phosphatidylserine
PTFE	polytetrafluoroethylene
PVA	polyvinyl alcohol
PVC	polyvinyl chloride
PVF	poly(vinyl flouride)
PVP	poly(vinyl pyrrolidone)
Pyr (or Py)	pyridine
RNA	ribonucleic acid
SDS	sodium dodecyl sulfate
Ser	serine
SLS	sodium lauryl sulfate
TAB	trimethylammonium bromide
TBE	tetrabromoethane
TCA	trichloroacetic acid
TCNQ	tetracyanoquinodimethane
TEA	triethylamine
TFA	trifluoroacetic acid
THE	tetrahydrocortisone
THF	tetrahydrofuran
Thr	threonine
TMB	N,N,N′,N′-tetramethylbenzidine
TMM	trimethylenemethane
TMS	tetramethylsilane
TMU	tetramethylurea
TNM	tetranitromethane
TNT	2,4,6-trinitrotoluene
TP	thymolphthalein
TPC	thymolphthalein complexone
TPE	tetraphenylethylene
Tr	trityl
Triglyme	triethylene glycol dimethyl ether
Trp	tryptophan
Ts	tosyl (or p-toluenesulfonyl-)
Tyr	tyrosine
U	uridine
UTP	uridine triphosphate
Val	valine
VTC	vinyltrichlorosilane
Xy	xylene
ZDBC	zinc dibutyldithiocarbamate

9. Equilibrium constants at 25°C/concentration units for solutions

Equilibrium constants at 25°C

Dissociation constants of weak acids

Formula	Name	K_a
H_3BO_3	boric acid	6.0×10^{-10}
HCN	hydrogen cyanide (hydrocyanic acid)	4.0×10^{-10}
$HC_2H_3O_2$	acetic acid	1.8×10^{-5}
H_2CO_3 (H_2O+CO_2)	carbonic acid	4.2×10^{-7} (K_1) 5.6×10^{-11} (K_2)
$H_2C_2O_4$	oxalic acid	5.4×10^{-2} (K_1) 5.0×10^{-5} (K_2)
$H_2C_6H_6O_6$	ascorbic acid	5.0×10^{-5} (K_1) 1.5×10^{-12} (K_2)
HNO_2	nitrous acid	5.0×10^{-4}
H_3PO_3	phosphorous acid	1.6×10^{-2} (K_1) 7×10^{-7} (K_2)

Formula	Name	K_a
H_3PO_4	phosphoric acid	7.6×10^{-3} (K_1) 6.3×10^{-8} (K_2) 4.4×10^{-13} (K_3)
H_2S	hydrogen sulfide (hydrosulfuric acid)	1.1×10^{-7} (K_1) 1.0×10^{-14} (K_2)
H_2SO_3 $H_2O + SO_2$	sulfurous acid	1.3×10^{-2} (K_1) 6.3×10^{-8} (K_2)
HSO_4^-	hydrogen sulfate ion (bisulfate ion)	1.2×10^{-2}
HF	hydrofluoric acid (hydrogen fluoride)	6.7×10^{-4}
HOCl	hypochlorous acid	3.2×10^{-8}
$HClO_2$	chlorous acid	1.1×10^{-2}

Equilibrium constants at 25°C

Dissociation constants of weak bases

Formula	Name	K_b
NH_3	ammonia	1.8×10^{-5}
NH_2OH	hydroxylamine	9.1×10^{-9}
CH_3NH_2	methylamine	4.4×10^{-4}

Formula	Name	K_b
$C_{10}H_{14}N_2$	nicotine	7.4×10^{-7} (K_1) 1.4×10^{-11} (K_2)
PH_3	phosphine	1×10^{-14}

Concentration units for solutions

Name	Symbol	Definition
Weight percent	%	(Grams of solute per grams of solution) × 100
Mole fraction[a]	X_A	Moles of A per total number of moles
Molarity	M	Moles of solute per litre of solution[b]
Normality	N	Equivalents of solute per litre of solution[b]
Formality[c]	F	Formula weights of solute per litre of solution[b]
Molality	m	Moles of solute per kg of solvent
Weight formality[c]	f	Formula weight of solute per kg of solvent

[a] The symbol Y_A is often used for the mole fraction of A in a gas phase that is in equilibrium with a liquid solution.

[b] Note that 1 litre = 1 dm³. The SI concentration unit of mol m⁻³ is not very attractive for most work in chemistry, and non-SI concentrations based on the litre are widely used.

[c] These units are infrequently used but are of great convenience in expressing the overall compositions of a solution when the solute is partially associated or dissociated.

10. Acronyms and abbreviations in quantum chemistry and related fields

AM1	Austin Method 1, a semiempirical procedure
AO	Atomic Orbital
CI	Configuration Interaction
CNDO	Complete Neglect of Differential Overlap, a semiempirical procedure
DZ	Double Zeta, a basis set consisting of two STOs for each atomic orbital
EH	Extended Hückel theory
EPR	Electron Paramagnetic Resonance (see ESR)
ESCA	Electron Spectroscopy for Chemical Analysis (see photoelectron spectroscopy)
ESR	Electron Spin Resonance
FEMO	Free-Electron Molecular Orbitals
GTO	Gaussian Type atomic Orbital
HF	Hartree-Fock method for self-consistent fields
HMO	Hückel Molecular Orbitals
HOMO	Highest Occupied Molecular Orbital
INDO	Intermediate Neglect of Differential Overlap, a semiempirical procedure
IR	Infrared Radiation, as in vibrational spectroscopy
JWKB	Jeffreys-Wentzel-Kramers-Brillouin, a semiclassical approximation to quantum mechanics
LCAO	Linear Combination of Atomic Orbitals
LUMO	Lowest Unoccupied Molecular Orbital (see HOMO)
MINDO	Modified Intermediate Neglect to Differential Overlap, a semiempirical procedure
MNDO	Modified Neglect of Differential Overlap, a semiempirical procedure
MO	Molecular Orbital
NMR	Nuclear Magnetic Resonance
PES	Photoelectron Spectroscopy
PPP	Pariser-Parr-Pople method, a semiempirical procedure
RHF	Restricted Hartee-Fock method (for open-shell molecules)
SCF	Self-Consistent Field
STO	Slater Type Orbital
STO-NG	A Slater type orbital expressed as the sum of N guassian orbitals
TFD	Thomas-Fermi-Dirac method, a statistical treatment of electron density
UHF	Unrestricted Hartee-Fock method for SCF calculations on open-shell molecules
UV	Ultraviolet Radiation
VB	Valence Bond theory
VSEPR	Valence-Shell Electron Pair Repulsion, a theory of molecular geometry
WKB	Wentzel-Kramers-Brillouin method, a semiclassical approximation to quantum mechanics
ZDO	Zero Differential Overlap, a semiempirical approximation

11. Standard potentials in aqueous solutions

Selected half-reaction potentials are listed in order of increasingly positive (anodic) assignments, providing a convenient guide to redox couples in a given range of standard potentials. The synopsis consists of two tables, one each for acid and basic solutions, which have been advisedly restricted to selected half-reactions.

Standard potentials in acid solutions

Couple	$E°$ (V)
$(3/2)N_2 + H^+ + e \rightarrow HN_3$	−3.10
$Li^+ + e^- \rightarrow Li$	−3.045
$K^+ + e^- \rightarrow K$	−2.925
$Rb^+ + e^- \rightarrow Rb$	−2.925
$Cs^+ + e^- \rightarrow Cs$	−2.923
$Ba^{2+} + 2e^- \rightarrow Ba$	−2.92
$Ra^{2+} + 2e^- \rightarrow Ra$	−2.916
$Sr^{2+} + 2e^- \rightarrow Sr$	−2.89
$Ca^{2+} + 2e^- \rightarrow Ca$	−2.84
$Na^+ + e^- \rightarrow Na$	−2.714
$No^{2+} + e^- \rightarrow No$	−2.5
$Md^{2+} + 2e^- \rightarrow Md$	−2.4
$Fm^{2+} + 2e^- \rightarrow Fm$	−2.37
$La^{3+} + 3e^- \rightarrow La$	−2.37
$Y^{3+} + 3e^- \rightarrow Y$	−2.37
$Ce^{3+} + 3e^- \rightarrow Ce$	−2.34
$Nd^{3+} + 3e^- \rightarrow Nd$	−2.32
$Sm^{3+} + 3e^- \rightarrow Sm$	−2.30
$Gd^{3+} + 3e^- \rightarrow Gd$	−2.29
$Mg^{2+} + 2e^- \rightarrow Mg$	−2.356
$Lu^{3+} + 3e^- \rightarrow Lu$	−2.30
$1/2H_2 + e^- \rightarrow H^-$	−2.25
$Cf^{2+} + 2e^- \rightarrow Cf$	−2.2
$Es^{2+} + 2e^- \rightarrow Es$	−2.2
$Am^{3+} + 3e^- \rightarrow Am$	−2.07
$AlF_6^{3-} + 3e^- \rightarrow Al + 6F^-$	−2.067
$Cm^{3+} + 3e^- \rightarrow Cm$	−2.06
$Sc^{3+} + 3e^- \rightarrow Sc$	−2.03
$Bk^{3+} + 3e^- \rightarrow Bk$	−2.01
$Cf^{3+} + 3e^- \rightarrow Cf$	−2.0
$Es^{3+} + 3e^- \rightarrow Es$	−2.0
$Be^{2+} + 2e^- \rightarrow Be$	−1.97
$Fm^{3+} + 3e^- \rightarrow Fm$	−1.96
$Th^{4+} + 4e^- \rightarrow Th$	−1.83
$Np^{3+} + 3e^- \rightarrow Np$	−1.79
$Md^{3+} + 3e^- \rightarrow Md$	−1.7
$Zr^{4+} + 4e^- \rightarrow Zr$	−1.70
$Al^{3+} + 3e^- \rightarrow Al$	−1.67
$U^{3+} + 3e^- \rightarrow U$	−1.66
$Ti^{2+} + 2e^- \rightarrow Ti$	−1.63
$Hf^{4+} + 4e^- \rightarrow Hf$	−1.56
$No^{3+} + 3e^- \rightarrow No$	−1.2
$SiF_6^{2-} + 4e^- \rightarrow Si + 6F^-$	−1.2
$TiF_6^{2-} + 4e^- \rightarrow Ti + 6F^-$	−1.191

Continued

Couple	$E°$ (V)
$Mn^{2+} + 2e^- \rightarrow Mn$	−1.18
$V^{2+} + 2e^- \rightarrow V$	−1.13
$Nb^{3+} + 3e^- \rightarrow Nb$	−1.1
$H_3BO_3 + 3H^+ + 3e^- \rightarrow B + 3H_2O$	−0.890
$SiO_2(vit) + 4H^+ + 4e^- \rightarrow Si + 2H_2O$	−0.888
$TiO^{2+} + 2H^+ + 2e^- \rightarrow Ti + H_2O$	−0.882
$Ta_2O_5 + 10H^+ + 10e^- \rightarrow 2Ta + 5H_2O$	−0.81
$Zn^{2+} + 2e \rightarrow Zn$	−0.7626
$TlI + e^- \rightarrow Tl + I^-$	−0.74
$Te + 2H^+ + 2e^- \rightarrow H_2Te$	−0.740
$TlBr + e^- \rightarrow Tl + Br^-$	−0.658
$Nb_2O_5 + 10H^+ + 10e^- \rightarrow 2Nb + 5H_2O$	−0.65
$TlCl + e^- \rightarrow Tl + Cl^-$	−0.5568
$Ga^{3+} + 3e^- \rightarrow Ga$	−0.529
$U^{4+} + e^- \rightarrow U^{3+}$	−0.52
$Sb + 3H^+ + 3e^- \rightarrow SbH_3(g)$	−0.510
$H_3PO_2 + H^+ + e^- \rightarrow P(w) + 2H_2O$	−0.508
$H_3PO_3 + 2H^+ + 2e^- \rightarrow H_3PO_2 + H_2O$	−0.499
$Fe^{2+} + 2e \rightarrow Fe$	−0.44
$Cr^{3+} + e^- \rightarrow Cr^{2+}$	−0.424
$Cd^{2+} + 2e \rightarrow Cd$	−0.4025
$Ti^{3+} + e^- \rightarrow Ti^{2+}$	−0.37
$PbI_2 + 2e^- \rightarrow Pb + 2I^-$	−0.365
$PbSO_4 + 2e^- \rightarrow Pb + SO_4^{2-}$	−0.3505
$Eu^{3+} + e^- \rightarrow Eu^{2+}$	−0.35
$In^{3+} + 3e^- \rightarrow In$	−0.3382
$Tl^+ + e^- \rightarrow Tl$	−0.3363
$PbBr_2 + 2e^- \rightarrow Pb + 2Br^-$	−0.280
$Co^{2+} + 2e^- \rightarrow Co$	−0.277
$H_3PO_4 + 2H^+ + 2e^- \rightarrow H_3PO_3 + H_2O$	−0.276
$PbCl_2 + 2e^- \rightarrow Pb + 2Cl^-$	−0.268
$Ni^{2+} + 2e^- \rightarrow Ni$	−0.257
$V^{3+} + e^- \rightarrow V^{2+}$	−0.255
$2SO_4^{2-} + 4H^+ + 4e^- \rightarrow S_2O_6^{2-} + 2H_2O$	−0.253
$SnF_6^{2-} + 4e^- \rightarrow Sn + 6F^-$	−0.25
$N_2 + 5H^+ + 4e^- \rightarrow N_2H_5^+$	−0.23
$As + 3H^+ + 3e^- \rightarrow AsH_3$	−0.225
$Mo^{3+} + 3e^- \rightarrow Mo$	−0.2
$CuI + e^- \rightarrow Cu + I^-$	−0.182
$CO_2 + 2H^+ + 2e \rightarrow HCOOH(aq)$	−0.16
$AgI + e^- \rightarrow Ag + I^-$	−0.1522
$Si + 4H^+ + 4e^- \rightarrow SiH_4$	−0.143
$Sn^{2+} + 2e^- \rightarrow Sn$	−0.136
$Pb^{2+} + 2e^- \rightarrow Pb$	−0.1251

Standard potentials in acid solutions

Couple	$E°$ (V)
$WO_3(c) + 6H^+ + 6e^- \rightarrow W + 3H_2O$	−0.090
$P(w) + 3H^+ + 3e^- \rightarrow PH_3$	−0.063
$O_2 + H^+ + e \rightarrow HO_2$	−0.046
$Hg_2I_2 + 2e^- \rightarrow 2Hg + 2I^-$	−0.0405
$Se + 2H^+ + 2e^- \rightarrow H_2Se$	−0.028
$GeO_2 + 4H^+ + 4e^- \rightarrow Ge(hex) + 2H_2O$	−0.009
$2H^+ + 2e \rightarrow H_2$	0.000
$CuBr + e^- \rightarrow Cu + Br^-$	0.033
$HCOOH(aq) + 2H^+ + 2e^- \rightarrow HCHO(aq) + H_2O$	0.056
$AgBr + e^- \rightarrow Ag + Br^-$	0.0711
$TiO^{2+} + 2H^+ + e^- \rightarrow Ti^{3+} + H_2O$	0.100
$CuCl + e^- \rightarrow Cu + Cl^-$	0.121
$C + 4H^+ + 4e^- \rightarrow CH$	0.132
$Hg_2Br_2 + 2e^- \rightarrow 2Hg + 2Br^-$	0.13920
$S + 2H^+ + 2e \rightarrow H_2S$	0.144
$Np^{4+} + e^- \rightarrow Np^{3+}$	0.15
$Sn^{4+} + 2e^- \rightarrow Sn^{2+}$	0.15
$Sb_4O_6 + 12H^+ + 12e^- \rightarrow 4Sb + 6H_2O$	0.1504
$SO_4^{2-} + 2H^+ + 2e^- \rightarrow H_2SO_3 + H_2O$	0.158
$Cu^{2+} + e^- \rightarrow Cu^+$	0.159
$UO_2^{2+} + e^- \rightarrow UO_2^+$	0.16
$BiOCl + 2H^+ + 3e^- \rightarrow Bi + H_2O + Cl^-$	0.1697
$2H_2SO_3^- + 3H^+ + 2e^- \rightarrow HS_2O_4^- + 2H_2O$	0.173
$ReO_2 + 4H^+ + 4e^- \rightarrow Re + 2H_2O$	0.22
$AgCl + e^- \rightarrow Ag + Cl^-$	0.2223
$HCHO(aq) + 2H^+ + 2e^- \rightarrow CH_3OH(aq)$	0.232
$(CH_3)_2SO_2 + 2H^+ + 2e^- \rightarrow (CH_3)_2SO + 2H_2O$	0.238
$HAsO_2(aq) + 3H + 3e^- \rightarrow As + 2H_2O$	0.248
$UO_2^{2+} + 4H^+ + 2e^- \rightarrow U^{4+} + 2H_2O$	0.27
$HCNO + H^+ + e^- \rightarrow 1/2C_2N_2 + H_2O$	0.330
$VO^{2+} + 2H^+ + e^- \rightarrow V^{3+} + H_2O$	0.337
$ReO_4^- + 8H^+ + 7e^- \rightarrow Re + 4H_2O$	0.34
$Cu^{2+} + 2e^- \rightarrow Cu$	0.340
$AgIO_3 + e^- \rightarrow Ag + IO_3^-$	0.354
$Fe(CN)_6^{3-} + e^- \rightarrow Fe(CN)_6^{4-}$	0.3610
$C_2N_2 + 2H^+ + 2e^- \rightarrow 2HCN(aq)$	0.373
$UO_2^+ + 4H^+ + e^- \rightarrow U^{4+} + 2H_2O$	0.38
$H_2N_2O_2 + 6H^+ + 4e^- \rightarrow 2NH_3OH^+$	0.387
$2H_2SO_3 + 2H^+ + 4e^- \rightarrow S_2O_3^{2-} + 3H_2O$	0.400
$Ag_2CrO_4 + 2e^- \rightarrow 2Ag + CrO_4^{2-}$	0.4491
$Ag_2MoO_4 + 2e^- \rightarrow 2Ag + MoO_4^{2-}$	0.486
$PdBr_4^{2-} + 2e^- \rightarrow Pd + 4Br^-$	0.49
$RhCl_6 + 3e^- \rightarrow Rh + 6Cl^-$	0.5
$H_2SO_3 + 4H^+ + 4e^- \rightarrow S + 3H_2O$	0.500
$2H_2SO_3 + 4H^+ + 6e^- \rightarrow S_4O_6^{2-} + 6H_2O$	0.507
$ReO_4^- + 4H^+ + 3e^- \rightarrow ReO_2 + 2H_2O$	0.51
$Cu^+ + e^- \rightarrow Cu$	0.520
$TeO_2(c) + 4H^+ + 4e^- \rightarrow Te + 2H_2O$	0.53
$I_2 + 2e^- \rightarrow 2I^-$	0.5355

Continued

Couple	$E°$ (V)
$I_3^- + 2e^- \rightarrow 3I^-$	0.536
$AgBrO_3 + e^- \rightarrow Ag + BrO_3^-$	0.546
$Cu^{2+} + Cl^- + e^- \rightarrow CuCl$	0.559
$TeOOH^+ + 3H^+ + 4e^- \rightarrow Te + 2H_2O$	0.559
$H_3AsO_4 + 2H^+ + 2e^- \rightarrow HAsO_2 + 2H_2O$	0.560
$MnO_4^- + e^- \rightarrow MnO_4^{2-}$	0.56
$S_2O_6^{2-} + 4H^+ + 2e^- \rightarrow 2H_2SO_3$	0.569
$CH_3OH(aq) + 2H^+ + 2e^- \rightarrow CH_4 + H_2O$	0.59
$Sb_2O_5 + 6H^+ + 4e^- \rightarrow 2SbO^+ + 3H_2O$	0.605
$Au(SCN)_4^- + 3e^- \rightarrow Au + 4SCN^-$	0.636
$PdCl_4^{2-} + 2e^- \rightarrow Pd + 4Cl^-$	0.64
$AgC_2H_3O_2 + e^- \rightarrow Ag + C_2H_3O_2$	0.643
$Cu^{2+} + Br^- + e^- \rightarrow CuBr$	0.654
$Ag_2SO_4 + 2e^- \rightarrow 2Ag + SO_4^{2-}$	0.654
$NpO_2^+ + 4H^+ + e^- \rightarrow Np^{4+} + 2H_2O$	0.66
$O_2 + 2H^+ + 2e^- \rightarrow H_2O_2$	0.695
$HN_3 + 11H^+ + 8e^- \rightarrow 3NH_4^+$	0.695
$PtBr_4^{2-} + 2e^- \rightarrow Pt + 4Br^-$	0.698
$2NO + 2H^+ + 2e^- \rightarrow H_2N_2O_2$	0.71
$H_2SeO_3 + 4H^+ + 4e^- \rightarrow Se + 3H_2O$	0.739
$PtCl_4^{2-} + 2e^- \rightarrow Pt + 4Cl^-$	0.758
$Rh^{3+} + 3e^- \rightarrow Rh$	0.76
$(SCN)_2 + 2e^- \rightarrow 2SCN^-$	0.77
$Fe^{3+} + e^- \rightarrow Fe^{2+}$	0.771
$Hg_2^{2+} + 2e^- \rightarrow 2Hg$	0.7960
$Ag^+ + e^- \rightarrow Ag$	0.7991
$2NO_3^- + 4H^+ + 2e^- \rightarrow N_2O_4 + 2H_2O$	0.803
$IrBr_6^{2-} + e^- \rightarrow IrBr_6^{3-}$	0.805
$AmO_2^+ + 4H^+ + e^- \rightarrow Am^{4+} + 2H_2O$	0.82
$OsO_4(c) + 8H^+ + 8e^- \rightarrow Os + 4H_2O$	0.84
$AuBr_4^- + 3e^- \rightarrow Au + 4Br^-$	0.854
$2HNO_2 + 4H^+ + 4e^- \rightarrow H_2N_2O_2$	0.86
$IrCl_6^{3-} + 3e^- \rightarrow Ir + 6Cl^-$	0.86
$Cu^{2+} + I^- + e^- \rightarrow CuI$	0.861
$IrCl_6^{2-} + e^- \rightarrow IrCl_6^{3-}$	0.867
$2Hg^{2+} + 2e^- \rightarrow Hg$	0.9110
$Pd^{2+} + 2e^- \rightarrow Pd$	0.915
$NO_3^- + 3H^+ + 2e^- \rightarrow HNO_2 + H_2O$	0.94
$NO_3^- + 4H^+ + 3e^- \rightarrow NO + 2H_2O$	0.957
$AuBr_2^- + e^- \rightarrow Au + 2Br^-$	0.960
$PtO + 2H^+ + 2e^- \rightarrow Pt + H_2O$	0.980
$HNO_2 + H^+ + e^- \rightarrow NO + H_2O$	0.996
$AuCl_4^- + 3e^- \rightarrow Au + 4Cl^-$	1.002
$Pu^{4+} + e^- \rightarrow Pu^{3+}$	1.01
$PuO_2^{2+} + e^- \rightarrow PuO_2^+$	1.02
$PuO_2^{2+} + 4H^+ + 2e^- \rightarrow Pu^{4+} + 2H_2O$	1.03
$N_2O_4 + 4H^+ + 4e^- \rightarrow NO + 2H_2O$	1.039
$PuO_2^+ + 4H^+ + e^- \rightarrow Pu^{4+} + 2H_2O$	1.04
$Sb_2O_5 + 2H^+ + 2e^- \rightarrow Sb_2O_4 + H_2O$	1.055

Standard potentials in acid solutions

Couple	$E°$ (V)
$Br_2(l) + 2e^- \rightarrow Br^-$	1.065
$ICl_2^- + e^- \rightarrow 2Cl^- + 1/2I_2$	1.07
$N_2O_4 + 2H^+ + 2e^- \rightarrow 2HNO_3$	1.07
$Cu^{2+} + 2CN^- + e^- \rightarrow Cu(CN)_2^-$	1.12
$H_2O_2 + H^+ + e^- \rightarrow OH + H_2O$	1.14
$SeO_4^{2-} + 4H^+ + 2e^- \rightarrow H_2SeO_3 + H_2O$	1.151
$ClO_3^- + 3H^+ + 2e^- \rightarrow HClO_2 + H_2O$	1.181
$ClO_2 + H^+ + e^- \rightarrow HClO_2$	1.188
$S_2Cl_2 + 2e^- \rightarrow 2S + 2Cl^-$	1.19
$IO_3^- + 6H^+ + 5e^- \rightarrow 1/2I_2 + 3H_2O$	1.195
$ClO_4^- + 2H^+ + 2e^- \rightarrow ClO_3^- + H_2O$	1.201
$O_2 + 4H^+ + 4e^- \rightarrow 2H_2O$	1.229
$MnO_2 + 4H^+ + 2e^- \rightarrow Mn^{2+} + 2H_2O$	1.23
$N_pO_2^{2+} + e^- \rightarrow NpO_2^+$	1.24
$N_2H_5^+ + 3H^+ + 2e^- \rightarrow 2NH_4^+$	1.275
$PdCl_6^{2-} + 2e^- \rightarrow PdCl_4^{2-} + 2Cl^-$	1.288
$2HNO_2 + 4H^+ + 4e^- \rightarrow N_2O + 3H_2O$	1.297
$NH_3OH^+ + 2H^+ + 2e^- \rightarrow NH_4^+ + H_2O$	1.35
$Cl_2 + 2e^- \rightarrow 2Cl^-$	1.3583
$Cr_2O_7^{2-} + 14H^+ + 6e^- \rightarrow 2Cr^{3+} + 7H_2O$	1.36
$2NH_3OH^+ + H^+ + 2e^- \rightarrow N_2H_5^+ + 2H_2O$	1.41
$HO_2 + H^+ + e^- \rightarrow H_2O_2$	1.44
$PbO_2(\alpha) + 4H^+ + 2e^- \rightarrow Pb^{2+} + 2H_2O$	1.468
$BrO_3^- + 6H^+ + 5e \rightarrow 1/2Br_2 + 3H_2O$	1.478
$Mn^{3+} + e^- \rightarrow Mn^{2+}$	1.5
$MnO_4^- + 8H^+ + 5e^- \rightarrow Mn^{2+} + 4H_2O$	1.51
$Au^{3+} + 3e^- \rightarrow Au$	1.52
$AmO_2^{2+} + e^- \rightarrow AmO_2^+$	1.59
$NiO_2 + 4H^+ + 2e^- \rightarrow Ni^{2+} + 2H_2O$	1.593
$H_5IO_6 + H^+ + 2e^- \rightarrow IO_3^- + 3H_2O$	1.603
$HBrO + H^+ + e^- \rightarrow 1/2Br_2 + H_2O$	1.604
$HClO + H^+ + e^- \rightarrow 1/2Cl_2 + H_2O$	1.630
$Bk^{4+} + e \rightarrow Bk^{3+}$	1.67
$AmO_2^{2+} + 4H^+ + 3e^- \rightarrow Am^{3+} + H_2O$	1.67
$HClO_2 + 2H^+ + 2e^- \rightarrow HClO + H_2O$	1.674
$PbO_2(\alpha) + SO_4^{2-} + 4H^+ + 2e^- \rightarrow PbSO_4 + 2H_2O$	1.698
$MnO_4^- + 4H^+ + 3e^- \rightarrow MnO_2 + 2H_2O$	1.70
$Ce^{4+} + e \rightarrow Ce^{3+}$	1.72
$AmO_2^+ + 4H^+ + 2e^- \rightarrow Am^{3+} + 2H_2O$	1.72
$H_2O_2 + 2H^+ + 2e^- \rightarrow 2H_2O$	1.763
$Au^+ + e^- \rightarrow Au$	1.83
$Co^{3+} + e^- \rightarrow Co^{3+}$	1.92
$HN_3 + 3H^+ + 2e^- \rightarrow NH_4^+ + N_2$	1.96
$S_2O_8^{2-} + 2e^- \rightarrow 2SO_4^{2-}$	1.96
$Ag^{2+} + e^- \rightarrow Ag^+$	1.980
$O_3 + 2H^+ + 2e^- \rightarrow O_2 + H_2O$	2.075
$F_2O + 2H^+ + 4e^- \rightarrow 2F^- + H_2O$	2.153
$OH + H^+ + e^- \rightarrow H_2O$	2.38

Continued

Couple	$E°$ (V)
$O(g) + 2H^+ + 2e^- \rightarrow H_2O$	2.430
$Am^{4+} + e^- \rightarrow Am^{3+}$	2.62
$F_2 + 2e^- \rightarrow 2F^-$	2.87
$F_2 + 2H^+ + 2e^- \rightarrow 2HF(aq)$	3.053

Standard potentials in basic solutions

Couple	$E°$ (V)
$Ca(OH)_2 + 2e^- \rightarrow Ca + 2OH^-$	−3.026
$Ba(OH)_2 + 2e^- \rightarrow Ba + 2OH^-$	−2.99
$Sr(OH)_2 + 2e^- \rightarrow Sr + 2OH^-$	−2.88
$Y(OH)_3 + 3e^- \rightarrow Y + 3OH^-$	−2.85
$Ho(OH)_3 + 3e^- \rightarrow Ho + 3OH^-$	−2.85
$Er(OH)_3 + 3e^- \rightarrow Er + 3OH^-$	−2.84
$Tm(OH)_3 + 3e^- \rightarrow Tm + 3OH^-$	−2.83
$Lu(OH)_3 + 3e^- \rightarrow Lu + 3OH^-$	−2.83
$Gd(OH)_3 + 3e^- \rightarrow Gd + 3OH^-$	−2.82
$Tb(OH)_3 + 3e^- \rightarrow Tb + 3OH^-$	−2.82
$La(OH)_3 + 3e^- \rightarrow La + 3OH^-$	−2.80
$Sm(OH)_3 + 3e^- \rightarrow Sm + 3OH^-$	−2.80
$Dy(OH)_3 + 3e^- \rightarrow Dy + 3OH^-$	−2.80
$Pr(OH)_3 + 3e^- \rightarrow Pr + 3OH^-$	−2.79
$Ce(OH)_3 + 3e^- \rightarrow Ce + 3OH^-$	−2.78
$Nd(OH)_3 + 3e^- \rightarrow Nd + 3OH^-$	−2.78
$Pm(OH)_3 + 3e^- \rightarrow Pm + 3OH^-$	−2.76
$Yb(OH)_3 + 3e^- \rightarrow Yb + 3OH^-$	−2.74
$Mg(OH)_2 + 2e^- \rightarrow Mg + 2OH^-$	−2.687
$Sc(OH)_3 + 3e^- \rightarrow Sc + 3OH^-$	−2.60
$ThO_2 + 2H_2O + 4e^- \rightarrow Th + 4OH^-$	−2.56
$Am(OH)_3 + 3e^- \rightarrow Am + 3OH^-$	−2.53
$Cm(OH)_3 + 3e^- \rightarrow Cm + 3OH^-$	−2.5
$Pu(OH)_3 + 3e^- \rightarrow Pu + 3OH^-$	−2.46
$Al(OH)_4^- + 3e^- \rightarrow Al + 4OH^-$	−2.310
$Al(OH)_3(g) + 3e^- \rightarrow Al + 3OH^-$	−2.300
$Np(OH)_3 + 3e^- \rightarrow Np + 3OH^-$	−2.2
$TiO + H_2O + 2e^- \rightarrow Ti + 2OH^-$	−2.13
$U(OH)_3 + 3e^- \rightarrow U + 3OH^-$	−2.10
$H_2PO_2^- + e^- \rightarrow P + 2OH^-$	−2.05
$Ti_2O_3 + H_2O + 2e^- \rightarrow 2TiO + 2OH^-$	−1.95
$B(OH)_4^- + 3e^- \rightarrow B + 4OH^-$	−1.811
$SiO_3^{2-} + 3H_2O + 4e^- \rightarrow Si + 6OH^-$	−1.7
$HPO_3^{2-} + 2H_2O + 2e^- \rightarrow H_2PO_2^- + 3OH^-$	−1.57
$Mn(OH)_2 + 2e^- \rightarrow Mn + 2OH^-$	−1.56
$ZnS + 2e^- \rightarrow Zn + S^{2-}$	−1.44
$PuO_2 + 2H_2O + e^- \rightarrow Pu(OH)_3 + OH^-$	−1.4
$2TiO_2 + H_2O + 2e^- \rightarrow Ti_2O_3 + 2OH^-$	−1.38
$Zn(CN)_4^{2-} + 2e^- \rightarrow Zn + 4CN^-$	−1.34
$Cr(OH)_3 + 3e^- \rightarrow Cr + 3OH^-$	−1.33
$Cr(OH)_4^- + 3e^- \rightarrow Cr + 4OH^-$	−1.33
$Zn(OH)_4^{2-} + 2e^- \rightarrow Zn + 4OH^-$	−1.285
$CdS + 2e^- \rightarrow Cd + S^{2-}$	−1.255
$Zn(OH)_2 + 2e^- \rightarrow Zn + 2OH^-$	−1.248

Standard potentials in basic solutions

Couple	$E°$ (V)
$H_2GaO_3^- + H_2O + 3e^- \rightarrow Ga + 4OH^-$	−1.22
$Te + 2e^- \rightarrow Te^{2-}$	−1.14
$PO_4^{3-} + 2H_2O + 2e^- \rightarrow HPO_3^{2-} + 3OH^-$	−1.12
$WO_4^{2-} + 4H_2O + 6e^- \rightarrow W + 8OH^-$	−1.074
$ZnCO_3 + 2e^- \rightarrow Zn + CO_3^{2-}$	−1.06
$Zn(NH_3)_4^{2+} + 2e^- \rightarrow Zn + 4NH_3$	−1.04
$HGeO_3^- + 2H_2O + 4e^- \rightarrow Ge + 5OH^-$	−1.03
$MnO_2 + 2H_2O + 4e^- \rightarrow Mn + 4OH^-$	−0.980
$CNO^- + H_2O + 2e^- \rightarrow CN^- + 2OH^-$	−0.97
$Cd(CN)_4^{2-} + 2e^- \rightarrow Cd + 4CN^-$	−0.943
$SO_4^{2-} + H_2O + 2e^- \rightarrow SO_3^{2-} + 2OH^-$	−0.94
$PbS + 2e^- \rightarrow Pb + S^{2-}$	−0.923
$MoO_4^{2-} + 4H_2O + 6e^- \rightarrow Mo + 8OH^-$	−0.913
$Sn(OH)_6^- + 2e^- \rightarrow HSnO_2^- + H_2O + 3OH^-$	−0.91
$P + 3H_2O + 3e^- \rightarrow PH_3 + 3OH^-$	−0.89
$2H_2O + 2e^- \rightarrow H_2 + 2OH^-$	−0.828
$Cd(OH)_2 + 2e^- \rightarrow Cd + 2OH^-$	−0.824
$VO + H_2O + 2e^- \rightarrow V + 2OH^-$	−0.820
$HFeO_2^- + H_2O + 2e^- \rightarrow Fe + 3OH^-$	−0.8
$MoO_4^{2-} + 2H_2O + 2e^- \rightarrow MoO_2 + 4OH^-$	−0.780
$CdCO_3 + 2e^- \rightarrow Cd + CO_3^{2-}$	−0.734
$Co(OH)_2 + 2e^- \rightarrow Co + 2OH^-$	−0.733
$Ni(OH)_2 + 2e^- \rightarrow Ni + 2OH^-$	−0.72
$CrO_4^{2-} + 4H_2O + 3e^- \rightarrow Cr(OH)_4^- + 4OH^-$	−0.72
$Ag_2S + 2e^- \rightarrow Ag + S^{2-}$	−0.691
$FeO_2^- + H_2O + e^- \rightarrow HFeO_2^- + OH^-$	−0.69
$AsO_2 + 2H_2O + 3e^- \rightarrow As + 4OH^-$	−0.68
$Se + 2e^- \rightarrow Se^{2-}$	−0.67
$AsO_4^{3-} + 2H_2O + 2e^- \rightarrow AsO_2^- + 4OH^-$	−0.67
$Sb(OH)_4^- + 3e^- \rightarrow Sb + 4OH^-$	−0.639
$Cd(NH_5)_4^{2+} + 2e^- \rightarrow Cd + 4NH_3$	−0.622
$ReO_4^- + 2H_2O + 7e^- \rightarrow Re + 8OH^-$	−0.604
$ReO_4^- + 2H_2O + 3e^- \rightarrow ReO_2 + 4OH^-$	−0.594
$2SO_3^{2-} + 3H_2O + 4e^- \rightarrow S_2O_3^{2-} + 6OH^-$	−0.58
$ReO_2 + H_2O + 4e^- \rightarrow Re + 4OH^-$	−0.564
$Cu_2S + 2e^- \rightarrow 2Cu + S^{2-}$	−0.542
$HPbO_2^- + H_2O + 2e^- \rightarrow Pb + 3OH^-$	−0.502
$Ni(NH_3)_6^{2+} + 2e^- \rightarrow Ni + 6NH_5$	−0.476
$Sb(OH)_6^- + 2e^- \rightarrow Sb(OH)_4^- + 2OH^-$	−0.465
$Bi_2O_3 + 3H_2O + 6e^- \rightarrow Bi + 6OH^-$	−0.452
$S + 2e^- \rightarrow S^{2-}$	−0.45
$NiCO_3 + 2e^- \rightarrow Ni + CO_3^{2-}$	−0.45
$Cu(CN)_2^- + e^- \rightarrow Cu + 2CN^-$	−0.44
$TeO_3^{2-} + 3H_2O + 4e^- \rightarrow Te + 6OH^-$	−0.42
$Cu_2O + H_2O + 2e^- \rightarrow 2Cu + 2OH^-$	−0.365
$SeO_3^{2-} + 3H_2O + 4e^- \rightarrow Se + 6OH^-$	−0.36
$Tl(OH) + e^- \rightarrow Tl + OH^-$	−0.343
$O_2 + e^- \rightarrow O_2^-$	−0.33
$Ag(CN)_2^- + e^- \rightarrow Ag + 2CN^-$	−0.31

Continued

Couple	$E°$ (V)
$Cu(SCN) + e^- \rightarrow Cu + SCN^-$	−0.310
$CuO + H_2O + 2e^- \rightarrow Cu + 2OH^-$	−0.29
$Mn_2O_3 + 2H_2O + 2e^- \rightarrow 2Mn(OH)_2^- + 2OH^-$	−0.25
$2CuO + H_2O + 2e^- \rightarrow Cu_2O + 2OH^-$	−0.22
$Cu(NH_3)_2^+ + e^- \rightarrow Cu + 2NH_3$	−0.100
$O_2 + H_2O + 2e^- \rightarrow HO_2^- + OH^-$	−0.0649
$Tl(OH)_3 + 2e^- \rightarrow TlOH + 2OH^-$	−0.05
$MnO_2 + H_2O + 2e^- \rightarrow Mn(OH)_2 + 2OH^-$	−0.05
$AgCN + e^- \rightarrow Ag + CN^-$	−0.017
$NO_3^- + H_2O + 2e^- \rightarrow NO_2^- + 2OH^-$	0.01
$SeO_4^{2-} + H_2O + 2e^- \rightarrow SeO_3^{2-} + 2OH^-$	0.03
$Co(NH_3)_6^{3+} + e^- \rightarrow Co(NH_3)_6^{2+}$	0.058
$TeO_4^{2-} + H_2O + e^- \rightarrow TeO_3^{2-} + 2OH^-$	0.07
$HgO(red) + H_2O + 2e^- \rightarrow Hg + 2OH^-$	0.0977
$N_2H_4 + 2H_2O + 2e^- \rightarrow 2NH_5(aq) + 2OH^-$	0.1
$VO_4^{3-} + 4H_2O + 5e^- \rightarrow V + 8OH^-$	0.120
$Co(OH)_3 + e^- \rightarrow Co(OH)_2 + OH^-$	0.17
$HO_2^- + H_2O + e^- \rightarrow OH + 2OH^-$	0.184
$O_2^- + H_2O + e^- \rightarrow HO_2^- + OH^-$	0.20
$PbO_2(\beta) + H_2O + e^- \rightarrow HPbO_2 + OH^-$	0.208
$PbO_2(\beta) + H_2O + 2e^- \rightarrow PbO(red) + 2OH^-$	0.247
$IO_5^- + 3H_2O + 6e^- \rightarrow I^- + 6OH^-$	0.257
$ClO_3^- + H_2O + 2e^- \rightarrow ClO_2^- + 2OH^-$	0.295
$PuO_2(OH)_2 + e^- \rightarrow PuO_2OH + OH^-$	0.3
$PbO_3^{2-} + 2H_2O + 2e^- \rightarrow HPbO_2^- + 3OH^-$	0.330
$Ag_2O + H_2O + 2e^- \rightarrow 2Ag + 2OH^-$	0.342
$Ag(NH_3)_2^+ + e^- \rightarrow Ag + NH_3$	0.373
$ClO_4^- + H_2O + 2e^- \rightarrow ClO_3^- + 2OH^-$	0.374
$O_2 + 2H_2O + 4e^- \rightarrow 4OH^-$	0.401
$NH_2OH + H_2O + 2e^- \rightarrow NH_3 + 2OH^-$	0.42
$Ag(SO_3)_2^{3-} + e^- \rightarrow Ag + 2SO_3^{2-}$	0.43
$Ag_2CO_3 + 2e^- \rightarrow 2Ag + CO_3^{2-}$	0.47
$IO^- + H_2O + 2e^- \rightarrow I^- + 2OH^-$	0.472
$NiO_2 + 2H_2O + 2e^- \rightarrow Ni(OH)_2 + 2OH^-$	0.490
$FeO_4^{2-} + 2H_2O + 3e^- \rightarrow FeO_2^- + 4OH^-$	0.55
$BrO_3^- + 3H_2O + 6e^- \rightarrow Br^- + 6OH^-$	0.584
$RuO_4^- + e^- \rightarrow RuO_4^{2-}$	0.593
$MnO_4^{2-} + 2H_2O + 2e^- \rightarrow MnO_2 + 4OH^-$	0.62
$2AgO + H_2O + 2e^- \rightarrow Ag_2O + 2OH^-$	0.640
$H_3IO_6^{2-} + 2e^- \rightarrow IO_3^- + 3OH^-$	0.656
$PbO_4^{4-} + 3H_2O + 2e^- \rightarrow HPbO_2^- + 5OH^-$	0.680
$ClO_2^- + H_2O + 2e^- \rightarrow ClO^- + 2OH^-$	0.681
$Ag_2O_3 + H_2O + 2e^- \rightarrow 2AgO + 2OH^-$	0.739
$BrO^- + H_2O + 2e^- \rightarrow Br^- + 2OH^-$	0.766
$HO_2^- + H_2O + 2e^- \rightarrow 3OH^-$	0.867
$ClO^- + H_2O + 2e^- \rightarrow Cl^- + 2OH^-$	0.890
$ClO_2 + e^- \rightarrow ClO_2^-$	1.041
$O_3 + H_2O + 2e^- \rightarrow O_2 + 2OH^-$	1.246
$OH + e^- \rightarrow OH^-$	1.985

12. Typical UV absorptions of unconjugated chromophores

Values are typically those found in nonpolar solvents

Chromophore	Transition	λ_{max} (nm)	log E_{max} (approx)
C_2H_6	$\sigma \to \sigma^*$	135	
H_2O	$n \to \sigma^*$	167	3.85
ROH	$n \to \sigma^*$	180–185	2.70
RSH	$n \to \sigma^*$	190–200; 225–230	3.18; 2.18
RCl	$n \to \sigma^*$	170–175	2.48
RBr	$n \to \sigma^*$	200–210	2.60
RI	$n \to \sigma^*$	255–260	2.70
R_2O	$n \to \sigma^*$	180–185	3.48
R_2S	$n \to \sigma^*$	210–215; 235–240	3.10; 2.00
RSSR	$n \to \sigma^*$	250	2.60
Amines	$n \to \sigma^*$	190–200	3.40–3.60
C_2H_4	$\pi \to \pi^*$	163; 174	4.18; 3.74
C_2H_2	$\pi \to \pi^*$	173	3.78 (vapour)
R–C≡CH	$\pi \to \pi^*$	185; 223	3.34; 2.08
R–C≡C–R	$\pi \to \pi^*$	178; 196–223	4.00; 3.30; 2.20
C=C=C	$\pi \to \pi^*$	170–185; 230	3.70–4.00; 2.78
C=C=O	$n \to \pi^*$	380	1.30
	$\pi \to \pi^*$	225	2.60
RCOCl	$n \to \pi^*$	280	1.00–1.18
RCO_2H, RCO_2R'	$n \to \pi^*$	195–210	1.60–2.00
$RCONH_2$	$n \to \pi^*$	175	3.85
RCN		<170	
RCONHCOR	$n \to \pi^*$	230–240	1.90–2.00
	$\pi \to \pi^*$	190–200	4.00–4.18
RNO_2	$n \to \pi^*$	270–280	1.30–1.48
	$\pi \to \pi^*$	200–210	4.18
RONO	$n \to \pi^*$	350	2.18
	$\pi \to \pi^*$	220	3.00
RNO (monomer)	$n \to \pi^*$	600–650; 300	1.30; 2.00
RN=NR	$n \to \pi^*$	350–370	1.00–1.18
	$\pi \to \pi^*$	<200	
RCHO	$n \to \pi^*$	290	1.18
	$\pi \to \pi^*$	185–195	
R_2CO	$n \to \pi^*$	270–290	1.00–1.30
	$\pi \to \pi^*$	180–190	3.30–4.00
RCOCOR	$n \to \pi^*$	420–460; 280–285	1.00; 1.30
RSOR	$n \to \pi^*$	210–230	3.18–3.40
RSO_2R		<190	

13. Typical UV absorption maxima of substituted benzenes

Bands arising from the same transition type and grouped together				
Substituent	**Solvent**	**λ_{max} nm (log E in brackets)**		
–OH	Water	210.5 (3.78)	270 (3.16)	
$–O^-$	Water	235 (3.97)	287 (3.42)	
$–OCH_3$	Water	217 (3.81)	269 (3.17)	
–SH	Hexane	236 (4.00)	269 (2.85)	
$–NH_2$	Water	230 (3.93)	280 (3.16)	
$–NH_3^+$	Water	203 (3.88)	254 (2.20)	
$–NO_2$	Hexane	252 (4.00)	280 (3.00)	330 (2.10)
–CHO	Ethanol	244 (4.18)	280 (3.18)	328 (1.30)
$–COCH_3$	Ethanol	240 (3.11)	278 (3.04)	319 (1.70)
$–CO_2H$	Water	230 (4.00)	270 (2.90)	
$–CO_2^-$	Water	224 (3.94)	268 (2.75)	
–CN	Water	224 (4.11)	271 (3.00)	
–F	Ethanol	204 (3.80)	254 (3.00)	
–Cl	Water	209.5 (3.87)	263.5 (2.28)	
–Br	Water	210 (3.90)	261 (2.28)	
–I	Water	207 (3.85)	257 (2.85)	
$–CH_3$	Water	207 (3.85)	261 (2.35)	
$–CH{=}CH_2$	Ethanol	244 (4.08)	282 (2.65)	
$–C{=}C–C_6H_5$	Hexane	236 (4.10)	278 (2.81)	
$–C_6H_5$	Ethanol	246 (4.30)		

14. Typical UV absorption maxima of aromatic and heteroaromatic compounds

Bands for different compounds arising from the same transition type are grouped together. For bands with fine structure, only λ_{max} of the band centre is given.						
Compound	**Solvent**	**λ_{max} nm (log E in brackets)**				
Benzene	Cyclohexane			183 (4.66)	204 (3.90)	256 (2.30)
Napthalene	Ethanol	167 (4.48)	190 (4.00)	220 (5.12)	286 (3.97)	312 (2.46)
Anthracene	Cyclohexane	186 (4.51)	221 (4.16)	256 (5.26)	375 (3.95)	(buried)
Napthacene	Benzene	187 (4.20); 211 (4.64)	230 (3.23)	272 (5.26)	474 (4.10)	
Azulene	Cyclohexane	700 (2.48)	193 (4.26)	236 (4.34)	269 (4.67)	357 (3.60)
Phenanthrene	Cyclohexane		222 (4.38)	252 (4.82)	292 (4.20)	345 (2.32)
Quinoline	Cyclohexane			228 (4.60)	270 (3.50)	315 (3.40)
Isoquinoline	Cyclohexane			218 (4.80)	265 (3.62)	313 (3.26)
Acridine	Ethanol			250 (5.30)	358 (4.00)	
Pyridine	Hexane	195 (3.88)		251 (3.30)		270 (2.65)
Pyrimidine	Cyclohexane			243 (3.31)		298 (2.48)
Pyrazine	Cyclohexane			260 (3.80)		327 (2.00)
Pyridazine	Cyclohexane			246 (3.11)		340 (2.50)
Purine	Water	<220 (3.48)		263 (3.90)		
Pyrrole	Hexane	210 (3.71)				240 (2.48)
Furan	Hexane	205 (3.81)				
Thiophene	Hexane	231 (3.85)				
Imidazole	Ethanol	207 (3.70)				
Pyrazole	Ethanol	210 (3.50)				
Isoxazole	Ethanol	211 (3.60)				
Thiazole	Ethanol	240 (3.60)				

15. Common isotopes for Mössbauer spectroscopy

Isotope	Natural abundance (%)	Spin states (gd)	(ex)	E_γ (keV)	Source isotope	Half-life	Line width (mm s^{-1})	Sign of $\Delta R/R$
^{57}Fe	2.2	1/2	3/2	14.4	^{57}Co	270 d	0.2	−ve
^{119}Sn	0.63	1/2	3/2	23.8	^{119m}Sn	240 d	0.8	+ve
^{121}Sb	2.1	5/2	7/2	37.2	^{121m}Sb	77 y	1.8	−ve
^{127}I	100	5/2	7/2	57.6	^{127m}Te	105 d	2.0	−ve
^{129}I	0.6	7/2	5/2	27.7	^{129m}Te	33 d	0.8	+ve
^{99}Ru	12.8	3/2	5/2	89.4	^{99}Rh	16 d	0.3	+ve
^{193}Ir	61.5	1/2	3/2	73.0	^{193m}Os	30 h	1.4	+ve
^{197}Au	100	3/2	1/2	77.3	^{197m}Pt	18 h	1.9	+ve

16. NMR frequency table

#	Isotope	Spin	Nat. Abundance%	Sensitivity rel.*	Sensitivity abs.**	NMR-frequency (MHz) at a field (T) of 2.3488	4.6975	5.8719	7.0463	9.3950	11.7440	14.0926	17.6157	18.7900
1	H	1/2	99.98	1.00	1.00	100.000	200.000	250.000	300.000	400.000	500.000	600.000	750.000	800.000
2	H	1	1.5×10^{-2}	9.65×10^{-3}	1.45×10^{-6}	15.351	30.701	38.376	46.051	61.402	76.753	92.102	115.128	122.804
3	H	1/2	0	1.21	0	106.663	213.327	266.658	319.990	426.664	533.317	639.980	799.974	853.328
3	He	1/2	1.3×10^{-4}	0.44	5.75×10^{-7}	76.178	152.355	190.444	228.533	304.710	380.888	457.066	571.332	609.420
6	Li	1	7.42	8.50×10^{-3}	6.31×10^{-4}	14.716	29.431	36.789	44.146	58.862	73.578	88.292	110.367	117.724
7	Li	3/2	92.58	0.29	0.27	38.863	77.727	97.158	116.590	155.454	194.317	233.180	291.474	310.908
9	Be	3/2	100	1.39×10^{-2}	1.39×10^{-2}	14.053	28.106	35.133	42.160	56.213	70.267	84.320	105.399	116.426
10	B	3	19.58	1.99×10^{-2}	1.39×10^{-3}	10.746	21.493	26.886	32.239	42.986	53.732	64.478	80.598	85.972
11	B	3/2	80.42	0.17	0.13	32.084	64.167	80.209	96.251	128.335	160.419	192.502	240.627	256.670
13	C	1/2	1.108	1.59×10^{-2}	1.76×10^{-4}	25.144	50.288	62.860	75.432	100.577	125.721	150.864	188.580	201.154
14	N	1	99.63	1.01×10^{-3}	1.01×10^{-3}	7.224	14.447	18.059	21.671	28.894	36.118	43.342	54.177	57.788
15	N	1/2	0.37	1.04×10^{-3}	3.85×10^{-6}	10.133	20.265	25.332	30.398	40.531	50.664	60.796	75.996	81.062
17	O	5/2	3.7×10^{-2}	2.91×10^{-2}	1.08×10^{-5}	13.557	27.113	33.892	40.670	54.227	67.784	81.340	101.676	108.454
19	F	1/2	100	0.83	0.83	94.077	188.154	235.192	282.231	376.308	470.385	564.462	705.576	752.616
21	Ne	3/2	0.257	2.50×10^{-3}	6.43×10^{-6}	7.894	15.788	19.736	23.683	31.577	39.472	47.366	59.208	63.154
23	Na	3/2	100	9.25×10^{-2}	9.25×10^{-2}	26.451	52.902	66.128	79.353	105.805	132.256	158.706	198.384	211.610
25	Mg	5/2	10.13	2.67×10^{-3}	2.71×10^{-4}	6.1195	12.238	15.298	18.358	24.477	30.597	36.716	45.894	48.954
27	Al	5/2	100	0.21	0.21	26.057	52.114	65.143	78.172	104.229	130.287	156.344	195.429	208.458
29	Si	1/2	4.7	7.84×10^{-3}	3.69×10^{-4}	19.865	39.730	49.662	59.595	79.460	99.325	119.190	148.986	159.280
31	P	1/2	100	6.63×10^{-2}	6.63×10^{-2}	40.481	80.961	101.202	121.442	161.923	202.404	242.884	303.606	323.846
33	S	3/2	0.76	2.26×10^{-3}	1.72×10^{-5}	7.670	15.339	19.174	23.009	30.678	38.348	46.018	57.522	61.356
35	Cl	3/2	75.53	4.70×10^{-3}	3.55×10^{-3}	9.798	19.596	24.495	29.395	39.193	48.991	58.790	73.485	78.386
37	Cl	3/2	24.47	2.71×10^{-3}	6.63×10^{-4}	8.156	16.311	20.389	24.467	32.623	40.779	48.934	61.167	65.246
39	K	3/2	93.1	5.08×10^{-4}	4.73×10^{-4}	4.667	9.333	11.666	13.999	18.666	23.333	27.998	34.998	37.332
41	K	3/2	6.88	8.40×10^{-5}	5.78×10^{-6}	2.561	5.122	6.403	7.684	10.245	12.806	15.368	19.209	20.490
43	Ca	7/2	0.145	6.40×10^{-3}	9.28×10^{-6}	6.728	13.456	16.820	20.184	26.913	33.641	40.368	50.460	53.826
45	Sc	7/2	100	0.30	0.30	24.290	48.588	60.735	72.882	97.176	121.470	145.764	182.205	194.352
47	Ti	5/2	7.28	2.09×10^{-3}	1.52×10^{-4}	5.637	11.273	14.092	16.910	22.547	28.184	33.820	42.276	45.094
49	Ti	7/2	5.51	3.76×10^{-3}	2.07×10^{-4}	5.638	11.276	14.095	16.914	22.552	28.191	33.828	42.285	45.104
50	V	6	0.24	5.55×10^{-2}	1.33×10^{-4}	9.970	19.940	24.926	29.911	39.881	49.852	59.822	74.778	79.762
51	V	7/2	99.76	0.38	0.38	26.289	52.576	65.720	78.864	105.152	131.440	157.728	197.160	210.304
53	Cr	3/2	9.55	9.03×10^{-4}	8.62×10^{-3}	5.652	11.304	14.130	16.956	22.608	28.260	33.912	42.390	45.216

* at constant field for equal number of nuclei.   ** product of relative sensitivity and natural abundance.

NMR frequency table (continued)

	Isotope	Spin	Nat. Abundance%	Sensitivity rel.*	Sensitivity abs.**	NMR-frequency (MHz) at a field (T) of 2.3488	4.6975	5.8719	7.0463	9.3950	11.7440	14.0926	17.6157	18.7900
55	Mn	5/2	100	0.18	0.18	24.664	49.328	61.661	73.993	98.657	123.322	147.986	184.983	197.314
57	Fe	1/2	2.19	3.37×10^{-5}	7.38×10^{-7}	3.231	6.462	8.078	9.693	12.925	16.156	19.386	24.234	25.850
59	Co	7/2	100	0.28	0.28	23.614	47.228	59.035	70.842	94.457	118.071	141.684	177.105	188.914
61	Ni	3/2	1.19	3.57×10^{-3}	4.25×10^{-5}	8.936	17.872	22.340	26.808	35.744	44.681	53.616	67.020	71.488
63	Cu	3/2	69.09	9.31×10^{-2}	6.43×10^{-2}	26.505	53.010	66.262	79.515	106.020	132.525	159.030	198.786	212.040
65	Cu	3/2	30.91	0.11	3.52×10^{-2}	28.394	56.788	70.986	85.183	113.577	141.972	170.366	212.958	227.154
67	Zn	5/2	4.11	2.85×10^{-3}	1.17×10^{-2}	6.254	12.508	15.635	18.762	25.160	31.271	37.524	46.905	50.320
69	Ga	3/2	60.4	6.91×10^{-2}	4.17×10^{-2}	24.003	48.006	60.008	72.009	96.012	120.016	144.018	180.024	192.024
71	Ga	3/2	39.6	0.14	5.62×10^{-2}	30.495	60.990	76.238	91.485	121.980	152.476	182.970	228.714	243.960
73	Ge	9/2	7.76	1.4×10^{-3}	1.08×10^{-4}	3.488	6.976	8.721	10.465	13.953	17.442	20.930	26.163	27.906
75	As	3/2	100	2.51×10^{-2}	2.51×10^{-2}	17.126	34.253	42.817	51.380	68.507	85.634	102.760	128.451	137.014
77	Se	1/2	7.58	6.93×10^{-3}	5.25×10^{-4}	19.067	38.135	47.669	57.203	76.270	95.338	114.406	143.007	152.540
79	Br	3/2	50.54	7.86×10^{-2}	3.97×10^{-2}	25.053	50.107	62.633	75.160	100.214	125.267	150.320	187.899	200.428
81	Br	3/2	49.46	9.85×10^{-2}	4.87×10^{-2}	27.006	54.012	67.515	81.018	108.025	135.031	162.036	202.545	216.050
83	Kr	9/2	11.55	1.88×10^{-3}	2.17×10^{-4}	3.847	7.695	9.619	11.543	15.391	19.238	23.086	28.857	30.782
85	Rb	5/2	72.15	1.05×10^{-2}	7.57×10^{-3}	9.655	19.310	24.138	28.965	38.620	48.276	57.930	72.414	77.240
87	Rb	3/2	27.85	0.17	4.87×10^{-2}	32.721	65.442	81.803	98.163	130.885	163.606	196.326	245.409	261.770
87	Sr	9/2	7.02	2.69×10^{-3}	1.88×10^{-4}	4.333	8.667	10.834	13.001	17.335	21.669	26.002	32.502	34.670
89	Y	1/2	100	1.18×10^{-4}	1.18×10^{-4}	4.899	9.798	12.248	14.697	19.596	24.496	29.394	36.744	39.192
91	Zr	5/2	11.23	9.48×10^{-3}	1.06×10^{-3}	9.330	18.660	23.325	27.991	37.321	46.651	55.982	69.975	74.642
93	Nb	9/2	100	0.48	0.48	24.442	48.885	61.107	73.328	97.771	122.214	146.656	183.321	195.542
95	Mo	5/2	15.72	3.23×10^{-3}	5.07×10^{-4}	6.514	13.029	16.287	19.544	26.059	32.574	39.088	48.861	52.118
97	Mo	5/2	9.46	3.43×10^{-3}	3.24×10^{-4}	6.652	13.304	16.630	19.957	26.609	33.261	39.914	49.890	53.218
99	Ru	3/2	12.72	1.95×10^{-4}	2.48×10^{-5}	3.389	6.779	8.474	10.169	13.559	16.949	20.338	25.422	27.118
101	Ru	5/2	17.07	1.41×10^{-3}	2.40×10^{-4}	4.941	9.882	12.353	14.824	19.765	24.707	29.648	37.059	39.530
103	Rh	1/2	100	3.11×10^{-5}	3.11×10^{-5}	3.147	6.295	7.868	9.442	12.590	15.737	18.884	23.604	25.180
105	Pd	5/2	22.23	1.12×10^{-3}	2.49×10^{-4}	4.576	9.152	11.440	13.728	18.305	22.881	27.456	34.320	36.610
107	Ag	1/2	51.82	6.62×10^{-5}	3.43×10^{-5}	4.046	8.093	10.116	12.139	16.186	20.233	24.278	30.348	32.372
109	Ag	1/2	48.18	1.01×10^{-4}	4.86×10^{-5}	4.652	9.304	11.630	13.956	18.608	23.260	27.912	34.890	37.216
111	Cd	1/2	12.75	9.54×10^{-3}	1.21×10^{-3}	21.205	42.410	53.013	63.616	84.821	106.027	127.232	159.039	169.642

* at constant field for equal number of nuclei.   ** product of relative sensitivity and natural abundance.

NMR frequency table (continued)

Isotope		Spin	Nat. Abundance%	Sensitivity		NMR-frequency (MHz) at a field (T) of								
				rel.*	abs.**	2.3488	4.6975	5.8719	7.0463	9.3950	11.7440	14.0926	17.6157	18.7900
113	Cd	1/2	12.26	1.09×10^{-2}	1.33×10^{-3}	22.182	44.365	55.457	66.548	88.731	110.914	133.096	166.371	177.462
113	In	9/2	4.28	0.34	1.47×10^{-2}	21.866	43.733	54.666	65.600	87.466	109.333	131.200	163.998	174.932
115	Ub	9/2	95.72	0.34	0.33	21.914	43.828	54.785	65.742	87.656	109.570	131.484	164.355	175.312
115	Sn	1/2	0.35	3.5×10^{-2}	1.22×10^{-4}	32.699	65.399	81.749	98.099	130.799	163.498	196.198	245.247	261.598
117	Sn	1/2	7.61	4.52×10^{-2}	3.44×10^{-3}	35.625	71.250	89.063	106.875	142.501	178.126	213.750	267.189	285.002
119	Sn	1/2	8.58	5.18×10^{-2}	4.44×10^{-3}	37.272	74.544	93.181	111.817	149.089	186.362	223.634	279.543	298.178
121	Sb	5/2	57.25	0.16	9.16×10^{-2}	23.930	47.860	59.826	71.791	95.721	119.652	143.582	179.478	191.442
123	Sb	7/2	42.75	4.57×10^{-2}	1.95×10^{-2}	12.959	25.918	32.398	38.878	51.837	64.796	77.756	97.194	103.674
123	Te	1/2	0.87	1.80×10^{-2}	1.56×10^{-4}	26.207	52.415	65.519	78.623	104.831	131.039	157.246	196.557	209.662
125	Te	1/2	6.99	3.15×10^{-2}	2.20×10^{-3}	31.596	63.193	78.992	94.790	126.387	157.984	189.580	236.976	252.774
127	I	5/2	100	9.34×10^{-2}	9.34×10^{-2}	20.007	40.014	50.018	60.021	80.029	100.036	120.042	150.054	160.058
129	Xe	1/2	26.44	2.12×10^{-2}	5.60×10^{-3}	27.660	55.321	69.151	82.981	110.642	138.302	165.962	207.453	221.284
131	Xe	3/2	21.18	2.76×10^{-3}	5.84×10^{-4}	8.199	16.399	20.499	24.598	32.798	40.998	49.196	61.497	65.596
133	Cs	7/2	100	4.74×10^{-2}	4.74×10^{-2}	13.117	26.234	32.792	39.351	52.458	65.585	78.702	98.376	104.916
135	Ba	3/2	6.59	4.90×10^{-3}	3.22×10^{-4}	9.934	19.868	24.835	29.802	39.736	49.670	59.604	74.505	79.472
137	Ba	3/2	11.32	6.86×10^{-3}	7.76×10^{-4}	11.113	22.226	27.783	33.339	44.452	55.566	66.678	83.349	88.904
138	La	5	0.089	9.19×10^{-2}	8.18×10^{-5}	13.193	26.386	32.982	39.579	52.772	65.965	79.158	98.946	105.544
139	La	7/2	99.91	5.92×10^{-2}	5.91×10^{-2}	14.126	28.252	35.315	42.378	56.404	70.631	84.756	105.945	112.808
141	Pr	5/2	100	0.29	0.29	29.291	58.582	73.227	87.872	117.163	146.454	175.744	219.681	234.326
143	Nd	7/2	12.17	3.38×10^{-3}	4.11×10^{-4}	5.437	10.875	13.594	16.313	21.750	27.188	32.626	40.782	43.500
145	Nd	7/2	8.3	7.86×10^{-4}	6.52×10^{-5}	3.345	6.690	8.364	10.036	13.381	16.727	20.072	25.092	26.762
147	Sm	7/2	14.97	1.48×10^{-3}	2.21×10^{-4}	4.128	8.256	10.320	12.384	16.512	20.640	24.768	30.960	33.024
149	Sm	7/2	13.83	7.47×10^{-4}	1.03×10^{-4}	3.289	6.578	8.224	9.868	13.156	16.446	19.736	24.672	26.312
151	Eu	5/2	47.82	0.18	8.51×10^{-2}	24.801	49.601	62.001	74.401	99.202	124.002	148.802	186.003	198.404
153	Eu	5/2	52.18	1.52×10^{-2}	7.98×10^{-3}	10.951	21.903	27.378	32.854	43.805	54.757	65.708	82.134	87.610
155	Gd	3/2	14.73	2.79×10^{-4}	4.11×10^{-5}	3.819	7.639	9.549	11.458	15.278	19.097	22.916	28.647	30.556
157	Gd	3/2	15.68	5.44×10^{-4}	8.53×10^{-5}	4.774	9.548	11.935	14.323	19.097	23.871	28.646	35.805	38.194
159	Tb	3/2	100	5.83×10^{-2}	5.83×10^{-2}	22.678	45.357	56.695	68.035	90.713	113.391	136.070	170.085	181.426
161	Dy	5/2	18.88	4.17×10^{-4}	7.87×10^{-5}	3.294	6.588	8.236	9.883	13.177	16.471	19.766	24.708	26.354
163	Dy	5/2	24.97	1.12×10^{-3}	2.79×10^{-4}	4.583	9.166	11.458	13.750	18.333	22.917	27.500	34.374	36.666
165	Ho	7/2	100	0.18	0.18	20.513	41.026	51.282	61.538	82.051	102.564	123.076	153.846	164.102
167	Er	7/2	22.94	5.07×10^{-4}	1.16×10^{-4}	2.890	5.780	7.226	8.671	11.560	14.451	17.342	21.678	23.120

* at constant field for equal number of nuclei.   ** product of relative sensitivity and natural abundance.

NMR frequency table (continued)

Isotope		Spin	Nat. Abundance%	Sensitivity		NMR-frequency (MHz) at a field (T) of								
				rel.*	abs.**	2.3488	4.6975	5.8719	7.0463	9.3950	11.7440	14.0926	17.6157	18.7900
169	Tm	1/2	100	5.66×10^{-4}	5.66×10^{-4}	8.271	16.543	20.679	24.814	33.086	41.358	49.628	62.037	66.172
171	Yb	1/2	14.31	5.46×10^{-3}	7.81×10^{-4}	17.613	35.226	44.032	52.839	70.452	88.065	105.678	132.096	140.904
173	Yb	5/2	16.13	1.33×10^{-3}	2.14×10^{-4}	4.852	9.704	12.130	14.556	19.409	24.261	29.112	36.390	38.818
175	Lu	7/2	97.41	3.12×10^{-2}	3.03×10^{-2}	11.407	22.815	28.518	34.222	45.629	57.036	68.444	85.554	91.258
176	Lu	7	2.59	3.72×10^{-2}	9.63×10^{-4}	7.928	15.858	19.822	23.786	31.715	39.644	47.572	59.466	63.430
177	Hf	7/2	18.5	6.38×10^{-4}	1.18×10^{-4}	3.120	6.240	7.801	9.361	12.481	15.602	18.722	23.403	24.962
179	Hf	9/2	13.75	2.16×10^{-4}	2.97×10^{-5}	1.869	3.739	4.674	5.609	7.479	9.349	11.218	14.022	14.958
181	Ta	7/2	99.98	3.60×10^{-2}	3.60×10^{-2}	11.970	23.940	29.925	35.910	47.880	59.850	71.820	89.775	95.760
183	W	1/2	14.4	7.20×10^{-4}	1.03×10^{-5}	4.161	8.322	10.402	12.483	16.644	20.805	24.966	31.206	33.288
185	Re	5/2	37.07	0.13	4.93×10^{-2}	22.513	45.027	56.284	67.541	90.055	112.569	135.082	168.852	180.110
187	Re	5/2	62.93	0.13	8.62×10^{-2}	22.744	45.488	56.861	68.233	90.977	113.722	136.466	170.583	181.954
187	Os	1/2	1.64	1.22×10^{-5}	2.00×10^{-7}	2.303	4.606	5.758	6.909	9.212	11.515	13.818	17.274	18.424
189	Os	3/2	16.1	2.34×10^{-3}	3.76×10^{-4}	7.758	15.517	19.397	23.276	31.035	38.794	46.552	58.191	62.070
191	Ir	3/2	37.3	2.53×10^{-5}	9.43×10^{-6}	1.718	3.437	4.296	5.156	6.875	8.593	10.312	12.888	13.750
193	Ir	3/2	62.7	3.27×10^{-5}	2.05×10^{-5}	1.871	3.743	4.678	5.614	7.486	9.357	11.228	14.034	14.972
195	Pt	1/2	33.8	9.94×10^{-3}	3.36×10^{-3}	21.499	42.998	53.747	64.497	85.996	107.495	128.994	161.241	171.992
197	Au	3/2	100	2.51×10^{-5}	2.51×10^{-5}	1.712	3.425	4.281	5.138	6.850	8.563	10.276	12.843	13.700
199	Hg	1/2	16.84	5.67×10^{-3}	9.54×10^{-4}	17.827	35.654	44.568	53.481	71.309	89.136	106.962	133.704	142.618
201	Hg	3/2	13.22	1.44×10^{-3}	1.90×10^{-4}	6.599	13.199	16.499	19.799	26.399	32.998	39.598	49.497	52.798
203	Tl	1/2	29.5	0.18	5.51×10^{-2}	57.149	114.298	142.873	171.448	228.597	285.747	342.896	428.619	457.194
205	Tl	1/2	70.5	0.19	0.13	57.708	115.416	144.270	173.124	230.832	288.540	346.248	432.810	461.664
207	Pb	1/2	22.6	9.16×10^{-3}	2.07×10^{-3}	20.921	41.843	52.304	62.765	83.687	104.609	125.530	156.912	167.374
209	Bi	9/2	100	0.13	0.13	16.069	32.139	40.174	48.208	64.278	80.348	96.416	120.522	128.556
235	U	7/2	0.72	1.21×10^{-4}	8.71×10^{-7}	1.790	3.580	4.475	5.371	7.161	8.951	10.742	13.425	14.322

* at constant field for equal number of nuclei. ** product of relative sensitivity and natural abundance.

17. ¹⁹F and ³¹P NMR chemical shifts

Some representative ¹⁹F NMR chemical shifts referenced to CFCl₃

	δ / ppm		δ / ppm		δ / ppm
MeF	−271.9	CFBr₃	7.4	FCH=CH₂	−114
EtF	−213	CF₂Br₂	7	F₂C=CH₂	−81.3
CF₂H₂	−1436	CFH₂Ph	−207	F₂C=CF₂	−135
CF₃R	−60 to −70	CF₂Cl₂	−8	C₆F₆	−163
AsF₅	−66	[AsF₆]⁻	−69.5	[BeF₄]⁻	−163
BF₃	−131	ClF₃	116; −4	ClF₅	247; 412
IF₇	170	MoF₆	−278	ReF₇	345
SeF₆	55	[SbF₆]⁻	−109	SbF₅	−108
[SiF₆]²⁻	−127	TeF₆	−57	WF₆	166
XeF₂	258	XeF₄	438	XeF₆	550

Some representative ³¹P NMR chemical shifts referenced to 85% H₃PO₄

(a) Phosphorus (III) compounds				(b) Phosphorus (V) compounds			
	δ / ppm		δ / ppm		δ / ppm		δ / ppm
PMe₃	−62	PMeF₂	245	Me₃PO	36.2	Me₃PS	59.1
PEt₃	−20	PMeH₂	−163.5	Et₃PO	48.3	Et₃PS	54.5
PPrⁿ₃	−33	PMeCl₂	192	[Me₄P]⁺	24.4	[Et₄P]⁺	40.1
PPrⁱ₃	19.4	PMeBr₂	184	[PO₄]³⁻	6.0	[PS₄]³⁻	87
PBuⁿ₃	−32.5	PMe₂F	186	PF₅	−80.3	[PF₆]⁻	−145
PBuⁱ₃	−45.3	PMe₂H	−99	PCl₅	−80	[PCl₄]⁺	86
PBuˢ₃	7.9	PMe₂Cl	96.5	MePF₄	−29.9	[PCl₆]⁻	−295
PBuᵗ₃	63	PMe₂Br	90.5	Me₃PF₂	−158	Me₂PF₃	8.0

18. Chemical shift ranges and standards for selected nuclei

Nucleus	Spin	Chemical shift range δ (ppm)	Standard
¹H	1/2	12 to −1	SiMe₄
⁶Li	1	5 to −10	1M LiCl in H₂O
⁷Li	3/2	5 to −10	1M LiCl in H₂O
¹¹B	3/2	100 to −120	BF₃ · OEt₂
¹³C	1/2	240 to −10	SiMe₄
¹⁵N	1/2	1200 to −500	MeNO₂
¹⁷O	5/2	1400 to −100	H₂O
¹⁹F	1/2	100 to −300	CFCl₃
²³Na	3/2	10 to −60	1M NaCl in H₂O
²⁷Al	5/2	200 to −200	[Al(H₂O)₆]³⁺
²⁹Si	1/2	100 to −400	SiMe₄
³¹P	1/2	230 to −200	H₃PO₄
⁴³Ca	7/2	40 to −40	CaCl₂
⁵¹V	7/2	0 to −2000	VOCl₃
⁶⁷Zn	5/2	100 to −2700	ZnClO₄
⁷⁷Se	1/2	1600 to −1000	SeMe₂
⁹³Nb	9/2	0 to −2000	NbCl₆⁻
⁹⁹Ru	3/2	3000 to −3000	RuO₃/CCl₄

Nucleus	Spin	Chemical shift range δ (ppm)	Standard
^{119}Sn	1/2	5000 to −3000	$SnMe_4$
^{121}Sb	5/2	1000 to −2700	Et_4NSbCl_6
^{129}Xe	1/2	2000 to −6000	$XeOF_4$
^{133}Cs	7/2	300 to −300	CsBr
^{195}Pt	1/2	9000 to −6000	Na_2PtCl_6
^{199}Hg	1/2	500 to −3000	$HgMe_2$*

* highly poisonous (see Chem. & Engin. News 6/1997).

19. Abbreviations and acronyms used in magnetic resonance

ACCORDION	2D technique, simultaneous incrementing of evolution and mixing times
ADA	Alternated Delay Acquisition
ADC	Analog-to-Digital Converter
ADLF	Adiabatic Demagnetization in the Laboratory Frame
ADRF	Adiabatic Demagnetization in the Rotating Frame
AEE	Average Excitation Energy approximation
AJCP	Adiabatic J Cross Polarization
APT	Attached Proton Test
AQ	Acquire
ARP	Adiabatic Rapid Passage
ASIS	Aromatic Solvent-Induced Shift
ASTM	American Society for Testing and Materials
BB	Broadband, as in decoupling
BIRD	Bilinear Rotation Decoupling
BLEW	A windowless multiple-pulse decoupling sequence
BPP	Bloembergen/Purcell/Pound (theory)
BR-24	Burum & Rhim (pulse sequence)
BURP	Band-selective Uniform Response Pure-phase
BWR	Bloch/Wangsness/Redfield (theory)
CAMELSPIN	Cross-relaxation Appropriate for Minimolecules Emulated by Locked -SPINs
CCPPA	Coupled Cluster Polarization Propagator Approximation
CH-COSY	Carbon-Hydrogen Correlation Spectroscopy
CHESS	Chemical Shift Selective Imaging Sequence
CHF	Coupled Hartree-Fock molecular orbital calculations
CIDEP	Chemically Induced Dynamic Electron Polarization
CIDNP	Chemically Induced Dynamic Nuclear Polarization
COCONOESY (CONOESY)	Combined COSY/NOESY
COLOC	Correlated Spectroscopy via Long Range Coupling
COSY	Correlated Spectroscopy

COSY-45	COSY with 45° mixing pulse
COSYDEC	COSY with F_1 Decoupling
COSYLR	COSY for Long-Range couplings
CP	Cross Polarization
CPD	Composite-Pulse Decoupling
CPMAS	Cross Polarization Magic-Angle Spinning
CPMG	Carr-Purcell-Meiboom Gill Sequence
CRAMPS	Combined Rotational and Multiple Pulse Spectroscopy
CRAZED	Correlated Spectroscopy Revamped by Asymmetric Z-gradient Echo Detection
CSA	Chemical Shift Anisotropy
CSCM	Chemical Shift Correlation Map
CSI	Chemical Shift Imaging
CT	Constant Time
CW	Continuous Wave
CYCLOPS	Cyclically Ordered Phase Sequence
CYCLPOT	Cyclic Polarization Transfer
DAC	Digital-to-Analog Converter
DANTE	Delay Alternating with Nutation for Tailored Excitation
DAS	Dynamic Angle Spinning
DCNMR	NMR in Presence of an Electric Direct Current
DD	Dipole-Dipole
DECSY	Double-quantum Echo Correlated Spectroscopy
DEFT	Driven Equilibrium Fourier Transform
DEPT	Distortionless Enhancement by Polarization Transfer
DEPTH	Spin-echo sequence for spatial localization
DFT	Discrete Fourier Transformation
DIGGER	Discrete Isolation from Gradient-Governed Elimination of Resonances
DIPSI	Composite-pulse Decoupling in the presence of Scalar Interactions
DISCO	Differences and Sums within COSY
DLB	Differential Line Broadening
DNMR	Dynamic NMR
DNP	Dynamic Nuclear Polarization
DOPT	Dipolar Order Polarization Transfer
DOR	Double-Orientation Rotation
DOUBTFUL	Double Quantum Transition for Finding Unresolved Lines

Abbreviations and acronyms used in magnetic resonance (continued)

DQ	Double Quantum		**HECTOR**	Heteronuclear Correlation Spectroscopy
DQC	Double Quantum Coherence		**HEHAHA**	Heteronuclear Hartmann Hahn
DQF	Double Quantum Filter		**HMBC**	Heteronuclear Multiple-Bond Correlation
DQF-COSY	Double Quantum Filtered COSY		**HMQ**	Heteronuclear Multiquantum
DQSY	Double Quantum COSY		**HMQC**	Heteronuclear Multiple Quantum Coherence
DRESS	Depth Resolved Spectroscopy		**HNCO**	triple resonance experiments involving ^{15}N, ^{13}C and ^{1}H
DSA	Data-Shift Acquisition		**HOESY**	Heteronuclear Overhauser Effect Spectroscopy
ECOSY	Exclusive Correlation Spectroscopy		**HOHAHA**	Homonuclear Hartmann-Hahn Spectroscopy
EFG	Electric Field Gradient		**HR**	High Resolution
EHT	Extended Hückel Molecular Orbital Theory		**HRPA**	Higher Random Phase Approximation
ELD	Energy Level Diagram		**HSQC**	Heteronuclear Single Quantum Coherence
ENDOR	Electron-Nucleus Double Resonance		**IDESS**	Improved Depth Selective single surface coil Spectroscopy
ENMR	Electrophoretic NMR		**IGLO**	Individual Gauge for different Localized Orbitals
EOM	Equations of Motion			
EPI	Echo Planar Imaging		**INADEQUATE**	Incredible Natural Abundance Double Quantum Transfer Experiment
ESR	Electron Spin Resonance		**INDO**	Intermediate Neglect of Differential Overlap
EXORCYCLE	4-step phase cycle for spin echoes		**INDOR**	Internuclear Double Resonance
EXSY	Exchange Spectroscopy		**INDO/S**	Intermediate Neglect of Differential Overlap Calculations for Spectroscopy
FC	Fermi Contact or Field Cycling		**INEPT**	Insensitive Nuclei Enhanced by Polarization
FFT	Fast Fourier Transformation		**INVERSE**	H, X corelation via ^{1}H detection
FID	Free Induction Decay		**IR**	Inversion-Recovery
FIRFT	Fast Inversion-Recovery Fourier Transform		**ISIS**	Image-Selected in vivo Spectroscopy
FLASH	Fast Low-angle Shot Imaging		**IST**	Irreducible Spherical Tensor
FLOPSY	Flip-Flop Spectroscopy			
FOCSY	Foldover-Corrected Spectroscopy		**JCP**	J Cross-Polarization
FOV	Field of View		**JR**	Jump-and-Return sequence $(90_y{-}\tau{-}90._y)$
FPT	Finite Perturbation Theory			
FT	Fourier Transform		**LAS**	Laboratory Axes System
FUCOUP	Fully Coupled Spectroscopy		**LIS**	Lanthanide Induced Shift
FWHM	Full (line) Width at Half Maximum		**LORG**	Local Origin
			LOSY	Localized Spectroscopy
GARP	Globaly Optimized Alternating Phase Rectangular Pulse		**LP**	Linear Prediction
GE	Gradient Echo		**LSR**	Lanthanide Shift Reagent
GES	Gradient-Echo Spectroscopy		**MAGROFI**	Magnetization Grid Rotating-Frame Imaging
GIAO	Gauge Included Atomic Orbitals		**MARF**	Magic Angle in the Rotating Frame
GRASS	Gradient-Recalled Acquisition in the Steady State		**MAS**	Magic-Angle Spinning
GRASP	Gradient-Accelerated Spectroscopy		**MASS**	Magic-Angle Sample Spinning
GRECCO	Gradient Enhanced Carbon Coupling		**MEDUSA**	Technique for the Determination of Dynamic Structures
GROPE	Generalized compensation for Resonance Offset and Pulse Length Errors		**MEM**	Maximum Entropy Method
GS	Gradient Spectroscopy		**MINDO**	Modified INDO
H, C-COSY	^{1}H, ^{13}C chemical-shift correlation spectroscopy		**MLEV**	M. Levitt's CPD sequence
H, X-COSY	^{1}H, X-nucleus chemical-shift correlation spectroscopy		**MP**	Multi Pulse
			MQ	Multiple-Quantum

Abbreviations and acronyms used in magnetic resonance (continued)

MQC	Multiple-Quantum Coherence		RIDE	Ring Down Elimination
MQF	Multiple-Quantum Filter		RODI	Rotating-frame relaxation Dispersion Imaging
MQS	Multi Quantum Spectroscopy		ROESY	Rotating Frame Overhauser Effect Spectroscopy
MREV	Mansfield-Rhim-Elleman-Vaughan sequence for dipolar line narrowing		ROTO	ROESY-TOCSY Relay
MRI	Magnetic Resonance Imaging		RPA	Random Phase Approximation
MRS	Magnetic Resonance Spectroscopy			
MRSI	Magnetic Resonance Spectroscopic Imaging		SA	Shielding Anisotropy
MSPGSE	Multiple-Stepped PGSE		SC	Scalar Coupling
			SCPT	Self Consistent Perturbation Theory
NMR	Nuclear Magnetic Resonance		SD	Spin Dipolar
NOE	Nuclear Overhauser Effect		SDDS	Spin Decoupling Difference Spectroscopy
NOESY	Nuclear Overhauser Effect Spectroscopy		SE	Spin Echo
NQCC	Nuclear Quadrupole Coupling Constant		SECSY	Spin-Echo Correlated Spectroscopy
NQR	Nuclear Quadrupole Resonance		SEDOR	Spin-Echo Double Resonance
			SEDUCE	Selective Decoupling Using Crafted Excitation
ODMR	Optically Detected Magnetic Resonance		SEFT	Spin-Echo Fourier Transform Spectroscopy (with J modulation)
OSIRIS	modification of ISIS		SELINCOR	Selective Inverse Correlation
			SELINQUATE	Selective INADEQUATE
PAR	Phase-alternated Rotation of magnetization		SELRESOLV	Selective Resolution of C,H Coupling
PAS	Principal Axes System		SEMUT	Subspectral Editing Using a Multiple-Quantum Trap
P.COSY	Purged COSY			
P.E.COSY	Primitive E.COSY		SERF	Selective Refocussing
PENDANT	Polarization Enhancement During Attached Nucleus Testing		SFORD	Single Frequency Off-Resonance Decoupling
			SGSE	Steady-Gradient Spin-Echo
PFG	Pulsed Field Gradient		SKEWSY	Skewed Exchange Spectroscopy
PGSE	Pulsed Gradient Spin Echo		SL	Spin-Lock pulse
PMFG	Pulsed Magnetic Field Gradient		SLITDRESS	Slice interleaved Depth Resolved Surface coil-Spectroscopy
POF	Product Operator Formalism			
ppm	Parts per million		SLOPT	Spin-Locking Polarization Transfer
PRE	Proton Relaxation Enhancement		SNR or S/N	Signal-to-noise Ratio
PRESS	Point-Resolved Spectroscopy		SOPPA	Second Order Polarization Propagator Approach
PRFT	Partially Relaxed Fourier Transform			
PSD	Phase-sensitive Detection		SPACE	Spatial and Chemical-Shift Encoded Excitation
PW	Pulse Width			
			SPI	Selective Population Inversion
QF	Quadrupole moment/Field gradient (interaction or relaxation mechanism)		SPT	Selective Population Transfer
			SQC	Single-Quantum Coherence
QPD	Quadrature Phase Detection		SQF	Single-Quantum Filter
			SR	Saturation-Recovery
			SSFP	Steady-State Free Precession
RARE	Rapid Acquisition Relaxation Enhanced		SSI	Solid State Imaging
RCT	Relayed Coherence Transfer		STE	Stimulated Echo
RE-BURP	Refocused Band selective Uniform Response Pure phase		STEAM	Stimulated Echo Acquisition Mode for imaging
RECSY	Multistep Relayed Coherence Spectroscopy		TANGO	Testing for Adjacent Nuclei with a Gyration Operator
REDOR	Rotational Echo Double Resonance			
RELAY	Relayed Correlation Spectroscopy		TART	Tip Angle Reduced T_1 Imaging
REX	Relativistically Extended Hückel molecular orbital theory		TCF	Time Correlation Function
RF	Radio Frequency		TE	Time delay between excitation and Echo maximum

Abbreviations and acronyms used in magnetic resonance (continued)

TMR	Topical Magnetic Resonance		WATERGATE	Water suppression pulse sequence
TOCSY	Total Correlation Spectroscopy		WEFT	Water Eliminated Fourier Transform
TOE	Truncated NOE			
TORO	TOCSY-ROESY Relay		XCORFE	H, X Correlation using a Fixed Evolution time
TOSS	Total Suppression of Sidebands		X-FILTER	Selection of ^{1}H-^{1}H correlation when both H are coupled to X
TPPI	Time-proportional Phase Incrementation		X-HALF-FILTER	Selection of ^{1}H-^{1}H correlation when one H is coupled to X
TQ	Triple Quantum			
TQF	Triple-Quantum Filter		Z-COSY	COSY with z-Filter
TR	Time for Repetition of excitation		Z-FILTER	pulse sandwich for elimination of signal components with dispersive phase
TRCF	Tilted Rotating Coordinate Frame		ZECSY	Zero Quantum Echo-Correlation Spectroscopy
UE	Unpaired Electron (relaxation mechanism)		ZQ	Zero-Quantum
			ZQC	Zero-Quantum Coherence
VAS	Variable Angle Spinning		ZQF	Zero-Quantum Filter
VOSING	Volume-selective Spectral editing		ZZ-Spectroscopy	Selection of coherences involving ZZ or longitudinal two-spin order
VOSY	Volume-Selective Spectroscopy		β-COSY	COSY with small flip angle mixing pulse
WAHUHA	Waugh-Huber-Haeberlen Sequence		Ψ-COSY	psuedo-COSY using incremented frequency-selective excitation
WALTZ	CPD Sequence Containing the Elements 1-2-3			

20. Symbols used in magnetic resonance

B	magnetic-flux density		Q	nuclear quadrupole moment
B_{loc}	local magnetic-flux density contribution by secular interaction		ρ	density operator
			R_1	spin-lattice relaxation rate
B_o	stationary main magnetic-flux density in z direction		R_2	transverse relaxation rate
			R_{IS}	cross-relaxation rate
η	asymmetry parameter (or anisotropy constant) or nuclear Overhauser enhancement factor		σ	chemical shift shielding tensor
δ	chemical shift (ppm)		T_1	longitudinal (spin-lattice) relaxation time for M_z
G	gradient of the main magnetic-flux density		T_2	transverse (spin-spin) relaxation time for M_{xy}
γ_n	gyromagnetic ratio of nucleus n		T_2^*	time constant of the FID in the presence of B_o inhomogeneities
I	spin quantum number			
J	(indirect) spin-spin coupling constant (in Hz)		T_{2e}	transverse relaxation time effective under multiple-pulse irradiation
M	magnetization in the laboratory frame			
M'_x, M'_y, M'_z	magnetization components in the rotating frame along the x', y' and z' axes, respectively		T_{2p}	transverse relaxation time during a spin-lock pulse
M_o	equilibrium (Curie) magnetization		t_1, t_2	time domains in multi-dimensional experiments
μ	magnetic dipole moment		T_{1p}	T_1 of spin-locked magn. in rotation frame
μ_o	magnetic field constant		T_{2p}	T_2 of spin-locked magn. in rotation frame
μ_B	Bohr magneton		T_E, t_e	echo time
μ_N	nuclear magneton		t_n	time domain of the n-th dimension
ω_L	Larmor (angular) frequency		F_n	frequency domain of the n-th dimension
ω_m	modulation angular frequency (in rad s^{-1})		t_m	mixing time
ω_0	Larmor or resonance (angular) frequency		T_R, t_r	repetition time
ω_r	sample rotation frequency (rad s^{-1})		τ_c	correlation time
ω_1	Larmor (angular) frequency in B_1		τ_{coll}	mean time between molecular collision in the liquid state

Symbols used in magnetic resonance (continued)

τ_j	angular momentum correlation time		Ω	frequency offset
$\tau, \tau_1, \tau_2,...$	intervals in pulse sequences		χ	magnetic susceptibility
W_0, W_1, W_2	transition probabilities for zero-, single-, and double-quantum transitions		$Y_{l,m}$	spherical harmonics

21. EPR/ENDOR frequency table

Z	Isotope[a]	Natural Abundance percent	Spin	ENDOR Frequency in MHz, for 3.5 kG Field	g_n	$\dfrac{g_n\beta_n}{g_e\beta_e}$	Electric Quadrupole Moment Q in multiples of $|e| \times 10^{-24}$ cm^2
0	n^{1}*	...	1/2	10.20769	−3.82608422	1.040674×10^{-3}	...
1	H^1	99.985	1/2	14.90218	5.5856912	1.519278×10^{-3}	...
	H^2	0.0148	1	2.287575	0.8574376	2.332185×10^{-4}	0.002875
	H^3*	...	1/2	15.89525	5.957920	1.62052×10^{-3}	...
2	He^3	0.000138	1/2	11.35266	−4.255248	1.157405×10^{-3}	...
3	Li^6	7.5	1	2.193167	0.8220514	2.235937×10^{-4}	−0.000644
	Li^7	92.5	3/2	5.791950	2.170961	5.904900×10^{-4}	−0.040
4	Be^9	100.	3/2	2.094	−0.7850	2.135×10^{-4}	0.053
5	B^{10}	19.8	3	1.60133	0.600216	1.63256×10^{-4}	0.08608
	B^{11}	80.2	3/2	4.782043	1.792424	4.875299×10^{-4}	0.040
6	C^{13}	1.11	1/2	3.74795	1.40482	3.82104×10^{-4}	...
7	N^{14}	99.63	1	1.077201	0.4037607	1.098208×10^{-4}	0.0193^v
	N^{15}	0.366	1/2	1.511052	−0.5663784	1.540519×10^{-4}	...
8	O^{17}	0.038	5/2	2.02099	−0.757516	2.06040×10^{-4}	−0.026
9	F^{19}	100.	1/2	14.02721	5.257732	1.430075×10^{-3}	...
10	Ne^{21}	0.27	3/2	1.17708	−0.441197	1.20003×10^{-4}	0.1029
11	Na^{22}*	...	3	1.553	0.5820	1.583×10^{-4}	?
	Na^{23}	100.	3/2	3.944228	1.478391	4.021146×10^{-4}	0.108
12	Mg^{25}	10.00	5/2	0.91291	−0.34218	9.3071×10^{-5}	0.22^v
13	Al^{27}	100.	5/2	3.886094	1.456601	3.961878×10^{-4}	0.150
14	Si^{29}	4.67	1/2	2.9630	−1.1106	3.0208×10^{-4}	...
15	P^{31}	100.	1/2	6.03804	2.26320	6.15579×10^{-4}	...
16	S^{33}	0.75	3/2	1.1448	0.42911	1.1672×10^{-4}	−0.064
17	Cl^{35}	75.77	3/2	1.461795	0.5479157	1.490302×10^{-4}	−0.08249
	Cl^{36}*	...	2	1.71477	0.642735	1.74820×10^{-4}	−0.0180
	Cl^{37}	24.23	3/2	1.216790	0.4560820	1.240519×10^{-4}	−0.06493
18	Ar^{39}*	...	7/2	0.99	−0.37	1.0×10^{-4}	?
19	K^{39}	93.26	3/2	0.6963030	0.2609909	7.098816×10^{-5}	0.054
	K^{40}*	0.0117	4	0.86582	−0.32453	8.8270×10^{-5}	−0.067
	K^{41}	6.73	3/2	0.3821910	0.1432542	3.896439×10^{-5}	0.060
20	Ca^{41}*	...	7/2	1.215641	−0.4556514	1.239348×10^{-4}	?
	Ca^{43}	0.135	7/2	1.00424	−0.376414	1.02983×10^{-4}	<0.23^v
21	Sc^{45}	100.	7/2	3.62586	1.35906	3.69657×10^{-4}	−0.22^u
22	Ti^{47}	7.4	5/2	0.84144	−0.31539	8.5784×10^{-5}	0.29^u
	Ti^{49}	5.4	7/2	0.841667	−0.315477	8.58081×10^{-5}	0.24^u
23	V^{50}	0.250	6	1.48495	0.556593	1.51390×10^{-4}	0.209^u
	V^{51}	99.750	7/2	3.91747	1.46836	3.99386×10^{-4}	−0.0515^u
24	Cr^{53}	9.50	3/2	0.8396	−0.3147	8.560×10^{-5}	−0.0285/+0.022

Z	Isotope[a]	Natural Abundance percent	Spin	ENDOR Frequency in MHz, for 3.5 kG Field	g_n	$\dfrac{g_n \beta_n}{g_e \beta_e}$	Electric Quadrupole Moment Q in multiples of $|e| \times 10^{-24}$ cm^2
25	Mn^{53}*	...	7/2	3.828	1.435	3.903×10^{-4}	?
	Mn^{55}	100.	5/2	3.6868	1.3819	3.7587×10^{-4}	0.33[u]
26	Fe^{57}	2.15	1/2	0.4818	0.1806	4.912×10^{-5}	...
27	Co^{59}	100.	7/2	3.516	1.318	3.585×10^{-4}	0.42
	Co^{60}*	...	5	2.025	0.7589	2.064×10^{-4}	0.44[u]
28	Ni^{61}	1.13	3/2	1.3340	−0.50001	1.3600×10^{-4}	0.162
29	Cu^{63}	69.2	3/2	3.959	1.484	4.036×10^{-4}	−0.222
	Cu^{65}	30.8	3/2	4.237	1.588	4.319×10^{-4}	−0.195
30	Zn^{67}	4.10	5/2	0.934604	0.350312	9.52830×10^{-5}	0.150[u]
31	Ga^{69}	60.1	3/2	3.58673	1.34439	3.65667×10^{-4}	0.168
	Ga^{71}	39.9	3/2	4.55729	1.70818	4.64616×10^{-4}	0.106
32	Ge^{73}	7.8	9/2	0.5214100	−0.1954371	5.315787×10^{-5}	−0.19[u]
33	As^{75}	100.	3/2	2.56026	0.959647	2.61019×10^{-4}	0.29[u]
34	Se^{77}	7.6	1/2	2.8528	1.0693	2.9084×10^{-4}	...
	Se^{79}*	...	7/2	0.7758	−0.2908	7.910×10^{-4}	0.8[u]
35	Br^{79}	50.69	3/2	3.746469	1.404266	3.819530×10^{-4}	0.293[u]
	Br^{81}	49.31	3/2	4.038446	1.513706	4.117201×10^{-4}	0.27
36	Kr^{83}	11.5	9/2	0.575481	−0.215704	5.86704×10^{-5}	0.260[u]
	Kr^{85}*	...	9/2	0.5957	0.2233	6.074×10^{-5}	0.45[u]
37	Rb^{85}	72.17	5/2	1.44402	0.541253	1.47218×10^{-4}	0.273
	Rb^{87}	27.83	3/2	4.89369	1.83427	4.98912×10^{-4}	0.130
38	Sr^{87}	7.0	9/2	0.64806	−0.24291	6.6070×10^{-5}	0.15
39	Y^{89}	100.	1/2	0.7332410	−0.2748361	7.475398×10^{-5}	...
40	Zr^{91}	11.2	5/2	1.39118	−0.521448	1.41831×10^{-4}	?
41	Nb^{93}	100.	9/2	3.6583	1.3712	3.7296×10^{-4}	−0.28[u]
42	Mo^{95}	15.9	5/2	0.9754	−0.3656	9.944×10^{-5}	−0.019[u]
	Mo^{97}	9.6	5/2	0.9962	−0.3734	1.016×10^{-4}	0.2[v]
43	Tc^{98}*	...	9/2	3.3701	1.2632	3.4358×10^{-4}	0.34[v]
44	Ru^{99}	12.7	5/2	0.664	−0.249	6.77×10^{-5}	0.076[u]
	Ru^{101}	17.0	5/2	0.744	−0.279	7.59×10^{-5}	0.44[u]
45	Rh^{102}*	...	(6)	1.83	0.685	1.86×10^{-4}	?
	Rh^{103}	100.	1/2	0.4717	−0.1768	4.809×10^{-5}	...
46	Pd^{105}	22.2	5/2	0.683	−0.256	6.96×10^{-5}	0.66[u]
47	Ag^{107}	51.83	1/2	0.606282	−0.227249	6.18105×10^{-5}	...
	Ag^{109}	48.17	1/2	0.698309	−0.261743	7.11927×10^{-5}	...
48	Cd^{111}	12.8	1/2	3.17597	−1.19043	3.23791×10^{-4}	...
	Cd^{113}*	12.2	1/2	3.3226	−1.2454	3.3874×10^{-4}	...
49	In^{113}	4.3	9/2	3.27791	1.22864	3.34184×10^{-4}	0.846
	In^{115}*	95.7	9/2	3.28498	1.23129	3.34904×10^{-4}	0.861
50	Sn^{115}	0.38	1/2	4.9028	−1.8377	4.9984×10^{-4}	...
	Sn^{117}	7.75	1/2	5.34139	−2.00208	5.44555×10^{-4}	...
	Sn^{119}	8.6	1/2	5.58812	−2.09456	5.69709×10^{-4}	...
51	Sb^{121}	57.3	5/2	3.5897	1.3455	3.6597×10^{-4}	−0.33[u]
	Sb^{123}	42.7	7/2	1.9443	0.72876	1.9822×10^{-4}	−0.68
	Sb^{125}*	...	7/2	2.005	0.7514	2.044×10^{-4}	?
52	Te^{123}	0.89	1/2	3.9314	−1.4736	4.0081×10^{-4}	...
	Te^{125}	7.0	1/2	4.7398	−1.7766	4.8323×10^{-4}	...
53	I^{127}	100.	5/2	3.00221	1.12530	3.06076×10^{-4}	−0.789[v]
	I^{129}*	...	7/2	1.9979	0.74886	2.0369×10^{-4}	−0.553[v]

Z	Isotope[a]	Natural Abundance percent	Spin	ENDOR Frequency in MHz, for 3.5 kG Field	g_n	$\dfrac{g_n \beta_n}{g_e \beta_e}$	Electric Quadrupole Moment Q in multiples of $\|e\| \times 10^{-24}$ cm^2
54	Xe^{129}	26.4	1/2	4.15115	−1.55595	4.23210×10^{-4}	...
	Xe^{131}	21.2	3/2	1.23055	0.461240	1.25455×10^{-4}	−0.120[u]
55	Cs^{133}	100.	7/2	1.968518	0.7378477	2.006907×10^{-4}	−0.003
	Cs^{134}*	...	4	1.9967	0.74842	2.0357×10^{-4}	0.389
	Cs^{135}*	...	7/2	2.0828	0.78069	2.1234×10^{-4}	0.051
	Cs^{137}*	...	7/2	2.1658	0.81180	2.2081×10^{-4}	0.052
56	Ba^{133}*	...	1/2	4.11	−1.54	4.19×10^{-4}	...
	Ba^{135}	6.59	3/2	1.4909	0.55884	1.5200×10^{-4}	0.20[u]
	Ba^{137}	11.2	3/2	1.6679	0.62515	1.7004×10^{-4}	0.34[u]
57	La^{137}*	...	7/2	2.054	0.7700	2.094×10^{-4}	0.26
	La^{138}*	0.089	5	1.9817	0.74278	2.0203×10^{-4}	0.51
	La^{139}	99.911	7/2	2.1215	0.79520	2.1629×10^{-4}	0.20
59	Pr^{141}	100.	5/2	4.3	1.6	4.4×10^{-4}	−0.041[u]
60	Nd^{143}	12.2	7/2	0.8207	−0.3076	8.367×10^{-5}	−0.56
	Nd^{145}	8.3	7/2	0.507	−0.190	5.17×10^{-5}	−0.29
61	Pm^{147}*	...	7/2	2.01	0.752	2.05×10^{-4}	0.66[u]
62	Sm^{147}*	15.1	7/2	0.6195	−0.2322	6.316×10^{-5}	−0.18[u]
	Sm^{149}	13.9	7/2	0.5109	0.1915	5.209×10^{-5}	0.056[u]
	Sm^{151}*	...	5/2	0.379	0.142	3.86×10^{-5}	0.52[v]
63	Eu^{151}	47.9	5/2	3.706	1.389	3.778×10^{-4}	1.53[u]
	Eu^{152}*	...	3	1.7265	0.64713	1.7602×10^{-4}	3.16[u]
	Eu^{153}	52.1	5/2	1.637	0.6134	1.668×10^{-4}	3.92[u]
	Eu^{154}*	...	3	1.783	0.6683	1.818×10^{-4}	3.9[u]
	Eu^{155}*	...	5/2	2.06	0.772	2.10×10^{-4}	?
64	Gd^{155}	14.8	3/2	0.4597	−0.1723	4.686×10^{-5}	1.30[u]
	Gd^{157}	15.7	3/2	0.6011	−0.2253	6.128×10^{-5}	1.34
65	Tb^{157}*	...	3/2	3.5	1.3	3.5×10^{-4}	?
	Tb^{158}*	...	3	1.563	0.5860	1.594×10^{-4}	2.7
	Tb^{159}	100.	3/2	3.580	1.342	3.650×10^{-4}	1.34
66	Dy^{161}	19.0	5/2	0.504	−0.189	5.14×10^{-5}	2.47
	Dy^{163}	2.49	5/2	0.710	0.266	7.24×10^{-5}	2.51
67	Ho^{165}	100.	7/2	3.180	1.192	3.242×10^{-4}	2.73[u]
68	Er^{167}	22.9	7/2	0.4317	−0.1618	4.401×10^{-5}	2.827[u]
69	Tm^{169}	100.	1/2	1.24	−0.466	1.27×10^{-4}	...
	Tn^{171}*	...	1/2	1.229	0.4606	1.253×10^{-4}	...
70	Yb^{171}	14.4	1/2	2.637	0.9885	2.689×10^{-4}	...
	Yb^{173}	16.2	5/2	0.72554	−0.27195	7.3969×10^{-5}	2.8
71	Lu^{173}*	...	7/2	1.78	0.669	1.82×10^{-4}	?
	Lu^{174}*	...	(1)	5.18	1.94	5.28×10^{-4}	?
	Lu^{175}	97.39	7/2	1.7059	0.63943	1.7392×10^{-4}	5.68[u]
	Lu^{176}*	2.61	7	1.21	0.452	1.23×10^{-4}	8.0[u]
72	Hf^{177}	18.6	7/2	0.6048	0.2267	6.166×10^{-5}	4.5
	Hf^{179}	13.7	9/2	0.3799	−0.1424	3.873×10^{-5}	5.1
73	Ta^{181}	99.9877	7/2	1.8070	0.67729	1.8422×10^{-4}	3.44[u]
74	W^{183}	14.3	1/2	0.6284800	0.2355694	6.407365×10^{-5}	...
75	Re^{185}	37.40	5/2	3.4011	1.2748	3.4674×10^{-4}	2.33[u]
	Re^{187}*	62.60	5/2	3.4357	1.2878	3.5027×10^{-4}	2.22[u]
76	Os^{187}	1.6	1/2	0.3498	0.1311	3.566×10^{-5}	...
	Os^{187}	16.1	3/2	1.30	0.488	1.33×10^{-4}	0.8
77	Ir^{191}	37.3	3/2	0.259	0.097	2.64×10^{-5}	0.78[u]
	Ir^{193}	62.7	3/2	0.285	0.107	2.91×10^{-5}	0.70[u]

Z	Isotope[a]	Natural Abundance percent	Spin	ENDOR Frequency in MHz, for 3.5 kG Field	g_n	$\dfrac{g_n \beta_n}{g_e \beta_e}$	Electric Quadrupole Moment Q in multiples of $|e| \times 10^{-24}$ cm^2
78	Pt^{195}	33.8	1/2	3.2522	1.2190	3.3156×10^{-4}	...
79	Au^{197}	100.	3/2	0.261371	0.097968	2.66468×10^{-5}	0.594[u]
80	Hg^{199}	16.8	1/2	2.699321	1.011770	2.751961×10^{-4}	...
	Hg^{201}	13.2	3/2	0.996423	−0.373483	1.01585×10^{-4}	0.42
81	Tl^{203}	29.5	1/2	8.656103	3.244514	8.824909×10^{-4}	...
	Tl^{204}*	...	2	0.1211	0.0454	1.235×10^{-5}	?
	Tl^{205}	70.5	1/2	8.7385	3.2754	8.9089×10^{-4}	...
82	Pb^{207}	22.1	1/2	3.1343	1.1748	3.1954×10^{-4}	...
83	Bi^{207}*	...	9/2	2.59	0.970	2.64×10^{-4}	−0.50[v]
	Bi^{209}	100.	9/2	2.50	0.938	2.55×10^{-4}	−0.46[u]
84	Po^{209}*	...	1/2	4.0	1.5	4.1×10^{-4}	...
89	Ac^{227}*	...	3/2	1.9	0.73	2.0×10^{-4}	1.7[v]
90	Th^{229}*	...	5/2	0.43	0.16	4.4×10^{-5}	4.4[v]
91	Pa^{231}*	...	3/2	3.58	1.34	3.64×10^{-4}	?
92	U^{233}*	...	5/2	0.69	0.26	7.1×10^{-5}	3.5/7.9[v]
	U^{235}*	0.720	7/2	0.27	−0.10	2.7×10^{-5}	4.3[v]
93	Np^{237}*	...	5/2	3.2	1.2	3.3×10^{-4}	4.5[v]
94	Pu^{239}*	...	1/2	1.08	0.406	1.10×10^{-4}	...
	Pu^{241}*	...	5/2	0.755	−0.283	7.70×10^{-5}	5.6[u]
95	Am^{241}*	...	5/2	1.72	0.644	1.75×10^{-4}	4.9[u]
	Am^{243}*	...	5/2	1.72	0.644	1.75×10^{-4}	4.9[u]
96	Cm^{243}*	...	5/2	0.43	0.16	4.4×10^{-5}	?
	Cm^{245}*	...	7/2	0.3	0.1	3×10^{-5}	?
	Cm^{247}*	...	9/2	0.21	0.08	2.2×10^{-5}	?

[a] Stable isotopes and those with half-lives >1 yr.
* Means radioactive
[u] Means polarization correction not made
[v] Means unclear whether the correction was made

Table by courtesy of Prof. J. A. Weil
and P. S. Rao, Dept. of Chemistry,
University of Saskatchewan,
Saskatoon, Canada S 7N OWO

22. Some useful conversion factors in EPR

1) To obtain the magnetic field B of a resonance line in teslas, the klystrom frequency ν_e expressed in GHz is multiplied by the factor 0.07144775 and divided by the g factor, and 2) a hyperfine coupling constant A expressed in the units cm^{-1} is multiplied by 2.9979246 $\times 10^4$ to convert it to MHz.

ν_e (GHz)	$= 13.9962\ g\,B$ (T) $= 1.39962\ g\,B$ (kG)		A (MHz)	$= 2.997\,9246 \times 10^4 A$ (cm^{-1}) $= 13.9962\ g\,A$ (mT) $= 1.399\,62\ g\,A$ (G)
B (T)	$= \dfrac{0.071\,447\,75}{g} \nu_e$ (GHz)		A(cm^{-1})	$= 0.333\,564\,10 \times 10^{-4}\ A$ (MHz) $= 4.668\,63 \times 10^{-4}\ g\,A$ (mT)
ν_p(MHz)	$= 42.576\,38\ B$ (T)	(H$_2$O)	1 mT	$= 10$ G
B (T)	$= 0.023\,487\,20\ \nu_p$ (MHz)	(H$_2$O)		
g	$= 0.071\,447\,75 \dfrac{\nu_e\ (\text{GHz})}{B\,(\text{T})}$ $= 3.041\,987 \dfrac{\nu_e\ (\text{GHz})}{\nu_p\ (\text{MHz})}$			

23. Mass spectrometry: atomic weights

Atomic weights of the isotopes of all elements based on ^{12}C = 12.00000000

Isotope	Exact Mass	Abundance (%)	Isotope	Exact Mass	Abundance (%)
^{1}H	1.00782504	100.0000	^{46}Ca	45.9536890	.0041
^{2}H	2.01410179	.0150	^{48}Ca	47.9525320	.1929
D	2.01410179	100.0000	^{45}Sc	44.9559136	100.0000
^{3}He	3.01602929	.0001	^{46}Ti	45.9526330	10.8401
^{4}He	4.00260325	100.0000	^{47}Ti	46.9517650	9.8916
^{6}Li	6.01512320	8.1081	^{48}Ti	47.9479470	100.0000
^{7}Li	7.01600450	100.0000	^{49}Ti	48.9478710	7.4526
^{9}Be	9.01218250	100.0000	^{50}Ti	49.9447860	7.3171
^{10}B	10.0129380	24.8439	^{50}V	49.9471610	.2510
^{11}B	11.0093053	100.0000	^{51}V	50.9439630	100.0000
^{12}C	12.0000000	100.0000	^{50}Cr	49.9460460	5.1915
^{13}C	13.0033548	1.1122	^{52}Cr	51.9405100	100.0000
^{14}N	14.0030740	100.0000	^{53}Cr	52.9406510	11.3379
^{15}N	15.0001090	.3673	^{54}Cr	53.9388820	2.8166
^{16}O	15.9949146	100.0000	^{55}Mn	54.9380460	100.0000
^{17}O	16.9991306	.0381	^{54}Fe	53.9396120	6.3236
^{18}O	17.9991594	.2005	^{56}Fe	55.9349390	100.0000
^{19}F	18.9984033	100.0000	^{57}Fe	56.9353960	2.3986
^{20}Ne	19.9924391	100.0000	^{58}Fe	57.9332780	.3053
^{21}Ne	20.9938453	.2983	^{58}Ni	57.9353470	100.0000
^{22}Ne	21.9913837	10.1867	^{60}Ni	59.9307890	38.2305
^{23}Na	22.9897697	100.0000	^{61}Ni	60.9310590	1.6552
^{24}Mg	23.9850450	100.0000	^{62}Ni	61.9283460	5.2585
^{25}Mg	24.9858392	12.6598	^{64}Ni	63.9279680	1.3329
^{26}Mg	25.9825954	13.9380	^{59}Co	58.9331980	100.0000
^{27}Al	26.9815413	100.0000	^{63}Cu	62.9295991	100.0000
^{28}Si	27.9769284	100.0000	^{65}Cu	64.9277921	44.5710
^{29}Si	28.9764964	5.0634	^{64}Zn	63.9291450	100.0000
^{30}Si	29.9737717	3.3612	^{66}Zn	65.9260350	57.4074
^{31}P	30.9737634	100.0000	^{67}Zn	66.9271290	8.4362
^{32}S	31.9720718	100.0000	^{68}Zn	67.9248460	38.6831
^{33}S	32.9714591	.7893	^{70}Zn	69.9253250	1.2346
^{34}S	33.9678677	4.4306	^{69}Ga	68.9255810	100.0000
^{36}S	35.9670790	.0220	^{71}Ga	70.9247010	66.3890
^{35}Cl	34.9688527	100.0000	^{70}Ge	69.9242500	56.1644
^{37}Cl	36.9659026	31.9780	^{72}Ge	71.9220800	75.0685
^{39}K	38.9637079	100.0000	^{73}Ge	72.9234640	21.3698
^{40}K	39.9639988	.0130	^{74}Ge	73.9211790	100.0000
^{41}K	40.9618254	7.2170	^{76}Ge	75.9214030	21.3698
^{36}Ar	35.9675456	.3380	^{75}As	74.9215960	100.0000
^{38}Ar	37.9627322	.0630	^{79}Br	78.9183360	100.0000
^{40}Ar	39.9623831	100.0000	^{81}Br	80.9162900	97.2776
^{40}Ca	39.9625907	100.0000	^{74}Se	73.9224770	1.8145
^{42}Ca	41.9586218	.6674	^{76}Se	75.9192070	18.1451
^{43}Ca	42.9587704	.1393	^{77}Se	76.9199080	15.3226
^{44}Ca	43.9554848	2.1518	^{78}Se	77.9173040	47.3790

Isotope	Exact Mass	Abundance (%)	Isotope	Exact Mass	Abundance (%)
^{80}Se	79.9165210	100.0000	^{114}Cd	113.903361	100.0000
^{82}Se	81.9167090	18.9516	^{116}Cd	115.904758	26.0703
^{78}Kr	77.9203970	.6140	^{113}In	112.904056	4.4932
^{80}Kr	79.9163750	3.9474	^{115}In	114.903875	100.0000
^{82}Kr	81.9134830	20.3509	^{112}Sn	111.904823	3.0864
^{83}Kr	82.9141340	20.1754	^{114}Sn	113.902781	2.1605
^{84}Kr	83.9115060	100.0000	^{115}Sn	114.903344	1.2346
^{86}Kr	85.9106140	30.3509	^{116}Sn	115.901744	45.3704
^{85}Rb	84.9117996	100.0000	^{117}Sn	116.902954	23.7654
^{87}Rb	86.9091840	38.5710	^{118}Sn	117.901607	75.0000
^{84}Sr	83.9134280	.6781	^{119}Sn	118.903310	26.5432
^{86}Sr	85.9092730	11.9399	^{120}Sn	119.902199	100.0000
^{87}Sr	86.9088900	8.4766	^{122}Sn	121.903440	14.1975
^{88}Sr	87.9056250	100.0000	^{124}Sn	123.905271	17.2840
^{89}Y	88.9058560	100.0000	^{121}Sb	120.903824	100.0000
^{90}Zr	89.9047080	100.0000	^{123}Sb	122.904222	74.5201
^{91}Zr	90.9056440	21.9048	^{127}I	126.904477	100.0000
^{92}Zr	91.9050390	33.3722	^{120}Te	119.904021	.2840
^{94}Zr	93.9063190	33.6832	^{122}Te	121.903055	7.6923
^{96}Zr	95.9082720	5.4033	^{123}Te	122.904278	2.6864
^{93}Nb	92.9063780	100.0000	^{124}Te	123.902825	14.2485
^{92}Mo	91.9068090	61.5002	^{125}Te	124.904435	21.1243
^{94}Mo	93.9050860	38.3340	^{126}Te	125.903310	56.0651
^{95}Mo	94.9058380	65.9760	^{128}Te	127.904464	93.7574
^{96}Mo	95.9046760	69.1256	^{130}Te	129.906229	100.0000
^{97}Mo	96.9060180	39.5773	^{124}Xe	123.906120	.3717
^{98}Mo	97.9054050	100.0000	^{126}Xe	125.904281	.3346
^{100}Mo	99.9074730	39.9088	^{128}Xe	127.903531	7.1004
^{96}Ru	95.9075960	17.4684	^{129}Xe	128.904780	98.1413
^{98}Ru	97.9052870	5.9494	^{130}Xe	129.903510	15.2416
^{99}Ru	98.9059370	40.1899	^{131}Xe	130.905080	78.8104
^{100}Ru	99.9042180	39.8734	^{132}Xe	131.904148	100.0000
^{101}Ru	100.905581	53.7975	^{134}Xe	133.905395	38.6617
^{102}Ru	101.904348	100.0000	^{136}Xe	135.907219	33.0855
^{104}Ru	103.905422	59.1772	^{133}Cs	132.905433	100.0000
^{103}Rh	102.905503	100.0000	^{130}Ba	129.906277	.1478
^{102}Pd	101.905609	3.7322	^{132}Ba	131.905042	.1409
^{104}Pd	103.904026	40.7611	^{134}Ba	133.904490	3.3710
^{105}Pd	104.905075	81.7051	^{135}Ba	134.905668	9.1939
^{106}Pd	105.903475	100.0000	^{136}Ba	135.904556	10.9540
^{108}Pd	107.903894	96.8167	^{137}Ba	136.905816	15.6625
^{110}Pd	109.905169	42.8833	^{138}Ba	137.905236	100.0000
^{107}Ag	106.905095	100.0000	^{138}La	137.907114	.0901
^{109}Ag	108.904754	92.9050	^{139}La	138.906355	100.0000
^{106}Cd	105.906461	4.3508	^{136}Ce	135.907140	.2147
^{108}Cd	107.904186	3.0978	^{138}Ce	137.905996	.2825
^{110}Cd	109.903007	43.4737	^{140}Ce	139.905442	100.0000
^{111}Cd	110.904182	44.5527	^{142}Ce	141.909249	12.5226
^{112}Cd	111.902761	83.9888	^{141}Pr	140.907657	100.0000
^{113}Cd	112.904401	42.5339	^{142}Nd	141.907731	100.0000

Isotope	Exact Mass	Abundance (%)	Isotope	Exact Mass	Abundance (%)
[143]Nd	142.909823	44.8949	[176]Hf	175.941420	14.7727
[144]Nd	143.910096	87.7258	[177]Hf	176.943233	52.8409
[145]Nd	144.912582	30.5934	[178]Hf	177.943710	76.9886
[146]Nd	145.913126	63.3616	[179]Hf	178.945827	39.0341
[148]Nd	147.916901	21.2311	[180]Hf	179.946561	100.0000
[150]Nd	149.920900	20.7888	[180]Ta	179.947489	.0120
[144]Sm	143.912003	11.6540	[181]Ta	180.948014	100.0000
[147]Sm	146.914907	56.7670	[180]W	179.946727	.4239
[148]Sm	147.914832	42.4810	[182]W	181.948225	85.7515
[149]Sm	148.917193	52.2560	[183]W	182.950245	46.6254
[150]Sm	149.917285	27.8200	[184]W	183.950953	100.0000
[152]Sm	151.919741	100.0000	[186]W	185.954377	93.2507
[154]Sm	153.922218	84.9620	[185]Re	184.952977	59.7444
[151]Eu	150.919860	91.5709	[187]Re	186.955765	100.0000
[153]Eu	152.921243	100.0000	[184]Os	183.952514	.0488
[152]Gd	151.919803	.8052	[186]Os	185.953852	3.8537
[154]Gd	153.920876	8.7762	[187]Os	186.955762	3.9024
[155]Gd	154.922629	59.5813	[188]Os	187.955850	32.4390
[156]Gd	155.922130	82.4074	[189]Os	188.958156	39.2683
[157]Gd	156.923967	63.0032	[190]Os	189.958455	64.3902
[158]Gd	157.924111	100.0000	[192]Os	191.961487	100.0000
[160]Gd	159.927061	88.0032	[191]Ir	190.960603	59.4896
[159]Tb	158.925350	100.0000	[193]Ir	192.962942	100.0000
[156]Dy	155.924287	.2128	[190]Pt	189.959937	.0296
[158]Dy	157.924412	.3546	[192]Pt	191.961049	2.3373
[160]Dy	159.925203	8.2979	[194]Pt	193.962679	97.3373
[161]Dy	160.926939	67.0212	[195]Pt	194.964785	100.0000
[162]Dy	161.926805	90.4255	[196]Pt	195.964947	74.8521
[163]Dy	162.928737	88.2978	[198]Pt	197.967879	21.3018
[164]Dy	163.929183	100.0000	[197]Au	196.966560	100.0000
[165]Ho	164.930332	100.0000	[196]Hg	195.965812	.5059
[162]Er	161.928787	.4167	[198]Hg	197.966760	34.0641
[164]Er	163.929211	4.7917	[199]Hg	198.968269	57.3356
[166]Er	165.930305	100.0000	[200]Hg	199.968316	77.9089
[167]Er	166.932061	68.3036	[201]Hg	200.970293	44.5194
[168]Er	167.932383	79.7619	[202]Hg	201.970632	100.0000
[170]Er	169.935476	44.3452	[204]Hg	203.973481	22.9342
[169]Tm	168.934225	100.0000	[203]Tl	202.972336	41.8922
[168]Yb	167.933908	.4088	[205]Tl	204.974410	100.0000
[170]Yb	169.934774	9.5912	[204]Pb	203.973037	2.6718
[171]Yb	170.936338	44.9686	[206]Pb	205.974455	45.9923
[172]Yb	171.936393	68.8679	[207]Pb	206.975885	42.1756
[173]Yb	172.938222	50.6918	[208]Pb	207.976641	100.0000
[174]Yb	173.938873	100.0000	[209]Bi	208.980388	100.0000
[176]Yb	175.942576	39.9371	[232]Th	232.038054	100.0000
[175]Lu	174.940785	100.0000	[234]U	234.040947	.0055
[176]Lu	175.942694	2.6694	[235]U	235.043925	.7253
[174]Hf	173.940065	.4545	[238]U	238.050785	100.0000

Table by courtesy of Prof. J. Seibl (ETH Zürich, Switzerland)

24. Conversion table of transmittance and absorbance units

Transmittance [%]	Absorbance	Transmittance [%]	Absorbance
1.0	2.000	51.0	.292
2.0	1.699	52.0	.284
3.0	1.523	53.0	.276
4.0	1.398	54.0	.268
5.0	1.301	55.0	.260
6.0	1.222	56.0	.265
7.0	1.155	57.0	.244
8.0	1.097	58.0	.237
9.0	1.046	59.0	.229
10.0	1.000	60.0	.222
11.0	.959	61.0	.215
12.0	.921	62.0	.208
13.0	.886	63.0	.201
14.0	.854	64.0	.194
15.0	.824	65.0	.187
16.0	.796	66.0	.180
17.0	.770	67.0	.174
18.0	.745	68.0	.167
19.0	.721	69.0	.161
20.0	.699	70.0	.155
21.0	.678	71.0	.149
22.0	.658	72.0	.143
23.0	.638	73.0	.137
24.0	.620	74.0	.131
25.0	.602	75.0	.125
26.0	.585	76.0	.119
27.0	.569	77.0	.114
28.0	.553	78.0	.108
29.0	.538	79.0	.102
30.0	.523	80.0	.097
31.0	.509	81.0	.092
32.0	.495	82.0	.086
33.0	.481	83.0	.081
34.0	.469	84.0	.076
35.0	.456	85.0	.071
36.0	.444	86.0	.066
37.0	.432	87.0	.060
38.0	.420	88.0	.056
39.0	.409	89.0	.051
40.0	.398	90.0	.046
41.0	.387	91.0	.041
42.0	.377	92.0	.036
43.0	.367	93.0	.032
44.0	.357	94.0	.027
45.0	.347	95.0	.022
46.0	.337	96.0	.018
47.0	.328	97.0	.013
48.0	.319	98.0	.009
49.0	.310	99.0	.004
50.0	.301	100.0	.000

25. Conversion table of energy and wavelength units

Wavenumber [cm⁻¹]	Wavelength [Micron]	Wavelength [nm]	Frequency [GHz]	Electron Volt [eV]
2.0	5 000.00	5 000 000	60	.00 025
4.0	2 500.00	2 500 000	120	.00 050
6.0	1 666.67	1 666 667	180	.00 074
8.0	1 250.00	1 250 000	240	.00 099
10.0	1 000.00	1 000 000	300	.00 124
12.0	833.33	833 333	360	.00 149
14.0	714.29	714 286	420	.00 174
16.0	625.00	625 000	480	.00 198
18.0	555.56	555 556	540	.00 223
20.0	500.00	500 000	600	.00 248
22.0	454.55	454 545	660	.00 273
24.0	416.57	416 667	719	.00 298
26.0	384.62	384 615	779	.00 322
28.0	357.14	357 143	839	.00 347
30.0	333.33	333 333	899	.00 372
32.0	312.50	312 500	959	.00 397
34.0	294.12	294 118	1 019	.00 422
36.0	277.78	277 778	1 079	.00 446
38.0	263.16	263 158	1 139	.00 471
40.0	250.00	250 000	1 199	.00 496
50.0	200.00	200 000	1 499	.00 620
60.0	166.67	166 667	1 799	.00 744
70.0	142.86	142 857	2 099	.00 868
80.0	125.00	125 000	2 398	.00 992
90.0	111.11	111 111	2 698	.01 116
100.0	100.00	100 000	2 998	.01 240
110.0	90.91	90 909	3 298	.01 364
120.0	83.33	83 333	3 597	.01 488
130.0	76.92	76 923	3 897	.01 612
140.0	71.43	71 429	4 197	.01 736
150.0	66.67	66 667	4 497	.01 860
160.0	62.50	62 500	4 797	.01 984
170.0	58.82	58 824	5 096	.02 108
180.0	55.56	55 556	5 396	.02 232
190.0	52.63	52 632	5 696	.02 356
200.0	50.00	50 000	5 996	.02 480
220.0	45.45	45 455	6 595	.02 728
240.0	41.67	41 667	7 195	.02 976
260.0	38.46	38 462	7 795	.03 224
280.0	35.71	35 714	8 394	.03 472
300.0	33.33	33 333	8 994	.03 720
320.0	31.25	31 250	9 593	.03 967
340.0	29.41	29 412	10 193	.04 215
360.0	27.78	27 778	10 792	.04 463
380.0	26.32	26 316	11 392	.04 711
400.0	25.00	25 000	11 992	.04 959
500.0	20.00	20 000	14 990	.06 199
600.0	16.67	16 667	17 987	.07 439

Conversion table of energy and wavelength units (continued)

Wavenumber [cm⁻¹]	Wavelength [Micron]	Wavelength [nm]	Frequency [GHz]	Electron Volt [eV]
700.0	14.29	14 286	20 985	.08 679
800.0	12.50	12 500	23 983	.09 919
900.0	11.11	11 111	26 981	.11 159
1 000.0	10.00	10 000	29 979	.12 398
1 100.0	9.09	9 091	32 977	.13 638
1 200.0	8.33	8 333	35 975	.14 878
1 300.0	7.69	7 692	38 973	.16 118
1 400.0	7.14	7 143	41 971	.17 358
1 500.0	6.67	6 667	44 968	.18 598
1 600.0	6.25	6 250	47 966	.19 837
1 700.0	5.88	5 882	50 964	.21 077
1 800.0	5.56	5 556	53 962	.22 317
1 900.0	5.26	5 263	56 960	.23 557
2 000.0	5.00	5 000	59 958	.24 797
2 200.0	4.55	4 545	65 954	.27 276
2 400.0	4.17	4 167	71 950	.29 756
2 600.0	3.85	3 846	77 945	.32 236
2 800.0	3.57	3 571	83 941	.34 716
3 000.0	3.33	3 333	89 937	.37 195
3 200.0	3.13	3 125	95 933	.39 675
3 400.0	2.94	2 941	101 929	.42 155
3 600.0	2.78	2 778	107 924	.44 634
3 800.0	2.63	2 632	113 920	.47 114
4 000.0	2.50	2 500	119 916	.49 594
5 000.0	2.00	2 000	149 895	.61 992
6 000.0	1.67	1 667	179 874	.74 390
7 000.0	1.43	1 429	209 853	.86 789
8 000.0	1.25	1 250	239 832	.99 187
9 000.0	1.11	1 111	269 811	1.11 586
10 000.0	1.00	1 000	299 790	1.23 984
11 000.0	.91	909	329 769	1.36 382
12 000.0	.83	833	359 748	1.48 781
13 000.0	.77	769	389 727	1.61 179
14 000.0	.71	714	419 706	1.73 578
15 000.0	.67	667	449 685	1.85 976
16 000.0	.62	625	479 664	1.98 374
17 000.0	.59	588	509 643	2.10 773
18 000.0	.56	556	539 622	2.23 171
19 000.0	.53	526	569 601	2.35 570
20 000.0	.50	500	599 580	2.47 968
22 000.0	.45	455	659 538	2.72 765
24 000.0	.42	417	719 496	2.97 562
26 000.0	.38	385	779 454	3.22 358
28 000.0	.36	357	839 412	3.47 155
30 000.0	.33	333	899 370	3.71 952
32 000.0	.31	312	959 328	3.96 749
34 000.0	.29	294	1 019 286	4.21 546
36 000.0	.28	278	1 079 244	4.46 342
38 000.0	.26	263	1 139 202	4.71 139
40 000.0	.25	250	1 199 160	4.95 936

26. Optical components used in FT-IR-Spectroscopy

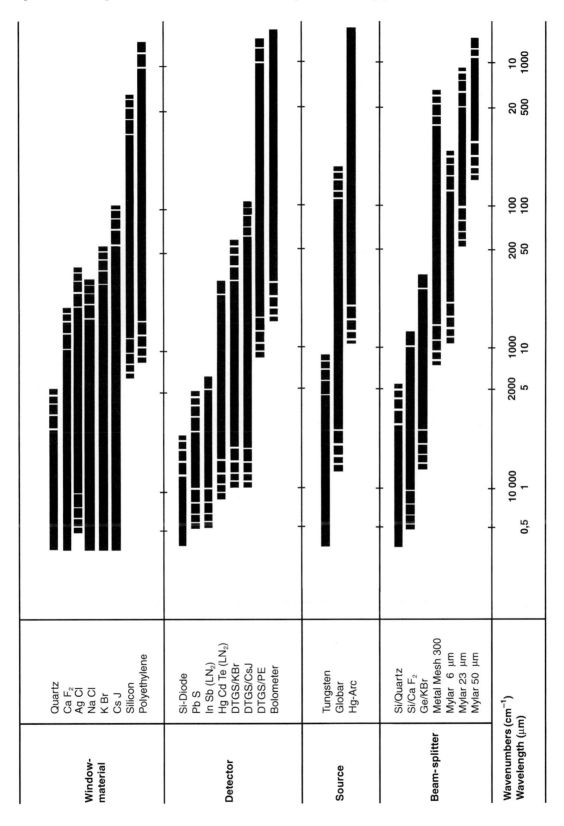

27. Infrared and Raman tables

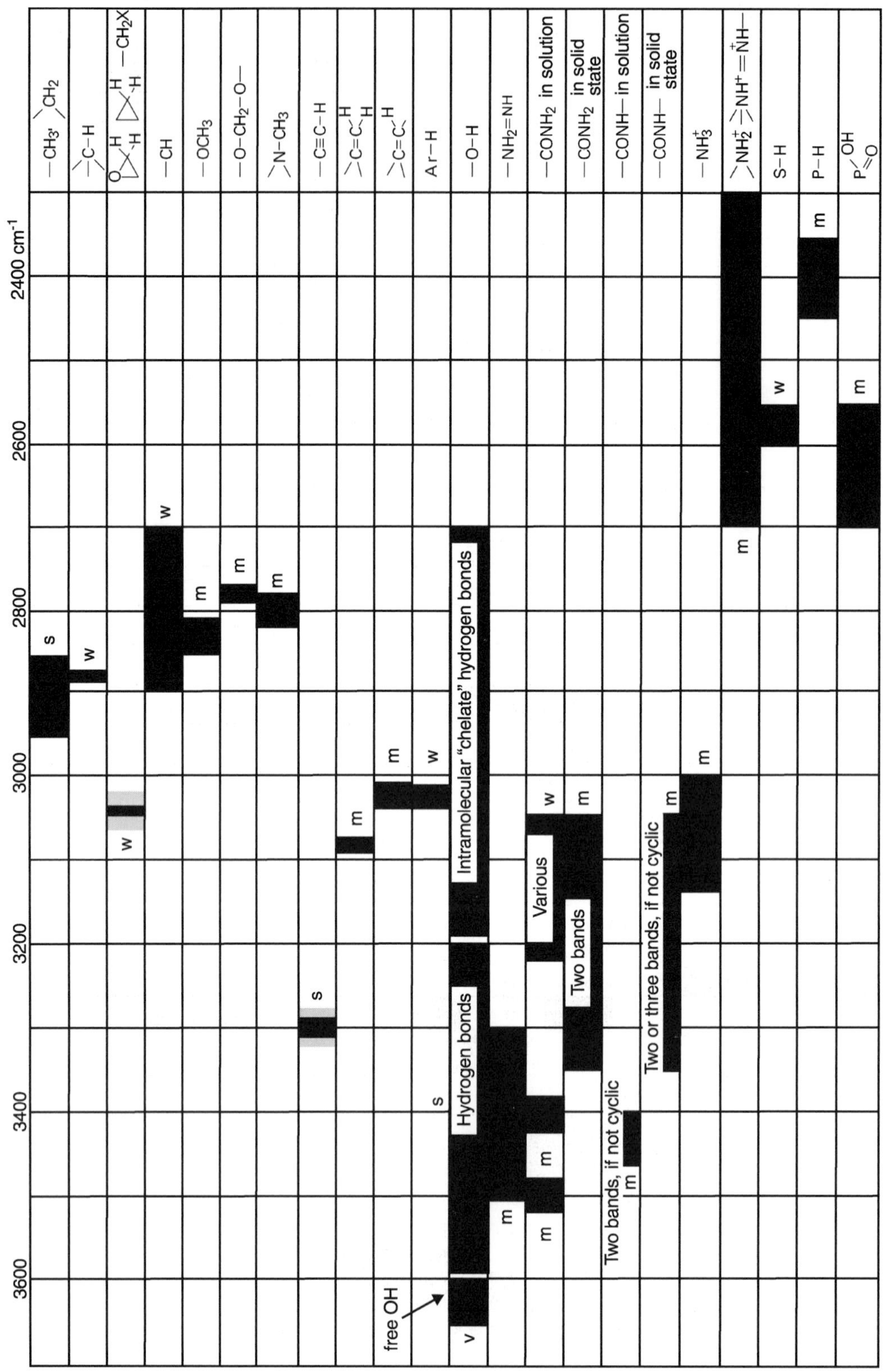

2400 cm⁻¹
2600
2800
3000
3200
3400
3600
—CH₃, CH₂
C—H
—CH₂X
—CH
—OCH₃
—O—CH₂—O—
N—CH₃
—C≡C—H
C=C—H
Ar—H
—O—H
—NH₂=NH
—CONH₂ in solution
—CONH₂ in solid state
—CONH— in solution
—CONH— in solid state
—NH₃⁺
NH₂⁺ NH⁺ =NH⁺—
S—H
P—H
P—OH =O
free OH
Hydrogen bonds
Intramolecular "chelate" hydrogen bonds
Various
Two bands
Two bands, if not cyclic
Two or three bands, if not cyclic

Positions of Stretching Vibrations of Hydrogen (in the hatched ranges the boundaries are not well defined);
Band intensity: s = strong, m = medium, w = weak, v = varying.

Infrared and Raman tables (continued)

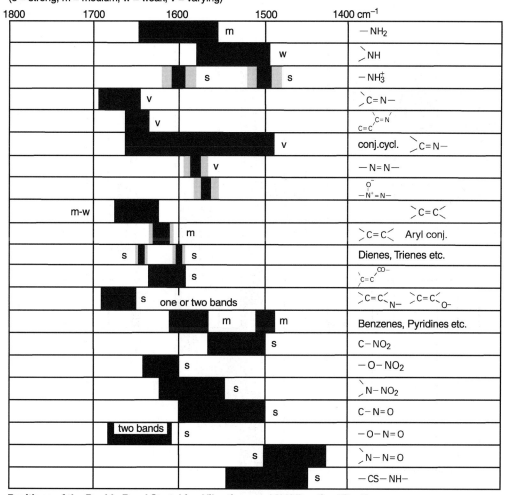

Positions of Stretching Vibrations of Triple Bonds and Cumulated Double Bonds
(s = strong, m = medium, w = weak, v = varying)

Positions of the Double Bond Stretching Vibrations and N-H Bending Vibrations
(s = strong, m = medium, w = weak, v = varying)

Infrared and Raman tables (continued)

Positions of Carbonyl Stretching Vibrations (all bands are strong)

Infrared and Raman tables (continued)

Group	1500	1400	1300	1200	1100	1000	900	800	700 cm⁻¹
Alkanes	m	m							w
—OCOCH₃ and —COCH₃			s						
—C(CH₃)₃		m	s						
C(CH₃)₂ (Double band)		s							
—CH=CH— trans						s			
C=C—H Alkenes						s/m			
—O—H				s					
C—O			s						
5 neighbouring aromatic C–H								s	
4 neighbouring aromatic C–H								s	
3 neighbouring aromatic C–H								s	
2 neighbouring aromatic C–H								s	
1 isolated aromatic C–H							w		
C—NO₂		s							
O—NO₂				s					
N—NO₂			s						
N—N=O		s							
>N⁺—O⁻					s		s		
>C=S						s			
—CSNH—	s								
>SO					s				
>SO₂					s				
—SO₂N<		s			s				
—SO₂O—		s			s				
P—O—Alkyl		s		s					
P—O—Aryl				s					
>P=O				s					
>P<O/OH			s						
C—F				s					
C—Cl						s		s	

Characteristic Absorptions in the Fingerprint Region (s = strong, m = medium, w = weak)

Stokes shifts (0–3500 cm⁻¹) of various laser sources

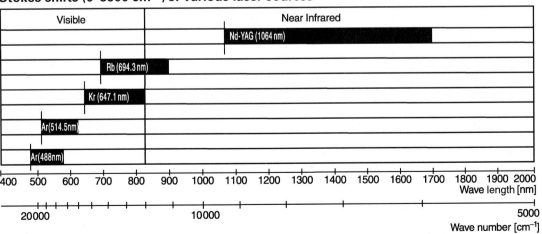

Nd-YAG: Neodynium YAG Laser   Rb: Ruby Laser   Kr: Krypton Ion Laser   Ar: Argon Ion Laser

28. Selected force constants and bond orders (according to Siebert) of organic and inorganic compounds

Bond A-B	Force Const. f (N cm⁻¹)	Bond Order	Compound
H–H	5.14	0.77	H_2
Li–Li	1.24	1.2	Li_2
B–B	3.58	1.2	B_2
C–C	16.5	3.2	HCCH
N–N	22.42	3.2	N_2
O–O	11.41	1.4	O_2
F–F	4.45	0.58	F_2
Na–Na	0.17	0.24	Na_2
Si–Si	4.65	2.0	Si_2
Si–Si	~1.7	~0.9	Si_2H_6
P–P	5.56	2.1	P_2
P–P	2.07	0.95	P_4
S–S	4.96	1.7	S_2
S–S	2.5	0.99	S_8
Cl–Cl	3.24	1.1	Cl_2
Ni–Ni	0.11	0.2	Ni solid
As–As	3.91	1.8	As_2
Se–Se	3.61	1.6	$^{80}Se_2$
Br–Br	2.36	1.1	Br_2
Rb–Rb	0.08	0.2	Rb_2
Cd–Cd	1.11	1.0	Cd_2^{2+}
Sb–Sb	2.61	1.9	Sb_2
Te–Te	2.37	1.7	Te_2
I–I	1.70	1.2	I_2
Hg–Hg	1.69	1.5	Hg_2^{2+}
Pb–Pb	4.02	3	Pb_2
Bi–Bi	1.84	1.6	Bi_2
H–B	2.75	0.68	BH_3
H–C	5.50	1.0	CH_4
H–N	7.05	1.1	NH_3
H–O	8.45	1.1	H_2O
H–O	7.40	1.0	HO^-
H–F	8.85	1.1	HF
H–Al	1.76	0.60	AlH_4^-
H–Si	2.98	0.84	SiH_4
H–P	3.11	0.82	PH_3
H–S	4.29	1.0	H_2S
H–Cl	4.81	1.0	HCl
H–Ge	2.81	0.82	GeH_4
H–As	2.85	0.81	AsH_3
H–Se	3.51	0.93	H_2Se
H–Br	3.84	0.98	HBr
H–Sn	2.03	0.76	SnH_4
H–Sb	2.09	0.77	SbH_3
H–I	2.92	0.97	HI
C–H	5.50	1.0	CH_4
C–B	3.82	1.1	$B(CH_3)_3$

Bond A-B	Force Const. f (N cm⁻¹)	Bond Order	Compound
C–C	16.5	3.2	HCCH
C–C	9.15	1.9	H_2CCH_2
C–C	7.6	1.7	C_6H_6
C–C	4.4	1.1	H_3CCH_3
C–N	18.07	3.0	HCN
C–N	11.84	2.1	CN_2^{2-}
C–N	6.54	1.3	$NNCH_2$
C–O	18.56	2.8	CO
C–O	15.61	2.4	CO_2
C–O	12.76	2.0	OCH_2
C–O	7.86	1.3	CO_3^{2-}
C–O	5.1	0.96	$O(CH_3)_2$
C–F	6.98	1.1	CF_4
C–P	8.95	2.4	HCP
C–S	7.67	2.0	CS_2
C–S	3.3	1.0	$S(CH_3)_2$
C–Cl	3.12	0.93	CCl_4
C–Ni	2.91	1.2	Ni_4CO
C–Ni	1.43	0.68	NiCO
C–Se	5.94	1.8	CSe_2
C–Br	2.42	0.86	CBr_4
C–Rh	2.4	1.2	$(Rh(CN)_6)^{3-}$
C–Ag	2.0	0.99	$(Ag(CN)_2)^-$
C–I	1.69	0.79	Cl_4
N–H	7.05	1.1	NH_3
N–B	7.2	1.6	BN_2^{3-}
N–C	18.07	3.0	HCN
N–N	22.42	3.2	N_2
N–N	16.01	2.4	N–NNH
N–N	13.15	2.0	$N–N–N^-$
N–O	25.07	3.1	$N–O^+$
N–O	17.17	2.3	NO_2^+
N–O	15.49	2.1	NO
N–O	15.18	2.0	ONCl
N–O	11.78	1.7	NNO
N–F	4.16	0.66	NF_3
N–Si	3.8	1.1	$((CH_3)_3Si)_2NH$
N–S	12.54	2.5	NSF_3
N–S	8.3	1.9	HNSO
N–S	3.1	0.87	$H_3N–SO_3$
O–Li	1.58	0.66	LiO
O–Be	7.51	1.8	BeO
O–B	13.66	2.5	BO
O–B	6.35	1.3	BO_3^{3-}
O–O	16.59	2.0	O_2^+
O–O	11.41	1.4	O_2

Selected force constants and bond orders
(according to Siebert) of organic and inorganic compounds (continued)

Bond A-B	Force Const. f (N cm^{-1})	Bond Order	Compound
O–O	6.18	0.89	O_2^-
O–O	5.70	0.83	O_3
O–Na	~3.2	~1.1	Na–OH
O–Mg	3.5	1.1	MgO
O–Al	5.66	1.5	AlO
O–Al	3.8	1.1	$Al(OH)_4^-$
O–Si	9.25	2.1	SiO
O–Si	4.75	1.2	SiO_4^{-4}
O–P	9.41	2.0	PO
O–P	6.16	1.4	PO_4^{3-}
O–S	10.01	2.0	SO_2
O–Cl	4.26	1.0	ClO_2^-
O–Cl	3.30	0.82	ClO^-
O–Ca	2.85	1.2	CaO
O–Ti	7.19	2.4	TiO
O–Cr	5.82	1.9	CrO

Bond A-B	Force Const. f (N cm^{-1})	Bond Order	Compound
O–V	7.36	2.3	VO
O–Mn	5.16	1.6	MnO
O–Fe	5.67	1.7	FeO
O–Cu	2.97	0.93	CuO
O–Ge	7.53	1.8	^{74}GeO
O–Se	6.45	1.5	SeO
O–Mo	3.05	1.2	Ba_2CaMoO_6 (solid)
O–Ru	6.70	2.2	RuO_4
O–Ag	2.00	0.79	AgO
O–Sn	5.53	1.7	SnO
O–Te	5.31	1.6	TeO
O–Ba	3.79	1.8	BaO
O–Ce	6.33	2.6	CeO
O–Pr	5.68	2.4	PrO
O–Nd	3.5	1.6	$NdAc_3 \cdot H_2O$ (polymer)

29. Fundamental physical constants

Physical constant	Symbol	Numerical value (SI units)	Physical constant	Symbol	Numerical value (SI units)
Proton rest mass	M_p	$1.6726231 \times 10^{-27}$ kg	Proton Lande factor	γ_H	5.58569
Electron rest mass	m_e	9.109389×10^{-31} kg	Proton-electron ratios[a]	m_p/m_e μ_e/μ_p γ_e/γ_p	1836.152701 658.2106881 658.2275841
Elementary charge	e	$1.6021773 \times 10^{-19}$ As	Electron Lande factor	g_e	2.0023193043
Charge-to-mass ratio for electron	e/m_e	1.758819×10^{11} As kg^{-1}	Electron magnetic moment	μ_e	$-9.2847701(31) \times 10^{-24}$ J/T
Atomic mass unit	amu	1.660540×10^{-27} kg	Free electron gyromagnetic ratio	$\gamma_e = g_e \mu_N / h$	$1.7608592(18) \times 10^{11}$ 1/s T
Bohr radius	a_0	5.291772×10^{-11} m	Neutron mass	m_n	$1.6749286(10) \times 10^{-27}$ kg
Electron radius	r_e	2.817940×10^{-15} m	Fine structure constant	$1/\alpha = 2h/\mu_0 c e^2$	137.0359895(61)
Planck constant	h $\hbar = h/2\pi$	6.62607×10^{-34} Js 1.054572×10^{-34} Js	Velocity of light in a vacuum	c	2.99792458×10^8 m s^{-1}
Boltzmann constant	k_B	1.38065×10^{-23} J K^{-1}	Permeability of a vacuum	$\mu_0 = 4\pi \times 10^{-7}$	$12.566370614 \times 10^{-7}$ T^2 m^3/J
Gas constant	R	8.3145 J mol^{-1} K^{-1}	Permittivity of a vacuum	$E_0 = 1/\mu_0 c^2$	$8.854187817 \times 10^{-12}$ c^2/Jm
Avogadro number	N_A	6.0221367×10^{23}	Gravitational constant	G	6.6725×10^{-11} m^3 kg^{-1}s^{-2}
Molar volume	V_{mol}	22.41383 m^3 kmol^{-1}	Acceleration due to gravity	g	9.80665 m s^{-2}
Faraday constant	$F = N_A \cdot e$	9.64853×10^4 C mol^{-1}	Compton vavelength of the electron ($h/m_e c$)	λ_c	2.426310×10^{-12} m
Bohr magneton	β_e	9.27401×10^{-24} J T^{-1}			
Nuclear magneton	β_N	5.05078×10^{-27} J T^{-1}			
Proton magneto gyric ratio	γ_H	2.67522×10^8 T^{-1} s^{-1}			

[a] The magnetic moment μ_p is for a bare proton and the gyromagnetic ratio γ_p is for protons in a spherical water sample which has diamagnetism and shape factor corrections, so the ratios μ_e/μ_p and γ_e/γ_p differ slightly.

$\pi = 3.14159$	$e = 2.71828$	$1/e = 0.36788$	$\ln(10) = 2.30259$

30. List of suppliers

Magnetic Resonance

Bruker Analytik GmbH Silberstreifen D-76287 Rheinstetten Germany	NMR, EPR
Bruker Medizintechnik GmbH Rudolf-Plank-Strasse 23 D-76275 Germany	MRI
Varian Inc 3120 Hansen Way Palo Alto CA 94304 USA	NMR
Jeol Ltd 1-2 Musashino 3-chome Akishima Tokyo 196 Japan	NMR, EPR
Oxford Instruments Ltd Head Office Old Station Way Eynsham Witney Oxon OX8 1TL UK	Magnets for NMR and MRI
Oxford Magnet Technology Ltd Wharf Road Eynsham Witney Oxon OX8 1BP UK	Magnets for MRI
G E Medical Systems 47697 Westinghouse Drive Fremont CA 94539 USA	MRI systems and PET systems
MR Resources Inc PO Box 880 158 R Main Street Gardner MA 01440 USA	Reconditioned NMR spectrometers and components
Fonar Corporation 110 Marcus Drive Melville NY 11747 USA	MRI
Picker International, Inc World Headquarters 595 Miner Road Cleveland Ohio 44143 USA	MRI

Doty Scientific, Inc 700 Clemson Road Columbia South Carolina 29229 USA	NMR Accessories
Magnex Scientific Ltd 21 Blacklands Way Abingdon Oxon OX14 1DY UK	MRI and NMR magnets
Resonance Instruments Ltd Unit 13 Thorney Leys Business Park Witney OX8 7GE UK	NMR Spectrometers
Tecmag, Inc 6006 Bellaire Blvd Houston TX 77081 USA	NMR Spectrometers
Goss Scientific Ltd 100 Vicarage Lane Great Baddow Essex CM2 8JB UK	NMR accessories, deuterated solvents, sample tubes
Martek Biosciences Corp 6480 Dobbin Road Columbia MD 21045 USA	Labelled compounds for NMR

Mass Spectrometry

Micromass Ltd Floats Road Wythenshawe Manchester M23 9LF UK	Sector mass spectrometers Time-of-flight mass spectrometers Quadrupole time-of-flight mass spectrometers Triple quadrupole mass spectrometers Quadrupole mass spectrometers
AMD Intectra GmbH Königsberger Strasse 1 D-27243 Harpstedt Germany	Sector mass spectrometers
Bruker Daltonics 15 Fortune Drive Billerica MA 01821 USA	Fourier transform ion cyclotron resonance mass spectrometers Ion trap mass spectrometers Time-of-flight mass spectrometers

Hitachi Instruments, Inc Separation Systems 3100 North First Street San Jose CA 95134 USA	Ion trap mass spectrometers
IonSpec Corporation 16 Technology Drive Suite 122 Irvine CA 92618 USA	Fourier transform ion cyclotron resonance mass spectrometers
Jeol Ltd 1-2 Musashino 3-chome Akishima Tokyo 196 Japan	Sector mass spectrometers
Kratos Analytical Shimadzu Wharfside Trafford Wharf Road Manchester M17 1GP UK	Sector mass spectrometers Time-of-flight mass spectrometers
PE Sciex Applied Biosystems Division 850 Lincoln Center Drive Foster City CA 94404-1128 USA	Time-of-flight mass spectrometers Triple quadrupole mass spectrometers
PerSeptive Biosystems, Inc 500 Old Connecticut Path Framingham MA 01701 USA	Time-of-flight mass spectrometers
ThermoQuest Finnigan Corporation 355 River Oaks Parkway San Jose CA 951134 USA	Sector mass spectrometers Ion trap mass spectrometers Time-of-flight mass spectrometers Fourier transform ion cyclotron resonance mass spectrometers Triple quadrupole mass spectrometers Inductively coupled plasme mass spectrometers; isotope ratio mass

Spatially Resolved Spectroscopic Analysis

Bruker AXS GmbH D-76181 Karlsruhe Germany	X-ray diffraction equipment
Digital Instruments GmbH Janderstrasse 9 D-68100 Manneheim Germany	Scanning probe microscopes
ThermoMicroscopes 1171 Borregas Avenue Sunnyvale CA 94089 USA	Scanning probe microscopes

Electronic Spectroscopy

Perkin-Elmer Ltd Post Office Lane Beaconsfield Bucks HP9 1QA UK	Fluorescence, fluorescence accessories, UV-Vis-NIR Absorption, UV-Vis-NIR Accessories
Jasco International 4-21 Sennin-cho 2-chome Hachioji Tokyo 193 Japan Jasco (UK) 18 Oak Industrial Park Great Dunmow Essex CM6 1XN UK	Fluorescence, fluorescence accessories, UV-Vis-NIR Absorption, UV-Vis-NIR Accessories, CD Spectrometers
Edinburgh Instruments Ltd Riccarton, Currie Edinburgh EH14 4AP UK	Fluorescence, fluorescence accessories
Instruments SA UK Ltd 2-4 Wigton Gardens Stanmore Middlesex HA7 1BG UK	Fluorescence, fluorescence accessories, UV-Vis-NIR Absorption, UV-Vis-NIR Accessories, CD Spectrometers, Ellipsometers
Molecular Probes Inc PO Box 22010 Eugene, OR 97402-0469 4849 Pitchford Ave. Eugene, OR 97402-9165 USA	Fluorescence accessories
Molecular Probes Europe BV For customers in Europe, Africa and the Middle East, contact: PoortGebouw Rijnsburgerweg 10 2333 AA Leiden The Netherlands	Fluorescence accessories
Hellma GmbH & Co. Postfach 1163 D-79371 Muellheim Germany Hellma (UK) Cumberland House 24-28 Baxter Avenue Southend on Sea Essex SS2 6HZ UK	Fluorescence accessories, UV-Vis-NIR Absorption
Specac, Inc. 500 Technology Ct. Southeast Smyrna, GA 30082 USA	Fluorescence accessories
Oxford Cryosystems 3 Blenheim Office Park Lower Road Long Hanborough Oxford OX8 8LN UK	Fluorescence accessories, UV-Vis-NIR Accessories

Varian (Cary) 3120 Hansen Way Palo Alto CA 94304 USA	UV-Vis-NIR Absorption, UV-Vis-NIR Accessories
Hitachi Instruments, Inc Separation Systems 3100 North First Street San Jose CA 95134 USA	UV-Vis-NIR Absorption, UV-Vis-NIR Accessories
Aviv Associates Inc. 750 Vassar Avenue Lakewood NJ 08701-6907 USA	CD Spectrometers
European Crystal Laboratories Unit 3 Pinnacle Business Centre Gordon Street Stockport SK4 1RS UK	UV-Vis-NIR Accessories
L.O.T – Oriel Ltd 1 Mole Business Park Leatherhead Surrey KT22 7AU UK	UV-Vis-NIR Accessories
Newport Corporation 1791 Deere Ave. Irvine, CA 92606 USA	UV-Vis-NIR Accessories
Hammamatsu Photonics UK Ltd Lough Point 2 Gladbeck Way Windmill Hill Enfield Middlesex EN2 7JA UK	UV-Vis-NIR Accessories
Spectroscopy Central Maple House Green Lane Warrington WA1 4JN UK	UV-Vis-NIR Accessories
Osram Hellabrunner Straße 1 81543 München Germany	UV-Vis-NIR Accessories

Vibrational, Rotational Spectroscopy

ASI SenseIR Technologies 15 Great Pasture Road Danbury CT 06810 USA	Standard equipment for reflection spectroscopy
Perkin-Elmer Ltd Post Office Lane Beaconsfield Bucks HP9 1QA UK	Infra-red spectrometers, Raman and IR accessories

Bomem 450 Saint-Jean-Baptiste Ave Quebec Canada G2E 5S5	Infra-red spectrometers
Midac Corporation 17911 Fitch Avenue Irvine CA 92714 USA	Infra-red spectrometers, Raman and IR accessories
Analect Instruments 2771 North Garey Ave Pomona CA 91714 USA	Infra-red spectrometers
Nicolet 5225 Verona Road Madison WI 53711-4495 USA	Infra-red spectrometers, Raman and IR accessories
Bruker Analytische Messtechnik GmbH Wikingerstrasse 13 D-76189 Germany	Infra-red spectrometers, Raman and IR accessories
Bio-Rad (Digilab division) 237 Putnam Ave Cambridge MA 02139 USA	Infra-red spectrometers, Raman and IR accessories
Innova Air Tech Instruments A/S Energivej 30 2750 Ballerup Denmark	Infra-red spectrometers
Block Engineering 164 Locke Drive Marlborough MA 01752-1178 USA	Infra-red spectrometers
Optical Solutions Inc 9477 Greenback Lane Suite 521 Folsom CA 95630 USA	Infra-red spectrometers
Jasco Inc 8649 Commerce Drive Easton MD 21601 USA	Infra-red spectrometers, Raman and IR accessories
Instruments SA (Yobin Yvon – Spex) 16018 rue du Canal 91165 Longjumeau Cedex France and 3880-T Park Avenue Edition NJ 08820 USA	Raman spectrometers

Renishaw (Raman Group) Old Town Wotton-under-Edge Glos GL12 7DH UK	Raman spectrometers, Raman and IR accessories
Kaiser Optical Systems Ltd PO Box 983 371 Parkland Plaza Ann Arbor MI 48106 USA	Raman spectrometers, Raman and IR accessories
Graseby Specac Ltd 97 Cray Avenue Orpington Kent BR5 4AA UK	Raman and IR accessories
Coherent Inc 5100 Patrick Henry Drive Santa Clara CA 95054 USA	Raman and IR accessories
The Svedberg Laboratory (TSL) Thunbergsvägen 5A Box 533 S-75121 Uppsala Sweden	Raman and IR accessories
Specac Inc 500 Technology Ct. Southeast Smyrna, GA 30082 USA	Raman and IR accessories
Spectroscopy Central Maple House Green Lane Warrington WA1 4JN UK	Raman and IR accessories
European Crystal Laborotories Unit 3 Pinnacle Business Centre Gordon Street Stockport SK4 1RS UK	Raman and IR accessories
Analytical Spectral Devices Boulder, CO USA	Near IR spectrometers
Bran & Luebbe Buffalo Grove, IL USA	Near IR spectrometers
Brimrose Corp of American Baltimore, MD USA	Near IR spectrometers
Bühler Inc Minneapolis, MN USA	Near IR spectrometers
Control Development Inc South Bend, IN USA	Near IR spectrometers

CVI Spectral Instrument Division Putnam, CT USA	Near IR spectrometers
Focus Engineering Budapest Hungary	Near IR spectrometers
Foss NIR Systems Höganäs Sweden	Near IR spectrometers
Infrared Engineering Concord, MA USA	Near IR spectrometers
Kett Villa Park, CA USA	Near IR spectrometers
Leco Saint Joseph, MI USA	Near IR spectrometers
LT Industries Rockville, MD USA	Near IR spectrometers
Mattson Instruments Madison, WI USA	Near IR spectrometers
Midac Irvine, CA USA	Near IR spectrometers
Ocean Optics Dunedin, FL USA	Near IR spectrometers
OLIS: On-line Instruments Bogart, GA USA	Near IR spectrometers
Oriel Instruments Stratford, CA USA	Near IR spectrometers
Perten Instruments Reno, NV USA	Near IR spectrometers
Rosemount Analytical Orrville, OH USA	Near IR spectrometers
OUP Guided Wave El Dorado Hills, CA USA	Near IR spectrometers
Wilks South Norwalk, CT USA	Near IR spectrometers
Zeiss Jena Jena Germany	Near IR spectrometers
Zeltex Hagerstown, MD USA	Near IR spectrometers

Optics, Detectors & Miscellaneous

Wright Instruments Ltd Unit 10 26 Queensway Enfield Middlesex EN3 4SA UK	CCD detectors
Spectra-Physics 1250 W.Middlefield Road PO Box 7013 Mountain View CA 94039-7013 USA	Lasers
Coherent Inc Laser Group 5100 Patrick Henry Drive Santa Clara CA 95054 USA	Lasers
Isotec Inc 3858 Benner Road Miamisburg OH 45342 USA	Deuterated solvents for NMR; labelled biochemicals
Optispec Rigistrasse 5 CH-8173 Neerach Switzerland	Special equipment for SBSR-ATR and ME-ATR Spectroscopy
ORC 1300 Optical Drive Azusa CA 91702-3251 USA	Light sources
Hamamatsu Corp 360-T Foothill Road PO Box 6910 Bridgewater NJ 08807-0910 USA	Light sources
Omega Optical, Inc 3 Grove Street PO Box 573 Brattleboro VT 05301 USA	Dielectric filters
Princeton Instruments Inc 3660-T Quakerbridge Road Trenton NJ 08619 USA	Detectors
Schott Glass Technologies Inc 400 York Avenue Duryea PA 18642 USA	Glass production
Heraeus Amersil, Inc 3473 Satellite Blvd Duluth GA 30136-5821 USA	Glass production

Polymicro Technologies Inc 18019N 25th Avenue Phoenix AZ 85023 USA	Fibers
Fiberguide Industries Inc 1-T Bay St Stirling NJ 07980 USA	Fibers
Ceramoptec GmbH Siemenstrasse 8-12 D-53121 Bonn 1 Germany	Fibers
Zeus Industrial Products Inc 620 Magnolia Street PO Box 2167 Orangeburg SC 29116 USA	Tubing
Kaiser Optical Systems Inc PO Box 983 371 Parkland Plaza Ann/Arbor MI 48106-0983 USA	Imaging spectrographs
Breault Research Organization Inc 6400 E Grant Ste 350 Tucson AZ 85715 USA	Illumination analysis
MicroQuartz Sciences 4420-T S 32nd St Phoenix AZ 85040-2804 USA	Custom fibreoptic probes
Rare Earth Medical Inc 126B Mid-Tech Drive West Yarmouth MA 02673 USA	Custom fibreoptic probes
Visionex Inc 151-T Osigian Blvd Dept 50 Warner Robins GA 31088 USA	Custom fibreoptic probes
QLT PhotoTherapeutics Inc 520 West 6th Avenue Vancouver British Columbia Canada V5Z 4H5	Photodynamic therapy
Labsphere PO Box 70A North Sutton NH 03260-0070 USA	Reflectance and scattering standards

NSG, Nippon Sheet Glass Co Ltd Fiber Optics Division 5-11-3, Shimbashi Minato-ku Tokyo 105 Japan	GRIN lenses
Swiss Jewel Via del sole CH-6598 Tenero	Sapphire products
Sapphire Engineering Inc 53 C Portside Dr Pocasset MA 02559 USA	Sapphire products
Saphirwerk Industrieprodukte AG Erlenstrasse 36 CH-2555 Brügg, Biel Switzerland	Sapphire products
Rubis Precis F-25140 Charquemont France	Sapphire products
Sandoz SA Clos de la Gare CH-1482, Cugy Switzerland	Sapphire products
Comadur SA Bernstrasse 11 CH-3608, Thun Switzerland	Sapphire products
Wilmad PO Box 688 1002 Harding Highway Buena New Jersey 08310 USA	NMR and EPR glassware
AMT 2570 E Cerritos Avenue Anaheim CA 92806 USA	

High Energy Spectroscopy

ARL Eclubens, CH-1024 Switzerland	X-ray fluorescence systems
ASOMA Austin TX 78759 USA	X-ray fluorescence systems
Bruker AXS GmbH D-76181 Karlsruhe Germany	X-ray fluorescence and diffraction systems
Amersham Bucks HP7 9NA UK	X-ray fluorescence systems

Atomika Oberschleissheim D-85764 Germany	X-ray fluorescence systems
BAIRD Bedford MA 01730-1468 USA	X-ray fluorescence systems
CIANFLONE Pittsburg PA 15275-1002 USA	X-ray fluorescence systems
EDAX Prairie View IL 60069 USA	X-ray fluorescence systems
EG & G Ortec	X-ray fluorescence systems
Meteorex Espoo, FIN 02201 Finland	X-ray fluorescence systems
N V Philips Eindhoven The Netherlands	X-ray fluorescence systems
Niton Bedford MA 01730-0368 USA	X-ray fluorescence systems
Oxford Instruments Oxon, OX14 1TX UK	X-ray fluorescence systems
Kratos Analytical Wharfside Trafford Wharf Road Manchester M17 1GP UK	X-ray fluorescence systems
Eraf-Nonius	X-ray diffractometers, X-ray accessories
Siemens	X-ray diffractometers, X-ray accessories
Oxford Cryosystems 3 Blenheim Office Park Lower Road Long Hanborough Oxford OX8 8LN UK	X-ray accessories

Atomic Spectroscopy

Cetac Technologies, Inc. 5600 South 42nd St. Omaha, NE 68107 USA	
Edinburgh Instruments Ltd Riccarton, Currie Edinburgh EH14 4AP UK	
GBC Scientific Equipment, Inc. 3930 Ventura Dr., Suite 350 Arlington Heights, IL 60004 USA	

Perkin-Elmer Ltd Post Office Lane Beaconsfield Bucks HP9 1QA UK	
Spectro Analytical Instruments GmbH Boschstr. 10 D-47533 Kieve Germany	
Thermo Jarrell Ash Corp 27 Forge Parkway, Franklin, MA 02038 USA	
ThermoQuest Corporation San Jose CA 95134 USA	

Software

ACD (Advanced Chemistry Development) 133 Richmond Street West Suite 605 Toronto Ontario Canada M5H 2L3	Chemical structure and spectroscopic manipulation

Acorn NMR Inc 46560 Fremont Blvd #418 Fremont CA 94538 USA	NMR data processing
Galactic Industries Corporation 395 Main Street New Hampshire 03079 USA	
Leeman Labs Inc 6 Wentworth Drive Hudson NH 03051 USA	
Shimadzu Corporation 1, Nishinokyo-Kuwabaracho Nakagyo-kur Kyoto 604-8511 Japan	
CIC Photonics Inc 3825 Osuna, NE Ste.6 Albuquerque, NM 87109 USA	
Bio-Rad Laboratories, Sadtler Division 3316 Spring Garden Street Philadelphia, PA 19104-2596 USA	
Ocean Optics Inc 380 Main Street Dunedin, FL 34698 USA	

Index